THE ATLAS OF AUSTRALIAN BIRDS

The Royal Australasian Ornithologists Union (RAOU) was founded in 1901 to advance knowledge and conservation of the birds of the Australasian region. It is the internationally recognized organization of ornithologists of Australia and nearby countries. Its members are people from all walks of life who have an interest in birds, as well as professional ornithologists. Projects and activities include the annual Congress and Campout, the annual Scientific Day, the Nest Record Scheme and research on waders, waterbirds, birds of prey and endangered species such as the Plains-wanderer. Donations to support the Union's research activities are tax deductible in Australia.

The RAOU publishes the *Emu*, a quarterly journal of world standing, and holds a comprehensive ornithological library. Bird observatories are maintained at Eyre in Western Australia, Rotamah Island in Victoria and Barren Grounds in New South Wales. With *The Atlas of Australian Birds* completed, the RAOU is now compiling *The Handbook of Australian Birds*.

Membership of the Union is open to anyone. Enquiries should be sent to: The Secretary, GPO Box 5236BB, Melbourne, Vic. 3001, Australia.

THE ATLAS OF AUSTRALIAN BIRDS

M. Blakers
S. J. J. F. Davies
P. N. Reilly

Royal Australasian Ornithologists Union

MELBOURNE UNIVERSITY PRESS
1984

First published 1984

Printed in Australia by
Globe Press Pty Ltd, Brunswick, Victoria, for
Melbourne University Press, Carlton, Victoria 3053
U.S.A. and Canada: International Scholarly Book Services, Inc.,
P.O. Box 1632, Beaverton, OR 97075
United Kingdom, Europe, Middle East, Africa:
HB Sales
Enterprise House, Ashford Road, Ashford, Middlesex,
England TW15 1XB

National Library of Australia Cataloguing-in-Publication entry

Blakers, M. (Margaret).
The atlas of Australian birds.

Bibliography.
Includes index.
ISBN 0 522 84285 2.

1. Birds—Australia—Maps. I. Davies, S. J. J. F. II. Reilly, Pauline, 1918–
III. Title.

598.2994

The Atlas of Australian Birds has been a co-operative project. It could not have been carried out without the help of thousands of people: from the 3000 atlassers who collected the information to the individuals, companies and institutions who provided financial and other support. The RAOU is indebted to them all and gives its sincere thanks.

ATLAS PATRONS

Dr M. Adena

Mr Tony Allingham

Mr R. C. Baird

Mrs M. Beurteaux

Dr J. M. Bonnin

P. A. Bourke

John and Edie Bransbury

S. R. F. Brown

Mr Anthony Bushell

R. G. H. Campbell

E. L. Carthew

Mr N. Chaffer

Miss A. W. Clucas

R. and P. Cundell

Dr Douglas D. Dow

H. M. Doyle

J. E. Friend

Mr H. N. Fretwell

Mrs I. Geddes

Gould League Bird Study Camp Club

Gould League of New South Wales

Mr C. N. Gove

P. W. Holbeach

Mrs A. D. Hosking

Mrs K. Hough

Mr M. Howlett

Mrs H. Laybourne-Smith

Colin F. Lewis

Mr Murray Lord

G. M. McArthur

Mrs Gwen McBurney

Dr J. M. McCracken

W. Manning

Mr S. Marchant

Mr Deane Morgan

Judith M. Moyes

Mr J. Nevinson

Juris and Melita Ozols

Mrs L. Perry

Mr W. Potts

Mr and Mrs G. D. Rogers

R. J. Rogers

Mrs L. M. Russell

Mr A. R. Sefton

Mr V. Serventy

Dr M. J. Seyfort

Ian Sinclair

L. H. Smith

Mr D. J. Stewart

Miss J. Strudwick

Mrs G. I. Surman

P. H. Sutton

Mr G. H. Watts

Dr H. N. B. Wettenhall

Mrs Jean Whatman

Miss M. J. Whitmore

Ida Willis

Edna M. Wilson

ATLAS SPONSORS

Production

Alcoa of Australia Ltd

ANZ Bank

Australian Biological Resources Study

Australian National Parks and Wildlife Service

Conservation Commission of the Northern Territory

Conzinc Riotinto of Australia Ltd

Dalgety Australia Ltd

Department of Fisheries and Wildlife, Western Australia

Estate of the late Mrs Sunday Ebbott

Electrolytic Zinc Company of Australasia Ltd

Esso Australia Ltd

Fisheries and Wildlife Division, Ministry for Conservation, Victoria

H. and L. Hecht Trust

M. A. Ingram Trust

Estate of the late Mrs Marjorie B. Murray

The Myer Foundation

National Parks and Wildlife Service, New South Wales

National Parks and Wildlife Service, Queensland

Publication

National Parks and Wildlife Service, Department of Environment and Planning, South Australia

National Parks and Wildlife Service, Tasmania

Perpetual Executors and Trustees Association of Australia Ltd

Reader's Digest Services Pty Ltd

The Trustees Executors and Agency Company Ltd

Western Mining Corporation Ltd

William Buckland Foundation

Alcoa of Australia Ltd

The George Alexander Foundation

National Australia Bank

The Ian Potter Foundation

The Shell Company of Australia Ltd

Westpac Banking Corporation

CONTENTS

FIGURES

TABLES

INTRODUCTION

A bird atlas to cover vast Australia? Impossible! Too many remote inaccessible areas, too few birdwatchers, no funds. That was the first reaction to an idea discussed in late 1972. But faith in it caused the idea to be pursued. By 1977 the Atlas was under way, to run for five years.

People who had passively watched birds in their own backyards from Cairns to Perth, Wyndham to Oodnadatta, caught the spirit of excitement. With the more densely populated areas covered, people trekked into the outback. To each there was a sense of achievement, of belonging. The justification for their labour is *The Atlas of Australian Birds.*

Background

The Atlas of Australian Birds maps the distribution and breeding distribution of birds in Australia from 1977 to 1981. It includes also what is known of bird distribution from the time of European settlement, a period of less than 200 years.

A knowledge of the distribution of organisms is vital to zoology and a powerful tool in managing populations, either to control them or to promote their conservation. In Australia the time since European settlement is brief and patterns of land use are still changing fast. The consequent effects on the distribution of birds can be measured by comparing atlases covering successive time periods.

The preparation of an atlas draws together the knowledge of amateur and professional. It increases the skills of participants and educates them in co-operative study, enabling people from different backgrounds to meet and share a common interest in birds.

The Botanical Society of the British Isles was the first to show that new information about distribution could be collected with the help of a large group of amateurs. By organizing them, the Society was able to publish in 1962 the *Atlas of the British Flora.*[1418] There followed, between 1968 and 1972, the collection of information about the breeding distribution of birds in Britain and Ireland. An estimated 10 000–15 000 people participated and the results were published in 1976.[1660] Since then, bird atlas projects have been developed in many other countries.[446,507,1924]

In October 1972, during the Annual Congress of the RAOU, its Field Investigation Committee (FIC) discussed tentatively the idea of an Australian bird atlas. The magnitude of the task of covering the Australian continent (7 682 300 sq. km) with the few hundred known field ornithologists seemed daunting but it was decided that a pilot atlas ought to be organized to test the feasibility of the project. This was carried out on the southern coast of New South Wales between March 1973 and September 1974 and involved 168 observers covering an area of 13 600 sq. km.[488] Then, in August 1974, the 16th International Ornithological Congress was held in Canberra, providing an opportunity for discussions between members of the FIC and overseas ornithologists involved in running atlas schemes. The outcome was a decision by the RAOU to go ahead with the project.

The Australian Government was approached and a meeting arranged with the Secretary of the Department of the Environment. The RAOU proposed to organize an Atlas of Australian Birds if the Australian Government would fund it. This proposal was accepted in principle in January 1975.

Other pilot atlases followed: the Adelaide Region,

April 1974 to December 1975,[1693] the Melbourne Region, December 1975 to December 1976[75] and Tasmania, which included data from 1900 to 1976.[1747] These served to test different procedures and train participants for the national project as well as to produce regional atlases of value in themselves.

In February 1976 the RAOU received the first grant from the Australian Government, which enabled the appointment of a full-time staff member whose task initially was to extract records for the Historical Atlas. Though funding for a 'field' atlas, mapping the current distribution of birds, had not been secured, the RAOU nevertheless decided to carry out the project and planning proceeded during 1976. In October grants were received enabling both the Historical and Field Atlases to go ahead, in the first instance until the end of June 1977.

In November 1976 the FIC decided how the Field Atlas would be organized, in particular: that the scheme would be computer-based with birds represented by numbers rather than names on the recording sheets; that species as defined in the *Checklist of the Birds of Australia, Part 1, Non-Passerines*[388] and *Interim List of Australian Songbirds, Passerines*[1576] would be the level of accuracy sought in the project; and that the Field Atlas would begin on 1 January 1977 and run for five years.

Immediate publicity was instigated and, by Christmas, Atlas kits had been designed, produced and distributed to 300 participants. By November 1981 all 1° blocks (p. xvii) had been visited and the project finished on 31 December 1981.

Scope

Geographical Limits

Observers were instructed to submit information from Australia, including islands and waters separating them from the mainland, and seas up to 50 km offshore and Bass Strait. In practice, records were also accepted from large embayments, e.g. Gulf of Carpentaria, and some seas and islands considerably more than 50 km from the mainland, e.g. Ashmore Reef (12/123). The outer limits were between 8°S and 44°S and between 112°E and 154°E. The RAOU recognized that money would not be available from the Australian Government for fieldwork outside Australia's political boundaries, so that New Guinea and New Zealand were excluded. Australian Antarctic Territory, Lord Howe, Norfolk, Christmas, Cocos-Keeling, Macquarie and Heard Is were excluded because they are remote and in some cases difficult of access. Their avifaunas differ from that of the Australian continent.

Taxonomy

Recording was at species level with provision for information about subspecies to be recorded if observers wished.

Periods

Five years was considered an adequate time in which to obtain information throughout Australia for the Field Atlas. It was long enough to cover some of the annual variation in distribution brought about by weather and short enough for large permanent changes in range not to take place during it. The distribution of those species that occasionally irrupt and reach places where populations do not persist cannot be adequately described in an atlas. The period of Atlas fieldwork, 1977 to 1981, is referred to as the Field Atlas throughout the text.

For the Historical Atlas, three time periods were chosen. These represent three distinct phases in the development of Australian ornithology: before 1901, exploration; 1901–50, ornithological development; 1951–76, popularization of ornithology.

BEFORE 1901. There was no specifically ornithological journal published in Australia. Results of expeditions and observations were published in books, reports, overseas journals and some local natural history journals. Only a small part of the continent had been cleared for agriculture and some parts were still unexplored.

1901–1950. Publication of the *Emu* began in 1901 so that ornithological work was now recorded in an Australia-wide journal. Works by Campbell[299] and North[1343] were published, the first attempts since Gould's[723] to present popular comprehensive accounts of the Australian avifauna. Mathew's *The Birds of Australia*[1219] also appeared during the period but was generally available only to professionals or serious and wealthy amateurs. Leach's *An Australian Bird Book*[1050] in 1911 and Cayley's *What Bird Is That?*[320] in 1931 were the first field guides within reach of many people. Most of Australia had been explored and agricultural and mining settlements extended into remote areas.

1951–1976. The first regional handbook, Serventy and Whittell's *Birds of Western Australia*,[1637] published in 1948, ushered in the third phase. It was followed by many more field guides giving ornithologists, for the first time, adequate tools with which to work. Skilled field ornithologists migrated to Australia following the 1939–45 war, boosting the field force and accelerating the pace of discovery. Agricultural expansion was rapid as bulldozers replaced horses to clear the scrub and trace elements were added to increase soil productivity. Conservation ideas began to be promoted, a fact partly responsible for the willingness of government to support the Atlas.

Scale

Atlases in Britain and the local pilot atlases used the Transverse Mercator Grid in 10 000-yard or 10-km units or a grid based on latitude and longitude. In 1976 the Australian Government, through the Australian Biological Resources Study (ABRS), was beginning to organize a biological data base for Australia. ABRS resolved to use the lines of latitude and longitude as the basic grid[287a] because it was universal and maps were readily available; the RAOU followed this recommendation. 'Fixed' grid lines 1° and 10′ apart were used for the Field Atlas and a programme to visit, at least once, all 812 of the 1° blocks in Australia, was set as the target. It was considered that this could be achieved within five years and that it should provide a sufficiently detailed picture of distribution on a continental scale for most purposes. Observations in the *Atlas* would be published in 1° blocks but observers were asked to record in 10′ blocks wherever possible. This smaller area (280 sq. km as against 10 000 sq. km for 1° blocks) would be more useful for conservation and management.

Many historical records could not be fitted into the 'fixed' grid used for the Field Atlas. For them, the accuracy with which the sighting was located was defined by adjusting the block size (p. xxv).

Computer

Previous atlas schemes used manual methods to collate data. More than three times as many species were involved in Australia as in Britain and Ireland[1660] and funding provided for only a small staff to co-ordinate the project. The RAOU therefore decided to automate, as far as possible, the sorting and analysis of data and to ask observers to submit data in a form compatible with computer processing. Without access to computer facilities the project could well have foundered under a mound of paper.

The use of a computer subsequently proved to have a further advantage: all data—every list of every observer—could be used to calculate the Reporting Rate of a species in different places or at different times of the year giving a comparative measure of recording frequency.

Information Required

Information was sought only on presence and, if possible, breeding of a species in a grid block. The main aim of the Atlas was to define the ranges of species by observations throughout Australia. These were solicited from birdwatchers and accepted unless further investigation showed a particular record to be doubtful or wrong. For the acceptance of a breeding record the observer had to see eggs or chicks in the nest or downy or dependent young. Empty nests were excluded because many Australian birds use nests for roosting and loafing. Activities such as display, courtship feeding, territorial fighting, copulation and nest building may follow environmental changes, e.g. rainfall, but in Australia they do not always lead to breeding and are inadequate evidence on which to define the breeding range.

Information was not sought on the habitat in which a bird was recorded nor on its abundance within the recording block. Even using 10′ blocks, the area was too great for such detail to be useful nor could it be handled efficiently. It would distract from the main aim of determining distribution.

The length of time to be covered by each set of observations was not specified and ranged from one day to five years for data in the Field Atlas or longer for the Historical Atlas. Observers were encouraged to collect data seasonally in the hope that large-scale movements of species, resulting in their absence from a district for part of the year, could be detected. This entailed repetition on record sheets of all species seen during each period and not just the addition of new ones. There was no attempt to collect data on the number of times a species was seen within the seasonal period. The periods were: spring, September to November; summer, December to February; autumn, March to May; winter, June to August. These are southern hemisphere seasons. The limitations of these periods were recognized: there is great variation from the tropical north to temperate south; springtime movement and activity often begin in August; and summer limits are often blurred in spring and autumn.

Administration

The Council of the RAOU has the ultimate responsibility for the affairs of the Union. Officers are elected for limited terms. The Field Investigation Committee (title changed to Research Committee in 1982) is directly responsible to the Council for the conduct of all field research, including the Atlas. An Atlas Committee was appointed in November 1976 to oversee the operation of the Atlas along the lines determined by the FIC.

The Atlas office in Melbourne co-ordinated the project. The staff of four, led by a Field Officer, was responsible for co-ordinating fieldwork, collecting records for the Historical Atlas, analysing data, producing newsletters and preparing the information for publication in the *Atlas*. Computer programming was carried out by a volunteer and volunteers were essential in handling much of the clerical work. Part-time typists were employed during preparation of the book.

Volunteer organizers were responsible for co-ordinating work and checking observations in nine regions covering Australia, again with the help of numerous other volunteers.

The Vignettes Committee solicited illustrations for the published *Atlas*. Payments to artists were made from a fund generated by the sale of the original drawings, delivery to be made after publication.

Financial support for the Atlas was forthcoming from many sources: the Australian National Parks and Wildlife Service, the Australian Biological Resources Study, state governments, philanthropic organizations, industry, individuals and the RAOU itself. Major donors are acknowledged on pp. vii-ix.

Field Atlas

Regions and Organizers

In order to decentralize the administration of the Atlas, Australia was divided into nine recording regions (Fig. 1) each with a Regional Organizer (RO). The boundaries of these regions were such that each could be conveniently administered from a major city. These recording regions are different from bioclimatic Atlas Regions used in the presentation of the maps (p. xl).

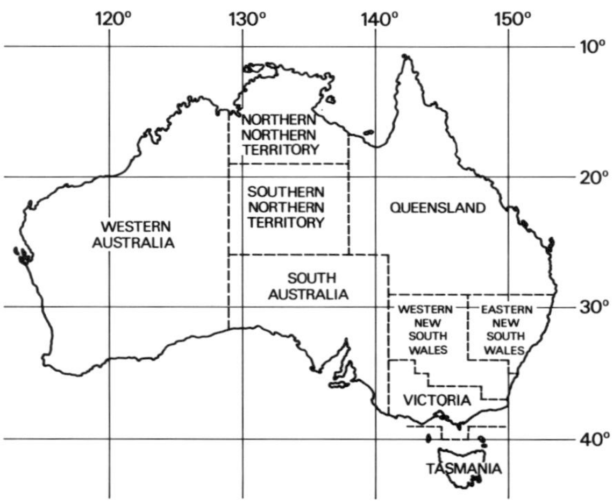

Fig. 1. Recording regions used for the collection of data

The tasks of the RO included processing record sheets, checking and assessing records, answering queries from atlassers, organizing and co-ordinating activities in their regions, e.g. camps. Sub-Regional Organizers (Sub-ROs) gave invaluable service and advice to the ROs. They were recruited by the ROs and worked with them in the way best suited to each of them. In some cases Sub-ROs took responsibility for a small area within the region; in others they acted as a committee. Usually data for the Historical Atlas were dealt with by a separate group of helpers.

Materials

A person who wanted to help collect information for the Atlas (an atlasser) was sent an Atlas Kit which contained:

a) an instruction folder (Fig. 2) which included the names and addresses of ROs to whom inquiries should be directed;

b) a set (usually 25) of record sheets (Fig. 3). The record sheet was based on a standard 80-column computer data sheet; all information was coded numerically;

c) a key listing the common name of each species and the three-digit code number assigned to it. The codes used were those of the Australian Bird Banding Scheme. The key lacked space to include the codes assigned to rare or vagrant species, which the observer was required to write at the foot of the record sheet;

d) a 'translation sheet' listing the common names of species and equivalents used in popular field guides;

e) an Unusual Record Report Form (URRF) (Fig. 4);

f) copies of past and current Atlas Newsletters;

g) a request card for record sheets with the RAOU's address on one side and space for the observer's name and address and the number of sheets required on the other; also used for enrolling new participants.

Processing Record Sheets

Atlassers were encouraged to record field observations in notebooks, taking descriptions of unusual species, breeding evidence etc. Records were grouped either in 10′ or 1° blocks. The flimsy Atlas record sheet was unsuitable for field use and was usually completed at home (Figs 2 and 3). A locality within the grid block provided a check for its co-ordinates. The minimum period of observation that could be entered was one day, for which a single date only was required. A three-digit code number represented each species observed followed by '1' for breeding and '2' if a Nest Record Card had been completed. Information confirming breeding, the names of rare or vagrant species observed but not included in the key and interesting observations were entered in *Other Species and Breeding Notes* or on the back of the record sheet.

Completed sheets were sent to the RO of the recording region in which the observations were made. ROs processed them as quickly as possible to shorten the time between observation and query should this be necessary. Co-ordinates of latitude and longitude were checked against the locality given by the ob-

ATLAS KIT

Each atlasser needs access to maps (see back page).

From us you receive quarterly Newsletters, 25 record sheets (replaced as required) and a list of common names with code numbers of most Australian species. Some names have been abbreviated to fit into the space available and others omitted because of rarity or strictly localised occurrence. These names can be entered in full at the bottom of the Record Sheet.

Names are taken from the "Checklist of the Birds of Australia, Part 1 (non-Passerines)" by H. T. Condon (1975) and from the "Interim List of Australian Songbirds (Passerines)" by R. Schodde (1975), both published by the RAOU and available from RAOU Headquarters.

FIELD ATLAS
HOW TO FILL IN RECORD SHEETS

(see Fig. 1)

1. Use one sheet for each grid block (defined below).
2. Ignore all green areas.
3. Write clearly using non-smudging ink.
4. **NAME, ADDRESS AND PHONE NUMBER.** Fill in your name, present address and phone number, including STD code. If the information has been collected by a group, the name and address of the leader, who accepts responsibility for the accuracy of the data, should be given.
5. **GRID REFERENCE AND BLOCK SIZE.** A grid based on the lines of latitude and longitude is used for the Atlas. A portion of the NSW coast with the grid superimposed upon it is shown in Fig. 2. Records from either a one degree (1°) or a ten minute (10') grid block are acceptable. A 1° grid block is that block bounded by **full degree** lines of latitude and longitude, each side measuring 1°. Fig. 2, for example, contains two complete 1° blocks.

Ten minute grid blocks are those within the 1° grid block bounded by lines of latitude and longitude which are multiples of 10 minutes. One of the 1° blocks in Fig. 2 is shown divided into its 36 component 10' blocks.

Under the heading **Grid Reference,** give the co-ordinates of the **centre point** of the block and under **Block Size** indicate whether it is a 1° or a 10' block.

Note: The centre point of a 1° block will always be x° 30' South, y° 30' East, while the co-ordinates of the centre point of a 10' grid block will always end with 05, 15, 25, 35, 45 or 55', e.g. The centre point of the 10' block containing Ulladulla in Fig. 2 (shaded) is 35° 25' S, 150° 25' E.

6. **PERIOD OF OBSERVATION.** Give the first and last days of the period of observation in the order day, month, year. If all observations are on the same day, write this date under **First Day** and leave the other blank. If you are resident in or frequently visit the same block, one sheet is adequate for each seasonal period, i.e., December to February, March to May, June to August, September to November. If visits are intermittent, use a separate sheet for each visit.
7. **LOCALITY.** Give the name of an identifying feature which is wholly within the grid block and **which is shown on the map** (e.g., Ulladulla in Fig. 1 below). It may be a town, hill, lake, etc., but not a river or creek. If there is no suitable feature within the grid block, select one nearby and specify the distance and direction of the centre of the grid block from the feature, e.g., 80 km SW of Tennant Ck for example B in Fig. 3.
8. **STATE.** This is the State in which the observations were made.
9. **MAP USED.** Please define as fully as possible (see Fig. 1; see back of this folder for suggested maps).
10. **SPECIES OBSERVED.** This section should be filled in starting at the left hand end of the top row and entering the data **across** the sheet. Complete each row before starting the next (the numbers at the end of each row are the cumulative species totals for the sheet). You will see that the rows are divided into 14 sections, each section into 4 spaces. In the first 3 spaces headed **Sp. No.** enter the 3 digit number of the bird (from the list of species' names and numbers). There is no need to enter species' numbers in numerical order.

The fourth space labelled **B** is for breeding data. In it:
enter 1 for confirmed breeding (nest with eggs or young; or downy or dependent young);
enter 2 if a Nest Record Card has been completed for this observation (see the section entitled **Notes on Recording);**

Note: The eggs or young of a cuckoo should be entered as a 1 or a 2 for the cuckoo species (if known) as well as the fostering species.

* If you record more than 126 species in one grid block, start a new sheet, completing all the information at the top as on the previous sheet.

11. **OTHER SPECIES AND BREEDING NOTES.** Write the name of any species not listed in the Key (e.g., Paradise Parrot).

Give notes on identification of confusing species (e.g., Crows, Ravens) and sub-species.

Give details of confirmed breeding records and probable breeding (e.g., birds carrying nesting material or food; or incubating in or on inaccessible holes or nests).

Enter any other notes. Continue on the back of the sheet if necessary.

* It cannot be overstressed that accurate observation and recording are essential. IF YOU ARE NOT ABSOLUTELY CERTAIN OF YOUR IDENTIFICATION, DO NOT RECORD IT.

Fig. 2. FIELD ATLAS. Map section showing two 1° blocks (grid references 35° 30' S, 149° 30' E; 35° 30' S, 150° 30' E), one of which is sub-divided into its 36 component 10' blocks. The grid reference for the shaded 10' block is 35° 25' S, 150° 25' E.

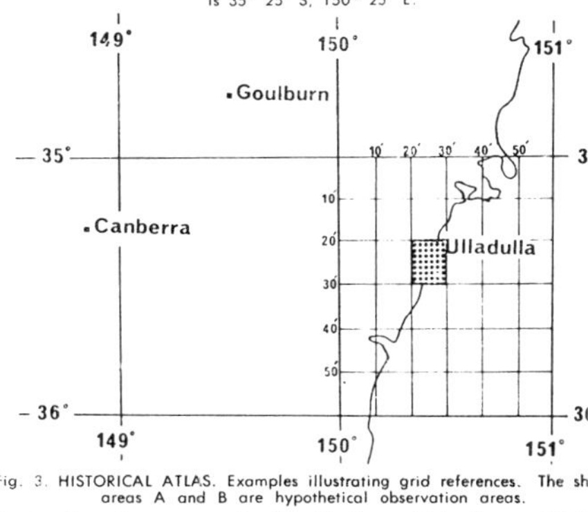

Fig. 3. HISTORICAL ATLAS. Examples illustrating grid references. The shaded areas A and B are hypothetical observation areas.

A: An observation area overlapping 10' lines. Grid reference 19° 30' S, 134° 23' E, Block Size 10'.

B: An observation area overlapping a whole degree line. Grid reference 20° 58' S, 133° 35' S. Block Size 1°.

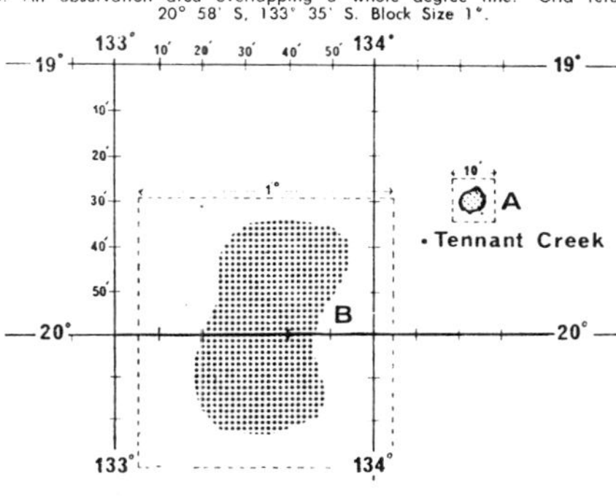

HISTORICAL ATLAS
HOW TO FILL IN RECORD SHEETS

Records should be entered in the same way as for the Field Atlas, with the following variations:

GRID REFERENCE AND BLOCK SIZE. The Historical Atlas differs from the Field Atlas in that data will have been collected from a locality rather than a grid block. Some records will be from small areas, e.g., a town, farm or lake; while others will be from large areas overlapping several Field Atlas grid blocks, e.g., a National Park, Shire.

Where the area of observation falls wholly within a 1° or a 10' grid block, as defined for the Field Atlas follow the Field Atlas procedure.

Where the area of observation overlaps a degree line or a ten minute line, find the approximate centre of the area and enter its co-ordinates (latitude and longitude) under the heading **Grid Reference.** Next, determine the size of the block, whose sides, measured in degrees and minutes, must be equal, within which the whole observation area can be contained. This block may be of any size between 1' and 2°, preferably 1° or less. Enter this figure under the heading **Block Size.** Examples are given in Fig. 3.

PERIOD OF OBSERVATION. Wherever possible, give the exact dates as for the Field Atlas. If you have records for a particular locality extending over a large number of years, these can be lumped into 10-year spans, i.e., 1921-1930, 1931-1940 . . . 1961-1970, 1971-1976.

If the date is known only to the nearest month, leave the day spaces blank; if only to the nearest year, leave both the month and day spaces blank.

Note: Records extracted from notebooks for the Historical Atlas may previously have been published. Information transcribed directly from notebooks is preferred, to avoid duplication, but give the complete reference of any articles at the bottom of the Record Sheet.

COMPLETED SHEETS

Send completed sheets to the **Atlas Regional Organiser of the State in which the observations were made.** Sheets may be folded. Where a grid block overlaps the boundaries of two States, send your Atlas Record Sheet to the Regional Organiser of the State in which the greater part of the block lies.

After some preliminary checking, the Regional Organiser will forward the sheets to the central Atlas Office, where the data will be transferred to a computer. At intervals, distribution maps and grid block species lists will be produced and sent to the Regional Organiser, who will inspect them for unusual or apparently erroneous records. At this stage, you may be asked to verify unusual observations on the Atlas Unusual Record Report Form.

* If you have any problems with filling in the Record Sheets, contact your Regional Organiser or the Atlas Office for help.

Fig. 2. Inside pages of the instruction folder

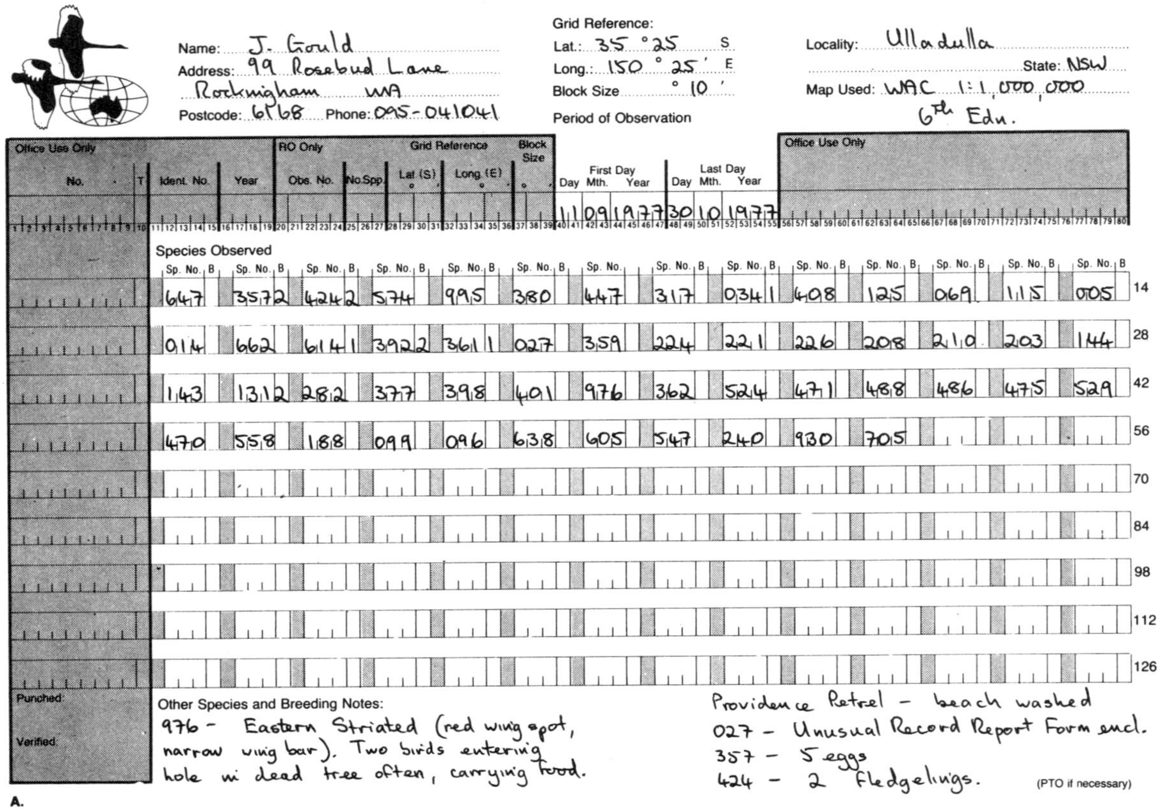
Name: J. Gould
Address: 99 Rosebud Lane
Rockingham WA
Postcode: 6168 Phone: 095-041041

Grid Reference:
Lat.: 35 °25 S
Long.: 150 °25 ' E
Block Size ° 10 '
Period of Observation

Locality: Ulladulla
State: NSW
Map Used: WAC 1:1,000,000
6th Edn.

Office Use Only No.	T	Ident. No.	Year	RO Only Obs. No.	No.Spp.	Grid Reference Lat. (S) °	Long. (E) °	Block Size °	First Day Day Mth. Year	Last Day Day Mth. Year	Office Use Only
									11 09 1977	30 10 1977	

Species Observed (Sp. No. B)

647	3572	4242	574	995	380	447	317	0341	408	125	069	115	005	14
014	662	6141	3922	3611	027	359	224	221	226	208	210	203	144	28
143	1312	2812	377	398	401	976	362	524	471	488	486	475	529	42
470	558	188	099	096	638	605	547	240	930	705				56
														70
														84
														98
														112
														126

Punched:
Verified:

Other Species and Breeding Notes:
976 - Eastern Striated (red wing spot, narrow wing bar). Two birds entering hole in dead tree often, carrying food.
Providence Petrel - beach washed
027 - Unusual Record Report Form encl.
357 - 5 eggs
424 - 2 fledgelings.
(PTO if necessary)

A.

Fig. 3. Record sheet A; completed by atlasser

server and entered on the line below. An identifying code number was allocated to each observer ('observer number' prefixed with a code for the RO's region) and entered on the sheet. Dates and breeding information were checked. Information about subspecies and URRFs or other data (e.g. feathers, photos, drawings) supporting the identification of a species were assessed and a code entered on the sheet accordingly next to the species' number:
'6': confirmed (URRF or equivalent accepted)
'7': hybrid
'8': subspecies and supporting description
'9': subspecies name only.
At this point some ROs also checked the species listed on each record sheet and contacted the observer to obtain additional information about unusual sightings. The RO assessed the information provided and amended the record sheet if necessary, e.g. deleted the species, marked it confirmed (6).

The last step for the RO was to count the number of species on the record sheet and enter it in the space provided before sending the completed sheet to the Atlas office in Melbourne.

At the Atlas office the number of record sheets submitted by each observer was tallied and entered on his or her address card which also contained a record of the person's 'observer number' in each region (allocated by the RO) and the number of record sheets supplied.

Record sheets were sorted into 1° blocks and filed by latitude and longitude. At intervals of four to six months the accumulated record sheets were individually numbered ('sheet number'), coded according to whether the locality was an island or not and whether the records derived from specimens, literature or observations and given a final general check. Information on the sheets was then punched on to a computer-compatible medium, initially cards, later magnetic tape. Data were punched and verified producing a punching accuracy of at least 99.75%.

Data were processed on CSIRONET. As each batch of new data was entered, the computer was programmed to point out inconsistencies, e.g. coordinates outside Australia, impossible dates, numbers for which there were no birds, duplicated species numbers, missing data, cases where the 'number of species' written on the top line of the sheet did not tally with the actual number on the sheet. The errors that had given rise to these inconsistencies were corrected and the batch of new data combined with that already in the data bank. The whole was then processed to obtain up-to-date maps and 'listings'.

Maps showed the 1° blocks in which each species was recorded ('O'—observed, 'B'—breeding) and covered the whole of Australia. Records confirmed by an acceptable URRF or equivalent were marked

Atlas of Australian Birds

Unusual Record Report Form

Name: A. BIRDWATCHER

Address: TWITCHERS LANE, EMU BOTTOM. 2345 **Phone Number:**

Species' Name: SQUARE-TAILED KITE. **Species' Number:** 230

Date of Observation: 27-28 NOV. 1980

Grid Reference: **Lat:** 37 ° 05 ' S **Long:** 141 ° 15 ' E **Block Size:** 0° 10'

Location (Be as precise as possible): MUNDARRA SWAMP, 12 km. S.W. of EDENHOPE.

Physical Description of Bird (Describe size and shape, plumage, colour of legs, bill and eyes, call, number present, sex, juvenile or adult, etc.; please describe only what you saw and include comparisons with other species present or familiar to you. If you have a sketch or field notes, include a copy.)

LONG WINGED RAPTOR ABOUT THE SIZE OF A MARSH HARRIER. VERY PROMINANT "FINGERS" ON THE WINGS AND BROAD SQUARE TAIL. UPPERPARTS GENERALLY REDDISH BROWN EXCEPT FOR PALE BUFF BAR ALONG WING COVERTS, WHITE FOREHEAD AND DARK BROWN PRIMARIES.

UNDERPARTS, BROWN STREAKED REDDISH, UNDERWING COVERTS BRICK RED; DARK TERMINAL BAND TO TAIL; PRIMARIES WHITE, TIPPED AND BANDED WITH BLACK GIVING THE IMPRESSION OF A LARGE WHITE AREA AT THE BASE OF THE PRIMARIES.

TWO BIRDS WERE PRESENT, PRESUMABLY A PAIR, ALTHOUGH DIRECT COMPARISONS OF SIZE WERENOT POSSIBLE.

DISTINGUISHED FROM WHISTLING KITE, LITTLE EAGLE AND MARSH HARRIER BY THE DISTINCT UNDERWING PATTERN, WHITE FOREHEAD AND FACE, DARK RUMP AND BROAD SQUARE TAIL.

SEE PHOTOGRAPH.

Behaviour (Describe where and how feeding, movements, flight pattern, interaction with other birds, etc.)

OBSERVED FLYING OVER RIVER RED GUMS SURROUNDING THE SWAMP. FLIGHT CONSISTED OF SLOW WINGBEATS INTERSPERSED WITH GLIDES ON UPSWEPT WINGS. OCCASIONALLY CIRCLED HIGHER BUT QUICKLY RETURNED TO TREE TOP LEVEL, BANKING AND TWISTING TO AVOID EMERGENT BRANCHES.

PTO

Fig. 4. Unusual record report form (URRF); completed front

Habitat (describe, together with any neighbouring habitats).

RIVER RED GUM WOODLAND AROUND AN EXTENSIVE SWAMP AND SURROUNDED BY PASTURE WITH SCATTERED RIVER RED GUMS AND YELLOW GUMS.

Where was the bird, e.g. ground, trees, flying? FLYING

Distance and height of the bird from you.

VARIED. CAME AS CLOSE AS ≃ 40m WHEN THE PHOTOGRAPH WAS TAKEN

For how long did you watch it?

40 MINUTES ON 27th AND ABOUT 15 MINUTES ON 28th.

Sighting conditions (time of day, weather, visibility, etc.).

SUNNY, GOOD VISIBILITY, 1200 - 1240 hrs.

Optical aids used.

ZEISS 10 x 40 BINOCULARS

What experience have you had with

(1) the species in question?

NONE UNTIL NOW.

(2) other species with which you think it may be confused?

AM FAMILIAR WITH BLACK KITES FROM QUEENSLAND; WHISTLING KITES, LITTLE EAGLES AND MARSH HARRIERS ARE VERY FAMILIAR TO ME.

What books have you used?

IMMEDIATELY CONSULTED SLATER.
LATER CHECKED PIZZEY AND READERS DIGEST.

Were other observers present (give names)?

YES. ANOTHER BIRDWATCHER

Does anybody agree or disagree with your identification?

SHE AGREES.

How confident are you of your identification, e.g. 70%, 100%? If not 100%, why not?

100%.

Other details (e.g. was it a roadkill, have you tapes or photos?).

SEE ACCOMPANYING PHOTOGRAPH TAKEN ON 27th NOV.

COPY OF FIELD NOTES ATTACHED.

(If more space is needed for any section, use extra paper and attach to this sheet.)

Fig. 4 (cont.). Unusual record report form (URRF); completed back

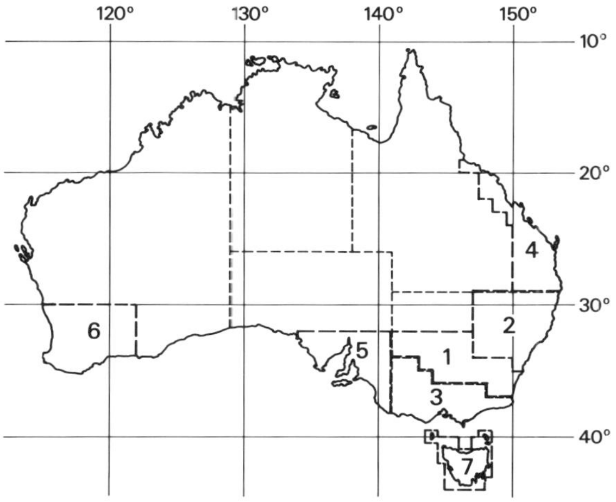

Fig. 5. Areas for which 10′ distribution maps were obtained

C (or A if breeding). The hierarchy was A>B>C>O. The first time a record appeared in a grid block it was marked '*'; additional records were marked '+'; doubtful ones '?'. From time to time maps at the 10′ scale were obtained for those areas sufficiently well-covered (Fig. 5). The same codes were used as on the 1° maps.

'Listings' for 1° blocks, in order by latitude and longitude, consisted of a cumulative list, by number and name, of the species recorded. Using the same codes as the maps, breeding records, confirmed records, new and additional data were all indicated for each species and, in later listings, the number of records was shown. In addition, records from the latest batch of data were listed in detail under the 10′ block where possible or under the 1° block if the sheet had been submitted for the 1° block as a whole. The detail included for data from the latest batch consisted of the sheet number only if five or more records of the species had been received for the 1° block; if there were fewer than five records, or if the record included breeding, then sheet number, observer number, first date, codes for breeding (1,2) and other information (6,7,8,9) were printed. The listings enabled any record on the map to be traced to the record sheet on which it was entered and provided extra details for records most likely to be queried (species reported fewer than five times in a 1° block and breeding records).

The cumulative listing for each 1° block also included a summary, by 10′ blocks and for the 1° block as a whole, showing the number of record sheets received, number of records, number of species recorded, number of species recorded breeding, average number of species per record sheet. On later printouts an aggregated summary was provided giving the same information for each region and for Australia as a whole and also showing the total number of records for each species in each region and in Australia as a whole.

Towards the end of the period of data collection records were sorted in different ways to test methods of analysing the information and to help with checking. These were:

1) 1° 'numbers' maps: showed the number of records of each species in each grid block and indicated where planning was required to overcome poor coverage;
2) 1° frequency map: showed the proportion of record sheets in each 1° block on which the species was recorded (Reporting Rate);
3) seasonal 1° frequency map: as for frequency map except that a separate map was produced for each season (December to January, March to May, June to August, September to November); record sheets falling into two or more seasons were ignored;
4) 'complete listing': gave sheet numbers for every record in every 10′ and 1° block.

ROs were supplied with a complete set of 1° maps and with the listings pertaining to their region each time new records were added to the data bank. Sub-ROs were sent listings for their areas.

Maps and listings were obtained on both paper and microfiche. Most ROs did not have access to microfiche readers and used the paper printouts.

Vetting Records

The system of processing data meant that most records were entered into the data bank without checking. A record was assumed to be accurate until queried and, within the expected range of a species, errors would not be detected.

ROs checked the maps and Sub-ROs the listings for records that needed verification. Often ROs asked local ornithologists to check the data too. Some maps were sent to ornithologists knowledgeable about the species concerned. They identified records needing verification but, in general, the most critical evaluations were made by people with good local knowledge, specially the Sub-ROs.

When a record needed verification, the observer was queried, usually by a letter, which specified the date and place of the sighting; these details were obtained directly from the listing. Many doubtful records at this stage turned out to be errors, e.g. wrong entry by observer, transcription error by computer operator, wrong grid reference. Genuinely unusual records, thought by the observer to be correct, were examined further. The observer filled in an URRF or provided whatever information was available in support of the record. This information was examined by the RO and others called upon for an opinion either in a committee or as individuals. The assessment took into account how unusual was the record, as well as the information actually supplied by the observer. Local observers familiar with their areas tended to be more accurate than visiting observers influenced by their expectations of what might be present. The decision to accept, reject or mark the record doubtful was taken by the RO and the data bank amended accordingly.

Towards the end of the Field Atlas, a National Vetting Committee at the Atlas office re-examined all URRFs previously accepted in order that there might be consistency within Australia as a whole. This committee was able to take account of what had been learned about species' distribution during the Atlas itself and also considered the skill and reliability of observers on the basis of their Atlas contributions.

Records doubted or rejected by the national committee were returned to the RO concerned for comment and final decision. A few significant records were sent to the RAOU's Record Appraisal Committee for consideration, e.g. new species for Australia, large extensions of range. Rejection of a record did not necessarily mean that it was wrong; often records were rejected or made doubtful because there was insufficient evidence to confirm the identification beyond doubt.

In these ways, the data were subjected to thorough scrutiny in an attempt to bring the Atlas to the highest possible standard of accuracy.

Recruiting Observers

When the Atlas started, the RAOU had fewer than 1000 members. The pilot atlases had trained a large workforce but a substantial recruiting programme was instituted to make contact with other potential contributors. Natural history societies were approached but there remained many competent birdwatchers not affiliated with any natural history organization. The Atlas staff provided articles for publication and gave interviews to the media. Staff and volunteers went on tour and gave talks, heralded by preliminary publicity, in regions where there were no atlassers. There was little tradition of co-operative ornithological research in Australia apart from that carried on in association with the Australian Bird Banding Scheme and the RAOU's own Nest Record Scheme.

Information was sent to potential contributors followed by an Atlas kit if the enquirer still wished to enrol. New contributors were still offering to help until the last months of the Field Atlas.

Newsletters

The importance of the quarterly Atlas Newsletter in co-ordinating the undertaking cannot be overemphasized. It acted as a bond between atlassers all over Australia, informing them of progress and enabling them to learn of the trials and triumphs of their colleagues. Maps at 10′ and 1° level indicated where coverage was needed; the distribution of interesting species was shown; and articles covered identification of difficult species, finding of nests and other field techniques. With some Newsletters additional detailed material was sent relating to the observer's own area: coverage maps showing the number of species recorded in each 10′ block and lists of species observed and breeding in the observer's home 1° block. Atlassers could also obtain this information, on request, for areas that they planned to visit. Meetings and camps were used to disseminate information about the project and what was most important to do next.

Field Methods

There was no restriction on the operations of an atlasser. Records were welcomed from any part of Australia as well as from the atlasser's own residential area. Atlassers were exhorted to be accurate: 'when in doubt, leave it out', and to enjoy themselves. The Atlas office and ROs arranged meetings, day trips and camps to train participants and keep them informed as well as to cover new country. Maps showing lack of coverage encouraged many atlassers to visit the blank areas. At first these were close to population centres but increasingly it became necessary to make expeditions to remote areas, inaccessible to ordinary vehicles. Individuals, companies and public bodies supported the expeditions, which included:

August	1978	Tambo (24/146)
August-September	1978	Tibooburra (29/142)
August-September	1978	Southern Pilbara Region
December	1978	Strahan (42/145)
August-September	1979	Nullarbor Plain
August-September	1979	Great Victoria Desert
September	1979	Gregory Downs (18/139)
December	1979	Temma (41/144)
July	1980	Great Sandy Desert
August	1980	Davenport Downs (24/141)
August-September	1980	Eastern Pilbara Region
September	1980	North-western South Australia
June	1981	Northern Simpson Desert
July	1981	Southern Tanami Desert
August	1981	Barkly Tableland
August-September	1981	Tanami Desert
August-September	1981	Top Springs (16/131)
November	1981	Poeppel's Corner (26/138)
December	1981	South-western Tasmania.

Co-ordination of the World Wildlife Fund's *Search for Australia's Rarest Birds* with the Atlas fieldwork improved coverage. Durham University's expedition to the Great Sandy Desert, *Shark Bay 81*, covered areas that would not otherwise have been surveyed. The last 1° block was Lake Amadeus (24/130) visited in November 1981 by the Northern Territory Conservation Commission.

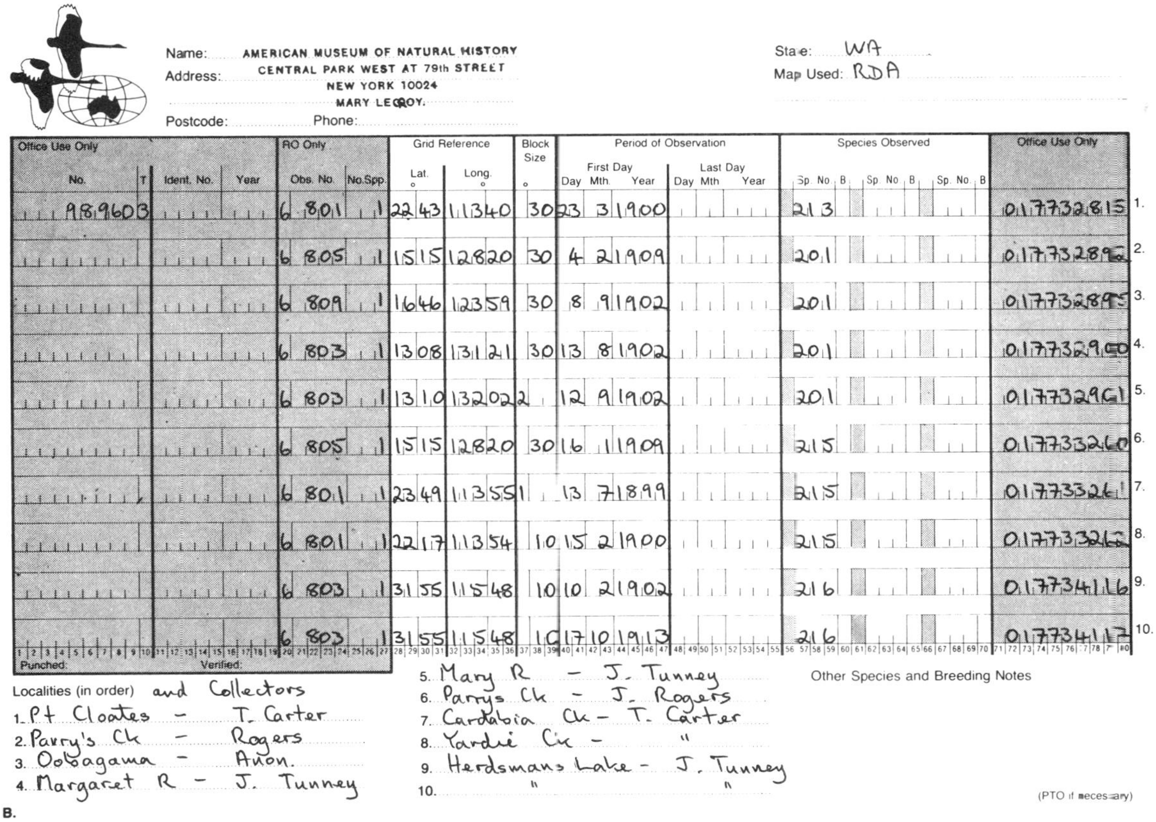

Name: AMERICAN MUSEUM OF NATURAL HISTORY
Address: CENTRAL PARK WEST AT 79th STREET
NEW YORK 10024
MARY LECROY.
Postcode: Phone:

State: WA
Map Used: RDA

Office Use Only No.	T	Ident. No.	Year	RO Only Obs. No.	No. Spp.	Grid Reference Lat. °	Long. °	Block Size °	First Day Day	Mth	Year	Last Day Day	Mth	Year	Species Observed Sp. No.	B	Sp. No.	B	Sp. No.	B	Office Use Only	
989603				6 801	1	22 43	113 40	30	23	3	1900				213						017732815	1.
				6 805	1	15 15	128 20	30	4	2	1909				201						01773289[illegible]	2.
				6 809	1	16 46	123 59	30	8	9	1902				201						01773289[illegible]	3.
				6 803	1	13 08	131 21	30	13	8	1902				201						0177329[illegible]	4.
				6 803	1	13 10	132 02	2	12	9	1902				201						017732961	5.
				6 805	1	15 15	128 20	30	16	1	1909				215						017733260	6.
				6 801	1	23 49	113 55	1	13	7	1899				215						017733261	7.
				6 801	1	22 17	113 54	10	15	2	1900				215						01773326[illegible]	8.
				6 803	1	31 55	115 48	10	10	2	1902				216						017734116	9.
				6 803	1	31 55	115 48	10	17	10	1913				216						01773411[illegible]	10.

Punched: Verified:

Localities (in order) and Collectors
1. Pt Cloates – T. Carter
2. Parry's Ck – Rogers
3. Oobagama – Anon.
4. Margaret R – J. Tunney
5. Mary R – J. Tunney
6. Parrys Ck – J. Rogers
7. Cardabia Ck – T. Carter
8. Yardie Ck – "
9. Herdsmans Lake – J. Tunney
10. "

Other Species and Breeding Notes

(PTO if necessary)

B.

Fig. 6. Record sheet B; completed

Historical Atlas

Sources

The main sources were observers' notebooks, published records and specimens in museums and private collections. Observers extracted their own records; staff and volunteers extracted data from books, journals and museums. The Historical Atlases represent a comprehensive catalogue of the distribution of Australian birds from the time of European settlement. Sources are listed on p. 671.

Processing Records

Records were entered on record sheet A (Fig. 3) as for the Field Atlas or on record sheet B (Fig. 6) designed to allow records from up to ten different localities to be entered on one sheet, each line independent of other lines. Records grouped by species rather than by locality, e.g. museum specimens, could thus be entered efficiently. Dates and other information could be different on each line.

Records deriving from observation, literature (including some unpublished archival material) and specimens were treated differently. Observations were entered in the same way as Field Atlas records (p. xviii). Literature records included codes for the book or journal, date of publication, volume and page number and a separate set of codes for the observers. Specimen records included codes for the collection, specimen number (if possible the number used by the institution holding the specimen), collector and identifier (rarely used) of the specimen. The same series of numbers was used for collectors and identifiers in each collection but a different series for each collection.

Completed sheets were checked, sorted and numbered in the same way as for the Field Atlas. Grid references proved particularly difficult because many place names had changed and some were so obscure or vague as to defy location.

Computer processing of historical records was similar to that for the Field Atlas. Maps (1° only) and listings were obtained for each time period. Records needing verification were referred to the original observer if possible or checked against the source. Assessment of the record and amendment of the data bank were carried out in the same manner as for the Field Atlas except that no literature or specimen record was actually removed from the data bank; unlikely records, including apparent errors were marked doubtful but retained.

Many historical records could not be located as precisely as those in the Field Atlas. The area containing a record was defined by giving the co-ordinates of its centre point together with the length of the side of a grid block which would enclose it. To map the record it was arbitrarily assigned to the 'fixed' grid block within which the centre point fell; records from grid blocks of greater size than those being

Table 1. Summary of Atlas statistics by recording region

Time Period	Recording Region	No. Record Sheets	No. Records	No. Species	No. Breeding Species	Average Species/Sheet
Field Atlas	Australia	89 134	2 715 413	716	581	30
	Western NSW	9 528	323 965	411	300	34
	Eastern NSW	11 958	425 062	452	321	35
	Vic.	16 968	615 060	442	313	36
	Qld	14 527	510 689	550	430	35
	SA	11 988	315 749	417	292	26
	WA	13 349	282 313	467	325	21
	Tas.	6 436	135 449	260	138	21
	Southern NT	1 770	28 038	222	109	15
	Northern NT	2 610	79 088	351	179	30
1951–76	Australia	51 443	594 139	727	598	11
	Western NSW	4 017	51 233	434	295	12
	Eastern NSW	7 014	71 743	482	339	10
	Vic.	7 265	125 155	448	313	17
	Qld	6 717	135 340	599	427	20
	SA	12 847	89 083	442	308	6
	WA	6 420	65 794	484	320	10
	Tas.	4 308	33 239	271	146	7
	Southern NT	952	6 619	229	95	6
	Northern NT	1 903	15 933	328	163	8
1901–50	Australia	34 652	149 607	680	590	4
	Western NSW	905	4 730	355	233	5
	Eastern NSW	4 177	17 616	428	320	4
	Vic.	4 603	20 472	406	294	4
	Qld	6 291	27 291	527	412	4
	SA	7 169	42 361	387	299	5
	WA	7 058	20 713	448	313	2
	Tas.	2 455	7 706	225	138	3
	Southern NT	478	2 176	185	70	4
	Northern NT	1 516	6 542	310	205	4
Before 1901	Australia	15 294	30 432	635	523	1
	Western NSW	787	1 300	264	162	1
	Eastern NSW	2 386	5 028	383	288	2
	Vic.	2 394	4 993	360	259	2
	Qld	3 660	8 823	493	334	2
	SA	1 417	2 839	304	207	2
	WA	2 201	3 406	366	199	1
	Tas.	1 602	2 445	181	125	1
	Southern NT	229	399	106	47	1
	Northern NT	618	1 199	224	46	1

mapped were excluded, e.g. a historical record at 32°51′ 147°25′, block size 50′ would be printed on a 1° map in the 1° block whose centre point was 32°30′ 147°30′ but excluded from a 10′ map. The assigned block size was an estimate of the accuracy with which the location was defined. The maximum allowed was 2°.

Records which could not be dated were omitted.

Results

Statistics

Between 1977 and 1981 records were received from a total of 885 different 1° blocks. These included the major land masses in Australia as well as many smaller islands but some contained only sea. In calculations, the 812 blocks that cover the mainland, Tasmania and large islands are taken to constitute 'Australia'. Table 1 summarizes by recording regions the number of record sheets submitted and species recorded in Field and Historical Atlases.

Figs 7–13 summarize in different ways the coverage achieved during the Field Atlas. Maps for individual species can be assessed by comparison with them. The number of record sheets submitted for each 1° block is shown symbolically in Fig. 7; Fig. 8 shows similar data for 10′ blocks in Tasmania; Fig. 9, the number of species recorded in each 1° block; Fig. 10, similar data for 10′ blocks in Tasmania; Fig. 11, the number of species recorded breeding in each 1° block; Fig. 12, similar data for 10′ blocks in Tasmania. Corresponding figures for the Historical Atlas are shown on p. 660.

In Fig. 13 a-d the number of record sheets received

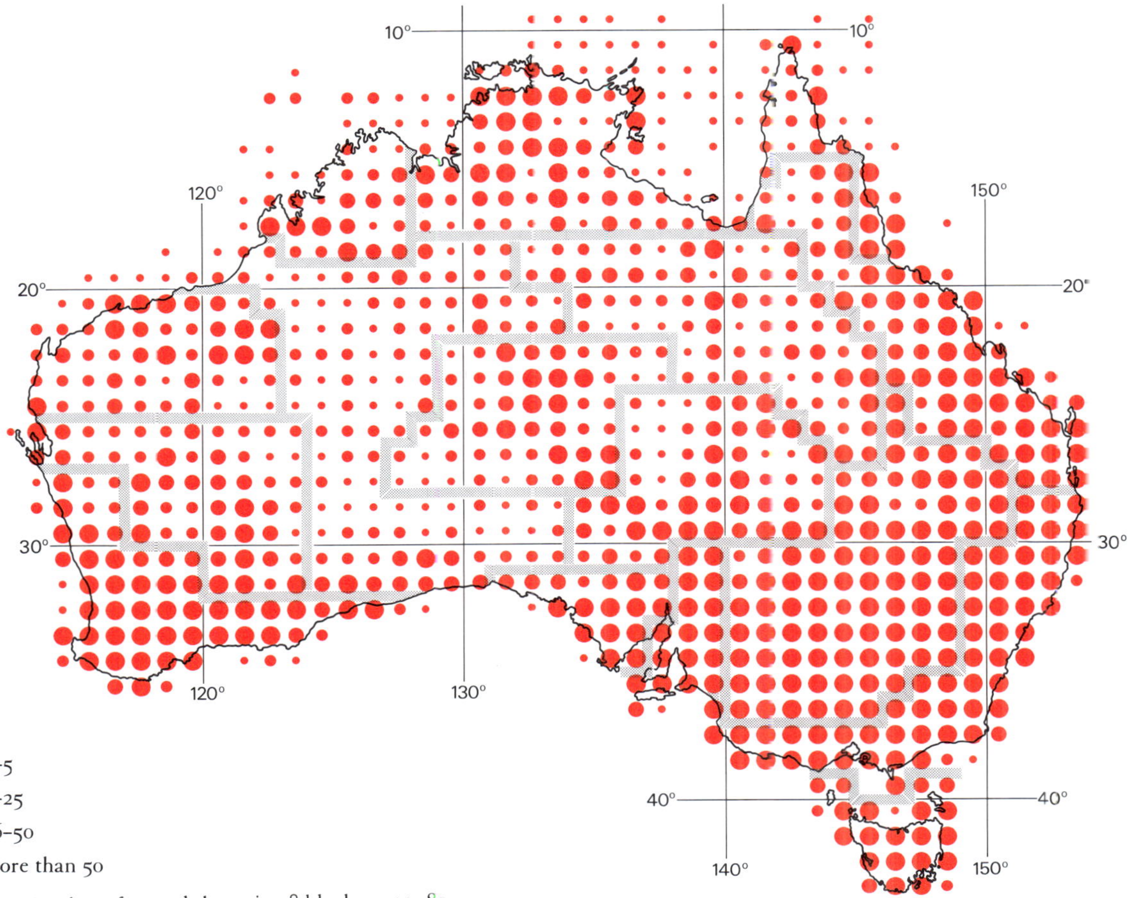

- 1–5
- 6–25
- 26–50
- more than 50

Fig. 7. Number of record sheets in 1° blocks, 1977–81

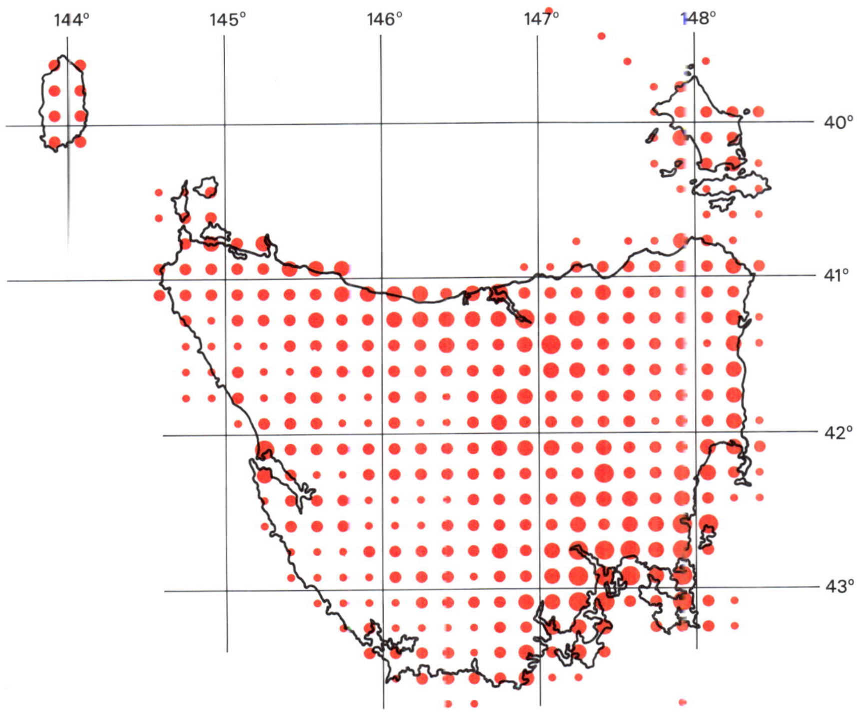

- 1–5
- 6–25
- 26–50
- more than 50

Fig. 8. Number of record sheets in 10′ blocks in Tasmania, 1977–81

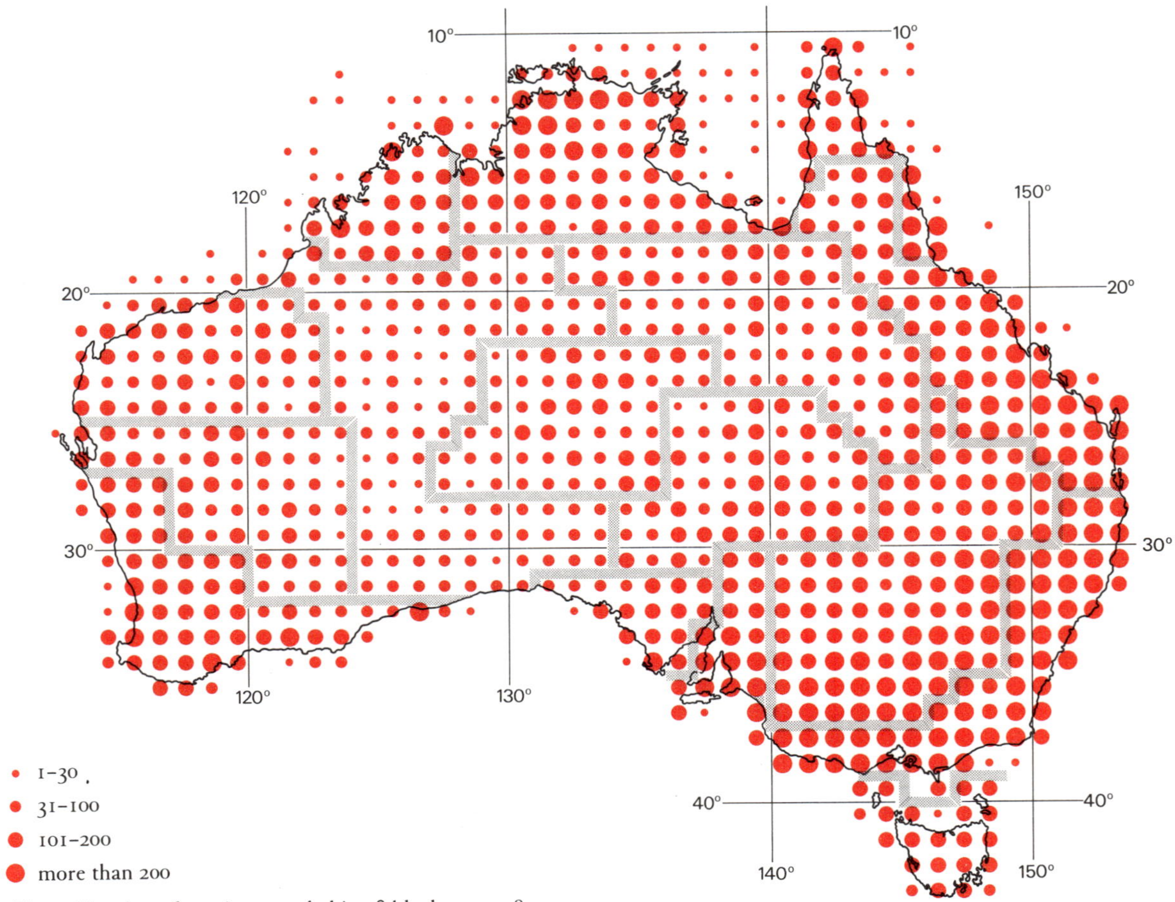

Fig. 9. Number of species recorded in 1° blocks, 1977–81

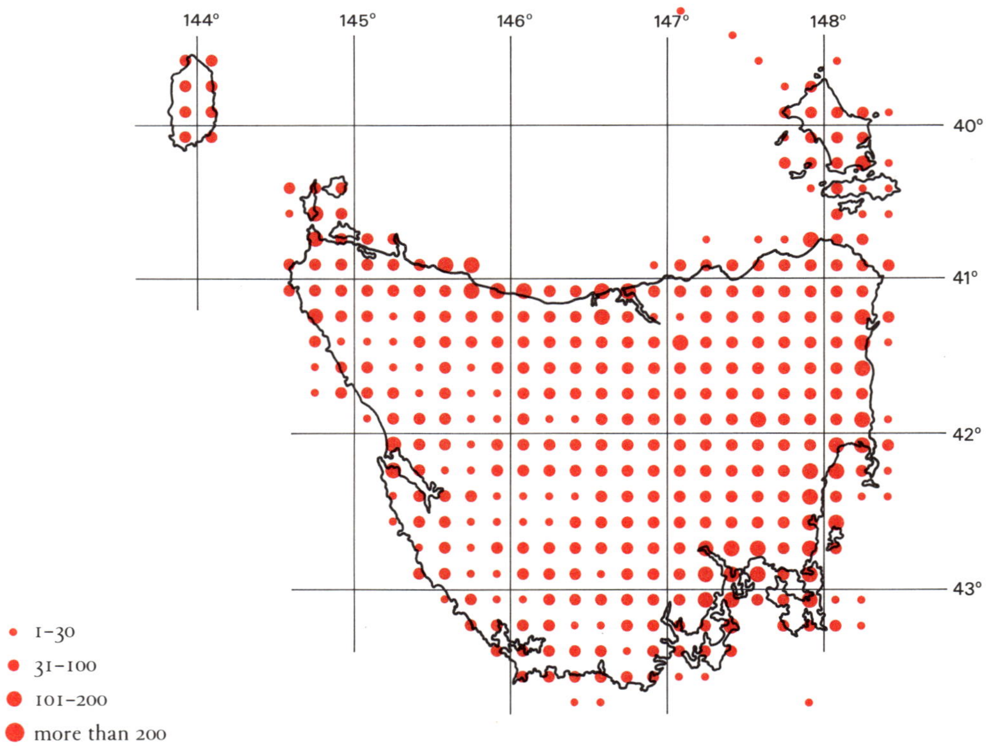

Fig. 10. Number of species recorded in 10′ blocks in Tasmania, 1977–81

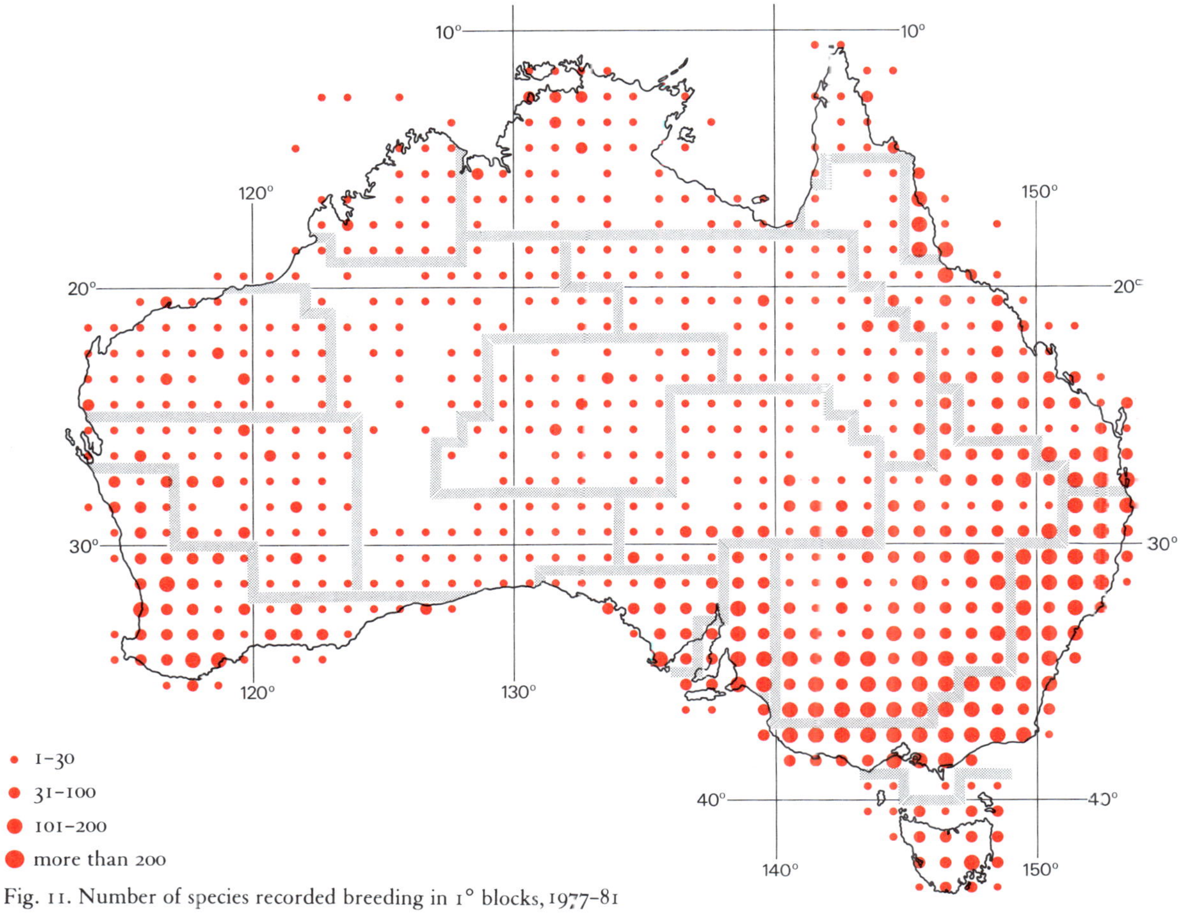

Fig. 11. Number of species recorded breeding in 1° blocks, 1977–81

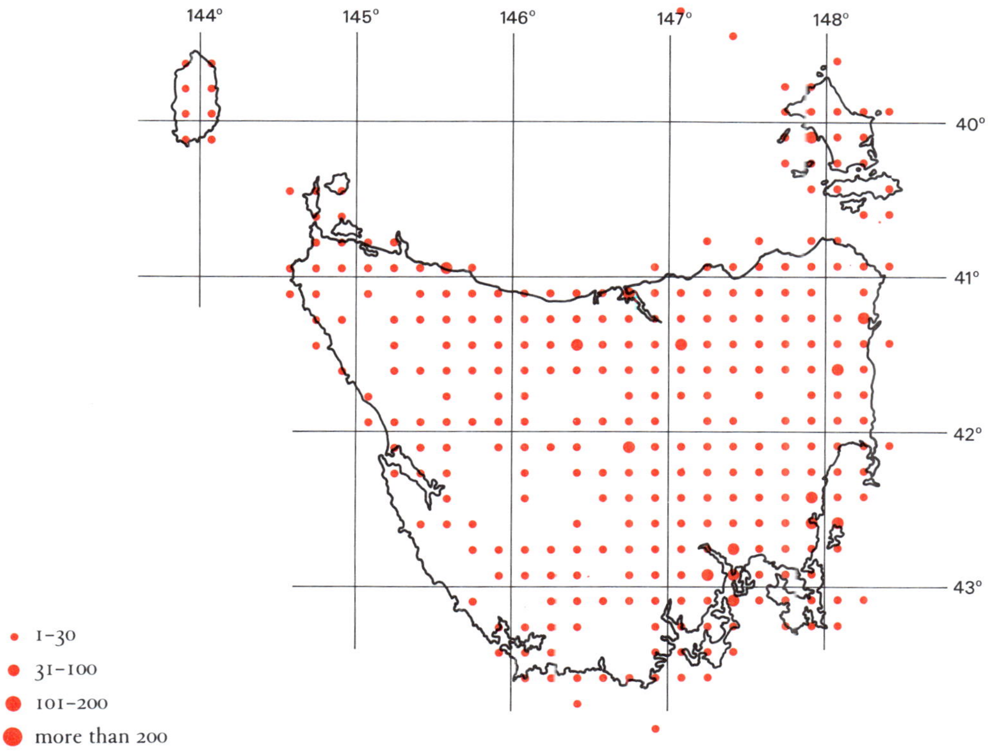

Fig. 12. Number of species recorded breeding in 10′ blocks in Tasmania, 1977–81

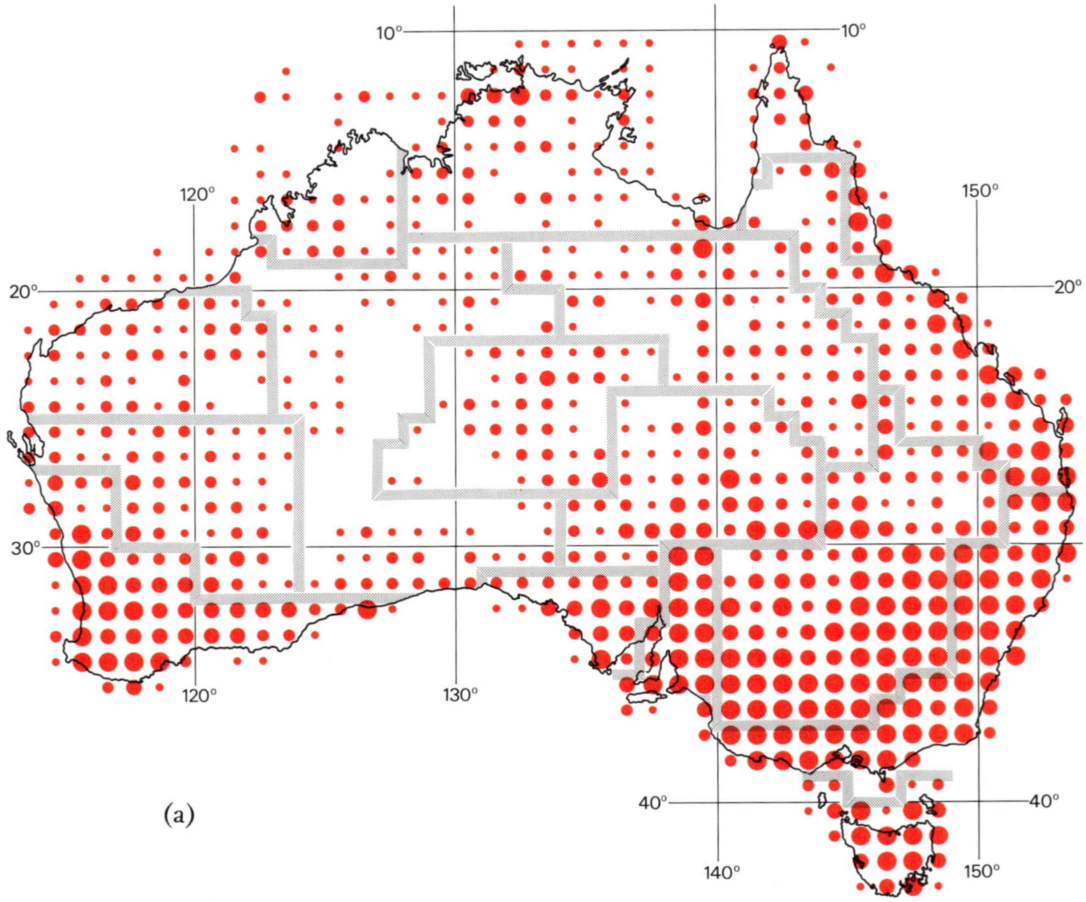

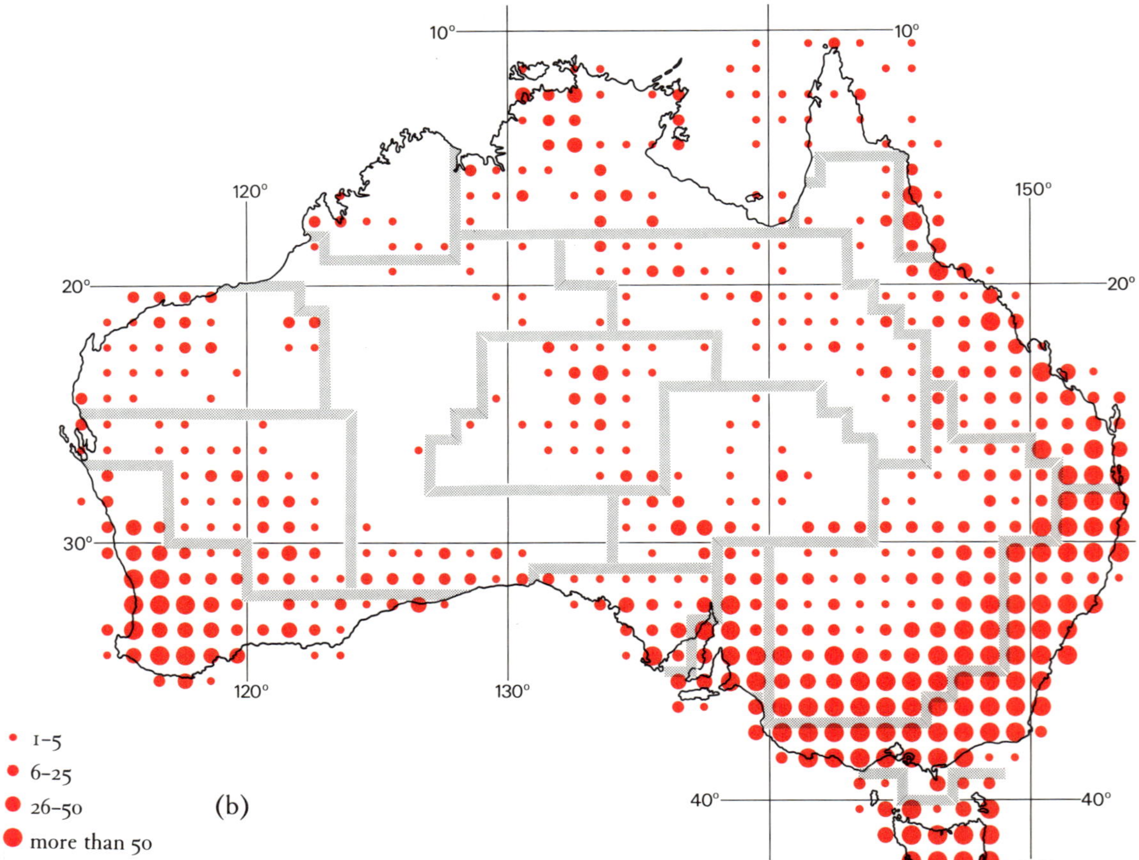

Fig. 13. Number of record sheets in 1° blocks in (a) spring, (b) summer, (c) autumn and (d) winter

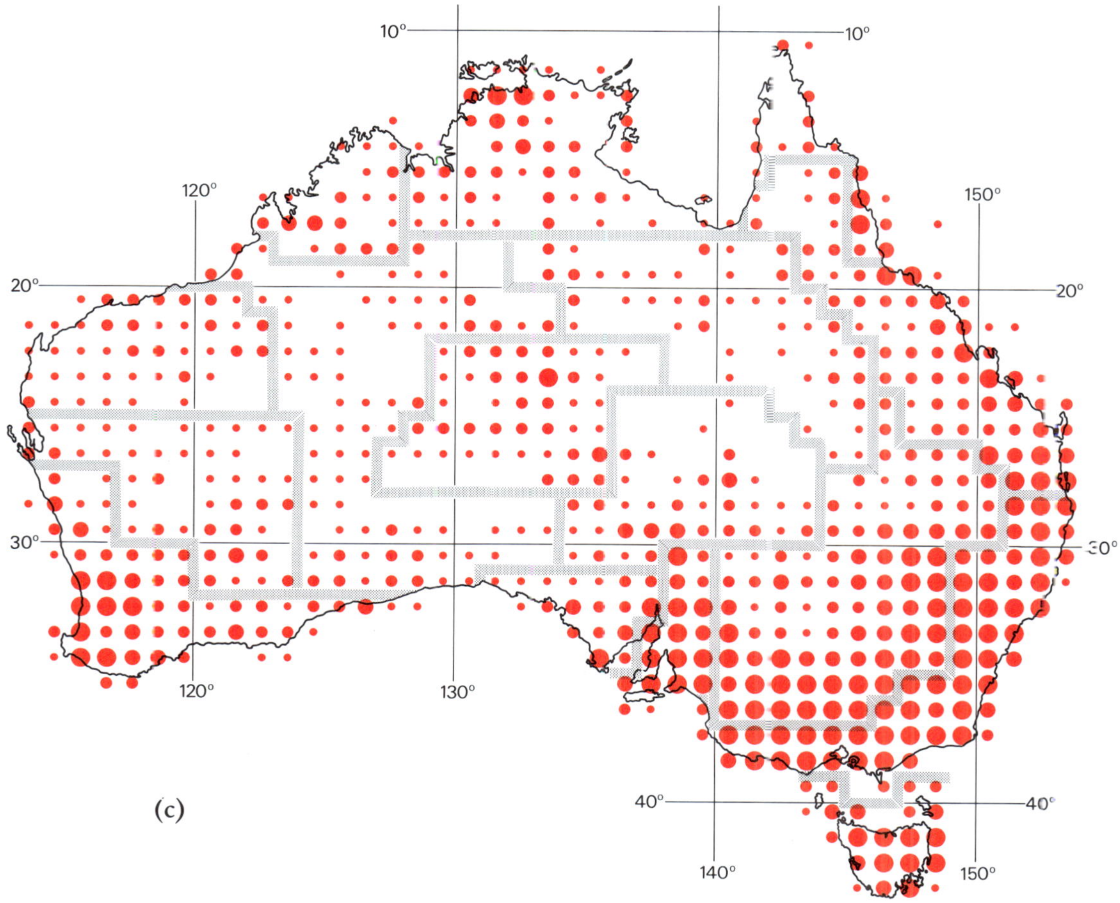
10°
10°
120°
150°
20°
20°
30°
30°
120°
130°
40°
40°
140°
150°

(c)

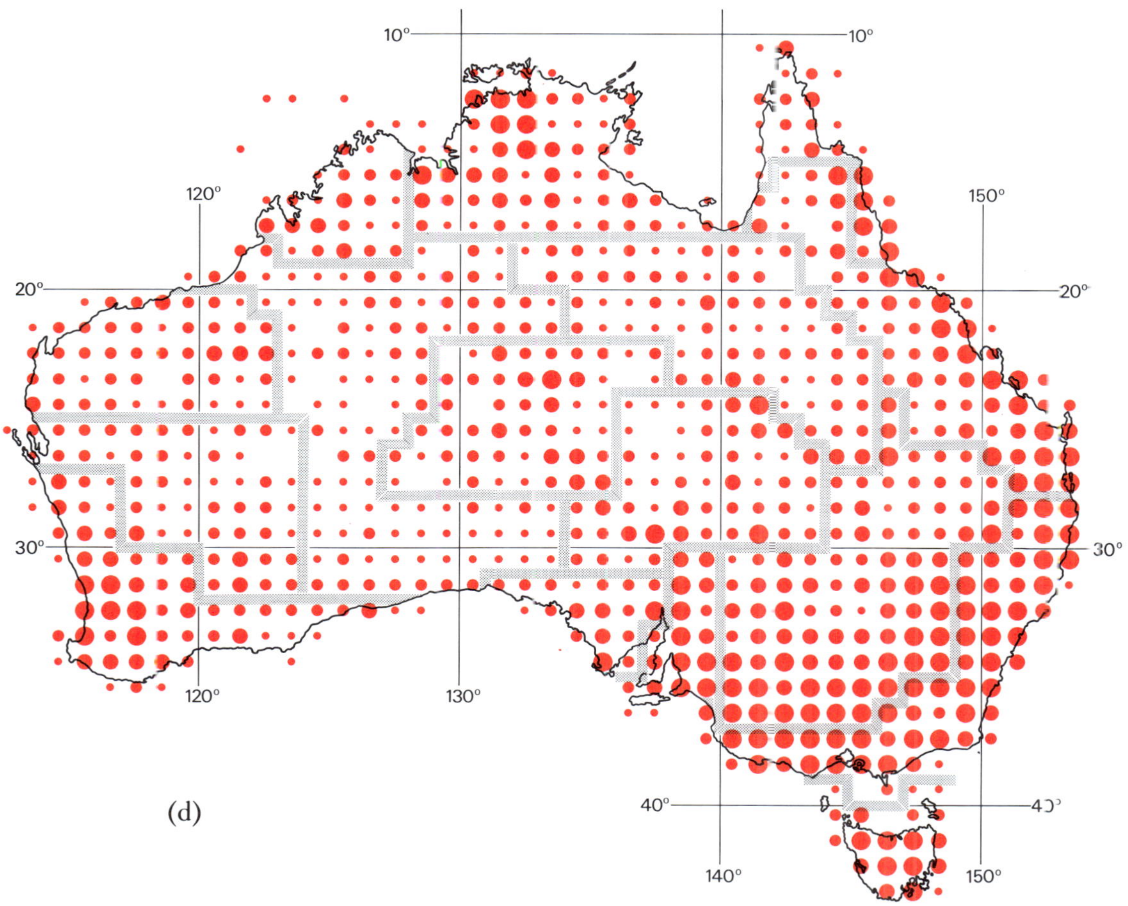
10°
10°
120°
150°
20°
20°
30°
30°
120°
130°
40°
140°
150°

(d)

for each 1° block in each season during the Field Atlas is shown. Record sheets that overlap into two or more seasons are omitted.

The greatest number of species recorded in any 1° block during the Field Atlas was 365 in the block (27/153) in which Brisbane lies. Breeding was less well recorded and the number of species reported breeding in any 1° block represents the effort put into finding them rather than the number that might actually breed there, e.g. in Brisbane only 137 species were recorded breeding. The greatest number of species reported breeding in a 1° block was 188 in 36/146, which includes Albury-Wodonga and part of the Murray R, in a total of 265 species recorded.

During the Field Atlas, there were breeding records from 88% of 1° blocks but many of these contained reports of fewer than ten breeding species. For non-passerines, excluding non-breeding migrants and vagrants but including species that breed colonially, breeding records were, on average, received from 21% of 1° blocks in which the species was recorded. For passerines the figure was 30%.

The number of 1° blocks in which a species was recorded is an indication of how widespread it is. In the Field Atlas the top ten in order were:

Species	Number of 1° blocks with records
Brown Falcon	769
Australian Kestrel	768
Willie Wagtail	762
Black-faced Cuckoo-shrike	756
Richard's Pipit	720
Wedge-tailed Eagle	702
Galah	678
White-winged Triller	676
Australian Magpie-lark	674
Black-faced Woodswallow	672

The species recorded breeding in the greatest number of 1° blocks were not necessarily the most widespread but rather those most easily proved breeding. The top ten were:

Species	Number of 1° blocks with breeding records
Australian Magpie-lark	340
Willie Wagtail	313
Australian Magpie	305
Galah	271
Striated Pardalote	269
Fairy Martin	267
Black-faced Cuckoo-shrike	265
Emu	258
Welcome Swallow	254
Australian Kestrel	247

The most commonly recorded species during the Field Atlas (as opposed to most widespread) were those that live in or near settlements (where most atlassers live) and are abundant, conspicuous and easy to identify. The top ten in each region and in Australia as a whole are listed in Table 2.

The number of people recorded as contributors to the Atlas was 2969. They are listed on pp. 683-91. Many of these names represent families or groups such as bird clubs or schools. In other cases observers passed on their information to atlassers without themselves being registered. The total number of contributors was probably well over 3000. The number of record sheets sent in by individual atlassers varied from 1 to 1669. Of the recorded contributors, 66% sent in up to 20 record sheets each and 8% sent in more than 100.

Coverage

There is no direct way of knowing how many species might be expected in an area. The number of species actually recorded during the Field Atlas cannot therefore be used to assess how well covered were different parts of Australia. Instead, an attempt was made to assess coverage by considering the number of record sheets received for a given time and area as a measure of the 'effort' expended in the survey: the greater the number of record sheets received, the greater the number of species recorded.

Australia was divided into 2° blocks and curves were generated of the number of species recorded against the number of record sheets received. Samples are shown in Fig. 14. When a large number of record sheets was received, the rate at which new species were added slowed, i.e. the curve generally flattened out. The curves did not appear to approach an asymptote—new species were still being added even in 2° blocks with up to 4000 record sheets. This meant that no 'maximum' possible number of species present in a 2° block could be calculated against which to compare the actual number recorded. Instead, 2° blocks that had been 'well-covered' were examined to determine the number of record sheets required to reach given proportions of the final number of species recorded.

Graphs of number of species against number of record sheets were obtained for the 251 different 2° blocks in Australia. Of these, 203 contained more than 20 record sheets. For each of these, the rate of addition of new species was calculated once 50% of record sheets had been plotted, i.e. on the second half of the curve. This rate varied from 0.02 new species/record sheet to 5.1 new species/record sheet.

The 50 2° blocks in which the rate was less than or equal to 0.1 new species/record sheet were considered to be 'well-covered'. In a sample of 11 of these, the number of record sheets needed to reach given percentages of the final number of species was calcu-

Table 2. Species most frequently reported during the Field Atlas in Australia as a whole and in recording regions

Recording Region	Species	Number of Records
Australia	Australian Magpie	52 945
	Willie Wagtail	48 133
	Australian Magpie-lark	47 200
	Welcome Swallow	39 074
	Black-faced Cuckoo-shrike	37 775
	Galah	37 359
	White-faced Heron	36 173
	Laughing Kookaburra	35 150
	Australian Kestrel	32 192
	Common Starling	31 764
Western NSW	Australian Magpie	7 437
	Galah	5 375
	Australian Magpie-lark	5 370
	Willie Wagtail	5 214
	Welcome Swallow	4 848
	Australian Raven	4 786
	Common Starling	4 764
	Laughing Kookaburra	4 682
	Australian Kestrel	4 481
	Yellow-rumped Thornbill	4 219
Eastern NSW	Australian Magpie	8 548
	Australian Magpie-lark	7 772
	Willie Wagtail	7 332
	Laughing Kookaburra	7 037
	Welcome Swallow	6 488
	Black-faced Cuckoo-shrike	6 232
	Galah	6 104
	Common Starling	5 928
	White-faced Heron	5 813
	Australian Raven	5 746
Vic.	Australian Magpie	13 348
	Welcome Swallow	10 875
	Australian Magpie-lark	10 742
	Willie Wagtail	10 553
	Common Starling	10 280
	Laughing Kookaburra	9 813
	White-faced Heron	9 623
	Superb Fairy-wren	9 566
	Masked Lapwing	8 847
	Grey Shrike-thrush	8 783
Qld	Australian Magpie-lark	9 056
	Australian Magpie	8 256
	Willie Wagtail	8 001
	Laughing Kookaburra	7 159
	Pied Butcherbird	6 920
	Australian Kestrel	6 036
	Masked Lapwing	5 724
	White-faced Heron	5 442
	Peaceful Dove	5 435
	Crested Pigeon	5 354
SA	Australian Magpie	7 751
	Galah	7 444
	Willie Wagtail	7 049
	Australian Kestrel	5 806
	Welcome Swallow	5 569
	Crested Pigeon	5 550
	Australian Magpie-lark	5 520
	Common Starling	5 374
	House Sparrow	5 082
	Richard's Pipit	4 652
WA	Willie Wagtail	6 804
	Black-faced Cuckoo-shrike	5 607
	Australian Magpie	5 568
	Australian Magpie-lark	5 542
	Australian Kestrel	4 824
	Richard's Pipit	4 677
	Australian Raven	4 455
	Singing Honeyeater	4 449
	Black-faced Woodswallow	4 241
	Welcome Swallow	4 164
Tas.	Forest Raven	4 273
	Masked Lapwing	3 473
	Superb Fairy-wren	3 458
	Common Starling	3 374
	Grey Fantail	3 221
	Green Rosella	3 133
	Welcome Swallow	2 958
	Grey Shrike-thrush	2 924
	Yellow-throated Honeyeater	2 663
	European Goldfinch	2 611
Southern NT	Willie Wagtail	918
	Australian Kestrel	810
	Black-faced Woodswallow	796
	Zebra Finch	793
	Brown Falcon	747
	Australian Magpie-lark	692
	Black-faced Cuckoo-shrike	675
	Crested Pigeon	653
	Yellow-throated Miner	626
	White-plumed Honeyeater	622
Northern NT	Australian Magpie-lark	1 501
	Peaceful Dove	1 410
	Rainbow Bee-eater	1 283
	Willie Wagtail	1 257
	Whistling Kite	1 228
	Red-collared Lorikeet	1 172
	Black Kite	1 165
	Black-faced Cuckoo-shrike	1 154
	Torresian Crow	1 134
	Blue-winged Kookaburra	1 110

lated. Table 3 gives the results. The median number of record sheets required to reach 50% of the final number of species was 25 and to reach 60% was 50. In all the well-covered 2° blocks, 50 was the maximum number of record sheets required to reach 50% of the final number of species recorded.

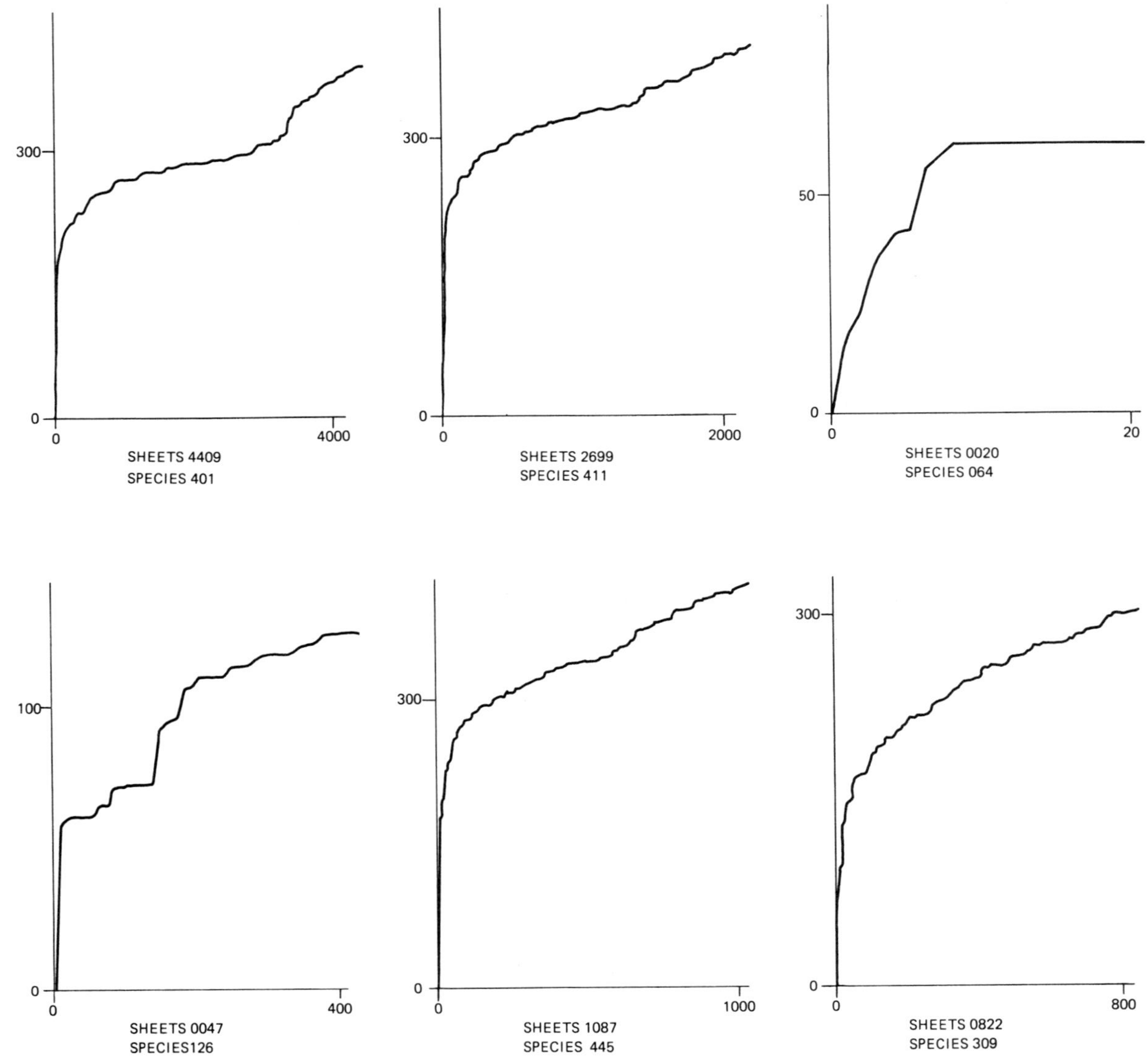

Fig. 14. Number of species (vertical axis) plotted against number of record sheets (horizontal axis) in 2° blocks. The grid blocks, identified by the latitude of their northern boundary and the longitude of their western boundary, are (left to right), top: 36/145, 34/149, 28/153; bottom: 20/131, 16/145, 12/131.

Table 3. Number of record sheets required to reach the given percentages of the total number of species recorded in 11 'well-covered' 2° blocks. Grid blocks are identified by their northern and western boundaries.

2° block	Total no. of species	Total no. of record sheets	No. of record sheets by which the given percentage of the total no. of species was reached			
			50%	60%	75%	90%
28/137	182	421	10	25	40	200
30/145	227	531	12	27	50	205
32/117	208	794	30	50	150	640
34/117	242	610	20	30	70	300
36/139	328	1031	12	50	225	625
36/141	308	1862	25	50	250	1500
36/143	370	2293	25	125	550	1875
40/145	203	1488	50	150	350	750
40/147	206	1245	25	75	225	650
42/145	182	1086	25	125	275	550
42/147	230	2020	25	150	450	1150
Median			25	50	225	640

Reporting Rate

The 'Reporting Rate' is the percentage of record sheets in the Field Atlas on which a given species was recorded. It is calculated for a selected area: an individual 10′ or 1° block or any combination of grid blocks including regions, States or Australia as a whole. For comparative purposes, record sheets may be grouped into different time periods, e.g. seasons, years, or the grid blocks may be selected to represent particular classes of country, e.g. different altitudes, coastal compared with inland grid blocks.

The Reporting Rate is determined by a variety of factors including the abundance and ubiquity of the species, its conspicuousness, accessibility of its range and ease of identification. Because of this, comparison of Reporting Rates between species is generally not justifiable except in special cases such as closely-related allopatric species. For an individual species, large differences in Reporting Rate from one area to another are assumed to be primarily related to differences in abundance and ubiquity. This assumption is not justified if the species' behaviour differs in such a way as to affect its conspicuousness, e.g. one that flocks in parts of its range but not in others. This lack of comparability is emphasized when Reporting Rates are calculated for each season because some species behave differently at different times of the year. Cuckoos call and are therefore conspicuous and frequently recorded in the breeding season but are silent at other times of the year. Robins move into more open country in winter and are more conspicuous then than in the breeding season.

Because Reporting Rates represent ubiquity as well as abundance, the area suitable for a species within a grid block will affect the rate. This is particularly obvious on the coast where seabirds will generally have a high Reporting Rate in a 1° block with a given length of coast and little land but a lower Reporting Rate in another 1° block with the same length of coast but a greater land area. Many of the biases introduced by observer behaviour, which affect distribution (see below), will also affect Reporting Rates.

The method used to calculate Reporting Rates for the maps in this book is given on p. 679.

Biases

The Atlas distribution maps incorporate a number of biases that need to be taken into account when interpreting them. First, the Field Atlas consists of records collected during a five-year period when particular weather and other environmental conditions prevailed. Distribution records collected at any other time will differ from those in the Field Atlas to the extent that environmental variables have changed. Second, most of the distribution maps are incomplete: not every species present in every grid block was recorded. A means of assessing the coverage is provided by the number of record sheets submitted (p. xxvii).

Breeding was poorly recorded for most species, specially in the northern half of Australia (see Fig. 11). No map of breeding distribution should be considered complete.

Several aspects of observer behaviour have affected the maps to varying degrees. Some areas, specially inland, were visited once only or for a short time, not necessarily when the majority of species would have been present. Parts of the country were consistently surveyed in some seasons only, e.g. much of northern Australia was not visited at all in the wet summer season (see Fig. 13b) leading to under-recording of summer migrants. Some areas of a 1° block were consistently visited and others avoided. Thus, in the inland, observers tended to make for any site where there was water in preference to spinifex plain or other inhospitable or inaccessible parts of the grid block.

The distribution shown for some rare species is biased because atlassers went to locations where they knew the birds were to be seen. Many people, for instance, visited Iron Ra. (12/143) and recorded species endemic to the area such as Eclectus and Red-cheeked Parrots but few made the more difficult journey to the Rocky R (13/143) where both species are probably just as common.

The Historical Atlas suffers from the same biases as the Field Atlas but they are accentuated because of the *ad hoc* nature of the records. There is possibly an additional factor in that common species may have been ignored in the literature and museum collections in favour of more unusual species. Comparison of Historical and Field Atlas maps needs to take into account this factor as well as the different levels of coverage and the different lengths of time of the four periods. The longer time span of the earlier historical periods increases the likelihood that relatively rare events, such as irruptions or visits by vagrants, will happen within them.

Access to Information

This book contains only part of the information that may be extracted from the Atlas data bank. Additional material is being prepared for publication in the form of regional atlases. Other data are available to anyone interested. Examples of types of information that may be useful for particular purposes include: maps at the 10′ scale, seasonal maps, lists of species recorded in a given area, access to part or all of the data bank to carry out biogeographical and other analyses, details of unusual records and so on. The Historical Atlas can act as an index to the literature, museums and other sources and may be sorted by species, time period, source, observer or location.

To obtain access to Atlas information, contact the RAOU (GPO Box 5236BB, Melbourne, 3001, Australia). Charges will be levied to cover the cost of extracting data.

The Next Atlas

Co-operative collection of data could be used to determine the distribution of a variety of organisms in Australia. When sufficient time has elapsed, another bird atlas might be organized to identify changes in distribution since this one.

Organization of the Atlas of Australian Birds was difficult because Australia is a large country with many bird species but a limited number of birdwatchers, most living in the major cities. Lack of funds also determined some aspects of the organization. The strategy adopted aimed to decentralize administration by using volunteer Regional Organizers and to use a computer to sort and store data. Except for rare species, vetting was limited to records outside the expected range of the bird. This system worked reasonably well but the administrative load of Regional Organizers and Atlas staff was heavy, particularly that involved in verifying records. It left little time for developing contacts with atlassers and keeping them fully informed, which, in turn, would have improved the Atlas.

Simon Bennett (RAOU) has developed a system that uses a micro-computer to handle many of the administrative chores in running an Atlas. It depends on a comparison of the 'observed' range of a species with the 'expected'.

Before data collection begins, the expected range of each species is defined (by 'experts', from published information etc.). Grid blocks outside this range are flagged to show that a record of the species in them requires verification. The requirement for verification may be refined to apply only to certain records, e.g. records in spring or breeding records.

Observers supply records in a computer-compatible medium. Each species recorded is compared with the expected range and, if verification is required, a letter to the observer is automatically generated giving details of the sighting and requesting substantiation. When the URRF is returned and assessed, the data bank is amended according to whether the record is accepted, rejected or made doubtful.

In the data bank, a summary of the number of records of each species in each grid block is maintained in whatever categories are required, e.g. present, breeding, confirmed, doubtful, seasonal. In time, enough observations will have been collected to enable the expected range to be altered. If sufficient confirmed observations of a species have been received from a grid block outside the previously expected range, then the expected range can be enlarged and verification is no longer required for records in that grid block. Alternatively, the expected range may have included grid blocks where the species is in fact rarely recorded. Grid blocks where the Reporting Rate is less than a selected threshold can be excluded from the expected range and query letters generated for any earlier observations and all subsequent ones. In this way the expected range is amended to fit the observed range.

The Book

Taxonomy

For all but a few species the taxonomy and nomenclature of the list of *Recommended English Names for Australian Birds*[1581] have been followed. Exceptions have been made for some species that are difficult to identify in the field or whose taxonomic status is so disputed that it seemed better to publish one combined map. They are:

Emu and Dwarf Emu
Broad-billed, Lesser Broad-billed and Antarctic Prions
Fulmar and Fairy Prions
Fluttering and Hutton's Shearwaters
Swamp and Brown Quails
Latham's, Pin-tailed and Swinhoe's Snipe
Great and South Polar Skuas
Little and Gould's Bronze-Cuckoos

For the Crimson, Yellow and Adelaide Rosellas there is only one text with the maps of the three birds.

The English and scientific names appear with the vignette and on each map the Atlas number appears at the left. Bird names used during the Atlas are listed in the index (pp. 730-8).

Species

Most species have been allocated a page containing a map, text and vignette. All Australian breeding species (except some introduced birds with limited ranges, e.g. Feral Chicken) and regular migrants have been so treated making a total of 656 species (including combined maps and texts). There are historical maps for 38 of them on pp. 660-70. A further 102 uncommon and vagrant species are treated briefly on pp. 651-9.

Maps

The Field Atlas map of each species (pp. 2–648) shows the grid blocks in which it was recorded (1° blocks for most species; 10′ blocks for the 11 Tasmanian endemics). An open circle indicates that the bird was reported one or more times and a closed circle that it was found breeding at least once. The circles are placed in the centre of the grid block.

The size of the circle in each grid block shows how frequently the bird was recorded (Reporting

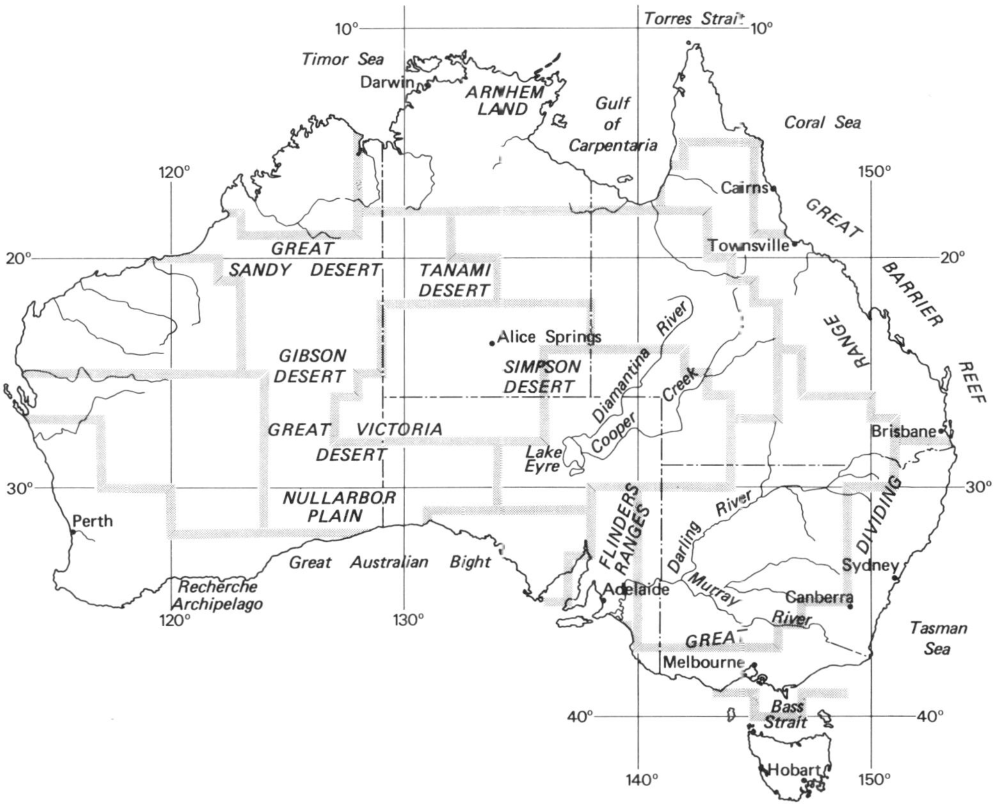

Fig. 15. Major geographical features of Australia

Rate: p. xxxv). The larger the circle the higher the Reporting Rate. The maximum Reporting Rate (Max. Rep. Rate) for the species is shown to the left of the map together with the grid block or blocks in which it occurred. All grid blocks with Reporting Rates greater than 40% of this maximum are allocated large circles; those with Reporting Rates between 11% and 40% of the maximum have medium-sized circles, and those with Reporting Rates less than 11% of the maximum have small circles. The method of calculating the Reporting Rates is given on p. 679.

The Historical Atlas maps (pp. 660-70) indicate only the grid blocks in which the species was recorded (open circles) and those in which it was found breeding (closed circles). The Reporting Rate does not apply to historical records.

Records considered doubtful, including some previously published, have been omitted from the maps. Reports of presumed escapees have been omitted, unless a feral breeding population is known to have been established.

The outline maps of Australia and Tasmania on which the species' maps are superimposed use an equal area projection. On the 1° maps, lines of latitude and longitude have been drawn at 10° intervals and, on the 10′ maps of Tasmania, at 1° intervals. Maps are subdivided into Atlas Regions (p. xl, Fig. 21) that represent bioclimatic zones. Fig. 15 shows the main geographical features mentioned in the text.

Overlays are available showing those environmental features mapped in Figs 15 to 18 (geographical features, rainfall, altitude, vegetation).

Text

The text for each species is intended to explain and complement the map. It deals mainly with range and movements but also identifies the species' known requirements and, if possible, factors on which these depend. It does not attempt to deal comprehensively with the bird's biology though many of the works to which reference is given do so. The subjects are always covered in the text in the same order though many categories are blurred and run into each other.

RANGE. Range outside Australia is indicated on a broad scale. Close to Australia, broad groupings are used for convenience and these may be separated into smaller groupings or into individual places, e.g. south-eastern Asia includes Burma, Malaysia, Indonesia, Philippines; Indonesia includes Sumatra, Java, Sunda Is, Borneo, Sulawesi (Celebes), Timor. New Guinea is always mentioned as a separate entity and includes all geographical New Guinea. Islands adjacent to New Guinea include Aru, Moluccas, Bismarck Archipelago.

The Australian range will be evident from the maps and will not be mentioned in the text unless there is good reason for it, such as historical comparison or to draw attention to gaps. Subspecific ranges are given because they often coincide with divisions apparent on the maps and are relevant to discussion of the distribution and behaviour of populations within a species. Confusing species are mentioned where these may cause problems in defining range.

MOVEMENT. Sedentary, nomadic and migratory species are identified where possible. A species was regarded as sedentary only on the basis of studies of marked birds; in many cases species are known to be present in a locality throughout the year but there is no information to show that the same individuals remain. Evidence about movements derived from analysis of Atlas data is included, particularly changes in Reporting Rate in an area from season to season. Altitudinal Reporting Rates were calculated by selecting 10′ blocks, each of which fell mostly into one of five altitude classes: 0–200, 200–500, 500–1000, 1000–1500 and above 1500 m. These 10′ blocks were not necessarily adjacent but were not farther away than in the next 1° block. Seasonal Reporting Rates were calculated on a 10′ block basis but have sometimes been averaged over two or more altitude classes or seasonal classes.

ENVIRONMENT AND REGULATING FACTORS. This includes the type of country where the species lives; whether it is solitary or in groups; communal and colonial breeding; presence of vegetation necessary for feeding or breeding; competition; predation; and the effects of European settlement.

STATUS. If there is clear evidence of a change in abundance, compared with the past, this is stated.

ABUNDANCE. Density measures or population counts are given where possible as an indication of the numbers of the species that might be expected in different areas and for comparison with the Reporting Rates on the map. Some density measures are assessed from short visits and others from long-term studies; some during breeding, others throughout the year.

To enable places to be located quickly on the map, a grid reference in brackets follows the first mention of the place in each text. The grid reference defines the 1° block in which the place is situated, giving the latitude and longitude of its northern and western boundaries. Thus 'Darwin (12/130)' means that Darwin lies in the 1° block whose centre point is 12°30′S 130°30′E. Grid references for places outside Australia, other than countries, have north or south and east or west indicated.

Atlas Regions (p. xl) and localities are usually covered in order from north to south, then from east to west, i.e. Cape York Region south to the Tas. Region, Top End Region south to the SA Ra. Region, Kimberley Region south to the South-West Region.

Abbreviations are listed on p. xlv; the scientific names of plants are on p. 698; and some specialized terms are listed in the Glossary (p. 694).

The text attempts to bring together references which may be consulted for more information. References are indicated by a small printed number in the text and are listed on pp. 701-29. Unreferenced matter is derived from Atlas sheets and URFFs and is on file at the RAOU.

At the foot of each text is a summary of the results for each species. It gives the total number of records of the species received during the Field Atlas. On the next line is the number of 1° blocks in which the species was recorded, followed in brackets by the percentage that this represents of the 812 1° blocks that cover Australia. The percentage is omitted for some seabirds for which it would be an unrealistic calculation. On the third line is the number of 1° blocks in which breeding was reported, followed in brackets by the percentage that this represents of the 1° blocks in which the species was recorded. The equivalent information is given for the Tasmanian endemics in 10′ blocks.

Vignettes

The purpose of using vignettes was twofold: first, to provide a pleasing complement to the text and an accurate representation of each species; and secondly, to involve as many bird artists as possible, specially those not yet established.

More than 100 artists responded to advertisements and submitted examples of their work, indicating preferences for specific birds. The Vignettes Committee (p. 681) assessed the submissions and about half the applicants were allotted species.

The size and requirement to draw in a black and white medium were strictly specified but design and style were left to the artist. All drawings had to be original. Finished work was examined for artistic merit and accuracy. Inevitably, the medium did not suit all artists. Others, though producing pleasing drawings, were unfamiliar with the anatomy of a bird and with the essential characteristics or 'jizz' of their species. The Vignettes Committee reluctantly rejected some submissions and the number of artists whose drawings were finally accepted was reduced to 23. The same fee was paid for each drawing.

Some species, mostly rare, had not been sought by any artist. Fortunately for the *Atlas*, some artists were prepared to expend a great deal of effort in examining skins and photographs to complete the 646 vignettes required. A few drawings had been completed before a decision was made to reduce some species from full to short texts.

The names of the artists and the pages on which their drawings appear are listed on pp. 692-3.

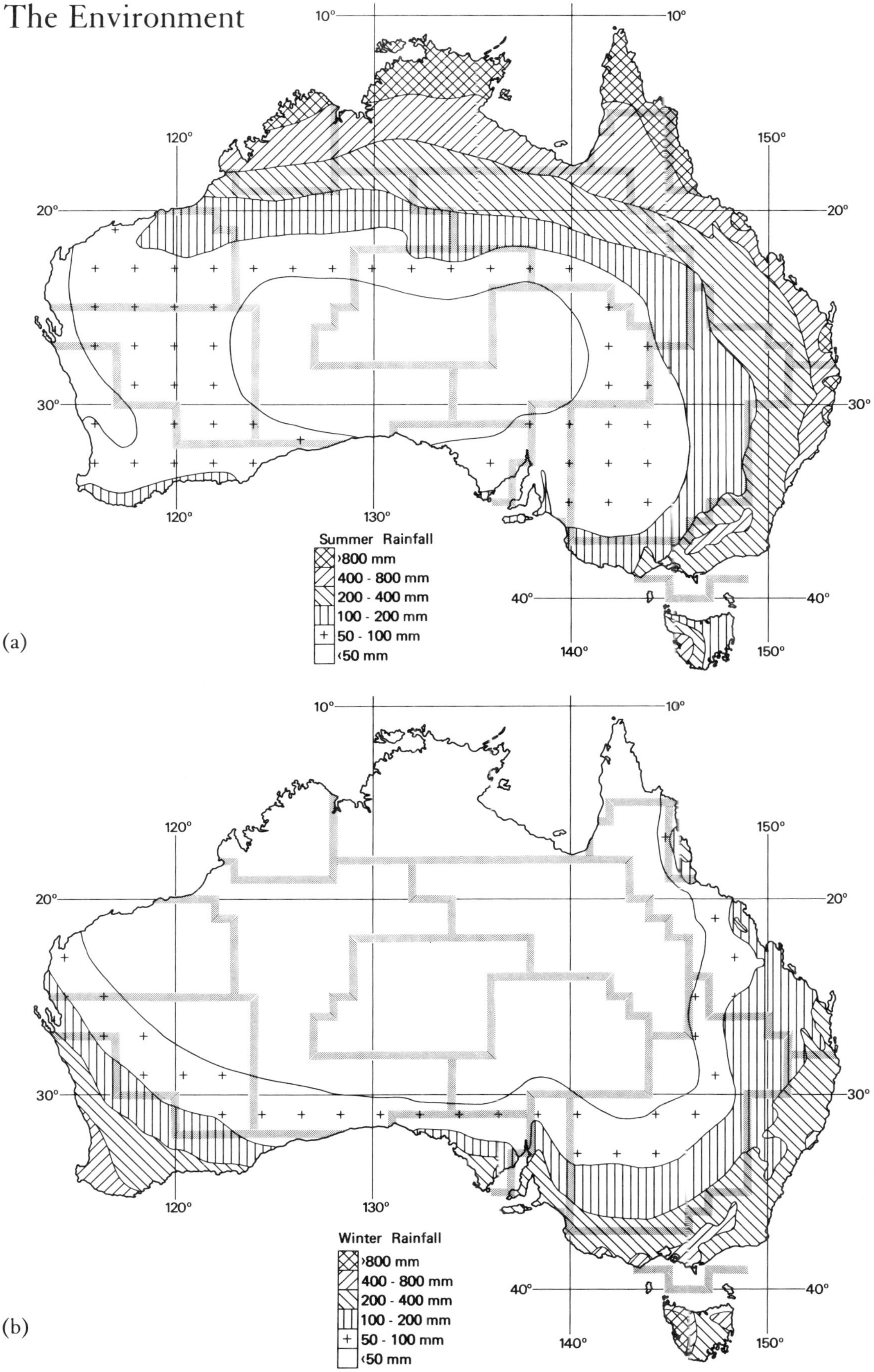

Fig. 16. Average annual rainfall received in eight years out of ten in (a) summer and (b) winter. After Nix and Austin.[1333]

Climate

Australia lies between 9°S and 44°S. Its ancient land surface, eroded in many places to a flat alluvial plain, is devoid of large, high mountain ranges that trap rainfall and spawn great rivers. The highest range, the Snowy Mts, does not rise above 2300 m. The climatic consequence of this physiography is described in the accompanying rainfall maps (Fig. 16 a, b). Instead of depicting annual rainfall, they show it separately for summer and winter.[1333] The isohyets join places receiving more than a specified amount of rain in eight years out of ten. This method of presenting the rainfall maps smoothes the curves by making the assumption that most vertebrates, as well as many invertebrates and plants, live more than a year and populations are able to survive droughts. So long as seasons with low rainfall are not too frequent, the survivors will breed up in good seasons and the population will not die out completely in dry seasons. Eight years out of ten is a level that can be tolerated by many bird species, accepting that some are adapted to arid and some to wet environments.

The maps show that northern and eastern Australia receive most rain in the summer, southern and western Australia most rain in the winter. South-western Australia receives little rain in summer and a huge area of northern Australia little rain in winter. Inland Australia receives little rain but may have falls in summer or winter. Eastern and particularly north-eastern Australia are areas of high rainfall. These characteristics are reflections of the latitudinal position of Australia, which means that the north is influenced by tropical summer cyclones and the south by winter storms from the Antarctic. Eastern Australia has, in addition, most of the significant mountain ranges (Fig. 17) and these promote extra rain.

Vegetation

Fig. 18 is a simplified map of the natural vegetation of Australia. Vegetation reflects not only the pattern of rainfall but also the distribution of soil types and physiography. Before Europeans arrived, the wet fringes of Australia were clothed in forest. In the less wet areas woodland dominated, giving way gradually to mallee vegetation in the dry inland. The low rainfall areas were covered with grasslands and shrublands. Three genera dominate the Australian vegetation: *Eucalyptus* in the forest and woodland; *Acacia* in the shrublands and *Triodia* (spinifex) in the grasslands. Only on the northern and eastern coasts are there patches of broadleaved forest that in very wet places may be called rainforest but are sometimes described as monsoon forest.

Three works classify the vegetation of Australia.[1695,1696,1896] Ornithologists face difficulties in using these classifications unchanged because floristics as well as structure and canopy density are important determinants of the suitability of the environment to bird species. Descriptive names for vegetation associations that are used and recognized by ornithologists are shown in Table 4, together with the approximate equivalents in published classifications.

Atlas Regions

Some means of dividing Australia into regions was needed to enable comparison of the distribution of birds and to study seasonal shifts of populations. Traditionally the country was divided into Eyrean, Torresian and Bassian regions when discussing avian distribution.[1636] A bioclimatic classification of Australia confirms the reality of the earlier division and identifies two zones of overlap where species diversity is enhanced (Fig. 19).[1332] Another recent definition of faunal regions identifies five zoogeographical barriers (Fig. 20), important in the distribution of birds.[1588] These two classifications provided a basis from which the Atlas Regions were developed with boundaries that followed lines of the 1° grid (Fig. 21). They are outlined below.

Cape York. High summer rainfall and dry winter, with rainforest near the coast. Many endemic bird species and subspecies with close relatives in New Guinea. Abundant water.

Atherton. A plateau sloping steeply to the coast, characterized by high summer and winter rainfall. Many endemic bird species and subspecies. Rainforest at all altitudes. Abundant water.

Eastern Qld. A heterogeneous region of land receiving much rain in summer and little in winter; warm throughout the year. Some patches of rainforest but mainly eucalypt forest and woodland. Extensive mangroves. Stock waters widespread.

South-East. Temperate region. Narrow coastal lowlands rising to Great Dividing Ra. Rain in winter and in summer. Characterized by eucalypt forest and woodland; some temperate rainforest and alpine conditions in the Snowy Mts (36/148). Abundant natural and artificial water.

Tas. An isolated region. Includes the islands of Bass Strait. Many endemic bird species and subspecies. Predominantly eucalypt forest; some temperate rainforest in the west; heath (buttongrass) in the south-west. Abundant natural water.

Murray-Darling. Catchment of Australia's two longest rivers. Watercourses lined with eucalypt woodland; patches of mallee in the south; elsewhere acacia scrub and saltbush. Low rainfall; most in winter. Stock waters widespread.

Top End. Sandstone escarpment (Arnhem Land) in north-east. Summer rainfall with pronounced dry winter season. Mangroves, coastal swamps and patches of rainforest in the north; predominantly

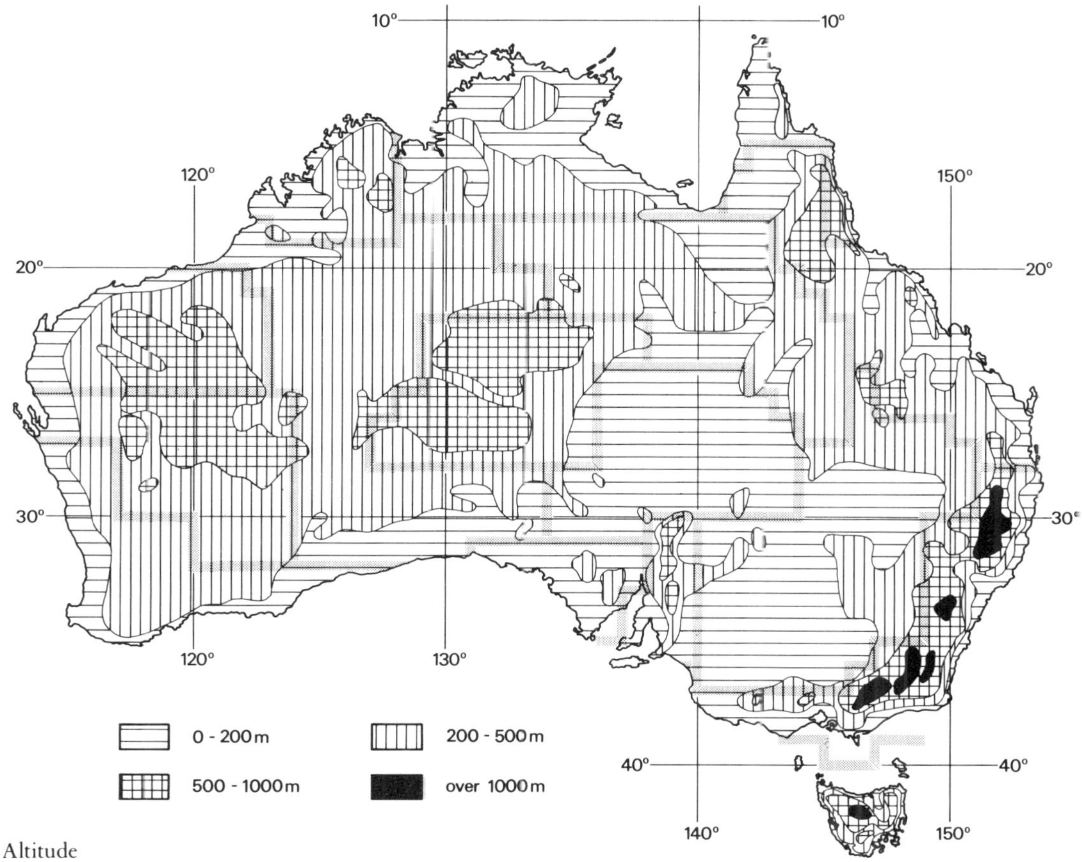

Fig. 17. Altitude

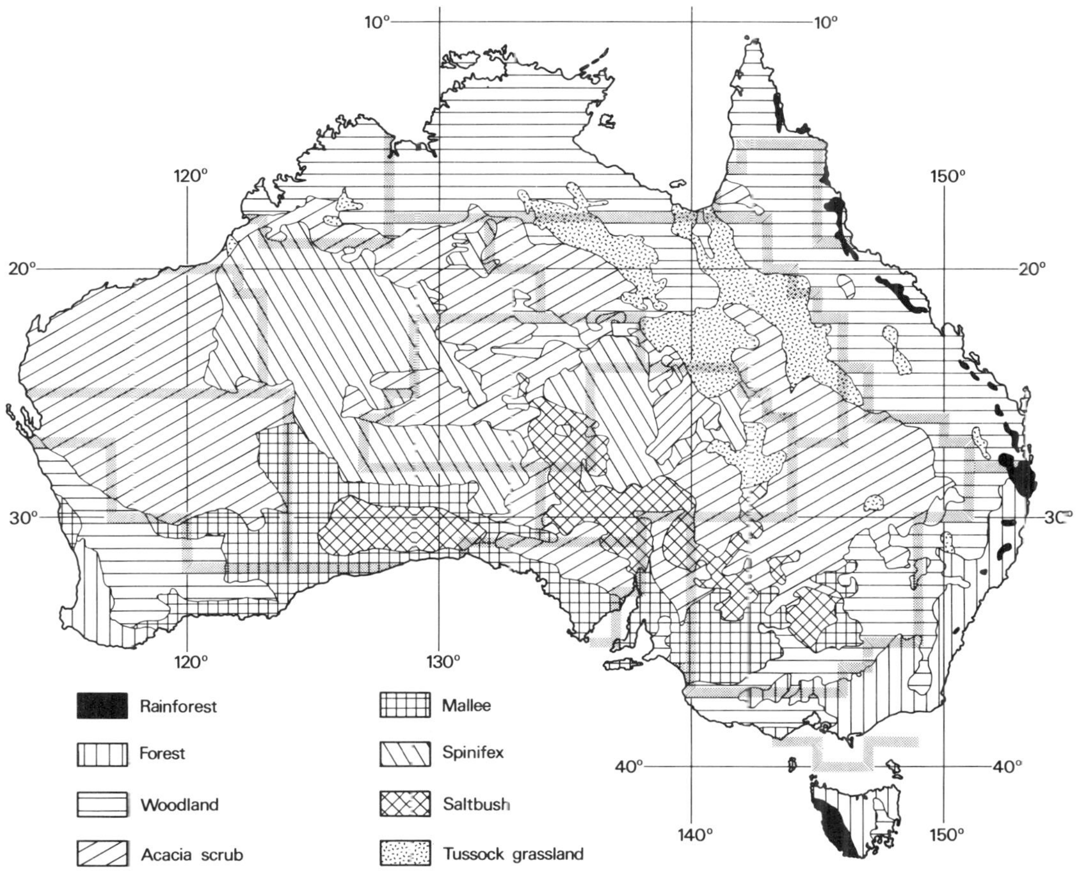

Fig. 18. Distribution of vegetation associations referred to in the *Atlas*. See also Table 4.

Table 4. Vegetation associations in Australia. The table shows the approximate equivalence between vegetation classifications. The names in the first column are those used on the vegetation map (Fig. 18) and throughout the text. Column 2 gives names of vegetation commonly used by ornithologists and sometimes added as descriptive terms in the text. The next four columns show the approximate equivalents from two well known classifications.[1695,1896] The foliage projective cover and height classes are taken from one of these.[1695]

			after Specht[1695] classification		
Atlas Map Name	Includes	Williams[1896] Name	Name	Foliage Projective Cover (%)	Height (m)
rainforest	nothofagus forest	tropical rainforest	tall closed-forest	70–100	>30
	monsoon forest	subtropical rainforest	closed-forest	70–100	10–30
	vine thicket	temperate rainforest	low closed-forest	70–100	<10
	riverine forest		closed shrub	70–100	>2
	gallery forest				
forest		wet sclerophyll	tall open-forest	50–70	>30
		dry sclerophyll	open-forest	50–70	10–30
			low open-forest	50–70	<10
woodland	pindan	tropical deciduous woodland	open-forest	30–50	10–30
	brigalow	tropical mixed woodland	low open-forest	30–50	<10
	river red gum	tropical layered woodland	woodland	10–30	<30
	riverine forest	mixed coastal woodland	low woodland	10–30	<10
	timbered country	temperate mixed woodland	open-woodland	<10	10–30
	paperbark swamp	tropical layered forest	low open-woodland	<10	<10
		tropical tree savannah			
		temperate tree savannah			
mangroves*	littoral complex				
acacia scrub	mulga	layered scrub	open-scrub	30–50	<10
	gidgee	arid scrub	tall shrubland	10–30	<10
	shrubland	tropical tall shrub savannah	tall open-shrubland	<10	<10
	myall	sclerophyll shrub savannah	low open-shrubland	<10	<10
		semi-arid shrub savannah			
		low arid woodland			
mallee	mallee sandplain	sclerophyll mallee	open-scrub	30–70	>2
	mallee heath	semi-arid mallee			
heath*	sandplain	heath	closed-heathland	70–100	0.25–2
	buttongrass	alpine complex	open-heathland	30–70	0.25–2
	alpine heath		dwarf open-heathland	<30	<0.25
saltbush	saltbush plain	shrub steppe	low shrubland	10–50	0.25–2
	bluebush		low open-shrubland	<10	0.25–2
	saltlake		very open sedgeland	<10	<2
spinifex		low shrub woodland	hummock grassland	<30	<2
		low arid woodland	open (tussock) grassland	10–30	<2
		sand plain desert	very open (tussock) grassland	<10	<2
		sclerophyll low tree savannah	low open-woodland	<10	<10
		semi-arid low tree savannah			
		sclerophyll hummock grassland			
		sandhill desert			
		stony desert			
tussock grassland	mitchell grassland	tropical tussock grassland	(tussock) grassland	10–100	<2
		temperate tussock grassland			
		semi-arid tussock grassland			

*not on map (Fig. 18)

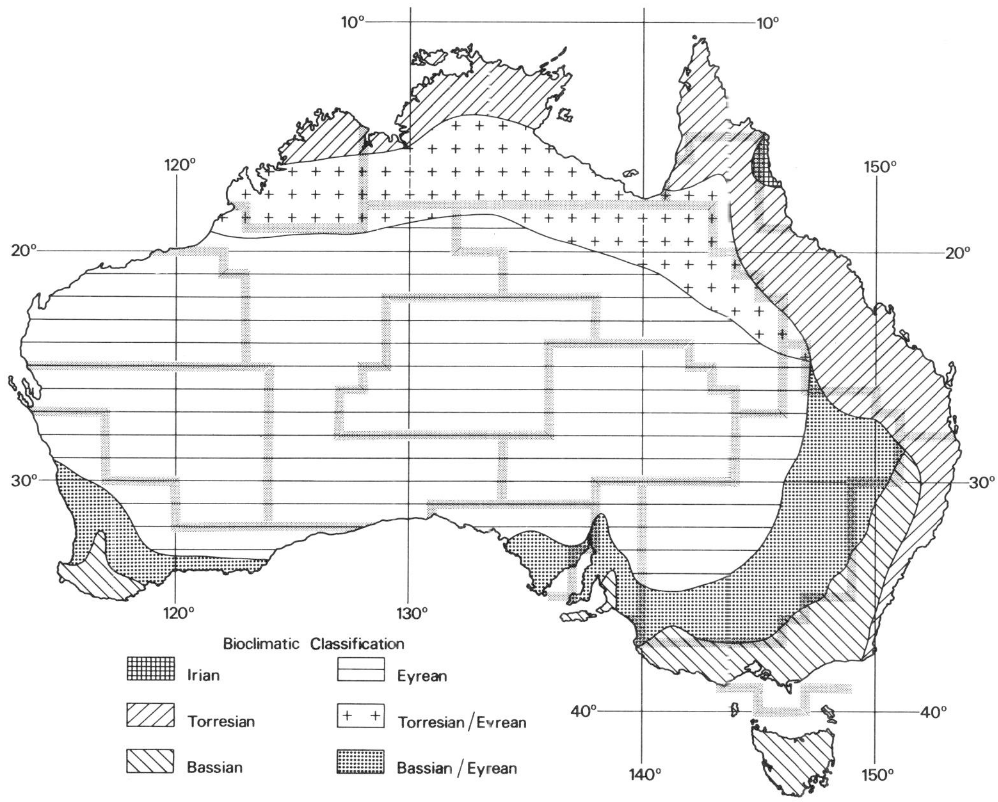

Fig. 19. Bioclimatic regions.[1332] These regions are defined in terms of the growth response of dominant plants to temperature. They closely parallel the fauna divisions of the early avian zoogeographers but add a new zone, the Irian, to indicate the links between the Atherton Region and New Guinea. In the Torresian zone plants show a significant growth response to summer conditions; in the Bassian they show a significant growth response to winter conditions. The remainder is the arid Eyrean zone with interzones (Torresian/Eyrean and Bassian/Eyrean) between it and the other two.

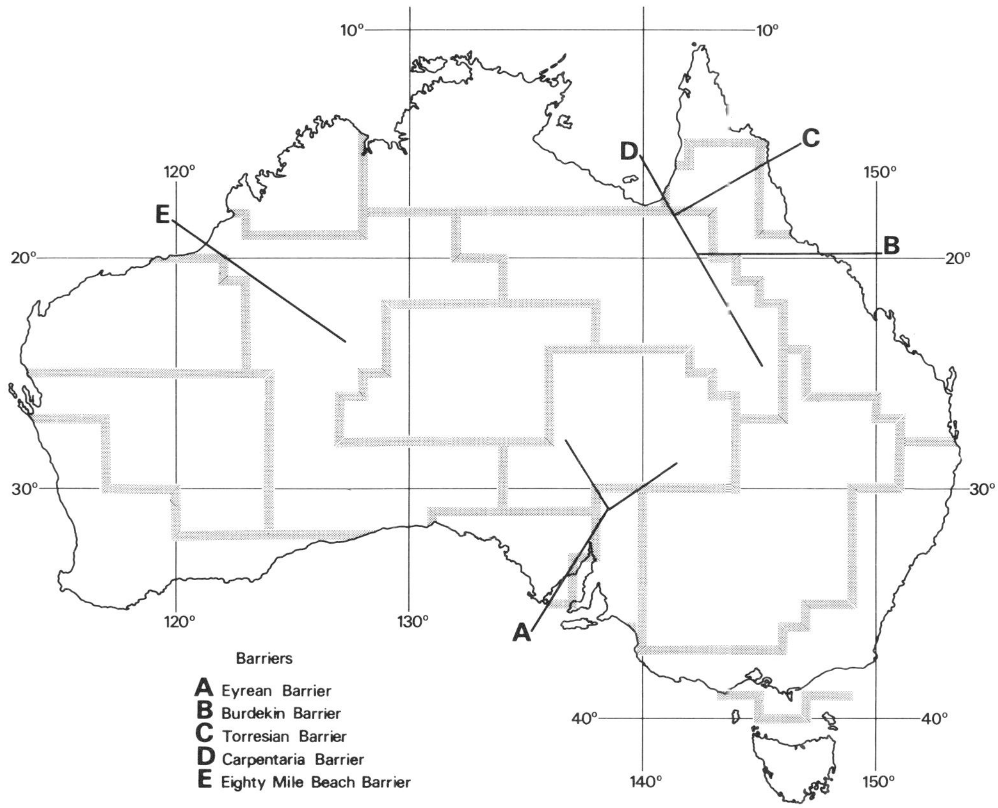

Fig. 20. Zoogeographical barriers defined by Schodde and Mason.[1588]

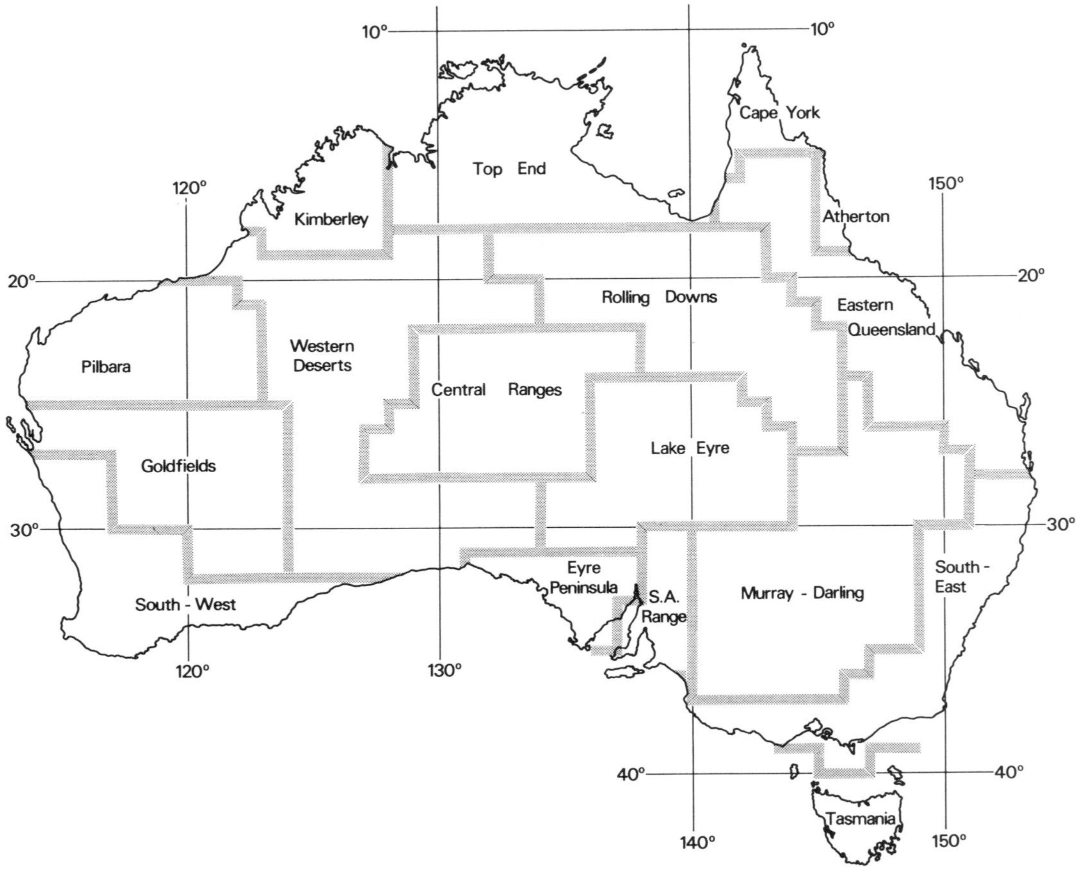

Fig. 21. Atlas Regions: biogeographical regions used to describe the distribution of birds

eucalypt woodland in the south. Natural water abundant only in the north.

Rolling Downs. Predominantly summer rainfall. Tussock grassland in the north; acacia scrub and spinifex in the south. Little surface water.

L. Eyre. The catchment of L. Eyre. Very low rainfall. Acacia scrub and spinifex intersected by large watercourses lined with eucalypt woodland. Includes sandy Simpson Desert. Little surface water.

SA Ra. Spine of ranges. Includes lower Murray R., Yorke Pen. and Kangaroo I. Predominantly winter rainfall, low in the north. Eucalypt woodland and forest in the Mt Lofty Ra. (34/138). Several isolated subspecies. Stock waters widespread.

Eyre Pen. Winter rainfall exceeding 100 mm but little summer rain. Eucalypt woodland and mallee; extensive clearing. Birds show relations to those of the South-West. Many stock waters.

Central Ra. Many low ranges. Runoff from these creates areas on the slopes below where vegetation is richer than would be expected from the climatic characteristics. Many stock waters and some natural rockholes.

Kimberley. Rugged ranges. Summer rainfall with long dry winter season. Wide mangrove belt in places; rainforest on the coast; eucalypt woodland inland. Many natural rockholes.

Western Deserts. Sandy (Tanami, Great Sandy, Great Victoria), stony (Gibson) and limestone (Nullarbor) deserts. The extension to the western coast follows a tongue of sand dunes separating the Kimberley and Pilbara Regions. Little surface water except after rain.

Pilbara. Rugged ranges. Low rainfall, predominantly in summer. Large watercourses lined with eucalypt woodland intersecting acacia scrub and spinifex. Stock waters widespread.

Goldfields. Flat. Saltlakes in the east. Low rainfall, predominantly in winter. Acacia scrub and spinifex in the north; eucalypt woodland in the south. Stock waters widespread.

South-West. Mainly flat. Almost exclusively winter rainfall. Eucalypt forest in the south-west; eucalypt woodland, mallee or heath elsewhere. Many endemic bird species and subspecies.

SYMBOLS AND ABBREVIATIONS

°	degrees (latitude, longitude and temperature)
′	minutes (latitude and longitude)
%	percentage
ACT	Australian Capital Territory
ANARE	Australian National Antarctic Research Expedition
Apr.	April
Aug.	August
cc	cubic centimetre
Ck	Creek
cm	centimetre
DDE	p,p'-dichlorophenyl-dichloroethylene
Dec.	December
dept	department
E	East
Feb.	February
g	gram
ha	hectare
I.	Island
Is	Islands
Jan.	January
km	kilometre
L.	Lake
m	metre
Mar.	March
mm	millimetre
Mt	Mount
Mt.	Mountain
Mts	Mountains
N	North
Nov.	November
NSW	New South Wales
NT	Northern Territory
Oct.	October
Pen.	Peninsula
pers. comm.	personal communication
Prom.	Promontory
Pt	Point
Qld	Queensland
R	River
Ra.	Range
S	South
SA	South Australia
Sept.	September
sp.	species (singular)
spp	species (plural)
sq. km	square kilometre
sq. m	square metre
St	Saint
Tas.	Tasmania
Vic.	Victoria
W	West
WA	Western Australia

HOW TO USE THIS BOOK

For quick reference, this is a summary of *The Book* (pp. xxxvi-xxxviii).

The English and scientific names appear with the vignette; the Atlas number appears at the left of each map and the bird names used during the Atlas are listed in the Index (pp. 730-8).

Maps

Open circle—observed.
Closed circle—breeding.
Size of circle indicates Reporting Rate (p. xxxv); the larger the circle, the more often the species was reported.
'Max. Rep. Rate'—Maximum Reporting Rate.
'Grid Block(s)'—those in which the maximum Reporting Rate was recorded.
Names of the Atlas Regions, shown on the map, are given in Fig. 21.

Text

Gives range ex-Australia, subspecies, historical range, movement, environment, factors influencing distribution, status and abundance where known.
Grid references follow place names.
Atlas Regions are usually referred to from north to south, then east to west.
Small numbers refer to references in the numbered bibliography (pp. 701-29).
'Blocks with records'—gives percentage that this is of 1° blocks in Australia (812) or 10′ blocks in Tasmania (368).
'Blocks with breeding records'—gives percentage that this is of blocks with records. Some unconfirmed breeding records are included in these figures but omitted from the maps.

Vignettes

The names of the artists and the pages on which their drawings appear are listed on pp. 692-3.

Definitions of Often-used Terms

Field Atlas. Data collected during the five years 1977–81.
Historical Atlas. Data collected before 1977, subdivided into three main periods: 1951–76, 1901–50, before 1901.
Record. A report of a bird in a place at a time entered on a record sheet.
Reporting Rate. The percentage of record sheets in an area on which a given species is reported.
1° block. A four-sided area whose sides measure 1° of latitude by 1° of longitude. 1° is approximately 100 km. A 1° block contains 36 10′ blocks.
10′ block. A four-sided area whose sides measure 10′ of latitude by 10′ of longitude. 10′ is approximately 17 km. There are 36 10′ blocks in a 1° block.
Grid reference. Latitude and longitude of the northern and western boundaries of a 1° block, e.g. 27/152 is the 1° block whose centre point is 27°30′S 152°30′E.
Seasons. Spring: Sept.-Nov., summer: Dec.-Feb., autumn: Mar.-May, winter: June-Aug.

Action

Fully-documented records (URRF) which suggest different ranges from those shown in the *Atlas* should be collected and sent for publication in the bird report of the relevant State. The RAOU will pass on any such information sent to it.

Main Species

Emu

Dromaius novaehollandiae

Dwarf Emu

Dromaius minor

When Europeans arrived, Australia was inhabited by at least four isolated populations of the endemic emus: one each on the mainland and Tas. (Emu) and one each on King (39/144) and Kangaroo (35/137) Is (Dwarf Emu). The Tas. Emu may have been distinct from the mainland bird.[1391] Populations became extinct on King I. about 1805, Kangaroo I. about 1827 and in Tas. by 1865.[734,866,1391] The mainland Emu was introduced to Maria I. (42/148) in 1968, and has become established there;[734] and to Kangaroo I. in 1957 where it is also breeding.[1391] It has been reintroduced locally to parts of Qld, NSW and Vic. as well.

Except for the eight weeks when the male is incubating, the Emu is completely mobile and moves to maintain contact with its food supply. It moves to places where substantial rain has fallen recently. Where populations have multiplied greatly, spectacular 'suicide' movements may result, specially from the Goldfields into the South-West Region.[459] The longest movement recorded for an individual bird is 990 km.[465]

The Emu avoids rainforest and areas that are waterless or highly developed but is otherwise present wherever trees and shrubs remain.

The Emu eats young green shoots, insects, fruit and seeds.[463] It has probably benefited from the provision of water for domestic stock because it drinks each day. The availability of such water leads to breeding in areas not otherwise habitable and thus to overpopulation which in turn may result in suicide movements.

Island populations of the Emu were apparently small and exterminated by sealers, whalers and settlers to provide food. Despite predictions that numbers of the mainland Emu would decline[723] the species remains widespread and, in many places, abundant.[459]

The density of the Emu in the Goldfields Region is about one bird to every 10–15 sq. km.[459]

Field Atlas Records: 10 352

1° blocks–with records: 580 (71%)
–with breeding records: 258 (44%)

Atlas No.: 001,848

Max. Rep. Rate: 100%
In block(s): 32/143

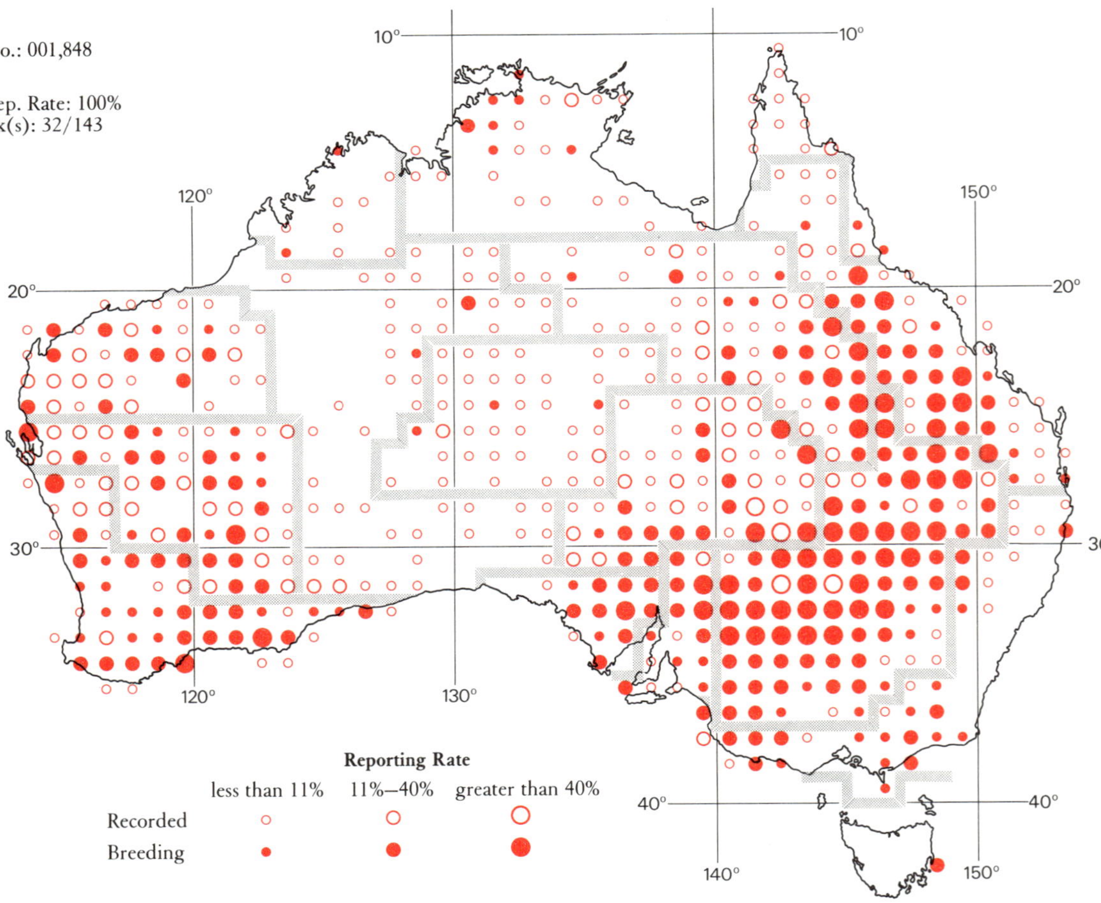

Southern Cassowary

Casuarius casuarius

The Southern Cassowary lives in New Guinea and adjacent islands and in Australia. Reports in the Field Atlas have come from all the 1° blocks where it was known historically. The two Australian populations are isolated but views about their subspecific status differ.[388,1709] It is unlikely that the species undertakes substantial movements though the birds are seldom seen during the summer.[432,1856]

In New Guinea the Cassowary inhabits lowlands and foothills. In Australia it lives in pockets of rainforest at all altitudes and also in canefields. It tends to be solitary and territorial, six being the largest group reported.[1856]

The Cassowary moves round the rainforest to take advantage of the most recent crops of fallen fruit. Snails are also eaten. The maintenance of extensive tracts of untouched rainforest is essential for the survival of the species because only by ranging over large areas can it find fruit throughout the year.[432]

There are no measures of abundance.

Field Atlas Records: 140

1° blocks–with records: 11 (1%)
–with breeding records: 6 (55%)

Atlas No.: 002

Max. Rep. Rate: 32%
In block(s): 17/146

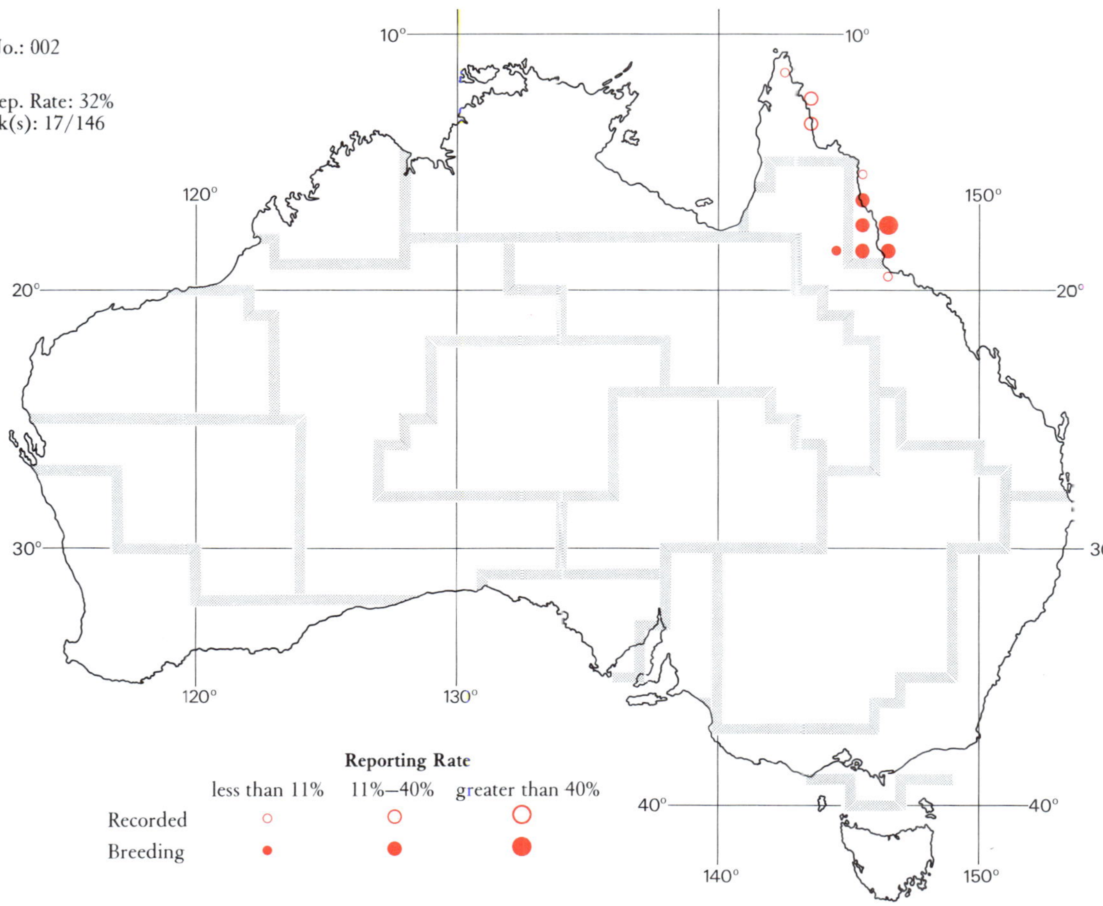

Great Crested Grebe

Podiceps cristatus

The Great Crested Grebe ranges throughout Europe, Asia, Africa, New Zealand and Australia, generally avoiding tropical areas. Australian and New Zealand birds have been regarded as the same subspecies *australis*[388] but may be distinct.[1391] Comparison with the historical maps (p. 661) shows that the range of the Great-crested Grebe has expanded since 1950, specially in the Atherton and Eastern Qld Regions. All but one (17/133) of the Top End records in the Field Atlas were made during an influx in June-Dec. 1978. One farther south (21/134) was made in Sept. 1977. Throughout the Field Atlas, records came from the Pilbara Region where the species had not previously been reported.

The breeding range has expanded also. The species was not recorded breeding at Mt Isa (20/139) until after 1949; before then the northernmost breeding reported was at 29°S. Breeding was first reported in the Tas. Region in 1971.[1307]

The Great Crested Grebe moves frequently but erratically about the continent. It generally inhabits fresh water but sometimes gathers in flocks in estuaries, bays and on inland lakes. Non-breeding winter gatherings have been recorded, e.g. 380 on Port Phillip Bay (38/144)[1839] and 600–800 on the Lachlan R swamps (33/146). At Rotamah I. Bird Observatory (37/147) the largest flocks of up to 200 are in Oct.[285] The Great Crested Grebe in Australia nests either alone or colonially. In Europe nests are usually well dispersed each pair requiring at least 1 ha of water; rarely, colonies form with nests as little as 2 m apart.[422] The Great Crested Grebe gathers its food by diving, generally for less than 30 secs.[422] One Australian gut sample of seeds[1049] is exceptional in comparison with overseas results where a small number of seeds was noted in many guts (89%) but the bulk of the diet was fish, greater than 10 cm long, and aquatic insects.[422]

On L. Bathurst (35/149) 400 birds have bred at the same time.[653] On L. Cooper (36/144) 100 pairs bred on a patch of floating weed 300 sq. m in area.[237]

Field Atlas Records: 1886

1° blocks–with records: 187 (23%)
–with breeding records: 22 (12%)

Atlas No.: 060

Max. Rep. Rate: 28%
In block(s): 24/153

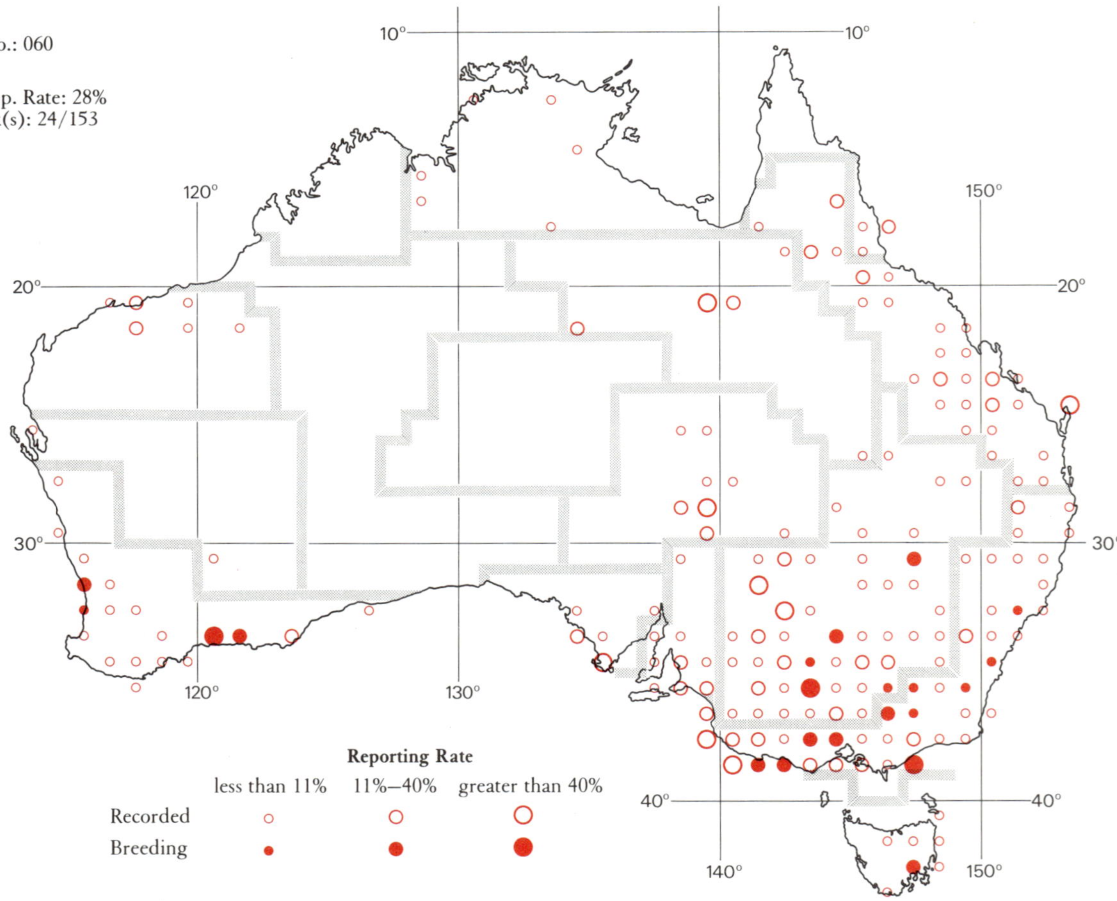

Hoary-headed Grebe

Poliocephalus poliocephalus

The endemic Hoary-headed Grebe had not been reported to breed north of 25°S before the Field Atlas. Knowledge of its breeding distribution previously was poor and the new breeding localities probably do not represent an extension of range. The species is often confused in the field with the Australasian Grebe in non-breeding plumage.

In some localities where there is a highly seasonal climate, the Hoary-headed Grebe is said to be present only seasonally, e.g. on the Avon R (31/116) in summer and autumn[1216] and at Kellerberrin (31/117) in winter.[627] Elsewhere, its movements are erratic, typified by those observed on the coast of the South-West Region. Many birds arrived in Apr. 1937 and stayed until Nov. 1938, after which few were seen until at least early 1940.[1601] Similar movements have been recorded in the southern Murray-Darling Region.[833] In that district and also at Canberra (35/149) and Laverton (38/143) some birds were always present, occasionally augmented by the arrival of many more.[1013,1822] Breeding was rarely reported in the Tas. Region (1% of records were of breeding compared with 2.4% in Vic.) but there are occasional influxes of thousands of birds. The Hoary-headed Grebe travels mainly at night:[1131] one pair landed at night on a boat crossing Bass Strait. The longest recorded movement of a banded bird is 572 km.[85]

The Hoary-headed Grebe occupies lakes, lagoons and swamps. During non-breeding periods large flocks form in estuaries and bays. At L. Illawarra (34/150) 1400 were counted in Aug. 1980; at Rotamah I. Bird Observatory (37/147) up to 400 in winter; and over 500 at Adelaide (34/138) in Mar. 1978. Hundreds have been seen in Port Phillip Bay (38/144) 2 km offshore,[1836] on the Swan R (31/115)[1638] and in other western estuaries. Except in Tas. where pairs nest singly, breeding is usually colonial. The species feeds on insects, which become abundant after seasonal or occasional flooding. The few gut contents examined have included the remains of water insects, spiders and crustaceans.[1049]

Breeding colonies of 400 nests are recorded on L. Bathurst (35/149);[653] 100 nests in the southern Murray-Darling Region;[831] and near Clunes (37/143) there were 10 breeding birds/ha.[1552]

Field Atlas Records: 8529

1° blocks–with records: 431 (53%)
–with breeding records: 102 (24%)

Atlas No.: 062

Max. Rep. Rate: 48%
In block(s): 0031,10,0 005128/139

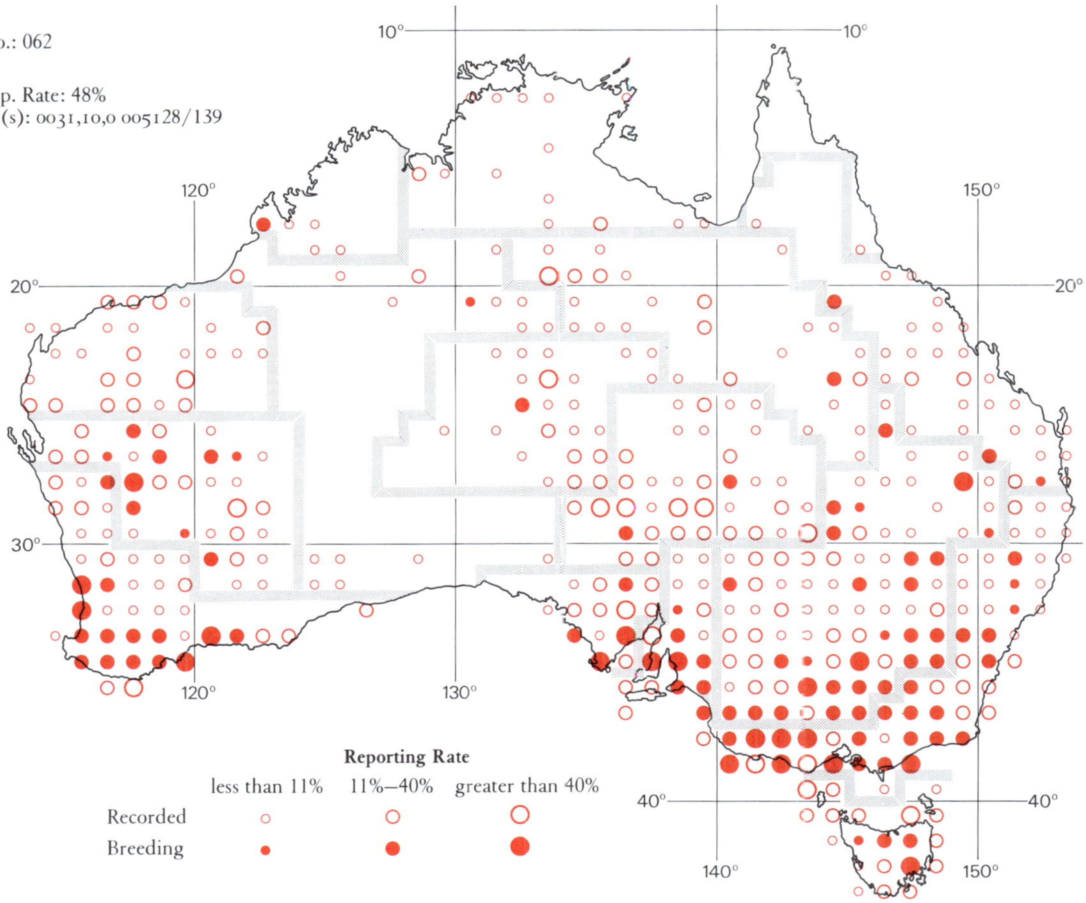

Australasian Grebe

Tachybaptus novaehollandiae

The Australasian Grebe ranges throughout Indonesia, New Guinea and adjacent islands, New Zealand and Australia. In Australia the subspecies is *novaehollandiae*.[388] The species is often confused in the field with the Hoary-headed Grebe in non-breeding plumage.

In the Tas. Region the Australasian Grebe is a rare species. There are no actual records before 1901 in the Historical Atlas though a writer in 1904 described it as 'universally distributed'.[1805] Subsequent reports are also scarce until the first record of breeding was made in 1965 and sightings became increasingly frequent. During the Field Atlas there were 87 reports in the Tas. Region compared with 861 for the Hoary-headed Grebe.

The Field Atlas does not provide evidence of seasonal movements on a continental scale but they have been reported from some localities. In the wet summer season the Australasian Grebe visits islands in Torres Strait.[503] In the Atherton Region it is present Apr.-Nov.[231] and in Darwin (12/130) it is present only May-Jan.[423] In the southern Murray-Darling Region it is present all year.[833] In the South-West it is present all year on the Avon R (31/116)[1216] but a winter visitor to the Middle Swan R (31/116) on the coastal plain nearby.[783] The bird apparently moves locally to find suitable water for each season. The longest movement recorded is of an adult from Lucindale (36/140) to L. Murdeduke (38/143), 338 km.[1436]

The Australasian Grebe occupies freshwater pools, farm dams and swamps and is rarely recorded on salt water. Flocks form in the non-breeding season, often mixed with other grebes. The bird usually nests singly, though at times in small colonies, and has once been reported to breed communally.[501] It may breed on any water providing the level is stationary for a short time, because the chicks leave the nest before they are dry, travelling on their parents' backs, and can thus move away from the edges of shrinking pools.[74] The Australasian Grebe feeds on water insects.[1049] It has been observed to follow Eurasian Coots closely as they move through weedy areas, gleaning insects disturbed by the passage of the larger birds.[831]

Six to seven nests may be found on ponds a few hectares in area.[1646]

Field Atlas Records: 13 117

1° blocks–with records: 457 (56%)
–with breeding records: 190 (42%)

Atlas No.: 061

Max. Rep. Rate: 49%
In block(s): 35/147

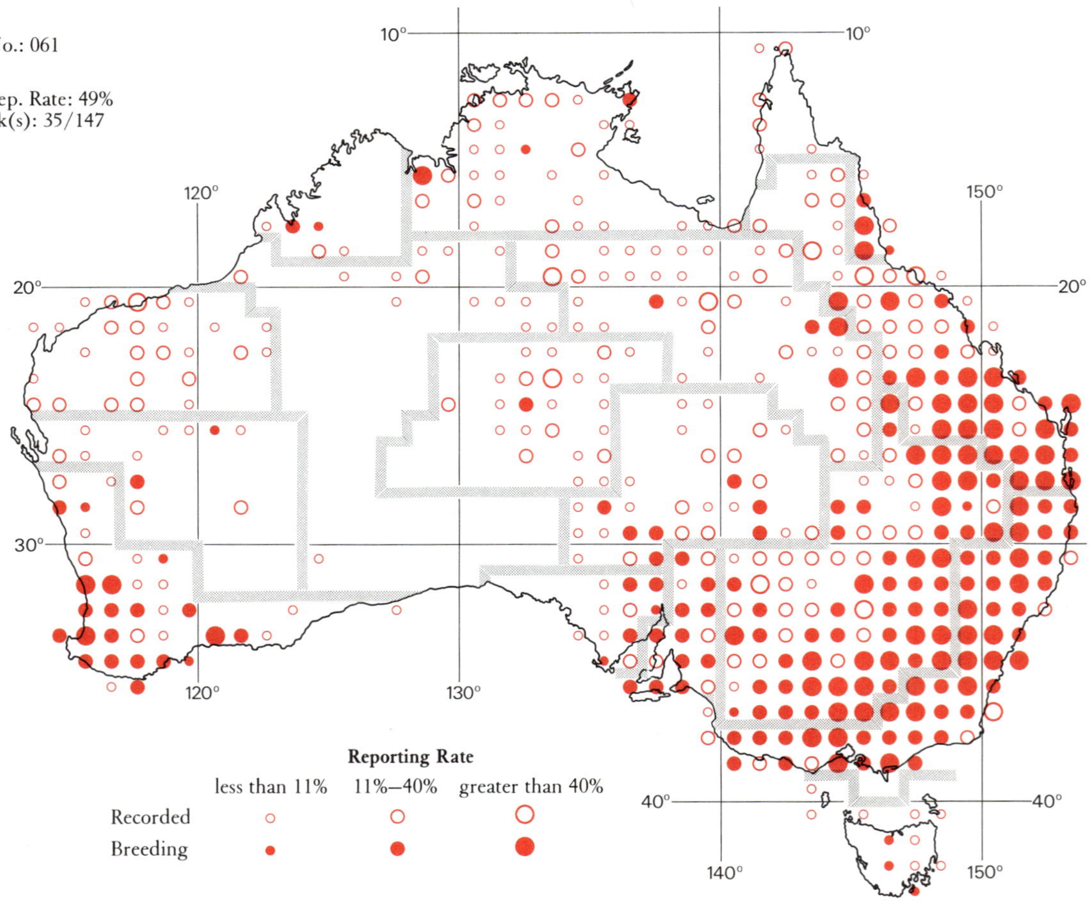

Little Penguin

Eudyptula minor

The Little Penguin breeds in New Zealand including Stewart (47S/168E) and Chatham (44S/176E) Is as well as in Australia. Breeding was confirmed during the Field Atlas in all but five of the 1° blocks in which it has been recorded historically. Of these five sites, three probably still exist: two along the Nullarbor cliffs (32/124, 32/125) and one on St Alouarn I. (34/115), which were visited only in the 1951–76 historical period. Penguins may still breed on Cheyne I. (34/118) and Marts I. (33/122) where chicks were recorded in the 1901–50 period.

Though adults may be absent at sea for weeks at a time when not breeding, they return to the same breeding area year after year.[1478] In contrast, the young disperse over great distances: two immatures were recovered more than 1000 km by sea from their banding place.[1479] In south-eastern NSW Penguins have been observed in groups as far out to sea as the edge of the continental shelf.[1475]

The Little Penguin has been watched landing in daylight at the base of the Nullarbor cliffs (32/125)[1472] but at Phillip I. (38/145) it comes ashore only at night. Distribution is influenced by availability of nesting sites that must be accessible from the sea and be either on islands or at the base of cliffs on the mainland.

The Little Penguin feeds on small fish and squid.[1275] Ashore it is vulnerable to attack by foxes, dogs and cats. Wherever human habitation has encroached on and radically altered breeding grounds, populations have declined or disappeared.

The estimation of numbers is influenced greatly by the timing of breeding, which varies annually, and by the stage of breeding when counts are made. The largest colonies recorded are: Gabo I. (37/149) where there were estimated to be 40 000 birds (including non-breeders) in 1977[1473] and 5000–10 000 burrows in 1978;[768] Phillip I., where 1.44 ha were occupied in a reserve, the active burrows numbering 617 in 1980–81, 582 in 1981–82 and 481 in 1982–83;[450] and St Helens I. (41/148) which had an estimated 30 000 birds in 1978.[1182]

Field Atlas Records: 1391

1° blocks–with records: 78 (10%)

–with breeding records: 46 (59%)

Atlas No.: 005

Max. Rep. Rate: 53%
In block(s): 39/146

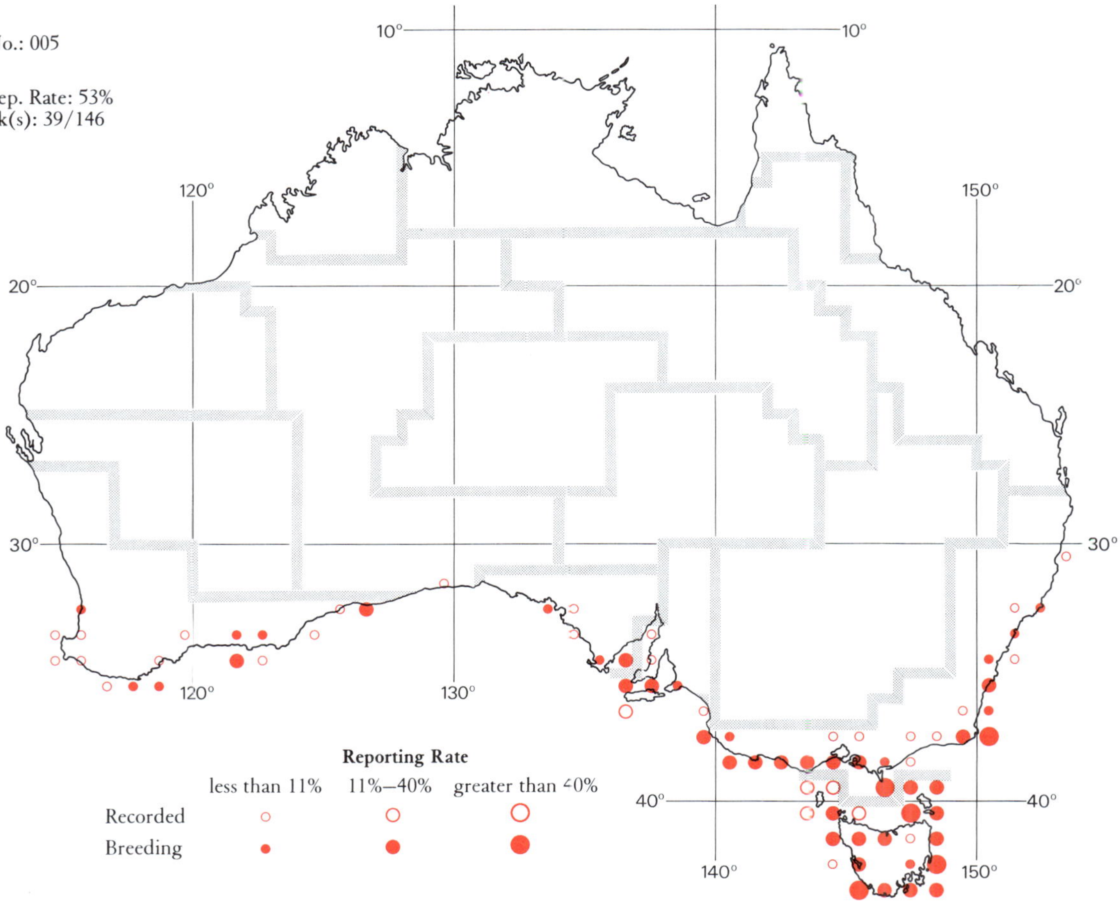

Wandering Albatross

Diomedea exulans

The Wandering Albatross breeds on islands in the sub-Antarctic zone. The subspecies most often in Australian waters, nominate *exulans*, breeds on Antarctic and sub-Antarctic islands of the Indian and Atlantic Oceans, including Macquarie I. (54S/158E).[362,1636] A second population, possibly subspecifically separable, breeds on New Zealand sub-Antarctic islands. The third, *dabbenena*, breeds at Tristan da Cunha (38S/12W) and Gough I. (40S/10W) and has not yet been reported in Australia. The Wandering Albatross was most often recorded by ANARE between the sub-tropical and Antarctic Convergences.[937] In Australia it has been reported as far north as the Whitsunday Passage (20/148) in the east and Fremantle (32/115) in the west.[20,1636]

The Wandering Albatross uses most of the Southern Ocean. It can travel 100–200 km a day as evidenced by birds that follow ships and recoveries of marked birds. Many have been recovered more than 10 000 km from where banded. Birds banded on the NSW coast have been found in breeding colonies on Bird I. at South Georgia (54S/35W) and at Marion (46S/37E), the Crozet (46S/50E) and Kerguelen (49S/70E) Is. Likewise, birds banded at Bird, Kerguelen and the Crozet Is as well as Macquarie I. have been recaptured off NSW. The species is most frequent in Australian waters in winter[1195,1760] (Reporting Rates: spring 0.23%, summer 0.13%, autumn 0.24% and winter 0.31%).

In NSW four times as many birds are seen offshore as inshore.[1760] Up to ten were seen on each of eight trips from Hobart (42/147) to the edge of the continental shelf in May-June 1979.[1314] Large congregations are unusual though up to 548 birds were recorded on one day in winter at offal from a sewerage outlet at Malabar (33/151)[1636] before improved processing removed this source of food. The Albatross scavenges during the day but feeds mainly at night on cephalopods and some fish.[113,896] Populations may be adversely affected by fishing operations in Australian waters; off southern Tas. birds drown when caught on long lines and are shot by fishermen.

The world population was estimated in the 1960s to be about 40 000 breeding birds.[1821] Breeding populations on Macquarie and Heard (53S/73E) Is are very small.[248]

Field Atlas Records: 245

1° blocks–with records: 40 (5%)

–with breeding records:

Atlas No.: 086

Max. Rep. Rate: 38%
In block(s): 37/150

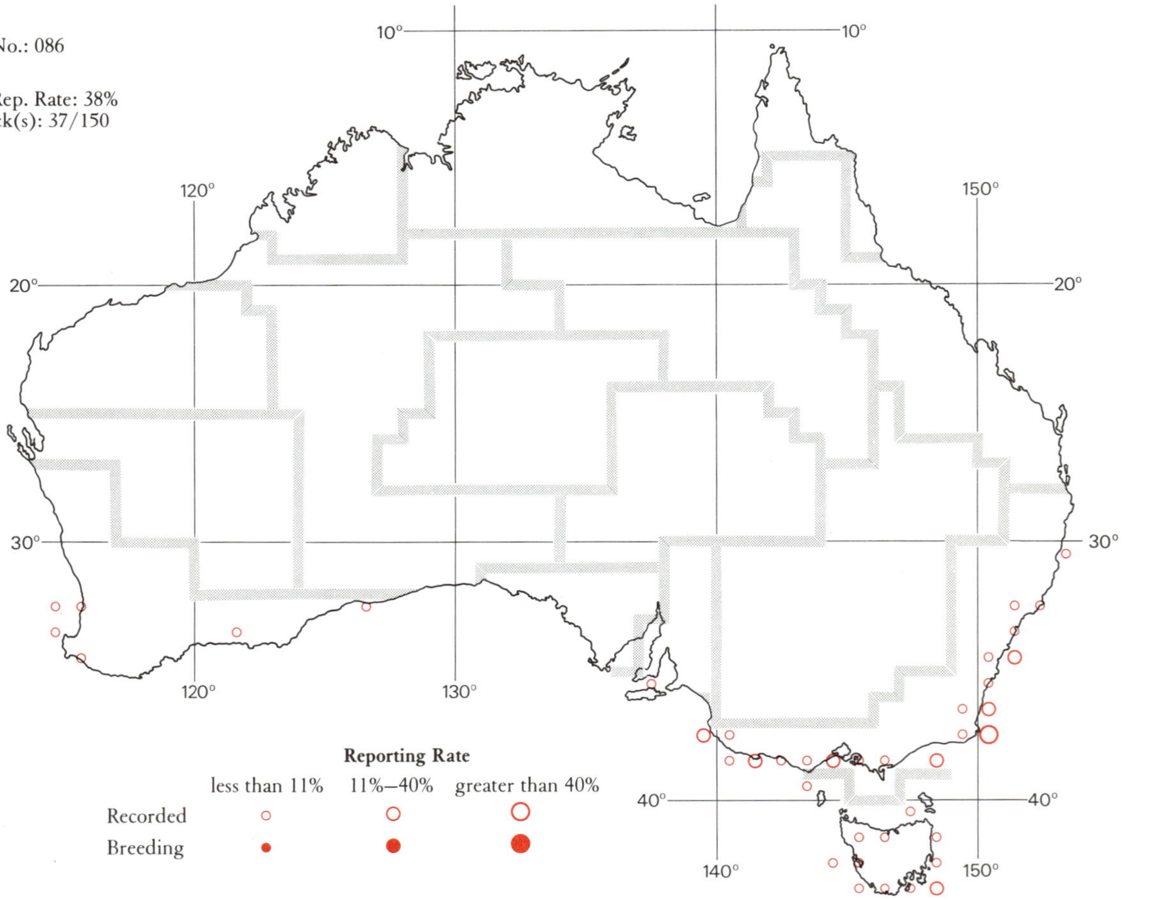

Royal Albatross

Diomedea epomophora

There are two subspecies of the Royal Albatross both of which breed in the New Zealand area: *epomophora* on Auckland I. (50S/166E) and Campbell I. (52S/169E) and *sanfordi* in New Zealand itself and on the Chatham Is (44S/176W). The Royal may be confused with the Wandering Albatross at sea.[1636]

The Royal is more frequent off southern Australia than the number of Field Atlas records indicates and both subspecies are thought to reach Australian waters at all times of the year. The species has been reported as far north as Red Rock (29/153) on the eastern coast[1292] and there is at least one record (31/115) from the South-West Region, in Mar. 1966.[1635]

Immatures of the Royal Albatross, and possibly non-breeding adults, are thought to circle the southern oceans with the prevailing westerly winds, probably north of the Antarctic Convergence.[1503] Banded birds have been recovered off South America and in the Indian Ocean as well as between Australia and New Zealand. Whether birds reach Australia by circumnavigating the South Pole or move west from New Zealand is not known. One newly-fledged bird from Campbell I. reached Ledge Pt (31/115) in WA in four months and an adult travelled at an average speed of at least 145 km per day from Campbell I. to Chile where it was recovered alive.[1636] Adults appear to leave the breeding area for at least 12 months after successfully rearing a chick but only one has been recovered far from its breeding ground during such a 'holiday' and it is possible that adults do not move as far as immatures.[1503]

The main reason for the scarcity of Royal Albatross records in the Atlases is that the species is most often encountered at the edge of the continental shelf. Of 17 sightings in the Field Atlas, for which details are available, only two were of birds seen from the shore and two others of beachwashed individuals. At sea the Royal is regularly seen in ones and twos off south-eastern Australia and sometimes in greater numbers. Unlike the Wandering, it is not attracted inshore to sewage outfalls. The Royal Albatross feeds on cephalopods and fish.

There are no measures of abundance.

Field Atlas Records: 22

1° blocks–with records: 11 (1%)
–with breeding records:

Atlas No.: 974

Max. Rep. Rate: 6%
In block(s): 37/150

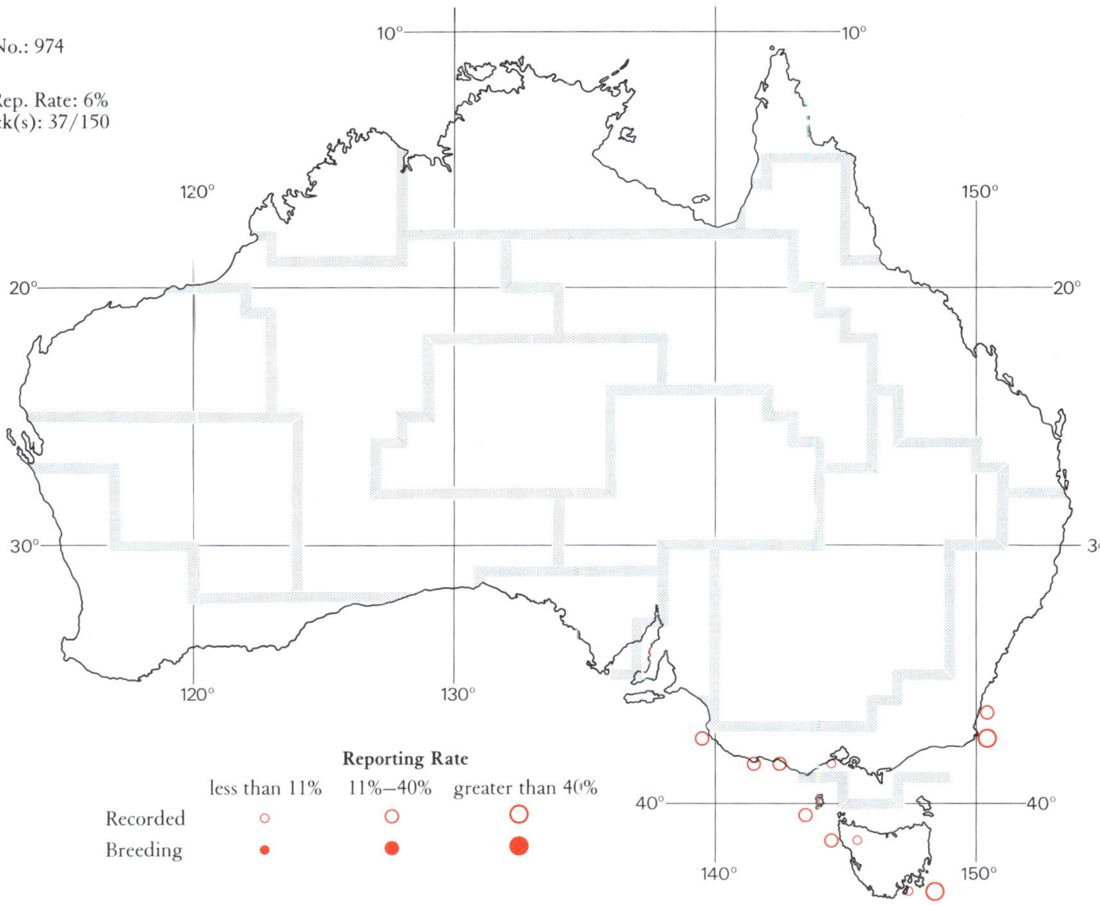

Black-browed Albatross

Diomedea melanophrys

There are two subspecies of the Black-browed Albatross: *impavida* breeds near New Zealand on Campbell I. (52S/169E) and possibly on the Antipodes Is (49S/178E); and *melanophrys* breeds on other sub-Antarctic islands including Macquarie (54S/158E) and Heard (53S/73E).[1636,1759,1818] Both subspecies reach Australian waters and are regularly reported on the eastern coast.[1292] Towards the west *impavida* becomes rarer, the westernmost report at 139°E.[375] The northern limits of the Black-browed Albatross, identified during the Field Atlas, are also those of the Historical Atlases. There was an unusual inland record at Hamilton (37/142) in July 1977 when a live bird was picked up, presumably storm-blown.

The Black-browed Albatross reaches the edge of the Antarctic ice but is commoner north of the Antarctic Convergence.[937] It is more often recorded at sea south of Tas. and New Zealand than off WA.[1821]

Counts off Sydney (33/151), Eden (37/149) and Adelaide (34/138) show that the species is present throughout the year but least abundant in summer.[113,413,1262] In the South-East and Tas. Regions Reporting Rates were correspondingly lowest in summer (spring 1.63%, summer 0.48%, autumn 1.23%, winter 2.59%). Recoveries in Australia of the subspecies *melanophrys* have been of birds banded on Macquarie and Heard Is.[1636] Few birds from the large colonies on the Falkland Is (51S/59W) appear to undertake the circumpolar movement that would bring them to Australian waters.[1821]

The species feeds on cephalopods[1759] and will follow ships to feed on offal. Off Chile the ratio of Black-browed to Grey-headed Albatrosses was 90:1 when the surface water temperature was 12°C but 2:3 when the temperature was 9.1°C.[921,1759] Off NSW the Black-browed was rarely seen when the surface water temperature was above 18°C.[113] It may therefore be an inhabitant of waters with surface temperatures between 9°C and 18°C.

Breeding colonies of the Black-browed Albatross in the Falkland Is, which contain 0.5–3.0 million birds, constitute perhaps 90% of the world's population.[1759] In comparison, numbers breeding on Campbell, Macquarie, Heard and Kerguelen (49S/70E) Is are reckoned only in hundreds or thousands.

Field Atlas Records: 495

1° blocks–with records: 65 (8%)
–with breeding records:

Atlas No.: 088

Max. Rep. Rate: 54%
In block(s): 37/150

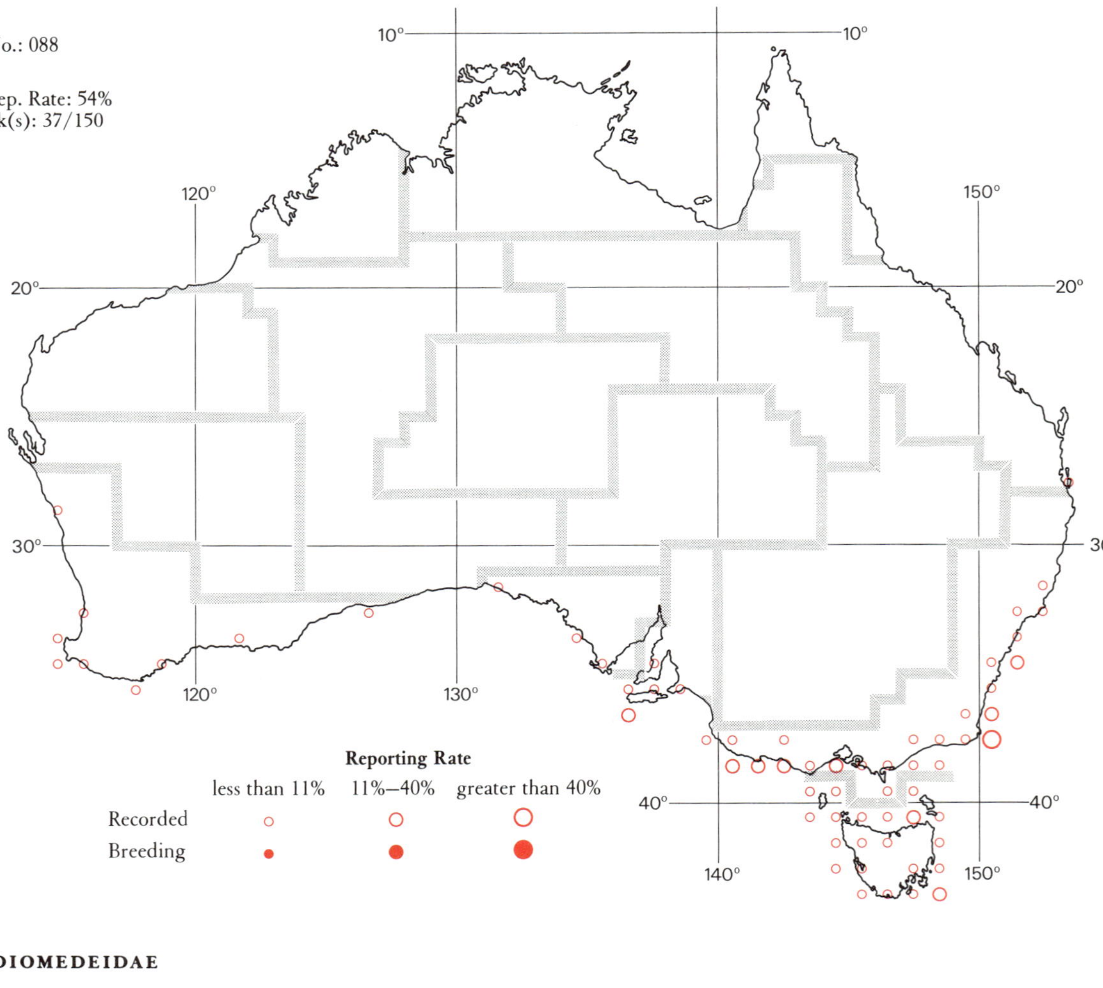

Yellow-nosed Albatross

Diomedea chlororhynchos

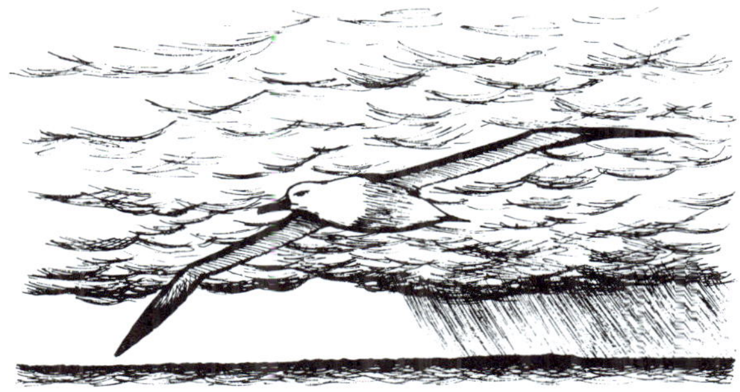

The Yellow-nosed Albatross breeds on northern sub-Antarctic islands in the Indian and Atlantic Oceans. Two subspecies have been recognized: *chlororhynchos* breeding on Tristan da Cunha (38S/12W) and Gough I. (40S/10W) and *bassi* breeding on St Paul (38S/77E), Prince Edward (46S/37E) and Amsterdam (37S/77E) Is. All SA specimens whose identity could be determined belonged to *bassi*.[1393]

Historical records show that, in Australian waters, the Yellow-nosed Albatross ranges north to Breaksea Spit (24/153)[20] in the east and to Barrow I. (21/115) in the west. A sighting east of Eucla (31/128) in 1973[1477] fills the gap in the Field Atlas round the Great Australian Bight and suggests that the species may be expected anywhere along the southern coast.

Recoveries of birds banded at the breeding grounds indicate a north-easterly movement of young away from the colonies in their first winter[1636] which would bring them first to the seas west of Australia. The Yellow-nosed is said to be the most plentiful albatross there.[1638] In the east it is generally less abundant than the other common albatross species[113,1195,1262] but off Coffs Harbour (30/153) it was more numerous than other species.[849]

The presence of the Yellow-nosed Albatross in parts of Australia is strongly seasonal: May-Oct. off Eastern Qld[20] and Apr.-Nov. in the west.[1638] Reporting Rates for the South-East, SA Ra., Eyre Pen. and South-West Regions were uniformly highest in winter. Combined they were: spring 0.57%, summer 0.34%, autumn 0.62%, winter 1.32%. In the Tas. Region Reporting Rates were highest in autumn (spring 0.30%, summer 0.08%, autumn 1.82%, winter 0.50%).

Oceanic observations indicate that the Yellow-nosed Albatross inhabits inshore waters rather than the open ocean.[415] It feeds on cephalopods, fish and crustaceans.

During an aerial survey from Shark Bay (25/113) to Geraldton (28/114) (330 km) on one day in July 1946, 175 birds were counted in flocks of up to 40. In 1972, also in July, 300 were counted on the water at the eastern end of Shark Bay.[1638]

Field Atlas Records: 327

1° blocks–with records: 73 (9%)

–with breeding records:

Atlas No.: 089

Max. Rep. Rate: 44%
In block(s): 37/150

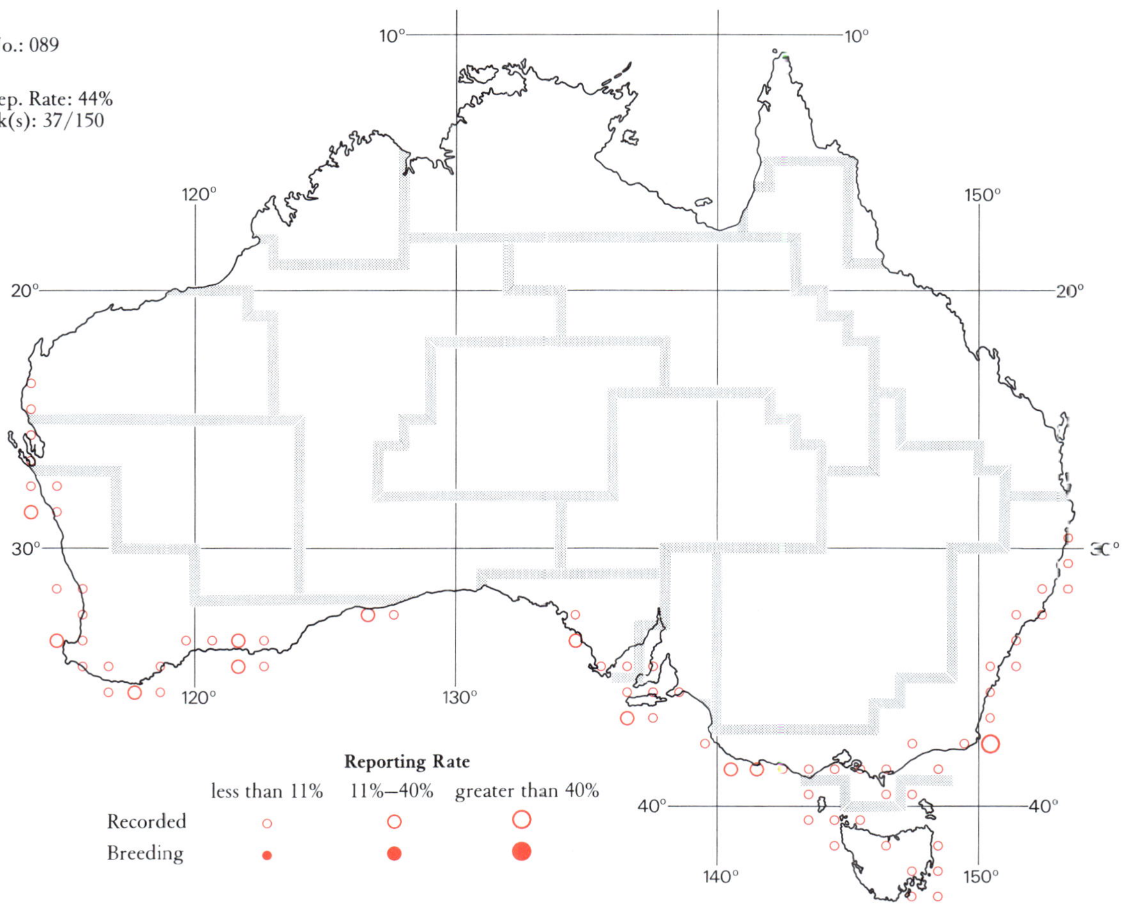

Shy Albatross

Diomedea cauta

The Shy Albatross is the only albatross to breed in Australian waters. There are three colonies of the subspecies *cauta* on islands in the Tas. Region: Albatross (40/144), Pedra Branca (43/146) and Mewstone (43/146). Other subspecies nest on islands round New Zealand. One of these, *salvini*, has also been recorded in Australian waters.

In eastern Australia specimens have been beachwashed as far north as Fraser I. (25/123). West of Eyre Pen. the species is rarely recorded; the report in the Field Atlas (28°S) is the northernmost on the western coast.

Some birds that breed in New Zealand are known to have a circumpolar distribution[415,1636] but movements of the Australian subspecies *cauta* are less clear. Many immatures and adults are seen close to Albatross I. and most appear to feed nearby. Of 1302 nestlings banded there in 1960 and 1973, 20 had been recovered by July 1973, most within the Tas. Region.[733] The longest movements were 885 km north-east to Wollongong (34/150) and 563 km north-west to Robe (37/139). A further 3000 nestlings were banded in 1980–83 and 12 have been recovered along the southern coast, one near Yanchep (31/115), 2800 km away.[248] One adult specimen of *cauta* was collected in the eastern North Pacific. Other birds, apparently *cauta*, have been photographed off Amsterdam I. (37S/77E) in the Indian Ocean[1636] and one was recorded near Macquarie I. (54S/158E).[930]

The Shy Albatross shows a seasonal pattern of movement in the Tasman Sea.[113] It moves south in late Nov., then reappears in the Tasman in large numbers in June. The Reporting Rate in the South-East and Tas. Regions was higher in winter (2.36%) than in summer (0.67%).

The Shy Albatross is usually inshore from the edge of the continental shelf. It feeds almost exclusively on mackerel and squid.[248]

The Albatross I. colony was raided by sealers after its discovery in 1798 and reduced to an estimated 250–300 nests by 1909.[939] By 1960 there were 600–700 nests; in 1973, 1460 nestlings were recorded; and in 1983 it was estimated that at least 2000 eggs were laid annually.[248] In 1978, 3000–4000 birds were estimated to breed on Mewstone[252] and 200 on Pedra Branca.[251]

Field Atlas Records: 493

1° blocks–with records: 57 (7%)
–with breeding records: 2 (4%)

Atlas No.: 091

Max. Rep. Rate: 40%
In block(s): 37/150

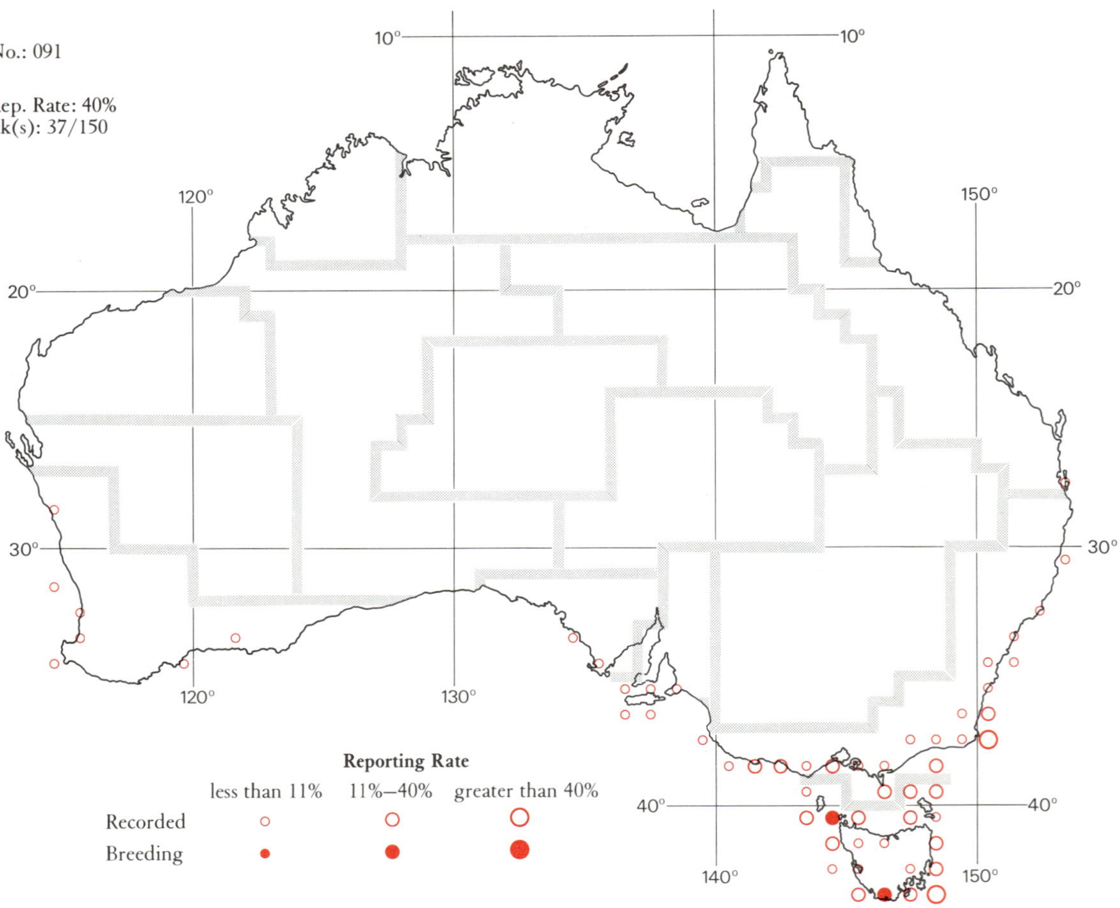

Southern Giant-Petrel

Macronectes giganteus

The Southern Giant-Petrel breeds from the Antarctic continent north to islands near the Antarctic Convergence (50–55°S).[935] The Southern and Northern Giant-Petrels were not distinguished as separate species until 1966.[217] Both species breed on Macquarie (54S/158E), Marion (46S/37E), Prince Edward (46S/37E), the Crozet (46S/50E) and South Georgia (54S/35W) Is.[878] Mixed breeding pairs have been reported; there is no evidence of successful hatching in them at Macquarie I.[934] or the Crozet Is[286] but at South Georgia about 1.5% of pairs are mixed and successfully raise hybrid young.[888] The northern records in the Field Atlas are as far north as the Southern Giant-Petrel has been reported in Australia but a nestling banded on South Georgia was recovered in New Guinea.[878]

Adult birds appear to remain close to their breeding grounds[1636] and most birds reaching Australia are immatures that disperse northward from their breeding colonies, then travel eastward with the prevailing westerly winds.[1761] Some cover long distances rapidly. A fledgeling banded at Macquarie I. was recovered in Chile three months later and a bird from South Orkney I. (60S/45W) reached SA in 10–12 weeks.[1636]

The Southern Giant-Petrel nests about six weeks later than the Northern and most chicks fly in Mar.-Apr.[931,1636] In Australia the Southern is predominantly a winter visitor, arriving later than the Northern and remaining into the spring (Reporting Rates: spring 0.15%; summer 0.04%; autumn 0.06%; winter 0.38%). Observations support the suggestion that the Southern arrives in Australian waters about a month later than the Northern.[115]

At sea the Southern Giant-Petrel feeds on fish, cephalopods, crustaceans, birds, floating refuse and carrion.[933,935] It does not follow ships to the same extent as the Northern[932] but large numbers gather at sewage outfalls, e.g. 50 were feeding near the Malabar (33/151) outfall in July 1973.[1262]

The Southern Giant-Petrel is more common than the Northern near Sydney (33/151)[1262] and probably also in SA.[415] On Macquarie I. there are an estimated 4000 breeding pairs of the Southern.[286,932]

Field Atlas Records: 197

1° blocks–with records: 43 (5%)

–with breeding records:

Atlas No.: 929

Max. Rep. Rate: 20%
In block(s): 37/150

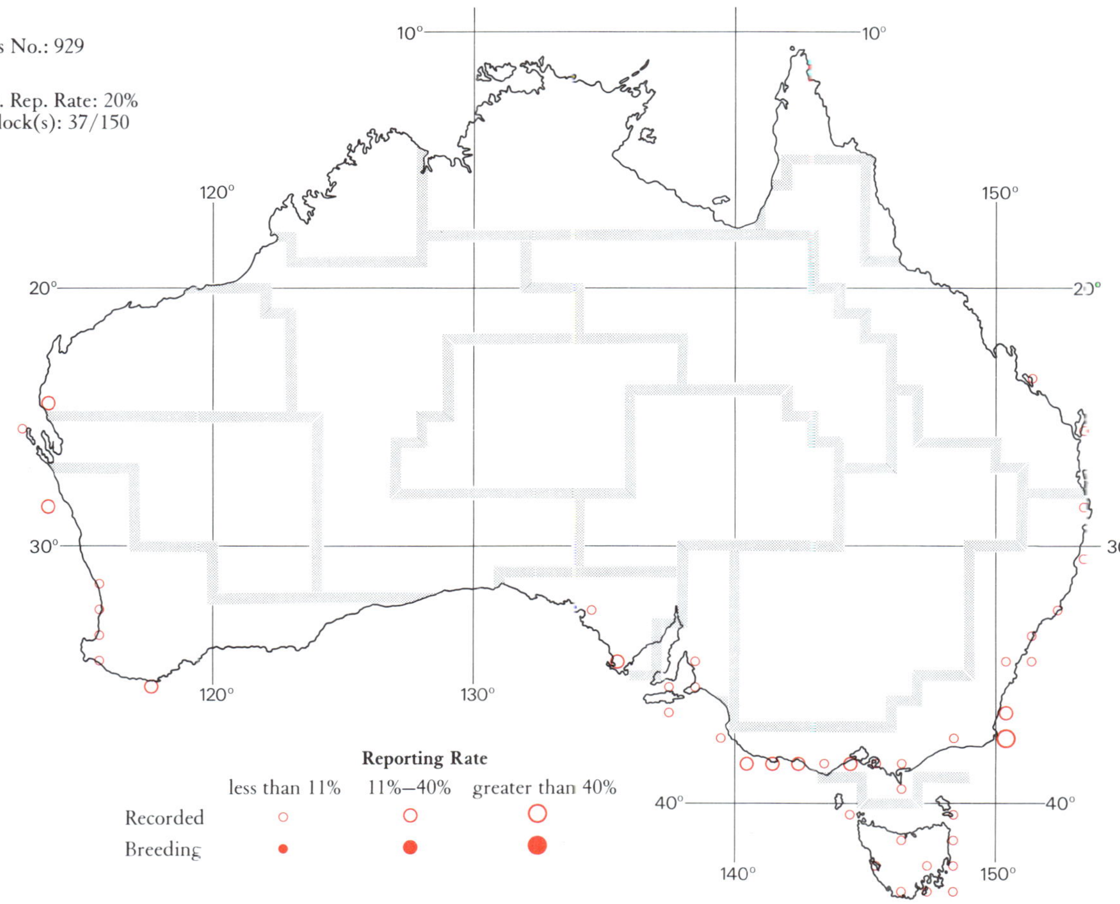

Northern Giant-Petrel

Macronectes halli

The Northern Giant-Petrel breeds on islands close to the Antarctic Convergence (50–55°S) and as far north as 35–45°S.[935] The Northern and Southern Giant-Petrels were not distinguished as separate species before 1966.[217] Both breed on Macquarie (54S/158E), Marion (46S/37E), Prince Edward (46S/37E), the Crozet (46S/50E) and South Georgia (54S/35W) Is.[878] Mixed pairs have been reported; there is no evidence of successful hatching on Macquarie I.[934] or the Crozet Is[286] but at South Georgia about 1.5% of pairs are mixed and successfully raise hybrid young.[888]

A record from Scarness (25/152) in Oct. 1975 indicates that the Northern Giant-Petrel may wander north of the range indicated by the Field Atlas. Like the Southern Giant-Petrel, immatures of the Northern species disperse north and east of their breeding colonies, moving with the prevailing winds.[932] A young bird banded on Macquarie I. in Apr. was seen at Fremantle (32/115) in July and two chicks banded on the Crozet Is reached WA.[1636] Adults banded at Macquarie I. have been recaptured off Sydney (33/151)[950] but immatures predominate in Australian waters.[415] They leave the colonies earlier than the Southern Giant-Petrel,[217] fledging in Feb. at Macquarie I.[931] and most Field Atlas records were in autumn and winter (Reporting Rates: spring 0.03%; summer 0.04%; autumn 0.07%; winter 0.13%). Observations support the suggestion that the Northern Giant-Petrel arrives in Australian waters about a month earlier than the Southern.[115]

In the southern Indian Ocean the Northern Giant-Petrel is more likely to follow ships than the Southern. It feeds on offal and takes carrion more than the Southern[933,935] but both species also feed on fish, cephalopods and crustaceans[1636] and are attracted to sewage outfalls.[1262]

The Northern Giant-Petrel is less abundant than the Southern near Sydney[1262] and probably also in SA.[415] On Macquarie I. there are an estimated 1000 breeding pairs.[932]

Field Atlas Records: 67

1° blocks–with records: 22 (3%)
–with breeding records:

Atlas No.: 937

Max. Rep. Rate: 16%
In block(s): 37/150

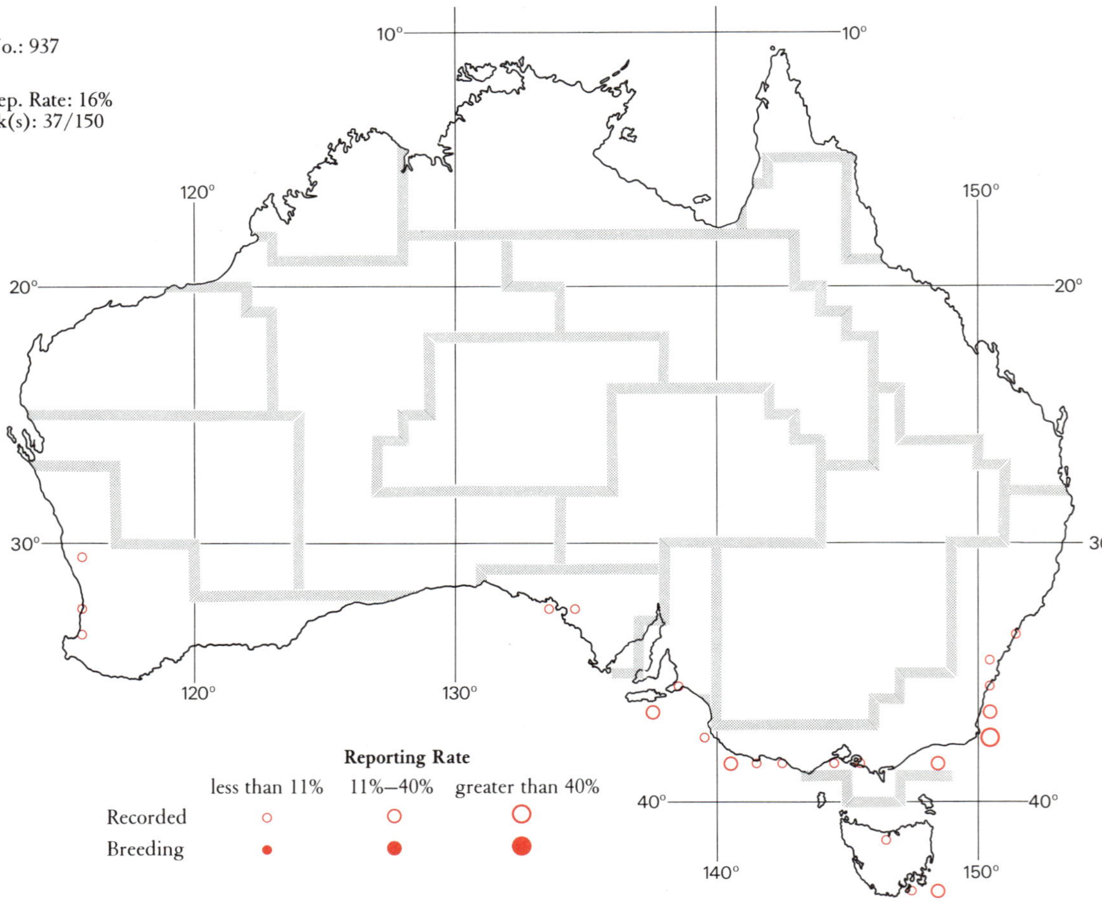

Southern Fulmar

Fulmarus glacialoides

The Southern Fulmar breeds on the coast and islands of Antarctica and is commonly recorded in the pack-ice north to the Antarctic Convergence.[937] In Australia specimens were collected at Queenscliff (38/144) in 1881 but there are no regular sightings in the Atlases until the 1951–76 period. In the east beachwashed specimens were recovered as far north as Moreton I. (27/153) in Oct. 1972.[1787] In the west, between 1945 and 1974, there were records from Jurien Bay (30/115) south to the Warren R (34/116). In the Field Atlas 92% of the reports are attributable to an irruption between June 1978 and Jan. 1979. There was a corresponding irruption in New Zealand.

The Southern Fulmar probably undertakes a regular migration northward in winter.[415,1821] In Australia, as with the 1978 invasion, most records have been made June–Jan.[415,1292] The species feeds on cephalopods, crustaceans, fish and offal.[1636]

There are no measures of abundance.

Field Atlas Records: 72

1° blocks–with records: 30 (4%)

–with breeding records:

Atlas No.: 074

Max. Rep. Rate: 3%
In block(s): 38/140,38/141,42/145

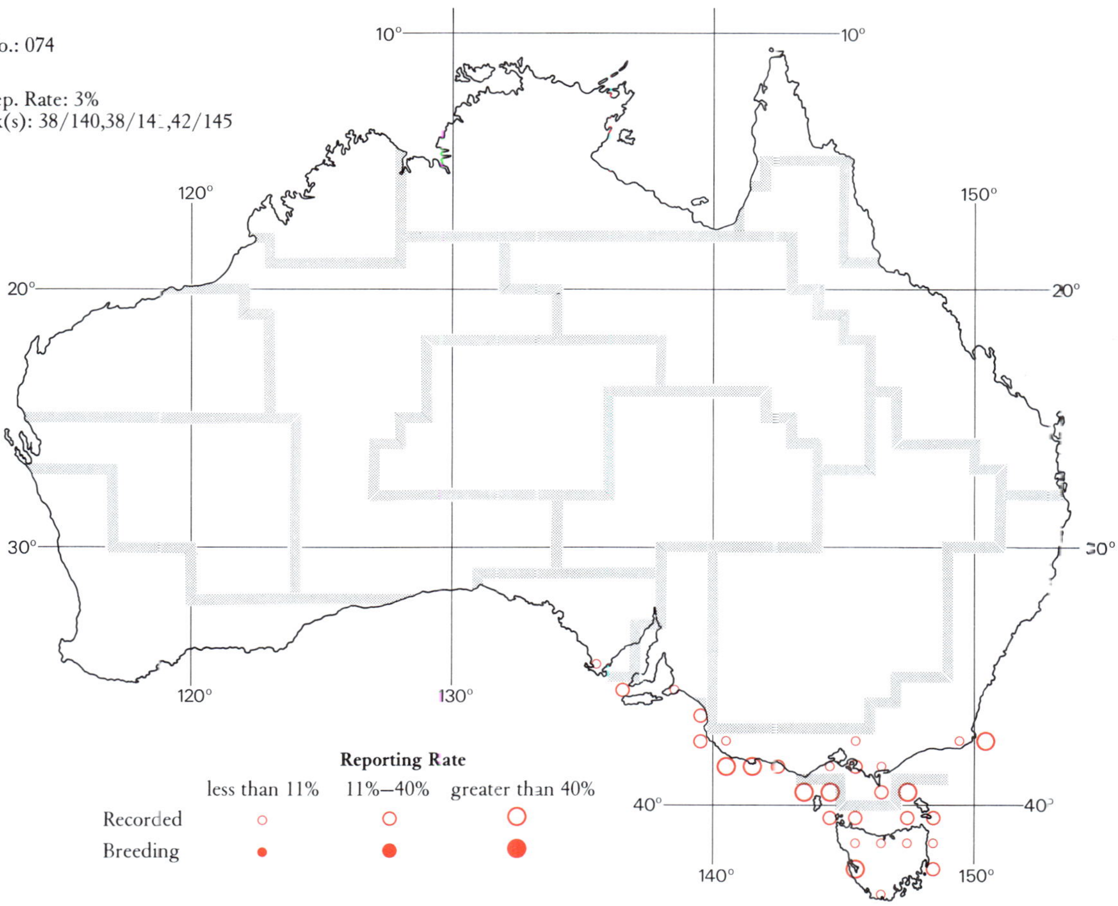

Cape Petrel

Daption capense

The Cape Petrel breeds on the mainland and islands of Antarctica and on sub-Antarctic islands, including Heard (53S/73E) and some off New Zealand. It might breed on Macquarie I. (54S/158E) but there is no conclusive evidence.[1535] New Zealand birds have been distinguished as a separate subspecies *australe*[1820] but subspecific identification is difficult and the origins of birds reaching Australian waters are unknown.

In eastern Australia the Atlases contain more or less continuous records north to 27°S and one as far north as Palm Beach (16/145) in Oct. 1975. In the west the species was regarded as rare before 1920 but records exist north to 21°S.[1622] There are only two sightings in the Atlases along the southern coast between 116°E and 133°E, both near Albany (35/117). Records of ANARE between Australia and Antarctica show the Cape Petrel to have a disjunct distribution at sea, which may represent the races *australe* of the New Zealand region and *capense* of the Antarctic:[937] one group of sightings is in the southern Tasman Sea and the other farther south and west near the edge of the pack-ice.

During the winter the Cape Petrel wanders farther north than many other Antarctic petrels, crossing the equator and possibly reaching Europe.[422] Australian records are correspondingly most frequent in winter (Reporting Rates: spring 0.10%, summer 0.08%, autumn 0.07%, winter 0.24%). The Cape Petrel is a winter visitor in the western Tasman Sea also, appearing in Apr.-May and reaching maximum abundance in Aug.-Sept.[339] Circumpolar movements are indicated by recoveries of South Orkney (60S/45W) birds in the New Hebrides (16S/166E), New Zealand and Chile.[422]

The Cape Petrel is often washed up on Australian beaches after storms. At sea it is seen singly, in pairs or in flocks. A flock of 250–300 birds was recorded near Sydney (33/151) in Sept. 1973.[1262] On seven out of eight recent sea trips to the edge of the continental shelf off Tas., small numbers were recorded well offshore. The Cape Petrel feeds on krill, other crustaceans, squid, fish and some offal from seal colonies. It is attracted to sewage outfalls such as those near Geelong (38/144) and the old whaling station at Albany (35/117) and follows ships.[215]

There are no measures of abundance.

Field Atlas Records: 129

1° blocks–with records: 42 (5%)
–with breeding records:

Atlas No.: 080

Max. Rep. Rate: 12%
In block(s): 37/150

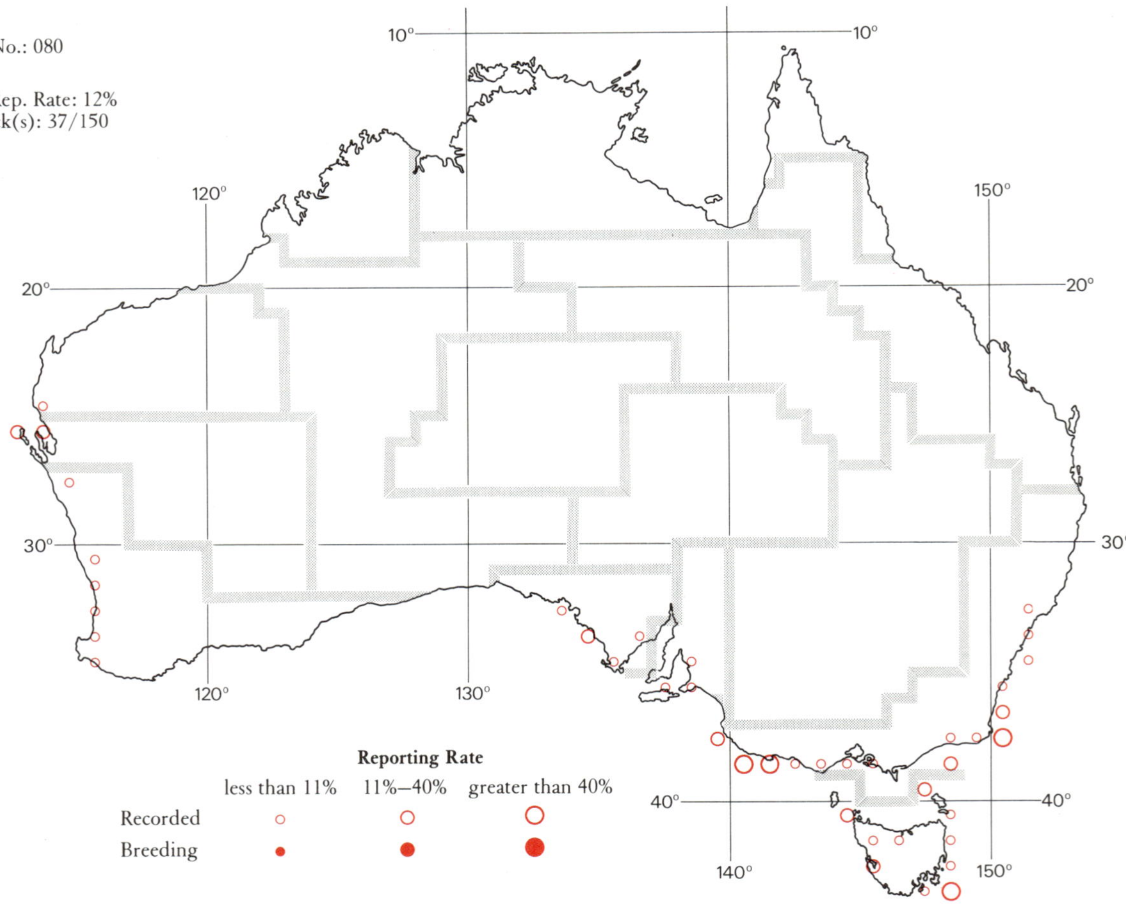

Great-winged Petrel

Pterodroma macroptera

There are breeding colonies of the Great-winged Petrel on sub-Antarctic islands, in New Zealand and in the South-West Region of Australia. Views about subspecies differ[415,1636] but birds of the New Zealand breeding populations can be distinguished from those of the South-West and both are seen in Australian waters. Birds from sub-Antarctic breeding islands in the Indian and Atlantic Oceans may also reach Australia. In NSW birds of the New Zealand subspecies predominate;[1292] in Vic. both subspecies have been recorded. Specimens in SA are thought to be from either the South-West colonies or the sub-Antarctic islands.[415] The origin of Tas. birds has not been determined.

In addition to the breeding areas shown on the map, there are colonies, visited in the 1951–76 period, on Bald I. (34/118) and islands in the Recherche Archipelago including Middle I. (34/123). Breeding was also confirmed on Remark I. (34/121) in 1981.[1031]

The Great-winged Petrel is occasionally blown inland specially in WA. Wrecks are associated with winter gales.

In the Field Atlas records of the Great-winged Petrel east of 129°E were made at all times of the year. West of 129°E there were also reports in all seasons but they were most widespread in spring, including a sighting in Nov. of 4–10 birds as far north as 22/113. The species is absent from its colonies in the South-West for only about two months each year and breeding adults are thought to remain nearby.[1636]

The Great-winged Petrel breeds in winter and some of its burrows are used in spring by breeding Flesh-footed Shearwaters. Occupation of colonies by the two species overlaps but it is not known whether they affect each other.

The only information about food is that cephalopod beaks have been found in guts.[1636]

On Eclipse I. (35/117) the breeding population was estimated at 10 000–15 000 in 1973.[675] In 1976, 2000–3000 Great-winged Petrels were seen at dusk circling Coffin I. (35/118).[1687]

Field Atlas Records: 70

1° blocks–with records: 28 (3%)
–with breeding records: 3 (11%)

Atlas No.: 075

Max. Rep. Rate: 18%
In block(s): 37/150

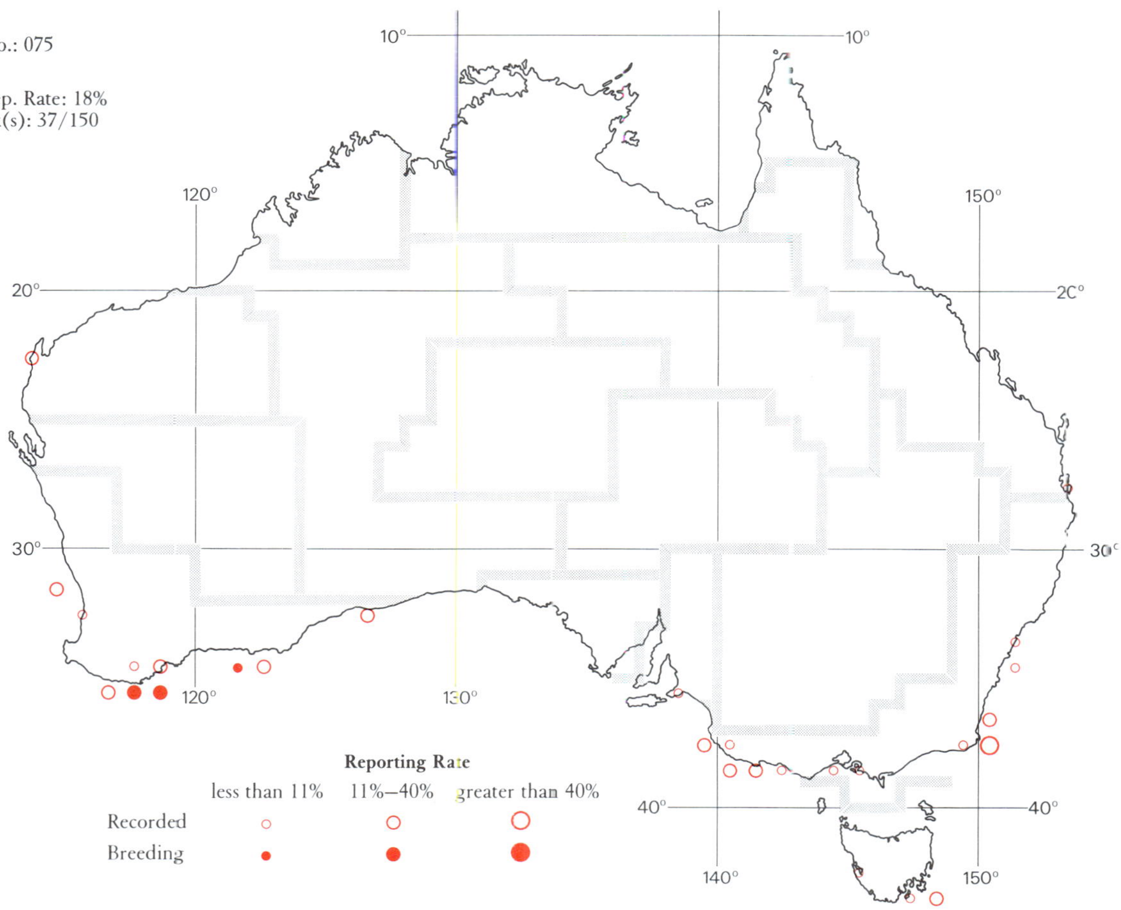

White-headed Petrel

Pterodroma lessonii

The White-headed Petrel breeds only on Auckland (50S/166E), Antipodes (49S/178E), Macquarie (54S/158E) and Kerguelen (49S/70E) Is.[1636] The origin of birds reaching Australia is unknown. The northernmost record in eastern Australia is of a beachwashed specimen found in June 1974 on North Stradbroke I. (27/153)[1492] and in the west the species is said to range north to the Dampier Archipelago (20/116).[1638] The White-headed Petrel is occasionally blown inland. Such a sighting was made west of Gunbar (33/145) in June 1977.[1272]

The White-headed Petrel is absent from its breeding colonies on Macquarie I. in June-July.[248] During the Field Atlas the species was least often reported in summer but the total number of records was small and seasonal differences were not great (Reporting Rates: spring 0.09%, summer 0.04%, autumn 0.06%, winter 0.07%).

Most Atlas records are of beachwashed birds and sightings of live birds are well out to sea, usually at the edge of the continental shelf. The White-headed Petrel feeds on cephalopods.[1636]

The breeding population on Macquarie I. was said, in the 1960s, to be declining as a result of predation by cats and skuas and disturbance by rabbits.[1814] The decline has continued[248] but the species is no longer considered seriously threatened following the control of rabbits by myxomatosis begun in 1978.[1537] The breeding population on Macquarie I. is estimated to be about 16 000 birds.[1537]

Field Atlas Records: 55

1° blocks–with records: 23 (3%)

–with breeding records:

Atlas No.: 077

Max. Rep. Rate: 4%
In block(s): 37/150

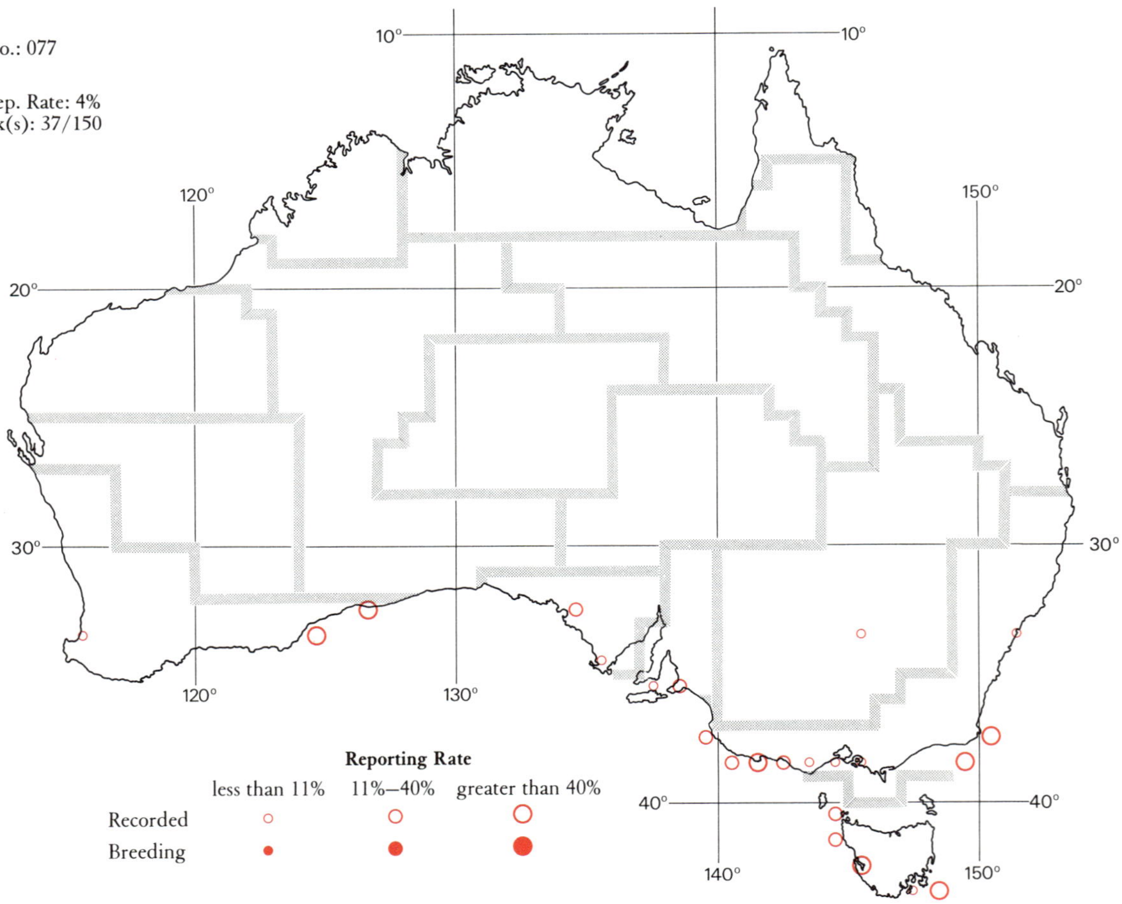

Kerguelen Petrel

Pterodroma brevirostris

The Kerguelen Petrel is washed up on southern Australian coasts in winter and spring. It breeds on sub-Antarctic islands in the Atlantic and Indian Oceans.[763] Since its discovery in Australia in 1926[1636] the species has been recorded in the South-West Region and from the coasts of Tas. and south-eastern Australia as far north as Mermaid Beach (28/153).[1786]

Breeding birds generally remain near the breeding islands from Sept.-Mar. though summer sightings of the species have been made as far north as 24°S on the Brazilian coast and as far south as the Antarctic pack-ice.[763] Sightings at sea indicate a nearly circumpolar movement.[763] All but one of the records in the Field Atlas were made in winter and spring, which suggests that the species may generally range farther north in winter than in summer.

In New Zealand the Kerguelen Petrel is regularly washed ashore.[763] Individuals and flocks apparently become trapped on lee shores by bad weather and succumb. Almost all Australian records are of wrecked birds, sometimes in large numbers.[1393] Between Aug. and Nov. 1981, 36 were picked up on beaches near Portland (38/141) and many were washed up elsewhere in Vic. as well as in SA and Tas. Records at this time account for 53% of Field Atlas reports. The number of reports partly depends on the frequency of beach-walks by observers[415] and lack of such systematic searches in the South-West is probably why there were few records there during the Field Atlas. The scarcity of sightings at sea in winter, both in Australian waters and elsewhere,[763] leaves open the question of where the bulk of the population winters.

Cephalopods and some fish remains have been found in guts.[1393]

There are no measures of its abundance but at Gough I. (40S/110W) thousands were attracted to a light switched on at night near the breeding grounds.[1729]

Field Atlas Records: 34

1° blocks–with records: 16 (2%)
–with breeding records:

Atlas No.: 935

Max. Rep. Rate: 2%
In block(s): 38/140

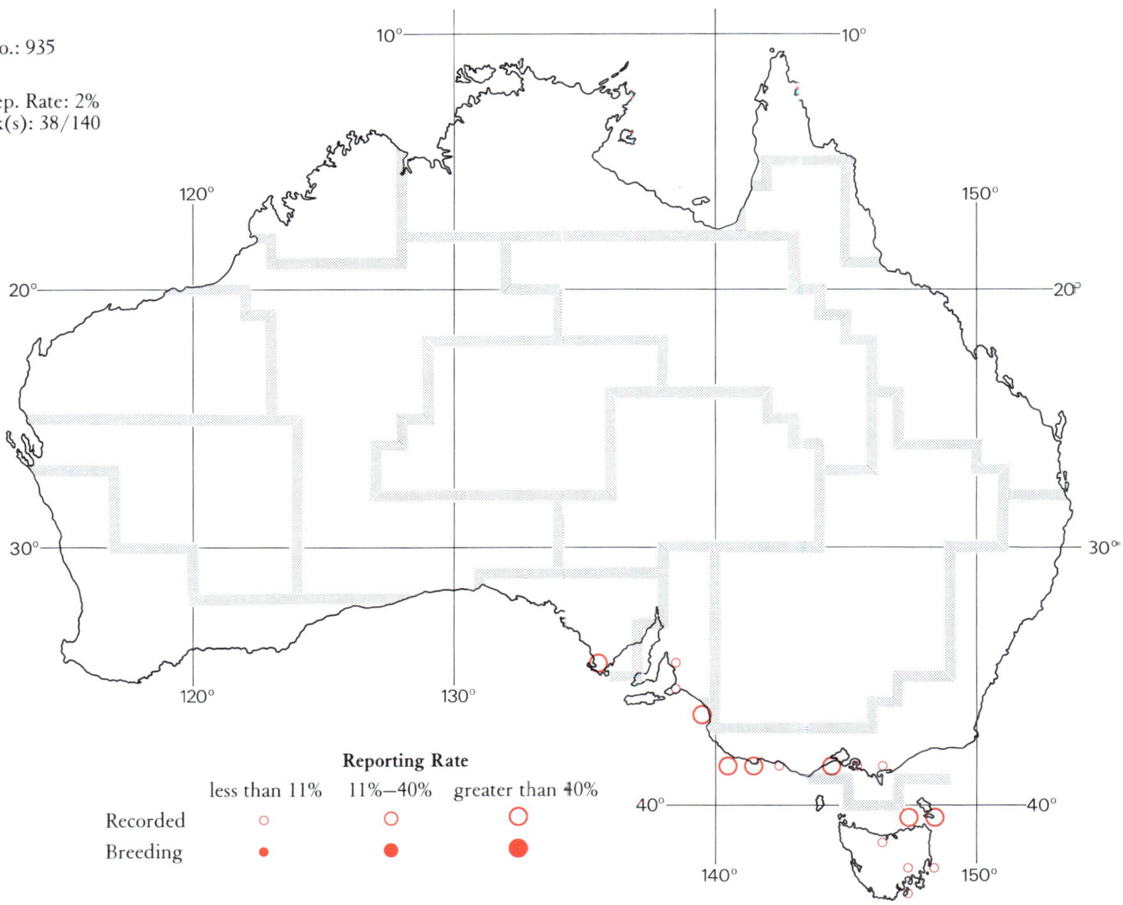

Gould's Petrel

Pterodroma leucoptera

Gould's Petrel lives in the south tropical Pacific Ocean. The Australian subspecies *leucoptera* breeds only on Cabbage Tree (Gould's) I. (32/152).[1636] A second subspecies *caledonica* nests in New Caledonia (21S/165E)[895] and others, sometimes regarded as separate species, breed in the New Hebrides (16S/166E) and Fiji (18S/175E). Birds breeding on Mas Afuera (36S/81W), previously identified as Gould's Petrel, are now considered a different species.[776]

The recently described subspecies, *caledonica*, has been recorded in Australia: there are specimens from Corindi (30/153) in Nov. 1975,[197] Middleton (35/138) in May 1976 and Southend (37/140) in Oct. 1979. Gould's Petrels regularly recorded in the Tasman Sea Dec.-Apr. are probably also *caledonica*.[895]

There are no confirmed records of *leucoptera* far from the breeding island yet the bird is absent from its colony May-Oct.[1292] Whether, like *caledonica*, it migrates eastward and northward in winter is not known. The longest movement recorded for *leucoptera* is from Cabbage Tree I. to Jervis Bay (35/150), 293 km.[85]

Gould's Petrel is seen at sea off the continental shelf usually singly. Away from the breeding grounds most records in the Field Atlas are of beachwashed birds. Cephalopods are the only known food of Gould's Petrel.[1636]

The total number of breeding birds on Cabbage Tree I. was estimated at 500–1000 in Dec. 1970.[673,1029] The small total population of *leucoptera* probably accounts for the lack of records away from the breeding ground.

Field Atlas Records: 15

1° blocks-with records: 9 (1%)
-with breeding records: 1 (11%)

Atlas No.: 078

Max. Rep. Rate: 2%
In block(s): 32/152,43/148

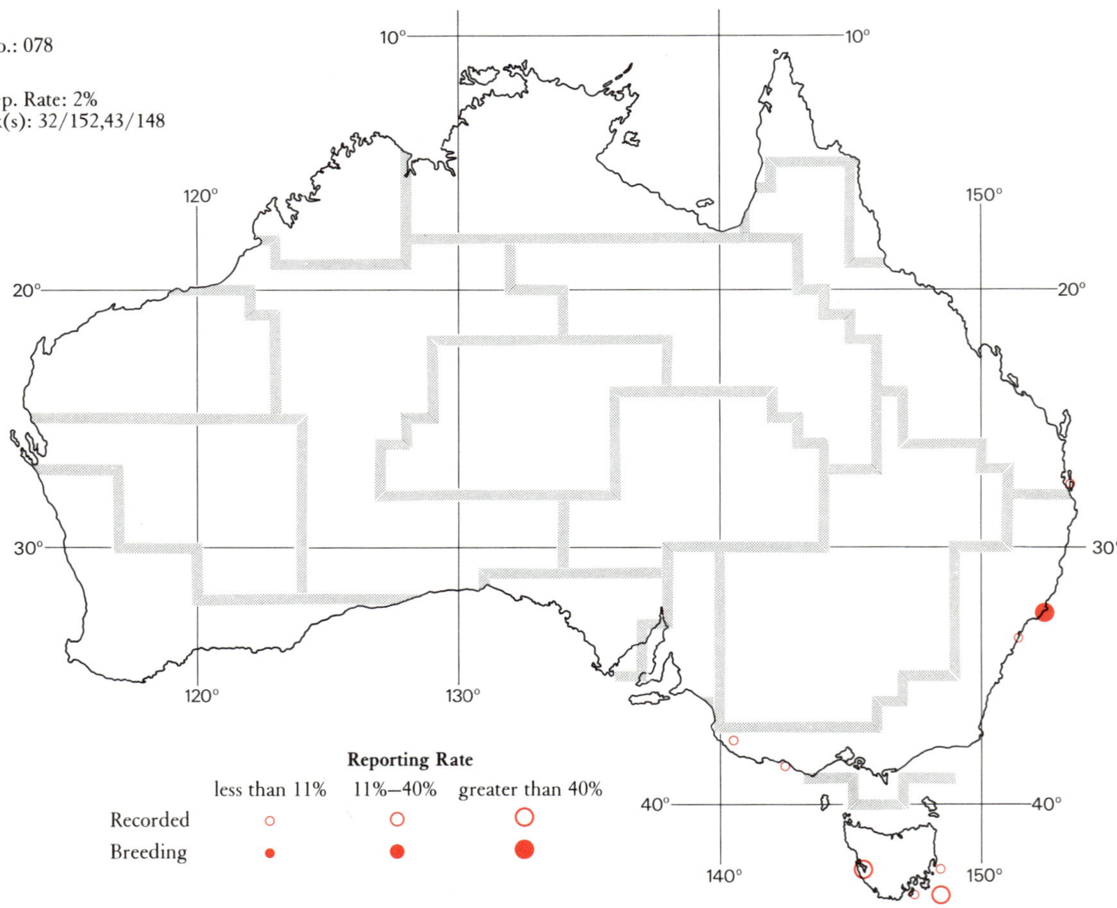

Broad-billed Prion

Pachyptila vittata

Lesser Broad-billed Prion

Pachyptila salvini

Antarctic Prion

Pachyptila desolata

These Prions proved almost impossible to differentiate so records are combined on the map. The Broad-billed breeds to the north of the other two species on islands west of Australia (Ile Amsterdam (34S/78E), Ile St Paul (38S/77E), Tristan da Cunha (38S/12W), Gough (40S/10W)), off the south island of New Zealand and in the Chatham Is (44S/176W). The Lesser Broad-billed breeds only on Indian Ocean islands (Prince Edward (46S/37E), Marion (46S/32E), Crozet (46S/50E)). The Antarctic breeds on islands in the Indian and Atlantic Oceans and south of Australia (Auckland (50S/156E), Macquarie (54S/158E) and Scott (67S/179W) Is)[419] and possibly at Cape Denison (67S/142E).[762]

Adult Antarctic and Lesser Broad-billed Prions, as well as immatures, are absent from the colonies Apr.-Sept. or Oct. Many Broad-billed Prions remain about the colonies all year though fewer are present Feb.-July.[762] All three prions are reported on the southern coast of Australia in winter (Reporting Rates: spring 0.09%, summer 0.01%, autumn 0.04%, winter 0.17%). The Antarctic is the most often reported and widespread, reflecting its circumpolar breeding distribution; there are confirmed records round the coast from southern Qld to the South-West Region. The Lesser Broad-billed is often reported on the southern coast but rarely in the east; one confirmed report was as far north as Brisbane (27/153).[1709] The rarest of the three, the Broad-billed, reaches the South-West and occasionally SA and is also reported on the eastern coast. The relative abundance of the three species in SA is indicated by the number of specimens lodged in the SA Museum: three Broad-billed, 41 Lesser Broad-billed and 72 Antarctic.[415] All Australian records of these three prions are of beach-washed birds. The birds eat mainly krill, also taking cephalopods, pteropods and small fish.[762,1636]

Field Atlas Records: 90

1° blocks–with records: 28 (3%)
–with breeding records:

Atlas No.: 082,941,084

Max. Rep. Rate: 5%
In block(s): 38/141

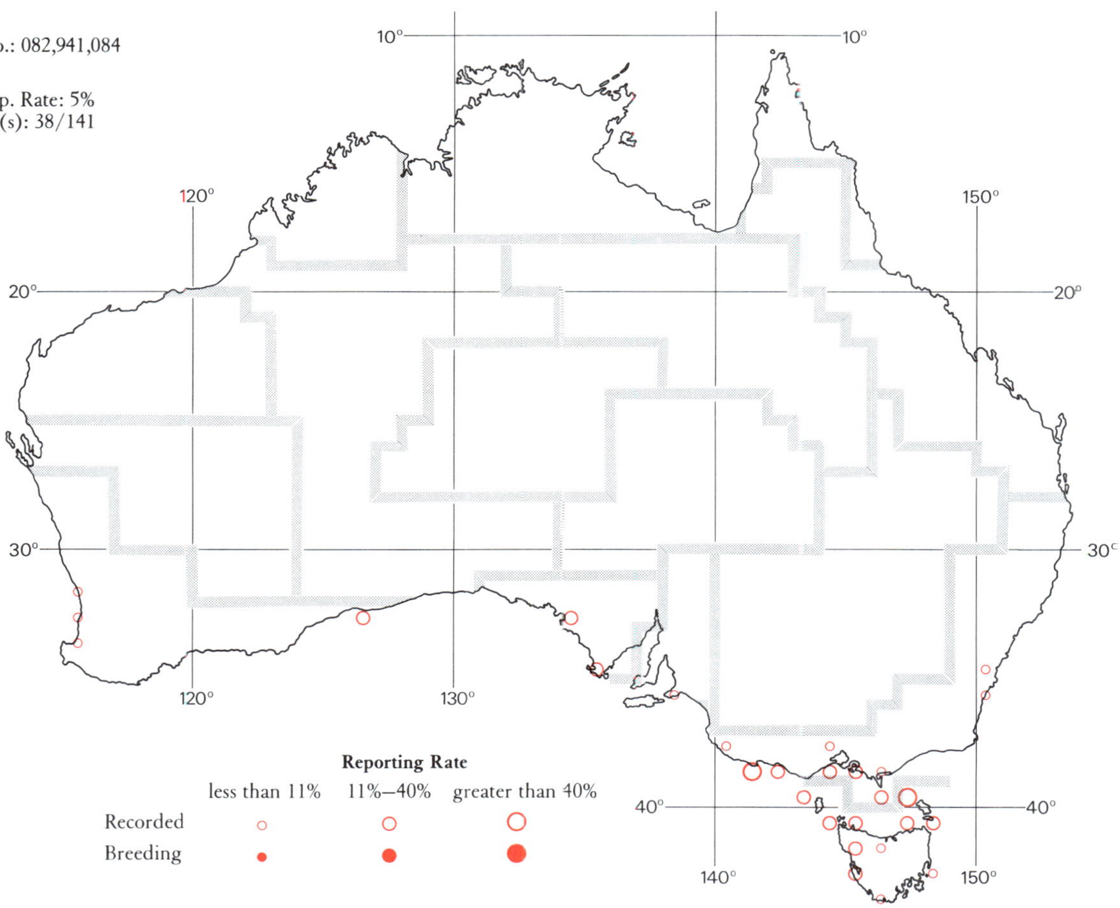

Slender-billed Prion

Pachyptila belcheri

The Slender-billed Prion visits the coasts of southern Australia in the winter non-breeding season. Breeding colonies are on sub-Antarctic islands to the west of Australia. The largest are on the Falkland Is (51S/59W)[319] and smaller ones are on the Crozet (46S/50E) and Kerguelen (49S/70E) Is and possibly on other islands off South America. The Slender-billed Prion is difficult to differentiate from other prions in the field.

In the Falkland Is the Slender-billed Prion is absent from the breeding grounds Apr.-Aug.[319] Reporting Rates in Australia were highest in winter (0.20%) and spring (0.10%) and the species was virtually absent in summer (0.02%) and autumn (0.01%). It is most abundant on the southern coasts of Australia, carried there by the prevailing westerly winds. The species is rarely recorded on the eastern coast though one specimen has been picked up as far north as Fraser I. (25/153) in Sept. 1973.[1788] Most Australian records are of birds beachwashed after winter storms.

In one area the Slender-billed Prion was found to feed mainly at night on an amphipod *Parathemisto gaudichaudii* and less often on small cephalopods.[761]

In the Falkland Is breeding colonies are being established on new sites and numbers seem to be increasing despite disturbance and predation.[319,419]

There are no measures of abundance.

Field Atlas Records: 109

1° blocks–with records: 25 (3%)
–with breeding records:

Atlas No.: 942

Max. Rep. Rate: 4%
In block(s): 38/141,39/146

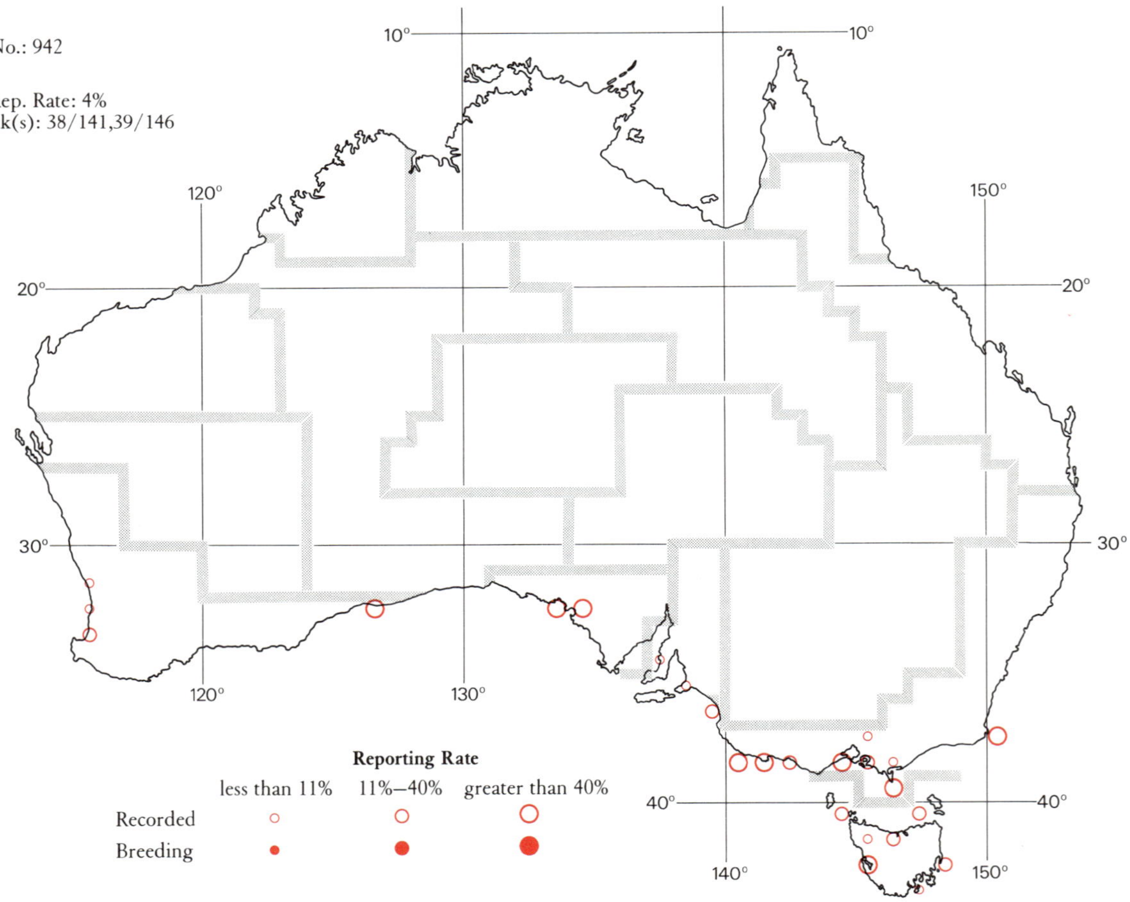

Fulmar Prion

Pachyptila crassirostris

Fairy Prion

Pachyptila turtur

The Fairy and Fulmar Prions are difficult to differentiate in the field or from beachwashed material so records of the two species are combined on the map. There is doubt about the definition of subspecies.[419,762] The Fairy Prion breeds in Australia on islands in Bass Strait and round the Tas. coast[1636] and on Macquarie I. (54S/158E);[248] it also breeds on islands near New Zealand, on Marion I. (46S/37E) and in the Falkland Is (51S/59E).[419,762,1636] In eastern Australia it has been recorded as far north as Wreck I. (23/151) where a juvenile was beachwashed in Sept. 1970.[205] It is rare west of 138°E. The Fulmar Prion breeds on islands near New Zealand and on Heard I. (53S/73E).[762] It remains near the colonies all year[762] and there are no confirmed records in Australia.

Fairy Prions breeding round New Zealand are migratory[762] but in Australia large numbers visit the colonies throughout the year.[248] Away from the colonies prions are recorded all year east of 140°E though more frequently June-Aug. on the NSW coast.[1292] West of 140°E Field Atlas records were predominantly in summer (Reporting Rates: summer 0.15%, winter 0.03%). It has been suggested that birds reaching the South-West Region and SA come from the Marion I. breeding colonies but that those in eastern Australia are from the local or New Zealand populations.[415,1638]

The Fairy Prion feeds on crustaceans, cephalopods and marine plankton.[1785] Colonies round Tas. appear to be stable but at Lawrence Rock (38/141) there is evidence of a decline attributed to erosion making the surface soils less suitable for burrows.[1419]

Estimates of breeding colonies are: Tasman I. (43/148) and Ile de Golfe (43/146) 50 000 birds each in 1978;[250,1853] Flat Witch I. (43/146) 20 000 birds in 1979;[1850] Flat I. (43/145) 5000 birds in 1977;[1851] Round Top I. (43/146) 2400 birds in 1980.[1854] There are also big colonies in the Hunter (40/144) and Kent (39/147) Groups[248] and on Lady Julia Percy I. (38/142).[768]

Field Atlas Records: 356

1° blocks–with records: 43 (5%)
–with breeding records: 10 (23%)

Atlas No.: 947,083

Max. Rep. Rate: 31%
In block(s): 39/146

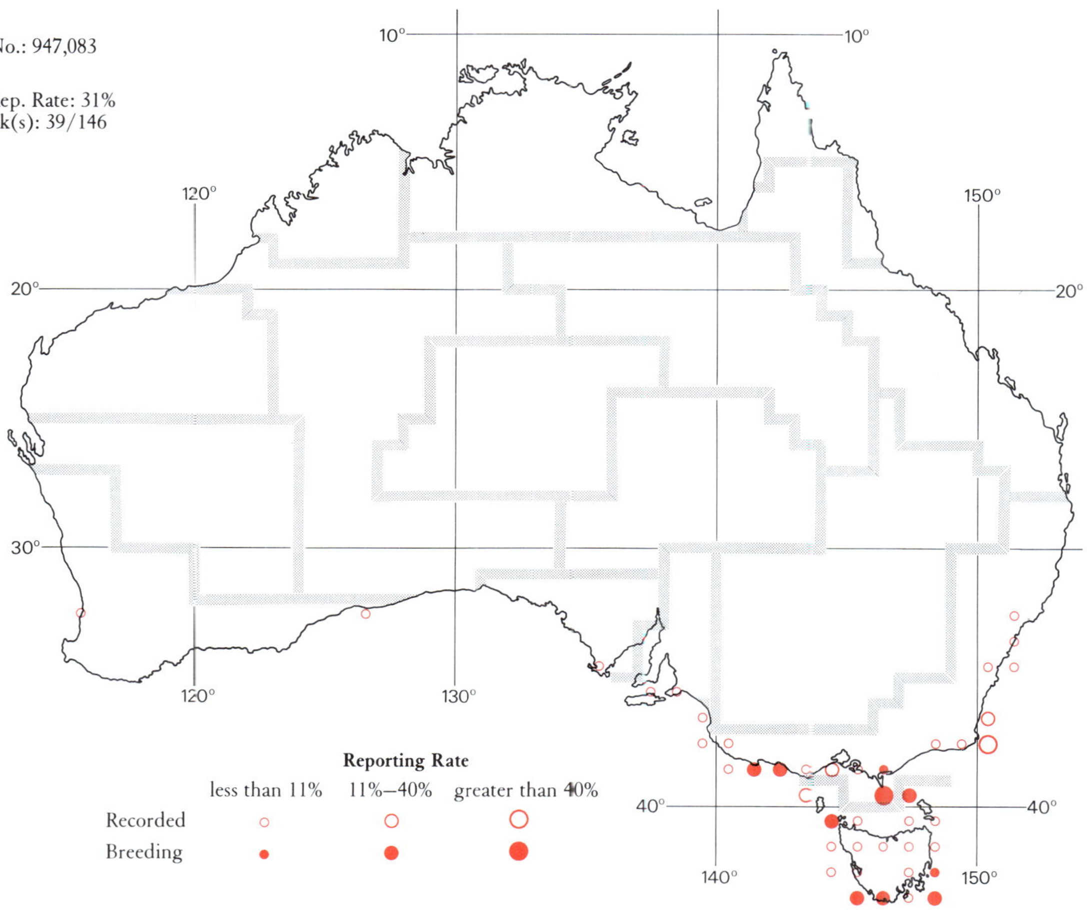

Streaked Shearwater

Calonectris leucomelas

The Streaked Shearwater, which breeds on islands in the north-western Pacific Ocean, was added to the Australian list in 1978,[1773] by three beachwashed specimens on Stradbroke I. (27/153) in Mar. 1974.[1789] The only previously published record was of a specimen collected in 'N. Australian seas' last century.[1636] In Dec. 1974 (incorrectly stated as 1976 in publication) more than 100 were seen in western Torres Strait.[338] Others were recorded 1978–80 on the Sahul Shelf (12/122), at Great Detached I. (11/143), near Ballina (28/153) and off Gabo I. (37/149).[112,316,1167] In Nov.-Dec. 1980 it became apparent that the species was more than a vagrant to Australia when it was recorded at sea round northern Australia from the Kimberley Region to Torres Strait.[316] Subsequently there have been sightings north-north-west of Geraldton (28/114) in May 1981[947] and on the eastern coast in 1982.[316]

The Streaked Shearwater migrates south and is seen in the non-breeding season off New Guinea, Malaysia, Ceylon and Australia.[947] Australian sightings have all been Nov.-May.

Individuals, groups and flocks of Streaked Shearwaters have been reported. Flocks of 70 were recorded on two occasions during a sea voyage round northern Australia and 300 were counted during one ten-minute period.[316] Groups congregate when feeding, sometimes with terns and other seabirds. The Streaked Shearwater has been observed feeding over schools of small fish, specially when these were being harried by predatory species such as mackerel and tuna.[112,316,947]

If 1980 was a typical year in northern Australia, it is surprising that the species was not recorded earlier. Lack of observers may be the reason but there is some evidence that the species may recently have expanded its range southward: it was not recorded in about 90 trips to sea off Ballina in the late 1970s[846] but has been recorded there most years since 1980.[315]

A total of about 1500 individuals was recorded on the voyage round northern Australia.[316] There are no other measures of abundance.

Field Atlas Records: 35

1° blocks–with records: 34 (4%)
–with breeding records:

Atlas No.: 853

Max. Rep. Rate: 4%
In block(s): 12/122

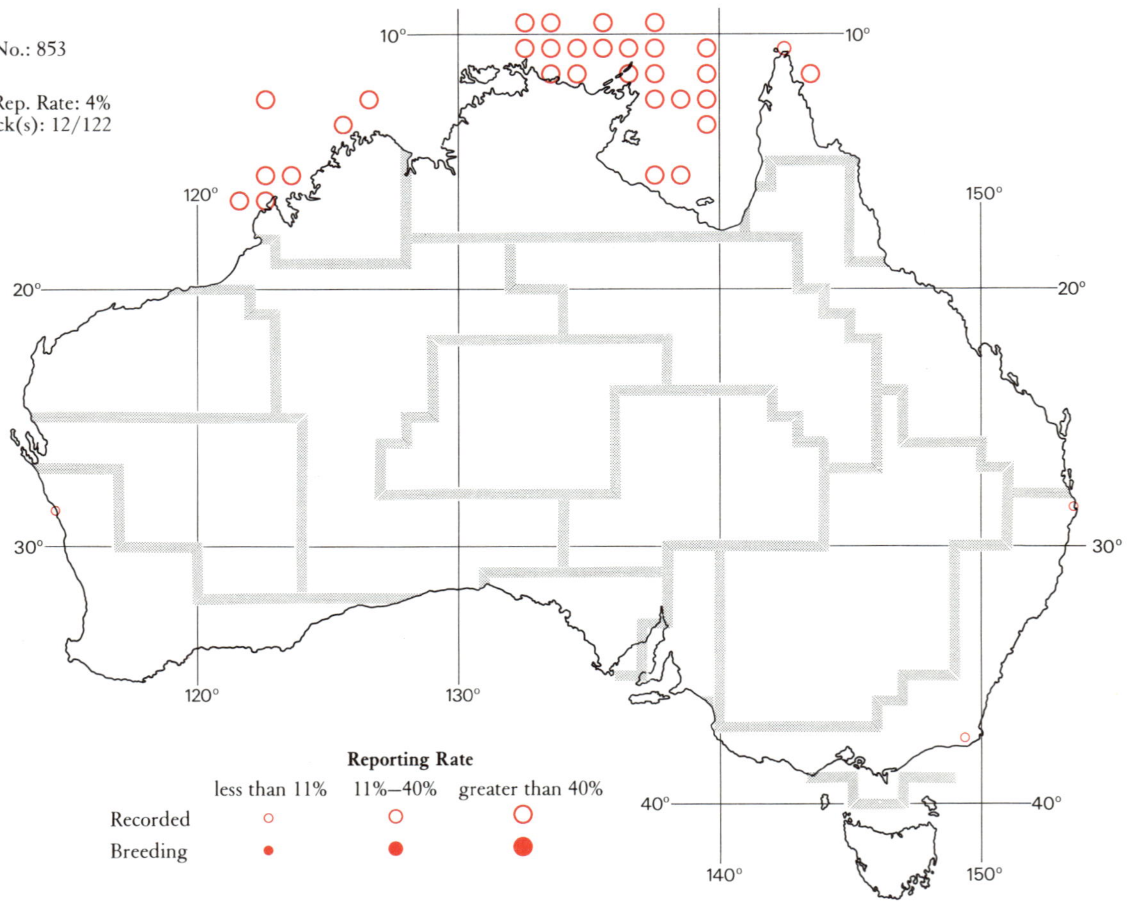

Flesh-footed Shearwater

Puffinus carneipes

The Flesh-footed Shearwater breeds in New Zealand and Australia including Lord Howe I. (31/159).[1636] Records from the South-East Region are known to include birds bred on Lord Howe I. and records west of 140°E are presumably birds of the population that breeds on islands off the South-West Region.

The Flesh-footed is the common breeding shearwater of the Recherche Archipelago. It also breeds on other islands west to Cape Leeuwin (34/115) including Coffin (35/118), Breaksea (35/118), Michaelmas (35/118), Eclipse (35/117), Stanley (35/117), Sandy (34/116) and St Alouarn (34/115).

The species migrates north in winter. Banded birds from Lord Howe I. have been recovered from as far north as Japan and Korea. Flesh-footed Shearwaters from colonies in the South-West are thought to migrate to the Indian Ocean in winter.[1636] The species is common in the Gulf of Oman (24N/58E) then, arriving about May. In the Field Atlas, the seasonal distribution of records differed in the east and west. East of 141°E Reporting Rates were highest in summer and autumn (spring 0.02%, summer 0.18%, autumn 0.09%, winter 0.02%) and near Sydney (33/151) regular counts recorded the species Dec.-Mar.[1262] Banding has shown that birds from the Lord Howe I. colony range as far as eastern Australia in the breeding season, Sept.-May.[676] In the west the species was most frequently reported in spring and summer (spring 0.51%, summer 0.60%, autumn 0.27%, winter 0.01%). All records on the western coast north of 34°S were made in spring when migrating birds were presumably returning to their breeding colonies.

The Flesh-footed Shearwater breeds in late spring. Some of its burrows are used in winter by breeding Great-winged Petrels. Occupation of colonies by the two species overlaps but whether they have any effect on each other is unknown. The Flesh-footed Shearwater feeds on small fish, cephalopods and crustaceans. It will congregate round fishing boats for offal.[1262]

Estimates of colony sizes include: Sandy I. 600 000 birds;[7] Eclipse 12 000–16 000 birds;[675] Breaksea 2000–10 000 birds;[2] and Lord Howe I. 35 000 birds in 1971.[376]

Field Atlas Records: 130

1° blocks–with records: 48 (6%)

–with breeding records: 6 (13%)

Atlas No.: 072

Max. Rep. Rate: 14%
In block(s): 37/150

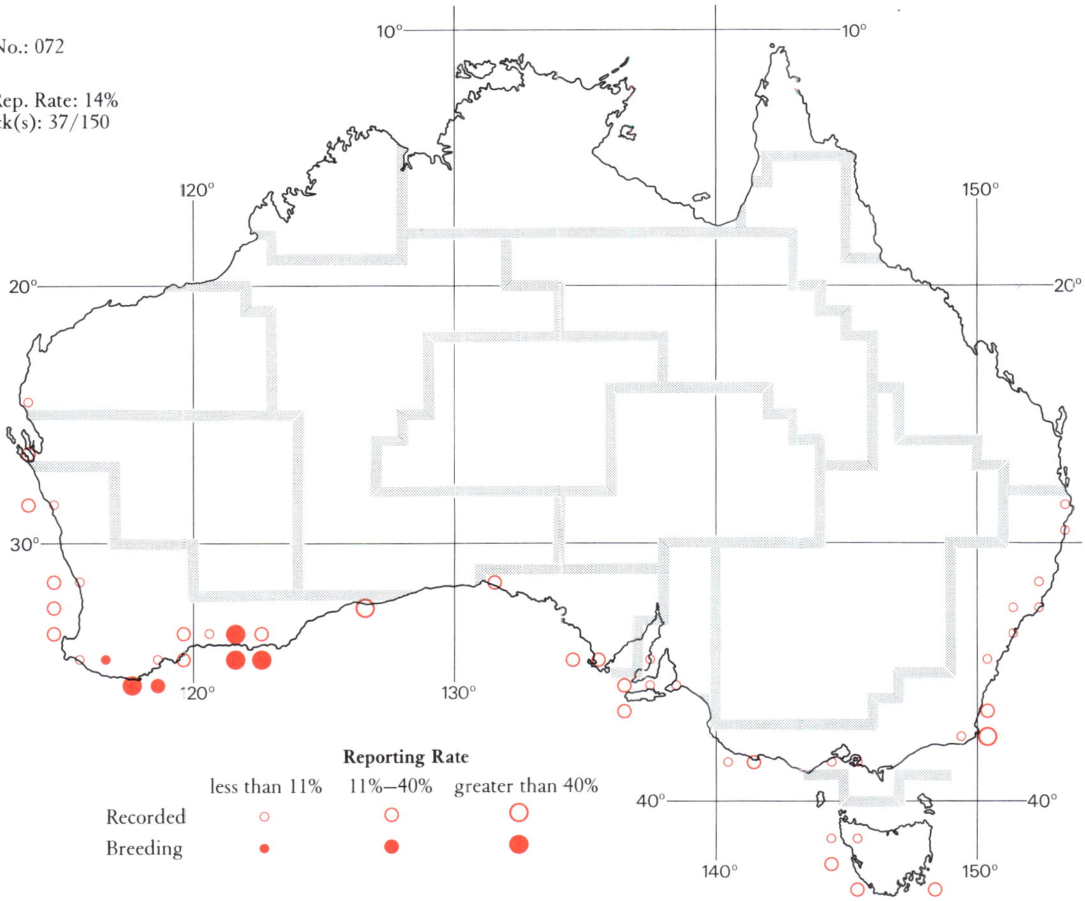

Wedge-tailed Shearwater

Puffinus pacificus

Two subspecies of the Wedge-tailed Shearwater are recognized in Australia: *royanus* breeding in the east and *chlororhynchus* in the west.[388] The species also breeds on tropical and sub-tropical islands in the Indian and Pacific Oceans. In Australia the bird is rare on northern and southern coasts. The Top End record (12/130) is the first in the Atlases from the northern coast between 124°E and 142°E; and on the southern coast between 115°E and 149°E the Historical Atlases contain few records to add to that on the map.

The Wedge-tailed Shearwater breeds on islands off the eastern coast from Raine (11/144) to Montagu (36/150), including some in the Coral Sea, and off the western coast from Forestier (20/117) to Carnac (32/115).[1636] There are brown- and white-bellied plumages. In Australia white-bellied birds are known to breed only at Shark Bay (26/113) where they constitute 20–30% of the breeding population; there are other mixed colonies in the Pacific.[1632] White-bellied individuals have rarely been seen in eastern Australia. Their origin is unknown though some may breed there.[1028]

Some populations of the Wedge-tailed Shearwater migrate.[925] In NSW the species is present only during the breeding season, Aug.-May,[1523] and the Field Atlas suggests that the pattern is similar in other parts of Australia (Reporting Rate: summer 0.84%, winter 0.07%). At least 13 individuals banded in NSW have been recovered in the Philippines but the wintering grounds of the bulk of the population are unknown.

On 14 islands off the NSW coast the species breeds in mixed colonies with Sooty and Short-tailed Shearwaters. The Wedge-tailed Shearwater feeds on cephalopods, crustaceans and small fish.[1636] Construction work on Slope I. (26/113) enabled cats and foxes to reach a colony there and it is declining.[1632]

Breeding colonies were estimated at 10 000–12 000 birds on Masthead I. (23/151) in 1972;[914] 16 000 on Heron I. (23/151) in 1978;[1357] 90 000 on 30 islands off the NSW coast in the 1960s and 1970s;[1029] and 300–400 on Carnac I. in 1979.[505] Large colonies on Norfolk I. (29/167) were estimated in 1978–79 to hold several hundred thousand breeding birds;[1731] 60 000 are estimated to breed on Lord Howe I. (31/159).[676]

Field Atlas Records: 344

1° blocks–with records: 67 (8%)

–with breeding records: 15 (22%)

Atlas No.: 069

Max. Rep. Rate: 24%

In block(s): 24/153,28/113,37/150

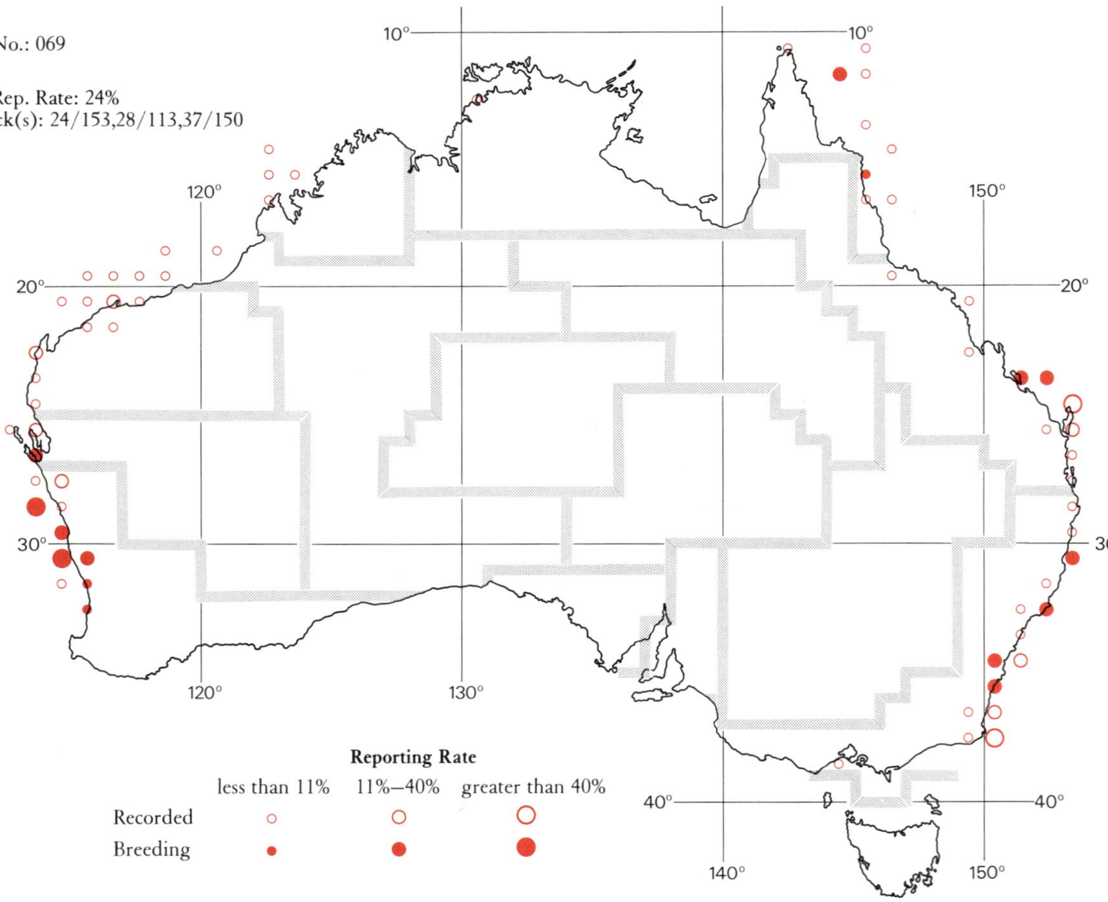

Sooty Shearwater

Puffinus griseus

The main breeding grounds of the Sooty Shearwater are on islands round New Zealand and Cape Horn.[1636] In Australia the species breeds on nine islands in the South-East Region from Little Broughton I. (32/152) to Montagu I. (36/150);[1029] and, in Tas., on Tasman I. (43/148) and Courts I. (43/147). In 1980–81 a pair nested on Hippolyte Rocks (43/148) also.[1036] Early reports of breeding from some islands in south-western Tas., e.g. Flat Witch I. (43/146) and Flat I. (43/145), are thought to have arisen from misidentification of the species[1850,1851] and breeding on any of the islets in Port Davey (43/145) remains unconfirmed.[1849]

Sightings outside the South-East and Tas. Regions are rare. There are several records from Encounter Bay (35/138).[1391] In the South-West Region, the only reports are of small flocks seen off Albany (35/117) in Aug. 1973 and July 1980.[677,1494]

The Sooty Shearwater migrates from its colonies during the non-breeding winter season but there are some winter records in Australia throughout its range. The Reporting Rate was 0.17% in summer and 0.08% in winter; these figures may be biased by the finding of old beachwashed specimens. Birds from Australia travel north towards Japan where a bird banded on Lion I. (33/151) was recovered. Several birds banded in New Zealand and one from Macquarie I. (54S/158E) have also been recovered off Japan.[1636] Large flocks have been reported from the edge of the Antarctic pack-ice during summer but they may include some Short-tailed as well as Sooty Shearwaters. It is thought that they are non-breeding birds and that the flocks rarely travel west of 70°E.[937,1423]

The Sooty Shearwater breeds in colonies always mixed with Short-tailed or Wedge-tailed Shearwaters.[1029] It feeds on small cephalopods, which are abundant where nutrients are brought to the surface or concentrated by ocean currents.[1423]

The breeding population on NSW coastal islands in the 1970s was estimated at 500 birds.[1029] On Tasman I. 2000 birds bred in 1978[250] and on Macquarie I. the breeding population was estimated at 3540 birds in 1979.[1537]

Field Atlas Records: 104

1° blocks–with records: 31 (4%)

–with breeding records: 3 (10%)

Atlas No.: 070

Max. Rep. Rate: 16%
In block(s): 37/150

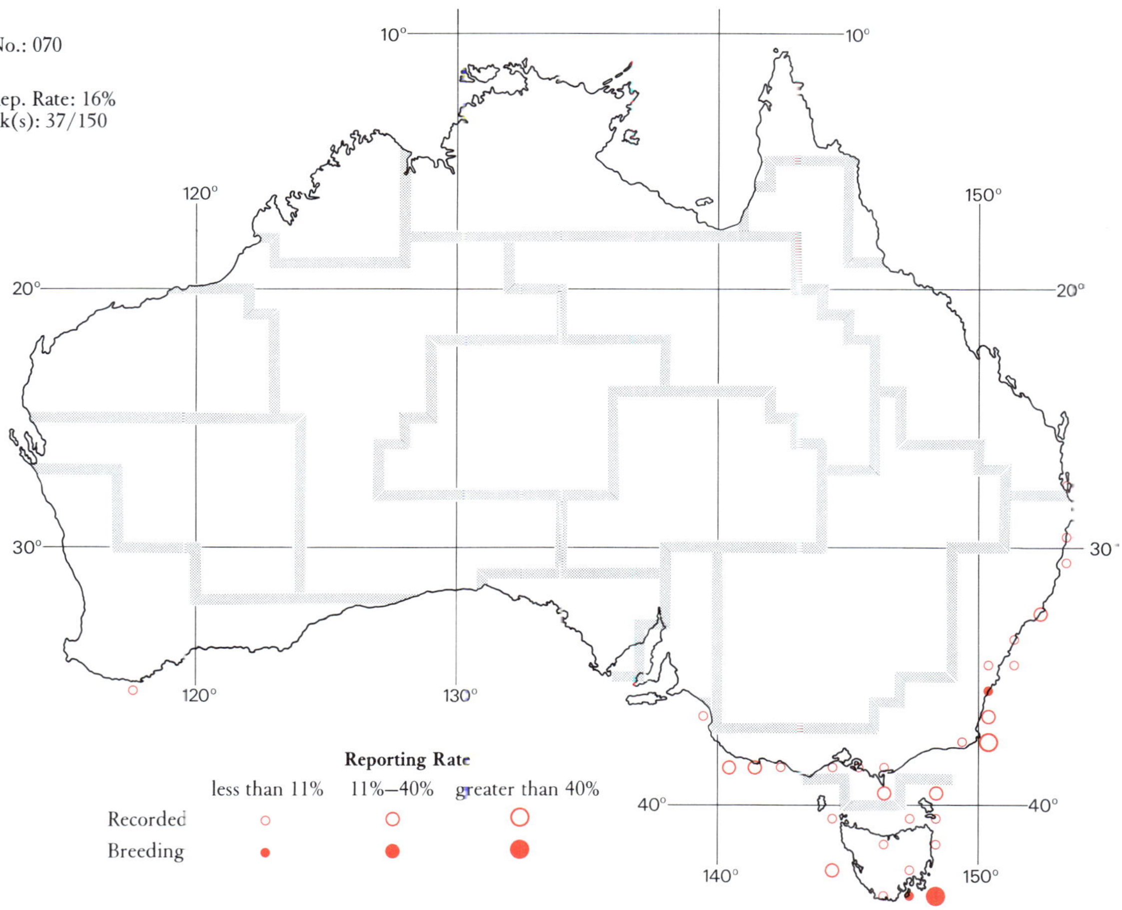

Short-tailed Shearwater

Puffinus tenuirostris

The Short-tailed Shearwater breeds only in Australia. Colonies were not discovered in NSW until 1958 and have now been found on 13 islands from Broughton (32/152) to Montagu (36/150).[1029] In Nov. 1981, 19 birds were discovered in burrows on Figure-of-Eight I. (34/121) and breeding was confirmed in Dec. 1982,[1032] extending the known breeding range by more than 1000 km. Beachwashed specimens have also been found farther west recently with the first record from the western coast in 1976 at Yallingup (33/115).[1715]

Additional 1° blocks from which breeding was reported in 1951-76 but not during the Field Atlas, because there were few visits, are: 32/152, 33/151, 34/150, 36/150, 38/143, 38/141, 35/136 and 34/135. In SA colonies on Wedge I. (35/136) and N. Gambier I. (35/136) are extinct and possibly also one at Douglas Pt (38/140).[418,556]

The Short-tailed Shearwater migrates from breeding colonies in Apr.-May and arrives off Japan in May-June. In Australia no sightings of live birds have been reported from mid-May to late Aug.[415,1533] Most winter records in the Field Atlas (Reporting Rates: summer 3.07%, winter 0.28%) were probably of old beachwashed specimens. The species' return to Australia is sometimes marked by spectacular wrecks on beaches in the South-East Region, with the greatest mortality in Dec.[674] Not all birds remain round the breeding colonies in summer. Large numbers of unknown age and breeding status were seen in Antarctic seas in Feb. 1980.[992]

In Tas. waters the Shearwater feeds on krill and small squid.[1533] Colonies on Phillip I. (38/145) appear to be declining[766] because housing encroaches and dogs and foxes prey on the birds. Licensed operators harvest about half a million chicks each year in Bass Strait. Despite this exploitation, some colonies appear to be enlarging and occupying new sites[1305] though this does not necessarily mean that the number of birds is increasing.

Most Short-tailed Shearwaters breed in the Tas. Region where there are 150 colonies with a total area of about 1050 ha, estimated in 1980 to hold 15-16 million birds.[1305] In Vic. burrows were estimated at 1.45 million in 1978[768] and in NSW there were estimated to be 51 000 birds in the 1970s.[1029]

Field Atlas Records: 1455

1° blocks-with records: 77 (9%)
-with breeding records: 29 (38%)

Atlas No.: 071

Max. Rep. Rate: 43%
In block(s): 39/146

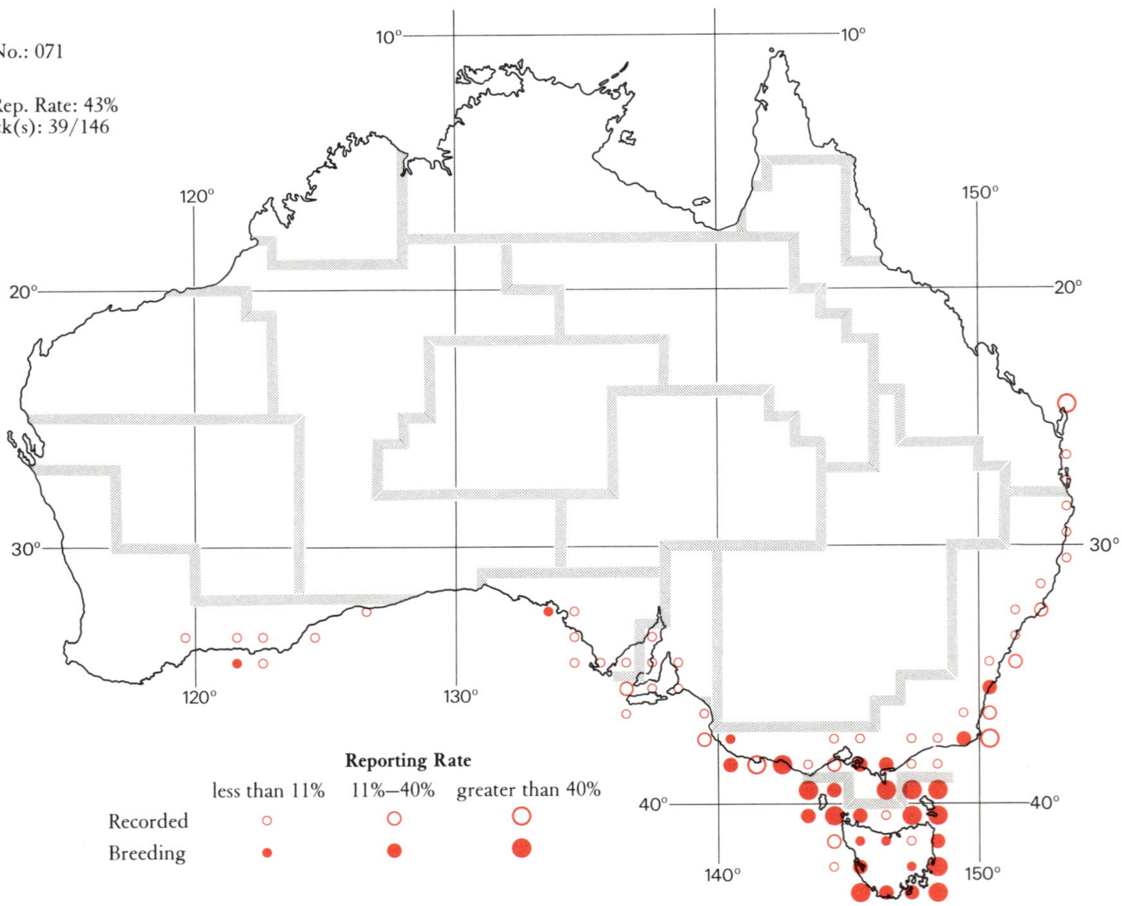

Fluttering Shearwater

Puffinus gavia

Hutton's Shearwater

Puffinus huttoni

Fluttering and Hutton's Shearwaters are treated together in the Atlas because of the great difficulty of separating them in the field and their disputed taxonomic status.[388,754,1636,1817] The Manx Shearwater *P. puffinus* is sometimes also considered conspecific with them but is treated separately here (p. 654).

The Fluttering Shearwater breeds on many islands in New Zealand but Hutton's Shearwater is known to breed only at high altitude on the Seaward Kaikoura Mts (42S/173E) in New Zealand. The Field Atlas map includes sightings of both species. All but three of the South-West records could be assigned to Hutton's. Fluttering records were of birds beachwashed at Esperance (33/121) in Feb. 1981 and near Eyre Bird Observatory (32/126), also in Feb. 1981. In the east, most records were of the Fluttering but Hutton's was recorded from eight 1° blocks in the South-East Region, two in the Tas. Region and one each in the Eastern Qld (27/153) and SA Ra. (35/138) Regions. There were also records of Hutton's from Booby I. (10/141) in Apr.-May[503] and of a live but exhausted bird well inland at Katherine (14/132) in June 1981.[1463] Several hundred were seen daily in July-Aug. 1978 and 1979 off Pt Cloates (22/113).[754]

Hutton's Shearwater may migrate widely, perhaps circumnavigating Australia counter-clockwise in the non-breeding season or as pre-breeders.[754,1817] Northern records are all Apr.-Sept. and the birds seen off Pt Cloates in July-Aug. were all moving south. An observer travelling by boat from Fremantle (32/115) north to Cairns (16/145) Nov.-Dec. saw none.[315]

Both shearwaters are present off the South-East Region and Tas. at all times of the year. Near Sydney (33/151) the greatest numbers were present in Oct.[1262] and at Burrewarra Pt (35/150) June-Oct.[1195] At Waitpinga (35/138) they were abundant Apr.-May and again July-Oct.[415] They are generally close inshore, within the continental shelf, feeding on crustaceans and pelagic fish.[1636] A flock of about 1000 Fluttering Shearwaters was seen in Oyster Bay (42/148) in Apr. 1981[1314] and flocks of up to 1500 have been recorded in SA.[415]

Field Atlas Records: 331

1° blocks–with records: 52 (6%)
–with breeding records:

Atlas No.: 068,913

Max. Rep. Rate: 38%
In block(s): 37/150

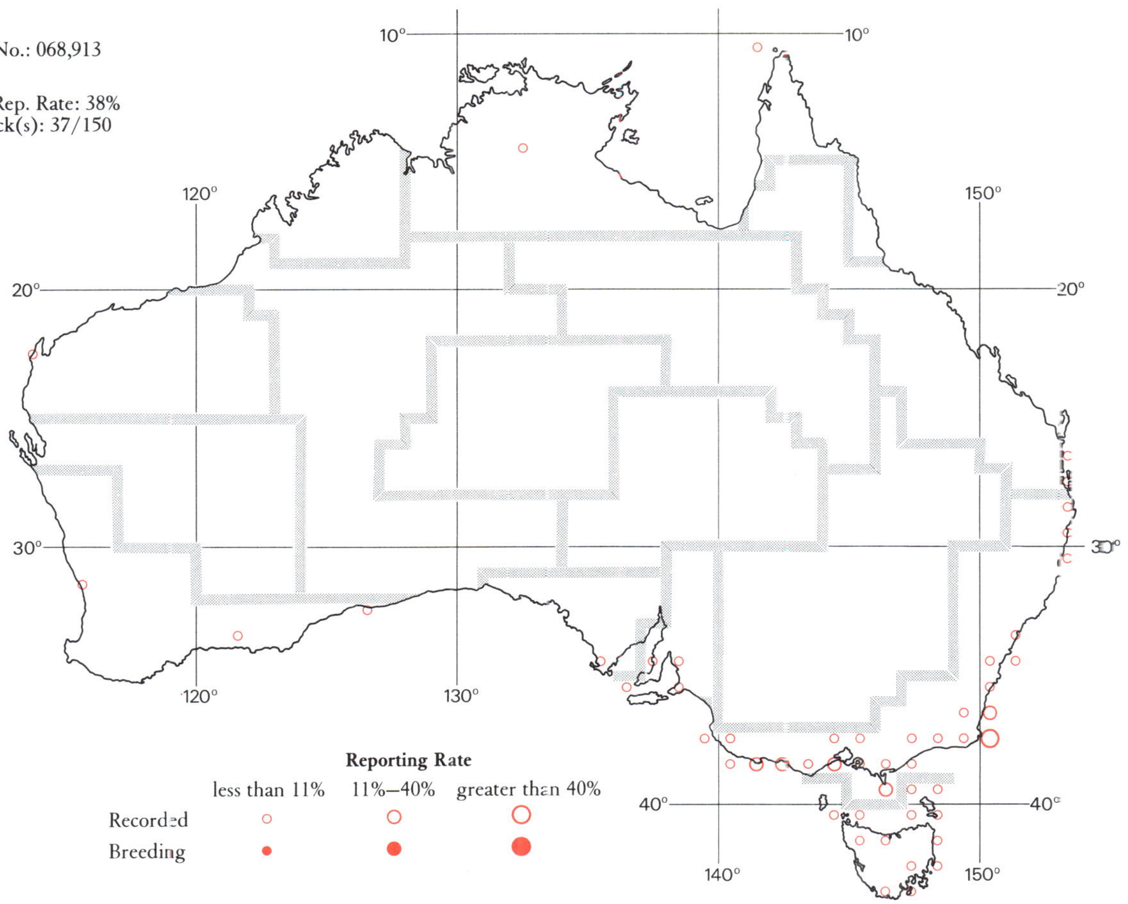

Little Shearwater

Puffinus assimilis

Breeding colonies of the Little Shearwater are distributed round the Southern Ocean and in the Atlantic north to the Azores Is (38N/28W). There are six subspecies of which two breed in Australia: *assimilis* on Lord Howe I. (31/159) and Norfolk I. (29/167); and *tunneyi* on many islands in the South-West Region.[1636] The first SA record (37/140), a specimen beachwashed in 1978, is of the western subspecies *tunneyi*,[1393] as is another from Barwon Heads (38/144) in Vic. in 1973.[1681] Records from the eastern coast may be of *assimilis*.[388,1615,1709]

Because breeding takes place in winter and visits to breeding islands were mostly in summer during the Field Atlas, only two colonies were reported. At other known nesting sites there was evidence of recent breeding and the breeding range includes the following 1° blocks in addition to those on the map: 28/113, 30/114, 30/115, 31/115, 32/115, 34/115, 35/116, 35/117, 34/121, 34/122.[1636]

The Little Shearwater is present near its breeding colonies for ten months of the year[1636] and appears to be rather sedentary.[422] Some dispersal is indicated by sightings up to 500 km seawards from the coast and records in the South-East and Tas. Regions which are made in all seasons.

During the day the Little Shearwater is either at sea or in the nesting burrow which it may occupy for up to four months before laying. All Australian breeding colonies are on islands, but some overseas are not only on coasts but up to 15 km inland and at 500 m altitude.[422] The Little Shearwater feeds on small cephalopods and fish.[422]

Fewer than 4000 birds bred on Eclipse I. (34/117) in 1973;[675] 100 on St Alouarn I. (34/115) in 1976[1022] and there were hundreds of burrows on Hood I. (34/122) in 1977.[1766] An estimated 8000 birds breed on Lord Howe I.[676] and possibly up to 2000 on Norfolk I.[1580]

Field Atlas Records: 26

1° blocks–with records: 17 (2%)
–with breeding records: 2 (12%)

Atlas No.: 067

Max. Rep. Rate: 6%
In block(s): 28/113

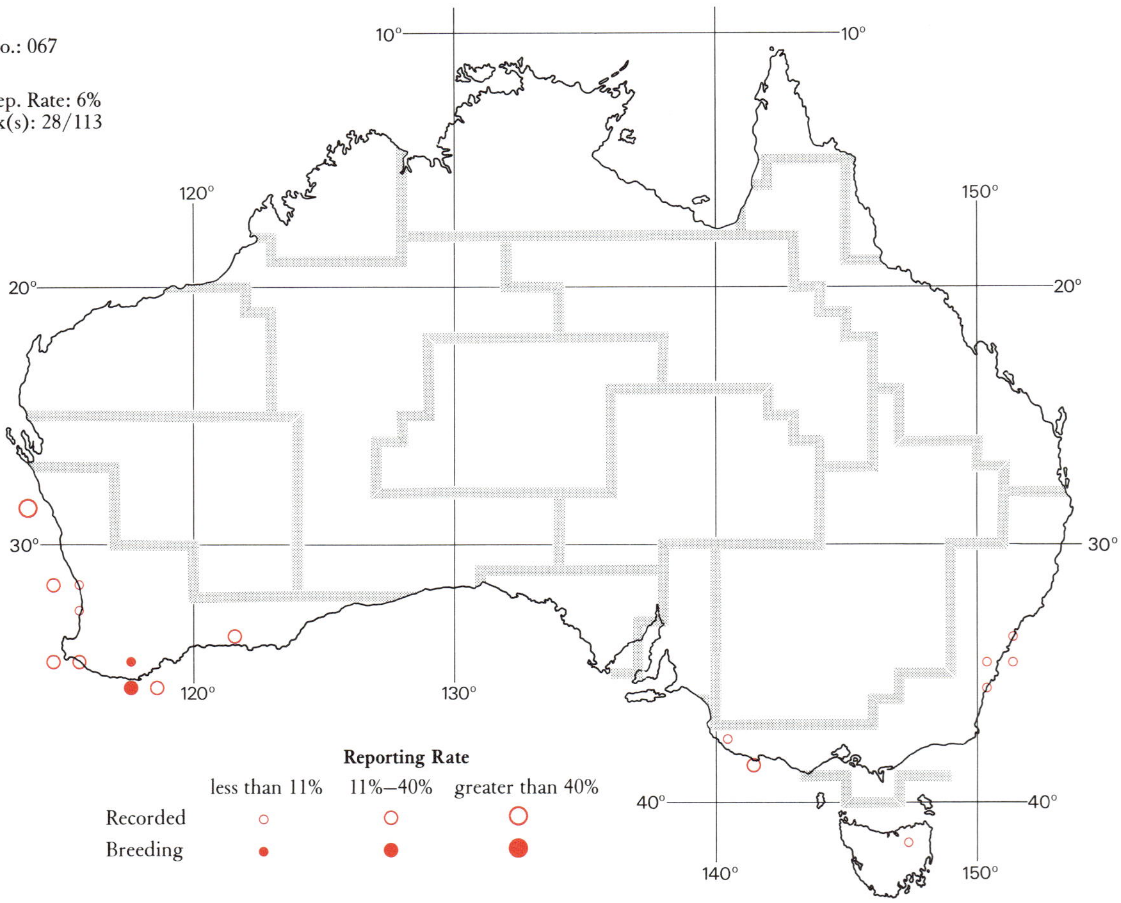

Wilson's Storm-Petrel

Oceanites oceanicus

Wilson's Storm-Petrel is possibly the most numerous species of bird breeding in Antarctica.[123] It has colonies at many places on the Antarctic continent as well as on some sub-Antarctic islands including Heard I. (53S/73E).[1636] In Australia the species is regularly reported from all parts of the coast except within the Gulf of Carpentaria and off Qld. There is an old record on the Qld coast between 13°S and 23°S in the Atlases: a specimen taken 40 km off Port Bowen (20/148) in May 1875.[11] But there are unpublished sightings of the species in the Coral Sea and inside the Great Barrier Reef between Gladstone (23/151) and Townsville (19/146) in May[401] so the gap may represent lack of observers rather than lack of birds.

Wilson's Storm-Petrel arrives at its breeding colonies Oct.-Dec. and leaves Mar.-May,[937,938] travelling north into the Atlantic and Indian Oceans, rarely into the Pacific.[1820] Large flocks have been seen off Sri Lanka[1636] and in the Red Sea (17N/40E).[1626]

It is thought that waters north of Australia may be a wintering ground also[1626] but visits during the Field Atlas were too few to confirm this supposition.

In autumn there is a passage of birds moving northward on the eastern and western coasts of Australia. At this time records are made on southern coasts also: the Field Atlas contains autumn records from six 1° blocks off the southern coast between 115°E and 145°E but only one during the period of the southward movement in spring. In NSW greatest numbers are seen Apr.-June and Sept.-Nov. corresponding with the northward and southward passages.[1292] Most of the records on the western coast in the Field Atlas were made by an observer travelling by boat north from Fremantle (32/115) to Cairns (16/145) between 1 Nov. and 15 Dec. 1980.[315] Only one sighting (10/137) was made east of 130°E on this trip which probably reflects completion of the main movement of birds through the region on the southward passage to breeding grounds.

Most records in Australia come from the edge of the continental shelf. Though the bird may be sometimes encountered farther inshore, it is seldom beachwashed. Wilson's Storm-Petrel feeds on cephalopods, crustaceans, including krill, fish and offal.[1636]

There are no measures of abundance.

Field Atlas Records: 1[illegible]3

1° blocks–with records: 61 (8%)

–with breeding records:

Atlas No.: 063

Max. Rep. Rate: 10%
In block(s): 37/150

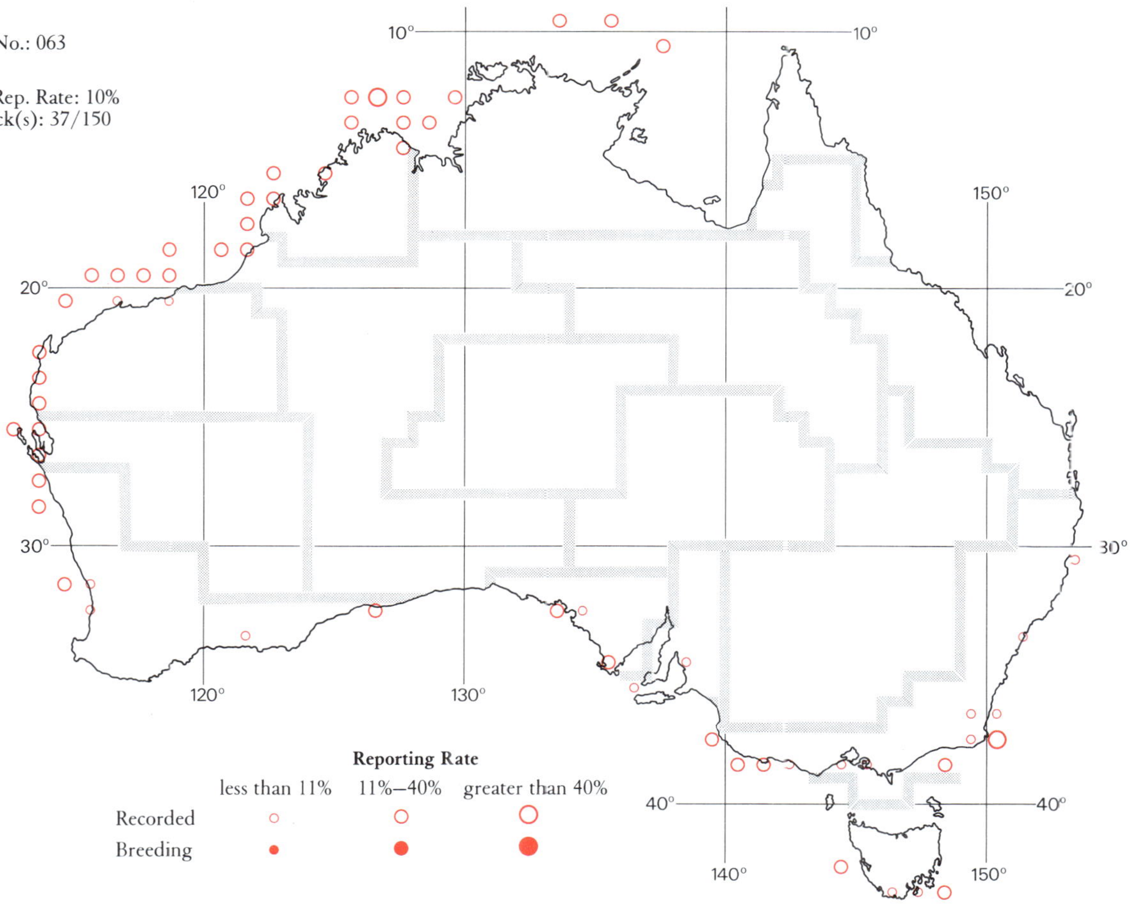

White-faced Storm-Petrel

Pelagodroma marina

The White-faced is the only storm-petrel to breed in Australia.[388] It also breeds on islands round New Zealand and in the Atlantic Ocean. The Australian subspecies *dulciae* breeds on many islands off southern Australia.[1636] The northernmost are Boondelbah I. (32/152) in eastern Australia, active in 1970; and, in the west, Pelsart I. (28/114) active in 1980.[678] The southernmost Australian breeding colony is that shown on the map at The Nuggets (42/148).[254] Though no records from the western Tas. Region appear in the Field Atlas, the Historical Atlas contains records from there, including breeding on Penguin I. (40/144).

The White-faced Storm-Petrel migrates north in winter. Reporting Rates were 0.28% in summer and 0.05% in winter. No birds banded in Australia have been recovered overseas but birds of the Australian subspecies appear in winter in the Arabian Sea (15N/65E) and northern Indian Ocean and as far east as Java (7S/110E).[1636] The July recovery of a beachwashed bird at Pt Cloates (22/113)[1219] suggests that some may winter not far from Australia. There is one observation of large numbers of White-faced Storm-Petrels in the Coral Sea in May[401] but whether this is a wintering ground is unknown.

Beachwashed specimens are frequent. Sightings at sea include a scattered flock of many thousands reported in Bass Strait in Mar. 1961.[1636] Australian breeding colonies suffer from the presence of sheep or goats, which trample the exposed burrows so that only those in the vegetation have any chance of success on grazed islands.[1914] The bird is a nocturnal visitor to the breeding colonies, even moonlight resulting in a decrease in activity.[1636] The White-faced Storm-Petrel feeds on plankton.

Breeding colonies were estimated at 21 400 birds on 11 islands off the NSW coast in the 1970s;[1029] 35 000 birds on Chalky I. (40/147) in 1981;[1501] and there were 5000–8000 fresh burrows on Frederick I. (34/122) in 1981.[1030]

Field Atlas Records: 161

1° blocks–with records: 49 (6%)
–with breeding records: 17 (35%)

Atlas No.: 065

Max. Rep. Rate: 16%
In block(s): 37/150

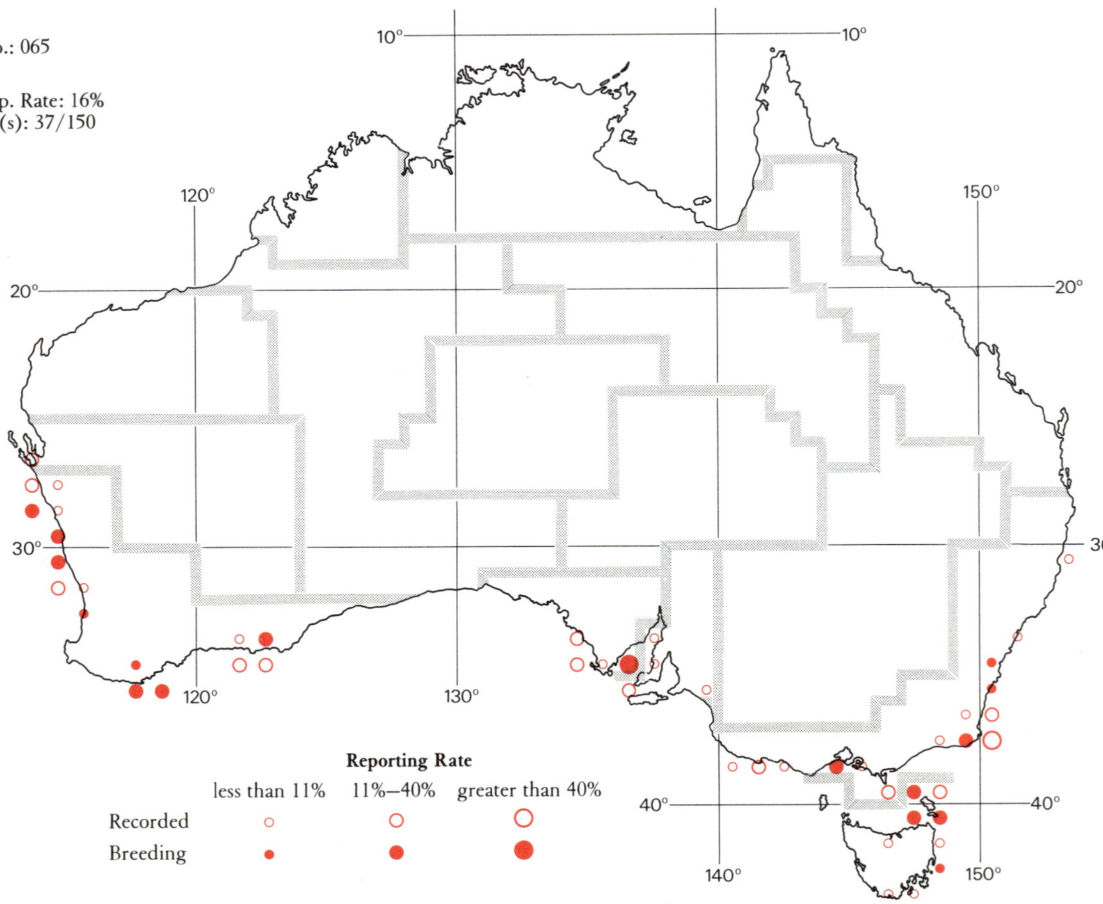

Common Diving-Petrel

Pelecanoides urinatrix

The Common Diving-Petrel breeds on islands round New Zealand, in the sub-Antarctic and in the South-East and Tas. Regions of Australia.[216] Most Australian records are of the subspecies *urinatrix* that breeds in Australia and New Zealand. Beachwashed specimens from WA have been identified as the subspecies *exsul* that breeds on many sub-Antarctic islands.[944] Six of seven specimens examined from SA are *urinatrix*; and it is now doubtful whether the tentative identification of the seventh, from Sellick's Beach (35/138) in May 1956, as *exsul* is correct.[415,1369,1393] A beachwashed specimen of one of the New Zealand breeding subspecies, *chathamensis*, has been picked up at The Nuggets (42/148) in Tas.[254] The Common Diving-Petrel has been recorded as far north as N. Stradbroke I. (27/153) where one was beachwashed in Aug. 1961.[819] It is rarely recorded west of 138°E. The 1951–76 Historical Atlas contains breeding records from colonies not visited during the Field Atlas at 39/144, 39/147, 40/144.

The Common Diving-Petrel is present in the breeding colonies throughout the year. Some birds may be absent for 1–2 months after breeding[248] but there is no evidence of large-scale seasonal movement.

The bird obtains most of its food by diving for crustaceans and small fish. Though usually recorded in small numbers, in 1977 and 1980 flocks of about 1000 birds were recorded in the Derwent Estuary (43/147). In some species of diving-petrel there is a simultaneous moult of all the flight feathers so that for a short time the birds are flightless.[1757] These unusually large flocks were not flightless but showed a tendency to settle very quickly.

The breeding population on five Vic. islands, estimated in 1978–80, probably amounted to no more than a few thousand pairs.[768] In Tas. 800 birds bred on Maatsuyker I. (43/146) in 1975[1263] and the islands of south-west Tas. were estimated in 1980 to hold 16 000 birds.[1852]

Field Atlas Records: 192

1° blocks–with records: 34 (4%)
–with breeding records: 6 (18%)

Atlas No.: 085

Max. Rep. Rate: 19%
In block(s): 39/146

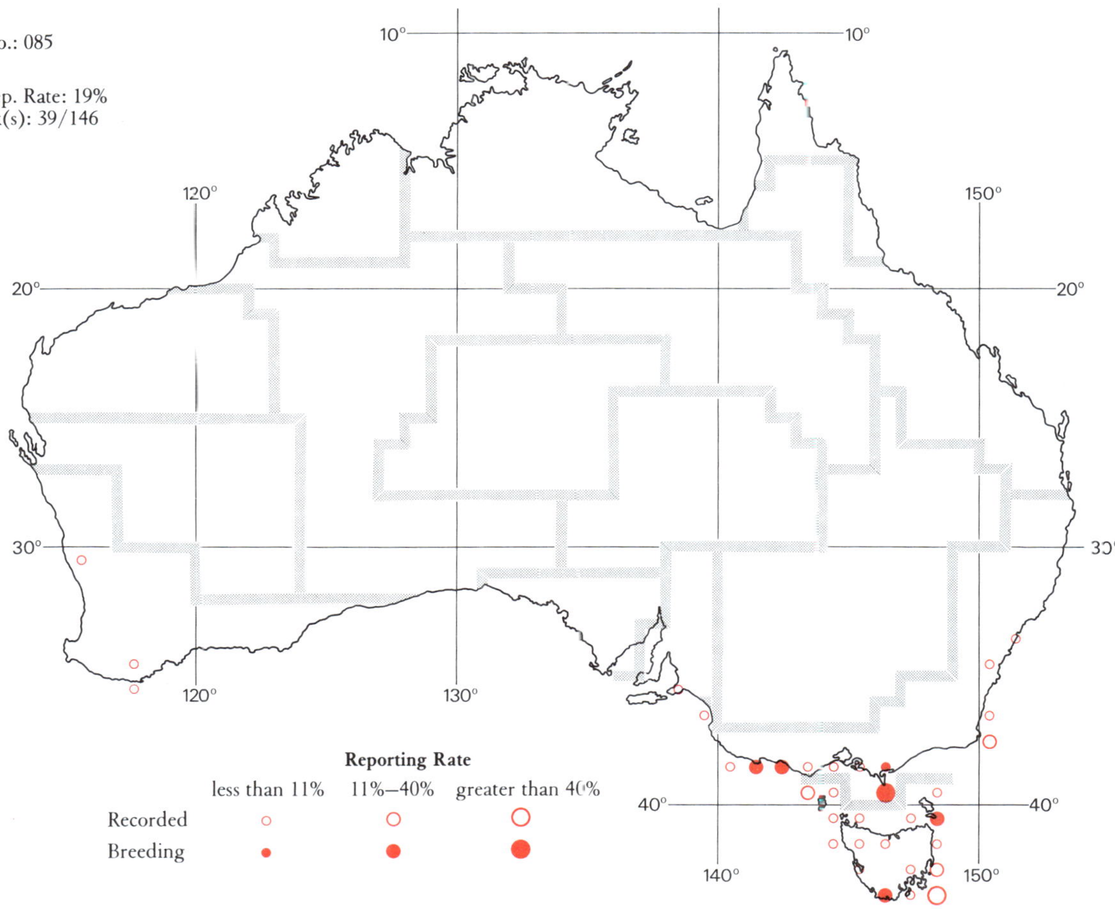

Australian Pelican

Pelecanus conspicillatus

The Australian Pelican lives in New Guinea and adjacent islands having become established there this century, particularly after the irruption of birds from Australia in 1977.[129] The Pelican is accidental to Indonesia and New Zealand. In Australia breeding colonies are few in number, some used regularly, others only occasionally. This is reflected in the Field Atlas, where the species bred in only 2% of the 1° blocks in which it was recorded. All breeding sites in WA are coastal: on offshore islands or islets in estuaries or inlets.

The Pelican can move great distances. Its wide distribution reflects its search for food. This was illustrated during 1977–78 when Pelicans turned up in unexpected places all over the country after successful breeding on inland waters. A colony of 1000 was recorded at L. Eyre (29/137) in 1974 and breeding probably continued in 1975–76. Large numbers were seen leaving the district in 1977 as the water dried up.[86] Many of those appearing in coastal districts in 1978 were emaciated and died but one flock of 150 reached Palau I. (7N/134E).[1779] Similar dispersals have been recorded from the L. Eyre Region in 1922[109] and from the Murray-Darling Region as floods receded in 1956.[833] The Pelican needs an assured supply of fish and an undisturbed area for three months in order to breed successfully.[1792] Breeding colonies are usually on islands either on the coast or in inland lakes. They suffer from falling water levels that allow predators access and from rising water levels that flood nests. Courting Pelicans, trampling in pursuit of each other, are also a source of disturbance, eliminating almost all peripheral nests at one colony studied for six years; people and dogs disturb accessible sites.[333,1792] At L. Cowal (33/147) and L. George (35/149) chicks were reared in only three of six breeding seasons.[1792]

Colonies reported include: 1500 birds on L. Hindmarsh (35/141) in 1967; 1600–3000 at L. Cowal in 1964, 144 in 1971 and 67 in 1975;[1793] 427 at L. George in 1968.[1792]

Field Atlas Records: 17 168

1° blocks-with records: 536 (66%)
-with breeding records: 12 (2%)

Atlas No.: 106

Max. Rep. Rate: 72%
In block(s): 36/150

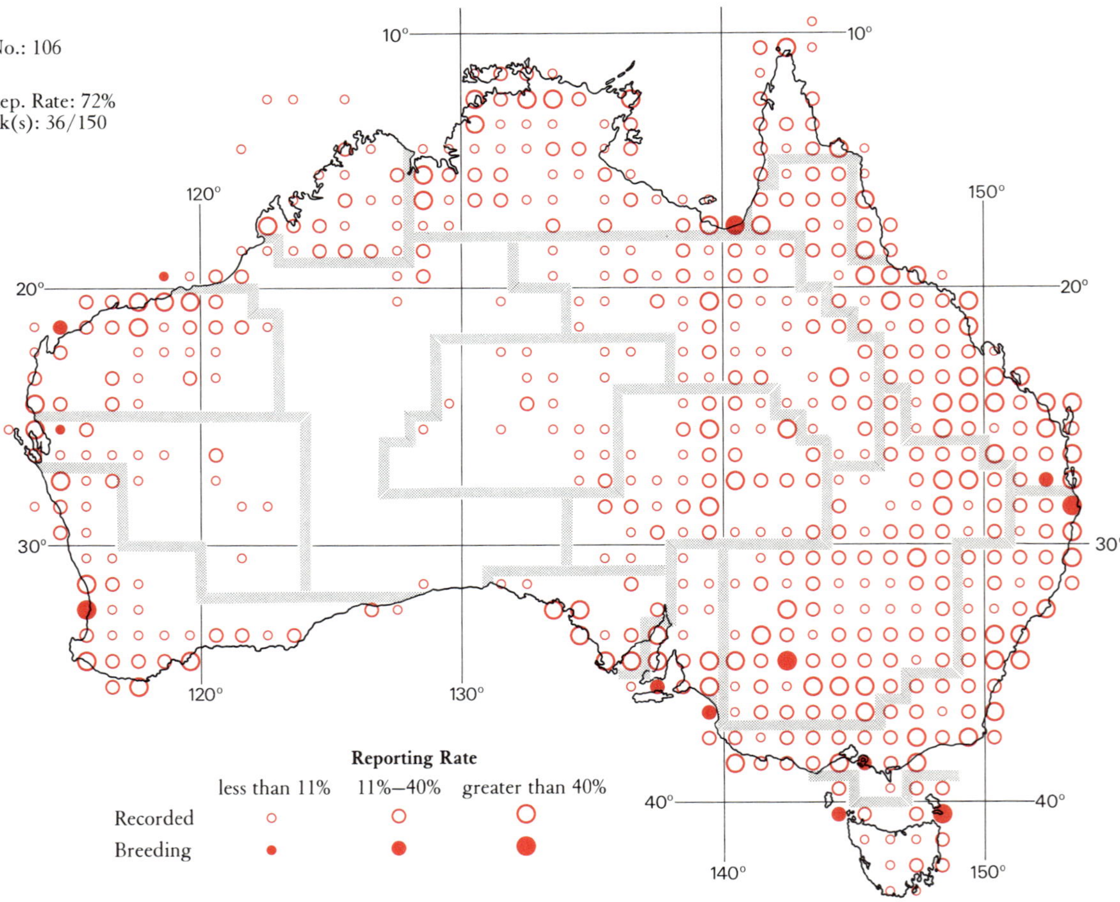

Australasian Gannet

Morus serrator

The Australasian Gannet breeds in New Zealand and Australia. Historical records show that, in Australia, it ranges north to Townsville (19/146) in the east[1906] and to Pt Cloates (22/113) in the west.[317] There are six breeding colonies in Australia, one of which, on Black Pyramid (40/144), was not visited during the Field Atlas.

Birds of breeding age may not wander far but young Gannets disperse widely from the colonies. Many birds bred in New Zealand migrate to Australian coasts for the first three years of life. One bird banded as a chick in New Zealand was recovered on the WA coast at 28°S nine months later.[1906] Within Australia the longest movement recorded is of a bird from Black Pyramid to Mandurah (32/115), 2719 km.[26] Off Caloundra (26/153) a regular southward movement of Gannets was noted in May-June each year for three years. One year, 1250 birds were counted in an hour.[1115] At Burrewarra Pt (35/150) monthly counts revealed no such striking passage but there were peak numbers moving north in May-June and south in Sept.[1195] The Gannet is said to be most numerous in NSW Mar.-Sept.[1292] and in SA May-Aug.[417]

The Gannet nests colonially. It feeds in deep water, either diving singly on to fish or herding shoals as a flock.[1636] The Gannet has few enemies ashore, other than man. The colony on Cat I. (39/148) was almost exterminated by cray fishermen seeking bait for their pots, leading over the years to a failure to recruit young breeding birds. It is now actively protected.

Numbers estimated at the six Australian colonies are: Wedge Light (38/144) 40 birds in 1980;[1552] Lawrence Rocks (38/141) 1200 birds in 1960, 2900 in 1978;[1419] Cat I. 5000–10 000 in 1908, but only 14 adults and six young in 1977;[1816] Black Pyramid 900 in 1961;[738] Eddystone Rock (43/147) 46 birds present in 1978, at least some on nests;[253] and Pedra Branca (43/146) 1000–2000 in 1978.[25] This gives an Australian breeding population of 5000–6000. The total Australian and New Zealand breeding population was estimated at 60 000 in 1971.[1312] Many birds in Australian waters probably come from New Zealand.

Field Atlas Records: 2623

1° blocks–with records: 109 (13%)

–with breeding records: 5 (5%)

Atlas No.: 104

Max. Rep. Rate: 84%
In block(s): 37/150

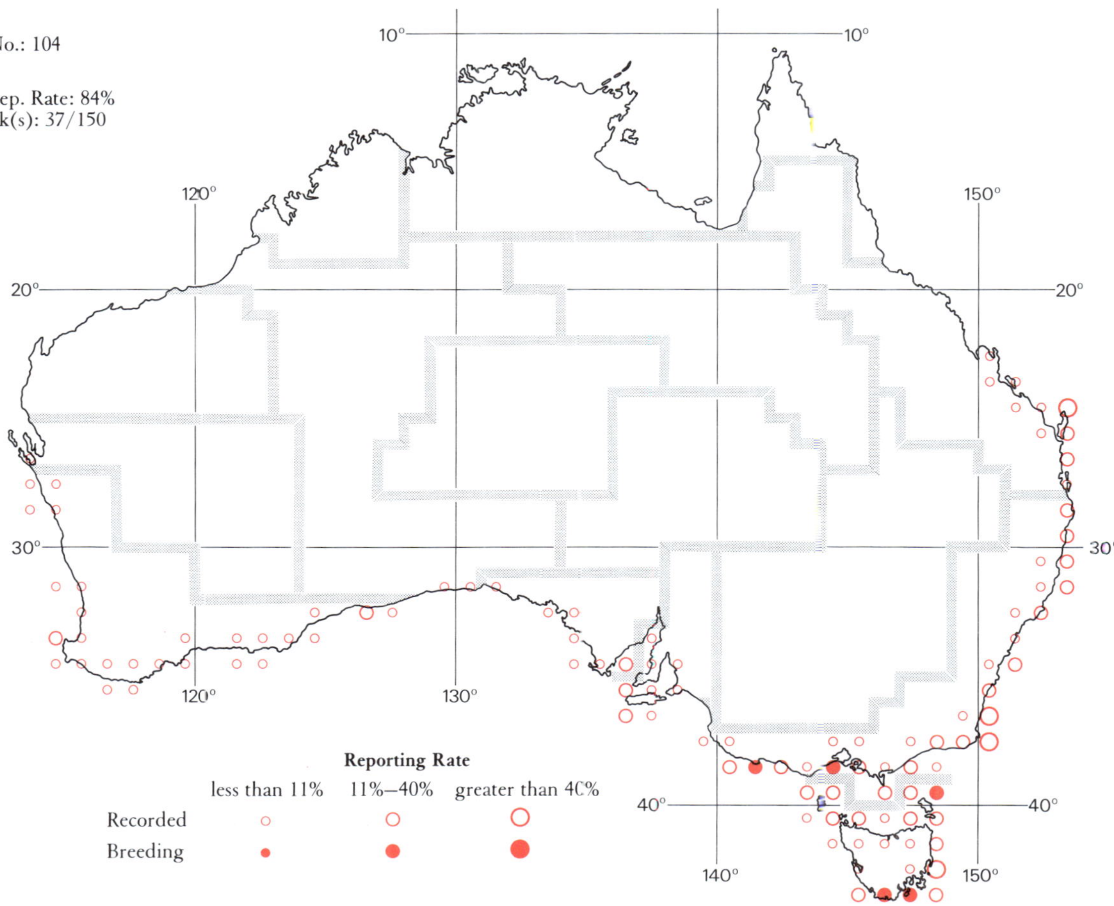

Red-footed Booby

Sula sula

The Red-footed Booby is a tropical species breeding in the Indian, Atlantic and Pacific Oceans.[1312] In Australia it breeds on Raine I. (11/144) and Pandora Cay (11/144) and on a number of islands in the Coral Sea. At least four different plumages have been seen on Raine I.[1811] and variation is great throughout the bird's range. Australian birds have been assigned to the subspecies *rubripes*.[1312] There are few records away from the seas and coasts adjoining the breeding colonies, the farthest south being at Ballina (28/153) in Jan. 1976.[1524]

Red-footed Boobies are to be found near the breeding colony throughout the year though overseas they are known to forage up to 100–150 km from the colony.[1312] All Atlas records south of 20°S have been made Sept.-Mar., which suggests some dispersal or wandering at this time, after breeding[1312] and also when cyclones are likely to drive birds south. Birds banded on Raine I. have been recovered to the north in southern New Guinea.[1000]

The Red-footed Booby nests on vegetated islands. It fishes in deep water for flying fish and squid,[1312] sometimes at dusk and dawn when other gannets are inactive.[1636]

The Red-footed is thought to be the world's most abundant booby[1312] but the breeding colony on Raine I. is small. It contained an estimated 300 birds in Feb. 1959[1811] and about 350 in Nov. 1979.[1000] Up to 20 birds were present on Pandora Cay on recent visits but only one or two pairs were breeding.[1000]

Field Atlas Records: 9

1° blocks–with records: 6 (1%)
–with breeding records: 1 (17%)

Atlas No.: 103

Max. Rep. Rate: 8%
In block(s): 11/144

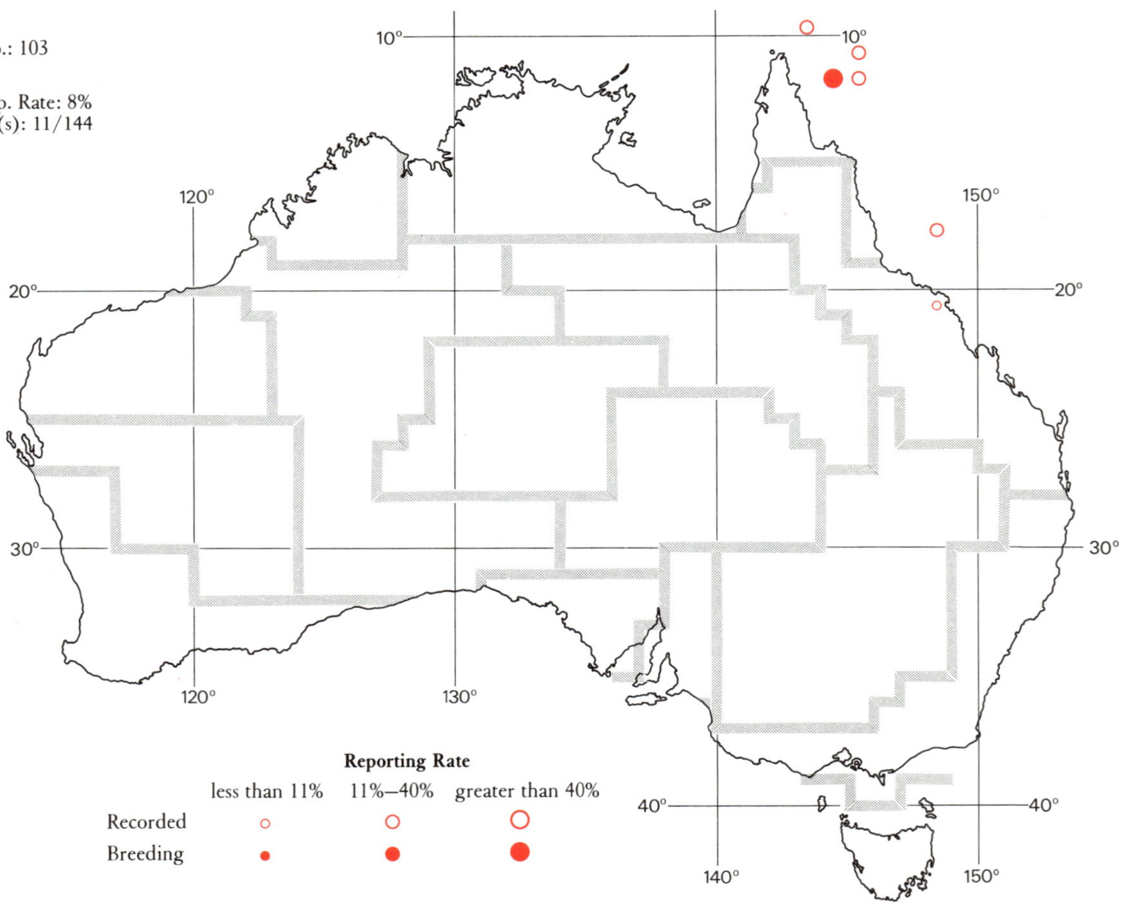

Masked Booby

Sula dactylatra

The Masked Booby, a tropical species of deep warm waters, is widely distributed in the equatorial parts of all the oceans of the world.[1312] It is divided into a number of subspecies, some only doubtfully separable. There may be morphological differences within the Australian population between birds breeding in the east and those in the north-west. In eastern Australia there are breeding colonies in the Coral Sea and at Raine I. (11/144), Pandora Cay (11/144) and Swain Reefs (21/151). In the west the species breeds only on Adele I. (15/123) and Bedout I. (19/119). On the northern coast, between the two breeding areas, reports are rare. There is only a single unconfirmed record from Melville Bay (12/136) between 125°E and 142°E.[884] The southernmost records in the Atlases are at Newcastle (32/151) in Sept. 1979[1292] in the east and Barrow I. (21/115) in the west. A sighting at Granite I. (35/138) in Dec. 1966 has been doubted.[417]

Birds up to three years old may move between neighbouring islands, establishing themselves as breeders on islands other than their natal one.[1312] A nestling banded on Norfolk I. (29/167) was recovered in the Tonga Is (20S/174W), 1900 km away and another from Lord Howe I. (31/159) was recovered at Willis I. (16/150), 1780 km distant.[67] Five birds banded on Raine I. have been recovered to the north on the southern coast of New Guinea.[1000] Movements of adults are little known. At colonies overseas breeding birds may be absent for up to three months of the year.[1312]

The Masked Booby feeds mainly on flying-fish caught over deep water.

Colonies are broken into groups with hundreds of metres and sometimes several kilometres between them. Within groups, the density of nests is very variable in one colony it was one pair to 42 sq. m.[1312] Colonies on Raine I. contained 1500 breeding birds in 1979;[1000] on Swain Reefs (21/151) 500 in 1976;[1073] on Bedout I. 560 birds in 1975[1008] and 240 in 1982;[280] and on Adele I. 540 in 1982.[280] On Lord Howe I., the southernmost breeding colony in the world, the population numbers 'hundreds'[676] and on Norfolk I. about 600 birds breed.[1731]

Field Atlas Records: 35

1° blocks–with records: 17 (2%)
–with breeding records: 2 (12%)

Atlas No.: 105

Max. Rep. Rate: 10%
In block(s): 16/146

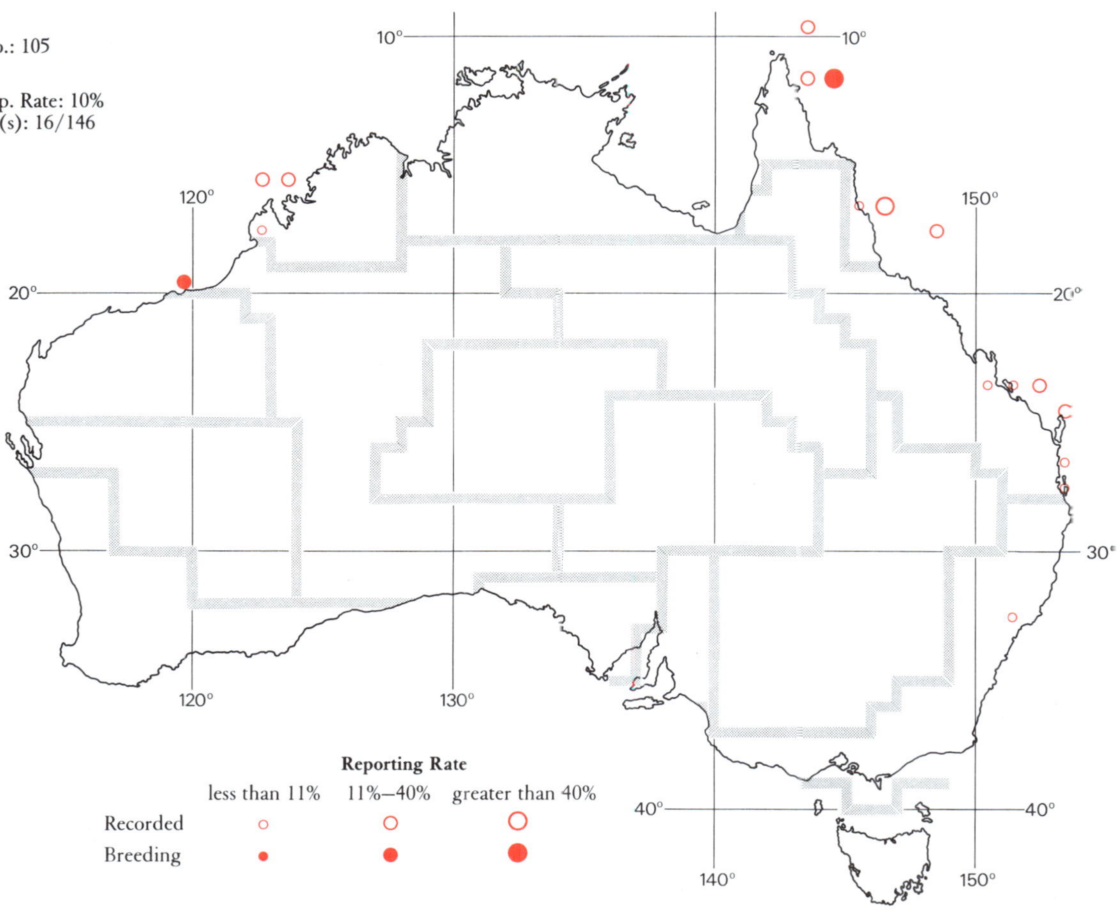

Brown Booby

Sula leucogaster

The Brown Booby nests on islands in tropical seas round the world. Australian populations belong to the subspecies *plotus* which ranges across the Pacific and into the Indian Ocean.[1312] In eastern Australia there are records as far south as Bowen I. (35/150) in Jan. 1978, Montagu I. (36/150) in Apr. 1976 and Altona (37/144) in May 1965.[1524,1636] In WA there is none south of those shown on the map.

Australian breeding colonies not visited during the Field Atlas are on islands in the Coral Sea and on Swain Reefs (21/151), Ellis I. (16/139), Rocky I. (16/139), Adele I. (15/122) and Lacepede Is (16/122).[1636]

Young Brown Boobies travel long distances after fledging. A nestling banded on Fairfax I. (23/152) was recovered three months later in New Guinea, 1600 km away.[34] Others banded on Raine I. (11/144) and No. 8 Sandbank (13/143) have travelled over 1000 km to New Guinea and adjacent islands.[1000] The movements of adult birds are not well known. Some appear off the coast of southern Qld[508] hundreds of kilometres from the nearest breeding colony. In overseas studies, adults at one colony were absent during the non-breeding season but in others the adults visited breeding sites throughout the year.[1312]

The Brown Booby takes fish in shallow seas, often flying-fish. Hunting and egg-collecting were traditionally practised by Torres Strait islanders at colonies off the eastern coast.[1145] In some parts of the world colonies have been exterminated by such steady harvesting and those in north-western Australia are still frequently harvested illegally by fishermen.

Reported colony sizes are: Raine I. 7000–9000 birds in 1959, 4000 in 1961[1815] and 12 000 in 1979;[1000] Maclennan Cay (11/143) 1840 in 1977;[1072] Rocky I. 20 000–30 000 in 1965;[1552] Swain Reefs 1040 in 1976;[1073] Adele I. 32 000 in 1982;[280] West and Middle Is in the Lacepede Group 14 700 and 20 600 respectively in 1982;[280] and Bedout I. (19/119) about 10 000 in 1972,[288] 1300 in 1978[6] and 10 000 in 1982.[280]

Field Atlas Records: 387

1° blocks–with records: 91 (11%)
–with breeding records: 8 (9%)

Atlas No.: 102

Max. Rep. Rate: 54%
In block(s): 24/153

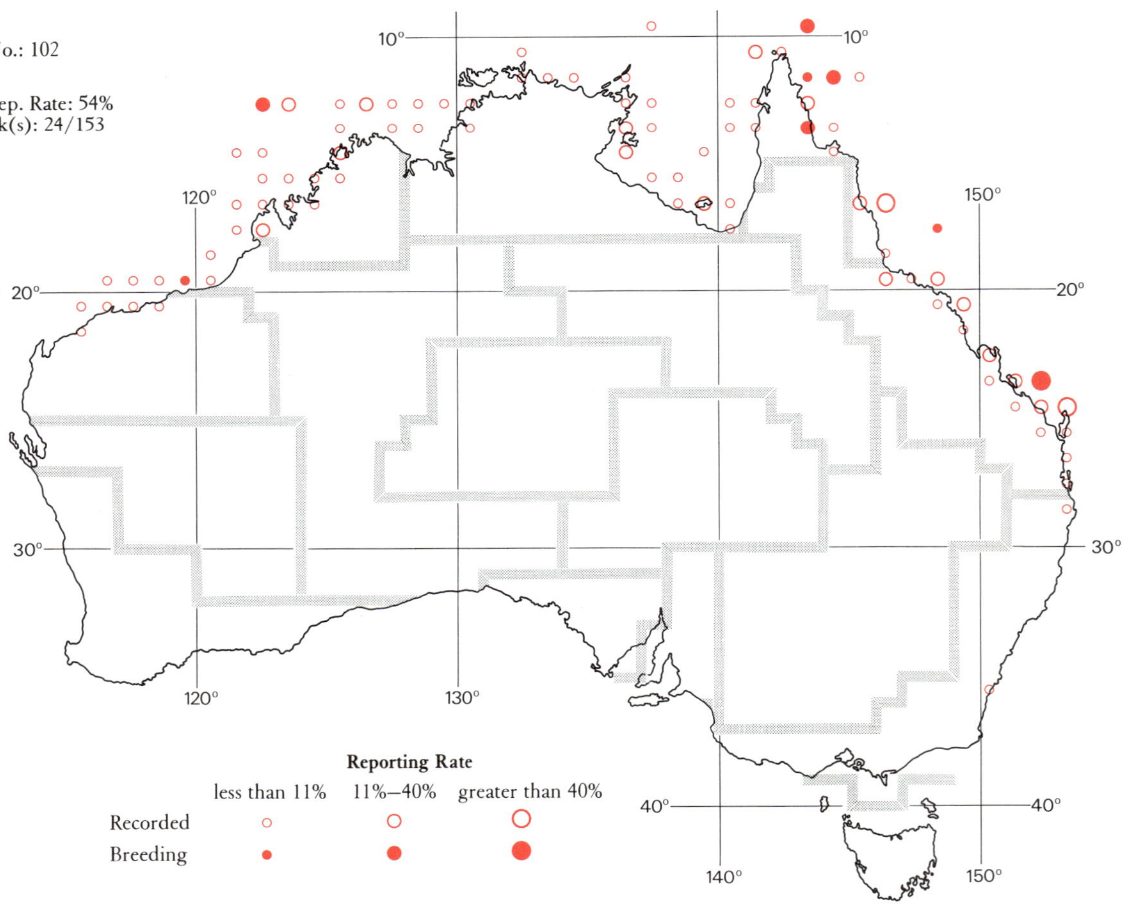

Darter

Anhinga melanogaster

The Darter lives in southern Africa, southern Asia, New Guinea and Australia and is accidental to New Zealand. Australian populations are sometimes considered a separate subspecies.[1391] In July-Aug. 1980 the first record of a Darter was made in the Tas. Region: a single bird, probably immature, was in company with cormorants at Brumby Ck (41/147). In the Pilbara, Goldfields and South-West Regions there are no records east of 123°E and this population may be isolated.

Populations overseas are migratory or sedentary.[422] In Australia the recovery of a banded bird 1008 km from its banding site and sightings such as that in Tas. or one at Leonora (28/121) in about 1974 confirm that the species moves long distances.

There is no suggestion of large-scale seasonal movement in the Field Atlas. In Torres Strait the Darter is nomadic.[503] At L. Cowal (33/147) numbers vary, fewest birds being present in June and most Dec.-Feb.[1790]

The Darter is most frequent on tropical and sub-tropical waters. It requires some open water and a perch on which to rest while drying its wings.[422] Breeding colonies are usually inland or on sheltered estuaries. Colonies and even individual nests may be used for several successive seasons. The Darter catches fish underwater by spearing them with its bill and also takes water insects.[1790] Human disturbance on the opening day of the duck-shooting season at L. Cowal accounted for two-thirds of all clutch failures.[1790]

Colonies have been reported of 60 birds at Block Waterhole (17/137) in 1967; 60 at L. Muncoonie West (25/138) in 1974; 80 near Goondiwindi (28/150) in 1975; 100 at Cawndilla Ck (32/142) in 1974 and 100 at Mystic Park (35/143) in 1976.[1552]

Field Atlas Records: 6707

1° blocks–with records: 432 (53%)
–with breeding records: 52 (12%)

Atlas No.: 101

Max. Rep. Rate: 48%
In block(s): 19/145

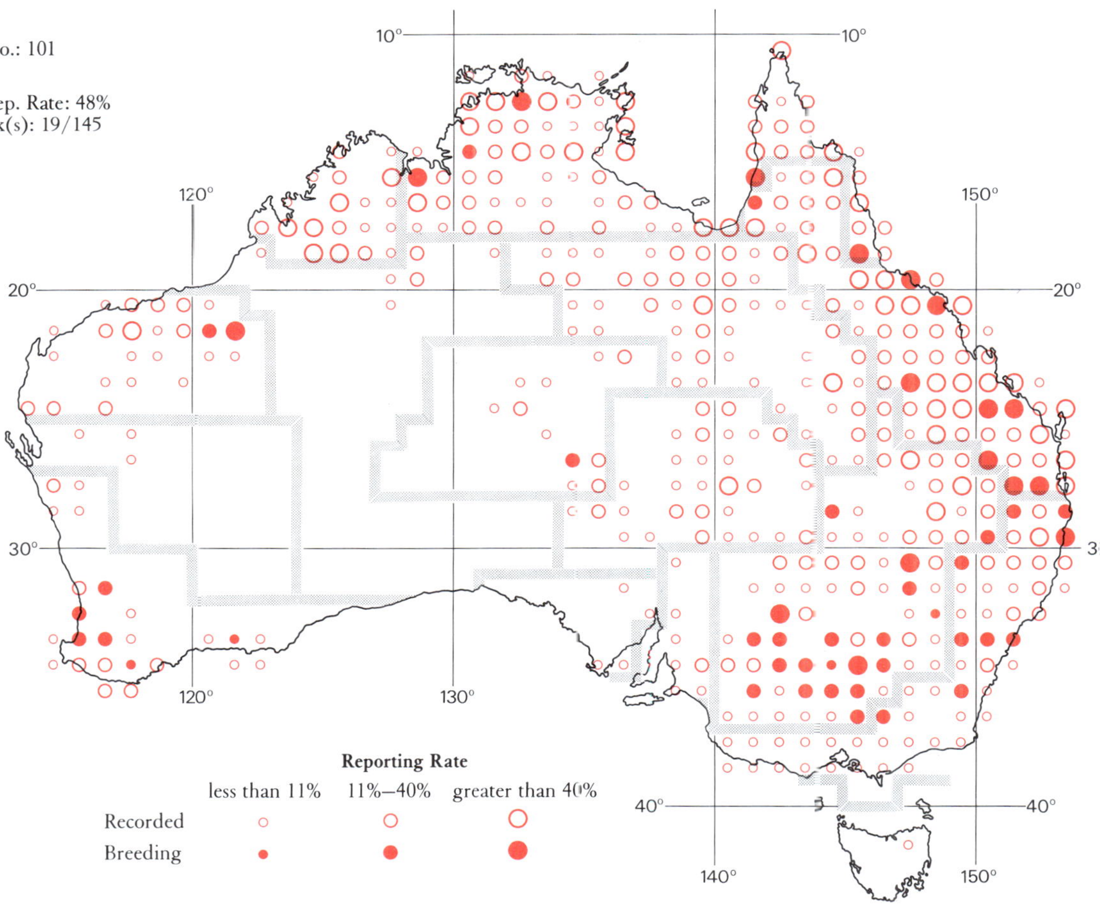

Black-faced Shag

Leucocarbo fuscescens

The Black-faced Shag is an endemic marine species of southern Australia. In the east it has once been recorded as far north as Montagu I. (36/150).[1292] A group of ten birds seen on Eucla jetty (31/128) in 1968 suggests that the species may be present in the Great Australian Bight, where topography permits, linking the eastern and western populations. Augusta (34/115), where a single bird was seen in company with Little Pied and Little Black Cormorants in Jan. 1980, is as far west as the species has been recorded. Records on the Swan R (31/115) during the Field Atlas were of one escapee from the zoo and do not appear on the map.

A study of banded birds in SA showed some dispersal from breeding colonies: 16% of recoveries were outside the banding district.[1778] The longest recorded movement is that of an immature from English I. (34/136) to Cape Leeuwin (34/115), 1930 km.[28]

The Shag inhabits rocky coasts and nests close to the seashore. Young birds gather into creches. The species feeds on fish, mostly in shallow water or on reefs.[1636] Persecution by fishermen round Tas. and probably Vic. may be leading to a gradual decline in numbers. In Tas. fishermen were seen shooting the last three breeding pairs on Flat I. (43/145) in Jan. 1977.[1851] On Granite I. (38/146) the Black-faced Shag was said to have been destroyed 'by order of the authorities' before 1918.[1338] On Lawrence Rocks (38/141) there are now no birds breeding but thousands were reported in 1907.[1419]

The largest breeding colonies reported are off Eyre Pen.: Dangerous Reef (34/136) 5000 birds in 1975;[1775] Winceby I. (38/136) 2000 birds in 1979; English I. (34/136) 2000 birds in 1974.[1552] The largest colonies in the Tas. Region are in the Furneaux Group (39/148).[248] Colonies in other parts of Australia are small: Notch I. (38/146) 200 birds in 1978;[767] Chalky I. (40/147) 130 birds in 1980;[1501] Wright and Egg Is (41/146) 350 birds in 1970;[1772] Maatsuyker I. (43/146) 400 birds in 1975;[1263] Hippolyte Rocks (43/143) 800 birds in 1980.[248] In the Recherche Archipelago the colony on Middle I. (34/123), first recorded in 1889, was still in use in 1976[1765] and on Remark I. (34/121) there were 20 nests in 1981.[1031]

Field Atlas Records: 1180

1° blocks–with records: 62 (8%)

–with breeding records: 17 (27%)

Atlas No.: 098

Max. Rep. Rate: 63%
In block(s): 34/136

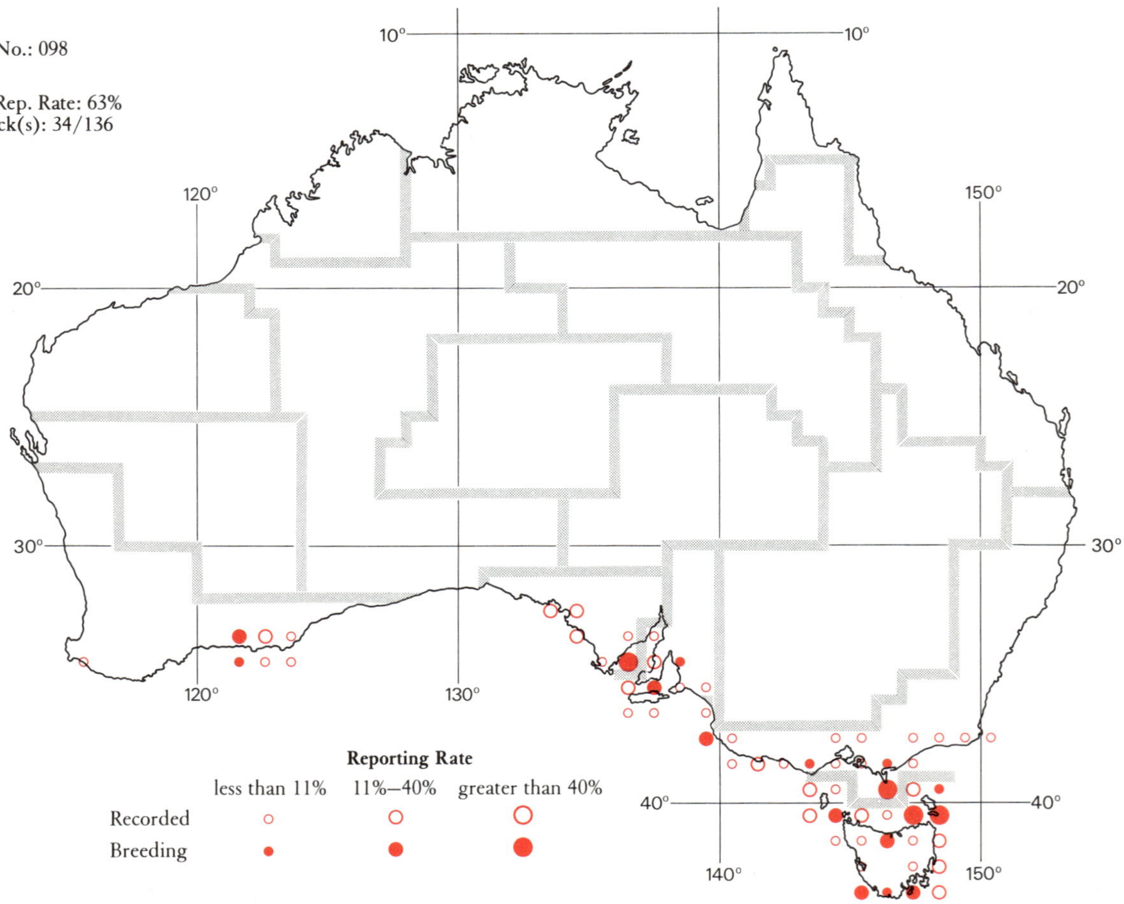

Great Cormorant

Phalacrocorax carbo

The Great Cormorant is a cosmopolitan species. The Australian subspecies *novaehollandiae* breeds only south of 20°S with the exception of breeding colonies reported on L. Argyle (16/128) in 1975.[1713] The species had not been reliably recorded from this district or the Kimberley Region before 1973 but the frequency of reports in the Field Atlas suggests that it has become established in the area.

The Great Cormorant can move long distances. Individuals bred either in Australia or New Zealand have reached Macquarie I. (54S/158E), birds banded in SA have been recovered in Tas.[1778] and one banded in NSW has been recovered in WA.[246] The Great Cormorant tends to disperse farther than other Australian cormorants. In a study of banded birds in SA, 56% of recoveries were outside the banding district.[1778]

Though the Great Cormorant breeds in Tas., there are, from time to time, far larger numbers present than could have been bred locally.[1636] During an irruption that began in 1976, 3000 were recorded at Launceston (41/147)[734] and large numbers were present at other sites. Flocks of 200–250 were still being recorded the following year on the Derwent R (42/147) and at Marion Bay (42/147). Most birds were juveniles and individuals picked up during the invasions were weak, some dying of malnutrition.[1533] Probably the birds were dispersing from breeding areas that had dried up on the mainland. The Great Cormorant was also noted in Torres Strait in 1977 and 1978.[503] Apparent movements of a more seasonal nature have been recorded on L. George (35/149) where maximum numbers are present in summer.[1013]

The Great Cormorant usually keeps to water several metres deep, both fresh and salt. On the Australian mainland it breeds colonially in lakes and estuaries;[1552] colonies in the Tas. Region are small. In coastal waters the Cormorant feeds on slow-moving fish;[1636] in the inland it will take insects and tadpoles.[246] Fishermen consider the Great Cormorant a serious competitor.

A colony of 40 000 birds at L. Menindee (32/142) in 1974 is much larger than any others reported. There were 1100 breeding at Mystic Park (35/143) in 1976 and 1200 at Moorook (34/140) in 1973.[1552]

Field Atlas Records: 15 112

1° blocks–with records: 434 (53%)

–with breeding records: 44 (10%)

Atlas No.: 096

Max. Rep. Rate: 70%
In block(s): 37/150

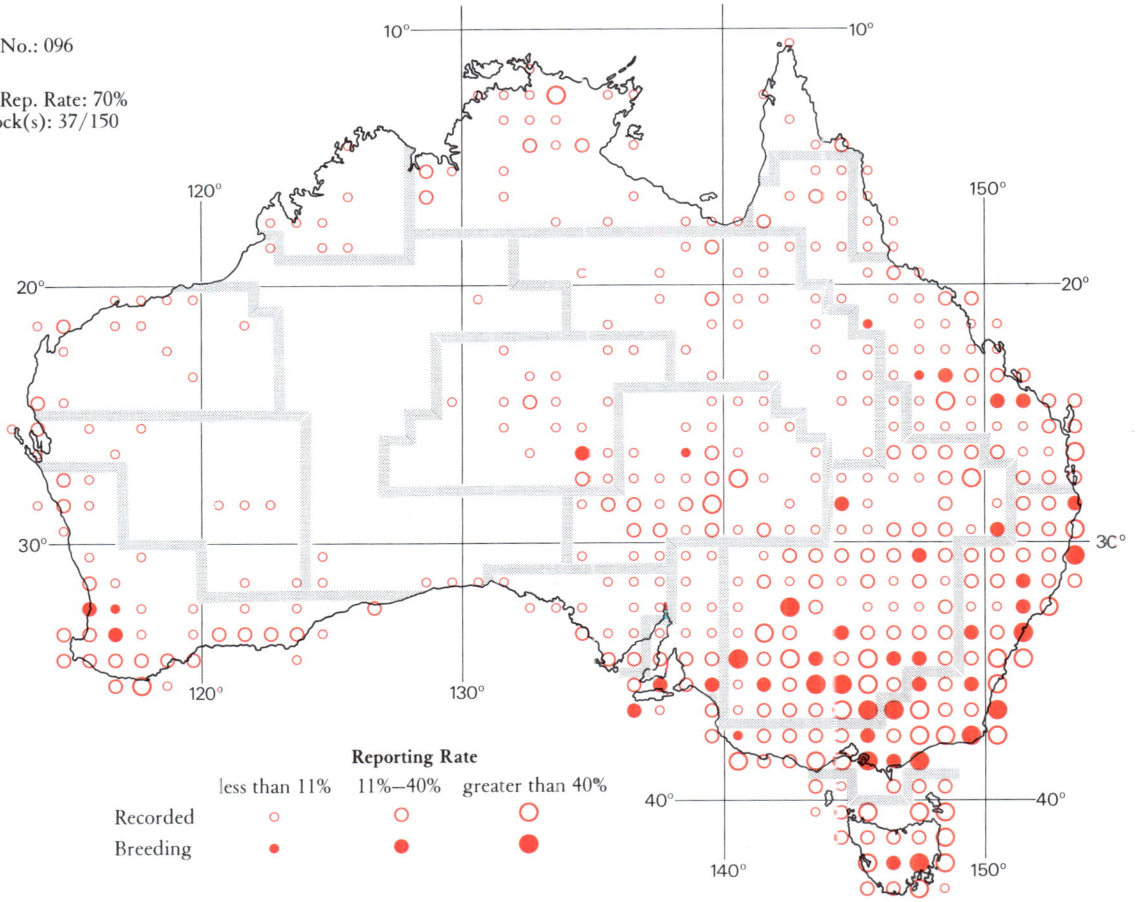

Pied Cormorant

Phalacrocorax varius

The Pied Cormorant is restricted to Australia and New Zealand. In the Tas. Region it became extinct in prehistoric times.[1778] Modern records in the region are of vagrants and the species was not reliably recorded there until 1971;[734] between Nov. 1977 and Sept. 1981 there were four records (a single bird was reported in several blocks in the south-west).

East of 135°E the Pied Cormorant breeds at inland and coastal sites but in the west it breeds only on the coast. It has bred in the past at a number of sites on the coasts of the Top End, Kimberley and Pilbara Regions which were not visited during the Field Atlas (12/135, 13/126, 14/125, 15/123, 15/136, 20/115, 22/113). Colonies also became established on L. Argyle (16/128) in 1975.[1713] The Atlases contain no breeding records from the Cape York Region.

A study of banded birds in SA showed that the Pied Cormorant tended to disperse less far than other Australian cormorants: only 7% of recoveries were outside the banding district.[1778] Nevertheless, individual birds can move long distances as the Tas. sightings demonstrate. The longest movement recorded is that of a bird banded as an immature from Price (34/138) to Condobolin (33/147), 869 km.[1436] The Field Atlas provides no evidence of large-scale seasonal movement.

The Pied Cormorant inhabits fresh and salt water. In the southern half of WA it is almost entirely marine. Breeding is colonial. It eats predominantly fish but also takes crustaceans and sometimes molluscs. Even though Pied Cormorants are protected, fishermen shoot them, in the belief that they compete for fish. The birds are also used as bait in crayfish pots.

An inland colony at L. Menindee (32/142) held 1000 birds in 1974.[1552] Colonies in coastal mangroves at Port Gawler (34/138) and Price (34/138) held 2000 and 3000 birds respectively in 1973.[1552] Two large breeding colonies in Shark Bay (25/113) were estimated in 1920 to hold 5000 birds each[1877] and a census of part of this area (the east side of Dirk Hartog I.) in 1948 recorded 15 000–20 000 birds.[1638] Recent WA colonies recorded are: 800 birds on Adele I. (15/123) and 950 on Mary Anne Reef (21/115) in 1978.[6]

Field Atlas Records: 7603

1° blocks–with records: 399 (49%)
–with breeding records: 32 (8%)

Atlas No.: 099

Max. Rep. Rate: 76%
In block(s): 30/114

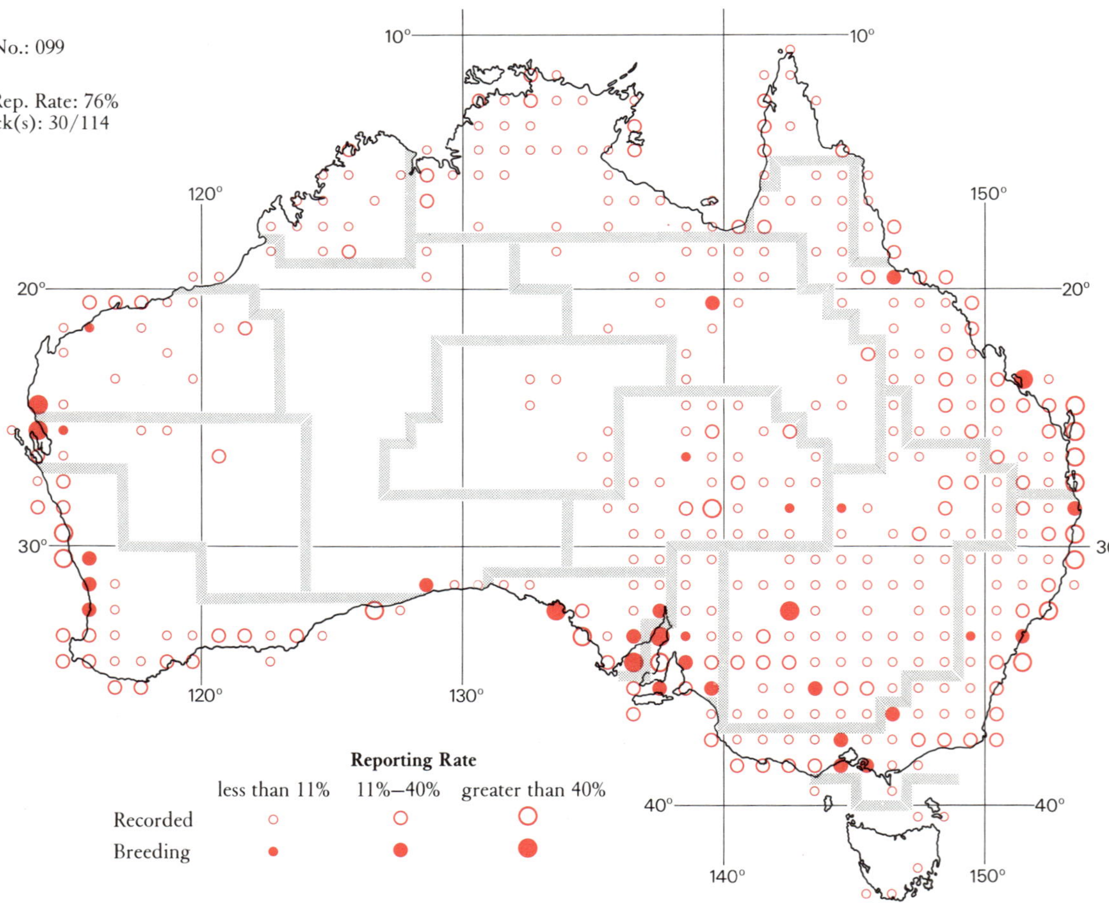

The Little Black Cormorant ranges from Indonesia to New Guinea, New Zealand and Australia. On the Australian mainland its breeding range is wide but it rarely nests in Tas. Single nests at Bridgewater (42/147) in 1976 and 1977 were the first reported since 1914.[570] There appears to have been a general increase in the range and abundance of the species in the Tas. Region since the mid-1970s. Reports came from only 8.9% of 10′ blocks visited in 1951–76 compared with 30.4% of 10′ blocks visited during the Field Atlas. The rarity of local breeding reports suggests that most birds have migrated from the mainland.

A study of the movements of birds banded in SA found that the Little Black Cormorant was intermediate among the Australian cormorants in its tendency to disperse: 31% of recoveries were outside the banding district.[1778] The longest movement recorded is that of an immature from L. Brewster (33/145) to L. Paringa (43S/169E) in New Zealand, 2200 km away.[49] Nomadic flocks reach islands in Torres Strait.[503] At L. George (35/149), at altitude 600 m, counts over three years recorded on average nine times as many Little Black Cormorants in summer as in winter.[1013]

The Little Black Cormorant inhabits fresh or salt water and in mainland Australia is more often inland than on estuaries. Even small pools and farm dams are used. In Tas. the species is most common on estuaries and brackish water. Sometimes large flocks form as at Mt Isa (20/139) in July-Aug. 1976 when 'thousands' were seen. The species usually breeds in colonies along rivers or near swamps, often with other species. It often fishes socially, groups of birds herding a school of fish. The number is critical: 30 were unable to hold a school but 100 could do so easily.[1636] On small bodies of water, insects and crustaceans are taken as well as fish. Fishermen treat the Cormorant as an enemy and some duck shooters regard it as game but the species is fully protected.

Most colonies contain fewer than 100 nests.[1552] Larger ones have been reported at L. Muncoonie (25/138), which had 1000 nests in 1974,[1552] and at Yunderup (32/115), which was estimated in 1980 to have 1000 Little Black Cormorants mixed with Little Pied.[1635]

Little Black Cormorant

Phalacrocorax sulcirostris

Field Atlas Records: 14 012

1° blocks–with records: 494 (61%)

–with breeding records: 56 (11%)

Atlas No.: 097

Max. Rep. Rate: 49%
In block(s): 32/142

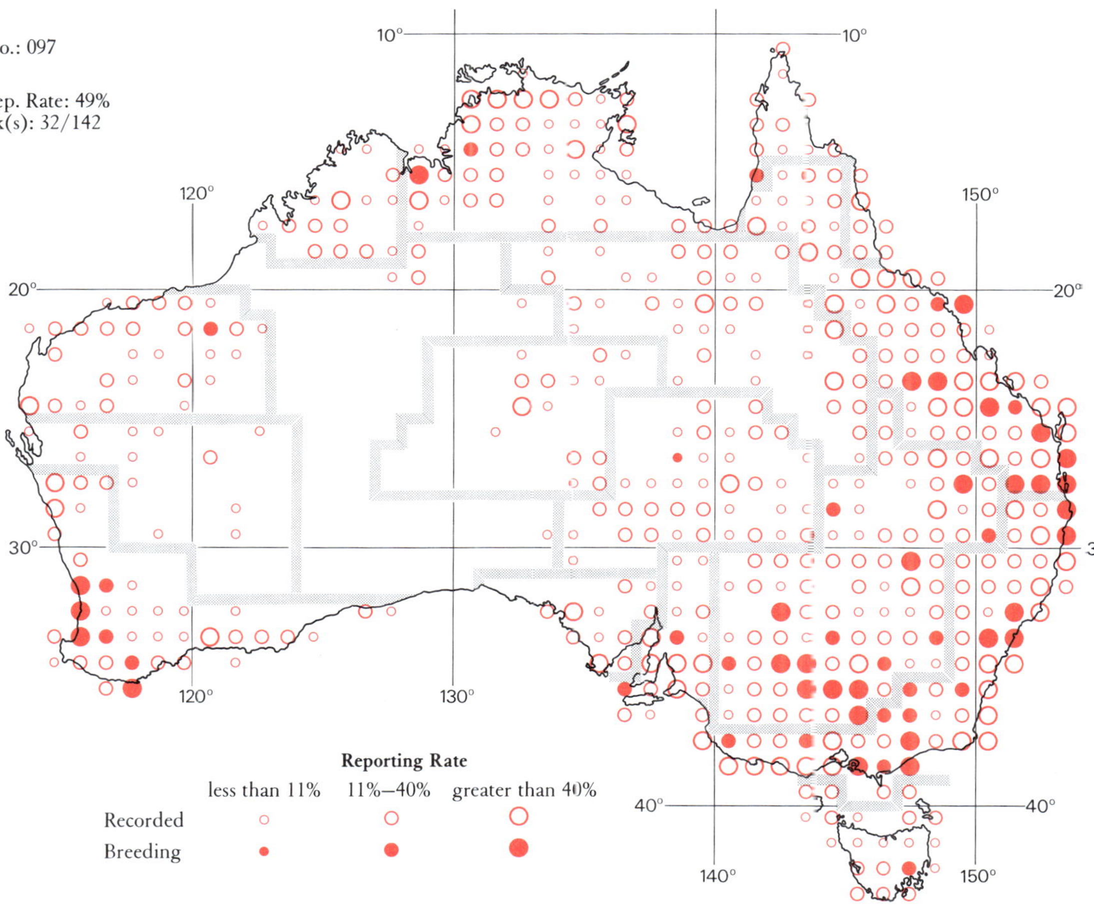

Little Pied Cormorant

Phalacrocorax melanoleucos

The Little Pied Cormorant lives in eastern Indonesia, the south-western Pacific, New Guinea and New Zealand south to Campbell I. (52S/169E), with the subspecies *melanoleucos* in Australia.

Like the other cormorants, the Little Pied is a scarce breeding species in the Tas. Region. Only two breeding reports were submitted during the Field Atlas (0.14% of records for the Little Pied Cormorant in the Tas. Region). For comparison, there were 64 breeding reports in Vic. (1.2% of Little Pied Cormorant records). This suggests that many birds in Tas. have migrated from the mainland.

A study of banded birds in SA showed that the Little Pied Cormorant was intermediate among the cormorants in its tendency to disperse: 20% of recoveries were outside the banding district.[1778] The longest movement recorded is that of an immature banded at Hattah (34/142) and recovered two years later at Arufi (8S/141E) in New Guinea, 2900 km north.[56] The species is regarded as nomadic on islands in Torres Strait.[503] On L. George (35/149) its numbers seem to remain stable throughout the year.[1013] At Rotamah I. Bird Observatory (37/147) numbers are greatest in autumn and winter.[285] The Field Atlas does not suggest large-scale seasonal movement.

The Little Pied Cormorant keeps mainly to fresh water but also lives on estuaries and coasts. Single birds will fish in almost any pool a metre or more deep. On larger bodies of water flocks of hundreds will form, depending on the quality of the feeding at the site. Pairs nest singly or in colonies, usually small. Unlike the other cormorants, the Little Pied feeds mainly on crustaceans[1778] but it is nevertheless persecuted by fishermen in the same way as other cormorants. At Northam (31/116) increased salting of the river has been correlated with the decline of colonies, originally of 5–50 nests.[1216]

Large colonies reported include: 460 birds at Kapalga (12/132) in 1979; two of 1000 birds near Barham (35/144) in 1974; 4000 near Yunderup (32/115) in 1980;[1635] and 1400 at Binningup (33/115) in 1972.[1552]

Field Atlas Records: 20 939

1° blocks–with records: 528 (65%)
–with breeding records: 94 (18%)

Atlas No.: 100

Max. Rep. Rate: 66%
In block(s): 36/150

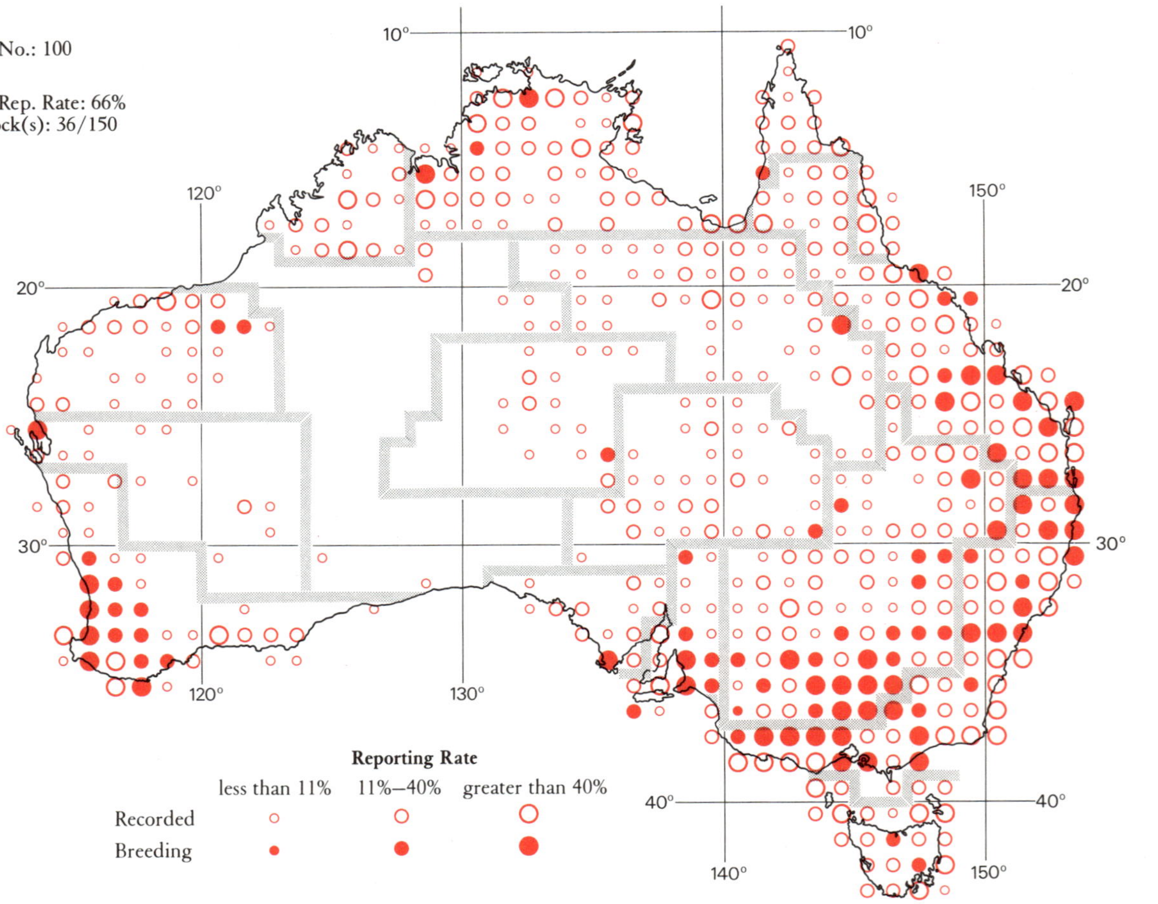

There are differing views about subspecies of the Great Frigatebird, which is one of a closely-related group of five similar species.[388,422] It breeds on islands in the tropical Pacific, western Atlantic and Indian Oceans; these include islands in the Coral Sea,[1636] Raine I. (11/144) and Quoin I. (12/143). The only previous evidence of breeding on Raine I. was an unconfirmed report last century[1811] and Quoin I. is a new location.

The usual distribution of the Great Frigatebird was said to include northern and western Australia as far south as Pt Cloates (22/113)[1636] but this is not borne out by the Atlases. A few sightings have been made in the Timor Sea[1713] but in coastal waters west of 130°E the Atlases contain only three records, all of birds far south of normal range: Swan R (31/115) May 1917; Fremantle (32/115) May 1951; and Walpole (34/116) June 1972.[1638] The Swan R bird was of the subspecies that breeds on Christmas I. (10S/105E).

In eastern Australia the species is distributed south to 29°S. Vagrants are occasionally recorded farther south: Mornington (38/145) 1902; Brighton (37/145) 1861.[21,388]

Though not undertaking regular migration, immature Frigatebirds, and some adults, disperse widely.[422] Individuals are occasionally blown far south by cyclones and lengthy cyclonic activity sometimes adversely affects the species, e.g. a wreck of exhausted birds was reported from Elcho I. (12/135).[1171]

The Great Frigatebird feeds on fish, squid and offal, taking all its food in flight. It is adept at catching flying-fish, hovers to take offal thrown overboard from fishing boats and sometimes pirates food from other seabirds or raids their breeding colonies.[1597]

Great Frigatebird

Fregata minor

Breeding colonies overseas vary in size from fewer than 100 birds up to 4000.[311] There were five nests on Raine I. in 1980 and on Quoin I., where a few birds are usually present, there was definitely one nest in June 1980.[1000]

Field Atlas Records: 64

1° blocks–with records: 26 (3%)
–with breeding records: 2 (8%)

Atlas No.: 094

Max. Rep. Rate: 19%
In block(s): 12/143

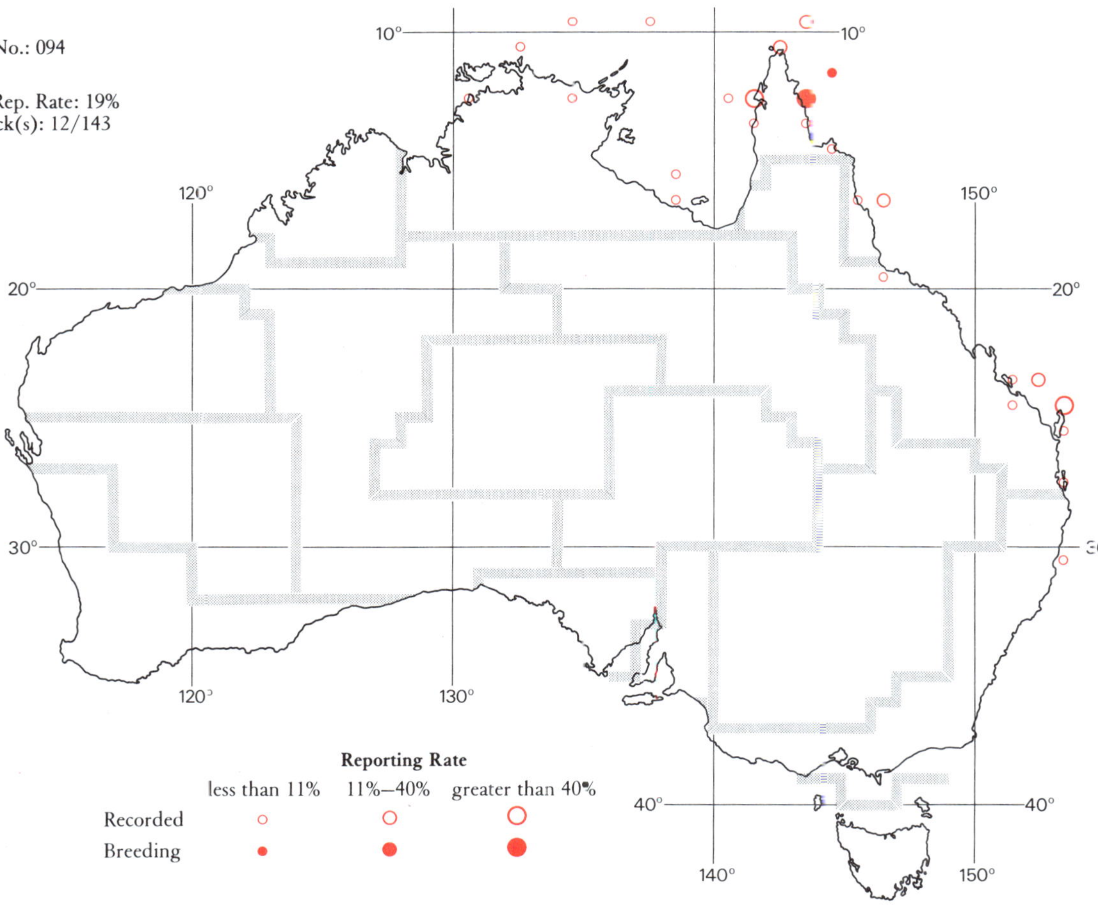

Least Frigatebird

Fregata ariel

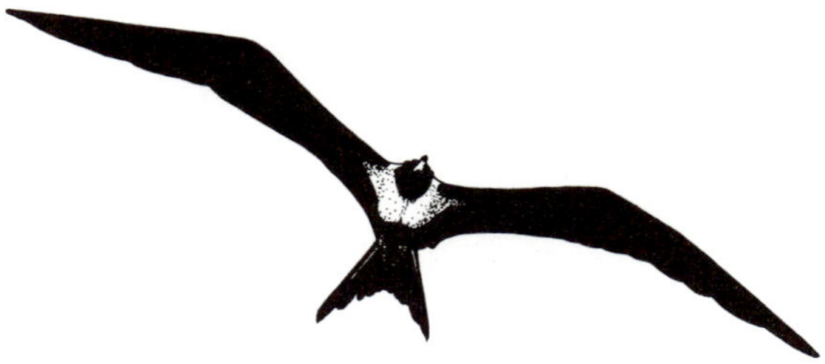

The Least Frigatebird, one of a closely related group of five similar species, breeds in the Atlantic, Pacific and Indian Oceans. It has three subspecies; one, *ariel*, nests on islands scattered round the coast of northern Australia and in the Coral Sea. Quoin I. (12/143) and Bell Cay (21/151) are new breeding locations and there are also colonies on Raine I. (11/144), Swain Reef (21/151), Rocky I. (16/139), Ashmore Reef (12/123), Adele I. (15/123), Lacepede Is (16/122), King Sound (16/123) and Bedout I. (19/119). The Least Frigatebird is occasionally reported well south of its usual range, vagrants reaching Vic. in Apr. 1929 (38/145) and, extraordinarily, an inland site (36/146) in June 1952.[820]

Recoveries from 300 000 birds banded in central Pacific colonies indicate that the young disperse great distances with the prevailing winds, moving west towards New Guinea and north to the Philippines.[1665] The adults also leave the colonies when breeding is completed in Nov. Whether Australian birds disperse over similar distances is not known. A seasonal pattern of movement is suggested by the Field Atlas, which shows that the species is more frequently reported in the Eastern Qld Region in summer (1.2%) than in winter (0.2%).

The Least Frigatebird catches fish and squid or pirates food from other seabirds.

The Ashmore Reef breeding colonies are often raided by Indonesian fishermen and most islands contain piles of bird remains. At Willis I. (16/150) birds were rendered down for their blubber and the colony destroyed.[1640]

Vast numbers of Least Frigatebirds breed in the central Pacific: Phoenix Group (3S/171W) 56 000; Howland I. (0N/176W) 16 000.[1665] Australian colonies are much smaller. In 1980 the Raine I. colony was estimated at 2670 birds and that on Quoin I. at more than 1000.[1000] Nest counts in 1982 at some colonies in WA recorded breeding populations of 11 400 birds on Adele I., 2700 birds on the Lacepede Is and 4580 birds on Bedout I.[283]

Field Atlas Records: 264

1° blocks–with records: 69 (8%)
–with breeding records: 4 (6%)

Atlas No.: 095

Max. Rep. Rate: 25%
In block(s): 10/142

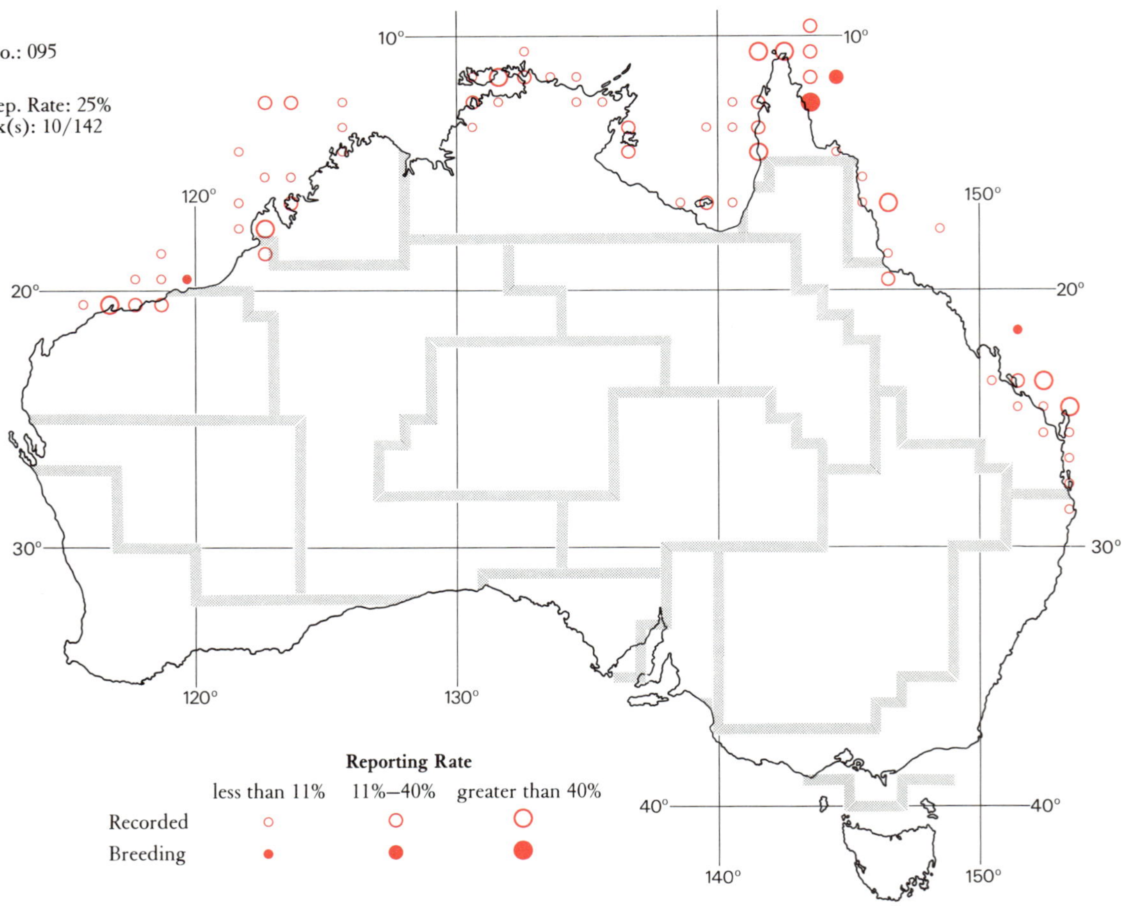

Red-tailed Tropicbird

Phaethon rubricauda

The Red-tailed Tropicbird lives in the Indian and Pacific Oceans. Two subspecies breed round Australia: *roseotincta* on two atolls in the Coral Sea and on Raine (11/144), Norfolk (29/167) and Lord Howe (31/159) Is; and *westralis* on Cocos-Keeling (12S/96E) and Christmas (10S/105E) Is and at a colony established in 1963 on Sugarloaf Rock (33/115).[388,1730] Since 1894 a few pairs have nested on the Houtman Abrolhos (28/114) but the last time seems to have been in 1954 when at least eight nests were counted.[1808] Single nests found at Busselton (33/115) in 1938, Rottnest I. (32/115) in 1957–59 and repeatedly on the mainland opposite Sugarloaf Rock all failed.[1638] The species has been observed right round the coast of Australia. The southernmost observations are from Guildford (37/144) in Jan. 1945 and from the Derwent Estuary (42/147) in Feb. 1917.[798]

In eastern Australia summer cyclones sometimes blow the Tropicbird far south and inland and sightings south of 24°S in the Eastern Qld, South-East and Tas. Regions show a seasonal pattern: all records in the Atlases were Nov.-Mar. except for one inland at Tamworth (36/149) in June 1945. Cyclonic weather caused a wreck in Feb.-Mar. 1978 when nine specimens, adult as well as immature, were found in the South-East and Murray-Darling Regions.[1288] The presence of adults suggests that they wander a considerable distance at times. At Norfolk and Lord Howe Is the species is almost absent in winter[676,1580] but at Sugarloaf Rock some birds remain near the colony at all times of the year.[1730] Dispersal of young birds is suggested by the failure of any chick banded at Sugarloaf Rock in 1967–75 to return and breed there.[1730]

Most of the food recorded at Sugarloaf Rock is fish but in other parts of the world squid is a prominent item of diet.[1730] The Sugarloaf Rock colony is sometimes harassed by Ospreys; and Silver Gulls apparently compete for nest sites.[1730]

The colony at Sugarloaf Rock had 26 occupied nests by 1974[1730] with at least 70 birds present. On Raine I. there is no record of more than 44 occupied nests on any one visit. One of the colonies on Norfolk I. numbered at least 400 breeding birds in 1979[1731] and hundreds of birds breed on Lord Howe I.[676]

Field Atlas Records: 46

1° blocks–with records: 22 (3%)
–with breeding records: 2 (9%)

Atlas No.: 107

Max. Rep. Rate: 8%
In block(s): 11/144

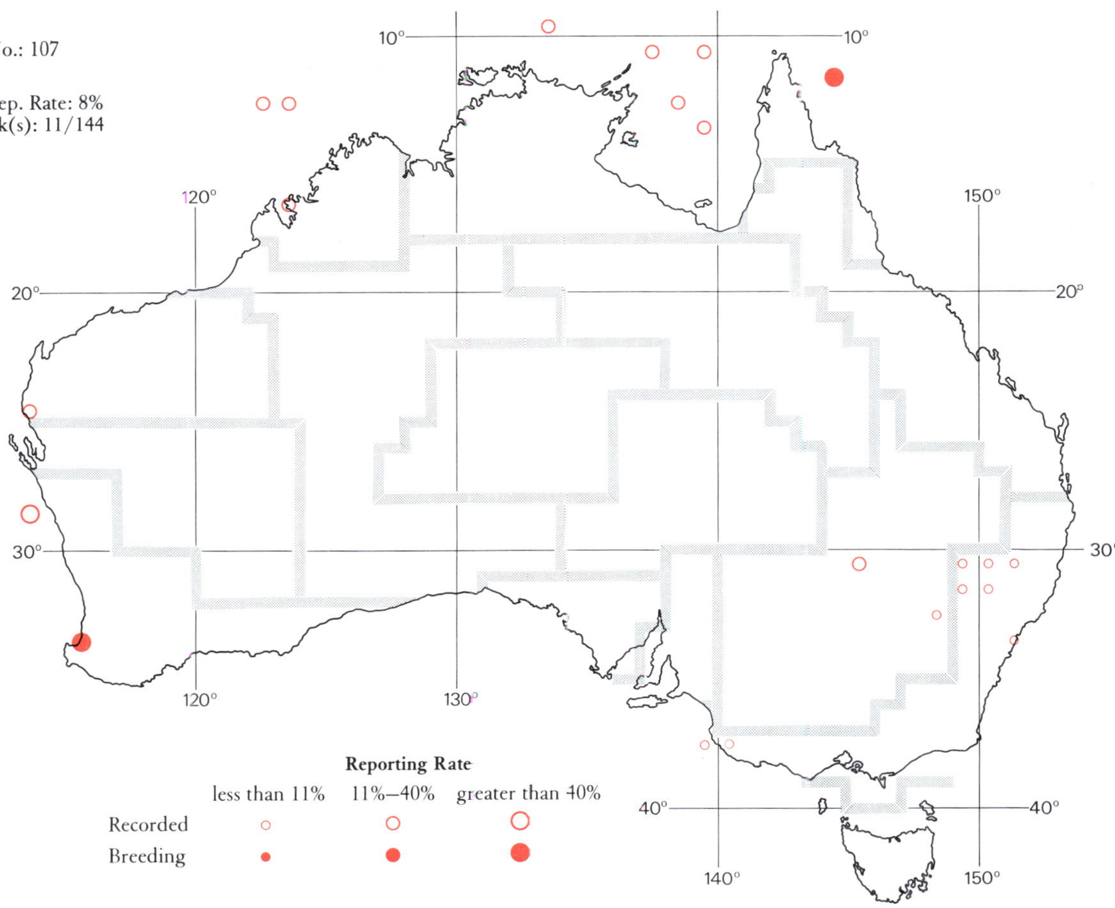

Great-billed Heron

Ardea sumatrana

The Great-billed Heron ranges through south-eastern Asia to New Guinea, where it is rare, and to Australia which is at its southern limit. In northern Australia the Heron's distribution is disjunct. The species is absent from the area south of the Gulf of Carpentaria except for the population shown on the map on the Gregory R (17/139, 18/139). The Top End and Kimberley populations are probably continuous west from the Macarthur R (15/136).

Two clutches of eggs collected on Fraser I. (25/153) in the 1950s[1784] represent the Heron's southern breeding limit but there are few sightings and no other breeding records south of Broad Sound (22/150) in the Atlases. One bird was seen near Yamba (29/153) in Apr. 1980.[687a] In WA, Derby (17/123) is as far south as the Great-billed Heron has been recorded.

The difficulty of finding nests is indicated by the single breeding record in the Field Atlas and the lack of authenticated breeding reports from the Kimberley Region.[1713]

The Field Atlas suggests no large-scale seasonal pattern of movement though the occasional southern records indicate that individuals may wander long distances at times. Near Jambin (24/150) the Heron appears only in wet years.

The Great-billed Heron is usually seen singly and at the same places year after year. It lives in mangroves and along small creeks, flying out to the main river to feed at dusk. The same nest sites may be used repeatedly but not necessarily in successive years. An area inhabited by a pair of Herons will have several nests within it in various states of repair.[404,1642] The food of this Heron is little known: a chick regurgitated three mullet 15–20 cm long[1642] and adults are thought to feed on large fish.[756]

Observations suggest that the species is at low density with pairs distributed 1–2 km apart along suitable creeks; the creeks themselves are widely separated.[1642]

Field Atlas Records: 183

1° blocks–with records: 47 (6%)
–with breeding records: 1 (2%)

Atlas No.: 184

Max. Rep. Rate: 14%
In block(s): 14/141

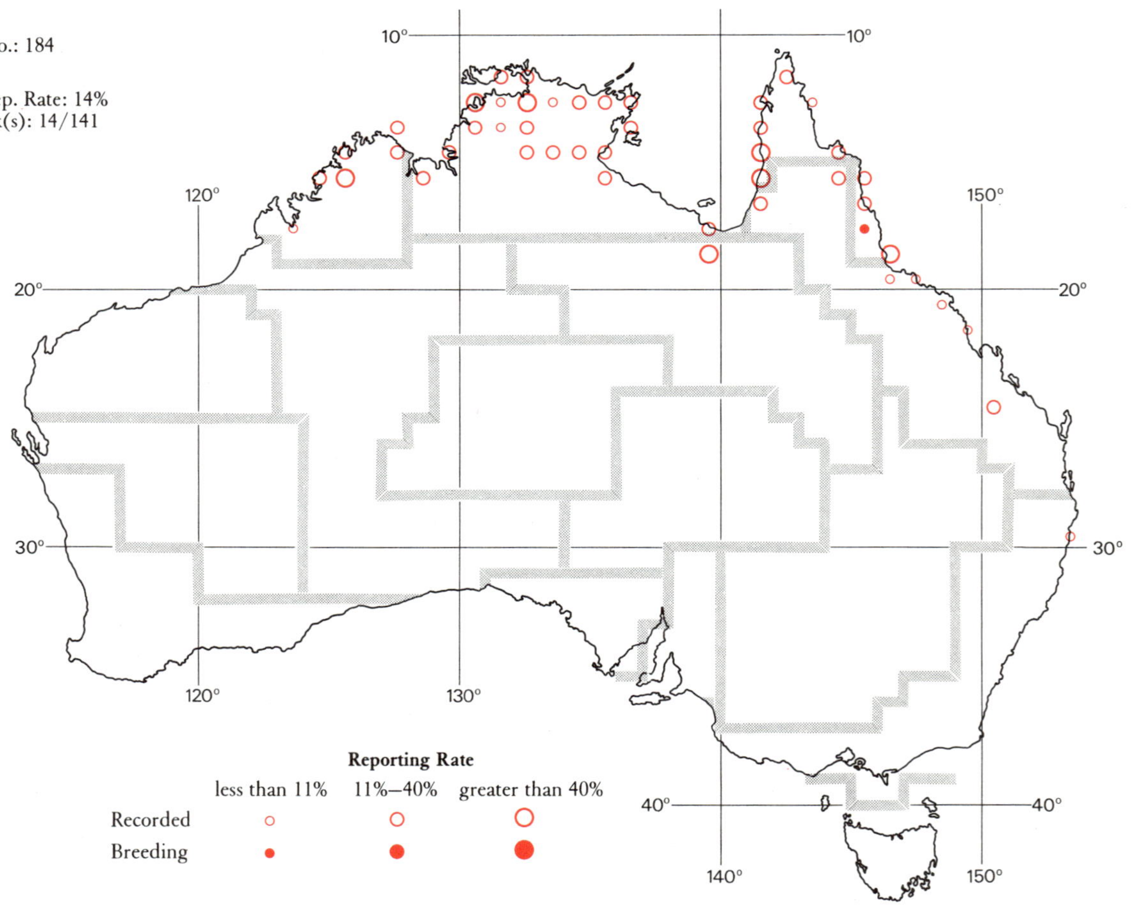

Pacific Heron

Ardea pacifica

The Pacific Heron lives mainly in Australia but occasionally appears in New Guinea and New Zealand.[388] In Australia its breeding range has expanded into the South-West Region, nesting being first confirmed there in 1964 at L. Toolibin (32/117).[1638] During the Field Atlas there was an irruption of Pacific Herons into the Tas. Region on an unprecedented scale. The species had been previously reported from only seven localities[734] but during 1978–79, when all but three of the Tas. Region reports were made, it was recorded in 37 different 10′ blocks. The only records in Torres Strait were made during Mar.-May, also in 1978 and 1979.

Regular movements have been noted as well as irruptions. At Townsville (19/146) the Heron arrives Mar.-June and departs at the end of the year[857] and round Darwin (12/130) it is most often seen June-Aug.[423] The Field Atlas reflects these seasonal changes in abundance in the Atherton Region and the Top End with Reporting Rates in winter (22.2%; 28.2% respectively) almost double those of summer (11.0%; 16.9%). At Portland (38/141) a spring invasion has been documented;[1052] in the Tas. Region all records are Apr.-Nov., including irruptions. Irruptions also take place in winter or spring in the South-West.[1216,1638]

The Pacific Heron forages in water or wet paddocks, often alone though flocks of 200 have been recorded in the Murray-Darling Region[833] and on Karakin L. (31/115) and one of 500–1000 birds was reported near Mystic Park (35/143) in 1976.[1552] The Pacific is more likely to breed inland than the White-faced Heron. In the Field Atlas only 20% of 1° blocks in which breeding Pacific Herons were reported include the coast compared with 46% for the White-faced Heron. The nest is usually solitary but sometimes in a loose colony or among the nests of other species of waterbird. Individuals defend a feeding territory[756,1458] and forage for fish, frogs, insects and aquatic invertebrates. These are abundant in inland pools while the water is drying up in spring and early summer.

If colonies form they usually have fewer than 40 nests though one of 250 nests was recorded at Mystic Park (35/143) in 1976.[1552]

Field Atlas Records: 18 955

1° blocks–with records: 603 (74%)

–with breeding records: 79 (13%)

Atlas No.: 189

Max. Rep. Rate: 83%
In block(s): 27/149

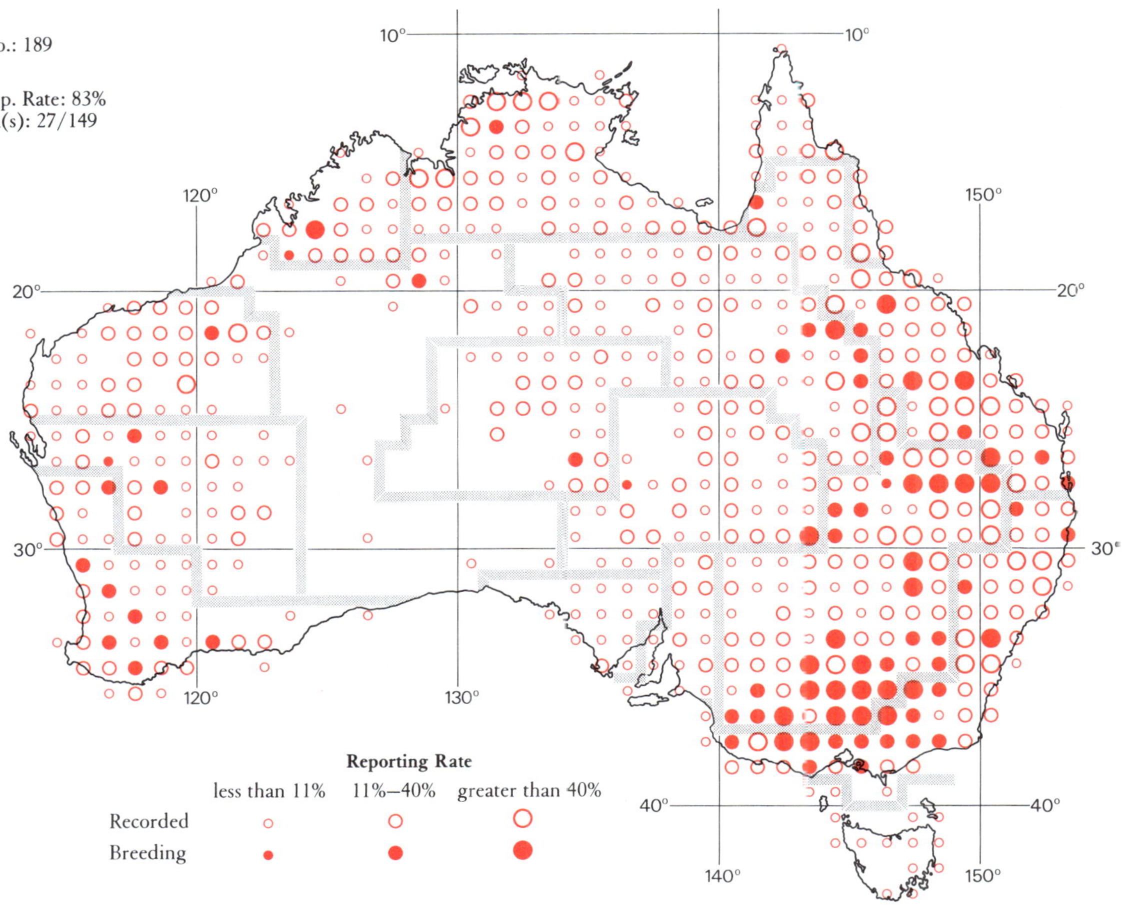

White-faced Heron

Ardea novaehollandiae

Outside Australia there are small populations of the White-faced Heron in New Zealand, New Caledonia and parts of Indonesia. In New Guinea the species is thought to be a non-breeding vagrant from Australia, though sometimes numerous.[129] Its colonization of New Zealand is recent, started in 1940 by birds from Australia such as those which still occasionally reach nearby Pacific and sub-Antarctic islands.[756]

In Australia the absence or rarity of breeding reports in some regions where the species is present implies considerable movement. Breeding has not been recorded in the Cape York Region, which may be populated largely by birds bred farther south. In the Tas. Region breeding is less frequent than in adjacent parts of Vic. (breeding reported in Tas. in 4.2% of 10′ blocks in which the species was recorded compared with 16.9% in the parts of Vic. south of 37°S). Movements take place between the mainland and Tas.[1801] but they are probably nomadic rather than regular; Reporting Rates in the Tas. Region changed little with the seasons (summer 37%, winter 39%). The White-faced Heron is a dry winter season visitor to islands in Torres Strait[503] and, near Darwin (12/130) also, the Reporting Rate was higher in winter (40.6%) than in summer (11.6%). Elsewhere the species is present throughout the year.

The White-faced Heron hunts in shallow water round pools, dams, lakes and irrigated crops, as well as in paddocks far from water. The species is often alone though a group as large as 400 has been reported in New Zealand.[756] When many are together, the birds are usually well dispersed, each defending a hunting beat. Nests are usually built singly. The White-faced Heron is more likely than the Pacific to breed in districts near the coast. In the Field Atlas, 46% of the 1° blocks in which the White-faced Heron was reported breeding include the coast, compared with 20% for the Pacific Heron.

The White-faced Heron feeds on small fish, tadpoles, frogs, lizards and aquatic invertebrates and several observers reported it feeding on road-kills. Much vegetable material has been found in its guts.[756] The frequency and persistence with which it uses farm dams and irrigated crops for feeding suggests that it has benefited from agriculture.

There are no measures of abundance.

Field Atlas Records: 36 173

1° blocks–with records: 644 (79%)

–with breeding records: 155 (24%)

Atlas No.: 188

Max. Rep. Rate: 77%
In block(s): 35/144

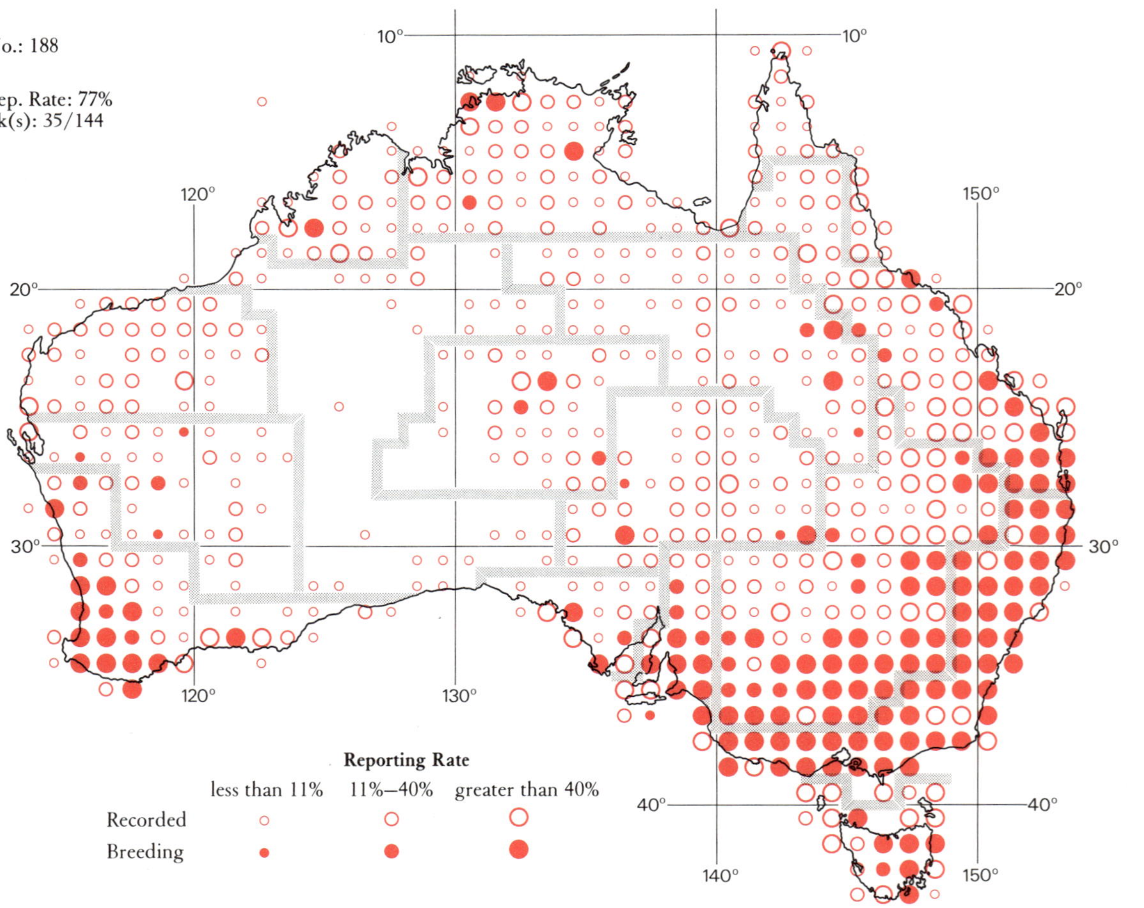

Pied Heron

Ardea picata

The Pied Heron lives in Indonesia and New Guinea as well as in Australia. The Australian range has been described as uneven[756] but the Field Atlas shows a continuous distribution round the northern coast from Townsville (19/146) to Wyndham (15/128). The species is occasionally seen in the western Kimberley Region, with Field Atlas records Apr.-May (17/123, 17/124). Four individuals have been seen even farther from the normal range: at Barham (35/144) in the early 1970s; Robertson Lagoon (34/139) in Nov. 1979; Leigh Ck (30/138) in Feb. 1979; and at Rabbit Flat (20/130) in 1975. A breeding colony on the Roper R (14/135) discovered in 1916[1859] was not visited during the Field Atlas.

It has been suggested that birds from Australia migrate to New Guinea where the Pied Heron is observed throughout the year but has not been found breeding.[756] This seems unlikely in view of the large numbers in New Guinea where 20 000–30 000 birds have been reported along one 15-km stretch of river.[555] Neither is there any evidence of passage through Torres Strait. Adults and juveniles are dry winter season visitors to islands in the Strait and may breed there.[503] A few Pied Herons appear each summer in the Atherton Region[231,698] but the species is most abundant near Townsville Mar.-May.[756] Changes in abundance round Darwin (12/130) have been interpreted to be the result of post-breeding dispersal from local colonies, the birds moving from freshwater lagoons and swamps to the coast in spring.[423]

The Pied Heron nests colonially in mangroves among other species of waterbird. It mostly feeds in swamps, lagoons and on muddy banks where the water is shallow, catching small fish, aquatic insects and their larvae.[1218] It will also feed away from water in rather dry places. Immature birds forage in small flocks but adults tend to be solitary.[1458] In New Guinea the species gathers on freshly burned ground and scavenges near settlements in garbage tips and sewage ponds.[756]

In Australia the Pied Heron is normally in flocks that number hundreds and the largest breeding colony reported is of 1200 nests at Kapalga (12/132) in 1980.[1552]

Field Atlas Records: 517

1° blocks–with records: 51 (6%)

–with breeding records: 5 (10%)

Atlas No.: 190

Max. Rep. Rate: 45%
In block(s): 12/132

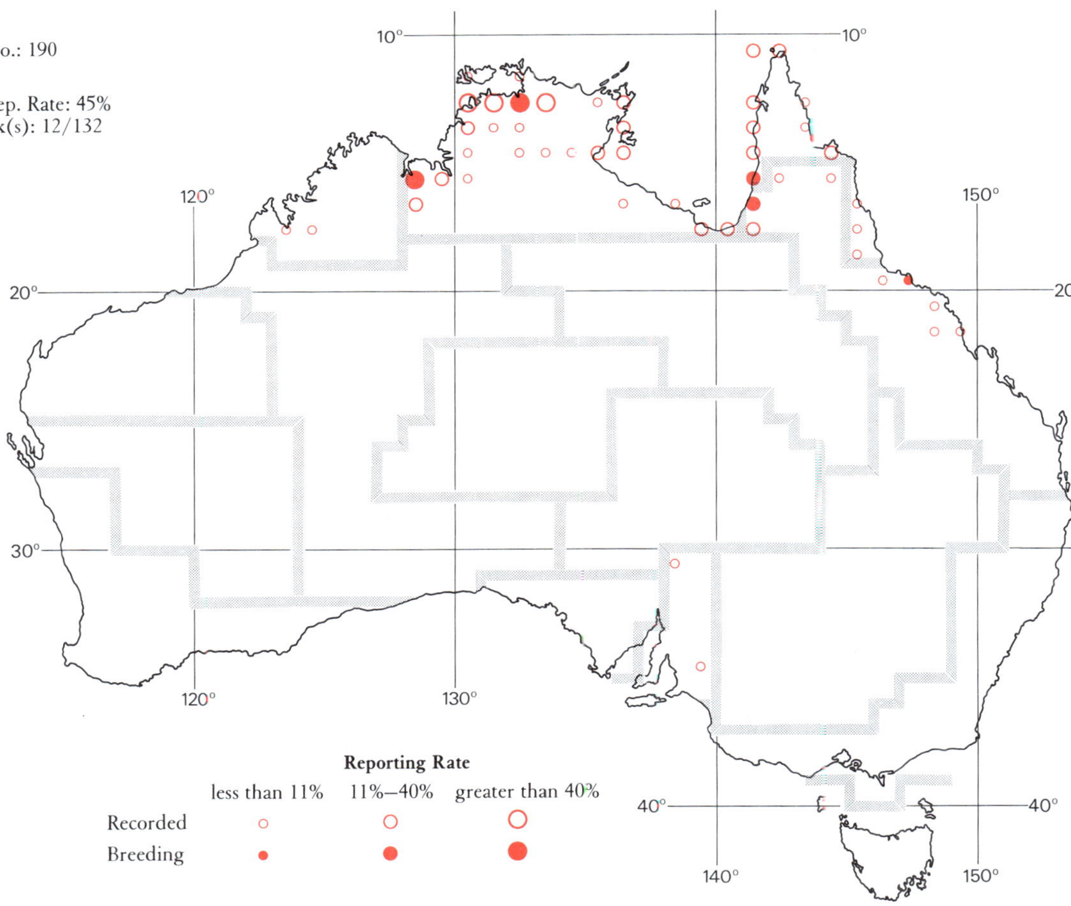

Cattle Egret

Ardeola ibis

The Cattle Egret is native to southern Europe, Africa and Asia but has colonized other areas. In Australia it is unlikely that 18 birds introduced to Kimberley Downs Station (17/124) in 1933[923] gave rise to the present population. Probably birds entered from the north as part of a world-wide expansion. Its spread in Australasia is well recorded (p. 661). First reports in each region were: western New Guinea, before 1941;[1075] Top End, presumed breeding at Oenpelli (12/133) in 1948,[475] certainly breeding by 1959;[786] SA Ra., 1948,[1056] bred once in 1971;[204,1391] South-East, near Cororooke (38/143) in 1949[329] and bred near Ulmarra (29/153) in 1954;[715] Pilbara, single bird in 1949;[1638] South-West, 1952;[923] Rolling Downs, at Mt Isa (20/139) in 1952;[1833] Atherton and Eastern Qld, 1960s and bred in Brisbane (27/153) in 1963;[1279] New Zealand, 1963;[779] Tas., 1965;[734] eastern New Guinea, 1969;[1075] Murray-Darling, bred on Murray R at 35/144 in 1974; Macquarie I. (54S/158E), single bird in 1975.[734] Breeding has not yet been reported in the lower South-East, Tas. or South-West Regions though many birds are regularly present, mostly in autumn and winter.

The Cattle Egret migrates.[1391] In New Guinea it is regarded as a dry winter season non-breeding migrant from Australia.[554] It also migrates to New Zealand, where it has not yet bred.[61,780] Within Australia a nestling from Murwillumbah (28/153) was recovered at Stratford (37/147) 1250 km south and another from Carr's Ck (29/152) 1550 km south at Granton (42/147).[1289] The Field Atlas suggests that birds move south in the winter non-breeding season (Reporting Rates south of 26°S: summer 2.4%, winter 4.8%). At L. Gutheridge (38/147) the bird is present Apr.-Nov.[329]

The Cattle Egret lives in flocks and breeds colonially. It feeds on insects and other invertebrates disturbed by grazing cattle, occasionally taking ecto-parasites from the animals themselves.[423,756,1653] It also eats small skinks[1533] and will feed in shallow water and drains. The breeding population in NSW was estimated at 4600 birds in 1979.[1289] A colony on the East Alligator R (12/132) was estimated to contain 3000–3400 birds in 1980.[1552]

Field Atlas Records: 3950

1° blocks-with records: 172 (21%)
-with breeding records: 14 (8%)

Atlas No.: 977

Max. Rep. Rate: 48%
In block(s): 27/153

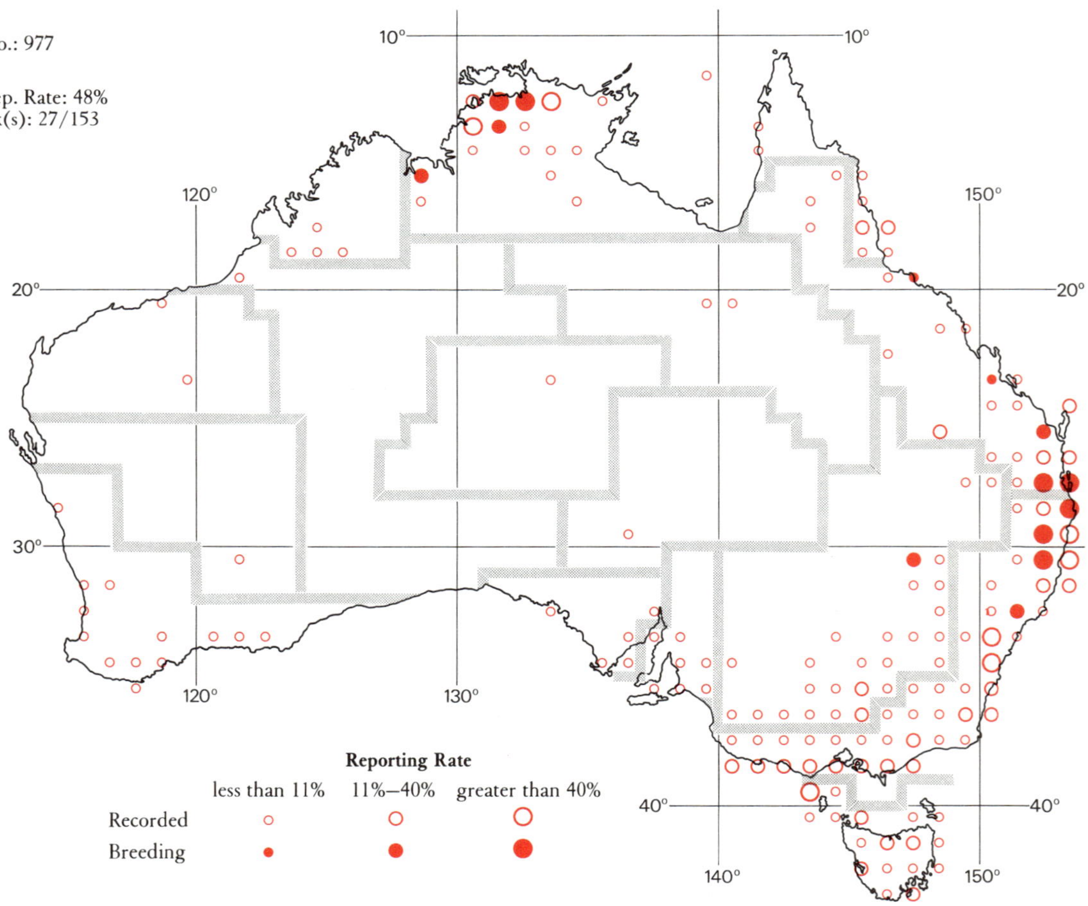

Great Egret

Egretta alba

The cosmopolitan Great Egret is widespread in coastal Australia and in the inland. Its range may have extended this century. Breeding was not reported in the SA Ra. Region until the 1951–76 period. In the South-West Region there are no records in the Atlases before 1901 and the species was considered rare until the 1950s.[10,1216] Since 1939, when a colony was discovered at Bunbury (33/115), the Egret has bred on the Collie R (33/116) and at Toolibin L. (32/117). The colonies at L. Chandala (31/115) and near Carnarvon (24/113) were first reported during the Field Atlas.

In northern Australia colonies recorded on the Roper R (14/135) 1901–50 were not visited during the Field Atlas; those on the Nassau (15/141) and Gilbert (16/141) Rs and on Waterbank Station (17/122) have not previously been reported.

Overseas, parts of some populations are migratory.[1117] The longest movement recorded in Australia is that of a bird from Balranald (34/143) to New Guinea, 3228 km.[85] The Great Egret is most frequently seen in Torres Strait in the dry winter season[503] though some birds are present throughout the year; in the Atherton Region,[231] it is absent Jan.-Mar. In the southern Murray-Darling Region, where it breeds, the bird disappears almost completely May-July.[833] At Laverton (38/143) in 1950–53 it was present irregularly for varying periods.[1822] The Field Atlas suggests no large-scale pattern of movement on the mainland but the Reporting Rate in the Tas. Region is lower in summer (2.8%) than in winter (11.6%), presumably because birds return to the mainland to breed.

Though mainly seen singly,[1859] flocks of up to 150 have been recorded in the southern Murray-Darling Region,[833] at Bool Lagoon (37/140) and other localities. The Great Egret nests in colonies with other waterbirds.[1552] It feeds mainly on fish, frogs and invertebrates in shallow swamps, flooded fields and along the edges of ditches and streams. Occasionally it will eat birds.[422,1004] Clearing and the development of irrigated agriculture have probably benefited the species.

Colonies usually contain fewer than 100 nests but larger ones have been reported: L. Menindee (32/142) 500 nests in 1974, near Koondrook (35/144) 200 in 1974 and Kapalga (12/132) 300 in 1979.[1552]

Field Atlas Records: 13 849

1° blocks–with records: 484 (60%)

–with breeding records: 24 (5%)

Atlas No.: 187

Max. Rep. Rate: 64%
In block(s): 13/130

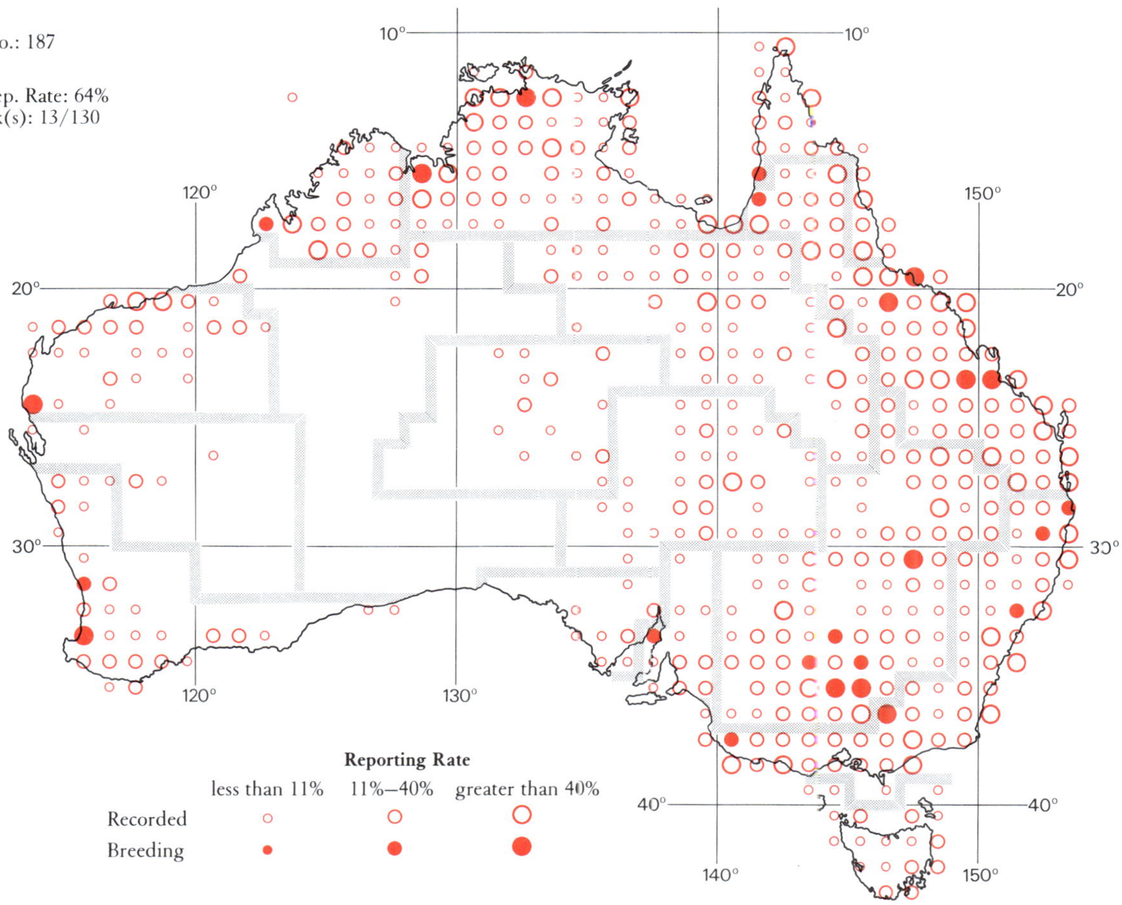

Little Egret

Egretta garzetta

The Little Egret ranges throughout southern Asia, parts of Europe and Africa, New Guinea and Australia. Its range in Australia has grown, particularly since 1901 (p. 661). The species was not recorded breeding as far south as the Murray R until 1939[955] and was still considered a summer visitor to the southern Murray-Darling Region in the 1950s.[833] In the SA Ra. Region sight records were made in the 1950s and have continued with increasing frequency since; the only report of breeding was in 1967. The first Tas. record was made in 1957[203] and the Egret is now a regular visitor in small groups of up to 10; it has not yet been reported to breed there. The Little Egret had spread to the South-West by the 1960s with a record from L. Yangebup (32/115) in 1965.[593] In 1980 a pair bred near Carnarvon (24/113) and in 1981 near Australind (33/115). The spread of the Little Egret has been unspectacular, one or two birds turning up here and there, so that it is not documented as well as that of the conspicuous Cattle Egret.

Overseas populations of the Little Egret make long migrations. In Australia numbers fluctuate at any given site but there is no indication in the Field Atlas of a large-scale seasonal pattern of movement. Small numbers migrate to Torres Strait in the dry winter season[503] and birds banded at Balranald (34/143) have been recovered in New Guinea and New Zealand so that there may be some interchange between Australian and Asian populations.[1117]

The Little Egret lives along streams, lake shores, the seashore and in mangroves but not usually in pastures or swamps. It nests colonially, often with other waterbirds.[1552] The species rarely hunts in company, though big concentrations of the small fish on which it feeds attract flocks of 30 or more.[1859] It has been observed feeding with Sacred Ibis, each Ibis accompanied by only one Egret. As the Ibis dabbles in the mud, the Egret darts its bill and neck about gathering prey.[1294]

Southern colonies are smaller than those in the north. The largest colonies reported in the south are of 30 nests near Newcastle (32/151) in 1982 and 40 nests near Koondrook (35/144) in 1974 compared with 554 at Kapalga (12/132) in 1981.[1552]

Field Atlas Records: 4098

1° blocks–with records: 284 (35%)
–with breeding records: 11 (4%)

Atlas No.: 185

Max. Rep. Rate: 54%
In block(s): 13/130

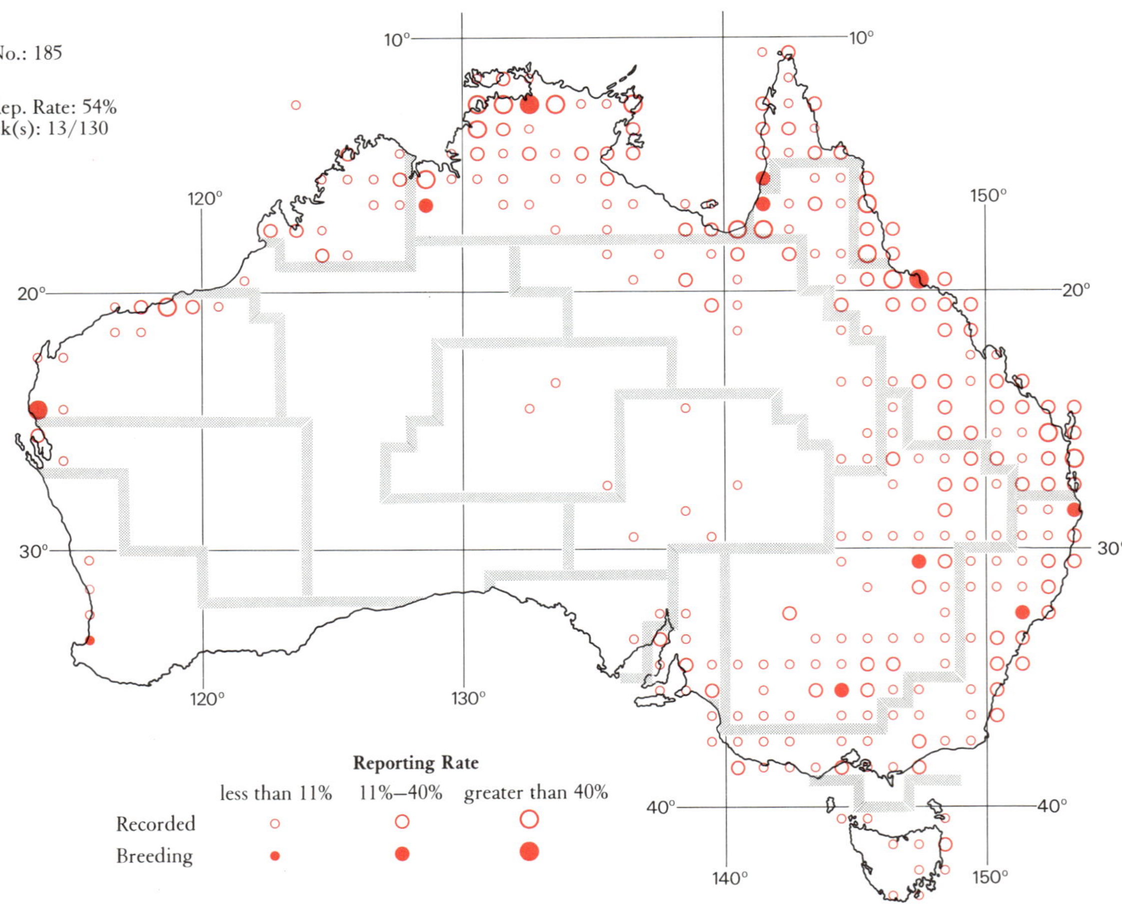

Intermediate Egret

Egretta intermedia

The Intermediate Egret has a discontinuous distribution through Africa, southern and eastern Asia, New Guinea and Australia. The species is intermediate in appearance and behaviour between the Great and Little Egrets and can be confused with them in the field. This has hindered interpretation of records, specially historical ones. The Intermediate Egret occasionally reaches the SA Ra. Region and breeds irregularly at Bool Lagoon (37/140); it has been recorded in the Tas. Region only in 1958 and during the Field Atlas.

Large breeding colonies on the Roper R (14/135) described by early ornithologists[1859] were not visited during the Field Atlas. No breeding colonies are known in the Kimberley Region, though there is an unconfirmed report of nesting on the Behn R (16/128) in 1972.[1713]

The Intermediate Egret is thought to be nomadic in Africa and it migrates regularly through south-eastern Asia.[422] Individuals from Australian breeding colonies move long distances, the longest recorded being by an immature from Barham (35/144) to New Guinea, 3700 km.[52] Sightings distant from the colonies are made most often in autumn and winter. Thus the species is a winter visitor to Torres Strait[503] and the Pilbara Region[1638] and the four Tas. records were made Mar.-June. In the southern Murray-Darling Region, where it breeds, the Egret is absent May-July.[833] Near Atherton (17/145) it is absent Feb.-Mar.[231]

Alone or in groups, the Intermediate Egret forages in shallow margins of swamps and sometimes follows feeding cattle. It feeds on frogs and aquatic insects.[748] Agricultural development of low-lying areas, formerly forested, has extended its feeding area and it has benefited from irrigation. The Intermediate Egret nests colonially, usually among other waterbirds.[1552] At the height of the plume trade a single day's sale in London in 1909 accounted for 24 000 birds taken at their breeding colonies.[1050] Perhaps twice as many unfledged chicks died as a result of the destruction of their parents and the disturbance to their colonies. The colonies survived and flourish now that the plumes are no longer in demand.

The largest southern colony reported is of 500 nests near Koondrook (35/144) in 1974 compared with a northern colony of 1845 nests at Kapalga (12/132) in 1981.[1552]

Field Atlas Records: 3000

1° blocks–with records: 239 (29%)

–with breeding records: 14 (6%)

Atlas No.: 186

Max. Rep. Rate: 50%
In block(s): 19/147

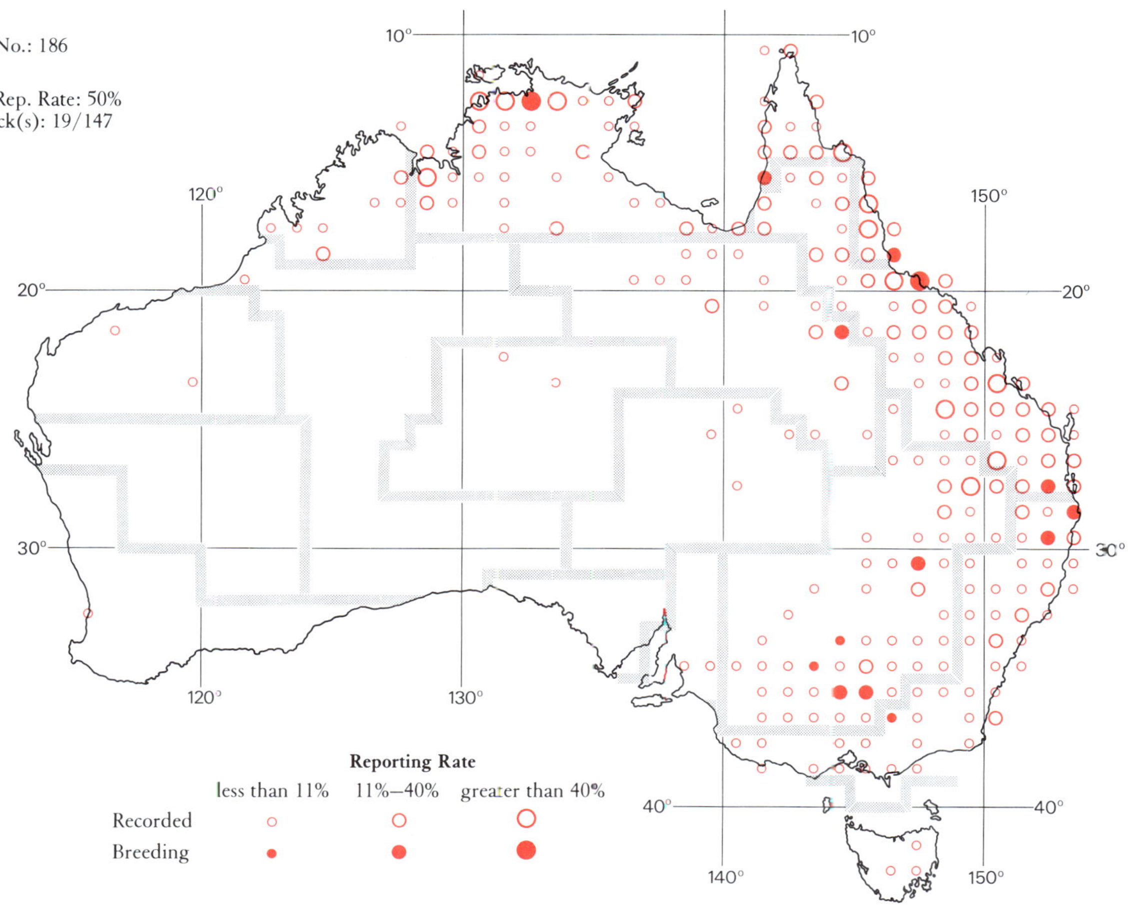

Eastern Reef Egret

Egretta sacra

The Eastern Reef Egret is widespread throughout south-eastern Asia, New Guinea and New Zealand with the subspecies *sacra* round the Australian coast. Throughout its range dark and white plumaged birds interbreed. White birds predominate in most places but not in southern latitudes. The population in the Eyre Pen. and SA Ra. Regions is isolated and only dark birds are known there.[387]

Confusion with the immature White-faced Heron casts doubt on many sightings, particularly in the south where the Reef Egret is rarely reported from the lower South-East Region between 141°E and 149°E or from the Tas. Region. Records from these regions that seem reliable were almost all made 1880–1909 when breeding of both dark and white birds was reported from Bass Strait islands including the Hunter Group (40/144), Sister I. (40/145), Rocky Cape (40/145), Clarke I. (40/148) and Church Rock (41/144).[734,1343] During this period there was also a sighting in the South-East Region at Phillip I. (38/145).[299] Since then the only confirmed reports from either region have been at Wilsons Prom. (39/146) in Apr. 1976 and Cox Bight (43/146) in Jan. 1975.[734]

A banded bird has been recovered 560 km from its banding site[85] and some evidence for post-breeding dispersal has been noted within SA[1391] but the Field Atlas suggests no large-scale seasonal movement.

The Reef Egret usually lives on the seashore in Australia but in New Guinea inhabits river banks well away from the sea and up to 600 m above sea level.[485] It breeds on islands, sometimes in colonies but not among other species. The Reef Egret feeds on small fish, marine invertebrates and insects.[1459] It is a predator of the eggs and nestlings of the Common and Black Noddies.[880,1459]

On the Great Barrier Reef a mated pair defended a feeding area of about 5 ha, one side facing the ocean and the other abutting the territories of conspecifics.[1456] At times the territory was occupied by other Reef Egrets, often as the tide first fell or just as it was about to re-cover the reef. The largest colony reported is of 21 nests on One Tree I. (23/152) in 1976.[1552]

Field Atlas Records: 1682

1° blocks–with records: 145 (18%)
–with breeding records: 12 (8%)

Atlas No.: 191

Max. Rep. Rate: 65%
In block(s): 10/142

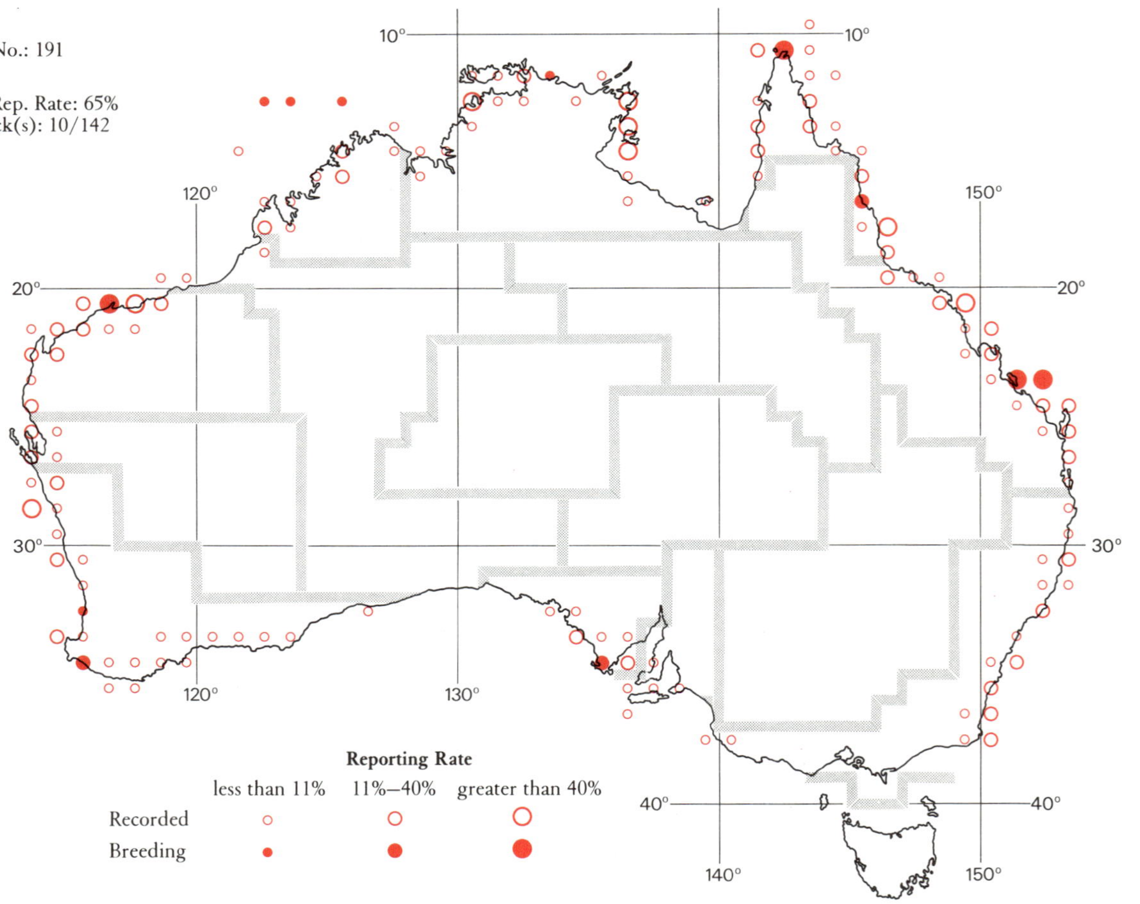

Striated Heron

Butorides striatus

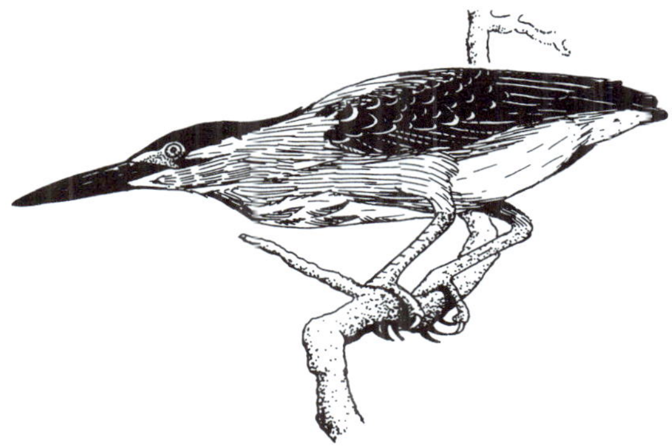

The pantropical distribution of the Striated Heron appears to have resulted from an expansion of range in the past similar to that which the Cattle Egret is now undergoing.[756] The generally sedentary habits of the Striated Heron and its dependence on mangroves led to the establishment of isolated populations following this expansion and hence to the development of numerous subspecies, some so distinct that they were previously considered full species. In Australia there are distinctive plumages from shades of grey to rufous but these are irregular in distribution and views about subspecies differ.[622,1589] Those sometimes recognized are separated by gaps in distribution where mangroves are absent. Taking Field and Historical records together, gaps are evident, on the scale of the Atlas, at only three points: the western corner of the Gulf of Carpentaria (15/137, 16/138); the northern end of Eighty Mile Beach (18/121); and 23/113 north of Carnarvon. The southern limits of distribution and the southernmost breeding records have not changed between the Historical and Field Atlases.

Some Asian birds undertake a regular migration[1117] but in Australia no large-scale seasonal movements have been reported and the Field Atlas gives no indication of them. A specimen from L. Cargelligo (33/146) taken in 1908 represents an exceptional inland record.[790]

The Striated Heron lives in mangroves. It feeds on small fish and crabs and will sometimes dive for food.[790] The nest is usually solitary in Australia but there is a record of a colony of 30 nests at Port Essington (11/132) last century.[790]

Field Atlas Records: 1105

1° blocks–with records: 100 (12%)
–with breeding records: 14 (14%)

Atlas No.: 193

Max. Rep. Rate: 33%
In block(s): 23/151

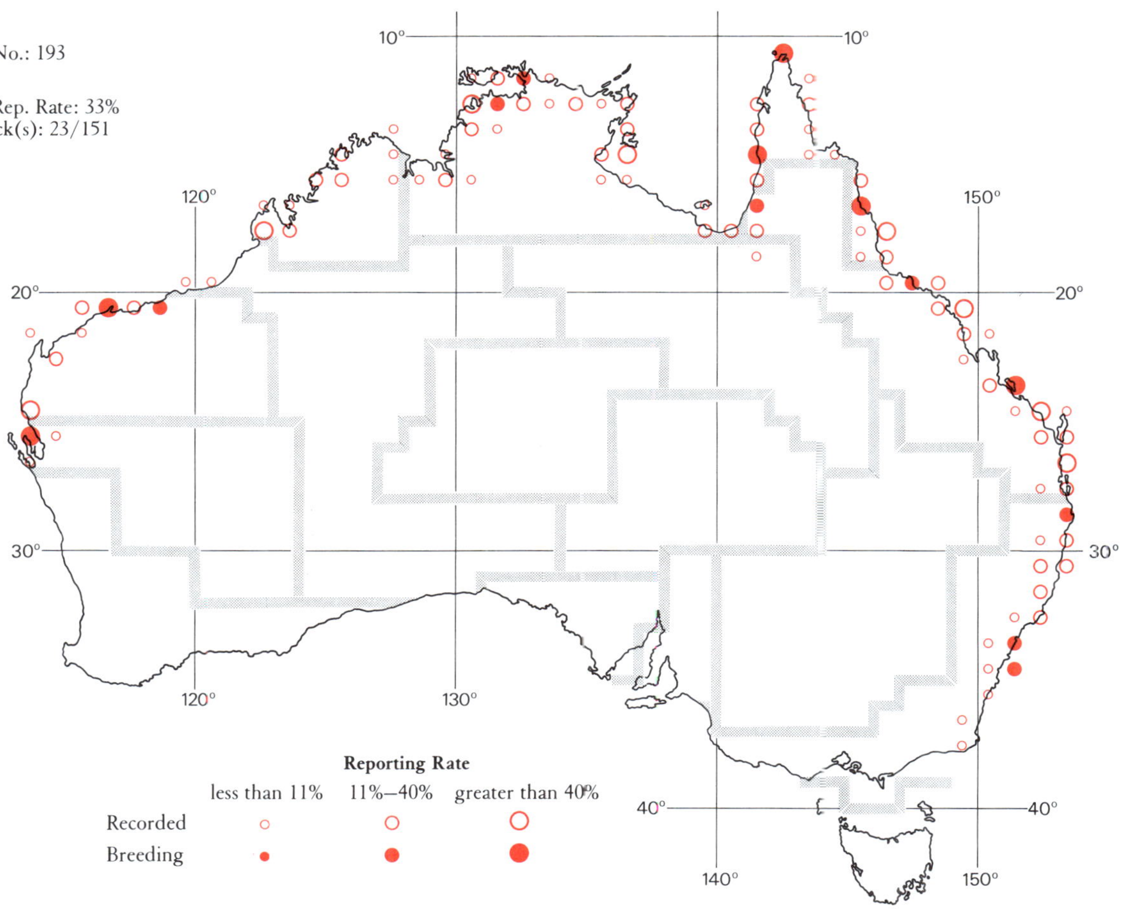

Rufous Night Heron

Nycticorax caledonicus

The range of the Rufous Night Heron includes south-eastern Asia, New Guinea, the Pacific islands and Australia. It is a vagrant to New Zealand. In Australia the species is rare in the Tas. Region but breeds regularly on King I. (39/144) and was located at five sites in Tas. itself during the Field Atlas, some used in more than one year by more than one bird. Breeding was reported from the Kimberley, Pilbara and Goldfields Regions in the 1951–76 period and earlier, though not during the Field Atlas.

Movements of the Rufous Night Heron appear to be governed by flooding of wetlands and availability of food.[1588] A seasonal regularity is imparted in some places where flooding is seasonal such as the southern Murray-Darling Region. There the bird is present each summer in varying numbers but almost entirely absent in winter.[833] In other places birds arrive when flooding occurs[759] or occupy roosts irregularly.[285] Many were present in the Macquarie Marshes (30/147) in the flood years of 1978–79 but in drought years there are few.[1211] Comparison of Reporting Rates north and south of 26°S shows the difference between the two to be greater in winter (6.8%:2.1%) than in summer (13.5%:11.0%). This suggests a general tendency for birds to move north in winter but the extent of the movement varies greatly from year to year.[1211] The number of birds in New Guinea fluctuates greatly[1588] and the species is considered nomadic in Torres Strait.[503] A nestling banded near Barham (35/144) was recovered in New Guinea (8S/141E), 2900 km distant.[53]

The Rufous Night Heron is nocturnal, roosting by day in trees, including pine and willow, near feeding grounds which are occasionally several kilometres from water.[759] It nests colonially.[1588] The sites of some colonies change from year to year; others are occupied annually; and some only during flood years. The Rufous Night Heron feeds on small fish, frogs, crustaceans and insects, both aquatic and terrestrial.[1211,1777,1793]

Breeding colonies containing more than 4000 birds have been reported at Gunbower I. (35/147) in Nov. 1973, Barmah Forest (36/144) in Jan. 1969 and near the mouth of the Alligator R (12/132) in Mar. 1981.[1552]

Field Atlas Records: 4914

1° blocks–with records: 384 (47%)
–with breeding records: 46 (12%)

Atlas No.: 192

Max. Rep. Rate: 38%
In block(s): 12/133

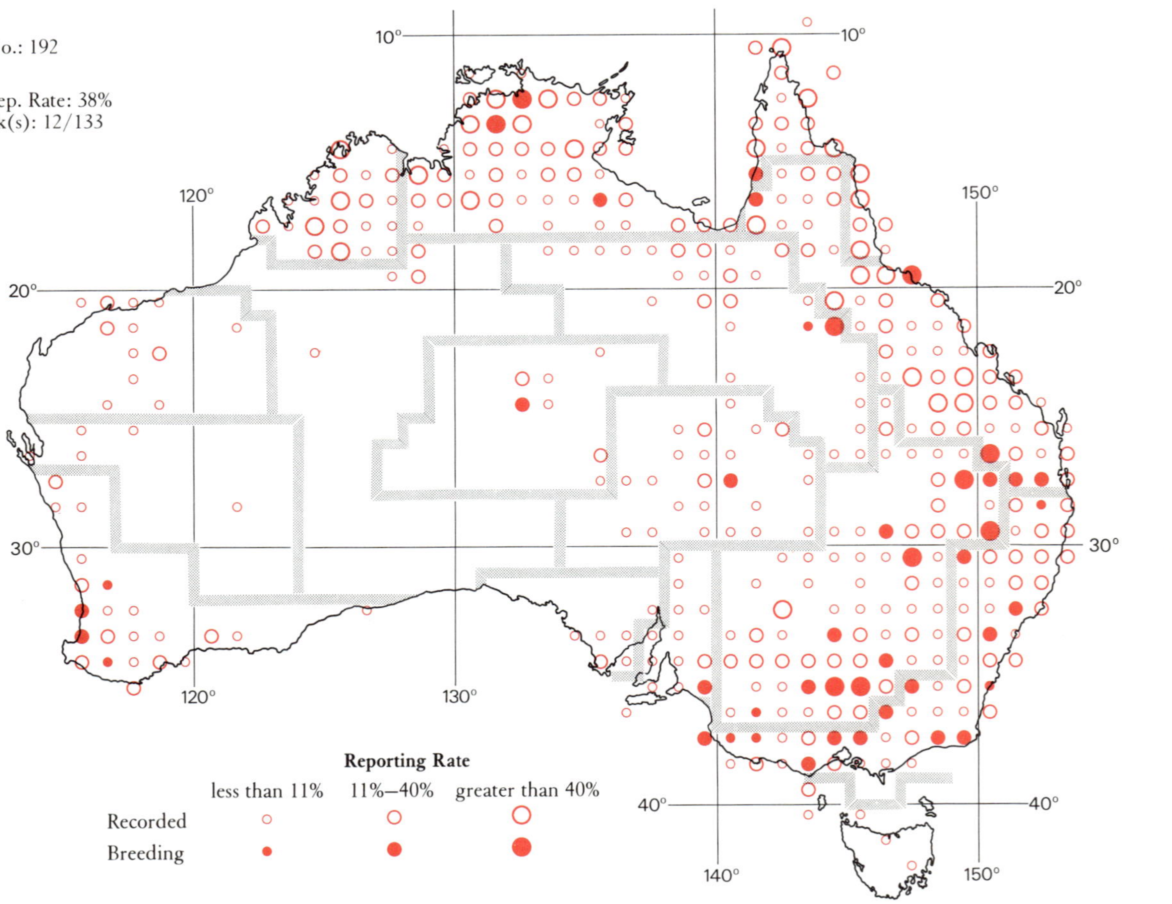

Little Bittern

Ixobrychus minutus

The Little Bittern lives in Africa, Europe, Asia, New Guinea, New Zealand and Australia. The Australian subspecies is *dubius*. The species is reported sporadically in the SA Ra. and Eyre Pen. Regions and has bred at Bool Lagoon (37/140).[712] There is no record from the Tas. Region. Sightings at Kununurra (15/128) in Jan. 1982 and Mar. 1983,[508,911] after the Atlas finished, extend the range of the Little Bittern in northern Australia.

The pattern shown on the map of inland breeding in the east and coastal breeding in the west is also evident in the Historical Atlases. The single coastal breeding record in the east was at Kooragang I. (32/151) in 1972.[1520] There are no confirmed breeding records in the Atlases north of 29°S.

Some overseas populations are migratory, others sedentary.[756] The Field Atlas showed the Little Bittern to have a Reporting Rate, in the regions in which it was recorded, of 0.42% in summer compared with 0.05% in winter, which suggests that it may leave Australia in winter. Several birds have been recorded at Booby I. (10/141) on nights of low cloud, probably migrating,[503] and in New Guinea the species is regarded as a wintering migrant from Australia.[129]

Its absence in winter has been noted from some southern Australian sites. In the Murray-Darling Region it is present Aug.-Mar.[833] and in SA Aug.-May.[712] In contrast it is recorded in all seasons at Groote Eylandt (13/136). On a more local scale an apparent movement to the coast brought unusually large numbers to Sydney (33/151) in 1958[1135] and the Illawarra (34/150) district in 1957[1616] perhaps in association with dry conditions inland. In the South-West at Herdsman L. (31/115) and Moora (30/116) the Bittern arrives in Sept. and departs when the wetlands dry out in summer.[442,1563]

The Little Bittern inhabits the dense vegetation of swamps, lakes and rivers. In the South-West Region dense, mature beds of cumbungi or baumea are preferred.[440] The species sometimes nests in loose colonies.[712,1552] At Herdsman L. in 1980, two singing males were within 100 m of each other.[442] The Little Bittern feeds from perches on emergent vegetation taking insects, small fish and frogs.[756]

There are no measures of abundance.

Field Atlas Records: 161

1° blocks–with records: 56 (7%)
–with breeding records: 8 (14%)

Atlas No.: 195

Max. Rep. Rate: 7%
In block(s): 13/136

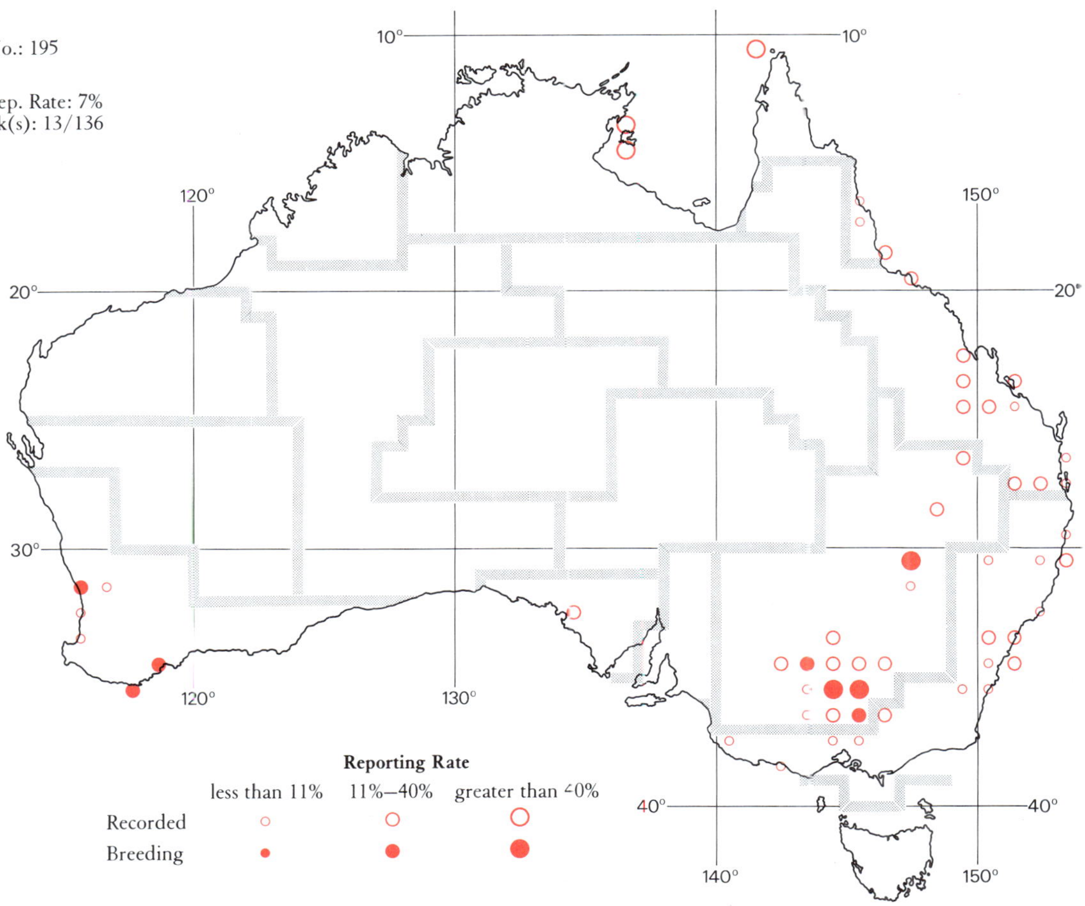

Black Bittern

Dupetor flavicollis

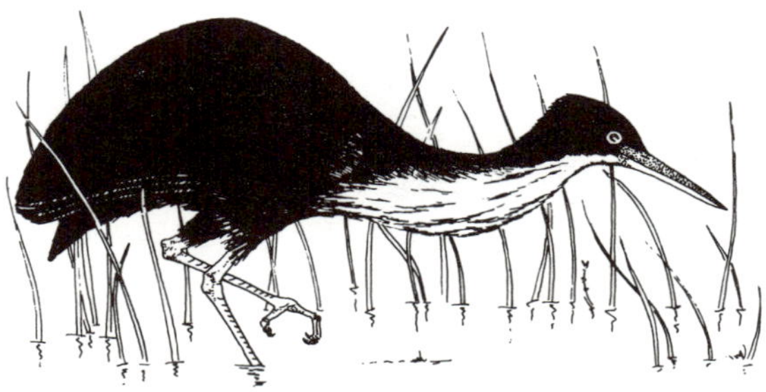

The Black Bittern lives in south-eastern Asia, New Guinea and Australia. The subspecies *gouldi* breeds in Australia. The isolated South-West population has not been reported breeding since the 1901–50 period and has declined. Before 1950 it bred at sites from Herdsman L. (31/115) and Northam (31/116) southward to the areas shown on the Field Atlas map (35°S).[1216,1638]

Records in the Historical Atlases suggest that the population is continuous from the Kimberley Region east and south to the South-East Region. The southernmost breeding report is at 34°S. The isolated population in the Pilbara Region has been recorded both summer and winter and is probably sedentary, though not yet reported breeding.

Birds breeding in China are known to migrate south in the winter; some other overseas populations are sedentary and some may have a more complex movement pattern.[756] In Torres Strait the Black Bittern is locally nomadic.[503] It may be only a summer visitor to Eastern Qld (south of 19°S)[1709] and the South-East Region where summer Reporting Rates were 2.6% and 0.3%, respectively, compared with 0.5% and 0.1% in winter. The Black Bittern's distinctive summer calling may account in part for this difference. In the South-West Region the Black Bittern is present all year.[1216]

The Black Bittern is solitary, living near water in mangroves and other trees which need to form only a narrow fringe of cover. In the South-West Region it lives in thicker vegetation such as paperbark woodland surrounding running water or coastal swamps. Pairs may be within hearing distance along a watercourse.[492] The Black Bittern feeds on small fish and invertebrates. Its decline in the South-West Region coincided with increases in the rate of clearing for agriculture and in the salinity of streams.

There are no measures of abundance.

Field Atlas Records: 441

1° blocks–with records: 109 (13%)
–with breeding records: 5 (5%)

Atlas No.: 196

Max. Rep. Rate: 17%
In block(s): 15/145

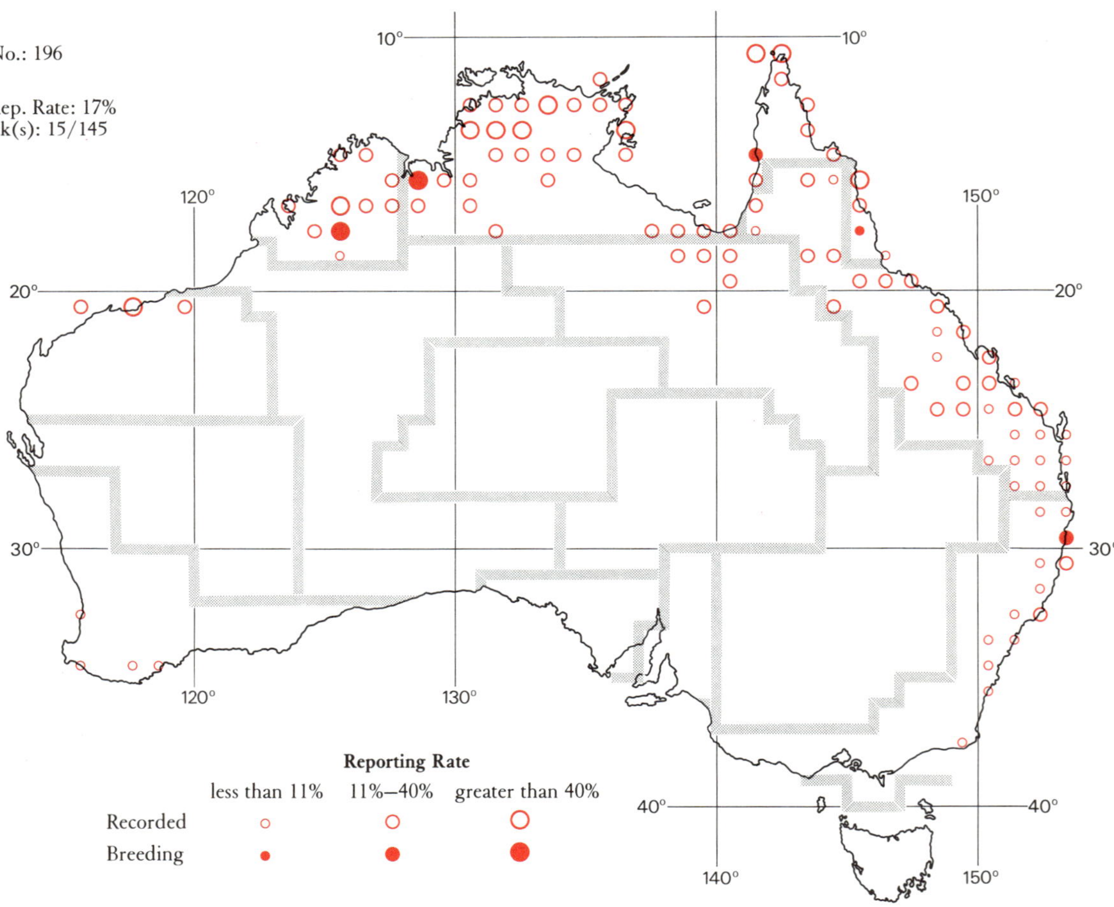

Australasian Bittern

Botaurus poiciloptilus

The Australasian Bittern is confined to Australia and New Zealand. Occasional records in the 1951–76 period in eastern Australia extend distribution north to Baralaba (24/149) and west to the Eyre Pen. Region. In the Kimberley Region, a specimen was taken on Argyle Downs (16/128) in Oct. 1971.[1713] Another specimen, taken in Dec. 1880, was labelled 'Shark Bay' (25/113), though this location may be open to doubt.[1638] In the South-West Region the Bittern's range may have retracted from the Moora (30/116) area where eggs were collected on at least two occasions before 1922 but there are no more recent records. In the field the Bittern may be confused with the immature Rufous Night Heron.

The Bittern seems to be nomadic. Sometimes when conditions are suitable many appear and at other times the bird disappears altogether: nine nests were built close together in 1956 at a site in the southern Murray-Darling Region following an influx of many birds;[833] also in 1956, considerable numbers appeared in the Illawarra (34/150) district;[1617] during a three-year survey of L. George (35/149) the species was often recorded in the winter of 1961 but not at other times;[1013] at Rochester (36/144) three nests were close together in 1932;[236] and at Fisherman's Bend (37/144) one bird stayed at least three years from 1947.[1837] The Bittern is present at Bool Lagoon (37/140) at most times of the year unless it is dry.[911]

The Bittern lives alone or in loose groups in extensive dense reedbeds and often on swamp edges. In the South-West it lives in tall reedbeds with short sedge. Breeding is sometimes loosely colonial[833] but in other cases pairs have been observed to maintain territories when several are present in a reedbed.[1363] The Bittern feeds on insects, small fish and other aquatic life, sometimes in rice-fields.[833]

At Bool Lagoon in 1981 there were groups of ten or more birds in 1-ha areas of cutting grass, i.e. 10 birds/ha.[911]

Field Atlas Records: 488

1° blocks–with records: 90 (11%)
–with breeding records: 5 (6%)

Atlas No.: 197

Max. Rep. Rate: 11%
In block(s): 40/147

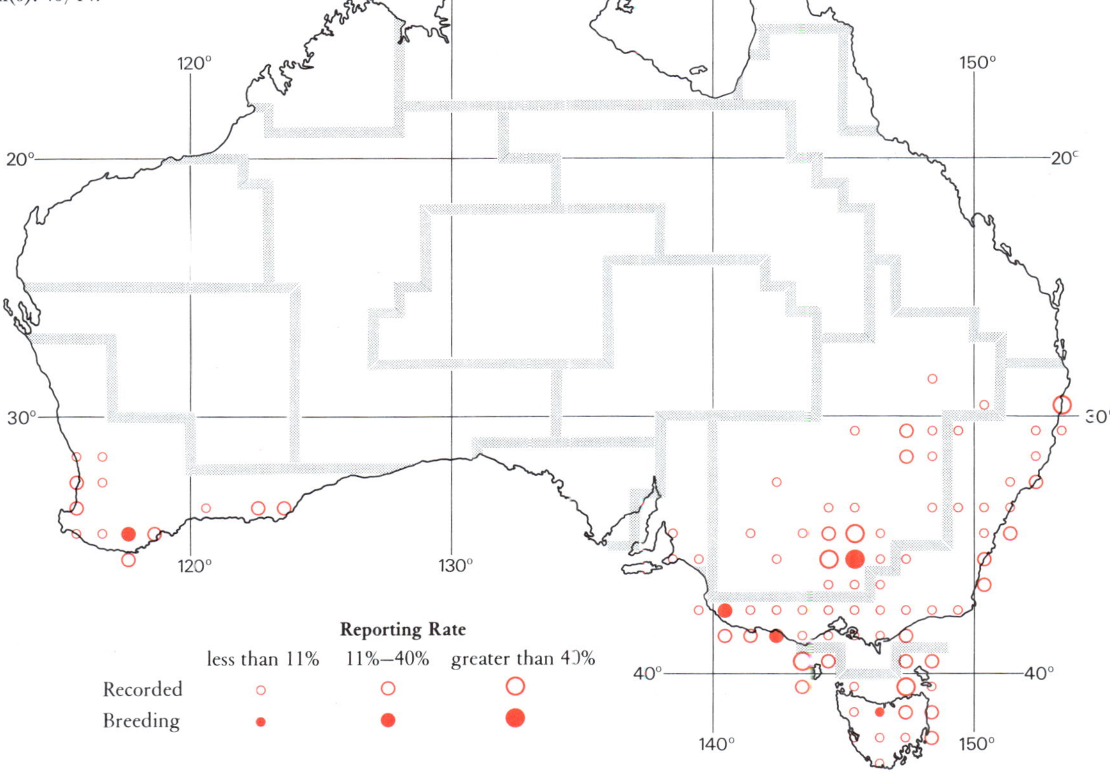

Black-necked Stork

Xenorhynchus asiaticus

The Black-necked Stork ranges through India, south-eastern Asia, southern New Guinea and into Australia. It was known in the Sydney (33/151) district when settlement was established, disappeared by the 1850s and began to recolonize from about 1930.[137] By the 1950s three nest sites were known in Ulmarra Swamp (29/153)[1653] and in 1940–60 two were known on the Hawkesbury R (33/151) only one of which was used each year.[1561] In the 1960s breeding was reported as far south as the Shoalhaven R (34/150). The Black-necked Stork was not recorded in the Pilbara Region until 1952 with breeding first reported in 1980.

There is nothing to suggest that the Black-necked Stork in Australia makes seasonal movements, though it is capable of moving long distances and turning up far from breeding grounds. No seasonal pattern is evident from such records, those in the Field Atlas having been made in Jan. (23/133), May (37/142), June (21/135), July (36/146), Aug. (25/142) and Nov. (24/132). In Torres Strait the Stork is considered nomadic.[503]

The Black-necked Stork is often in pairs and occasionally in parties of up to 18.[423] It usually inhabits swamps associated with river systems and large permanent pools but sometimes appears on the coast or in estuaries. There are no studies of banded birds but it appears that an individual, having once settled in a locality, remains for many years.[1561] The bird feeds in shallow water, stalking its prey of fish, frogs[231] and possibly rats.[423]

The NSW population was estimated at 37–43 birds in 1965.[1561] There are no other measures of abundance.

Field Atlas Records: 2420

1° blocks–with records: 224 (28%)
–with breeding records: 21 (9%)

Atlas No.: 183

Max. Rep. Rate: 64%
In block(s): 13/130

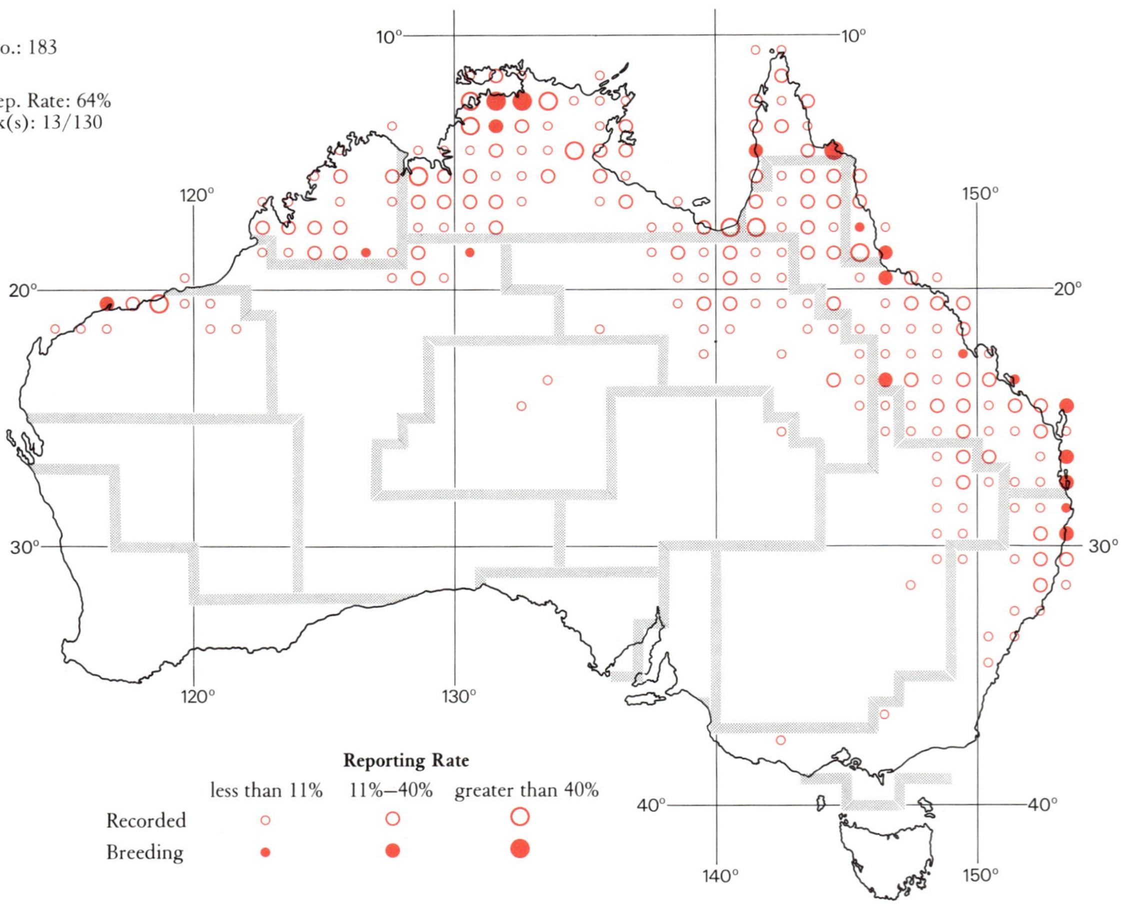

Glossy Ibis

Plegadis falcinellus

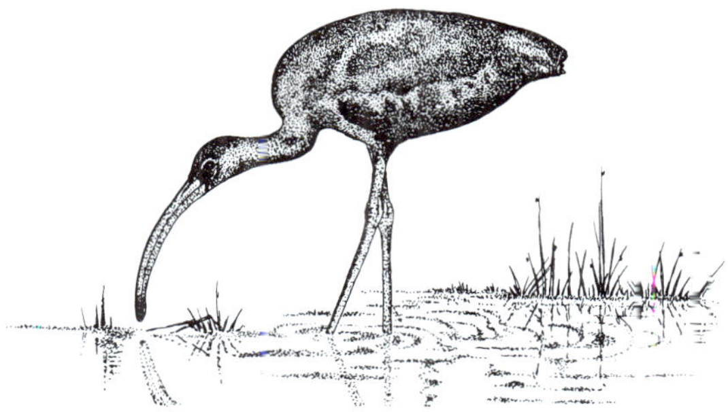

The Glossy Ibis ranges through southern Europe, Africa, America, southern Asia, New Guinea and Australia. The species is a vagrant in the Tas. Region mostly recorded Oct.-Jan. but, in 1975, birds were seen in Apr. and May and two were present until Sept.[734] The Field Atlas record (41/145) was of a flock of 40 in Oct. 1979. Despite the much greater frequency with which the Glossy Ibis is recorded in northern Australia compared with the south, all breeding records in the Atlases are south of 26°S and east of 137°E except for one at Kununurra (15/128).

Some overseas populations of the Glossy Ibis are migratory, birds banded in Europe having been recovered in northern Africa and India.[422] In Australia the species is an uncommon and irregular visitor to Torres Strait[503] and considered less frequent round Darwin (12/130) Nov.-May than during the dry winter season.[423] In the Kimberley Region it inhabits the floodplain of the Ord R (15/128) in the wet summer season and congregates on lakes and irrigated areas farther inland late in the dry winter season.[724] In the southern Murray-Darling Region the Glossy Ibis is a summer visitor.[833] The Field Atlas supports this observation of regular movement: the Reporting Rate in the South-East, Murray-Darling and SA Ra. Regions combined was 3.5% in summer and 0.6% in winter.

In the non-breeding season the Glossy Ibis lives in shallow lake margins and flooded places, usually freshwater. Breeding colonies form in trees and shrubs standing in water and in reedbeds, sometimes near colonies of other waterbirds.[724,1552] Overseas studies found that the Glossy Ibis eats mostly invertebrates.[422] At Kununurra it takes rice seeds, generally feeding in flocks of 20–50 and sometimes up to 500.[724]

About 500 Glossy Ibis bred at L. Bullawarra (27/143) in 1974, 1000 near Balranald (34/143) in 1981,[1105] 500 at Tragowel (35/143) in 1968 and more than 400 at L. Alexandrina (35/139) in 1964.[1552] The total breeding population suggested by colony counts in south-eastern Australia does not seem large enough to account for the numbers of birds seen in northern Australia. On the floodplain of the Ord R alone, the peak population in the dry winter season of 1980 was estimated to be 9000 birds.[724] Perhaps northern colonies are yet to be found.

Field Atlas Records: 2393

1° blocks–with records: 320 (39%)

–with breeding records: 11 (3%)

Atlas No.: 178

Max. Rep. Rate: 41%
In block(s): 12/132

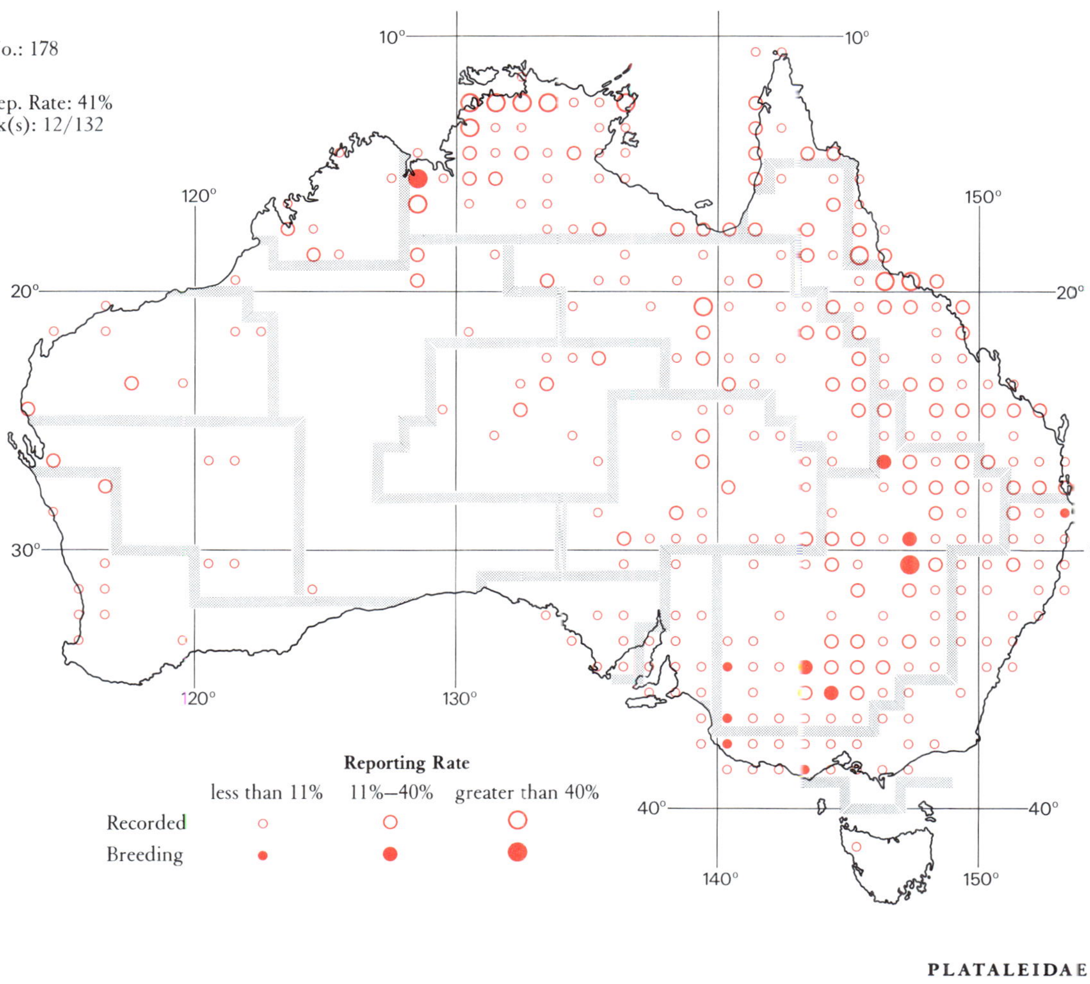

Sacred Ibis

Threskiornis aethiopica

The Sacred Ibis ranges throughout Africa, parts of the Middle East, south-eastern Asia, New Guinea and Australia. It is accidental to New Zealand. In Australia the species' range has expanded since 1950 (p. 661). The first reports in the South-West Region followed an irruption in 1952[1638] but breeding was not reported in the wild until 1979, near Ludlow (33/115); free-flying birds had bred in the zoo earlier. Field Atlas records in the Tas. Region were of vagrants but there were major irruptions in 1957–58 and in 1975.

It has been shown at two sites in Vic. that most adults are sedentary but juveniles leave the natal area and move northward.[1101] Several nestlings banded in the South-East Region were recovered in New Guinea.[310] Migrating flocks of 20–500 birds are observed in Torres Strait[503] but no seasonality of movements is evident in the Field Atlas. Moderate numbers arrived at Sturt Ck (19/128) with the wet summer season. Irruptions into the Tas. Region follow flood years in the Murray-Darling and, in 1975, birds in one flock of about 40 were all immature.[1316]

The Sacred Ibis lives near the shallow margins of lakes and swamps, in fields near water and on coastal mudflats. In the South-East it is the most common ibis near the coast.[1101] It breeds colonially, often with other waterbirds.[1552] North of 26°S the only inland breeding records are of five pairs, which nested in a lignum swamp at Brunette Downs (18/135) in 1913,[107] and of a small colony with at least eight young at Sturt Ck in 1978. Many breeding sites are used only in flood years.[310]

The Sacred Ibis feeds mainly on aquatic invertebrates and also takes some small vertebrates.[309,1102] At Phillip I. (38/145) it will fly up to 15 km each day from breeding colonies to feed.[1103] It has benefited from the development of irrigated agriculture, though the breeding colonies are sensitive to human disturbance.[310]

In Vic. in 1979–80 two colonies exceeded 10 000 birds and the total number in the main colonies was estimated at 33 000.[412] On Barren Box Swamp (34/145) in 1967 there were up to 1000 nests.[1552] Colonies in the north are smaller: an estimated 184 nests at Kapalga (12/132)[1552] and up to 1000 nests on the Fitzroy (17/123) and Ord (15/128) Rs.[1713]

Field Atlas Records: 17 976

1° blocks-with records: 370 (46%)
-with breeding records: 40 (11%)

Atlas No.: 179

Max. Rep. Rate: 72%
In block(s): 36/145

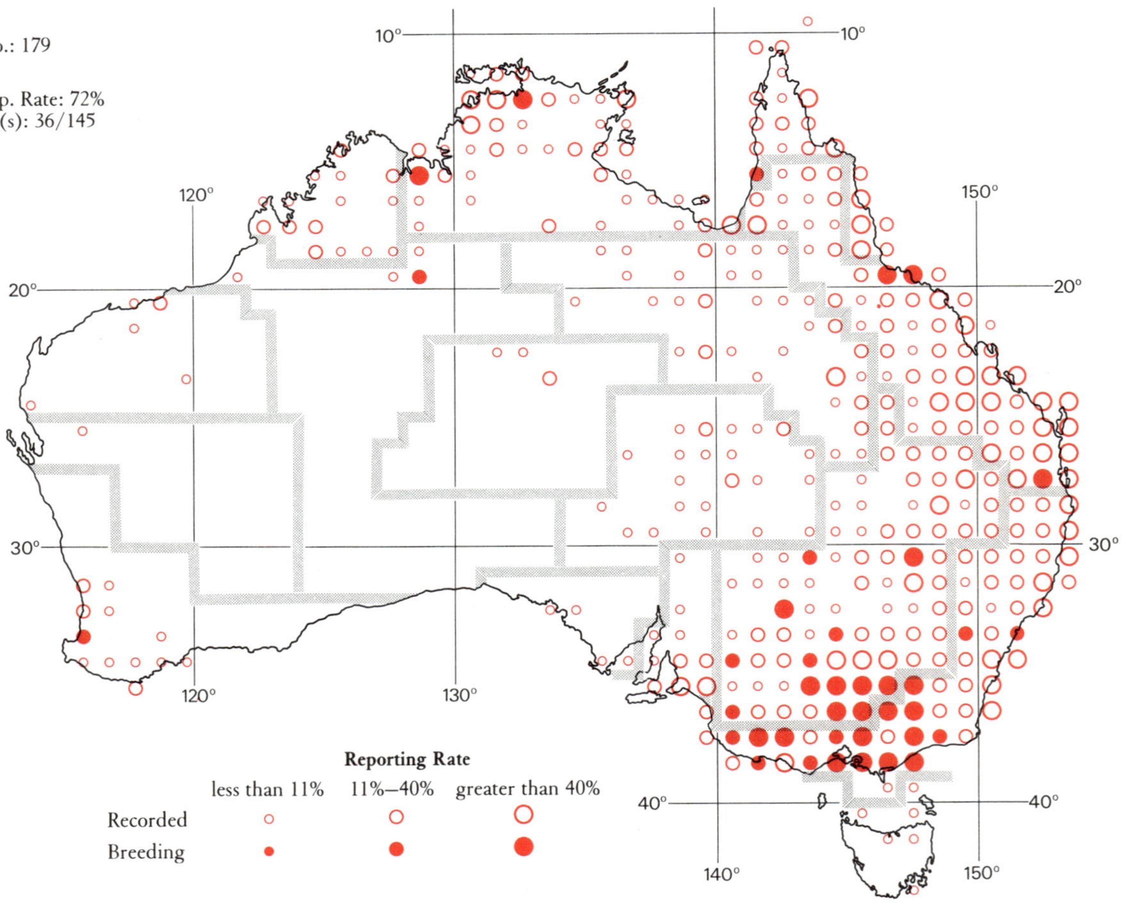

Straw-necked Ibis

Threskiornis spinicollis

The Straw-necked Ibis occasionally strays outside Australia to New Guinea. In Australia its range has increased. It was rare in the South-West until large numbers arrived in 1892 and breeding was not reported there until after 1950. In the Tas. Region the first record was not made until 1895[1343] and breeding has not yet been reported.

North of 26°S the Atlases and a recent review[1177] have established only ten sites where breeding has been attempted. All involved only small numbers of nests except for one at Cardabia Ck (23/114) where some hundreds nested in 1900.[1638] The implication is that most birds in northern Australia have been bred in the south. Banding shows that immatures disperse widely and rapidly[1177] and recoveries include: from Gingin (31/115) to Arnhem Land;[310] from the Murray-Darling Region to the Cape York and Top End Regions; two from the South-West to eastern Australia;[1638] and two from Muchea (31/115) to Beagle Bay (16/122) and Forrest R (15/127).[1713]

In Torres Strait the Straw-necked Ibis is a dry winter season passage migrant in flocks of 50–100.[503] Birds are scarce near Darwin (12/130) Jan.-Feb. suggesting seasonal movement[423] and the Reporting Rates of 61.1% in winter compared with 21.0% in summer correspond with this observation. The number of birds breeding at sites in southern Vic. fluctuates widely, which also suggests movement,[1103] but no large-scale seasonal pattern is evident in southern Australia from the Field Atlas. Sightings in Tas. are made in all seasons.

The Straw-necked Ibis inhabits shallow flooded areas and dry land where insects are abundant. Away from the coast it usually far outnumbers the Sacred Ibis.[1101] The Straw-necked Ibis breeds colonially, often with other waterbirds.[1552] Food includes locusts, beetles, other invertebrates and frogs[309] and the species has benefited from the development of irrigated agriculture.

A colony at L. Cowal (33/147) had 20 000 birds in 1964 and near Ivanhoe (32/144) 5300 young were counted in two colonies in 1973.[1552] In Vic. in 1979–80 four colonies exceeded 10 000 adults each and a total of 76 000 was estimated in the 16 main colonies.[412] The largest colonies in the South-West have numbered only a few hundred birds.[1638]

Field Atlas Records: 18 015

1° blocks–with records: 478 (59%)
–with breeding records: 26 (5%)

Atlas No.: 180

Max. Rep. Rate: 73%
In block(s): 19/147

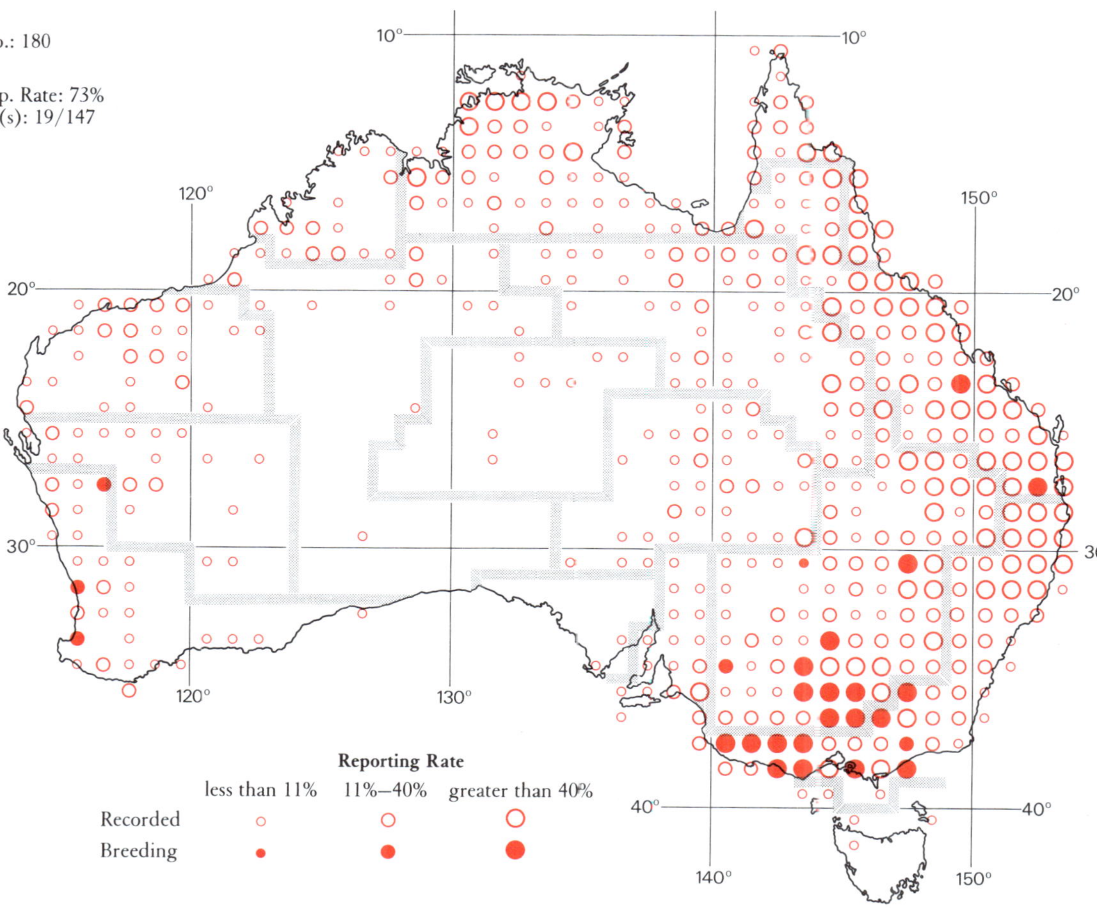

Royal Spoonbill

Platalea regia

The Royal Spoonbill ranges through Indonesia, New Guinea and New Zealand to Australia. It was first reported in the Tas. Region in 1957 with sightings in several places between May 1957 and Apr. 1958 and subsequent records in Sept. 1959, July 1966 and Sept. 1975.[1804] In the South-West Region it is rare and not known to breed.

Breeding has been reported from a number of northern localities including: Mitchell R (15/141) in 1981, Nicholson R (17/139) in 1898, Kapalga (12/132) in 1979–80, Humpty Doo (12/131) in 1977, Daly R (13/130) in 1902 and Fitzroy Crossing (18/125) in 1960. It seems likely that breeding is regular on northern rivers from the Cape York to the Kimberley Region, though rarely reported because colonies are small and difficult of access.

Movements of the Spoonbill have been recorded, including one of 1440 km.[85] A few visit Torres Strait in the dry winter season.[1791] Round Western Port Bay (38/145) the population is sedentary but augmented in summer and autumn by birds from the inland.[1103] The Field Atlas does not provide evidence of any large-scale seasonal movement. Local movements probably take the bird from one good feeding site to another.

The Spoonbill is usually in groups and inhabits shallow waters, inland and coastal. It breeds in colonies, often among nests of other species of waterbird.[1552] The Spoonbill feeds in water less than 40 cm deep. At L. Cowal it takes fish and invertebrates in the ratio of 3:4 and the species' absence from there June-Sept. is thought to be a direct result of a shortage of these foods.[1791] On intertidal mudflats at Western Port Bay the Spoonbill feeds mainly on shrimps and fish. One shrimp species composed 73% by weight of the food in stomach samples and the Spoonbill depends on it.[1104]

Numbers of the Royal Spoonbill are usually fewer than those of the Yellow-billed.[86,833] It was estimated that the Royal was outnumbered 20:1 by the Yellow-billed in the southern Murray-Darling Region. Colonies of the Royal Spoonbill are small, the largest reported containing about 200 breeding birds at Kerang (35/143) in 1980.[1552]

Field Atlas Records: 6780

1° blocks–with records: 342 (42%)
–with breeding records: 25 (7%)

Atlas No.: 181

Max. Rep. Rate: 49%
In block(s): 19/147

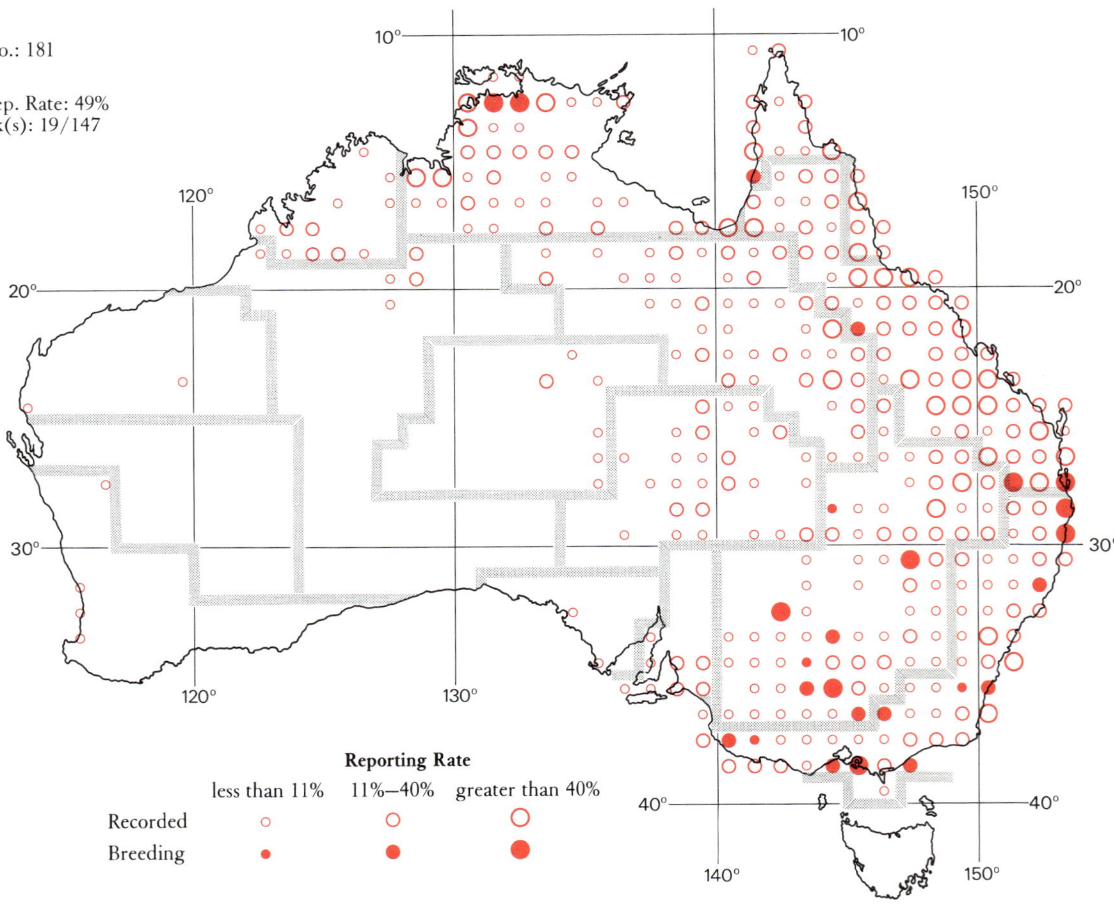

Yellow-billed Spoonbill

Platalea flavipes

There are only small numbers of the endemic Yellow-billed Spoonbill in the Top End[1755] and few breeding records from northern Australia though it nested at Sturt Ck (19/128) in 1978. The Spoonbill was first recorded in the Tas. Region in 1965 but has not bred there. Its range has extended into the South-West Region since 1900 with breeding first reported in 1921 at Moora (30/116) (p. 662). During the Field Atlas single pairs bred at L. Eganu (29/115) and Toolibin L. (32/117). In 1973 a nest was found inland near Carnegie (25/122).

The Field Atlas supports the observation that the Yellow-billed Spoonbill is a dry winter season visitor to floodplains of the coastal rivers in the Top End;[1755] Reporting Rates near Darwin (12/130) were 14.5% in winter compared with 3.0% in summer. Farther south there are sightings in all seasons, though numbers in the southern Murray-Darling Region are reported to be halved May-July compared with the rest of the year.[833] In the Tas. Region the Yellow-billed Spoonbill has been seen only May-Oct.;[1804] the two Field Atlas sightings were Aug.-Sept. 1977.

The Yellow-billed Spoonbill is typically an inland bird, appearing on the coast infrequently. Single birds or pairs are usual but a gathering of 250 has been reported at L. Cowal (33/147).[1791] The species roosts in trees, usually in or near water. It seldom forms breeding colonies with other members of its own species but solitary nests are often among those of other waterbirds.[1552] The Yellow-billed Spoonbill feeds in fresh or brackish water less than 40 cm deep[1791] and will use even small bodies of water.[86] It preys mainly on aquatic invertebrates, taking four times as many of these as of small fish.

Numbers of the Yellow-billed Spoonbill seem usually to be greater than those of the Royal[86,833] and in the southern Murray-Darling the Yellow-billed was estimated to outnumber the Royal 20:1. The largest colonies reported are: one with about 40 birds near Dingee (36/144) in 1973 and one with 70 near Kerang (35/143) in 1981.[1552]

Field Atlas Records: 10 405

1° blocks–with records: 423 (52%)
–with breeding records: 34 (8%)

Atlas No.: 182

Max. Rep. Rate: 57%
In block(s): 27/149

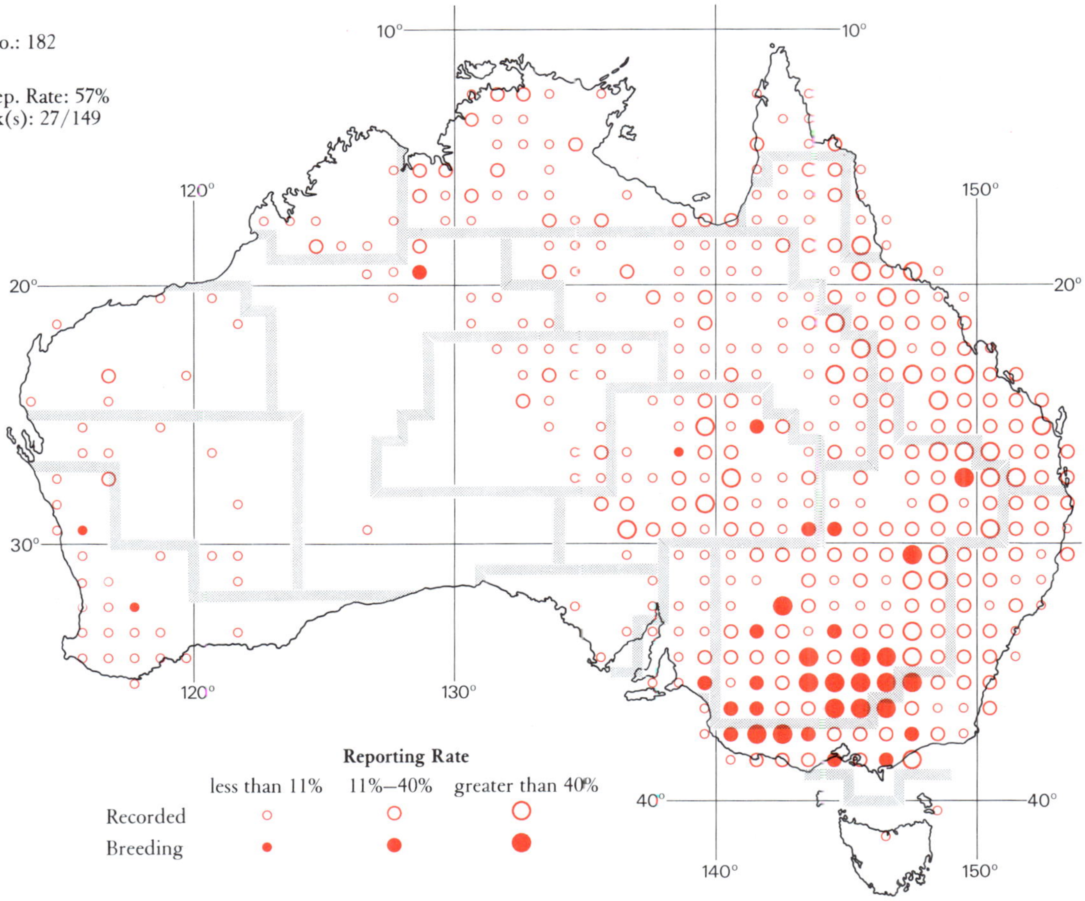

Magpie Goose

Anseranas semipalmata

The Magpie Goose lives in southern New Guinea and Australia. It was never more than a vagrant in the South-West Region but at the time of European settlement was distributed round eastern Australia to the SA Ra. Region (p. 662). Its numbers in western Vic. last century were described as 'incalculable'.[240] By 1911 it had gone from most of the South-East and Murray-Darling Regions and only a few bred near Moree (29/149). Since then wild birds have not bred south of 21°S.

Magpie Geese in the southern Murray-Darling and South-East Regions during the Field Atlas probably had escaped or been released. In Sydney (33/151) birds fly between the zoo and the botanic gardens. In Vic. about 300 birds have been released at Sale (38/147), Kerang (35/143) and Tower Hill (38/142) since 1973 but most have not survived. Some bred at Tower Hill where the largest group, about 45 birds, remains. No introductions have been made to Tas. but two birds that appeared in Oct. 1979 near Elizabeth Town (41/146) and near Ross (42/147) are presumed to have originated in Vic. Attempts are being made to re-establish the Magpie Goose at Bool Lagoon (37/140).

Movements show no seasonal pattern. Large numbers of Magpie Geese fly over the western islands of Torres Strait at all times of the year[503] possibly crossing to New Guinea. Sightings in southern WA have been made at all times of the year. In the Top End and Kimberley Regions most of the population may be forced into a few refuges round remaining water during the dry winter season. The longest movement recorded is that of an immature from Berrimah (15/132) to Milingimbi (12/134), 499 km.[1436]

The Magpie Goose lives in swampy country, nesting colonially.[1552] It feeds on grass seeds and the bulbs of rushes and sedges.[668] Predation at the nest is high (70–100% in one three-year study) and often the whole colony is destroyed by floods or premature drying of swamps.[457,668] Low productivity is compensated for by longevity and low adult mortality.[656] An increase in adult mortality through poisoning, shooting and netting at the turn of the century is probably why the species disappeared from the South-East.[668] In 1956–58 there were estimated to be 350 000 birds between Darwin (12/130) and Arnhem Land.[668]

Field Atlas Records: 1078

1° blocks-with records: 82 (10%)
-with breeding records: 14 (17%)

Atlas No.: 199

Max. Rep. Rate: 52%
In block(s): 13/130

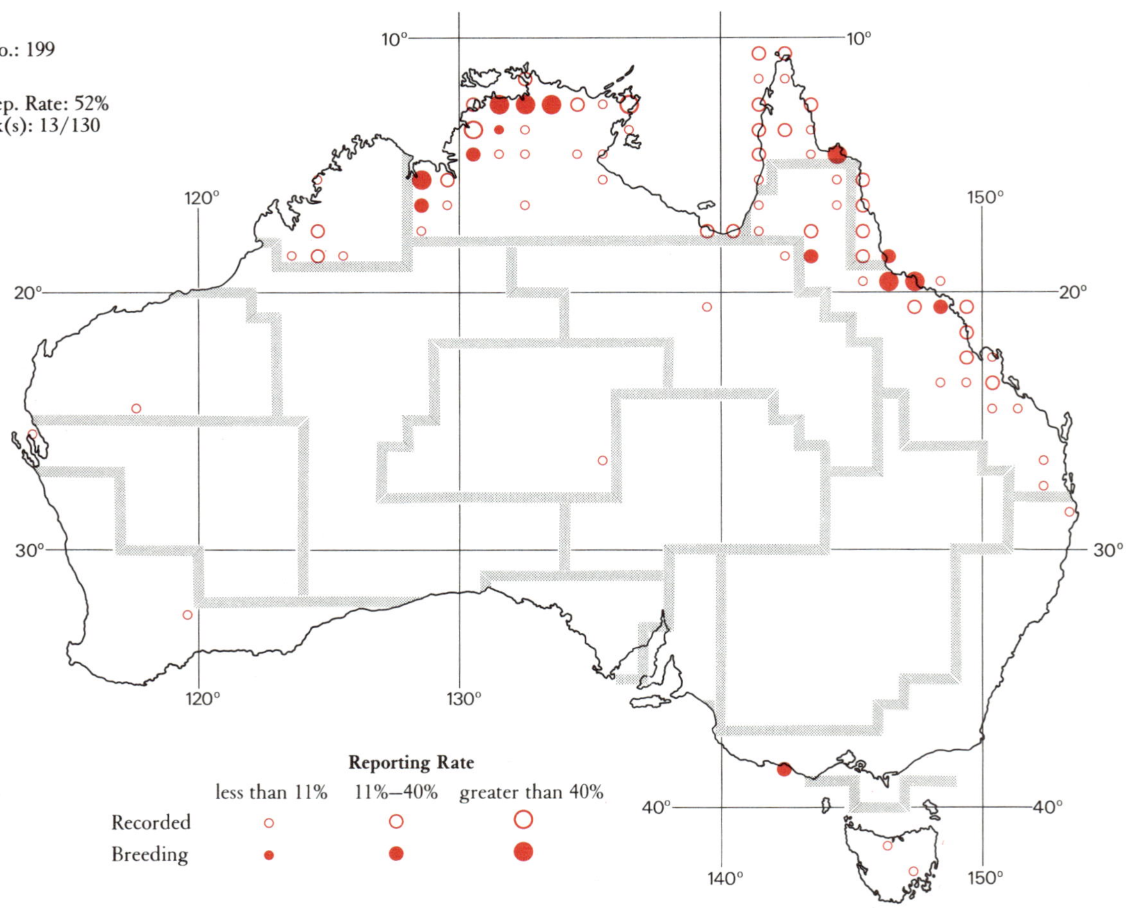

Wandering Whistling-Duck

Dendrocygna arcuata

The range of the Wandering Whistling-Duck includes northern Australia and New Guinea as well as many Pacific, Indonesian and Philippine islands. Populations in Timor (9S/125E) and New Britain (5N/150E) are subspecifically distinct from each other and from the subspecies *australis* that breeds in New Guinea and Australia. The Atlases suggest a gap in the distribution of the Australian population south of the Gulf of Carpentaria between 137°E and 139°E; and the western Kimberley population has been recorded only as far east as L. Gladstone (17/126). Sightings near Healesville (37/145) of probable escapees are omitted from the map but a group of about seven near Jandakot (32/115) in winter 1981 were apparently wild vagrants.

The range of the Wandering Whistling-Duck may have shrunk but perhaps not as greatly as has been suggested.[658] In the east the species bred in the Clarence R (29/153) district in the 1880s and was sufficiently abundant to be shot for the market.[1343] Only small numbers appear irregularly in this district now. South of 30°S, there are only two reports in the Atlases of the species before 1901[751,1301] so that it was probably no more than an occasional visitor even then.

The bird moves in response to rainfall, flooding and food.[658] Flocks congregate on permanent water in the dry winter season. A seasonal pattern has been noted in the Townsville (19/146) district where it is usually absent July-Nov.[1044] In Eastern Qld south of 25°S Reporting Rates (summer 3.0%, winter 0.5%) suggest that some birds may move north in winter. The Wandering Whistling-Duck flies across Torres Strait[658] and is probably a regular migrant, travelling at night.[503] The morphological differences between New Guinea specimens and those taken on other islands north of Australia (Timor and New Britain) indicate that these movements take the ducks no farther than New Guinea.

The Wandering Whistling-Duck lives in flocks. It forages on flooded pastures feeding on the recently submerged grasses and, when these are unavailable, moves to deeper water and dives for aquatic vegetation such as waterlilies.[658]

Huge flocks have been reported: in winter 1964 there were an estimated 30 000 on the Mary R (13/132) and 40 000 on the Daly R (13/130).[658]

Field Atlas Records: 1111

1° blocks-with records: 121 (15%)
-with breeding records: 28 (23%)

Atlas No.: 204

Max. Rep. Rate: 38%
In block(s): 13/130

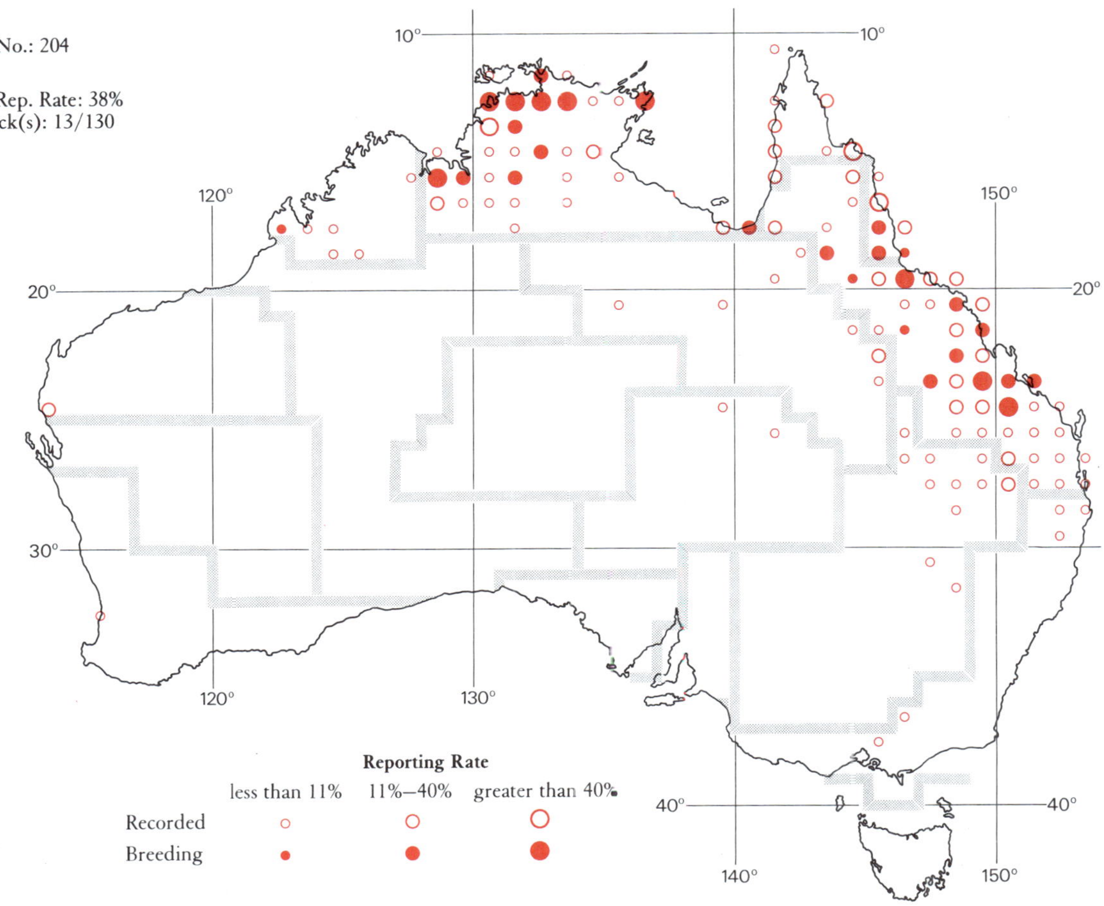

Plumed Whistling-Duck

Dendrocygna eytoni

The range of the Plumed Whistling-Duck does not normally extend beyond Australia but vagrants reach New Guinea[140] and New Zealand.[388] Historical records suggest that its range in northern Australia is continuous westward to the Kimberley Region. Thought early last century to be confined to northern Australia,[723] the Plumed Whistling-Duck is now widespread in the Murray-Darling Region and was recorded in the SA Ra. Region in the 1951–76 period. It bred in the Macquarie Marshes (30/147) before 1901[1343] and a single pair bred near Deniliquin (35/144) in 1900.[1062] Other early southern records include a specimen taken at L. Buloke (36/142) in 1890 and Tas. records in May 1871 and June 1872,[1082] the only ones from that region. The 1901–50 Atlas contains no breeding records south of 30°S and the breeding population now established along the Murray R and into the South-East Region apparently originated during floods in 1955–56.[833]

The Plumed Whistling-Duck forms large flocks, often of thousands, in the non-breeding season.[658] These flocks congregate round water and remain fairly stable in composition until the arrival of the wet summer season allows birds to disperse. Banding recoveries show movements of over 1900 km but no tendency to move in any particular direction. The fact that the six WA records south of the Kimberley Region were all made Sept.-Oct., though in different years, suggests a seasonal factor.

The non-breeding roosts are often in different places in successive years and sometimes far from the food being used at the time.[698,1044] The Plumed Whistling-Duck may fly 20 km or more to feed on dry land at night, leaving its roost in the evening and returning at dawn to rest by water during the day. Its distribution, specially during the breeding season, lies in the richer grasslands of Australia where it feeds largely on grass seed.[658]

There are no measures of abundance.

Field Atlas Records: 1876

1° blocks–with records: 207 (25%)
–with breeding records: 41 (20%)

Atlas No.: 205

Max. Rep. Rate: 37%
In block(s): 26/150

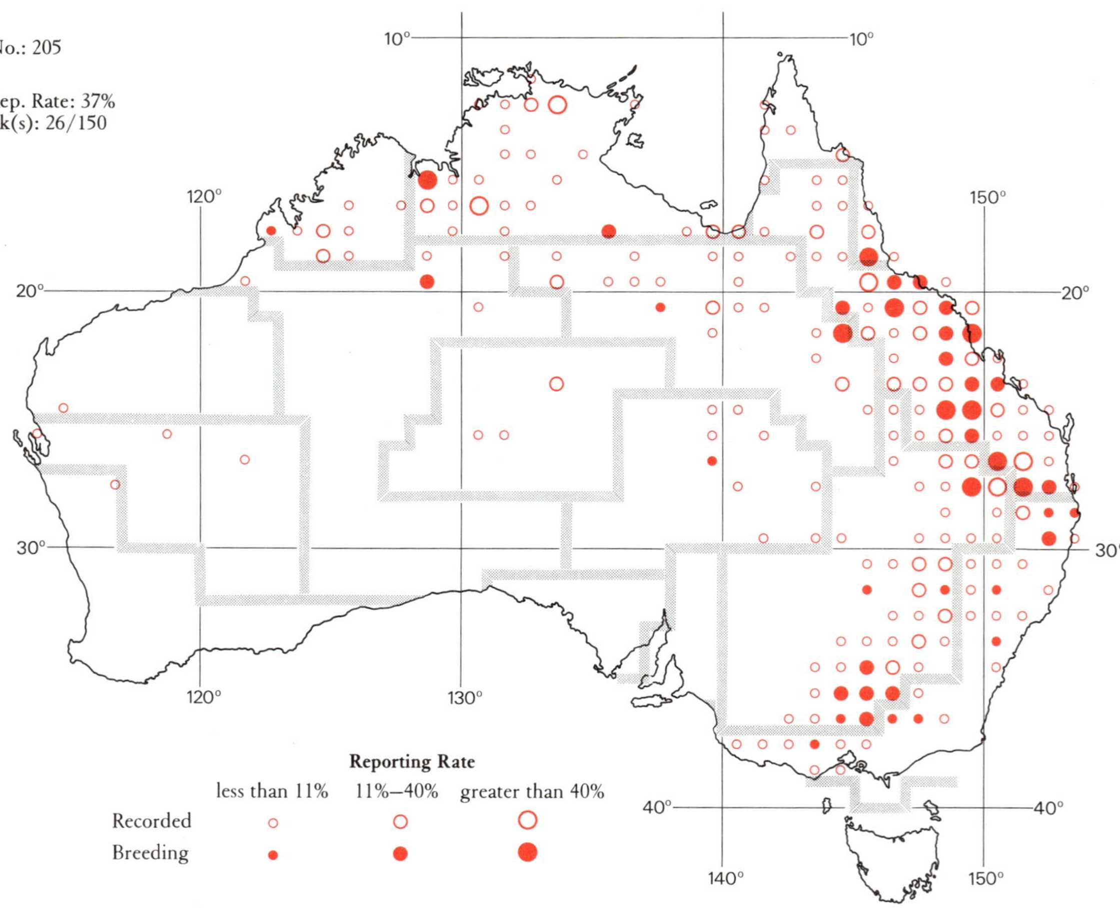

Black Swan

Cygnus atratus

Black Swans are often present in 'tens of thousands'.[23] There was a colony of about 120 nests on L. George (35/149) in summer 1962[220,224] and 110 used nests were counted on Sabinas I. (42/148) in Oct. 1980.[961]

The Black Swan lived only in Australia at the time of European settlement. It was introduced to New Zealand in 1864[1085] and first recorded in New Guinea in 1979.[127] In Australia its range has expanded northwards since 1951. The earliest record in the Central Ra. Region was of a single bird on the Finke R (23/132) in 1923. After good rains in 1966–67, the Black Swan bred in at least four localities in the Central Ra. Region[1370] and in 1980 it was reported breeding for the first time in the Top End Region. There is a record from Argyle Downs (16/128) in 1944 followed by breeding near Cape Bossut (18/121) in 1974 and at L. Eda (17/122) in 1976.[1713]

The composition of a flock of Swans changes continually. At L. Brewster (33/145) 381 members of a flock of about 1200 were marked over a short period. Three months later, about 1800 Swans were present of which only 20 were marked.[658] No large-scale seasonal pattern is suggested by the Field Atlas. Movements of more than 600 km have been recorded, the longest that of a banded adult from L. Cowal (33/147) to Moulting Lagoon (42/148), 965 km.[1437]

The Black Swan lives on fresh, brackish and salt water. It gathers on large bodies of water including marine estuaries and bays, specially when flightless during moult. Up to 3000 Swans use Corner Inlet (38/146) to complete their moult.[266] Flocks form when food becomes abundant and breed where sufficient vegetation can be uprooted and dragged to form a substantial platform for the nest.[222,223] The Black Swan eats vegetation, mainly submerged aquatic plants, including algae. It also grazes on pasture.[658,661]

On Moulting Lagoon and in The Coorong (36/139)

Field Atlas Records: 17 667

1° blocks–with records: 457 (56%)
–with breeding records: 190 (42%)

Atlas No.: 203

Max. Rep. Rate: 70%
In block(s): 38/147

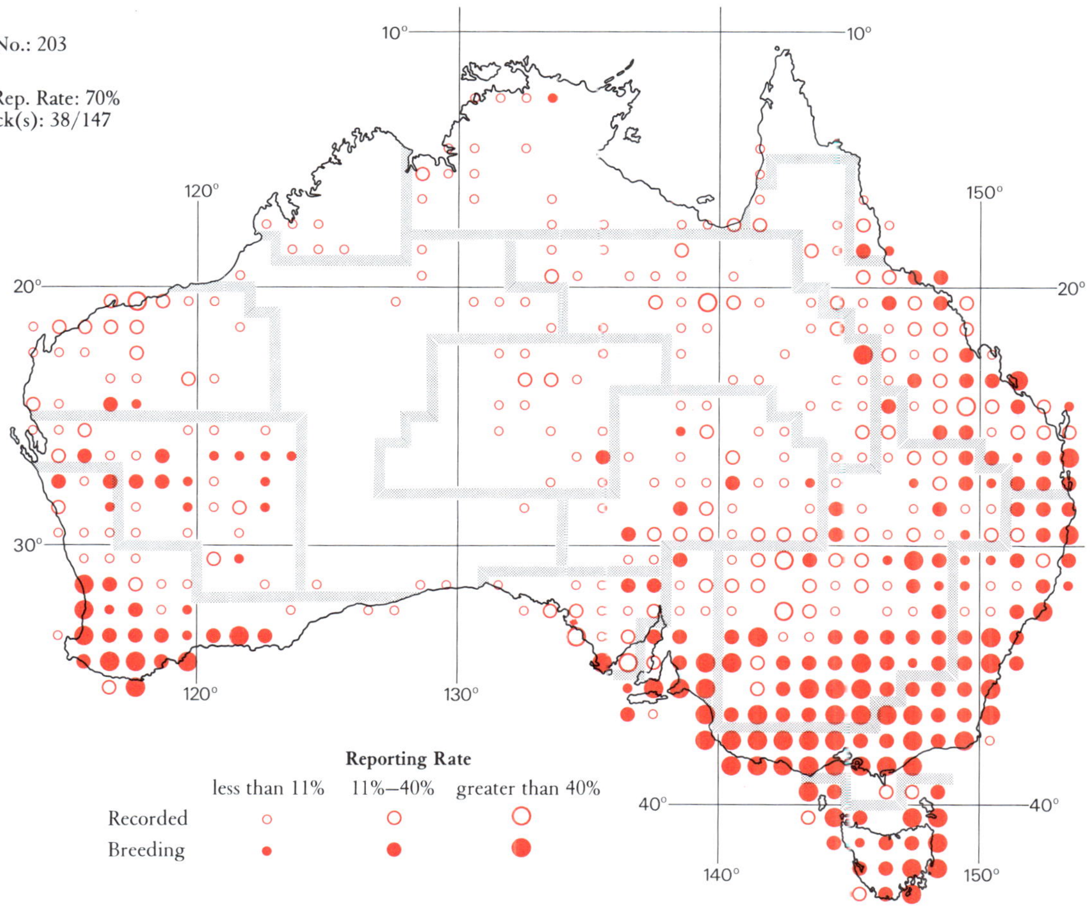

Freckled Duck

Stictonetta naevosa

The endemic Freckled Duck has been recorded at some time in all regions of Australia except the Cape York Region. In contrast, reports of breeding are clustered in the Murray-Darling and South-West Regions. Of the 1° blocks in which breeding has been reported since 1900, ten (53%) are in the Murray-Darling Region and six (32%) in the South-West Region.

Little is known about movements of the Freckled Duck. When inland waters dry up, many Freckled Duck gather in a few refuge areas. At Brunette Downs (18/135) the species is said to be a winter visitor, present in small numbers most years.[1370]

The Freckled Duck inhabits both open lakes and wetlands surrounded by thick vegetation, specially swamps in which lignum, cumbungi or paperbarks grow.[221,652] Permanent or temporary wetlands of varying salinity are used.[406] Most sightings during the Field Atlas were of six or fewer birds but flocks of hundreds have been reported. One study of food found that the bulk (89% by volume) was vegetable material;[652] another that 61% (dry weight) was invertebrates.[235] The two studies used different methods and sites differed in the relative amounts of plant and animal food available.

Though protected, the Freckled Duck suffers high mortality when large numbers in south-eastern Australia coincide with the duck-shooting season. In Mar. 1980 about 500 of an estimated 700 present were shot on Bool Lagoon (37/140)[1467] and in Mar. 1981 about 800 of 3000 estimated to be in Vic. were shot.[407,1341]

The total number of breeding reports historically is small, never involving more than a few pairs[658] and during the Field Atlas only four breeding reports were received. Breeding on a large scale may take place only when floods create extensive and inaccessible wetlands in the Channel Country of the L. Eyre Region; these floodplains have large areas of lignum in which the species appears commonly to nest.[234] The observation in Sept. 1977 of birds in moult and unable to fly on L. Muncoonie (25/138) suggests that the area may be important to the species.

A national count of Freckled Duck in Jan. 1983, during severe drought, recorded a total count of 7926, the largest gathering, 900 birds on L. Hungerford (28/144).[1210]

Field Atlas Records: 417

1° blocks–with records: 106 (13%)
–with breeding records: 4 (4%)

Atlas No.: 214

Max. Rep. Rate: 10%
In block(s): 28/139,29/141

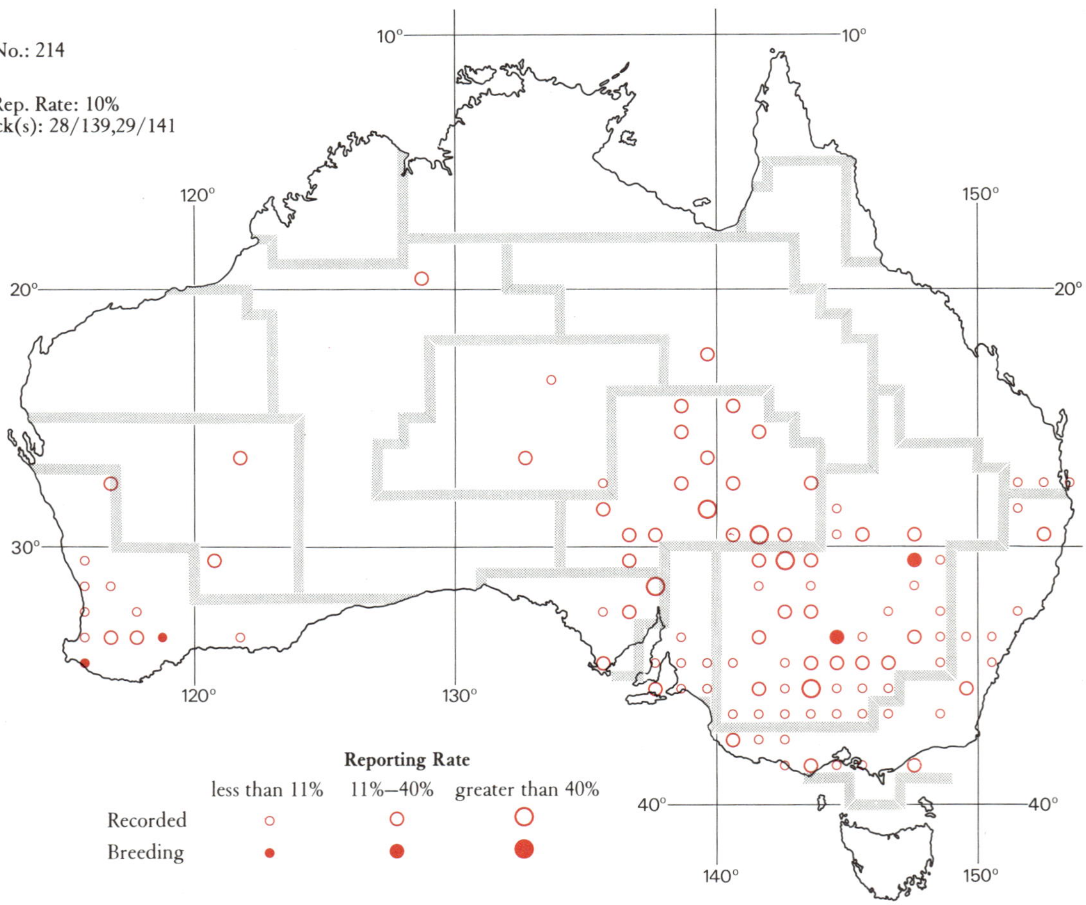

Cape Barren Goose

Cereopsis novaehollandiae

Eastern (*novaehollandiae*) and western (*grisea*) populations of the endemic Cape Barren Goose are recognized as subspecifically distinct.[1712] The breeding population near Tidbinbilla (35/148) resulted from an introduction and the species was also introduced successfully to Maria I. (42/148) and Three Hummock I. (40/144) in 1968. Birds reached King I. (40/144) in 1969, apparently from a group liberated on nearby Hunter I. (40/144) in 1968.[734,739] The population established on Kangaroo I. (35/137) also results from an introduction.[1369] In 1974 groups of birds, apparently wild, were seen near Ivanhoe (32/144) and several times between 1951 and 1976 appeared as far north as the Murray R in the Murray-Darling Region.[1521] The breeding range in the Field Atlas is as great as has been determined historically, except that islands in the northern Furneaux Group (39/147) and the Eyre Pen. Region west to St Francis I. (32/133) were not visited during the Field Atlas.

In the South-West Region the Cape Barren Goose was occasionally seen as far west as Busselton (33/115) before 1950. Since then a record at Narrikup (34/117) in 1952 is the only one not near the Recherche Archipelago and this population may have declined.

Local post-breeding movements take place in Bass Strait and SA. The young geese form flocks by Oct. and, during the summer, leave the breeding islands to join other immatures on the larger islands and adjacent mainland.[509] A movement of 258 km by an immature has been recorded within SA.[48]

The Cape Barren Goose breeds on small islands vegetated with tussock grassland and heath. The same birds use the same breeding site each winter. The species grazes and much of the food of the eastern population now consists of introduced grasses and legumes. The western bird is known to eat the fruit of boobialla.[185]

In eastern Australia the Cape Barren Goose has benefited from agriculture and is now at least as abundant, if not more so, than before European settlement.[509] Breeding density ranges from 0.007 to 4 birds/ha, depending on the island censused.[743] The numbers of Geese have been estimated: Tas. 5000 (750 breeding pairs); South-East 500; SA Ra. and Eyre Pen. Regions 10 000; and South-West 1000.[509]

Field Atlas Records: 750

1° blocks–with records: 53 (7%)
–with breeding records: 15 (28%)

Atlas No.: 198

Max. Rep. Rate: 38%
In block(s): 40/147

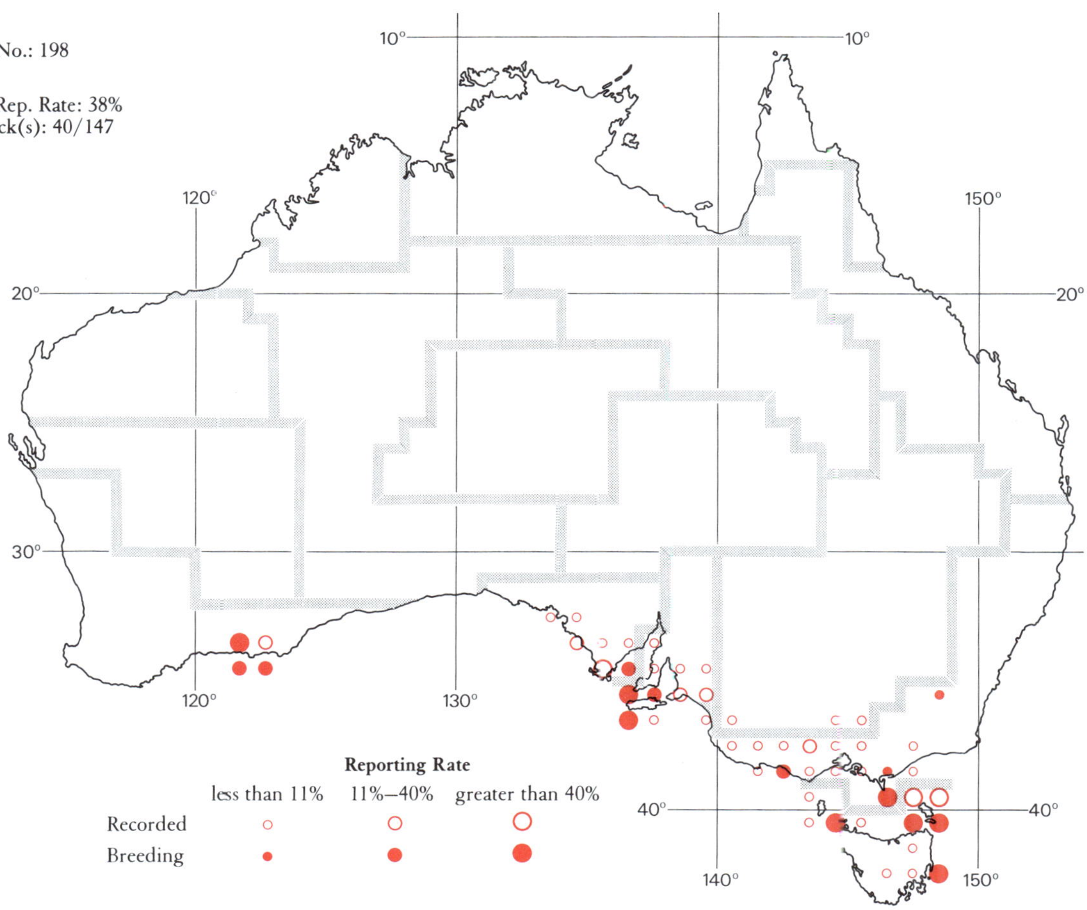

Australian Shelduck

Tadorna tadornoides

The endemic Australian Shelduck has two populations separated by the Nullarbor Plain. Comparison of the Field and Historical Atlases suggests that its range and abundance may have increased since European settlement particularly in the west where it was recorded in only three 1° blocks in the Goldfields Region before 1950. It is occasionally reported well north of its breeding range, the northernmost being two pairs near Wyndham (15/128) in June 1970.[1638]

During summer, before it moults its primaries and becomes flightless, the Shelduck moves to a few large freshwater lakes where moulting has traditionally taken place. At L. George (35/149) 1000–2000 birds gather to moult where only 20–30 pairs breed.[658] Birds banded at the moulting sites disperse widely and in all directions when the moult is complete. The longest movement recorded is that of an adult from Moorook (34/140) to Valleyfield (41/147), 1094 km.[43] Records north of 25°S have been made in all seasons and no large-scale seasonal pattern of movement is suggested by the Field Atlas.

The Shelduck needs fresh water and, if not using freshwater swamps, selects estuaries and saline lakes adjacent to freshwater springs or soaks. The nest is often in hollow trees, not always close to water and there can be substantial mortality when ducklings are led to the water.[1487] In the long term, clearing will reduce the number of hollow trees,[1566] limiting the usefulness of farm dams, which at present are excellent breeding sites. The Shelduck feeds on algae, aquatic invertebrates and plants. It grazes on pasture and takes the seeds and shoots of pasture plants, crops and legumes, a practice that has led to its legal destruction by farmers.

Each pair of ducks maintains a breeding territory, either a whole farm dam up to about 1 ha in area, or at least 200 m of the fringe of a large body of water.[658] Other waterfowl are driven out as well as members of its own species.

In a survey of 56 000 ha of wetlands in western Vic. in Feb. 1982, 23 200 Shelduck were counted (0.41 birds/ha).[267]

Field Atlas Records: 8581

1° blocks–with records: 218 (27%)
–with breeding records: 103 (47%)

Atlas No.: 207

Max. Rep. Rate: 56%
In block(s): 37/141

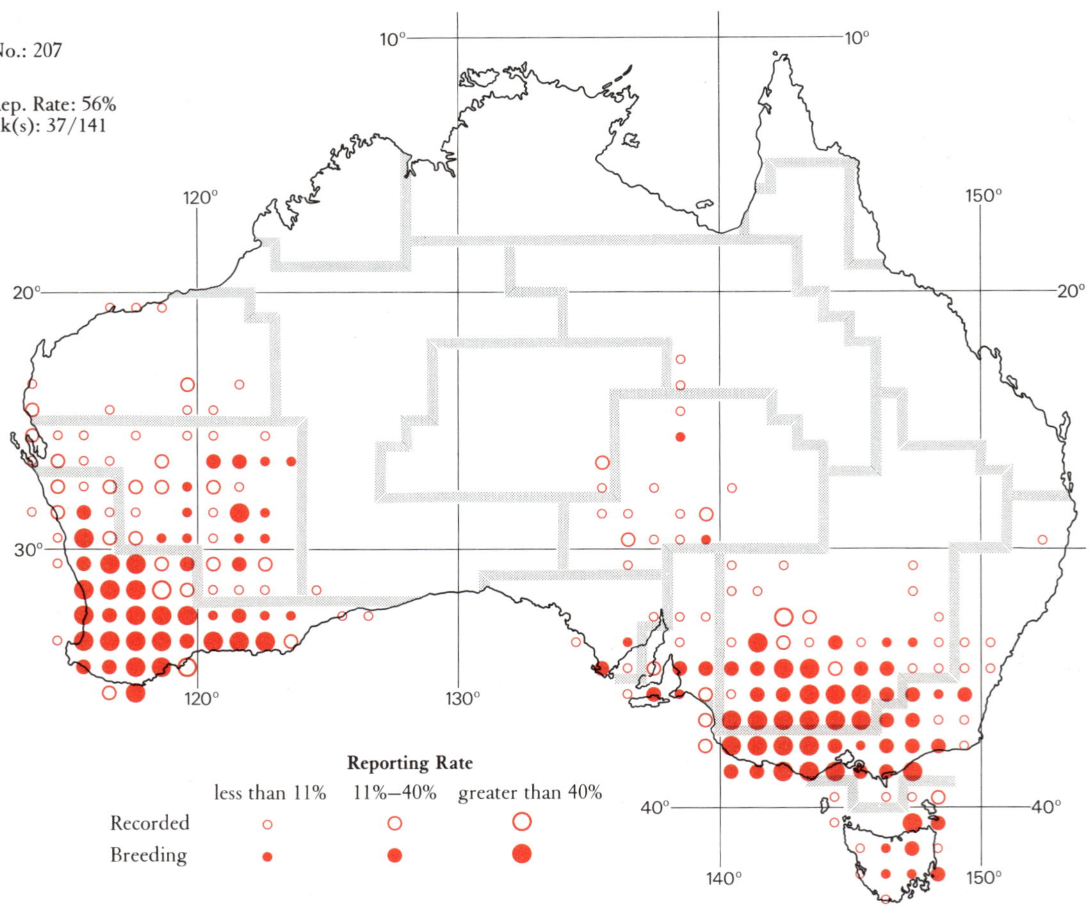

Radjah Shelduck

Tadorna radjah

The Radjah Shelduck ranges from New Guinea and adjacent islands to Australia. The Australian subspecies *rufitergum* has a disjunct distribution with no records in the Atlases at the Gulf of Carpentaria (137–139°E) linking the eastern and Top End populations. In the Kimberley Region there were records last century, including breeding on the Margaret R (18/126), but none since 1900. Before 1901 the Shelduck was seen as far south as Moreton Bay (27/153) and eggs were collected in 1851 labelled at that locality, though possibly collected elsewhere. In 1897, 15 Shelducks were seen near Moree (29/149)[1343] and the species appears on lists for Melbourne (37/145) in 1885 and 1907 and Adelaide (34/138) in 1930. These few records do not suggest that it was ever frequent south of Brisbane (27/153). The 1960s brought a spate of sight records in southern Australia: Katanning (33/117) two in Feb. 1962; Warwick (28/152) 20 birds in July 1964; Hamilton (37/142) two in Aug. 1964; Glencoe East (37/140) one in Aug. 1964; near Coward Springs (29/136) seven in Mar. 1966; L. Brewster (33/145) one in Sept. 1968;[397,706,708,718] and Purnong (34/139) one in Oct. 1968. The only southern record since then was an unconfirmed report in May 1977 of 10–12 at Lipscombe Dam (34/146).

As the dry winter season progresses, flocks of Shelducks gather on the coast or large rivers and inland lagoons. They disperse when the rains come. A few birds visit Torres Strait irregularly but specially at the end of the dry winter season.[503]

The Shelduck often occupies inland waters but is most abundant in mangroves lining tropical rivers.[658] In Eastern Qld it uses artificial lakes and swamps.[1039] In the dry winter season, flocks of 50–60 sometimes form but the observation of 'immense flocks' on the Burdekin R (19/147) in the 1850s[1125] seems remarkable and may be wrong. The Shelduck feeds on invertebrates at the water's edge.[658] In the non-breeding season pairs will roost at least 6 km from feeding sites.

Of the 11 northern ducks, the Shelduck has been thought the least secure because it is tame and vulnerable to hunting[658] and perhaps to competition from the Grey Teal and Black Swan whose ranges have expanded northward.[1036] On the Adelaide R (12/131) pairs defended on average 3 km of river frontage.[658]

Field Atlas Records: 825

1° blocks–with records: 70 (9%)
–with breeding records: 11 (16%)

Atlas No.: 206

Max. Rep. Rate: 52%
In block(s): 13/130

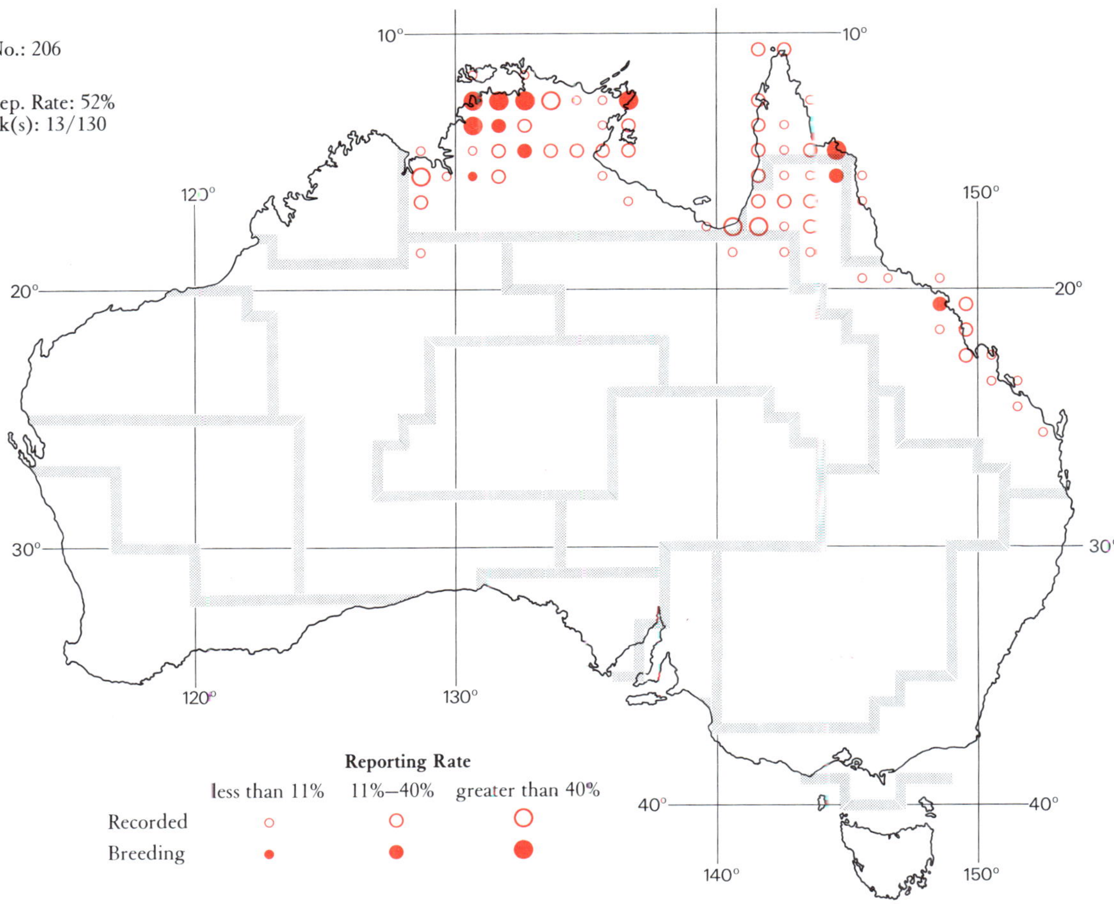

Pacific Black Duck

Anas superciliosa

The Pacific Black Duck inhabits the Pacific islands, New Zealand, New Guinea and Australia. The subspecies *superciliosa* breeds in Australia, southern New Guinea and parts of Indonesia.[658] The introduced Mallard hybridizes with the Black Duck.

The movements of an individual Black Duck are unpredictable.[658] Banding has shown that some remain on the same waterhole for months or years and others move long distances across the continent[1638] or to New Zealand.[658] Banding in eastern Australia has also shown a general trend for Black Ducks to move north in winter and south in summer, though individual ducks may move in any direction in any season.[651] In addition, the seasonal drying out of wetlands, in summer in southern Australia and in winter in northern Australia, results in the ducks moving to more permanent waters and dispersing when the rains come.[658]

The Black Duck can be found wherever there are swamps, rivers and lakes. Deep permanent wetlands with low salinity and heavy vegetation are preferred. The species tends to breed in the north in autumn and in the south in spring. It feeds by filtering in shallow water for the seeds of grasses, sedges and other aquatic plants and also takes some water insects.[658,661]

Even though the Black Duck has a high reproductive potential,[651] surveys of shooting in Vic. have shown that its representation in the bag is declining. Between 1972 and 1976 it composed 32% to 43% of the harvest but from 1977 to 1981 made up only 3% to 13% of the total bag.[267] In a survey of 56 000 ha of wetlands in western Vic. in Feb. 1982, 5700 Black Ducks were counted (0.10 birds/ha). Many of the wetlands surveyed are not well suited to the species but the low count, together with declining representation in hunters' bags, suggests that the number of Black Ducks may have decreased.

Field Atlas Records: 26 862

1° blocks–with records: 577 (71%)
–with breeding records: 223 (39%)

Atlas No.: 208

Max. Rep. Rate: 74%
In block(s): 27/149

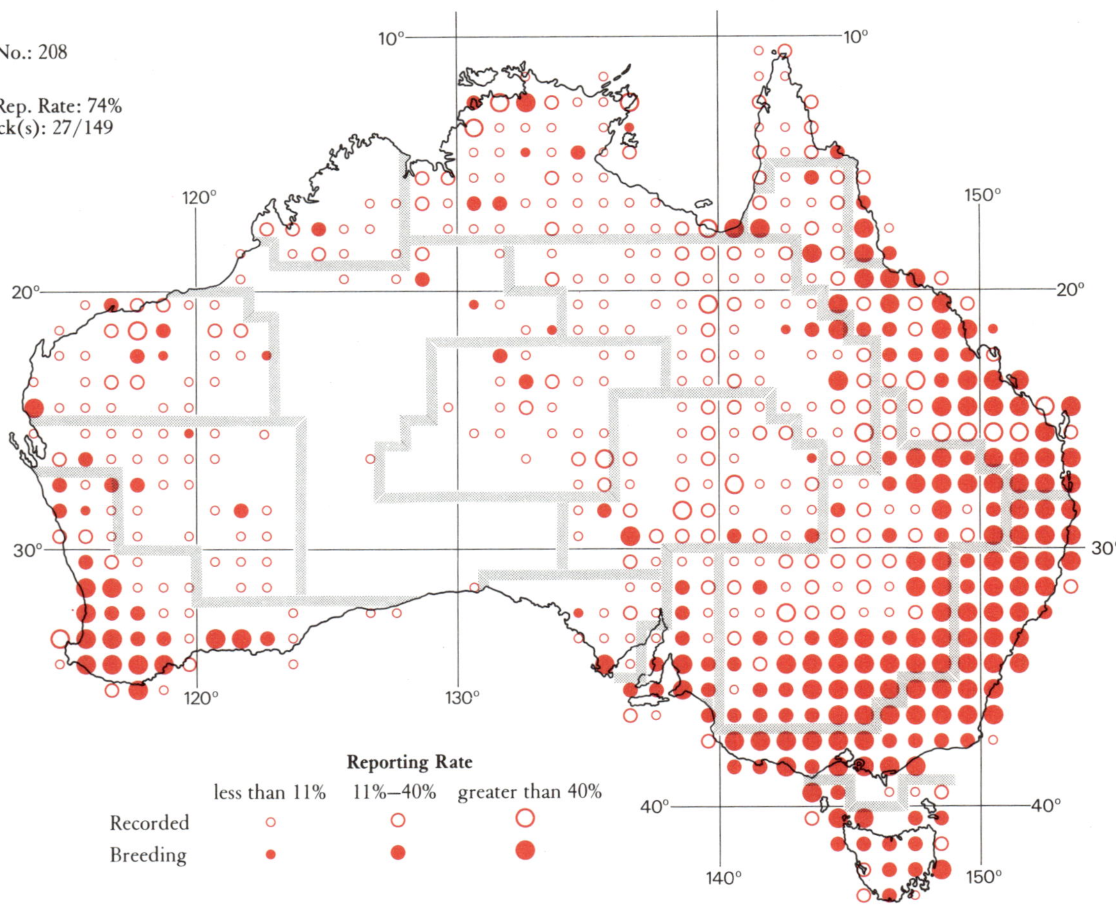

Mallard

Anas platyrhynchos

The Mallard is native to the Holarctic.[1085] In the 1860s it was introduced from England to New Zealand and probably Australia.[97] The first recorded introduction in Australia was to the Melbourne Botanic Gardens (37/145) in 1871 and another 80 were released there in 1872.[1553] These apparently had almost disappeared by 1905. Later introductions are not recorded but the Mallard has often been used as an ornamental species in urban parks and gardens. The main expansion of its range has taken place since 1950. Before then the Atlases contain records only from Sydney (33/151), Melbourne, Adelaide (34/138) and Perth (31/115). The species was breeding on a number of ornamental waters in Perth before 1912 and has remained there without spreading.[1085]

In the eastern States breeding appears still to be mainly confined to semi-domestic situations. Breeding records in the Field Atlas were all in or near settlements, mainly the capital cities and towns along the Murray R.

In the northern hemisphere the Mallard is mostly migratory but in New Zealand seldom moves far from its breeding area.[658] In Australia Mallard are sometimes recovered in wetlands away from settlements but how far or how freqently they move is not known.

The Mallard lives on coastal and inland waters, specially in city parks and gardens. In Europe it is omnivorous and opportunistic, foraging on dry land as well as in water. Its diet includes seeds, buds, aquatic and terrestrial invertebrates, amphibians, fish and occasionally birds and mammals.[422] Australian birds presumably are equally adaptable.

The ability of the Mallard to hybridize with the native Pacific Black Duck and produce fertile offspring is a cause for concern. Hybrids live in semi-domestic situations in some towns[658] and have occasionally been found among wild Pacific Black Duck shot in Tas.[1533] In New Zealand the Mallard has flourished and is now more commonly shot by hunters than the native Pacific Black Duck.[658] How much of the Mallard's expansion in Australia since 1950 is due to natural increase and how much to new releases cannot be determined.

There are no measures of abundance.

Field Atlas Records: 931

1° blocks–with records: 31 (10%)
–with breeding records: 13 (16%)

Atlas No.: 948

Max. Rep. Rate: 9%
In block(s): 33/151,34/138

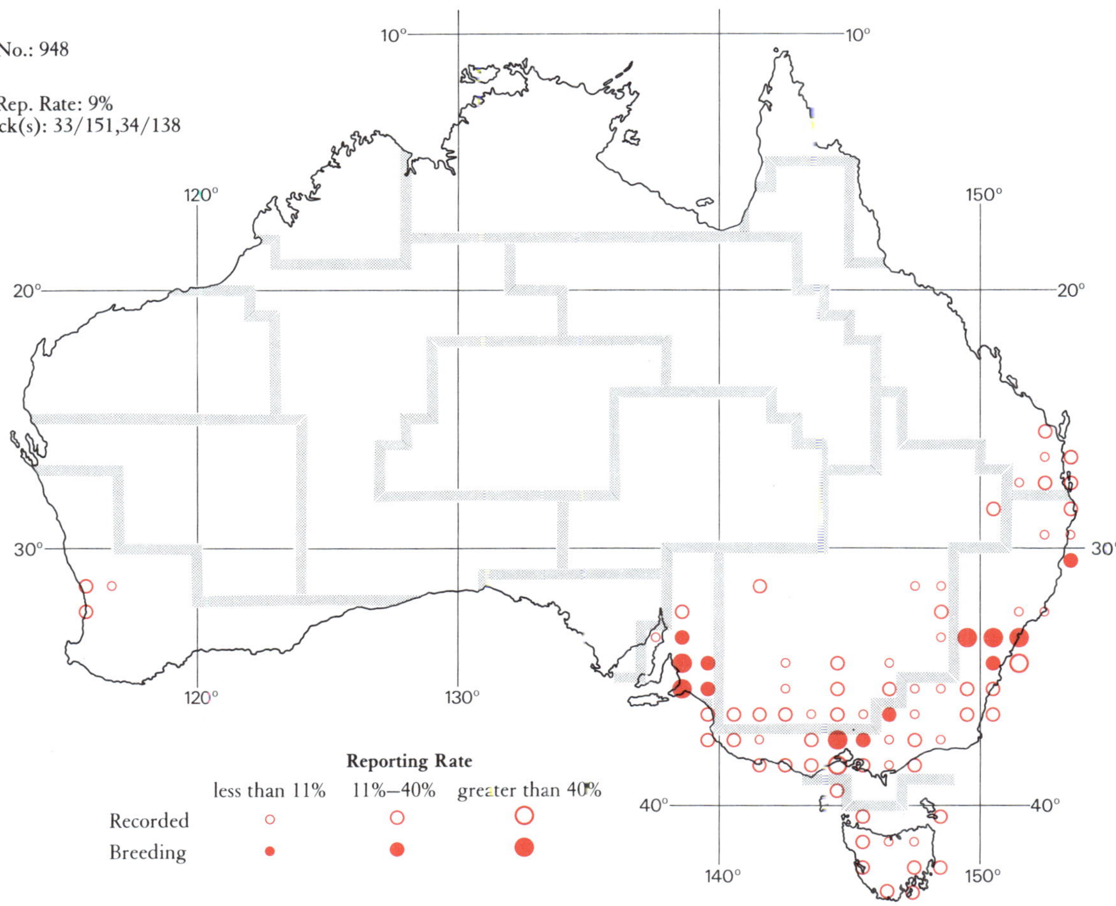

Grey Teal

Anas gibberifrons

Grey Teal of the Australian subspecies *gracilis* also inhabit New Zealand and New Guinea and adjacent islands. Other subspecies breed in Indonesia and the Andaman Is (12N/92E).[658] In Australia the Grey Teal rarely breeds in the north or in the Tas. Region. North of 20°S there are breeding reports in the Atlases from only 9% of the 1° blocks in which the species is recorded compared with 49% for 1° blocks south of 20°S. In the Tas. Region only nine breeding reports were submitted during the Field Atlas.

The Grey Teal is nomadic. The composition of flocks changes continually as birds arrive and leave though, on permanent wetlands, some individuals are sedentary and have been repeatedly recaptured.[658] Neither individuals nor populations show any regularity in the direction of movement apart from gathering on remaining wetlands as inland waters dry up in summer. Banded birds recovered on one wetland, Hume Weir (35/147), had come from Griffith (34/146), Dareton (34/142), Sydney (33/151), Darwin (12/130), Hobart (42/147), Naracoorte (36/140), Adelaide (34/138), Melbourne (37/145) and Perth (31/115). Conversely, during a drought-induced irruption in 1957–58, birds banded in the Murray-Darling Region were recovered throughout Australia and in New Guinea and New Zealand.[658] The large numbers of Teal that arrived in the Top End and New Zealand following this irruption established breeding populations. That in the Top End did not persist but breeding has been regular in New Zealand since.

The Grey Teal inhabits wetlands of any shape, size and salinity but most often fresh. Flocks gather in bays and estuaries, specially in the non-breeding season.[1038] The birds congregate on spreading floodwaters where the seeds of pasture plants lie beneath shallow water.[649] This suddenly increased food supply may contribute to the rapid onset of breeding.[661] A typical dabbling duck, the Grey Teal feeds in the shallows on seeds, vegetative material, bulbs and insects.[480,649,661] In bays and estuaries the species may rely almost entirely on molluscs and other marine invertebrates.[1038] The Grey Teal is a game species.

In a survey of 56 000 ha of wetlands in western Vic. in Feb. 1982, 34 850 Grey Teal were counted (0.62 birds/ha).[267]

Field Atlas Records: 17 529

1° blocks–with records: 599 (74%)
–with breeding records: 212 (35%)

Atlas No.: 211

Max. Rep. Rate: 67%
In block(s): 27/148

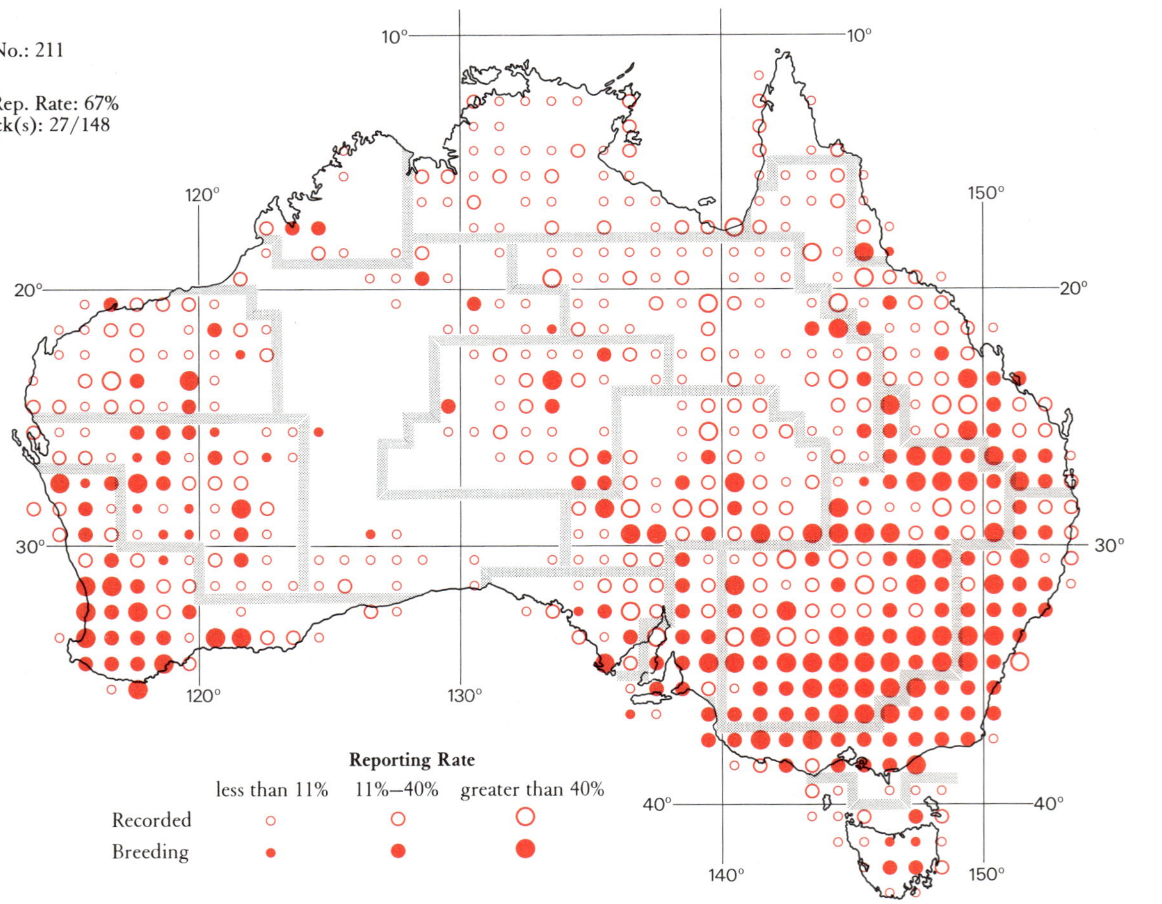

Chestnut Teal

Anas castanea

The Chestnut Teal is most abundant on coastal wetlands of southern and eastern Australia. A few birds reach northern Australia and the species was recorded in New Guinea for the first time in 1980.[1721] In the Tas. Region it is the most abundant duck. Early taxonomic confusion between the Chestnut and Grey Teal makes the historical distribution of the Chestnut Teal uncertain but its historical breeding range falls within that determined by the Field Atlas except for breeding reports before 1901 from North West Cape (21/114).[1343] There are no historical records linking the eastern and western ranges but the species' capacity for long-distance movement indicated by sightings as far north as Darwin (12/130)[658] and New Guinea means that there may still be interchange between the populations. Most banded birds have been recovered close to the banding place and adults appear to be fairly sedentary though gathering into flocks in autumn and dispersing in spring.[658] The birds that move greater distances are mostly juvenile. Some are known to have crossed Bass Strait and others have been recovered within eastern Australia up to 500 km from the banding site. Vagrants, recorded in northern Australia in all seasons, have probably been caught up with moving flocks of Grey Teal.

The Chestnut Teal inhabits large and small bodies of water, including saline coastal wetlands, bays and estuaries. It gathers in the non-breeding season on large bodies of water. At this time there are flocks of 300–400 on L. Reeve (37/147),[285] up to 2000 at Corner Inlet (38/146)[1335] and 500–1000 at several sites in the Derwent Estuary (42/147).[1314] The Chestnut Teal feeds in and near the shallows or on mudflats and, when it follows feeding Australian Pelicans, may be collecting food disturbed by them.[124] On the Gippsland Lakes (38/147) in 1971–73, gizzards of Chestnut Teal contained mainly plant material (74% by volume) with the remainder insects. Some molluscs and crustaceans were also recorded.[1340] In saline areas it feeds on invertebrates or plant tubers.[479,1335] The Chestnut Teal is a game bird.

There are no measures of abundance.

Field Atlas Records: 4956

1° blocks–with records: 199 (25%)

–with breeding records: 67 (34%)

Atlas No.: 210

Max. Rep. Rate: 43%
In block(s): 34/135

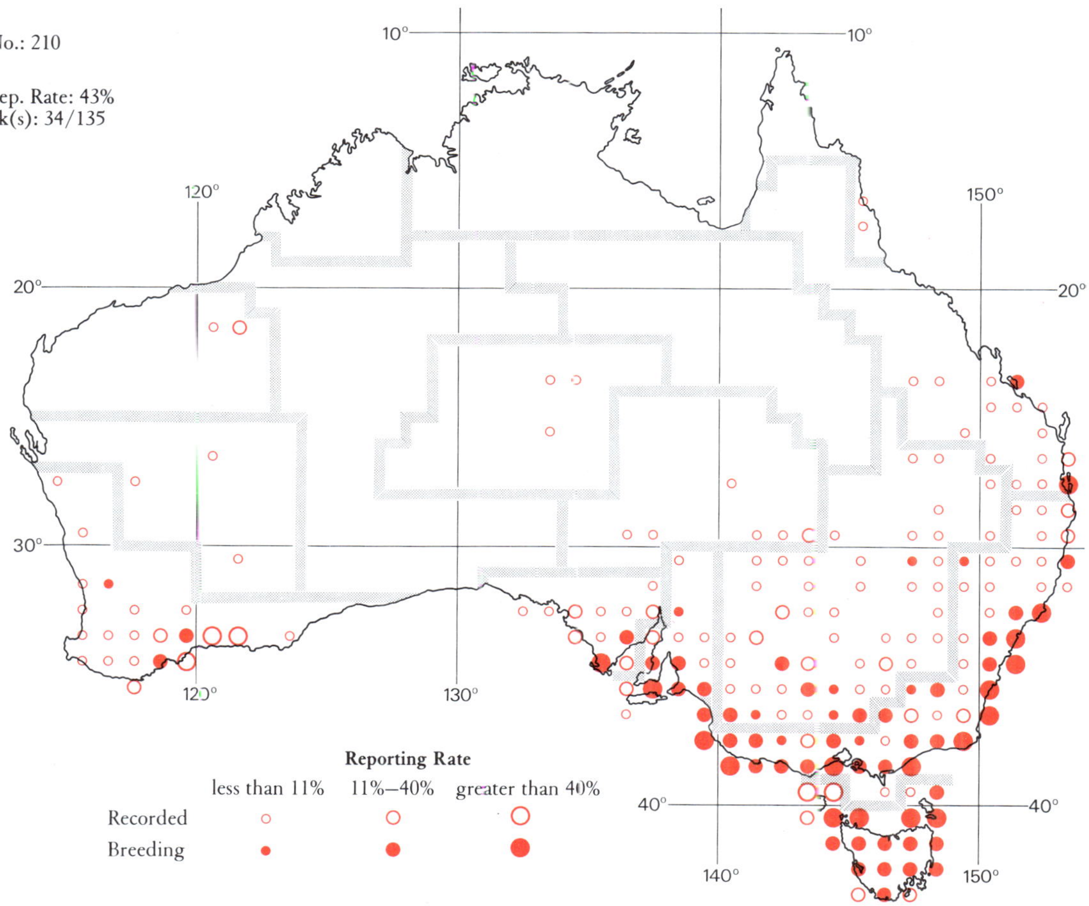

Australasian Shoveler

Anas rhynchotis

The Australasian Shoveler has two subspecies: *variegatus* in New Zealand and *rhynchotis* in Australia. No records in the Atlases link the eastern and western populations in Australia. The breeding range of the eastern population may have declined this century: breeding has not been reported north of 30°S since 1920; earlier breeding records came from as far north as 22°S, though the Shoveler was described as 'rather rare' in the area.[102] In WA, in contrast, the Atlases suggest that the species' range and numbers may have expanded. There were reports from only four 1° blocks in WA and no breeding records in the 1901–50 period, increasing to 28 in 1951–76, including six breeding records. Breeding in the Goldfields Region during the Field Atlas followed heavy rains and a subsequent irruption in 1980.

Movements of the Shoveler are little known. Vagrants often turn up in the tropics and inland, specially after a good breeding season and subsequent drought. In 1956–58 there was such an exodus from the Murray-Darling Region.[658] At other times thousands of Shovelers gather on a single wetland. Autumn flocks have been noted occasionally at L. Reeve (37/147);[285] at L. Dulverton (42/147) flocks of up to 150 are present in autumn and winter but the bird is rare Oct.-Feb.[1900] The longest recorded movement is 960 km.[85]

Pairs and flocks of Shovelers inhabit wetlands of all types. In eastern Australia they are most often in the more permanent, heavily-vegetated swamps[658] but in WA are more characteristically in flocks of up to several hundred on open water.[1021] The Shoveler feeds by filtering surface water through its bill, sometimes upending and dredging on the bottom. One study near Griffith (34/146) found that it ate mainly food of animal origin, specially insects, and also took a variety of plant material.[649,661] The Shoveler is a game species in some states. Its numbers are generally thought to have declined since European settlement.[658]

In a survey of 56 000 ha of wetlands in western Vic. in Feb. 1982, 4980 Shovelers were counted (0.09 birds/ha).[267] In some years Bool Lagoon (37/140) holds more than 10 000 Shovelers; thousands gather at L. Alexandrina (35/139) at times; and 2000 have been recorded at Thompsons L. (32/115).[911]

Field Atlas Records: 2947

1° blocks–with records: 225 (28%)
–with breeding records: 41 (18%)

Atlas No.: 212

Max. Rep. Rate: 21%
In block(s): 34/145

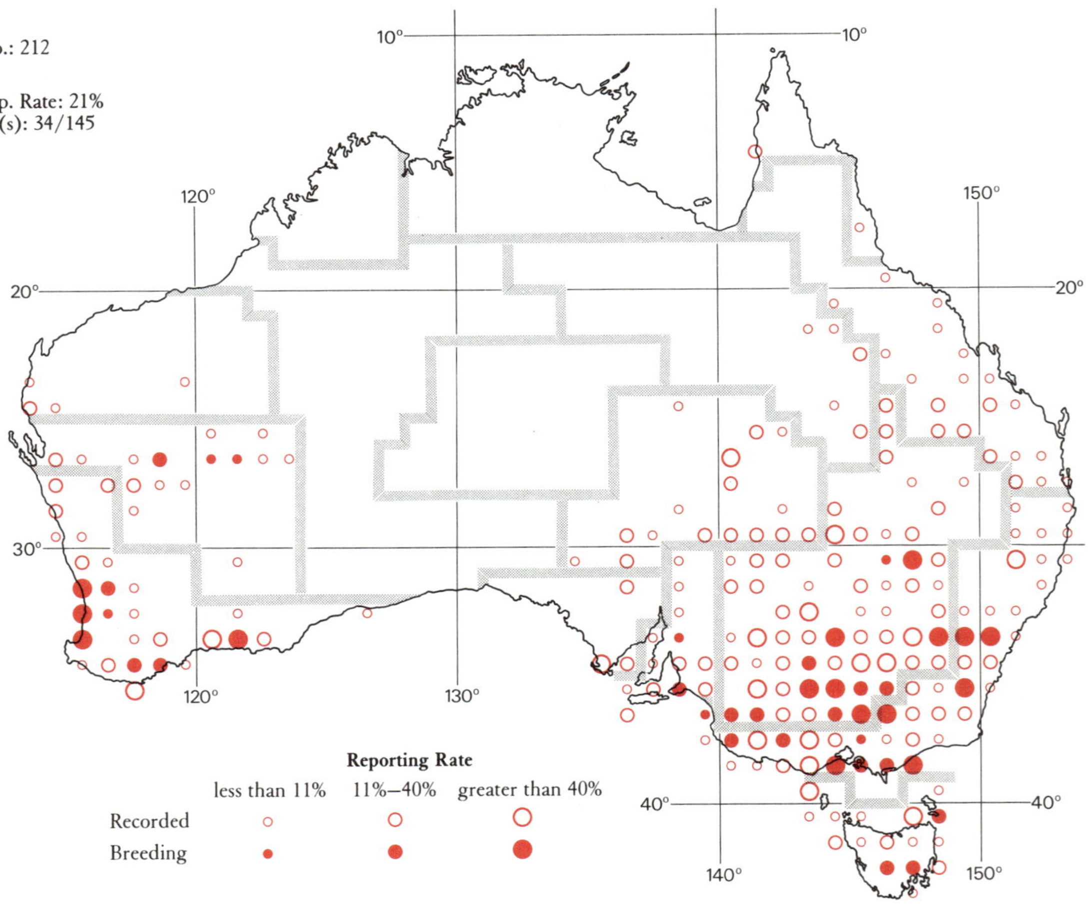

Pink-eared Duck

Malacorhynchus membranaceus

The endemic Pink-eared Duck is characteristic of the wetlands of inland Australia. It does not often breed near the coast, specially in eastern Australia, and seldom north of 26°S. In the Atlases breeding has been reported in only eight 1° blocks north of 26°S compared with 103 to the south. There are few records of the Pink-eared Duck from the Tas. Region and no breeding reports.

The Pink-eared Duck moves round the continent in search of suitable feeding conditions. It seldom visits the eastern coast except when most inland swamps have dried up. The longest movement recorded is that of an immature from Booligal (33/144) to Skipton (37/143), 486 km.[1440]

The Pink-eared Duck will breed at any time of the year.[658] Its distribution is closely controlled by the need to maintain contact with shallow water, supporting dense populations of the aquatic invertebrates on which it feeds by filtering the water.[649,661] Food requirements are best met in warm shallow water, such as when waters are receding or drying up after extensive flooding; hence the species' more inland distribution. Some artificial sites, such as sewage ponds, also provide suitable conditions. When the bird is forced to use deep-water refuges in dry periods, seeds make up a greater proportion of its diet.

In a survey of 56 000 ha of wetlands in western Vic. in Feb. 1982, 23 000 Pink-eared Duck were counted (0.41 birds/ha), 13 000 of them on the sewage-polluted coastal L. Borrie (38/144).[267]

Field Atlas Records: 2977

1° blocks-with records: 379 (47%)
-with breeding records: 72 (19%)

Atlas No.: 213

Max. Rep. Rate: 34%
In block(s): 28/139

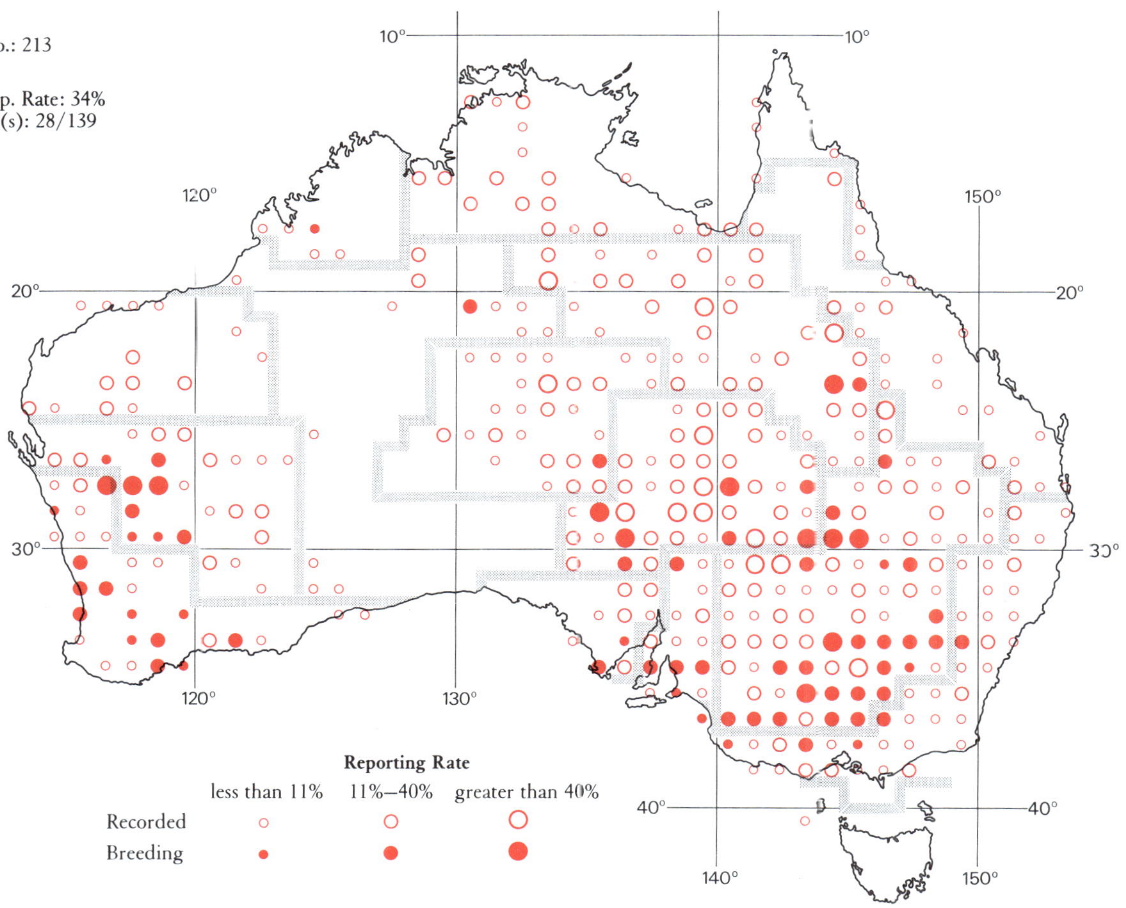

Hardhead

Aythya australis

The Hardhead breeds mainly in Australia. Individuals reach New Zealand, New Guinea, Indonesia and the Pacific islands and breeding has occasionally been reported but colonies thus established seem not to have survived.[658] In Australia breeding reports from regions outside those shown on the map have been few. They include some breeding at Mt Isa (20/139)[862] and old breeding reports from the Fitzroy R (17/123).[1713] In Mar. 1982 a bird was found on eggs near Broome (17/122) and northern breeding may be more widespread than these instances suggest because wetlands are inaccessible to observers in the wet summer season.[911] The Hardhead is said to breed on King I. (39/144)[739] but there is no report of breeding in Tas. itself.

Hardheads banded in Vic. have been recovered in Qld.[1709] The longest recorded movement of a banded adult was from Booligal (33/144) to Charters Towers (20/146), 1768 km.[41] The species leaves wetlands before they become shallow but the Field Atlas suggests no large-scale pattern of movement. In Tas. the number of birds varies, e.g. at L. Dulverton (42/147) the species is usually present[1900] but the number counted in autumn for four successive years ranged from 31 to 324.[1409]

The Hardhead usually lives in deep permanent freshwater lakes and swamps and the distribution of breeding records reflects the distribution of such waters. Its food contains much vegetable matter (57–74% by volume of gizzards sampled) most from deep-water plants.[649,661] The remainder is invertebrate material, specially molluscs. More animal material is taken during the spring than at other times. The Hardhead searches the bottom but feeds also at the surface and in shallow water where it will dredge as do typical dabbling ducks.[658]

Formerly abundant in the deep swamps of the northern coast of NSW (29/152), it is much rarer there now because flood mitigation has been undertaken, specially where this results in the draining of swamps and lagoons.[658] Farther south flocks of 2000–10 000 were reported on a single lagoon in the Myall Lakes (32/152) during the Field Atlas and similar numbers were reported on wetlands near Mt Isa and Kerang (35/143).[219] There were more than 5000 Hardheads on flooded coastal plains east of Wyndham (15/128) in Mar. 1983 and 6000 at Anna Plains lagoon (19/121) in Sept. 1981.[911]

Field Atlas Records: 7767

1° blocks–with records: 439 (54%)
–with breeding records: 56 (13%)

Atlas No.: 215

Max. Rep. Rate: 38%
In block(s): 18/145,27/140

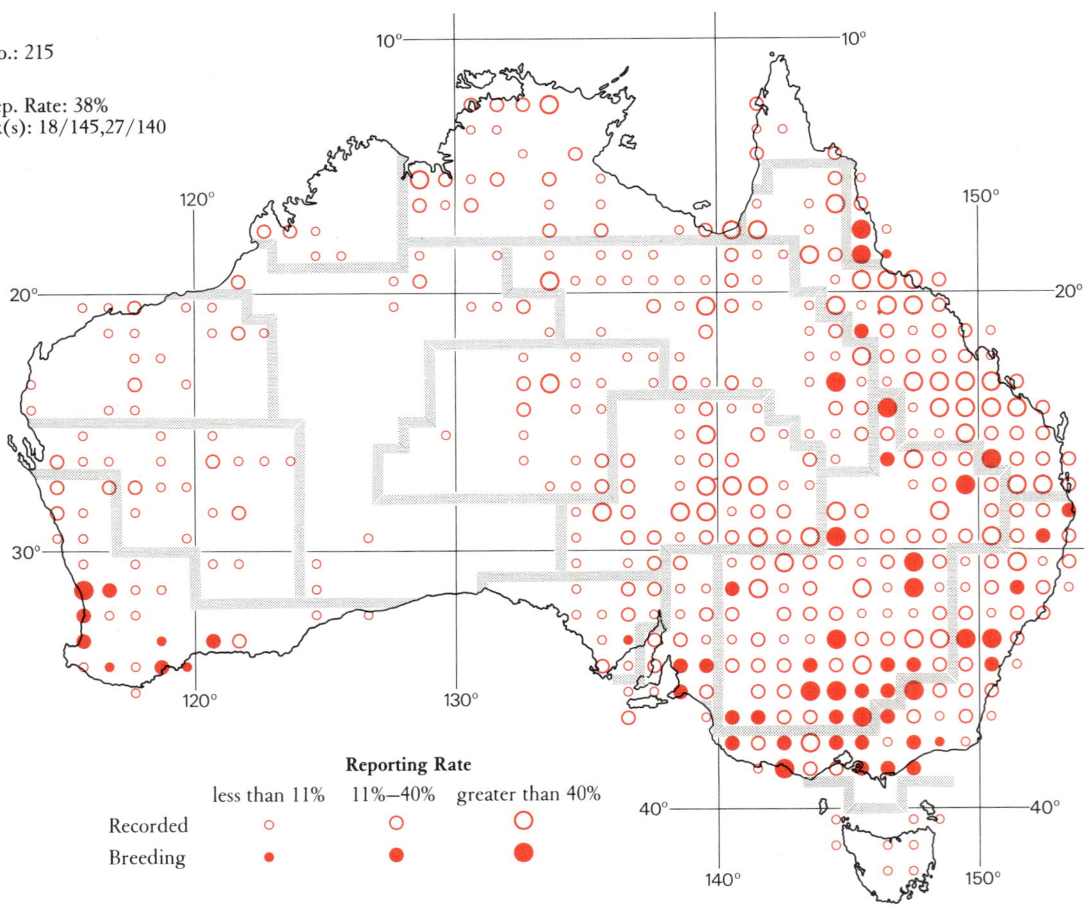

Maned Duck

Chenonetta jubata

The range of the endemic Maned Duck has expanded since European settlement. Comparison with the historical maps (p. 662) shows its spread into newly-cleared areas, specially in the Eastern Qld, Tas. and South-West Regions. In Tas. the species was regarded as rare before 1970. Breeding was first reported near Ross (42/147) in 1973[1315] and the Maned Duck has since become widespread. During the Field Atlas it bred near Gretna (42/146), Ravensdale (42/147) and Broadmarsh (42/147).

The outlying northern records were of a single adult at Wildman R (12/131) in July 1979 and of an unspecified number of birds at Weipa (12/141) in June 1978.

There is a seasonal pattern of movement. The Maned Duck gathers into flocks round fresh water once breeding is over.[649,1337] It roosts at these sites throughout the summer then moves out again when pastures become green. Roosts are used year after year but recoveries of birds banded in them show that the individual Maned Duck, though faithful to a roost during one dry spell (often the whole summer), may not return there next summer. Two studies of banded birds suggest that the Maned Duck does not disperse widely. In one, 75% of recoveries were within 160 km of the banding place[649] and, in the second, 60% of recoveries were within 200 km of the banding site.[1337] One bird banded at Serendip (38/144) was recovered at Prairie (20/144), 1950 km distant.[1337]

The Maned Duck inhabits pastures near fresh water.[658] It is usually in groups and non-breeding flocks of up to 2000 have been recorded. Improved pastures and farm dams have provided the Maned Duck with many new suitable pools of fresh water. It feeds almost entirely on vegetation, chiefly the grasses and legumes of pastures. Feeding sites may be up to 10 km from roosts.[649] Annual mortality is high, 57–80% in Vic., depending apparently on shooting pressure.[1337]

There are no measures of abundance.

Field Atlas Records: 17 623

1° blocks–with records: 456 (56%)
–with breeding records: 200 (44%)

Atlas No.: 202

Max. Rep. Rate: 77%
In block(s): 27/149

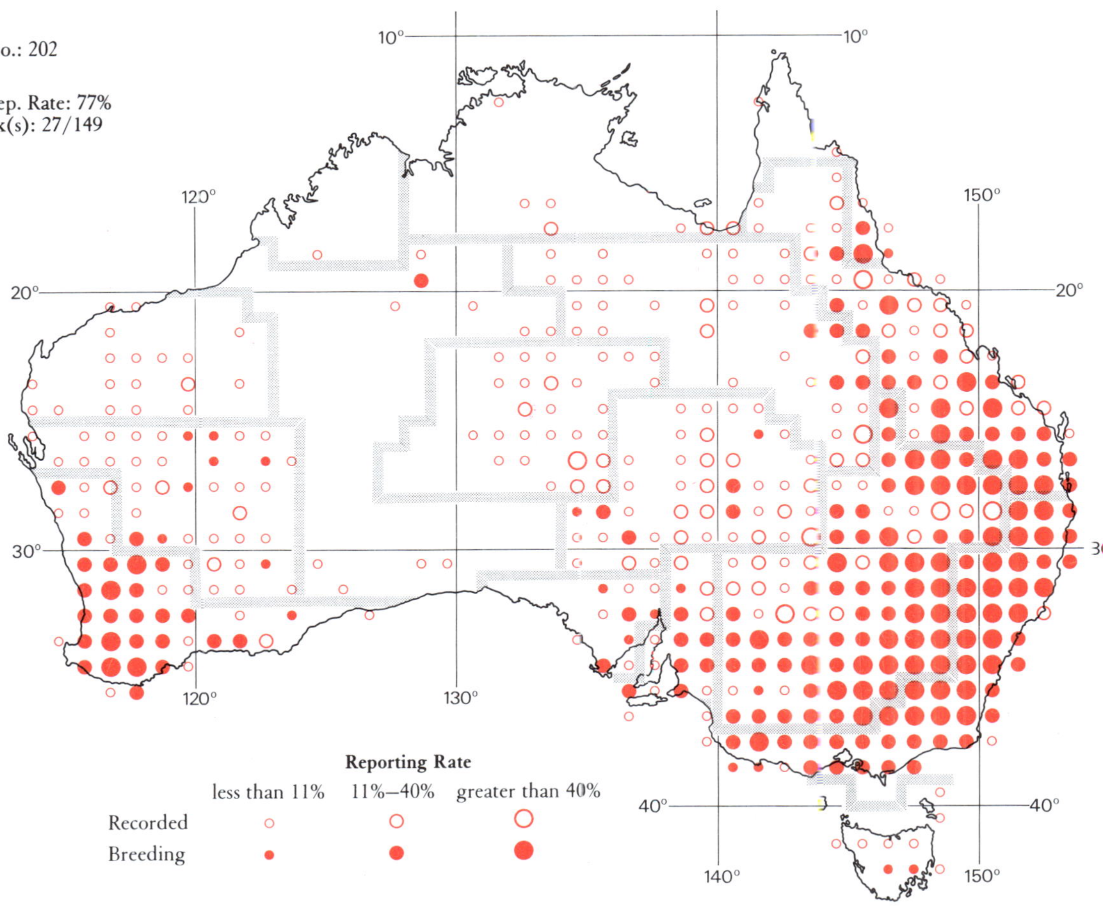

Cotton Pygmy-Goose

Nettapus coromandelianus

Subspecies of the Cotton Pygmy-Goose live in south-eastern Asia, Indonesia and northern New Guinea. In Australia the subspecies is *albipennis*. Early records of the Cotton Pygmy-Goose from the Claudie R (12/143) and Watson R (13/141), in the Cape York Region, are probably attributable to the Green Pygmy-Goose.[958] If the Cotton Pygmy-Goose occurs in the region, north of 14°S, it is probably only as a vagrant. The species has been observed at Lotus Vale (17/141).[1036]

The Hunter R (32/151) was given as the southern limit of the species' distribution last century[723] but there is no supporting evidence for a substantial population so far south.[658] Vagrants still reach this area[1292] and some come even farther south, e.g. L. Mokoan (36/146) in Mar. 1978. There was certainly a population on the Richmond R (29/153) and Clarence R (29/153) last century and until about the 1930s, which has subsequently disappeared[658] (see maps p. 662). Possibly the Cotton Pygmy-Goose declined in other parts of its range at this time: in the Duaringa (23/149) district it was described in 1924 as formerly common and then scarce[102] and it was not recorded during a campout in the Rockhampton (23/150) area in 1924.[1911] Any such decline has been reversed and the species was regularly reported in central Eastern Qld during the Field Atlas.

Some Asian populations are locally migratory.[1117] Seasonal changes in abundance have been observed near Townsville (19/146), birds gathering on water that remains in the dry winter season.[658]

Pairs and small flocks usually occupy deep lagoons vegetated with waterlilies and submerged and floating aquatics.[658] Many farm dams provide similar conditions and may have increased the number of wetlands suitable for the species.[1036] The Cotton Pygmy-Goose feeds mainly on seeds, inflorescences and other parts of aquatic plants, specially the seeds of waterlilies and hydrilla.[658] Dabbling on the surface of the water, it strips the seed heads of water plants and is almost always on the water, rarely resting on logs in the water or on the edges of pools.

The total Australian population in the early 1960s was estimated at 1500 birds.[1037]

Field Atlas Records: 481

1° blocks–with records: 41 (5%)
–with breeding records: 7 (17%)

Atlas No.: 200

Max. Rep. Rate: 36%
In block(s): 18/145

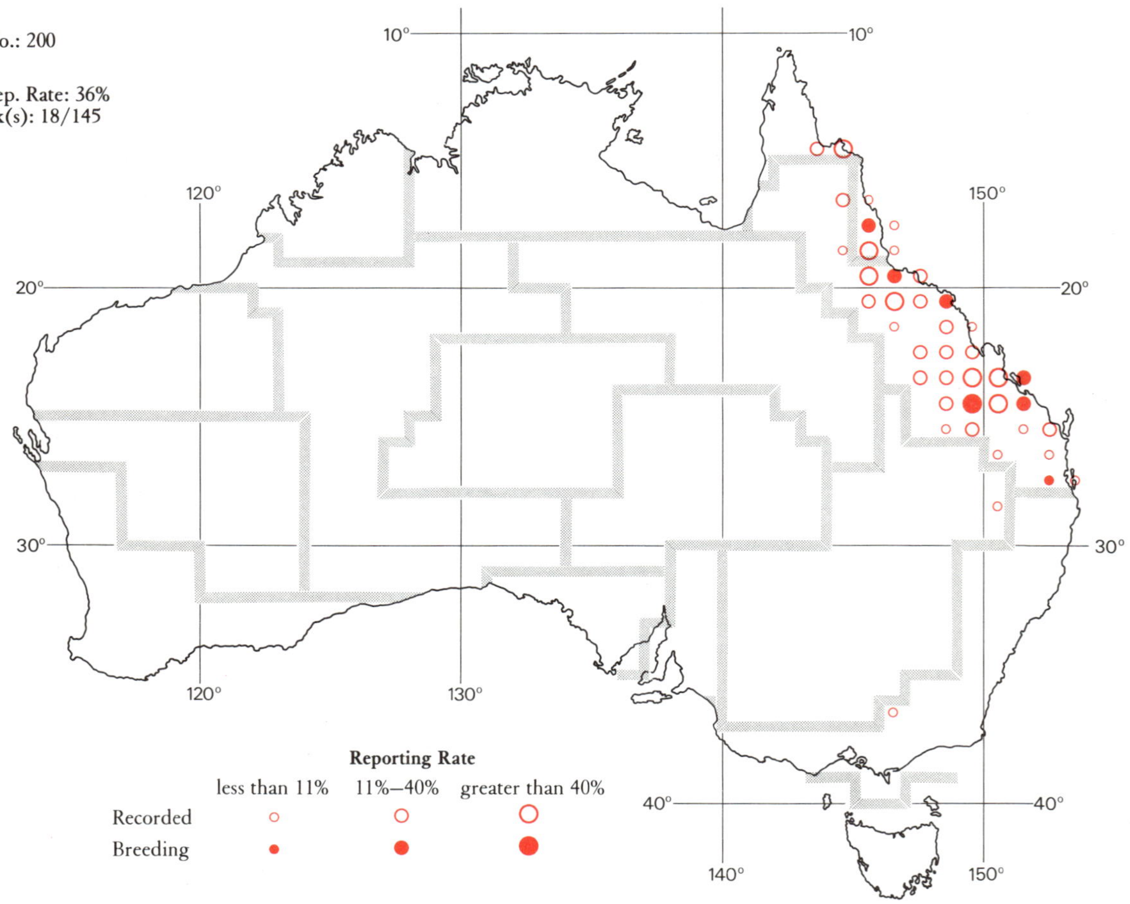

Green Pygmy-Goose

Nettapus pulchellus

The Green Pygmy-Goose lives in Indonesia, southern New Guinea and Australia. In Australia its range is continuous between Rockhampton (23/150) and Kununurra (15/128). Its distribution in eastern Australia seems unchanged from the historical periods but in the Kimberley Region its range and numbers appear to have declined this century. In the 1890s flocks of up to a dozen were recorded on 'all the swamps or lagoons along the course of the Fitzroy R' (17/123)[1343] and in 1886 flocks of up to 100 populated its lower reaches.[1713] In 1902 the species was collected at Oobagama (16/123) where it has not been recorded since. Isolated populations persist on the Fitzroy R: a group of eight was observed near Camballin (18/124) in 1979 and parties of 20–30 on L. Gladstone (17/126) in 1968.[748]

The Green Pygmy-Goose has not been reported breeding south of 21°S. It is rare south of 25°S but vagrants have even been reported south of 30°S: once in eastern Australia and four times in WA.[1292,1638]

When the rains of the wet summer season flood the northern plains, Green Pygmy-Geese spread out but during the dry winter season the bird is restricted to deep permanent pools where aquatic vegetation persists. It is a dry winter season visitor to parts of Torres Strait.[503] Long flights must be made at times because the species is occasionally reported in southern areas. Nine dated records south of 25°S are all Aug.-Mar. and mostly in summer.

The Green Pygmy-Goose lives in pairs or small flocks on deep permanent freshwater lagoons and man-made lakes. Large flocks sometimes form, specially in the dry winter season, some containing as many as 600 birds.[658] The bird feeds on waterlily flowers, as well as on the seeds of swamp grasses and on submerged vegetation. It does not come out of the water to feed and will take the seeds of fringing grasses only when it can float to do so.[658] The bird rapidly invades man-made reservoirs in the north, e.g. at Mt Isa (20/139) and Kununurra. The Green Pygmy-Goose is little sought as a game species and its numbers, in Qld at least, appear not to have changed since early European settlement.[1037] Destruction of aquatic vegetation by cattle is blamed for the decline in the Kimberley Region.[1713]

There are no measures of abundance.

Field Atlas Records: 660

1° blocks–with records: 73 (9%)

–with breeding records: 6 (8%)

Atlas No.: 201

Max. Rep. Rate: 46%

In block(s): 13/130

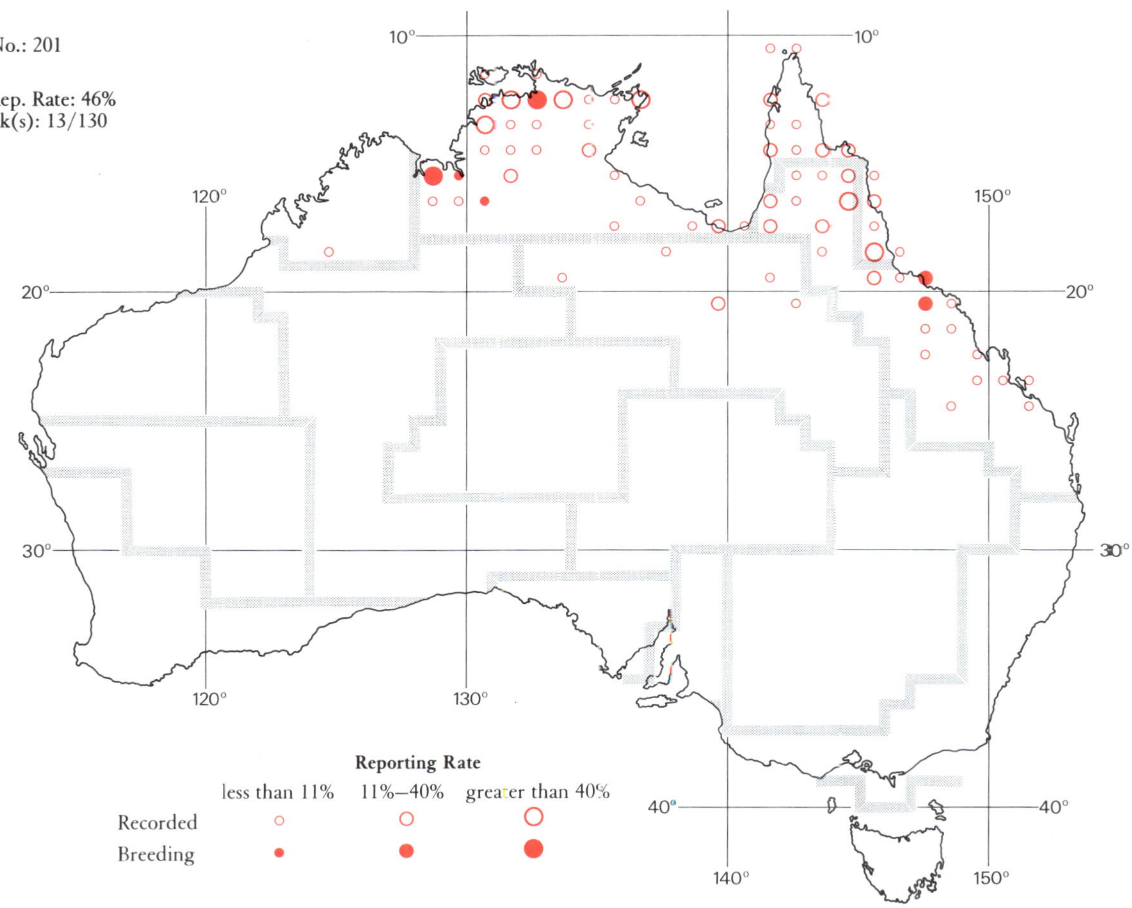

Blue-billed Duck

Oxyura australis

The eastern and western populations of the endemic Blue-billed Duck appear isolated but no subspecies are recognized.[658] Breeding records in the east are mostly south of 32°S but in 1974 when L. Eyre was full, the Blue-billed Duck bred on Frome Ck (29/137)[421] and at L. Cootabarlow (30/140). The species has only occasionally been recorded breeding in the Tas. Region, most recently in the 1970s.[734]

A regular movement of the Blue-billed Duck has been described. Flocks gather on large bodies of open water in winter, augmenting local resident populations. The migratory birds move out to smaller, densely vegetated swamps to breed in spring.[588,658,1832] Despite this, records show no seasonal pattern at the edges of the species' range in eastern Australia, that is north of 29°S and in the Tas. Region. The Field Atlas suggests that a small irruption occurred about 1978. The species was recorded north of 32°S and in the Tas. Region more often between July 1977 and Dec. 1978 than throughout the remainder of the Field Atlas period (52% and 59% respectively of Blue-billed Duck records in these areas were made during this time). The Mt Isa (20/139) sighting was in Mar. 1978.

The Blue-billed Duck inhabits large deep permanent lakes and swamps, natural or artificial, but is rarely on salt water or small farm dams. The bird feeds both by dabbling at the surface and by diving, taking about equal proportions of animal and plant food.[661] Midge and stonefly larvae are abundant in the diet during the breeding season.

In winter flocks numbering 500–1000 are present on lakes in the southern Murray-Darling[833] and South-East Regions[1107] and of up to 300 in the South-West Region.[911] These flocks may form jointly with flocks of Eurasian Coots.

There are no measures of abundance.

Field Atlas Records: 1050

1° blocks–with records: 109 (13%)
–with breeding records: 21 (19%)

Atlas No.: 216

Max. Rep. Rate: 14%
In block(s): 31/115,39/143

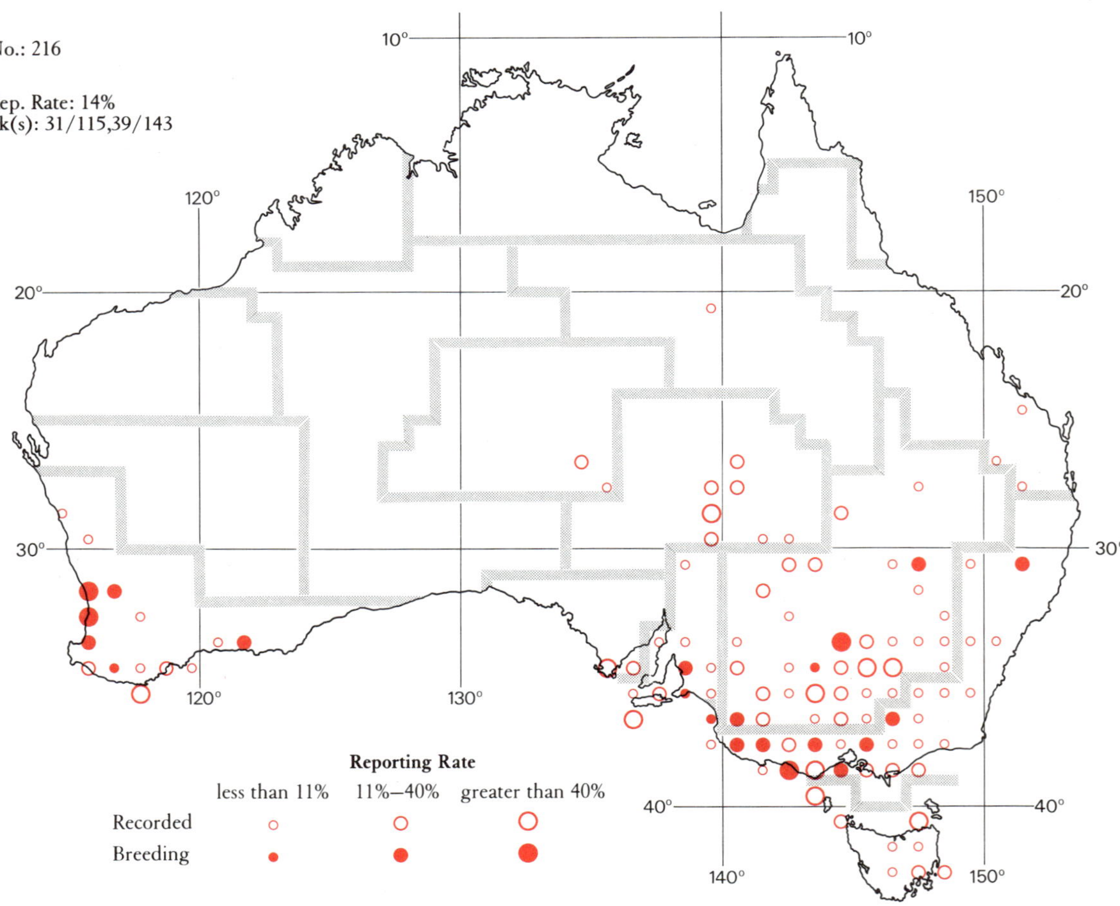

Musk Duck

Biziura lobata

The eastern and western populations of the endemic Musk Duck have not been separated subspecifically.[658] Musk Ducks were seen on the sea near Eyre Bird Observatory (32/126) during the Field Atlas so that interchange may take place between eastern and western populations. Vagrants sometimes wander well north of the usual southern range and have been reported as far north as the Herbert R (18/146) in 1888[239] and Cardabia Ck (23/114) in 1902.[317] There are no reports of breeding in the Eastern Qld or Goldfields Regions and only one in the L. Eyre Region though sightings are made there regularly.

The Musk Duck rapidly occupies deep waters that form occasionally in inland Australia, e.g. L. Eyre.[658] Flocks gather in winter, apparently mainly of local birds. There is thought to be a movement by juveniles from swamps in inland NSW to areas farther south in winter.[658] The Field Atlas suggests no large-scale seasonal pattern of movement.

The Musk Duck lives on lakes, swamps and farm dams provided the water is deep enough for diving and is often seen on the sea round the southern coast and in bays and estuaries. It feeds mainly on aquatic animals (72–78%) including large crayfish, molluscs and insects.[661] Because they feed by diving, Musk Duck are taken accidentally in fishing nets in estuaries and inland waters.[1638]

In a survey of 56 000 ha of wetlands in western Vic., in Feb. 1982, 660 were counted (0.01 birds/ha).[267] Round Manjimup (34/116), where the Musk Duck breeds, one pair to a dam is usual.[265]

Field Atlas Records: 4930

1° blocks–with records: 203 (25%)
–with breeding records: 54 (27%)

Atlas No.: 217

Max. Rep. Rate: 32%
In block(s): 24/153,37/139,39/143

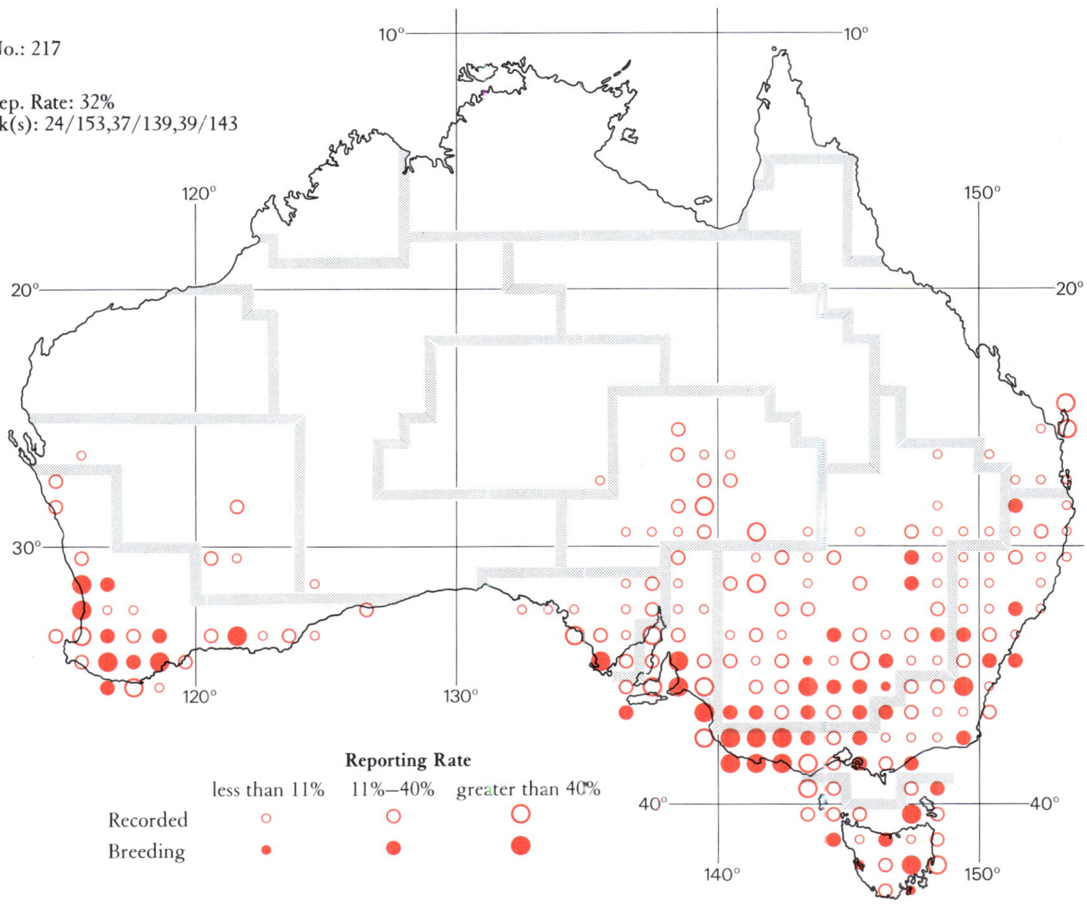

Osprey

Pandion haliaetus

The Osprey breeds in the northern parts of America, Europe and Asia and migrates south in Sept. to spend the southern summer in South America, Africa and southern Asia respectively.[422] A single subspecies, *cristatus*, breeds in Australia and New Guinea. The combined observations of the Atlases suggest that western and eastern populations are separated at the Eighty Mile Beach (19/121) where the Western Deserts reach the coast; and that the population in the Eyre Pen. and SA Ra. Regions is isolated. The Osprey reaches the Tas. Region rarely but at least one specimen was taken there last century;[15] two unconfirmed sightings were made during the Field Atlas.

Southern and inland records show that the Osprey in Australia is able to travel long distances. There is a seasonal pattern in records inland along the Murray R, four out of the six being in autumn. The fifth was a breeding report from near Nildottie (34/139) in Nov. 1980 and the sixth a sighting two months later near Berri (34/140).

The Osprey fishes in fresh, brackish or salt water.[439] Individuals are sometimes seen inland along the larger northern rivers and the Murray R though breeding is usually confined to the coast and islands. The Osprey feeds on live fish, usually 20–40 cm in length.[422] It will submerge itself to at least 1 m when fishing.

In many parts of the world the Osprey population declined during the 19th and 20th centuries. Human disturbance was blamed for these declines which have now been reversed in some places, specially northern Europe.[422] Elsewhere pollution has adversely affected reproductive success and could do so in Australia near large industrial centres.

In some parts of its range overseas nests are as little as 10 m apart but in Australia most are separated by at least 1 km.[456] On Rottnest I. (32/115) three pairs nested each year from 1961 to 1963; the island has an area of 23.5 sq. km.[852]

Field Atlas Records: 2267

1° blocks–with records: 173 (21%)
–with breeding records: 60 (35%)

Atlas No.: 241

Max. Rep. Rate: 72%
In block(s): 24/153

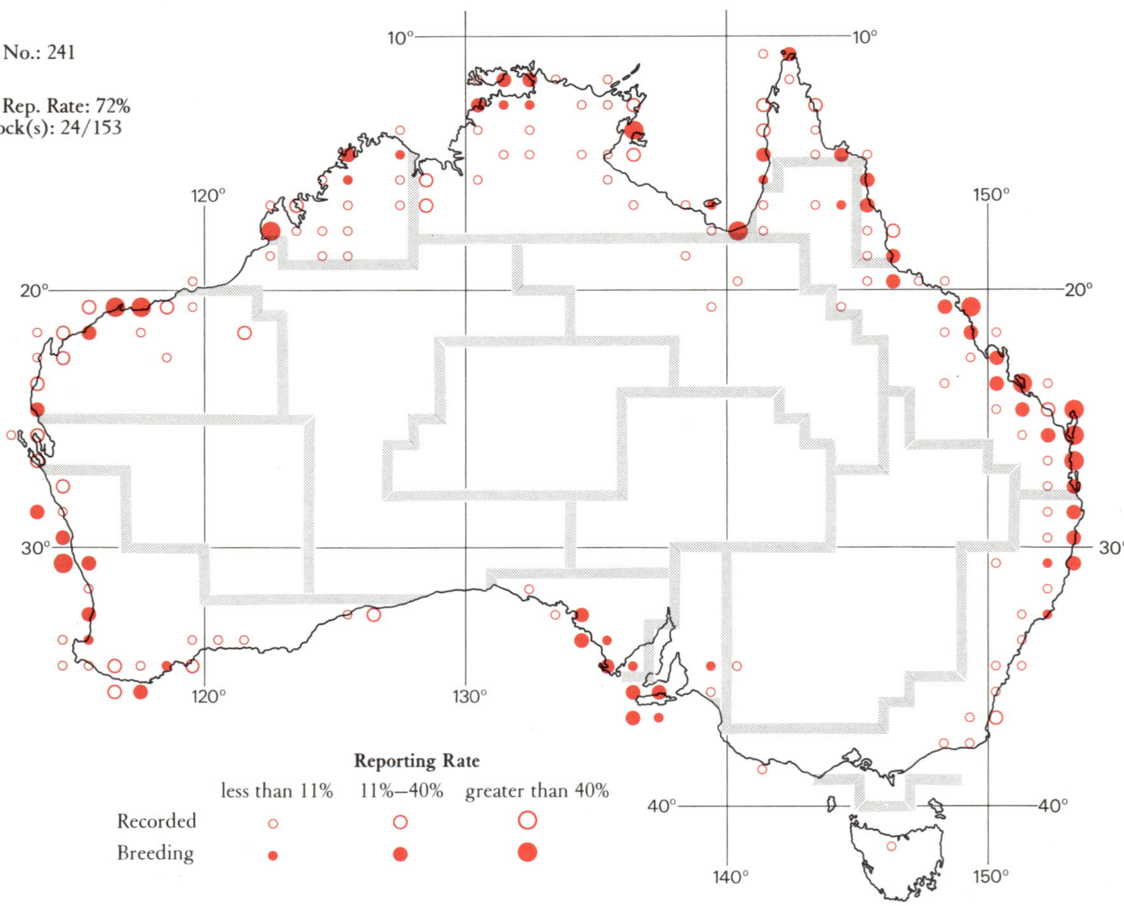

Black-shouldered Kite

Elanus notatus

The endemic Black-shouldered Kite has expanded its range. Until the early 1950s it was reported only irregularly from south-western Vic.[79,1052] but is now an established resident. It has also spread into the South-West Region:[1216,1638,1888] at Northam (31/116) the first two were recorded in Apr. 1935, breeding was recorded in Dec. 1948 and the species is now seen every year. In the Tas. Region it is a rare vagrant to Flinders I. (40/148) and the north-western coast, immature birds only having been recorded.[1276] Unlike the closely related Letter-winged Kite, most Black-shouldered Kites breed in coastal and sub-coastal districts.

A seasonal pattern of movement has been reported in three areas. The bird appears on the coast of the Atherton Region only May-Oct.[698] Near Sydney (33/151) it is always present but there is a marked influx in autumn and winter;[473] and in the Mildura (34/142) district a few arrive to breed each year from autumn to spring but the numbers vary greatly.[439] The Field Atlas shows no evidence of large-scale seasonal movement. In many places fluctuating numbers or irruptions are reported: in parts of the Atherton Region the Black-shouldered Kite is present all year but numbers vary; in the upper South-East Region 'many dozen pairs' stayed for a few months from May 1897;[906] in the southern Murray-Darling Region a small resident population was augmented after the 1956 floods by many breeding Kites which left by 1958;[833] and at Laverton (37/144) there were occasional small invasions.[1841] The longest movement recorded is that of a bird, banded as a nestling, from Redbanks (35/137) to Lithgow (33/150), 998 km.[46]

The Black-shouldered Kite hunts over plains, crops or pasture 0.3–1.5 m high. Its prey is usually small mammals, predominantly the house mouse, and grasshoppers.[91] Irruptions are often associated with an abundance of rodents and grasshoppers,[396] the bird staying to breed if the abundance of prey is great enough.[439,1414] Some coastal gatherings have been related to dry conditions inland.[906] Agricultural clearing has probably favoured the Kite by providing good hunting grounds and good conditions for the development of plagues of the house mouse. There are no measures of abundance.

Field Atlas Records: 13 641

1° blocks–with records: 421 (52%)

–with breeding records: 89 (21%)

Atlas No.: 232

Max. Rep. Rate: 71%
In block(s): 27/151

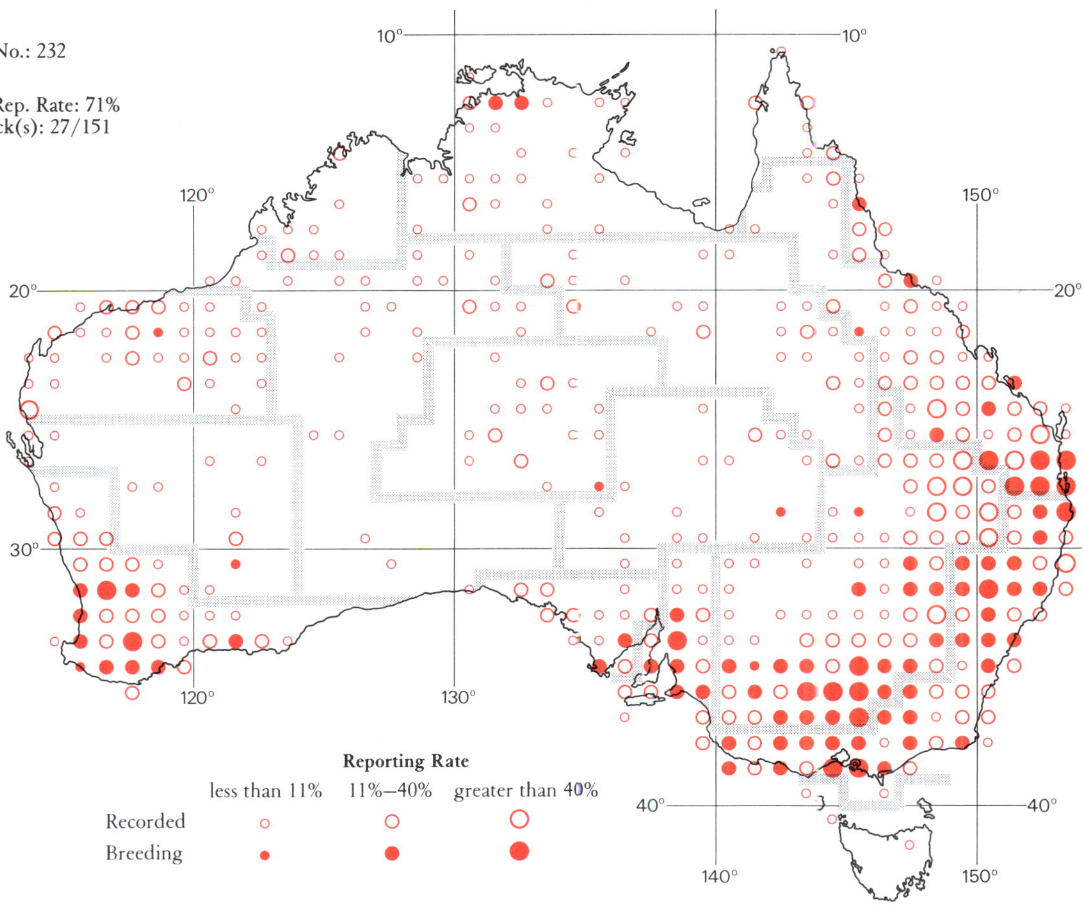

Letter-winged Kite

Elanus scriptus

Breeding of the endemic Letter-winged Kite takes place predominantly in the Rolling Downs and L. Eyre Regions: in the Atlases 61% of 1° blocks containing breeding records are in these regions. From them, the Kites disperse during irruptions. In 1976–77 such an irruption brought large numbers of the Kites to coasts round Australia but particularly the lower South-East Region;[845] a few remained into 1978. A second, much smaller, influx took place in 1980–81.

In its breeding range the Letter-winged Kite lives along watercourses and near bores, nesting colonially though seldom with more than one pair to a tree. Up to several hundred birds will roost and breed together.[910,1588] The Letter-winged Kite usually hunts at night. Rodents are its main prey, particularly the rat *Rattus villosissimus*.[439] When rats are scarce it eats insects and there is one record of it eating honeyeaters. Recently the house mouse has been found to be prominent in its diet.[91,439] Breeding is continuous whenever rats have reached plague proportions after heavy rains.[910,1385] When the land dries up and the rats disappear, a random dispersal follows leading to spectacular irruptions.[1374,1627] During these the Letter-winged Kite may turn up in unusual places such as Ashmore Reef (12/123) where a flock of 16 was reported in 1977 and at 38/144 near Werribee where up to 95 were counted in May 1977.[845] Birds reaching the coast are often in poor condition and sometimes die. Many are immature. Sufficient birds remain in the breeding range to form colonies again when food is abundant. The pattern is one of excessive reproduction when resources are abundant followed by an almost suicidal scatter when food supplies diminish, a boom and bust strategy of survival.

On the Diamantina R (26/140) colonies of 20–40 birds were distributed 7–10 km apart, though plenty of trees were available between colonies.[910]

Field Atlas Records: 280

1° blocks–with records: 117 (14%)
–with breeding records: 7 (6%)

Atlas No.: 233

Max. Rep. Rate: 8%
In block(s): 29/140

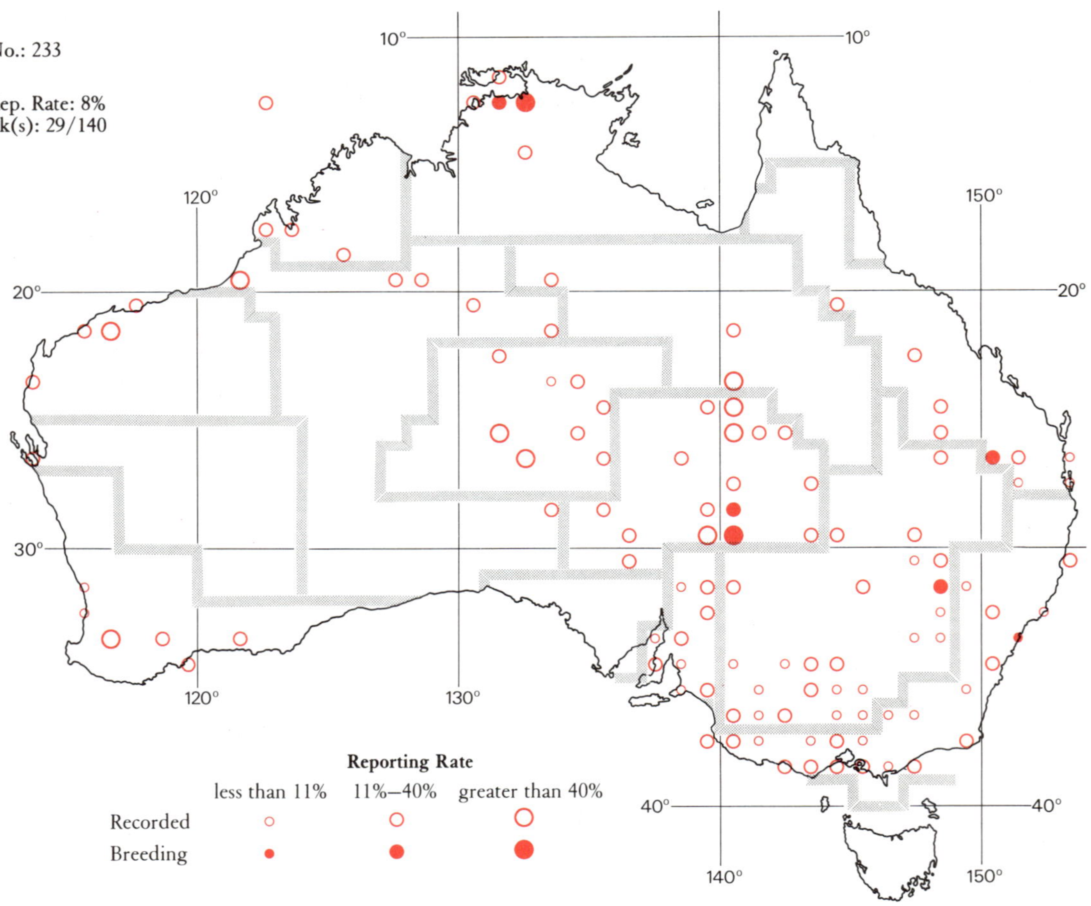

Pacific Baza

Aviceda subcristata

The Pacific Baza lives in New Guinea and adjacent islands and Australia. Subspecies have been described but views about them conflict.[388,1711,1713] The Historical Atlases contain no records from Torres Strait, though breeding has been reported near the tip of Cape York (10/142). Neither are there any records in the gap round the Gulf of Carpentaria. Records are too few to determine whether any gap exists between the Kimberley population and that of the Top End. Except for the northern Cape York Region, Field Atlas records come from throughout the range determined in the Historical Atlases. Breeding has not been reported from the Kimberley Region.

The Pacific Baza is apparently sedentary in some places, maintaining a territory throughout the year.[1099] Near Darwin (12/130) a lack of records May-Aug. may indicate local seasonal movement[423] but the Field Atlas does not suggest any large-scale movement.

Though the Pacific Baza is usually in pairs or family parties, 17 adults and juveniles were seen together on one occasion near Thinoomba (25/152). The species usually lives in forest. In Brisbane (27/153) it also inhabits gardens and orchards. Insects such as stick insects and mantids are taken from the outer foliage of trees and lizards from cliffs.[1094,1099] Observers have recorded frogs and beetles in the species' diet and it sometimes feeds on grasshoppers on the ground.[439,1099]

There are no measures of abundance.

Field Atlas Records: 1042

1° blocks–with records: 92 (11%)
–with breeding records: 26 (28%)

Atlas No.: 234

Max. Rep. Rate: 25%
In block(s): 15/145

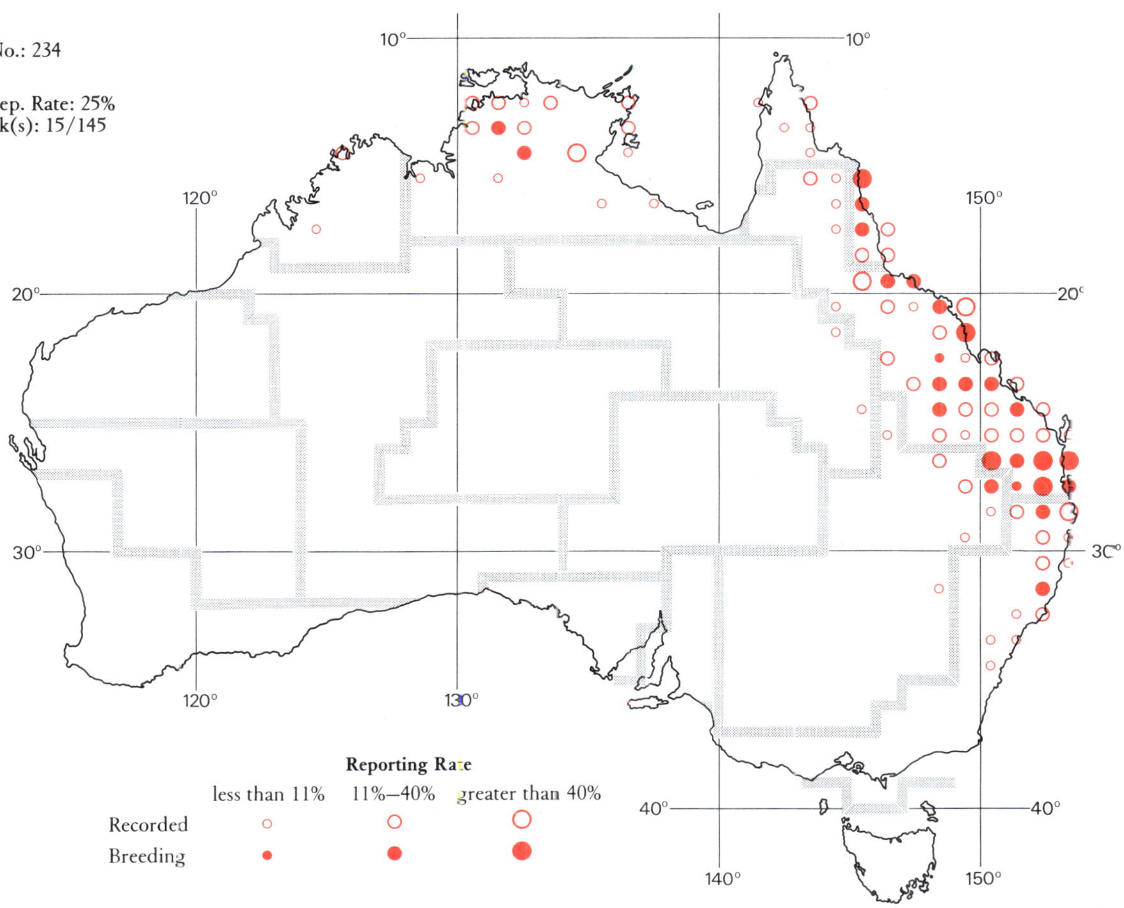

Black Kite

Milvus migrans

The Black Kite is widespread throughout the world apart from the Americas. In Australia there is one subspecies *affinus*. During the Field Atlas two confirmed records were made in Tas. where the species was first seen in 1966.[734] The Atlases show an expansion of range into the South-West Region where, before 1951, there was only one record: in 1943 at Northam (31/116) after a cyclone.[1216] An irruption in 1952 brought many Black Kites as far south as Perth (31/115) where they have since appeared regularly though not breeding.

In the northern hemisphere the Black Kite is mostly a regular migrant but there are also sedentary populations.[422] In Australia seasonal movements have been noted near Darwin (12/130) where the species is scarce or absent in the wet summer season[423] but at Mt Isa (20/139) recognizable individuals have stayed for at least nine months in one place.[1066] In Qld populations are probably rather sedentary and the greatly increased numbers occasionally present in some places are likely to be the result of local irruptions.[1045] In coastal NSW the species is a vagrant.

The Black Kite lives in a wide variety of country, characteristically gathering in groups at abattoirs, garbage dumps, shearing sheds and other sources of food. Sometimes huge flocks assemble such as one of more than 2800 birds recorded at Mt Isa.[1066] Though primarily a scavenger, the Black Kite catches mammals, specially rabbits and birds, reptiles, invertebrates both larval and adult, and frogs.[439,1045,1157,1343] Large numbers built up during an irruption of the rat *Rattus villosissimus* in 1974–76 in the Murray-Darling Region and many birds died when the plague subsided.[439] Lack of water may force the Black Kite from an area.[1066] The use of the inland for grazing, the establishment of waterpoints and increased road-kills have benefited the Black Kite.

In Sept. 1981, 47 pairs nested in less than 20 km of the Strzelecki Ck (28/140) floodplain (no more than 3 km wide).[1186]

Field Atlas Records: 9188

1° blocks–with records: 554 (68%)
–with breeding records: 51 (9%)

Atlas No.: 229

Max. Rep. Rate: 79%
In block(s): 17/141

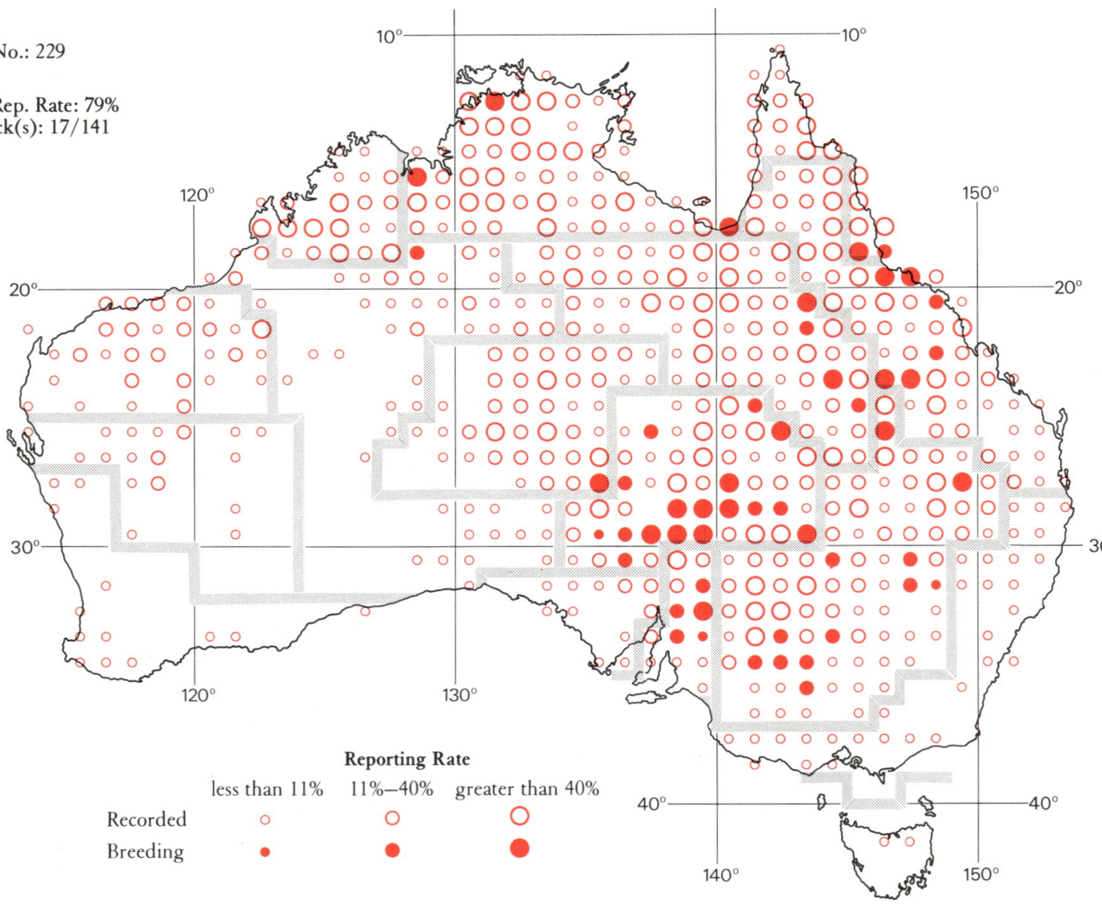

Square-tailed Kite

Lophoictinia isura

The endemic Square-tailed Kite has not been recorded from the Tas. Region and, on the mainland, is very rarely reported in the inland and in SA.[1162,1369] Breeding reports are confined to coastal and sub-coastal districts north of 35°S, except for a nest recorded at Bondi State Forest (36/149) in 1981. The species may be confused in the field with the Black Kite.

The Square-tailed Kite appears to undertake a regular seasonal movement, south in summer and north in winter, in south-eastern Australia[474] and probably in south-western Australia as well.[476] Reporting Rates on the mainland south of 30°S showed a greater difference between seasons in the east than in the west (east of 129°E: summer 0.33%, winter 0.11%; WA: summer 2.45%, winter 1.72%).

Typically, the Square-tailed Kite is seen alone soaring above or just below the tree canopy. There was no report of more than two Square-tailed Kites together during the Field Atlas, except at the nest. In the north and east most sightings came from forest and woodland and sometimes rainforest along rivers; breeding was recorded in forest of eucalypt, angophora and native pine.[929,1598] In the South-West the bird inhabits yate and wandoo woodland, mallee and heath.

Birds, mainly fledgeling passerines, rabbits, crickets and beetles have been recorded as food at the nest.[1598] Other food recorded includes grasshoppers[126] and lizards.[827] One bird has been seen on a carcass[126] and one chasing a rabbit. At Grafton (29/152) the Square-tailed Kite raids nests of the House Sparrow in street trees. In the South-West it has been observed to quarter heath systematically when passerines are breeding, taking such small prey as nestling Silvereyes.[1528]

There are no measures of abundance.

Field Atlas Records: 747

1° blocks–with records: 225 (28%)
–with breeding records: 10 (4%)

Atlas No.: 230

Max. Rep. Rate: 26%
In block(s): 13/130

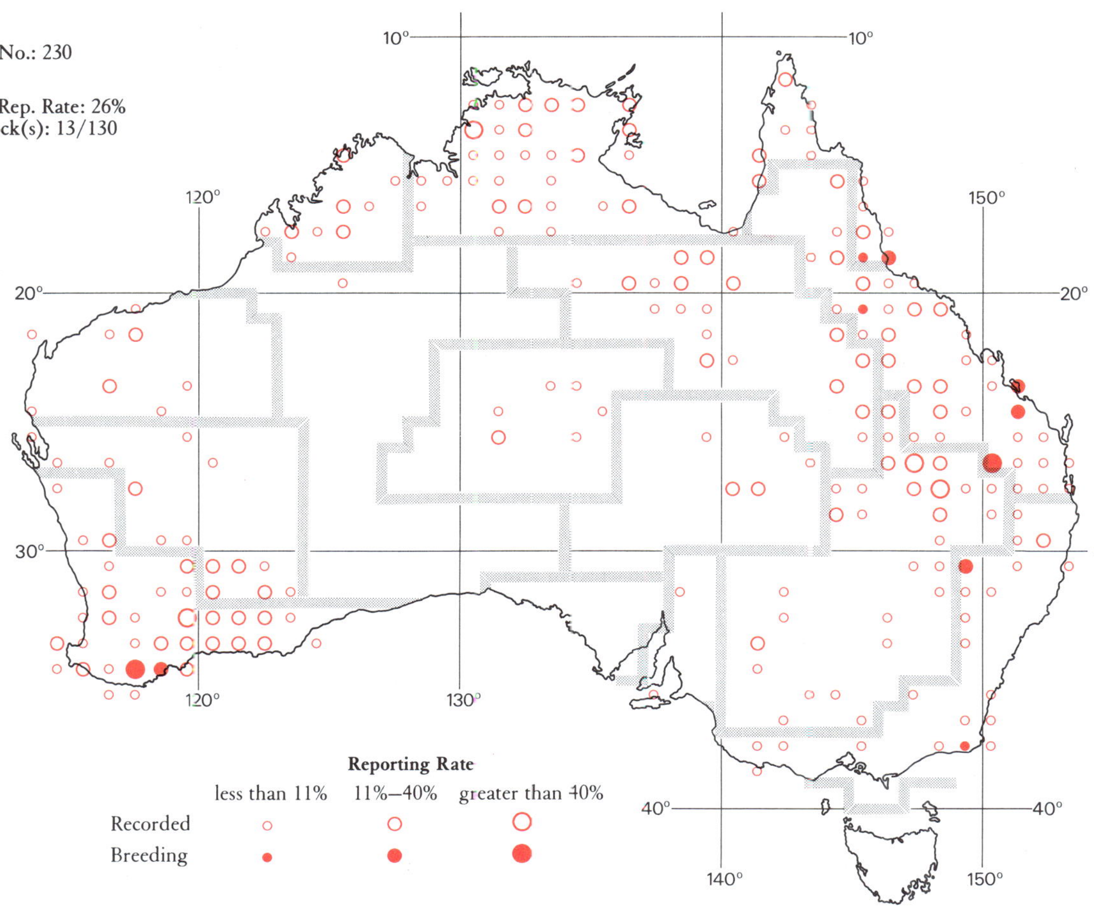

Black-breasted Buzzard

Hamirostra melanosternon

The endemic Black-breasted Buzzard is recorded in the Field Atlas throughout its historical range except for the Tas. Region. There, one was seen at Cullenswood (41/148) in Nov. 1916 and another doubtfully recorded on King I. (39/144) in Apr. 1923.[23,734] Sightings in the southern Murray-Darling Region were sporadic during the Field Atlas; until 1912 breeding was reported regularly there, once as far south as Moulamein (35/144).[299] An observer in the district in 1891 noted that the Black-breasted Buzzard was becoming rarer and the Atlases confirm its decline in this region. It has never been reported breeding at equivalent latitudes in the west and has remained an irregular visitor to the South-West Region.

Observations in the L. Eyre and Murray-Darling Regions suggest that adult Black-breasted Buzzards are sedentary.[299,439,1343] The same pairs probably nest year after year in the same area, if not in the same nests, and only during prolonged dry conditions do the birds disappear. A seasonal pattern of movement is suggested by the Field Atlas in the south of the species' range: all but one of the records south of 31°S was in Aug.-Jan. A minor irruption at the end of 1977 provided six of the ten records from the southern Murray-Darling Region.

The Black-breasted Buzzard inhabits woodland and pasture. It feeds on lizards, adult birds and nestlings including those of hawks and falcons and, in the south, on rabbits.[439] In captivity Black-breasted Buzzards presented with fresh Emu eggs broke them with one blow of their bills and ate the contents, contrary to some reports that the species uses stones to break eggs.[1638,1698]

Two pairs of Black-breasted Buzzards nested in less than 20 km of the Strzelecki Ck (28/140) floodplain (no more than 3 km wide) in Sept. 1981.[1186]

Field Atlas Records: 706

1° blocks–with records: 248 (31%)
–with breeding records: 8 (3%)

Atlas No.: 231

Max. Rep. Rate: 26%
In block(s): 28/140

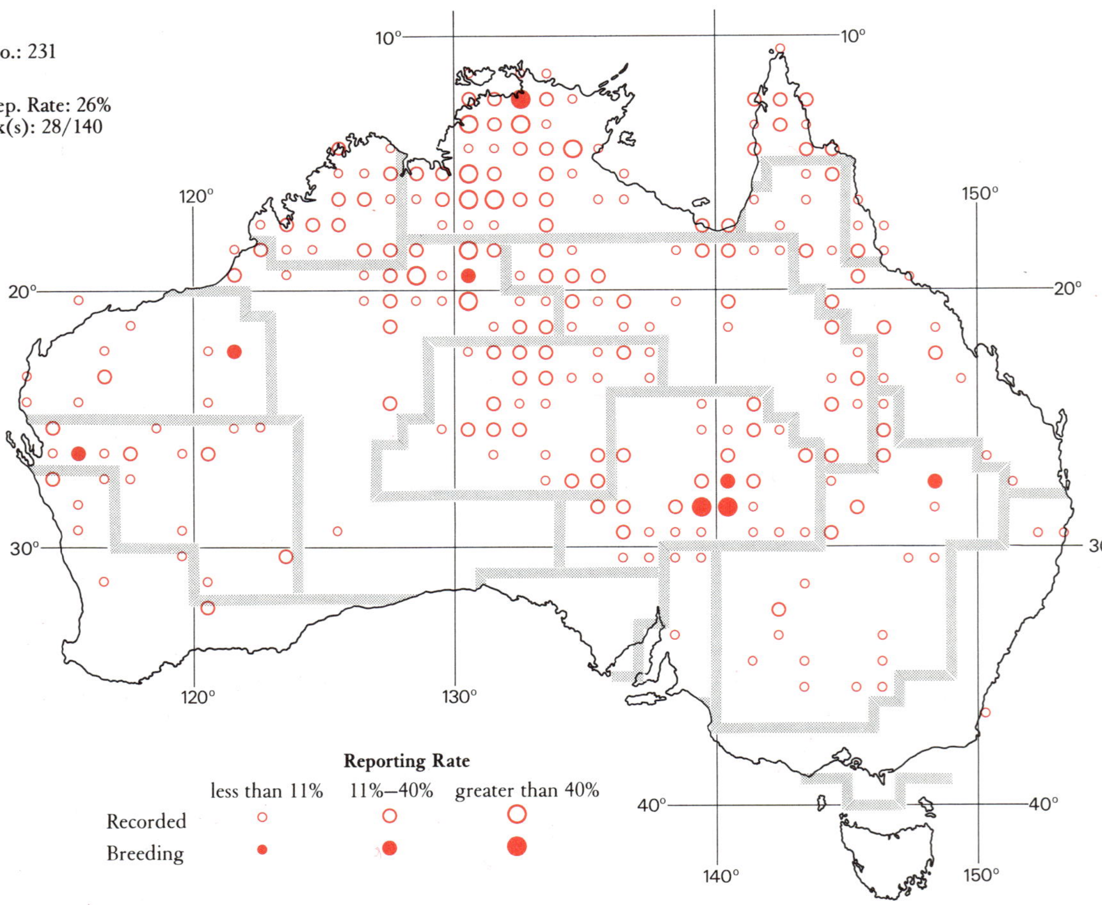

Brahminy Kite

Haliastur indus

The Brahminy Kite lives in India, south-eastern Asia, New Guinea, the Pacific islands and Australia. There are four subspecies only one of which, *girrenera*, lives in Australia.[388] The southern limits of distribution in the Field Atlas are the same in the Historical Atlases. Inclusion of historical records suggests that the species' distribution is continuous round the northern coasts.

The Brahminy Kite is a dry winter season visitor to parts of Torres Strait, numbers varying markedly from year to year.[503] Elsewhere there is no evidence of seasonal movement except for dispersal of the young from the breeding area; pairs remain near the nest throughout the year. Occasional inland records show that individuals can move far.

The Brahminy Kite lives singly or in family parties in woodland, mangroves, along beaches, estuaries, coastal lakes and on the lower reaches of rivers. In New Guinea the species ranges inland to altitudes of 1000 m feeding on lizards, possibly birds and human refuse.[148,1824] In Australia it feeds on fish and carrion and there are records of it taking insects, crabs and cephalopods.[439] It often acts as a scavenger along the shore and even along the streets of towns in northern Qld. When the tide drops it collects fish stranded in pools and shallow creeks.

There are no measures of abundance.

Field Atlas Records: 2557

1° blocks–with records: 112 (14%)

–with breeding records: 32 (29%)

Atlas No.: 227

Max. Rep. Rate: 74%
In block(s): 24/153

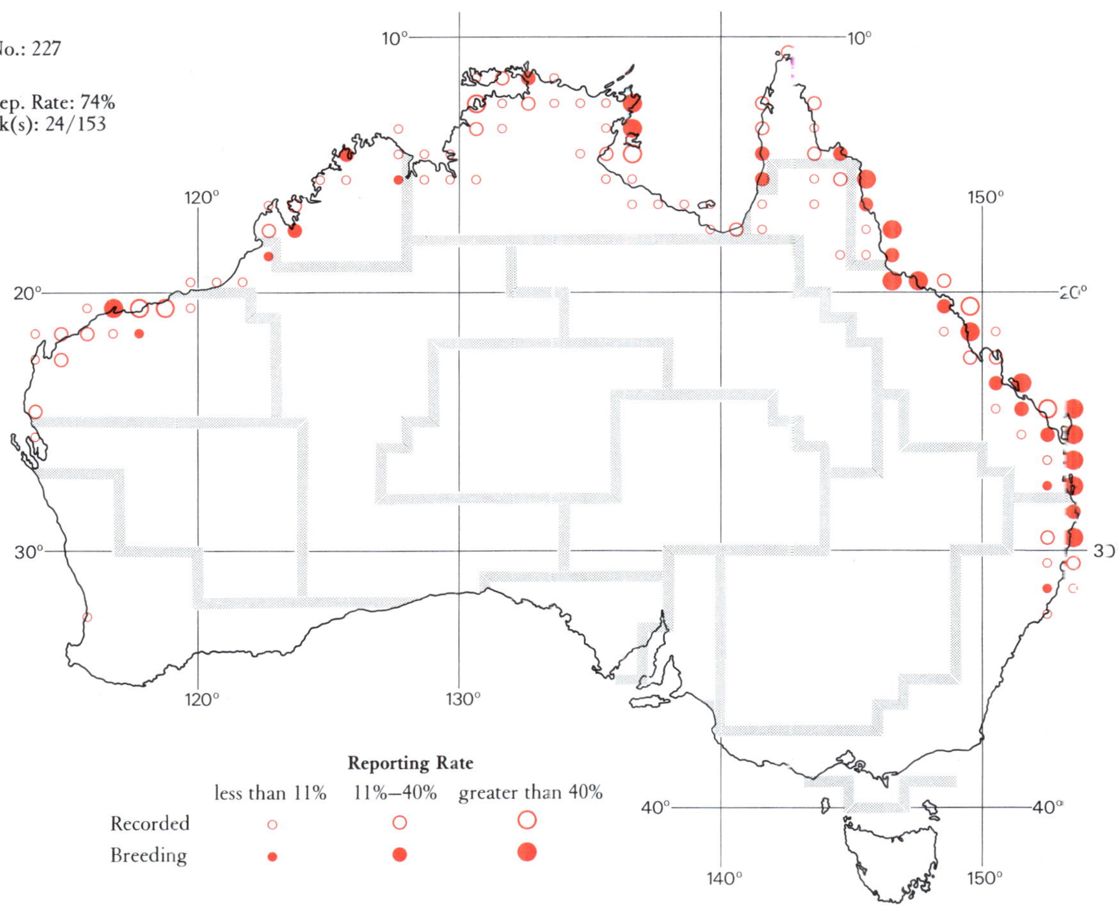

Whistling Kite

Haliastur sphenurus

The Whistling Kite lives in New Guinea, western Pacific islands and Australia. In the Tas. Region, where the first confirmed sighting was made in Nov. 1964 near Falmouth (41/148), the Whistling Kite is usually a vagrant. An early breeding report from Tas. was almost certainly erroneous[1802] but recently a pair was found resident at one site and from one to three birds were present at another for two years during the Field Atlas. Pairs have probably bred at both places intermittently.

Though the Field Atlas suggests no large-scale seasonal pattern of movement, an autumn and winter increase in numbers has been observed at several coastal sites in northern and eastern Australia. The species is a scarce winter visitor to the western islands of Torres Strait;[503] near Darwin (12/130) it is also slightly more frequent in winter than in summer;[423] an autumn and winter influx takes place round Sydney (33/151);[473] and at Laverton (37/144) numbers are greatest in autumn and least in spring and summer.[1841] Inland at Armidale (30/151) fewer are present in winter than in other seasons.[470] In the Tas. Region the only winter reports in the Field Atlas came from a site where the species was thought to be resident and breeding. The longest movement recorded is that of a banded adult from Humpty Doo (12/131) to near Bourke (30/145), 2430 km.[62]

The Whistling Kite inhabits eucalypt forest and woodland, generally near water.[439] An opportunistic species, it can breed throughout the year in response to abundant food[653] and form flocks, sometimes numbering tens of birds.[833,1099,1216] At Armidale its diet in 1980 was 64% mammal, 29% bird, 6% reptile and 1% crustacean.[470] At a study site (34/141) west of Mildura birds in the breeding season appeared to take mainly live prey[89] but the species also feeds extensively on carrion from abattoirs, roads and seashores. As with other carrion eaters, the Whistling Kite has probably benefited greatly from agriculture, farming providing carcasses and clearing making them visible.

Near Fogg Dam (12/131) three pairs nested less than 1 km apart and near Inverell (29/151) there was one active nest per 1.6 km of river.[885] In Sept. 1981 two pairs were nesting in less than 20 km of the Strzelecki Ck (28/140) floodplain (no more than 3 km wide).[1186]

Field Atlas Records: 15 785

1° blocks–with records: 639 (79%)

–with breeding records: 195 (31%)

Atlas No.: 228

Max. Rep. Rate: 78%
In block(s): 13/130

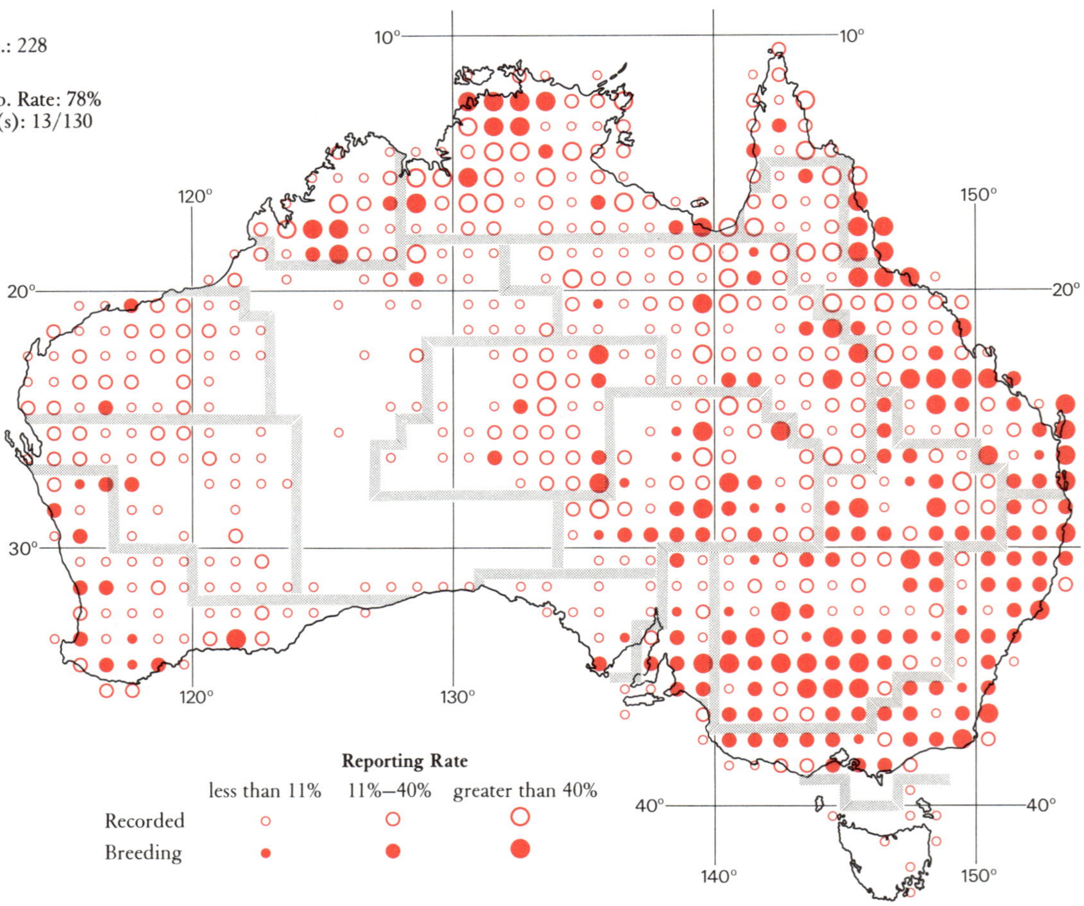

Brown Goshawk

Accipiter fasciatus

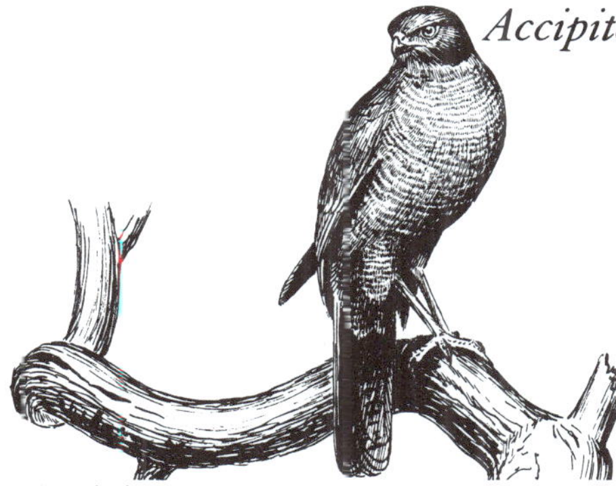

The Brown Goshawk inhabits islands of the Indian and Pacific Oceans, New Guinea and Australia. Two subspecies are recognized in Australia: *didimus* on the northern coast from about Cooktown (15/145) north and west to Derby (17/123) and *fasciatus* everywhere else. Pale Brown Goshawks are sometimes confused with the Grey Goshawk and the two species occasionally hybridize.[439]

The Brown Goshawk is a dry winter season visitor to parts of Torres Strait[503] but is seen all year in the Atherton Region.[231,698] It appears to leave the Armidale (30/151) area in winter;[470] near Sydney (33/151) it is more numerous in autumn and winter than in spring and summer.[473] The species was not recorded near Moruya (35/150) May-Aug. 1975–78.[1197] At a study site (38/144) near Werribee most birds arrive in Mar. and leave in Aug.-Sept., a small number remaining to breed.[90] In the Tas. Region banding has shown adults to be permanent residents in most nesting territories; in some the adults are absent Apr.-Aug. perhaps because of local food shortage.[1276] The Field Atlas does not suggest large-scale seasonal movement. The longest recorded is 912 km.[85]

The Brown Goshawk lives wherever there are groves of trees but is most abundant in forest and woodland. It still breeds in areas near Melbourne (37/145) where up to 80% of trees have been cleared[89] but in the South-West its population decreases markedly where clearing is extensive.[1216] The Brown Goshawk feeds on ground-living mammals, reptiles and insects; birds as large as the Australian Magpie are often eaten.[231,439,1531,1670] The staple diet throughout much of southern Australia now seems to be the rabbit. In the Tas. Region the species eats mainly small birds.[1276] The Brown Goshawk is widespread despite illegal destruction by poultry keepers and the owners of racing pigeons. It has probably benefited from the introduction of the rabbit and the partial clearing of forest

Near Armidale active nests are 1 km apart in eucalypt woodland.[470] In the southern Murray-Darling three nests were within 1100 m of each other.[439] In suitable country near Hobart (42/147) pairs often nest 1.5 km apart.[1276] Three pairs nested in a small valley about 8 km long in the South-West Region.[1670]

Field Atlas Records: 8358

1° blocks–with records: 548 (67%)
–with breeding records: 112 (20%)

Atlas No.: 221

Max. Rep. Rate: 56%
In block(s): 13/130

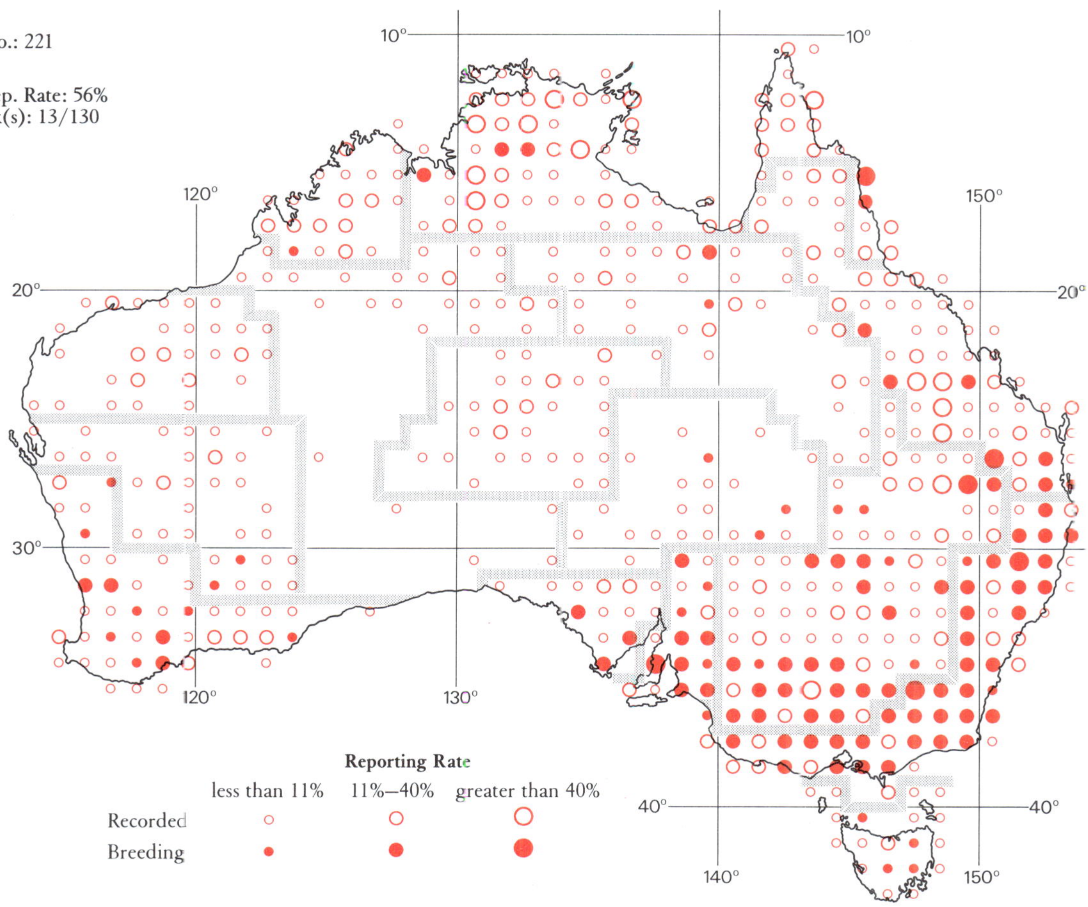

Collared Sparrowhawk

Accipiter cirrhocephalus

The Collared Sparrowhawk lives in New Guinea and Australia. Two subspecies are recognized in Australia: *quaesitandus* in the Cape York Region and coastal areas of northern Australia and *cirrhocephalus* elsewhere.[388]

The Sparrowhawk is a scarce dry winter season visitor to parts of Torres Strait.[503] In the Atherton Region it is rarely recorded and then only Mar.-May.[231] In Sydney (33/151) it is more frequently reported in summer and autumn than in winter or spring.[473] The Field Atlas does not suggest large-scale seasonal movement.

The Sparrowhawk inhabits eucalypt woodland and forest. In Tas. the species is generally uncommon but locally more abundant in forest of low height or in rainforest canopy.[1276] It feeds mainly on birds, including swift-flying lorikeets[318] and Budgerigars[439,1049] as well as nestlings. It often takes insects as well but rarely mammals and lizards. It can exhibit great skill in hunting. An atlasser at Daly Waters (16/133) reported a single bird regularly diving at flocks of feeding finches, sending them flying into the netting round a tennis court. The Sparrowhawk would then pick up a stunned finch and eat it. In the southern Murray-Darling Region the species flies low over towns in the evenings hunting sparrows.[833]

In the Top End the Sparrowhawk is considered to be widely and evenly distributed with moderate frequency throughout the forest.[1602] In the Avon Valley (31/116) the bird is said to have become scarcer since 1930 possibly as a result of clearing.[1216] Because sparrows and starlings have not been introduced the wheatlands of WA are devoid of flocks of small birds on which the Sparrowhawk can feed.

Two nesting pairs were located in a day's excursion on foot along a watercourse at Elsey (14/133).[1602] There are no other measures of abundance.

Field Atlas Records: 3462

1° blocks–with records: 470 (58%)
–with breeding records: 85 (18%)

Atlas No.: 222

Max. Rep. Rate: 43%
In block(s): 27/148

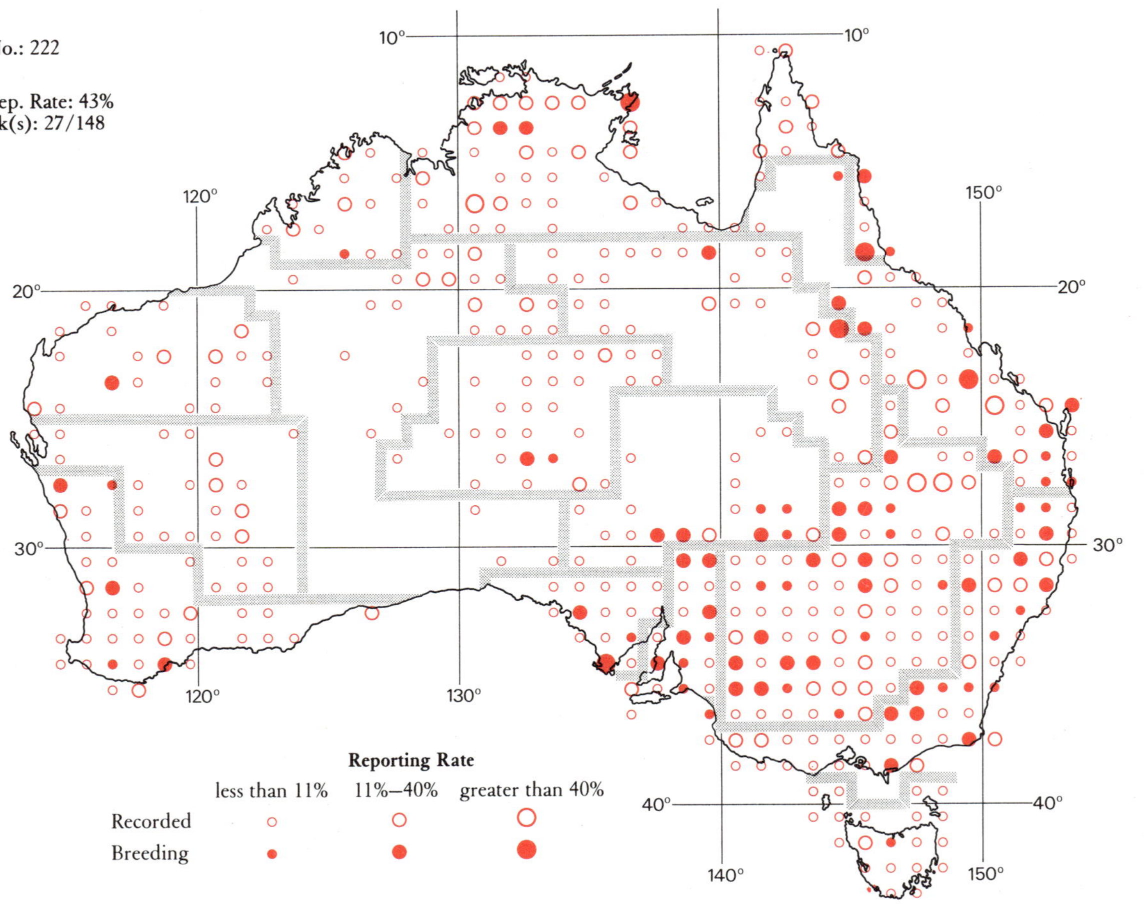

Grey Goshawk

Accipiter novaehollandiae

The Grey Goshawk lives in Australia and in New Guinea and adjacent islands. The Australian subspecies is *novaehollandiae*, which has a white or a grey plumage. In different areas the proportion of white birds to grey ones changes: white predominates in the south and north-west and grey elsewhere.[1694] Birds of each plumage interbreed but intermediates are rare.[561]

In parts of Australia pale Brown Goshawks are easily misidentified as Grey Goshawks. Few records from the coast of the Gulf of Carpentaria are unequivocal. There are sightings from the Archer R (13/141)[1147] and near Weipa (12/141) and one set of eggs collected near Burketown (17/139) in 1898 is of this species.[533] In the Pilbara Region eggs said to have been collected in 1908[1871] and a sighting in the Barlee Ra. (23/116) in 1933[1510] are unsupported by description. Possible sightings during the Field Atlas from high-rainfall country in the region (20/116, 20/117, 22/118) have been omitted though some observers maintain that they saw pure-white raptors. In the Tas. Region the Grey Goshawk has not been recorded on Flinders I. (40/148) and reports from King I. (39/144) in 1980–82 were the first since 1912. In the SA Ra. Region the species has been recorded only rarely but again confusion with Brown Goshawks makes many early sightings doubtful.[387]

The Field Atlas does not show any evidence of large-scale seasonal movement. In Tas. adults are sedentary and confined to the west; birds recorded in the east seem to be all juveniles, apparently vagrants.[1276]

The Grey Goshawk breeds in forest including the mangrove fringe in northern Australia; non-breeders may live in more open country. Prey is mainly birds and sometimes insects and carrion.[439,447,448,891] In undeveloped parts of Tas. the female takes medium to large birds such as herons, currawongs and parrots and the smaller male takes smaller ones including parrots and insectivorous birds.[1277] When available both male and female take rabbits and introduced birds; these may have enabled the Goshawk to survive in some cleared rural areas.[1276] Clearing of the forest, in which it breeds, and human persecution adversely affect the species which, though widespread, is at low density.

There are no measures of abundance.

Field Atlas Records: 1261

1° blocks–with records: 129 (16%)
–with breeding records: 15 (12%)

Atlas No.: 220

Max. Rep. Rate: 38%
In block(s): 12/143

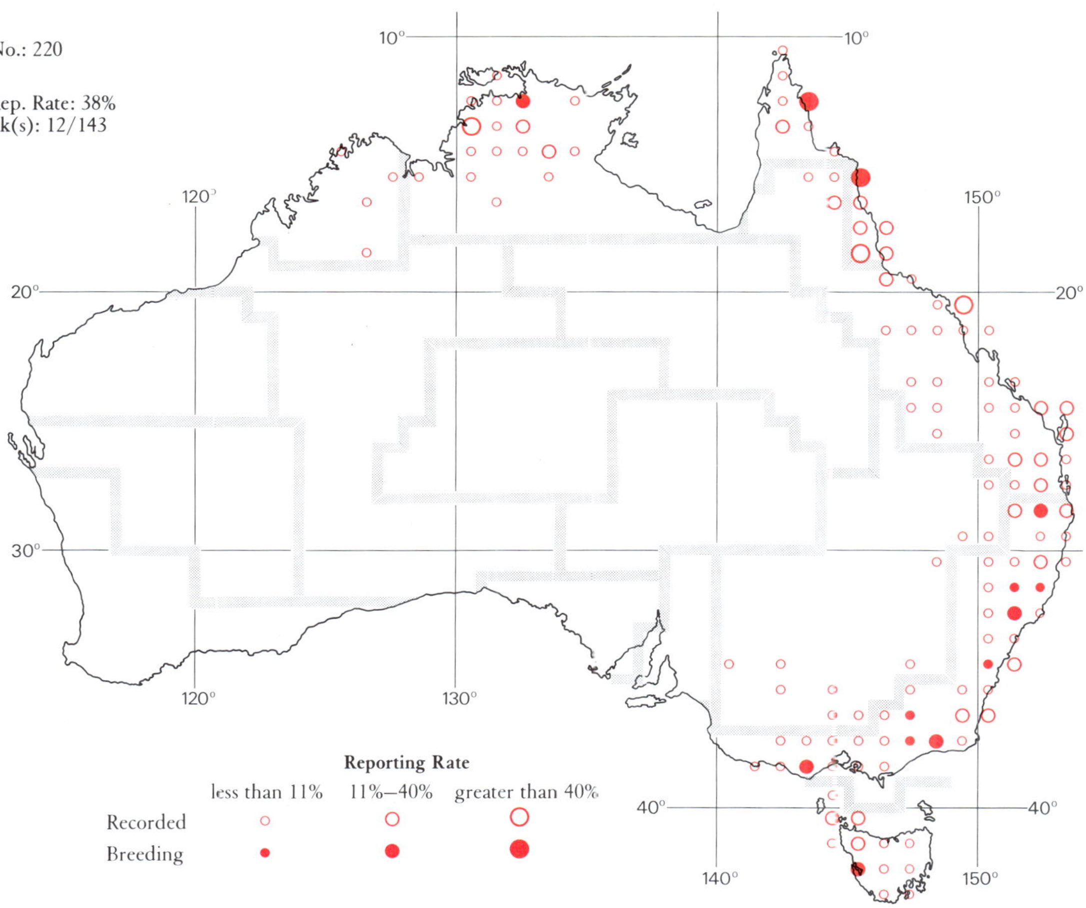

Red Goshawk

Erythrotriorchis radiatus

The endemic Red Goshawk was recorded in the Field Atlas from most of its historical range; that its distribution extends to the western Kimberley Region is shown by a sighting near Barred Ck (17/122) in Apr. 1976. In the northern Rolling Downs Region an observer saw the Goshawk several times in 1909 and it was reported at Mt Isa (20/139) in July 1958.[862,1145] These records suggest that the eastern and western populations are connected round the Gulf of Carpentaria. In the eastern Top End Region the most recent record in the Atlases is of a nest found on the Macarthur R (15/136) in 1913.[107]

The Red Goshawk was described from a painting of a bird collected soon after European settlement, possibly as far south as Sydney (33/151). Other southern records about this time were from the Manning R (31/152) and perhaps Bourke (30/145) (the locality of this record is open to doubt).[472,723] There were no other reports south of 30°S until the species was seen at Ben Lomond (30/151) and Tilbuster (30/151) in the 1960s, at Scone (32/150) in Jan. 1968 and doubtfully at Narromine (32/148) and Canberra (35/149) in Dec. 1969. The clumping of records at this time, and lack before and since, may indicate a minor irruption.[472] Breeding has not been reported in the South-East Region which, together with the irregularity of sightings, suggests that the area is peripheral to the species' main range.

Breeding reports in the Atlases have come from ten sites between Murphy's Ck (27/152)[1098] and Napier Broome Bay (14/126).[542,787] Adults have been found breeding in the same locality one or more years apart[439,542] but there is little information on the species' movements. It appears in rainforest country in the Atherton Region only when there are dry conditions to the north and west.[231]

The Red Goshawk lives mainly in forest and woodland. It takes large prey, usually birds, among which cockatoos, parrots, pigeons, ducks, kookaburras, Rufous Night Herons and crows have been recorded.[16,107,110,439,1145] The species has once been observed to take a snake and to pursue a Black-shouldered Kite.[1098]

The Red Goshawk was reported only 66 times during the Field Atlas which, given its wide range, suggests it to be at low density but there are no measures of abundance.

Field Atlas Records: 66

1° blocks–with records: 28 (3%)
–with breeding records: 2 (7%)

Atlas No.: 223

Max. Rep. Rate: 6%
In block(s): 13/142

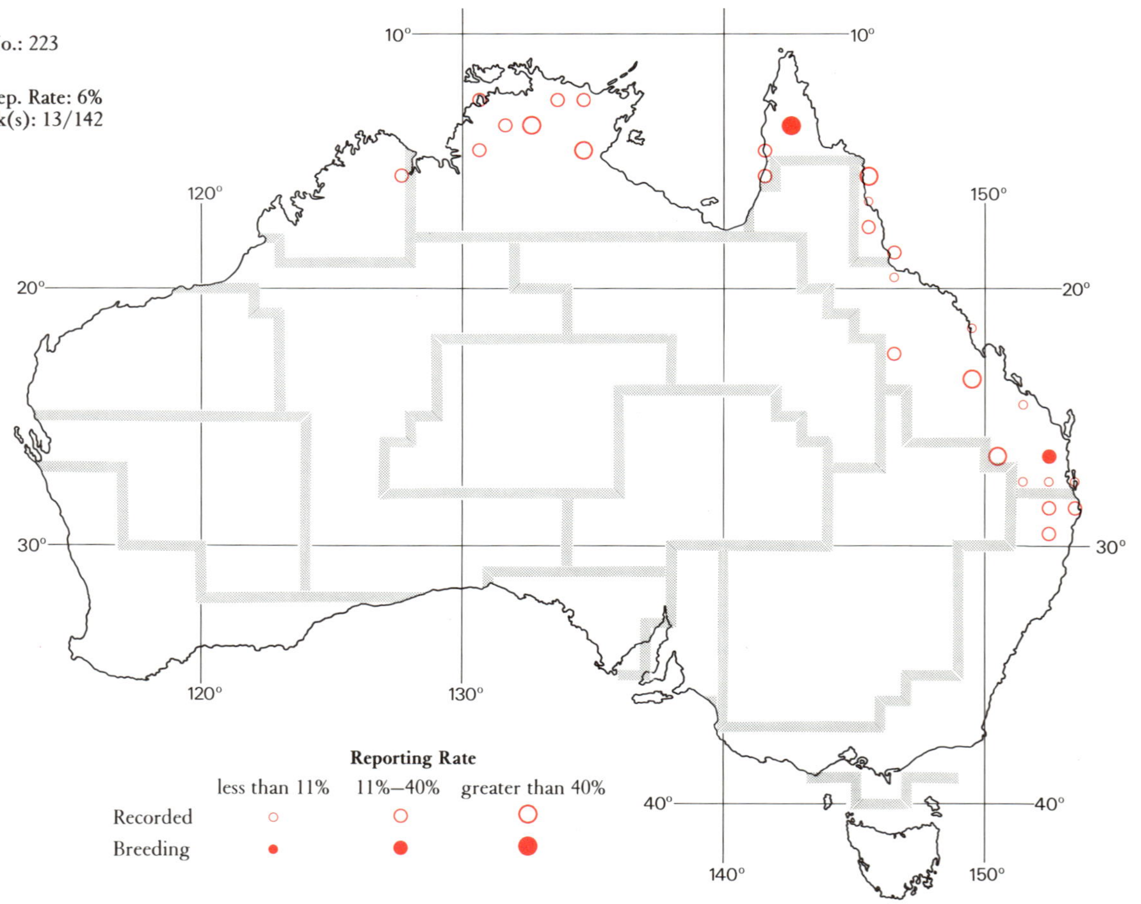

White-bellied Sea-Eagle

Haliaeetus leucogaster

The White-bellied Sea-Eagle ranges from India, south-eastern Asia and New Guinea to Australia but no subspecies are recognized.[388] Breeding pairs are sedentary in the Gippsland Lakes (37/147) area[176] and probably are usually so elsewhere[439] though some movements of adults, as well as juveniles, have been reported. At Mataranka (14/133) the Sea-Eagle was reported to arrive at the same time as bands of flying foxes on which it appeared to prey.[858] Along the Murray R both adults and young disperse when breeding is finished.[540,833] A single adult visited Cardinia Reservoir (37/145) in July of 1979, 1980 and 1981.[1768] The longest recorded movement is of a nestling from Cowell Harbour (33/136) to Fraser I. (25/153), 3000 km.[44] The Sea-Eagle lives on coasts and islands and along inland rivers and lakes. The bird has colonized and breeds on most large man-made reservoirs within its range.[176] It lives singly or in pairs or family parties. One group of seven juveniles has been recorded near Port Davey (43/145).[1276] Breeding areas are traditional, though several nest sites may be used in each area.[175,540,729]

The diet of the White-bellied Sea-Eagle varies from place to place. The species is a predator of birds including penguins, shearwaters, terns and Cape Barren Geese;[729,743,882,1474,1707] live lambs, domestic turkeys, reptiles and carrion have also been recorded.[1276] Along the Murray R at least one nest was associated with a cormorant colony.[540] On Franklin I. (32/133) an estimate of the composition of prey found at a nest site was: shearwaters (60%), bandicoots (20%) and fish (20%).[512] In the Gippsland Lakes most fish taken are carrion, live fish being caught only in shallow water or from fishermen's nets.[176] The Sea-Eagle may have increased its range inland because European carp have become abundant.[176]

In the 1940s there was a nesting pair on average every 65 km along the Murray R between Wentworth (34/141) and Echuca (36/144), measured in a direct line rather than following the river.[560] Eight birds have been seen together here.[540] There are about ten pairs in the Furneaux Group (39/148) and about 20 pairs in the Hunter Group (40/144).[1276] In the Houtman Abrolhos (28/113) ten birds have been seen in the air at once.[1707]

Field Atlas Records: 5487

1° blocks–with records: 310 (38%)
–with breeding records: 77 (25%)

Atlas No.: 226

Max. Rep. Rate: 74%
In block(s): 37/150

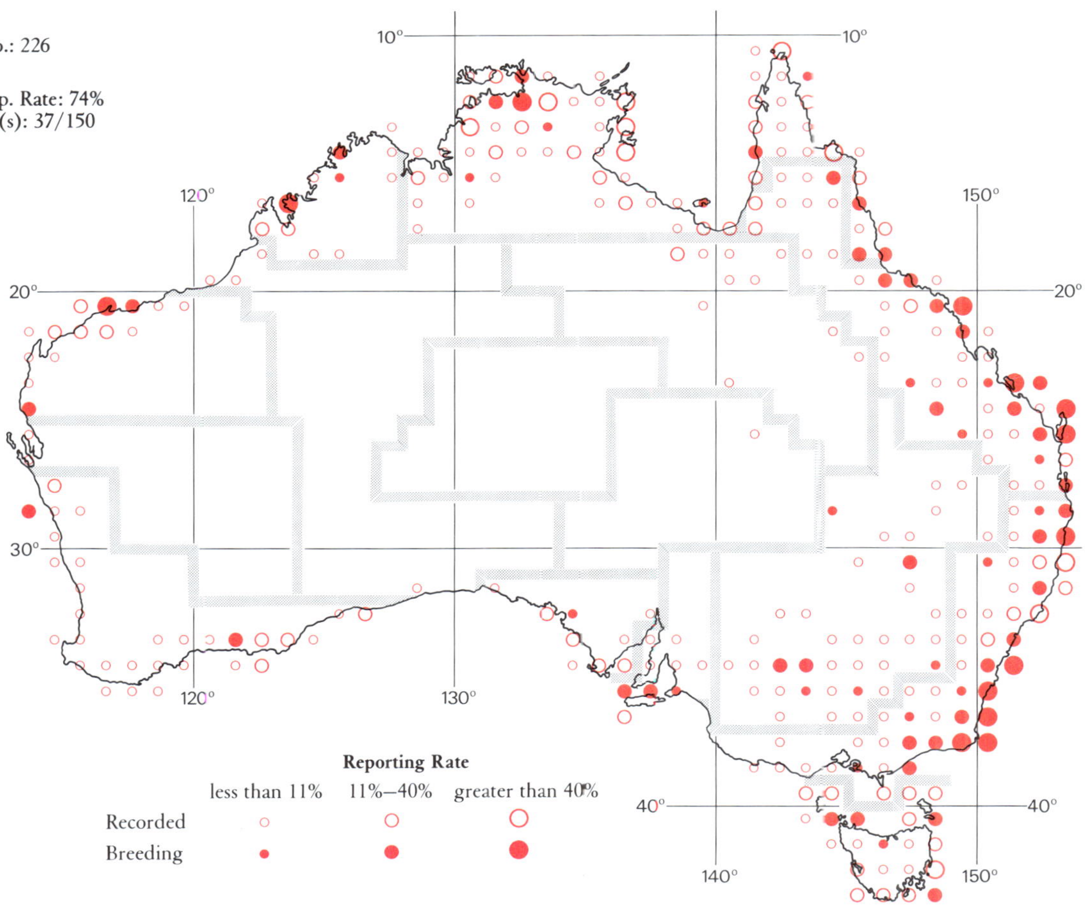

Wedge-tailed Eagle

Aquila audax

The Wedge-tailed Eagle of southern New Guinea is the same subspecies *audax* as that on mainland Australia, though there are no Torres Strait records linking the populations.[503] The other subspecies *fleayi* lives in the Tas. Region.[388] The species was not recorded on King I. (39/144) during the Field Atlas and is a rare visitor there.[734]

The Field Atlas provides no evidence of large-scale seasonal movement. Young birds disperse and may move from the inland when it dries up after a series of wet years, banding returns showing that some travel hundreds of kilometres. The longest movement recorded is 856 km.[85]

The Wedge-tailed Eagle inhabits all types of vegetation and will even hunt in coastal areas, though reports from there are less common than from the inland. It is rare in rainforest but will breed there in emergent eucalypts, travelling to more open areas to hunt. The bird lives singly or in family parties. During the Field Atlas there was a record of 15 birds congregated on a rocky outcrop north of Andamooka (30/137). The Wedge-tailed Eagle takes different and varied food in different areas.[247,1059] Where rabbits are abundant, they dominate the diet;[439] carrion is sometimes a major food; and the bird also takes kangaroos, introduced mammals, lizards, birds and sometimes lambs, usually sick or weak.[247,1059] Clearing of forests and the introduction of rabbits has probably provided more food in some places but the Wedge-tailed Eagle is sometimes killed by poison at carcasses of animals. Though in most places the bird is protected, it is sometimes shot by farmers who regard it as a predator of sheep.

Pairs occupied about 30–35 sq. km each near Armidale (30/151)[470] and near Canberra (35/149).[1059] In a year when rabbits were in plague numbers on Strzelecki Ck (28/140) two pairs of Eagles nested within 700 m of each other and farther along the creek no more than 2 km separated each of four nests from its nearest neighbour.[439] At three sites in different parts of WA the number of Wedge-tailed Eagles varied from 3–6/100 sq. km.[247]

Field Atlas Records: 19 403

1° blocks–with records: 702 (86%)
–with breeding records: 227 (32%)

Atlas No.: 224

Max. Rep. Rate: 78%
In block(s): 27/147,31/139

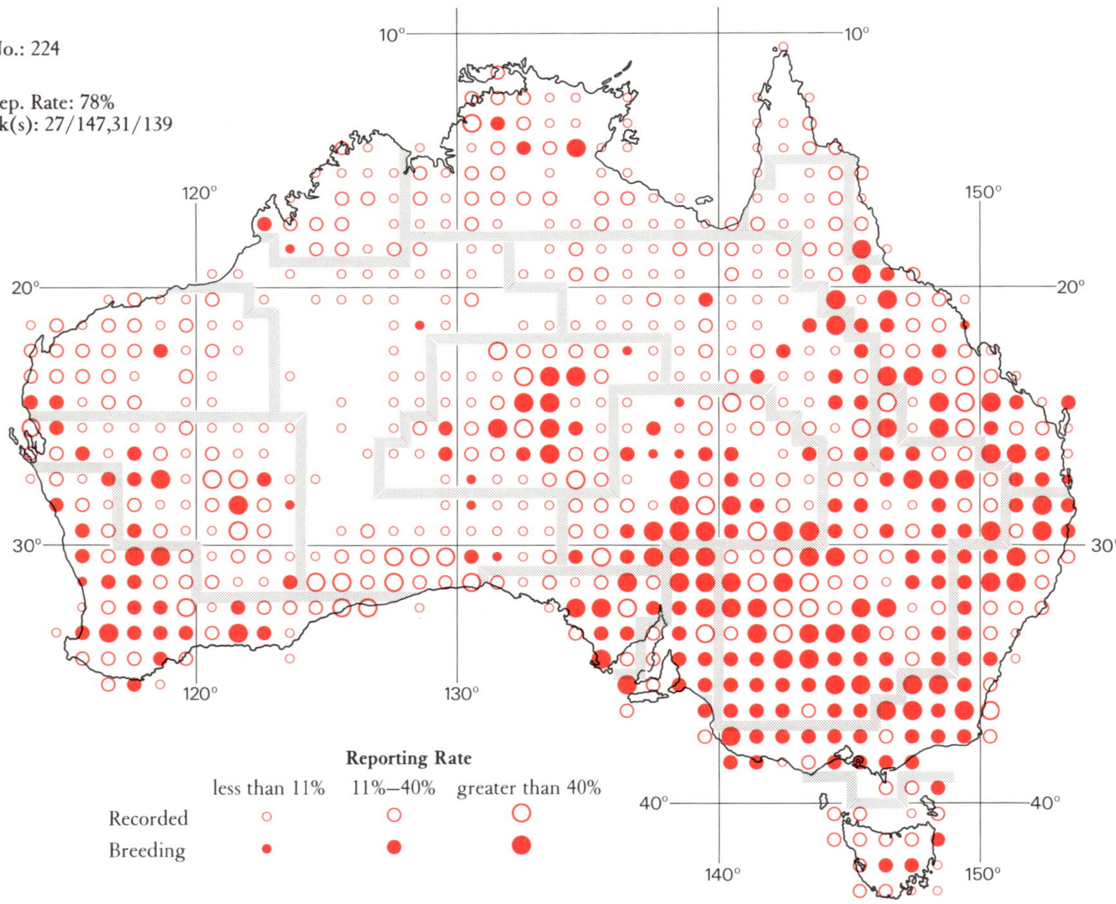

Little Eagle

Hieraaetus morphnoides

The Little Eagle lives in New Guinea and Australia. The subspecies *morphnoides* breeds in Australia. There are no records from Torres Strait or the northern tip of the Cape York Region and the species has never been reported from the Tas. Region.

The Little Eagle is present all year in some districts[562,833] though there is evidence that some birds wander.[439,1128,1216] At Warren (31/147) the number of Little Eagles increased from one to 32 in six months during a plague of house mice.[836] The Little Eagle is a winter visitor to Rockhampton (23/150)[1089] and Sydney (33/151);[473,1521] and a partial movement from the Armidale (30/151) area was noted in winter 1980.[470] The Field Atlas provides no evidence of large-scale seasonal movement.

The Little Eagle lives singly or in family parties where there are tall trees to use as lookouts. It hunts by soaring and will hawk to catch and eat prey on the wing. Prey is usually taken live but the bird is known sometimes to take carrion. The Little Eagle prefers young rabbits under 650 g in weight but, when they are not available, will eat adults and other small mammals. Reptiles were rarely recorded in the diet of birds studied in Vic.[562] but formed 15% of the items taken near Armidale.[470] The remainder was 60% rabbits and 25% birds. The diet of one pair watched at a nest near Duaringa (23/149) was mostly reptiles.[1343] One observer reported a Little Eagle carrying a goanna.[748] Near Brewarrina (29/146) six Little Eagles were watched hawking low for grasshoppers and eating them on the wing within an area of 100 sq. m.[1724]

At Armidale in 1980 Little Eagles bred at a density of one pair/1600 ha.[470] At a study site (34/141) west of Mildura five pairs bred in 134 sq. km (one pair/2860 ha) in 1980–81.[90] Three pairs of Little Eagles nested in less than 20 km of the Strzelecki Ck floodplain (28/140) (no more than 3 km wide) in Sept. 1981.[1186]

Field Atlas Records: 5799

1° blocks–with records: 524 (65%)
–with breeding records: 93 (18%)

Atlas No.: 225

Max. Rep. Rate: 41%
In block(s): 25/148

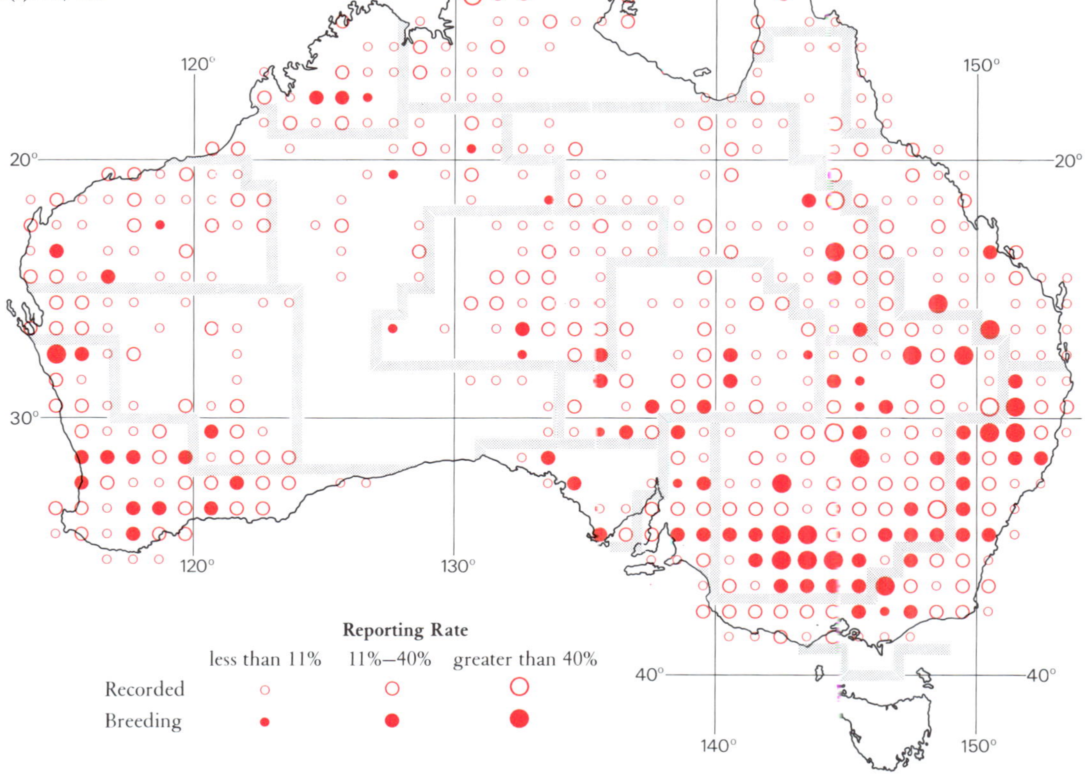

Spotted Harrier

Circus assimilis

The Spotted Harrier lives in the Lesser Sunda Is (9S/120E) and Sulawesi (2S/121E) as well as in Australia. It rarely reaches the Tas. Region, the Field Atlas record near Elizabeth Town (41/146) in Feb. 1977 being only the second published sighting.[734]

Observations suggest that the Spotted Harrier tends to be farther south in summer and north in winter though records were made throughout its range in all seasons in the Field Atlas and Reporting Rates south of 30°S did not differ greatly between seasons (summer 3.4%, winter 2.7%). A few Harriers visit Darwin (12/130) in winter[423] (Reporting Rates: summer 1.0%, winter 3.3%). In the Atherton Region it is an irregular visitor throughout the year.[231] It is a scarce summer migrant near Armidale (30/151).[470] In the southern Murray-Darling Region some birds are present all year but the number increases Oct.-Dec. over wheatfields and Feb.-Apr. over rice crops.[833] At Booligal (33/144) it was considered migratory, arriving at the end of Aug. and departing in Feb.[1343] The pattern at a study site (34/141) west of Mildura was similar but the number of birds arriving was very variable and in one year none came.[90] In the Avon R valley (31/116) single birds arrive each summer and establish a hunting area, traverse it regularly each day, then disappear about May.[1216] In arid areas, such as at Wanjarri (27/120), the Spotted Harrier will appear after summer rain.[1282] The longest movement recorded is that of an immature from Calomba (34/138) to L. Clarendon (27/152), 1650 km.[68]

The Spotted Harrier lives singly or occasionally in small groups and ranges over plains and open wooded grassland. It hunts low, usually taking its prey from the ground. Large insects, birds, reptiles and mammals as big as bandicoots and rabbits are taken. The Spotted Harrier gathers where food is abundant and sometimes stays to breed.[439,1282]

In the Atherton Region the Spotted Harrier was estimated to be at a density of 1 bird/25 sq. km.[231] In the Murray-Darling Region 17 pairs defended territories in 134 sq. km (0.0030 birds/ha) in 1980 but only two pairs were present in 1981 (0.0003 birds/ha).[90]

Field Atlas Records: 3587

1° blocks–with records: 544 (67%)
–with breeding records: 55 (10%)

Atlas No.: 218

Max. Rep. Rate: 39%
In block(s): 21/122

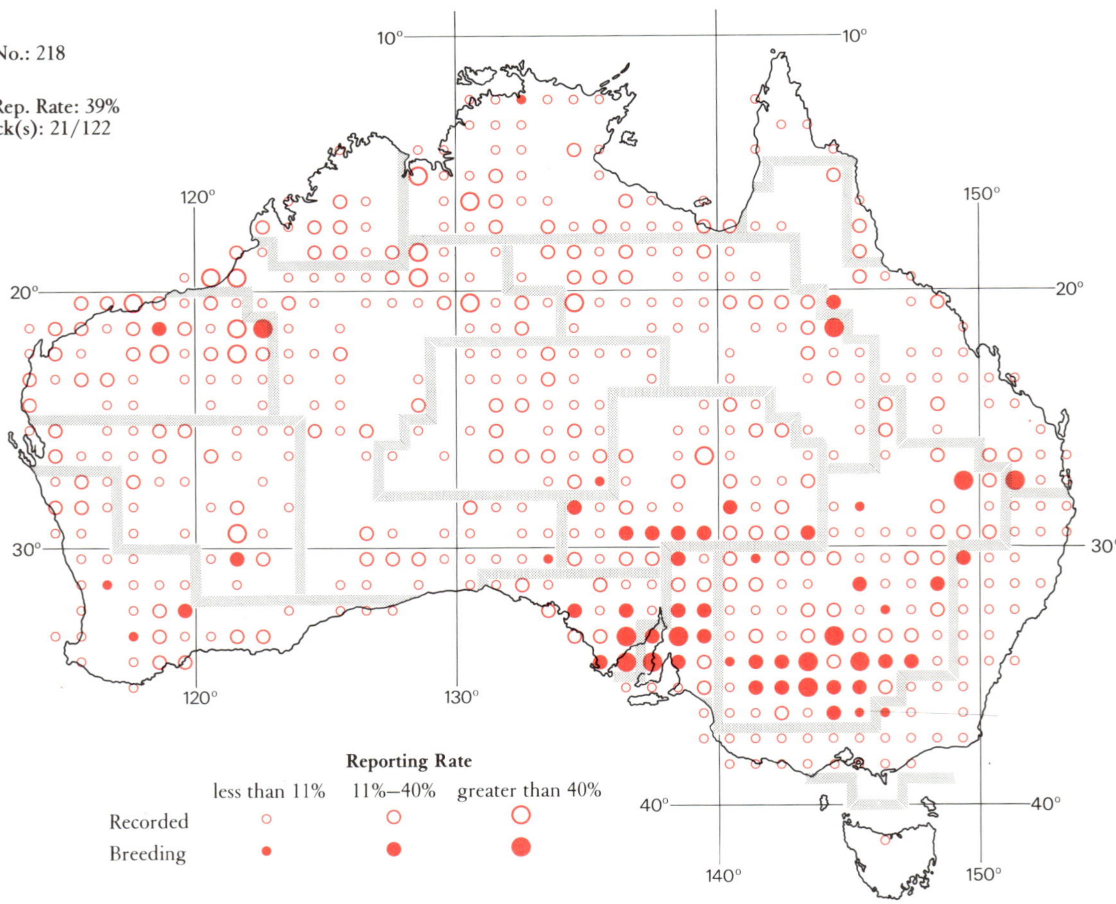

Marsh Harrier

Circus aeruginosus

The Marsh Harrier is cosmopolitan in distribution with the exception of the American continent. The subspecies *gouldi* breeds in Australia. In eastern Australia all but five breeding sites in the Atlases are south of 33°S and in the west all are south of 31°S. This means that most Marsh Harriers in northern Australia migrate there from the south. The species is a winter visitor to New Guinea,[1449] where there are also resident populations, and to Torres Strait[503] and the Top End.[423] Near Sydney (33/151) autumn and winter sightings exceed those of spring and summer by 2–2.5 times.[473] Spring and autumn movements have been reported in the southern Murray-Darling Region[832] and many birds cross Bass Strait to breed in summer in Tas.[1654] (Reporting Rates, Tas. Region: summer 34%, winter 14%). The winter rate is inflated by early arrivals and late departures and few birds actually remain in Tas. during June-July. In late Feb. 1953 about 200 were recorded crossing Bass Strait by an observer on a three-hour sea passage[832] and birds banded in Tas. disperse to the mainland. Six of 78 birds banded at 38/144 near Werribee were seen in Tas.[90,1442] and a further 15 seen again more than 10 km west or south-west of Werribee, one as far away as Bool Lagoon (37/140). The longest movement recorded is that of an adult from Antill Ponds (42/147) to Chatsworth (29/153), 1520 km.[826] Movement at times is in response to the changing availability of food and hunting areas such as hay and grain crops.[89]

The Marsh Harrier inhabits open swampy country, including pasture and crops, and in the inland lives near bores.[439] It preys on native and introduced birds and mammals, frogs, reptiles, large insects and carrion.[439] In the Tas. Region rabbits are a staple item of diet.[1276]

Near Werribee density of Marsh Harriers in favoured areas of long grass was 0.008–0.017 birds/ha in autumn and winter; only one pair remained to breed.[90] The Field Atlas indicates that the Tas. Region is the stronghold of the Marsh Harrier: 26% of records came from the region, despite its small area (2.7% of 1° blocks in Australia) and the absence of the birds in winter. At one site where breeding density was high nests were placed every 2 km along a creek.[1276]

Field Atlas Records: 6564

1° blocks–with records: 283 (35%)

–with breeding records: 38 (13%)

Atlas No.: 219

Max. Rep. Rate: 46%
In block(s): 40/147

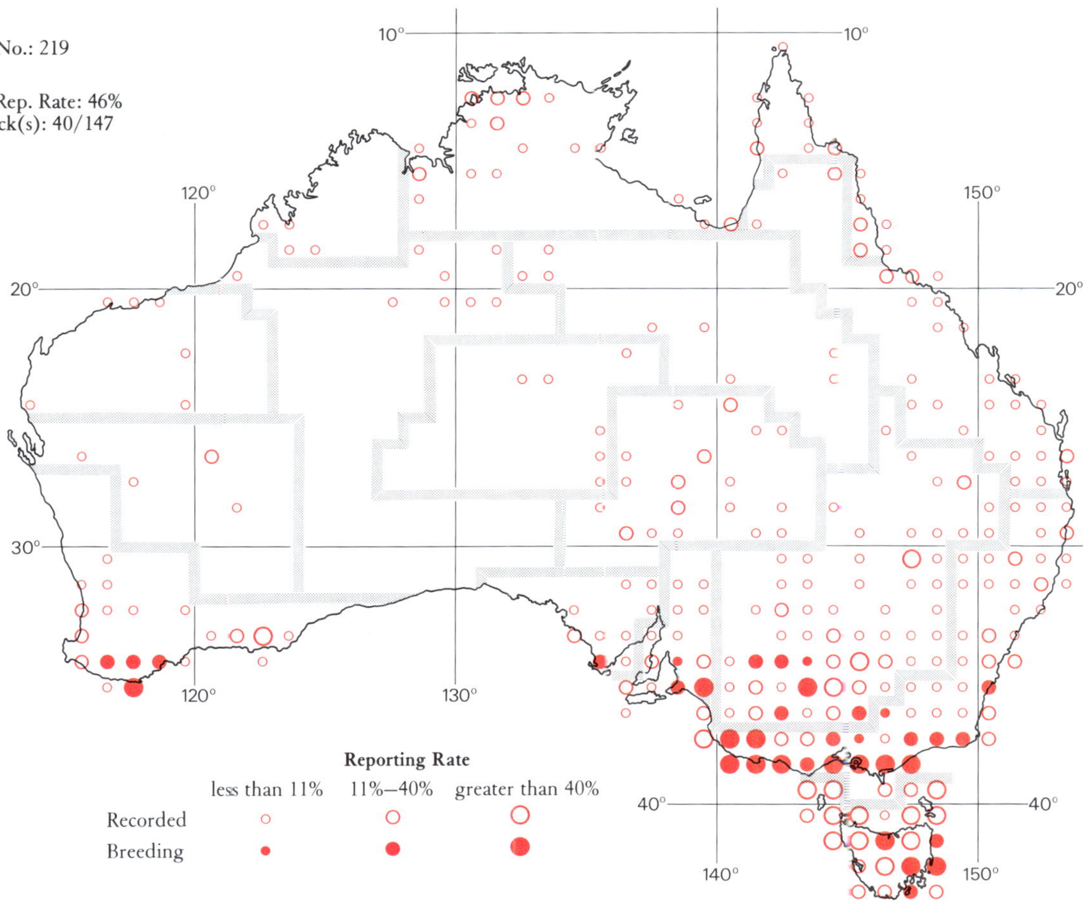

Black Falcon

Falco subniger

The endemic Black Falcon is a vagrant to the Pilbara, Goldfields and South-West Regions and there is only a single record, unconfirmed, from the Tas. Region.[734] Breeding is almost confined to the Murray-Darling, L. Eyre and SA Ra. Regions which together contain 81% of the 1° blocks where breeding has been reported in the Atlases. There are no confirmed breeding reports at all from the Top End and Kimberley Regions. References by one observer to breeding at L. Way (26/120) in the Goldfields Region last century and on the Fitzroy R (17/123) in the Kimberley Region[1343] may be errors because there are no later records.

The western records show that the Black Falcon can move far. Though the species is often thought to be nomadic and irruptive[439] there are several observations of regular movement. In the Mossgiel (33/144) district it was regarded as a migrant arriving in Sept. to feed on quail and departing about Feb.[1343] Records at Armidale (30/151) 1979–82, though few, were all spring-autumn.[470] In the southern SA Ra. Region the Black Falcon appears each year in late spring and at Bool Lagoon (37/140) in Jan. Peak numbers in both areas are in Mar. but by late winter the species is usually absent.[911] The Field Atlas also suggests a seasonal pattern of movement, at least on the southern edges of the species' range: in the South-East Region Reporting Rates were 0.4% in winter and 2.2% in summer. In the Darwin (12/130) area a small irruption took place in 1977, perhaps resulting from the conditions that caused Letter-winged Kites to arrive at the same time.[1755]

The Black Falcon inhabits eucalypt woodland, mostly breeding along watercourses. It lives singly or in pairs, feeding mainly on birds taken in the air. Its diet during the breeding season at a study site (34/141) west of Mildura was 57% birds, 40% rabbits and 3% house mice (data from 335 prey items).[90] It will take pipits and button-quail disturbed by tractors during farming.[911] There is one observation, near Burra (33/138), of six Black Falcons together feeding on locusts, which were in plague numbers;[639] and twice the Falcon has been recorded eating carrion.[640,1881]

On the Strzelecki Ck (28/140) in Sept. 1981 four pairs of Black Falcons nested in less than 20 km of the floodplain (no more than 3 km wide).[1186]

Field Atlas Records: 1758

1° blocks–with records: 305 (38%)
–with breeding records: 16 (5%)

Atlas No.: 238

Max. Rep. Rate: 21%
In block(s): 28/140

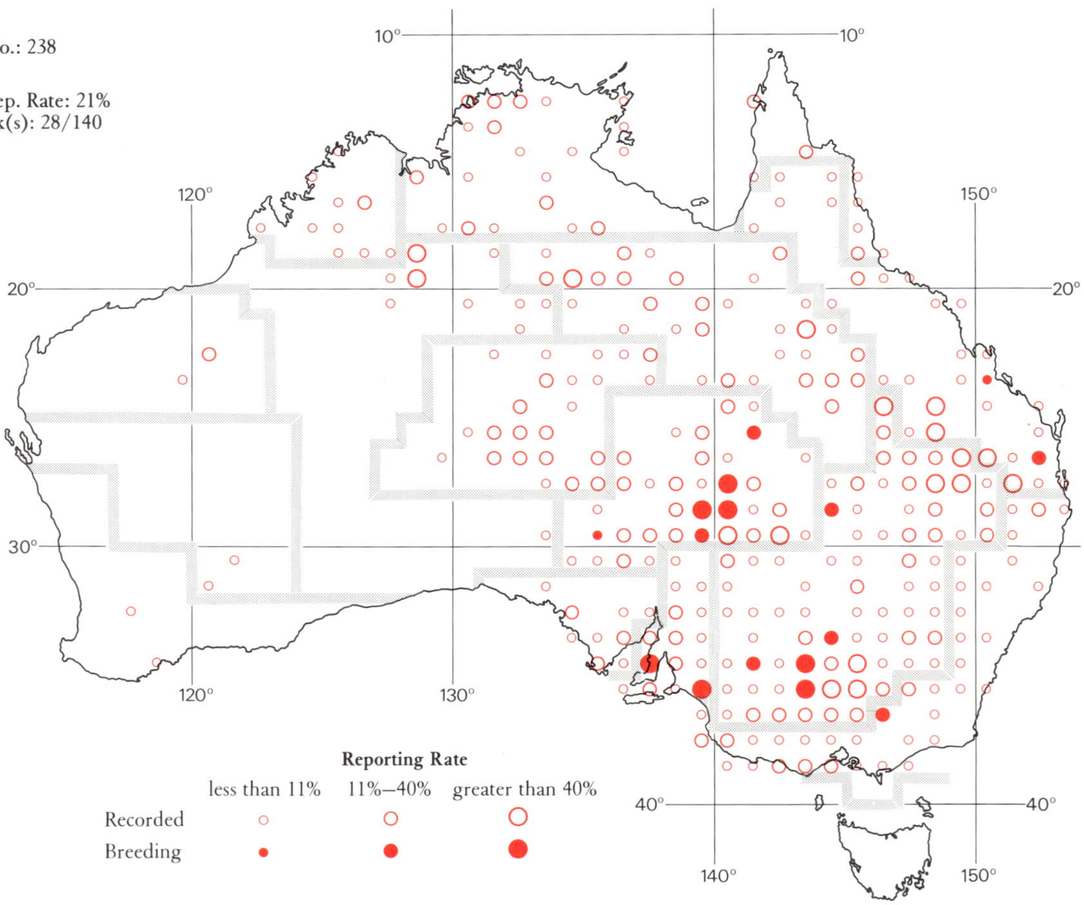

Peregrine Falcon

Falco peregrinus

The Peregrine Falcon is almost cosmopolitan. In Australia two subspecies are recognized: *submelanogenys* in the South-West and *macropus* elsewhere. No distributional gap is evident in the Atlases. Breeding records are virtually absent from the Rolling Downs, L. Eyre, Western Deserts and Pilbara Regions. These together contain less than 2% of the 1° blocks in which breeding has been reported in the Atlases.

The Peregrine Falcon is generally sedentary[439,960] though reported to be a dry winter season visitor in small numbers to Torres Strait.[503] In Vic. young Peregrines apparently disperse randomly.[525] Banding does not suggest interchange of birds between Vic. and Tas. though individuals have been seen at sea in Bass Strait.[1276] In Tas. adults are resident all year in good sites but may be absent during winter from poorer breeding sites.[1276] The Field Atlas does not suggest any large-scale seasonal movement.

The Peregrine Falcon inhabits inland tree-lined watercourses, woodland, pasture, swamps and eucalypt forest near the coast. The Peregrine's reproductive capacity is reduced by ingestion of pesticide residues concentrated in the food chain.[1450] Since the introduction of chlorinated hydrocarbons in the 1940s there has been a 20.4% reduction in egg-shell thickness in the Vic. population and pesticides are now present in all eggs. In 1976–77, 99% of the prey items in 73 eyries in Vic. were birds[1434] and most of the common prey analysed contained DDE residues.[1435] The residue level is high in two species of prey, the Feral Pigeon and Common Starling, which compose respectively 20% and 16% of the food.[525,1434,1435,1845] A study in Tas. found the Common Starling to constitute 60% of prey items; small parrots, waterbirds, cuckoos and honeyeaters accounted for 20% and introduced pigeons, 10%; seabirds were important on the coast.[1276]

The Peregrine in Vic. is at a density of one pair to 650–750 sq. km, about the same as in Europe before the use of chlorinated hydrocarbons.[525,1434,1435,1845] In Tas. there are an estimated 80 breeding pairs, on average 15–20 km apart.[1276] There are estimated to be 3000–5000 pairs of Peregrines in Australia and perhaps only 12 000–18 000 pairs in the world, making Australia one of the strongholds of the species.[290]

Field Atlas Records: 2719

1° blocks–with records: 365 (45%)
–with breeding records: 73 (20%)

Atlas No.: 237

Max. Rep. Rate: 24%
In block(s): 24/153

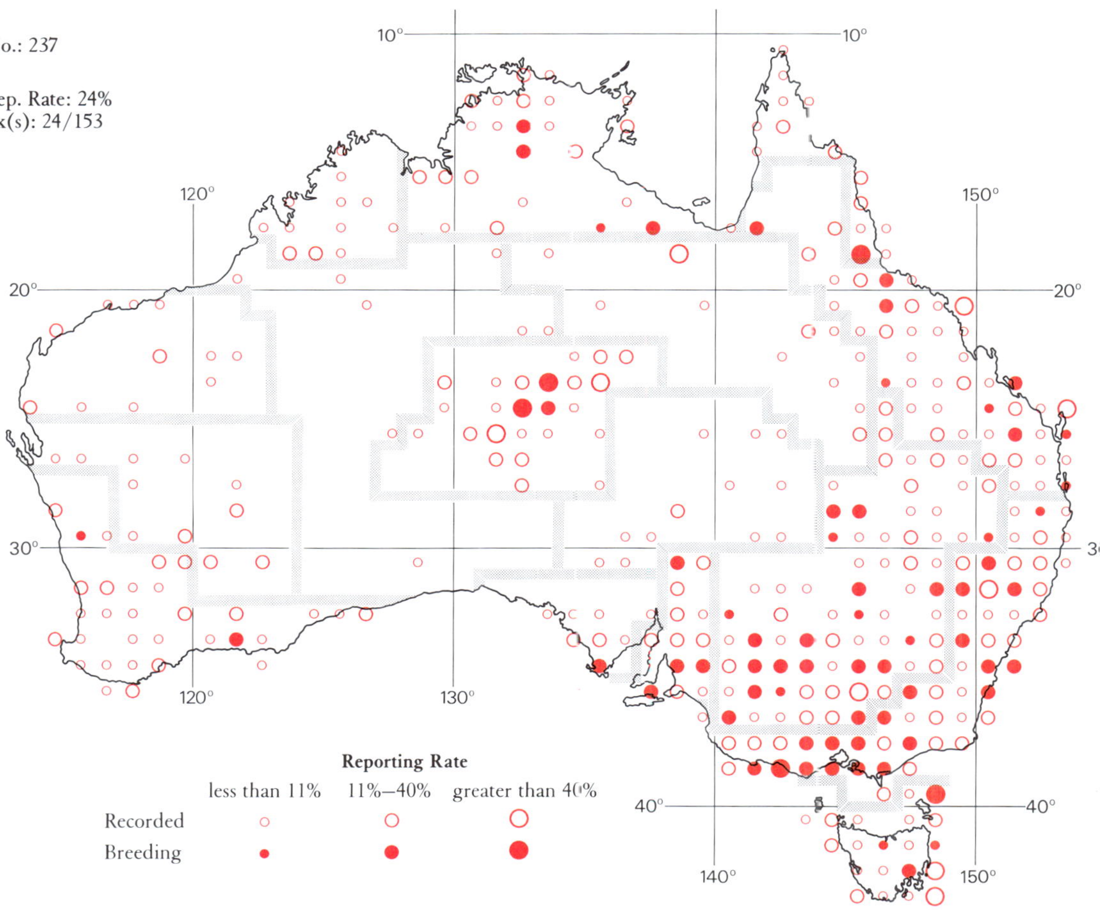

Australian Hobby

Falco longipennis

The Australian Hobby lives in Australia and parts of Indonesia and visits New Guinea. In Australia two subspecies are recognized: *murchisonianus* in the drier parts extending throughout northern Australia; and *longipennis* in southern Australia.[388] There is no gap between them. In the Tas. Region the 10′ maps show the Hobby to be scarce except on Flinders I. (40/148) where it is regularly reported. There is no record from King I. (39/144) and only two breeding reports from the Tas. Region before the Field Atlas.

Small numbers visit Torres Strait in the dry winter season[503] and the Hobby is recorded in New Guinea only as a rare winter visitor.[1449] In the Atherton Region an observer on the tableland regards it as present all year[231] and another on the coast as present only June-Oct.[698] At Inverell (29/151) and in the southern Murray-Darling Region the Hobby is present all year[96,833] but in the high country at Thredbo (36/148) it is seen only on passage in summer and autumn.[682] In Tas. there are a few resident pairs, the numbers augmented by vagrants;[1276] Reporting Rates show that the Hobby was recorded there most often in spring (0.6%), summer (0.8%) and autumn (1.1%) and least often in winter (0.2%). At Northam (31/116), where it is rare, most sightings were in autumn and of single birds.[1216] The longest movement recorded is that of a bird banded as a nestling in Canberra (35/149) to Brisbane (27/153), 1000 km.[64]

The Australian Hobby inhabits eucalypt woodland and often suburban parks and gardens.[439] It is a predator of small birds and able to live wherever these are in any number. The number of Hobbies in Tas. appears to have increased recently possibly because drought has caused more migrant passerines to remain during winter.[1323] The Hobby sometimes enters towns at dusk to catch sparrows on their way to roost[833] and may visit fowlyards and zoological gardens to catch pigeons and doves attracted to spilt grain.[1638] It catches bats[449] and will also gather at flights of insects, e.g. the nuptial flights of termites[1807] and plagues of grasshoppers,[107] both of which it catches on the wing.

There are no measures of abundance.

Field Atlas Records: 4944

1° blocks–with records: 520 (64%)

–with breeding records: 60 (12%)

Atlas No.: 235

Max. Rep. Rate: 30%
In block(s): 26/150

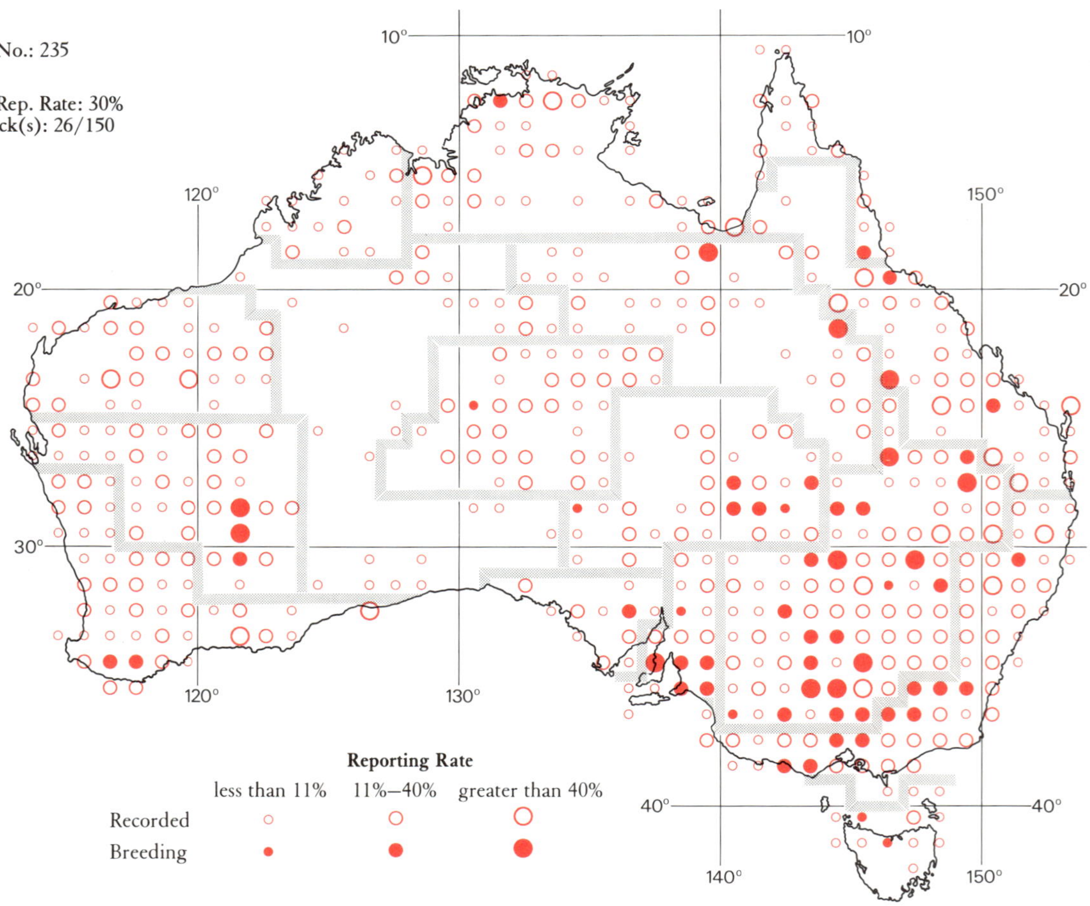

Grey Falcon

Falco hypoleucos

The Grey Falcon breeds only in Australia. In Nov. 1980 it was recorded for the first time in New Guinea.[552,1722] In Australia, of the 1° blocks in which breeding is reported in the Atlases, 67% are in the L. Eyre and Murray-Darling Regions. The remainder are scattered throughout the species' range in the Eastern Qld and Kimberley Regions, as well as the Central Ra., SA Ra. and Eyre Pen. Regions. In the Goldfields and South-West Regions, where only two sightings were made during the Field Atlas, the bird bred at Carnegie (25/122) in 1973 and near Perth (31/115) in 1929.

The Grey Falcon is rarely seen. Outside its main inland range reports are intermittent and years apart. The paucity of Field Atlas records in the Eastern Qld, South-East, Goldfields and South-West Regions may therefore represent the short period over which observations were made rather than a decline of the species.

Pairs are known to be resident at several sites in the L. Eyre and Murray-Darling Regions[833,1159,1292] though the species is stated by other observers to be largely nomadic.[439] The Field Atlas suggests no seasonal pattern in the appearance of birds at the periphery of its range.

The Grey Falcon lives in acacia scrub, spinifex and tussock grassland.[439,1162] The birds are usually in pairs; more than two are seen together only for a short time after breeding if the young remain near the adults. The Grey Falcon eats mainly birds and sometimes small mammals and reptiles, sweeping low over open country, hunting at pools[1369] or perching in trees to wait for prey. It also feeds on plagues of grasshoppers.[833,1162]

No more than one pair of Grey Falcons has been recorded to breed on the Strzelecki Ck (28/140).[89]

Field Atlas Records: 160

1° blocks–with records: 99 (12%)
–with breeding records: 6 (6%)

Atlas No.: 236

Max. Rep. Rate: 10%
In block(s): 28/140

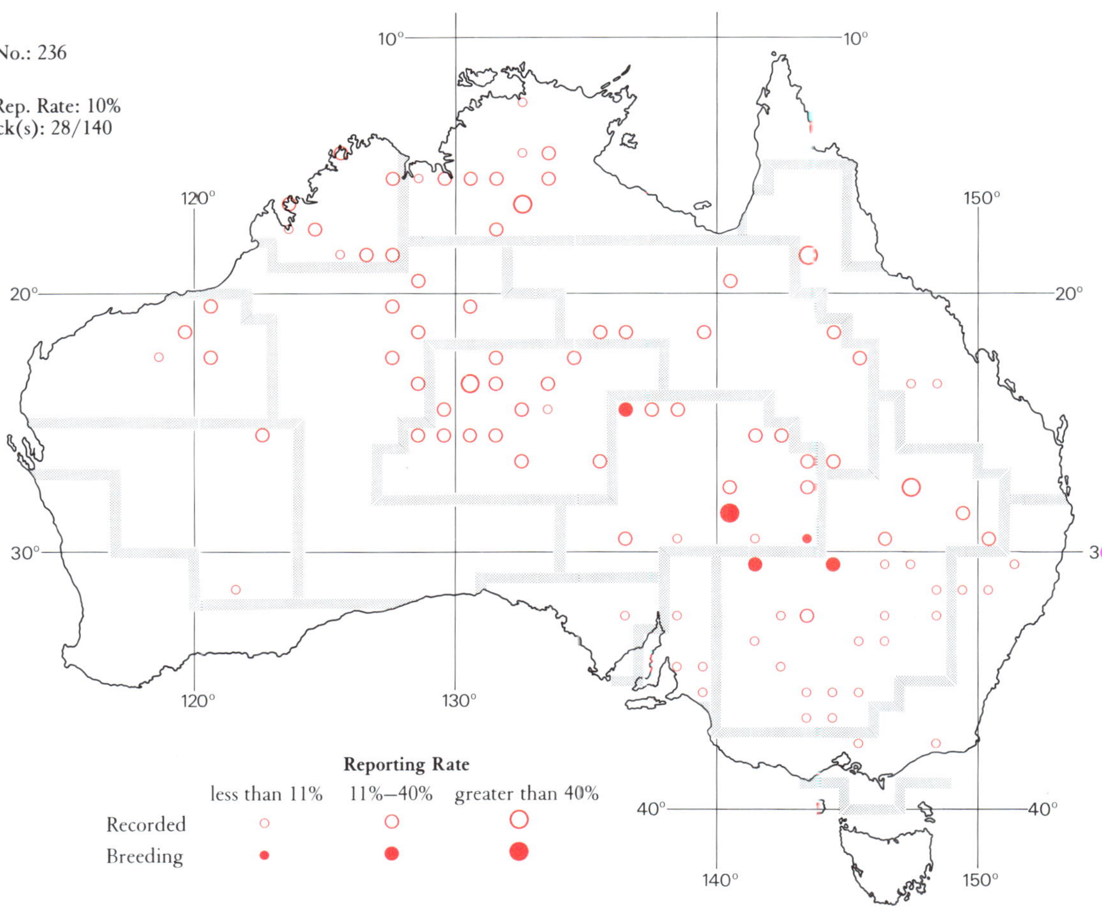

Brown Falcon

Falco berigora

The Brown Falcon lives in New Guinea and Australia. Views differ about subspecies in Australia.[385,388,748] Light, dark and red plumages exist, often in the same locality, and may indicate polymorphism of a single population. The Brown Falcon was the species most widely reported during the Field Atlas. Breeding is recorded historically in both the Cape York and Kimberley Regions though not during the Field Atlas.

Adult pairs of the Brown Falcon seem to be rather sedentary.[1276,1360] In the Tas. Region adult territories may shift a short distance during winter; juveniles disperse soon after fledging.[1276] Though there are observations of seasonal movement, no large-scale pattern is suggested by the Field Atlas. The Brown Falcon is commoner at Darwin (12/130) in winter than in summer.[423] It is a scarce winter visitor to Torres Strait[503] and to the coast of the Atherton Region.[698] On the Atherton (17/145) tableland it is present Feb.-Dec. but numbers fluctuate from year to year.[231] Numbers were reduced near Armidale (30/151) in winter 1980.[470] Near Sydney (33/151) and Griffith (34/146) the species is recorded much more often in autumn-winter than in spring-summer[473,1360] but at Laverton (37/144) it tends to build up to large numbers in summer.[1841] In the southern Murray-Darling Region numbers fluctuate but not seasonally;[833] similarly at Northam (31/116) where influxes are related to dry periods.[1216] The longest movement recorded is that of a banded immature from Calomba (34/138) to Streaky Bay (32/134), 410 km.[1436]

The Brown Falcon is absent only from rainforest and some eucalypt forest. It perches on exposed branches and fence posts, often near roads, which makes it conspicuous. It is a highly opportunistic feeder: in Tas. the diet of adults changed according to the season from 50% mammals in winter (birds, reptiles, insects and carrion making up the remainder) to 50% insects in summer-autumn.[1276] Immatures may gather in loose flocks of up to 40 to feed on plagues of grasshoppers and crickets.[70,1276] House mice are taken in preference to rabbits;[291] and crustaceans and frogs have also been recorded.

Six pairs of Brown Falcons nested in less than 20 km of Strzelecki Ck (28/140) floodplain (no more than 3 km wide) in Sept. 1981.[1186]

Field Atlas Records: 20 167

1° blocks–with records: 769 (95%)
–with breeding records: 167 (22%)

Atlas No.: 239

Max. Rep. Rate: 63%
In block(s): 35/145

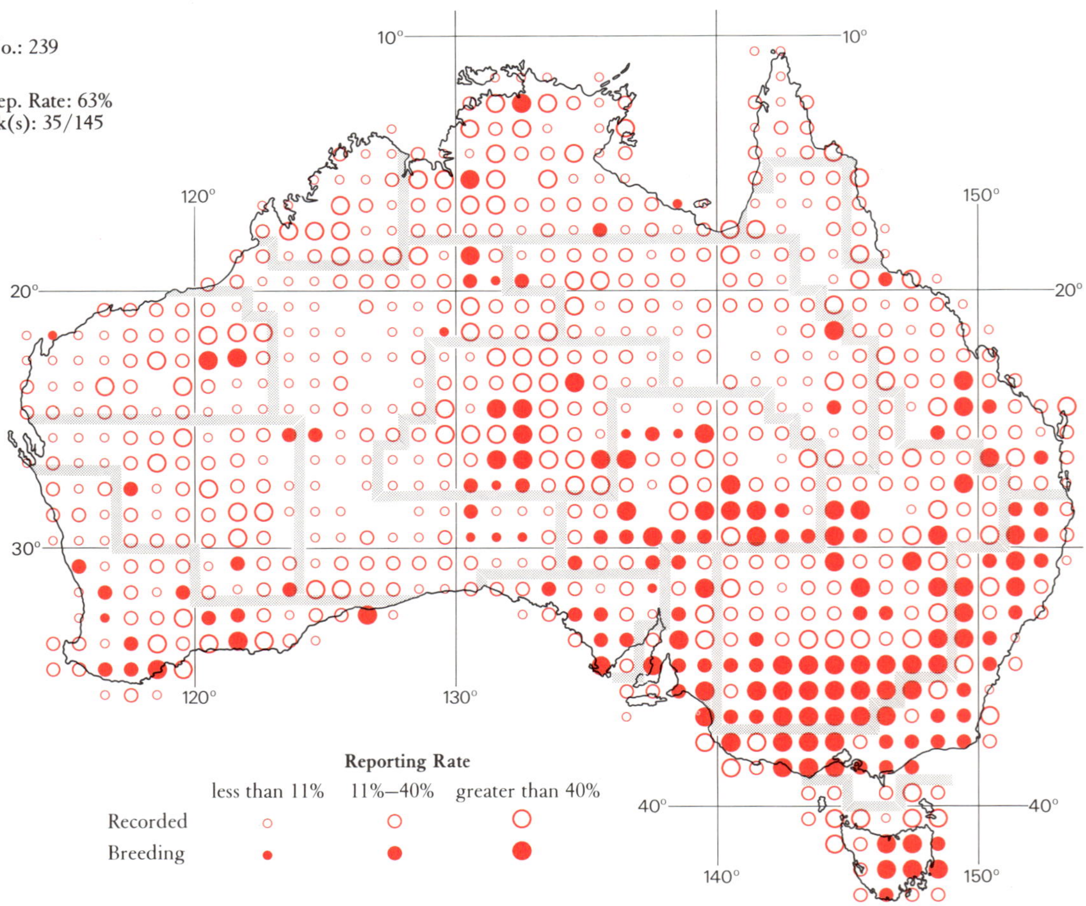

Australian Kestrel

Falco cenchroides

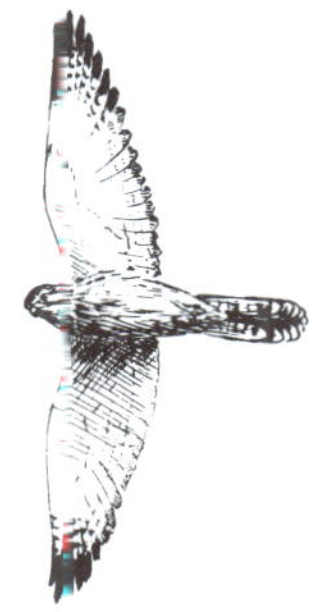

The Australian Kestrel also lives in Indonesia and New Guinea but is a vagrant to New Zealand. The Australian subspecies *cenchroides* breeds on Lord Howe I. (31/159) and Norfolk I. (29/167) as well as on the Australian mainland. In the Tas. Region it breeds on King I. (39/144) and the Furneaux Group (39/148); in Tas. itself a few birds are present throughout the year but breeding has not been recorded.[1276] The Kestrel was one of the most widely reported birds in the Field Atlas, despite its rarity in Tas.

In Torres Strait the Kestrel is a winter visitor and passage migrant[503] probably visiting New Guinea.[484] Near Armidale (30/151) observations from Feb.-June suggested that adult pairs maintained a territory; groups of immatures wandered, disappearing steadily from Mar. onward.[687] In 1980 at Armidale numbers were lowest in winter[470] but near Sydney (33/151) the population in winter was more than double that in spring-summer.[473] At a study site (34/141) west of Mildura most birds left in Feb.; they began to return in Aug.; many more came in Sept. but by Oct. only the breeding population remained.[90] Mainly breeding pairs are present at Griffith (34/146) Sept.-Apr. followed by an increase in autumn and winter.[1360] The Field Atlas suggests altitudinal movement in the mountains of the South-East (35/148–9, 36/148–9) (Reporting Rates: summer 45.4%, winter 21.6%). A winter influx into south-western Vic. is also suggested (Reporting Rates: summer 16.8%, winter 49.3%). Vagrants reach Tas. in all seasons. On the Nullarbor the Kestrel appeared after heavy rains and bred.[246] The longest movement recorded is from near Sydney to Moggill (27/152), 758 km.[50]

The Kestrel inhabits any country where it can hunt over open space, often urban areas, even town centres. It feeds on insects, birds, mammals and reptiles.[1361] Any particularly abundant prey may dominate but vertebrates are the most important.

In the Armidale district a pair, watched for four months in winter, occupied a territory of at least 200 ha (0.01 birds/ha).[687] At Boola Boola (38/146) in patches of clear-felled eucalypt forest density was 0.06 birds/ha.[1113] Seven pairs nested in less than 20 km of the Strzelecki Ck (28/140) floodplain (no more than 3 km wide) in Sept. 1981.[1186]

Field Atlas Records: 32 192

1° blocks–with records: 768 (95%)

–with breeding records: 247 (32%)

Atlas No.: 240

Max. Rep. Rate: 94%
In block(s): 21/122

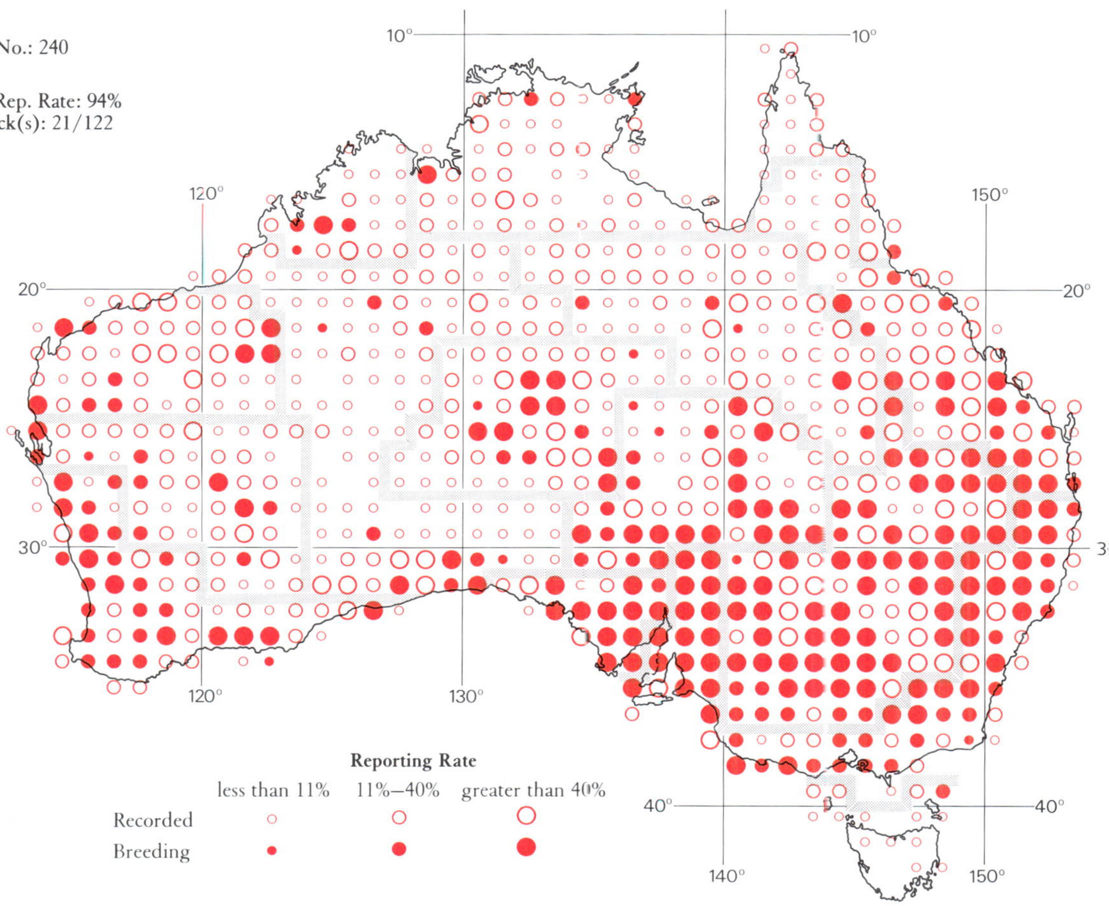

Orange-footed Scrubfowl

Megapodius reinwardt

The Orange-footed Scrubfowl lives in western Indonesia, southern New Guinea and Australia.[1362,1773] Four subspecies are recognized in Australia: *duperreyii* on islands in Torres Strait; *yorki* in the Cape York Region; *castanonotus* in the Atherton Region and *tumulus* in the Top End and Kimberley Regions. Birds in the Eastern Qld Region, south of 20°S, are isolated and possibly a separate subspecies.[1709] The Atlases show that any gap between *yorki* and *castanonotus* must be narrow. They suggest also that the Kimberley population is separated from that in the Top End though the populations have not differentiated.[608]

In Australia the Orange-footed Scrubfowl maintains a territory throughout the year.[435] It lives in patches of rainforest, not necessarily large. In the northern parts of its range, specially in New Guinea and the islands, it lives in mangroves. One population near Port Moresby (9S/147E) in New Guinea lives on a coral island where the only trees, mangroves, occupy no more than 5 ha and are no taller than 2 m.[145] This population was thought to have been established less than ten years previously.

The Orange-footed Scrubfowl builds a mound in which to lay its eggs. The mound is placed on flat or sloping ground usually in the shade, rarely in full sun.[435] The Scrubfowl does not lay if the season is dry or if the mound is very wet. It requires an environment where humidity and rainfall are sufficiently great for the vegetation in the mound to ferment. More than one pair of Scrubfowl may use a single mound.[435]

The Orange-footed Scrubfowl feeds on insects and snails scratched from the forest floor.[1145] It is preyed upon by the Grey Goshawk[100] and its eggs are taken by people, pigs and lizards. Pigs greatly disturb nests in some places[1355] but eggs may be protected by being deeply buried.[104] At the southern extremity of its range the Scrubfowl has disappeared from some localities it formerly occupied.

At Sweet Ck (16/145) three different pairs used the same mound in three successive years. The territory occupied by each pair was about 1.5 ha, 2.0 ha and 1.3 ha in the three years.[435]

Field Atlas Records: 707

1° blocks–with records: 43 (5%)
–with breeding records: 19 (44%)

Atlas No.: 006

Max. Rep. Rate: 55%
In block(s): 12/143

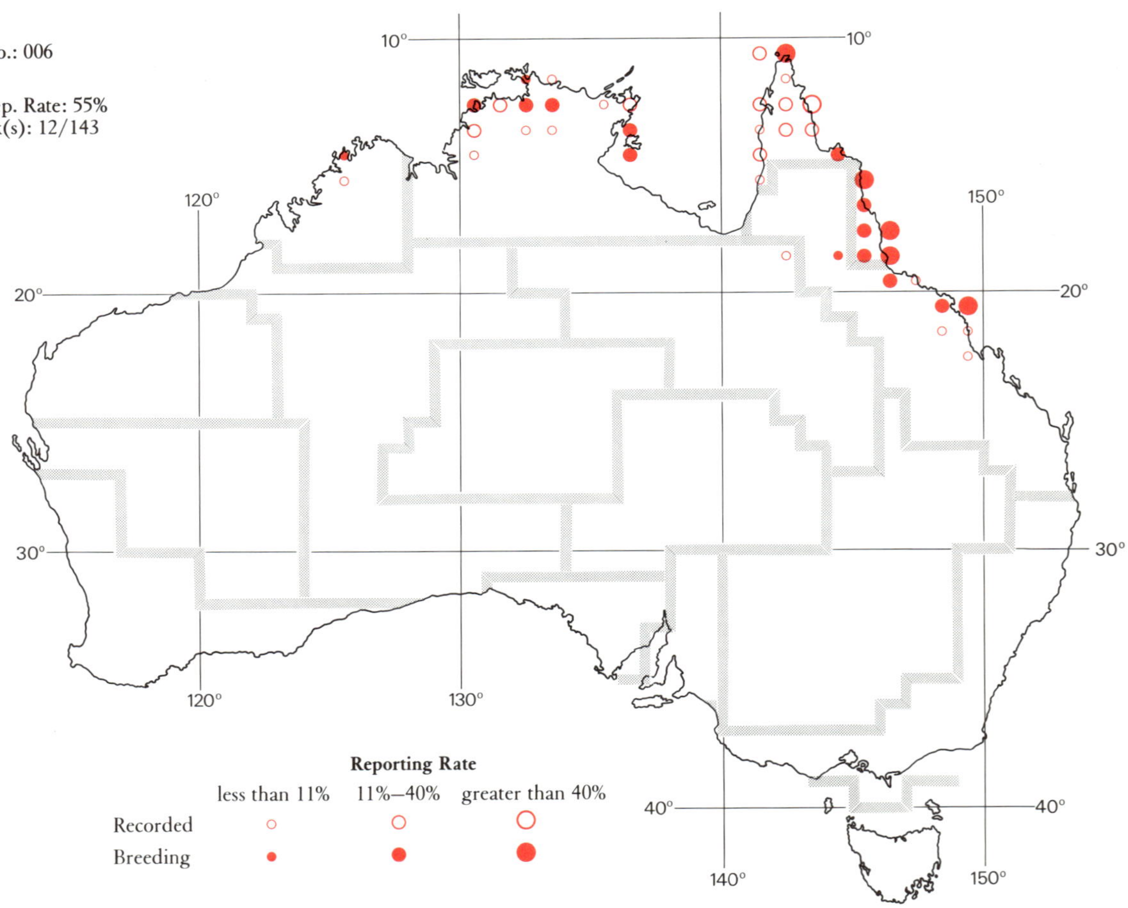

Malleefowl

Leipoa ocellata

The distribution of the endemic Malleefowl before 1901 is better defined than that of most inland species because it provided food for the early explorers. Changes in distribution are evident on the historical maps (p.663). By 1950 the species was gone from central Vic. and from the forests of the South-West Region. By 1977 it had almost gone from the arid inland also. The only records (29/131, 30/131) made during the Field Atlas in the Western Deserts Region were of two recently used mounds. A survey of the Great Victoria Desert in 1980 found evidence that the species is still sparsely distributed there.[180]

The Malleefowl, a sedentary and territorial species, lives in mallee and eucalypt woodland growing on poor sandy soil.[650] Many records in the Atlases are of the conspicuous nesting mounds of the Malleefowl which may remain for years after their last use.

The Malleefowl depends upon the shoots and flowers of herbs for food in winter and spring and the fruit and seeds of acacias and other shrubs in summer and autumn.[650] Grazing by rabbits and sheep in uncleared mallee destroys these foods. Clearing deprives the bird of the litter needed to build its mound. Spilled wheat is an important food where mallee or woodland survives in association with wheatfields.

At Griffith (34/146) a pair of Malleefowl, during a breeding life of about six years, produced about 120 eggs.[650] Foxes, long known as predators of eggs,[163] accounted for about half but the high reproductive rate has enabled many more populations of Malleefowl to survive than might have been expected.

The bird's density at Griffith varied from 0.03 birds/ha in poor mallee to 0.05 birds/ha in dense mallee.[650]

Field Atlas Records: 487

1° blocks–with records: 71 (9%)
–with breeding records: 39 (55%)

Atlas No.: 007

Max. Rep. Rate: 18%
In block(s): 30/117,30/118

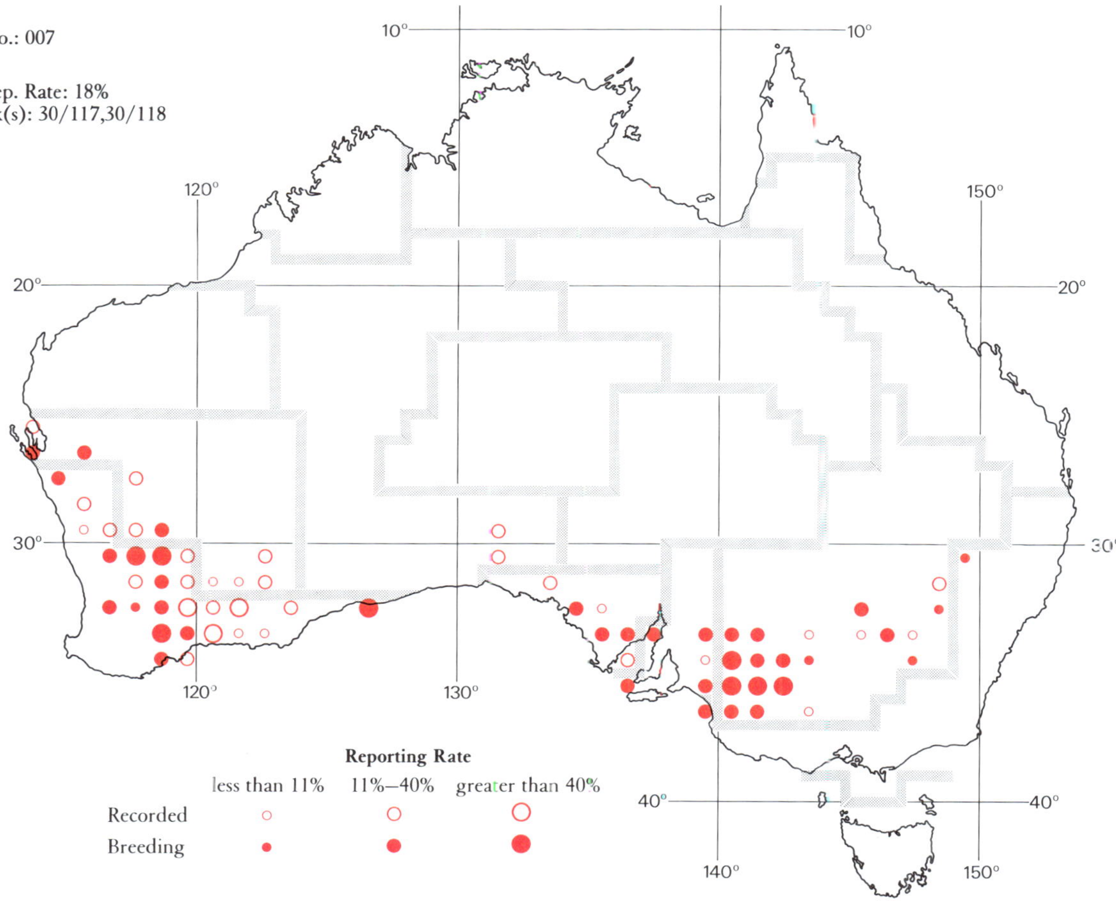

Australian Brush-turkey

Alectura lathami

There are two subspecies of the endemic Australian Brush-turkey: *purpureicollis* in the Cape York Region and *lathami* elsewhere. The subspecies *lathami* was introduced to Kangaroo I. (35/137) in 1936 and has persisted there.[1,78]

The species is sedentary.

In the Atherton Region it lives in rainforest or the introduced lantana.[231,698] In other places it tolerates open and barren surroundings,[972] e.g. at Chillagoe (17/144) it occupies woodland on the tops of isolated hills that are separated by grassland.[755] In the Field Atlas breeding was reported in semi-arid country at Barakula (26/150) and Bingara (29/150). Many records in the Atlases are of the conspicuous nesting mound of the Brush-turkey.

The species eats fruit, seeds, insects, snails and other invertebrates and will take the young sprouts of sugar cane[101] and potatoes.[1227] It usually feeds on the ground but has been seen eating the fruit at the top of a tree.[698] The early settlers hunted the bird for both food and sport. At Murphy's Ck (27/152) it was common until 1902 when hunting led to its disappearance.[1099] In 1934 it reappeared and began to establish itself in thickets of lantana which offered shelter, abundant litter and edible fruit. The pattern at Murphy's Ck has probably been common elsewhere because there are many accounts linking the Brush-turkey with lantana.[231,1099]

The spread of the prickly pear into southern Qld and northern NSW also allowed the Brush-turkey to expand its range.[647] During the 1901–50 period it was present as far inland as Rio (23/144) and a set of eggs was taken at Blackall (24/145) in 1925. The bird fed on the fruit and used the succulent plant segments as nesting material in the mound. When the prickly pear was eliminated, the Brush-turkey retreated from some semi-arid areas.

There are no measures of abundance.

Field Atlas Records: 1486

1° blocks–with records: 79 (10%)

–with breeding records: 34 (43%)

Atlas No.: 008

Max. Rep. Rate: 70%
In block(s): 20/149

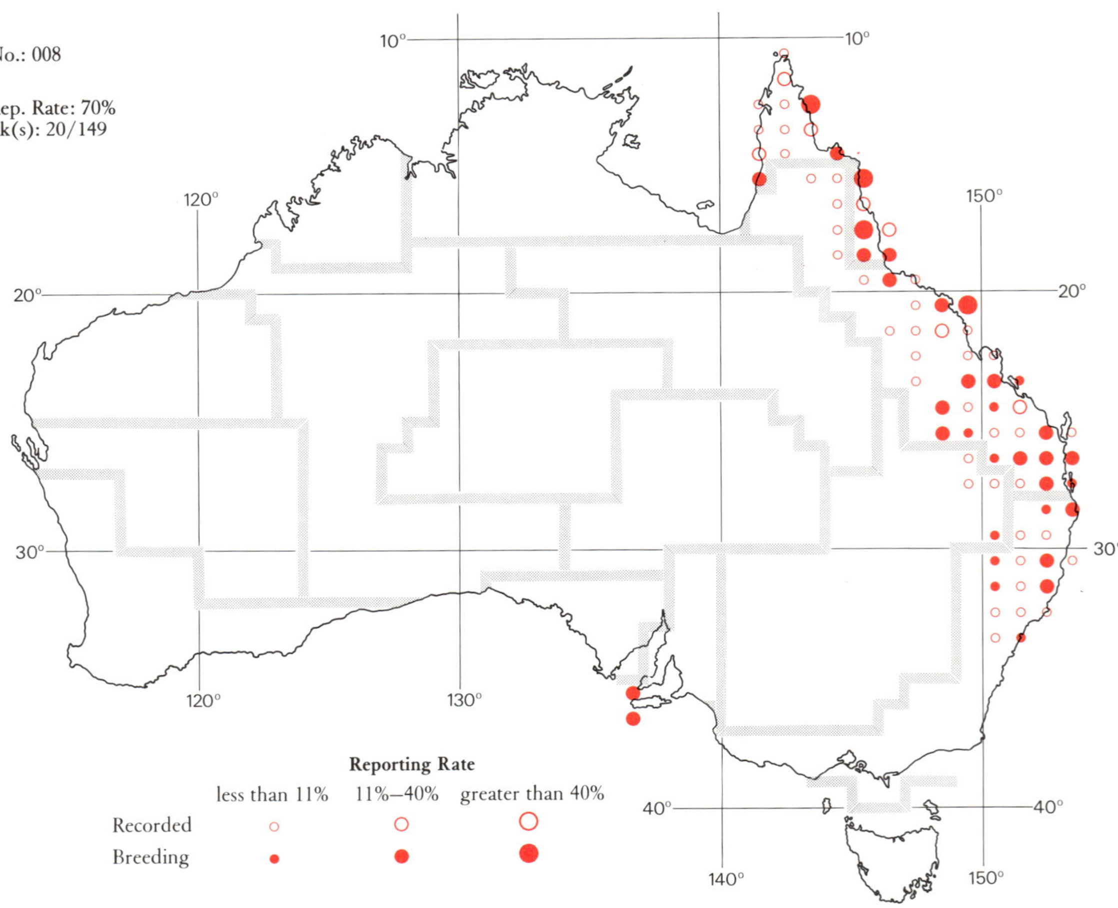

Stubble Quail

Coturnix novaezealandiae

The Stubble Quail became extinct in New Zealand about 1870.[1003] In Australia it was common last century in Tas. and, even in 1910, was considered an important game species throughout the State's agricultural areas. Between 1940 and 1960 it disappeared almost entirely from Tas. So complete and quick was its demise that some comprehensive reviews[653,734] overlook its former abundance. All but two Field Atlas records from the Tas. Region were from King I. (39/144).

In WA, in contrast, the range of the Stubble Quail seems to have expanded. There are no records of it before 1901 in that State, which suggests that it was rare.

The Stubble Quail can move great distances. The longest movement recorded of a banded bird is from Mallala (34/138) to Gunnedah (30/150), 1142 km.[40] When food becomes scarce the bird disperses.[662,671] Many also invade the inland after substantial rain.[671] The Quails do not seem to travel in groups but numbers gradually build up at a favourable locality as wandering birds arrive. The Field Atlas shows no large-scale seasonal change in Reporting Rate.

The Stubble Quail lives in all types of country except forest[653] but mainly in well-drained plains and agricultural land. It will also enter gardens. The nest is often in crops about to be harvested, which leads to its destruction. The bird eats seeds, mainly of introduced plants, and also takes insects.[662]

On the mainland numbers of Stubble Quail fluctuate greatly at any one place. Movements contribute to this but many occurrences of large numbers of Stubble Quail are probably the result of successful local breeding. When conditions are good, two or three broods may be raised in succession.[665] The Stubble Quail is becoming more abundant on the mainland, taking advantage of newly irrigated agricultural lands. In the Atherton Region, for instance, it was rare and irregular before 1960 but is now resident.[31] The disappearance of the species from Tas. correlated with a change in agricultural methods to improved pasture. Because the bird was sedentary in Tas.[734] it may have been more vulnerable to changing conditions.

There are no measures of abundance.

Field Atlas Records: 3497

1° blocks–with records: 265 (33%)
–with breeding records: 91 (34%)

Atlas No.: 009

Max. Rep. Rate: 26%
In block(s): 34/143,35/143

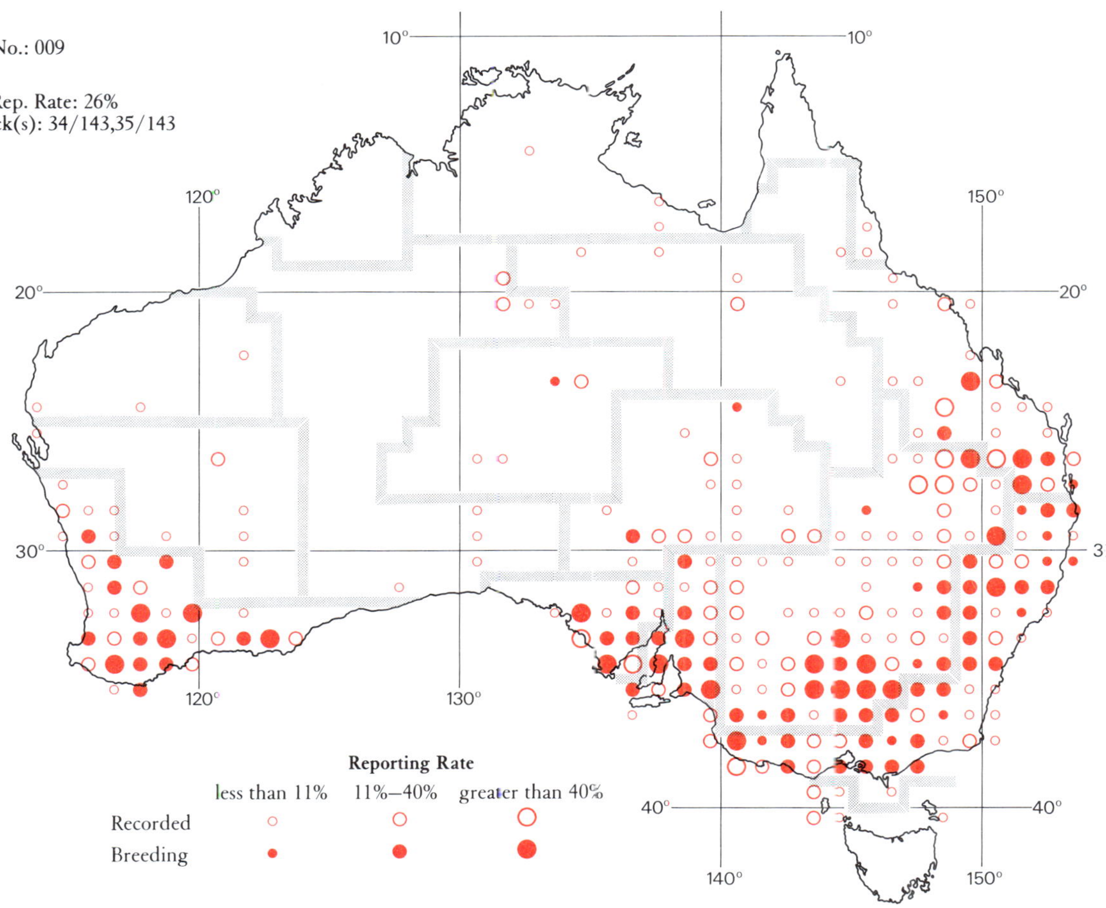

Swamp Quail

Coturnix ypsilophora

Brown Quail

Coturnix australis

The Brown Quail inhabits Indonesia, New Guinea and Australia and was introduced to New Zealand and Fiji. The Swamp Quail is now thought to be a subspecies of the Brown restricted to Tas.[1574] The two are treated together because observers could not separate them in the field. Historical records from Inkerman (16/141) and Normanton (17/141) suggest that the range of the Brown Quail is continuous round the Gulf of Carpentaria.

The Brown Quail is present all year in the wetter parts of its range but historical reports in the Western Deserts Region suggest that the bird comes and goes in the inland. A report early this century mentions it crossing Torres Strait to land 'in numbers' on Booby I. (10/141)[1145] and a recent review considers it nomadic in Torres Strait as well as a breeding resident of some islands.[503] Round Canberra (35/149) and in the southern Murray-Darling Region, the Quail is sometimes seen often and sometimes not at all, as if it were moving into and out of the districts.[657,833] These observations point to the bird moving at times but the Field Atlas suggests no large-scale seasonal pattern.

The Brown Quail mostly lives in rank, moist vegetation.[1099] Backwaters of reservoirs and headlands in irrigated fields are rapidly occupied by the bird because it can use very small areas. It will also live in crops such as lucerne[231,653] and in the Tas. Region sometimes breeds on small islands devoid of swamps.[1314] The bird is usually in parties of fewer than 30 but at times is extremely numerous. Thousands seen on the 200-ha Passage I. (40/148) may have been migrants but were thought to have been bred on the island.[1178] The Brown Quail feeds on small grass seeds, insects and green leaves of grass and clover.[787,1269,1531] It may have declined this century in the lower South-East Region.[583,657] There are no measures of abundance.

Field Atlas Records: 1925

1° blocks–with records: 219 (27%)
–with breeding records: 54 (25%)

Atlas No.: 011,010

Max. Rep. Rate: 21%
In block(s): 40/144

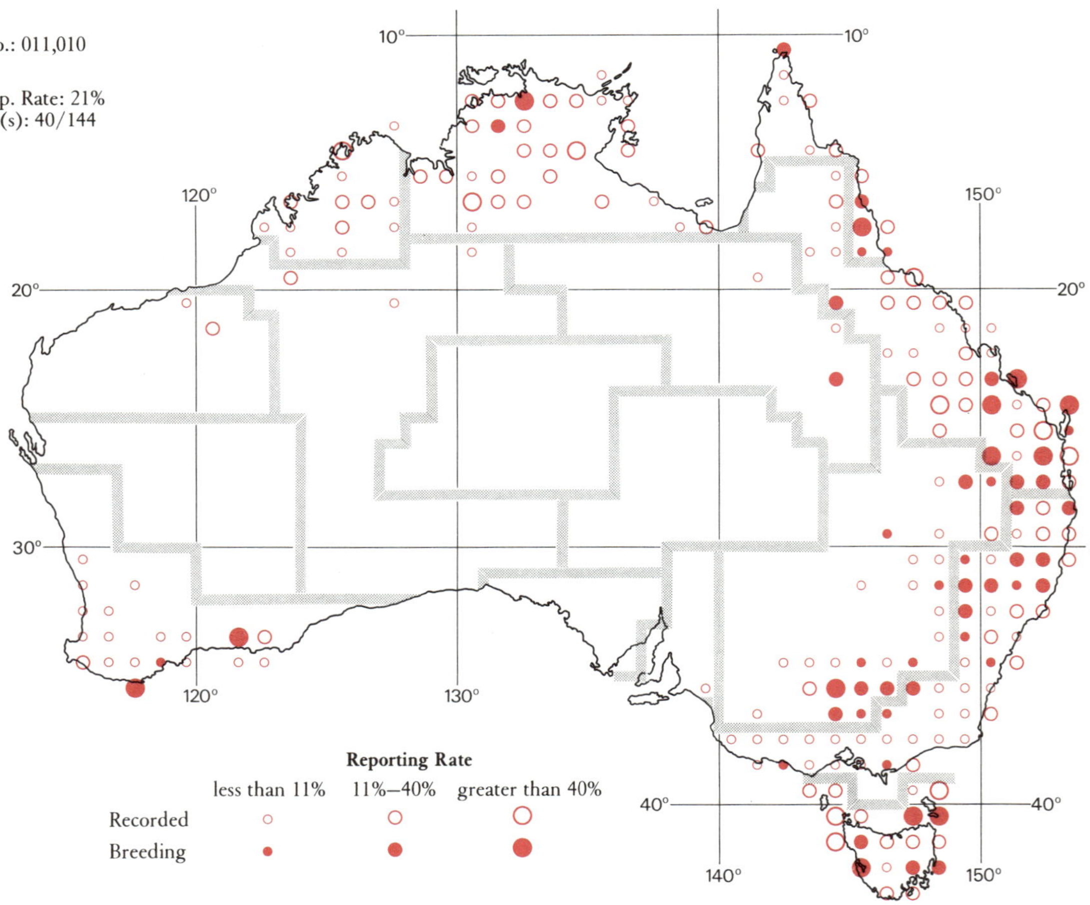

King Quail

Coturnix chinensis

The King Quail has an extensive distribution throughout south-eastern Asia.[1230] Eastern and southern Australia are inhabited by a population that may differ morphologically from that of the Kimberley and Top End Regions but views about subspecies differ.[388,1574] The King Quail has never been reported from the Tas. Region except for sightings on King I. (39/144) in the 1960s, possibly of escapees.[739] Its range in northern and eastern Australia south to Sydney (33/151) seems not to have changed since European settlement. South of Sydney it has declined. Historically, breeding records and observations extend throughout the South-East Region and there was an apparently isolated population in the Mt Lofty Ra. (34/138) as well, last reported breeding in the early 1960s. There is one recent breeding report south of 35°S at French I. (38/145) in 1975.

The King Quail is probably present all year in northern Australia. In the south it was regarded at the turn of the century as a spring-summer visitor.[299] Observations at Murphy's Ck (27/152) suggest local movement: the Quail is abundant in good seasons, moderately common in average ones and not seen at all in poor ones.[1099] The Field Atlas shows no evidence of large-scale seasonal movement.

In undeveloped land, the King Quail lives in rank vegetation beside streams, wet heaths and swamps. In agricultural areas it often invades and nests in summer crops, specially lucerne.[231,746,891,1099]

The King Quail feeds on insects as well as small seeds.[1733] It can reproduce rapidly and changes in abundance at one locality could therefore be explained by local breeding as well as movements.

There are no measures of abundance.

Field Atlas Records: 149

1° blocks–with records: 42 (5%)

–with breeding records: 5 (12%)

Atlas No.: 012

Max. Rep. Rate: 8%
In block(s): 12/143,14/136,18/145

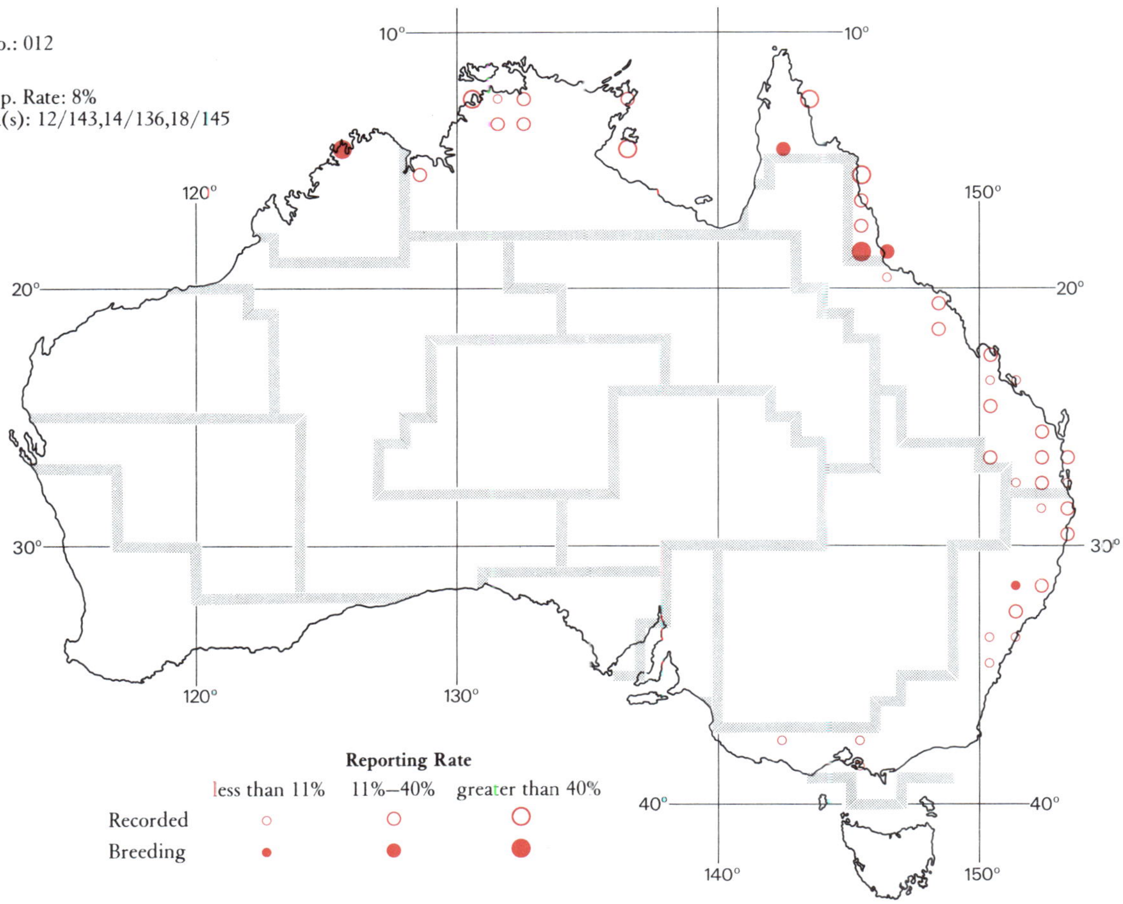

Red-backed Button-quail

Turnix maculosa

The Red-backed Button-quail lives in Indonesia, New Guinea and adjacent islands and Australia.[388] The northern and eastern populations in Australia were considered morphologically distinct[388] but views about subspecies differ.[748] The populations do seem to be isolated because there are no records in the Atlases from the southern coast of the Gulf of Carpentaria. Though not recorded in the Kimberley Region during the Field Atlas, the species' range extends there.[1713]

In the Eastern Qld and South-East Regions the Quail's range seems to have fluctuated. The bird was virtually unrecorded from Eastern Qld in the 1901–50 period, the only report in the Atlases being of eggs collected at Duaringa (23/149) in 1905. A population present round Sydney (33/151) in the 1890s[1343] was last recorded about 1904.[816] In the 1951–76 period and during the Field Atlas the species has again been reported from many localities in Eastern Qld and as far south as Finley (35/145) where breeding was recorded in 1954.[270]

There may be more movement of the species than is realized because a bird, apparently migrating, was killed when it flew into a wall at night.[303] If there is movement, isolated populations may still be in contact and changes in abundance may be related to movements as well as to change in status. The Field Atlas shows no evidence of large-scale seasonal movement.

The Red-backed Button-quail lives in rank grass and herbs beside northern watercourses; in eucalypt woodland and rainforest; and in crops and gardens.[212,303,1767] In the Atherton Region, it is said to be well distributed in cleared areas up to 900 m altitude, apparently breeding in them as well.[231] The Red-backed Button-quail feeds on small seeds and insects as well as green vegetation.[748]

The species was described as 'numerous' in the Cape York Region in 1935.[1758] A sighting during the Field Atlas of four birds in as many days at Iron Ra. (12/143) suggests that, in parts, it may still be so but there are no measures of abundance. South of the Atherton Region its status seems to have varied. To suggest that it has declined generally[657] may be an exaggeration.

Field Atlas Records: 97

1° blocks–with records: 24 (3%)
–with breeding records: 1 (4%)

Atlas No.: 013

Max. Rep. Rate: 17%
In block(s): 15/145

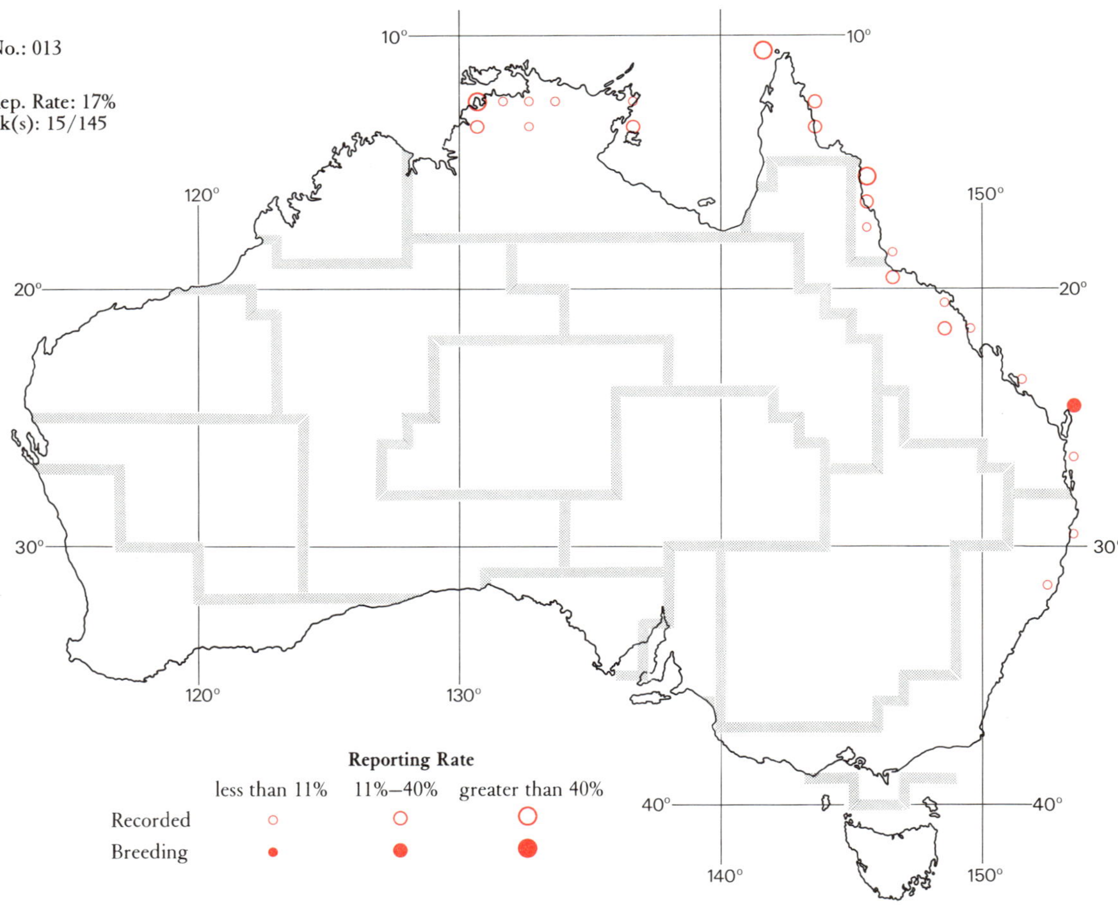

Painted Button-quail

Turnix varia

The Painted Button-quail lives in Australia with an outlying population in New Caledonia (21S/165E). A population on the Houtman Abrolhos (28/113), not recorded during the Field Atlas, is regarded as the subspecies *scintillans* distinct from the nominate *varia* elsewhere.[388,1707] Otherwise no morphological differences have been detected among the several isolated populations in Australia. Breeding has been reported from all these populations either historically or during the Field Atlas. The isolated sighting (27/114) on the Murchison R was of a pair in Mar. 1979.

In the middle of last century the Painted Button-quail was common and widespread in Tas.[723] but since 1950 it has rarely been recorded there[657] and only 17 times during the Field Atlas.

The Painted Button-quail is thought to be present all year in some places[214] but its capacity to move long distances is evident from records such as that on the Murchison. Any movements are usually considered erratic though near Rockhampton (23/150) it is said to be present Aug.-Oct.[1089] The Field Atlas shows no seasonal pattern except possibly south of 37°S in the South-East Region where the Reporting Rate in summer (1.6%) is slightly greater than in winter (1.1%) This implication of regular movements is supported by observations at Jamieson (37/146) where the Quail bred and was present only Oct.-Apr. each year for 14 years.[1469]

The Painted Button-quail lives in eucalypt forest, woodland and heath where there is a layer of leaf and twig litter. Cleared or grazed areas are shunned.[657,1099,1216] When feeding, the Quail makes characteristic circular scrapes and its presence in an area can often be detected by these.[1413,1415]

The Painted Button-quail feeds on seeds and takes many insects as well.[748,1049,1531] It is now protected.

Its abundance in the 1850s near Sydney (33/151) can be gauged by the fact that 15 pairs could be shot in a morning.[793] Density in eucalypt woodland near Armidale (30/150) was 0.01 birds/ha.[582]

Field Atlas Records: 1061

1° blocks–with records: 144 (18%)
–with breeding records: 37 (26%)

Atlas No.: 014

Max. Rep. Rate: 19%
In block(s): 25/148

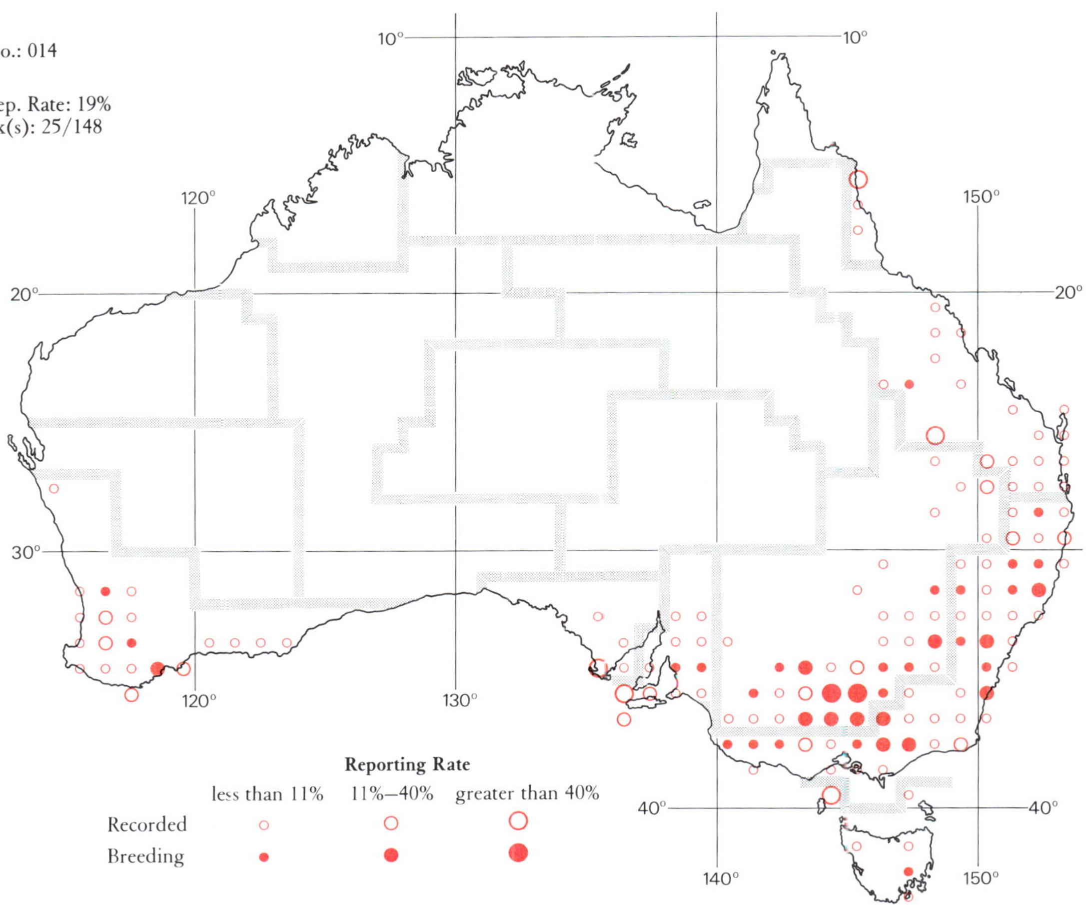

Chestnut-backed Button-quail

Turnix castanota

Populations of the endemic Chestnut-backed Button-quail are restricted to the Top End and Kimberley Regions but there is doubt whether they are morphologically distinct.[388,1574,1713] Additional localities from which the species has been reported before the Field Atlas are Melville I. (11/130) and as far south-east as Borroloola (16/136).[107] It is absent from the Darwin (12/130) district. The Kimberley population has been recorded as far east as 128°E.

There is no information about movements. The Chestnut-backed Button-quail lives in eucalypt woodland including that with a grassy understorey. In the Top End it tends to be found on the ridges rather than in the intervening flat lands and in the Kimberley Region it lives in sandstone and lateritic country[1713] and is said to be locally common.[664,787,1711,1713] It lives in coveys of 6–20 birds and tends to run rather than fly when flushed. It feeds on the seeds of grasses and also takes insects.[1860]

There are no measures of abundance.

Field Atlas Records: 27

1° blocks–with records: 9 (1%)

–with breeding records:

Atlas No.: 015

Max. Rep. Rate: 13%
In block(s): 12/133

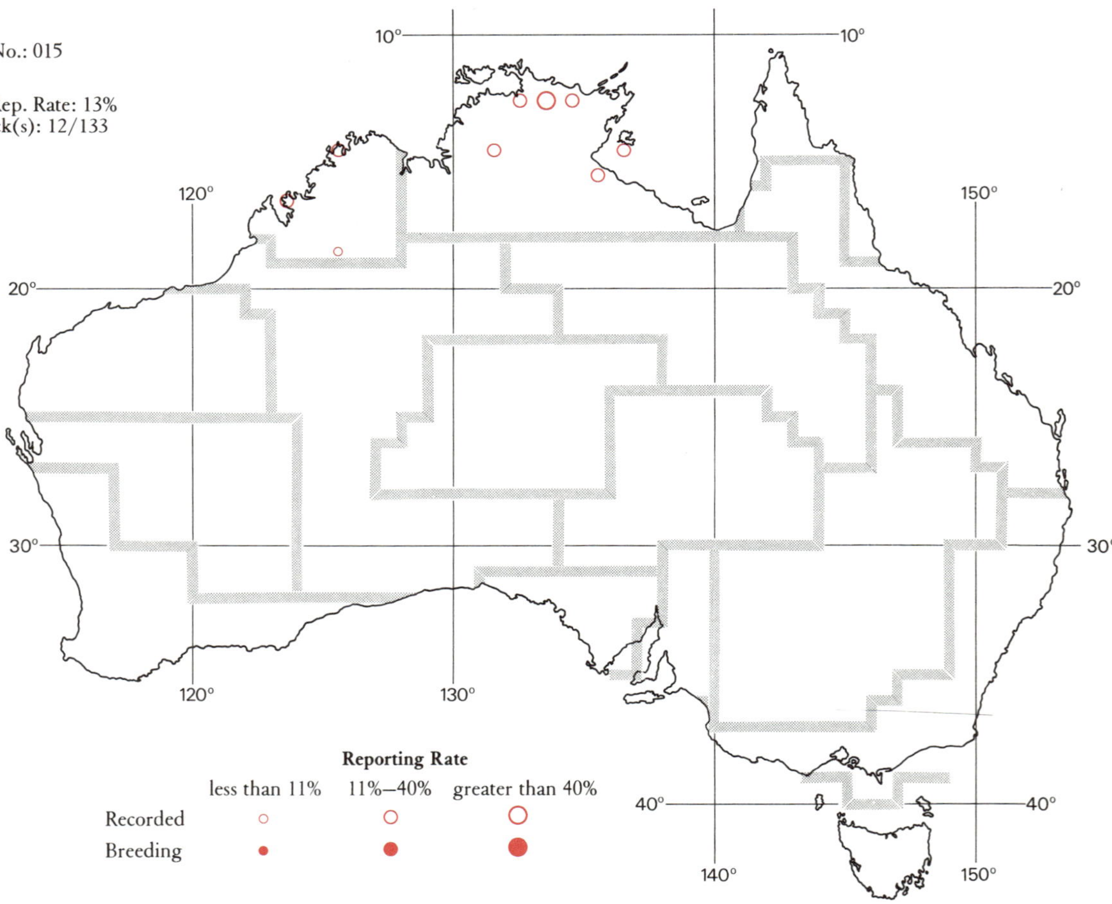

Buff-breasted Button-quail

Turnix olivei

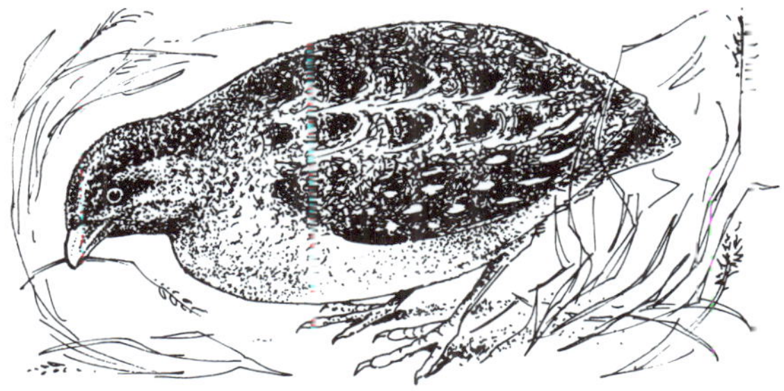

The endemic Buff-breasted Button-quail has been recorded by few observers, none during the Field Atlas. The species was first collected in 1894 at a locality cited as 'Cooktown' (15/145) but the specimen probably did not come from near the town.[1123] Additional specimens and eggs were not obtained until 1921–22 when the species was rediscovered near Coen (13/143).[1861] Subsequent records have come from Iron Ra. (12/143), some as recently as the 1970s.

Nothing is known of movements of the Buff-breasted Button-quail. It inhabits clearings where grasses dominate, including patches in rainforest, woodland and swamps. The only consistent series of sightings of the species was made during the wet summer season when the bird was calling and breeding.[1861] Nothing is known of the Button-quail's food or other requirements.

Despite the lack of records during the Field Atlas, there is no reason to doubt that the Buff-breasted Button-quail still exists.

There are no measures of abundance.

Field Atlas Records:

1° blocks–with records:

–with breeding records:

Atlas No.: 016

Max. Rep. Rate:

In block(s):

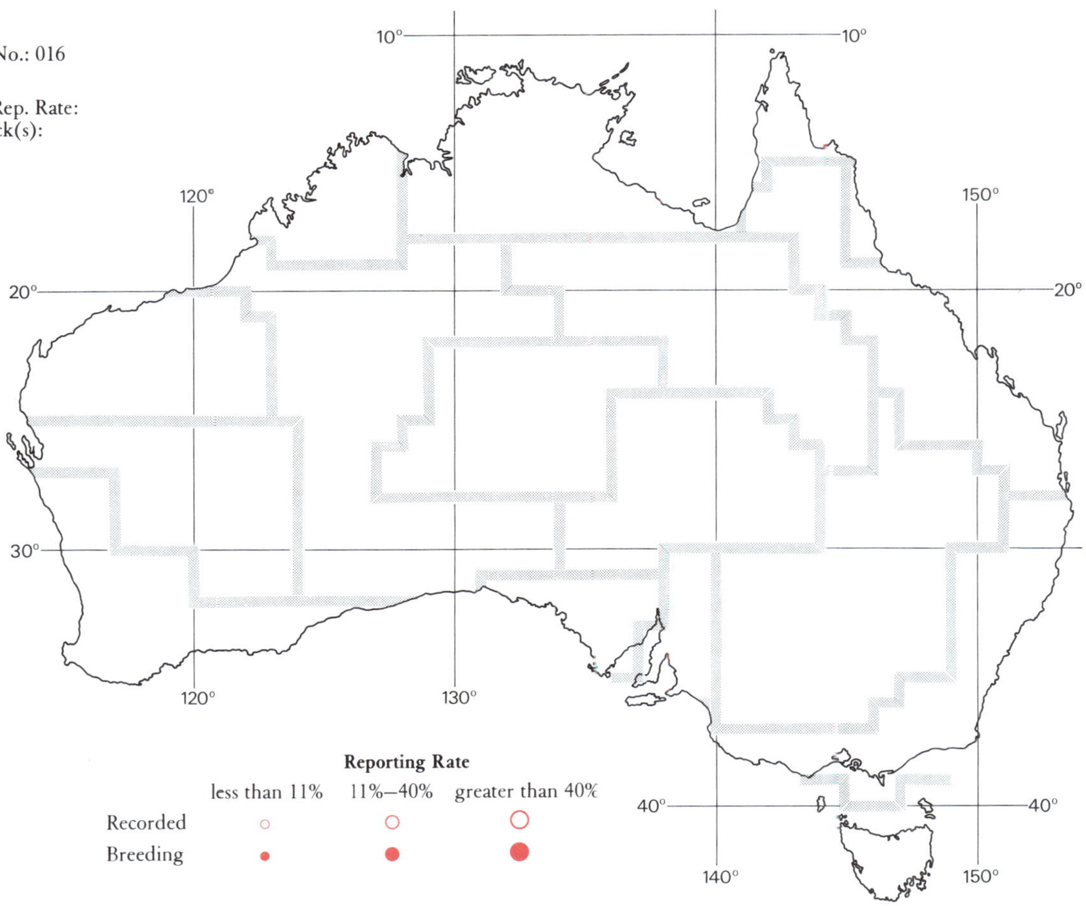

Black-breasted Button-quail

Turnix melanogaster

The endemic Black-breasted Button-quail is a cryptic bird, easily overlooked, whose range has declined greatly. Before 1901 it was recorded in the Atherton Region and on the eastern coast between 23 and 29°S. The Atherton population was not recorded again until 1956[1656] and subsequently not since 1966.[231] Most records in its southern range were from the Nanango (26/152)-Yarraman (26/151) district with a few from the Macpherson Ra. (28/152) and Bunya Mts (26/151). The few early accounts indicate that it was common, e.g. in the thick vegetation in the Dawson R (23–25°S) area in 1883–84, but it had completely disappeared by 1900.[1343] Since 1968 reports have come from Fraser I. (25/153) indicating a recent expansion of range.

The species appears to be present all year in a district, living in small parties of 4–5 birds in rainforest that is dry rather than wet. It needs dense overhead cover and a thick, dry, litter layer. Most recent records show the bird to be localized and associated with the introduced lantana and plantations of hoop pine. The Black-breasted Button-quail moves slowly about the forest floor scratching characteristic oval depressions in the litter but nothing is recorded of its food.

Though the bird is still rare, recent observations of it in lantana and hoop pine suggest that it is adapting to man-modified environments and may be increasing in numbers.

There are no measures of abundance.

Field Atlas Records: 26

1° blocks–with records: 4 (1%)

–with breeding records: 1 (25%)

Atlas No.: 017

Max. Rep. Rate: 6%
In block(s): 24/153

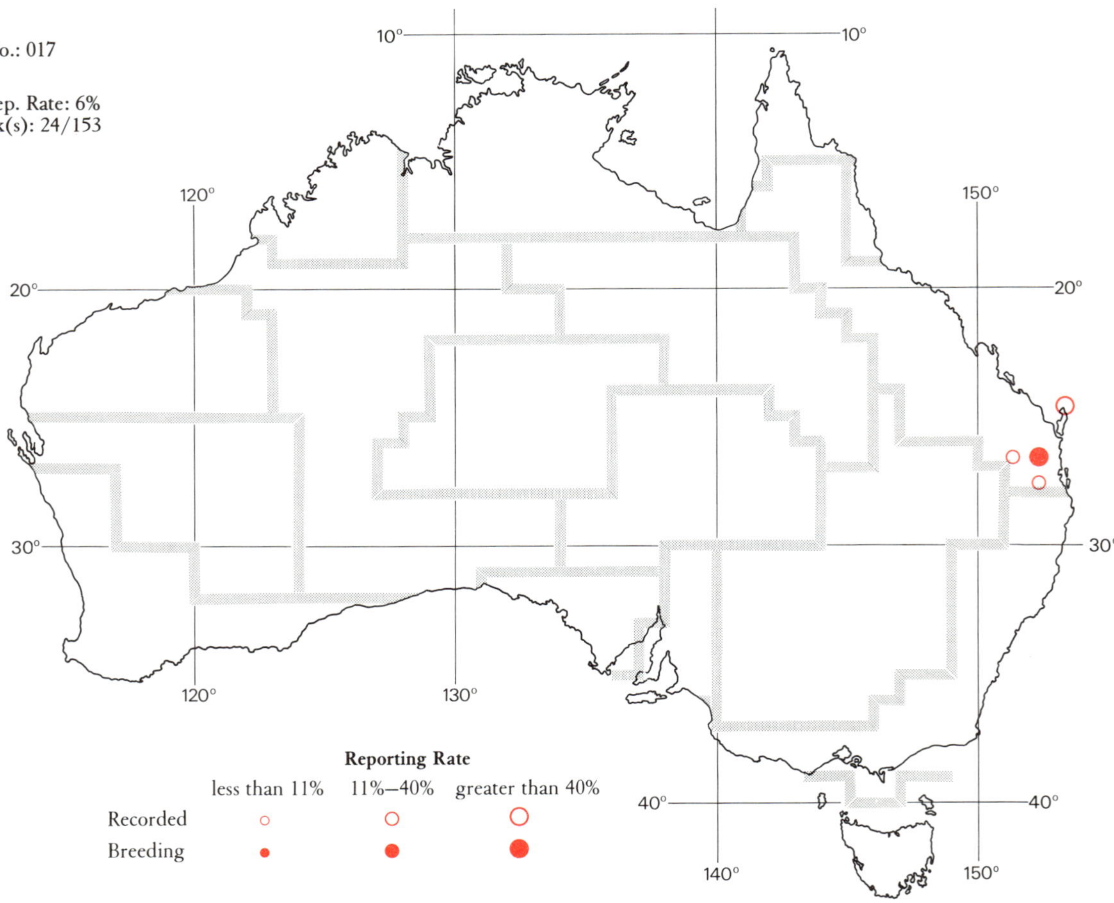

Little Button-quail

Turnix velox

The Little Button-quail is endemic. Historical records fill many of the apparent gaps in its distribution, particularly in the Western Deserts, Pilbara and Goldfields Regions. In 1964 reports came from the Atherton Region also.

The Little Button-quail is known to move erratically: large numbers sometimes appear at a locality, stay to breed and as rapidly vanish.[1413] The numbers seen during an irruption can run into thousands.[1216] The species is regarded as nomadic in the southern Murray-Darling Region[833] but other observations and the Atlas suggest a seasonal component in movements in southern Australia. Reporting Rates south of 26°S were 0.010% in summer compared with 0.005% in winter. At Rupanyup (36/142) the Little Button-quail arrives each year when the crops are nearly in ear. In southern SA it is a regular migrant in spring and summer though numbers may vary a hundredfold from year to year.[214,513] In WA there were no winter records in the Field Atlas south of 30°S and observers at Kellerberrin (31/117) have seen the species only in spring and summer.[627]

The Little Button-quail inhabits woodland, acacia scrub, spinifex and tussock grassland. It takes readily to cereal crops from which it may be flushed during harvesting. The species feeds mainly on seeds, specially of grass, and its movements seem closely related to the presence of abundant seed. At Garah (29/149) in 1918 good rains led to heavy seeding of the local grasses. A plague of house mice followed and when, later, many Little Button-quail arrived, almost all the seed was gone and the birds rapidly disappeared.[1295]

On about 270 ha of farmland cropped in the Strathalbyn (35/138) district estimates were made of the number of Button-quail present on migration each summer. They varied between 2–4 and 500–800 (0.01–2.76 birds/ha), the latter during an excellent season.[513]

Field Atlas Records: 877

1° blocks–with records: 293 (36%)

–with breeding records: 55 (19%)

Atlas No.: 018

Max. Rep. Rate: 26%
In block(s): 20/144

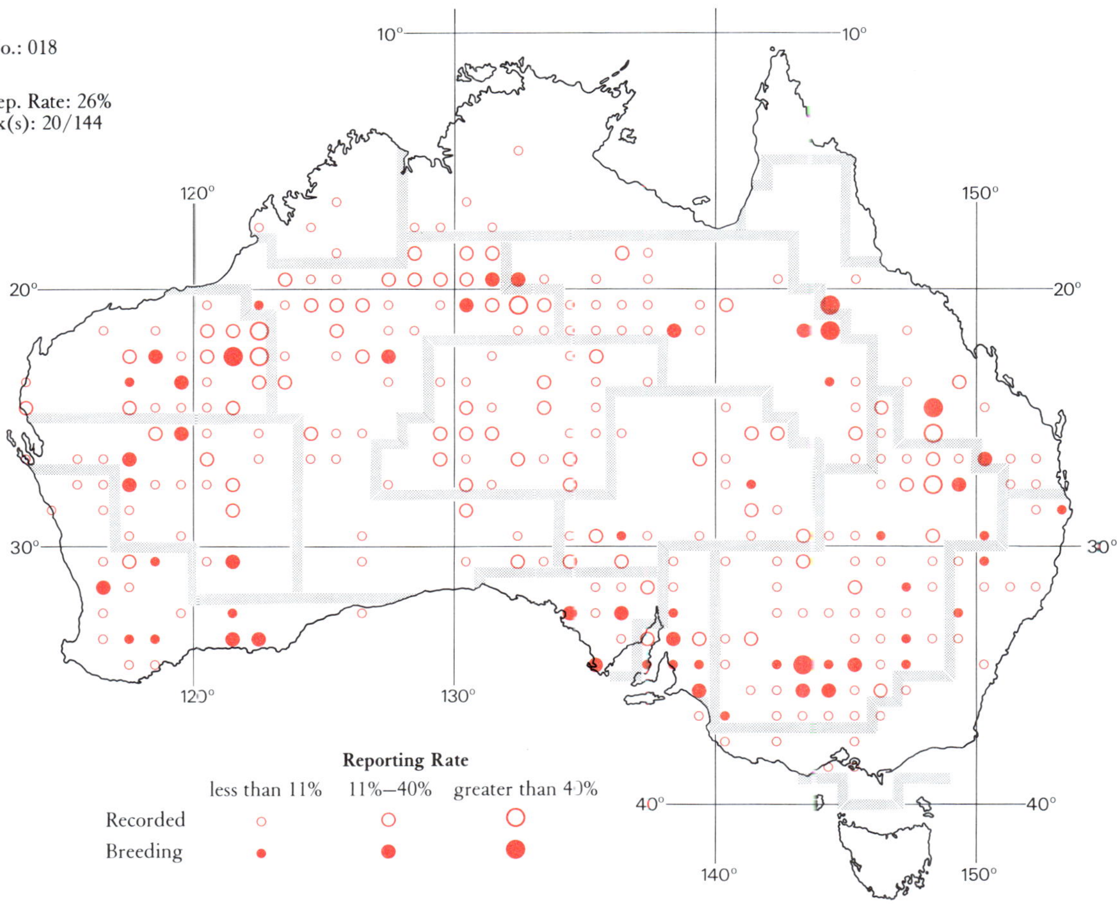

Red-chested Button-quail

Turnix pyrrhothorax

The endemic Red-chested Button-quail is sparsely distributed across northern Australia to the western side of the Kimberley Region[1713] but has not been reported farther south in WA.

An analysis in 1971 of NSW records suggested that the species may have undergone substantial fluctuations of range and abundance[1284] but observations at Cooyal (32/149) cast doubt on this conclusion. The species was always present but difficult to detect and easily confused with the Little Button-quail.[1291] The Historical Atlases do not suggest any long-term change in range.

Reports of seasonal movement are widespread as are records of erratic increases and decreases. In the Richmond (20/143) district, the Red-chested Button-quail is seen mainly Jan.-June, numbers varying from year to year.[168] It is an erratic visitor round Atherton (17/145), most often present Nov.-Mar.[231] It is also regarded as a spring-summer visitor to the Cobar (31/145) district;[1573] to Rupanyup (36/142); and probably to south-eastern SA.[214] Near Cobar it was locally abundant during one summer (1974–75) in a ten-year period of observation. Reporting Rates south of 26°S were greater in summer (0.20%) than in winter (0.03%).

At Cooyal there is no evidence of movement: a few birds are present all year but more easily located during spring and summer when crops are harvested and in the resulting stubble.[1291] There are also winter records from other parts of NSW so that part or all of the difference between summer and winter Reporting Rates may be accounted for by the difficulty of observation in winter.

Local fluctuations in abundance probably often result from successive good breeding seasons rather than movements[1291] but this does not appear to have been so at Richmond where the bird was abundant in a dry year but hard to find in a much wetter one.[168]

The Red-chested Button-quail lives in crops and stubble as well as native pasture with and without light tree cover.[231,1284,1573] It is recorded singly, in pairs and in small coveys of three to five birds. The Red-chested Button-quail feeds on insects and small seeds including spilled wheat.[748,1049,1291]

There are no measures of abundance.

Field Atlas Records: 123

1° blocks–with records: 48 (6%)
–with breeding records: 16 (33%)

Atlas No.: 019

Max. Rep. Rate: 10%
In block(s): 21/144

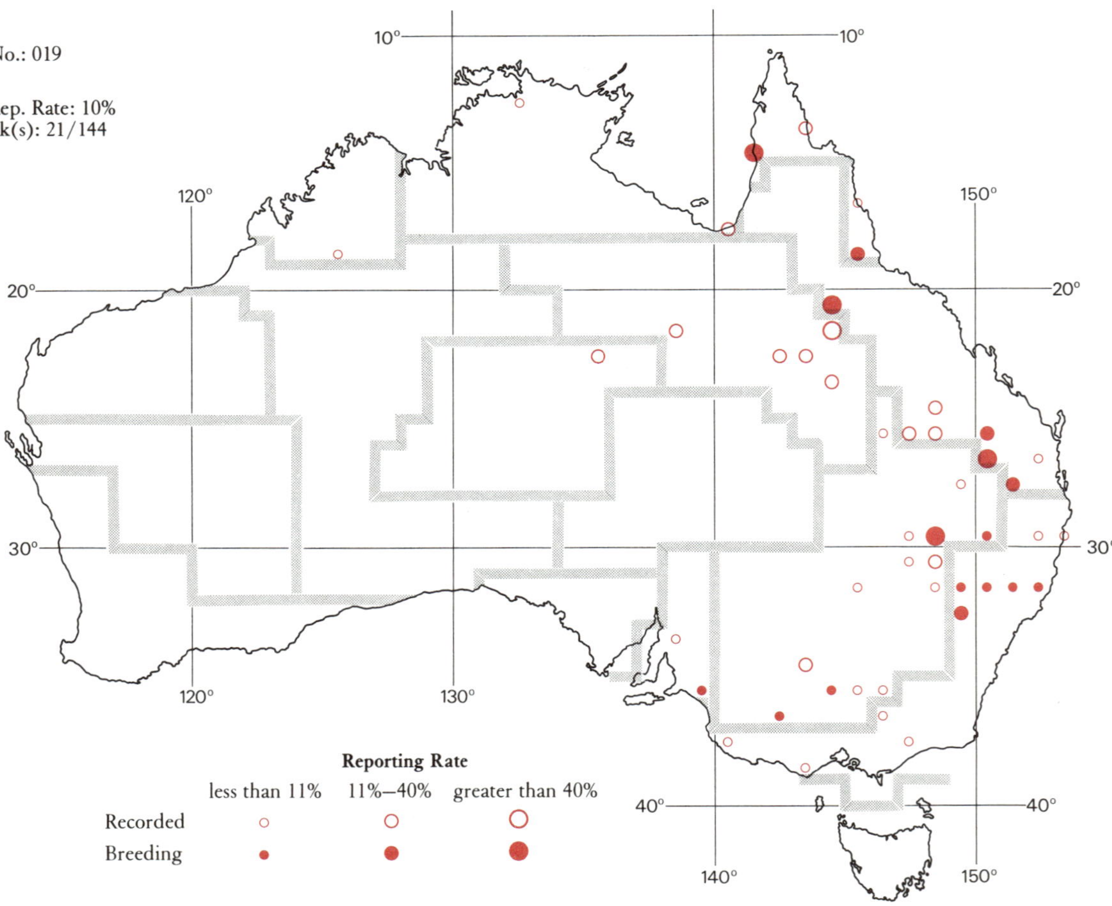

Plains-wanderer

Pedionomus torquatus

The Plains-wanderer is not just an endemic Australian species: it is the only member of a separate avian family. Its range has shrunk since European settlement, specially in the South-East Region (see maps p. 663). Most reports recently are from the Riverina in the southern Murray-Darling Region and from north-central Vic.[165]

The individual Plains-wanderer probably remains in an area as long as conditions are suitable but the bird is capable of moving long distances as sporadic records in central Australia demonstrate.

To maintain populations the Plains-wanderer requires extensive areas vegetated with a varied community of grasses and other herbs.[165] It will tolerate some forms of agriculture. Large expanses of suitable vegetation are provided where there is broad-acre grazing on unimproved pasture, such as in the Hay (34/144) and Narrandera (34/146) districts and also in cereal-growing areas near Swan Hill (35/143). Round Swan Hill only one cereal crop in three years is grown in any one paddock. After harvesting, the paddocks are left for a year or more, allowing a varied plant community to develop.

The Plains-wanderer is generally encountered singly, in pairs or in family parties. When disturbed, it tends to run away slowly, not attempting to fly.

The Plains-wanderer feeds on insects and the seeds of a variety of exotic and native grasses and other herbs. Such seeds were found in the gut contents of two specimens.

The Plains-wanderer has disappeared from most areas where more intensive forms of agriculture are practised. The clearing of mallee for marginal farmland has enabled it to move into some areas which it would not previously have inhabited.

Generally it is present at low density though periods of abundance have been reported.

There are no measures of abundance.

Field Atlas Records: 18

1° blocks–with records: 9 (1%)

–with breeding records:

Atlas No.: 020

Max. Rep. Rate: 2%
In block(s): 24/139

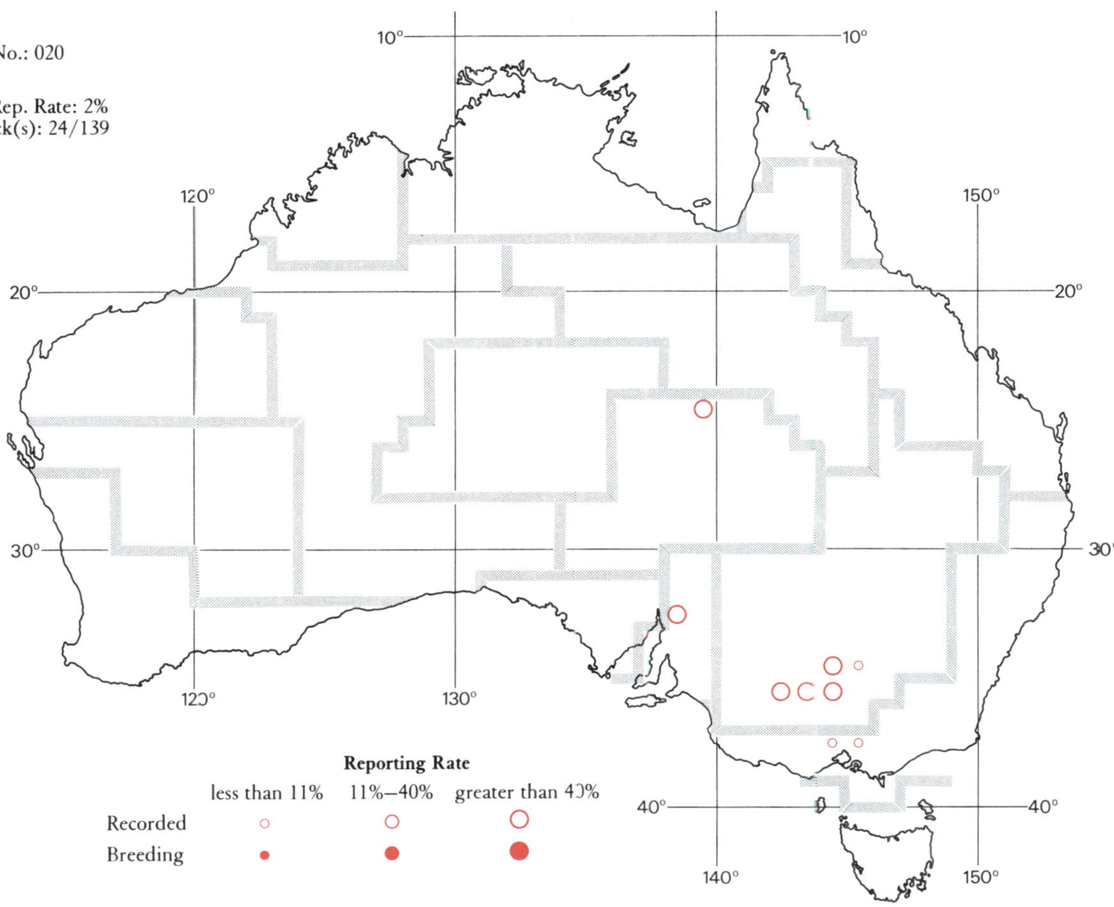

Buff-banded Rail

Rallus philippensis

The Buff-banded Rail inhabits islands of the Indian Ocean and south-western Pacific, Indonesia, New Guinea and Australia. Australian populations show no geographically consistent variation and are considered to belong to the single subspecies *mellori*.[1579] In the Tas. Region, the Rail is seen occasionally on King I. (39/144)[734] but reports from Tas. itself are unconfirmed.

The Buff-banded Rail is widely regarded as a spring-summer visitor to sites in southern Australia such as Moree (29/149),[1296] the southern Murray-Darling Region[833] and Melbourne (37/145).[1343] Near Brisbane (27/153) a few birds remain in winter but most leave in Apr.[506] At Murphy's Ck (27/152)[1099] and farther north in the Rockhampton (23/150) district,[1089] Atherton Region[231,698] and Darwin (12/130)[423] the species is present all year. During the Field Atlas, records came from throughout the species' range in all seasons, except for the SA Ra. and Eyre Pen. Regions where there were no reports in autumn and winter. Reporting Rates showed a general drop from 1.1% in summer to 0.4% in winter. This may be partly because the species is more conspicuous when calling in the breeding season but there is evidence that some birds leave Australia and migrate through Torres Strait.[503]

The Buff-banded Rail usually occupies rank vegetation in swamps and along creeks or in paperbark woodland; it also enters crops.[231,1343] In SA some birds inhabit islands with little cover and no surface water.[911] On islands of the Great Barrier Reef, it is often flushed from petrel burrows.[1147] The Buff-banded Rail feeds on small insects and aquatic life, foraging round the edges of reedbeds specially in the evening. At Eyre Bird Observatory (32/126) a single bird watched closely for three weeks was observed to hoard food after the fashion of a rodent.[440]

There are no measures of abundance.

Field Atlas Records: 809

1° blocks-with records: 127 (16%)
-with breeding records: 39 (31%)

Atlas No.: 046

Max. Rep. Rate: 18%
In block(s): 23/152

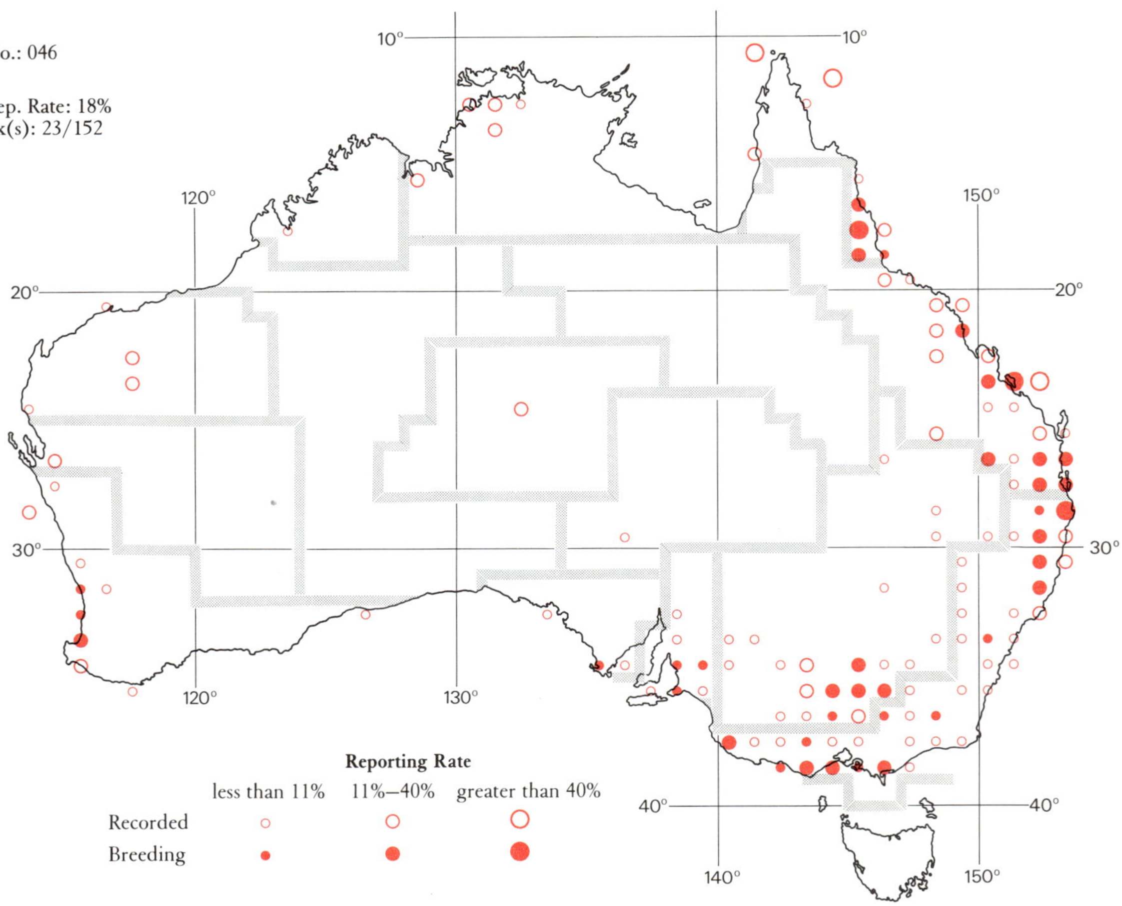

Lewin's Rail

Rallus pectoralis

Lewin's Rail lives in Indonesia and New Guinea as well as Australia. Three Australian subspecies are recognized: *pectoralis* in the eastern mainland, *brachipus* in the Tas. Region and *clelandi* in the South-West Region.[773] The last sighting of *clelandi* was made in 1932 at 34/115 near Bridgetown.[1884] This subspecies is known from only four specimens: two from Margaret R (33/115), taken in 1907, one from King George's Sound (35/117) and one from an unknown location. Despite organized searches there are no records in the Field Atlas and it seems possible that *clelandi* is extinct.

The distribution of Lewin's Rail in eastern Australia determined by the Field Atlas is similar to that obtained by adding in all the Historical Atlas records. The combined map suggests that there may be a gap in the species' range in Eastern Qld between 22°S and 26°S. A record on Eyre Pen. in 1951[1702] is unconfirmed.

Movements are suggested by observations such as at Laverton (37/144) in the South-East Region where Lewin's Rail was present only intermittently.[1822] At some sites in Tas. as well numbers change from year to year.[249,571] In southern SA a lack of records in winter suggests that the Rail is absent but this could also reflect the greater difficulty of observing it in winter. Reporting Rates generally were lower in winter, presumably for the same reason, and suggest no large-scale seasonal movement.

Lewin's Rail lives in dense vegetation.[1343] In Tas. it occupies swamps[1054] and on Maatsuyker I. (43/146) tree fern and cutting grass under woodland.[1261] Nests are often at high density along drainage areas and river flats.[1668] Lewin's Rail feeds on invertebrates, including insects, probing with its bill into crevices in dry ground, mud and soft soil. Birds brought in by cats accounted for a number of Field Atlas records. The species' decline in the South-West Region resembles that of the Noisy Scrub-bird and the bristlebirds. The decline of these passerines has been attributed to grazing by domestic stock and frequent firing of the country.[1684]

There are no measures of abundance.

Field Atlas Records: 214

1° blocks–with records: 54 (7%)
–with breeding records: 13 (24%)

Atlas No.: 045

Max. Rep. Rate: 6%
In block(s): 19/147,39/143

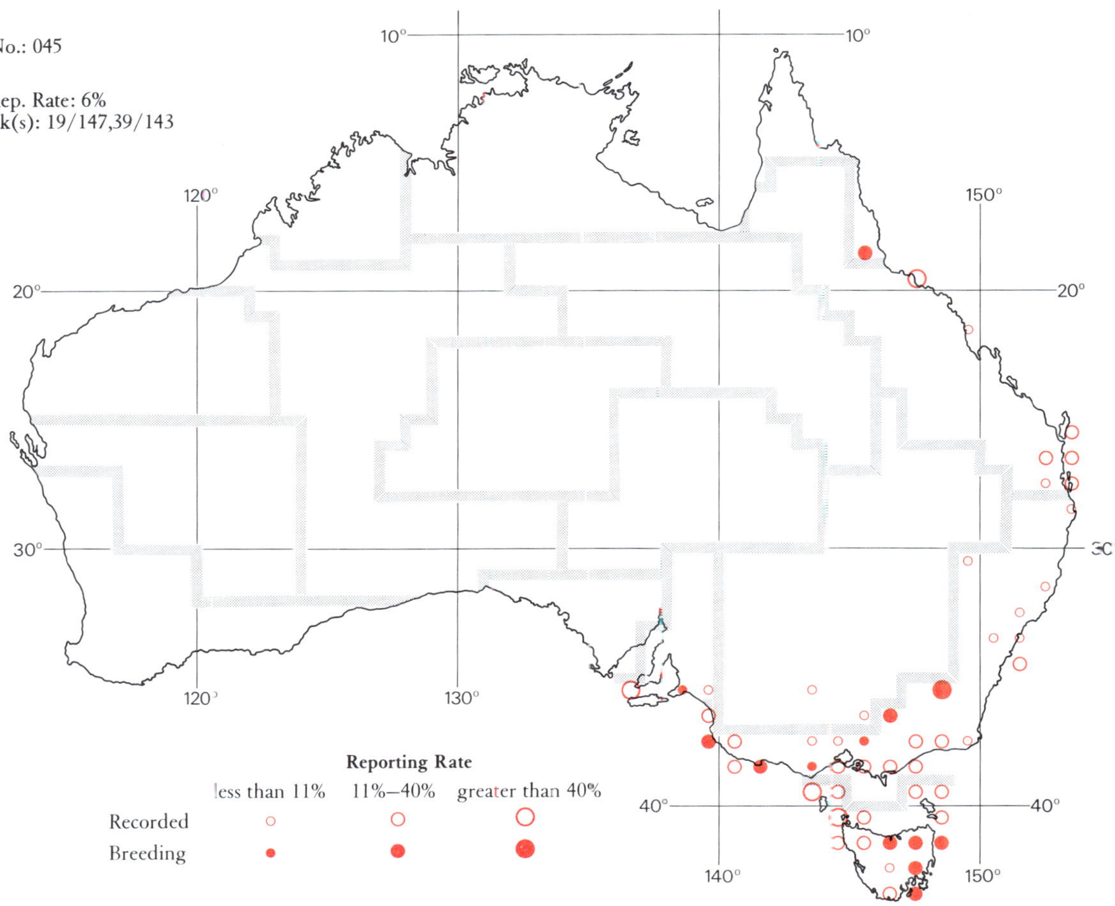

Chestnut Rail

Eulabeornis castaneoventris

The Chestnut Rail lives only on the Aru Is (6S/134E) and in Australia. Combined Historical and Field Atlas records show it to have been recorded as far east as Disaster Inlet (17/139) on the Gulf of Carpentaria and as far west and south as Derby (17/123) on the Kimberley coast. Gaps in range suggested by this combined map are probably attributable to lack of coverage rather than lack of birds. The Atlases contain no records this century from the southern coast of the Gulf of Carpentaria, the easternmost being from Centre I. (15/136). This may also represent lack of coverage but one recent expedition failed to find the bird there.[748]

There is no information about movements of the Chestnut Rail. Though capable of flight, it usually runs through the mangroves where it invariably lives. Except where accustomed to powerboats, it is shy and difficult to see. Despite the number of observers who have visited Darwin (12/130), the species was not recorded there until 1972.[1446] Its presence can be detected by its trumpeting duet. The Chestnut Rail feeds on crabs.[1859]

There are no measures of abundance.

Field Atlas Records: 35

1° blocks-with records: 10 (1%)
-with breeding records:

Atlas No.: 047

Max. Rep. Rate: 11%
In block(s): 12/130

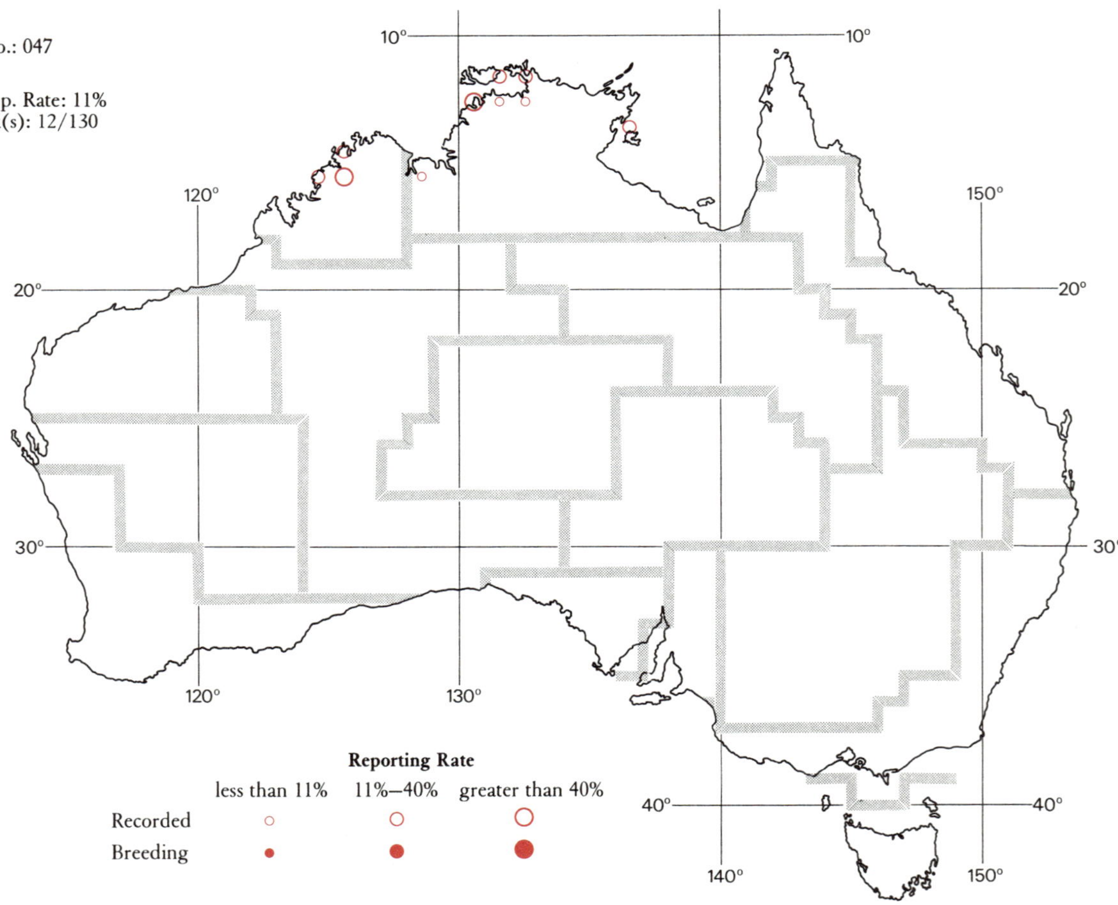

Red-necked Crake

Rallina tricolor

The Red-necked Crake lives in Indonesia, New Guinea and, in Australia, only in the Cape York and Atherton Regions. Combined Historical and Field Atlas records give a continuous distribution on the eastern side of the Cape York Region with a gap to the Atherton population.

Some Red-necked Crakes apparently leave Australia in winter. The species is thought to move through Torres Strait travelling at night[503] and has been noted arriving on Cape York (10/142) at the beginning of summer.[1343] It probably leaves the Cape York Region when the rainforest floor becomes dry in winter.[129] In the Atherton Region where rain falls throughout the year observers have recorded the Red-necked Crake in all months of the year and conclude that it may be sedentary.[1213] The Field Atlas contains insufficient records from the Cape York Region to compare seasonally. In the Atherton Region the Reporting Rate in summer was 6.0% compared with 1.7% in winter, which may reflect only greater conspicuousness in the summer breeding season.

The Red-necked Crake lives in thick vegetation, seasonally rather than permanently wet. In the south it is in thicker and moister areas than in the Cape York Region[700] living in the densest parts of the rainforest. In the north it is found also on the ridges. Many records are based on the bird's distinctive calls. The Red-necked Crake eats aquatic invertebrates, frogs, tadpoles, insects, earthworms and snails, feeding mostly at night.[104,1145,1213]

There are no measures of abundance.

Field Atlas Records: 82

1° blocks–with records: 9 (1%)
–with breeding records: 1 (10%)

Atlas No.: 048

Max. Rep. Rate: 10%
In block(s): 15/145

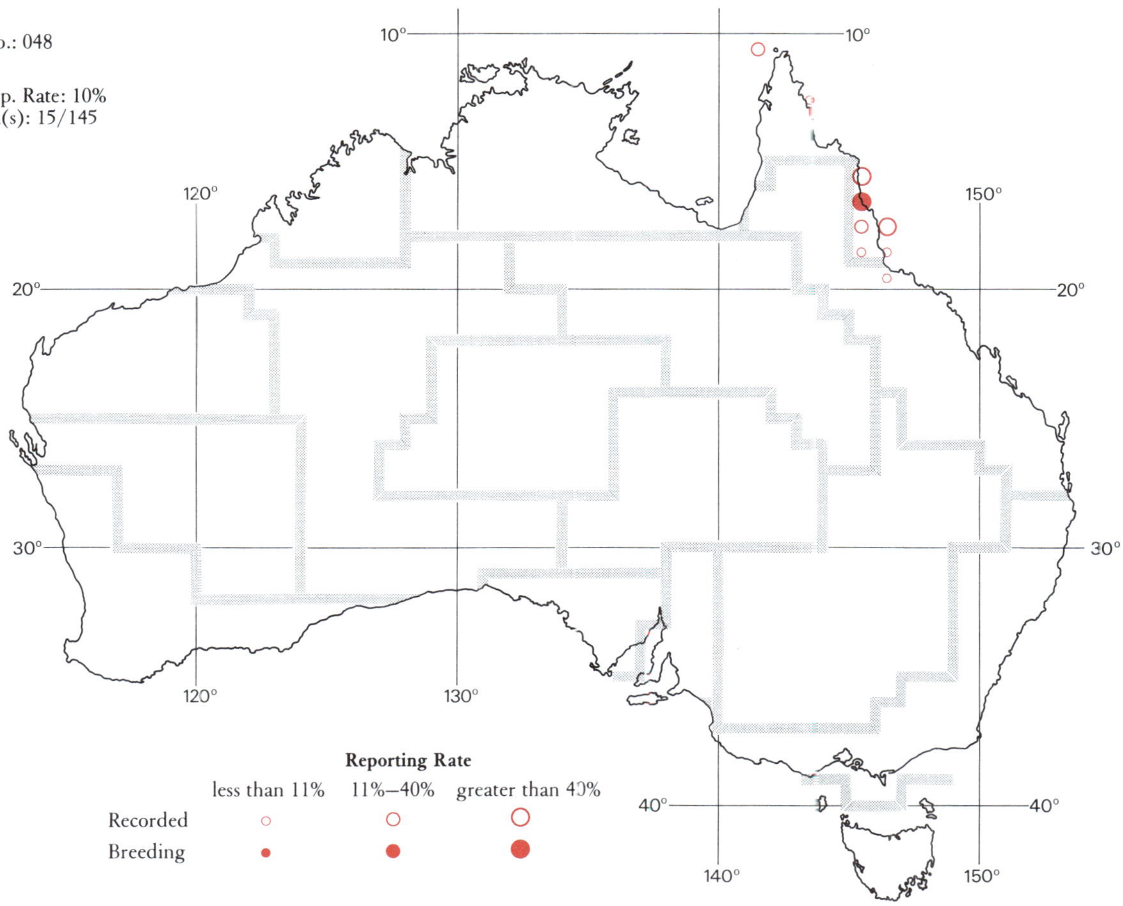

Baillon's Crake

Porzana pusilla

Baillon's Crake lives in Europe, Africa, Asia, New Guinea and Australia. In Australia it is represented by the subspecies *palustris*.[388,422] There are no breeding records in the Atlases north of 26°S though sightings are made throughout northern Australia. In the Tas. Region, the species has not been recorded from Flinders I. (40/148); there is one undocumented reference to eggs taken on King I.;[299] and in Tas. itself it is rarely reported but has bred. It has been recorded as a vagrant on Macquarie I. (54S/158E).[734]

A regular north-south winter-summer migration is likely. Seasonal movements are reported to take the bird out of the southern parts of its range Apr.-Sept. in both east and west.[276,1216,1822] Two birds were caught at Booby I. (10/141) in June 1979 indicating some movement in Torres Strait[503] and specimens taken from flocks in the Top End in July 1974 had fat stored in the manner of migratory birds.[1215] The species is rarely recorded in the Top End and probably winters there only when conditions are suitable.

Reporting Rates showed marked seasonal changes. In the South-East and Murray-Darling Regions respectively they were 0.8% and 1.2% in summer, dropping to 0.06% and 0.4% in winter. These changes may be due to emigration but other rails and crakes show similarly decreased Reporting Rates in winter apparently because the birds are not calling and are therefore less conspicuous. The nominate subspecies *pusilla*, breeding in central Asia east to Japan, migrates as far south as the Philippines, so perhaps an equivalent movement takes birds out of Australia in winter.

Baillon's Crake lives in swamps, along the edges of reed-beds and in moderately dense cover where floating vegetation is abundant. It is rarely if ever in salty water. In Europe, Baillon's Crake feeds on insects and seeds in shallow water[422] and in Australia it has been reported to eat insects.

Thirty Crakes were found in two pools each of 2.5 ha (6 birds/ha) near Melbourne (37/145) in a year when they were said to be very common.[276] In most years in large areas of suitable vegetation the density is 1 bird/ha and in smaller isolated reed clumps 2 birds/ha.[911]

Field Atlas Records: 352

1° blocks-with records: 99 (12%)

-with breeding records: 9 (9%)

Atlas No.: 050

Max. Rep. Rate: 8%

In block(s): 15/128

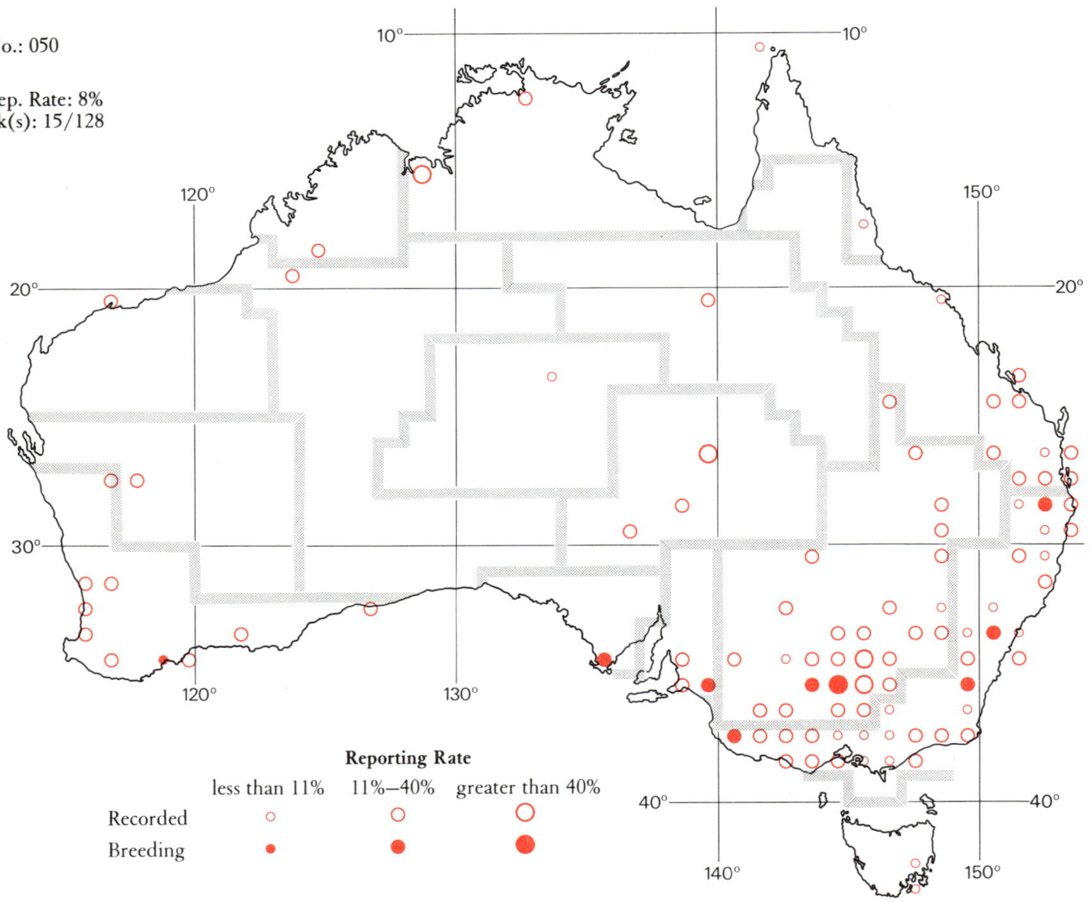

Australian Crake

Porzana fluminea

All breeding records of the endemic Australian Crake in the Atlases were south of 27°S but sightings were made as far north as Kununurra (15/128) and Innisfail (17/146). There were records throughout Eastern Qld in the 1951–76 period, though none was made during the Field Atlas. In the past the species has bred in Tas. and during the Field Atlas was recorded for the first time on King I.: at Tatham's Lagoon (39/143) in Feb. 1979.

In parts of the southern Musray-Darling and western L. Eyre Regions the Australian Crake is always present[86,833] but there are movements. There were 80–90 Crakes at Laverton (38/143) in Apr. 1951 that had gone by the end of May.[1822] At Coongie L. (27/140) the bird was 'almost in hundreds' in Aug.-Sept. 1979. Tens and twenties were reported at sites in the Kimberley, Western Deserts and Pilbara Regions as wetlands dried up following good rains in 1980–81. Reporting Rates were uniformly higher in summer than in winter presumably because the bird is more conspicuous when calling in summer. No seasonality in the pattern of movement is suggested by the Field Atlas.

The Australian Crake lives near pools about 1 ha in extent. It inhabits lagoons and areas of mangrove and samphire, feeding among the reeds in water up to 5 cm deep. It follows receding wetlands in preference to the deeper, flooded conditions of winter. Gatherings of 15 or more birds are common on bore drains or swamps that are drying out. The Australian Crake runs to cover when disturbed but, when undisturbed, will fly over open water between stands of reeds. Boldest of the crakes when feeding, it takes aquatic insects and their larvae, hatching-flies and sometimes tadpoles.[277] It appears to be more tolerant of salt conditions than other crakes and small rails.

There are no measures of abundance.

Field Atlas Records: 802

1° blocks–with records: 128 (16%)

–with breeding records: 18 (14%)

Atlas No.: 049

Max. Rep. Rate: 30%
In block(s): 29/136

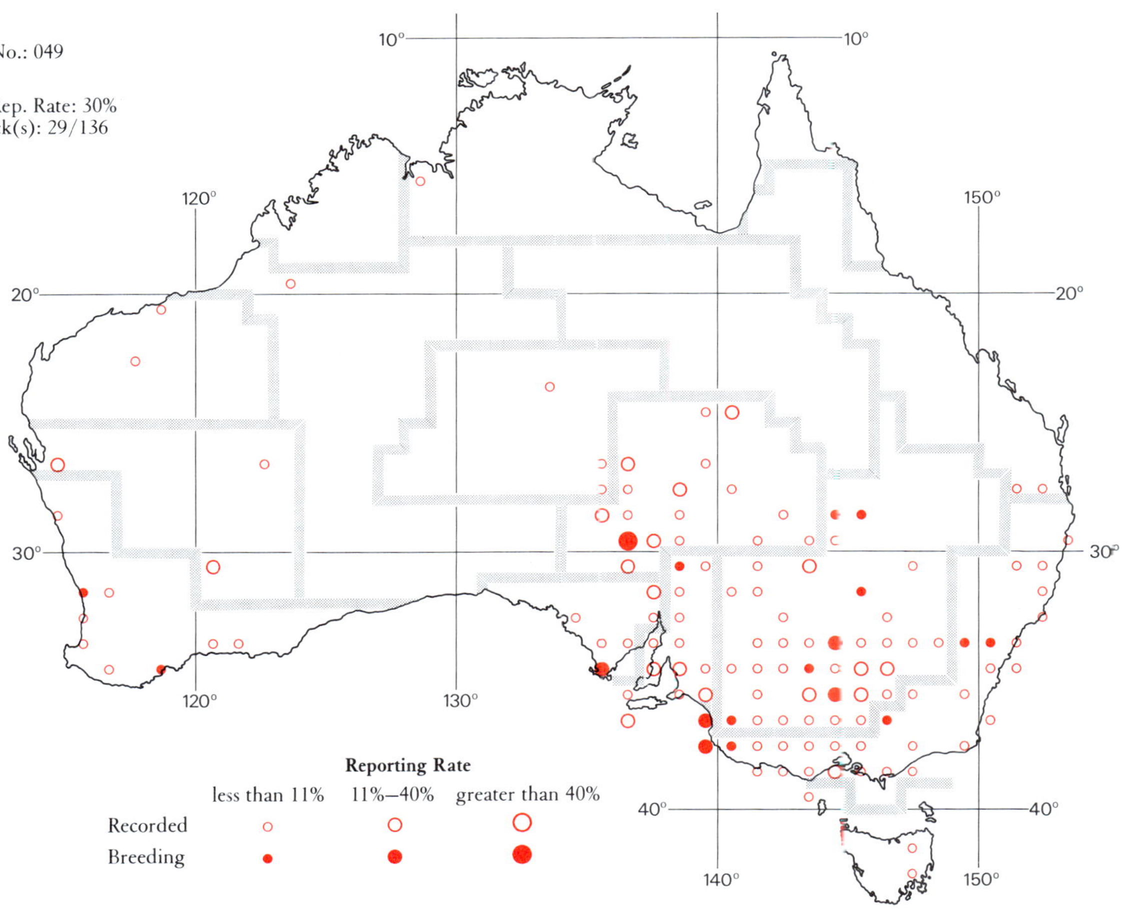

Spotless Crake

Porzana tabuensis

The Spotless Crake ranges throughout south-eastern Asia, New Guinea, the south-western Pacific, New Zealand and Australia. The Australian subspecies is *plumbea*.[388] There are no records from Flinders I. (40/148) in the Atlases. The northernmost confirmed breeding records are those in the Field Atlas.

Records from the Top End and Pilbara indicate the Crake's ability to move long distances beyond its normal range. There are reports from Booby (10/141) and Thursday (10/142) Is in Torres Strait where the bird is thought to be a regular nocturnal migrant.[503] Reporting Rates in the South-East, Murray-Darling, SA Ra., Tas. and South-West Regions taken together were 0.5% in summer and 0.2% in winter; there were no winter records at all in the Tas. Region. This suggests there may be some seasonal movement, though the bird is more difficult to find in winter when it is likely to be in deep water with some cover and calling less frequently.

The Spotless Crake usually lives in swamps, often those dominated by cumbungi,[277] as well as along streams, in mangroves and occasionally in saltmarshes at the edges of lagoons. On islands it is sometimes in less sheltered places including on dry land.[1638] The Spotless Crake feeds on insects and has also been observed to take carrion and ripe blackberries.[277] It is probably common in many places but overlooked because it is the shyest of the crakes and always feeds within cover.

At one locality, in the southern Murray-Darling Region, five pairs usually bred along a watercourse where there was a strip of suitable rushes 800 m long and 25 m wide. It is not known whether the areas used by these pairs were contiguous.[1297] Near Melbourne (37/145) individual pairs were thought to occupy home ranges of 0.04 ha.[277]

Field Atlas Records: 299

1° blocks–with records: 79 (10%)

–with breeding records: 10 (13%)

Atlas No.: 051

Max. Rep. Rate: 20%
In block(s): 28/113

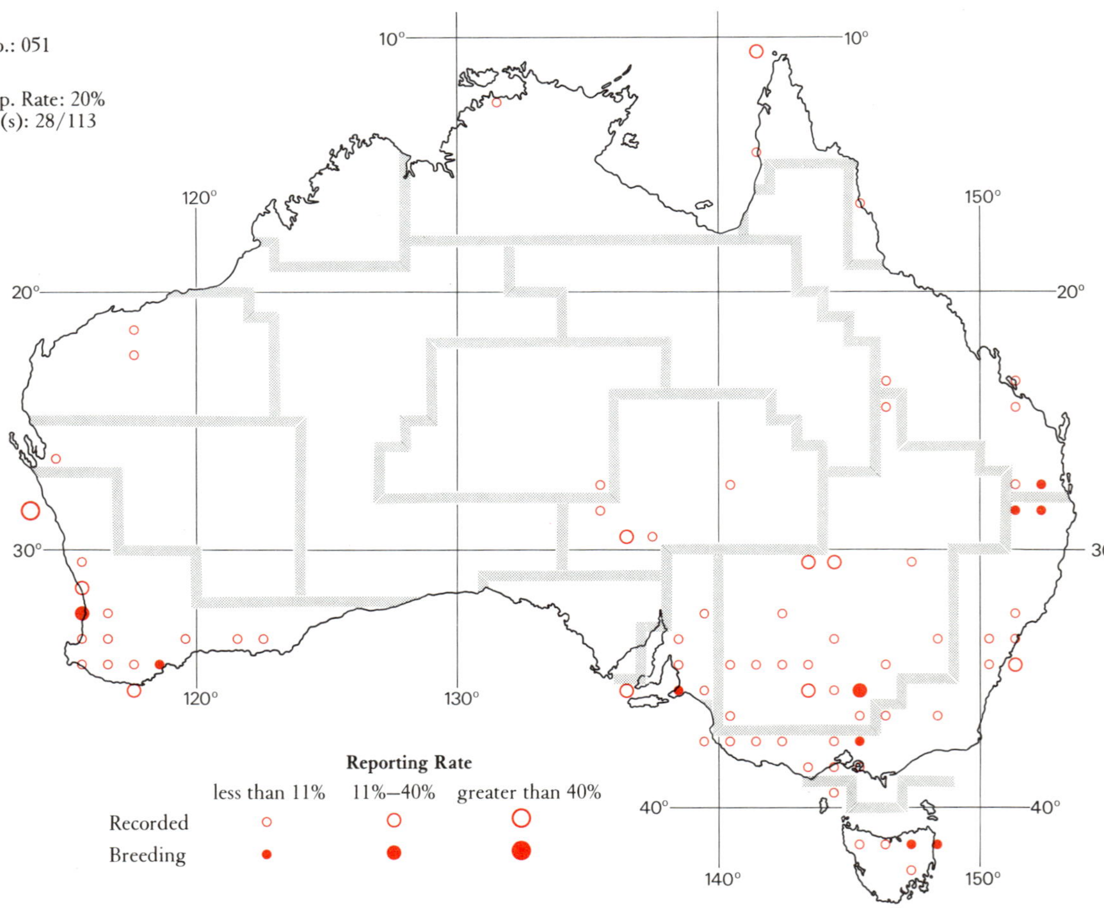

White-browed Crake

Poliolimnas cinereus

The White-browed Crake is distributed beyond Australia into south-eastern Asia, the Pacific and New Guinea. The Australian subspecies is *leucophrys*.[388] Records of eggs collected before 1901 on the Nicholson R (about 17/139) south of the Gulf of Carpentaria[299] and other historical records help to fill gaps in the Field Atlas map and suggest that the distribution of the White-browed Crake may be more continuous than appears on the map.

There are two accounts of the Crake arriving on the northern coast of Australia for the wet summer season, breeding, then departing[107,423] and near Darwin (12/130) the species is most abundant during the wet summer season.[423] Surveys in the late 1960s failed to find the bird at Port Essington (11/132)[664] where specimens were collected in 1840,[723] perhaps because the visits in the 1960s were in late Feb.-Sept., outside the wet summer season. Reporting Rates in the Atherton Region were higher in summer than winter (2.9% compared with 0.8%) as they were round Darwin (3.0% compared with 0.4%). In other rails and crakes similar differences are at least partly attributable to greater conspicuousness when the birds are calling during the breeding season but movements of this species do take some birds through Torres Strait on what is thought to be a regular migration.[503]

The White-browed Crake inhabits lagoons and moves over the aquatic vegetation, picking insects from the surface.[701] Gut contents have also included earthworms, slugs and vegetable matter.[1343] The Crake will move to the side of a lagoon when observed but tends to keep out of the surrounding forest, staying on the edge of the open surface of the pool. It rarely swims but will sometimes run up the roots and limbs of trees, as well as flying. It is noisy and often gives its presence away by calling in response to sudden loud noises.[231,701]

Eight pairs were counted in about 1 km along the edge of a lagoon on Bougainville I. (5S/155E).[701] There are no other measures of abundance.

Field Atlas Records: 76

1° blocks-with records: 15 (2%)
-with breeding records: 5 (33%)

Atlas No.: 052

Max. Rep. Rate: 10%
In block(s): 10/141

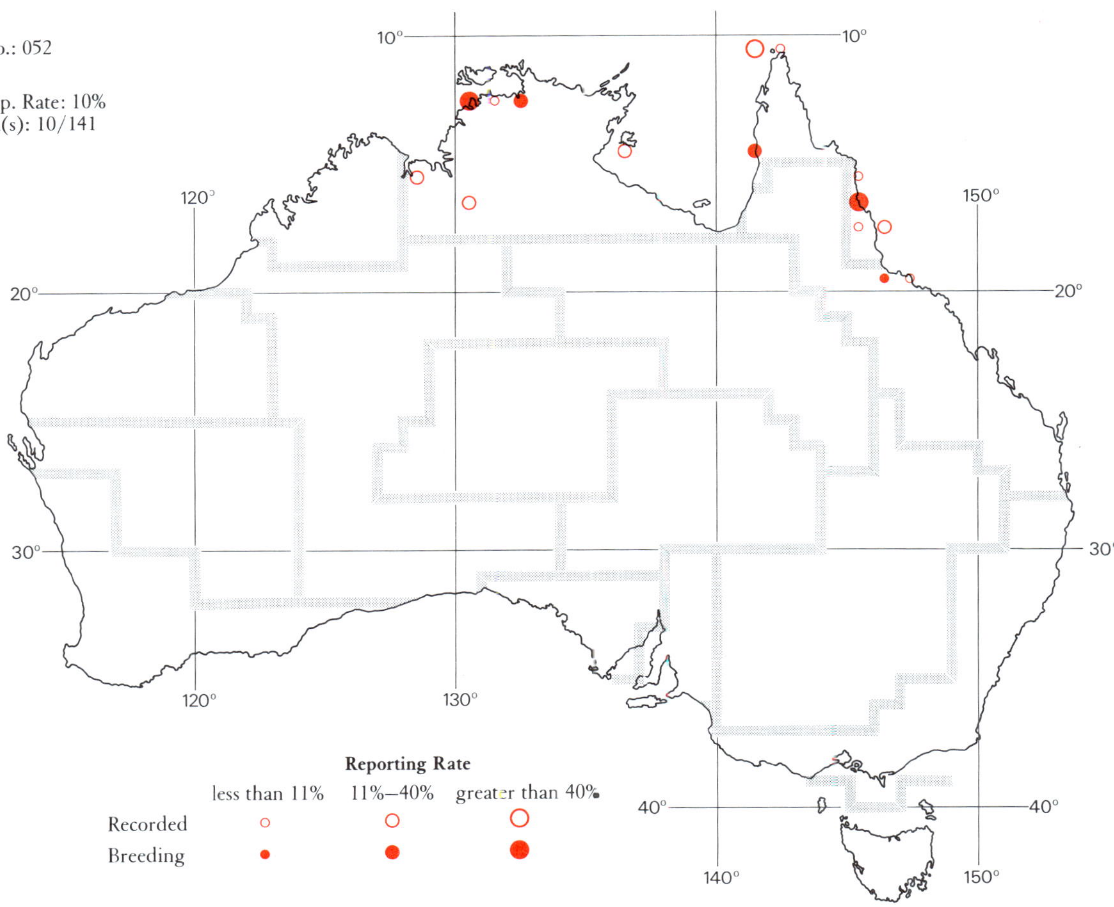

Bush-hen

Gallinula olivacea

The Bush-hen ranges from south-eastern Asia to New Guinea and Australia. The known range of the Australian subspecies *ruficrissa* was extended in 1977 from the eastern coast to the Top End.[1174] Early records of the Bush-hen in this region have been doubted[1711] but seem likely to have been correct, e.g. Aborigines at Millingimbi (12/134) had a name for the Bush-hen.[1828]

Though the Bush-hen was found breeding at Grafton (29/152) as early as 1864,[199] there are no records in the 1901–50 period south of 24°S. Its discovery near Brisbane (27/153) in 1964[1280] may simply reflect the bird's secretiveness but, considering the extent of coverage of the Qld coast in 1900–64, it seems unlikely to have been overlooked. More likely the Bush-hen was absent from these areas and returned to them when conditions were favourable.[638] The Top End birds may also be part of such an expansion but the lack of resident observers there makes it probable that the species was simply overlooked.

It has been suggested that the Bush-hen is nomadic, retreating to areas of permanent water during dry conditions and spreading out when rains come.[171] The arrival of the bird at Cardwell (18/146) after rain and its subsequent disappearance when rain fell farther north is consistent with this interpretation.[108] Part of the population may undertake a regular seasonal movement. The Bush-hen is thought to cross Torres Strait regularly[503] and in the Cape York Region birds have been recorded flying south with the onset of the wet summer season.[1145,1147,1343,1861] The Field Atlas contains Bush-hen records throughout its range in all seasons. At Mackay (21/149) the bird is present all year. It is secretive and generally located by its call, more frequent and prolonged in summer than in winter. This may be the main reason for the lower Reporting Rate in winter (2.4%) than in summer (10.5%) in the Atherton Region.

The Bush-hen lives in long, dense vegetation adjacent to fresh water and has also been recorded in lantana 0.5 km from water. It feeds on insects and some vegetation.[231,1145] It can survive clearing only where small patches of scrubby vegetation are left on swampy ground that retains fresh water.

There are no measures of abundance.

Field Atlas Records: 190

1° blocks–with records: 26 (3%)
–with breeding records: 9 (35%)

Atlas No.: 053

Max. Rep. Rate: 11%
In block(s): 15/145

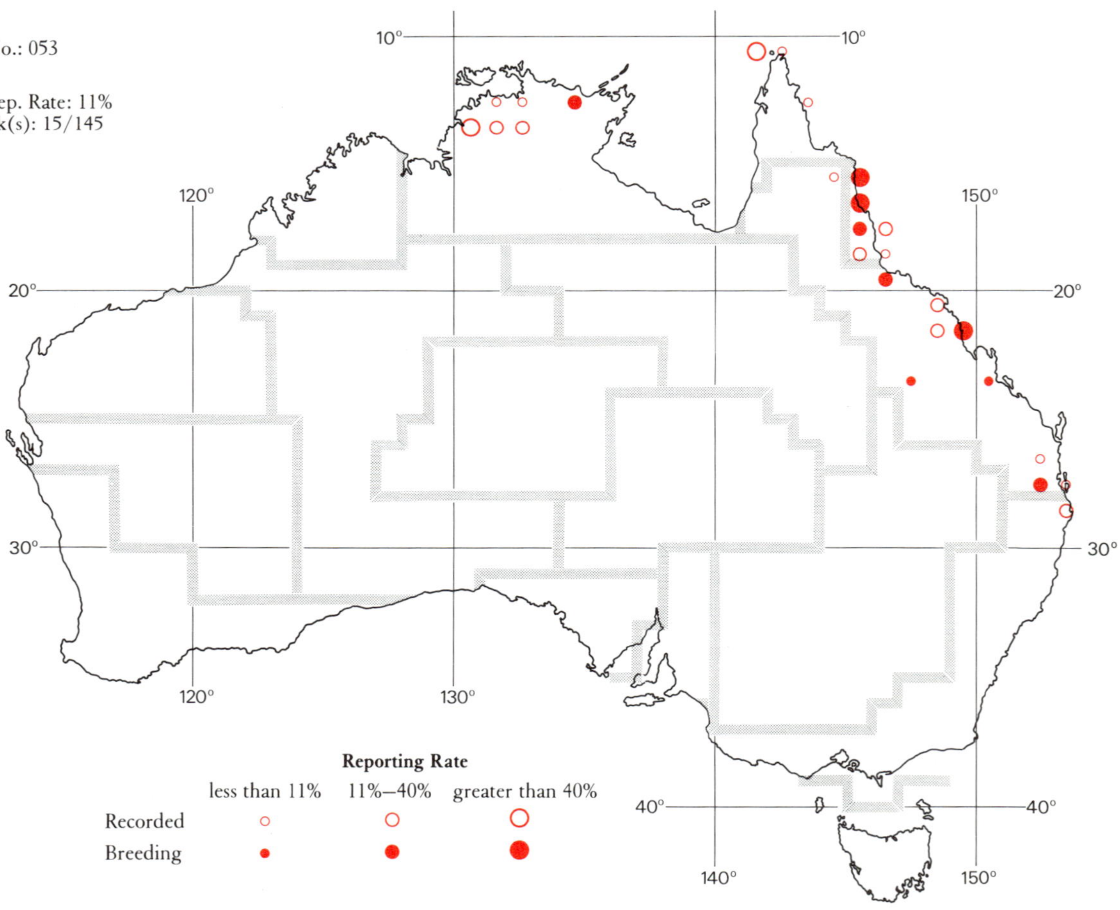

Tasmanian Native-hen

Gallinula mortierii

The Tasmanian Native-hen is endemic to the Tas. mainland but was introduced to Maria I. (42/148) in 1969.[734] Comparison of the Field and Historical Atlases indicates that its range has not changed substantially though it has expanded into newly opened land.

There is little movement; individuals disperse at the end of their first year but 15 km was the farthest movement recorded during a detailed study.[1485] A movement of 40 km has since been recorded.[85]

The Native-hen lives throughout the agricultural areas of Tas., specially where farming is less intense and vegetation along creek banks provides cover.[1485] It is absent from the heath dominated by button-grass in south-western Tas. The bird requires short, grazed pasture throughout the year and damp pastures adjacent to streams with cover for nesting. Breeding is often communal.[1485] An analysis of the gut contents of 447 specimens showed that the Native-hen feeds principally on plants (90% by volume) but that the parents feed many insects to young chicks.[1485]

A decline was reported in the number of Native-hens in the Midlands (42/147) after the introduction of myxomatosis and the consequent reduction in the number of rabbits.[730] Probably the bird originally lived on 'lawns' maintained by the grazing of marsupials and benefited from the introduction of the rabbit as well as the clearing of land for sheep grazing.[1485] The clearing of country for other purposes also enables the Native-hen to expand its range. Road construction provided a corridor along which the bird entered the L. Gordon-New L. Pedder (42/146) area during the Field Atlas.[1533] Farmers regard the bird as a pest and it is not protected.[1533] Its disappearance from the Australian mainland roughly coincided with the arrival of the dingo but in Tas. it survives predation by cats as well as by native species such as the Tasmanian Devil.

There are no estimates of abundance.

Field Atlas Records: 1869
10′ blocks–with records: 202 (55%)
–with breeding records: 107 (53%)

Atlas No.: 054

Max. Rep. Rate: 82%
In block(s): 41°05′/145°45′

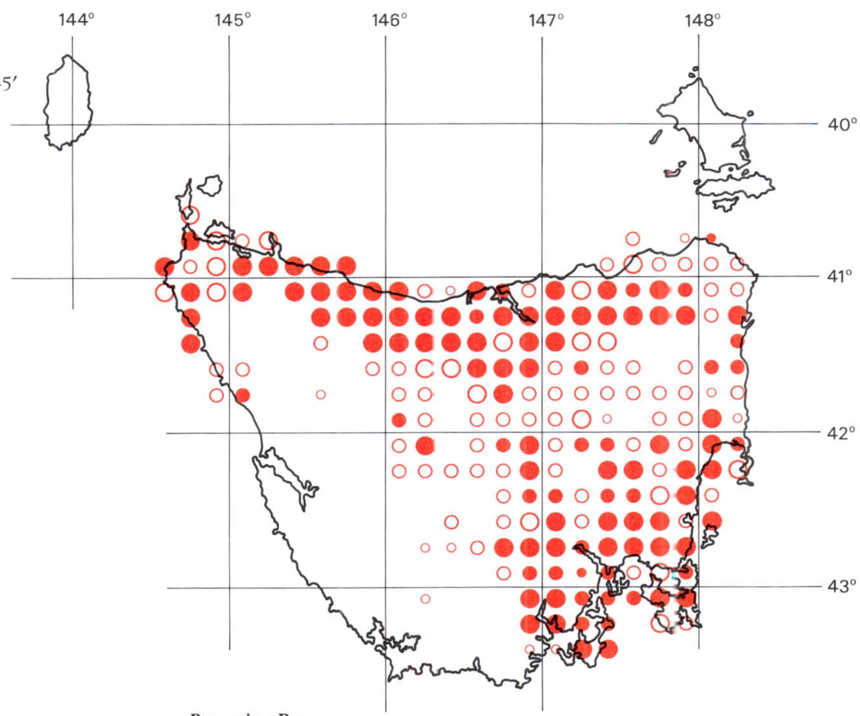

Reporting Rate

	less than 11%	11%–40%	greater than 40%
Recorded	○	○	○
Breeding	●	●	●

Black-tailed Native-hen

Gallinula ventralis

The endemic Black-tailed Native-hen is almost confined to mainland Australia. The only Tas. record is a specimen taken at Epping (41/147) in June 1916[734] but not identified until 1970.[1314] It has not been reported breeding north of 22°S in the Atlases.

The Black-tailed Native-hen is apt to arrive in or disappear from a locality in a matter of days or even hours.[1343] In May 1973 more than 10 000 arrived near Manara (32/143) within 14 hours.[1594] Such irruptive movements brought large numbers into the South-East and SA Ra. Regions in 1941–42, 1968–69, 1972–73 and 1975.[1217] Similar irruptions into the South-West were reported in 1833–35, 1853, 1872, 1886, 1897, 1919, 1952, 1964 and 1975.[1638] The Field Atlas suggests a seasonal tendency in movements of the bird taking it farther north in summer. The Reporting Rate north of 26°S was 5.3% in the wet summer season and only 0.3% in the dry winter season.

The Black-tailed Native-hen lives in the inland near water. During irruptions there are large gatherings. One site near Adelaide (34/138) that usually held 30 birds contained 4000 during the 1975 irruption.[1217] Sometimes small breeding colonies form, one at Mystic Park (35/143) containing nine nests.[1552] The Black-tailed Native Hen feeds on vegetation and insects. The cause of irruptions is thought to be the exodus of large numbers from the inland as it dries up after a wet year when there is extensive breeding.[1217] In the SA Ra. Region the bird was observed to leave as soon as heavy rains fell inland, even though green feed remained in the south.

There are no measures of abundance.

Field Atlas Records: 2700

1° blocks–with records: 308 (38%)
–with breeding records: 42 (14%)

Atlas No.: 055

Max. Rep. Rate: 41%
In block(s): 29/136

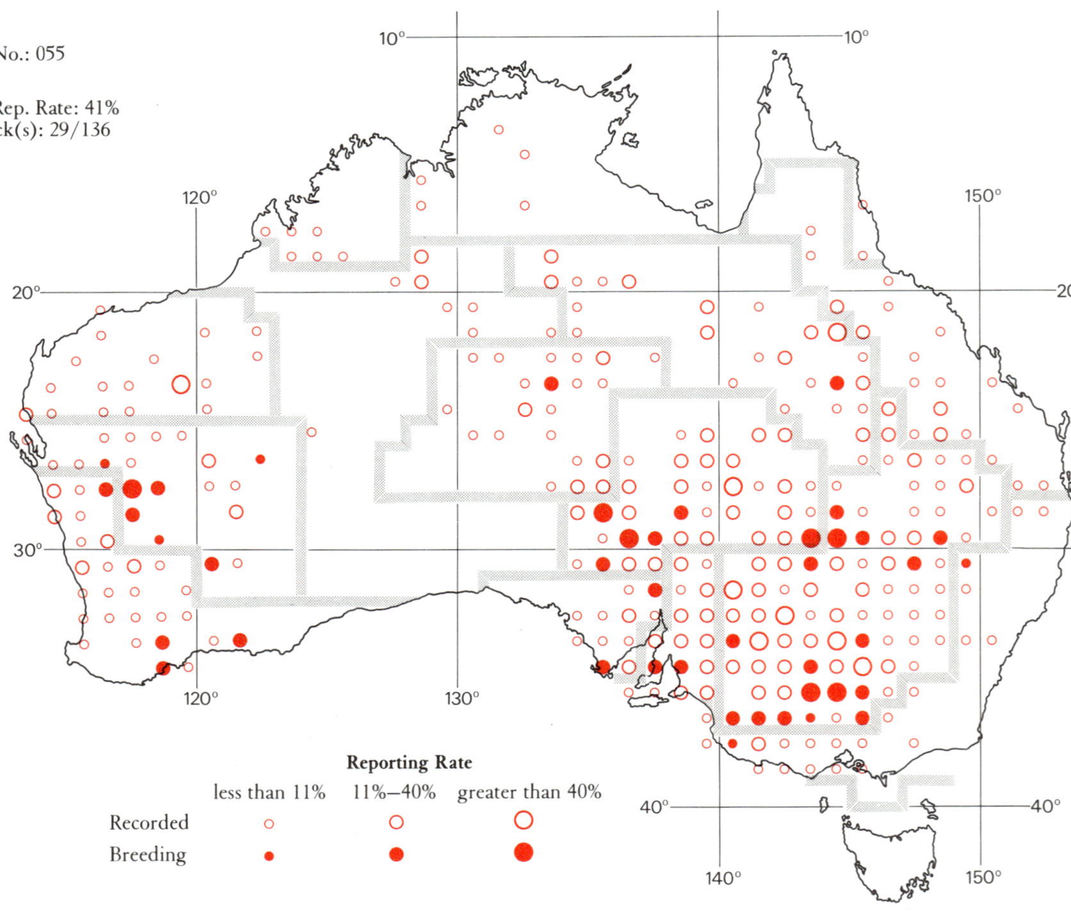

Dusky Moorhen

Gallinula tenebrosa

The Dusky Moorhen lives in Indonesia, New Guinea and Australia. The range of the Australian subspecies *tenebrosa*[388] is expanding. The first Top End record was in 1976[1754] and the Field Atlas contains the first breeding reports from the Rolling Downs, L. Eyre and Eyre Pen. Regions.

The earliest record in the Tas. Region is of eggs taken on Flinders I. (40/148) in 1935.[732] The Moorhen has been regularly present and breeding on King I. (39/144) at least since the 1960s[739] and its range has recently extended into Tas. where the first record was in 1976. In 1977 the Moorhen bred near Launceston (41/147) and has continued to do so there and at other sites during succeeding seasons.

Throughout the Field Atlas, a small population was reported near Port Hedland (20/118) where the bird had been previously only a vagrant.

The Field Atlas does not suggest a large-scale seasonal pattern of movement though observations of movement are reported in the literature. At Canberra (35/149) a colour-banded population built up to a peak each Dec. (the end of the breeding season) after which many birds departed leaving only those holding territories.[686] There was no indication where the rest may have gone. Dispersal must take place from permanent breeding sites to intermittent waters in the inland such as form in the L. Eyre Region.[86] Populations in the Tas. Region fluctuate with movements to and from the South-East Region.[734]

The Dusky Moorhen inhabits reed-lined pools of fresh water, including bore drains in the inland, as well as ornamental pools in parks and reserves within cities. It is likely to appear wherever there are pools associated with reed-beds. Social groups jointly defend a territory and breed communally.[686] The Dusky Moorhen feeds on aquatic vegetation.

At Canberra breeding density varied from 53–39 birds/ha of lakeside reeds.[686]

Field Atlas Records: 10 139

1° blocks-with records: 227 (28%)

-with breeding records: 118 (52%)

Atlas No.: 056

Max. Rep. Rate: 42%
In block(s): 27/152

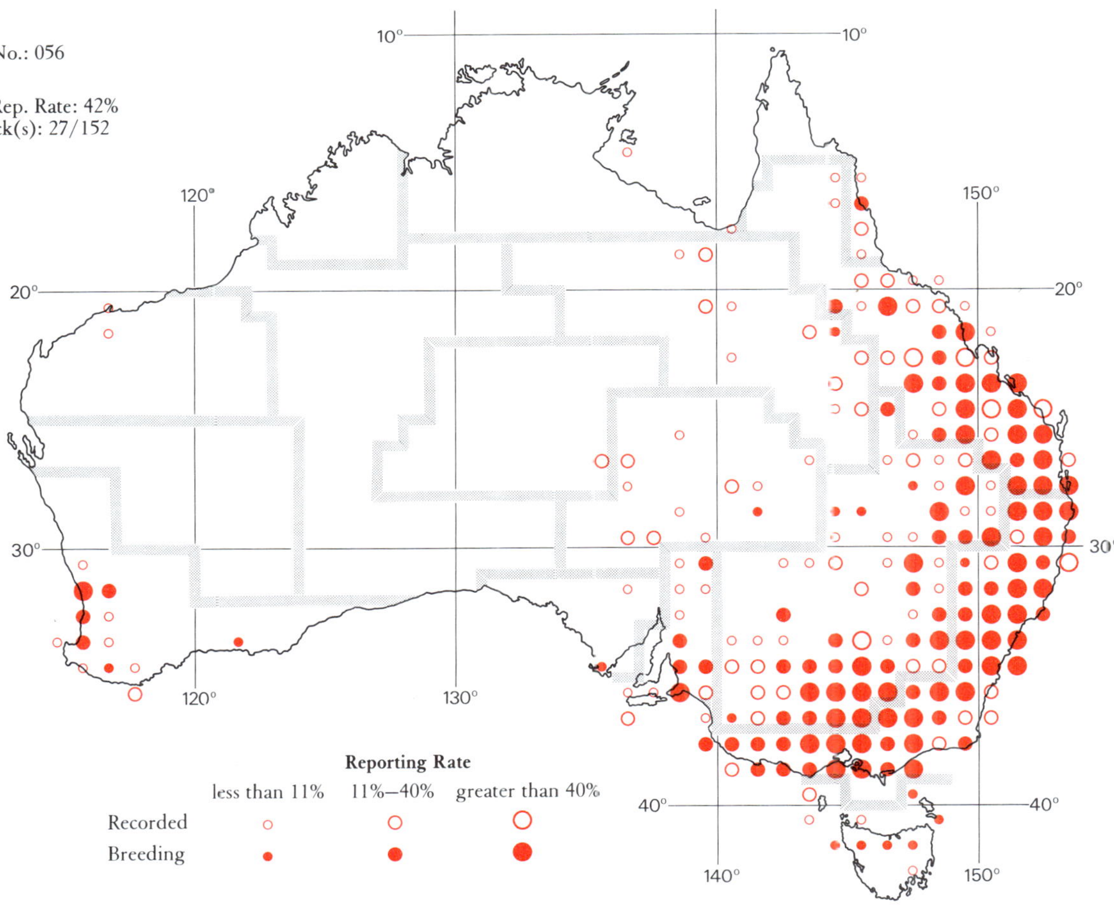

Purple Swamphen

Porphyrio porphyrio

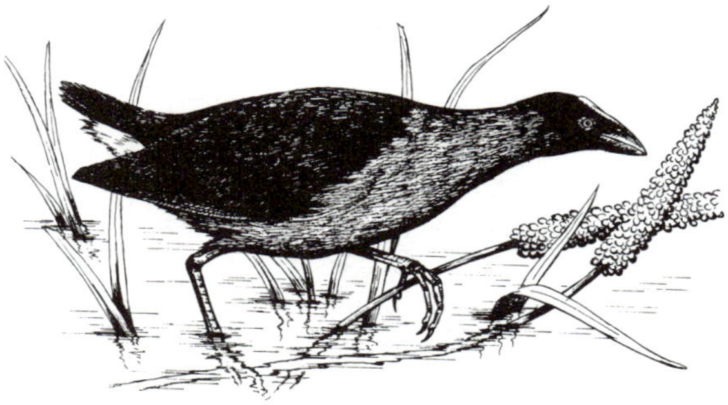

The distribution of the Purple Swamphen stretches from Africa, southern Europe, Asia, Pacific islands, New Zealand and New Guinea to Australia. In Australia two subspecies are recognized: *bellus* in the South-West Region and *melanotus* elsewhere.[1488] The subspecific identity of birds in the Pilbara is not known.

The Field Atlas does not reflect any large-scale seasonal movement but the species is said by one observer to migrate across Torres Strait[1811] and by others to be a wet summer season visitor there.[503] Birds banded in Townsville (19/146) have been recovered in Merauke (8S/140E) in New Guinea[129] and huge fluctuations in numbers of birds in ricefields there are attributed to migration.[855] The longest movement recorded for a banded bird is 1420 km.[85]

The Swamphen is always near water but often grazes 100 m or more from the edge of the swamp in which it lives. In the non-breeding season flocks of several hundred individuals form. Breeding is often communal.[501] During years of heavy rainfall the Swamphen breeds on swamps in the inland.[833]

The Swamphen feeds mainly on vegetation, including seeds and fruit.[1049,1542] At times its grazing of young crops is a nuisance to farmers. As a result, it is only partly protected in the Tas. Region and permits can be issued for the destruction of the bird.[1533] It feeds on the eggs of many species and some birds raid fowlhouses.[177] The bird has also been seen killing ducklings[1771] and frogs[274] as well as taking stranded fish.

A maximum of 174 was observed on about 45 ha of L. George (35/149) during regular counts for three years.[1013]

Field Atlas Records: 10 186

1° blocks–with records: 270 (33%)
–with breeding records: 100 (37%)

Atlas No.: 058

Max. Rep. Rate: 45%
In block(s): 36/150

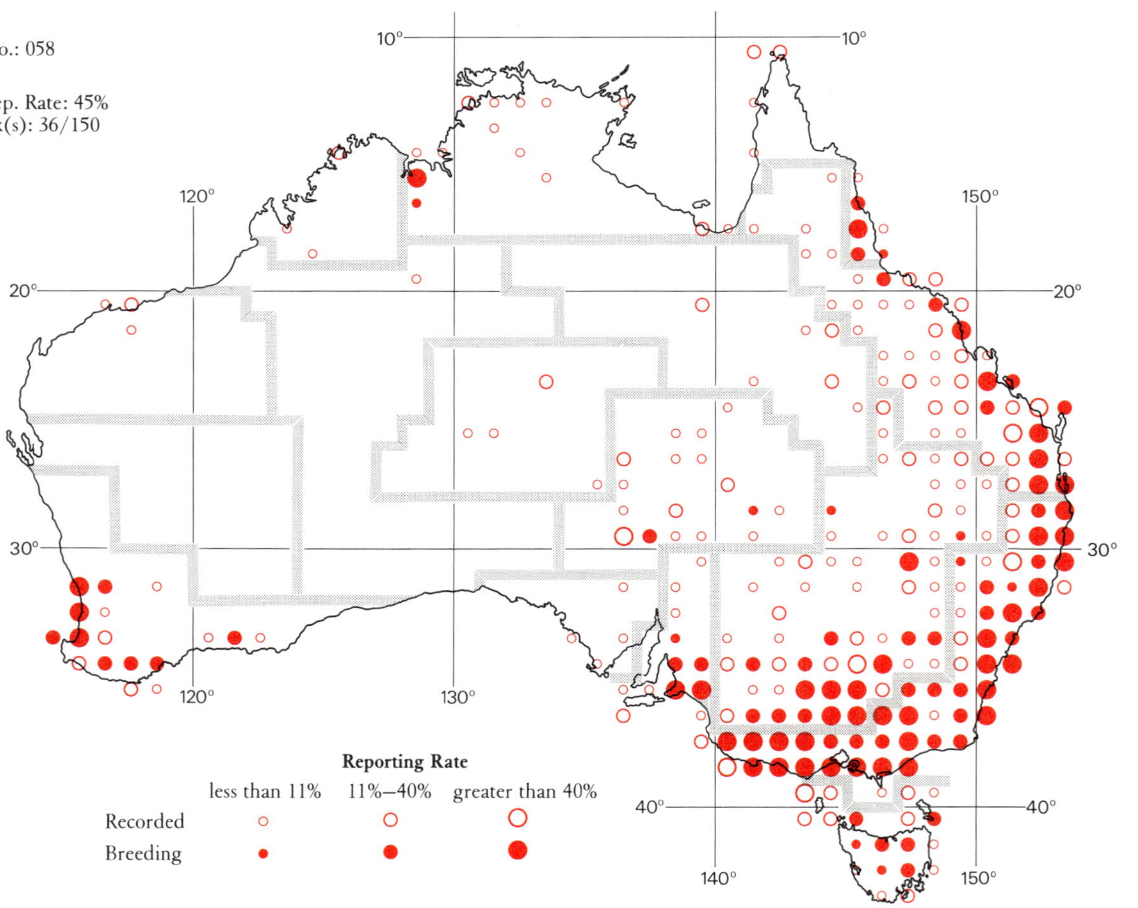

Eurasian Coot

Fulica atra

The Eurasian Coot ranges from northern Europe, northern Africa, India, China, south-eastern Asia and New Guinea to Australia. In Australia the subspecies *australis* has been recognized.[388] The rarity of breeding in the Tas. Region has been noted previously[1657] and confirmed during the Field Atlas. Breeding was reported in only 4.5% of 10′ blocks in which the Coot was recorded in the Tas. Region. The corresponding figure in the South-East Region was 12.2%.

Throughout its world range great changes in the number of Coots at any one site are characteristic but the Field Atlas suggests no large-scale seasonal pattern of movement in Australia. Typical fluctuations were observed on L. George (35/149) in 1961–63 when numbers in a 45-ha section varied from less than 100 to more than 5000.[1013] At Rotamah I. Bird Observatory (37/147) in 1980–81 up to 1600 birds were present in winter and none in summer, the birds presumably departing to breed elsewhere.[285] A spectacular exodus from Tas. was noted overnight in Apr. 1956 when the whole Coot population left synchronously and did not reappear in numbers for 11 months.[1657] An increase and decrease of flocks in autumn in northern Tas. has been related to the arrival of birds from the mainland.[395]

Studies in Vic. show that only five out of 3177 banded birds were recovered at the same site more than six months later.[1339] Of those recovered elsewhere, the mean distance moved was 295 km and four covered more than 500 km.

The Coot gathers on large bodies of water soon after they form. Where nest sites are available, some may breed. Flocks, some exceeding several thousand, were seen in 1977–78 on Cooper Ck (28/137) in the L. Eyre Region[86] but breeding was reported from only one block in the region and its associated rivers. It has been suggested that some big flocks may consist of moulting birds.[1335] The Coot also breeds in isolated pairs on small bodies of water, inland waterholes and farm dams.[264]

The Coot is said to feed mainly on vegetation in open water, sometimes near other feeding waterfowl. Food is gathered either by diving or by pecking it from the surface.[422] The Coot also grazes on grass near wetlands.

There are no measures of abundance.

Field Atlas Records: 14 314

1° blocks–with records: 455 (56%)

–with breeding records: 118 (26%)

Atlas No.: 059

Max. Rep. Rate: 52%
In block(s): 35/143

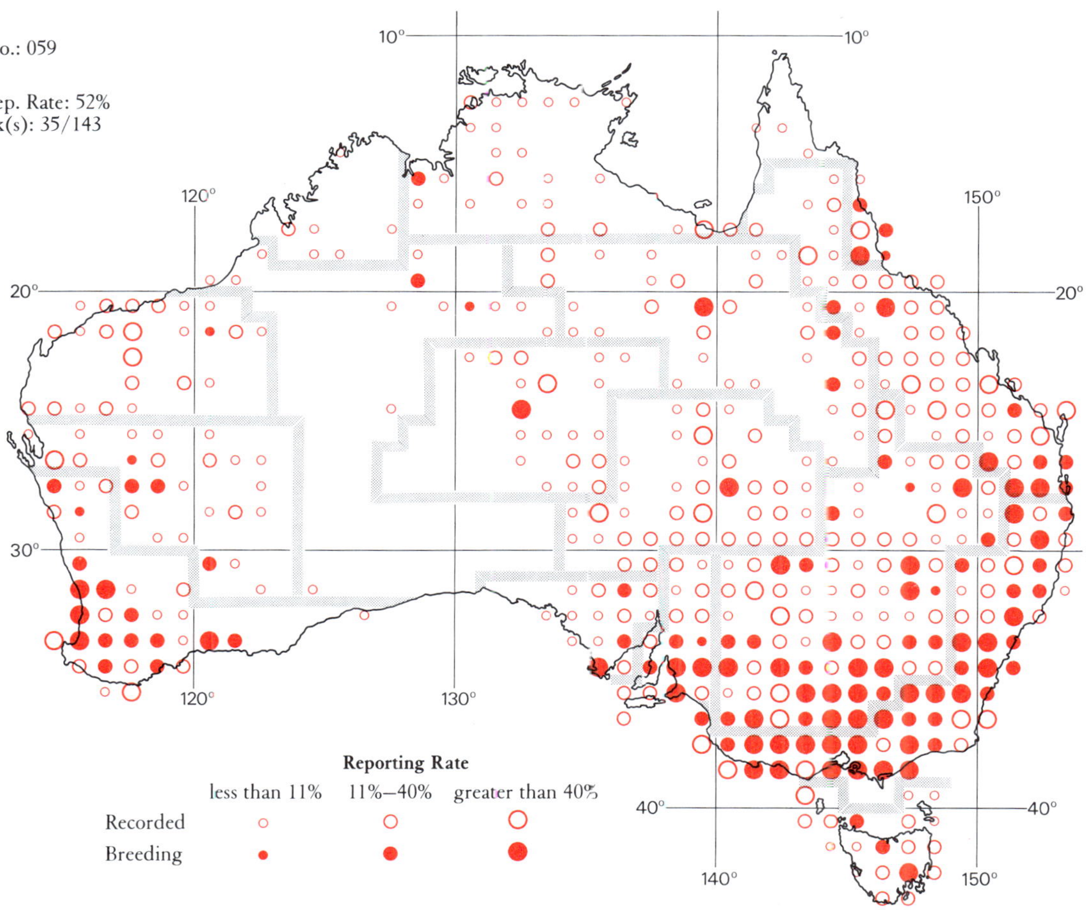

Brolga

Grus rubicundus

The Brolga lives in Australia and New Guinea. It probably breeds in New Guinea though there is no direct confirmation of this.[550] In Australia the Brolga was more widely distributed in the South-East Region last century than now. Before 1901 there are records from several independent sources of the Brolga on the eastern coast between the Hunter R (32/151) and Ulladulla (35/150) and in the country to the east of Melbourne (37/145) where it also bred. The only subsequent sightings from these districts were in 1938 at Port Macquarie (31/152) and in 1957 on the Hunter R and near Wollongong (34/150).[1292]

The Brolga has not been reported breeding in the Pilbara Region and is a vagrant to other regions on the mainland outside the breeding range indicated on the map. There are no records from the Tas. Region.

The Field Atlas suggests no large-scale seasonal pattern to the Brolga's movements. Flocks, sometimes large, form during the non-breeding season.[1800] In the Darwin (12/130) district and the Kimberley Region the birds converge on remaining wetlands during the dry winter season.[423,764] No movement across Torres Strait is reported.[503]

The Brolga lives near swamps, shallow lakes, flooded grassland, saltmarsh and also on dry land. It feeds on large invertebrates, small vertebrates, seeds, green shoots, rhizomes and bulbs.[1144,1800] Tubers of bulkuru sedge are particularly important in northern Australia. In the Kimberley Region dry season flocks of up to 1000 birds gather on irrigated grain and nut crops causing damage locally and there are places where farmers poison the birds to try to prevent damage to crops.[764]

The closely related Sarus Crane was discovered near Normanton (17/141) in 1966 and in the Atherton Region in 1967 and its range appears to be expanding. The two species interbreed in captivity[257] where the Sarus Crane is dominant[268] but there is no evidence yet that it is displacing the Brolga in the wild. Both species were present near Gregory Downs (18/139) in Aug.-Sept. 1979 but were never seen together.

There are no measures of abundance.

Field Atlas Records: 3710

1° blocks–with records: 330 (41%)
–with breeding records: 66 (20%)

Atlas No.: 177

Max. Rep. Rate: 59%
In block(s): 23/144

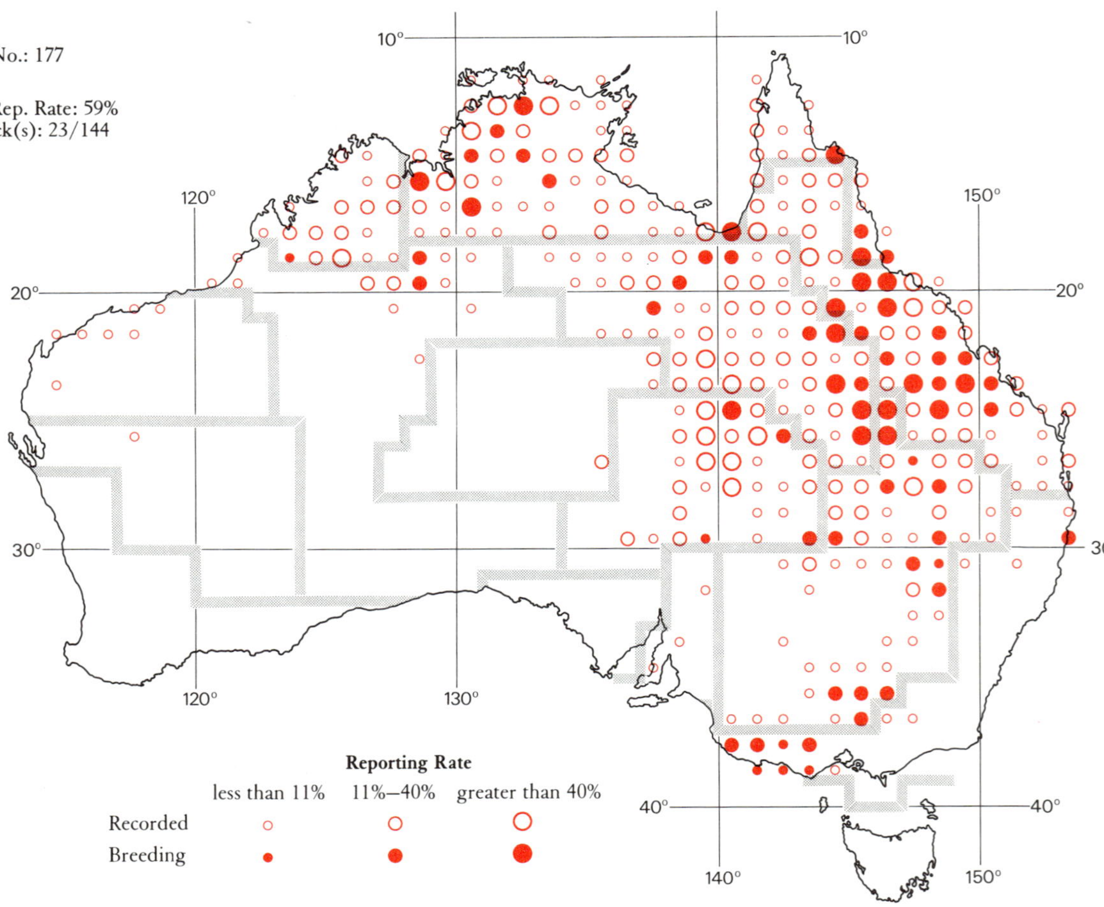

Sarus Crane

Grus antigone

The nominate subspecies of the Sarus Crane is more or less confined to India; the other *sharpei* is distributed in south-eastern Asia and Australia, though not reported from Indonesia or New Guinea.[1800] The Sarus Crane was first recorded in Australia near Normanton (17/141) in Oct. 1966 and found breeding in the area in Apr. 1967.[697] This discovery on the Gulf of Carpentaria was followed, in July 1967, by the first recognition of the species in the Atherton Region.[230] These two areas are connected by a flyway regularly used by waterbirds.[182] The spread of the Sarus Crane has been to the north and south of that flyway, with scattered reports in the Top End and Kimberley Regions since 1977.

Some overseas populations of the Sarus Crane are migratory[1800] and birds in the Atherton Region appeared to be so too in the 1960s, arriving in July and leaving in Dec.[699] The Field Atlas supports this observation of movement. Reporting Rates in the Atherton Region were: spring 5.7%, summer 1.1%, autumn 1.9%, winter 1.6%.

The Sarus Crane lives in swamps, tussock grassland and farmland. During the non-breeding season birds gather at night into large roosts. In Aug. 1970, 350 were counted going to roost at Willets Swamp (19/145).[699] The Sarus Crane eats vegetable matter, including bulbs and seeds. It is also known to eat grasshoppers[1800] and in captivity will take small animals.[257] Since its recognition in Australia the Sarus has spread rapidly and the population appears to be still expanding. In captivity the Sarus Crane and Brolga interbreed and produce fertile offspring[268] and the Sarus dominates the Brolga.[268] If this happens in the wild, the Sarus may gradually displace the Brolga from parts of northern Australia.

More than 100 breeding pairs were estimated to be on the Atherton Tableland in 1970.[699] There are no other measures of abundance.

Field Atlas Records: 183

1° blocks–with records: 39 (5%)
–with breeding records: 3 (8%)

Atlas No.: 898

Max. Rep. Rate: 34%
In block(s): 17/140,17/141

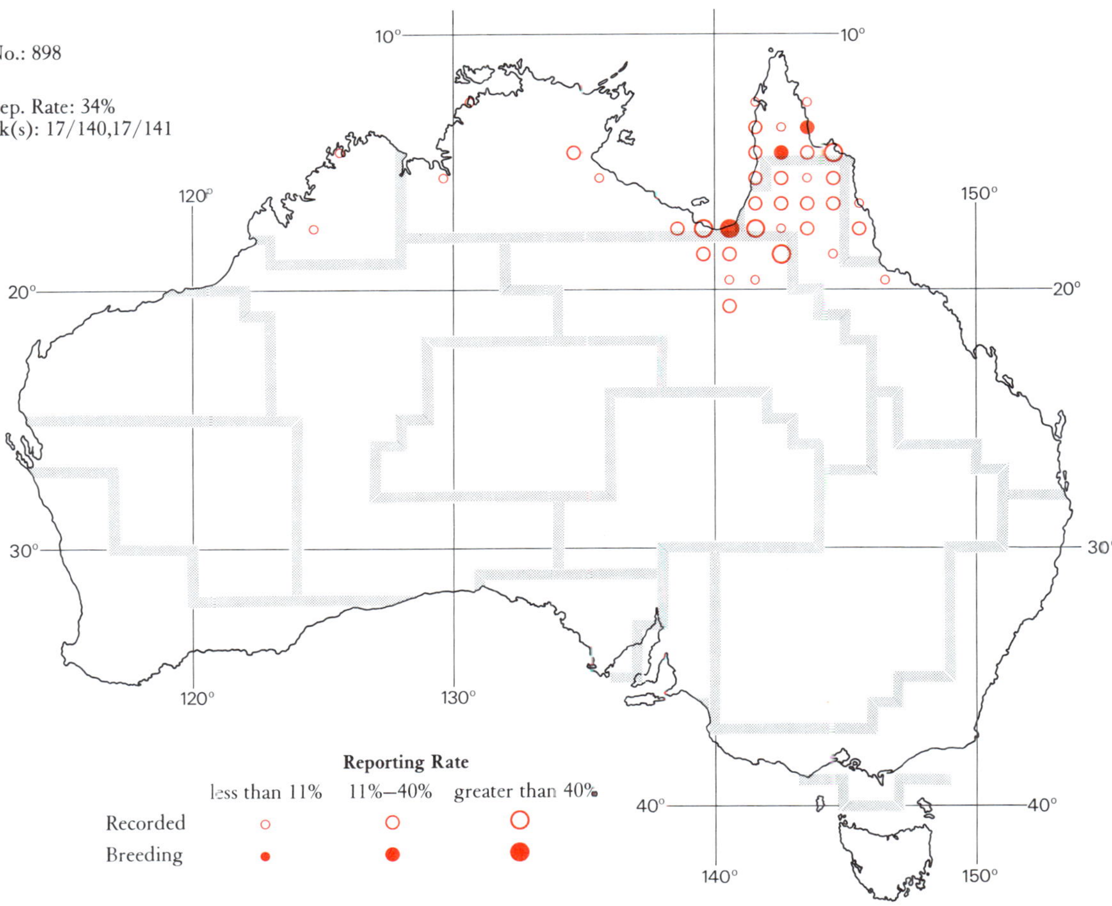

Australian Bustard

Ardeotis australis

The Australian Bustard also lives in southern New Guinea. In Australia it has generally retreated from the Murray-Darling and South-East Regions. Before 1900 the Bustard was widespread in both regions, mainly on the inland side of the Great Dividing Ra., reaching the coast west of Melbourne and breeding there (38/144, 38/145). Historically, breeding has been recorded throughout the mainland, including the Cape York Region and the Top End as far north as Darwin (12/130).

The Bustard is thought to move between Australia and New Guinea but how regularly is not known.[503] Large-scale seasonal movements are not suggested by the Field Atlas but the bird is probably nomadic in response to increases in food supply after rain.[1427] Observations early this century of increased abundance of Bustards following clearing[103,318] probably represent local immigration.

The Bustard is usually encountered singly or in small groups of one sex, sometimes in flocks of 30 or greater. One near Hay (34/144) in 1897 was estimated at more than 1000.[1343] The species lives in open country, timbered or completely treeless, and in crop stubble. It feeds on insects, small vertebrates, seeds and fruit.[1049] Large numbers gather at plagues of mice and locusts[184,999] and feed near bushfires on dead and injured insects. Some of the changes that accompanied settlement by Europeans have harmed the species, e.g. it is hunted illegally and often inadvertently poisoned;[83] the introduced fox has replaced the dingo as a predator of eggs and chicks;[83,169,1064] and human disturbance during breeding adversely affects the population.[559]

Comparison of the Atlases and comments of observers suggest that, outside the South-East Region, the range and abundance of the Bustard may recently have increased. In Eastern Qld in particular, the Bustard was often reported during the Field Atlas as common in places from which it had been absent for many years or known only as a rarity. In the five years of the Field Atlas breeding was reported in 11% of 1° blocks in which the Bustard was seen. This suggests that breeding was more widespread than in the 1951–76 period when breeding was reported in only 6% of 1° blocks where Bustards were seen.

There are no measures of abundance.

Field Atlas Records: 3259

1° blocks–with records: 518 (64%)
–with breeding records: 57 (11%)

Atlas No.: 176

Max. Rep. Rate: 62%
In block(s): 24/148

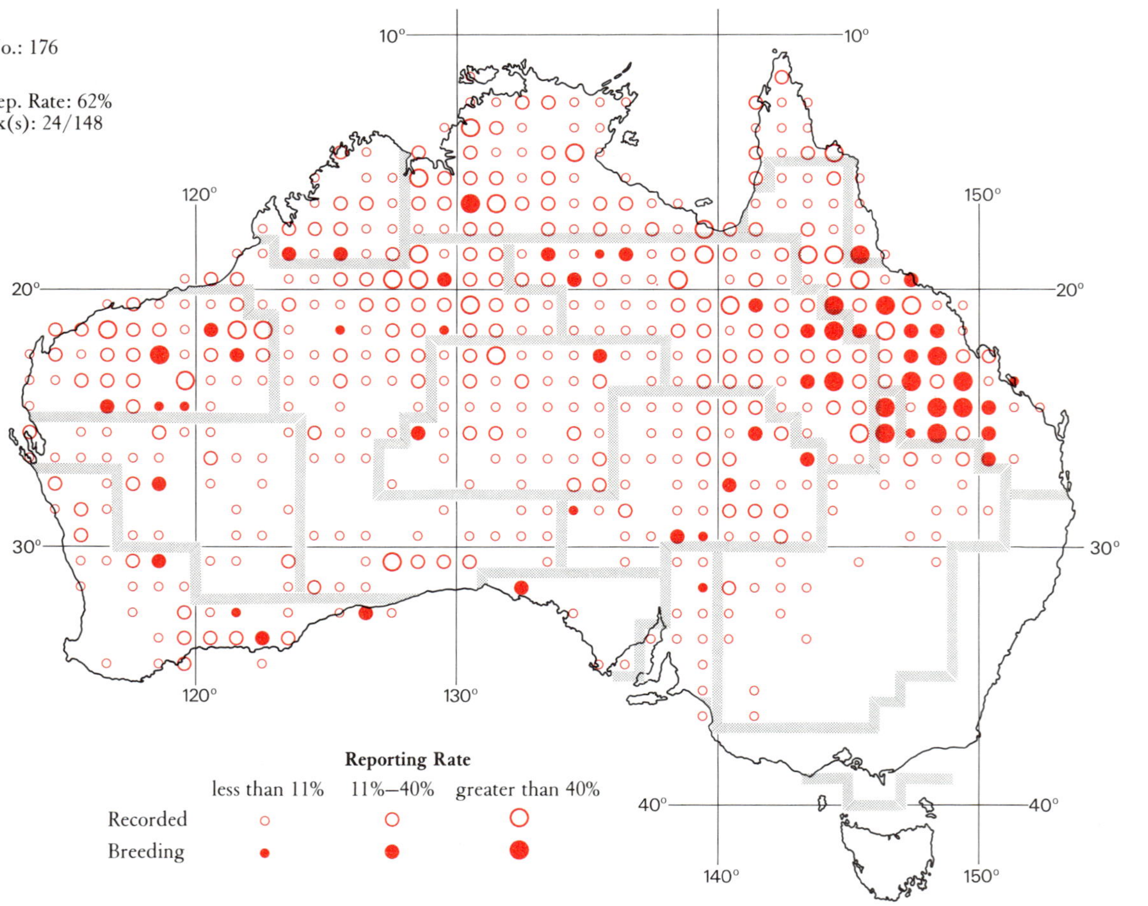

Comb-crested Jacana

Irediparra gallinacea

The Comb-crested Jacana has three subspecies in south-eastern Asia, the Pacific and New Guinea, one of which *novaehollandiae* lives in northern and eastern Australia. The Field Atlas record at Bilpin (33/150) is as far south as the bird has been reported. The species previously bred in this district;[794,812] it appeared about 1930 on the Hawkesbury R (33/151) and bred in substantial numbers until 1940 when there was a decline coincident with a prolonged dry spell. The swamps dried out completely and surviving birds stayed along the main river bank. By 1948 numbers had increased again but a severe flood in 1949, 6 m above the normal level, coincided with a decrease in numbers, from which the population has not recovered.

The Jacana has been recorded from three isolated localities in the Kimberley Region: near Derby (17/123) in 1886, Townsend R (16/123) in 1902 and most recently in 1974 in the Prince Regent R Reserve (15/125). The gaps round the Gulf of Carpentaria on the Field Atlas map are filled by historical records.

The Jacana can move far as demonstrated by occasional inland records such as at Mudgee (32/149) in 1889 and at Mt Isa (20/139) in 1976 and 1978. Near Darwin (12/130) it is restricted during the dry winter season to 2% of the area it occupies in the wet summer season.[425] The species is present all year near Innisfail (17/146),[698] also in the Rockhampton (23/150) district though numbers there decrease in summer.[1089] The Field Atlas does not show any evidence of large-scale seasonal movement.

The Jacana lives on deep permanent lagoons where conditions are stable, its long toes enabling it to walk with ease on floating vegetation.

It gathers insects and vegetable matter both from the water and from the flowers of waterlilies and other aquatic plants.[1428]

Up to 150 Jacanas have been counted on Fogg Dam (12/131) (about 0.6 birds/ha of water).

Field Atlas Records: 1163

1° blocks–with records: 82 (10%)
–with breeding records: 27 (33%)

Atlas No.: 171

Max. Rep. Rate: 39%
In block(s): 18/145

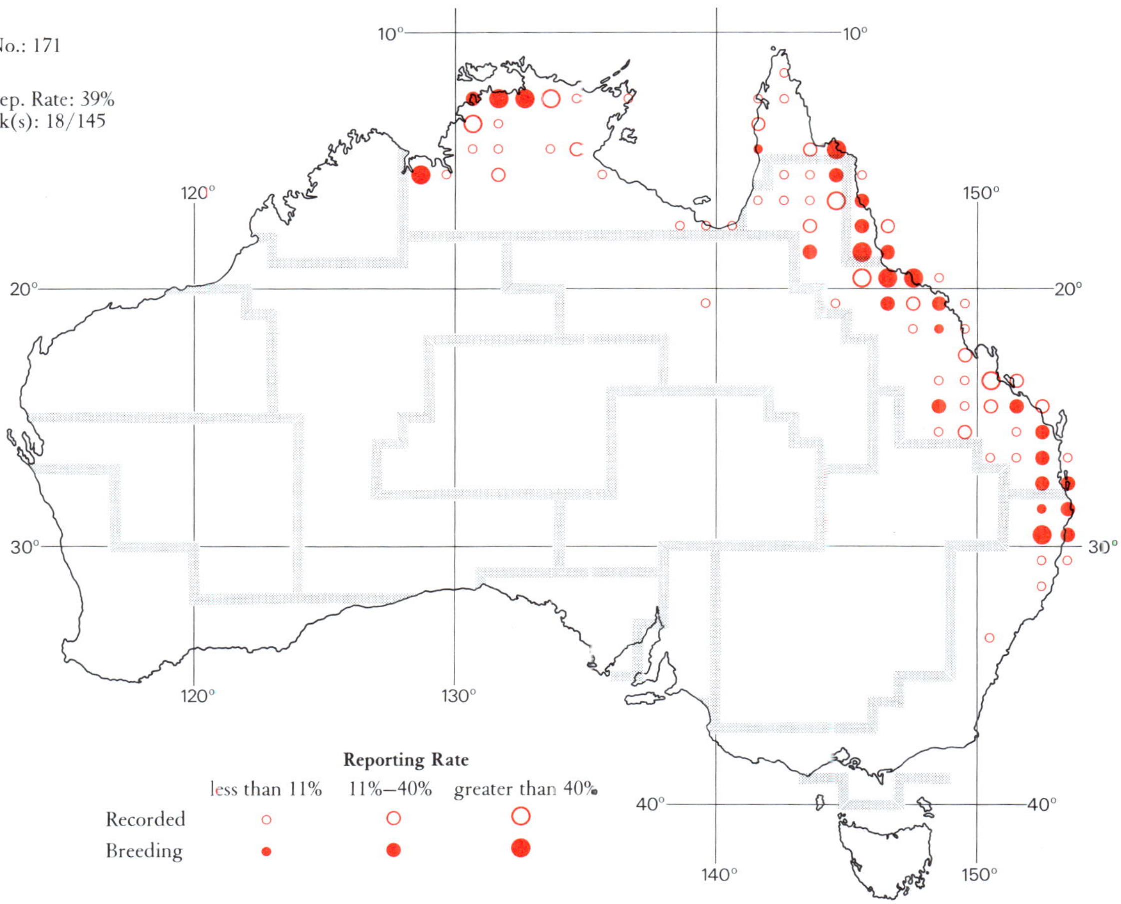

Bush Thick-knee

Burhinus magnirostris

The only record of the Bush Thick-knee outside Australia is from southern New Guinea.[1075] Within Australia its distribution has changed over the historical periods (p. 663). It has almost disappeared from the South-East Region, except for a small population north of Sydney (33/151), and from the SA Ra. Region except for a population on Kangaroo I. (35/137). Most records in the Eyre Pen. Region are now from islands and the species has also gone from the environs of Perth (31/115). Specimens confirm the presence of the Bush Thick-knee in Tas. in 1895 but there have been no confirmed reports since.[734]

There is no evidence of large-scale movement. After breeding, birds from a district gather into loose flocks of 5–10 individuals in southern areas, and sometimes as many as 30–40 in the north. They move about locally so that they may come and go erratically from a particular spot, even though they have not gone far.[1588] Flocks disband during breeding, pairs tending to nest each year at the same spot. The Bush Thick-knee inhabits woodland with short or scattered grass. Occasionally it will be present in mangroves, probably only as a daytime refuge. In central Australia it is tied closely to the ranges, large lakes and watercourses. It has not been recorded from saltbush and spinifex. The Bush Thick-knee persists in many wheat-growing areas despite their intense agricultural development.

The food of the Bush Thick-knee is primarily insects, spiders and other invertebrates; occasionally small vertebrates are taken and fruits as they become available.[1588] Many Atlas records were based on the species' distinctive call, given at night.

Despite predictions that the introduced fox would destroy populations of the Bush Thick-knee throughout Australia[186] the bird has survived even in some places where foxes are common, e.g. South-West and Goldfields Regions. Its decline has been caused mainly by urban development, intense cultivation, burning and overgrazing.

Anecdotal evidence suggests that pairs probably occupy 10–20 ha when breeding but loose flocks or clans may roam over 100 sq. km in the non-breeding season.[1588]

Field Atlas Records: 2437

1° blocks-with records: 271 (33%)
-with breeding records: 53 (20%)

Atlas No.: 174

Max. Rep. Rate: 48%
In block(s): 20/149

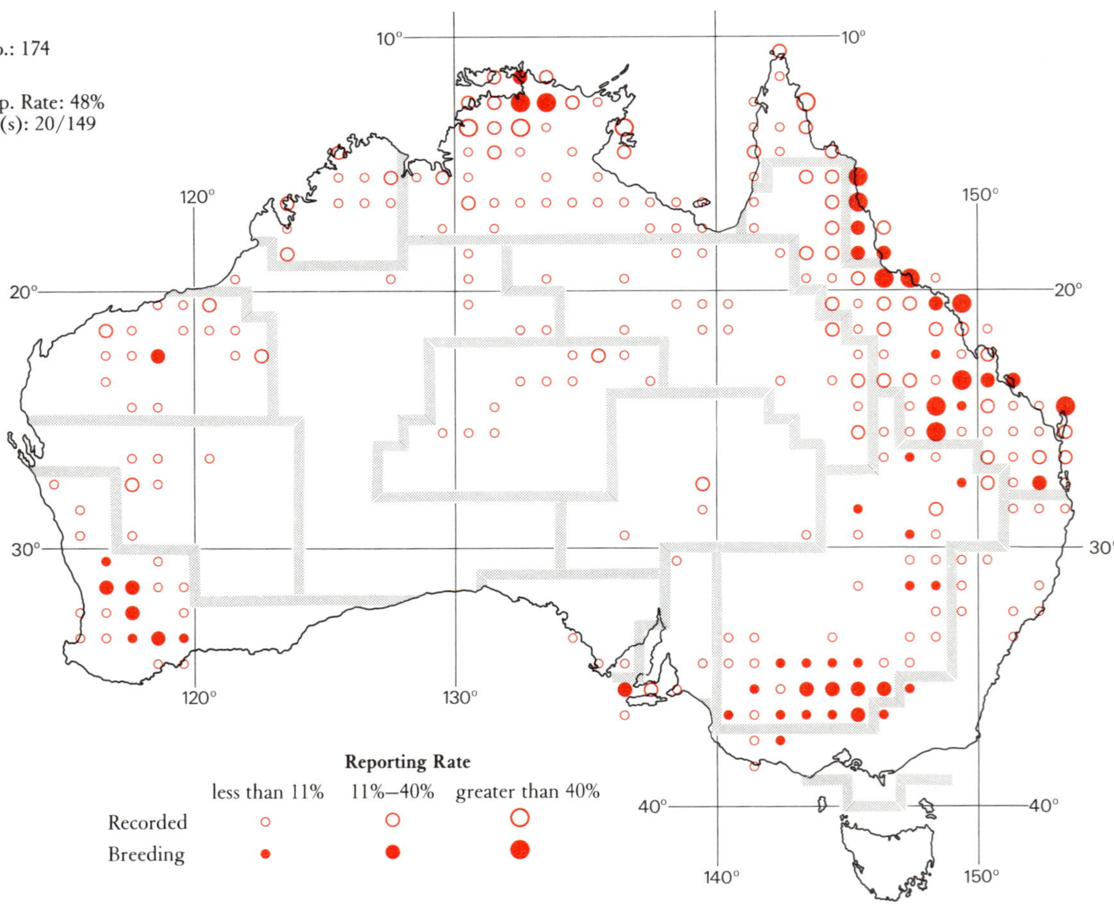

The Beach Thick-knee ranges through south-eastern Asia, New Guinea and some Pacific islands to Australia. Birds in Australia are of the subspecies *neglectus*.[388] The Atlases show that the Beach Thick-knee is distributed round the northern coast from Red Rock (29/153) to Pt Cloates (22/113) with gaps on the southern coast of the Gulf of Carpentaria and at Eighty Mile Beach (19/121). Breeding is recorded throughout this range.

Vagrants are recorded as far south as Mallacoota (37/149), which indicates that it can move far. The southerly records have been made Dec., Jan., Feb., July and Aug. One observer reported several birds at sea off the Qld coast, flying past his boat singly or in pairs.[1063] The Field Atlas suggests no large-scale seasonal pattern of movement and the species is generally regarded as sedentary.

The Beach Thick-knee lives on beaches, mudflats and reefs, specially on islands. Crabs are its principal prey.[1498] At one place the numbers of Beach Thick-knees declined after the introduction of a cat and recovered when it had died.[1498]

The species is thinly scattered along the Great Barrier Reef. It was seen on 10 of 41 islands visited in 1958, no more than four individuals being recorded from any one island.[1813] There are no other measures of abundance.

Beach Thick-knee

Burhinus neglectus

Field Atlas Records: 459

1° blocks-with records: 59 (7%)
-with breeding records: 6 (10%)

Atlas No.: 175

Max. Rep. Rate: 58%
In block(s): 24/153

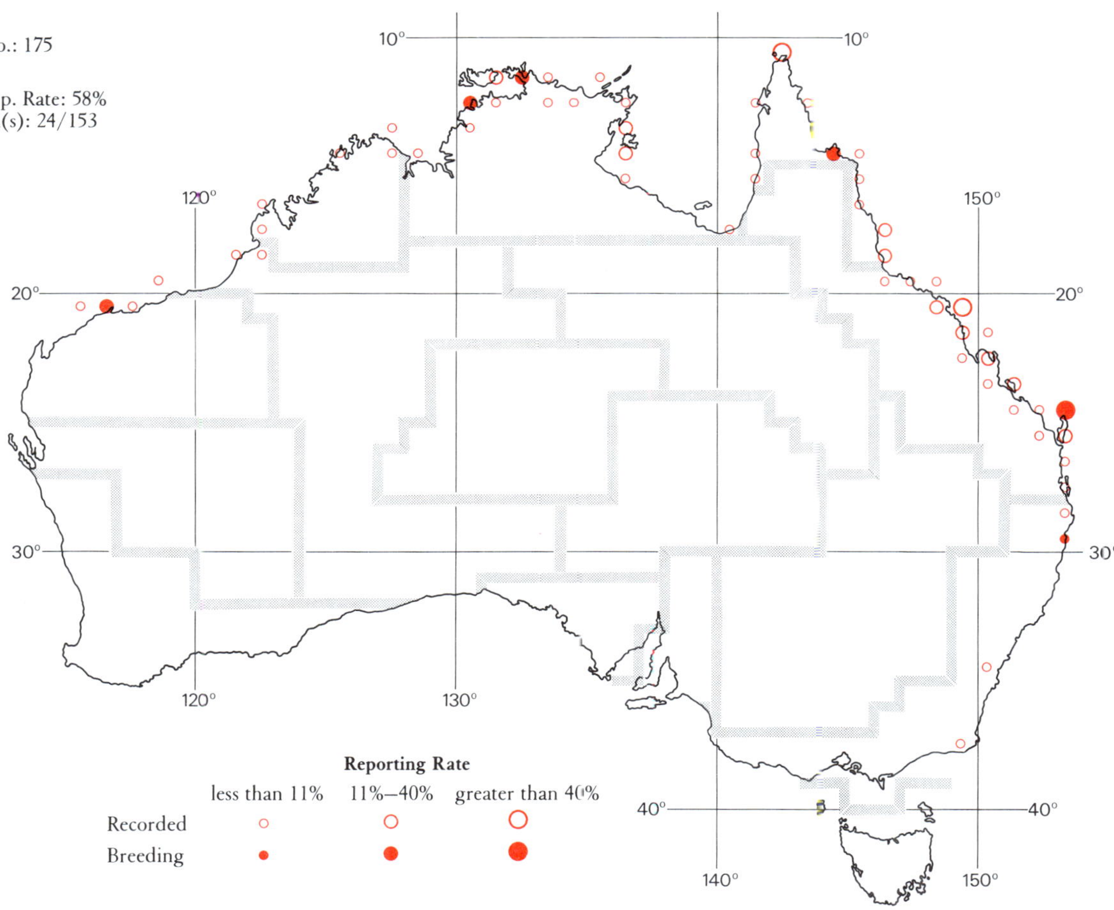

Painted Snipe

Rostratula benghalensis

The Painted Snipe ranges through Africa, India, China, south-eastern Asia and the Pacific. Two subspecies are recognized, one, *australis*, breeding in Australia. In the Tas. Region the only record is of a specimen from Sandford (42/147) in July 1910. In WA the species has not been recorded since a sighting near Derby (17/123) in the 1950s.[1638] Other records, all before 1950, include specimens taken near Derby in July 1886[1713] and Jan. 1896; one taken in Sept. 1907 on the Strelley R (20/119) in the Pilbara Region; and sightings in the Goldfields Region at Brockman Ck (25/122) in Aug. 1896; and at a number of places in the South-West.

Breeding reports are confined to the Murray-Darling, South-East and the SA Ra. Regions except for one in the 1950s of the bird arriving to breed at Ayr (19/147) in Mar.-Apr. each year.[1108] Otherwise the northernmost breeding records are at 28°S. Occasional records from remote places show that the Painted Snipe can move far. This and its apparently erratic appearances, common in some years, scarce or absent in others, led earlier writers to consider it nomadic,[653,1108,1638] e.g. birds at Laverton (37/144) May-Sept. 1951 were the first recorded there since 1897.[1838] In southern Australia the Painted Snipe was reported to be present only during floods at long intervals[1108] but the Field Atlas and other observations suggest that movements are mainly north in winter and south in summer (Reporting Rates in the Eastern Qld, South-East and Murray-Darling Regions combined: summer 0.50%, winter 0.04%). The Painted Snipe is a spring-summer visitor to Cunnamulla (28/145), Mossgiel (33/144)[1343] and Sale (38/147). The Reporting Rate did not increase in the north in winter, which suggests that the species may even migrate out of the country though there is little evidence of overseas populations migrating[13,1117] or of a possible wintering ground for Australian birds.

The Painted Snipe requires shallow fresh water for breeding, though the nest is not deserted if the water dries up.[1108,1161] It probes in mud along the shore to gather snails and water insects.[119] Nests are usually in groups, once recorded at a density of 24 nests/ha.[1108] Other smaller colonies have had distances of 15-50 m between nests. The largest flock reported is of 25 birds near a nesting colony on the Murray R.

Field Atlas Records: 132

1° blocks–with records: 64 (8%)
–with breeding records: 5 (8%)

Atlas No.: 170

Max. Rep. Rate: 2%
In block(s): 20/139, 23/147, 24/140, 24/152, 25/138, 25/149, 31/145, 34/151, 37/140

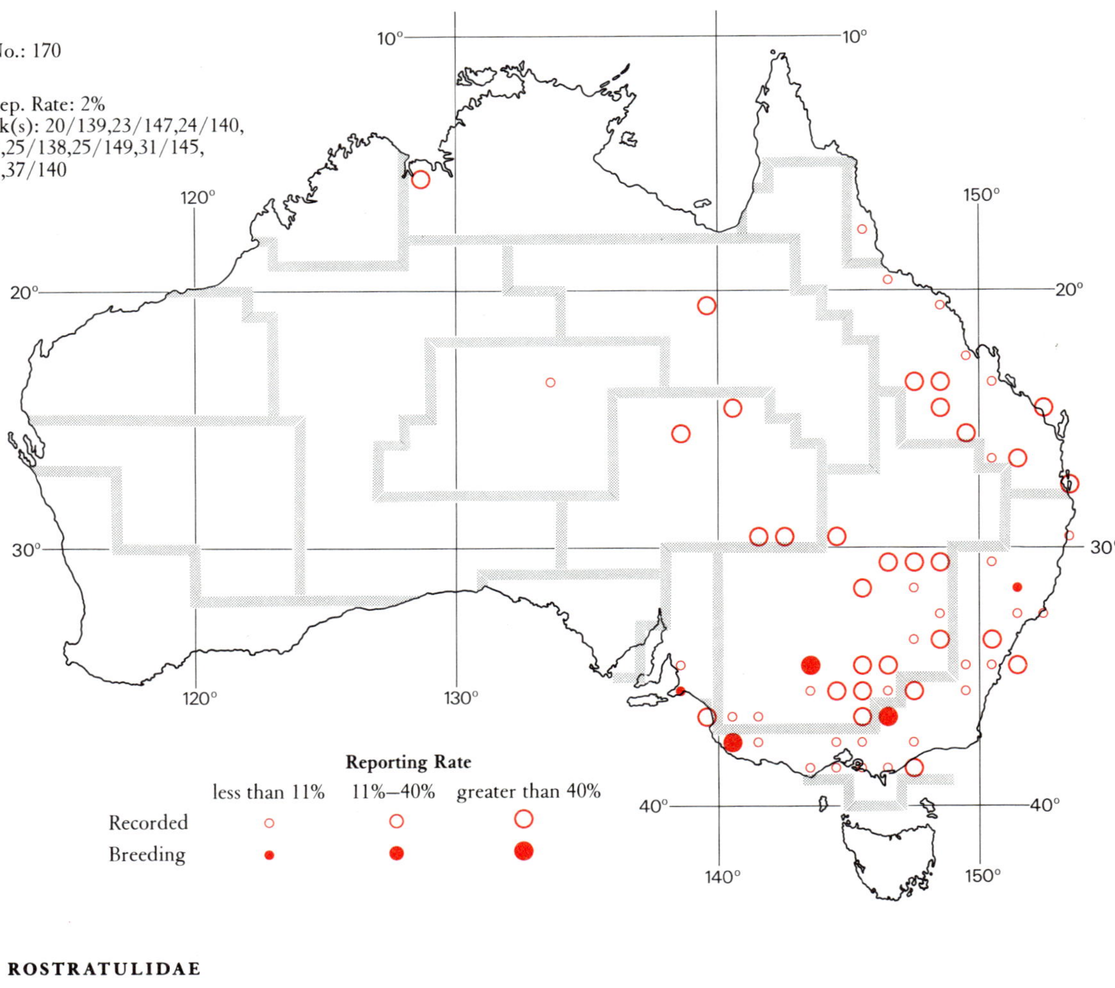

Pied Oystercatcher

Haematopus longirostris

The Pied Oystercatcher lives in Australia, New Zealand and southern New Guinea. Only one subspecies *longirostris* is recognized in Australia.[1773] On occasions mixed pairs of Pied and Sooty Oystercatchers have been observed[1131] but no successful hatching of hybrid progeny has been reported.

The Pied Oystercatcher is continuously distributed round Australia except where cliffs replace beaches such as on the Nullarbor coast. Occasionally vagrants appear on saline wetlands in the inland, e.g. L. Tutchewop (35/143) where one was present among other waders in Sept. 1977.

A five-year study of banded birds near Hobart (42/147) showed that breeding adults were completely sedentary. Thirteen immatures less than one year old banded in the same area were more mobile. They were often re-sighted up to at least 12 months after banding but moved round within the local complex of bays and beaches; two were seen 35 km distant but later returned.[1320] Some birds move farther: one colour-banded at 38/144 near Werribee was seen on King I. (39/144).[1017]

In Australia the Pied Oystercatcher lives on sandy rather than rocky beaches and breeds on the coast, mainly on sandy beaches, dune systems and shallow muddy bays. In New Zealand it also breeds inland. The Pied Oystercatcher does not breed until about four years old. Non-breeding birds form flocks in bays, augmented by adults outside the breeding season.[1314] It feeds on molluscs that live in sand and on polychaete worms.

Like other shore-nesting birds, the Pied Oystercatcher suffers disturbance from people and dogs using beaches for recreation. The bird has gradually disappeared from Moreton Bay (27/153)[19] and in NSW it is considered endangered.

Because the species is exclusively coastal the Australian population is small by comparison with New Zealand where in one area alone, between Kaipara Harbour (36S/174E), Manukau Harbour (37S/174E) and the Firth of Thames (37S/175E), 50 000 Pied Oystercatchers have been counted.[1781] The greatest number recorded at any one site in Australian National Wader Counts was 932 at Corner Inlet (38/146) in Feb. 1982. A count of 50 km of ocean beach north of Salt Ck (36/139) in Feb. 1983 recorded 385 Pied Oystercatchers (7.7 birds/km).[226]

Field Atlas Records: 5048

1° blocks–with records: 177 (22%)

–with breeding records: 59 (33%)

Atlas No.: 130

Max. Rep. Rate: 78%
In block(s): 24/153

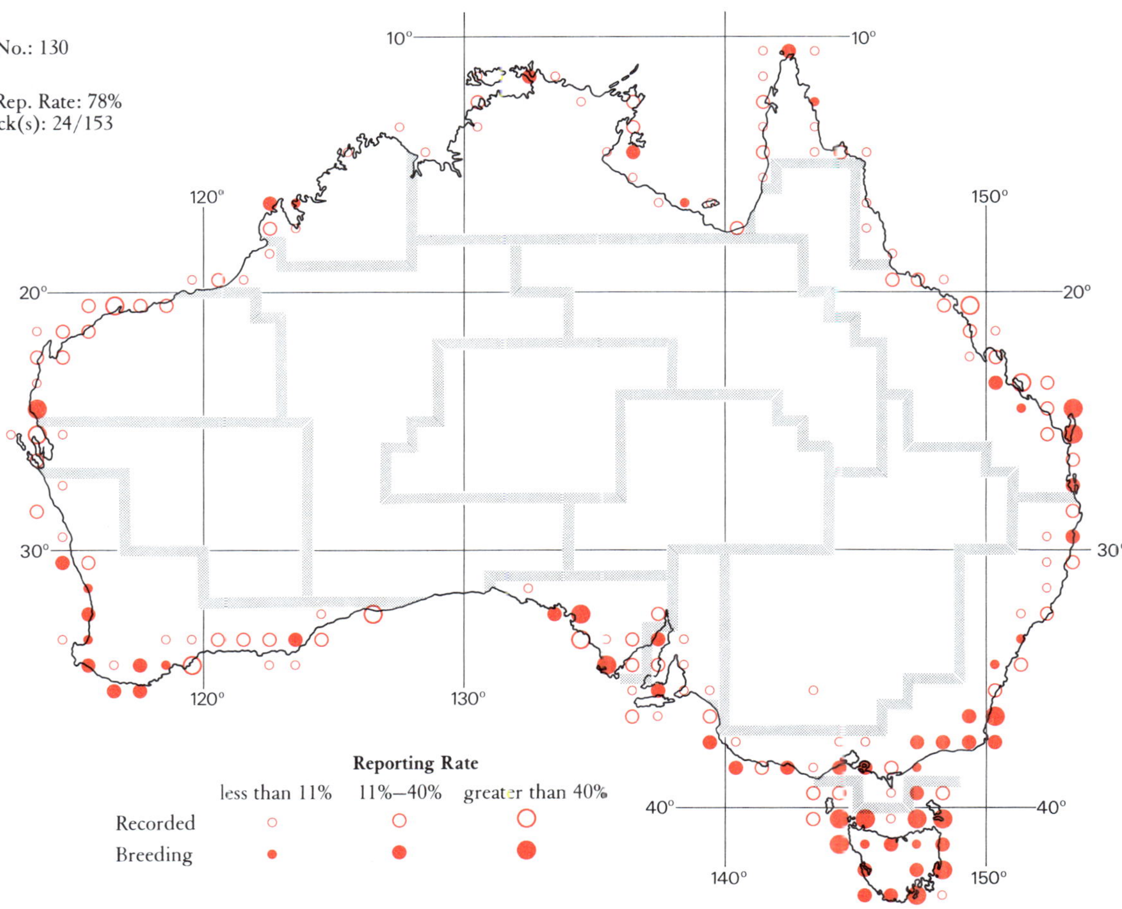

Sooty Oystercatcher

Haematopus fuliginosus

Two subspecies of the endemic Sooty Oystercatcher are recognized: *ophthalmicus* in the north and *fuliginosus* everywhere else.[388] The species is rare on the southern and eastern coasts of the Gulf of Carpentaria but has been recorded on a number of islands in the Gulf. It is also rare on Eighty Mile Beach (19/121) and along the Nullarbor cliffs. Despite the absence of breeding records from northern coasts in the Field Atlas, breeding was reported in the historical periods and there is no reason to doubt that the species breeds throughout its range.

At Wilsons Prom. (39/146) and on Eyre Pen. the Sooty Oystercatcher breeds on islands; flocks form in the non-breeding season and move to nearby mainland sites in autumn and winter.[84] Movements in the Hobart (42/147) area appear to follow a similar pattern though some immature birds may stay on the mainland during the summer breeding season.[1324] The species' status as a scarce visitor in Torres Strait[503] and round Darwin (12/130)[1755] indicates some movement in these areas also. The longest recorded movement of a banded bird is 48 km.[85]

The Sooty Oystercatcher lives on rocky coasts within 50 m of the shore line. It almost always breeds on islands. In the Top End its distribution is thought to be largely determined by the availability of the coral reefs on which it prefers to forage.[1755] A study of diet at five sites on Phillip I. (38/145) and Wilsons Prom. showed that various molluscs were eaten, specially limpets and the dogwhelk, as well as crustaceans and polychaete worms. The importance of different foods varied from site to site.[393]

The greatest numbers of Sooty Oystercatchers recorded at individual sites in National Wader Counts were: 242 at Corner Inlet (38/146) in Feb. 1981 and 450 on Flinders I. (40/148) in Feb. 1983.[84]

Field Atlas Records: 2735

1° blocks–with records: 153 (19%)
–with breeding records: 27 (18%)

Atlas No.: 131

Max. Rep. Rate: 67%
In block(s): 39/146

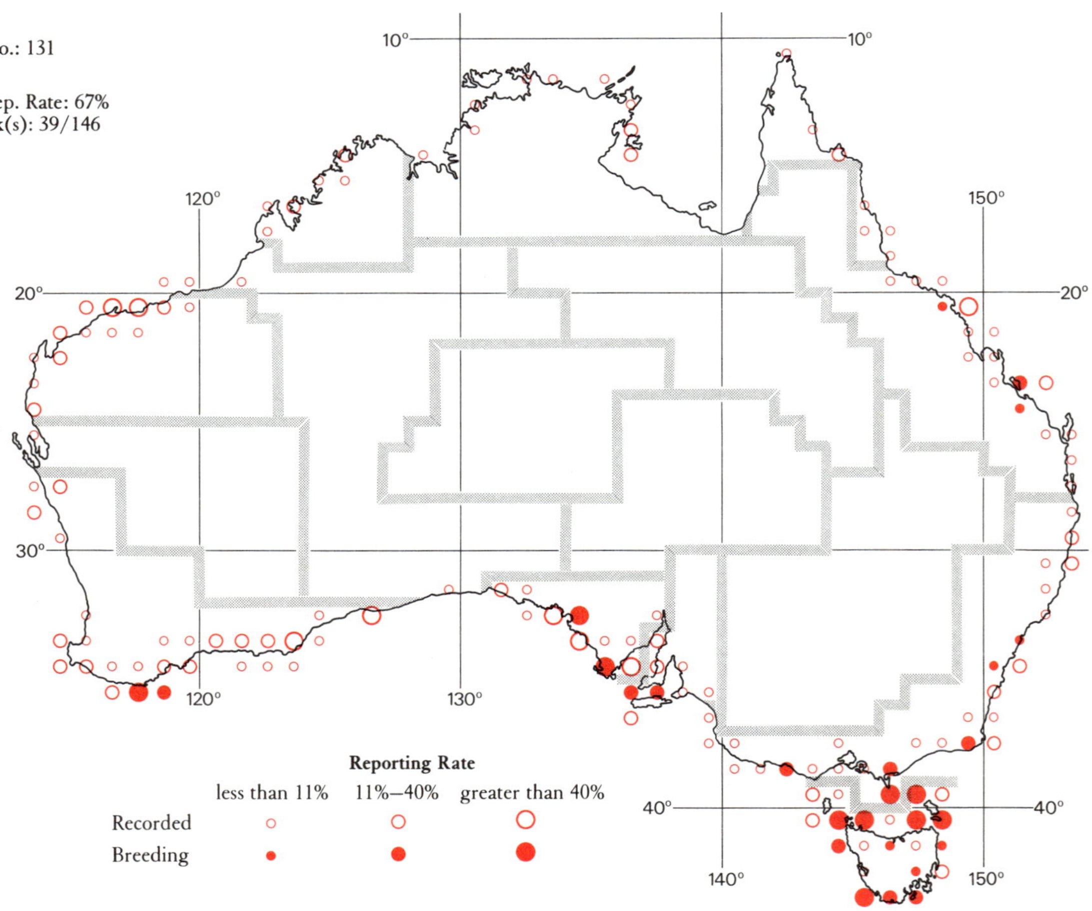

Masked Lapwing

Vanellus miles

The Masked Lapwing lives in Indonesia and New Guinea as well as Australia. It colonized New Zealand, first breeding there in the 1940s.[1003] There are two subspecies in Australia: *miles* in northern Australia south to about Townsville (19/146) and *novaehollandiae* elsewhere.[388] The first record for the South-West Region was in 1956 (subspecies *novaehollandiae*)[1638] and the species remains a vagrant to the region.

The Lapwing appears to be expanding its range into the Central Ra. Region. Before the Field Atlas it was regarded as a scarce winter visitor to Alice Springs (23/133)[1370] and there were records from only two other 1° blocks in the region. Field Atlas records were made throughout the region in all seasons.

Individuals can move far as is shown by the arrival of the species in New Zealand and the South-West Region[388,1774] and occasional records of the subspecies *novaehollandiae* in New Guinea.[129] No regular seasonal movement is reported nor is any suggested by the Field Atlas. The Lapwing is always present on islands in Torres Strait though not recorded breeding there.[503] Composition of non-breeding flocks on the Derwent R (42/147) appeared to vary from day to day.[1742] Most movements take place at night.[539]

The Lapwing is usually near water. It lives in short grass pastures and on the margins of lakes, lagoons, beaches and irrigated crops. Of 255 nests recorded in Tas., 147 were in grazed paddocks, 18 on airfields, 28 on the verges of lagoons and 38 on ploughed land; most others were in short vegetation and only a few (3%) in growth taller than 10 cm.[1742] Pairs breed solitarily but flocks form in the non-breeding season.[1742] In dry weather birds may congregate on remaining moist ground. The Lapwing eats insects and invertebrates, specially earthworms,[453] as well as seeds and other vegetable matter.[748] It uses man-made environments and, since European settlement, has extended its range.[453] In Tas. it has become more abundant.[1742]

At Laverton (37/144) there were 16 birds in 8 sq. km (0.02 birds/ha) in spring and up to 308 in the non-breeding season.[1838] At Barilla Bay (42/147) 11 pairs held territories in about 24 ha (1 bird/ha).[1742] On 25 km of ocean beach near Kingston (36/139) 300 were counted in Jan. 1983 (12 birds/km).[226]

Field Atlas Records: 31 320

1° blocks–with records: 445 (55%)
–with breeding records: 213 (48%)

Atlas No.: 133

Max. Rep. Rate: 82%
In block(s): 27/149

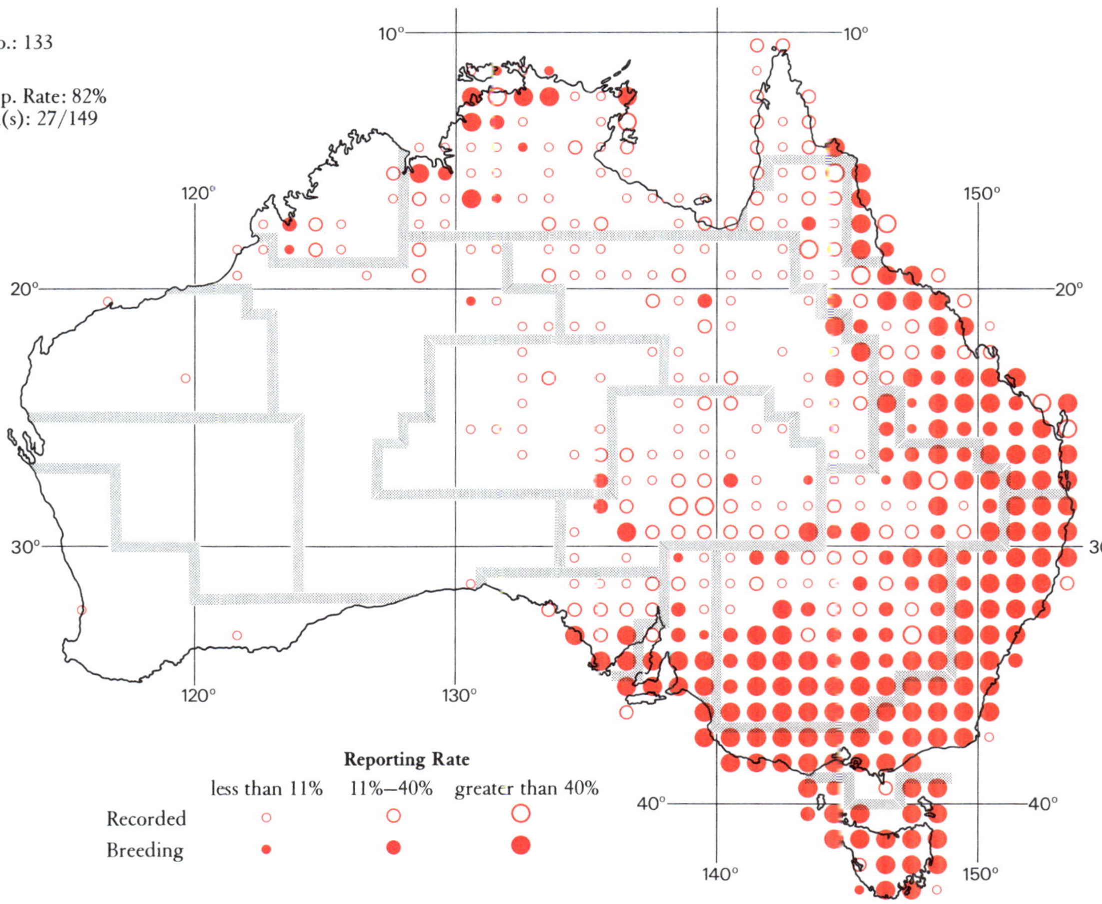

Banded Lapwing

Vanellus tricolor

Over most of its range the endemic Banded Lapwing has changed in distribution and abundance only locally but it has expanded its range in the South-West. Before 1901 the Atlases contain only a single record in WA.

The Lapwing is a vagrant in northern parts of its range[862] and records in the north of the Western Deserts Region show it to be capable of long-distance movement. A regular movement that appeared to take birds east in autumn and west in spring was noted at Murphy's Ck (27/152) but the species was no longer present after 1937.[1099] Nomadic movement is suggested in the southern Murray-Darling Region where large numbers arrive in favourable seasons.[539] The Field Atlas suggests no large-scale pattern of seasonal movement.

The Banded Lapwing lives on short grass pastures[453,1774] or bare paddocks. Unlike the Masked Lapwing, it does not need to be near water but, in hot weather, it will travel many kilometres to obtain shade.[539] The species spreads out after rain over a larger area than it occupies during dry periods[1216,1740] but for most of the year it remains in flocks, foraging in small parties of 5–6 birds; pairs breed solitarily.[539] It feeds on surface-living invertebrates, seeds and green shoots.[453,539]

In Tas. numbers of the Banded Lapwing have declined this century[1740] but the decrease has not been steady and at least a small increase took place about 1950.[1649] In the SA Ra. Region a decrease has taken place since 1940 associated with reduced wheat farming in marginal land.[191] In the South-West there has been a general increase in abundance since European settlement.[1638] Round Northam (31/116) a local decline was associated with a change to taller vegetation as a result of pasture improvement.[1216]

When the Banded Lapwing is abundant in the southern Murray-Darling Region, there are often 4–5 nests within 50 m of each other; once six nests were found within 1 ha.[539] There are no other measures of abundance.

Field Atlas Records: 6025

1° blocks–with records: 403 (50%)
–with breeding records: 212 (53%)

Atlas No.: 135

Max. Rep. Rate: 42%
In block(s): 25/113,30/117

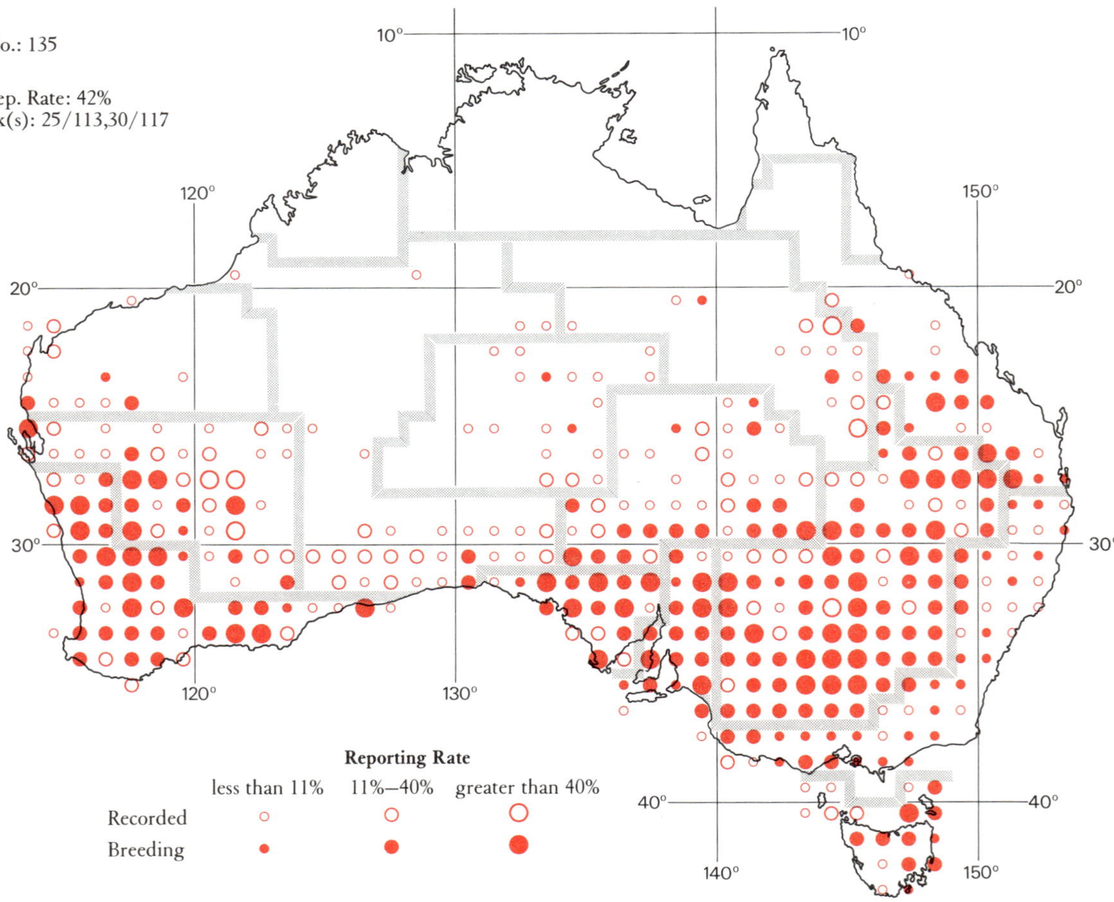

Grey Plover

Pluvialis squatarola

The Grey Plover breeds on the tundra in the Holarctic. Migration takes the species to Africa and South America as well as Australia where it is a visitor in summer, some birds remaining during the winter (Reporting Rates: summer 1.43%, winter 0.34%). The three inland records in the Field Atlas, one of which (35/142) was of a flock of 14, and an earlier one from Sutherlands (34/139)[193] were all made Sept.-Nov. This suggests that on their southward migration some birds cross the continent. Others move round the coast: the species is a passage migrant in Torres Strait[503] and at Eyre Bird Observatory (32/126) numbers in four successive spring passages were 36, 16, 44 and 19 compared with none recorded in autumn.[389]

The Grey Plover is a coastal species in Australia but in the Americas and Africa it is regularly present inland, specially on the shores of large lakes.[1905] In Australia it lives on mud flats on tidal shores and feeds on marine invertebrates.[1049] The inland specimen from Sutherlands had mosses and a few insects in its gut.[193] In the breeding season in the northern hemisphere the Grey Plover takes many insects.

Most Grey Plovers in Australia in summer are concentrated in a few areas. The coast between Broome (17/122) and Port Hedland (20/118) held an estimated 3000 birds in Nov. 1982; Eyre Pen. and Spencer Gulf (34/136) and Gulf St Vincent (34/138) held 1650 in Feb. 1982; Corner Inlet (38/146) held 850 in Feb. 1983.[84] Large numbers are also known to occupy the Gulf of Carpentaria.

Field Atlas Records: 916

1° blocks–with records: 111 (14%)

–with breeding records:

Atlas No.: 136

Max. Rep. Rate: 23%

In block(s): 24/113

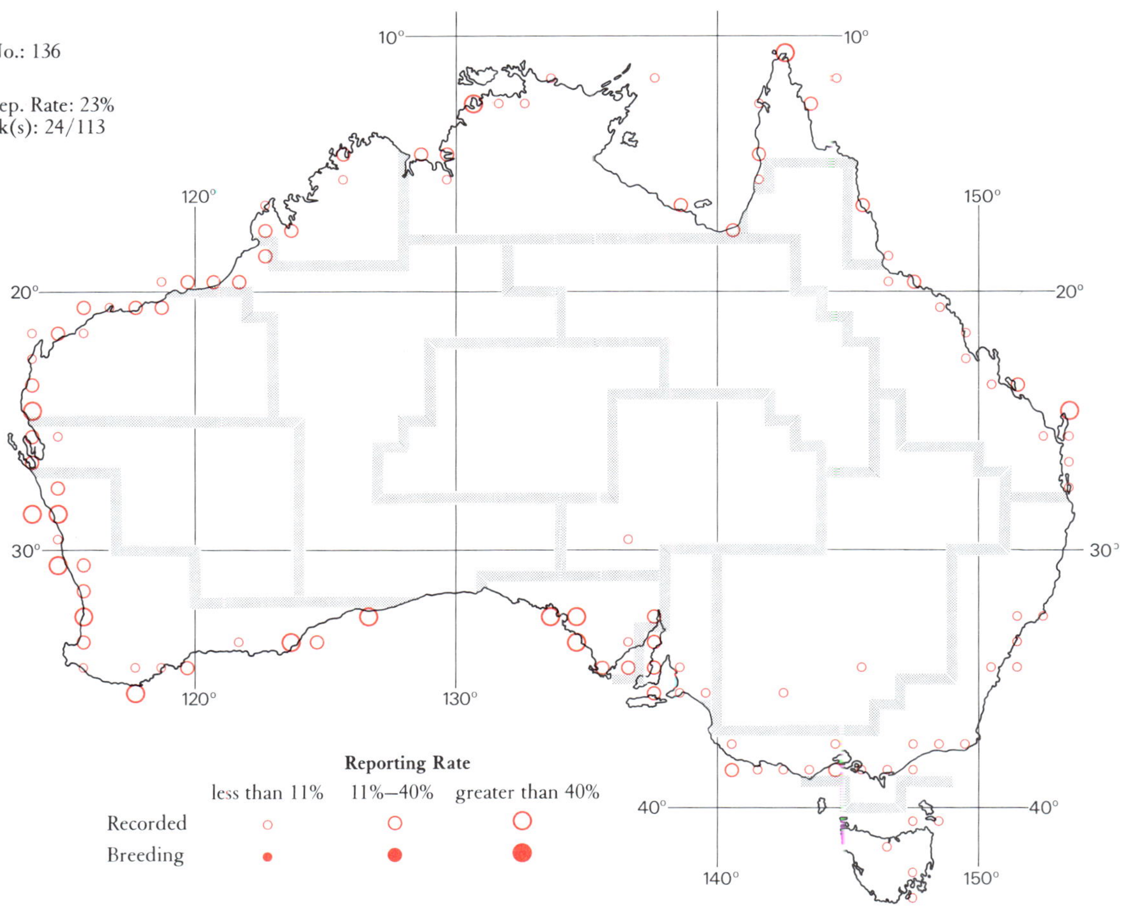

Lesser Golden Plover

Pluvialis dominica

The Lesser Golden Plover breeds in northern Asia and North America and migrates south in the non-breeding season. Some pass along the coasts and islands of eastern Asia to spend the summer in Australia.[1117] A few remain to winter in Australia[503] (Reporting Rates: summer 2.33%, winter 0.49%). One bird banded in Alaska was recovered in Australia at Urunga (30/153).[85]

That some birds migrate southward across the continent is indicated by records in the Field Atlas from 13 inland 1° blocks in spring compared with three in summer and one each in autumn and winter. A few have been recorded on the coast at Eyre Bird Observatory (32/126) in spring and autumn but not every year.[389]

The Lesser Golden Plover lives on rocky coasts,[1838] mud flats, estuaries and coastal marshes,[19] often where there is vegetation in shallow water.[1706] It also visits, and roosts on, pasture and paddocks well away from water.[231,653,698] It is usually in flocks and feeds on small invertebrates picked from the surface of shallow water and from dry vegetation.[1905]

Sites where large numbers have been recorded in National Wader Counts include, in Feb. 1983, 500 on the Hunter R (32/151) estuary, 300 at Shallow Inlet (38/146), 250 on the NSW north coast, 220 in the east Derwent, Pittwater (42/147) area and 200 at Botany Bay and Boat Harbour (34/151).[84]

Field Atlas Records: 1444

1° blocks-with records: 151 (19%)

-with breeding records:

Atlas No.: 137

Max. Rep. Rate: 43%
In block(s): 10/142

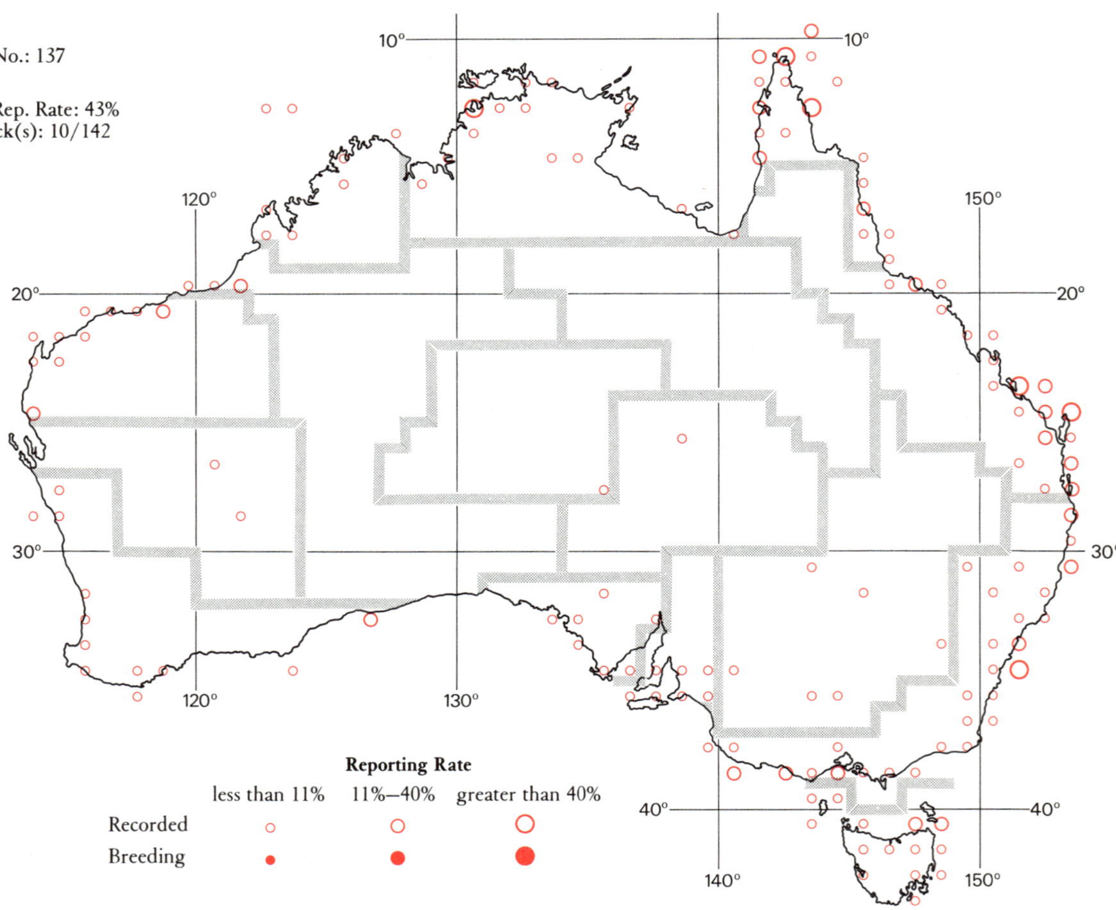

Red-kneed Dotterel

Erythrogonys cinctus

The Red-kneed Dotterel lives in southern New Guinea[129] and Australia. It has become more frequently reported on the coast of the South-East Region since 1950.[1838] It was first reported in the Tas. Region in 1967[734] and Field Atlas records there were made in Feb. 1977 (39/144) and Apr. 1978 (42/147).

The Red-kneed Dotterel moves far but unpredictably. Numbers are reduced in the southern Murray-Darling Region in winter.[833] The Field Atlas suggests no pattern of movement but Reporting Rates generally were lower in winter than in summer. Presumably the birds disperse and become less conspicuous.

The Red-kneed Dotterel lives round shallow, vegetated lakes and is rarely on river banks or pools.[833,1713] It is gregarious and, even when breeding, lives in loose colonies, each pair defending a small nesting territory.[1181] Breeding colonies are small, containing up to 14 nests.[1552]

The Red-kneed Dotterel feeds both in water and on muddy shores. It probes into the mud as well as taking insects from the surface. It may actually require the salt content of the water to be low because it does not have the salt-excreting nasal glands common to waders using saline lakes.[1181]

There are no measures of abundance.

Field Atlas Records: 2674

1° blocks–with records: 351 (43%)
–with breeding records: 55 (16%)

Atlas No.: 132

Max. Rep. Rate: 48%
In block(s): 29/136

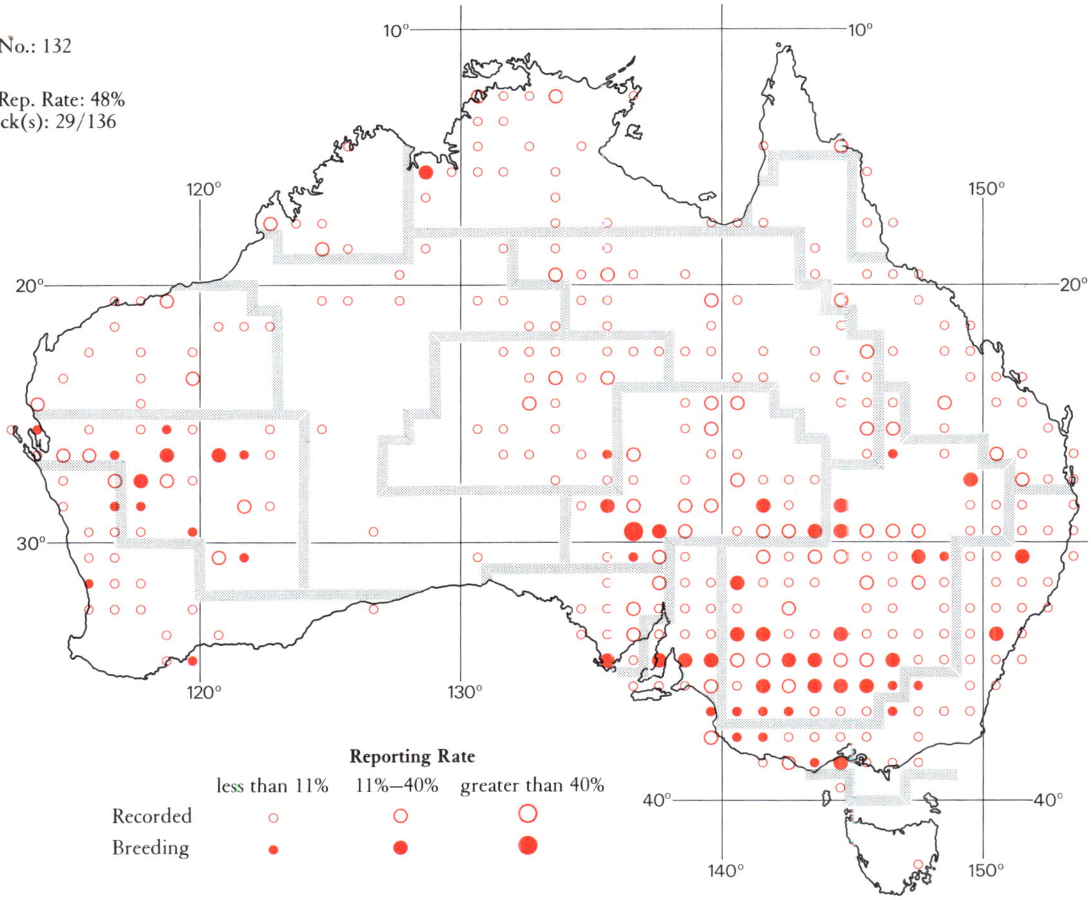

Hooded Plover

Charadrius rubricollis

The range of the endemic Hooded Plover has declined. Before 1950 stragglers were recorded as far north as Port Stephens (32/152) but the species was last recorded on Sydney (33/151) beaches in the 1940s.[812] There are no breeding records north of 35°S in eastern Australia or 30°S in the west.

In south-eastern SA there is some movement in the non-breeding season: birds were absent or sparse Apr.-Aug. in several sites where they bred in spring-summer.[226] They may move to nearby lakes and to the inner shores of The Coorong (36/139). In Vic. and Tas. flocks form in winter, indicating at least local movement.[1314,1599] In the South-West birds move in summer from inland salt lakes to the coast[1706] and the Field Atlas correspondingly contains no inland records in summer. The longest recorded movement of a banded bird is 130 km.[85]

In eastern Australia the Hooded Plover is coastal, living on wide beaches backed by sand dunes and sometimes on lakes close to the coast. Dunes as well as the shore are used for nesting.[1019,1552] Counts in summer and winter showed that about 70% of the Plovers were in pairs. Most others were alone or in groups of three, rarely more,[226,1019] though flocks of 30–40 sometimes form in winter in Vic. and Tas.[1019,1314] In the South-West the bird behaves differently. It lives on beaches but also inland where it will breed round salt lakes. Much larger flocks form than in the east: several of more than 100 have been recorded and in Feb. 1983 one of 393 occupied about 4 km of shore line at L. Gore (33/121).[911]

The Hooded Plover has been observed to feed on insects trapped on sand left wet by receding waves;[1469] nothing else is known of its food. It is easily disturbed and increasing human use of beaches must be a factor in its declining status.[1019] In Tas. storms may destroy early nests.[1131]

In Vic. in Oct. 1980 there were 475 Plovers on 708 km of beach (0.7 birds/km);[1019] in Tas. in Oct. 1982 there were 257 birds in 155.5 km of beach (1.7 birds/km);[1409] and in SA in July-Aug. 1982 there were 156 birds in 400 km (0.4 birds/km).[226] On Guichen Bay (37/139) breeding density was 1 bird/km in Sept.-Nov. 1982.[226]

Field Atlas Records: 1644

1° blocks–with records: 83 (10%)
–with breeding records: 41 (49%)

Atlas No.: 138

Max. Rep. Rate: 34%
In block(s): 42/148

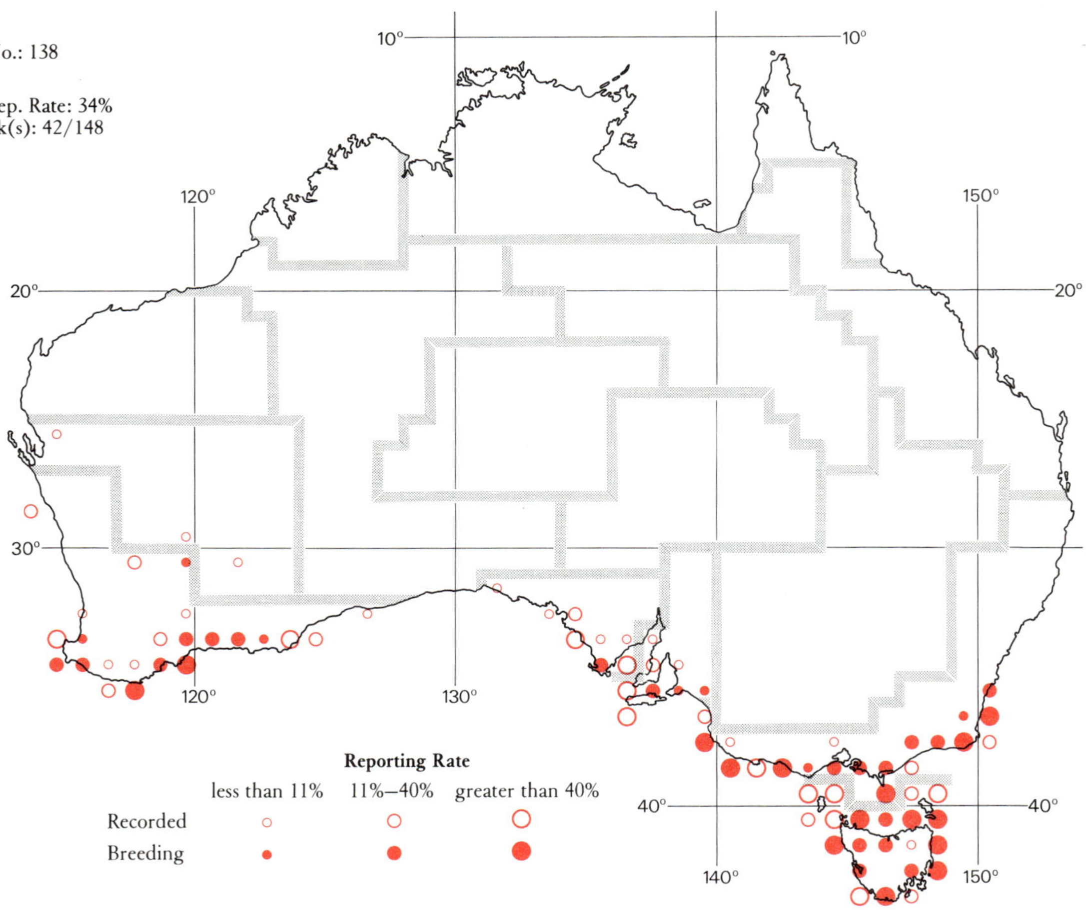

Mongolian Plover

Charadrius mongolus

Two subspecies of the Mongolian Plover visit Australia: *mongolus* breeds in eastern Siberia and is recorded throughout Australia; *stegmanni* breeds in north-eastern Siberia and has been recorded definitely only once, in the Darwin (12/130) area, in 1948.[388] In the field the Mongolian Plover is sometimes confused with the Large Sand Plover.

The Mongolian Plover visits Australia in summer and some stay throughout the winter[423,503,1044] (Reporting Rates: summer 1.17%, winter 0.52%). The only inland record in the Field Atlas was of a bird south-west of Clifton Hills (27/138) in Dec. 1981. At Eyre Bird Observatory (32/126) the species is recorded only in spring and then only one or two birds at a time.

The Mongolian Plover lives on muddy and sandy shores, usually in isolated flocks of up to 100 birds, but associates with other species when roosting. Little is known of its diet nor in what way it differs from the crab- and shrimp-eating Large Sand Plover.

The Mongolian Plover is far more abundant in Qld than anywhere else in Australia. Counts there include: 1770 in Moreton Bay (27/153) in Feb. 1983, 1434 in the Great Sandy Strait and Hervey Bay (25/152) area in Feb. 1983, 653 in the Mackay (21/149) area in Feb. 1983, and 705 on three islands in Torres Strait in Feb. 1982.[84] There is also a record of 500 at Port Hedland (20/118) in Mar. 1982.

Field Atlas Records: 816

1° blocks–with records: 101 (12%)
–with breeding records:

Atlas No.: 139

Max. Rep. Rate: 37%
In block(s): 10/142

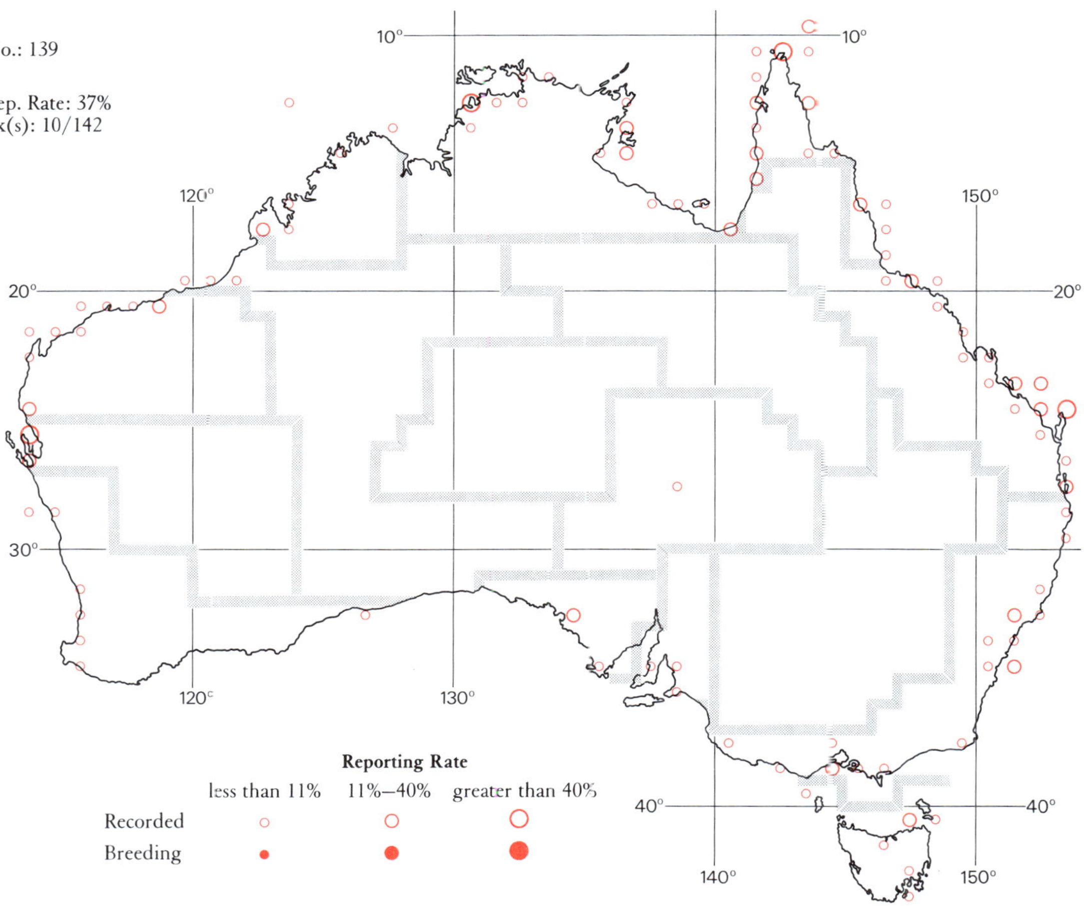

Double-banded Plover

Charadrius bicinctus

The Double-banded Plover breeds in New Zealand and part of the population visits Australia in the non-breeding season each year. Some birds remain in New Zealand in winter and others straggle as far as Fiji (18S/175E) and Vanuatu (16S/167E). Cairns (16/145) is the farthest north that the species has been regularly recorded in eastern Australia. There is a record in 1875 from Warraber I. (10/142) in Torres Strait.[503] In WA claimed specimens from Pt Cloates (22/113) were misidentified,[595] leaving West Wallabi I. (28/113) as the northernmost locality.

Most birds arrive in Australia in Jan.-Feb. and depart in June-July, though individuals have been recorded in both Qld and Tas. in Nov.[19,1745] Reporting Rates were highest in autumn and lowest in spring (spring 0.31%, summer 0.54%, autumn 1.65%, winter 1.18%). The bird is present at Eyre Bird Observatory (32/126) in May-June with a few as late as Aug.; there have been counts of 73, 22 and 31 in successive years.[389]

The Double-banded Plover usually lives on ocean beaches and estuarine mudflats. Paddocks and ploughed fields may be used as roosts during high tide. A small proportion of the population winters inland on salt and freshwater lakes. A flock of 36 appeared at Wanganella (35/144) during floods in 1956.[833] The Plover is sometimes in flocks of hundreds and associates with the Red-capped Plover.[1838] It feeds on dry or damp, recently wet ground but there is no specific information about its food in Australia or New Zealand.[1424]

Australia is the main wintering ground of the species. The most important site known is Corner Inlet (38/146) which held 1460 in July 1982. At the same time Port Phillip Bay (38/144) held 1075; beaches on the north-eastern coast of Tas. 677; and Anderson Inlet (38/145) 550.[84]

Field Atlas Records: 1017

1° blocks–with records: 90 (11%)
–with breeding records:

Atlas No.: 140

Max. Rep. Rate: 36%
In block(s): 24/153

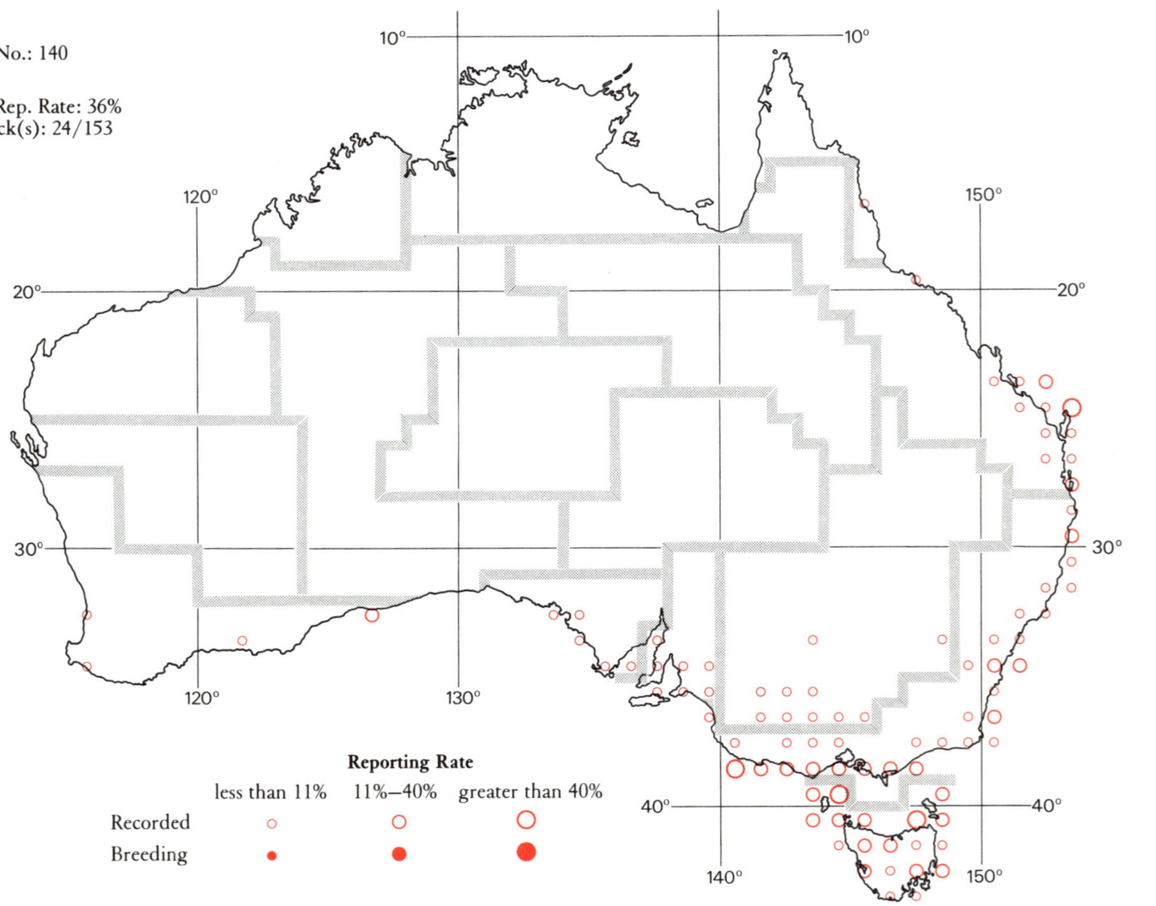

Large Sand Plover

Charadrius leschenaultii

The Large Sand Plover breeds from Caucasia to Mongolia, mainly in the USSR, and visits Australia in summer. A bird banded in Sept. 1981 at Broome (17/122) was recovered in Aug. 1982 at Shanghai (31N/121E) in China.[85] Some remain in Australia during the winter (Reporting Rates: summer 0.73%, winter 0.23%). In the field the Sand Plover is sometimes confused with the Mongolian Plover.

Small numbers migrate through Torres Strait, usually among flocks of Mongolian Plover.[503] At Eyre Bird Observatory (32/126) counts of 0, 1, 6 and 0 were recorded in four successive spring passages and none in autumn.[389]

The Large Sand Plover lives on the coast, most commonly on tidal sandbanks. There are no inland records in the Atlases. Single birds are usual in the south and parties of 10–20 in the north, often in flocks with other species. The Large Sand Plover feeds on insects, specially beetles, and crustaceans.[422]

In Nov. 1982, 58 000 Large Sand Plover were counted at Roebuck Bay (18/122) and Eighty Mile Beach (19/121).[84] This is a stronghold for the species, which generally is far more abundant in north-western Australia than elsewhere in the country.

Field Atlas Records: 413

1° blocks–with records: 86 (11%)
–with breeding records:

Atlas No.: 141

Max. Rep. Rate: 20%
In block(s): 12/130

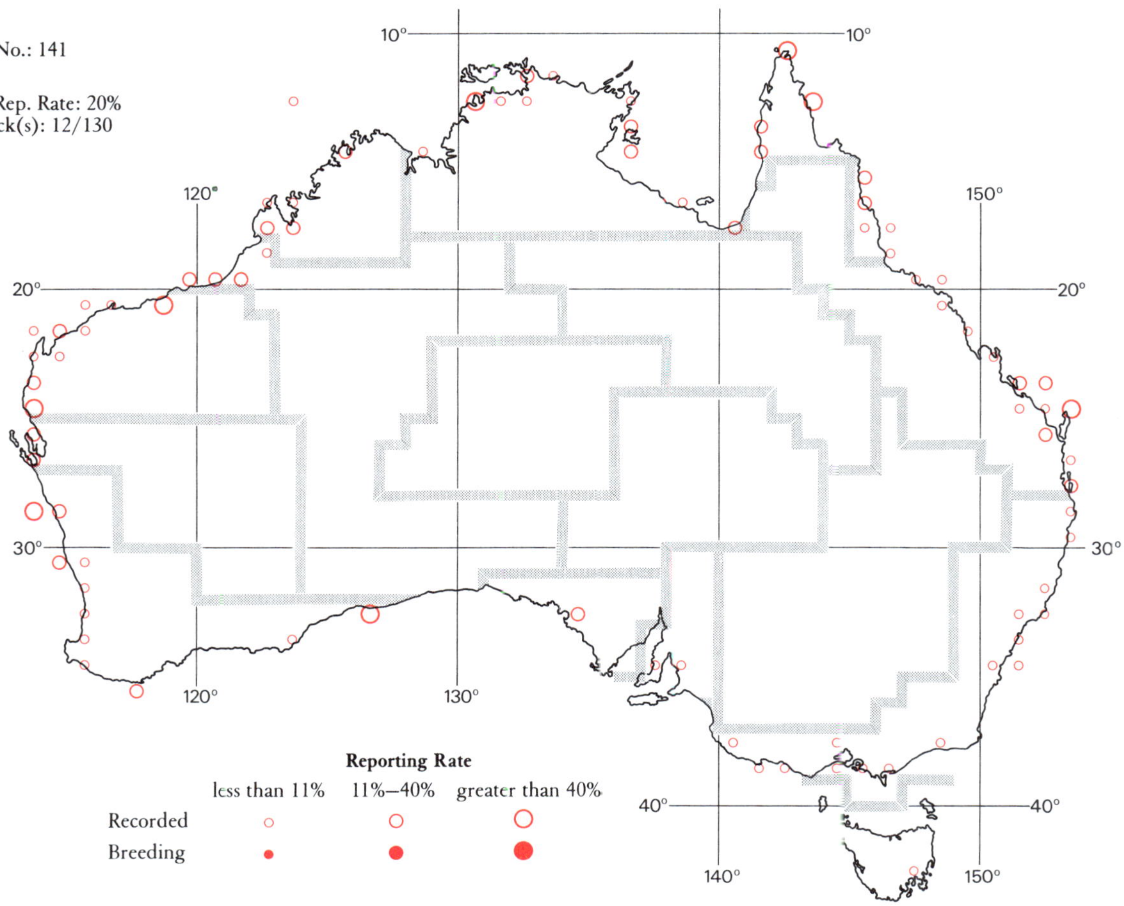

Oriental Plover

Charadrius veredus

The Oriental Plover breeds in northern China and some birds migrate to Australia for the non-breeding season. Few remain in Australia during winter (Reporting Rates: summer 0.11%, winter 0.04%).

Within Australia the species is very mobile. Oriental Plover at Mt Isa (20/139) disappeared immediately heavy rain fell.[312] In southern Australia the species rarely remains more than a few days at any site.[374]

The Oriental Plover lives on open plains and on the seashore. Round Darwin (12/130) flocks have been recorded in recently burnt woodland.[423] Sightings in southern Australia during the Field Atlas were made on sandy, rocky and muddy shores, on a salt lake, near water lying in a paddock and in a dry paddock. The birds are seen in ones and twos or flocks of up to 50 or more. There is no information about the food of the species in Australia.

In Nov. 1982 there were more than 21 000 Oriental Plover between Broome (17/122) and Port Hedland (20/118).[84] How this compares with numbers in other parts of Australia is not known.

Field Atlas Records: 131

1° blocks-with records: 60 (7%)
-with breeding records:

Atlas No.: 142

Max. Rep. Rate: 14%
In block(s): 17/139

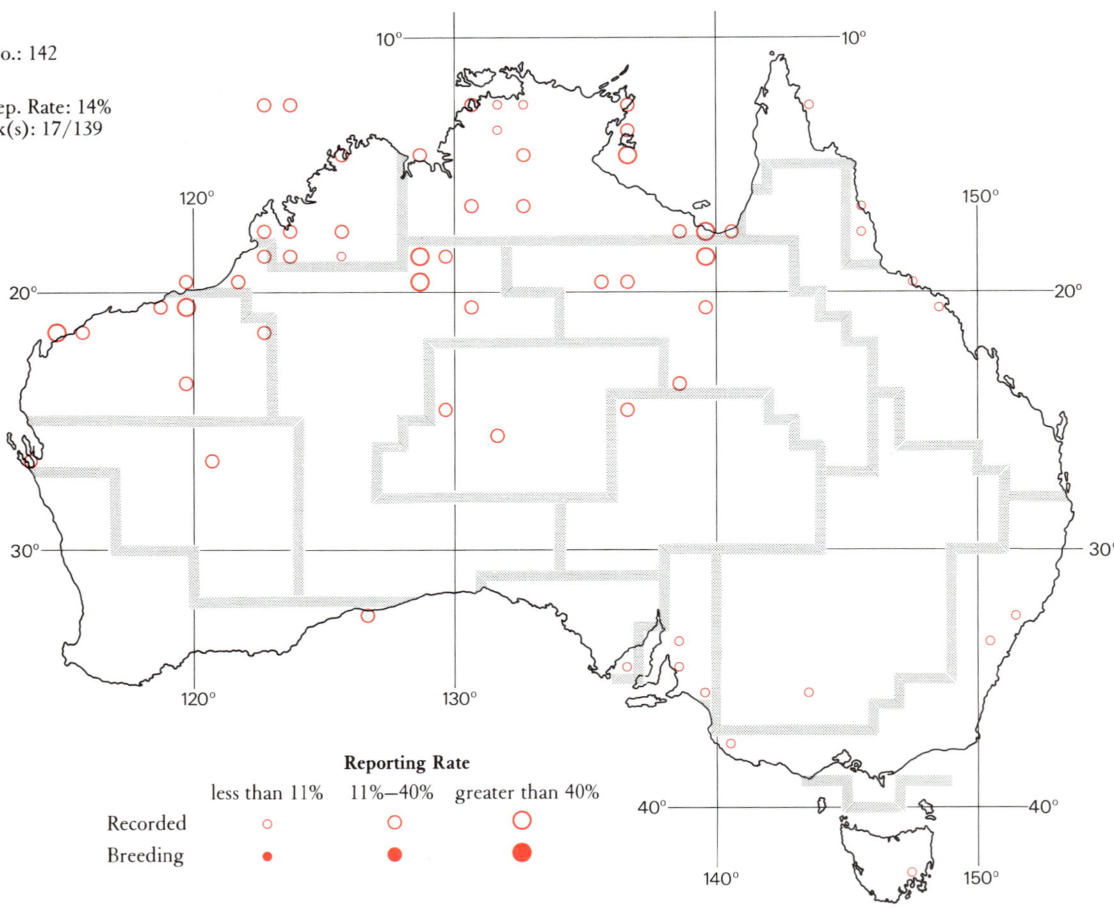

Red-capped Plover

Charadrius ruficapillus

The Red-capped Plover breeds in Australia and occasionally straggles to New Zealand. Its presence in New Guinea has not been confirmed.[1449]

Flocks form in the non-breeding season and disperse to breed[812,1601,1838] but the Field Atlas suggests no large-scale pattern of seasonal movement. In the inland the bird arrives and breeds in large numbers when flooding occurs.[541,728,838] An adult banded at L. George (35/149) was recovered at Broken Hill (31/141), 832 km away.[27]

The Red-capped Plover lives on sand or shingle beaches along the coast or inland waters, on saltpans, dried out saltmarsh where there are bare patches and on muddy or sandy river edges.[748] In the South-West birds remain on coastal lakes for up to two months after they have dried out in autumn, e.g. L. Forrestdale (32/115).[911]

At Eyre Bird Observatory (32/126) highest numbers live on those parts of the beach where the gradient is least.[1209] The population of hundreds shelters in the sand dunes on windy days, moving back on to the beach when it is calm.

Comprehensive counts conducted in Vic. show that inland sites hold more birds than coastal ones: in Feb. 1982 nearly 75% of birds counted were on inland lakes and, in Feb. 1983, 72%. More than 85% of those in the inland were on saline and brackish waters (compared with 57% of all waders counted).[1020]

The Plover feeds on beetles, insect larvae and vegetable matter.[748,1430] Human disturbance, specially on beaches, may cause nests to fail.

Because much of the population lives inland, counts outside Vic. are incomplete. Samples are: Spencer Gulf (34/136) 2000 in Feb. 1983, Coorong (36/139) 4820 in Feb. 1982, Wimmera lakes (36/142) 1160 in Feb 1983, Kerang (35/143) area 1340 in Feb. 1982.[84]

Field Atlas Records: 6315

1° blocks–with records: 350 (43%)

–with breeding records: 129 (37%)

Atlas No.: 143

Max. Rep. Rate: 78%
In block(s): 24/153

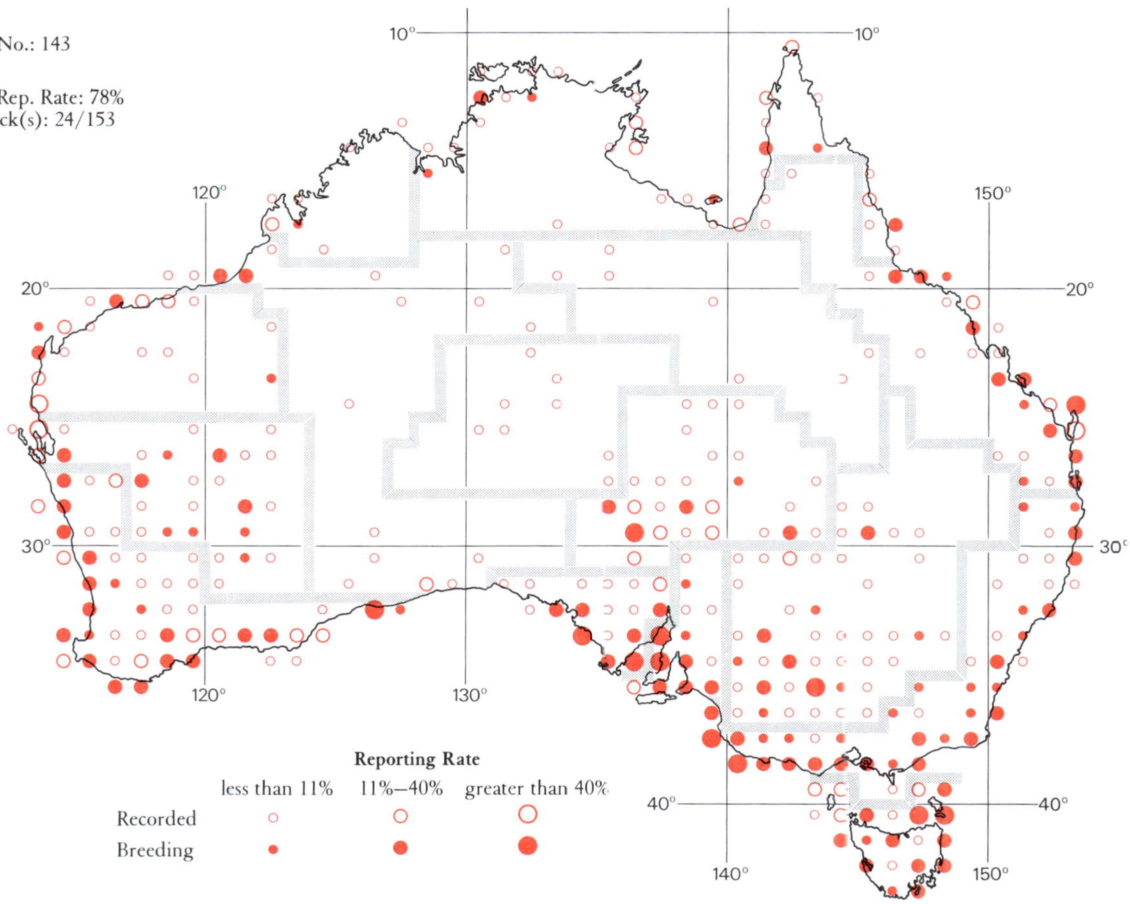

Black-fronted Plover

Charadius melanops

The Black-fronted Plover colonized New Zealand from Australia and was first recorded breeding there in 1954.[388,1424] It has also expanded its range into Tas. where, as late as 1968, the species was considered rare.[653] During the Field Atlas it had become sufficiently widespread to be recorded in 57 different 10′ blocks, breeding in seven. There are still no records from the Furneaux Group (39/148).

Having colonized New Zealand, the Black-fronted Plover can clearly move far. Within Australia the bird occupies wetlands when they form but the Field Atlas suggests no large-scale pattern of seasonal movement.

In the non-breeding season flocks of 20 or more Black-fronted Plover may congregate but pairs are solitary in the breeding season. At Broken Hill (31/141) they occupy a territory on a freshwater shore line with a gravelly beach.[1181] Typically it includes part or all of the shore of a freshwater pool backed by low pasture or is on a river bank which rises steeply from the beach. Nests are usually built close to water: 15 were on average 14.4 m from the water's edge.[1181] The Black-fronted Plover feeds on insects, small crustaceans and a few seeds.[1424] Since European settlement the establishment of numerous farm dams has multiplied by many times the number of places meeting the requirements of the species.

There are no measures of abundance.

Field Atlas Records: 11 361

1° blocks–with records: 568 (70%)
–with breeding records: 172 (30%)

Atlas No.: 144

Max. Rep. Rate: 58%
In block(s): 21/144,29/136

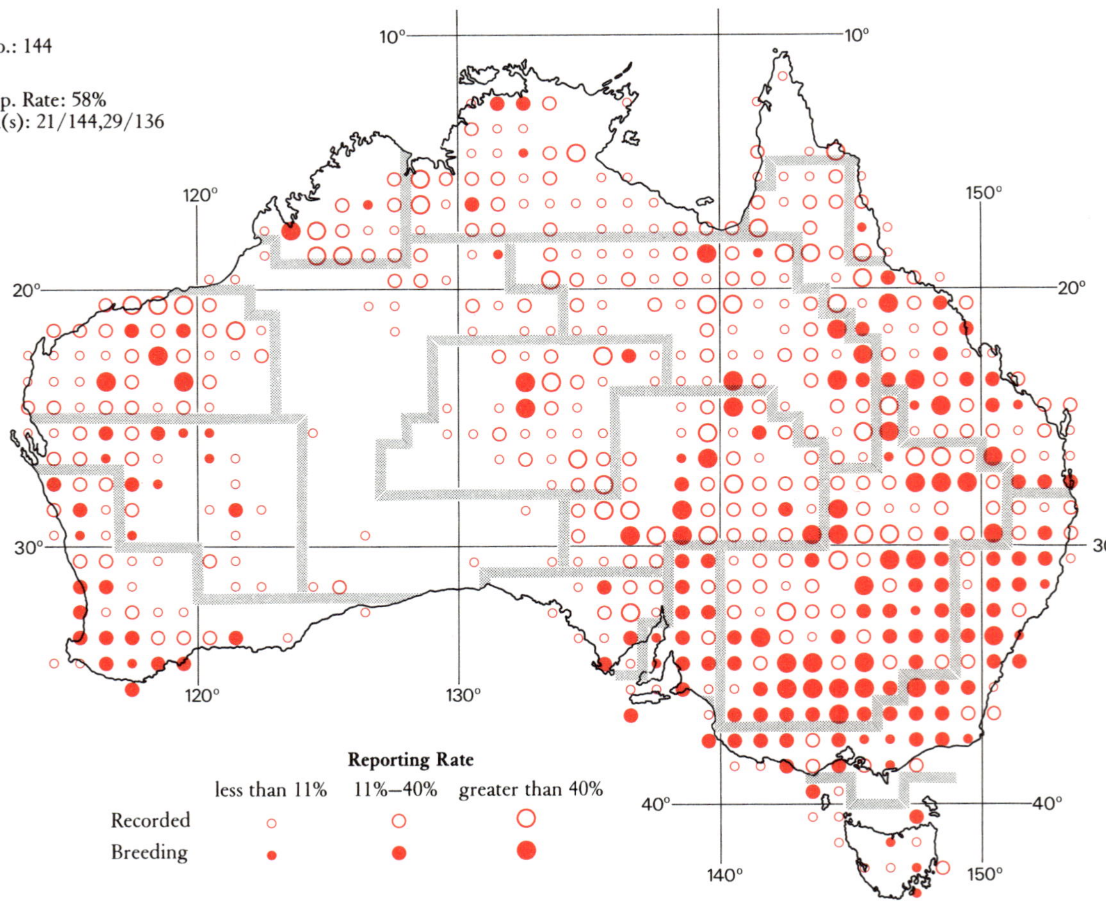

Inland Dotterel

Peltohyas australis

Seasonal movements of the endemic Inland Dotterel have been suggested in the southern Murray-Darling Region: south in spring and north in summer.[273] Usually the bird is reported to arrive in a locality irregularly,[1109] and the Field Atlas suggests no seasonal pattern to its movements.

The Inland Dotterel always inhabits plains, specially those with scattered shrubs or grass tussocks 20–40 cm high. It seems able to live wherever there are open plains of stones, clay or sand, mainly south of exclusively summer rainfall areas (p. xxxix). In the non-breeding season groups are usually of 10–50 birds[1180] but as many as 400 have been seen together in Nov. The nest is sometimes in a loose colony[1109,1155] and sometimes solitary. The Inland Dotterel feeds at night on insects including swarming grass-hoppers.[1155,1180,1185] During the day it feeds on the flowers and leaves of succulent plants from which it is thought to derive most of its water,[1180] though it is also known to drink both from natural claypans and artificial water points.[1158] Large salt glands in its head presumably enable it to excrete salts absorbed from the succulent vegetation.

There are no measures of abundance.

Field Atlas Records: 542

1° blocks–with records: 164 (20%)
–with breeding records: 56 (34%)

Atlas No.: 145

Max. Rep. Rate: 35%
In block(s): 30/128

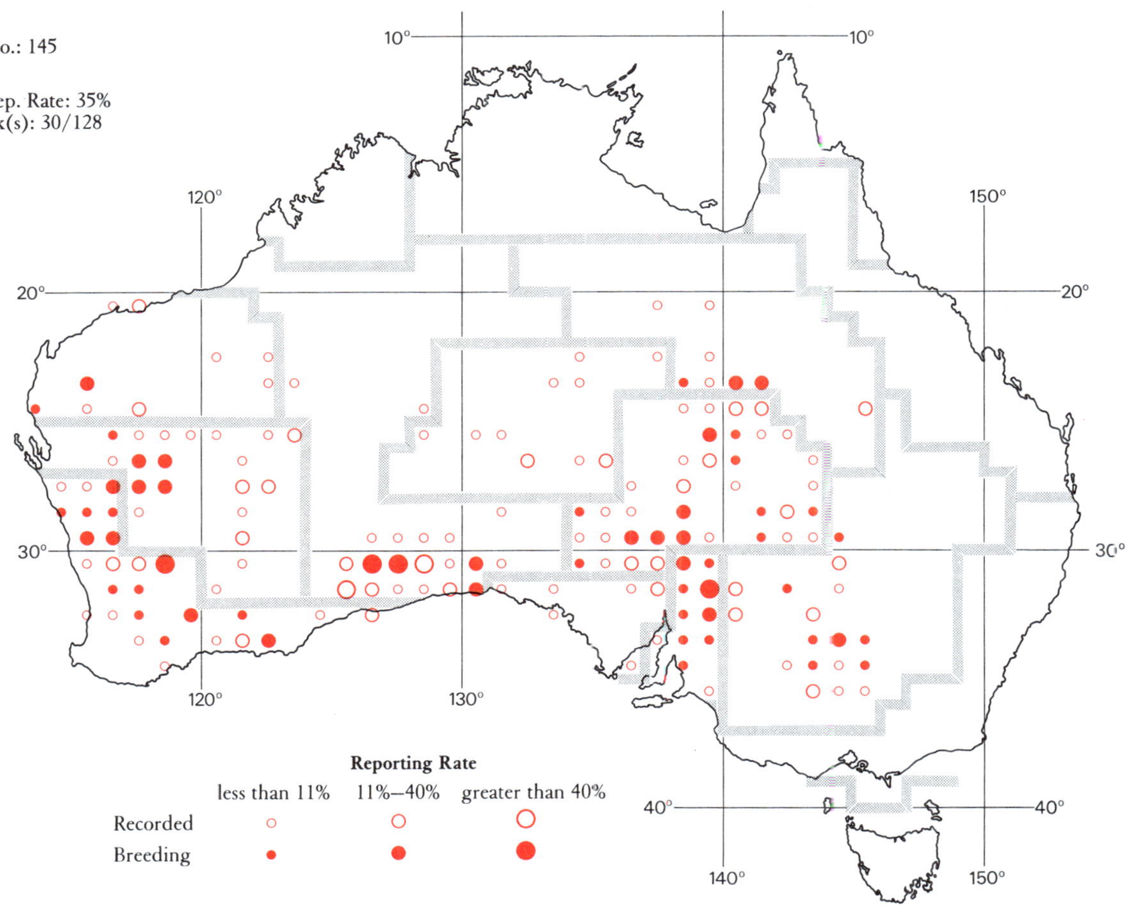

Black-winged Stilt

Himantopus himantopus

The Black-winged Stilt is almost cosmopolitan in distribution. The Australian subspecies *leucocephalus* ranges throughout the mainland but is only an occasional visitor to the Tas. Region.

Until recently the Black-winged Stilt in New Guinea was thought to be a winter visitor from Australia[388] but it is now known to be a breeding resident there.[129,554] This, together with lack of evidence of migration through Torres Strait,[503] suggests that most Australian birds do not move outside the country. In Torres Strait the Black-winged Stilt appears to be sedentary on northern islands and a rare winter visitor to southern ones. At several localities in southern Australia it has been noted to be more common in summer than in winter. These include the southern Murray-Darling Region,[833] Laverton (37/144)[1838] and Perth (31/115).[1601] Reporting Rates were higher in summer than in winter, presumably because the birds are more conspicuous when forced to congregate in flocks as wetlands dry out in summer.

The Black-winged Stilt lives in shallow fresh or saline water and nests colonially or singly. Its toes are only partly webbed and food is gathered while wading, unlike the Banded Stilt, which swims. The Black-winged Stilt has been recorded feeding on molluscs, insects and diatoms.[366,1860] The guts of a bird at Laverton contained brine shrimps and brine flies.[927] Irrigation and clearing have increased the salinity of previously fresh lakes and rivers in the Northam (31/116) district and the Black-winged Stilt is now more abundant there than it was before 1960 when the first signs of salting appeared.[1216]

Colonies are usually small, up to about 60 pairs. The largest on record contained an estimated 200 nests, in a flooded area of 400 ha, at Mystic Park (35/143) in 1973.[1552] Because most Black-winged Stilts live inland, National Wader Counts do not produce representative population estimates except in Vic. where the largest number recorded in one area was 1450 on the Western District (38/143) lakes in Feb. 1982.[84]

Field Atlas Records: 7941

1° blocks–with records: 485 (60%)
–with breeding records: 95 (20%)

Atlas No.: 146

Max. Rep. Rate: 64%
In block(s): 29/136

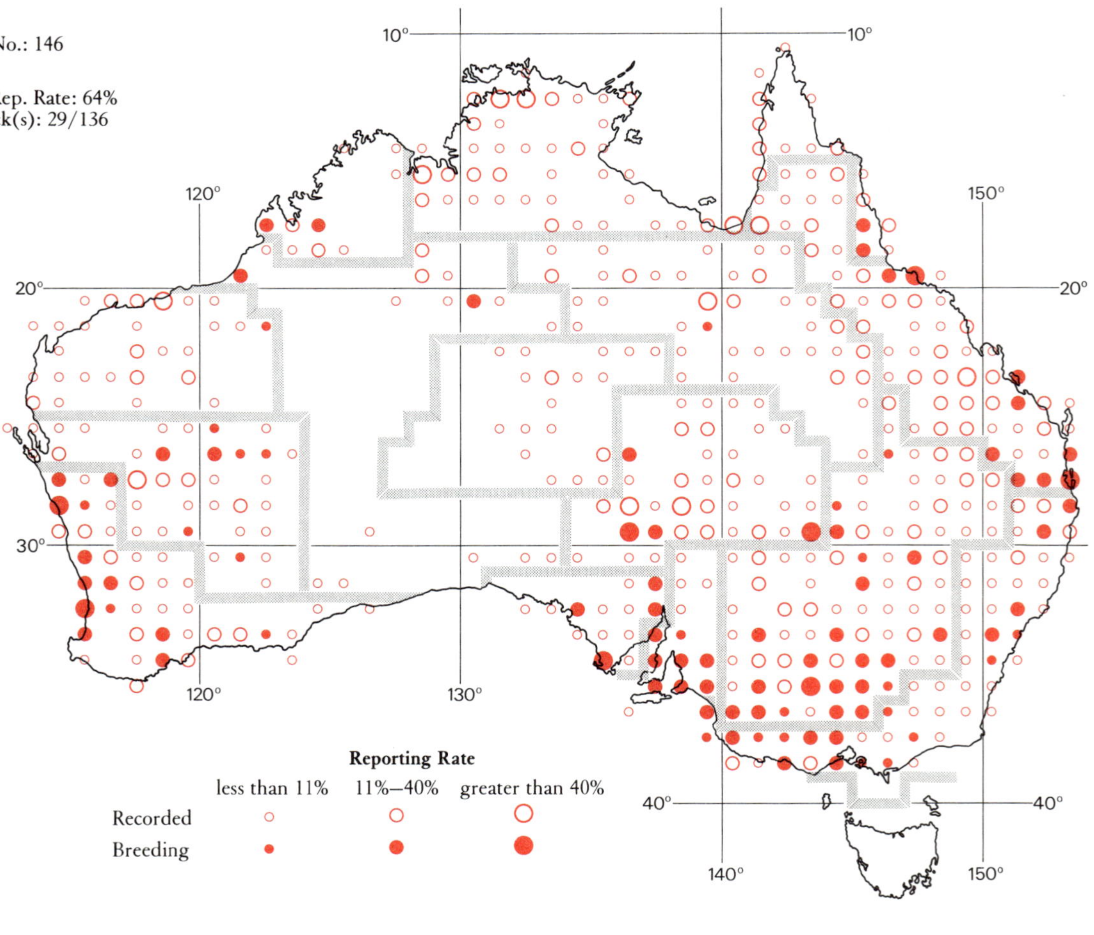

Banded Stilt

Cladorhynchus leucocephalus

It was not until 1930 that a breeding colony of the endemic Banded Stilt was recorded, 114 years after the species was described. Some earlier breeding reports became known subsequently[957] but by 1981 there were still only 19 reports of actual or probable breeding in the Atlases. Dates and places were: L. Cowan (31/121) 1904; Nannine (26/118) 1923; L. Grace (33/118) 1930, 1945, 1946; L. King (33/119) 1930; Menzies (29/121) 1930; L. Callabonna (29/140) 1930, 1936; Wagga Wagga L. (28/116) 1960; L. Disappointment (23/122) 1971; L. Ballard (29/120–121) 1973, 1975, 1980; L. Marmion (29/121) 1975; Percival L. (21/124) 1975; L. Barlee (29/119) 1980; L. Goongarrie (30/121) 1980; and Esperance (33/121) 1980.[1007,1638]

The Stilt's breeding distribution is so small in comparison with its range that the species can clearly move far. It tends to be more abundant on the edges of its range in summer than in winter, presumably because inland waters dry up. Thus at Laverton (37/144) many Banded Stilts appear in summer but a few may be recorded in any month.[1838] At Moolap Saltworks (38/144) there are usually about 2000 present in summer and up to about 300 in winter. At Rottnest I. (32/115) it usually arrives in Sept.[1706] The Stilt is seldom recorded in the Tas. Region but in summer 1980–81 an influx brought birds, most apparently immatures, to several places on both eastern and western coasts and King I. (39/144). They included groups of seven and eight near Sandy Cape (41/144) in Jan. 1981.[781]

The Banded Stilt occupies salt lakes, saltmarsh, saltfields, some coastal mudflats and occasionally shallow freshwater lakes. The large salt lakes on which breeding colonies usually form are filled only after very heavy rains such as drenched the Goldfields Region in 1980. When full, salt lakes are densely populated by brine shrimps and the bird feeds mostly on them. Nesting may continue for several months. It is sometimes curtailed by flooding but more often the lake dries out and thousands of chicks have been recorded to walk away from the colony, many dying. Mammalian and avian predators take heavy toll of eggs and young.[282] A colony at L. Barlee in 1980 contained an estimated 179 000 nests.[282] The National Wader Count recorded 70 000 Banded Stilts on The Coorong (36/139) in Feb. 1982.[84]

Field Atlas Records: 1034

1° blocks–with records: 136 (17%)
–with breeding records: 5 (4%)

Atlas No.: 147

Max. Rep. Rate: 19%
In block(s): 33/137

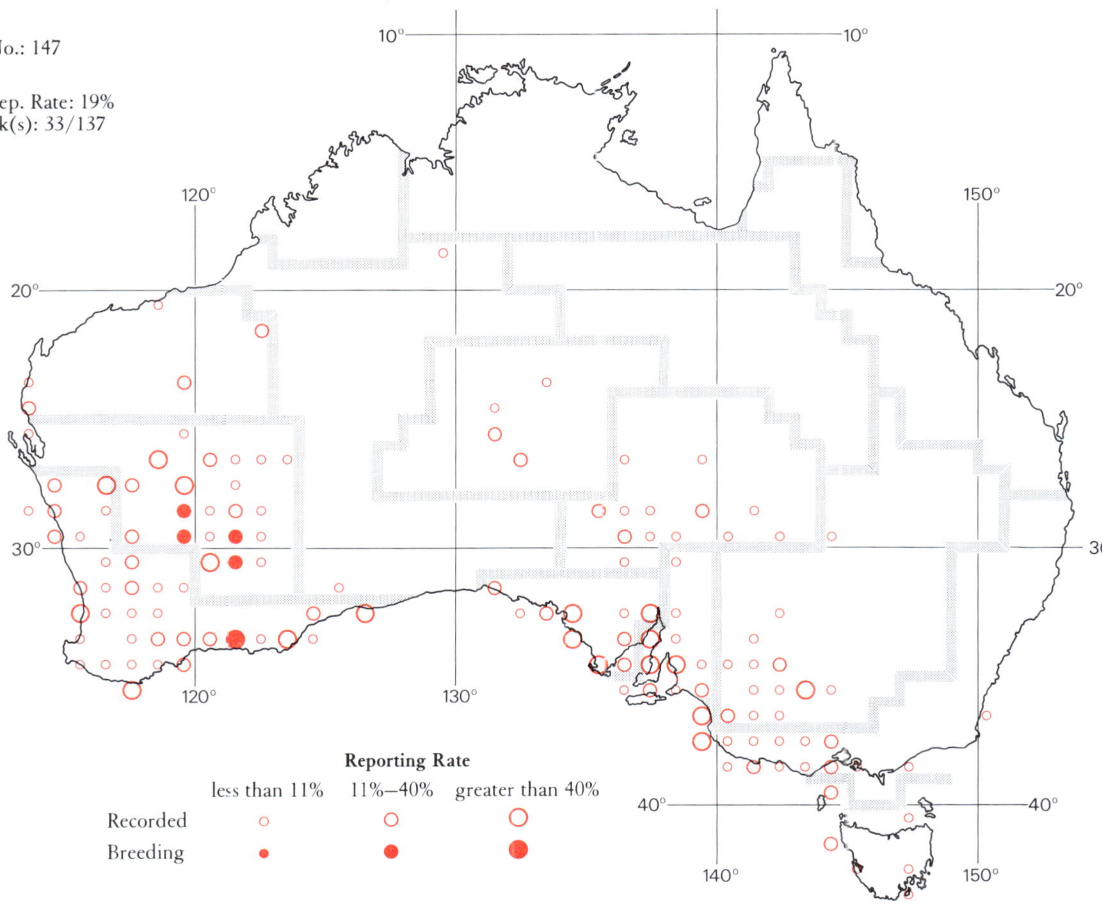

Red-necked Avocet

Recurvirostra novaehollandiae

The Red-necked Avocet breeds only in Australia. It is a vagrant to New Zealand, widely reported last century but rarely since 1892.[971,1003] In the Tas. Region it is also a vagrant; two of the three records there were made last century, in 1880 at George Bay (41/148) and in 1887 at Black R (40/145). The third was at 39/148 on Flinders I. in 1973.

The Avocet is rarely reported east of the Great Dividing Ra. in mainland Australia, though there are some historical records from the Atherton and Cape York Regions. There are no breeding records in the Atlases north of 23°S or east of 148°E.

Movements generally are not regular or seasonal but may be extensive as the New Zealand records indicate. They bring large numbers of birds to shallow lakes formed after heavy rain.[1159] In the southern Murray-Darling Region the species is present all year.[833] Round Kellerberrin (31/117) it usually arrives in June-July of wet years.[627]

During the non-breeding season the Avocet can be found in small groups on the edges of estuaries and rivers, often associated with stilts. Breeding colonies are sometimes large but there may be intervals of many years between breeding attempts when few birds are present in a locality.[88,627,1159] Flocks of 1000 or more form after successful breeding.[833] Flocks of a similar size have also been reported on lakes in Perth (31/115) as they dry out in summer.[911]

The Red-necked Avocet feeds in shallow water from which it filters crustaceans and aquatic insects and their larvae. Seeds and vegetable matter have also been found in guts.[748] At Laverton (37/144) the largest numbers of birds are present when the brine shrimp is most abundant.[1838]

There are no measures of abundance.

Field Atlas Records: 1804

1° blocks–with records: 220 (27%)

–with breeding records: 28 (13%)

Atlas No.: 148

Max. Rep. Rate: 35%
In block(s): 29/136

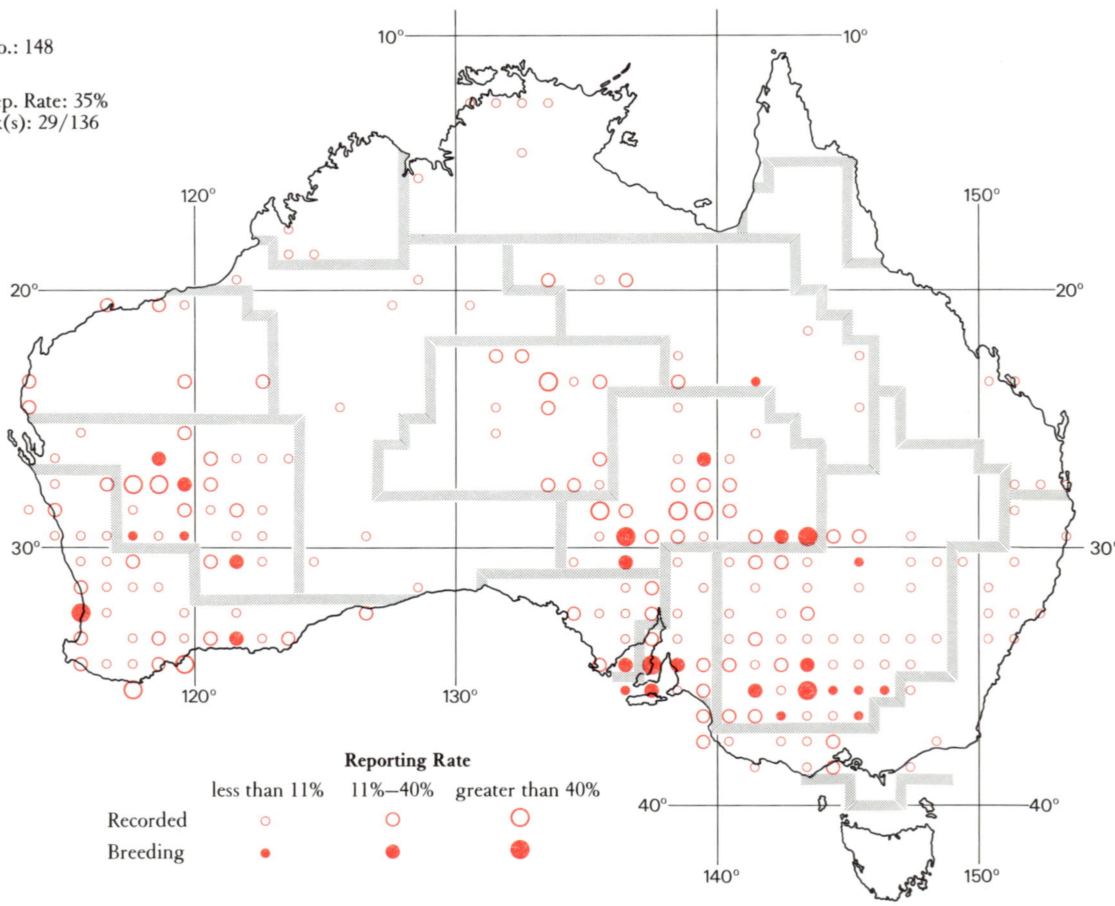

Ruddy Turnstone

Arenaria interpres

The Ruddy Turnstone is a summer visitor to Australia, mainly birds of the subspecies *interpres* which breeds in Siberia. A second subspecies *morinella*, which breeds in eastern North America, has been recorded in the Darwin (12/130) area.[388] Birds of other Asian populations winter in south-eastern Asia and India[1117] and could appear occasionally in Australia. A bird banded at Newcastle (32/151) has been recovered in the USSR (43N/131E)[85] and birds colour-dyed in Alaska have been sighted in Australia.

A proportion of the Turnstone population remains in Australia during winter.[19,812] Most are in eclipse plumage and are probably immature. Field Atlas records from inland 1° blocks were made only Sept.-Nov., which suggests that some Turnstones cross Australia on their southward migration. Other birds move round the coast. The species is a passage migrant through Torres Strait.[503] At Eyre Bird Observatory (32/126) there is a regular passage Sept.-Oct. each year but no return passage in autumn. Counts there in four successive springs were 79, 39, 33 and 13.[389]

The Ruddy Turnstone inhabits shores where stones or stony pavements are exposed as well as sandier shores, where seaweed has accumulated, and mudflats. The sites at which it is abundant in southern Australia are all characterized by rocky intertidal areas.

The Turnstone appears round inland waters only on migration and in small numbers: one inland Field Atlas record was of about six birds and the remainder were of single birds.

The Ruddy Turnstone feeds on a wide range of small invertebrates, turning over pebbles, shells and seaweed with its bill to expose them. It often takes carrion and has been recorded eating maggots from a carcass.[1756]

The Turnstone is generally commoner in northern than southern Australia. It is particularly abundant in a few areas, e.g. counts in Feb. 1983 recorded 2060 at Roebuck Bay (18/122), 784 at Eyre Pen. and Spencer Gulf (34/136), 754 in south-eastern SA, 287 in the Darwin area and 28[illegible] on King I. (39/144).[84]

Field Atlas Records: 1774

1° blocks–with records: 162 (20%)

–with breeding records:

Atlas No.: 129

Max. Rep. Rate: 53%
In block(s): 25/113

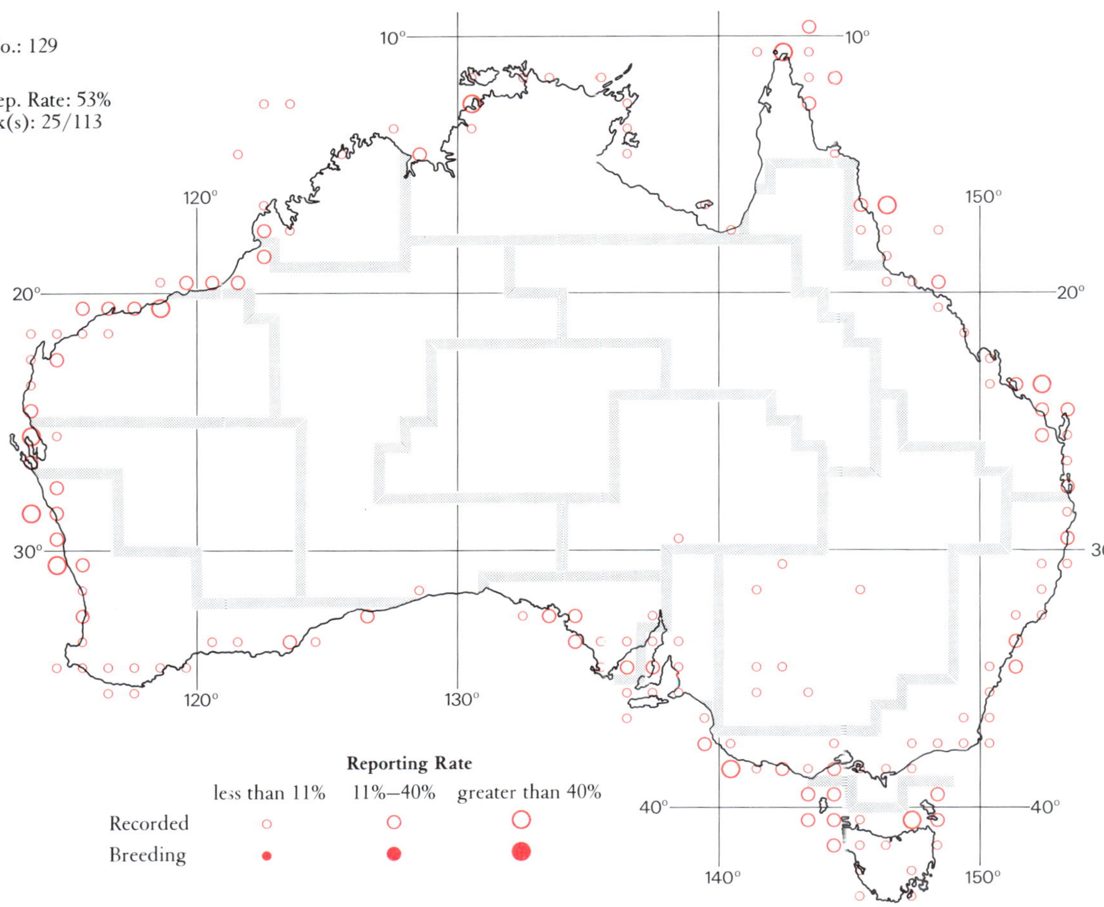

Eastern Curlew

Numenius madagascariensis

The Eastern Curlew breeds in eastern USSR and northern China and migrates south in the non-breeding season.[1238] The Curlew is transient in Indonesia[1846] and the Malay Pen.[1238] and probably the bulk of the world population visits Australia for the summer. Reporting Rates suggest that a significant proportion of the population remains in Australia during winter, mainly in the north. The summer Reporting Rate was 4.2%, which dropped in winter to 1.2% south of 26°S, but 3.2% north of 26°S. In Tas., birds that have not left by May remain for the whole winter.[1745]

Roebuck Bay (18/122) is an important area for birds arriving in Australia on migration. It held 2160 Curlew in Sept. 1982. By mid-Nov. this had dropped to 960.[84] In Torres Strait, in comparison, the species is regarded as an uncommon passage migrant.[503] The only inland record during the Field Atlas was of two birds at L. Tutchewop (35/143) in Nov. 1977. The scarcity of inland records suggests that the bulk of the population may travel along the coast. The longest recorded movement of a banded bird is 8160 km from Botany Bay (34/151) to South Korea.[85]

The Curlew lives on the coast in muddy, tidal estuaries. Sometimes hundreds gather and remain in one site throughout the summer. It eats crustaceans, specially crabs, extracted from deep in the mud.[812]

Annual counts round Hobart (42/147) reveal a decline from 500–600 birds in 1948 to an average summer population of 265 in the 1960s and 192 between 1974 and 1981.[1319] The summer 1982 count of 140 is the lowest since counts began in 1948. There is also evidence of widespread decline in SA.[377] Both areas are at the limit of the species' range in Australia and held only 8% of the 5788 birds counted in the National Wader Count of summer 1982. Human activities that disturb feeding and roosting areas may cause the decline.

National Wader Counts show the Curlew to be more abundant in northern and eastern Australia than in the west or south. In particular, numbers are low in SA despite the huge numbers of other wader species there. In Feb. 1983 four areas held substantial numbers: Corner Inlet (38/146) 2280, Hervey Bay and Great Sandy Strait (25/152) area 1970, Moreton Bay (27/153) 1080, Western Port Bay (38/145) 763.[84]

Field Atlas Records: 2680

1° blocks–with records: 139 (17%)
–with breeding records:

Atlas No.: 149

Max. Rep. Rate: 54%
In block(s): 24/153

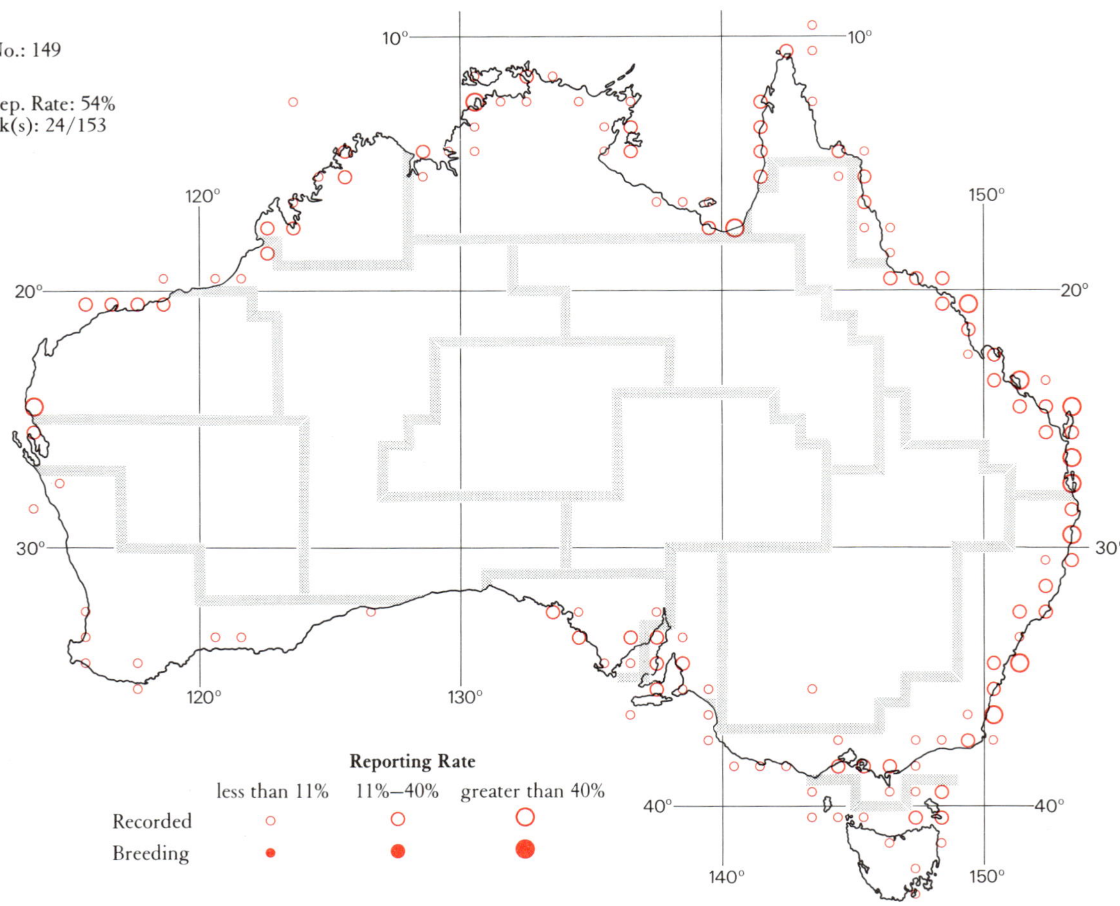

Whimbrel

Numenius phaeopus

The Whimbrel breeds in the northern Holarctic. The subspecies *variegatus* that migrates to Australia in the non-breeding season breeds in eastern Siberia.[388] Some birds cross the mountains of New Guinea at high altitude on migration.[1591]

At Cairns (16/145) there is a regular passage of up to 100 birds south in spring and north in autumn. Inland Atlas records in spring (32/142, 33/141) and autumn (26/140) suggest that some birds travel across Australia on migration.

The Whimbrel is scarce in southern Australia even in summer and only occasionally reaches Tas., usually singly. A substantial non-breeding population remains in winter, mostly in northern Australia, though a wintering flock of more than 20 was recorded from Western Port Bay (38/145).[1112] Reporting Rates south of 26°S were 0.6% in winter and in summer 1.4%.

The Whimbrel lives along the coast, foraging on tidal mudflats and roosting on rocky islets and coral cays, often in mangroves and, occasionally, in other trees. In the Top End birds are sometimes inland along mangrove-lined rivers. Flocks of up to 50, such as are seen in Qld,[19] are rare in the south where most sightings are of single birds or small parties.

Little is known of the food requirements of the Whimbrel in Australia but in Britain a variety of invertebrates is taken, including insects, worms and crustaceans, as well as some plant material.[422]

Preliminary aerial surveys of the Gulf of Carpentaria suggest that this may be an important area for the species in Australia.

The largest counts from individual localities in the National Wader Survey of Feb. 1983 were: 1080 at Roebuck Bay (18/122) and 380 at Moreton Bay (27/153).[84]

Field Atlas Records: 1843

1° blocks–with records: 144 (18%)
–with breeding records:

Atlas No.: 150

Max. Rep. Rate: 57%
In block(s): 10/142

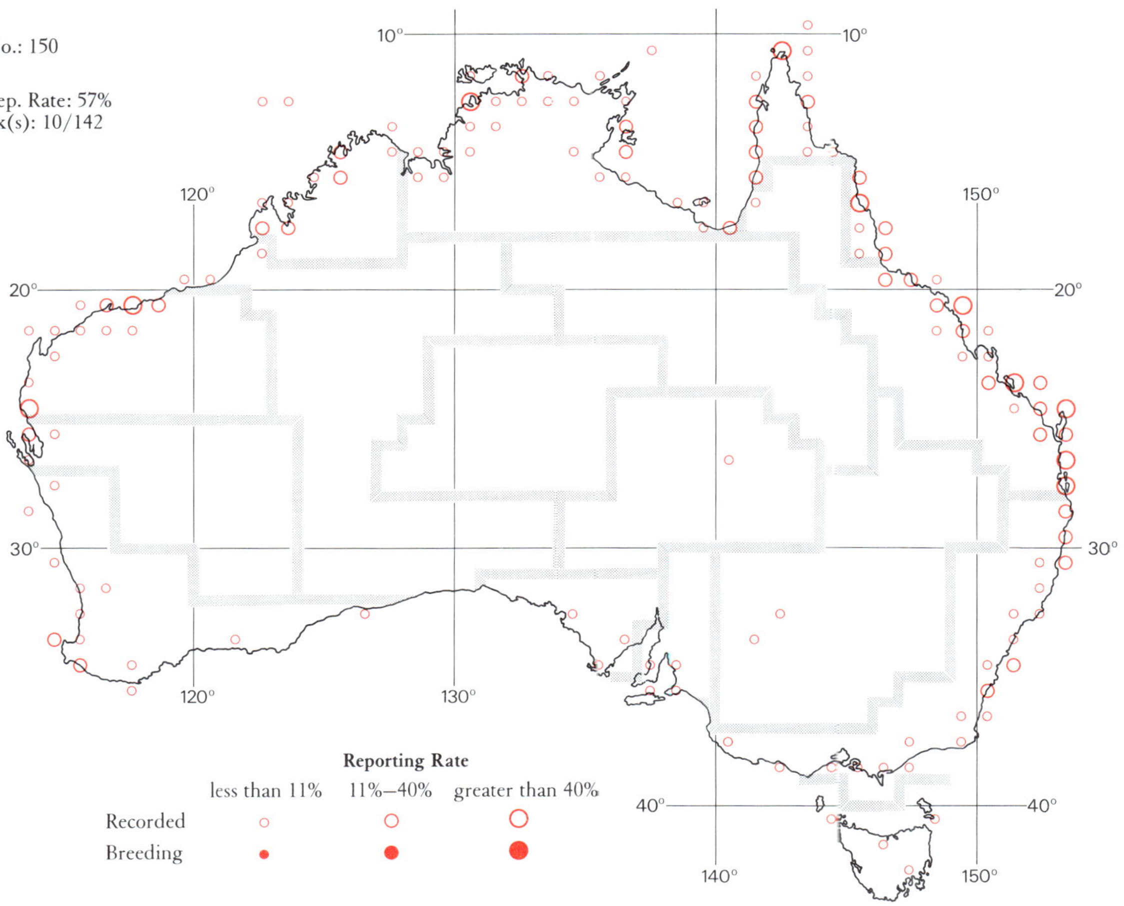

Little Curlew

Numenius minutus

The Little Curlew breeds in Siberia and Mongolia,[1011] migrating south for the non-breeding season. There are few summer records outside Australia and New Guinea and it appears that Australia may hold the bulk of the Little Curlew population in summer.[1138]

At Darwin (12/130) flocks are present Oct.-Nov. and again in Mar.-Apr. but not during the height of the wet summer season[423] when the birds probably disperse inland to the extensive damp areas formed then. There have been observations of large numbers of birds at inland sites in summer, including 183 at L. Moondarra (20/139) in Feb. 1982[84] and 2000–3000 in the Richmond (20/143) district in Feb. 1903.[1138] The Field Atlas also shows that the bird is widely distributed inland in summer. North of 26°S there are summer records from eight inland 1° blocks compared with three in autumn and only one in spring.

The Little Curlew is an uncommon visitor to southern Australia. There have been two records in Tas. and occasional reports in the South-West Region, though not during the Field Atlas. South of 26°S there were records from four inland 1° blocks in spring and one each in summer and autumn, which suggests that part of the population, at least, flies overland to reach southern sites. There are few winter records of Little Curlews in Australia; most of the population apparently leaves in Apr.

The Little Curlew feeds on dry or damp ground. It deserts flooded areas and has even been recorded from recently burnt woodland.[423] A small flock at Sydney (33/151) in 1958 was feeding mainly on beetles living in short grass.[1138]

An estimated quarter of a million Little Curlews were recorded on migration at Fogg Dam (12/131) on two days in Oct.[1745] In the National Wader Count of Feb. 1983, 5000 were recorded at Karumba (17/140).[84]

Field Atlas Records: 275

1° blocks–with records: 72 (9%)
–with breeding records:

Atlas No.: 151

Max. Rep. Rate: 26%
In block(s): 24/153

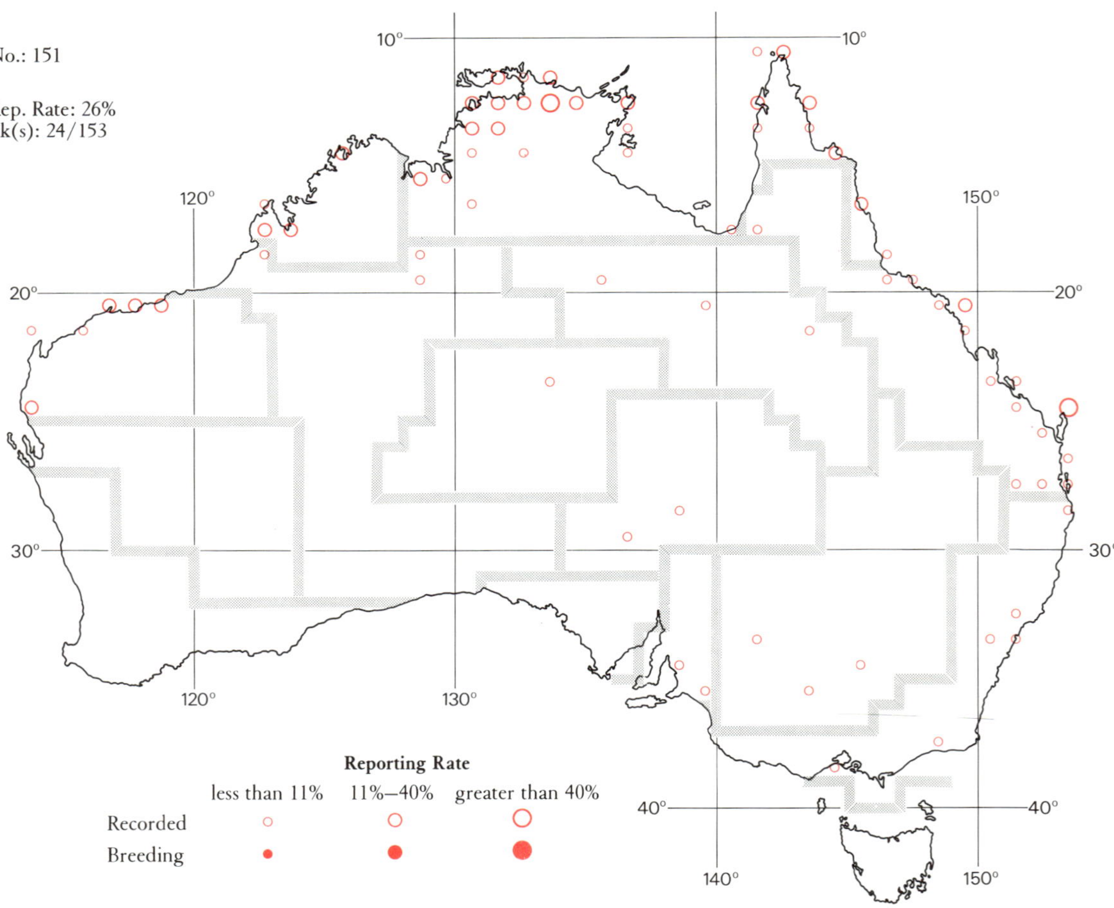

Wood Sandpiper

Tringa glareola

The Wood Sandpiper breeds in the Palearctic. Birds visiting Australia in the non-breeding season are probably from eastern Siberia. The species is more frequently reported from northern Australia than southern and in the Tas. Region had only once been recorded before the Field Atlas records (42/148) in Mar. and Dec. 1978. Few birds winter in Australia.

The Wood Sandpiper lives on shallow margins of freshwater lagoons and swamps, specially those fringed with river red gum. More birds are recorded inland than on the coast. The species is usually solitary or in parties of up to ten.[815,956,1609] Large flocks have occasionally been recorded including one of about 80 at Camballin Irrigation Area (18/124) in Sept. 1982 and one of 95 on flooded, grassy coastal plain at Twelve Mile Lagoon (15/128) in Mar. 1983.[911]

In southern Asia this species shows a strong tendency to return to the same sites in the non-breeding season year after year.[1117] In Australia, as well, it is a regular visitor to some sites. A vagrant was recorded in Tas. at the same place in two consecutive seasons. The Wood Sandpiper also appears erratically in summer at storm pools throughout the inland. It feeds on insects, including beetles, ants and the larvae of flies.[815] This species benefits from irrigation in the tropics and from open sewage farms and ponds throughout the continent.

There are no measures of abundance.

Field Atlas Records: 348

1° blocks–with records: 99 (12%)
–with breeding records:

Atlas No.: 154

Max. Rep. Rate: 17%
In block(s): 29/136

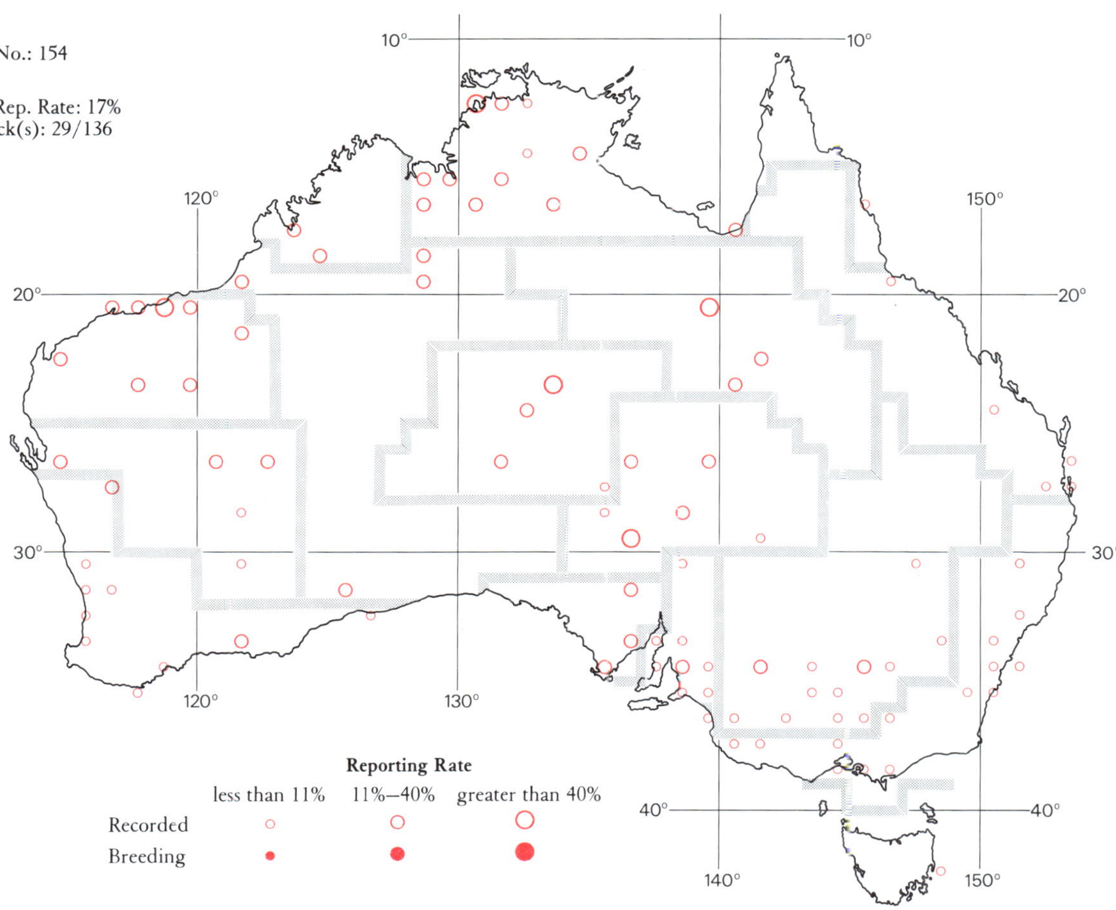

Grey-tailed Tattler

Tringa brevipes

The Grey-tailed Tattler breeds in Siberia but its breeding range is incompletely known.[388] Birds banded in Australia have been recovered in Japan and the USSR. The longest movement recorded is that of a banded bird from Newcastle (32/151) to USSR (62N/174E), 9300 km.[55]

The Grey-tailed and Wandering Tattlers are sometimes considered conspecific and are difficult to distinguish in the field.[1078] The Grey-tailed Tattler is a non-breeding visitor to Australia in summer, some birds remaining during the winter (Reporting Rates: summer 1.5%, winter 0.7%). On the estuary of the Hunter R (32/151) it is most common Nov.-Apr. but some are present May-Oct.[975]

The Field Atlas contains only one record from an inland 1° block, 34/139 near the Murray R in Sept. 1980, which suggests that most Tattlers move round the coast to reach southern sites. At Cairns (16/145) substantial numbers are present only during southward and northward migration. At Eyre Bird Observatory (32/126) on the southern coast the Tattler appears only in spring and only in some years with counts of 2, 0, 2, 17 in four successive spring passages.[389]

The Grey-tailed Tattler occupies coastal mudflats and reefs, including oyster leases, and only rarely sandy beaches. It is commonly associated with mangrove-fringed shores and will roost at high tide in mangrove trees. It was recorded eating small crabs in the Hunter R estuary[975] but usually probes into mud, presumably for intertidal invertebrates.

Counts and Atlas data suggest that the Grey-tailed Tattler is most common on the coasts of northern Australia with smaller numbers widely dispersed in the south. Numbers estimated at individual locations include: 675 in the area between Broome (17/122) and Port Hedland (20/118) in early Aug. 1982 rising to 8410 in Sept. and 11 900 in mid-Nov.; and 3190 at Moreton Bay (27/153) in Feb. 1983.[84] In the Hobart (42/147) area counts from 1965-71 recorded only one to three birds annually. There was an increase to a maximum of 17 in 1974 followed by a decline to its former status between 1976 and 1983.[1409]

Field Atlas Records: 1124

1° blocks-with records: 114 (14%)
-with breeding records:

Atlas No.: 155

Max. Rep. Rate: 34%
In block(s): 10/142

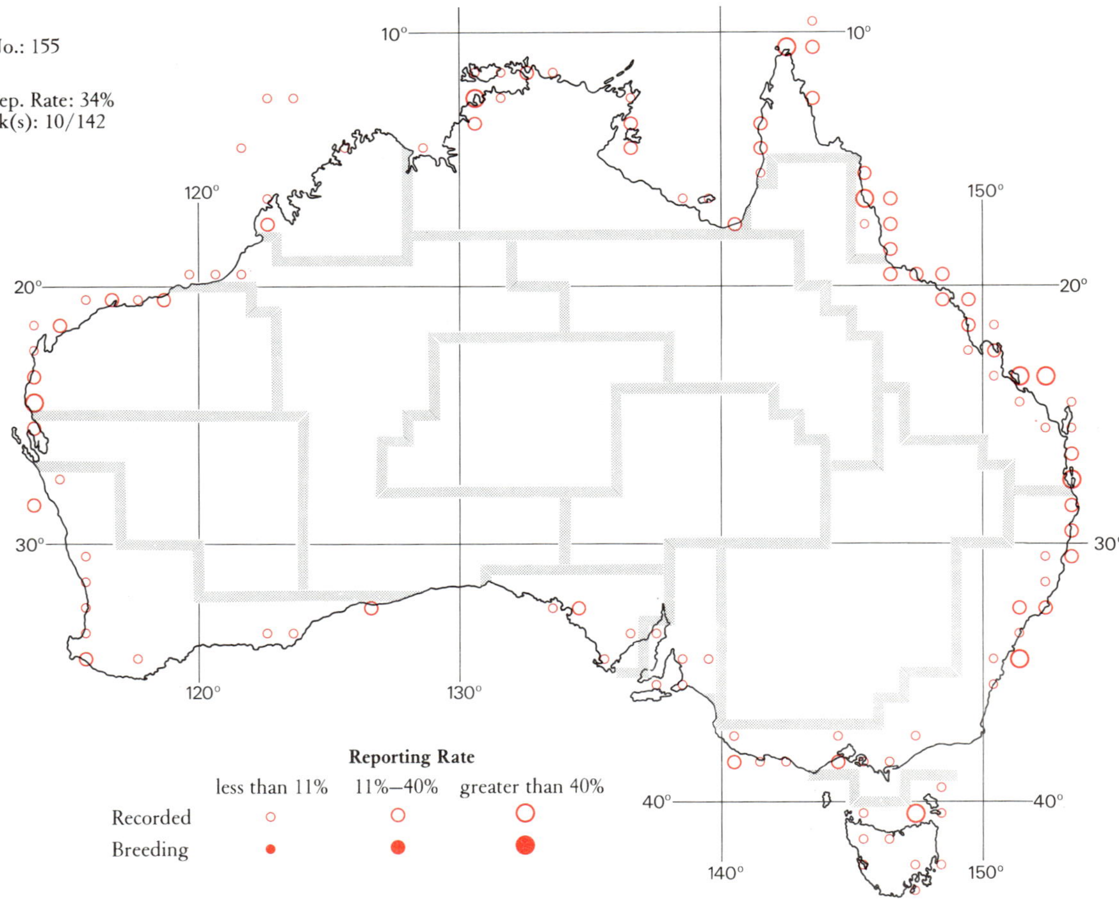

Wandering Tattler

Tringa incana

The Wandering Tattler breeds in eastern Siberia and Alaska, farther east than the Grey-tailed but its breeding range is incompletely known.[388] It is difficult to distinguish the Wandering from the Grey-tailed Tattler in the field[1078] and being in much smaller numbers the Wandering is probably often overlooked. A bird banded in Tokyo and recovered at Moreton Bay (27/153) was reported to be a Wandering[1117] but at the time the two species were not separated. Its identity is uncertain but it was probably the Grey-tailed.

The Wandering Tattler is a summer visitor to Australia, few birds remaining over winter. It usually lives on the open shore, often on wave-washed reefs and rocky coasts. Field Atlas records were of single birds and small parties of up to about four. The species feeds on crabs and marine worms in Australia and has been observed to catch and break up a crab on rocks.[812]

There are no measures of abundance.

Field Atlas Records: 47

1° blocks–with records: 15 (2%)
–with breeding records:

Atlas No.: 156

Max. Rep. Rate: 4%
In block(s): 10/142

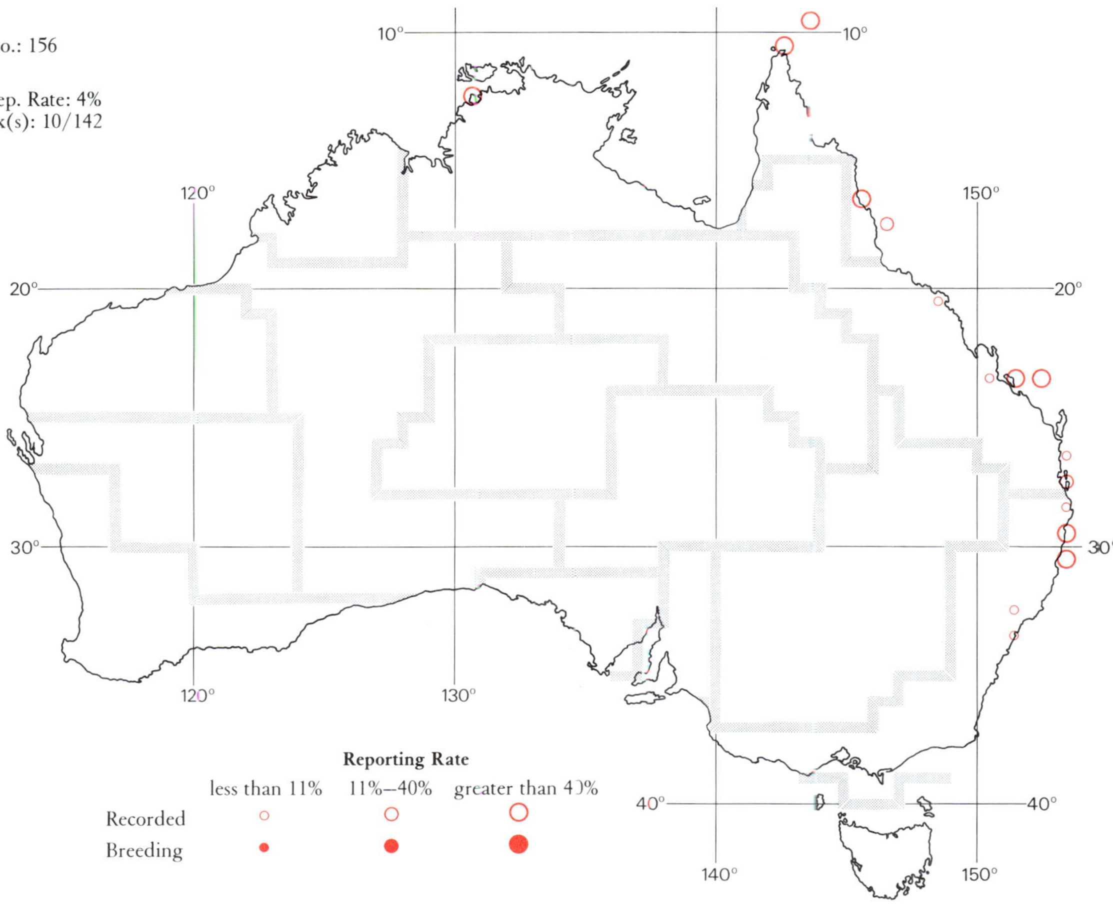

Common Sandpiper

Tringa hypoleucos

The Common Sandpiper is a summer visitor to Australia from breeding grounds that stretch from the British Isles to eastern Siberia south of the Arctic Circle.[388] Some birds remain in Australia in winter.[503,1346] South of 26°S the Reporting Rate in summer was 2.2% and in winter 0.3%. Some of the winter records are probably early arrivals: birds reaching Australia in Aug.

Inland records suggest movement across the continent and many birds probably remain on inland waters throughout the summer. On the coast at Eyre Bird Observatory (32/126) the species appears only in spring with counts of 3, 9, 8, 10 in four years.[389]

The Common Sandpiper is usually alone or in loose groups of less than five in Australia. It inhabits rocky shores, mangroves and the tidal course of rivers as well as inland lakes, streams, pools and dams. It is seldom in mixed flocks of waders and in the Top End feeds on river margins not used by other species.[423] It feeds by probing in shallow water taking aquatic invertebrates and insects and will stalk flies to catch them with a final pounce.[1462]

Counts and observations show the Common Sandpiper to be more abundant in northern than southern Australia but there are no measures of abundance.

Field Atlas Records: 1873

1° blocks–with records: 276 (34%)
–with breeding records:

Atlas No.: 157

Max. Rep. Rate: 35%
In block(s): 25/113

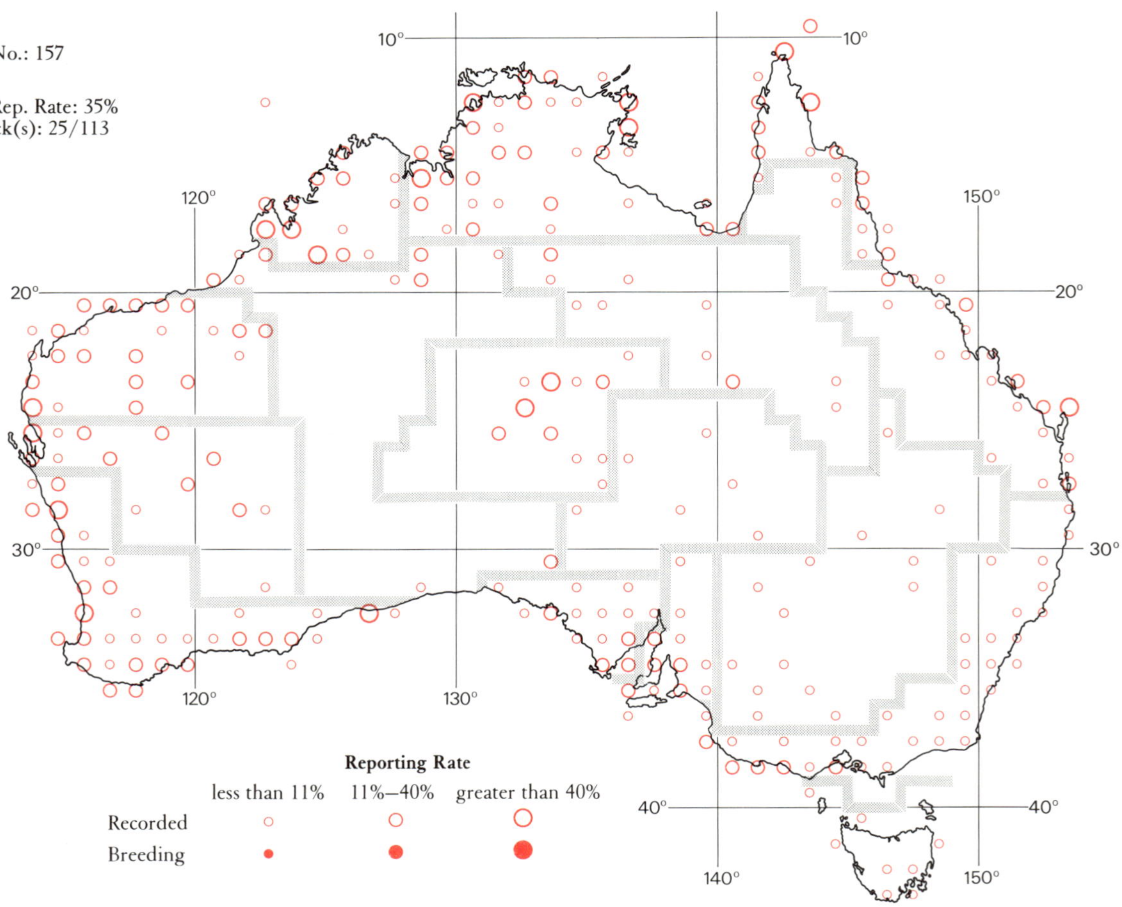

Greenshank

Tringa nebularia

The Greenshank has breeding grounds stretching from the British Isles to eastern Siberia, south of the Arctic Circle.[388] Some, at least, of the Asian birds come to Australia each summer.[1313] A few birds remain in Australia during the winter (Reporting Rates: summer 5.5%, winter 1.1%). In NSW birds are reported throughout the year but most commonly Sept.-Apr.[396]

Birds are recorded on the coast near Eyre Bird Observatory (32/126) only in spring with counts of 22, 4, 4, 21 in four years.[389] At Laverton (37/144) the Greenshank is most abundant in spring and autumn but some are present all summer.[1838]

The Greenshank occupies the shores of lagoons, swamps and large rivers as well as muddy and sandy coasts.[812] In the Mt Isa (20/139) district it inhabits areas too rocky and steep to attract other waders.[1067] The Greenshank tends to feed in dispersed groups. It sometimes forms mixed flocks with the Marsh Sandpiper, which it resembles. The Greenshank feeds on fish and aquatic invertebrates, occasionally chasing them through shallow water.[1313,1838]

The Greenshank population in Australia is widely dispersed, no one site holding large numbers. Numbers estimated in National Wader Counts include 670 on The Coorong (36/139) in Feb. 1982, 540 at Roebuck Bay (18/122) in Feb. 1983, 370 at Spencer Gulf (34/136) in Feb. 1983, 310 at Swan Bay (38/144) in Feb. 1983 and 290 at Corner Inlet (38/146) in Feb. 1982.[84] At Hobart (42/147) the population has been almost constant at about 70 each summer between 1965 and 1981.[1322]

Field Atlas Records: 3253
1° blocks–with records: 321 (39%)
–with breeding records:

Atlas No.: 158

Max. Rep. Rate: 39%
In block(s): 25/113

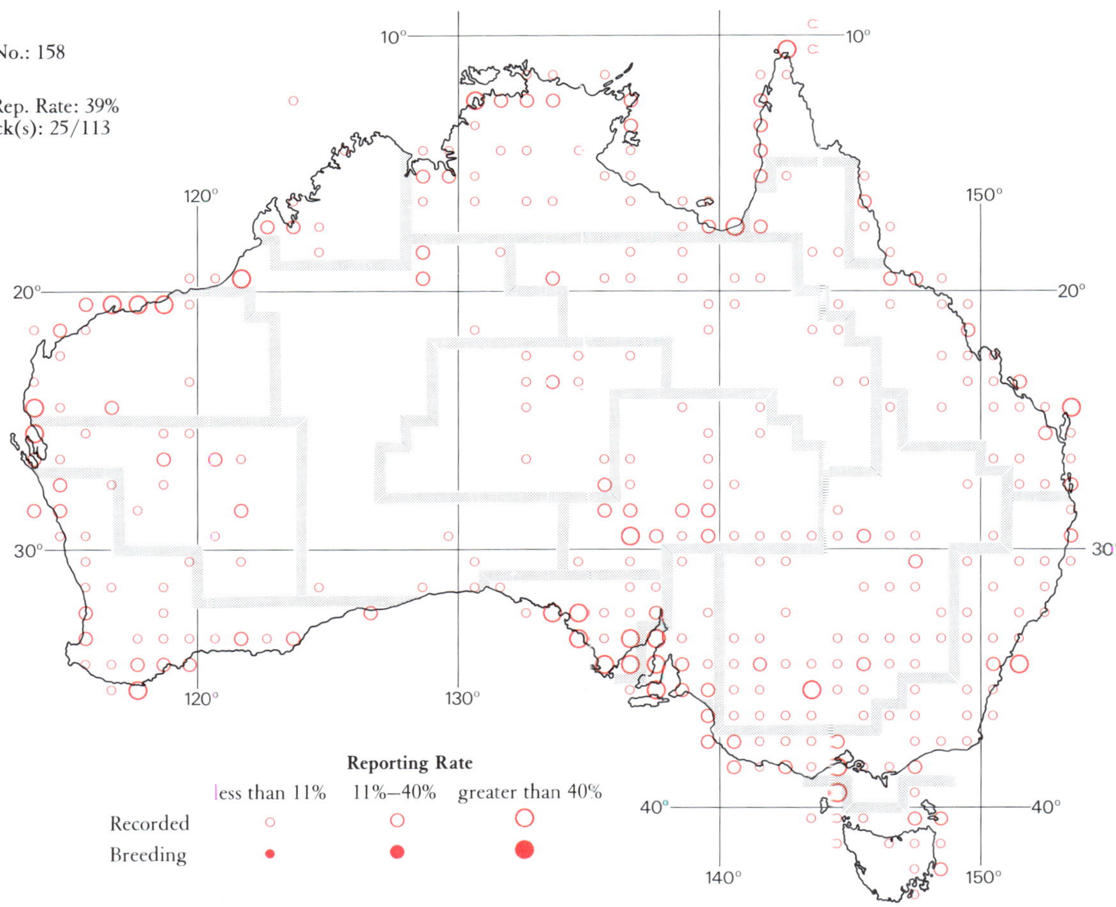

Marsh Sandpiper

Tringa stagnatilis

The Marsh Sandpiper breeds from Bulgaria to northern Mongolia and some birds visit Australia each summer. The species is a vagrant to the Tas. Region. The Reporting Rate in autumn (0.5%) was lower than in spring (0.9%) or summer (1.3%), which suggests that the Marsh Sandpiper may leave Australia at or before the end of summer. Few appear to remain during the winter (Reporting Rate 0.1%) and only in the north.

In Australia the Marsh Sandpiper is sometimes in mixed flocks with the Greenshank, which it resembles. Flocks of up to 100 Marsh Sandpipers have been recorded near Darwin (12/130)[423] but usually the species is solitary or in small groups. It feeds near the water's edge in the upper parts of estuaries, on the margins of inland waters and in artificial salt works. In Europe the species feeds on invertebrates, chiefly insects, molluscs and crustaceans.[422]

There is no information about its diet in Australia. The largest count of Marsh Sandpipers at a single location is 500 at Port Hedland Saltworks (20/118) in Nov. 1982.[84]

Field Atlas Records: 653

1° blocks–with records: 141 (17%)

–with breeding records:

Atlas No.: 159

Max. Rep. Rate: 13%
In block(s): 12/130

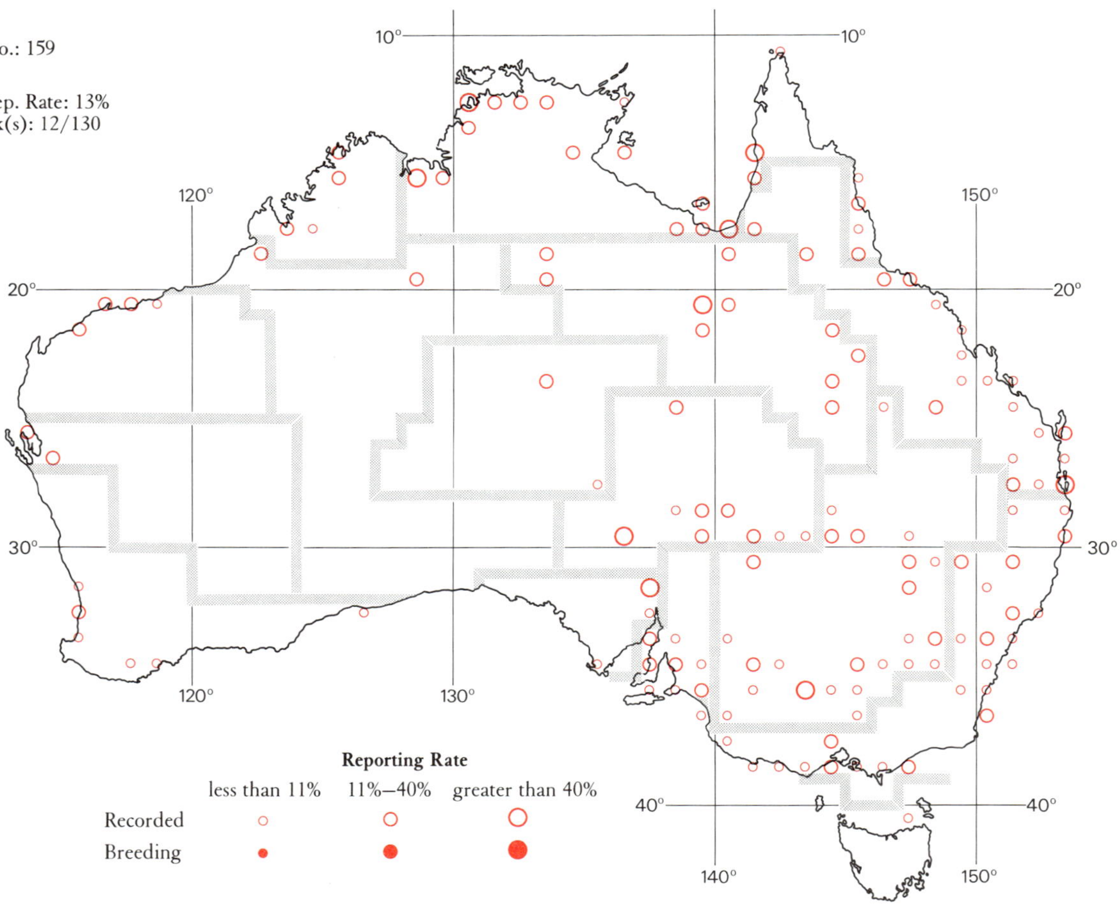

Terek Sandpiper

Tringa terek

The Terek Sandpiper breeds south of the Arctic Circle from Finland to Siberia.[388] It visits Australia each summer and some birds remain for the winter in the north.[503] South of 26°S the Reporting Rate in summer was 0.40% and in winter 0.04%.

The only inland record was of a single bird at Coward Springs (29/136) in Oct. 1978, presumably moving across the continent on the southward migration. Single birds have appeared at Eyre Bird Observatory (32/126) in three springs out of four but none in autumn.[389] A bird banded in Japan (35N/139E) in 1978 was recovered at Kooragang I. (32/151) in 1980.[65]

The Terek Sandpiper occupies the coast, specially river estuaries. It perches in low mangroves or roosts on flat shores with other waders. Large roosting flocks, brought together by rising tides, break up into singles or small parties when feeding.[423] Flocks of more than 100 have been reported from Kooragang I.[1519] and Darwin (12/130)[423] and 400 were present near Broome (17/122) in Mar.-Apr. 1982. Overseas the bird feeds on invertebrates and, at its breeding grounds, on seeds.[422] There is no information about food in Australia.

The Terek Sandpiper is much more abundant in northern than southern Australia. Counts include 510 at Roebuck Bay (18/122) in Nov. 1982, 130 at Moreton Bay (27/153) in Feb. 1982, 110 at Mackay (21/149) in Feb. 1983 and 70 at the Hunter R (32/151) in Feb. 1983.[84]

Field Atlas Records: 533
1° blocks–with records: 74 (9%)
–with breeding records:

Atlas No.: 160

Max. Rep. Rate: 21%
In block(s): 12/130

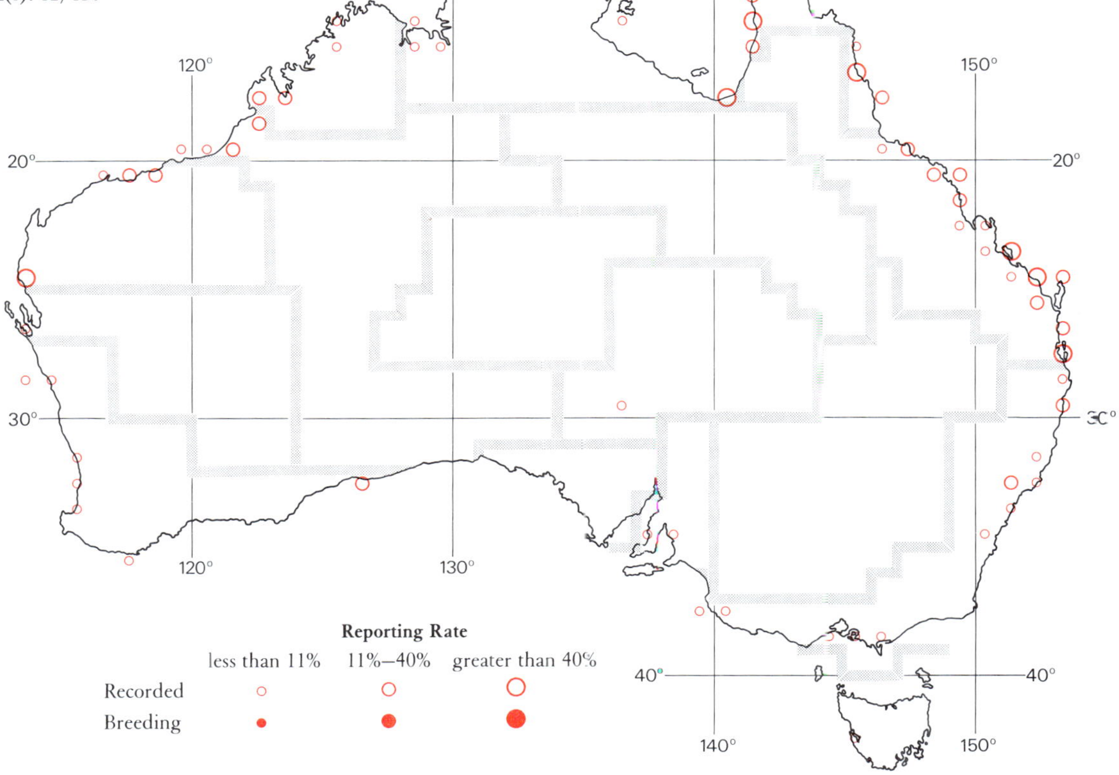

Latham's Snipe
Gallinago hardwickii
Pin-tailed Snipe
Gallinago stenura
Swinhoe's Snipe
Gallinago megala

Latham's Snipe breeds in Japan and spends the summer in southern Australia; Swinhoe's and the Pin-tailed breed in Siberia and migrate to Africa, Asia and Australia, the Pin-tailed with a more westerly distribution than Swinhoe's.[1117,1716] In Australia their distribution is uncertain because they can only be identified in the hand so the map combines records of all three species. Latham's is the common snipe in eastern Australia; claimed records in the NT are disputed.[666] In the Top End and Kimberley Regions Swinhoe's is probably the main species. Only the Pin-tailed has been confirmed to be in the Pilbara Region[1716] and it has been recorded nowhere else. Snipe, of whichever species, are probably commoner in northern Australia than the map suggests, records being limited by the lack of observers there in summer.

Latham's Snipe arrives in Qld from Aug. onward and migrates south. Small flocks pass through King I. (39/144) to Tas.[231,654,1306] While in Australia Latham's Snipe is highly mobile, probably in response to rainfall.[1306] Northward movement begins in Feb. in the south and continues to Apr. in the north.

Latham's Snipe lives singly or in small groups at all altitudes, occupying muddy edges of fresh, and sometimes saline, swamps and pools, wet grass and heath.[1306] It probes in soft mud mainly for worms and also eats aquatic larvae of insects and some seeds.[666] In Qld it is a game bird. Numbers declined this century[1306] presumably because wetlands have been drained and cleared and hunting has been excessive.[666] Swinhoe's Snipe occupies freshwater swamps and pools but the Pin-tailed inhabits drier grasslands.[1716]

Unusually high densities of Latham's Snipe include: 32 in 1 ha of swamp in the Atherton Region in Mar. 1964[228] and 50 birds/ha at Seaford Swamp (38/145) in Feb. 1983.[1306]

Field Atlas Records: 2257

1° blocks–with records: 182 (22%)
–with breeding records:

Atlas No.: 168,852,169

Max. Rep. Rate: 19%
In block(s): 26/150

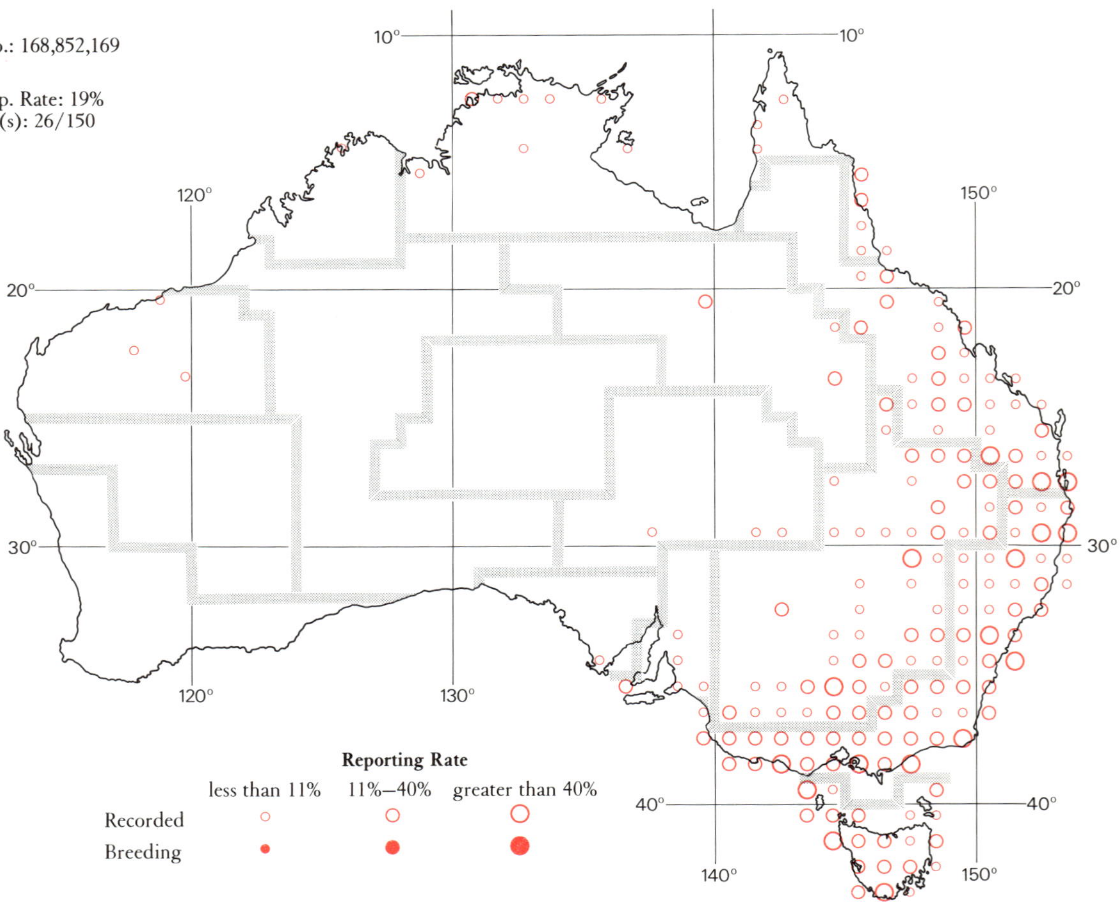

Black-tailed Godwit

Limosa limosa

The Black-tailed Godwit breeds in the Palearctic. The subspecies *melanuroides*, which breeds in Mongolia and eastern Siberia, visits Australia in summer. Most remain in northern Australia but those moving south may travel across the continent.[1745] The Field Atlas contains records from 15 inland 1° blocks in spring compared with six in autumn. There were also reports from 12 inland 1° blocks in summer, which suggests that some birds remain inland on suitable waters, unlike the more strictly coastal Bar-tailed Godwit.

Some birds stay in Australia during the winter (Reporting Rates: summer 0.8%, winter 0.2%). In Darwin (12/130) the Black-tailed Godwit has been recorded in every month of the year.[423]

The Black-tailed Godwit occupies coral islets, coastal spits, lagoons, mudflats and inland fresh water.[19] It is frequently with the Bar-tailed Godwit.[1226] The Black-tailed Godwit feeds in water, often as deep as it can stand in.[1067] Food requirements of the bird in Australia are not known but the European subspecies eats invertebrates and some plant material.[422]

Some migrating Black-tailed Godwits pass through the Roebuck Bay (18/122) area where 1945 were counted in Sept. 1982 decreasing to 370 by mid-Nov.[84] but the species' stronghold in Australia appears to be the south-eastern corner of the Gulf of Carpentaria. In Dec. 1981, 60 000 were counted there and, in Dec. 1983, 49 400.[84]

Field Atlas Records: 497

1° blocks–with records: 10 (12%)
–with breeding records:

Atlas No.: 152

Max. Rep. Rate: 22%
In block(s): 24/153

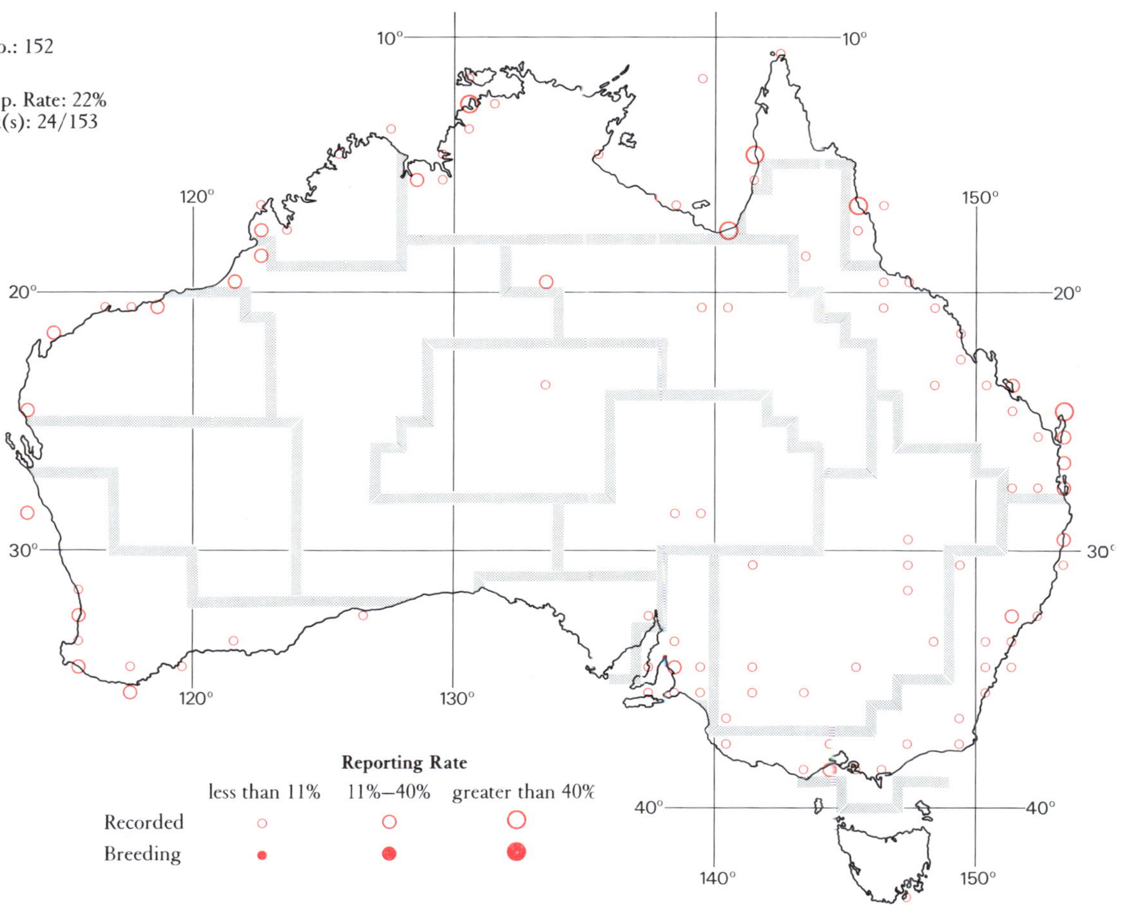

Bar-tailed Godwit

Limosa lapponica

The Bar-tailed Godwit breeds in eastern Siberia and northern Alaska, the subspecies *baueri* visiting Australia in the non-breeding season. An adult banded at Botany Bay (34/151) has been recovered in South Korea, 7900 km distant.[69]

Although there are many more Bar-tailed than Black-tailed Godwits in southern Australia in summer, the number of inland Bar-tailed records was much lower, which suggests that most migrants move round the coast (number of inland 1° blocks in the Field Atlas in which Bar-tailed Godwits were reported: spring seven, summer three, autumn three and winter one). Movement has been observed at Eyre Bird Observatory (32/126) where 22 were counted during the spring passage and three in autumn during four years.[389]

Many Bar-tailed Godwits spend the winter in Australia (Reporting Rates: summer 3.7%, winter 1.5%). The Field Atlas indicates that only a few of these remain on the southern coast or in the Tas. Region.

The Bar-tailed Godwit occupies coastal sandy shores, mudflats and adjacent marshes.[1124] Twelve were recorded on an inland lagoon near Darwin (12/130) in stormy weather[423] but the species seldom visits inland lakes. It is frequently with the Black-tailed Godwit.

The Bar-tailed Godwit feeds in shallow water at the edge of the sea. In Europe it takes chiefly invertebrates, specially insects, molluscs, crustaceans and annelid worms.[422]

The area between Broome (17/122) and Port Hedland (20/118) appears to be the most important site for the Bar-tailed Godwit in Australia. Counts in 1982 estimated the population there to be 9750 in Aug., increasing to 41 900 in early Sept. and 58 400 by mid-Nov.[84] On the coast of Eastern Qld the species is the most abundant wader. Counts in Feb. 1983 recorded 7670 in the Hervey Bay and Great Sandy Strait (25/152) area and 6280 in Moreton Bay (27/153). Though the Bar-tailed Godwit is generally most abundant in northern Australia, a population of about 7000 uses Corner Inlet (38/146) each summer.

Field Atlas Records: 2461

1° blocks–with records: 150 (18%)

–with breeding records:

Atlas No.: 153

Max. Rep. Rate: 51%
In block(s): 25/113

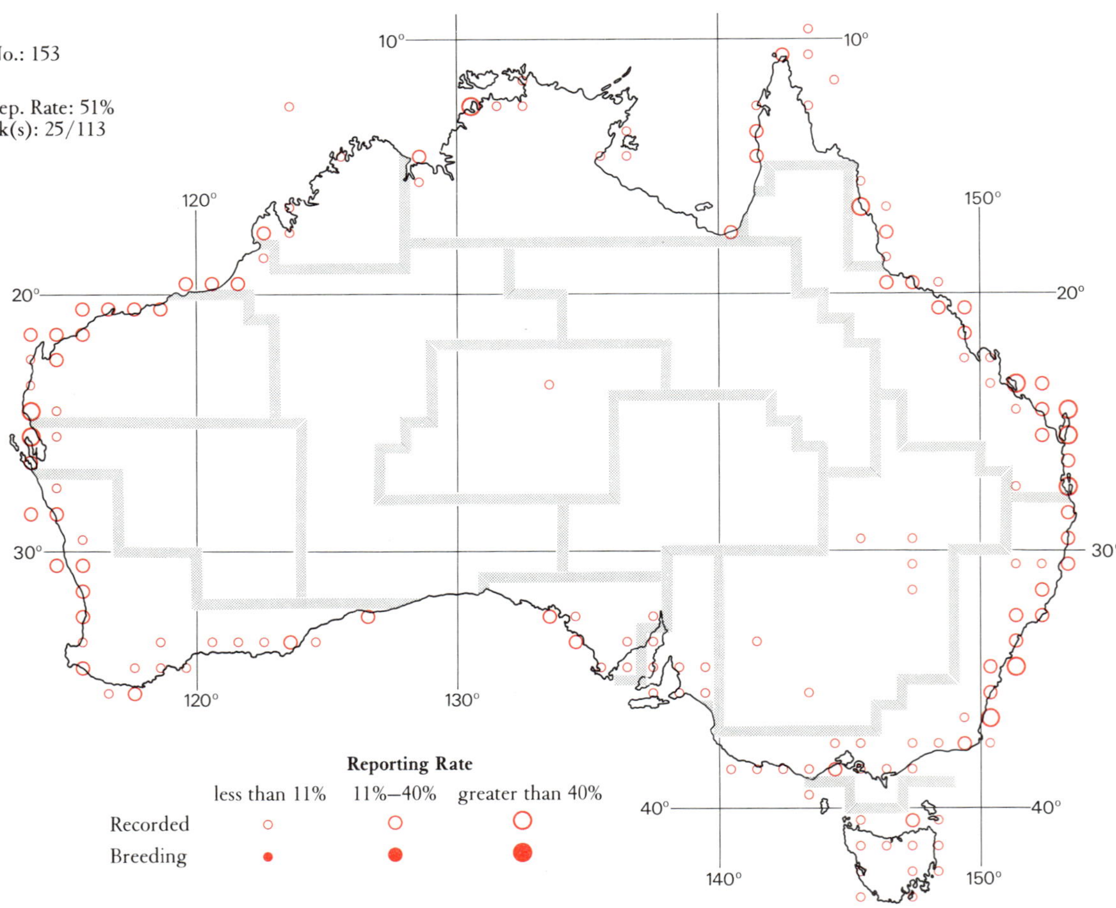

Red Knot

Calidris canutus

The Red Knot breeds in the Holarctic. The subspecies *canutus* is a summer visitor to Australia, some birds remaining for the winter, mainly in the north (Reporting Rates: summer 0.72%, winter 0.22%).

The bulk of the population appears to arrive in Australia in the north-west, many remaining there through the summer. In Torres Strait the species is only an irregular migrant.[503] There are few inland records: all those in the Field Atlas were of 1–2 birds and dated Sept.-Nov. when the birds are migrating southward. A few are recorded on the coast at Eyre Bird Observatory (32/126) with counts of 25, 4, 15 and 23 in four successive springs and 0, 0 and 2 in autumn.[389]

Banding recoveries include one bird from Perth (31/115) to Okara Bay (36S/174E) in New Zealand, 5200 km; and another banded at 38/144 near Werribee in Jan. 1979 to Kaipara Harbour (36S/174E) in New Zealand in Mar. 1980.[85] The Red Knot keeps to open mudflats and wide sandy shores.[423,812] It probes rapidly up and down, sometimes with a sideways movement, in soft sand or mud covered with shallow water.[1132] On its breeding grounds it feeds on invertebrates, chiefly molluscs, and some plant material.[422]

The area between Broome (17/122) and Port Hedland (20/118) appears to be the stronghold of the species in Australia. Counts in 1982 estimated the population in this area at 18 300 in Aug., 90 300 in early Sept. and 52 500 in mid-Nov. but the proportion moving on into southern Australia is unknown. Other counts include 4750 in Spencer Gulf (34/136) in Feb. 1982 and 2700 in Corner Inlet (38/146) in Feb. 1982.[84] The species is uncommon on the eastern coast of Qld.

Field Atlas Records: 648

1° blocks–with records: 90 (11%)
–with breeding records:

Atlas No.: 164

Max. Rep. Rate: 13%
In block(s): 12/130,32/126

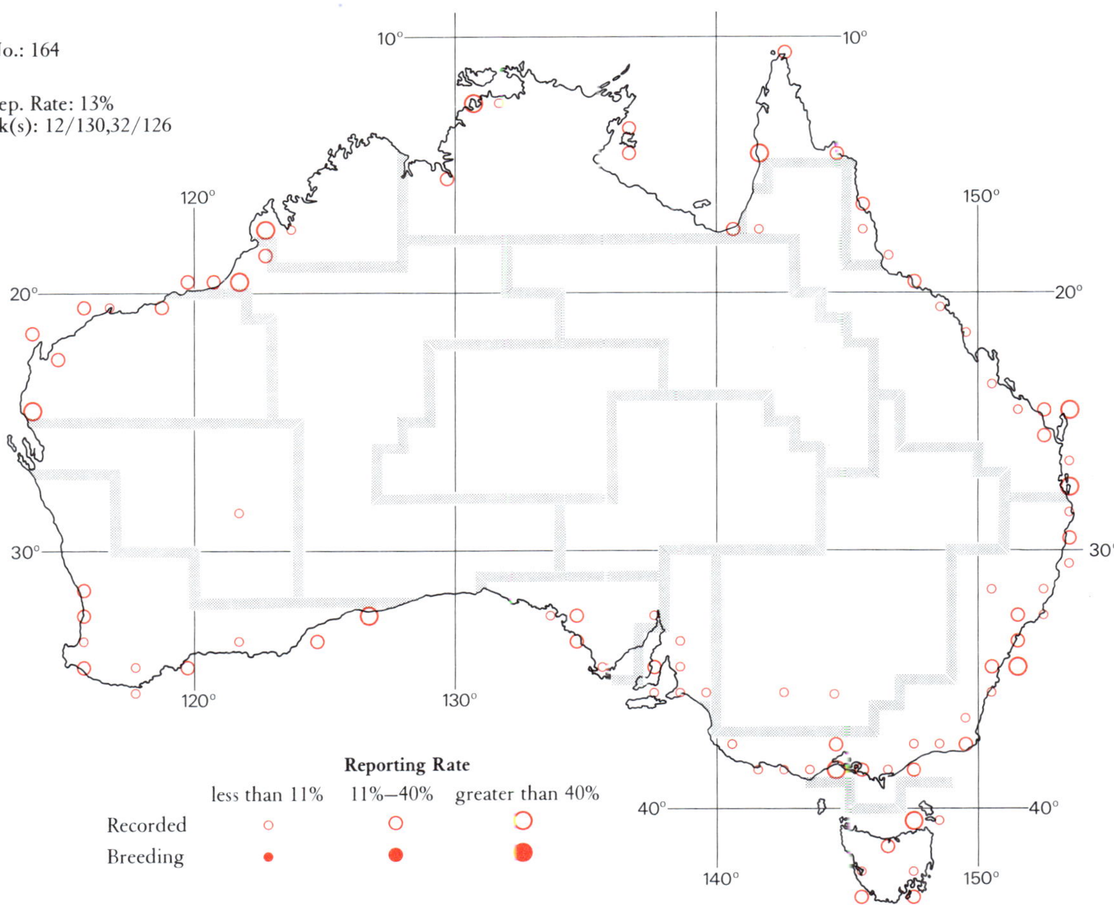

Great Knot

Calidris tenuirostris

The Great Knot breeds in eastern Siberia and within the Arctic Circle. It visits Australia in summer, some birds remaining in winter mostly in northern Australia (Reporting Rates: summer 0.57%, winter 0.16%).

There have been few reports from the inland: the Field Atlas record from Woorinen (35/143) was of a single bird in Nov. 1980, presumably migrating southward. In Torres Strait the Great Knot is a passage migrant through western and central island groups.[503] It is recorded at Eyre Bird Observatory (32/126) only in spring with counts of 3, 0, 2, 8 in four successive years.[389]

The Great Knot lives on coastal mudflats and feeds in shallow water. Flocks of up to 20 000 have been recorded at roosts during high tide in northern Australia where it is more abundant than in the south. In Tas. it is a straggler, only recorded some years, one or two birds at a time.[1409] Minute gastropods were found in the guts of two specimens from Bunbury (33/115).[1619]

The Great Knot is predominantly a northern Australian species and the north-west, Top End and, probably, the Gulf of Carpentaria, are its principal strongholds. Large numbers were recently discovered in the area between Broome (17/122) and Port Hedland (20/118), which held an estimated 107 000 in Sept. 1982. Counts at Darwin (12/130) recorded 8400 in Feb. 1983 as well as a substantial winter population: 6600 in July 1982.[84]

Field Atlas Records: 428

1° blocks–with records: 74 (9%)
–with breeding records:

Atlas No.: 165

Max. Rep. Rate: 22%
In block(s): 12/130

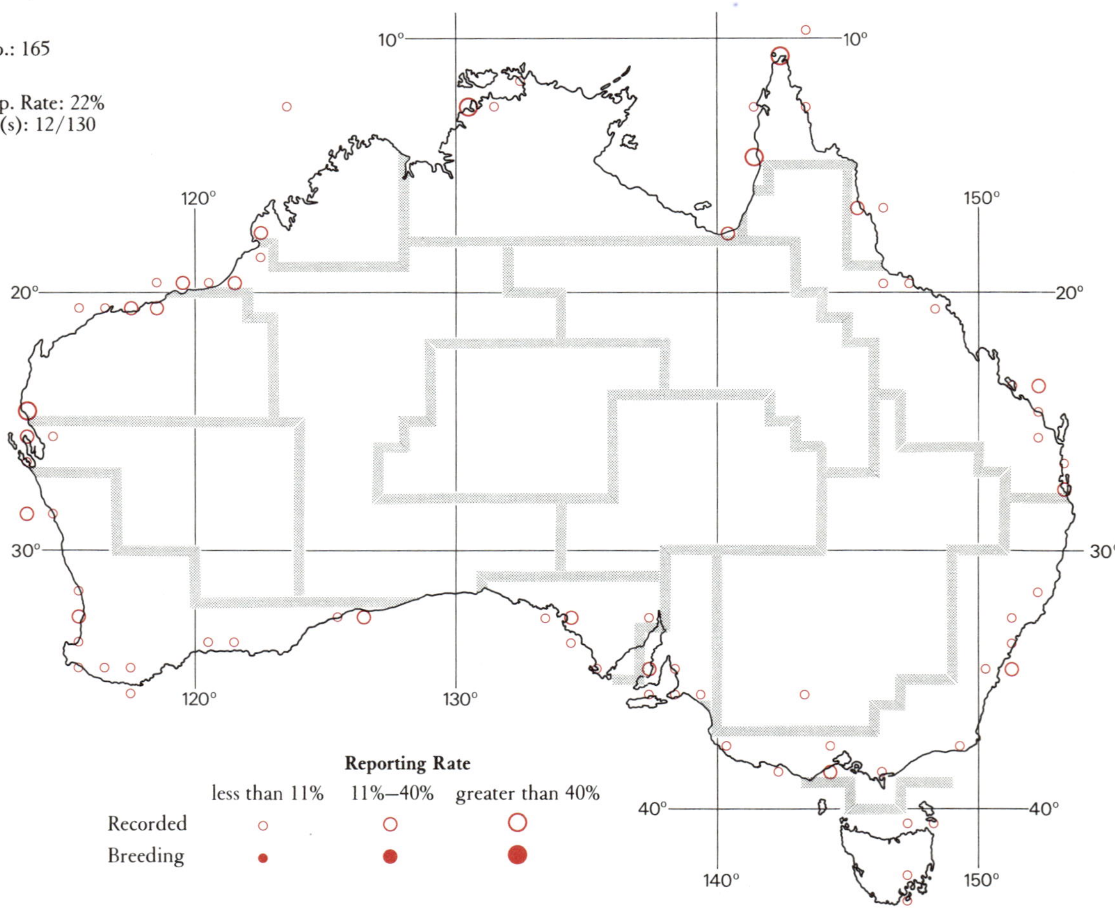

Sharp-tailed Sandpiper

Calidris acuminata

The Sharp-tailed Sandpiper breeds in arctic Siberia and visits Australia each summer. A bird banded near Perth (31/115) has been recovered near the Yana R (67N/135E) in the USSR and was presumably close to its breeding grounds.[1326] Another banded at Woolooware Bay (34/151) in Feb. 1977 was recovered at 25N/119E in China in Apr. the same year.[57]

Only a few Sharp-tailed Sandpipers pass through Torres Strait.[503] At Mt Isa (20/139) the species is present Aug.-Apr., often in large numbers, which decrease in Dec.-Feb.[862] The spring and autumn peaks presumably represent the passage of migrating birds through the area. At Eyre Bird Observatory (32/126) there were counts of 23, 37, 301 and 63 in the spring of four successive years and 0, 4 and 1 in autumn.[389] It is one of the first of the abundant waders to leave in autumn, few remaining after Mar.

The Sharp-tailed Sandpiper occupies fresh water more than the sea coast[1745] and also frequents brackish water and estuaries. In Vic. in Feb. 1982 a comprehensive survey of both inland and coastal wetlands recorded 32% of Sharp-tailed Sandpipers inland (compared with 7% of Curlew Sandpipers and 6% of Red-necked Stints).[84]

The Sharp-tailed Sandpiper forms mixed flocks with other waders, specially when roosting, but also during feeding periods. In Tas. flocks are always small compared with those on the mainland because suitable wetlands are scarce.[1314] At the mouth of the Little R (38/144), where Sharp-tailed and Curlew Sandpipers and the Red-necked Stint co-exist, the Sharp-tailed feeds mostly in shallow water.[452] Its main prey is polychaete worms. At L. Reeve (37/147) it takes gastropods and amphipods in Oct. and, in Mar., pupae of midges.[1430] Because it is an inland wader, counts, which are concentrated on the coast, underestimate its numbers greatly. In Sept. 1982 there were 25 000 on a freshwater lagoon near Anna Plains (19/121) while on the beach nearby flocks of several thousand each were arriving. Counts of individual sites include 54 700 on The Coorong (36/139) in Feb. 1982 and 13 300 at Port Phillip Bay (38/144) in Feb. 1982.[84]

Field Atlas Records: 2912

1° blocks–with records: 296 (36%)
–with breeding records:

Atlas No.: 163

Max. Rep. Rate: 29%
In block(s): 29/136

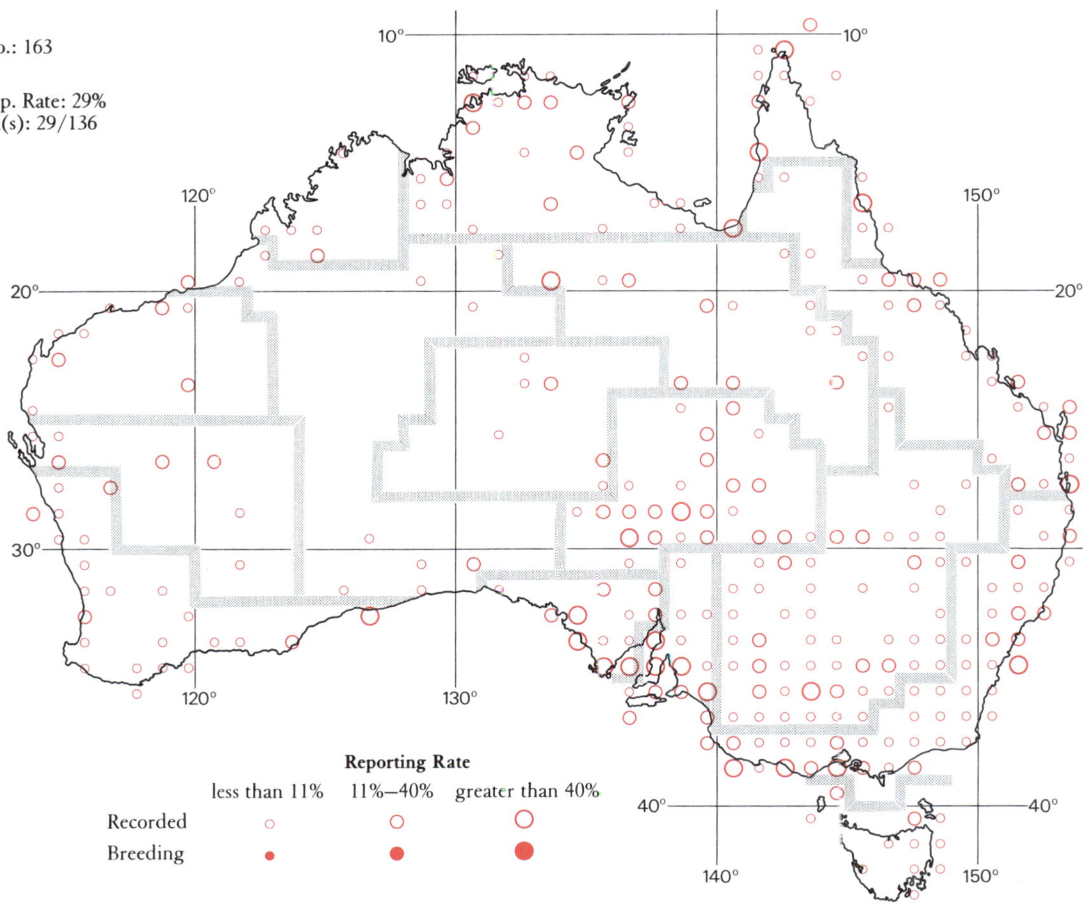

Pectoral Sandpiper

Calidris melanotos

The Pectoral Sandpiper breeds in arctic Siberia and Alaska[388] and small numbers migrate to Australia. The bulk of the population migrates through the Americas, spending the non-breeding period in the lands between Peru and Patagonia.

Since the late 1950s the Pectoral Sandpiper has been recognized as a regular visitor to Australia and probably has always visited in small numbers. The sparseness of records in northern Australia may reflect the few observers there, specially during the wet summer season when the bird would be present.

The Pectoral Sandpiper inhabits shallow water, both fresh and saline. It usually occupies wetlands where there is mud with some superficial vegetation; rarely the seashore. Many Field Atlas records were from sewage ponds. The species is recorded singly or in parties of up to ten birds. It is often in close association with flocks of Sharp-tailed Sandpipers, with which it can be confused in the field. It eats insects and some crustaceans, including crabs. Algae have been found in guts overseas[1905] but it is not clear whether they are an important item of food.

There are no measures of abundance.

Field Atlas Records: 113

1° blocks–with records: 32 (4%)
–with breeding records:

Atlas No.: 978

Max. Rep. Rate: 4%
In block(s): 26/120

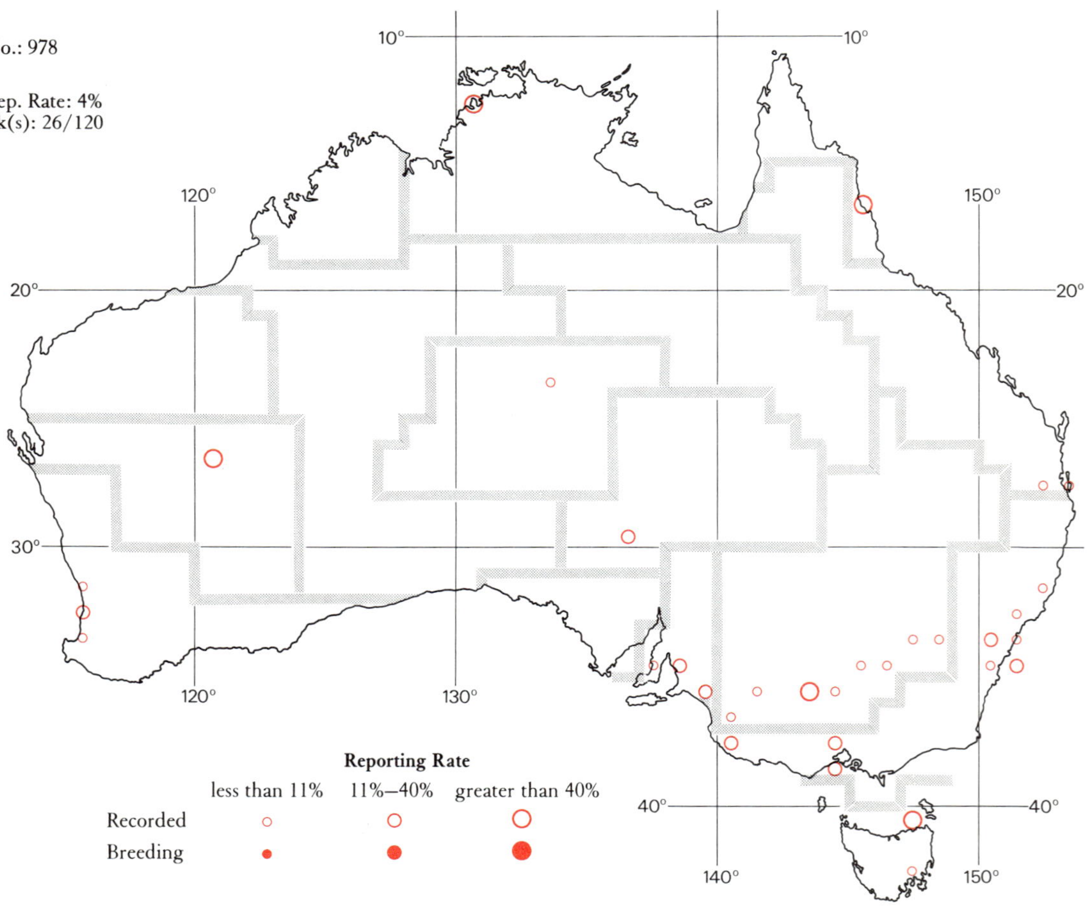

Red-necked Stint

Calidris ruficollis

The Red-necked Stint breeds in arctic Siberia and migrates to Australia each summer. Many remain in Australia during winter, mostly first year birds[1404] (Reporting Rates: summer 5.5%, winter 1.5%).

The Field Atlas suggests that many birds move across the continent on their southward migration in spring. There were records from 57 inland 1° blocks then compared with 33 in summer, 25 in autumn and 14 in winter. At Mt Isa (20/139) the Stint is a regular migrant in small numbers Aug.-Jan.[862] In Torres Strait it is an abundant passage migrant;[503] and on the coast at Eyre Bird Observatory (32/126) counts of 1057, 332, 765 and 943 were recorded in spring passages over four years; and 577, 338 and 889 in autumn.[389]

Birds banded overseas and recovered in Australia include one from 66N/169W in eastern USSR to Perth (32/115)[66] and others from Hong Kong (22N/114E) to Fraser I. (25/153), Queenscliff (38/144) and Pipeclay Lagoon (42/147).[66,85] Once arrived at the non-breeding grounds in Australia, movements are only local.[452] Banding near Hobart (42/147) showed that birds were more likely to be recaptured in their banding area than elsewhere though there was a continuous interchange between locations up to 10–15 km apart.[1314] In the Perth area there is local movement between the Swan R estuary and lakes, dictated by tides.[911] Within Port Phillip Bay (38/144) there is very little movement between Werribee and Queenscliff, 25 km apart.[452]

The Stint lives mainly by the sea or on estuaries. At the mouth of the Little R (38/144), where it co-exists with Sharp-tailed and Curlew Sandpipers, the Stint tends to forage above the water line taking food mostly from the surface of the mudflats and to a lesser extent by jabbing and probing into the substrate.[452] It eats invertebrates including polychaete worms, amphipods, snails and insects.[1749] At L. Reeve (37/147) in Oct. it ate the seeds of sea tassel.[1430]

The Stint is the most abundant Palearctic wader in Australia though uncommon on the eastern coast of Qld. Counts include an estimated 97 500 in the area between Broome (17/122) and Port Hedland (20/118) in Nov. 1982, 53 800 on The Coorong (36/139) in Feb. 1982, 20 600 in Port Phillip Bay in Feb. 1983 and 20 300 at Corner Inlet (38/146) in Feb. 1983.[84]

Field Atlas Records: 3610

1° blocks–with records: 259 (32%)
–with breeding records:

Atlas No.: 162

Max. Rep. Rate: 36%
In block(s): 32/126

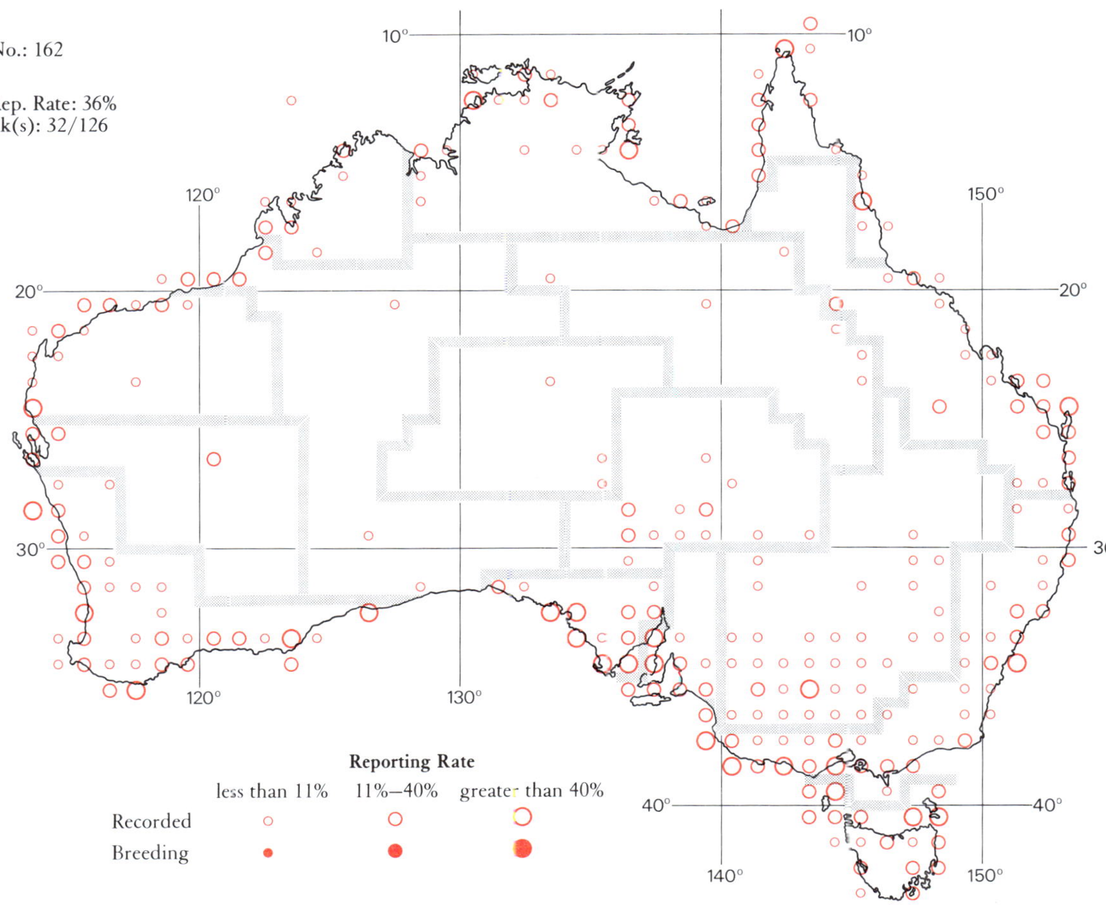

Long-toed Stint

Calidris subminuta

The Long-toed Stint breeds in Siberia and migrates southward in the non-breeding season. It was first collected in Australia in 1886 on the Fitzroy R (17/123) but it remained a rare vagrant until the 1960s when it began to be seen fairly regularly in Vic.[1676] This seems less likely to represent a change in the pattern of movement than to reflect an increasing awareness by observers. The species is now recognized as a regular summer visitor and has been recorded throughout Australia, including Qld, though no sightings were confirmed there during the Field Atlas.

Field Atlas sightings indicate that the Long-toed Stint begins arriving in Australia in Aug. There was only one record in June-July, which suggests that very few birds remain through the winter. The species is most abundant in WA where Field Atlas reports were mostly of parties of 4–13 birds. On L. Forrestdale (32/115) up to 80 have been seen together[440] and 92 were reported from Carnarvon (24/113) in Feb. 1981.[1491] In eastern Australia reports were generally of one or two birds.

The Long-toed Stint inhabits the drying margins of shallow freshwater lakes and, less often, beaches or short grass where there are freshwater soakages. Characteristically it is on mud margins, with or without emergent vegetation. Even its most favoured localities in the South-West attract parties for only a few weeks as lake beds approach dryness.[440] It has been reported near brackish water but not salt. After a summer thunderstorm it appeared at an intermittent lake at Wiluna (26/120).[441]

The Long-toed Stint feeds by probing in mud and hunts flying insects.

There are no measures of abundance.

Field Atlas Records: 108

1° blocks–with records: 34 (4%)
–with breeding records:

Atlas No.: 965

Max. Rep. Rate: 5%
In block(s): 20/118

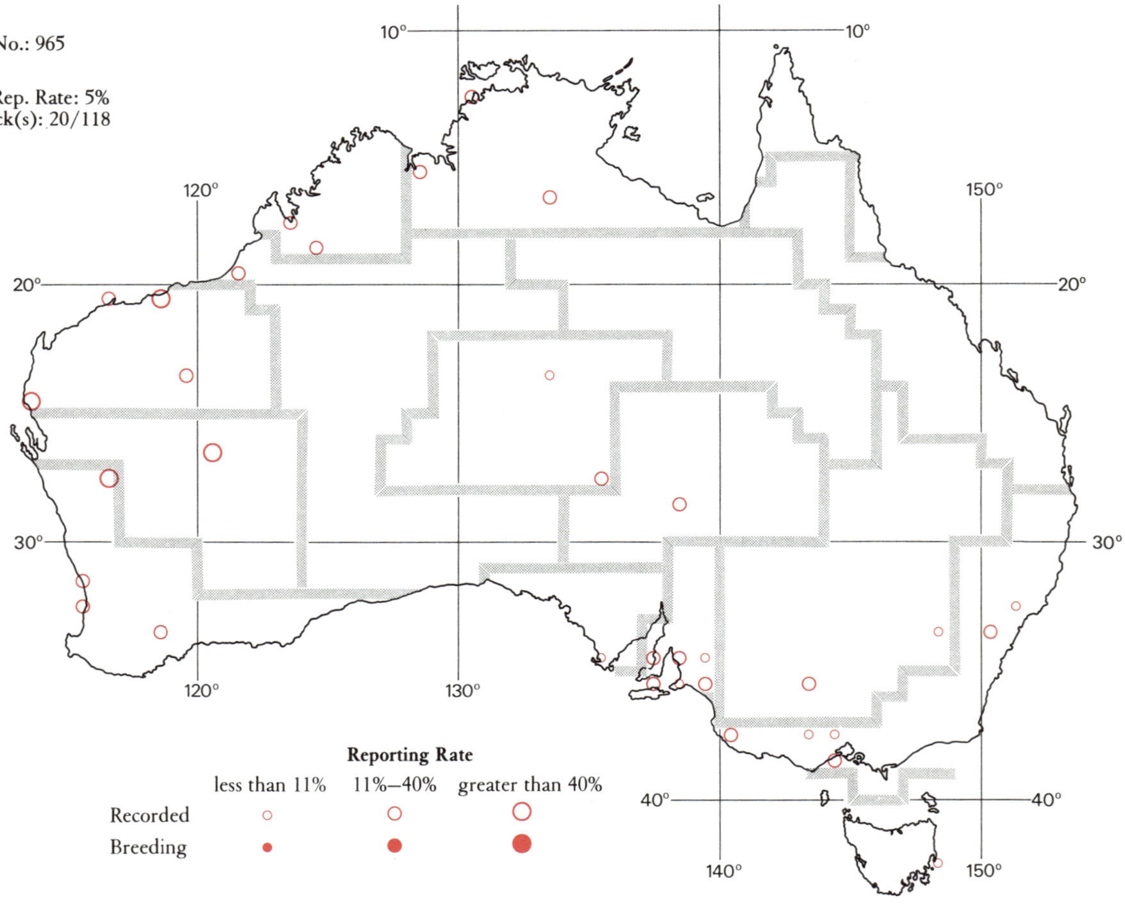

Curlew Sandpiper

Calidris ferruginea

The Curlew Sandpiper breeds in arctic Siberia and visits Australia in summer, many birds remaining for the winter (Reporting Rates: summer 3.1%, winter 0.8%).

Banding at 38/144 near Werribee has resulted in several recoveries overseas in India, Malaysia, USSR and China. A bird banded in Singapore has been retrapped in Australia.

The Curlew Sandpiper probably travels across Australia at least on its southward migration. The Field Atlas suggests it to be widespread across the continent in spring when there were reports from 38 inland 1° blocks compared with 14 in summer, 12 in autumn and eight in winter. At Mt Isa (20/139) it is a regular migrant from Aug.-Oct. in flocks of 6–50.[862] At Eyre Bird Observatory (32/126) on the coast there were counts of 32, 8, 7 and 11 in four successive spring passages and 2, 2 and 4 in autumn.[389]

In Australia the Curlew Sandpiper tends to remain in one place throughout the summer while moulting its flight feathers[1750] and to return to the same place in subsequent years.[451] It occupies coastal mudflats, often in mixed flocks with the Red-necked Stint and Sharp-tailed Sandpiper. The Curlew Sandpiper feeds usually belly-deep in water at the edge of the tide, probing into the mud to depths of 20–40 mm.[1749] At the mouth of the Little R (38/144) it took mainly polychaete worms.[452] A variety of molluscs, crustaceans and insects has also been recorded in its guts.[1749]

The Curlew Sandpiper is uncommon on the eastern coast of Qld. Elsewhere, sites at which large numbers have been recorded include: 96 800 in the area between Broome (17/122) and Port Hedland (20/118) in Nov. 1982, 21 600 on The Coorong (36/139) in Feb. 1982, 13 600 in Port Phillip Bay (38/144) in Feb. 1983 and 12 200 at Corner Inlet (38/146) in Feb. 1982.[84] In the Hobart (42/147) area the summer population has increased significantly from an average of 589 in 1965–69 to an average of 973 in 1974–81.[1322] Birds have begun to occupy mudflats in the Derwent R estuary (42/147) previously unused by this species.

Field Atlas Records: 2140

1° blocks–with records: 189 (23%)
–with breeding records:

Atlas No.: 161

Max. Rep. Rate: 28%
In block(s): 25/113

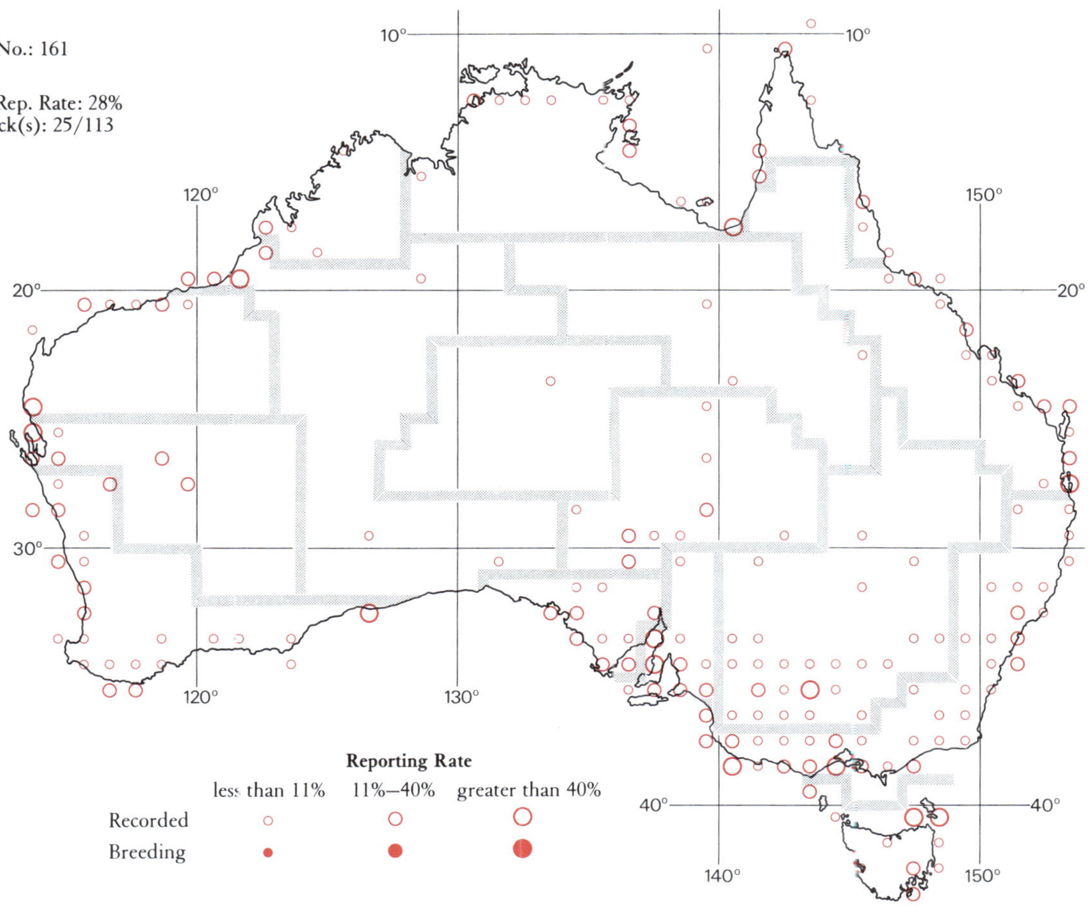

Sanderling

Calidris alba

The Sanderling breeds in the Holarctic and visits Australia in the summer. Some birds remain in the winter, flocks regularly being reported at sites in the western South-East Region and in the South-West Region then.[911] Its Reporting Rate in summer was 0.76% and 0.17% in winter. The inland observation at Wiluna (26/120) was made in Dec. 1977 after a heavy thunderstorm.[441] On the coast at Eyre Bird Observatory (32/126) there were counts of 21, 12, 16 and 19 in spring of four successive years and 161, 1 and 20 in autumn.[389]

The Sanderling usually lives on ocean beaches and occasionally on some coastal lagoons, e.g. on Rottnest I. (32/115).[1706] In Tas. it is restricted to a few coastal sites, usually near an estuary or river mouth, and is generally absent from long stretches of sandy beach.[1314] Birds congregate in sheltered sites during storms, e.g. round pools in dunes at the Henty R mouth (41/145), and disperse along the beach in good weather. The Sanderling feeds in the ebb and flow of waves on flat beaches where the water is not too turbulent.[802] Many sorts of insects have been recorded in its guts.[1049]

Flocks of 300–500 are known from the western South-East Region and South-West Region[911] and the coast between Warrnambool (38/142) and Kingston (36/139) regularly holds about 1000 Sanderling in summer.[84] On 45 km of ocean beach between Carpenters Rocks (37/140) and Southend (37/140) 400 Sanderlings were counted in Nov. 1982, 240 in Dec. 1982 and 220 in Apr. 1983.[226] At Roebuck Bay (18/122) 1500 were counted in Nov. 1982.[84]

Field Atlas Records: 421

1° blocks–with records: 83 (10%)
–with breeding records:

Atlas No.: 166

Max. Rep. Rate: 22%
In block(s): 32/126

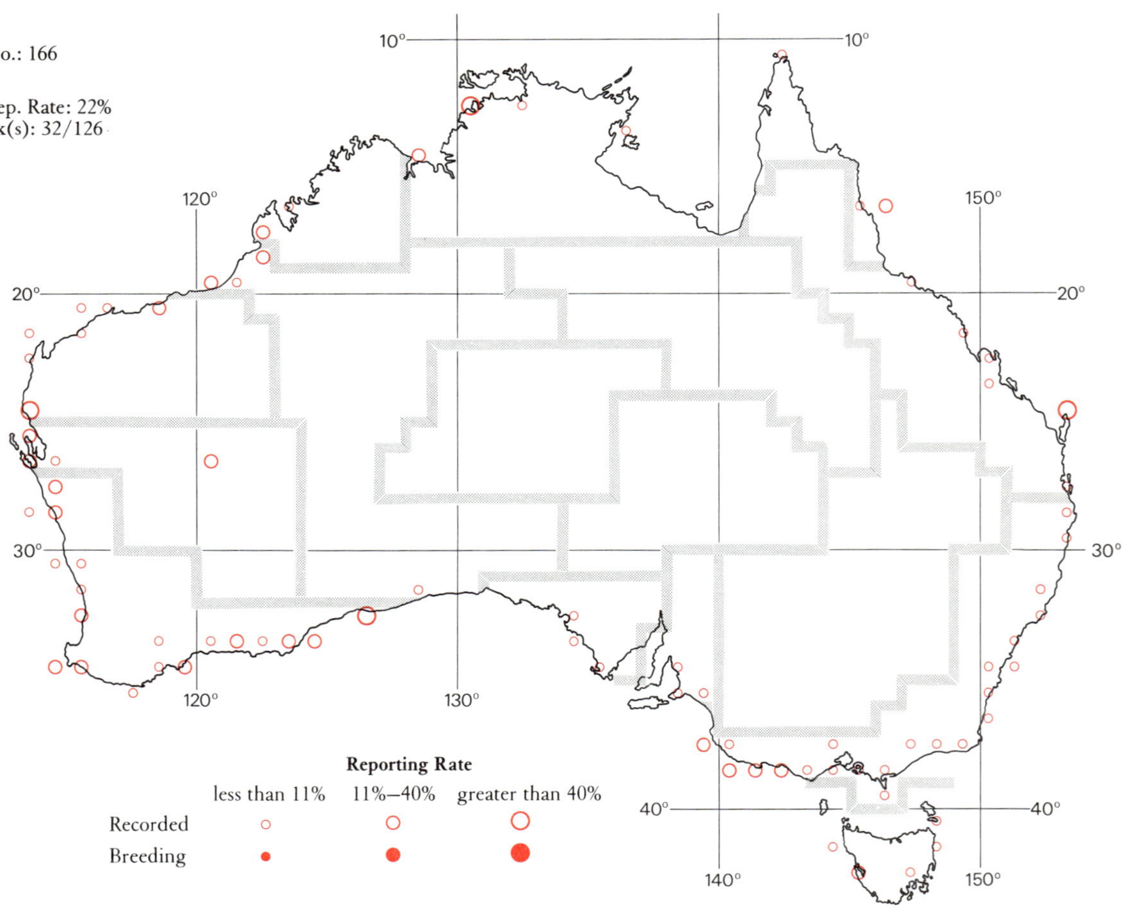

Broad-billed Sandpiper

Limicola falcinellus

Two subspecies of the Broad-billed Sandpiper breed in the Palearctic. One, *sibericus*, breeds in northern Siberia and visits Australia in summer. It is much more frequent in northern than in southern Australia, e.g. it is reported regularly from Darwin (12/130).[423]

In southern Australia single birds, and occasionally parties, appear only irregularly and in many years few or no birds are seen.[19,1745] In the South-East Region unusually large groups of 17 and 15 respectively were reported at Botany Bay (34/151) and Altona (37/144) in Jan.-May 1953.[812] A similar influx during the Field Atlas brought 24 birds to Augusta (34/115) in Jan. 1980. The single inland record was at L. Daley (18/124) in Nov. 1979.

The Broad-billed Sandpiper usually occupies muddy shores and spits on the coast where there is some mud mixed with sand. It will feed on exposed sites, reefs and in pools so deep that it has to submerge its head and neck in water. At Darwin it is often on muddy banks on the seaward side of mangroves.[423] Overseas the Broad-billed Sandpiper feeds on invertebrates and seeds.[422] In Australia the only record identifying prey is of a bird taking an annelid worm on a mudbank in Vic.[1838]

At Port Hedland Saltworks (20/118) in Mar.-Apr. 1982, 500 Broad-billed Sandpipers were present, the largest gathering yet reported in Australia.

Field Atlas Records: 74

1° blocks–with records: 24 (3%)
–with breeding records:

Atlas No.: 167

Max. Rep. Rate: 6%
In block(s): 24/153

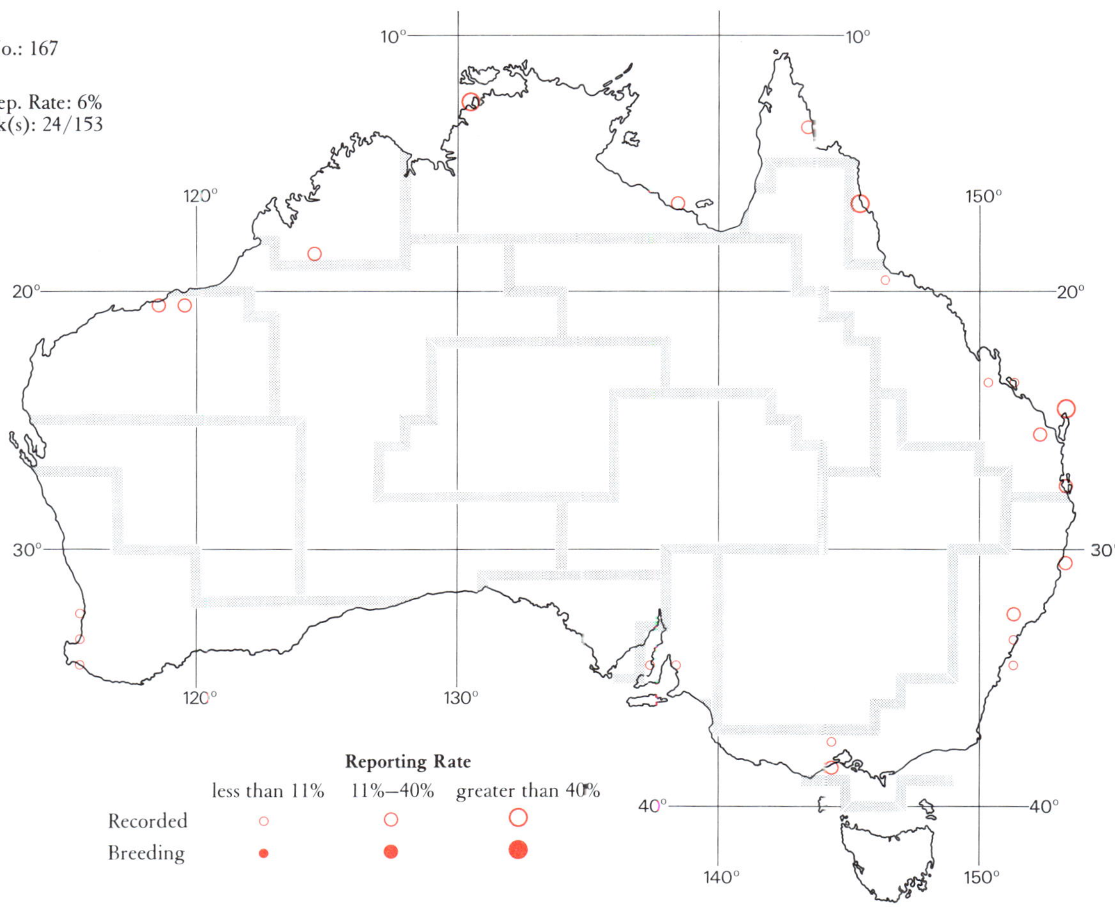

Ruff

Philomachus pugnax

The Ruff visits Australia in the non-breeding season from its northern Asian and European breeding grounds. It is a rare visitor to the Malay Pen.[1238] and its principal non-breeding grounds are in India and eastern Africa. The first record in Australia was from Port Gawler (34/138) in 1962[388] and the species is now recorded from scattered localities throughout Australia in most summers.

During the Field Atlas Ruffs were found on freshwater lagoons whose edges were muddy or vegetated with sedges as well as in saltmarsh, saltponds and sewage works. Most sightings were of single birds and one of three together. A probable record of 40 or more at Cape Portland (40/147) in Mar. 1966[1751] was of an unusually large flock for Australia and was also unusual in being by the sea. In non-breeding areas overseas, flocks of several hundred thousand have been recorded.[422]

In Europe the Ruff feeds on insects and their larvae.[422] The development of irrigated agriculture in northern Australia may provide favourable conditions for it.

There are no measures of abundance.

Field Atlas Records: 65

1° blocks-with records: 14 (2%)

-with breeding records:

Atlas No.: 934

Max. Rep. Rate: 3%

In block(s): 12/130

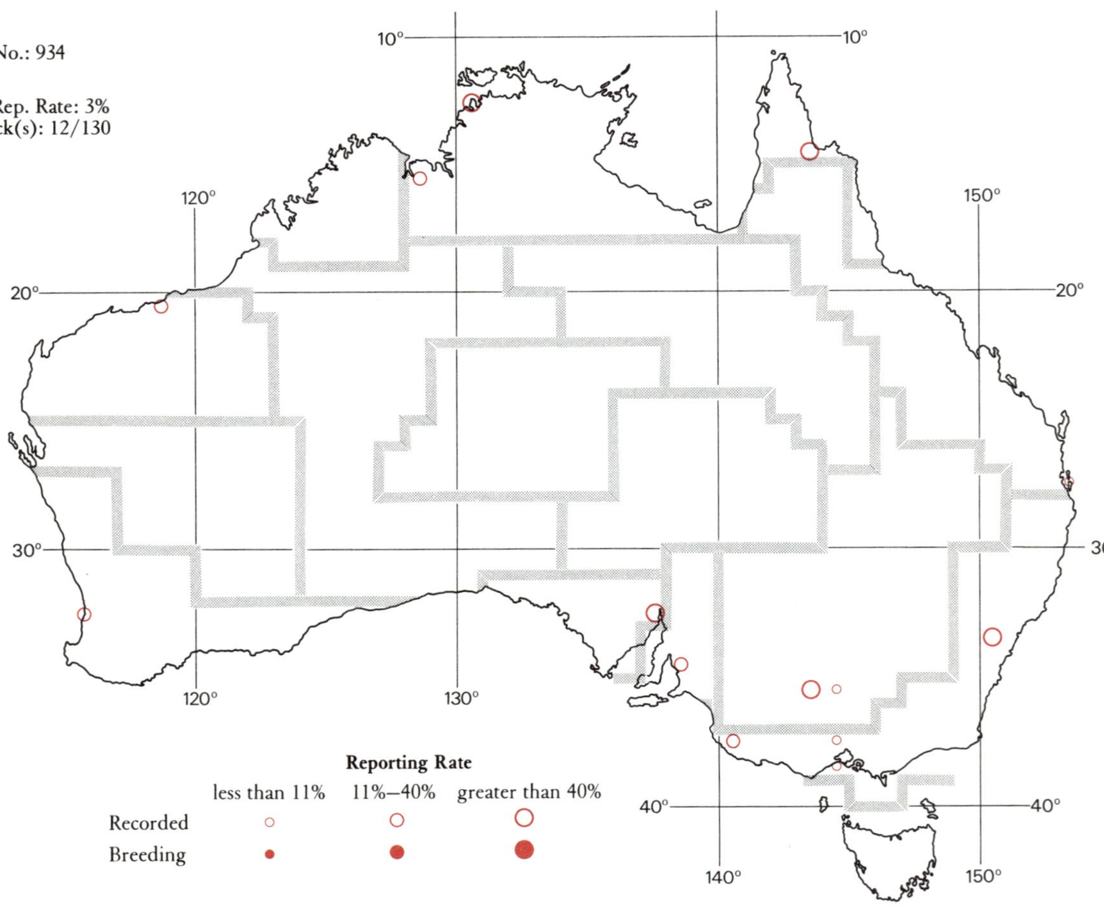

Oriental Pratincole

Glareola maldivarum

The Oriental Pratincole breeds in eastern Asia from southern USSR through China to India and visits Australia in summer.[388] It is probably more widespread and abundant in the Top End and Kimberley Regions than the Field Atlas suggests because few observers visit these areas in summer when the bird is present. In other parts of Australia the species appears erratically but there are records from all regions except Tas. in places as far apart as Moa I. (10/142) in Torres Strait[302] and the Nullarbor.[1364]

Within Australia the Oriental Pratincole moves about during the summer. Resident observers report that birds stay for varying periods in any one district.[423] A flock of about 5000 remained near Mt Isa (20/139) for several weeks in Dec. 1967 then disappeared completely after a rainstorm.[312] The Field Atlas suggests that few if any birds remain during winter though sightings have been reported as late as May.

The Oriental Pratincole inhabits open plains, usually near water. Southern records are of ones, twos and small groups but flocks of thousands are reported in the north.

The Oriental Pratincole feeds on insects, many taken on the wing. It also gathers where there are groups of grasshoppers, catching them on the ground.[999,1218] It is known to feed on the winged forms of termites, which sometimes hatch in large numbers before rain. A flock of about 30 hunted insects near a bushfire on Moa I.[302]

In Mar. 1980, following a series of cyclones, a large flock appeared along the highway north of the De Gray R (20/119). It stretched for about 300 km and the density of birds counted within the vehicle's headlights alone was 10–50 per km.

Field Atlas Records: 70

1° blocks–with records: 31 (4%)
–with breeding records:

Atlas No.: 172

Max. Rep. Rate: 6%
In block(s): 18/133

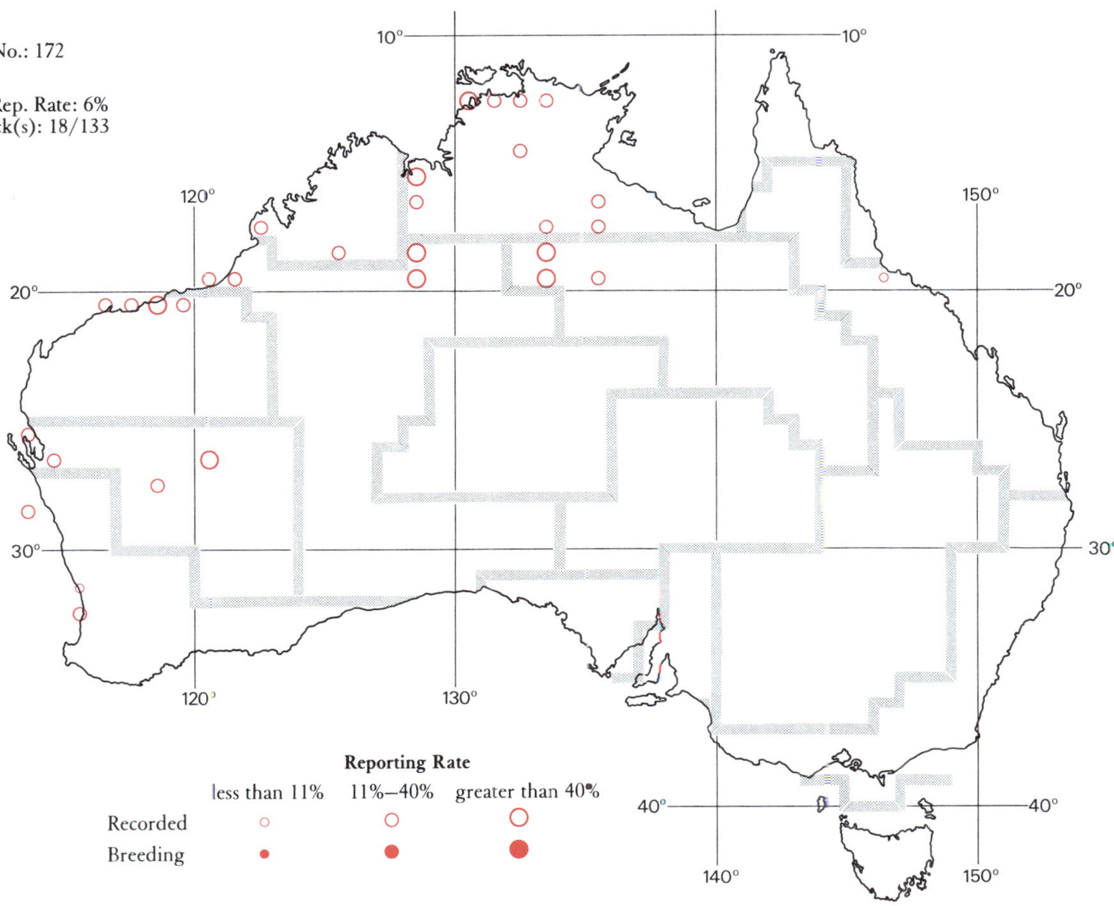

Australian Pratincole

Stiltia isabella

The Australian Pratincole lives mainly in Australia but visits Indonesia and New Guinea in the non-breeding season. In New Guinea there is one possible old breeding record.[129] In Australia it is rarely recorded in the Goldfields and South-West Regions, the first report being from Mileura (26/117) in June-July 1947.[924] Breeding is not recorded in these regions and is seldom reported north of about 18°S.

The Australian Pratincole is migratory, moving south-east in spring and summer to breed and returning to northern areas in winter. In southern Australia Reporting Rates were accordingly low in winter and high in summer (Murray-Darling 0.2%-3.0%, SA Ra. Region 0.1%-5.7%). Near Darwin (12/130) the trend was reversed (winter 40.3%, summer 6%). The species is known as a winter visitor to Torres Strait[503] as well as to Darwin.[423] Round Mt Isa (20/139) it is present Aug.-Apr.;[862] in the L. Eyre Region it is common Sept.-Mar. with occasional winter records;[88] and it is recorded Oct.-Feb. in the southern Murray-Darling Region.[833]

The Australian Pratincole lives on open plains and lagoon edges, almost always near water.[1179] There are several reports from southern parts of its range of greatly increased numbers arriving in very wet years.[833,1159] On the eastern side of the Flinders Ra. flocks of 50 or more arrived continually, totalling up to 1000 birds per day in Sept. of the wet years 1971–75.[88] Field observations show that the Pratincole drinks frequently during the day. It has a salt gland and is able to drink saline water. Nests are in loose groups, with an average of 55 m between nests at one colony in the L. Eyre Region.[1179] They are often in places with scattered low shrubs that are used as refuges by the chicks. The Australian Pratincole feeds on insects,[1179] which it finds on the ground or catches in flight after flushing them.

There are no measures of abundance.

Field Atlas Records: 1631

1° blocks–with records: 267 (33%)

–with breeding records: 43 (16%)

Atlas No.: 173

Max. Rep. Rate: 42%
In block(s): 17/139

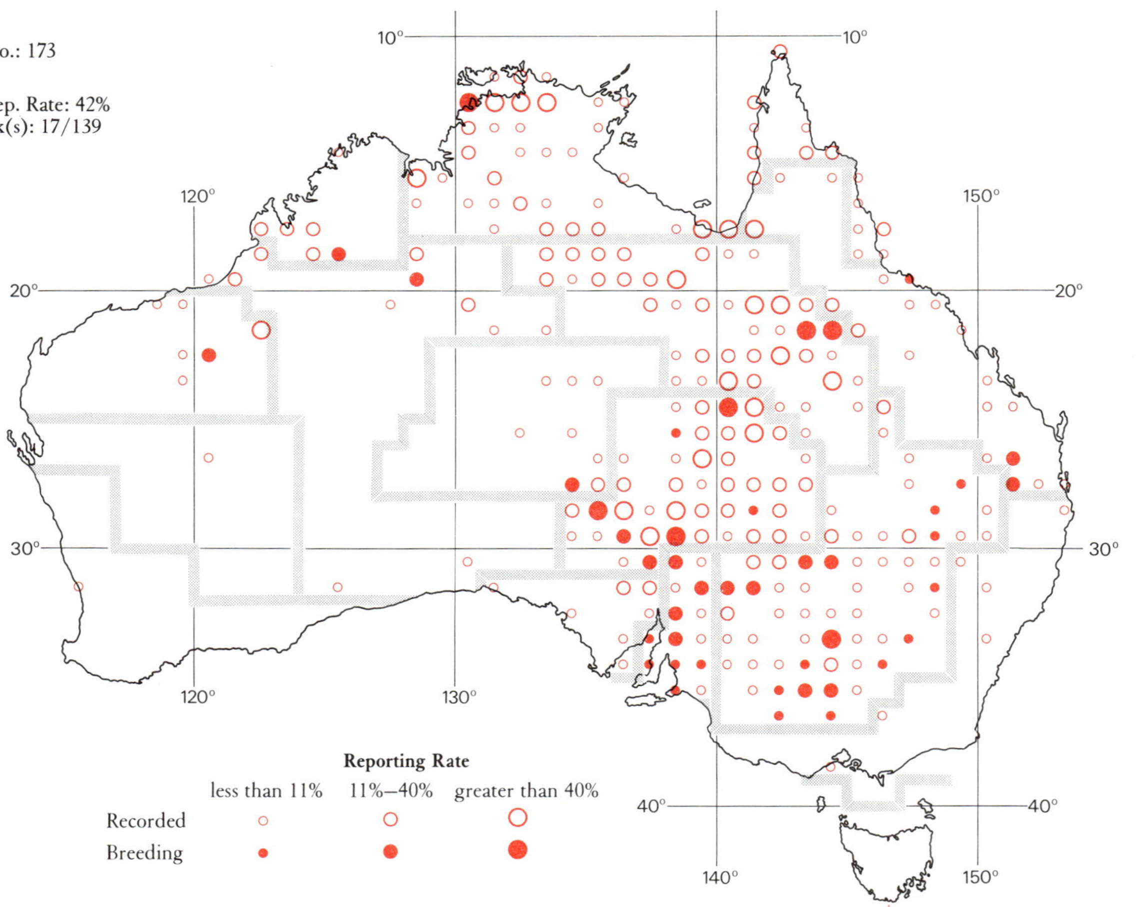

Great Skua

Stercorarius skua

South Polar Skua

Stercorarius maccormicki

The Great Skua and the South Polar Skua have been treated together in the Atlas because their taxonomic status is doubtful and they are hard to identify in the field.[388,1636] The Great Skua is regularly observed in Australian waters in winter and breeds on sub-Antarctic islands including Macquarie I. (54S/158E).

The South Polar Skua breeds farther south on the coast of the Antarctic continent and migrates farther north in winter, crossing the equator in the Pacific Ocean.[482] There are few confirmed Australian records. During the Field Atlas the South Polar Skua was recorded at 37/150 off Tathra in Nov. 1977 and off Geraldton (28/114) in Nov. 1980. One adult banded in Antarctica was recovered at Hardwicke Bay (34/137) in May 1958.[388,1636]

The Great Skua has been recorded as far north in Australia as Sandy Cape (24/153) in the east[12] and Houtman Abrolhos (28/113) in the west.[1638]

Both species are most frequently observed in winter in Australian waters but there are some sightings in summer (Reporting Rates: summer 0.06%, winter 0.23%). The Great Skua is absent from its breeding grounds on Macquarie I. Apr.-Aug.[954] Near Eden (37/149) it arrives in May and departs in Nov.[114]

In eastern Australian waters the Great Skua is usually seen singly or in pairs.[114,1262] Gatherings of up to 50 birds, such as have been observed at Fremantle (32/115) and Albany (35/117), are not reported from eastern Australia.[1636]

The Skuas spend most of their time at sea in the non-breeding season, roost on the water and often follow ships for days at a time. Food is scavenged from the oceans, from ships and from the shore.[114,954,1481,1636]

On Macquarie I. there are approx. 550 breeding pairs of the Great Skua (0.09 birds/ha).[954]

Field Atlas Records: 115

1° blocks–with records: 34 (4%)
–with breeding records:

Atlas No.: 980,979

Max. Rep. Rate: 14%
In block(s): 37/150

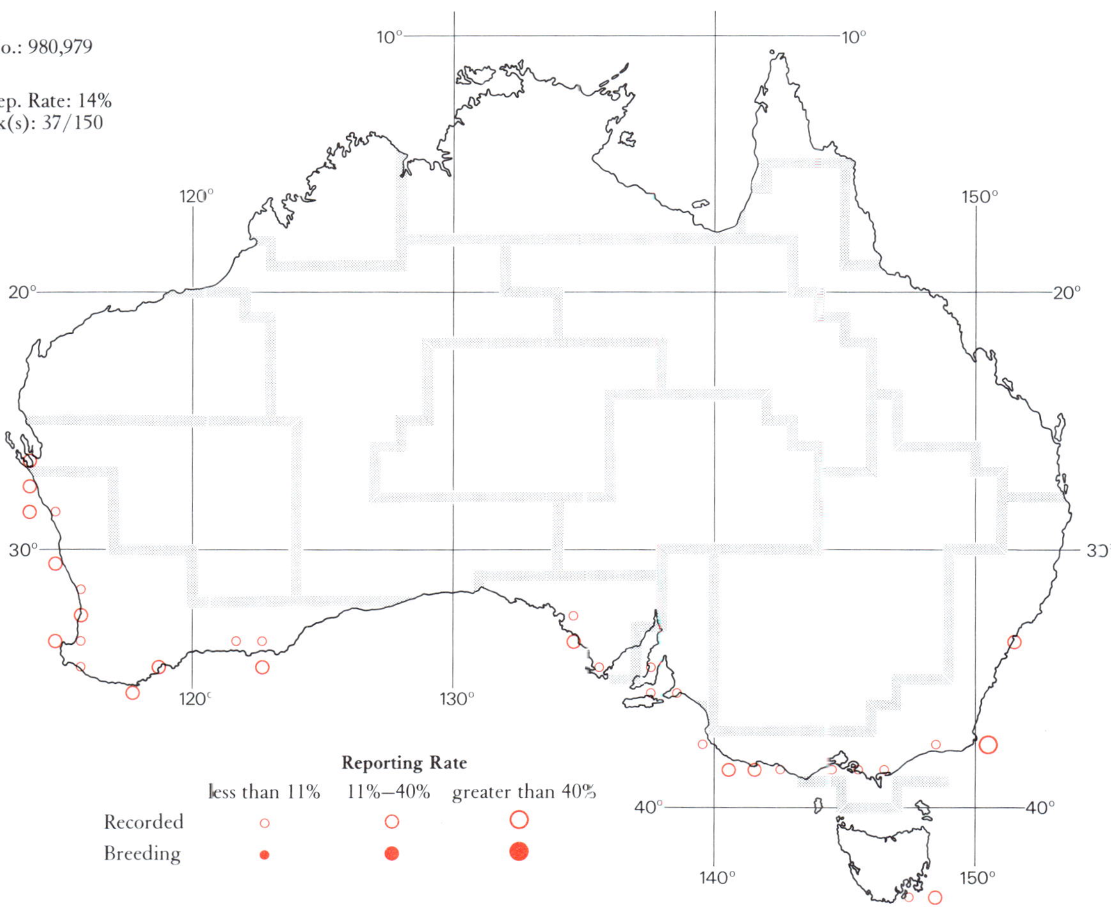

Arctic Jaeger

Stercorarius parasiticus

The Arctic Jaeger has a circumpolar breeding distribution in the northern hemisphere and migrates south across the equator in the non-breeding season. There are dark- and light-plumaged birds, the dark-plumaged ones breeding farther south than the light. In Australia the ratio of dark to light birds was about 2:1 off Eden (37/149)[114] and in the Fremantle (32/115) area 87 dark birds were counted compared with 20 light.[1705]

The Jaeger is a summer visitor to Australia recorded no farther south than the Tas. coast on voyages to Antarctica.[937] Northern Australian records in the Field Atlas were made in Nov. (15/122), Dec. (10/145, 14/146) and Feb. (12/130). The species arrives in southern waters in Oct.-Nov., which coincides with the rise in water temperature and strongest flow of the East Australian Current.[113] It departs in late Apr.[114] Reporting Rates for the four seasons in the South-East Region, where most sightings were made, were: spring 0.4%, summer 1.0%, autumn 0.8%, winter 0.1%.

Except on migration, the Jaeger is primarily a bird of harbours and bays in Australia. Near Sydney (33/151) twice as many were seen in the harbour as outside the heads.[1262] In late Apr. small parties (rarely more than 6-7 birds) are recorded 25-60 km offshore in the Eden area.[114] The Jaeger tends to hunt singly, possibly remaining in the same area throughout the summer.[114] In Australia it feeds mainly on the wing, robbing food from terns, gulls and Fluttering Shearwaters.[114] It also takes small pieces of offal near fishing boats; plague locusts, presumably blown out to sea, were once recorded in the guts of birds taken in SA.[1867] On the breeding grounds it also takes small passerines and berries.[1660]

The Jaeger is expanding its breeding range in northern Europe.[1660]

There are no measures of abundance.

Field Atlas Records: 254
1° blocks-with records: 53 (7%)
-with breeding records:

Atlas No.: 128

Max. Rep. Rate: 16%
In block(s): 37/150

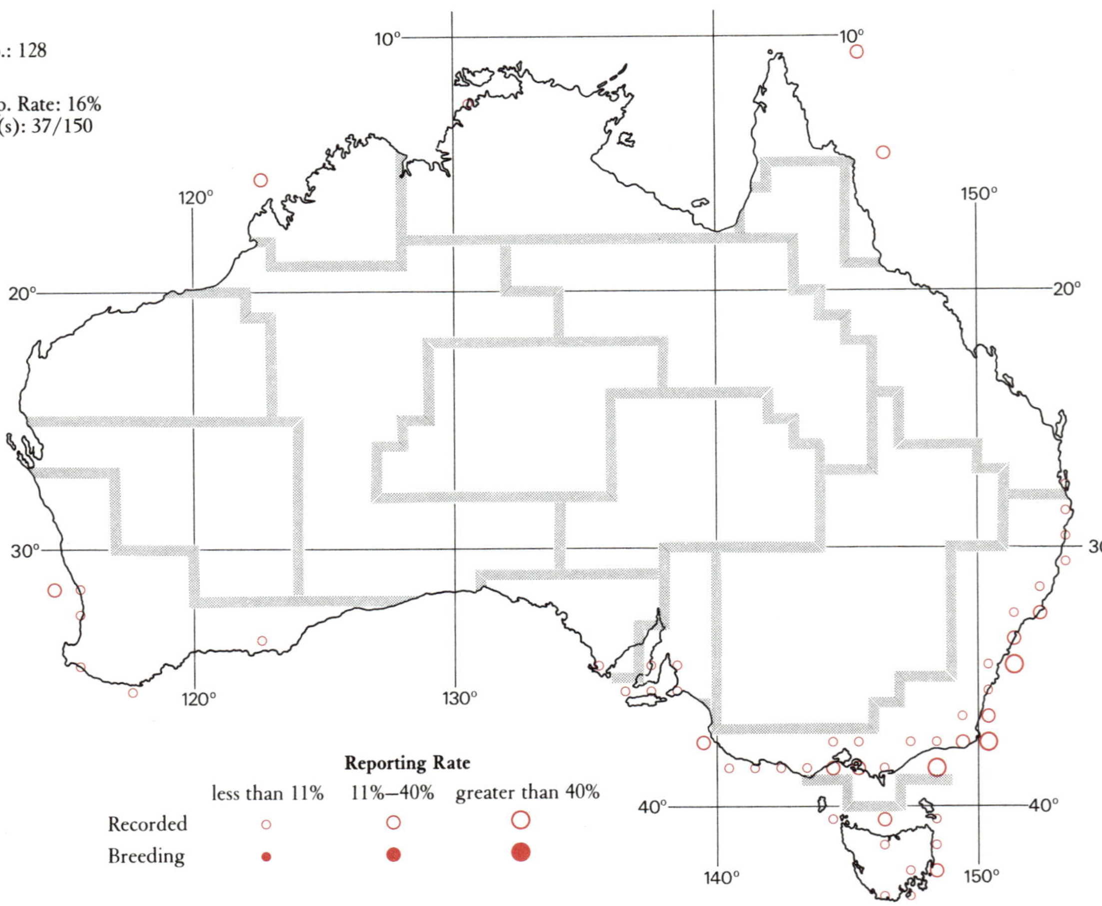

Pomarine Jaeger

Stercorarius pomarinus

The Pomarine Jaeger is a summer visitor to Australia. It breeds north of the Arctic Circle, most birds spending the non-breeding season in the Atlantic off North America and Africa. A smaller number visit the Pacific, a few going as far south as Antarctica.[388,558] There are pale- and dark-coloured adult plumages. In the northern hemisphere the pale plumage outnumbers the dark about 20:1.[1636] In Australia in eight years up to 1963 one observer saw 25 Jaegers off Fremantle (32/115), all pale-coloured birds.[1705] Off Eden (37/149) the ratio of dark to light was estimated at 60:40 and adult to immature at 65:35[114] but another observer found most Pomarine Jaegers off south-eastern Australia to be pale.[508]

In Australian waters the Pomarine Jaeger is most abundant off the coast of the South-East Region.[1262] Few were seen in northern Australian waters by an observer travelling by ship north from Fremantle to Cairns (16/145) in Nov.-Dec. 1980.[315] The species is most frequently reported in summer and the few winter records in the Field Atlas were made at the beginning and end of the season, apparently birds arriving early or leaving late. Off the South-East coast the species appears to move south as the warm surface water flows south in summer.[114]

The Pomarine Jaeger is a bird of offshore waters in Australia but is primarily pelagic when on migration. It is often seen singly but sometimes in parties of up to 30–40.[508] It readily takes offal and follows ships to obtain it.[114,1262] The Jaeger also robs food from gulls and terns and will take its own prey of fish and small birds.

There are no measures of abundance.

Field Atlas Records: 96

1° blocks–with records: 31 (4%)

–with breeding records:

Atlas No.: 945

Max. Rep. Rate: 16%
In block(s): 37/150

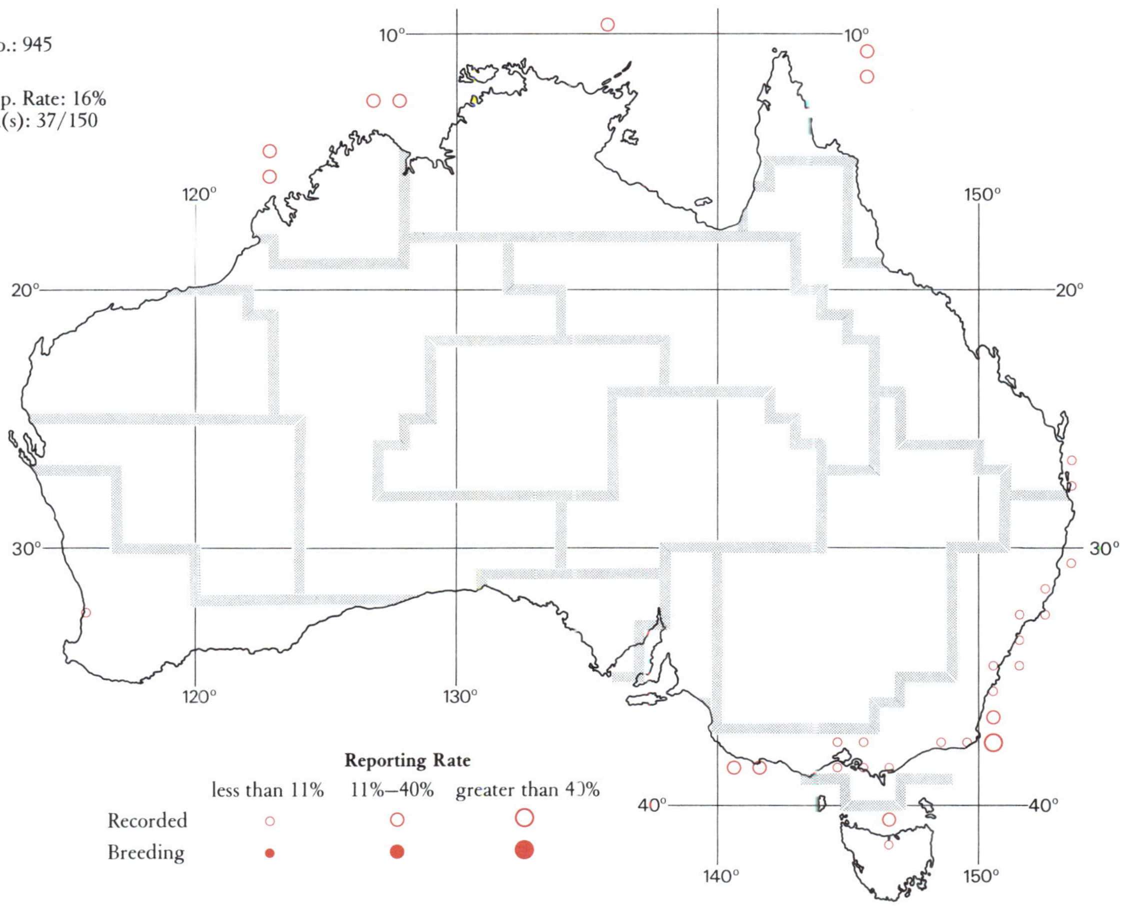

Silver Gull

Larus novaehollandiae

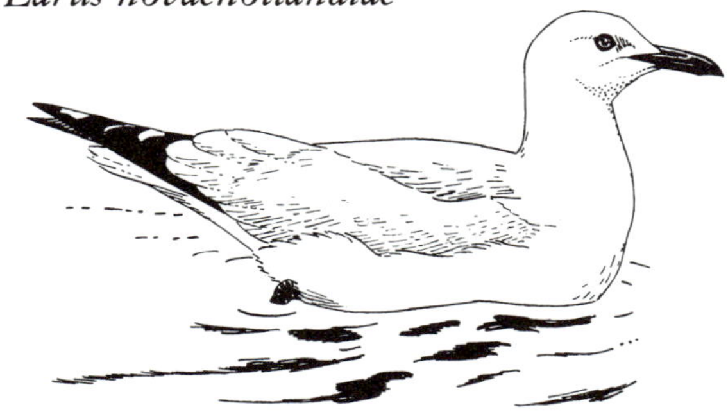

The Silver Gull lives in South Africa, New Zealand, New Caledonia and Australia. Two subspecies are recognized in Australia: *forsteri* in the Top End, Cape York and Atherton Regions and *novaehollandiae* elsewhere.[1422] Some northern breeding colonies were not visited during the Field Atlas. Breeding has been reported in the Top End from Haul Round I. (11/134), Melville Bay (12/136) and Groote Eylandt (13/136) and in the Kimberley Region only from Jones I. (13/126).[1711,1713]

After breeding, young Gulls from all colonies and adults from inland colonies gather at sites where food is plentiful,[311,1303] e.g. Port Phillip Bay (38/144); St Vincent (34/138) and Spencer Gulfs (34/136). Round Darwin (12/130) a drop in numbers in the dry winter season has been correlated with departure to breeding grounds.[1169] Observations of Silver Gulls in the Western Deserts Region indicate that it will wander inland even to uninhabited regions. The longest movement recorded is that of an immature from Five Is (34/150) to Alice Springs (23/133), 2043 km.[45]

The Silver Gull nests colonially on islands, coastal or inland, and on artificial constructions such as salt works. There are permanent breeding colonies on many inland waters, notably in the South-East Region, and opportunistic colonies on intermittent lakes. The only site at which inland breeding is reported north of 28°S is Mt Isa (20/139).[862] The Silver Gull was originally a coastal scavenger[311,1303] and a persistent predator at the colonies of other nesting seabirds.[881] Some of the largest colonies are now near cities because most urban areas have rubbish tips on which the Gull forages. Near Melbourne (37/145) hundreds commute daily 25–30 km inland for this purpose.

The Silver Gull is becoming more abundant in some places. It was rarely recorded in Darwin until the 1970s but has since increased rapidly.[1169] At Mud I. (38/144) numbers rose steadily from 400 birds in 1959 to 3000–4000 in 1979–80.[768] Other colonies in Vic. have fewer than 1000 birds each.[768] In NSW an estimated 60 000 birds bred on 13 islands in the 1970s.[1029] Other colonies include: 500 birds on Flat I. (43/145) in 1977;[1851] 500 on St Helens I. (41/148) in 1978;[1182] 6000–8000 on Carnac I. (32/115) in 1979;[505] and 8000–10 000 on Green I. (34/117) in 1977.[3]

Field Atlas Records: 20 042

1° blocks–with records: 426 (52%)
–with breeding records: 85 (20%)

Atlas No.: 125

Max. Rep. Rate: 86%
In block(s): 30/114,39/146

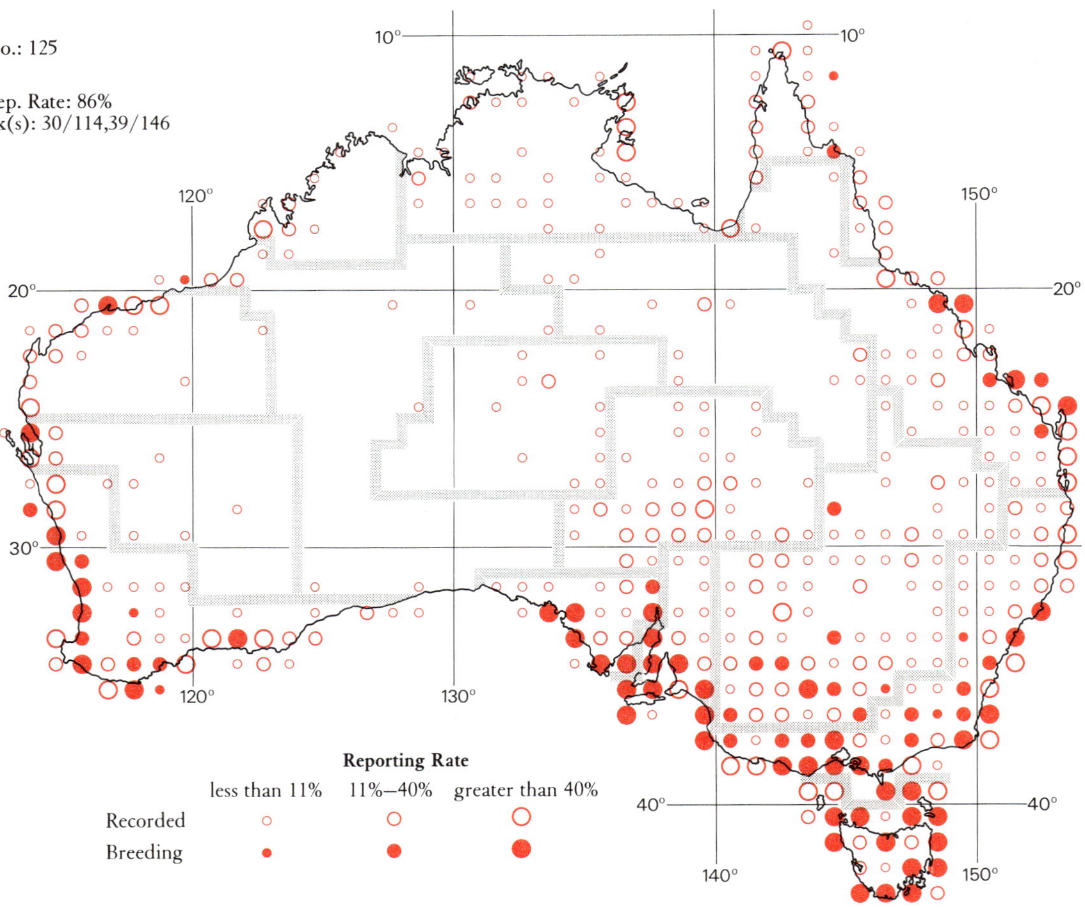

Pacific Gull

Larus pacificus

The gap at 140°E in the distribution of the endemic Pacific Gull is consistent in the Historical Atlases: birds to the east are subspecies *pacificus* and those to the west are *georgii*.[1219] The Pacific Gull does not breed east of Wilsons Prom. (39/146) on the Australian mainland and early records from Eastern Qld are questionable.[1134]

Young Pacific Gulls may fly some distance from their breeding grounds soon after fledging. At Laverton (37/144) there are influxes of young birds in early summer and in winter.[1822] A nestling banded in SA moved 260 km east-north-east within five months but had returned to its breeding place 18 months later.[1636] The longest movement recorded is that of a bird from Goose I. (40/147) to Lauderdale (42/147), 295 km.[1499] In the non-breeding part of its range, east of 147°E on the mainland, birds are recorded at all times of the year.

The Pacific Gull lives on ocean beaches and offshore islands, only exceptionally appearing far inland. Sometimes it will feed at rubbish dumps, specially in Tas. where flocks of 500 have been recorded at the Launceston (41/147) tip.[1799] The bird also forages on sub-coastal pastures such as at Wilsons Prom. and on Flinders I. (40/148). The Pacific Gull feeds on fish, a variety of marine invertebrates, carrion, sometimes at seal colonies, and will take seabird chicks and sea-urchins. Shellfish are broken by being dropped on to rocks. Particular rocks are used for many years and large 'middens' of broken shells accumulate.[1636]

The Kelp Gull has recently begun to breed in the Tas. Region and has now established mixed colonies with the Pacific Gull on Visscher I. (42/147), Lachlan I. (42/147) and Green I. (43/147).[1499] In these colonies the Kelp Gull attacks young Pacific Gulls, causing about 10% mortality on Green I.[1799]

Colonies of the Pacific Gull seldom number more than 100. Recorded sizes are: 40–60 birds on Cat I. (39/148) in 1957;[1816] 46 birds on Mile I. (40/147) in 1979;[1499] 42 birds on Chalky I. (40/147) in 1979;[1501] 74 on Little Chalky I. (40/147) in 1979;[1499] and 32 on Green I. in 1978.[1552] More than 800 birds were breeding on islands off Wilsons Prom. in 1978–80.[768]

Field Atlas Records: 4878

1° blocks-with records: 99 (12%)

-with breeding records: 32 (32%)

Atlas No.: 126

Max. Rep. Rate: 92%
In block(s): 39/146

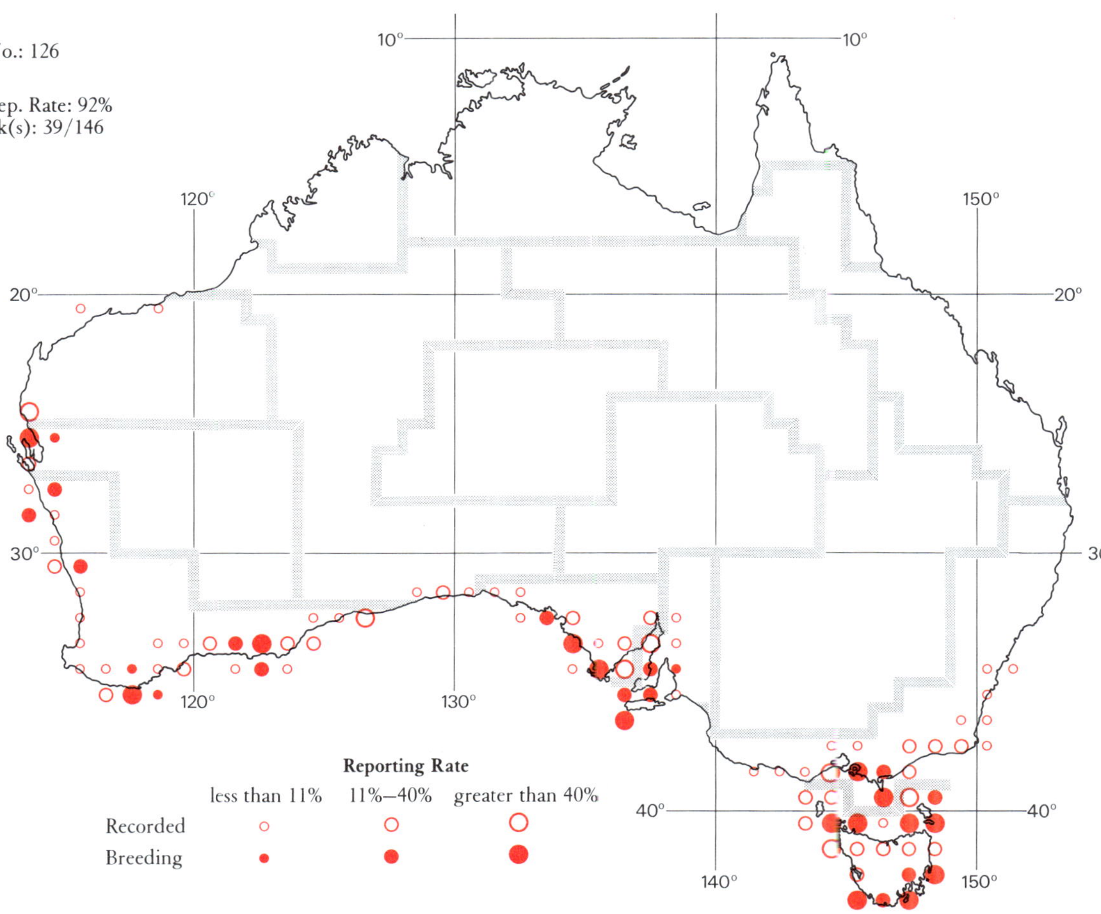

Kelp Gull

Larus dominicanus

The Kelp Gull inhabits South America, South Africa, New Zealand and Australia, including Macquarie I. (54S/158E).[388] It has increased its range and abundance in New Zealand since European colonization.[1502] A similar expansion appears to be taking place in Australia where it was first reported in 1943, at Botany Bay (34/151), but breeding was not recorded until 1958. Some early arrivals may have been overlooked, because a specimen collected near Perth in 1924 and misidentified as a Pacific Gull was not recognized until 1964.[591] The presence or absence of a tail band, formerly thought diagnostic, is not constant.[373,1500]

The Kelp Gull has been recorded in Australia as far north as Cairns (16/145)[909] and round the southern coast to Jurien Bay (30/115). Breeding is confined to Moon I. (33/151), the Five Is (34/150) and Seal Rocks (38/145) in the South-East Region; and in the Tas. Region mainly to three colonies mixed with Pacific Gulls on Lachlan I. (42/147), Visscher I. (42/147) and Green I. (43/147). As numbers increase in south-east Tas. the Kelp Gull is starting to establish colonies by itself on other islands in Frederick Henry Bay (42/147) and Pitt Water (42/147).[1499]

Kelp Gulls breeding on sub-Antarctic islands are mostly sedentary[1636] but, in Australia, both adults and immatures are observed away from breeding grounds[373] and a nestling banded at Moon I. has been recovered near Fremantle (32/115), 3328 km away.[1636] No seasonal pattern is evident in records from SA and WA where the species has been recorded in all months except July. The few Qld records were Feb.-Aug.[373]

The Kelp Gull is coastal in Australia but in New Zealand has invaded inland sites since European settlement.[1502] Like other large scavenging gulls it eats fish, marine invertebrates and carrion;[1636] in New Zealand it sometimes attacks weak sheep and lambs, poultry and tuataras. In Tas. it attacks and often kills immature Pacific Gulls on the sea and in flight over the colony and the two species compete for food.[1499]

The Australian breeding population was estimated at 400 pairs in 1982–83[1499] and is increasing. Most are in south-eastern Tas. Colonies in NSW held an estimated 13 pairs in 1981.[1292] At Seal Rocks one pair has bred each year since 1970–71.[1111]

Field Atlas Records: 599

1° blocks-with records: 34 (4%)
-with breeding records: 5 (15%)

Atlas No.: 981

Max. Rep. Rate: 35%
In block(s): 43/147

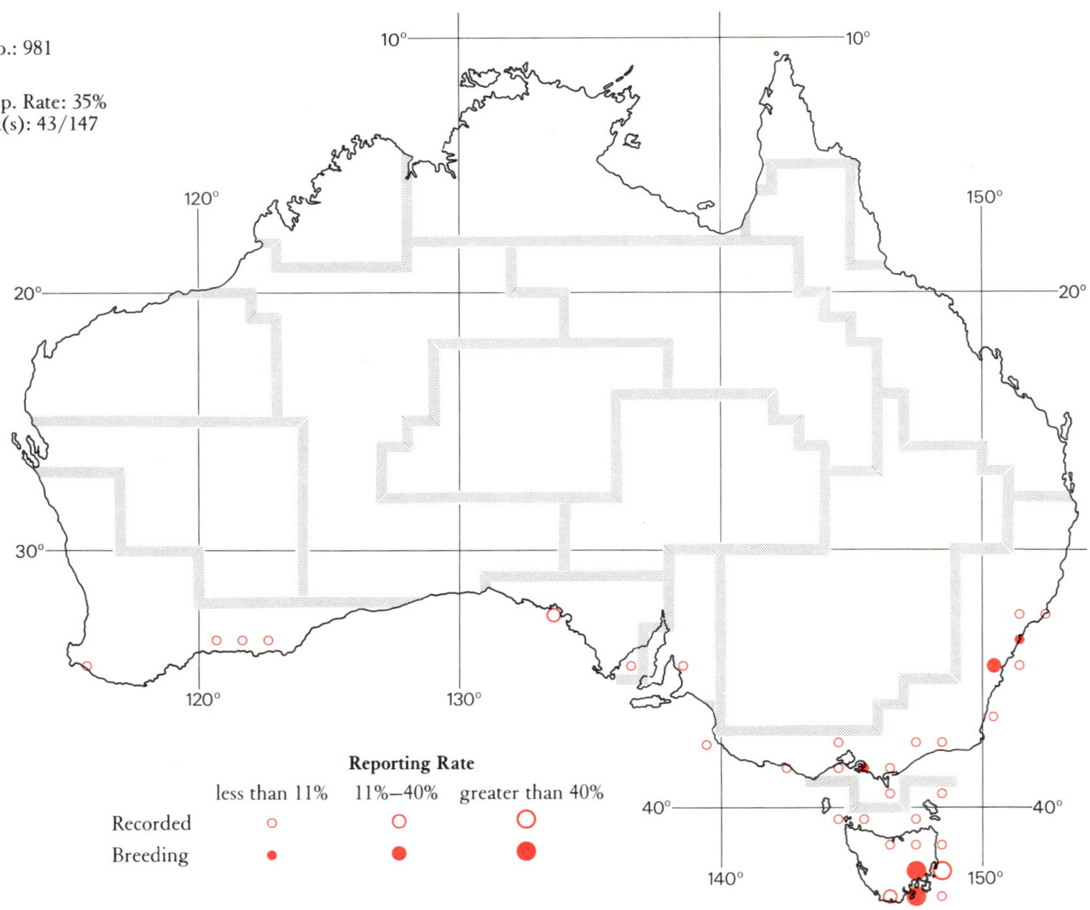

Whiskered Tern

Chlidonias hybrida

The Whiskered Tern breeds in Eurasia, Africa and Australia and not, as is sometimes stated,[388] in south-eastern Asia or New Guinea.[1249] The subspecies *javanica* breeds only in Australia.[1249] The Whiskered Tern is widespread on the mainland but in the Tas. Region there are only three records in the Atlases: Ouse (42/146) in Sept. 1967, Cape Portland (40/147) in Apr. 1968[734] and Cameron Inlet (40/148) in Nov. 1981.

Breeding of the Whiskered Tern has not been confirmed in northern inland Australia though an early unconvincing reference[903] has been used to imply that it is usual there.[427,1636,1711] Of the 47 1° blocks in which breeding has been reported in the Atlases, 47% are in the Murray-Darling Region, 34% in the South-East, 13% in the Goldfields and South-West and the rest in the L. Eyre and SA Ra. Regions. All are south of 25°S and most in eastern Australia.

Part of the population appears to migrate north out of Australia in the non-breeding season. Banded birds have been recovered overseas, e.g. Goose Lagoon (38/142) to New Guinea, Ivanhoe (32/144) to Java (7S/110E); and specimens of *javanica* were collected in Sulawesi (2S/121E), Luzon (2N/121E) and Java.[1249,1395] Whiskered Terns are not recorded in Torres Strait[503] but in Darwin (12/130) numbers vary seasonally,[427,1169] which suggests that migrants pass through the Top End rather than the Cape York Region. Counts at Darwin are lowest Feb.-Apr. and highest Oct.-Nov. In the South-East, Murray-Darling and SA Ra. Regions Reporting Rates dropped from 5.1% in summer to 1.4% in winter but some Whiskered Terns remain in southern Australia in winter; some also remain in southern Asia in summer.[1249]

The Whiskered Tern inhabits swamps and lagoons, fresh or salt, intermittent or permanent. Colonies form quickly when floods fill swamps in summer[210] but breeding sites are not used every year and many birds, presumably non-breeders, frequent coastal and other areas during the summer breeding season. The Tern feeds on insects taken on the wing (dragonflies, flying ants), from the water (beetles) or from dry ground (grasshoppers).[210,42-] It also eats small fish and crustaceans and fed on house mice during a plague.[840]

Many colonies contain fewer than 100 nests but some with 500–2000 nests have been reported.[1552]

Field Atlas Records: 2683

1° blocks–with records: 301 (37%)
–with breeding records: 11 (4%)

Atlas No.: 110

Max. Rep. Rate: 28%
In block(s): 12/133

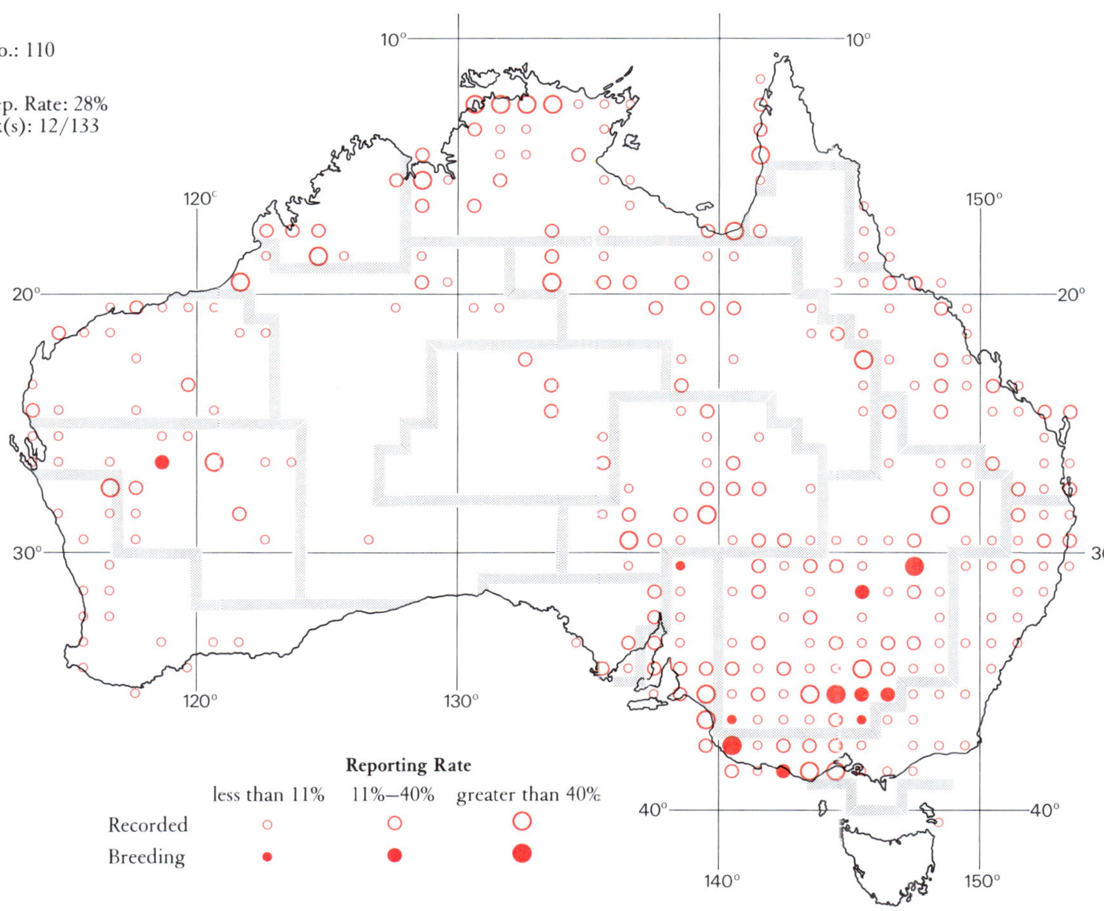

White-winged Tern

Chlidonias leucoptera

The White-winged Tern breeds in the swamps of southern Europe, across Asia and perhaps in Africa.[1425,1905] It migrates south for the non-breeding season to Africa and as far as Australia and New Zealand. One bird banded in eastern USSR has been recovered in New Guinea.[63]

The White-winged Tern is a regular wet summer season visitor round Darwin (12/130) with maximum numbers in Jan.[423,1169] and a passage migrant in small numbers through Torres Strait.[503] An observer travelling by sea round northern Australia, Nov.-Dec. 1980, regularly recorded small flocks, mostly fewer than ten birds, between 120°E and 140°E, presumably migrating.[315] In Mar.-Apr. 1982 an estimated 15 000, some in breeding plumage, massed at Port Hedland (20/118) presumably preparatory to migration.[911]

Normally small numbers reach south-eastern Australia. In the South-West Region few were recorded during the Field Atlas but occasionally there are irruptions. These have been linked with the occurrence of tropical cyclones.[1638] An inland record at Alice Springs (23/133) in Dec. 1978 suggests that some birds move overland to southern coasts or are displaced inland by cyclones.

The White-winged Tern lives near swamps and shallow coastal waters in Australia. It feeds on insects, including termites, locusts, dragonflies and moths, taking food from the water surface and slightly below it as well as on land.[424,1636] Large congregations of White-winged Terns are sometimes associated with abundances of insects, e.g. 700 at Herdsman L. (31/115) in Mar.-Apr. 1917;[9] thousands in Apr. 1979 at Anna Plains (19/121) feeding on swarms of locusts. About 300 birds were reported at sea off the Atherton Region in Mar. 1955, also feeding on unusual concentrations of insects.[18]

There are no measures of abundance.

Field Atlas Records: 306

1° blocks–with records: 86 (11%)
–with breeding records:

Atlas No.: 109

Max. Rep. Rate: 13%
In block(s): 12/130

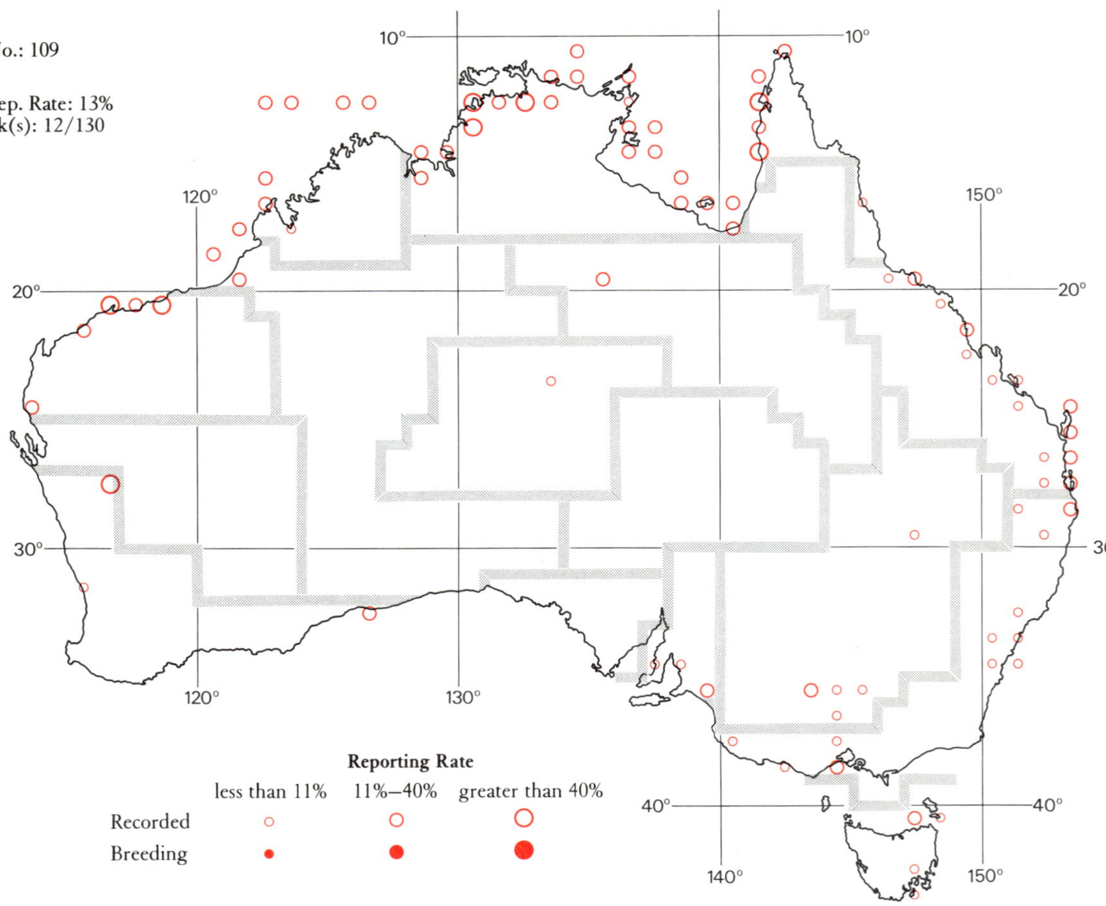

Gull-billed Tern

Gelochelidon nilotica

The Gull-billed Tern lives on all the continents. The subspecies *macrotarsa* breeds in Australia.[388] An Asian race *affinis* is recorded round Darwin (12/130) in small numbers Sept.-Apr.[1169] and was also reported in the Kimberley Region in Oct. 1976.[941] In the Tas. Region there is a reported sighting of the Gull-billed Tern south of Richmond (42/147) in 1949[1658] but single birds on the Tamar Estuary (41/146) in Apr.-July 1979 and Feb. 1980 are the first confirmed records.

Only two breeding records in the Atlases are north of 25°S: Maud's Landing (23/133) in 1900[317] but not since; L. Moondarra (20/139), 12 nests in 1963.[1552]

The lack of breeding records from northern Australia suggests that birds disperse over long distances from southern breeding sites. Reports show some seasonal regularity of movement: in New Guinea the species is regarded as a dry winter season migrant from Australia,[554] with small numbers throughout the year in Torres Strait;[503] in northern Australia the species occurs mainly in winter;[313,423,1169] and in the southern Murray-Darling it arrives each summer to breed regularly at a few sites and occasionally at many.[233] Reporting Rates in the South-East, Murray-Darling and SA Ra. Regions combined were lowest in autumn (spring 1.1%, summer 1.2%, autumn 0.5%, winter 0.9%). In the South-West Region visits are infrequent and irregular.[629]

The Gull-billed Tern lives near fresh or saline wetlands, inland or on the coast. Inland it may sometimes be far from water as well. No coastal breeding sites are recorded. Breeding may begin because of an abundance of food rather than from any direct effect of rainfall or water conditions, e.g. at Ivanhoe (32/144) in 1974 a colony (or colonies) bred in response to a plague of house mice.[840] The Gull-billed Tern is also recorded to take grasshoppers,[18] fish,[313] crabs,[1840] food from freshly ploughed ground[698] and insects in the air and from the water surface.[233]

Most colonies contain about 50 nests, a few over 100, e.g. 200 birds at L. Tutchewop (35/143) in 1971;[213] 300 birds at Ivanhoe (35/128) in 1973; and several colonies with more than 300 birds at L. Bancannia (30/141) in 1974.[1552]

Field Atlas Records: 1734

1° blocks–with records: 213 (26%)
–with breeding records: 6 (3%)

Atlas No.: 111

Max. Rep. Rate: 44%
In block(s): 24/153

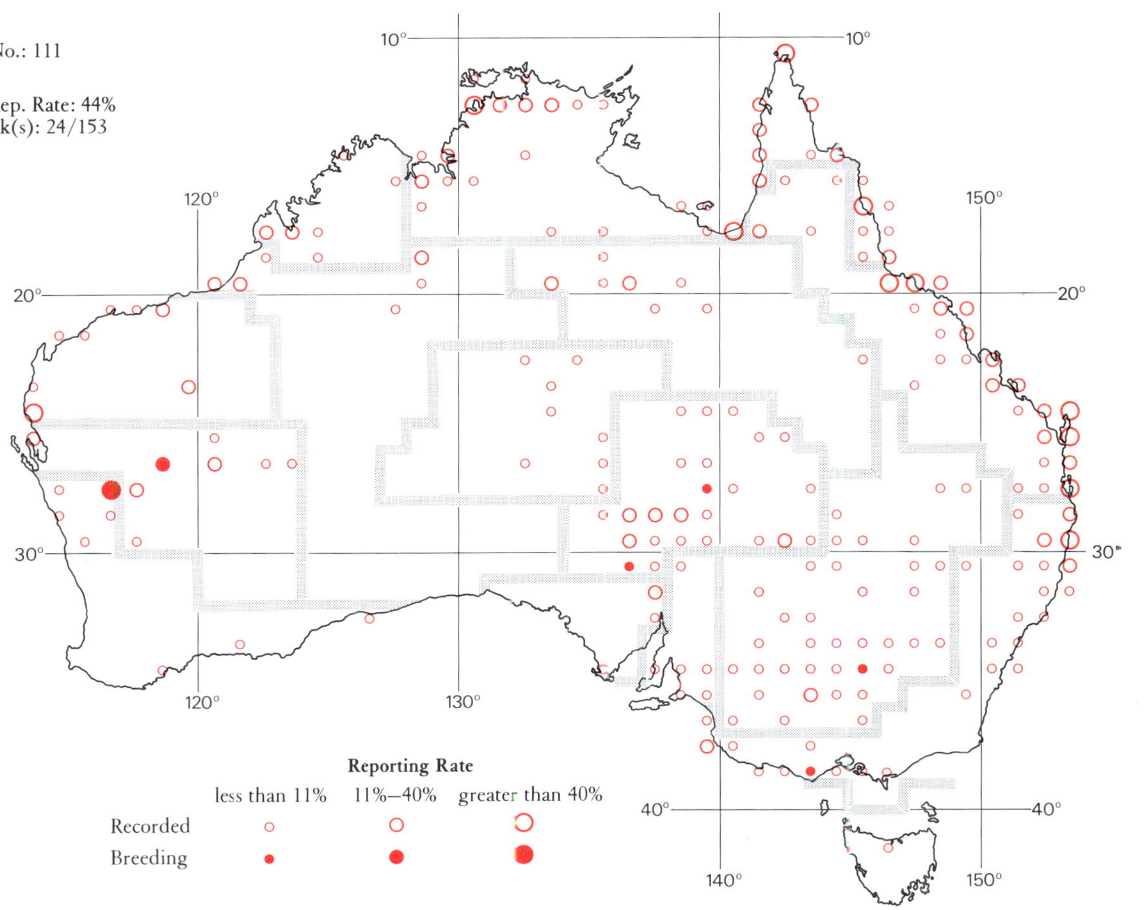

Caspian Tern

Hydroprogne caspia

The Caspian Tern lives in North America, Europe, Asia, Africa, New Zealand, New Guinea and Australia. In the Top End it is reported to breed in the Sir Edward Pellew Group (15/136) and perhaps on other islands.[1169] On the coast of the South-East Region there are no breeding reports between 29°S and 37°S. Inland breeding is rare, sites reported in the Atlases being L. Moondarra (20/139), L. Dynevor (28/144), Menindee Lakes (32/142), L. Eyre (29/137) and L. Watherston (30/138).

The Caspian Tern was previously considered to move only locally.[1636] Though it is present all year in some places, there is evidence of seasonal movement in others. In Torres Strait it is seen throughout the year.[503] Round Darwin (12/130) small numbers are always present[1169] but in some years the species may be more abundant in the dry winter season.[423] At L. Moondarra, where up to 400 birds breed May-Sept., few are present in the wet summer season, Dec.-Mar.[862] In the southern Murray-Darling the bird is most numerous Jan.-Aug. though some may be seen in all months; presumably it leaves the area to breed.[833] In the Tas. Region the Reporting Rate was 1% in winter compared with 10% in summer, which suggests that there may be movement out of the region in winter. The longest movement recorded is from Mt Isa (20/139) to Murray Bridge (35/139), 1633 km.[47]

The Caspian Tern lives on the seashore and round some inland waters. Many breeding colonies are small and sometimes the bird nests singly among other species of seabirds. In the Tas. Region breeding is not usually colonial.[1533] The Caspian Tern feeds on fish up to about 18 cm in length, which it catches by diving, sometimes submerging completely.[1636]

There are colonies of up to 400 birds on West I. (35/138)[1406] and 200 at Houtman Abrolhos (28/113).[1636]

Field Atlas Records: 4860

1° blocks–with records: 271 (33%)
–with breeding records: 43 (16%)

Atlas No.: 112

Max. Rep. Rate: 58%
In block(s): 25/113

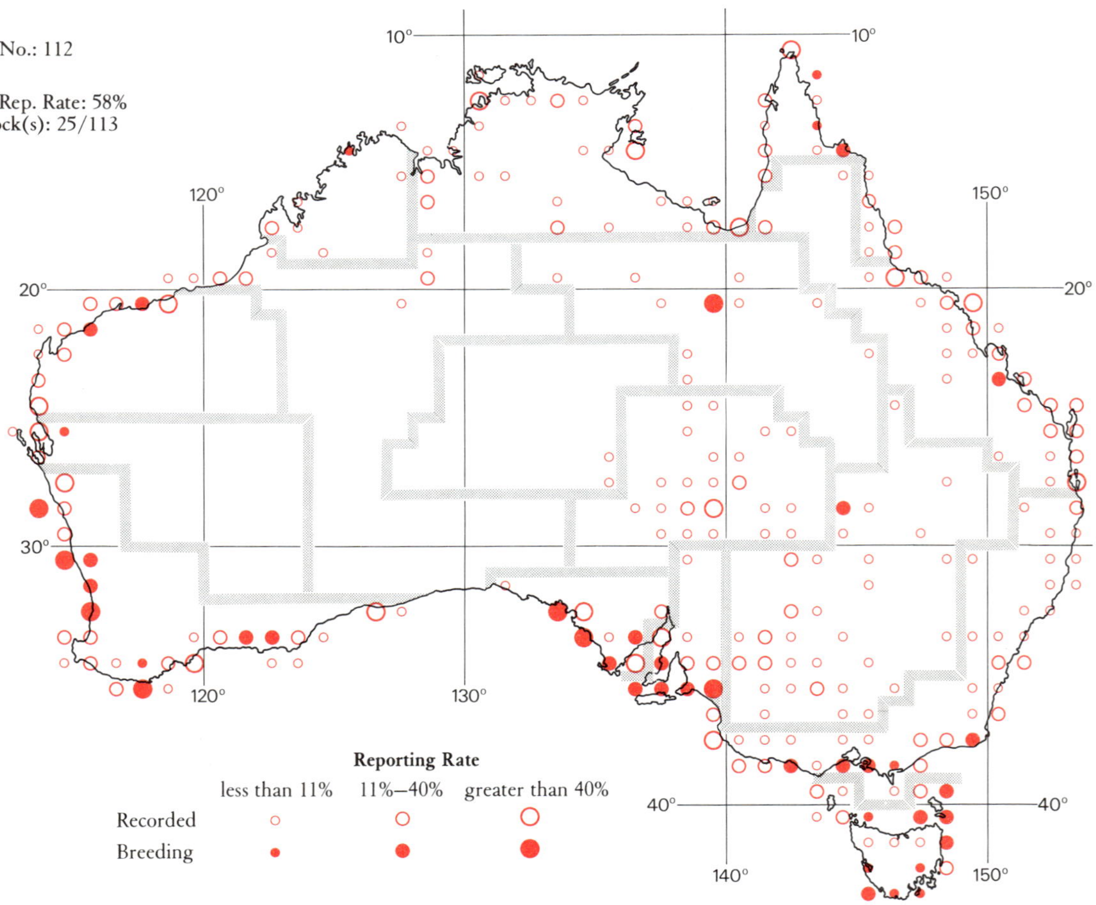

Common Tern

Sterna hirundo

The Common Tern breeds in North America, Europe, temperate Asia and Africa[388] and is a regular summer visitor to Australia. Most birds in northern and eastern Australia, from Port Hedland (20/118) to SA, appear to be referable to the subspecies *longipennis* that breeds in northern and eastern Siberia.[824,1583,1638] Banded birds of the nominate subspecies that breeds in America, Europe and northern Africa have also been recovered in Australia: one from Sweden to Fremantle (32/115); another from Ireland to Kow Swamp (35/144).[85,1638] A specimen from Goolwa (35/138) has been identified as *minussensis* whose breeding range lies between those of *longipennis* and *hirundo*.[388,511] The Common Tern has not yet been confirmed to be pesent in the Tas. Region and is rare in the South-West Region.

Though primarily a summer visitor to Australia some Common Terns remain throughout the winter (Reporting Rates: summer 1.2%, winter 0.2%).

The bird lives on shallow estuaries with sand banks as well as tidal flats, harbours and beaches. It is usually seen in small groups or flocks but a flock of 1000 has been observed at Boat Harbour (34/151),[1724] one of 150 was recorded on L. Alexandrina (35/139) in 1976[1465a] and flocks of 100–200 were present at Ocean Grange (37/147) on several days in Jan. and Dec. 1981.[284] Between Port Hedland and Cape Keraudren (19/119) flocks of 800–1000 gather in Apr. preparatory to migration.[1638] Guts of eight specimens contained insects as well as fish.[824]

There are no measures of abundance.

Field Atlas Records: 635

1° blocks–with records: 71 (9%)
–with breeding records:

Atlas No.: 953

Max. Rep. Rate: 34%
In block(s): 24/153

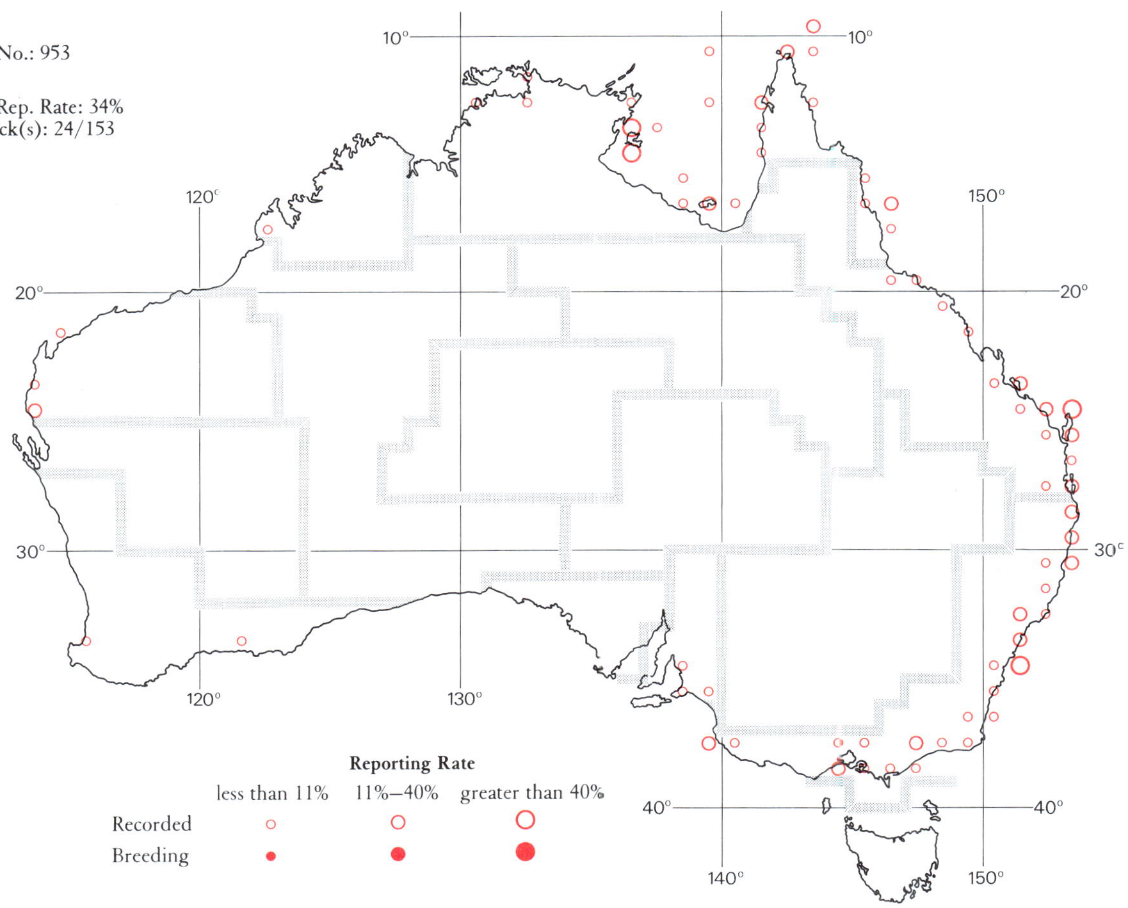

Roseate Tern

Sterna dougallii

The Roseate Tern has an uneven coastal distribution round the world, the subspecies *gracilis* breeding in Australia[388] mainly in the north. The southernmost records in the Atlases are at Ballina (28/153) in the east and Mandurah (32/115) in the west. There is one doubtful record from Sutherlands (34/139) in winter 1928.[821] Some Arctic Terns have been mis-identified as Roseates.[1638]

In addition to sites shown on the map, the Roseate Tern breeds in Torres Strait and on Haul Round I. (11/134); there were no breeding records during the Field Atlas because these localities were not visited. Breeding was rare south of Geraldton (28/114) until 1960 when colonies spread to islands near Perth (31/115). In 1978 the Roseate Tern bred on Carnac I. (32/115) even farther south. The colony on Bedout I. (19/119) may have been abandoned. It was in use in 1901; flocks of some hundreds were near it in 1972[1008,1638] but there were no nests in 1975.[288]

The Field Atlas suggests some movement out of the South-West Region in winter with no records south of 30°S then.

The Roseate Tern nests on islands. As overseas, some nests are within colonies of other species. The Roseate Tern feeds on small fish. It often tries to rob fish from other species of birds but rarely successfully.[881]

On the Great Barrier Reef it fishes in deep and shallow water.[880] In Peel Inlet (32/115) it fishes in shallow inshore waters, which is unusual for this species.[911] On One Tree I. (23/152), at the southern limit of its breeding range in eastern Australia, breeding was attempted each year between 1971 and 1976 but no chick was reared. Predation by gulls of small colonies of 20–30 nests was severe.[882] This contrasts with the expansion of range taking place in the west.

On Haul Round I. 4000 or more birds nested in 1972; and, in 1947, 5000 birds nested on Pelsart I. (28/114).[1638] Most other reported colonies contain fewer than 200 birds.

Field Atlas Records: 168

1° blocks–with records: 47 (6%)
–with breeding records: 7 (15%)

Atlas No.: 113

Max. Rep. Rate: 28%
In block(s): 28/113

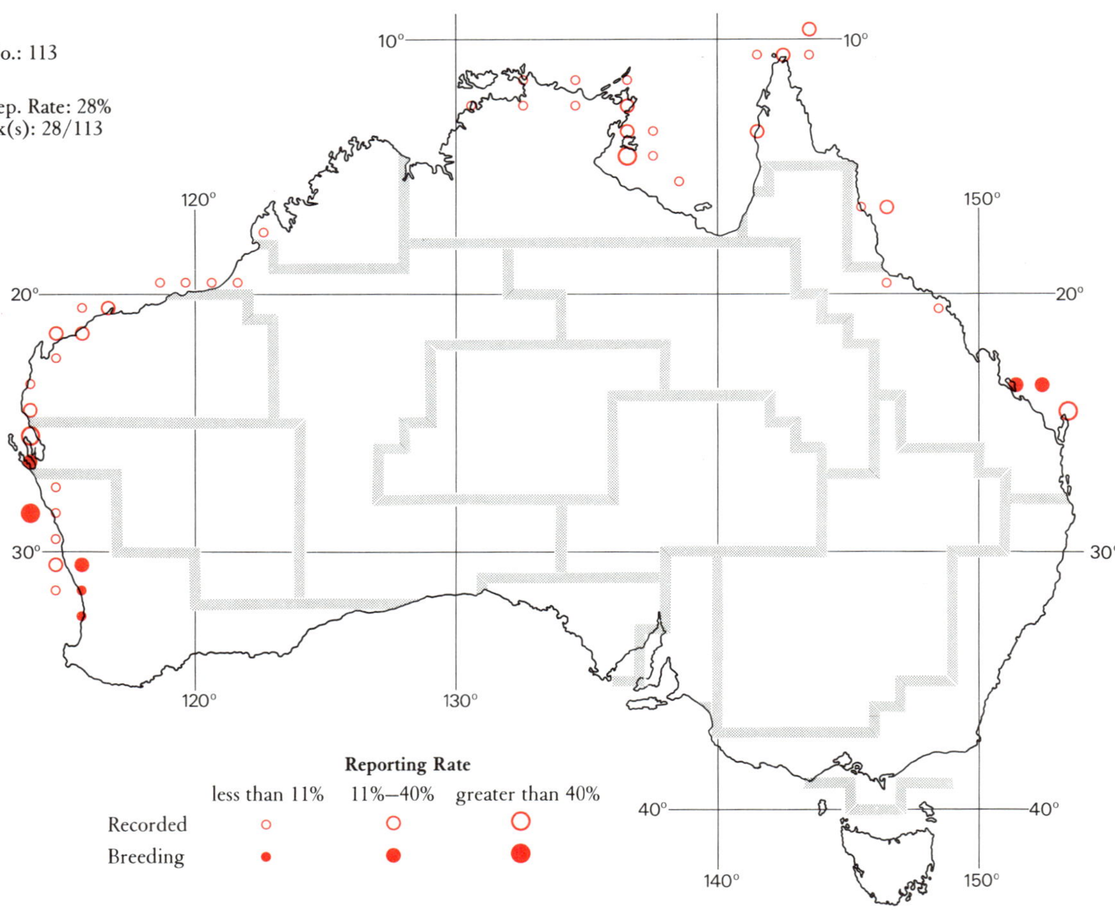

White-fronted Tern

Sterna striata

In 1979 the White-fronted Tern was proved to breed in Australia. Though some reports imply that it may have bred in Tas. last century,[1844] it was previously confirmed to breed only in New Zealand. In 1979, 4–6 pairs bred successfully on Battery I. (40/148) in Bass Strait.[1844] Subsequently the species attempted to breed there and on Mile I. (40/147) without success but it bred successfully on Fisher Reef (40/148) in 1979, 1980 and 1981. On Billy Goat Reef (40/148) in Nov. 1981 six nests had eggs or young and 14 birds were in attendance.[1310] It remains to be seen whether the White-fronted Tern can maintain itself in Australia as a breeding species. The limits of distribution established by the Field Atlas are the same as in the Historical Atlases.

The Field Atlas confirms that the White-fronted Tern is predominantly a winter visitor to Australia, mainly May-Nov.,[795] few birds remaining over the summer. The visiting birds appear to be mainly immatures. One estimate in the South-East Region put the proportion of birds in 'adult' plumage at less than 5%.[1166]

The White-fronted Tern will fish in the surf zone[1636] but observations round Tas. and in Bass Strait during the Field Atlas show it to be more common offshore than inshore. It feeds only on fish and catches them by diving from heights of 6–10 m above the water.

Numbers visiting Australia vary from year to year. There was an unusually large influx in May-Sept. 1959 with heavy mortality in storms in Aug.: on 9 Aug. 21, including five dead, were seen round Phillip I. (38/145); in Aug.-Sept. at Port Fairy (38/142) over 20 were found dead; one bird was blown inland to Wyperfeld (35/141), 320 km from the coast; and at least 1000 were counted from Windang I. (34/150) in Aug. 1959.[1166] These larger than usual numbers included many adult birds (more than 50%).

Field Atlas Records: 170

1° blocks–with records: 33 (4%)
–with breeding records: 2 (6%)

Atlas No.: 114

Max. Rep. Rate: 22%
In block(s): 37/150

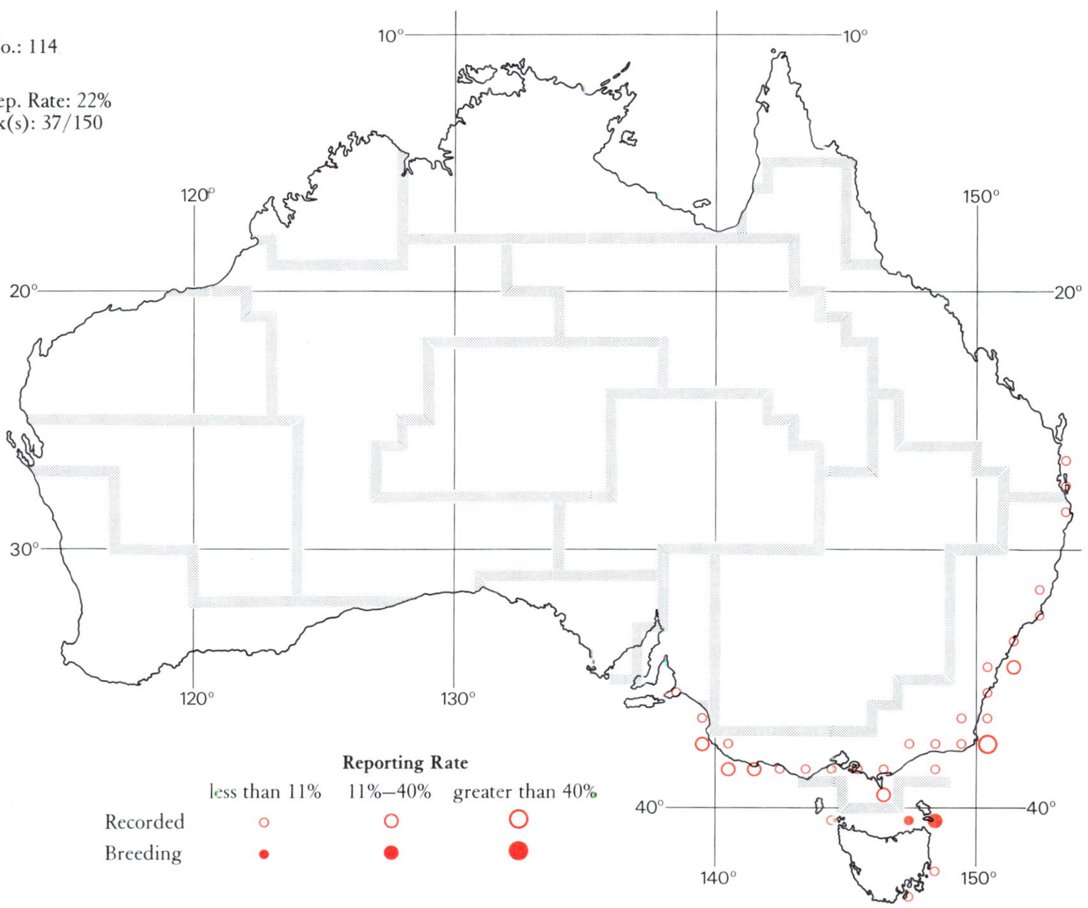

Black-naped Tern

Sterna sumatrana

The Black-naped Tern is present throughout the tropics of the Indian and Pacific Oceans. In Australia it breeds on islands off the Qld coast from Lady Elliott (24/152) north to Torres Strait; on Coral Sea islands; and on at least two islands off the Top End coast: Sandy I. (11/132) visited in 1978 and Turtle I. (15/136) visited in 1964. The species has been sighted as far south as Stradbroke I. (27/153). In the Top End there are no confirmed records west of 132°E though reports have come from as far west as Darwin (12/130).[1490] The Black-naped Tern may be confused in the field with the Little Tern in non-breeding plumage.[880]

The Black-naped Tern is mostly absent from colonies off the Qld coast for several months in the winter non-breeding season.[495,843,880]

Colonies are not large and are usually close to the high-tide line.[1148] On One Tree I. (23/152), where the Black-naped Tern feeds on small fish, it is not attacked much in the air by gulls, which select terns carrying larger fish, but is severely harassed by gulls on the breeding grounds.[881,882] Silver Gulls (300 g) can push this small tern (98 g) off its nest to steal the eggs. In this way up to 62% of eggs laid are lost. Predation is less severe when Crested Terns nest in large numbers and attract many gulls away from the Black-naped colony. Flooding, specially from high tides, may destroy up to 50% of the eggs. Disturbance by people also causes losses. All these factors lead to low breeding success in the Capricornia section of the Great Barrier Reef Marine Park and the species may be unable to maintain its population.[882]

On One Tree I. there were 102 birds in 1973–74, 276 birds in 1974–75[882] and 146 birds in 1982–83.[880] The colony on Sandy I. contained 100 birds in Dec. 1978. On Swain Reefs (21/151) there were 400 birds breeding in 1976[1073] and on Turtle I. 200 in 1964.[1552]

Field Atlas Records: 192

1° blocks–with records: 30 (4%)
–with breeding records: 5 (17%)

Atlas No.: 119

Max. Rep. Rate: 35%
In block(s): 10/142

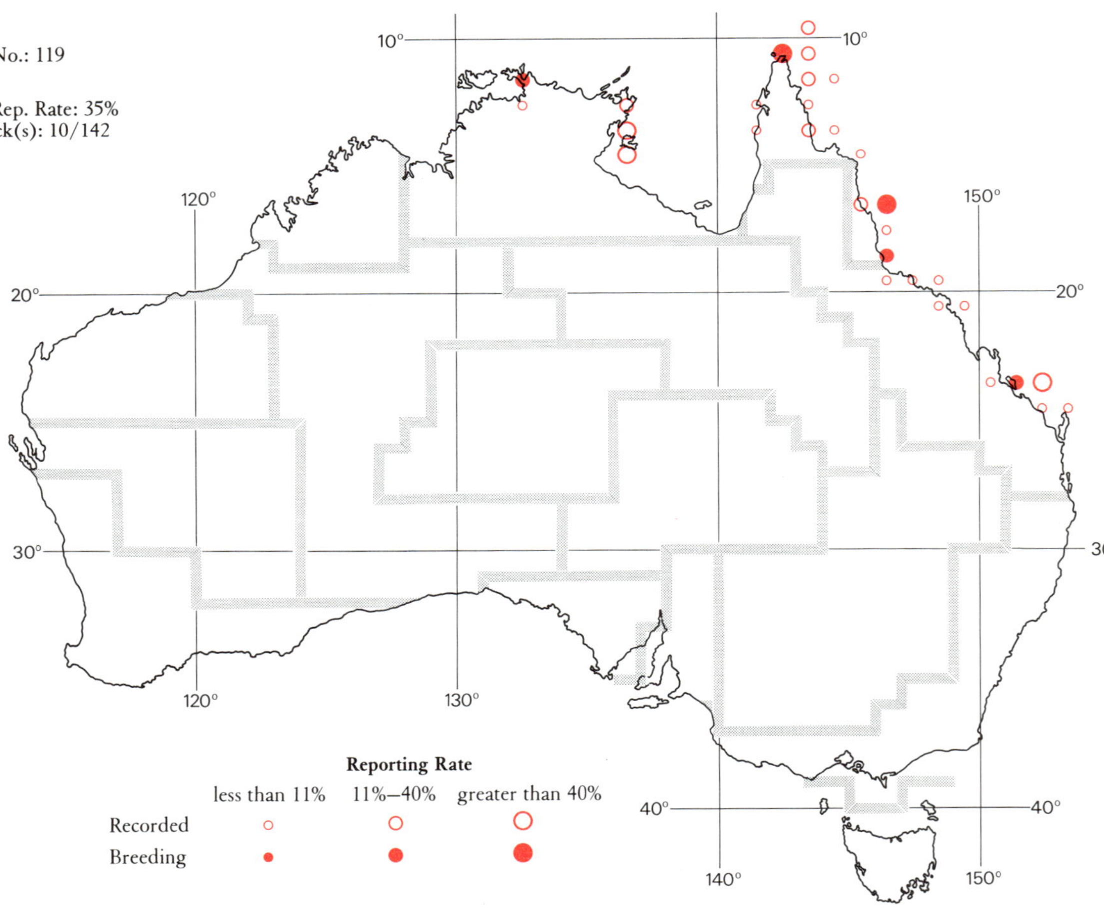

Sooty Tern

Sterna fuscata

The Sooty Tern is present throughout the tropics with substantial breeding colonies in Australia of the subspecies *serrata*. In eastern Australia it breeds on islands from Torres Strait south to Cairns (16/145), in the Coral Sea and on Lord Howe (31/159) and Norfolk (29/167) Is; in the west on Ashmore Reef (12/123), Bedout I. (19/119) and Pelsart I. (28/114).

During the Field Atlas colonies in the Coral Sea and Bedout I. were not visited. Vast colonies on Rat I. (28/113) and in the Wallabi Group (28/113) have vanished since 1913. Breeding has never been reported from the Top End. There has been one breeding record from SA: on Kangaroo I. (35/137) in 1884, and occasional sightings since.[1386]

When breeding is over the Sooty Tern leaves the area of the colony.[1638] Banding in the USA showed that only juveniles appeared to migrate, the adults simply dispersing into surrounding seas.[1636] In NSW waters both adults and juveniles are seen, which indicates considerable movement from Australian colonies. The species is absent from Lord Howe I. Apr.-Aug.[676] and from Norfolk I. Apr.-Sept.[1580] Birds banded on Lord Howe I. have been recovered in Australia[503,1636] and in the Philippines.[31] Another from the Seychelles (4S/55E) was recovered in Darwin (12/130).[1711] Sometimes wrecks occur, e.g. gales in Nov. 1964 were followed by records of Sooty Terns in WA up to 100 km inland and in the Field Atlas the two southernmost records on the eastern coast were made during storms in Jan. 1978.

The Sooty Tern feeds mainly at dusk, when small cephalopods surface, and takes fish of species that are known only to surface then.[1636] Eggs from Sooty Tern colonies have been much harvested by man but this does not account for the disappearance of colonies on Rat I. and in the Wallabi Group.

Estimates of the size of colonies are difficult because vegetation obscures them. Huge numbers nest round Lord Howe I., including an estimated 100 000 birds on Roach I. (31/159) alone;[676] 80 000–140 000 breed on Norfolk I.[731] Other estimates of Australian colonies include: 40 000 birds on Bramble Cay (9/143);[503] 2000 birds on Raine I. (11/144) in 1980 though there were only three nests;[1000] 15 000–20 000 birds on Michaelmas Cay (16/145) in 1982 of which 60–70% were nesting.[1826] The Ashmore Reef colony held about 1000 birds in 1949.[1625]

Field Atlas Records: 131

1° blocks–with records: 62 (8%)
–with breeding records: 9 (15%)

Atlas No.: 120

Max. Rep. Rate: 30%
In block(s): 16/146

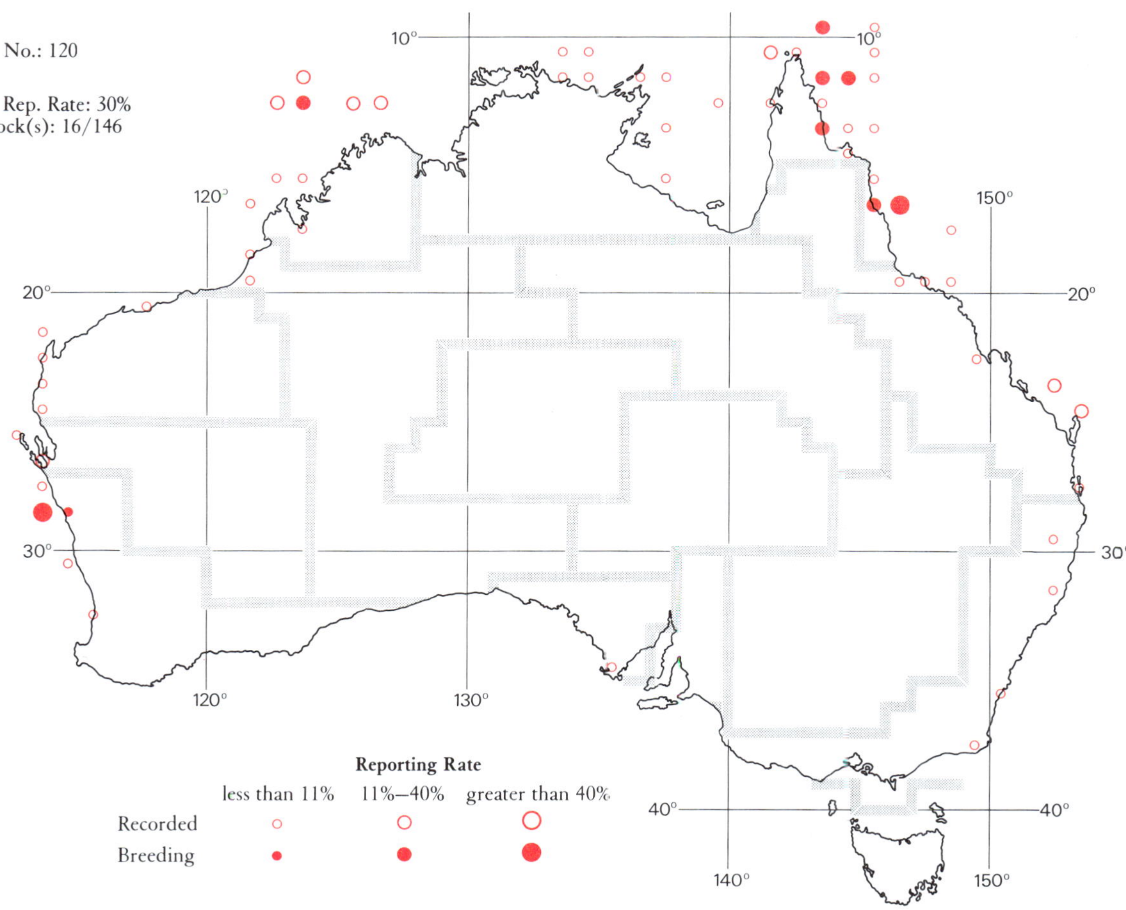

Bridled Tern

Sterna anaethetus

The Bridled Tern lives in the Indian, Pacific and western Atlantic Oceans, with the subspecies *anaethetus* breeding in Australia. As well as the breeding sites mapped during the Field Atlas, there are others on islands off the Qld coast; in the Top End on Haul Round I. (11/134) and an island in the Sir Edward Pellew Group (15/136); and on islands round the Kimberley coast.[1636] In WA the Bridled Tern has extended its range southward. In 1839–43 it was present no farther south than the Houtman Abrolhos (28/113). In 1920, 32°S was given as the southern limit. In 1957 it was discovered nesting on Seal I. (34/115) and continues to do so.[1636] The broken shell of a hatched egg was collected from Baudin Rocks (37/139) in SA in 1969 and the species was reported again in SA in 1973.[711]

The Bridled Tern migrates to islands in Torres Strait Aug.-Jan. to breed[503] and is also a spring-summer breeding migrant on the Qld coast and in the South-West Region. In the Top End it has been recorded breeding June-July and Oct. In the Kimberley the breeding season is given as spring-summer only[1713] or autumn as well.[1636]

The Field Atlas suggests that most of the spring-summer breeding birds leave coastal waters in autumn and winter. In the regions where the bird was most often seen (Atherton, Eastern Qld, Top End and South-West) the combined Reporting Rates were: spring 1.1%, summer 3.5%, autumn 0.2%, winter 0.1%. One bird banded at N. Fisherman I. (30/114) was recovered north of the equator in Borneo (3N/117E).[942]

The Bridled Tern is a bird of the open sea, rarely fishing inshore, even near breeding colonies. It hunts in daylight taking fish rather than cephalopods.

The main breeding islands are: Channel I. (10/142) 160 birds in 1979;[1552] Swain Reefs (21/151) 1100 birds in 1976;[1073] Tryon I. (23/151) 762 birds;[880] One Tree I. (23/152) 652 birds in 1974;[880] N. Fisherman I. 4000 birds in 1976;[942] S. Fisherman I. (30/114) 1000 birds in 1976;[943] Lancelin I. (31/115) 400–1000 birds in 1976;[5] Carnac I. (32/115) 100–200 birds in 1980;[505] Hamelin I. (34/115) 1000 birds in 1977.[4]

Field Atlas Records: 259

1° blocks–with records: 57 (7%)
–with breeding records: 19 (33%)

Atlas No.: 121

Max. Rep. Rate: 38%
In block(s): 30/114

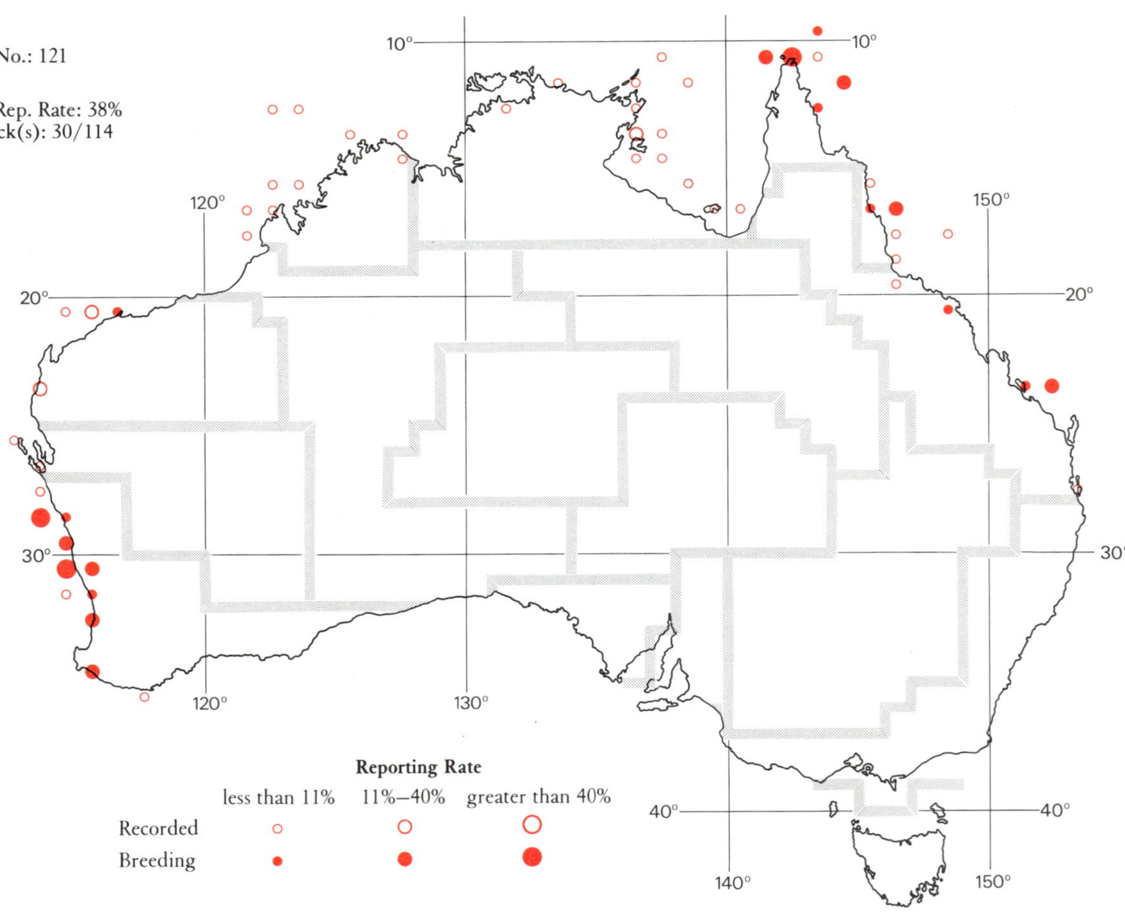

Little Tern

Sterna albifrons

The Little Tern has a world-wide distribution; the Australian subspecies *sinensis* is also distributed throughout Asia. In south-eastern Australia Little and Fairy Terns overlap in range, sometimes breed in mixed colonies, may hybridize[420] and can be hard to differentiate in the field. The Historical Atlases contain breeding records of the Little Tern along the coasts of the Atherton and Eastern Qld Regions but none west of 135°E. Small outlying breeding populations are established from time to time, e.g. on The Coorong (36/139) and Price Saltfields (34/138).[420] In the Tas. Region birds nested regularly near St Helens (41/148) between 1968 and 1978 and one pair nested at Marion Bay (42/147) in 1976. In WA the Little Tern was not recorded south of Derby (17/123) until 1975 when it was collected at Port Hedland (20/118);[1638] it was seen as far south as Monkey Mia (25/113) in 1979. There is no evidence of breeding on the western coast and it is thought that many of the birds there are migrants from north-eastern Australia.[1638]

The Little Tern appears to be mainly a summer visitor to northern Australia, e.g. present near Darwin (12/130) Sept.-June[423,1169] and in Torres Strait in summer.[503] There is, however, a population in the Gulf of Carpentaria breeding in autumn (14/141, 15/141) and winter.[1169] Many of the birds in the north in summer may be migrants from populations breeding north of the equator, as may the occasional individuals in non-breeding plumage in the south in summer.[822,1709] One nestling banded in Japan was recovered at Kerema (8S/145E) in New Guinea, 4800 km distant.[54] In southern Australia the Little Tern is most abundant in summer, the breeding season[1292,1709] (Reporting Rates south of 26°S: summer 1.8%, winter 0.2%).

The Little Tern feeds mainly on fish taken in a dive from above the surface of the water. Nesting colonies, usually small and dispersed, are established on sand or shingle beaches. Encroaching vegetation, human activity, industrial and recreational, and predation threaten them and have led to declines in the breeding populations of NSW and Vic.[1897] In NSW the number of breeding birds fell from about 680 in the 1950s to 64 in 1978–79.[1287] The Vic. breeding population was estimated at fewer than 200 birds in 1982–83.[1795]

Field Atlas Records: 1103

1° blocks–with records: 88 (11%)

–with breeding records: 15 (17%)

Atlas No.: 117

Max. Rep. Rate: 36%
In block(s): 24/153

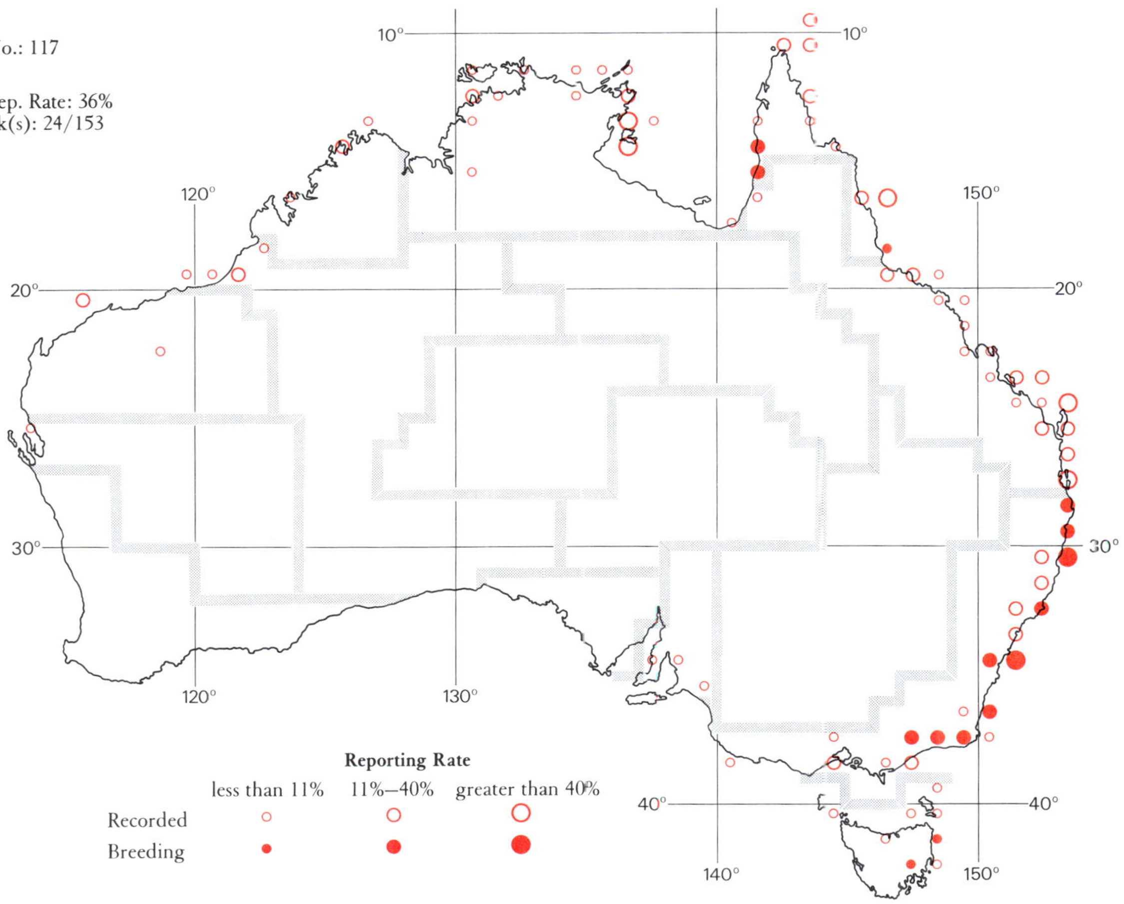

Fairy Tern

Sterna nereis

There are small populations of the Fairy Tern breeding on the coasts of New Zealand and New Caledonia but the largest are in Australia where there is one breeding subspecies *nereis*. Three specimens from Heron I. (23/151) in Dec.-Jan. are apparently of the New Caledonian subspecies *exsul*.[401] The northernmost records of *nereis* are from Botany Bay (34/151) in the east in Dec. 1962[864] and Admiralty Gulf (14/126) in the west in July 1973.

On the mainland the Fairy Tern rarely breeds east of Corner Inlet (38/146); small numbers, usually one or two pairs, have occasionally bred at the Gippsland Lakes (38/147) and Marlo (37/148).[1795] In WA colonies are reported north to the Lacepede Is (16/122).[1636] Fairy and Little Terns overlap in distribution, sometimes breed in loose mixed colonies, may hybridize[420] and are difficult to differentiate in the field. Mixed breeding colonies have been recorded at St Helens (41/148), Marion Bay (42/147), on The Coorong (36/139) and at Price Saltfields (34/138).

The Field Atlas suggests movement out of the Tas. and South-West Regions in winter (Reporting Rates: summer 5.5% and 4.3% respectively, winter 1.1% and 0.6%). It does not reflect suggested gatherings in winter on The Coorong nor in Spencer Gulf (34/136) and Gulf St Vincent (34/138).[1704] The longest movement reported is that of an immature from Beachport (37/140) to 38/144 near Werribee, 410 km.[1439]

The Fairy Tern lives along the coast, rarely out of sight of land and seldom inland. Fish, the main food, are taken by plunge-diving in shallow water. Most nesting colonies are on islands, the shores of lagoons or coastal beaches.[1636] Breeding colonies suffer from encroachment of vegetation, storm damage, human disturbance and expansion of Silver Gull colonies.[565,1018,1318,1407,1897]

The average size of ten colonies examined 1974–80 in Vic. was 53 birds[1018] and of 29 colonies examined 1955–82 in Tas. was 40 birds.[1533] The largest reported is of 800 birds on Flinders I. (40/148) in 1971.[1260] Probably fewer than 500 birds were present in Vic. in 1981–82.[1897]

Field Atlas Records: 1314

1° blocks–with records: 88 (11%)
–with breeding records: 32 (36%)

Atlas No.: 118

Max. Rep. Rate: 34%
In block(s): 34/136

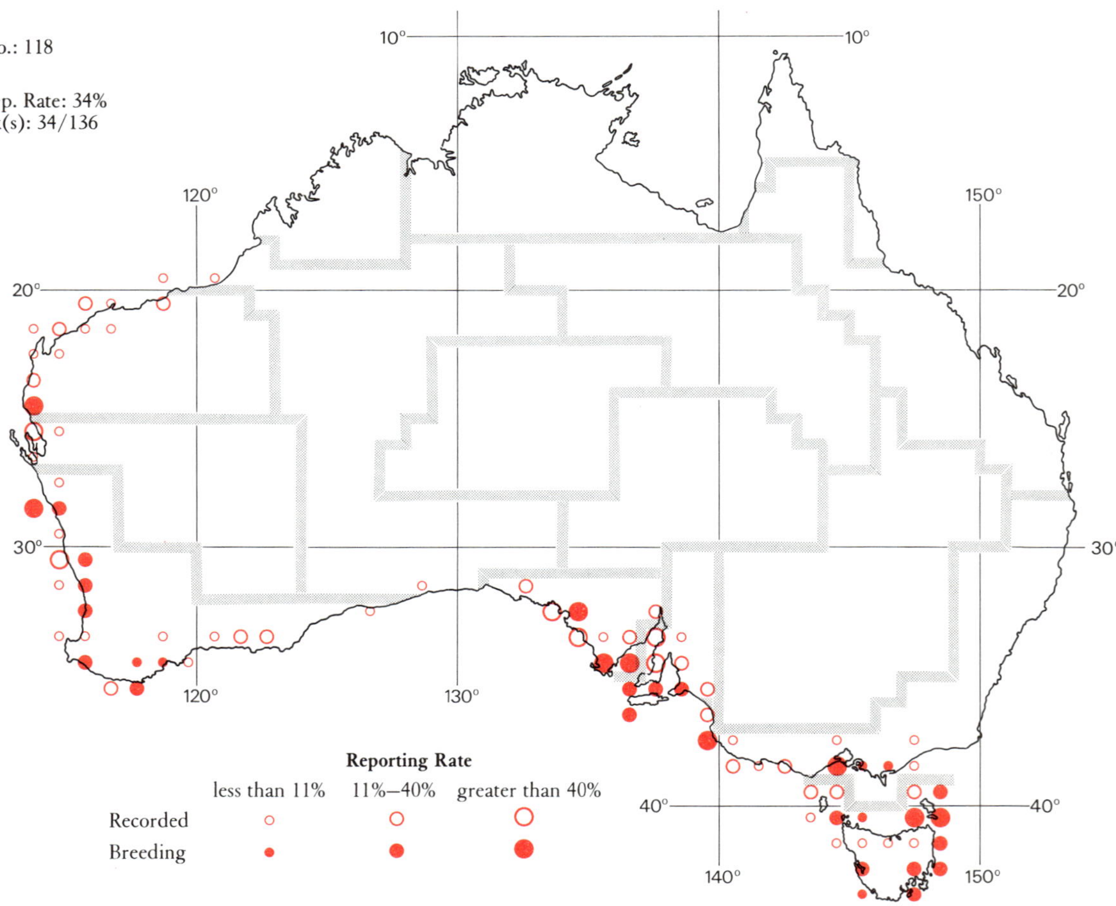

Crested Tern

Sterna bergii

The Crested Tern lives in south-eastern Asia, some Pacific islands, New Guinea and countries bordering the Indian Ocean including Australia. Opinions differ about the subspecific status of birds in Australia.[388,1250] Though the species breeds round the Top End coast, breeding was not reported there in the Field Atlas because few observers visited at the right time.

There is a seasonal fluctuation in the number of Crested Terns present in Darwin (12/130) with few recorded Apr.-July. The pattern may result from a local movement to breeding grounds or it may represent movement in and out of Australia. Some south-eastern Asian populations are known to undertake regular migrations.[427,1169] Banding studies have shown long movements for adults and young of the Crested Tern, the longest recorded being that of an immature from Beachport (37/140) to Hervey Bay (25/152), 1850 km.[35] Eastward movement from colonies in SA has been determined but so far none to the west.[1636]

The Crested Tern may be seen on any coast and even far out to sea. It fishes by plunge-diving. Nesting is colonial. Sites of colonies change and numbers vary year by year, e.g. on West I. (35/138) no terns nested 1964–69; about 1000 in 1970–71, 2800 in 1972–73 and 2000 in 1975–76.[1406] Small scattered colonies suffer predation from Silver Gulls, which not only take eggs but rob parents bringing fish to the chicks.[881] It is thought that the predatory activities of gulls caused a colony on One Tree I. (23/152) to decline from 181 nests on 8 Dec. 1973 to none on 25 Dec.[882]

Major breeding islands are: Lizard I. (14/145) 2000 birds in 1974;[494] Michaelmas Cay (16/145) 2500–4000 birds in 1982;[1826] Swain Reefs (21/151) 1960 birds in 1976;[1073] One Tree I. 1600 birds in 1980;[880] on nine NSW islands 26 000 birds (summary of the 1970s);[1029] Mud I. (38/144) 2000 birds in 1978;[768] Georges I. (40/148) 4000 birds in 1977;[1309] Penguin I. (37/140) 1100 birds in 1976;[1552] Fifth I. (36/139) 2000 birds in 1964;[1552] Lipson Cove I. (34/136) 2000 birds in 1980;[1552] Gull Rock (35/117) 2000 birds in 1977.[1009]

Field Atlas Records: 7253

1° blocks-with records: 219 (27%)

-with breeding records: 52 (24%)

Atlas No.: 115

Max. Rep. Rate: 72%
In block(s): 24/153

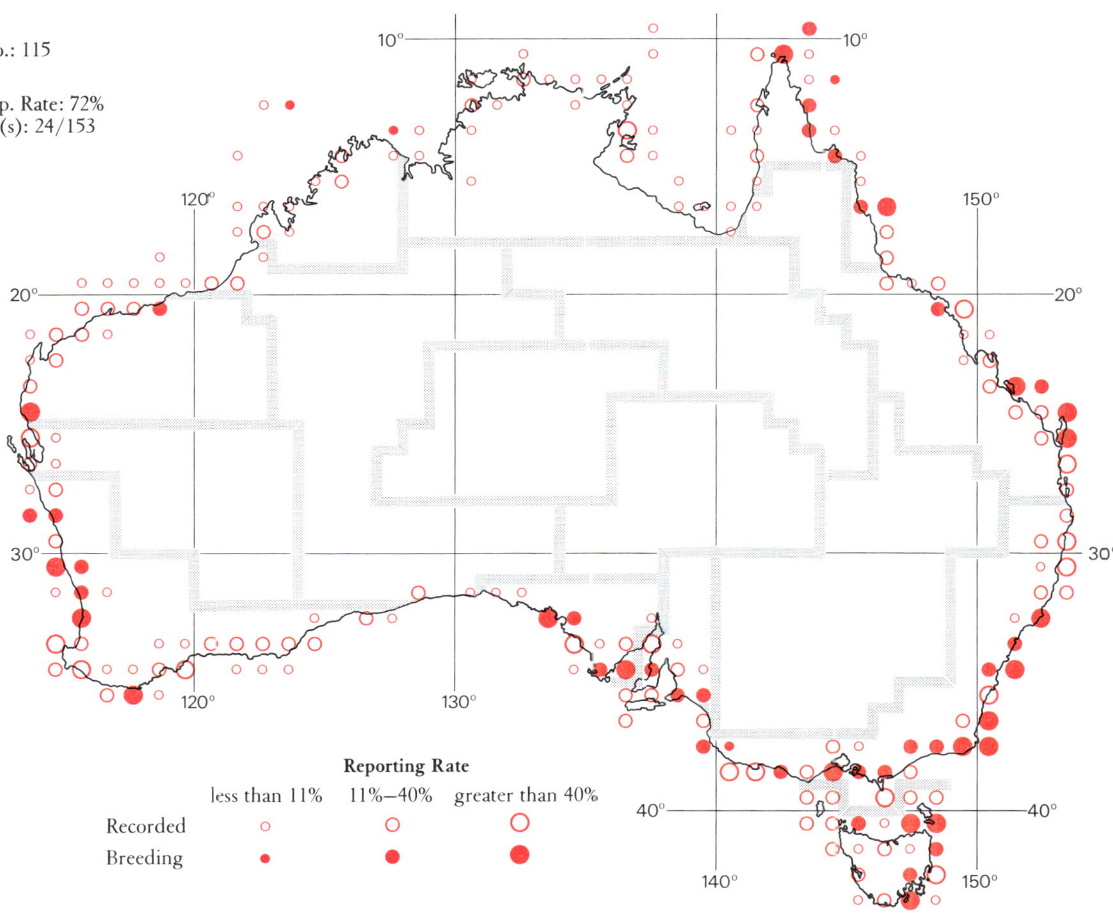

Lesser Crested Tern

Sterna bengalensis

The Lesser Crested Tern lives throughout the tropics of the Indian Ocean, the Indonesian and Malaysian archipelagos, New Guinea and Australia. The subspecies *bengalensis* breeds in Australia. There are no records in the Atlases farther south than those shown on the map. Colonies not recorded in the Field Atlas exist on Adele (15/123) and Bedout (19/119) Is.[1636] An unconfirmed report last century has the Lesser Crested Tern breeding on islands near Port Essington (11/132)[299] but otherwise there are no breeding records in the Atlases between 128°E and 142°E.

The Lesser Crested Tern wanders from its breeding colonies and a few pairs can be met with on any Australian tropical beach. Round Darwin (12/130) greatest numbers are present July-Sept. perhaps representing local movement to and from breeding grounds or, because it is known to migrate in parts of south-eastern Asia, possibly reflecting movement in and out of Australia.[427,1169] The species is a winter visitor to Torres Strait.[503]

The Lesser Crested Tern feeds on small fish caught in shallow water or the open sea. Silver Gulls attack the species both on the feeding ground and at the creches, converging mainly on terns carrying fish more than 6 cm long. Though at one foraging site the Lesser Crested Tern lost 12.5% of its fish, 65% of eggs laid resulted in chicks that fledged in 1973–74.[881,882] On the Qld coast colonies suffer severely from cyclones.[305]

The Lesser Crested Tern is usually in company with other species of tern. On One Tree I. (23/152) in 1973–74 there were 890 pairs of terns of five species, of which Lesser Crested made up 13.5%. In 1974–75 there were 1192 pairs of terns of the same five species; 3.4% were Lesser Crested.[881] Most accounts suggest that the 1974–75 figure reflects the usual abundance of the species.[288,1008] At One Tree I. the density of the Lesser Crested Tern nests in the colony was 2.76 nests/sq. m. On Michaelmas Cay (16/145) in 1982 there were 600–700 birds of which 80% were nesting.[1826] On Taylor Cay (17/146) a colony of 4000 birds nested in 1973.[1904]

Field Atlas Records: 394

1° blocks–with records: 71 (9%)
–with breeding records: 9 (13%)

Atlas No.: 116

Max. Rep. Rate: 48%
In block(s): 24/153

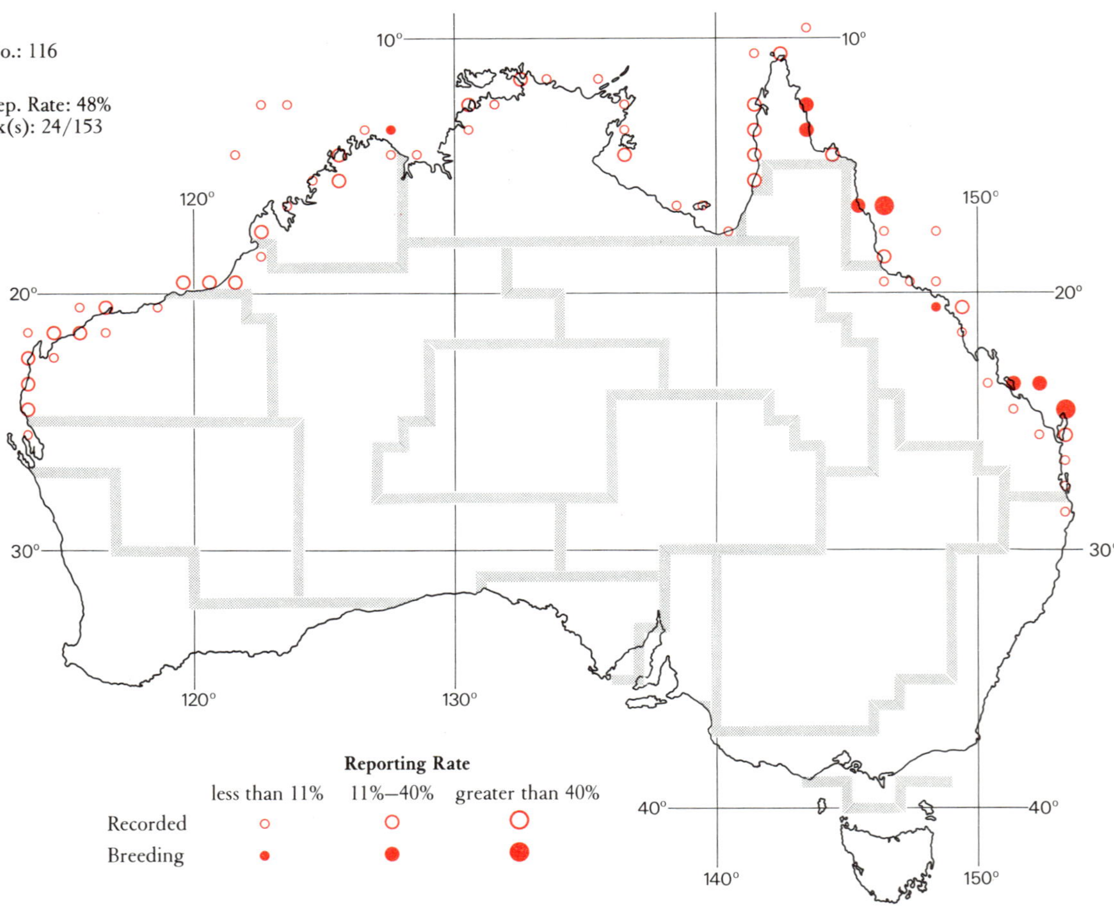

Common Noddy

Anous stolidus

The Common Noddy lives on tropical and sub-tropical islands round the world, with the subspecies *pileatus* breeding in Australia on many islands, including Lord Howe (31/159) and Norfolk (29/167) Is, islands in the Coral Sea and in Eastern Qld south to Lady Elliott (24/152). In the west, in addition to those on the map, it breeds on the Lacepede Is (16/122).[1636] Breeding has not been reported on Bedout I. (19/119) since 1901.[288]

Top End records of this species are rare. The first was made on Wessel I. (11/136) in Oct. 1972 and two birds were blown into Darwin (12/130) during storms in Feb. 1973.[867,1171] Field Atlas records in the Top End were made May, June, July and Nov. The Common Noddy can be confused in the field with the Black and Lesser Noddies where ranges overlap.

The Common Noddy is known to be capable of swift flight, some individuals returning to their nests in five days after release 1920 km away.[1823] At Norfolk I. birds seem to disperse to sea in the non-breeding season, Apr.-Aug.,[1580] and at the Houtman Albrolhos (28/113) they disappear May-July.[1636]

Food is mainly fish and cephalopods taken from the surface. Nesting Noddies are preyed upon by lizards, Silver Gulls, Eastern Reef Egrets[1459] and people.[1776] In the Atlantic distribution has been correlated with water temperature, nesting colonies being confined to the warmer seas.[1302]

Estimates of abundance must be treated with caution because colony occupation is erratic and breeding seasons are ill-defined. Some of the larger breeding colonies are: more than 10 000 birds on Bramble Cay (9/143);[503] more than 10 000 birds on Raine I. (11/144) in July 1980, many

breeding;[1000] 15 000 birds present on Michaelmas Cay (16/145) in Mar. 1978;[1826] and more than 2000 birds on Swain Reefs (21/151) in 1976.[1073] Early visitors to Houtman Albrolhos found the Common Noddy breeding in vast numbers.[1887] In 1889 they estimated 1.5 million birds on Rat I. (28/113) and Wooded I. (28/113).[1636] These colonies have disappeared to be replaced by one on Pelsart I. (28/114), where 40 000 bred in the 1940s;[1639] a very large colony was still present in 1981.[678]

Field Atlas Records: 132

1° blocks–with records: 43 (5%)

–with breeding records: 13 (30%)

Atlas No.: 122

Max. Rep. Rate: 28%
In block(s): 16/146

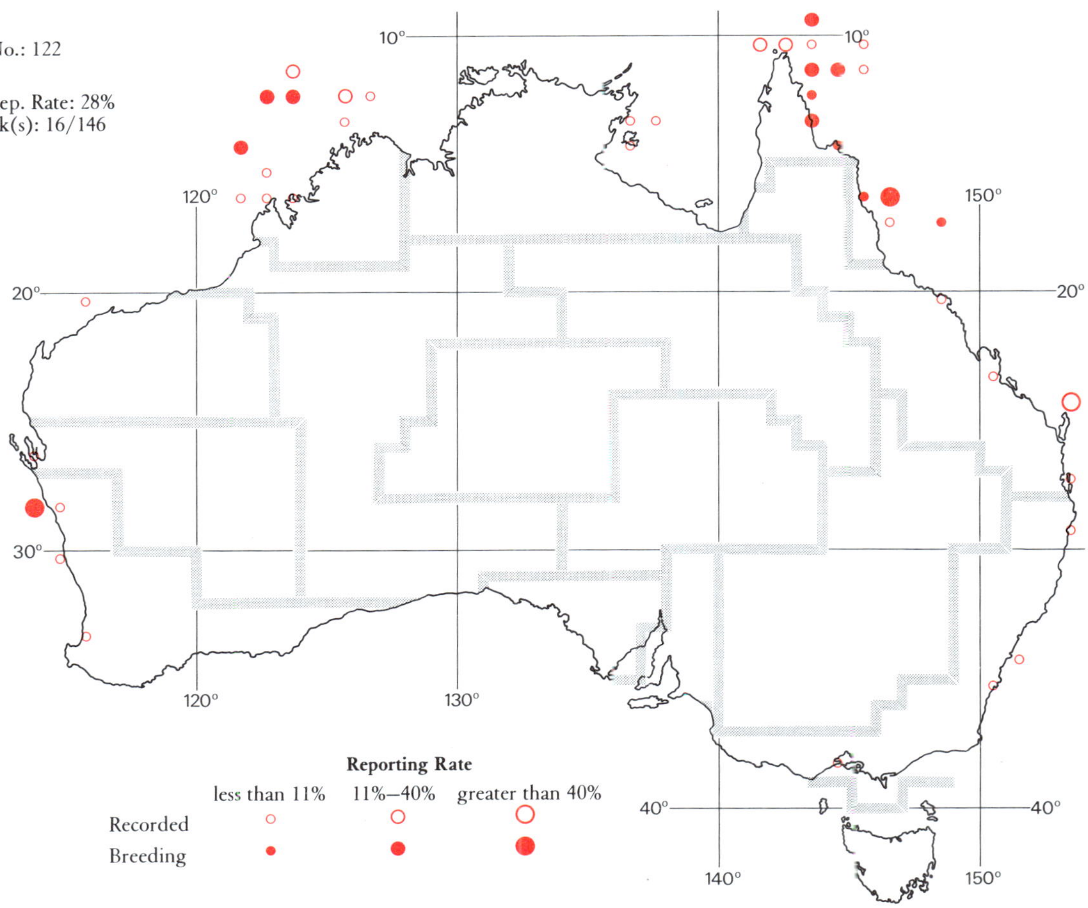

Lesser Noddy

Anous tenuirostris

On a world scale the Lesser Noddy has a restricted breeding range: one subspecies, *melanops*, breeds only on Houtman Albrolhos (28/113) and the other, *tenuirostris*, in the Seychelles (4S/55E), Madagascar (19S/46E) and the Mascarene Is (21S/57E).[1417] The species can be confused in the field with the Common Noddy where ranges overlap. The Lesser Noddy is seen about the Houtman Albrolhos in winter[1494] as well as in summer and the species appears to be present all year. Wrecks occur as far south as 34°S after storms.[1636] The longest movement recorded is that of a bird from Pelsart I. (28/113) recovered about 250 km south.

The Lesser Noddy feeds on fish taken near the surface. Colonies have changed in size and position since their discovery. In the 1830s the only colony was on Pelsart I. and the Lesser Noddy was reported to be 'even much more numerous than the Common Noddy'.[1887] Guano deposits on the island were worked from 1847. Between 1899 and 1907, when the Lesser Noddy ceased to breed on Pelsart I., a colony estimated to contain hundreds of thousands was established on Wooded I. (28/114).[688] By 1913 the Lesser Noddy in this colony was less numerous than either the Sooty Tern or Common Noddy and in 1936 the Lesser Noddy had again established a colony on Pelsart I.[1562] The colony on Wooded I. persisted with about 5000 birds in 1940. A colony of perhaps 10 000 on adjacent Morley I. had dispersed by 1945 but was re-established by 1970.[726] The colony on Wooded I. was reported flourishing in 1969[726] and that on Pelsart I. contained 'tens of thousands of nests' in 1981.[678]

Field Atlas Records: 10

1° blocks–with records: 2 (1%)
–with breeding records: 1 (50%)

Atlas No.: 123

Max. Rep. Rate: 18%
In block(s): 28/113

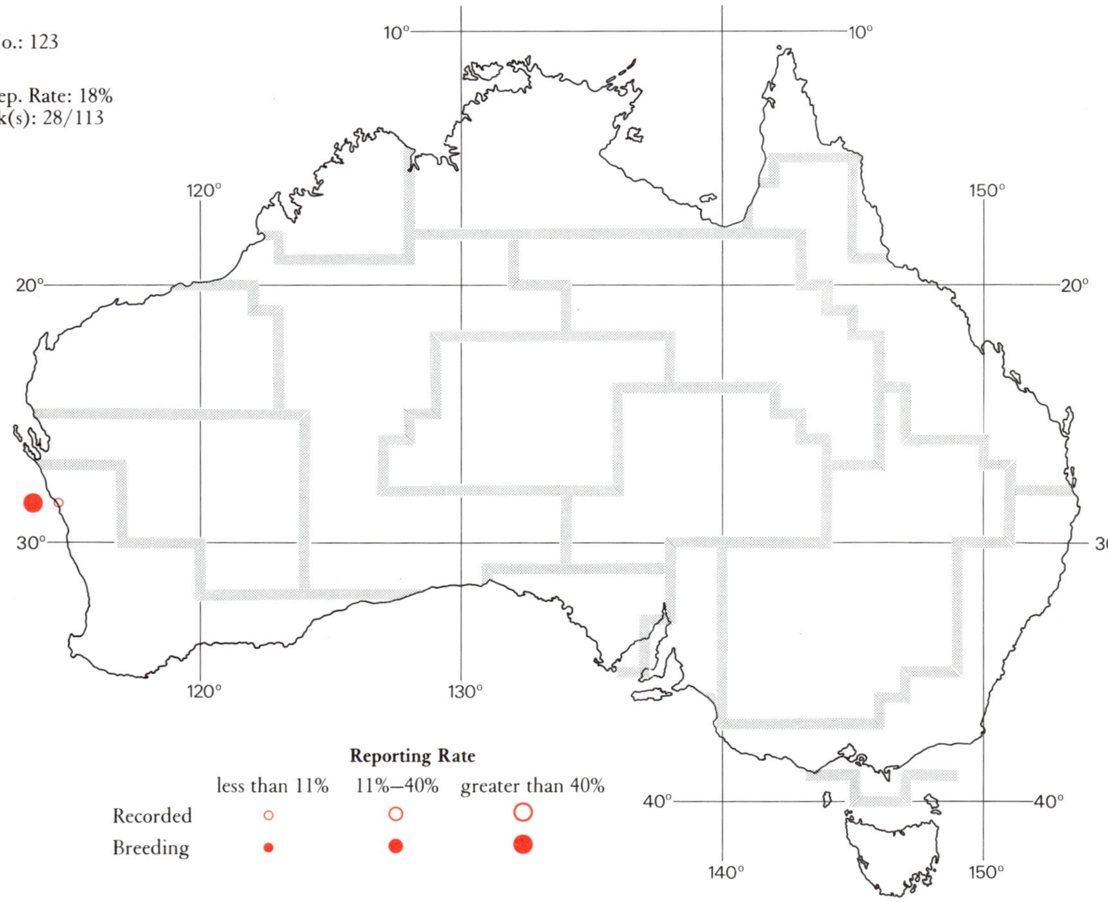

Black Noddy

Anous minutus

The Black Noddy lives in the tropical Pacific and Atlantic Oceans. In Australia the subspecies *minutus* breeds on islands in the Coral Sea and on Norfolk I. (29/167)[1731] as well as in the colonies shown in the map. Small numbers are recorded on Lord Howe I. (31/159) and possibly breed there.[676] An early report of breeding near Darnley I. (9/143) has not been confirmed[1145] but breeding on Quoin I. (12/143) has been. One bird was caught at sea (9/133) off the Top End in Nov. 1980, the first record west of Cape York Pen. (10/142). The Black Noddy can be confused in the field with the Common Noddy where ranges overlap.

The Black Noddy is present throughout the year at its breeding colonies,[996] though some individuals wander. Records south of 30°S have been made Dec.-May.[1292] The longest movement recorded is 2500 km.[85]

The Black Noddy rarely hunts alone and is usually in flocks of 200–500 birds, sometimes up to 3000. It feeds on small fish, euphausids and cephalopods and hunts near islands, hovering over a school of fish and plunging feet first on to its prey.[880] Eggs and chicks are preyed on by Silver Gulls and Eastern Reef Egrets. Some adults are eaten by the White-breasted Sea-Eagle.[880]

The colony on Heron I. (23/151) is expanding: 53 nests were occupied in 1910, increasing to 17 000 birds breeding in 1965,[1662] 26 000–50 000 in 1978[1357] and 60 000 in 1983.[880] On Masthead I. (23/151) there were an estimated 170 000 breeding birds in 1972.

Field Atlas Records: 84

1° blocks–with records: 17 (2%)
–with breeding records: 3 (18%)

Atlas No.: 124

Max. Rep. Rate: 34%
In block(s): 23/152

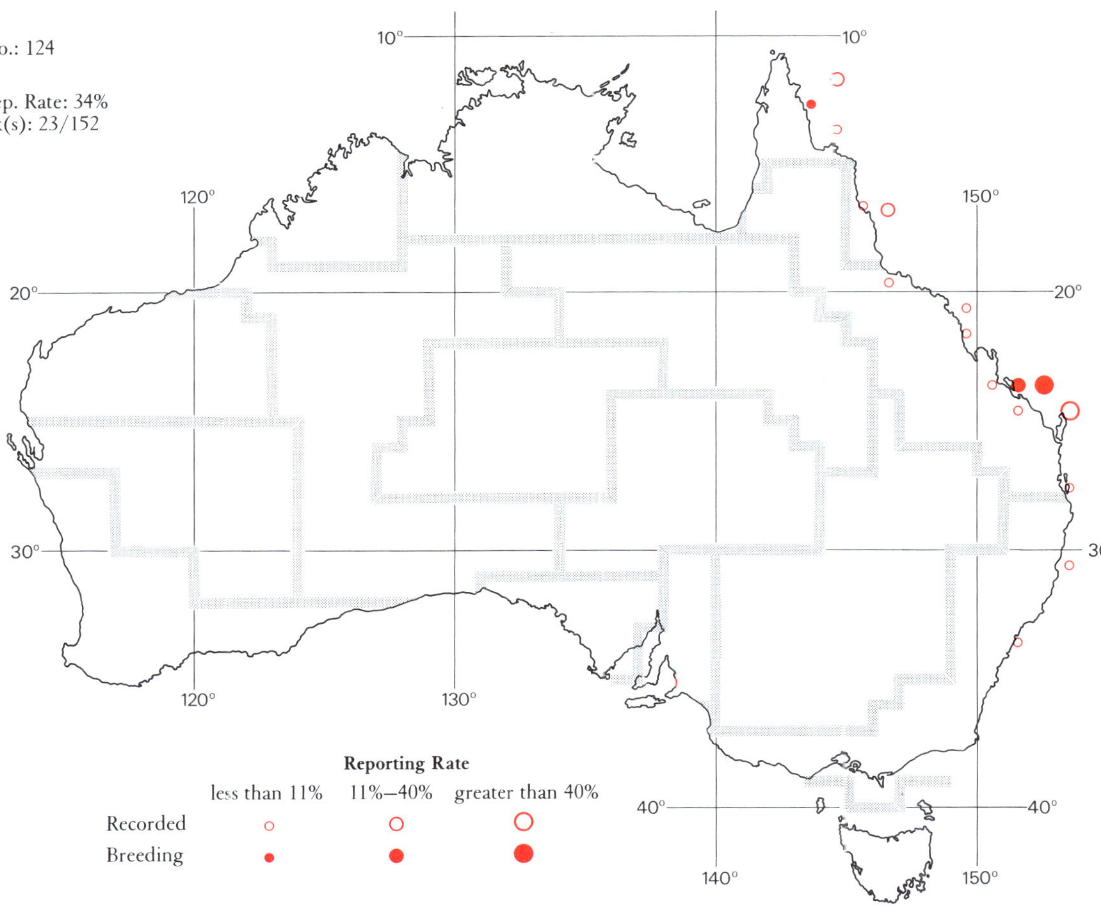

Banded Fruit-Dove

Ptilinopus cinctus

The Banded Fruit-Dove lives in parts of Indonesia and Australia, Australian birds being regarded as the subspecies *alligator*.[659] There are a few historical records to the west of the Arnhem Land escarpment, to which the bird is mostly confined,[659] including one at Acacia Gap (12/131) in 1975. Whether these represent a separate population or are vagrants is not known. A record from Melville Bay (12/136) in 1944[884] needs confirmation.

There is nothing to suggest that the species undertakes any substantial movement.

The Banded Fruit-Dove is rarely away from the escarpment of the sandstone plateaux of Arnhem Land. It inhabits patches of rainforest in gullies, living singly, in pairs or in small parties.[82,748] It feeds early in the morning and may spend much of the day in the dense foliage of trees where it is difficult to see. In the morning and evening, it will visit pools and watching at these is perhaps the best way of determining whether the bird is in an area.

The Banded Fruit-Dove feeds on the fruit of rainforest trees, specially figs, and has been observed taking cultivated mulberries.

At Mt Brockman (12/132) on the edge of Arnhem Land breeding pairs were found at a mean distance apart of 1.7 km.[655] There are no other measures of abundance.

Field Atlas Records: 67

1° blocks–with records: 6 (1%)
–with breeding records: 1 (17%)

Atlas No.: 024

Max. Rep. Rate: 14%
In block(s): 13/132

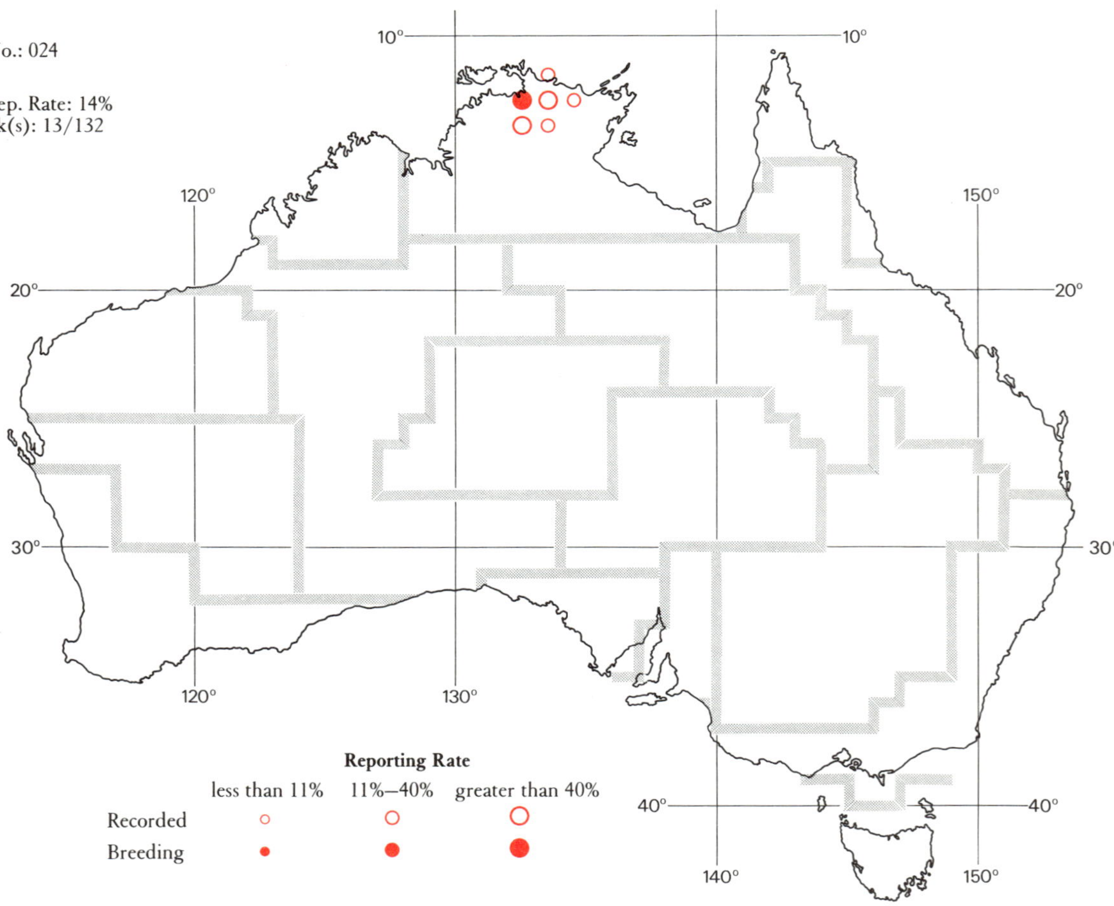

Superb Fruit-Dove

Ptilinopus superbus

The Superb Fruit-Dove lives in Indonesia and New Guinea as well as Australia. No subspecies are recognized.[659] Historical records show the species' distribution within the Cape York and Atherton Regions to be continuous. Breeding reports are almost confined to these two regions but the species regularly ranges farther south and there are records of stragglers as far away as Tas.[810] The Tas. record in the Field Atlas (41/147) was of an immature in May 1981.

Reporting Rates were 14.3% in summer and 4.3% in winter. Some of the difference may be attributed to inconspicuousness in the non-breeding season[659] but observations support the suggestion that part of the population may leave Australia in winter. The bird is reported from Booby I. (10/141) Sept.-Apr. during its passage through Torres Strait.[503,1813] Other observations indicate that it migrates at night and suggest that it winters in New Guinea[1041] though numbers of locally breeding birds there are so great that any influx from Australia has not been evident.[129] Despite the apparent regularity of its movements in Australia, southern sightings of the species are made at all times of the year.

The Superb Fruit-Dove lives mainly in rainforest but will feed in adjacent mangroves or eucalypt forest. In New Guinea it is fairly common in the lowlands and lives in secondary growth as well as rainforest.[667] In the Innisfail (17/146) district it is widespread at all altitudes.[698] At Mission Beach (17/146), the Superb Fruit-Dove feeds mainly on the fruit of tall trees (72% of its time) and on fruit of lower vegetation for the rest of its feeding time.[431] The species may have one of the shortest nestling periods of any pigeon,,perhaps no more than seven days.[429]

There are no measures of abundance other than an index used to estimate relative numbers of rainforest pigeons at Mission Beach.[431] On a series of four-hour walks each month for three years, the average number contacted was 15 during the month when abundance was greatest.

Field Atlas Records: 273

1° blocks-with records: 31 (4%)
-with breeding records: 6 (19%)

Atlas No.: 023

Max. Rep. Rate: 37%
In block(s): 12/143

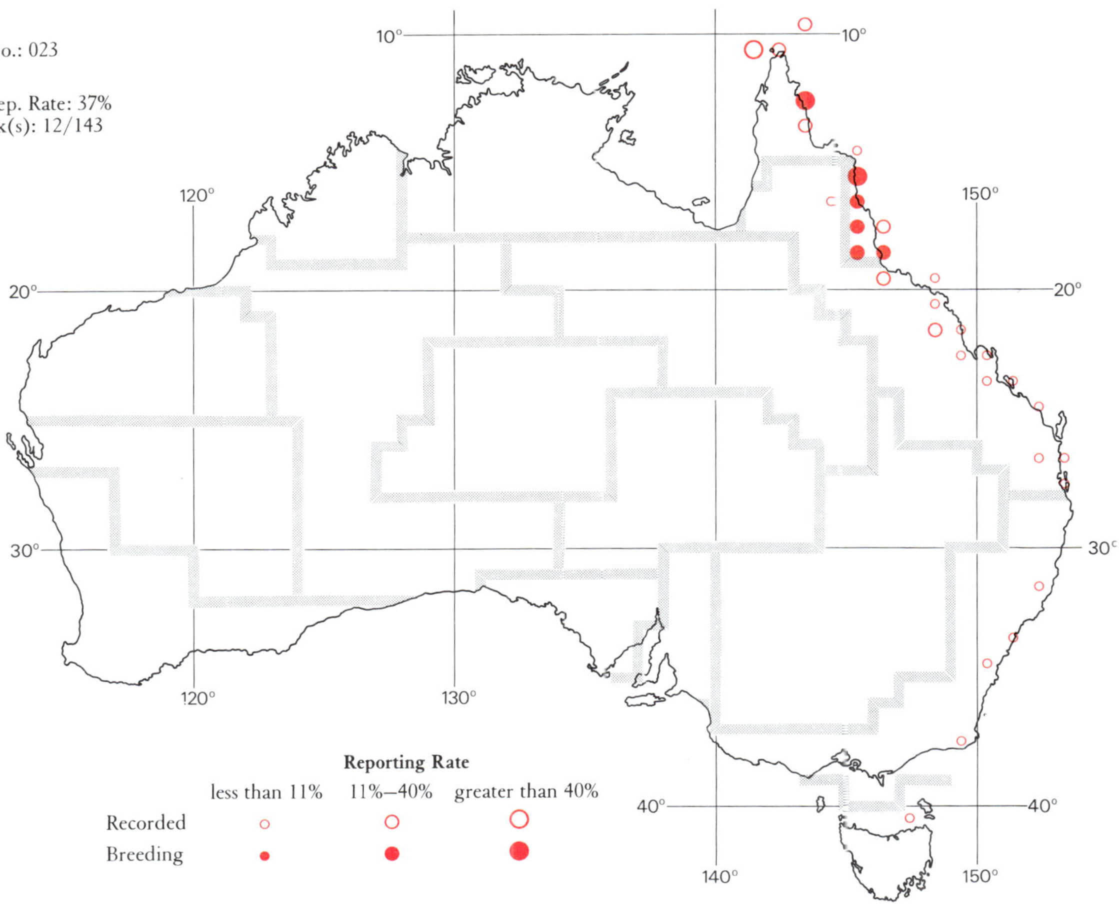

Rose-crowned Fruit-Dove

Ptilinopus regina

The Rose-crowned Fruit-Dove lives on islands adjacent to New Guinea and in Australia. Its status in New Guinea itself is uncertain. There are records from Daru I. (9S/143E) and Port Moresby (9S/147E)[659] but any small local population may be hidden among abundant populations of other species of *Ptilinopus*. In Australia the Rose-crowned Fruit-Dove has three isolated populations. Those of the Top End and Kimberley Regions *ewingii* were once separated as a different species but are now considered conspecific with the eastern population *regina*.[945] Field Atlas records came from throughout the range determined in the Historical Atlases.

The Field Atlas suggests that part of the eastern population may move seasonally: the Reporting Rate south of 27°S was lower in winter (0.08%) than in summer (0.36%). Observations at some sites also show a seasonal pattern. At Mission Beach (17/146) the Dove is a regular visitor June-Sept. and rare at other times.[431] In the Innisfail (17/146) district it is present all year[698] but near Atherton (17/145), at altitudes greater than 1000 m, it is present rarely and only Feb.-June.[231] In the Richmond R (29/153) district 20 or more pigeons annually visited a single fruiting tree Oct.-Mar.[645] The pattern of movements in the eastern population seems to take birds from the southern end of the range north in winter; others move seasonally in relation to availability of crops of fruit,[431] distance and direction varying from one year to the next.

On most islands in Torres Strait the species is present all year; it is a vagrant to those in the north-west[503] and perhaps also to New Guinea. There is no evidence of seasonal movements in the Top End and Kimberley populations.

The Rose-crowned Fruit-Dove generally lives in rainforest, sometimes in patches as small as 2 ha.[645] It feeds mainly in tall trees (92% of the time) on fruit 5.0–7.5 mm long and black or purple in colour (90% of fruits seen eaten).[431]

There are no measures of abundance other than an index used to estimate relative numbers of rainforest pigeons at Mission Beach.[431] On a series of four-hour walks each month for three years, the average number contacted was 5.6 during the month when abundance was greatest.

Field Atlas Records: 572

1° blocks–with records: 64 (8%)
–with breeding records: 4 (6%)

Atlas No.: 021

Max. Rep. Rate: 32%
In block(s): 20/149

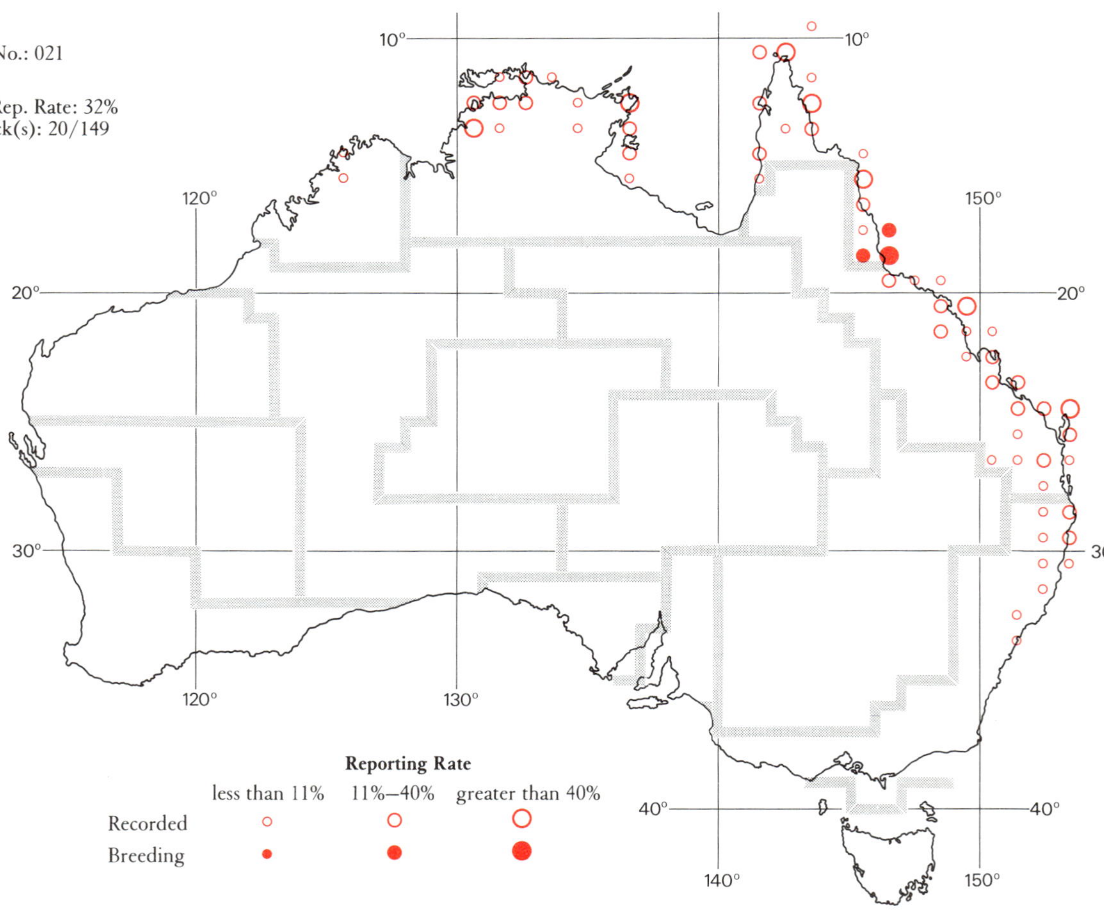

Wompoo Fruit-Dove

Ptilinopus magnificus

The Wompoo Fruit-Dove lives in New Guinea and adjacent islands and in three isolated populations in Australia: *assimilis* in the Cape York Region; *keri* in the Atherton Region; and *magnificus* in the Eastern Qld and South-East Regions.[659] The species formerly bred as far south as Port Kembla (34/150) but has not been reported there since the 1920s.[690]

There is known to be local movement by the bird in response to an abundance of food, often generated by the fruiting of one or a few trees.[431] There may also be movement out of the South-East Region in winter where its Reporting Rate was lower in winter (0.04%) than in summer (0.20%).

The Wompoo Fruit-Dove lives in large undisturbed patches of rainforest and feeds on fruit. At Mission Beach (17/146) it feeds mainly in large forest trees, specially laurel and olive-berry, and moves from species to species because they fruit at different times of the year.[431] There are several observations of the bird eating the fruit of at least one introduced species, the loquat.[345]

Reporting Rates in the Cape York, Atherton, Eastern Qld and South-East Regions were, respectively, 22.5%, 18.2%, 3.9% and 2.8%, which agree with statements that the northern populations are more abundant than the southern population.[645,657] There are no measures of abundance.

Field Atlas Records: 684

1° blocks–with records: 41 (5%)
–with breeding records: 9 (22%)

Atlas No.: 025

Max. Rep. Rate: 55%
In block(s): 12/143

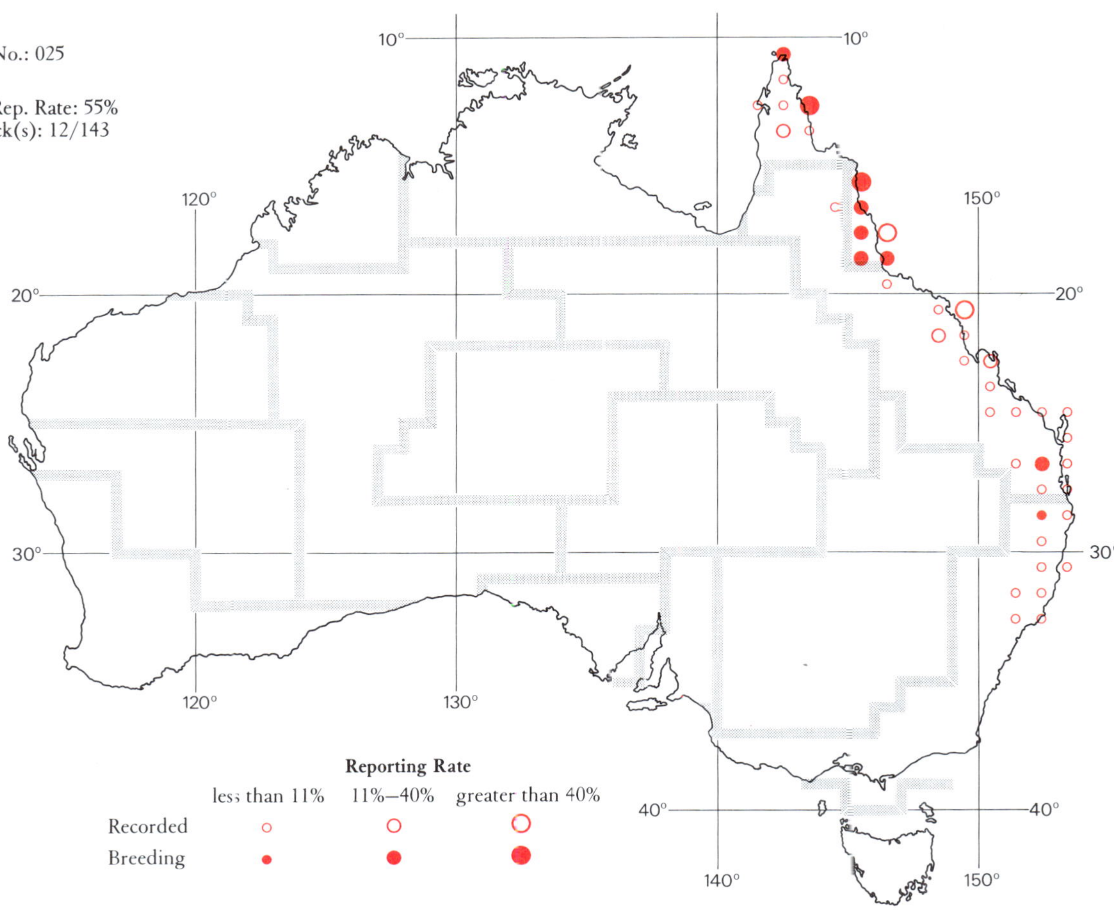

Torresian Imperial-Pigeon

Ducula spilorrhoa

The range of the Torresian Imperial-Pigeon includes Indonesia and New Guinea as well as Australia. Australian populations are all subspecies *melvillensis*[659] though Top End birds are isolated from those of the Kimberley Region. Records south of the Gulf of Carpentaria suggest that the populations in the Cape York and Top End Regions are in contact. Comparison of Field and Historical Atlases shows no substantial change in the bird's range but large colonies on the Barnard (17/146) and Family (18/146) Is have disappeared.[430]

The population in Qld and the Top End is largely migratory. Birds banded at Low I. (16/145) were recovered in western New Guinea (5S/138E) 1300 km distant[58] and on the Fly R (8S/141E) in New Guinea where large flocks, apparently of migrants from Australia, have been reported.[140] At Low I. the date of arrival differed by only 12 days between 1968 and 1972 with a median date of 17 Aug.;[430] the bird departed in Jan.-Feb. before the cyclone season was at its worst. In Darwin (12/130) it arrives in Aug. and leaves in Mar.-Apr.[423] The Kimberley population seems to remain in Australia throughout the year.[945] It ranges widely and is known to visit offshore islands in the winter.

In the Kimberley Region the Pigeon is confined to pockets of rainforest. Elsewhere it lives and breeds in eucalypt woodland as well. No evidence of colonial breeding, typical of birds on the eastern coast, has been found in the Kimberley Region.

The Torresian Imperial-Pigeon feeds on fruit.[430] Its arrival on Torres Strait islands and in the Cape York Region coincides with the ripening of the wongai.[503,1758] In Qld birds from island colonies fly each day to rainforest on the mainland to feed.[1147] Visitors and shooters on the islands, rather than destruction of the rainforest in which the bird feeds, appear to have caused the decline of the southern colonies.[430] Some southern colonies are now recovering.

Nesting density on the island colonies is high, e.g. up to 0.64 nests/sq. m on Low I.[430] An estimated 20 000–25 000 Pigeons nested on Low I. in 1971–72. Other large colonies are on Hope (15/145), Low Woody (15/145), Turtle (14/145), Howick (14/144), King (14/144), Clack (14/144) and Night (13/143) Is and on mangrove islands near Portland Roads (12/143).

Field Atlas Records: 711

1° blocks–with records: 61 (8%)
–with breeding records: 11 (18%)

Atlas No.: 026

Max. Rep. Rate: 51%
In block(s): 12/143

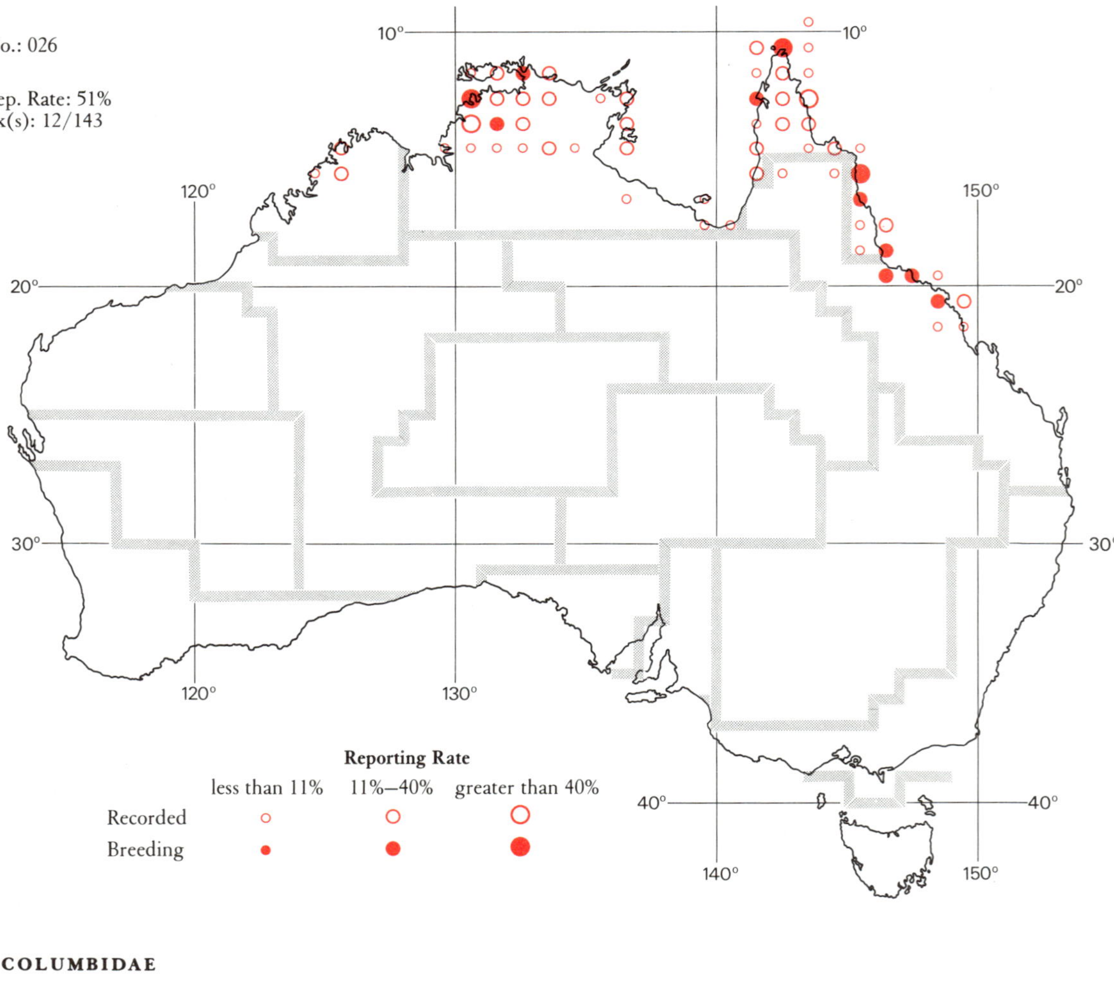

Topknot Pigeon

Lopholaimus antarcticus

The endemic Topknot Pigeon was recorded during the Field Atlas throughout its main historic range of 15–37°S. In the Cape York Region, where only one record was made during the Field Atlas, the Topknot Pigeon has always been rarely recorded[1147] while south of 37°S vagrants occasionally appear, such as the single bird seen in Tas. at Paradise Lagoon (43/145) in Nov. 1979. No subspecific variation is recognized[659] though the species' breeding range appears disjunct with breeding records in the Atlases confined to the Atherton Region and the eastern coast between 25–35°S. The Topknot Pigeon can be confused in the field with the White-headed.

The Topknot Pigeon is nomadic, moving about in search of fruit.[648] When a crop is found, many birds gather to eat until it is exhausted and then leave to seek another. In some places the Pigeon's movements may take on a seasonal regularity, e.g. round Innisfail (17/146) where it is less frequent May-June[698] and Moruya (35/150) where small parties fly north in Dec.[1192] The Field Atlas shows no large-scale seasonal change in Reporting Rate and vagrants south of the usual range arrive at all times of the year. Those reaching Tas. have been reported in Mar., May, Nov. and Dec.

The Topknot Pigeon inhabits rainforest and sometimes eucalypt forest. The nest is difficult to find as evidenced by the low proportion of breeding records in the Field Atlas (0.01% of sightings) compared with other pigeons with a similar distribution. The Topknot Pigeon feeds on the fruit of rainforest trees, including lemonwood.[431] In 1955, on the Richmond R (29/153), it was forced to seek another food supply when its natural one failed.[645] Many died but in that year it started to eat the fruit of the introduced camphor laurel and has done so ever since.[657] It is possible that the episode brought the camphor laurel into the suite of 'traditional' foods.

There have been several assertions that the species is heading for or close to extinction.[645,648,693] A recent assessment was more optimistic, but considered its numbers to be greatly reduced.[657]

There are no measures of abundance.

Field Atlas Records: 808

1° blocks–with records: 51 (6%)
–with breeding records: 5 (10%)

Atlas No.: 027

Max. Rep. Rate: 20%
In block(s): 28/153

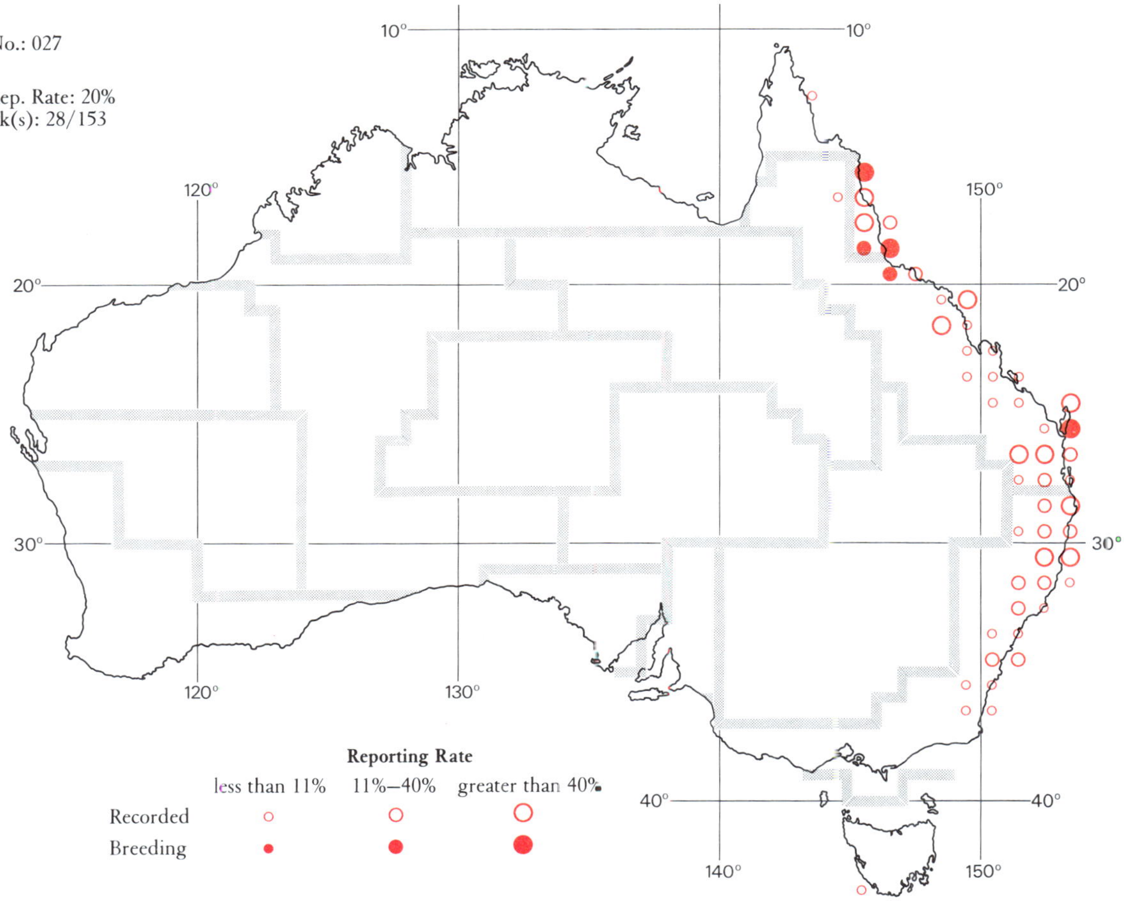

White-headed Pigeon

Columba leucomela

Records of the endemic White-headed Pigeon come from three discrete areas: the Atherton Region, the Eungella (21/148) district and south of Byfield (22/150) but no subspecies have been recognized.[659] In the Atlases breeding has been recorded only in the Atherton Region and in the South-East Region between 28–32°S. The species can be confused in the field with the Topknot Pigeon.

The White-headed Pigeon is said to be nomadic, coming and going erratically, sometimes present in great numbers, sometimes absent from a district season after season.[431,648,891] The report of a single bird at Tuross Falls (36/149) in Nov. 1980 is the only one in the Atlases of the Pigeon moving outside its usual southern limit of 35°S.

The White-headed Pigeon lives in rainforest and partly cleared land from sea level to 1500 m. It has been observed feeding in isolated trees and even nesting in them.[657] The Pigeon eats the fruit of laurels, figs and pittosporum among others. It will feed on the ground, sometimes in the stubble of coarse grains. More recently it has been recorded taking fruit of exotic trees, e.g. the camphor laurel and privet.[657,891,1531]

Last century the White-headed Pigeon was commonly seen by settlers clearing the rainforest. It became rare but recently has begun to increase again.[657] These changes in abundance are attributed to changes in agricultural practice. At first, settlers clear-felled the forest, leaving no trees to provide fruit for the Pigeons, so that their numbers declined. Subsequently shelter trees were planted, often allowing pockets of rainforest to grow again. As these trees began to fruit the Pigeons increased in numbers, feeding on the fruit of native and exotic species.

There are no measures of abundance other than an index used to estimate relative numbers of rainforest pigeons at Mission Beach (17/146).[431] On a series of four-hour walks each month for three years, the average number contacted was two during the month when abundance was greatest.

Field Atlas Records: 623

1° blocks–with records: 34 (4%)

–with breeding records: 7 (21%)

Atlas No.: 028

Max. Rep. Rate: 22%
In block(s): 15/145

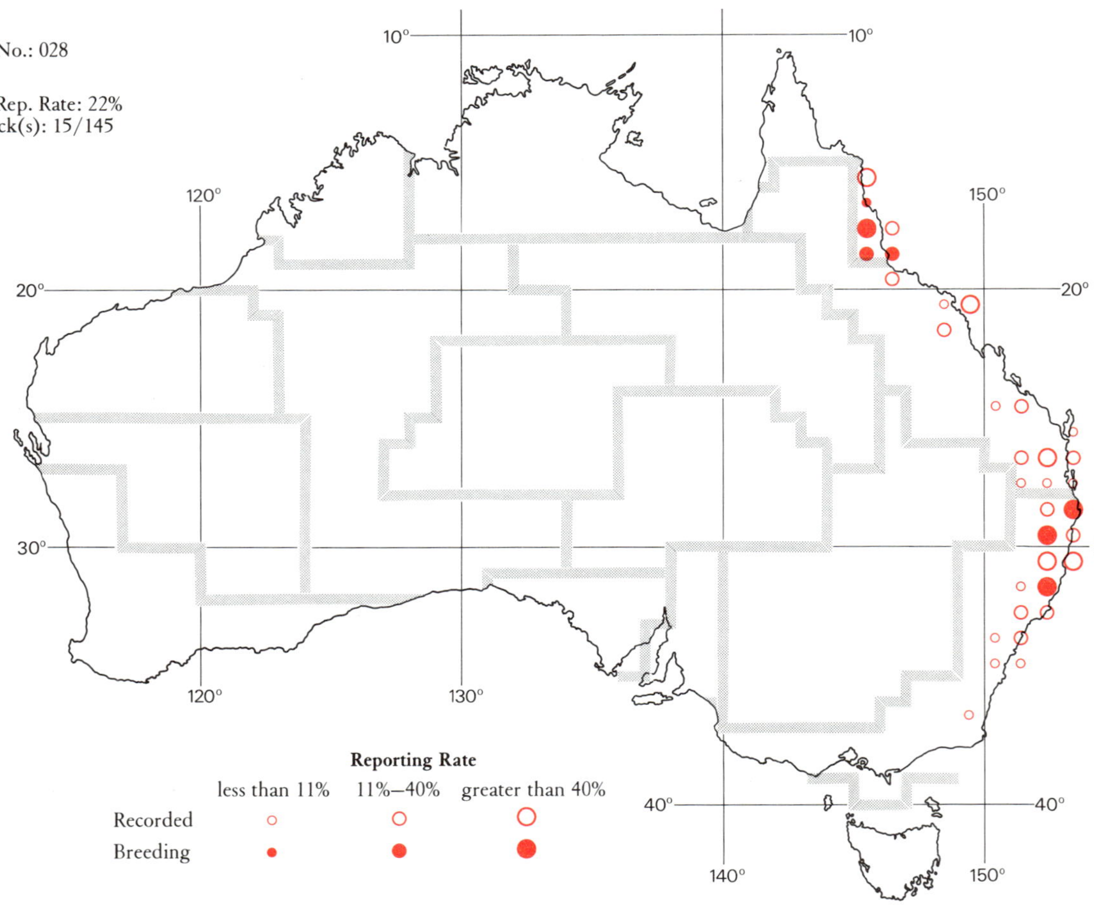

Feral Pigeon

Columba livia

The original range of the Feral Pigeon overseas is difficult to reconstruct because crossing with domestic stock has obscured the matter.[1085] Probably the species was distributed from the British Isles, northern Africa and southern Europe to India and Sri Lanka.[388] Few deliberate introductions to Australia are recorded. The earliest liberation was at Cape Liptrap (38/145) before 1873[97] but birds probably escaped from lofts before that. Additions to the feral population are made each year when birds, taking part in pigeon races, become lost; others stray from badly managed lofts or are abandoned to fend for themselves.[1084] Though most perish, some survive on farms and in townships and the Pigeon's range continues to expand.

Racing pigeons are able to travel great distances but the Field Atlas suggests no seasonal pattern of movement by the Feral Pigeon in Australia. The Feral Pigeon lives most commonly in urban environments but occasionally becomes truly wild, nesting in hollow trees away from settlements.[1085] Breeding is usually associated closely with towns and homesteads. The Pigeon feeds on seeds, grain, shoots, berries, worms, slugs and snails.[1085]

There are no measures of abundance.

Field Atlas Records: 10 422

1° blocks–with records: 304 (37%)

–with breeding records: 89 (29%)

Atlas No.: 957

Max. Rep. Rate: 53%
In block(s): 34/138

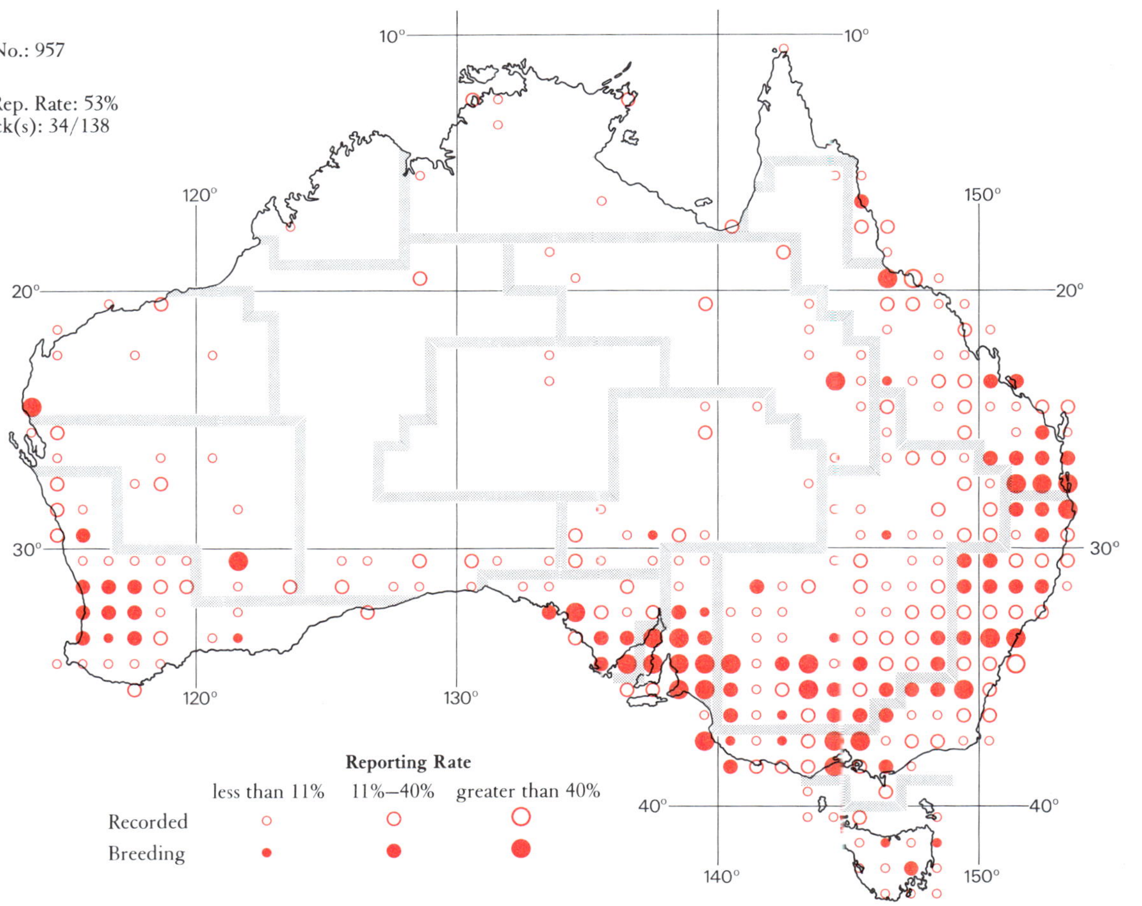

Spotted Turtle-Dove

Streptopelia chinensis

The Spotted Turtle-Dove is native to those parts of Asia south and east from southern China and India.[388] In Australia numerous escapes and releases, still continuing, have given rise to the feral population (p. 663). The first birds were brought, probably from China and Indo-Malaysia, to Melbourne (37/145) in the 1860s. They became established and spread slowly, supported by numerous other importations and releases.[669,1085,1607] Other known dates of early introductions are Adelaide (34/138) in 1881[192] and Perth (31/115) in 1898.[1607] By 1898 the bird was in Sydney (33/151), by 1913 in Brisbane (27/153)[1085] and by 1926 in Townsville (19/146).[1150] It also became established in Hobart (42/147) early this century but has not spread in the Tas. Region away from the urban areas of Hobart and Launceston (41/147).

The species was first seen round Cairns (16/145) and Innisfail (17/146) in 1945 and this was probably a separate importation. Unlike all the other Australian populations, which are hybrids of the subspecies *tigrina* and *chinensis*, the Cairns birds are almost pure *tigrina*, the Indo-Malaysian subspecies.[669]

The spread of the Spotted Turtle-Dove has been slow, involving advances and retreats rather than a steady expansion. In the South-West the bird was deliberately taken out into the country and released. Its continuing spread in other regions has been helped by additional releases and escapes up to the present day such as took place at Port Lincoln (34/135) in the early 1970s and Tamworth (31/150) in 1975 or 1976.

The Spotted Turtle-Dove is present throughout the year in Australian cities. It inhabits suburban parks and gardens, feeding in small flocks on open ground. It eats the small seeds of weeds as well as bread and other food provided by people.[670]

The absence of suitable food probably limits expansion of the Spotted Turtle-Dove away from urban areas.[1367] It has been suggested that increases in range in the South-West were facilitated by grain, specially wheat, spilled during transit.[1607] This may have been an important factor but the bird is still far from common even in farmland close to its points of introduction.

There are no measures of abundance.

Field Atlas Records: 7189

1° blocks–with records: 89 (11%)
–with breeding records: 31 (35%)

Atlas No.: 989

Max. Rep. Rate: 54%
In block(s): 33/151

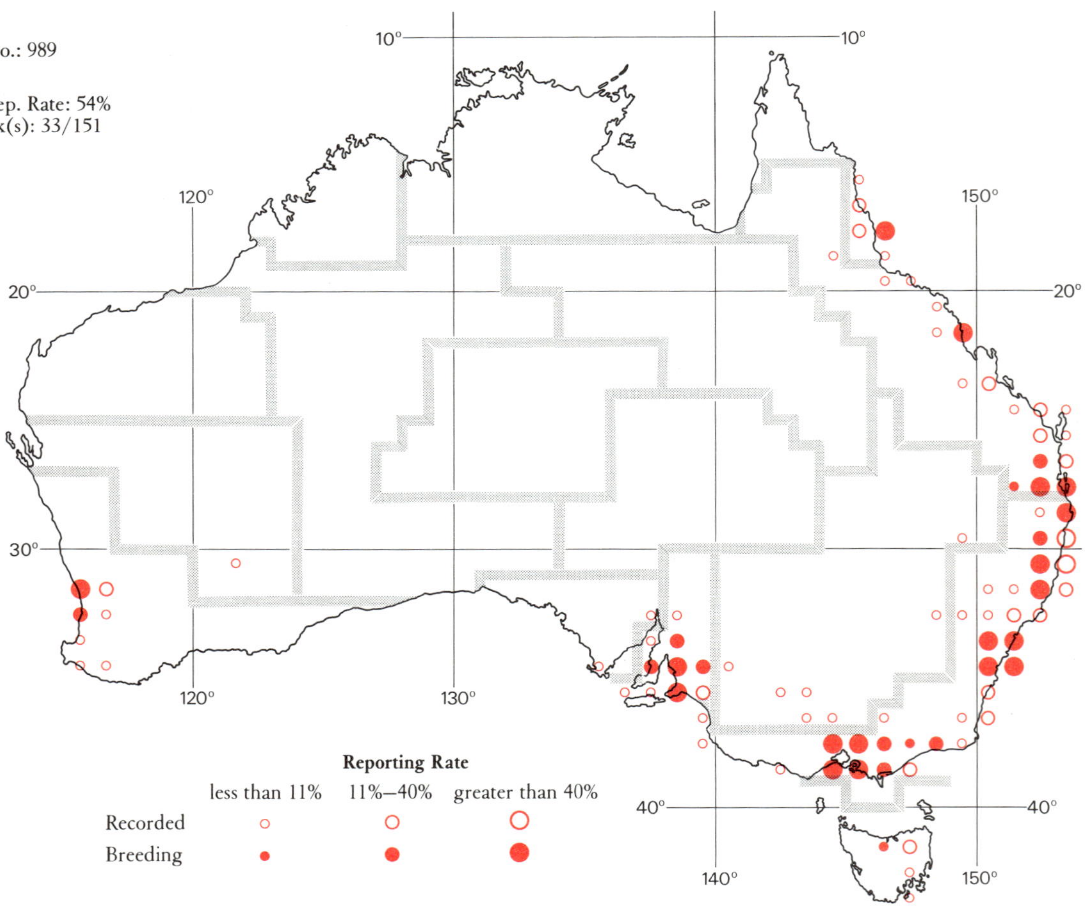

Laughing Turtle-Dove

Streptopelia senegalensis

The Laughing Turtle-Dove, a native of Africa and southern Asia, was first released in Australia at the Perth Zoological Gardens (31/115) in 1898.[1085,1607] It was possibly also released in Sydney (33/151) and Toowoomba (27/151) in the 1940s[1735] but has not been recorded in either place since. During the 1930s it expanded into the wheatbelt east of Perth (p. 664). It is now well established and apparently still expanding down the coast, west of the Darling Ra., and to the north. Whether the isolated population between Kalgoorlie (30/121) and Esperance (33/121) resulted from independent releases is not known. The species was certainly in Kalgoorlie by 1949 and Esperance by 1955.[1607]

Banding in Perth has shown adults to be mainly sedentary, occupying a territory throughout the year.[1720] Juveniles may disperse and colonize new areas.

The Laughing Turtle-Dove lives mainly round settlements and is rare in the jarrah forest and in the wet coastal parts of the South-West. It feeds on seeds and grain including crops of small grain. Its spread seems mainly to have been along road and, specially, railway reserves where spilt grain provides a ready source of food. Its distribution between Kalgoorlie and Esperance is an example of this. Field Atlas records were made only in or near four towns on the railway which links the two centres. The coastal expansion north of Perth (31/115) is not associated with railways. The Laughing Turtle-Dove may gradually spread through pastoral areas of the state where permanent water is available for stock. In the north it may compete with the Diamond and Peaceful Doves, with which it is spreading into contact.

There are no measures of abundance.

Field Atlas Records: 1421

1° blocks–with records: 34 (4%)
–with breeding records: 13 (38%)

Atlas No.: 988

Max. Rep. Rate: 49%
In block(s): 31/115

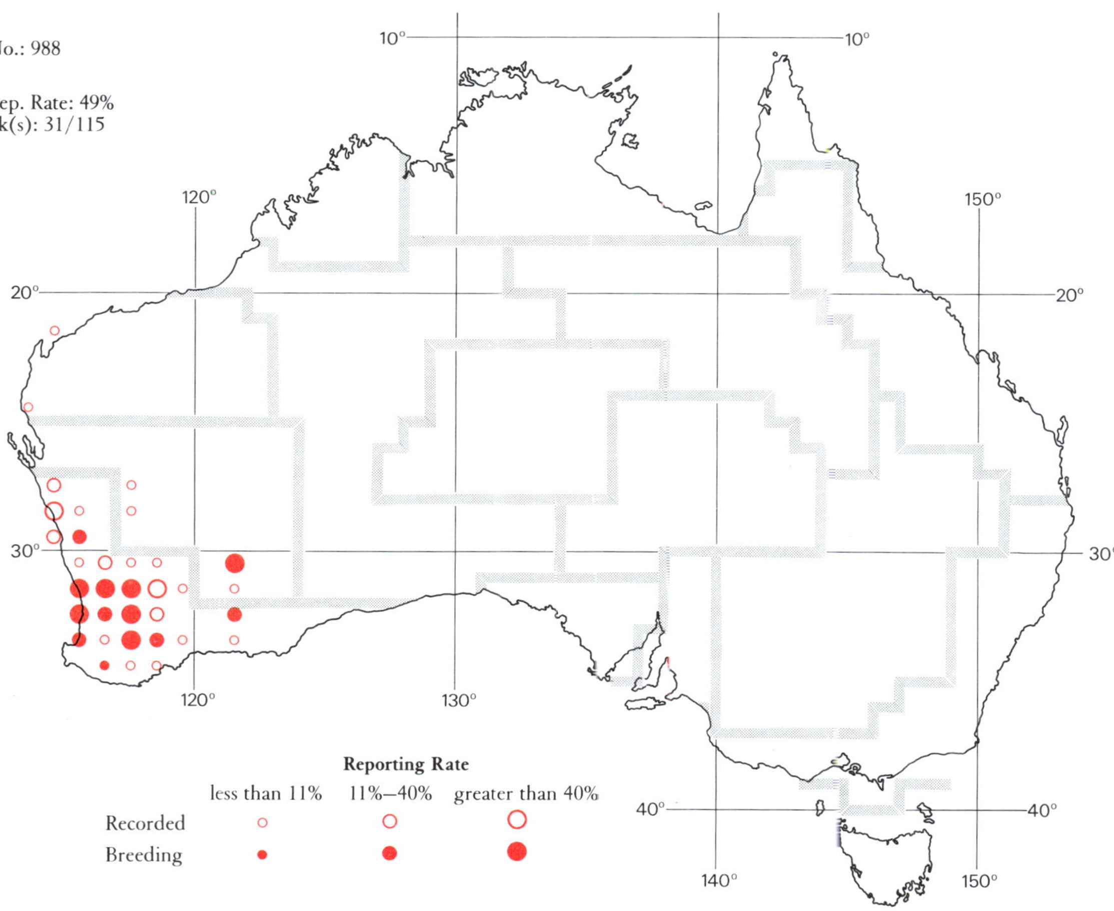

Brown Cuckoo-Dove

Macropygia amboinensis

The Brown Cuckoo-Dove ranges into south-eastern Asia. There is doubt about the specific and subspecific relationships of the different forms.[659] In Australia there were Field Atlas records from all 1° blocks where the species has been reported historically, except for the tip of Cape York (10/142). A specimen claimed to have been taken in the Rolling Downs Region has been shown to be wrongly located.[398]

The Brown Cuckoo-Dove is widely reported as present all year or, at most, locally nomadic.[231,398,645,659,698] At Mission Beach (17/146) numbers varied greatly in counts made over three years, reaching peaks in winter.[431] In the Rockhampton (23/150) district it is reported to be a summer migrant, present Sept.-Feb.[1089] The species visits Royal National Park near Sydney (33/151) when the austral mulberry is fruiting and is seldom present at other times.[1724] The Field Atlas suggests no large-scale seasonal shift of the population.

In the South-East Region the Brown Cuckoo-Dove inhabits pockets of rainforest almost entirely near the coast. To the north it lives at higher altitudes.[1709] Reporting Rates in the Atherton Region were 27% below 500 m and 49% above. The Cuckoo-Dove feeds mainly on seeds and fruit[431] gathered on the ground[129] and in large trees and shrubs both native and introduced. Most seeds eaten are about 8 mm long. Plants carrying seeds of this type grow in profusion at the edges of tracks and round secondary growth.

There are no measures of abundance other than an index used to estimate numbers of rainforest pigeons at Mission Beach.[431] On a series of four-hour walks each month for three years, the average number contacted was ten during the month when abundance was greatest.

Field Atlas Records: 2163

1° blocks–with records: 63 (8%)
–with breeding records: 15 (24%)

Atlas No.: 029

Max. Rep. Rate: 53%
In block(s): 12/143

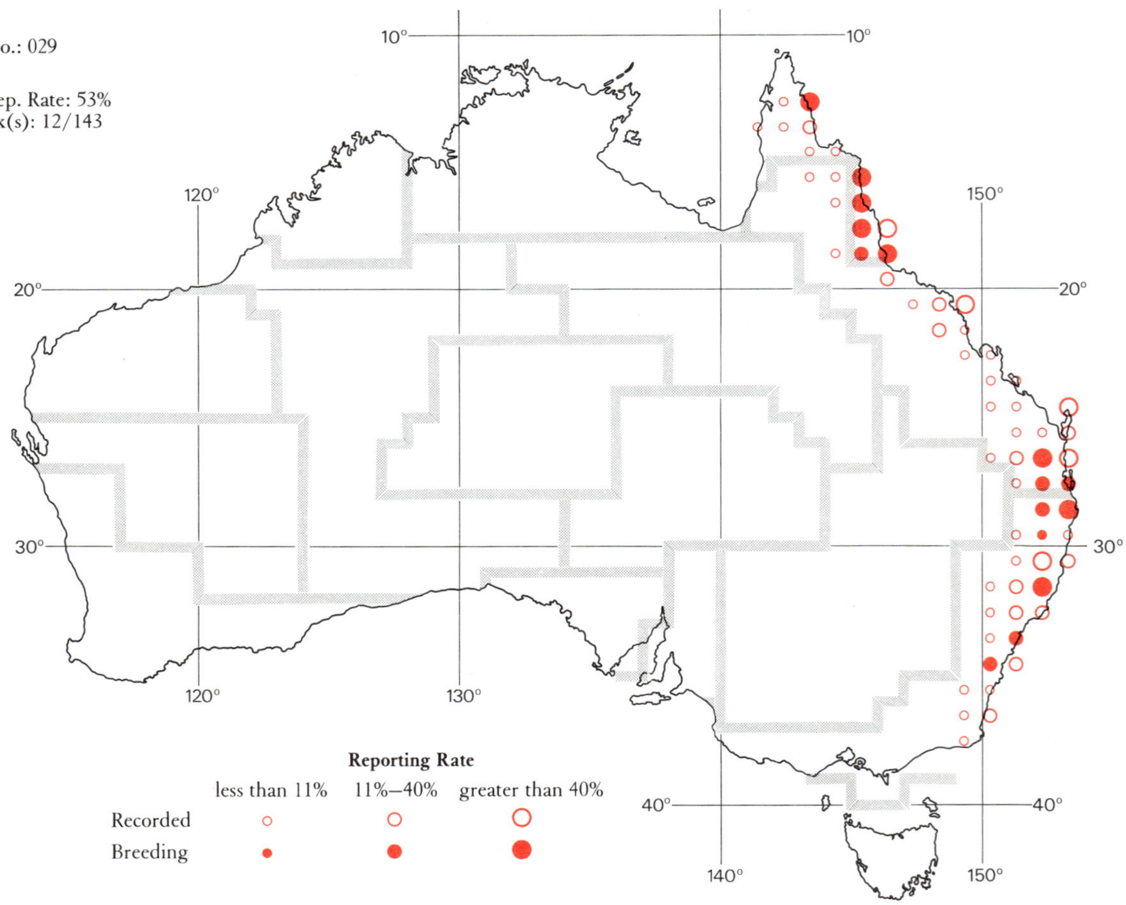

Peaceful Dove

Geopelia placida

The Peaceful Dove is widely distributed in south-eastern Asia, New Guinea and adjacent islands and Australia. In Australia, three subspecies are recognized: *placida* in tropical northern Australia; *tranquilla* to the south in eastern Australia, variation between these two being clinal;[748] and *clelandi* in the Pilbara Region.[388,659] The Peaceful Dove was introduced to Kangaroo I. (35/137) in 1937 and 1940[388] but was not recorded there during the Field Atlas. Probable aviary escapees seen at Kambalda (31/121) and Eyre Bird Observatory (32/126) are not shown on the map.

Round Melbourne (37/145) and in coastal districts to the east birds occasionally sighted appear to be wild, perhaps augmented by escapees such as those known to have been present at Healesville Fauna Park (37/145) in 1978. The species may be expanding its range in this area. It is also becoming more frequently reported in the Central Ra. Region. The earliest record in the Historical Atlas was of eggs taken at Simpson's Gap (23/133) in 1911. In the 1901–50 period the Dove was recorded from two 1° blocks in the region and from three in the 1951–76 period compared with 11 during the Field Atlas.

Generally the Peaceful Dove is present all year in a district. Winter flocking has been observed in the southern Murray-Darling Region in a locality where a small number of adults that were banded proved to be sedentary.[1106] In Darwin (12/130) there is some evidence of movement towards the coast in the dry winter season.[423] The longest movement recorded for a banded bird is of an adult from Kingston (34/140) to Tintinara (35/140), 193 km.[1437]

The Peaceful Dove lives in forest and woodland, often in verges adjacent to open grassland. In the southern Murray-Darling it is more frequent in woodland dominated by river red gum than in mallee.[833] In WA it is confined to the neighbourhood of large watercourses where summer rainfall is regular. It is always near a supply of water.[748] The Peaceful Dove feeds on seeds of native and introduced grasses.[672]

Density in eucalypt woodland at Armidale (30/151) was 0.08 birds/ha.[582] The number of pairs nesting in orchards at Mystic Park (35/143) varied from year to year, ranging from 1–4 birds/ha.[1106]

Field Atlas Records: 14 735

1° blocks–with records: 476 (59%)
–with breeding records: 108 (23%)

Atlas No.: 030

Max. Rep. Rate: 82%
In block(s): 16/130

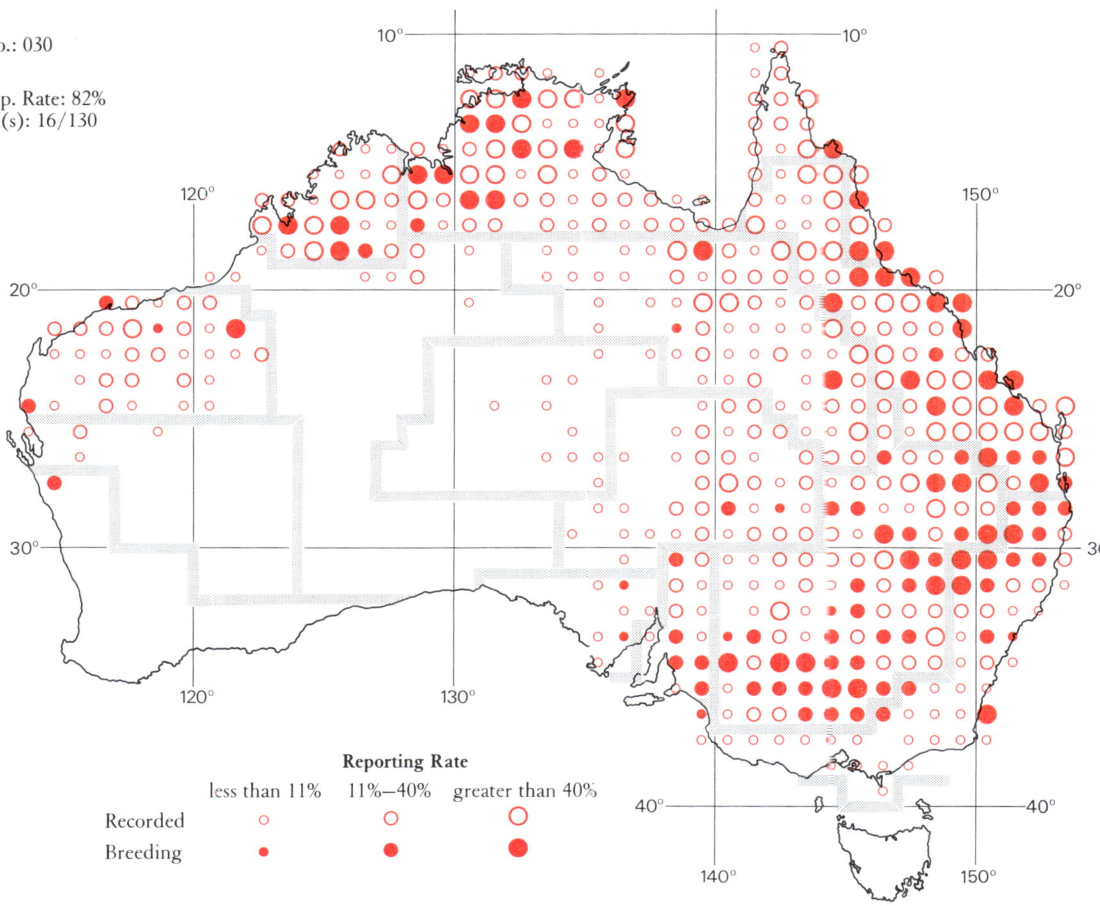

Diamond Dove

Geopelia cuneata

No subspecies of the endemic Diamond Dove are recognized.[659] The species is rarely recorded in the Cape York Region but has been found breeding near the Escape R (10/142).[1144] Groups of two or three birds that were seen at 34/138 near Nuriootpa in the summer of 1979–80 were almost certainly wild birds wandering south of their normal range. Records near Melbourne (37/145, 38/144) and one in Tas. (41/147), not shown on the map, are probably of aviary escapees.

At times Diamond Doves irrupt into country south of the usual range or, in eastern Australia, closer to the coast.[659,1638] The visitors breed for a season or so but do not remain, either returning to the dry country or dying out. Records throughout the SA Ra. Region in the 1901–50 period and south to 34°S in the South-West Region in the 1951–76 period represent such influxes.

There is, in addition, a seasonal pattern of movement suggested by the Field Atlas and by observation. In the southern Murray-Darling Region the Diamond Dove is a summer breeding migrant Nov.-Mar.[833] Reporting Rates in the region as a whole correspond. They were higher in summer (2.2%) than in winter (0.8%). The reverse was true near Darwin (12/130) where the Dove appears only in winter (June-Oct.).[423] Reporting Rates were 1% in summer and 3% in winter.

The Diamond Dove lives in woodland or open country with scattered trees or bushes.[748] Water must be available. The contents of guts sampled during 20 months at Alice Springs (23/133) included grass seeds (30% by volume) and the seeds of other herbs and shrubs.[672] During the Field Atlas, the Diamond Dove was recorded feeding on ants.

There are no measures of abundance.

Field Atlas Records: 2693

1° blocks-with records: 433 (53%)
-with breeding records: 33 (8%)

Atlas No.: 031

Max. Rep. Rate: 72%
In block(s): 22/118

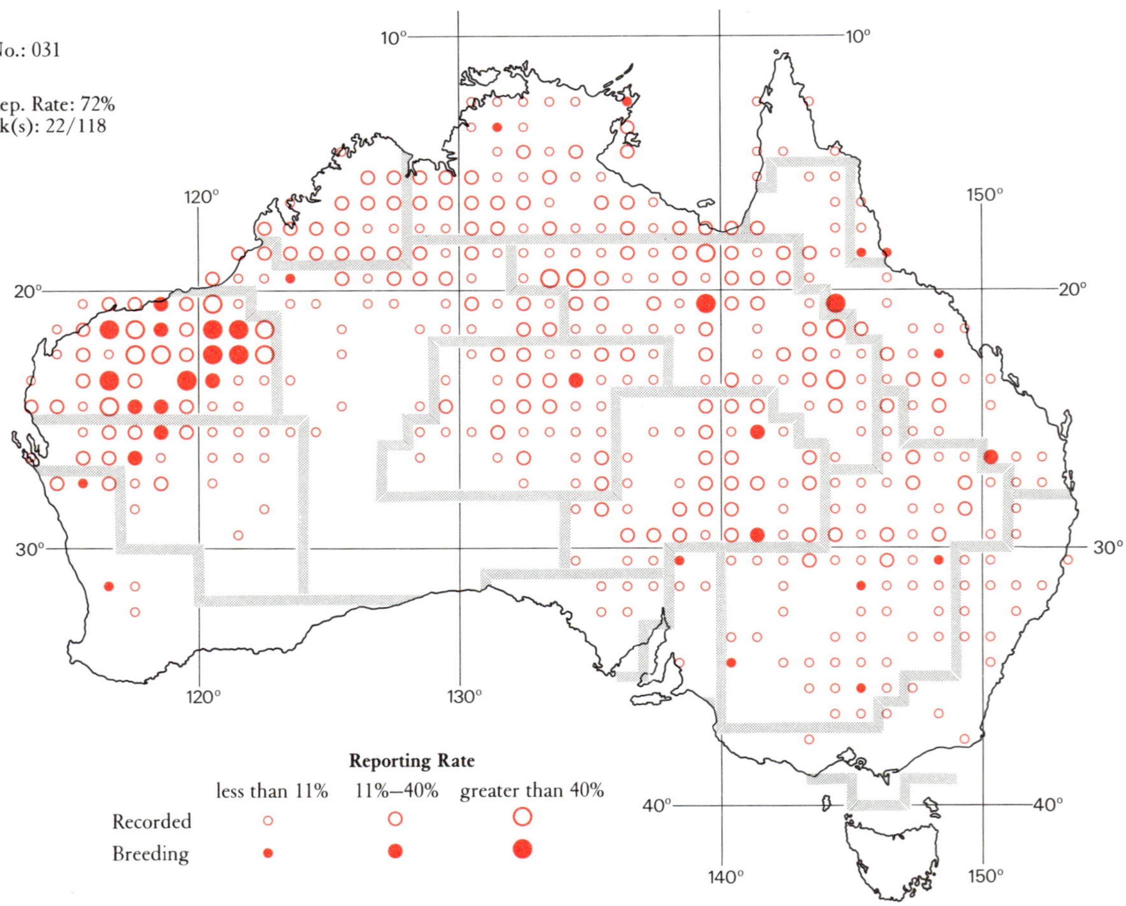

Bar-shouldered Dove

Geopelia humeralis

The Bar-shouldered Dove lives in southern New Guinea and Australia. Comparison with the historical maps suggests that none of the four proposed subspecies in Australia is isolated and there is doubt whether they should continue to be recognized.[388,659,748] Records near Melbourne (37/145) of presumed aviary escapees have been omitted.

The range of the Bar-shouldered Dove has expanded south and west in the Murray-Darling and South-East Regions. Before 1951 there are no records in the Historical Atlas south of 32°S, except for a pair at Penshurst (33/151) in 1928.[1726] Neither is there any record in NSW west of 148°E.

The Field Atlas shows no evidence of large-scale seasonal movements. Birds banded at Murwillumbah (28/153) have been recovered at varying intervals up to ten years later suggesting that part of the population at least is sedentary.[1432]

The Bar-shouldered Dove lives near water in forest, woodland and acacia scrub. In northern Australia it is often in mangroves and on islands. The Dove is commoner near the coast than inland and rarely more than 300 m from a watercourse lined with moderately dense vegetation.[1145,1602,1661] In southern parts of its range, high country in the Great Dividing Ra. is avoided. The Bar-shouldered Dove feeds on the seeds of grasses and other herbaceous plants. When clearing opened up the forest following settlement its distribution expanded and it began to feed round farms and in suburban gardens.[231,645,1099,1703] Counteracting this in some places is the effect of competition from the introduced Spotted Turtle-Dove. Near Lismore (28/153), the Spotted Turtle-Dove arrived in 1935, by 1952 was in equal numbers to the Bar-shouldered and by 1970 was much more common.[645,657]

There are no measures of abundance.

Field Atlas Records: 5844

1° blocks–with records: 244 (30%)
–with breeding records: 59 (24%)

Atlas No.: 032

Max. Rep. Rate: 73%
In block(s): 20/149

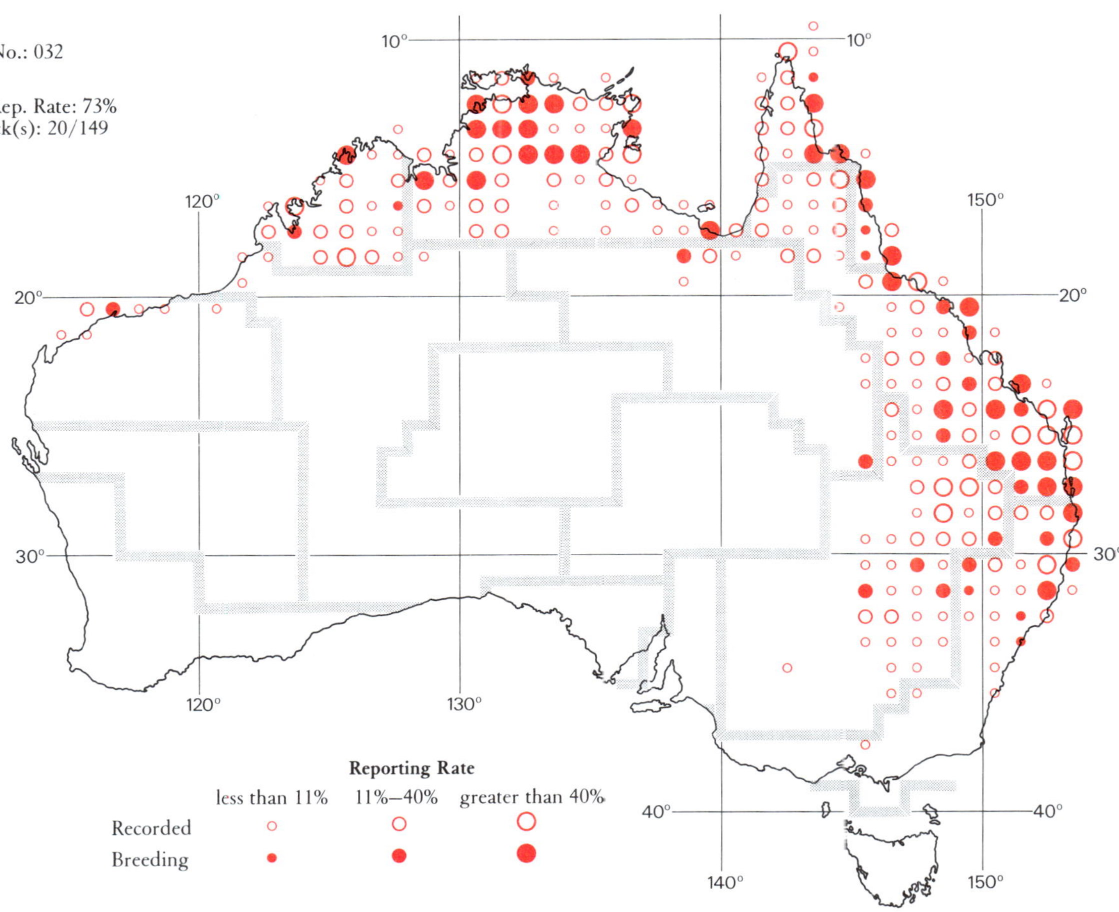

Emerald Dove

Chalcophaps indica

The Emerald Dove ranges through south-eastern Asia to New Guinea, the New Hebrides, New Caledonia and Australia. In Australia, Top End and Kimberley birds are separated as *longirostris* and eastern ones as *chrysochlora*.[659] The Emerald Dove has never been recorded from the western side of the Cape York Region. The Kimberley population may not be widely separated from that of the Top End[608] because the Atlases contain records from a continuous sequence of adjoining 1° blocks connecting the two. At the southern end of the species' range, vagrants have reached the Dandenongs (37/145) about 1882,[299] Heathmere (38/141) in 1931 and 1932[1053] and Genoa (37/149) in 1961.[136]

Many northern populations occupy the same small areas of rainforest throughout the year.[659] The Field Atlas suggests no large-scale seasonal movement but the Vic. sightings show that the species can disperse over long distances.

The Emerald Dove lives mainly at the edges of rainforest, including that along streams; it also occupies farmland with remnants of forest, banana plantations, areas overrun by lantana and even city parks.[645] It is usually in pairs but occasionally in small parties.[748] The Emerald Dove forages on the forest floor, sometimes, in the Top End, on burnt ground;[748] on islands in Torres Strait it also forages on beaches and among mangrove roots.[503] It eats seeds, including those of introduced weeds, fallen fruit and a few invertebrates.[659,1141] In Qld the bird frequents colonies of Metallic Starlings to eat seeds voided by the nesting birds.[1147]

In strips of rainforest passing through eucalypt woodland near Port Moresby (9S/147E) in New Guinea density was 0.01 birds/ha.[158]

Field Atlas Records: 1211

1° blocks–with records: 69 (8%)
–with breeding records: 8 (12%)

Atlas No.: 033

Max. Rep. Rate: 49%
In block(s): 12/143

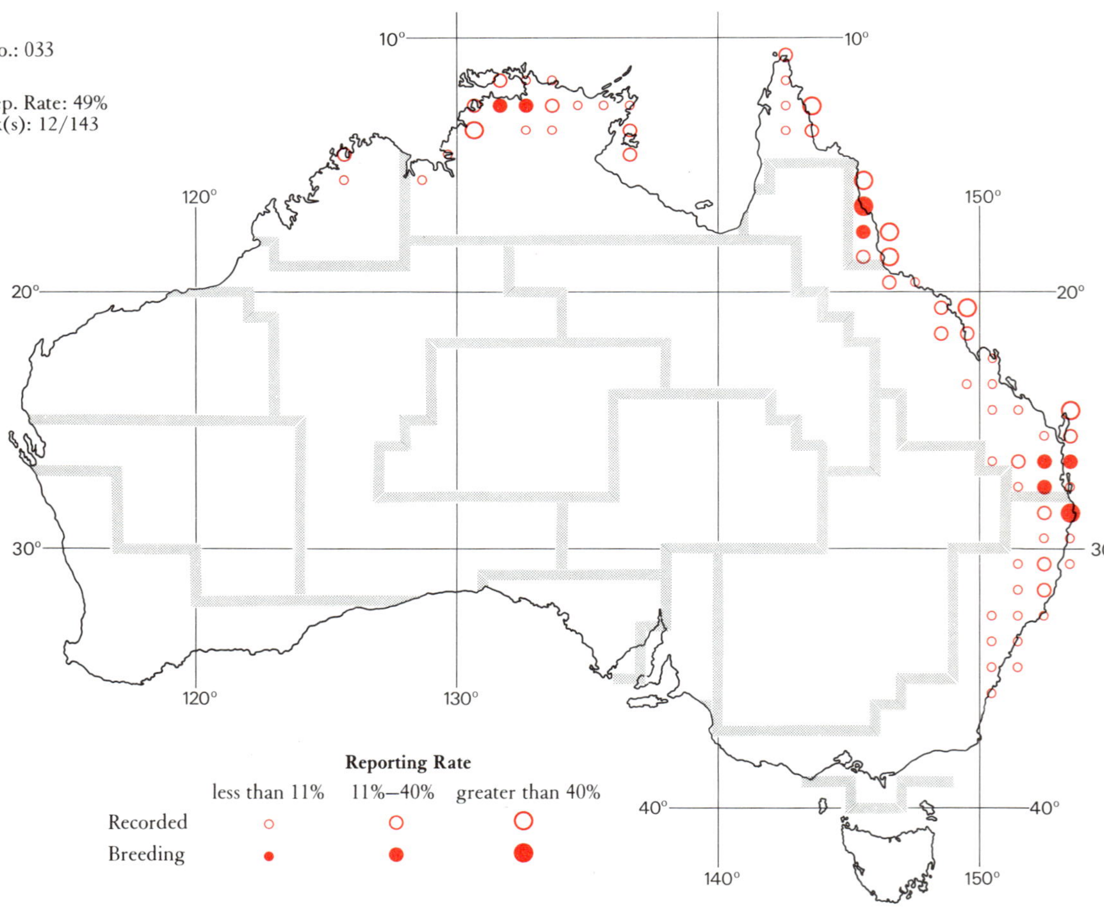

Common Bronzewing

Phaps chalcoptera

Three subspecies of the endemic Common Bronzewing are sometimes recognized but they are not isolated and morphological differentiation among them is clinal.[388,748] There is only one record from Cape York (10/142) in the Historical Atlas, made in 1912.[1145] The species was introduced to Kangaroo I. (35/137) in 1937[388] though it has been listed as a natural resident there.[1] The Common Bronzewing can be mistaken in the field for the Brush Bronzewing.

Seasonal movements are reported from the South-East Region where the bird appears to leave the mountains in winter[536] but the Field Atlas showed no seasonal change in Reporting Rate related to altitude. The longest movement recorded is of an adult from Eremophila Flat (34/140) to St Arnaud (36/143), 362 km.[32]

The Common Bronzewing lives in forest, woodland and mallee provided surface water is available for drinking. It avoids woodland dominated by river red gum in the southern Murray-Darling Region.[833] Though generally encountered singly or in pairs, flocks of more than 200 Common Bronzewings will gather to feed on abundant food.[833] The bird eats seeds, particularly those of acacias and other legumes. It also takes those of some introduced plants and cereals[657,1099] and will feed on spilt grain.[748,1638] It can safely metabolize the seeds of plants in the South-West Region that contain the poison sodium fluoroacetate.

In eucalypt forest at Jamieson (37/146) density was 0.2 birds/ha.[1469]

Field Atlas Records: 11 310

1° blocks–with records: 484 (60%)
–with breeding records: 116 (24%)

Atlas No.: 034

Max. Rep. Rate: 65%
In block(s): 27/147

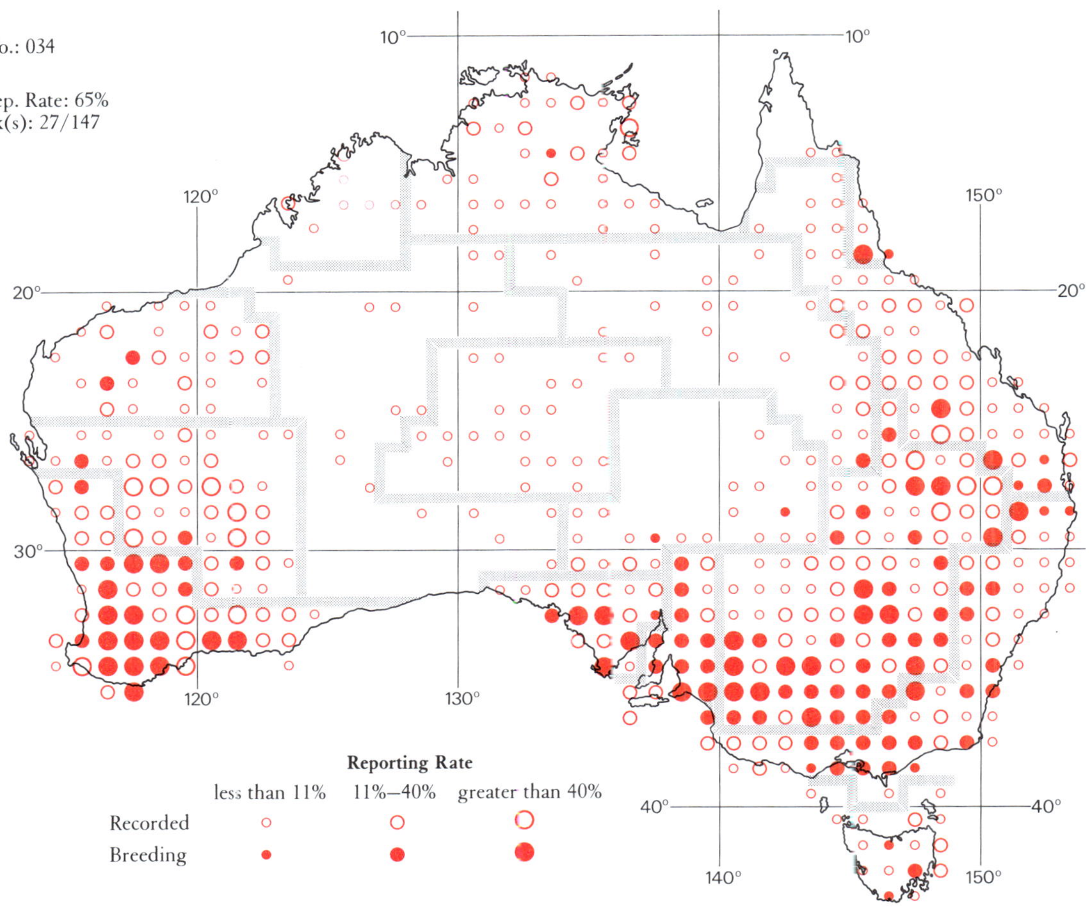

Brush Bronzewing

Phaps elegans

The gap separating the eastern and western populations of the endemic Brush Bronzewing is as great in the Historical Atlases as in the Field Atlas but no subspecies are recognized.[659] From Eyre Bird Observatory (32/126) the species' range is probably continuous west along the coast to Israelite Bay (33/123). It breeds on the Houtman Abrolhos (28/113), though not recorded there during the Field Atlas, and there were occasional records in the 1951-76 period on the coast south of the Houtman Abrolhos.

In the South-West Region the Brush Bronzewing is said to have declined in abundance[1638] and it disappeared from Rottnest I. (32/115) last century.[1706]

The Brush Bronzewing can be mistaken in the field for the Common.

Some movement of the Brush Bronzewing is suggested by the fact that near Canberra (35/149) most records are made Dec.-Feb.[1575] but the Field Atlas suggests no large-scale seasonal pattern of movement.

The Brush Bronzewing is generally a bird of forest and heath, often on islands. In Qld it is restricted to the coast. In the Snowy Mts (36/148) it lives mainly in forest growing in gullies with thick undergrowth[1575] and in similar country in the Mt Lofty Ra. (34/138). In Tas. its distribution is complementary to that of the Common Bronzewing. The Common lives in dry forest while the Brush occupies wet forest, is dominant in heath near the coast in western Tas. and is almost exclusive on islands in Bass Strait.[1314] On the mainland, in both east and west, the Brush Bronzewing is sparsely distributed in mallee. It feeds on seeds, berries and cultivated grain[1049] and requires water to drink. Clearing has been blamed for its decline in the South-West Region.

There are no measures of abundance.

Field Atlas Records: 1942

1° blocks–with records: 116 (14%)
–with breeding records: 27 (23%)

Atlas No.: 035

Max. Rep. Rate: 32%
In block(s): 39/144

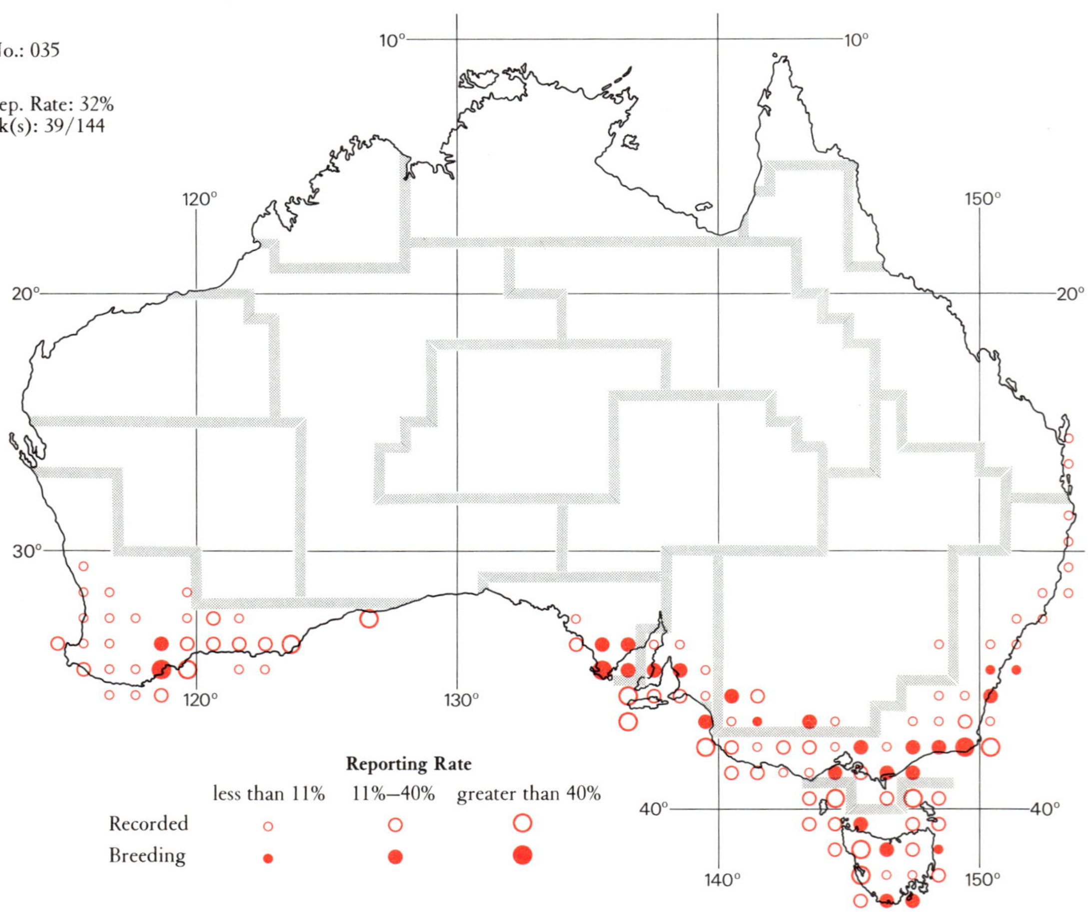

Flock Bronzewing

Phaps histrionica

No subspecific variation of the endemic Flock Bronzewing is recognized.[659] At times its range has extended beyond that shown on the map to include Eastern Qld north to the Mitchell R (15°S) and east to 25/150 on the Dawson R; the South-East Region north and west of the Mooki R (31/150); the Murray-Darling Region south to 34°S; and the Central Ra. and Pilbara Regions. It seems to have almost disappeared from the Eastern Qld, South-East and Murray-Darling Regions from which it has been reported on only four occasions since 1950 in the Historical Atlas. Elsewhere its range has expanded and contracted at different times being greatest last century and in the 1951–76 period.

The unpredictable appearance of large numbers of Flock Pigeons in widely separated localities[602,1153,1343] has led to the assumption that the bird is highly nomadic. This may not always be so. Resident observers have noted the birds building up in numbers slowly during a prolonged breeding period.[168,1151] The large flocks could be birds bred locally during a lengthy period of favourable conditions.

The Flock Bronzewing inhabits tussock grassland. When alarmed it squats on the ground until flushed at close range. It is mostly observed in flocks round waterholes, wheeling in tight formation, dropping briefly to drink and then departing. Away from water, it lives in pairs or small flocks. The Flock Bronzewing has always been said to feed on grass seeds,[1153] but analysis of the contents of two guts showed that seeds of a variety of herbs were eaten.[672,1218] Seeds of herbs, specially legumes, would provide a diet of higher quality than grass seed alone.

Reports last century speak of Flock Bronzewings in 'countless multitudes', 'immense numbers' and 'countless myriads'.[1343] In 1931 flocks estimated at 20 000 and 80 000–100 000 were reported from central Qld.[1153] Since then the largest flocks reported have numbered less than 5000.[1208,1370] In Aug. 1980, 3000 were seen at 18/134 south-east of Elliott.[1484]

Flock Pigeons were hunted by Aborigines and Europeans alike,[22] but the effects of grazing stock have probably been more significant in reducing their range and numbers. There are no measures of abundance.

Field Atlas Records: 84
1° blocks–with records: 45 (6%)
–with breeding records: 2 (4%)

Atlas No.: 036

Max. Rep. Rate: 22%
In block(s): 24/140

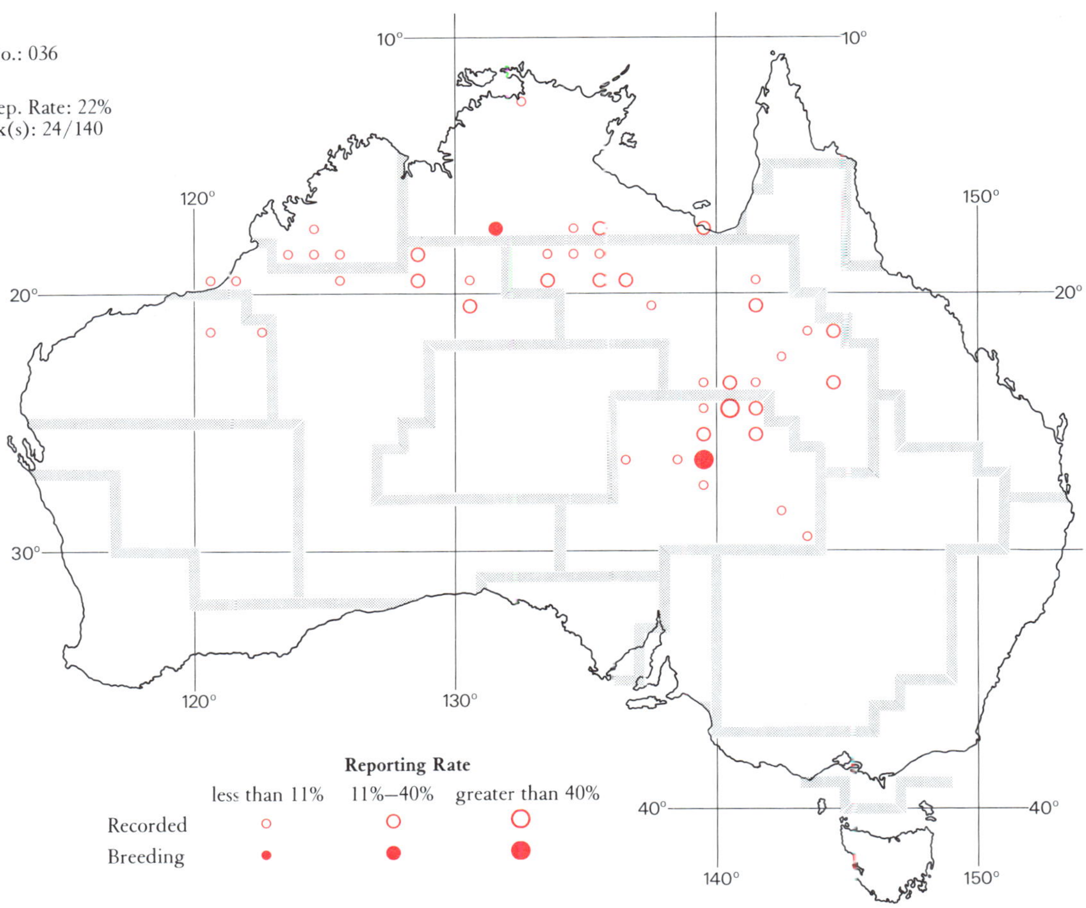

Crested Pigeon

Ocyphaps lophotes

Two subspecies of the endemic Crested Pigeon are recognized: *lophotes* in eastern and central Australia and *whitlocki* in WA.[659] The populations are continuously distributed and intermediates are found where they meet.[748,1231] The expansion of range of the Crested Pigeon is well-documented and can be seen on the historical maps (p. 664). Before 1920 there were records from only two places in Eastern Qld: on the Dawson R (23/149) last century and near Normanton (17/141) in 1913. By 1950 the Pigeon was established from Townsville (19/146) to Brisbane (27/153). In the South-East Region there were isolated records from Canberra (35/149)[467] and Geelong (38/144)[1219] last century followed by rapid expansion in the 1920s and 1930s. The bird reached Cowra (33/148) in 1925,[208] Canberra again in 1937[1015] and the Richmond R (29/153) in 1938.[645] The population at Canberra did not become established but the Crested Pigeon has since re-invaded the environs of the city successfully. It was deliberately released at Ingleburn (34/150) in the 1930s[799] and a feral population is breeding at Healesville (37/145). In the Top End a spread northward has taken place, mainly since 1950. In the South-West the species has been recorded only since 1915[1634,1638] and reached Perth (31/115) in 1956.

The Field Atlas shows no evidence of large-scale seasonal movement and none is reported.

The Crested Pigeon is ubiquitous, requiring only a ready supply of seeds, fresh water and, usually, dense cover for nesting.[659,1372] In the southern Murray-Darling Region it lives in open woodland and near human habitation, avoiding mallee and woodland dominated by river red gum.[833] A tendency to associate with human habitation is also evident at Murphy's Ck (27/152)[1099] and in the Goldfields Region.[458] The Crested Pigeon eats a variety of seeds and other plant material[748] and some insects and their larvae.[659] Its dependence upon water[458,748] suggests that the provision of artificial water points with the gradual extension of pastoral activities must have benefited it. It has also invaded land cleared for agriculture.

There are no measures of abundance.

Field Atlas Records: 25 039

1° blocks–with records: 615 (76%)
–with breeding records: 203 (33%)

Atlas No.: 043

Max. Rep. Rate: 92%
In block(s): 27/149

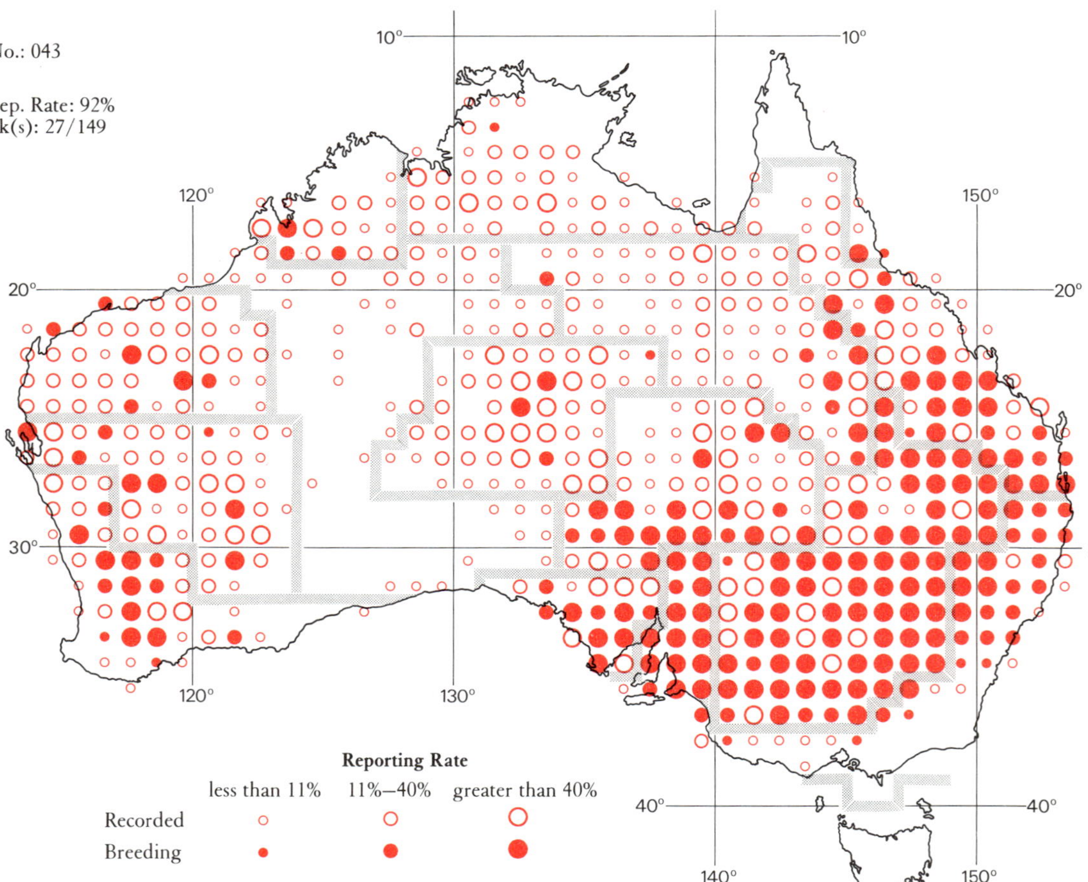

Squatter Pigeon

Petrophassa scripta

Subspecies of the endemic Squatter Pigeon have been recognized but views about their validity differ because variation is clinal.[388,659] The species' range has decreased and it is now rarely reported south of 26°S. There were reports last century from the South-East and Murray-Darling Regions south to 34°S. Since 1901 the only records south of 30°S in the Historical Atlas are of eggs taken near Bourke (30/145) in 1922, a single bird near West Wyalong (33/147) in 1975[1088] and two sightings in the Cobar area (30/145, 31/145) in the 1970s.

The species has not been reported from the Cape York Region except for an unconfirmed sighting in 1981 north of Bamaga (10/142). Its range in Qld seems otherwise unchanged. Reports from the Rolling Downs and upper Murray-Darling Regions have always been rare and the Historical Atlas includes some in the 1970s though none appears in the Field Atlas.

The Field Atlas does not suggest any large-scale seasonal movement and the species is thought to be only locally nomadic.[659]

The Squatter Pigeon is essentially terrestrial, running rather than flying and perching on vegetation only when flushed.[433] It is usually in pairs though a flock of 30 has been reported.[748] Most birds live in sandy sites near permanent water in woodland rather than in more open country. The Squatter Pigeon eats mainly grasses and other herbs, its diet varying with the seasons. In the Atherton Region grass seeds and insects are eaten in autumn and the seeds of Townsville lucerne during the rest of the year.[433]

The decline of the Squatter Pigeon in southern parts of its range has been attributed to competition for food from stock and rabbits[1343] and to its being regarded as a game bird by European settlers.[1060] It was described as 'plentiful' by settlers in the lower Murray-Darling Region[1343] but has disappeared from this part of its former range. It remains common in northern Qld though it is sometimes shot illegally by hunters.[433]

There are no measures of abundance.

Field Atlas Records: 602

1° blocks–with records: 70 (9%)
–with breeding records: 5 (7%)

Atlas No.: 039

Max. Rep. Rate: 48%
In block(s): 19/145

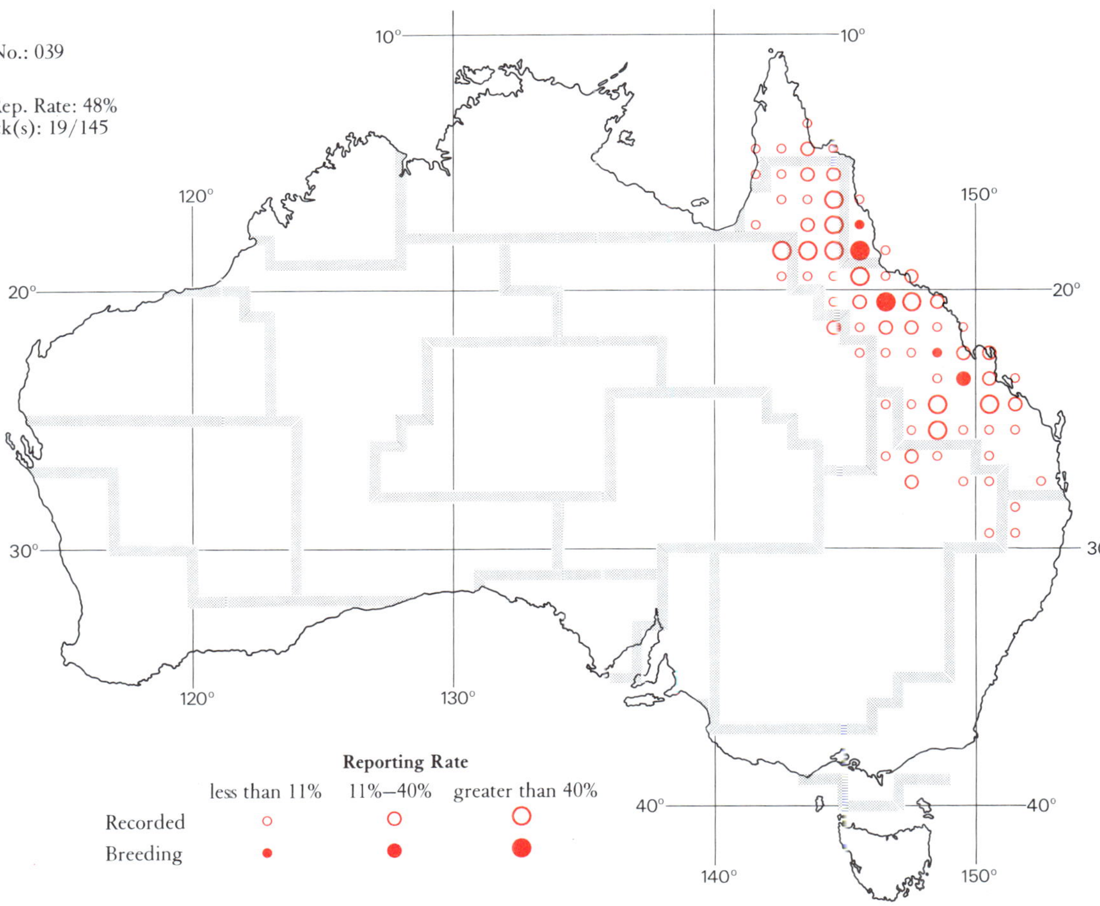

Partridge Pigeon

Petrophassa smithii

Two subspecies of the endemic Partridge Pigeon are recognized: *smithii* in the Top End has red bare skin round the eyes, differentiating it from *blaauwi* in the Kimberley Region with yellow skin.[1244] A pigeon sighted at Moonlight Ck (17/139) in 1964 is possibly a Partridge Pigeon[1709] though well east of any other record of the species.

The Historical Atlas contains records of *smithii* near Kununurra (15/128) in 1902[1713] and the easternmost records of *blaauwi* are from Napier Broome Bay (14/126). This suggests that the two populations may have been isolated.

The range of *smithii* has shrunk. It was recorded round Borroloola (16/136) in 1911 and 1913, on the Roper R (14/134) in 1910 and 1916 and on the Chambers R (14/133) in 1924 but has not since been reported in any of these 1° blocks.

There are no reports in the Historical Atlas from the Kununurra area later than 1911.

The western population *blaauwi* has been reported since 1960 from all 1° blocks where it was recorded before then. These are, in addition to the Field Atlas sighting, Kalumburu (14/126) in 1960,[1244] the Prince Regent R Reserve (15/125) in 1974[1257] and Kimbolton Springs (16/123) in 1975.[289]

The Partridge Pigeon is said to be nomadic[388,659,1343] but the Field Atlas gives no evidence of large-scale seasonal movement. Pairs and parties of 15–20 are usual and a flock of more than 100 has been recorded.[659] The species lives in woodland with a grass understorey where there are moist areas.[788] It is often on recently burnt ground[723,748] where seeds of the grasses and other herbs on which it feeds are conspicuous.[672,748]

In the Top End *smithii* has declined in abundance as well as range. In 1840 many were recorded at Port Essington (11/132)[723] but few were seen on surveys in the early 1970s.[664] Round Katherine (14/132) where it was reported only three times during the Field Atlas, 87 were counted at one pool early this century.[785] It may now be most numerous where there is least cattle-grazing.[659] The status of *blaauwi* is probably unchanged.

There are no measures of abundance.

Field Atlas Records: 214

1° blocks–with records: 12 (1%)
–with breeding records: 2 (17%)

Atlas No.: 040

Max. Rep. Rate: 43%
In block(s): 12/133

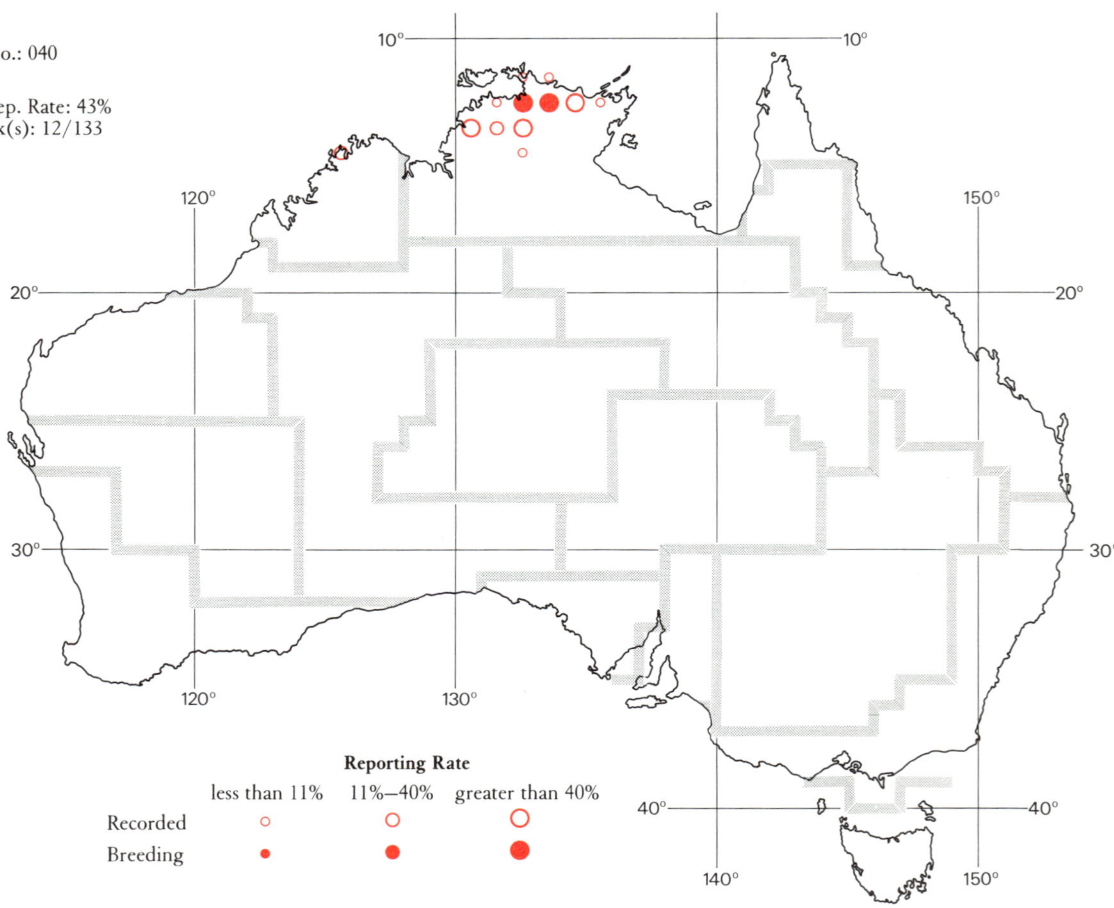

White-quilled Rock-Pigeon

Petrophassa albipennis

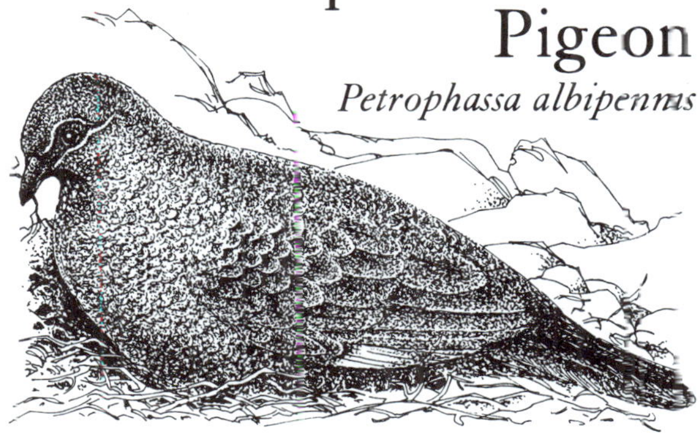

Two subspecies of the White-quilled Rock-Pigeon are recognized: *boothi* in the western part of the Top End and *albipennis* in the Kimberley Region.[659] Records of the Rock-Pigeon from within the range of *boothi* come from pockets as isolated from each other as they are from the eastern edge of the range of *albipennis*. Whether the two subspecies are in contact is not known.

The range of the species defined by the Field Atlas includes most 1° blocks where it has been reported historically. In the Top End additional records in the 1951–76 period came from the Dee R (14/130), Auvergne Station (15/130) and south of Waterloo Station (16/129).

The White-quilled Rock-Pigeon appears to be sedentary[659] and is confined to sandstone gorges with permanent water. Unlike some pigeons it will visit water during the middle of the day, usually in pairs but sometimes in small flocks.[748] Early observers found the nests plentiful and they have remained so.[787,999,1713]

The seeds on which the White-quilled Rock-Pigeon feeds are so small that the volume of 50–100 is 1 cc. They come from a variety of plants. Food is sought from plants growing in shady, moist places round the gorges.[672]

There are no measures of abundance.

Field Atlas Records: 107

1° blocks–with records: 19 (2%)
–with breeding records: 1 (5%)

Atlas No.: 037

Max. Rep. Rate: 46%
In block(s): 16/125

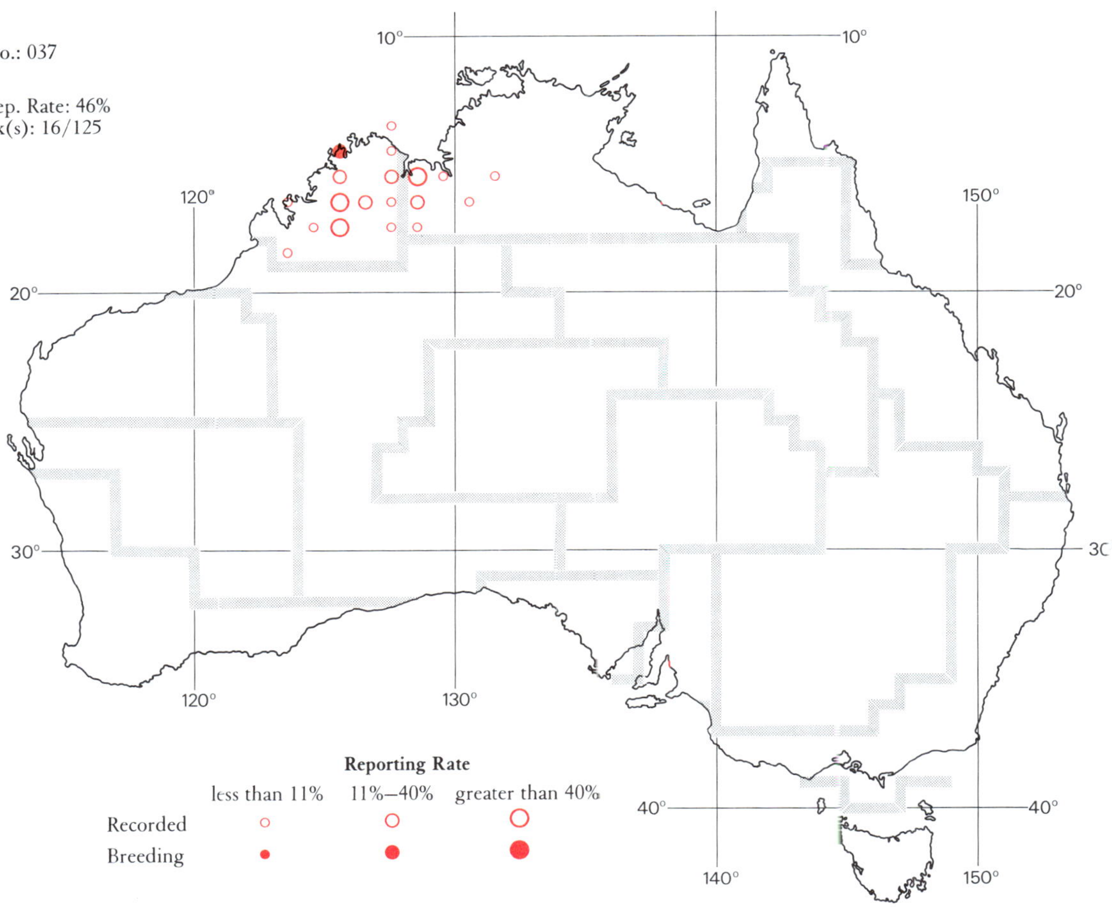

Chestnut-quilled Rock-Pigeon

Petrophassa rufipennis

No subspecific variation of the endemic Chestnut-quilled Rock-Pigeon is recognized.[659] It lives in the Top End on the western escarpment of Arnhem Land. There is an unconfirmed record from Melville Bay (12/136) in 1944–45.[884] Otherwise historical records have come only from 1° blocks where the species was observed during the Field Atlas.

The Field Atlas does not suggest any large-scale seasonal movement and local changes in abundance appear to result from movement over short distances in response to changing availability of food and water.[659]

The Chestnut-quilled Rock-Pigeon shelters in places where there are permanent pools and where the cliff faces of deeply dissected gorges are eroded into horizontal ledges. It is seldom more than 100 m from the rocks and cliffs.[655] Most birds observed in Sept.-Oct. were in shelter during the hot part of the day.[748] During the non-breeding season the bird is commonly in groups of five or six.[659] The Chestnut-quilled Rock-Pigeon takes many seeds of peas and of other herbs including grasses.[672]

It remains common within its range[659] but there are no measures of abundance.

Field Atlas Records: 123

1° blocks–with records: 7 (1%)
–with breeding records: 2 (29%)

Atlas No.: 038

Max. Rep. Rate: 28%
In block(s): 13/132

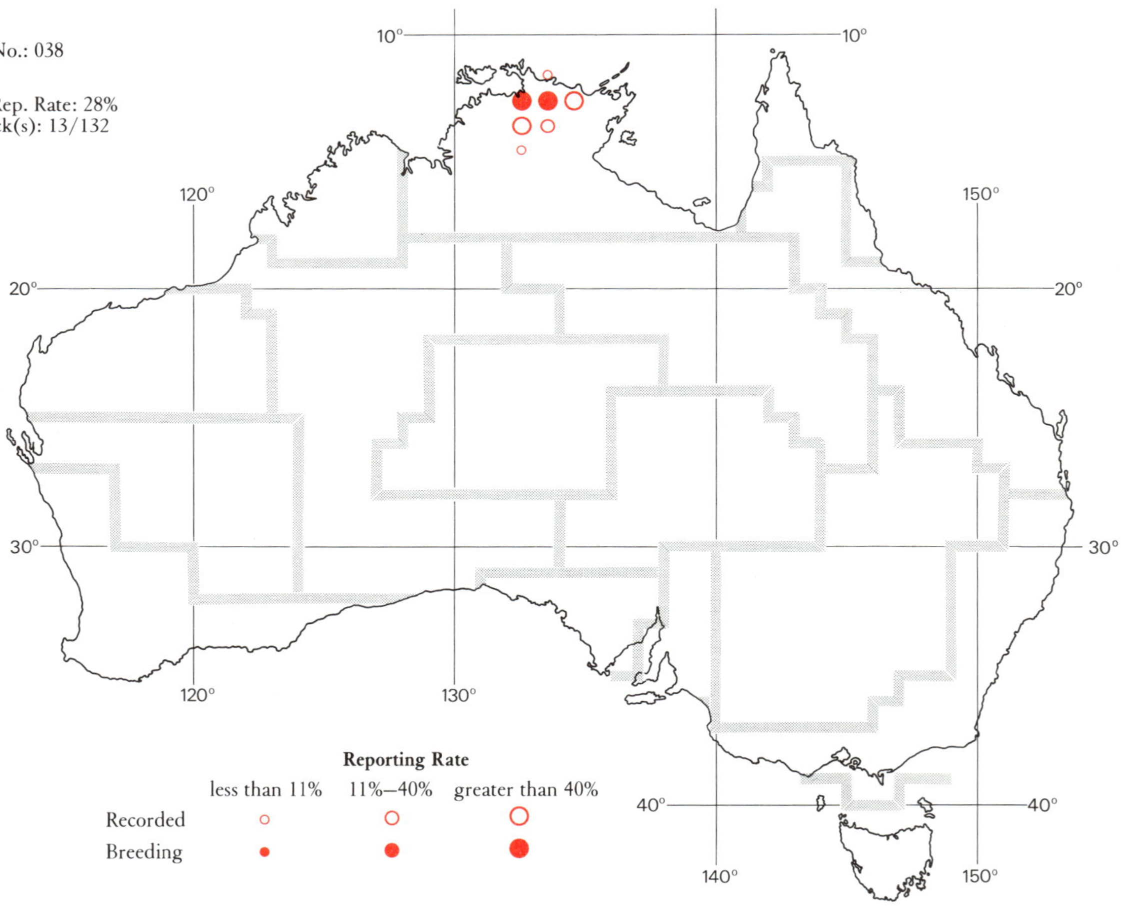

There are different views about subspecies of the endemic Spinifex Pigeon[388,659,946] and the Atlases show only the population in the Pilbara and Goldfields Regions to be isolated. In 1900 this population was claimed to have been observed as far south as the Champion Bay (28/114) district.[299] The Historical Atlas contains no records of actual sightings farther south than 26°S in the Goldfields Region. The Field Atlas sightings south to 28°S probably represent a small recent expansion of range.

Nothing is recorded of movements and the species is thought to be generally sedentary.[659]

The Spinifex Pigeon lives in acacia scrub and spinifex growing in rocky or stony country with permanent water. It is usually in pairs or small parties but occasionally in flocks of up to 100 near water.[946] Though associated with spinifex, the pigeon is not dependent on it and feeds also on the seeds of grasses and other herbs.[660,946]

At Blanchewater Station (29/139) and in the Musgrave Ra. (26/131) the disappearance of the Spinifex Pigeon has been correlated with the introduction of sheep[1163] but the species has survived in country grazed by sheep since the 1860s east of Gascoyne Junction (25/115). The Pigeon may have benefited by the establishment of bores.

There are no measures of abundance.

Spinifex Pigeon

Petrophassa plumifera

Field Atlas Records: 734

1° blocks–with records: 120 (15%)
–with breeding records: 12 (10%)

Atlas No.: 042

Max. Rep. Rate: 54%
In block(s): 21/116

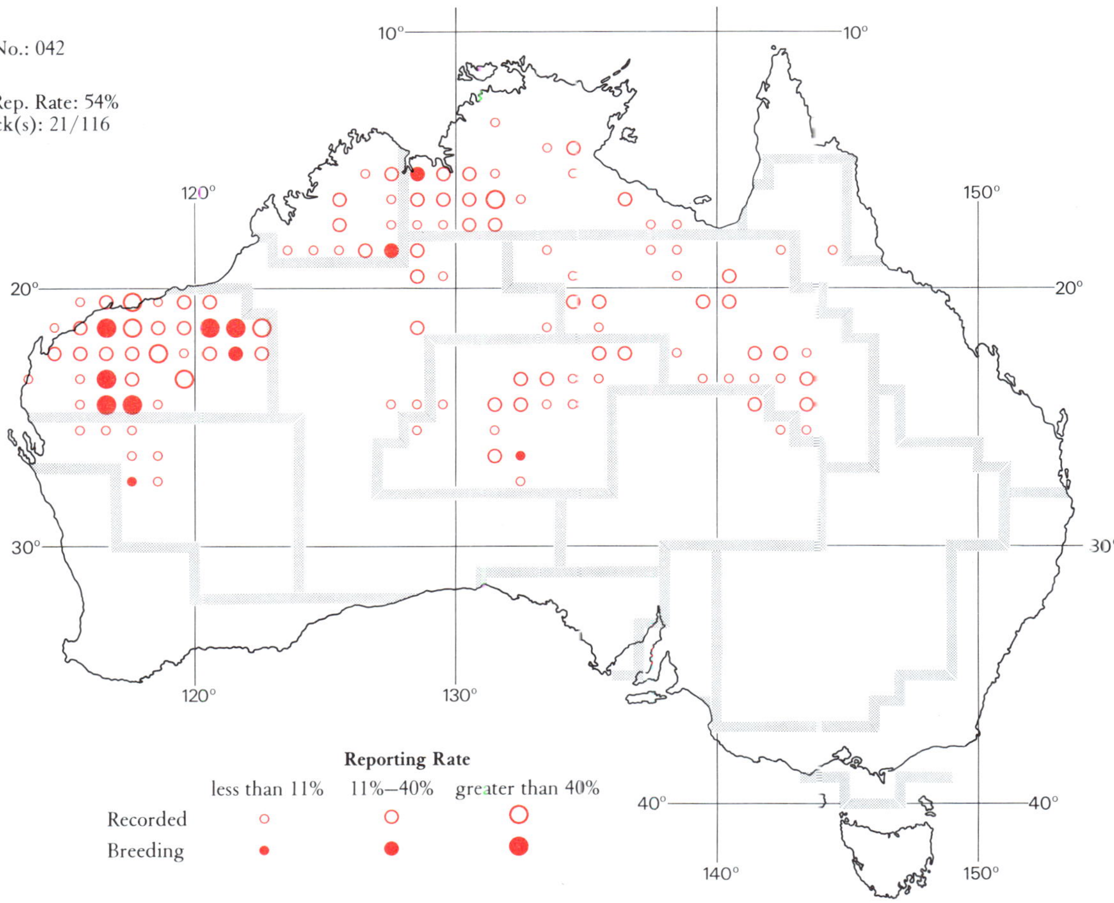

Wonga Pigeon

Leucosarcia melanoleuca

No subspecific variation of the endemic Wonga Pigeon is recognized.[659] Its range has possibly been reduced since European settlement. The Historical Atlas contains records of eggs taken near Cairns (16/145) in 1896 and references to sightings last century from Boar Pocket (16/145), Herbert R (18/146)[1343] and south to the present range. The Wonga Pigeon is also included in a list for Finch Hatton (21/148) in 1908. Since then, all records in the Historical Atlas have come from south of 22°S. In 1980 three possible sightings were made in the Eungella (21/148) district by two different observers but the records are omitted from the map because they are unconfirmed.

The southern limits of the species' range have remained unchanged. A sighting in the Field Atlas at Mt Macedon (37/144) was probably of an escapee and is not shown on the map.

No movements have been reported[659] and the Field Atlas suggests none. The Wonga is a ground-feeding pigeon that lives mainly in rainforest and eucalypt forest. It is particularly abundant where there are pockets of dense growth in gullies but it is often in open forest and clearings.

The Wonga Pigeon feeds on the fruit and seeds of native and introduced plants, specially acacias and inkweed.[659,1049] Its distribution is probably determined by that of uncleared forest and its disappearance from some places has been linked to land clearing.[1343] It used to be shot as a pest of sprouting maize when large numbers of birds gathered to uproot the young seedlings on newly sown plots.[645] It was also hunted for food.[1343] At Murphy's Ck (27/152)[1099] and King R (36/146)[1644] declines were attributed to predation by foxes and cats.

The Wonga Pigeon was at one time so abundant that 1000 birds were taken round Highfield (30/149) in 11 months.[299] Density at Moruya (35/150) in eucalypt woodland was 0.13–0.20 birds/ha.[1192]

Field Atlas Records: 1819

1° blocks–with records: 56 (7%)
–with breeding records: 13 (23%)

Atlas No.: 044

Max. Rep. Rate: 49%
In block(s): 36/150

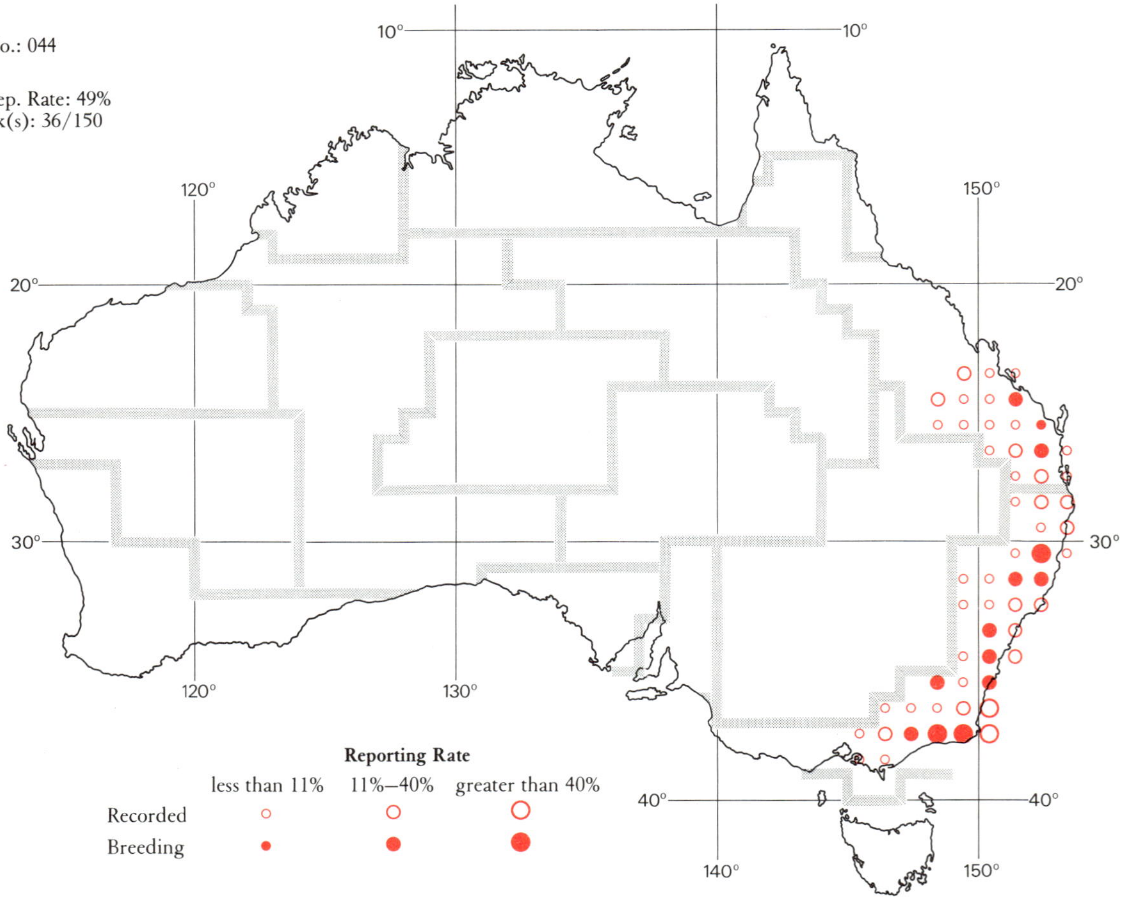

Palm Cockatoo

Probosciger aterrimus

The Palm Cockatoo inhabits lowland New Guinea and adjacent islands as well as the Cape York Region. There are different opinions about the subspecific status of the Australian population.[635,1250] The distribution shown in the Field Atlas includes all 1° blocks where the species was recorded historically.

The Palm Cockatoo is present all year throughout its range and lives where rainforest and woodland are close to each other.[635] Often it feeds in eucalypt and paperbark woodland and retreats to the rainforest in the middle of the day apparently to escape the hot sun.[635] Living in pairs and small family parties, it indulges in noisy social behaviour which generally betrays its presence.[632] Fruit, specially of pandanus, as well as seeds and nuts are eaten, both from trees and on the ground.

Near Somerset (10/142) the population apparently decreased between 1896 and 1910, the difference attributed to shooting.[104] In New Guinea the species has disappeared from near human settlements.[635] Because it is long-lived[491] there is a risk that the population may seem to be stable and surviving well and yet be failing to recruit young at a sufficient rate to replace the aging breeding population.

There are no measures of abundance.

Field Atlas Records: 81

1° blocks–with records: 9 (1%)
–with breeding records: 1 (11%)

Atlas No.: 263

Max. Rep. Rate: 55%
In block(s): 12/143

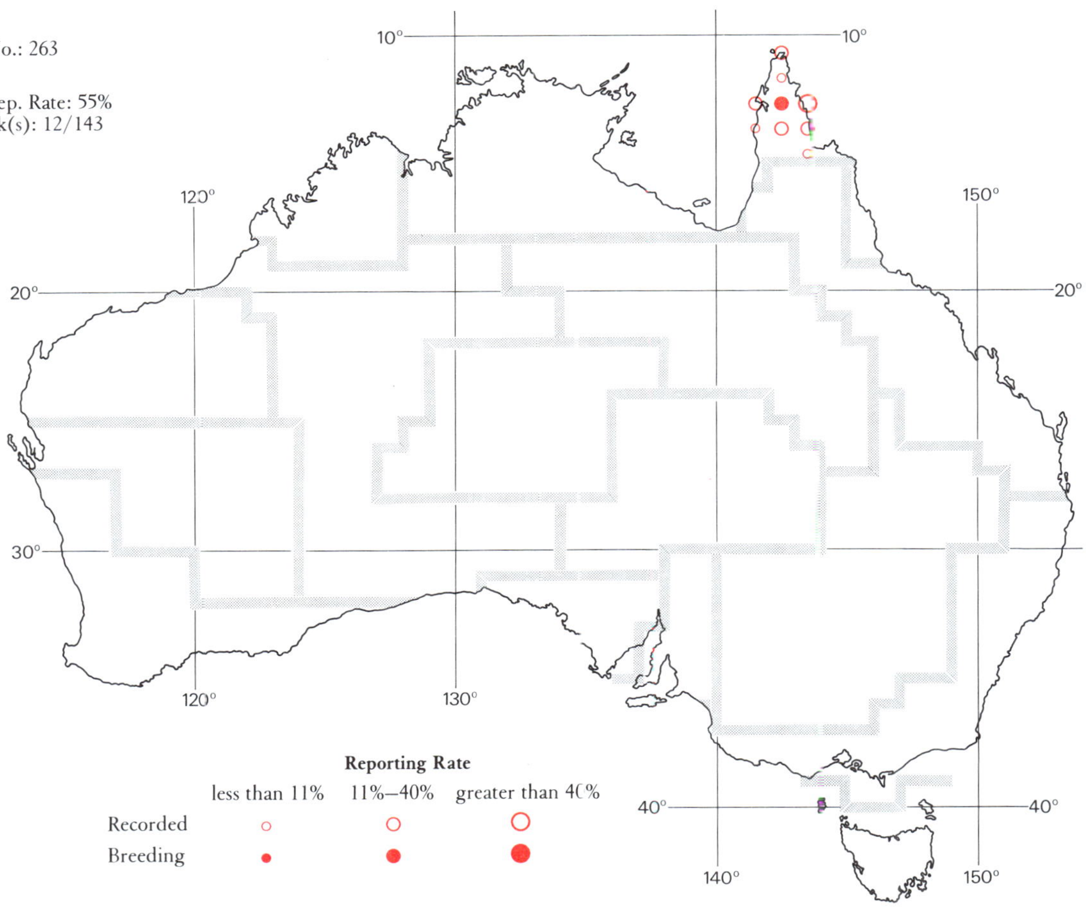

Red-tailed Black-Cockatoo

Calyptorhynchus magnificus

The endemic Red-tailed Black-Cockatoo has a range broken into seven and possibly eight populations whose subspecific relationships have yet to be determined.[617] That with the largest distribution extends from the Kimberley to the Eastern Qld Region and into northern parts of the South-East Region. Apparently connected with it are two associated with river systems: one on the Diamantina R and Cooper Ck in the Rolling Downs Region extending into the L. Eyre Region and one on the Darling R and its tributaries in the Murray-Darling Region. Populations in the lower South-East and Central Ra. Regions are isolated, as is that on the De Grey R in the eastern Pilbara Region. Two in the South-West are separated at 32°S,[1567] the Field Atlas 10′ maps showing a gap of about 50 km.

Where ranges overlap, observers sometimes confuse the Red-tailed and Glossy Black-Cockatoos. In addition the range of the Red-tailed seems to have changed. It bred as far south as the Clarence R (29/153) in 1900[617] but is now only an occasional visitor to the district, last reported in 1975.[1433] This population also bred in the Mt Isa (20/139) district before 1950. In the lower South-East Region birds were reported last century as far east as Ballarat (39/143) and perhaps within 50 km of Melbourne (37/145).[967] There was an unconfirmed record in 1901 on Flinders I. (40/148).

Seasonal movement is suggested by the Field Atlas in the Darwin (12/130) district (Reporting Rate: summer 15%, winter 60%). Elsewhere movements appear to be nomadic,[635] perhaps with a seasonal element imposed in some districts by regular fruiting of some plants.[617,967] The Red-tailed Black-Cockatoo lives mainly in eucalypt woodland and along watercourses.[635]

The lower South-East population feeds on fruit of the brown stringybark and buloke and clearing of these trees threatens its existence.[967] A flock of 450–650 seen during drought in 1967–68 may have included all birds in the Casterton (37/141) district. The largest flock recently reported in the lower South-East is 75–100 birds.[967] In the northern South-West Region the bird has benefited from introduction of the double-gee, which now composes 99% of its diet.[1565] There, flocks exceed 2000 and range appears to be expanding.

There are no measures of abundance.

Field Atlas Records: 3607

1° blocks–with records: 294 (36%)
–with breeding records: 18 (6%)

Atlas No.: 264

Max. Rep. Rate: 64%
In block(s): 13/132

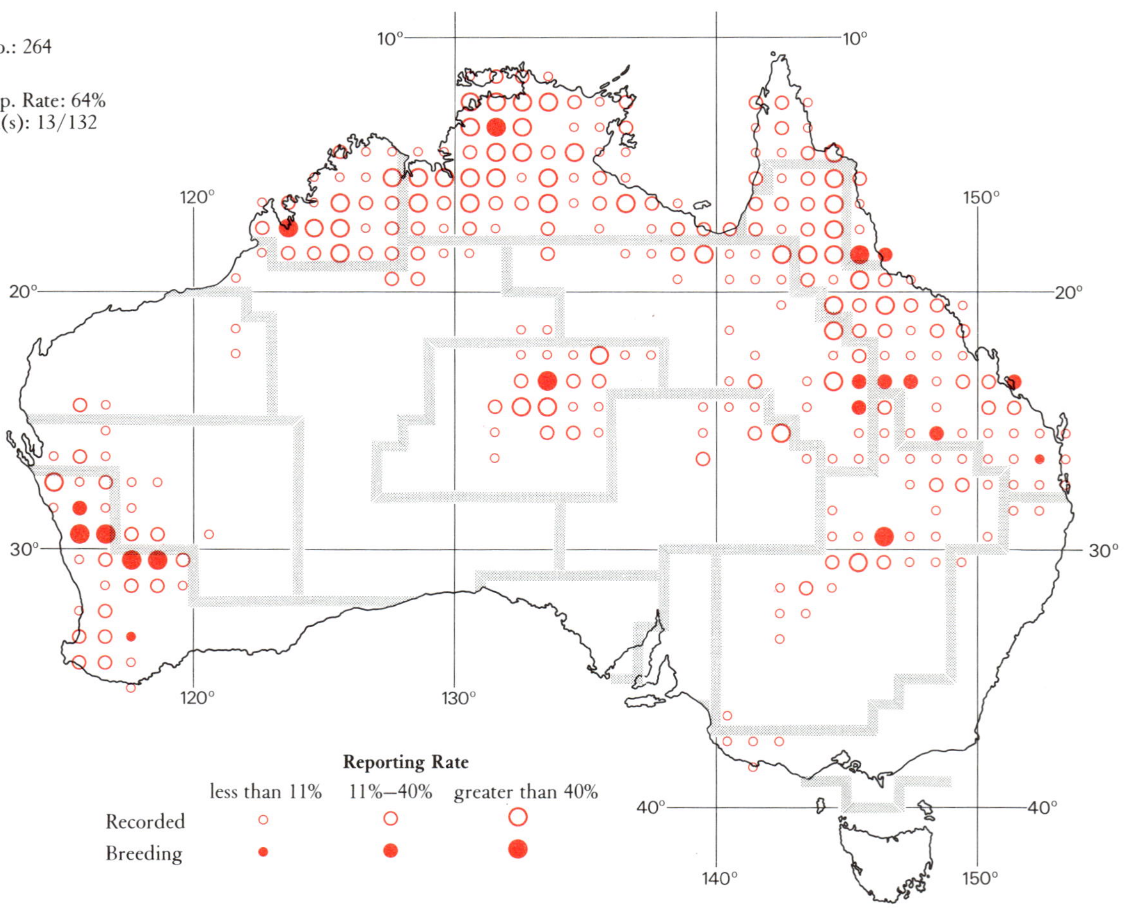

Glossy Black-Cockatoo

Calyptorhynchus lathami

The distribution of the endemic Glossy Black-Cockatoo appears to reflect three distinct populations. The main one inhabits the Great Dividing Ra. and eastern coast north to Shoalwater Bay (22/150) and the Eungella (21/148) district. This population has been reported farther west than the map indicates: in 1899 birds bred at Tarwin (38/145) and again in Oct. 1974 four birds were seen in the same area. In the Murray-Darling Region, Glossy Black-Cockatoos in the Pilliga (30/149) and Round Hill (33/146) area form a second population. The third breeds on Kangaroo I. and birds may move across to the adjoining mainland.[1369] The Glossy may be confused in the field with the Red-tailed Black-Cockatoo.

Groups of Glossy Black-Cockatoos seem to occupy an area permanently though individuals and sub-groups may move round within it.[396] Birds that occasionally appear away from the main range of the species[635] may have been forced to wander more widely by dry conditions.

The Glossy Black lives in loose groups of 2–20 birds.[396] Pairs or family parties roost by themselves but during the day join other birds of the group at food or water. Of 396 sightings in the Field Atlas for which data are available, 84% were of 2–9 birds (mode 2). Generally groups on the ranges and coast are smaller than those of the inland and Kangaroo I. populations. The Glossy Black-Cockatoo lives in eucalypt forest and woodland. It feeds almost exclusively on casuarina fruit,[372] ripping and crushing the cones to extract the seeds. *Casuarina torulosa* and *C. littoralis* are the predominant food trees of the main population; *C. cristata* is significant in the diet of inland birds, which use a more diverse range of casuarinas; Kangaroo I. birds appear to be restricted to *C. stricta*. The Glossy Black-Cockatoo has also been recorded eating angophora fruit,[590] wood-boring grubs from casuarinas[635] and perhaps acacias[396] and, near Sydney (33/151), sunflower seeds.

The Glossy Black-Cockatoo appears to be as widely distributed now as in the recent past but its dependence on one type of food makes it vulnerable. The population on Kangaroo I. may have declined and was estimated in 1980 to number 115–150 individuals, probably with fewer than 60 breeding birds.[967a]

Field Atlas Records: 862

1° blocks–with records: 65 (8%)
–with breeding records: 21 (32%)

Atlas No.: 265

Max. Rep. Rate: 25%
In block(s): 27/149

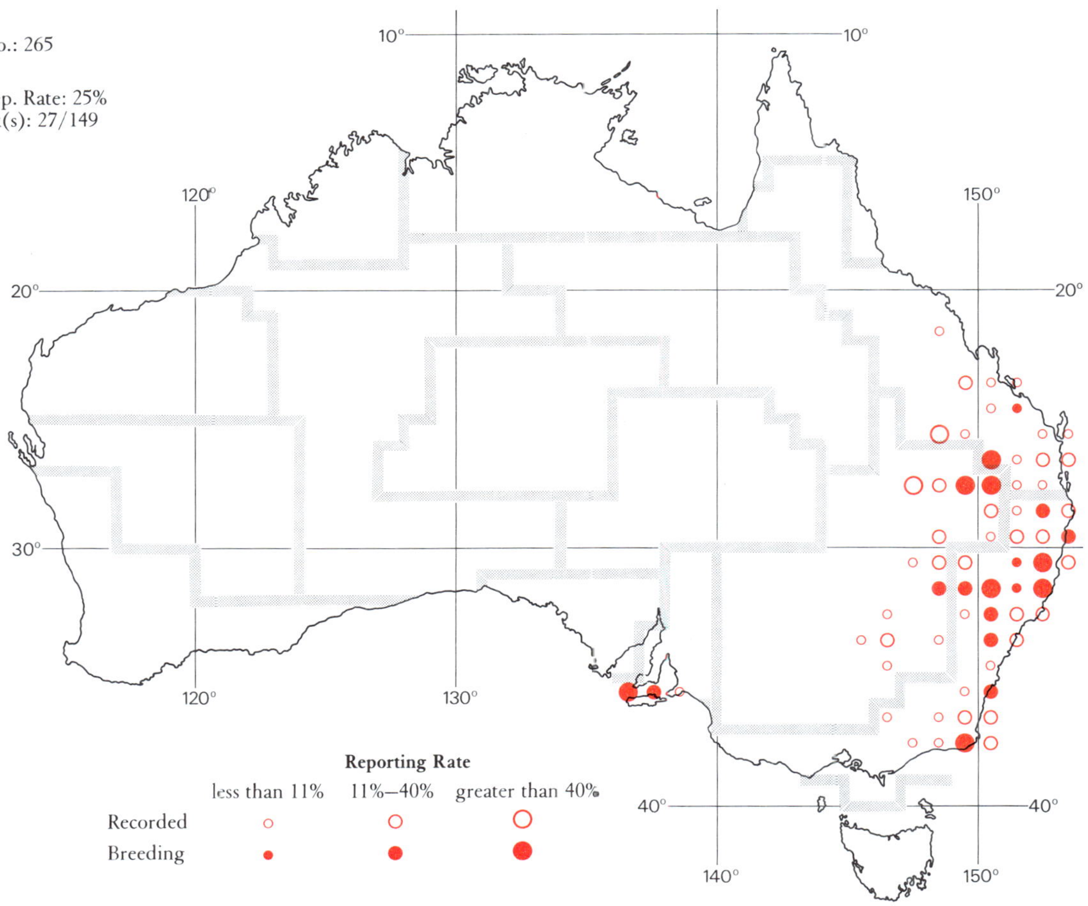

Yellow-tailed Black-Cockatoo

Calyptorhynchus funereus

It has been suggested that the yellow-tailed population of the endemic Black-Cockatoo should be united with one or both of the western white-tailed populations,[635,1569] though it is here presented as a separate species. Two subspecies of the Yellow-tailed Black-Cockatoo have been recognized: *xanthanotus* from western parts of the South-East Region and the SA Ra. and Tas. Regions; and *funereus* from the remainder of the range. Birds which are definitely *funereus* have been collected from Iguana Ck (37/147) and definite *xanthanotus* from the Grampians (37/142) so the two populations must meet in between.[1569] The Atlases show a gap in breeding records east of the Grampians (143–144°E) and the 10′ maps a gap in distribution at 144°30′–145°E, the latter probably a recent result of the growth of Melbourne (37/145). In the SA Ra. Region the Yellow-tailed Black-Cockatoo inhabits Kangaroo I. (35/137) but is absent from Yorke Pen. (35/137). The Eyre Pen. population is isolated.

Both local and seasonal movements of the Yellow-tailed Black-Cockatoo have been observed.[635] At Murphy's Ck (27/152) the species is an accidental visitor[1099] but in northern parts of the South-East Region birds are said to move to the ranges to breed, returning in Mar.[1344] Groups of up to 50 birds visit Sydney (33/151) suburbs in late autumn.[396] Post-breeding flocks visit King I. (39/144) and other offshore islands in the Tas. Region and birds have been observed crossing between Kangaroo I. and the SA mainland.

The Yellow-tailed Black-Cockatoo lives in flocks, one of more than 200 being recorded in woodland with a grass understorey during the Field Atlas and another of 300 in southern Tas.[635] It feeds on a wide range of food, mostly seeds of forest trees including eucalypts, banksias, acacias, hakea and pines. It also eats the larvae of wood-boring insects: near Coffs Harbour (30/153) it visits young hardwood forests to excavate cossid moths from saplings.[1164] In Tas. it has been observed tearing bark from casuarinas and cutting open and eating the heartwood of regenerating eucalypts.[259]

At Boola Boola (38/146) in eucalypt forest its density was 0.26 birds/ha in the gullies.[1113]

Field Atlas Records: 7154

1° blocks–with records: 128 (16%)
–with breeding records: 32 (25%)

Atlas No.: 267

Max. Rep. Rate: 52%
In block(s): 43/146

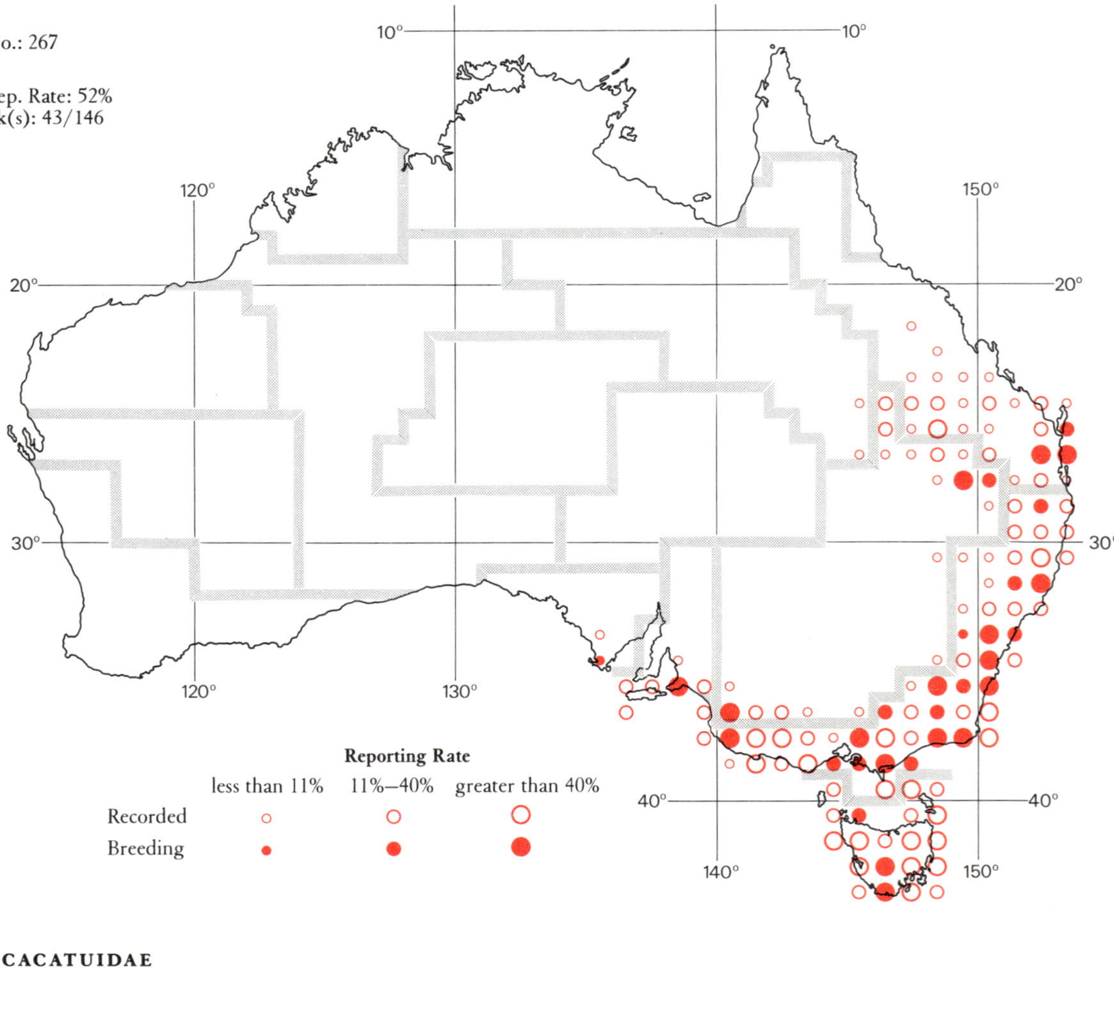

White-tailed Black-Cockatoo

Calyptorhynchus baudinii

The White-tailed Black-Cockatoo is endemic to the South-West Region. It has two populations: one, *latirostris*, breeds in country receiving less than 750 mm annual average rainfall; and the other, *baudinii*, breeds in the extreme south-west in areas receiving more than 750 mm. The two differ in morphology and behaviour and their breeding ranges evidently do not overlap. Though recorded as one species during the Atlas, it may be more accurate to regard them as separate: *baudinii* as a species in its own right and *latirostris* as a subspecies of the Yellow-tailed Black Cockatoo *funereus* of eastern Australia.[1569]

During the non-breeding season birds of the population *baudinii* may forage widely; *latirostris* moves to higher rainfall areas, inland birds undertaking a definite migration to the coast.[1571]

The hard fruits of proteaceous shrubs, nectar from their flowers and seeds of pines are eaten by *latirostris*;[1571] *baudinii* feeds chiefly on the seeds of marri, wood-boring grubs and the pips of pome fruit.[1569]

The numbers of *latirostris* have evidently been greatly reduced. One breeding population declined in eight years from 23 pairs to zero; yet the species was still seen occasionally in the district.[1570] The decline was attributed to the failure of the population to reproduce successfully, probably as a consequence of lack of food; lack of nest hollows was not the limiting factor. Clearing has removed much of the heath (sandplain) flora and the birds now have to go far for food, with a consequent increase in time spent away from eggs and young. Individual birds are long lived and remain in a district for many years after effective breeding has ceased. In this way *latirostris* appeared to be maintaining itself at a time when it was in fact declining.

The populations of *latirostris* and *baudinii* together were estimated at 14 000–60 000 birds in 1977.[1565]

Field Atlas Records: 1853

1° blocks–with records: 38 (5%)

–with breeding records: 10 (26%)

Atlas No.: 266

Max. Rep. Rate: 50%
In block(s): 34/116

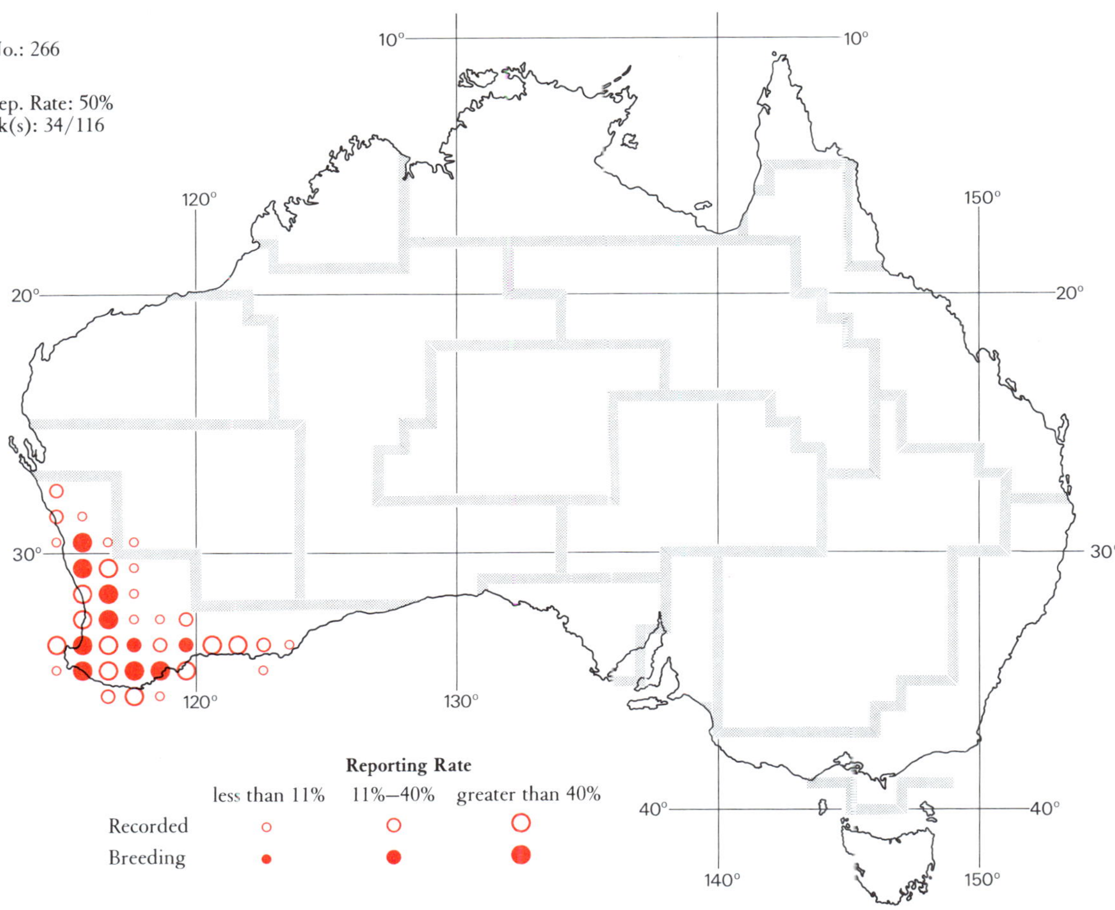

Gang-gang Cockatoo

Callocephalon fimbriatum

In the north the range of the endemic Gang-gang Cockatoo is unchanged from the historical periods. In the south the species has disappeared from King I. (39/144) where specimens were collected last century and early this century. Occasional sightings in Tas. before 1914[734] were presumably of birds from King I. The western part of the Gang-gang's range has become fragmented leaving three populations virtually isolated: in the Otway Ra. (38/143), Grampian Ra. (37/142) and the Portland (38/141) district. The Gang-gang was introduced to Kangaroo I. (35/137) in 1947 and 1957.[635] Some birds survive but are very rarely reported.

Individual Gang-gangs can move far as occasional records outside the main range demonstrate, e.g. near Ebor (30/152) in Nov. 1967.[635] Some birds move seasonally, leaving the mountains to spend the winter at lower altitudes. The Reporting Rate in the Snowy Mts (36/148) above 500 m dropped from 47% in spring-summer to 20% in autumn-winter. Below 500 m it was 78% in spring-summer and 75% in autumn-winter. In the Brindabella Ra. (35/148) the Gang-gang is present Sept.-May;[1016] nearby at Canberra (35/149), at a lower altitude, some birds are present all year but there is a marked influx in Apr. and exodus in Sept.-Oct.[635] In some places the Gang-gang remains all year at high altitudes, e.g. Blue Mts (33/150),[396] Thredbo (36/140)[682] and Jamieson (37/146).[1469] Near Colac (38/143) it is a frequent autumn and winter visitor.[256]

The Gang-gang lives in eucalypt forest, some birds moving in winter into woodland, parks and gardens. During the breeding season it is usually in pairs or family parties but flocks of up to 60 congregate at other times.[635]

The Gang-gang feeds on green acacia seeds and eucalypt fruit in the summer.[518,586,635,1865] In the winter it also takes fruit and sometimes buds of cultivated plants including hawthorn, pine, quince, pyracantha and cotoneaster often in suburban streets and near farms.[256,635] The bird tends to return to one source of food until that is exhausted then move on to another.[635] During the Field Atlas one party of birds was seen to demolish a clump of sawfly grubs.

At Boola Boola (38/146) in eucalypt forest its density was 0.14–0.44 birds/ha.[1113]

Field Atlas Records: 4252

1° blocks–with records: 40 (5%)
–with breeding records: 25 (63%)

Atlas No.: 268

Max. Rep. Rate: 44%
In block(s): 35/148

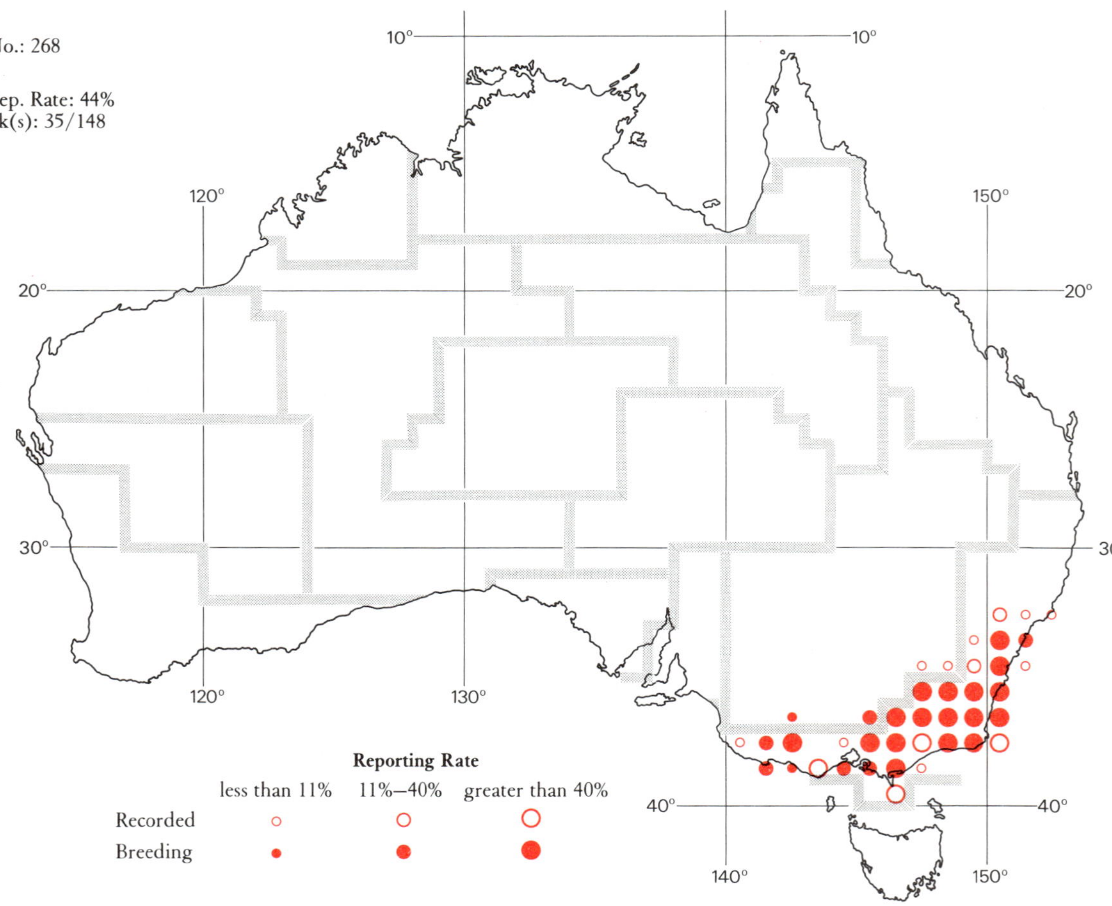

Galah

Cacatua roseicapilla

Recent taxonomic treatment separates two populations of the endemic Galah on morphological characters, specially the colour of the bare skin round the eye.[635] These populations lie east and west of the Western Deserts Region and the Field Atlas indicates them to be in contact.

The range of the Galah has expanded since European settlement to include most of the continent. Before 1901 the Atlases contain records from only five 1° blocks south of 20°S in the Eastern Qld and South-East Regions, and in the west there is no record south of 29°S before 1901. Partly perhaps because aviary escapees have established themselves round population centres, the main expansion has taken place this century. Even in the 1950s the species was rare on the tablelands south of Uralla (30/151) where it is now frequently recorded in flocks of up to 100.[396]

In the Tas. Region the Galah was first recorded in 1842 then not again until after 1900. The first breeding report was in 1925. The species is still rather scarce but flocks of 20–30 have been reported from several localities.[1409]

Breeding pairs of Galahs are effectively sedentary but may range many kilometres in search of food. Immature birds flock and wander widely.[635] There are no breeding records in the Atlases from the Western Deserts Region west of 130°E so sightings in this region may be of flocks wandering after heavy rains. The Field Atlas does not suggest any large-scale pattern of seasonal movement. The longest movement reported is of a bird banded at Lower Light (34/138) recovered at Pira (35/143), 473 km.[635]

The Galah lives along watercourses in the inland and in eucalypt woodland. Since European settlement it has colonized open country of all types including suburban gardens and parks. Large flocks are common, sometimes containing over 1000 individuals. The Galah feeds on seeds, fruit and insects. In cereal growing areas it can live almost exclusively on spilled grain. It also feeds on storks-bill, a weed of improved pastures.

In south-western Vic. the minimum density of Galahs has been estimated to be 0.002 birds/ha.[526]

Field Atlas Records: 37 359

1° blocks–with records: 678 (83%)
–with breeding records: 271 (40%)

Atlas No.: 273

Max. Rep. Rate: 93%
In block(s): 27/149

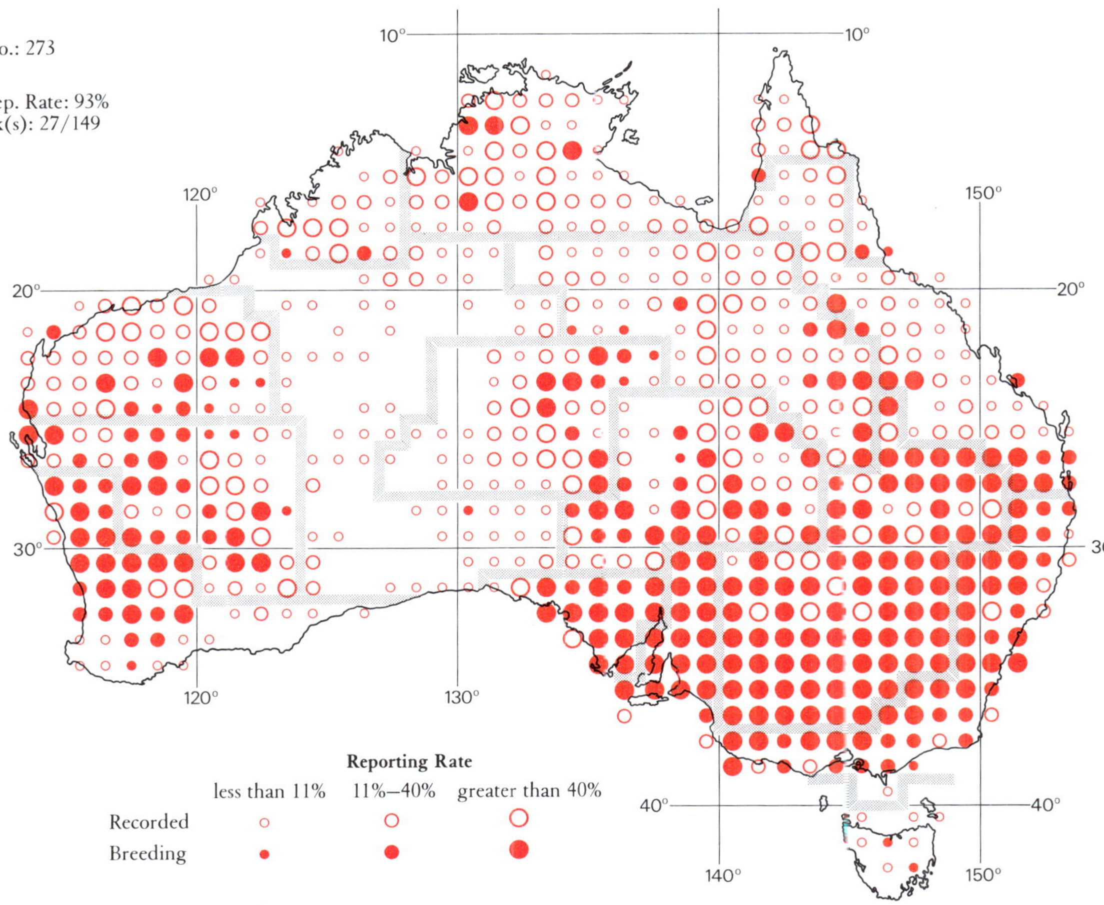

Long-billed Corella

Cacatua tenuirostris

A recent review regards the 'Long-billed' Corella of the South-West Region as a subspecies *pastinator* of the Little Corella *Cacatua sanguinea*.[1590] When the Field Atlas began, the eastern and western 'Long-billed Corellas' were considered a single endemic species and as a result both are displayed on the map on this page. The vignette illustrates a bird from Vic., the true Long-billed Corella.

In the South-West there are two isolated populations of *pastinator* separated at 32°S. These appear to be relics of the formerly more widely distributed population of the coastal plains and woodlands.

At the time of European settlement the range of the Long-billed Corella in eastern Australia extended along the Darling and Murrumbidgee Rs[1343] and east to the Mornington Pen. (38/145).[1843] A decline beginning last century greatly reduced its range and numbers. The species has recently begun to recover but its present distribution is still small compared with that last century.

Records from the SA Ra. Region, omitted from the map, probably originated from 25–30 birds that escaped from captivity in 1976 and remained round McLaren Flat (35/138). Sightings of probable escapees in Sydney (33/151) are also omitted from the map.

Movements seem to be local; there is no evidence in the Field Atlas of large-scale seasonal movement.

Flocks of Long-billed Corellas live along tree-lined watercourses in farmland.[635] The species sometimes forms mixed flocks, particularly with Sulphur-crested Cockatoos and occasionally with Galahs. It feeds mainly on onion grass and emerging crops as well as other corms and bulbs of sedges, rushes and pasture weeds.

The decline of the species on the Murrumbidgee R has been linked with a change from cattle to sheep grazing.[1343] The range of the species in WA shrank as a result of persecution by farmers who regarded the bird as a major pest of grainfields. It also suffers from bird dealers illegally taking fledglings for the bird trade. In Vic. it may be trapped under permit.[526]

Recent aerial and ground surveys indicate that the population in Vic. and SA exceeds 250 000 birds.[524]

Field Atlas Records: 1882

1° blocks–with records: 44 (5%)
–with breeding records: 14 (32%)

Atlas No.: 272

Max. Rep. Rate: 57%
In block(s): 37/142

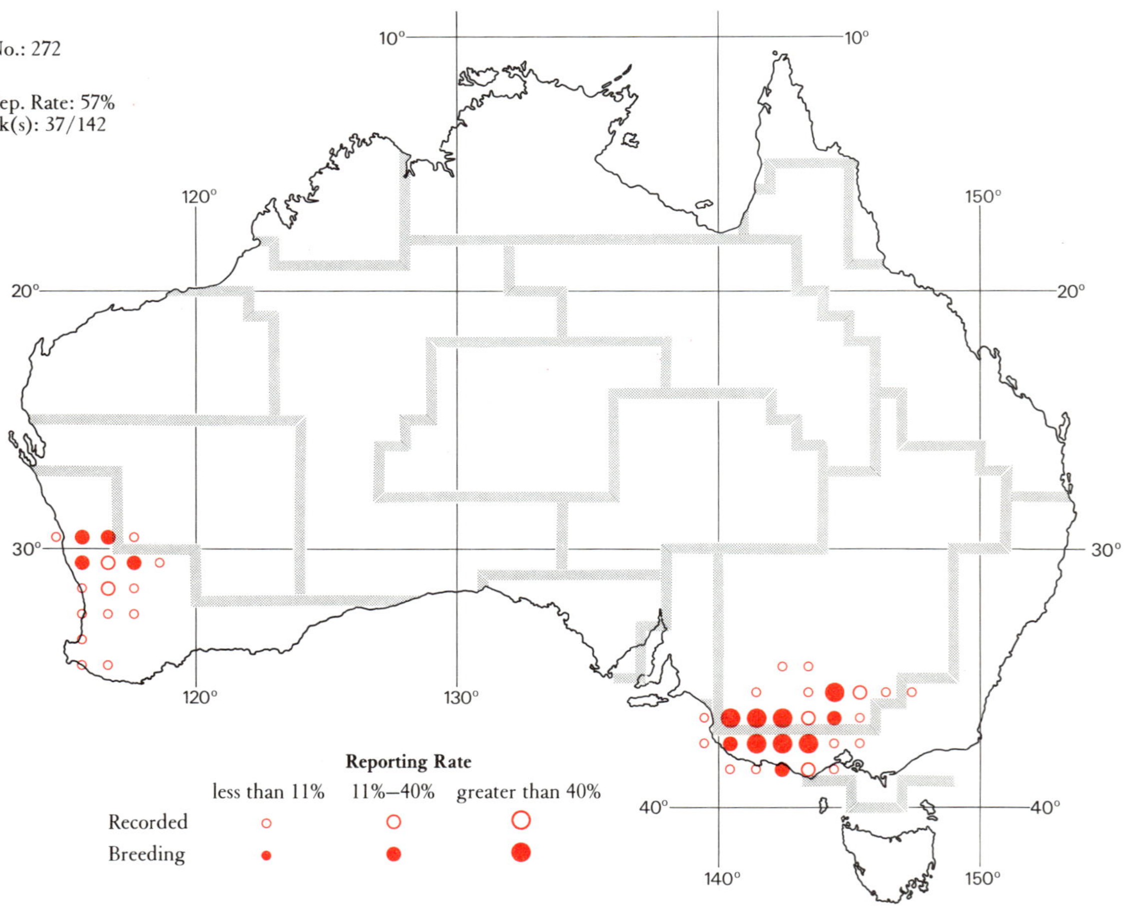

Little Corella

Cacatua sanguinea

Outside Australia the Little Corella lives only in southern New Guinea. The taxonomy of both the Little and Long-billed Corellas has been revised recently and in Australia four subspecies of the Little are now recognized.[1590] One of these, *pastinator*, was previously classified as a western subspecies of the Long-billed Corella *C. tenuirostris* and is treated under that species in the text. Birds of the western Cape York Region and southern New Guinea are *normantoni*. The most widespread subspecies, *gymnopis*, has a disjunct distribution. The eastern population occupies the Murray-Darling, South-East, Rolling Downs, Central Ra. and L. Eyre Regions. It has expanded into the Murray-Darling and South-East Regions since the 1901–50 period (p. 664) and appears to be well separated from its western counterpart in the Pilbara and Goldfields Regions. This western population of *gymnopis* is also spreading southwards. The fourth subspecies, *sanguinea*, of the Kimberley and Top End Regions may be an isolated population.

Sightings at a distance from the main range, such as the four birds near the Wingan R (37/149) in Dec. 1977, seem to be part of the general advance of the Little Corella in the South-East Region. The populations round Sydney (33/151) and Perth (31/115) originated from aviary escapees, as also perhaps some birds round Melbourne (37/145).

Tagged birds at Three Springs (29/115) moved 150–250 km north-west to near Dongara (29/114) and Geraldton (28/114) after breeding, returning in Mar.-Apr.[635,1682] Seasonal movements have been observed in the Kimberley Region also but at Coomallo Ck (30/115) birds join flocks in the non-breeding season and move only locally.[635]

The Field Atlas does not suggest a large-scale pattern of seasonal movement.

The Little Corella inhabits a variety of woodland, acacia scrub and pasture provided fresh water is available. Birds are usually in flocks. In the Kimberley Region flocks of 60 000–70 000 birds have been reported.[635] Single birds and small groups sometimes join flocks of other cockatoos. The Little Corella feeds mainly on seeds, specially those of grasses, as well as eating fruit, shoots and insects.

There are no measures of abundance.

Field Atlas Records: 5708

1° blocks–with records: 400 (49%)

–with breeding records: 56 (14%)

Atlas No.: 271

Max. Rep. Rate: 74%
In block(s): 27/140

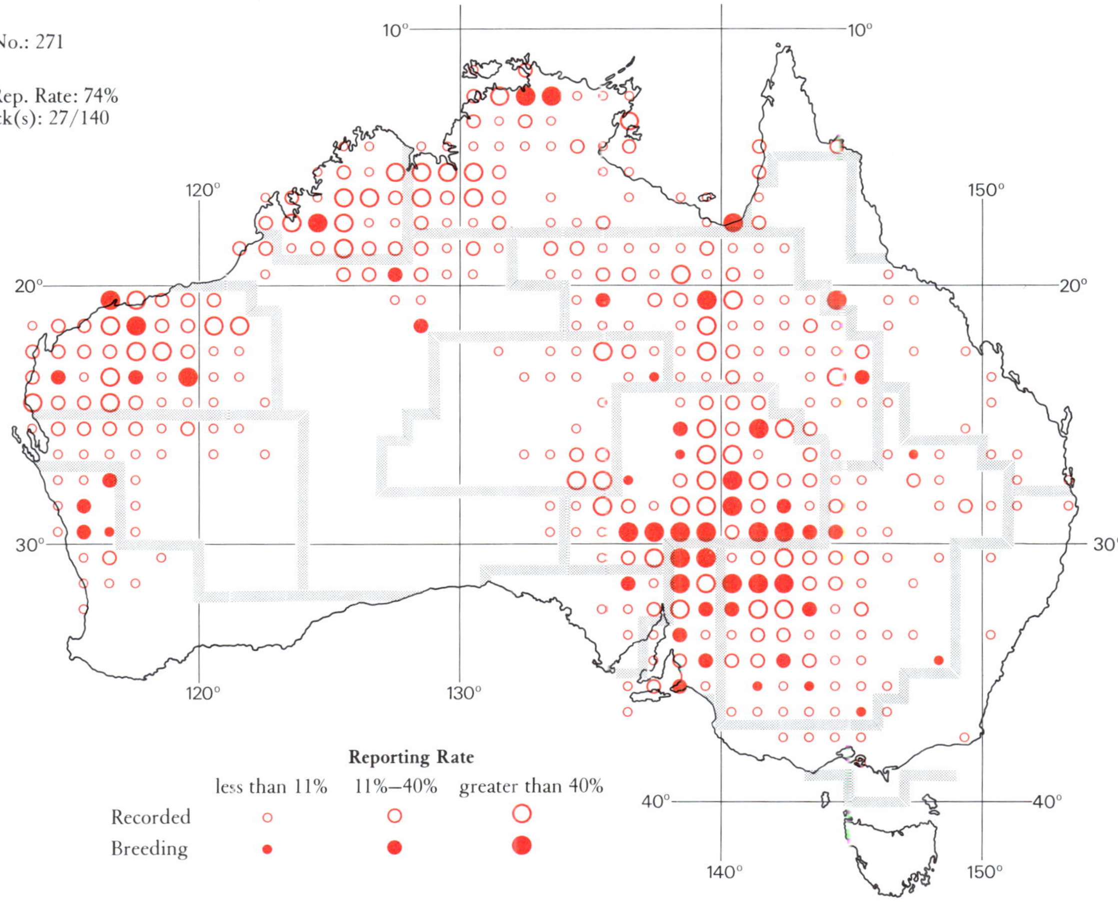

Pink Cockatoo

Cacatua leadbeateri

The population of the endemic Pink Cockatoo in southern WA has been described as a distinct subspecies *mollis* on the basis of specimens from Warburton (26/126) and Carnamah (29/115). Another subspecies *superflua* has been suggested for SA birds.[748] Both these are doubtfully distinct from nominate *leadbeateri*, which takes in birds from the rest of the range.[635] The Atlases show no breaks in distribution corresponding to this subspecific arrangement. The ranges of the Pink and Sulphur-crested Cockatoos are largely complementary.

No decline in the range of the Pink Cockatoo is evident except in the SA Ra. Region, where it bred near Adelaide (33/138) last century but had disappeared from Mt Mary Plains (34/139) by the 1950s. There is only one later breeding report in the Atlases from this region: near Waikerie (34/139) in 1964.

Field Atlas records of probable escapees have been omitted from the map, notably of the Sydney (33/151) population which is not known to breed but is replenished from escapees or releases.

For the most part, the Pink Cockatoo is sedentary. When food is short, breeding birds may join wandering flocks of immatures and appear in unusual places.[635] Vagrants often turn up many kilometres from the population in which they were bred and single birds are reported in flocks with other species of cockatoo.

The Pink Cockatoo inhabits a variety of country in the inland providing there is fresh surface water and large hollow trees for nesting.[1552] It is usually in pairs or small parties but flocks of 300 birds have been recorded.[635] It feeds on fruit, including that of the native fig, and seashore plants and the seeds of native and introduced pines. Seeds are excavated from dried paddy melons.

Though its total range has altered little, local declines in numbers have been noted after clearing of woodland, trapping[635] and nest robbing.[1540] Small populations (5–30 birds) may persist over long periods, e.g. McLarty Hills (19/123) and Windy Corner (23/125).

Each breeding pair occupies an area of about 500 ha which it rarely leaves.[1540]

Field Atlas Records: 2327

1° blocks-with records: 194 (24%)
-with breeding records: 33 (17%)

Atlas No.: 270

Max. Rep. Rate: 73%
In block(s): 27/147

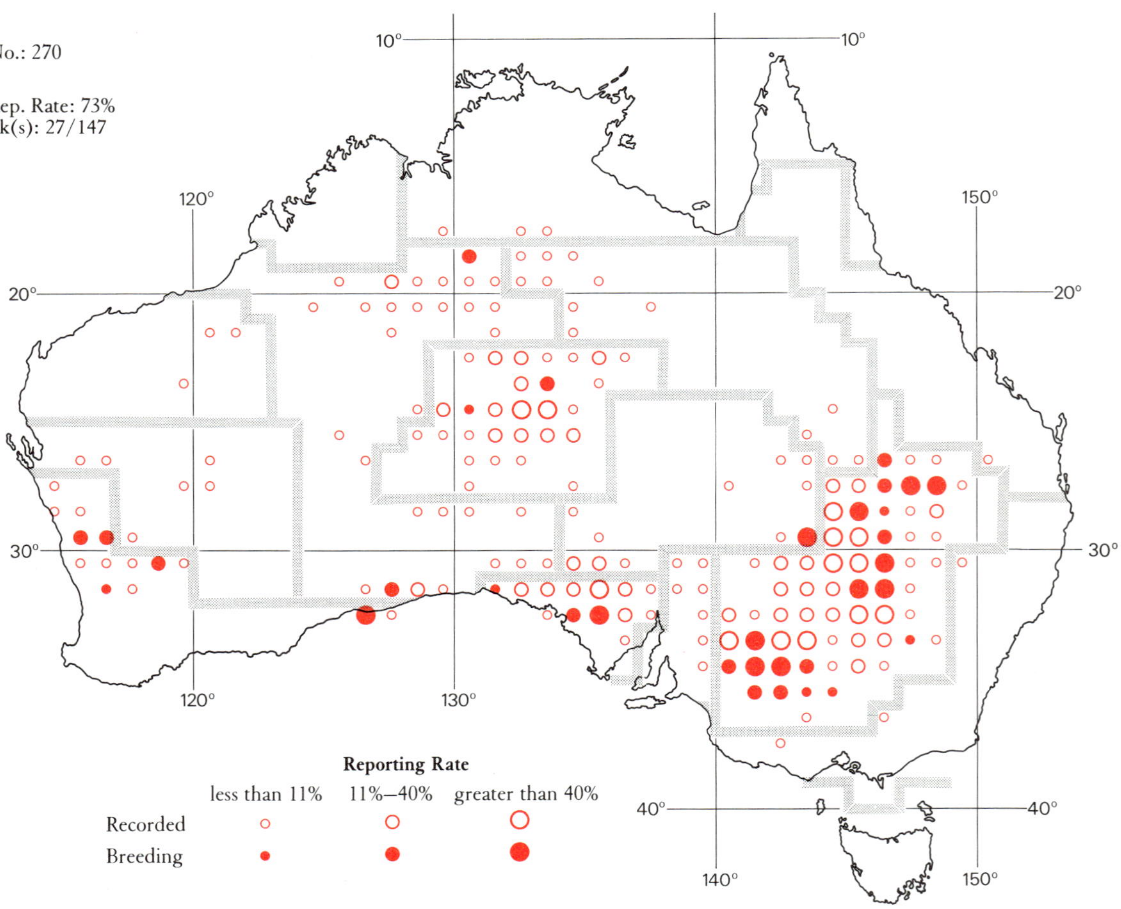

Sulphur-crested Cockatoo

Cacatua galerita

As well as living in parts of Australia, the Sulphur-crested Cockatoo inhabits New Guinea and was introduced to some south-eastern Asian islands and to New Zealand. Two Australian populations that are morphologically distinct meet south of the Gulf of Carpentaria: *galerita* to the east and *fitzroyi* to the west.[635] In western Tas. the species is widespread and breeds[260] but it is absent from Flinders I. (40/148) and only an occasional visitor to King I. (39/144), where it was more common until the 1920s. The ranges of the Sulphur-crested and Pink Cockatoos are largely complementary.

The Sulphur-crested Cockatoo was introduced to or escaped in the South-West Region before 1935.[1085] In other parts of Australia escapees are not breeding and records of them are omitted from the map.

In the alpine regions of southern NSW there is said to be some altitudinal movement of flocks to lower country in winter.[635] Reporting Rates did not reflect it (below 500 m: spring-summer 14.8%, autumn-winter 15.0%; above 500 m: spring-summer 15.8%, autumn-winter 13.3%). Elsewhere movements appear to be local.[635] In the Atherton Region there are invasions by flocks of a few hundred when crops are being harvested.[231] In Tas. the species is faithful to roosting and breeding sites.[259] The Field Atlas suggests no large-scale pattern of seasonal movement.

The Sulphur-crested Cockatoo generally lives along watercourses. It feeds in open areas away from watercourses but returns to them for nesting, roosting and loafing. Highlands are avoided: in Qld the species tends to stay below 1000 m[1709] and in the South-East Region below 1500 m. Flocks of several hundred birds are often reported; some inland records were of a few Sulphur-cresteds among flocks of Little Corellas. The food of the Sulphur-crested Cockatoo varies from nectar and wood-boring grubs to fruit, grain[635] and rhizomes.[259] The bird causes damage to crops and is sometimes shot illegally by farmers.

In south-western Vic. the minimum density of Sulphur-crested Cockatoos has been estimated at 0.007 birds/ha.[52]

Field Atlas Records: 18 174

1° blocks–with records: 332 (41%)
–with breeding records: 86 (26%)

Atlas No.: 269

Max. Rep. Rate: 75%
In block(s): 12/136

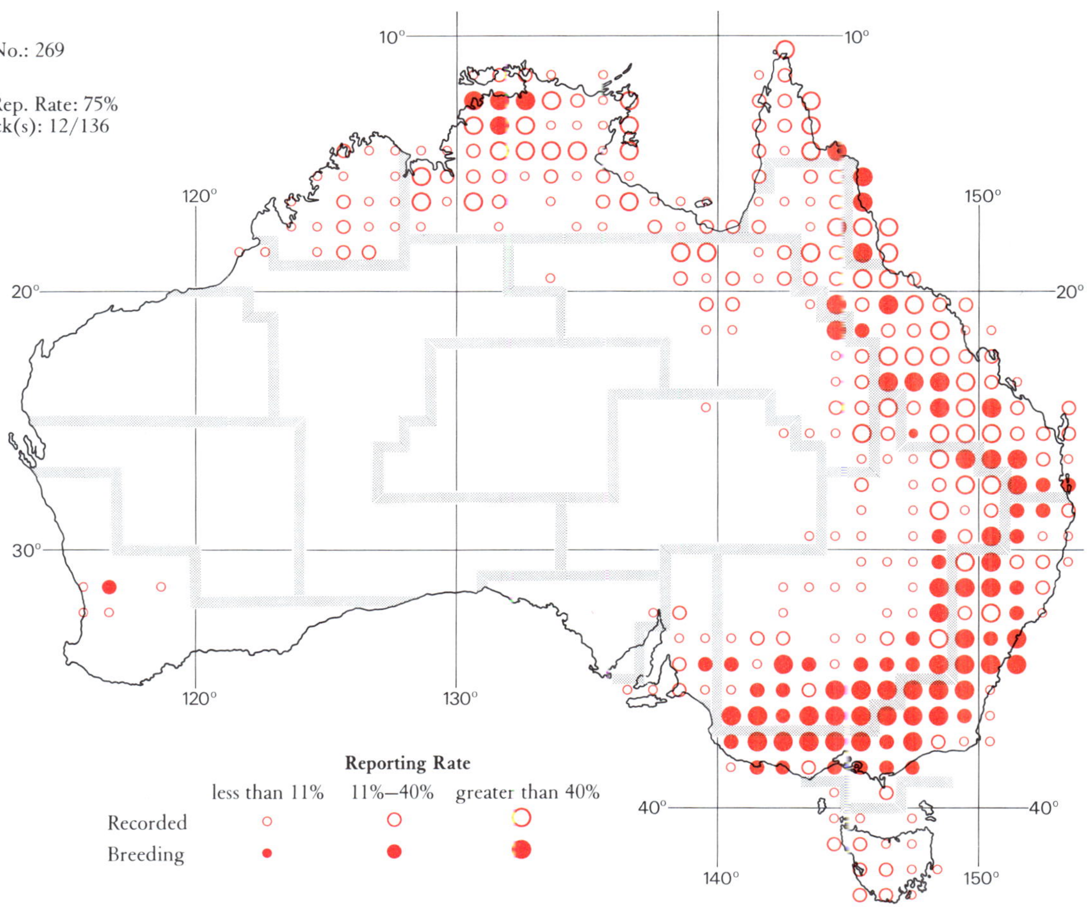

Eclectus Parrot

Eclectus roratus

The Eclectus Parrot ranges through parts of Indonesia, New Guinea, some Pacific islands and Australia. In Australia the subspecies *macgillivrayi*, discovered in 1913,[1145] is restricted to the eastern side of Cape York Region from the Pascoe R (12/143) south to Massy Ck (13/143). It breeds there though not reported to do so during the Field Atlas.

The Eclectus Parrot was reported in 1953 in the Atherton Region at L. Barrine (17/145) and in 1955 about 55 km away.[80] There have been no subsequent sightings and it seems more likely that the birds were escapees than that they arrived naturally. No movements have been reported.[635]

The Eclectus Parrot lives in rainforest and occasionally in less dense areas of eucalypt woodland. It forages in groups, usually in the tops of the trees. One group of 80 roosted together.[635] The Eclectus Parrot seems to be exceptional among the cockatoos and parrots in that it breeds communally. Six to eight individuals have commonly been observed attending one nest.[635]

The Electus Parrot feeds on fruit, specially of pandanus, nuts, leaf buds and blossom, taking the nectar as well as the whole blossom.[635]

At a lowland rainforest site near Port Moresby (9S/147E) in New Guinea the density of the Eclectus Parrot has been estimated to be 0.1 birds/ha.[160]

Field Atlas Records: 39

1° blocks-with records: 2 (1%)

-with breeding records:

Atlas No.: 275

Max. Rep. Rate: 49%
In block(s): 12/143

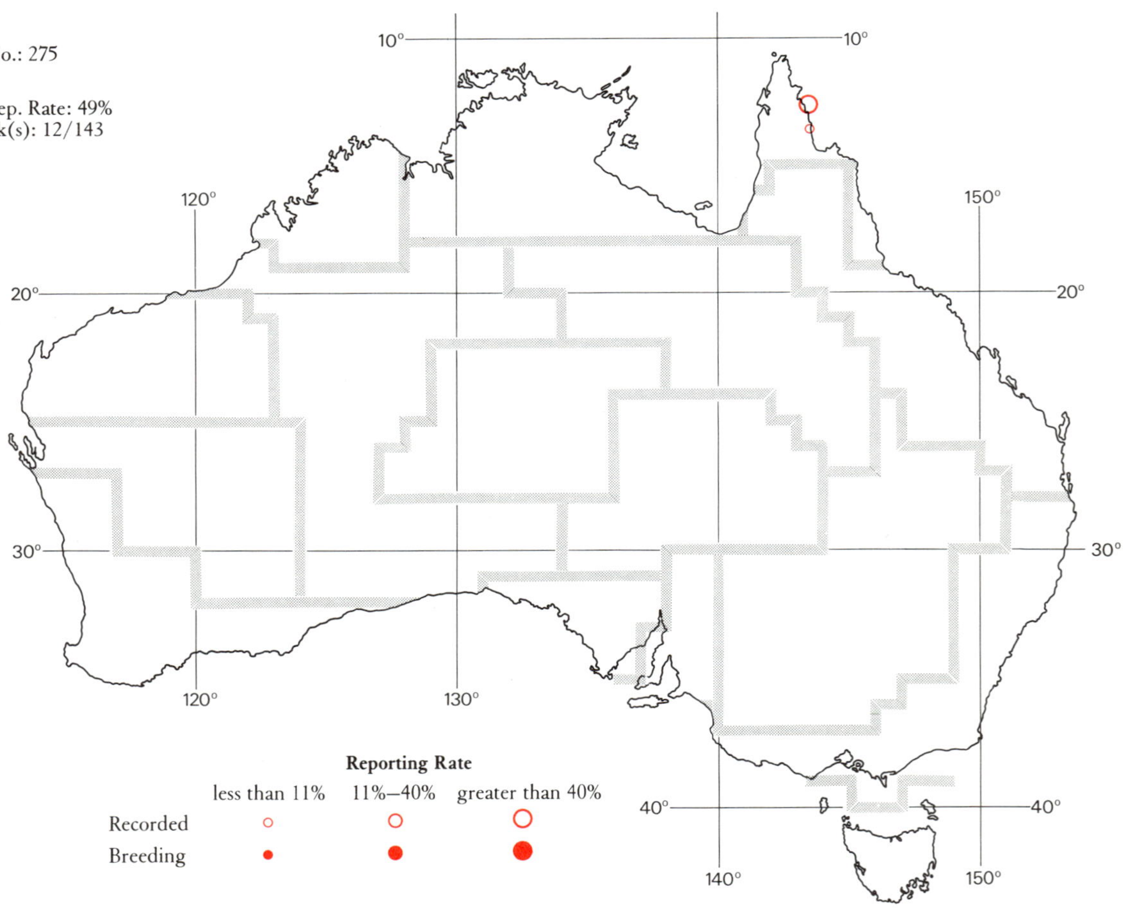

Red-cheeked Parrot

Geoffroyus geoffroyi

Many subspecies of the Red-cheeked Parrot are described from its wide range in Indonesia, New Guinea and Australia.[635] In Australia the subspecies *aruensis* is restricted to the Cape York Region from the Pascoe R (12/143) south to the Rocky R (13/143) and inland to Kann Ck and Peach Ck (143°E). Most records come from round Iron Ra. (12/143) but the species is said also to be distributed through the less accessible country to the south. Since its discovery in Australia in 1913 the species' range does not appear to have changed.

The Red-cheeked Parrot was present throughout the year at a lowland rainforest site near Port Moresby (9S/147E) in New Guinea.[160] In the Cape York Region local movements related to food have been suggested but the species does not wander far.[635]

In Australia the Red-cheeked Parrot lives in rainforest, particularly along watercourses[635] but in New Guinea it is also present in mangroves.[149] It lives in pairs and small parties but will congregate in larger groups where a tree is bearing a heavy crop of suitable fruit. The species is conspicuous, perching in the tops of trees, often on bare branches, where it calls loudly. There seems to be little difficulty in determining whether it is in an area and its late addition to the Australian list was because it is local rather than rare. It feeds on blossom, nectar, seeds and a wide range of soft and hard forest fruit, taking both flesh and kernel.

At a lowland rainforest site near Port Moresby in New Guinea its density was estimated at 0.3 birds/ha.[160]

Field Atlas Records: 41

1° blocks–with records: 2 (1%)
–with breeding records: 1 (50%)

Atlas No.: 276

Max. Rep. Rate: 51%
In block(s): 12/143

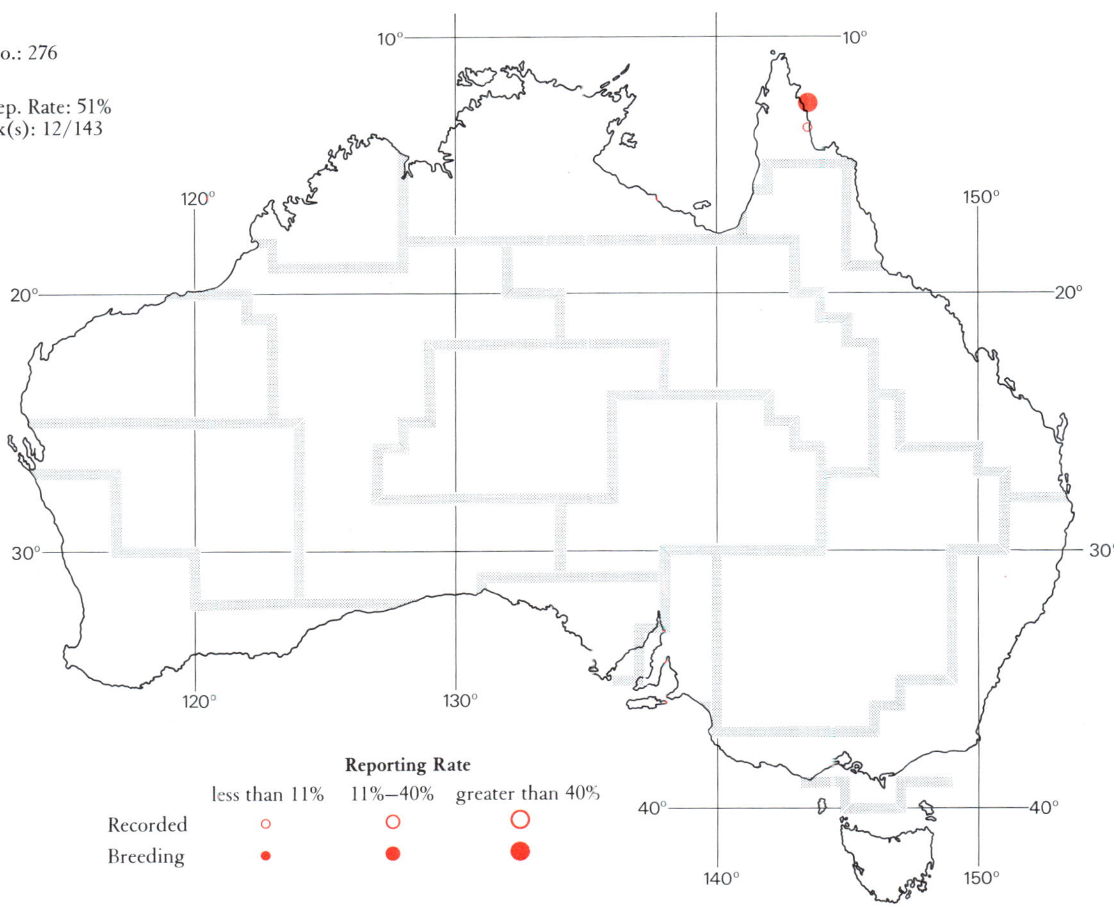

Rainbow Lorikeet

Trichoglossus haematodus

The Rainbow and Red-collared Lorikeets are probably conspecific, both subspecies of the species *haematodus*, which is distributed through southern Indonesia, south-western Pacific islands, New Guinea and adjacent islands and Australia.[635] The Rainbow Lorikeet was treated as a separate species during the Field Atlas.[1581] The Atlases suggest that the gap between the Rainbow and the Red-collared at the eastern end of the Gulf of Carpentaria is narrow, if it occurs at all. There are no records in 10′ blocks between 139°30′E and 140°50′E, an area where difficulty of access has precluded many visits.

There were occasional records of the Rainbow Lorikeet from the Tas. Region last century.[734] In 1969 a flock of about 16 was recorded on King I. (39/144) where it remained for several months before disappearing.[259] Other recent Tas. records are probably of escaped birds.[259] In Perth (31/115) the Rainbow Lorikeet escaped from captivity in the late 1960s.[635] It now breeds there, though not reported to do so during the Field Atlas.

It occasionally flocks with the Scaly-breasted Lorikeet with which it has interbred.[747]

At any site numbers of the Rainbow Lorikeet vary erratically and the bird may be completely absent in some years.[96,285] In Torres Strait it breeds on the larger islands and visits the smaller ones when plants are flowering.[503] The Field Atlas does not suggest any large-scale seasonal movement but the species can move far as occasional records in the Tas. Region show.

Flocks of the Rainbow Lorikeet gather on flowering eucalypts, paperbarks and other tall native trees as well banksias, bottlebrushes[306] and some introduced species such as coral trees. The bird regularly visits artificial feeding sites in a wildlife park at Currumbin (28/153) and in towns and urban areas elsewhere. It feeds on nectar, pollen,[306] native and cultivated fruit and occasionally insects.[635]

The average density of the Rainbow Lorikeet at 12 sites in suburban Townsville (19/146) was 4.1 birds/ha.[953]

Field Atlas Records: 7791

1° blocks–with records: 172 (21%)
–with breeding records: 60 (35%)

Atlas No.: 254

Max. Rep. Rate: 77%
In block(s): 18/145,23/147

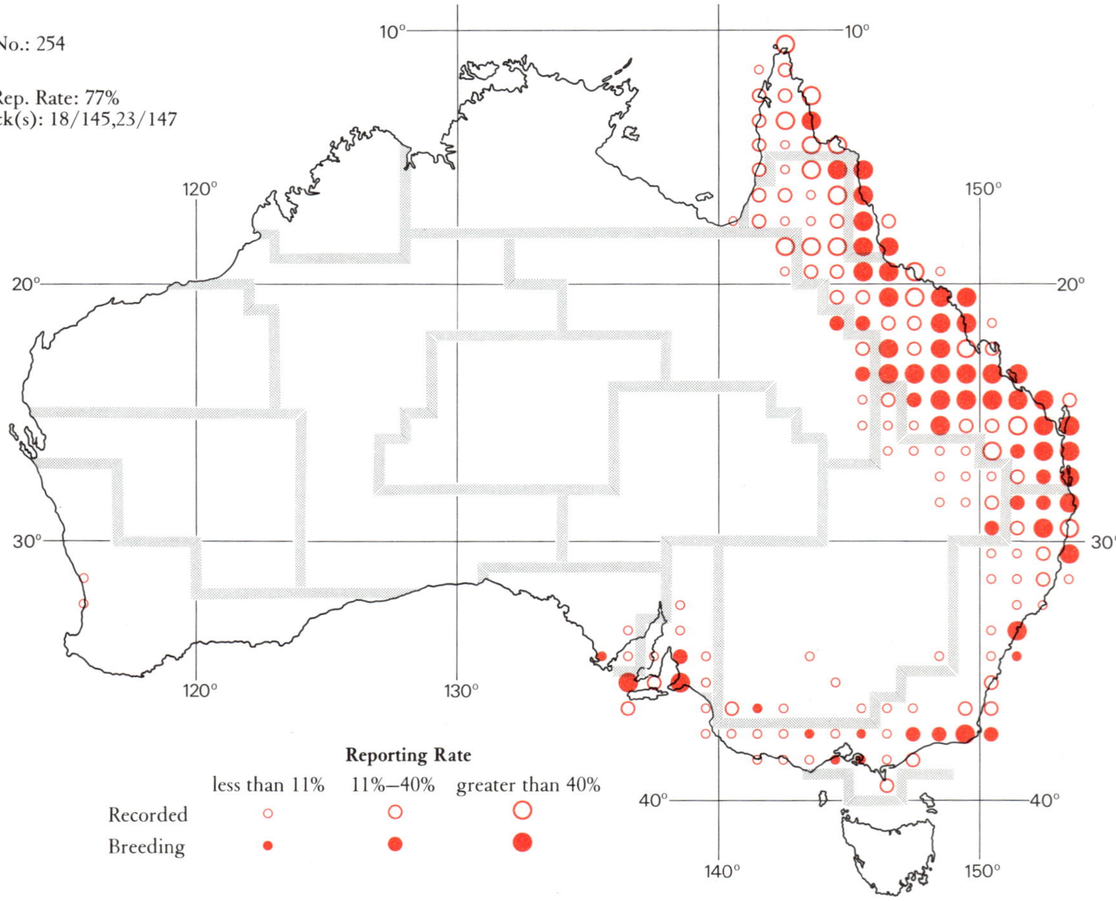

Red-collared Lorikeet

Trichoglossus rubritorquis

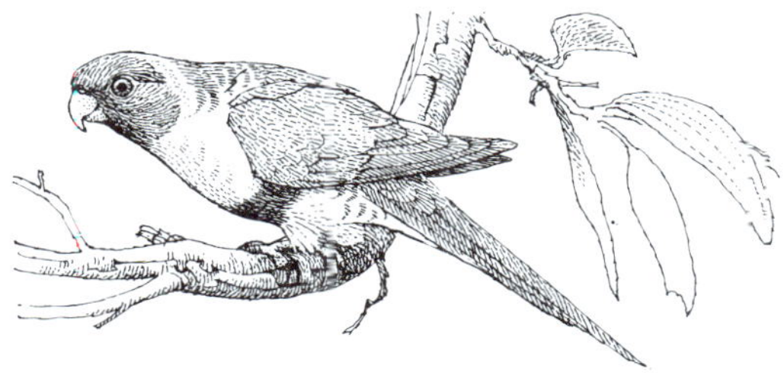

The Red-collared and Rainbow Lorikeets are probably conspecific, both subspecies of the species *haematodus*, which is distributed through southern Indonesia, south-western Pacific islands, New Guinea and adjacent islands and Australia.[635] The Red-collared Lorikeet was treated as a separate species during the Field Atlas.[1581] The Atlases suggest that the gap between the Red-collared and the Rainbow at the eastern end of the Gulf of Carpentaria is narrow, if it occurs at all. There are no records in 10′ blocks between 139°30′E and 140°50′E, an area where difficulty of access has precluded many visits. Breeding is reported throughout the range of the Red-collared Lorikeet in the Historical Atlases though rarely reported during the Field Atlas. A record of a single bird near Connewarre (38/144) in 1978, in company with a Rainbow Lorikeet, has been omitted from the map because it had probably escaped from captivity.

The Red-collared Lorikeet will move long distances. One reached Booby I. (10/141) in Feb. 1979[503] and another was seen on the De Grey R (20/119) in June 1979. The Field Atlas suggests no large-scale seasonal movement but local shifts in population may correspond with flowering of trees. Such movements are most conspicuous at isolated sites such as Groote Eylandt (13/136): some birds stay there throughout the year but large numbers visit Apr.-May when banksias are flowering.[635]

The Red-collared Lorikeet usually lives along watercourses in eucalypt forest, moving into eucalypt and paperbark woodland when the trees flower. It is also present in Darwin (12/130) and other northern towns. The Red-collared Lorikeet takes a variety of fruit and insects as well as nectar[748] often feeding in flocks with the Varied Lorikeet. The Red-collared is one of the commonest parrots in the Top End[635] but in parts of the Kimberley Region is less common than the Varied.

There are no measures of abundance.

Field Atlas Records: 1399

1° blocks-with records: 76 (9%)
-with breeding records: 5 (7%)

Atlas No.: 255

Max. Rep. Rate: 80%
In block(s): 12/133

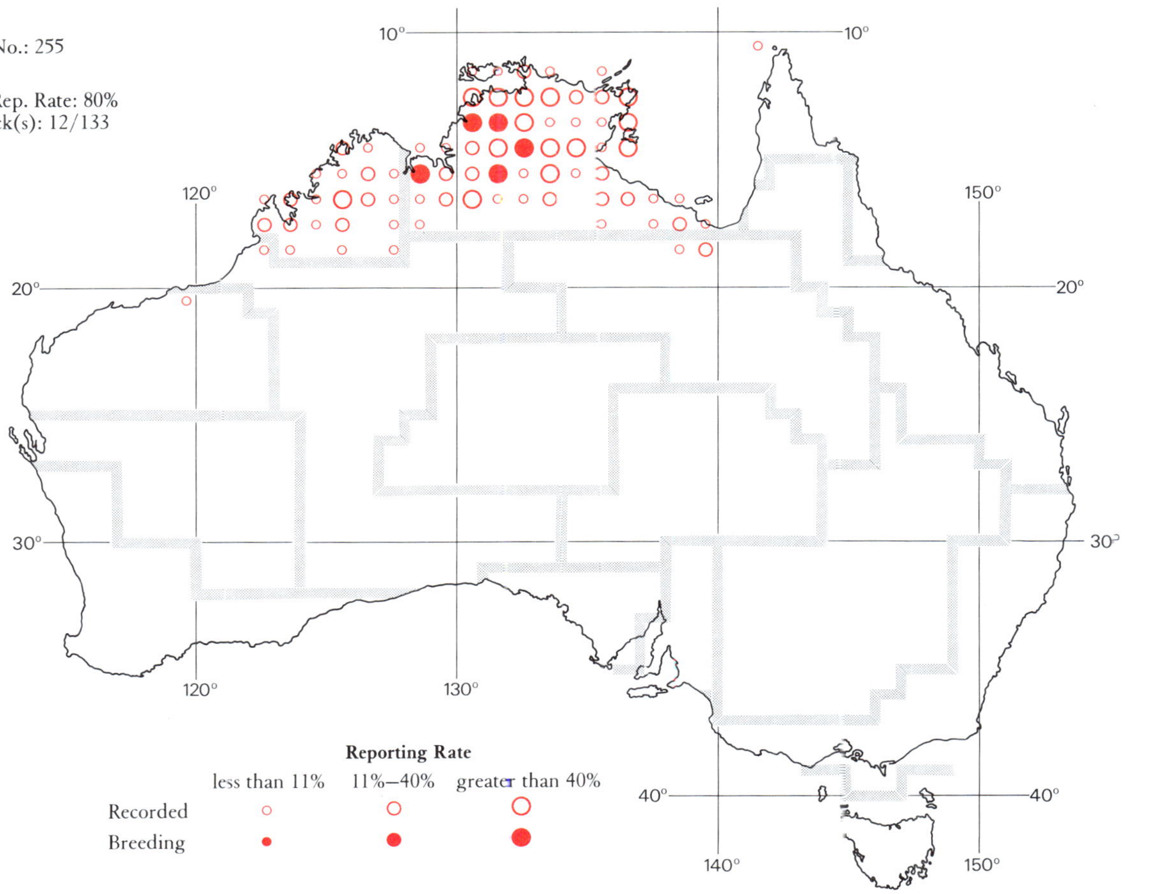

Scaly-breasted Lorikeet

Trichoglossus chlorolepidotus

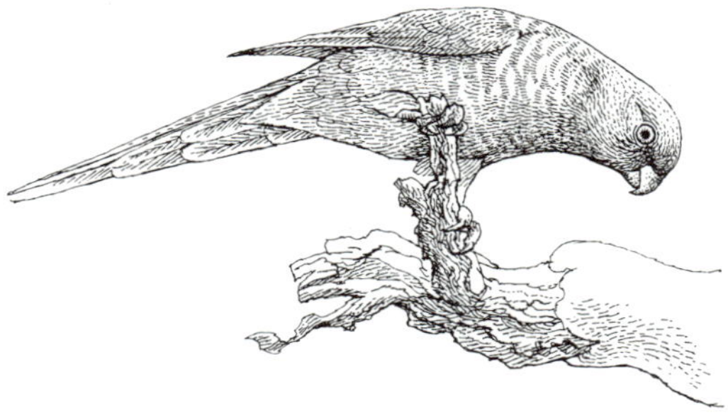

The endemic Scaly-breasted Lorikeet historically has not been recorded north of 15°S. During the Field Atlas an irruption appears to have taken place and from 1978 the species was recorded throughout most of the Cape York Region. There were several unconfirmed records from Groote Eylandt (13/136) at the same time. The species' range is otherwise unchanged from the historical periods except for the small breeding colony round Melbourne (37/145) that became established in the 1970s, probably from aviary escapees. A sighting of Scaly-breasted Lorikeets, presumably escapees, in Perth (31/115) in 1978 has been omitted from the map. The species is sometimes in mixed flocks with the Rainbow Lorikeet with which it has interbred in Sydney (33/151).[747]

The Field Atlas does not indicate any substantial north-south seasonal movement, though the birds move to where there are flowering trees. Inland its distribution extends along river systems where the species is always rare compared with coastal sites. It will live in cultivated lands, including orchards and sorghum crops and suburban gardens.[635] Though described as a lowland bird,[635] in the Atherton Region it was commoner above 600 m than on the coast.[698] The Field Atlas suggests a difference in altitudinal distribution of northern and southern populations. At 17°S the Reporting Rate above 500 m was 24% and below 500 m was 19%, whereas at 30°S the corresponding Reporting Rates were 0% and 55%. The bird feeds on nectar, pollen, fruit, seeds and insects.[635]

There are no measures of abundance.

Field Atlas Records: 3843

1° blocks–with records: 102 (13%)
–with breeding records: 27 (26%)

Atlas No.: 256

Max. Rep. Rate: 59%
In block(s): 26/150

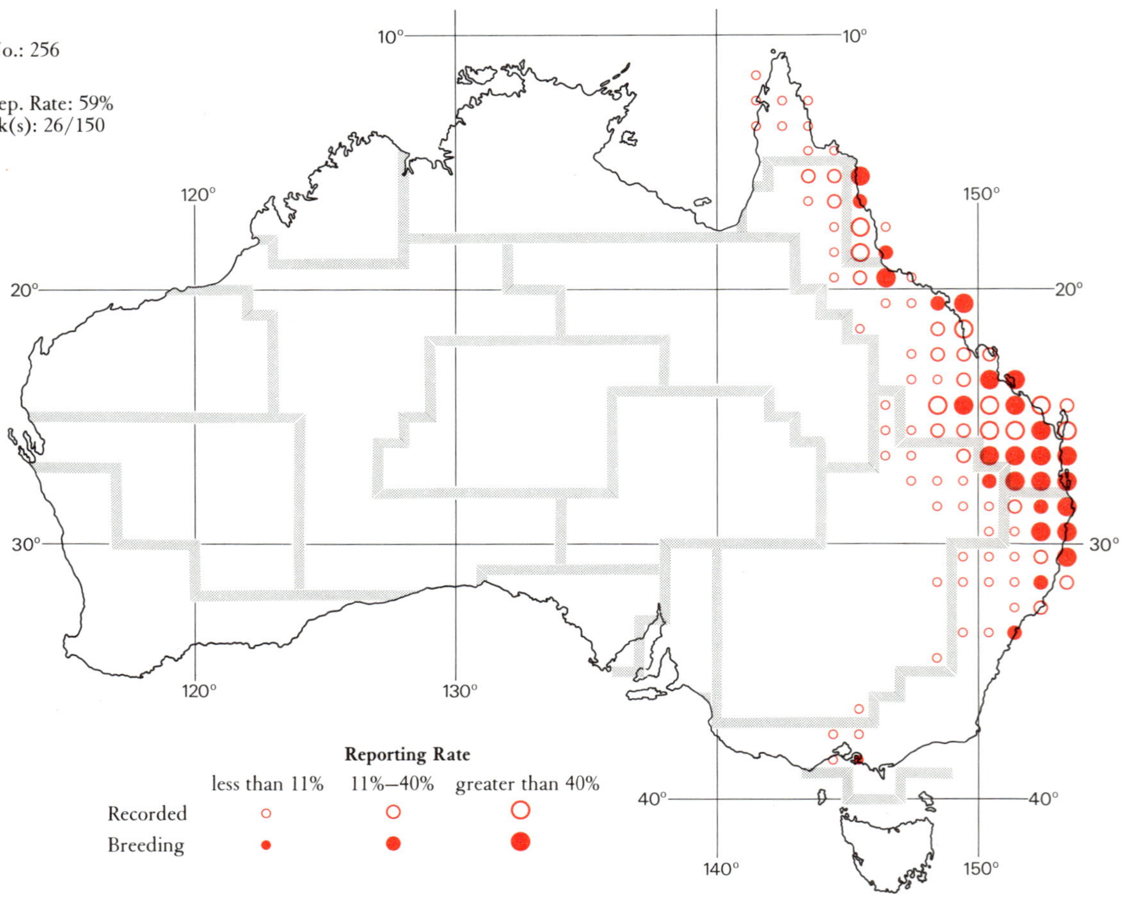

The Historical Atlases fill the gaps in the range of the endemic Varied Lorikeet on the southern coast of the Gulf of Carpentaria and in the Cape York Region. There have been occasional unconfirmed records of the species from the Atherton Region and it is said to have been seen as far south as Gin Gin (25/151).[322]

Movements of the Varied Lorikeet are unpredictable. Sometimes large flocks gather on flowering paperbarks, eucalypts and other species. In Darwin (12/130) it appears irregularly, abundant at times when trees are in flower and usually in greater numbers Apr.-May and Sept.-Oct.[423,475] There is insufficient seasonal information in the Field Atlas to describe such nomadic movements.

The Varied Lorikeet lives in forest and woodland, rarely in mangroves. It is often in mixed feeding flocks with Red-collared Lorikeets and takes nectar, pollen and at times fruit. It is not known to take seeds and it is only inferred that it takes insects. Aviary experience has indicated that it succumbs to dampness and cold draughts,[635] a factor which may account for its absence from the coast of the Eastern Qld Region where the climate can be cold and wet, whereas in the Top End the cool season is dry.

There are no measures of abundance.

Varied Lorikeet

Psitteuteles versicolor

Field Atlas Records: 1010

1° blocks–with records: 121 (15%)
–with breeding records: 4 (3%)

Atlas No.: 257

Max. Rep. Rate: 66%
In block(s): 12/133

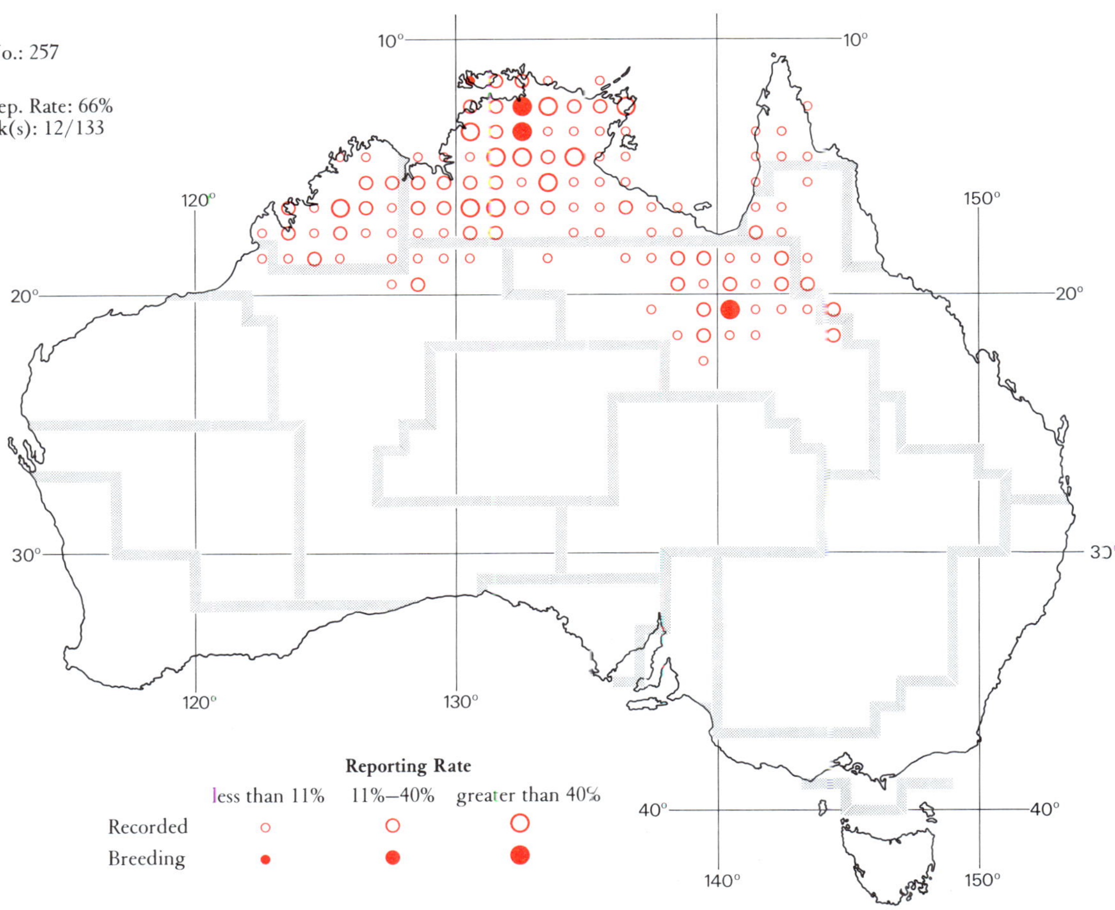

Musk Lorikeet

Glossopsitta concinna

The distribution of the endemic Musk Lorikeet seems to have changed this century. A specimen was collected at Cape Upstart (19/147) in 1843 and in 1924 the species was regarded as an irregular visitor to the Duaringa (23/149) district.[102] There is no breeding record north of 26°S in the Atlases and its appearance farther north must be regarded as intermittent. The southern end of its range seems unchanged. In the Tas. Region, it has not been recorded from Flinders I. (40/148) and rarely on King I. (39/144) but it breeds in Tas. where it is restricted to the east.

Though subspecies are not recognized[388] some apparently isolated populations exist: on the inland side of the Great Dividing Ra. north from Sydney (33/151); on the coast between Sydney and Melbourne (37/145); on the inland side of the Great Dividing Ra. in the lower Murray-Darling Region; in the Mt Lofty Ra. (34/138); Kangaroo I. (35/137); Eyre Pen.; and Tas. There is a small feral population in Perth (31/115).

Movements are erratic. At Murphy's Ck (27/152) the Musk Lorikeet was a regular visitor 1936–42 and then not seen again until 1952.[1099] It was not known to breed there. At Boree (33/148) it appeared when eucalypts were in flower[1344] as it does at Rotamah I. Bird Observatory (37/147).[285] In the Tas. Region the Field Atlas showed little difference between summer and winter Reporting Rates (2.5% and 1.9% respectively) and, though numbers in any one locality fluctuate erratically,[1314,1659] there is no evidence of significant migration across Bass Strait.[734] At least on the fringes of its range, the Musk Lorikeet can be called nomadic.

The Musk Lorikeet lives in eucalypt forest and woodland as well as orchards, parks and gardens. Flocks of more than 200 gather in flowering trees and shrubs, feeding on pollen, nectar, native and cultivated fruit, seeds and insects.[285,635] It is sometimes illegally shot when it feeds on orchard fruit.

The lack of records of the Musk Lorikeet north of 25°S since the 1920s suggests some decline in abundance in northern parts of its range. It is also reported to be no longer as common near to and south of Sydney as formerly.[635]

There are no measures of abundance.

Field Atlas Records: 2453

1° blocks–with records: 86 (11%)
–with breeding records: 22 (26%)

Atlas No.: 258

Max. Rep. Rate: 31%
In block(s): 34/138

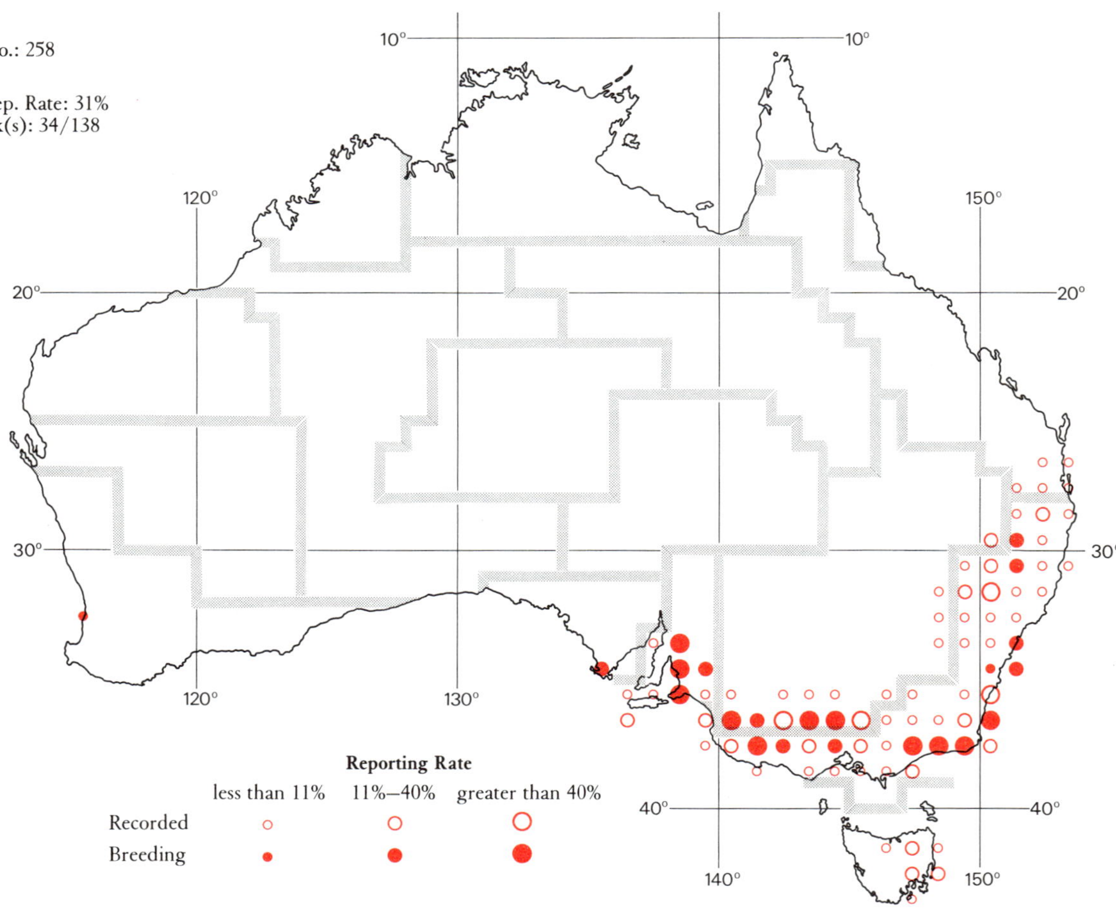

Purple-crowned Lorikeet

Glossopsitta porphyrocephala

The range of the endemic Purple-crowned Lorikeet is disjunct. There is no record in the Atlases between 129°E and 131°E though the species' mobility means that the eastern and western populations may not be isolated and no morphological differentiation between them is recognized.[599] The eastern population is rarely recorded north of the Murray R. When it is, large numbers of birds appear, as at Rand (35/146) in winter 1958[211] and Bega (36/149) in Apr. 1958.[1136] There were also reports of the Purple-crowned Lorikeet in Eastern Qld near Brisbane (27/152) in Jan. 1946 and autumn 1970[1709] and doubtfully from the Tweed R (28/153) district in 1907.[1251] No records link this isolated group of sightings to the main range of the species.

The western population moves north and into the Western Deserts Region at times, with several records as far north as 27°S in May 1969.[599] Near the coast the Field Atlas record at Three Springs (29/115) is the farthest north.

The pattern of distribution of the Purple-crowned Lorikeet varies from year to year, great numbers gathering in places where eucalypts are flowering heavily. Just as spectacular is the total disappearance of the species in some years. Its movements do not seem to be seasonal.

The Purple-crowned Lorikeet inhabits mallee, eucalypt woodland and forest. Dense breeding congregations have been reported in east and west.[308,331] The bird specializes in feeding on eucalypts, extracting pollen from young flowers and nectar from older ones. It can work so fast that food requirements for a day may be satisfied in two hours' feeding, roughly equal quantities of pollen and nectar being taken.[860,861] Though eucalypt flowers form the main food source, it feeds on other blossoms, fruit, seeds[635] and male flowers of casuarina.[370]

At Eyre Bird Observatory (32/126), when the eucalypts flower, the Purple-crowned Lorikeet reaches a density of 6 birds/ha.[464] In a breeding congregation at Dumbleyung (33/117) 55 nests were occupied simultaneously within 2.5 ha.[635]

Field Atlas Records: 2481

1° blocks–with records: 103 (13%)
–with breeding records: 21 (20%)

Atlas No.: 259

Max. Rep. Rate: 46%
In block(s): 34/119

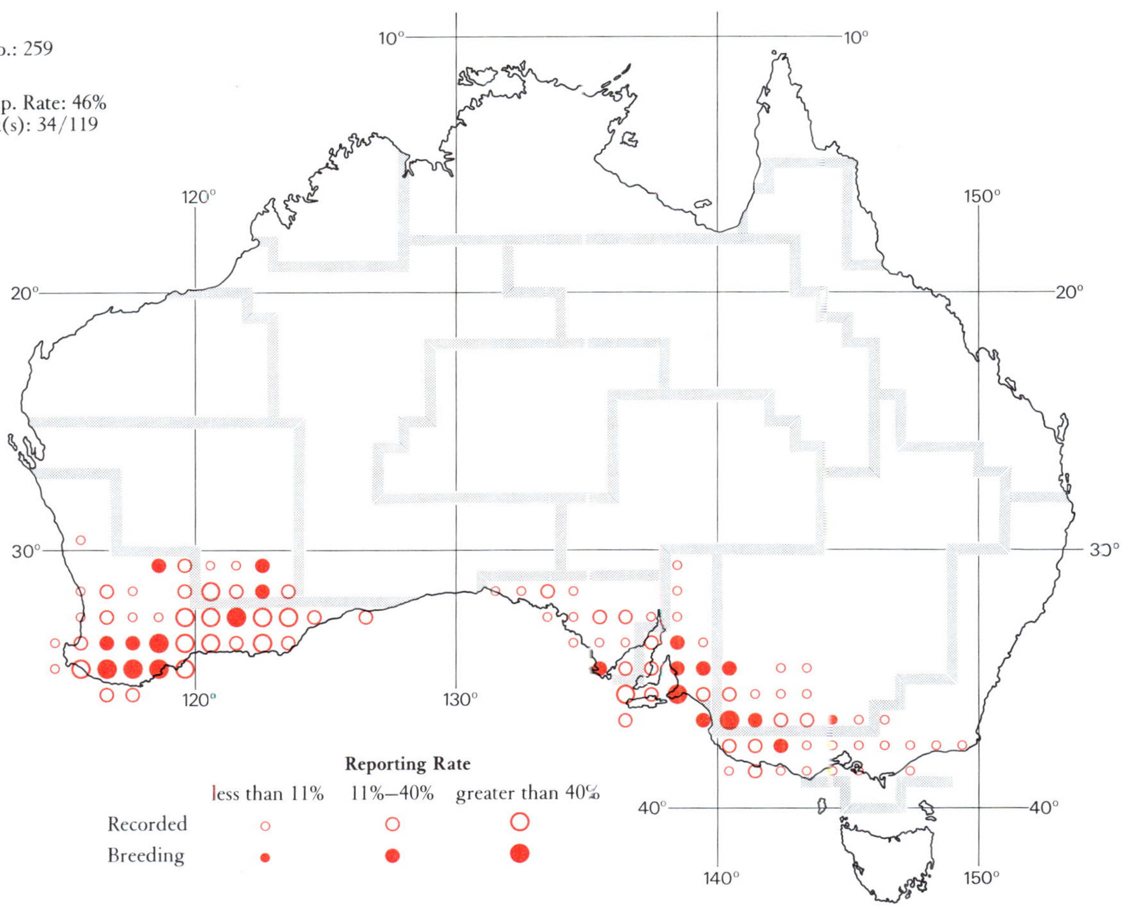

Little Lorikeet

Glossopsitta pusilla

The endemic Little Lorikeet may have declined at the southern end of its range. Formerly a non-breeding visitor to the Mt Lofty Ra. (34/138), the species was last reliably reported there in 1957.[635] In adjacent parts of the South-East Region (west of 144°E) the most recent breeding record in the Atlases was in 1951 at Naracoorte (36/140).

Some early records may have confused Little and Purple-crowned Lorikeets,[635] e.g. the sole record claimed for the Little Lorikeet on Kangaroo I. (35/137) is probably of the Purple-crowned.[1369] In the Tas. Region there are several unconfirmed sightings of the Little Lorikeet but only one appears to be reliable.[1654]

The Field Atlas shows no evidence of seasonality in movements within the species' present range though visits to the Mt Lofty Ra. historically were confined to autumn and winter.

The Little Lorikeet is nomadic, moving about to maintain contact with flowering trees.

The Little Lorikeet stays in the tops of trees, merging with the canopy, and is inconspicuous apart from its chattering calls.[635] South of about 30°S the species is rarely recorded from the higher country of the Great Dividing Ra. It feeds mainly on nectar and pollen, principally from eucalypts.

There are no measures of abundance.

Field Atlas Records: 2162

1° blocks–with records: 107 (13%)
–with breeding records: 20 (19%)

Atlas No.: 260

Max. Rep. Rate: 33%
In block(s): 26/150

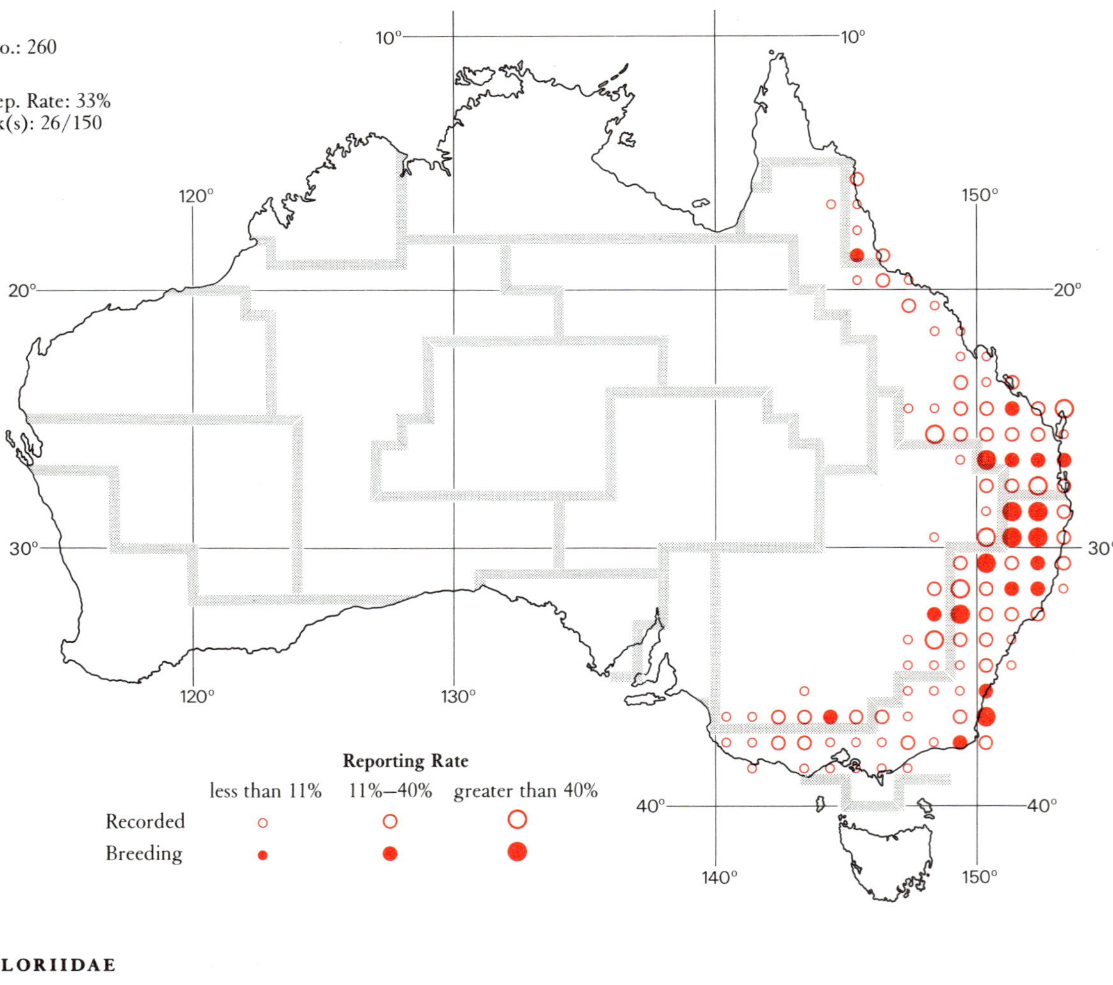

Double-eyed Fig-Parrot

Psittaculirostris diophthalma

The Double-eyed Fig-Parrot lives in New Guinea and adjacent islands and Australia. Three subspecies are recognized in Australia: *marshalli* in the Cape York Region; *macleayana* in the Atherton Region; and *coxeni* from the Conondale Ra. (26/152) and Blackall Ra. (26/152) south to the Macleay R. (30/152).

A set of eggs taken near Maryborough (25/152), claimed to be of *coxeni*, was misidentified but the bird possibly inhabited the area.[633]

The ranges of the two northern populations are unchanged from the Historical Atlases (p. 664) but there were no confirmed records for *coxeni* during the Field Atlas. Reports of this subspecies since 1950 have been only near the Macpherson Ra. (28/152), most recently Koreelan Ck (28/152) in Feb. 1976 and Lamington National Park (28/153) in Dec. 1976.[402,1524] Two birds only were seen on each occasion. Breeding of this subspecies has never been recorded.

Nothing is known of the movements of the Double-eyed Fig-Parrot. It lives where annual rainfall is highest and the forests are most dense. The population *marshalli* inhabits lowland rainforest; *macleayana* lives in both highland and lowland rainforest, sometimes round human habitation, and even breeds in Cairns (16/145).[635] The subspecies *coxeni* also occupied both lowland and highland rainforest but lowland rainforest within its range has almost all been cleared.

The Fig-Parrot will feed in isolated trees but roosts in rainforest, rarely staying long in eucalypt forest. It does indeed principally eat figs, selecting the seeds rather than pulp.

It has also been observed to feed on bark or fungal growth and nectar from flowering trees, sometimes gathering in mixed flocks with lorikeets.[635]

There are no measures of abundance.

Field Atlas Records: 250

1° blocks–with records: 10 (1%)
–with breeding records: 3 (3%)

Atlas No.: 261

Max. Rep. Rate: 44%
In block(s): 12/143

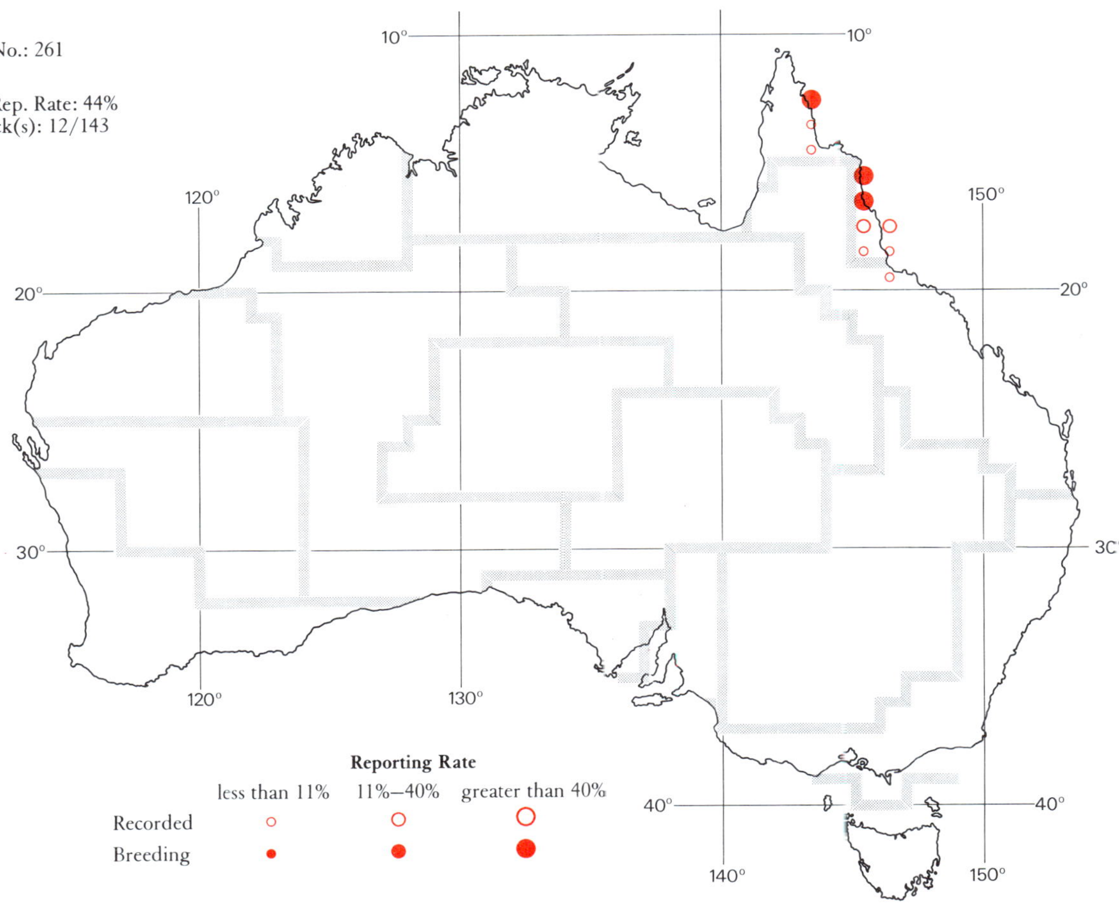

Australian King-Parrot

Alisterus scapularis

There are two subspecies of the endemic Australian King-Parrot: *minor* in the Atherton Region and Eungella (21/148) district; *scapularis* everywhere else.[635] The Eungella population appears to be isolated.

The Field Atlas contains records throughout the historical range of the King-Parrot. No confirmed reports have been made west of 143°E in the lower South-East Region apart from presumed escapees seen round Adelaide (34/138) historically. Other presumed escapees seen along the Murray R (34/142, 35/144) and in the Little Desert (36/141) during the Field Atlas are omitted from the map.

There is evidence of post-breeding dispersal from the highlands in autumn when young birds wander in search of food.[635] Large numbers visit Canberra (35/149) in autumn and winter[653] and flocks of more than 50 are present at Jamieson (37/146) throughout the winter.[1469] A bird banded at Grafton (29/152) has been recovered 270 km south near Krambach (32/152).[635]

The King-Parrot lives in eucalypt forest and rainforest. In the non-breeding season some birds visit more open country including gardens and parks. The species is usually in pairs or small parties; in autumn immatures may band together in flocks.[635]

The King-Parrot feeds on the fruit and seeds of a variety of introduced and native plant species including eucalypts, acacias, angophora, maize and cultivated fruit. Nectar has also been recorded in gut contents.[635,1049] Damage by immature birds to fruit in orchards and crops has led to the bird being shot illegally in some districts.[358,1099]

At Boola Boola (38/146) in eucalypt forest the density of the King-Parrot in the gullies was 0.22 birds/ha.[1113]

Field Atlas Records: 4993

1° blocks–with records: 89 (11%)
–with breeding records: 40 (45%)

Atlas No.: 281

Max. Rep. Rate: 57%
In block(s): 25/148

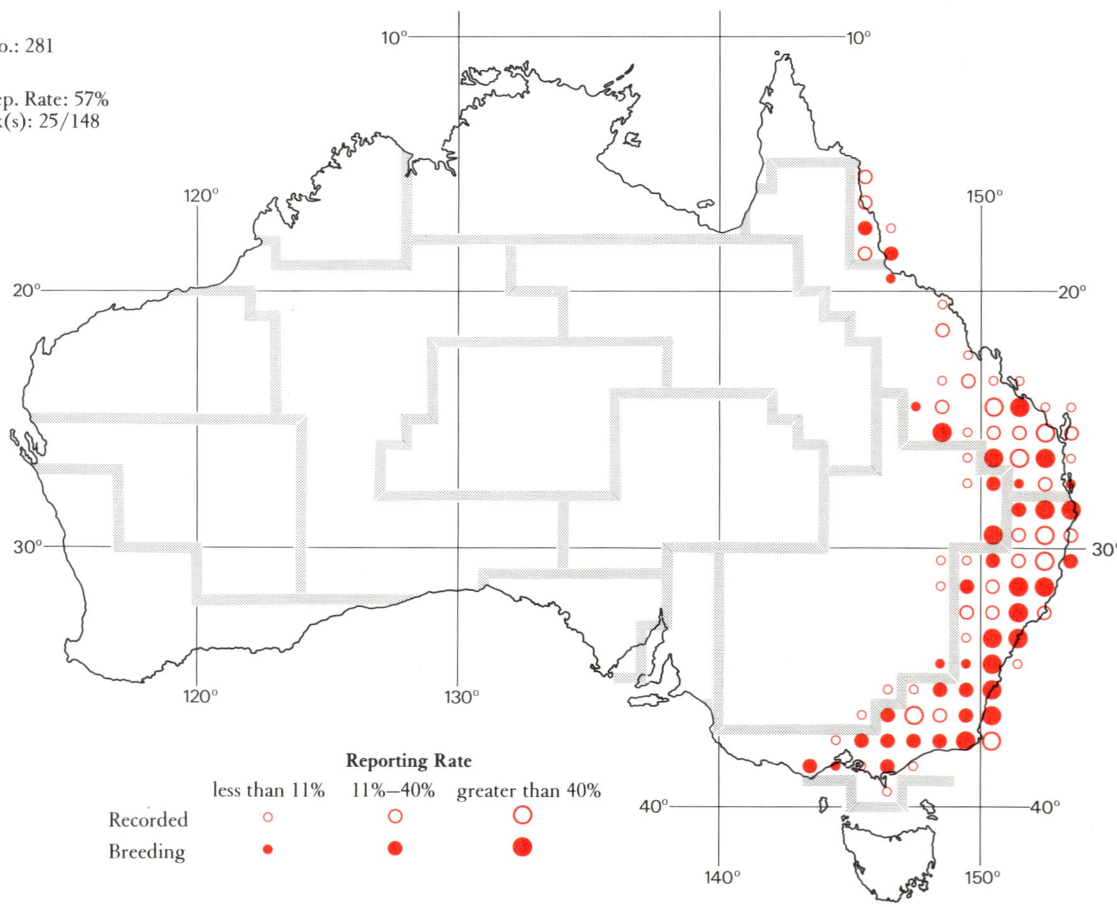

Red-winged Parrot

Aprosmictus erythropterus

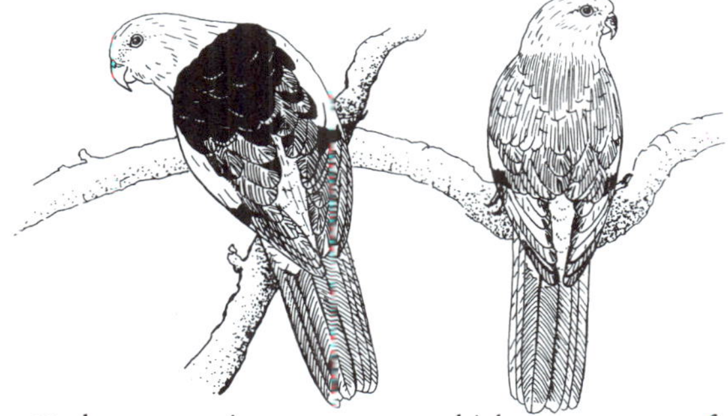

The Red-winged Parrot lives in southern New Guinea and in Australia. Two subspecies are recognized in Australia: *erythropterus* south of about Cooktown (15/145) merging into *coccineopterus* in the Cape York Region and across northern Australia to the west.[635] No gap is evident on the Field Atlas map to separate the northern from the eastern subspecies.

The species is present in the northern L. Eyre Region from time to time and bred near Innamincka (27/140) and Coongie L. (27/140) before 1926.[1464] Twice it has been recorded farther south: in 1919–20, during a drought, Red-winged Parrots appeared near L. Frome (30/139) and Broken Hill (31/141); and in 1966 two were seen near Mutooroo (33/140).[707] The Historical Atlases do not otherwise suggest any change in the species' range. Records of escapees in Sydney (33/151) and Melbourne (37/145) are omitted from the map as is a sighting near Victory Downs (26/132) presumed to be of an escapee.

The Red-winged Parrot appears to be nomadic[635] and there is no evidence in the Field Atlas of large-scale seasonal movement.

In northern Australia the Red-winged Parrot lives in eucalypt woodland and mangroves.[635] At the southern edge of its range, where temperatures tend to be cooler, it inhabits forest along watercourses as well as woodland. In some places it has benefited by partial clearing and has been recorded in large clearings in rainforest.[635] The Red-winged Parrot is usually in pairs or small family groups but flocks of about 60 have been reported during the non-breeding season.[635] It feeds on fruit and seeds of trees and shrubs, including pigface and sometimes crops, e.g. soya. It has also been recorded feeding on lerp[368] and on mistletoe[423] that grows in mangroves, which may account for the bird's presence in coastal areas.

There are no measures of abundance.

Field Atlas Records: 4281

1° blocks–with records: 254 (31%)

–with breeding records: 42 (17%)

Atlas No.: 280

Max. Rep. Rate: 80%
In block(s): 27/149

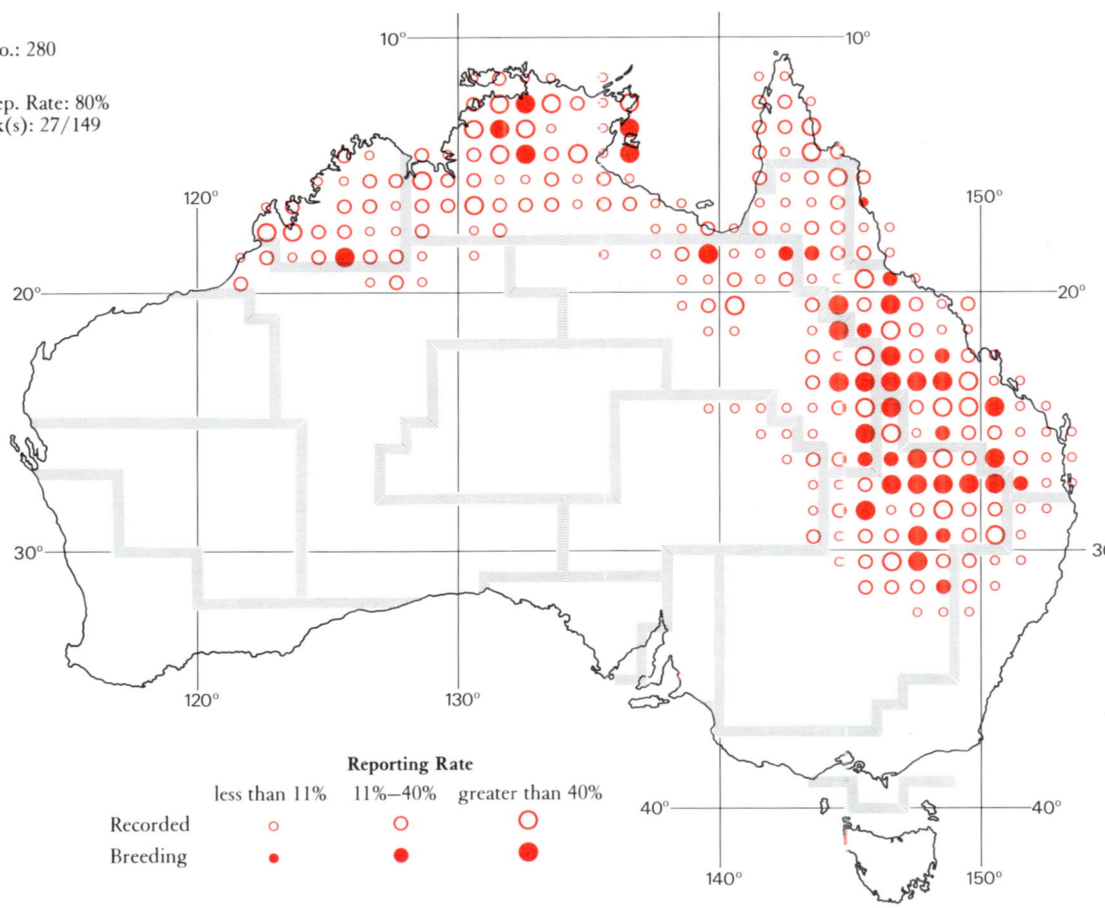

Superb Parrot

Polytelis swainsonii

Reports of the endemic Superb Parrot near Melbourne (37/145) last century are not supported by description or specimens but suggest that the species' distribution may then have extended farther south than the Field Atlas map shows.[120,1129,1843] It is now rarely recorded south of the Murray R. Records of escapees near Sydney (33/151), Greenpatch (35/150) and Melbourne (37/145) have been omitted from the map.

The Superb Parrot has not been reported breeding north of 33°S though sightings are made as far north as 30°S. This relates to a regular seasonal movement apparent in the Field Atlas: in summer no sightings were made farther north than 32°S. Observations show that the species is a visitor to northern parts of its range Apr.-Aug.[396,1596] but Atlas reports show that some birds remain in the breeding quarters throughout the year.

The Superb Parrot usually lives in parties of about 20 birds but post-breeding flocks sometimes number 100 or more.[1593] It inhabits woodland dominated by river red gum. In the west of its range it is confined to near watercourses because the intervening plains are devoid of large trees; in the east, where much of the land between rivers is lightly timbered, the parrot may range much farther from rivers. It forages mainly on the ground on the seeds of grasses and other herbs, crops and weeds, which it sometimes takes green. Flocks have been recorded feeding on spilt grain along roads and railway lines. The species also feeds on fruit and blossom of acacias and eucalypts including the sugar gum, introduced within the bird's range from SA.[1596]

Early agricultural clearing,[1060] laying of poisoned baits for rabbits[1061] and trapping for the bird trade may have reduced the species' abundance. It was said to have declined early in this century and subsequently recovered.[663] More recent changes in land use may be adversely affecting the species again and numbers in the Parkes (33/148) district have declined since the 1950s.[1596]

Near Darlington Pt (34/146), where it was common, 30 were counted in a transect of 3 km in summer.[663] There are no other measures of abundance.

Field Atlas Records: 352

1° blocks–with records: 25 (3%)
–with breeding records: 7 (28%)

Atlas No.: 277

Max. Rep. Rate: 21%
In block(s): 34/148

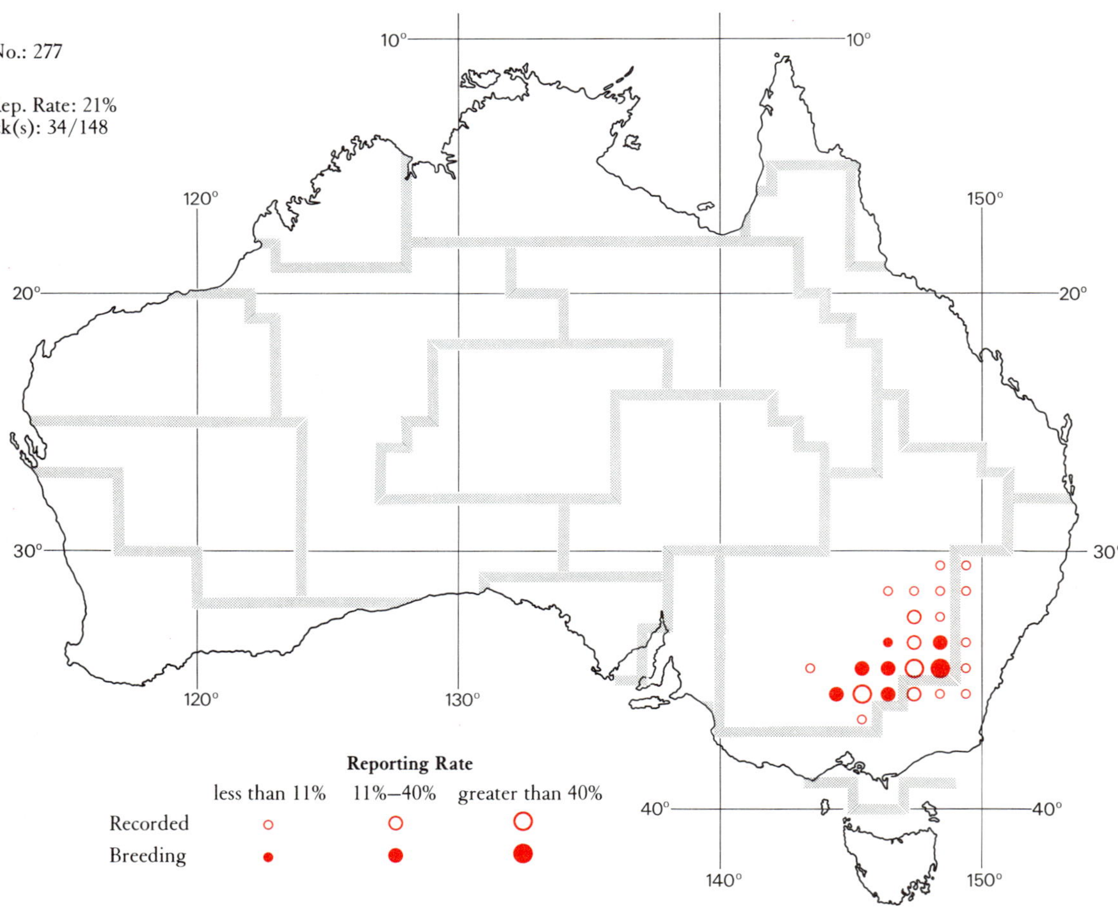

Regent Parrot

Polytelis anthopeplus

The endemic Regent Parrot is considered to be separable into two subspecies though the morphological differences between them are not great. They are *anthopeplus* in the east and *westralis* in the west.[635] The range of *anthopeplus* has declined since the end of last century when breeding was regularly reported on the Wimmera R at least as far south as Dimboola (36/142). It used to be recorded north-east of Murray Bridge (35/139) before 1950.[284] The bird may also have ranged closer to Adelaide (34/138) than it now does.[966]

The Field Atlas suggests no large-scale seasonal movement of the nominate subspecies, though a northward post-breeding dispersal may take place.[635] There are three reports, none confirmed, of the bird on the Murray R (36/146) east of the present distribution (36/146).[125,296,337] The most recent was in 1957 and they suggest that the bird may sometimes move out of its main range. North-east of Murray Bridge it was present seasonally but did not breed.[284]

In the South-West Region a regular pattern of movement is suggested by the Field Atlas. During spring *westralis* is widespread throughout the South-West but as summer progresses congregations are apparent in the southern portion of the range and by autumn the species is mostly south of 32°S. In winter the population spreads out again and the highest Reporting Rates are in the north-east of the region. Winter rainfall enables the bird to feed on fruiting saltbush[334] but its range is reduced when summer dries out the vegetation.

In the east the Regent Parrot inhabits eucalypt woodland along the Murray R and adjacent mallee. In the west it also inhabits mallee but its stronghold is in eucalypt woodland. It is scattered through farmland as well. The Regent Parrot eats seeds, berries, buds, blossom and rarely insects.[1084] It will eat spilt wheat and, south of Hattah (34/142), as many as 300 have been recorded along 3 km of roadway feeding on spilt grain.[635] In so doing many are killed.

The abundance of *westralis* has fluctuated.[1638]

There are no measures of abundance.

Field Atlas Records: 770

1° blocks–with records: 53 (7%)
–with breeding records: 12 (23%)

Atlas No.: 278

Max. Rep. Rate: 31%
In block(s): 34/142

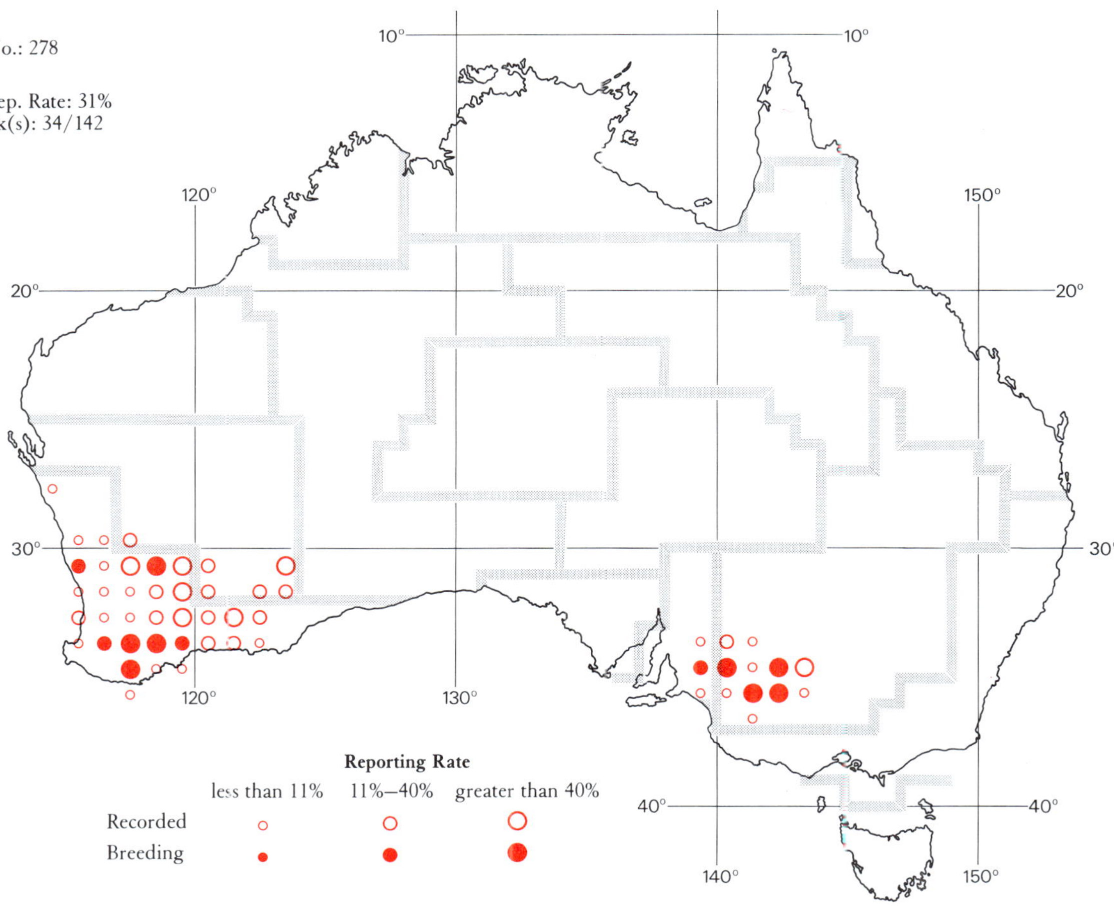

Alexandra's Parrot

Polytelis alexandrae

The range of the endemic Alexandra's Parrot is indicated in a combination of the Historical (p. 665) and Field Atlas maps. Doubtful records from the northern L. Eyre Region have been omitted. Breeding was first recorded in 1889 at Crown Pt (25/134) and widespread in 1894 when numbers nested on the Hale (23/134) and Hugh (24/133) Rs and other watercourses in the district.[1343] Breeding records have come from widely scattered places and many birds are usually involved. Breeding was reported in the Central Ra. Region in 1889, 1894, 1903, 1912, 1920, 1933 and 1968 and in the Goldfields Region in 1902, 1903 and 1909. The breeding record (20/130) in the Western Deserts Region was of a single pair in 1968.

Alexandra's Parrot is highly nomadic. It appears in a district, breeds, leaves soon after the young have fledged, then is not seen again for many years.[635]

Sightings in the Field Atlas were of single birds, pairs, two groups of three, one of seven and one of 15 adults at water (22/121). The species inhabits desert vegetated with casuarinas and spinifex;[532] and nests in eucalypts along watercourses and in casuarina woodland far from water. Five or more nests have been found in one tree.[1343] Alexandra's Parrot feeds on seeds of spinifex, acacias and desert herbs, green casuarina leaflets[532,635] and perhaps also takes blossom and insects as do other *Polytelis* parrots.[635] It is apparently tame and feeds much on the ground.[1343]

Because Alexandra's Parrot is highly nomadic and lives in areas infrequently visited its status is difficult to determine. Only one breeding episode involving many birds has been recorded since 1933, which supports a recent suggestion that the species may be declining.[1685]

There are no measures of abundance.

Field Atlas Records: 16

1° blocks–with records: 11 (1%)
–with breeding records:

Atlas No.: 279

Max. Rep. Rate: 4%
In block(s): 20/124,20/125,24/128

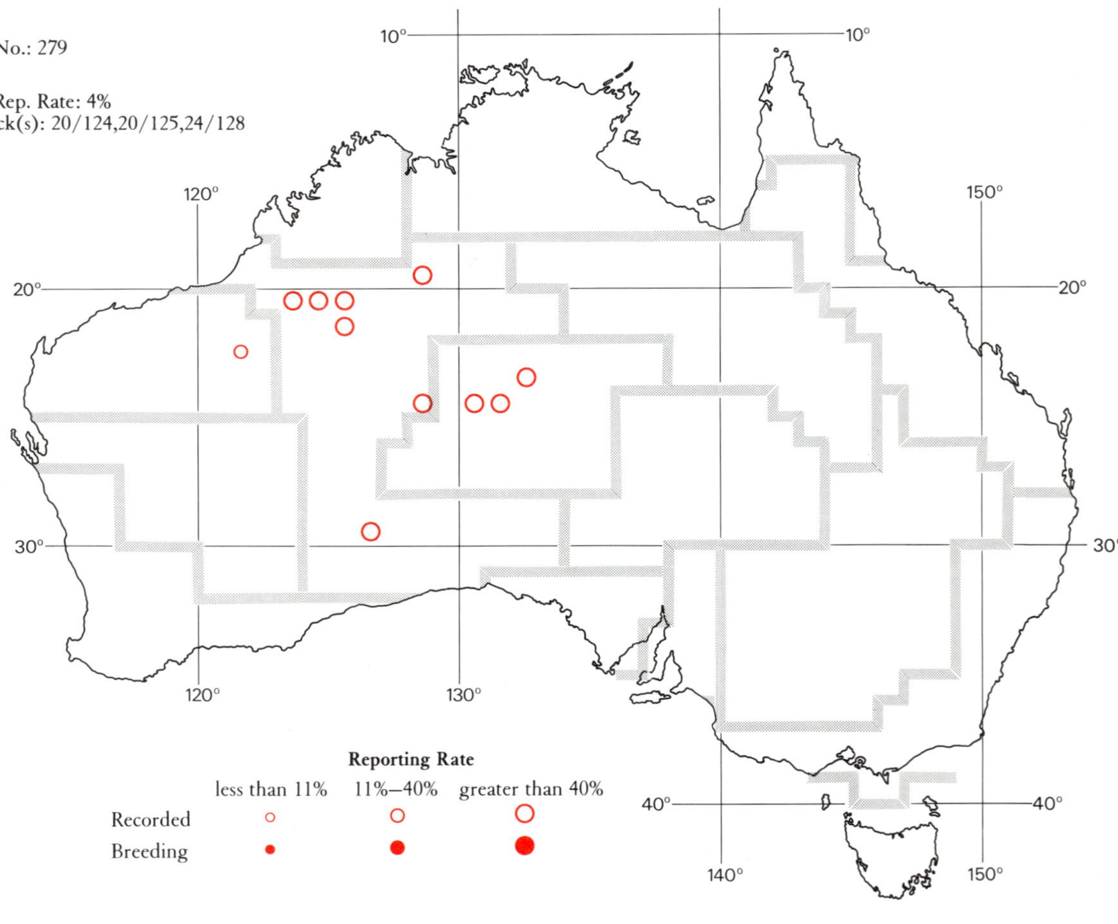

Cockatiel

Nymphicus hollandicus

Populations of the endemic Cockatiel are morphologically uniform[602] and have an unbroken distribution; records in the 1970s from the lower Western Deserts Region link the populations of the Central Ra. and Goldfields Regions. The Atlases suggest no alteration of range since European settlement.

Reports of the Cockatiel in the Tas. Region, including one during the Field Atlas, are probably of escapees rather than vagrants. Other sightings of probable escapees including some round Sydney (33/151) have been omitted from the map.

The Field Atlas suggests that the Cockatiel moves south in summer and north in winter in both eastern and western Australia. South of 36°S in eastern Australia the Reporting Rate in summer was 7% and in winter 1%. In WA south of 30°S the Reporting Rate in summer was 8% and in winter 4%. In the Cobar (31/145) district the species is a regular spring-summer visitor in varying numbers; it often arrives and breeds after heavy rain.[1573] In the southern Murray-Darling also, the species is regarded as a spring-summer visitor.[833] In the Top End the Cockatiel moves north each winter, some years coming as far as Darwin (12/130) and in others no farther north than Pine Ck (13/131).[423] During the Field Atlas there were no reports of the Cockatiel in the Darwin district in summer. In several inland localities there are reports of the species becoming common, after rain or during good seasons, in places where normally it is rare or absent.[635]

The Cockatiel lives in acacia scrub, eucalypt woodland, pasture, wheatland and irrigated orchards. Usually it is in pairs or flocks, rarely exceeding 100 birds, but flocks of 1000 or more gather at dumps of rice husks, e.g. near Colleambally (34/145).[390] The Cockatiel feeds on the seeds of grasses, trees and shrubs, particularly of acacias.[635] It will occasionally eat fruit in orchards and is a serious pest in sorghum and sunflower crops. It may have increased in abundance since permanent waters for domestic stock have been established across the inland.

There are no measures of abundance.

Field Atlas Records: 8597

1° blocks-with records: 515 (63%)
-with breeding records: 88 (17%)

Atlas No.: 274

Max. Rep. Rate: 89%
In block(s): 27/149

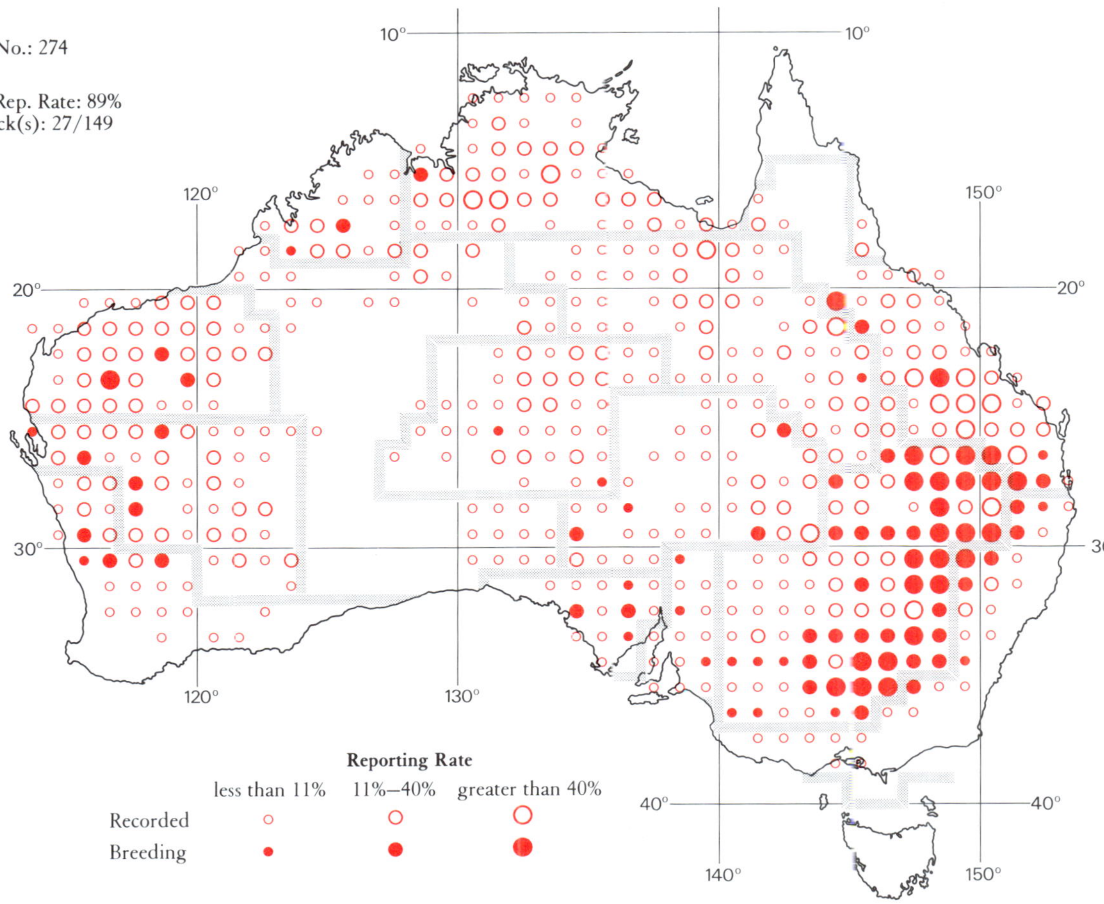

Ground Parrot

Pezoporus wallicus

There are two subspecies of the endemic Ground Parrot: *wallicus* in the east and *flaviventris* in the west. The range of both subspecies has contracted to isolated pockets (p. 665). The main loss of *wallicus* has been near the urban centres of Sydney (33/151) and Melbourne (37/145) and along the coast between Melbourne and Adelaide (34/138). It has disappeared entirely from the SA Ra. Region and from King (39/144) and Flinders (40/148) Is and the Kent Group (39/147). It appears to have become scarce in eastern Tas. but south-western Tas. remains a stronghold. The range of *flaviventris* has contracted to isolated pockets along the coast of the South-West Region. It is known still to be near West Cape Howe (35/117) though sightings were not submitted to the Atlas[1189] and a systematic search may reveal other small populations surviving in this region.

Ground Parrots probably live as sedentary territorial pairs but there is substantial post-breeding dispersal, presumably of the year's young. Individuals have moved at least 120 km[1253] and, in winter, records have been made far from breeding grounds.

The Ground Parrot inhabits only heath and is restricted to that providing thick cover and a high density of food plants.[1253] It eats a variety of small seeds and an adequate supply of food may be provided either by a high density of one or a few seeding sedges or by a diverse array of heath plants, none very common but all producing suitable seeds.

In heath the frequency and intensity of fire determines whether the vegetation cover and food supply are suitable for the Ground Parrot.[1253] Different types of heath respond differently and are suited to the species at different ages after fire. In sedgelands fire is infrequent and unimportant to Ground Parrot populations. Some populations have been elimated by drainage and clearing of swamps and trampling by cattle.[384,635]

By listening for calls at dusk, the Ground Parrot can be located even in dense vegetation. In Vic. densities are 0.05–0.30 birds/ha; peaks occur 5–10 years after fire in heath dominated by shrubs such as casuarina and banksia and 15–18 years after fire in heath dominated by sedges and grass trees.[1252] Near Strahan (42/143) density was 0.22 birds/ha.[1317]

Field Atlas Records: 262

1° blocks–with records: 28 (3%)
–with breeding records: 6 (21%)

Atlas No.: 311

Max. Rep. Rate: 18%
In block(s): 37/150

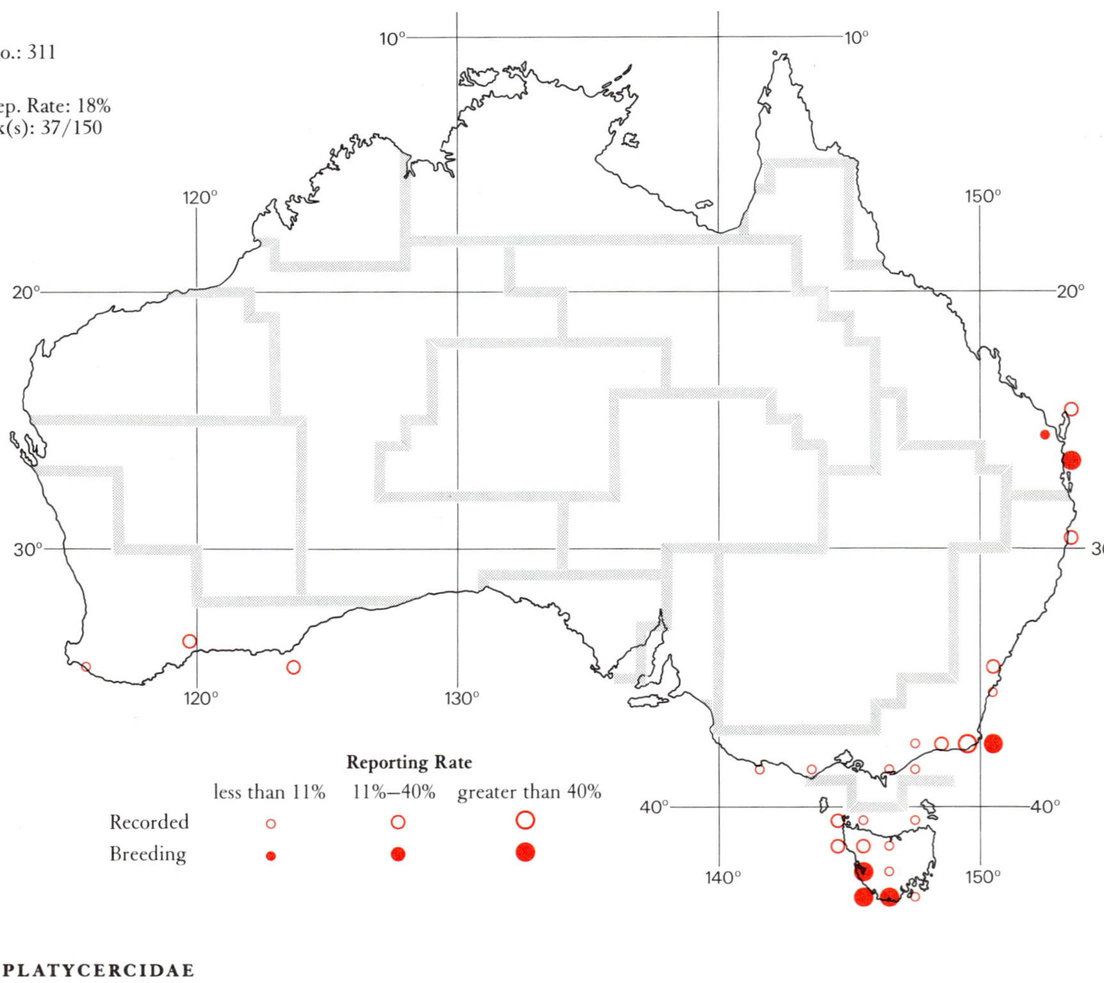

Night Parrot

Geopsittacus occidentalis

Sightings of the endemic Night Parrot are very rare. Its possible range is indicated by the combined records of the Field and Historical Atlases (p. 665). The greatest number of reports was made in the 1870s when up to 16 specimens were collected in the L. Eyre Region.[635] The Atlases contain no reliable records between 1935 and 1978 except for two in 1967. All the confirmed Field Atlas reports as well as some additional unconfirmed sightings were made in 1979–80: one bird south of Muncoonie L. (25/138), four near L. Perigundi (27/139), one near Meentheena (21/120) and one north of Glenayle (25/122).

The Night Parrot is thought to be nomadic.[1588] An observer familiar with the species in the L. Eyre Region last century, when it was more abundant, noted that it came and went according to the seasons.[299] A study of the records in SA concludes that samphire flats associated with lake systems may be important to the survival of the Night Parrot, perhaps specially between years of heavy rain; when rain brings the spinifex into flower and seeds ripen, the parrot may move out from the lakes to the spinifex sandplains.[1369] A survey of all locations in WA where the Night Parrot had been recorded showed that there were large freshwater pools at each site, often surrounded by mulga with an understorey of saltbush and other shrubs. There was always spinifex nearby.[164] The contents of the only crop examined included seeds of spinifex and of herbage growing on limestone.[1899]

The Night Parrot has certainly declined in abundance but may still be sparsely distributed throughout much of its former range. There are no measures of abundance.

Field Atlas Records: 4

1° blocks–with records: 4 (1%)
–with breeding records:

Atlas No.: 312

Max. Rep. Rate: 4%
In block(s): 25/138

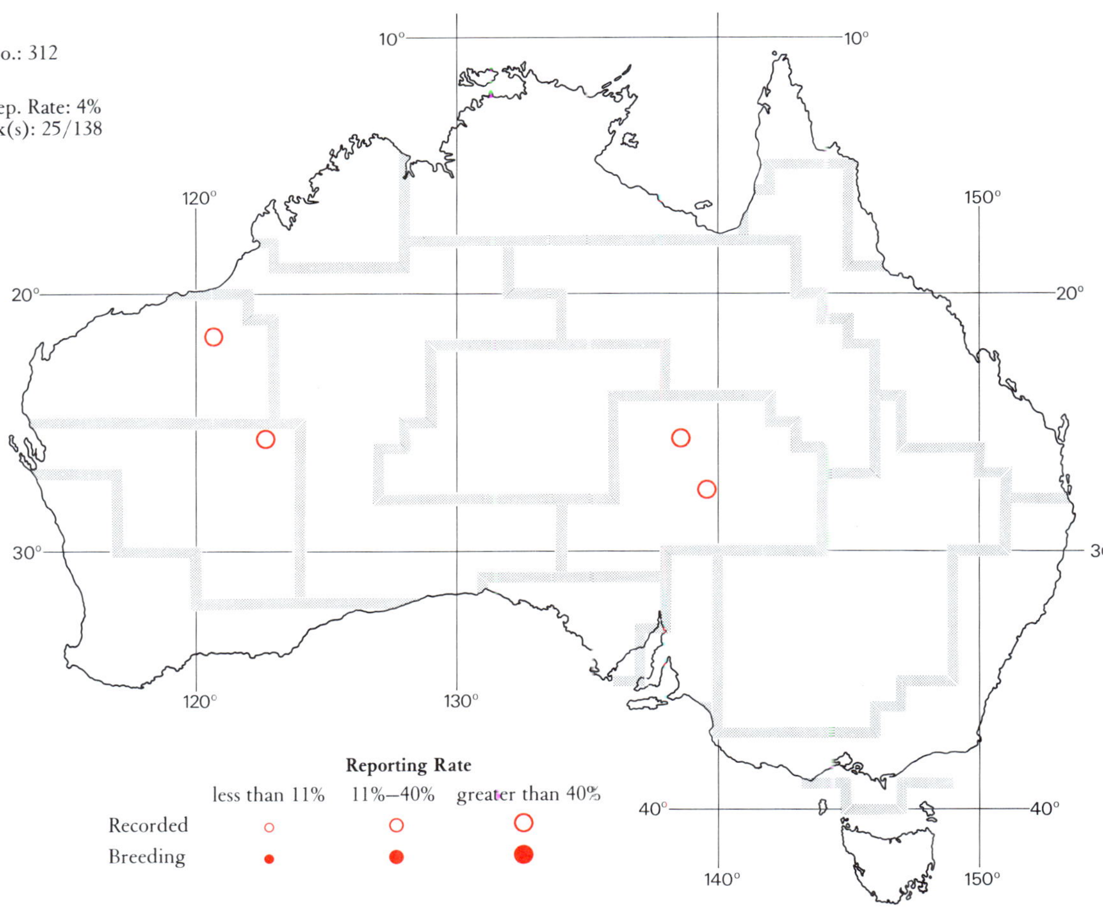

Budgerigar

Melopsittacus undulatus

Assessment of the range of the endemic Budgerigar is complicated by the presence of escapees of this widely-kept aviary bird. Some records on the map round capital cities and, in NSW, those east of the Great Dividing Ra., may be of escapees. There are no records of other than escapees from the Tas. Region.

The Field Atlas suggests that there is a seasonal shift of the Budgerigar population towards the south in summer and away from it in winter (Reporting Rates south of 26°S: summer 2.9%, winter 1.4%). Consistent with this are the reports of several observers who have noted the species to be a spring-summer visitor to their areas, e.g. Cobar (31/145),[1573] lower Murray-Darling Region[833] and Swan Hill (35/143) district.[635]

Local weather may disturb the pattern: at Cobar birds did not arrive one summer until drought-breaking rain had fallen; in the lower Murray-Darling birds appear only in good seasons.

The Budgerigar lives in acacia scrub, tussock grassland, spinifex, eucalypt woodland, mallee and farmland usually in or near woodland along watercourses. It feeds mostly on the seeds of grasses. In inland eastern Australia the most important species was mitchell grass; seeds of chenopods also formed a significant part of the diet.[1922] Breeding takes place in winter in northern Australia, in the hot months (Oct.-Mar.) in the middle latitudes and in spring in the south.[1923] In the north rain falls in summer leading to an autumn crop of grass seeds but in the south the main rains in winter lead to a spring crop.[1333] The Budgerigar follows these crops, rarely roosting more than a few kilometres from a feeding ground and breeding where seeds are abundant.[1923] The Budgerigar is rarely far from surface water and during drought flocks of thousands gather at available water. In Dec. 1972, 10 000–15 000 were visiting Ellery Ck Big Hole (23/133) daily to drink.[635] In those parts of the continent developed for grazing the open stock waters provided may have enabled the bird to extend its range.[635]

There are no measures of abundance.

Field Atlas Records: 3463

1° blocks–with records: 521 (64%)
–with breeding records: 97 (19%)

Atlas No.: 310

Max. Rep. Rate: 48%
In block(s): 22/122

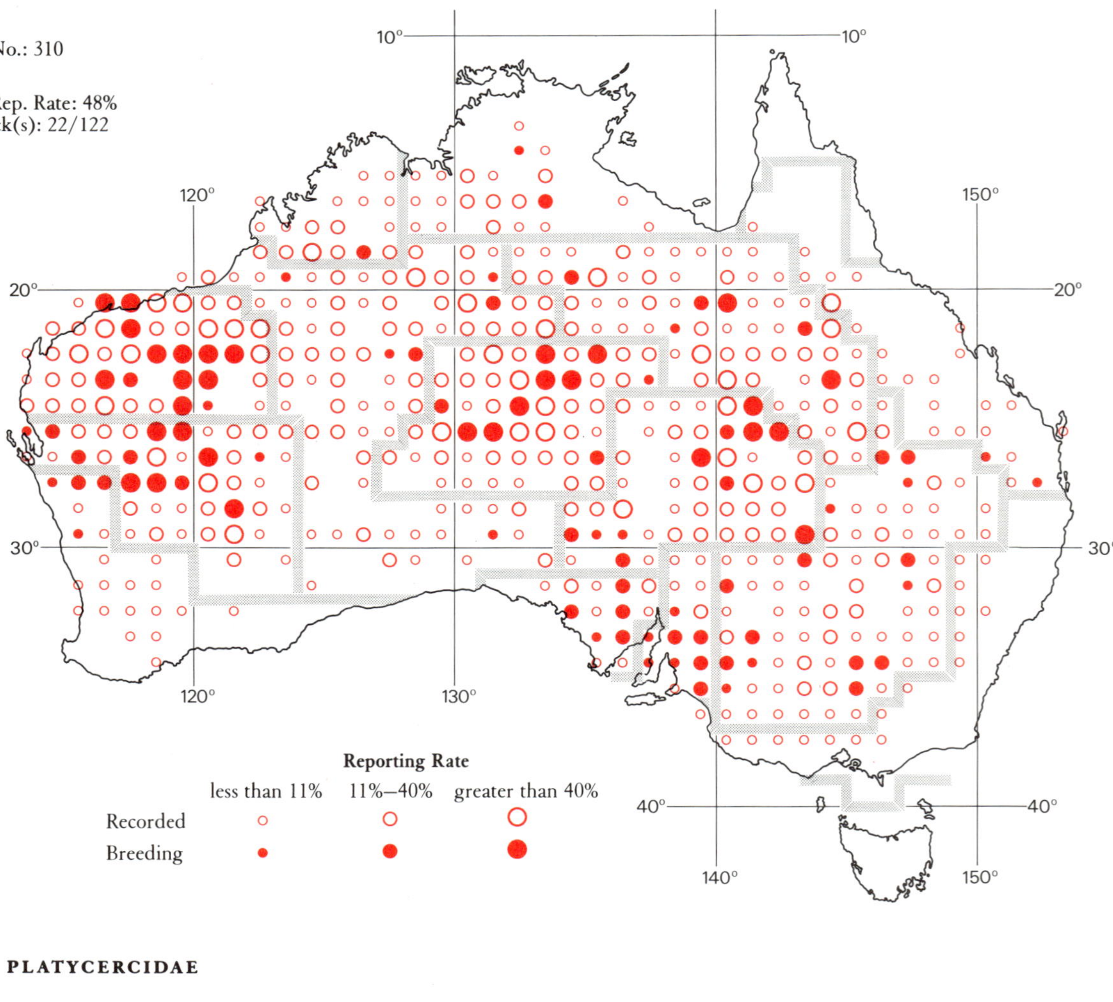

Swift Parrot

Lathamus discolor

The endemic Swift Parrot is known to breed only in Tas. in the east and midlands. After breeding there appears to be a general dispersal into western Tas. where the bird is not recorded before Jan.[259] Then in autumn the bulk of the population migrates across Bass Strait to the mainland[635,818] (Reporting Rates in the Tas. Region: summer 9.8%, winter 0.6%; South-East Region: summer 0.1%, winter 0.8%).

Flocks of 20–30 Swift Parrots are reported to fly north over King I. (39/144) in Mar., returning south in Sept.[739] In some years the species crosses the Murray R in considerable numbers. During an irruption in May-Nov. 1958 several localities near Sydney (33/151) were estimated to hold thousands of Swift Parrots: one flock of 500–600 was observed; and more than 1000 individuals moved past a point at Lane Cove (33/151) within 20 minutes.[818] In the same year the species also appeared as far north as Tenterfield (29/152) and west to Mildura (34/142). It has moved as far north as Duaringa (23/149)[102] and appears on a list for Port Denison (20/148) in 1871.[487] Before 1950 the Swift Parrot regularly visited the Mt Lofty Ra. (34/138)[322] but it is now only intermittently reported there.

The Swift Parrot inhabits eucalypt forest and woodland and is almost invariably in small flocks. Many appear where eucalypts are flowering in profusion. Because these abundances rarely occur at the same place in successive years, neither do the birds. The Swift Parrot feeds mainly on nectar, pollen and lerp and some of the largest irruptions of the Swift Parrot into NSW coincide with a great abundance of lerp.[818] In Tas. the blue gum is the major food source during the breeding season; the dominant eucalypt of western Tas., *E. nitida*, does not flower until middle to late summer, perhaps explaining the Parrot's absence from this area early in summer.[259] The Swift Parrot also eats fruit and seeds of native and exotic plants and is common in suburbs when plants are flowering or in fruit. In Tas. fruit such as plums and apricots are eaten.[217]

There are no measures of abundance.

Field Atlas Records: 736

1° blocks-with records: 5 (6%)
-with breeding records: 3 (6%)

Atlas No.: 309

Max. Rep. Rate: 22%
In block(s): 42/148

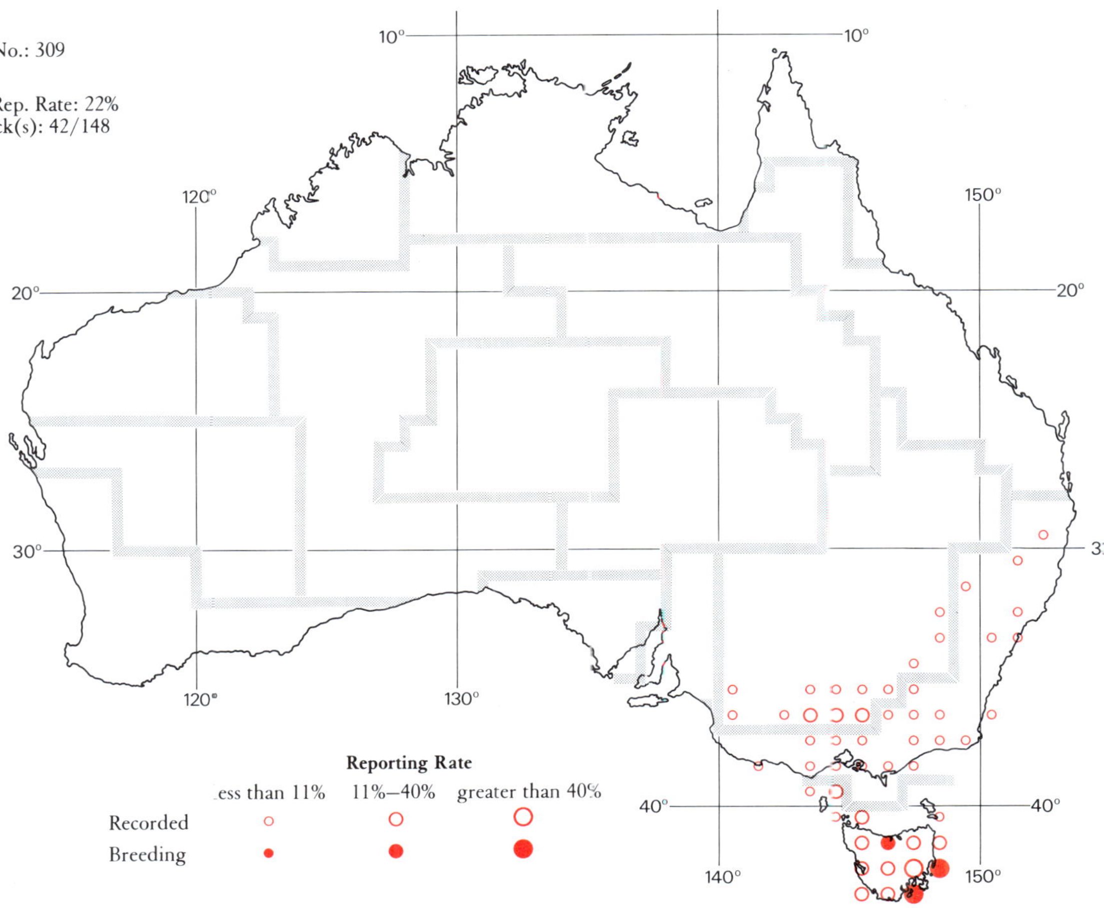

Red-capped Parrot

Purpureicephalus spurius

In the 1930s the endemic Red-capped Parrot bred as far north as Dandaragan (30/115)[592] and was seen in that area in the 1970s. Otherwise the Field Atlas range is unchanged from historical periods. The bulk of the population lives within 100 km of the coast, though small parties may be found inland. Adults are sedentary but immatures may wander, flocks of more than 20 being reported.[635]

An important factor in the distribution of this Parrot is the marri, the seed of which is the mainstay of its diet in the central part of its range. The marri nut contains large seeds which are easily extracted by the specially shaped bill of the Red-capped Parrot, providing it with reliable food all the year.[1514] The marri is now more widespread and abundant than at the time of European settlement. After firing or logging of the jarrah forest, marri, regarded as a weed species, grows thickly and fruits when quite small. The spread of marri has probably favoured the Red-capped Parrot. It does take seeds of other plants, particularly of eucalypts and casuarinas and of introduced weeds.[1084] Fruit, blossom, lerp and insects are also eaten.[635] The Red-capped Parrot is regarded as a nuisance in orchards and is often shot, legal only after a permit has been obtained.

There are no measures of abundance.

Field Atlas Records: 1220

1° blocks–with records: 20 (2%)
–with breeding records: 12 (60%)

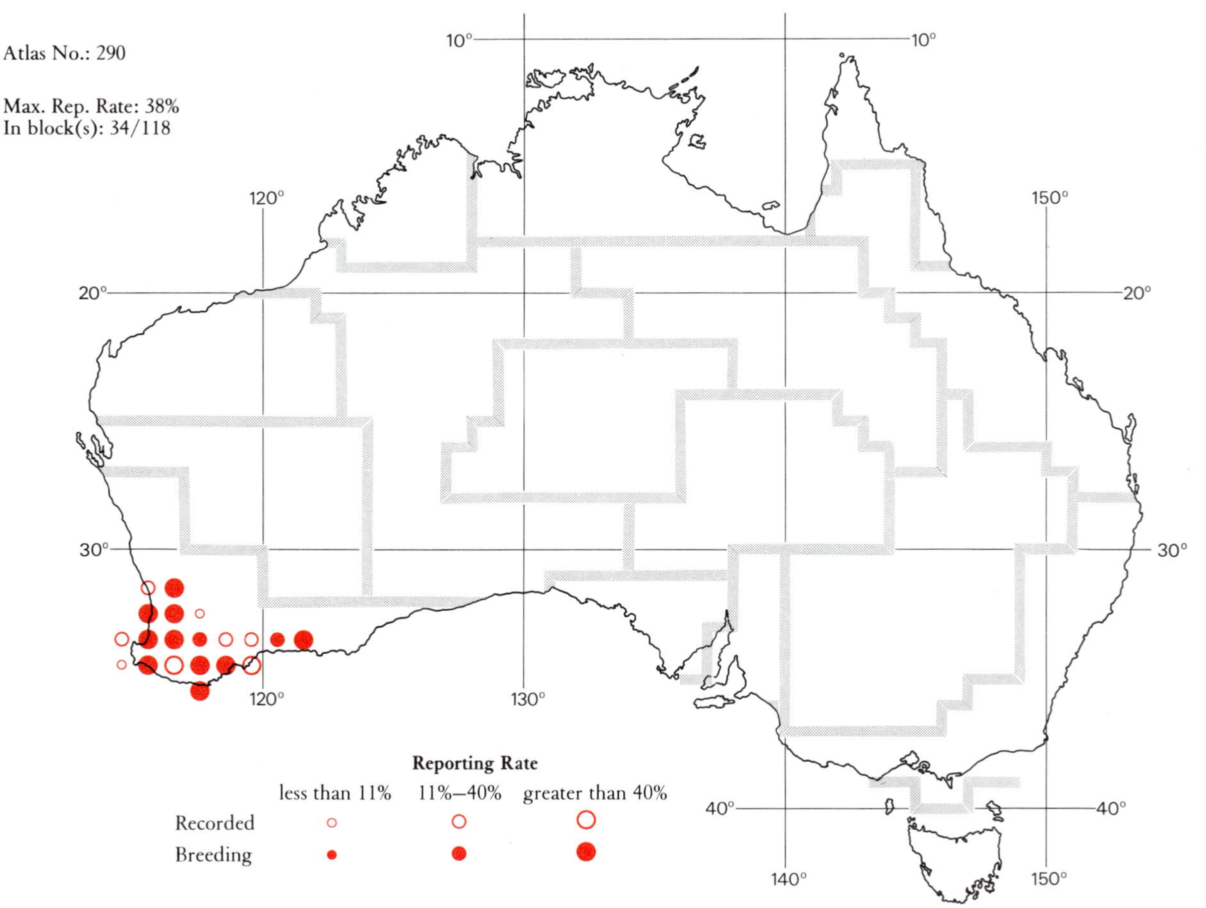

Green Rosella

Platycercus caledonicus

The Green Rosella is endemic to and ubiquitous in the Tas. Region, breeding on both King (39/144) and Flinders (40/148) Is, as well as Tas., though no breeding reports came from the islands during the Field Atlas. As with many parrots, aviary escapees are common and Green Rosellas have been seen round Sydney (33/151).

There is no evidence of large-scale seasonal movement.

The Green Rosella lives and breeds in eucalypt forest and visits rainforest, heath with isolated patches of trees, cleared country, orchards and gardens. During spring and summer it lives in pairs but in autumn and winter birds of all ages form flocks; one of 80 birds has been recorded.[635] It is present in winter near the top of Mt Wellington (42/147)[1452] and in western coastal rainforest,[504] tolerating severe wet and cold.

The Green Rosella feeds on seeds of grasses, shrubs and trees, including eucalypt, myrtle and sassafras,[259] native and introduced fruit, blossom and insects.[635]

On King I. numbers are said to have declined this century as a result of extensive forest clearing.[739] In the Tas. region as a whole the Green Rosella is common and was the fourth most frequently recorded species during the Field Atlas. There are no measures of abundance.

Field Atlas Records: 3133
10′ blocks–with records: 324 (88%)
–with breeding records: 23 (7%)

Atlas No.: 285

Max. Rep. Rate: 90%
In block(s): 42°05′/146°45′

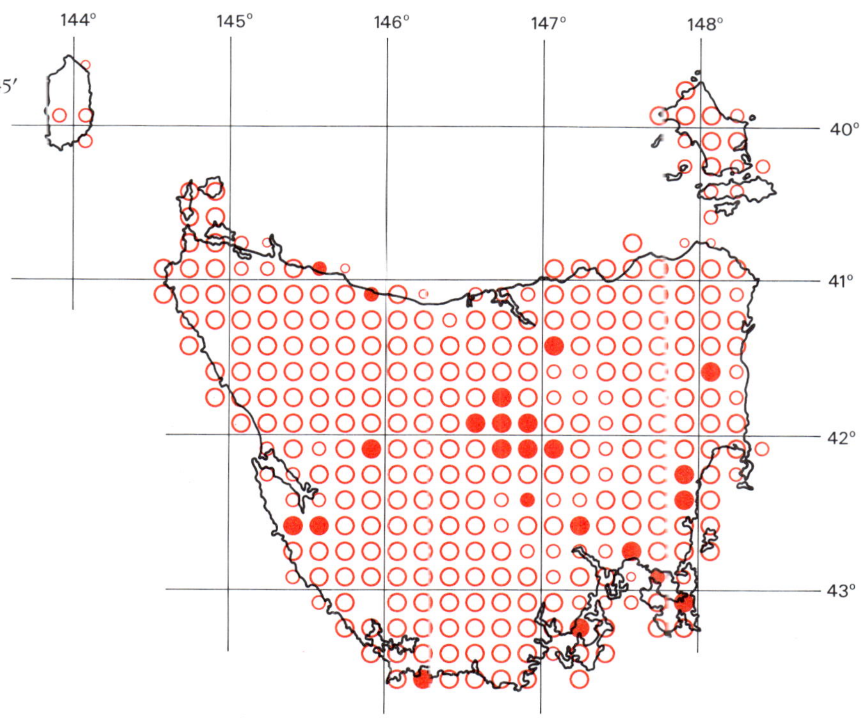

Reporting Rate

	less than 11%	11%–40%	greater than 40%
Recorded	○	○	○
Breeding	●	●	●

Crimson Rosella
Adelaide Rosella
Yellow Rosella

Platycercus elegans

The endemic Crimson, Adelaide and Yellow Rosellas are closely related. Views differ about whether they should be regarded as species or subspecies.[388,635,1581] Records of the three were collected separately for the Atlases. The Crimson Rosella was introduced to New Zealand and Norfolk I. (29/167); in Australia it has at least four isolated populations: in the Atherton Region; the Eungella (21/148) district; southward to the extreme south-east of SA; and on Kangaroo I. (35/137).[635] The Adelaide Rosella has northern and southern populations said to be separated by a gap of about 60 km.[635] No such gap is evident on the 10′ maps, though breeding records are absent from 33°0′-33°40′S. The Yellow Rosella lives along the Murray R and its tributaries including the Darling R upstream to about Tilpa (30/144). Any of these rosellas beyond its usual range is likely to be an aviary escapee, e.g. Adelaide Rosella in Sydney (33/151). Only local wandering by immatures and some nomadic movement at the edges of the range have been observed.[635] The Atherton population of the Crimson Rosella lives at 600–1500 m altitude in eucalypt forest.[1709] Farther south in Qld and NSW the Crimson is rare on the coast and mostly lives in the ranges above 450 m.[635] In the south it is common at all altitudes.[488] It will enter suburban gardens and parks where large trees are plentiful. The Yellow Rosella is primarily a bird of eucalypt woodland along rivers but may move out into surrounding lands when not breeding. The Adelaide Rosella inhabits eucalypt woodland and forest, keeping to watercourses in the drier parts of its range. All three rosellas eat native and introduced seeds and fruit. The Crimson and Adelaide Rosellas feed on the ground but the Yellow Rosella most often feeds in trees. Densities of the Crimson Rosella were: near Bathurst (33/149) 3.5 birds/ha and 0.75 birds/ha respectively in wet and dry eucalypt forest;[489] 0.5–1.0 birds/ha in eucalypt forest at Jamieson (37/146);[1469] and 0.12–0.58 birds/ha at Boola Boola (38/146).[1113]

Field Atlas Records: 13 807

1° blocks–with records: 89 (11%)
–with breeding records: 58 (65%)

Atlas No.: 282 Crimson Rosella

Max. Rep. Rate: 80%
In block(s): 37/146

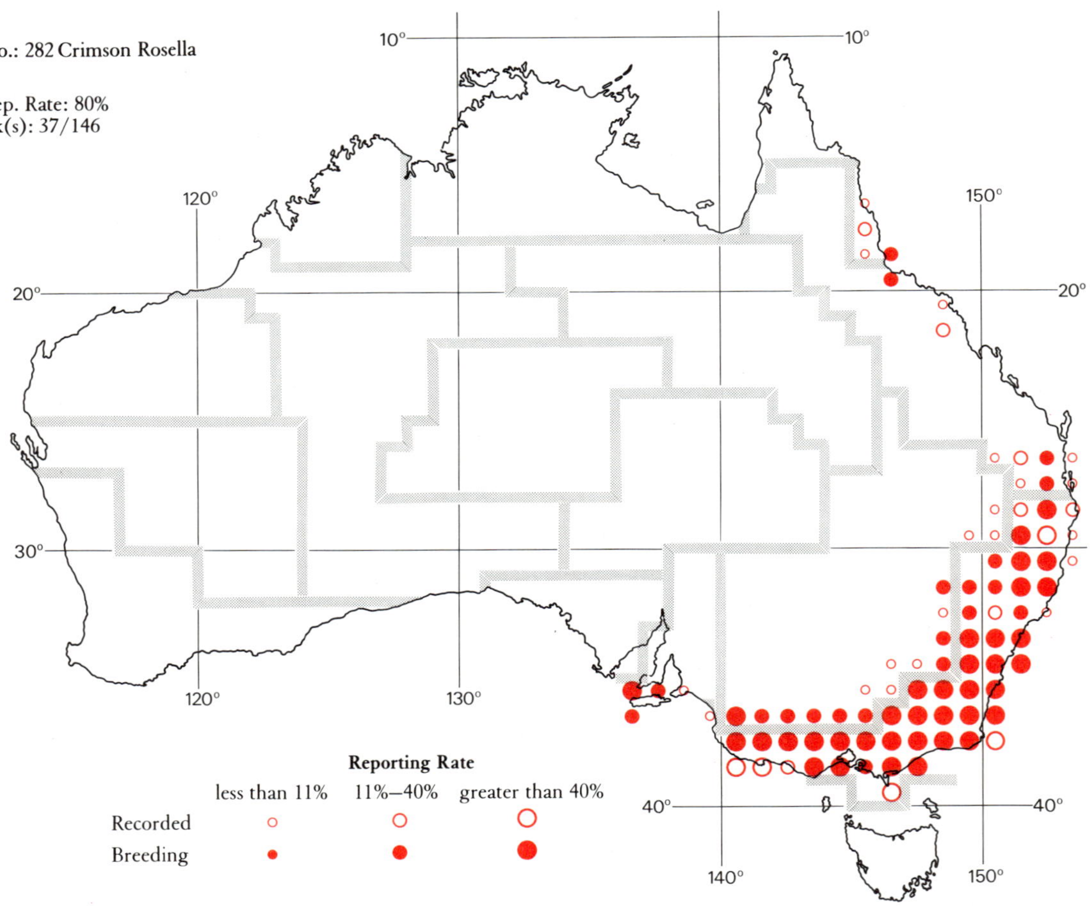

Atlas No.: 283 Adelaide Rosella

Max. Rep. Rate: 64%
In block(s): 35/138

Field Atlas Records: 1691
10′ blocks–with records: 8 (1%)
–with breeding records: 6 (75%)

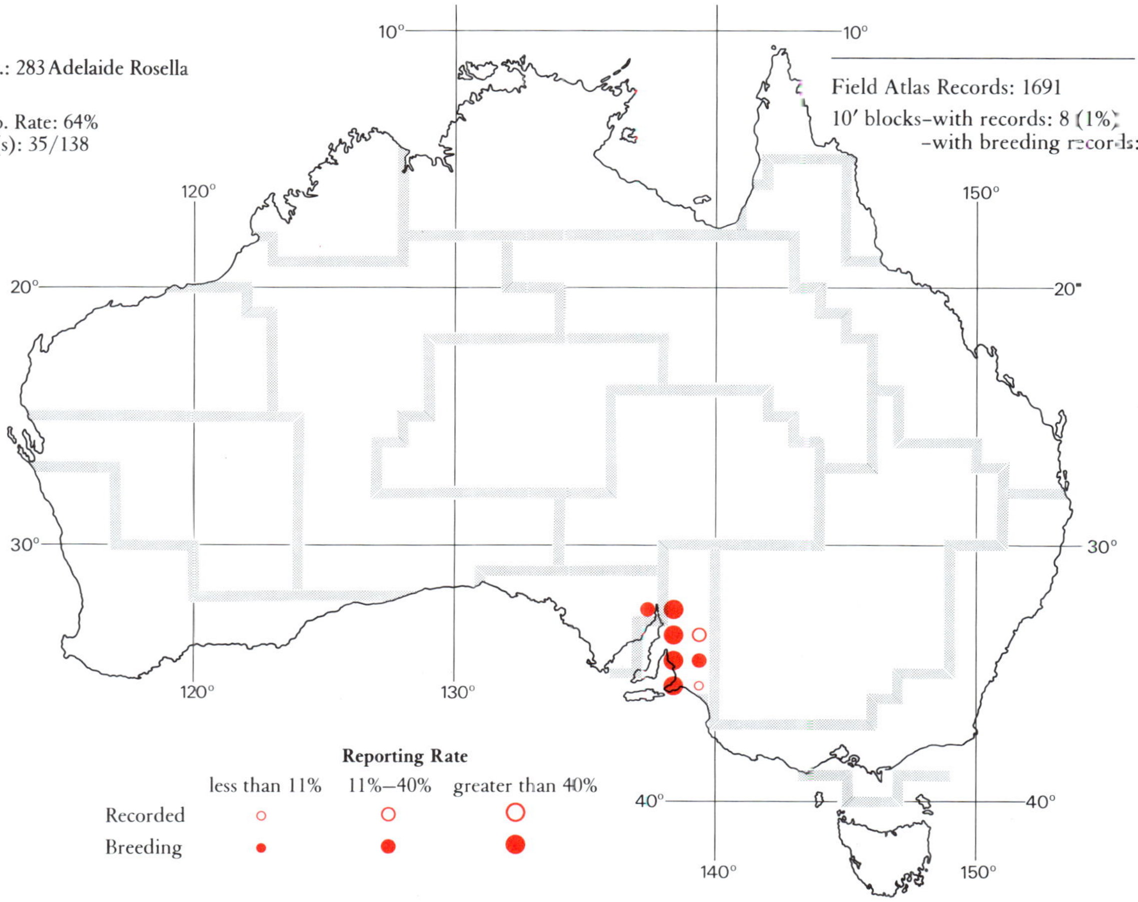

Atlas No.: 284 Yellow Rosella

Max. Rep. Rate: 51%
In block(s): 35/144

Field Atlas Records: 1518
10′ blocks–with records: 26 (3%)
–with breeding records: 14 (54%)

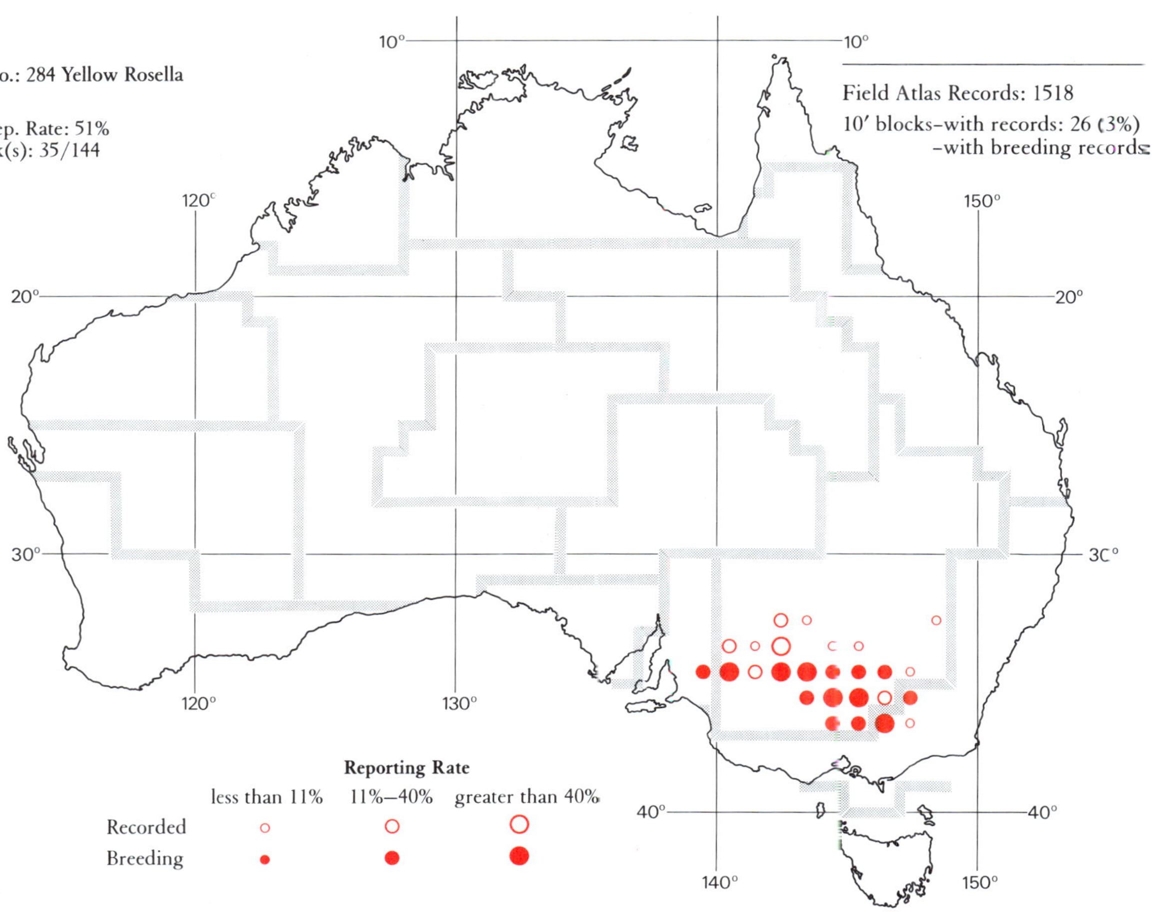

Eastern Rosella

Platycercus eximius

The Eastern Rosella is native to Australia and was introduced to New Zealand. Three subspecies are recognized.[635] The Field Atlas shows no discontinuity of abundance or distribution at the Hunter R (32/151) which is said to be the boundary between the mainland populations of *cecilae* to the north and *eximius* to the south. The population in the Tas. Region is *diemenensis*. On King I. (39/144) a pair arrived after strong winds in 1979 and bred but may have been escapees. There is no previous confirmed record from there or from Flinders I. (40/148). The population in the SA Ra. Region, shown to be isolated on the 10′ maps, is thought to have become established from escapees. It was not reported before 1901 and the earliest record of breeding for that area in the Atlases is in 1934.[1399] The Eastern Rosella hybridizes with Crimson and Pale-headed Rosellas.

Adult Eastern Rosellas are sedentary round Armidale (30/151) though immatures may be more mobile.[1920] On Rotamah I. (37/147) numbers vary little throughout the year.[284] In SA there is pronounced post-breeding movement.[1369] The Field Atlas suggests no large-scale seasonal movement.

The Eastern Rosella inhabits eucalypt woodland, clearings in forest, parks, gardens and orchards. It is absent from rainforest. Usually it lives in small flocks but up to 100 birds have been recorded in non-breeding flocks. It feeds on seeds of grasses, shrubs and trees, fruit, vegetable matter, nectar and occasionally insects. There is competition for nest sites with the Common Starling and in Tas. the Rosella is most common in areas inhabited by the Noisy Miner. The Noisy Miner characteristically drives many other species away from the territory held by a group.[499] Miners are so vigorous that even the Common Starling is discouraged and this may give the Rosellas some protection during the nesting season.[1314]

There are no measures of abundance.

Field Atlas Records: 17 282

1° blocks–with records: 111 (14%)
–with breeding records: 73 (66%)

Atlas No.: 288

Max. Rep. Rate: 78%
In block(s): 35/147

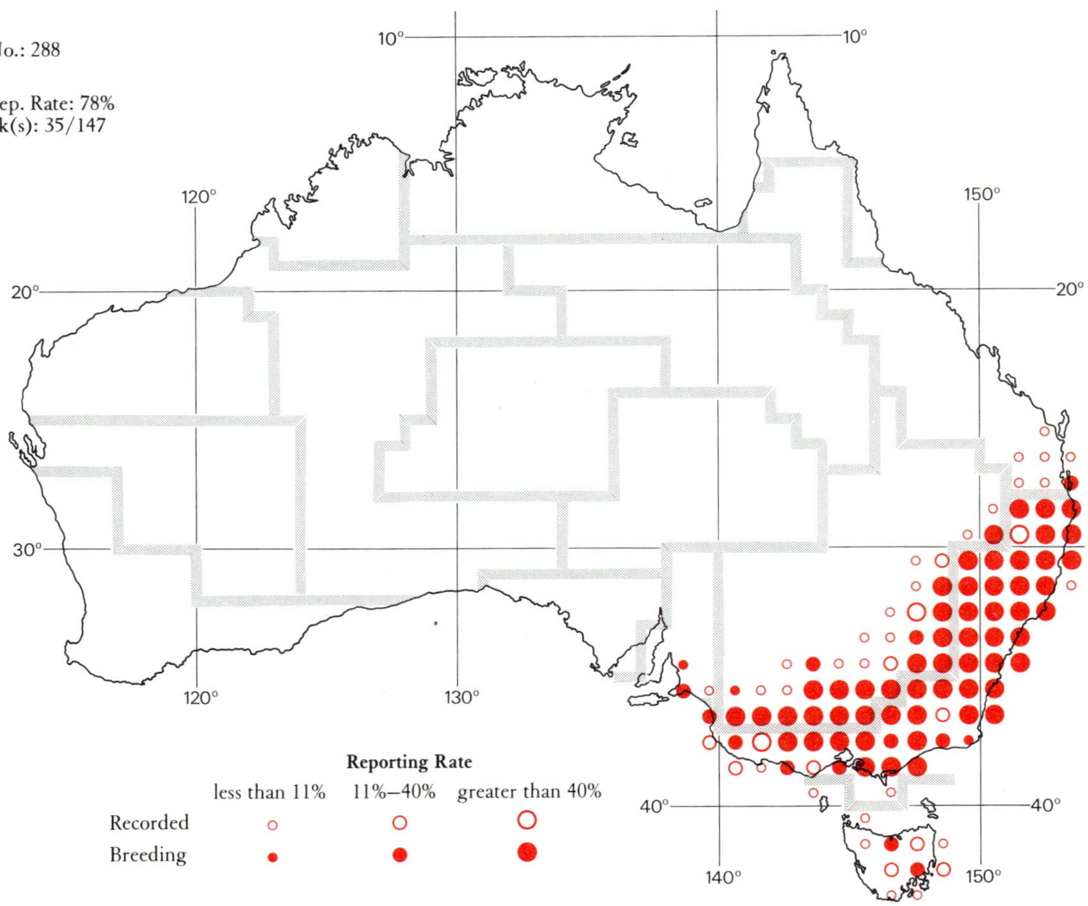

Up to four subspecies of the endemic Pale-headed Rosella have been described but views about them differ.[388,635,1709] No gaps in distribution are evident in the Field Atlas. There is a population of escapees at Kurnell (34/151) and records in Melbourne (37/145) during the Field Atlas and in the 1951–76 period are presumably of aviary escapees also but breeding in the wild has not yet been reported and the records are therefore omitted from the map. The Historical Atlases suggest that the species' range is otherwise unchanged. There are hybrids between the Pale-headed and Eastern Rosellas where their ranges overlap.

The Field Atlas shows no evidence of large-scale seasonal movement and the only reports of movement have come from the edges of the species' range and appear to be associated with drought or other climatic extremes.[635] The Pale-headed Rosella seldom penetrates deeply into rainforest, though it lives on the fringes, in coastal heath, on islands and in paperbark woodland. Though often in trees along watercourses, the species does not penetrate along them into the arid zone. The western limit of its range coincides quite well with the line west of which less than 200 mm of rain falls in eight summers out of ten.[1333] The Pale-headed Rosella feeds on seeds, fruit, insects and occasionally flowers. The bird will take the seeds of many introduced weeds that have spread with cultivation[635] and has been observed eating the red fruit of peppercorn trees.

There are no measures of abundance.

Pale-headed Rosella

Platycercus adscitus

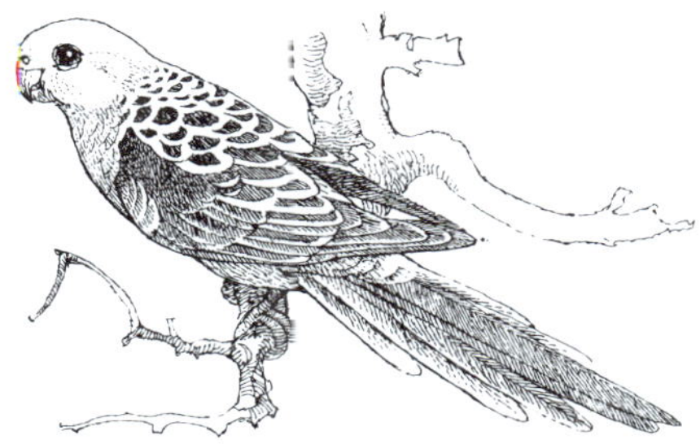

Field Atlas Records: 5417

1° blocks-with records: 116 (14%)
-with breeding records: 39 (34%)

Atlas No.: 286

Max. Rep. Rate: 84%
In block(s): 27/149

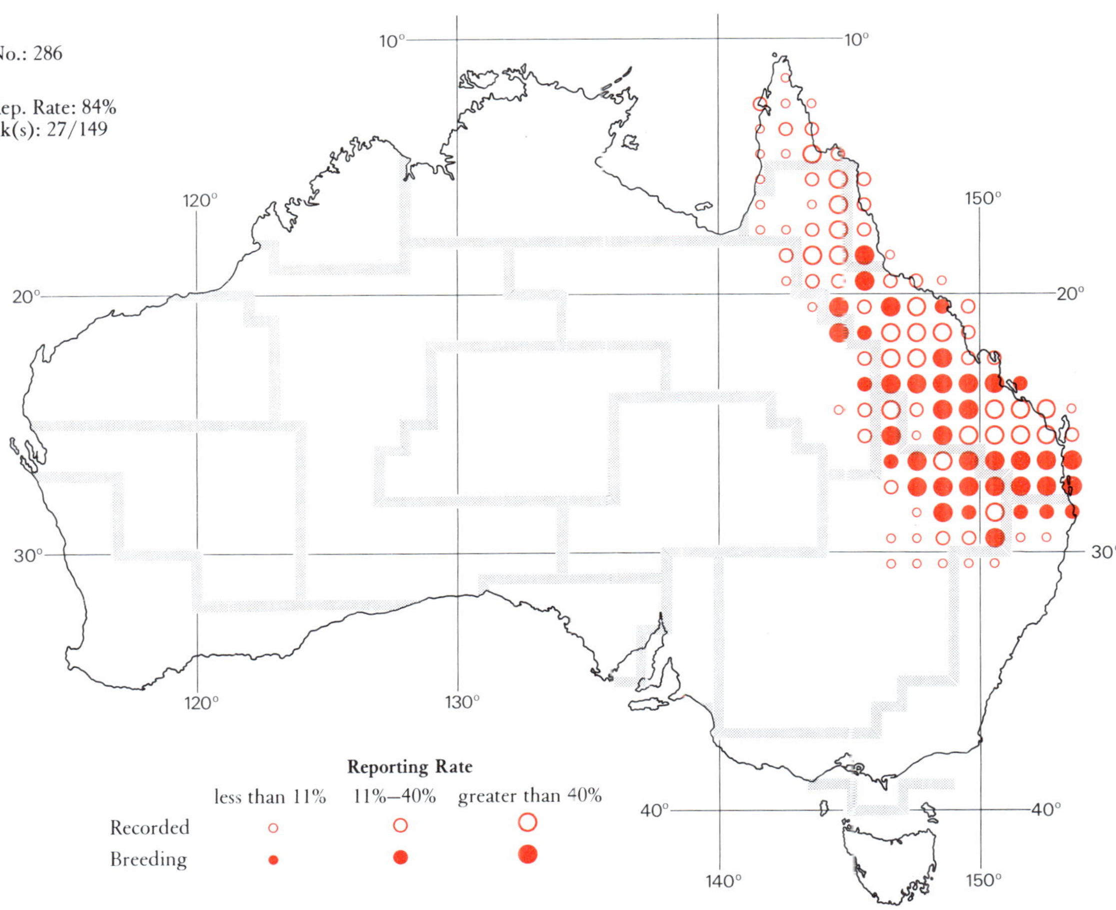

Northern Rosella

Platycercus venustus

Views about subspecies of the endemic Northern Rosella differ.[388,635] Its plumage varies considerably and there is at least some difference between birds of the Top End and Kimberley Regions,[475,748] so that the zone of low Reporting Rate in the western Top End may represent a real gap in range. The species is unevenly distributed within its range even in remote Kimberley reserves, which have been little disturbed by European development.[1257] It was recorded on only five of 27 islands off the north-western Kimberley coast visited in 1971–73.[635] Field Atlas reports have come from throughout the species' historical range.

There is little information about movements of the Northern Rosella, which, like other rosellas, is probably mainly sedentary. Most birds are encountered singly or in pairs[748] but parties of 12–15 have been reported and once a flock of 50.[635] The species usually inhabits eucalypt or paperbark woodland, often along watercourses; it is rare in rainforest and suburban areas. Some birds also live in mangroves and feed on the shore line and along creek banks.[635] The Rosella eats fruit, blossom, nectar and seeds of eucalypts, paperbarks and acacias.[635] It is more arboreal than many rosellas but has been observed feeding on seeds on the ground. It requires access to surface water and, perhaps for this reason, lives mostly in hilly country where pools remain in the dry winter season.

There are no measures of abundance.

Field Atlas Records: 399

1° blocks–with records: 48 (6%)
–with breeding records: 4 (8%)

Atlas No.: 287

Max. Rep. Rate: 31%
In block(s): 13/131

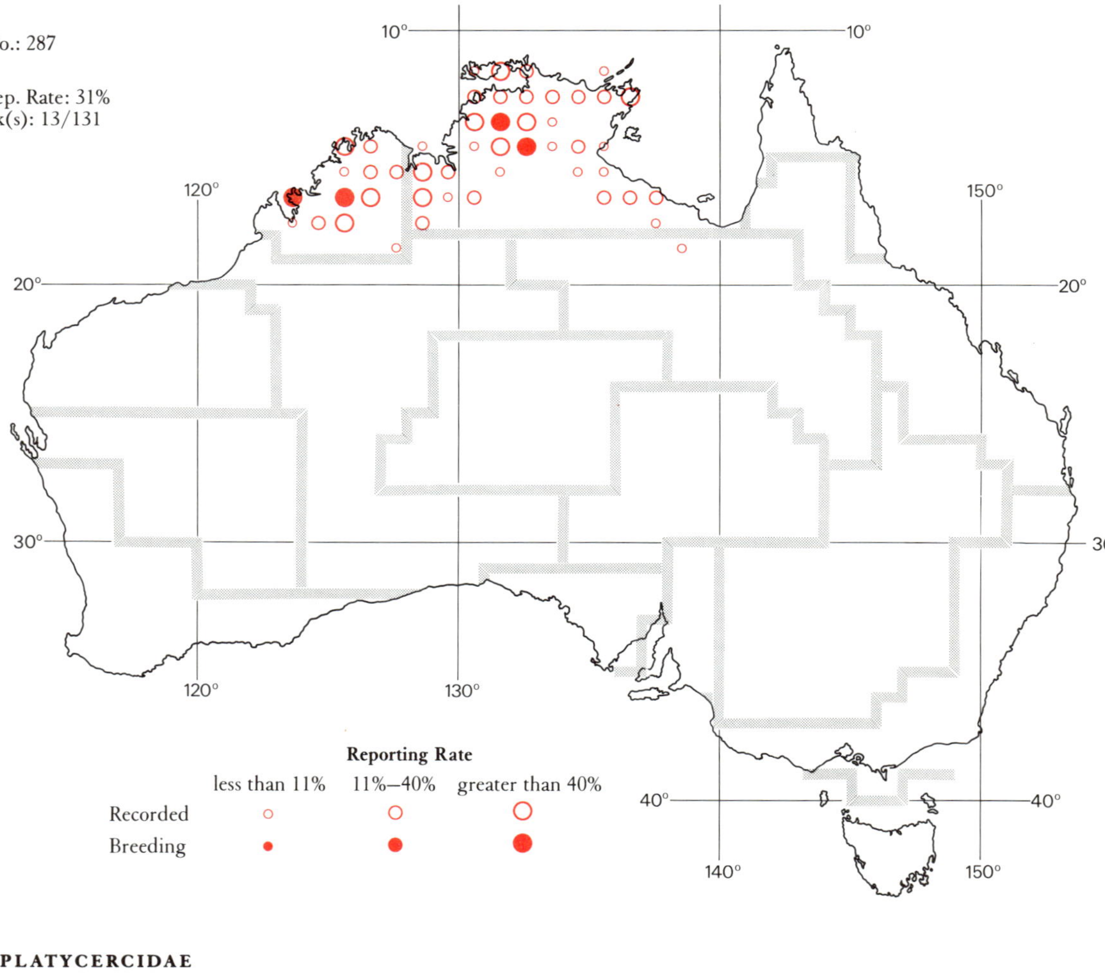

Western Rosella

Platycercus icterotis

The Western Rosella is isolated from all other rosellas. Two subspecies have been recognized: *icterotis* confined to eucalypt forest of the western and southern coasts and *xanthogenys* in areas of eucalypt woodland to the east. Doubts have been expressed about the validity of these subspecies[635,1084] and no gap is apparent on the 10′ maps. The species was recorded in the Field Atlas throughout its historical range.

In northern parts of the species' range it has been suggested that post-breeding movements may take the birds towards the coast in summer.[635] The Field Atlas shows no evidence of large-scale seasonal movement.

The Western Rosella inhabits open and partly cleared eucalypt woodland and forest but not heath. The absence of records from the jarrah forest, apparent on the 10′ map, may be correlated with the scarcity of hollows in trees used for nesting. The Western Rosella does much feeding in trees as well as on seeds of grasses and other herbaceous plants.[635] Seeds of casuarina, capeweed and clover are prominent in its diet and insects are also eaten.[1084] The Rosella will forage on roadsides and in pastures and invades orchards to eat fruit. Shooting in orchards, legal only after obtaining a permit, may eliminate local birds.

The species reappeared after 15 years at Grass Patch (33/121) where it was common in the late 1950s. No other changes in abundance have been documented and there are no measures of abundance.

Field Atlas Records: 1249

1° blocks–with records: 32 (4%)
–with breeding records: 10 (31%)

Atlas No.: 289

Max. Rep. Rate: 63%
In block(s): 34/116

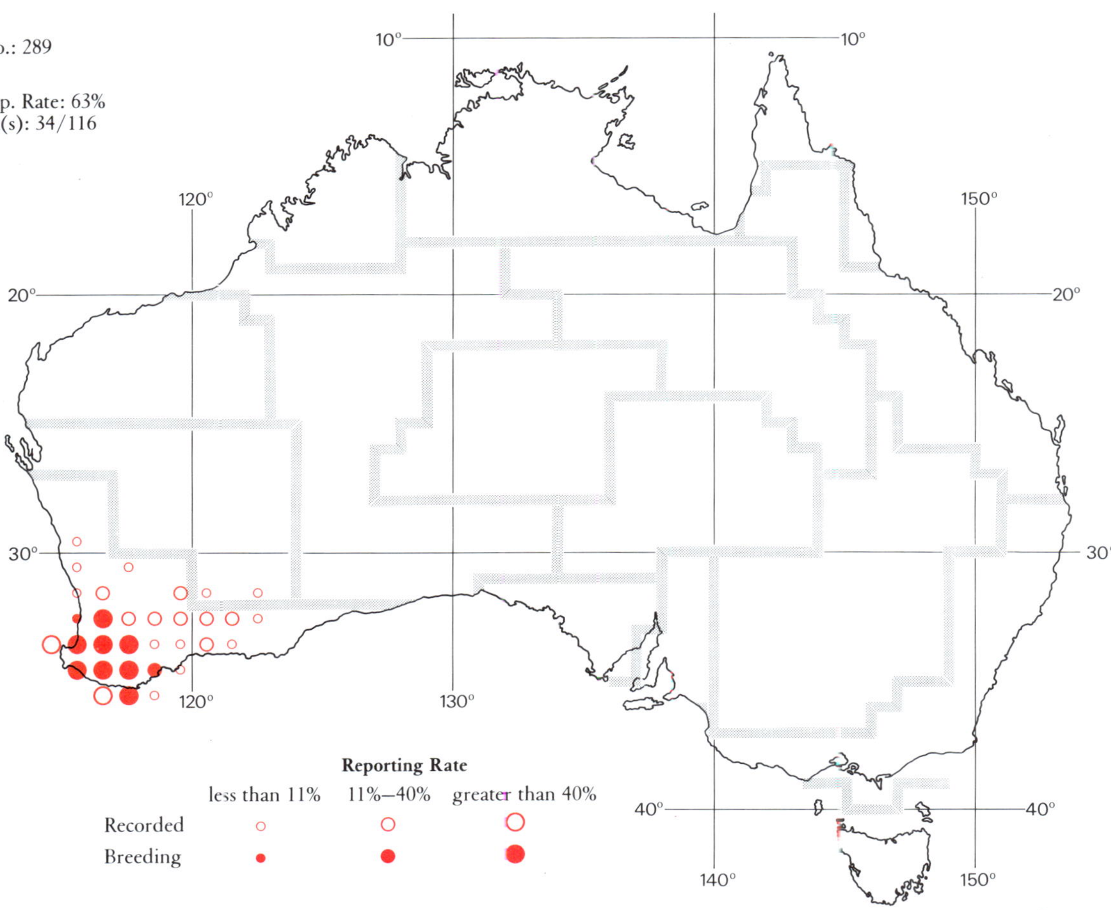

Mallee Ringneck

Barnardius barnardi

At least two and sometimes more subspecies of the endemic Mallee Ringneck are recognized.[626,635] Northwestern birds belong to the subspecies *macgillivrayi*, which appears not to be isolated from the main population. In the Field Atlas records of *macgillivrayi* were made at Cadell Ck (22/141), a tributary of the Diamantina R, less than 60 km from where possible hybrids were reported on the Diamantina itself (21/142).

The SA Ra. Region is populated by hybrids between the Mallee Ringneck and Port Lincoln Ringneck.[1383] There is another area (22/136–7) in the eastern Central Ra. Region from which both species were reported during the Field Atlas.

The distribution of the Mallee Ringneck determined by the Field Atlas covers that known historically. Records of presumed escapees in Brisbane (27/153), Sydney (33/151) and Melbourne (37/145) have been omitted from the map. The record from Tweed Heads (28/153) was of a small flock in Nov.-Dec. 1981 behaving like wild birds.

The Mallee Ringneck is generally present all year in a district though local movements, possibly in response to climatic conditions, have been reported from some places in the north of the species' range.[635]

The Mallee Ringneck occupies the mallee belt of eastern Australia and extends into the adjacent eucalypt woodland and acacia scrub wherever there are suitable nesting trees, e.g. along large watercourses. It competes for nesting hollows with the Common Starling whose distribution almost completely overlaps that of the Mallee Ringneck. Though unusual, flocks of up to 200 have been reported.[635] The Mallee Ringneck feeds on fruit and seeds, including seeds excavated from dried paddymelons.[635] It is a persistent visitor to flowers of the emu-bush, blue gum and mistletoe,[1403] as well as taking caterpillars and insect galls.

There may have been local decreases in abundance caused by illegal shooting, trapping and extensive clearing; the bird apparently avoids settlement.[190,635]

There are no measures of abundance.

Field Atlas Records: 4996

1° blocks–with records: 155 (19%)
–with breeding records: 51 (33%)

Atlas No.: 291

Max. Rep. Rate: 76%
In block(s): 27/147

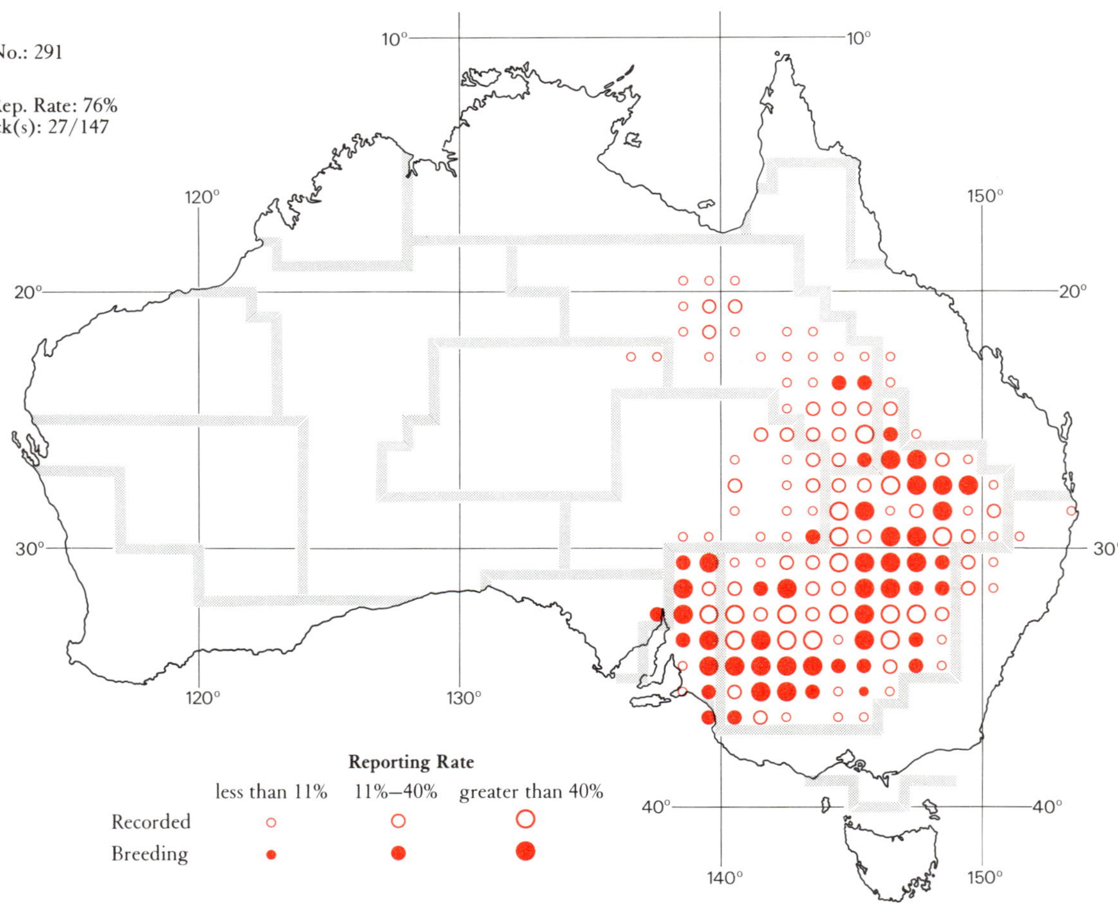

Port Lincoln Ringneck

Barnardius zonarius

There are three subspecies of the endemic Port Lincoln Ringneck: *occidentalis* in the Pilbara and northern Goldfields Region; *semitorquatus* in the southern part of the South-West Region; and *zonarius* in between, merging with both and extending east to take in the rest of the species' range.[635,1638]

The population in the upper Western Deserts Region may be isolated. Additional records from the 1951–76 period and earlier extend its distribution west to Nicholson (18/128) and the Gardiner Ra. (19/128).[1713] The easternmost record is from Winnecke Ck (18/131) on the western edge of the Tanami Desert, which probably forms a barrier between the two populations.

The SA Ra. Region is populated by hybrids between the Port Lincoln Ringneck and the Mallee Ringneck.[635,1383] The two species are also in contact in the eastern Central Ra. Region (22/136,22/137). Aviary escapees reproduce naturally in Melbourne (37/145) and probably near Sydney (33/151). A population in the Mt Lofty Ra. (34/138) was destroyed to prevent interbreeding with the Mallee Ringneck. This and other sightings of probable escapees near Darwin (12/130), Canberra (35/147) and Hobart (42/148) have been omitted from the map. Birds at Pine Ck (32/141) appeared to be wild.

Little is known about movements of the Port Lincoln Ringneck.[635] At Callagiddy (25/114) it was present only Mar.-June in three of five years; in the years when the parrot appeared the previous summer had been dry, which suggests a movement to the coast in these conditions.[245] The development of distinctive morphological types, even if clinal, argues against random mixing of the population.

The Port Lincoln Ringneck inhabits eucalypt forest and woodland and any arid area where river red gums grow. It eats a wide range of foods: nectar, insects, seeds, fruit, including that of the introduced peppercorn, and native and introduced bulbs.[635,1384] It probably requires water daily.[458] Because it damages cultivated fruit, orchardists sometimes shoot it, legal only after a permit is obtained.

There are no measures of abundance.

Field Atlas Records: 7009

1° blocks–with records: 273 (34%)

–with breeding records: 63 (23%)

Atlas No.: 294

Max. Rep. Rate: 84%
In block(s): 32/117

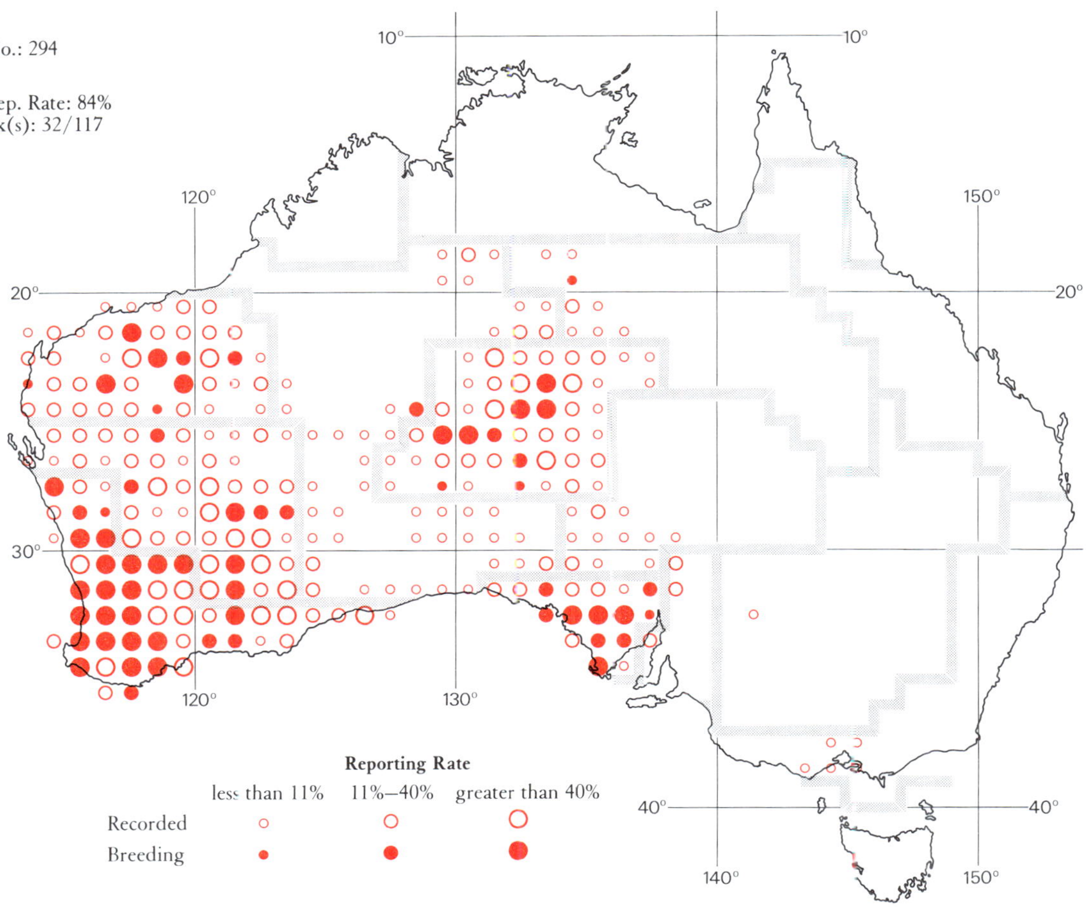

Red-rumped Parrot

Psephotus haematonotus

Two subspecies of the endemic Red-rumped Parrot are recognized. The pale population *caeruleus* in the L. Eyre Region may be isolated from the nominate subspecies *haematonotus*, which inhabits the remainder of the range.[635] Taking earlier records into account, Red-rumped Parrots, presumably *caeruleus*, have been recorded along the Cooper Ck system from the Cheviot Hills (25/143) to Etadunna (28/138). At the northern end of this range Red-rumped Parrots have been seen at L. Bullawarra (27/143) possibly indicating a link between the Cooper Ck population and that in the northern Murray-Darling Region. At the southern end, small flocks seen at 29/137 west of Marree during the Field Atlas[86] are midway between the Cooper Ck population and that of the Flinders Ra., which comes no farther north than Leigh Ck (29/138) in the Atlases.

The Red-rumped Parrot has expanded its range towards the coast in the South-East Region. Before 1901 there was only one record near the coast east and north of Melbourne (37/145), a specimen taken at Lithgow (33/150) in 1871. By 1950 the species had reached coastal districts near Brisbane (27/153) and Sydney (33/151) and by 1976 had spread along the coast east of Melbourne to Wilsons Prom. (39/146).

The Red-rumped Parrot seems to be sedentary except on the edges of its range where some local movement has been recorded.[635] It lives in woodland and penetrates along timbered watercourses into acacia scrub. Dense vegetation is avoided. The species' range has expanded following agricultural clearing and the bird is common in farmland, parks and gardens. It feeds on the seeds of grasses and exotic weeds.

There are no measures of abundance.

Field Atlas Records: 15 058

1° blocks–with records: 156 (19%)
–with breeding records: 94 (60%)

Atlas No.: 295

Max. Rep. Rate: 80%
In block(s): 35/146

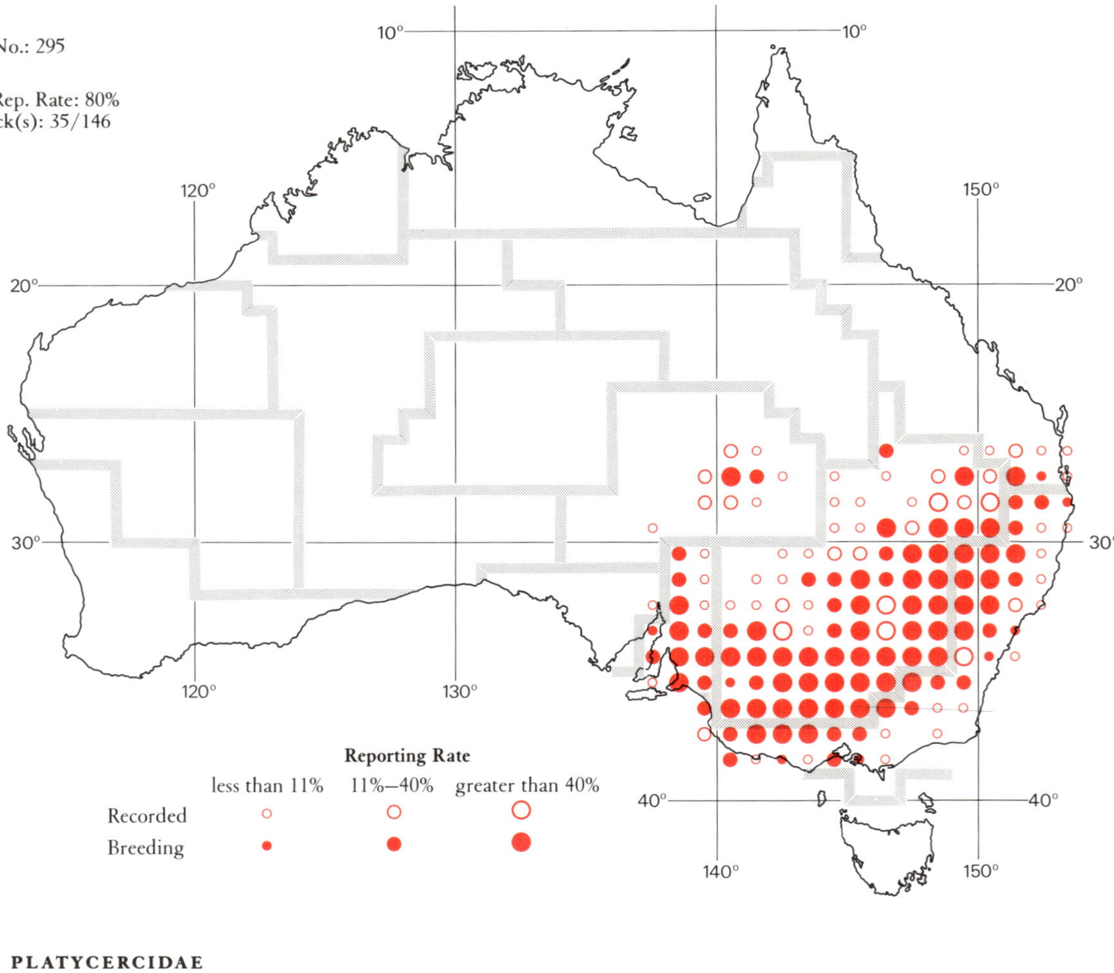

Mulga Parrot

Psephotus varius

No subspecies of the endemic Mulga Parrot are recognized. Its distribution is continuous across the continent and appears unchanged from the historical periods.

The Field Atlas suggests no large-scale seasonal movement. Pairs have been noted visiting nesting hollows throughout the year[456] but some movement is indicated by the species' intermittent presence round Wilpena Pound (31/138).[635]

The Mulga Parrot is aptly named because its distribution closely follows that of the mulga. But the association is probably fortuitous because the bird is not dependent on mulga for food or nesting. The mulga is distributed largely within the zone in which both summer and winter precipitation, treated separately, are below 100 mm in eight years out of ten.[1333] The range of the Mulga Parrot crosses this line only in the South-West where there is a drought of five months each year between Nov. and Mar.

The Mulga Parrot is generally encountered in pairs or small groups. Larger gatherings have been reported, such as on the edges of the Nullarbor where a flock of 100 was observed.[246] The bird feeds on the seeds of grasses and other herbs, trees and shrubs including mulga, mistletoe and saltbush.[635]

It has been suggested that the Mulga Parrot has declined in some places since European settlement[635] but the Atlases show no evidence of a general decrease in range.

There are no measures of abundance.

Field Atlas Records: 3714

1° blocks–with records: 284 (35%)
–with breeding records: 66 (23%)

Atlas No.: 296

Max. Rep. Rate: 47%
In block(s): 27/147,33/140

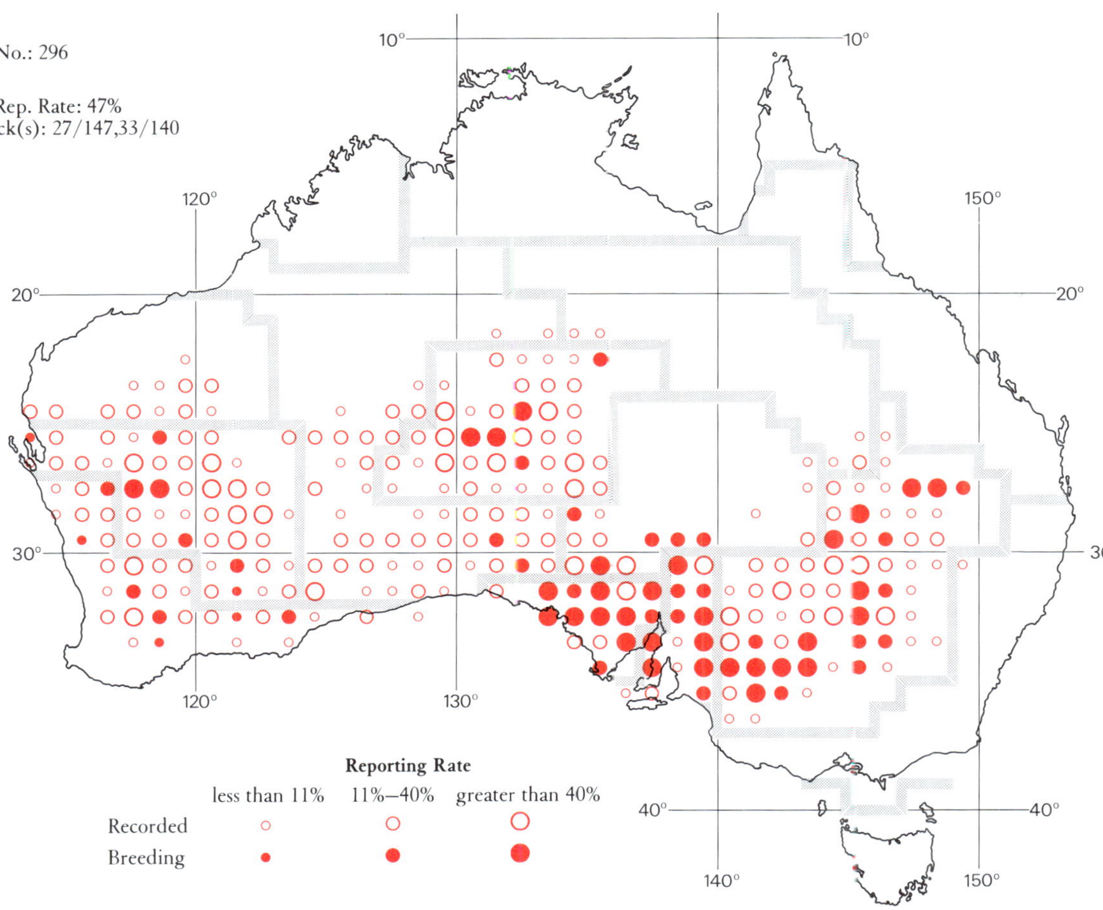

Golden-shouldered Parrot

Psephotus chrysopterygius

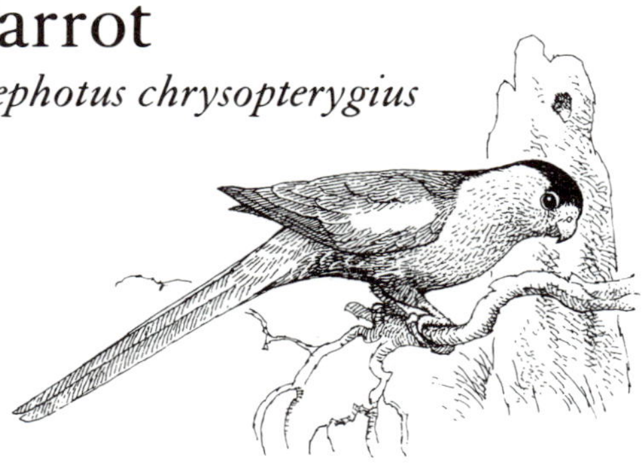

The range of the endemic Golden-shouldered Parrot has declined since European settlement. It was formerly distributed in the Cape York Region as far south as Byerstown (16/144) in the eastern part of its range and, in the west, south along the coast of the Gulf of Carpentaria to the Norman R (17/140). No records have been made in this south-western sector since specimens were collected there in 1856. Neither are any reports confirmed from the western coast of the Cape York Region since the original sightings on the Watson R (13/141) in 1915.[1147] An unconfirmed sighting of a single bird was made near Edward R (14/141) in Apr. 1981.

All earlier records have been south of 13°S and a sighting near the Pascoe R (12/143) in Dec. 1980 is a significant extension to the north-east of the Parrot's range.[1826] Confirmed Field Atlas reports otherwise came only from the eastern edge of the species' original range between Musgrave (14/143) and Koolburra (15/143). A survey in 1975–79 also confirmed this area to be the centre of the Golden-shouldered Parrot's present breeding distribution.[1827] The only movement recorded is post-breeding dispersal.[635]

The Golden-shouldered Parrot is generally in pairs or small, apparently family, parties[635] and is not particularly wary. It is most often in eucalypt or paperbark woodland with a grass understorey[1827,1861] but also in mangroves.[1147] It excavates nesting hollows in termitaria in drainage depressions that are water-logged most of the year. A significant preference is shown for conical rather than meridian termitaria. Several partly completed nest chambers may be found in the same mound in addition to the single completed one.[1827] The Golden-shouldered Parrot feeds on grass seeds picked from the ground or stripped from seed heads.

The Golden-shouldered Parrot was subjected to heavy trapping in the 1950s and 1960s. Possibly the damage done by trappers has been overstated[635] and that caused by the grazing of domestic stock or dry winter season burning overlooked.

During 71 days in the field over a five-year period three groups of 6–10 birds, eight solitary pairs and seven single birds were recorded by one person engaged in searching for the species.[1827]

There are no other measures of abundance.

Field Atlas Records: 7

1° blocks–with records: 3 (1%)
–with breeding records: 1 (33%)

Atlas No.: 300

Max. Rep. Rate: 8%
In block(s): 14/143

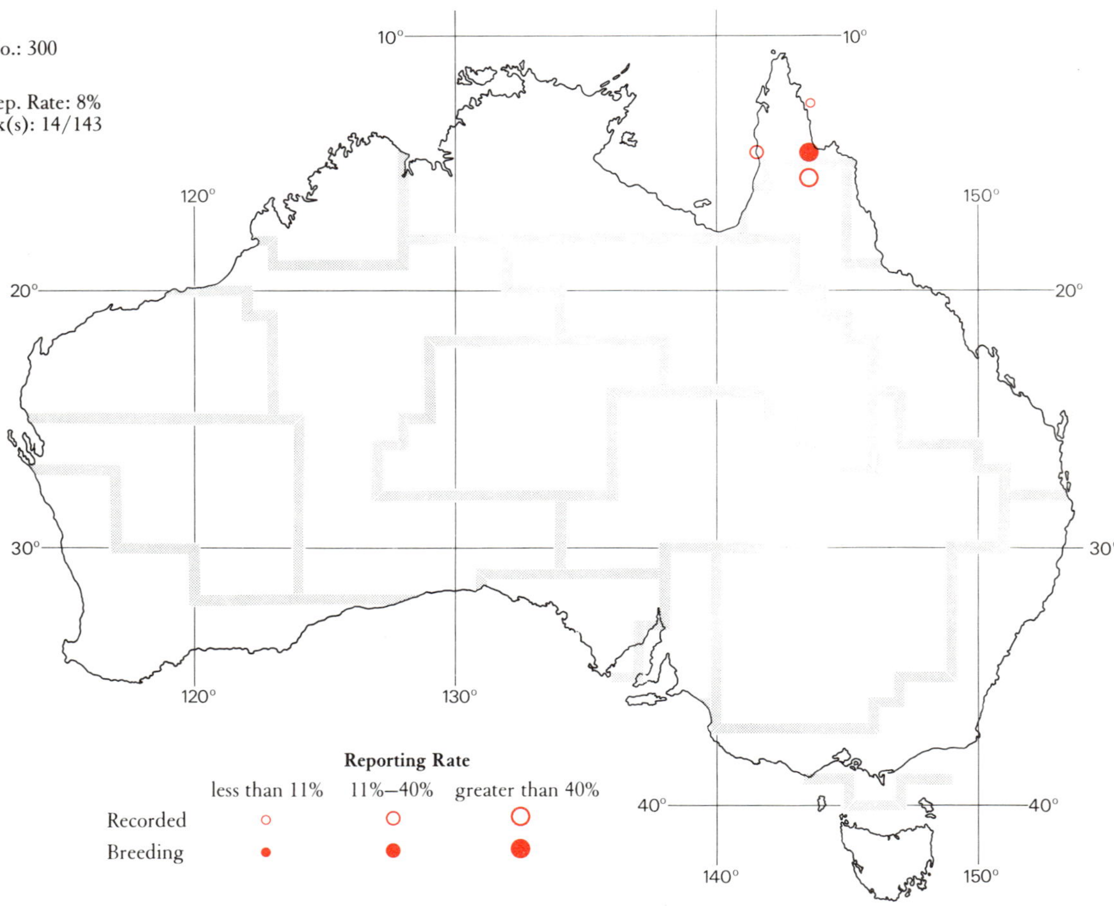

Hooded Parrot

Psephotus dissimilis

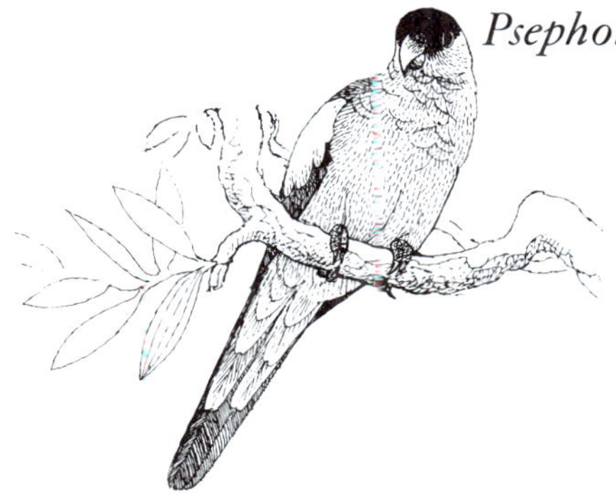

The endemic Hooded Parrot is still reported throughout most of its historic range though declines have taken place in the north and south-east. It was recorded before 1950 on Melville I. (11/130) and Banyan I. (12/135)[635] and eggs labelled Port Darwin (12/130) were collected in 1904 though this is probably not where they were actually collected. The species was also fairly plentiful along the lower McArthur R (15/136) in 1913[107] but has not since been recorded there. The Field Atlas otherwise includes reports from all 1° blocks where the Hooded Parrot was recorded historically.

In the early literature there is a suggestion of seasonal movement, e.g. to Melville I.,[635] but there is no recent evidence of this probable post-breeding dispersal.

The Hooded Parrot spends a good deal of the middle of the day near water, usually shallow pools and soaks where sand spits allow the bird to walk out to the water in the open.[635] Flocks of up to 100 birds form in the non-breeding season round sources of food and water. In 1979, in 30 minutes, 32 parrots visited one pool near Pine Ck (13/131).[635] The nest is excavated by the bird in a termitarium. The Hooded Parrot feeds on seeds of grasses and shrubs. Several observers, e.g.[1602] have reported gatherings of parrots along roadsides, apparently feeding on the seeds of plants growing in the moist verges. As a result casual observers may think that the bird is commoner than it actually is, though recent observations away from the roads indicate that it is still numerous.

Illegal trapping has been blamed for the species' decline but grazing and alteration in the timing and frequency of fires, which adversely affect the ground cover, particularly seeding grasses, may be more important.[635]

There are no measures of abundance.

Field Atlas Records: 58

1° blocks–with records: 10 (1%)
–with breeding records: 4 (40%)

Atlas No.: 301

Max. Rep. Rate: 12%
In block(s): 14/134

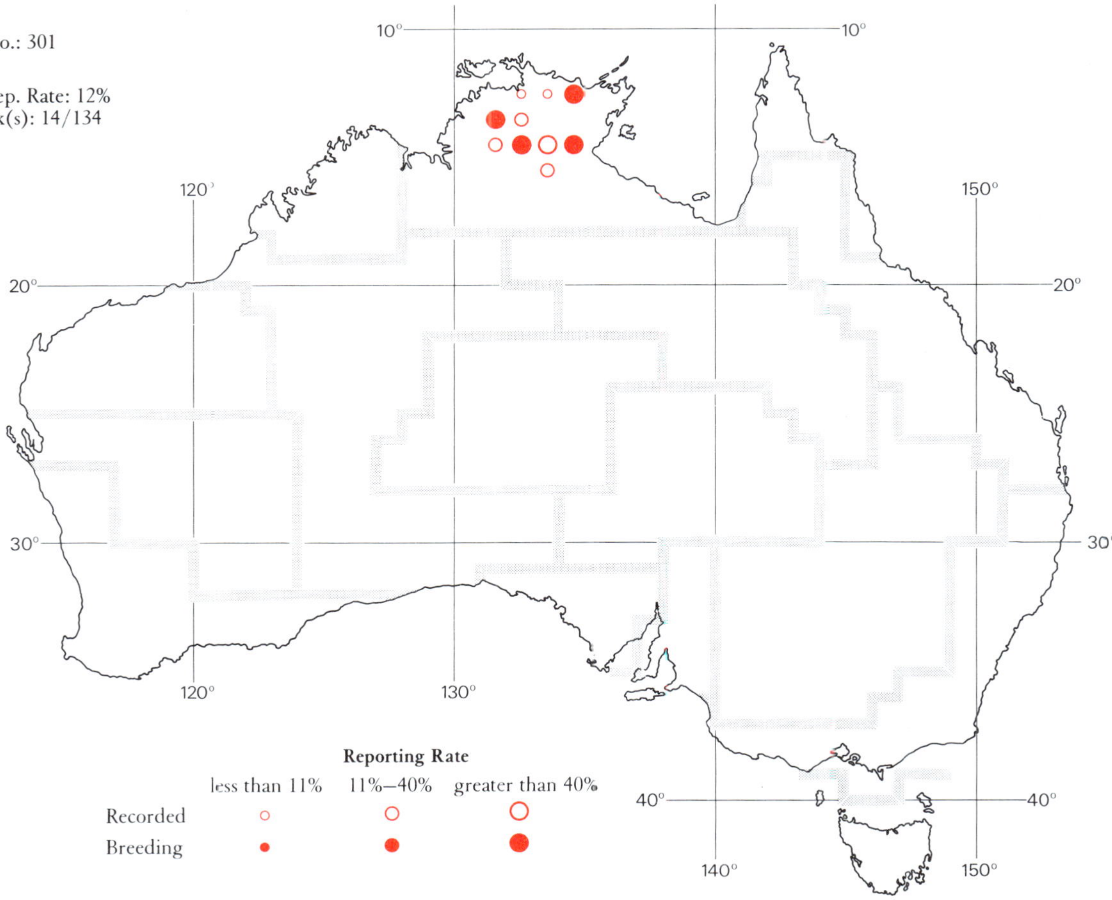

Paradise Parrot

Psephotus pulcherrimus

The endemic Paradise Parrot was not reported during the Field Atlas. Its former range and decline are illustrated by the historical maps (p. 665). Commonly reported in the lower Eastern Qld Region until the 1880s, it declined rapidly, e.g. it had gone from the Duaringa (23/149) district by 1902.[344] A publicity campaign in 1918, after a number of years during which the species had not been reliably recorded, elicited several reports. The last confirmed sightings came from opposite ends of the Paradise Parrot's southern range: near Casino (28/153) in Nov. 1926, on the Burnett R (24/152) in Nov. 1927 and near Ipswich (27/152), also in 1927.[322,1895] Reports of the Paradise Parrot from northern Qld are unconfirmed and possibly refer to the Golden-shouldered Parrot but they cannot be discounted. The most convincing comes from the Mitchell R (15/141) in 1845 and there are also references to the possible presence of the Paradise Parrot on the Archer R (13/142) and near Coen (13/143) in the 1920s.[344,347]

There is no information about movements of the Paradise Parrot. It was seen only in pairs or family parties. One pair was located regularly for a month in less than 2 ha.[635] It lived in eucalypt woodland with grassy ground cover, mainly along river valleys, nesting in a hole excavated in a termitarium. Some accounts suggest that birds nested in the same termitarium year after year. It may also have nested in holes tunnelled into river banks.[904] The Paradise Parrot fed on grass seed on the ground and on seed heads. As European settlement spread through the species' range, grazing and firing to provide a green pick for stock destroyed seeds.[344] A bad drought in 1902 probably hastened the Paradise Parrot's demise in Eastern Qld.

Of the Australian parrots this is the only one likely to have become extinct. Searches have failed to trace countless rumours to their source. If the Paradise Parrot still exists it is as small remnant populations in remote country, perhaps in northern Qld where many rumours are centred.

Field Atlas Records:

1° blocks–with records:

–with breeding records:

Atlas No.: 299

Max. Rep. Rate:

In block(s):

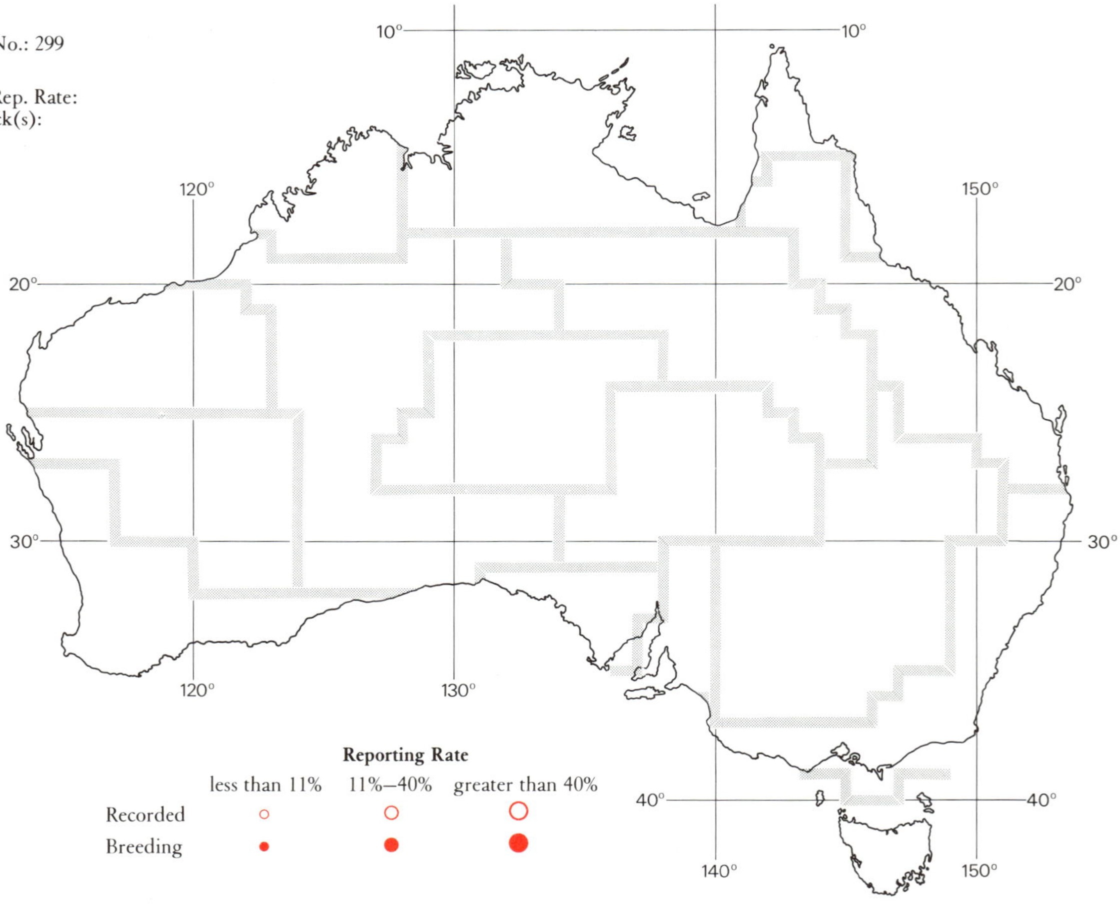

Blue Bonnet

Northiella haematogaster

Four subspecies of the endemic Blue Bonnet are recognized.[635] One, *haematorrhous* in the north-east Murray-Darling Region, is separated from *haematogaster* by a zone along the Warrego R (146°E) where Reporting Rates are low. The southern and western Murray-Darling Region is occupied by *haematogaster* whose distribution extends west through the SA Ra. Region to the southern L. Eyre Region. Birds from the eastern L. Eyre Region are also of this subspecies but, in the centre of the region, the zone of high Reporting Rate represents *pallescens* distributed along watercourses flowing into L. Eyre. The subspecific status of birds from Muncoonie L. (25/138) is unknown; the Blue Bonnet has been reported several times from there but not from surrounding 1° blocks. The fourth subspecies *narethae* is apparently isolated in the lower Western Deserts Region. The main population of *narethae* has not been reported farther east than appears on the map. Records north of Colona (31/132) are also of this subspecies but whether this group of birds is connected to populations farther west or originated from escapees is not known. Apart from a mention of Blue Bonnets in the Musgrave Ra. (26/131) in the 1930s,[1163] records in the Field Atlas have come from throughout the historic range of the species.

There is no record of movements.[635] The Field Atlas suggests no large-scale seasonal movement.

The subspecies *haematorrhous* and *haematogaster* live in woodland. Near L. Eyre *pallescens* inhabits open country with scattered trees, not always close to water.[86,1387] Normally *narethae* is common in the mallee belt on the northern and north-western edges of the Nullarbor; in good seasons it lives in most dongas as well. All Blue Bonnets eat fruit and seeds of saltbush, mistletoe[748] and other arid zone plants but take rather little grass seed. The subspecies *narethae* commonly eats acacia seed green but will take dry seeds, unripe fruit of storksbill and the seeds of composites.[246,635,1878] Food preferences do not appear to control distribution, whereas the availability of nesting hollows may limit breeding range, specially on the Nullarbor. The subspecies *narethae* lives in remote places where illegal trapping may cause local extinction.[635]

On the Nullarbor Plain density in dongas in saltbush was 0.13 birds/ha.[246]

Field Atlas Records: 4812

1° blocks–with records: 161 (20%)

–with breeding records: 58 (36%)

Atlas No.: 297

Max. Rep. Rate: 80%
In block(s): 27/149

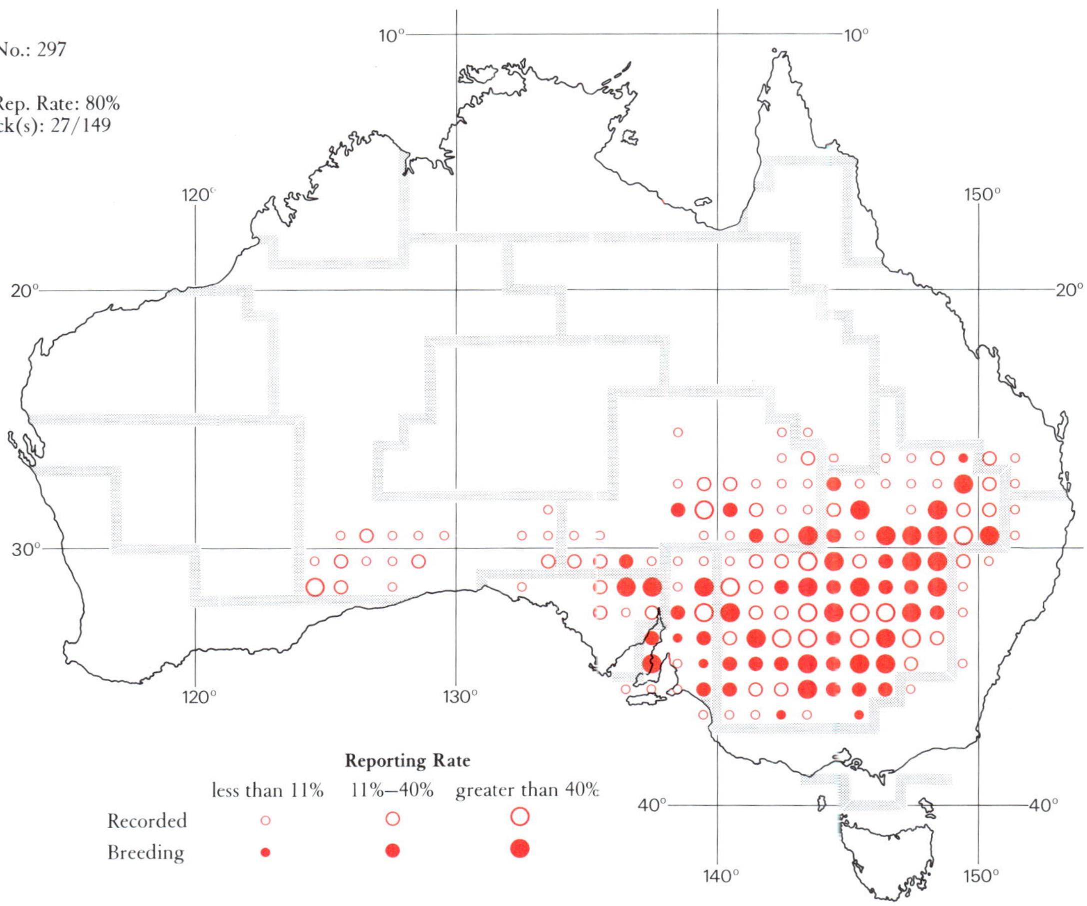

Bourke's Parrot

Neophema bourkii

No subspecies of the endemic Bourke's Parrot are recognized but the population at the eastern end of the species' range may be isolated because the Historical Atlases also contain no records between 139°E and 140°E. The two populations approach each other most closely on either side of the Flinders Ra. Farther north, in the L. Eyre Region, there is only one other record (28/138) on the eastern side of the lake to add to those in the Field Atlas. The western population is continuous across the Western Deserts Region.[87]

The Field Atlas map shows the usual distribution of the Bourke's Parrot. During the 1960s several reports came from outside this range: at Peak Hill (32/148) and Narrandera (34/146) in the Murray-Darling Region and in the NT as far north as 20°S. As well there is an area (34/140) in the lower SA Ra. Region where the species is regarded as an irregular visitor; the last report there was in 1971.[710] Reports of probable escapees in Brisbane (27/153) are omitted from the map.

Bourke's Parrot was present at Thundelarra (28/117) throughout a four-year study[1076] but in central Australia it is regarded as nomadic.[1369] It inhabits acacia scrub, specially where mulga is dominant, though it does not occupy the dense mulga woodland of southern Qld. During drought flocks of up to 100 may gather at watering places, which are visited morning and evening in summer and throughout the day in winter.[458] Bourke's Parrot feeds in the grass or sits quietly in trees so that large resident populations can persist unnoticed. Studies at Thundelarra showed that it ate small seeds, specially of wind grass;[1076] it did not feed on seeds of acacia and cassia, which it has been observed to eat elsewhere.[635]

Bourke's Parrot was considered to have declined in the Murray-Darling and eastern L. Eyre Regions by the 1920s.[1149] The decline has been attributed to pastoral development and the introduction of cats, rabbits and sheep.[1149] In WA, in contrast, Bourke's Parrot was regarded as rare before 1930 then seems to have become more abundant.[461,589] The increase may be correlated with a reduction in sheep numbers since 1936[589] or with the change in ground vegetation from saltbush to herb dominance.[461]

There are no measures of abundance.

Field Atlas Records: 444

1° blocks–with records: 123 (15%)
–with breeding records: 13 (11%)

Atlas No.: 304

Max. Rep. Rate: 42%
In block(s): 27/118

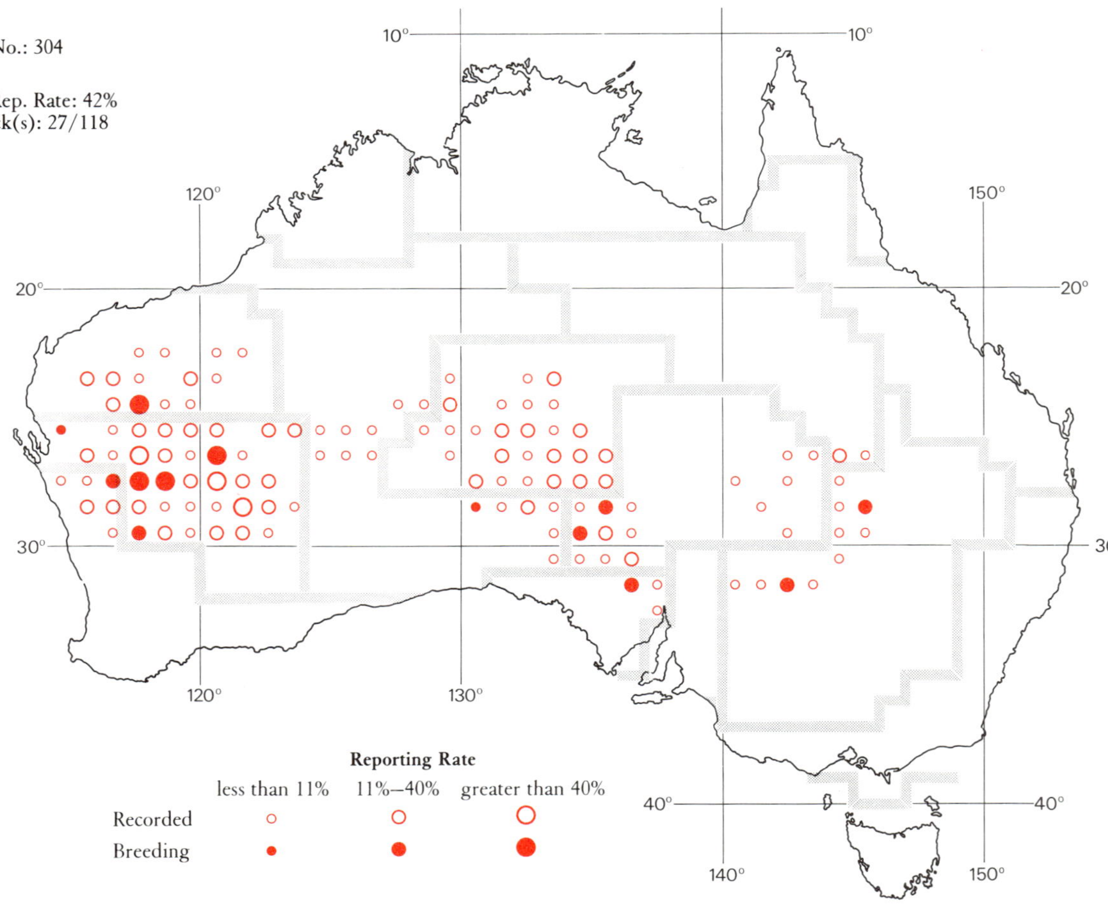

Blue-winged Parrot

Neophema chrysostoma

The endemic Blue-winged Parrot undertakes a regular migration. In summer no record was made north of 34°S in the Field Atlas; the species' winter range extends north generally as far as 26°S with one record in May 1980 of a single bird near Mt Morris (22/129). Breeding is confined to the Murray-Darling and South-East Regions south of 36°S and to the Tas. Region. An early breeding report near Wilmington (32/138) in the Flinders Ra.[225] is probably attributable to confusion with the Elegant Parrot.

In Tas. Blue-winged Parrots are frequently reported in summer from the midlands and north-western coast as far south as the Pieman R (41/145). The Field Atlas shows them to be sparsely distributed down the western coast to Melaleuca Inlet (43/146)[260] but south of Macquarie Harbour (42/145) there is no evidence of breeding.[259] A few birds winter in Tas., e.g. round Orielton Lagoon (42/147) where winter flocks number up to 100,[259] but the bulk of the population migrates (Reporting Rates: summer 6.0%, winter 1.2%). Flocks, mainly of juveniles, gather in autumn in north-western Tas. An estimated 2000 birds were present at Woolnorth (40/144) in Mar., apparently preparatory to migrating. By July, when coasts of north-western Tas. were searched, only 14 birds were counted.[260] There are few reports from King I. (39/144), which suggests that most birds fly directly over Bass Strait to the mainland.[259] Near Colac (38/143) the Blue-winged Parrot is not seen on the autumn movement north but passes through on its southerly migration.[635] In the lower Murray-Darling Region both autumn and spring movements have been noted.[833]

In winter on the mainland the Blue-winged Parrot inhabits eucalypt woodland, saltbush and open places where there are grasses, including coastal and mountain heaths and lignum swamps. It is common on the coastal salt-marshes and dune systems of the lower South-East Region.[259] In Tas. the Blue-winged Parrot lives mainly in eucalypt woodland visiting pasture, clearings and orchards.[635] It feeds on seeds of grasses and other herbs, blossom, fruit and insects.[635]

On a forest block at Maggs Mt (41/146) 20 parrots were found living on 104 ha of regenerating grasses shortly after the area was clear-felled.[635] There are no other measures of abundance.

Field Atlas Records: 1205

1° blocks–with records: 88 (11%)
–with breeding records: 12 (14%)

Atlas No.: 306

Max. Rep. Rate: 31%
In block(s): 41/144

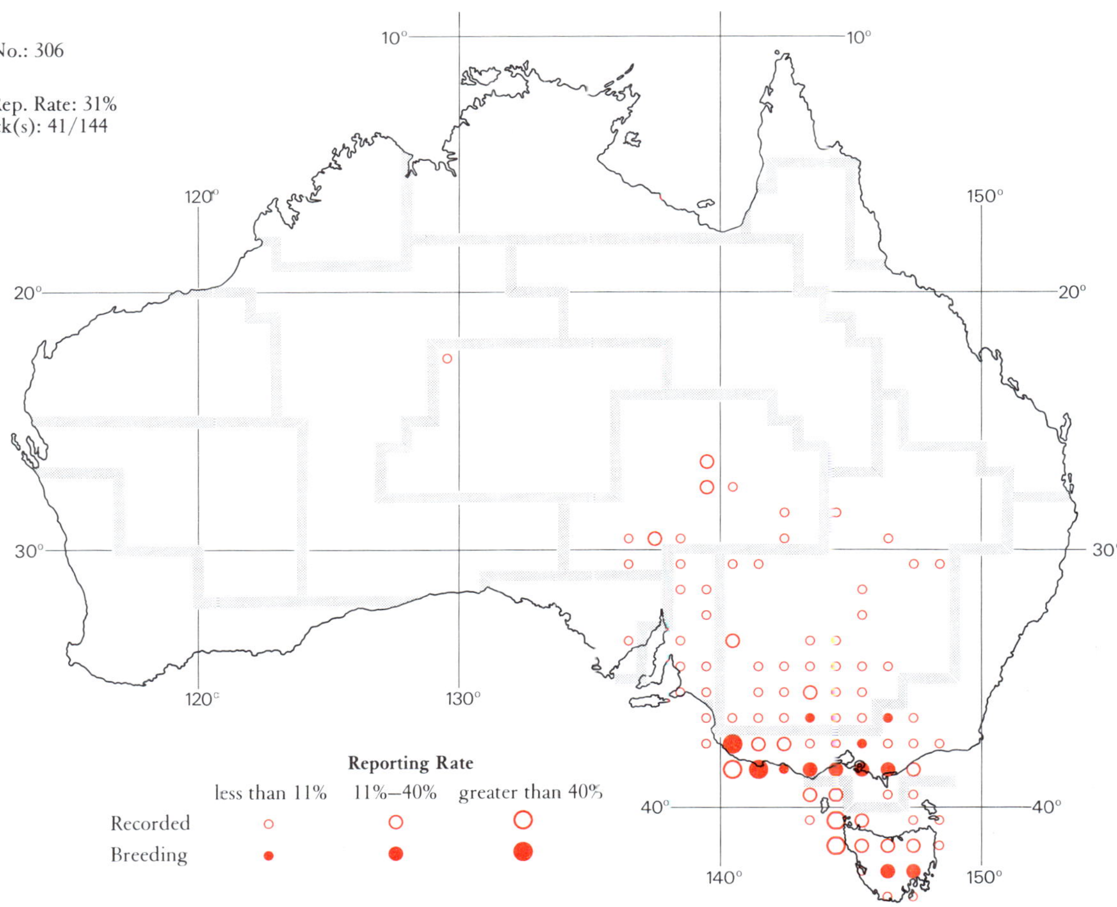

Elegant Parrot

Neophema elegans

The eastern and western populations of the endemic Elegant Parrot seem to have lost contact with each other but there is no evidence of morphological differentiation. Confusion with the Blue-winged Parrot makes it difficult to determine the historical range of the Elegant Parrot in eastern Australia. The few unequivocal records suggest that it has never been more than an irregular visitor to the Murray-Darling and South-East Regions east of 140°E.

In the South-West the Elegant Parrot appears to have extended its range since the late 1920s particularly to the east and south-west.[1638] The historical maps suggest that, in addition to this expansion within the South-West Region, the species moves far from its stronghold under favourable conditions. It was recorded at Pt Cloates (22/113) in the 1890s and several times north of 30°S between 1957 and 1975.

At the outer limits of its range local movements of the Elegant Parrot have been reported.[635] Records on Kangaroo I. (35/137), including those in the Field Atlas, are confined to Nov.-Apr., which suggests that the species is a visitor, probably breeding, moving to the mainland opposite in winter.[122]

The Elegant Parrot usually lives where there is grass, both native and introduced, in eucalypt woodland, mallee, acacia scrub and tussock grassland. It is often flushed from roadsides in farmland and is present on coastal sand dunes. Flocks of 20–100 or more have been recorded.[635] It feeds on the seeds of grass and other herbs and shrubs.

There are no measures of abundance.

Field Atlas Records: 1115

1° blocks–with records: 57 (7%)
–with breeding records: 14 (25%)

Atlas No.: 307

Max. Rep. Rate: 40%
In block(s): 32/117

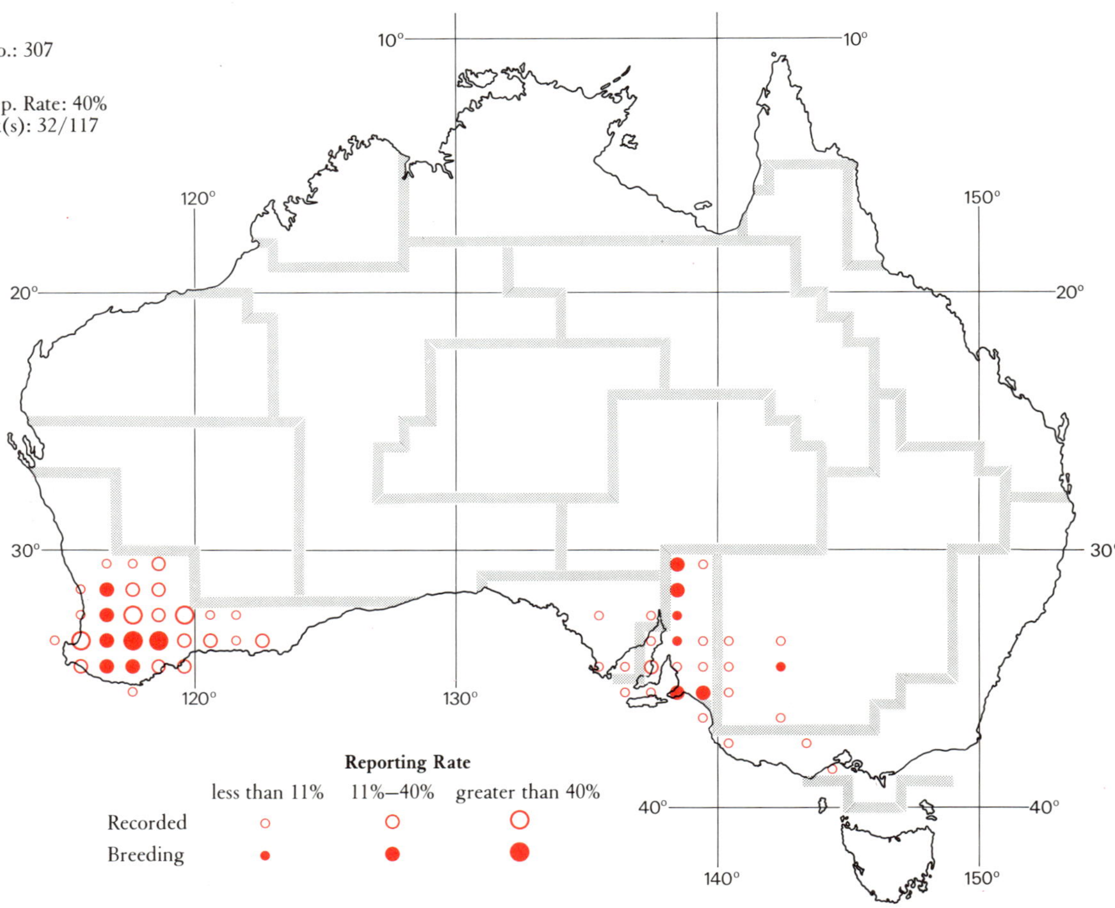

Rock Parrot

Neophema petrophila

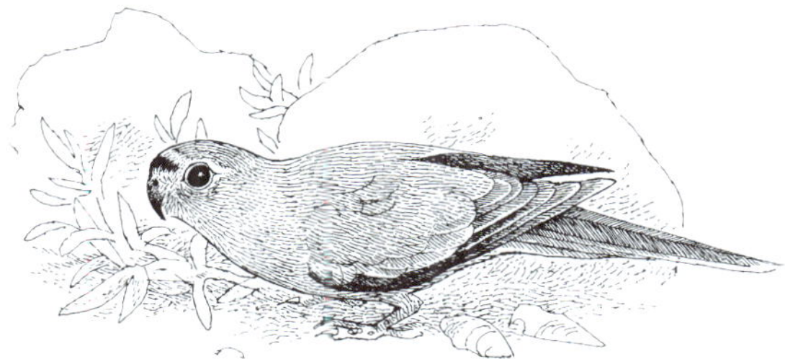

The endemic Rock Parrot has two populations, which appear to be isolated but without morphological differentiation. Records of the two populations that are closest to each other are from near Nullarbor (31/130) and Twilight Cove (32/126) in the 1970s.[1477] The 10′ maps show all sightings to be close to the coast or on islands. Breeding of the eastern population has been recorded only on islands of the SA Ra. Region and Eyre Pen. Region west to 133°E. In the South-West Region there are two records of breeding on the mainland: at Margaret R (33/115) before 1900 and an undated record near Pt Malcolm (33/123);[916] all others are on islands.

After breeding some Rock Parrots disperse to other islands or to the adjacent mainland.[635] Flocks may form then, particularly at feeding and watering sites.

The Rock Parrot lives on the seashore or rocky islands, usually where there is pigface. It feeds on grass seeds and the fruit of shrubs and succulents that grow on the seashore.[635] East of Esperance (33/121) the Rock Parrot feeds on the mainland and can be watched flying out to roost on the islands in the evening, travelling at least 20 km twice a day. It is not known whether it requires water daily but it visits freshwater soaks regularly at Rottnest I. (31/115). A factor limiting its distribution is its habit of nesting on the ground in crevices and in the burrows of seabirds. Such nest sites on the mainland are vulnerable to predation by rats, foxes, goannas and cats and the Rock Parrot seems able to nest securely only on islands. Towards the end of the 1970s, when illegal trapping ceased on Rottnest I. (32/115), the numbers of the bird increased rapidly, indicating its high reproductive potential.

There are no measures of abundance.

Field Atlas Records: 434

1° blocks–with records: 41 (5%)
–with breeding records: 5 (12%)

Atlas No.: 308

Max. Rep. Rate: 38%
In block(s): 34/136

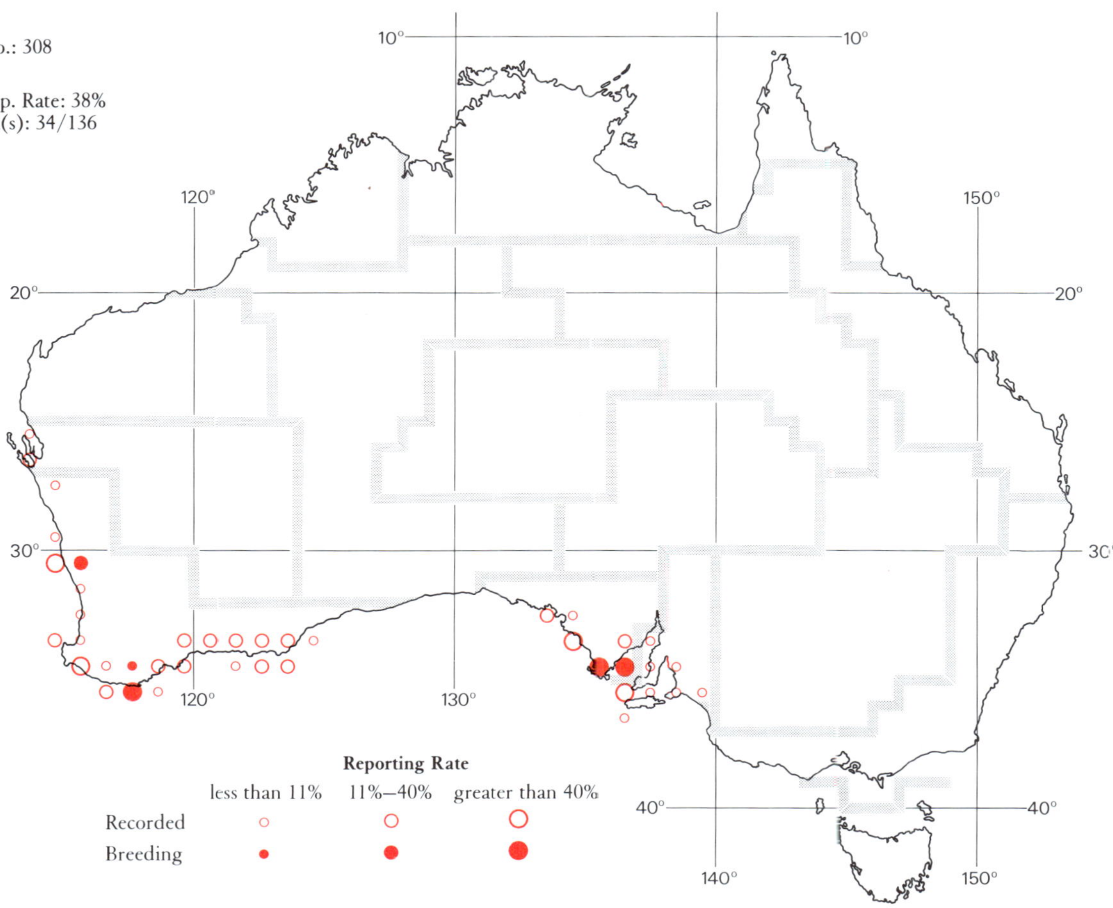

Orange-bellied Parrot

Neophema chrysogaster

The endemic Orange-bellied Parrot now breeds only on the western coast of Tas. south of Macquarie Harbour (42/145) but its breeding range before 1901 may have been more extensive. Its non-breeding distribution is restricted to the South-East and Tas. Regions.

After completion of breeding in Mar.-Apr. there is a northward movement of the population through western Tas., the Hunter Group (40/144) and King I. (39/144). In 1982 three birds were recorded at Port Davey (43/145) in June-July and a few birds remain through the winter on King I. Most birds winter in the South-East Region between Gippsland in the east and The Coorong in the west. There may have been a withdrawal from The Coorong in recent years. The largest numbers of Orange-bellied Parrots now winter in the saltmarshes of Port Phillip Bay (38/144) and the coastal dune systems of SA between Kingston (36/139) and the Vic. border. Records of birds returning to Tas. are few but it appears that the same western Bass Strait route is used and migration is less protracted than in autumn. The Orange-bellied Parrot returns to breeding locations in early Oct. to nest in hollows in eucalypts. Previously successful birds use the same nest site.[261]

In the breeding season the Orange-bellied Parrot forages on heath dominated by buttongrass, mostly in areas burned within the last 15 years, seldom in older vegetation. It takes seeds of heath and other plants including buttongrass and boronia. On migration along the western coast of Tas. seeds of grasses, pasture weeds and strand vegetation predominate in the diet; on islands in Bass Strait and in Vic. saltmarsh species provide most food. In SA birds feed mainly in dunes taking seeds of sea-rocket and sheep's burr.[261]

The present population of the Orange-bellied Parrot is probably about 120 pre-breeding and about 170 post-breeding.[261] There is little doubt that the population was much larger at the turn of the century and, as recently as the 1960s, there is evidence that the species was considerably more numerous than now. The causes of the decline in numbers have been destruction and alteration of vegetation combined with competition from other species. Regular trapping for aviculture of birds flocking during winter, specially in SA, has accelerated the decline.[259]

Field Atlas Records: 152

1° blocks–with records: 18 (2%)
–with breeding records: 2 (11%)

Atlas No.: 305

Max. Rep. Rate: 14%
In block(s): 43/145

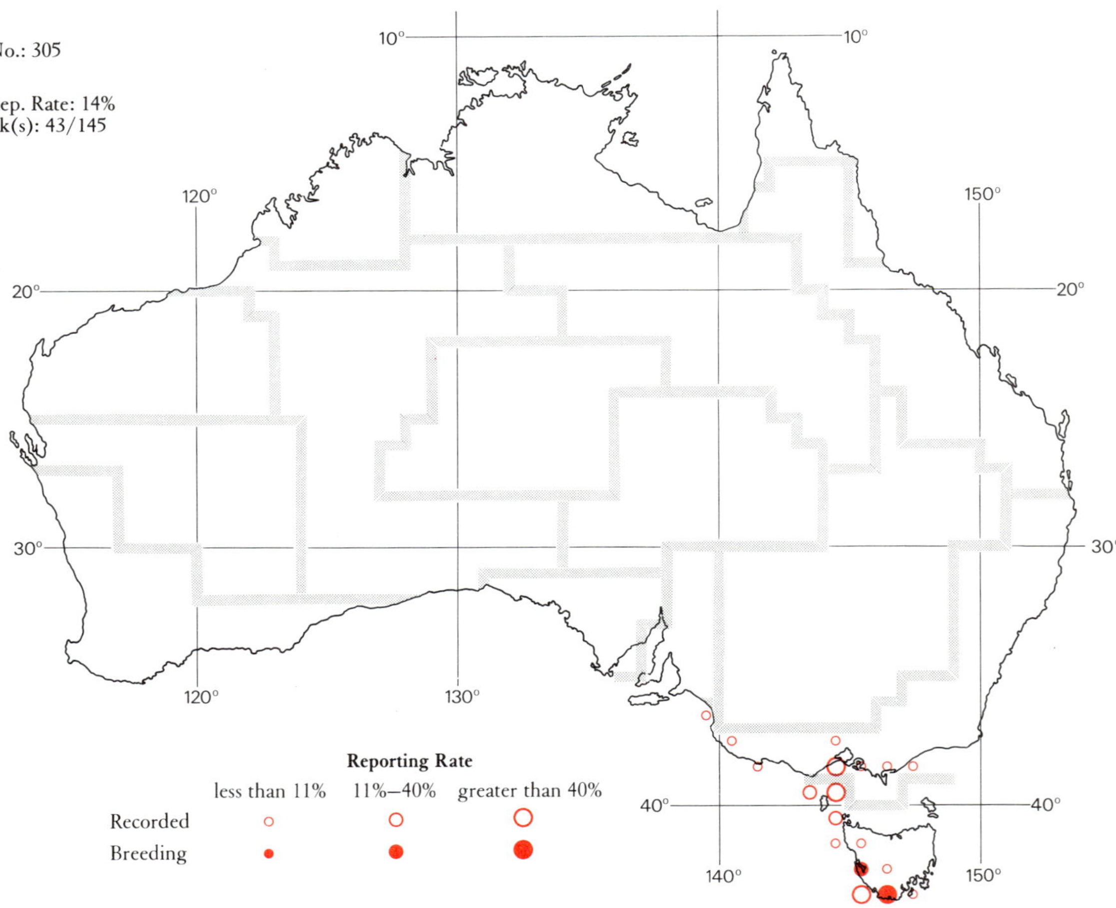

Turquoise Parrot

Neophema pulchella

The range and abundance of the endemic Turquoise Parrot have fluctuated. Some of the changes are evident on the historical maps (p. 666). The species was recorded as far north as the Suttor R (20/146) in 1844 but not since[347] and, in the south, it once extended along the Murray R to its junction with the Darling R.

A reduction in range and abundance, which took place between 1880 and 1920, is not evident on the maps but is reflected in the numbers of records. Between 1901 and 1918 there is only a single record in the Atlases: from Belltrees (33/147) in 1907. The disappearance of the species was noted by observers at the time.[1343] From 1920 onwards records became more frequent with a noticeable increase in the 1940s.[646]

There is no evidence of large-scale seasonal movement by the Turquoise Parrot though local changes in abundance have been noted. Birds may move from eucalypt woodland to pasture after the breeding season in some places.[635]

The Turquoise Parrot lives on the edges of eucalypt woodland adjoining clearings and on timbered ridges and creeks in farmland. Sometimes flocks of up to 50 birds form but usually the species is in groups of up to six, presumably families.[635,1290] The Turquoise Parrot forages on the ground for the seeds of grasses and other herbs, both native and introduced.

It is likely that its earlier decline was caused by domestic livestock competing for food in time of drought at the turn of the century. The species probably survived in refuge areas such as the MacPhersons Ra. (34/146).[646] Trapping, blamed in part for the species' decline, may have contributed. Now re-established over much of its former range, the Turquoise Parrot is regularly reported and seems secure.

There are no measures of abundance.

Field Atlas Records: 563

1° blocks–with records: 40 (5%)

–with breeding records: 12 (30%)

Atlas No.: 302

Max. Rep. Rate: 19%
In block(s): 29/150

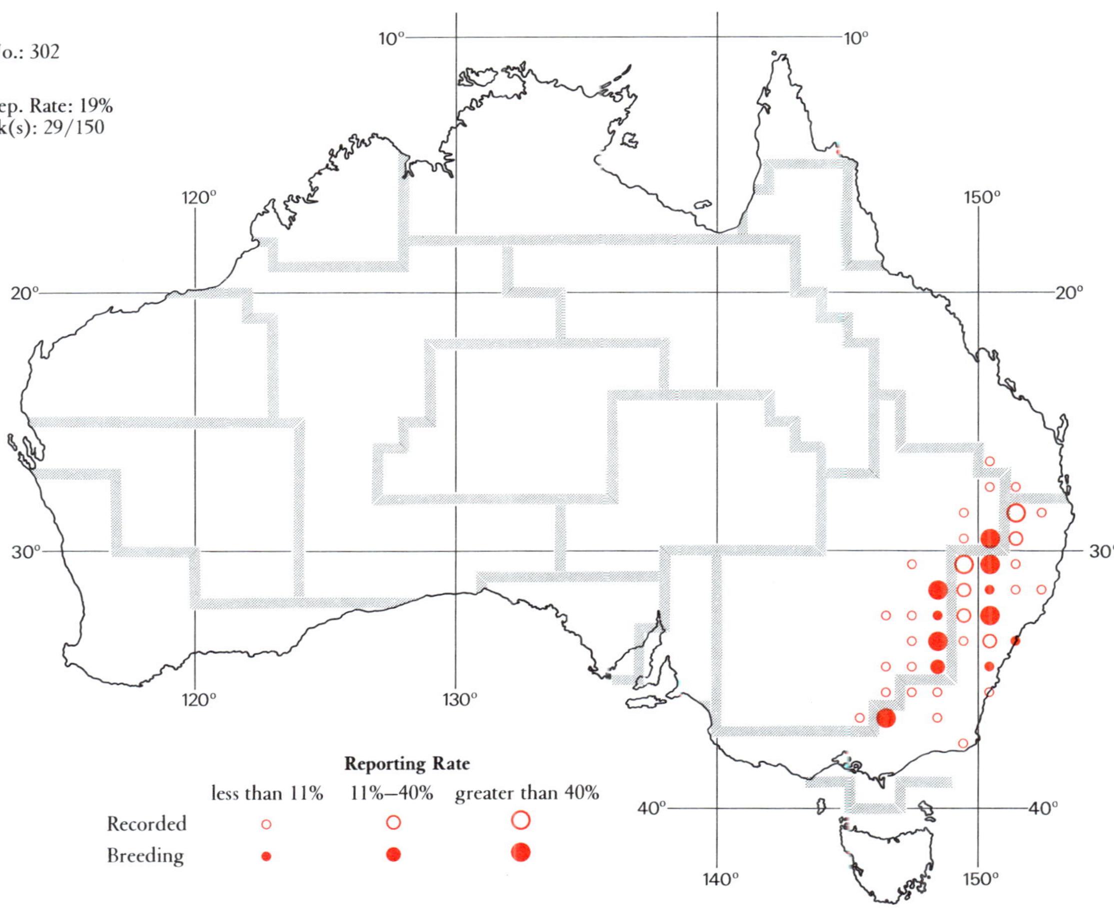

Scarlet-chested Parrot

Neophema splendida

The endemic Scarlet-chested Parrot is at times abundant and at others rarely reported. The Field Atlas was a period of comparative rarity. Documented periods of abundance include 1939 when many Scarlet-chested Parrots were caught on the Nullarbor Plain, particularly round Wynbring (30/133).[1638] From then until the 1960s the species was considered very rare. In 1960–66 large numbers visited the mallee north of the Murray R in SA[782] and in 1969 many nested on the SA-WA border north of the Nullarbor.[599] The species was common then in the better-vegetated southern parts of the Great Victoria Desert and was thought to spread into the desert in favourable seasons.

Though records were few during the Field Atlas, they cover most of the range determined historically. The Scarlet-chested Parrot has been recorded near Bourke (30/145)[1343] but claimed sightings farther east must be doubted.[1137] Breeding was reported near Pinnaroo (35/140) in 1941.[874] In the South-West Region, a specimen was taken on the Moore R (31/115) in 1845 and a small party appeared near Kweda (32/117) in 1949.

At Hawks Nest Station (34/140), during the irruption of the 1960s, flocks appeared regularly Apr.-Aug. then disappeared presumably to breed.[782] Taking all records from the area into account there are some in all months except Sept.-Oct.[965]

The Scarlet-chested Parrot inhabits mallee and acacia scrub usually with a ground cover of spinifex. In the south-east of its range it is often in recently burnt country:[635] both Field Atlas records in the Sunset Country (34/141) were in recently burnt mallee but 12 sightings elsewhere, which give details of vegetation, were not. Except for breeding records, Field Atlas sightings that gave details were of individuals or pairs only, though flocks of at least 100 have sometimes been recorded.[782] When breeding, the bird may be described as semi-colonial. Under suitable conditions it apparently breeds rapidly and becomes common, only to disperse and decline until the next favourable season. The Scarlet-chested Parrot eats seeds of grasses, including spinifex, and other herbs and acacias.

During a good breeding season in 1969 five pairs and two nests were found in 8 ha at one site; and at another ten pairs and four nests in 30 ha.[599]

Field Atlas Records: 25

1° blocks–with records: 23 (3%)
–with breeding records: 2 (9%)

Atlas No.: 303

Max. Rep. Rate: 4%
In block(s): 31/133

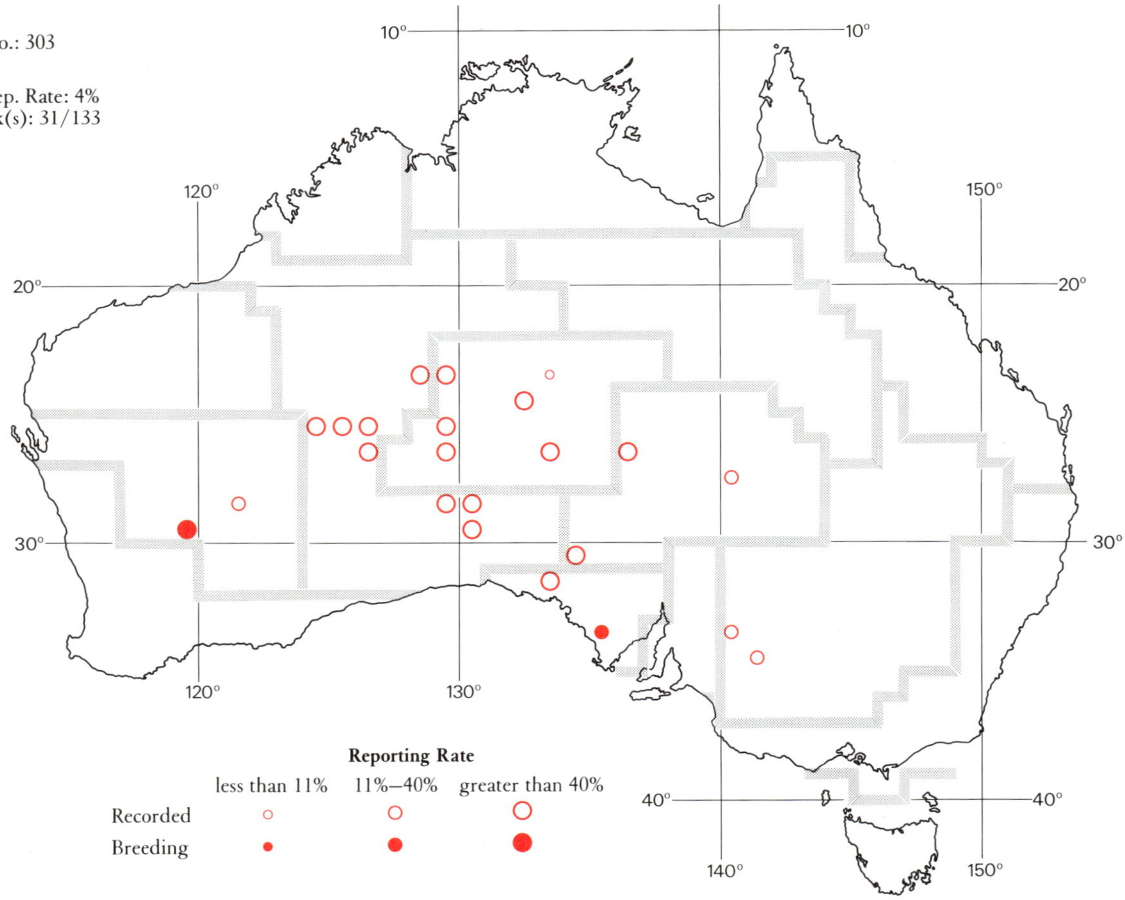

Oriental Cuckoo

Cuculus saturatus

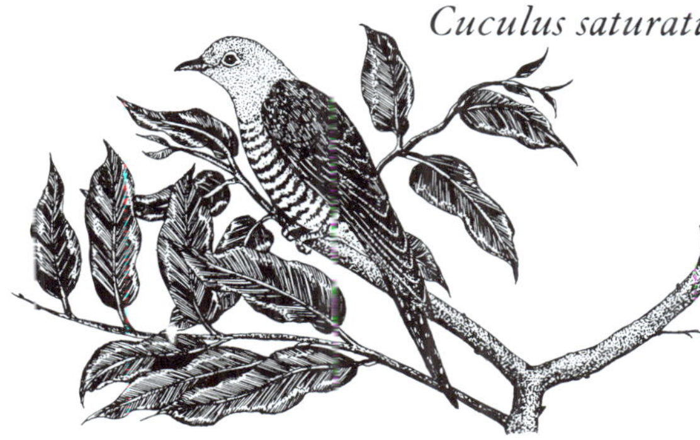

The Oriental Cuckoo breeds in Asia and migrates across the equator for the southern summer. Considerable numbers reach Indonesia[1847] but the species is regarded as rare in New Guinea.[1449] Australia is at the limit of its normal range though single birds occasionally turn up in New Zealand.[1359] The lack of records in northern Australia, including the apparent gap in range south of the Gulf of Carpentaria, probably reflects scarcity of observers there in summer rather than of birds. Within Australia the Oriental Cuckoo is present mainly Nov.-May[1211] but has been recorded in the Kimberley Region June-Aug.[1713] Its appearance at any one locality is unpredictable. The southernmost record in the Atlases is at Narooma (36/150) in Dec. 1972.

The Oriental Cuckoo lives in rainforest, eucalypt forest and woodland, paperbark woodland and mangroves. It is usually absent from more open areas where it is replaced by the Pallid Cuckoo.[1211] It appears to require large trees to act as vantage points from which to pounce on prey, both on the foliage and on the ground. The Oriental Cuckoo is usually solitary, less often in parties of up to five birds. Of 15 Field Atlas records with details, only one was of two birds together. A flock of 12 birds has been seen in the Cape York Region[1703] and in New Guinea up to 300 have been recorded congregating each year to feed on swarms of moths.[153] The species feeds mainly on large insects including caterpillars and mantids.[229,302]

A measure of abundance is given in Arnhem Land where 10–30 individuals were recorded regularly on a 20-km transect traversed daily in late Mar. 1979.[1211]

Field Atlas Records: 174

1° blocks–with records: 57 (7%)

–with breeding records:

Atlas No.: 336

Max. Rep. Rate: 13%
In block(s): 15/145

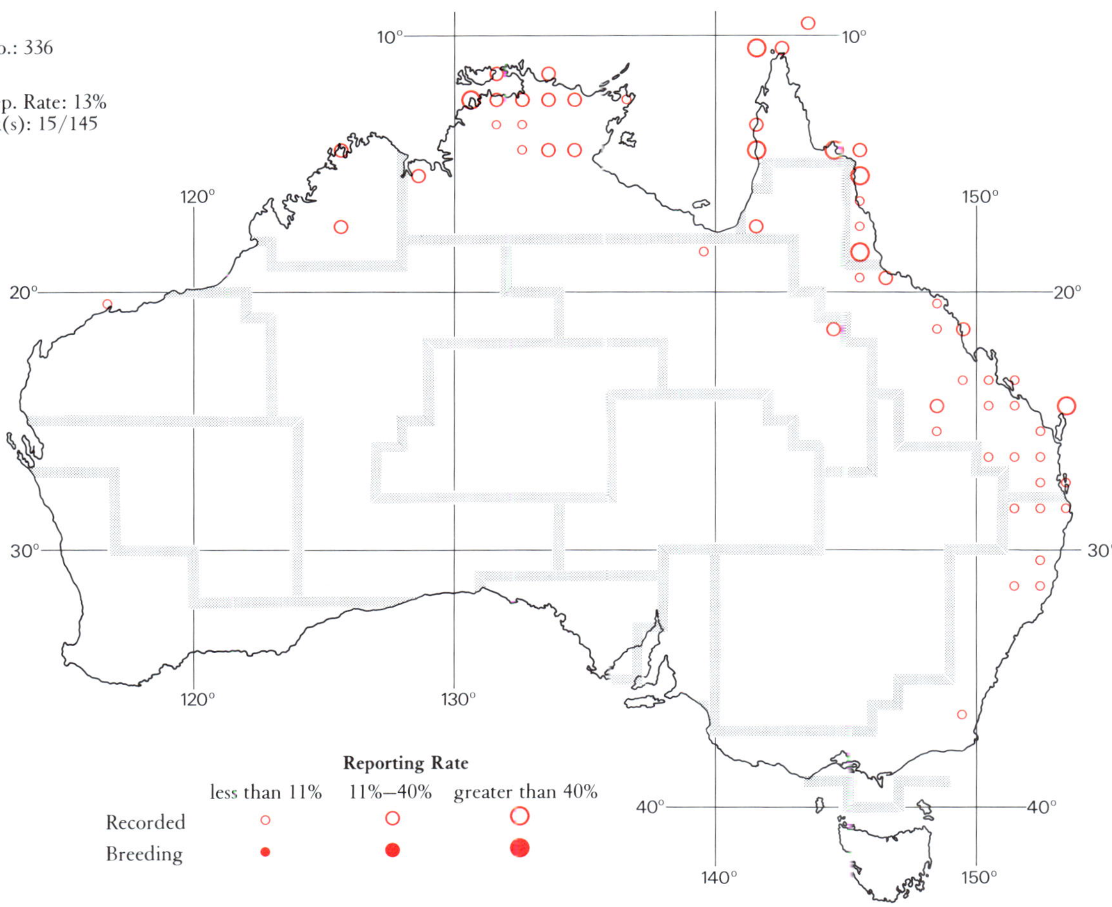

Pallid Cuckoo

Cuculus pallidus

The Pallid Cuckoo has been recorded outside Australia in Timor[1173,1247] and occasionally in southern New Guinea[1250,1723]. There are few breeding reports in the Atlases north of 20°S; of those that are dated, three are in Mar.-June and one in Sept.

Observers in some southern areas regard the Pallid Cuckoo as a migrant. At Sydney (33/151) it is present and breeding Aug.-Jan.[692] and it is a passage migrant in the southern Murray-Darling Region[833] and farther north at Murphy's Ck (27/152).[1099] In southern Vic. large numbers arrive in dry springs but, whatever the weather conditions, only immatures remain later than Dec.[1110] The Pallid Cuckoo leaves Tas. entirely in winter.[1314] At Northam (31/116) it is a common breeding migrant that varies in number and time of appearance but is absent Jan.-Mar.[1216]

There was little difference north of 26°S between the summer (7.5%) and winter (6.9%) Reporting Rates, which suggests that there is no general north-south shift in the population with the seasons. The Reporting Rate was highest in southern localities in the breeding season: spring in the South-East (20.5%), Murray-Darling (22.0%) and Tas (30.0%) Regions and winter in the South West (23.7%). It was lowest in autumn in the South-East (1.5%), Murray-Darling (4.7%) and South-West (4.0%) but in the Tas. Region it was lowest in winter (0.9%). Some of this variation may be attributable to increased conspicuousness when calling in the breeding season but differences from region to region, specially in Tas., suggest that movements are involved also.

The Cuckoo inhabits all types of vegetation but is least often in rainforest. At least 80 species have been recorded as hosts of the Pallid Cuckoo, mainly Meliphagidae and Muscicapidae.[757,1552] This wide choice may partly explain the seemingly erratic movement pattern, e.g. some resident honeyeaters breed in winter in WA but late spring in NSW. The Cuckoo feeds on large insects including caterpillars and grasshoppers.[653]

At Inverell (29/151) density was 0.5 birds/ha.[96] At Boola Boola (38/146) in eucalypt forest density was 0.22 birds/ha.[1113]

Field Atlas Records: 9198

1° blocks–with records: 586 (72%)
–with breeding records: 115 (20%)

Atlas No.: 337

Max. Rep. Rate: 53%
In block(s): 27/149

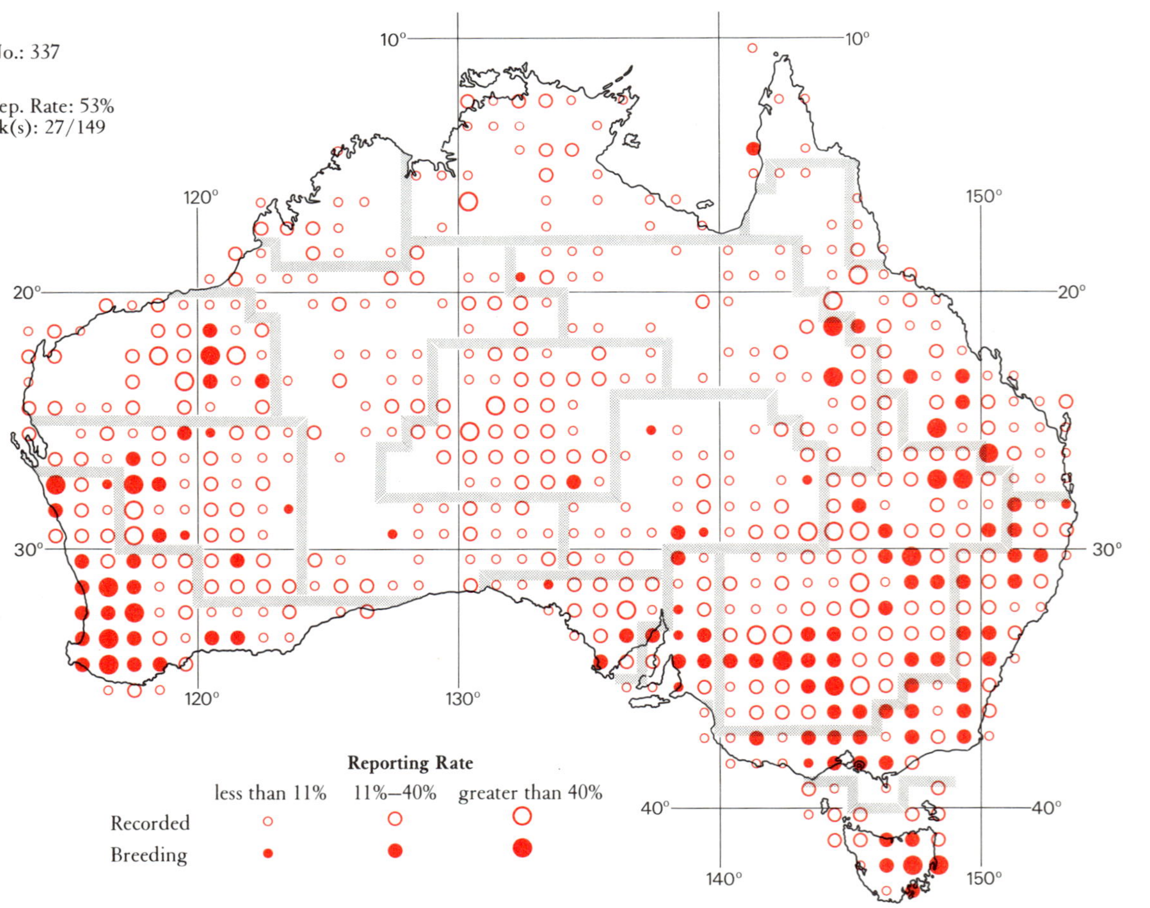

Brush Cuckoo

Cuculus variolosus

About 18 subspecies of the Brush Cuckoo are recognized in south-eastern Asia, New Guinea and adjacent islands and Australia. Two breed in Australia, meeting and intergrading on the eastern coast at the Burdekin R (about 20°S): *dumetorum* to the north and west and nominate *variolosus* to the south.[1211]

The southern subspecies *variolosus* migrates north to winter in southern New Guinea and islands in the Banda Sea (7S/124E).[1211,1449] The winter range does not seem to include south-eastern New Guinea where no change in numbers was detected in woodland or rainforest.[158,160] In Torres Strait the Brush Cuckoo is a breeding summer migrant and possible passage migrant.[503] In the South-East Region it is a summer visitor, Oct.-Feb.,[692] young birds leaving later than adults to return north.[653] The northern subspecies *dumetorum* is not migratory[1211] but may move locally.[423] Reporting Rates were higher in summer than winter partly because the species is more conspicuous when calling during the breeding season. The difference in the South-East Region, where the species is known to migrate, was much greater than in other regions, e.g. South-East Region: summer 7.9%, winter 0.9%; Eastern Qld: summer 9.3%, winter 3.0%; Atherton Region: summer 21.2%, winter 5.3%; Top End: summer 13.1%, winter 3.3%.

The Brush Cuckoo inhabits a variety of country. In the north it lives in eucalypt forest and woodland, paperbark woodland and, less frequently, rainforest and mangroves.[423, 748,1147,1211] In the south it inhabits eucalypt woodland and forest and sometimes paperbark woodland.[653,692,1211]

About 50 species are recorded as hosts of the Brush Cuckoo, mainly Muscicapidae, Maluridae and Meliphagidae.[231,653,872,1552] Host preference varies regionally: Brown-backed and Bar-breasted Honeyeaters in the north; flycatchers in Eastern Qld and the northern South-East Region; and fantails and robins in the South-East Region.[172,1211] The Brown-backed Honeyeater hatched 11 cuckoos in one sample of 39 nests.[1265] The few reports of diet indicate that the Brush Cuckoo feeds on caterpillars and small insects of the foliage. Densities were: 0.1 birds/ha in eucalypt woodland at Wollomombi (30/152);[129] 0.5 birds/ha in a gully in eucalypt forest at Boola Boola (38/146).[1113]

Field Atlas Records: 2279

1° blocks–with records: 140 (17%)
–with breeding records: 31 (22%)

Atlas No.: 339

Max. Rep. Rate: 26%
In block(s): 15/145,19/146

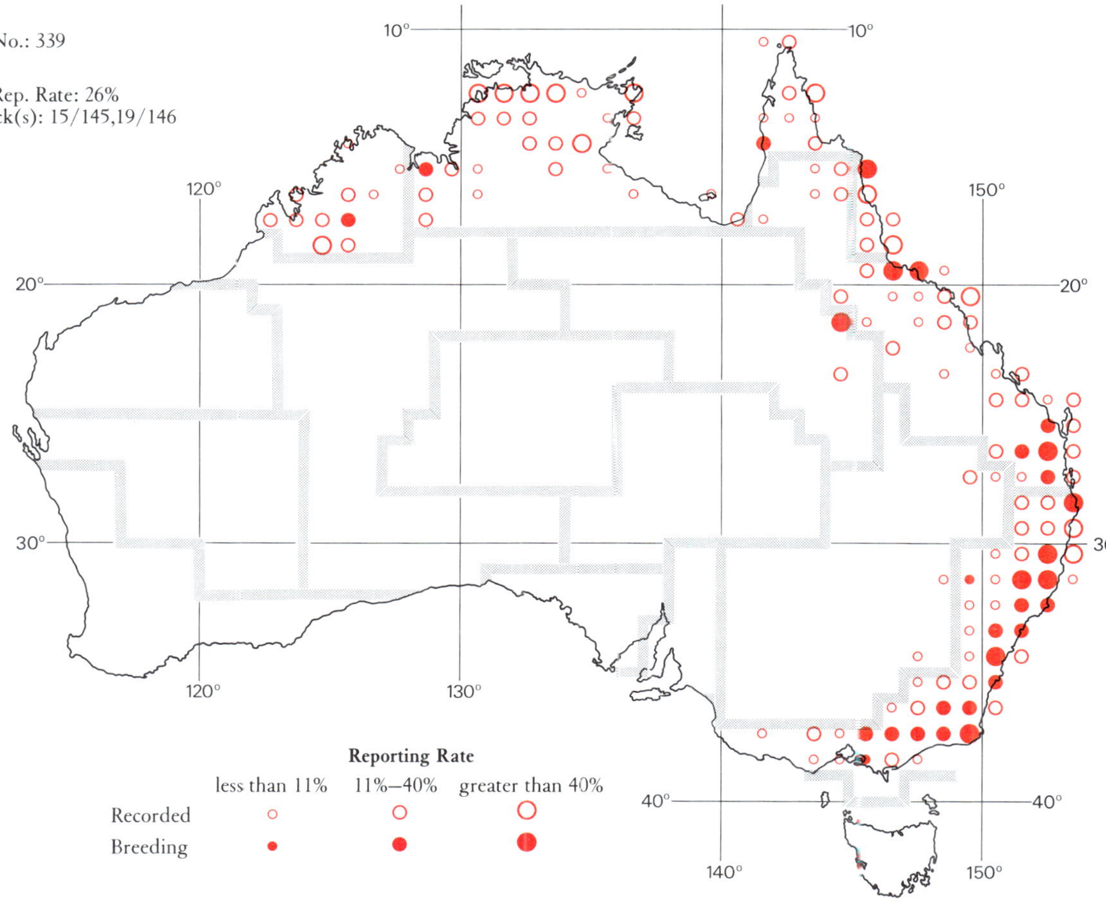

Chestnut-breasted Cuckoo

Cuculus castaneiventris

The Chestnut-breasted Cuckoo has three subspecies in New Guinea and adjacent islands, one of which *castaneiventris* inhabits the Cape York Region. Its distribution is continuous from the tip of Cape York Pen. (10/142) south to 16°S on the eastern coast when records from the Historical and Field Atlases are combined. The Shipton's Flat (15/145) area is a new location for the species. Reports farther south in the Atherton Region (17/145, 19/146) in the 1960s are unconfirmed.[81,231] Two breeding records in Australia are accepted, both from the Claudie R (12/143).[1211]

The Chestnut-breasted Cuckoo is considered to have a small resident population in the Cape York Region[1211] and the species is recorded at Shipton's Flat throughout the year. Some movement through Torres Strait is suggested by the sighting from a boat of a pair west of Booby I. (10/141) in Oct. 1979 and of another on Booby I. in Dec. 1978.[503] The Field Atlas contains records in all seasons suggesting that some birds, at least, are present all year. Reporting Rates in the Cape York Region were highest in summer (13.2%) and lowest in winter (3.4%).

The Chestnut-breasted Cuckoo lives in rainforest[1355] sometimes moving up to a kilometre out into adjacent eucalypt forest. It feeds on insects, particularly caterpillars.[1147] It is uncommon in Australia where there seems to be no more than a small outlier of the main population; in New Guinea it has been described as abundant.[148] There are no measures of abundance.

Field Atlas Records: 69

1° blocks–with records: 5 (1%)

–with breeding records:

Atlas No.: 340

Max. Rep. Rate: 35%
In block(s): 12/143

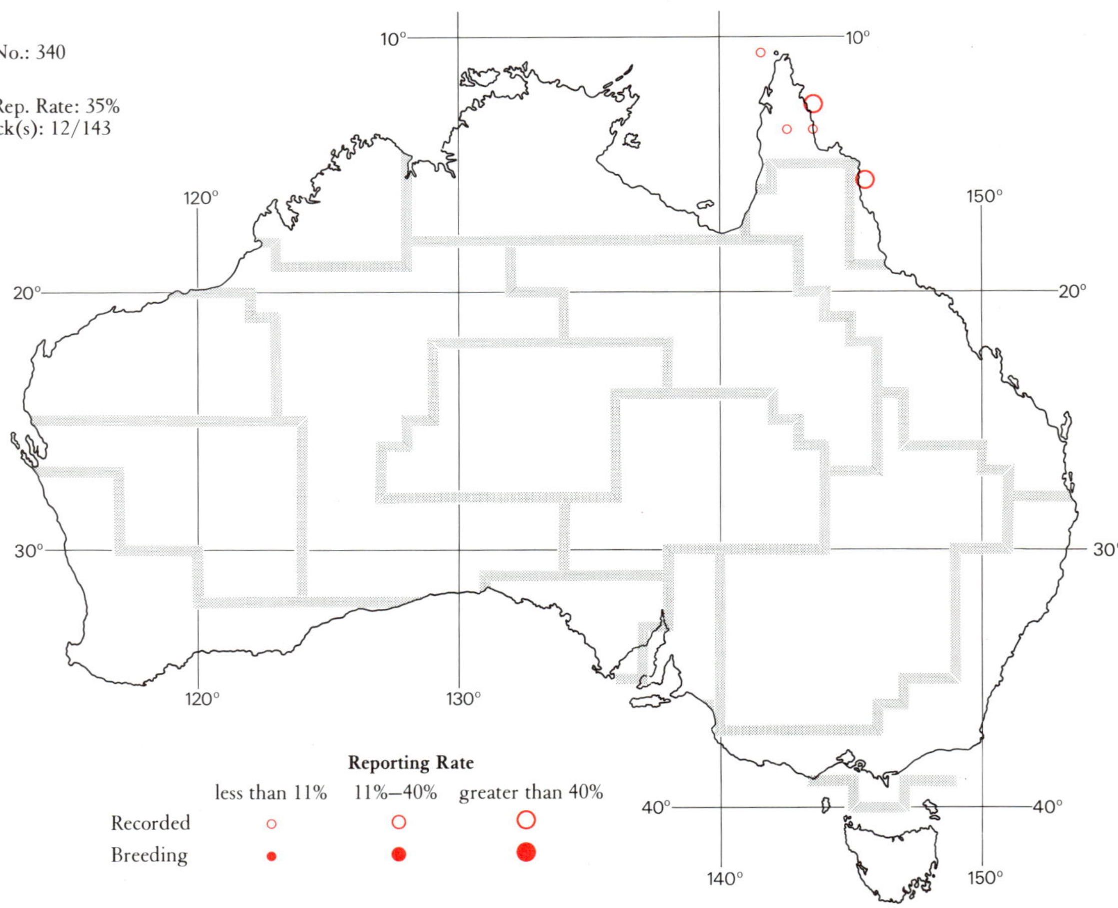

Fan-tailed Cuckoo

Cuculus pyrrhophanus

The six subspecies of the Fan-tailed Cuckoo range through New Guinea and the South Pacific islands with one, *prionurus*, in Australia.[388] This subspecies also occurs in New Guinea, mainly in the Fly R (8S/141E) area.[1449] In the Atlases there are breeding reports from throughout the species' range except for the Cape York Region.

Movements are known to take place, e.g. at Wollomombi (30/152) fewer Fan-tailed Cuckoos are present in winter than in other seasons and at Canberra (35/149) the species is present Aug.-Mar.[154] The pattern suggested by the Field Atlas is complicated by the greater conspicuousness of the bird in the breeding season and by its tendency to breed early, before the end of winter. In the Atherton, Murray-Darling and SA Ra. Regions, Reporting Rates in spring (12.1%, 7.5%, 5.4% respectively) and winter (11.4%, 6.6%, 6.1%) were similar and markedly higher than in summer (4.3%, 1.9%, 2.6%) and autumn (8.1%, 2.8%, 3.1%). In contrast, in the South-East Region the spring Reporting Rate (32.3%) was higher than for any other season (summer 14.7%, autumn 12.0%, winter 17.1%) and similarly in the Tas. Region (spring 41.3%, summer 23.4%, autumn 7.0%, winter 8.8%). The different pattern of Reporting Rates in different regions suggests movement north and inland in winter, taking birds out of the South-East and Tas. Regions.

In the South-West Region reports were restricted to areas south of 31°S in summer, which suggests seasonal movements taking the birds to wetter parts of the region in summer.

The Fan-tailed Cuckoo lives in rainforest, eucalypt forest and woodland, mallee and heath. It parasitizes species of Maluridae and Acanthizidae,[1552] choice of host depending on availability.[1199] It feeds on caterpillars and other insects taken on the ground and in vegetation.

Densities in eucalypt woodland were: Armidale (30/151) 0.02–0.04 birds/ha;[582] Wollomombi 0.2 birds/ha; Canberra 0.08 birds/ha.[154] In eucalypt forest at Boola Boola (38/146) it was 0.2–1.1 birds/ha.[1113] At Moruya (35/150) six cuckoos were seen in 1 ha of eucalypt woodland in Aug.-Sept. when hosts were breeding.[1199]

Field Atlas Records: 10 345

1° blocks–with records: 272 (33%)

–with breeding records: 58 (21%)

Atlas No.: 338

Max. Rep. Rate: 43%
In block(s): 34/150

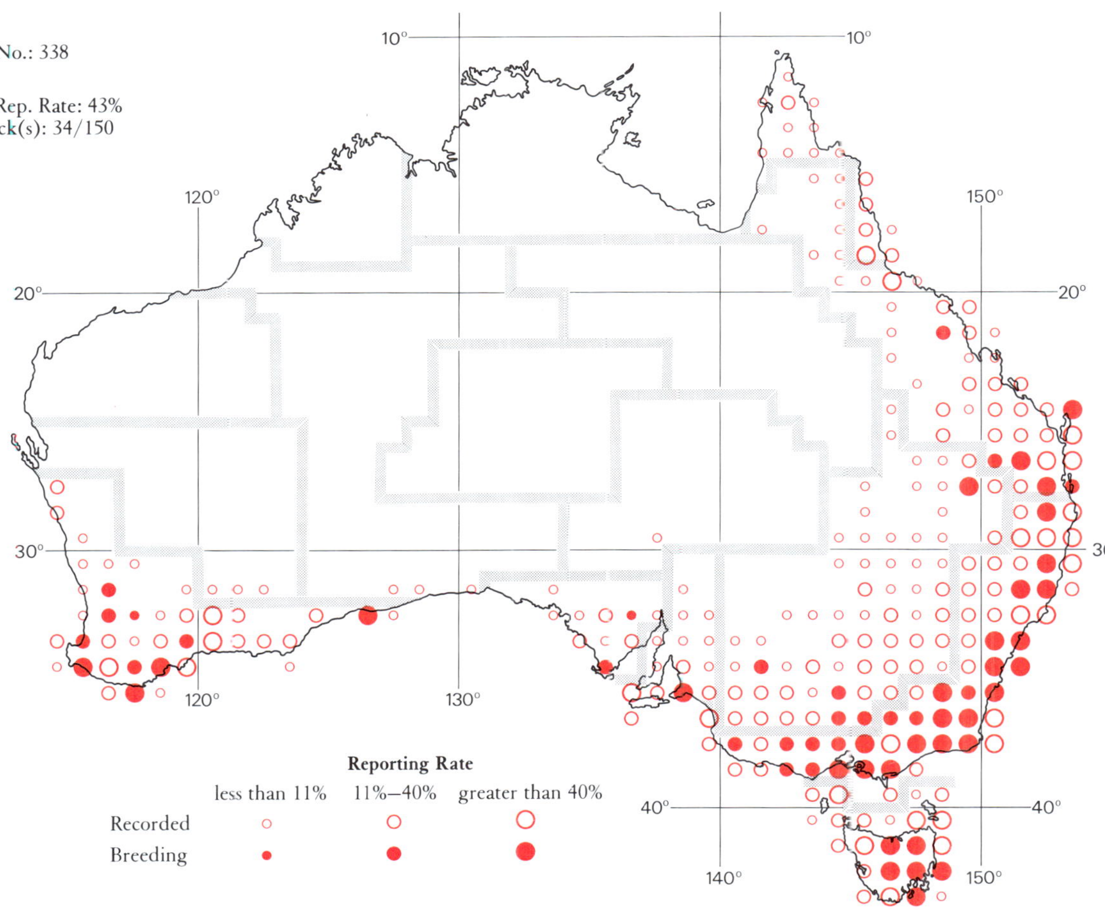

Black-eared Cuckoo

Chrysococcyx osculans

The Black-eared Cuckoo breeds only in Australia and mainly in the southern part of its range (south of 26°S). There is only one breeding report north of 23°S in the Atlases. The species was first reported in Tas. during the Field Atlas in Dec. 1977.

Movements of the Black-eared Cuckoo are not understood. Some birds reach islands in the Banda Sea (7S/124E) and southern New Guinea[1211] and the species is reported to be an uncommon migrant across Torres Strait.[503] In northern Australia, where it seldom breeds, observers have reported that: at Atherton (17/145), single birds arrived in June of most years;[231] at Mt Isa (20/139) the species was recorded Apr., May and Aug. and was thought to be a regular winter visitor;[862] at Darwin (12/130) birds arrive Aug.-Sept. and return to wintering grounds Apr.-June;[1211] and in the Kimberley Region birds are said mainly to move north in Oct. and south in Apr.-May.[1713] In the Field Atlas the Cuckoo was reported in northern Australia in all seasons (Reporting Rates north of 26°S: spring 0.61%, summer 0.91%, autumn 0.85%, winter 0.99%).

On the southern edge of its range the Black-eared Cuckoo appears to be absent Jan.-Aug.[627,653] Reporting Rates in southern regions were highest in spring but this probably reflects, at least partly, increased conspicuousness in the breeding season.

The Black-eared Cuckoo lives in open areas in many types of country: eucalypt woodland, specially where there are also paperbarks; dongas on the Nullarbor Plain; mangroves on the coast; acacia scrub; and spinifex. It parasitizes hosts that build dome-shaped nests, usually the hylacolas, Redthroat and Speckled Warbler;[1211] the White-browed Scrubwren, Superb and White-winged Fairy-wrens have occasionally been recorded as hosts also.[684,692,694] The Black-eared Cuckoo feeds on insects, chiefly moths and caterpillars, and sometimes seeds.[748,1049]

There are no measures of abundance.

Field Atlas Records: 903

1° blocks–with records: 272 (33%)
–with breeding records: 20 (7%)

Atlas No.: 341

Max. Rep. Rate: 23%
In block(s): 25/113

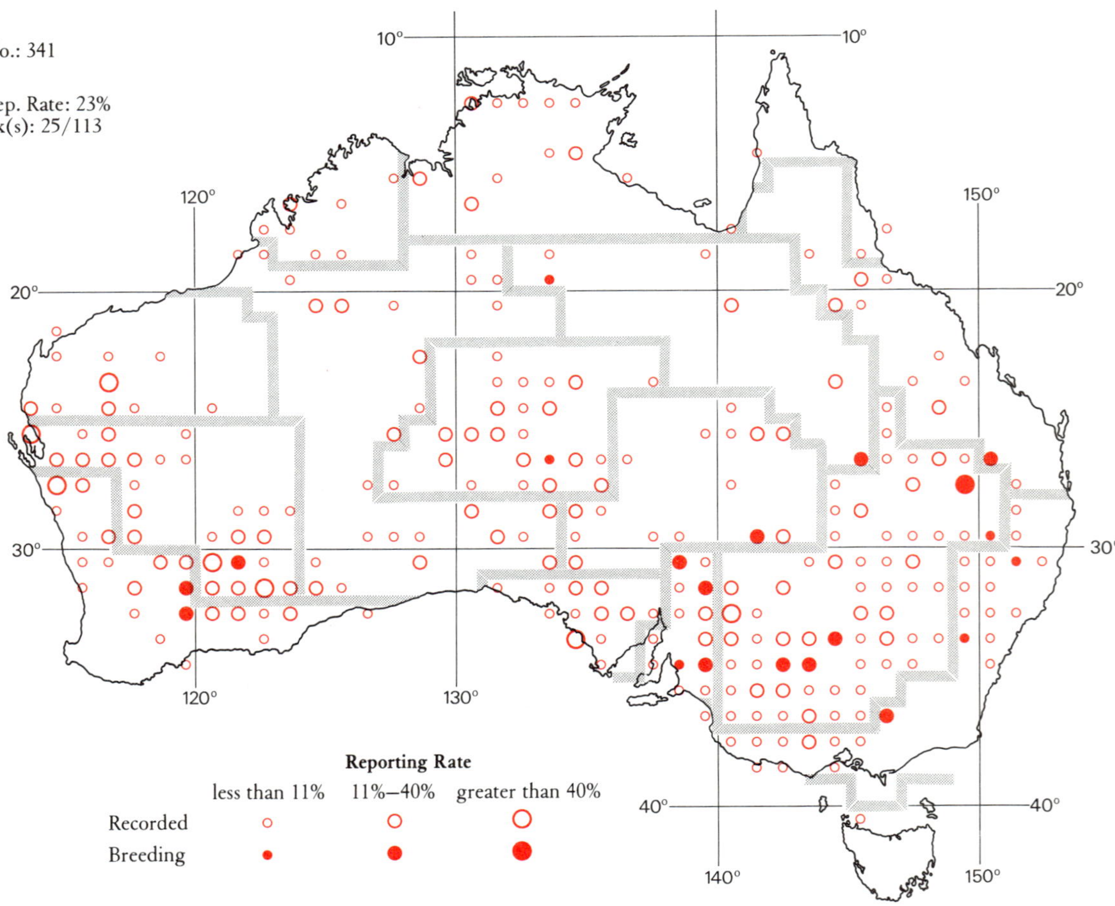

Horsfield's Bronze-Cuckoo

Chrysococcyx basalis

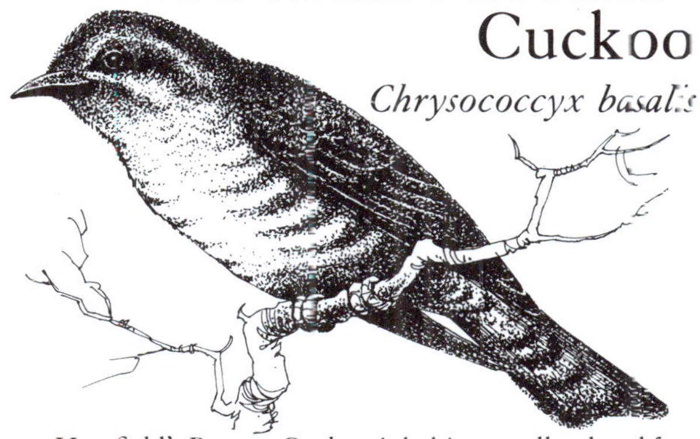

Horsfield's Bronze-Cuckoo breeds in Australia and some birds migrate in winter to New Guinea, Indonesia and islands of the eastern Indian Ocean. A few pass through Torres Strait.[503] In the Atherton Region birds are present and call all year but sometimes many arrive in Aug.;[231,698] Reporting Rates were higher in summer (5.2%) than in winter (1.2%), which suggests that some birds leave in winter. To the south, in districts on the coastal side of the Great Dividing Ra., Horsfield's Bronze-Cuckoo is known as a breeding migrant, though in most places some birds remain in winter. Thus it is common in spring and summer at Rockhampton (23/150);[1089] present all year but in varying numbers at Murphy's Ck (27/152);[1097,1099] most common Aug.-Feb. near Sydney (33/151)[396,692] and recorded June-Dec. at Laverton (37/144).[1822] Reporting Rates in the South-East and Tas. Regions were high in spring (15.0% and 5.7%) and low in winter (3.9% and 0.5%) reflecting conspicuousness in the breeding season and movement away in winter; the bird is almost completely absent from Tas. in winter.

To the west of the Great Dividing Ra. movements also take place. Birds migrate through the southern Murray-Darling in spring and autumn, sometimes staying to breed or to winter.[833] At Cobar (31/145) the species is usually present Aug.-Mar.[1573] Round Darwin (12/130) the Reporting Rate dropped from 8.0% in summer to 4.0% in winter and a resident observer recorded the species as a visitor, most often Sept.-Dec.[423]

In the South-West the Reporting Rate was high in winter (8.7%) and spring (9.6%) and low in the dry summer (2.8%) and the species is regarded as a migrant, arriving mid-June.[1638]

Horsfield's Bronze-Cuckoo inhabits woodland and forest, specially edges and clearings within it; mallee; acacia scrub; coastal saltmarsh; and spinifex. It parasitizes the nests of Muscicapidae, Maluridae and Acanthizidae.[1552] In south-eastern Australia it seems to be most successfully raised by fairy-wrens and thornbills.[410,748,1193] The bird feeds on insects including hairy caterpillars.

At Armidale (30/151) near eucalypt woodland density was 0.04 birds/ha.[582] At Boola Boola (38/146) in eucalypt forest density Aug.-Feb. was 0.1–0.2 birds/ha.[1113]

Field Atlas Records: 7070

1° blocks–with records: 581 (72%)
–with breeding records: 125 (22%)

Atlas No.: 342

Max. Rep. Rate: 36%
In block(s): 21/144

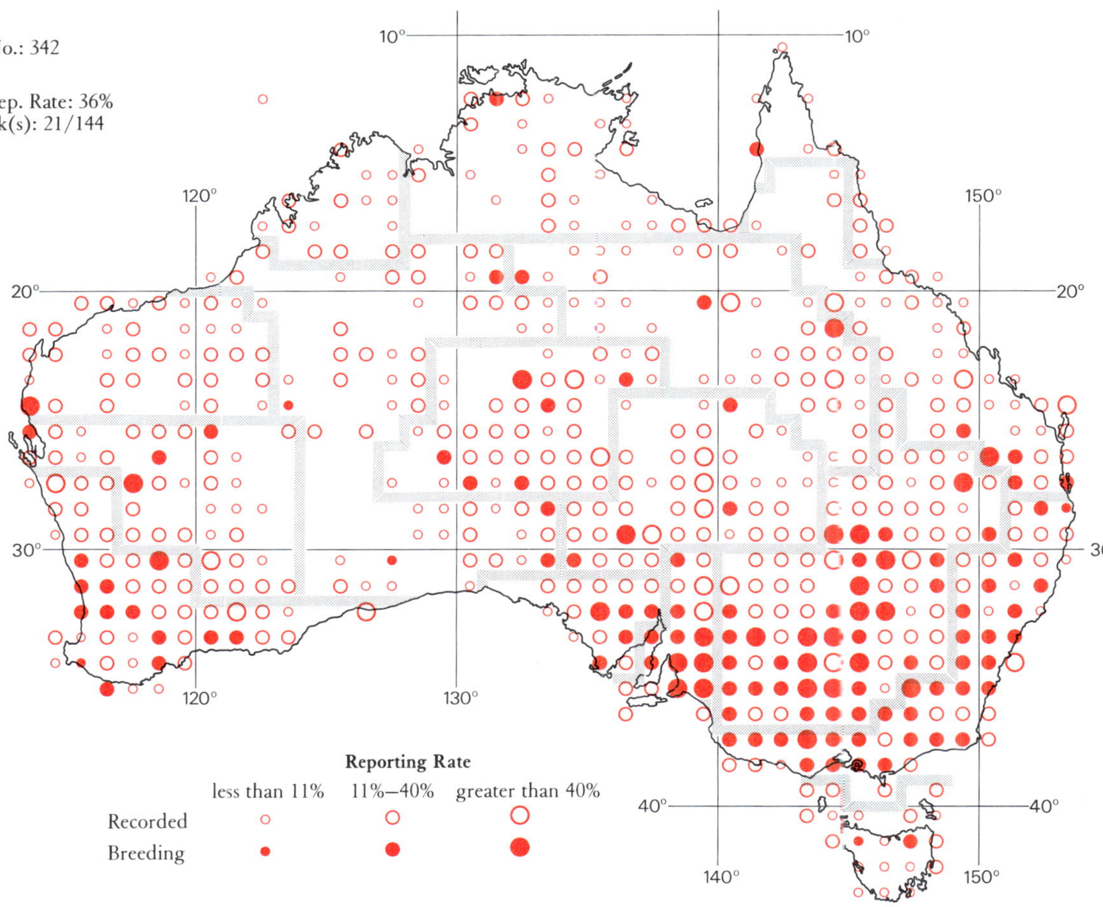

Shining Bronze-Cuckoo

Chrysococcyx lucidus

One subspecies of the Shining Bronze-Cuckoo *lucidus* breeds in New Zealand and another *plagosus* in Australia. Both are migratory, wintering in New Guinea and adjacent islands and parts of Indonesia. There are also sedentary populations in the Philippines and western Pacific islands.[388] In Australia eastern and western populations of *plagosus* are separated. Breeding has not been recorded in the Cape York Region.

The New Zealand subspecies *lucidus* migrates in winter through Lord Howe (31/159) and Norfolk (29/167) Is to the Solomons; a few birds are recorded in Eastern Qld and the South-East Regions on migration, mostly in autumn, occasionally in spring. They have been reported as far west as Glen Innes (29/151), Munghorn Gap (32/149) and L. George (35/149).[1292]

Part of the eastern Australian population of *plagosus* moves north in winter.[692,734] Small numbers pass across Torres Strait.[503] Reporting Rates were highest in spring in the South-East (14.2%) and Tas. (17.7%) Regions, which presumably reflects conspicuousness in the breeding season as well as migration, and lowest in winter (3.4% and 0.1% respectively). The Cuckoo is present at Wollomombi (30/152) mainly Oct.-Mar.;[129] and at Canberra (35/149) Oct.-Apr.[154]

The western population of *plagosus* migrates to Indonesia.[1638] The Field Atlas suggests it to be absent in autumn rather than winter (Reporting Rates: spring 12.4%, summer 4.1%, autumn 1.0%, winter 4.9%). Birds, presumably on passage, are occasionally recorded in WA north of the breeding range.[1713]

The Shining Bronze-Cuckoo inhabits eucalypt forest,[653] mainly areas receiving over 380 mm of rain a year.[194] It parasitizes the Acanthizidae, mostly thornbills but also gerygones.[1552] The Yellow-rumped Thornbill is commonly used.[1193] The frequency of parasitism varies: in a 40-ha site south of Perth (31/115) 21 of 65 Yellow-rumped Thornbill nests were parasitized in 1956 when the Cuckoo was common but only five out of 24 in 1959 and one out of 23 in 1961.[590] The Cuckoo feeds on caterpillars, other insects and spiders.

Densities in eucalypt woodland were: Armidale (30/151) 0.02–0.42 birds/ha;[582] Wollomombi 0.1 birds/ha;[129] Canberra 0.02 birds/ha;[154] and in eucalypt forest at Boola Boola (38/146) 0.1–0.9 birds/ha.[1113]

Field Atlas Records: 4649

1° blocks–with records: 218 (27%)
–with breeding records: 58 (27%)

Atlas No.: 344

Max. Rep. Rate: 24%
In block(s): 30/151

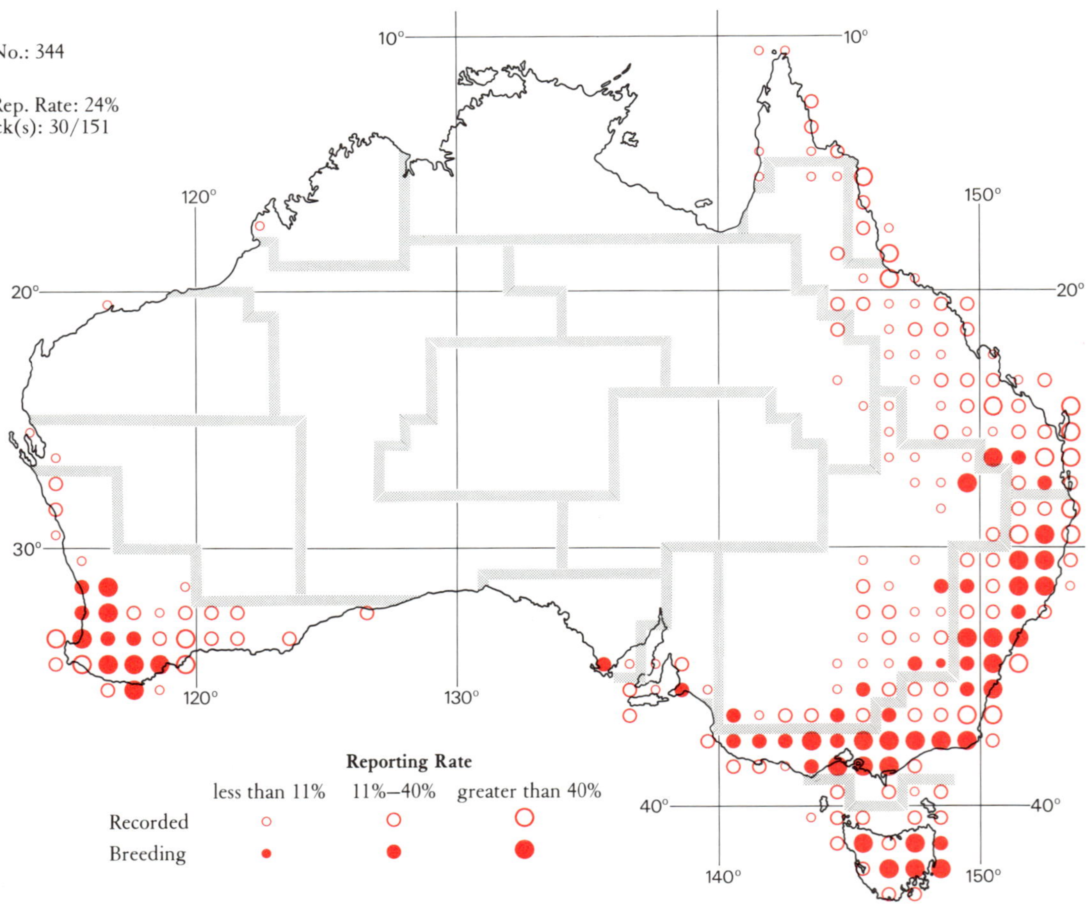

Little Bronze-Cuckoo

Chrysococcyx malayanus

Gould's Bronze-Cuckoo

Chrysococcyx russatus

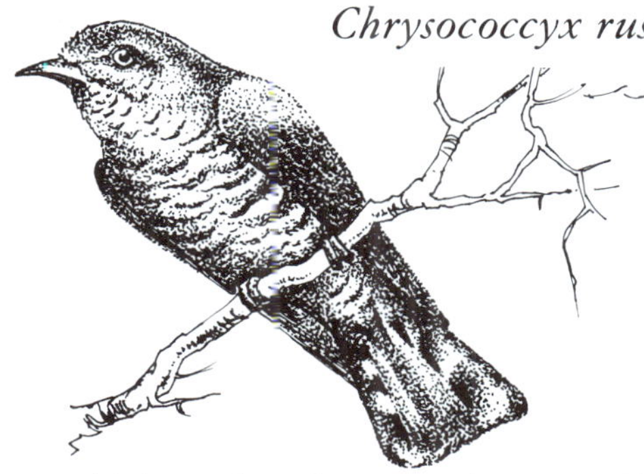

Little and Gould's Bronze-Cuckoos belong to a group of species living in south-eastern Asia, New Guinea and Australia.[1388] Whereas in Sulawesi the two species overlap without apparent interbreeding, in Australia they apparently interbreed and form intermediates.[621,1388] They are therefore treated together here. Gould's Bronze-Cuckoo breeds in the Cape York and Atherton Regions and in Eastern Qld south to about Bowen (20/148). The Little Bronze-Cuckoo has two subspecies and a disjunct range: *minutillus* inhabits the Top End and Kimberley Regions, meeting Gould's Bronze-Cuckoo at the eastern end of this range; *barnardi* breeds in Eastern Qld and the South-East Region south of Byfield (22/150). These ranges are established from specimens. Atlas reports suggest little gap between the breeding ranges of Gould's Bronze-Cuckoo and *barnardi*. Birds identified as Gould's seem to be resident near Pleystowe (21/149) and birds identified as *barnardi* were present in Nov., the breeding season, in the Eungella (21/148) district.

The subspecies *barnardi* appears to migrate north in winter.[621,1388,1709] The Field Atlas had no winter records south of 22°S and specimens of *barnardi* have been taken in the Cape York Region in winter. A few Bronze-Cuckoos migrate across Torres Strait each year,[503] apparently both species since specimens of *barnardi* and of individuals resembling the Australian subspecies of Gould's have been identified from New Guinea.[1388] Gould's Bronze-Cuckoo may be migratory at the southern end of its range[1388,1709] though some birds were present at Townsville (19/146) and Pleystowe all year. Reporting Rates in the Atherton Region were lower in winter (7.5%) than in summer (12.0%). Those in the Darwin (12/130) district suggest that some birds may leave in summer (summer 7.6%, winter 11.5%).

Both species live in mangroves, edges of rainforest, paperbark woodland and gardens.[621] Birds are usually alone or in pairs; a group of more than six was seen in winter at Bellenden Ker (17/145).[1469] The two species parasitize gerygones[621,1552,1859] and feed on insects.[748] There are no measures of abundance.

Field Atlas Records: 460

1° blocks–with records: 58 (7%)
–with breeding records: 8 (14%)

Atlas No.: 345,346

Max. Rep. Rate: 33%
In block(s): 12/143

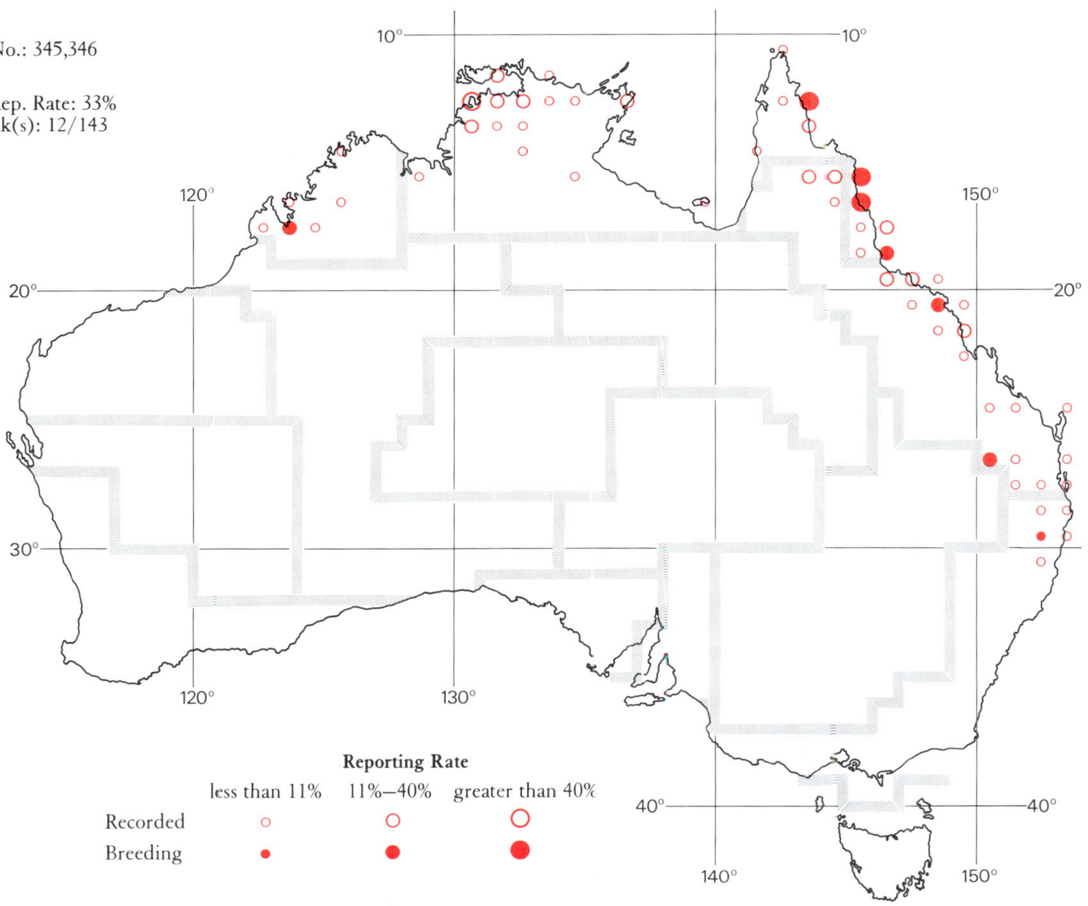

Common Koel

Eudynamis scolopacea

The Common Koel breeds in Asia, New Guinea, western Pacific islands and Australia. The two subspecies recognized in Australia are migratory: *subcyanocephala* breeds across the north; *cyanocephala* breeds in eastern Australia south of about 20°S.[1211] The Koel is more widespread in northern Australia, particularly the Kimberley Region, than the map indicates. Lack of observers in summer, when most Koels are present, accounts for the difference. Breeding is reported throughout the Koel's range except in the far south where the species is only a vagrant. The Koel has once been recorded in southern WA: an immature female appeared near Quairading (32/117) after a cyclone in Apr. 1971.[1633]

The subspecies *subcyanocephala* winters in southern New Guinea[1211,1449] though no change in numbers is evident near Port Moresby (9S/147E);[158] *cyanocephala* is thought to winter in Indonesia and possibly as far north as the Philippines.[1211] The Koel is an uncommon migrant through Torres Strait.[503] The Field Atlas contains no winter records of the Koel in northern Australia and the few in the east are probably mostly accounted for by early arrivals.

At Grafton (29/152) and Sydney (33/151) *cyanocephala* arrives in Sept.[396,692,721] and leaves late Jan.-Mar., with stragglers as late as May.[721] Within the breeding range of *subcyanocephala* extreme dates of arrival and departure are Aug.-May in the Innisfail (17/146) district,[698] Oct.-Apr. at high altitudes near Atherton (17/145),[231] Nov.-Apr. at Mt Isa (20/139),[862] Sept.-Mar. at Darwin (12/130).[423] In the Kimberley Region it is present Oct.-Apr.[1211]

The Koel inhabits eucalypt forest, woodland and rainforest along streams. Host preference varies regionally but friarbirds are often used as well as the Magpie Lark, Figbird, Blue-faced Honeyeater and wattlebirds.[1211,1552,1857] The Koel eats fruit, including mulberries,[396] occasionally taking other food in forests as well as feeding in street trees.

Near Grafton one pair occupied a territory covering those of about five pairs of Magpie Larks. One pair of Magpie Larks may rear up to three Koels in a season.[721]

Field Atlas Records: 2811

1° blocks–with records: 141 (17%)
–with breeding records: 23 (16%)

Atlas No.: 347

Max. Rep. Rate: 29%
In block(s): 24/148

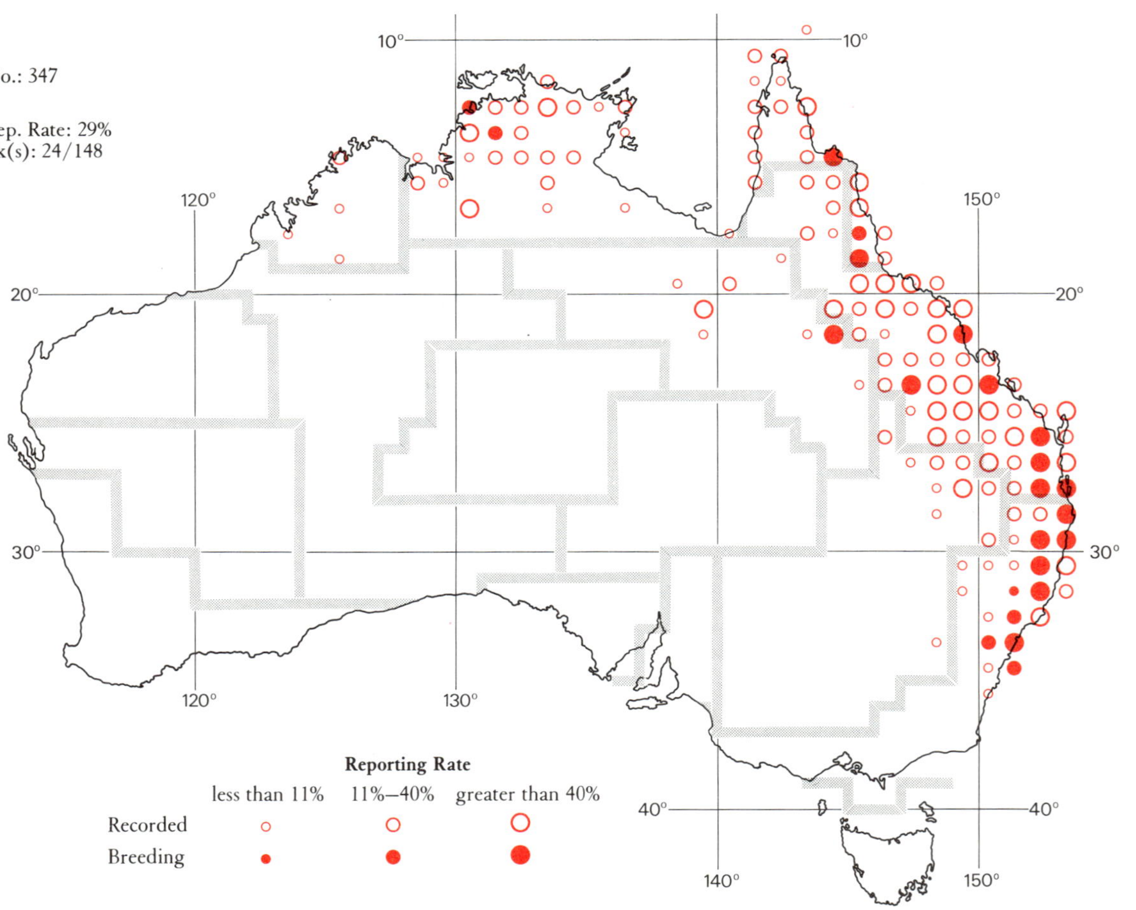

Channel-billed Cuckoo

Scythrops novaehollandiae

The Channel-billed Cuckoo breeds in Australia and possibly New Guinea.[129] It migrates in winter to southern Indonesia, the Bismarck Archipelago[1212] and New Guinea.[129] It was recorded once in New Caledonia and occasionally reaches New Zealand.[804] On the map the lack of records in northern Australia is partly because observers rarely went there in summer. The bird is common round the Gulf of Carpentaria[1211] but the low Reporting Rates in the Top End and Kimberley Regions suggest that it is less abundant there than in the east, which confirms the opinion of local observers.[423,1713] There are scattered historical records in inland Australia, particularly the L. Eyre Region where the Cuckoo breeds when the Diamantina R floods.[1465] It was also recorded in Tas. in Nov. 1867 and Feb. 1943.[734]

The Channel-billed Cuckoo is a passage migrant through Torres Strait;[503] only a few birds are seen moving south in Aug.-Sept. but there is a clear movement northward in Mar.[503] Exhausted birds arrive in Port Moresby (9S/147E) from the direction of the sea, which suggests that some cross the Coral Sea as well.[129] In the Atherton Region birds arrive Aug.-Sept. and depart Apr.-May (at higher altitudes they are present Oct.-Mar.).[231,698] Farther south at Rockhampton (23/150) and Murphy's Ck (27/152) the species arrives in Aug. each year and departs in Mar.[1089,1099]

The Channel-billed Cuckoo usually lives singly or in pairs; at the end of the breeding season it forms small flocks, one of which at Innisfail (17/146) contained as many as 32 birds.[698] The species inhabits eucalypt forest, specially along watercourses, and rainforest along streams, places where both hosts and figs abound.[1211] Hosts are varied and include crows, currawongs[1552] and the Australian Magpie, all of which range much more widely than the Cuckoo. Two eggs are often laid in the same nest and up to five cuckoo chicks have been reared together in one nest.[1145]

The Channel-billed Cuckoo feeds mainly on figs[953] but will also take other fruit and occasionally insects including stick insects.[928] One of the few to reach Tas. had mulberries in its guts.[1647] There are no measures of abundance.

Field Atlas Records: 1777

1° blocks–with records: 147 (18%)
–with breeding records: 30 (20%)

Atlas No.: 348

Max. Rep. Rate: 30%
In block(s): 21/144

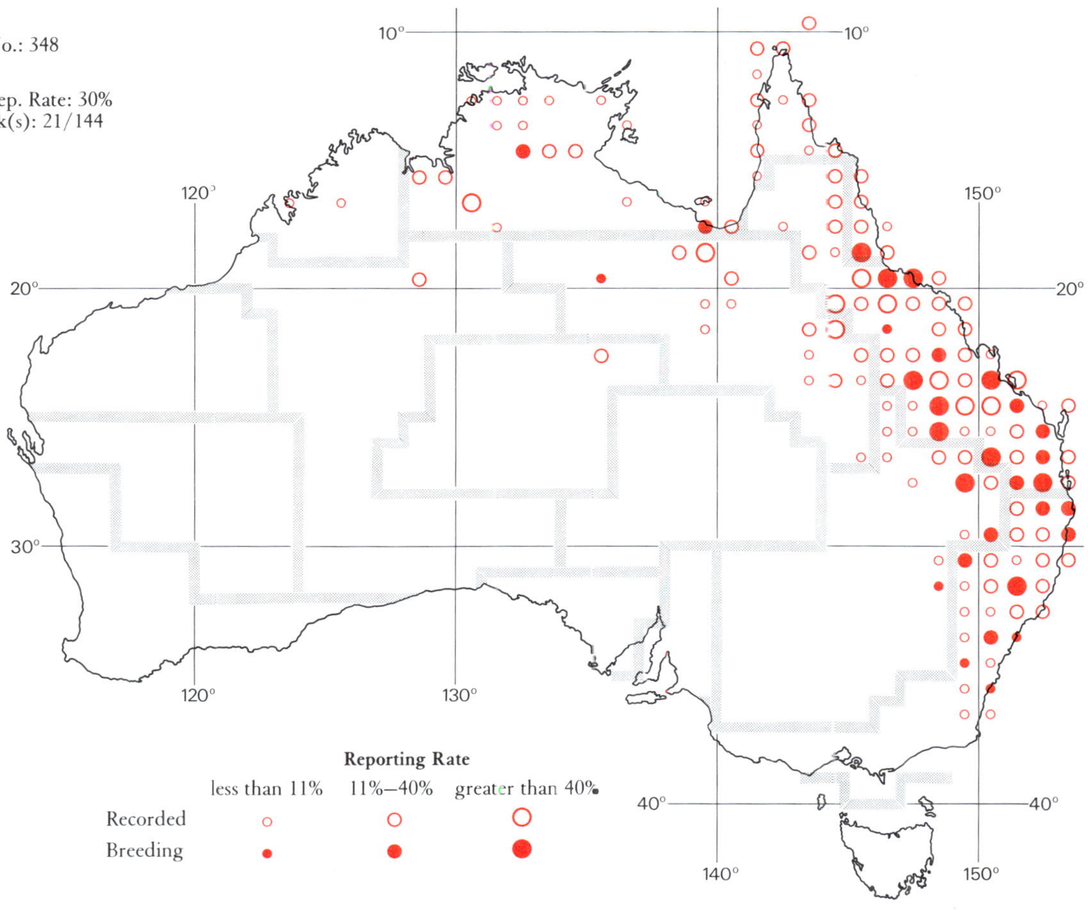

Pheasant Coucal

Centropus phasianinus

The Pheasant Coucal lives in Timor, New Guinea and Australia. The two subspecies recognized in Australia are divided on the eastern coast at the Burdekin R (about 20°S): *phasianinus* to the south and *macrourus* to the north and west.[1214] The southernmost record of *phasianinus* is from Conjola (35/150) in 1953.[805] Historical records fill the gaps in the Field Atlas map on the southern coast of the Gulf of Carpentaria. The population in the Pilbara Region is isolated by the Great Sandy Desert. Before 1950 it was recorded on the De Grey R (20/119) and Oakover R (20/120) east to 121°E and as far south as Minilya (23/113).[805]

The Pheasant Coucal is territorial when breeding but during non-breeding months will move locally.[1211] The occasional inland record such as at Bingara (29/150) shows that some birds wander long distances but the Field Atlas does not suggest any large-scale seasonal movement.

In the north the Coucal inhabits eucalypt woodland with an understorey of rank grass, paperbark woodland, rainforest and pandanus along streams; in the south it lives in eucalypt forest with a grass understorey, heath and swamps.[1709,1711,1713] In the South-East Region it is confined to low altitudes, e.g. the Reporting Rate in the New England (30/152) district was higher below 500 m (50%) than above 500 m (14%).

Unlike other cuckoos, the Coucal builds its own nest and does not parasitize other species. It feeds on large insects and some small vertebrates[1188] and spends much of its time on the ground. At Murphy's Ck (27/152) the arrival of the fox was followed by a decrease in the number of Coucals but the population later recovered.[1099] The Coucal has adapted well to living in canefields but, specially after the breeding season, is often killed on roads in sugar-growing districts.

In New Guinea density in woodland near Port Moresby (9S/147E) was about 0.12 birds/ha.[158]

Field Atlas Records: 4621

1° blocks–with records: 176 (22%)
–with breeding records: 33 (19%)

Atlas No.: 349

Max. Rep. Rate: 63%
In block(s): 20/149

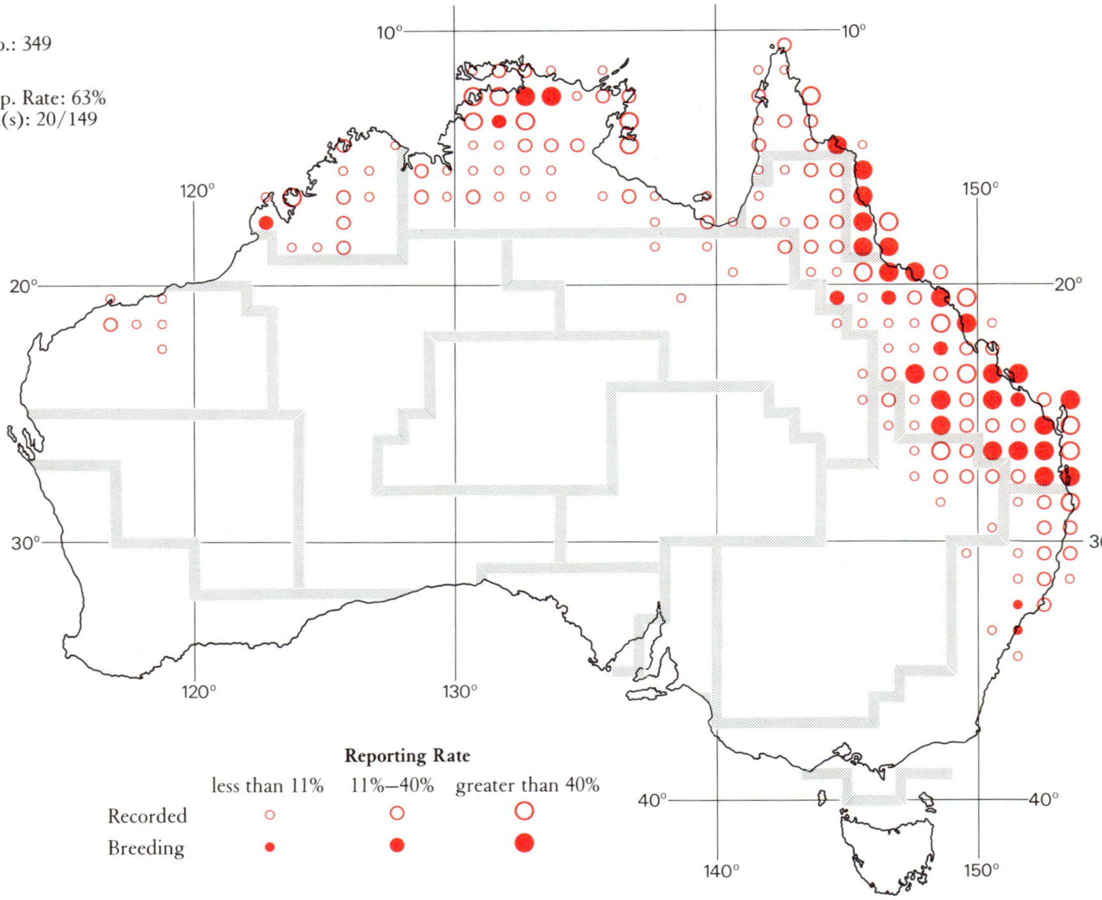

Rufous Owl

Ninox rufa

The Rufous Owl lives in New Guinea and adjacent islands as well as in Australia. Three subspecies are recognized in Australia: *meesi* in the Cape York Region; *queenslandica* in the Atherton Region; and *rufa* in the Top End and Kimberley Regions.[1588] In the Cape York Region there are historical records north to Somerset (10/142) and on the western coast at the Archer R (13/141). In the Kimberley Region the species has been recorded from Napier Broome Bay (14/126) west to the Prince Regent R (15/125) as recently as 1974. The lack of records from these regions in the Field Atlas reflects the difficulty of access rather than any change in distribution.

The Rufous Owl is present all year within its range. Pairs live in rainforest, including that along rivers, paperbark woodland and occasionally enter mangroves. The diet of the Rufous Owl varies according to what is available. It preys on birds as large as the Australian Brush Turkey and Pacific Heron and also takes mammals, including flying foxes, and insects.[1588]

Nests can be 3–4 km apart and near Ingham (18/146) a foraging territory may cover 400–800 ha.[1588]

Field Atlas Records: 77

1° blocks–with records: 18 (2%)
–with breeding records: 1 (6%)

Atlas No.: 247

Max. Rep. Rate: 13%
In block(s): 15/145

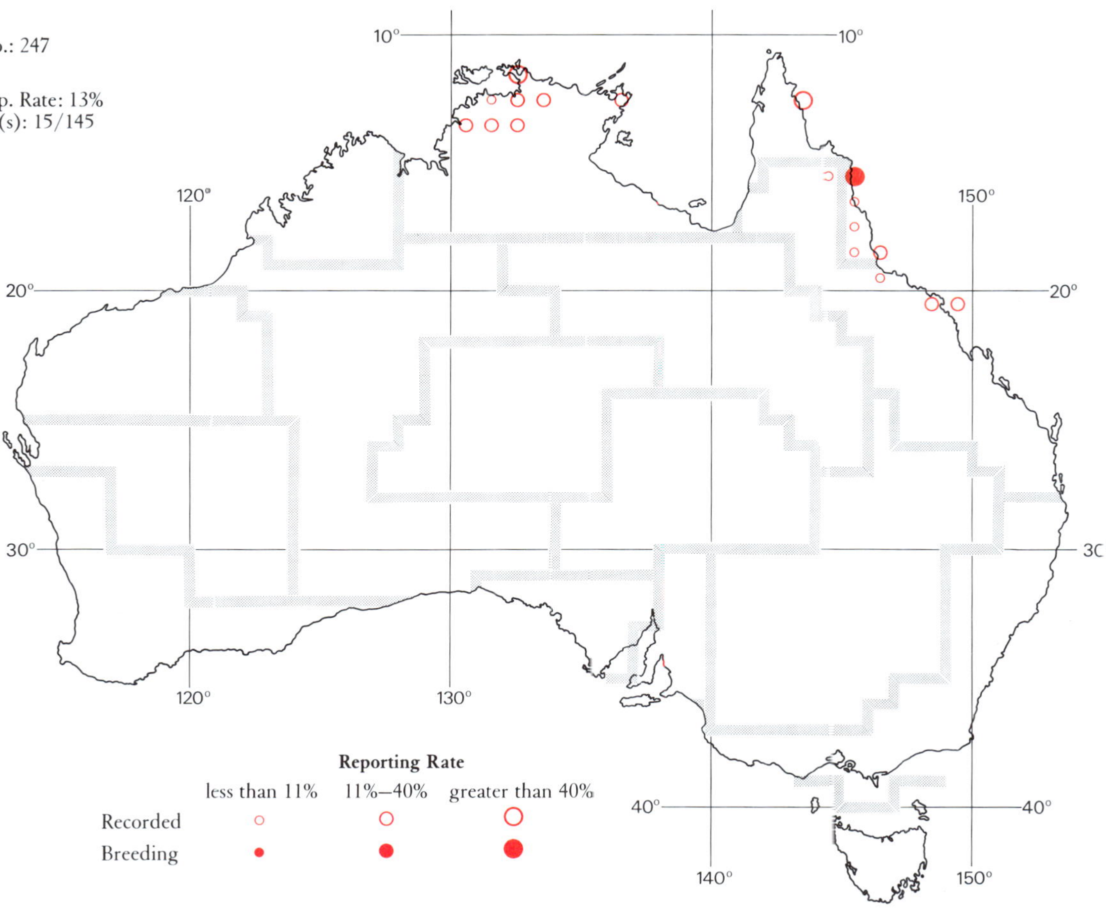

Powerful Owl

Ninox strenua

The distribution of the endemic Powerful Owl determined by the Field Atlas is similar to that in the Historical Atlases. The species was recorded a little farther north at Duaringa (23/149)[102] in the 1901–50 period and at the southern end of its range it was seen once, in 1968, near Mt Burr (37/140). Reports north of this range probably refer to the Rufous Owl; a specimen of a supposed Powerful Owl taken near Adelaide (34/138) was misidentified and is actually a Barking Owl.[1577] The Powerful Owl is generally confined to the coastal side of the Great Dividing Ra. but it is also recorded on the inland slopes. Occasional records well inland are of young vagrants dispersing in search of territory.[1588]

The Powerful Owl is sedentary.[894] It lives alone or in pairs and occupies a permanent territory containing a number of roost sites. The Owl roosts by day in dense foliage, often along streams between ridges covered with eucalypt forest.[1588] The open structure of the forest suits its hunting method.[1588] It hunts nocturnally. Arboreal or semi-arboreal mammals form about 80% of its diet.[1763] Birds account for about 18%; insects and some terrestrial mammals are also taken. The prey varies from place to place depending on seasonal availability.[1614] The ring-tailed possum was the most common prey species at study sites in Vic. and the sugar glider was also important at some times in some places.[1763] One study found the greater glider to be important but fire and clearing have reduced the glider's range and abundance and may have altered its relative importance in the Owl's diet.[1763]

The Powerful Owl is adversely affected when populations of arboreal marsupials are reduced by bush fires or by conversion of native forest to plantations of exotic or native trees.[1613] Pairs of Powerful Owls occupy permanent territories of 800–1000 ha in eastern Vic.[563] but larger in western Vic. where less prey is available.[1614] Territories are usually well separated with pairs at intervals of 5–20 km.[1588]

Field Atlas Records: 287

1° blocks–with records: 45 (6%)
–with breeding records: 8 (18%)

Atlas No.: 248

Max. Rep. Rate: 4%
In block(s): 38/146

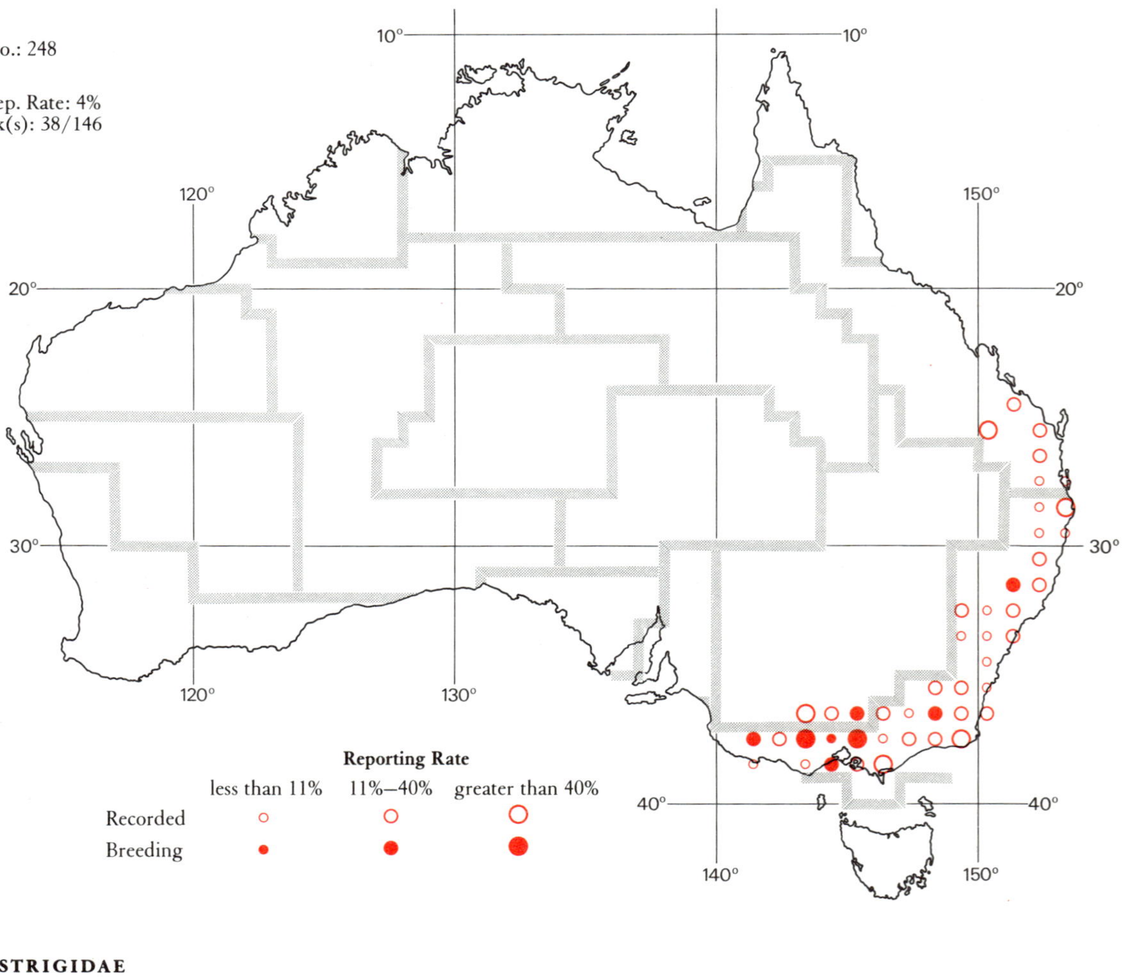

Southern Boobook

Ninox novaeseelandiae

The Southern Boobook inhabits New Zealand, eastern Indonesia, New Guinea and adjacent islands and Australia, including Lord Howe I. (31/159). In Australia recent taxonomic treatment recognizes four subspecies: *lurida* in the Atherton Region; *leucopsis* in the Tas. Region; and, in the remainder, *boobook* in the east and *ocellata* in the west, with a boundary along the west of the Great Dividing Ra.[1588] Records in the 1951–76 period link the Central Ra. and Goldfields populations across the lower Western Deserts and fill the apparent gap isolating the South-West population. The Historical Atlases contain breeding records from all regions.

Immature birds disperse locally. A few Boobooks arrive in parts of Sydney (33/151) each autumn and winter.[363] Early specimens of the Tas. population taken in Vic. and NSW Apr.-Oct. and one taken at sea in Bass Strait (in May 1906) indicate that some Tas. birds migrate north in winter but the last specimen was taken 60 years ago.[1241,1588] A bird on passage was recorded on Albatross I. (40/144) in Dec. 1982.[248]

The Boobook, having great ecological tolerance, ranges through eucalypt forest and woodland, rainforest and the inland wherever there are suitable nesting hollows in living or dead trees.[1588] Most Atlas records resulted from its easily recognized calls given repeatedly in long bouts all year round but more sustained when breeding. The Boobook hunts nocturnally in clearings and is often seen by the side of roads or flying under street lights after moths.[1588] Unlike other members of the genus, the Boobook is mainly insectivorous, though it also takes small birds and mammals, joining other raptors to exploit plagues of the house mouse. In Tas. the population may have been reduced by clearing and through competition with starlings and kookaburras for nest holes.[1276]

The area covered by a hunting pair in one night was estimated to average about 8 ha in the South-East Region though probably less in eucalypt forest and more in arid country.[1588] At Boola Boola (38/146) in eucalypt forest the Boobook's density was 0.1 birds/ha.[1113]

Field Atlas Records: 8344

1° blocks-with records: 537 (66%)
-with breeding records: 67 (12%)

Atlas No.: 242

Max. Rep. Rate: 40%
In block(s): 13/130

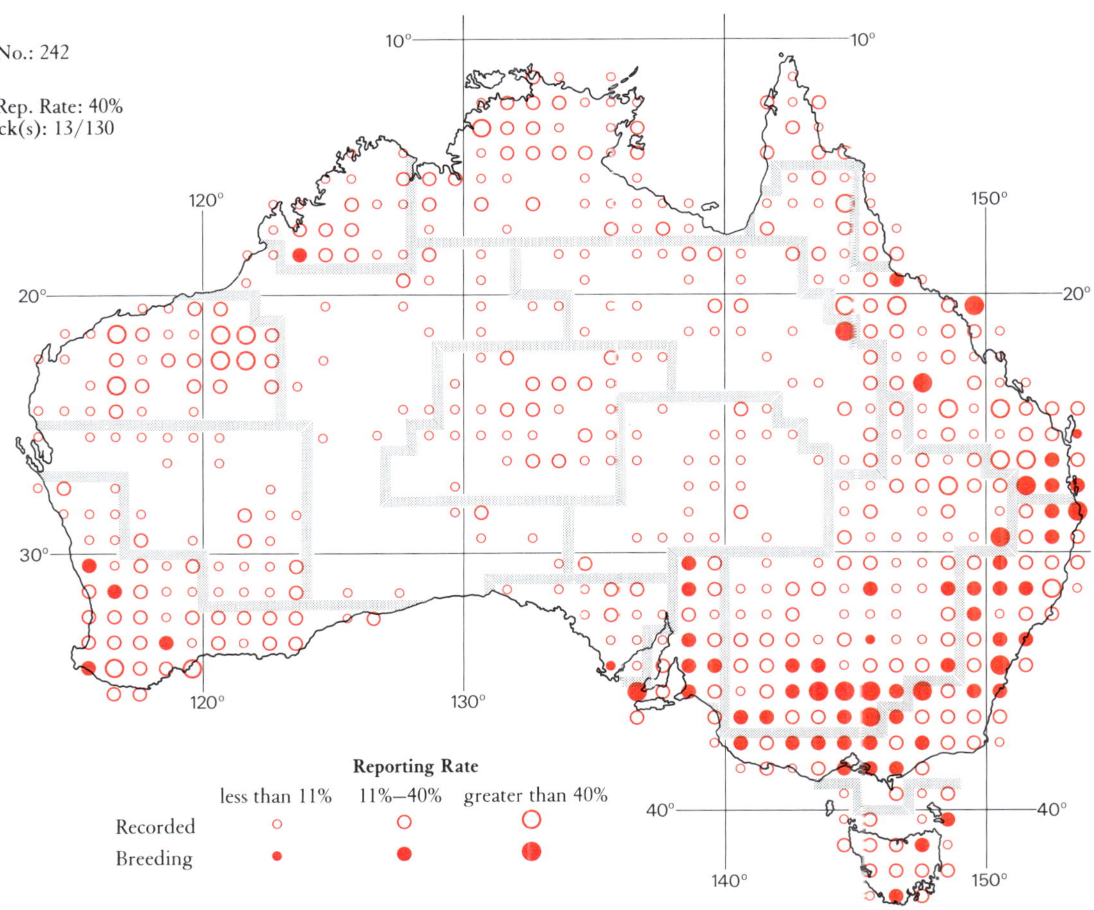

Reporting Rate

	less than 11%	11%–40%	greater than 40%
Recorded	○	○	○
Breeding	●	●	●

Barking Owl

Ninox connivens

The Barking Owl lives in New Guinea and adjacent islands and in Australia. The morphologically distinct South-East population *connivens* grades into the northern subspecies *peninsularis* at about latitude 20°S.[1588] The isolated Pilbara population also belongs to *peninsularis* whose distribution extends across the Top End. In the South-West, where the Reporting Rate was low, there is an isolated population of *connivens*. The Barking Owl is absent from most islands[1588] and is absent from the Tas. Region. In the SA Ra. Region it was recorded at the Reedbeds (34/138) until 1906 and round Wilmington (32/138) in 1944–50.[1380] Later records have come only from northern parts of the Flinders Ra. including Arkaroola (30/139) in 1972. In adjacent regions the Barking Owl has been recorded near Naracoorte (36/140), most recently in 1952.[1380] The Atlases contain breeding records from the Top End eastwards but none from the WA populations. The Barking Owl is sedentary[1588] and recorded in all seasons in WA and so undoubtedly breeds throughout its range.

The Barking Owl lives in pairs in eucalypt woodland, its range extending inland along drainage systems in the L. Eyre and Murray-Darling Regions. The species is more often recognized by its distinctive call at dusk and dawn than by sight. Confusion of this call with the bark of a fox has produced some erroneous records.[1380] The Barking Owl feeds on mammals, birds and insects.

Breeding and roosting territories may cover 200 ha in southern Qld[1927] and some are as small as 30 ha.[1588] The bird ranges over a greater area to feed.

Field Atlas Records: 724

1° blocks–with records: 172 (21%)
–with breeding records: 11 (6%)

Atlas No.: 246

Max. Rep. Rate: 27%
In block(s): 12/133

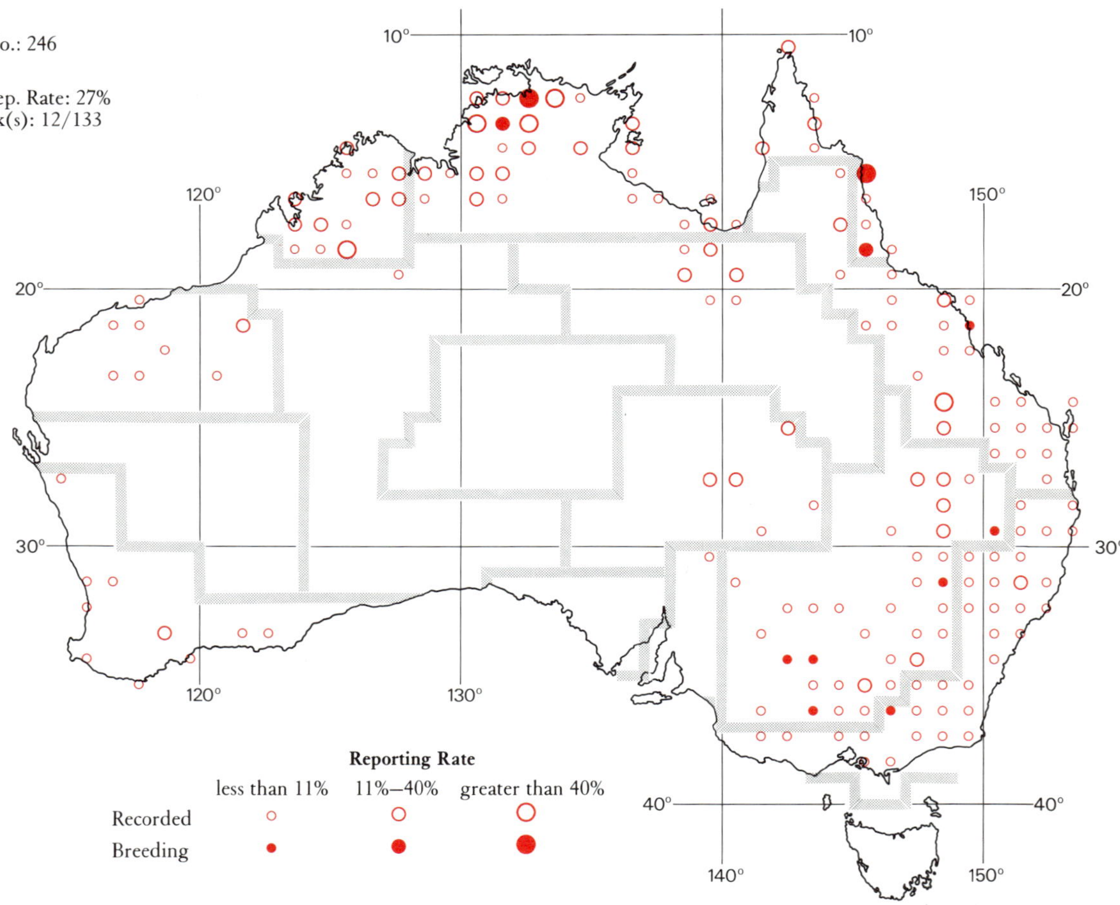

Barn Owl

Tyto alba

The cosmopolitan Barn Owl is represented in Australia by the subspecies *delicatula*, distributed throughout the mainland but rare in the Tas. Region.[736] The only breeding report in Tas., from Devonport (41/146) in the 1940s, is unconfirmed.[1276] Breeding is not recorded in the Atlases in the Cape York and Kimberley Regions and is infrequent in coastal 1° blocks of the South-East and South-West Regions. The Barn Owl is a vagrant in the arid inland.[1381,1588]

When small mammals breed up to plague proportions, Barn Owls congregate and breed at a high density to disperse when the plague subsides.[1588] Such an irruption occurred in south-eastern Australia in 1980 after a plague of house mice; large numbers of Barn Owls appeared on the coasts of the South-East Region and some reached Tas. All but one of the Tas. Region records in the Field Atlas was made Aug.-Dec. 1980. All recent Tas. specimens appear to be of immatures that have died from starvation.[1276]

Seasonal movements have been reported: near Sydney (33/151), where the population appears to decline markedly in summer,[1672] and in the Kimberley Region where the species is thought to be mainly a dry winter season visitor.[1713] The Field Atlas suggests no large-scale seasonal pattern of movement. The longest recorded in Australia is that of a bird banded as a nestling from Moolap (38/144) to Baroota (32/137), 840 km [441] but some birds from Australia have reached New Zealand.[1588]

The Barn Owl lives in woodland, mallee, saltbush, spinifix, heath and pasture, rarely in eucalypt forest or rainforest. It is solitary and hunts by gliding over the ground or perching, waiting to pounce. Small vertebrates including rodents, birds, young rabbits and frogs are its main prey.[1298,1588,1866] The rat *Rattus villosissimus* and house mouse were the most important prey species at several sites.[1299,1770] In some places the house mouse has become the staple item, presumably replacing the native rodents, which have disappeared from much of Australia since European settlement. Most plagues of house mice are in the wheatbelts.

There are no measures of abundance.

Field Atlas Records: 2919

1° blocks–with records: 343 (42%)
–with breeding records: 37 (11%)

Atlas No.: 249

Max. Rep. Rate: 26%
In block(s): 35/145

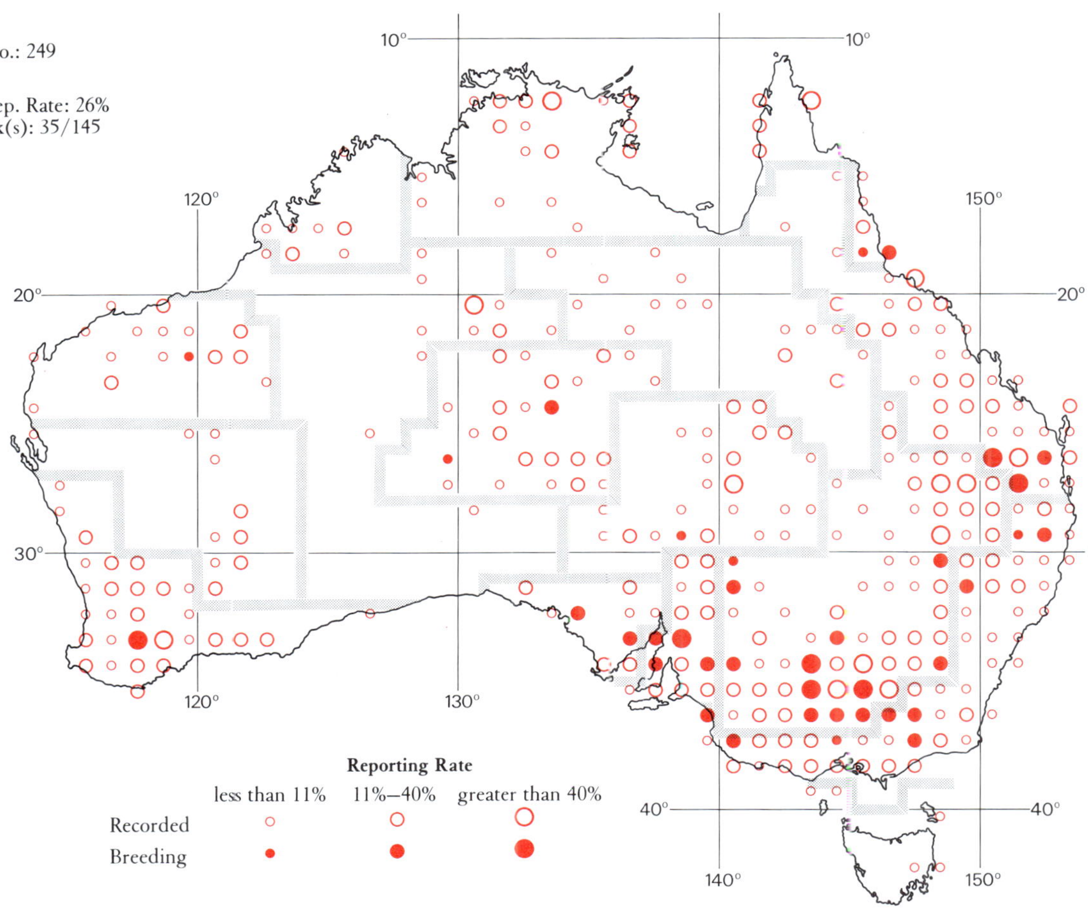

Masked Owl

Tyto novaehollandiae

The Masked Owl inhabits southern New Guinea and adjacent islands as well as Australia where four subspecies are recognized: *melvillensis* on Melville (11/130) and Bathurst (11/130) Is (not recorded in the Field Atlas); *kimberli* from about Cooktown (15/145) north and west to the Kimberley Region; *novaehollandiae* south of Cooktown and west to the Pilbara; and *castanops* in the Tas. Region. The subspecies *novaehollandiae* and *kimberli* are separated in the west where the Great Sandy Desert reaches the coast at Eighty Mile Beach (19/121); whether they actually meet near Cooktown is unknown.[1588]

The Masked Owl is so rarely observed on the Australian mainland and its call so often unrecognized that the Field Atlas map incompletely represents its distribution. In addition it has probably disappeared from some areas, though many early records resulted from confusion with the Barn Owl.[1381] Historical records, some recent, from the Kimberley, Pilbara and SA Ra. Regions suggest that the Masked Owl is sparsely distributed in coastal regions of Australia. The Nullarbor population was known with certainty only from two specimens collected in the 1920s and a third found dead in a cave in 1962.[1240,1381] The record (30/127) of a single bird in June 1980 is confirmation that this population still exists, though probably in reduced numbers.

The Masked Owl has been said to range no more than 300 km from the coast.[1588] Except for Field Atlas records in the northern Murray-Darling Region, reports from farther inland[596,1240,1522] are unconfirmed.

The Masked Owl is sedentary. It lives in eucalypt forest and woodland, requiring partial clearing or forest edges for hunting but roosting in the dense cover of gullies or caves. Prey varies and includes insects, mammals and birds. In Tas. the most common items are rodents (native and introduced), bandicoots, sugar gliders, ring-tailed possums and rabbits; bats and birds, including raptors and domestic poultry, have also been recorded.[1276] Its probable decline, apart from in the Tas. Region, has been attributed to the decline of native mammals since European settlement.[244,1588] The Tas. Region, with 65% of Field Atlas records, is clearly the stronghold of the species.

The Masked Owl occupies a territory of about 500–1000 ha in the South-East Region.[1588]

Field Atlas Records: 168

1° blocks–with records: 42 (5%)
–with breeding records: 2 (5%)

Atlas No.: 250

Max. Rep. Rate: 3%
In block(s): 27/147,41/146,42/147, 43/146

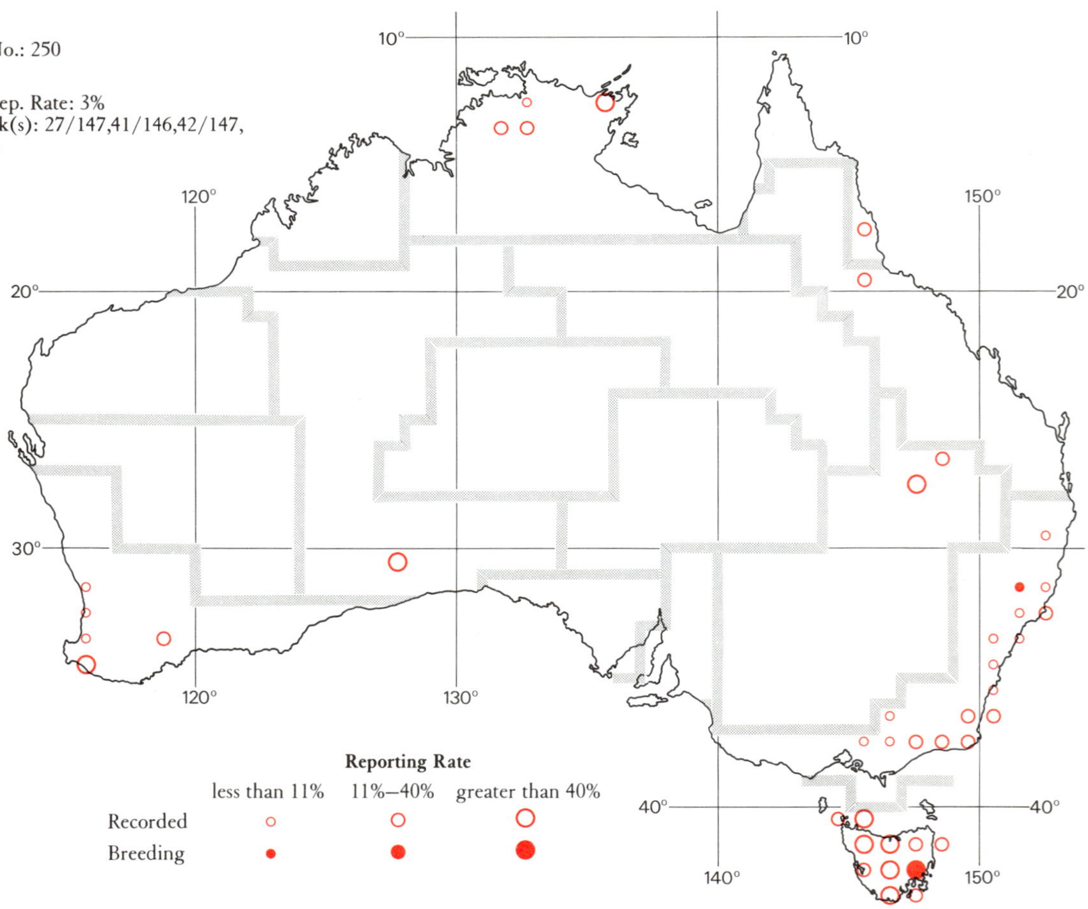

Eastern Grass Owl

Tyto longimembris

The Eastern Grass Owl ranges from India eastward to Fiji, excluding New Guinea. In Australia the scarcity of records perhaps partly reflects confusion with the Barn Owl and means that the distribution of the Grass Owl has not yet been satisfactorily determined. A recent review recognizes two Australian populations.[1588] One includes the eastern coast from the Cape York Region to Harrington (31/152), the southernmost breeding record.[1520] The other has a stronghold in the Rolling Downs and L. Eyre Regions from which birds disperse in years of plenty to all parts of the continent, most often to the South-East, Murray-Darling, Top End and Kimberley Regions and only once to the South-West.[703] Vagrants have also been recorded in Torres Strait.[503]

Movements are nomadic. Inland birds appear to breed up in response to plagues of small mammals, specially the rat *Rattus villosissimus*, then disperse, often to die, as the plague subsides.[1381] During such an episode in 1905, 17 dead or dying Grass Owls were found in Vic. within a few weeks.[380] The pattern is like that of the Letter-winged Kite. The Owls follow the direction of rat plagues and, during those of the mid-1970s, appeared as far north as Darwin (12/130) and as far south as Wilsons Prom. (39/146).[399,530] They appear able to remain longer in the Top End and Kimberley Regions, where rats *R. colletti* are available for longer periods than in the south.

Coastal populations of the Grass Owl are considered fairly sedentary though numbers have been observed to fluctuate.[1588] At the Chapman R (14/141) the Grass Owl is said to be a wet summer season visitor.

The Grass Owl lives in a loose community, several hundred metres usually separating solitary roosting birds or breeding pairs. The birds roost and nest on the ground in crops, thick grass and tussock often associated with swamps, forming extensive tunnels and trampling the area. The species usually hunts by night but when under stress will hunt by day.[416,530,1211] Up to 30 have been observed in the air at the same time, sweeping over the grassland at heights of 1–4 m and dropping directly on to prey. Rodents, including the house mouse, are taken and apparently grasshoppers and other insects.[174,243,530,725,1588] At Ingham (18/146) 80–100 birds occupied 40 ha of fallow rice while breeding (2–2.5 birds/ha).[1588]

Field Atlas Records: 49

1° blocks–with records: 22 (3%)
–with breeding records: 3 (14%)

Atlas No.: 252

Max. Rep. Rate: 9%
In block(s): 12/133

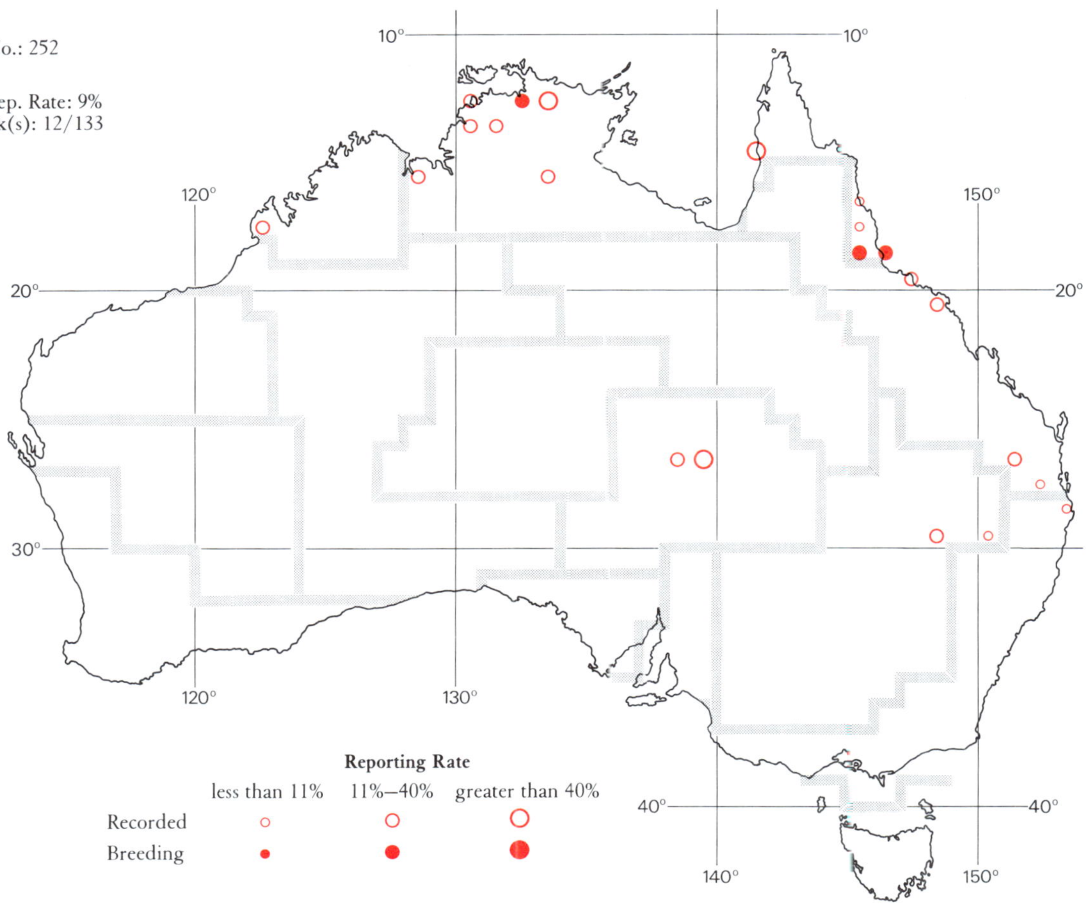

Sooty Owl

Tyto tenebricosa

The Sooty Owl lives in New Guinea and in Australia. The Australian population has recently been divided into two species: the Lesser Sooty Owl *T. multipunctata* of the Atherton Region and the Sooty Owl *T. tenebricosa* of the South-East Region.[1588]

The Lesser Sooty Owl was reported in the Field Atlas throughout its historical range. The Sooty Owl is probably distributed continuously within the range shown on the map. An 1887 record from the Fitzroy R (23/149) is unconfirmed[238] as are records from the Grampians (37/142)[1588] at the southern end of its range. A possible sighting near Blackwood (37/144) during the Field Atlas is omitted from the map because it was unconfirmed. An outlying population, rarely reported, apparently lives on Flinders I. (40/148) but was not recorded during the Field Atlas.[732]

Both Sooty Owls are sedentary and territorial. Members of a pair maintain their bond but do not roost together. The Sooty inhabits pockets of rainforest and wet eucalypt forest, roosting in hollows in the trunks of trees. The Lesser Sooty inhabits extensive areas of rainforest, roosting in crevices in tree trunks, passages in the aerial roots of figs and overhanging banks.[1448,1588] Because the habitat has been altered, both Sootys can be seen hunting along roadsides and on adjoining grassed areas.[1211] Recently, increased familiarity with its distinctive whistling call has resulted in detection of the Sooty Owl in many places from which it was believed to be absent. Both species feed on mammals: mainly terrestrial (rodents and bandicoots) and, to a lesser extent, arboreal (possums and gliders). They also take birds, usually pouncing on prey from a perch.[1211]

The Sooty may have a territory of 200–800 ha or more[1588] but only about 50–60 ha seems to be required by the Lesser Sooty[1211] and its correspondingly greater density is reflected in the frequency of recording in the Field Atlas.

Field Atlas Records: 130

1° blocks–with records: 21 (3%)

–with breeding records: 5 (24%)

Atlas No.: 253

Max. Rep. Rate: 15%
In block(s): 15/145

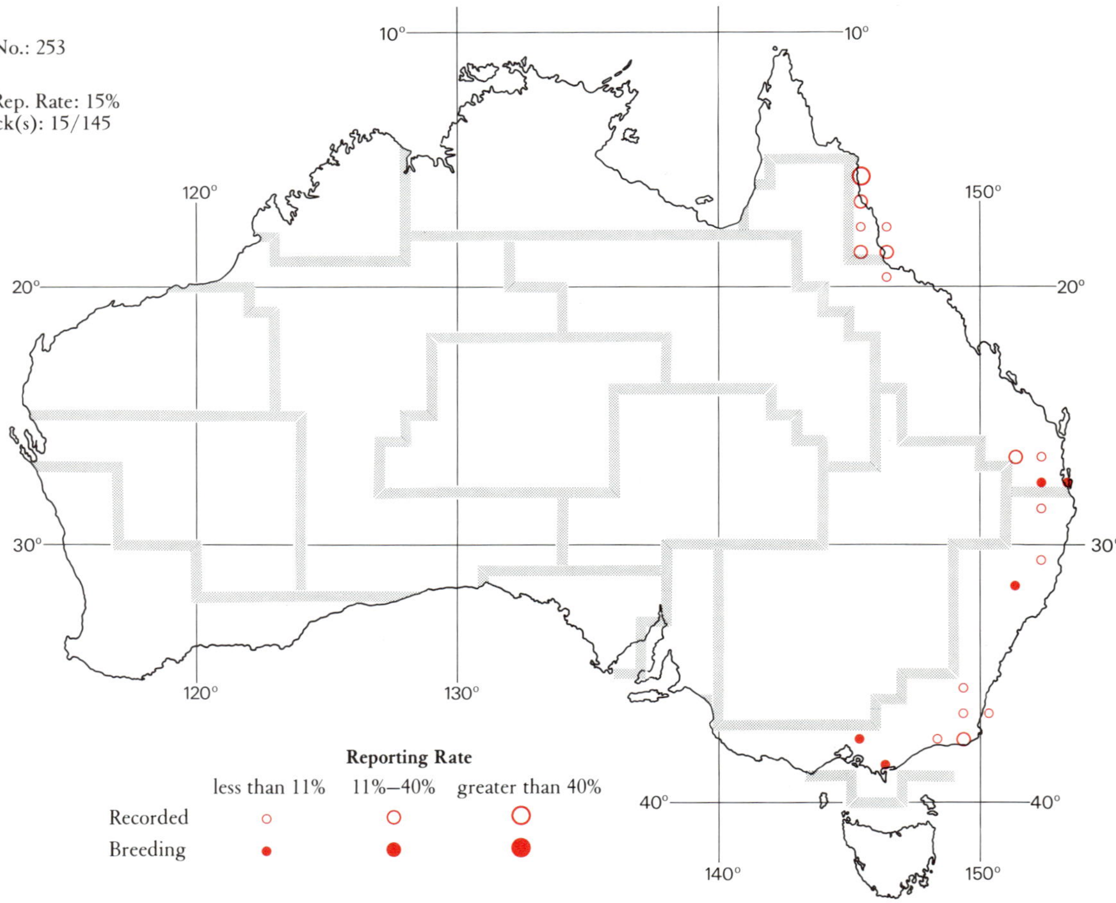

Tawny Frogmouth

Podargus strigoides

The Tawny Frogmouth is essentially confined to Australia though it has recently been recorded in southern New Guinea.[1588] Numerous colour variants, sexual dimorphism and latitudinal clines of size have given rise to the description of many subspecies. Only two seem to be warranted: *strigoides* of coastal eastern Australia including the Tas. Region, meeting *phalaenoides* along the western edge of the Great Dividing Ra.[1588] There is no gap apparent in the Field Atlas. The species has been recorded to the tip of Cape York Pen. (10/142) but is absent from King (39/144) and Flinders (40/148) Is.

The Field Atlas suggests no large-scale seasonal movement. The frequency of colour variants matching the local tree bark and soil colour attest its sedentary status.

The Tawny Frogmouth lives in pairs and appears to occupy a permanent territory.[1588] It generally requires trees or shrubs in which to roost and perches from which it can pounce on prey at night. The Frogmouth therefore inhabits most eucalypt forests and woodlands. In the inland it lives in mulga groves between the sand dunes and among breakaways, far from the nearest eucalypt-lined watercourse. It is frequently reported from the cleared farmlands of the south but is largely absent from the wet south-west of Tas. It does not generally inhabit rainforest but is quick to colonize along roadways in rainforest where it can forage.[846,1276] It feeds on a variety of invertebrates, mainly scorpions, spiders and centipedes, and also takes many frogs and occasionally small mammals.[1276] Like some other predatory birds it will exploit house mice plagues but it does not feed on carrion.[1276] The Tawny Frogmouth has benefited from clearing, commonly feeds along roadsides and is often killed as a result.

Sometimes it roosts on the ground, where it is vulnerable to foxes and cats.

Territories typically range in size from 40 to 80 ha but are sometimes as small as 20 ha.[1588]

Field Atlas Records: 6030

1° blocks–with records: 445 (55%)
–with breeding records: 196 (44%)

Atlas No.: 313

Max. Rep. Rate: 47%
In block(s): 27/149

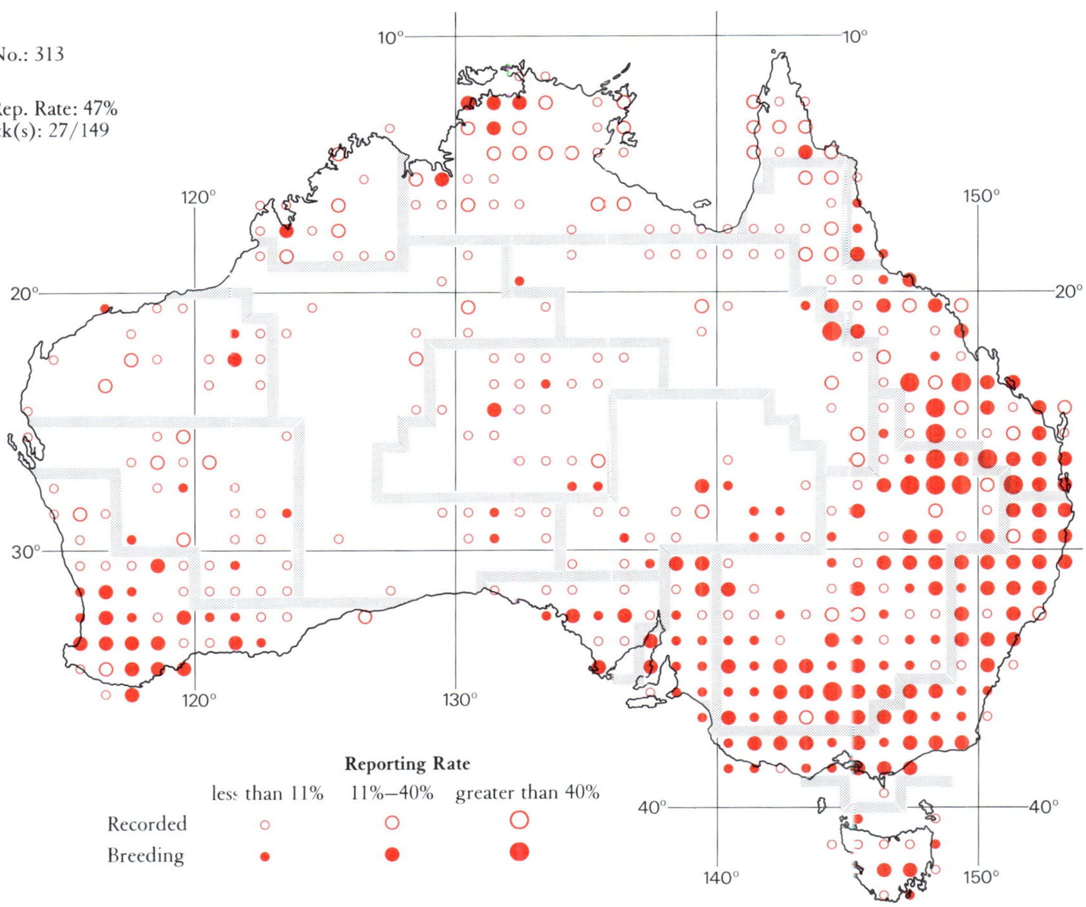

Papuan Frogmouth

Podargus papuensis

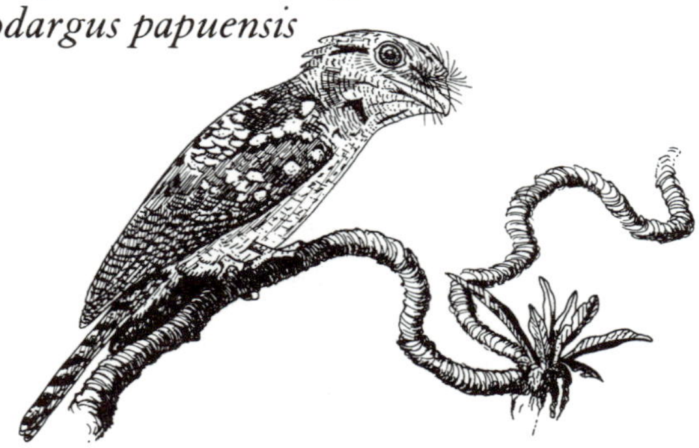

The Papuan Frogmouth inhabits New Guinea and adjacent islands and Australia. In Australia there are two subspecies: *rogersi* in the Cape York Region and *baileyi* in the Atherton Region.

The Papuan Frogmouth has been reported crossing Torres Strait at night[104,1145] but recent observers regard it as a breeding resident on islands there and reports of it migrating appear to be mistaken.[503] The species probably maintains a territory all year.[1588]

The Papuan Frogmouth lives in eucalypt forest and woodland, on the edges of rainforest and in mangroves.[748] In the Atherton Region it is most abundant in lowland areas.[698] It eats nocturnal beetles and other insects and small vertebrates.[1588] Probably most feeding is near the ground because the species is often killed on roads in the Atherton Region.[698] It can tolerate some clearing and agricultural development and is still present over its former Australian range.

In mangroves nests may be as little as 100 m apart.[1588] In the Cape York Region nesting and roosting territories may be 5–7 ha and foraging range 20–30 ha.

Field Atlas Records: 115

1° blocks–with records: 18 (2%)

–with breeding records: 4 (22%)

Atlas No.: 316

Max. Rep. Rate: 22%

In block(s): 12/143

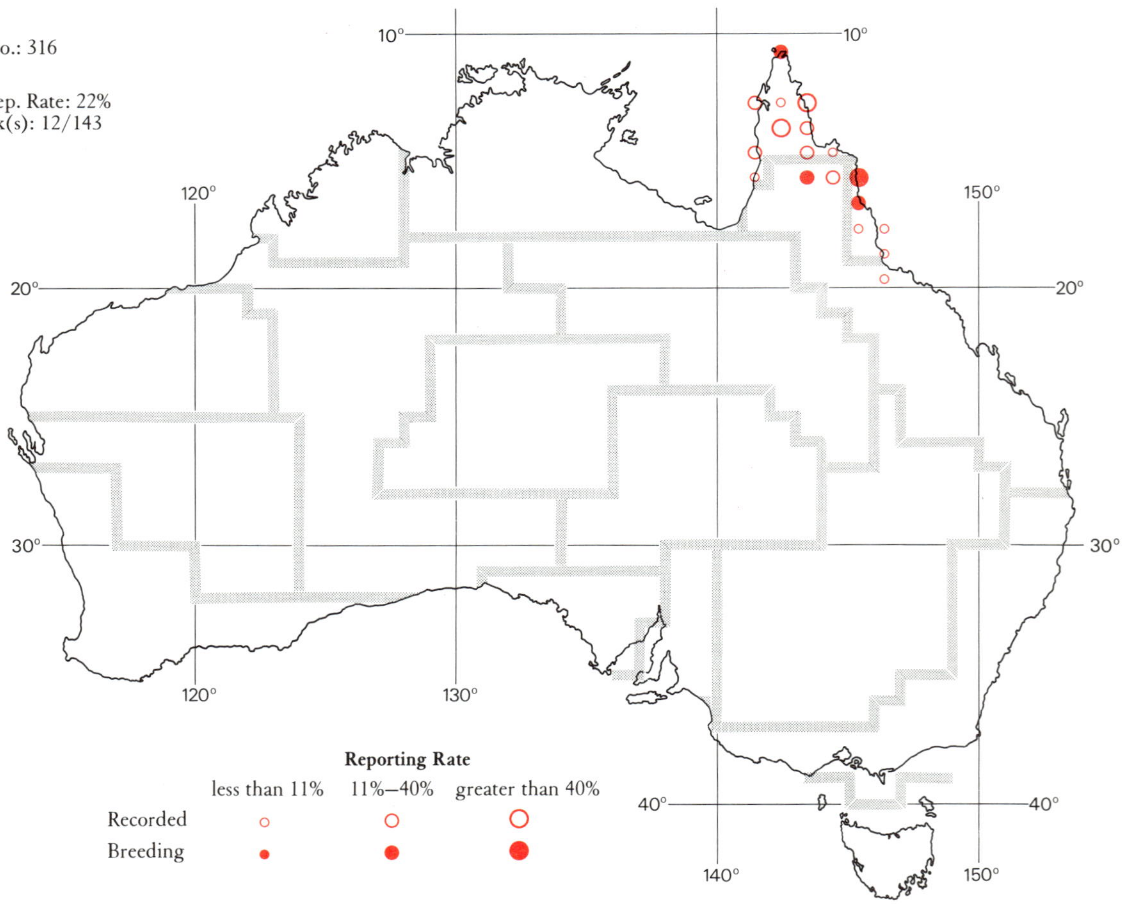

Marbled Frogmouth

Podargus ocellatus

The Marbled Frogmouth lives in the Solomons, New Guinea and adjacent islands and Australia. In Australia it has two widely separated populations, which are considered subspecifically distinct: *marmoratus* in the north and *plumiferus* in the south.[1588] The Historical Atlas shows that the range of *marmoratus* extends from the tip of Cape York Pen. (10/142) south along the eastern coast to the McIlwraith Ra. (13/143). There are unconfirmed sightings of *plumiferus* as far south as Taree (31/152)[1142] but specimens have not been collected south of Grafton (29/152). The Marbled Frogmouth is often confused with rufous Tawny Frogmouths.

The Marbled Frogmouth seems to be permanently territorial. Birds were recorded in the same place in the Conondale Ra. (26/152) in Oct.-Nov. 1976 and Apr. 1977.[1496] Others were recorded at two localities in the Tweed Ra. (28/153) and Night Cap Ra. (28/153) in Feb.-Apr. and Dec. 1980 and Jan.-Feb. 1981.[850] Subsequent observations confirmed that the species was sedentary.[846]

The Marbled Frogmouth lives only in rainforest, often along flowing creeks, and usually roosts in thick vegetation.[164,1588] The southern population has been reported at altitudes up to 700 m.[850] Beetles and grasshoppers and, once, a frog have been reported in the diet.[1147,1449,1588] The crop of a specimen taken at the Rocky R (13/143) contained only the remains of arthropods such as cockroaches, crickets, beetles, moth larvae and spiders; most of the prey items would be expected to live on foliage, trunks or branches rather than on the ground.[76]

The southern population is said to have declined because lowland rainforest has been cleared.[1588]

There are no measures of abundance.

Field Atlas Records: 14

1° blocks–with records: 5 (1%)
–with breeding records:

Atlas No.: 314

Max. Rep. Rate: 8%
In block(s): 12/143

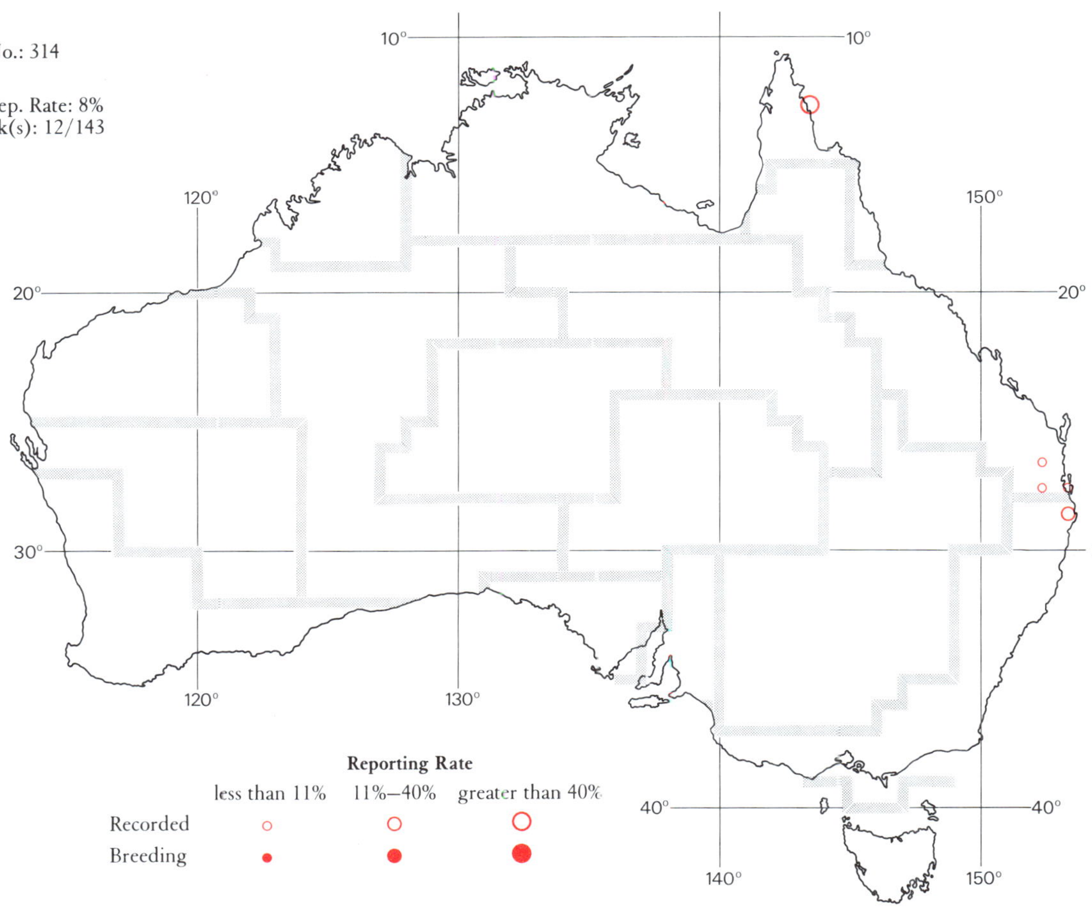

Australian Owlet-nightjar

Aegotheles cristatus

The Australian Owlet-nightjar lives in New Guinea and Australia. Two subspecies are recognized in Australia. The mainland subspecies *cristatus* is absent from the tip of Cape York Pen. (10/142) and Torres Strait islands. The subspecies *tasmanicus* in the Tas. Region is sparsely distributed mainly in the east with isolated records in the south-west. It is absent from King (39/144) and Flinders (40/148) Is.

The Owlet-nightjar is apparently sedentary[1588] and lives in eucalypt forest and woodland as well as in much of the inland. It roosts in hollow trees or in hollows in banks, crevices, cliffs or breakaways and termite mounds. It will also use roof cavities and abandoned babbler nests. The Owlet-nightjar can be located by its call, given mainly at dusk and dawn but also at intervals during the night and day.[846]

The Owlet-nightjar feeds on ground-dwelling and often wingless insects.[1531,1620] Possibly the birds collect these by swooping to the ground but observers during the Field Atlas and at other times have often found the Owlet-nightjar perched on the road, apparently feeding on insects struck by passing traffic. It probably does take flying insects when it can but the bulk of its food comes from the ground, unlike the true nightjars, which feed mainly in flight. In some areas it may suffer predation from goannas.[239]

At Moruya (35/150) in eucalypt woodland density was about 0.2 birds/ha.[1192]

Field Atlas Records: 2700

1° blocks–with records: 450 (55%)
–with breeding records: 57 (13%)

Atlas No.: 317

Max. Rep. Rate: 41%
In block(s): 27/149

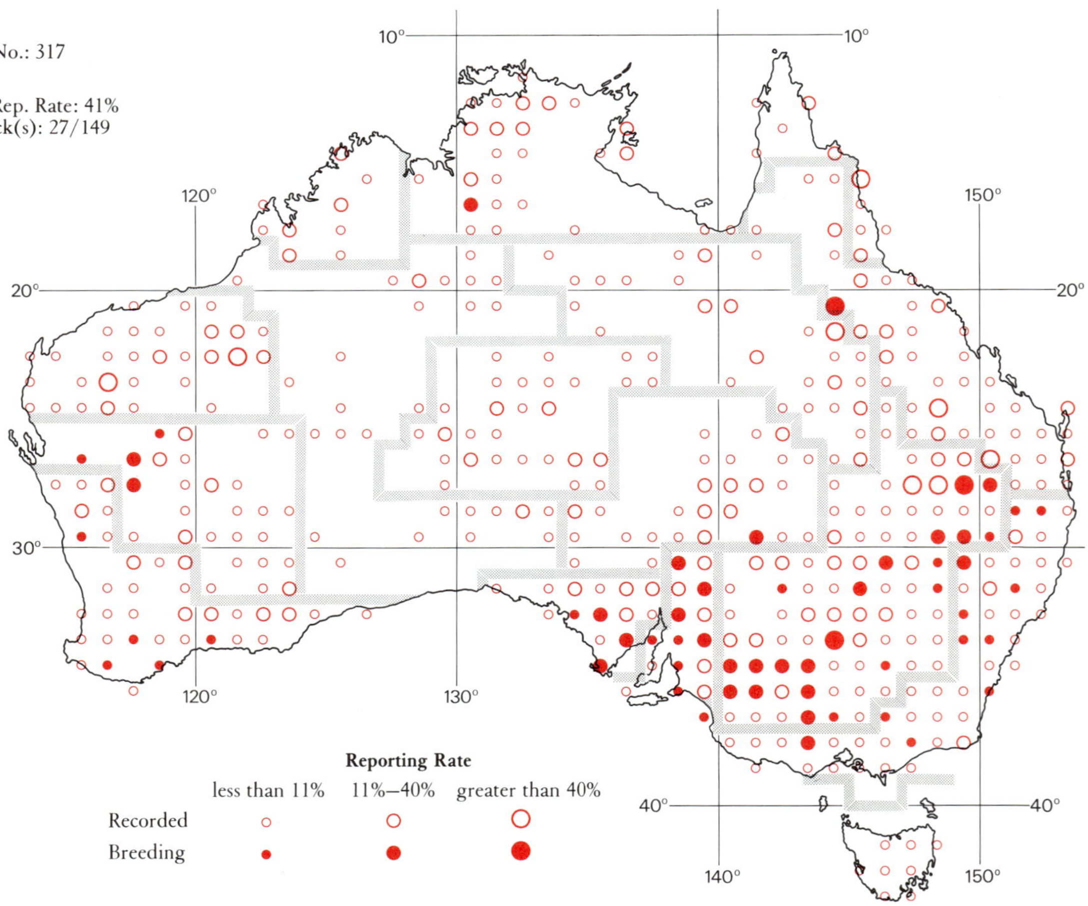

White-throated Nightjar

Caprimulgus mystacalis

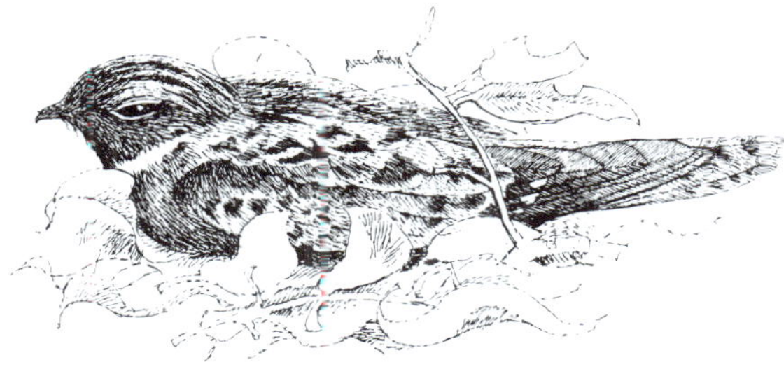

The White-throated Nightjar inhabits New Guinea, the Solomon Is and New Caledonia as well as Australia. All Australian birds are considered to belong to a single subspecies *mystacalis*[1588] whose range in the South-East Region extends west to Melbourne (37/145) with a few sightings and a possible breeding record as far west as the Otway Ra. (38/143).[576] The species has never been recorded in the Tas. Region. The ranges of the White-throated and Spotted Nightjars overlap along the Great Dividing Ra. Pairs of each species were found nesting within 50 km of each other during the Field Atlas.

Southern birds, at least, migrate. The Reporting Rate in the South-East Region in summer (1.2%) was much higher than that in winter (0.2%) and suggests that the migration involves most of the southern population. During 14 years of observations at Jamieson (37/146) the Nightjar was recorded only Nov.-Mar. each year[1469] and in eastern Vic. was said to arrive each Jan.[902] Migrating birds move north, some passing through Torres Strait to winter in New Guinea[129,503] but part of the population remains in northern Australia throughout the year. North of Townsville (19/146) permanent territories are maintained.[1588] On the Atherton Tableland a more irregular movement is suggested by the observation that large numbers sometimes appear, stay for a few months and then depart.[231]

The White-throated Nightjar lives in pairs in eucalypt forest and woodland, specially along dry, sparsely-forested ridges where the understorey is not continuous.[1192,1588] It hunts above the canopy and along forest margins and streams. The bird breeds in the same spot year after year, successive nests being within a few metres of each other.[521,871,891] The White-throated Nightjar takes many large nocturnal insects disturbing them from the treetops. It is often observed when attracted to a concentration of moths round an artificial light.

Each pair requires a territory of about 1 ha or less in which it roosts and breeds but it forages over an area of 100 ha or more and feeding ranges may overlap.[1588]

Field Atlas Records: 425

1° blocks–with records: 74 (9%)
–with breeding records: 17 (23%)

Atlas No.: 330

Max. Rep. Rate: 28%
In block(s): 24/153

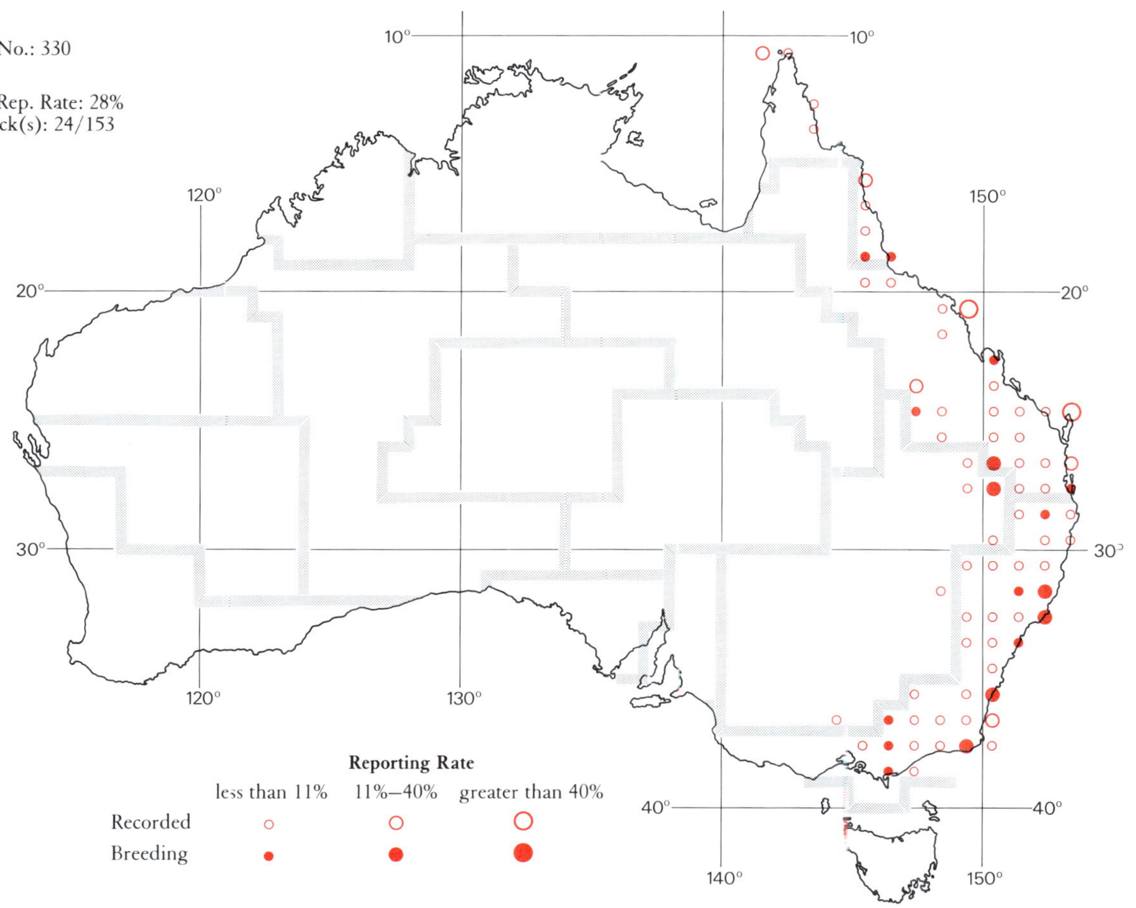

Spotted Nightjar

Caprimulgus guttatus

The Spotted Nightjar occasionally reaches islands north of Australia during the non-breeding period but breeds only in Australia.[1588] The ranges of the Spotted and White-throated Nightjars overlap along the Great Dividing Ra. Pairs of each species were found nesting within 50 km of each other during the Field Atlas.

Reporting Rates in the Murray-Darling Region (summer 2.4%, winter 0.7%) support the suggestion that some southern populations of the Spotted Nightjar migrate north for the winter[1588] but no comparable change in Reporting Rate was evident in the South-East Region (summer 0.13%, winter 0.11%) and South-West Region (summer 0.6%, winter 0.9%). The non-migratory population of the north is said to undertake local nomadic movements.

The Spotted Nightjar lives in forest, woodland, acacia scrub, spinifex and tussock grassland. It spends the day roosting in open sites exposed for at least part of the time to full sunshine. Physiological studies have shown it to be remarkably resilient to temperatures in excess of 40°C.[468] At night it feeds on flying insects, specially moths and beetles, hawking over about 1 sq. km of airspace.[1588]

Breeding pairs seem to occupy only 1–2 ha often having common territorial boundaries with adjacent nesting pairs.[1588] There are no other measures of abundance.

Field Atlas Records: 1027

1° blocks–with records: 305 (38%)
–with breeding records: 25 (8%)

Atlas No.: 331

Max. Rep. Rate: 29%
In block(s): 21/121

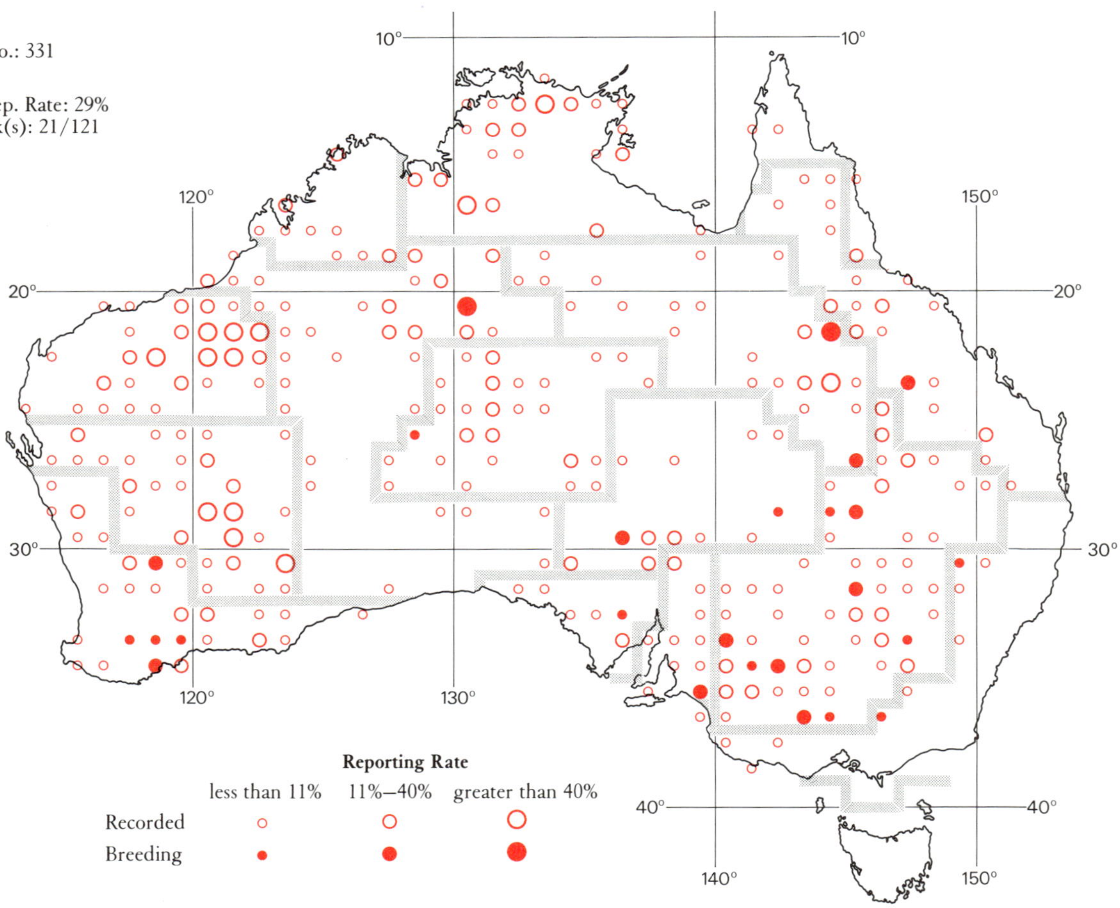

Large-tailed Nightjar

Caprimulgus macrurus

The range of the Large-tailed Nightjar extends from southern Asia to the Pacific.[1248] Though the two Australian populations are widely separated, they are not morphologically distinct from each other, which suggests that the bird may have arrived in Australia in recent geological time.[1588] Records in the Historical Atlases extend the species' range down the western coast of the Cape York Region to the Bynoe R (17/140) and it has also been recorded on Melville I. (11/130). The range in the Field Atlas is otherwise consistent with that determined historically.

The Large-tailed Nightjar is sedentary, calls thoughout the year and appears to maintain a permanent territory.[1588] It lives at the edges of rainforest in paperbark woodland and mangroves.[129] It preys on large nocturnal insects, including moths and beetles, and there is no evidence that it feeds other than on the wing.

Each pair appears to hunt over about 50 ha. As many as eight individuals may be found in 500 m of forest fringe, separated from each other and apparently occupying different territories.[1588]

Field Atlas Records: 290

1° blocks–with records: 39 (5%)

–with breeding records: 10 (26%)

Atlas No.: 332

Max. Rep. Rate: 45%
In block(s): 12/143

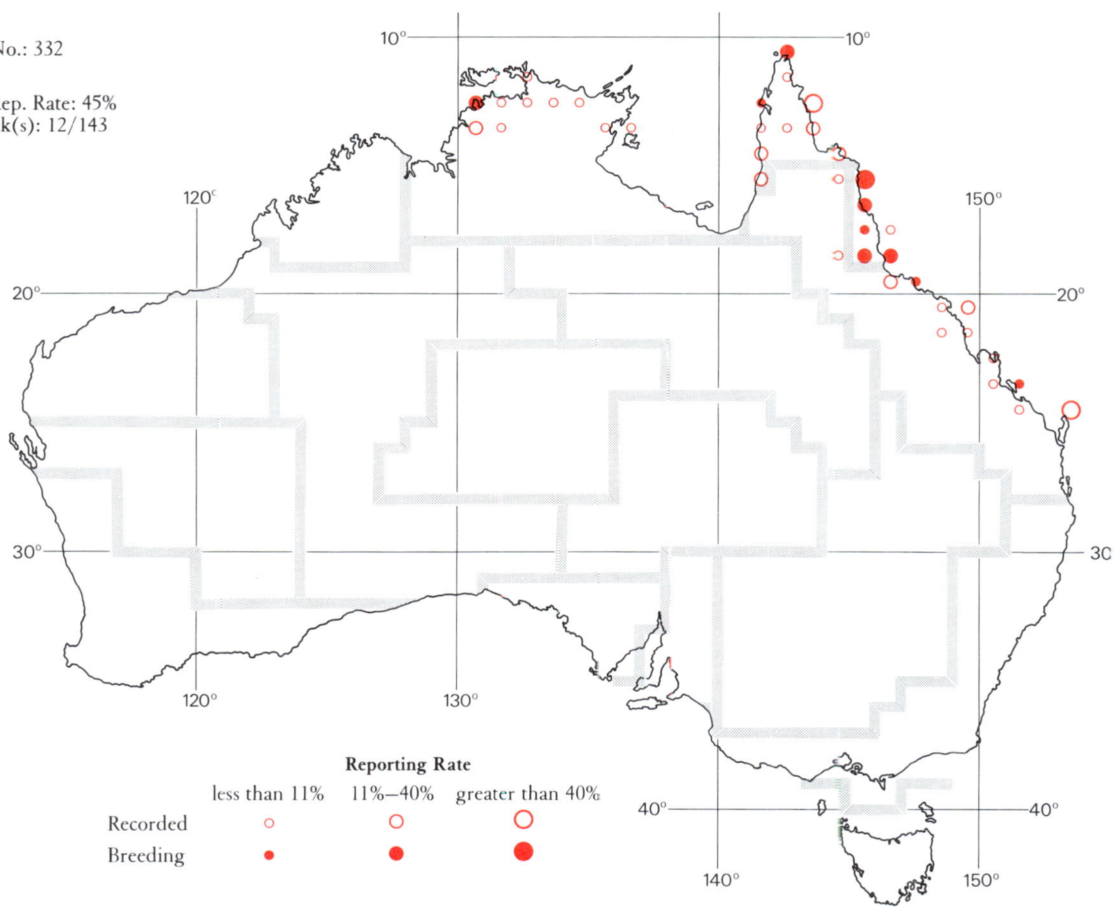

White-rumped Swiftlet

Collocalia spodiopygia

The distribution of the White-rumped Swiftlet includes islands adjacent to New Guinea and some western Pacific islands as well as Australia. Two subspecies are recognized in Australia: *terraeregina* on the coast and *chillagoensis* inland.[1412] Apart from records of vagrants in the South-East Region, the species' range is unchanged from the historical periods. Six breeding sites have been located: upper Mitchell R (16/144), not discovered until after the Field Atlas,[1412] Chillagoe (17/144), Tully Gorge (17/145), Mt Peter Mines (17/146), the Family Is (18/146) and Finch Hatton (21/148). The frequency of sightings in other areas during the Field Atlas, notably between Cooktown (15/145) and Cairns (16/145) and at Iron Ra. (12/143), suggests that other colonies are yet to be discovered.

The White-rumped Swiftlet roosts in the breeding cave throughout the year and there is no evidence of seasonal movement.[1692] Vagrants have occasionally been reported well to the south of the breeding range. Two such sightings during the Field Atlas were of single birds among swallows, martins or woodswallows: one at Canungra (28/153) in Mar. 1980 and the other at Iluka (29/153) in June 1980.

An aerial feeder, the White-rumped Swiftlet forages over all sorts of country both on the mainland and on islands. It will forage up to 35 km from the breeding colony[1412] and is often in company with the Welcome Swallow and Fairy Martin.[437] It feeds on insects, mainly flies[1691] taken in the air, and has been reported hawking at all heights from a few feet above the ground or along river gorges to above the forest canopy and high in the air. At each of the six known breeding sites there are from one to ten colonies situated in caves, rock overhangs and old mine shafts. Size of colony varies from one or a few pairs to greater than 1000 birds and fluctuates from year to year. Breeding success is reduced by human disturbance of breeding colonies, flooding, cave-ins, and immersion in seaspray during storms.

The largest colonies are two at Tully Gorge, each with more than 1000 birds in 1976, two at Chillagoe also with more than 1000 birds in 1983, and one at Wheeler I. (18/146) with 500 nests in 1977.[1692]

Field Atlas Records: 571

1° blocks–with records: 21 (3%)
–with breeding records: 5 (24%)

Atlas No.: 333

Max. Rep. Rate: 44%
In block(s): 12/143

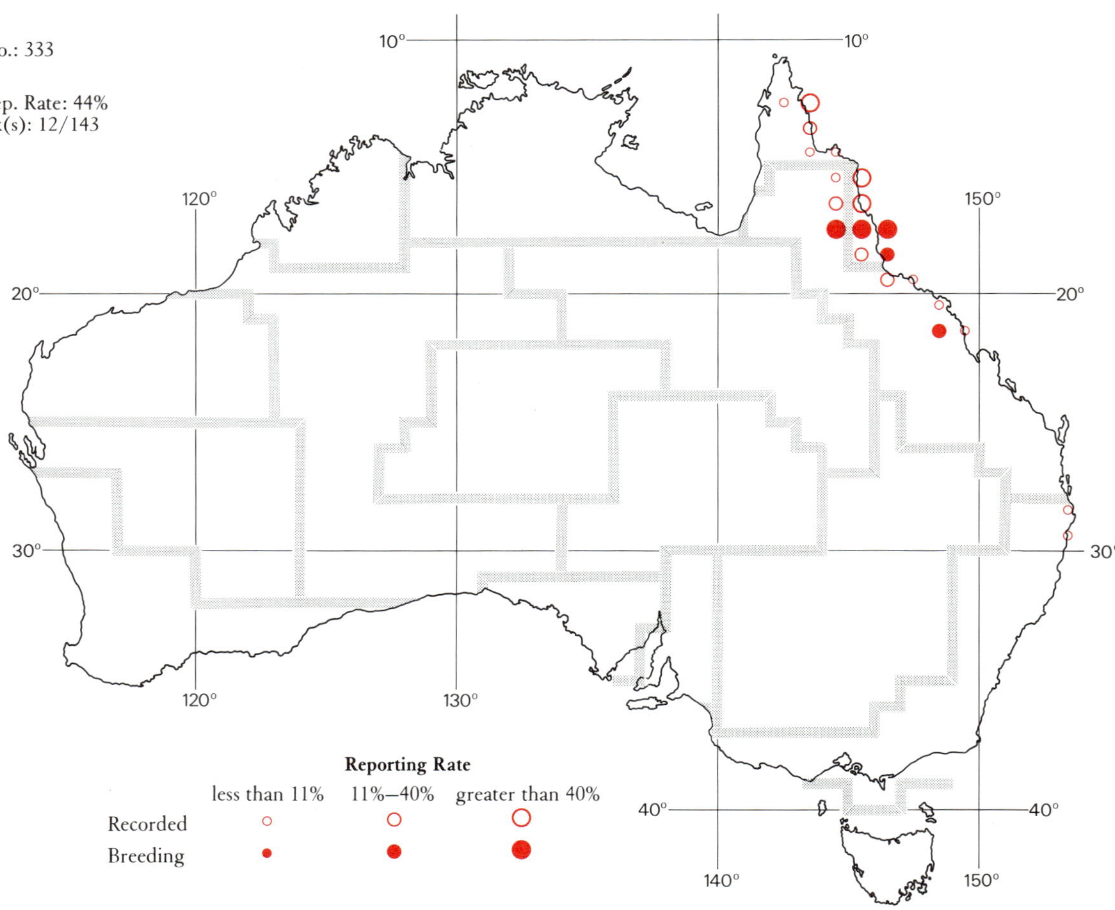

White-throated Needletail

Hirundapus caudacutus

The White-throated Needletail breeds in Asia and migrates to Australia in summer. Records in the Cape York Region are lacking in the Field Atlas because few observers went there in summer when the birds are present. West of the Cape York Region confirmed sightings are rare; many claimed records result from confusion with the Fork-tailed Swift. The Needletail is commoner on the coast than the Fork-tailed Swift.

The Needletail migrates across Torres Strait, south in Oct. and north in Apr., sometimes in flocks numbering hundreds.[503] At Atherton (17/145) there is a well-defined southward movement Oct.-Nov. but no obvious return in autumn.[231] At Murphy's Ck (27/152) first sightings over five years varied from 15 Oct. to 21 Nov. and last from 16 Mar. to 11 Apr.[1096] Arrival in NSW is also in Oct.-Nov.[396] The Field Atlas indicates that few birds reach the Tas. Region or areas west of Melbourne (37/145) before Dec., there being spring records from only five 1° blocks in these areas.

The Needletail spends most of the day high in the air,[1667] often out of sight from the ground, coming low in unsettled weather. It is regularly seen in association with storms at Atherton in Jan.-Feb.[231] The species is thought to remain in the air during its time in Australia but there are some instances of birds roosting.[1443] At Dartmouth Dam (36/147) one perched in a small tree over water and was caught by hand. It was uninjured and flew off after release. About 40 were reported clinging to the stringy bark of eucalypts at Lower Glenelg National Park (38/141) in late morning in Mar. 1982 and others were on the ground under a sprinkler.[455]

The Needletail feeds on flying insects, gathering over bushfires to catch fleeing insects,[1096] and on swarms, e.g. a farmer ploughing near Chinchilla (26/150) caused grass-hoppers from 50 ha to be concentrated into the last 4 ha of cover from which they exploded as a swarm. Needletails had not been in evidence during the day but about 20 appeared and began to feed. More arrived over the next hour until an estimated 1000 were present. Parties of about 20 returned briefly several times next day, swept over the padddock, found no grasshoppers and left.[294]

There are no measures of abundance for this totally migratory, aerial species.

Field Atlas Records: 4663

1° blocks–with records: 187 (23%)
–with breeding records:

Atlas No.: 334

Max. Rep. Rate: 33%
In block(s): 26/150

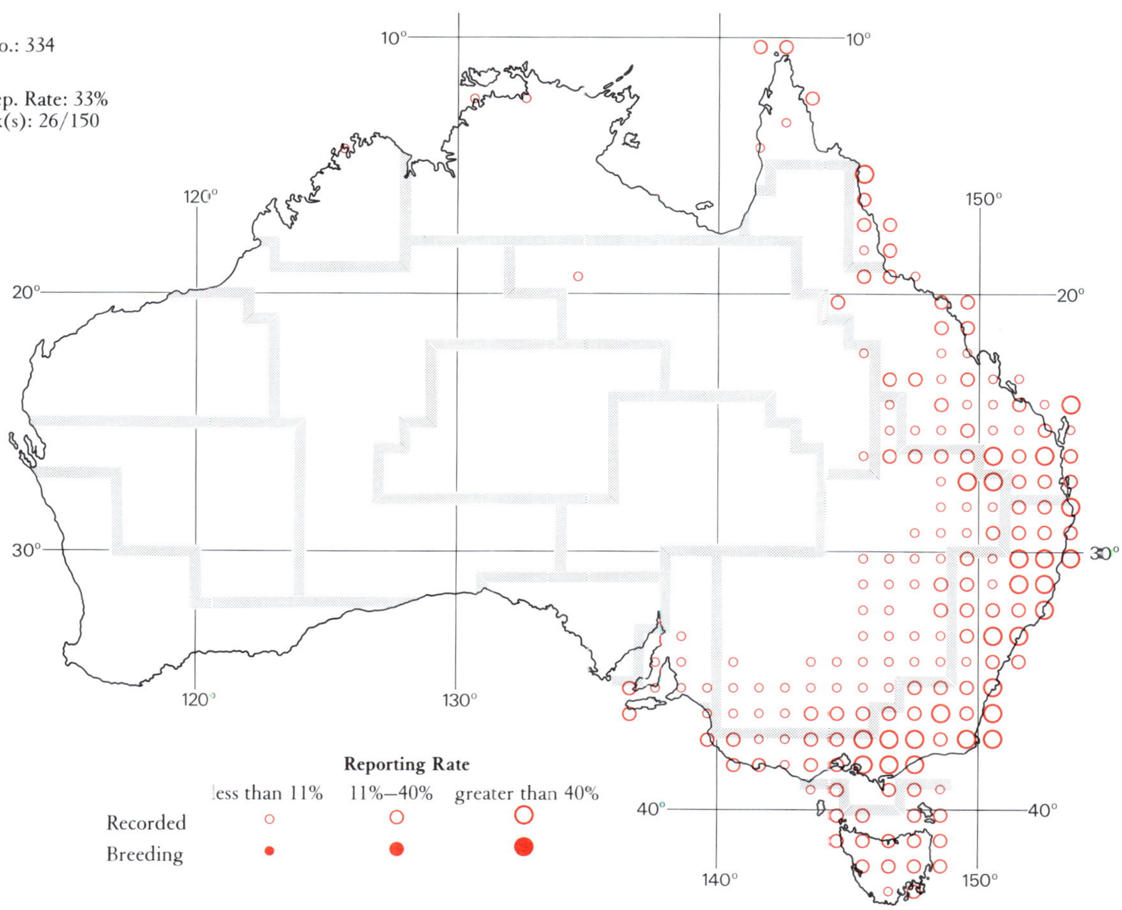

Fork-tailed Swift

Apus pacificus

The Fork-tailed Swift breeds in Asia and arrives in Australia in Oct. each year. Only occasionally does it reach Tas.[178] The lack of records in northern Australia is probably due more to the absence of observers in summer than to lack of birds. The Fork-tailed Swift is sometimes in mixed flocks with the White-throated Needletail. The Fork-tailed Swift is more common inland than the Needletail.

There is no obvious passage of the Fork-tailed Swift through Torres Strait where it is known as a nomadic wet summer season visitor.[503] Birds, apparently migrating, have been seen in parties of 1–6 in the Timor Sea in Oct.[1624] The Field Atlas suggests no differences in time of arrival in different parts of southern Australia, unlike the Needletail. The species normally leaves southern Australia by mid-Apr.[188] and Darwin (12/130) by the end of the month.[423] The Field Atlas indicates that the migration is almost total though there have been occasional winter records.[181] Birds are usually either singly or in small flocks, though huge flocks are sometimes reported.[396]

There are four reliable reports of the Fork-tailed Swift roosting in reedbeds, cliffs and large trees[187] but it may not normally roost. It feeds on flying insects, conspicuously so at nuptial flights of termites when it may hawk within a metre of the ground. Bees, beetles, bugs and flies have been found in guts.[371] Swifts are often seen before a change of weather. Weather systems are preceded by masses of humid air which is of lower density than dry air. Insects that habitually fly at an altitude corresponding with a certain air density tend to fly lower when relative humidity is high. Consequently the birds that prey upon them must seek them at a lower altitude. At other times most insects are higher and so are the Swifts.[293,1012] The Fork-tailed Swift has been seen to congregate near a bushfire, flying through the smoke to gather insects flushed by the fire.[1119]

There are no measures of abundance for this totally migratory, aerial species.

Field Atlas Records: 812

1° blocks–with records: 215 (26%)
–with breeding records:

Atlas No.: 335

Max. Rep. Rate: 26%
In block(s): 21/144

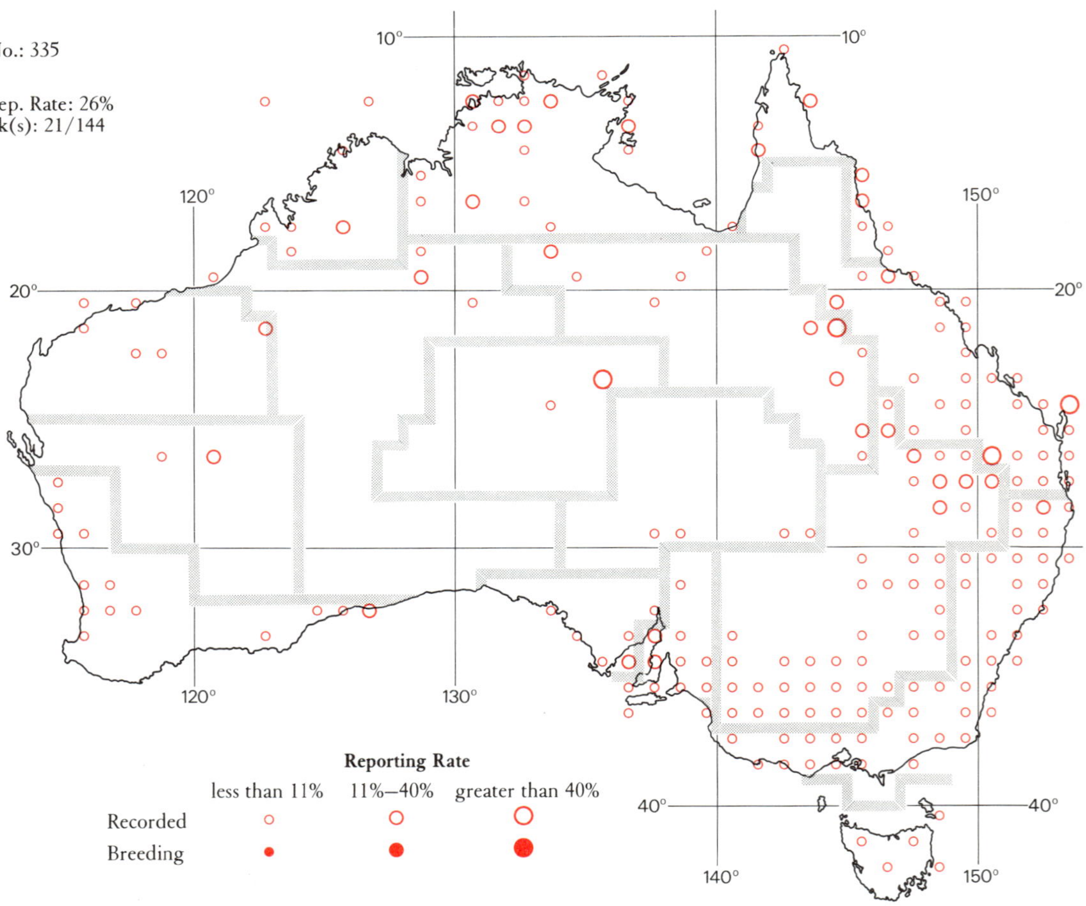

Azure Kingfisher

Ceyx azurea

The range of the Azure Kingfisher extends through Indonesia and New Guinea to Australia. Three subspecies are recognized in Australia.[1586] The range of the northern one, *ruficollaris*, appears to be broken by a gap at the eastern end of the Gulf of Carpentaria (140°E), which is also evident in the Historical Atlases. The second subspecies *azurea* takes in birds on the mainland south of about Cooktown (15/145) and the third, *diemenensis*, is confined to Tas. The Azure Kingfisher has not been recorded in WA south of the Kimberley Region. In the South-West Region orange-breasted Sacred Kingfishers may be misidentified as Azure Kingfishers.

In Eastern Qld the Azure Kingfisher is now seen farther inland than historically but, at the southern end of its range, its distribution has contracted. The Field Atlas suggests that in Tas. the Azure Kingfisher is now virtually confined to the western coast with the stronghold in the south-west. Before 1950 it was recorded from a number of sites in eastern Tas. but never from King (39/144) or Flinders (40/148) Is. Reports near Lagoon of Islands (42/146) in central Tas. in 1977 and 1979 represent a new locality for the species. The Azure Kingfisher bred in the south of the SA Ra. Region before 1950 but the most recent record in the Atlases was in 1965. This population may be extinct and was apparently isolated.

The Azure Kingfisher is sedentary in many, perhaps most, areas[979] but in the Rutherglen (36/146) district it is considered a summer migrant.[1128] In the non-breeding season it will use different parts of its territory for some days and then move elsewhere.

The Azure Kingfisher lives along rivers, streams and tidal estuaries and in mangroves.[792] In the lower Murray-Darling Region it occupies bore drains.[833] It feeds on small fish and also takes frogs, insects and crustaceans, including crabs.[792,1099,1201,1618]

At Murphy's Ck (27/152) a larger territory is occupied during the non-breeding season (1.6 km of creek) than when breeding (1 km of creek).[1099] Near Sydney (33/151) territories occupy about 200 m of creek.[1201]

Field Atlas Records: 3165

1° blocks–with records: 188 (23%)

–with breeding records: 32 (17%)

Atlas No.: 319

Max. Rep. Rate: 47%
In block(s): 12/143

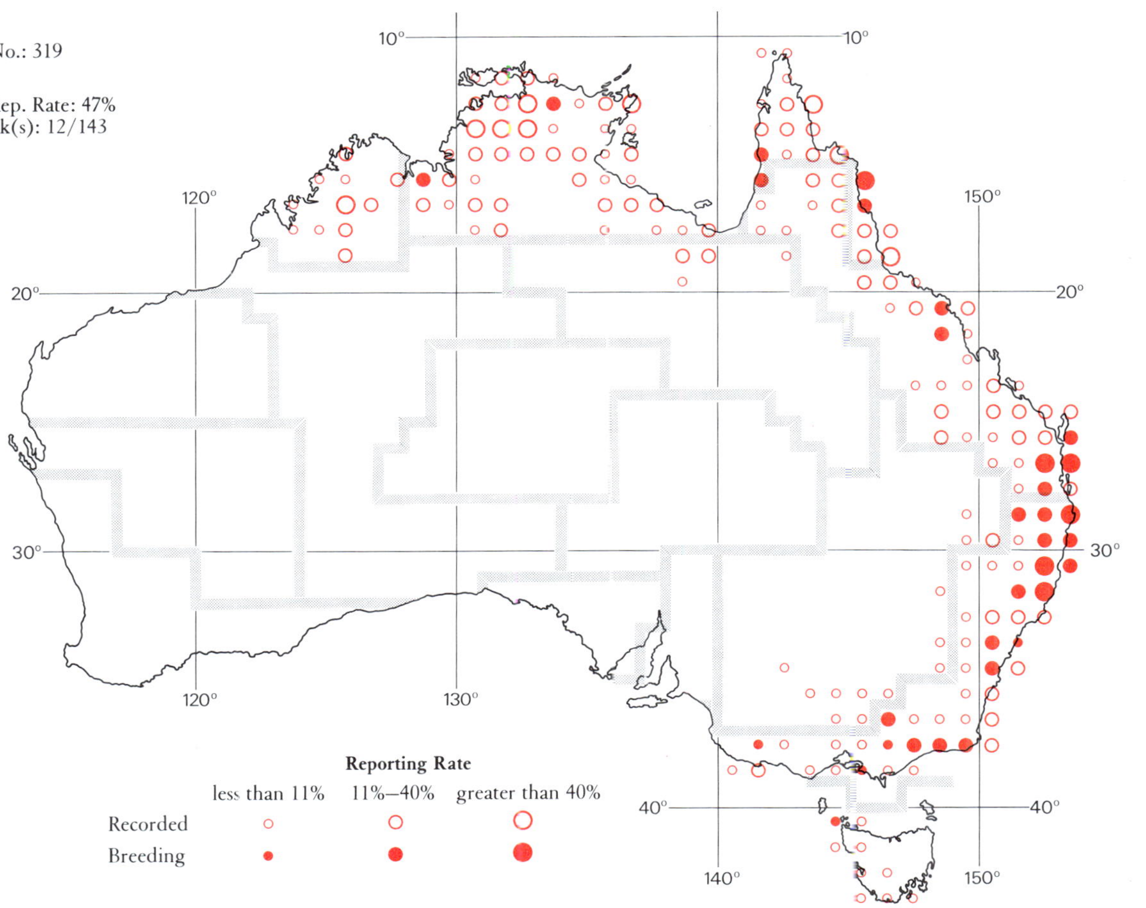

Little Kingfisher

Ceyx pusilla

Outside Australia the Little Kingfisher is recorded in New Guinea[149] and adjacent islands.[1411] Views differ about the subspecies recognized in eastern Australia.[388,1709] The isolated Top End population is regarded as subspecifically distinct and the Historical Atlas contains no confirmed records from the southern coast of the Gulf of Carpentaria to narrow the gap shown on the map. Historically, the Little Kingfisher has been recorded on some Torres Strait islands and from additional sites on the coast of the Cape York Region. The Field Atlas shows it to range farther south than the coast of the Atherton Region, its previously accepted southern limit.

Recent observers in the Atherton Region suggest that the Little Kingfisher may be sedentary[231,698] but one early author considered that it moved from the coast in spring to breed in the upper reaches of coastal streams.[239]

The Little Kingfisher lives along rivers near the coast, in mangroves, in forest upstream and on the highlands of the Atherton Region. It feeds on small fish, obtained by diving.[1049,1145,1264] Eight fish an hour were brought to a nest throughout the day.[1264]

There are no measures of abundance.

Field Atlas Records: 177

1° blocks-with records: 28 (3%)

-with breeding records:

Atlas No.: 320

Max. Rep. Rate: 17%
In block(s): 12/143

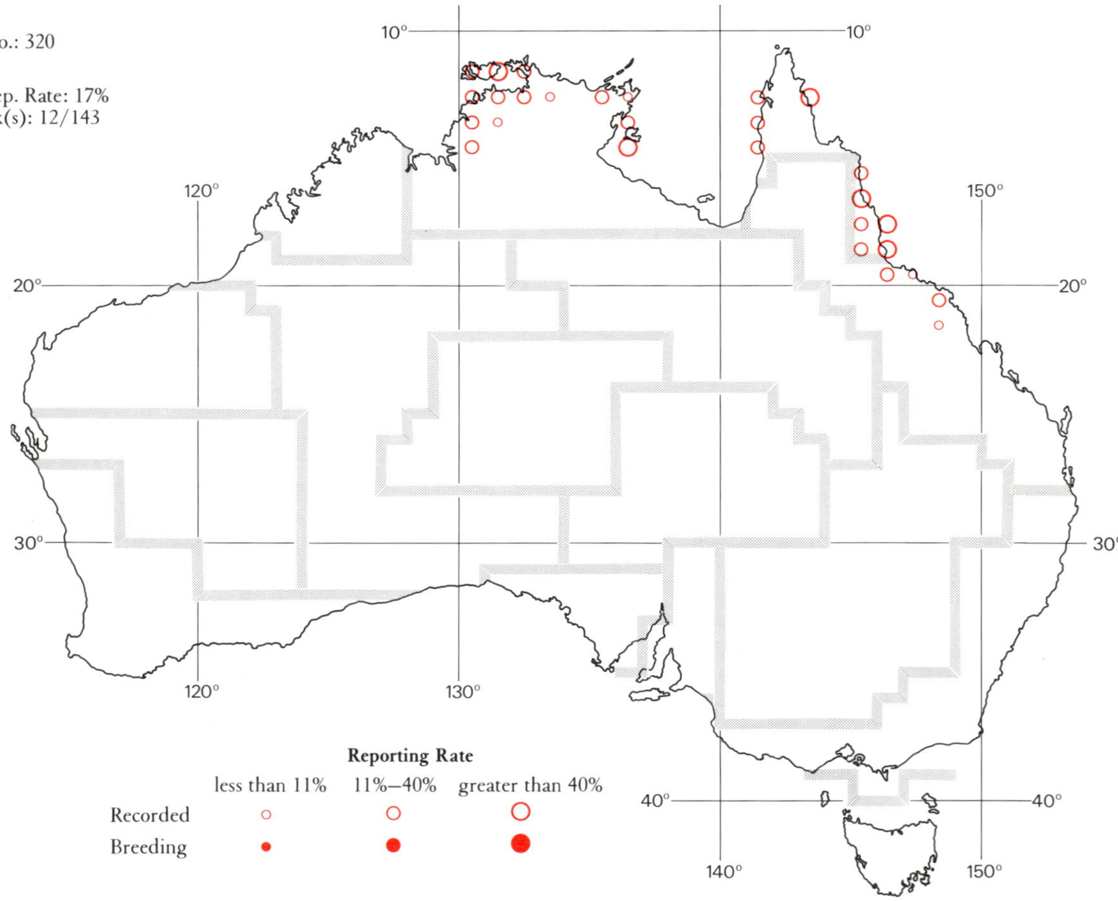

Laughing Kookaburra

Dacelo novaeguineae

The Laughing Kookaburra is endemic to Australia from where it has been introduced to New Zealand. There are two subspecies: *minor* in the Cape York Region and *novaeguineae* everywhere else.[388] The species' distribution at the time of European settlement in Australia included only the eastern part of the mainland. It was successfully introduced to the South-West in 1897, Tas. about 1905, Kangaroo I. (35/137) in 1926 and Flinders I. (40/148) about 1940.[732,1085] In northern Tas. and in the South-West there were multiple introductions and the Kookaburra is now widespread. It is recorded throughout Flinders I. but on Kangaroo I. survives only in a small area. Distribution elsewhere in Australia does not appear to have changed. The range of the Laughing Kookaburra overlaps that of the Blue-winged in Qld.

There is no evidence of movement of the Laughing Kookaburra. It lives in eucalypt forest and woodland, extending inland along watercourses in the Murray-Darling Region, and also survives well in cleared farmland and suburbia, provided there are suitable tree hollows for nesting.[1552] About half of its diet is invertebrate, including insects, earthworms and crayfish and the remainder is reptiles, frogs and other small vertebrates.[1397]

Groups of Kookaburras live in a territory. Each group has a dominant breeding pair and from none to four or more auxilliaries.[1398] In the Dandenong Ra. (37/145) each adult required 1.3 ha and the group's territory size varied accordingly. The requirement of 1.3 ha per bird did not differ between seasons when food was plentiful and those when it was not. The size seemed to be determined by the area the bird could defend rather than by the amount of food the territory contained. Densities measured elsewhere include, in eucalypt woodland: Wollomombi (30/152) 0.05 birds/ha;[125] Armidale (30/151) about 0.1 birds/ha;[582] Canberra (35/149) about 0.1 birds/ha;[154] and Moruya (35/150) 0.23–0.32 birds/ha.[1197] In eucalypt forest they were: 0.5 birds/ha at Jamieson (37/146)[1469] and 0.1–0.5 birds/ha at Boola Boola (38/146).[1113]

Field Atlas Records: 35 150

1° blocks–with records: 280 (34%)
–with breeding records: 159 (57%)

Atlas No.: 322

Max. Rep. Rate: 89%
In block(s): 27/149

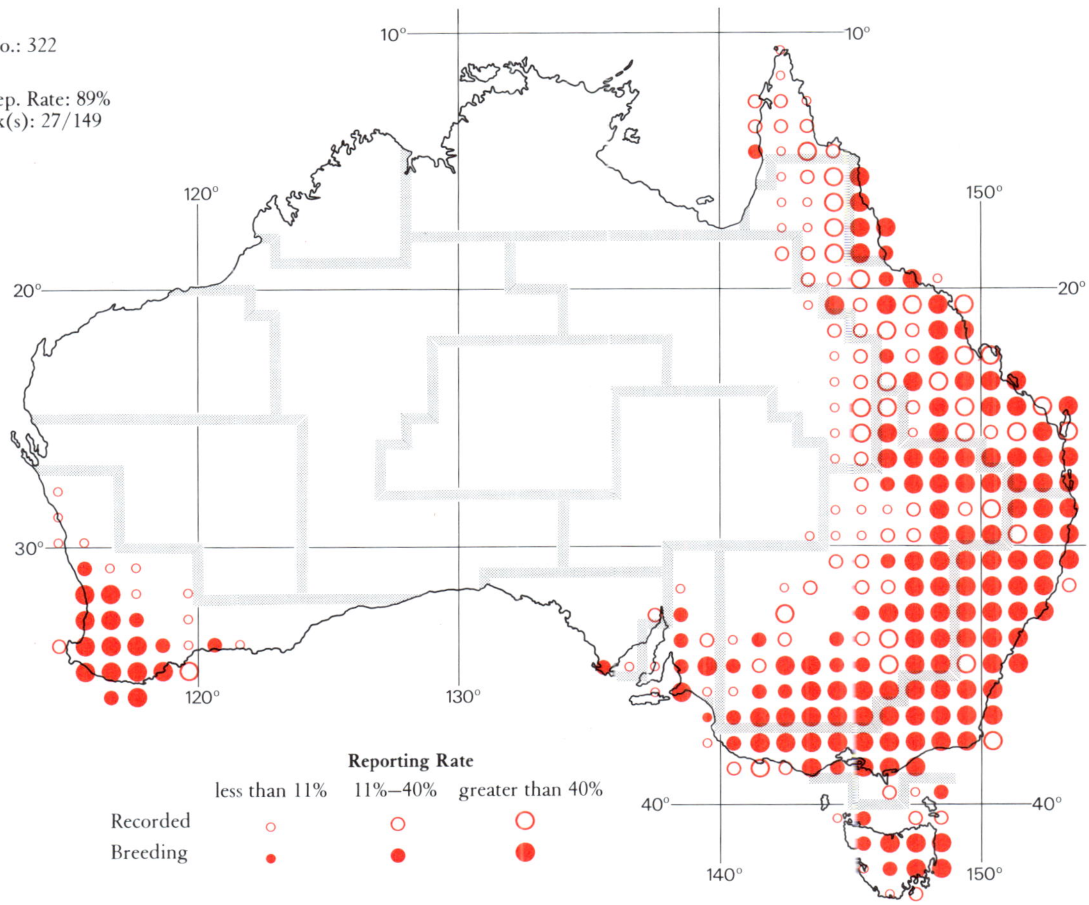

Blue-winged Kookaburra

Dacelo leachii

The Blue-winged Kookaburra lives in southern New Guinea and Australia. The four subspecies in Australia are: *kempi* in the Cape York Region south to Cooktown (15/145); *leachii* from Derby (17/123) east across northern Australia to southern Qld, omitting the Cape York Region; *cervina* on Melville I. (11/130); and *cliftoni* from the De Grey R (20/119) south to the Wooramel R (25/116).[388] Of these, only *cliftoni* in the Pilbara Region appears as an isolated population in the Field Atlas. The narrow gap south of the Gulf of Carpentaria (141°E) is maintained in the Historical Atlases. No change in distribution is evident. In eastern Australia the range of the Blue-winged overlaps that of the Laughing Kookaburra.

There is no suggestion of large-scale seasonal movement in the Field Atlas but in the Top End the Blue-winged Kookaburra tends to desert wet creekside vegetation in the wet summer season and move to better-drained forest.[426]

In the Atherton Region, though present in all vegetation types other than rainforest, the Blue-winged Kookaburra is most often in pastures near streams lined with trees.[231,698] Round Darwin (12/130) it is common in eucalypt forest and was specifically noted to avoid areas with dense undergrowth close to the ground.[426] In the Kimberley Region it inhabits rainforest in gorges and eucalypt woodland.[1257] In the Pilbara it lives along large rivers lined with tall eucalypts and paperbarks resembling the forest and woodland occupied in northern Australia. It feeds on ground-dwelling insects, particularly grasshoppers, as well as reptiles, frogs and scorpions.[787,859,1861]

In New Guinea near Port Moresby (9S/147E) in woodland with a grass understorey density was 0.05 birds/ha.[129]

Field Atlas Records: 3927

1° blocks–with records: 211 (26%)
–with breeding records: 22 (10%)

Atlas No.: 323

Max. Rep. Rate: 78%
In block(s): 12/133

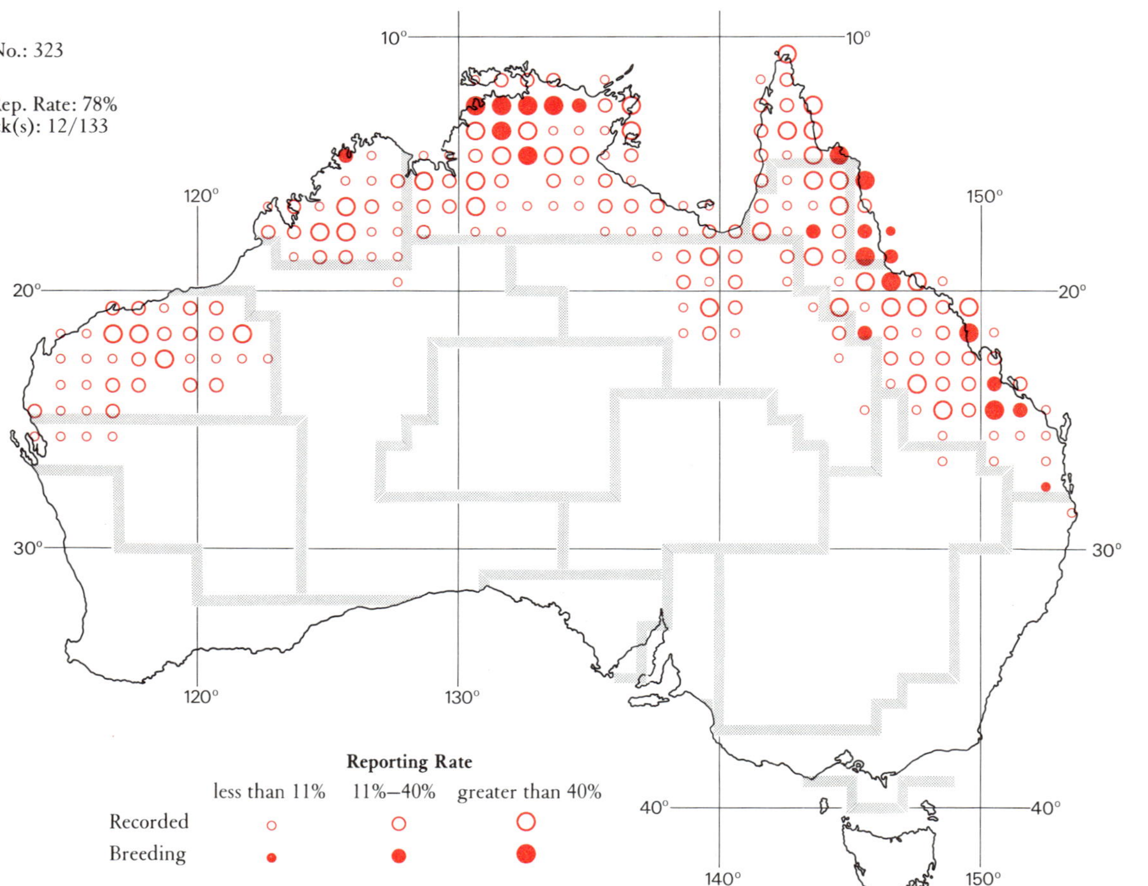

Forest Kingfisher

Halcyon macleayii

The Forest Kingfisher lives in New Guinea and Australia. In Australia the Top End subspecies *macleayii* is isolated from that of the eastern coast *incincta*.[388,748] There are no confirmed records of the species in WA despite one description of its range as south to Derby (17/123).[1428]

Migrants from Australia winter in New Guinea at altitudes up to 1500 m.[156,1449] The birds pass through Torres Strait but there is great variation in their number and in the extent of movement from year to year.[503] In Australia the pattern of movements varies. There is no information from the Cape York Region but in the Atherton Region the species is reported to be present all year.[231,698] Farther south it is described as a partial or complete migrant or nomad;[1292,1711] at Murphy's Ck (27/152) and Casino (28/153) the migrants arrive in Sept. to breed[905,1099] and at Inverell (29/151) in Oct.[96] Near Casino some birds remain as late as June in many years.

Reporting Rates increased slightly in the Cape York Region in winter (summer 21.0%, winter 31.5%), showed little change in the Atherton (summer 50.0%, winter 46.2%) and Eastern Qld (summer 23.3%, winter 22.2%) Regions and decreased slightly in winter in the South-East Region (summer 1.7%, winter 1.1%). Contrary to the general pattern of movement northward in winter the record near Melbourne (37/145) was made in winter 1977. Reporting Rates near Darwin (12/130) varied little (summer 23.7%, winter 17.5%).

The Forest Kingfisher lives in pairs, generally near water, at the edges of rainforest and eucalypt forest, in eucalypt and paperbark woodland and in mangroves.[1428,1449] In northern parts of its range, it has a wide altitudinal distribution but in the south is restricted to lowlands. The Reporting Rate below 200 m in the New England (30/152) district was 33.0% compared with 6.2% above 200m. The Forest Kingfisher feeds on insects, pouncing on them on the ground.[1703] It often takes dragonflies, presumably catching them as they rest.

During a traverse along the eastern coast in winter of 1960, single birds were seen northwards from Ballina, with a concentration of 12 birds/km from Ingham (18/146) for 160 km northwards.[134] There are no other measures of abundance.

Field Atlas Records: 3699

1° blocks–with records: 122 (15%)

–with breeding records: 37 (30%)

Atlas No.: 324

Max. Rep. Rate: 66%

In block(s): 15/145

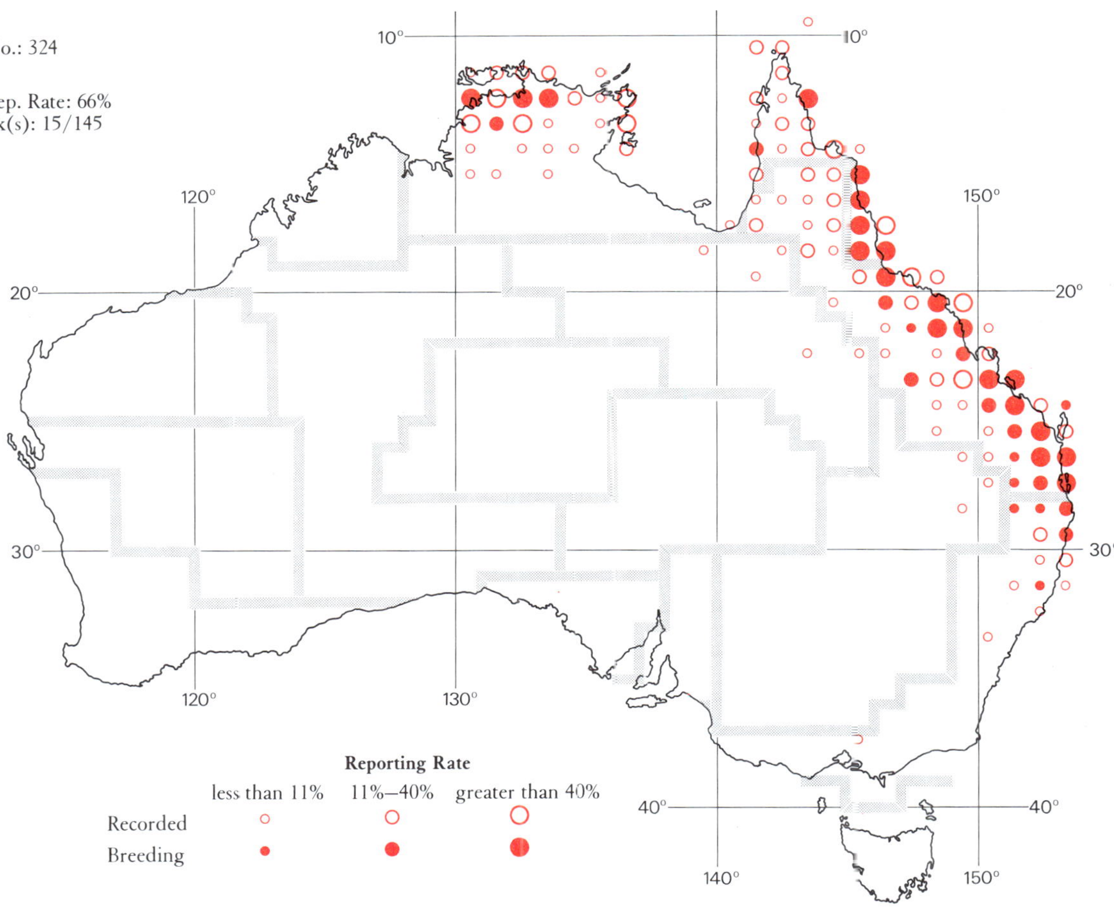

Red-backed Kingfisher

Halcyon pyrrhopygia

The endemic Red-backed Kingfisher is widespread but without significant morphological variation.[388,748] The range shown by the Field Atlas does not differ from that of the Historical Atlases.

Movements have been reported from some places and the Field Atlas suggests a regular north-south trend. South of 26°S the Reporting Rate in winter (1.7%) was lower than in summer (2.4%). The difference was most evident in Eastern Qld where the winter rate (4.1%) was nearly twice that in summer (2.2%); it may partly reflect movement towards the coast in winter[1709] such as has been reported at Innisfail (17/146) where the Red-backed Kingfisher is a scarce non-breeding winter visitor.[698] In NSW it is also regarded as a non-breeding visitor to the coast May-Aug.[396] It is a summer breeding migrant at Murphy's Ck (27/152) from late Aug. to Apr.,[1099] in the Cobar (31/145) district Sept.-Mar.[1573] and in the southern Murray-Darling Region Aug.-Mar.[833]

At Darwin (12/130), where a few pairs breed, the species is present throughout the year but most commonly June-Oct.[423] It may be absent from some areas on the edge of its range for several years: at Murphy's Ck the bird was not recorded 1940–56, having been regular before;[1099] at Glen Innes (29/151) it was not recorded 1944–57.[409]

Inland, the Red-backed Kingfsher is often present far from water but always where there are trees in spinifex and tussock grassland or in acacia scrub. It is commonly seen perching on telegraph wires. It feeds on insects and their larvae, spiders, frogs and reptiles.[1049] Factors detrimental to breeding success include heavy rain, which can seep into the nesting burrow, damaging its contents;[1095] when the burrow entrance is near the top of a bank, the chamber itself is close to the surface. Both dingoes and foxes have been observed to excavate nest chambers and eat the contents, sometimes catching an adult on the nest.[409,1095]

There are no measures of abundance.

Field Atlas Records: 3655

1° blocks–with records: 551 (68%)
–with breeding records: 76 (14%)

Atlas No.: 325

Max. Rep. Rate: 71%
In block(s): 21/122

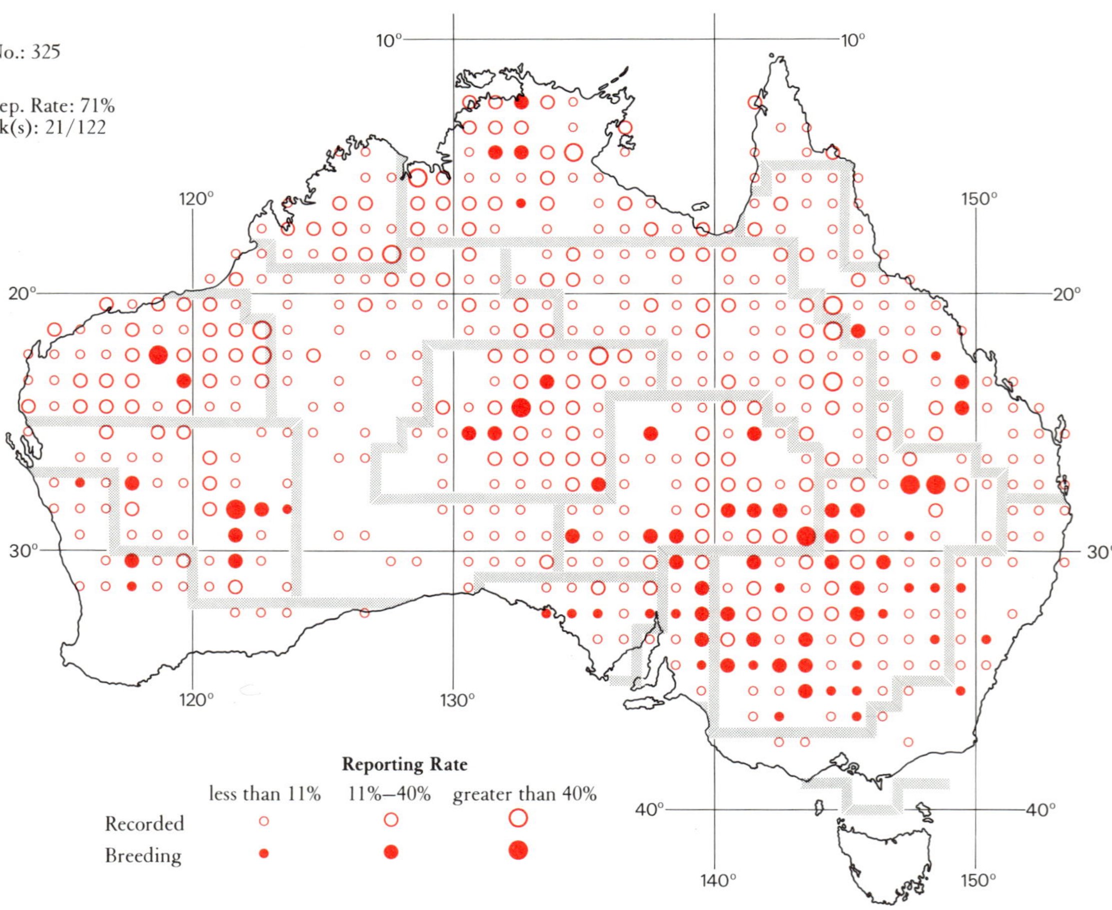

Sacred Kingfisher

Halcyon sancta

The Sacred Kingfisher lives in the Pacific islands, New Zealand, Indonesia, New Guinea and Australia. Inland in Australia an increase in range is greater than would be expected from increased coverage alone: in the Central Ra. Region, during the 1901–50 period, there were records from one 1° block, in 1951–76 from three and in the Field Atlas from ten, with breeding first reported in 1976.[340] In mangroves the Sacred Kingfisher may be confused with the Collared.[396]

The Sacred Kingfisher is sedentary on Lord Howe (31/159) and Norfolk (29/167) Is and in New Zealand, New Caledonia and the Solomons.[631] In southern Australia most birds migrate (Reporting Rates south of 26°S: summer 23.5%, winter 2.6%); a few remain in winter, e.g. in mangroves near Sydney (33/151).[1724] In the north the species is present all year. Australian birds winter in New Guinea, Mar.-Oct., and parts of Indonesia.[156,1117,1449] They migrate through Torres Strait, where some birds remain in winter,[503] and almost certainly cross the Coral Sea as well because many exhausted birds arrive on islets off Port Moresby (9S/147E).[129] Round Darwin (12/130) numbers increase Apr.-May and again Aug.-Sept.,[423] which suggests passage there also. Some birds migrate at night and can be heard calling.[692]

The Sacred Kingfisher is a scarce non-breeding visitor to the Tas. Region, mostly in winter: eight records in the Atlases were in winter, five in autumn, four in spring and only two in summer. Winter migrants were invariably on the coast; those at Dunalley (42/147) and Margate (43/147) remained some months and seemed to take up feeding territories[1314] as wintering birds in Torres Strait have been shown to do.[503]

The Sacred Kingfisher inhabits eucalypt woodland, sea-shores, mangroves and islands.[1428] It feeds mainly on insects but it will take marine animals, house mice and small lizards.[748,840,1420]

In New Guinea 80–100 birds fed in 4 ha on the Nomad R airstrip;[147] and at Port Moresby ten pairs were along 600 m of road.[156] In Australia densities in eucalypt woodland were: Armidale (30/152) 0.17–0.19 birds/ha;[582] Wollomombi (30/152) 0.4 birds/ha;[129] Canberra (35/149) 0.12 birds/ha;[154] and Moruya (35/150) 0.05–0.18 birds/ha.[1197] In eucalypt forest at Jamieson (37/146) it was 0.31 birds/ha.[1469]

Field Atlas Records: 13 218

1° blocks–with records: 483 (59%)
–with breeding records: 173 (36%)

Atlas No.: 326

Max. Rep. Rate: 54%
In block(s): 27/148

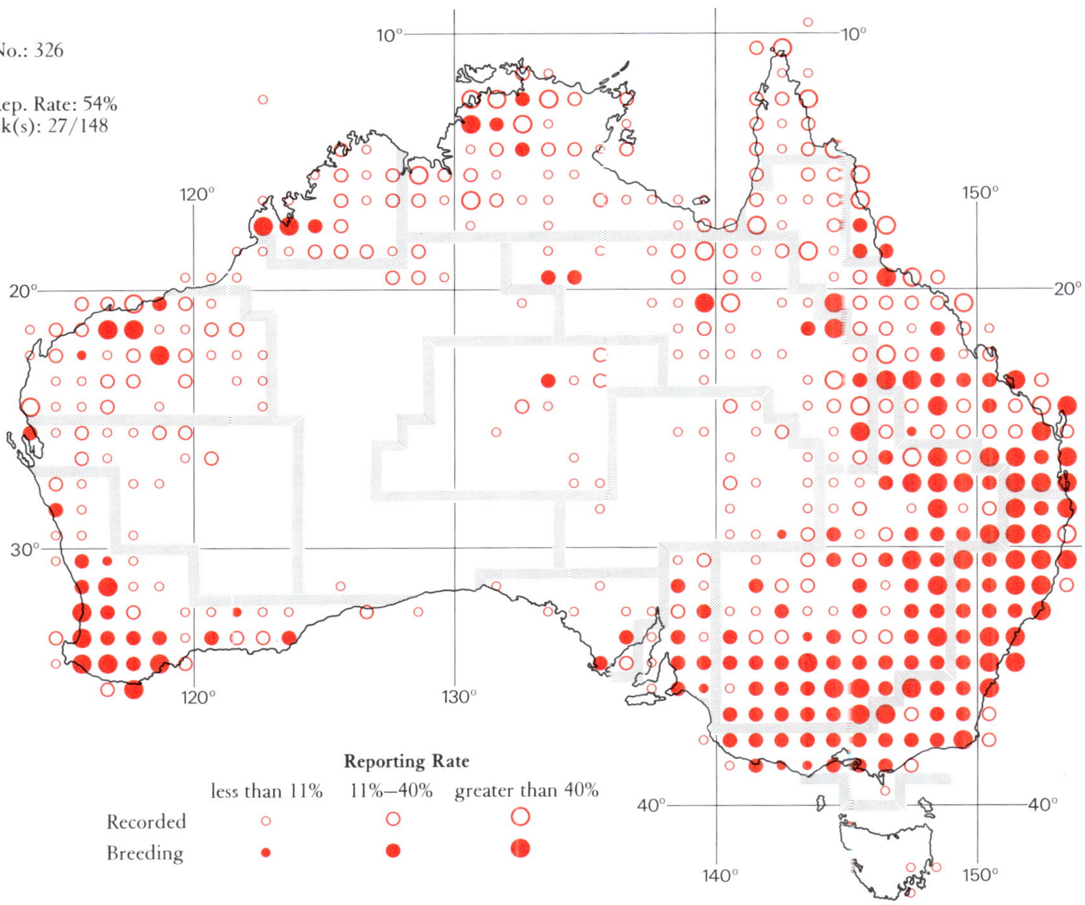

Collared Kingfisher

Halcyon chloris

The Collared Kingfisher has more than 40 described subspecies stretching from the Pacific islands to the Red Sea, with only one, *sordida*, living in Australia.[808] Historical records fill some of the gaps in the Field Atlas map but records are lacking at the western end of the Gulf of Carpentaria (137–139°E), between the Top End and Kimberley populations (126–130°E), along the Eighty Mile Beach (120–122°E) and near Warroora (23/113). There are no mangroves in the gaps except for the one between the Kimberley and Top End Regions.[622] The species may be confused in the field with the Sacred Kingfisher which also inhabits mangroves[396].

The Collared Kingfisher may migrate. In the Cape York Region it was thought to arrive in Sept.[1145] and in southern Qld at the end of Sept. At Pt Cloates (22/113) it was most frequent Feb.-Apr.[1266] More recent observers have concluded that in Torres Strait it is nomadic and uncommon but shows no evidence of migration.[503] It is stated to be present all year at Innisfail (17/146),[698] Brisbane (27/153) and Darwin (12/130).[423] The Field Atlas does not suggest large-scale seasonal movement.

In Australia the Collared Kingfisher usually lives in mangroves or woodland close to salt water. In Brisbane it sometimes appears in gardens up to 1 km from mangroves. In countries where it is in contact with fewer species of kingfisher than in Australia it inhabits many types of vegetation though it is better adapted to mangroves than other kingfishers.[485,1797] The Collared Kingfisher swoops on tadpoles, crabs, lizards and insects from vantage points such as dead limbs or telegraph posts but rarely takes fish.

Each pair occupied an area of about 7–10 ha near Brisbane in 1934, including teatree woodland and mangrove-lined creeks.[1266] In 1956–57 pairs each defended 1.5 ha of mangroves near the mouth of the Brisbane R where the fringe is particularly wide and undisturbed.[133]

Field Atlas Records: 629

1° blocks–with records: 63 (8%)
–with breeding records: 4 (6%)

Atlas No.: 327

Max. Rep. Rate: 30%
In block(s): 12/136

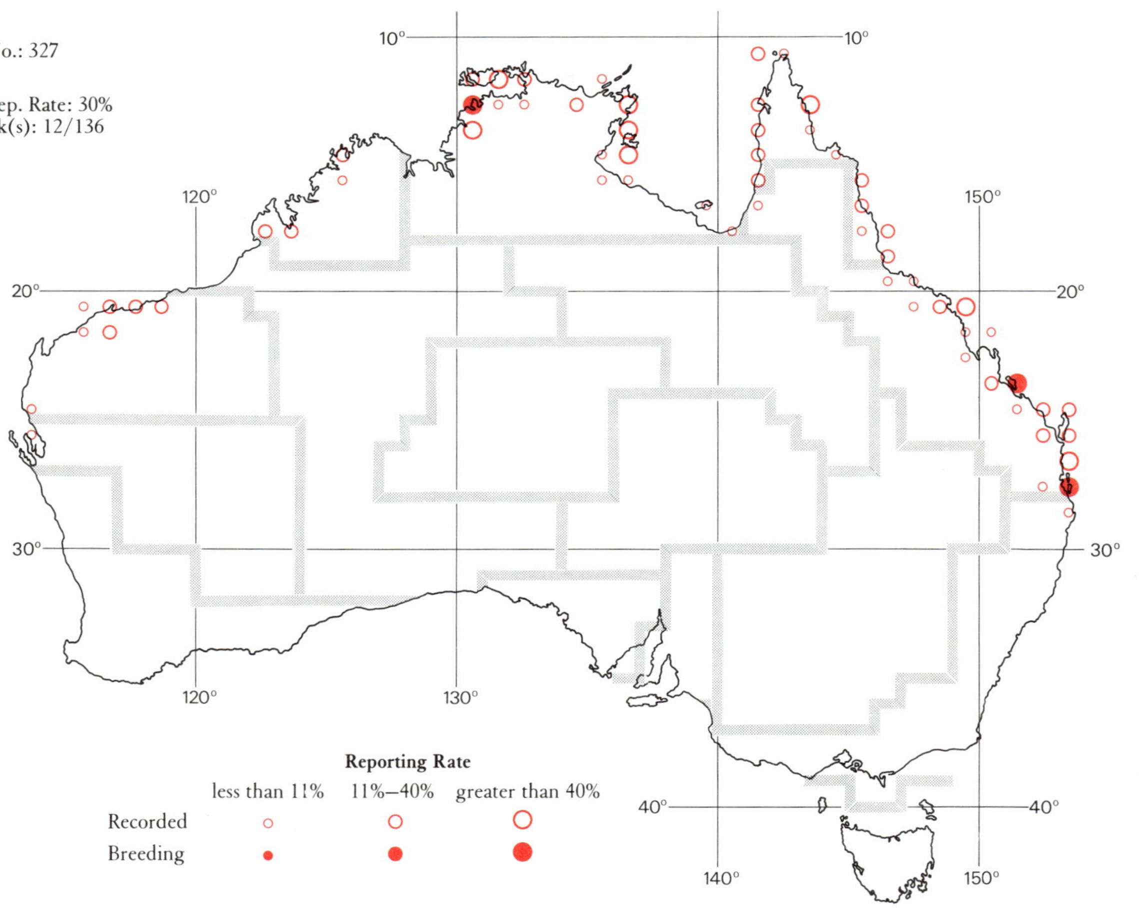

Yellow-billed Kingfisher

Syma torotoro

The Australian population *flavirostris* of the Yellow-billed Kingfisher is a subspecies of this predominantly New Guinean bird.[388]

During a two-year study in New Guinea banded individuals were repeatedly recaptured within 100 m of their banding site.[156] The Australian population is probably also sedentary and is confined to pockets of undisturbed rainforest in the Cape York Region. The bird keeps below the canopy in the forest[160] and uses arboreal termitaria up to 15 m above the ground as nest sites.[1154,1552] These termitaria and the hollow tree limbs in which it also nests are only in mature forest. During the breeding season the Yellow-billed Kingfisher is noisy, bold and aggressive near the nest but at other times of the year it is much less conspicuous and easily overlooked.

The Yellow-billed Kingfisher hunts from perches low down in rainforest, often close to its margin with eucalypt forest, and makes sallies from its vantage point to capture small lizards, dragonflies, beetles, grasshoppers and butterflies. Despite its serrated bill, it apparently does not take fish.[1154]

In New Guinea in rainforest near Port Moresby (9S/147E) density was 1 bird/ha.[160]

Field Atlas Records: 44

1° blocks-with records: 5 (1%)
-with breeding records: 1 (20%)

Atlas No.: 321

Max. Rep. Rate: 46%
In block(s): 12/143

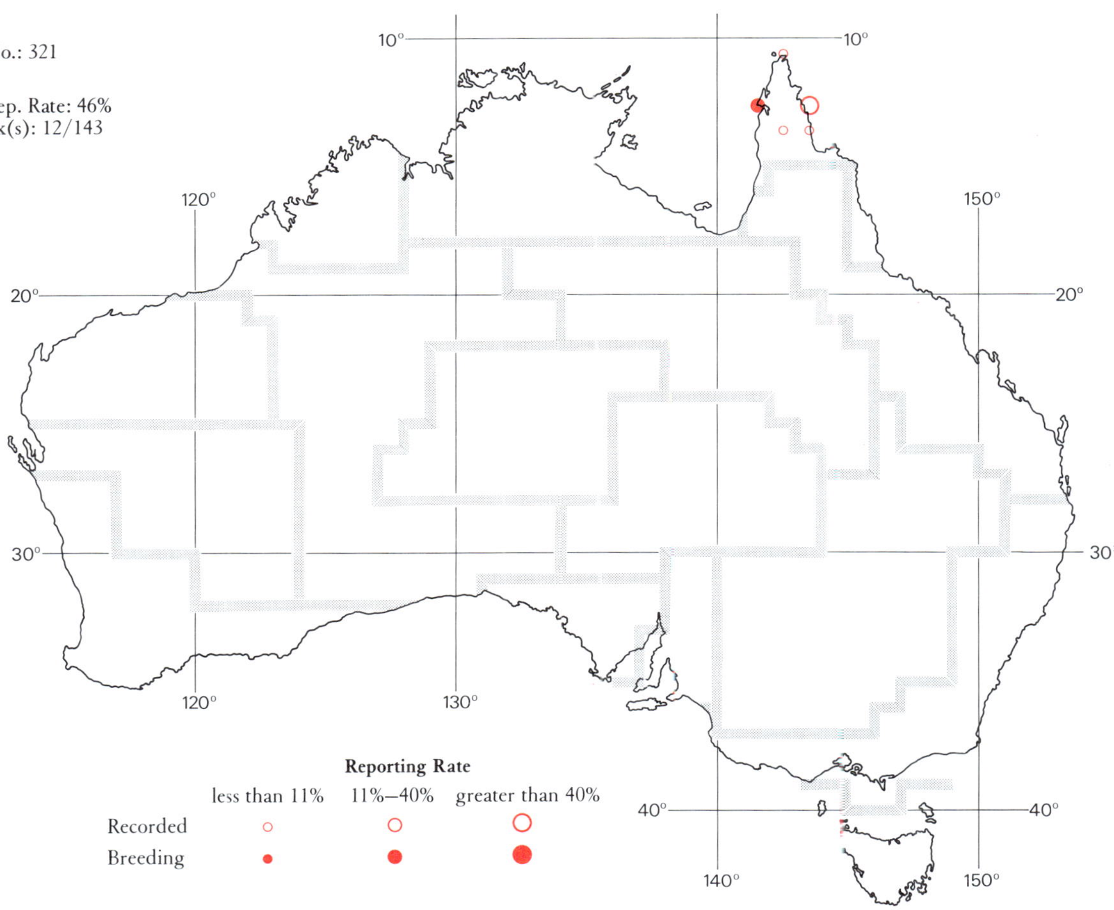

Buff-breasted Paradise-Kingfisher

Tanysiptera sylvia

Two subspecies of the Buff-breasted Paradise-Kingfisher breed in New Guinea and a third, *sylvia*, breeds in Australia and migrates to New Guinea.[388,631] Though the Australian range appears to be disjunct on the map, records from the 1951–76 period fill the gaps and show that the bird breeds in the Cape York Region. The record in the Eungella (21/148) district in Nov. 1981 is an extension of known range.

The Field Atlas shows that the Paradise-Kingfisher is frequently recorded in summer in the Cape York and Atherton Regions but not at all in winter. It is known as a passage migrant across Torres Strait[503] and some birds may fly direct across the Coral Sea because records have come from islets there and on the Great Barrier Reef,[156] including birds killed when they struck lighthouses. Possibly the populations from the Cape York Region and the Atherton Region take different routes and have different wintering grounds in New Guinea. On their return to Australia birds arrived in the Atherton Region in Nov. but not until Dec. in the Cape York Region.[129,994] They appear to migrate in flocks.[346] At the northern tip of Cape York Pen. (10/142) a flock of exhausted birds was observed one morning where none was before and large numbers were found resting on York I. (10/142).[1144] On the northward migration adults appeared to leave first as a group, the young of the year following a fortnight later.[696]

The Buff-breasted Paradise-Kingfisher inhabits only rainforest and can live in small patches at high density. When breeding, the bird is noisy. It feeds insects, spiders and lizards to its young as well as snails, the shells of which are first removed.[696] There is no record of adult food.

At Innisfail (17/146) at least eight pairs occupied 10 ha and more were thought to be present.[696]

Field Atlas Records: 119

1° blocks-with records: 12 (1%)
-with breeding records: 4 (33%)

Atlas No.: 328

Max. Rep. Rate: 23%
In block(s): 15/145

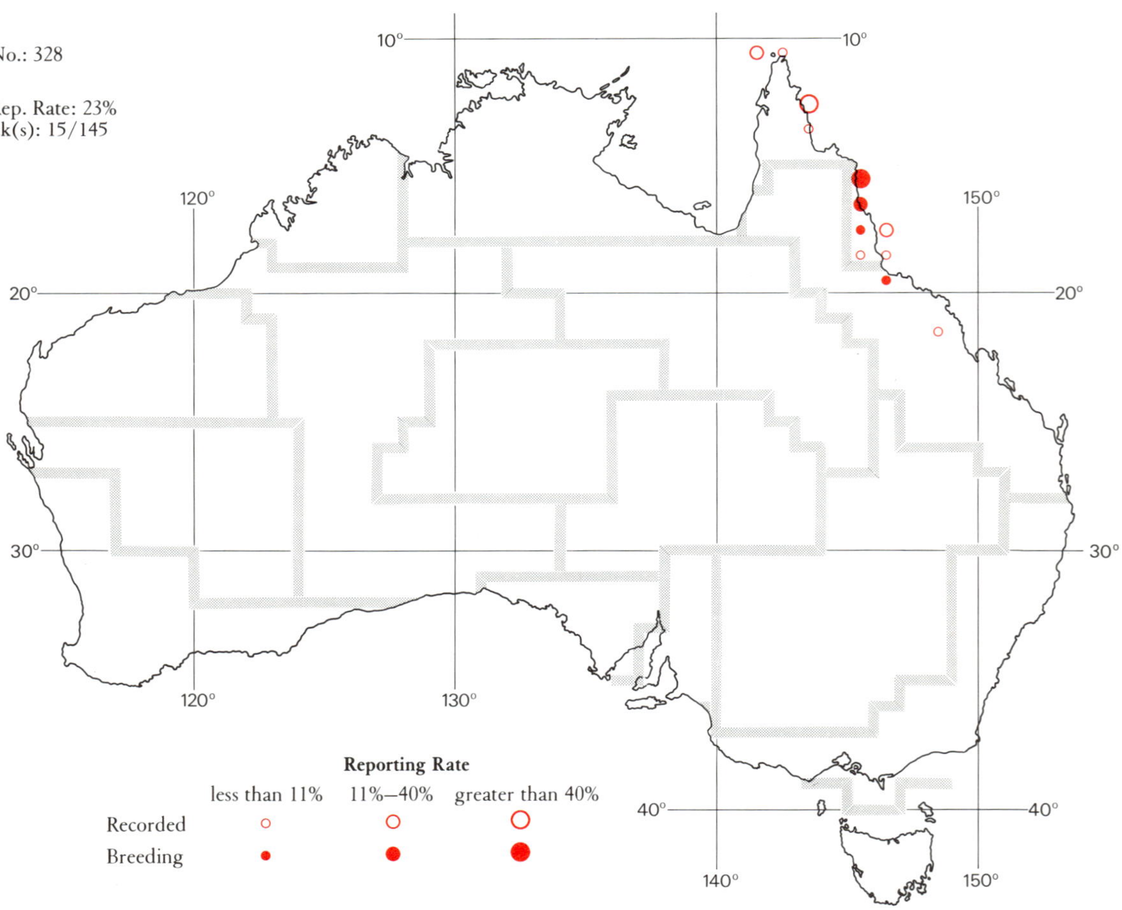

Rainbow Bee-eater

Merops ornatus

The Rainbow Bee-eater breeds in New Guinea at Port Moresby (9S/147E) and in Australia. Australian birds winter throughout New Guinea and some overshoot to Micronesia (round 5N/170E) and to the Ryuhu Is of Japan.[129] In Australia the bird is absent from the Tas. Region. Distribution appears unchanged from that shown in the Historical Atlases.

Most birds migrate north for the winter (Reporting Rates south of 26°S: summer 22.9%, winter 1.6%). During late summer groups in the south coalesce into flocks of up to 200 birds and begin to drift northward, moving faster as autumn advances.[833,1099] In Mar.-Apr. flocks of several thousand travel north and pass Innisfail (17/146), often at night.[698] In the Cape York Region and at Darwin (12/130) many birds occupy the mangroves in May-June and depart during July.[423,1147] Large numbers pass through Torres Strait on migration. Generally the Bee-eaters travel in small flocks but, in Sept. 1978, 864 birds per hour were recorded passing over Thursday I. (10/142).[503] During the southward migration in Aug.-Oct. they have been seen at sea from Darwin east to Thursday I., the greatest numbers being in Torres Strait.[301] Groups of 5–50 birds were seen flying at heights from just above the waves to 100 m or more.[301] Some flocks were close together, others widely separated.

The Rainbow Bee-eater lives almost anywhere suitable for hawking insects. It nests either solitarily or colonially and has been recorded to breed communally.[557] One site at Jindabyne (36/148) had 55 nests. The distribution of the Rainbow Bee-eater is determined largely by the availability of flying insects on which it feeds. It takes insects of many orders and will gather at swarms. As the vignette implies, there is some justice in its name and the gut contents of a large proportion of those sampled included the honeybee[1049] though these are not fed to the young.[1420] Apiarists shoot the Bee-eater illegally, regarding it as a pest. There is heavy mortality on the southward migration through Torrres Strait when little water is available [503]

Density near Port Moresby in woodland with a grass understorey was 0.12–0.33 birds/ha.[158]

Field Atlas Records: 15 830

1° blocks–with records: 608 (75%)

–with breeding records: 181 (30%)

Atlas No.: 329

Max. Rep. Rate: 76%
In block(s): 13/130,24/153

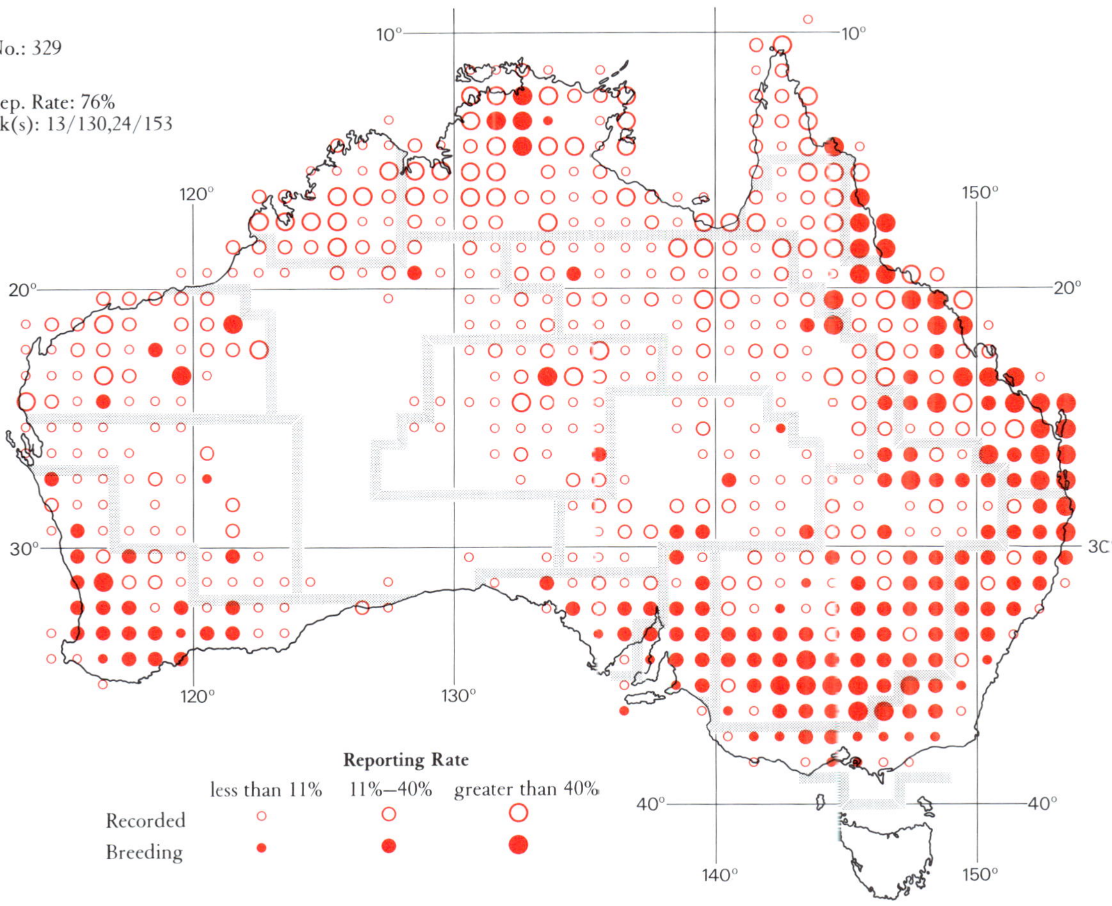

Dollarbird

Eurystomus orientalis

The Dollarbird ranges from India and China south-east through the Pacific islands and New Guinea to Australia. In Australia the apparent gap in distribution round 138°E is probably an artifact caused by lack of coverage in summer when the birds are present. The isolated inland and Pilbara records are of single birds Nov.-May. There is a winter record from the Pilbara Region in June 1953[1638] and one specimen from western Tas. in 1918, season unknown.[1091]

The Dollarbird is a migrant, breeding in Australia Sept.-Mar. and spending the winter months in New Guinea.[1591] Adult Dollarbirds leave Australia soon after the young are fledged and the young follow later. Birds arrived at a site in south-eastern New Guinea in Apr. and departed in Nov.[160] The main passage through Torres Strait is north in Mar.-May and south in Oct.-Nov.[503] though some birds arrive earlier in Australia. The species' time of arrival is remarkably regular. At Urangan (25/152) the dates varied over 14 years from 7-30 Sept.[361] with similar regularity reported at other sites. Records of the heights at which Dollarbirds migrate are 4500 m in New Guinea[1591] and 2500 m in mid-eastern Qld;[183] in Torres Strait ones and twos have been reported travelling 20-30 m above the sea.

The Dollarbird lives on the edge of rainforest, along rivers, in areas of regrowth, in eucalypt forest and woodland and in cleared areas with some old trees. It swoops from vantage points high on dead limbs and twigs to take insects in flight. It will come down to ground level to take grasshoppers when they are in plagues[1099] as well as to feed on the crepuscular flights of beetles and other large insects.

Pairs aggressively defend territories round their nests. On the Ord R (15/128) nests were about 0.75 km apart along creeks and streams.[1218] Densities in eucalypt woodland were: Armidale (30/151) 0.14 birds/ha[582] and Moruya (35/150) about 0.05-0.10 birds/ha.[1192]

Field Atlas Records: 6625

1° blocks–with records: 235 (29%)
–with breeding records: 92 (39%)

Atlas No.: 318

Max. Rep. Rate: 38%
In block(s): 12/136,16/130,24/148, 27/152

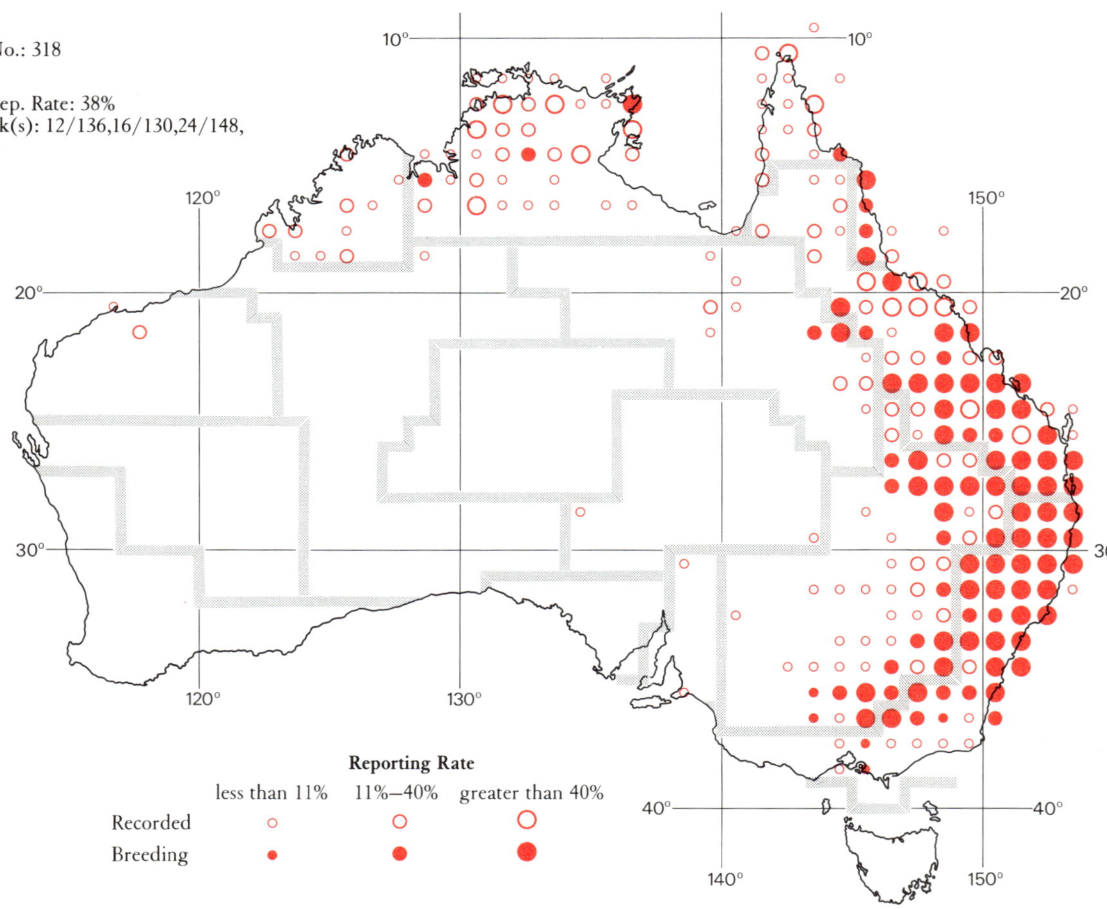

Red-bellied Pitta

Pitta erythrogaster

The Red-bellied Pitta has a wide range overseas through New Guinea, Indonesia and the Philippines. Many subspecies have been described.[1422,1449] The Australian one, *macklotii*, is confined to the eastern coast of the Cape York Region from the tip (10/142) south to the McIlwraith Ra. (13/143). Breeding has been reported throughout its Australian distribution.

The Pitta migrates to spend the dry winter season in New Guinea and is a rarely seen passage migrant in Torres Strait.[503] It returns to Australia to breed, some birds arriving as early as Oct. and most leaving in Mar.-Apr.[269] At the Claudie R (12/143) it arrived in Dec.[1147]

The Red-bellied Pitta in Australia is confined to rainforest.[1147] It searches the litter for its food of ground-dwelling invertebrates, including snails, which it breaks on anvils. It also spends some time in the trees calling and possibly feeding.

In New Guinea density in lowland rainforest near Port Moresby (9S/147E) was 0.3 birds/ha.[160]

Field Atlas Records: 9

1° blocks–with records: 3 (1%)
–with breeding records:

Atlas No.: 353

Max. Rep. Rate: 8%
In block(s): 12/143

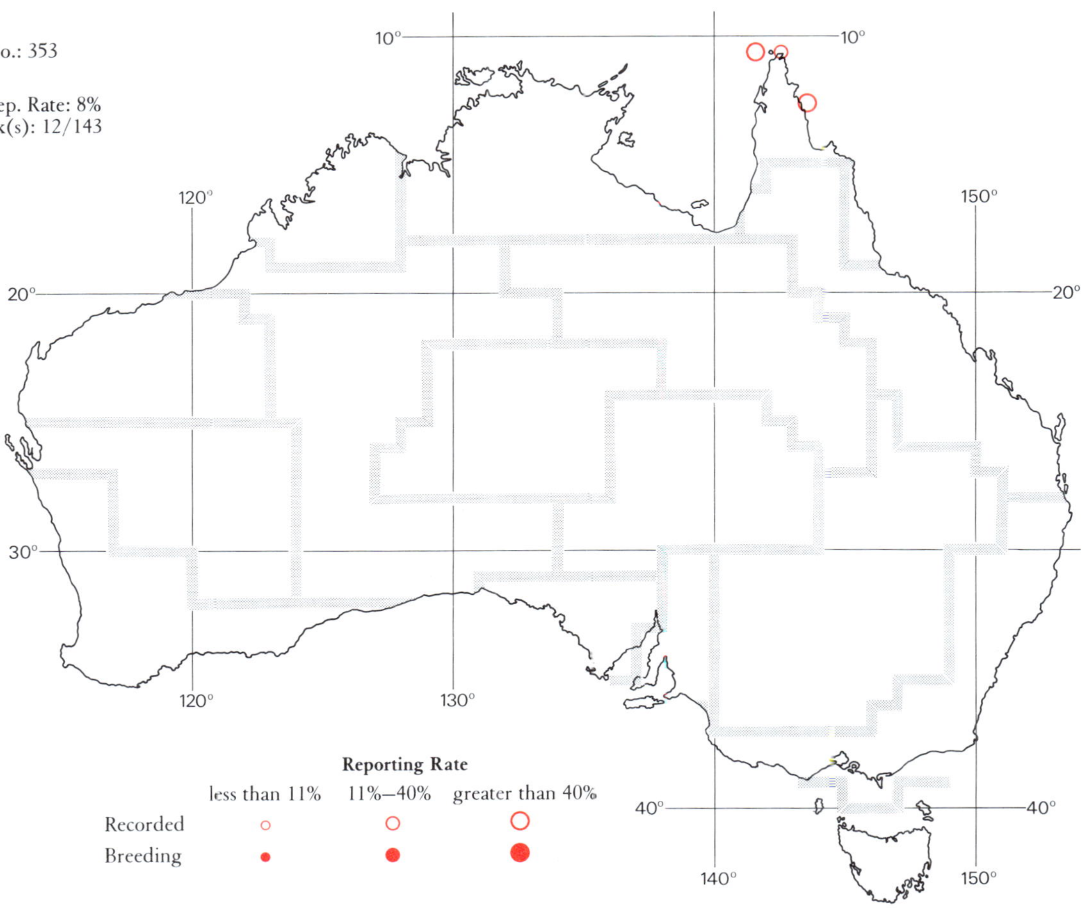

Noisy Pitta

Pitta versicolor

The Noisy Pitta lives in New Guinea and Australia. The two subspecies in Australia are *simillima* in the Cape York Region and *versicolor* southwards. There are no records in the Atlases linking the Cape York Region, Atherton, Eungella (21/148) and southern populations.

The Noisy Pitta is a passage migrant throughout Torres Strait where it also breeds.[503] In the Cape York Region it is regarded as a summer visitor[104,143] and the assumption has sometimes been made that it is a migrant elsewhere in Australia.[1709] Migration of the population in the Cape York Region is reflected in Reporting Rates of 20.6% in summer and 6.7% in winter. The Field Atlas suggests that birds in the Atherton and South-East Regions do not undertake large-scale movements. Seasonal Reporting Rates in the Atherton Region varied from 8.4 to 10.6% and in the South-East from 0.5 to 0.7%.

The Noisy Pitta is usually reported from rainforest, from near sea level to 1500 m. It also inhabits woodland adjacent to rainforest and mangroves. In the Tooloom Scrub (28/152) it is in eucalypt forest.[1342] The Noisy Pitta feeds on ground-dwelling invertebrates in damp places.[1049,1145] It is best known for anvils on which it cracks the shells of large forest snails.[536]

An average of two individuals was recorded on each of six transects, 3 km long, through Tooloom Scrub in Dec. 1962.[1342] There are no other measures of abundance.

Field Atlas Records: 513

1° blocks–with records: 40 (5%)

–with breeding records: 12 (30%)

Atlas No.: 352

Max. Rep. Rate: 46%
In block(s): 12/143

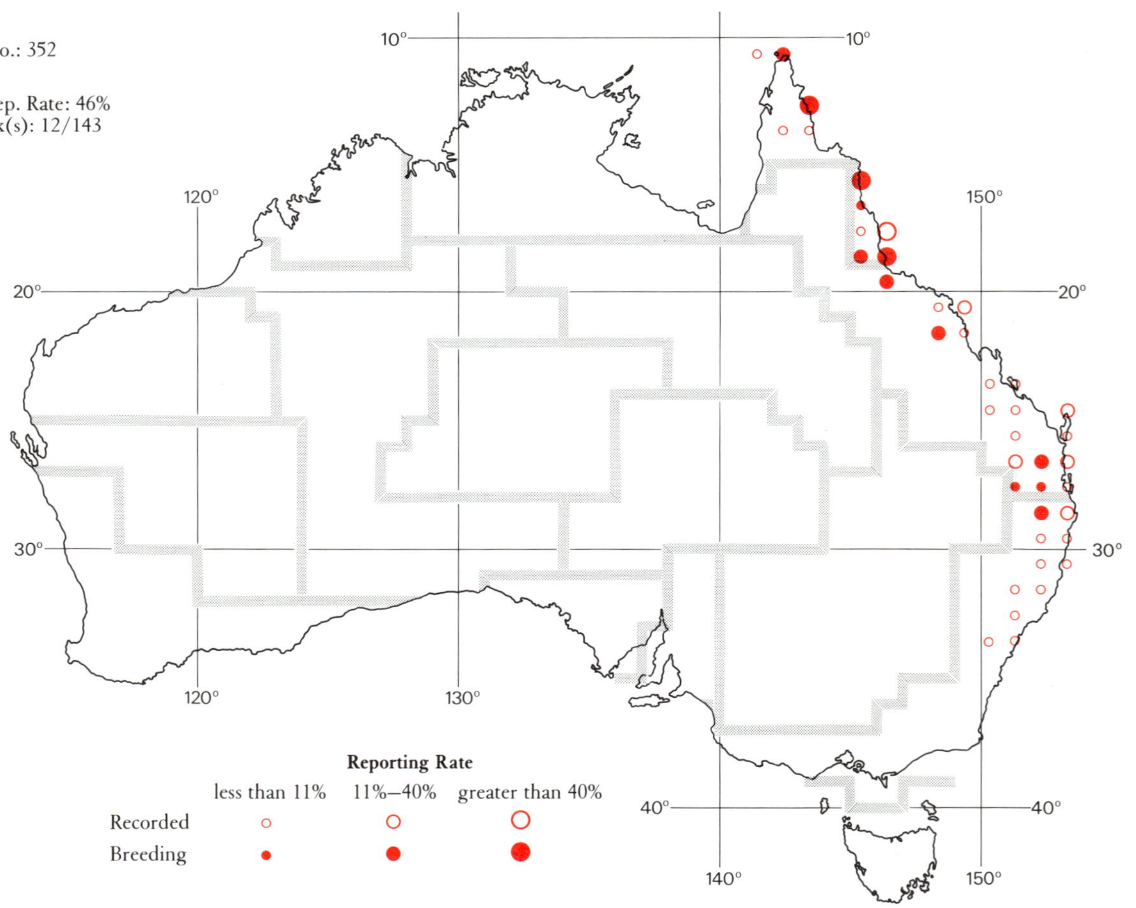

Rainbow Pitta

Pitta iris

The Atlases show two main populations of the endemic Rainbow Pitta, one in the Top End Region and one in the Kimberley Region, though no morphological differences are recognized.[608,1711,1713] The Top End population has been recorded from Port Bradshaw (12/136) west to Port Keats (14/129) and on Melville I. (11/130). In the Kimberley Region additional historical reports have come from Parry Harbour (13/126) and Napier Broome Bay (14/126) south and west to the Prince Regent R Reserve (15/125), including some from islands off the coast. The species breeds throughout its range and is thought to be sedentary.

The Rainbow Pitta lives in patches of rainforest, mangroves, eucalypt forest adjacent to the shore and rainforest along streams.[423,787,1257] Its noisy foraging among fallen leaves forewarns of the bird's approach.[423] The Pitta feeds on ground-dwelling insects and small snails, breaking them on a branch or root; no anvils have been found.[664,1859] It will forage on mounds of the Orange-footed Scrubfowl.

On the Cobourg Pen. (11/132) there was a pair in every patch of rainforest larger than 1 ha. In one patch of 5 ha seven pairs were located.[664]

Field Atlas Records: 96

1° blocks–with records: 12 (2%)
–with breeding records: 2 (17%)

Atlas No.: 354

Max. Rep. Rate: 20%
In block(s): 13/130

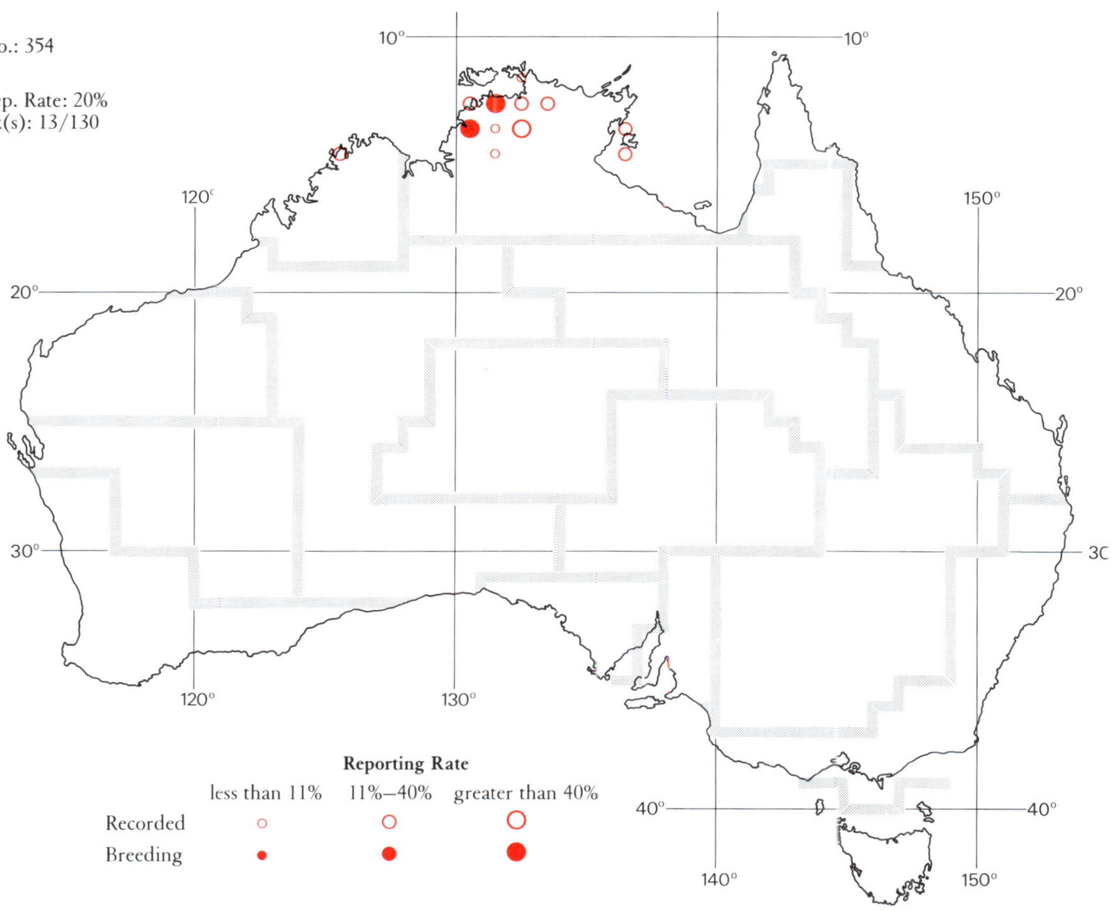

Albert's Lyrebird

Menura alberti

The range of the endemic Albert's Lyrebird stretches from the Mistake Mts (27/152) south along the Great Dividing Ra. to Killarney (28/152), east along the McPherson Ra. to the Springbrook Plateau (28/153)[445] and south to Bangalow (28/153). An isolated population still survives on Tamborine Mt. (27/153). Records from the Blackall Ra. (26/152)[322,1122,1328] have not been confirmed and may have been in error.

Albert's Lyrebird is sedentary, though birds may leave their territories and wander locally after the breeding season. The species lives in rainforest or mixed eucalypt forest and rainforest with adequate ground cover, sometimes that of the introduced lantana.[444] The bird breeds in mid-winter.[1515]

Only insect remains have been found in guts.[723] Albert's Lyrebird is also said to feed on large land snails[322] but these and earthworms would rapidly be digested to an unrecognizable mush.

The population on Tamborine Mt. seems unlikely to survive because clearing is isolating it.[444]

There are no measures of abundance.

Field Atlas Records: 64

1° blocks–with records: 4 (1%)
–with breeding records: 2 (50%)

Atlas No.: 351

Max. Rep. Rate: 7%
In block(s): 28/153

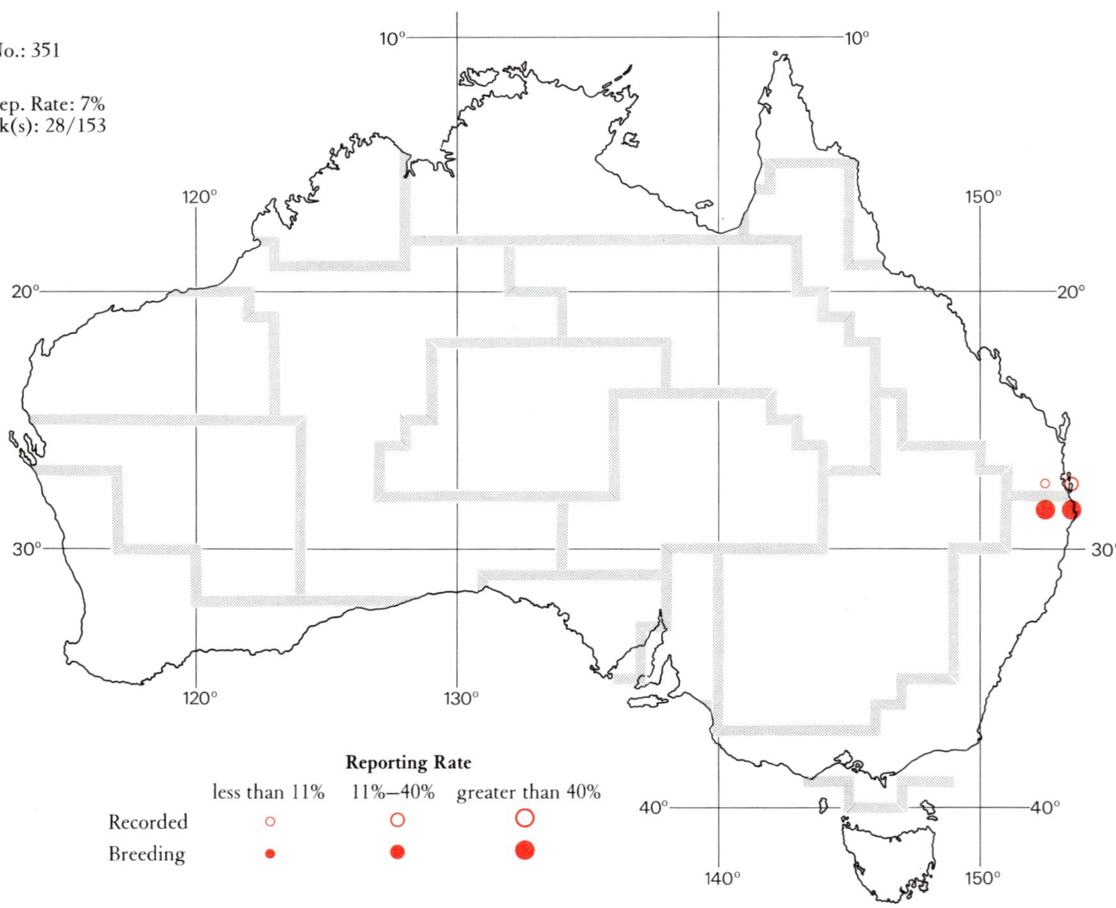

The natural range of the endemic Superb Lyrebird is within the South-East Region. In 1934 it was introduced at Mt Field (42/146) and Hastings (43/146) in Tas. It has bred and is slowly expanding its range there. No large-scale change in distribution on the mainland is suggested by the Atlases. The Lyrebird appears on a list for Mt Macedon (37/144) in 1856 but supporting records are lacking and the possibility that it was introduced or misidentified there cannot be discounted. There are also recent unconfirmed reports from Macedon, Lerderderg Gorge (37/144) and in the Otway Ra. (38/143) but all records west of 145°E are extremely doubtful.

The Superb Lyrebird is sedentary and lives singly or in small parties in the forests of south-eastern Australia from the mountains down to sea level.[488] On the mainland it lives in rainforest and eucalypt forest. In Tas. it has become established in rainforest. The male seeks an elevated area for his territory in which he builds display mounds, while the female is more often in cool humid gullies. Territories of the two sexes overlap.[991,1071,1517] Breeding takes place in winter.

The Superb Lyrebird feeds on insects, spiders, seeds, earthworms and crustaceans.[1049,1517] As a ground-foraging bird, it is the prey of native cats and dingoes; corvids and goannas prey on eggs and young. Dogs and foxes are also predators in Sherbrooke Forest (37/145), which is in a closely settled area. Settlement has narrowed the corridors between adjoining forests necessary for the dispersal of young and this may be the cause of the noticeable decline to few if any birds in Olinda State Forest (37/145). Nevertheless in Sydney (33/151) the Lyrebird survives in gullies in sandstone areas surrounded by suburbs.[129]

Superb Lyrebird

Menura novaehollandiae

At Tidbinbilla (35/148) the average area of nine male territories was 2.4 ha.[1517] Density at Moruya (35/150), in eucalypt woodland, was 0.15 birds/ha,[1197] and at Boola Boola (38/146) in eucalypt forest 0.1–0.5 birds/ha.[1113] During an eight-year period in Sherbrooke Forest an average of one nest was found in each 18 ha of the two gullies searched though not all nests were completed.[1471] Females ranged over about 5 ha and the male in the most studied site included the foraging areas of at least four females in his territory.[991]

Field Atlas Records: 2767

1° blocks–with records: 39 (5%)
–with breeding records: 21 (54%)

Atlas No.: 350

Max. Rep. Rate: 38%
In block(s): 37/146

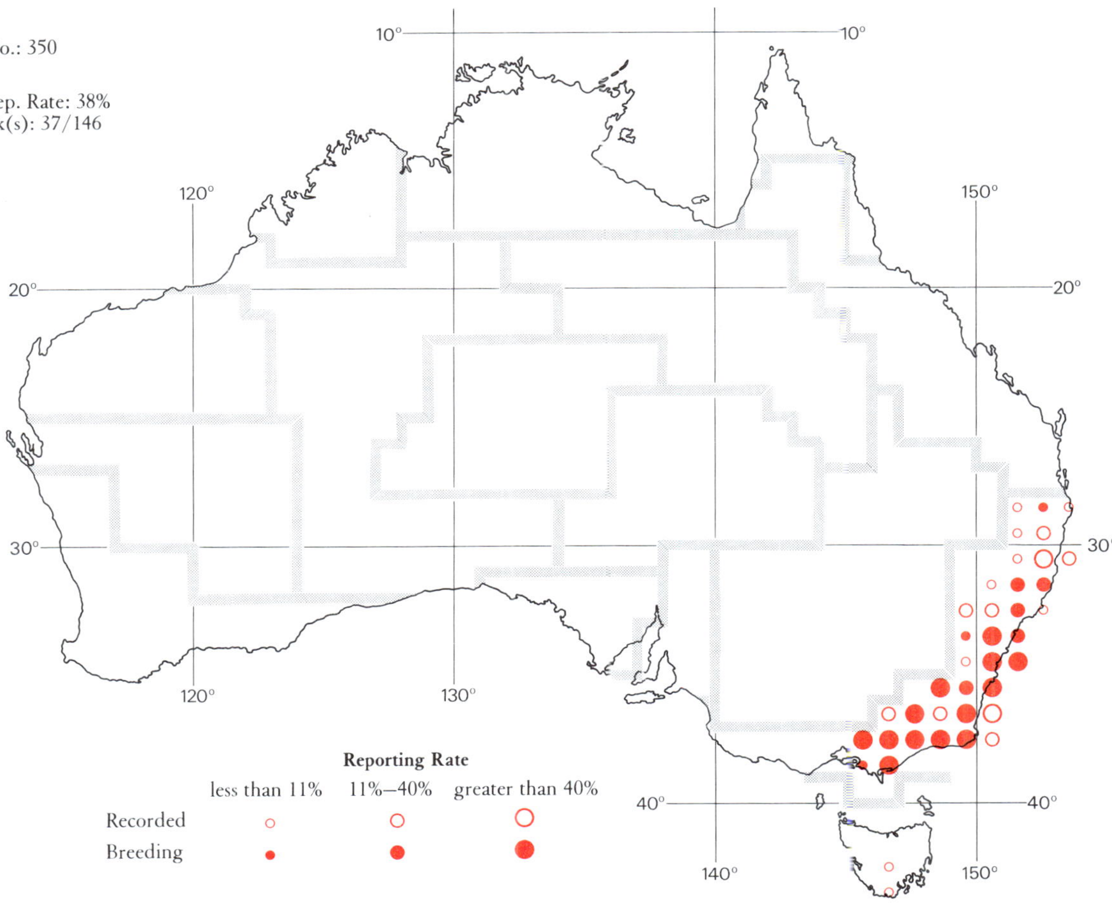

Rufous Scrub-bird

Atrichornis rufescens

The endemic Rufous Scrub-bird is distributed in isolated populations in the highlands of the Great Dividing Ra. The northernmost record is at the junction of the Mistake Mts and the Little Liverpool Ra. (27/152). Other populations inhabit the McPherson Ra. (28°S), the Gibraltar Ra. (29/152), New England Tableland (30/152), upper Hastings R (31/152), and Barrington Tops (32/151). Before 1900 the Rufous Scrub-bird was known from lowland sites round Lismore (28/153)[351] but it is now restricted to highland forest.

The Rufous Scrub-bird is sedentary and lives in rainforest and adjacent eucalypt forest where undergrowth is specially thick.[1516,1683] Early naturalists found it common round natural clearings within lowland rainforest.[351] It requires a dense layer of ground cover at least 1 m high, a moist microclimate at ground level, abundant leaf litter and rainforest near but not necessarily in the territory.[545] Its food has not been recorded.

Even when Europeans arrived, the distribution of the Rufous Scrub-bird was that of a relict species, fragmented populations surviving in patches of suitable country.[1684] Clearing of rainforest, specially in the lowlands, has destroyed some of these populations, leaving the species isolated in small pockets in the ranges. Populations in eucalypt forest are at risk from fire.[544] On the other hand there is a suggestion that selective logging, specially in rainforest, might lead to an increase in the density of undergrowth to the benefit of the Scrub-bird.[544]

During a study of the Rufous Scrub-bird where all known records were followed up, 172 territories were located, 93 of them within the McPherson Ra. in the Lamington-Wiangaree (28/152–3) area. A detailed census at Barrington Tops and Wiangaree gave an average density of 4 territories per sq. km.[545]

Field Atlas Records: 29

1° blocks–with records: 5 (1%)
–with breeding records: 2 (40%)

Atlas No.: 355

Max. Rep. Rate: 3%
In block(s): 28/153

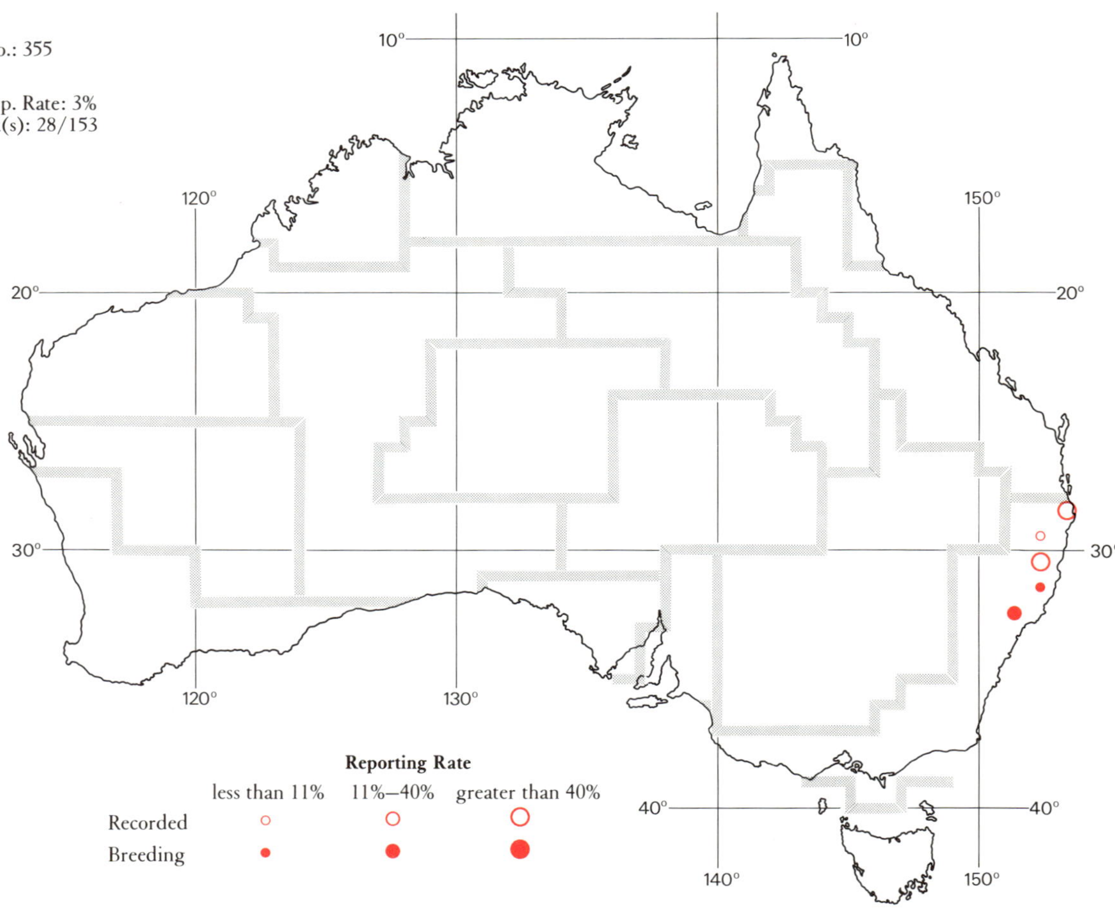

Noisy Scrub-bird

Atrichornis clamosus

The endemic Noisy Scrub-bird has become one of Australia's legendary birds, at present known only from Two People Bay Nature Reserve (34–35/118) in the South-West Region. At the time of European settlement it also lived at Drakes Brook (32/115), Margaret R (33/115), Augusta (34/115), Torbay (35/117) and Mt Barker-Albany (34–35/117).[1684,1686] The species declined rapidly and was not reported at all between 1889 and 1961. Since 1976 its range has begun to expand in the Two People Bay Nature Reserve.[1682]

The Scrub-bird shows no seasonal movements other than dispersal of the young.

The male holds a territory of up to 6 ha in the gullies of Mt Gardner (34/118) where there is eucalypt forest. New territories are being established round L. Gardner (1 km north-west), presumably by young bred in the main colony. Formerly the Noisy Scrub-bird was more common where swamps fringed eucalypt forest, specially near Albany.[1684] It feeds on large insects, including crickets and cockroaches, soft-bodied invertebrates and slaters, which it finds by turning over the litter.[1688]

Fires started by early settlers and forest managers became too frequent to allow the bird's invertebrate food supply to survive in the forest, though nesting areas were little damaged. This was probably the chief factor in reducing the range of the Noisy Scrub-bird; clearing for agriculture and climatic oscillation, including long dry periods, have also contributed. The Scrub-bird apparently survived on Mt Gardner because rock outcrops prevented fires from burning over the whole mountain at any one time. Repopulation from surviving colonies was therefore possible. Experience suggests that the heath and forest needs to be free from fire for 8–10 years to regenerate a sufficiently deep litter layer to support a Noisy Scrub-bird population.[1683]

At present the Mt Gardner area holds about 74 territories.[1686]

Field Atlas Records: 19

1° blocks–with records: 2 (1%)
–with breeding records: 2 (100%)

Atlas No.: 356

Max. Rep. Rate: 7%
In block(s): 34/118

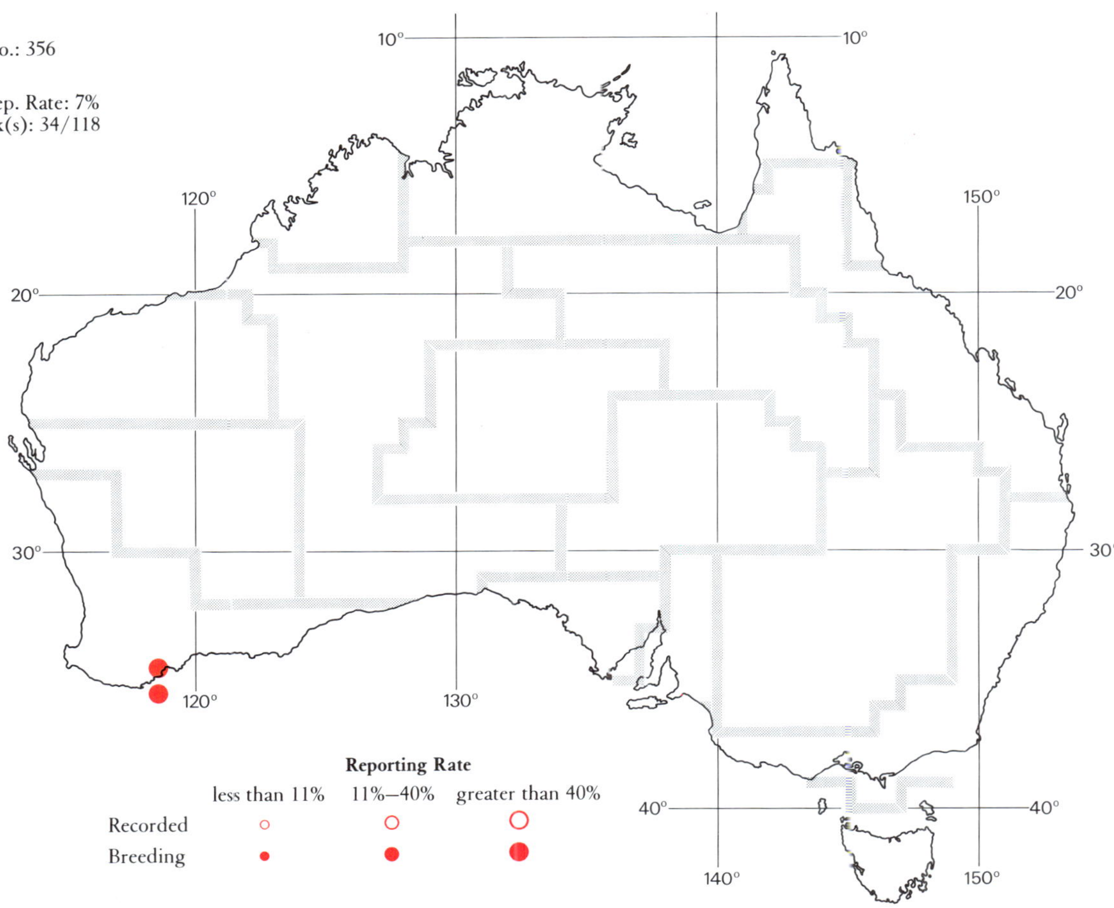

Singing Bushlark

Mirafra javanica

The range of the Singing Bushlark extends from Africa and India to Australia. Twelve subspecies have been described in Australia,[748,1234] one of which, *melvillensis* from Melville I. (11/130), was not recorded during the Field Atlas. In the Tas. Region Bushlarks have been reported only from Big Green I. (40/147) in Aug.-Oct. 1967.[1336] A record from Duketon (27/122) in Dec. 1974 seems to be better evidence of the capacity of Bushlarks to wander long distances than of a general expansion of range.

In the Atherton Region the Singing Bushlark is present all year but in the non-breeding season becomes secretive, moving from open fields to the seclusion of rank herbage.[227] In Eastern Qld it is regarded as a partial migrant.[1709] In the southern Murray-Darling Region movements are sudden and erratic.[833] Reporting Rates in the Eastern Qld, South-East and Murray-Darling Regions were greater in summer than in winter (summer: 11.9%, 2.7%, 7.8% respectively; winter: 3.6%, 0.9%, 3.0% respectively). The difference may simply reflect a change in the bird's behaviour because in the summer breeding season frequent and lengthy song-flights make it conspicuous.[207] The differences are similar in all regions and do not suggest large-scale seasonal movement.

The Singing Bushlark lives in open country such as tussock grassland, cereal crops and irrigated lucerne. Numbers encountered vary from single birds to small or large groups. In the Murray-Darling Region 50 were in the air at once in autumn.[1234] In 1955 'countless thousands' appeared after heavy rain on grass and saltbush plains near Conargo (35/145)[833] and non-breeding flocks of many hundreds were observed during the Field Atlas.

The diet of the Singing Bushlark includes much seed, mostly of grasses, and some insects including ants, grasshoppers and caterpillars.[748,1049] The Bushlark has taken advantage of clearing, rapidly establishing populations on farmland.[227]

There are no measures of abundance.

Field Atlas Records: 3008

1° blocks–with records: 350 (43%)
–with breeding records: 50 (14%)

Atlas No.: 648

Max. Rep. Rate: 46%
In block(s): 21/144

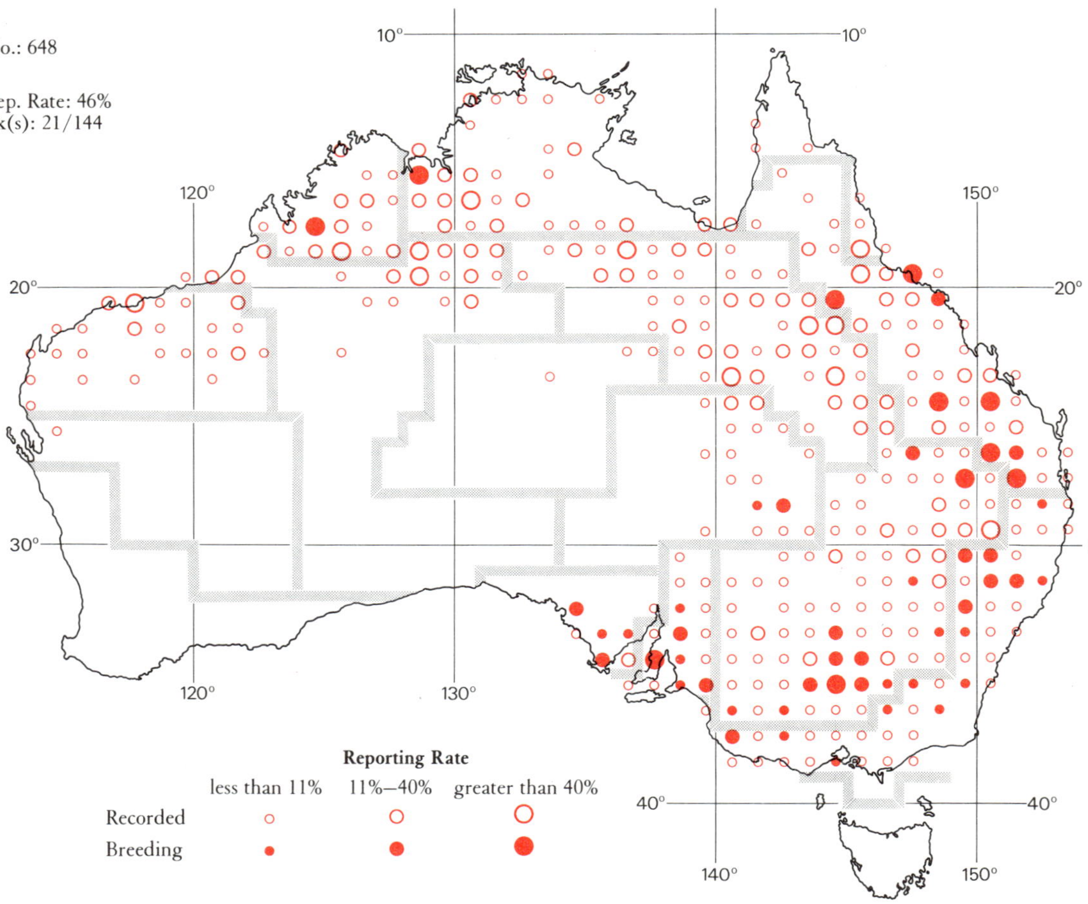

Skylark

Alauda arvensis

The Skylark, which is native to Eurasia and northern Africa, was introduced to Australia. Its spread is well recorded (p. 666) following liberations in many places: Geelong (38/144) in 1850 or 1854;[97,760] Melbourne (37/145) in 1855;[97,1553] Adelaide (34/138) in 1862; Sydney (33/151) in 1866;[1553] Qld in 1869 (unsuccessful);[343] Tas. in 1862 or 1872; WA before 1912 (unsuccessful).[1085] In most places the initial release was followed by others up to the turn of the century. The spread of the Skylark has been generally slow but is continuing, particularly into the Murray-Darling Region and west to Eyre Pen. It is often confused in the field with Richard's Pipit.

The Skylark is migratory in most overseas countries. It is present throughout the year over its range in Australia though there is evidence of flocking in Tas.[575] It lives in heath, pasture, farmland, swamps and sometimes on the seashore. It is generally in areas modified by man and absent from those that are undisturbed, e.g. western and south-western Tas.[1314] The Skylark feeds on seeds, shoots, insects and other invertebrates.[1085] It damages some young crops in England, specially sprouting seeds and seedlings of market garden varieties. Similar damage has been done to oil poppy seedlings in southern Tas.[1533] Possible competition with native species has not been studied.

In Europe estimates of the Skylark's density vary. On downland pasture in summer it was 0.31 birds/ha, similar to the density on heath; in mixed farmland it was 0.34 birds/ha; and in arable land it was 1.43 birds/ha.[1304] In Australia there are no measures of abundance.

Field Atlas Records: 5461

1° blocks–with records: 88 (11%)
–with breeding records: 37 (43%)

Atlas No.: 993

Max. Rep. Rate: 57%
In block(s): 40/144

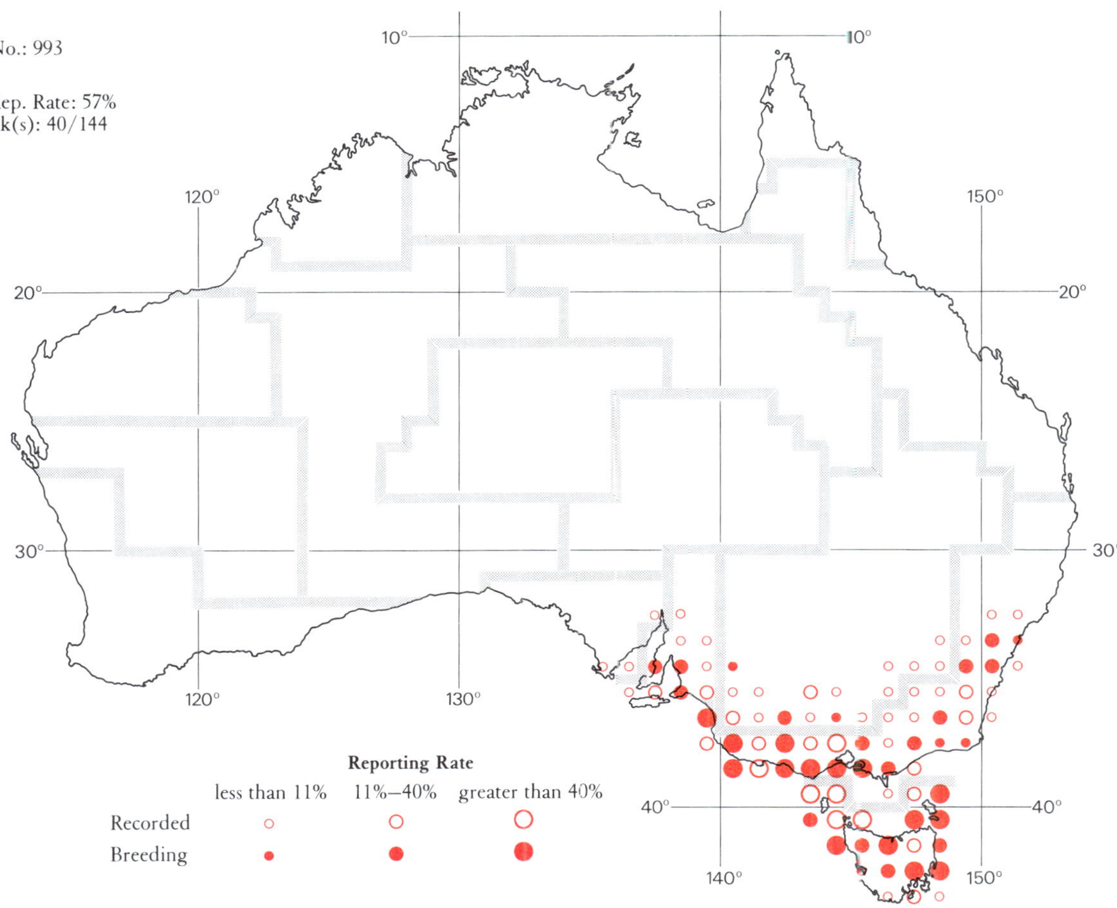

White-backed Swallow

Cheramoeca leucosternum

In eastern Australia the range of the endemic White-backed Swallow has expanded towards the coast since 1900. Before then the only record in the Atlases in the Eastern Qld and South-East Regions is of birds breeding at Narrabri (30/149) in 1896.[1343] In the north, where the species' range appears unchanged, it is an occasional non-breeding visitor to the Kimberley and Rolling Downs Regions. In the South-West there is some evidence of it extending south along the Swan coastal plain (32/115)[281,1638] and west along the southern coast to at least Denmark (34/117).[478]

There are no observations of large-scale movement of the White-backed Swallow. In the southern Murray-Darling Region individual colonies may be deserted after the breeding season[833] but the summer (7.3%) and winter (4.6%) Reporting Rates in the region as a whole differed little, which suggests that many birds remain throughout the year. On the edges of its range the Swallow may be present only intermittently, as at Mt Isa (20/139),[862] so that some movements do take place; the longest recorded is that of a banded adult from Moonta (34/137) to Angle Vale (34/138), 121 km.[85]

The White-backed Swallow nests either singly or colonially in sandy banks[1343] and was found in the early days of European settlement in the South-West in burrows of bandicoots and rat-kangaroos.[723] One colony in an old quarry contained about 80 birds but there is no evidence of communal breeding. The Swallow feeds by hawking along watercourses and other areas for flying insects. It has been reported to roost communally at night and 27 birds have been found in one burrow.[1819] The species can become torpid in winter, groups of birds clustering at the end of a roosting tunnel.[391,1630] One observer suggests that it may demonstrate territorial behaviour, a family apparently driving away strange conspecifics from a hunting area.[517]

There are no measures of abundance.

Field Atlas Records: 3479

1° blocks–with records: 360 (44%)
–with breeding records: 65 (18%)

Atlas No.: 358

Max. Rep. Rate: 52%
In block(s): 21/122

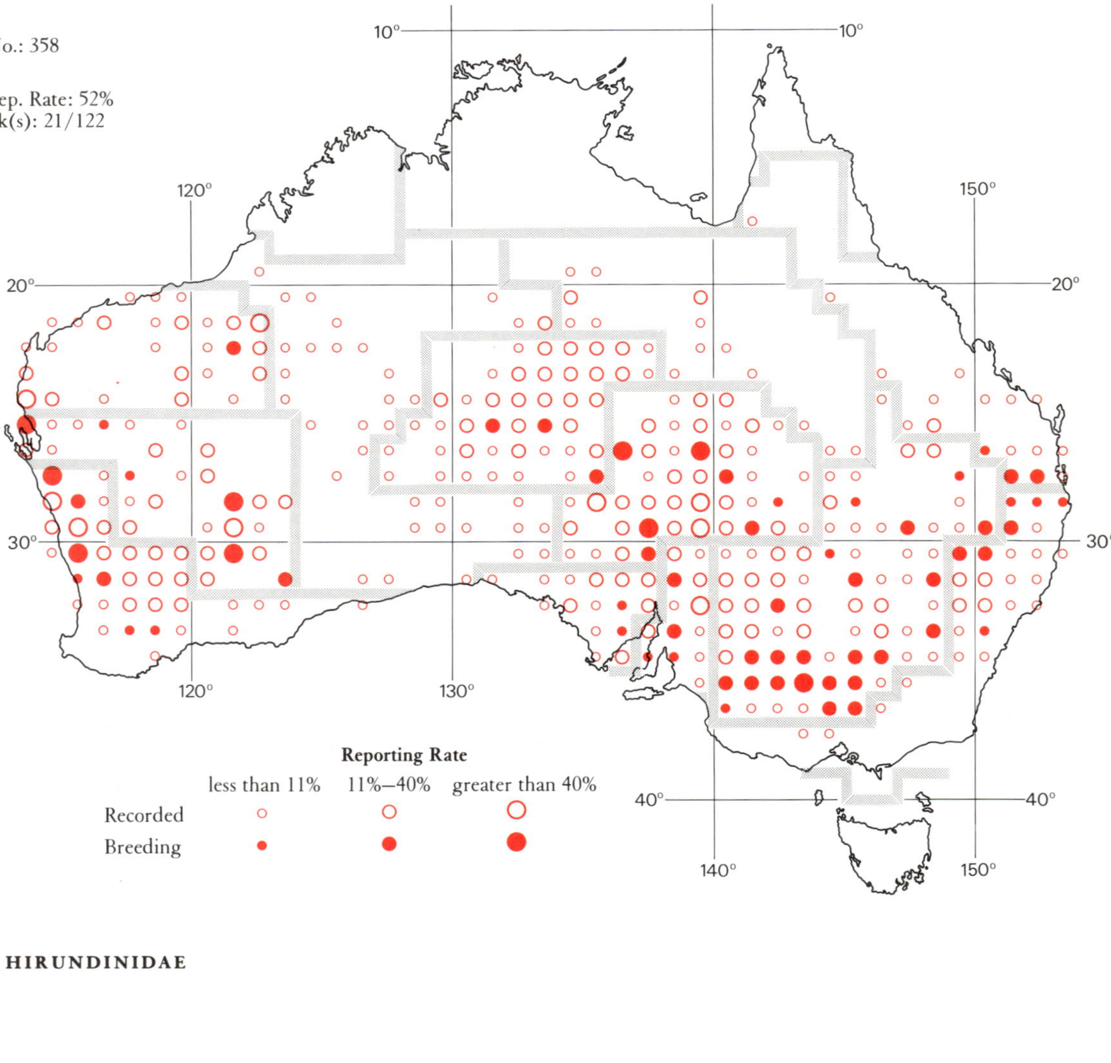

Barn Swallow

Hirundo rustica

One hundred years separate the first two records of the Barn Swallow in Australia. The first was in Torres Strait in 1860[723] and the next at Derby (17/123) in 1960,[1671] though a record on Moa I. (10/142) in Dec. 1919 is probably of this species.[503] The Barn Swallow breeds across North America, Europe, northern Africa and Asia. The subspecies *gutturalis* breeds in eastern Asia south to China and spends the non-breeding season in south-eastern Asia.[1173,1848] It is now recognized as a regular migrant to northern Australia.

Though the Barn Swallow is a regular visitor to the western Kimberley Region and near Innisfail (17/146)[1006] there are no reports in the Atlases from northern Australia between 132°E and 145°E. The lack probably partly reflects the few observers there in summer but even in Darwin (12/130) only occasional sightings of a few birds are reported.[1755] In southern Australia (south of about 25°S) the species is a vagrant and only single birds have been recorded including one at Koolunga (33/138) in Mar.-Apr. 1977 and one at McGrath's Hill (33/150) in Dec. 1981.

The Barn Swallow lives near water usually in open country and, in towns, is often seen perched on overhead wires. Frequently it is in company with other swallows, martins or swifts. Flocks as large as 300 have been seen at Derby and at least 220 were estimated to be near Innisfail in Dec. 1975.[1005] The Barn Swallow feeds on insects taken in the air.

There is a suggestion that the arrival of the Barn Swallow in Australia is recent and that increasing numbers will visit New Guinea and Australia in future.[549] In South Africa such a change was recorded in 1962–63 when the wintering range of the Barn Swallow suddenly increased by several hundred kilometres southward and eastward,[1237] possibly because the weather was abnormal.[1539]

There are no estimates of abundance in Australia.

Field Atlas Records: 30

1° blocks–with records: 12 (1%)
–with breeding records:

Atlas No.: 879

Max. Rep. Rate: 4%
In block(s): 15/128

Welcome Swallow

Hirundo neoxena

From Australia the Welcome Swallow spread naturally to New Zealand where it was first recorded in the 1920s and found breeding in 1958.[1254] In 1980 it was first recorded in New Guinea.[550] Confusion with the Barn Swallow makes some early records uncertain but the range of the Welcome Swallow seems not to have increased greatly in Australia. It was not reported in the Top End before the Field Atlas but there were previous reports in the Kimberley Region. Breeding has not been reported in the Kimberley, Top End or Cape York Regions.

Populations in eastern Australia undertake a partial migration. In Tas. Swallows arrive Aug.-Sept. and depart Feb.-Mar.[1368,1738] Many pass over King I. (39/144) on their northward passage[1737] and flocks of 60–100 fly over Swan I. (40/148) in Oct. and Apr. Reporting Rates in the Tas. Region (summer 56.2%, winter 20.0%) suggest that the migration is not complete. No exodus from southern areas of the mainland has been observed nor is it suggested by the Field Atlas but there is evidence of movement. There are flocks of up to 2000 migrating birds at Laverton (37/144) in Feb.[1822] and an autumn-winter passage is reported in the southern Murray-Darling Region.[833] In Sydney (33/151) banding showed that different individuals were present in winter than in summer.[1026] Farther north near Inverell (29/151) more birds are present in winter and at Murphy's Ck (27/152) fewer.[96,1099] In the Atherton Region the winter Reporting Rate (26.1%) was slightly higher than in summer (18.6%) and in Torres Strait the Welcome Swallow is known as a winter visitor.[503] The longest movement recorded for a banded bird is from St Kilda (34/138) to Kimba (33/136), 270 km.[25] Western populations appear not to undertake extensive movements though at Northam (31/116) large flocks form in Feb.[1216,1622]

The Welcome Swallow breeds singly or in loose colonies, e.g. there were 20–25 nests under one jetty near Rotamah I. Bird Observatory (37/147) and 500 birds occupied a deserted farmhouse near Moonta (34/137). Before European settlement it nested in such places as cliffs and hollow trees. Now it uses buildings, bridges and the eaves of houses and may consequently have increased in numbers and range. It feeds on insects caught in the air, specially near water.

There are no measures of abundance.

Field Atlas Records: 39 074

1° blocks–with records: 491 (60%)
–with breeding records: 254 (52%)

Atlas No.: 357

Max. Rep. Rate: 80%
In block(s): 35/147

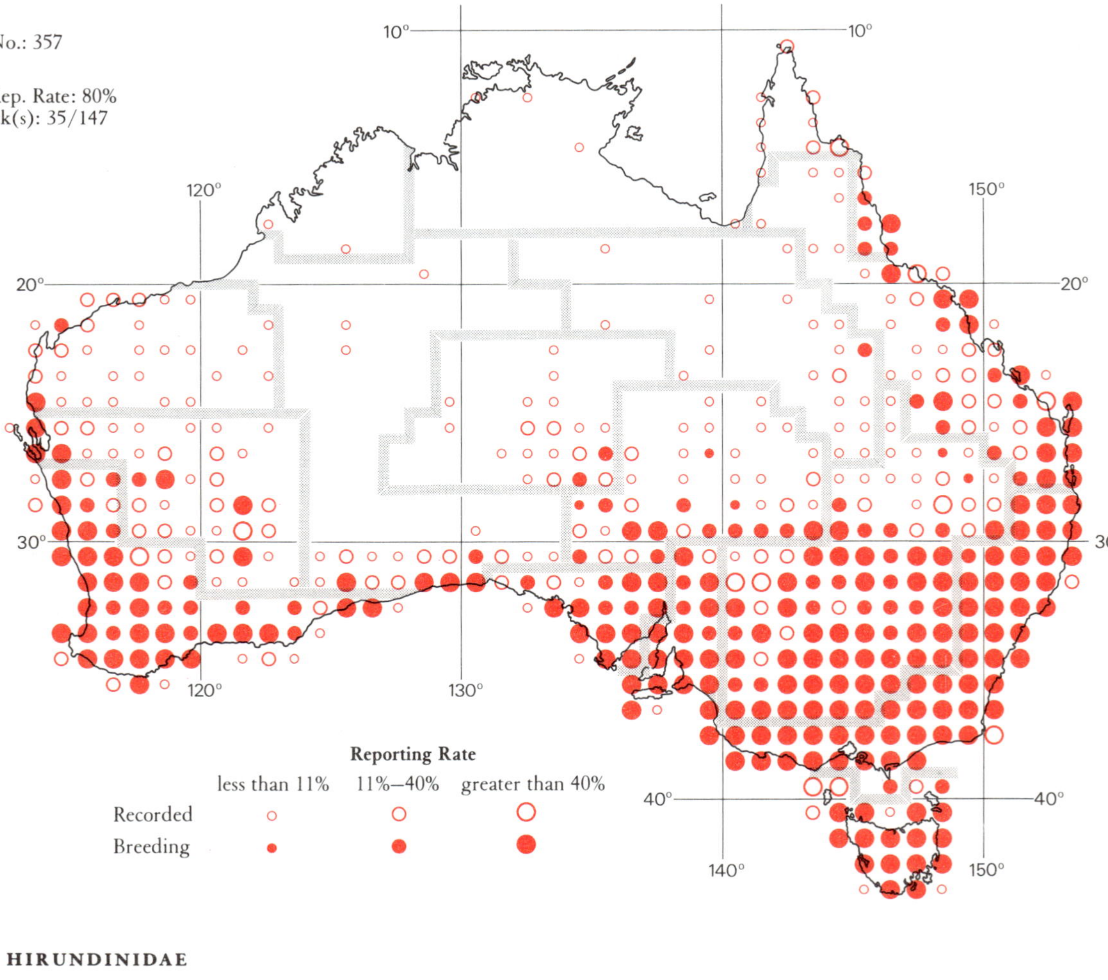

Tree Martin

Cecropis nigricans

The Tree Martin lives in southern Indonesia, New Caledonia and Australia. Australian birds migrate to New Guinea and adjacent islands.[1422,1428] Birds of northern and western Australia are ascribed to the subspecies *neglecta* and the remainder to *nigricans*. Most breeding takes place in southern Australia with only two breeding records north of 20°S in the Atlases, both in the Atherton Region.

Many Tree Martins from southern Australia migrate north in winter, some to New Guinea.[140] Observers in Torres Strait describe the species only as a winter visitor in small numbers[503] and it may be that migrating birds travel at high altitude. Their arrival in New Guinea is sudden and more are seen in the north than in the south in winter.[129] The Tree Martin is present all year near Innisfail (17/146)[698] but a migrant Dec.-Sept. above 1000 m altitude on the tableland.[231] Reporting Rates in the Atherton Region of 5.4% in summer and 11.7% in winter suggest an influx. At Coonabarabran (31/149) it was present for all of 1981.[396] Movement out of the South-East and Tas. Regions in winter is reflected in Reporting Rates of 20.3% and 31.3% respectively in summer and 10.6% and 2.4% in winter. Birds leave Tas. Feb.-Mar. and return in Aug., many travelling via Flinders I. (40/148).[1274,1534,1738] In Darwin (12/130) the Tree Martin is common May-Aug., with a few birds remaining in some summers.[423]

Near Perth (31/115) the Tree Martin is present all year though a movement north in autumn and south in spring has been noticed.[1622] The Field Atlas suggests movement out of the South-West Region in winter (Reporting Rates: summer 34.0%, winter 20.3%).

The Tree Martin lives near water. It will nest and roost in the same hollow trees year after year[1648] but is sufficiently adaptable to use buildings and bridges for nesting in Brisbane (27/153), Sydney (33/151), Adelaide (34/138) and Perth.[1648,1369,1638] Flocks sometimes make spectacular roosts, e.g. in reedbeds and river red gums; some during the non-breeding season have been estimated to contain hundreds or thousands of birds.[833,1216] The Tree Martin hawks for insects above the treetops and sometimes over the sea.[153] At Boola Boola (38/146) in gullies in eucalypt forest its density was 0.5 birds/ha.[1113]

Field Atlas Records: 16 187

1° blocks–with records: 584 (72%)
–with breeding records: 198 (34%)

Atlas No.: 359

Max. Rep. Rate: 50%
In block(s): 27/140

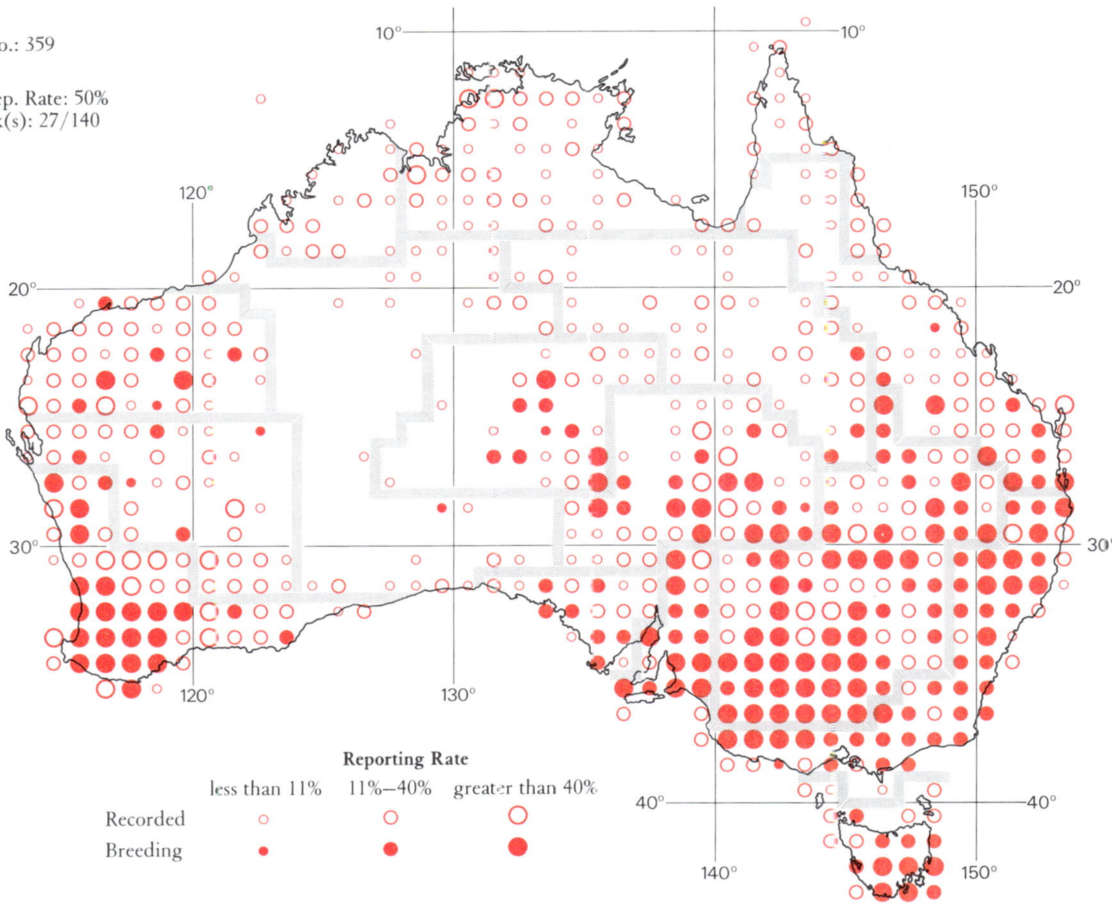

Fairy Martin

Cecropis ariel

The Fairy Martin occasionally straggles to New Guinea and is also a vagrant in Torres Strait.[503] Breeding is not reported from the Cape York Region nor from the Top End north of 14°S. There were breeding records from Tas. last century[299] and an unoccupied nest was found in 1968.[1796] The few other Tas. records are of vagrants in spring and summer.

A north-south pattern of migration is observed in eastern Australia. At Murphy's Ck (27/152) the Fairy Martin is common in summer, gathering into large flocks in autumn to leave in Apr., though a few remain for the winter.[1099] In the southern Murray-Darling Region it arrives in Aug. and departs Jan.-Mar.[833] Reporting Rates increased in winter in the Cape York and Atherton Regions and round Darwin (12/130) (summer: 2.5%; winter 11.1%) matched by a decrease in the South-East Region (summer: 17.2%; winter 6.3%). Reporting Rates suggest no such regular pattern of movement in the South-West Region and none has been recorded.[1216,1638] The longest movement recorded is that of an adult from 34/140 near Waikerie to Salters Springs (34/138), 129 km.[1437]

The Fairy Martin lives in open country, often near water. It nests colonially. One colony under a bridge at Bedourie (24/139) had 700 nests. Natural nest sites are under overhanging cliffs and rocks and in tree trunks.[1552] Artificial sites include bridges, buildings and culverts. Only culverts with a minimum clearance of 0.4 m above the soil or water contained nests in one survey.[1480] Banding of individuals at a colony in a culvert near Singleton (32/151) showed that adults returned to breed there again in a subsequent year.[1024] Site tenacity was great also in the Goulburn Valley (32/150)[1035] and at Moruya (35/150)[1192] but at Northam (31/116) Fairy Martins apparently seldom used the same site in successive years.[1216]

The Fairy Martin feeds on insects. It is subject to competition for its nest from House Sparrows, pardalotes and Zebra Finches and to predation by kingfishers, which break open the nest to get at the contents.[653]

There are no measures of abundance.

Field Atlas Records: 10 425

1° blocks–with records: 529 (65%)
–with breeding records: 267 (50%)

Atlas No.: 360

Max. Rep. Rate: 61%
In block(s): 23/144

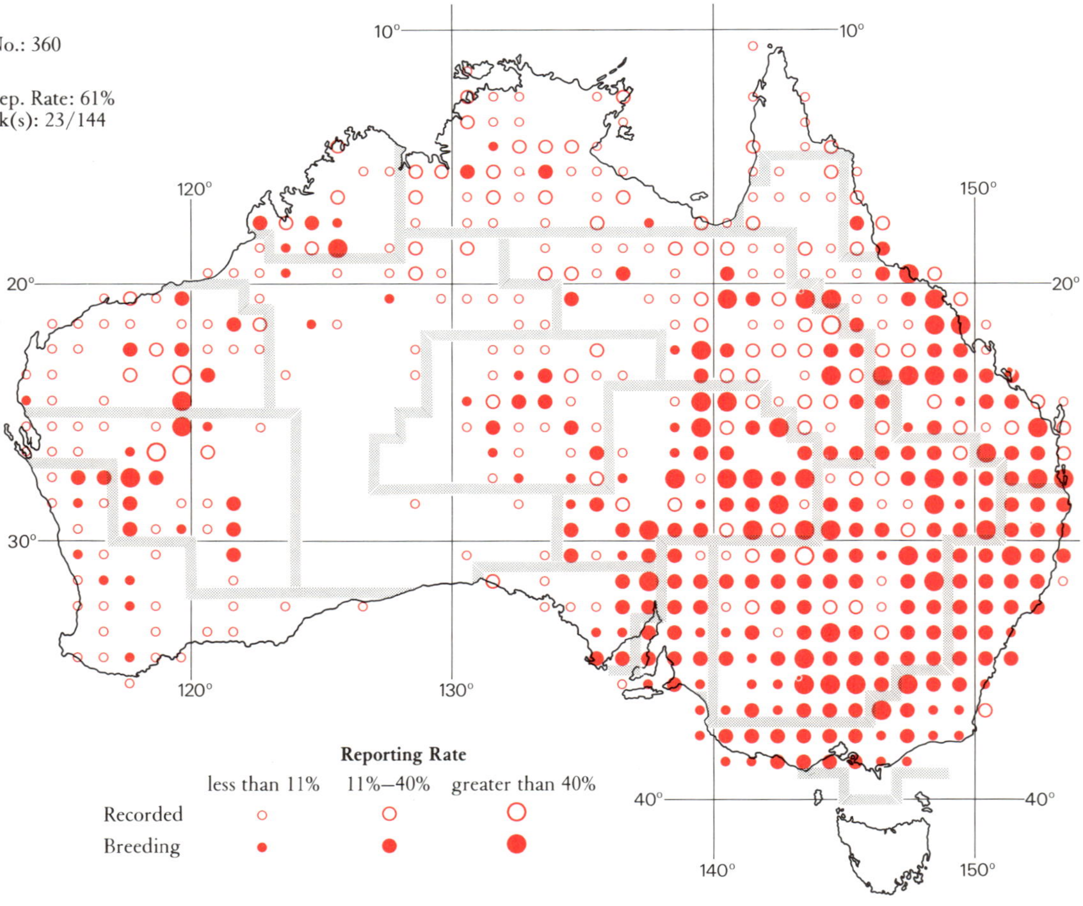

Richard's Pipit

Anthus novaeseelandiae

Richard's Pipit ranges from Africa, Europe and Asia to New Guinea and Australia. The subspecies in Australia are under review. The Field Atlas does not indicate any separation between populations on the mainland.

Near Nareen (37/141) birds are present all year but form flocks in winter.[1126,1127] In Tas. some redistribution of the population takes place, birds becoming more common on beaches in winter though not in flocks.[1533] There have also been observations suggesting migration by part of the population.[469,497,740] Reporting Rates support the suggestion that some birds leave the Tas. Region in winter (summer 26.5%, winter 10%). On the adjacent mainland they changed little (South-East south of 35°S: summer 34.9%, winter 29.3%). Single birds appeared on the coast near Melbourne (37/145) June-Oct.[1476] At Eyre Bird Observatory (32/126), where the Pipit is rare, large numbers arrived in Jan. 1980. In the north the species is reported to be present all year near Darwin (12/130)[423] but a winter visitor to the Kimberley Region.[1713]

Richard's Pipit lives in clearings and pasture, including gibber plains in the inland, beaches and their backing dunes and bare or cleared areas in woodland or forest. It rapidly invades newly cleared areas.[189,212,1622] It is abundant on the high plains of the Great Dividing Ra.; Reporting Rates in the Snowy Mts (36/148) were 47.0% above 1000 m and 15.6% below. The Pipit has been observed feeding over and round the edge of snow patches[1401] but is inactive in wet weather.[1126,1127] Flocks of up to 100 birds form in the non-breeding season[1127] congregating round a scarce resource, e.g. food and water at Mt Mary Plains (34/139),[189] or in shade in summer.[627] Richard's Pipit feeds chiefly on insects, catching them among vegetation on the ground and in crevices of logs and rocks.[1049,1222] One was observed feeding on crabs.[1226] It takes some seeds[748] and may damage young vegetables.[740]

In two successive years a breeding pair at Nareen repelled intruders from, and foraged within, an area of about 9 ha (0.22 birds/ha).[1126,1127] Density in pasture at Armidale (30/151) was 0.22 birds/ha[582] and on the Nullarbor Plain was 0.07 birds/ha.[246]

Field Atlas Records: 27 015

1° blocks–with records: 720 (89%)
–with breeding records: 202 (28%)

Atlas No.: 647

Max. Rep. Rate: 88%
In block(s): 31/125

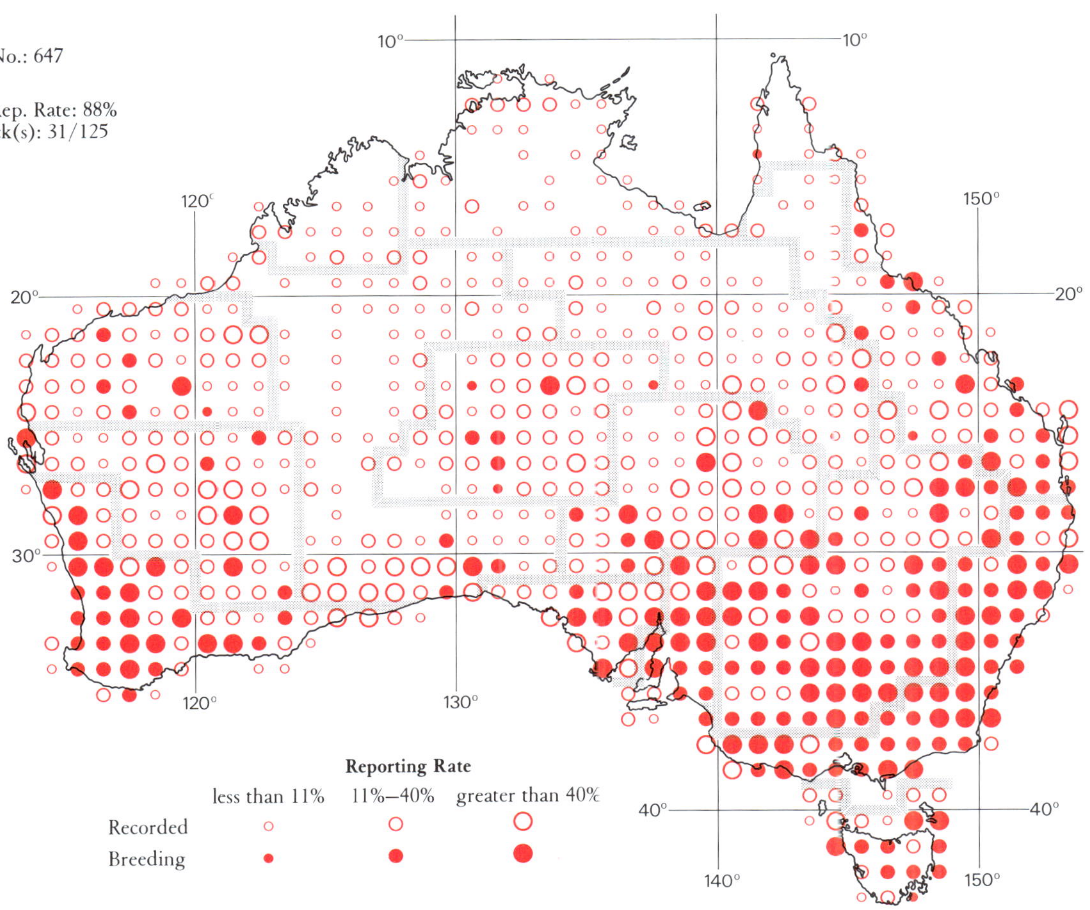

Yellow Wagtail

Motacilla flava

The Yellow Wagtail visits Australia in summer from breeding grounds in Asia and Alaska. It was first identified in Australia in June 1905 when a specimen was taken at Duaringa (23/149). Subsequent records have come from Innisfail (17/146),[698] Heron I. (23/151), Fogg Dam (12/131), Broome (17/122), Mt Goldsworthy (20/119) and Shay Gap (20/120)[948] in addition to the localities on the map. The species is a regular visitor to northern Australia, occasionally straying farther south. Southern vagrants recorded during the Field Atlas included one bird at Baker's Lagoon (33/150) in Apr. 1979 and one at Herdsman L. (31/115) in Feb. 1981.

Specimens taken at Harrison Dam (12/131) and Duaringa are of the subspecies *tschutschensis* that breeds in the north-east of the USSR and in Alaska; that taken at Kununurra (15/128) is of the subspecies *simillima* that breeds in eastern Russia and Korea. This subspecies and another, *taivana*, not yet recorded in Australia,[948] spend the non-breeding season mainly in south-eastern Asia. The main non-breeding grounds of *tschutschensis* are unknown.[948]

Most Australian sightings have been Nov.-Feb. In eastern Australia a few records have been made outside this period: two in June and one in Apr.[948]

The Yellow Wagtail lives in wet places such as where shallow water lies at the edges of swamps, in irrigated crops and along bore drains. Most sightings have been of three or fewer birds but at Kununurra a flock of 50 was seen in Feb. 1980.[948] The Yellow Wagtail feeds on small insects caught on the ground and in the air.

Many more sightings have been made since 1960 than earlier. The increasing number of ornithologists visiting northern areas during the wet summer season has probably contributed to this.

There are no measures of abundance.

Field Atlas Records: 19

1° blocks–with records: 8 (1%)

–with breeding records:

Atlas No.: 877

Max. Rep. Rate: 4%

In block(s): 12/130,17/123

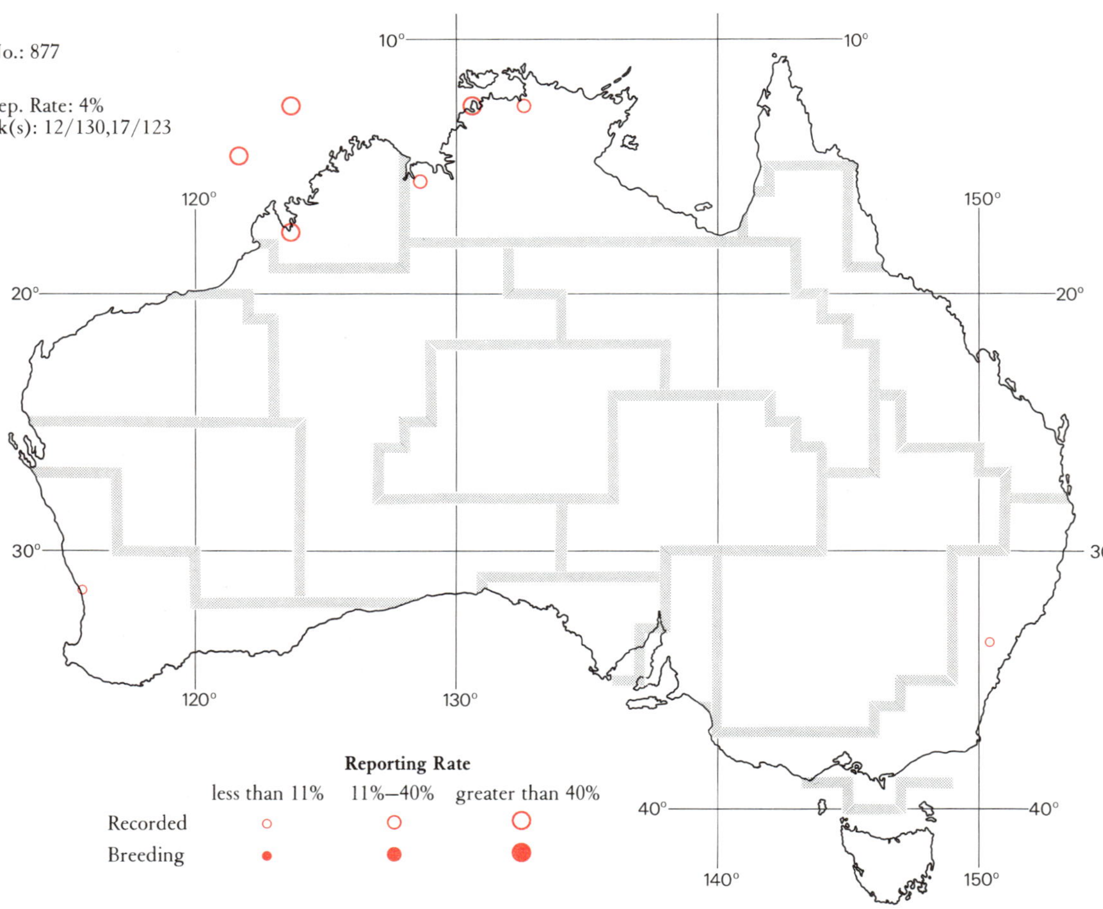

Black-faced Cuckoo-shrike

Coracina novaehollandiae

The Black-faced Cuckoo-shrike is distributed from Australia to New Guinea, south-eastern Asia and India. There are different views about subspecies in Australia.[748,985,1239] The Black-faced Cuckoo-shrike was the fourth most widely reported bird in the Field Atlas.

In eastern Australia the Cuckoo-shrike makes seasonal movements, part of the population migrating north each winter. At least part of the Australian population migrates to New Guinea. At Bensbach (9S/141E) the species is thought to be present only in the dry winter season and birds were reported massing for migration in Oct.[550] Possibly, the Cuckoo-shrike flies directly across the Coral Sea.[158] No large movements have been observed on islands in Torres Strait[503] but gatherings are reported on Cape York Pen. (10/142) in winter.[1145] In the Atherton Region the local population is augmented in winter by migrants.[231] There are records of movements in autumn and spring from Inverell (29/151),[96] the southern Murray-Darling Region[833] and Hobart (42/147).[1452] A few birds winter in these places but at Wollomombi (30/152), Canberra (35/149) and Jamieson (37/146) the bird is present only Sept.-Apr. each year.[129,154,1469] Reporting Rates were higher in winter in the Cape York and Atherton Regions and lower in summer (Cape York: summer 13%, winter 44%; Atherton: summer 21%, winter 34%). The reverse was true in the South-East and Tas. Regions (South-East: summer 45%, winter 35%; Tas.: summer 30%, winter 8%). Because winter Reporting Rates were high the migration is probably only partial. In Darwin (12/130) the Cuckoo-shrike is absent during the wet summer season.[423] No evidence of movement is shown in the west.

The Cuckoo-shrike inhabits all types of country. It feeds on grasshoppers, caterpillars and beetles, vegetable matter and seeds.[748,778]

Densities in eucalypt woodland were: Armidale (30/151) 0.03–0.25 birds/ha;[582] Wollomombi 0.2 birds/ha;[19] Canberra 0.2 birds/ha;[154] and Moruya (35/150) 0.03 birds/ ha.[1197] Densities in eucalypt forest were: Jamieson 0.5–1.0 birds/ha[1469] and Boola Boola (38/146) 0.01–0.53 birds/ha.[1113]

Field Atlas Records: 37 775

1° blocks–with records: 756 (93%)
–with breeding records: 265 (35%)

Atlas No.: 424

Max. Rep. Rate: 81%
In block(s): 19/145,27/149

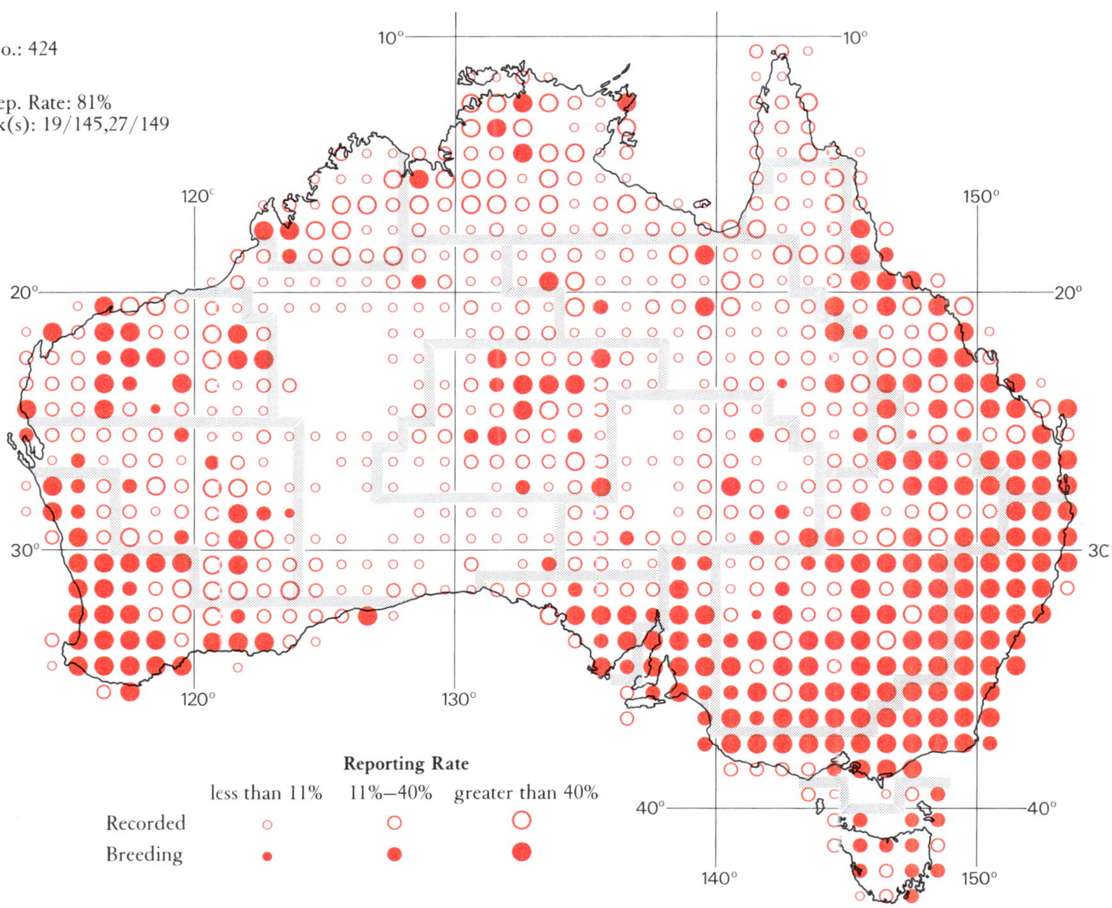

White-bellied Cuckoo-shrike

Coracina papuensis

The White-bellied Cuckoo-shrike ranges through New Guinea and adjacent islands to Australia. Three subspecies have been recognized in Australia: *hypoleuca* west of the Cape York Region; *stalkeri* in the Cape York and Atherton Regions: and *robusta* to the south.[680,748] The morphological differences between these subspecies are complex and the only discontinuity in eastern Australia is at the Burdekin R (19/147).[680] The gap in distribution south of the eastern end of the Gulf of Carpentaria (141°E) is consistent in the Historical Atlases except for records from Cabanda (20/141) in 1970.[150] In the south the White-bellied Cuckoo-shrike is only an occasional visitor to the SA Ra. Region.

It is locally nomadic during the dry winter season on islands throughout Torres Strait,[503] though previously thought to be sedentary in the north.[680] The subspecies *robusta* is probably a spring-summer visitor to south-eastern SA.[912] Other movements have been reported[680] but not on a large scale or with any regularity and the Field Atlas shows no seasonal changes in Reporting Rate.

The White-bellied Cuckoo-shrike lives in many types of vegetation. Throughout the tropics, it is abundant in urban areas.[129] It is present in Qld in eucalypt forest and rainforest as well as in woodland, mangroves and sub-urban gardens; across the north it is in eucalypt woodland with a grassy understorey; and in the north-west near rivers and in gorges. The White-bellied Cuckoo-shrike feeds on large insects and fruit. Clearing in the Wau Valley (7S/146E) of New Guinea has been followed by a great increase in abundance.[126a] There are no similar observations in Australia.

In New Guinea density near Port Moresby (9S/147E) in eucalypt woodland was 0.08–0.19 birds/ha.[158]

Field Atlas Records: 3914

1° blocks–with records: 261 (32%)
–with breeding records: 42 (16%)

Atlas No.: 425

Max. Rep. Rate: 68%
In block(s): 15/145

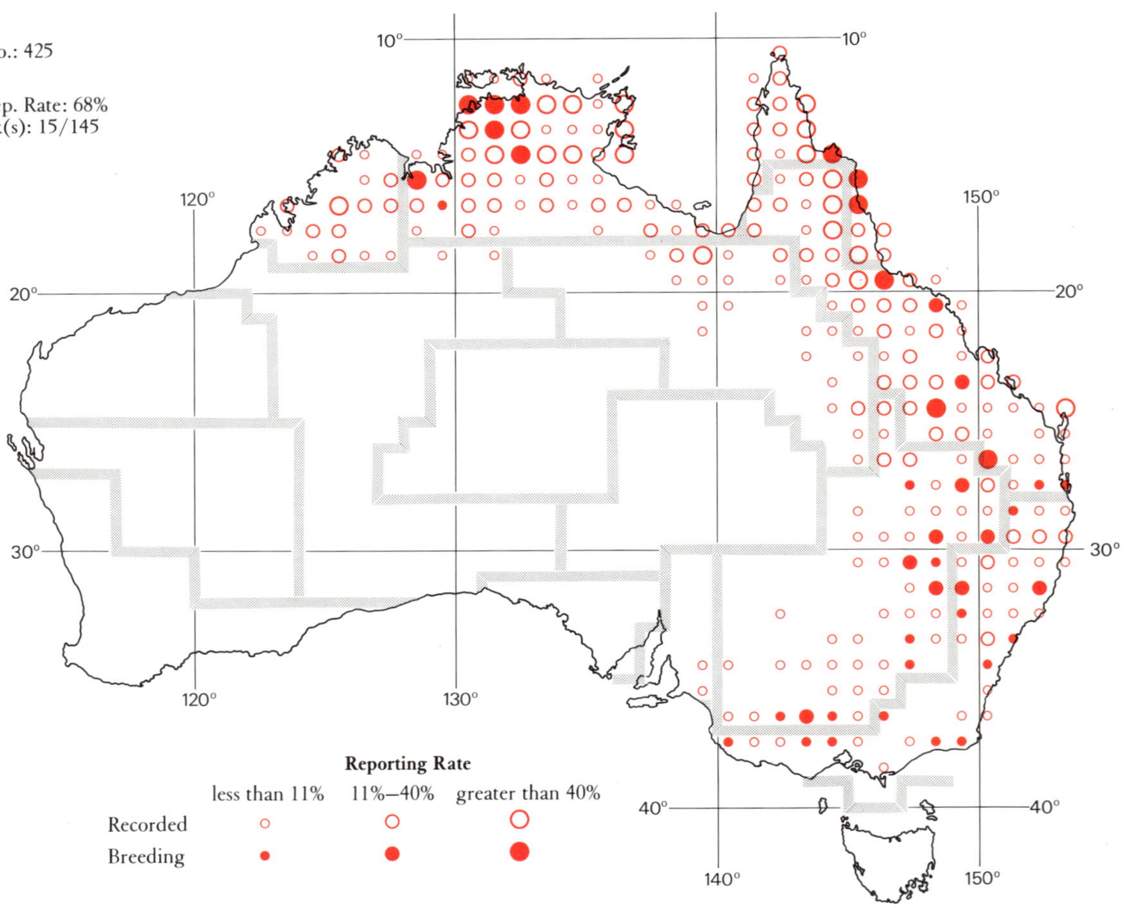

Yellow-eyed Cuckoo-shrike

Coracina lineata

Two subspecies of the Yellow-eyed Cuckoo-shrike live in New Guinea and adjacent islands; another, *lineata*, lives in Australia.[1449] The disjunct distribution shown in the Field Atlas, with the Cape York Region and Atherton populations isolated, is also evident in the historical maps, where there were records north to the tip of Cape York Pen. (10/142). Breeding has been reported only from the Atherton Region and south of 22°S.

The Yellow-eyed Cuckoo-shrike is reported to be present all year near Innisfail (17/146)[698] but has also been considered locally nomadic moving in search of fruit.[985] Small parties have been reported from the Murray R (18/145) moving south in the spring.[303]

Reporting Rates were generally much lower in winter than in summer (Atherton Region: summer 16%, winter 5%; Eastern Qld: summer 1%, winter 0.1%; South-East Region: summer 0.13%, winter 0.03%). There are no records of the species from Torres Strait and only one, of *lineata*, from New Guinea;[1449] so, it seems unlikely that a regular migration takes birds out of the country. The difference in Reporting Rates is presumably a result of decreased conspicuousness in the non-breeding season.

The Yellow-eyed Cuckoo-shrike inhabits rainforest and eucalypt forest, including margins and regrowth.[698,1703] It lives in pairs or parties of up to 50 individuals. The parties roost together in the non-breeding season; 36 have been recorded roosting in one tree.[698] Despite this sociability, communal breeding has not been reported. The Yellow-eyed Cuckoo-shrike feeds on fruit and insects.[748,1147,1732]

There are no estimates of abundance.

Field Atlas Records: 280

1° blocks–with records: 30 (4%)
–with breeding records: 5 (17%)

Atlas No.: 428

Max. Rep. Rate: 33%
In block(s): 15/145

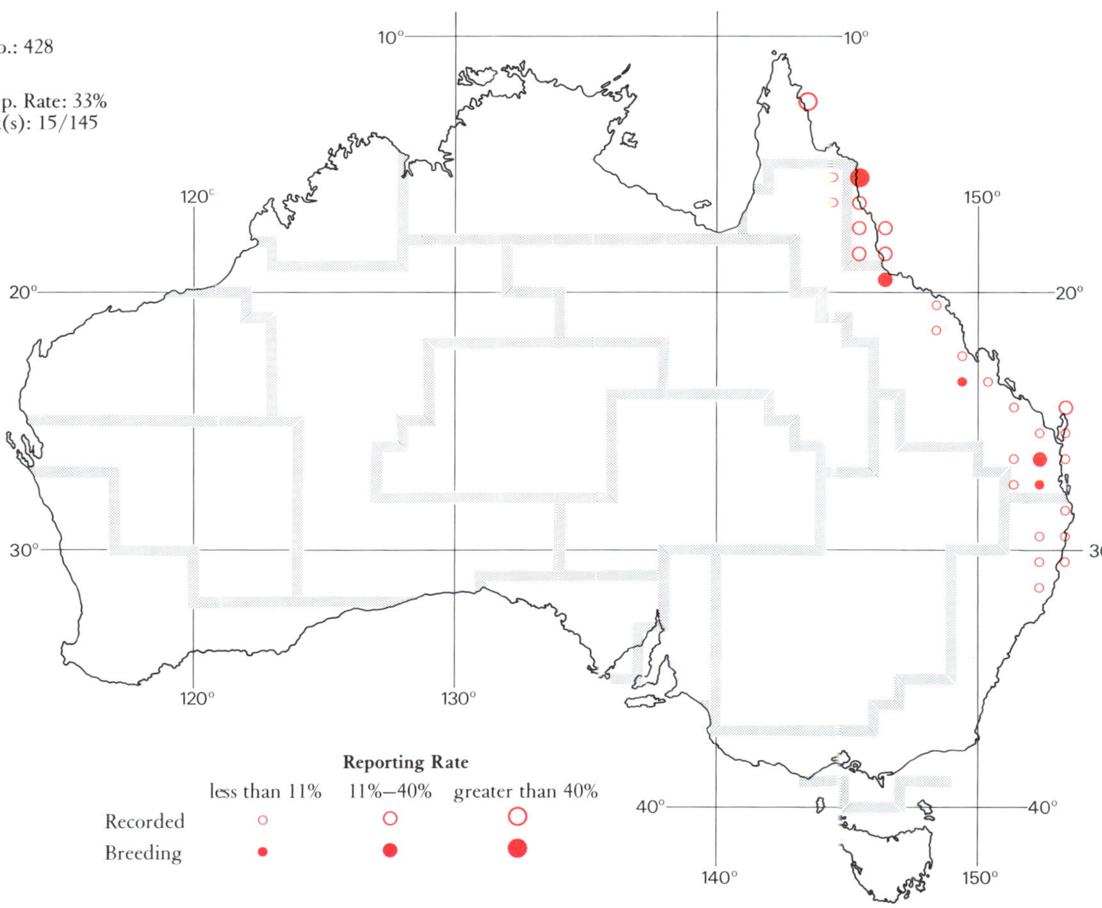

Cicadabird

Coracina tenuirostris

The Cicadabird has a wide distribution through the south-western Pacific, New Guinea and adjacent islands and Australia. There are different views about Australian subspecies.[748,985,1257,1709] The Atlases show consistent gaps between records in the Cape York Region and Top End and between the Top End and Kimberley Regions.

In the Top End, where historical records fill the gap in the Field Atlas map at 133–135°E, the Cicadabird is more widely distributed than in the Kimberley Region where it is known only between the Mitchell Plateau (14/125) and Napier Broome Bay (14/126). Before 1950 the Cicadabird bred as far south and west as 38/145 on the outskirts of Melbourne[299] but it is now a vagrant there. The record (35/138) in the SA Ra. Region was made in Adelaide in Nov. 1978.

The southern part of the eastern population of the Cicadabird is migratory. In winter the species moves through the western islands of Torres Strait[503] and Australian birds are recorded in New Guinea.[158] Dates of arrival at places in Australia are: Innisfail (17/146) 20 Aug. to 12 Oct.;[698] Urangan (25/152) 2 Sep. to 2 Nov.;[361] Wollomombi (30/152) late Oct.;[129] Sydney (33/151) regularly about 4 Oct.;[692] Moruya (35/150) 4–28 Oct.[1196] Last sighting dates at Moruya are usually from 10–24 Mar.[1196] but in dry years as early as Jan.[1192]

In the Top End the Cicadabird is reported throughout the year though easily overlooked.[423] In the Kimberleys it is said to be confined to rainforest in the dry winter season, dispersing to eucalypt forest in the wet summer season.[1713]

In northern Australia the Cicadabird lives in rainforest along creeks and watercourses, in paperbark and eucalypt woodland and at the inner edge of mangroves. In the South-East it is in eucalypt forest and woodland.[653] The Cicadabird feeds on large insects collected from the leaves and bark.[1196]

Density in eucalypt woodland at Wollomombi was 0.3–0.5 birds/ha[129] and at Moruya 0.08–0.11 birds/ha.[1197] At Boola Boola (38/146) in eucalypt forest it was 0.1 birds/ha.[1113]

Field Atlas Records: 1366

1° blocks–with records: 108 (13%)
–with breeding records: 18 (17%)

Atlas No.: 429

Max. Rep. Rate: 33%
In block(s): 15/145

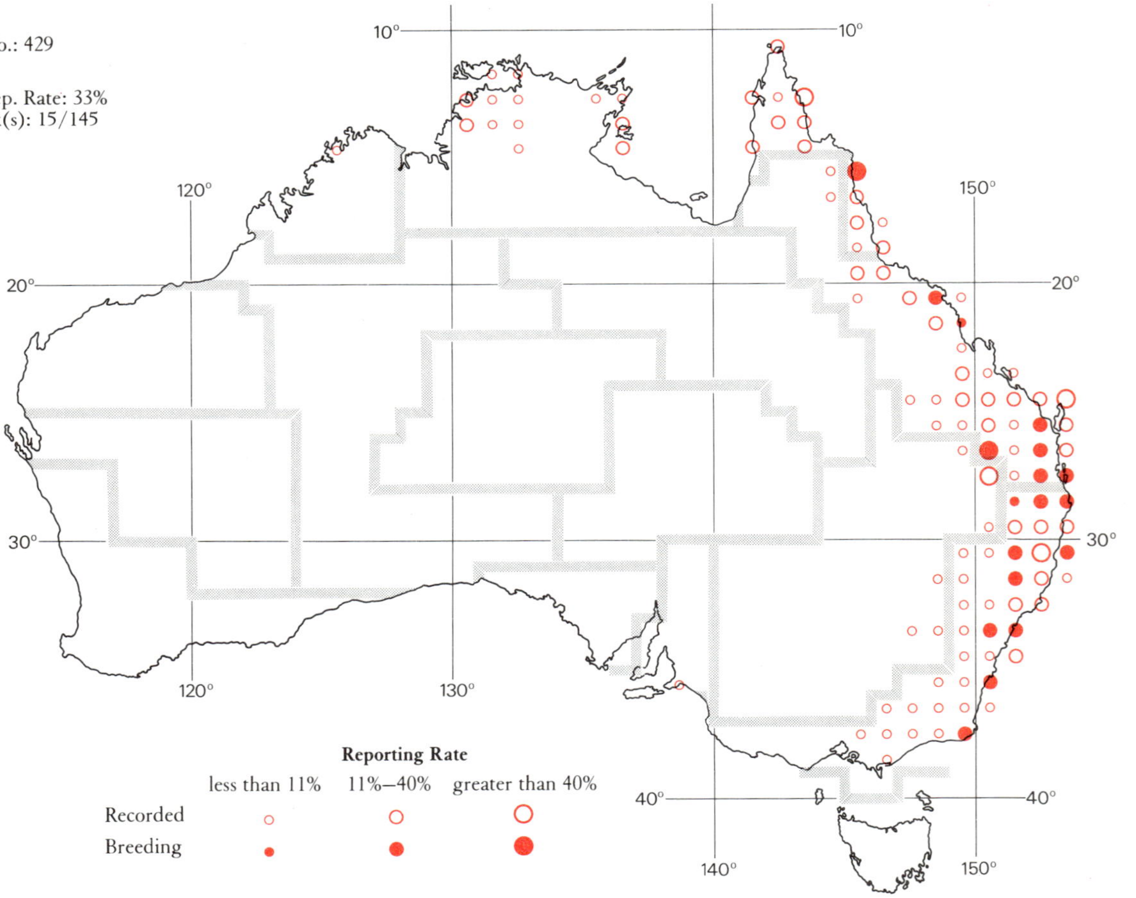

Ground Cuckoo-shrike

Coracina maxima

There are different views about subspecies of the endemic Ground Cuckoo-shrike.[748] The species has apparently moved into Eastern Qld since 1950. Until then the Atlases contain records from only four 1° blocks in the Eastern Qld and Atherton Regions. In one of these blocks, Murphy's Ck (27/152), it visited as a winter migrant for some years from 1935 before it bred.[1099]

The Field Atlas gives no indication of large-scale seasonal movement but in some places movements have been reported. In the southern Murray-Darling Region the Ground Cuckoo-shrike is an irregular visitor, not apparently influenced by local rain or dry spells, high or low temperatures.[833] At Coolup (32/115) winter flocks visit fairly regularly and some stay most of the year.[1622]

The Ground Cuckoo-shrike lives in small groups of three or more birds in eucalypt woodland, acacia scrub and spinifex, usually in open places. The bird breeds communally.[292,1547] It feeds on the ground on grasshoppers and other insects.[271,748]

There are no measures of abundance.

Field Atlas Records: 1489

1° blocks–with records: 322 (40%)

–with breeding records: 56 (17%)

Atlas No.: 423

Max. Rep. Rate: 35%
In block(s): 26/150

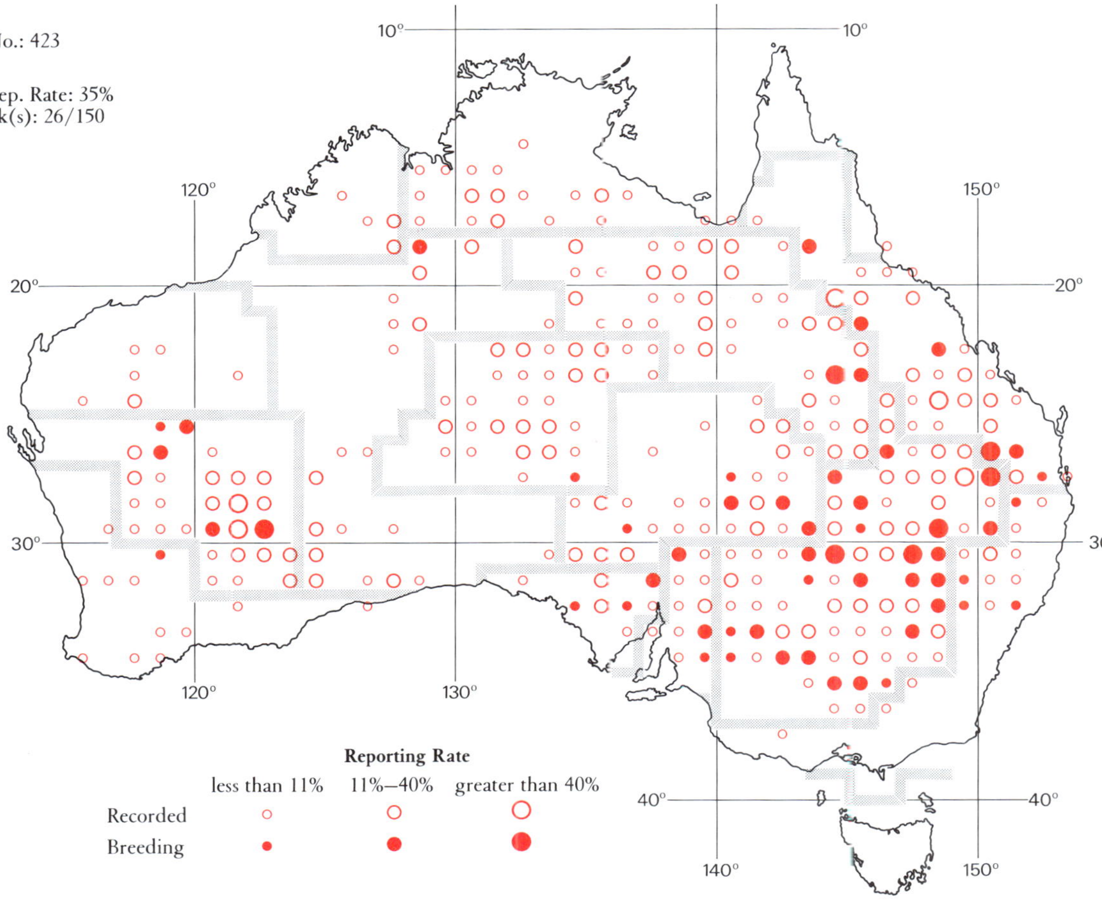

White-winged Triller

Lalage sueurii

The White-winged Triller lives in Indonesia and New Guinea with one subspecies, *tricolor*, in Australia. It is widespread except in the Tas. Region where there are only occasional records including one of breeding in 1968.[1426]

The Field Atlas shows the White-winged Triller to be recorded more frequently south of 26°S in summer (10.4%) than in winter (3.3%). There is only one record from Torres Strait[503] and the regular northward movement appears to take few birds out of Australia. There is a sedentary breeding population at Port Moresby (9S/147E) in New Guinea and occasionally Australian birds may reach there though no major migration is likely.[146,158] In Australia this Triller's appearance in many districts is irregular, despite the regularity of movement suggested by the Field Atlas. In the Murray-Darling Region in 1981 it was at its most abundant for ten years and in the same year near Sydney's Royal National Park it was reported for the first time in about 50 years.[396] Many birds bred at Jamieson (37/146) in 1968 but during the following years a few birds appeared only in 1973 and 1980.[1469] In good seasons the Triller can be common in desert areas such as the Nullarbor.[246]

The White-winged Triller lives in eucalypt woodland, mallee, acacia scrub and spinifex in wetter northern and drier southern regions.[748] It often nests in loose colonies.[138,1552] The Triller eats insects, including winged termites, ants, grasshoppers and caterpillars, and feeds both on the ground and in the air.

At Murphy's Ck (27/152) six pairs nested within 12 ha.[1093] At Macleod (37/145) four pairs, whose nests were no more than 20 m apart, occupied 50 ha of the surrounding woodland.[138] At Kununurra (15/128) 20 pairs nested in an area of a few sq. km.[899] Density in eucalypt woodland at Armidale (30/151) was 0.04 birds/ha.[582]

Field Atlas Records: 8738

1° blocks–with records: 676 (83%)

–with breeding records: 192 (28%)

Atlas No.: 430

Max. Rep. Rate: 52%
In block(s): 21/144,27/149

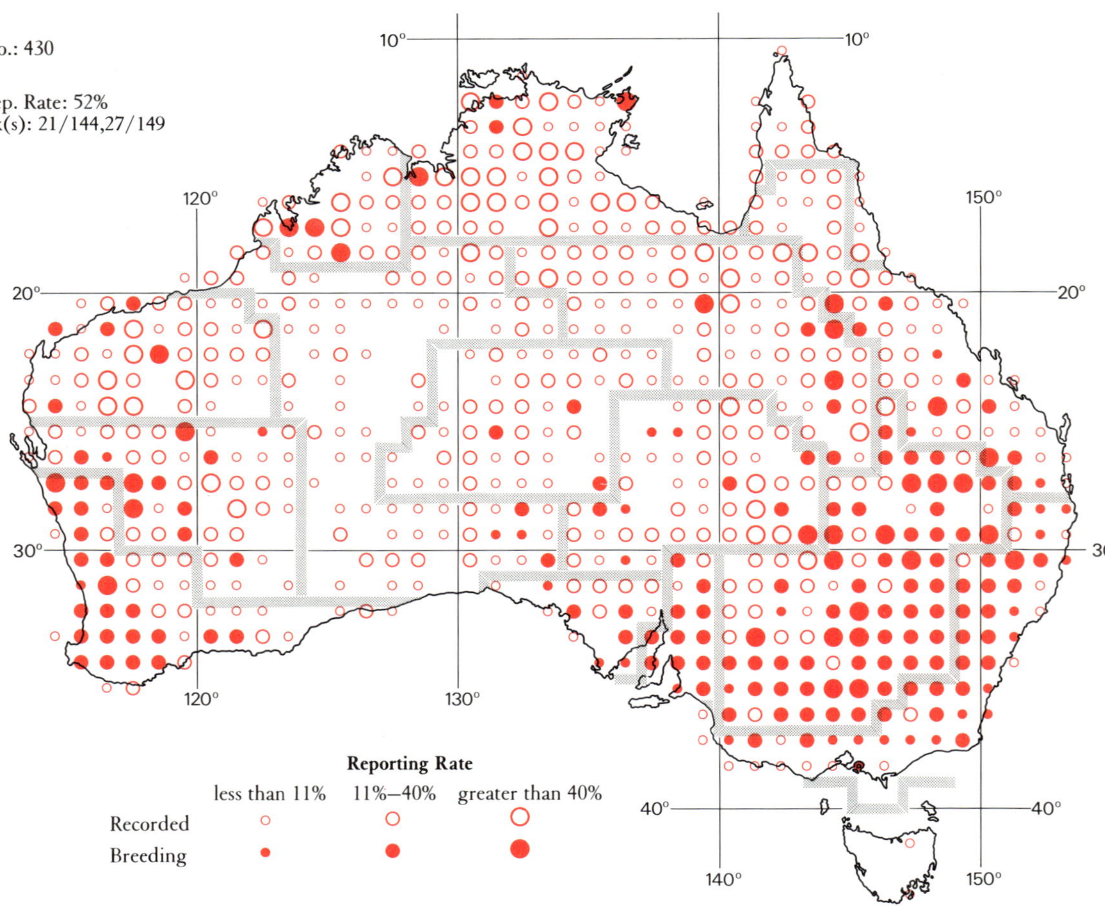

Varied Triller

Lalage leucomela

The Varied Triller lives in New Guinea and adjacent islands and Australia. In Australia it is separated into three subspecies: *yorki* in the Cape York Region; *leucomela* to the south; and *rufiventris* in the Top End and Kimberley Regions.[608,748,1713] An earlier review of the species suggested that the population of *leucomela* might have isolates within it[1229] but there is no evidence of any gaps in distribution on the 10′ maps. Historical ranges of the three subspecies are similar to those shown but include a record on the western coast of the Cape York Region at Inkerman (16/141)[748] and records of the Kimberley population as far north and east as Parry Harbour (13/126).[787]

The Varied Triller is present all year in the Atherton Region[231,698] but reported to decrease in numbers Nov.-Feb. near Rockhampton (23/150).[1089] The bird inhabits rainforest, eucalypt forest and woodland, paperbark woodland and mangroves adjacent to rainforest. In NSW it is absent above 850 m.[396] It lives in pairs, sometimes threes, and seldom associates with other species. In New Guinea in lowland rainforest, where it is always in pairs, it usually stays in the canopy but is lower in regrowth.[160] The population in the Atherton Region lives near lantana thickets as well as in rainforest along streams and that in the Kimberley Region occupies the bottom of sandstone gorges where vegetation is dense and figs abound. The Varied Triller feeds on insects, fruit and seeds but is particularly attracted to figs.[1713]

There are no measures of abundance.

Field Atlas Records: 1901

1° blocks–with records: 79 (10%)

–with breeding records: 19 (24%)

Atlas No.: 431

Max. Rep. Rate: 67%

In block(s): 20/149

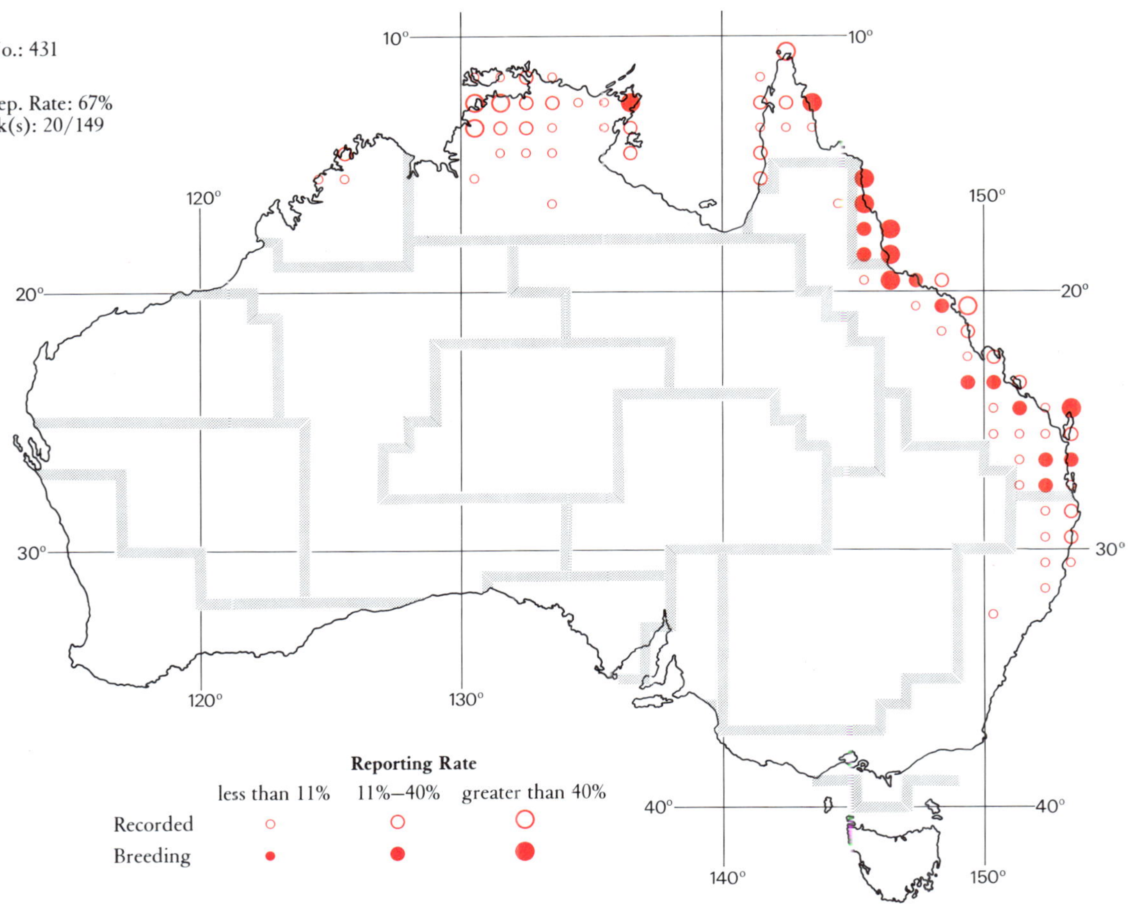

Red-whiskered Bulbul

Pyconotus jocosus

The Red-whiskered Bulbul, a native of south-eastern Asia, was introduced to Sydney (33/151) in 1880.[1085] It spread slowly but is now present on the coast from L. Macquarie (32/151) to the Shoalhaven R (35/150) and west to the Blue Mts (33/150). A second population, presumably derived from escapees, had become established round Coffs Harbour (30/153) by 1972[1292] and is expanding.

In Melbourne (37/145) the first definite record of the species was in 1915.[1085] It has lived in the Melbourne Botanic Gardens since then and occasionally appears in other suburbs. Few sightings have been made recently and this population may be dying out.

Other Field Atlas records in Deniliquin (35/144) and Morwell (38/146) and historical records in Adelaide (34/138)[1055] were of individual birds presumably escaped or released and are omitted from map.

The Red-whiskered Bulbul is present throughout the year in Sydney and Coff's Harbour.[1085] It is closely associated with human habitation and lives mainly in gardens and parks. In Sydney it occupies some native woodland, specially privet-infested gullies. The Bulbul is a nuisance in orchards where it eats fruit. It also eats berries of the introduced lantana,[326] widespread in coastal NSW, specially in the north, and this may enable the bird to extend its range away from settlement.

There are no measures of abundance.

Field Atlas Records: 1140

1° blocks–with records: 11 (1%)
–with breeding records: 4 (36%)

Atlas No.: 990

Max. Rep. Rate: 44%
In block(s): 33/151

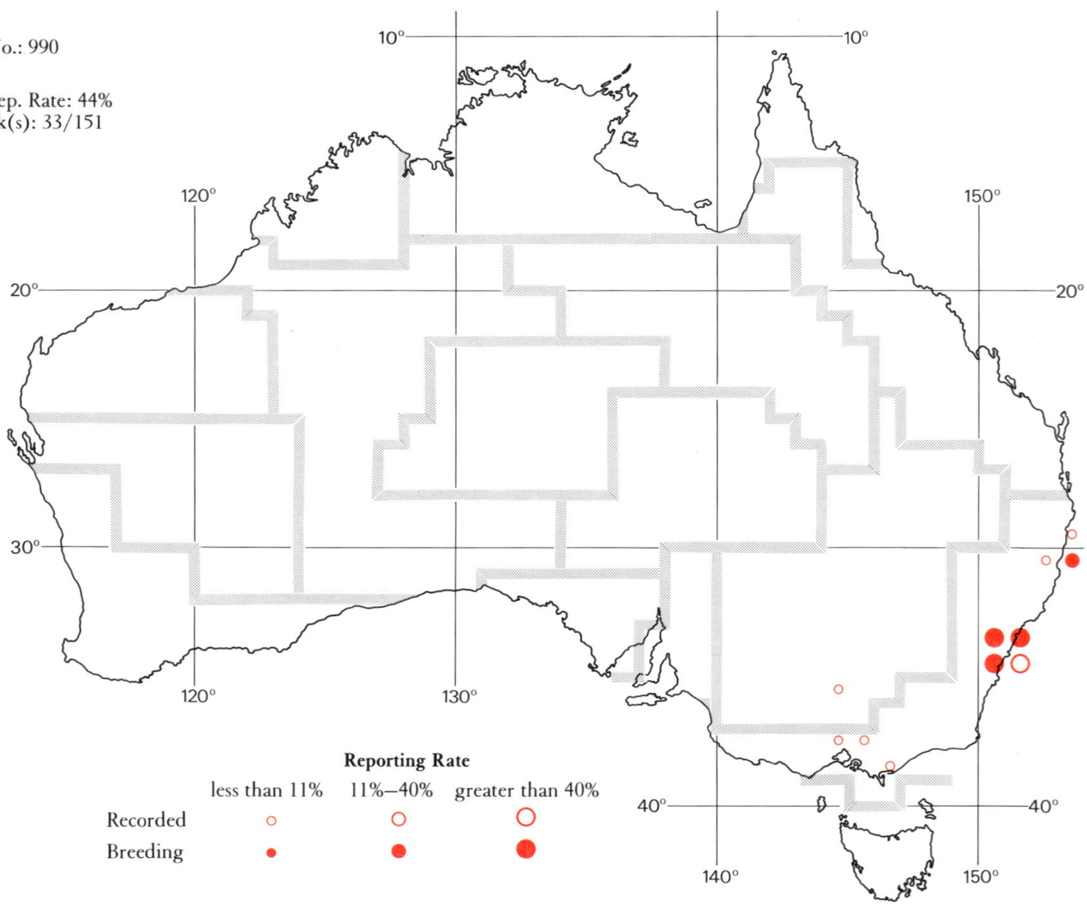

White's Thrush

Zoothera dauma

The range of White's Thrush includes Siberia, eastern Europe, south-eastern Asia, New Guinea and Australia. Recent taxonomic work shows that White's Thrush in Australia is divisible into two separate species.[624] One, *lunulata*, has an isolated population in the highlands of the Atherton Region and adjacent parts of Eastern Qld; its main distribution extends from the Bunya Ra. (26/151) south to Tas. and west to the SA Ra. Region, which also has an isolated population. The second species, *heinei*, extends from the Bowen R (20°S) south to Ourimbah (33/151). The two have ranges in the same latitudes in parts of the Eastern Qld and South-East Regions but on opposite sides of the Great Dividing Ra. In the Atlases there are no breeding records in the Eastern Qld Region between 19°S and 26°S and only scattered sightings.

The Field Atlas shows no evidence of large-scale seasonal movement but observations suggest that it may take place. At Moruya (35/150) White's Thrush is present only Apr.-Oct. to breed.[1197] At Rotamah I. Bird Observatory (37/147) it is less frequently reported in summer than in winter but perhaps is less conspicuous in summer; at Jamieson (37/146) it was netted only July-Nov.[1469] Inland, in the southern Murray-Darling Region, birds seen irregularly in July-Aug. are non-breeding visitors.[833,1558] The longest movement recorded is that of a bird from Lynch's Ck (33/150) to Katoomba (33/150), 32 km.[1438]

White's Thrush lives in eucalypt forest and woodland, mainly in gullies in the South-East Region.[1081,1113] It will breed in plantations of the introduced pine.[1552] Outside the breeding season, White's Thrush appears in gardens and fields on King I. (39/144).[1131] It feeds on insects, earthworms and crustaceans.[1049]

At Moruya density during breeding was 0.11–0.15 birds/ha in eucalypt woodland.[1197] At Boola Boola (38/146) density was 0.3–0.5 birds/ha in eucalypt forest undisturbed for 15 or more years after logging.[1113]

Field Atlas Records: 2519

1° blocks–with records: 95 (12%)
–with breeding records: 36 (38%)

Atlas No.: 447

Max. Rep. Rate: 36%
In block(s): 37/150

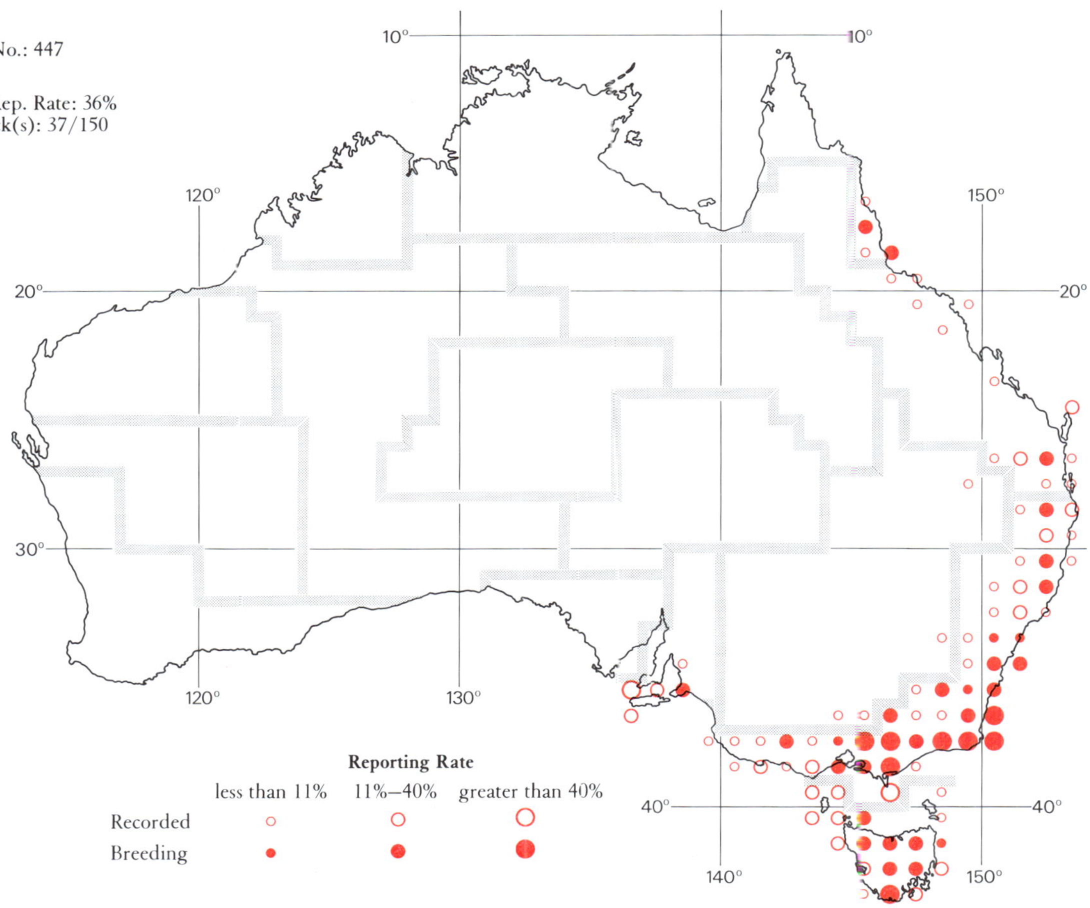

Blackbird

Turdus merula

The Blackbird, a native of Europe and Asia, was introduced to Australia before 1862 by which time the species was already established in Melbourne (37/145).[97] Releases of additional birds followed in Melbourne and other centres including Adelaide (34/138) in 1863 (successful), possibly Qld about 1869 (unsuccessful) and Sydney (33/151) in 1872 (unsuccessful).[1085] The species was in Albury (36/146) in 1926, apparently introduced there separately, and also in Sydney at about the same time[357] though the present Sydney population is thought to be derived from birds released in 1940.[1085] The first breeding record from Tas. was just before 1919[1085] and the Blackbird was introduced to Flinders I. (40/148) about 1930.[732] From all these points of release, and probably many others, the species has spread rapidly and continues to do so (p. 666). The first birds reached the Canberra (35/149) district in 1949, Deniliquin (35/144) in 1954, Dareton (34/142) in 1959,[833] Broken Hill (31/141) in 1975-76, Cobar (31/145) in 1976 and Armidale (30/151) in 1977.[1292] From Adelaide the species had reached Kangaroo I. (35/137) by 1947.

Though most recorded movements are short, less than 10 km,[1070] there is one of 500 km by an immature from Canberra to Castlemaine (37/144).[36] The Blackbird lives in parks and gardens, orchards and disturbed land and has spread slowly into woodland and forest. The bird feeds both on and above the ground on earthworms, snails, spiders, insects, fruit, seeds[1085] and tadpoles. In Vic. its diet includes fruit of two native species, *Pittosporum undulatum* and *Exocarpos cupressiformis*, which appear to be spreading because the Blackbird voids their seeds in new areas.[450]

The Blackbird's range is still expanding in Australia specially north along the main watercourses of the Murray-Darling Region. The species has colonized settlements in the mallee of north-western Vic. and has become ubiquitous in the Tas. Region in all but the south-west where it is confined to coastal districts. The Blackbird probably competes with some native species such as White's Thrush but as yet there is no evidence.[450]

Breeding density at the University of Melbourne in 1977 was 5.90 birds/ha,[454] which is within the range (0.02–14.08 birds/ha) recorded for Europe.[1666]

Field Atlas Records: 12 663

1° blocks–with records: 117 (14%)
–with breeding records: 78 (67%)

Atlas No.: 991

Max. Rep. Rate: 74%
In block(s): 37/145

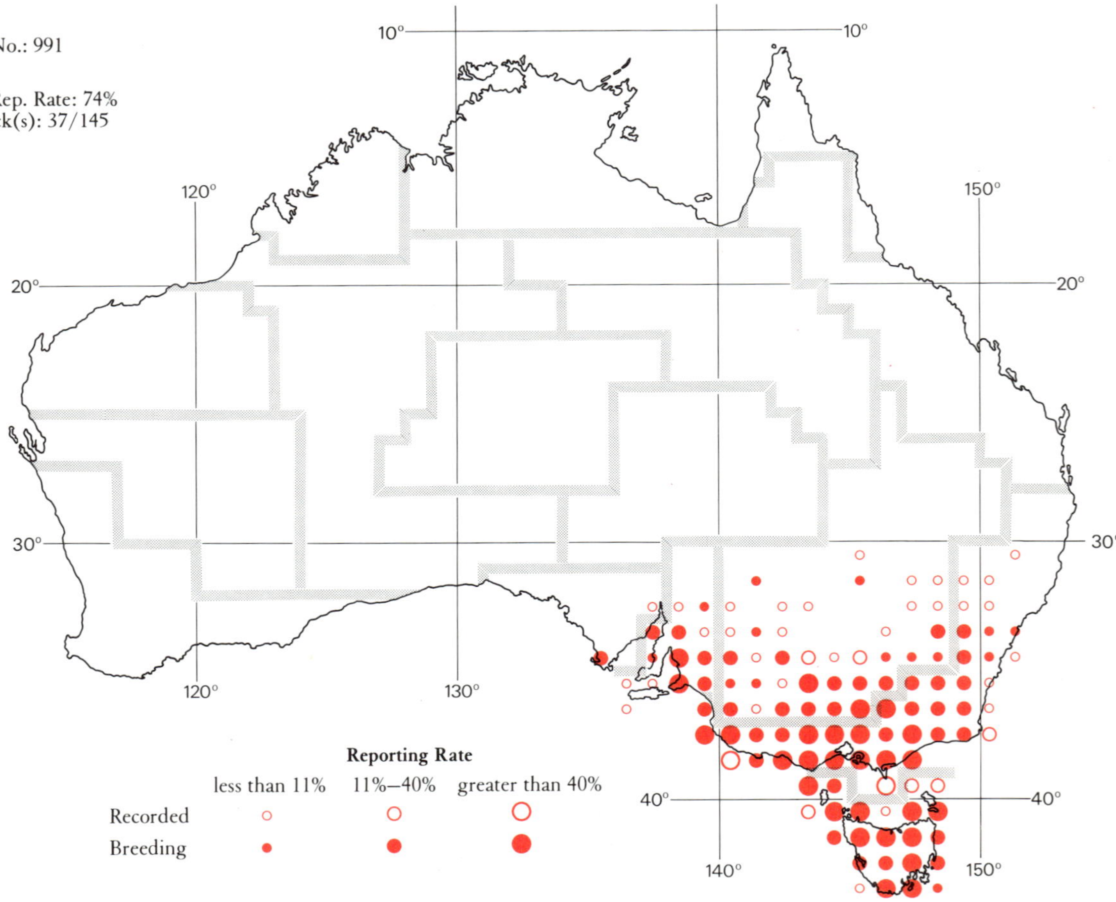

Song Thrush

Turdus philomelos

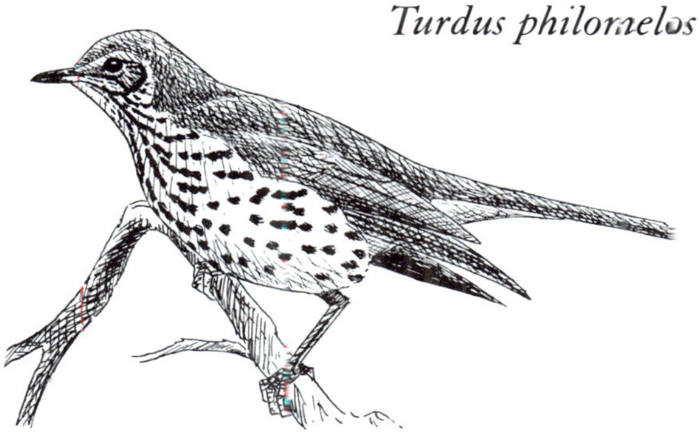

Compared with the Blackbird, the Song Thrush, native to Europe and central Asia, has made little headway in Australia. Introductions at Phillip I. (38/145) about 1857, Brisbane (27/153) 1869, Sydney (33/151) 1872, Adelaide (34/138) 1879–80 and Canberra (35/149) 1935 all failed.[1085] Only round Melbourne (37/145), where the species was released in the Botanic Gardens and Royal Park from 1857, has a population become established,[75] confined to the suburbs and the Geelong (38/144) district. No confirmed sightings have been made away from these areas. Records elsewhere seem all to have been made in autumn and winter when the immature Blackbird has a plumage resembling the Song Thrush.

The Song Thrush is migratory in some overseas countries but there is no evidence of movement by the Australian population.

The Song Thrush lives in gardens and parks; in Melbourne mainly in those vegetated with deciduous trees. It feeds on earthworms, snails, grubs, slugs, insects, berries and fruit.[1085] The species seems to be sensitive to hot conditions, keeping to the shade during the Australian summer. It has done much better in New Zealand where it is known as a pest of soft fruit and berry crops.

In British farmland in 1980 its density was 0.15 birds/ha[1191] but in Australia there are no measures of abundance.

Field Atlas Records: 823

1° blocks–with records: 4 (1%)

–with breeding records: 4 (100%)

Atlas No.: 992

Max. Rep. Rate: 26%
In block(s): 37/145

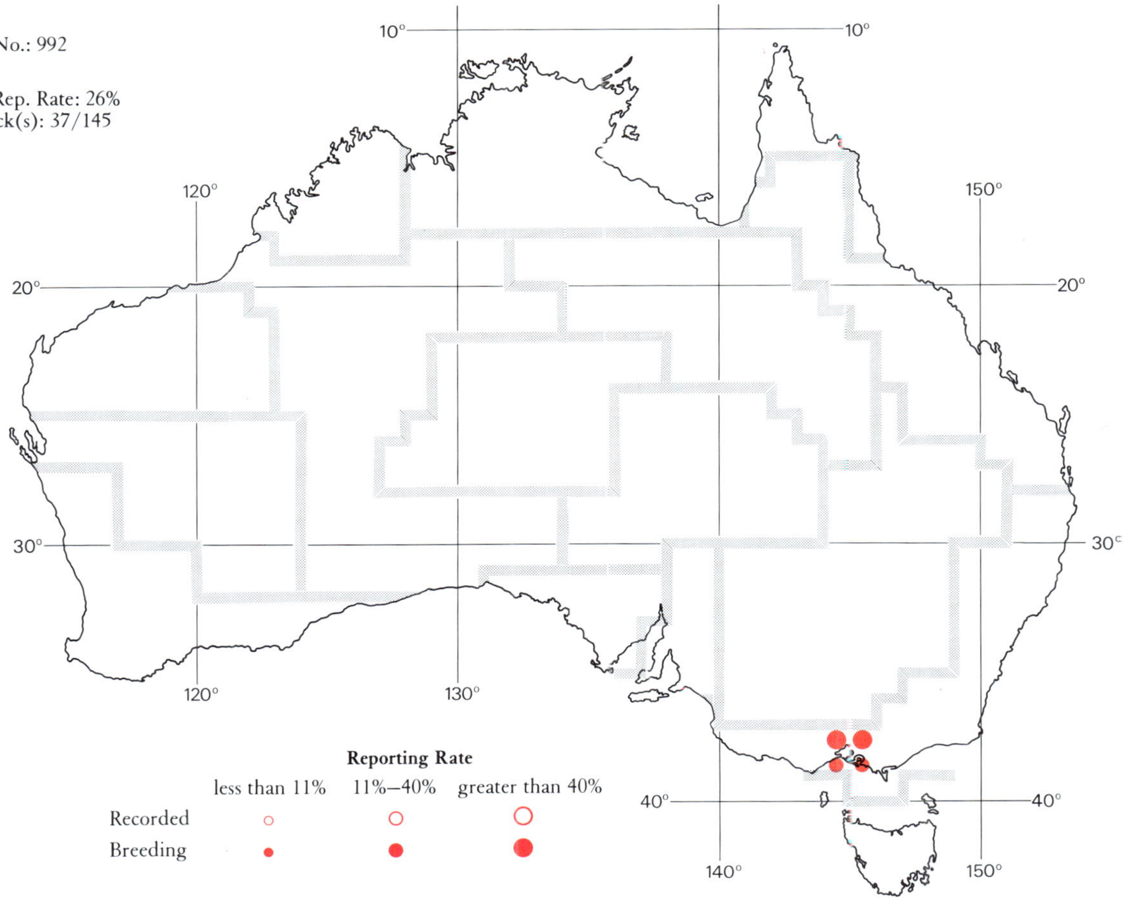

Northern Scrub-robin

Drymodes superciliaris

The Northern Scrub-Robin is a widespread New Guinean bird, also inhabiting the Aru Is and reaching northern Australia. One of the two Australian subspecies, *superciliaris*, lives in the Cape York Region. The second, *colcloughi*, is known only from two specimens collected at 14/134 on the Roper R in 1909. Intensive searches during the Field Atlas failed to find any trace of *colcloughi* where the original population was recorded but small numbers may survive in more remote parts of the region.[166]

The subspecies *superciliaris* has been recorded on the western coast of the Cape York Region at the Archer R (13/142)[1147] and the Watson R (13/141) and on the eastern coast from Cape York Pen. (10/142) south to Coen (13/143). Reports in the Atherton Region and farther south are unconfirmed.

The Northern Scrub-Robin is thought to be sedentary; *superciliaris* lives in pairs in rainforest; *colcloughi* occupied narrow stands of similar vegetation that grow along the Roper R. The Scrub-robin feeds on insects in the leaf litter and on land snails, leaving conspicuous broken shells on the forest floor.[1861] The disappearance of *colcloughi* from the Roper R is attributed to burning. The fire regime of aborigines left unburnt vegetation suitable for the Scrub-Robin whereas present practice burns indiscriminately.[166]

There are no measures of abundance but in the Cape York Region *superciliaris* is still locally common.

Field Atlas Records: 35

1° blocks–with records: 2 (1%)

–with breeding records:

Atlas No.: 442

Max. Rep. Rate: 44%
In block(s): 12/143

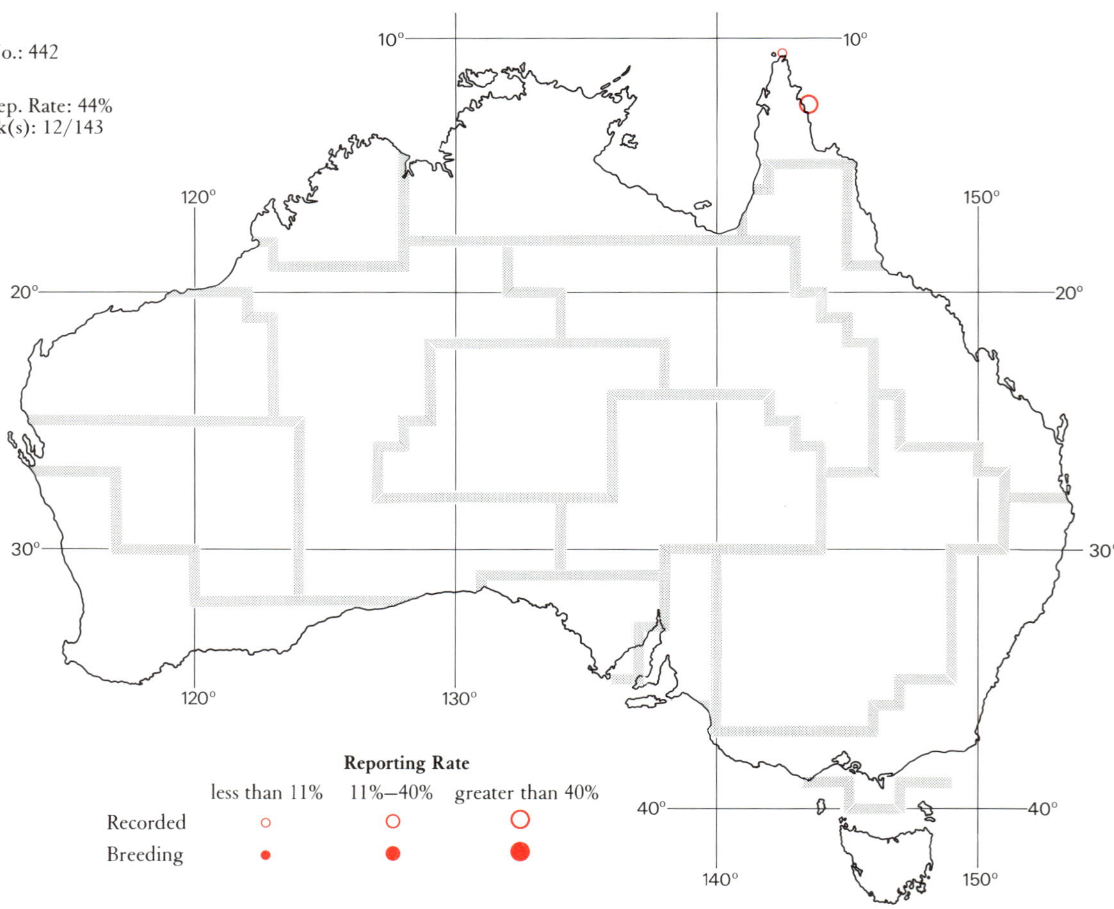

Southern Scrub-robin

Drymodes brunneopygia

The gap between the eastern and western populations of the endemic Southern Scrub-robin is wide yet there is no evidence of morphological differentiation.[599] Within the two main populations are a number of others. Those in and near Round Hill Nature Reserve (33/146), in the northern SA Ra. Region, near Eyre Bird Observatory (32/126) and at Shark Bay (25/113) are clearly isolated. In addition, the 10′ maps show birds in the Wedderburn (36/142) area to be isolated.

Comparison with the historical maps does not suggest any significant change in range but there is an indication on the Peron Pen. (25/113) that abundance may have fluctuated. The species was recorded there in 1858 then not again until the 1940s, despite visits by competent observers.[466] In the 1940s it was conspicuous and abundant.

The Field Atlas gives no indication of large-scale seasonal movement. The bird is strictly territorial and sedentary throughout the year at Eyre Bird Observatory.[48]

The Southern Scrub-robin lives in acacia scrub and mallee, specially where broombush is common. It spends most of its time on the ground where it feeds on beetles and other invertebrates.[743,1049] The population on the Peron Pen. survives despite light grazing of the area by sheep.[466]

At Eyre Bird Observatory the Southern Scrub-robin has a disjunct distribution within the mallee; where there are several territories together its density is 3 birds/ha.[390]

Field Atlas Records: 902
1° blocks–with records: 75 (9%)
–with breeding records: 14 (19%)

Atlas No.: 441

Max. Rep. Rate: 31%
In block(s): 32/120

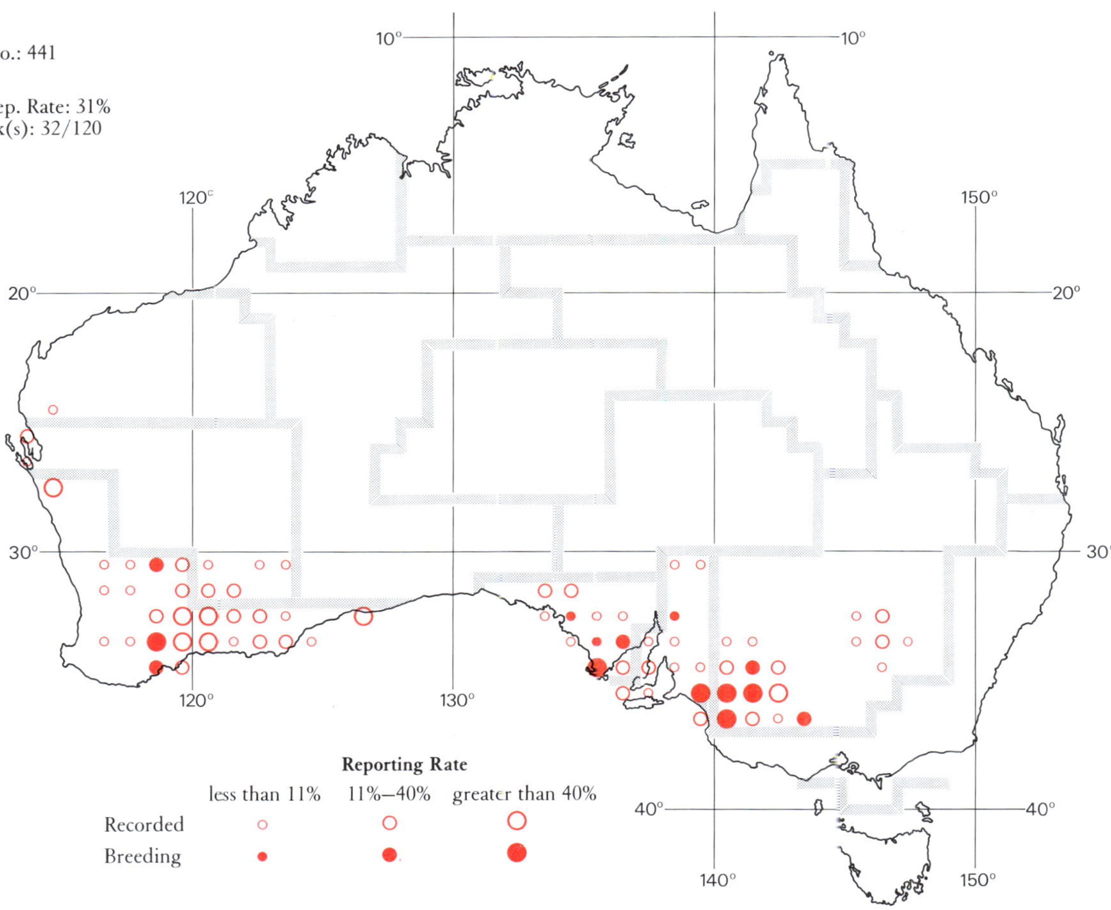

Rose Robin

Petroica rosea

The endemic Rose Robin is not reported breeding north of 28°S in the Atlases. In the south the record near Portland (38/141) in Dec. 1977-Jan. 1978 is a westward extension of breeding range and the species was first reported in the SA Ra. Region in Aug. 1967.[1408]

The Field Atlas shows the Rose Robin to be virtually confined to the South-East Region in summer with no records north of 26°S then. In winter there is a northward movement and few birds remain in Vic. This is only partly reflected in the Reporting Rate (south of 35°S: summer 3.2%, winter 2.6%) because the birds arrive back at the end of winter.[1110] Near Canberra (35/149) the bird may be seen on passage in spring and autumn.[653] In northern parts of its range, high country appears to be deserted in winter: in summer the Rose Robin breeds on the McPherson Ra. (28/152) at 2000 m but not on the lower Tamborine Mt. (27/153) at 1000 m where it appears in winter.[354] Dated records in the SA Ra. Region are all May-Oct., which suggests some westward movement in winter.

In New England National Park (30/152) the Rose Robin breeds in rainforest.[129] In Vic. it breeds in eucalypt forest, using wet gullies in which large acacias are growing, specially silver wattle but also blackwood;[1110,1113] the few remaining over winter move into drier country such as forest dominated by narrow-leaved peppermint.[1110] Pairs, perhaps the same birds, often return to nest in the same place year after year.[653,871] The Rose Robin feeds on insects mainly flies and beetles taken in the air or from the foliage of acacias and sometimes from the ground in late summer and autumn.[1110]

In eucalypt woodland at Armidale (30/151) density was 0.06 birds/ha[582] and at Moruya (35/150) 0.03-0.10 birds/ha.[1197] At Boola Boola (38/146) in eucalypt forest densities were 0.5-1.0 birds/ha, the greater density being in the gullies with abundant mature silver wattles.[1113]

Field Atlas Records: 1912

1° blocks-with records: 79 (10%)
-with breeding records: 19 (24%)

Atlas No.: 384

Max. Rep. Rate: 20%
In block(s): 37/149

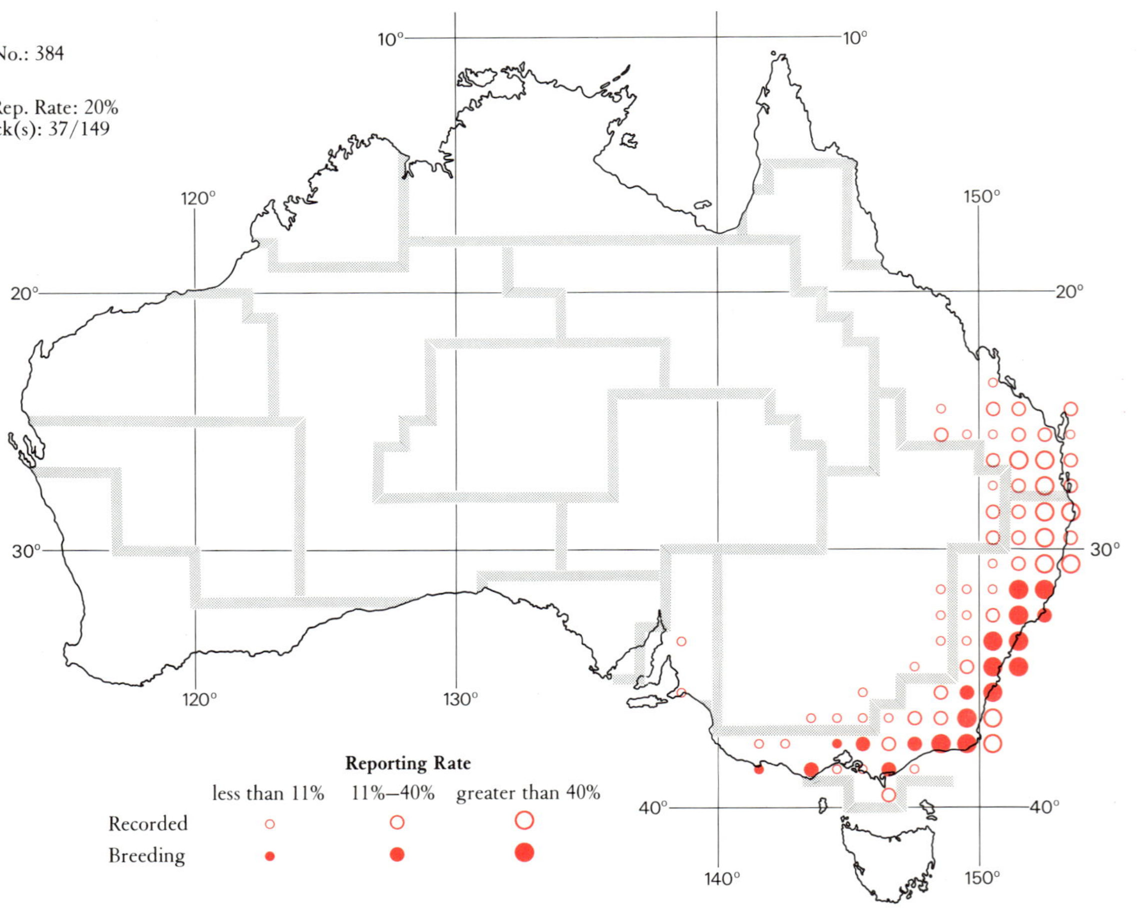

Pink Robin

Petroica rodinogaster

On the mainland the endemic Pink Robin breeds only in the Alps south of 36°S, the Yarra R catchment area (37/145) and the Otway Ra. (38/143, 38/144). In Tas. breeding is widespread, even on offshore islands. Occasionally the Pink Robin is recorded as far north as Sydney (33/151) and Corowa (36/146) and as far west as Adelaide (34/138). Females and immatures are sometimes misidentified as those of the Flame Robin.

Movements take place. Pink Robins have been seen north of the breeding range in all seasons but, at Canberra (35/149), mainly in winter.[1903] The Field Atlas suggests a movement westward in winter: it contains no sightings in summer west of 143°E and all dated reports from the SA Ra. Region are Mar.-Sept. In Tas. colour-banding has shown that males tend to remain near the breeding territory throughout the year though they may forage over a greater area in the non-breeding season. Juveniles are driven out and wander widely.[1314] Probably the same happens in Vic. where birds disperse into a variety of country in winter but males are seldom seen away from breeding areas.[1110] In both Vic. and Tas. the Reporting Rate increases in winter because resident birds tend to forage in open parts of their territories then[1314] and wandering birds are often also in open country.[734,1452] Movement across Bass Strait has been suggested[653,1739] but there is no direct evidence for it and the Field Atlas provides none.

The Pink Robin breeds in rainforest and eucalypt forest, usually in gullies.[354,1110] The materials from which it builds its nest are found only in these gullies possibly limiting places where it can breed.[1321] The Pink Robin feeds on insects in the air and on the ground.[1261] It forages more on the ground than the Rose Robin.[1314] Clearfelling of rainforest adversely affects the breeding of the Pink Robin.[127]

At Ferntree (42/147) on average five pairs bred each year along about 300 m of creekside.[1314]

Field Atlas Records: 1043

1° blocks–with records: 49 (6%)

–with breeding records: 16 (33%)

Atlas No.: 383

Max. Rep. Rate: 26%
In block(s): 41/145

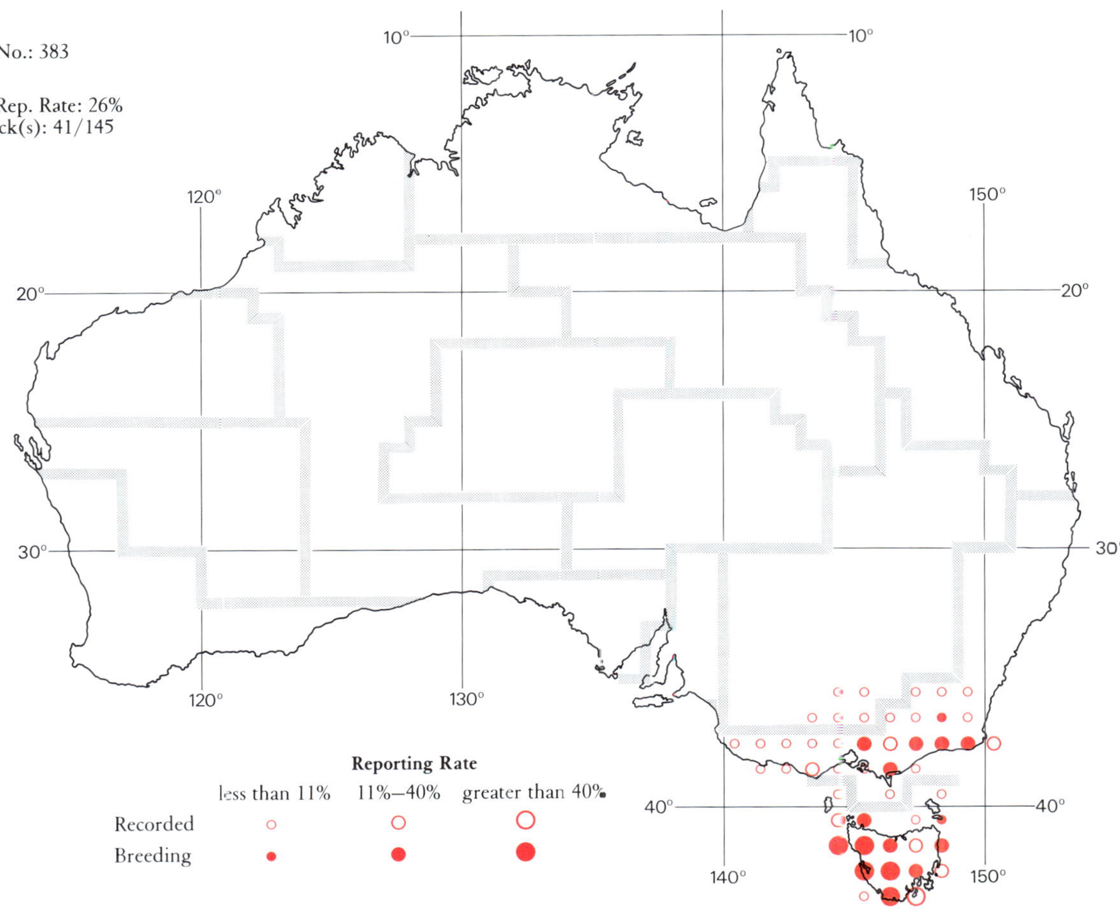

Flame Robin

Petroica phoenicea

Part of the population of the endemic Flame Robin migrates. Some movements are altitudinal; birds that breed in the Alps in summer visit the lowlands in winter, e.g. at Jamieson (37/146) and in the southern lowlands of Vic. the bird is present Mar.-Sept.[1469,1743,1822] There is also a pronounced westward movement in winter taking birds into the SA Ra. Region: in summer the Flame Robin was not recorded west of 142°E in the Field Atlas. Another part of the population moves north into the Murray-Darling Region were the Flame Robin appears regularly Mar.-Sept. at Rutherglen (36/146)[1128] and Apr.-Sept. at Shepparton (36/145)[630] and Cobram (35/145).[117] It returns to the same sites near Cobram and Shepparton each winter, some birds remaining all winter, others passing through.[116,630] At Cobram there is no difference in time of arrival and departure of birds of either sex[116] but in southern Vic. males arrive after and depart before birds in female or immature plumage.[1743,1822]

Some Flame Robins probably migrate across Bass Strait from Tas.[354,480,1131] The Field Atlas suggests that a large part of the population remains in Tas. where there is also a dispersal in winter specially of birds that breed at high altitude.

The Flame Robin lives in eucalypt forest and woodland and heath and may show a preference for recently burnt areas.[1700] It also inhabits plantations of pine. In the winter it lives in flocks mainly in pasture,[833,1128]which increases its conspicuousness. There are areas of overlap with the Scarlet Robin but the two species occupy mutually exclusive territories. The Flame is generally in wetter areas.[1113] It roosts at night in trees with dense foliage including those in orchards. It feeds on insects taken either in sallies from vantage points or when hopping along the ground.

At Bathurst (33/149) breeding densities in wet and dry eucalypt forest and pine plantation were 1.25, 0.50 and 2.00 birds/ha respectively.[489] At Boola Boola (38/146) in eucalypt forest densities ranged from 0.1–1.2 birds/ha with the greatest density in mature forest.[1113] In wintering roosts in orchards near Cobram, 6641 individuals have been banded since 1964.[118]

Field Atlas Records: 6675

1° blocks–with records: 101 (12%)
–with breeding records: 45 (45%)

Atlas No.: 382

Max. Rep. Rate: 47%
In block(s): 37/146

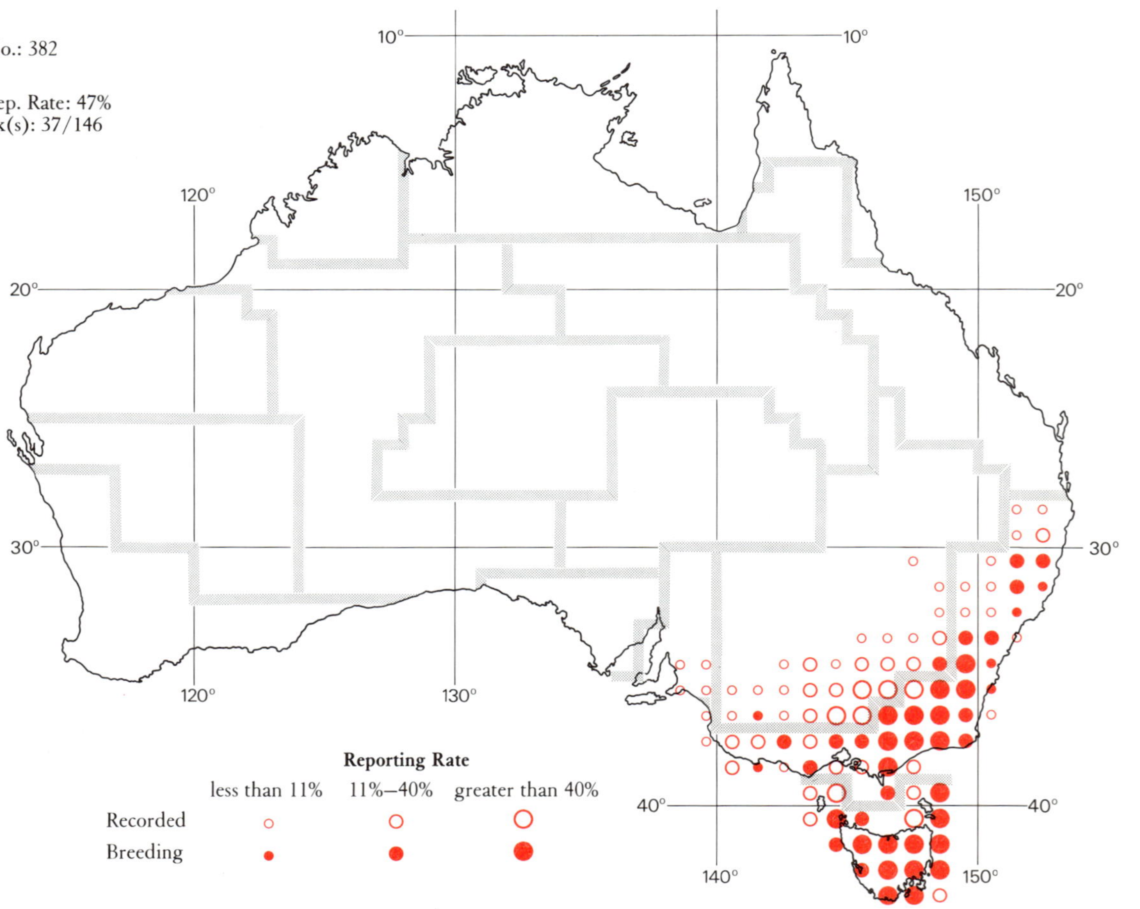

Scarlet Robin

Petroica multicolor

Outside Australia there are subspecies of the Scarlet Robin on many Pacific islands, including Norfolk (29/167), and in Australia two: *boodang* to the east and *campbelli* to the west of the Great Australian Bight.[986] Despite the presence of subspecies at latitudes as far north as 5°S in the Solomon Is, the Scarlet Robin in Australia is confined to areas south of 25°S. Reports last century from Rockhampton (23/150) and Rockingham Bay (18/146)[354] are unconfirmed. In the Tas. Region the species was first reported on King I. (39/144) during the Field Atlas.

At Wollomombi (30/152) a study of colour-banded birds showed that the Scarlet Robin lives in pairs in a permanent territory though it may forage in adjoining open country in winter. The young wander in autumn[129,877] and this may give a false impression of nomadism previously reported in the literature.[354,734,833] A similar pattern has been observed near Hobart (42/147) where adults remain near their breeding territory throughout the year but young are driven out in autumn.[1314] The Field Atlas shows the species to be recorded up to two degrees farther north in winter in the Eastern Qld Region.

The Scarlet Robin breeds in eucalypt forest with an open understorey. There are areas of overlap with the Flame Robin but the two species occupy mutually exclusive territories. The Scarlet is generally in drier forest.[1113] In autumn and winter it feeds on small insects, taken on the ground after a sortie from a perch up to 3 m above the forest floor,[77] but stays in the canopy in spring and summer.[877] In the non-breeding season, it joins mixed species flocks.[155] In Vic. clearing and bushfires have attracted Robins to places where the original understorey was too dense but conversely they are excluded from young regrowth.[1110] In Tas. the bird is reported less frequently where forests have been cleared for woodchipping.[1901]

Densities in eucalypt woodland were: Armidale (30/151) 0.04–0.65 birds/ha;[582] Wollomombi 0.3 birds/ha;[29] and Canberra (35/149) 0.1 birds/ha.[154] In eucalypt forest they were: Jamieson (37/146) 0.1 birds/ha[1469] and Boola Boola (38/146) 0.1–0.8 birds/ha.[1113]

Field Atlas Records: 10 852

1° blocks–with records: 148 (18%)
–with breeding records: 77 (52%)

Atlas No.: 380

Max. Rep. Rate: 58%
In block(s): 34/116

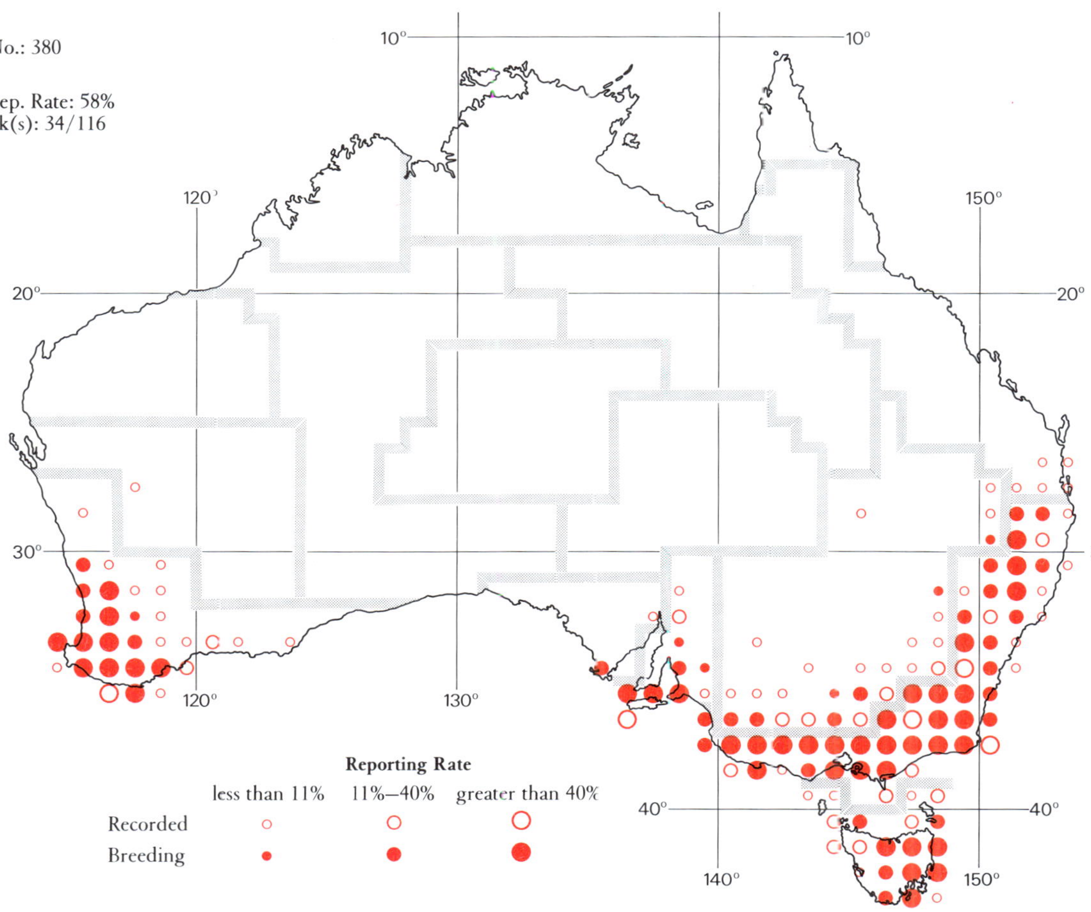

Red-capped Robin

Petroica goodenovii

Breeding of the endemic Red-capped Robin is not reported north of 21°S. There are records historically from the Mt Isa (20/139) district where it is regarded as an uncommon winter visitor.[862] It has declined locally near Perth (31/115).[1638]

The Red-capped Robin is a winter visitor to northern parts of its range.[1709,1713] In the South-East and the southern part of the Eastern Qld Region the bird is sedentary though it may change its hunting grounds slightly with the seasons.[96,189,833] In the South-West it visits the Swan R coastal plain (32/115) only in winter and is seldom reported near the coast except for the isolated population on Rottnest I. (32/115).[1638]

The Red-capped Robin lives in most inland vegetation associations where there are trees, specially in acacia scrub and native pine woodland. It is often in flocks of mixed species.[1606] It feeds on insects caught in the air, on the ground and on trees, generally perching low to make its sallies or even hopping about on the ground.[748]

The only measure of density is a report that a pair near Naracoorte (36/140) occupied about 0.5 ha.[77]

Field Atlas Records: 8718

1° blocks-with records: 462 (57%)

-with breeding records: 179 (39%)

Atlas No.: 381

Max. Rep. Rate: 66%
In block(s): 28/121

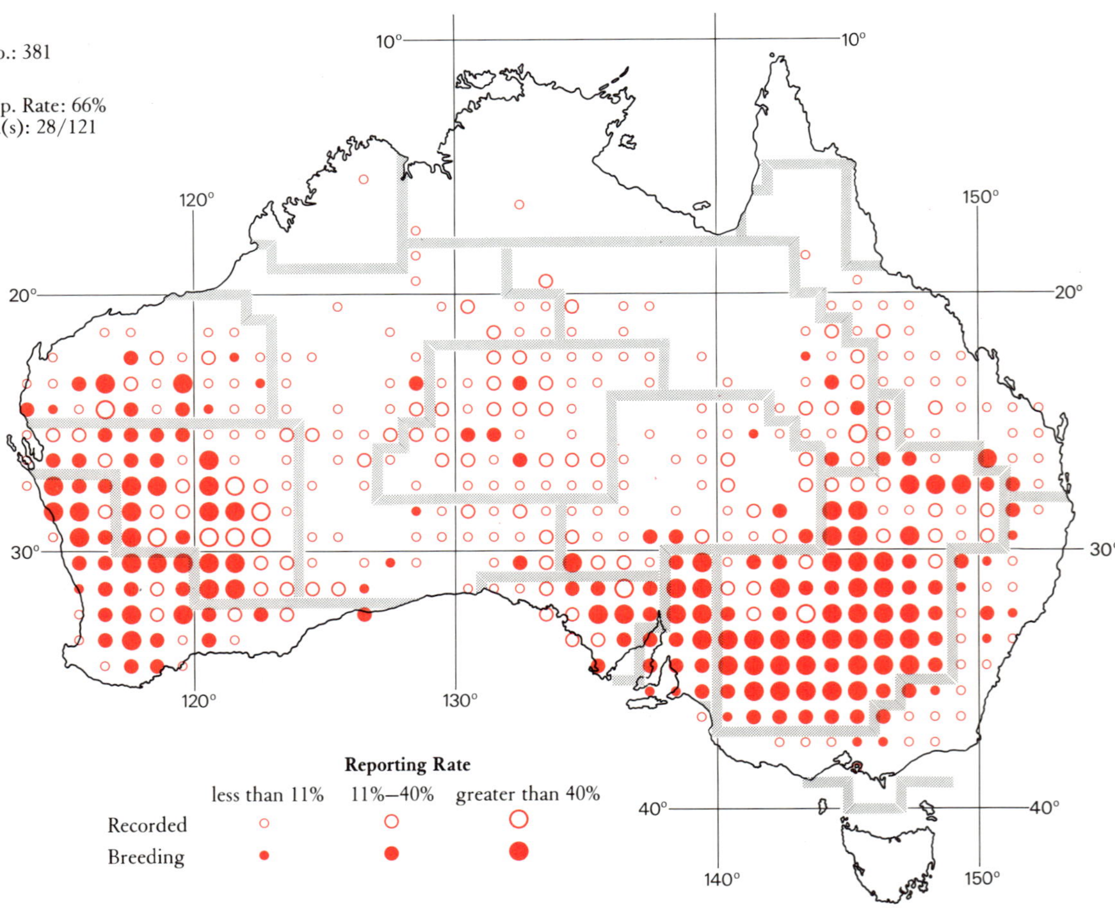

Hooded Robin

Melanodryas cucullata

Individuals of the southern population of the endemic Hooded Robin are distinctly larger than those of the northern[748] but the difference is clinal and no subspecies are recognized. The species has not been reported in Tas. It breeds throughout its range.

The Field Atlas gives no evidence of large-scale seasonal movements. The species is sedentary at Wollomombi (30/152)[129] and probably in most places, though observers at three widely separated localities (Canberra (35/149),[653] the Central Ra. Region[1880] and the Fitzroy R[1661]) have reported it to be more common in winter than summer. Local seasonal movements are possible, e.g. the increased numbers in Canberra in winter may mean that birds leave the surrounding high country and move to lower areas then. On the Fitzroy R a local dispersal may occur in the wet season. There is no evidence of nomadism though such behaviour has been suggested.[1356]

The Hooded Robin lives in acacia scrub, eucalypt woodland and mallee, often in or near clearings. It joins mixed species flocks.[129] Breeding is often communal.[129,1547] The Hooded Robin darts down from stumps and bare limbs to collect insects, some as large as 2.5 cm in length.

The population declined at Murphy's Ck (27/152) though no reason could be suggested.[1099]

At Wollomombi in eucalypt woodland the Robin's territory in the breeding season is about 9 ha, increasing to about 15 ha in winter.[129] Density in eucalypt woodland at Armidale (30/151) was 0.03 birds/ha[582] and on the Nullarbor Plain 0.05 birds/ha.[246]

Field Atlas Records: 6001

1° blocks–with records: 466 (57%)
–with breeding records: 134 (29%)

Atlas No.: 385

Max. Rep. Rate: 78%
In block(s): 27/147

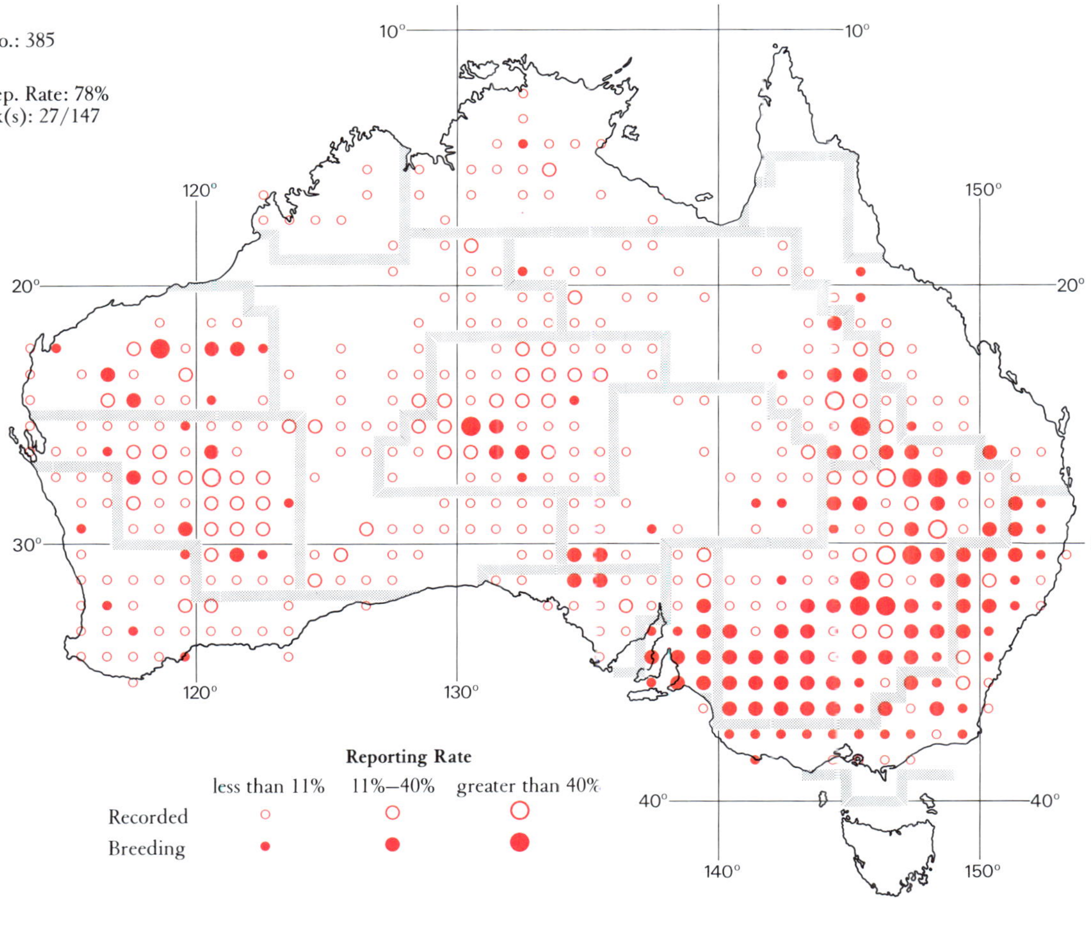

Dusky Robin

Melanodryas vittata

The Dusky Robin is endemic to the Tas. Region. It has bred on King I. (39/144) and probably still does though not reported to do so during the Field Atlas.

There is little information about movements. At Woodsdale (42/147) one banded bird was recaptured three times within a year at the same spot.[1899a] Nests may be built in the same bush for several successive seasons[1343] but the bird also seems able to colonize areas as they become suitable, e.g. country that is regenerating after clearing or fire.[496,1450a] Outside the breeding season the Dusky Robin forages in family groups, often moving into more open country.[1131]

The Dusky Robin's wide distribution reflects its ability to live in a range of vegetation from eucalypt forest to heath, including buttongrass plains in the south-west.[740] It requires some open ground to forage in and is therefore restricted to the edges of undisturbed rainforest but benefits from clearing for roads and tracks. Pairs and groups live at altitudes up to 1200 m.[1451] The Dusky Robin usually hunts by perching motionless on low objects such as stumps and fences, darting down to gather prey. Occasionally it will hop along the ground when foraging[1314] and near houses often attends gardeners at work. Guts of the Dusky Robin contain a great number of beetle remains. It also feeds on earthworms[1645] and has been known to take flies in a house.

Densities range up to three groups/sq. km.[1899a]

Field Atlas Records: 1505
10′ blocks–with records: 271 (74%)
–with breeding records: 60 (22%)

Atlas No.: 386

Max. Rep. Rate: 78%
In block(s): 42°05′/146°45′

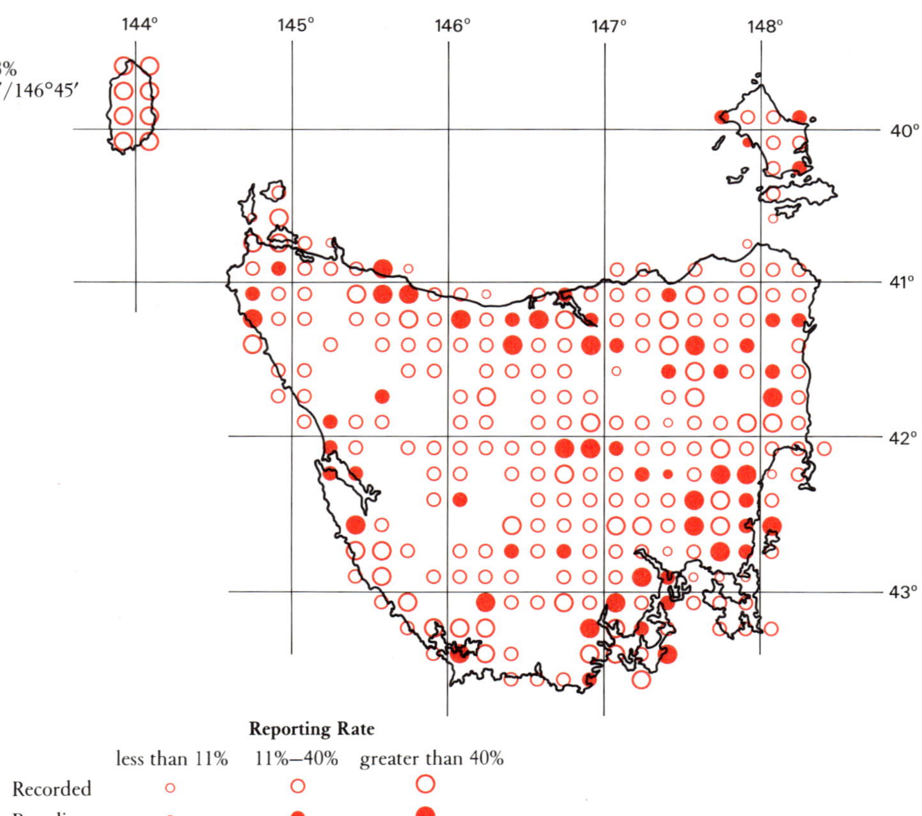

Mangrove Robin

Eopsaltria pulverulenta

The Mangrove Robin inhabits New Guinea and adjacent islands as well as the coast of northern Australia. There is a slight morphological difference between populations east and west of 127°E[622] but there is doubt whether the populations warrant subspecific status.[748,1709,1713] In the Atlases as a whole, the species' distribution is more continuous than it appears on the map, though there are gaps along the southern coast of the Gulf of Carpentaria (137–139°E) and at 129°E, 127°E, 16/123, Eighty Mile Beach (19/119–120) and 20/114. Breeding takes place throughout the species' range.

There is no information about movements of the Mangrove Robin. It lives in pairs and small parties of up to 15,[748] in Australia always in mangroves, usually in thickets near the edge.[1203] It is present in some patches of mangroves but not in others apparently equally suitable.[698,748] In New Guinea it has once been recorded in a freshwater environment: on the Sepik R (4S/142E) 300 km from the sea in reedbeds with a few shrubs.[553]

The Mangrove Robin feeds on insects, including ants, and on invertebrates caught on the ground below the mangroves at low tide.[748,788,1855]

There are no measures of abundance.

Field Atlas Records: 123

1° blocks–with records: 29 (4%)
–with breeding records: 1 (3%)

Atlas No.: 388

Max. Rep. Rate: 13%
In block(s): 12/143,20/117

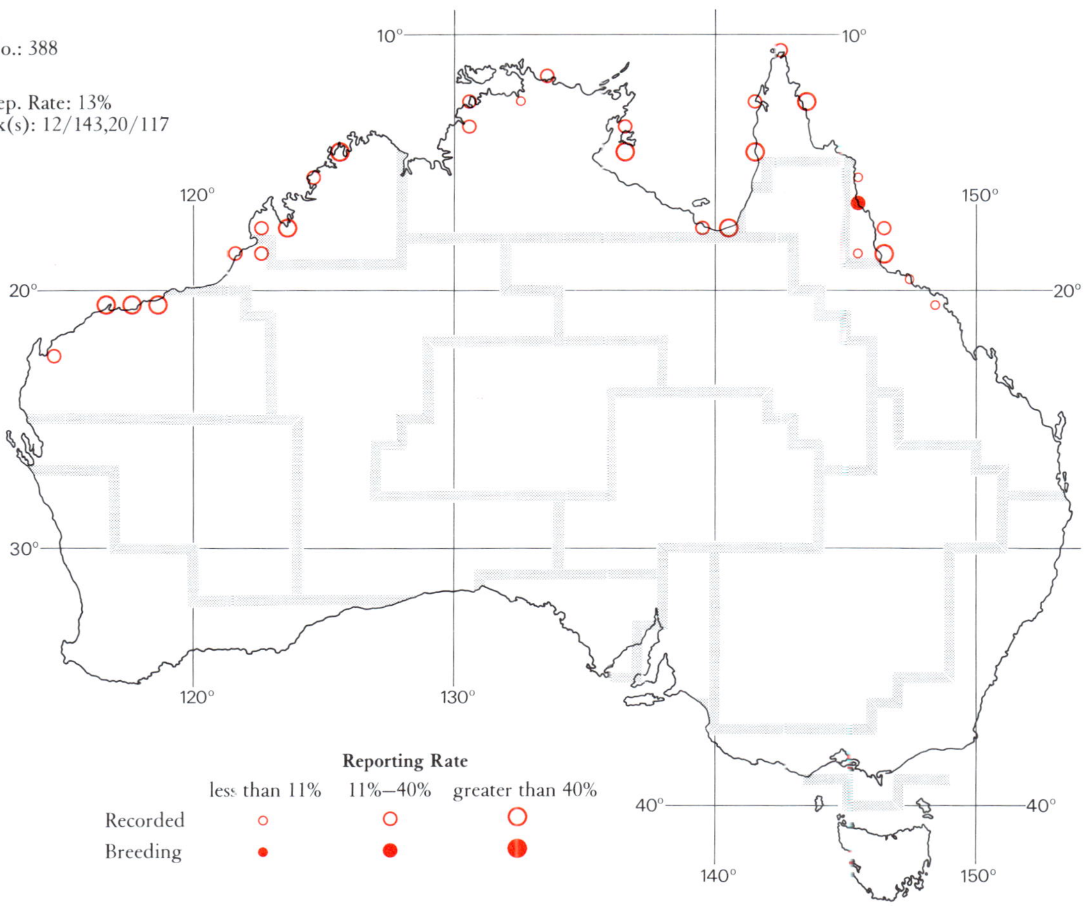

White-breasted Robin

Eopsaltria georgiana

The White-breasted Robin is endemic to the South-West Region. Before 1950 the only record north of 31°S was a doubtful one between New Norcia (30/116) and Wongan Hills (30/116) but the species is now frequently recorded in a narrow coastal strip south from 28°S.[600] The Field Atlas map indicates a continuous distribution from Geraldton (28/114) southwards but the 10′ map shows that there is a 100 km gap at metropolitan Perth (31/115). Individuals of the two populations have slight morphological differences, clinal rather than abrupt.[628]

The White-breasted Robin is sedentary.[265] It requires thickets with plenty of litter and a damp environment, either maritime or in a high rainfall district. Whether or not there is a tree layer seems to be less important. The species is usually thought of as inhabiting thickets along creeks in eucalypt forest dominated by karri but along the coast it lives in thickets of heath among dunes. It sometimes lives away from creeks in eucalypt forest[1719] or on the edges of forest bordering cleared land.[263] It sometimes breeds communally.[263] The White-breasted Robin feeds on insects collected in sallies from a bare twig or other vantage point[1222,1719,1885] and will forage on recently burnt ground. It often perches sideways on the boles of trees and rarely spends long on the ground.

At Smith Brook (34/116) in a 72-ha study area density was one breeding group (2–6 individuals)/ha.[262]

Field Atlas Records: 501

1° blocks–with records: 18 (2%)
–with breeding records: 11 (61%)

Atlas No.: 387

Max. Rep. Rate: 48%
In block(s): 34/116

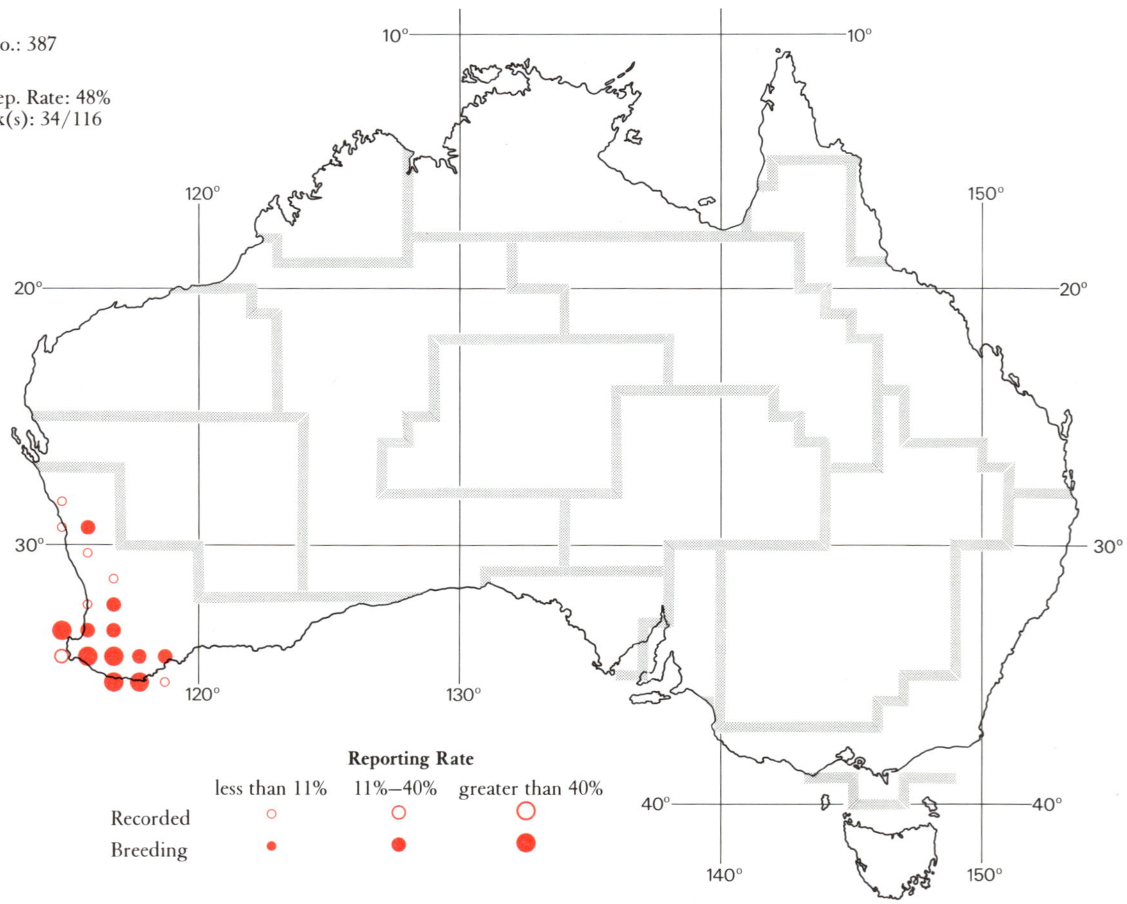

Eastern Yellow Robin

Eopsaltria australis

During the Field Atlas the endemic Eastern Yellow Robin was recognized as a separate species[1581] but a recent taxonomic opinion suggests that it and the Western Yellow Robin should be treated as one species despite their different plumages.[613] There are different opinions about the validity of subspecies of the Eastern Yellow Robin. Variation may be clinal.[613,748,986]

North of Rockhampton (23/150) breeding is reported only in the Eungella (21/148) district and the Atherton Region. The Field Atlas shows records of the Eastern Yellow Robin to be isolated in these two areas in summer, which suggests a winter dispersal to surrounding districts. There appears to be some altitudinal migration also and the Eastern Yellow Robin is considered a rare summer breeding migrant in the Thredbo Valley (36/148) at 1100–2000 m.[682] Reporting Rates in the South-East Region were higher below 500 m than above at all times of the year but suggested that the species was slightly more frequent above 500 m in spring-summer (26.3%) than in autumn-winter (13.3%).

The Eastern Yellow Robin inhabits rainforest, eucalypt forest, woodland, mallee and heath.[1575] The aspect common to these vegetational formations seems to be an understorey to 8 m high or a shrubbery up to 5 m high with little ground cover.[1192] Typically a pair holds an area for itself but there are records of more than one female laying in a nest[871,889] and more than two birds attending a nest.[501] The Eastern Yellow Robin feeds on grubs and insects that it sees in ground litter from vantage points on the sides of tree trunks.

Densities in eucalypt woodland were: Armidale (30/151) 0.68 birds/ha;[582] and Moruya (35/150) 0.97 birds/ha.[1197] In eucalypt forest they were: Bathurst (33/149) 0.75 and 0.25 birds/ha respectively in wet and dry forest;[489] Jamieson (37/146) 0.5–1.0 birds/ha;[1469] and Boola Boola (38/146) 0.3–3.0 birds/ha, the greatest densities being in gullies.[1113]

Field Atlas Records: 13 138

1° blocks–with records: 141 (17%)

–with breeding records: 89 (63%)

Atlas No.: 392

Max. Rep. Rate: 70%
In block(s): 36/150

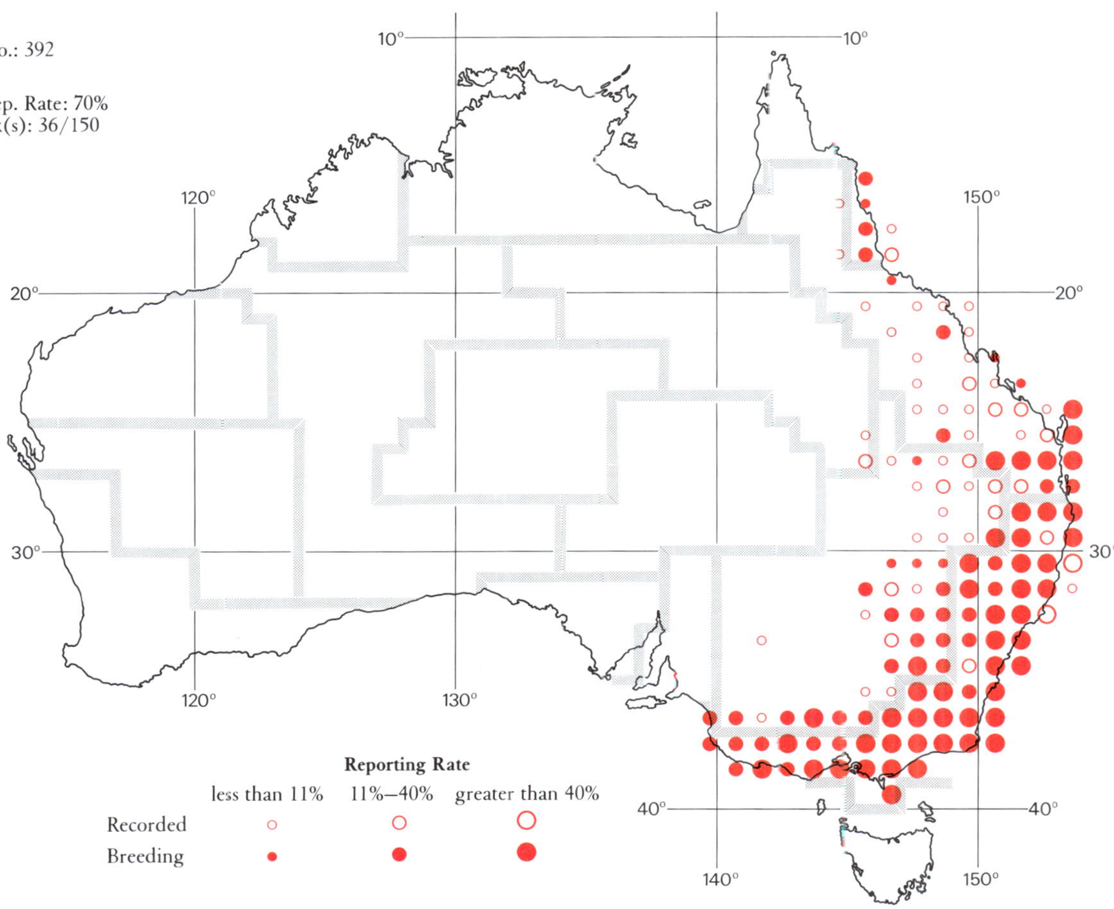

Western Yellow Robin

Eopsaltria griseogularis

During the Field Atlas the endemic Western Yellow Robin was recognized as a separate species from the Eastern Yellow Robin.[1576] A recent taxonomic opinion suggests that they should be treated as one species despite their different plumages.[613] There are two subspecies of the Western Yellow Robin: *rosinae* on Eyre Pen. ranges west to 131°E; the range of the South-West population *griseogularis* is probably continuous east to Eucla (31/128). In the west *griseogularis* is continuously distributed as far north as 25°S. Breeding is reported throughout the range of the Western Yellow Robin.

In the past a local but regular seasonal movement brought birds to the coastal plain near Perth (31/115) in winter but the last observation there was in 1935.[1638]

The Western Yellow Robin ranges right through the vegetational zones of the South-West: eucalypt forest, woodland, mallee and acacia scrub. Wherever the bird lives, there is a shrub layer 2–3 m tall. In the mallee and acacia scrub this forms the canopy as well but in woodland and forest the Western Yellow Robin is present only where a shrubby understorey grows.[600] It sometimes breeds communally.[263]

The Western Yellow Robin feeds on insects, usually alone but sometimes in mixed species feeding flocks.[1049,1606] It catches most of its prey on the ground, waiting on the twigs and branches of the understorey to pounce on it.

The Western Yellow Robin is most often reported from eucalypt woodland, a vegetational type that is being rapidly cleared for agriculture.[1216] The decline of winter visitors near Perth has been attributed to a decline in the quality of the forest on the adjacent Darling Scarp.

In 72 ha at Smith Brook (34/116) over three years, totals of 1, 13 and 16 individuals were banded or retrapped.[265] Searches in 1980 located two breeding groups of three and four birds in the area giving a breeding density of about 0.1 birds/ha.[262]

Field Atlas Records: 822

1° blocks–with records: 55 (7%)
–with breeding records: 21 (38%)

Atlas No.: 394

Max. Rep. Rate: 33%
In block(s): 34/117

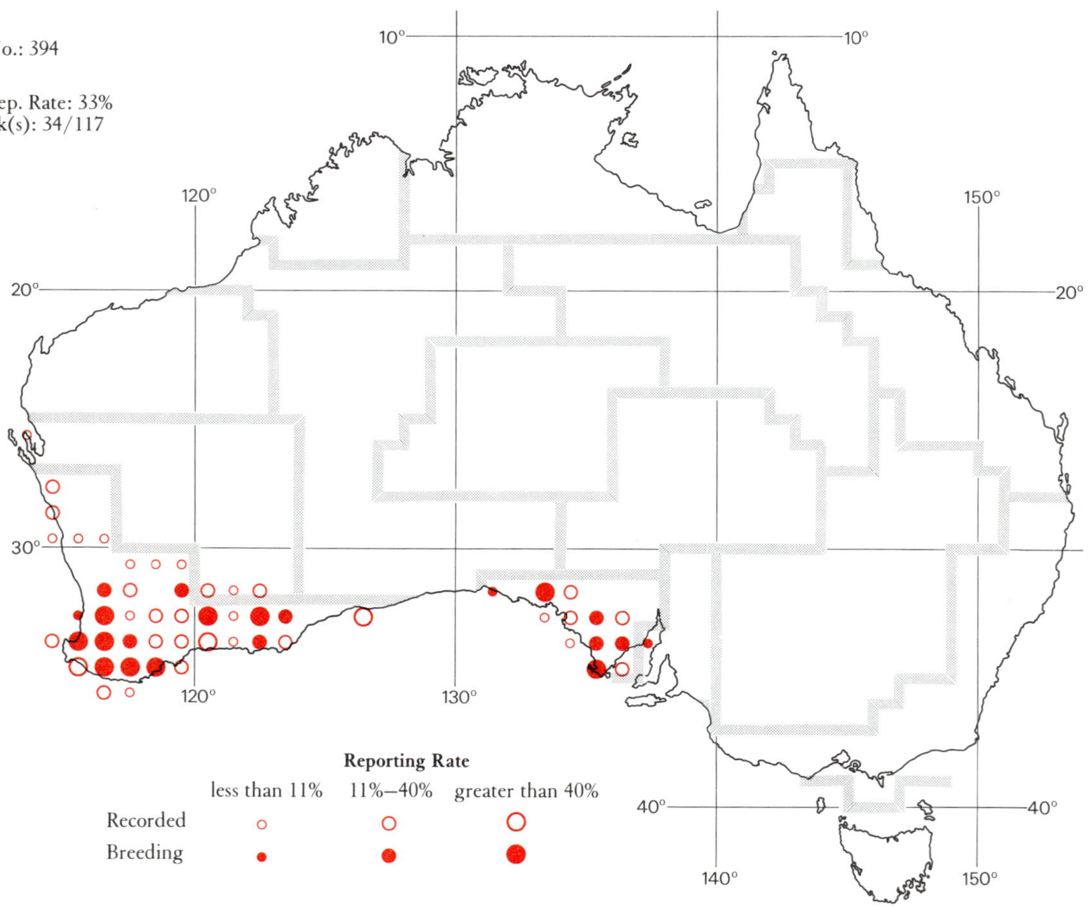

Yellow-legged Flycatcher

Microeca griseoceps

The Yellow-legged Flycatcher lives in New Guinea as well as Australia. Its presence in Australia was unknown until 1913 when it was found in the Cape York Region. The Field Atlas record from Weipa (12/141) was the first from the western coast of the region.

There are several unconfirmed sight records of the Yellow-legged Flycatcher in the Atherton Region[212,231,1869] and recently a previously overlooked specimen was discovered whose label gives the place and date of collection as the Herbert R (18/146) in 1874.[201] The possibility of a labelling error for the specimen cannot be dismissed and the species may be confused with the Lemon-breasted Flycatcher and Grey Whistler in the field; extension of the range of the Yellow-legged Flycatcher to the Atherton Region therefore remains unproved.

The Yellow-legged Flycatcher lives on the fringes of rainforest[1356] and is a species about which so little is known that its nest was not found in Australia until 1977.[1353] It feeds on insects, catching prey flushed from the outer foliage and from vine covered branches.[1356]

There are no measures of abundance.

Field Atlas Records: 34

1° blocks–with records: 5 (1%)

–with breeding records: 1 (20%)

Atlas No.: 395

Max. Rep. Rate: 33%
In block(s): 12/143

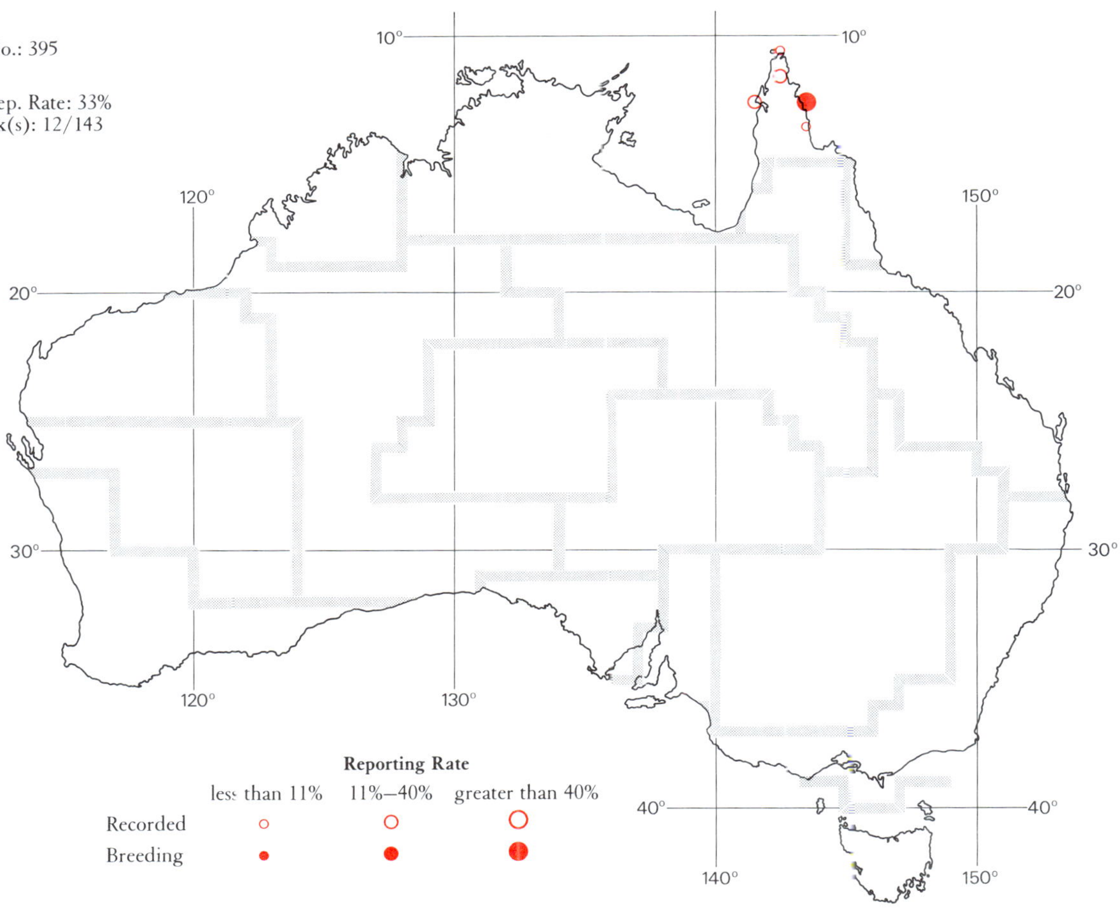

Lemon-bellied Flycatcher

Microeca flavigaster

Outside Australia the Lemon-bellied Flycatcher lives in New Guinea. In Australia it is divided at the Gulf of Carpentaria into two subspecies: *terraereginae* to the east; *flavigaster* to the west. The gap between them in the Field Atlas is consistent with that in the Historical Atlases. There is no confirmed record of the species in the Kimberley Region.

There is a suggestion that the Lemon-bellied Flycatcher is more abundant round Townsville (19/146) Apr.-Oct. each year.[1044] Other observers in the Atherton Region report no variation in numbers[231,698] and the Field Atlas does not suggest any large-scale seasonal movement.

At Darwin (12/130) the Lemon-bellied Flycatcher inhabits rainforest and occasionally mangroves; at Oenpelli (12/133) and Pine Ck (13/131) it lives along pool edges and in isolated trees in flower; on the coast of the Gulf of Carpentaria it is in the densest vegetation; and on the eastern Qld coast in eucalypt forest, woodland and clearings fringing rainforest.[748] It has also been seen in paperbark woodland and at times in gardens.[1703]

The Lemon-bellied Flycatcher feeds on ants, caterpillars and other sedentary insects[748] and is also frequently observed making sallies for flying insects from dead limbs[1356] and gleaning food from the outer leaves of forest trees.[1147,1483]

Density at Port Moresby (9S/147E) in New Guinea in eucalypt woodland was 0.3 birds/ha.[158]

Field Atlas Records: 915

1° blocks–with records: 69 (8%)
–with breeding records: 16 (23%)

Atlas No.: 379

Max. Rep. Rate: 48%
In block(s): 13/130

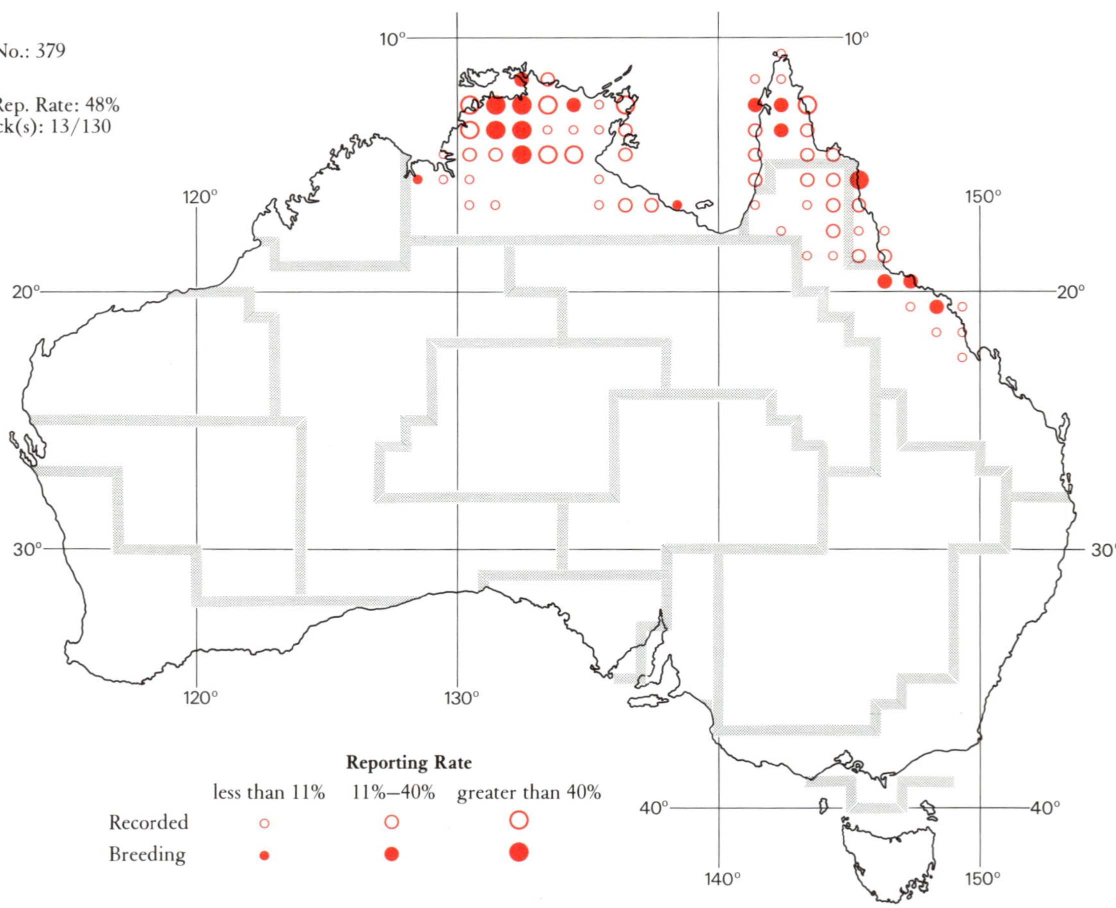

Kimberley Flycatcher

Microeca tormenti

The Kimberley Flycatcher is endemic to the Kimberley Region. The eight records in the Field Atlas are in 1° blocks at the southern end of its range which lies between Napier Broome Bay (14/126) and Barred Ck (17/122). The only breeding report comes from Napier Broome Bay where eggs were collected in 1909.[1858] Records in the Top End in historical periods have proved erroneous.[1377] Nothing is known of the species' movements.

The Kimberley Flycatcher lives singly or in small parties, keeping to the thickest patches of mangroves subject to regular flooding.[748,787] Its presence is betrayed by a loud call, said to be 'honeyeater-like' in sound.[787] The Kimberley Flycatcher feeds on crabs, flies[748] and other invertebrates of the tidal mangroves.

There are no measures of abundance.

Field Atlas Records: 3

1° blocks–with records: 3 (1%)

–with breeding records:

Atlas No.: 378

Max. Rep. Rate: 4%
In block(s): 17/123

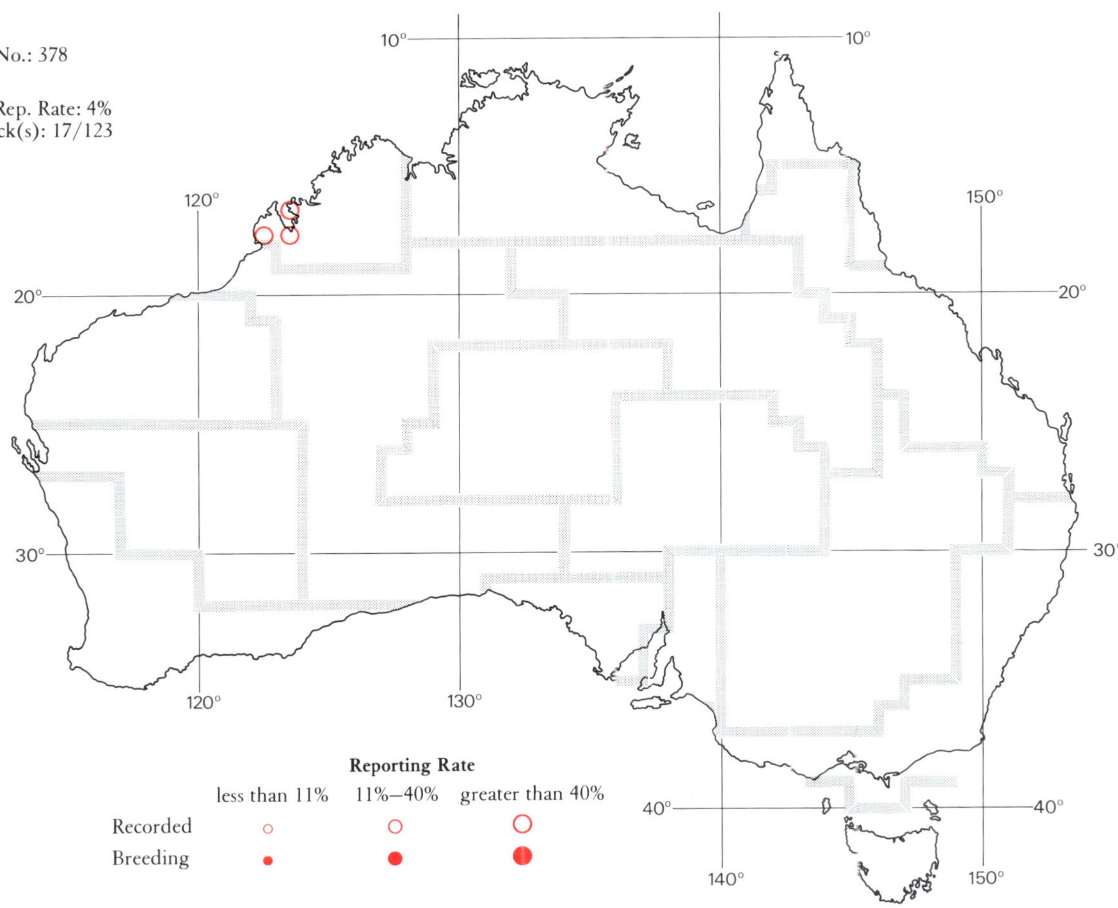

Jacky Winter

Microeca leucophaea

Outside Australia the Jacky Winter inhabits only a small area of New Guinea. In Australia it has been divided into four subspecies[986] but no breaks are apparent in its distribution on the Field Atlas map. The subspecies are: *pallida* in the north; *leucophaea* in the east; *assimilis* westward from the SA Ra. Region; and *barcoo* in the inland. The Jacky Winter has never been recorded in Tas. and rarely in the Central Ra. Region where sightings appear attributable to occasional influxes of a few birds. Comparison of Historical and Field Atlases indicates that it may have extended its range slightly towards the coast in the South-West Region.

Qld populations are thought to be augmented by southern birds in autumn and winter[1709] but at Wollomombi (30/152) colour-banding has shown the population to be sedentary.[129] At Jamieson (37/146) in eucalypt forest the Jacky Winter is present Sept.-Mar., moving to adjacent open lakeside areas for the winter.[1469] The Field Atlas does not suggest any large-scale seasonal movement and the bird is usually considered to be sedentary.

The Jacky Winter lives where cleared areas or heath provide open country against eucalypt forest. It is generally scarce along the coast all round the continent and is more often present in eucalypt woodland and mallee west of the Great Dividing Ra. It is uncommon even in the drier parts of the Atherton Region.[231] It inhabits sites with little undergrowth and is often round farms and gardens. The Jacky Winter lives in pairs or joins mixed species flocks and feeds by sallying after flying insects from a perch and by taking insects from the ground.

Density at Wollomombi was about 0.2 birds/ha[129] and at Jamieson in eucalypt forest 0.1 birds/ha.[1469]

Field Atlas Records: 13 413

1° blocks–with records: 477 (59%)
–with breeding records: 145 (30%)

Atlas No.: 377

Max. Rep. Rate: 79%
In block(s): 27/147

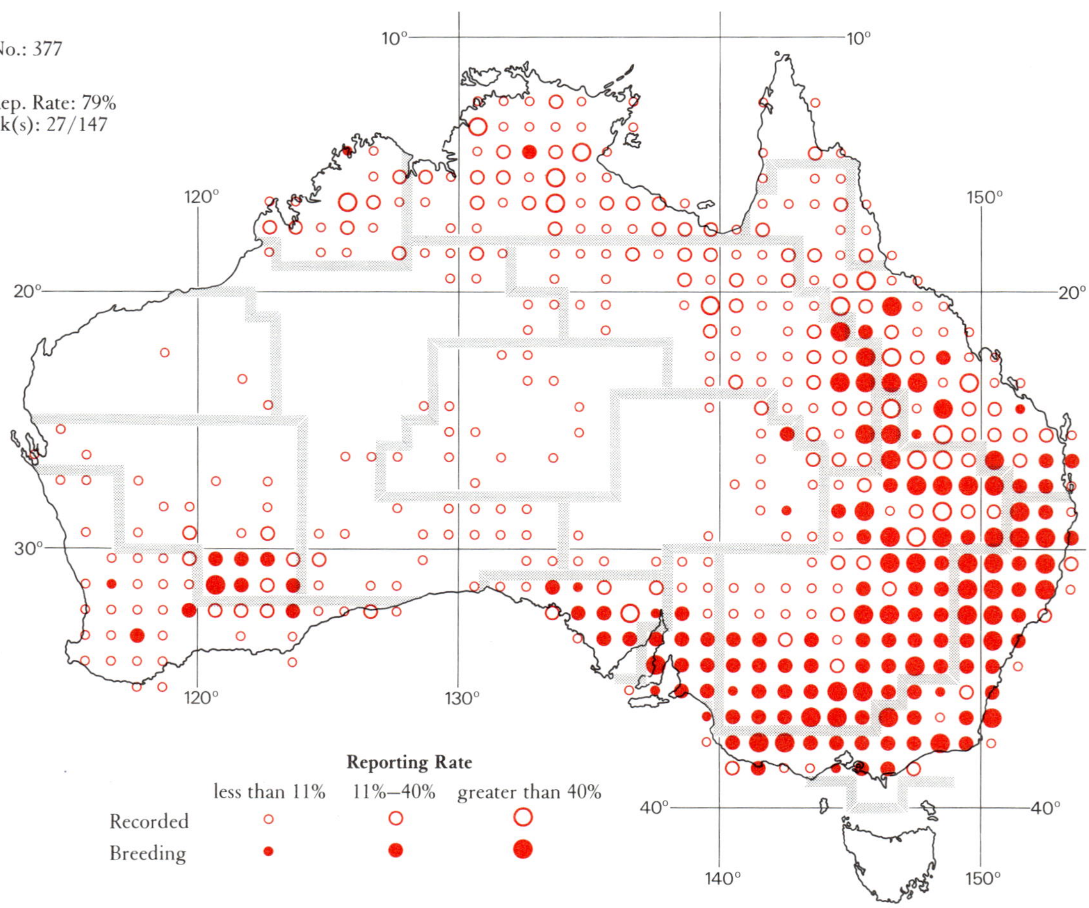

Pale-yellow Robin

Tregellasia capito

Two subspecies of the endemic Pale-yellow Robin are recognized: *nana* in the Atherton Region and *capito* in the Eastern Qld and the South-East Regions. The species' distribution has not changed over the historical periods. It breeds throughout its range.

The Pale-yellow Robin is reported to be present all year at several places and presumably is so throughout its range.[231,698,1292] It inhabits rainforest, often the very dense parts specially where lawyer vine grows.[354] The bird lives at all altitudes from the coast to as high as 1500 m in the Atherton Region.[231,239,698,997]

The Pale-yellow Robin feeds on insects, including ants and beetles. Prey is captured on the ground or in sorties from a vantage point.[354] Tiny red-brown seeds have been found in one gut[748] but whether seeds form a significant part of the diet is not known.

The Pale-yellow Robin is considered one of the commonest species in the rainforest of the Atherton Region[695] but there are no measures of abundance.

Field Atlas Records: 621

1° blocks–with records: 23 (3%)
–with breeding records: 12 (52%)

Atlas No.: 396

Max. Rep. Rate: 34%
In block(s): 15/145

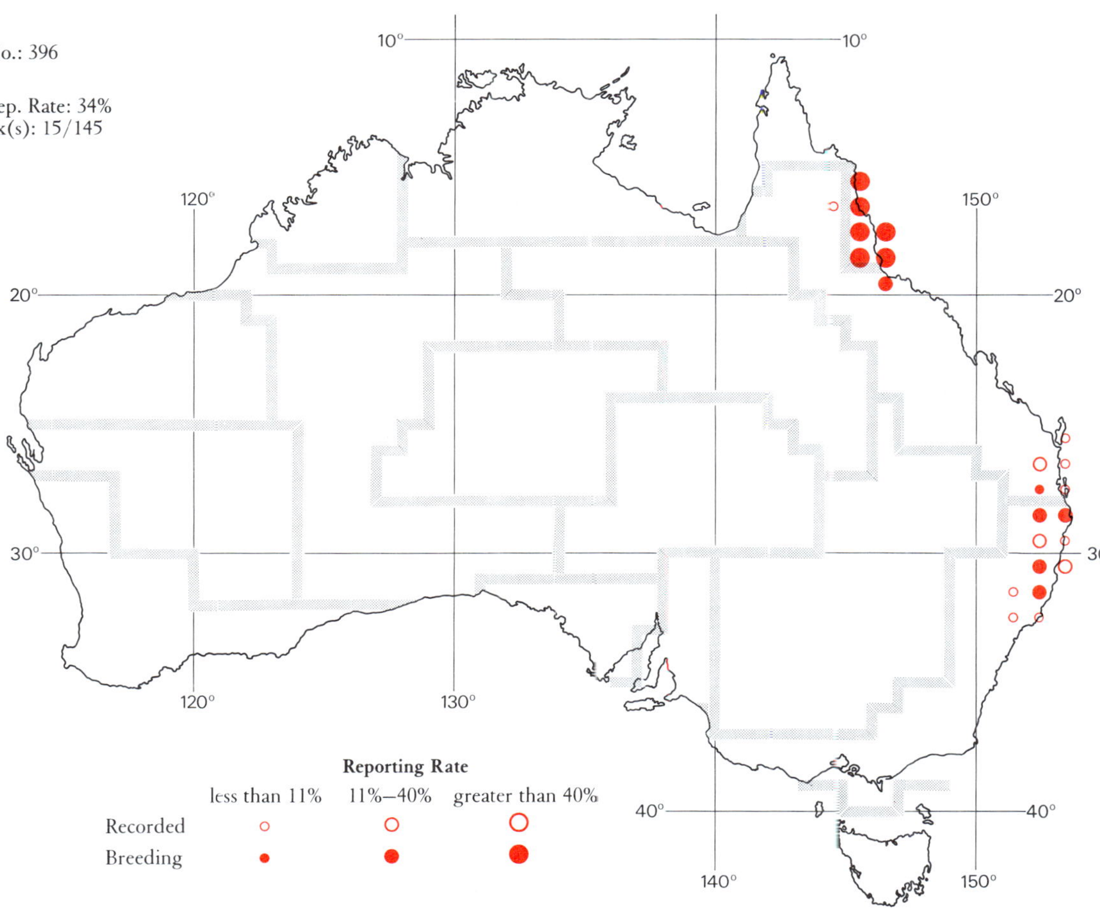

White-faced Robin

Tregellasia leucops

The White-faced Robin lives in New Guinea, where there are nine subspecies,[1449] and in Australia where there is a tenth, *albigularis*.[1709] In the Cape York Region the species has been recorded from the tip (10/142) down the eastern coast to the Rocky R (13/143). It breeds throughout its range but nothing is known of its movements.[1145] The bird lives in rainforest, including that along streams, and feeds on insects on the ground. It perches on the limbs, trunks and twigs of shrubs and on vines, making sallies for its prey.[1356]

The White-faced Robin is considered reasonably common[164] but there are no measures of abundance.

Field Atlas Records: 41

1° blocks-with records: 3 (1%)
-with breeding records: 1 (33%)

Atlas No.: 397

Max. Rep. Rate: 49%
In block(s): 12/143

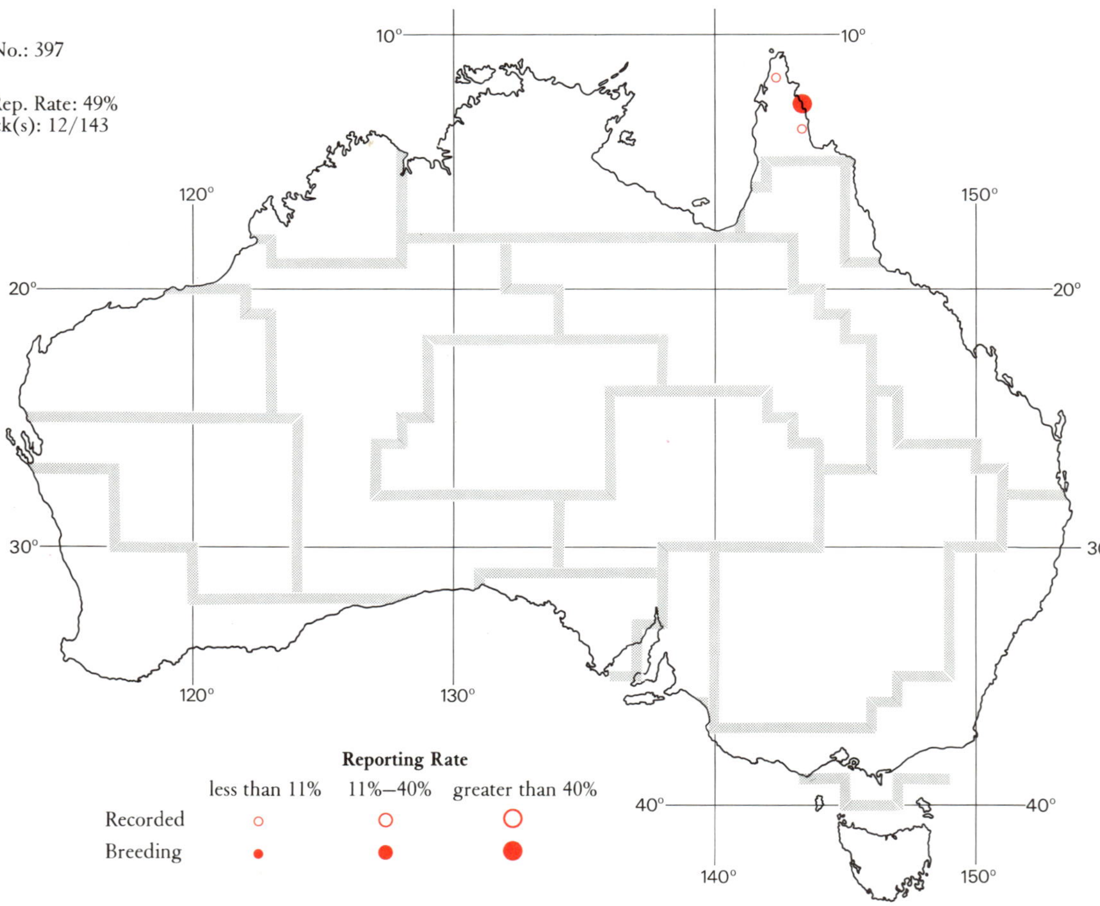

White-browed Robin

Poecilodryas superciliosa

There are two subspecies of the endemic White-browed Robin: *superciliosa* east of 140°E and *cerviniventris* to the west. Taking the Historical and Field Atlases together, no gaps are evident within the range of *superciliosa*, which appears unchanged from historical periods.

The distribution of *cerviniventris* is closely tied to river systems and is therefore disjunct. In the western part of the Top End Region it extends inland along the Victoria R (131°E) and, in the Kimberley Region, along the Fitzroy R (18°S). There are also records in the 1970s from rivers round the northern Kimberley coast. The species' range in the Kimberley Region has shrunk.[1690] There have been no records since 1904 on the Ord R upstream from Wyndham (15/128) where the bird was once described as common.[999] Since the turn of the century the White-browed Robin has also disappeared from the lower reaches of the Fitzroy R.

There is no evidence of movement. The White-browed Robin usually lives in pairs though one observer noted threes as common.[1602] It does not associate with other species, each group maintaining a discrete territory. In the east, it lives in rainforest along streams and in woodland near the coast.[741] In the west it occupies mangroves at the edge of river deltas as well.[1690]

The White-browed Robin feeds on caterpillars and tree-ants, as well as beetles, grasshoppers and other ground-dwelling insects.[748,1218] Much of its time is spent feeding on the ground, though it has been observed to take insects from tree trunks.[212] In mangroves it takes crabs, molluscs and other invertebrates.[1855] On the Roper R (14/135) the White-browed Robin is still often recorded though the riverside vegetation is heavily grazed and burnt. Its decline in the Kimberley Region is attributed to trampling of the understorey vegetation by cattle.[1690]

There are no measures of abundance.

Field Atlas Records: 340

1° blocks–with records: 48 (6%)

–with breeding records: 5 (10%)

Atlas No.: 390

Max. Rep. Rate: 36%
In block(s): 12/143

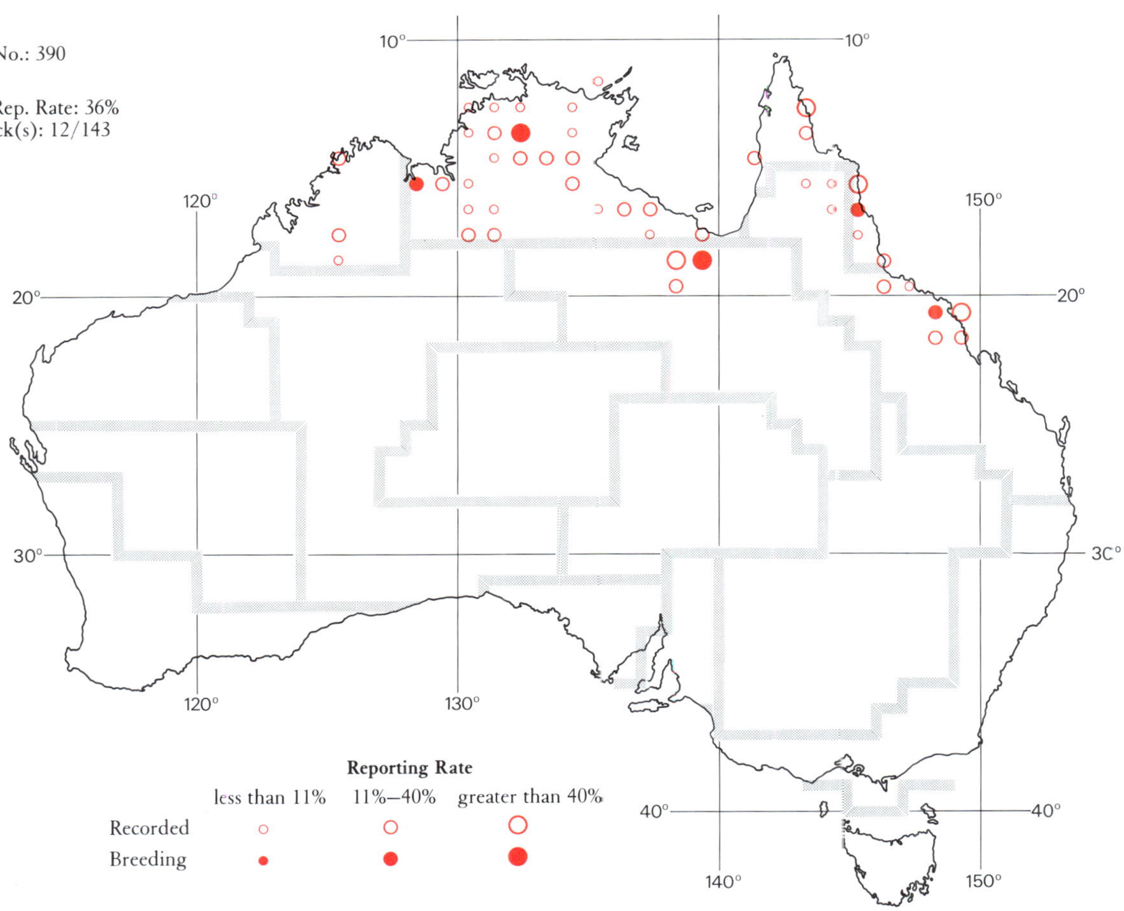

Grey-headed Robin

Poecilodryas albispecularis

For long known as *Heteromyias cinereifrons*, the endemic Grey-headed Robin has recently been grouped as a species with *H. albispecularis* of New Guinea and the genus *Heteromyias* sunk in *Poecilodryas*.[1576,1709] The species is present all year throughout its range.[231,698] It is confined to country above 200 m in the Atherton Region, inhabiting rainforest. The bird is most often seen feeding on the fringes, along tracks and at the edges of clearings. It feeds on insects and their larvae, caught on the ground.[748,1187] Generally the bird perches on a low branch or sideways on the trunk of a tree and darts down on its prey. Occasionally it hawks for insects such as termites on the wing.[643] It is abundant in the rainforest of northern Qld but there are no measures of its abundance.

Field Atlas Records: 299

1° blocks–with records: 8 (1%)
–with breeding records: 5 (63%)

Atlas No.: 389

Max. Rep. Rate: 27%
In block(s): 17/145

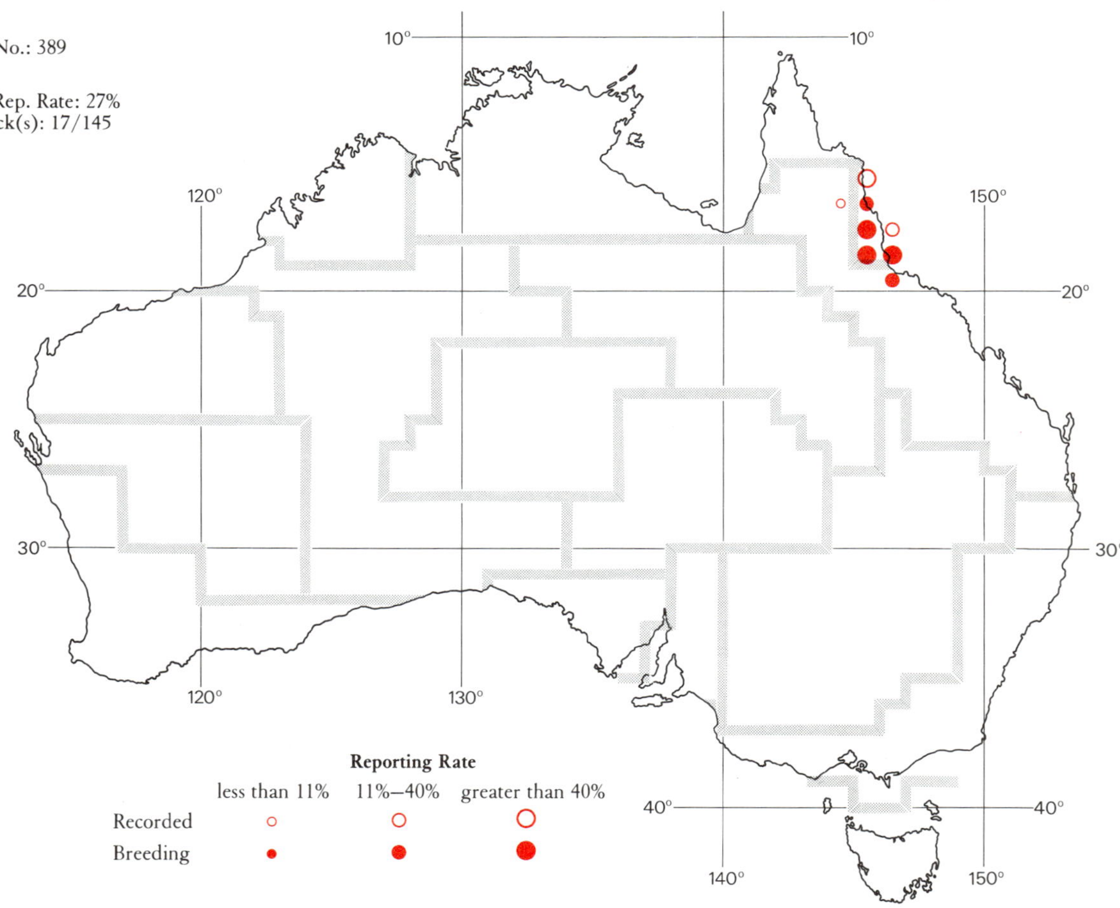

Crested Shrike-tit

Falcunculus frontatus

There are three subspecies of the endemic Crested Shrike-tit: the populations in the north, *whitei*, and west, *leucogaster*, are separated from each other and from *frontatus*, the main population in eastern Australia.[1232] The northern *whitei* has always been sparsely distributed. In the Top End there are records before 1950 from the additional localities of Larrimah (15/133) and the McArthur R (16/136).[1232] The species was also recorded from the Kimberley Region at Napier Broome Bay (14/126) in 1909–10,[787] the Ord R (16/128) in 1955[1669] and on the Drysdale R (15/126) in 1976.

The population in the Atherton Region may be isolated from the main range of *frontatus*, as is that in the SA Ra. Region. In the Murray-Darling Region *frontatus* extends inland along river systems but on the Murray R itself only as far downstream as Ponde (34/139).[969] The distribution of *leucogaster* has been said to extend north to Geraldton (28/114)[1428] and Morawa (29/116)[1638] but no actual records, either historical or recent, could be found north of 30°S.

The Crested Shrike-tit is sedentary.[341,871,1016,1099] Groups of 3–5,[876] often in mixed species flocks,[129] live in eucalypt forest and woodland, specially that dominated by river red gums that fringe major watercourses.[833,969] Communal feeding of young has been suggested.[876] The Crested Shrike-tit tears bark from the trees in its search for insects. In Vic. honeyeaters, particularly New Hollands, actively drive Shrike-tits away from manna gum in which both species feed.[1402]

In eucalypt woodland at Armidale (30/151) densities of 0.13–0.15 birds/ha[876] and 0.29 birds/ha[582] were recorded; at Wollomombi (30/152) in eucalypt woodland 0.2 birds/ha;[129] and at Moruya (35/150) in eucalypt woodland 0.08–0.13 birds/ha.[1197] Densities in eucalypt forest were: Jamieson (37/146) 0.31 birds/ha[1469] and Boola Boola (38/146) 0.3 birds/ha.[1113]

Field Atlas Records: 5970

1° blocks–with records: 139 (17%)
–with breeding records: 57 (41%)

Atlas No.: 416

Max. Rep. Rate: 39%
In block(s): 35/147

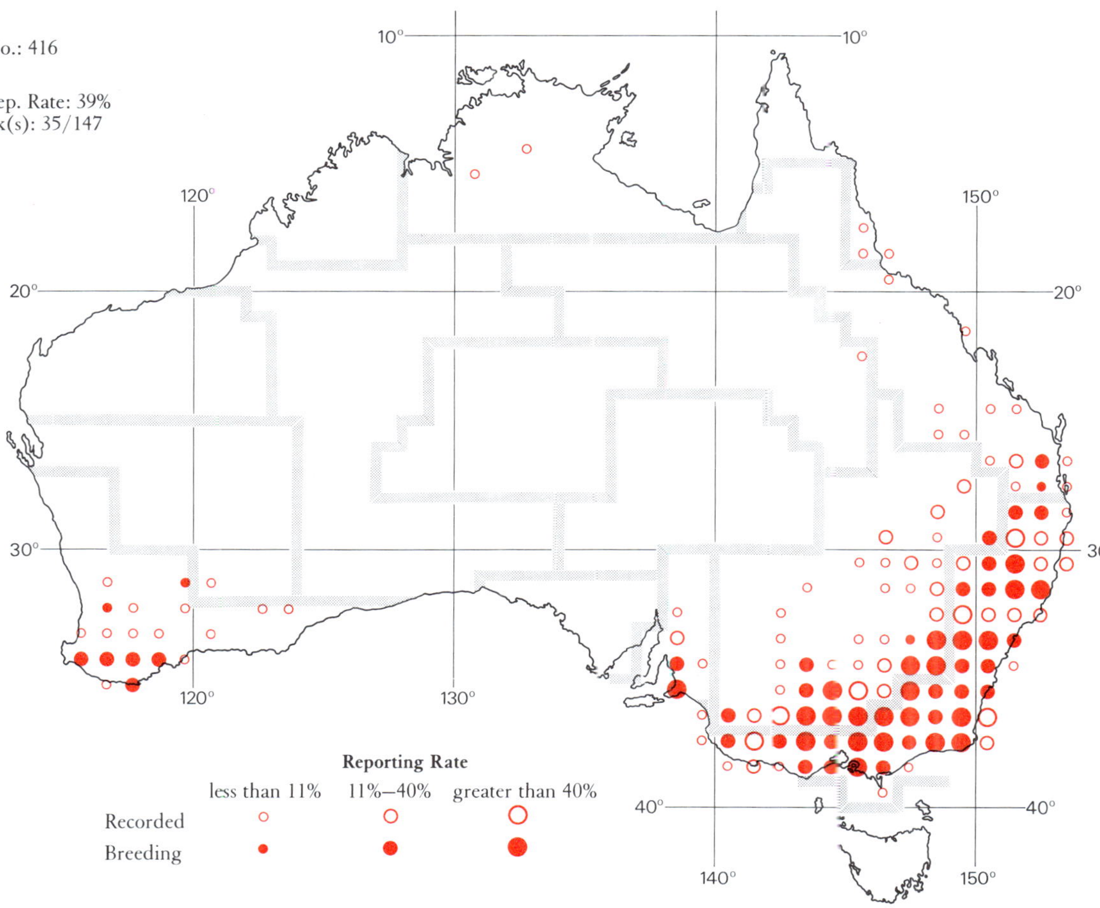

Olive Whistler

Pachycephala olivacea

The endemic Olive Whistler has been recorded as far north as Mistake Mts (28/152)[353] and breeding north to 28°S. It has not been seen near Sydney (33/151) since the 1920s. The 10′ maps show the population west of 142°E to be isolated.

A partial migration in winter, from the mountains to lower altitudes, is thought to take place near Canberra (35/149)[653] and the Field Atlas supports these observations. In the Snowy Mts (36/148) the Reporting Rate above 500 m was 29% in summer and 0% in winter; below 500 m it was 20% in summer and 36% in winter. A similar altitudinal movement is suspected in the McPherson Ra. (28/152).[1709] At Boola Boola (38/146) an increase in numbers Apr. and Sept. has been noted and is thought to represent passage of birds through the area.[1113]

In northern parts of its range, including the McPherson Ra., New England National Park (30/152) and Barrington Tops (32/151), the Olive Whistler lives mainly in rainforest and eucalypt forest above 500 m.[129,396,1356] Farther south it inhabits eucalypt forest and, on the Vic. coast, Flinders I. (40/148), King I. (39/144) and the western coast of Tas., it lives in woodland near the coast, specially that dominated by teatree.[1314] In Tas. there is a tendency for the bird to be outside the thickest vegetation in winter on Mt Wellington (42/147) but it was not observed during a detailed study of dry eucalypt forest nearby.[1453] The Olive Whistler feeds mainly on the ground where it forages for insects, specially large beetles and weevils.[1049]

At Boola Boola it was so rare that no density measure could be obtained.[1113]

Field Atlas Records: 1881

1° blocks–with records: 56 (7%)
–with breeding records: 20 (36%)

Atlas No.: 405

Max. Rep. Rate: 40%
In block(s): 43/146

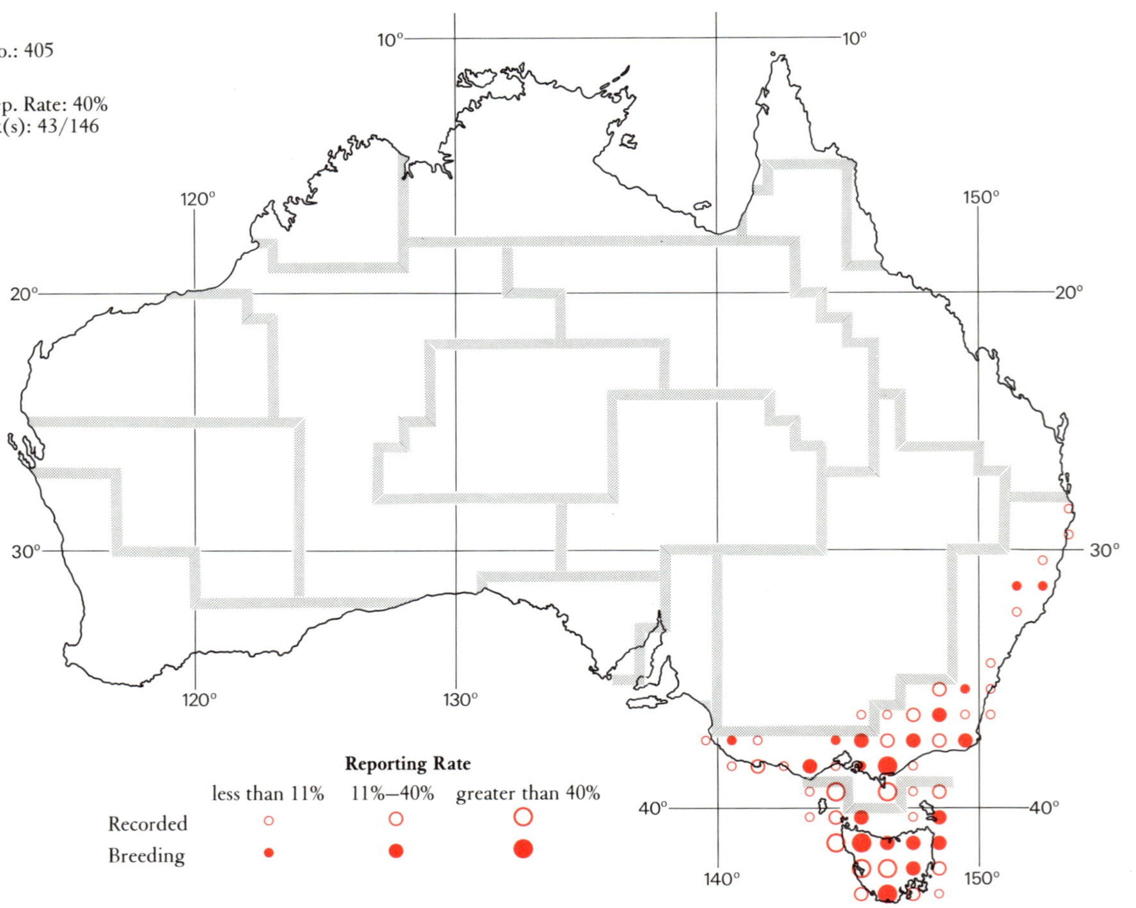

Red-lored Whistler

Pachycephala rufogularis

The range of the endemic Red-lored Whistler has contracted. An outlying population round Pulletop Nature Reserve (33/146,34/146), present in the 1951–76 period, was not recorded during the Field Atlas and may have disappeared. The species also bred at least as far south as Yanac (36/141) in 1935[348] but has not been recorded there since. The original specimens of the species were collected by Gould within 3 km of Adelaide (34/138)[1489] but the closest records in the Field Atlas were about 100 km distant.

The Red-lored Whistler may not be as sedentary as is usually thought. A bird found dead at Para Hills (34/138) in Oct. 1976 may have been a vagrant or a migrant.[1369]

The species is confined not only to mallee in south-eastern Australia but to characteristic vegetation formations within it. At Pulletop it lived in mallee where broombush and native pines, *Callitris verrucosa* or *C. canescens* dominate[1575] though there were similar patches of mallee from which the bird was absent. In the Big Desert (35/141) it is confined to mallee with low ground cover, specially the creeper *Cassytha*,[1369] whereas the Golden Whistler inhabits the taller and more open stands of broombush. The Red-lored Whistler feeds on insects and sometimes berries.[1049,1400] Most food is taken on the ground but, while in flight, the bird also picks insects from the vegetation. It apparently does not survive disturbance to the mallee. Its dependence on ground-dwelling insects would also make it extremely sensitive to fires, specially to regular burning off.

There are no measures of abundance.

Field Atlas Records: 85

1° blocks-with records: 10 (1%)

-with breeding records: 2 (20%)

Atlas No.: 402

Max. Rep. Rate: 13%
In block(s): 35/141

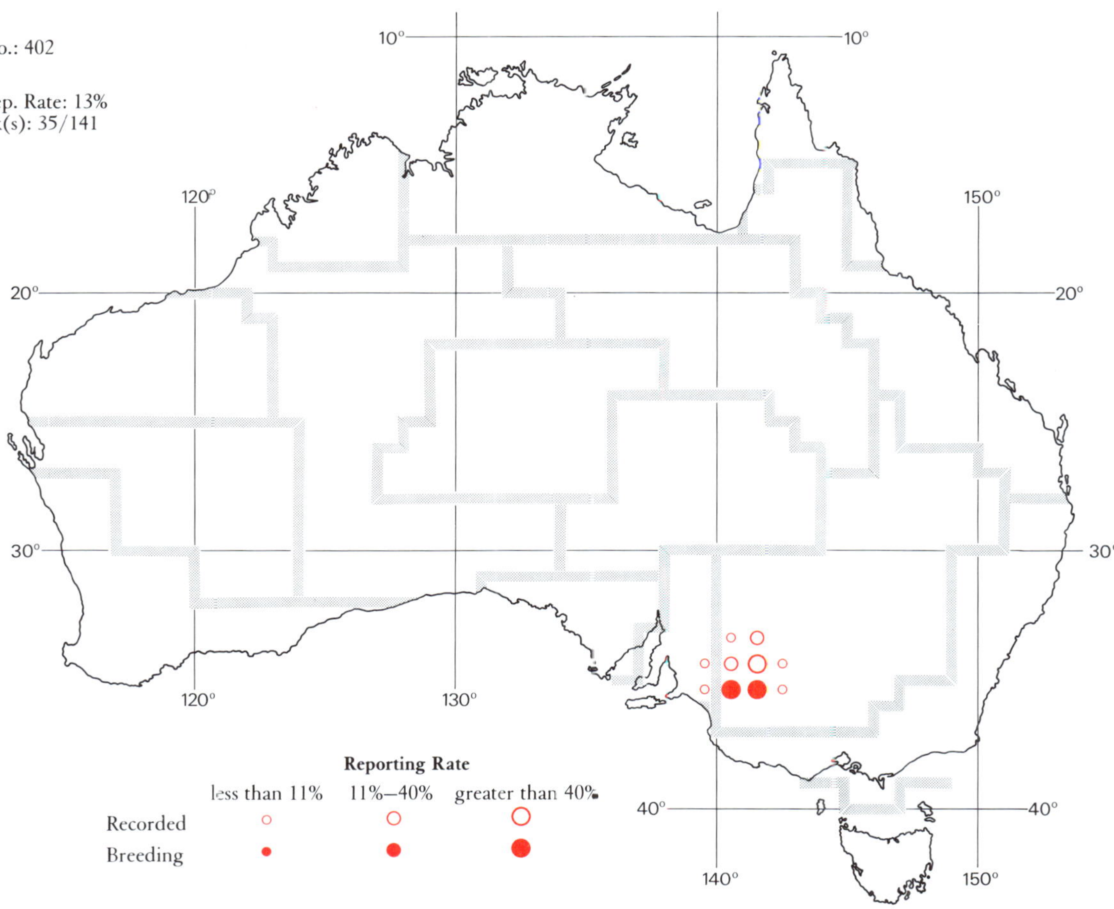

Gilbert's Whistler

Pachycephala inornata

The endemic Gilbert's Whistler is distributed almost continuously throughout the southern mallee, crossing the Great Victoria Desert through the fringe of mallee at its southern edge.[599] Lack of records from the Western Deserts during the Field Atlas reflects the few visits made to the area rather than change in distribution.

In 1947 two birds were collected on the coastal plain south of Perth (31/115)[599] and there were sightings to the east of Perth (31/116, 31/117) in the 1950s but none since despite good coverage.

The Field Atlas does not suggest any large-scale seasonal movement. Gilbert's Whistler usually lives in mallee and at times in taller eucalypt woodland or partly cleared country.[209] It feeds on large insects, specially caterpillars, beetles and weevils[1049] taken on the ground[1400] and sometimes from trees.[206]

Near Cowra (33/148) three pairs nested in about 25 ha (0.24 birds/ha) but there was a substantial distance between that group of nests and the next.[209]

Field Atlas Records: 553

1° blocks–with records: 78 (10%)
–with breeding records: 17 (22%)

Atlas No.: 403

Max. Rep. Rate: 26%
In block(s): 29/118

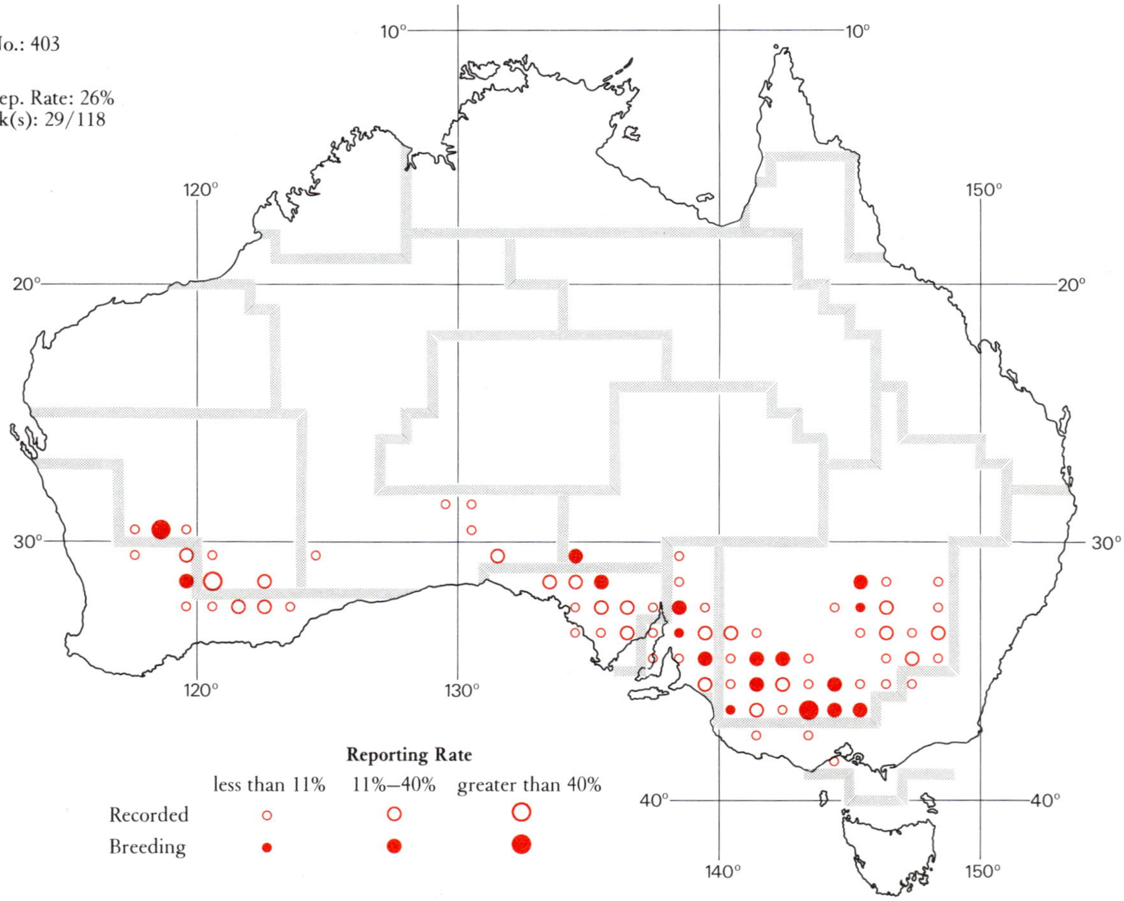

Golden Whistler

Pachycephala pectoralis

There are different views about the overseas range and subspecies of the Golden Whistler.[679,748,1124] In the South-West Region historical records show its range to be continuous east to 31/128. Ranges of the Golden and Mangrove Golden Whistlers adjoin on the eastern coast of Australia but they occupy different vegetation. The Mangrove Golden Whistler has been collected on the Qld coast at Cape Grenville (11/143), Cape Gloucester (20/148) and Springcliff (20/148).[679,748] In WA the Golden Whistler has been recorded as far north as Tamala (26/113) in the breeding season[600] more than 200 km from the southernmost sighting of the Mangrove Golden Whistler at Carnarvon (24/113).[748]

There appear to be both sedentary and migratory populations of the Golden Whistler. In the Atherton Region Reporting Rates differed little (summer 16.0%, winter 13.7%). In the Eastern Qld Region they rose from 6.2% in summer to 15.2% in winter and in the Murray-Darling Region from 2.4% in summer to 10.4% in winter. Some observations reflect this seasonal pattern: the species is present at Wollomombi (30/152) and Orange (33/149) Mar.-Sept.[129,784] and in the southern Murray-Darling Region Apr.-Oct.[833] At Moruya (35/150) and Jamieson (37/146) it is sedentary.[1192,1469] Some birds migrate altitudinally: they breed Sept.-Apr. in mountains near Canberra (35/149)[1016] but visit the city only Mar.-Nov.[154] (Reporting Rates above 500 m in the Alps of the South-East Region: spring-summer 32%, autumn-winter 0%). In the Tas. Region Reporting Rates differed little: summer 20.2%, winter 21.2%. In WA, inland records, e.g. near Tom Price (22/117) in July 1977, indicate that some birds migrate north and east out of the South-West Region in winter. In the Goldfields Region Reporting Rates rose from 0.5% in summer to 2.9% in winter.

The Golden Whistler lives singly or in pairs in eucalypt forest, woodland and heath usually where there is undergrowth. It feeds mainly on insects from the ground and trees.[748,1049]

Densities in eucalypt woodland were: Armidale (30/151) 0.43–0.79 birds/ha;[532] Wollomombi 0.20 birds/ha;[129] Canberra 0.15 birds/ha;[154] and Moruya (35/150) 0.28–0.42 birds/ha.[1197] In eucalypt forest they were: Jamieson 0.2–0.5 birds/ha[1469] and Boola Boola (38/146) 0.4–6.0 birds/ha.[111]

Field Atlas Records: 14 847

1° blocks–with records: 242 (30%)
–with breeding records: 87 (36%)

Atlas No.: 398

Max. Rep. Rate: 55%
In block(s): 37/149

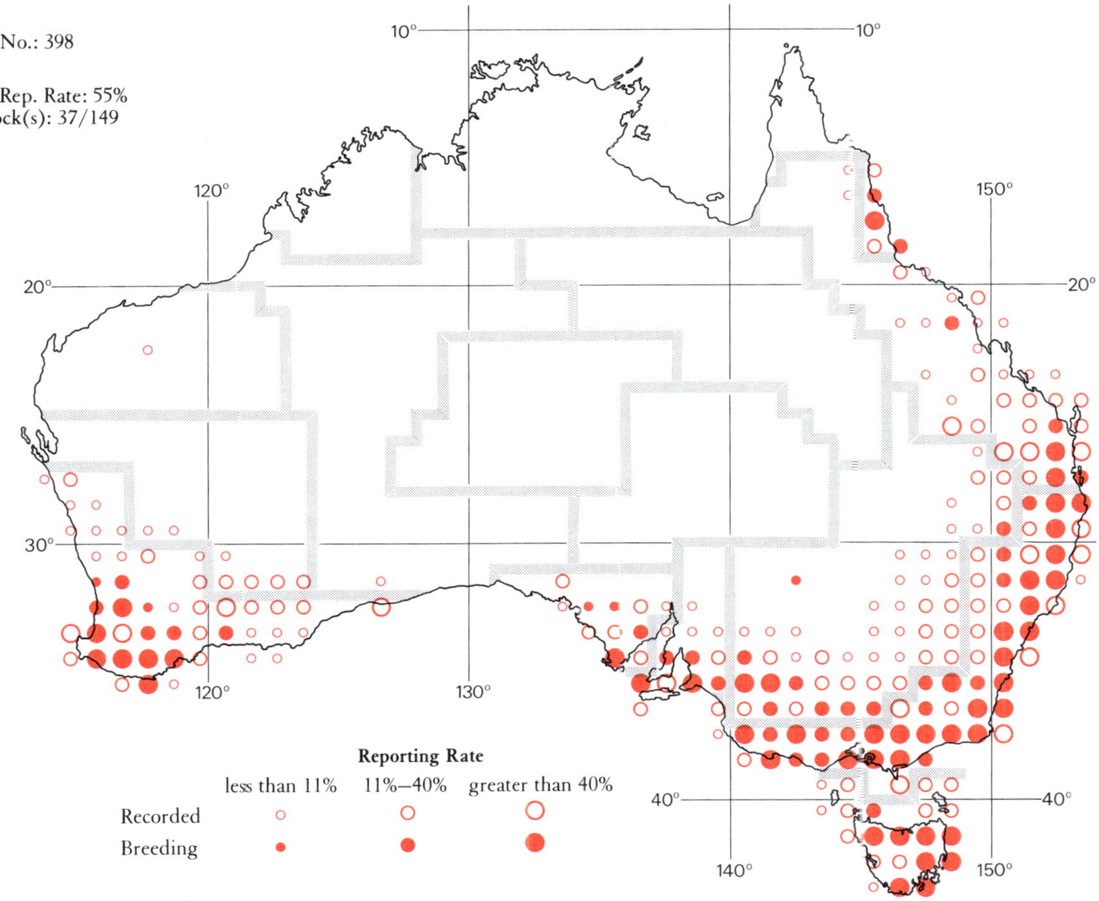

Mangrove Golden Whistler

Pachycephala melanura

Views differ about the overseas range and subspecies of the Mangrove Golden Whistler.[679,748] Historical records narrow but do not close the gaps in its range evident on the Field Atlas map. Some correspond with gaps in mangrove distribution round the coast;[622] others, particularly in the Cape York Region, probably represent lack of visits by observers. There are only two breeding records of the Mangrove Golden Whistler in the Historical Atlases and none at all in the Field Atlas but the species probably breeds throughout its range.

Specimens of the Mangrove Golden Whistler collected at Cape Grenville (11/143), Cape Gloucester (20/148) and Springcliff (20/148)[679,748] support Field Atlas observations of the species on the eastern coast of Qld. The relation between its range and that of the Golden Whistler in Qld is inadequately portrayed on the maps. Because they are produced on a 1° grid the two species are shown with overlapping ranges but they probably never meet. The Mangrove Golden Whistler lives in mangroves, rarely even in adjacent rainforest. The Golden Whistler inhabits eucalypt woodland and forest and rarely adjacent rainforest. The breeding records for the Golden Whistler shown in coastal 1° blocks (16/145, 17/145, 18/146) were all on or west of the Great Dividing Ra. The Golden Whistlers seen nearer the coast may have been non-breeding visitors to the lowlands. In WA a sighting of the Mangrove Golden Whistler near Carnarvon (24/113)[748] brings the species to about 200 km from the northernmost record of the Golden Whistler in the breeding season. During the Field Atlas two records were made of 'golden' whistlers at 16/125 and 16/126 near Mt Barnett; they are omitted from the map because there is doubt about which species they were.

The Mangrove Golden Whistler is a vagrant on some islands in Torres Strait;[503] nothing is known of its movements elsewhere. It usually lives singly or in pairs. On islands in the Great Barrier Reef it will live in vegetation other than mangroves, e.g. pisonia.[255,1709] The species feeds on insects gleaned from twigs and foliage and it will also take crabs, presumably from the mud.[748,1859] It is apparently absent from those larger islands in south-western Torres Strait inhabited by the Little Shrike-thrush.[503]

There are no measures of abundance.

Field Atlas Records: 99

1° blocks–with records: 30 (4%)
–with breeding records:

Atlas No.: 400

Max. Rep. Rate: 13%
In block(s): 10/142

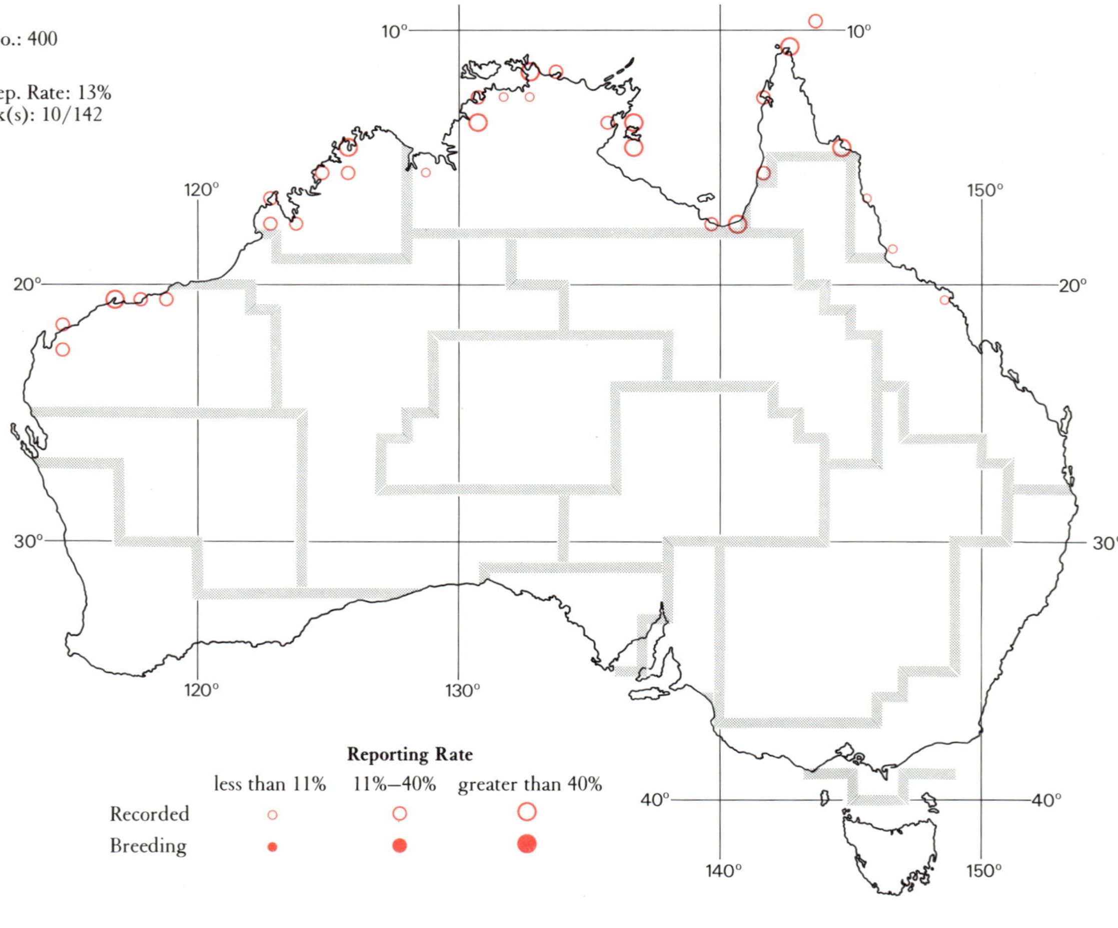

Grey Whistler

Pachycephala simplex

The Grey Whistler lives in New Guinea and adjacent islands and in Australia. The two subspecies in Australia are widely separated: *peninsulae* in Qld and *simplex* in the Top End.[748] Historical records link the three groups of Field Atlas records in Qld. In the Top End the distribution of *simplex* is also continuous.

The Grey Whistler is a lowland species (Reporting Rates in the Atherton Region were 13.5% below 500 m and 2.7% above). Birds that live at high altitude are thought to leave in winter[231] but the Reporting Rate above 500 m was so low that no seasonal pattern could be inferred. There is no other evidence of seasonal movement.

The Grey Whistler lives in rainforest, including regrowth, mangroves[1703] and occasionally eucalypt forest and woodland adjacent to rainforest.[423,748] The bird will stay motionless in dense cover for hours[856] but its call can be used to determine its presence in an area. The Grey Whistler takes a variety of insect food, seldom from the ground. It joins mixed flocks of small birds to forage.[108,160,748,1145]

In New Guinea in lowland rainforest near Port Moresby (9S/147E) density was 2 birds/ha.[160]

Field Atlas Records: 381

1° blocks–with records: 24 (3%)
–with breeding records: 1 (4%)

Atlas No.: 406

Max. Rep. Rate: 47%
In block(s): 12/143

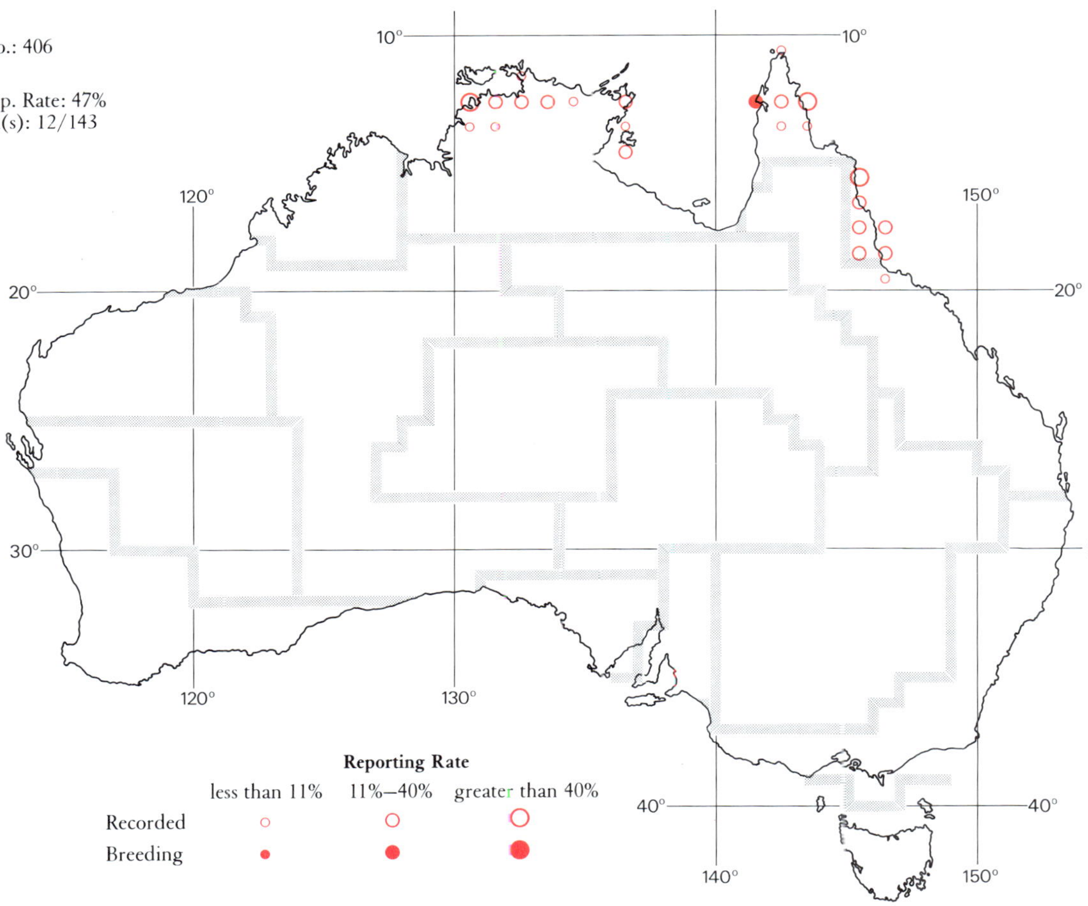

Rufous Whistler

Pachycephala rufiventris

Outside Australia the Rufous Whistler lives in New Caledonia and New Guinea and adjacent islands.[1124,1449] The New Guinean birds are sometimes considered a separate species.[129] In Australia the population varies clinally and there seems little justification for the recognition of subspecies.[748] The first record for King I. (39/144) was in 1968 and the only Field Atlas record for Tas. was at Green's Beach (41/146) in Oct. 1980. There are recent historical records from the western Nullarbor Plain though few in the Field Atlas.

There appears to be extensive movement by the Rufous Whistler in eastern Australia: north in autumn and south in spring. It leaves NSW in Apr.-May and returns in Aug.-Sept.[396] and is present at Canberra (35/149) Oct.-Apr.[155] In the South-East Region the Reporting Rate dropped from 38% in summer to 8% in winter. There was a corresponding increase in Eastern Qld from 21% in summer to 43% in winter and a suggestion of an increase in the Cape York Region also (spring 19%, summer 18%, autumn 31%, winter 26%). Birds from the south have been recorded round the Gulf of Carpentaria in winter[1709] but there is no record of the species from Torres Strait.[503] The longest movement recorded for a banded bird is that of an adult from Waikerie (34/139) to Red Cliffs (34/142), 201 km.[1436]

The Rufous Whistler lives singly or in pairs in most vegetational types apart from rainforest,[748] usually in thickets where there is dense cover beneath forest or woodland. At Wollomombi (30/152) migrants return to their previous territories each year.[129] The bird feeds on insects, mostly ground-dwelling, and occasionally joins mixed species flocks to forage.[155,748] In the South-West there is some evidence of its spreading as forest is cleared.[1638]

In eucalypt woodland densities were: Armidale (30/151) 0.15–0.71 birds/ha,[582] Wollomombi 0.8–1.4 birds/ha,[129] Canberra 0.3 birds/ha,[154] Moruya (35/150) 0.55 birds/ha[1197] and Bolgart (31/116) 0.06–0.28 birds/ha.[528] In eucalypt forest they were: Jamieson (37/146) 0.6–1.2 birds/ha[1469] and Boola Boola (38/146) 0.1–1.3 birds/ha.[1113] At Bathurst (33/149) in wet and dry eucalypt forest and pine plantation densities were 2.0, 0.25 and 1.25 birds/ha respectively.[489]

Field Atlas Records: 21 896

1° blocks–with records: 656 (81%)
–with breeding records: 151 (23%)

Atlas No.: 401

Max. Rep. Rate: 77%
In block(s): 27/149

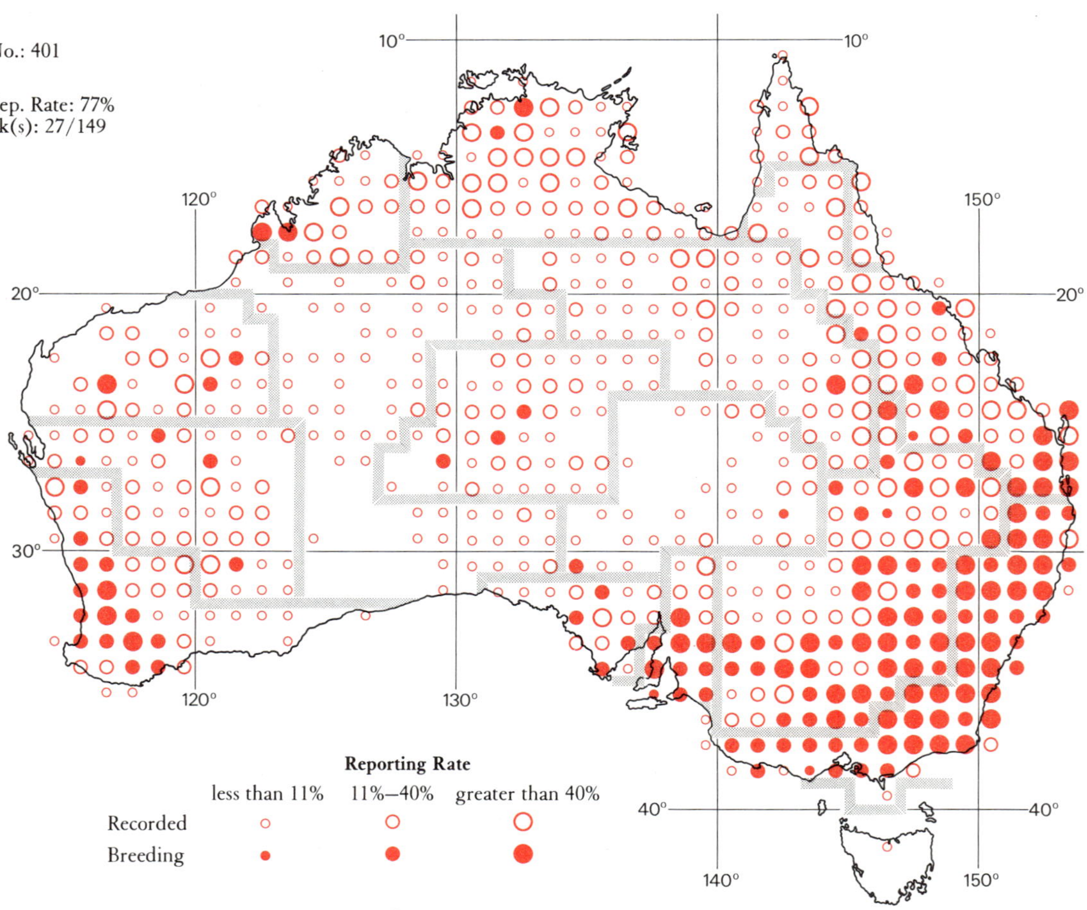

White-breasted Whistler

Pachycephala lanioides

The endemic White-breasted Whistler is divided into three subspecies: *carnarvoni* south of Eighty Mile Beach (19/121); *lanioides* between Eighty Mile Beach and Kalumburu (14/126); and *fretorum* to the east.[622,748] The Historical Atlases include records at Wyndham (15/128), the mouth of the Roper R (14/135) and Bing Bong Ck (15/136)[1711] partly closing some of the gaps on the map. Though there are no breeding records in the Field Atlas, the species is thought to be sedentary and to breed throughout its range.[748,1124]

All populations inhabit mangroves but in the Darwin (12/130) area the White-breasted Whistler is occasionally in adjacent rainforest.[423] It calls at dawn but is otherwise difficult to see because it tends to remain still for long periods, waiting for prey. The White-breasted Whistler feeds on crabs and large insects, specially grasshoppers.

The bird's territory occupies strips of mangroves about 50 m in length in WA but several hundred metres in Qld.[748]

Field Atlas Records: 94

1° blocks–with records: 23 (3%)

–with breeding records:

Atlas No.: 404

Max. Rep. Rate: 15%
In block(s): 20/117

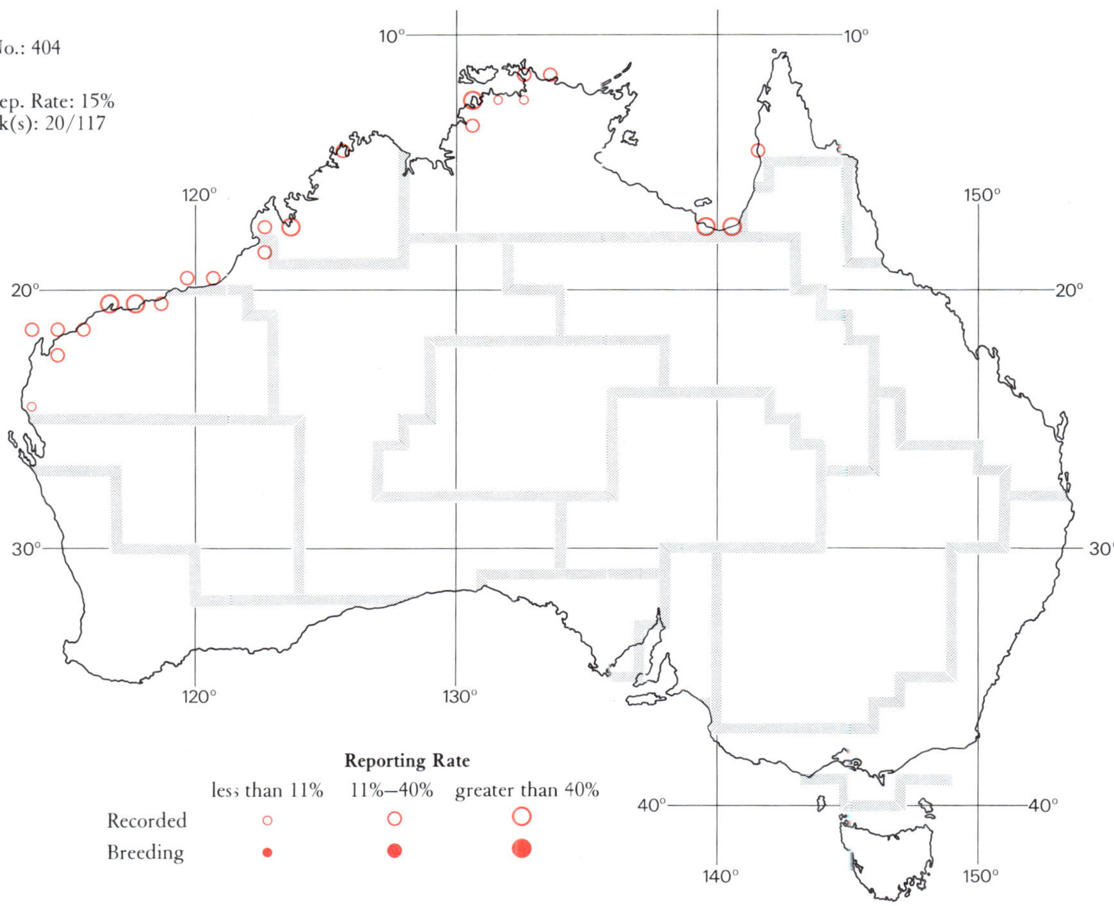

Little Shrike-thrush

Colluricincla megarhyncha

The Little Shrike-thrush inhabits New Guinea and adjacent islands and Australia. For many years the populations east and west of the Gulf of Carpentaria were regarded as separate species. Now, *parvula* to the east and *megarhyncha* to the west are considered conspecific.[615] Specimens taken at the mouth of the Macarthur R (15/136) have been described as an intermediate form, the extent of whose distribution is unknown.[1585] It was supposed that small populations along the coast of the Gulf of Carpentaria might link this intermediate form with birds in the Cape York Region. Mangrove and rainforest pockets in which such populations could live would need to be 5–10 ha in extent and close to each other.[1585] The Field Atlas sighting (17/139) presumably represents such a population.

The Little Shrike-thrush was present all year at a study site in New Guinea as well as on islands in Torres Strait.[160,503] It lives in the dense damp vegetation of rainforest and mangroves and is also reported in woodland near the coast and eucalypt forest.[104,1228] In NSW it is most often below 200 m and absent above 850 m.[1292] It eats large insects, spiders, beetles and crabs, feeding actively and tearing bark from trees to reveal prey.[78,748,1228] The species has been reported to tear open birds' nests to take the young. Where it occurs on large islands in south-western Torres Strait it appears to exclude the Mangrove Golden Whistler.[503]

Density in New Guinea in lowland rainforest near Port Moresby (9S/147E) was 4 birds/ha.[160]

Field Atlas Records: 1461

1° blocks–with records: 71 (9%)

–with breeding records: 17 (24%)

Atlas No.: 413

Max. Rep. Rate: 65%
In block(s): 12/143

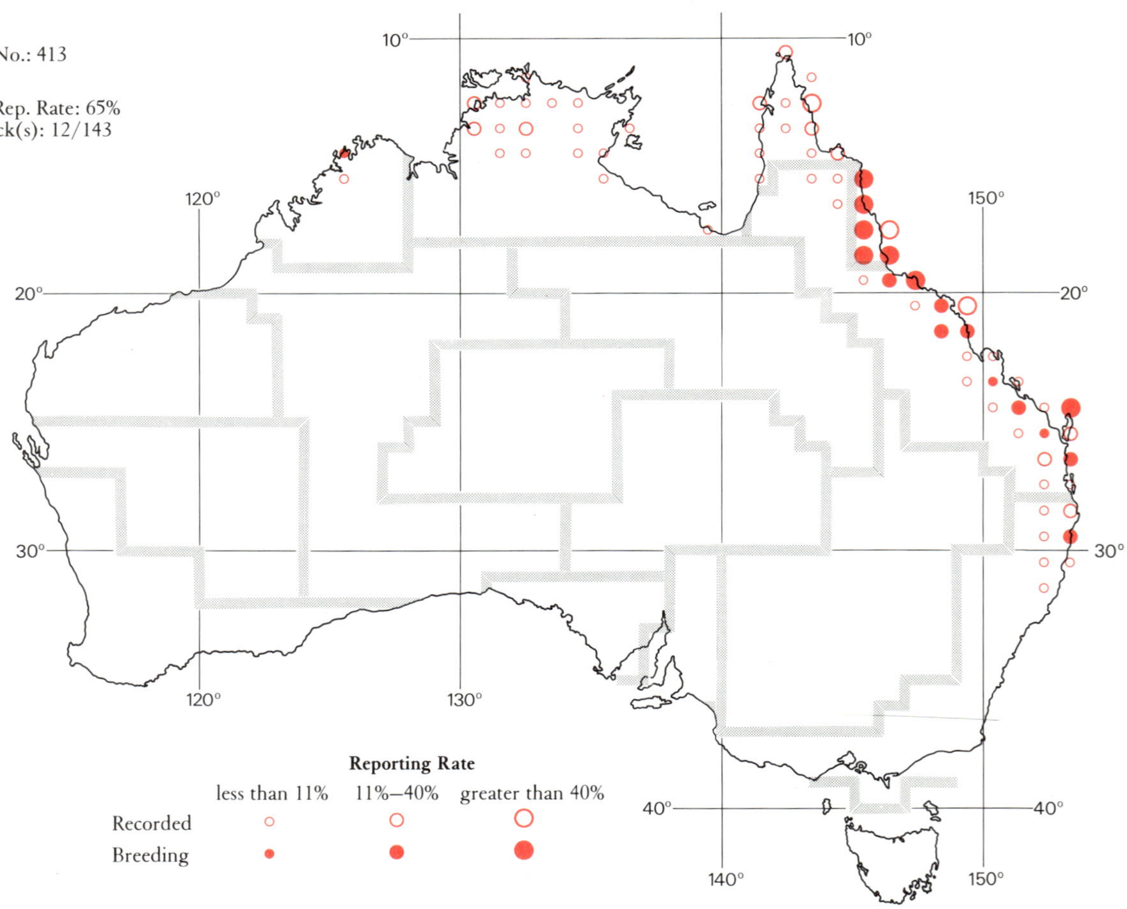

Bower's Shrike-thrush

Colluricincla boweri

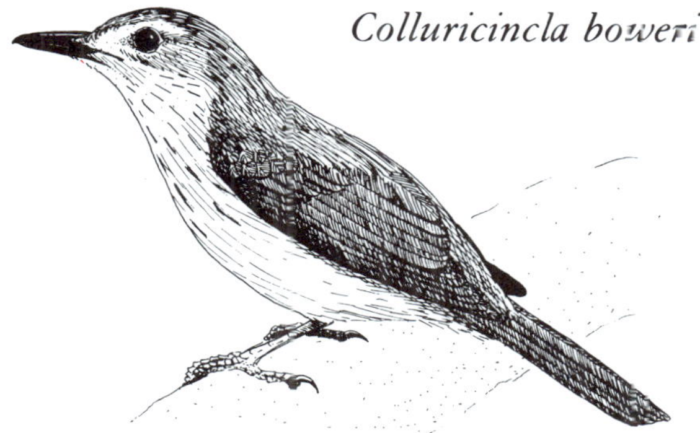

Bower's Shrike-thrush is endemic to the Atherton Region. It is usually in mountains above 250 m[231,698,1703] but observations of the bird near the sea at Tully (17/145) in winter[748] and near Cape Tribulation (16/145)[695] suggest that it may come lower in winter or where lowland rainforest is suitable for it. In the Field Atlas its Reporting Rate was 25% above 500 m and 0% below. It lives in rainforest and feeds on large insects at all levels of the forest.[1343] Occasionally it takes insects on the wing but usually waits to pounce on prey from a short distance above the ground.[748,1124,1356]

There are no measures of abundance.

Field Atlas Records: 191

1° blocks-with records: 7 (1%)

-with breeding records: 3 (43%)

Atlas No.: 414

Max. Rep. Rate: 17%
In block(s): 15/145

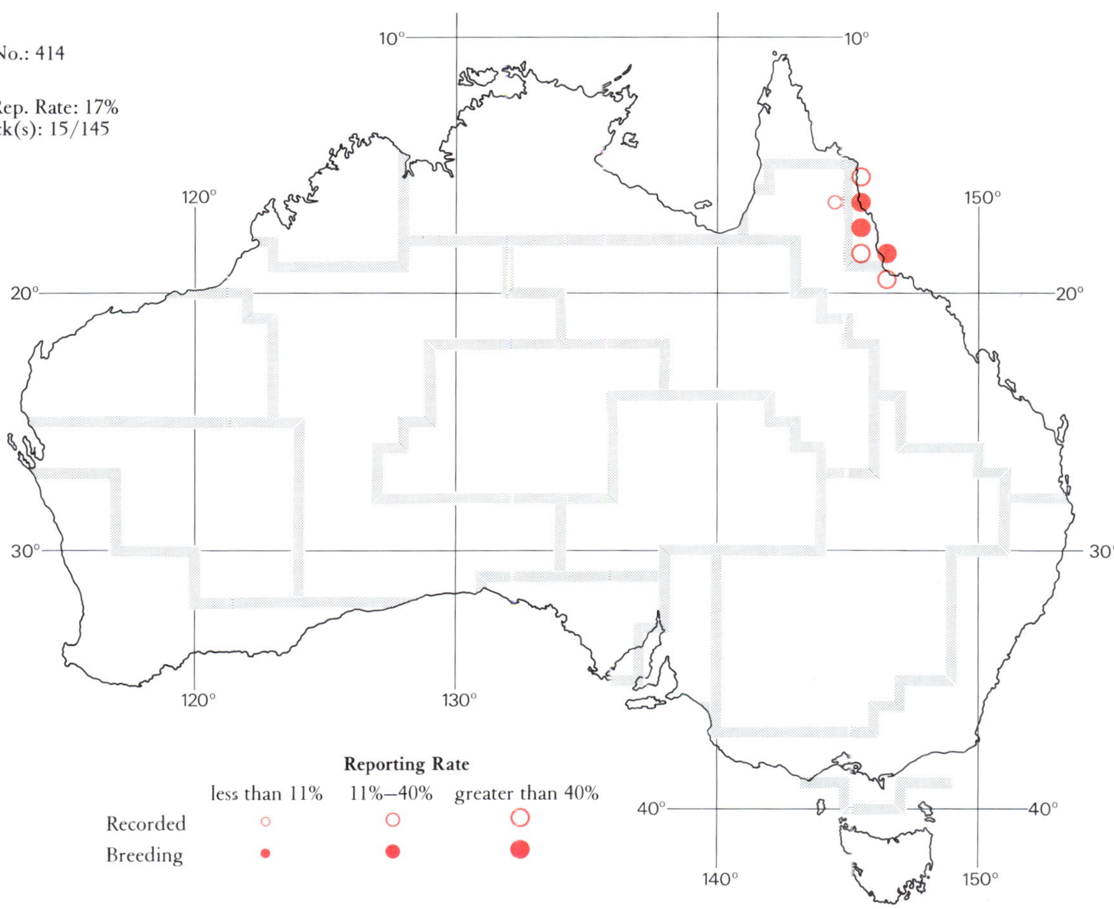

Sandstone Shrike-thrush

Colluricincla woodwardi

There are different views about subspecies of the endemic Sandstone Shrike-thrush.[608,748,1713] Atlas records at the 10′ scale show that there is a gap in distribution east of the Victoria R (132°E)[608] but there are insufficient data to determine where a similar gap, said to isolate the Arnhem Land population from that round the Macarthur R (15/136), might lie.

The Sandstone Shrike-thrush lives in coastal cliffs, gorges in the ranges, and on sandstone islands.[280] It is usually seen on the ground or perched on a conspicuous pinnacle of sandstone at the top of a cliff or gorge from which its song carries for some distance.[748,1860,1881] The Sandstone Shrike-thrush feeds on spiders, grasshoppers and probably other large insects, as well as small vertebrates.[748]

There are no measures of abundance.

Field Atlas Records: 157

1° blocks–with records: 32 (4%)
–with breeding records: 3 (9%)

Atlas No.: 411

Max. Rep. Rate: 25%
In block(s): 13/132

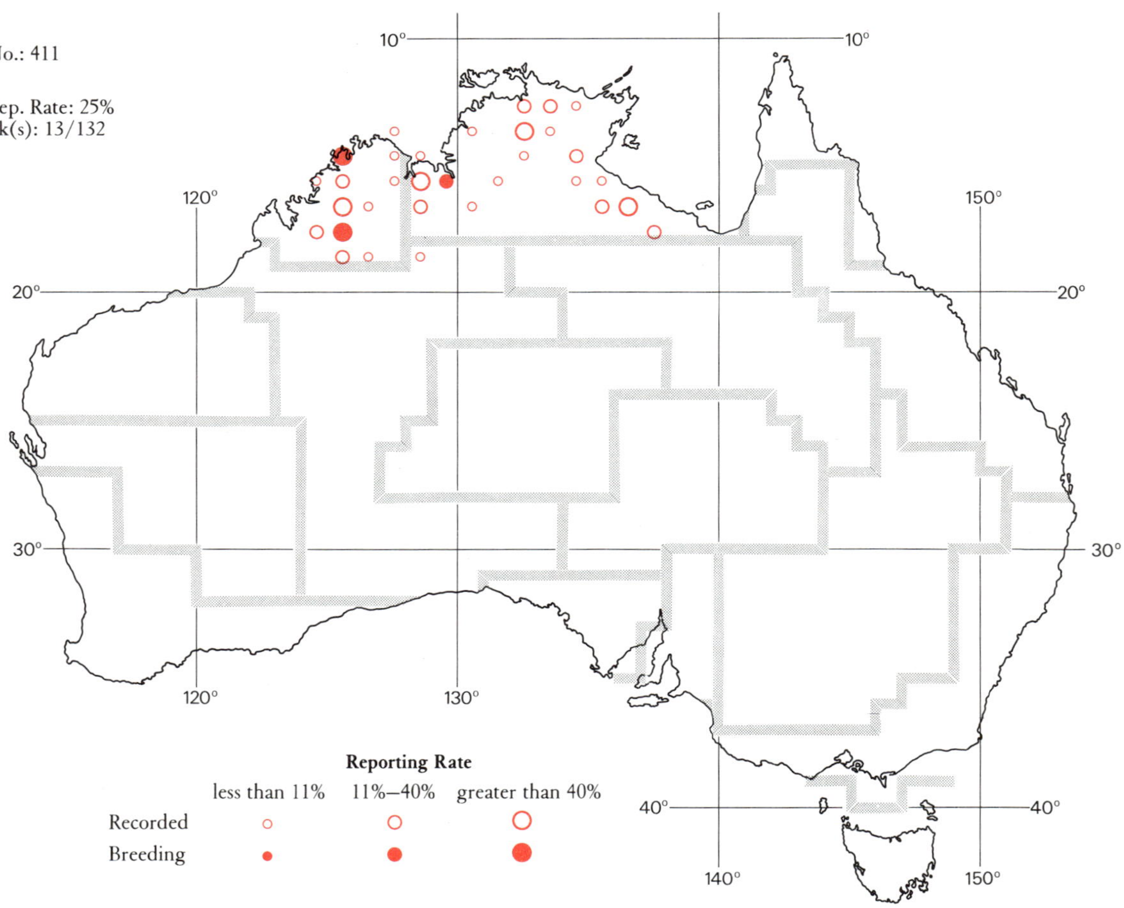

Grey Shrike-thrush

Colluricincla harmonica

The Grey Shrike-thrush lives in New Guinea and Australia. Australian birds vary clinally from north to south and east to west.[626,1121] The sharpest step in the cline is across the upper Western Deserts, L. Eyre and SA Ra. Regions. Birds north and east of an arc connecting these places are subspecies *harmonica* and those south and west of it *rufiventris*.

The Grey Shrike-thrush is sedentary. Individuals have been repeatedly retrapped throughout the year in their known territories at Moruya (35/150), Jamieson (37/146) and Eyre Bird Observatory (32/126).[481,1192,1469] The Field Atlas does not show evidence of any large-scale seasonal movement either.

The Grey Shrike-thrush inhabits many types of vegetation, specially places where water accumulates after rainfall in dry areas or where temperatures are mild in colder parts. It eats a variety of insects and will take small vertebrates, including nestlings, as well as carrion.[1470]

In eucalypt woodland densities were: Armidale (30/151) 0.5 birds/ha,[129] Wollomombi (30/152) 0.5 birds/ha[129] and Moruya (35/150) 0.23–0.32 birds/ha.[1197] In eucalypt forest they were: Bathurst (33/149) 0.25–1.25 birds/ha,[489] Jamieson 0.4–1.0 birds/ha[1469] and Boola Boola (38/146) 0.3–2.3 birds/ha, with the highest density in mature gullies.[1113] In mallee at Eyre Bird Observatory mapping recorded 0.2 birds/ha and a census estimated 0.03–0.44 birds/ha.[464] On the Nullarbor Plain in dongas in saltbush density was 0.06 birds/ha[246] and in partly cleared farmland south of Perth (32/115) 0.11 birds/ha.[1429]

Field Atlas Records: 30 497

1° blocks–with records: 628 (77%)
–with breeding records: 171 (27%)

Atlas No.: 408

Max. Rep. Rate: 73%
In block(s): 37/148

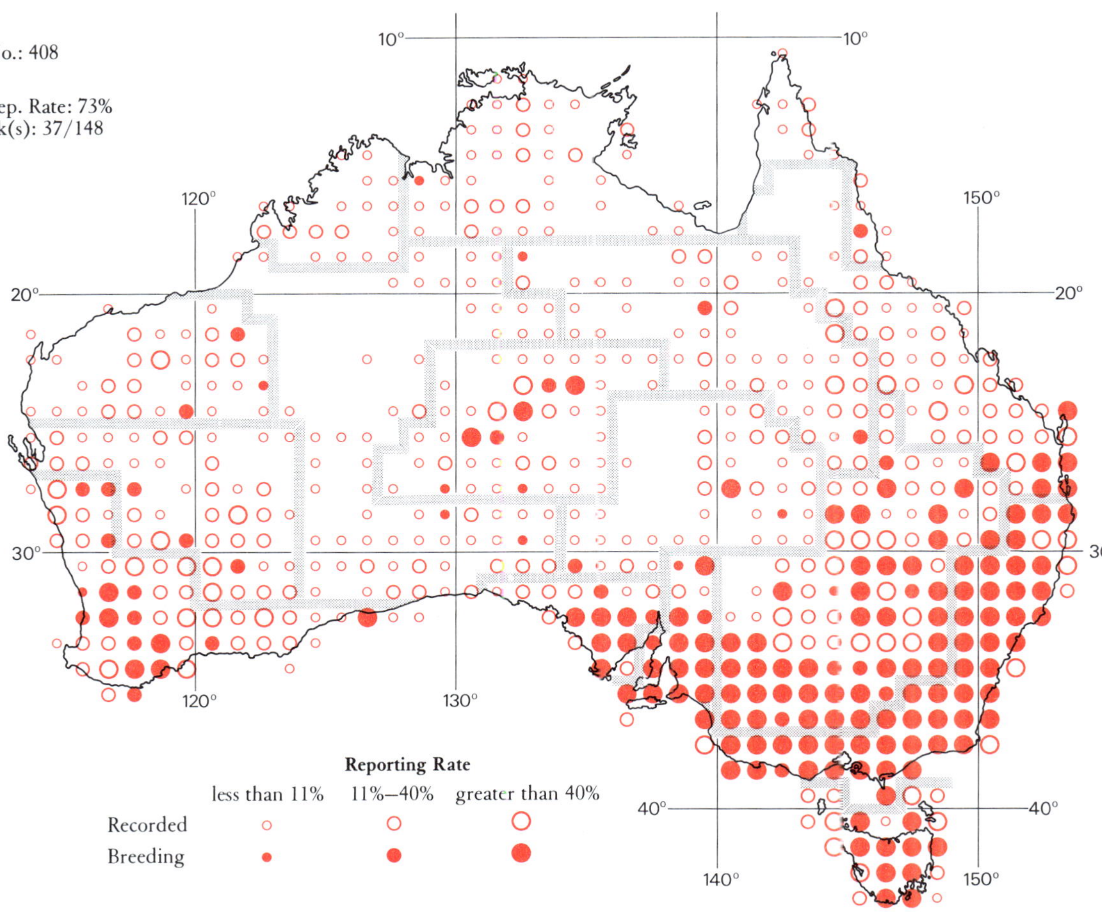

Crested Bellbird

Oreoica gutturalis

Up to five subspecies of the endemic Crested Bellbird have been described.[1233] Most of the variation seems to be individual or clinal and there is little justification for the separation of any subspecies.[748]

The Crested Bellbird no longer inhabits much of the Eastern Qld and Atherton Regions. In the 1840s it was distributed from the Mitchell R (16/142) to the Darling Downs (26/151).[347] By the 1920s it was gone from the Duaringa (23/149) district.[102] It has not been reported from the Atherton Region since the 1940s when it still survived near Wondecla (17/145).[212] It was recorded east of the Great Dividing Ra. in north-eastern NSW in 1900–05 but not since. In Vic. its range has shrunk and the 10′ maps show the populations at Chiltern (36/146) and north of Bendigo (36/144) to be isolated. Its range may also have shrunk in the Top End, where records were made round Borroloola (16/136) in 1914,[106] and in the Kimberley Region, where specimens were taken near Napier Broome Bay (14/126)[787] and the Forrest R (15/127) in 1910–11.

In the west the bird's range may be slightly more extensive than it was in the 1920s, having expanded down the coastal plain almost as far as Perth (31/115).

The Field Atlas suggests no large-scale seasonal movement. The Crested Bellbird was present in a donga on the Nullarbor Plain on every visit, whatever the time of year.[246] It lives in eucalypt woodland, mallee, acacia scrub, spinifex, saltbush and heath. Usually it occupies the thickest parts of the vegetation near the ground. The Bellbird feeds on caterpillars and keeps a larder of incapacitated but living caterpillars round its nest.[1727] It also eats many seeds.[748,1049]

On the Nullarbor Plain in dongas in saltbush density was 0.05 birds/ha.[246]

Field Atlas Records: 4995

1° blocks–with records: 447 (55%)
–with breeding records: 52 (12%)

Atlas No.: 419

Max. Rep. Rate: 63%
In block(s): 28/121

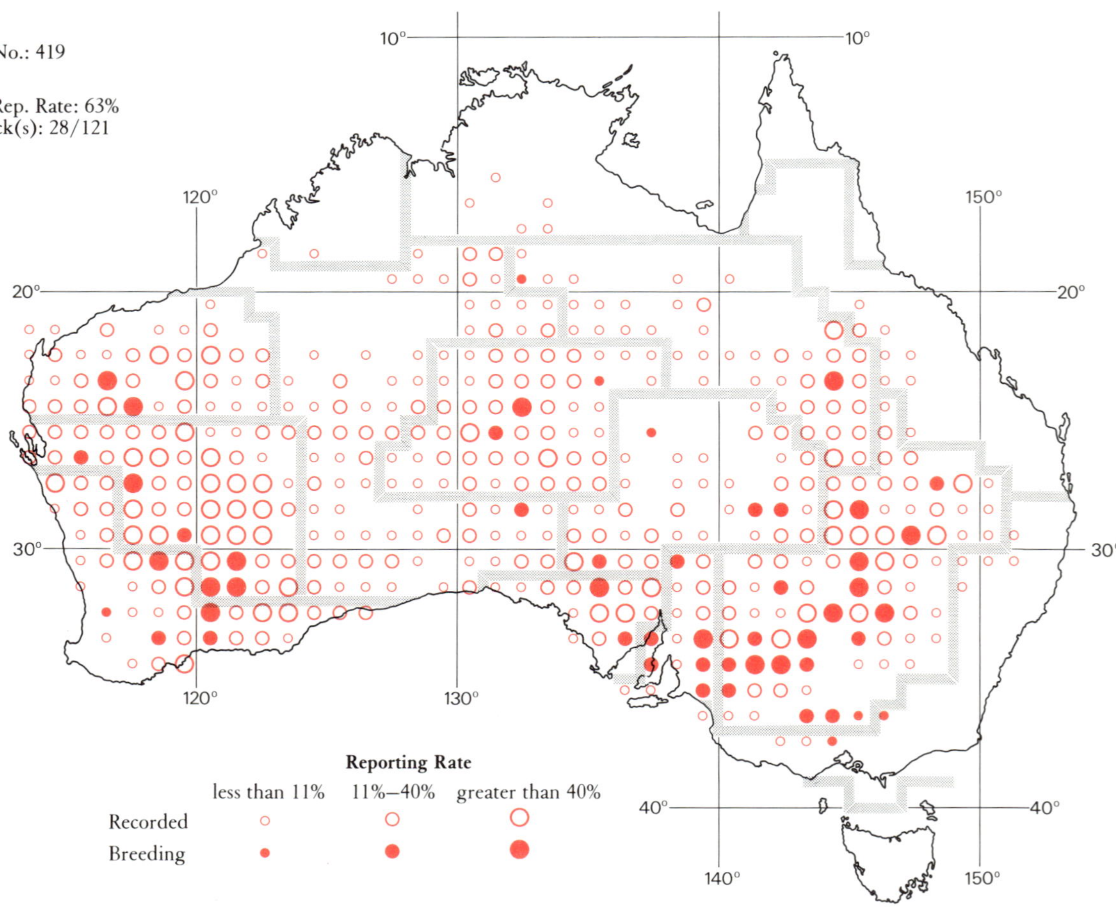

Yellow-breasted Boatbill

Machaerirhynchus flaviventer

There are four subspecies of the Yellow-breasted Boatbill in New Guinea[1449] and two in Australia: *flaviventer* in the Cape York Region and *secundus* in the Atherton Region.[748,1709] Taken in conjunction with the Historical Atlas, the distribution in the Cape York Region is not disjunct and extends down the western coast to 14°S. The Atherton population is isolated.

The Boatbill was present throughout the year at a study site in New Guinea.[160] It is confined to rainforest, rarely emerging even on the fringes. Observers have made varying reports about the extent to which it calls: it was remarkably silent in July[748] but early observers commented on the species' whistling calls when feeding and when nest building.[104,1145] An observer not familiar with its call, which is unlike that of other flycatchers, could assume the bird to be rare; in fact it is common.[695] The bird lives in scattered pairs in the canopy of rainforest trees, often in mixed species flocks,[160,748] but pairs rarely associate with others of the same species.[748] The Boatbill feeds on flying insects and by accompanying flocks of other species it may be able to take insects flushed by the flock.

Density in rainforest near Port Moresby (9S/147E) in New Guinea was 0.1 birds/ha.[160]

Field Atlas Records: 267

1° blocks–with records: 12 (1%)
–with breeding records: 3 (25%)

Atlas No.: 368

Max. Rep. Rate: 50%
In block(s): 12/143

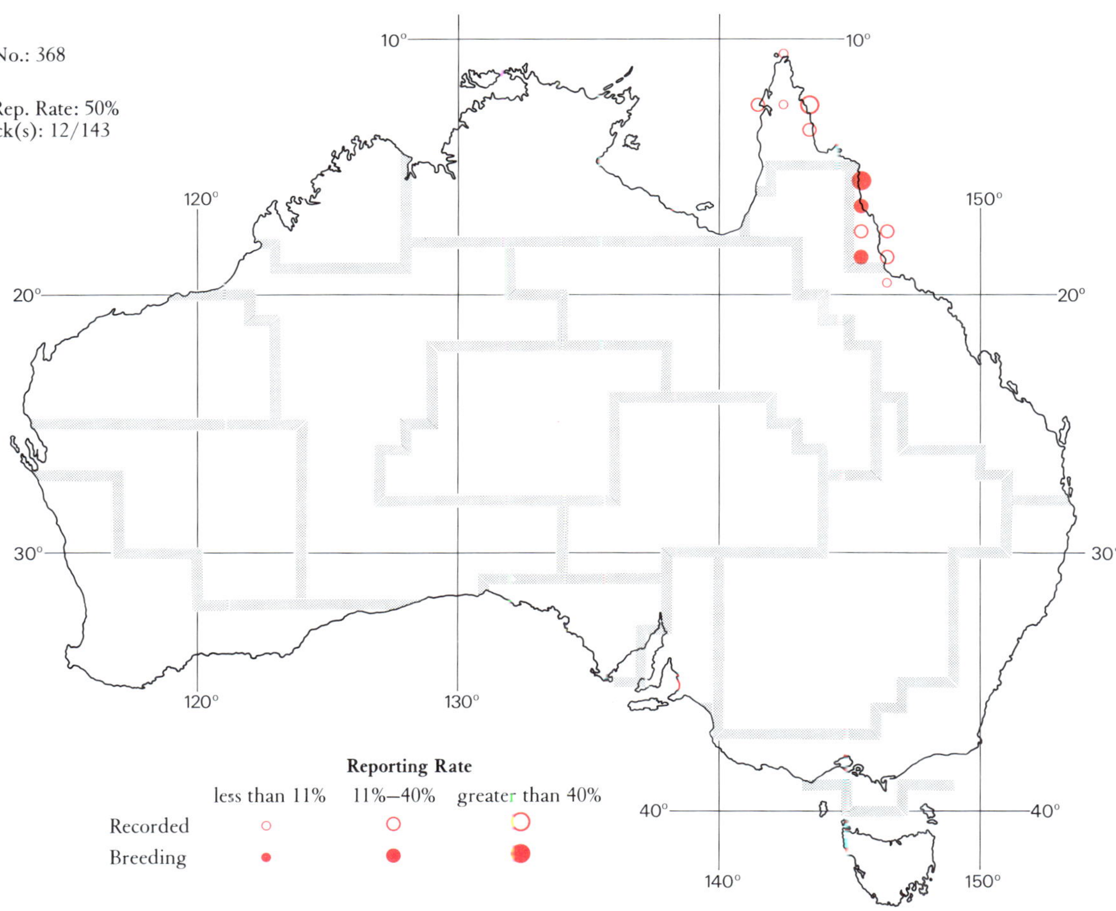

Black-faced Monarch

Monarcha melanopsis

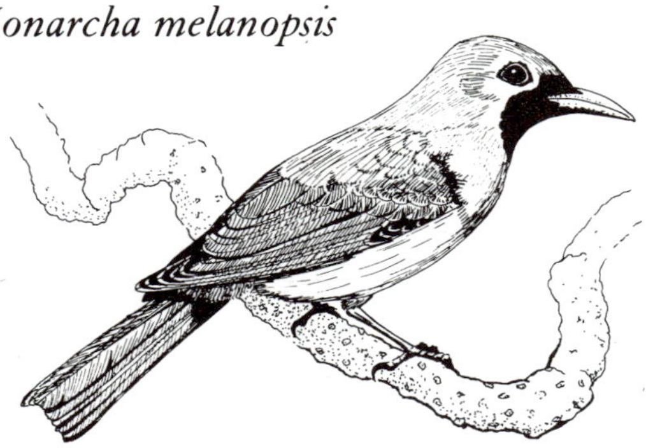

The Black-faced Monarch breeds only in Australia but some birds winter in New Guinea. It has two subspecies: *pallidus* in the Cape York Region south to Cardwell (18/146) and *melanopsis* to the south.[1709] At the southern end of its range three of four sightings west of 146°E were made early in 1981, which suggests a minor irruption that year.

The only record of breeding north of the Atherton Region, at Iron Ra. (12/143) in 1913, is doubtful. Breeding is confirmed in the Atherton Region and from places south of 26°S. The Field Atlas shows sightings in summer to be almost confined to this breeding range with additional records at Iron Ra., Eungella (21/148 and 21/149) and near Rockhampton (23/149, 23/150).

The Black-faced Monarch is migratory: *pallidus* has been reported to winter in New Guinea and *melanopsis* is thought to winter in Qld from Rockhampton north to the Bloomfield R (15/145) with some possibly going to New Guinea[1709]. The Field Atlas contains few sightings in winter in Australia: Reporting Rates in the Atherton Region were 19.2% in summer and 1.2% in winter. Therefore the bulk of the Australian population may winter in New Guinea where the species is present Mar.-Oct. on the Fly R (8S/141E) and in south-eastern New Guinea.[129,160] Small numbers appear on islands in Torres Strait during migration[503] and the bird has been recorded well out to sea, e.g. on Pine Islet Lightstation (21/150), during and after a rainy night in Sept.[1190] This islet is 65 km off the Qld coast but on the flight line to south-eastern New Guinea from the South-East Region. In NSW the Monarch arrives late Sept. and leaves in Feb.[396] Evidence from lighthouses indicates that it moves at night as well as by day.

The Black-faced Monarch lives in rainforest and eucalypt forest. In New Guinea it colonizes hardwood plantations unlike other New Guinean monarchs.[160] It feeds where there is tangled understorey, moving along branches and poking into crevices to take insects and caterpillars.[879,1049]

On its wintering grounds in New Guinea density was 0.2 birds/ha;[160] at Moruya (35/150) in eucalypt woodland it was 0.15 birds/ha.[1197]

Field Atlas Records: 1629

1° blocks–with records: 65 (8%)
–with breeding records: 20 (31%)

Atlas No.: 373

Max. Rep. Rate: 24%
In block(s): 15/145

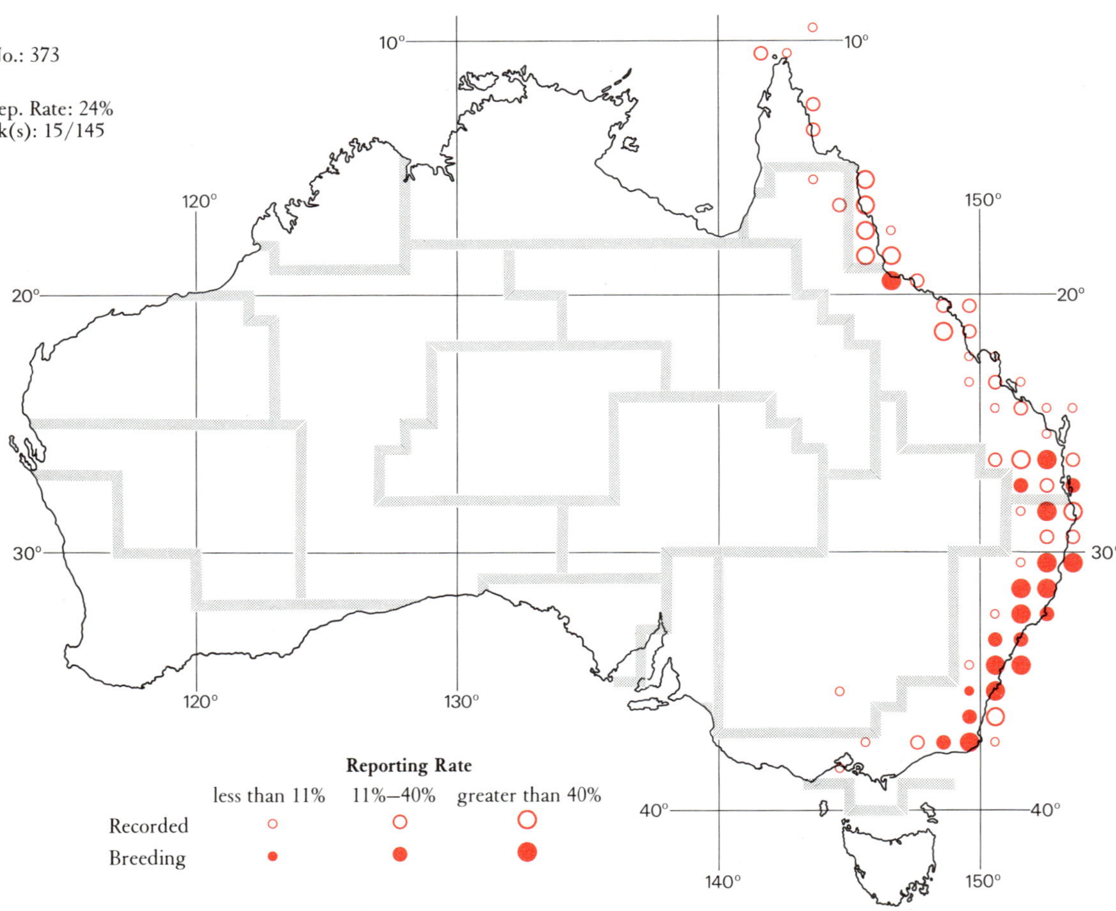

Black-winged Monarch

Monarcha frater

The Black-winged Monarch breeds in the Cape York Region and migrates to New Guinea in winter. The only confirmed record south of the Cape York Region was made at Edmonton (17/145) in Jan. Breeding has rarely been reported: eggs were collected on the Claudie R (12/143) in Jan. 1915 and a nest was watched at Iron Ra. (12/143) in Jan. 1966.[1552]

The Black-winged Monarch is a passage migrant through the western islands of Torres Strait[503] and the Atlases contain no winter records of it in Australia. The species lives in eucalypt forest and rainforest. The Black-winged Monarch seldom takes flight after airborne prey, as do many flycatchers, but rather searches twigs and leaves of trees and shrubs for its insect food.[1356,1824]

There are no measures of abundance.

Field Atlas Records: 25

1° blocks–with records: 5 (1%)

–with breeding records:

Atlas No.: 374

Max. Rep. Rate: 23%
In block(s): 12/143

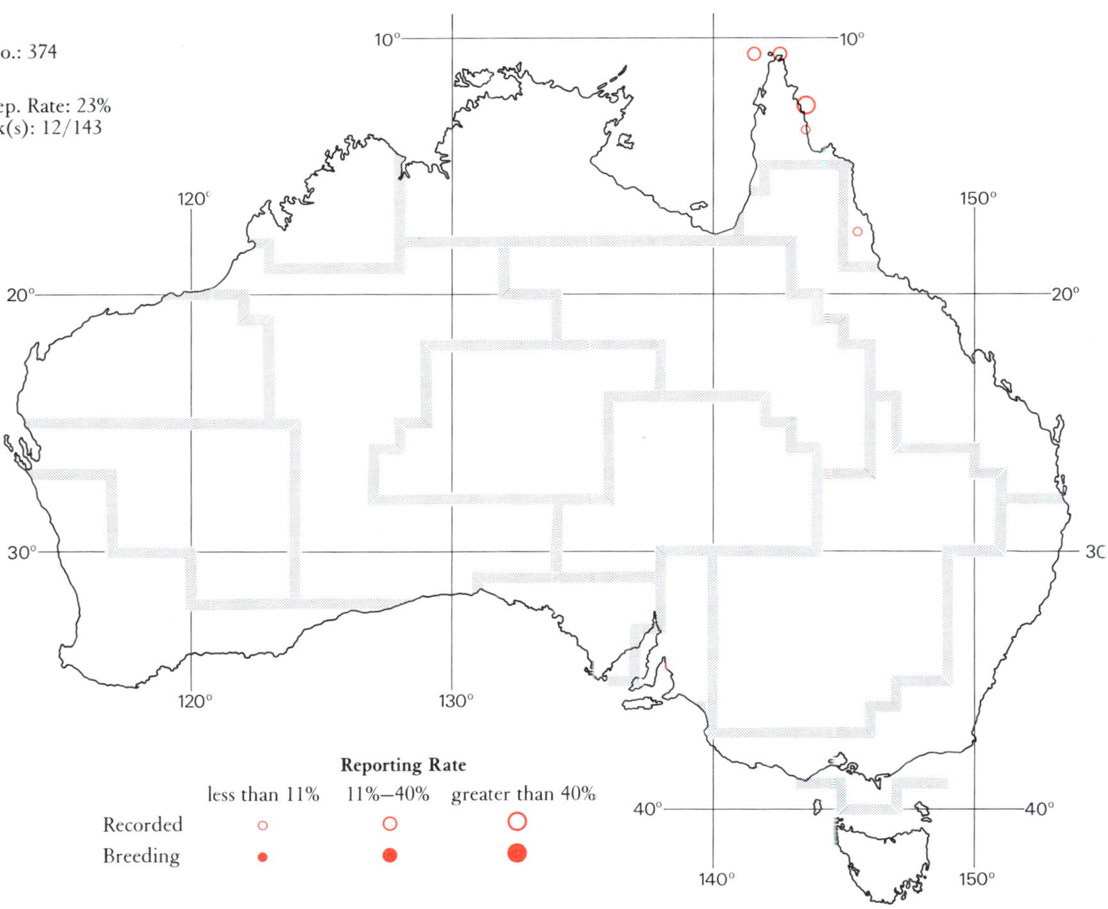

Spectacled Monarch

Monarcha trivirgatus

The Spectacled Monarch breeds in Australia and visits New Guinea in the winter.[1235] Some islands adjacent to New Guinea have sedentary populations.[129] In Australia there are two subspecies: *albiventris* in the Cape York Region and *gouldi* south of 15°S.[1709] The low Reporting Rate in Eastern Qld at 19/147 suggests there may be a gap in the distribution of *gouldi*. The southern record was of a single bird near Pt Addis (38/144) in Dec. 1977.

There are both sedentary and migratory populations of the Spectacled Monarch. In the Cape York Region *albiventris* is present all year[986] as is *gouldi* in the Atherton Region.[231,698] The Field Atlas suggests that the population south of 25°S migrates north in winter (Reporting Rates: summer 6.1% and winter 1.6%). The bird is mainly a passage migrant at Rockhampton (23/150), a few remaining to breed.[1089] In NSW it arrives late Sept. and leaves late Mar.-May.[396] It is present in the New England (30/152) district only in summer.[997] Birds identified as *gouldi* appear in winter in the Cape York Region[748] and the species is a passage migrant in Torres Strait.[503] The origin of birds that visit and breed in Torres Strait during summer is unknown. In the part of New Guinea closest to Australia only immature birds have been recorded. The Spectacled Monarch is sometimes seen well out to sea on migration, e.g. on Pine Islet Lightstation (21/150) during and after a rainy night in Sept.[1190] This islet is 65 km off the Qld coast but on the flight line to New Guinea from the South-East Region.

The Spectacled Monarch lives in rainforest, mangroves often remote from rainforest and occasionally in eucalypt forest adjacent to rainforest. It may use different localities at different times of the year, e.g. at Atherton (17/145) where it moves out of the rainforest in winter.[231] The Spectacled Monarch hunts in mixed flocks with other species, taking insects in the air, from the vegetation and the ground. It feeds on ants, cockroaches, snails[748] and termites when these are flying in quantities.[1147] At Widgee (26/152)[879] and in NSW[396] it usually feeds within 6 m of the forest floor, seeking insects under loose bark and on vine stems. It seldom visits the canopy. There are no measures of abundance.

Field Atlas Records: 1098

1° blocks–with records: 47 (6%)
–with breeding records: 15 (32%)

Atlas No.: 375

Max. Rep. Rate: 51%
In block(s): 12/143

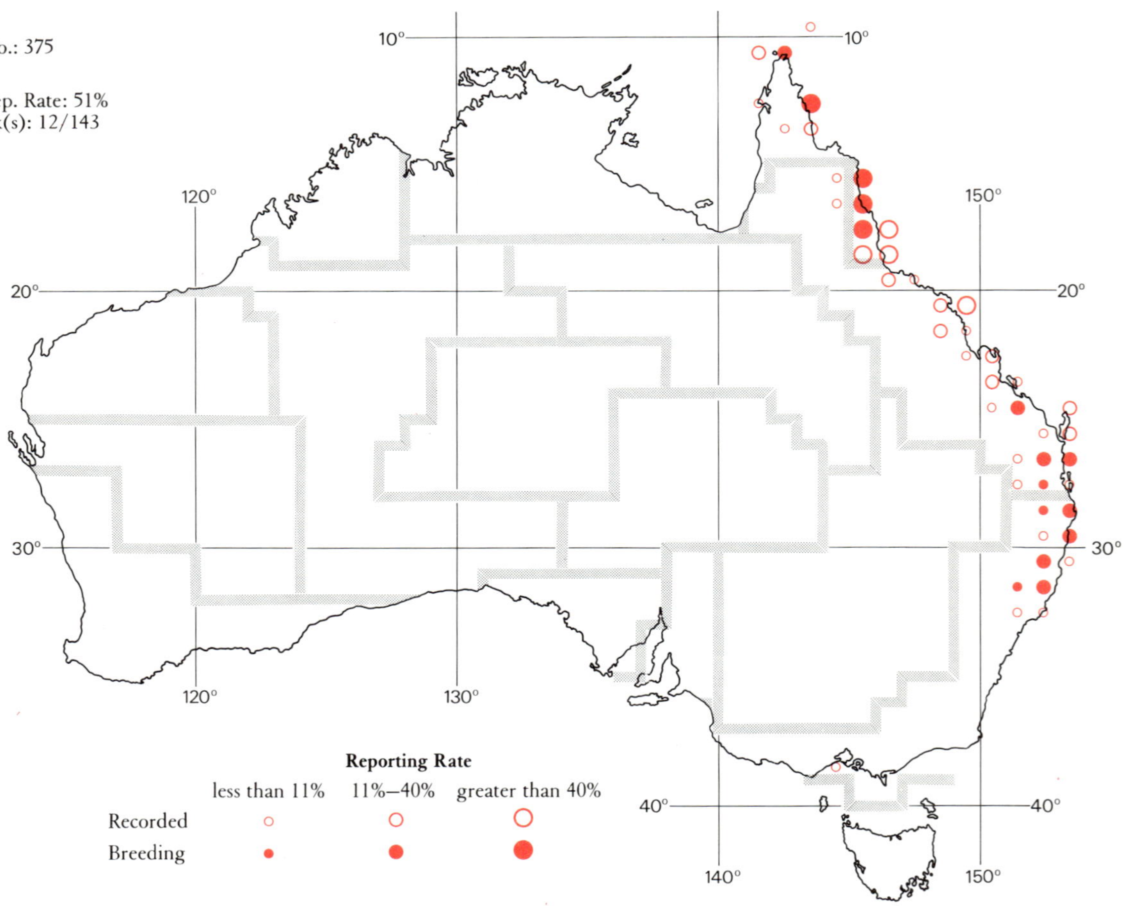

White-eared Monarch

Monarcha leucotis

The endemic White-eared Monarch is closely related to species on adjacent Pacific islands.[748] The species' distribution in the Atlases extends to the top of the Cape York Region but there is a gap between the Cape York and Atherton Regions. Its nest was not found until 1923[1224] and eggs were first collected in 1930.[534] Even since then, the Atlases contain breeding records from only a single 1° block in the Atherton Region and seven south of 23°S.

The movements of the White-eared Monarch are variously described as: apparently nomadic in autumn and winter;[1709] migratory at Widgee (26/152) arriving late Aug. to breed and leaving late Dec.;[879] sedentary at Noosa Head (26/153) and near Fernvale (27/152);[172] recorded throughout the year at Iluka (29/153).[396] The Field Atlas suggests no large-scale seasonal movement.

The White-eared Monarch lives in eucalypt forest including regrowth, at the edge of rainforest, along creeks and round clearings and sometimes in mangroves. It is often on islands. At Noosa Head pairs defend territories in the breeding season and range more widely the rest of the year.[172] The White-eared Monarch feeds on insects taken from the outer foliage of the forest, below the canopy and above the ground.[879] It frequently perches on leaves as well as twigs, fluttering to keep its balance, and takes food in the air. Insects including beetles have been reported in its guts and the bird is probably able to eat any insect disturbed by its feeding method.[748]

It persists in localities at low population densities but there are no measures of abundance.

Field Atlas Records: 288

1° blocks–with records: 28 (3%)
–with breeding records: 2 (7%)

Atlas No.: 376

Max. Rep. Rate: 25%
In block(s): 15/145

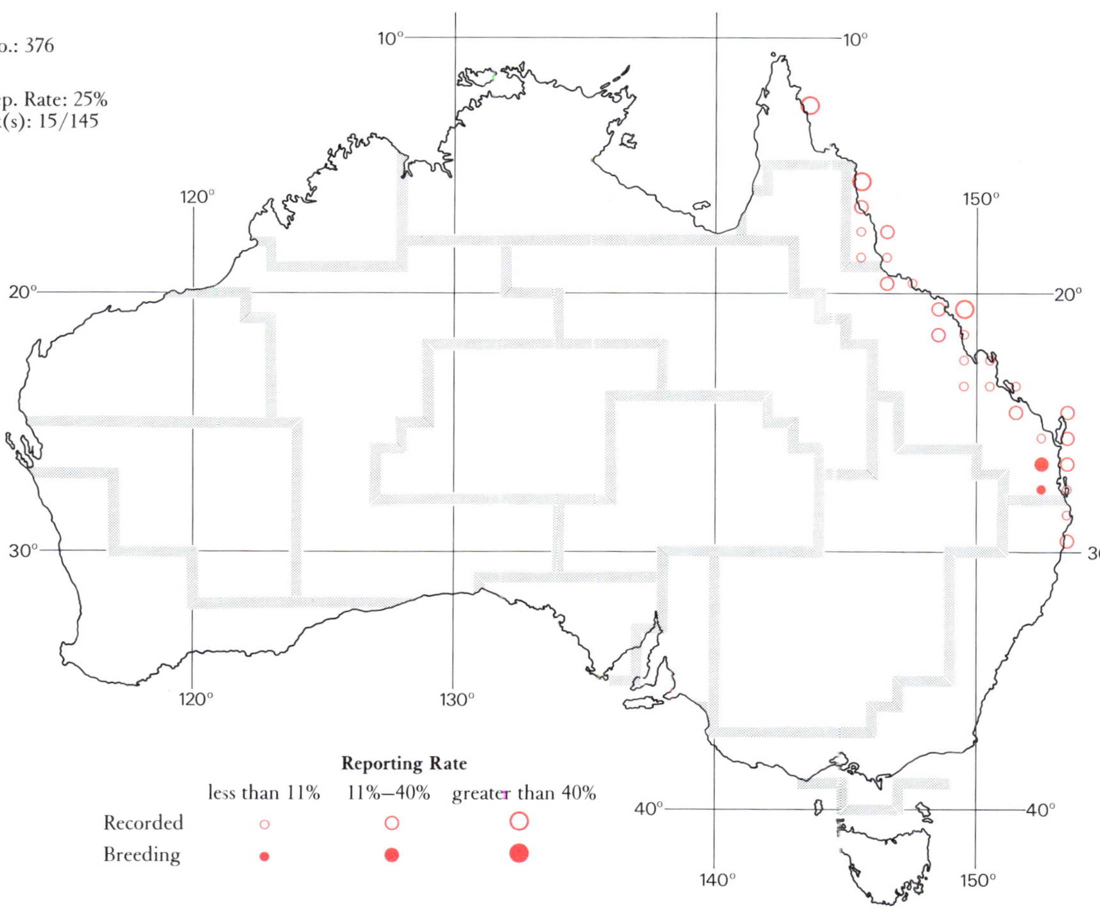

Frilled Monarch

Arses telescophthalmus

There are seven subspecies of the Frilled Monarch in New Guinea and an eighth, *lorealis*, in the Cape York Region.[1709] The distribution of the population in the Cape York Region is not disjunct when Historical and Field Atlas maps are examined. Published references to the Frilled Monarch in the Atherton Region[1225,1869] are probably of the Pied Monarch.[1369]

The Frilled Monarch was present all year in a study site in New Guinea.[160] It lives in rainforest with eucalypt forest adjacent and breeds in the wet summer season.[1145] The bird feeds on crevice-dwelling insects, searching the trunks of trees like a treecreeper, as well as logs, branches and leaves.[1147] It also feeds, in company with other passerine species, on swarms of flying termites.[157] The nest is liable to attack by shrike-thrushes.[104,1449]

Early observers described it as nowhere plentiful,[104,535] whereas recent visitors to Iron Ra. (12/143) recorded it frequently.

Density in lowland rainforest near Port Moresby (9S/147E) in New Guinea was 2 birds/ha.[160]

Field Atlas Records: 51

1° blocks-with records: 4 (1%)

-with breeding records: 1 (25%)

Atlas No.: 371

Max. Rep. Rate: 53%
In block(s): 12/143

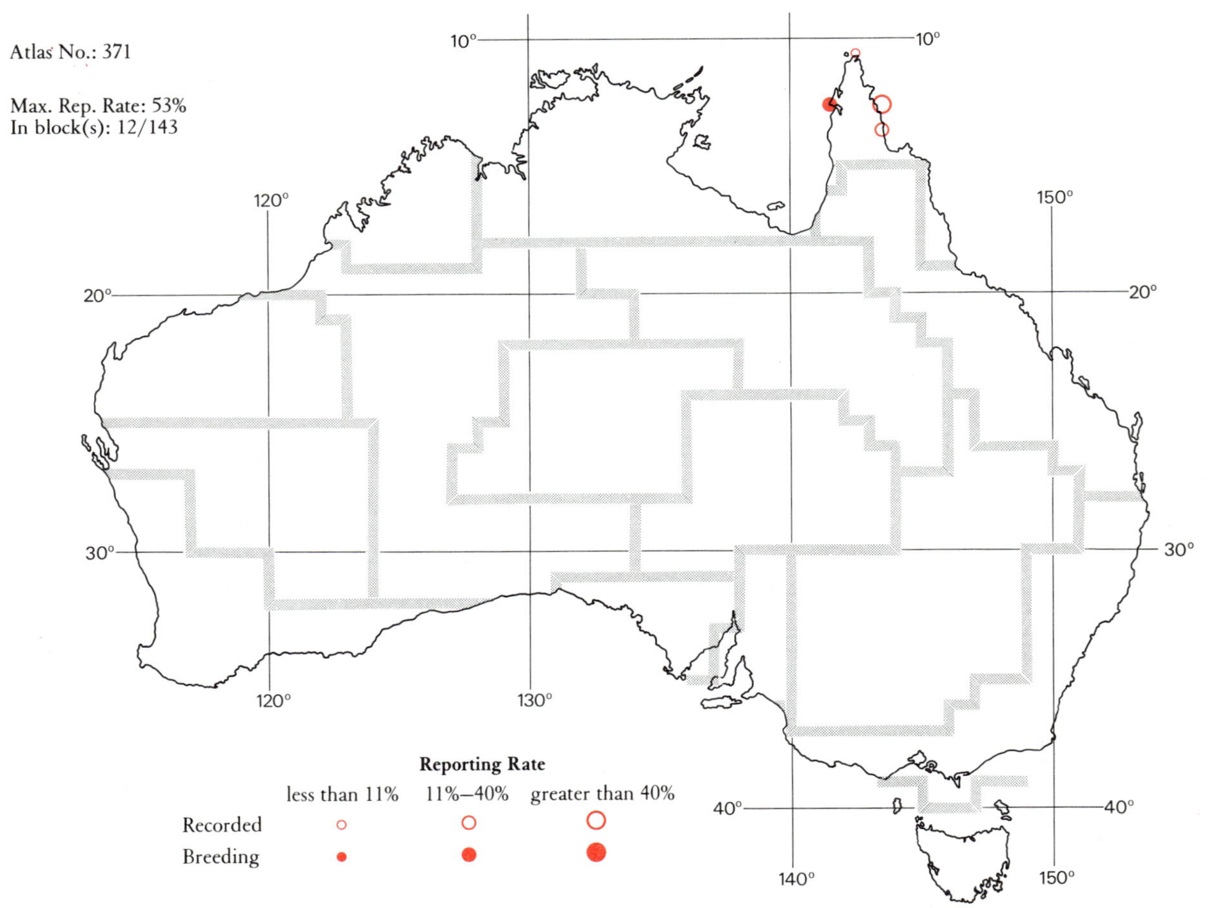

Pied Monarch

Arses kaupi

The endemic Pied Monarch is confined to the coastal side of the Great Dividing Ra. It appears to be present all year at all altitudes, living in rainforest and sometimes in nearby eucalypt forest.[108,231,698] It breeds in the wet summer season.[534,698]

The Pied Monarch has a treecreeper's habit of clinging to tree trunks and moving down them as well as up, searching under bark and in crevices for insects.[748] One, watched for more than an hour, chased small insects often using its wings to help it progress up tree trunks.[12] Several observers have remarked that it is reasonably common within its range[239,535,695] though clearing of the rainforest is reducing the area in which it can live.[695]

There are no measures of abundance.

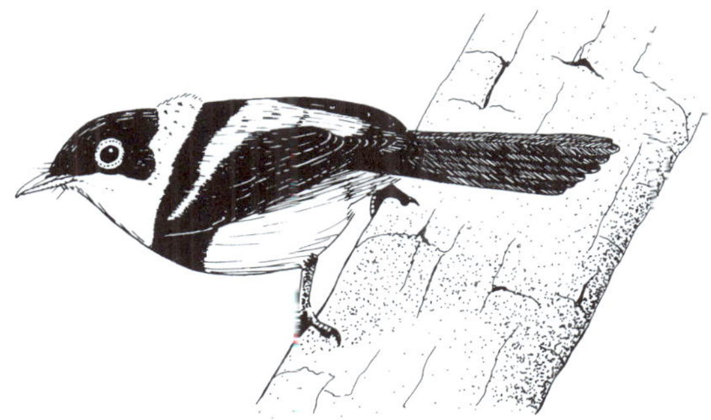

Field Atlas Records: 203

1° blocks–with records: 7 (1%)

–with breeding records: 2 (29%)

Atlas No.: 370

Max. Rep. Rate: 34%
In block(s): 15/145

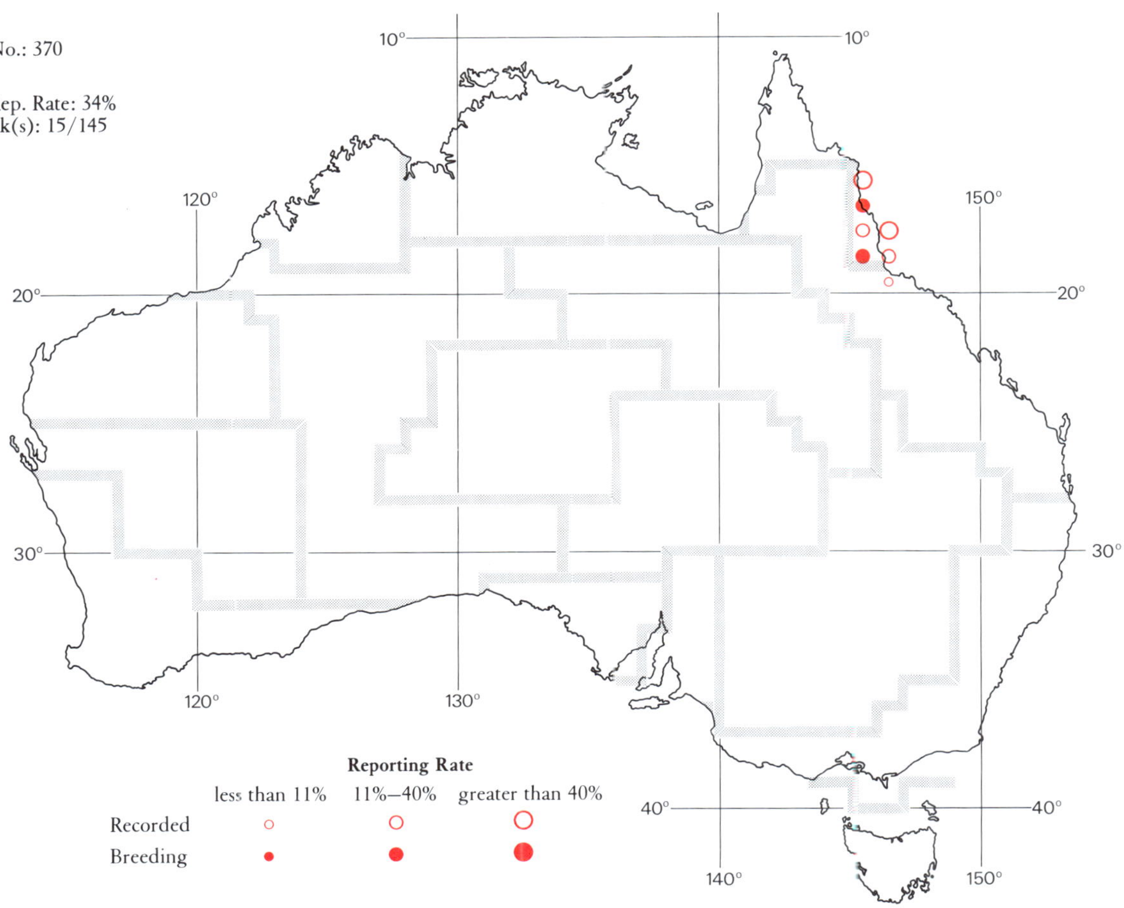

Broad-billed Flycatcher

Myiagra ruficollis

The Broad-billed Flycatcher lives in Indonesia, New Guinea and adjacent islands and Australia. All Australian populations are regarded as subspecies *mimikae*.[622,748] It is difficult in the field to differentiate between the Broad-billed and the female Leaden Flycatcher because white in the tail is not diagnostic.

There are few reliable records of the Broad-billed Flycatcher from the eastern coast of Qld. Haggerstone I. (12/143) was previously the southernmost confirmed locality[1709] but specimens collected at Cooktown (15/145) in 1938 and at Keppel Bay (23/150) in 1901 have now been confirmed as this species.[198] When Historical and Field Atlases are combined gaps in the records remain at 127°E, 129°E and 137–139°E, some corresponding with gaps in the distribution of mangroves.

Records of the Broad-billed Flycatcher are made throughout its range in all seasons and the species appears to be mainly sedentary. Some local movement in Torres Strait is suggested by records from Booby I. (10/141), which has no resident population.[503]

The Broad-billed Flycatcher lives singly or in pairs, mainly in mangroves but also in paperbark woodland and rainforest surrounding swamps at Darwin (12/130).[423] It searches the leaves for insects or catches them in short flights from a vantage point.

There are no measures of abundance.

Field Atlas Records: 142

1° blocks–with records: 31 (4%)
–with breeding records: 2 (6%)

Atlas No.: 367

Max. Rep. Rate: 14%
In block(s): 12/130

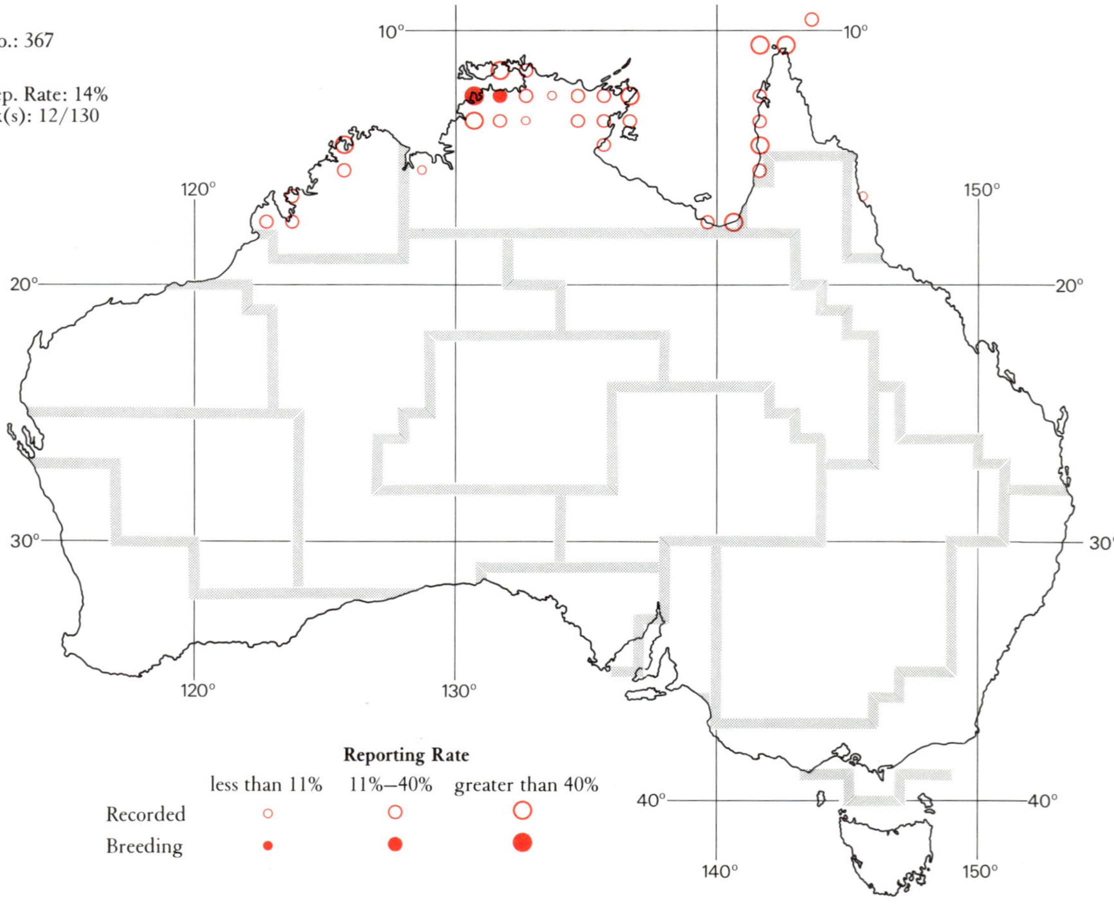

Leaden Flycatcher

Myiagra rubecula

The Leaden Flycatcher breeds in New Guinea and adjacent islands and in Australia. The gap between the subspecies *rubecula* in eastern Australia and *concinna* in the west is said to be between the Leichhardt R (139°40′E) and the Norman R (140°40′E)[1709] but the Atlases lack confirmed records between 138°E and 139°E. There are slight morphological differences between northern and southern breeding birds of *rubecula*.[748,986] The Leaden Flycatcher is a vagrant to the SA Ra. Region and a rare migrant to Tas. where there are breeding records, most recently in 1968.[734]

Individuals of the southern type of *rubecula* winter in New Guinea and northern Qld.[653,748] The species migrates through Torres Strait[503] and also winters on Barrier Reef islands.[1526] It is absent in winter from Murphy's Ck (27/152),[1099] Wollomombi (30/152), Canberra (35/149),[129] Moruya (35/150)[1192] and Jamieson (37/146)[1469] and, the Field Atlas suggests, from the rest of the South-East Region (Reporting Rates: summer 9.7%, winter 0.4%). Most birds visit south-eastern NSW Sept.-Mar.[1526] In the Atherton Region the Leaden Flycatcher is mainly a lowland bird (Reporting Rates: below 500 m 55%, above 500 m 4%), resident near Innisfail (17/146)[698] but some birds possibly move to higher altitudes Aug.-Dec.[231]

The Leaden Flycatcher is rarely in large parties, even on migration. It lives in eucalypt forest, rainforest along rivers and mangroves; in the north it also inhabits eucalypt woodland specially during the wet summer season. Individual birds return to the same place each year.[871,1192] It feeds on flying insects.

Territories of pairs are discrete and may be several kilometres apart[1526] but in some places, e.g. Canberra,[653] nests are clumped into groups one or more kilometres apart. Such clumping will affect measures of density. Densities in eucalypt woodland were: Armidale (30/151) 0.04–0.08 birds/ha;[582] Wollomombi 0.8 birds/ha;[129] Canberra about 0.03 birds/ha;[154] Moruya 0.21 birds/ha;[1197] and Eden (37/150) 0.05–0.40 birds/ha.[1526] At Bathurst (33/149) in wet and dry eucalypt forest it was 0.75 and 0.25 birds/ha respectively[489] and at Jamieson in eucalypt forest it was 0.1 birds/ha.[1469]

Field Atlas Records: 4255

1° blocks–with records: 194 (24%)

–with breeding records: 67 (34%)

Atlas No.: 365

Max. Rep. Rate: 65%
In block(s): 20/149

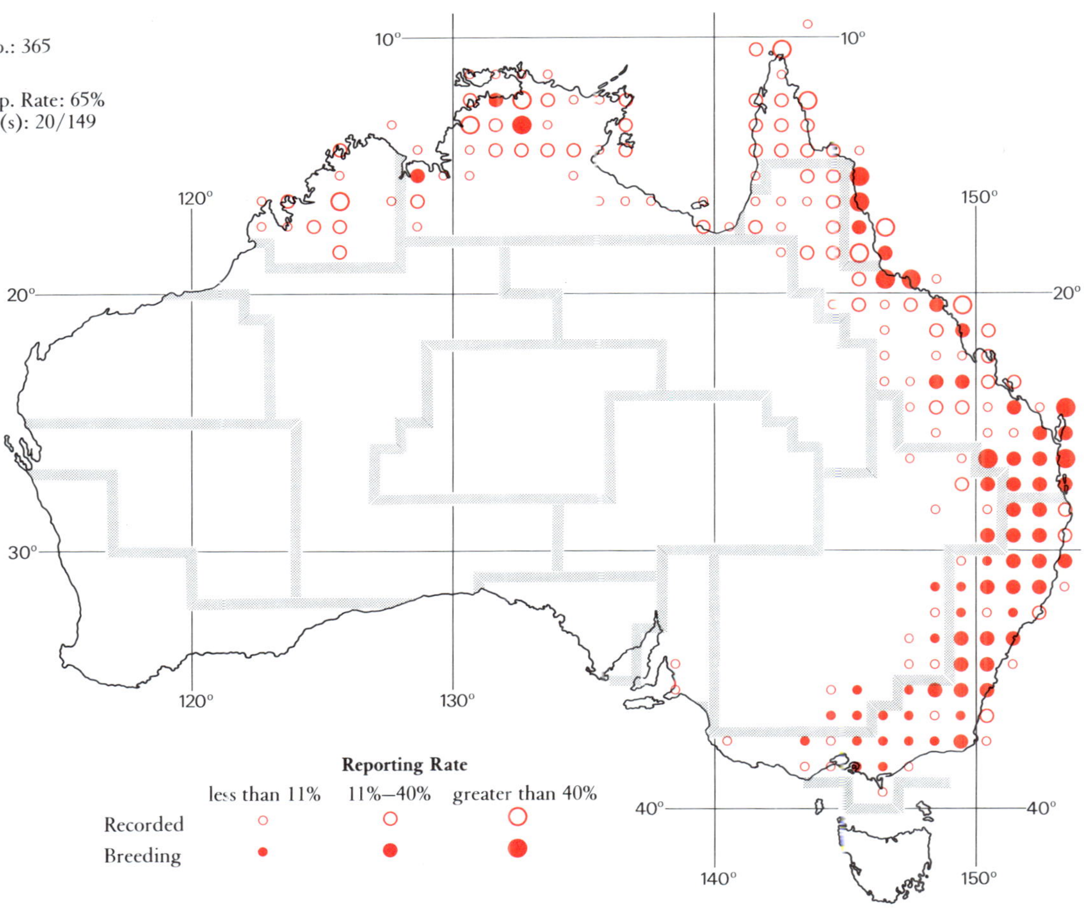

Satin Flycatcher

Myiagra cyanoleuca

The Satin Flycatcher is a winter non-breeding visitor to New Guinea and adjacent islands, though reports are few, and a vagrant to New Zealand. In Australia vagrants reach the SA Ra. Region[1468] and places even farther west, e.g. Victory Well (27/132) and Twilight Cove (32/126).[242] Both these records were in Apr. The species has not been reported to breed north of 23°S.

The Satin Flycatcher migrates north in winter. Small numbers pass through Torres Strait[503] but some winter on Barrier Reef islands[1526] and on the mainland from Rockhampton (23/150) north.[231,1089] It is present at Wollomombi (30/152) Sept.-Mar. and co-exists with the Leaden Flycatcher.[129] The South-East and Tas. Regions appear to be deserted almost entirely in winter with Reporting Rates of 7.8% and 13.6% respectively in summer and 0.3% in both in winter. The Satin Flycatcher does not migrate at all conspicuously, perhaps because it moves singly, in pairs or small loose groups through the treetops and perhaps at night.

It lives in eucalypt forest on flat to moderately undulating tableland country and on the coast west into SA. South of Sydney (33/151) it lives in lowland eucalypt forest as well as on the tablelands[488] but it is absent from rainforests including those of south-western Tas. On Flinders I. (40/148) it breeds in woodland near the coast.[1314] At Bombala (36/149) and Mt Werong (34/149) it nests in small colonies of 2–5 pairs with nests 20–50 m apart.[1526] In many parts of eastern Tas. loose nesting colonies are in densely vegetated gullies and sometimes nests have been noted close to those of the Australian Goshawk.[1314] Once chicks are out of the nest the family may move rapidly to drier and more open woodland not used for breeding. The Satin Flycatcher feeds on insects, specially large flies, and spiders, either hawked on the wing or snatched from the trees.[1526] Clearing and logging of forests in the South-East Region is reducing the population of the Satin Flycatcher.[1526]

Densities in eucalypt woodland were: Wollomombi 0.25 birds/ha[129] and Armidale (30/151) 0.08 birds/ha.[582] In eucalypt forest they were: Bathurst (33/149) 1.25 birds/ha;[489] 37/150 near Bombala 0.2–0.5 birds/ha;[1527] Boola Boola (38/146) 0.2–0.5 birds/ha;[1113] and Maria I. (42/148) 0.2–0.4 birds/ha.[1526]

Field Atlas Records: 2377

1° blocks–with records: 117 (14%)
–with breeding records: 53 (45%)

Atlas No.: 366

Max. Rep. Rate: 18%
In block(s): 12/143

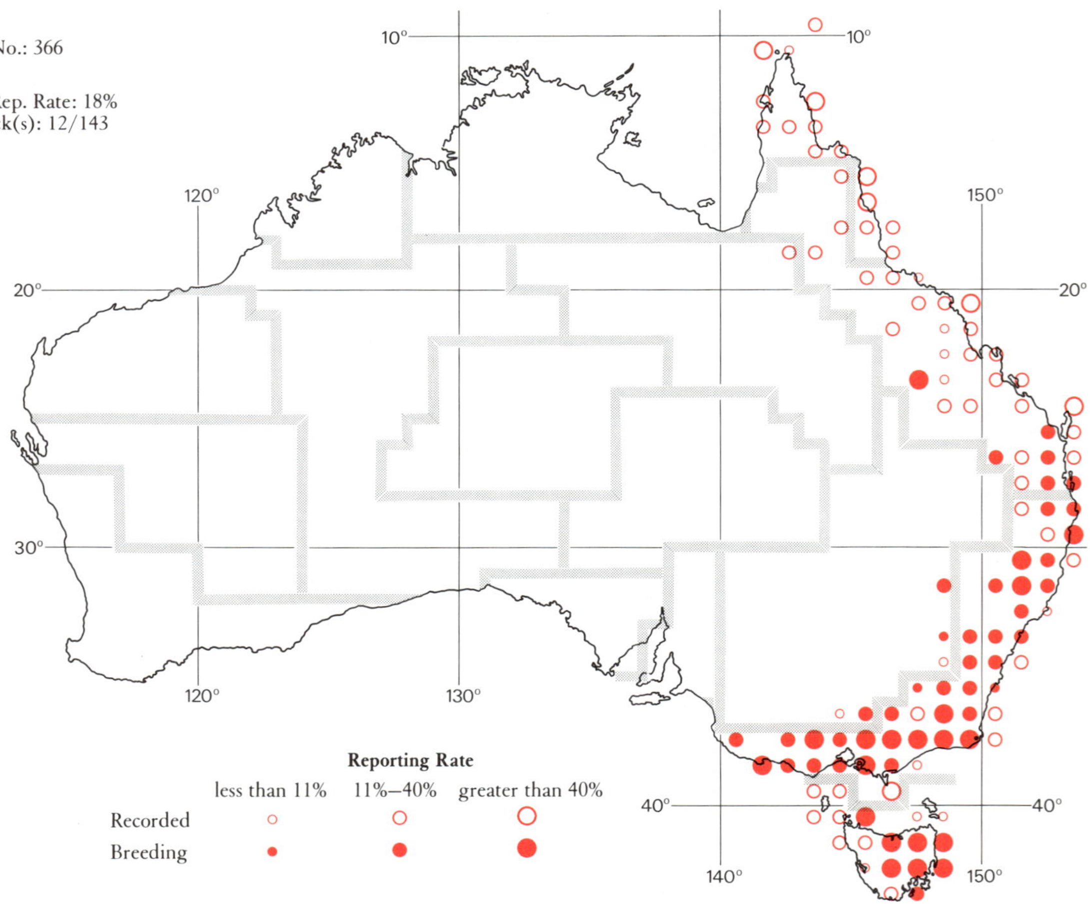

Shining Flycatcher

Myiagra alecto

The Shining Flycatcher lives in New Guinea and adjacent islands and in Australia. There are three morphologically distinct populations in Qld, the Top End and Kimberley Regions but views about their subspecific status differ.[748,1709,1711,1713] The Atlases suggest that populations in the Top End and Kimberley Regions are continuously distributed west from the Macarthur R (16/136). There is a single record of the species south of the Gulf of Carpentaria on the Nicholson R (16/139).[1709] No historical records extend the species' distribution on the eastern and western coasts beyond that shown on the map.

The Shining Flycatcher was present all year at a study site in New Guinea[160] and is always present on the coastal plain near Innisfail (17/146).[698] In New Guinea and on islands in Torres Strait it lives in mangroves, lowland rainforest and rainforest regrowth;[140,148,160,503] in Australia it lives mainly in rainforest along streams and in mangroves, rarely in eucalypt forest.[423,787,1152] It may exhibit lek behaviour in some places:[148,1300,1767] in New Guinea five males were seen displaying to one female, using the same site for display on successive days. The Flycatcher feeds on invertebrates, hopping about on the mud to take crustaceans and molluscs as well as catching some insects on the wing.

In New Guinea in roadside vegetation density was estimated at one pair to 100 m of road.[129] On Badu I. (10/142) in rainforest and mangroves density was 1.25 birds/ha.[685]

Field Atlas Records: 653

1° blocks–with records: 65 (8%)
–with breeding records: 12 (18%)

Atlas No.: 372

Max. Rep. Rate: 54%
In block(s): 12/143,13/130

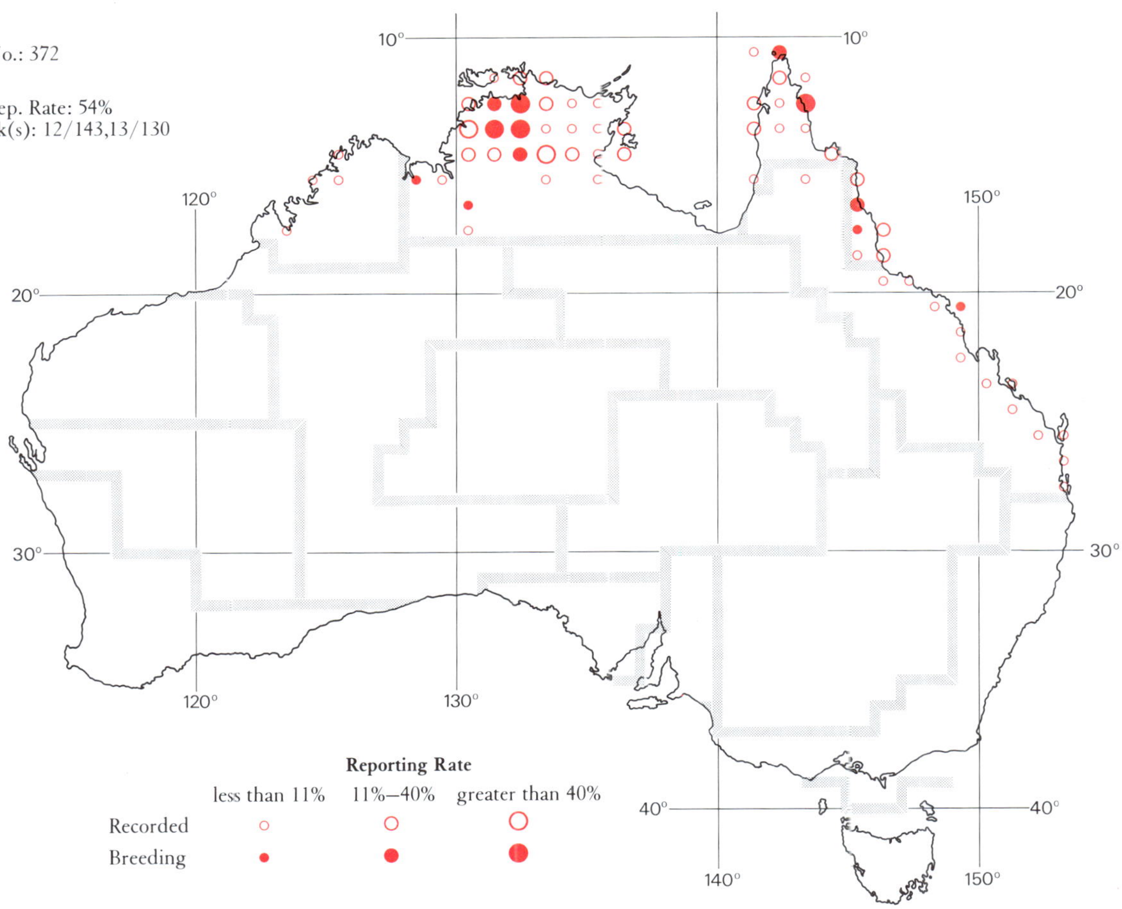

Restless Flycatcher

Myiagra inquieta

The range of the Restless Flycatcher extends beyond Australia to southern New Guinea. Few records have been made in New Guinea[550] but some birds may be resident.[129] In Australia there are two subspecies: *nana* in the north and *inquieta* in the south.[748] A gap is thought to separate them south of Normanton (17/141). A specimen from Julia Ck (20/141) belongs to *inquieta*. Many Field Atlas records came from the Julia Ck area but it is not known whether they were breeding birds or migrants from southern populations.[1369] There is a single record from Torres Strait[503] but otherwise few reports from the Cape York Region apart from those on the map. The population in the South-West Region is not morphologically different from the eastern one[748] and records in the 1960s on the Nullarbor suggest that there may be a connection between them.[600]

The Restless Flycatcher is thought to migrate north in Qld in winter.[1709] At Murphy's Ck (27/152) it appeared mainly in the warmer months[1099] and a slight winter invasion was detected in the southern Murray-Darling Region.[833] Reporting Rates varied little, e.g. Eastern Qld Region: summer 8.0%, winter 12.6%; South-East Region: summer 22.6%, winter 21.2%; Murray-Darling Region: summer 21.2%, winter 25.6%. The species occasionally appears in the L. Eyre Region in winter.[1369]

The Restless Flycatcher is less common on the coast than inland. It lives in eucalypt woodland and mallee, often near water. Generally eucalypt forest is avoided[488] except where clearings have been formed.[1693] The Restless Flycatcher is often in mixed species flocks.[129] It feeds in a characteristic way, hovering about a metre above the ground, calling continually and dropping to take insects, specially flies, that hide in grass or herbage.

At Armidale (30/151) in eucalypt woodland density was 0.03–0.28 birds/ha;[582] and at Jamieson (37/146) in eucalypt forest 0.2 birds/ha.[1469]

Field Atlas Records: 13 821

1° blocks–with records: 404 (50%)
–with breeding records: 136 (34%)

Atlas No.: 369

Max. Rep. Rate: 65%
In block(s): 27/147

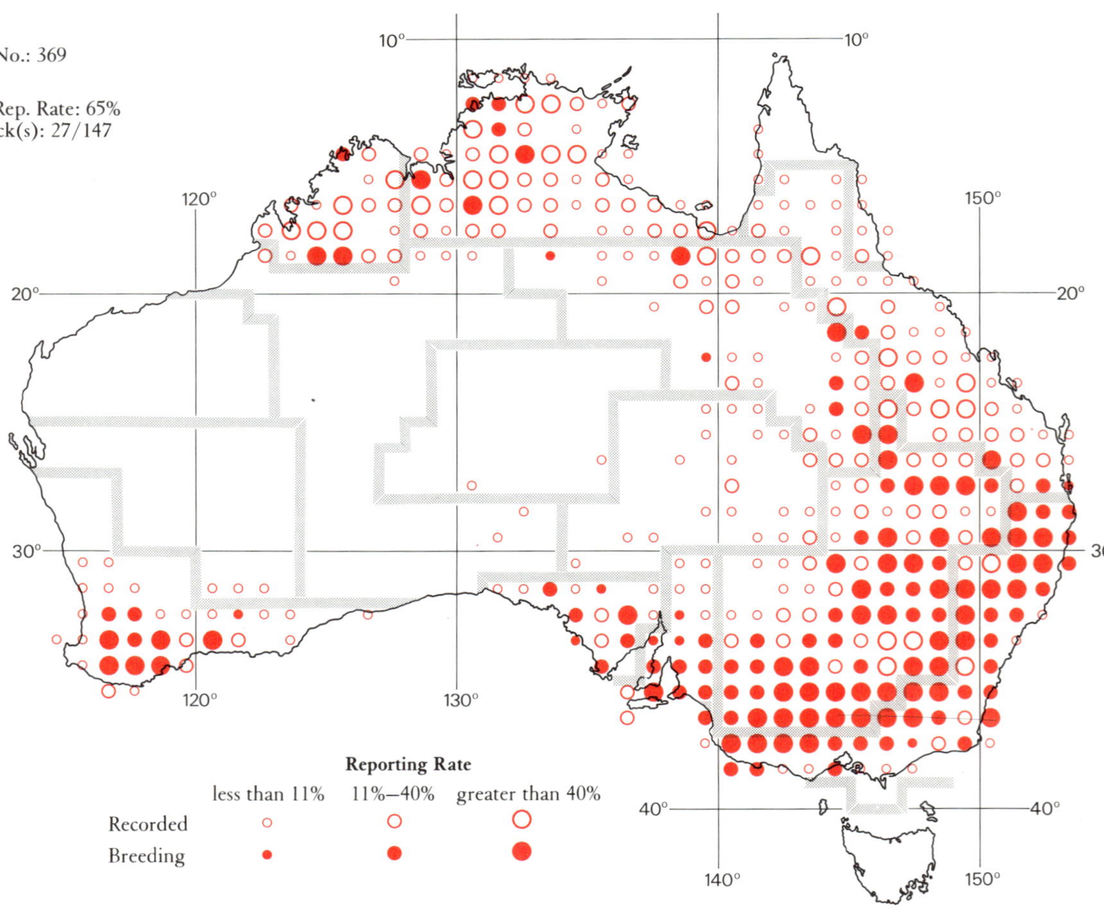

Rufous Fantail

Rhipidura rufifrons

The Rufous Fantail inhabits New Guinea, some Pacific and some Indonesian islands as well as Australia. There are three subspecies in Australia: *dryas* in the Kimberley and Top End Regions and on the western coast of the Cape York Region; *intermedia* breeding in the Atherton Region; and *rufifrons* breeding south of 20°S.[1709,1713] Most birds in the Cape York Region are winter visitors; *dryas* has not been recorded breeding there or round the southern coast of the Gulf of Carpentaria; nor has breeding been reported in the Kimberley Region though birds there are presumed to breed locally.

The three subspecies appear to have different patterns of movement. Round Darwin (12/130) the Reporting Rate of *dryas* in summer (1.0%) was much lower than in winter (6.7%) and a movement inland in the wet summer season has been suggested.[1711] In the Atherton Region an altitudinal movement of *intermedia* is reported, taking the birds to breed at altitudes above 900 m in summer.[698] The observations are supported by the Reporting Rates (above 500 m: summer 37%, winter 0%; below 500 m: summer 0%, winter 40%). A passage movement through the region has also been noted,[231] perhaps of *rufifrons* which breeds to the south and migrates, many birds passing through Torres Strait to New Guinea.[149,503] The Rufous Fantail is present in Sydney (33/151) between mid-Sept. and Mar.-Apr.;[692] near Canberra (35/149) and Melbourne (37/145) it is present Oct.-Apr.[75,653] The Field Atlas suggests an almost complete absence from the South-East Region in winter (Reporting Rates: summer 13.5%, winter 0.7%) though some birds winter in Eastern Qld (Reporting Rates: summer 6.2%, winter 4.6%). The species appears to migrate only in the twilight hours.[692]

The Rufous Fantail lives in eucalypt forest in the South-East, in rainforest in the north and in mangroves as well round the Gulf of Carpentaria. On migration it occasionally turns up in the confined spaces between city buildings.[800] It feeds on insects mostly gleaned from the foliage.[1049,1703] A few are taken on the wing.

At Moruya (35/150) in eucalypt woodland density was 0.03–0.05 birds/ha.[1197] At Boola Boola (38/146) in eucalypt forest it was 0.6–0.8 birds/ha.[1113]

Field Atlas Records: 3634

1° blocks–with records: 138 (17%)
–with breeding records: 34 (25%)

Atlas No.: 362

Max. Rep. Rate: 42%
In block(s): 12/143

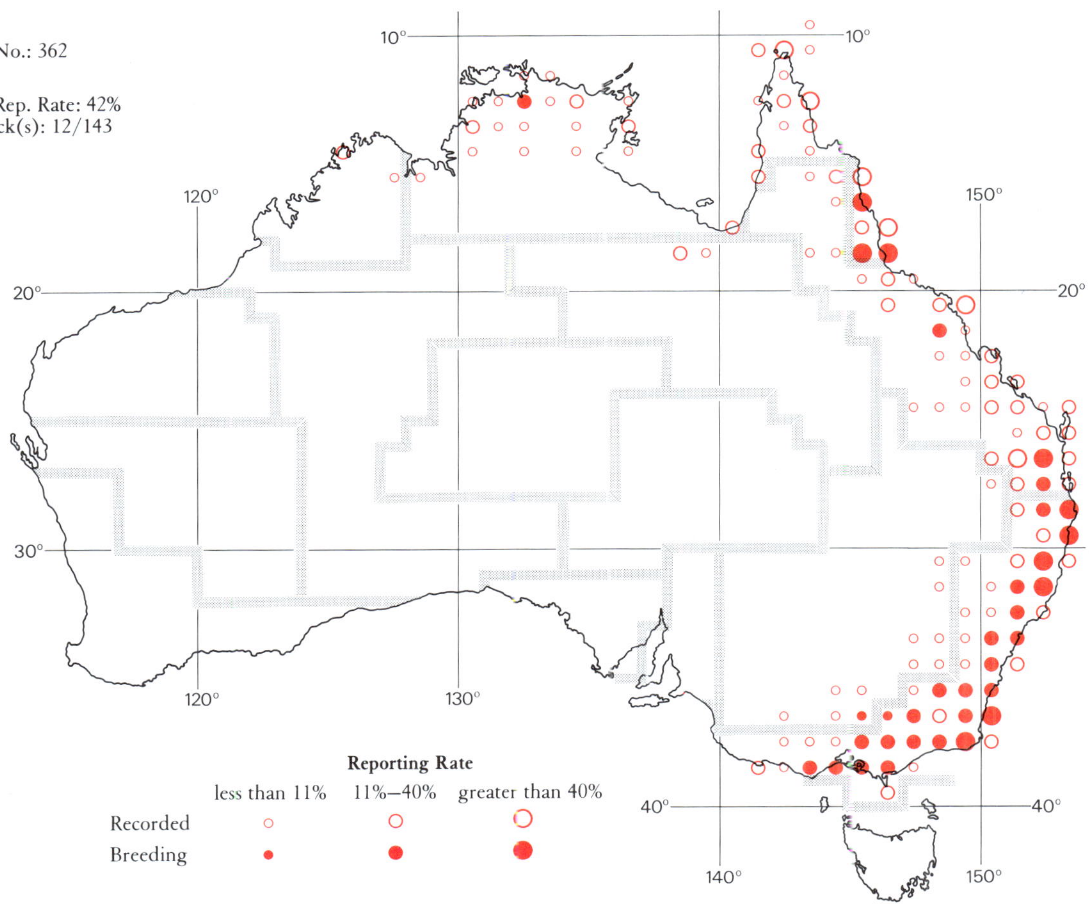

Grey Fantail

Rhipidura fuliginosa

The Grey Fantail has six subspecies in Australia and others in New Zealand, New Guinea and the Pacific islands.[619] The Australian subspecies have isolated breeding ranges: *phasiana* on the coasts of southern New Guinea and of Australia between the Norman R (17/140) and Shark Bay (25/113); *keasti* in the Atherton Region and Eungella (21/148) district; *alisteri* in the lower Eastern Qld, South-East, Murray-Darling, SA Ra. and lower Eyre Pen. Regions; *albiscapa* in Tas; *albicauda* in the inland from the Goldfields north-east to the Rolling Downs Region; and *preissi* in the South-West. Breeding records on the map at 23/141 and 23/144 are presumably *alisteri* though outside its known breeding range.

The subspecies *phasiana* and *keasti* are present all year throughout their range. Part of the population of *alisteri* migrates to northern Australia, perhaps as far as New Guinea, though records from Torres Strait are lacking. Thus in NSW the species is always present on the coast, though many fewer in winter than in summer; on the tablelands, including Canberra (35/149), it is present Sept.-Apr.; and farther west it is a non-breeding winter visitor.[154,396] Some *albiscapa* migrate in winter to the mainland though the Field Atlas suggests that many birds remain in Tas. (Reporting Rates: spring 60%, summer 52%, autumn 50%, winter 40%). The subspecies *albicauda* is nomadic and *preissi* moves north and east in winter. The Field Atlas map shows the result of the complex mixing of all the subspecies.

The Grey Fantail lives in eucalypt forest and woodland; *phasiana* is also in mangroves; *keasti* is confined to the mountains; *albiscapa* occupies all types of country including the rainforest of south-western Tas.[1314] The species is usually in pairs though flocks may form before migration. It feeds on insects caught on the wing.

Densities in eucalypt woodland were: Armidale (30/151) 0.92 birds/ha;[582] Wollomombi (30/152) 0.2–0.8 birds/ha; Moruya (34/150) 0.55–0.82 birds ha;[1197] and Rotamah I. Bird Observatory 7–9 birds/ha after breeding.[284] At Bathurst (33/149) in wet and dry eucalypt and pine forests respectively densities were 5.25, 0.25 and 0.25 birds/ha.[489] In eucalypt forest they were: Jamieson (37/146) 0.6–1.2 birds/ha[1469] and Boola Boola (38/146) 0.3–4.0 birds ha.[1113]

Field Atlas Records: 29 830

1° blocks–with records: 476 (59%)
–with breeding records: 132 (28%)

Atlas No.: 361

Max. Rep. Rate: 73%
In block(s): 37/146,38/146

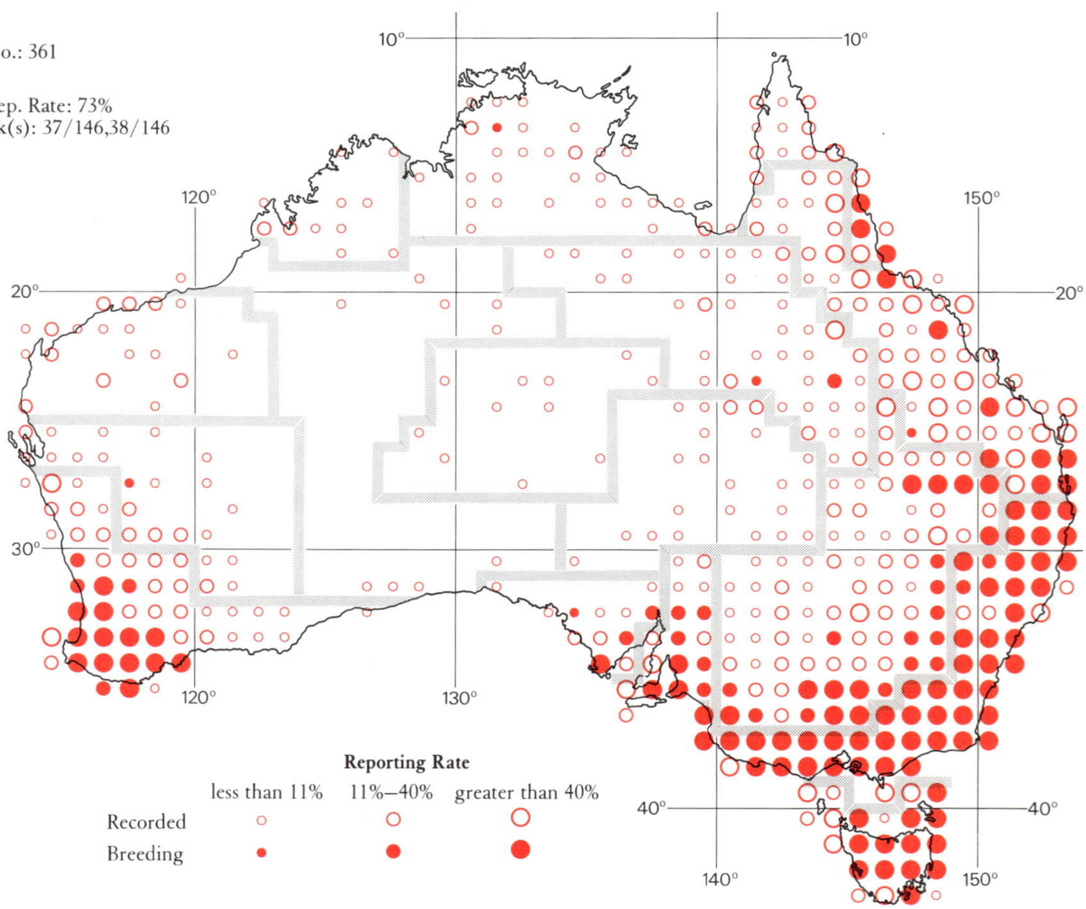

Northern Fantail

Rhipidura rufiventris

No subspecies of the Northern Fantail are recognized within Australia,[748] though populations in New Guinea and adjacent islands differ morphologically.[774] Records in the combined Atlases suggest that the Fantail's distribution is continuous across northern Australia. No change in range from the historical periods is evident and breeding has been recorded in all regions.

In New Guinea, banding studies have shown the Northern Fantail to be sedentary.[160] The Field Atlas gives no evidence of large-scale seasonal movements in Australia though there is one observation in the Top End of birds apparently migrating at Elsey (14/133).[1602] In the Atherton Region observers have variously stated that the Fantail: is seen in eucalypt forest only in winter;[1044] visits only in winter and spring;[231] moves to the hills to breed in the wet summer season;[239] is resident at Cooktown (15/145).[748,1703] Observations of colour-banded birds in New Guinea showed them to be active all day, descending nearly to ground level from 1200–1500 hours; observers visiting a forest early or late in the day could easily miss the bird.[129]

In New Guinea the Northern Fantail lives in rainforest mostly below the canopy[129] and in Australia in rainforest including that along streams; it also lives in eucalypt forest, mangroves and swamps.[1713] It is a regular member of mixed species flocks.[160] The Northern Fantail catches its prey in flight, mostly flies but also dragonflies.

In New Guinea in rainforest near Port Moresby (9S/147E) density was 2 birds/ha.[160]

Field Atlas Records: 864

1° blocks–with records: 83 (10%)

–with breeding records: 10 (12%)

Atlas No.: 363

Max. Rep. Rate: 49%
In block(s): 13/132

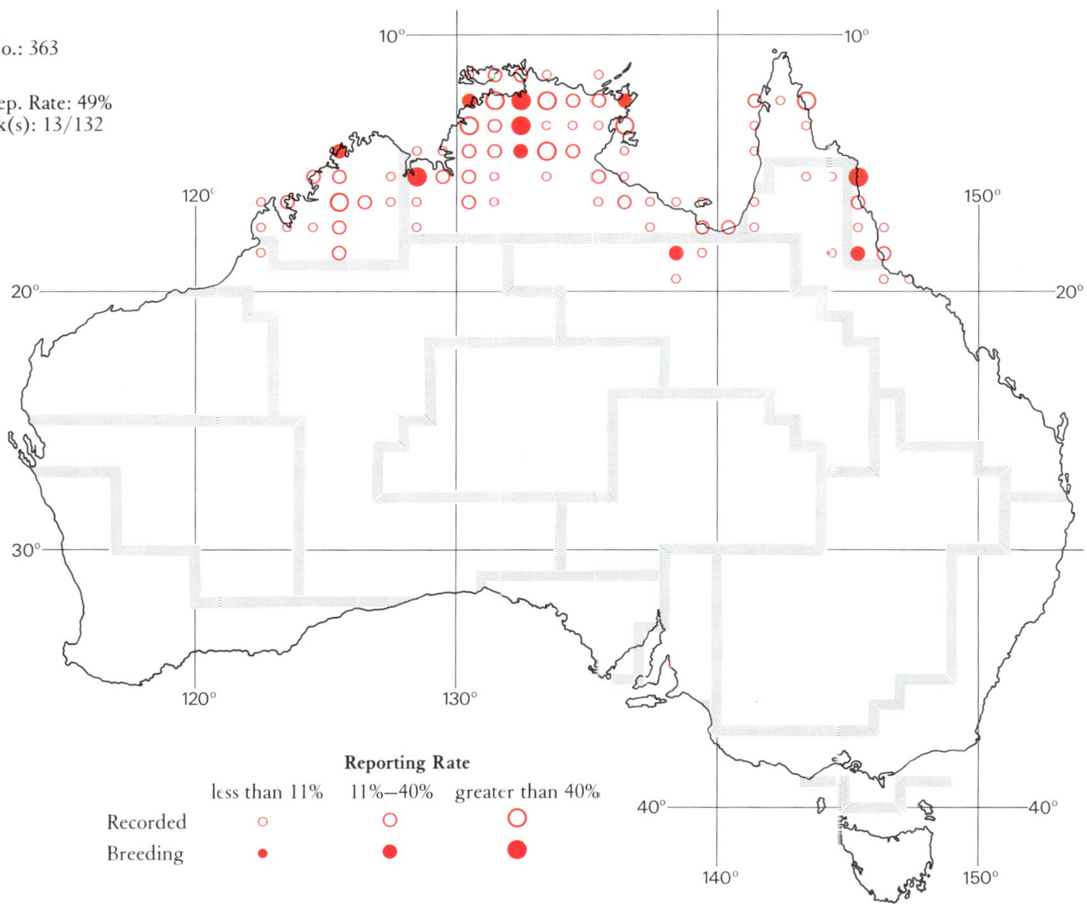

Willie Wagtail

Rhipidura leucophrys

The Willie Wagtail lives in New Guinea and adjacent islands and in Australia. It was one of the most widely recorded birds in the Field Atlas though rarely reported in the Cape York Region north of 12°S and not generally present in Tas. In the South-West the Willie Wagtail has spread with clearing into the Darling Ra.[1622] but no general expansion of range is evident.

The Willie Wagtail is present thoughout the year in the southern Murray-Darling Region,[833] at Jamieson (37/146)[1469] and on the Nullarbor Plain.[246] The Field Atlas provides no evidence of large-scale seasonal movements, though seasonal changes in distribution are reported. At Jandowae (26/151) it moves in Apr. to open farmland nearby, returning Aug.-Sept.; during winter the sorghum fields on the open plain are filled with Willie Wagtails, which feed and roost there.[1329] At Inverell (29/151) numbers are reduced in autumn and winter with small flocks forming in Feb.-Mar.[96] In Canberra (35/149) after breeding the population seems greatly reduced.[1192]

The Willie Wagtail lives in pairs or family parties at the margins of eucalypt woodland and in open country, including urban areas, where there are trees for nesting. Sometimes flocks of 15-20 birds form. The Willie Wagtail feeds on insects which it takes in the air, sometimes flushing them with its wings and tail. It will also rob food from other species when it meets a mixed feeding flock[719] and will feed its young on maggots. It has extended its range locally as a result of clearing and agricultural development, not only because it can feed over clearings but also because it is often closely associated with domestic stock. The bird rides on the backs of cattle and sheep making sallies to catch the insects they flush[774] and also takes ticks from cattle.[231,1099]

Density near Port Moresby (9S/147E) in New Guinea in eucalypt woodland was 0.15 birds/ha;[158] at Armidale (39/151) in eucalypt woodland 0.02–0.25 birds/ha;[582] at Jamieson (37/146) on the margin of eucalypt forest 0.31 birds/ha;[1469] in dongas in saltbush on the Nullarbor Plain 0.06 birds/ha;[246] and in eucalypt woodland at Wellard (32/115) one population had a density of about 1 bird/ha.[1429]

Field Atlas Records: 48 133

1° blocks-with records: 762 (94%)
-with breeding records: 313 (41%)

Atlas No.: 364

Max. Rep. Rate: 94%
In block(s): 27/148

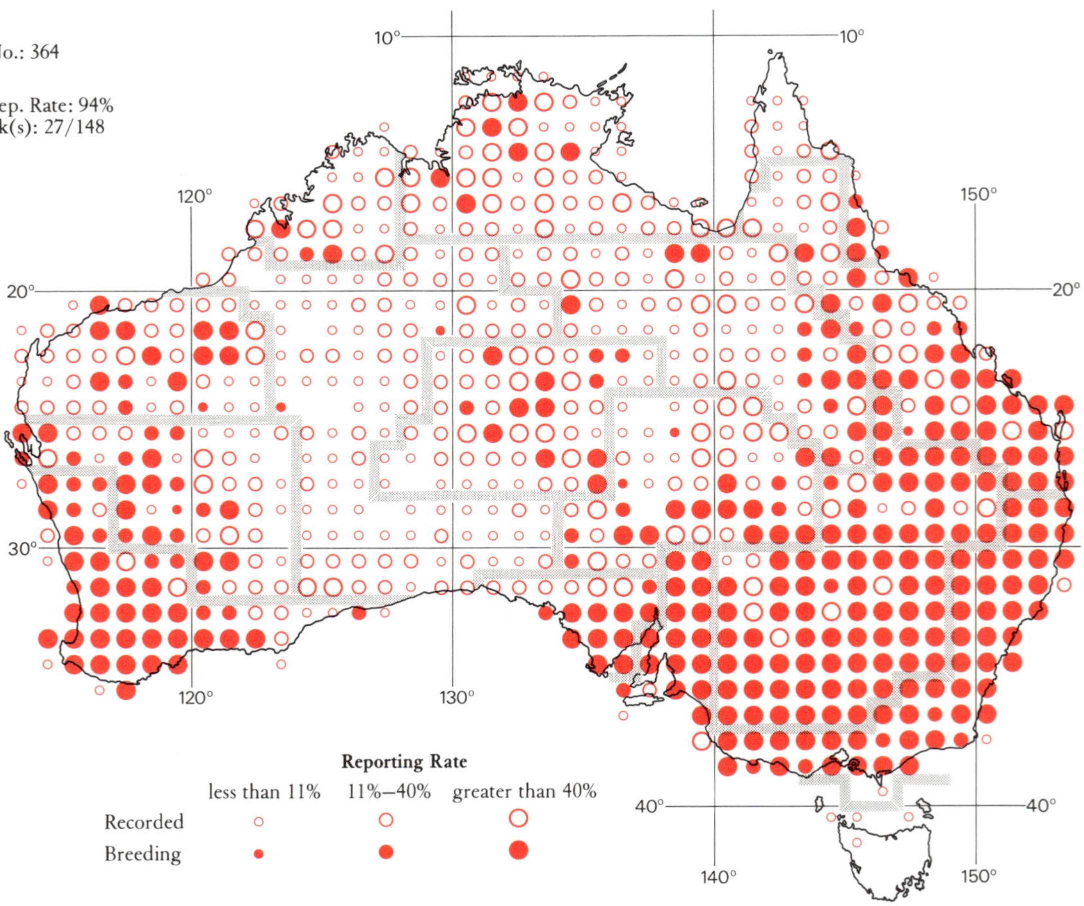

The endemic Logrunner is continuously distributed within its range except where it has disappeared from the environs of Sydney (33/151). No seasonal movements are reported.

The Logrunner usually lives in pairs but occasionally in family parties. It inhabits rainforest, including places where the understorey is the introduced lantana[974] or blackberry.[1183] By using the introduced species of weed it is able to move out from rainforest into cleared land and along watercourses. It has once been recorded as breeding communally.[501]

The Logrunner eats insects, specially their larvae. Often it feeds in company with the Yellow-throated Scrubwren and Eastern Whipbird.[791] To unearth its prey, the Logrunner scratches aside the litter, bracing itself on its spiny tail as it does so. The circular scratching sites are characteristic and enable the presence of the bird to be detected even if its loud calls are not heard.

At Dorrigo (30/152) one pair maintained itself in 2 ha of isolated rainforest surrounded by dairying country.[875]

Logrunner

Orthonyx temminckii

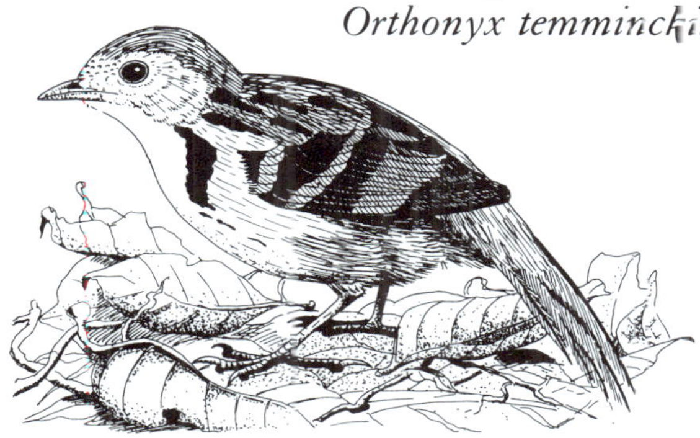

Field Atlas Records: 292

1° blocks–with records: 16 (2%)
–with breeding records: 7 (44%)

Atlas No.: 434

Max. Rep. Rate: 19%
In block(s): 28/153

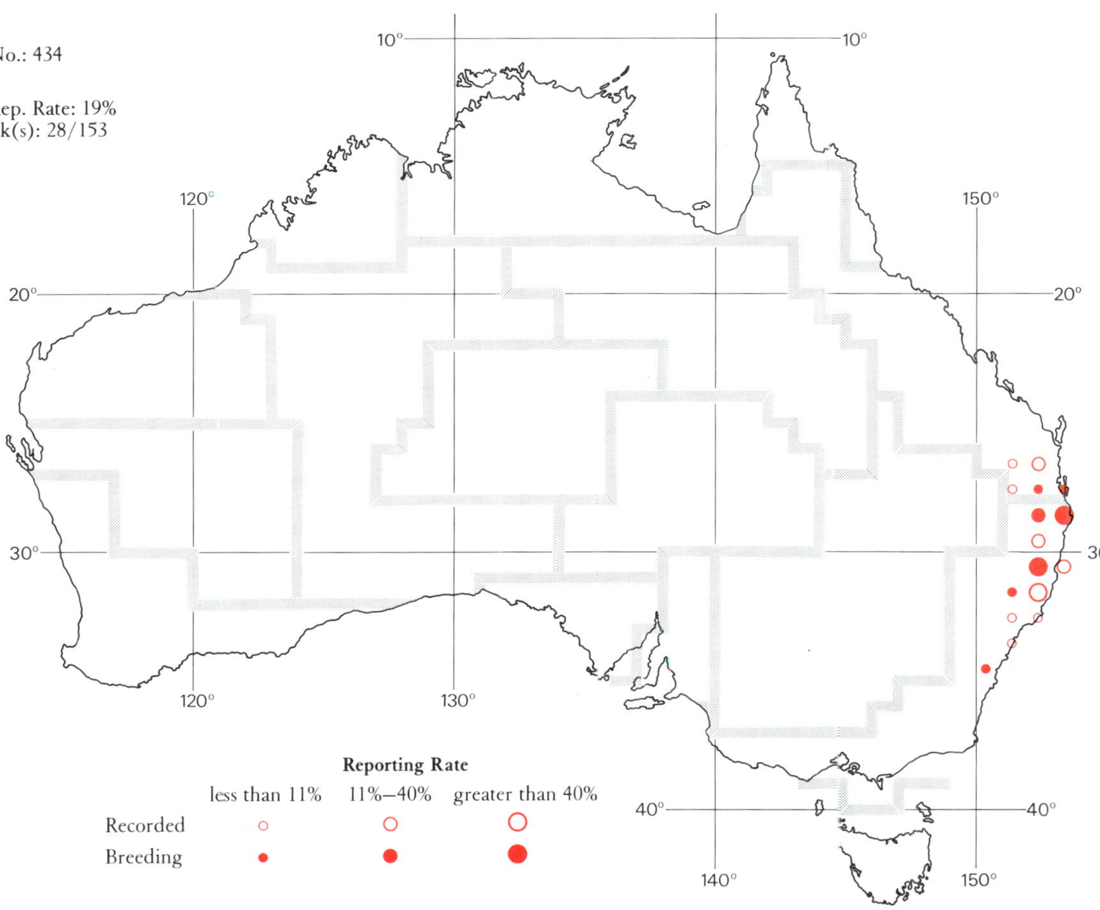

Chowchilla

Orthonyx spaldingii

The Chowchilla is endemic to the Atherton Region and recorded in the Field Atlas in all the 1° blocks from which it was reported historically. It is sedentary[1040] and inhabits rainforest.[748,1703] Breeding pairs usually occupy a territory but bands of 3-5 birds are often reported and as many as 16 immature birds were observed in one flock.[1036,1928] The Chowchilla feeds by scratching the forest floor to expose insects, earthworms and berries.[1928] When disturbed it has been known to hide under leaves on the forest floor, scratching litter over itself in the same way as it scratches when feeding.[231]

The average area occupied by breeding pairs is 3 ha.[1040]

Field Atlas Records: 209

1° blocks–with records: 8 (1%)
–with breeding records: 3 (38%)

Atlas No.: 435

Max. Rep. Rate: 20%
In block(s): 15/145

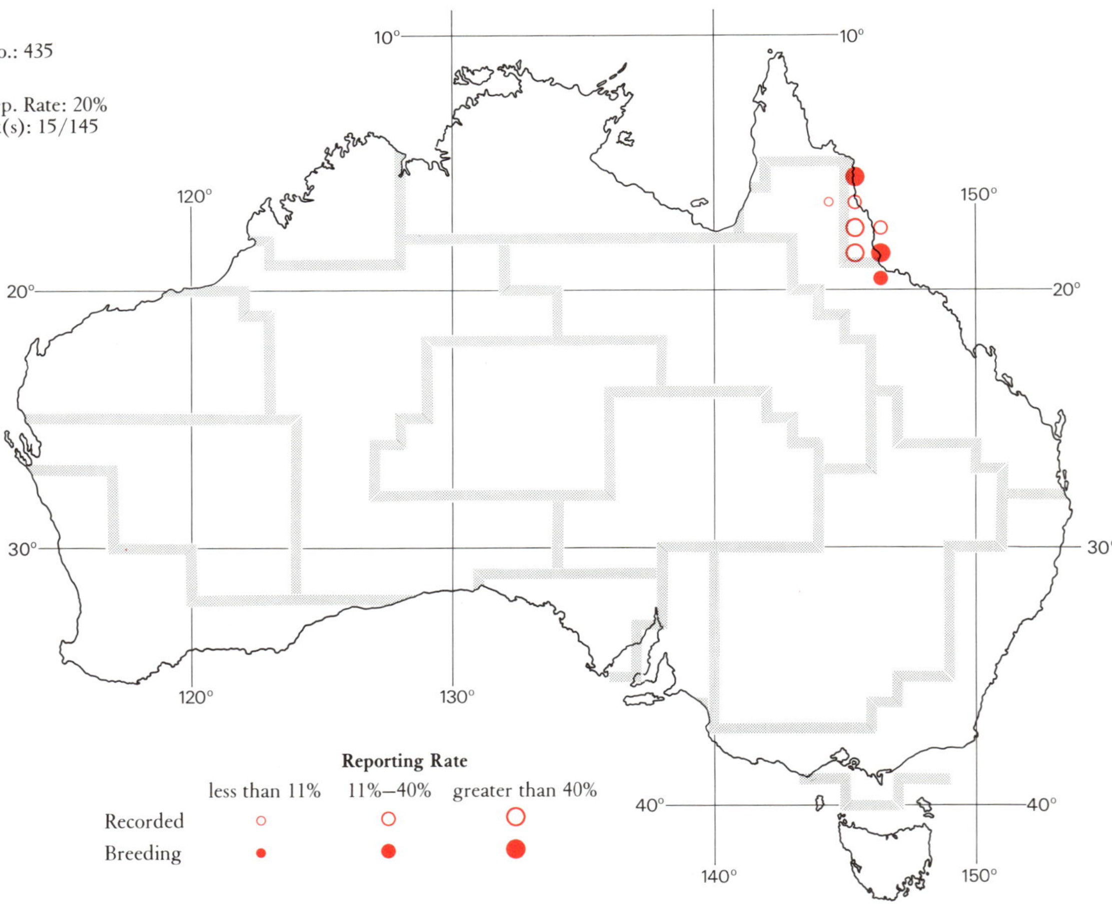

Eastern Whipbird

Psophodes olivaceus

There are two subspecies of the endemic Eastern Whipbird; *lateralis* of the Atherton Region is isolated from *olivaceus* to the south. The population in the Eungella (21/148) district is also isolated but not considered distinct from *olivaceus*.[983] The Whipbird's present range appears unchanged from the historical periods.

Data in the Field Atlas give no indication of large-scale movements and the species is known to be sedentary at Moruya (35/150).[1198] The resident nature of local populations and their isolation has been demonstrated by a study of their songs.[1825] Those of the male do not differ much throughout the bird's range but the antiphonal response from the female may vary from district to district.

The Eastern Whipbird lives in rainforest and eucalypt forest, specially in gullies where there is dense ground cover. It will also occupy heath, and thickets of the introduced lantana and blackberry.[974,1183] The Whipbird feeds on insects found under litter on the ground.[758,1049] Clearing adversely affects the species but where lantana and blackberries are established, the Eastern Whipbird is said to have increased in abundance.[974]

In eucalypt woodland at Moruya (35/150) density was 0.05–0.11 birds/ha[1197] and in eucalypt forest at Jamieson (37/146) 0.1 birds ha.[1469] At Boola Boola (38/146) in eucalypt forest density was 1.2 birds/ha[1113] in the gullies about 15 years after logging. Older regrowth was thought to be too open for the species. At Wandin (37/145) three pairs were recorded in 0.8 km of creek frontage.[758]

Field Atlas Records: 6328

1° blocks–with records: 65 (8%)
–with breeding records: 32 (49%)

Atlas No.: 421

Max. Rep. Rate: 58%
In block(s): 28/153

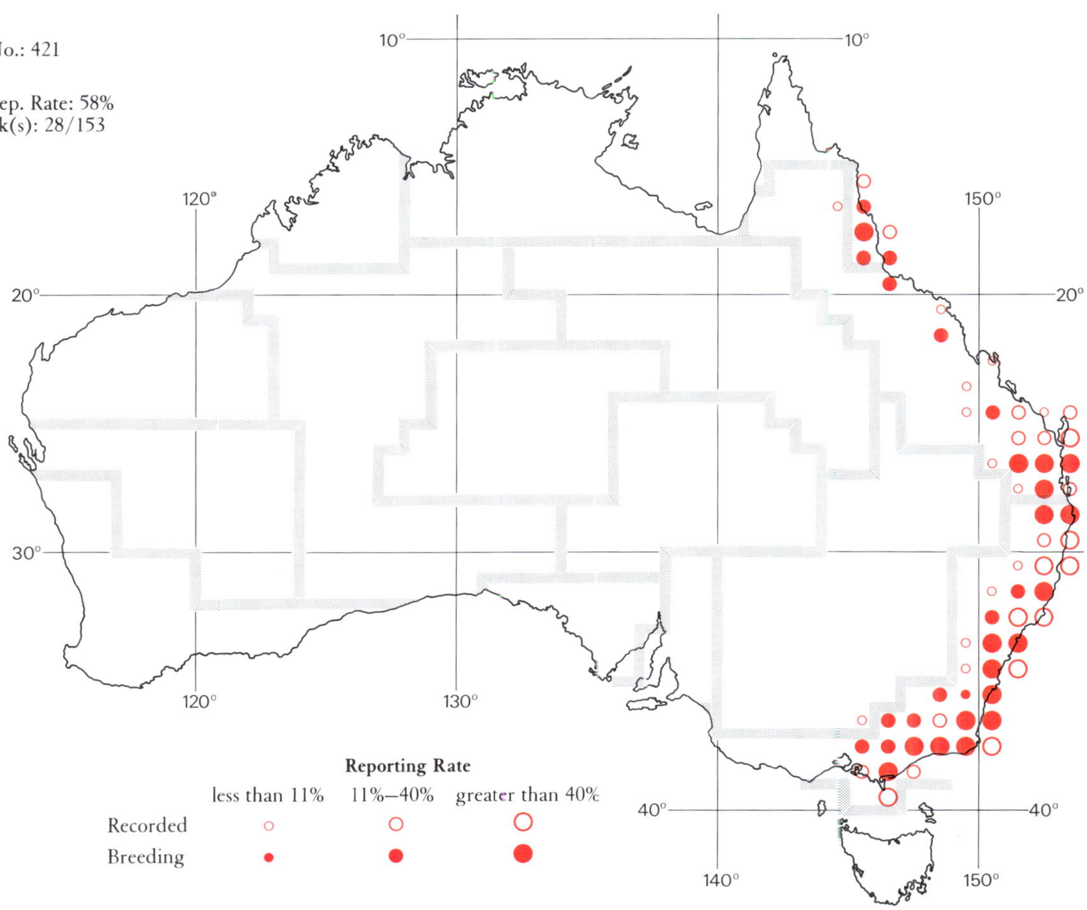

Western Whipbird

Psophodes nigrogularis

There are three subspecies of the endemic Western Whipbird: *leucogaster* in the Murray-Darling Region; *pondalowiensis* on the tip of Yorke Pen. (35/137) and perhaps including birds on Kangaroo I. (35/137); and *nigrogularis* of southern Eyre Pen. and the South-West Region.[387,1394] Eastern populations of the Western Whipbird were not discovered until fairly late: *leucogaster*, in 1932; *pondalowiensis* on Yorke Pen., in 1965; the population on Kangaroo I., in 1967; and *nigrogularis* on Eyre Pen., in 1966.[386,387,887] The historical maps (p. 666) show changes in distribution. Despite searches *leucogaster* has not been recorded in Vic. since 1969 and its range has probably declined. The western population of *nigrogularis*, which last century lived as far north as Wongan Hills (30/116) and west to near Perth (31/115) and Margaret R (33/115), has also declined.[1638,1684]

The Western Whipbird lives in mallee and heath. It is difficult to flush from the ground where it feeds on insects and small invertebrates, including land snails.[1270a]

The two main populations seem originally to have been isolated by increasing aridity[983] and there are small isolated populations within the range shown on the map. European settlement is the cause of a continuing shrinkage of range.[1684] Once the mallee or heath is cleared the Western Whipbird seems to have no ability to survive elsewhere. Burning has also eliminated it from many areas.

About 70 pairs lived in the reserve at Two People Bay (34/118) in 1976.[1684]

Field Atlas Records: 107

1° blocks–with records: 12 (1%)
–with breeding records: 4 (33%)

Atlas No.: 422

Max. Rep. Rate: 16%
In block(s): 35/136

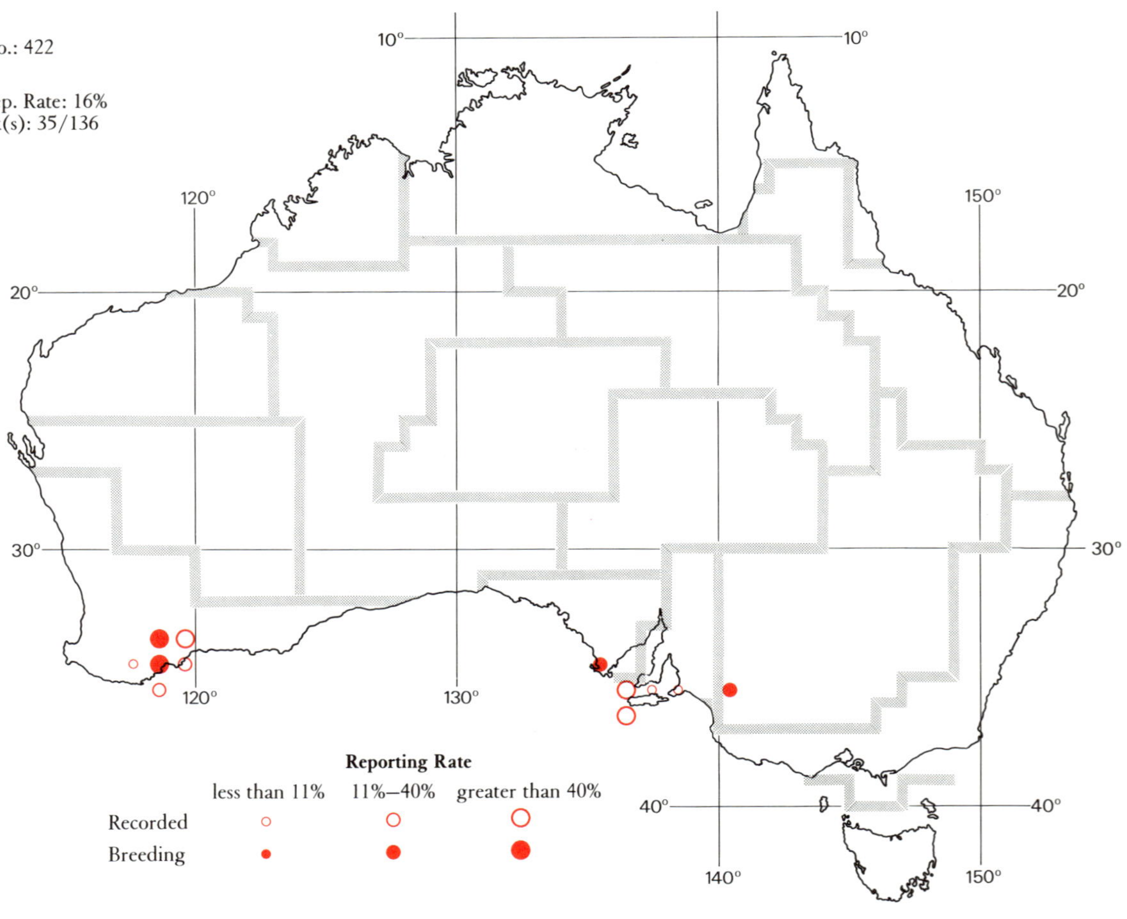

In 1973 the eastern and western populations of the endemic Wedgebill were recognized as probably being separate species.[625,1379] They approach each other most closely in the Oodnadatta (27/135) area. A specimen of the Chirruping has been collected at Oodnadatta, 80 km from where the Chiming was collected at Todmorden (27/134).[1379] During the Field Atlas both species were heard in the same 1° block 'near Oodnadatta' but not in the same place.

The Field Atlas suggests no large-scale seasonal movement but individuals have turned up well outside the normal range of the species: at L. Boga (35/143)[625] and, in Nov. 1979, south of Springsure (24/148).

The Chirruping Wedgebill lives in lower and more scattered vegetation than the Chiming, inhabiting saltbush, including that dominated by bluebush, acacia scrub, spinifex and lignum swamps. It is less skulking than the other species.[625,1160] It feeds on insects[1160] and seeds.[748]

There are no measures of abundance.

Chirruping Wedgebill

Psophodes cristatus

Field Atlas Records: 871

1° blocks–with records: 57 (7%)
–with breeding records: 12 (21%)

Atlas No.: 866

Max. Rep. Rate: 76%
In block(s): 31/140

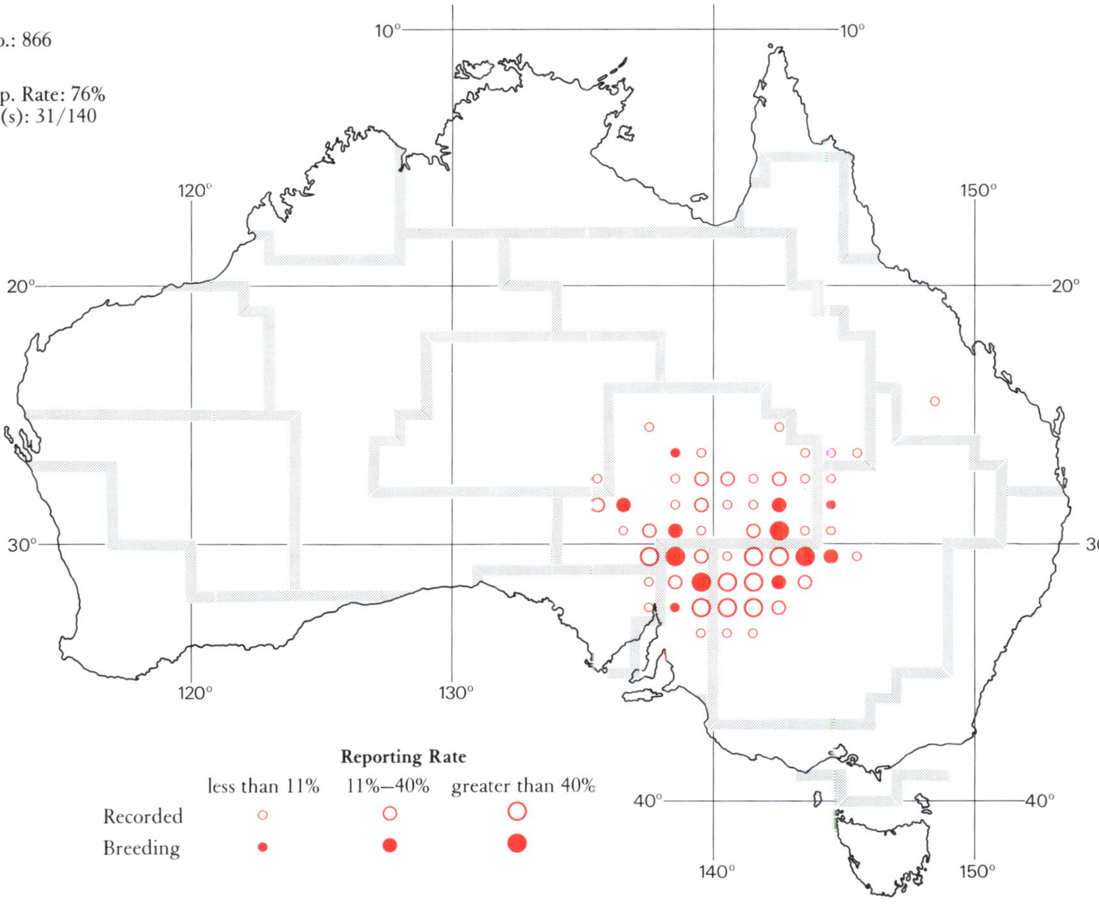

Chiming Wedgebill

Psophodes occidentalis

In 1973 the eastern and western populations of the endemic Wedgebill were recognized as probably being separate species.[625,1379] The principal difference between them is their song. They approach each other most closely in the Oodnadatta (27/135) area. Specimens of the Chiming have been taken at Todmorden (27/134), 80 km from where a specimen of the Chirruping has been collected at Oodnadatta.[1379] During the Field Atlas both species were heard in the same 1° block 'near Oodnadatta' but not in the same place. The range of the Chiming Wedgebill is continuous between the Goldfields and Western Deserts Regions and extends east to Urandangi (21/138) and to 23/138 in the Simpson Desert. The species seems to have disappeared from the Nullarbor Plain where there were records before 1935 from Naretha (31/124), Rawlinna (31/125) and Haig (31/126).[381,1878]

The Field Atlas suggests no large-scale seasonal movement though, in WA, inland birds are said to move locally.[587]

The Chiming Wedgebill lives in acacia scrub, mallee, specially that dominated by broombush, and in spinifex along watercourses[625] and is more often heard than seen. Presumably, like the Chirruping Wedgebill, it feeds on seeds and insects.[748,1160] In captivity it will take termites.[907]

There are no measures of abundance.

Field Atlas Records: 407

1° blocks–with records: 97 (12%)
–with breeding records: 4 (4%)

Atlas No.: 865

Max. Rep. Rate: 61%
In block(s): 25/113

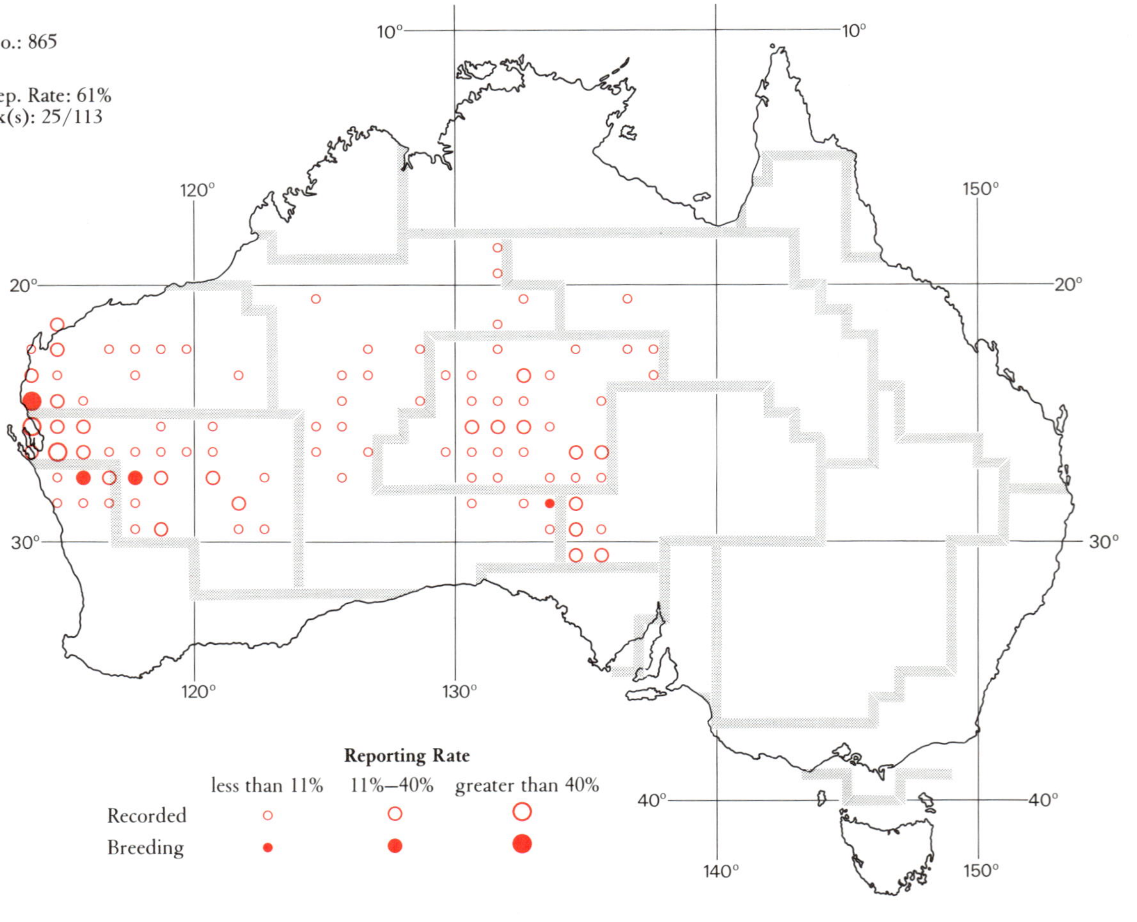

Spotted Quail-thrush

Cinclosoma punctatum

The Field Atlas shows the distribution of the endemic Spotted Quail-thrush to be fragmented but only Tas. birds have been separated as a subspecifically distinct form *dovei* from the mainland *punctatum*.[304,605,1422] The species is absent from islands in Bass Strait. Birds in the southern Mt Lofty Ra. (35/138) are isolated. This population was formerly distributed north to Angaston (34/139) and has declined in range and abundance. It was not recorded at all between 1972 and 1976.[1693]

The Spotted Quail-thrush is sedentary.[605] It lives in eucalypt forest, specially in high country and on ridges running off the high plateaux. It also lives at lower altitudes in eucalypt forest where there is an understorey of small shrubs, grasses and other herbs. At Barrington (31/151) it inhabits sunny, north-facing ridges.[892] The Spotted Quail-thrush feeds on insects, seeds and lizards.[605,1049]

Density in eucalypt woodland at Wollomombi (30/152) was 0.04 birds/ha[129] and at Moruya (35/150) 0.2–0.4 birds/ha.[1192] At Boola Boola (38/146) in eucalypt forest it was 0.4–0.5 birds/ha.[1113]

Field Atlas Records: 1243

1° blocks–with records: 74 (9%)
–with breeding records: 18 (24%)

Atlas No.: 436

Max. Rep. Rate: 22%
In block(s): 37/150

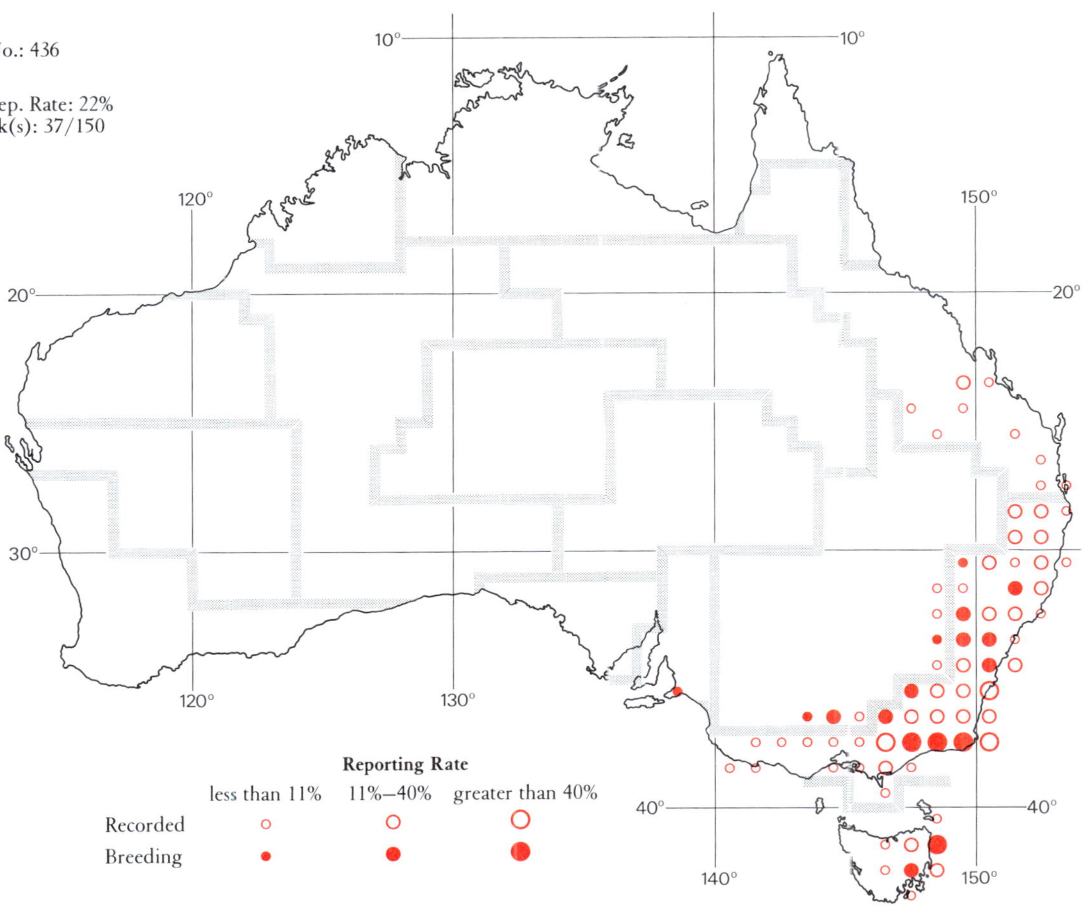

Chestnut Quail-thrush

Cinclosoma castanotum

The distribution of the endemic Chestnut Quail-thrush is not as disjunct as it appears in the Field Atlas. It is better illustrated in the 1951–76 map (p. 667). This shows the species to be more or less continuously distributed across southern Australia, though any connection between the population in and near Round Hill Nature Reserve (33/146) and birds farther west in the Murray-Darling Region appears tenuous. There is geographical variation within the species. The subspecies *castanotum* occupies the southern fringe of the range and areas east of Spencer Gulf (137°E). This is linked by intermediate populations to *clarum* in the inland.[620]

The range of the Chestnut overlaps that of the Cinnamon and is contiguous with that of the Nullarbor Quail-thrush.[597,599,601] Occasional hybrids are produced with the subspecies *marginatum* of the Cinnamon Quail-thrush.

The inland subspecies *clarum* of the Chestnut Quail-thrush is thought to be partly nomadic but birds inhabiting more coastal environments are sedentary.[189,331,620] The Chestnut Quail-thrush usually lives in pairs but is sometimes in small parties. It inhabits acacia scrub and mallee, including that where the understorey is dominated by spinifex. Usually the bird is associated with sandy rather than lateritic soil[599] and it therefore lives both north and south of the Nullarbor Plain where there are sandy soils in which mallee grows. It has disappeared from large areas of SA and WA cleared for agriculture.[587]

The Chestnut Quail-thrush feeds on insects and seeds.[1049]

At Eyre Bird Observatory (32/126) density was 0.1 birds/ha.[390] There were two pairs 90 m apart near Yalgoo (28/116).[602]

Field Atlas Records: 470

1° blocks–with records: 66 (8%)
–with breeding records: 11 (17%)

Atlas No.: 437

Max. Rep. Rate: 26%
In block(s): 32/126

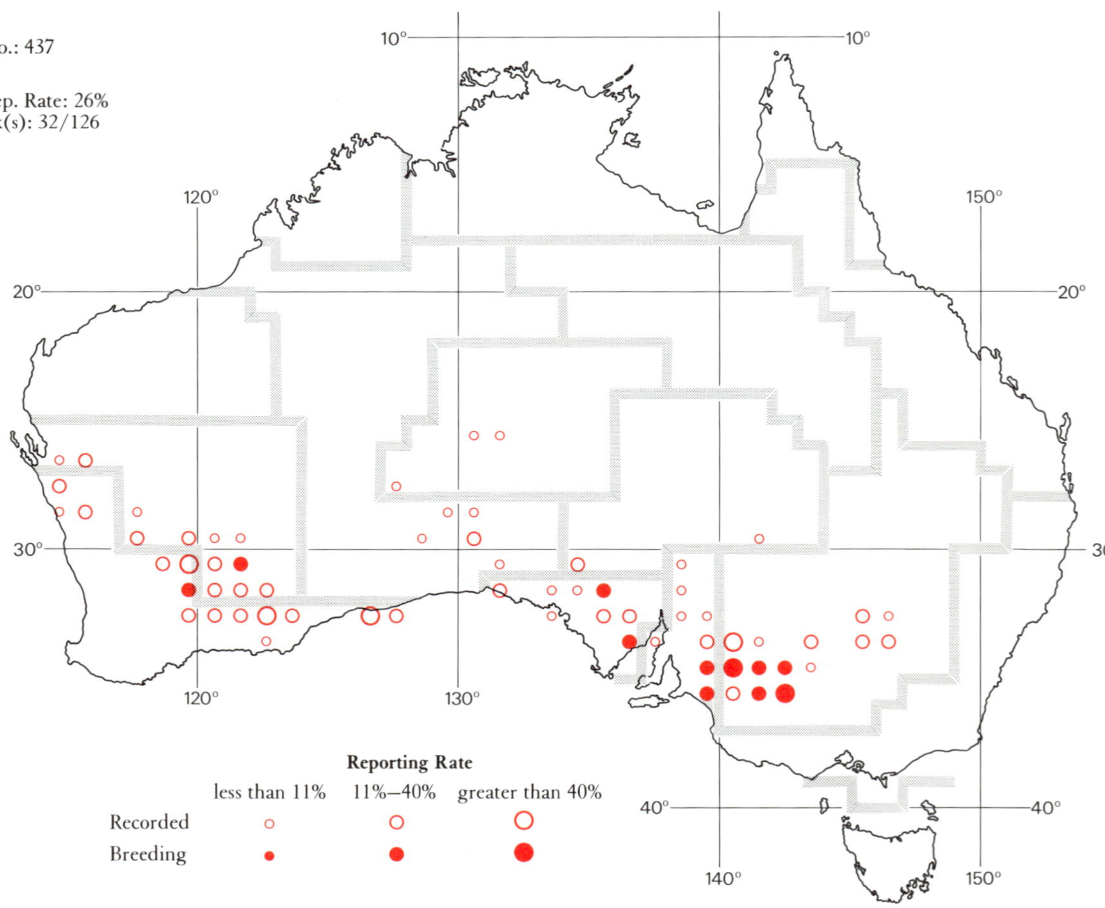

Cinnamon Quail-thrush

Cinclosoma cinnamomeum

The endemic Cinnamon Quail-thrush is separated into three subspecies: *marginatum* in the west, *cinnamomeum* in the centre and *castaneothorax* in the east.[601] The subspecies *marginatum* is sometimes called a full species.[1717] The Nullarbor Quail-thrush is treated as a separate species here though there are some grounds for considering it conspecific with the Cinnamon.[597] The subspecies *marginatum* of the Cinnamon Quail-thrush sometimes hybridizes with the Chestnut Quail-thrush.[605]

From Field and Historical Atlases, no break in the range of the Cinnamon Quail-thrush is evident, in particular between *cinnamomeum* and *marginatum*, which have been said to be separated.[605] The distribution of *marginatum* extends farther south in the Western Deserts than shown on the map. The southernmost records reach 30°S, the southern edge of the Great Victoria Desert.

The Field Atlas provides no evidence of large-scale seasonal movement.

Throughout its range the Cinnamon Quail-thrush lives in acacia scrub, saltbush or spinifex where this grows on stony country.[748] It avoids sandy areas. Loose groups have been reported[719,1605] and there is one record of communal breeding.[241] The Cinnamon Quail-thrush feeds on insects and seeds, turning over small stones with its bill.[587] It seems to have been unaffected by European pastoral development and is still reported in areas where stock have been grazed for many years, e.g. the Goldfields Region.

There are no measures of abundance.

Field Atlas Records: 792

1° blocks–with records: 139 (17%)
–with breeding records: 26 (19%)

Atlas No.: 439

Max. Rep. Rate: 41%
In block(s): 29/137

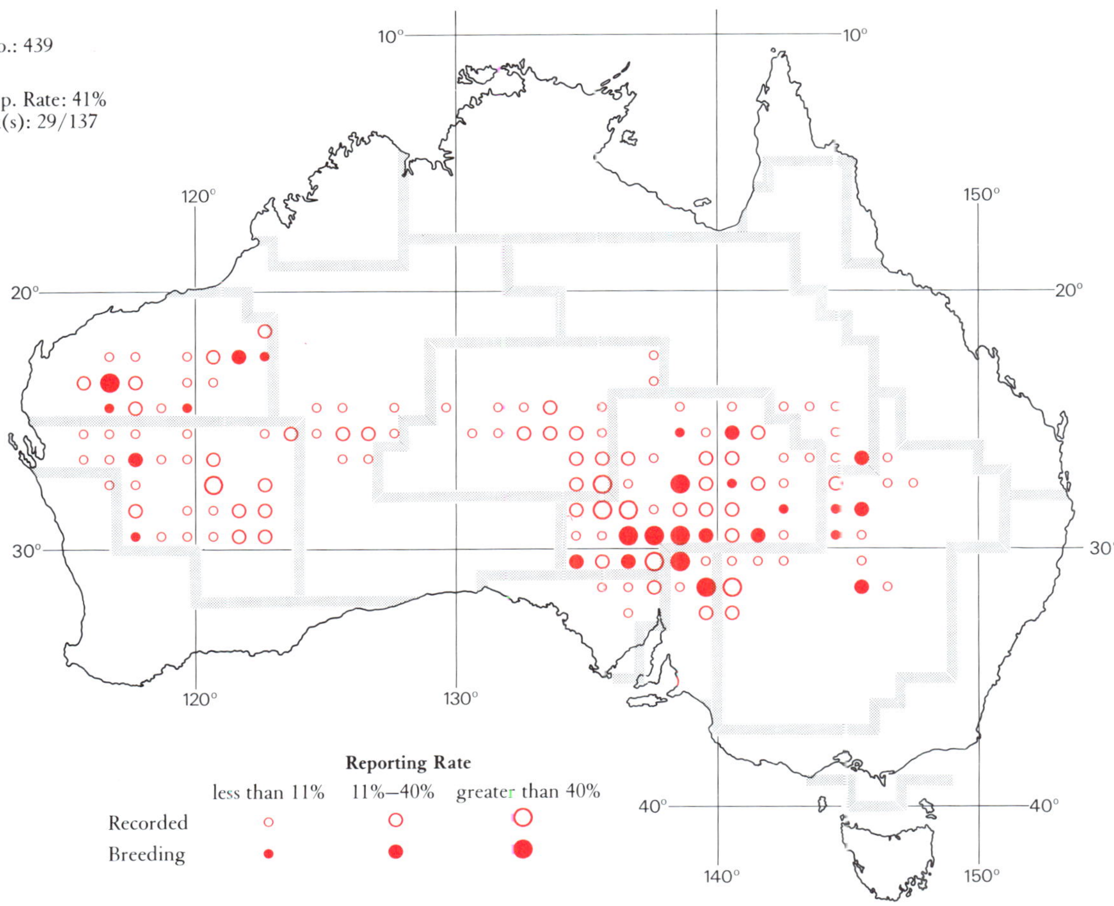

Nullarbor Quail-thrush

Cinclosoma alisteri

The Nullarbor Quail-thrush is endemic to the Nullarbor Plain. Though treated here as a separate species, there are grounds for considering it conspecific with the Cinnamon Quail-thrush.[597] The two groups of sightings which are isolated on the map are connected by records in the Historical Atlases.

There is no evidence of large-scale seasonal movement in the Atlas, though the bird may move round the Nullarbor Plain depending on seasonal conditions.[587]

The Nullarbor Quail-thrush lives in pairs or small family parties. It is confined to saltbush plains where bluebush also grows. Dongas with their richer vegetation are avoided.[1878] The Quail-thrush inhabits stony country and is absent from the sand dunes at the edge of the Nullarbor Plain where the Chestnut Quail-thrush lives.[601] Its distribution is uneven, many apparently suitable areas being uninhabited by the Quail-thrush. The species feeds on seeds and insects[601] and round Rawlinna (31/125) survives in country lightly stocked with sheep.

During a 14-day search in the Rawlinna-Seemore Downs (30/125) area nine birds were counted. There are no other measures of abundance.

Field Atlas Records: 12

1° blocks–with records: 9 (1%)

–with breeding records:

Atlas No.: 440

Max. Rep. Rate: 8%
In block(s): 31/125

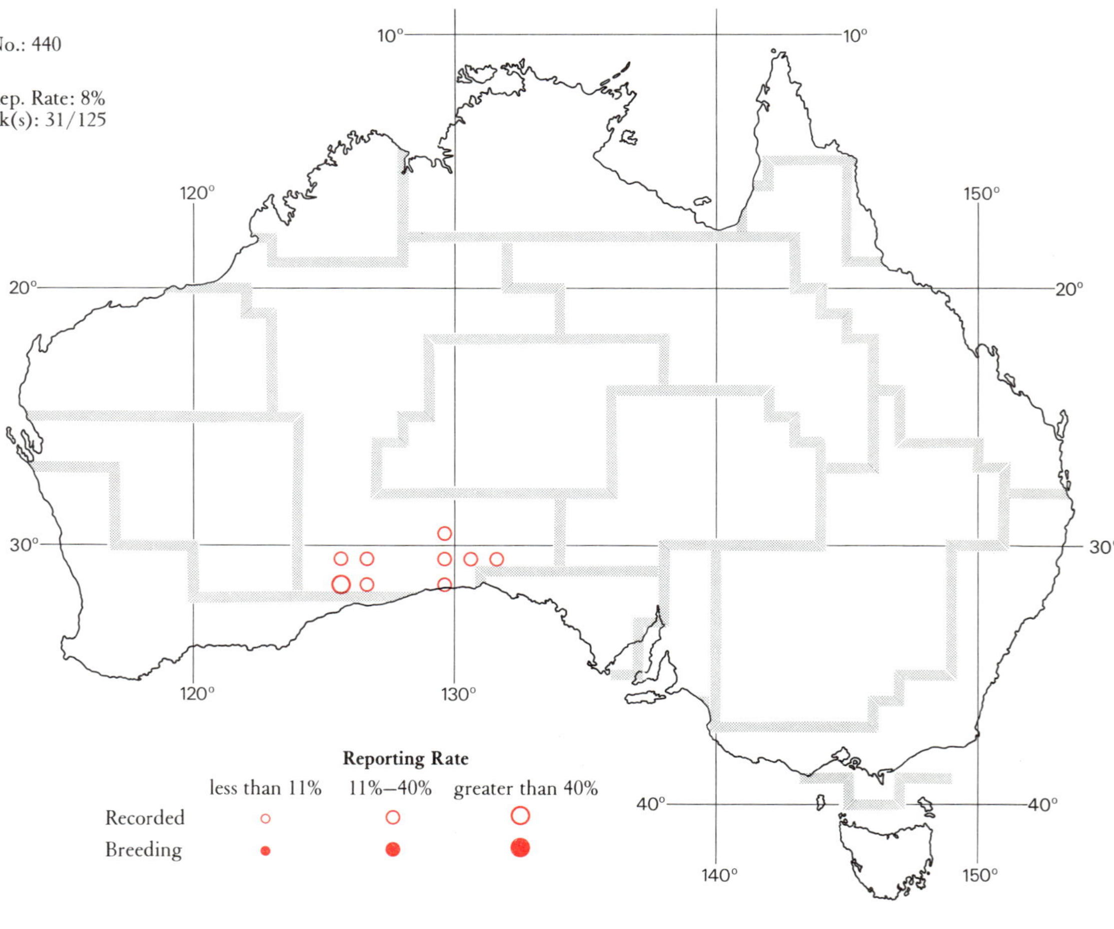

The Grey-crowned Babbler lives in southern New Guinea and Australia. Up to ten subspecies have been recognized but much of the variation is clinal and there is disagreement about them.[748,1124] The Western Deserts isolate the Pilbara-Goldfields population from the remainder of the species.

The Grey-crowned Babbler was once distributed along the coast from Melbourne (37/145) westwards into the SA Ra. Region. A specimen was collected as far west as the southern Mt Lofty Ra. (35/138) in 1887. The species has retreated from The Coorong (36/139) since the 1930s and disappeared from the Portland (38/141) district in the 1920s. Birds are only occasionally recorded near Bangham (36/140), now the western end of its range in southern Australia.[865] Round Melbourne the Grey-crowned Babbler has been reduced to an isolated population on the Mornington Pen. (38/145).

The Grey-crowned Babbler is sedentary.[1001] Groups maintain communal territories. The birds may nest as simple pairs but usually as a larger social group.[258] The species inhabits acacia scrub, woodland and farmland. Sandy deserts and very wet areas are avoided. In places it overlaps other species of babbler.[600,748] The Grey-crowned Babbler feeds on spiders, centipedes and other invertebrates, which it seeks under litter, in crevices of branches and tree trunks and in vegetation on and near the ground.

In a study that compared populations of the Grey-crowned Babbler in two areas with annual rainfall of 560 and 710 mm, group size varied from 4–11 and was larger in the drier area.[1001] Size of territory did not correlate with numbers of birds in a group and ranged from 1.7 to 52.7 ha. Larger territories were probably associated with lack of tall vegetation.[1001]

Grey-crowned Babbler

Pomatostomus temporalis

Field Atlas Records: 7072

1° blocks–with records: 394 (49%)
–with breeding records: 109 (28%)

Atlas No.: 443

Max. Rep. Rate: 75%
In block(s): 27/149

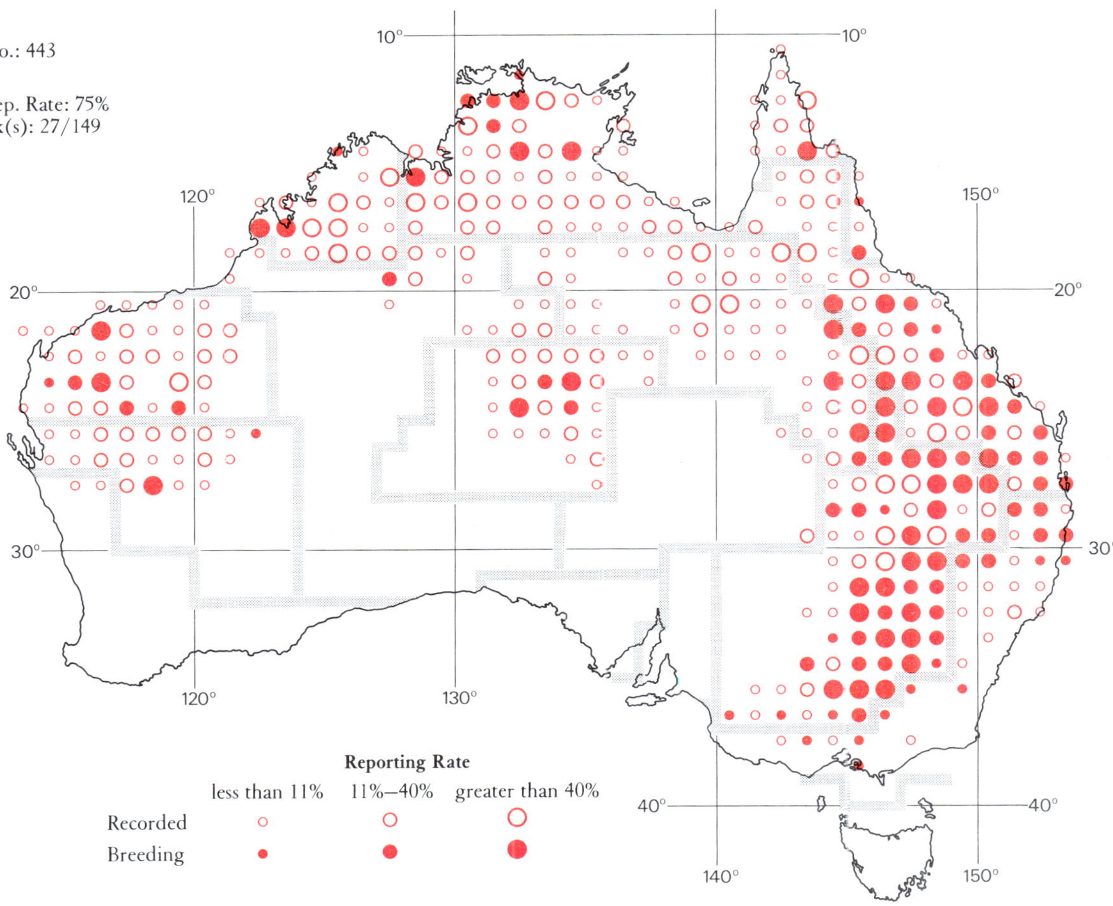

White-browed Babbler

Pomatostomus superciliosus

Up to four subspecies of the endemic White-browed Babbler have been recognized but some are now thought to be invalid.[411] The species' distribution has not changed over the historical periods. Records in 1897 from the Fitzroy R (18/123, 18/124) in the Kimberley Region[1343] are well separated from the northernmost record (20/129) in the Tanami and seem unlikely to be correct. The White-browed Babbler does not undertake regular seasonal movements but is nomadic in the non-breeding season at Inverell (29/151).[96]

The distribution of the White-browed Babbler overlaps those of the other three babblers: Grey-crowned, Chestnut-crowned and Hall's. The White-browed is closely related to the Chestnut-crowned and Hall's Babblers but has a far more extensive range; the reasons for this may be related to climatic changes in the past.[602,772]

In the Murray-Darling and South-East Regions the White-browed Babbler inhabits mainly eucalypt woodland. In the SA Ra. Region and parts of the South-West it occupies eucalypt forest as well[584] but most reports in the Adelaide Regional Atlas came from woodland and mallee.[1693] Inland the species lives in acacia scrub[607,626] and on the Nullarbor in dongas among saltbush.[246] Where it is sympatric with the Grey-crowned Babbler, the White-browed occupies the drier, more stony sites.

Groups of 3–15 White-browed Babblers live together and breed communally. The bird feeds on insects and seeds which it finds under litter and in crevices near the ground.

In dongas in saltbush on the Nullarbor Plain density was 0.15 birds/ha.[246] At Eyre Bird Observatory (32/126) densities measured by censusing were 0.1–1.5 birds/ha; when territories were mapped in Aug. of two successive years densities were 0.44 and 0.56 birds/ha.[464]

Field Atlas Records: 6912

1° blocks–with records: 345 (42%)
–with breeding records: 111 (32%)

Atlas No.: 445

Max. Rep. Rate: 59%
In block(s): 34/137

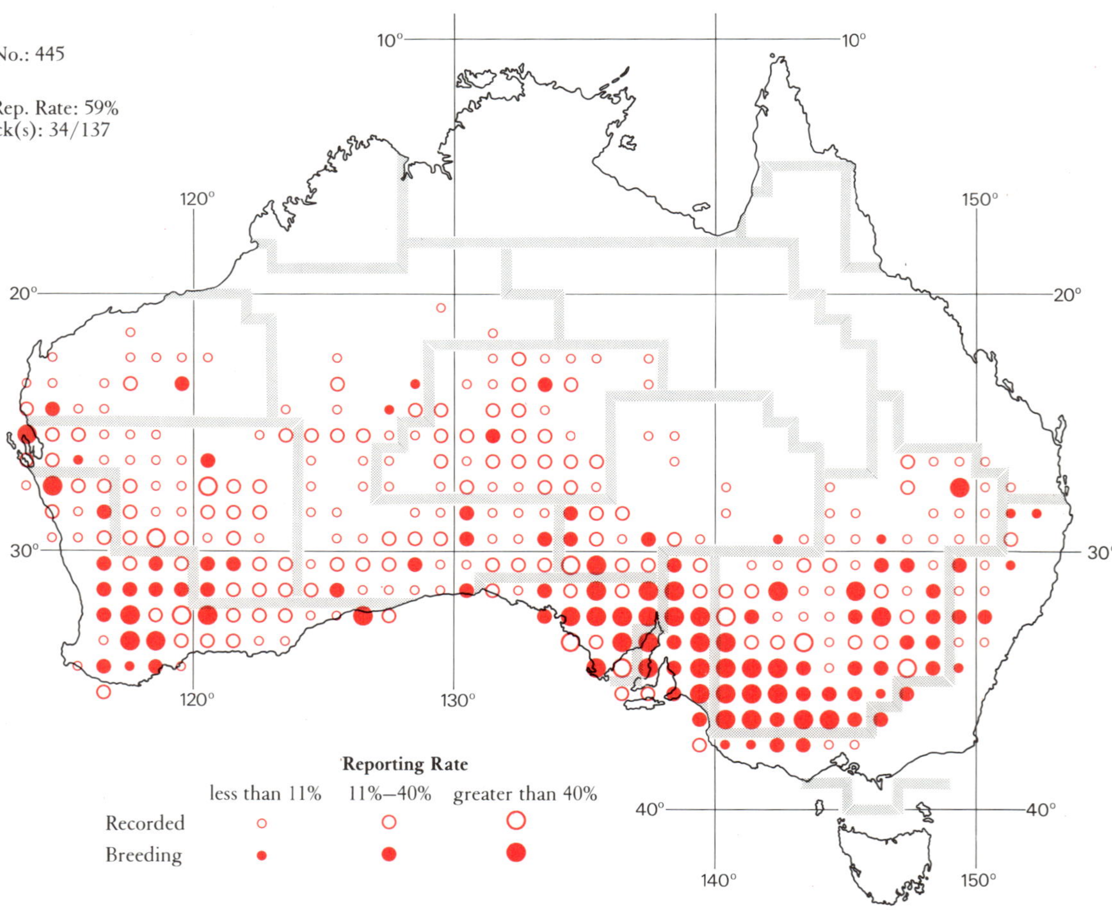

Hall's Babbler

Pomatostomus halli

Hall's Babbler was first collected in 1963 at Tyrone (26/145) and described as a new species in 1964.[411] In addition to the range shown on the map, there are earlier records from as far south as Mootwingee (31/142).

Hall's Babbler does not show seasonal movements. The species inhabits acacia scrub, mainly mulga with a grassy understorey. It is not limited to watercourses and is widespread throughout its range. Small groups roost and breed communally, as do other babblers, possibly coalescing to form larger groups in the non-breeding season.[498]

Hall's Babbler feeds on insects and other invertebrates which it finds in the grass or under debris on the ground.[92] It also probes in the bark of trees. The members of a group forage together, rarely spread over an area of more than 100 sq. m at any one time. When a large food item is found, it is shared by several members of the group.

During a week's study west of Cunnamulla (28/145) in 1974 home ranges were estimated to be 1.8–3.5 ha, with 5–15 birds in each group.[92] In the same place in Aug. 1980 home ranges were estimated at 13.7–17.3 ha, with 2–5 birds in each breeding group. These birds, some of which were colour-banded, had been observed in May 1980 in groups of 20–25.[502]

Field Atlas Records: 83

1° blocks–with records: 26 (3%)
–with breeding records: 6 (23%)

Atlas No.: 938

Max. Rep. Rate: 23%
In block(s): 28/145

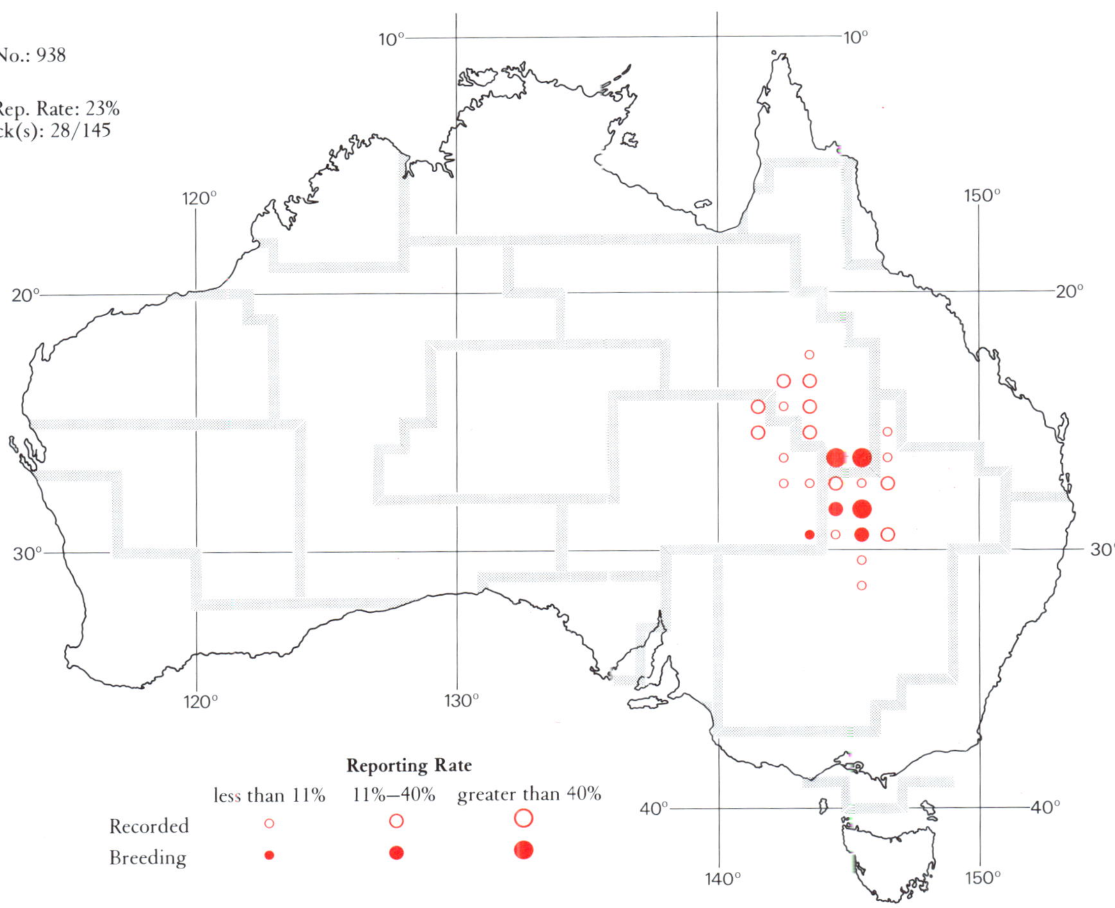

Chestnut-crowned Babbler

Pomatostomus ruficeps

The range of the endemic Chestnut-crowned Babbler appears unchanged over the historical periods. Its distribution overlaps those of the Grey-crowned and Hall's Babblers and is almost entirely within that of the White-browed Babbler. The reason that the White-browed Babbler now has a more extensive distribution than the closely-related Chestnut-crowned may be related to past climatic change.[602,772]

The Chestnut-crowned Babbler lives in mallee, woodland dominated by native pine and belah, acacia scrub, saltbush plains and along watercourses.[748,959] Groups of up to 12–15 birds live together. They use specific sites for social activities which take on the nature of playgrounds.[189] Communal breeding is likely but has not been proved.[501]

The Chestnut-crowned Babbler feeds on insects as well as seeds, collecting them on the ground and in trees. As many as 100 weevils were found in the guts of one bird.[1049]

There are no measures of abundance.

Field Atlas Records: 1242

1° blocks–with records: 80 (10%)

–with breeding records: 36 (45%)

Atlas No.: 446

Max. Rep. Rate: 36%
In block(s): 32/139

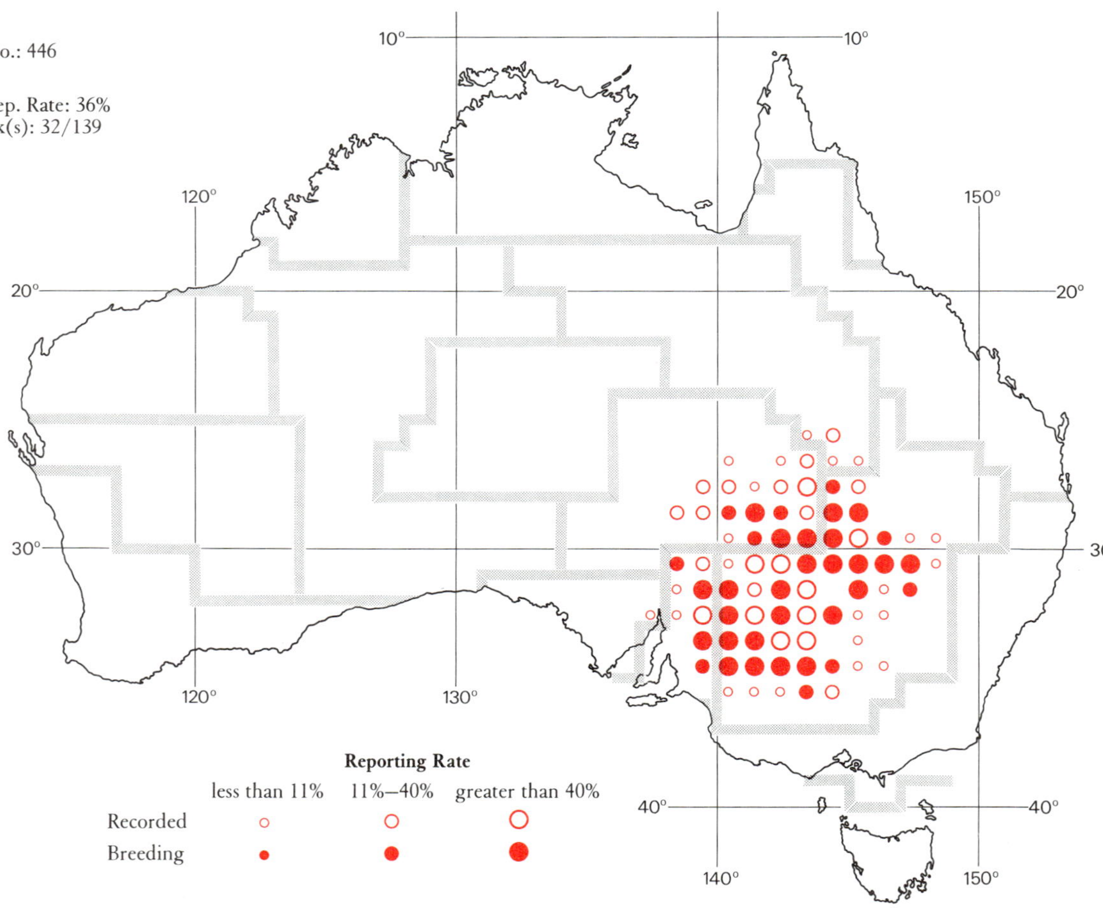

The range of the Clamorous Reed-Warbler is disjunct but reaches from north-eastern Africa, southern Asia and New Guinea to Australia.[1428] There are doubts about subspecies.[748,1124] In Tas. the Reed-Warbler is scarce but probably breeds regularly, though not reported to do so during the Field Atlas.

The Reed-Warbler is at least a partial migrant in eastern Australia. It visits Torres Strait and the Atherton Region in winter[231,503,698] and the lack of breeding reports in the Cape York and Atherton Regions suggests that most or all Reed-Warblers there are migrants. An adult banded at Willaroo (37/141) was recovered 2124 km to the north at Bambaroo (18/146) in winter.[37] The species is a summer visitor to the lower Murray-Darling Region[833] and at Moruya (35/150).[1192] There are no records in autumn and winter from Tas. Reporting Rates everywhere were lower in winter than in summer because the bird is much more conspicuous when calling during the breeding season. Movement north in winter is nevertheless indicated because the difference in Eastern Qld (summer 4.2%, winter 3.2%) was slight compared with the change in the South-East (summer 15.0%, winter 1.6%), Murray-Darling (summer 17.9%, winter 2.9%) and SA Ra. (summer 12.2%, winter 3.8%) Regions.

Whether migrants augment populations in the Top End and Kimberley Regions in winter is not known. In Darwin (12/130) the species is present only June-Jan., which indicates either local dispersal or movement inland.[423] In the South-West at Manjimup (34/116) the species arrives in two waves, Oct.-Nov. and Dec.-Jan., breeds, then departs again.[265]

The Reed-Warbler lives singly or in pairs usually in

Clamorous Reed-Warbler

Acrocephalus stentoreus

wetlands where reeds grow. It feeds on insects. Agriculture has enabled it to spread and invade newly formed reed-beds. It followed quickly on the establishment of artesian bores in the Richmond (20/143) district.[168] It did not appear in the Moora (30/116) district until there were two successive wet years.[1563] The bird did not colonize L. George (35/149) when it filled in 1950 though it did so immediately L. Burley Griffin (35/149) became suitable.[1013]

During the summer of 1979-80 at Manjimup 11 nests were used in 0.15 ha of cumbungi. At least five individually marked pairs were involved.[265]

Field Atlas Records: 5821

1° blocks–with records: 297 (37%)
–with breeding records: 93 (31%)

Atlas No.: 524

Max. Rep. Rate: 36%
In block(s): 28/148

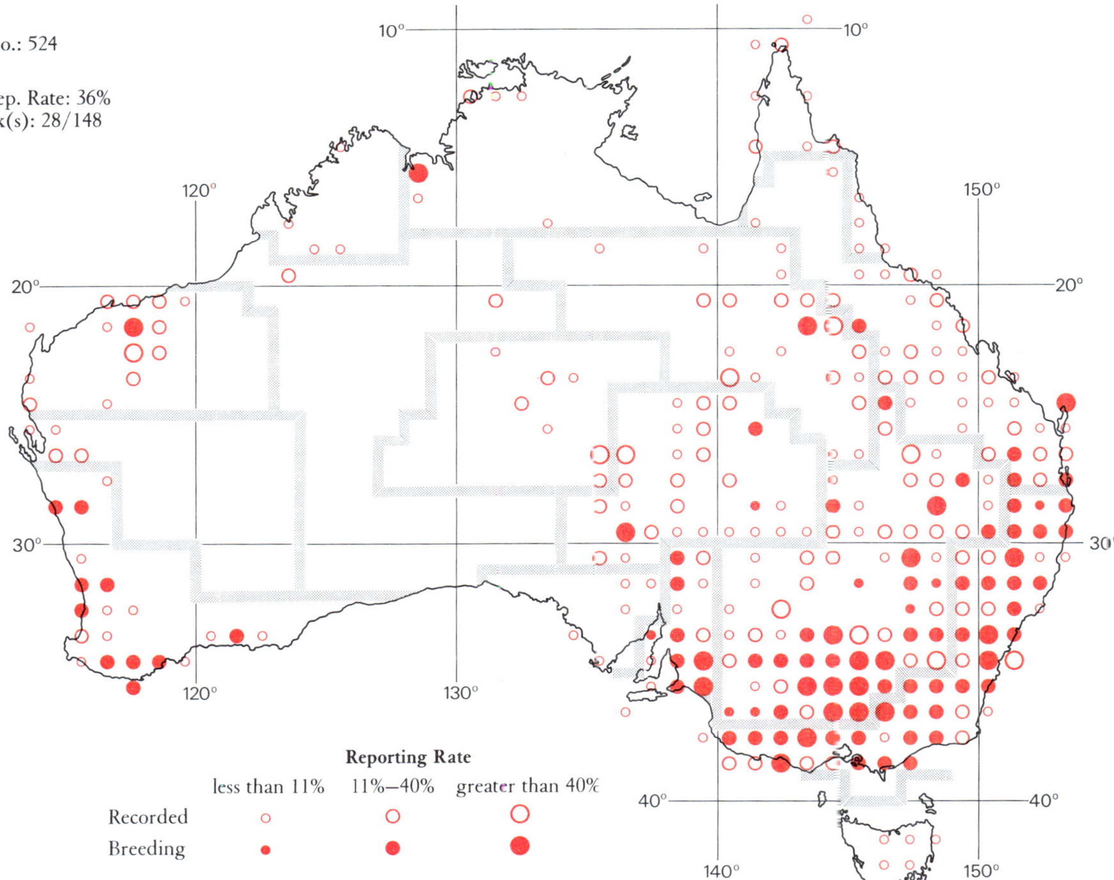

Tawny Grassbird

Megalurus timoriensis

The range of the Tawny Grassbird extends from the Philippines and New Guinea to Australia.[1124] In New Guinea eight subspecies are recognized[1449] but all Australian birds are considered to belong to the single form *alisteri*.[976] The first record for the Tas. Region was made on King I. (39/144) in May 1981.[1902] The Tawny Grassbird has been reported from some Torres Strait islands but is scarce.[503] Records round the Gulf of Carpentaria suggest contact between the eastern and Top End populations. Kimberley birds may be isolated, the most easterly record in that region being from Napier Broome Bay (14/126).

The Tawny Grassbird appears to be present all year in the Atherton Region,[231] whereas in the Rockhampton (23/150) district it is regarded as a scarce breeding migrant, present Aug.-Nov.[1089] At Murphy's Ck (27/152) it was said to arrive in Aug. each year for 15 years after which it was not seen for at least seven years.[1099] The Field Atlas does not suggest large-scale seasonal movement but occasional inland records indicate some wandering. The species often seems to be nomadic, occupying sites when they are suitable, leaving when they are not. Some established breeding pairs could be sedentary.[714]

The Tawny Grassbird lives in pairs, usually in heath, reedbeds or on the edge of large, deep wetlands that contain open water and reeds.[797] Sometimes it is recorded in drier environments, e.g. tall grass and weeds on the outskirts of a town, in rank grass bordering mountain eucalypt forest,[714] or in crops and fringes of gardens. The Tawny Grassbird feeds on insects.[748] There are suggestions that populations in some areas have declined as the result of the persistent grazing of tall grass by domestic stock[814] but, at a site in New Guinea, grass, reduced from 2 m high to dense regrowth 0.5 m high by grazing cattle, supported greater numbers of birds than neighbouring taller grass.[142]

There are no measures of abundance in Australia. In New Guinea at about 1000 m altitude density in pasture was at least 0.2 birds/ha.[150]

Field Atlas Records: 739

1° blocks–with records: 73 (9%)
–with breeding records: 9 (12%)

Atlas No.: 523

Max. Rep. Rate: 30%
In block(s): 24/153

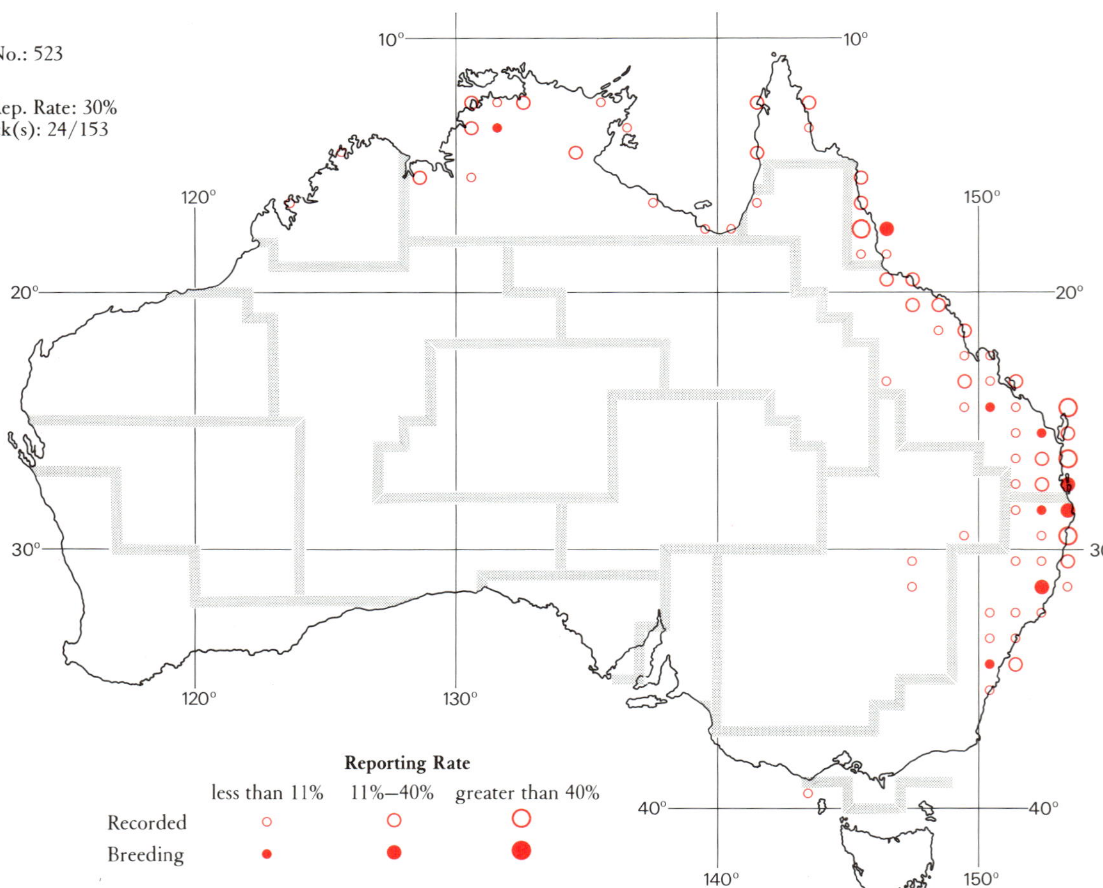

Little Grassbird

Megalurus gramineus

The Little Grassbird lives mainly in Australia. In New Guinea the single known specimen has been described as a separate subspecies.[1449] Three Australian subspecies are recognized: *goulburni* throughout eastern and southern Australia; *gramineus* confined to the Tas. Region; and *thomasi* in the South-West.[976] Sightings near Kununurra (15/128) during the Field Atlas were the first for the Top End.

All but three breeding reports in the Atlases are south of 26°S and the northernmost was at Alice Springs (23/133) in 1967.[1370] Birds carrying food have been seen in the Atherton Region.[231]

The literature contains conflicting statements about changes in abundance and movements of the Little Grassbird. The species breeds at L. George (35/149) and remains during the winter.[1013] Counts suggested that numbers were lowest there in summer, perhaps indicating a local movement to surrounding swamps. Round Melbourne (37/145) the Little Grassbird remains throughout the year, though silent in autumn and winter.[275] In Tas. a winter movement into heath has been noted.[564] At Herdsman L. (31/115), after breeding in cumbungi, the Grassbird moves to adjacent, greener sedge-covered swamps in dry autumns.[442] At another swamp in the South-West Region the bird left its breeding grounds at the end of summer.[1875] Records from scattered inland sites suggest a capacity for movement over long distances.

The Little Grassbird lives and breeds mainly in reedbeds and lignum swamps but it is sometimes present in saltmarsh,[275] on lake shores,[1013] in woodland dominated by teatree[275] and, on Bass Strait islands, in coastal grass tussocks[1048] and saltbush. Construction of artesian bores in much of northern Australia may have enabled it to colonize new areas, though there are reports from the northern Rolling Downs Region as early as 1909.[1374] The Little Grassbird feeds on insects, terrestrial and aquatic, molluscs and spiders.[1049]

At Herdsman L. (450 ha) estimates in 1980–81 put the population at 1200–1500 birds (2.67–3.33 birds/ha).[442] In another swamp in the South-West about 24 nests were found in 20 ha in 1911–12.[1875]

Field Atlas Records: 3173

1° blocks–with records: 233 (29%)
–with breeding records: 63 (27%)

Atlas No.: 522

Max. Rep. Rate: 40%
In block(s): 29/136

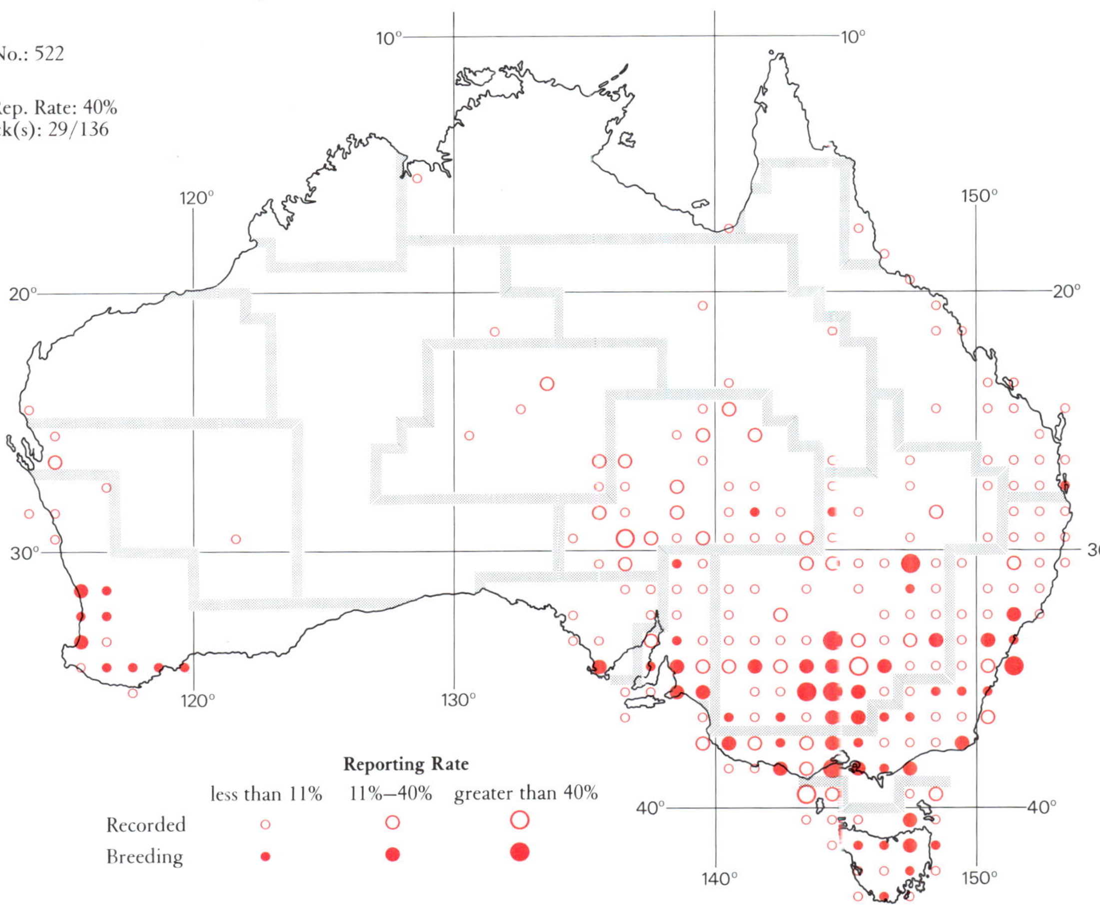

Spinifexbird

Eremiornis carteri

Morphological variation of the endemic Spinifexbird is not great enough to warrant separation into subspecies.[626] Northernmost records are from Wave Hill (17/130)[1370] and 26°S appears to be the southern limit. Records from Wanjarri (27/120) are unconfirmed. Historical Atlas records fill some gaps in the Great Sandy and Tanami Deserts and suggest a continuous distribution from the Central Ra. Region westward. The population in the eastern Rolling Downs Region appears isolated.

The Spinifexbird is apparently sedentary and maintains a small permanent territory.[1916] An early statement about the bird is all too true: '. . . . this species cannot be called rare; it is very local, and shows a decided preference for haunts where observation is difficult'.[1876] It lives in acacia scrub and spinifex, specially along small watercourses and at the bases of hills, where run-off of water causes dense spinifex growth.[626] On Barrow I. (20/115) much of the food of the Spinifexbird is insects.[1917] There is an early account of repeated observations of the parents carrying seeding heads of mulla mulla to their chicks, apparently to feed them.[1876] The gut of one adult contained the seeds of the plant.

There was estimated to be a total of 17 800 Spinifexbirds on Barrow I. in Aug. 1976 (0.32 birds/ha).[1611] It is the most common bird there.

Field Atlas Records: 124

1° blocks–with records: 48 (6%)

–with breeding records: 4 (8%)

Atlas No.: 507

Max. Rep. Rate: 20%
In block(s): 21/120

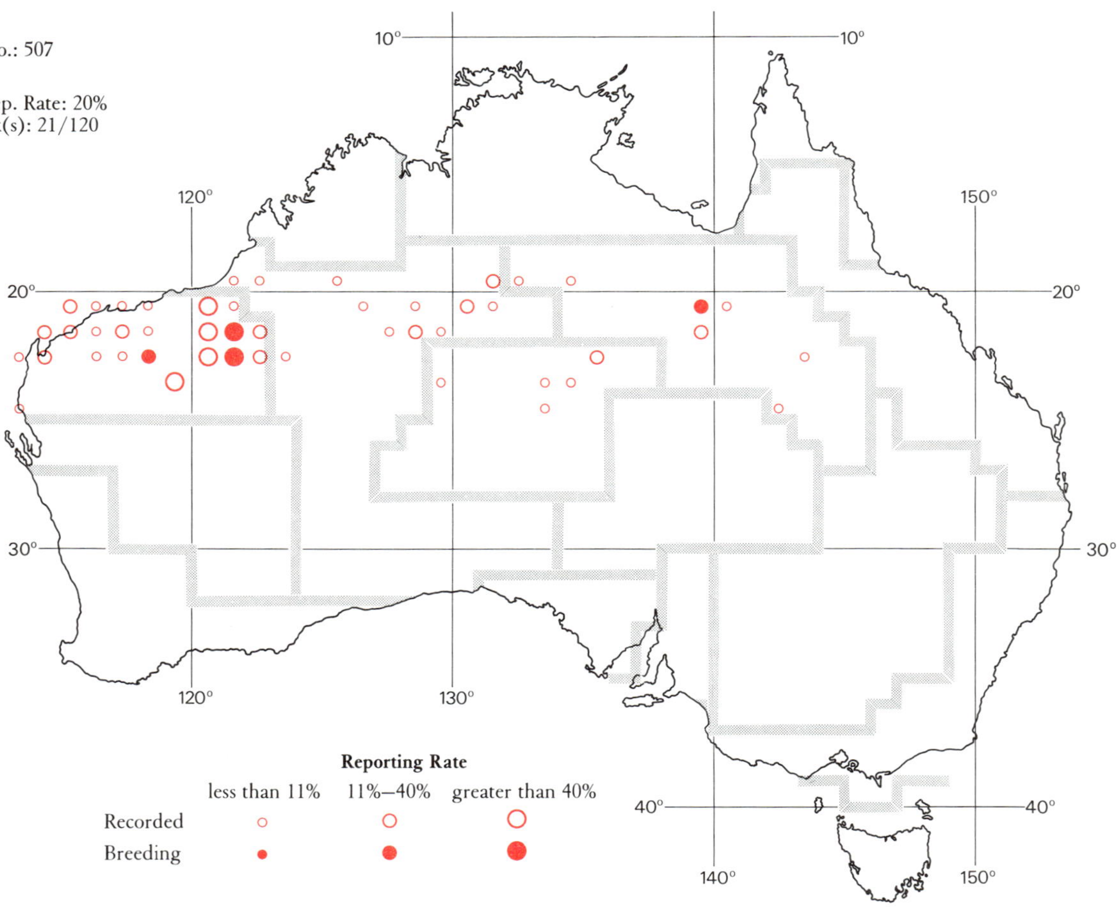

Zitting Cisticola

Cisticola juncidis

The Zitting Cisticola inhabits southern Europe, Africa, southern Asia, New Guinea[551] and Australia. In Australia three subspecies are recognized: *laveryi* in the Atherton and Eastern Qld Regions; *normani* at the eastern head of the Gulf of Carpentaria; and *leanyeri* in the Top End Region.[1587] The species is known only from isolated localities. It is difficult to differentiate between the two Australian cisticolas, the Zitting and Golden-headed,[1046,1114,1220] except in the breeding season when the males are singing.[702,1587]

There is no information about movements but the species appears to be present all year at Darwin (12/130)[702] and near Ayr (19/147).[531]

The Zitting Cisticola lives in short grass swamps almost permanently damp but seldom flooded and distinct from the seasonally flooded long-grass swamps used by the Golden-headed. Near Ayr the Zitting Cisticola lives in flat grassy country on the fringe of saltmarsh vegetated with couch, open fields and water meadows but not in cultivated areas.[531,1046] Vegetable matter and insects have been found in guts.[748]

Near Ayr four nests were found in a 100-m strip on the edge of a saltmarsh within the territory of one dominant male.[531] There are no other measures of abundance.

Field Atlas Records: 90

1° blocks–with records: 12 (1%)
–with breeding records: 3 (25%)

Atlas No.: 969

Max. Rep. Rate: 16%
In block(s): 12/130

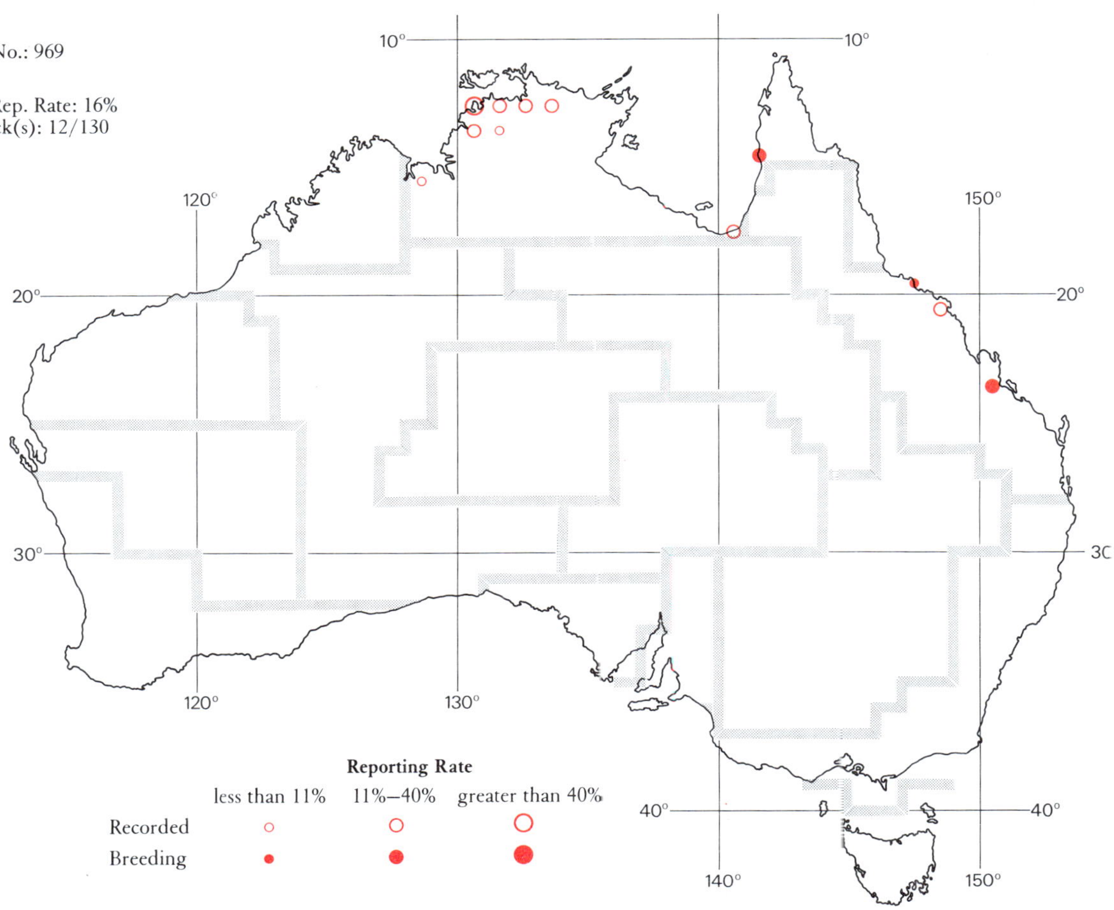

Golden-headed Cisticola

Cisticola exilis

The Golden-headed Cisticola ranges from India, China, south-eastern Asia and New Guinea to Australia.[1124] There are four Australian subspecies: *diminuta* in the Cape York Region; *exilis* to the south; *lineocapilla* in the northern Top End and *alexandrae* in the drier areas of the Rolling Downs, southern Top End and across to the Kimberley Region.[748,1114] Historical records link the Top End and Kimberley populations. Sporadic records from the Pilbara Region in different seasons, including one on the Oakover R (21/121) in May 1980, leave it uncertain whether the species is always there or an intermittent visitor. There is a population on King I. (39/144) in the Tas. Region but the Cisticola has been recorded only once (1913) in Tas. itself.[568]

The Cisticola has expanded rapidly into the Murray-Darling Region since 1950.[841] Before then, records in the region were confined to the south-western corner (36/140, 36/141) with the exception of a single report from Moree (29/149) in Nov. 1943, the first for that district.

The Cisticola is inconspicuous in winter but probably does not migrate.[1140] Birds reached Ivanhoe (32/144) during an irruption in 1976[841] and appear irregularly at other inland sites. Near Melbourne (37/145) a pair occupied a small swamp in dunes for a few days in May, the only record during 13 months.[1476]

The Cisticola lives near lowland swamps,[691] in dense stands of native grass along the borders of rivers and in grain crops. At Darwin (12/130) annual burning of swamps in the dry winter season forces it to retreat sometimes to the fringes of rainforest.[423] In the Rolling Downs Region birds were reported in tussock grassland well away from water.[1724] The Cisticola feeds mainly on the ground on invertebrates.[272] Establishment of irrigation in the lower Murray-Darling Region has enabled the Cisticola to expand its range and it now inhabits crops and rank vegetation along roads and drains in the area.[841]

In 1954, 38 nests were found adjacent to Dee Why Lagoon (33/151) which covers about 66 ha.[1140] In the Kimberley Region 15–20 individuals were estimated to be in an area of 0.75 ha (20–26.7 birds/ha).[748] In New Guinea in eucalypt woodland near Port Moresby (9S/147E) density was 0.6 birds/ha.[158]

Field Atlas Records: 5398

1° blocks–with records: 226 (28%)
–with breeding records: 52 (23%)

Atlas No.: 525

Max. Rep. Rate: 51%
In block(s): 19/147

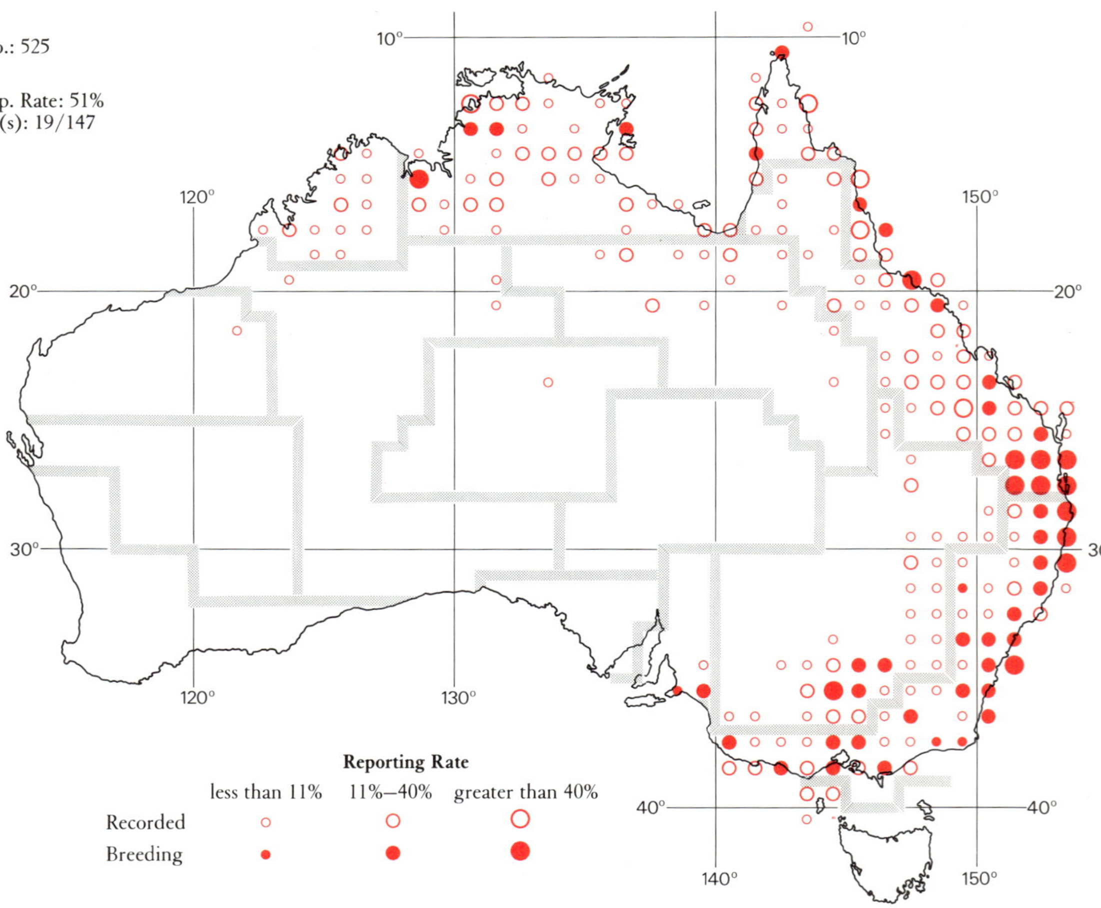

Rufous Songlark

Cinclorhamphus mathewsi

No geographical differentiation is evident in the population of the endemic Rufous Songlark.[748] Its distribution does not extend to the Tas. Region. Most breeding reports are south of 20°S but there are some in the Historical Atlases from as far north as Kununurra (15/128) and King R (1[illegible]/133).

The Rufous Songlark is migratory. It is a winter visitor near Edward R (14/141) in the Cape York Region and a spring-summer breeding visitor to places in southern Australia such as Murphy's Ck (27/152),[1099] Inverell 29/151)[96] and Bendigo (36/144).[1555,1557] Observers in the South-West also record the bird arriving in Aug. and departing in Dec.[1638] Some birds remain in the south during winter and the numbers migrating vary from year to year.[833,862] In the Field Atlas the greater conspicuousness of the Songlark when calling during the breeding season means that Reporting Rates were uniformly highest in spring and summer.

The Rufous Songlark lives on plains dominated by grass in woodland, acacia scrub and spinifex. On the Nullarbor Plain it occupies shrubby dongas.[246] It feeds on seeds and insects caught on the ground.[748,1049]

At Inverell there was apparent alternation in abundance with a two year periodicity between 1963 and 1971; one pair occupied 4 ha in one year and four pairs the same area in the next year of each two year period.[96] Density in eucalypt woodland at Armidale (30/151) was 0.04–0.14 birds/ha[582] and in eucalypt forest at Jamieson (37/146) was 0.1–0.4 birds/ha.[1469]

Field Atlas Records: 6048

1° blocks-with records: 526 (65%)
-with breeding records: 96 (18%)

Atlas No.: 509

Max. Rep. Rate: 44%
In block(s): 28/141

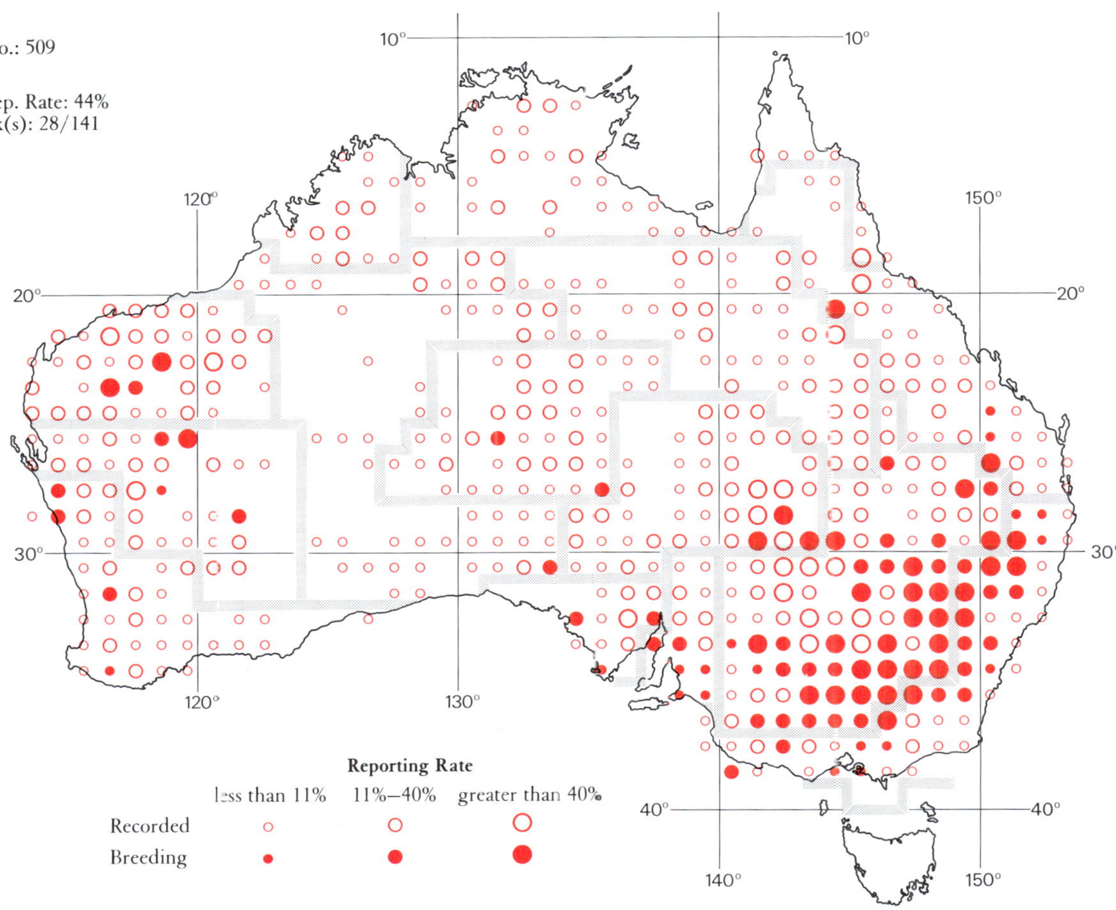

Brown Songlark

Cinclorhamphus cruralis

There is morphological differentiation among populations of the endemic Brown Songlark but subspecific names have yet to be allocated.[748] The species has not been recorded in the Tas. Region. Breeding is rarely reported from northern parts of the range, the northernmost breeding record being from Brunette Downs (18/135) in 1913.[106]

The Brown Songlark is known both as a migrant and a nomad. Long-distance movements are indicated by the collection of a bird on a ship 230 km east of the NSW coast.[811] In 1980 drought appeared to induce a significant movement of the Brown Songlark into eastern parts of NSW, including an invasion into the Nowra (34/150) district in Sept., many of the birds staying to breed.[1079] In the southern Murray-Darling Region sudden increases in abundance have been observed in a short period, so great that new individuals must have arrived.[833] On the Nullarbor Plain in two successive years birds arrived 2–4 days after heavy rain but did not breed for another 15 weeks.[246] In the South-West the literature indicates a consistent pattern of arrival in July and departure in Dec.[1216,1638]

At nesting time the song flights of the male and its calls make it conspicuous. When not breeding, the species is not easy to see and may often be overlooked. Reporting Rates appear to represent conspicuousness during the breeding season more than movement. They were highest in spring in the South-East and South-West and in summer in the Eastern Qld Region.

The Brown Songlark lives on plains vegetated with grass, crops or low shrubs in woodland, acacia scrub and spinifex. It has benefited from agricultural clearing. It forages on the ground, mainly taking insects, but also some seeds.[748,1049]

There are no measures of abundance.

Field Atlas Records: 5957

1° blocks–with records: 505 (62%)
–with breeding records: 70 (14%)

Atlas No.: 508

Max. Rep. Rate: 48%
In block(s): 30/127

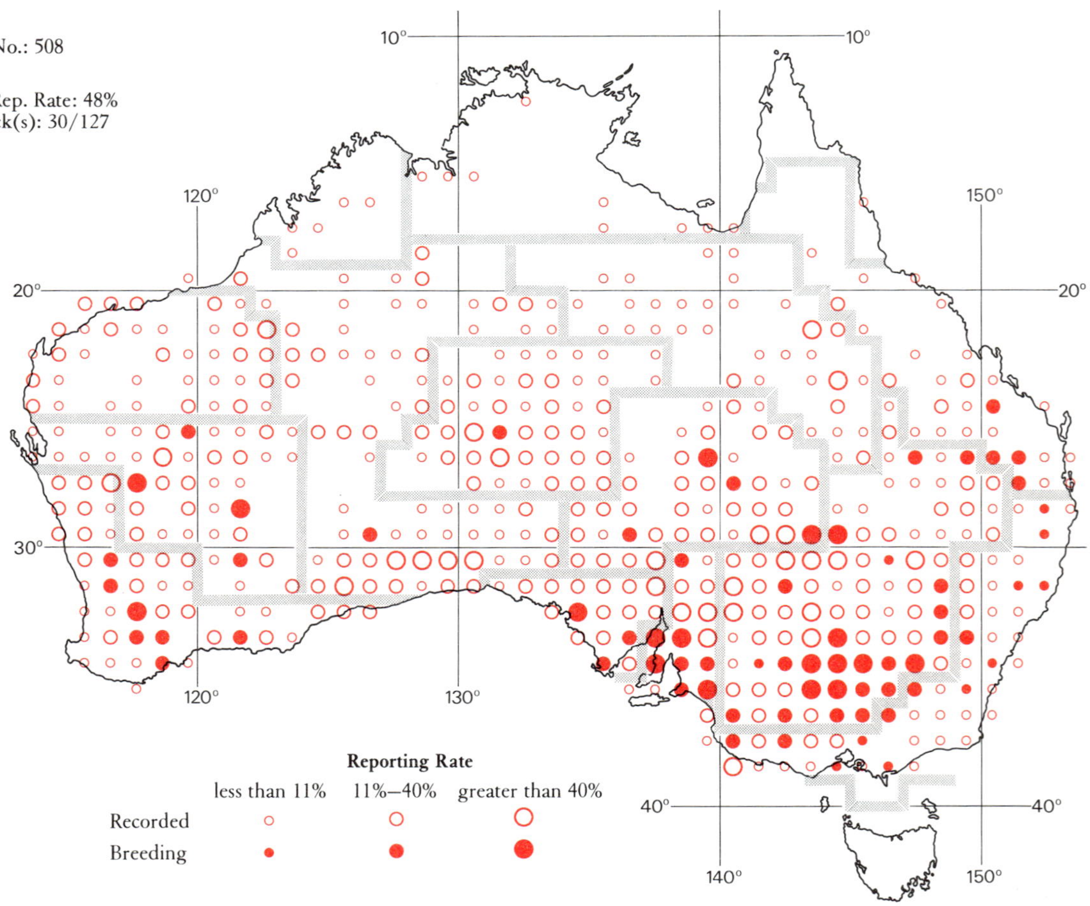

Purple-crowned Fairy-wren

Malurus coronatus

There are two subspecies of the endemic Purple-crowned Fairy-wren: *macgillivrayi* in the east and *coronatus* in the west.[1578] The range and numbers of *coronatus* have been reduced by pastoral development. It is all but gone from the Fitzroy R where it was abundant in 1886. Only one site (18/125) is now known on the Fitzroy itself together with small populations (16/125, 16/126) on some of its tributaries. Similarly, only relict populations remain on the Drysdale (15/126), Pentecost (15/127) and Ord (15/128) river systems and on the upper reaches of the Victoria R (131°E). No corresponding decline in range is evident for *macgillivrayi*.

At Victoria River Downs (16/131) the Purple-crowned Fairy-wren was consistently located in the same places in the dry winter season but not in the wet summer season when parts of the areas it occupied were inundated; possibly the birds had moved to higher ground.[196] There is no other evidence of movement.[1578]

The Purple-crowned Fairy-wren always lives close to water, often in pandanus.[748] It also inhabits swamp canegrass, paperbark and mangroves.[196,748,1834] There is no evidence of communal breeding except that the species has often been seen in small parties. It feeds mainly on insects, including beetles,[748,1578] and has been reported to take insects in the air in the manner of flycatchers.[321] The species' decline has been linked with the trampling of canegrass by cattle along the edges of streams.[196,1690] That the abundance of *coronatus* has declined as well as its range is evident from descriptions of its status up to 1904 as: 'abundant'; 'in groups of 5–7'; 'almost wherever there is water'.[1690] Field Atlas observations were rarely of more than two birds together and confined to isolated sites on a few rivers. Measures of abundance are few: four pairs on 1 km of the Drysdale R in June 1973;[1690] six breeding pairs and nine dependent young on 5 km of the Victoria R in 1976–77; and four breeding pairs and four dependent young on 3 km of the Wickham R (16/131) in 1976–77.[19] The last two places are separated by 20 km of river, some apparently suitable, but none with Fairy-wrens. Relict colonies seem unable to re-colonize the intervening stretches even when vegetation has regenerated. No data are available for *macgillivrayi*.

Field Atlas Records: 79

1° blocks-with records: 22 (3%)

-with breeding records: 2 (9%)

Atlas No.: 542

Max. Rep. Rate: 19%
In block(s): 18/139

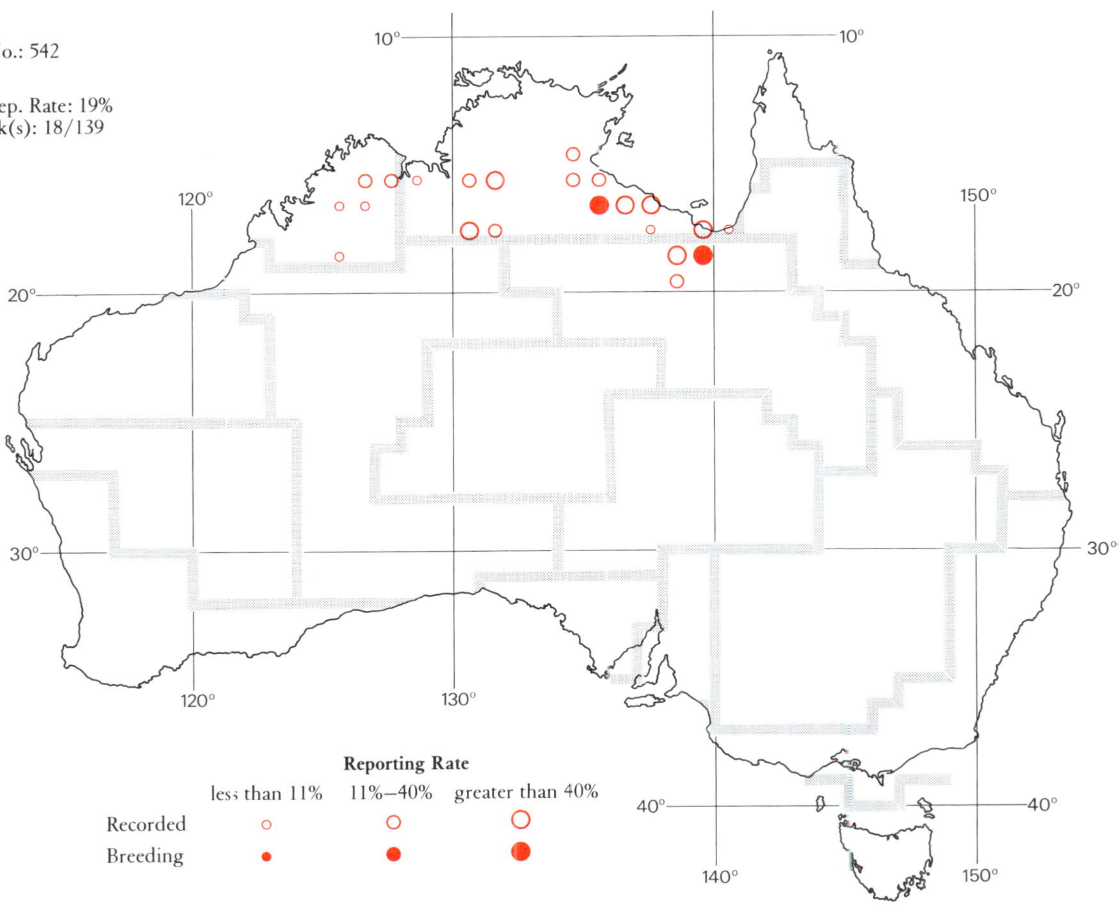

Superb Fairy-wren

Malurus cyaneus

The subspecies of the endemic Superb Fairy-wren on the mainland is *cyanochlamys* and, in the Tas. Region, *cyaneus*.[1578] The distribution of *cyanochlamys* is disjunct. On the Murray R, west of its confluence with the Murrumbidgee, is a gap of about 100 km to Mildura (34/142) where no Superb Fairy-wren was recorded during the Field Atlas. Expansion of range along the Murray was thought to be closing the gap in 1961[833] but has not progressed in the succeeding 20 years. West of Mildura the Fairy-wren follows the Murray down to about Swan Reach (34/139) where it becomes broadly distributed over the countryside. The 10′ maps show a small relict population on northern Yorke Pen. (34/137); another lives on the southernmost tip of Eyre Pen. (34/135).

Colour-banding of birds at Sydney (33/151)[218] and Canberra (35/149)[1541] showed that groups of the Superb Fairy-wren occupy a permanent territory. Young females leave their natal territory and move several kilometres before settling in a new site, whereas young males are more likely to remain in a breeding group for several years.[1541] The sighting of a male Superb Fairy-wren near Hattah (34/142),[1165] at least 40–50 km from the nearest population reported in the Field Atlas, shows that the species can disperse far.

The Superb Fairy-wren occupies many different types of vegetation, avoiding only saltbush and mallee. Where clearing has opened up the mallee the bird has moved in. The limits of its distribution coincide well with the area within which more than 100 mm of rain falls in eight summers out of ten. The Fairy-wren usually lives in groups of 6–12 birds and nests communally.[1541,1552] It feeds mainly on insects but also takes vegetable material.[1578] Cats may take a heavy toll. In one Canberra territory five successive males were taken by a cat before it was shot.[1541] Neither climate nor lack of suitable vegetation seems adequate to explain the absence of the Superb Fairy-wren from parts of the Murray R or the southern Yorke Pen.[1578]

Near Canberra density in a garden was 0.6–1.5 birds/ha.[1541] Densities in eucalypt woodland were: Armidale (30/151) 0.19–0.41 birds/ha[582] and Moruya (35/150) 0.21–0.87 birds/ha.[1197] In eucalypt forest they were: Jamieson (37/146) 0.5–1.0 birds/ha[1469] and Boola Boola (38/146) 0.2–1.7 birds/ha.[1113]

Field Atlas Records: 26 146

1° blocks–with records: 172 (21%)

–with breeding records: 126 (73%)

Atlas No.: 529

Max. Rep. Rate: 83%
In block(s): 27/149

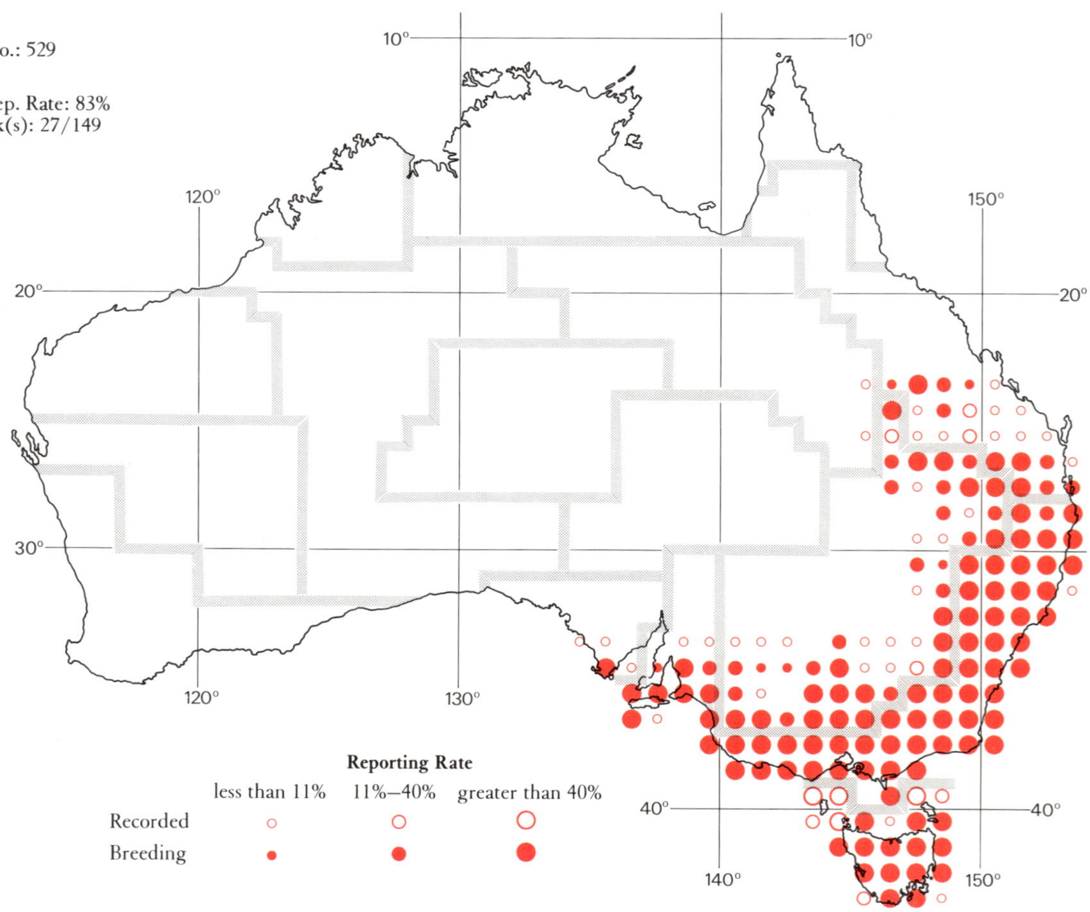

Splendid Fairy-wren

Malurus splendens

The endemic Splendid Fairy-wren has three morphologically recognizable subspecies, all of which hybridize across broad zones of contact.[602,603,1578] The three populations are roughly: eastern, *melanotus*; central, *callainus*; and western, *splendens*, the Simpson Desert and L. Eyre acting as one barrier and the Nullarbor Plain as the other.[1578] Historical records in the Western Deserts between 25°S and 30°S show that the Splendid Fairy-wren is continuously distributed across the continent. The Field Atlas shows that it is widespread in the Tanami Desert, several hundred kilometres north of its previously known range.

Study of colour-banded birds at Gooseberry Hill (31/116) showed that groups of the Splendid Fairy-wren occupy a territory permanently with negligible dispersal even of young birds.[1550] The species occupies a wide range of vegetation types: *melanotus* inhabits mallee with abundant ground cover, belah woodland with low cover and watercourses with associated vegetation; *callainus*, acacia scrub with an understorey of emu-bush;[603] and *splendens*, eucalypt forest, eucalypt woodland with a grass understorey and swamps.[1140] Though the Splendid Fairy-wren occupies the high-rainfall area of the South-West Region, it is replaced in such areas in the South-East by the Superb Fairy-wren. It appears to thrive in the dry summer of the South-West, which is usually rainless Nov.-Mar., but not in the wetter summer of the South-East.[1550] The Splendid Fairy-wren lives in family groups, breeding communally.[1550] It feeds mainly on insects, including small beetles and ants, and on some vegetable material.[748,1578]

The number of breeding birds occupying a 65-ha site at Gooseberry Hill varied from 25–31 during six years and the number of territories from 9–11. Family groups averaged 2.9 individuals per group, with a density of 0.53 birds/ha, but large apparently suitable areas surrounding the territories were unoccupied.[1550]

Field Atlas Records: 3203

1° blocks–with records: 230 (28%)
–with breeding records: 61 (27%)

Atlas No.: 532

Max. Rep. Rate: 61%
In block(s): 33/114

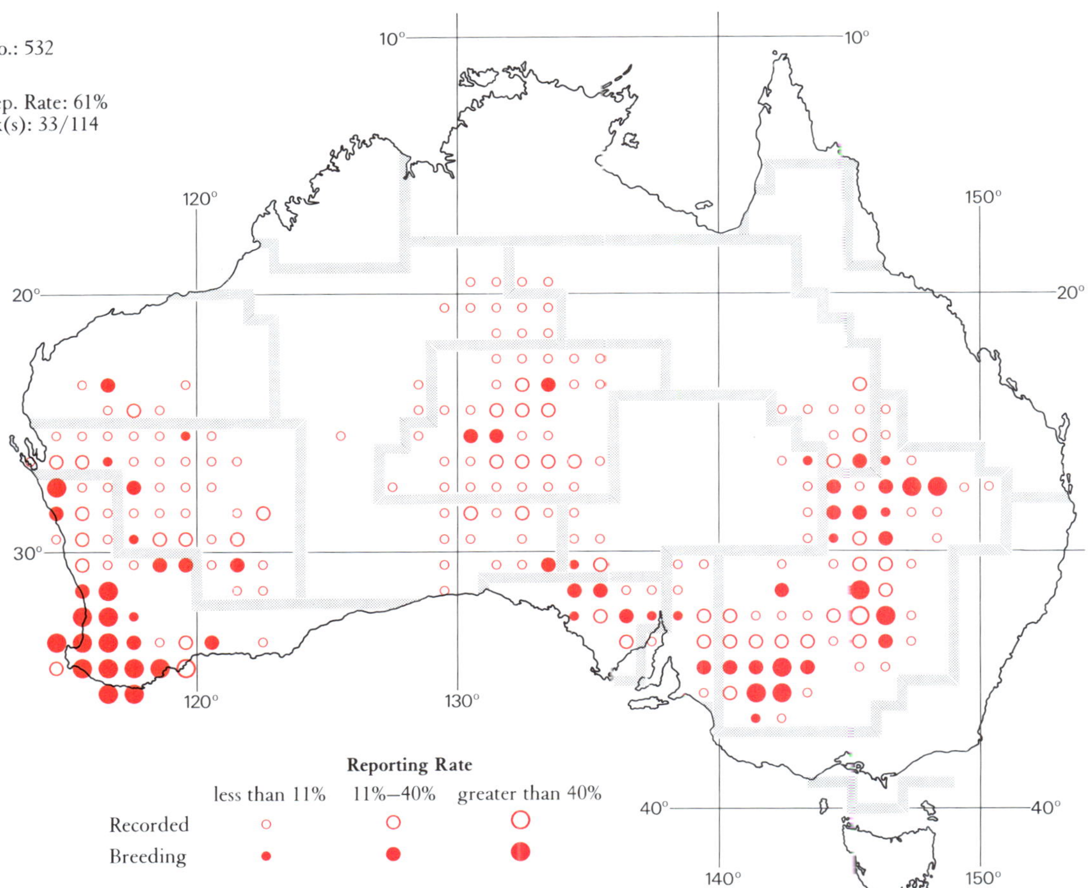

Variegated Fairy-wren

Malurus lamberti

The endemic Variegated Fairy-wren has been separated into a number of subspecies or, at different times, species. The Lovely Fairy-wren, *amabilis*, in the Cape York and Atherton Regions is now separated again as a species[1578] but is included on this map in conformity with the taxonomy in use during the Field Atlas.[1576] In addition to *amabilis*, four subspecies are recognized, some intergrading broadly: *lamberti* on the coast in the Eastern Qld and South-East Regions; *dulcis* in the northern Top End Region; *rogersi* in the Kimberley Region; and *assimilis* over the rest of the continent. South of about 30°S *lamberti* becomes isolated on the coastal side of the Great Dividing Ra.; its distribution was extended south to the Wallagaraugh R (37/149) by Field Atlas records. The subspecies *assimilis* has an outlying population in the Whipstick (36/144) but is absent from the Mt Lofty Ra. (34/138), Kangaroo I. (35/137) and southern Eyre Pen. (34/135). The Variegated Fairy-wren was not recorded in Perth (31/115) until 1975 when it was found in the heart of the city at Pelican Pt.[1634] It may have expanded its range there or have been previously overlooked because it looks so similar to the Red-winged Fairy-wren.

The Variegated Fairy-wren is thought to be territorial like other fairy-wrens, probably foraging more widely in the non-breeding season.[1578] It inhabits spinifex, acacia scrub, mallee, heath, rainforest margins, mangroves and eucalypt woodland wherever shrubby thickets are present. The species usually lives in small groups of 2–5 birds and breeds communally.[1578]

The Variegated Fairy-wren includes a high proportion of plant bugs and weevils in its diet[1049] and takes some seeds and vegetable material as well.[1578] It also eats caterpillars.

At Moruya (35/150) in eucalypt woodland density was 0.08–0.37 birds per ha.[1197]

Field Atlas Records: 8046

1° blocks–with records: 547 (67%)
–with breeding records: 135 (25%)

Atlas No.: 536

Max. Rep. Rate: 64%
In block(s): 21/144

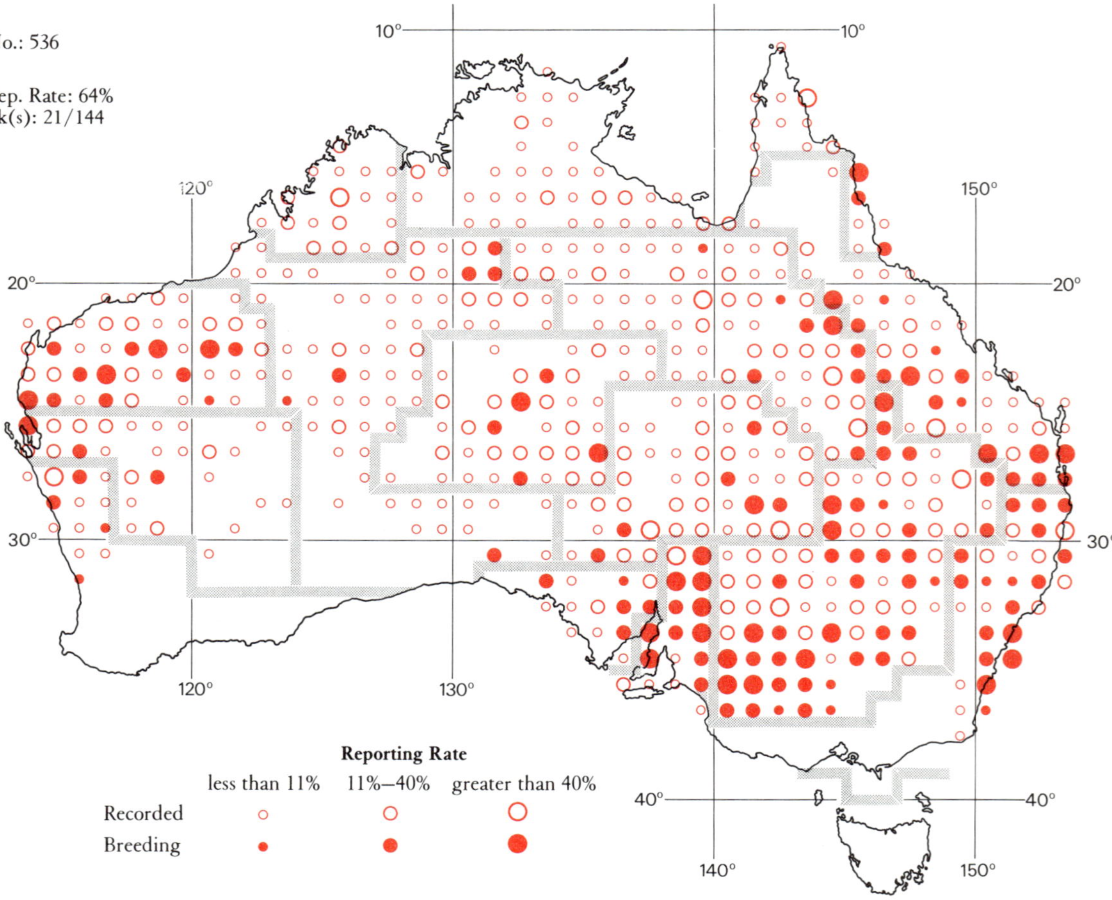

Blue-breasted Fairy-wren

Malurus pulcherrimus

The two populations of the endemic Blue-breasted Fairy-wren shown as isolated from each other on the map, appear not to be separable subspecifically.[1578] The population at Eyre Bird Observatory (32/126) is linked with that farther west by recent historical records.

Like other fairy-wrens, groups of the Blue-breasted appear to be territorial and sedentary.[1549] Groups of up to 5–7 birds live together and breed communally.[1549] The species' distribution closely follows that of mallee with a dense understorey.

The Blue-breasted Fairy-wren feeds on insects including ants.[1049] Large-scale clearing for farming on Eyre Pen. and in WA has left the species surviving only in islands of native vegetation in much of its former range.

In the wheatbelt of the South-West Region a relict population in 48 ha of uncleared land maintained itself in isolation for at least 30 years without even occupying the whole area.[1549] The seven or eight territories averaged 3.2 ha each, sizes ranging from 2.0–7.6 ha. The study, carried out in a dry period, suggested that the population was not at a maximum at the time. There were 22 adults in the 48 ha, a density of 0.46 birds/ha. At Eyre Bird Observatory in mallee and coastal heath territories were mapped in the area most used by the species for two years, giving a density of 1.56 birds/ha.[464] In a nearby area surveyed regularly by transect densities were 0.48–2.22 birds/ha depending on the season; the highest were in summer.

Field Atlas Records: 343

1° blocks–with records: 46 (6%)

–with breeding records: 9 (20%)

Atlas No.: 540

Max. Rep. Rate: 32%
In block(s): 32/126

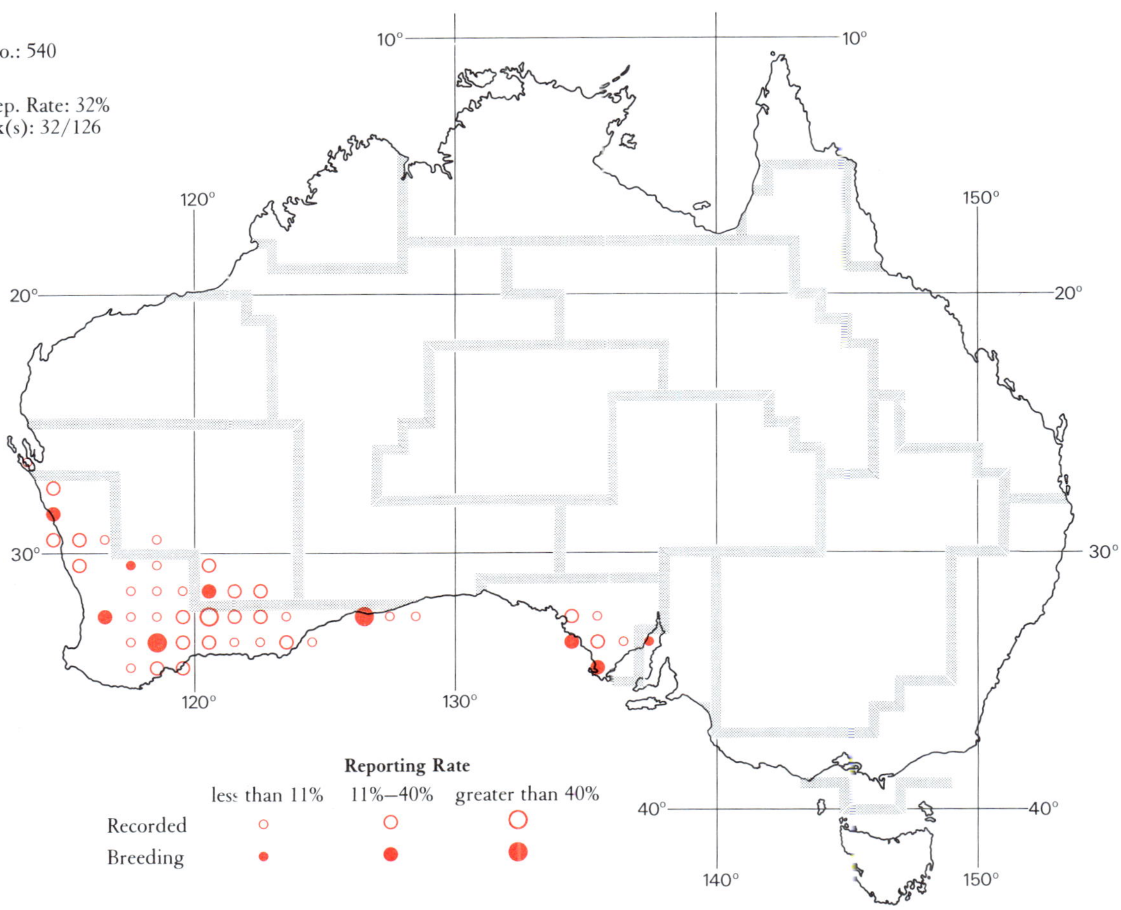

Red-winged Fairy-wren

Malurus elegans

The Red-winged Fairy-wren is endemic to the South-West Region and shows no subspecific variation.[594,1578,1623] It appears to be a relict species isolated by climatic changes. Apart from a single early reference to its presence in the Moora (30/116) district[1365] there are no records in the Historical Atlases north of those in the Field Atlas. Local declines are documented, such as its disappearance from Herdsman L. (31/115) about 1901,[1638] but the Atlases do not suggest a general decrease in range. Most Field Atlas records were within 50 km of the coast. There is an outlying population near Mt Barker (34/117).[748]

Groups of the Red-winged Fairy-wren live in a permanent territory. The species occupies gullies in jarrah and karri forests. Except in karri forest, where it can be found some distance from surface water,[1830] it is closely associated with streams and the rank vegetation that lines them. It also lives on the margins of freshwater lakes. The species is a communal breeder.[336,1830] It feeds mainly on insects.[748,1830] Drainage of swamps and increasing salinity of water, as well as predation, may seriously affect its numbers. At Manjimup (34/116) in the 1940s the numbers of the Fairy-wren were thought to increase over three years after 40 cats were shot nearby.[1830]

There are no measures of abundance.

Field Atlas Records: 286

1° blocks–with records: 14 (2%)
–with breeding records: 8 (57%)

Atlas No.: 538

Max. Rep. Rate: 29%
In block(s): 34/116

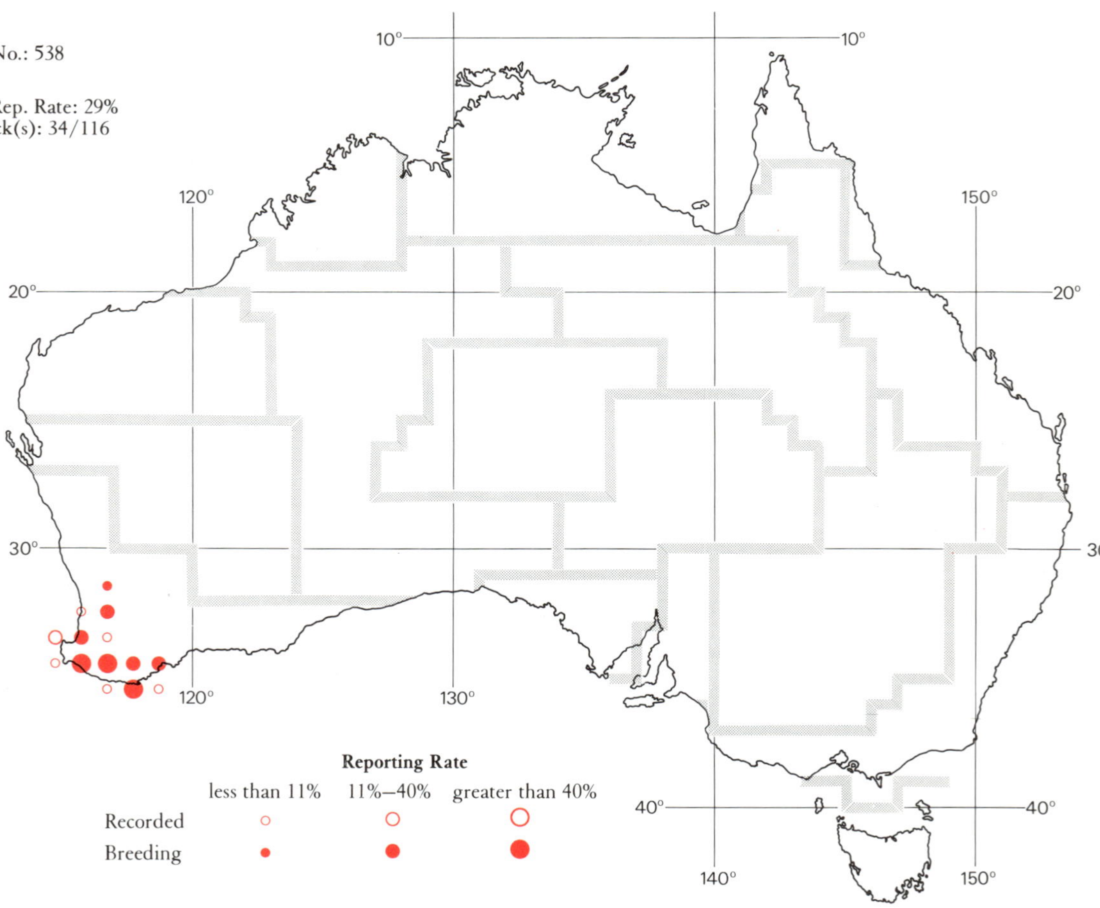

Black-plumaged populations of the endemic White-winged Fairy-wren on Dirk Hartog I. (25/113), *leucopterus*, and Barrow I. (21/115), *edouardii*, are differentiated subspecifically from *leuconotus* of the mainland.[1578] Birds in dark plumage have also been observed on the mainland: on the western coast; in the L. Eyre Region; and in the Central Ra. Region east of Alice Springs (23/133) to Dalhousie Springs (26/135).

Historical records of the White-winged Fairy-wren fill gaps across the lower Western Deserts where the species was not reported during the Field Atlas.

Near Booligal (33/144) groups occupied the same territories in successive breeding seasons but wandered over larger areas in the non-breeding season, overlapping into adjacent territories.[1762] Round Mt Mary Plains (34/139) large flocks form in autumn.

The White-winged Fairy-wren lives in saltbush, spinifex, heath, mallee and acacia scrub. It is absent from eucalypt forest but does enter some open woodland. The species lives in groups of up to 8–12 birds and breeding is communal.[1547,1762] The White-winged Fairy-wren feeds on arthropods, the prey gleaned mainly from bushes.[1049,1762,1917] Few ants are taken.

Near Booligal, where breeding territories ranged from 1.3–8.3 ha in area, density ranged from 1.5–5.0 birds/ha.[1762] This is denser than the estimate for Barrow I. where the population was estimated at 8150 birds in 27 000 ha, 0.3 birds/ha, spread evenly throughout the spinifex that covers most of the island.[1611] Both populations are much less dense than one at Redcliffs (34/142) that had 30 birds/ha[537] but these observations were made along an irrigation channel which may be atypical.

White-winged Fairy-wren

Malurus leucopterus

Field Atlas Records: 4638

1° blocks–with records: 404 (50%)
–with breeding records: 95 (24%)

Atlas No.: 535

Max. Rep. Rate: 74%
In block(s): 31/140

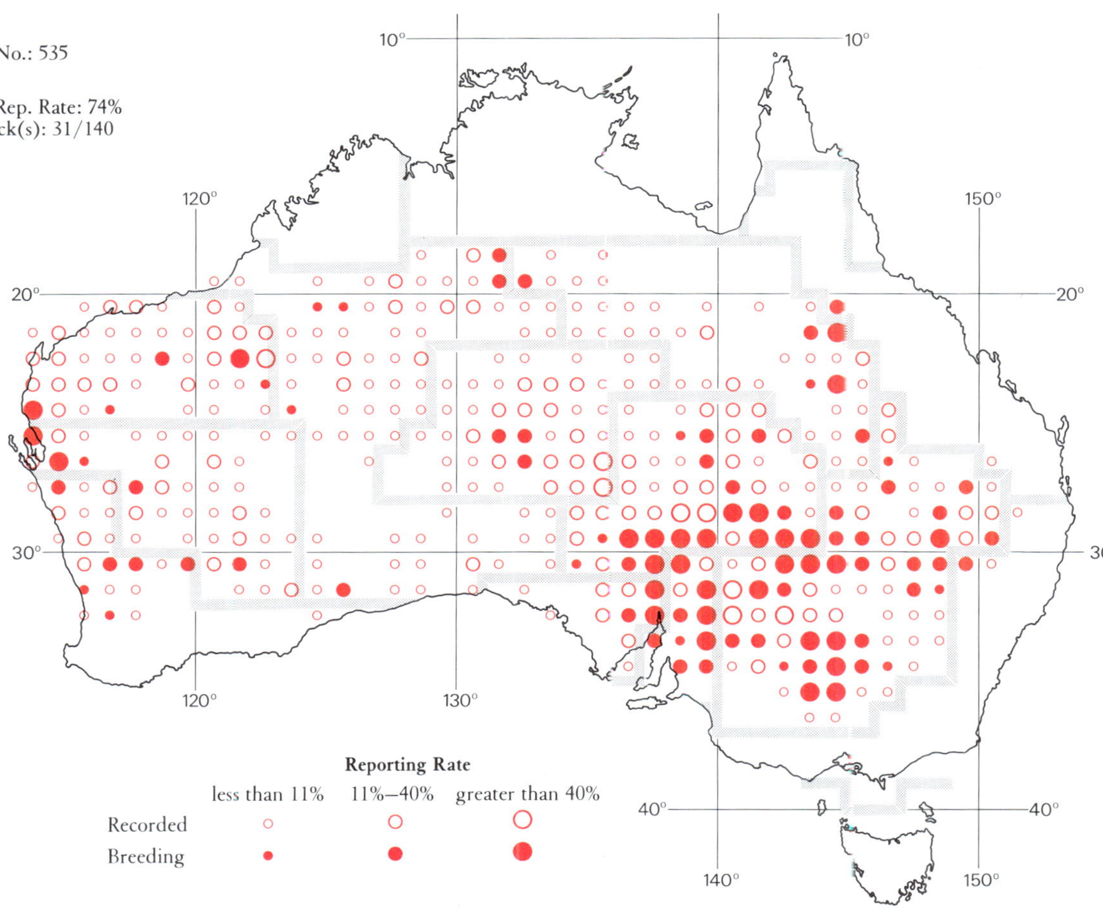

Red-backed Fairy-wren

Malurus melanocephalus

The northern population of the endemic Red-backed Fairy-wren is regarded as a single subspecies, *cruentatus*, that merges, in the Atherton Region, with *melanocephalus* of the eastern coast.[1578]

The Red-backed Fairy-wren may be less sedentary than other fairy-wrens, abandoning its territory after the breeding season to wander locally in nomadic flocks[1578] but in some places groups remain in one place for long periods.[321] Groups as large as 30 have been recorded at Murphy's Ck (27/152) after breeding.[1099] In the Top End, after fires in the dry winter season have burnt through the eucalypt woodland, the Fairy-wren retreats to damper places.[426] It lives in eucalypt woodland, specially where there is an understorey of grass or shrubs, and on the margins of rainforest.[748] The species breeds communally though apparently less frequently and with fewer helpers than other fairy-wrens.[1578] The Red-backed Fairy-wren feeds mainly on insects, gleaning them from the ground and low vegetation, and it will hunt over burnt ground.[426] No seeds or vegetable matter were found in any of nine guts examined.[1578]

There are no measures of abundance.

Field Atlas Records: 4449

1° blocks–with records: 201 (25%)
–with breeding records: 47 (23%)

Atlas No.: 541

Max. Rep. Rate: 72%
In block(s): 24/153

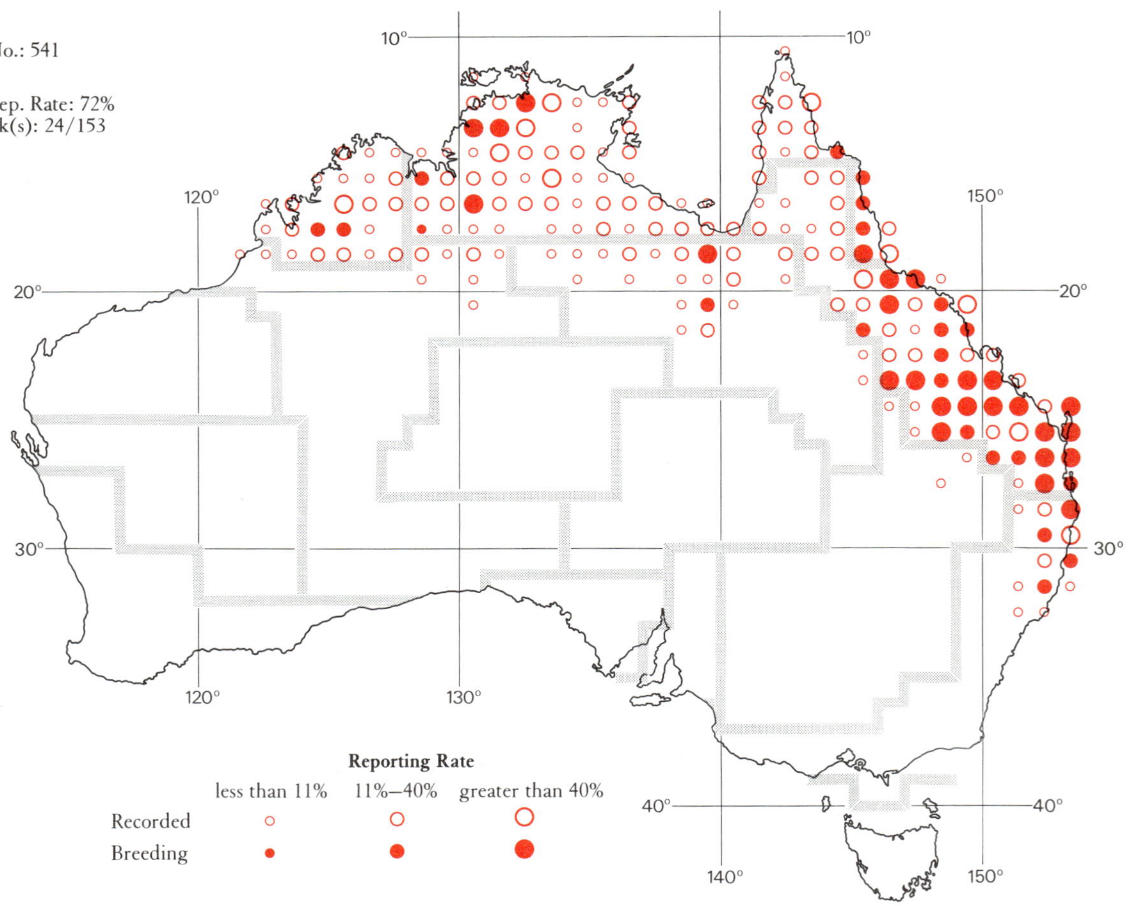

There are two subspecies, sometimes regarded as species,[1578] of the endemic Rufous-crowned Emu-wren: *mallee*, an isolated population in the southern Murray-Darling Region; and *ruficeps*, which ranges across central Australia. Records in the 1930s near Meningie (35/139) attributed to the Rufous-crowned are probably of the Southern Emu-wren whose range contacts that of the Rufous-crowned near Coombe (35/140).[514]

The Rufous-crowned Emu-wren was recorded on the southern edge of the Simpson Desert during the Field Atlas and on its eastern edge historically, which suggests that the species is present throughout the Desert.[1578] Farther east there is apparently a gap (140–142°E) to populations on the eastern side of the Georgina R.[626] In the Goldfields and Western Deserts Regions historical records extend the species' distribution south to 28°S and there is one report farther south on the northern edge of the Nullarbor Plain near Naretha (31/124).[381]

The Rufous-crowned Emu-wren may be found repeatedly in the same localities. Pairs or family parties appear to coalesce into larger groups to forage more widely in the non-breeding season.[1578]

Seven birds were reported in one group[873] and in another at least three adult males were recorded together during the Field Atlas.

The northern bird, *ruficeps*, is widespread in spinifex where there are few trees; *mallee* also inhabits spinifex with an overstorey of mallee eucalypts. The bird is secretive but, even in areas where it is well known, seems to be unevenly distributed, its dispersal blocked by the absence of spinifex.[833]

It is assumed to feed mainly on insects and *mallee* is known also to take some vegetable matter and seeds.[1578]

There are no measures of abundance.

Rufous-crowned Emu-wren

Stipiturus ruficeps

Field Atlas Records: 125

1° blocks–with records: 38 (5%)

–with breeding records: 2 (5%)

Atlas No.: 528

Max. Rep. Rate: 14%
In block(s): 21/114

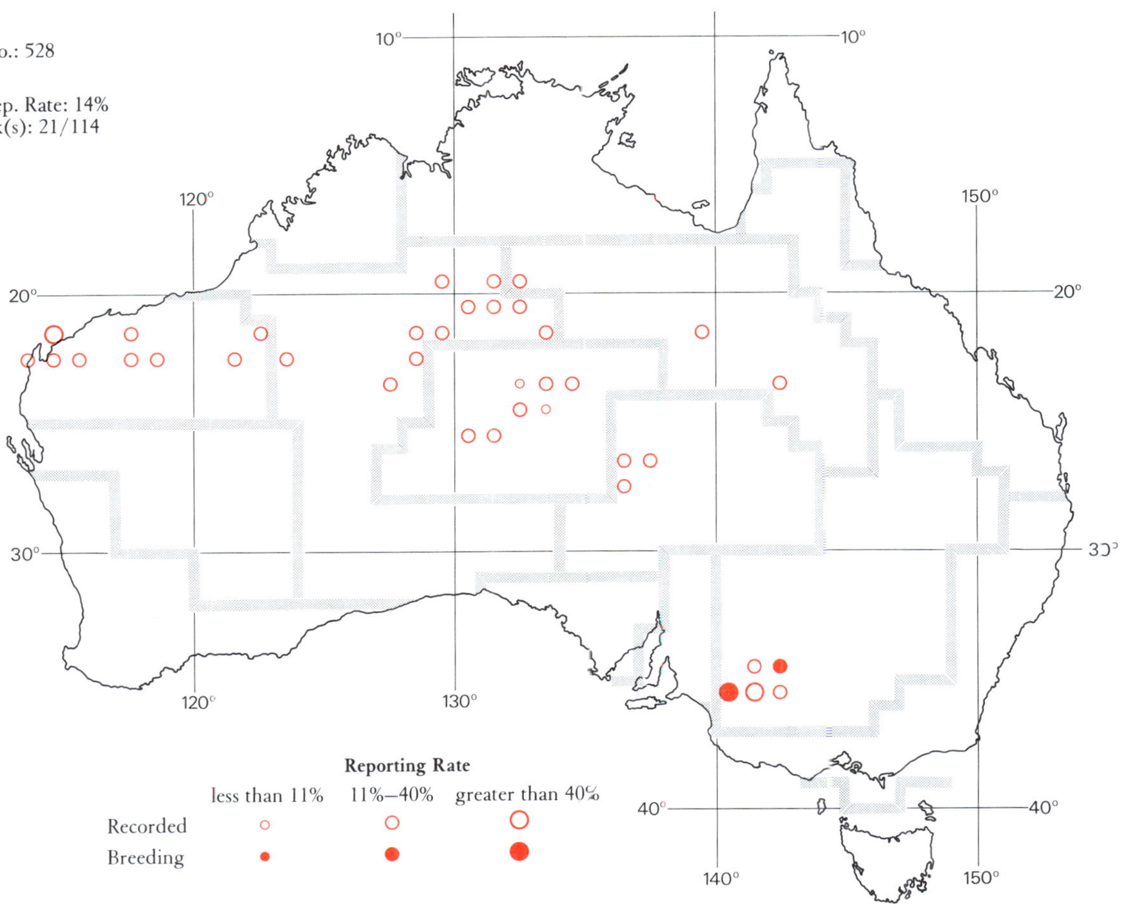

Southern Emu-wren

Stipiturus malachurus

The range of the endemic Southern Emu-wren is disjunct and six of the many isolated populations have been described as subspecies.[1578] All birds in the South-East Region are *malachurus*, which is distributed more or less continuously down the eastern coast to Melbourne (37/145) with outlying populations in the higher country round the Gibraltar Ra. (29/152) and Blue Mts (33/150). West of Melbourne there are enclaves of *malachurus* in the Otway Ra. (38/143) area and Grampians (37/142) and along the coast at the western end of the region, all apparently isolated. The subspecies *littleri* is in Tas. but not on the Bass Strait islands. The population at the southern end of the Mt Lofty Ra. (34/138) is *intermedius*; on Kangaroo I. *halmaturinus* and, on the tip of Eyre Pen. (34/135), *parimeda*. These three subspecies are all isolated. Western birds are all *westernensis* whose range along the coast is extended by records in the Field Atlas north to 26/113 near Denham. The population on Dirk Hartog I. (25/113) was not reported during the Field Atlas but the bird is so secretive that it may have been overlooked. The Field Atlas shows that the ranges of the Southern and the Rufous-crowned Emu-wrens do not overlap but they may make contact at Coombe (35/140).[514]

The Southern Emu-wren does not make large-scale seasonal movements but parties wander in the non-breeding season.[1578]

The species requires dense undergrowth.[569,977] It lives in swampy areas vegetated with species such as sawsedge, rushes, lignum and samphire;[510] in dunes near the coast; in heath at altitudes up to 950 m[835] including that dominated by buttongrass. The Southern Emu-wren feeds on a range of small arthropods in the undergrowth.[569,1049] It was abundant in the swamps on what is now the Sydney Domain (33/151) in the early days of settlement[789] but has disappeared from this and other settled or cleared areas.

Small parties of the Southern Emu-wren have been found in islands of suitable vegetation only 1–2 ha in extent remaining after recent clearing.[770] Whether populations of the species can survive indefinitely in such a small area is not known.

Field Atlas Records: 1138

1° blocks–with records: 75 (9%)
–with breeding records: 18 (24%)

Atlas No.: 526

Max. Rep. Rate: 46%
In block(s): 37/150

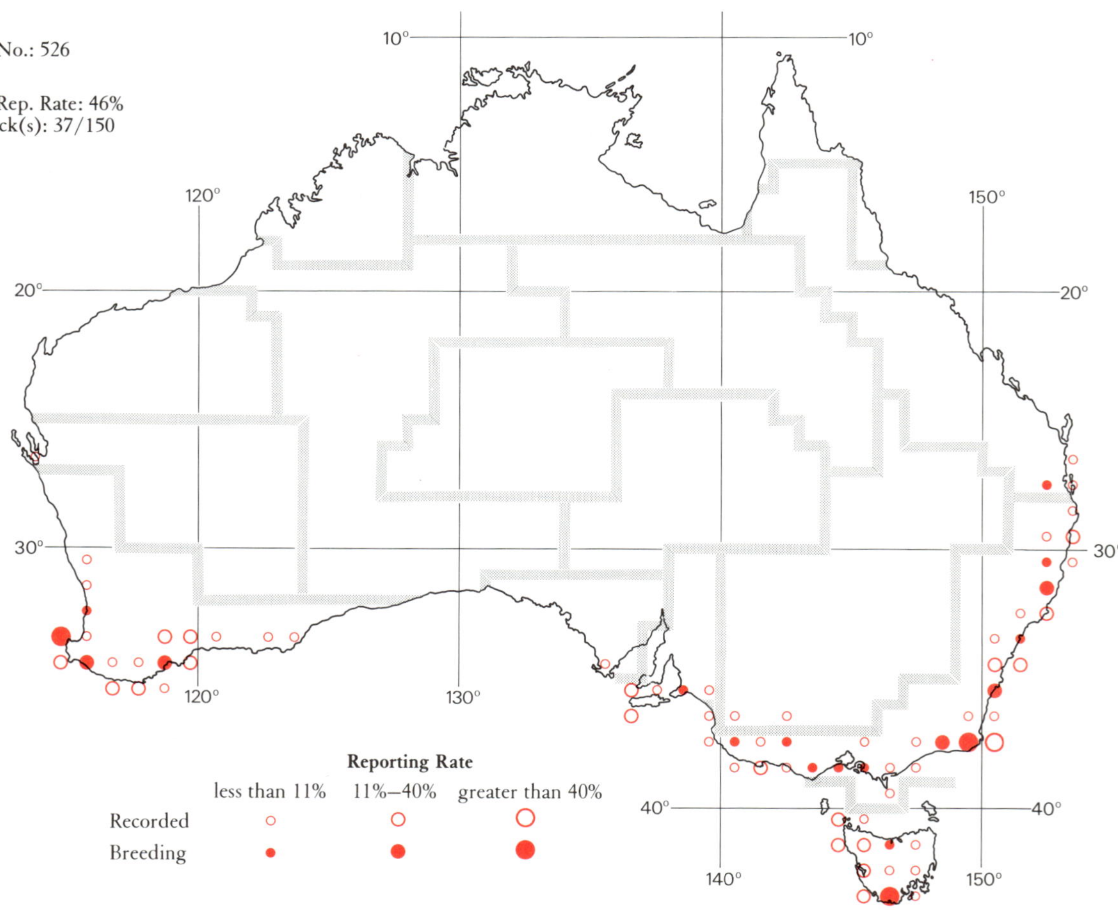

Black Grasswren

Amytornis housei

Until its rediscovery in 1968[641] the endemic Black Grasswren was known from only three specimens taken in 1901.[1578] Since 1968 the Grasswren has been found at a number of localities between the mouth of the Mitchell R (14/125) and Mt Barnett Station (16/125).

The Black Grasswren gathers into parties in the non-breeding season but the extent to which they wander is not known.[1578] Up to 12 birds have been seen in a group. The Black Grasswren lives in rough sandstone country among boulders and spinifex. It hides in rock crevices when disturbed and hops or runs over the ground rather than flies. Its eggs have not yet been described.[1578] The guts of six birds all contained insect remains and seeds[748].

There are no measures of abundance.

Field Atlas Records: 12

1° blocks–with records: 3 (1%)
–with breeding records:

Atlas No.: 518

Max. Rep. Rate: 16%
In block(s): 14/125

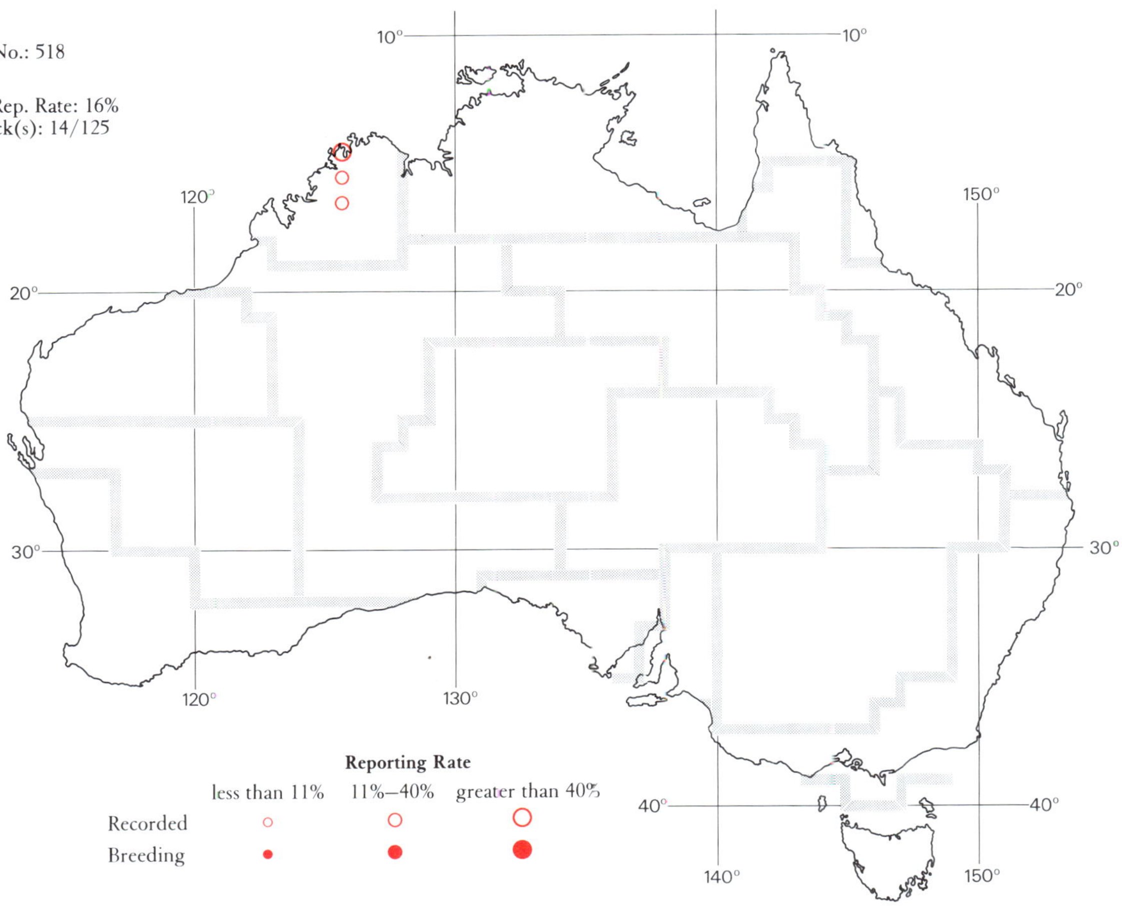

White-throated Grasswren

Amytornis woodwardi

The endemic White-throated Grasswren is confined to the sandstone country of the Arnhem Land escarpment.

In the non-breeding season small flocks form and may wander locally.[1578] Singly, in pairs and in small parties of up to seven, the Grasswren lives among boulders where there are large old spinifex tussocks on the upper slopes of valleys.[1584] When disturbed the bird runs rapidly over the ground with head and tail lowered to shelter in a rock crevice.[1584] Males sing loudly from a vantage point such as a prominent rock, the song carrying several hundred metres. There is one observation of communal breeding: three or four adults were watched feeding recently fledged young.[1578]

The White-throated Grasswren feeds on insects, including ants, and on seeds and vegetable matter.[748] Fire affects it adversely by destroying large old clumps of spinifex.[1584] There are no measures of abundance.

Field Atlas Records: 18

1° blocks–with records: 5 (1%)
–with breeding records: 2 (40%)

Atlas No.: 516

Max. Rep. Rate: 13%
In block(s): 13/132

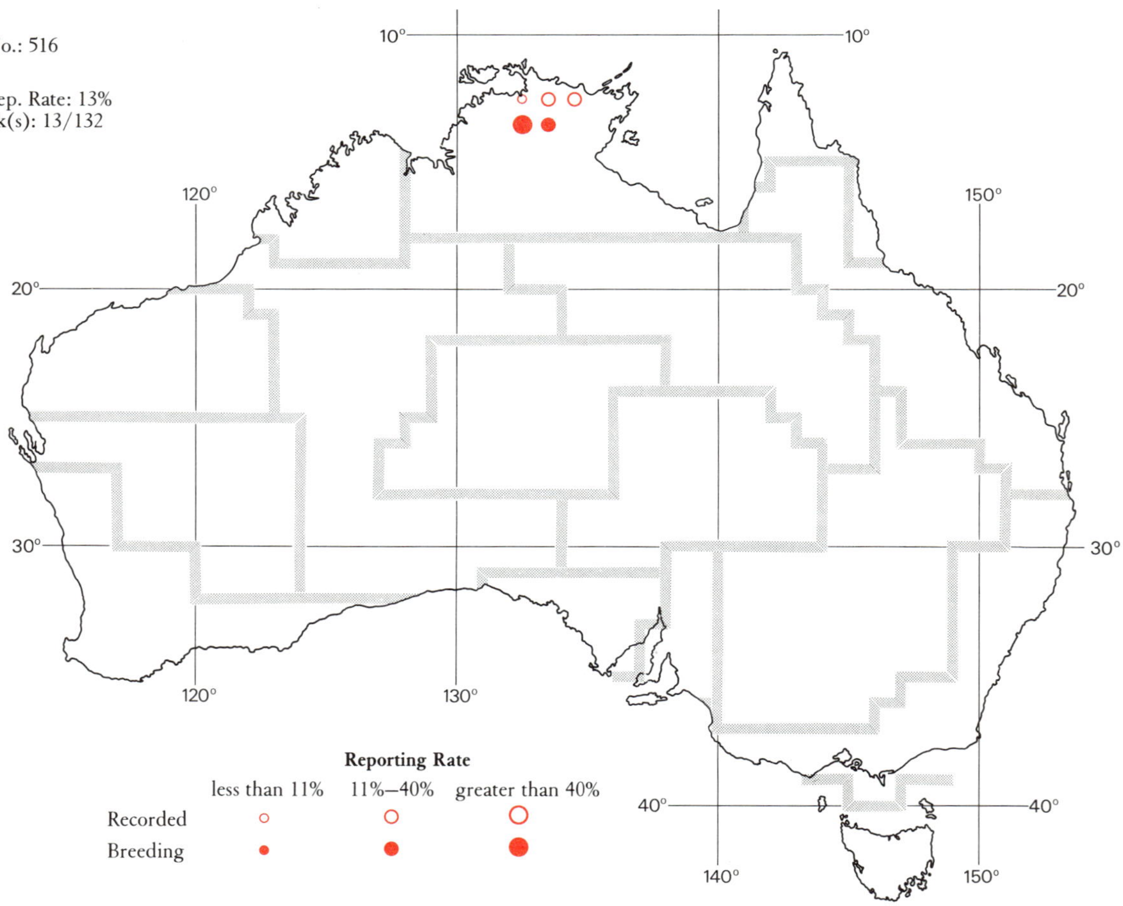

Carpentarian Grasswren

Amytornis dorotheae

The Carpentarian Grasswren is endemic to country south of the Gulf of Carpentaria and recorded from only two localities. Most reports are from the Macarthur R at 16/136 near Borroloola but the species is also known from the China Wall (17/137) area in the Nicholson R drainage and presumably is present in between.[1578]

The species is assumed to become only locally nomadic in the non-breeding season but apparently varies in abundance from time to time in any place.[1578] It lives in sandstone gorges and gullies vegetated with spinifex, using the rock crevices for shelter[105] and eating insects and seeds.[1578] The Carpentarian Grasswren seems to require dense stands of spinifex which take some time to regenerate after fire.[1578] Too frequent burning would therefore reduce the number of suitable places with a consequent reduction in the number of Grasswrens.

There are no measures of abundance.

Field Atlas Records: 2

1° blocks–with records: 1 (1%)
–with breeding records:

Atlas No.: 517

Max. Rep. Rate: 4%
In block(s): 16/136

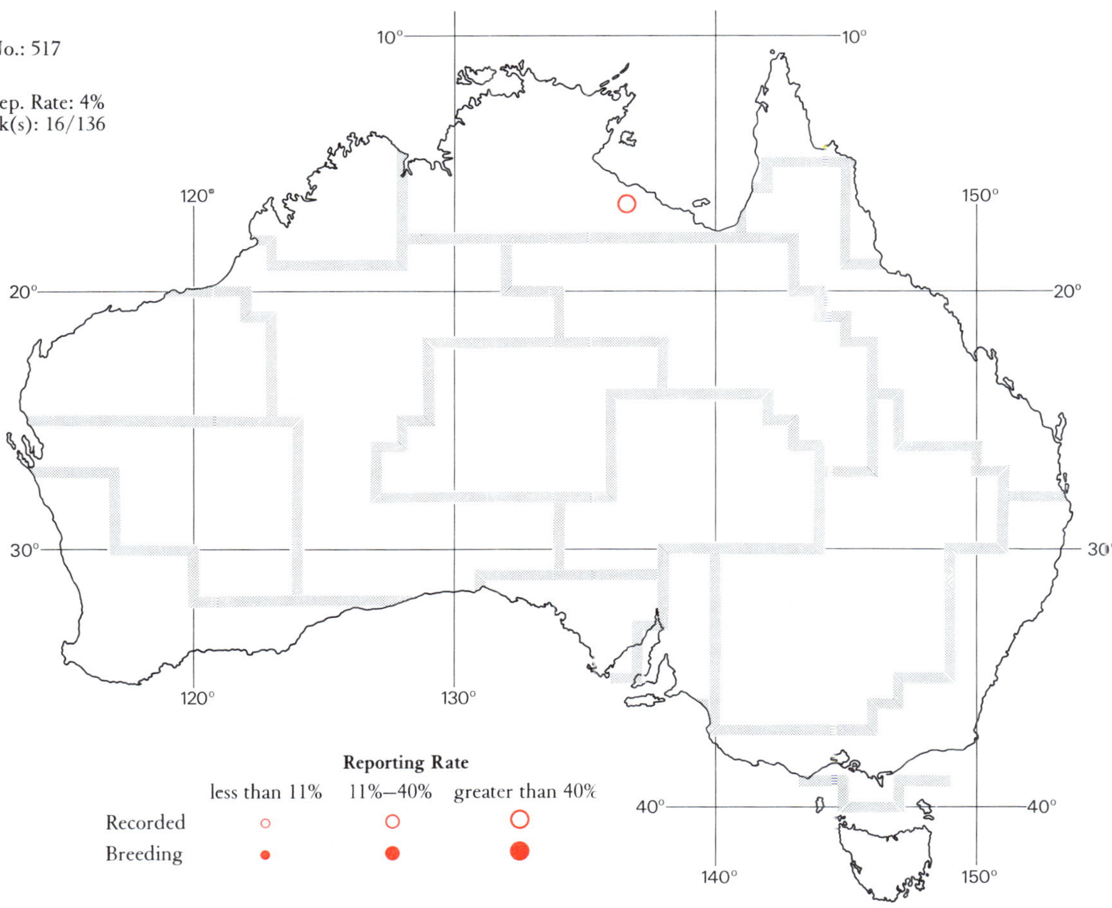

Striated Grasswren

Amytornis striatus

The most recent study of the endemic Striated Grasswren recognizes three subspecies.[1578] In the northern Flinders Ra. an isolated population, *merrotsyi*, [1389,1578] is represented in the Field Atlas only by records from the southern end of its distribution (33/138), which extends north through the ranges to Freeling Gorge (30/139). A second subspecies *whitei* in the Pilbara Region also appears, from the Atlases, to be isolated. The third subspecies *striatus* covers the remainder of the species' range.[1578] This appears to have extended last century east to the Namoi R (30/148) and through the southern half of the Murray-Darling Region.

The bird has disappeared from much of the eastern part of its range but records in the 1970s from Cobar (31/145) and south-westward suggest that it is still sparsely distributed there. No records in the Atlases link the Eyre Pen. population to that in the Western Deserts even though apparently suitable country lies between. Birds seen in the Rolling Downs Region in the 1970s at Fermoy (23/143) and Opalton (23/142)[626] appear to represent an isolated population. Because the species is elusive and perhaps varies greatly in abundance in a locality, it may yet be found in places where it now seems absent.

The extent of movements of the Striated Grasswren is unknown; groups form in the non-breeding season and may wander widely.[1578] The species generally lives where spinifex is growing, using it for cover. In the Big Desert (35/141) prickly hakea is also frequently used.[1334] The subspecies *striatus* lives in spinifex growing on sandy soil but *merrotsyi* and *whitei* inhabit stony country.[1578] The Grasswren eats seeds and arthropods including ants and beetles.[748,1049,1578] When fires destroy large areas of spinifex, observations suggest that, though regeneration is steady, it is many years before clumps of spinifex become really dense and suitable for the Grasswren; it may be many more years before they are actually re-populated though pairs are able to breed in patches of spinifex less than 1 ha in area.[1578]

There are no measures of abundance.

Field Atlas Records: 150

1° blocks–with records: 45 (6%)
–with breeding records: 8 (18%)

Atlas No.: 513

Max. Rep. Rate: 10%
In block(s): 33/140

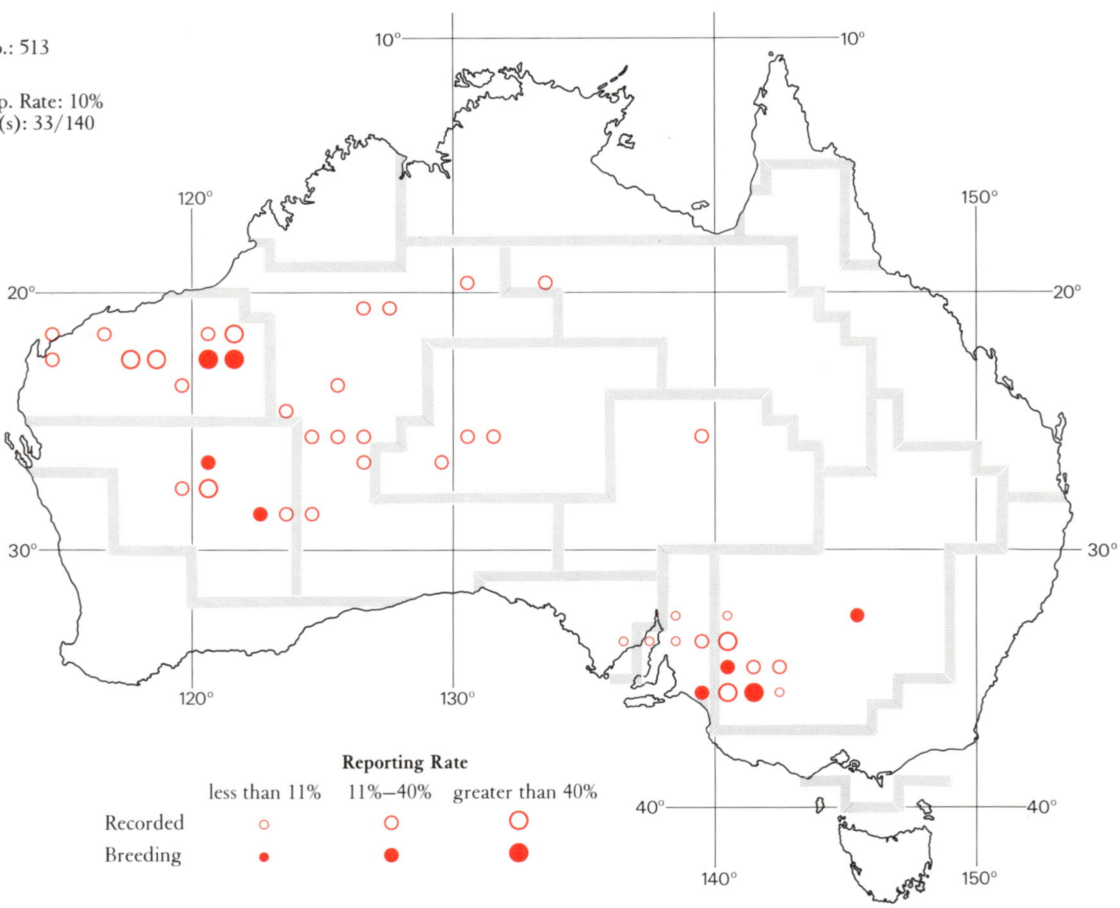

Eyrean Grasswren

Amytornis goyderi

There were no definite reports of the endemic Eyrean Grasswren between its discovery in the 1870s and apparent rediscovery in 1961 on the Macumba R (27/136),[1281] confirmed by collection of specimens in 1976 near Poeppel's Corner (25/138, 26/138).[1223,1392] It has subsequently proved to be widespread in the northern L. Eyre Region but all records in the Atlases are south of 25°S and it remains to be seen whether the species is present farther north in the Simpson Desert.

The Eyrean Grasswren is apparently at least locally nomadic,[1369] forming parties in the non-breeding season.[1578] It lives and nests mainly in dense tussocks of sandhill canegrass, the characteristic plant of the sandhills.[1392] Parties, usually of more than two birds, keep to the tussock and are hard to flush, even from one clump to the next. The bird hops or flutters over the ground rather than flies. Seeds, mainly of canegrass, were in all guts examined and most contained insects, principally bugs and ants.[1392]

Rabbits and exotic plants have infested areas within the species' range[1369] but the lack of historical records makes comparison of its status impossible. In any locality abundance varies greatly depending on the season.[1394,1578]

There are no measures of abundance.

Field Atlas Records: 18

1° blocks–with records: 7 (1%)
–with breeding records: 3 (43%)

Atlas No.: 515

Max. Rep. Rate: 12%
In block(s): 26/136

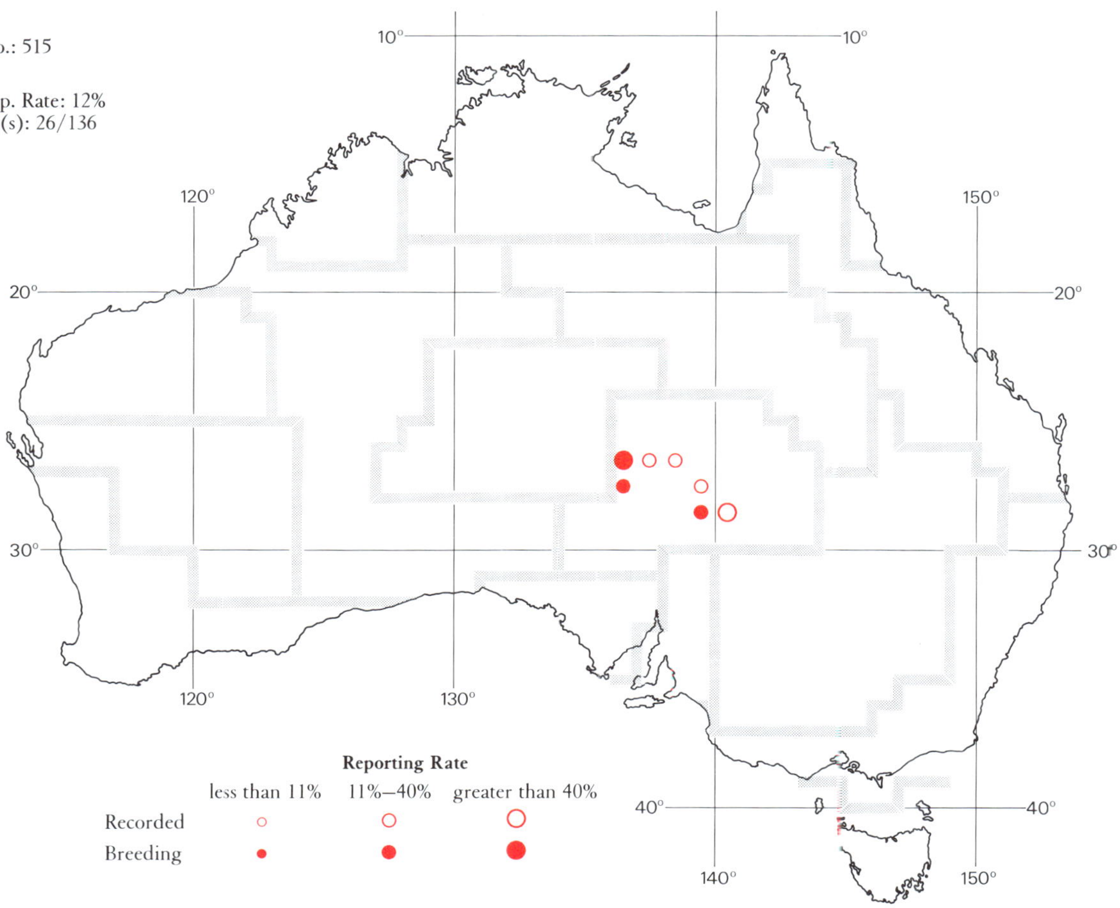

Grey Grasswren

Amytornis barbatus

The endemic Grey Grasswren was the last of the group to be described. It was first recorded in 1921 in the Bulloorine (29/142) area and again in 1942, but not collected until 1967.[543,1709] During the Field Atlas it was recorded from only two widely separated localities: the Bulloo R (142°E) on either side of the NSW-Qld border; and the Diamantina R at Goyder Lagoon (26/139).[416] Subsequently it has been reported at Embarka Waterhole (27/140); the junction of the Kallakoopah Ck and Warburton R (27/138); L. Cuddapan (25/141); and between L. Machattie (24/139) and L. Koolivoo (24/139).[968,1578]

The Grey Grasswren appears to leave its territory after the breeding season and gather into locally nomadic parties. Drought may force groups of 30–40 to congregate in remaining patches of lignum.[1578] It lives in lignum, apparently only in tall dense clumps with little space between them.[1578] On the Bulloo R there is swamp canegrass among the lignum and on the Diamantina, sedge. The bird eats seeds and insects and, at times, water snails.[1369]

At Goyder Lagoon three pairs occupied territories of approximately 200 sq. m.[416]

Field Atlas Records: 13

1° blocks–with records: 3 (1%)
–with breeding records: 1 (33%)

Atlas No.: 871

Max. Rep. Rate: 14%
In block(s): 28/142

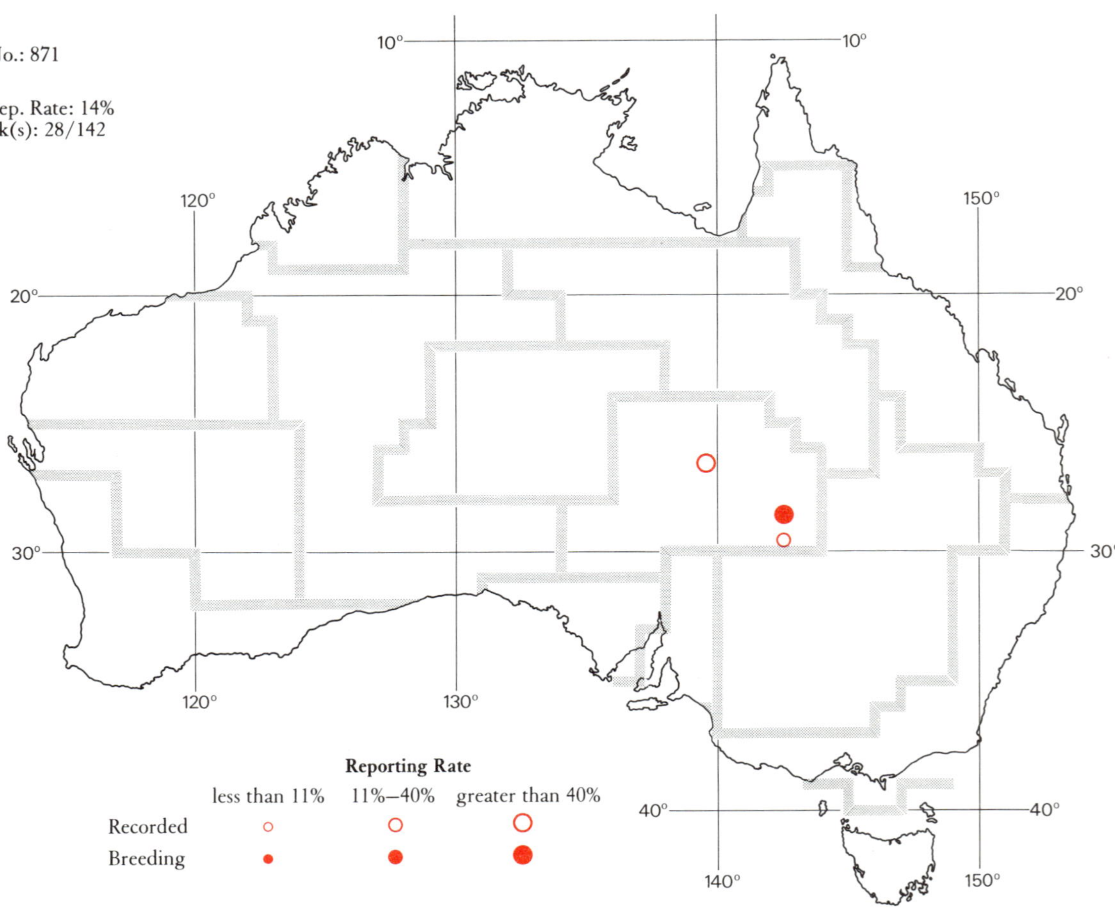

Confusion between the endemic Thick-billed Grasswren and the Dusky Grasswren in the Central Ra. Region was not fully resolved until 1972.[297,1376,1578] Of the three isolated populations of the Thick-billed Grasswren, two at least have declined greatly this century (p. 667).[1578] The Field Atlas map shows, in the L. Eyre Region and northern fringes of the SA Ra. Region, the remnants of *modestus* whose range last century stretched from the Namoi R (30/148) through the Murray-Darling Region and into the Central Ra., there overlapping the distribution of the Dusky Grasswren.

The population *myall* is represented in the Field Atlas by the single record (32/136) in the Eyre Pen. Region. This subspecies is probably still well distributed in the north of the peninsula though unobtrusive and hard to find.[414] Farther west a specimen was collected near Immarna (30/132) in 1909 but there have been no subsequent records.

The third subspecies is *textilis*, formerly distributed throughout the Goldfields Region and adjacent parts of the South-West Region. Its decline has been so great that it is now almost confined to the vicinity of Shark Bay (25/113). The Field Atlas record from 22/114 near Exmouth in Aug. 1979 extends the range of *textilis* northward.

Breeding adults of the Thick-billed Grasswren are thought to be sedentary.[1578] The species usually lives alone or in pairs in saltbush, often where there is much bluebush. In many places bluebush has been eliminated and saltbush degraded by grazing, preventing young plants from growing and maturing into the thick, old bushes required by the Grasswren. This appears to have been the major influence in the decline of *textilis* in WA.[440] The population that survives at Shark Bay inhabits acacia scrub with chenopods but usually no saltbush, vegetation dominated by mulla mulla and drainage sites vegetated with dense lignum and emu-bush. The common feature is large shrubs in which the birds can shelter at ground level.[440]

There are no measures of abundance but pairs are thought to remain within an area of 4–5 ha throughout the year.[1578]

Thick-billed Grasswren

Amytornis textilis

Field Atlas Records: 38

1° blocks–with records: 10 (1%)
–with breeding records: 2 (20%)

Atlas No.: 512

Max. Rep. Rate: 11%
In block(s): 25/113

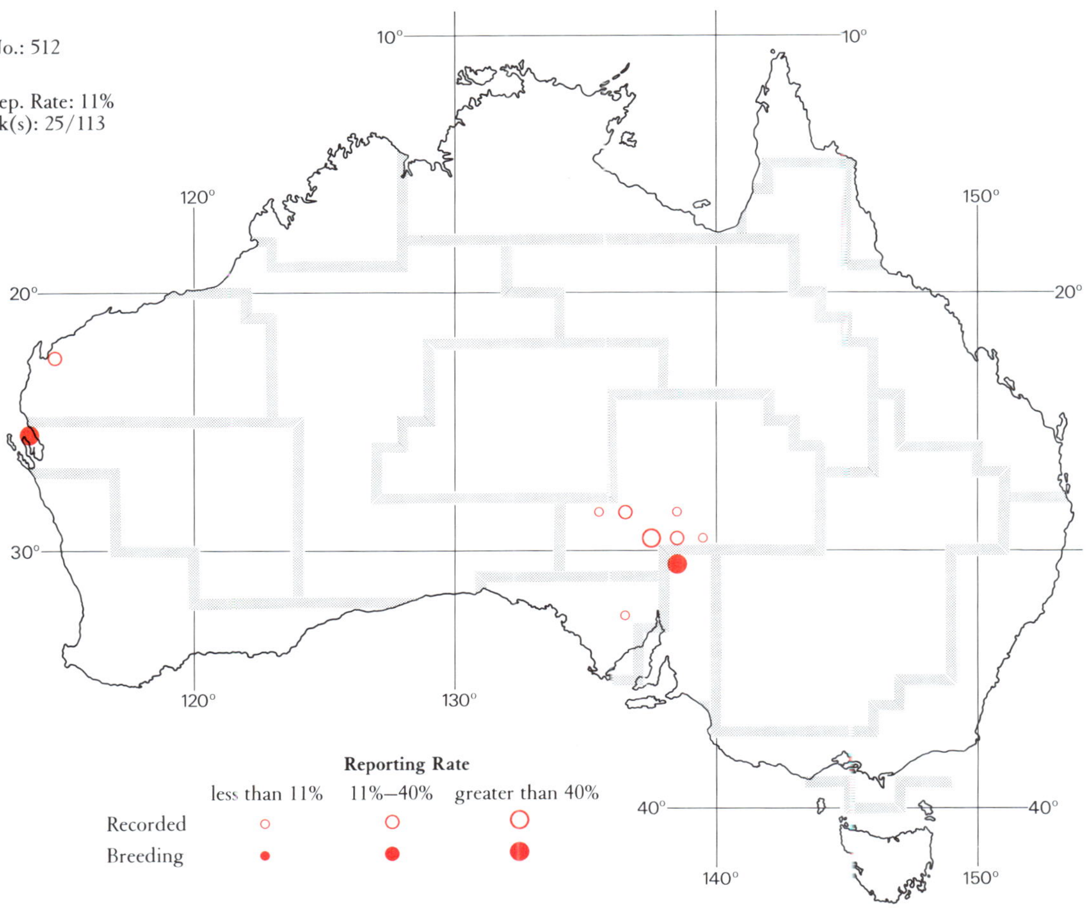

Dusky Grasswren

Amytornis purnelli

From 1894, when the endemic Dusky Grasswren was first collected, there has been confusion between it and the Thick-billed Grasswren whose range it overlaps in the Central Ra. Region. The problem was not fully resolved until 1972.[297,1376,1578]

The outlying population of the Dusky Grasswren, discovered in the Mt Isa (20/139) district in 1966,[314] is the subspecies *ballarae*, which extends north to Thorntonia (19/138)[1578] and south to Mt Ninmaroo (22/140). Other populations all belong to the nominate subspecies *purnelli* though many are probably isolated in outlying hills and ranges.[1578] The northernmost report of the species is from the Ashburton Ra. (18/133)[1376] and the southernmost from the Birksgate Ra. (27/129)[1578] and the Everard Ra. (27/132). The species has not been found in the Tanami Desert.

Nothing is known about movements of the Dusky Grasswren. It lives in groups of up to a dozen or more, specially in the non-breeding season,[1578] and is confined to the ranges of central Australia, characteristically on rock-strewn slopes among clumps of spinifex. The bird hops from boulder to boulder rather than flies.[719] Seeds and insects are eaten in roughly equal proportions at most times of the year.[1578]

There are no measures of abundance but breeding territories, assumed to be held by pairs, are often close together, each covering perhaps no more than 1 ha.[1578]

Field Atlas Records: 92

1° blocks–with records: 25 (3%)
–with breeding records: 3 (12%)

Atlas No.: 511

Max. Rep. Rate: 14%
In block(s): 24/131

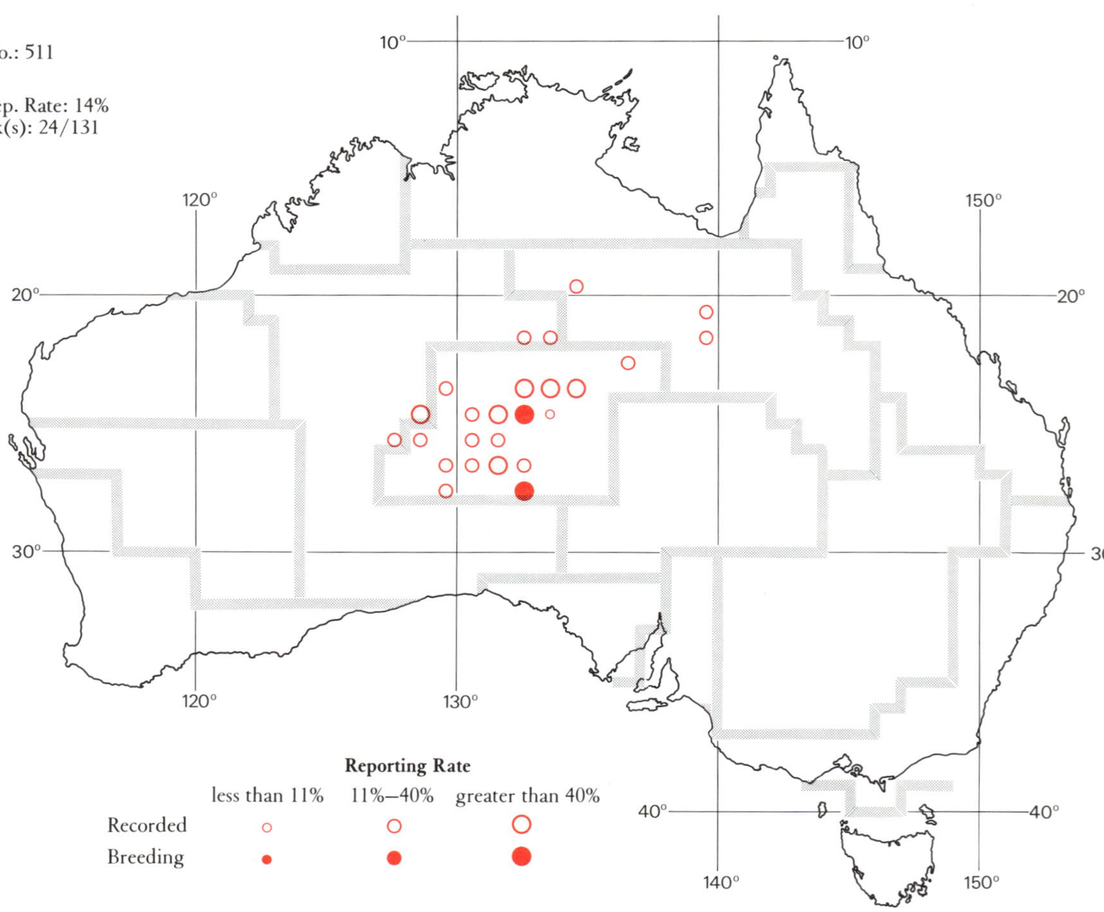

Eastern Bristlebird

Dasyornis brachypterus

The range of the endemic Eastern Bristlebird extends from Cunningham's Gap (28/152) to Marlo (37/148) in a series of isolated pockets. There are records from Cunningham's Gap in the 1970s so this population probably survives but others from there south to Sydney (33/151) have not been reported since before 1950. The last records from Sydney were made in the 1920s.[328]

The Eastern Bristlebird is sedentary. It lives in rank vegetation bordering on heath in coastal and mountain environments. Barren Grounds (34/150) and adjoining Budderoo Plateau and Tianjara (35/150) are the southernmost localities where it is present at high altitudes (above 600 m at Barren Grounds). Farther south all records are coastal. The bird is easy to find because of its loud distinctive calls. It also responds sometimes to a replay of its calls.[963] The Eastern Bristlebird feeds mainly on insects but also takes some epacris fruit. It scratches for food usually in the litter and on the ground, even bare gravel by a stream[1505] and has been observed feeding in low foliage.[963]

The causes of the Bristlebird's decline seem to relate to the spread of settlement and the frequent burning or grazing of areas where it lives.[328,1684] Infrequent fires may be advantageous because thick low vegetation is often a successional stage reached some years after a fire; provided small areas can be left unburnt, Bristlebirds will survive in them[1184] and re-populate the surrounding area after it has regrown. The Barren Grounds have a long history of fire, yet the birds are fairly common there.[328,963,1286]

In a survey of 24 ha at Barren Grounds 13 calling birds were located. If these represented pairs then density was 1.1 birds/ha. The maximum density recorded was about 2 birds/ha.[962]

Field Atlas Records: 53

1° blocks–with records: 6 (1%)
–with breeding records: 1 (17%)

Atlas No.: 519

Max. Rep. Rate: 4%
In block(s): 35/150

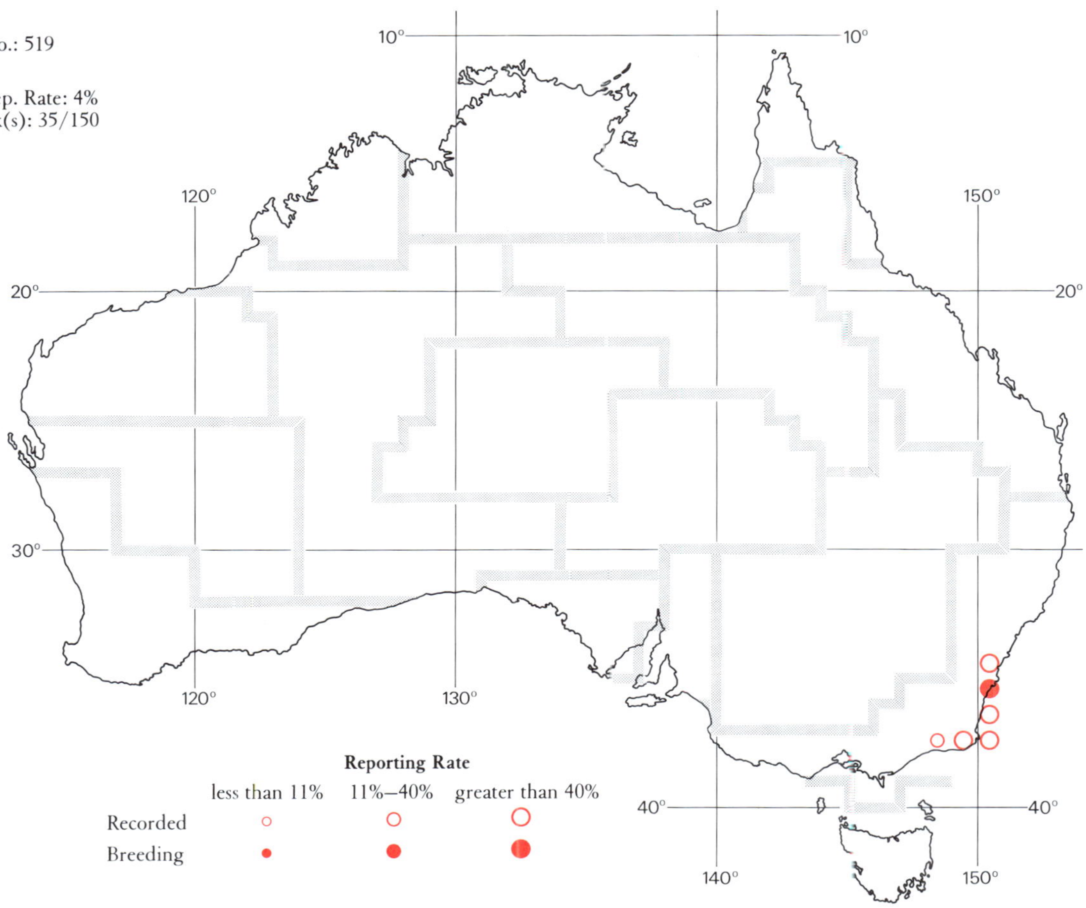

Western Bristlebird

Dasyornis longirostris

Even at the time of European settlement the endemic Western Bristlebird lived in relict populations isolated by past climatic change.[1684] Its range has shrunk further in the last 150 years (p. 667). In 1839 it inhabited the coastal plain near Perth (31/115) and presumably similar country right round the coast to Albany (35/117) where it survives. There are no subsequent reports from the Perth area and the only other population reported west of Albany, at Wilson Inlet (34–35/117), was discovered in 1907 and destroyed by fire in 1914. Field Atlas records came from the Fitzgerald R National Park (33/119), Waychinicup R (34/118) and Two People Bay (34–35/118). Studies at Two People Bay suggest that the Western Bristlebird is sedentary.[1682] It lives in heath, reedbeds and long grasses.[1682] The bird perches on a twig above the top of the vegetation but, at the slightest movement, ceases to sing and dives down out of sight. It feeds on insects and seeds.[278,1682] Settlement and burning have eliminated populations of the Western Bristlebird though it will survive even in small patches of unburnt country.

At Two People Bay Fauna Reserve the population was about 80 pairs in 1976[1684] and had increased to about 100 pairs in 1982.[1682]

Field Atlas Records: 15

1° blocks–with records: 4 (1%)
–with breeding records:

Atlas No.: 520

Max. Rep. Rate: 4%
In block(s): 34/118

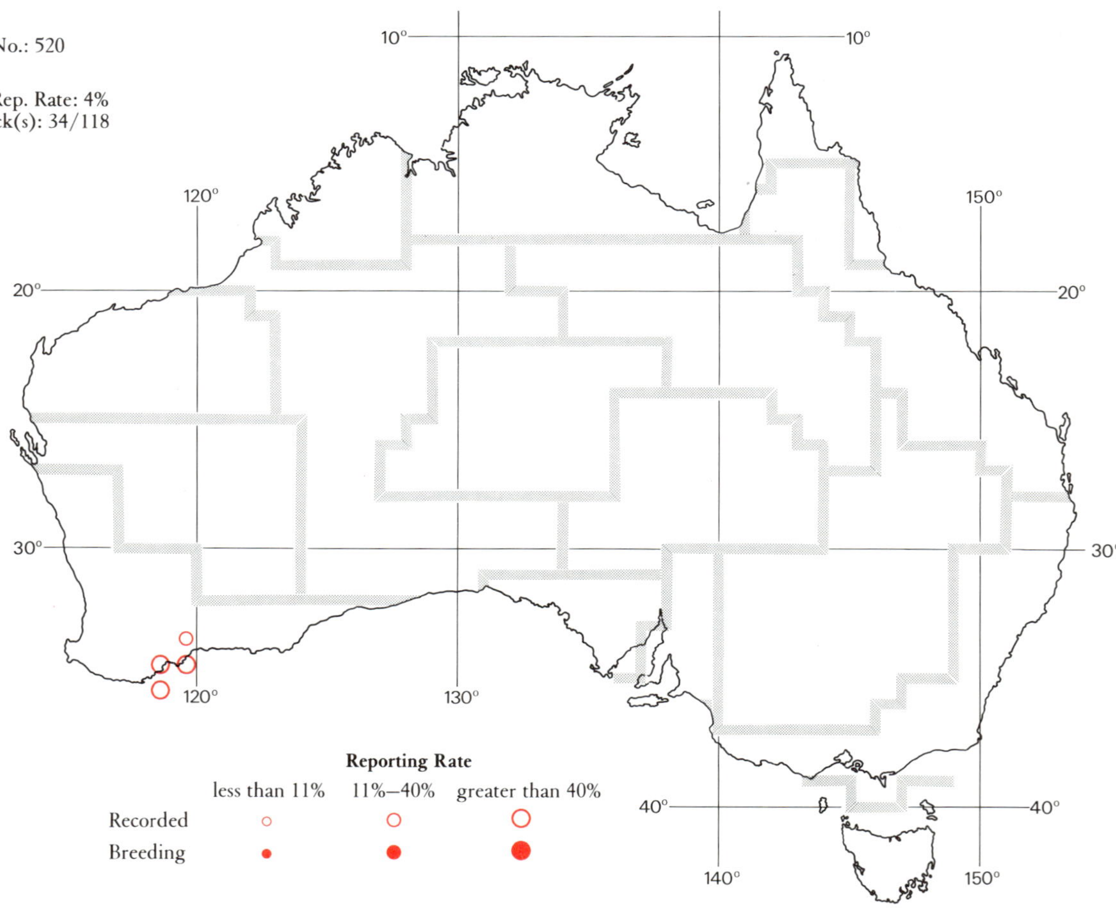

Rufous Bristlebird

Dasyornis broadbenti

Three subspecies of the endemic Rufous Bristlebird are recognized: *broadbenti* east of 141°E, *whitei* west of 141°E as far as the Murray Mouth (35/138) and *litoralis* in the South-West Region. The population in the South-West Region has not been reliably recorded since 1906[1684] when it occupied the coast between Cape Mentelle (33/114) and Cape Naturaliste (33/115).[270] Only unconfirmed reports resulted from intensive searches for it during the Field Atlas.

Both the eastern subspecies still occupy their historical range from near Anglesea (38/144) to the Murray Mouth where reports as recent as 1970[709] are from as far west as *whitei* has ever been recorded. Small populations are spread throughout this range except for a gap near Warrnambool (38/142).

The Rufous Bristlebird is present all year throughout its range. It lives mainly in heath near the coast but in the Otway Ra. (38/143) it penetrates a short distance inland along streams. It will forage in gardens near thick natural vegetation. In the South-West *litoralis* lived in heath about 40 cm high on the coast. The bird feeds on insects, notably beetles, and also takes considerable quantities of the seeds of seashore and heath plants.[1049] It has been observed to cross the open beach to reach banks of beachwashed seaweed where it feeds on the animal life.[1051] It feeds on the ground though it will perch low in trees when calling.[1469] In the South-West Region the only gut examined contained snails.[1270]

The Rufous Bristlebird is surviving in eastern Australia even in settled areas but, because it is confined almost exclusively to a thin strip of coastline, severe bushfires may fragment the population.[1469] In the South-West Region the disappearance of *litoralis* coincided with a period from 1906–16 of extensive firing of the country by settlers seeking grazing for their cattle. The fires destroyed many of the places where the Rufous Bristlebird lived[318,1684] and it may be extinct.

If calling birds represent a simple pair, there were 12 birds in a 3 km strip of coast 200 m wide at Airey's Inlet (38/144) in 1982.[1469]

Field Atlas Records: 265

1° blocks–with records: 9 (1%)

–with breeding records: 4 (44%)

Atlas No.: 521

Max. Rep. Rate: 28%
In block(s): 37/139

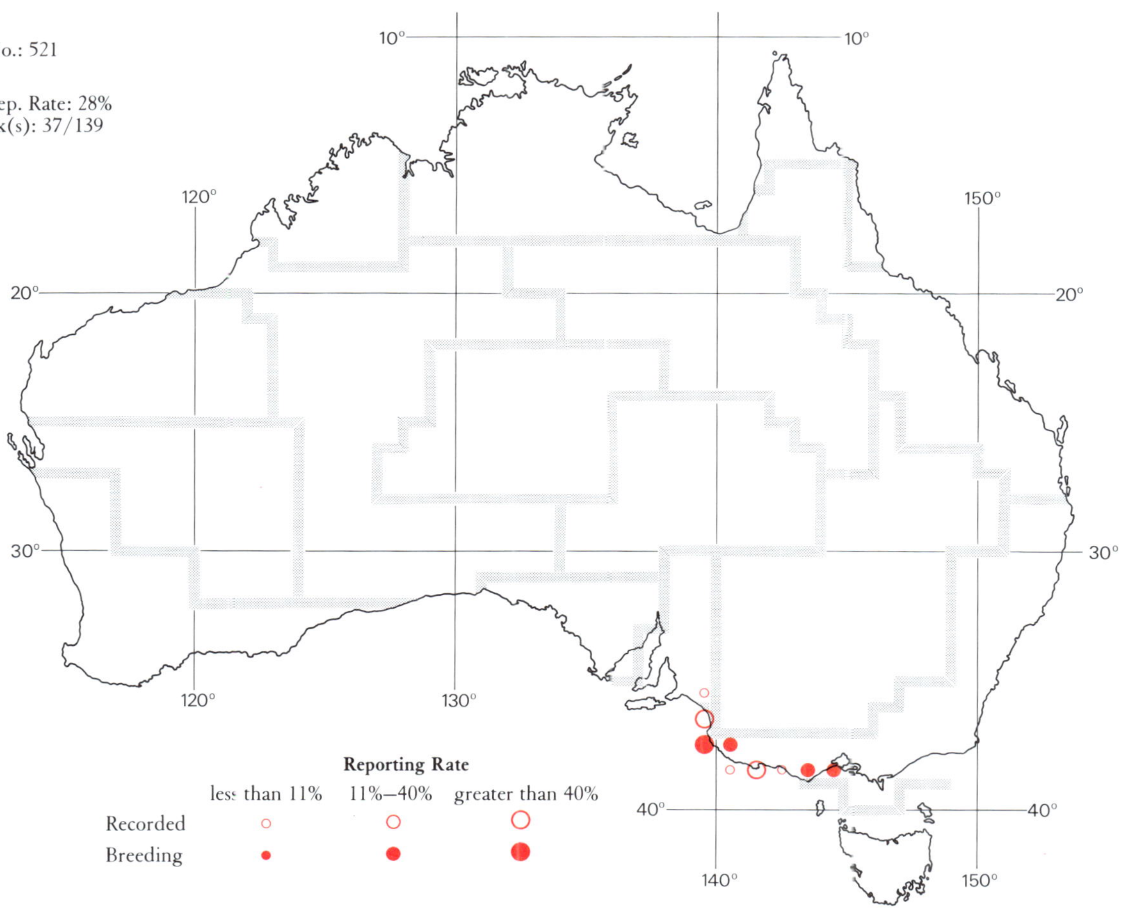

Pilotbird

Pycnoptilus floccosus

The Atlases indicate no change in range of the endemic Pilotbird. Numbers are reduced in winter in high mountain gullies near Canberra (35/149),[1016] possibly indicating some seasonal altitudinal movement but there were too few records for any such movement to be reflected in Reporting Rates. In Sherbrooke Forest (37/145) each pair occupies its territory permanently.[1929]

The Pilotbird lives in eucalypt forest as well as in woodland dominated by snowgum. In the foothills or on the coast it is in wet gullies. In eastern Vic. it has sometimes been recorded on dry slopes under tall forest with dense undergrowth of daviesia and uses regenerating gully sites as well as mature areas.[1113] The Field Atlas shows that in the Snowy Mts (36/148) it was more often recorded between 200 m and 500 m in altitude (28.6%) than above (5.9%) or below (2.5%).

The Pilotbird seeks insects, earthworms and other invertebrates beneath litter, which it turns over with its bill.[1929] It will also feed on wet ground near streams and birds feeding in such sites were found to have eaten small mussels.[868] The Pilotbird sometimes accompanies the Lyrebird as it feeds, taking invertebrates exposed by the Lyrebird's scratching.[1890] The range of the Pilotbird is completely within that of the Superb Lyrebird but they are rarely seen together. Locally each species is sometimes present in the absence of the other and the smaller bird is unlikely to be dependent on the larger.

In Sherbrooke Forest the Pilotbird lives at a density of 1.3–2.0 birds/ha.[1929]

Field Atlas Records: 580

1° blocks–with records: 19 (2%)
–with breeding records: 7 (37%)

Atlas No.: 506

Max. Rep. Rate: 11%
In block(s): 37/146

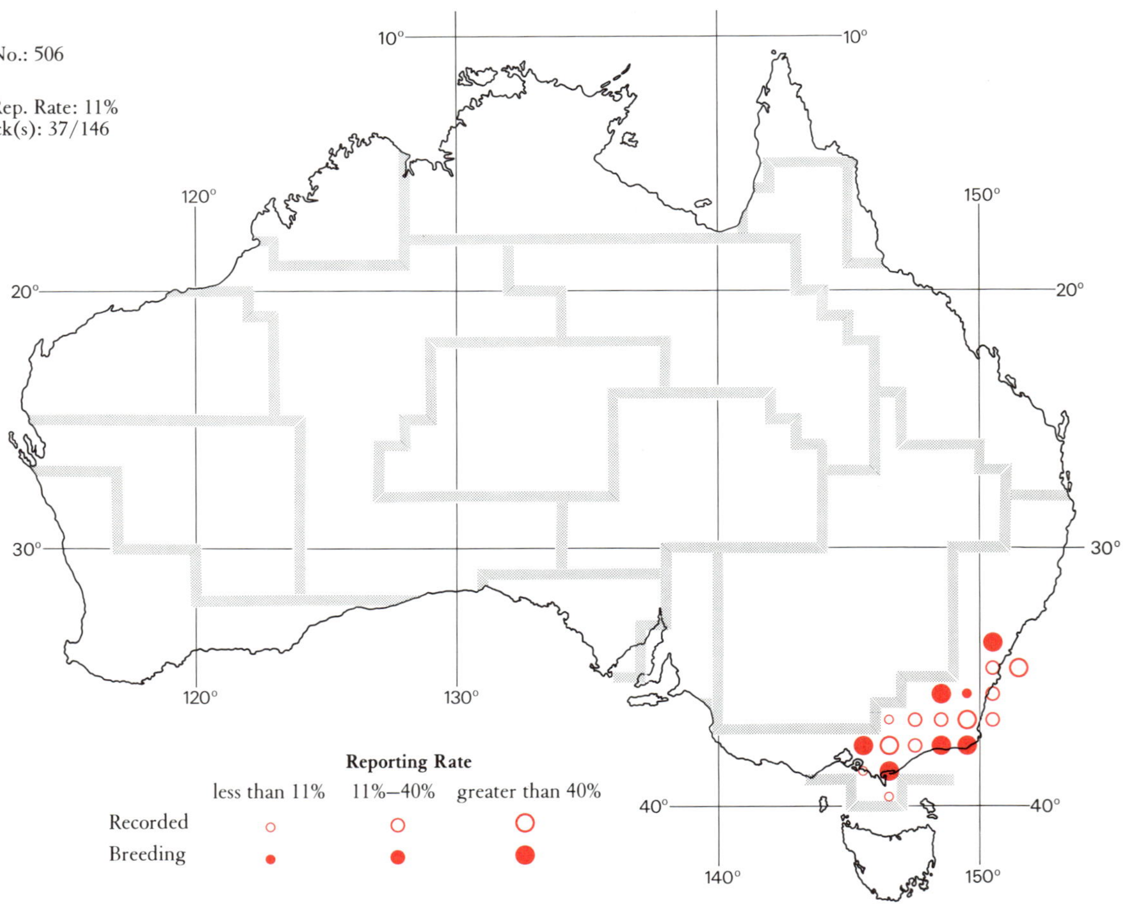

There have been several observations, some unconfirmed, of the endemic Origma outside the range defined by the Field Atlas. To the north there were sightings at Crawney Pass (31/151) and Timor (31/151) in 1951–76.[749] To the south, also in 1951–76, there were reports from Krawarree (35/149) and Big Badja (36/149) on the headwaters of the Shoalhaven R[488,863] and from the coast south of Bermagui (36/150) in about 1914.[520] Such sightings may represent relict populations or indicate an ability to disperse widely, perhaps along river systems in rugged terrain, but banding data suggest that breeding pairs are mainly sedentary and that the young do not move great distances.[749] Some movement is suggested by the observation that the Origma deserts dried up creeks in summer.[1343]

Though known as a characteristic bird of the Hawkesbury sandstones, the Origma is not limited to them as implied in some accounts.[684] Its distribution extends into adjacent limestone areas. Rocky ravines near water are its usual haunts. The bird is usually in pairs or family parties.[396,1343] An observation in 1924 of about 20 Origma in a loose flock feeding on the ground in the upper Colo (33/150)[1002] may have been a misidentification. Sometimes the Origma is recorded as exclusively insectivorous,[320,1343] though seeds were found in four guts examined and insects in only two.[365,1049] The species has suffered from urban development near Sydney (33/151).

There are no measures of abundance.

Origma

Origma solitaria

Field Atlas Records: 564

1° blocks–with records: 11 (1%)

–with breeding records: 8 (73%)

Atlas No.: 505

Max. Rep. Rate: 19%
In block(s): 34/151

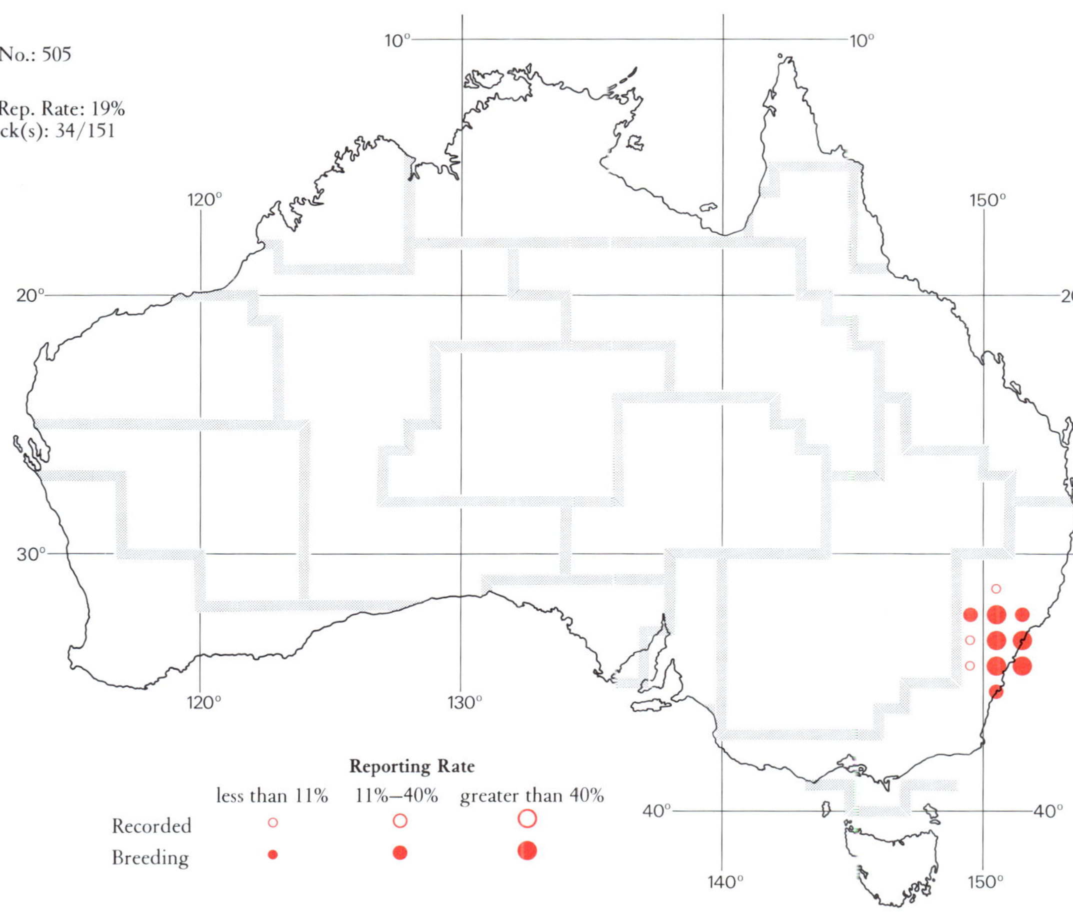

Australian Fernwren

Crateroscelis gutturalis

The endemic Australian Fernwren lives only in the Atherton Region. It is present throughout the year above 600 m altitude, living on the floor of rainforest in damp mountain gullies.[231,698]

The Fernwren feeds on small invertebrates that it uncovers by lifting and turning over litter on the ground[1582] often almost disappearing under the forest leaves.[1140] One bird was watched following an Orange-footed Scrubfowl, foraging in litter disturbed by the larger bird.[1077]

The species is said to be reasonably abundant near Atherton (17/145)[231] but there are no measures of abundance.

Field Atlas Records: 120

1° blocks–with records: 7 (1%)
–with breeding records: 4 (57%)

Atlas No.: 496

Max. Rep. Rate: 13%
In block(s): 15/145

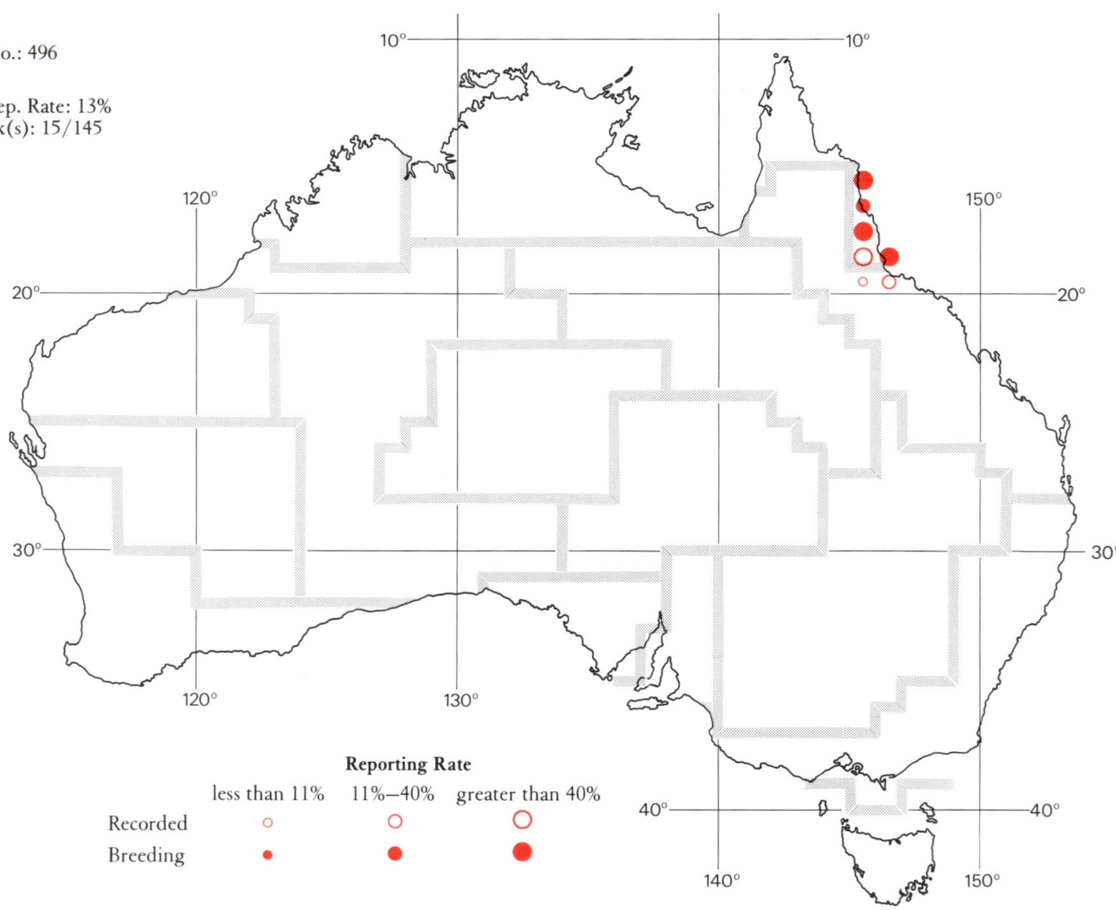

Atherton Scrubwren

Sericornis keri

Three specimens of the endemic Atherton Scrubwren were collected in 1889 and described as a new species in 1920 but it was subsequently regarded as an atypical Large-billed Scrubwren, until rediscovered in 1964.[681] The species is confined to the Atherton Tableland and is sympatric with the Large-billed Scrubwren from which it is most easily distinguished by behaviour. It shows no pattern of movement.

The Atherton Scrubwren is usually solitary and rarely even in pairs. It lives round the fringes of and within rainforest, mostly above 600 m.[681] In contrast to the Large-billed, it is largely terrestrial, nesting and feeding on or close to the ground.[1643] It forages in litter for snails and insects.[681,748,990]

There are no measures of abundance.

Field Atlas Records: 73

1° blocks–with records: 5 (1%)
–with breeding records: 3 (60%)

Atlas No.: 946

Max. Rep. Rate: 8%
In block(s): 17/145

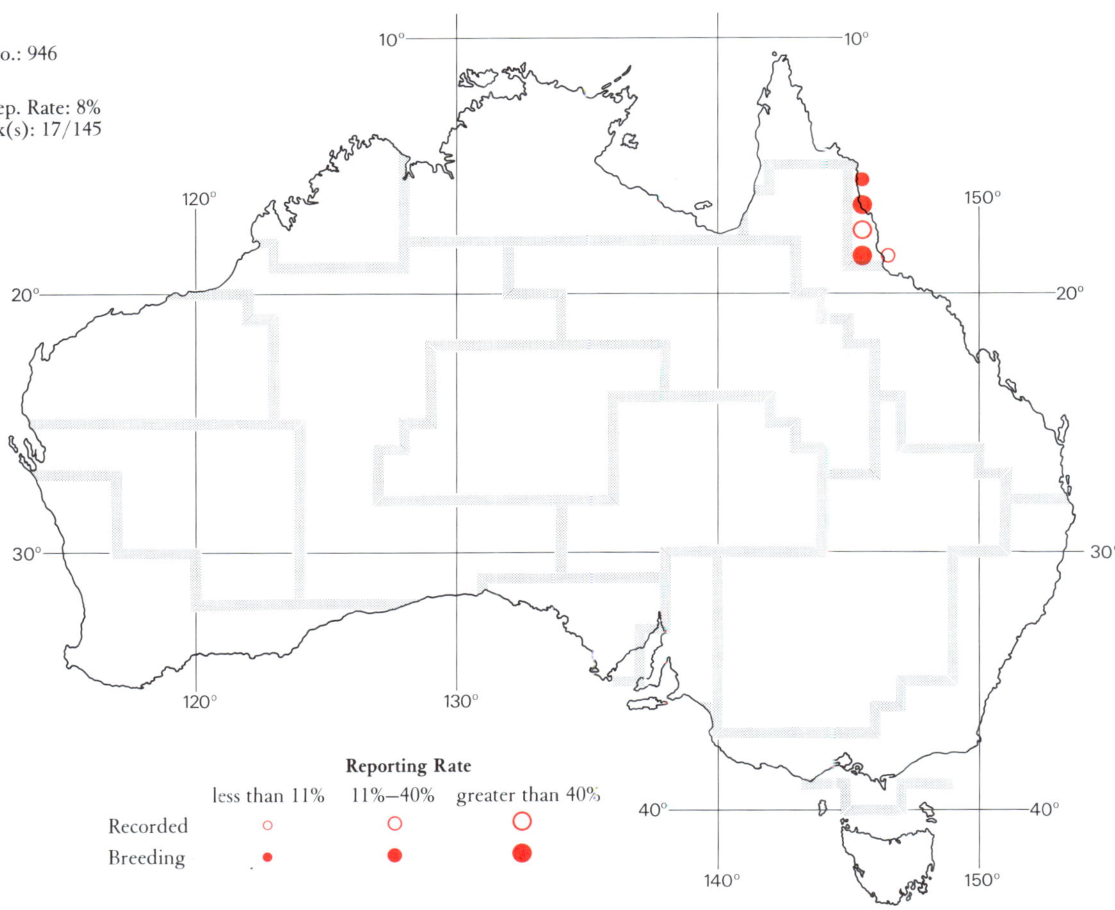

Large-billed Scrubwren

Sericornis magnirostris

Two subspecies of the endemic Large-billed Scrubwren are recognized: *viridior*, isolated in the Atherton Region, and *magnirostris* to the south. The Eungella (21/148) population is isolated, the nearest records to the south being at Shoalwater Bay (22/150) and to the north in the Paluma Ra. (19/146). In the Atherton Region *viridior* overlaps the range of the Atherton Scrubwren.[681] Sight records in the Atlases as far north as Cooktown (15/145) support the suggestion that it also overlaps the southern edge of the range of the Tropical Scrubwren.[200]

There is no evidence of large-scale seasonal movement. The Large-billed Scrubwren lives in rainforest and wet eucalypt forest, usually on the eastern slopes of the Great Dividing Ra. The farther north, the higher the altitude at which the species lives. The Reporting Rate of *viridior* above 500 m was 33.8% and below was 13.2%; in the New England (30/152) district the Reporting Rate of *magnirostris* above 500 m was 8.3% compared with 14.3% below. Parties of up to six birds are reported and the Scrubwren sometimes joins other species in a mixed foraging flock.[653] It has once been recorded breeding communally.[501] Unlike most scrubwrens, the Large-billed does not feed on the ground; it works upwards on the trunks and stems of trees and vines 3–17 m above the forest floor.[990] Its food includes insects and small snails.[748]

At Boola Boola (38/146) in eucalypt forest density in gullies 15 or more years after logging was 0.5 birds/ha or fewer.[1113]

Field Atlas Records: 1115

1° blocks–with records: 50 (6%)
–with breeding records: 14 (28%)

Atlas No.: 494

Max. Rep. Rate: 32%
In block(s): 15/145

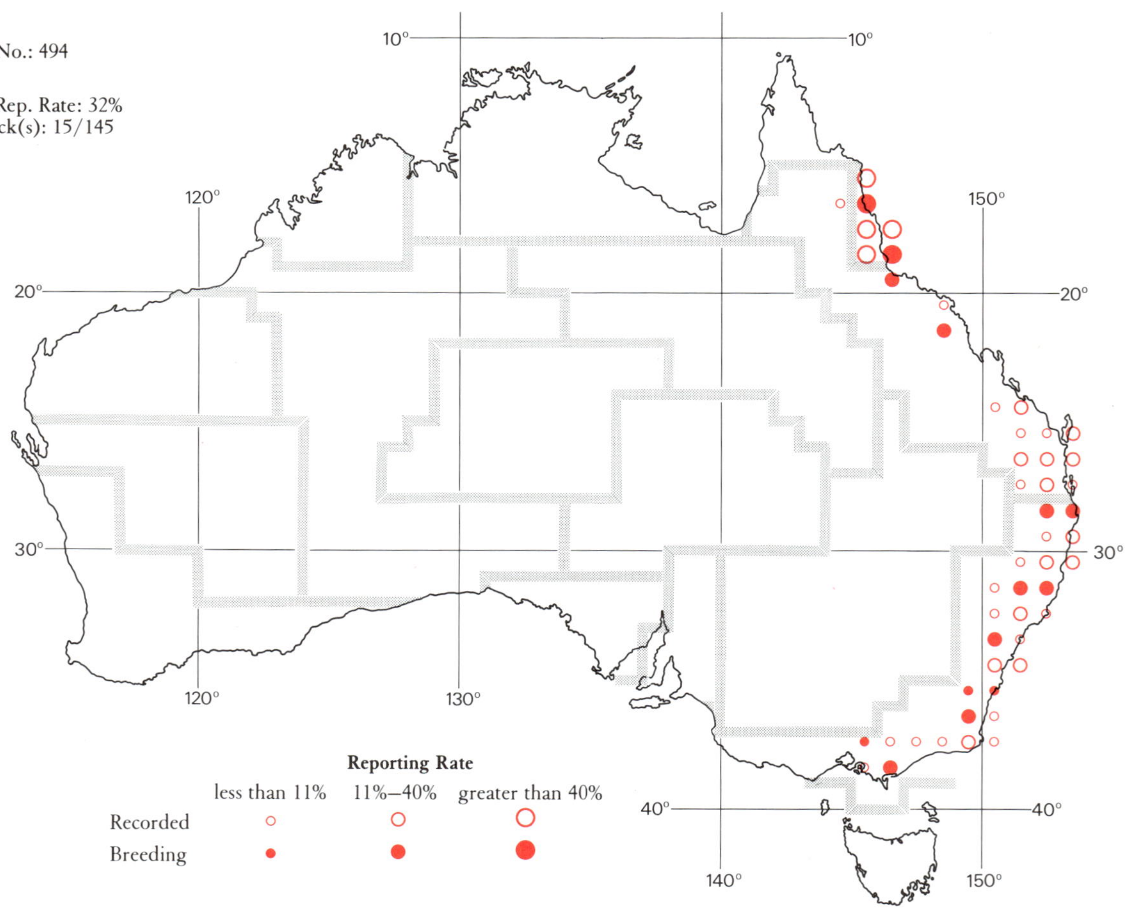

Reporting Rate

	less than 11%	11%–40%	greater than 40%
Recorded	○	○	○
Breeding	●	●	●

Yellow-throated Scrubwren

Sericornis citreogularis

The isolated populations of the endemic Yellow-throated Scrubwren in the Atherton Region and south of 26°S have been described as subspecies[748] but there are different views about their validity.[1709]

There is no evidence of movements by the Yellow-throated Scrubwren. The Atherton population lives at higher altitudes than southern birds. The Reporting Rate in the Atherton Region was 40.2% above 500 m and 0% below. The Yellow-throated Scrubwren is usually in pairs and often in company with other species. It lives in rainforest and eucalypt forest[1575] and hunts for insects and small seeds among fallen timber and debris on the forest floor.[748,1140,1200] Only 14–26% of its feeding time is spent in vegetation above the ground.[990] It is known to accompany the Logrunner foraging in the litter scratched aside by the larger bird.[1928]

The Yellow-throated Scrubwren is abundant in some rainforest. At Tooloom Scrub (28/152) in 1962 an average of 40 was seen or heard each time a 3.2 km transect was walked.[1342] At Lilyvale (34/151) 11 nests were found in a gully, approximately one each 60 m.[1200]

Field Atlas Records: 839

1° blocks–with records: 33 (4%)
–with breeding records: 19 (58%)

Atlas No.: 493

Max. Rep. Rate: 23%
In block(s): 31/152

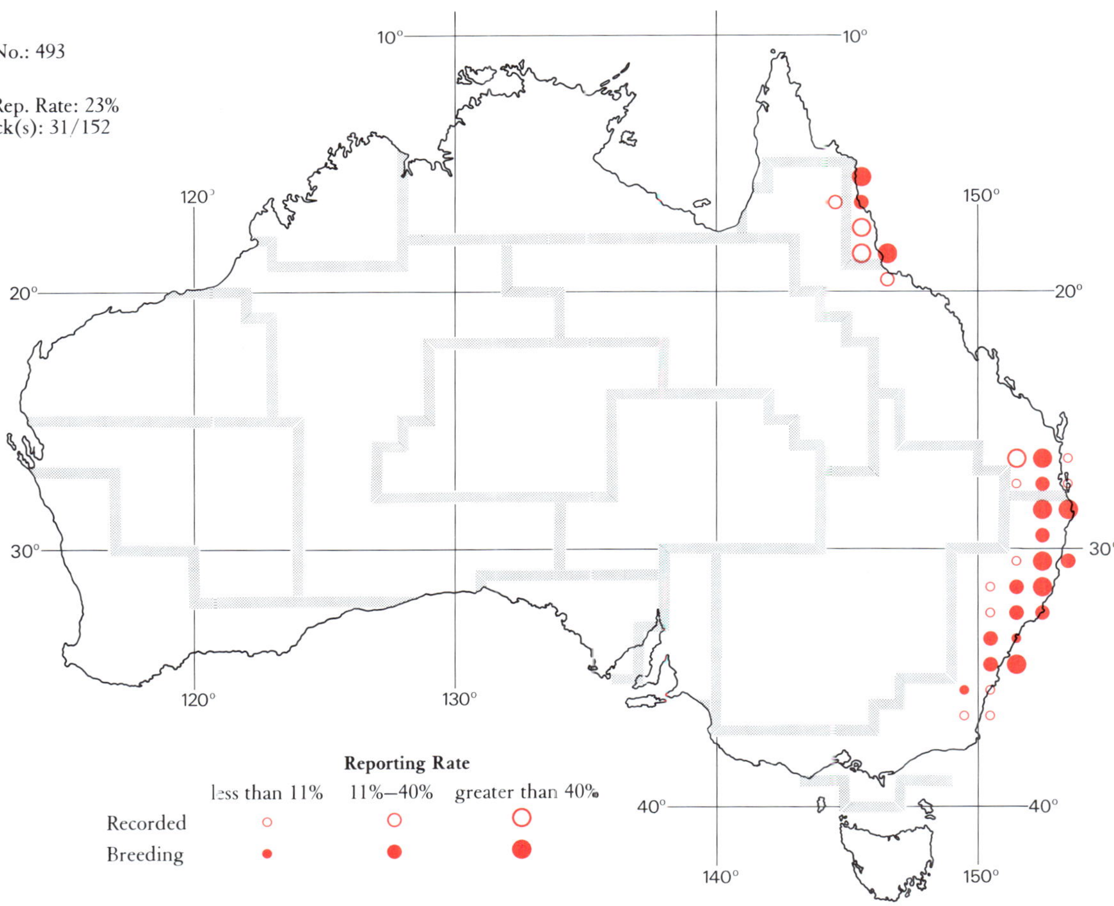

Tropical Scrubwren

Sericornis beccarii

The Tropical Scrubwren lives in New Guinea and adjacent islands and in Australia. The two subspecies in Australia are *minimus* on the tip of the Cape York Region and *dubius* to the south. They intergrade near Iron Ra. (12/143).[1371] On the eastern coast the species' range extends to just south of Cooktown (15/145) where sight records in the Atlases of the Large-billed Scrubwren support the suggestion that the two overlap.[200] An observation inland on the Archer R (13/142) gave rise to the hypothesis that the western Cape York Region was colonized by birds moving along corridors of suitable vegetation lining west-flowing rivers.[1371]

There is no evidence of large-scale movements by the species. Both subspecies live in rainforest. The northern *minimus* forages in the litter and debris of the forest floor.[1145] It may spend less time in the middle strata than *dubius* which, on the Claudie R (12/143), spends 70% of its foraging time in the middle levels of the forest, searching for insects on the trunks of trees and in vines and saplings.[990,1355]

There are no measures of abundance.

Field Atlas Records: 43

1° blocks–with records: 6 (1%)
–with breeding records: 1 (17%)

Atlas No.: 490

Max. Rep. Rate: 45%
In block(s): 12/143

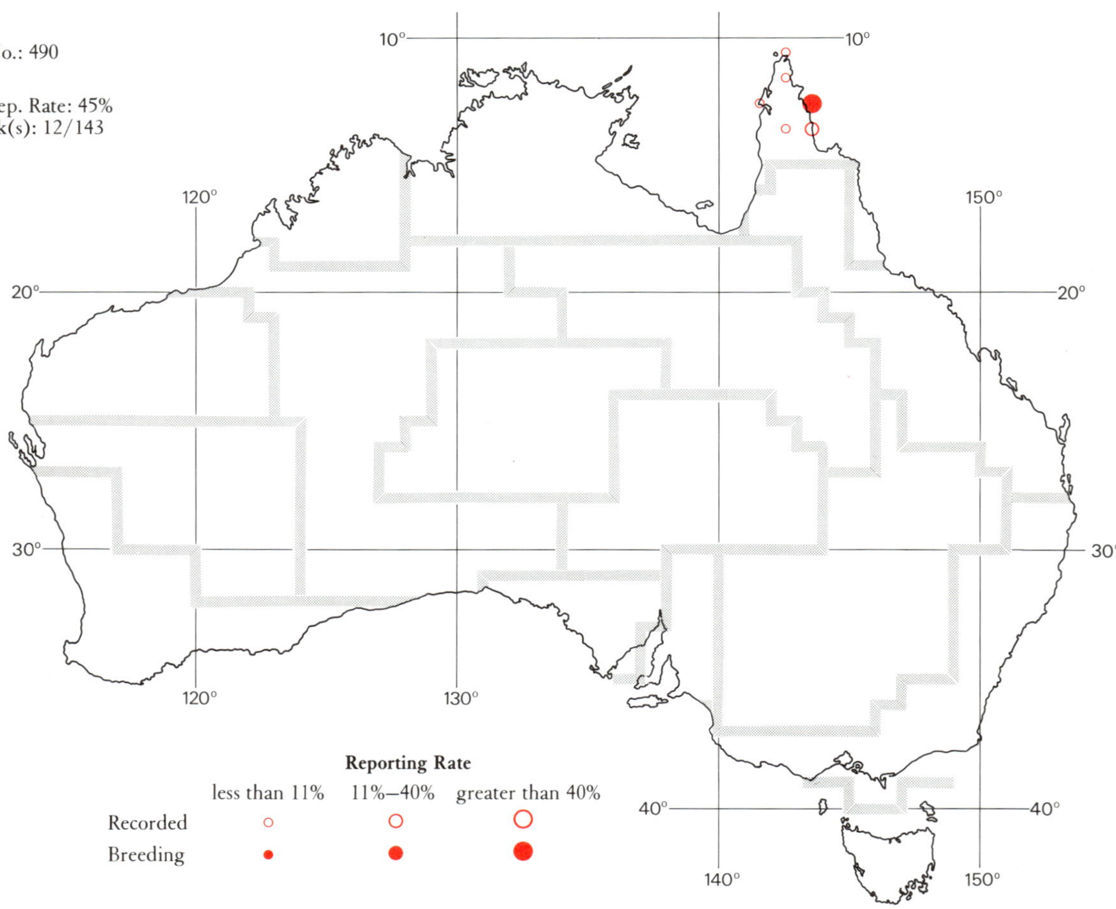

White-browed Scrubwren

Sericornis frontalis

Many subspecies of the endemic White-browed Scrubwren have been recognized. There are different views about their number and range[598,748,1124] but they fall broadly into eastern and western groups which appear to intergrade in the SA Ra. Region.[748] The 10′ maps suggest no gap in distribution there, though a population at the foot of Yorke Pen. (35/137) may be isolated. Distribution round the Great Australian Bight is probably continuous though records are lacking in the Atlases between 132°E and 133°E.

The Field Atlas suggests that populations at the northern end of the species' range in the Atherton Region and Eungella (21/148) district may be isolated because they live at higher altitudes (Reporting Rates: below 500 m 0%, above 500 m 10.4%). The northernmost records are at 16°S, within 50 km of southern populations of the Tropical Scrubwren.

At Rotamah I. (37/147) and Eyre (32/126) Bird Observatories adult Scrubwrens are sedentary and occupy a permanent territory.[17,284]

The White-browed Scrubwren lives in dense undergrowth in eucalypt forest and woodland, mallee and coastal heath. Often it persists in creekside vegetation or thickets of the introduced blackberry when surrounding country has been cleared. It lives in pairs or small parties, often in mixed flocks.[129] Sometimes it breeds communally.[17,161,765]

The White-browed Scrubwren feeds mainly on invertebrates taken on the ground or within 1 m of it.[990] Beetles, including weevils, bugs,[1049] adult moths and butterflies and spiders[17] have been recorded in its diet. In Tas. it takes a significant amount of seed.[735] In some places the White-browed Scrubwren has disappeared because agriculture has destroyed the native vegetation.[598]

Densities were: at Bathurst (33/149) in wet eucalypt and introduced pine forests 5.0 and 1.75 birds/ha;[489] at Moruya (35/150) in eucalypt woodland 0.39–0.82 birds/ha;[1197] in eucalypt forest at Jamieson (37/146) 0.5–1.0 birds/ha[1469] and Boola Boola (38/146) 0.3–4.4 birds/ha;[1113] at Eyre Bird Observatory in mallee 0.78–3.23 birds/ha;[464] and at Rockingham (32/115) in heath near the coast 1.43–4.83 birds/ha.[17]

Field Atlas Records: 13 509

1° blocks–with records: 202 (25%)
–with breeding records: 85 (42%)

Atlas No.: 488

Max. Rep. Rate: 57%
In block(s): 37/149

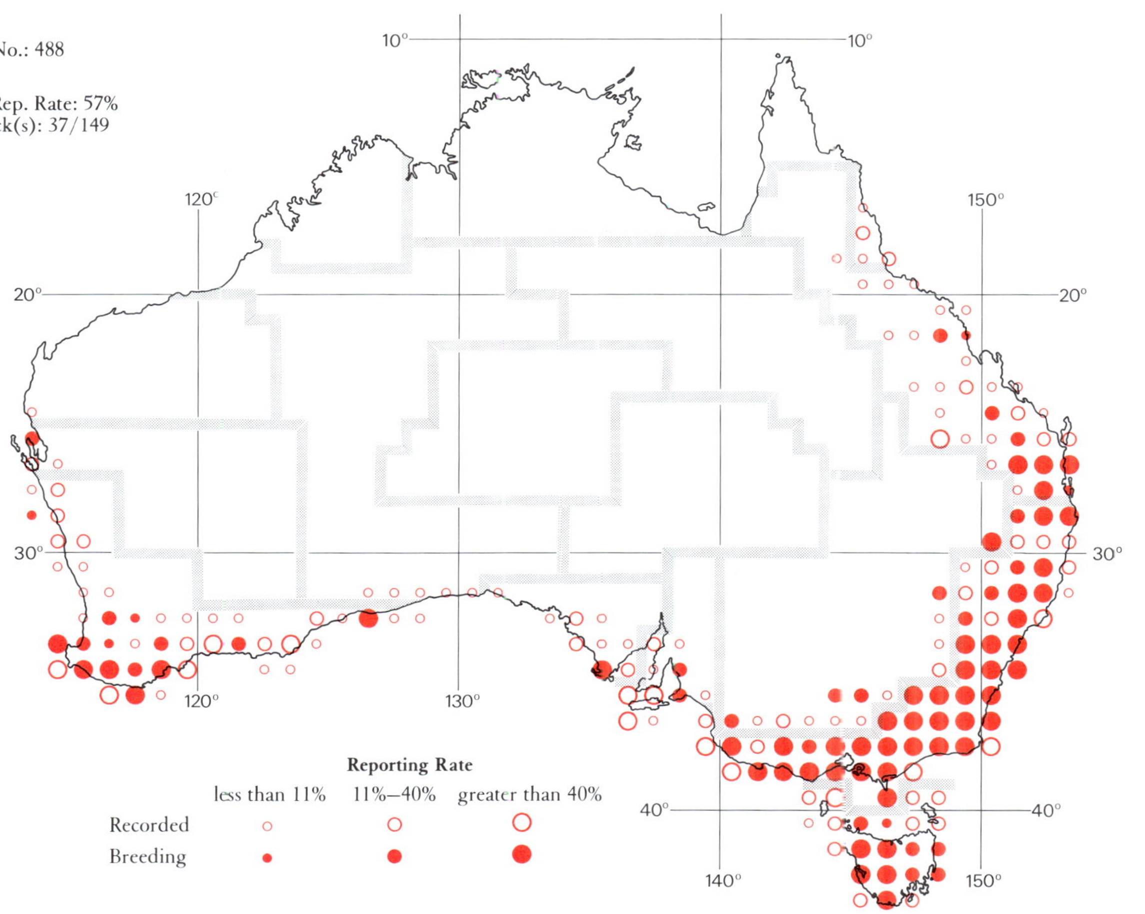

Scrubtit

Sericornis magnus

The Scrubtit is endemic to the Tas. Region, though absent from Flinders I. (40/148).

An early suggestion was made that it lived at higher altitudes in summer than in winter[72] but a study on Mt Wellington (42/147) recorded it at the same altitudes in both seasons though more frequently in winter.[1452]

The Scrubtit is a secretive and little-known species, which inhabits dense wet vegetation: fern gullies, eucalypt forest and rainforest in Tas.[1652,1746] and teatree scrub on King I. (39/144).[734] Usually it is seen singly or in pairs but there are observations in Apr. of groups of four birds and on one occasion of the Scrubtit in a mixed foraging flock with the Tasmanian Thornbill and the White-browed Scrubwren.[1746] The Scrubtit co-exists with these two species and, though less frequently reported than them during the Field Atlas, its wide distribution (reported in 155 different 10′ blocks) suggests it is in no danger of being eliminated by them as has been proposed.[988] The three species forage in different zones of vegetation, the Scrubtit behaving like a treecreeper. In one study it spent 90% of its time above the ground foraging on tree trunks and branches, fern trunks and fronds[1746] and, in another, 70%.[990] It feeds on small crevice-dwelling arthropods.[1049,1746]

There are no measures of abundance.

Field Atlas Records: 428

10′ blocks–with records: 155 (42%)
–with breeding records: 12 (8%)

Atlas No.: 487

Max. Rep. Rate: 65%
In block(s): 42°45′/146°35′

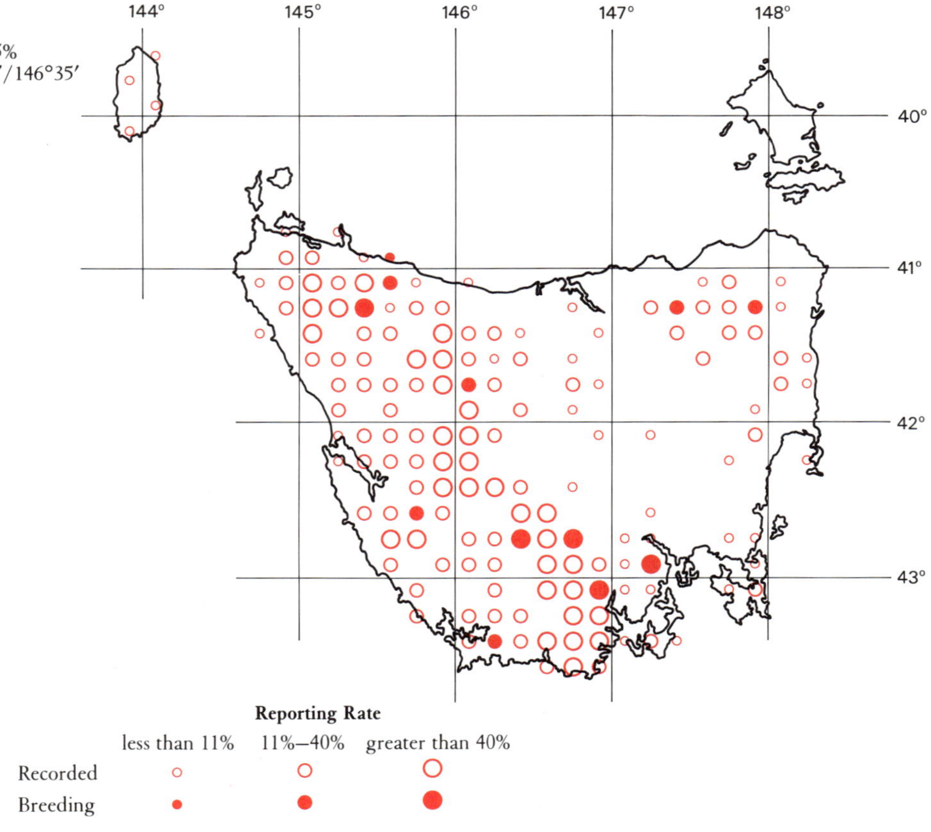

Two subspecies of the endemic Chestnut-rumped Hylacola have been recognized: *pyrrhopygius* near the coast and *magnus* inland.[748] Populations on either side of the Great Dividing Ra. are isolated as is that in the SA Ra. Region. Records at 140°E are linked with those farther east in the South-East Region by records in the Historical Atlas. The Field Atlas does not suggest any large-scale seasonal movement.

The Chestnut-rumped Hylacola lives in small parties which spend most of their time on the ground, generally where the undergrowth is thick.[869,1140] It is a bird of heath and eucalypt woodland and forest. In the Warrumbungle Mts (31/148) it inhabits heath covering rocky outcrops. North of Bendigo (36/144) there is a gap of only 5 km between the ranges of the Chestnut-rumped and Shy Hylacolas but the two live in completely different vegetation: the Chestnut-rumped in eucalypt forest and the Shy in mallee.[1559] The Chestnut-rumped Hylacola is usually absent from country at high altitude but has been recorded at 1500 m.[129] It feeds on small insects and seeds.[1049] Urbanization has displaced it from some localities where it was recorded 1901–50.[355]

There are no measures of abundance.

Chestnut-rumped Hylacola

Sericornis pyrrhopygius

Field Atlas Records: 471

1° blocks–with records: 54 (7%)
–with breeding records: 11 (20%)

Atlas No.: 498

Max. Rep. Rate: 8%
In block(s): 37/150

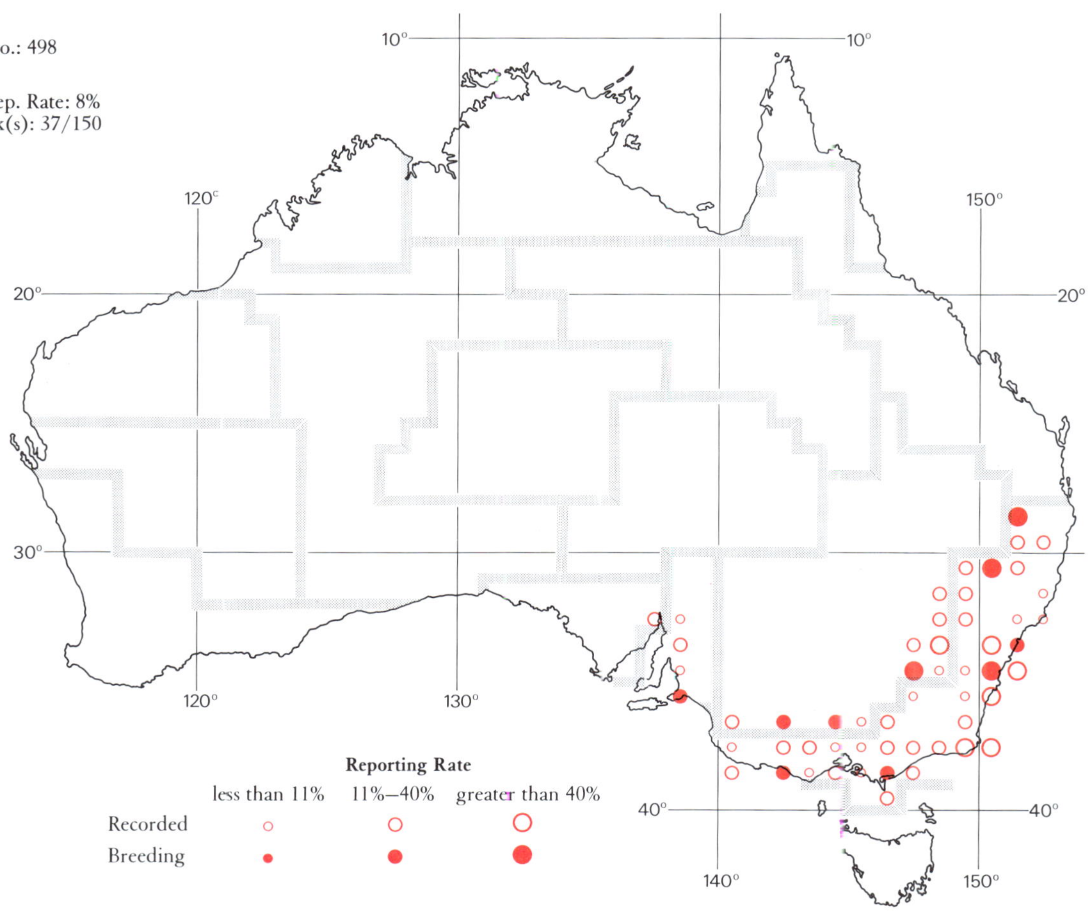

Shy Hylacola

Sericornis cautus

There are different views about subspecies of the endemic Shy Hylacola.[387,1124] The Field Atlas shows that a number of populations in the eastern part of its range are isolated: in and near Round Hill Nature Reserve (33/146); in the Whipstick north of Bendigo (36/144); east and west of the Mt Lofty (34/138) and Flinders Ra; on Kangaroo I. (35/137); and at the foot of Yorke Pen. (34/137). Distribution west along the coast from Eyre Bird Observatory (32/126) appears continuous as far as the Stirling Ra. (34/117) when historical records are included. In 1948 this population was recorded as far north as the Murchison R (27/114).[1603] It is absent from the forested south-western corner.

The Field Atlas does not suggest large-scale seasonal movement. Distribution of the Shy Hylacola follows that of the mallee; where mallee has been cleared the bird has gone too. In the SA Ra. Region the Shy Hylacola is more often present in regrowth of cleared mallee than in undisturbed areas because the cover is denser.[189] In the South-West Region it lives on well drained heaths that become dry in summer.

North of Bendigo there is a gap of only 5 km between the ranges of the Shy and Chestnut-rumped Hylacolas but they live in completely different vegetation: the Shy in mallee and the Chestnut-rumped in eucalypt forest.[1559] The Shy Hylacola has once been recorded in a mixed feeding flock of small passerines[1604] but normally forages alone on the ground, feeding almost entirely on insects.[1049] Rarely, its guts have included seeds.

Within apparently suitable mallee the Hylacola's distribution is not uniform.[523,1873,1874] Sometimes pairs are close together[1725] and others are isolated as near Dumbleyung (33/117) where one pair was at least 3 km from any other.[523]

There are no measures of abundance.

Field Atlas Records: 574

1° blocks–with records: 61 (8%)
–with breeding records: 16 (26%)

Atlas No.: 499

Max. Rep. Rate: 29%
In block(s): 35/141

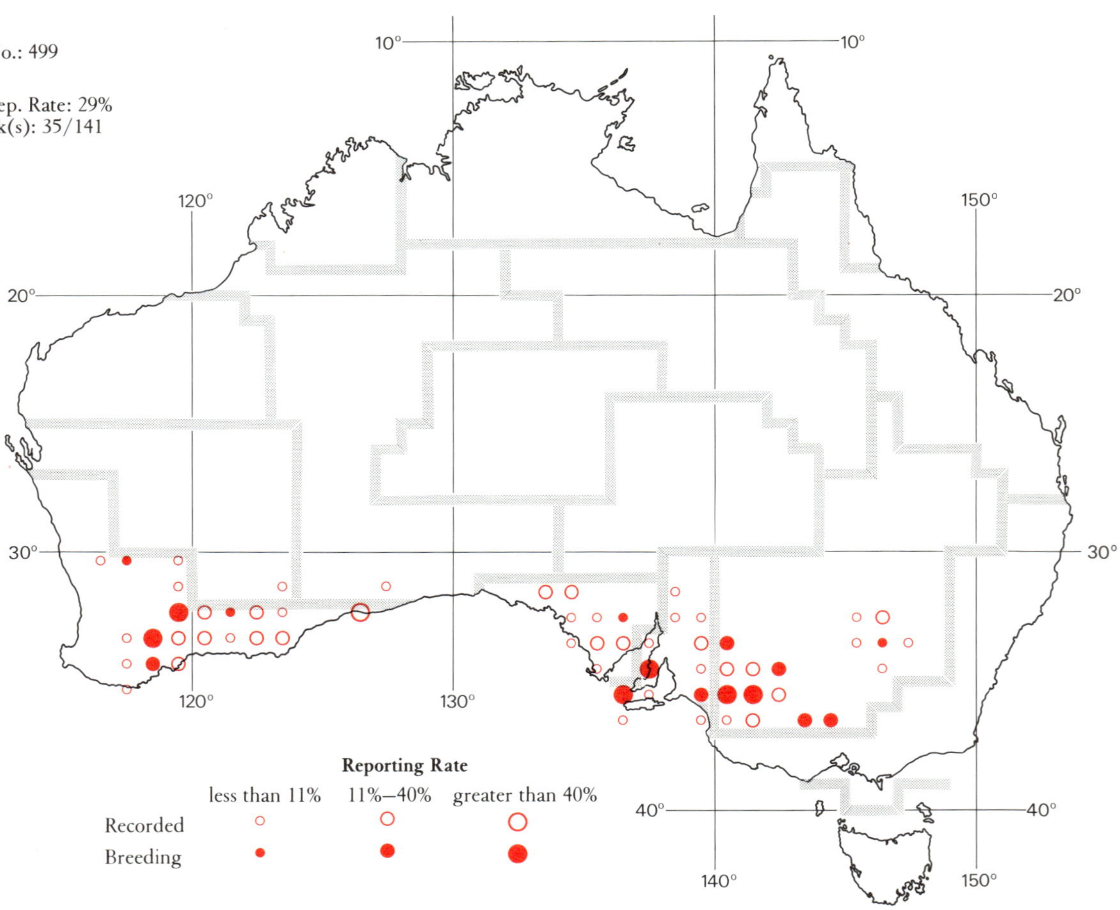

Redthroat

Sericornis brunneus

The endemic Redthroat varies geographically and a subspecific classification has been tentatively proposed.[298,748] Records in the 1951–76 period extend the species' range north to the Granites (20/130) and Barrow Ck (21/133).[626] It used to be more widely distributed: reports before 1950 came from as far east as Tyndarie (31/145)[1343] and south to 36°S in the Murray-Darling and SA Ra. Regions.

The Field Atlas suggests no large-scale seasonal movement and the bird is generally regarded as sedentary.[1140] The Redthroat lives in acacia scrub and saltbush, specially along watercourses, generally where rainfall is less than 250 mm per year. It feeds mainly on insects, including beetles and ants, gathered from the ground and among low foliage, specially saltbush; seeds are also taken.[748,1049] Conversion to farmland of country in the Murray-Darling Region has presumably been responsible for the reduction in the Redthroat's range. Recently the species has gone from parts of the Nullarbor Plain also. It still lives in mallee on the edge of the Nullarbor, usually where undergrowth is thick, but fire and grazing by rabbits have destroyed the vegetation of many dongas.[246] As a sedentary species it could no longer find food and shelter continuously in one place.

There are no measures of abundance.

Field Atlas Records: 824

1° blocks–with records: 148 (18%)
–with breeding records: 20 (14%)

Atlas No.: 497

Max. Rep. Rate: 36%
In block(s): 32/120

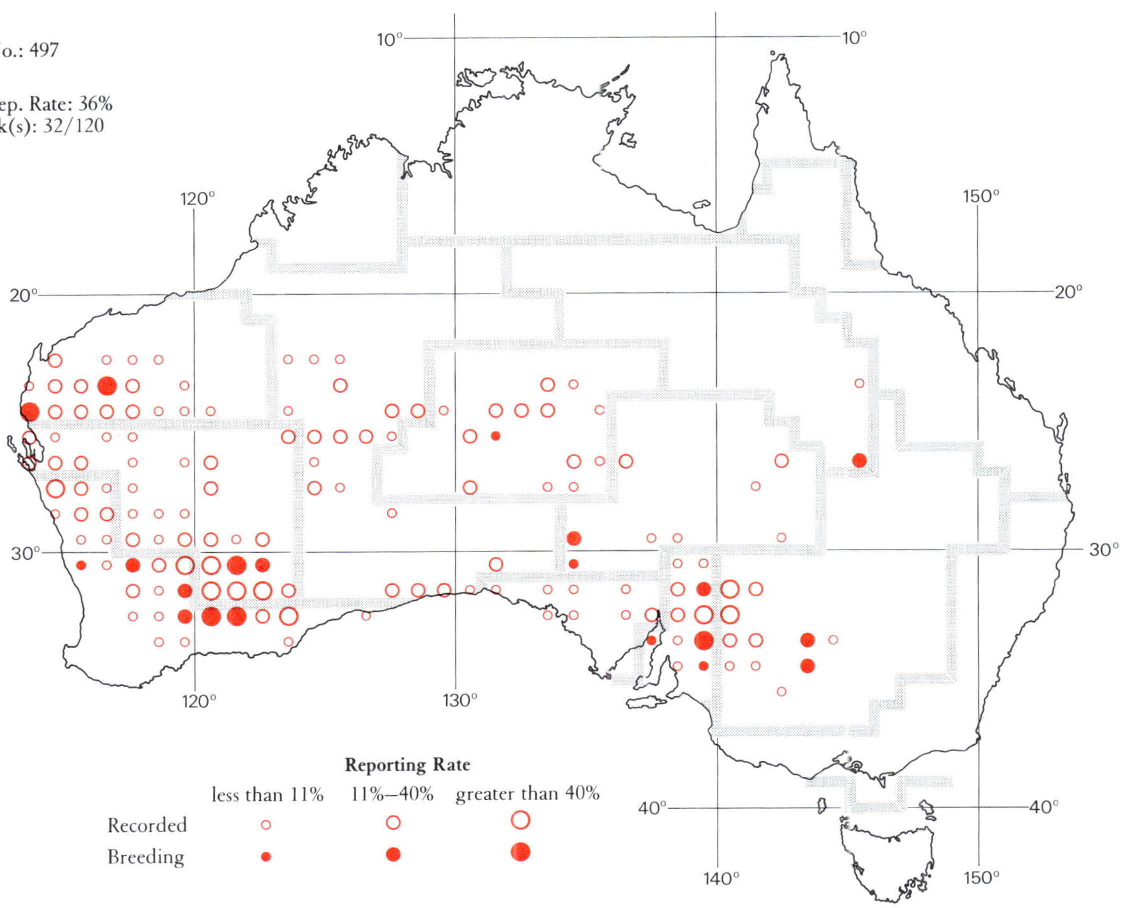

Calamanthus

Sericornis fuliginosus

Two species of the endemic Calamanthus were recognized originally. They were combined at the time when data were collected for the Atlas[1576] but a later review again proposes two species: one restricted to the South-East and Tas. Regions and the other elsewhere.[1369] Calamanthus is absent from King I. (39/144), possibly replaced there by the Golden-headed Cisticola. It is also absent from Flinders I.(40/148) though recorded frequently throughout Tas.

Historical records fill out the inland distribution. They include a number of reports in the 1951–76 period from the Western Deserts between the Central Ra. and Pilbara Regions. The Calamanthus thus appears to be sparsely distributed throughout the Western Deserts, possibly as far north as L. Doman (19/126) where an unconfirmed record was made during the Field Atlas. The gap in distribution where the L. Eyre and Western Deserts Regions meet is maintained in the Historical Atlases.

The Field Atlas does not suggest large-scale seasonal movement.

The Calamanthus appears able to inhabit any low vegetation, wet or dry. It lives in samphire, saltbush and heath where shrubs are well spaced and the canopy is incomplete. Whether trees grow above the low shrubs does not seem to matter. Though recorded on the snow in Tas.[1651] it is absent from altitudes above about 1000 m in the South-East Region. The Calamanthus feeds on insects and seeds.[1049]

At Laverton (37/144) there were at least 0.07 birds/ha restricted to those parts on which dense heath grew.[1822]

Field Atlas Records: 1666

1° blocks–with records: 144 (18%)

–with breeding records: 32 (22%)

Atlas No.: 500

Max. Rep. Rate: 30%
In block(s): 37/150,40/144

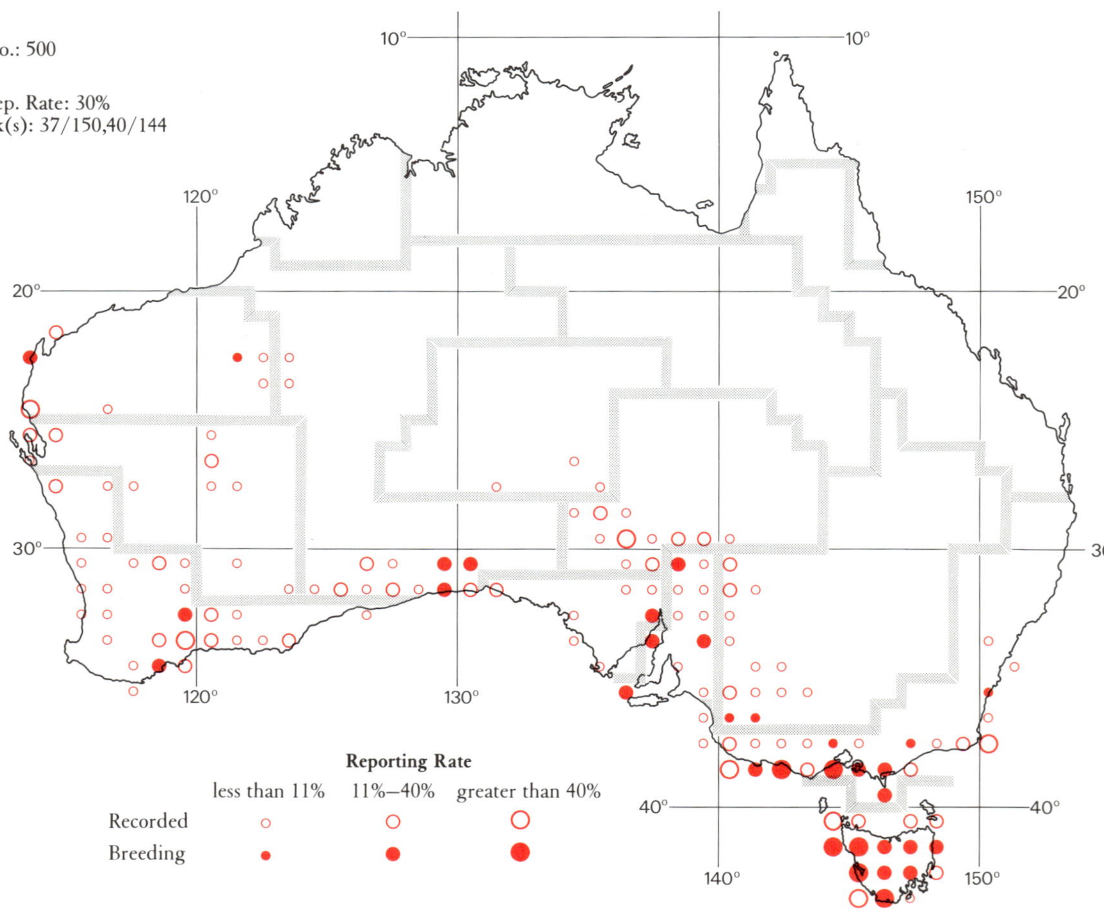

Speckled Warbler

Sericornis sagittatus

There are different views about subspecies of the endemic Speckled Warbler.[748,1709] The only claimed SA records were made before 1900 but there is doubt about their authenticity.[1384]

Banding at Wollomombi (30/152) and Canberra (35/149) has shown that the Speckled Warbler is sedentary.[130,653] It lives in woodland with a grassy understorey, often on rocky ridges or in gullies. It is most frequently reported from the hills and tablelands of the Great Dividing Ra. and rarely from the coast but can live in many vegetation and topographical systems. The species' social organization is not clear. Among a colour-banded population at Wollomombi, one male attended two nests and the female from one nest with her own young feeding independently helped feed the young at a second nest.[129] Speckled Warblers composed 9% of birds in a study of 1300 mixed species flocks mainly of Brown, Striated and Buff-rumped Thornbills.[129] Seeds and insects have been found in guts.[129]

In eucalypt woodland at Armidale (30/151) density was 0.49–0.54 birds/ha.[582] At Wollomombi in eucalypt woodland breeding territories occupied about 10 ha.[130]

Field Atlas Records: 2258

1° blocks–with records: 97 (12%)
–with breeding records: 39 (40%)

Atlas No.: 504

Max. Rep. Rate: 30%
In block(s): 26/150

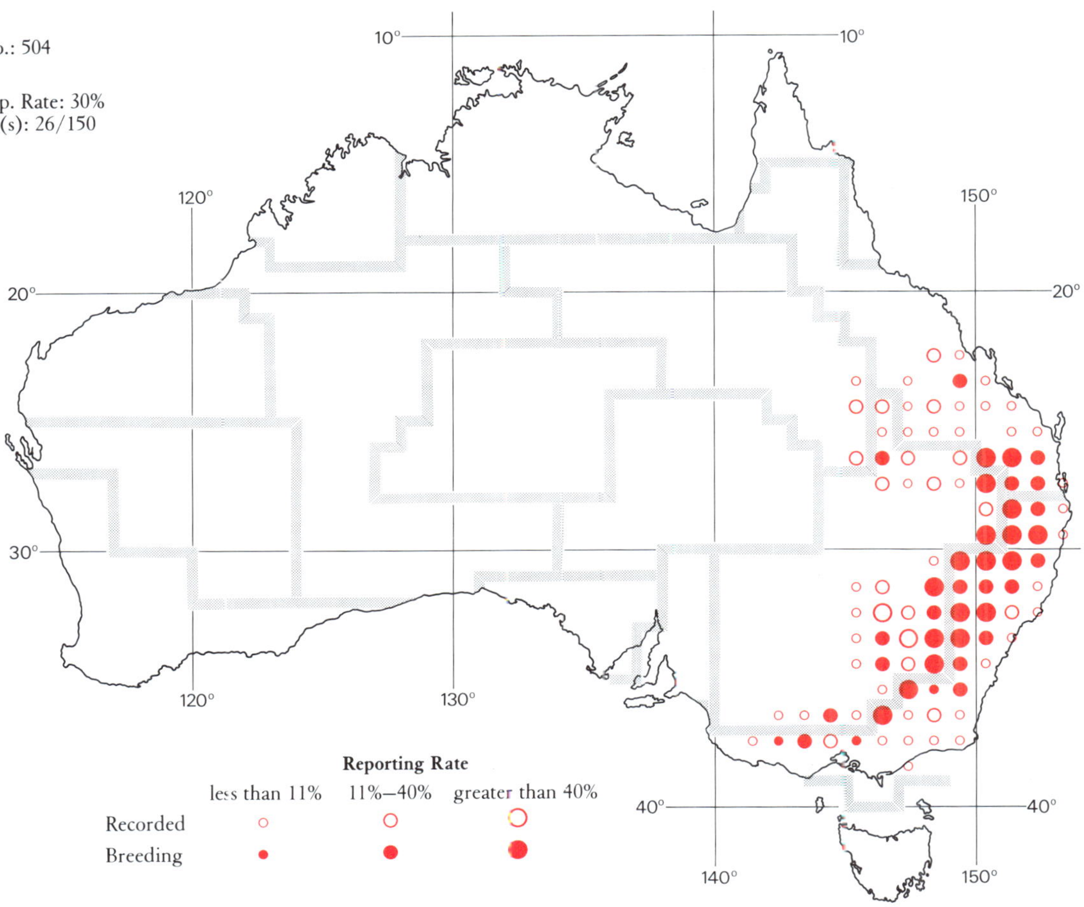

Weebill

Smicrornis brevirostris

Seven subspecies of the endemic Weebill have been recognized.[748] The validity of some of these is doubtful and the Field Atlas does not indicate that any population is isolated. Even birds of the Pilbara, Goldfields and South-West Regions are tenuously connected with eastern populations round the Nullarbor Plain and through the central Western Deserts. There are no records of the Weebill in the Tas. Region in the Atlases.

There is no evidence of seasonal movement. The Weebill lives in parties of up to ten birds[748] in mallee, eucalypt woodland and occasionally forest.[75,488,1693] In the inland its distribution reflects that of eucalypts and in mallee, where the climate is dry, it is particularly abundant. In the southern Murray-Darling Region it is the commonest bird in the mallee but avoids stands of river red gum.[833] There are a few records of communal breeding.[501] The Weebill frequently joins mixed feeding flocks[748] and eats small insects gleaned from the outer foliage of eucalypts.

Densities in Vic. were: 0.62–1.20 birds/ha in mallee and eucalypt woodland near Bendigo (36/144); 0.10–0.37 birds/ha in eucalypt woodland at the You Yangs (37/144); and 0.14–0.61 birds/ha in native pine at Beechworth (36/146).[1907] At Mt Mary Plains (34/139), where the Weebill is said to be the commonest bird, density was 0.8 birds/ha.[189] The species is also abundant at densities of 1.8–3.9 birds/ha at Eyre Bird Observatory (32/126).[464]

Field Atlas Records: 12 193

1° blocks–with records: 585 (72%)
–with breeding records: 121 (21%)

Atlas No.: 465

Max. Rep. Rate: 69%
In block(s): 32/146

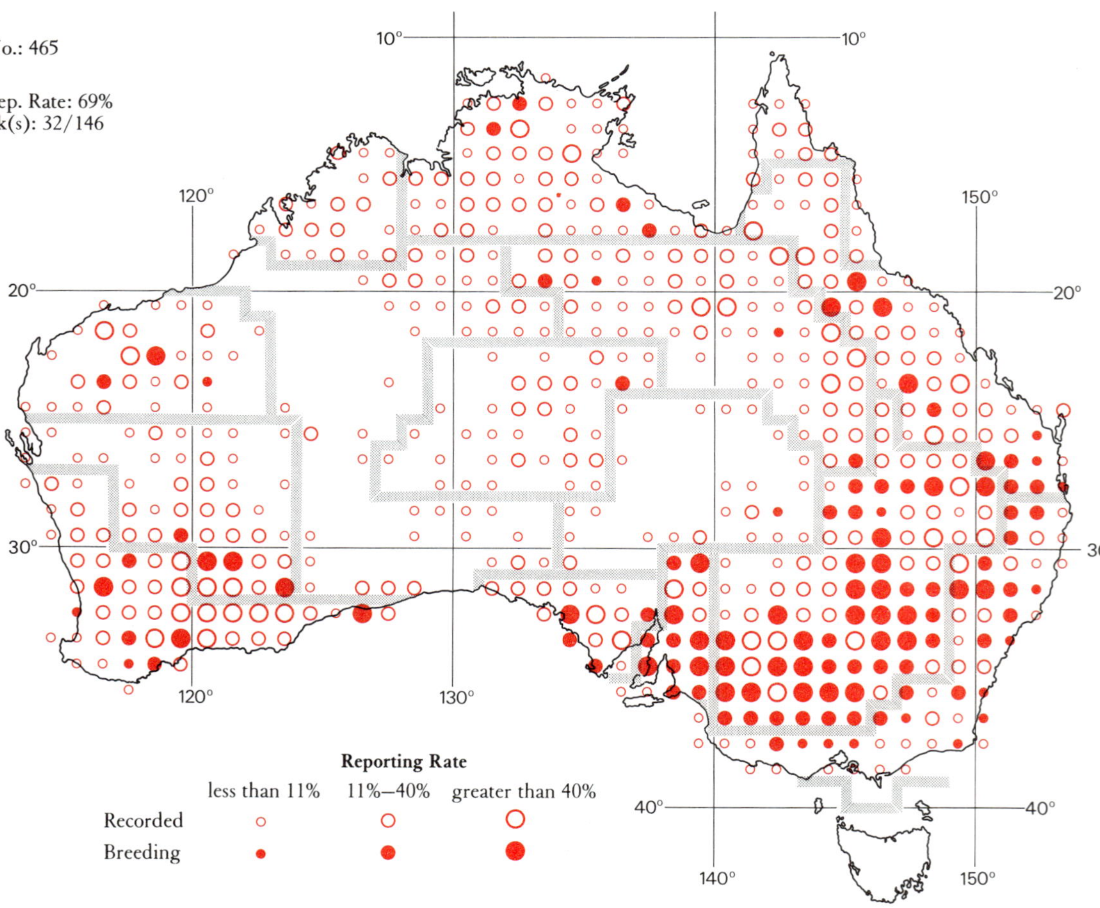

Brown Gerygone

Gerygone mouki

There are three subspecies of the endemic Brown Gerygone: *mouki* in the Atherton Region; *amalia* in the Eungella (21/148) district and southern Qld; and *richmondi* to the south.[748,1369] The population in the Atherton Region is isolated as is *amalia* in the Eungella district. Breeding records in the Historical Atlases are within the range determined by the Field Atlas with the exception of an early report from the Fitzroy R (23/150)[299] based only on a nest and open to doubt.[1369]

The Brown Gerygone can move long distances, e.g. it has colonized the botanic gardens of both Sydney (33/151) and Canberra (35/149).[129] Other reported movements are of local wandering in the Atherton Region;[231] movement by some birds near Sydney into more open country in winter[816] though another observer regarded the species as sedentary there;[323] and, in Vic., local movement in winter from rainforest to wet gullies in eucalypt forest.[1110]

The Brown Gerygone lives in rainforest in the Atherton Region[698] and farther south in rainforest along watercourses,[323] eucalypt forest where it grows in gullies and mangroves.[1079] The species has once been reported breeding communally.[501] Small groups of 3–10 Brown Gerygones will feed together during the non-breeding season and the species often feeds in company with other small, insectivorous birds. It takes insects from the outer foliage, hunting at all levels from the undergrowth to the canopy.[748]

At Moruya (35/150), in eucalypt woodland, density was 0.33 birds/ha[1197] and in rainforest at Glenaladale (37/147) 0.19–1.20 birds/ha.[1907]

Field Atlas Records: 1830

1° blocks–with records: 55 (7%)
–with breeding records: 32 (58%)

Atlas No.: 454

Max. Rep. Rate: 30%
In block(s): 15/145,31/152

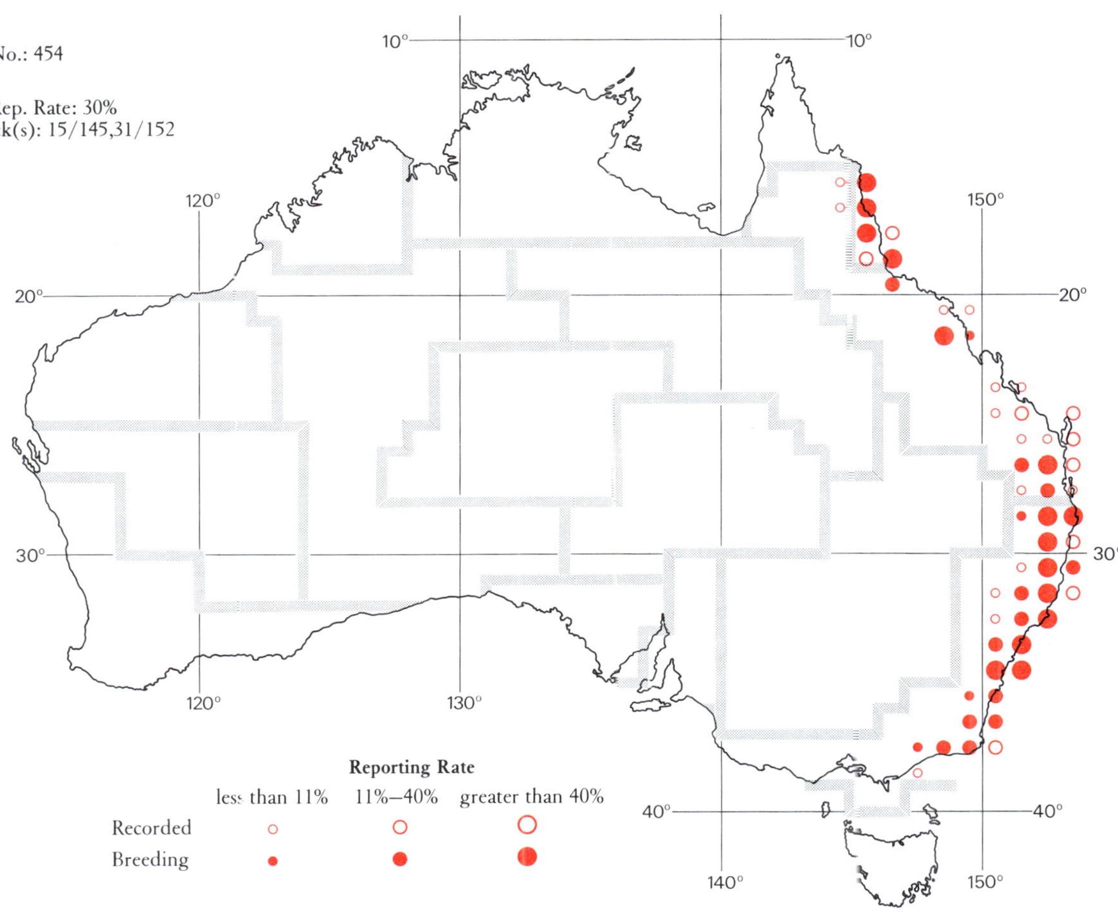

Large-billed Gerygone

Gerygone magnirostris

The Large-billed Gerygone lives in New Guinea and adjacent islands and in Australia. There are different views about subspecies in Australia.[748,1709,1711,1713] The separation into eastern, Top End and Kimberley populations shown in the Field Atlas is consistent in the Historical Atlas. It has been suggested that scattered populations round the coast of the Gulf of Carpentaria link birds in the Cape York and Top End Regions[1585] but the only record from here is a clutch of eggs reputed to have been collected on the Nicholson R (17/139) early this century.[1709]

The Large-billed Gerygone was once grouped with the Dusky. The ranges of the two species overlap at King Sound (17/123).

The Large-billed Gerygone is present throughout the year on islands in Torres Strait,[503] round Innisfail (17/146)[698] and Darwin (12/130).[423] Usually it lives near the coast in mangroves and other wet vegetation. It also lives in paperbark woodland and rainforest, including that along rivers though not at high altitude.[423,698,748] In New Guinea it is often in parties and leads mixed species flocks.[158] The Large-billed Gerygone feeds on insects, which it takes from the leaves of trees, mainly in the canopy.

There are no measures of abundance.

Field Atlas Records: 436

1° blocks–with records: 42 (5%)
–with breeding records: 14 (33%)

Atlas No.: 457

Max. Rep. Rate: 29%
In block(s): 12/143

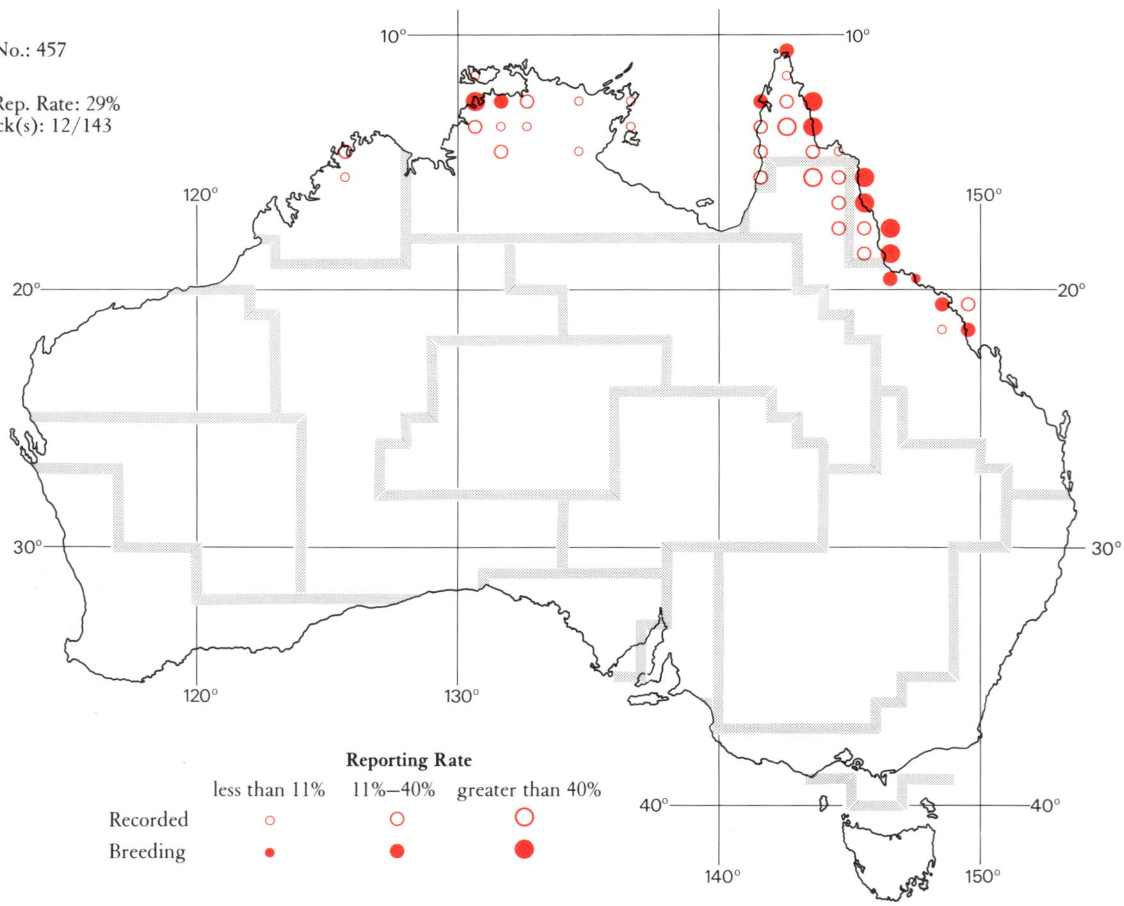

Dusky Gerygone

Gerygone tenebrosa

The Dusky Gerygone is endemic to WA and ranges from King Sound (17/123) south to Bush Bay (25/113). In King Sound it overlaps the similar Large-billed Gerygone. The two species were treated as one for many years.[940] The Field Atlas shows that at least three populations of the Dusky Gerygone exist, separated at the Eighty Mile Beach (19/120) and at the break in mangroves between Pt Cloates (22/113) and Carnarvon (24/113). Three corresponding subspecies have been described, namely *tenebrosa*, *whitlocki* and *christopheri* in order from north to south.[940]

The Dusky Gerygone is probably sedentary and lives singly, in pairs or in small parties. One group of five was observed feeding fairly close together in mangrove foliage. The Dusky Gerygone is confined to mangroves and observations during the Field Atlas suggested that it is more abundant on the seaward side of mangrove belts. The bird feeds on small insects gleaned from the leaves of the canopy.[748] It has been recorded feeding on the ground: several birds were pecking among dead mangrove leaves at Port Hedland (20/118).

One observer recorded at least six Dusky Gerygones in 1 km of mangroves at Bush Bay. There are no other measures of abundance.

Field Atlas Records: 43

1° blocks–with records: 10 (1%)
–with breeding records: 2 (20%)

Atlas No.: 461

Max. Rep. Rate: 11%
In block(s): 20/117

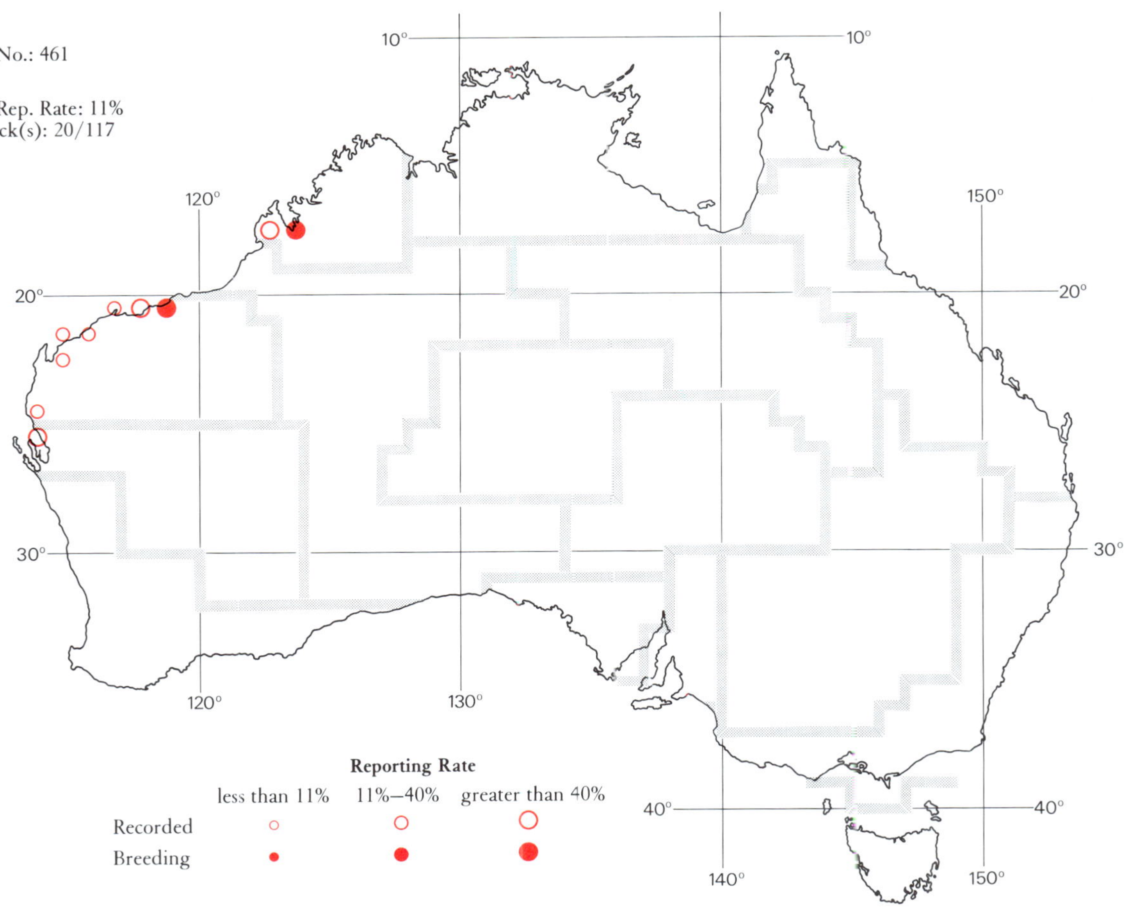

Mangrove Gerygone

Gerygone laevigaster

The Mangrove Gerygone lives in southern New Guinea and Australia. In Australia it is separated into two subspecies: *cantator* on the eastern coast and *laevigaster* westward from the Gulf coast of the Cape York Region.[618] The range of *cantator* is expanding southward. The first record south of the Tweed R (28/153) was at Nambucca (30/153) in 1944.[796] By 1952 the species was south of 31°S; by 1966 it was south of 32°S; and in 1978 it reached L. Macquarie (33/151).

A seasonal pattern of movement has been evident at the expanding edge of the range of *cantator*. The bird was regarded as a spring-summer visitor to the Nambucca area in 1944[796] and to Kooragang I. (32/151) in the 1970s. The six Field Atlas records from L. Macquarie were all spring-summer.

The Field Atlas provides no other suggestion of large-scale seasonal movement and the northern subspecies *laevigaster* is regarded as sedentary.[618]

The Mangrove Gerygone usually lives in mangroves but in Brisbane (27/153) it comes into parks and gardens near the river[342] and at Darwin (12/130) it lives in rainforest bordering mangroves.[423] There is also an observation of it in eucalypt woodland.[748] South of Cape Leveque (16/122) it extends slightly inland and lives in paperbark woodland and acacia scrub. At this locality the Dusky Gerygone is abundant in the mangroves and possibly competes successfully with the Mangrove Gerygone.[618] The guts that have been examined contained insects, including beetles and white grubs.[748]

There are no measures of abundance.

Field Atlas Records: 628

1° blocks-with records: 60 (7%)
-with breeding records: 9 (15%)

Atlas No.: 460

Max. Rep. Rate: 26%
In block(s): 27/153

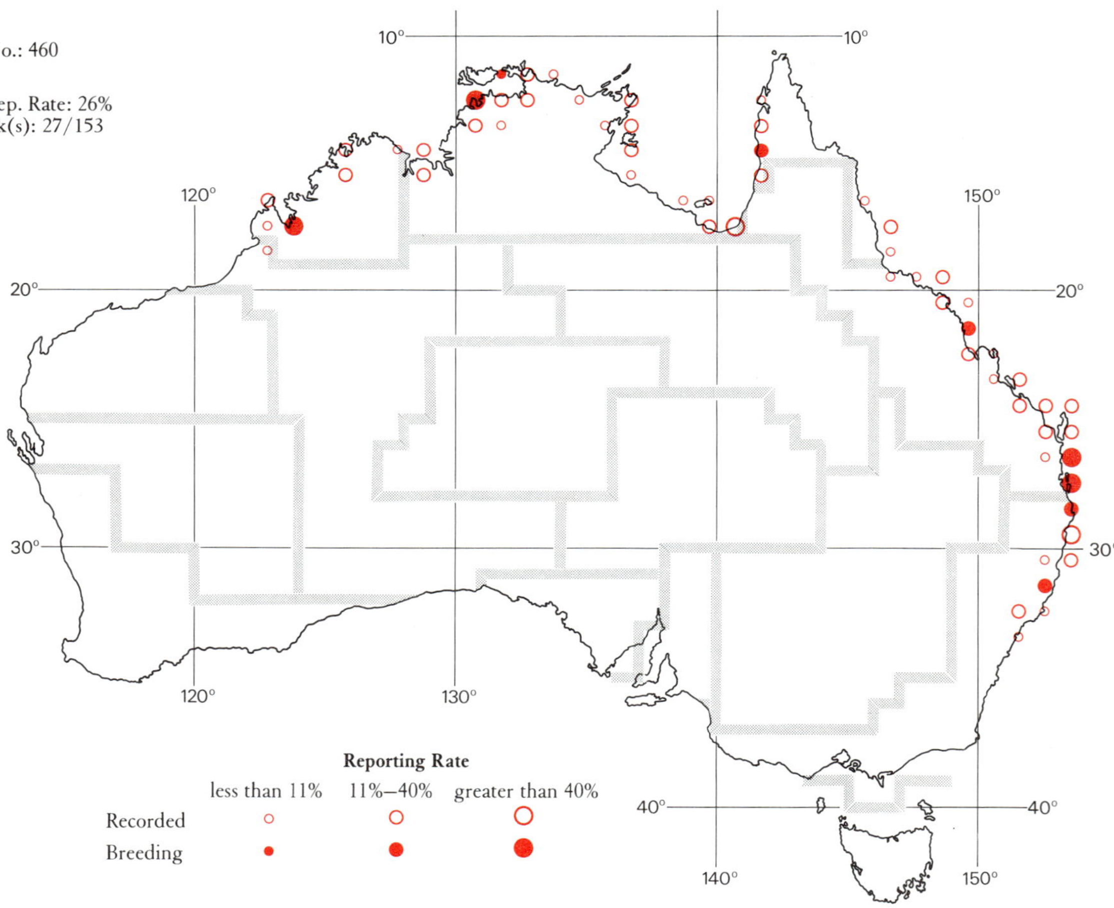

Western Gerygone

Gerygone fusca

Two subspecies of the endemic Western Gerygone are recognized: *mungi* breeding from the base of the Cape York Region through the Central Ra. Region and across to the Pilbara; and *fusca* breeding in the South-West Eyre Pen. (34/135) and throughout much of inland eastern Australia.[618] The Historical Atlases contain records from the Western Deserts linking the Central Ra. and Pilbara populations. The species has colonized Rottnest I. (32/115) since 1950.[1706]

In the eastern population of *fusca*, three types of movement have been suggested: altitudinal; inland dispersal in the spring of wet years; and northward movement in winter.[618] The Field Atlas shows no evidence of any large shift in this population and records were made throughout the species' range in all seasons. The population on Eyre Pen. is sedentary.[1466] Part of the western population of *fusca* migrates in winter north as far as the Kimberley Region and east to the Nullarbor, retreating to the South-West Region to breed in spring-summer.[618] The Field Atlas supports the idea that this movement is undertaken by only part of the population because some records were made throughout the South-West in winter. A winter movement eastward is consistent with the status of the species as a rare winter visitor to Eyre Bird Observatory (32/126).

The population of *mungi* in the NT may also move northward in winter[618] but there are too few summer records in the Field Atlas to be useful.

The Western Gerygone lives in eucalypt forest and woodland, mallee and acacia scrub from the coast inland. One of the few consistent aspects of its pattern of distribution is that it avoids wet areas, both temperate and tropical. In NSW it is absent above 850 m.[396] The Western Gerygone is insectivorous, seeking its prey in the outer foliage of almost any kind of tree or shrub. In some parts clearing is considered to have caused a decrease, e.g. in the Avon Valley (32/116),[1216] but in other places an increase, e.g. Canberra (35/149).[618]

There are no measures of abundance.

Field Atlas Records: 3687

1° blocks–with records: 255 (31%)
–with breeding records: 49 (19%)

Atlas No.: 463

Max. Rep. Rate: 48%
In block(s): 31/116

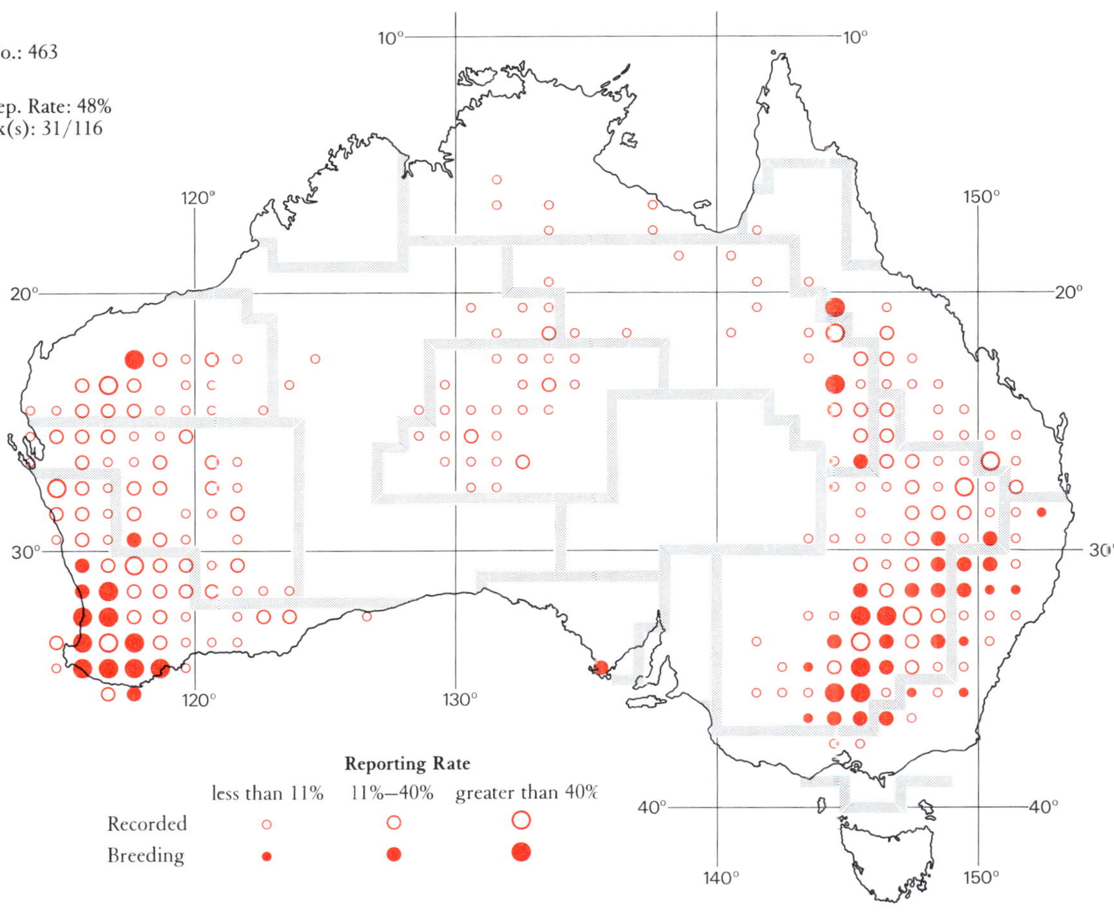

Green-backed Gerygone

Gerygone chloronota

The Green-backed Gerygone lives in New Guinea and adjacent islands and northern Australia. In conjunction with the Historical Atlas, the population appears to be continuous in the Top End and Kimberley Regions. It also inhabits Melville I. (11/130) and Groote Eylandt (13/136). Records in the 1951–76 period extend the species' range in the Kimberley south to Mt House (17/125) and Windjana Gorge (17/124).[587]

The Field Atlas shows no evidence of large-scale seasonal movement. The Green-backed Gerygone is mainly confined to wet areas in eucalypt forest; it also inhabits mangroves, bamboo and paperbark swamps.[587,970,1257] In the Top End the species ranges inland where large rivers maintain fringing forests. An inland record on Manning Ck (16/125) in the Kimberley Region was made in eucalypt woodland in a sandstone gorge at an altitude of 600 m.[748] The Green-backed Gerygone feeds on insects.[748]

There are no measures of abundance.

Field Atlas Records: 184

1° blocks-with records: 23 (3%)
-with breeding records: 4 (17%)

Atlas No.: 458

Max. Rep. Rate: 27%
In block(s): 12/130

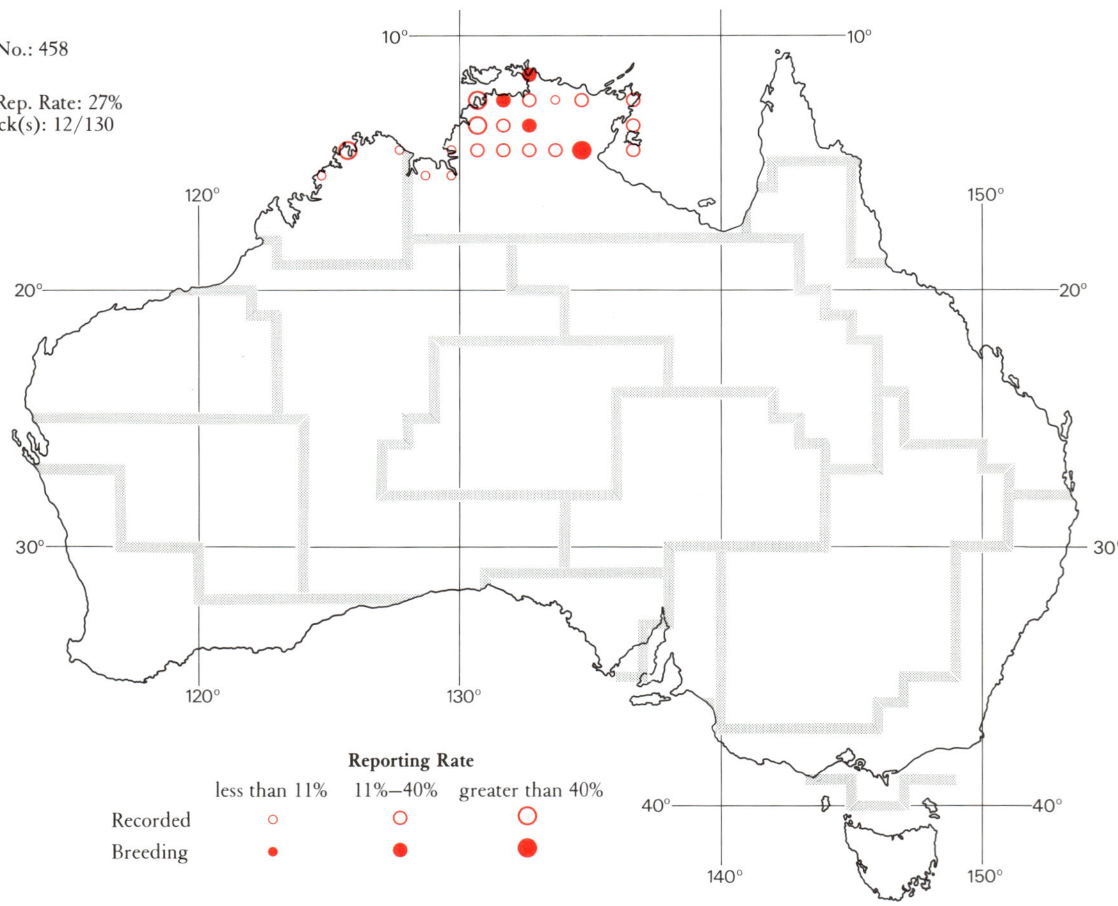

Fairy Gerygone

Gerygone palpebrosa

The Fairy Gerygone in Australia has been shown to be part of a clinally varying population that runs from New Guinea into eastern Australia. It may tentatively be broken into two subspecies in Australia: *personata* in the Cape York Region and *flavida* to the south.[610] There is only one record from islands in Torres Strait.[503] On the coast of the Gulf of Carpentaria the southernmost record is from Inkerman (16/141).[748]

There is no evidence of seasonal movement. In New Guinea the Fairy Gerygone is a bird of the canopy in rainforest and in Australia it lives in rainforest, including edges and that along streams.[610] The species is also common in mangroves, specially on the shore side, where it will penetrate short distances into rainforest. Near Cooktown (15/145) it forages in gardens, specially on mango trees.[1703] In the non-breeding season the Fairy Gerygone lives in family parties or in mixed flocks with other species.[160,748,1145] In the breeding season it is territorial and lives in pairs. It seeks insects among the leaves and twigs of the forest canopy.[1145]

The density measured in lowland rainforest near Port Moresby (9S/147E) in New Guinea was 0.1 birds/ha[160] but this is almost certainly an underestimate.[129]

Field Atlas Records: 460

1° blocks–with records: 39 (5%)
–with breeding records: 13 (33%)

Atlas No.: 456

Max. Rep. Rate: 49%
In block(s): 12/143

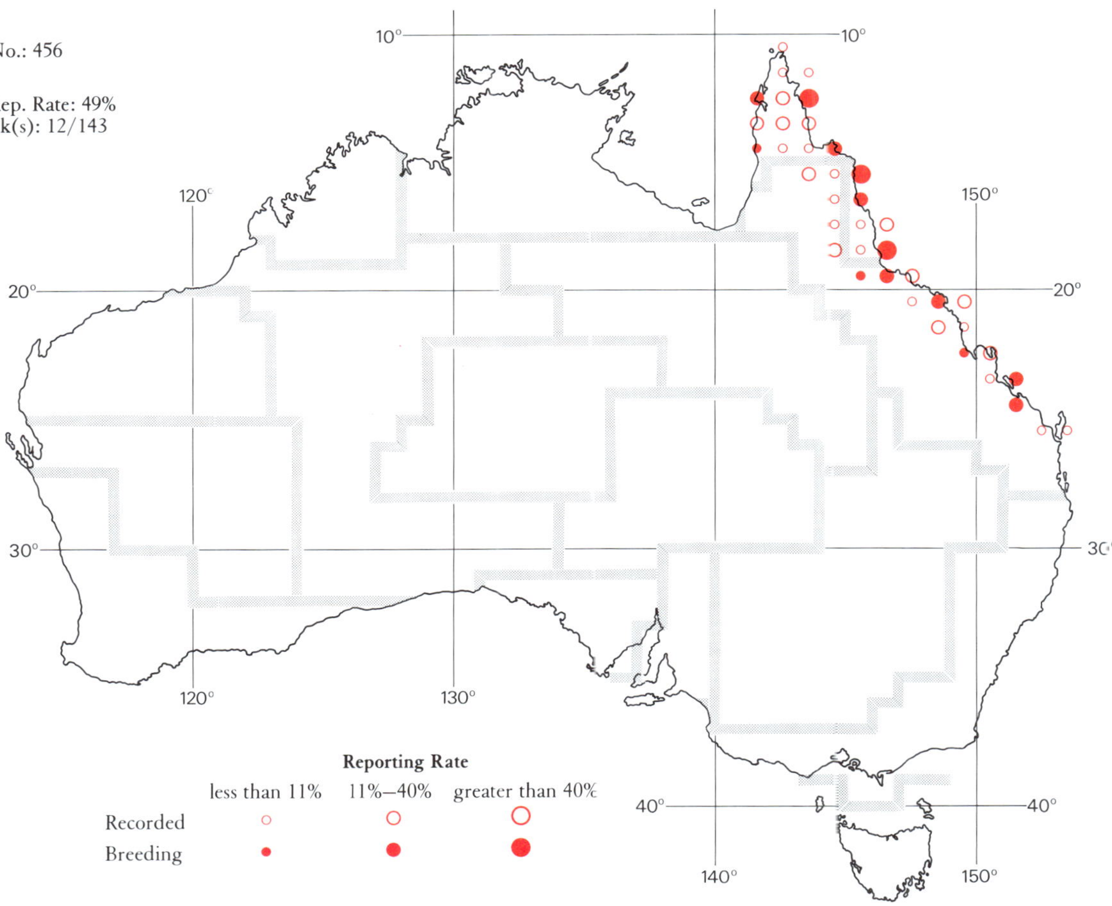

White-throated Gerygone

Gerygone olivacea

The White-throated Gerygone lives in south-eastern New Guinea and Australia. Two morphologically distinct populations in Australia intergrade near the Leichhardt R (140°E) south of the Gulf of Carpentaria. They are *olivacea* in the east and *rogersi* west to the Kimberley Region.[611]

The pattern of movement of the two subspecies differs: *olivacea* is migratory in southern parts of its range. It is present at Wollomombi (30/152) Sept.-Apr.;[129] and at Jamieson (37/146) Oct.-Mar.[1469] There were no Field Atlas reports west of Melbourne (37/145) in winter and the Reporting Rates reflect the exodus from the South-East Region (summer 11%, winter 2%). Correspondingly the Reporting Rate in Eastern Qld showed a small increase (summer 12%, winter 17%). There is no evidence of large-scale seasonal movement by the western subspecies *rogersi*.

The White-throated Gerygone is usually in pairs and occasionally in small groups during the non-breeding season. It is a bird of eucalypt woodland and forest but seems able to live in farmland, including gardens, where a few clumps of trees remain. Some records during the Field Atlas were made in mallee and woodland dominated by teatree or native pine. In southern parts of its range the species is absent from high altitudes. The White-throated Gerygone feeds on insects from the leaves of the canopy. In the few guts analysed, plant bugs were a common item.[1049] Ants and other insects are also recorded.[748]

In eucalypt woodland in the New England (30/152) district one pair occupied 4 ha (0.5 birds/ha).[96] Density at Armidale (30/151) in eucalypt woodland was 0.46 birds/ha;[582] at Wollomombi in eucalypt woodland 0.4–0.9 birds/ha;[129] and at Jamieson, in eucalypt forest, 0.31 birds/ha.[1469]

Field Atlas Records: 4195

1° blocks–with records: 215 (26%)
–with breeding records: 48 (22%)

Atlas No.: 453

Max. Rep. Rate: 45%
In block(s): 26/150

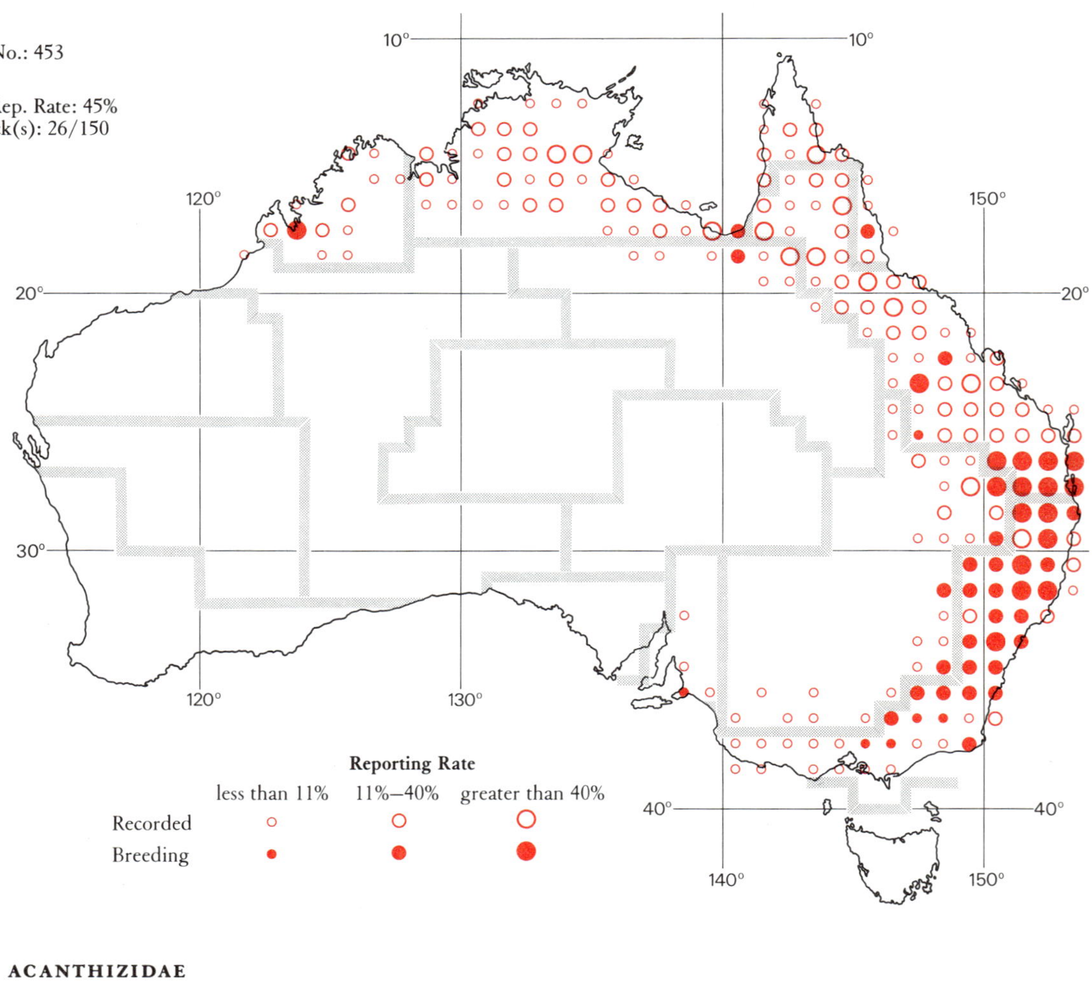

Mountain Thornbill

Acanthiza katherina

The Mountain Thornbill is endemic to the Atherton Region. Previously regarded as a subspecies of the Brown Thornbill, it was separated as a full species in 1969.[1172] It lives in small parties confined to rainforest above 450 m.[989] The Mountain Thornbill gleans invertebrates from the leaves of canopy trees 6.5–13.0 m high. Most (66%) of its feeding actions are made in the outer foliage of the canopy.[989]

The species appears to have been isolated in the Atherton Region by its need for a wet environment and inability to tolerate the increasing aridity of surrounding lowlands.[989]

There are no measures of abundance.

Field Atlas Records: 148

1° blocks–with records: 6 (1%)
–with breeding records: 2 (33%)

Atlas No.: 474

Max. Rep. Rate: 12%
In block(s): 15/145,18/146

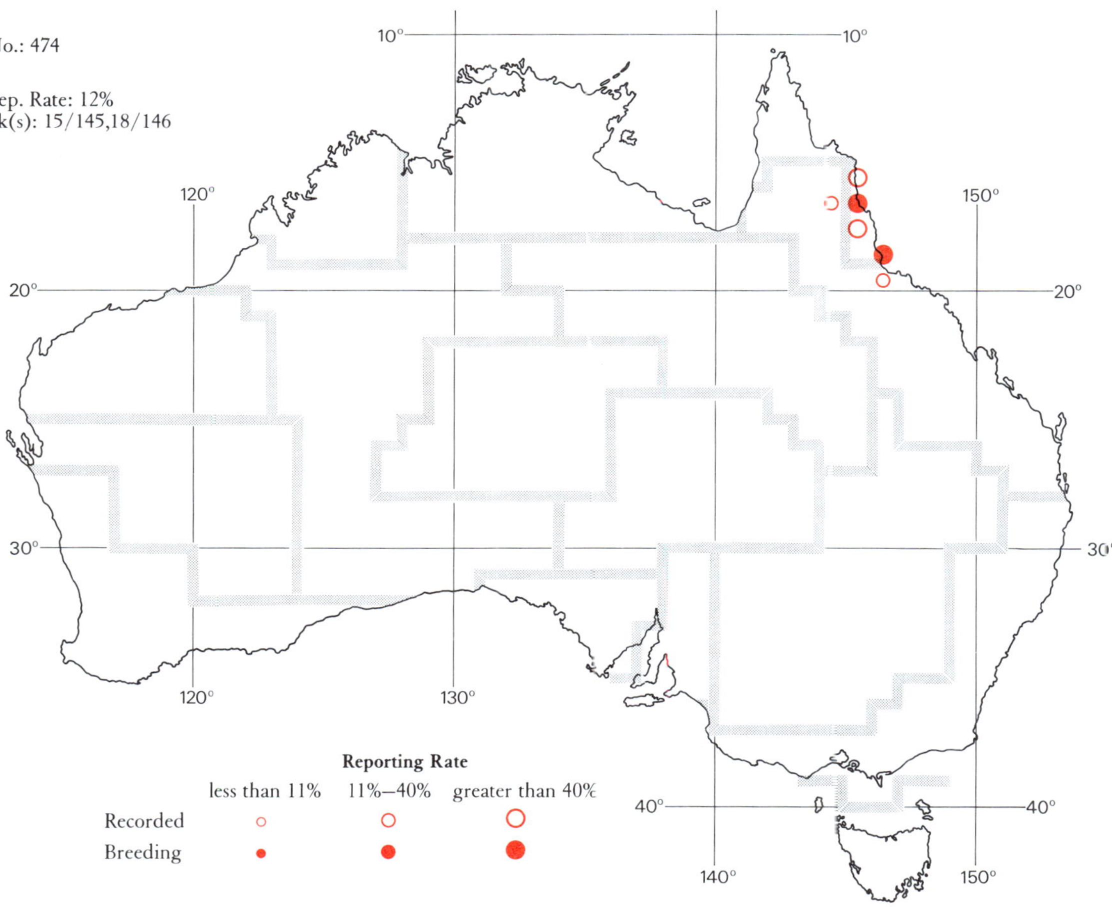

Brown Thornbill

Acanthiza pusilla

Five subspecies of the endemic Brown Thornbill are recognized: *pusilla* on the mainland except in the Eungella (21/148) district, which has a distinctive isolated population *mcgilli*; *diemenensis* in Tas.; *archbaldi* on King I. (39/144); and *zietzi* on Kangaroo I. (35/137).[202] The Brown Thornbill may be confused in the field with the Inland Thornbill: a zone of possible hybridization has been identified in the western part of the South-East Region[202] and sightings assigned to both species are reported along the inland side of the Great Dividing Ra. In the Tas. Region there may be confusion between the Brown and Tasmanian Thornbills. The Brown is virtually absent from south-western Tas. apart from the coastal belt, has never been recorded from islands in the Furneaux Group (39/148) and is rare on King I.

Near Armidale (30/151) pairs hold a territory all year and only the young leave it.[159] Banding in the ACT shows similarly that juveniles may undertake small movements but adults are sedentary.[653] Farther south, formation of flocks in the non-breeding season suggests at least local movement: flocks of up to ten are reported at Rotamah I. Bird Observatory (37/147)[284] and large flocks form in Tas. in autumn.[1314]

The Brown Thornbill inhabits forest with dense undergrowth, pine plantations, mangroves, saltmarsh and gardens.[486,792,1110] Where its range overlaps that of the Tasmanian Thornbill, the Brown lives in drier places. It forages preferentially in narrow-leaved shrubs and hence generally in the understorey of eucalypt forest but often high in the canopy in pines.[1110] It joins mixed species flocks[162] and feeds on insects and spiders[1049,1531] and also blackberries.[1183]

At Wollomombi (30/152) pairs held territories of 0.25–4.00 ha (smallest where the understorey was most dense).[162] Densities at Bathurst (33/149) in wet eucalypt, dry eucalypt and pine forest were 2.25, 0.75 and 0.50 birds/ha.[489] In eucalypt woodland densities were: Armidale 0.31 birds/ha,[582] Canberra (35/149) 0.3–1.4 birds/ha[154] and Moruya (35/150) 1.41–1.58 birds/ha;[1197] in eucalypt forest: Jamieson (37/146) 0.5–1.0 birds/ha,[1469] Tynong (38/145) 0.91–1.78 birds/ha[1907] and Boola Boola (38/146) 1.2–8.4 birds/ha;[1113] and in rainforest: Donna Buang (37/145) 0.91–1.34 birds/ha and Glenaladale (37/147) 0.38–1.23 birds/ha.[1907]

Field Atlas Records: 12 741

1° blocks–with records: 118 (15%)
–with breeding records: 64 (54%)

Atlas No.: 475

Max. Rep. Rate: 60%
In block(s): 35/150,37/146,37/150

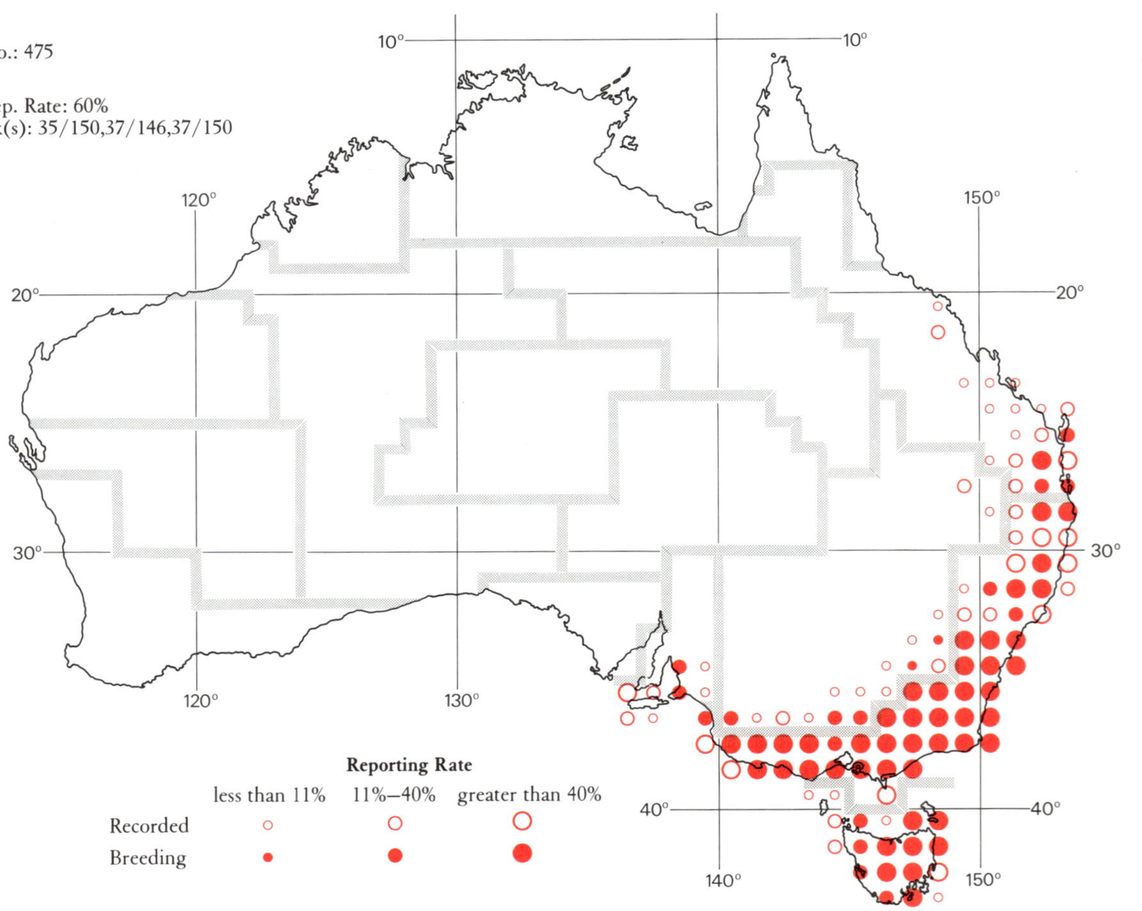

Inland Thornbill

Acanthiza apicalis

Four subspecies of the endemic Inland Thornbill have been described but their ranges have not been clearly defined[748,1124] and none is isolated in distribution. A small population in the Whipstick (36/144) is cut off from the main range of the species. There are historical records in the central Western Deserts Region where few were made during the Field Atlas.

The Inland Thornbill may be confused with the Brown in the field. Sightings assigned to both species are reported from a zone along the inland side of the Great Dividing Ra., e.g. in the Little Desert (36/141) the boundary seems to be at about 36°30′S with no reliable reports of the Inland Thornbill to the south or the Brown to the north. In 10′ blocks on the line 'brown' thornbills were identified as both species and twice as 'intermediates'.

At Wyperfeld (35/141) and Bendigo (36/144) the Inland Thornbill appeared to be present all year[1907] but it was thought not to be so in central Australia.[1369]

The Inland Thornbill lives singly or in pairs, rarely in small parties. Eucalypt woodland, mallee, acacia scrub and heath are occupied, specially near water. The Inland Thornbill forages in shrubs and the foliage of trees, gleaning small insects and spiders. It takes few seeds and little vegetable matter.[1049] The contents of only one of five guts examined included material other than insects.

Densities in mallee and woodland at Wyperfeld were 0.1–0.4 birds/ha and at Bendigo 0.02–0.08 birds/ha.[1907] At Eyre Bird Observatory (32/126), in mallee, density was 2.64–5.48 birds/ha along a transect but 0.33 birds/ha in a mapped site of 9 ha.[464] Three active nests were found in a 32-ha study site south of Perth (31/115).[1429]

Field Atlas Records: 4598

1° blocks-with records: 326 (40%)
-with breeding records: 82 (25%)

Atlas No.: 476

Max. Rep. Rate: 58%
In block(s): 32/120

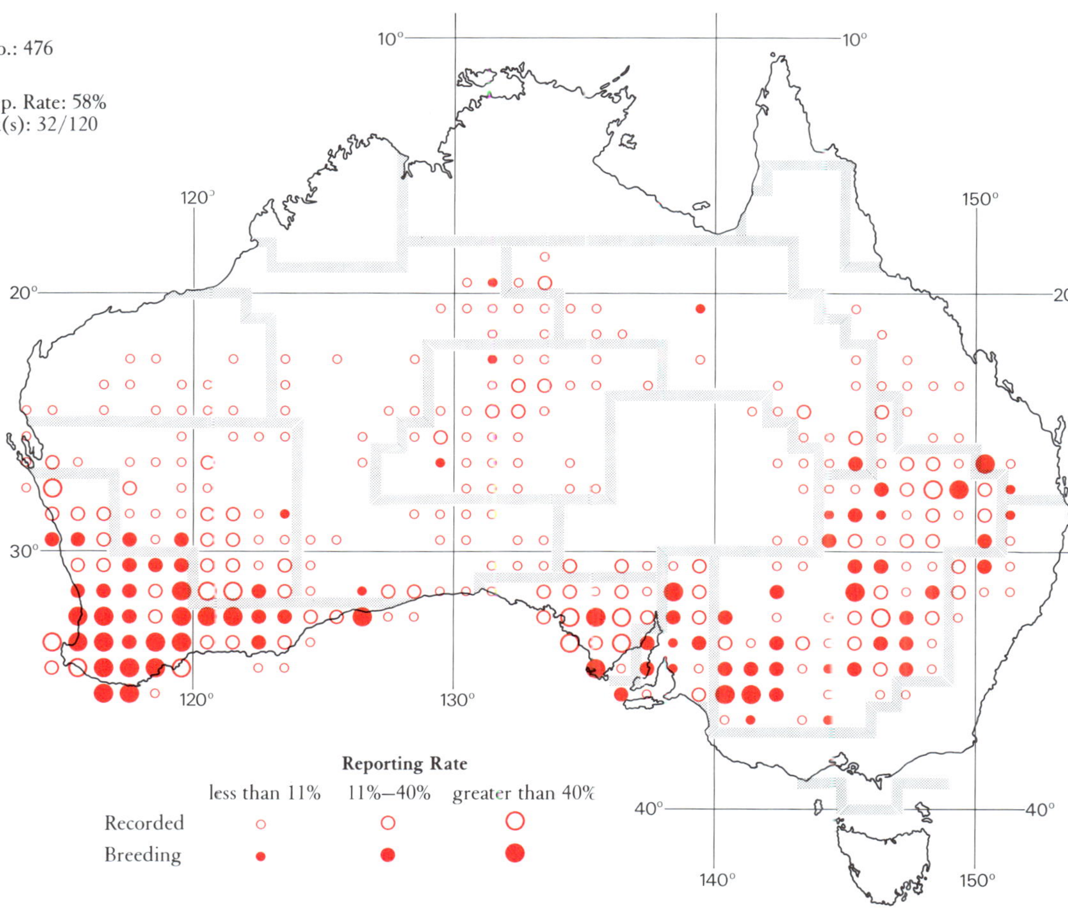

Tasmanian Thornbill

Acanthiza ewingii

The Tasmanian Thornbill is endemic to the Tas. Region. In the field it is difficult to distinguish from the Brown Thornbill.

There is no evidence of large-scale seasonal movement.

Generally the Tasmanian Thornbill inhabits wetter areas than the Brown and is consequently the dominant species in western Tas.; along the coast, where both are present, the Tasmanian Thornbill occupies teatree swamps and the Brown the drier heaths.[1314] In eastern Tas. both species are present and the Tasmanian Thornbill penetrates dry eucalypt forest only along the edges of creeks. In many areas the mosaic of wet and dry vegetation provides edges where the two species co-exist. At altitudes above 500 m, wet environments tend to predominate and the Tasmanian Thornbill may replace the Brown, as at Mt Field (42/146) in summer,[1092] or both may be present, e.g. on Mt Wellington (42/147).[1451] In contrast, the offshore islands, where heath predominates, support the Tasmanian Thornbill but the Brown is either absent (Furneaux Group (39/148) and Maria I. (42/148)) or rare (King I. (39/144)).[739]

The Tasmanian Thornbill is less often in flocks in the non-breeding season than the Brown[1314] but the two may sometimes be in loose nomadic feeding flocks.[727] The Tasmanian Thornbill forages at any height from a little over a metre above the ground to the tops of the trees.[1486] It feeds mainly on invertebrates.[1049]

There are no measures of abundance.

Field Atlas Records: 1273

10′ blocks-with records: 268 (73%)

-with breeding records: 26 (10%)

Atlas No.: 473

Max. Rep. Rate: 78%

In block(s): 41°05′/145°35′

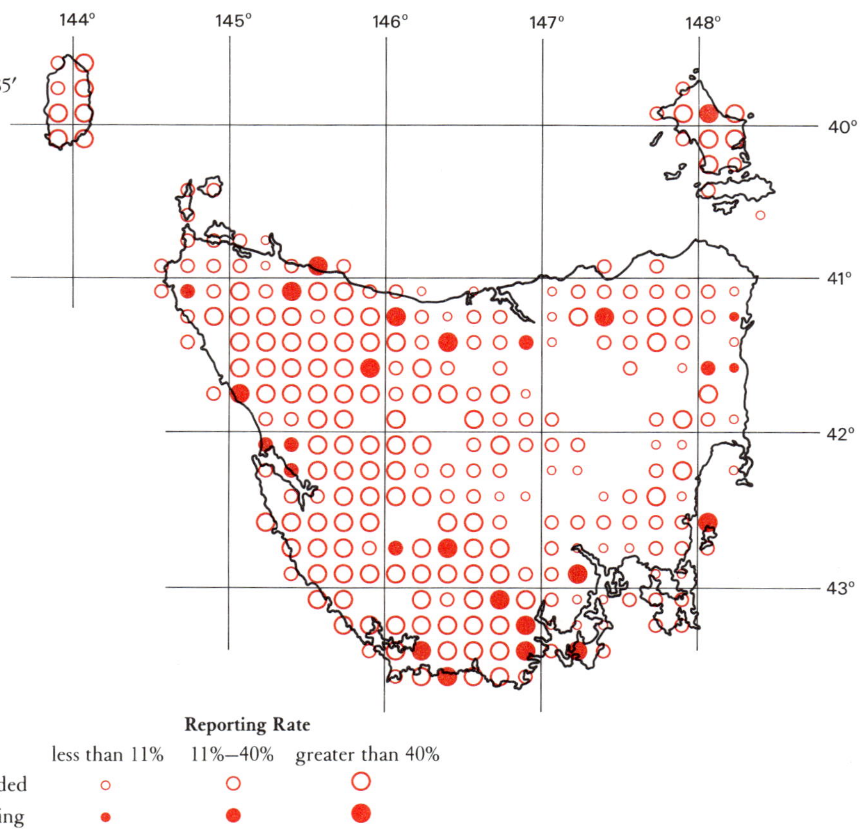

Chestnut-rumped Thornbill

Acanthiza uropygialis

There are records of the endemic Chestnut-rumped Thornbill in the 1951–76 Historical Atlas that eliminate much of the gap apparent in the Field Atlas in the lower Western Deserts Region. No change in distribution over the historical periods is suggested by the Atlases.

The Chestnut-rumped Thornbill is reported to be present throughout the year in some places, e.g. Cobar (31/145)[1573] and the southern Murray-Darling Region.[833] Round Wiluna (26/120) it forms mixed flocks with other thornbills and may be seasonally nomadic.[440] Vagrants are occasionally recorded beyond the species' normal range, which suggests that some individuals move long distances.[653] The Field Atlas suggests no large-scale seasonal movement.

The Chestnut-rumped Thornbill lives in eucalypt woodland, acacia scrub, saltbush, lignum and grassland[748] as well as near farm buildings and in partly cleared paddocks. Particularly in the non-breeding season it is often in small parties and joins other thornbills and small passerines in mixed flocks.[653,1140,1604,1606] Low bushes and tangles of undergrowth are used for roosting.[719] The Chestnut-rumped Thornbill is often in areas with dead trees, perhaps because it uses hollow timber in which to nest. Though most breeding records are of pairs, there is one mention of communal breeding[1870] and a recent observation of three adults attending a nest.[1083] Breeding territories may be defended against Buff-rumped Thornbills.[1907] The Chestnut-rumped Thornbill feeds mainly on small invertebrates collected in trees, undergrowth or on the ground.[1600] The contents of three of 16 guts examined included seeds;[367,1049] some contained buds as well.[748]

Densities in mallee and woodland in Vic. were: Wyperfeld (35/141) 0.21–0.74 birds/ha and Bendigo (36/144) 0.03–0.13 birds/ha.[1907]

Field Atlas Records: 4791

1° blocks–with records: 325 (40%)
–with breeding records: 118 (36%)

Atlas No.: 481

Max. Rep. Rate: 53%
In block(s): 33/143

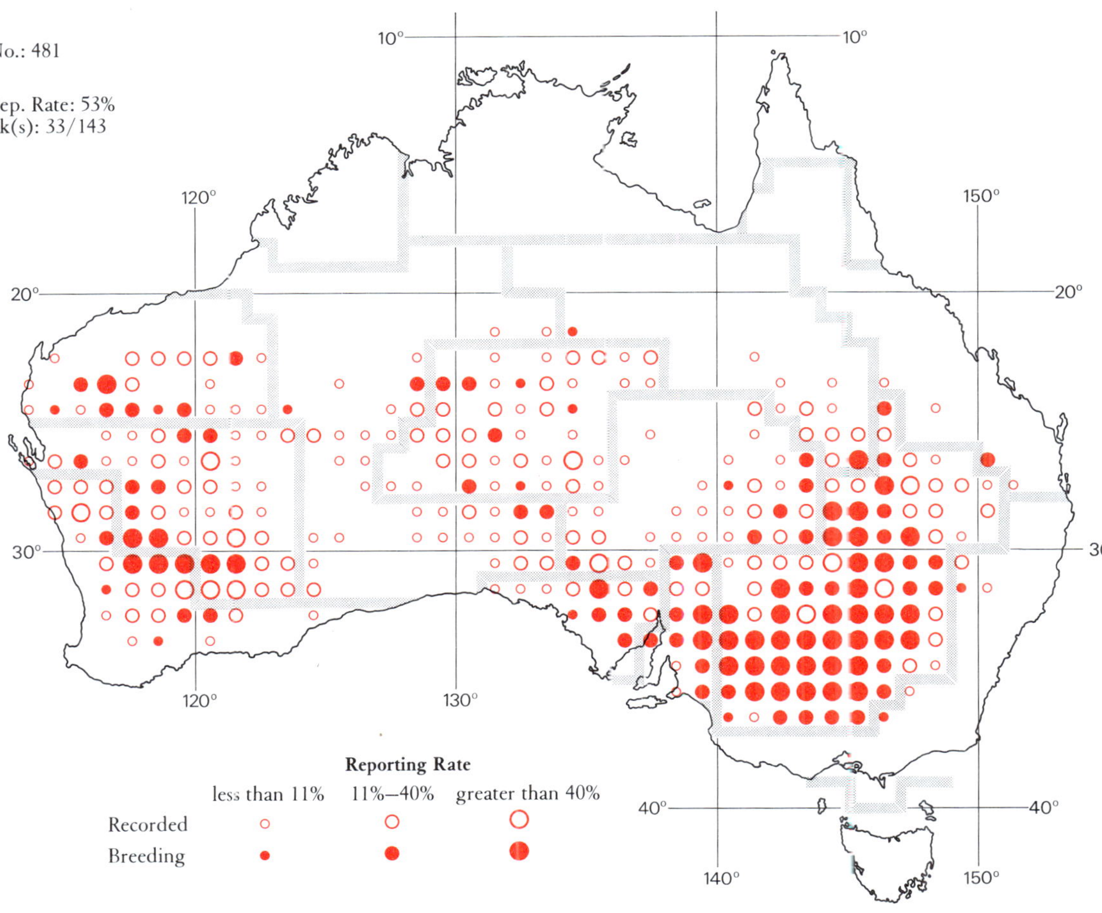

Slaty-backed Thornbill

Acanthiza robustirostris

A record of the endemic Slaty-backed Thornbill from near Eromanga (26/143) in 1971 extends its range eastward and suggests that the species may be sparsely distributed round the northern edge of the L. Eyre Region. The isolated Field Atlas record on the Mulligan R (23/138) was of a pair feeding in gidgee in Aug. 1980. The Slaty-backed Thornbill may be confused in the field with the Chestnut-rumped; their ranges are sympatric but the Chestnut-rumped is generally more abundant.

Round Wiluna (26/120) the Slaty-backed Thornbill sometimes forms mixed groups with the Chestnut-rumped and other thornbills and may be seasonally nomadic.[440] There is no other information about movements. It lives in that part of the dry interior where summer and winter rainfall are equally likely. It inhabits acacia scrub and spinifex, specially dense patches of shrubs characteristic of the foothills of the ranges,[1882] as well as groves of tall mulga. Such sites develop where water runs on after rain, so that plants there enjoy a favourable soil moisture balance for much of the year. In eastern parts of its range the species is usually low in dense mulga or on the ground, the belly feathers of some specimens being stained by the red soil.[1369] In WA, in contrast, the bird is known as a foliage and aerial gleaner.

Of 30 Slaty-backed Thornbills encountered by one expedition, six were in parties and the remainder solitary or in pairs.[748] Remains of insects and caterpillars were found in the guts of 20 birds examined.[748,1049,1872]

There are no measures of abundance.

Field Atlas Records: 207

1° blocks–with records: 94 (12%)
–with breeding records: 9 (10%)

Atlas No.: 480

Max. Rep. Rate: 24%
In block(s): 27/120

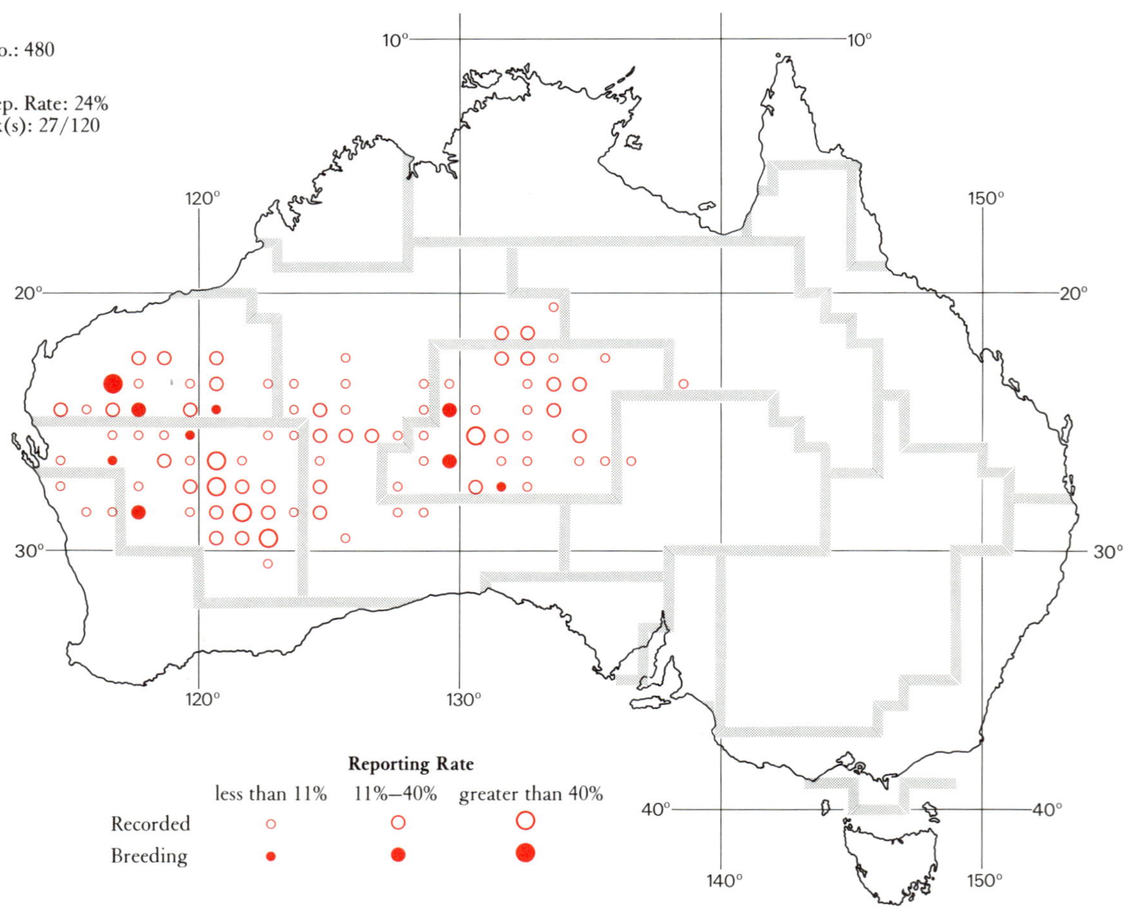

Western Thornbill

Acanthiza inornata

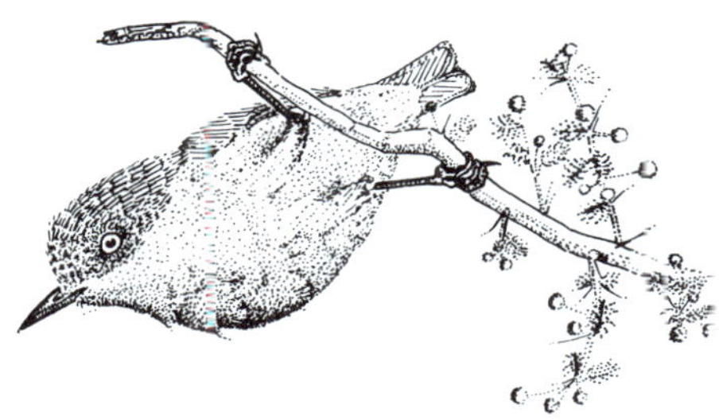

Subspecies have been recognized within the small range of the endemic Western Thornbill[1236] but their validity is disputed.[587] There is no break in distribution on the 10′ maps. The record at Galena (27/114) was of a small group of birds in July 1979.

The Western Thornbill is said to wander in small flocks in the non-breeding season.[638] Though primarily a woodland species, it also inhabits eucalypt forest and the fringes of agricultural land.[1429,1612] The bird lives in small parties and is frequently in mixed flocks with other passerines. Small insects, mostly beetles and bugs, are eaten, the bird foraging at any height from the treetops to the ground.[1222]

There are no measures of abundance.

Field Atlas Records: 644

1° blocks–with records: 22 (3%)

–with breeding records: 13 (59%)

Atlas No.: 472

Max. Rep. Rate: 25%
In block(s): 31/116

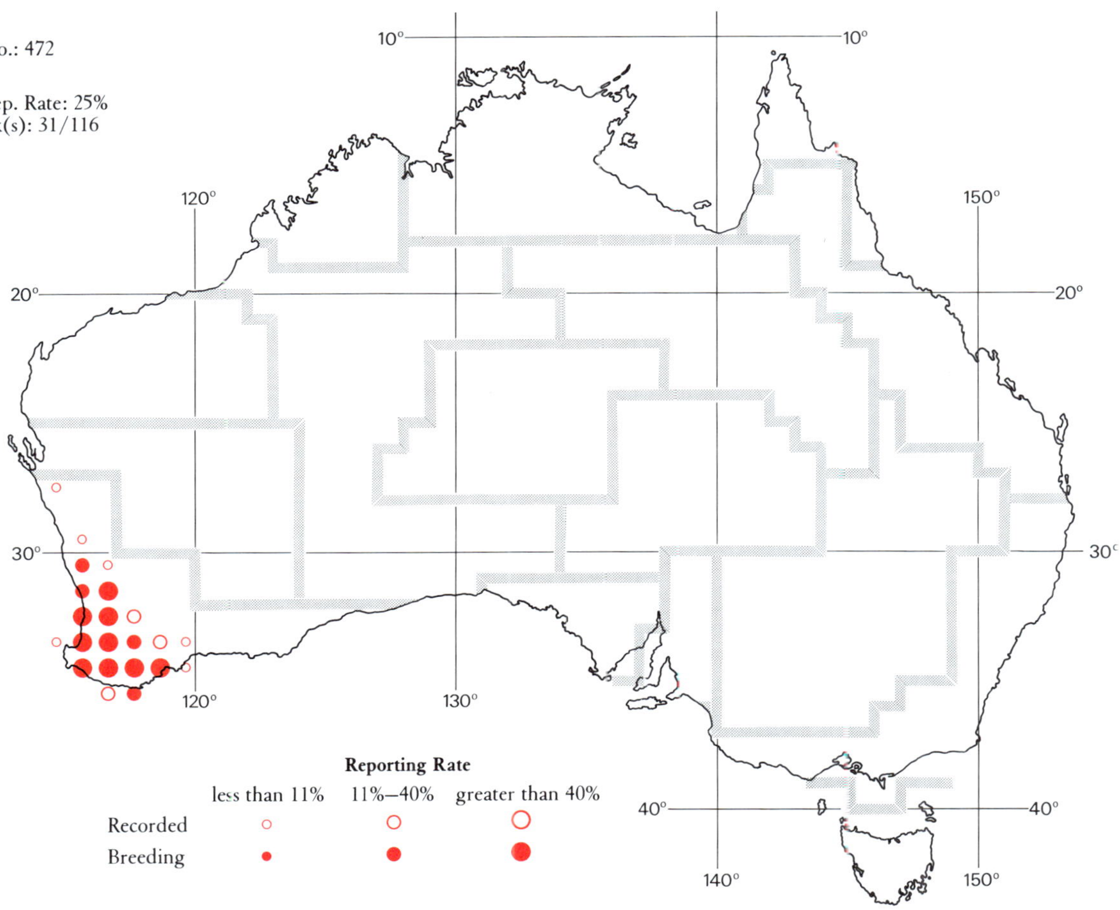

Buff-rumped Thornbill

Acanthiza reguloides

Several subspecies of the endemic Buff-rumped Thornbill have been described but views about them differ.[748,1124] The population *squamata* of the Atherton Region and Eungella district (21/148) has long been recognized as distinct[197] but there is no apparent break in distribution between it and southern subspecies as has sometimes been suggested.[1140] Breeding is reported north to the Atherton Region.

The 10′ map suggests that the population in the SA Ra. Region is only tenuously connected with that of the South-East Region.

The Buff-rumped Thornbill is sedentary near Armidale (30/151) where clans of 10–16 birds forage and roost within a group territory throughout the year. During the breeding season they split into smaller groups of one female and a number of males to breed communally.[159] The Buff-rumped Thornbill lives mainly in eucalypt forest or woodland with a grassy or shrubby understorey; it will enter orchards.[14] The northern population *squamata* lives along forest margins and occasionally in rainforest.[212] These birds glean from twigs and leaves rather than forage on the ground as do Buff-rumped Thornbills in most places. At Wollomombi (30/152) and Canberra (35/149) the Buff-rumped Thornbill is a prominent member of mixed species flocks.[155,162] It feeds on small arthropods.[1049]

Near Armidale density was 1.54 birds/ha[582] and clans of 10–16 birds held territories of about 15 ha.[159] Densities in eucalypt woodland were: Canberra 0.3–1.4 birds/ha,[154] Tooboorac (37/144) 0.59–1.10 birds/ha, You Yangs (37/144) 0.24–0.43 birds/ha and Barmah (36/144) 0.02–0.16 birds/ha;[1907] in native pines at Beechworth (36/146) 0.22–1.58 birds/ha;[1907] and in eucalypt forest at Jamieson (37/146) 0.42 birds/ha.[1469] At Boola Boola (38/146) the bird lived only in eucalypt regrowth forest at densities of 0.2–0.5 birds/ha.[1113]

Field Atlas Records: 4963

1° blocks–with records: 142 (17%)
–with breeding records: 67 (47%)

Atlas No.: 484

Max. Rep. Rate: 38%
In block(s): 30/151

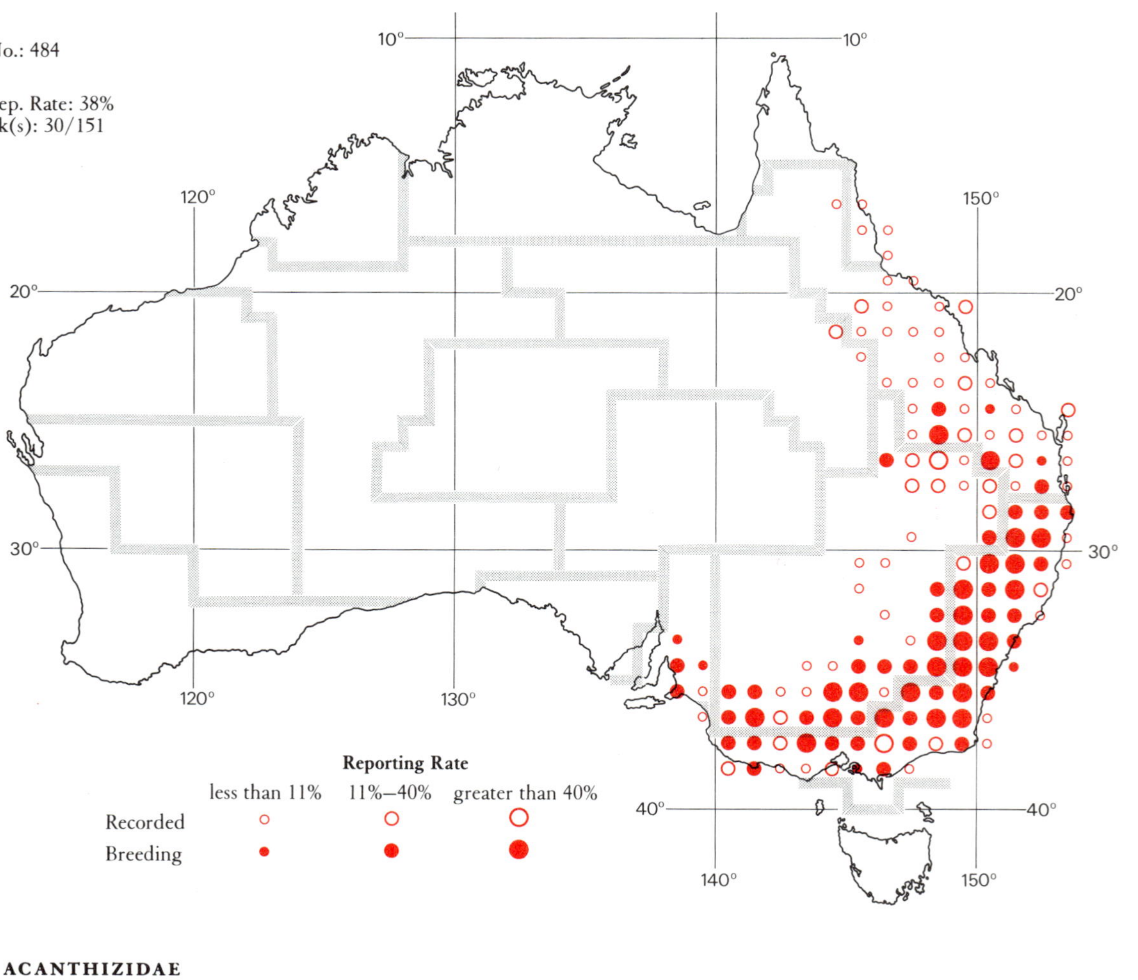

Though the distribution of the endemic Slender-billed Thornbill appears disjunct on the Field Atlas map, addition of historical records suggests that, with the exception of the apparently isolated population in the south-western Murray-Darling Region, the species is sparsely but continuously distributed across Australia from 139°E to the western coast. This accords with the fact that morphological differences are clinal. There are different views about subspecies.[748,1124] Skins collected in 1913–14 at Idracowra (25/133)[1374] and Wantapella Swamp (27/133) extend the species' distribution to the Central Ra. Region. No subsequent records are so far north in that region and the species may have disappeared from there.

In the Big Desert (35/141) the bird is present throughout the year.[886] There is no information about movements from other parts of the species' range.

In the main part of its range the Slender-billed Thornbill lives, usually in parties,[748] in areas dominated by saltbush and samphire. In the south-western Murray-Darling it inhabits heath. In the Big Desert the Thornbill is restricted to dense heath about 1 m high with no mallee; it is in ones and twos Aug.-Oct. and in groups of 4–10 Dec.-Apr.; it forages, apparently for insects, mostly inside bushes of banksia, teatree, casuarina and hakea.[886] Elsewhere it feeds on small spiders[1049] and insects found among samphire, saltbush and bluebush.[748]

On the Nullarbor Plain it was noted that, though widespread and often reported 1918–35, the Thornbill is now rarer and more local.[246] The change in status may be due to changes in the environment, particularly the destruction of bluebush brought about by rabbits and domestic stock.

Slender-billed Thornbill

Acanthiza iredalei

Density in the Big Desert in banksia-dominated heath averaged 2.0–2.3 birds/ha.[886]

Field Atlas Records: 138

1° blocks–with records: 34 (4%)
–with breeding records: 4 (12%)

Atlas No.: 482

Max. Rep. Rate: 12%
In block(s): 35/141

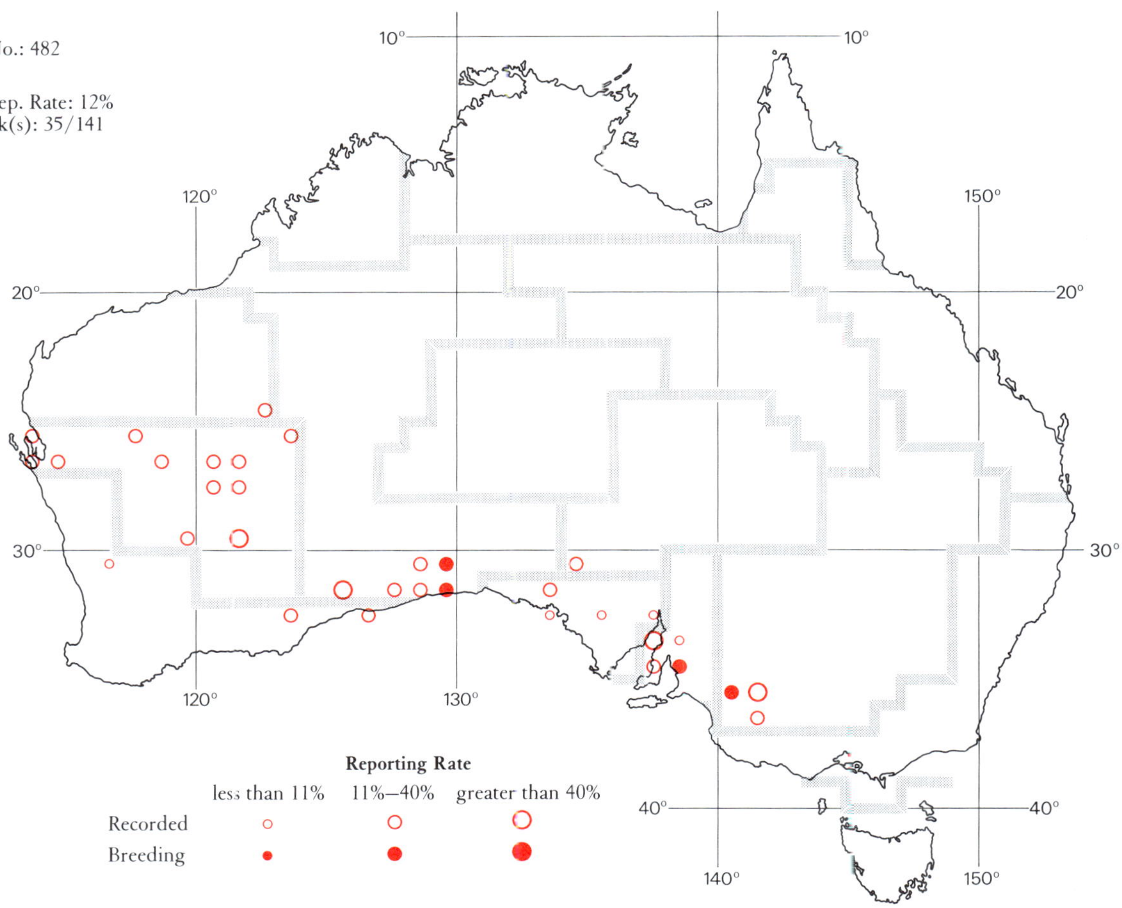

Yellow-rumped Thornbill

Acanthiza chrysorrhoa

There are different views about subspecies of the endemic Yellow-rumped Thornbill.[748,1124] Many have been recognized but the Atlases do not show any isolated populations on the mainland. The species is absent from King (39/144) and Flinders (40/148) Is and the west of Tas. The distribution determined by the Field Atlas is similar to that of the Historical Atlases and suggests no change in range.

There is no evidence of seasonal movements near Canberra (35/149)[1014] or from other places. Colour banding at Wollomombi (30/152) suggests that a flock occupies a territory in the non-breeding season and, during the breeding season, splits into pairs or communally breeding groups.[129] The Yellow-rumped Thornbill inhabits country ranging from eucalypt forest to eucalypt woodland with a grassy understorey. It shows a marked preference for fern-leaved acacias, pines and native pines and usually builds its nest in such plants if they are available. The nest may be heavily parasitized by cuckoos, e.g. 21 nests from a sample of 65 studied near Perth (31/115) were parasitized by the Shining Bronze-Cuckoo.[590]

The Yellow-rumped Thornbill is joined by other species to form mixed flocks during the non-breeding season.[129] It feeds on spiders and insects, specially caterpillars,[1222,1531] and occasionally takes seeds, buds and other vegetable material.[748]

Four nests were found within 100 m of each other at Cooyal (32/149).[1010] At Eyre Bird Observatory (32/126) in mallee density was 0.14–1.07 birds/ha.[464] In an area of about 40 ha south of Perth the Thornbill bred at a density of 0.6 nests/ha in two years; some nests were attended by more than two birds.[590]

Field Atlas Records: 24 657

1° blocks–with records: 437 (54%)
–with breeding records: 220 (50%)

Atlas No.: 486

Max. Rep. Rate: 73%
In block(s): 27/149

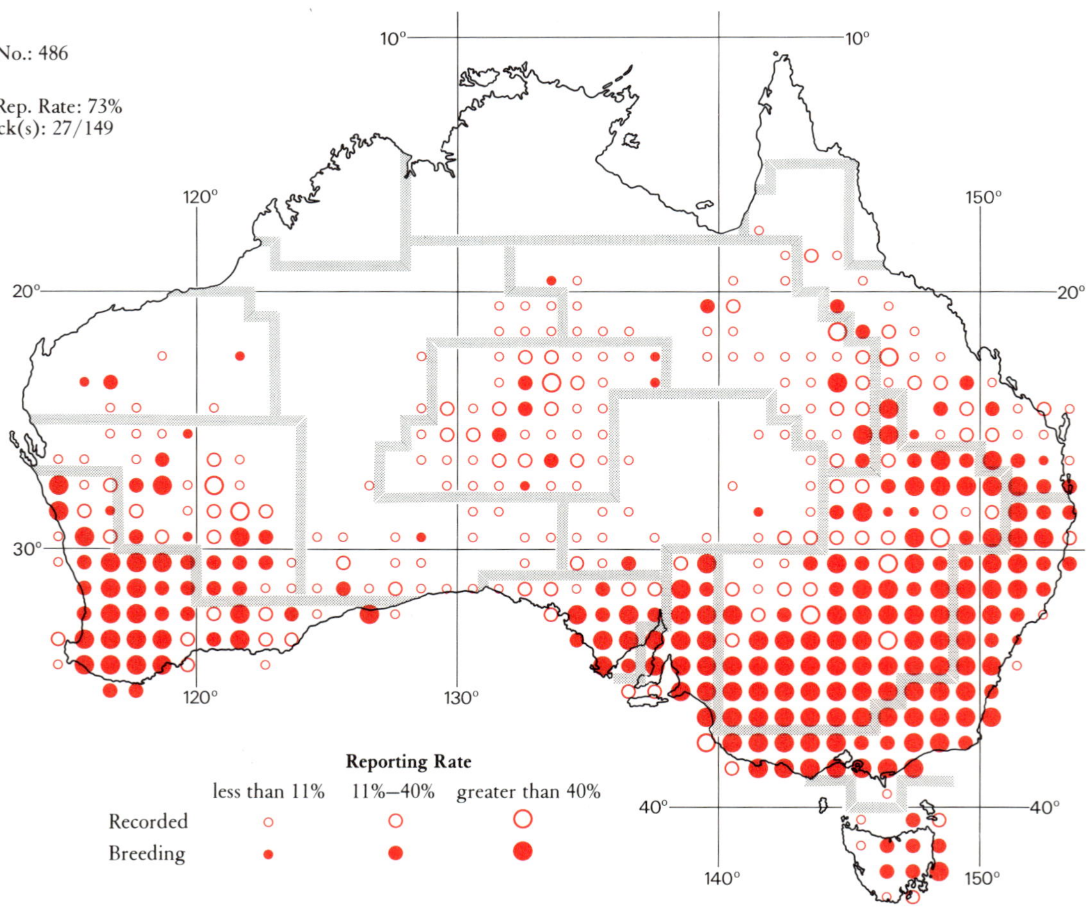

Yellow Thornbill

Acanthiza nana

Four subspecies of the endemic Yellow Thornbill are recognized: *modesta* reaching the coast in central Qld but mostly west of the Great Dividing Ra. and *flava*, *nana* and *mathewsi* east of the range from north to south.[1124] The population in the Atherton Region *flava* is apparently isolated but the Atlases do not show any other break in distribution.

The Field Atlas does not suggest large-scale seasonal movement. At Wangaratta (36/146) the bird visits farmland in spring.[337]

The Yellow Thornbill inhabits a range of vegetation: eucalypt forest and woodland, including regrowth, mangroves, relict timber on cleared land, homestead gardens and orchards. In the Atherton Region it is confined to the highlands. It was observed at 1000 m in the McPherson Ra. (28/152)[1205] but not reported above 300 m in ranges farther south near Barellan (34/146).[360] In the Snowy Mts (36/148) the Field Atlas Reporting Rate was much higher below 200 m (35%) than above (5%). The Yellow Thornbill has once been recorded to breed communally.[501]

The Yellow Thornbill eats small arthropods.[1049] Foraging tends to be concentrated in acacias, casuarinas and native pines. The bird will feed in introduced vegetation such as ornamental shrubs and orchards.[332,891]

Densities in Vic. were: in mallee and woodland at Wyperfeld (35/141) 0.24–0.56 birds/ha and Bendigo (36/144) 0.29–0.54 birds/ha; in eucalypt woodland at Tooboorac (37/144) 0.02–0.26 birds/ha and the You Yangs (37/144) 0.19–0.56 birds/ha; and in native pines at Beechworth 0.10–0.48 birds/ha.[1907]

Field Atlas Records: 7329

1° blocks–with records: 177 (22%)
–with breeding records: 72 (41%)

Atlas No.: 471

Max. Rep. Rate: 56%
In block(s): 27/149

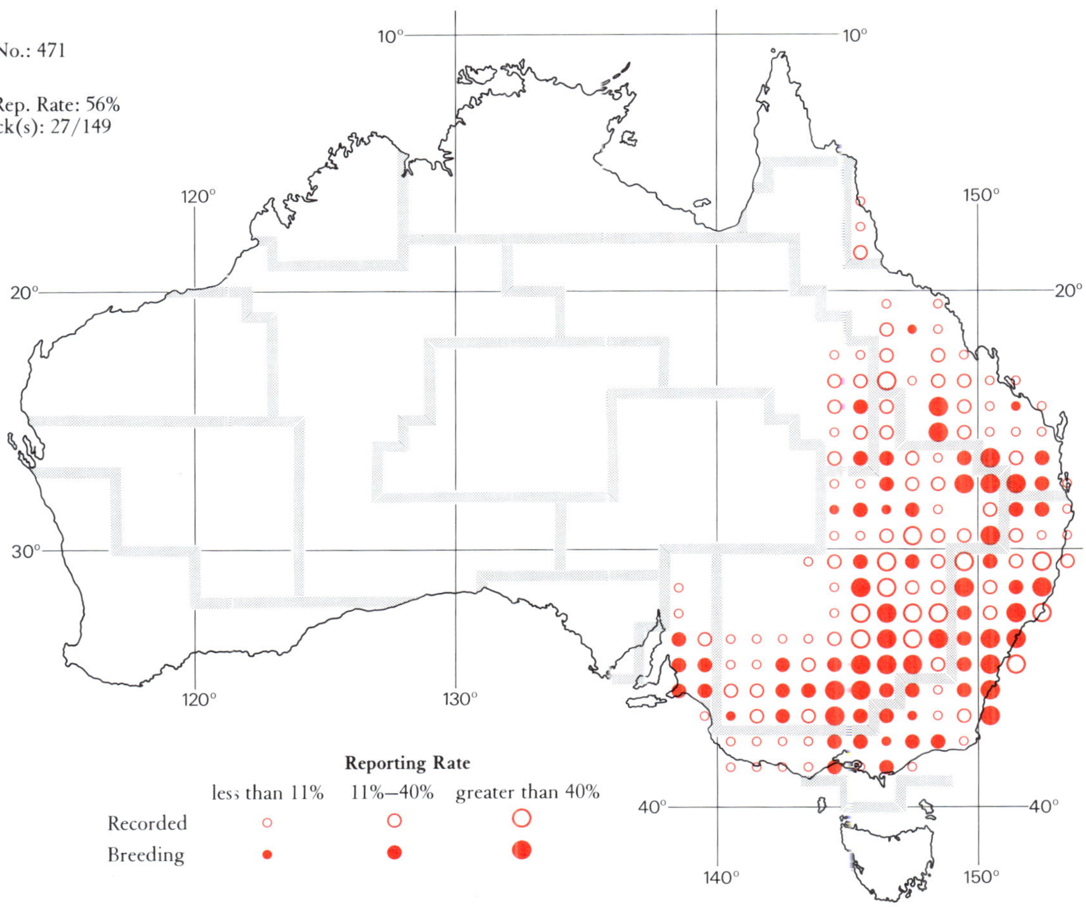

Striated Thornbill

Acanthiza lineata

Four subspecies of the endemic Striated Thornbill are recognized: *alberti* in Qld, *lineata* in NSW, *chandleri* in Vic. and on Kangaroo I. (35/137) and *clelandi* in the Mt Lofty Ra. (34/138).[1124] The Field Atlas does not show any break in the distribution of mainland populations. The Striated Thornbill is not known in the Eungella (21/148) district where Brown Thornbills were sometimes misidentified as Striated. The Striated has not been reported breeding north of 26°S.

Banding has shown the Striated Thornbill to be sedentary near Armidale (30/151),[159] Jamieson (37/146),[1469] Bogong (36/147), Healesville (37/145) and Cranbourne (38/145).[1907]

A suggestion that there is altitudinal movement in the Brindabella Ra. (35/148)[1016] is not supported by the Field Atlas and may derive from the fact that birds are in flocks in the non-breeding season, making them more conspicuous.[129]

Near Armidale clans of 15–20 birds forage and roost in a territory during the non-breeding season, dividing it amongst smaller communally breeding groups in the breeding season.[159] The Striated Thornbill lives in eucalypt forest and will enter gardens. It works systematically over mature trees or saplings, preferably eucalypts, often hovering beside the outer foliage as it searches for leaf-dwelling arthropods. In the non-breeding season it is often a member of mixed species flocks.[155] The Striated Thornbill feeds mainly on arthropods, specially spiders, beetles, weevils, bugs and caterpillars; lerp is also taken.[1049,1907] At Barrington (31/151) 23 Striated Thornbills were seen clustering in contact on a cold windy day.[893]

Near Armidale clans of 15–20 birds held territories of about 6 ha.[159] Densities in eucalypt woodland were: Armidale 0.25–1.58 birds/ha,[582] Moruya (35/150) 0.74–0.77 birds/ha,[1197] Tooboorac (37/144) 0.66–1.74 birds/ha, the You Yangs (37/144) 0.03–0.08 birds/ha and Cranbourne 0.69–1.33 birds/ha.[1907] In eucalypt forest they were: Corranderrk (37/145) 0.75–1.76 birds/ha,[1907] Jamieson 0.25–0.50 birds/ha,[1469] Tynong (38/145) 0.93–4.13 birds/ha[1907] and Boola Boola (38/146) 1.2–5.6 birds/ha (greatest in mature eucalypt forest).[1113] In rainforest densities were: Donna Buang (37/145) 0.29–0.69 and Glenaladale (37/147) 0.02–0.03 birds/ha.[1907]

Field Atlas Records: 9080

1° blocks–with records: 102 (13%)
–with breeding records: 60 (59%)

Atlas No.: 470

Max. Rep. Rate: 55%
In block(s): 36/150

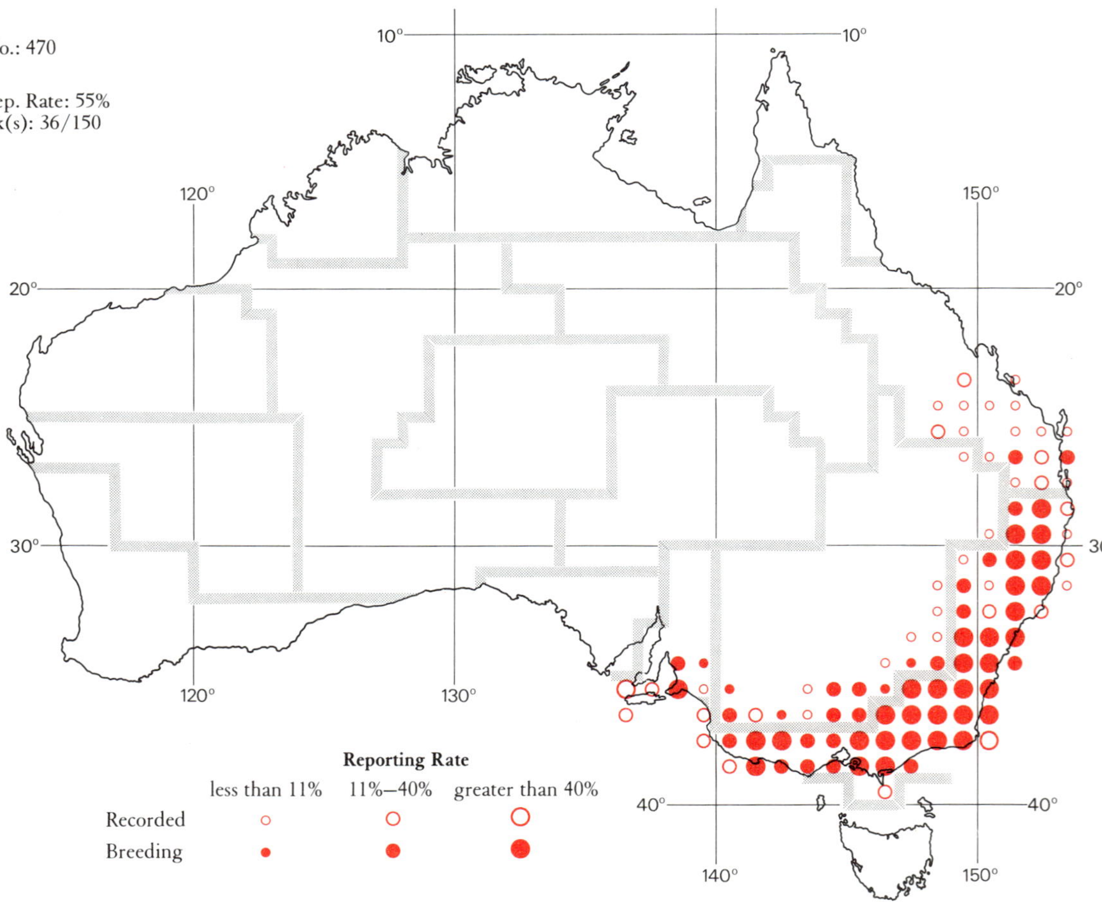

Three subspecies of the endemic Southern Whiteface are recognized: *leucopsis* in the east, *whitei* in the centre and *castaneiventris* in the west.[748] Two of these, *leucopsis* and *whitei*, intergrade north and east of the Great Victoria Desert.[1369] When historical records are added to the Field Atlas the gap suggested to separate *whitei* from *castaneiventris*[748] does not occur.

The Field Atlas suggests no large-scale seasonal movement and the species is resident throughout the year in the southern Murray-Darling Region.[833,1128] In the non-breeding season the Southern Whiteface lives in parties: groups of 10–70 birds have been recorded near Meekatharra (26/118).[456,460] The species is often in mixed flocks with other whitefaces and Yellow-rumped and Chestnut-rumped Thornbills.[1156,1606] It lives in acacia scrub, specially that dominated by mulga, and in spinifex.

The Southern Whiteface feeds on insects and seeds.[1049] Most feeding is done on the ground, often round farm and station homesteads. The bird turns over objects in its search for food, even tackling cow pats much larger than itself.[719] The House Sparrow has been observed pulling the nest of the Southern Whiteface to pieces.[337]

There are no measures of abundance.

Southern Whiteface

Aphelocephala leucopsis

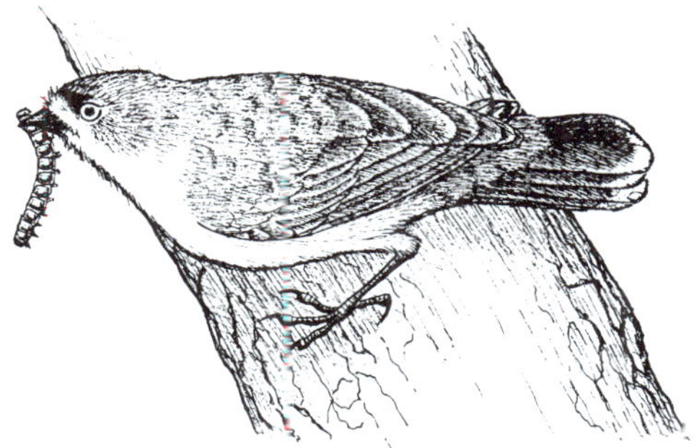

Field Atlas Records: 5537

1° blocks–with records: 323 (40%)
–with breeding records: 117 (36%)

Atlas No.: 466

Max. Rep. Rate: 59%
In block(s): 33/143

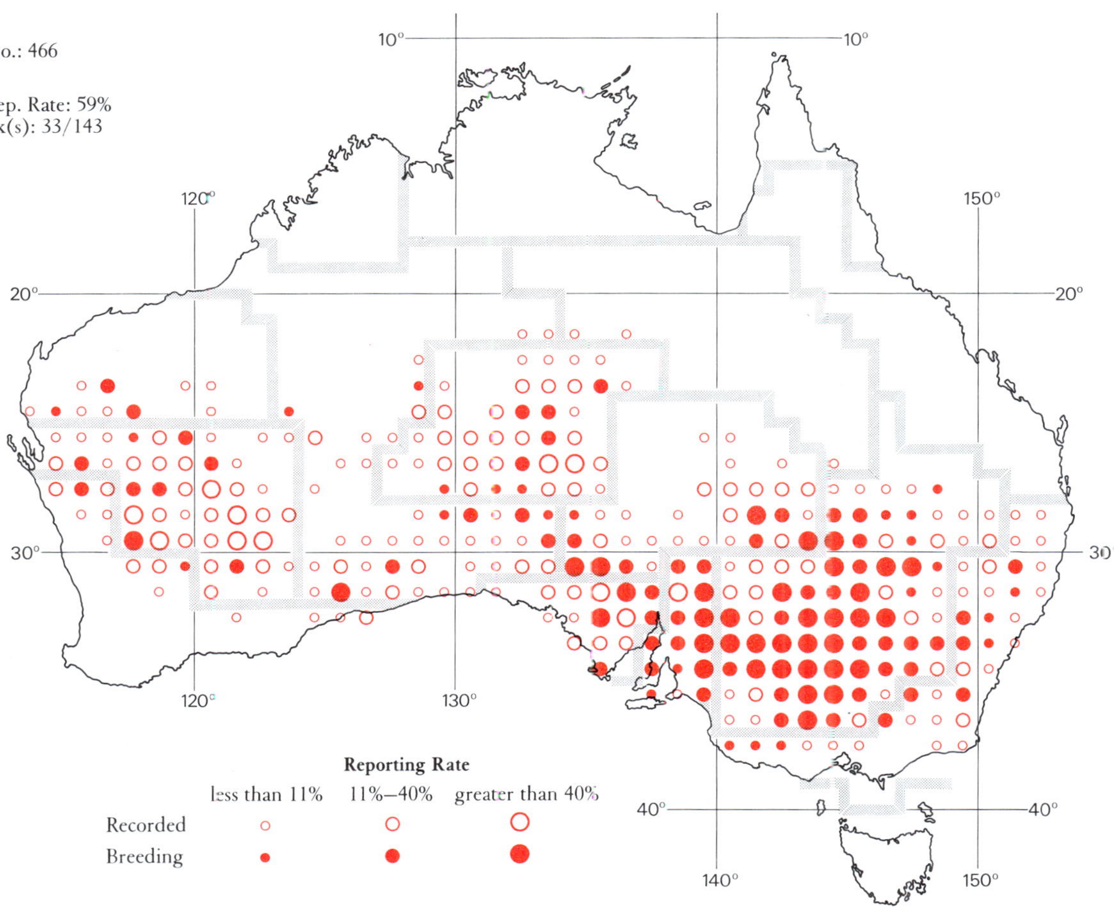

Chestnut-breasted Whiteface

Aphelocephala pectoralis

There was an early suggestion that the endemic Chestnut-breasted Whiteface was only the juvenile of the Banded Whiteface.[1864] The species is still little known, its nest and eggs undescribed until 1968. A combination of Historical (p. 667) and Field Atlas records gives the range over which it may be expected. Records are too few to provide evidence of any change in range since European settlement.

The Chestnut-breasted Whiteface is thought to be nomadic because it cannot predictably be found even in localities where it has bred.[1444,1864] A pair nested at Myrtle Springs (30/138) following good rains in July 1968; there were old nests nearby, possibly dating from another good season two years previously,[1445] but though the area has been visited many times subsequently, the bird has not been located again.[1444] A nest with young was found near Cannuwaukaninna Bore (28/138) in Sept. 1976 but no Whitefaces were present on a visit a month later.[1699]

The Chestnut-breasted Whiteface generally lives in small groups and sometimes in mixed flocks with other whitefaces.[1863] Field Atlas sightings were of one, two, three and four birds. The species lives where there are shrubs mostly on stony rises or gibber plains.[602,1699] It feeds on the ground presumably on seeds and insects as do the other whitefaces.

Its rarity is emphasized by the fact that only four sightings of a total of ten birds were reported to the Field Atlas.

Field Atlas Records: 4

1° blocks–with records: 4 (1%)
–with breeding records:

Atlas No.: 468

Max. Rep. Rate: 4%
In block(s): 30/134

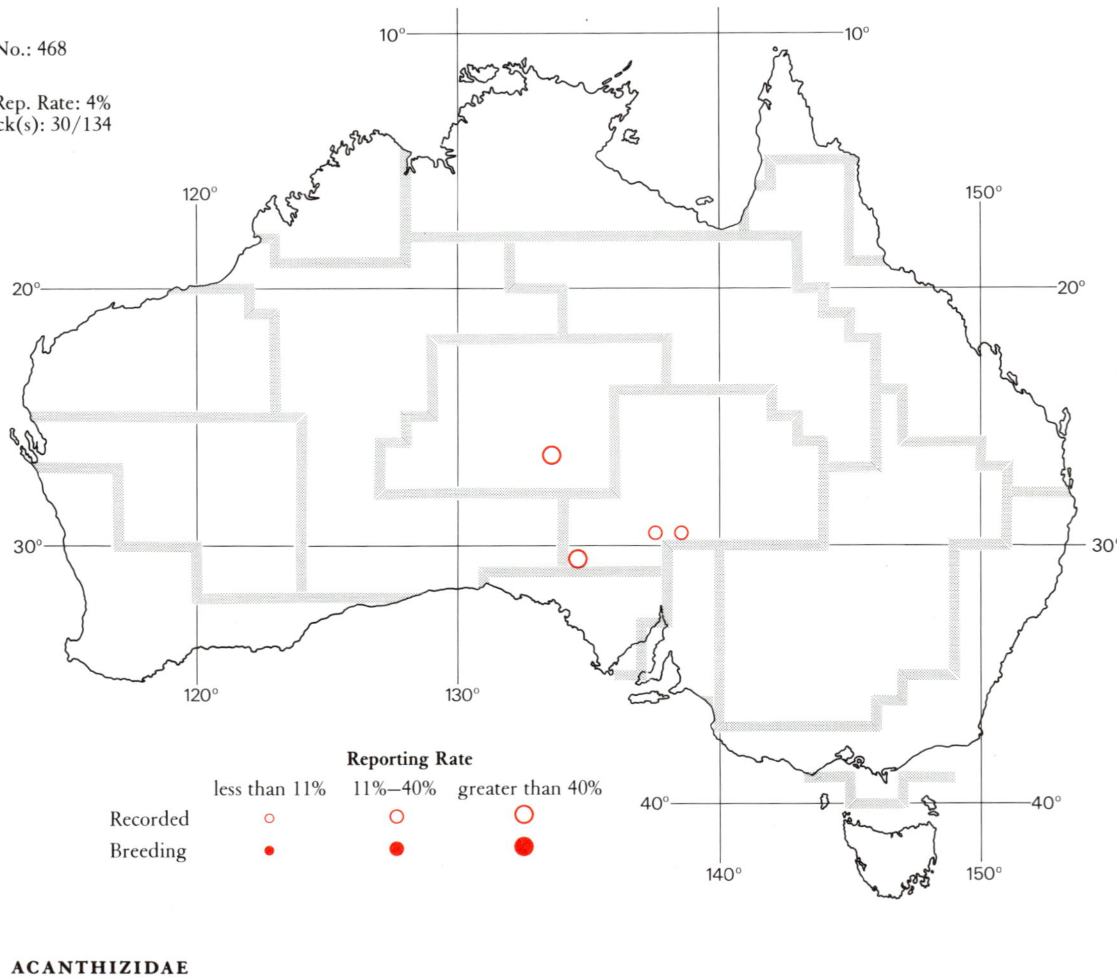

Banded Whiteface

Aphelocephala nigricincta

The range of the endemic Banded Whiteface in the Field Atlas is similar to that in the historical periods.

The Field Atlas suggests no large-scale seasonal movement. The Banded Whiteface lives in acacia scrub and spinifex, specially in open places[748] or where there has been a recent fire; and on the edges of small watercourses in the Western Deserts Region. Field Atlas sightings were of pairs or small parties, sometimes in mixed flocks with other whitefaces or the Crimson Chat. During the Field Atlas four nests, one active, were found in one small bush; two birds were present. At another site three birds were recorded building one nest. The Banded Whiteface feeds on the ground on seeds and insects.[719,748,1156]

There are no measures of abundance.

Field Atlas Records: 192

1° blocks–with records: 95 (12%)
–with breeding records: 15 (16%)

Atlas No.: 469

Max. Rep. Rate: 14%
In block(s): 25/124

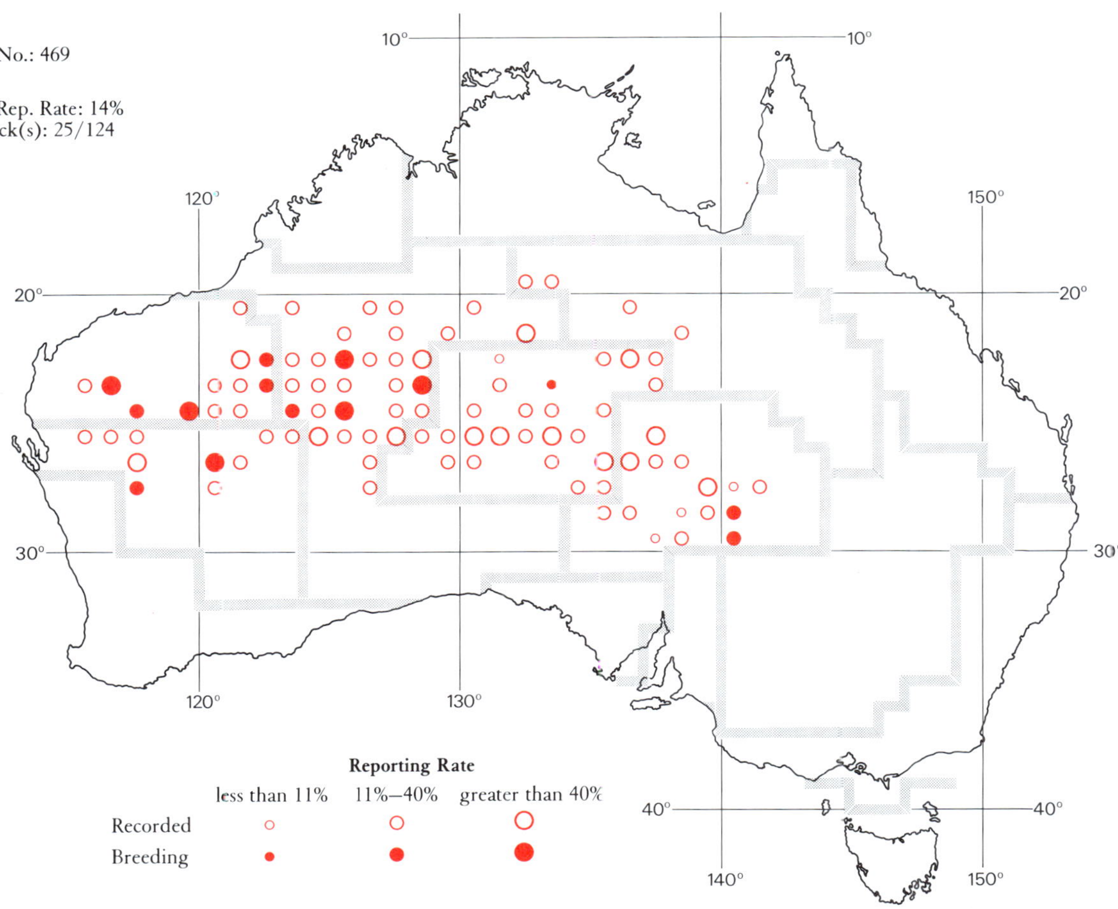

Varied Sittella

Daphoenositta chrysoptera

Five subspecies of the endemic Varied Sittella are recognized: *striata* in the Cape York and Atherton Regions south to 19°S; *leucocephala* in the Eastern Qld Region; *chrysoptera* in the South-East and Murray-Darling Regions east of approximately 144°E; *leucoptera* in the northern Rolling Downs, Top End and Kimberley Regions; and *pileata* elsewhere.[616,1124] Hybrid zones occur where ranges meet.[616,1372]

During the 1951–76 period the Varied Sittella was recorded across the lower Western Deserts Region and its absence during the Field Atlas is probably because few people visited the area. Other districts where the Varied Sittella was recorded 1951–76 but not during the Field Atlas are the eastern Rolling Downs, the area between Tibooburra (29/142) and Broken Hill (31/141) and the Pilbara coast. These were visited often during the Field Atlas and the lack of records suggests that the bird was rare or absent during that time. There are no records from the Tas. Region.

Local nomadic movements are reported in the non-breeding season but not large seasonal shifts of population. The Varied Sittella lives where there are rough-barked eucalypts in forest and woodland, irrespective of the height or density of the trees.[1345] In arid areas it inhabits mallee and acacia scrub. It lives in groups generally of 2–12 birds but up to 20 have been recorded.[748] It breeds communally.[501,1350] Insects are excavated from the bark of trees as the bird works its way along branches and up or down the trunks.

Density at Armidale (30/151) in eucalypt woodland was 0.2 birds/ha[582] and a group of 5–8 birds ranged over about 20 ha of semi-cleared woodland there.[1350] In eucalypt woodland at Wollomombi (30/152) density was 0.5 birds/ha,[129,1345] at Canberra (35/149) 0.2 birds/ha[154] and at Moruya (35/150) 0.05–0.15 birds/ha.[1197] In eucalypt forest at Jamieson (37/146) density was up to 0.5 birds/ha[1469] and at Boola Boola (38/146) 0.2–0.9 birds/ha.[1113]

Field Atlas Records: 8503

1° blocks–with records: 427 (53%)

–with breeding records: 134 (31%)

Atlas No.: 549

Max. Rep. Rate: 41%
In block(s): 26/150

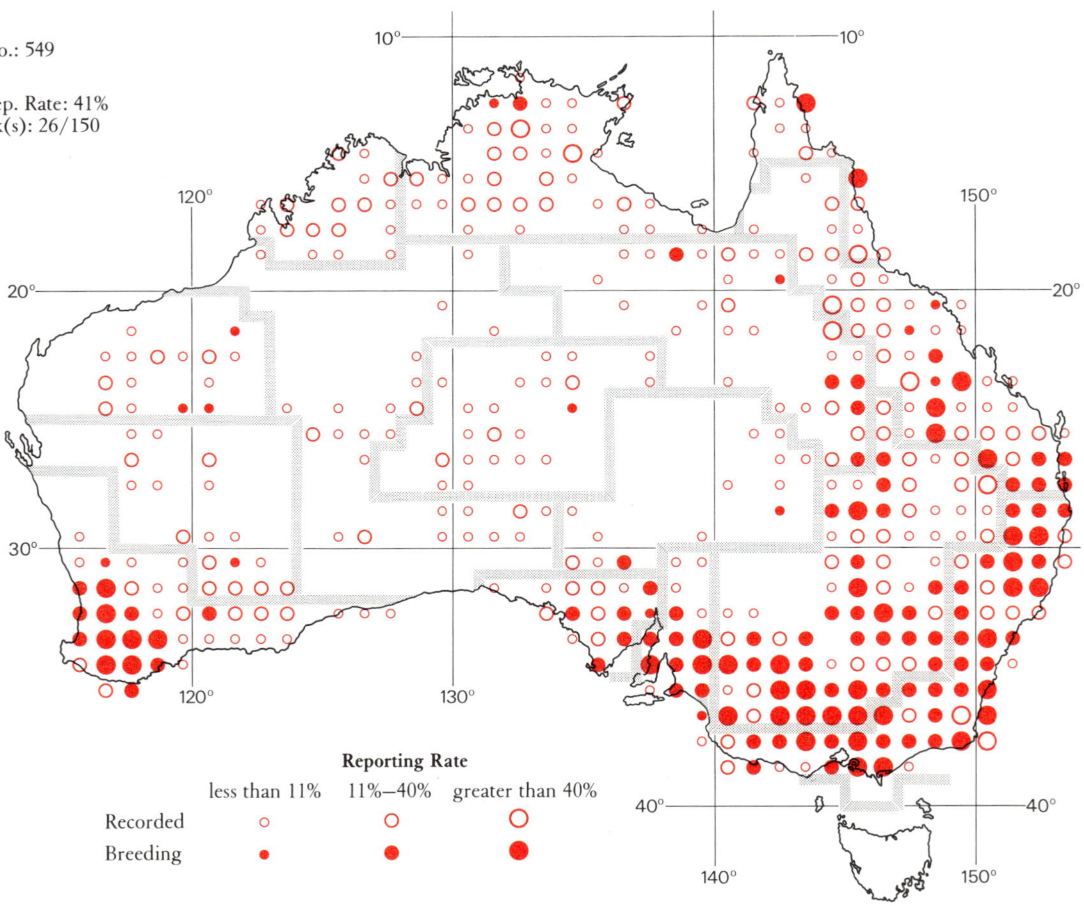

Little Treecreeper

Climacteris minor

The endemic Little Treecreeper, first described late last century as a species, was subsequently reduced to subspecific rank as an outlier of the White-throated Treecreeper,[978] and later reinstated as a full species.[1576] The isolated population in the Eungella (21/148) district is an undescribed intermediate between the White-throated and Little Treecreepers.[197]

The Little Treecreeper is present all year throughout its range. It lives singly or in pairs in the mountains in rainforest and on the edges of rainforest clearings.[748] It is known to favour feeding on epiphytic mosses. Its food is not recorded but, if similar to that of other treecreepers, would be mainly insects, including ants. The epiphytic mosses need light and are particularly common at the edge of rainforest, so that patches of clearing may provide more areas where the Little Treecreeper can feed than were available before European settlement.

There are no measures of abundance.

Field Atlas Records: 219

1° blocks–with records: 8 (1%)
–with breeding records: 1 (13%)

Atlas No.: 559

Max. Rep. Rate: 22%
In block(s): 15/145

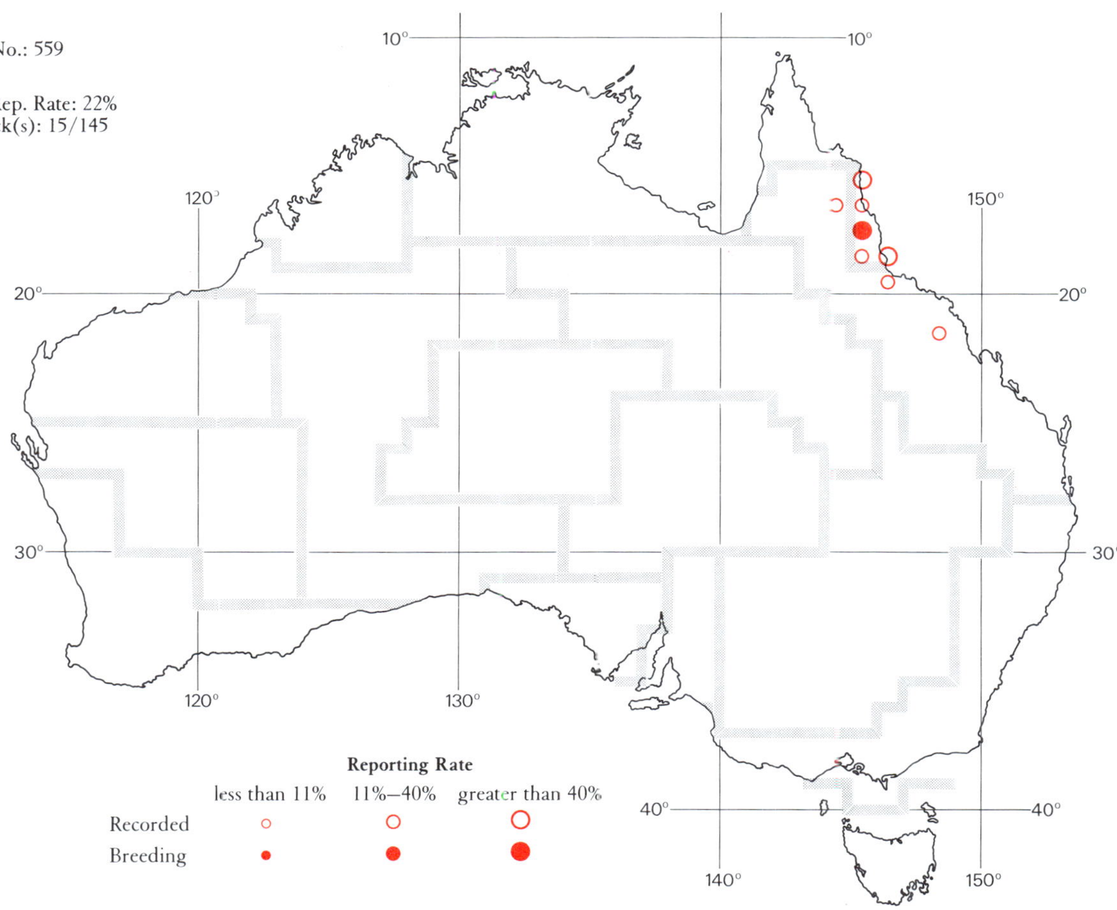

White-throated Treecreeper

Climacteris leucophaea

Within its main range the endemic White-throated Treecreeper displays no subspecific differentiation[748,978] though the population in the SA Ra. is isolated. The population in the Eungella (21/148) district is an undescribed intermediate between the Little and White-throated Treecreepers.[197]

The White-throated Treecreeper is sedentary. Pairs maintain a territory throughout the year.[1351] Only pairs have been recorded attending a nest and the species is unlikely to breed communally.[1349] It inhabits rainforest, eucalypt forest and woodland and extends into partly cleared farmland. The bird is usually solitary in the non-breeding season[1348] but sometimes joins mixed species flocks: 17% of mixed flocks recorded at Wollomombi (30/152) included it.[129]

The White-throated Treecreeper spirals up trees to forage, spending less than 1% of its time on the ground,[1348] and feeds on ants and other insects, which it finds by tugging and pecking at the bark mostly of rough-barked trees. At Wattle Ck (37/147) a bird was observed to catch a moth and cache it with others in a small split in a tree trunk.

Densities in eucalypt woodland were: Armidale (30/151) 0.03–1.13 birds/ha,[582] Wollomombi 0.47–0.56 birds/ha in three successive seasons,[1351] Canberra (35/149) 0.6 birds/ha[154] and Moruya (35/150) 0.42–0.46 birds/ha.[1197] In eucalypt forest densities were: Bathurst (33/149) 1.50 and 0.75 birds/ha in wet and dry forest respectively,[489] Jamieson (37/146) 0.2 birds/ha[1469] and Boola Boola (38/146) 0.3–2.0 birds/ha, highest in mature forest and gullies.[1113]

Field Atlas Records: 13 057

1° blocks–with records: 110 (14%)

–with breeding records: 53 (48%)

Atlas No.: 558

Max. Rep. Rate: 71%
In block(s): 35/148

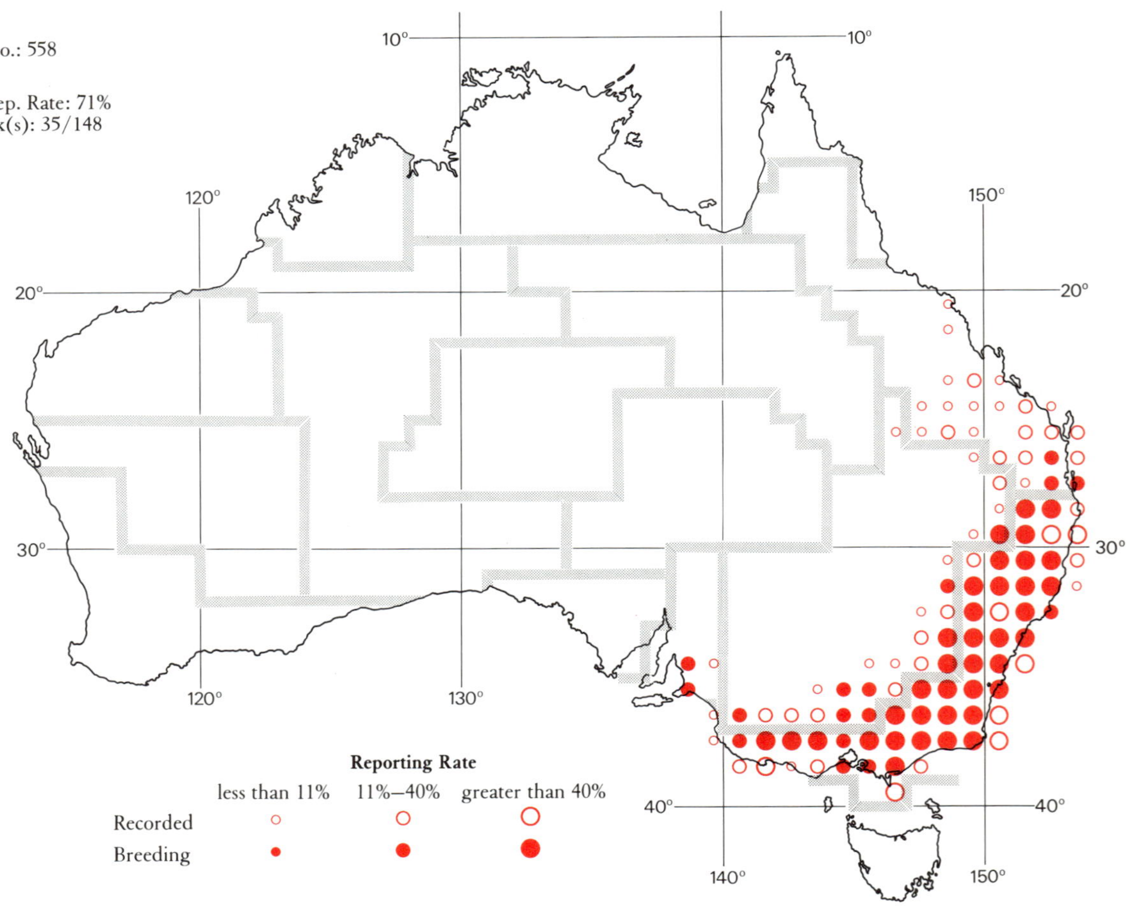

Red-browed Treecreeper

Climacteris erythrops

Two subspecies of the endemic Red-browed Treecreeper are recognized: *olinda* in the Dandenongs (37/145) and *erythrops* to the east and north.[978] The Atlases suggest no gap between them but on the 10′ maps a small population in the Macedon-Daylesford (37/144) area appears isolated and there is also a break in distribution at the Hunter R (32°S).

The Red-browed Treecreeper is sedentary, pairs or small groups occupying permanent territories.[1351] Breeding is sometimes communal.[1349] The species inhabits eucalypt forest[489,870,978] and is absent from rainforest except where eucalypts emerge above regenerating rainforest.[1345] It forages on many eucalypt species but prefers to probe through ribbons of bark hanging from the trunks and branches of smooth-barked species.[1348] Less than 1% of its time is spent on the ground but it often visits fallen logs.[1348] The Red-browed Treecreeper feeds on ants and other arthropods.

Density at Wollomombi (30/152) in eucalypt woodland was 0.28–0.32 birds/ha over three years;[1351] at Moruya (35/150) in eucalypt woodland 0.15 birds/ha;[1197] and at Boola Boola (38/146) in eucalypt forest, including the gullies where it was most abundant, 0.1–1.0 birds/ha.[1113]

Field Atlas Records: 1164

1° blocks-with records: 43 (5%)
-with breeding records: 20 (47%)

Atlas No.: 560

Max. Rep. Rate: 14%
In block(s): 28/152,30/151,36/149

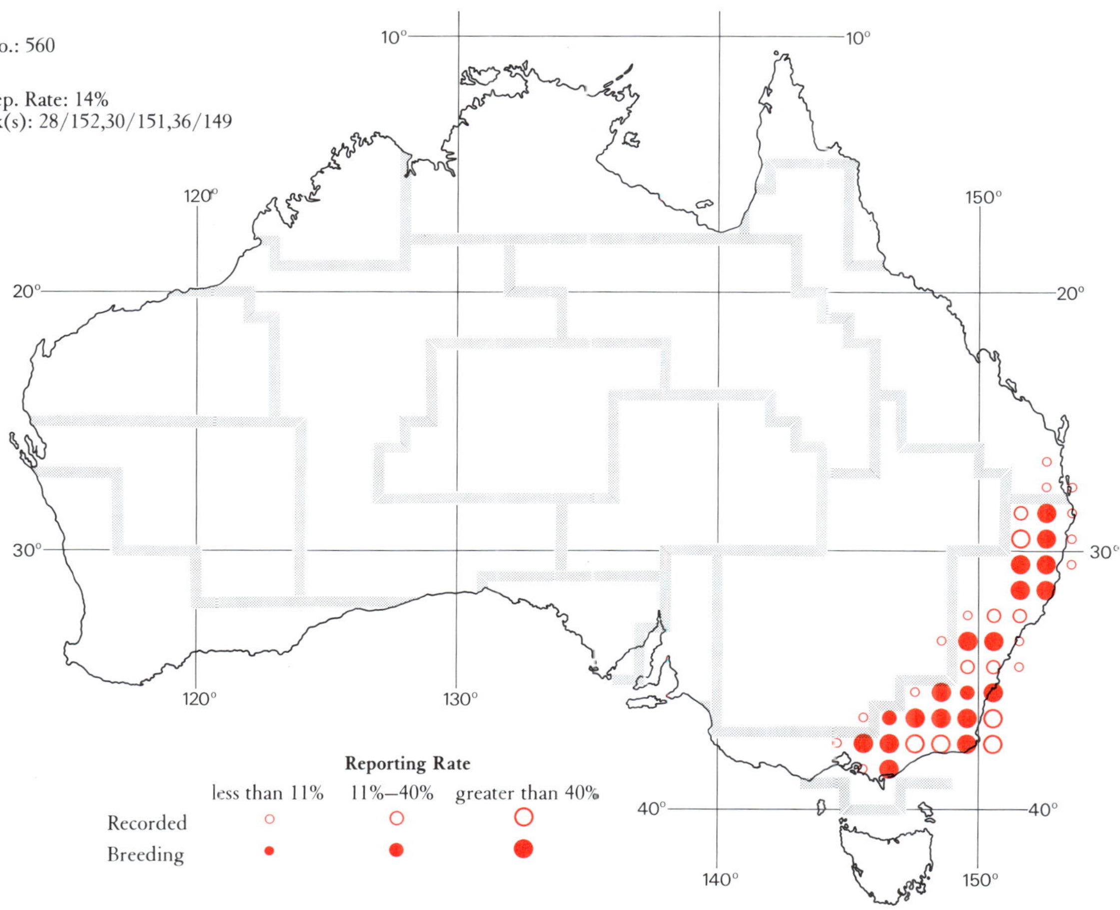

White-browed Treecreeper

Climacteris affinis

Two subspecies of the endemic White-browed Treecreeper are recognized: *affinis* in the east and *superciliosa* in the west, with a boundary north of Spencer Gulf (137°E).[602,978] Though this division coincides with a gap in Field Atlas records, the Historical Atlas contains records from a series of 1° blocks linking the two populations. Sightings north of the Simpson Desert (22/137, 23/137) suggest the possibility of contact between the two subspecies in northern parts of their ranges also. In addition, the Historical Atlas contains records from the lower Western Deserts Region, linking the central and western portions of the species' distribution, and extending it to the western coast.

The White-browed Treecreeper is present throughout the year in the same localities, inhabiting mainly acacia scrub and spinifex where mulga and sheoke are dominant.[599] It is usually solitary but has occasionally been reported in groups of two or three.[748] Communal breeding has not been recorded but is possible given the close relationship of the White-browed Treecreeper to the communally breeding Red-browed Treecreeper.[1349] The White-browed Treecreeper feeds on ants and other insects.[748,1049] It takes prey from the trunks of trees, as well as on the ground.[1345]

There are no measures of abundance.

Field Atlas Records: 455

1° blocks–with records: 119 (15%)
–with breeding records: 18 (15%)

Atlas No.: 561

Max. Rep. Rate: 26%
In block(s): 27/146

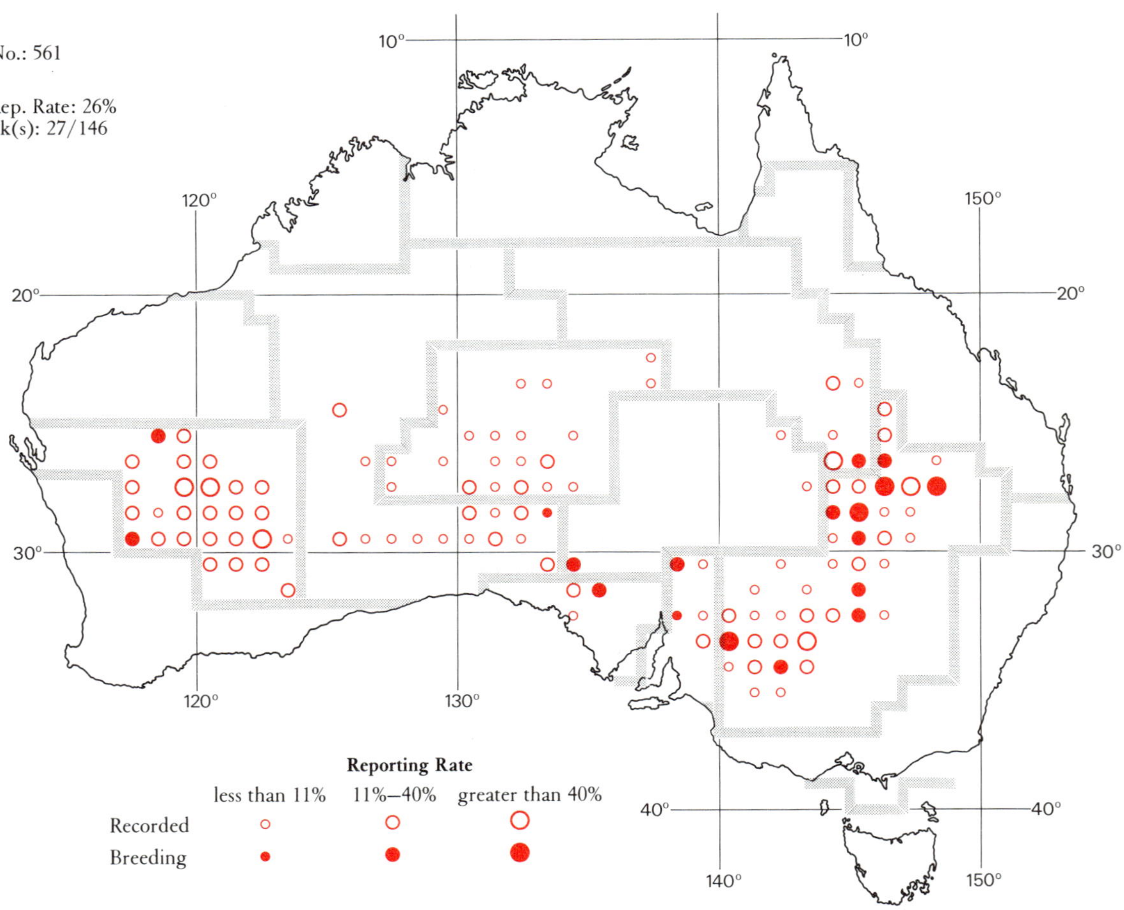

Brown Treecreeper

Climacteris picumnus

Two subspecies of the endemic Brown Treecreeper have been recognized: *melanota* to the north and *picumnus* to the south, intergrading in the region of a line joining Cairns (16/145) and Hughenden (20/144).[748,771,978] Records in the Historical Atlas show that the species is continuously distributed north to 15°S but in the Cape York Region it has been recorded only in 1° blocks on the western coast.

The Brown Treecreeper is sedentary.[189,1349] Groups of up to five individuals live together in permanent territories and breed communally.[501,1349,1351] The species occupies many different types of vegetation, commonly eucalypt woodland, sometimes adjacent forest and often where there is dead timber.[208,271,1348] The Brown Treecreeper forages by turning over leaves and litter,[870] spending about half of its time on the ground and on fallen logs.[1348] It also spirals up tree trunks both dead and living and will take insects on the wing. At Wollomombi (30/152) other species join the Brown Treecreeper to form foraging flocks.[129] The Brown Treecreeper feeds mainly on ants, beetles and larvae.[1049,1348] There is evidence of eviction from nest hollows by Common Starlings.[538,1128]

In 90 ha of semi-cleared stringybark-dominated farmland near Glen Innes (29/151) density varied from 0.20–0.27 birds/ha in three successive breeding seasons.[1351] At Armidale (30/151) in eucalypt woodland density was 0.04–0.10 birds/ha.[582] There were 0.25–0.50 birds/ha at Cowra (33/148)[208] and 2 birds/ha at Birchip (35/142).[1663] At Jamieson (37/146) in eucalypt forest density was 0.42 birds/ha.[1469]

Field Atlas Records: 8596

1° blocks–with records: 223 (27%)
–with breeding records: 113 (51%)

Atlas No.: 555

Max. Rep. Rate: 57%
In block(s): 29/146,35/144

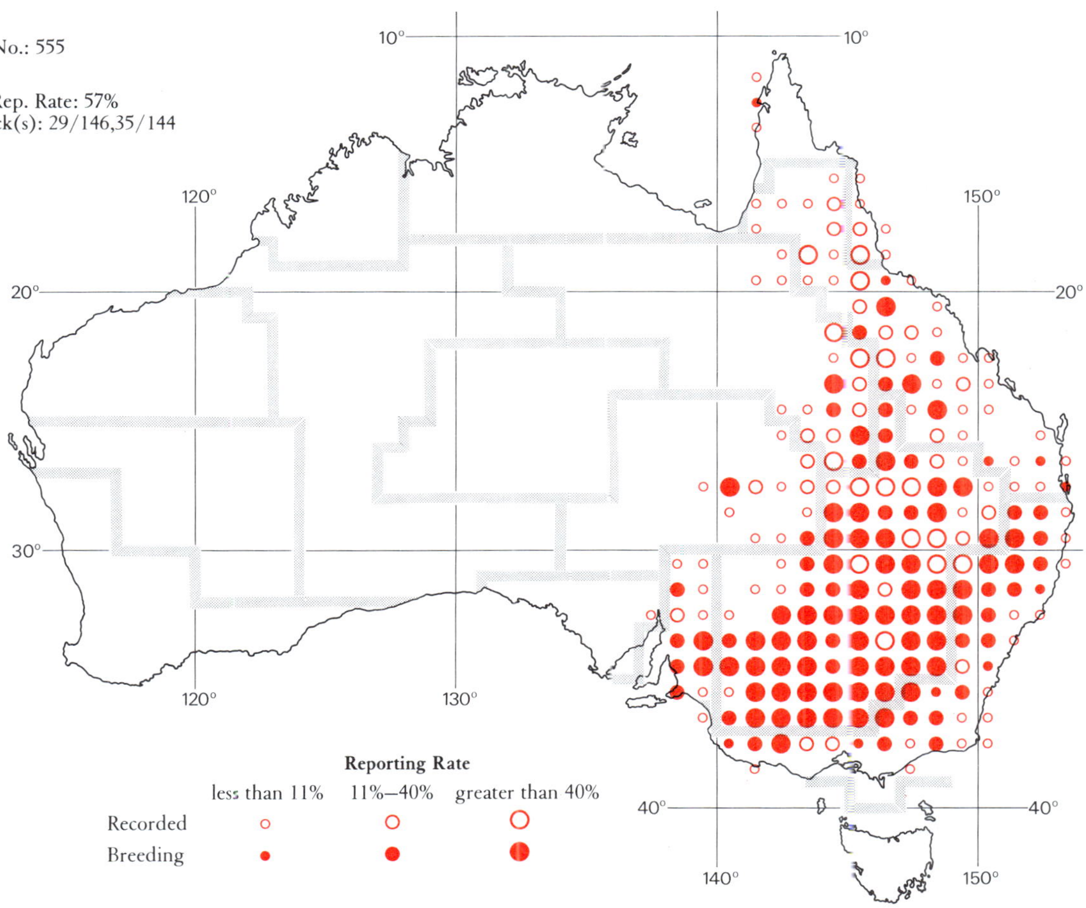

Rufous Treecreeper

Climacteris rufa

The South-West and Eyre Pen. populations of the endemic Rufous Treecreeper were once thought to be isolated and to differ morphologically.[978] The Atlases confirm that the population is continuously distributed in a narrow belt of country bordering the Great Victoria Desert and subspecies are no longer recognized.[599,748] In northern parts of the species' range it is sympatric with the White-browed Treecreeper.

The Rufous Treecreeper is present all year throughout its range and lives in eucalypt woodland, mallee[599] and forests dominated by karri and jarrah. In the Great Victoria Desert it occupies areas where eucalypts and casuarinas grow.[180] Where it overlaps with the White-browed Treecreeper the Rufous occupies mallee growing on sandy soil and the White-browed acacia scrub on loamy soil.[599] The Rufous Treecreeper has been recorded breeding communally.[1349] It feeds on insects, including ants, beetles and larvae, spiralling up the trees as it searches the bark and also foraging on the ground.[1862] In the South-West the bird no longer lives near settled areas.[1622]

There are no measures of abundance.

Field Atlas Records: 810

1° blocks-with records: 55 (7%)
-with breeding records: 22 (40%)

Atlas No.: 556

Max. Rep. Rate: 38%
In block(s): 32/120

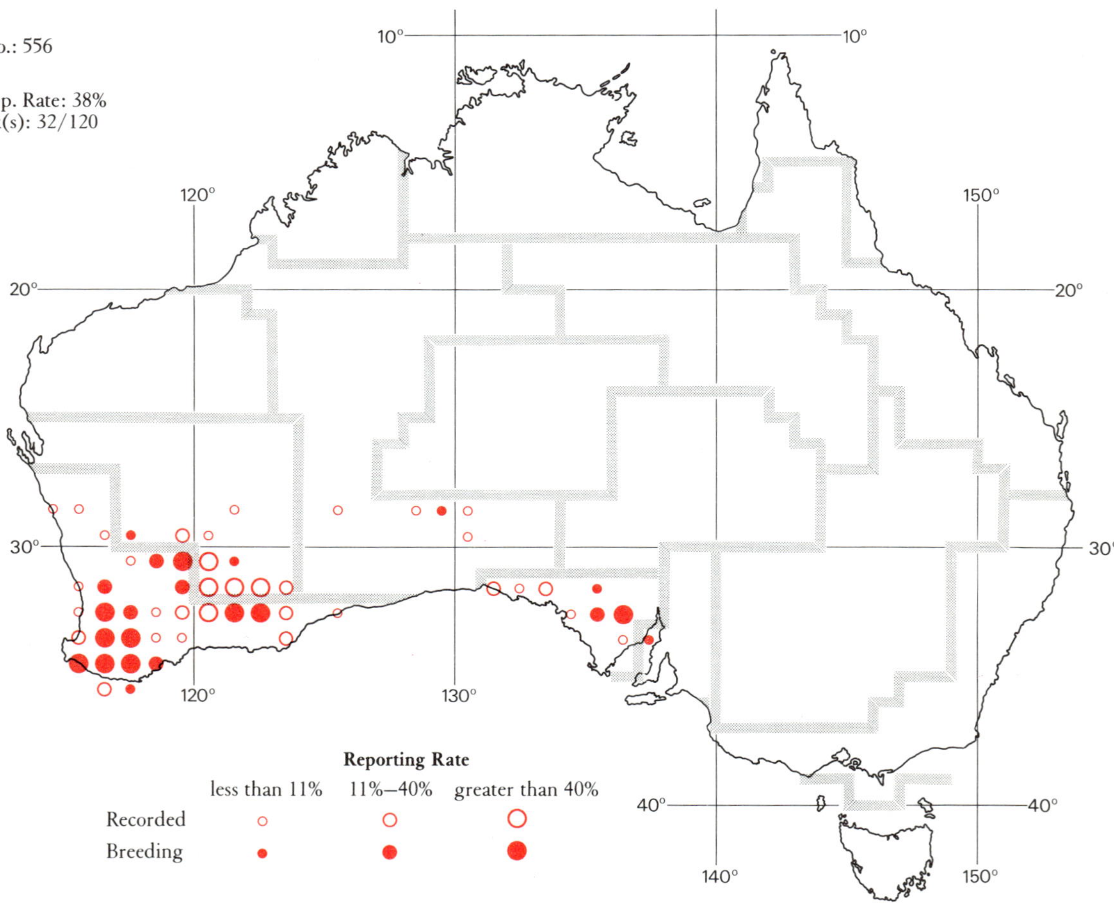

Black-tailed Treecreeper

Climacteris melanura

The Great Sandy Desert separates the two subspecies of the endemic Black-tailed Treecreeper: *melanura* in the north and *wellsi* in the Pilbara Region.[978,1124] Both populations have been recorded south of the range determined in the Field Atlas: *melanura* in several localities south to Alexandria (19/136) and Tennant Ck (19/134) in the western Rolling Downs Region; and *wellsi* at 24/114 near Carnarvon and at Gascoyne Junction (25/115).

The species is at the same localities throughout the year. It lives in woodland with a grassy understorey, in acacia scrub and, in the inland, along creeks lined with woodland.[608] It usually lives in pairs or small groups and communal breeding has been recorded.[501,1349]

The Black-tailed Treecreeper forages on the ground and in trees and searches live and dead timber.[748,1218,1602] Guts of specimens have contained ants, beetles, larvae and other insects but ants appear to be the main food.[748,787,1049]

There are no measures of abundance.

Field Atlas Records: 638

1° blocks–with records: 86 (11%)
–with breeding records: 6 (7%)

Atlas No.: 562

Max. Rep. Rate: 52%
In block(s): 16/130

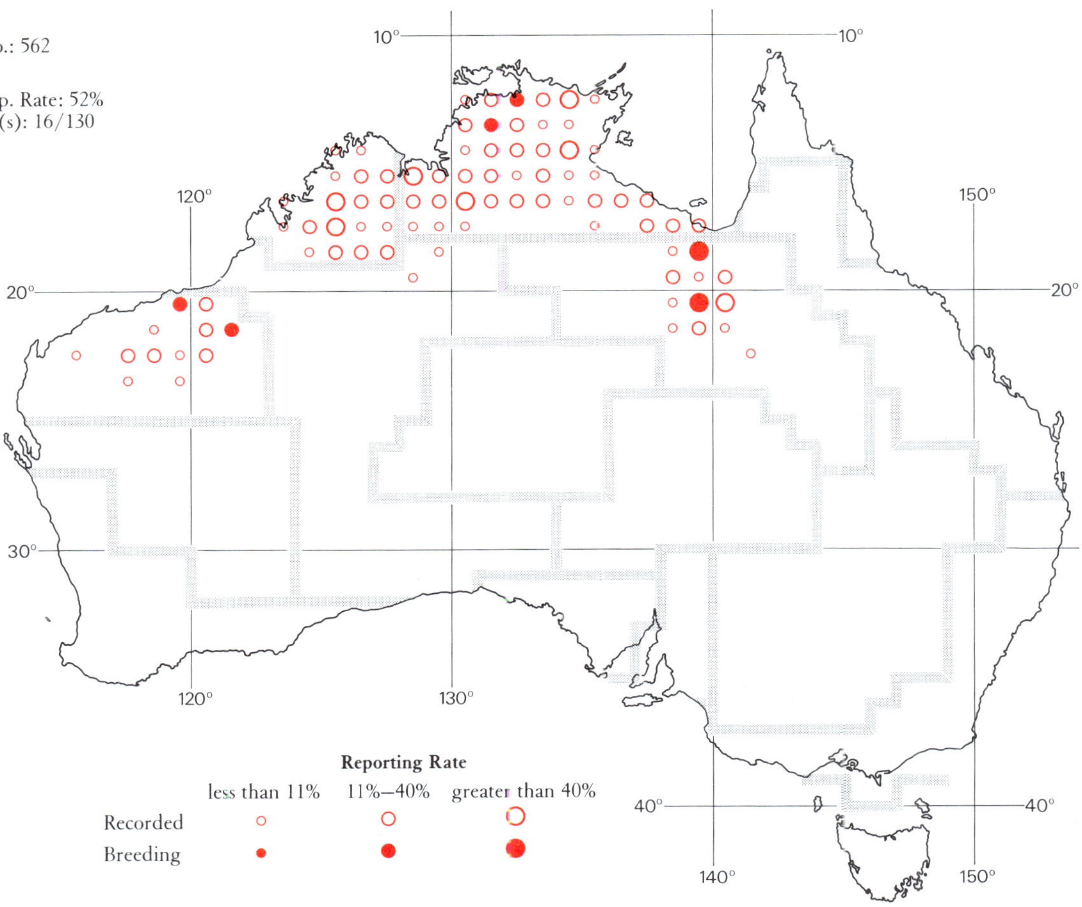

Red Wattlebird

Anthochaera carunculata

The Red Wattlebird lives mainly in Australia, occasionally straggling to New Zealand. In Australia, two subspecies are recognized: *carunculata* in the east and *woodwardi* in the Goldfields and South-West Regions.[1422] The Atlases combined show a continuous distribution round the Great Australian Bight. Early this century the species was unknown as a breeding bird near Sydney (33/151) where now it is abundant.[129]

Many observations suggest northward movement in winter in eastern Australia[284,987,1469] but this is not evident in the Field Atlas. Altitudinal movement is: in the Snowy Mts (36/148) above 500 m, Reporting Rates were 74% in spring-summer and 33% in autumn-winter; below 500 m there was little change (spring-summer 74%, autumn-winter 80%). The longest movement recorded is that of a banded adult from Berri (34/140) to Hazelwood Park (34/138), 200 km.[1438] In the South-West there are reports of movements north in autumn and south in spring near Perth (31/115).[783,1621] The Red Wattlebird is a winter visitor to Kellerberrin (31/117)[627] and the Avon Valley (32/116)[1216] and probably to places farther north as well; Reporting Rates north of 29°S in the South-West Region were 15% in winter and 7% in summer. At Eyre Bird Observatory (32/126) hundreds of Red Wattlebirds have been noted moving along the southern coast, east in summer and west in autumn.[390]

The Red Wattlebird lives usually in pairs in eucalypt forest, woodland, mallee, parks and gardens.[748,1124] It feeds on nectar, insects, fruit, manna and lerp[748,1049,1402] and several accounts mention birds feasting on insect abundances such as grasshopper swarms and soldier caterpillars. At Golton Vale (36/142) it feeds mainly on honeydew, taking much nectar in autumn but few insects.[1402]

Densities in eucalypt woodland were: Armidale (30/151) 0.05–0.92 birds/ha[582] and Wollomombi (30/152) 0.04–0.38 birds/ha.[129] In eucalypt forest they were: Jamieson (37/146) 0.2 birds/ha[1469] and Boola Boola (38/146) 0.44 birds/ha.[1113] In mallee at Eyre Bird Observatory density was 1.3 birds/ha in spring.[464]

Field Atlas Records: 18 047

1° blocks–with records: 192 (24%)
–with breeding records: 120 (63%)

Atlas No.: 638

Max. Rep. Rate: 72%
In block(s): 34/119

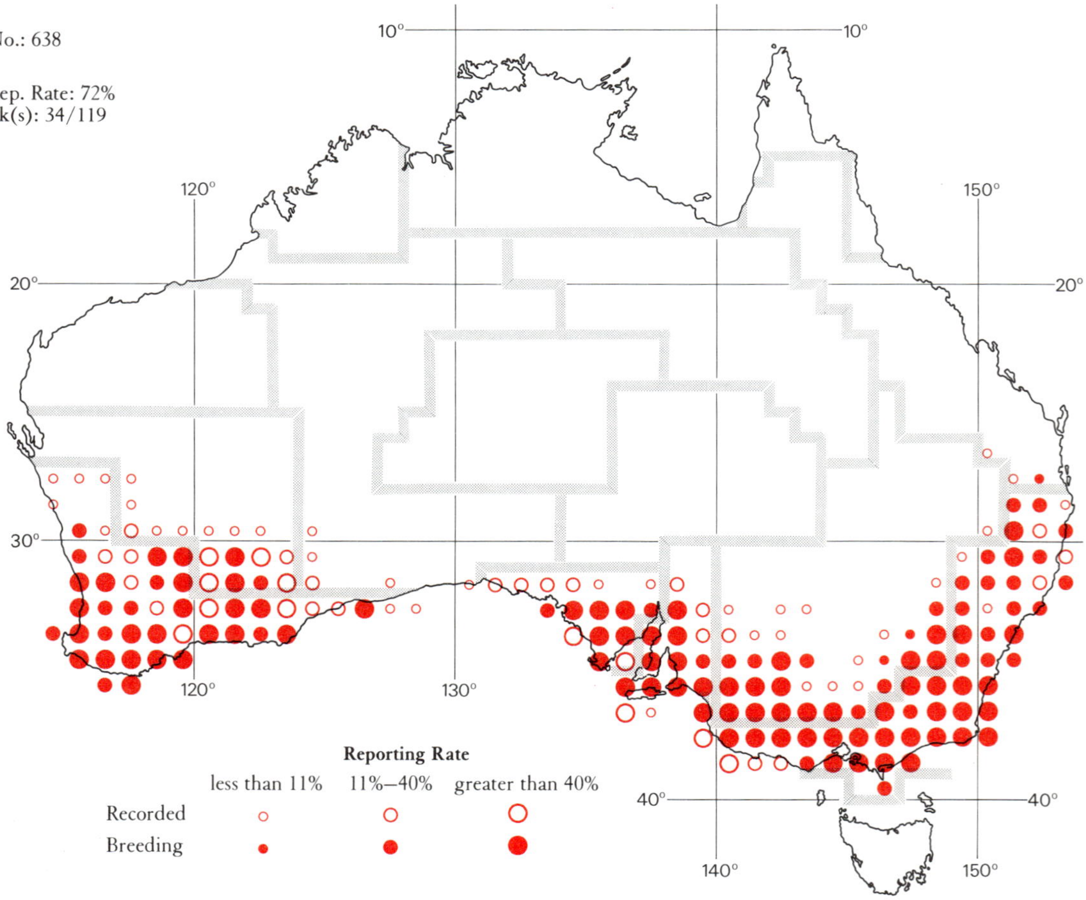

Yellow Wattlebird

Anthochaera paradoxa

The Yellow Wattlebird is endemic to Tas. and King I. (39/144) but absent from the Furneaux Group (39/148). Distribution has not changed over the historical periods.

Banding studies have shown that some birds are apparently sedentary, others undertake regular seasonal movement and yet others are nomadic.[1259] Birds have been recovered up to 155 km from the banding place.[1259] On Mt Wellington (42/147) the species is usually below 240 m but is present at higher altitudes in late summer.[1451] Near sea level at Green's Beach (41/146) birds are present throughout the year, numbers varying little.[636]

The Yellow Wattlebird lives in eucalypt forest and woodland, heath near the coast, gardens and orchards.[574,1124,1259] Distribution is less coastal than the Little Wattlebird and the Yellow is absent from the wet south-west.[1314]

The Yellow Wattlebird feeds on nectar, insects, spiders, fruit, both native and introduced, honeydew and manna.[574,1049,1259] It feeds mainly on insects obtained from the foliage and small branches of eucalypts.[1259] But the bird will forage anywhere from the tops of trees or under bark, to close to the ground on native fuchsia.[1259] Along with other honeyeaters, flocks converge on the cider gum in early autumn when manna is abundant.[1259] The Yellow Wattlebird is considered to be a game bird. Though on the partly-protected list, an open season may be declared, the most recent of which took place in 1972.

There are no measures of abundance.

Field Atlas Records: 1464

10′ blocks–with records: 212 (58%)

–with breeding records: 15 (7%)

Atlas No.: 639

Max. Rep. Rate: 87%

In block(s): 42°05′/147°45′

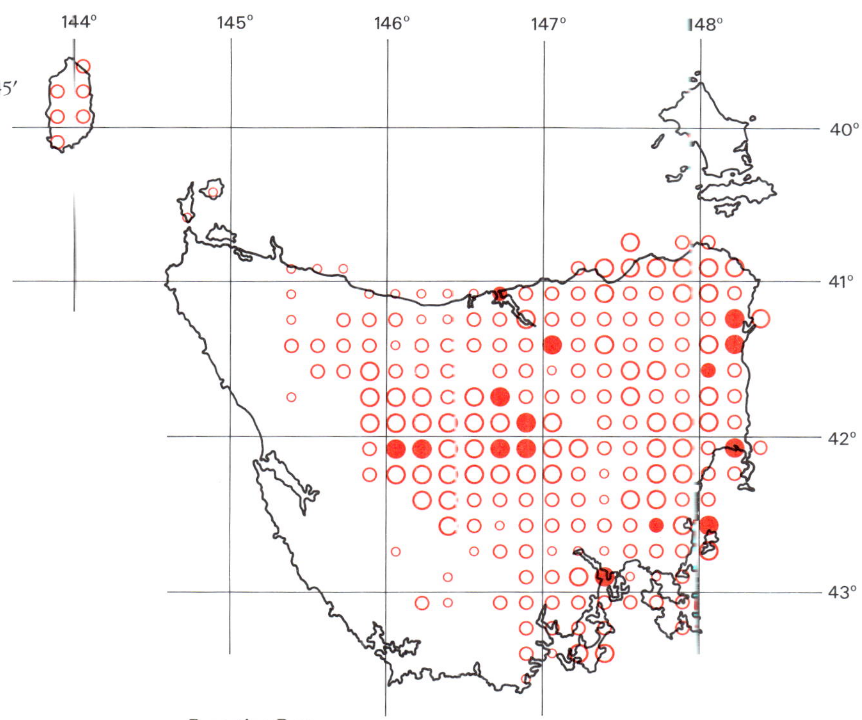

Reporting Rate

less than 11% 11%–40% greater than 40%

Recorded

Breeding

Little Wattlebird

Anthochaera chrysoptera

Four subspecies of the endemic Little Wattlebird are recognized: *chrysoptera* in eastern Australia; *halmaturina* on Kangaroo I. (35/137); *tasmanica* in Tas.; and *lunulata* in the South-West.[1422] The species is absent from the Bass Strait islands.

The Little Wattlebird is a winter visitor to Murphy's Ck (27/152)[1099] and the Field Atlas suggests that there may be some winter movement into Eastern Qld (Reporting Rates: summer 3.0%, winter 4.7%). Over much of its range the species is thought to be sedentary but in some places, e.g. Rockingham (32/115), it is reported to arrive when nectar-bearing plants are in flower.[987]

The Little Wattlebird lives in eucalypt forest, heath and gardens mostly at low altitude in coastal and sub-coastal districts. Individuals appear sporadically inland and at higher altitudes, e.g. in the Gibraltar Ra. (29/152) at 1000 m in Mar. 1978. In Tas. the distribution of the Little Wattlebird is more coastal than that of the Yellow Wattlebird.[1314]

The Little Wattlebird feeds on nectar, insects, manna, some honeydew[1402] and fruit, including that of lantana[1099] and inkweed.[483] Near Adelaide (34/138) it regularly visited three species of flowering eucalypt, two of banksia and one of astroloma, contributing to their pollination.[1403]

The population in Perth declined in the 1920s and 1930s[1622] but the species was frequently reported and found breeding there during the Field Atlas.

There are no measures of abundance.

Field Atlas Records: 6613

1° blocks–with records: 117 (14%)

–with breeding records: 55 (47%)

Atlas No.: 637

Max. Rep. Rate: 67%
In block(s): 34/151

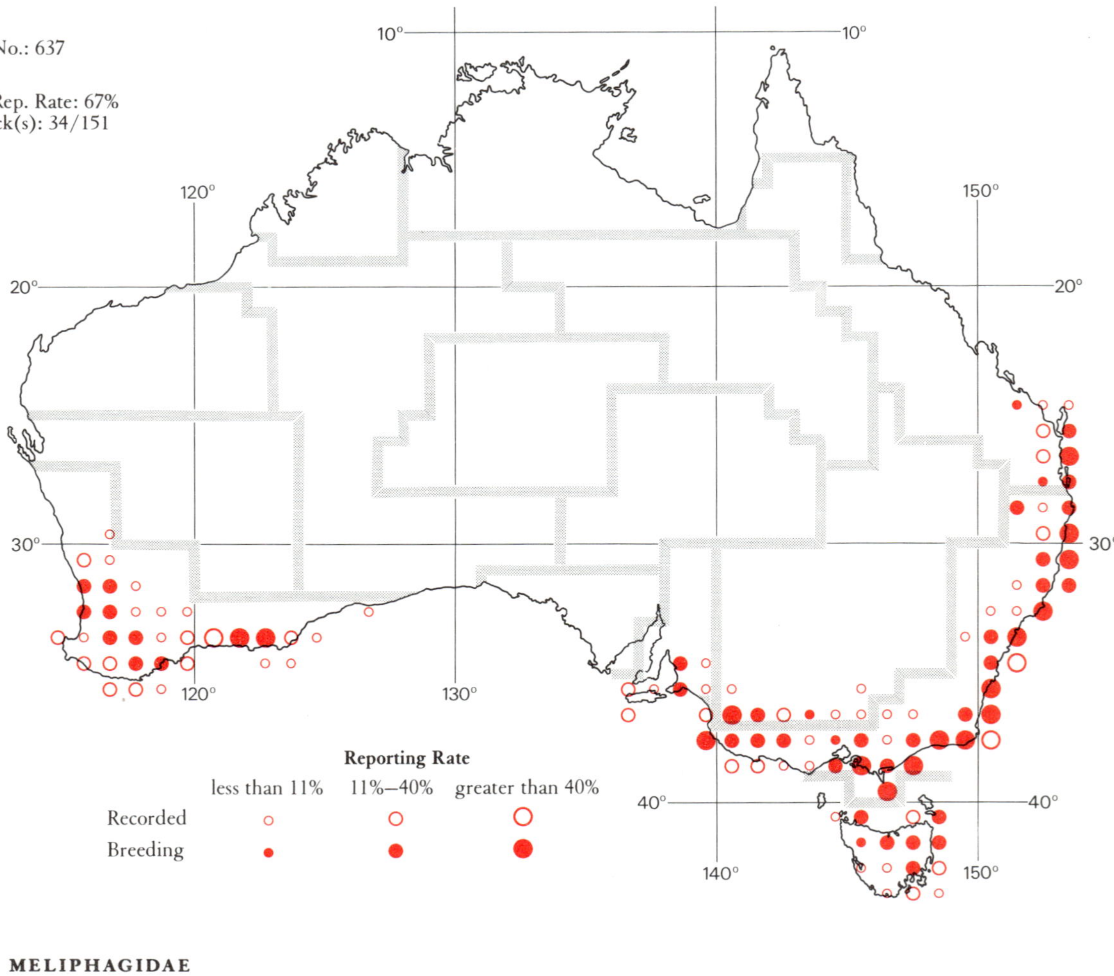

Spiny-cheeked Honeyeater

Acanthagenys rufogularis

No subspecies are recognized in the main population of the endemic Spiny-cheeked Honeyeater[748] but a specimen supposedly collected on Friday I. (10/142) in Torres Strait in 1883 has recently been described as a new subspecies.[1396] There has been no later record of the species from Torres Strait nor is it reported from the Cape York Region. In the South-East Region the population round Melbourne (37/145) appears only tenuously connected with the main range of the species.

The Spiny-cheeked Honeyeater is considered sedentary in southern parts of its range but there are observations of seasonal movement in the north.[987] At Mt Isa (20/139) it is more common in summer than in winter;[862] it is a winter visitor to the Macdonnell Ra. (23/133).[987] The longest movement recorded is that of an adult from Eremophila Flat (34/140) to Robertstown (33/139), 148 km.[30]

The Spiny-cheeked Honeyeater lives in eucalypt woodland, spinifex and acacia scrub,[1124] flocks of up to 100 gathering when food is abundant.[987] It takes a wide variety of food including insects, both on the wing and at rest, fruit, nectar, manna and mistletoe.[748,1049,1532] It will eat the fruit of such introduced species as grapes and peppercorns.

There are no measures of abundance.

Field Atlas Records: 10 244

1° blocks-with records: 515 (63%)
-with breeding records: 108 (21%)

Atlas No.: 640

Max. Rep. Rate: 74%
In block(s): 27/149

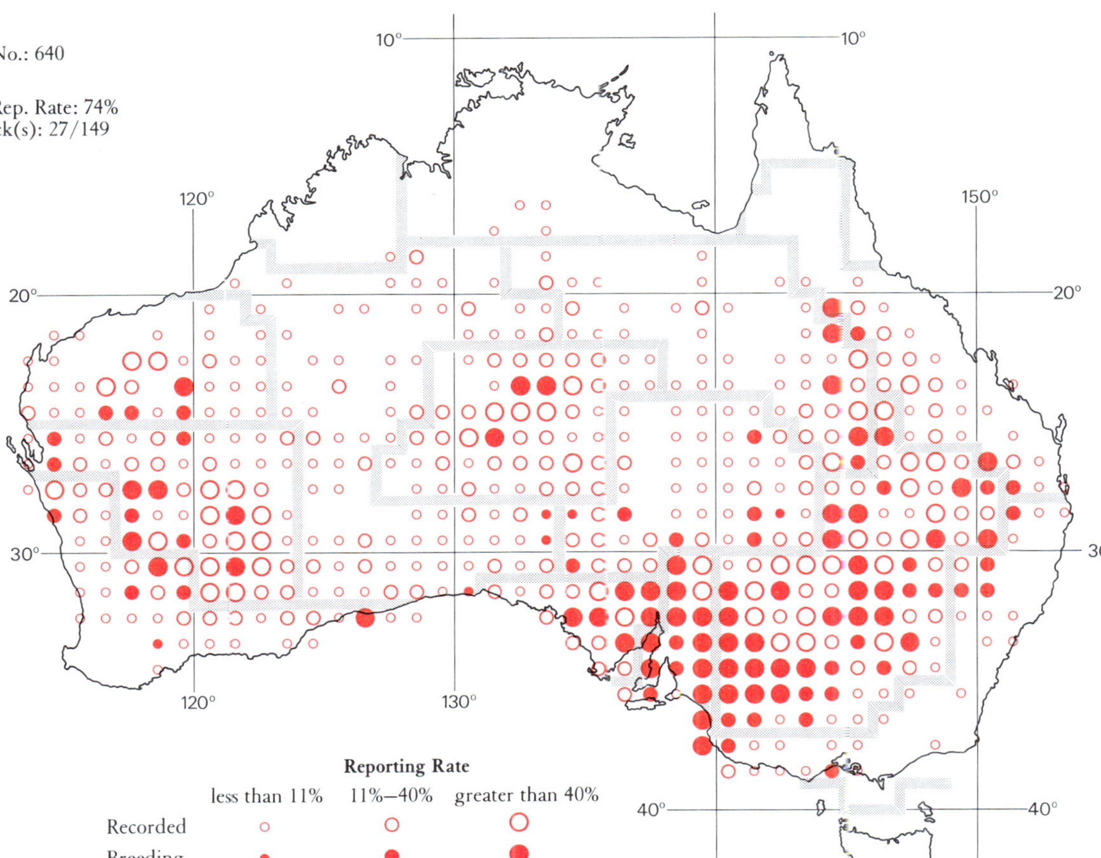

Striped Honeyeater

Plectorhyncha lanceolata

No subspecies of the endemic Striped Honeyeater have been recognized though the bird's size varies clinally and Qld birds are smaller than those in the south.[748] A coastal population north of Newcastle (32/151) is isolated by the Great Dividing Ra. but comes close to contacting an extension of the inland population down the Hunter R (31–32/151) valley.[842] Birds on Yorke Pen. (34/137) appear isolated from the main range of the species.

There are irruptions from time to time near the edges of the bird's range.[987] At Tuggerah (33/151) it is present each year in summer and autumn. The Field Atlas suggests no large-scale seasonal movement.

The Striped Honeyeater is usually encountered singly or in pairs, rarely in parties as large as eight. Inland it lives in mallee and eucalypt woodland. On the coast at Cape Gloucester (20/148) the bird was observed in mangroves[748] and on the NSW coast it inhabits dense stands of paperbarks and casuarinas.[842] The Striped Honeyeater feeds on insects of many kinds[1049] and takes some nectar.[1403] It has been observed eating prickly pear[1086] and will attack orchard fruit so that it is sometimes considered a pest by orchardists. It also eats mistletoe berries.[1099]

There are no measures of abundance.

Field Atlas Records: 3931

1° blocks–with records: 159 (20%)
–with breeding records: 76 (48%)

Atlas No.: 585

Max. Rep. Rate: 77%
In block(s): 27/149

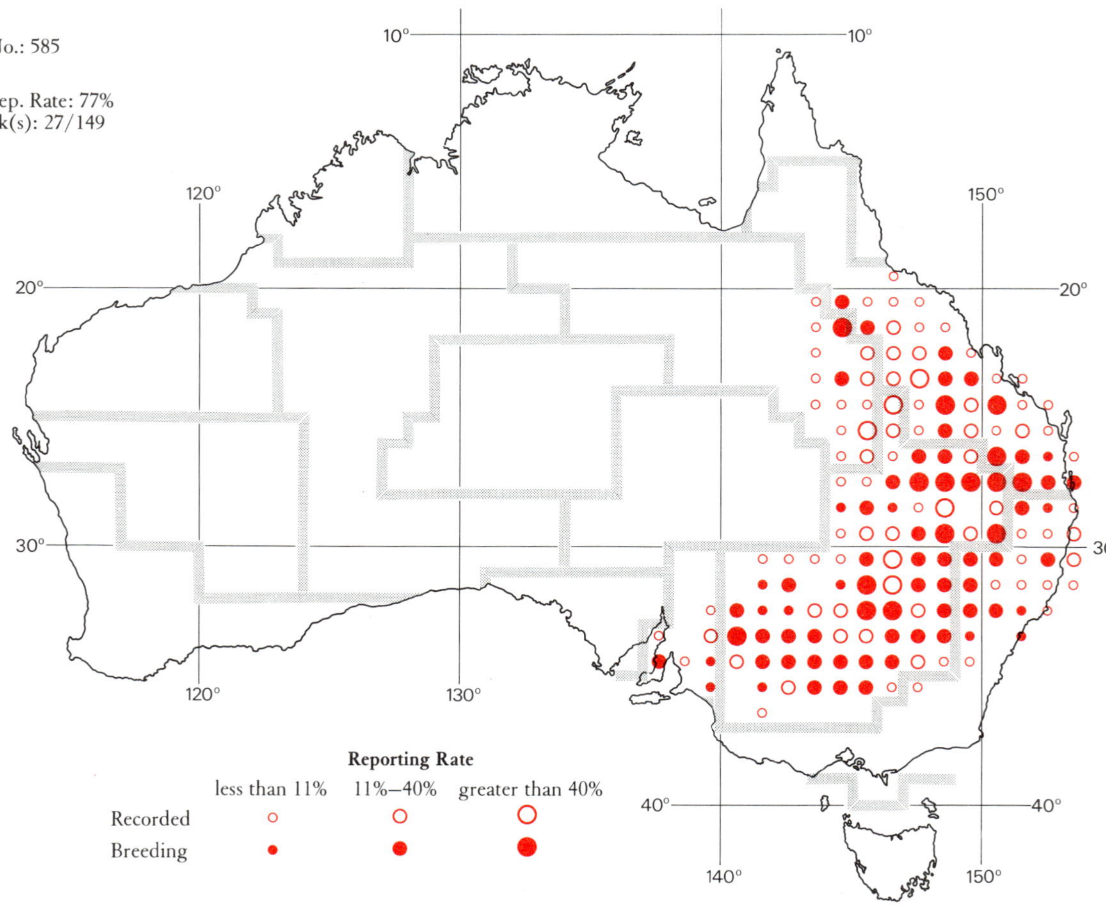

Helmeted Friarbird

Philemon buceroides

The range of the Helmeted Friarbird extends through southern Indonesia, New Guinea and adjacent islands to Australia. Geographical variation in morphology has been described but there is doubt about whether more than one species should be recognized in Australian populations.[1373,1576] The gap between Qld and Top End birds is wide. The population in the Cape York Region has not been reported farther south on the coast of the Gulf of Carpentaria than shown. In the Top End, distribution is continuous along the northern coast.

Nomadic movements related to the presence of flowering plants have been noted in Qld but there is no evidence of large-scale movement.[503,987] In Darwin (12/130) the Helmeted Friarbird is considered an irregular visitor, commonest Dec.-Jan.,[423] but Reporting Rates in the district varied little between summer (10.0%) and winter (9.4%).

The Helmeted Friarbird lives in rainforest on the sandstone escarpment of Arnhem Land, mangroves, open eucalypt woodland, gardens and orchards.[423,1267] Its nest is often near those of the Figbird, Spangled Drongo and Metallic Starling.[104,683,1145] The bird eats insects, spiders, fruit and nectar.[748,1124]

In New Guinea density near Port Moresby (9S/147E) in eucalypt woodland was 0.6 birds/ha throughout the year.[129]

Field Atlas Records: 984

1° blocks–with records: 40 (5%)
–with breeding records: 7 (18%)

Atlas No.: 642

Max. Rep. Rate: 58%
In block(s): 15/145

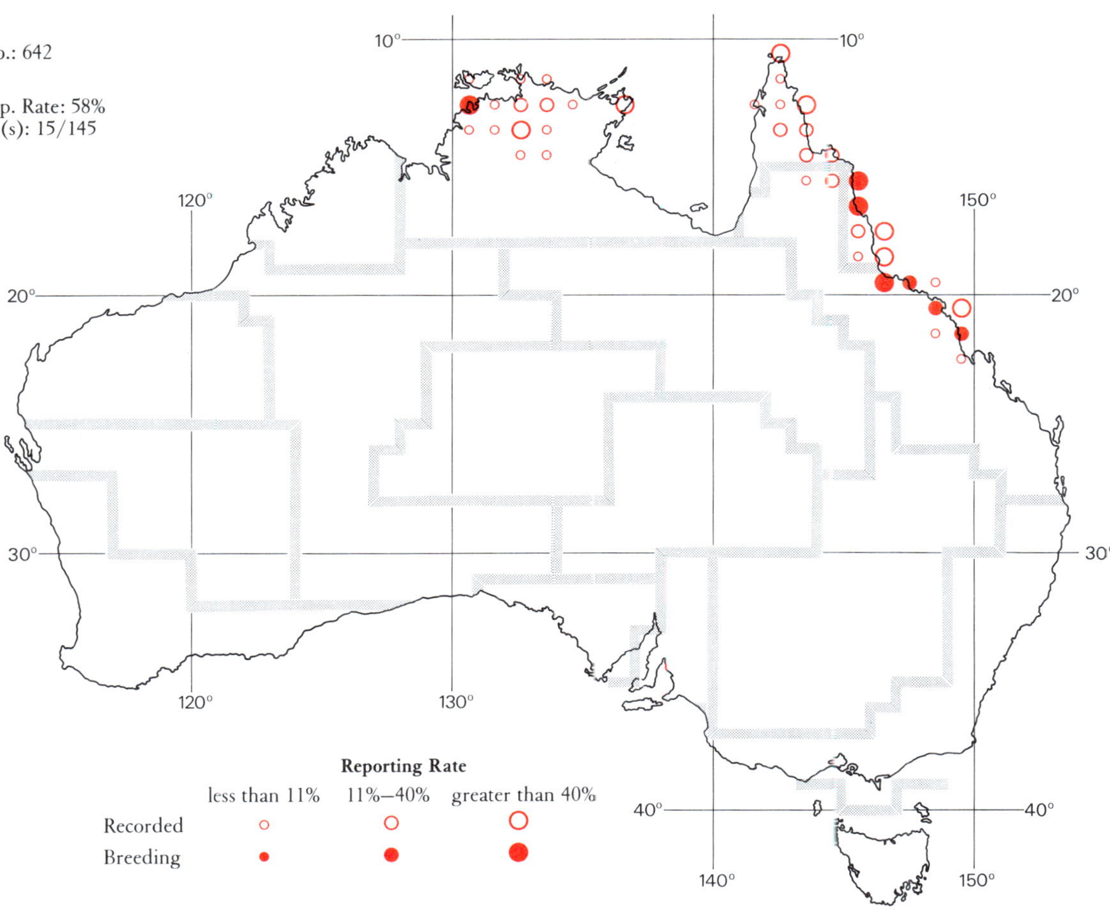

Silver-crowned Friarbird

Philemon argenticeps

Views about subspecies of the endemic Silver-crowned Friarbird differ.[748,1422] In the Atlases there is no gap between the Top End and Kimberley populations but there is one between birds in the Cape York Region and those in the Top End.

The species' distribution in Qld is uncertain. There are occasional historical reports on the coast as far south as Shoalwater Bay (22/150)[144] and Parkhurst (23/150).[1087] Inland, an unusual influx in Mar.-June 1974 brought the bird to Prairie (21/144) and Tiree (21/145), not far from where it had been reported in Mar.-Apr. 1964 at Glendower (20/144).[748]

Movements may be seasonal: at Mt Isa (20/139) numbers fluctuate but the species is most abundant in the wet summer season;[862] at Batchelor (13/131) it appeared to be absent in the dry winter season;[1602] and in the Kimberley Region it leaves the level forest country in Jan. to move to the sandstone hills where it breeds.[787] The sightings in southern and inland Qld suggest that nomadic or irruptive movements also take place.

The Silver-crowned Friarbird lives in eucalypt forest, woodland with an understorey of grass, acacia scrub and gardens.[748,1124] It feeds mainly on insects, usually on those attracted to flowering trees.[108,857,1145,1218,1859] It will also take seeds and figs and has been observed to catch insects on the wing.[787,1809]

There are no measures of abundance.

Field Atlas Records: 977

1° blocks-with records: 91 (11%)
-with breeding records: 9 (10%)

Atlas No.: 644

Max. Rep. Rate: 52%
In block(s): 12/136,13/130

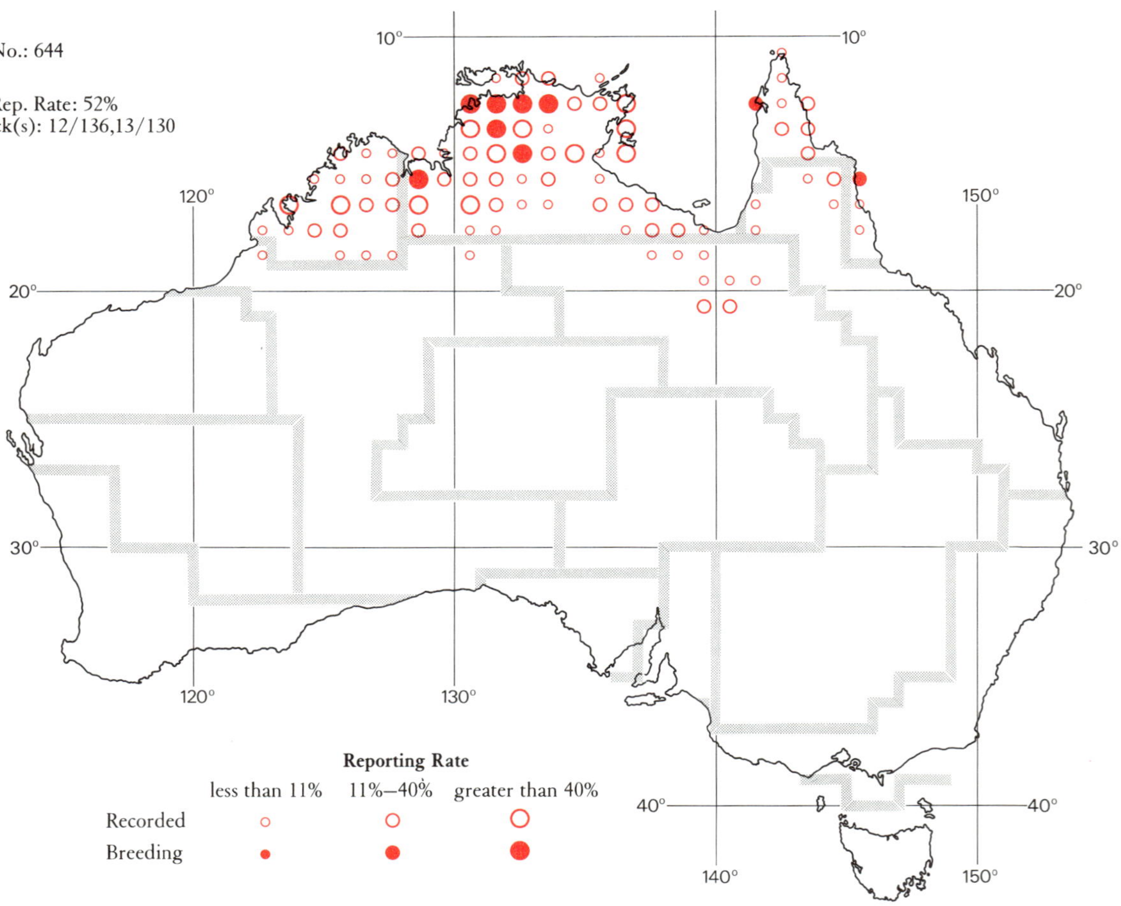

Noisy Friarbird

Philemon corniculatus

The Noisy Friarbird lives in southern New Guinea and eastern Australia. A gradient of increasing size from north to south has been used to define three subspecies but there is doubt about their geographical and taxonomic limits.[748,1422] The species is absent from Torres Strait and the tip of the Cape York Region. Birds sometimes wander west of the main range, usually along watercourses, e.g. there are records from the Murray R as far west as Overland Corner (34/140).[713]

Reported movements include: north through Kingscliff (28/153) in June and Hastings Pt (28/153) in Mar.-Apr.;[1069] out of the Armidale (30/151), Wollomombi (30/152) and Canberra (35/149) districts in Apr.-May with a return in spring;[154,653,987] and out of the northern[1344] and southern[833] Murray-Darling Region in winter. At Moruya (35/150) the Noisy Frairbird is a partial migrant.[1197] These observations indicate movement both to lower altitudes and northward in winter. Movement out of southern parts of the species' range in winter was reflected in Reporting Rates (south of 35°S in the South-East Region: summer 10.3%, winter 3.3%). No specimen of the southern population has been taken north of central Qld,[1086] which suggests this to be the limit of northward migration and Reporting Rates suggest no influx of birds into Eastern Qld in winter (summer 29.8%, winter 26.5%). In the Atherton Region the Noisy Friarbird moves about locally to visit flowering trees[231] (Reporting Rates: summer 23.3%, winter 14.3%). Elsewhere, movements are associated with good flows of nectar, e.g. at Murphy's Ck (27/152) the species is only present in winters when eucalypts flower heavily.[1099] The longest movement recorded is that of a bird from Mudgee (32/149) to Mitta Mitta (36/147), 510 km.[60]

The Noisy Friarbird lives singly, in pairs or flocks in eucalypt forest and woodland, avoiding dense vegetation.[488,748] It feeds on insects and nectar and will take nestlings.[327,1049,1531] In late summer and autumn it raids orchards and in the past was often illegally shot.[1128] At Rutherglen (36/146) it was common in the 1930s but the Little Friarbird gradually replaced it as clearing and development progressed.[1128]

Density in eucalypt woodland was: Armidale 0.67 birds/ha,[582] Wollomombi 0.1 birds/ha,[129] Canberra 0.06–0.20 birds/ha,[154] Moruya 0.21 birds/ha.[1197]

Field Atlas Records: 8670

1° blocks–with records: 180 (22%)
–with breeding records: 84 (47%)

Atlas No.: 645

Max. Rep. Rate: 70%
In block(s): 24/153,26/153

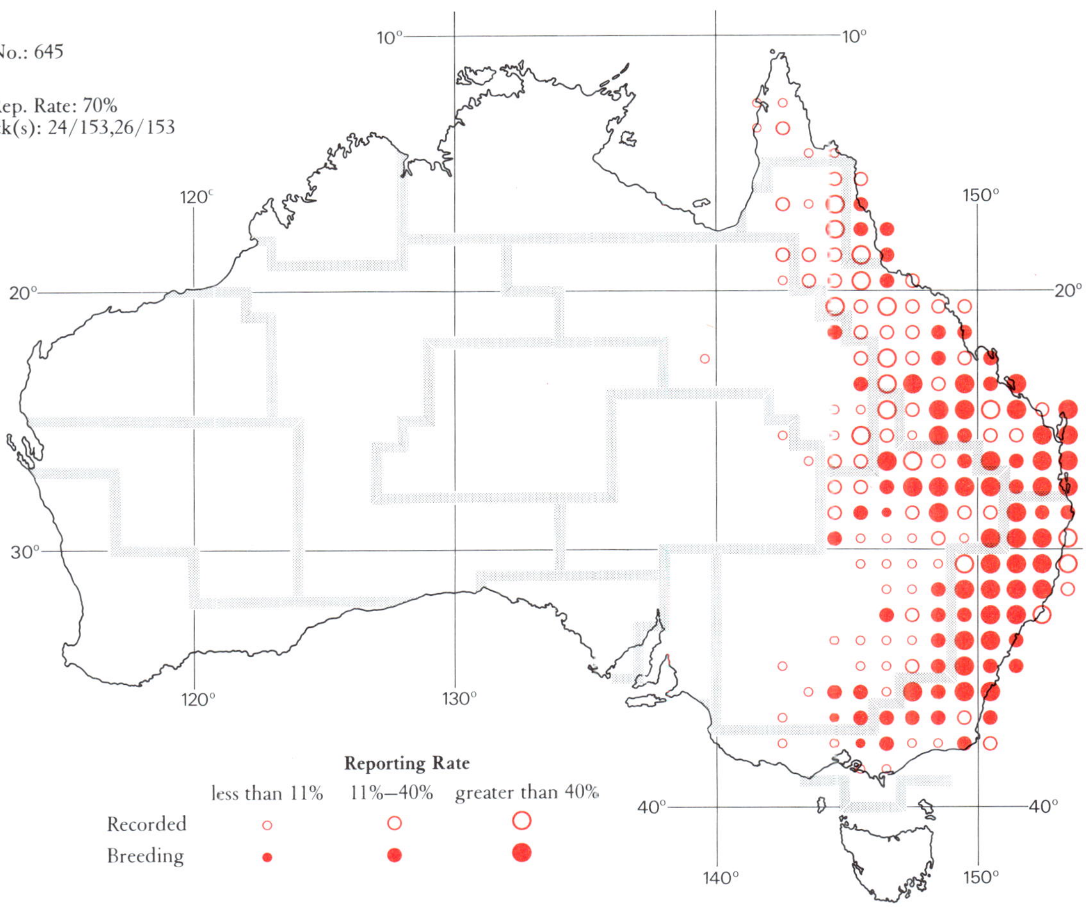

Little Friarbird

Philemon citreogularis

The Little Friarbird lives in southern New Guinea and adjacent islands and Australia. The Atlases show no gaps in distribution but six Australian subspecies are recognized: *johnstoni* in the Cape York Region; *citreogularis* in the rest of eastern Australia; *carpentariae* south of the Gulf of Carpentaria; *breda* on Melville I. (11/130); *sordidus* elsewhere in the Top End; and *occidentalis* in the Kimberley Region.[1422] There is only a single record from Torres Strait[503] though the species is abundant in the Cape York Region. In the South-East Region it appears sporadically on the coastal side of the Great Dividing Ra.

The Field Atlas suggests a northward movement in winter by the subspecies *citreogularis* with the Reporting Rate south of 26°S higher in summer (8.1%) than in winter (3.9%). The movement is confirmed by observations.[1709] In the Atherton Region the Little Friarbird is a visitor June-Dec.[231] At Murphy's Ck (27/152) it is rare in winter.[1099] It leaves the southern Murray-Darling Region in Mar.-Apr. and returns in Sept.[833]

South of the Gulf of Carpentaria at Mt Isa (20/139) birds are present throughout the year but most common Sept.-Mar.[862,987] Fluctuating numbers round Darwin (12/130)[423] suggest it to be nomadic there. The longest movement recorded is that of a bird from Gregory (35/147) to Boggabri (30/150), 644 km.[823]

The Little Friarbird lives in eucalypt forest and farmland and, in northern Australia, in eucalypt woodland near water and in the gorges of the Arnhem Land escarpment.[748] It congregates in small but noisy parties of 6–15 birds. The Little Friarbird feeds on nectar, insects and fruit, native and cultivated. At Rutherglen (36/146) it has gradually replaced the Noisy Friarbird since the 1930s with the progress of clearing and development.[1128]

There are no measures of abundance.

Field Atlas Records: 7486

1° blocks–with records: 309 (38%)
–with breeding records: 85 (28%)

Atlas No.: 646

Max. Rep. Rate: 68%
In block(s): 27/149

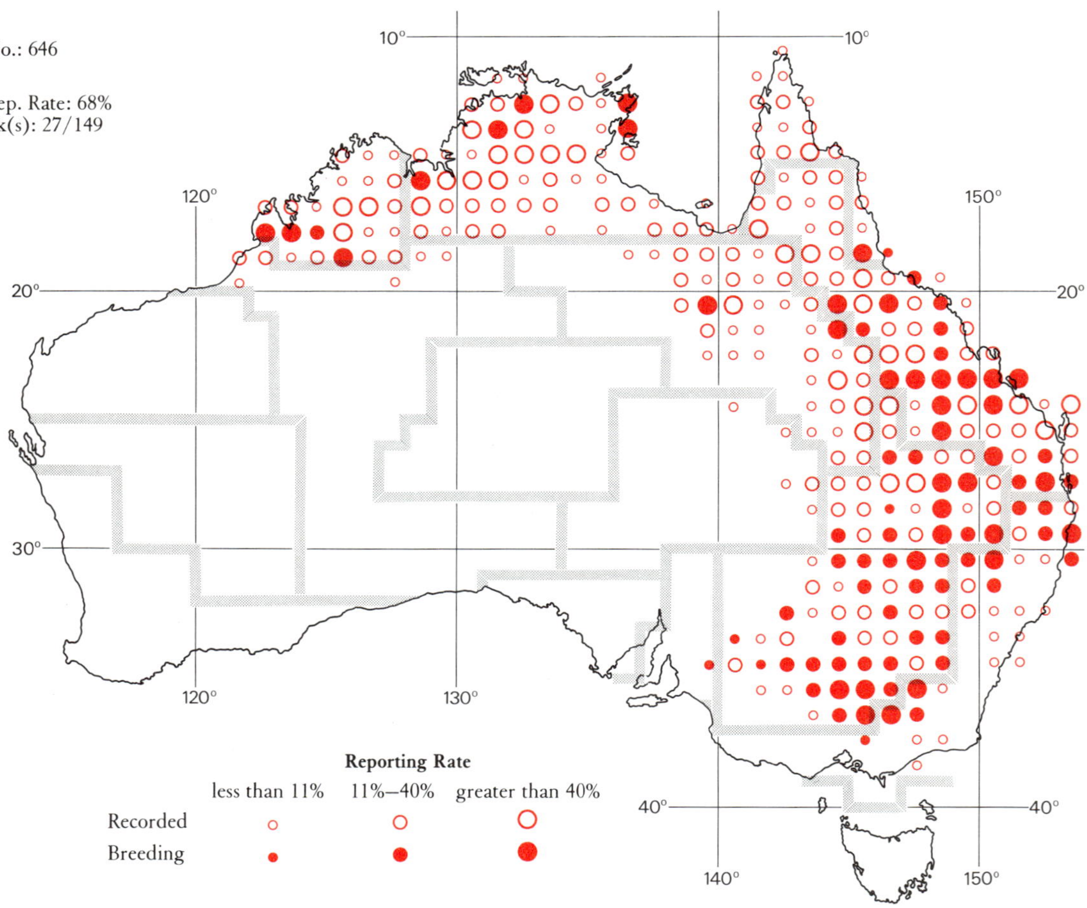

Regent Honeyeater

Xanthomyza phrygia

Compared with the Historical Atlas, the Field Atlas shows that the range of the endemic Regent Honeyeater has contracted. Before 1950 the species appeared sporadically in Eastern Qld as far north as Byfield (22/150) and the Dawson R (23/149).[102,1709] The northernmost breeding record is from Buckinbah (28/148) in 1891. To the west and south the Honeyeater's distribution extended to Kangaroo I. (35/137) and the Mt Lofty Ra. (34/138) where it bred in 1958[383] and again in the 1970s.[1369] The only confirmed sighting in this region during the Field Atlas was of a single bird at Rostrevor (34/134) in 1977.

Movements seem to be nomadic though the bird will regularly appear in some districts each year when banksias and eucalypts flower.[325,987,1343] Sometimes it is absent for years or arrives in reduced numbers in districts where it was formerly abundant. At Rutherglen (36/146) the Regent Honeyeater is said to be an irregular summer migrant.[1128] At Jamieson (37/146) it is present Nov.-Jan. but only in some years and has declined in numbers and frequency of appearance.[1469] Until 1977 small parties visited the Maryborough (37/143) district when the ironbark flowered, both summer and winter. The Field Atlas suggests no large-scale seasonal pattern of movement.

The Regent Honeyeater lives in eucalypt forest and woodland. Most Field Atlas sightings were of single birds or small groups of up to about six, nearly all high in the canopy of flowering eucalypts. Flocks gather in flowering trees, including those in street and garden.[973,987] At these times the Regent Honeyeater takes much nectar but guts contained many insects[1049] and the bird has been recorded to eat caterpillars.[637] It will raid orchards to take cultivated fruit.[1556]

The Regent Honeyeater is thought to be decreasing in numbers as well as range.[1113,1292] The frequency of records in the Atlases has declined during the three historical periods to its lowest point in the Field Atlas[1421] and fewer birds are now seen. Flocks of 50–100 were regularly reported in the past[1343] but in the Field Atlas there was only one account of large numbers: flocks of 30–40 appeared irregularly near Bermagui (36/150).

Field Atlas Records: 172

1° blocks–with records: 38 (5%)

–with breeding records: 12 (32%)

Atlas No.: 603

Max. Rep. Rate: 3%
In block(s): 29/151, 32/149, 36/146, 36/150

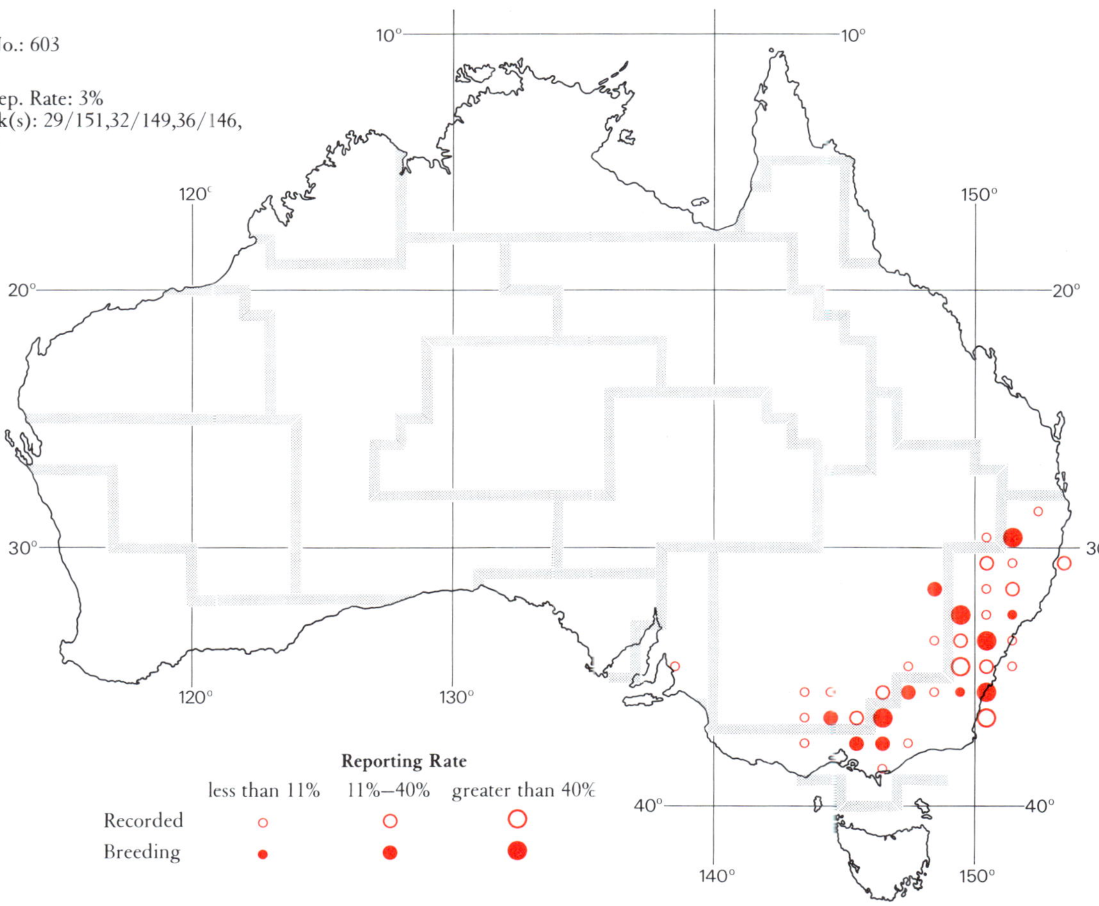

Blue-faced Honeyeater

Entomyzon cyanotis

The Blue-faced Honeyeater lives in southern New Guinea and Australia. Four Australian subspecies are recognized: *harterti* in the Cape York Region south to Cairns (16/145); *cyanotis* south of Cairns; *apsleyi* on Melville I. (11/130); and *albipennis* in the rest of the Top End and Kimberley Regions. Though *harterti* and *albipennis* appear from the maps to meet at the head of the Gulf of Carpentaria, no intergradation was found in specimens.[748]

Though the species is said to be sedentary over most of its range,[987] regular movements of small numbers of birds have been reported. At Jandowae (26/151) the species moves north-east in Mar.-June, returning south-west in July-Aug.;[1331] at Inverell (29/151) eastwards in Jan.-May and westwards in June-July;[96] and near Darwin (12/130) there is a regular increase in numbers in woodland in summer and a corresponding decrease in rainforest along creeks.[426] A report of a single bird near Cowell (33/136) in Dec. 1981 shows that the species can move far but the Field Atlas does not suggest large-scale seasonal movement.

The Blue-faced Honeyeater lives in eucalypt woodland, specially along watercourses, on the fringes of rainforest and in mallee but it is also well known in farmland. Sometimes it forms groups of up to 30 individuals but it is usually seen singly, in pairs or in small parties.[748,1124] It feeds on insects as well as nectar and fruit.

There are no measures of abundance.

Field Atlas Records: 6493

1° blocks–with records: 250 (31%)
–with breeding records: 81 (32%)

Atlas No.: 641

Max. Rep. Rate: 78%
In block(s): 27/149

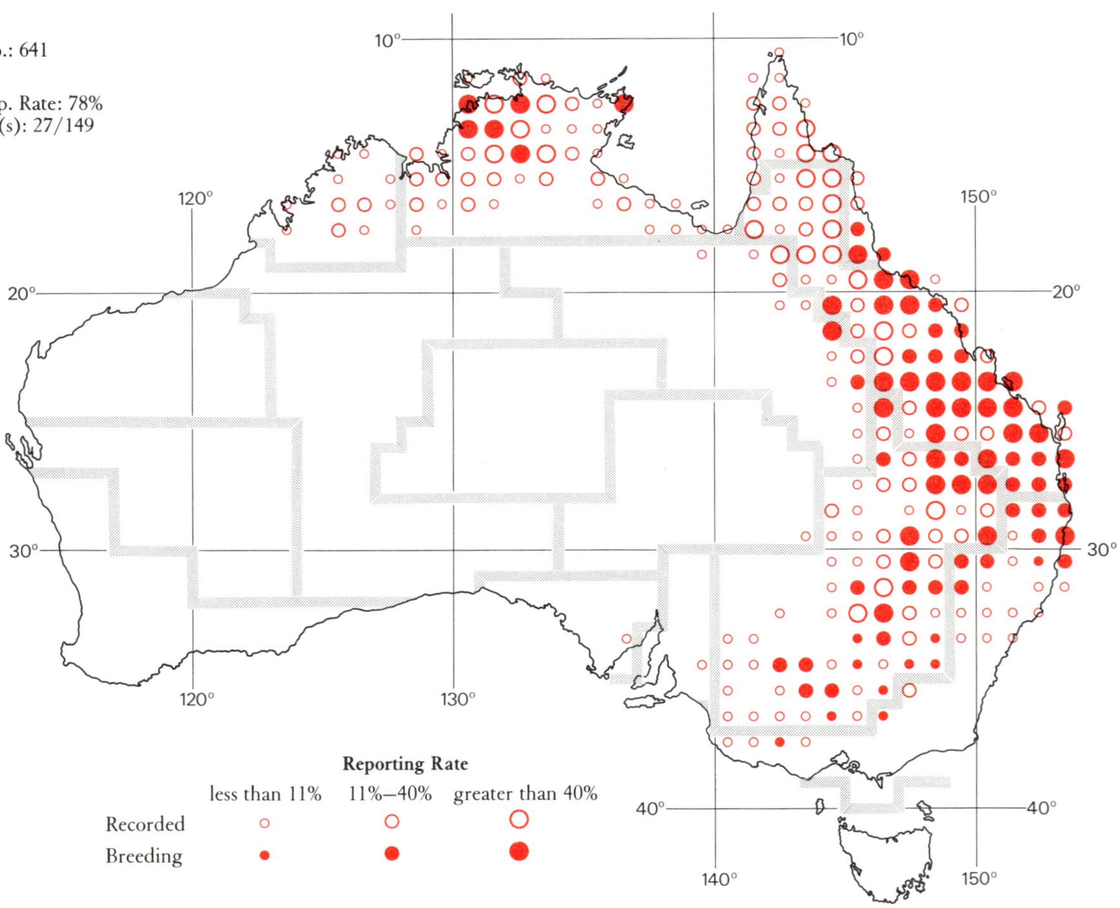

No subspecies of the endemic Bell Miner are recognized though there are apparent breaks in distribution near Ulladulla (35/150) and between the Dandenong Ra. (37/145) and populations to the east.

The species is no longer found west of Melbourne (37/145).[120,299] Until the 1850s it inhabited creeks in the Sunbury (37/144) area but there are no later records except for two birds seen drinking at a pool in the Lerderderg Gorge (37/144) during the drought of Mar. 1967. Last century an observer claimed to have heard Bell Miners on the Murray R at Moira L. (35/144)[299] but there are no other reports so far inland.

The Bell Miner is sedentary. It lives in 'clans' of one or more breeding pairs and numerous helpers.[364] Each clan has a foraging area outside which the members rarely move; within it, breeding pairs occupy loosely overlapping home ranges. A population has been observed to move at least 600 m, probably because an infestation of saw-fly larvae had reduced the food supply.[1674] The Bell Miner lives in eucalypt woodland and forest, often along gullies that have heavy undergrowth beneath the trees. In places the species inhabits gardens.

The Bell Miner feeds largely on lerp and occasionally on insects.[1728] Near Melbourne birds left 77% of the psyllids uneaten after consuming the lerp which is the cover protecting them.[1402] Wherever Bell Miners live the trees are found to be infested with psyllids which cause varying degrees of defoliation. The Bell Miner is often blamed for the dieback though the bird is actually eating the psyllids.[1918]

Recent evidence shows that the Bell Miner is less effective than other species in reducing the number of psyllids

Bell Miner

Manorina melanophrys

but it excludes these other birds from its territory.[1110,1769]

Near Melbourne groups occupy territories of 0.2–2 ha[1728] or 0.2–0.3 ha[1674] and densities are 13–23 birds/ha.[364]

Field Atlas Records: 2650

1° blocks–with records: 31 (4%)
–with breeding records: 15 (48%)

Atlas No.: 633

Max. Rep. Rate: 65%
In block(s): 36/150

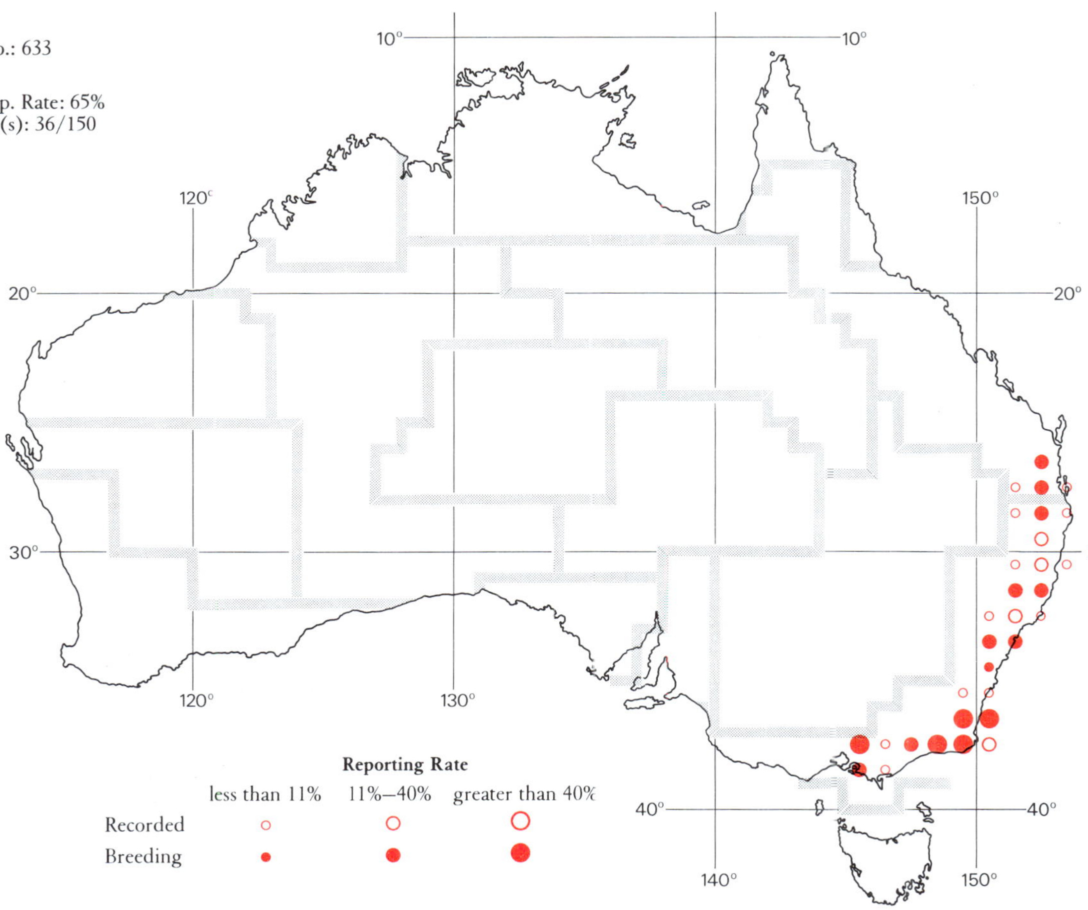

Noisy Miner

Manorina melanocephala

Two subspecies of the endemic Noisy Miner are recognized: *crassirostris* in Qld and *melanocephala* to the south[748] but no gap between them is evident in the Field Atlas. The species is absent from the Bass Strait islands and restricted to the eastern half of Tas. Hybrids have been recorded between the Noisy and Yellow-throated Miners.[581]

The Noisy Miner is sedentary, 10 km being the longest movement recorded for a banded bird.[85] It lives in loosely territorial groups of 6–30 individuals, which combine to form colonies numbering up to several hundred birds. Colonies attempt to exclude not only conspecifics but all other birds from their areas.[499,500] Breeding is communal. The Noisy Miner lives in eucalypt forest and woodland and can inhabit suburban parks and gardens. It is absent from high country in the lower South-East Region and, where mallee becomes the predominant vegetation in the western Murray-Darling Region, the Miner is confined to river red gum along the Murray R.

The Noisy Miner feeds on nectar, lerp, insects, vegetable matter and fruit including that of the peppercorn.[748,1124,1402] If a colony drives out all other species from an area, the range of food taken is wider than in sites where it cannot do so.[499]

Where some clearing has taken place and an open woodland results, the species becomes established.[1883]

Densities in a large colony near Laidley (27/152) were estimated at 7.8–10.4 birds/ha.[500]

Field Atlas Records: 19 327

1° blocks–with records: 201 (25%)

–with breeding records: 122 (61%)

Atlas No.: 634

Max. Rep. Rate: 86%
In block(s): 27/149

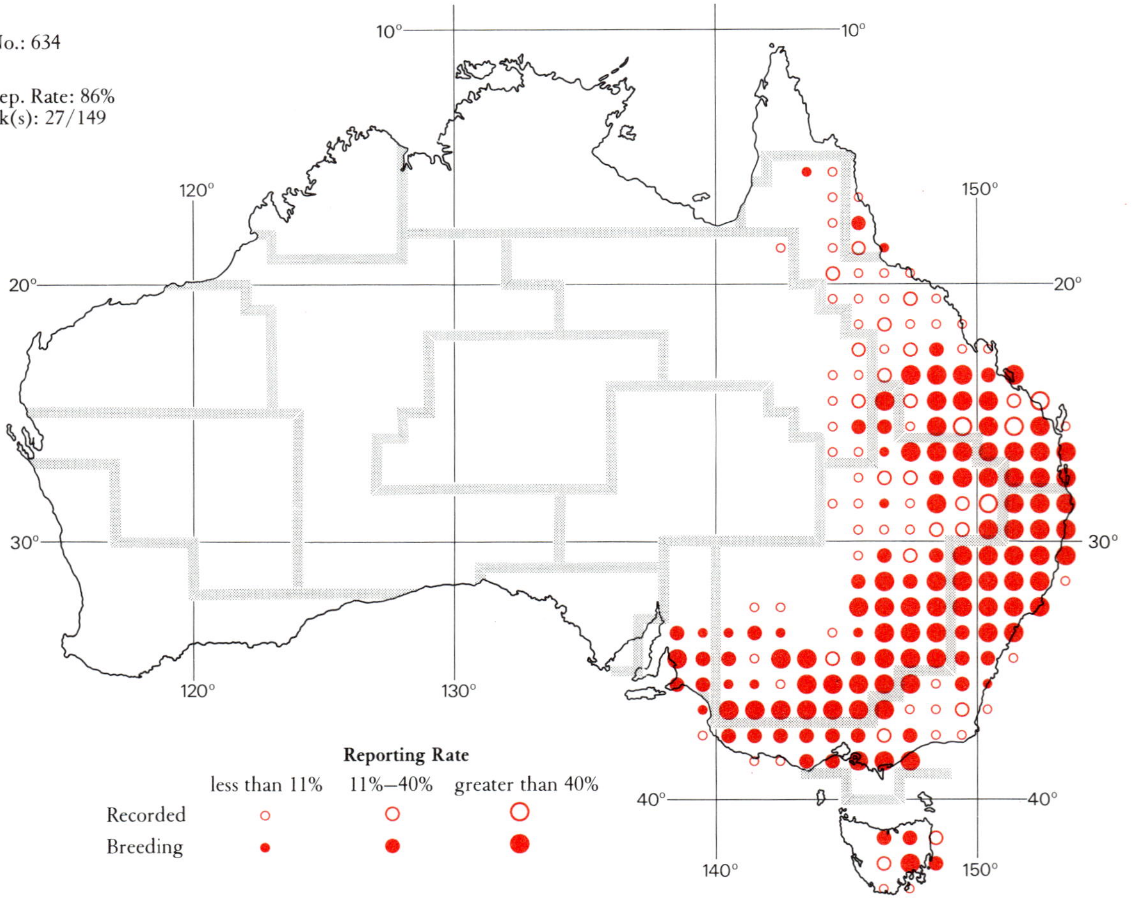

Yellow-throated Miner

Manorina flavigula

Up to eight subspecies of the endemic Yellow-throated Miner have been recognized but views about them differ.[748,1422] One population is from Melville I. (11/130) but this was not reported during the Field Atlas. The Field Atlas does not show any gaps in range. Hybrids have been recorded between the Yellow-throated and Noisy Miners.[581]

There is no suggestion from the Field Atlas of large-scale seasonal movement and the species is thought to be sedentary in most places.[987] Banding showed that it was probably so on the Nullarbor Plain: four of 23 birds banded in an isolated donga were recaptured in the same place 4, 12, 22 and 23 months later.[246]

The Yellow-throated Miner lives in scattered groups in woodland, mallee and cleared farmland if some trees and bushes remain. Flocks, which usually number fewer than ten birds, may contain up to 50 probably representing the coalescence of several adjoining groups. The species breeds communally.

The Yellow-throated Miner feeds on nectar[987] and seeks insects in crevices and under litter and cow-pats.[719] It also eats fruit and seeds have been found in guts.[748,1124] The Miner was once a pest of orchards south of Perth (31/115) but populations have retreated with the spread of suburbia.[1622]

Near Mt Isa (20/139) 4–8 individuals were seen in 1 km of transect.[1068] At Eyre Bird Observatory (32/126) densities ranged from 0.23–1.21 birds/ha.[464]

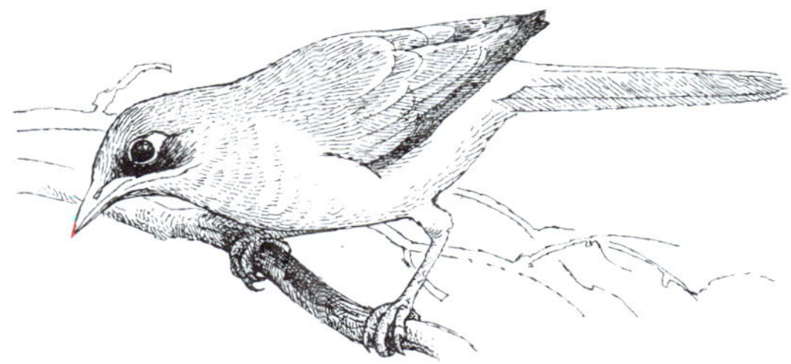

Field Atlas Records: 12 545

1° blocks–with records: 599 (74%)
–with breeding records: 194 (32%)

Atlas No.: 635

Max. Rep. Rate: 80%
In block(s): 29/143,29/144

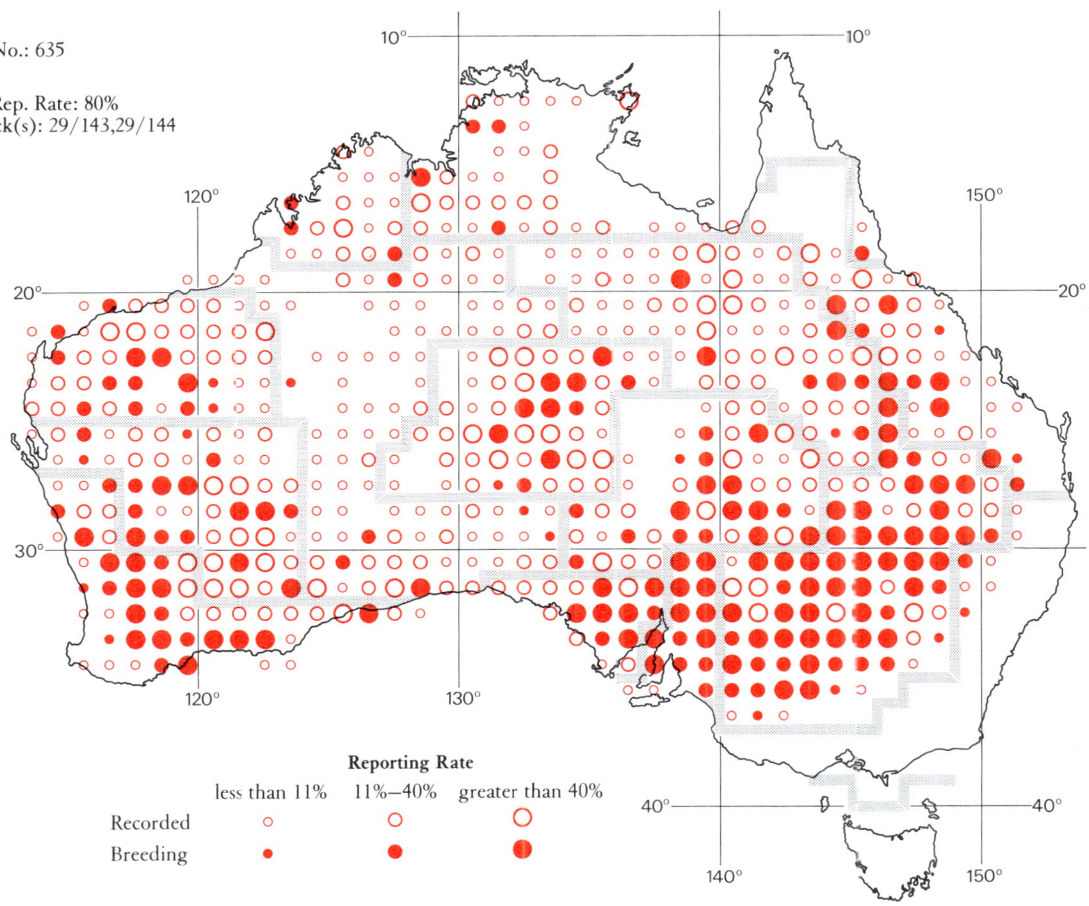

Black-eared Miner

Manorina melanotis

The taxonomic status of the endemic Black-eared Miner is uncertain;[581] no single plumage character distinguishes the Black-eared from the Yellow-throated Miner; only several characters considered together enabled a series of specimens to be divided into two groups representing the two species.[581] The difficulty of identifying the Black-eared Miner in the field means that the map may not represent its true distribution. As recognized in the Field Atlas, its range overlaps that of both the Yellow-throated and Noisy Miners.

Nothing is known of the bird's movements but it is presumably sedentary like other miners.

The Black-eared Miner is almost always reported from mallee and is thereby separated from the Noisy Miner of the eucalypt woodlands dominated by river red gums.[73] Colonies of the Black-eared and Yellow-throated Miners have been found close to each other. Considerable fighting has been seen between them[959] but aggression is characteristic of the Noisy Miner[499] and is perhaps general in the genus. The Black-eared Miner breeds in colonies but communal breeding has not been reported. It feeds on insects from under the bark of mallee, spends much time fossicking on the ground and also feeds on nectar.[331,577]

The Black-eared Miner is common in some areas but there are no measures of its abundance.

Field Atlas Records: 44

1° blocks–with records: 9 (1%)
–with breeding records: 2 (22%)

Atlas No.: 967

Max. Rep. Rate: 4%
In block(s): 35/140

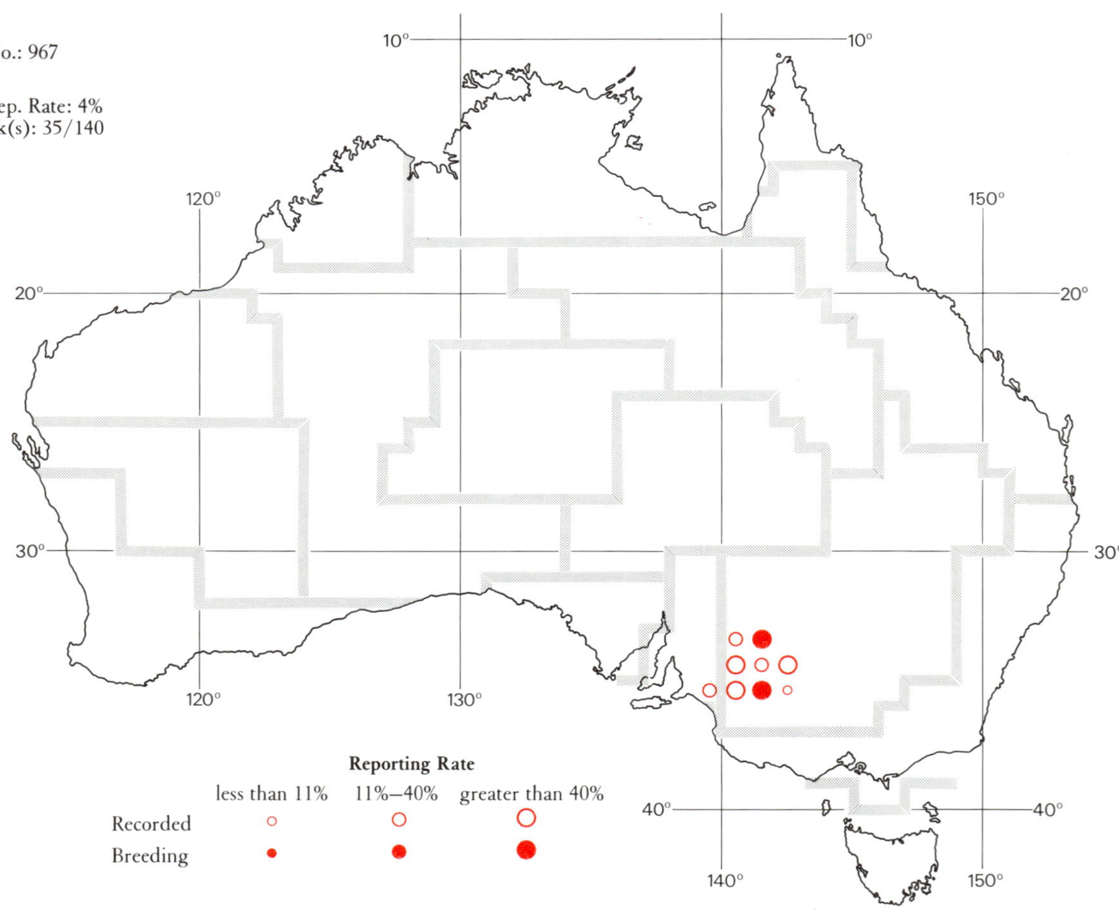

Macleay's Honeyeater

Xanthotis macleayana

Macleay's Honeyeater is endemic to the Atherton Region. It is present all year throughout its range.[231,698,742] It lives singly or in pairs in rainforest and eucalypt forest, at times entering orchards and gardens. Reporting Rates were 23.5% below 1000 m and 10.9% above. Some visiting observers have reported the species as shy and quiet and recorded only occasionally whereas resident observers consider it common and conspicuous because its calls are loud.[212,231,698,1354]

The diet of Macleay's Honeyeater is about 70% insects and spiders, 20% nectar and 10% fruit.[434,1354] The bird searches epiphytic growths and vines as well as gleaning from leaves. It will also take ripe orchard-fruit and various berries.[748]

There are no measures of abundance.

Field Atlas Records: 345

1° blocks–with records: 7 (1%)
–with breeding records: 3 (43%)

Atlas No.: 615

Max. Rep. Rate: 37%
In block(s): 15/145

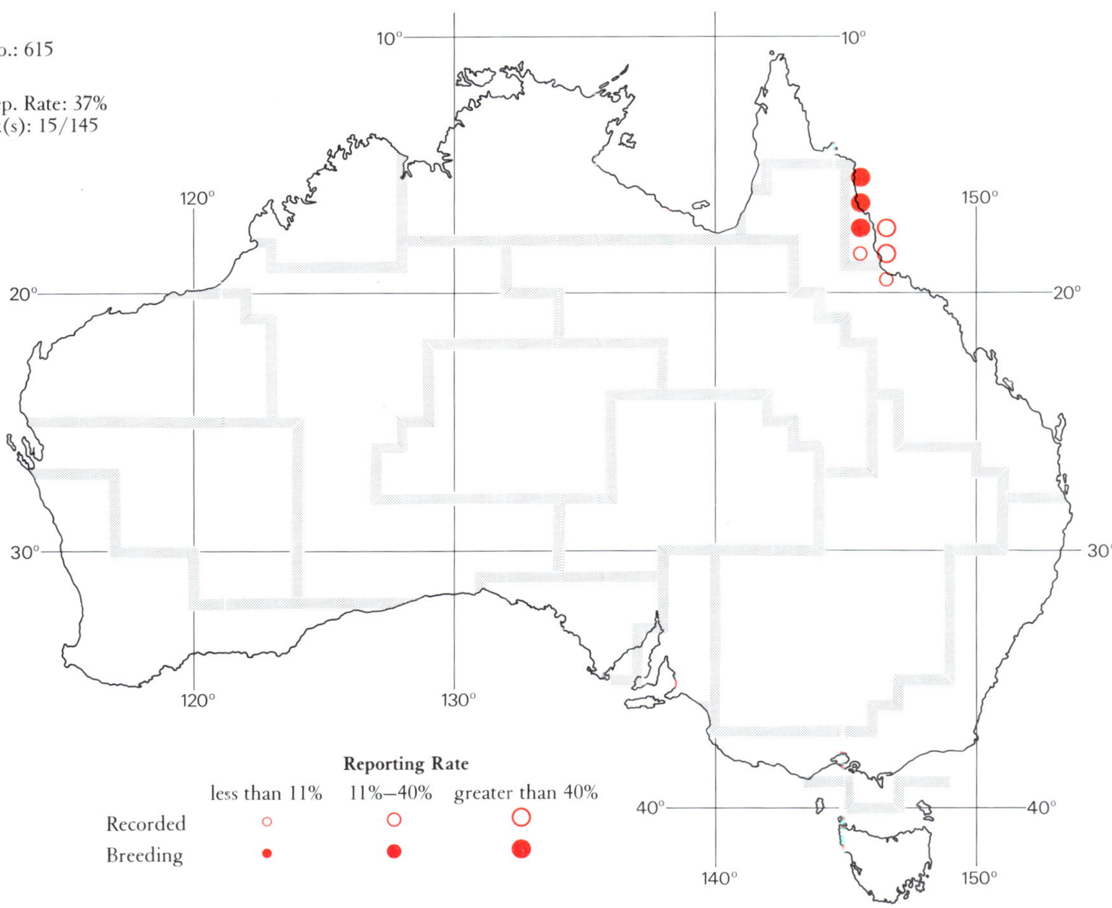

Tawny-breasted Honeyeater

Xanthotis flaviventer

The Tawny-breasted Honeyeater lives in New Guinea and adjacent islands and has one subspecies *filigera* in Australia.[1709] The sighting at the mouth of the Malaman Ck (15/141) in Apr. 1981 is farther south than the species has previously been reported.

The Tawny-breasted Honeyeater is present all year near Port Moresby (9S/147E) in New Guinea[160] and on islands in Torres Strait.[503] Pairs or small parties inhabit rainforest, including regrowth, mangroves and eucalypt woodland.[1354] The bird lives mainly in the canopy or just below it and is most conspicuous where the canopy descends to lower levels along streams and at the edge of rainforest stands.[1147]

The Tawny-breasted Honeyeater is generally seen gleaning insects from the foliage and trunks of tall forest trees and from dead leaves.[1145] It also feeds on nectar and some fruit.[129] In rainforest near Port Moresby density was 1 bird/ha.[160]

Field Atlas Records: 64

1° blocks–with records: 11 (1%)
–with breeding records: 1 (9%)

Atlas No.: 629

Max. Rep. Rate: 53%
In block(s): 12/143

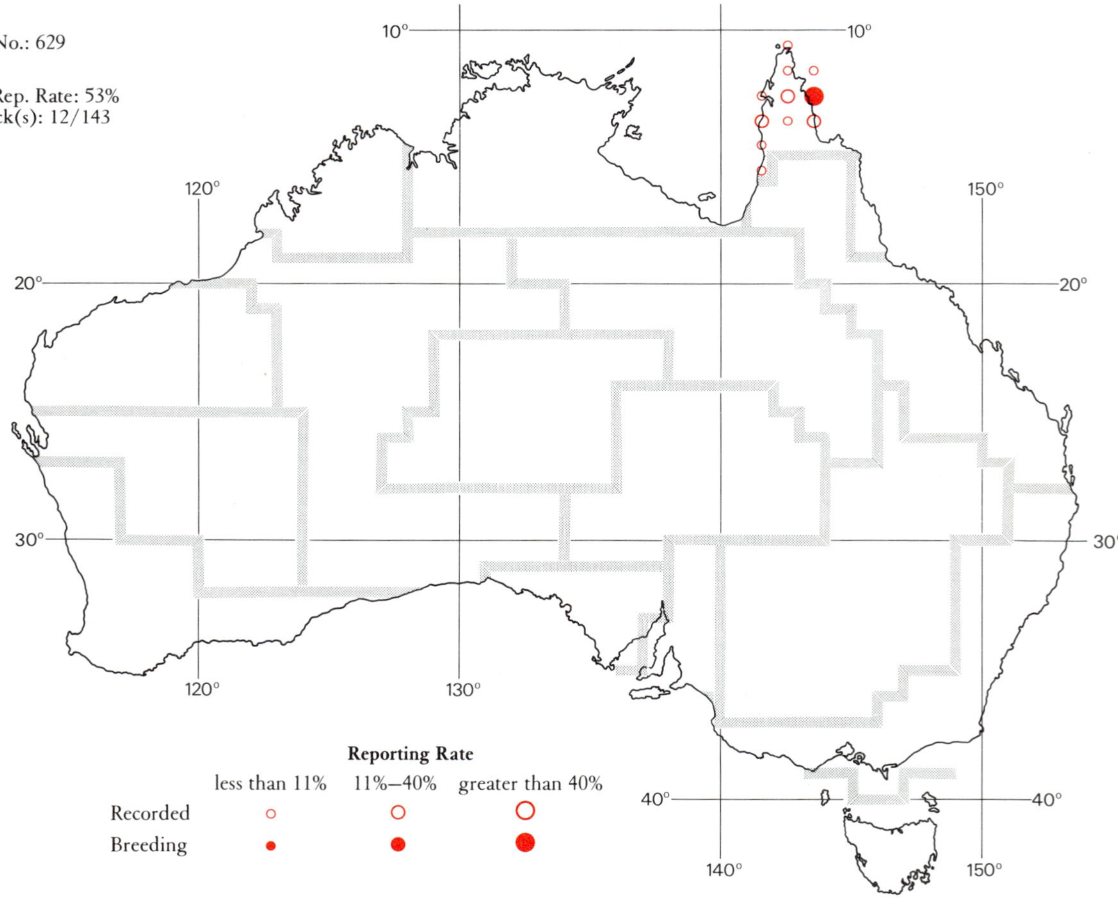

Populations of the endemic Lewin's Honeyeater in the lower South-East Region are separated as the subspecies *nea* from the nominate subspecies to the north[1422] though there is no break in distribution between them. An isolated population exists in the Nandewar Ra. (30/149); another in the McIlwraith Ra. (13/143)[998] was not recorded during the Field Atlas. Northern parts of the Lewin's range overlap those of two similar and closely related species: the Yellow-spotted and Graceful Honeyeaters and some observers experience difficulty in distinguishing between them.

Lewin's Honeyeater is present throughout the year in the Atherton Region[231,698] and round Rockhampton (23/150).[1089] Some birds in northern parts of the species' range may move to lower altitudes in winter[1709] and at Wollomombi (30/152) the species moves into eucalypt woodland in autumn and winter presumably from the adjacent gorge.[129] The Atlas suggests no large-scale seasonal movement.

Lewin's Honeyeater inhabits rainforest and eucalypt forest and sometimes eucalypt woodland, orchards and gardens.[1354] At the northern tip of its range the species lives in the highlands (Reporting Rate: 51% above 500 m and 12% below) and the Yellow-spotted and Graceful Honeyeaters predominate at lower altitudes. In the south Lewin's is a bird of the lowlands and coast, occasionally in localities as high as 600 m but more abundant in gullies at lower altitudes.[488,683,1113] Lewin's Honeyeater feeds on nectar, insects from the leaves and foliage and joins fruit-pigeons and orioles to feed on fruit in the tops of trees. It takes figs and lantana berries among a variety of native and introduced fruit.[748] At Wollomombi guts contained

Lewin's Honeyeater

Meliphaga lewinii

large numbers of hard chenopod seeds.[129]

Density at Wollomombi in eucalypt woodland was 0.05 birds/ha in autumn and winter;[129] at Moruya (35/150) in eucalypt woodland 0.18–0.21 birds/ha;[1197] and at Boola Boola (38/146) in eucalypt forest the species was most abundant in wet gullies 0.11–0.13 birds/ha.[1113]

Field Atlas Records: 5649

1° blocks–with records: 84 (10%)
–with breeding records: 32 (38%)

Atlas No.: 605

Max. Rep. Rate: 70%
In block(s): 24/153

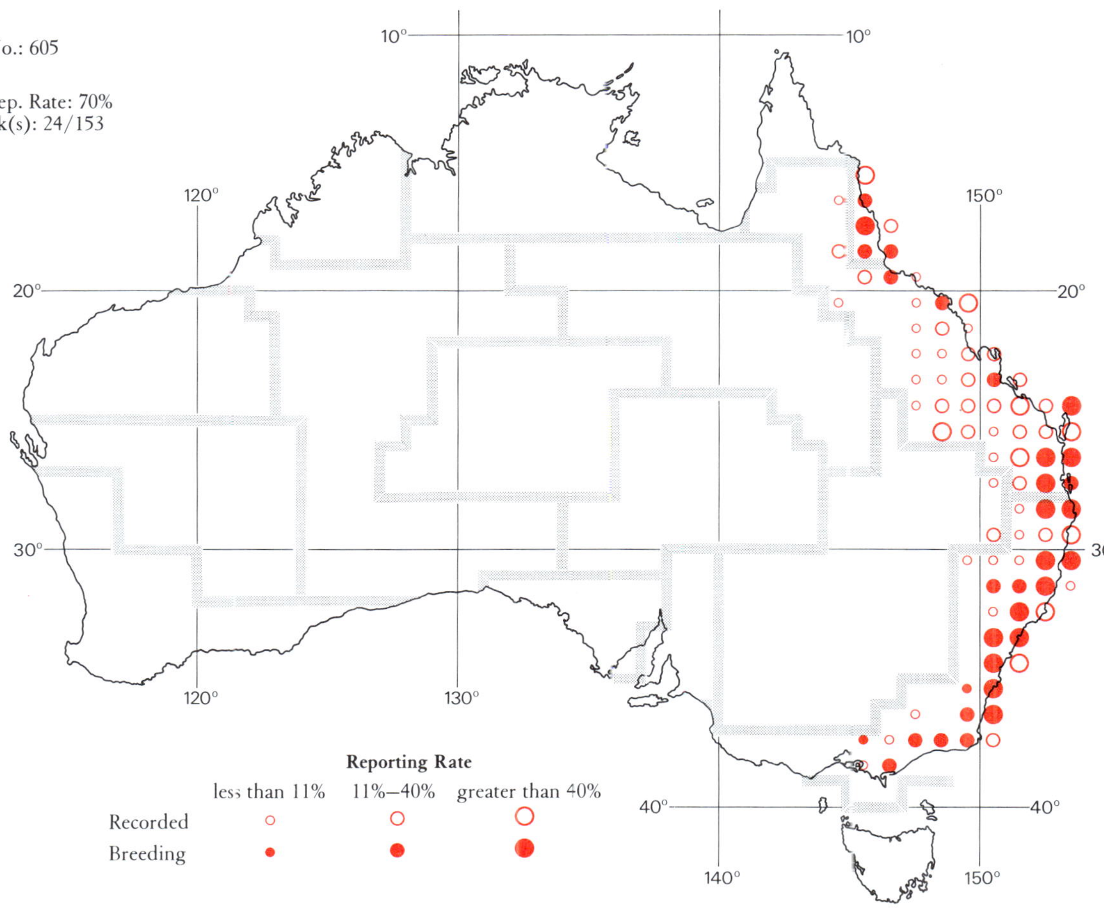

Yellow-spotted Honeyeater

Meliphaga notata

The Yellow-spotted Honeyeater is endemic. The subspecies *notata* occupies the Cape York Region and the larger islands of Torres Strait, and *mixta* areas to the south.[1422] Some observers experience difficulty in distinguishing between the closely-related Yellow-spotted, Lewin's and Graceful Honeyeaters whose ranges overlap in north-eastern Qld.

The Yellow-spotted Honeyeater is present all year in the Innisfail (17/146) district[698] but there is no other information about movements. It is a lowland species (Reporting Rate: 31% below 500 m altitude and 10% above 500 m) that lives in mangroves, rainforest along streams and rivers and among flowering trees in woodland with a grassy understorey. It also inhabits the edges of cultivation, including gardens. It feeds on native and cultivated fruit, including lantana berries, and insects, including ants.[748,1354] It lives singly or in flocks and is often in company with other honeyeaters at flowering trees.[748,1355]

There are no measures of abundance.

Field Atlas Records: 650

1° blocks-with records: 23 (3%)

-with breeding records: 7 (30%)

Atlas No.: 606

Max. Rep. Rate: 63%
In block(s): 12/143

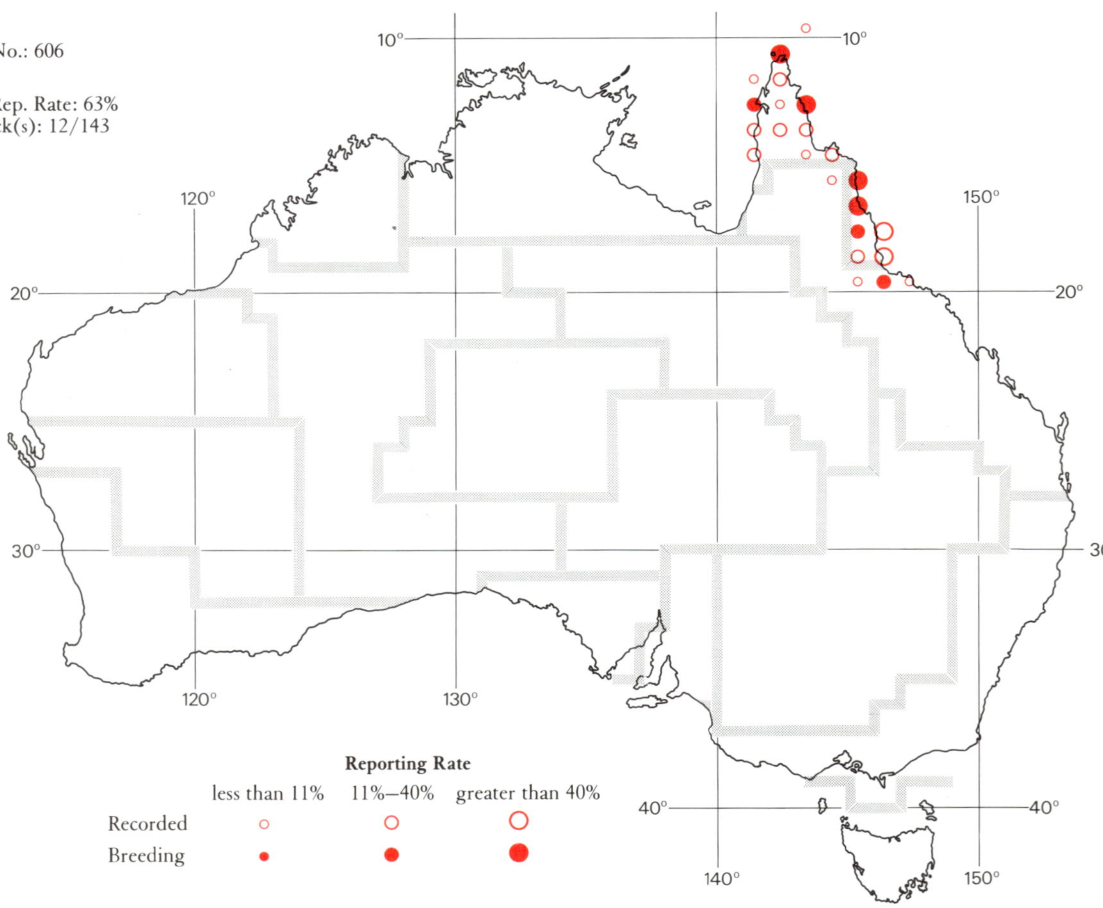

Graceful Honeyeater

Meliphaga gracilis

The Graceful Honeyeater ranges from New Guinea and adjacent islands to Australia. Two subspecies have been recognized: *gracilis* in the Cape York Region and *imitatrix* in the Atherton Region.[1709] Some observers experience difficulty in distinguishing between the Graceful, Lewin's and Yellow-spotted Honeyeaters where their ranges overlap in north-eastern Qld.

The Graceful Honeyeater is present all year near Innisfail (17/146);[698] there is no other information about movements. Though recorded above 600 m in the Atherton Region[748], it is usually a resident of the lowlands.[698,987] Reporting Rates in the Atherton Region were 16% below 500 m altitude and 4% above. Generally, the species lives in rainforest or its fringes, but also visits eucalypt forest, mangroves and gardens when trees are in flower.[303] In New Guinea it lives only on the edge of rainforest.[160] The Graceful Honeyeater feeds on insects, spiders and lantana berries,[303,748,1703] joining other honeyeaters round flowering trees.

There are no measures of abundance.

Field Atlas Records: 429

1° blocks–with records: 20 (2%)
–with breeding records: 3 (15%)

Atlas No.: 612

Max. Rep. Rate: 56%
In block(s): 12/143

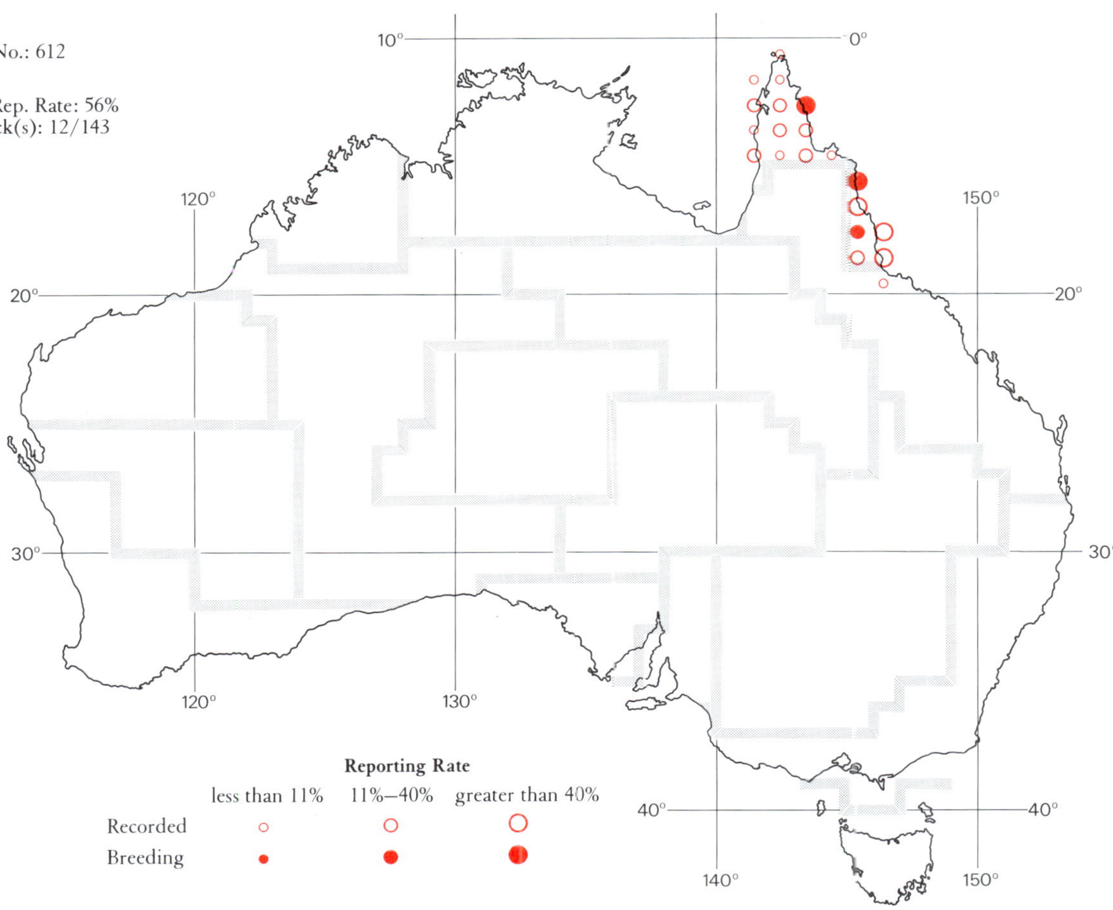

White-lined Honeyeater

Meliphaga albilineata

The two populations of the endemic White-lined Honeyeater are widely separated but not considered morphologically distinct.[1422,1711,1713] That in the Top End has once been reported farther south at Katherine Gorge (14/132) but the record needs confirmation. The Kimberley distribution includes the Drysdale R National Park (15/127) and the King Leopold Ra. (17/125).

The White-lined Honeyeater lives in scattered patches of rainforest in sandstone gorges[1689] and on floodplains near escarpments.[1086] Calls seem the best indication of its presence.[642,748] It forms loose flocks when foraging[1086] and eats a variety of fruit, seeds, vegetable matter, insects, spiders and probably nectar.[748,1584]

There are no measures of abundance.

Field Atlas Records: 97

1° blocks–with records: 6 (1%)
–with breeding records: 2 (33%)

Atlas No.: 607

Max. Rep. Rate: 27%
In block(s): 13/132

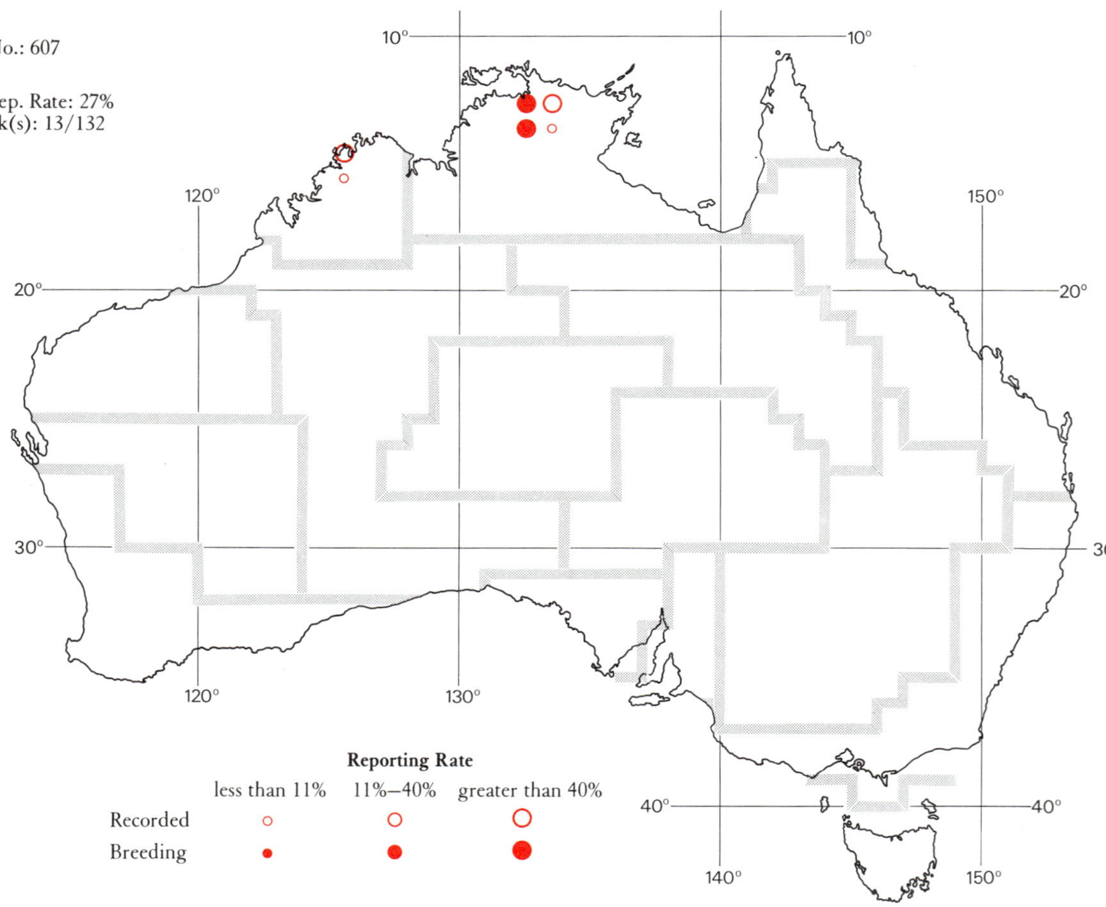

The population of the endemic Yellow-faced Honeyeater in the SA Ra. Region is doubtfully distinct from birds elsewhere.[1422] In Eastern Qld between 19°S and 26°S the Atlases show a gap in breeding records except for one unconfirmed report at Duaringa (23/149); birds in this area may therefore be mostly migrants.

The Yellow-faced Honeyeater has long been known to undertake extensive movements but there is no clear definition of where any southern population winters. In the Atherton Region the species is present all the year with a winter influx on the Atherton tableland.[231,698] Along the NSW coast northward migration in autumn is pronounced. Near Hastings Pt (28/153) up to 100 000 birds have been counted passing in a single day in May but flocks moved southward as well as northward.[1069] At Wollomombi (30/152) a sedentary colour-banded population did not join flocks of migrants present in winter.[129] Flocks of birds moving north in autumn have been reported as far south as Rotamah I. Bird Observatory (37/147)[285] and the number of Yellow-faced Honeyeaters is significantly reduced in Melbourne (37/145) in winter.[75] Reporting Rates reflect altitudinal movement (Snowy Mts (36/148) above 500 m: spring-summer 84%, autumn-winter 33%; below 500 m: spring-summer 81%, autumn-winter 50%). The longest movement recorded is that of a banded adult from Grafton (29/152) to Pt Hut (35/149), 756 km.[38] Near Adelaide (34/138) counts recorded more Yellow-faced Honeyeaters in winter than in summer[579] but the Field Atlas suggests only that there is some movement south in summer.

The Yellow-faced Honeyeater inhabits eucalypt forest, woodland and farmland. It lives singly, in pairs or in flocks.

Yellow-faced Honeyeater

Lichenostomus chrysops

Birds of the sedentary population at Wollomombi joined other species in mixed flocks.[129] In Vic. the Yellow-faced Honeyeater feeds mainly on manna in summer and autumn and on nectar in spring.[1402] It also takes insects and some honeydew.

Densities in eucalypt woodland were: Armidale (30/151) 0.03–0.16 birds/ha,[582] Wollomombi 1.4 birds/ha in the breeding season and 2.0 birds/ha in winter,[129] Canberra (35/149) 1.5 birds/ha[154] and Moruya (35/150) 0.9–1.76 birds/ha.[1197] In eucalypt forest at Bathurst density was 2.5 birds/ha[489] and at Boola Boola (38/146) 0.5–3.0 birds/ha.[1113]

Field Atlas Records: 11 588

1° blocks–with records: 132 (16%)
–with breeding records: 60 (45%)

Atlas No.: 614

Max. Rep. Rate: 66%
In block(s): 35/150

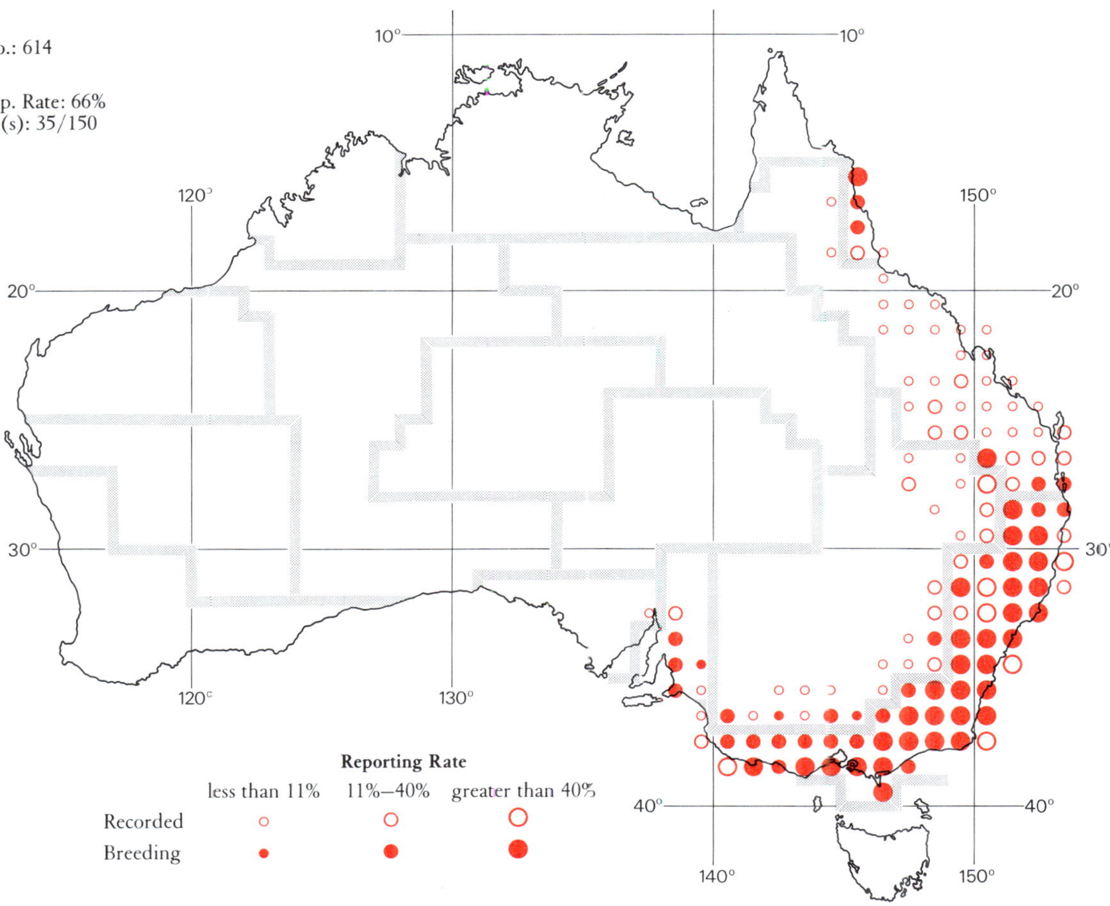

Bridled Honeyeater

Lichenostomus frenatus

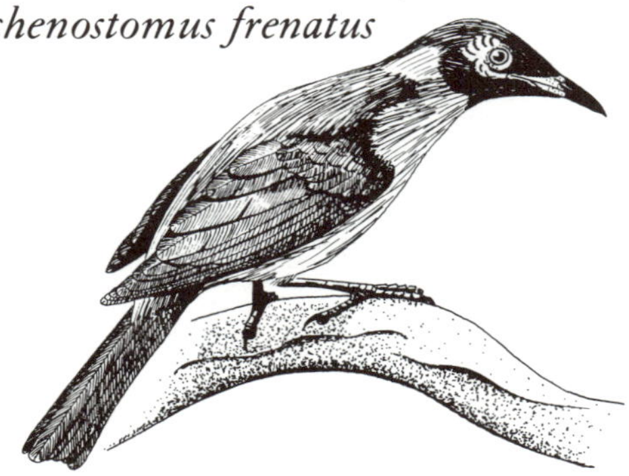

The Bridled Honeyeater is endemic to the Atherton Region. The honeyeater in the Eungella (21/148) district, formerly thought to be this species, is a new species: the Eungella Honeyeater.[1090]

The Bridled Honeyeater is present all year at altitudes above 450 m.[231,698] It is considered by one observer not to move to lower altitudes[698] but by others to descend to the foothills and coasts in winter in some districts.[683,1709] It lives in rainforest and eucalypt forest. Groups forage where there is an abundance of flowering plants in forests, along streams and in swamps.[212] The species feeds on nectar, specially from mistletoe,[1221] and on beetles and other insects.[748]

There are no measures of abundance.

Field Atlas Records: 234

1° blocks–with records: 8 (1%)
–with breeding records: 5 (63%)

Atlas No.: 611

Max. Rep. Rate: 25%
In block(s): 15/145

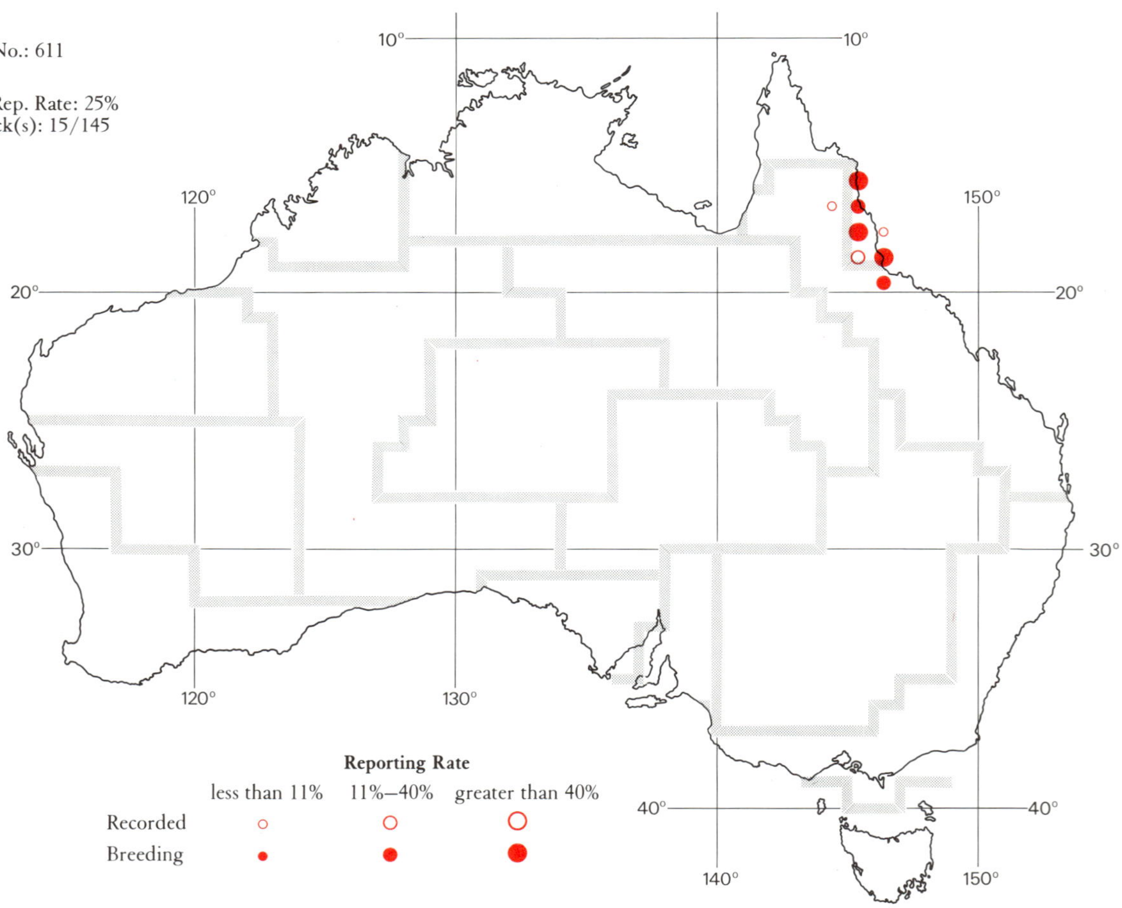

In 1977 it was realized that honeyeaters in the Eungella (21/148) district, formerly thought to be the Bridled,[1506,1709] belonged to a different and new species, the Eungella Honeyeater.[1090] At present it is known only from a small area near Eungella.

There is nothing published about movement.

The Eungella Honeyeater lives in rainforest and surrounding eucalypt forest at altitudes above 1000 m. Single birds, pairs and small parties have been recorded. The species has been observed feeding in mistletoe, among a bunch of palm seeds and in the outer foliage of trees.

There are no measures of abundance.

Eungella Honeyeater

Lichenostomus hindwoodi

Field Atlas Records: 13

1° blocks–with records: 2 (1%)

–with breeding records:

Atlas No.: 834

Max. Rep. Rate: 4%
In block(s): 21/148

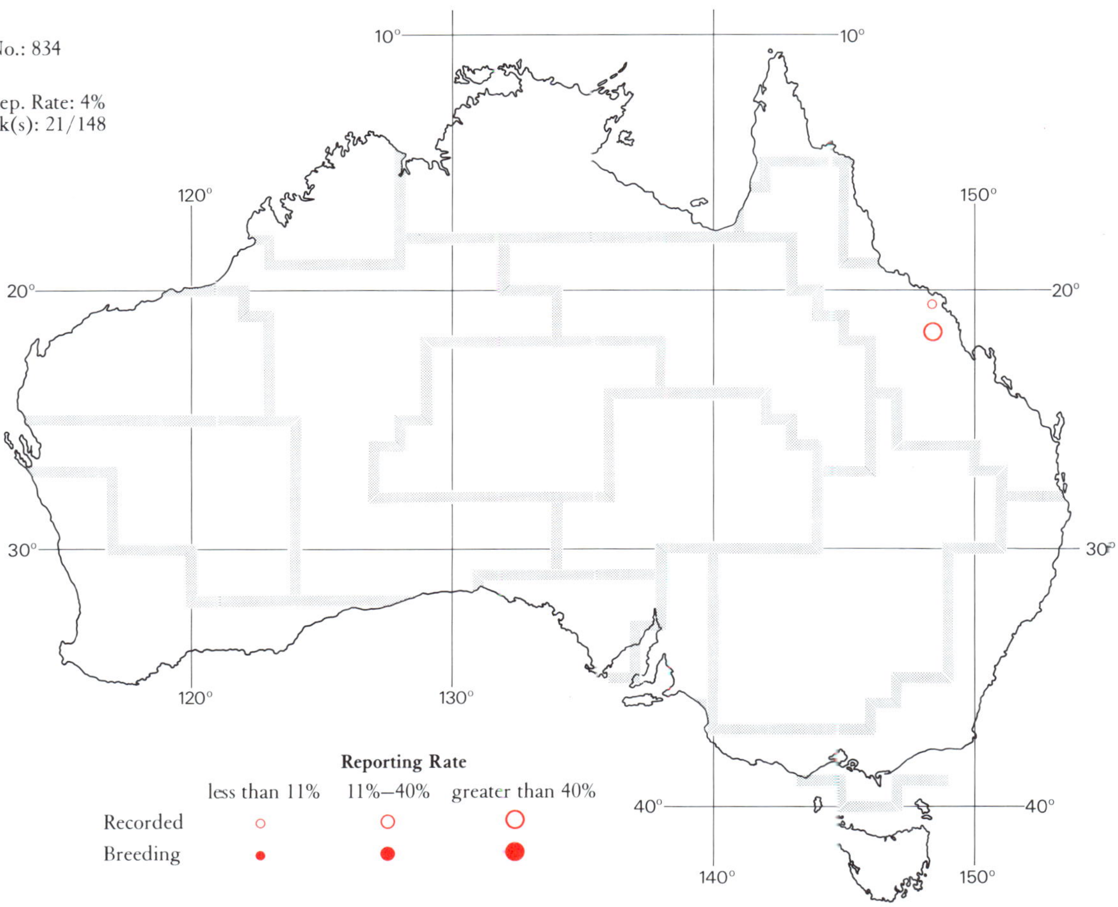

Singing Honeyeater

Lichenostomus virescens

The Singing Honeyeater lives in northern New Guinea[129] and Australia. Morphological variation is clinal and there is doubt whether application of subspecific names to some variants is justified.[748] The species is not recorded from the Tas. Region. The Singing, Varied and Mangrove Honeyeaters are closely related[609] and sometimes confused in the field by observers.

There is little information about the movements of the Singing Honeyeater. On the Nullarbor Plain 40 birds were colour-banded in one donga between 1970 and 1973 though the average number present at any one time was seven. Two of the banded birds were seen repeatedly over 25 months and five others for 6–10 months,[246] which suggests that the Singing Honeyeater is usually sedentary there. At Eyre Bird Observatory (32/126) numbers increase in spring and autumn, though there do not seem to be regular movements.[464] Sightings beyond the normal range of the Singing Honeyeater, such as that at Green Cape (37/150) in Sept.-Oct. 1977, indicate that it can move long distances but not in large numbers and the Field Atlas does not suggest large-scale seasonal movement.

The Singing Honeyeater lives in acacia scrub, spinifex, mallee, eucalypt woodland and, along the Vic. coast, in heath. It has been recorded in dwarf mangroves on the western coast.[748] In New Guinea it lives in association with man, e.g. in settlements and plantations.[129] Normally singly or in pairs, small flocks sometimes form, e.g. 20 birds together at Mt Isa (20/139).[1068] The Singing Honeyeater feeds on nectar[653,1403] and a variety of insects, including ants. Guts contained fruit, such as mistletoe berries, and some seeds[748,1049] and the bird is known to raid the nests of Zebra and Double-barred Finches, taking eggs and chicks.[1638] It often feeds from the ground.

On the Nullarbor Plain the mean density in a three-year census of a 10-ha plot was 0.2 birds/ha.[246] At Eyre Bird Observatory estimates of density in Aug., using two different methods, gave figures of 0.8–0.9 birds/ha.[464]

Field Atlas Records: 12 633

1° blocks–with records: 649 (80%)
–with breeding records: 149 (23%)

Atlas No.: 608

Max. Rep. Rate: 85%
In block(s): 28/121

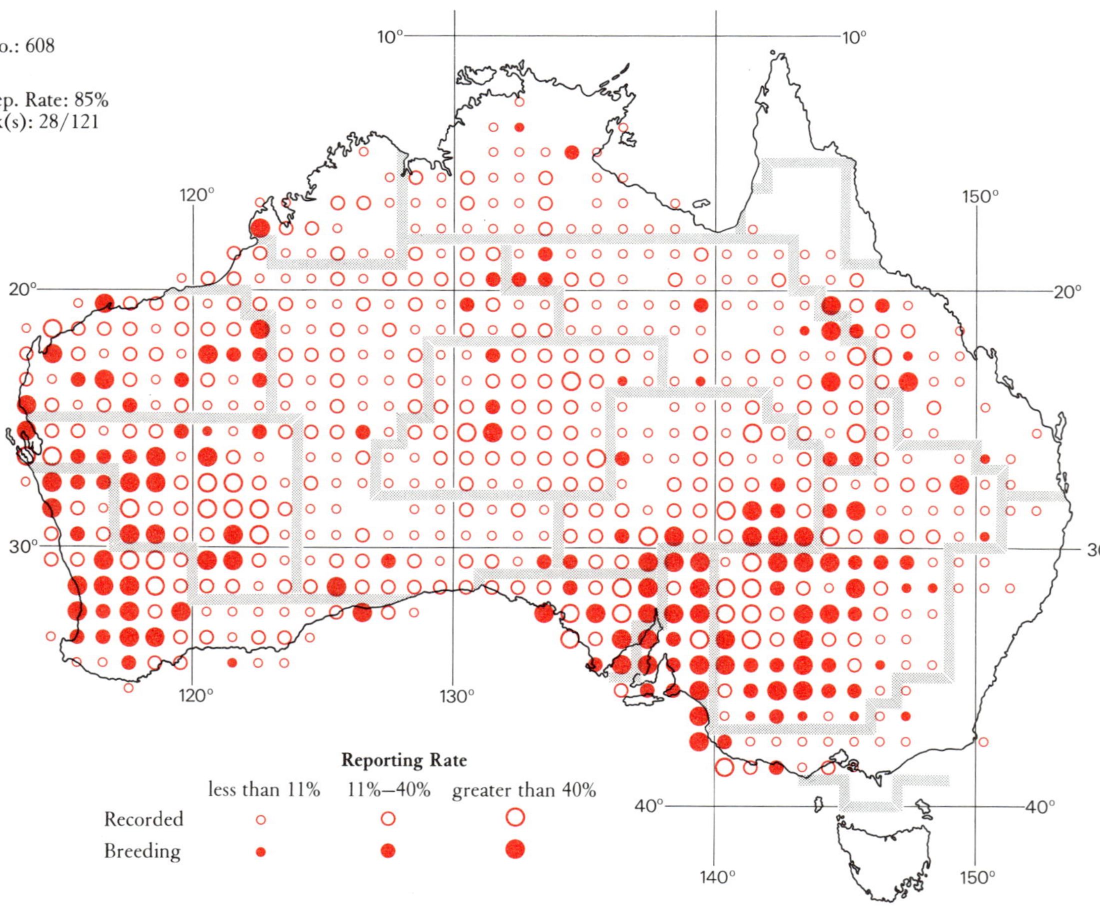

Varied Honeyeater

Lichenostomus versicolor

The Varied Honeyeater, which lives in southern New Guinea and adjacent islands as well as in Australia, is the counterpart of the more southerly ranging Mangrove Honeyeater. Intermediates have been recorded where the two species meet between Cardwell (18/146) and Townsville (19/146) and they are sometimes considered conspecific.[609,622] The Varied Honeyeater is sometimes confused in the field with the Mangrove and Singing Honeyeaters.

The Varied Honeyeater is present throughout the year in the Innisfail (17/146) district[698] but a local fluctuation in abundance from one year to the next has been reported from Cape York (10/142).[1145] There is no other information about movements. The Varied Honeyeater usually lives in mangroves. In Cairns (16/145) it frequents the street trees[698,1354] and on Daru I. (9S/143E) is often in gardens and settlements[140] as it is in New Guinea.[129] In Torres Strait it inhabits small islands where there is forest and larger ones dominated by mangroves.[503] It has been recorded feeding on nectar from the flowers of mangroves.[303] The guts of two birds taken in the Cape York Region held the remains of numerous small insects.[1758]

There are no measures of abundance.

Field Atlas Records: 150

1° blocks–with records: 10 (1%)
–with breeding records: 3 (30%)

Atlas No.: 609

Max. Rep. Rate: 20%
In block(s): 16/145

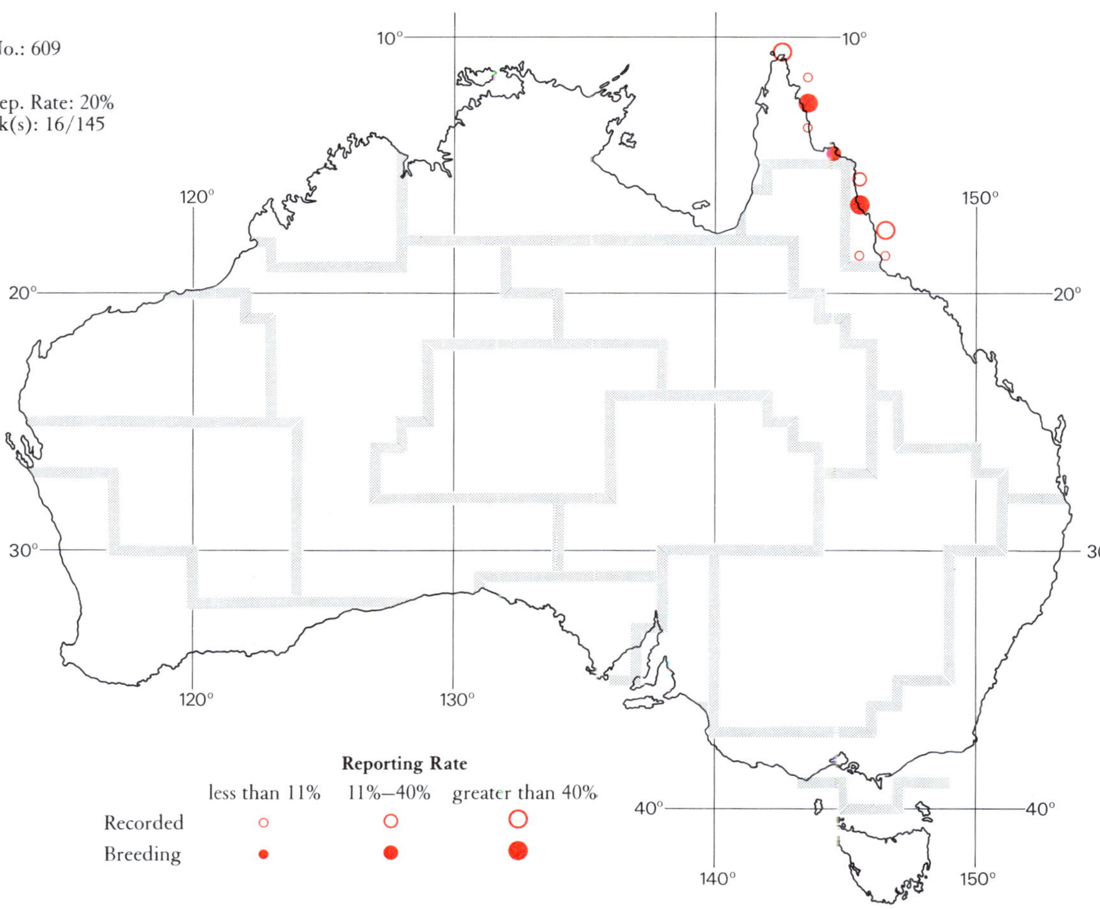

Mangrove Honeyeater

Lichenostomus fasciogularis

The endemic Mangrove Honeyeater is the counterpart of the more northerly Varied Honeyeater. Intermediates have been recorded where the two species meet between Cardwell (18/146) and Townsville (19/146) and they are sometimes considered conspecific.[609,622] The Mangrove, Varied and Singing Honeyeaters are sometimes confused in the field by observers.

The Mangrove Honeyeater is extending its range southward.[809] A thorough survey of birds at Stuarts Pt (30/153) on the Macleay R in 1944–45 failed to find the species, whereas it was common there in 1958. Its discovery even farther south in 1981 means that it is now known from three isolated localities at the southern end of its range: the mouths of the Clarence R (29/153), Macleay R (30/153) and Hastings R (31/152).

The species is said to be sedentary[987] but may wander locally seeking flowering plants.[1086] It lives in mangroves, frequently visiting flowering shrubs in towns adjacent to mangroves.[806] The bird is conspicuous and noisy, specially in the breeding season. It feeds on nectar, insects, molluscs and crabs,[748,1176] spending some of its time feeding low down in mangroves close to the mud[806] where it is more likely to find both insects and marine invertebrates.

There are no measures of abundance.

Field Atlas Records: 427

1° blocks-with records: 22 (3%)
-with breeding records: 5 (23%)

Atlas No.: 610

Max. Rep. Rate: 27%
In block(s): 27/153

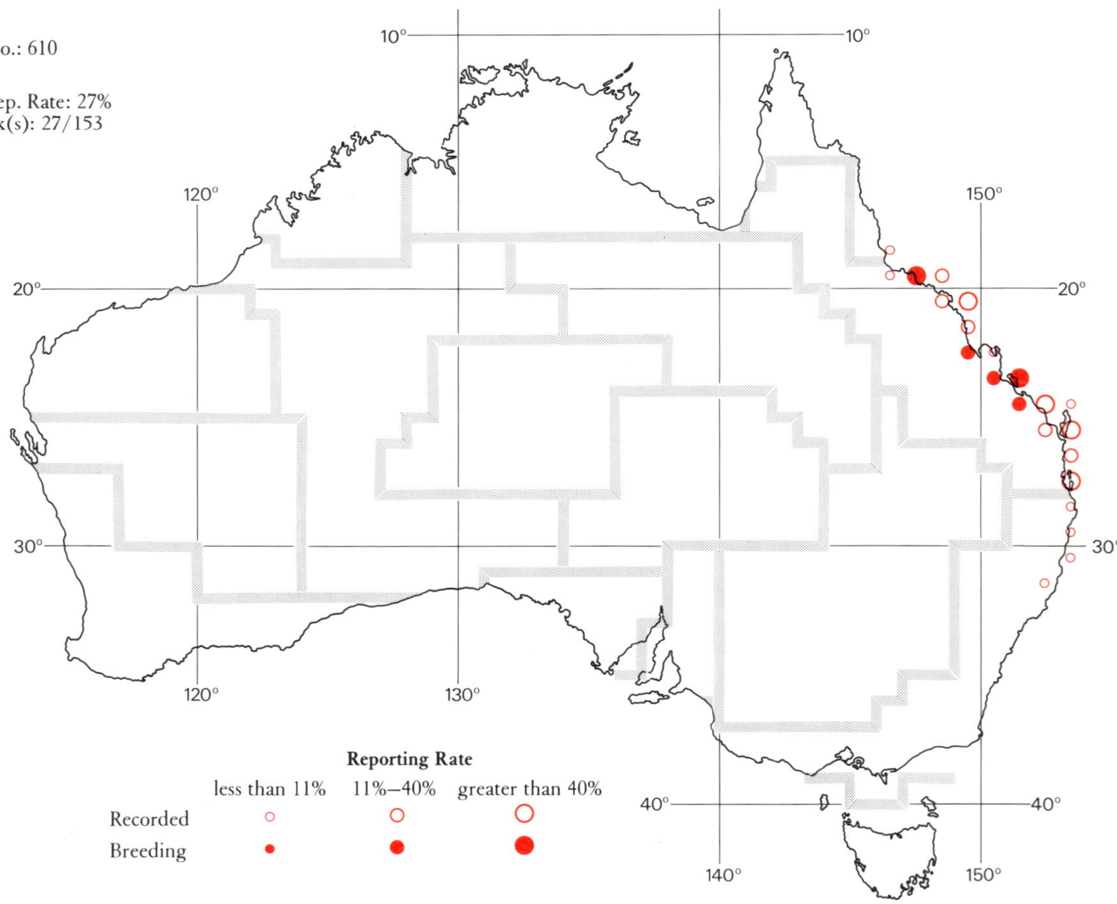

White-gaped Honeyeater

Lichenostomus unicolor

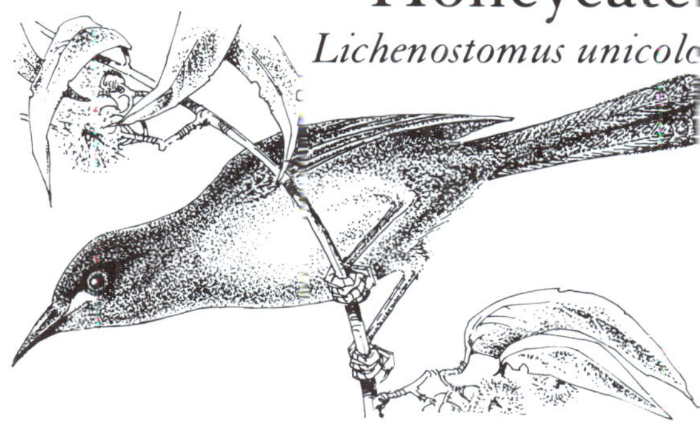

The Atlases show a consistent break in the distribution of the endemic White-gaped Honeyeater south of the Gulf of Carpentaria (140–41°E) but subspecies are generally not recognized.[748,1422,1709,1711,1713] The species may have retreated from the Mt Isa (20/139) district: it was reported to be common there in 1932[1204] but later observers have recorded it only some distance to the north.[862,1068] The single Field Atlas report from the district was from L. Julius (20/139), about 60 km north of Mt Isa.

The White-gaped Honeyeater is said to be sedentary.[898,987] It lives singly, in pairs or in flocks most often in eucalypt woodland with a dense understorey along watercourses. It also inhabits mangroves, gardens and orchards[748,898] and, in the most northerly parts of its range, it is present in rainforest in gorges and estuaries.[748]

The White-gaped Honeyeater gleans from bark and twigs. Near Kununurra (15/128) it eats mostly insects caught in spiders' webs as well as the spiders themselves, even entering houses to seek them.[898] Insects, nectar, fruit and seeds have been found in the guts of specimens.[748] At the end of the dry winter season the bird has been recorded sucking the contents from the fruit of the passionflower and eating berries.[898]

At Kununurra density was 0.3 birds/ha in gardens near houses but along the river, in vegetation 100 m wide, there was only one pair to each 1–3 km (0.07–0.20 birds/ha).[898]

Field Atlas Records: 1490

1° blocks–with records: 102 (13%)
–with breeding records: 17 (17%)

Atlas No.: 628

Max. Rep. Rate: 72%
In block(s): 13/130

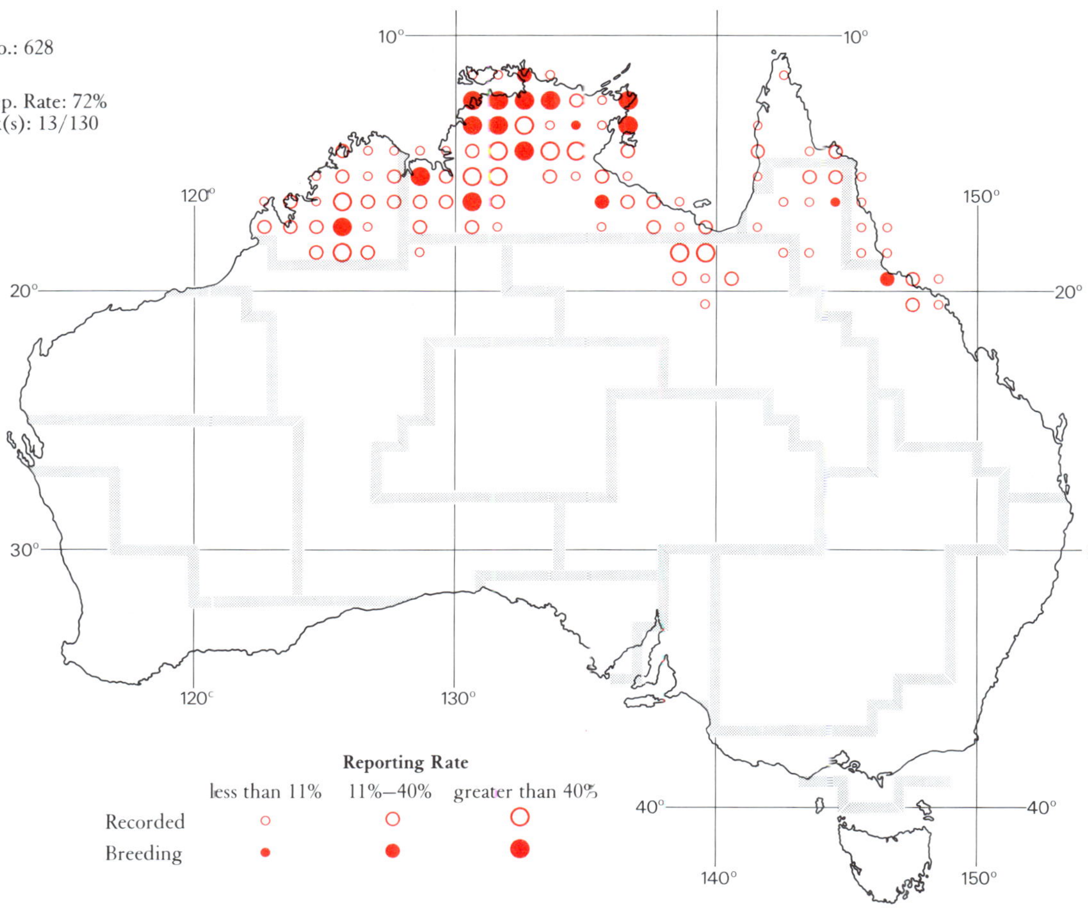

Yellow Honeyeater

Lichenostomus flavus

The range of the endemic Yellow Honeyeater is unchanged from the Historical Atlases. It is present all year in the Innisfail (17/146) area,[698] though numbers fluctuate on the tableland which suggests that there may be local nomadic movement in some places.[231] The species lives singly, in pairs or in flocks in eucalypt woodland and forest, often along rivers. It also inhabits many other sites including mangroves, swamps, gardens and orchards.[403,748,1703,1869] It will breed in suburbs and appears to thrive round human habitation. In Townsville (19/146) it is one of the most abundant birds, at higher density in the town than in woodland.[953] The Yellow Honeyeater feeds on nectar as well as the insects, spiders and some seeds which were found in guts.[748,1758]

There are no measures of abundance.

Field Atlas Records: 970

1° blocks-with records: 47 (6%)

-with breeding records: 10 (21%)

Atlas No.: 626

Max. Rep. Rate: 50%
In block(s): 19/147

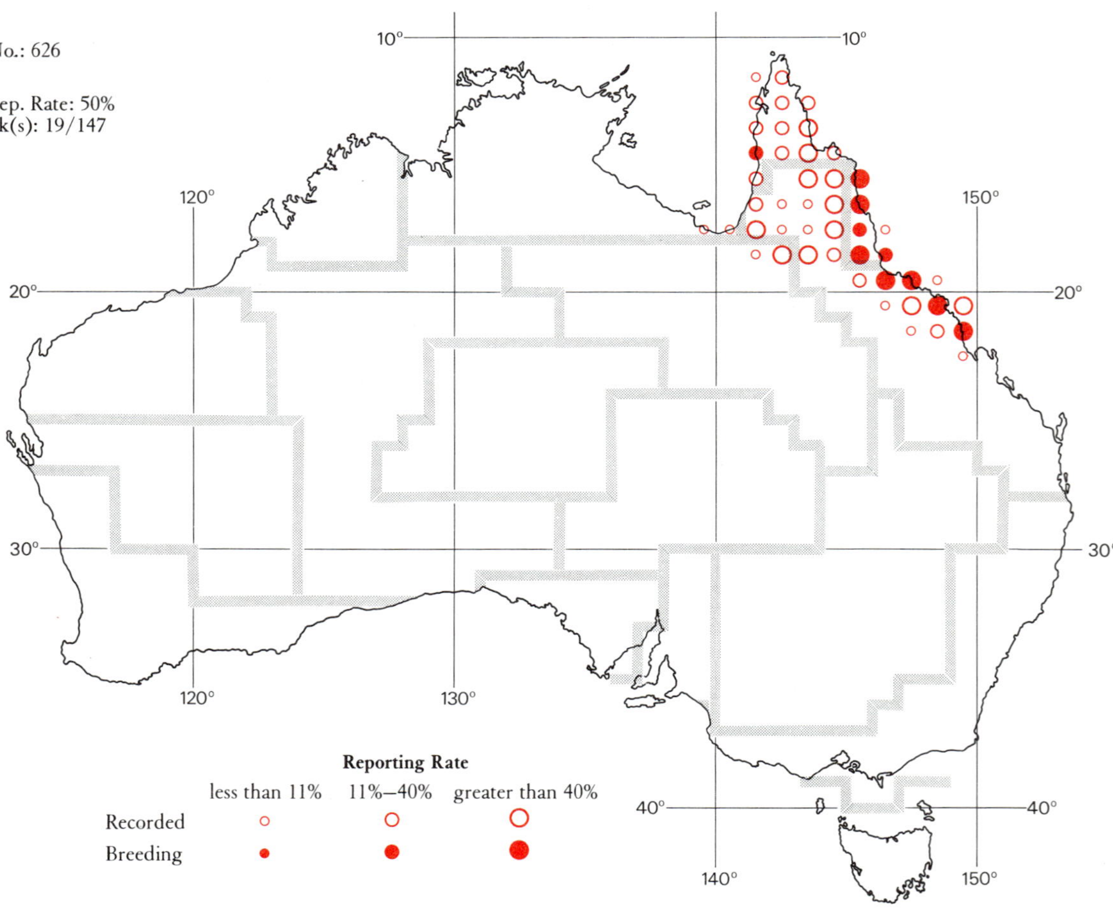

White-eared Honeyeater

Lichenostomus leucotis

The endemic White-eared Honeyeater has eastern and western populations, which approach each other but probably do not meet across the head of the Great Australian Bight. The two have been described as separate subspecies but views about them differ.[600,748] The western population has been reported no farther east than Eucla (31/128)[600] and is probably continuously distributed along the southern coast but there is a small, and perhaps isolated population, near Geraldton (28/114). A historical record of the species at Kalbarri (27/114) was dismissed[600] but confirmed records were made in the area during the Field Atlas.

The White-eared Honeyeater has been considered to be at most a local nomad[987] but there is evidence of altitudinal movement in the South-East Region. In the Thredbo Valley (36/148) birds are present all year but the population is augmented in summer by migrants returning from the lowlands.[682] At Canberra (35/149) and Moruya (35/150) it is present Apr.-Sept..[154,1192] Reporting Rates in the Snowy Mts (36/148) above 1000 m were 73% in spring-summer and 33% in autumn-winter; below 1000 m they were 34% in spring-summer and 52% in autumn-winter.

The White-eared Honeyeater lives in eucalypt forest and woodland, and heath. At Cranbourne (38/145) it feeds mostly on manna and honeydew, the proportions varying with the seasons.[1402] Some insects and a little nectar are also taken. At Golton Vale (36/142), where records were obtained only in autumn and winter, honeydew formed the bulk of the diet together with a few insects.[1402] Near Adelaide (33/138) the Honeyeater visits a few species of flowering plants persistently, including mistletoe.[1403] Many beetles and other insects were found in its guts.[748,1049]

In eucalypt woodland densities were: Armidale (30/151) 0.01 birds/ha,[582] Wollomombi (30/152) 0.1–0.3 birds/ha[129] and Canberra 0.13–0.36 birds/ha.[154] At Boola Boola (38/146), in eucalypt forest regrowth 5–10 years old density was 0.5–1.4 birds/ha[1113] and at Eyre Bird Observatory (32/126) in mallee 0.25–0.97 birds/ha.[464]

Field Atlas Records: 8834

1° blocks–with records: 197 (24%)

–with breeding records: 70 (36%)

Atlas No.: 617

Max. Rep. Rate: 55%
In block(s): 32/120

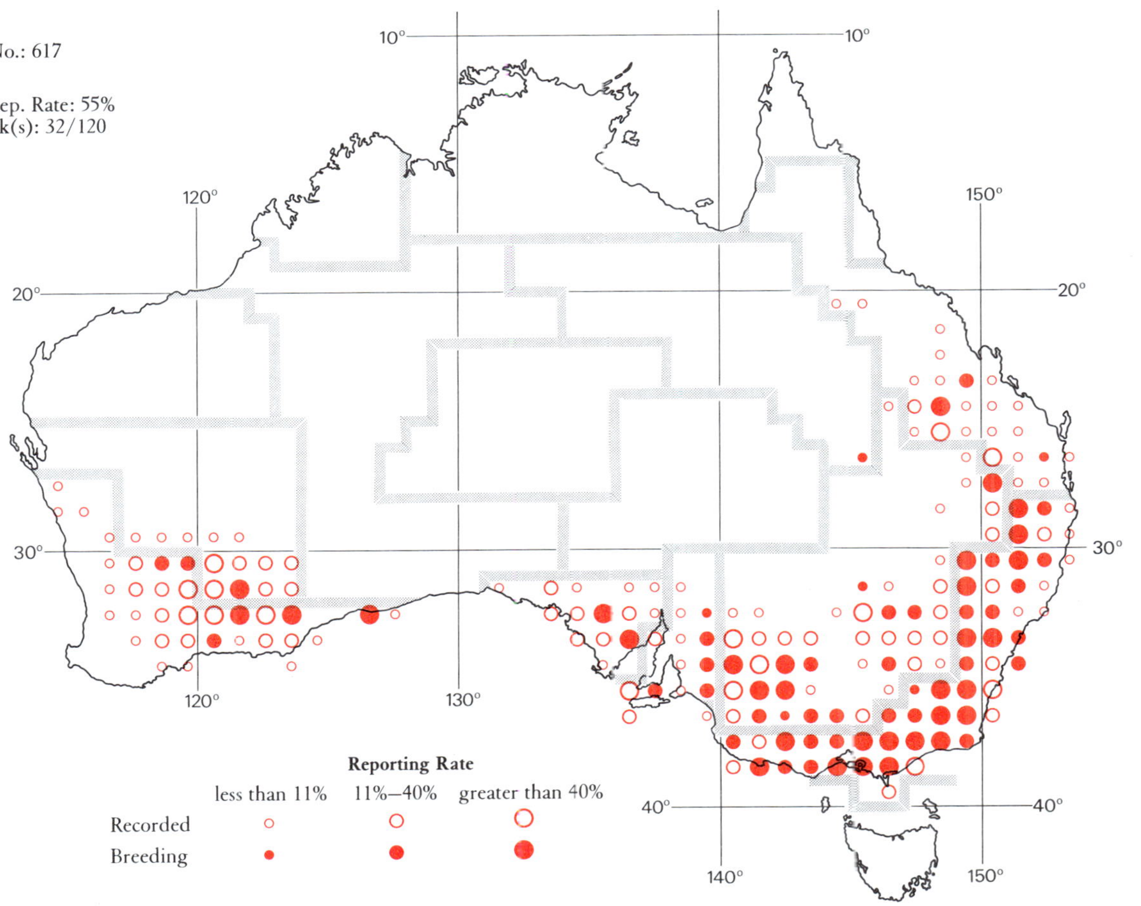

Yellow-throated Honeyeater

Lichenostomus flavicollis

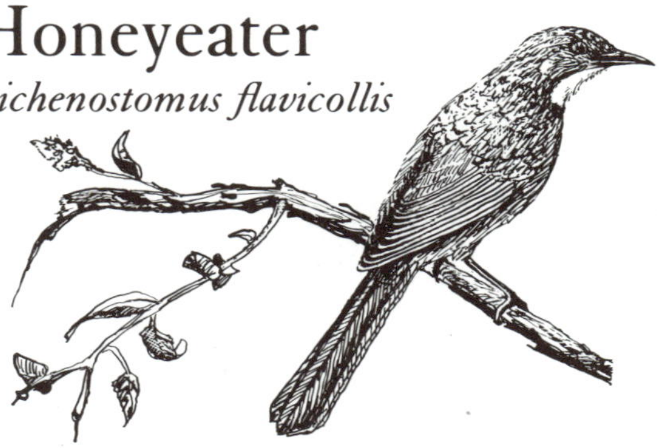

The Yellow-throated Honeyeater is endemic to the Tas. Region. The male is sedentary but the female and young disperse after the nesting is over. Sometimes the male will drive them off, re-mate and nest again in the same season.[740,987]

The species lives in eucalypt forest, woodland and heath, sometimes feeding in gardens. On Mt Wellington (42/147) it is widely distributed in eucalypt forest, wet and dry, but is not on the bare summit and rarely in cleared land.[1452] Nests may sometimes be close to each other.[567]

For most of the year the Yellow-throated Honeyeater forages for insects. South of Hobart (42/147) over 70% of its foraging time is spent on bark but it also hawks for insects, gleans from the foliage and ground layer and takes some nectar as well.[1748] At Port Davey (43/145) it has been recorded to take manna.[740] Along with other honeyeaters, flocks converge on the cider gum when manna is abundant in early autumn.[1259] Its preference for seeking insects on bark correlates well with its abundance in areas where large trees are common and provide a large surface to search.

During the Field Atlas the Yellow-throated was the most frequently recorded honeyeater in the Tas. Region, which reflects its ubiquity and conspicuousness as well as its abundance. There are no measures of abundance.

Field Atlas Records: 2663

10′ blocks–with records: 309 (84%)
–with breeding records: 34 (11%)

Atlas No.: 616

Max. Rep. Rate: 90%
In block(s): 42°05′/146°45′

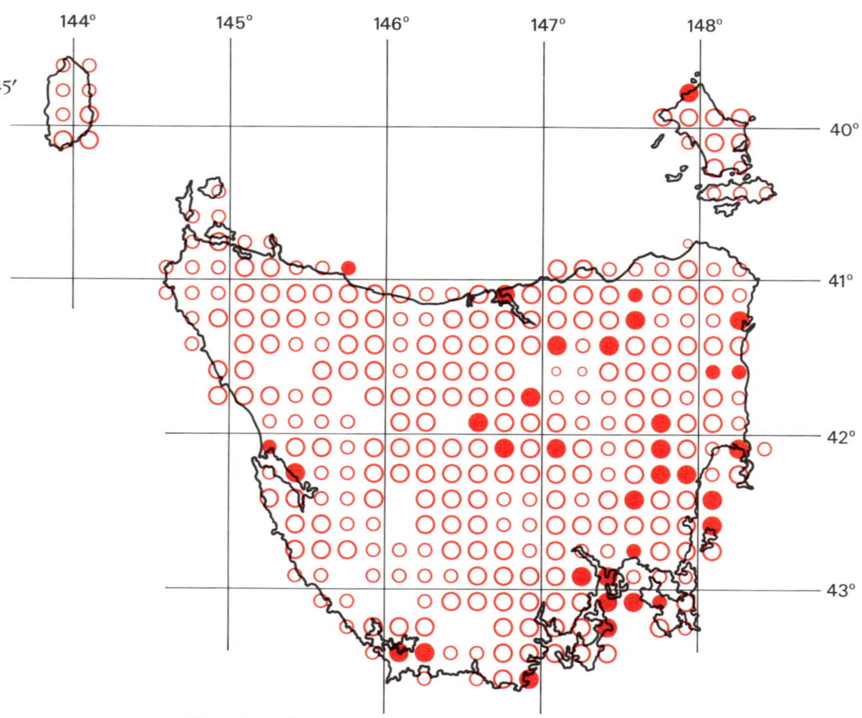

Reporting Rate

	less than 11%	11%–40%	greater than 40%
Recorded	○	○	○
Breeding	●	●	●

Yellow-tufted Honeyeater

Lichenostomus melanops

Four subspecies of the endemic Yellow-tufted Honeyeater are recognized: *melanops* on the coastal side of the Great Dividing Ra. south to Narooma (36/150); *gippslandicus* from Narooma to Noojee (37/146); *cassidix* round Yellingbo (37/145); and *meltoni* on the inland side of the Great Dividing Ra.[428] The subspecies *cassidix* (Helmeted Honeyeater) is now extremely rare.

Populations of the three subspecies in Vic. are mobile in autumn and winter, *meltoni* moving the greatest distance, specially in drought years. One *melanops* was captured with *cassidix* at Yellingbo in winter 1978.[1919]

The Yellow-tufted Honeyeater is restricted to eucalypt forest or woodland.[1919] The subspecies *meltoni* lives as isolated colonies or as large intermixing populations but the other subspecies are almost always in colonies.[1919] Groups or colonies may nest close together and remain attached to one locality for long periods, probably permanently. Breeding sites of all subspecies are traditional, some known to have been used for up to 20 years.[1798] The species has been recorded breeding communally.[501] It feeds on nectar, insects and spiders throughout the year; and on honeydew, manna and lerp in the breeding season.[1919] The fruit of mirrorbush is a major component of the food of nestling *cassidix*.[1919] Populations of *cassidix* are now mainly restricted to pockets of eucalypt forest that remain along creeks.[1919] Further reduction of vegetation by clearing and bushfires may lead to the extinction of these populations. Furthermore, psyllid infestations are killing the eucalypts and attracting Bell Miners that aggressively annex *cassidix* breeding areas.[1919]

At Sentry Box (30/151) 15 nests of the Yellow-tufted Honeyeater were recorded within a circle 1 km in diameter. At Koorawatha (34/148) one colony of about 50 breeding pairs occupied 8 ha (12.5 birds/ha).[1798] At Moruya (35/150) in eucalypt woodland density was 1.05 birds/ha.[1197]

Field Atlas Records: 2770

1° blocks–with records: 72 (9%)
–with breeding records: 40 (56%)

Atlas No.: 619

Max. Rep. Rate: 43%
In block(s): 36/143

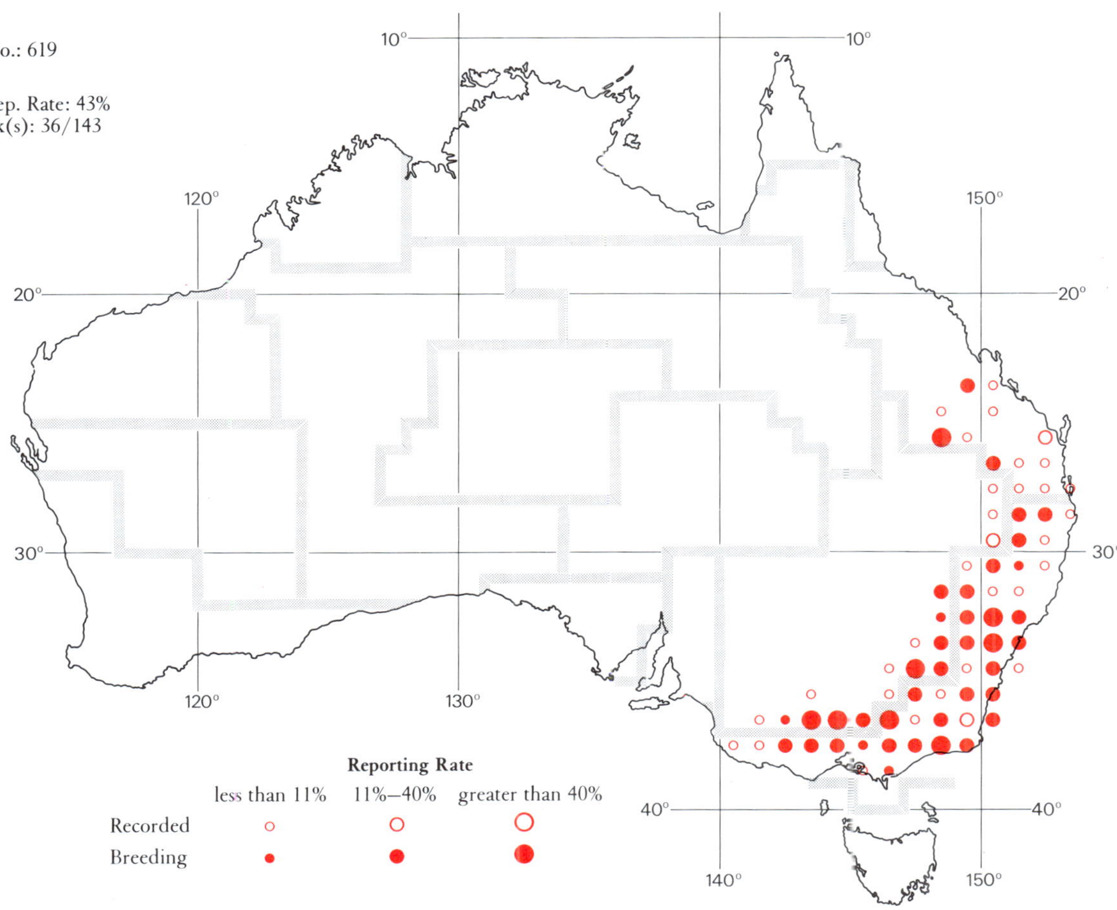

Purple-gaped Honeyeater

Lichenostomus cratitius

Two subspecies of the endemic Purple-gaped Honeyeater have been described but there is doubt about their validity.[578,1422] There is an isolated population in the Whipstick (36/144) and the species' distribution in eastern Australia is further split by the Flinders Ra. The large gap round the Great Australian Bight is confirmed in the Historical Atlases.

In the Big Desert (35/141) and at Eyre Bird Observatory (32/126) the Purple-gaped Honeyeater is present all year though the bird moves locally to areas of temporary food abundance.[390,886,1554] The Purple-gaped Honeyeater lives in mallee and heath, sometimes entering built-up areas. In the Big Desert 80% of sightings were of one or two birds; only 8.5% were of five or more birds, nearly all where mallee was in flower.[886]

The Purple-gaped Honeyeater feeds on insects gleaned from foliage and bark and caught by hawking.[748,886,1049] Much time is spent probing flowers of eucalypts and banksias, presumably for nectar.[886] On Kangaroo I. (35/137) several marked birds maintained territories for a number of days, feeding almost exclusively on honeydew.[1402]

In the Big Desert density in heath with no mallee was 0.15 birds/ha and in mallee about 5.0 birds/ha.[886] At Eyre Bird Observatory in mallee, density measured on a transect was 0.5–1.1 birds/ha; measured by mapping in a nearby area in Aug. it was 0.1 birds/ha.[464]

Field Atlas Records: 800

1° blocks–with records: 51 (6%)
–with breeding records: 15 (29%)

Atlas No.: 620

Max. Rep. Rate: 40%
In block(s): 35/136

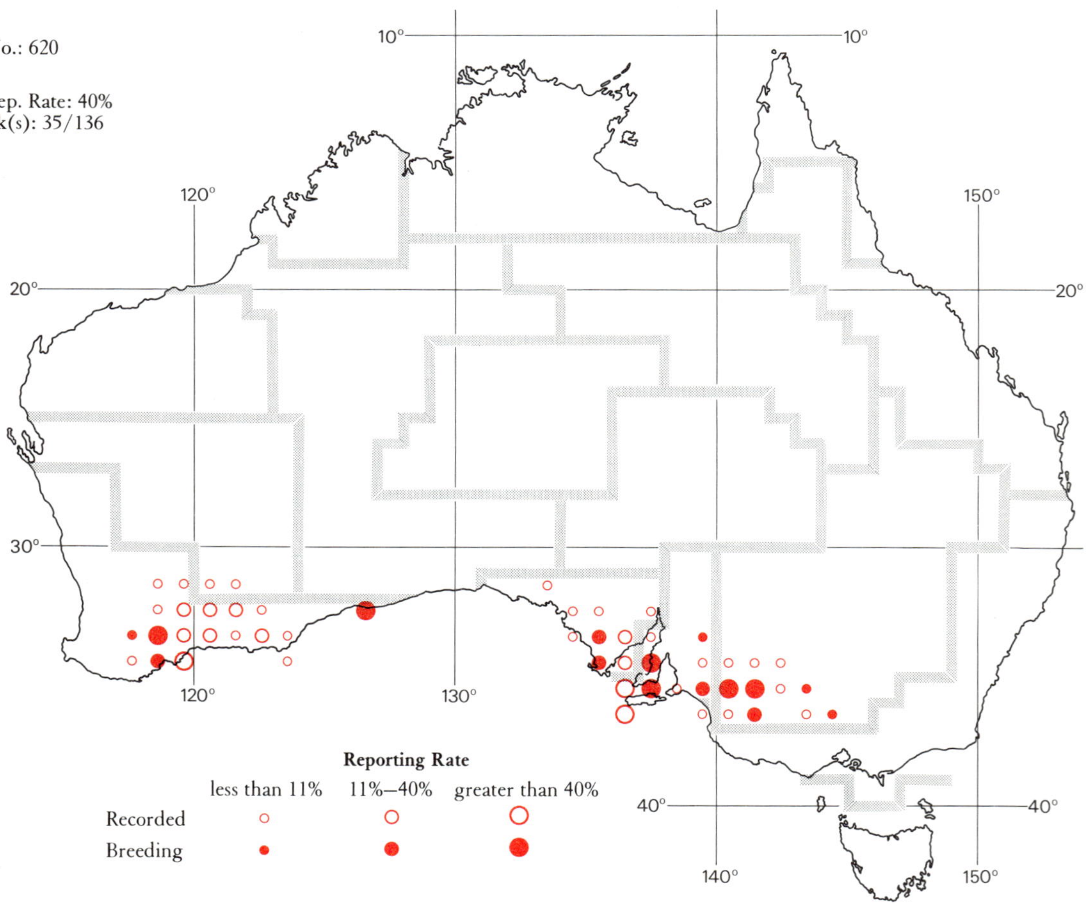

Grey-headed Honeyeater

Lichenostomus keartlandi

The range of the endemic Grey-headed Honeyeater lies in the driest part of northern Australia, receiving less than 200 mm of summer rain and less than 50 mm of winter rain. It is thought to be at most locally nomadic[987] but some birds occasionally stray far beyond the species' normal range, as in June 1977 when at least one was recorded at 28/145 west of Cunnamulla.[1347]

In the Great Sandy Desert the Grey-headed Honeyeater was found to be abundant wherever there were eucalypts, as described by other observers.[626,1714] It also occupied hills and gorges but was not only in rocky places as formerly suggested.[1354,1638] The species lives singly or in small groups, though flocks of 20–200 have been seen at Mt Isa (20/139).[1068] It feeds on insects and nectar.[1049,1086] Unidentified vegetable matter has been found in guts.[748]

There are no measures of abundance.

Field Atlas Records: 728

1° blocks–with records: 153 (19%)
–with breeding records: 13 (8%)

Atlas No.: 621

Max. Rep. Rate: 59%
In block(s): 22/118

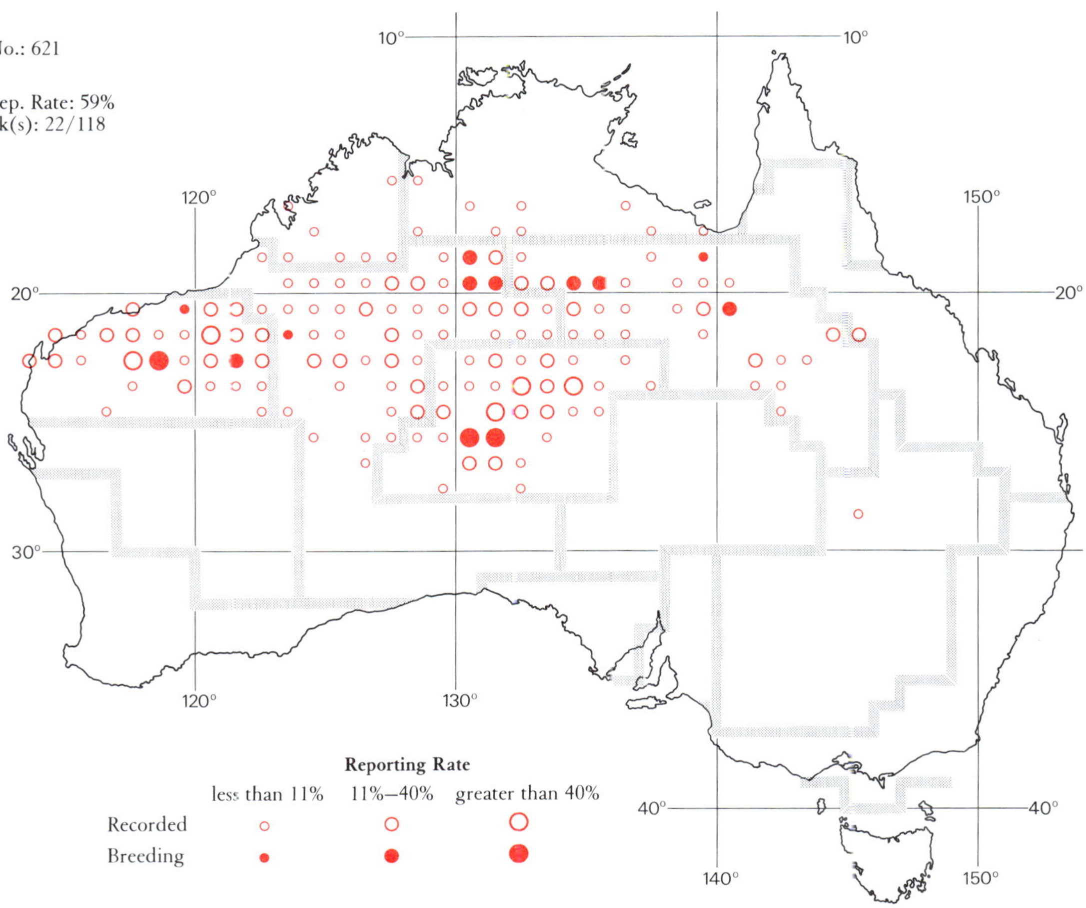

Yellow-plumed Honeyeater

Lichenostomus ornatus

Though no subspecies of the endemic Yellow-plumed Honeyeater have been recognized[599] there are barriers that at least partly isolate the population in the eastern half of the Murray-Darling Region and those on either side of the Flinders Ra. Historical and Field Atlases together indicate that the population is sparsely continuous round the Great Australian Bight and north of the Nullarbor Plain.[180]

In the Big Desert (35/141) the bird is present throughout the year with peak numbers in May-June.[886]

The Yellow-plumed Honeyeater lives in eucalypt woodland, mallee and acacia scrub.[748,833,886] Most records (80% in the Big Desert) are of one or two birds; parties of up to 12 gather on flowering trees.[886] The Yellow-plumed Honeyeater feeds on lerp, insects from the bark and foliage of eucalypts or caught in the air and nectar from eucalypts and mistletoe.[748,1402,1403,1919] In the South-West the species appears to be penetrating the forest country in some places following clearing.[1634]

In the Big Desert density varied from 0.04 birds/ha in heath with no mallee to 3.6 birds/ha in tall mallee (4-8 m high).[886]

Field Atlas Records: 2481

1° blocks–with records: 117 (14%)
–with breeding records: 50 (43%)

Atlas No.: 622

Max. Rep. Rate: 64%
In block(s): 32/122

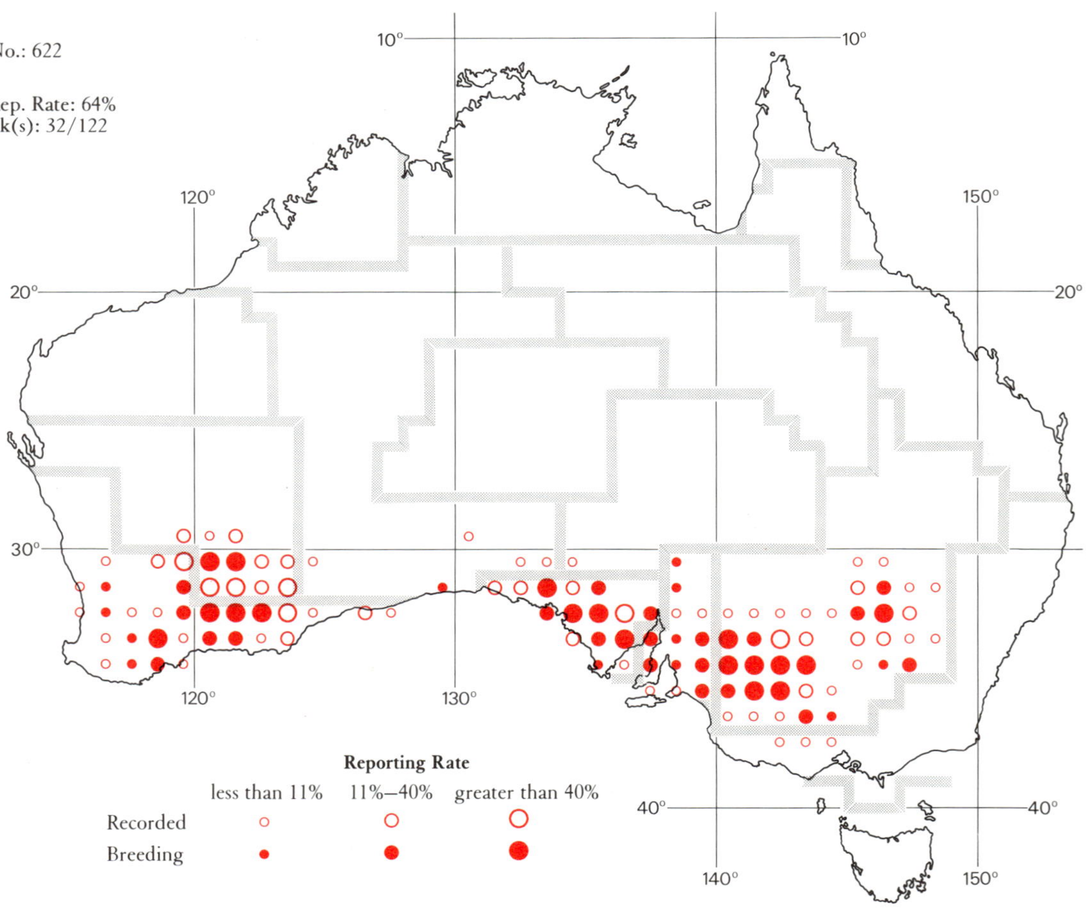

Grey-fronted Honeyeater

Lichenostomus plumulus

Subspecies of the endemic Grey-fronted Honeyeater have been described but views about them differ and variation is probably clinal.[748,1422] Historical Atlas records fill only a few of the gaps in the Field Atlas map so that there are large areas within the species' general distribution where it is rare or absent.

The Grey-fronted Honeyeater is thought to be sedentary over much of its range but periodically 'swarms' on flowering eucalypts.[987] At Mt Isa (20/139) it is recorded most often May-Dec. in parties of 20–30 birds and once up to 80 birds.[1068] The Field Atlas suggests no large-scale seasonal movement.

The Grey-fronted Honeyeater inhabits mallee, acacia scrub and eucalypt woodland, with a grassy understorey, often along creeks.[748] Near Mt Isa it is always to be found in short dense mallee clumps with plentiful small water-courses and sometimes in patches of flowering eucalypts.[1068] In the Great Victoria Desert it is widespread in mallee with spinifex.[180] Usually it lives in pairs or small groups. It feeds in the foliage on insects, including lerp and aphid, seeds, nectar and berries.[748] At Wanjarri (27/120) it was recorded feeding in mistletoe.[1282]

There are no measures of abundance.

Field Atlas Records: 756

1° blocks-with records: 215 (26%)
-with breeding records: 27 (13%)

Atlas No.: 623

Max. Rep. Rate: 39%
In block(s): 20/139

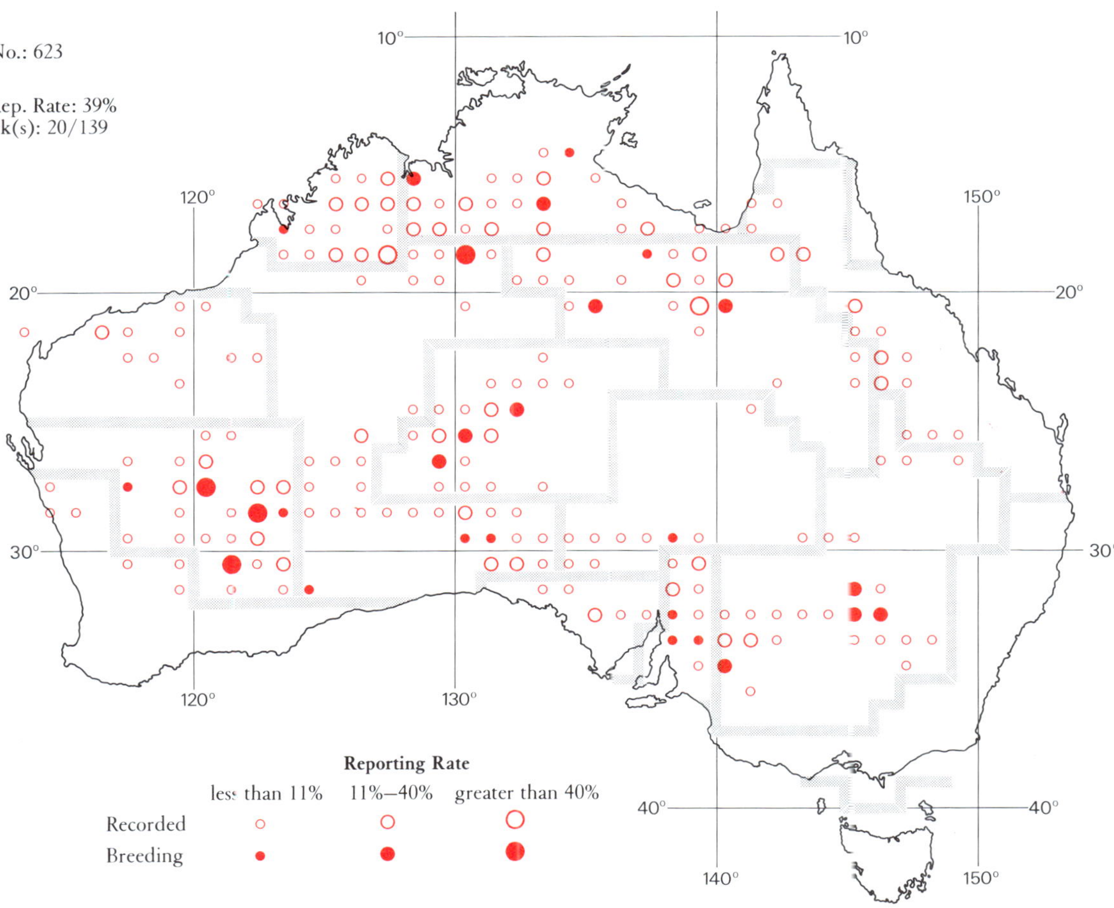

Fuscous Honeyeater

Lichenostomus fuscus

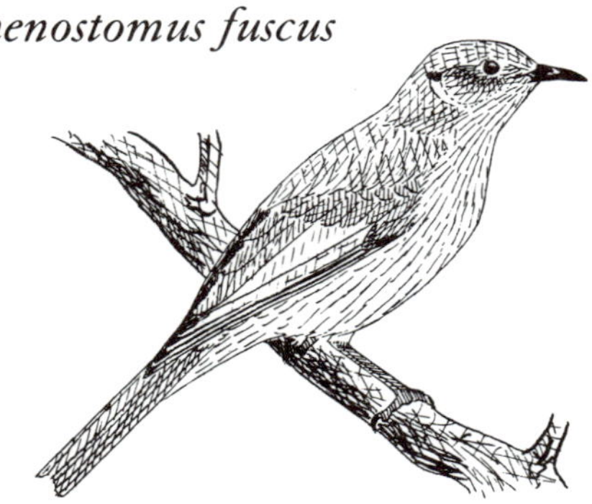

Three subspecies of the endemic Fuscous Honeyeater are recognized but there is doubt about their validity.[1124,1375,1422] In the South-East Region the Fuscous Honeyeater has occasionally been noted migrating in company with the Yellow-faced Honeyeater, e.g. at Hastings Pt (28/153), Barrington (31/151) and Blackheath (33/150).[987,1069] Only small numbers seem to be caught up in these mass honeyeater-migrations but in some places observations suggest regular seasonal movements. At Canberra (35/149) the Fuscous Honeyeater is a passage migrant: a few birds winter but there are no summer records.[653] Reporting Rates in the southern high country (35/148, 36/148) were 13.5% in summer and 0% in winter; they were reversed on the nearby coast (35/150, 36/150) (summer 0%, winter 12.4%), which suggests an almost complete altitudinal migration. In the Warby Ra. (36/146) the Fuscous Honeyeater is present only during winter.[1293] At Jamieson (37/146), where the bird breeds Aug.-May, there were only two winter records in 14 years.[1469]

The Fuscous Honeyeater lives in eucalypt forest and woodland, occasionally in mallee and in heath near the coast. It is abundant where native shrubs are flowering profusely including gardens and parks. The Fuscous Honeyeater feeds on honeydew, nectar and insects[1402] and has been observed eating blackberries.

At Armidale (30/151) in eucalypt woodland density was 0.94–5.46 birds/ha.[582] At Jamieson in 20 ha of eucalypt forest more than 100 birds have been recorded at the peak of breeding activity (5 birds/ha) and, in one weekend before the young had fledged, 75 individuals were banded.[1469]

Field Atlas Records: 2584

1° blocks–with records: 98 (12%)
–with breeding records: 36 (37%)

Atlas No.: 613

Max. Rep. Rate: 35%
In block(s): 30/151

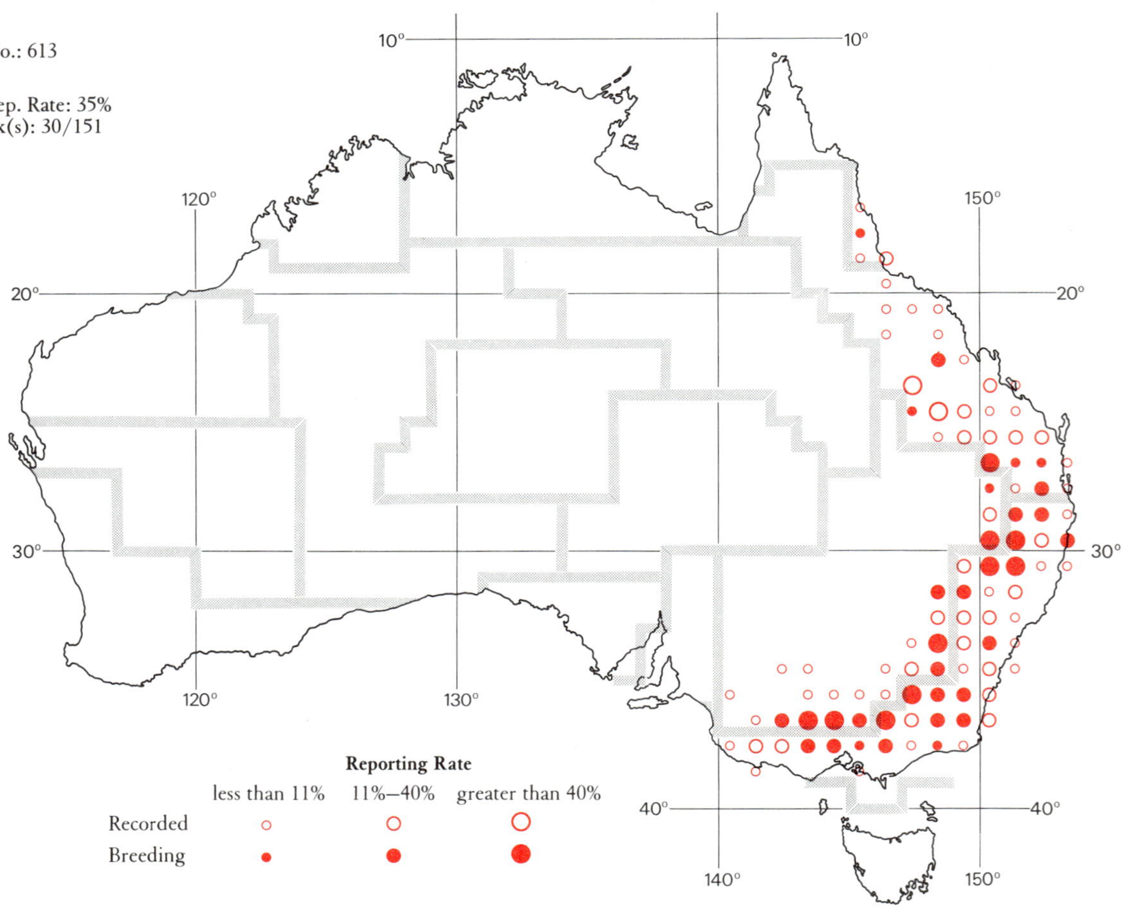

Yellow-tinted Honeyeater

Lichenostomus flavescens

The Yellow-tinted Honeyeater lives in New Guinea[1824] and Australia. Several subspecies have been described including one from Melville I. (11/130) not recorded during the Field Atlas. There is doubt about the validity of some.[748,1375,1422] The record from Weipa (12/141) is a northward extension of the species' range by about 400 km.[378] The birds were numerous there in June 1978.

The Yellow-tinted Honeyeater is said to be sedentary[898,987] though one observer regarded it as a rare summer visitor to the Mt Isa (20/139) district.[862] It lives in rainforest and eucalypt woodland lining watercourses and large tropical rivers[1068,1661] and in the Kimberley Region is also found in eucalypt woodland with an understorey of spinifex.[898]

The Yellow-tinted Honeyeater feeds on nectar and insects[748] moving singly and in small flocks from one flowering eucalypt to another. Insects are caught on leaves and by hovering, e.g. near a colony of winged termites.[898]

South of Wyndham (15/128) each pair of Honeyeaters defended a territory 100–200 m in diameter.[898] North of Mt Isa 6–20 individuals were recorded in each of a series of 30-minute walks along watercourses.[1068] In new Guinea in eucalypt woodland near Port Moresby (9S/147E) density was 0.12–0.64 birds/ha, probably fluctuating with the flowering of trees.[158]

Field Atlas Records: 810

1° blocks–with records: 90 (11%)
–with breeding records: 17 (19%)

Atlas No.: 624

Max. Rep. Rate: 62%
In block(s): 16/125

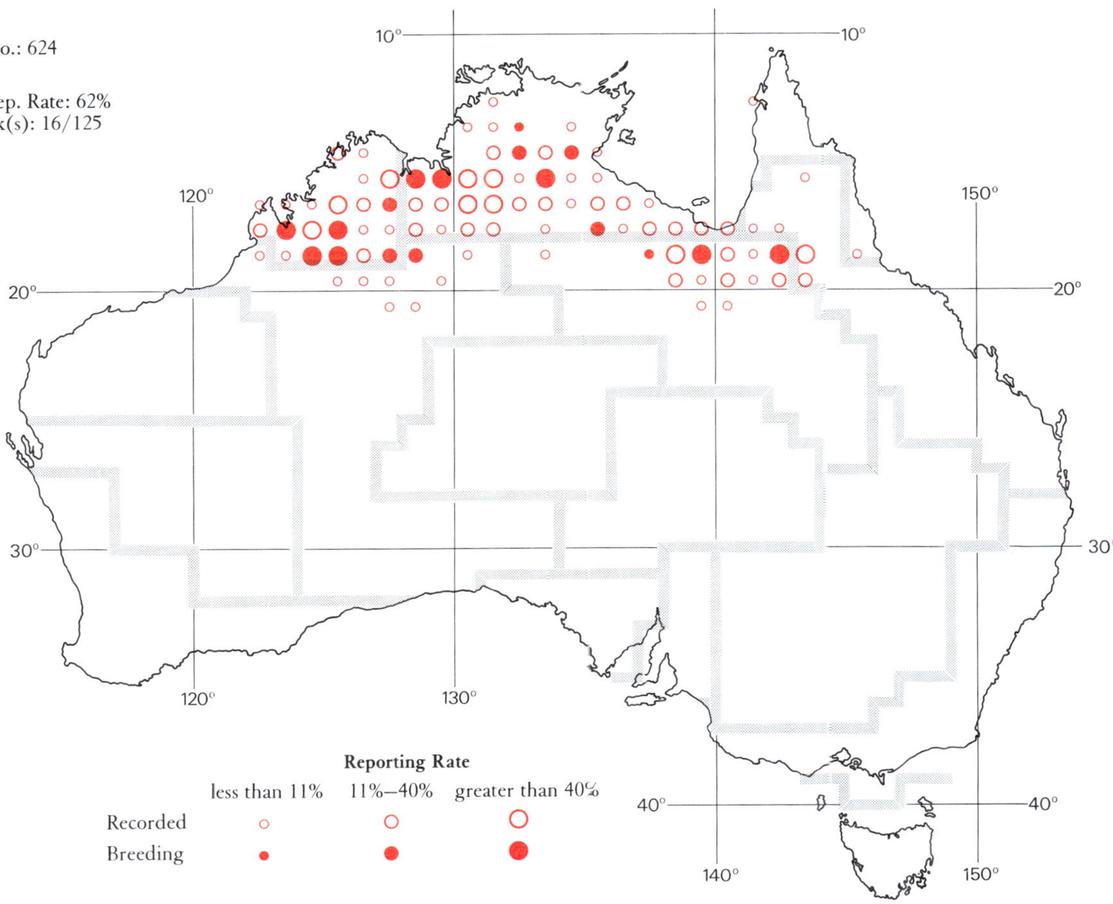

White-plumed Honeyeater

Lichenostomus penicillatus

Eight subspecies of the endemic White-plumed Honeyeater have been recognized but variation is clinal and there is doubt about how many should be retained.[748,1422] In a small extension of range the species was observed to establish itself in the Sydney (33/151) area gradually between 1900 and 1950.[803] It is occasionally recorded near Perth (31/115), seeming to penetrate down the coast from Dongara (29/114).

The species is regarded as mainly sedentary[987] but because the bird frequently appears at stock waters remote from watercourses it must undertake some nomadic wandering. The longest movements recorded are of an adult from Maryborough (37/143) to Maude (34/144), 290 km,[24] and of a bird from Narrandera (34/146) to Mt Piper (37/145), 330 km.[85] The Field Atlas does not suggest large-scale seasonal movement.

Distribution of the White-plumed Honeyeater follows that of large eucalypts, specially the river red gum, so it tends to be most abundant along watercourses.[684,833,1068] Individuals and small parties live in woodland near the coast and in eucalypts well away from watercourses in the desert. Though not as abundant there as close to pools, the bird obviously has the ability to survive without access to surface water. In Melbourne (37/145) and Adelaide (33/138) the White-plumed Honeyeater is one of the few native species frequently inhabiting parks and gardens.[75,1693] It has once been recorded to breed communally.[501] Observations at Cranbourne (38/145) show that manna forms a large proportion of the species' diet, amounting to over 90% in summer; some nectar is taken in spring and insects are eaten throughout the year.[1402] At Golton Vale (36/142) honeydew is the main food; only small amounts of nectar and insects are eaten, except in summer.[1402] In Adelaide the species was recorded persistently visiting a number of different flowering plants.[1403] Blackberries are also eaten.

Density at Wagga Wagga (35/147) in eucalypt woodland was 14.3 birds/ha.[952]

Field Atlas Records: 19 868

1° blocks–with records: 434 (53%)

–with breeding records: 200 (46%)

Atlas No.: 625

Max. Rep. Rate: 79%
In block(s): 27/147

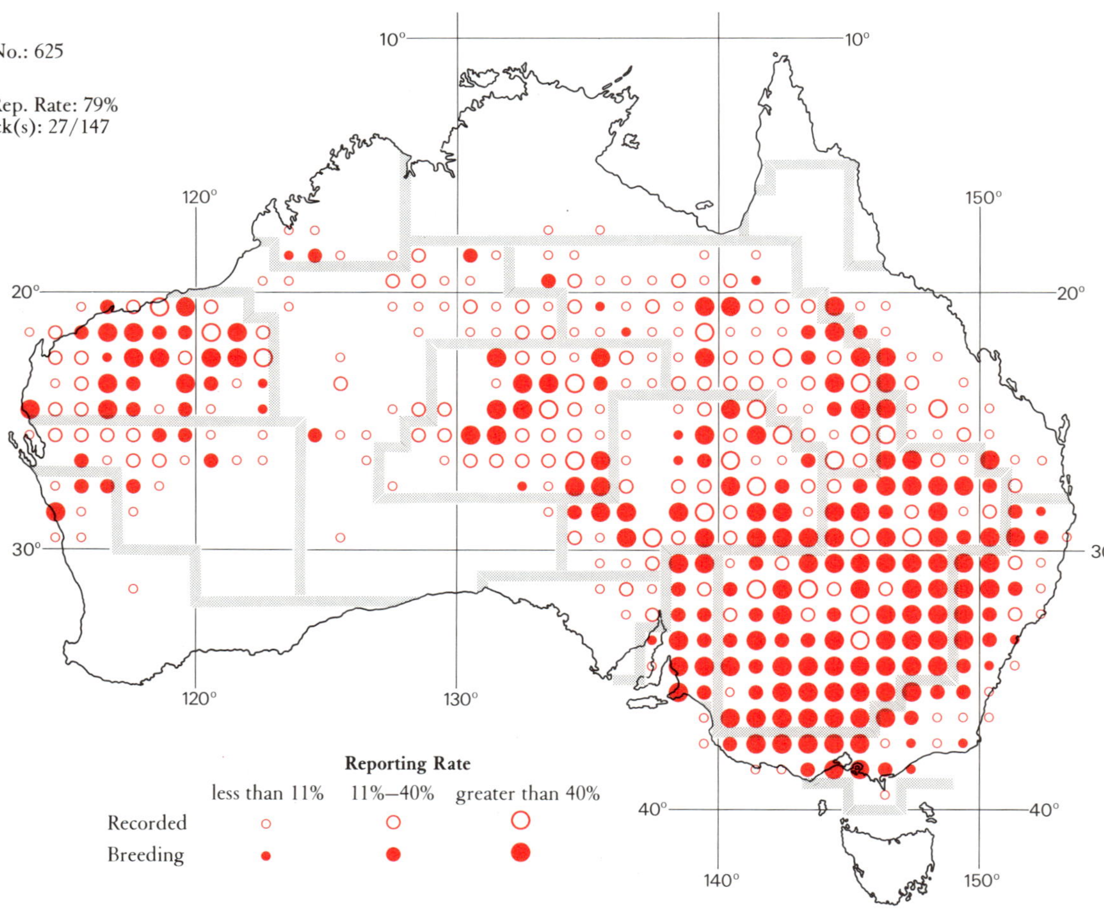

Black-chinned Honeyeater

Melithreptus gularis

Two subspecies of the endemic Black-chinned Honeyeater are recognized: *gularis* in Eastern Qld and areas to the south; *laetior* (Golden-backed) elsewhere.[1124,1709] They are sometimes treated as two species.[1422] The population in the SA Ra. Region appears isolated from birds farther east. The Black-chinned Honeyeater is generally present throughout the year in a locality, any movement being only local.[987] It is rarely found close to the coast. In the Atherton Region, where the distribution map suggests its range reaches the coast, it was seen by one observer only twice during 15 years, each time more than 1000 m above sea level.[698] In the south it is absent from higher altitudes and generally confined to the inland side of the Great Dividing Ra. except in the Grafton (29/152) district, where the 10′ maps show a population apparently isolated on the coastal side of the range, and between the Hawkesbury R (33/150) and the Shoalhaven R (34/150).

The Black-chinned Honeyeater lives in pairs or small groups in eucalypt and paperbark woodland, acacia scrub and spinifex, specially where there are patches of flowering shrubs. During the Field Atlas three adults were reported feeding a fledgling near Katherine (14/132) and there are a few other records of communal breeding.[501]

The Black-chinned Honeyeater feeds on nectar, honeydew and insects.[1402] In Vic. where detailed studies have been made, insects are most important as food in the spring. Throughout the year much honeydew is eaten, specially in summer and autumn. Nectar is gathered mainly from eucalypts but at times from banksias.[1403]

There are no measures of abundance.

Field Atlas Records: 1428

1° blocks–with records: 217 (27%)

–with breeding records: 33 (15%)

Atlas No.: 580

Max. Rep. Rate: 18%
In block(s): 16/125

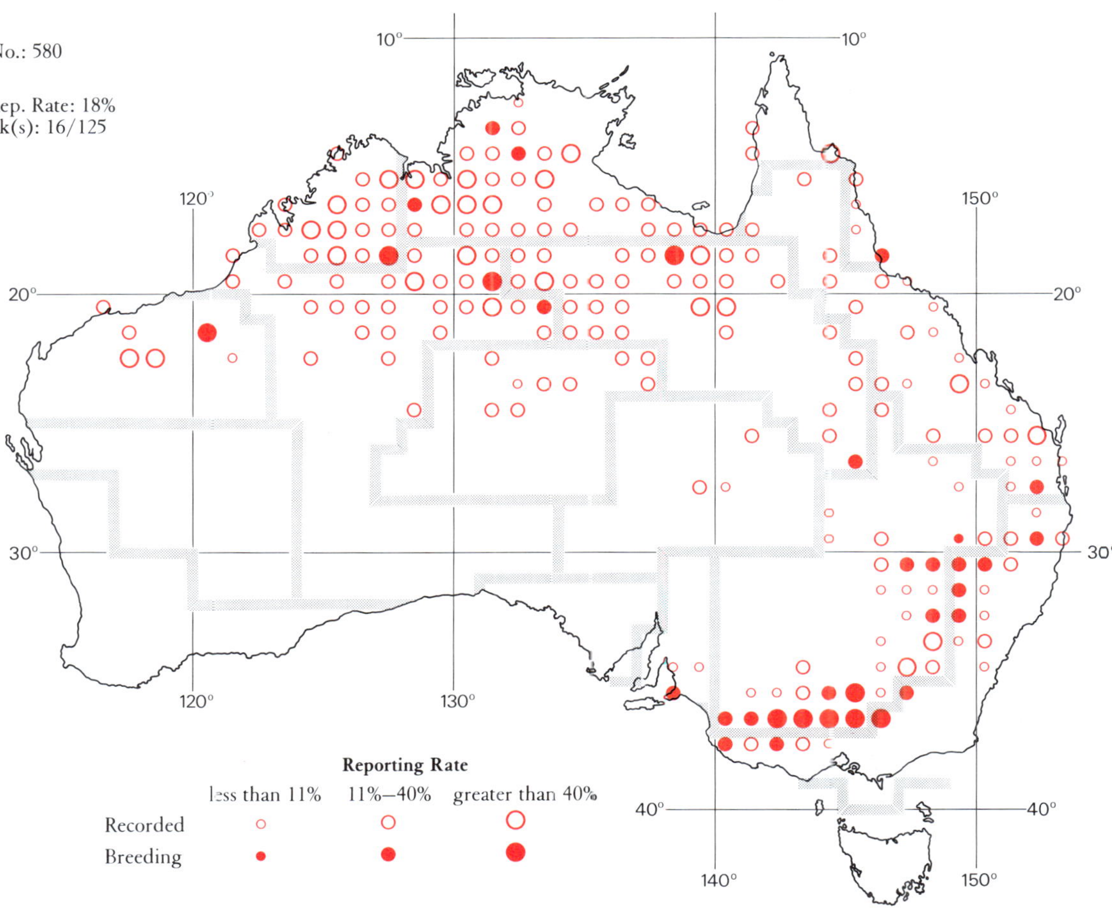

Strong-billed Honeyeater

Melithreptus validirostris

The Strong-billed Honeyeater is endemic to the Tas. Region. It is nomadic for much of the year, making no regular seasonal movements.

It lives in small parties in eucalypt forests, specially those in gullies. In winter on Mt Wellington (42/147) it may move to drier forest on the slopes to form mixed foraging flocks with the Black-headed Honeyeater.[1314,1452] It breeds semi-colonially.[1354,1748]

The Strong-billed Honeyeater feeds on insects from crevices in eucalypt bark, as well as nectar. In a region where well-known bark-gleaning birds are absent, e.g. treecreepers, Crested Shrike-tit, this member of a nectarivorous genus takes over the role of treecreeper.[1748] Recent work has shown that it lives up to its name and is able to extract insects from deep crevices that require more force to open than those from which other honeyeaters extract insects.[438] Observations at Hobart (42/147) showed that the Strong-billed Honeyeater spent over 80% of its time foraging on the bark of trees for most of the year; but can feed opportunistically as well, e.g. taking caterpillars from the ground.[1748] It also feeds on manna.

There are no measures of abundance.

Field Atlas Records: 1064

10′ blocks-with records: 230 (63%)

-with breeding records: 36 (16%)

Atlas No.: 582

Max. Rep. Rate: 70%
In block(s): 42°05′/147°45′

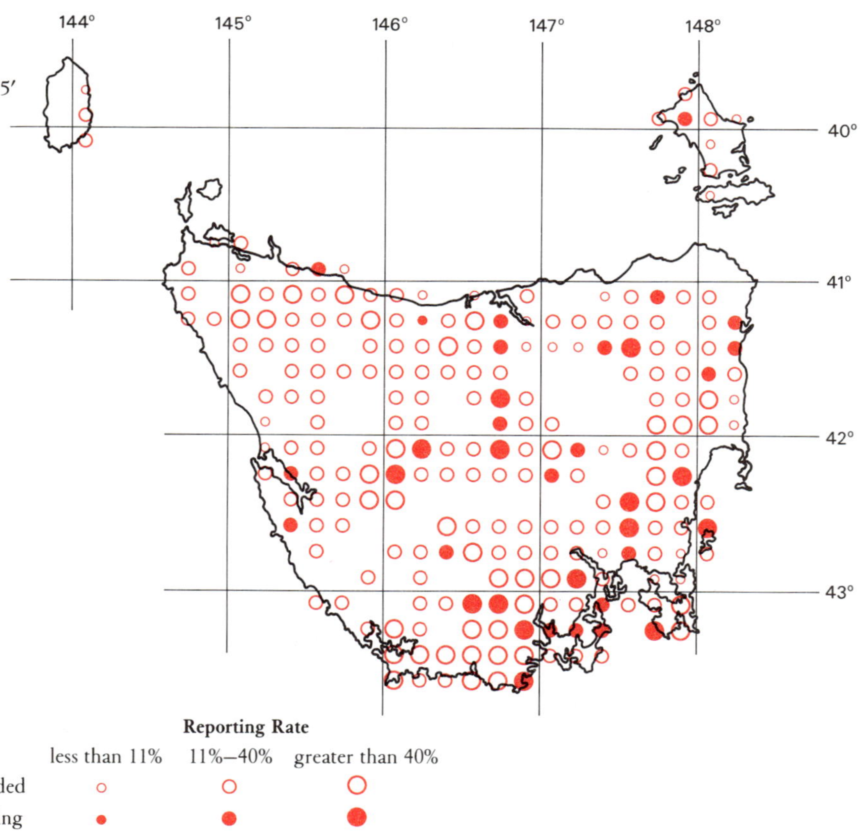

Brown-headed Honeyeater

Melithreptus brevirostris

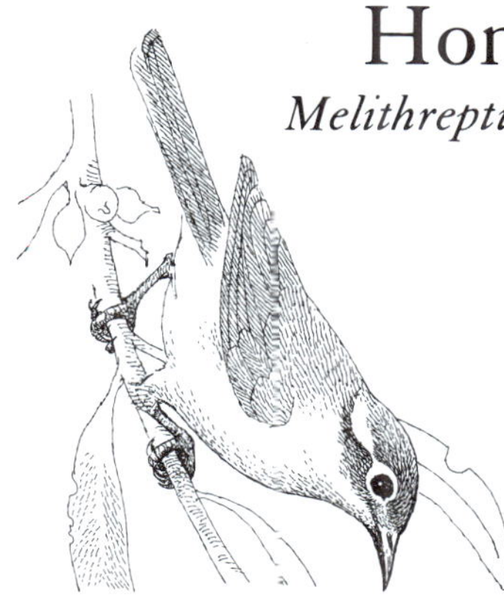

Eastern and western populations of the endemic Brown-headed Honeyeater are separated at the Great Australian Bight. Up to four subspecies have been recognized but views about them differ.[600,748,1124]

At Armidale (30/151) and Wollomombi (30/152) the Brown-headed Honeyeater is present all year but has a big home range.[130] Similarly, at Canberra (35/149), groups seem to wander over an area of about 3–5 sq. km.[653] At Boola Boola (38/146) more Brown-headed Honeyeaters are found in lowland forest in winter than in summer[1113] but altitudinal movement is not reflected in Reporting Rates. In areas in and near the Snowy Mts (36/148) few were seen in any season except summer, when the Reporting Rate was 18%, but birds were recorded at all altitudes then. At Eyre Bird Observatory (32/126) the Brown-headed Honeyeater is much more abundant in spring and autumn than in summer and winter, so that it presumably makes at least local movements.[464]

The Brown-headed Honeyeater inhabits eucalypt forest, woodland and mallee, reaching the coast only where there is heath and other moderately dry environments. Near Canberra banding has shown that the same birds stay together in small groups of up to 20.[653] Communal breeding has occasionally been recorded.[501,1352] The Brown-headed Honeyeater also joins mixed species flocks.[129] It feeds on nectar, insects and manna.[1402] In the inland it spends much time gleaning from the foliage as well as feeding on flowering mallee. At Cranbourne (38/145) its diet in winter is almost entirely manna.[1402] In the South-West in spring it feeds almost exclusively on nectar of the parrot bush.[753]

The large home range of groups of this species makes it difficult to census.[130] Densities in eucalypt woodland were: Armidale 0.08–0.65 birds/ha,[582] Wollomombi 0.03–0.99 birds/ha[129] and Canberra 0.04–0.22 birds/ha.[54] At Boola Boola in eucalypt forest density was 0.3–0.5 birds/ha.[1113] At Eyre Bird Observatory in mallee it was 0.1–3.8 birds/ha.[464]

Field Atlas Records: 7223

1° blocks–with records: 220 (27%)
–with breeding records: 83 (38%)

Atlas No.: 583

Max. Rep. Rate: 46%
In block(s): 35/141

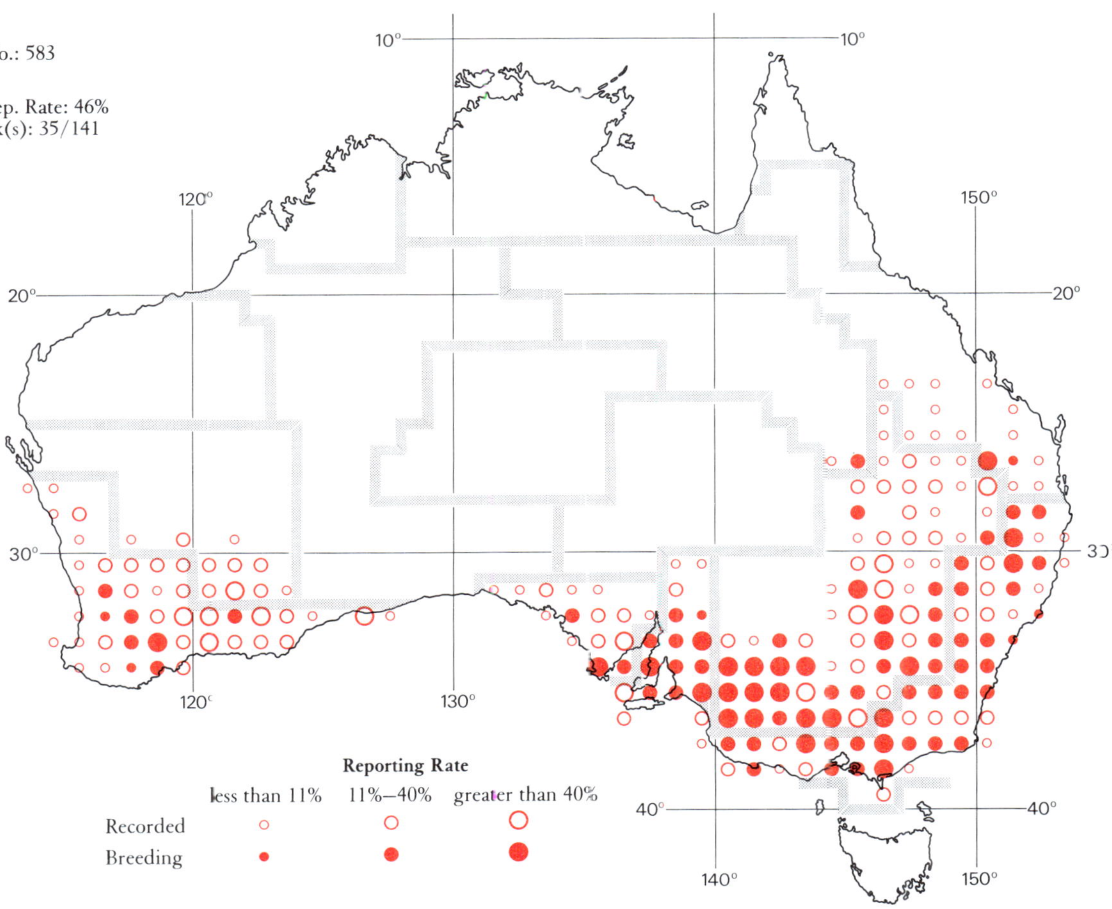

White-throated Honeyeater

Melithreptus albogularis

The White-throated Honeyeater lives in southern New Guinea and Australia. Views about the validity of subspecies in Australia differ.[748,1422]

The Field Atlas does not provide evidence of any large-scale seasonal movement though local movements have been reported from the edges of its range.[987,1089]

The White-throated Honeyeater lives singly or in small parties[748] in eucalypt forest and woodland and paperbark woodland. It is often in trees along watercourses[748] but uncommon in rainforest. In the east it is not present inland of the slopes of the Great Dividing Ra. The species has once been reported to breed communally.[501] The White-throated Honeyeater feeds on insects gathered from the foliage. It forages in flowering trees, specially paperbarks, probably taking some nectar as well as insects attracted to the blossoms.

In New Guinea in eucalypt woodland near Port Moresby (9S/147E) density ranged from 0.28–0.77 birds/ha, the greatest abundance occurring when trees flowered.[158]

Field Atlas Records: 3564

1° blocks–with records: 166 (20%)
–with breeding records: 31 (19%)

Atlas No.: 579

Max. Rep. Rate: 67%
In block(s): 13/132

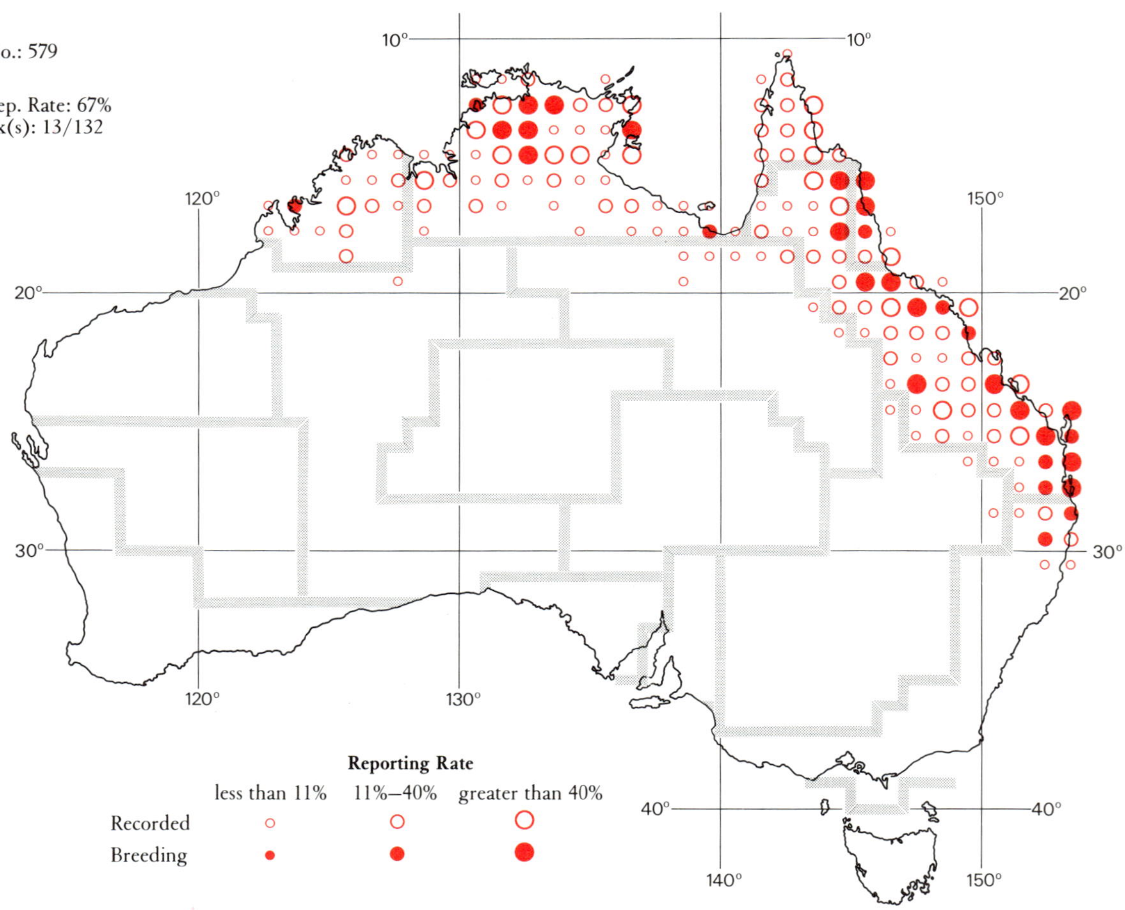

White-naped Honeyeater

Melithreptus lunatus

Two subspecies of the endemic White-naped Honeyeater are recognized: *lunatus* in the east and *chloropsis* in the South-West Region.[748,1124] The species has not been recorded in the Tas. Region.

On the eastern coast between the Goulburn Valley (32/150) and Mallacoota (37/149) spectacular movements involving thousands of White-naped Honeyeaters are observed, the flocks passing northward in autumn and south in spring.[807] The destination of the migrating flocks is not known. In the Field Atlas birds were recorded throughout the year south and west of Mallacoota, which suggests that the migrants may originate in the high country of the Great Dividing Ra. The White-naped Honeyeater is absent in winter from the tablelands near Canberra (35/149) at altitudes above 700 m.[653,682] The exodus takes place in Apr.-May and birds start to return in Aug. Reporting Rates in the Snowy Mts (36/148) above 500 m dropped from 57.9% in summer to 20.0% in winter. Below 500 m there was also a decrease from 70.4% in summer to 50.0% in winter but its magnitude was much less, which suggests that a large population remains at lower altitudes throughout the year. The longest movement recorded for the White-naped Honeyeater is 252 km.[85] In the Atherton Region the population is sparse and resident all year.[231,698] At Armidale (30/151) some birds are sedentary but the population is augmented by passage migrants in spring and autumn.[129] In Eastern Qld, inland areas and the SA Ra. Region, *lunatus* appears to be sedentary or locally nomadic as is *chloropsis* in the South-West Region.[987]

The White-naped Honeyeater lives in small flocks and has been reported breeding communally.[501] It inhabits eucalypt forest and woodland and feeds on insects, nectar, manna and honeydew. In a series of observations at Cranbourne (38/145) the species spent nearly 60% of its time feeding on manna throughout the year.[1402]

Densities in eucalypt woodland were: Armidale 0.81–4.63 birds/ha;[582] Wollomombi (30/152) 0.8 birds/ha while breeding but up to 2.3 birds/ha during migration;[129] Canberra up to 0.96 birds/ha.[154] In eucalypt forest they were: Bathurst (33/149) 14.25 birds/ha during breeding;[489] Jamieson (37/146) 0.5–1.0 birds/ha and occasionally 5.0 birds/ha;[1469] and Boola Boola (38/146) 0.3–2.8 birds/ha.[1113]

Field Atlas Records: 7743

1° blocks–with records: 143 (18%)
–with breeding records: 66 (46%)

Atlas No.: 578

Max. Rep. Rate: 51%
In block(s): 36/150

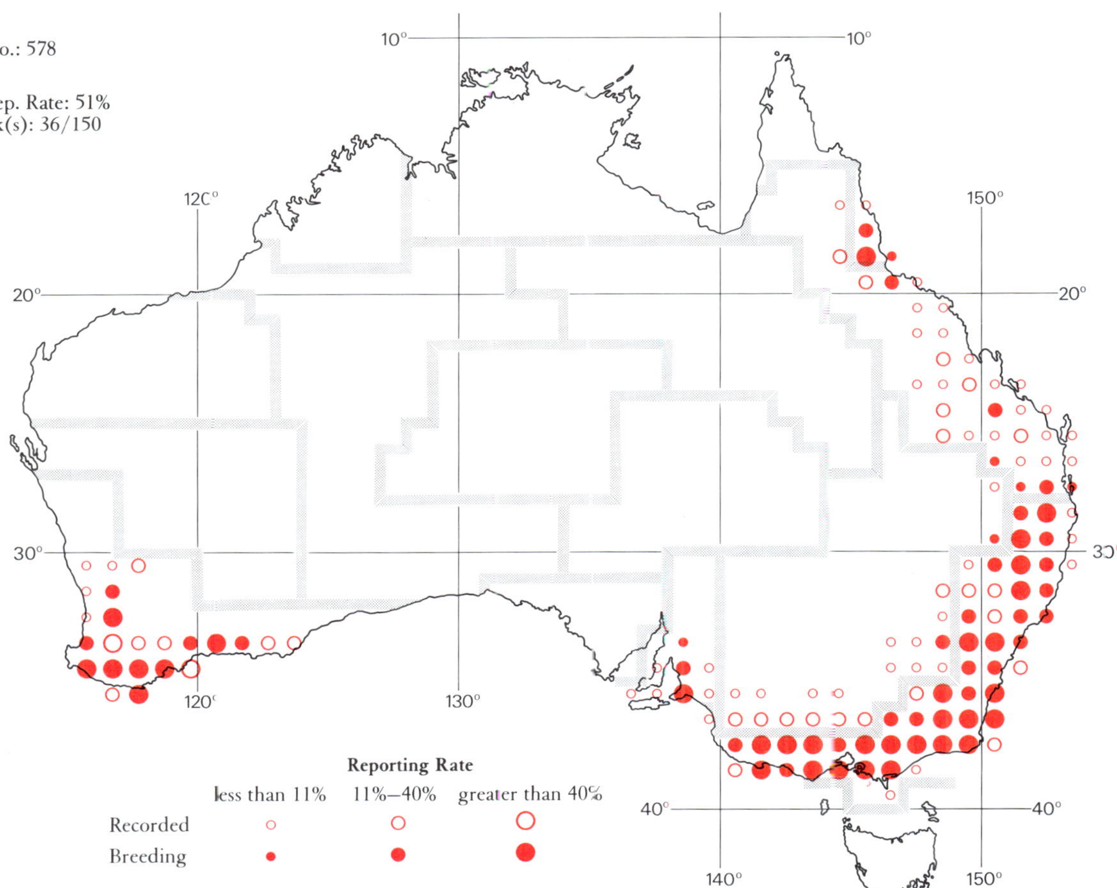

Black-headed Honeyeater

Melithreptus affinis

The Black-headed Honeyeater is endemic to the Tas. Region. There is no evidence of a seasonal altitudinal shift of the population on Mt Wellington (42/147)[1452] but post-breeding flocks wander and may be joined in winter by flocks of the Strong-billed Honeyeater.[1314] The Black-headed Honeyeater lives in pairs in summer[232] mainly in eucalypt forest.[1452] It is most often in dry forests and the 10′ maps show that few live in the wet forests of the western and south-western parts of the region. Several observers have noted it in places where there is young regrowth.[566,1461] The species also breeds in sites disturbed by development such as gardens and orchards[1354] though an earlier observer regarded it as uncommon in gardens.[1645]

The species has been recorded to breed communally.[1325] Near Hobart (42/147) the Black-headed Honeyeater spends more than 75% of its feeding time gleaning invertebrates, including ants, from leaves; some foraging is done on the bark, mainly of branches; nectar is taken only in summer when post-breeding flocks wander.[1748] The species is also recorded feeding on honeydew[232,1049] and will cause damage to fruit.[683]

Near Hobart there were four nests within 1 ha,[1645] which is the only measure of abundance recorded.

Field Atlas Records: 1140

10′ blocks-with records: 199 (54%)

-with breeding records: 28 (14%)

Atlas No.: 584

Max. Rep. Rate: 66%
In block(s): 42°05′/146°45′

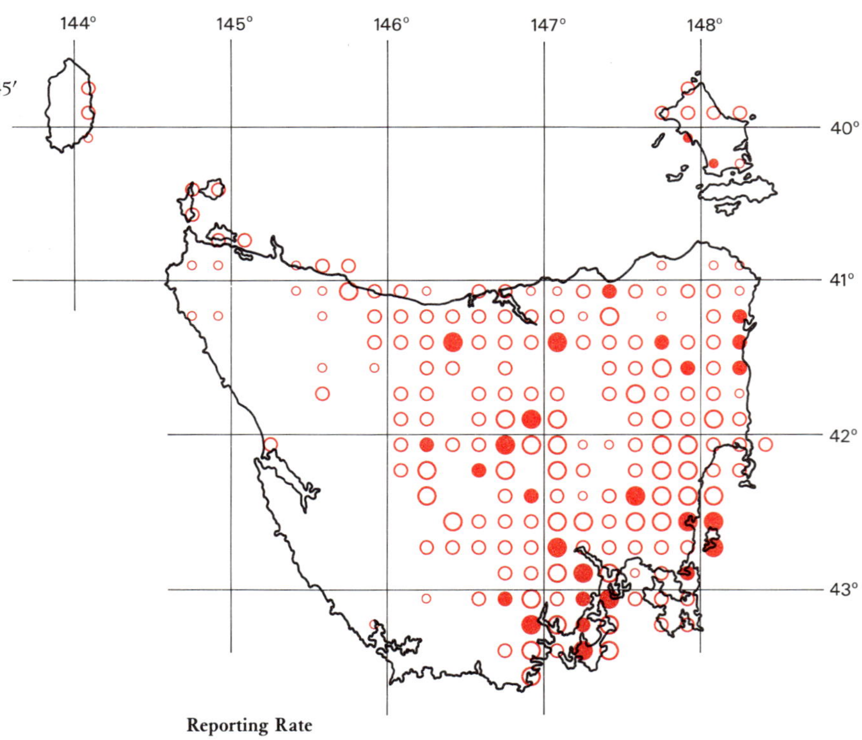

Reporting Rate

	less than 11%	11%–40%	greater than 40%
Recorded	○	○	○
Breeding	●	●	●

Green-backed Honeyeater

Glycichaera fallax

This predominantly New Guinean bird is represented in Australia by a population confined to a small area in the Cape York Region near the Claudie R (12/143). The species is sedentary and lives, usually in pairs, in the foliage of rainforest vegetation from the tree-tops to within 1 m of the ground.[844] In New Guinea it is most often in secondary vegetation.[160] The Field Atlas breeding report is of adults feeding dependent young in June 1981. The Green-backed Honeyeater feeds almost exclusively on insects gleaned from the outer foliage or caught while the bird flutters and hovers round the leaves.[844,1146] It has been seen visiting eucalypt blossom but whether for nectar, insects or both is unknown.[1354]

Within its restricted range, the Green-backed Honeyeater is common but there are no estimates of abundance.

Field Atlas Records: 29

1° blocks–with records: 1 (1%)
–with breeding records: 1 (100%)

Atlas No.: 604

Max. Rep. Rate: 37%
In block(s): 12/143

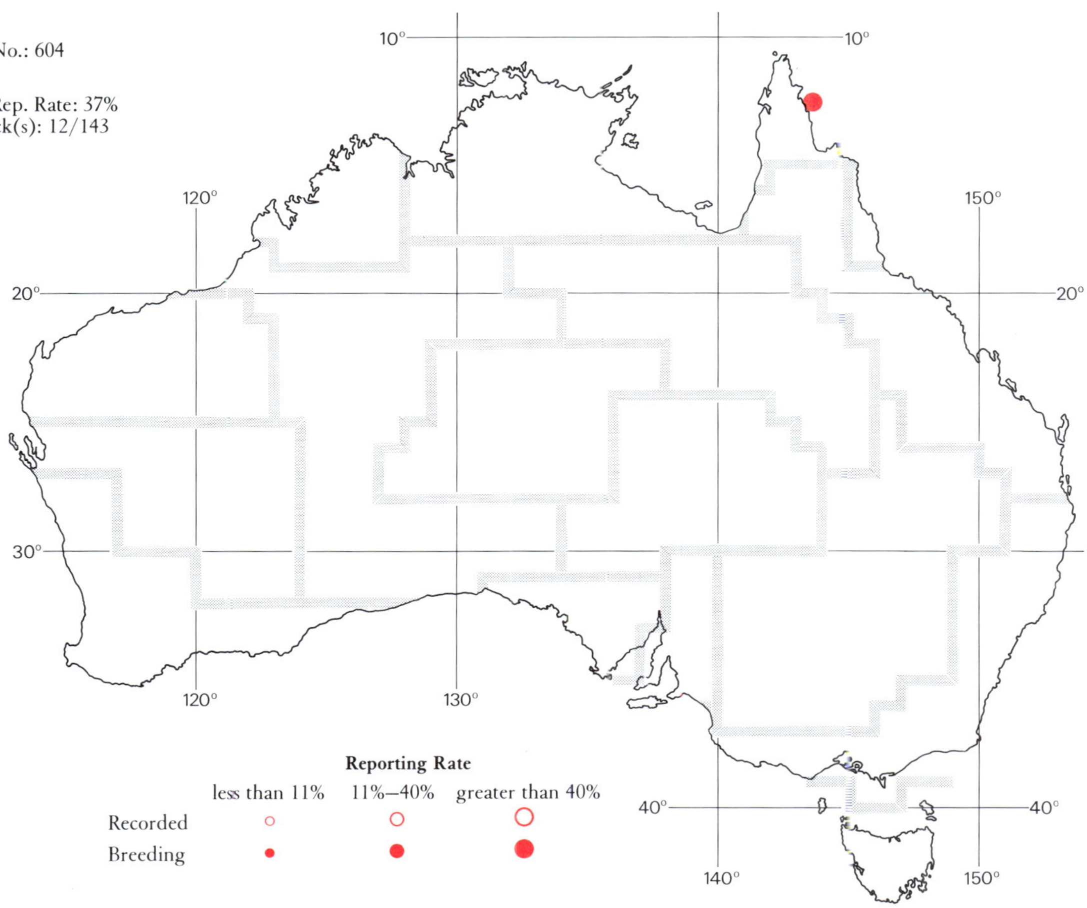

Brown Honeyeater

Lichmera indistincta

The Brown Honeyeater ranges from the Lesser Sunda Is, southern New Guinea and adjacent islands to Australia. The subspecies *indistincta* is the only one in Australia.[1239] On the eastern coast the species has been recorded as far south as 36/150.[1086,1454a] It is absent from those islands in Torres Strait occupied by the Dusky Honeyeater.[503]

Throughout its range the Brown Honeyeater is partly nomadic, gathering where shrubs are flowering.[987] In some places a population remains all year, augmented seasonally by immigrants. At Eyre Bird Observatory (32/126) the Brown Honeyeater is most abundant in spring and autumn; some birds are present in winter but few or none Dec.-Feb.[464,481]

The range of the Brown Honeyeater incorporates many different climates and types of vegetation from mangrove, forest, woodland and heath to suburban parks and gardens.[1124,1638] The bird feeds on nectar and insects. At Murphy's Ck (27/152) it is said to take insects throughout the year, catching them from a high perch.[1099] At Newcastle (32/151) it appeared to feed entirely on insects July-Aug. when mangroves were not in flower.[745] On the other hand, in a study near Pingelly (32/117), 99% of feeding time in Aug.-Sept. was spent on flowers.[753]

At Eyre Bird Observatory in mallee, density was 0.26–0.78 birds/ha.[464]

Field Atlas Records: 8563

1° blocks-with records: 440 (54%)
-with breeding records: 79 (18%)

Atlas No.: 597

Max. Rep. Rate: 70%
In block(s): 24/153,26/153

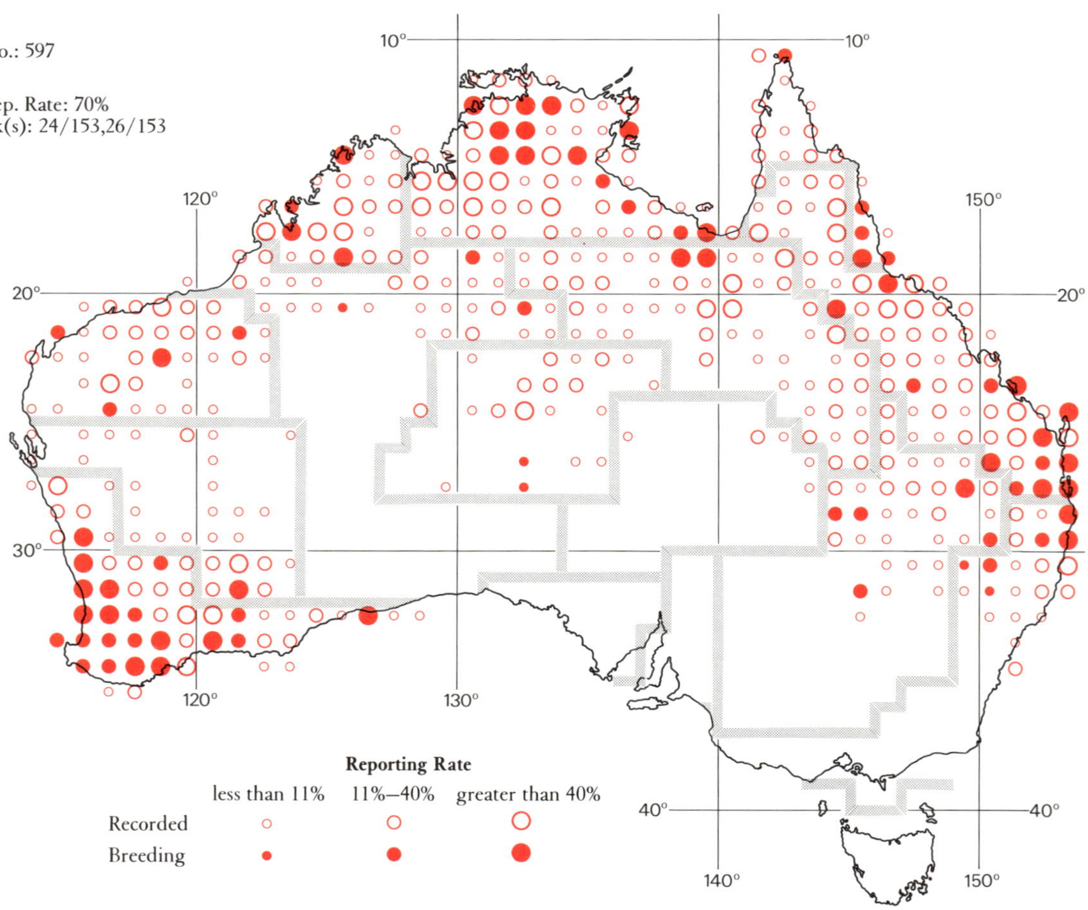

White-streaked Honeyeater

Trichodere cockerelli

The endemic White-streaked Honeyeater has previously only once been recorded outside the Cape York Region when specimens were collected near Shipton's Flat (15/145) in 1948.[1378] This isolated population has been overlooked but the species was recorded throughout the area during the Field Atlas.

The White-streaked Honeyeater disappeared within a month from a locality in the Cape York Region where it had been plentiful,[1145] which suggests at least local nomadic movement. It lives in heath, eucalypt woodland with a dense understorey, mangroves and rainforest along rivers and occasionally away from them. At Iron Ra. (12/143) the bird was common in eucalypt woodland with a heath understorey but not recorded in adjacent eucalypt-paperbark woodland.[164] Usually the bird is singly or in pairs. It is attracted to flowering plants such as paperbarks, grevilleas and banksias to feed on insects and nectar.[164,1145]

Near Iron Ra. singing birds were spaced 70–100 m apart and seemed to be strongly territorial, responding vigorously to tape-recorded calls.[164]

Field Atlas Records: 71

1° blocks–with records: 7 (1%)

–with breeding records:

Atlas No.: 627

Max. Rep. Rate: 37%
In block(s): 12/143

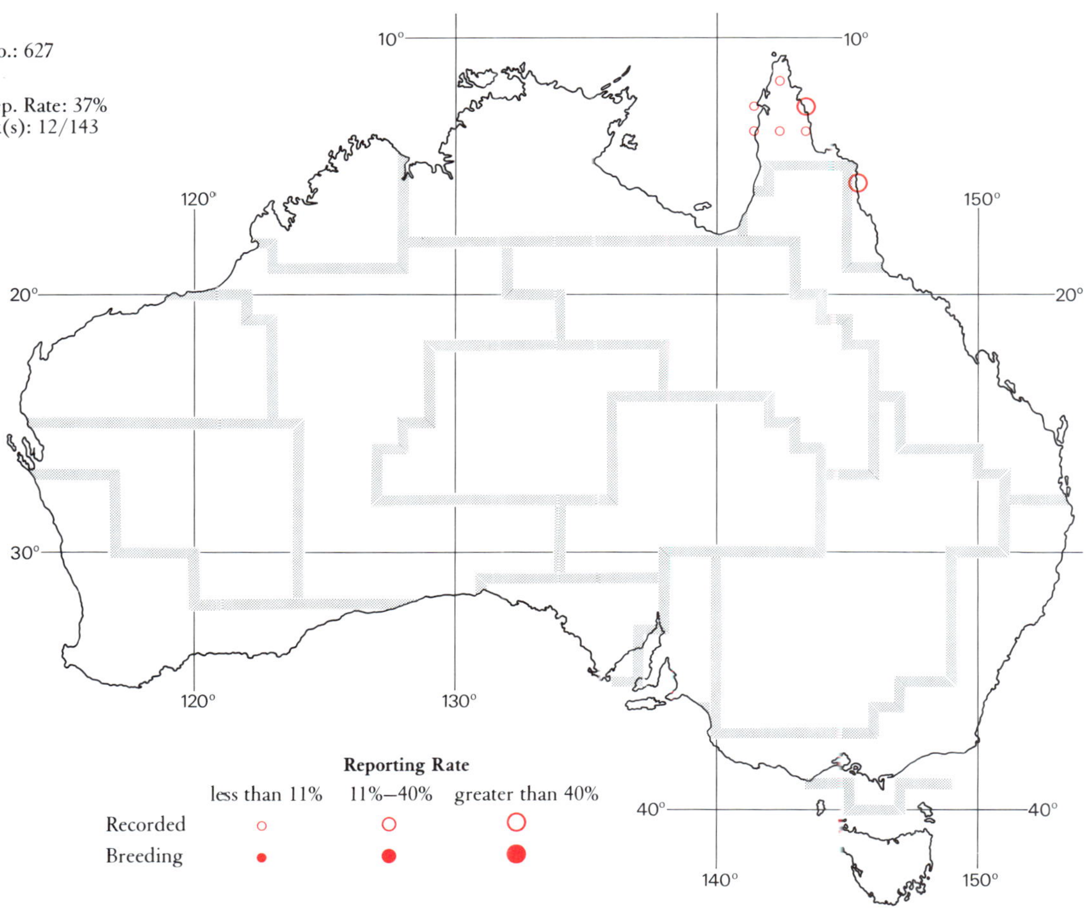

Painted Honeyeater

Grantiella picta

The endemic Painted Honeyeater is migratory, breeding in southern Australia, mostly south of 26°S, and moving north in winter. Reporting Rates south of 26°S were 1.3% in summer compared with 0.1% in winter. Observations confirm the north-south movement.[987] In the southern Murray-Darling Region it is present Dec.-Apr.;[833] near Canberra (35/149), where small numbers occasionally breed along the Murrumbidgee R, it is present Nov.-Feb.[653]

The Painted Honeyeater lives in forest and woodland usually in pairs or groups of up to six but with as many as 30 recorded once.[748] It feeds on mistletoe berries, sometimes exclusively, but will also take nectar from mistletoe flowers.[359,653,748] Some Field Atlas sightings were of it apparently taking nectar or insects in flowering shrubs but most also associated it with heavy infestations of mistletoe. The flowering and fruiting of mistletoe is said to govern the species' pattern of movement.[987] The Painted is unusual among honeyeaters in its preference for berries.

Twelve pairs of Painted Honeyeaters bred in an area of about 800 ha near Bendigo (36/144) (0.03 birds/ha).[515]

Field Atlas Records: 308

1° blocks–with records: 81 (10%)
–with breeding records: 20 (25%)

Atlas No.: 598

Max. Rep. Rate: 16%
In block(s): 27/149

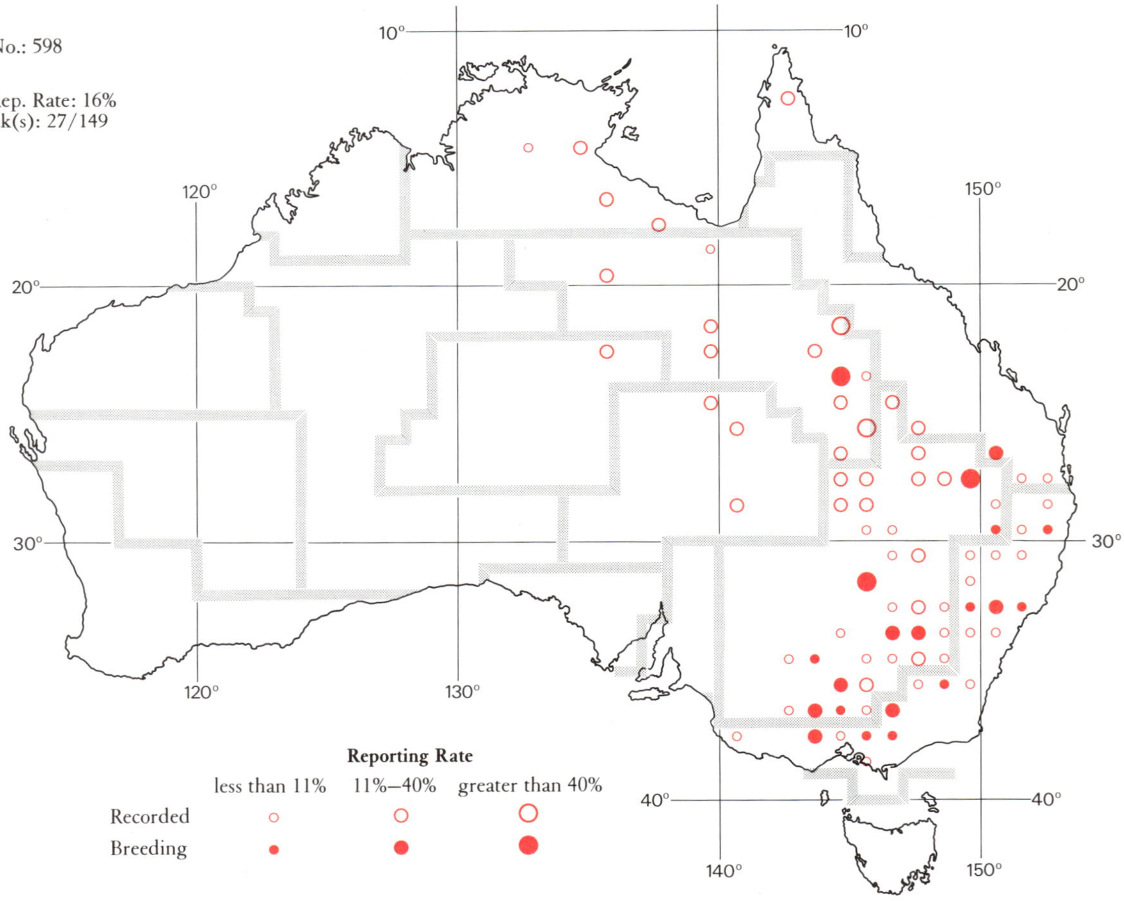

Crescent Honeyeater

Phylidonyris pyrrhoptera

Isolated populations of the endemic Crescent Honeyeater correspond with the five subspecies recognized. They are: *pyrrhoptera* in the South-East Region; *indistincta* in the Mt Lofty Ra. (34/138); *halmaturina* on Kangaroo I. (35/137); *rex* on the Bass Strait islands; and *inornata* in Tas.[1422] A population in the Grampian Mts (37/142) appears also to be isolated.

Movements have been reported in the South-East Region in autumn and winter.[987] An altitudinal component is evident near Canberra (35/149), where numbers decrease in the ranges in winter when the bird visits the suburbs of the city.[653] In the Snowy Mts (36/148) Reporting Rates above 500 m dropped from 26% in summer to 0% in winter. There was a corresponding increase below 500 m from 30% in summer to 45% in winter.

Two sightings out of the normal range in summer indicate that individuals can wander far. They were at Mystic Park (35/143) in Feb. and at Inbarendi (36/143) in Dec. Fluctuating numbers at one site in Hobart (42/147) showed at least some local movement.[1739,1748] The species is abundant in Hobart gardens in winter but moves to higher altitudes in the breeding season.[1314]

The Cresent Honeyeater lives in eucalypt forest and in heath near the coast. It takes nectar from many plants[1403] and sometimes honeydew.[1402] In Tas. it is an opportunistic feeder and will forage on bark and among foliage where it gleans insects.[1314,1748] There, along with other honeyeaters, flocks of the Crescent Honeyeater converge on the cider gum when manna is abundant in early autumn.[1259]

At Moruya (35/150), where the Crescent Honeyeater appeared one year out of six, in eucalypt woodland density was 0.5 birds/ha.[1197] At Boola Boola (38/146) density in eucalypt forest was 0.3–1.6 birds/ha.[1113] The Tas. Region appears to be the stronghold of the species, 47% of Field Atlas reports having come from there.

Field Atlas Records: 4609

1° blocks–with records: 64 (8%)
–with breeding records: 26 (41%)

Atlas No.: 630

Max. Rep. Rate: 68%
In block(s): 43/146

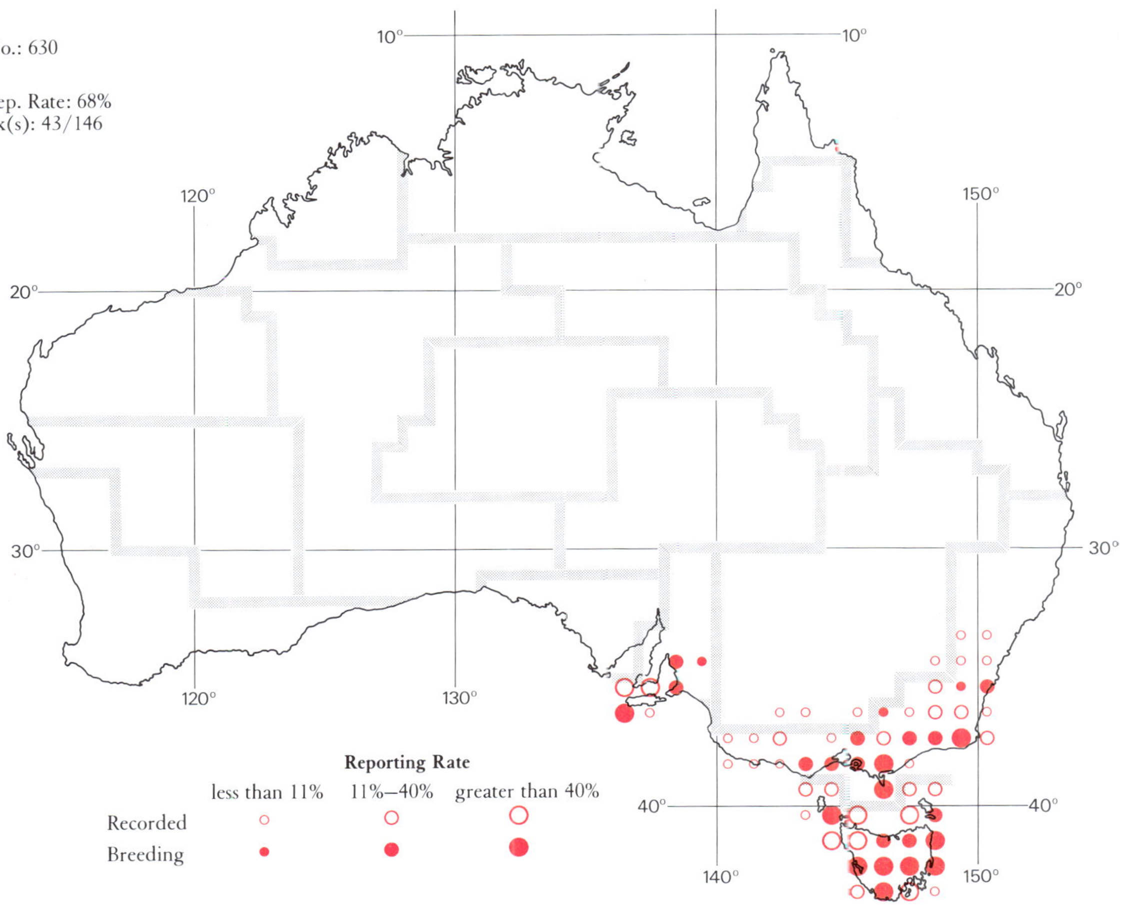

New Holland Honeyeater

Phylidonyris novaehollandiae

The five subspecies of the endemic New Holland Honeyeater correspond with geographically isolated populations: in eastern mainland Australia, *novaehollandiae*; Kangaroo I. (35/137), *campbelli*; Bass Strait islands, *caudata*; Tas., *canescens*; and the South-West, *longirostris*.[748,1422] Interbreeding with the White-cheeked Honeyeater has been recorded.[1124]

Movements, apparently coinciding with flowering of plants, have been observed in both east and west, though the species has been thought to be largely sedentary.[987] At Warrah (33/151) banding has shown that many New Holland Honeyeaters move through the area at times, most remaining only a few days.[1457] Near Manjimup (34/116) noticeable fluctuation in the numbers of this species, compared with other honeyeaters, is thought to reflect movement.[265] The longest movement recorded is that of a bird from Pondalowie Bay (35/136) to Clarendon (35/138), 161 km.[39]

The New Holland Honeyeater lives in heath, woodland and forest and, on the NSW coast, is most abundant where banksias grow.[1457] It is not as strictly coastal as the White-cheeked Honeyeater. North of Sydney (33/151) the bulk of the New Holland Honeyeater population lives in the ranges, separated from birds occupying a narrow coastal strip. The species sometimes nests in loose colonies. The New Holland Honeyeater feeds on nectar and insects. The proportion of feeding activity devoted to each differed in studies at Warrah (22% and 78% respectively)[1457] and in SA (51% and 49% respectively).[585] In summer, when nectar is scarce, the diet includes manna and honeydew. In Tas., along with other honeyeaters, flocks of the New Holland Honeyeater converge on the cider gum when manna is abundant in early autumn.[1259]

At Warrah, in heath, the highest nesting density recorded between 1967 and 1976 was 5.97 birds/ha in 1972.[1457] At Holsworthy (33/150) in heath density was 0.10–1.01 birds/ha.[139] At Moruya (35/150) in eucalypt woodland, density was 0.39 birds/ha.[1198] In WA, in a gully in eucalypt forest dominated by jarrah, six pairs nested close together at a density of 16 birds/ha.[1529] For three years the birds always nested near each other but ranged over a much larger area to feed.

Field Atlas Records: 12 032

1° blocks–with records: 139 (17%)
–with breeding records: 85 (61%)

Atlas No.: 631

Max. Rep. Rate: 78%
In block(s): 34/119

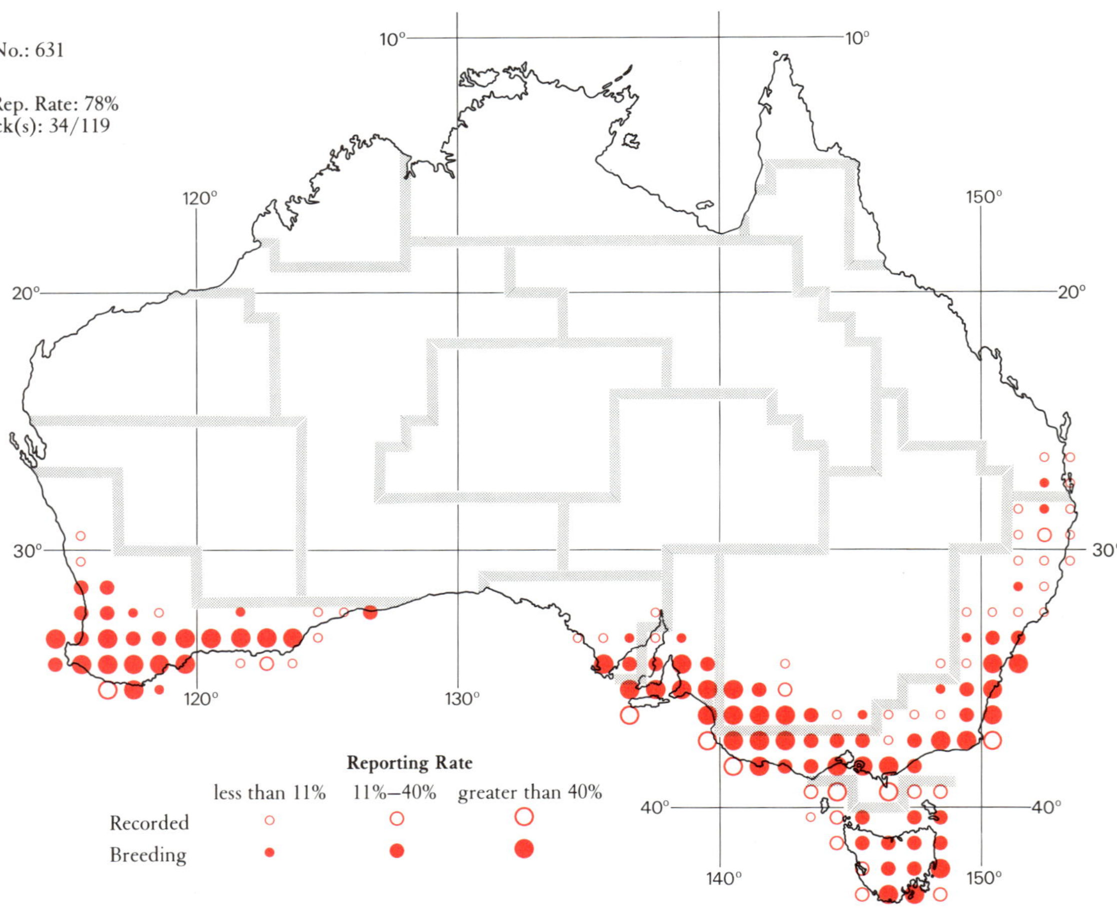

White-cheeked Honeyeater

Phylidonyris nigra

Two subspecies of the endemic White-cheeked Honeyeater are recognized: *nigra* in the east and *gouldi* in the west.[600,748] The isolated population in the Atherton Region is sometimes regarded as a third subspecies *herbertoni*.[1709] The distribution of *gouldi* is probably continuous westward along the coast from Eyre Bird Observatory (32/126). Interbreeding with the New Holland Honeyeater has been recorded.[1124]

The White-cheeked Honeyeater was thought to be sedentary in most places[987] but mist-netting at Warrah (33/151) has shown that, at times, many birds move through the area, most remaining only a few days;[1457] and at Holsworthy (33/150) the species was present only Nov.-June.[139] The longest movement recorded is that of an adult from Lennox Head (28/153) to near Kingscliffe (28/153), 53 km.[33]

The White-cheeked Honeyeater lives in eucalypt woodland and heath. The populations in the Atherton Region and Eungella (21/148) district are confined to higher altitudes. The subspecies *nigra* is otherwise more strictly coastal than its close relative, the New Holland Honeyeater. The White-cheeked Honeyeater nests colonially.[139] Many gather on flowering shrubs such as banksias and dryandras. Insects and nectar are the main foods, gathered mostly below 4.5 m;[600,753,1457] vegetable matter has also been found in guts.[748] At Warrah 22% of feeding time was expended on gathering nectar from many plants and the rest on catching insects.[1457]

At Pearl Beach (33/151) 27 pairs nested in heath at a density of 6.2 birds/ha.[1457] At Holsworthy in low heath 7-8 pairs nested colonially in an area of 9 ha (1.6-1.8 birds/ha).[139]

Field Atlas Records: 2012

1° blocks–with records: 79 (10%)

–with breeding records: 25 (32%)

Atlas No.: 632

Max. Rep. Rate: 70%

In block(s): 24/153

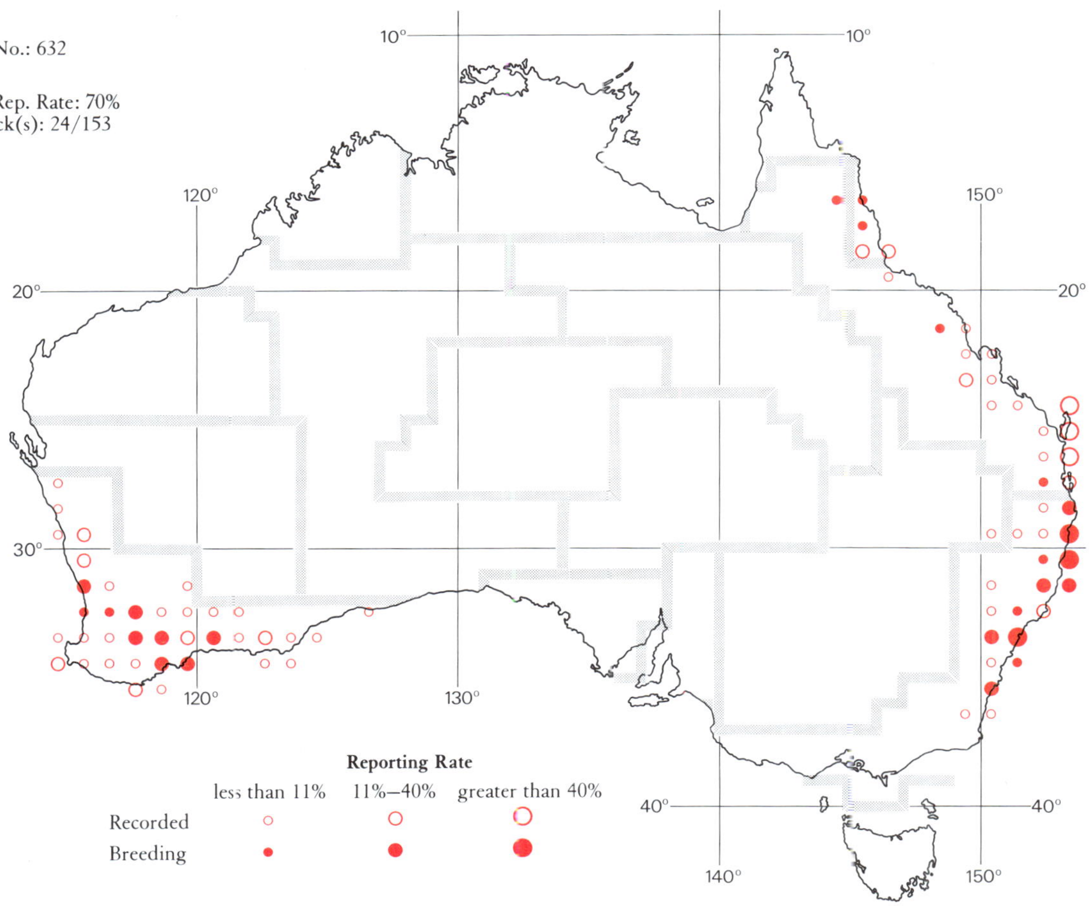

White-fronted Honeyeater

Phylidonyris albifrons

Movements of the endemic White-fronted Honeyeater are frequently reported, probably because it moves in large flocks.[987] It is said to visit the Kimberley Region in winter.[1713] In the southern Murray-Darling there was an eastward movement in Aug. with a return passage in autumn.[833] At Naretha (31/124) westward movements have been reported.[381] At Eyre Bird Observatory (32/126) 205 passed westward in 25 minutes in Apr. 1981.[392] At Mukinbudin (30/118) southward movements have been observed in Oct. and Jan.[335] Movements are erratic in timing and direction, yet consistent in any one region, so that the distances moved may be fairly small. The Field Atlas does not show any regular seasonal change in Reporting Rate.

The bird inhabits acacia scrub, eucalypt woodland and mallee.[748] It lives singly, in pairs or in flocks. Breeding sometimes takes a semi-colonial form, with numerous nests in a small area but communal breeding has not been reported. The White-fronted Honeyeater feeds on nectar, honeydew and insects. There are some reports of vegetable matter in its gut.[748,1049]

At Eyre Bird Observatory in mallee its density was 0.51–0.73 birds/ha.[464]

Field Atlas Records: 1681

1° blocks–with records: 266 (33%)

–with breeding records: 59 (22%)

Atlas No.: 594

Max. Rep. Rate: 35%
In block(s): 35/141

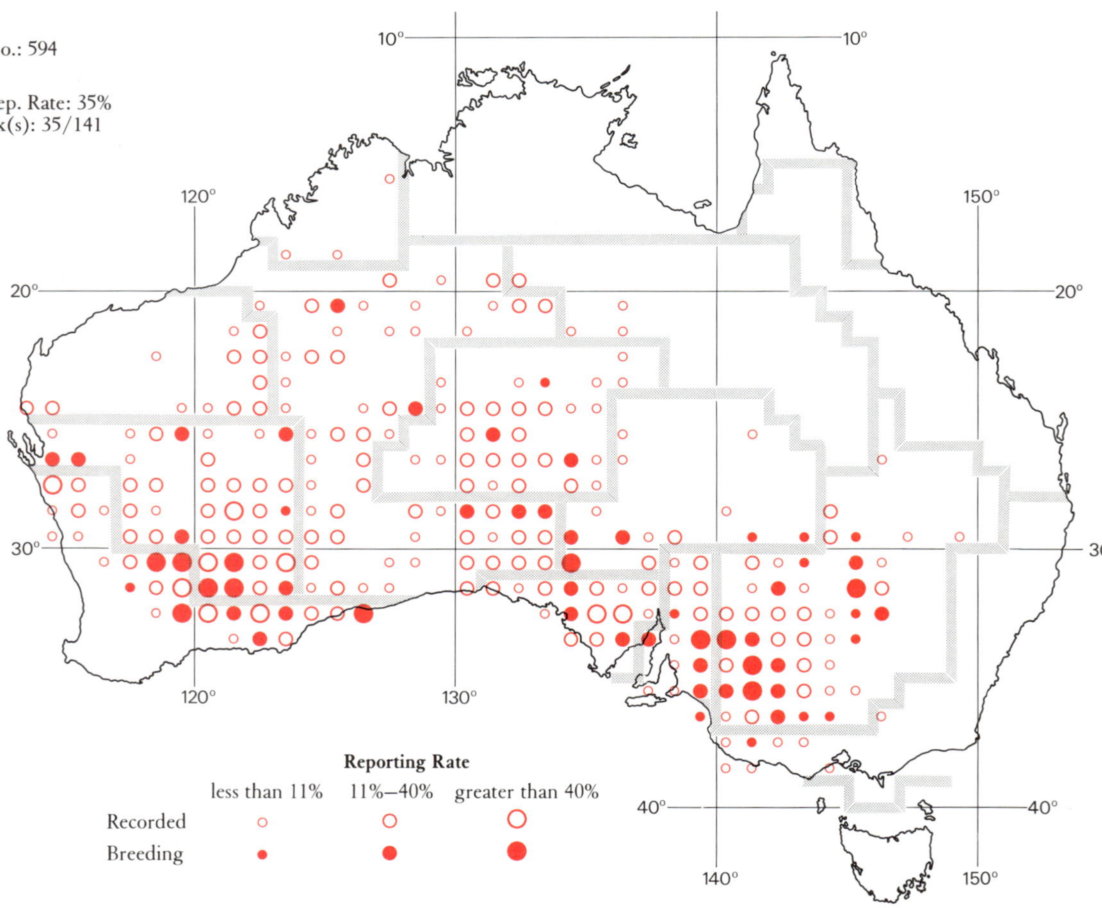

Tawny-crowned Honeyeater

Phylidonyris melanops

There is a gap in records of the endemic Tawny-crowned Honeyeater at 127–133°E in the Atlases but no morphological differentiation of eastern and western populations has been suggested. Within the eastern population, Kangaroo I. (35/137) birds are doubtfully distinct. Those of the Tas. Region have been recognized as a separate subspecies *crassirostris*.[1422]

The eastern population is said to be sedentary.[987] In the South-West the Tawny-crowned Honeyeater is sedentary in some heath but visits and breeds in others only seasonally.[1915] An irruption into forest has been documented: in Mar. 1945 large numbers arrived to feed on marri flowers near Perth (31/115) and at Smith Brook (34/116).[1829] At Smith Brook, which is 40 km from the nearest heath, they were abundant Mar.-May and had left by July. A study of birds in the same area in 1973–82, with mist-netting, recorded no Tawny-crowned Honeyeaters though in some years the marri flowered well.[262,265]

Distribution of the Tawny-crowned Honeyeater is essentially coastal. It inhabits heath and, in Vic. and SA, also lives in some areas of mallee. In Tas. it is restricted to heath within a few kilometres of the sea.[1314]

The Tawny-crowned Honeyeater feeds mainly on nectar. North of Sydney (33/151), it foraged on the ground to take nectar from flowers growing within the heath rather than near the tops of the shrubs and rarely hawked for insects.[1455] At Golton Vale (36/142) it also fed mainly on nectar, a little on manna and rarely hawked for insects.[1402] On heath in the WA wheatbelt the species behaved differently, spending less than 5% of its time feeding on the ground and over 50% hawking for insects above the vegetation.[753]

Density at Holsworthy (33/150) in low heath was 0.28–0.63 birds/ha;[139] and at Pearl Beach (33/151) in 1968 the density of nesting birds was 1.15 birds/ha and, in 1969, 1.38 birds/ha.[1455]

Field Atlas Records: 1973

1° blocks–with records: 115 (14%)

–with breeding records: 51 (44%)

Atlas No.: 593

Max. Rep. Rate: 62%
In block(s): 37/150

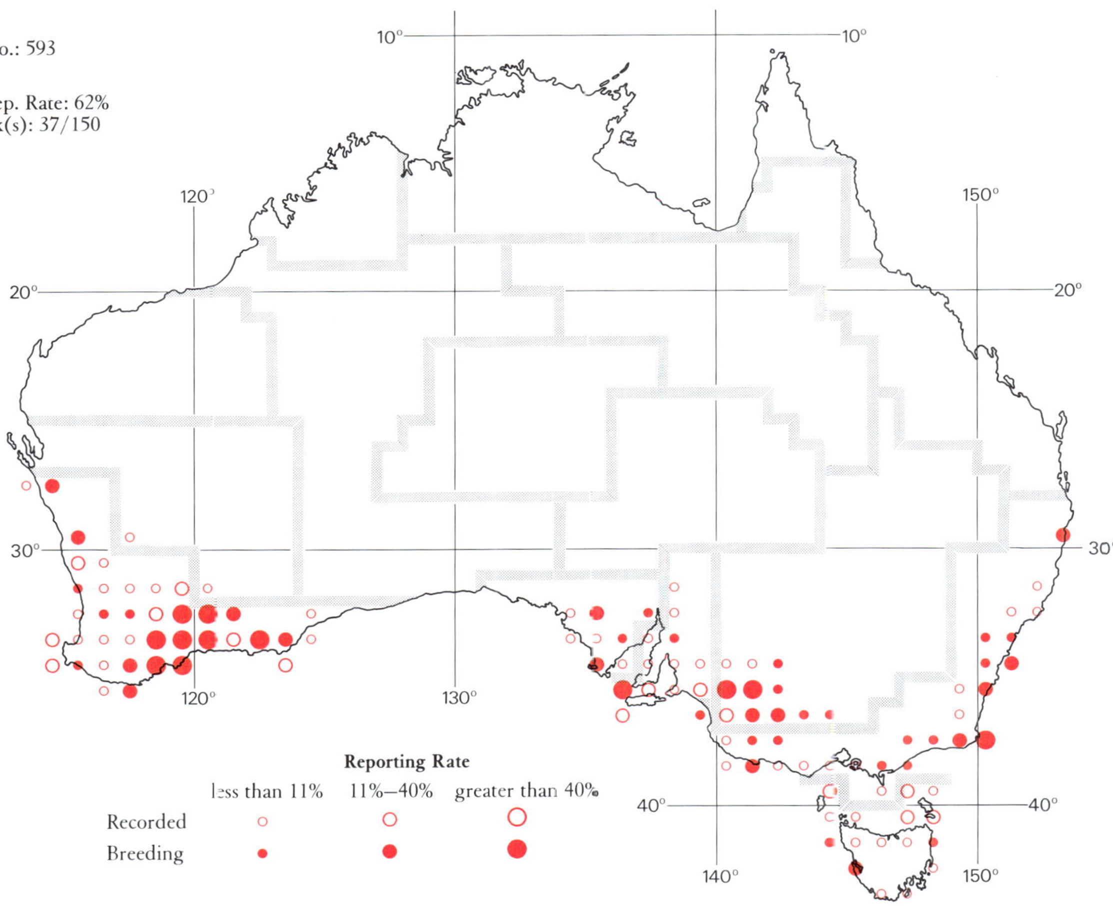

Brown-backed Honeyeater

Ramsayornis modestus

The Brown-backed Honeyeater lives in New Guinea and adjacent islands and in Australia. It has been regarded as conspecific with the Bar-breasted Honeyeater to which it is closely related.[1709]

It has been suggested that some Australian birds winter in New Guinea[1709] though the species is regarded there as an abundant breeding resident.[129] No movement through Torres Strait is reported and the bird's occurrence on the islands is related to temporary swamps in the wet summer season.[503] In the Cape York Region the bird is also reported to leave as swamps dry out in autumn[1354] and at Innisfail (17/146) it is regarded as a definite migrant arriving in Aug. and leaving in May.[698] Reporting Rates were lower in winter in both the Atherton (summer 21%, winter 8%) and Cape York (summer 18%, winter 7%) Regions, which gives some support to the suggestion of migration out of Australia. The Reporting Rate in the Cape York Region was highest in autumn (22%).

The Brown-backed Honeyeater inhabits mangroves and woodland dominated by paperbark and adjacent eucalypt woodland. It is also reported from the built-up area of Cairns (16/145). The species lives singly, in pairs or in flocks and nests are sometimes grouped in colonies. The Brown-backed Honeyeater feeds in flowering paperbarks and other trees and shrubs and also takes insects, sometimes on the wing.[1265]

There are no measures of abundance.

Field Atlas Records: 352

1° blocks–with records: 24 (3%)
–with breeding records: 7 (29%)

Atlas No.: 595

Max. Rep. Rate: 27%
In block(s): 12/143

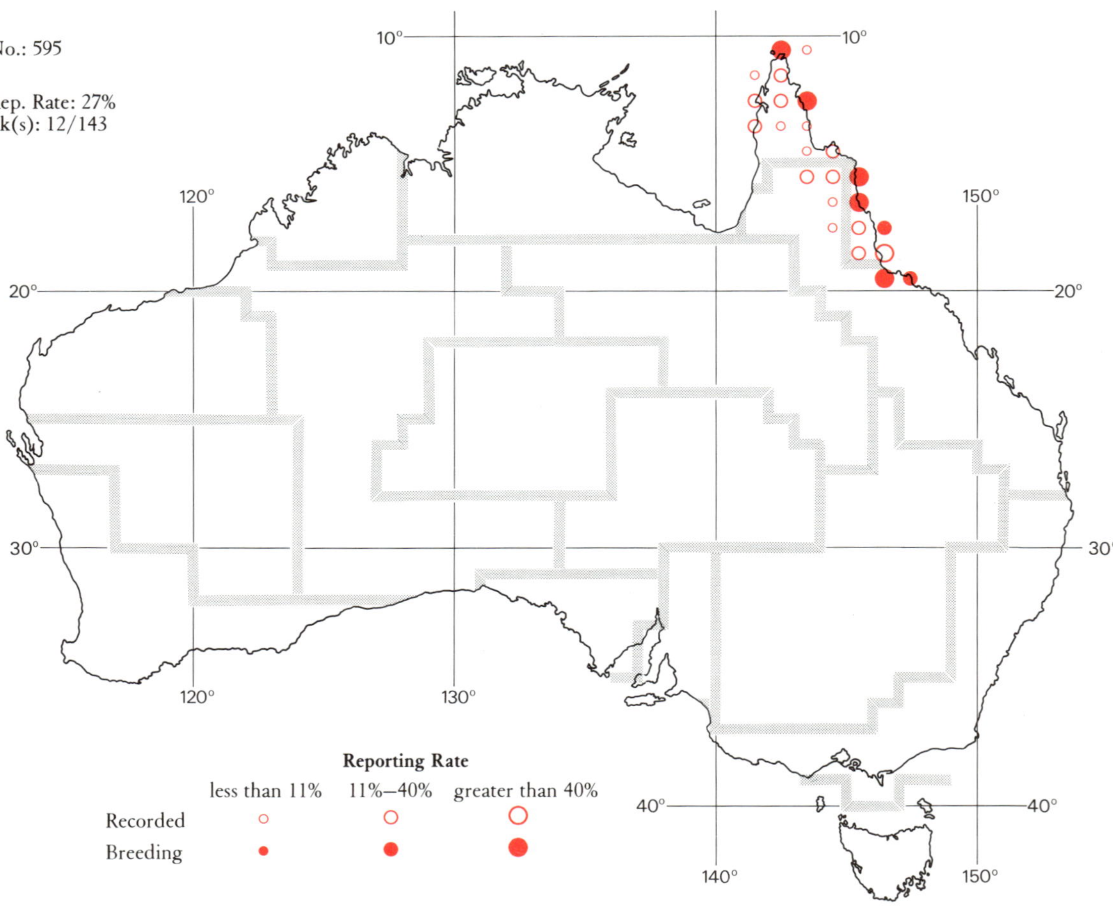

Bar-breasted Honeyeater

Ramsayornis fasciatus

Three subspecies of the endemic Bar-breasted Honeyeater have been recognized but views about them differ.[748,1422] The species has been regarded as conspecific with the Brown-backed to which it is closely related.[1709] Historical records from the Atherton Region fill the gap in the Field Atlas distribution and breeding has been reported as far south as Duaringa (23/149). There are historical reports in the Atlases from the northern Cape York Region.

The number of Bar-breasted Honeyeaters in a locality fluctuates as the birds gather where stands of paperbarks are flowering.[423,787] No seasonal pattern of movement is suggested by the Field Atlas.

The species lives singly, in pairs or in flocks in eucalypt forest and woodland close to water, as well as in paperbark swamps and rainforest along streams.

Nectar and insects have been recorded in the gut of the Bar-breasted Honeyeater.[748] The nectar is often from paperbarks. The bird has also been recorded feeding in grevilleas[1602] and hawking for insects over water.[748]

During the Field Atlas the Bar-breasted Honeyeater was reported to be abundant in the Kimberley Region, where in 1910 an observer found 19 nests in a day along one small creek.[787] There are no other measures of abundance.

Field Atlas Records: 555

1° blocks–with records: 76 (9%)
–with breeding records: 15 (20%)

Atlas No.: 596

Max. Rep. Rate: 48%
In block(s): 16/125

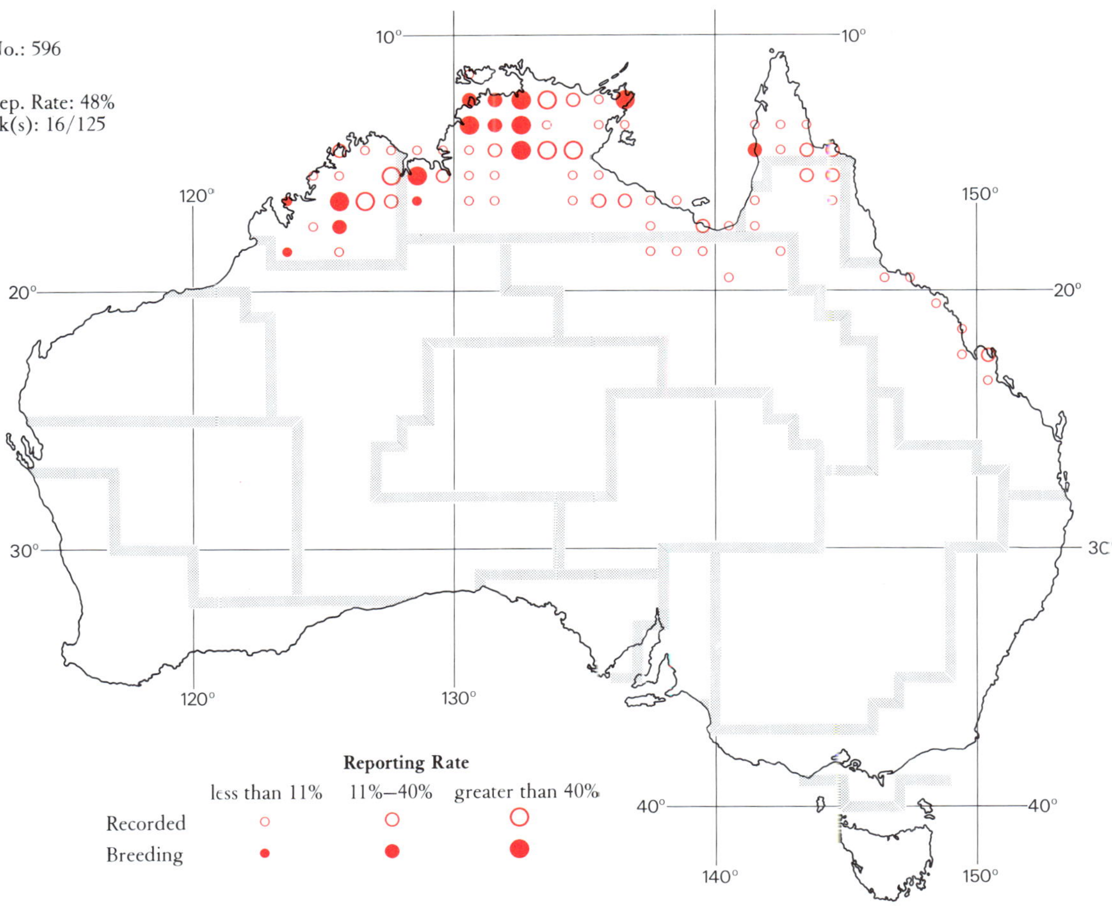

Rufous-banded Honeyeater

Conopophila albogularis

The Rufous-banded Honeyeater ranges through New Guinea and adjacent islands to Australia. In the Cape York and Top End Regions populations are apparently not morphologically distinct[1422] but the gap south of the Gulf of Carpentaria is consistent in the Atlases. The Field Atlas confirms that the species is present on the eastern side of the Cape York Region as well as the western.

The Rufous-banded Honeyeater is present all year on islands in north-western Torres Strait[503] but round Darwin (12/130) numbers fluctuate markedly.[423] Over much of its range the species appears to be nomadic in that it cannot reliably be found in the same spot each year.[987] The Rufous-banded Honeyeater lives singly, in pairs or in flocks in mangroves and paperbark woodland growing in swamps, as well as eucalypt woodland and forest.[423,748,1124] Flocks of 10–30 birds forage in and hawk above the tree-tops. Mixed flocks are formed with other species, specially the Rufous-throated Honeyeater. The Rufous-banded Honeyeater catches insects on the wing and also takes nectar.[723,748]

In New Guinea, though abundant in Port Moresby (9S/147E), it is much less common in adjacent eucalypt woodland[158] but there are no measures of abundance.

Field Atlas Records: 354

1° blocks–with records: 26 (3%)
–with breeding records: 3 (12%)

Atlas No.: 600

Max. Rep. Rate: 48%
In block(s): 12/136

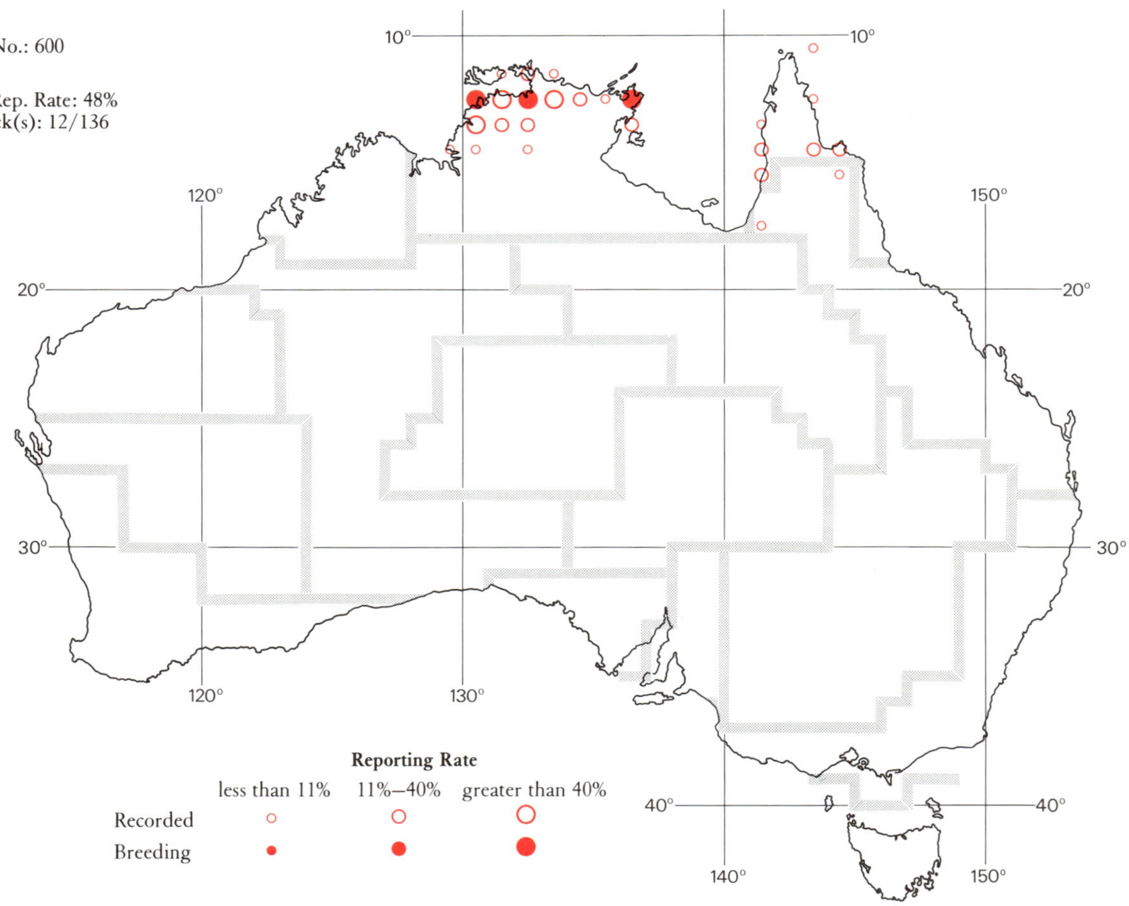

Rufous-throated Honeyeater

Conopophila rufogularis

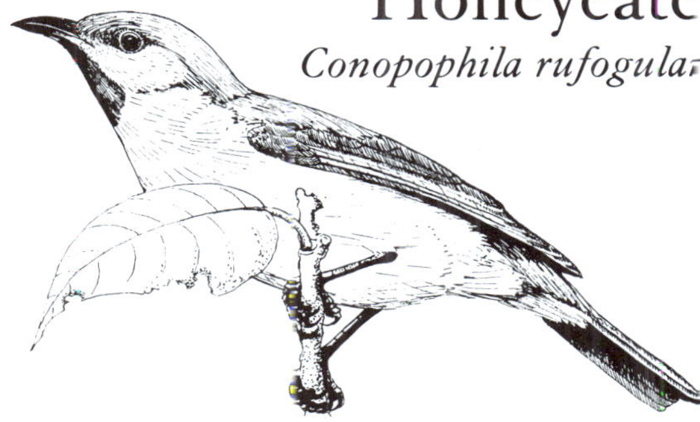

The Qld population of the endemic Rufous-throated Honeyeater has been doubtfully distinguished as a separate subspecies from birds in other parts of Australia.[748] The species is continuously distributed across northern Australia except for the gap on the coast of the Eastern Qld Region south of which records are sporadic and breeding is not recorded.

The Rufous-throated Honeyeater has been considered nomadic, timing of flowering imparting a seasonal bias to its movements.[987] In Qld it is thought to migrate south in Aug. and north in Apr.-May.[1709] It is present in Mt Isa (20/139) Aug.-Mar.[1068] Reporting Rates in Mt Isa and surrounding 1° blocks were 53.6% in summer and 20.3% in winter. At Kununurra (15/128) the species is considered sedentary.[898]

The Rufous-throated Honeyeater inhabits eucalypt forest and woodland, specially along rivers and creeks.[608,1124] It lives in groups and at times joins large mixed flocks of honeyeaters round flowering shrubs,[1507] specially bauhinias.[1724] Communal breeding has occasionally been recorded.[501]

The Rufous-throated Honeyeater feeds on nectar as well as insects, specializing in such ground-dwelling ones as grasshoppers and crickets.[898] It will also hawk for insects over water.

At Mt Isa 12–35 birds were observed per kilometre of river.[1068] At Kununurra breeding territories were about 50 m in diameter.[898]

Field Atlas Records: 996

1° blocks–with records: 131 (16%)
–with breeding records: 16 (12%)

Atlas No.: 601

Max. Rep. Rate: 52%
In block(s): 21/144

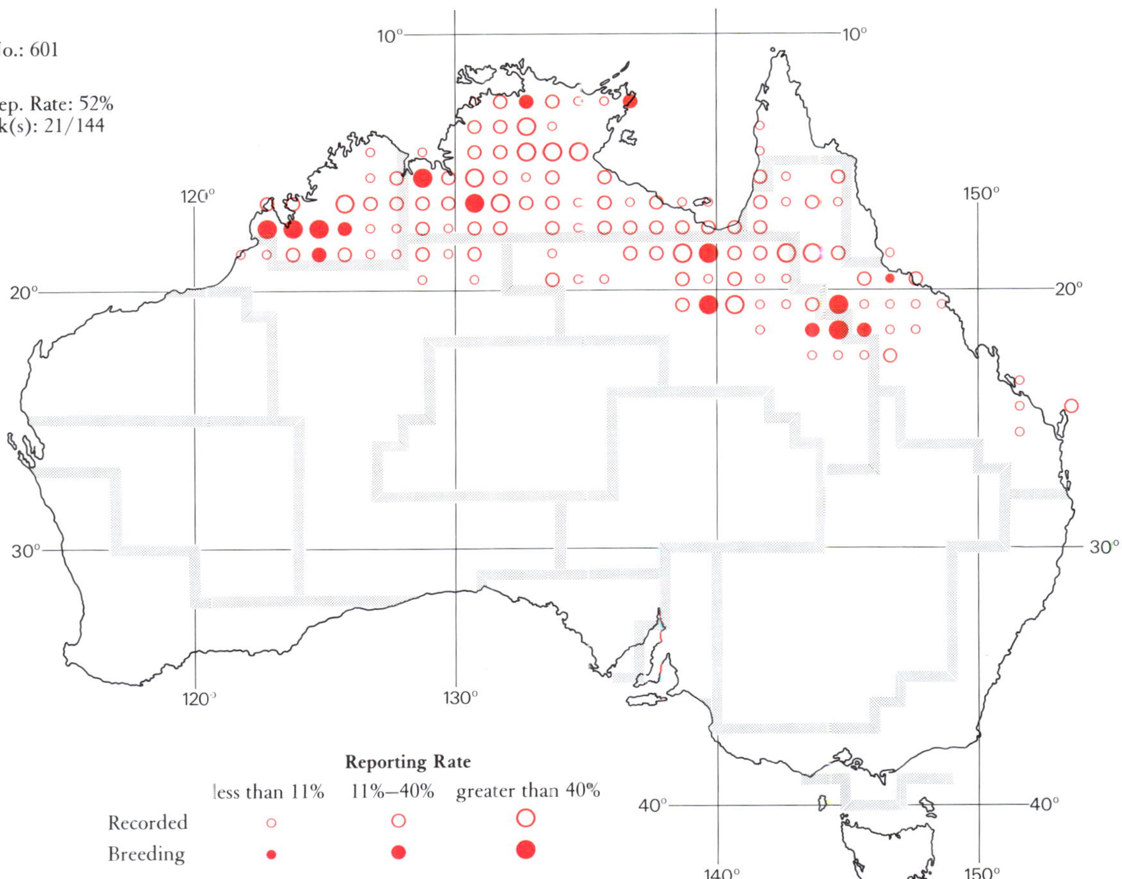

Grey Honeyeater

Conopophila whitei

Until the Field Atlas began only 16 records of the endemic Grey Honeyeater had been published.[1405] Descriptions suggest geographical variation in plumage but Atlas records, though sparse, do not suggest any break in distribution.

The extent of movements is unknown. One author considers that the species is nomadic[602] and another that it is sedentary.[987]

The Grey Honeyeater lives in acacia scrub dominated by dense mulga or in spinifex either with thickets or scattered trees and bushes.[602] The bird, in pairs or parties of up to about eight and sometimes in company with thornbills, calls frequently as it feeds.[443,1282,1872,1880]

The Grey Honeyeater gleans insects from leaves and flowers,[1282,1872] feeds on mistletoe berries[748,1831] and takes nectar.[1493]

It is inconspicuous and inhabits inaccessible places so that the rarity of reports may not reflect its true abundance. At Yalgoo (28/116) there were estimated to be five pairs within a radius of 1 km (0.3 birds/ha).[1831]

Field Atlas Records: 35

1° blocks-with records: 18 (2%)
-with breeding records: 2 (11%)

Atlas No.: 599

Max. Rep. Rate: 16%
In block(s): 23/116

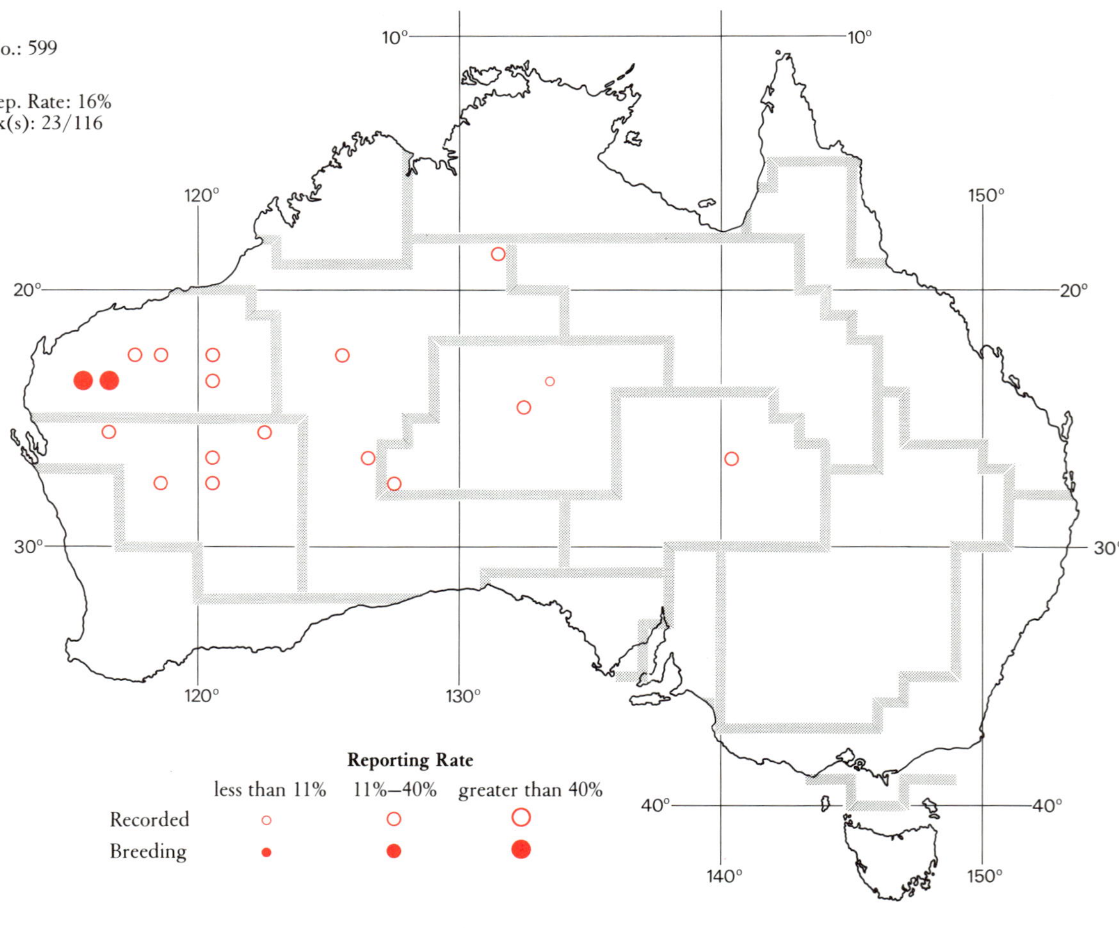

There is doubt about the validity of some of the seven subspecies of the endemic Eastern Spinebill that have been described.[748,1124,1422] The populations in the Atherton Region, and probably the Eungella (21/148) district, are isolated, as is another small group of birds in the Mt Remarkable (32/138) area of the SA Ra Region. The species was believed to be extinct on King I. (39/144)[734] but one was reported from Sea Elephant R (39/144) during the Field Atlas.

In the Atherton Region, where the species is confined to high altitudes (Reporting Rates: above 1000 m 23%, 500–1000 m 6%, below 500 m 0%), birds are present all year though numbers fluctuate.[231] There are no breeding records in the Atlases between the Atherton Region and 26°S. Birds in the intervening area, south of Eungella, may be winter visitors only; Reporting Rates in Eastern Qld were greater in winter (4.6%) than in summer (2.2%). A winter movement into the Hastings Pt (28/153) area of NSW has been recorded and in the New England National Park (30/152) at 1400 m the species was rare or absent in summer and most abundant in autumn and winter.[1130] Banding near Richmond (33/150) revealed that the population was constantly changing though no seasonal pattern was evident.[1026] Near Canberra (35/149) altitudinal movement has been suggested.[987] Reporting Rates in the Snowy Mts (36/148) above 500 m were higher in spring-summer (26.3%) than in autumn-winter (10%). The corresponding Reporting Rates below 500 m (81.5% and 85.0%) were much higher and showed no seasonal change. On Mt Wellington (42/147) in Tas. there is no evidence of such an alititudinal movement.[1452] The longest movement of a banded bird is 235 km.[85]

Eastern Spinebill

Acanthorhynchus tenuirostris

The Eastern Spinebill inhabits rainforest, eucalypt forest and woodland, heath and suburban gardens.[1183] It lives singly or in small groups and sometimes joins mixed species feeding flocks. It feeds on nectar and also takes insects, specially when feeding young.[369,1049,1403,1497]

Densities in eucalypt woodland were: Armidale (30/151) 0.08–0.11 birds/ha,[582] Wollomombi (30/152) 0.04–0.22 birds/ha[129] and Moruya (35/150) 0.46–0.55 birds/ha.[1197] In eucalypt forest they were: Bathurst (33/149) 1.25 birds/ha[489] and Boola Boola (38/146) 0.1–1.7 birds/ha (greatest in mature forest).[1113]

Field Atlas Records: 11 414

1° blocks–with records: 123 (15%)
–with breeding records: 55 (45%)

Atlas No.: 591

Max. Rep. Rate: 62%
In block(s): 34/150

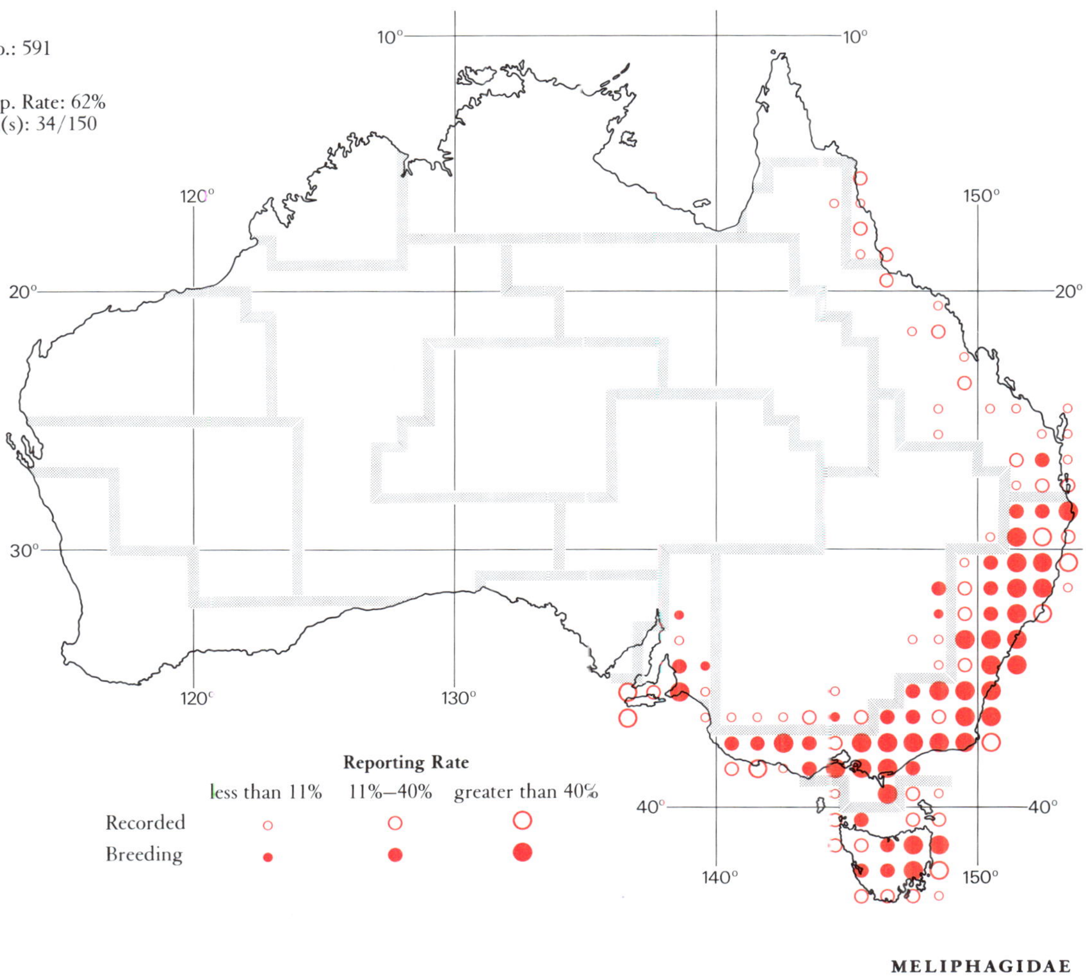

Western Spinebill

Acanthorhynchus superciliosus

The Western Spinebill is endemic to the South-West Region and recorded historically no farther north and east than in the Field Atlas.

The species has been described as 'swarming' into Broomehill (33/117) and Albany (35/117)[987] but this may simply be the gathering together of birds from the surrounding district. At Wembley (31/115) a few pairs appear Apr.-May each year. The Field Atlas does not suggest large-scale seasonal movement.

The Western Spinebill lives in eucalypt woodland and heath.[600,1124] It feeds on nectar from the kangaroo paw and other flowers,[753,1564] particularly dryandras and banksias. Insects are also taken,[753,897] mainly gleaned from bushes but also caught by hawking in the air.[1868] In taking nectar from the kangaroo paw, the Spinebill often picks up copious amounts of pollen on its back. At the same time the stigma picks up pollen from other flowers which the Spinebill has previously visited.[1243] Most of the heath in the wheatbelt has now been cleared for agriculture, which has reduced the range of the Western Spinebill.

There are no measures of abundance.

Field Atlas Records: 1070

1° blocks-with records: 31 (4%)
-with breeding records: 13 (42%)

Atlas No.: 592

Max. Rep. Rate: 36%
In block(s): 34/118

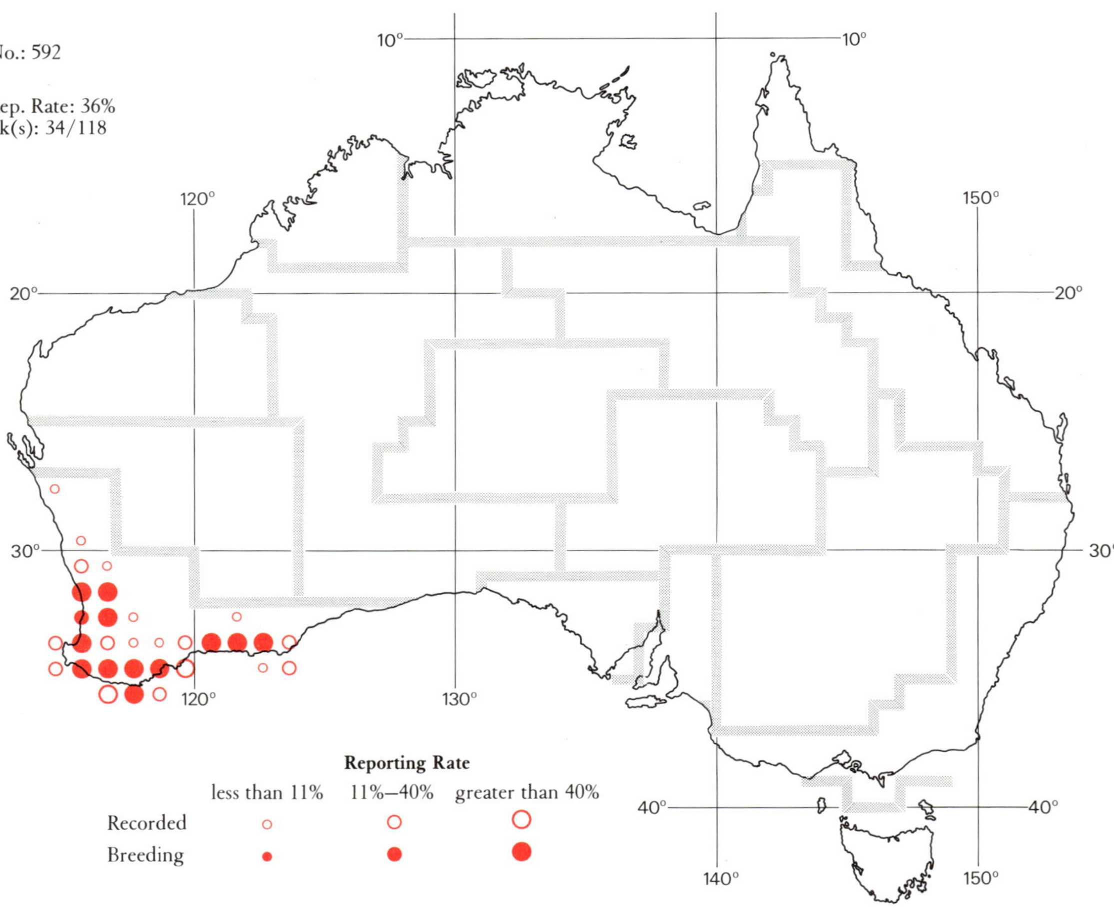

Banded Honeyeater

Certhionyx pectoralis

No subspecies of the endemic Banded Honeyeater are recognized though the Atlases suggest a discontinuity in its range at the eastern end of the Gulf of Carpentaria (140°E).

The species is nomadic and can move rapidly and in large numbers, e.g. in one locality on the Gregory R (18/138) there were none in May 1957, eight in May 1958 and an estimated 8000 per kilometre of river in Aug. 1958.[1068] Similar large invasions have been reported elsewhere usually associated with abundant flowering of paperbarks,[1068,1354,1881] eucalypts,[748,1145,1602] grevilleas[748,787] and other Proteaceae.[1703]

The Banded Honeyeater lives in eucalypt woodland where grass dominates the understorey and in paperbark woodland near water.[1369] Pairs appear to remain together throughout the year[231] even though the birds are usually in small parties. The Banded Honeyeater feeds on insects and nectar.[748]

There are no measures of abundance.

Field Atlas Records: 518

1° blocks–with records: 97 (12%)
–with breeding records: 2 (2%)

Atlas No.: 588

Max. Rep. Rate: 38%
In block(s): 14/134,16/130

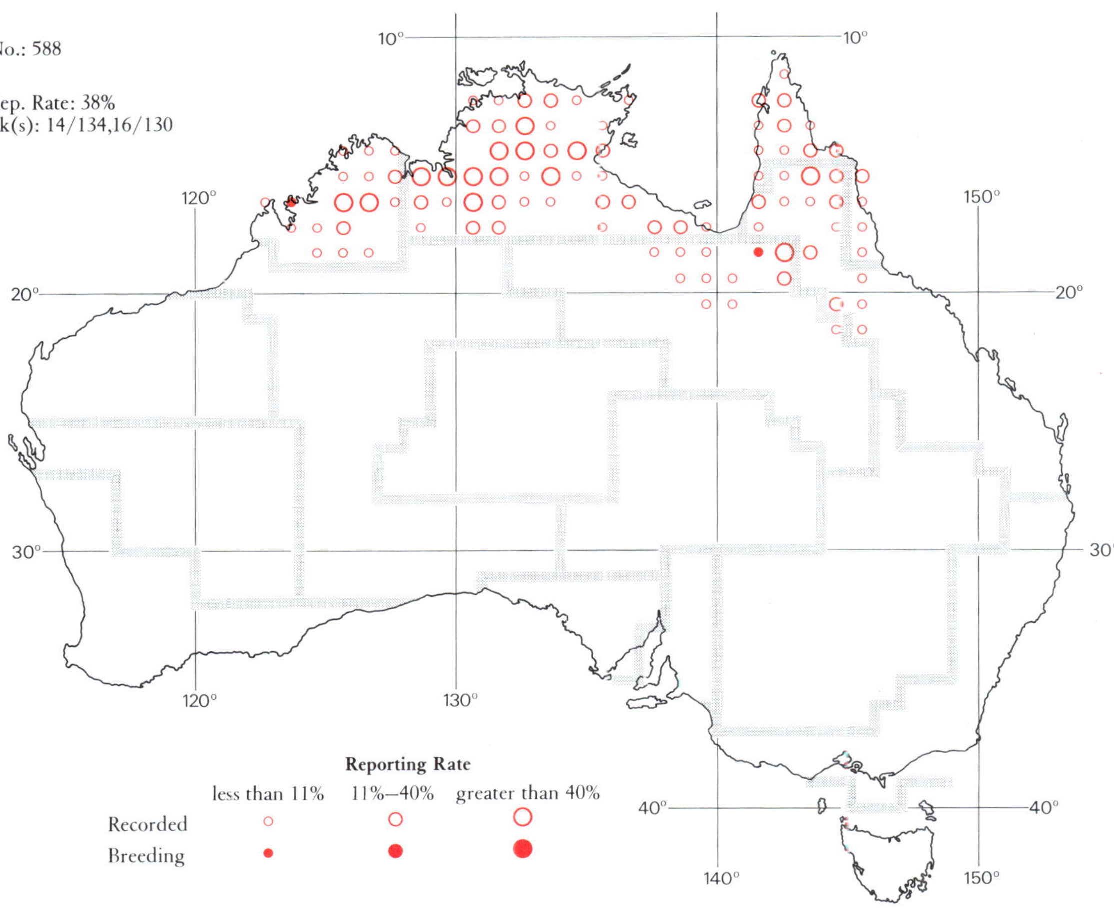

Black Honeyeater

Certhionyx niger

The endemic Black Honeyeater is sparsely but continuously distributed across the continent. The southern and eastern edges of its range fluctuate from year to year[580] with breeding reported as far south as Maryborough (37/143)[349] and east to Sydney (33/151).[816] There are no breeding reports north of 20°S.

There is some evidence of regular seasonal movement by the Black Honeyeater. An observer in the southern Rolling Downs Region at 24/143 concluded that the bird appeared in the area on passage, south in Aug.-Sept. and north in Apr.-May, rarely outside these months.[170] An analysis of records in SA between 1901 and 1975 showed that 75% were made in Sept.-Oct. and most of the remainder in Aug. and Nov.-Dec.[580] The Field Atlas also showed that the species was more often recorded in the south in spring and summer than in autumn and winter, e.g. in the Murray-Darling Region Reporting Rates were: spring 1.6%, summer 0.6%, autumn 0%, winter 0.2%. The winter records probably represent early arrivals. There were too few observers in central Australia in summer to determine whether birds leave northern parts of the range and move south. In the Great Victoria Desert the bird is nomadic and may appear wherever suitable conditions occur.[180]

The Black Honeyeater lives in pairs or flocks in acacia scrub, spinifex and mallee.[748] Field Atlas records in southern areas were of one or two birds but sometimes large numbers arrive and stay to breed.[1271] The species shows a preference for red-flowered emu-bushes[834] and flowering bloodwoods,[170] though it takes nectar from other species and feeds on insects.[580,748]

There are no measures of abundance.

Field Atlas Records: 319

1° blocks–with records: 157 (19%)
–with breeding records: 16 (10%)

Atlas No.: 589

Max. Rep. Rate: 18%
In block(s): 25/119

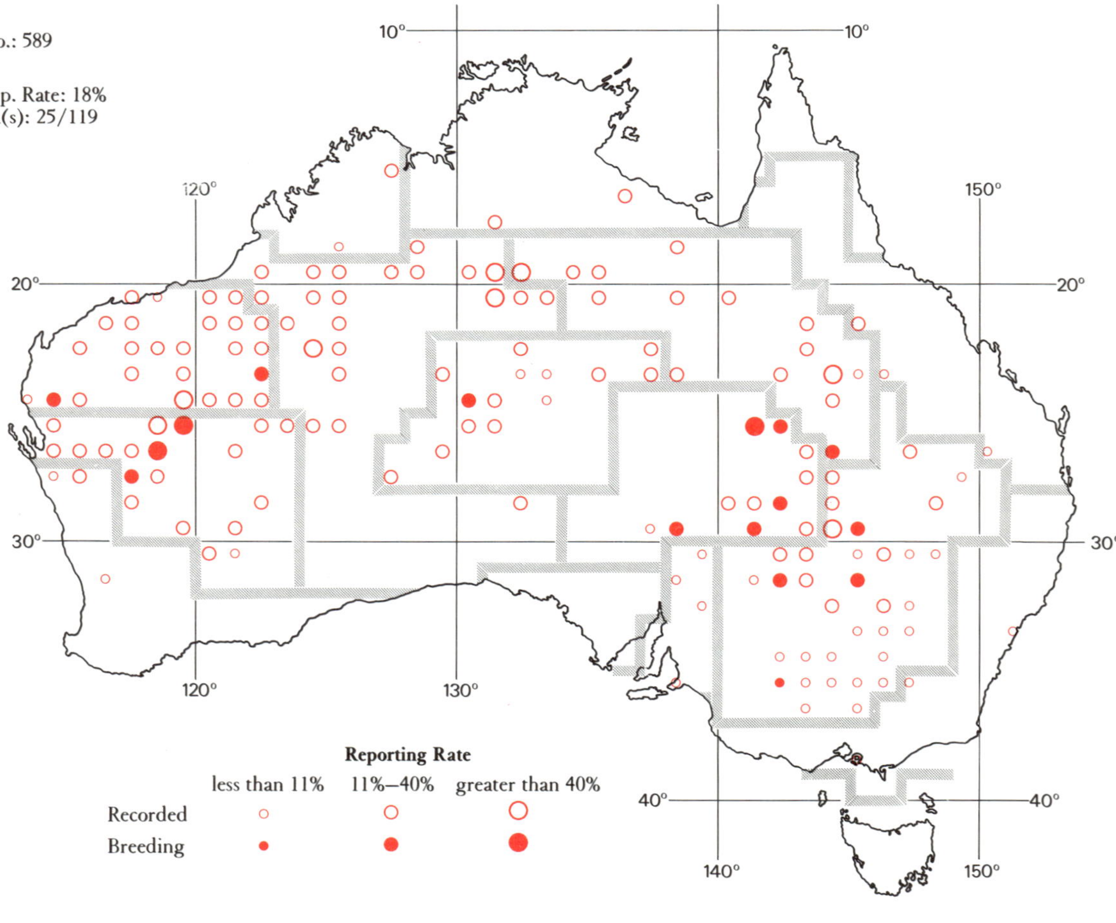

Pied Honeyeater

Certhionyx variegatus

The endemic Pied Honeyeater shows no morphological differentiation across the continent.[602] The population moves nomadically.[179,987] In places a seasonal regularity has been observed sometimes related to flowering times of local plants. Thus, near Burra (33/138), the species is often present in spring when emu-bushes are flowering. The Reporting Rates north and south of 26°S do not suggest a general movement north in winter.

Field Atlas sightings of Pied Honeyeaters were of single birds, pairs and groups of up to about six. The species inhabits acacia scrub, mallee, spinifex and eucalypt woodland usually where shrubs are flowering.[748,1124]

The Pied Honeyeater takes nectar from mistletoe and a variety of tubular flowers, e.g. emu-bush and brachysema. It also eats saltbush fruit, presumably digesting the flesh and voiding the seeds, and takes some insects.[748,1049]

There are no measures of abundance.

Field Atlas Records: 322

1° blocks–with records: 153 (19%)
–with breeding records: 17 (11%)

Atlas No.: 602

Max. Rep. Rate: 25%
In block(s): 27/117

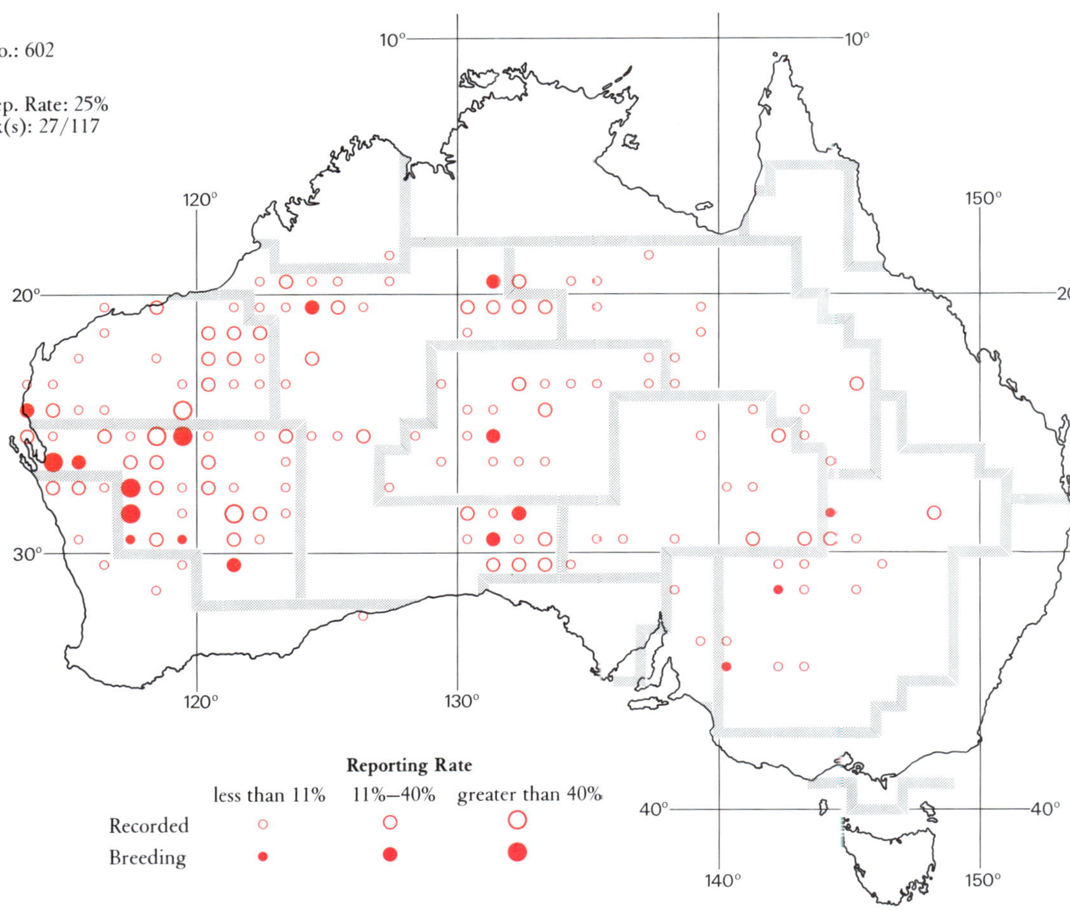

Dusky Honeyeater

Myzomela obscura

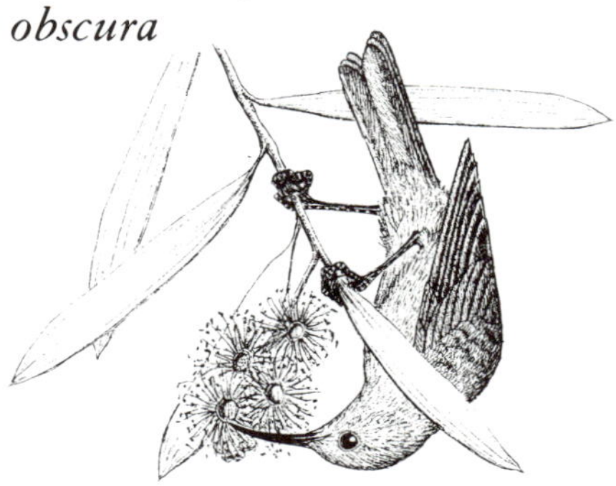

The Dusky Honeyeater lives in New Guinea and adjacent islands and in Australia. Two subspecies are recognized in Australia: *harterti* in the east; *obscura* in the Top End.[748] Early records from Borroloola (16/136) and south of Burketown (17/139)[107,1151] are probably errors; records from Groote Eylandt (13/136) are unconfirmed. The species is absent from those Torres Strait islands occupied by the Brown Honeyeater.[503]

The Dusky Honeyeater is nomadic in Torres Strait,[503] present all year in the Innisfail (17/146) district[698] and local movements have been noted at Rockhampton (23/150).[1086] The Field Atlas does not suggest any large-scale seasonal movement.

The range of the Dusky Honeyeater is limited to the humid areas of the north where it inhabits rainforest along streams, paperbark woodland and mangroves. Using creekside vegetation, it penetrates eucalypt woodland. It is sometimes in mixed flocks with other honeyeater species. The Dusky Honeyeater feeds on nectar, specially from grevilleas and paperbarks, and insects, including winged termites and wasps,[748,1049] which it catches in the air.

There are no measures of abundance.

Field Atlas Records: 1539

1° blocks–with records: 68 (8%)

–with breeding records: 13 (19%)

Atlas No.: 590

Max. Rep. Rate: 70%
In block(s): 20/149

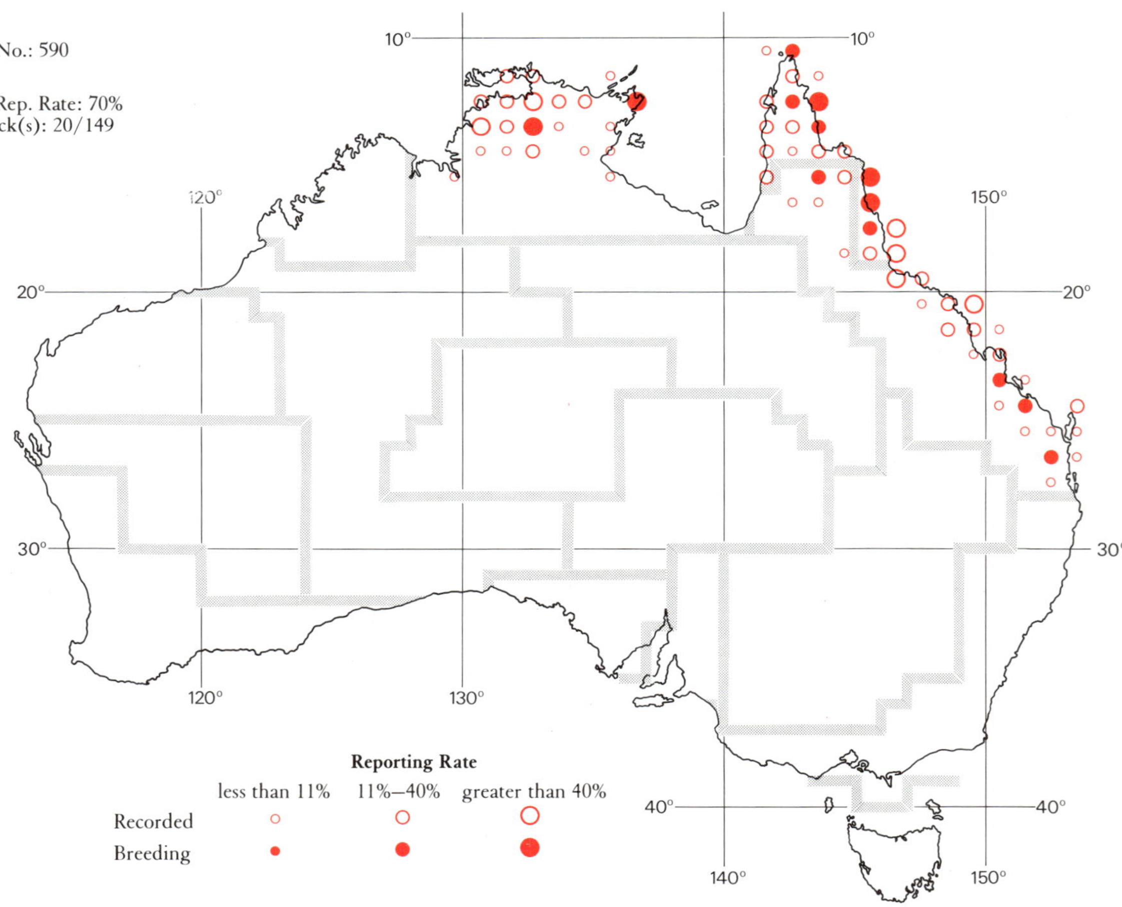

Red-headed Honeyeater

Myzomela erythrocephala

The Red-headed Honeyeater ranges from the Lesser Sunda Is and New Guinea to Australia.[1124] Two subspecies are recognized in Australia: *infuscata* in the Cape York Region; *erythrocephala* in the Top End and Kimberley Regions.[748] The combined Atlases suggest a continuous distribution, gaps at 129–130°E and 137–138°E probably reflecting lack of visits rather than of birds. The southern limits of range on both eastern and western coasts are as determined by the Field Atlas. Records from the Atherton Region[231,1842] are unconfirmed.

Fluctuations in abundance have been recorded, probably associated with the flowering of trees within a district. Birds flock and move round, at least locally, if not over large distances.[423,1145] The species visits the larger islands of western Torres Strait late in the dry winter season.[503]

The Red-headed Honeyeater lives singly, in pairs or in small flocks, sometimes mixed with other species. It inhabits mangroves and eucalypt woodland along streams, visiting places where eucalypts are in flower. It feeds on nectar and insects including ants.[748]

There are no measures of abundance.

Field Atlas Records: 212

1° blocks–with records: 38 (5%)
–with breeding records: 1 (3%)

Atlas No.: 587

Max. Rep. Rate: 30%
In block(s): 12/130

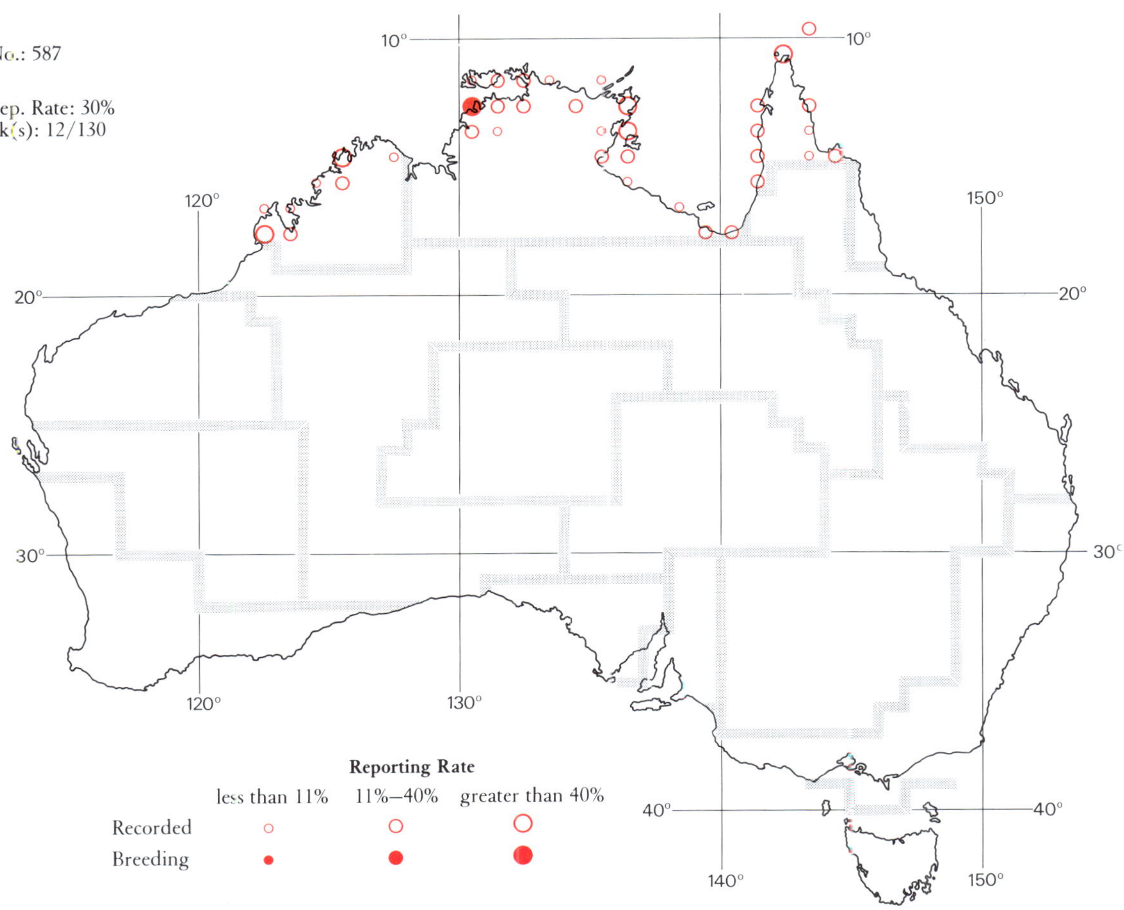

Scarlet Honeyeater

Myzomela sanguinolenta

The Scarlet Honeyeater lives in Sulawesi, New Caledonia and in Australia. A single subspecies, *sanguinolenta*, is recognized in Australia. In the Atherton Region the Scarlet Honeyeater is present all year[698] but there is evidence of movements farther south. At Murphy's Ck (27/152) it is plentiful in spring and summer when plants are flowering[1099] but in coastal south-eastern Qld, some birds are said to be winter visitors and others passage migrants moving north in winter and south in summer.[987,1709] Near Sydney (33/151) movements are also north-south but incomplete and irregular.[692,1912] The Field Atlas reflects an apparent winter influx into Eastern Qld, the Reporting Rate increasing from 4.8% in summer to 16.5% in winter, but shows little change in the South-East Region (summer 3.8%, winter 3.2%).

All records at the south-western end of its range (west of 149°E), where it is an irregular visitor, were made Sept.-Mar.

The Scarlet Honeyeater lives singly or in small groups in eucalypt woodland moving in response to the flowering of various plants, particularly paperbarks, bottlebrushes, turpentine and mistletoe and sometimes eucalypts.[987] In the Atherton Region it is said to be rare below 400 m;[683] in the south it is usually in coastal lowlands though large numbers have sometimes been recorded at high altitude, e.g. at 1300 m in New England National Park (30/152) and above 1000 m in the Moonbi Ra. (31/151).[129] Reporting Rates in the Atherton Region, New England district and Snowy Mts (36/148) were respectively 11.7%, 0%, 0% above 500 m and 12.5%, 30.5%, 6.7% below 500 m. The Scarlet Honeyeater feeds on nectar and insects caught in the air, some very small such as midges. Seeds have also been found in its gut.[1912]

At Moruya (35/150), near the southern limit of its range, in eucalypt woodland density was 0.18 birds/ha.[1197]

Field Atlas Records: 2217

1° blocks-with records: 79 (10%)
-with breeding records: 23 (29%)

Atlas No.: 586

Max. Rep. Rate: 40%
In block(s): 18/145

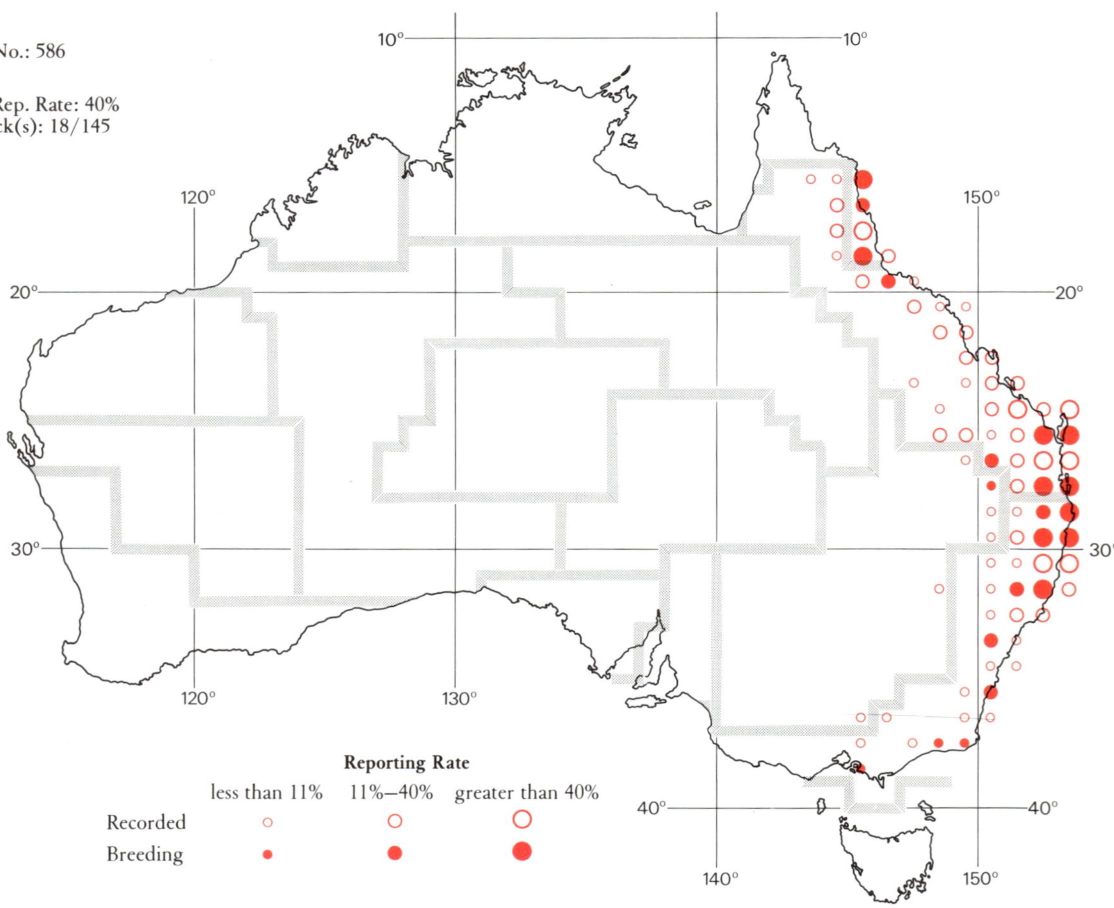

Crimson Chat

Ephthianura tricolor

The range of the endemic Crimson Chat fluctuates from year to year. In some years it includes the whole of the Rolling Downs Region and parts of the Top End and Kimberley Regions as far north as Wyndham (15/128). The species is an intermittent winter-spring visitor to these areas[862,1713] and has not been recorded breeding north of 19°S in the Atlases.

In southern Australia the Chat sometimes irrupts south of the usual range to breed in spring-summer in the east and winter-spring in the west. This happened at Rutherglen (36/146) in 1896[1128] and at Yallum (35/142) in 1923.[1725] Other early irruptions have been documented.[984] No such irruption occurred in eastern Australia during the Field Atlas but, in 1981, considerable numbers of Chats bred in the lower South-West Region from which the species is usually absent.[1258]

Reporting Rates were higher in winter than in summer both north and south of 26°S (north of 26°S: summer 2.3%, winter 5.0%; south of 26°S: summer 1.1%, winter 2.5%). This perhaps indicates that the bird is more conspicuous in winter than in summer but does not support the suggestion of a north-south shift of the population, winter and summer.[984] The Field Atlas record at Bemm R (37/148) was of two birds in Dec. 1978.

The Crimson Chat lives in acacia scrub and other open, sparsely-vegetated areas of the inland, specially near salt lakes. It sometimes forms mixed flocks with other chats. It breeds in loose colonies.[1638,1892] The Crimson Chat feeds almost entirely on insects and occasionally nectar and will drink when water is available, even as dew. Studies in WA related its biology to its need to conserve water,[1892] e.g. in inland WA it breeds in winter when evaporation is low and its behaviour shows other adaptations to water conservation. Young birds have been found in a torpid state on cold mornings.[904a]

When breeding, the male initially defends the female and later an area of about 20 m radius round the nest.[1892] There are no other measures of abundance.

Field Atlas Records: 2191

1° blocks–with records: 416 (51%)
–with breeding records: 99 (24%)

Atlas No.: 449

Max. Rep. Rate: 42%
In block(s): 28/121

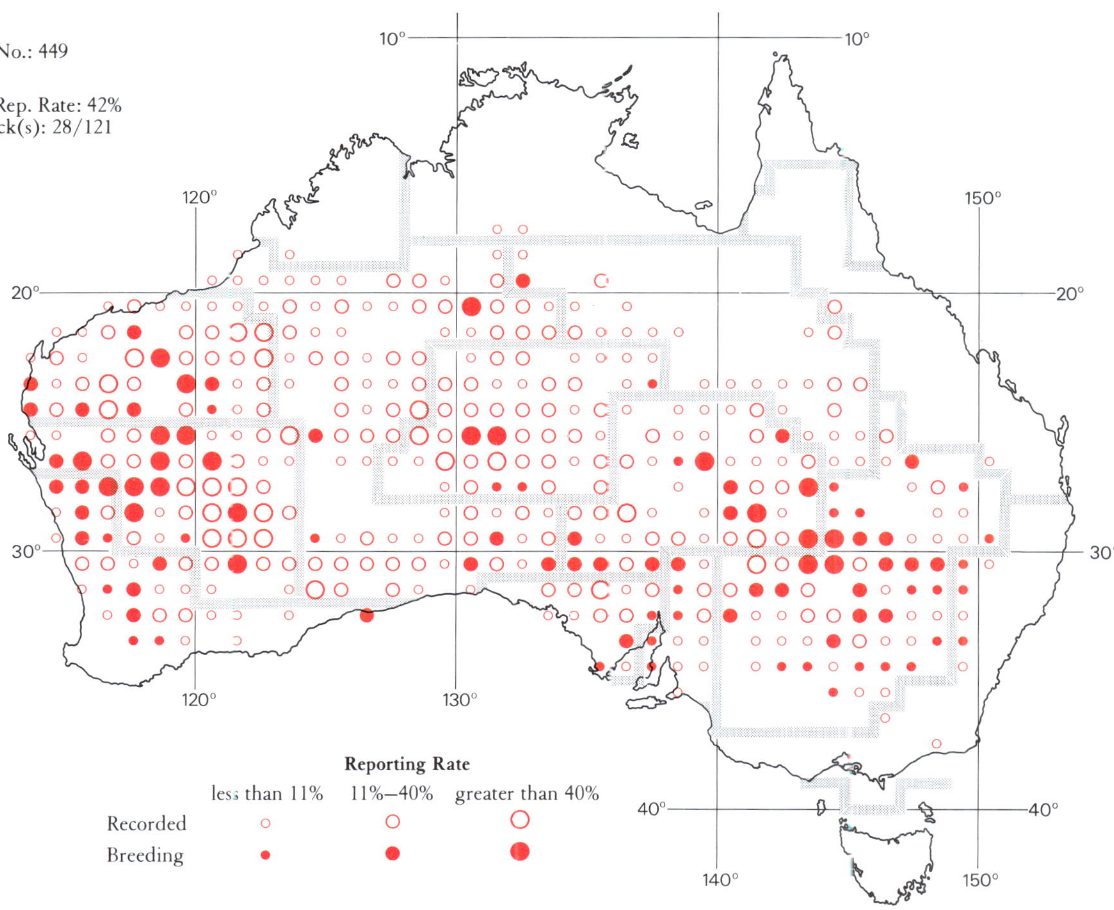

Orange Chat

Ephthianura aurifrons

The range of the endemic Orange Chat does not fluctuate markedly from year to year as does that of the Crimson Chat. Most records of the Orange Chat are south of 20°S and all breeding records in the Atlases are south of 22°S. The Field Atlas sighting at Burketown (17/139) was of a flock of about 20 birds in June 1980 and is the northernmost record in the Atlases. The outlying report, also in June 1980, from Coolmunda (28/151) was of a flock of 11–12 birds.

Reporting Rates of the Orange Chat were higher in winter both north of 26°S (summer 0.8%, winter 1.8%) and south of 26°S (summer 1.6%, winter 2.6%), which suggests that the bird is more conspicuous in winter than in summer. The differences do not suggest a regular seasonal movement but nomadic movements are reported. Flocks sometimes augment resident populations in the southern Murray-Darling[833] and after heavy rain in 1980 large numbers bred on the Nullarbor Plain.

The Orange Chat lives in acacia scrub, spinifex, tussock grassland and saltbush, specially round salt lakes. It has been reported to forage in sunflower crops and cotton fields. Flocks of up to 40 birds form in the non-breeding season, sometimes together with Crimson Chats. The Orange Chat forages on the ground and eats only insects. Physiological studies show it to be more tolerant of desiccation than the White-fronted and Crimson Chats.[1893,1894] This is perhaps the reason that its range fluctuates less from year to year than does that of the Crimson Chat.

There are no measures of abundance.

Field Atlas Records: 1775

1° blocks–with records: 275 (34%)
–with breeding records: 59 (21%)

Atlas No.: 450

Max. Rep. Rate: 44%
In block(s): 26/139

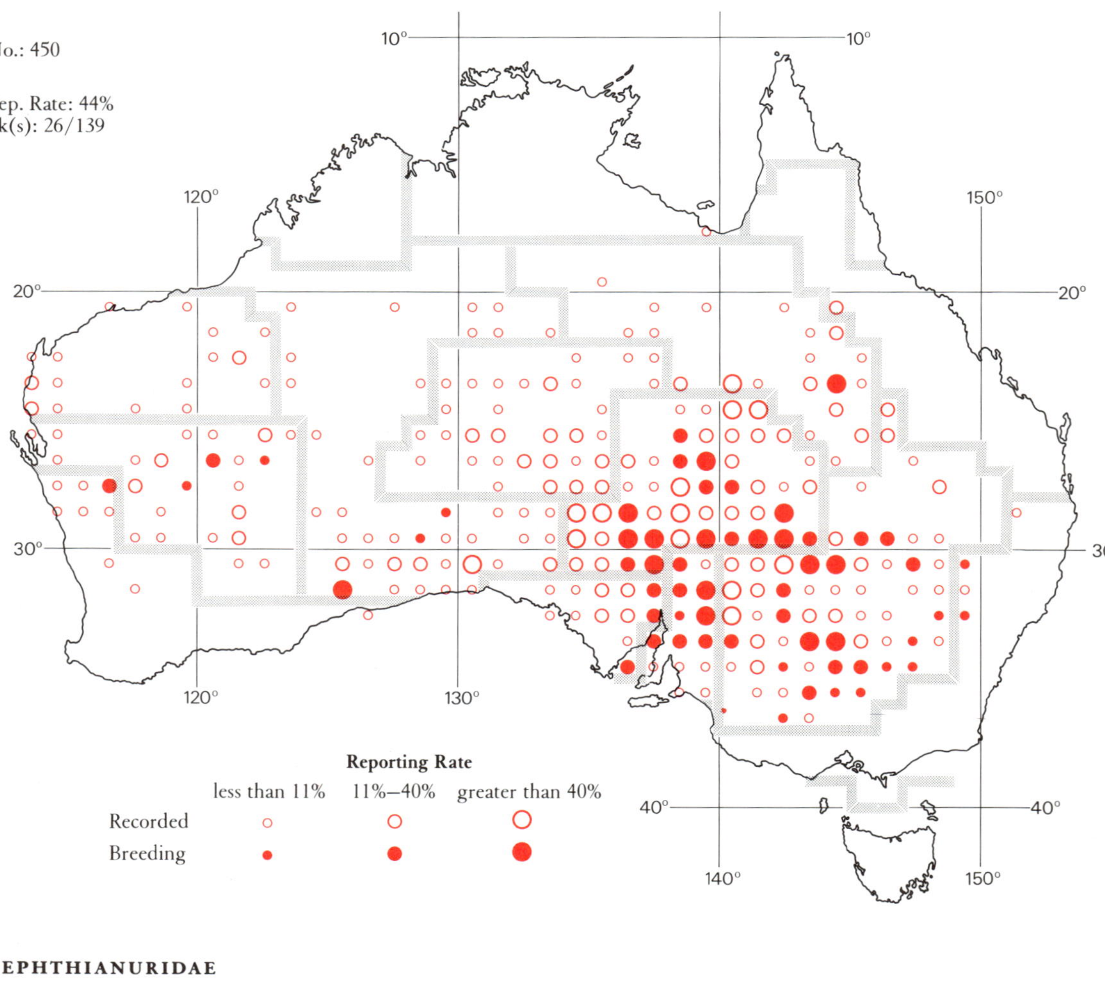

Yellow Chat

Ephthianura crocea

The range of the endemic Yellow Chat has always been poorly known because it inhabits remote areas and is often inconspicuous.[1891] Before 1950 it was known from a few places on northern coastal rivers between the Fitzroy R (17/123) and the Norman R (17/141) with inland records near Sedan (25/143) in 1910[626,984] and west of Davenport Downs (24/141) in 1918. There was also a population in Eastern Qld known from specimens taken at Fitzroy Vale (23/150) in 1859 and at Torilla (22/150) in 1917.[300,984] The species has not subsequently been reported in Eastern Qld but has now been recorded at places throughout the Rolling Downs Region. Breeding was confirmed at Coorabulka (23/140) in June 1974.[1892]

The Yellow Chat is nomadic.[1891] Local movements are recorded at Oenpelli (12/133) when marshes shrink seasonally[626] and are thought to take place at Coorabulka also.[1482]

The Yellow Chat lives close to bodies of water. In the Rolling Downs Region it inhabits swamps that develop by bore drains, most of which have been constructed since 1890.[626] It is often in lignum, cumbungi, sedge or other dense vegetation but there were Atlas records from such open places as tidal samphire, a dry lake bed with sparse ground cover dotted with low bushes and bare ground round a bore. Most sightings were of groups of two to ten birds. On the Roebuck Plains (17/122) in the Kimberley Region, in June 1979, more than 100 were present in two flocks of about 50 and several smaller groups. The Yellow Chat feeds on insects, often flies.[626]

There are no measures of abundance.

Field Atlas Records: 58

1° blocks–with records: 14 (2%)
–with breeding records: 1 (7%)

Atlas No.: 451

Max. Rep. Rate: 18%
In block(s): 24/140

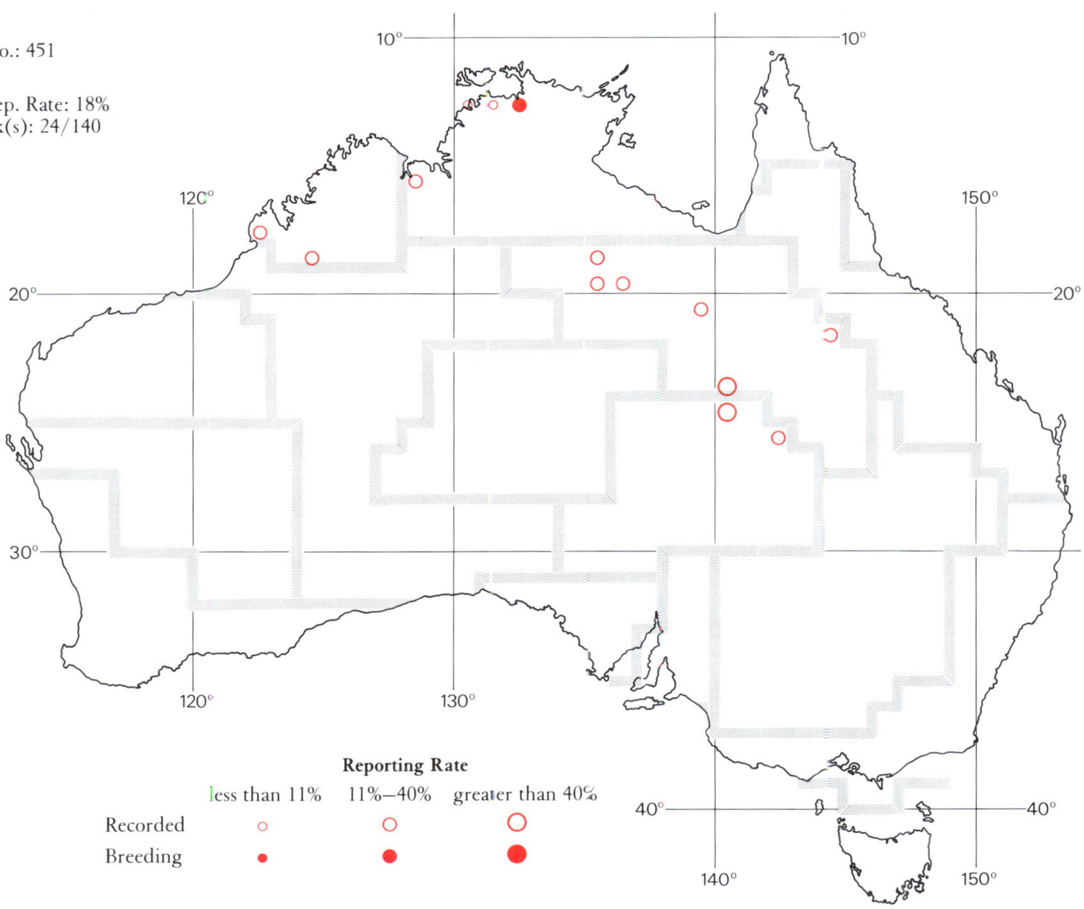

White-fronted Chat

Ephthianura albifrons

In the historical periods the endemic White-fronted Chat bred as far north as Peron Pen. (25/113) and there are references to sightings as far north as Wycliffe Ck (20/134).[1366] In Tas. the Chat's distribution is discontinuous in the eastern half; in the west only the coastal belt south to Strahan (42/145) is occupied.

There is a tendency for the White-fronted Chat to leave the inland during summer but the trend is not uniform and the movement is nomadic rather than regular and seasonal.[1892] The Field Atlas shows no evidence of large-scale seasonal movement. The longest movement recorded is that of an immature from Rottnest I. (32/115) to Herdsman L. (31/115), 31 km.[1438]

The White-fronted Chat is present most often in damp places. Along the southern coasts it is abundant in heath, near the shore and in pasture. Away from the coast it is also widespread wherever there are damp areas: along rivers and fringes of lagoons in south-eastern Australia and round the salt lakes of the inland. Above 1000 m it is rare. Flocks of varying size form in the non-breeding season. In the southern Murray-Darling Region one of over 1000 was recorded[833] and at Cramphorne (31/118) 400–500 birds perched on wires near the road. On King I. (39/144) and at Mt Mary Plains (34/139) flocks as large as 50 are notable.[189,1131] Sometimes the White-fronted joins flocks of other chats. It breeds semi-colonially.

The White-fronted Chat feeds mainly on ground-dwelling insects and guts examined have contained no seeds.[1049] One flock has been observed feeding in the outer foliage of trees. At Orange (33/149) before 1950 the Chat bred commonly in open country but now it is rarely recorded.[784]

At Fishermen's Bend (37/145) in 1948, 18 nests were found in 1 ha of saltmarsh.[1837]

Field Atlas Records: 9672

1° blocks–with records: 271 (33%)
–with breeding records: 126 (46%)

Atlas No.: 448

Max. Rep. Rate: 59%
In block(s): 40/147

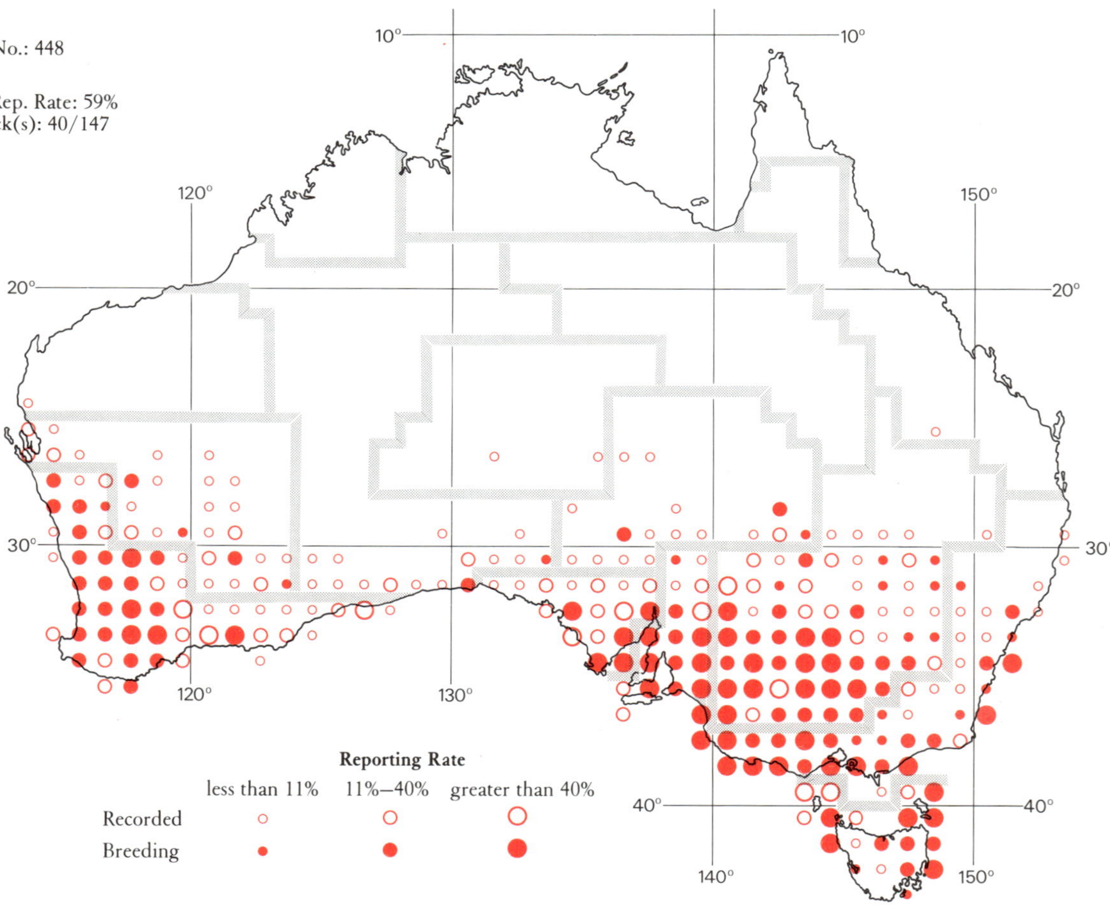

Gibberbird

Ashbyia lovensis

The Gibberbird is endemic. The record of birds breeding at Toko Ra. (22/137) extends to the north-west the species' previously known distribution.

Some movement is implied by the observation that the small flock from which the first specimens were taken in 1910, near Leigh Ck (30/138), was present only a short time in the district.[1100] The species is thought to be generally nomadic.[1369]

The Gibberbird lives on sparsely vegetated gibber plains, sometimes in small areas between dunes.[1921] Small flocks have been recorded,[1100] though the species is usually in ones and twos. There were beetles in the guts of the four wild birds that have been examined.[910] The Gibberbird has also been watched feeding on grasshoppers, grubs, moths and spiders, all taken on the ground. Blowfly larvae, left on the ground by infected sheep, are an important food in winter.[1595]

The late discovery of the Gibberbird in 1910, despite earlier work by competent ornithologists in regions where it now lives, has prompted the suggestion that the species was once much less abundant and widespread. Grazing cattle and sheep may have modified the environment to the advantage of the Gibberbird.[1387]

There are no measures of abundance.

Field Atlas Records: 201

1° blocks–with records: 51 (6%)

–with breeding records: 5 (10%)

Atlas No.: 452

Max. Rep. Rate: 30%

In block(s): 24/140

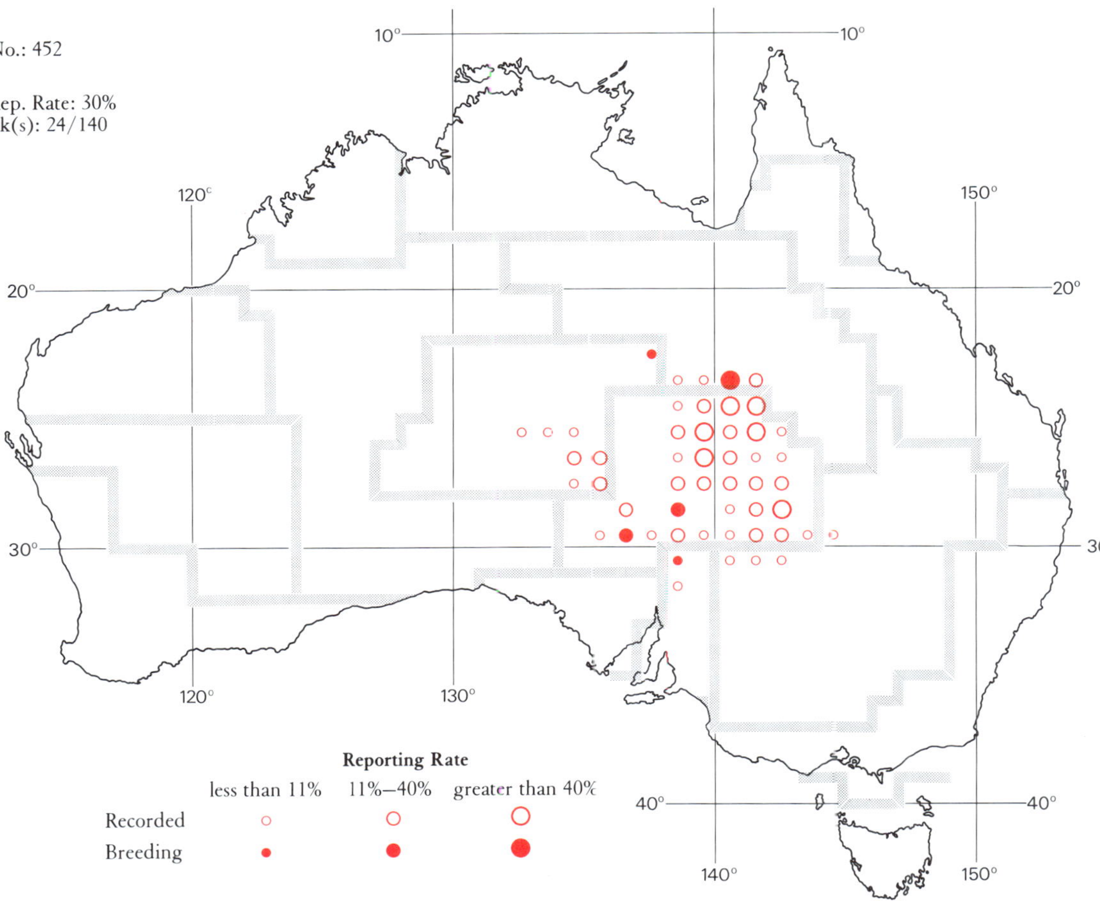

Yellow-bellied Sunbird

Nectarinia jugularis

The Yellow-bellied Sunbird lives in New Guinea and adjacent islands with one subspecies *frenata* in Australia.[1124] It has been suggested that the species' range is extending southward.[1124] The Atlases provide some support for this: the Sunbird was recorded only as far south as Yeppoon (23/150) in 1901–50[1911] and not on the well-settled coast to the south. The Field Atlas 10′ maps show a continuous series of records along the coast from Yeppoon (23°) to Round Hill (24°15′S) and one farther south at Mullett Ck (24°30′S). The inland record (24/149) in the Field Atlas was of two birds on the Dawson R.

The Sunbird is reported to be present all year in Torres Strait[503] and near Innisfail (17/146).[698] In Townsville (19/146) density was greatest in the wet summer season (up to 4 birds/ha) and least in the dry winter season (less than 0.5 birds/ha),[951] which suggests some movement.

The Sunbird lives in pairs near the coast in rainforest, including margins and regrowth, mangroves and homestead and suburban gardens, all areas where there are nectar-bearing flowers.[151,231,1703] It feeds on insects and the nectar of flowering plants, native and introduced.[1498,1758,1889] Cyclones can destroy local populations: Sunbirds disappeared from Mackay (21/149) after one in 1918 and did not reappear until 1921.[405]

The Sunbird was more abundant in suburban Townsville (average 0.8 birds/ha) than in adjacent eucalypt woodland (0.3 birds/ha).[951]

Field Atlas Records: 1338

1° blocks–with records: 42 (5%)

–with breeding records: 17 (40%)

Atlas No.: 572

Max. Rep. Rate: 80%
In block(s): 20/149

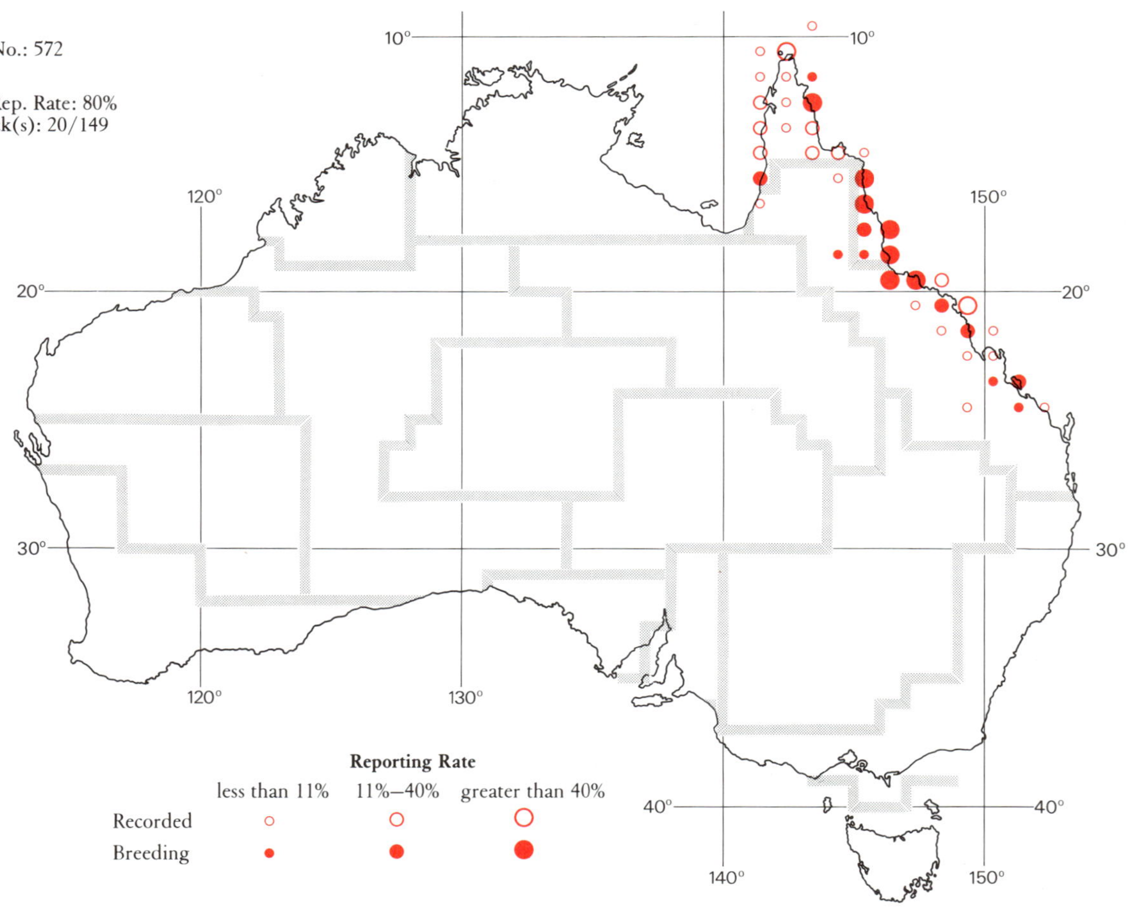

Mistletoebird

Dicaeum hirundinaceum

The Mistletoebird lives on Tanimbar (7S/131E), Kai (5S/133E), Aru (6S/134E) and Torres Strait islands[503] and in Australia.[1124] There is a slight clinal variation in the morphology of the Australian birds, the northern ones being smaller than the southern and the females in the Cape York Region slightly different in colour from other females, but no subspecies have been recognized.[748,980] The Mistletoebird is absent from the Tas. Region.

There are many reports of the Mistletoebird arriving seasonally[980] but the Field Atlas suggests no major population shift. Local nomadic movements keep the bird in contact with fruiting mistletoe, which is the movement pattern observed in the southern Murray-Darling Region.[833]

The Mistletoebird appears wherever mistletoe grows, be it rainforest, desert or mangrove though it is rare above 1000 m in the Snowy Mts. The bird is usually solitary or in small groups. Though feeding mainly on the berries of mistletoe the Mistletoebird also eats insects and the fruit of other plants, some introduced, including saltbush, privet, hawthorn and peppercorn.[653] The timing of nesting and the extent of nomadism of the Mistletoebird are governed by the fruiting of mistletoes. This varies with the species, the season and the locality[653] with the proviso that the bird avoids cold weather because it may become torpid in very cold periods.[1631] In few other Australian birds does co-evolution of bird and plant seem to be so clear.

Density in eucalypt woodland at Armidale (30/151) was 0.08–0.21 birds/ha[582] and at Moruya (35/150) 0.18 birds/ha.[1197] In eucalypt forest at Boola Boola it was 0.22–0.52 birds/ha.[1113]

Field Atlas Records: 13 525

1° blocks–with records: 615 (76%)
–with breeding records: 127 (21%)

Atlas No.: 564

Max. Rep. Rate: 66%
In block(s): 23/144,27/149

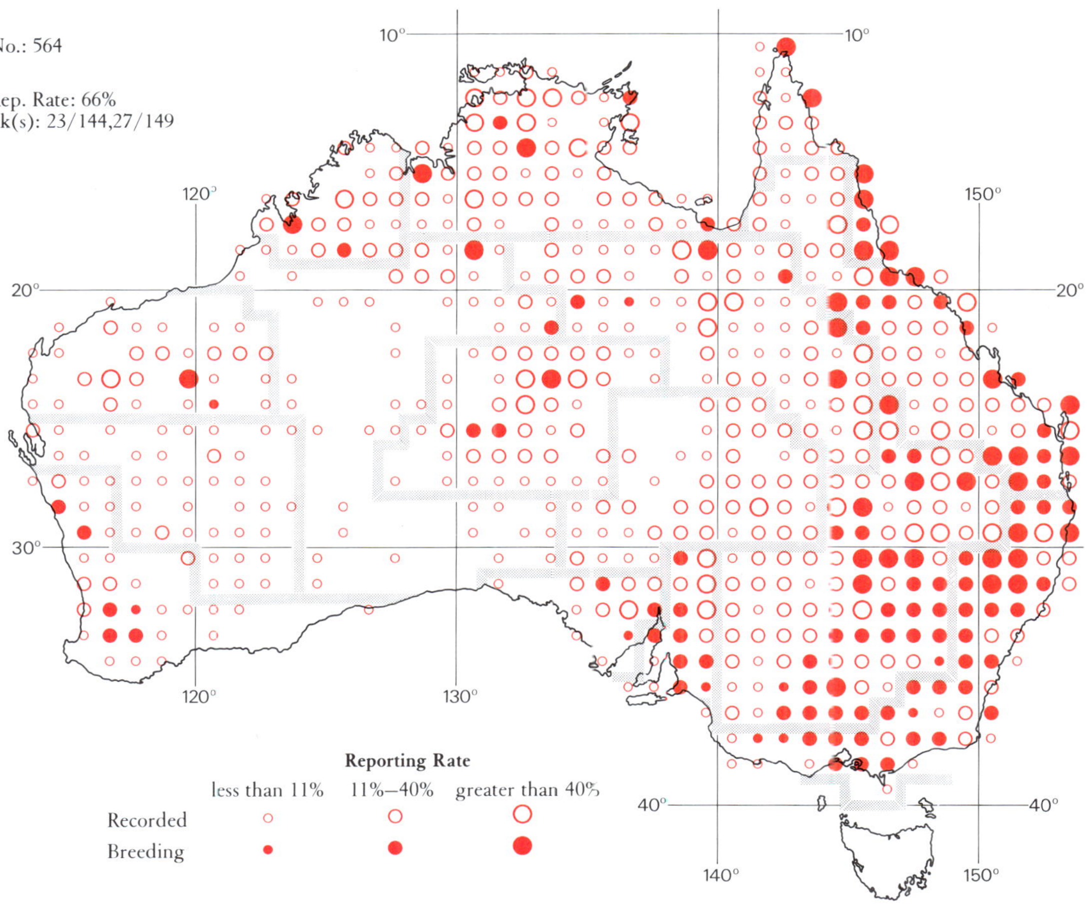

Spotted Pardalote

Pardalotus punctatus

Though populations of the endemic Spotted Pardalote are isolated in eastern Australia, the Tas. Region and the South-West Region, no subspecies are recognized.[748] The Spotted and Yellow-rumped Pardalotes interbreed in areas of contact[1907] and may be hard to differentiate in the field. They usually occupy different types of vegetation and replace each other rather than overlap in breeding distribution. Both species have been reported from Kangaroo I. (35/137). The only pardalote recorded in the Cave Hill (31/121) area could have been either species and is omitted from the map.

In eastern Australia Spotted Pardalotes appear to migrate in winter, some to lower altitudes, some north and some inland. There is a winter movement into the Eastern Qld Region (Reporting Rates: summer 3.4%, winter 6.8%). At Wollomombi (30/152) some birds are present all year but numbers increase in autumn and spring, which may represent birds on passage.[129] The species is regarded as a winter visitor to the lower Murray-Darling Region specially along the Murray R[833] (Reporting Rates: summer 3.2%, winter 5.3%). In the Snowy Mts (36/148) Reporting Rates above 500 m in autumn-winter (13%) were lower than in spring-summer (32%); below 500 m there was little difference (40% and 48%). No movement is apparent in the Tas. Region.[1533,1801] In the Perth (31/115) area some birds are present all year but there is an influx in May.[1610,1638] The longest movement recorded is that of an adult from Kangola (36/140) to Melbourne (37/145), 590 km.[59]

The Spotted Pardalote lives in pairs or small groups, forming flocks of several hundred in autumn and winter.[653] It inhabits eucalypt forest and woodland[131,1907] and feeds by gleaning insects, specially psyllids, from the leaves of the canopy.

Densities in eucalypt woodland were: Armidale (30/151) 0.33 birds/ha,[582] Wollomombi 0.1 birds/ha when breeding and up to 2.5 birds/ha in spring and autumn,[129] Canberra (35/149) 0.3 birds/ha,[154] Moruya (35/150) 0.69–0.82 birds/ha,[1197] Tooboorac (37/144) 0.42–0.83 birds/ha and the You Yangs (37/144) 0.06–0.62 birds/ha.[1907] In eucalypt forest they were: Lysterfield (37/145) 0.08–2.56 birds/ha,[1907] Jamieson (37/146) 0.5–1.0 birds/ha[1469] and Boola Boola (38/146) 0.2–1.5 birds/ha.[1113]

Field Atlas Records: 10 328

1° blocks–with records: 179 (22%)

–with breeding records: 95 (53%)

Atlas No.: 565

Max. Rep. Rate: 50%
In block(s): 33/150

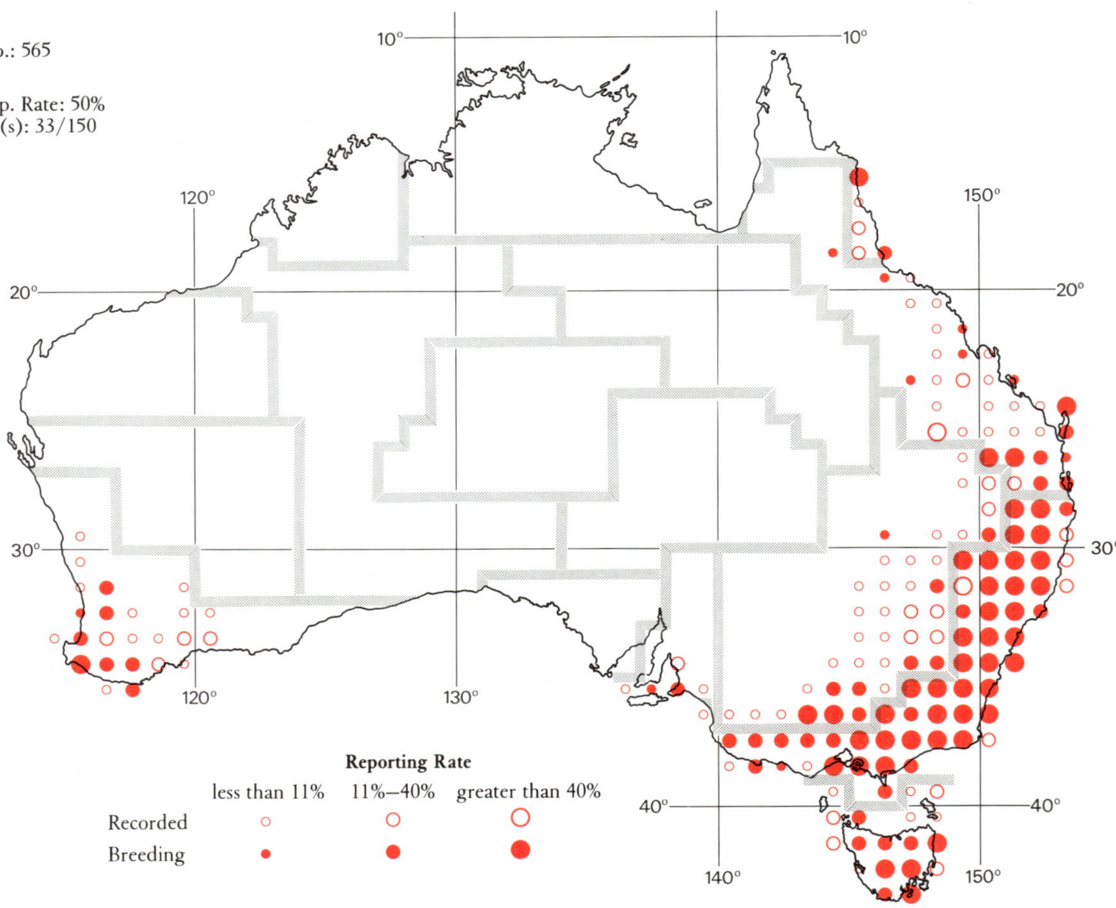

Yellow-rumped Pardalote

Pardalotus xanthopygus

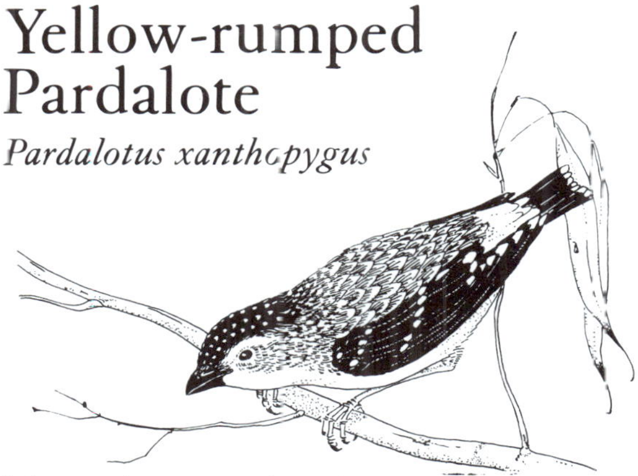

The endemic Yellow-rumped Pardalote is closely related to the Spotted Pardalote. They interbreed in areas of contact[1907] and are sometimes hard to differentiate in the field. The two usually occupy different types of vegetation and replace each other rather than overlap in breeding distribution. They come closest where there is an 'island' of one vegetation type within another such as in the Whipstick (36/144). Both species have been reported from Kangaroo I. (35/137). The only pardalote recorded in the Cave Hill (31/121) area could have been either species and is omitted from the map.

No subspecies of the Yellow-rumped Pardalote are recognized.[748] The population in and near Round Hill Nature Reserve (33/146) in the Murray-Darling Region is probably isolated. Those on either side of the Great Australian Bight are connected north, as well as south, of the Nullarbor Plain: in 1969 one pair was found breeding west of Vokes Hill (28/130) and others were seen north of the Nullarbor (28/128, 29/131).[600] Lack of records from this area during the Field Atlas is partly because there were few visits but may also indicate that the population is usually sparse. The species is known to irrupt occasionally onto the Nullarbor itself at Cook (30/130).[600]

Flocks form at Mt Mary Plains (34/139) in autumn and the species seems most numerous then, though present throughout the year.[189] There are no other reports of movements and the Field Atlas does not suggest any large-scale pattern of seasonal movement.

The Yellow-rumped Pardalote lives in pairs or small groups in mallee, remaining outside the eucalypt forest and woodland where the Spotted Pardalote lives.[833] The Yellow-rumped Pardalote feeds on insects gleaned from foliage and sometimes from the ground.[748]

Densities in mallee and woodland in Vic. were: Bendigo (36/144) 0.05–0.30 birds/ha and Wyperfeld (35/141) 0.02–0.37 birds/ha.[1907] At Eyre Bird Observatory (32/126) in mallee density in spring was 1.75 birds/ha.[464]

Field Atlas Records: 1690

1° blocks–with records: 78 (10%)
–with breeding records: 28 (36%)

Atlas No.: 566

Max. Rep. Rate: 41%
In block(s): 35/141

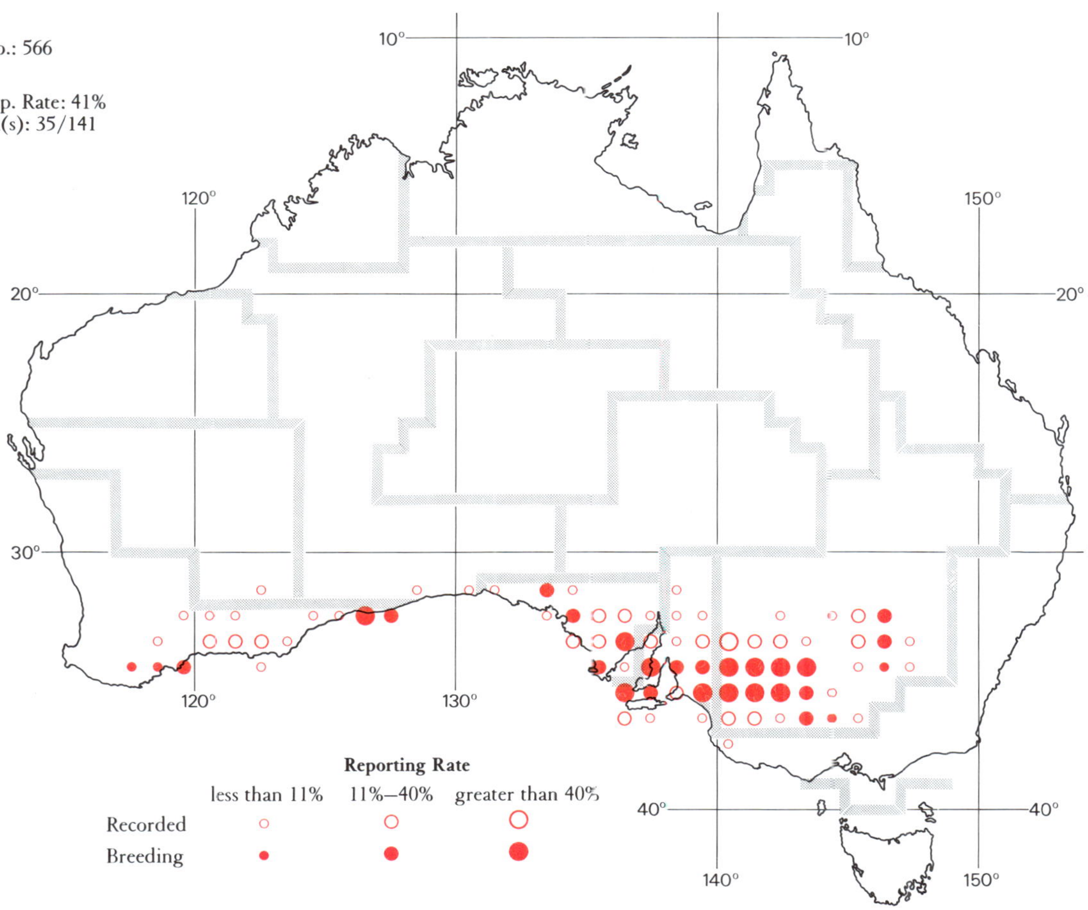

Forty-spotted Pardalote

Pardalotus quadragintus

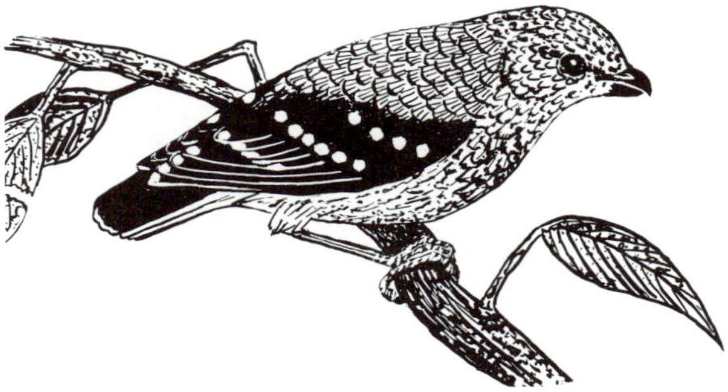

The Forty-spotted Pardalote is endemic to the Tas. Region. Its distribution and abundance have declined greatly this century and the species is now endangered. Until the turn of the century it was distributed on King I. (39/144) and Flinders I. (40/148) as well as in eastern Tas. at least as far north and west as Wynyard (41/145). It was locally common from sea level[1645] up to 1000 m.[574] Since 1950 there has been no report from King I.; the most recent from Flinders I. is a specimen collected in Nov. 1970; and, in Tas. itself, there has been no confirmed record outside the distribution shown on the map. The Forty-spotted Pardalote is now restricted to coastal south-eastern Tas. where eight small populations survive on islands and coastal peninsulas within 80 km of Hobart (42/147).[1538]

Movements of the Forty-spotted Pardalote appear to involve only local dispersal after breeding. The species inhabits forest and woodland undisturbed by human development, particularly that dominated by manna gum. Like the Spotted and Striated Pardalotes, with which it is sympatric in Tas., the Forty-spotted Pardalote gleans insects from leaves, mainly in the upper foliage.

It has been suggested that a major cause of the decline of the Forty-spotted Pardalote is competition from other pardalotes.[1910] The Spotted is at low density, each pair occupying a territory from which all its needs are derived but the Striated defends only the immediate vicinity of the nest and may reach high population densities where nest sites are abundant. Smaller colonies of the Forty-spotted Pardalote are vulnerable to competition during the breeding season. Broods fail because adults spend so much time defending the colony, specially from Striated Pardalotes.

Surveys in 1982 estimated the total population of the Forty-spotted Pardalote at about 850 birds.[1538] The largest population, containing several hundred pairs, is on Maria I. (42/148).

Field Atlas Records: 69

10′ blocks-with records: 11 (3%)
-with breeding records: 3 (27%)

Atlas No.: 571

Max. Rep. Rate: 58%
In block(s): 42°35′/148°05′

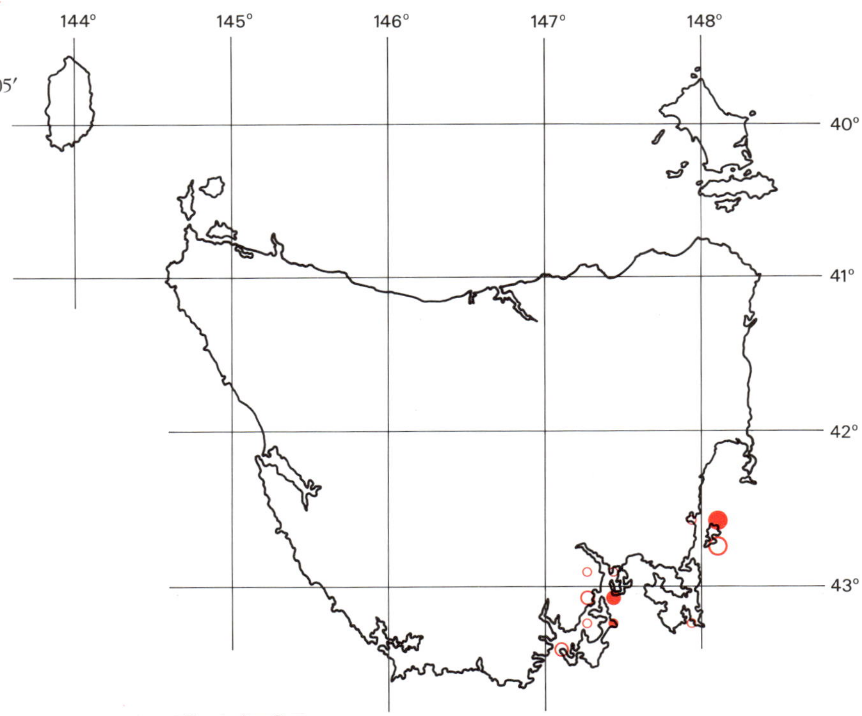

Reporting Rate

	less than 11%	11%–40%	greater than 40%
Recorded	○	○	○
Breeding	●	●	●

Red-browed Pardalote

Pardalotus rubricatus

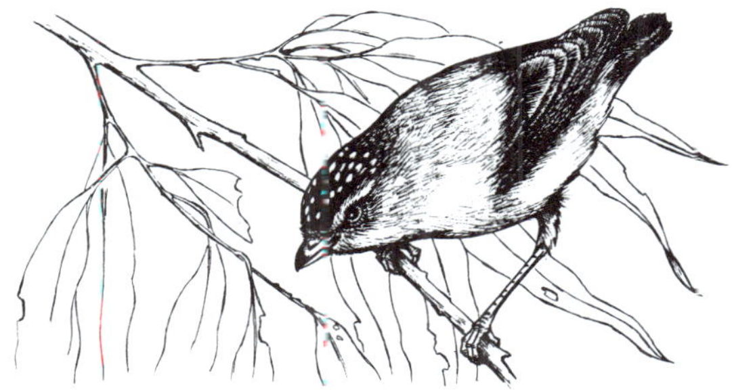

Four subspecies of the endemic Red-browed Pardalote are recognized: *yorki* in the Cape York Region; *carpentariae* on the southern coast of the Gulf of Carpentaria; *parryi* in the remainder of the north; and *rubricatus* in the southern part of the species' range.[748,1422] Historical records in the Western Deserts link the population in the Pilbara to that in the Central Ra. Region.

The Red-browed Pardalote is thought to be sedentary but the two isolated records in the Murray-Darling Region show that it can move long distances, perhaps along river systems, because both records are close to tributaries of the Darling R. One sighting was of a pair of birds south of the Moonie R (28/148) in Dec. 1980 and the other of a single bird near Pine Ridge (31/147) in Sept. 1977.

The Red-browed Pardalote lives singly or in pairs in eucalypt and paperbark woodland, including woodland along watercourses in acacia scrub and spinifex. It feeds on small insects from among the leaves and branches of trees and guts have also contained vegetable matter.[748]

There are no measures of abundance.

Field Atlas Records: 1182

1° blocks–with records: 291 (36%)

–with breeding records: 35 (12%)

Atlas No.: 570

Max. Rep. Rate: 40%
In block(s): 28/139

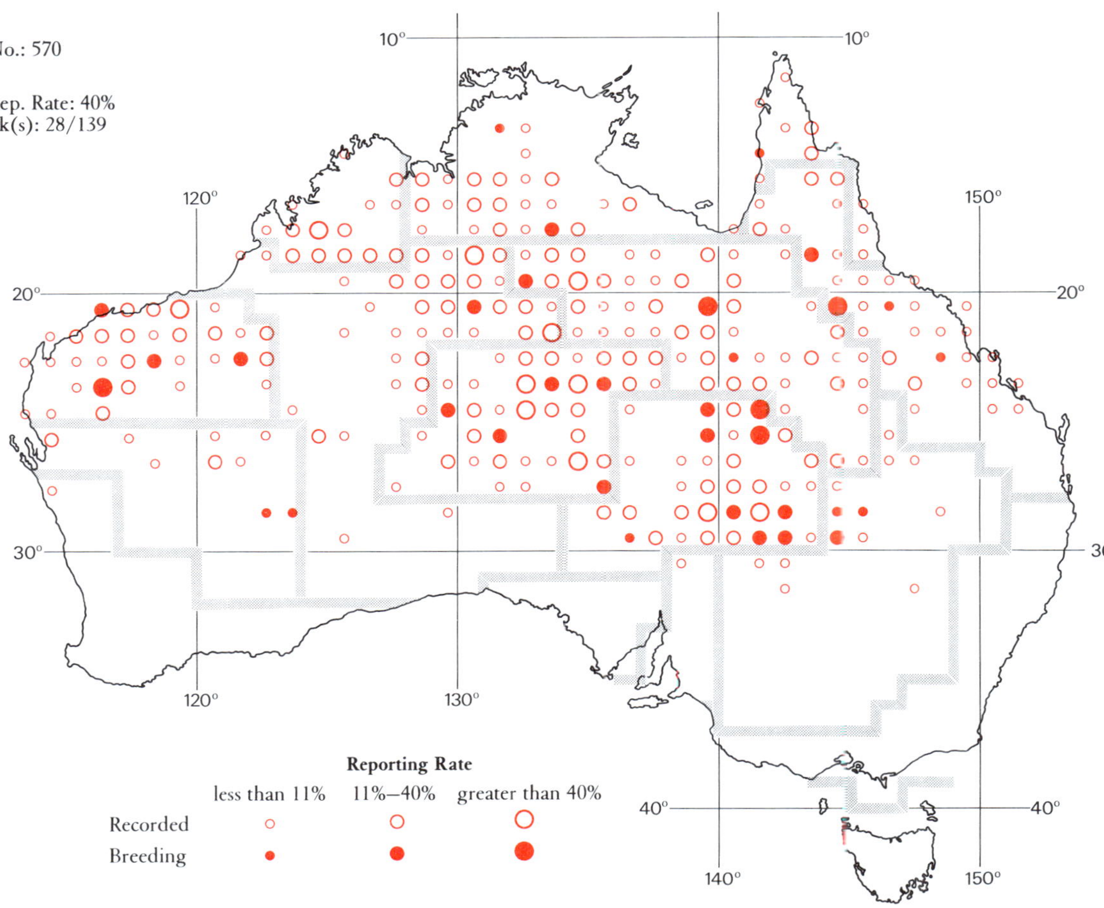

Striated Pardalote

Pardalotus striatus

The four forms of the endemic Striated Pardalote have been variously regarded as different species and as conspecific. Breeding ranges overlap but are broadly: *melanocephalus* from the Kimberley through Qld to the upper South-East Region; *substriatus* south of *melanocephalus* across the southern half of the continent; *ornatus* in eastern Australia south of the Atherton Region; and *striatus* in the Tas. Region and perhaps on the mainland.[817,1124,1242,1576] Individuals of *ornatus* have been recorded in breeding colonies of *striatus* in northern Tas.[737,1536] On the Australian mainland the only gap in range is between the Kimberley and Pilbara Regions.

In most places *melanocephalus* is recorded all year[231,698,1089] though near Darwin (12/130) it is apparently absent from the coast Oct.-Feb.[423] The other three subspecies migrate, sometimes in large flocks[330,1608] and in company with the Spotted Pardalote. Numbers increase in Qld in winter[1099,1709] (Reporting Rates in Eastern Qld: summer 22.6%, winter 44.1%). These are presumably *ornatus* and *striatus* from the south.[1089,1709] At Canberra (35/149) there appear to be sedentary birds, passage migrants and winter visitors from higher altitudes.[154,653] In the lower Murray-Darling Region birds are present all year with a marked winter influx, apparently mainly of *substriatus*.[833] Most birds from the Tas. Region probably migrate to the mainland in winter[1536] (Tas. Reporting Rates: summer 27.3%, winter 14.5%). In the South-West Region birds move inland in winter.[1608] Movements may be accompanied by massive mortality.[1909]

The Striated Pardalote lives mostly in eucalypts: in mallee, forest and woodland, including that along inland watercourses. Communal breeding has been reported.[501,1908] It feeds on insects and sometimes nectar.[748,1403] Birds converge on abundances of lerp[818] for which the bill is probably specialized.[1907]

Densities in mallee and woodland were: Bendigo (37/144) 0.16–3.00 birds/ha and Wyperfeld (35/141) 0.02–0.21 birds/ha.[1907] In eucalypt woodland they were: Armidale (30/151) 1.66 birds/ha,[582] Wollomombi (30/152) 0.3–2.54 birds/ha[129] and Canberra 0.24–2.07 birds/ha.[154] In eucalypt forest they were: Jamieson (37/146) 0.2–0.5 birds/ha[1469] and Boola Boola (38/146) 0.6–2.6 birds/ha.[1113]

Field Atlas Records: 23 581

1° blocks–with records: 570 (70%)

–with breeding records: 269 (47%)

Atlas No.: 976

Max. Rep. Rate: 84%
In block(s): 27/149

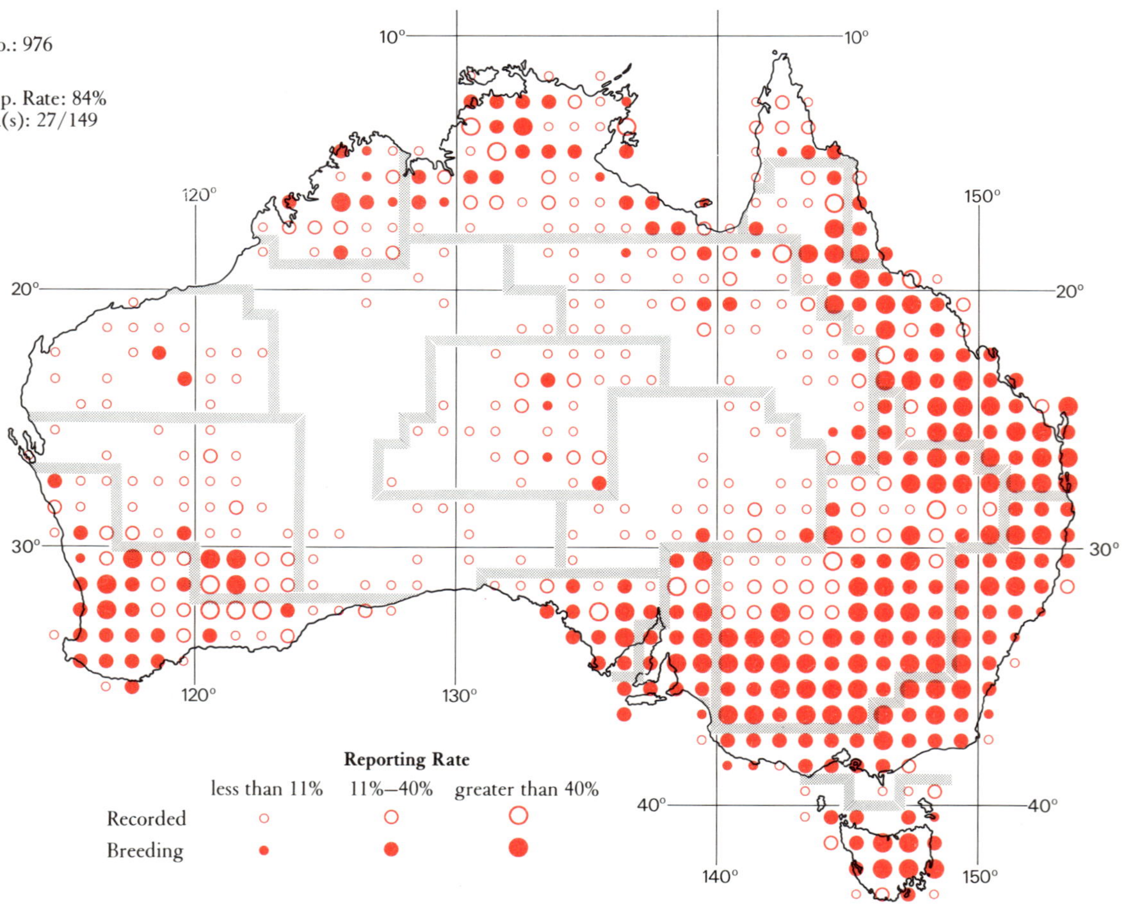

Pale White-eye

Zosterops citrinella

The Pale White-eye lives on the Lesser Sundas, on islands of Torres Strait and off the eastern side of the Cape York Region south to Eagle Islet (14/145).[1245] Its status and its relations with the Silvereye on the eastern coast are confused, as the size and colouration of the Pale White-eye at the southern end of its range resemble those of the northern Silvereye.[993] Sightings of the Pale White-eye on the mainland at Byfield (22/150) and Yeppoon (23/150) have never been confirmed.[1911]

A record from Booby I. (10/141), which has no vegetation suited to the species, is evidence of some local movement in Torres Strait.[503] The Pale White-eye normally lives on wooded islands. It has been recorded feeding on insects and berries.[1145,1147]

There are no measures of abundance.

Field Atlas Records: 28

1° blocks–with records: 4 (1%)
–with breeding records: 1 (25%)

Atlas No.: 577

Max. Rep. Rate: 21%
In block(s): 10/142

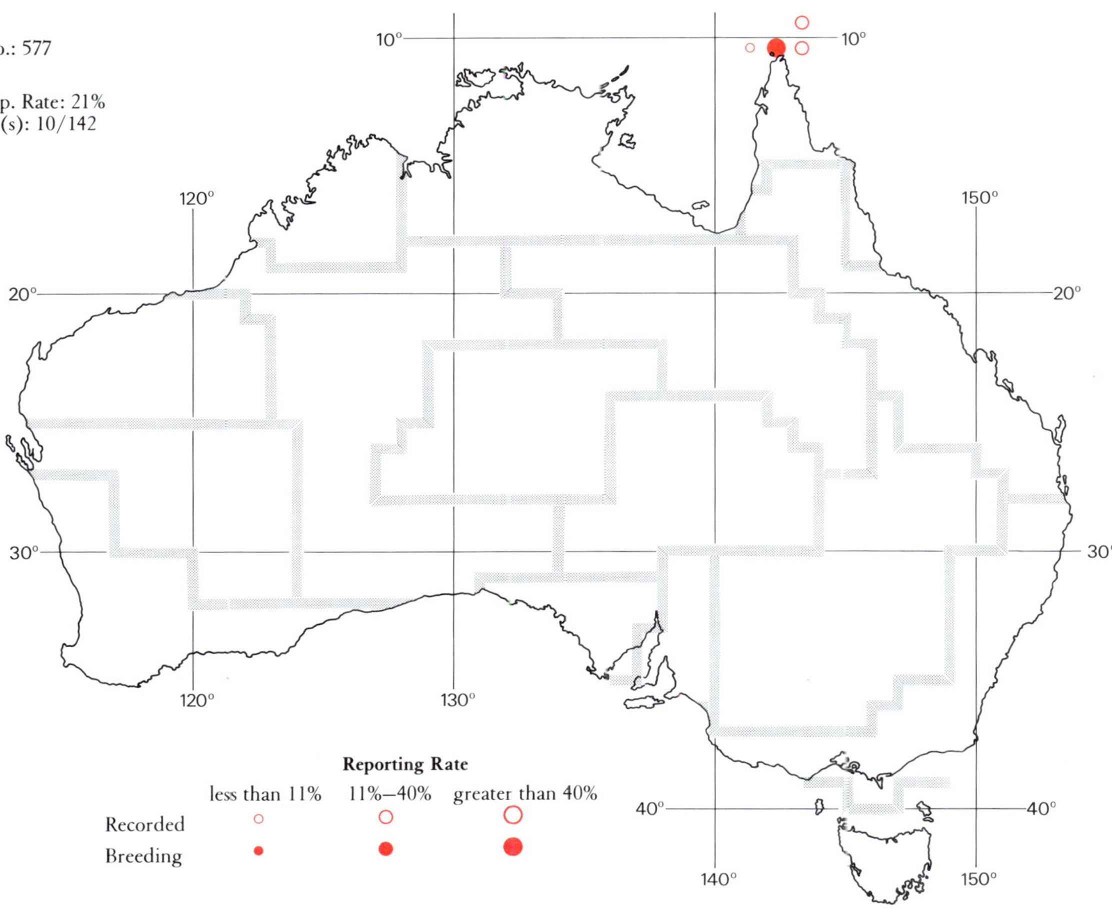

Yellow White-eye

Zosterops lutea

There are two subspecies of the endemic Yellow White-eye: *lutea* east of Wyndham (15/128) and *balstoni* to the west.[1124] There are no records from the western coast of the Cape York Region north of the limit determined by the Field Atlas but specimens collected in mangroves on the eastern coast at Barratta Ck (19/147) match typical Yellow White-eyes in plumage.[1043] An earlier unconfirmed record from Cairns (16/145)[1043] suggests that the species is distributed in mangroves along the coast of the Atherton Region. The Yellow White-eye must therefore closely approach the forests where Silvereyes occur. In WA the range of the Yellow White-eye overlaps that of the Silvereye between Exmouth (22/114) and Shark Bay (25/113).[769] The Atlases confirm that the Yellow White-eye's distribution is continuous from the Cape York Region westward.[608]

The Yellow White-eye is apparently nomadic at Darwin (12/130) because its numbers fluctuate irregularly.[423] There is no evidence of seasonal movement.

The Yellow White-eye lives near the coast in flocks, sometimes numbering up to 40 birds. It is usually in mangroves but not confined to them. On the Edward R (14/141) and near Anna Plains (19/121) flocks were recorded in paperbark woodland, up to several kilometres from mangroves. In the Pilbara Region, the species is sometimes well inland along rivers[622] and on islands of the western coast it is occasionally in eucalypt woodland adjoining mangroves.[748] Near Exmouth flocks may move out of the mangroves in winter to feed in acacia scrub.[1638] Because the Yellow White-eye is not confined to mangroves it can range across the gaps in mangroves round the coast that isolate populations of other species. The Yellow White-eye feeds on insects, including ants and larvae, fruit and nectar. Fruit is rare in mangroves and nectar seasonal, so probably the bird is mainly insectivorous.[748,769]

There are no measures of abundance.

Field Atlas Records: 328

1° blocks–with records: 55 (7%)
–with breeding records: 2 (4%)

Atlas No.: 576

Max. Rep. Rate: 30%
In block(s): 17/140

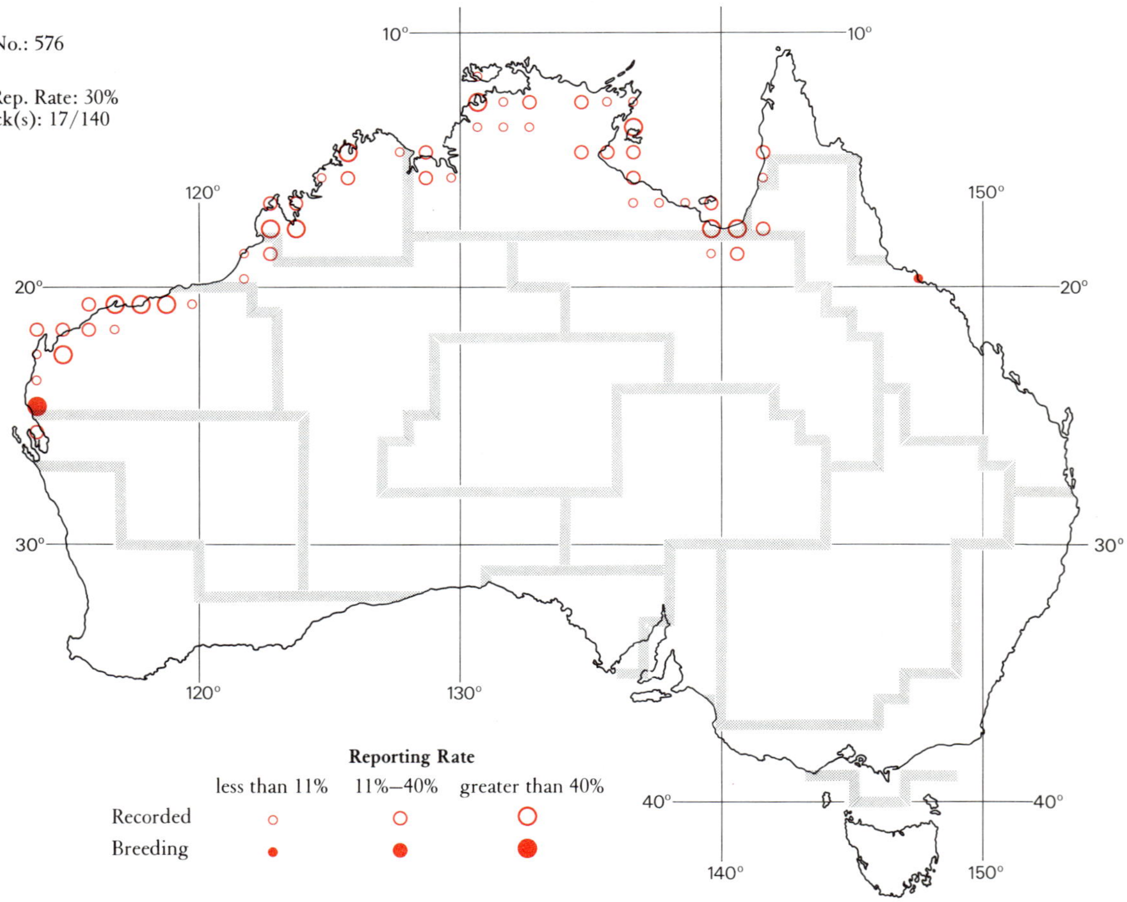

Silvereye

Zosterops lateralis

The Silvereye lives on south-western Pacific islands and in New Zealand and Australia. Six subspecies are recognized in Australia:[748,993,1124] *ramsayi* north of Mackay (21/149); *chlorocephala* on the Capricorn and Bunker Groups (23/152); *familiaris* from Rockhampton (23/150) to central Vic; *lateralis* in Tas; *halmaturina* from central Vic. west to 132°E, where the Atlases show a consistent break in distribution; and *gouldi* to the west. In the Cape York Region the Silvereye has been recorded on the mainland between the Pascoe R (12/143) and Coen (13/143); this population is probably isolated. Silvereyes on Barrier Reef islands north to about Cairns (16/145) are larger and differ in colour from mainland birds; some, on northern islands, resemble Pale White-eyes and the status and relationships of the two species are confused.[993] In WA the range of the Silvereye overlaps that of the Yellow White-eye between Exmouth (22/114) and Shark Bay (25/113),[769] historical records of the Silvereye extending north to Pt Cloates (22/113).

Movements of Silvereyes involve only parts of populations and individuals will migrate in some years and not in others.[1246] Banded birds have been recovered hundreds of kilometres from their banding place[265,653,1033,1192] but only birds from the Tas. Region have been identified as undertaking a regular seasonal migration[1530] (Reporting Rate: summer 36.7%, winter 18.2%). Part of this population winters in the South-East Region at least as far north as Murwillumbah (28/153), a bird banded near there having been recovered in Tas. at Don (41/146), 1600 km distant.[29,1026,1027] At Wollomombi (30/152) colour-banded breeding birds disappear in winter and are replaced by Tas. type birds.[129]

The Silvereye lives in eucalypt forest and woodland, mallee, heath and mangroves. It feeds on nectar, insects, earthworms and fruit. When native food is scarce, it will attack fruit in orchards and vineyards.[653,1638]

Densities were: in eucalypt woodland at Armidale (30/151) 0.03–0.88 birds/ha[582] and Moruya (35/150) 0.41–0.58 birds/ha;[1192] at Boola Boola (38/146) in eucalypt forest 0.3–1.8 birds/ha.[1113] In WA they were: Eyre (32/126) in mallee 1.3–1.5 birds/ha and Margaret R (33/115) in heath 3.5 birds/ha;[1528] in both places densities were up to four times greater after the breeding season.[464]

Field Atlas Records: 22 490

1° blocks–with records: 286 (35%)
–with breeding records: 135 (47%)

Atlas No.: 574

Max. Rep. Rate: 67%
In block(s): 34/151,38/140

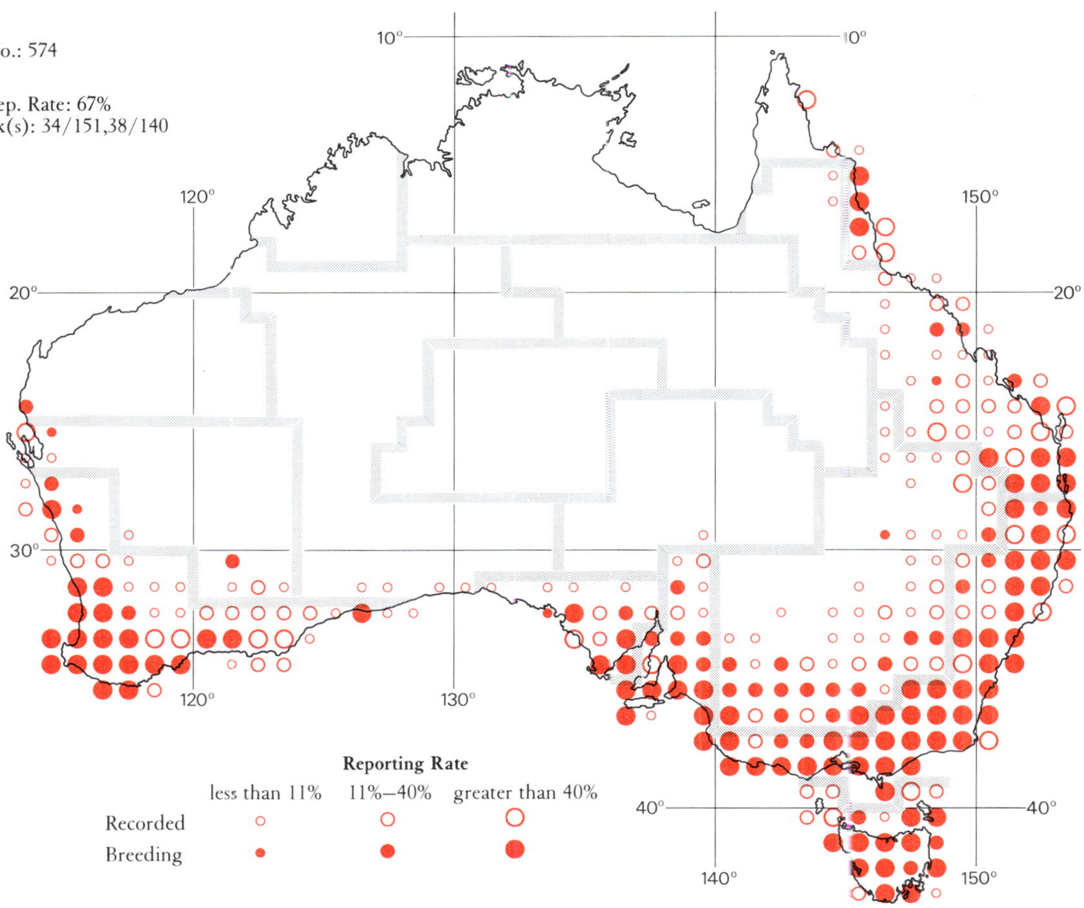

European Goldfinch

Carduelis carduelis

The European Goldfinch is native to Eurasia and northern Africa. Some may have been released in Melbourne (37/145) in 1857–58 but the first definite introduction was of 20 in 1863 near Melbourne.[1085] Similarly, birds were sent to Adelaide (34/138) in 1862 but the first release recorded was not until 1879. Other introductions followed at several centres in Vic. None is recorded in Qld or NSW but the species was established in Sydney (34/151) by 1886 and probably spread into Qld from NSW. The Atlases suggest (p. 668) that there may have been another centre of establishment in the northern South-East Region now merging with the Sydney population. Some Goldfinches arrived in Tas. in 1827 aboard ship. It is not known whether they were released but in 1902 the species was said to have been present for about 20 years.[1085] It is now absent only from undisturbed areas in south-western Tas.

The Goldfinch was liberated in Perth (31/115) sometime between 1899 and 1912 and had become common round Perth and Albany (35/117) by 1950–70. It has since declined and persists only in isolated pockets. During the Field Atlas one Goldfinch arrived in Darwin (12/130) but was shot.

The Field Atlas does not suggest large-scale seasonal movement in Australia where the longest recorded movement is by a banded adult from Berri (34/140) to Bordertown (36/140), 233 km.[1437] The species can disperse far because it colonized islands up to 700 km distant from New Zealand.[1084]

In Australia the Goldfinch is confined to areas where exotic plants have largely replaced native vegetation.[1255] Birds flock throughout the year, groups of several hundred being common in the non-breeding season. Seeds of composite weed species are its principal food and a few insects are taken. The range of the Goldfinch during the Field Atlas is similar to that for the 1951–76 period, which suggests that expansion has slowed or stopped recently. In the Wimmera round Dimboola (36/142) it is less common now than in the 1950s. Its food has been reduced by changed farming practices and the use of herbicides.[1256]

Density in eucalypt woodland at Armidale (30/150) was 0.07–0.31 birds/ha[582] and in pines at Bathurst (33/149) 0.75 birds/ha.[489] In farmland in England in 1980 it was 0.061 birds/ha.[1191]

Field Atlas Records: 14 556

1° blocks–with records: 123 (15%)
–with breeding records: 77 (63%)

Atlas No.: 996

Max. Rep. Rate: 60%
In block(s): 38/142

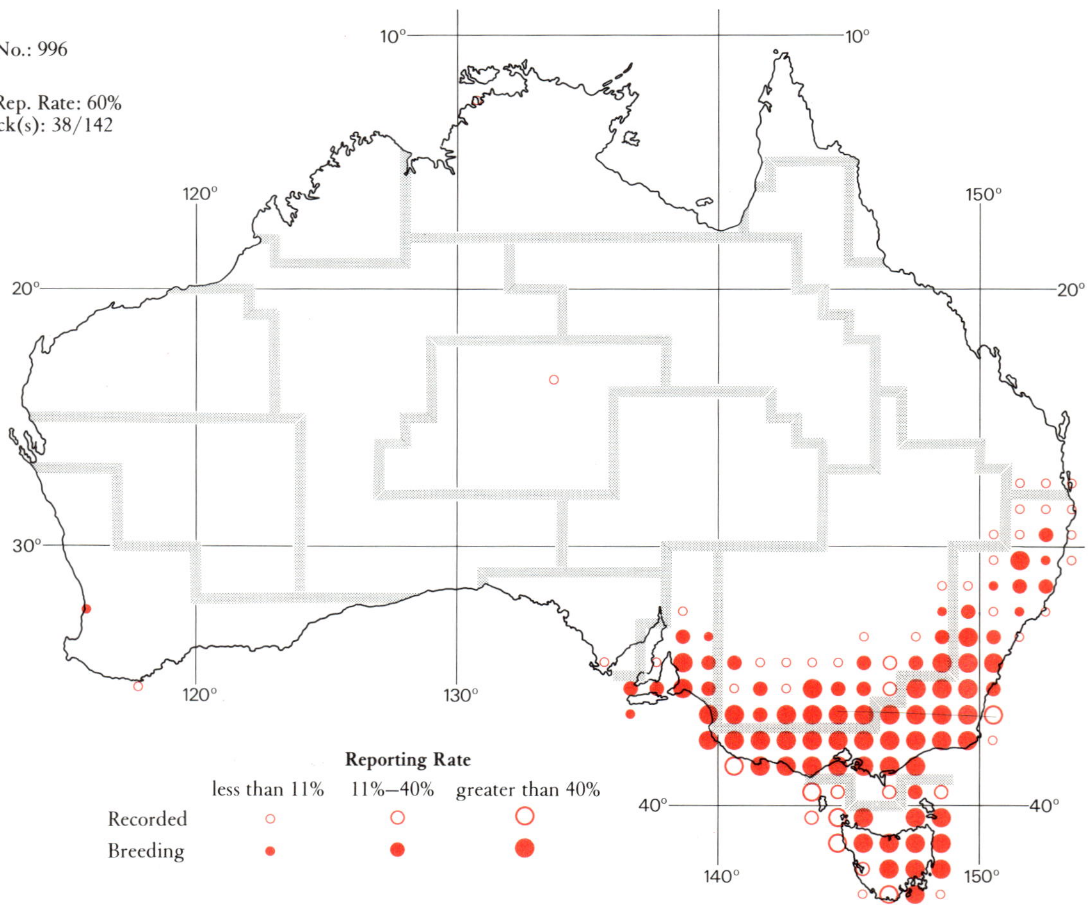

European Greenfinch

Carduelis chloris

The European Greenfinch is native to Europe, western Asia and northern Africa and was introduced to Australia but its release was poorly documented. In Melbourne (37/145) it was liberated one or more times between 1863 and 1873.[97,1553] It was first released in Adelaide (34/138) in 1879 or earlier and in Sydney (33/151) a population, presumably derived from escapees, was established by the 1890s.[343] The Greenfinch was recorded in Bathurst (33/149) and Albury (36/146) about 1910, whether from independent releases or natural spread is not known. Its arrival in Colac (38/143) in 1929[256] was part of its natural expansion of range, evidently fairly slow. The first record for Canberra (35/149) was in 1961.[653]

There is a continuous population of the Greenfinch in the lower South-East Region isolated from that in the Mt Lofty Ra. (34/138). There are also isolated populations in Sydney, Bathurst, Canberra, Albury and perhaps other urban centres where it has been reported historically but not during the Field Atlas (p. 668). In the Tas. Region the Greenfinch was first reported in 1941[734] and has spread throughout eastern Tas.; in the west it is confined to the coast.

Some movement is indicated by the observation that the species is absent from a locality near Geelong (38/144) in winter. The Field Atlas suggests no large-scale seasonal pattern.

The Greenfinch lives in farmland, coastal heath, eucalypt woodland and urban areas and is often found in conifers or native plants with similar foliage. It flocks in the non-breeding season and several pairs will often nest together in a loose colony. The species feeds on weed seeds round crops and stubble and apparently on sea rocket at Queenscliff (38/144) where large flocks gather. Its spread appears to be continuing.

There are no measures of abundance in Australia but in English farmland in 1980 density was 0.124 birds/ha.[1151]

Field Atlas Records: 2538

1° blocks–with records: 55 (7%)

–with breeding records: 23 (42%)

Atlas No.: 997

Max. Rep. Rate: 28%
In block(s): 38/142

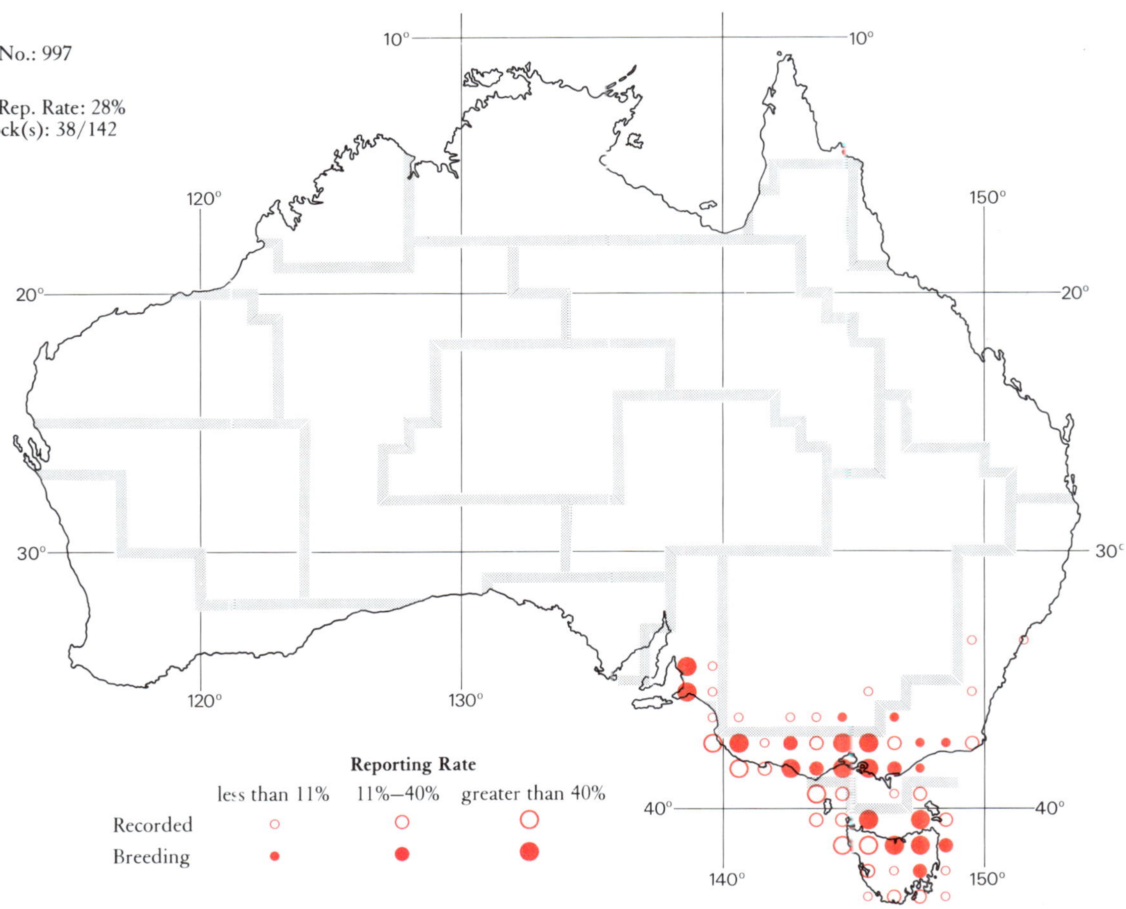

House Sparrow

Passer domesticus

The House Sparrow, native to Eurasia and northern Africa, was introduced into Australia in the middle of last century. In Melbourne (37/145) 19 were released in 1863. In the next ten years there were releases throughout Vic.; in Adelaide (34/138) in 1863, Hobart (42/147) in 1863–73 and Brisbane (27/153) in 1869–70.[1085] No release is recorded in Sydney (33/151) but by 1906 the Sparrow was well distributed in southern NSW and by 1950 had spread over the whole state (p. 668). The release in Brisbane was unsuccessful, judged by the lack of subsequent records until the NSW population had spread northward. In 1960 the Sparrow was absent north of Rockhampton (23/150) except for vagrants at Mackay (21/149).[129] By 1978 it was breeding on Torres Strait islands.[503] In the Tas. Region it spread rapidly where there was clearing but is still absent from forests.

The House Sparrow can disperse rapidly but the Field Atlas suggests no seasonal pattern. It is commensal with man and its world range closely follows human settlement. It is abundant in towns but largely absent from unsettled areas, e.g. at road houses on the Nullarbor Plain hundreds have sometimes bred from the original colonists but there are none between the settlements. The Sparrow lives in flocks and feeds on seeds, vegetable matter, fruit, insects, spiders and food scraps. It often drives native birds from nest sites.

In the USA its range expanded at 140 km/year in densely settled areas but at 8 km/year in the deserts of Utah.[949] There were similar differences in Australia. From Adelaide the Sparrow took about 90 years to reach Tarcoola (30/134) (6.7 km/year) through thinly settled country. In contrast, its spread in Qld through settled farmland was rapid. From being just north of Brisbane in 1950, it reached Mt Isa (20/139) by 1964[862] and Atherton (17/145) by 1965,[231] equivalent to minimum rates of travel of 103.6 and 85.3 km/year respectively. The westward spread of the Sparrow along the southern coast seems to have been checked by the Western Deserts. Authorities in the NT and WA are attempting to control it by destroying any birds detected.

Density in suburban Townsville (19/146) was 33.7–39.1 birds/ha[953] and in Wagga Wagga (35/147) was highest, 35.1 birds/ha, in a 20-year old suburb.[952]

Field Atlas Records: 26 980

1° blocks–with records: 330 (41%)
–with breeding records: 185 (56%)

Atlas No.: 995

Max. Rep. Rate: 82%
In block(s): 34/137

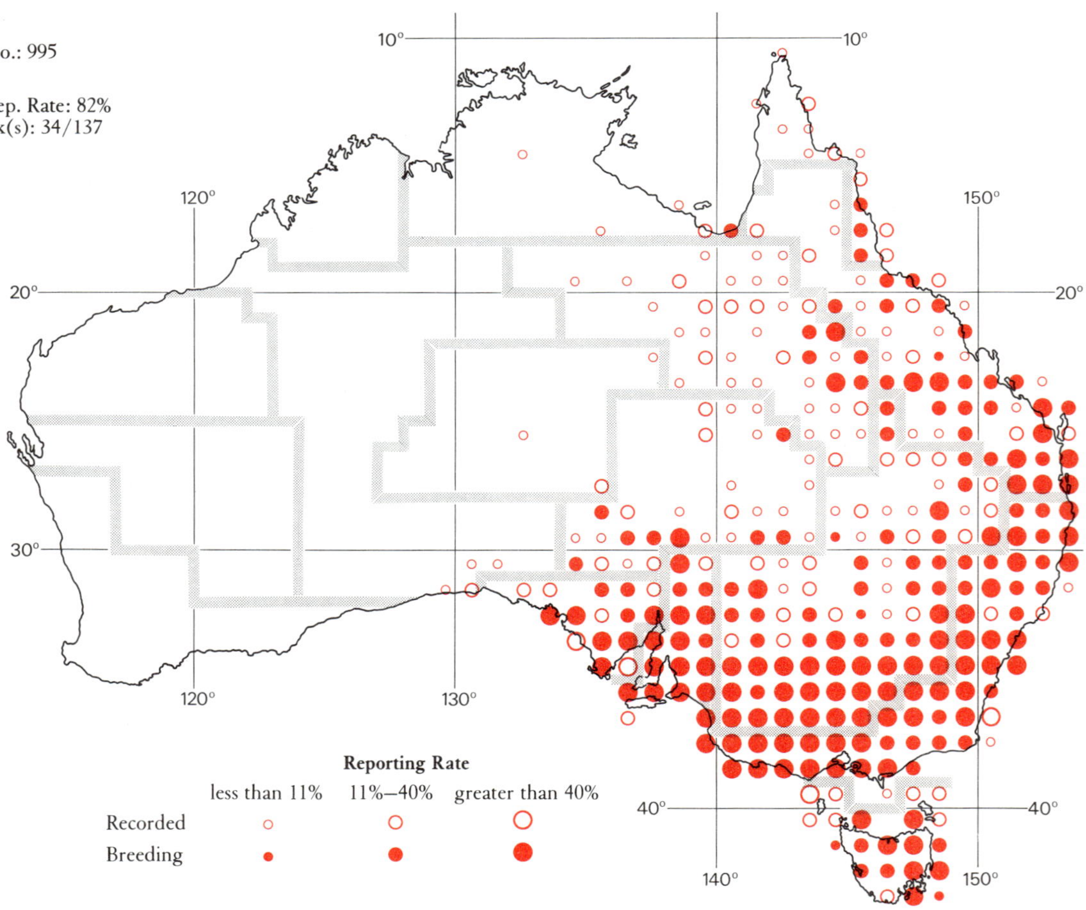

Tree Sparrow

Passer montanus

The Tree Sparrow is native to Eurasia. Birds of Chinese origin are known to have been introduced to Vic. in 1863–64[1560] but there is no information about subsequent introductions there or elsewhere apart from an unsuccessful attempt in Tas. in the 1860s. It is not always clear whether Tree or House Sparrows are being referred to in early records[1085] and the female House Sparrow is sometimes misidentified in the field as the Tree Sparrow.

The Melbourne (37/145) population spread north over the Murray R and reached Tocumwal (35/145) and Moama (36/144) in the 1950s.[833] The only other established population identified during the Field Atlas is round Cowra (33/148), where the birds were first recorded about 1950.[828] There are isolated records before 1950 from a number of places where the Tree Sparrow has not subsequently been recorded (p. 668).

Sparrows arriving on ships in northern and western Australia are likely to be Tree Sparrows, the 'house sparrow' of south-eastern Asia. Authorities are attempting to control their spread by destroying any birds detected. Two that arrived at Geraldton (28/114) in 1966 and another that arrived in Perth (31/115) in 1970 were shot[1085] as was one in Darwin (12/130) in 1981.

Overseas the Tree Sparrow does not migrate and the Field Atlas does not suggest any large-scale seasonal movement in Australia.

The Tree Sparrow lives round towns and in areas where agriculture has long been established. It sometimes breeds in colonies and usually forages in flocks, feeding on grain, fruit and insects in gardens as well as in more rural areas.

The relations of the Tree Sparrow with native species have not been studied but it probably seldom meets the few native finches that remain in areas of intense cultivation. In Asia its damage to crops has provoked active control measures but in Australia it does not appear to be a problem. In eastern Australia the bird has spread slowly[830] but in towns near the Murray R it is said to outnumber the House Sparrow.

Density in a 30-year-old suburb of Wagga Wagga (35/147) was 29.2 birds/ha.[952]

Field Atlas Records: 1136

1° blocks–with records: 20 (2%)

–with breeding records: 14 (70%)

Atlas No.: 994

Max. Rep. Rate: 27%
In block(s): 36/145

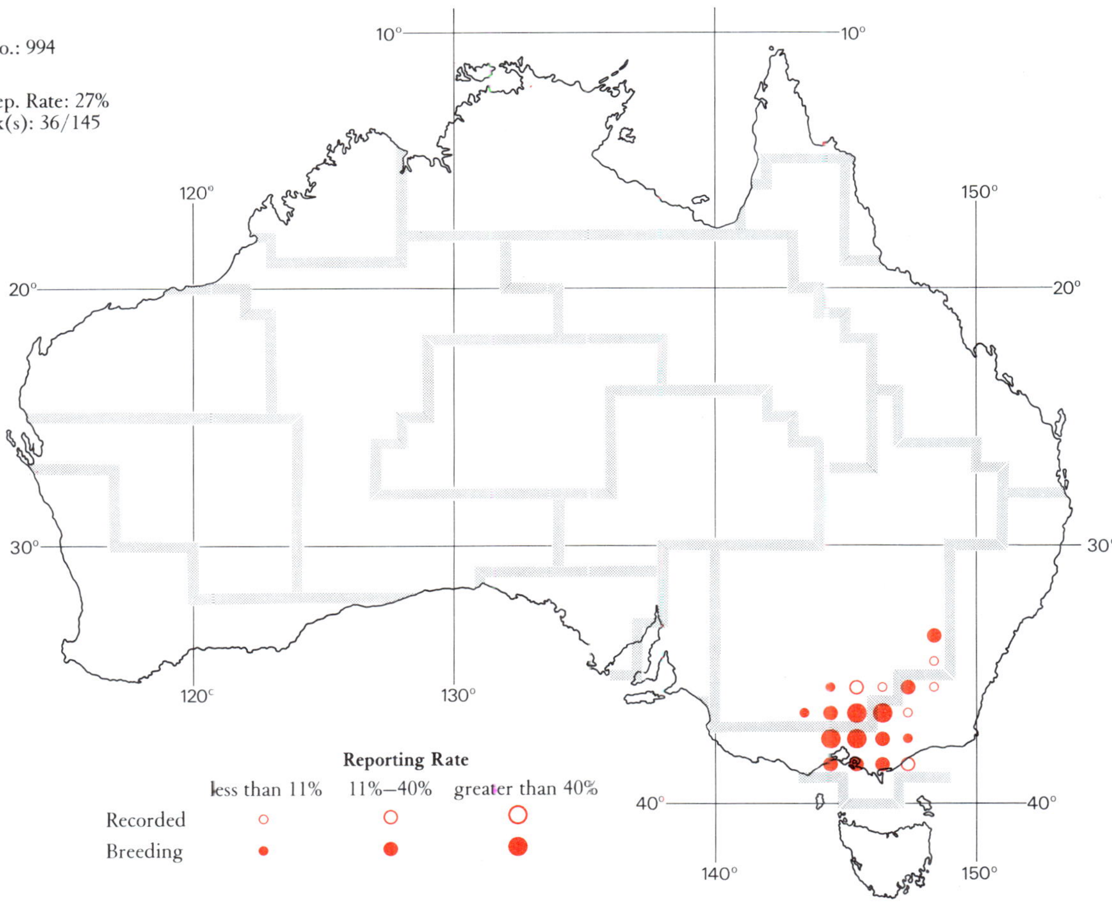

Red-browed Firetail

Emblema temporalis

The Red-browed Firetail is native to eastern Australia. It has been introduced to Tahiti and to the South-West Region.[477] Three subspecies are recognized: *minor* in the Cape York Region; *temporalis* in the Atherton, Eastern Qld and South-East Regions; and *loftyi* in the SA Ra. Region. The subspecies *minor* has been recorded throughout the Cape York Region south to the Mitchell R (16/141). The population introduced to the South-West persists in the Darling Ra. (32/116) but is spreading only slowly. Many claimed reports from other places result from confusion with young Red-eared Firetails.

The Red-browed Firetail seems to be locally nomadic when in loose non-breeding flocks.[901] Near Colac (38/143) the bird would be absent for many years then reappear in large numbers.[256] The longest movement recorded is 13 km.[85]

The Red-browed Firetail lives in clearings in rainforest,[129] as well as in eucalypt forest and woodland, heath, mangroves, gardens and orchards. It breeds close to houses in some places, particularly in orchards, but near Sydney (33/151) young born in orchards were always taken to the native vegetation lining adjacent creeks.[918,920] It has adapted itself to agricultural development and will feed on introduced weeds.[483,891] In the breeding season it is in pairs or small groups but forms flocks of up to 200–300 at other times of the year.[901] It forages in mixed species flocks.[129]

The Red-browed Firetail feeds on the seeds of small-seeded grasses, native and introduced, and many berries and other seeds including those of casuarinas. Insects are also taken.[1343] It soon began eating native sedge seeds once it became established near Perth (31/115).[477] Flocks numbered up to 300 birds at Wollomombi (30/152) but the population collapsed after two consecutive years of drought when grasses did not seed.[129] The Red-browed Firetail may have been adversely affected by competition with the introduced Nutmeg Mannikin.[657,901]

Densities in eucalypt woodland were: Armidale (30/151) 0.33–0.69 birds/ha,[582] Wollomombi 0.12–0.96 birds/ha[129] and Canberra (35/149) 0.07–0.22 birds/ha.[154] In eucalypt forest they were: Bathurst (33/149) 1.0 birds/ha[489] and Jamieson (37/146) 0.5 birds/ha.[1469]

Field Atlas Records: 12 753

1° blocks–with records: 134 (17%)
–with breeding records: 78 (58%)

Atlas No.: 662

Max. Rep. Rate: 65%
In block(s): 36/150

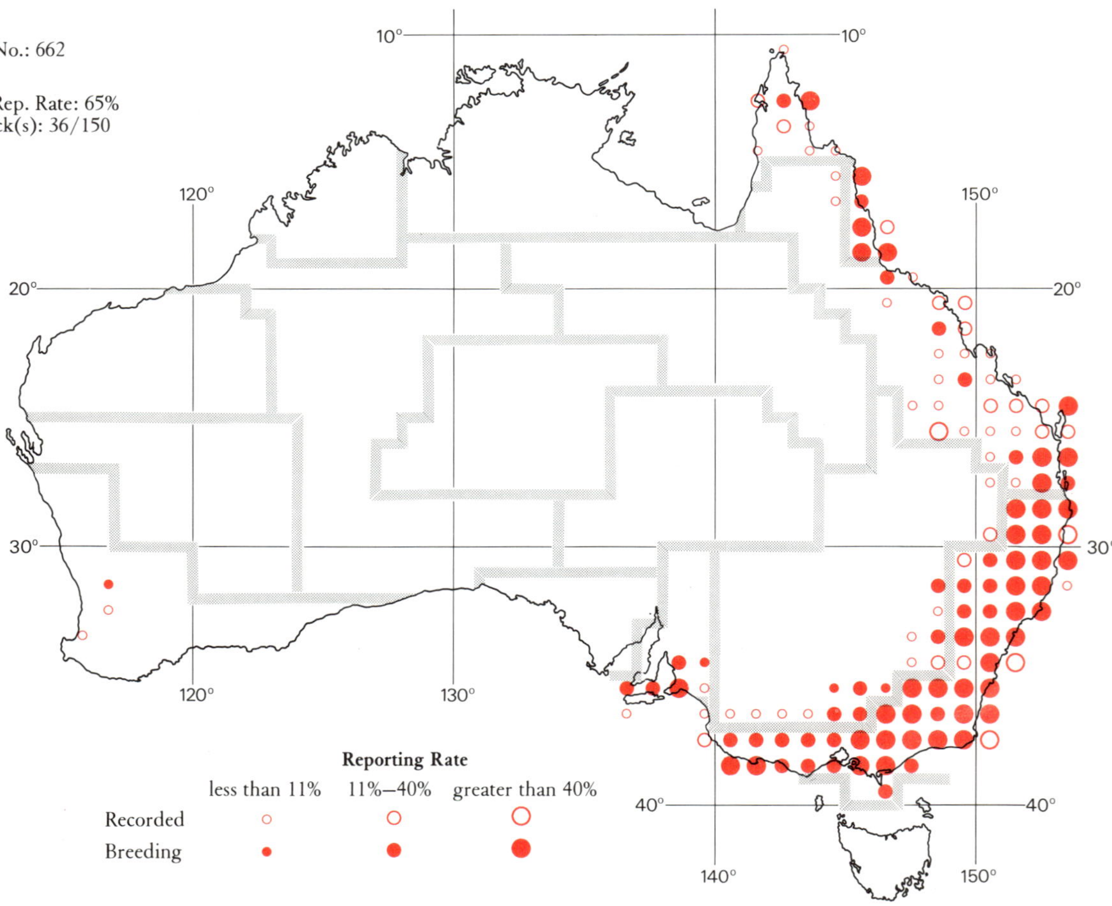

Beautiful Firetail

Emblema bella

The endemic Beautiful Firetail shows no significant morphological variation throughout its range[901] though gaps appear to isolate a number of populations round the coast of the South-East Region. These include birds between 139°E and 142°E, a population in the Mt Lofty Ra. (35/138) and one on Kangaroo I. (35/137). The species is absent from King I. (39/144).

Pairs of the Beautiful Firetail occupy a territory during the breeding season but may wander locally in the non-breeding season when groups form.[901] Flocks are rarely of more than 20 birds but at least 50 have been observed at an artificial feeding point. The Beautiful Firetail lives in thick vegetation in eucalypt forest, heath, swamps, farmland and gardens adjacent to forests. On the mainland it is usually confined to the coastal side of the ranges though there are unconfirmed historical records inland. In Tas., where it is the only native finch, it inhabits heath, woodland and forest[1486] to altitudes of at least 1000 m.[734] It forages on the ground or in vegetation eating the seeds of grass, casuarinas and other trees and shrubs.[901]

Historical records are too few to determine whether the range of the Beautiful Firetail has always been as fragmented as it now is. It is thought that there has been a decline in numbers caused by feral cats,[901] burning and clearing of heath near the coast.[1925]

The Field Atlas suggests that the species' stronghold is in the Tas. Region from which 63% of records came. There are no measures of abundance.

Field Atlas Records: 1244
1° blocks–with records: 47 (6%)
–with breeding records: 24 (51%)

Atlas No.: 650

Max. Rep. Rate: 32%
In block(s): 37/139

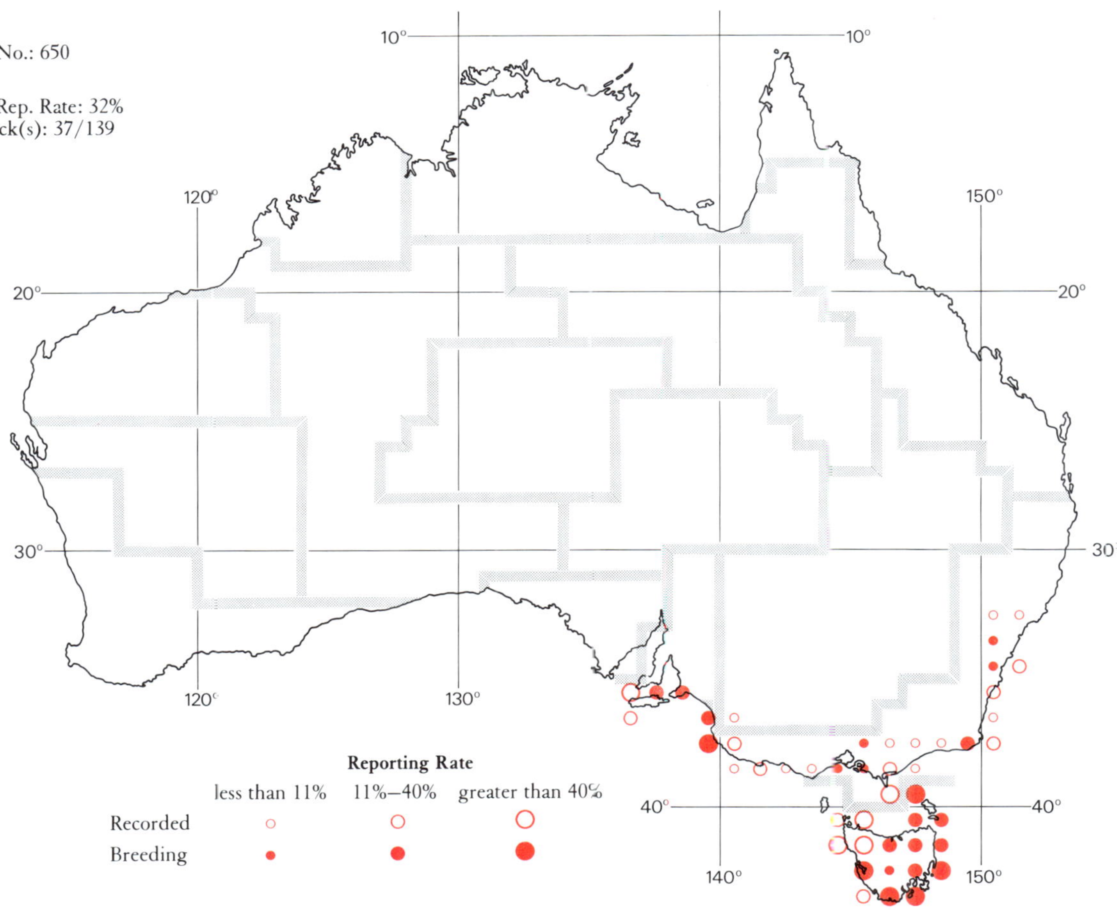

Red-eared Firetail

Emblema oculata

The Red-eared Firetail is endemic to the South-West Region. It has apparently disappeared from the coastal plain near Perth (31/115) where no sightings have been reported for at least 40 years.[1327,1622] Records were made during the Field Atlas in all other 1° blocks in which the species had earlier been reported. The population to the east of Esperance (33/121) appears to be isolated.

Each pair occupies a territory throughout the year[901] and there is no evidence in the Field Atlas of movement. The Red-eared Firetail inhabits streamside vegetation in eucalypt forest dominated by jarrah and karri as well as heath where there is thick damp vegetation such as near the coast and in the Stirling Ra. (34/117). The species lives singly or in pairs, rarely in groups of three or four and never in flocks.

During the non-breeding season it is quiet and inconspicuous but responds readily to tape recordings of its calls.[1327]

The Red-eared Firetail feeds in low cover above the ground on the seeds of grasses, sedges and related plants. It will take casuarina seeds directly from the cones and also eats some insects.[720]

Territories were estimated to be 90–180 m in diameter (1–2 birds/ha);[901] birds responded to taped calls in well-defined areas taking in 100–200 m of stream length.[1327]

Field Atlas Records: 297

1° blocks–with records: 21 (3%)
–with breeding records: 7 (33%)

Atlas No.: 651

Max. Rep. Rate: 27%
In block(s): 34/116

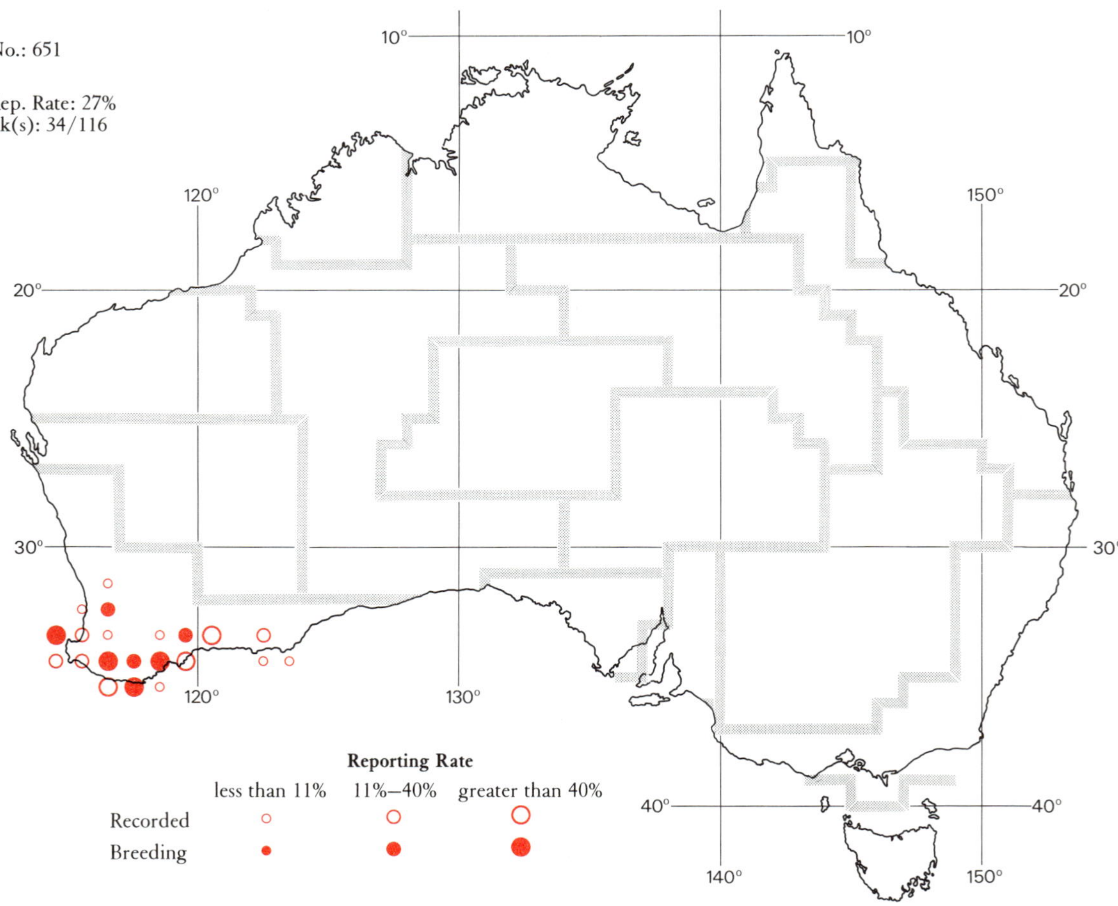

Painted Firetail

Emblema picta

No subspecific variation of the endemic Painted Firetail is recognized[901] though the Atlases suggest the Pilbara population to be isolated by the Great Sandy Desert. In the Kimberley Region there are records as far west as Broome (17/122) including one in 1968 at Mt Anderson (18/123).[748,1343] This population appears to be continuously distributed east to the Rolling Downs Region (143°E) and south to the Musgrave Ra. (26/131) in the Central Ra. Region. Outside this area are scattered records near Oodnadatta (27/135), Tibooburra (29/142) and in the SA Ra. and Eyre Pen. Regions as far south as Corunna (32/137).[964] No breeding has been recorded in any of these districts and the only one that might harbour a resident population is the northern SA Ra. Region.

The Painted Firetail sometimes irrupts to establish temporary populations beyond its usual range, e.g. at Yalgoo (28/116), Mt Hale (26/117), Laverton (28/122), Leonora (28/121) in the 1951–76 period[1638] and Woomera (31/136) in 1963.[152] The other scattered records show that groups and individuals can move far.

Normally the species lives throughout the year near permanent water in hills and gorges vegetated with acacia scrub and spinifex. Living in pairs or small groups, it rarely forms flocks even at waterholes though it sometimes joins other species to drink. Up to about 30 birds have been seen together. The Painted Firetail feeds on the ground on fallen seeds, mainly of spinifex, and requires fresh water for drinking.[901] Licensed trappers still catch some in WA for the finch trade.

There are no measures of abundance.

Field Atlas Records: 419

1° blocks–with records: 82 (10%)
–with breeding records: 12 (15%)

Atlas No.: 654

Max. Rep. Rate: 48%
In block(s): 22/118

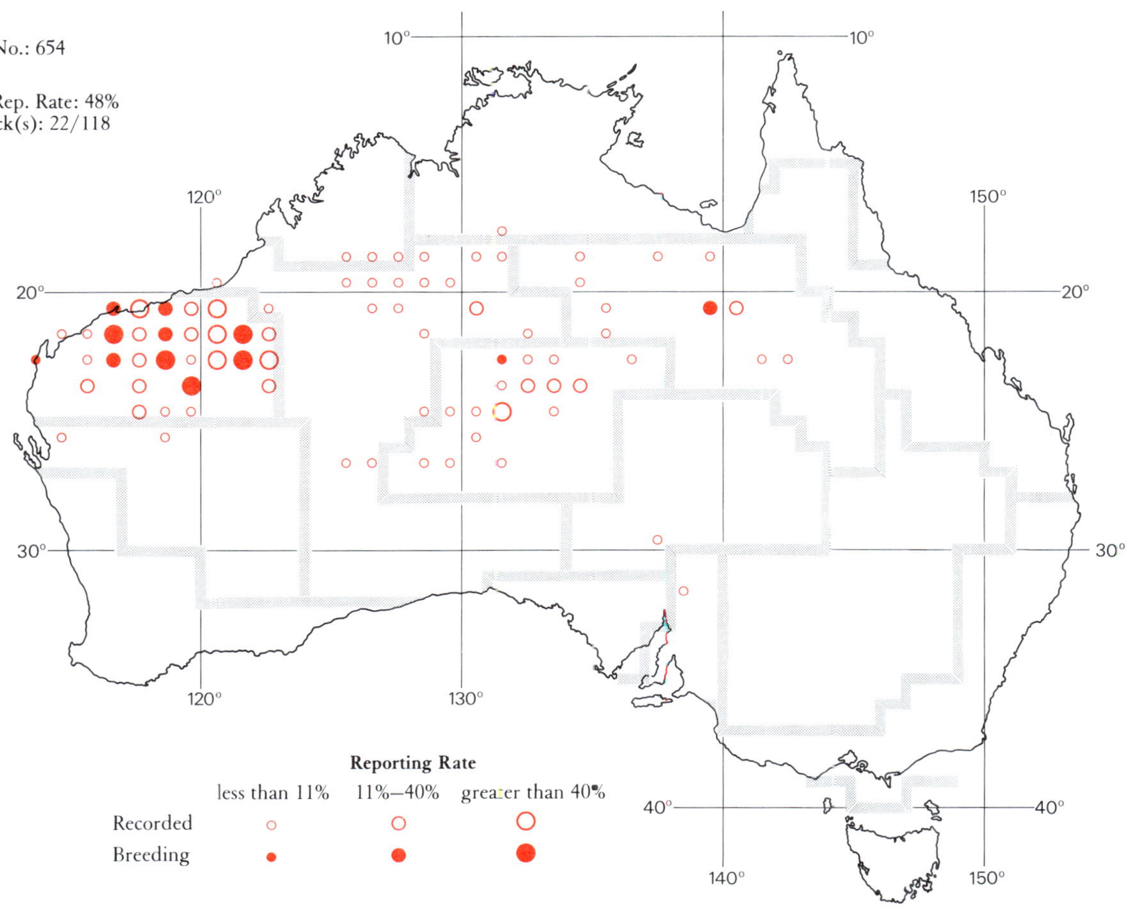

Diamond Firetail

Emblema guttata

No subspecies of the endemic Diamond Firetail are recognized.[901] The gap isolating the population in the SA Ra. and Eyre Pen. Regions in the Field Atlas is filled by historical sightings but not by breeding records. This population has been reported breeding as far north as 30°S. The Diamond Firetail appears to have declined in northern parts of its range. Between 1901 and 1950 it was recorded as far north as the Kirrama Tableland (18/145).[303] It has not since been reported farther north than 23°S. At Duaringa (23/149) it was regarded as more plentiful in the 1920s than previously[102] but there are no later records from this district. By the 1950s it had become rare at Murphy's Ck (27/152).[1099] Its numbers are believed to have decreased in Vic. also.[1358] At Jamieson it has not been seen at all since early 1977.[1469]

The Diamond Firetail has been thought to remain in the breeding area all year round[901] but in the northern Grampian Mts (37/142) there is a regular exodus to mallee farther north in winter.[1358] At Jamieson (37/146) it was regarded as a scarce breeding visitor Dec.-Mar.[1469] The Field Atlas suggests no large-scale seasonal movement.

The Diamond Firetail lives in mallee, eucalypt forest, eucalypt woodland, including that along rivers, and in farmland. The bulk of the population is on the inland side of the Great Dividing Ra. with only small pockets along the coast. Pairs breed singly or in colonies of up to 20. The species is usually in small groups which, after breeding, consolidate into flocks of 20–30 or up to 150.[720,1358] It regularly joins mixed species flocks.[129] The Diamond Firetail feeds on the ground on seeds, green plants and insects.[720,1358] Trappers have been blamed for its decline in Qld.[1099]

Density at Armidale (30/151) in eucalypt woodland was 0.25–1.18 birds/ha[582] and at Wollomombi (30/152) in eucalypt woodland 0.08–0.70 birds/ha.[129]

Field Atlas Records: 3638

1° blocks–with records: 123 (15%)
–with breeding records: 55 (45%)

Atlas No.: 652

Max. Rep. Rate: 34%
In block(s): 30/151

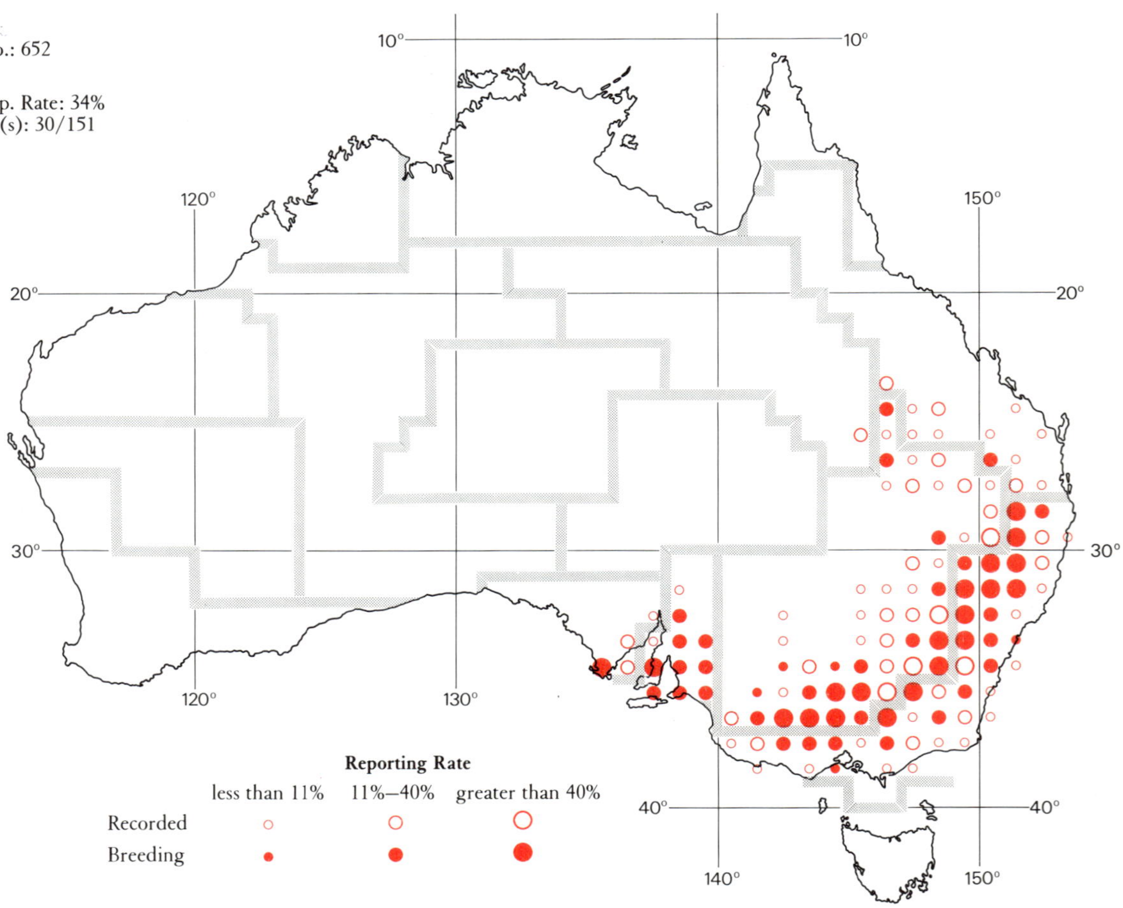

Star Finch

Neochmia ruficauda

Two subspecies of the endemic Star Finch are recognized: *ruficauda* south of about Rockhampton (23/150) and *clarescens* in northern Australia.[748,982] The population in the Pilbara appears to be isolated by the Great Sandy Desert. A flock of 20–25 birds, including immatures being fed, recorded at 33/150 in Sydney in 1980 is presumed to be of escapees.

The range of the Star Finch has decreased perhaps more than that of any other Australian finch (p. 669). The species was described in 1837 from a population 'thinly disposed' on the banks of the Namoi R (30/149)[723] but since 1900 few records have been made south of Townsville (19/146). They include an unconfirmed record in 1932 from Blackall (24/145)[1202] (omitted from the 1901–50 map); another unconfirmed sighting of about 50 birds was made in the same area, near Emmet (24/144) in Aug. 1978 after rain (omitted from map); and a group of about 20 at 29/150 near Inverell in Sept. 1964[96] (shown on 1951–76 map).

South of the Gulf of Carpentaria between Normanton (17/141) and Cloncurry (20/140), where many Star Finches were trapped last century,[1343] there have been no recent records.

There is no information about movements but the formation of flocks in the non-breeding season implies at least local movement. Flocks of 10–20, sometimes larger, are recorded[901] and one of more than 400 was reported during the Field Atlas. The Star Finch lives in eucalypt woodland along rivers where there are rushes along the banks[723] interspersed with a few bushes or low trees.[901]

The Star Finch feeds on the half-ripe and ripe seeds of grass, including spinifex, usually taken from the fruiting head.[748,901] It has now become numerous in northern irrigated farms but eats seeds of the surrounding grasses rather than those of the crop.[901] Insects, including flying ants and termites, are also eaten. Licensed trappers still catch some birds in the Kimberleys for the finch trade.

There are no measures of abundance.

Field Atlas Records: 181

1° blocks–with records: 45 (6%)
–with breeding records: 10 (22%)

Atlas No.: 663

Max. Rep. Rate: 21%
In block(s): 15/128

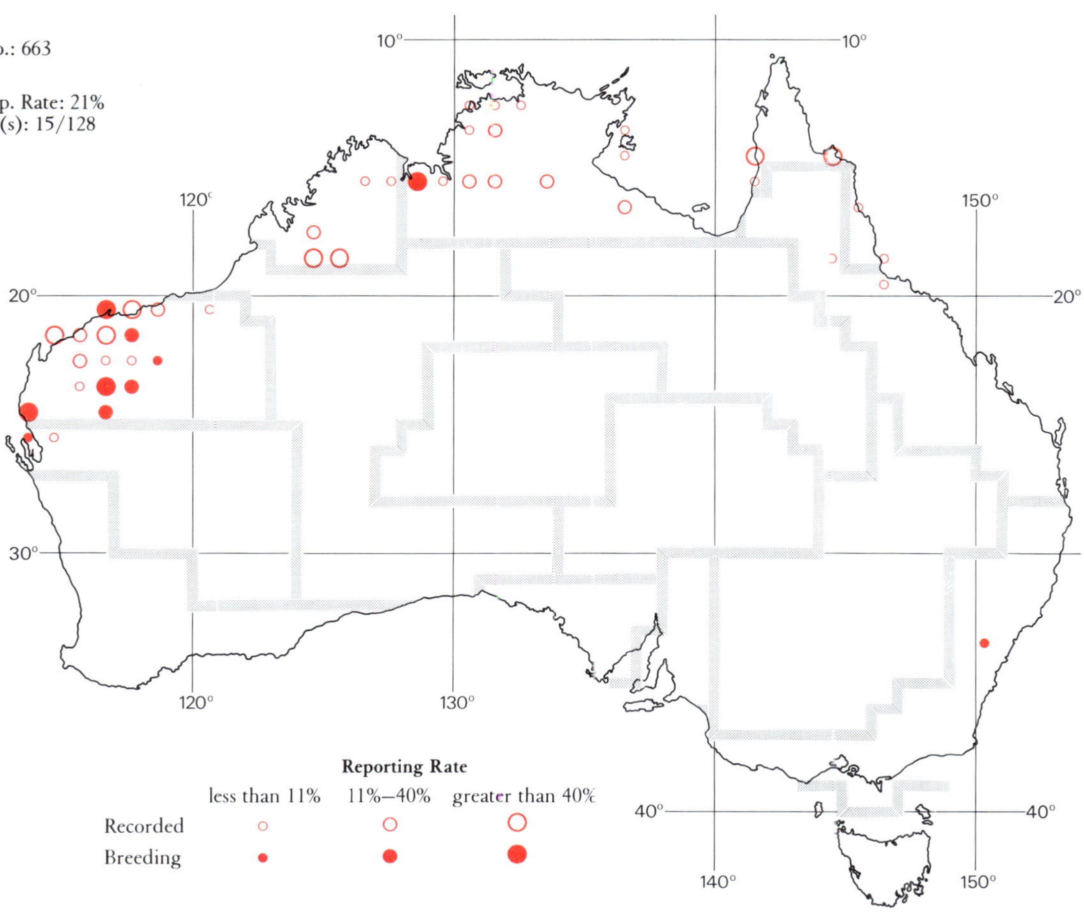

Crimson Finch

Neochmia phaeton

The Crimson Finch inhabits southern New Guinea and Australia. Three subspecies are recognized in Australia: *evangelinae* in the Cape York Region; *iredalei* in the Atherton and Eastern Qld Regions; and *phaeton* west of the Leichhardt R (140°E). A population of the subspecies *phaeton* in the Atherton Region apparently became established there only since the 1920s.[1709] On the western coast of the Cape York Region *evangelinae* has a small range from the Archer R (13/142) to the Mitchell R (15/141); on the eastern coast it has been recorded only near the Claudie R (12/143) but was not reported there during the Field Atlas. The population of *iredalei* has declined. It occurred as far south as 25/149 on the Dawson R in the 1840s but has not been recorded south of 21°S since 1900.

There is no information about movements of the Crimson Finch but flocking in the non-breeding season implies at least local movement. It is in pairs in the breeding season and in small family parties and flocks of up to 20 birds in the non-breeding season. It lives in eucalypt woodland where there is long grass, including the fringes dominated by pandanus, and in farmland such as pineapple groves and margins of canefields. It is always near a watercourse though apparently obtaining most of its water from rain drops or dew on the vegetation.[901]

The Crimson Finch feeds on the seeds of pandanus, grasses[748] and other herbaceous plants; on insects, including flying ants and termites; and on spiders.[901] It is aggressive and appears able to compete successfully with introduced species, e.g. Nutmeg Mannikin. Disappearance from its former southern range has been attributed to cattle grazing close to watercourses.[1709] Licensed trappers still catch some birds in the Kimberley Region for the finch trade.

There are no measures of abundance.

Field Atlas Records: 790

1° blocks–with records: 73 (9%)
–with breeding records: 22 (30%)

Atlas No.: 664

Max. Rep. Rate: 48%
In block(s): 16/125

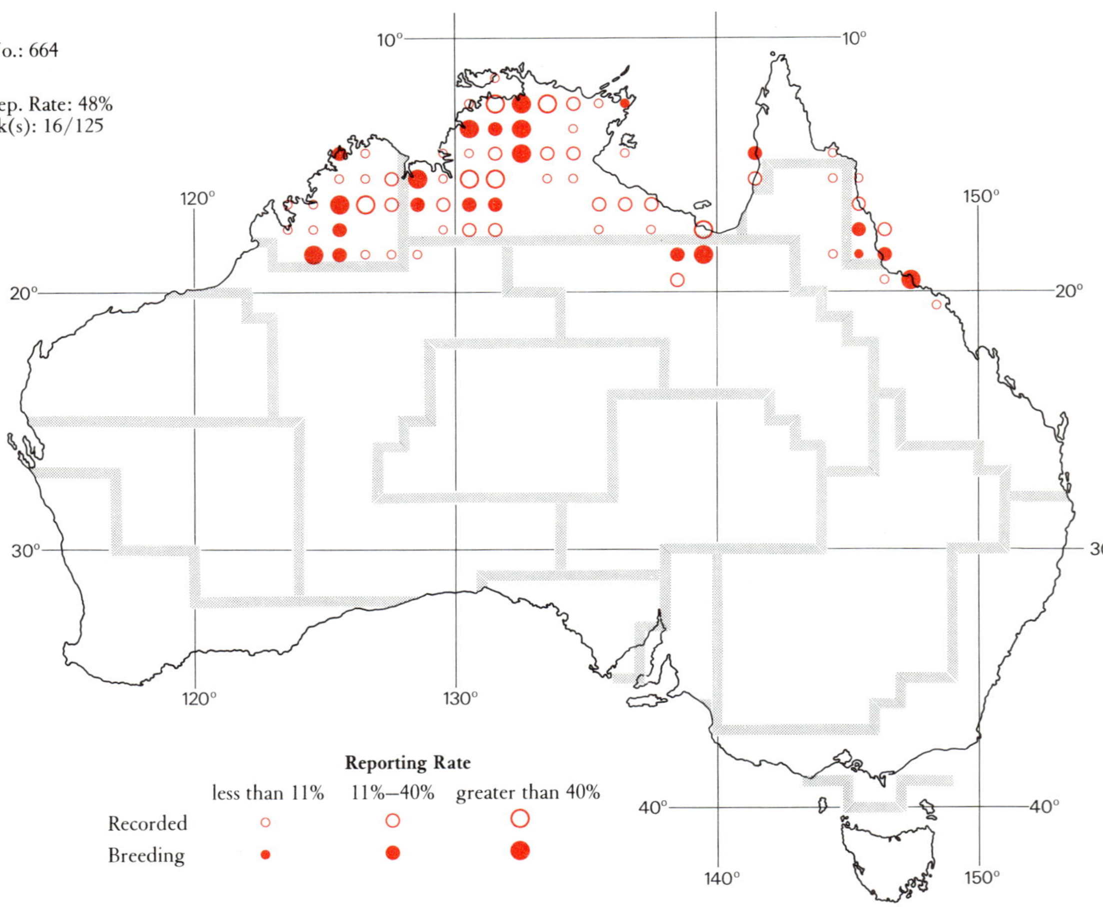

Zebra Finch

Poephila guttata

The nominate subspecies of the Zebra Finch lives in the Lesser Sundas (9S/120E) and the subspecies *castanotis* in Australia.[748,901] North of 15°S the Zebra Finch is rarely recorded and there are no breeding reports from there. The Atlases suggest some expansion of range towards the coast in the lower South-East Region. The species was introduced to Kangaroo I. (35/137) in 1937[1] but has not been recorded there subsequently.

The Field Atlas suggests no large-scale seasonal movement and the Zebra Finch is generally regarded as sedentary.[901] Only two of about 5000 birds banded at an isolated watering point in WA during ten years were recovered away from the site, both within 6 km.[456] The longest movement recorded is 24 km.[85]

The Zebra Finch lives in flocks near water in natural and cultivated grassland, needing only scattered trees or shrubs in which to nest. It usually breeds in colonies of 40–100 birds.[1925] Separate roosting nests are built. The Zebra Finch feeds on seeds of grasses, both native and introduced,[462,995] growing tips of grass shoots and insects, including flying termites.[720] Fresh water is essential, salty stock water being avoided. It has been suggested that the provision of watering points has enabled the Zebra Finch to live in places previously too far from water for it to travel each day, with a concomitant increase in the total population.[461] Young birds can breed when 90 days old and the build up of a population is fast.

During ten years the population living round one watering point at Mileura (26/117) varied between 1000 and 33 000 birds.[456] There are no other measures of abundance.

Field Atlas Records: 11 020

1° blocks–with records: 607 (75%)
–with breeding records: 272 (45%)

Atlas No.: 653

Max. Rep. Rate: 79%
In block(s): 22/118

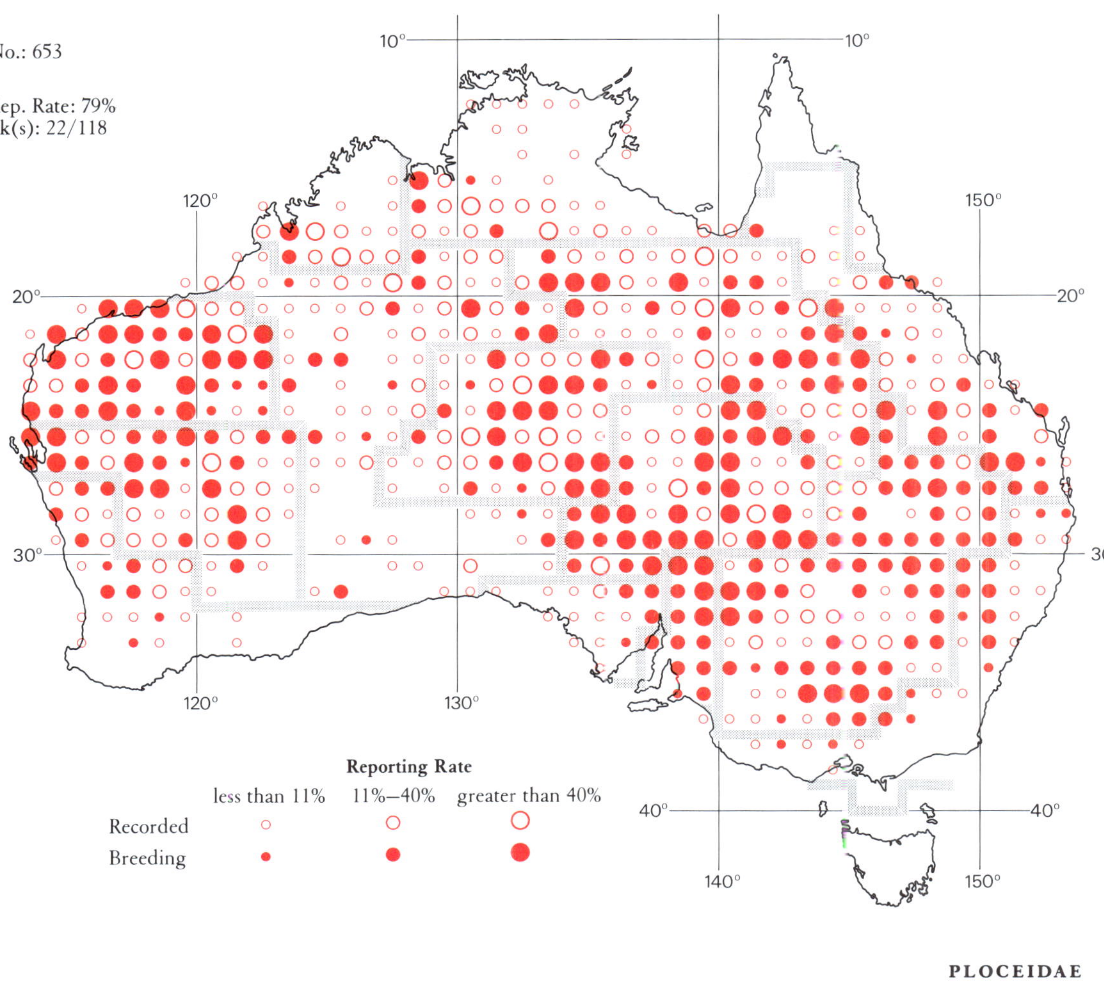

Double-barred Finch

Poephila bichenovii

Two subspecies of the endemic Double-barred Finch are recognized: *bichenovii* in the east, which has a white rump, and *annulosa* in the Top End and Kimberley Regions, which has a black rump.[982] The two interbreed in a broad area south of the Gulf of Carpentaria. The species' range has expanded towards the coast in the lower Eastern Qld and the South-East Regions since 1900 (p. 669). It has also expanded southward in the Murray-Darling and South-East Regions where, before 1900, there were no records south of 34°S. The species bred at Albury (36/146) during the Field Atlas but has not yet crossed the Murray R. Records from Melbourne (37/145) are presumed to be of escapees and are omitted from the map.

There is no information about movements of the Double-barred Finch. Observers living within its range report it to be present all year[862,1089] but at least local movement is implied by gatherings that develop at water in dry seasons.[423,901] The Double-barred Finch lives in close-knit groups, as the vignette shows, containing up to about 40 birds. It is always near water, usually in woodland with a grassy understorey but also in acacia scrub and cultivated land such as parks, gardens and canefields.[901]

The Double-barred Finch feeds mainly on the ground on the small seeds of grasses and many other plants. It is often recorded in cleared areas, which confirms the impression that it has benefited by settlement and clearing. Opinions differ as to whether it has been displaced in parts of northern Qld by the Nutmeg Mannikin.[653,901] Licensed trappers still catch some in the Kimberleys for the finch trade.

Density at Armidale (30/151) in eucalypt woodland was 0.08 birds/ha.[582]

Field Atlas Records: 7316

1° blocks–with records: 262 (32%)
–with breeding records: 98 (37%)

Atlas No.: 655

Max. Rep. Rate: 76%
In block(s): 27/147

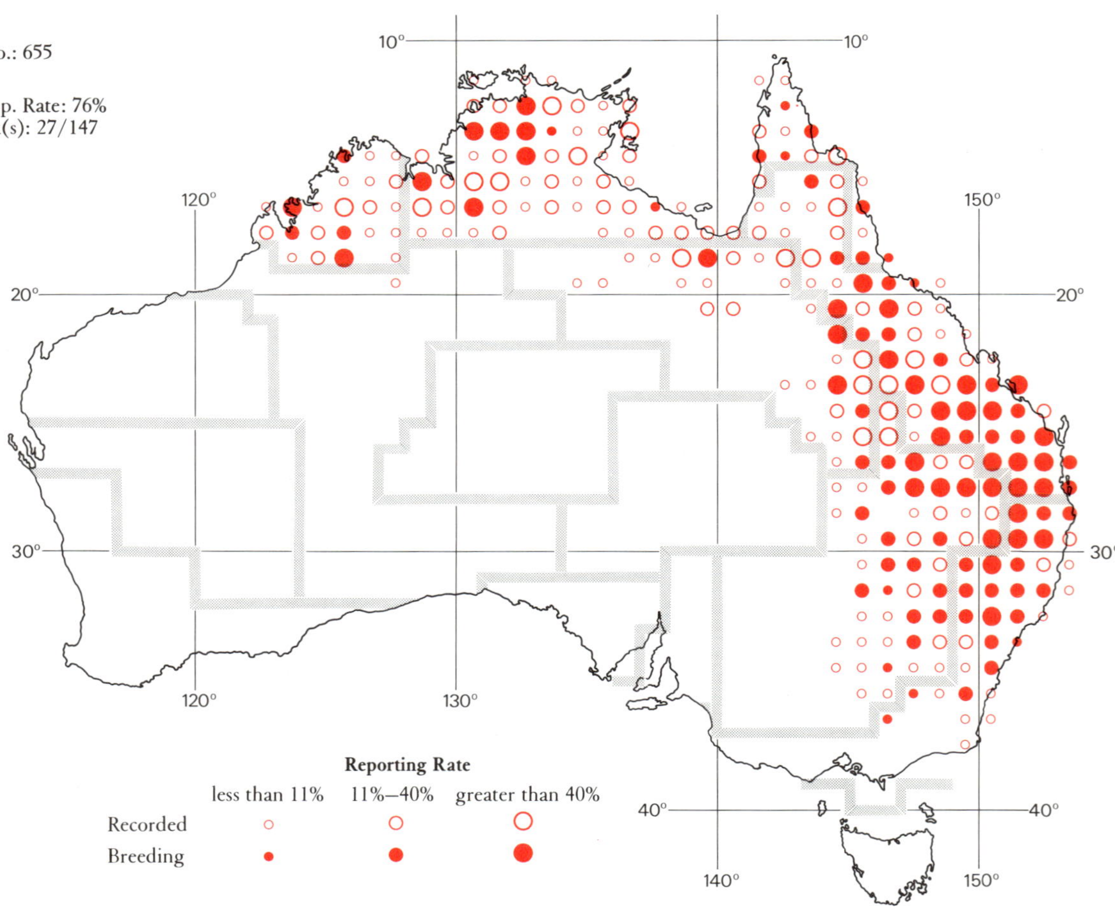

Masked Finch

Poephila personata

Two subspecies of the endemic Masked Finch are recognized: *leucotis* in the Cape York Region and areas to the south and *personata* in the Top End and Kimberley Regions.[901] The Atlas shows a consistent gap separating them at 140–141°E, south of the Gulf of Carpentaria. The species has not been recorded south of 19°S.

No movements are described but flocks of many thousands have been recorded round water during long droughts.[901]

The Masked Finch lives near water, along watercourses in eucalypt woodland where there is a grassy understorey and in nearby clearings.[982] It is usually in flocks of 20–30 birds, often with Long-tailed or Black-throated Finches. It breeds semi-colonially, though nests may be separated by 20–50 m.[475] The Masked Finch feeds on grass seeds, ripe and half-ripe, as well as insects, flies and, specially, flying termites and ants. It can drink from cattle troughs and so extend its range into dry country after the establishment of stock waters.[901] Licensed trappers still catch some birds in the Kimberley Region for the finch trade.

There are no measures of abundance.

Field Atlas Records: 593

1° blocks–with records: 60 (7%)
–with breeding records: 15 (25%)

Atlas No.: 669

Max. Rep. Rate: 48%
In block(s): 13/130

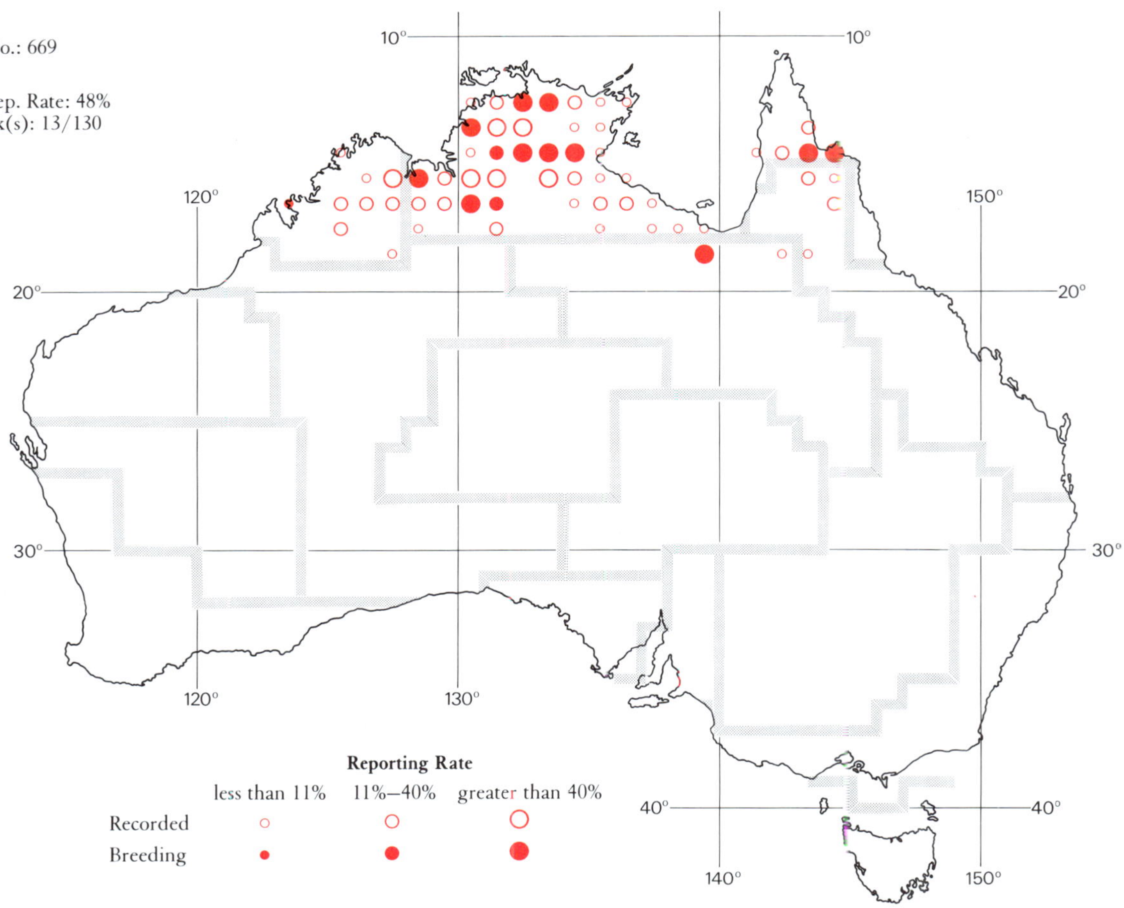

Long-tailed Finch

Poephila acuticauda

There are different views about subspecies of the endemic Long-tailed Finch.[748,901] The species is absent from Melville I. (11/130) and Groote Eylandt (13/136). On the mainland it ranges east to the Leichhardt R (140°E). A record of the closely related Black-throated Finch near Doomadgee (18/138) in Sept. 1977 suggests that the two species may be in contact.

The Field Atlas shows no evidence of large-scale seasonal movement and the species is present throughout the year at Mt Isa (20/139).[862]

The Long-tailed Finch lives in eucalypt woodland with a grassy understorey and other undisturbed sites including woodland fringe dominated by pandanus.[901] Usually it is in pairs or in flocks of 20–30 birds, often with the Masked Finch. Flocks of many thousands of Long-tailed Finches have been reported to congregate round water during extended drought.[901] It nests in loose colonies, usually one pair to a tree. It eats seeds, ripe and half-ripe. Most food is taken on the ground[901] but the bird sometimes flies upwards to catch flying ants and flies. Licensed trappers still catch some birds in the Kimberley Region for the finch trade.

There are no measures of abundance.

Field Atlas Records: 823

1° blocks–with records: 75 (9%)
–with breeding records: 16 (21%)

Atlas No.: 666

Max. Rep. Rate: 43%
In block(s): 18/139

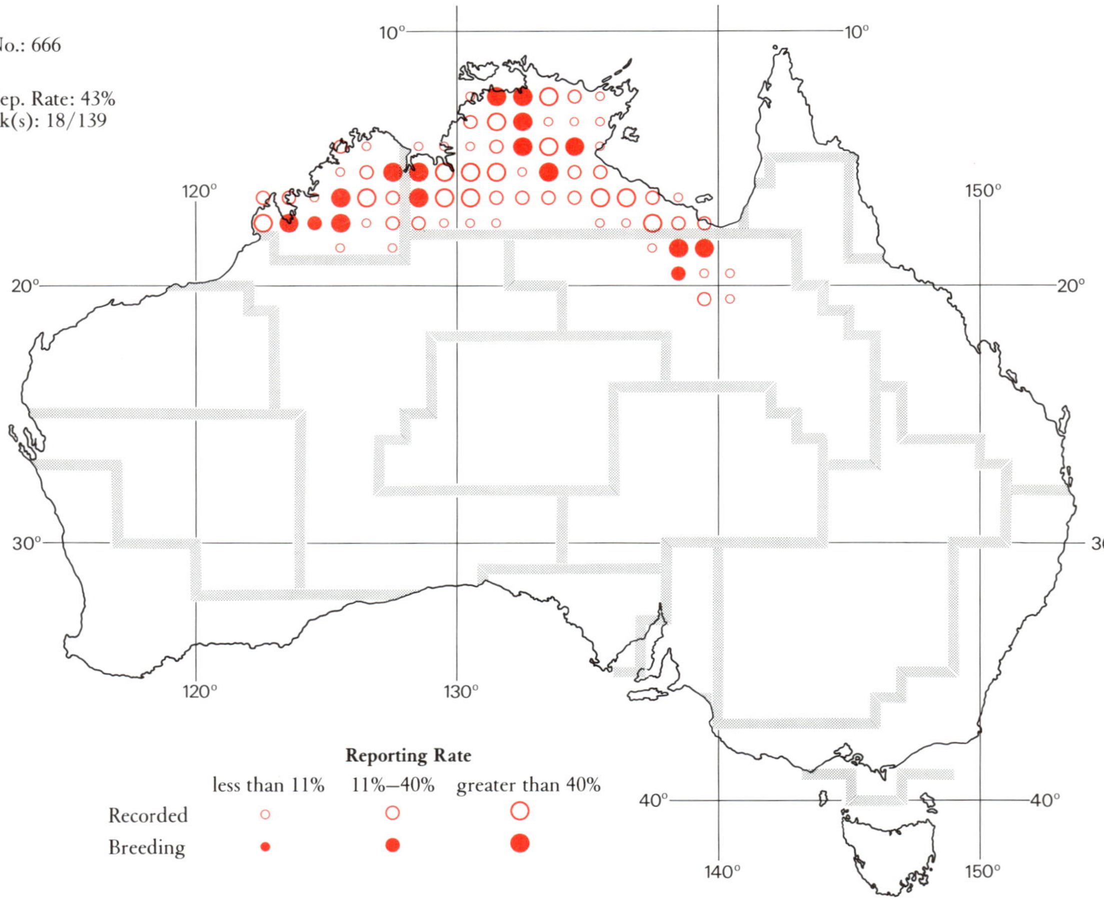

Black-throated Finch

Peophila cincta

Three subspecies of the endemic Black-throated Finch are recognized: *nigrotecta* in the northern half of the Cape York Region; *atropygialis* south to 17°S, where there are intermediates with the southern subspecies *cincta*.[1926] In 1952 when intermediates were first recorded it was thought that the range of *cincta* was extending north[982] but, by 1970, the opposite had occurred: the zone of interbreeding had moved 50 km south.[1926]

The records near Doomadgee (18/138) in Sept. 1977 and Gregory R (18/139) in Mar. 1978 suggest that the species may be in contact with the closely-related Long-tailed Finch.

The range and numbers of *cincta* have declined greatly since 1900. The subspecies was 'tolerably abundant' on the Liverpool Plains (30/148) in 1839[723] and distributed northward to its present stronghold. Since 1900 the number of records throughout this range has dwindled. The most recent NSW sightings have been at Tenterfield (29/152) in Apr. 1963,[748] about Inverell (29/151) between May 1965 and Oct. 1967 but not since and near Gilgandra (31/148) in 1968.[96]

The Black-throated Finch is probably sedentary but its erratic appearance in the few southern localities where it has recently been recorded suggests some movement. The species is highly social, living in flocks of 12–30 individuals.[1926] It inhabits eucalypt woodland with a grassy understorey, specially along watercourses, and sometimes cultivated fields where trees remain, though generally avoiding settled areas.[748,982] It nests in loose colonies.

It usually feeds on fallen seed of small size[1926] sometimes eating half-ripe seeds still in their seed heads and occasionally taking termites and other insects.[901] The causes of the species' decline in the southern half of its range are unknown.

Near Mareeba (16/145) two flocks of 21 and 30 individuals occupied nesting colonies 800 m apart.[1926] There are no other measures of abundance.

Field Atlas Records: 238

1° blocks–with records: 47 (6%)
–with breeding records: 6 (13%)

Atlas No.: 667

Max. Rep. Rate: 36%
In block(s): 18/143

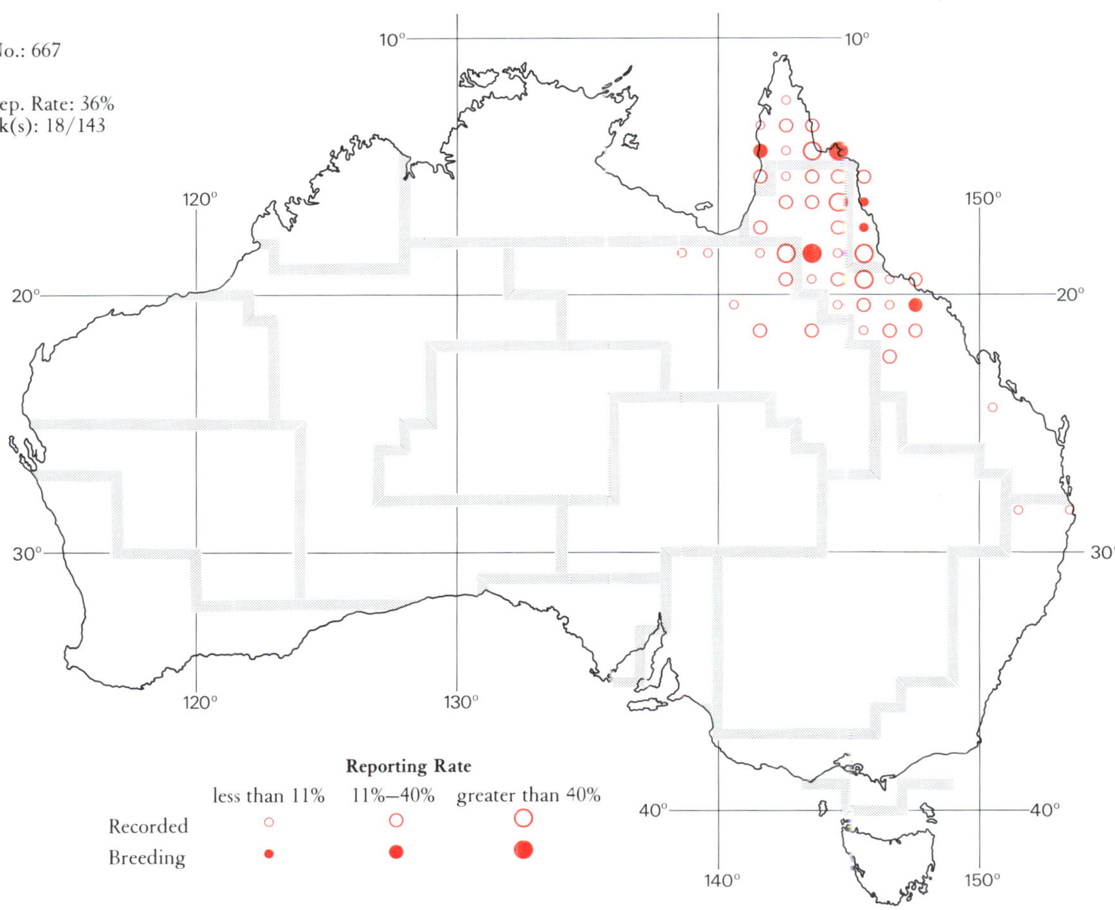

Plum-headed Finch

Aidemosyne modesta

The endemic Plum-headed Finch shows no subspecific variation.[901] It is occasionally reported as far north as Atherton (17/145), most recently in May 1965 and June 1966.[231] A small flock appeared in the far south of its range at Uriarra Crossing (35/148) in Jan. 1960. The dates of these records reflect the movements of the species: north in winter and south in summer. South of 30°S Reporting Rates were 4.3% in summer and 1.0% in winter. There appear to be nomadic movements also, the bird being present irregularly in many areas.[901]

The Plum-headed Finch lives usually in pairs or small flocks in eucalypt woodland where there are watercourses and swamps with tall vegetation.[102,1025,1343] It eats seeds on or near the ground and will take them from fruiting heads also.[813,901] The bird is said to drink at least every hour.[901] Because it was common round Townsville (19/146) in 1932 and not subsequently it is sometimes thought to have declined.[901] An irruption in winter 1972 brought at least 500 birds to Townsville Common, suggesting that the earlier record may have coincided with a similar invasion. In the Murphy's Ck (27/152) district the species was temporarily abundant and nested where maize, millet and sorghum were grown.[1099] Competition with the introduced Nutmeg Mannikin is thought to have adversely affected the Plum-headed Finch in parts of northern Qld but this inference has been disputed.[657,901] The Atlases do not indicate any decrease in the range of the Plum-headed Finch though numbers may have changed.

There are no measures of abundance.

Field Atlas Records: 760

1° blocks–with records: 93 (11%)
–with breeding records: 18 (19%)

Atlas No.: 661

Max. Rep. Rate: 36%
In block(s): 24/148

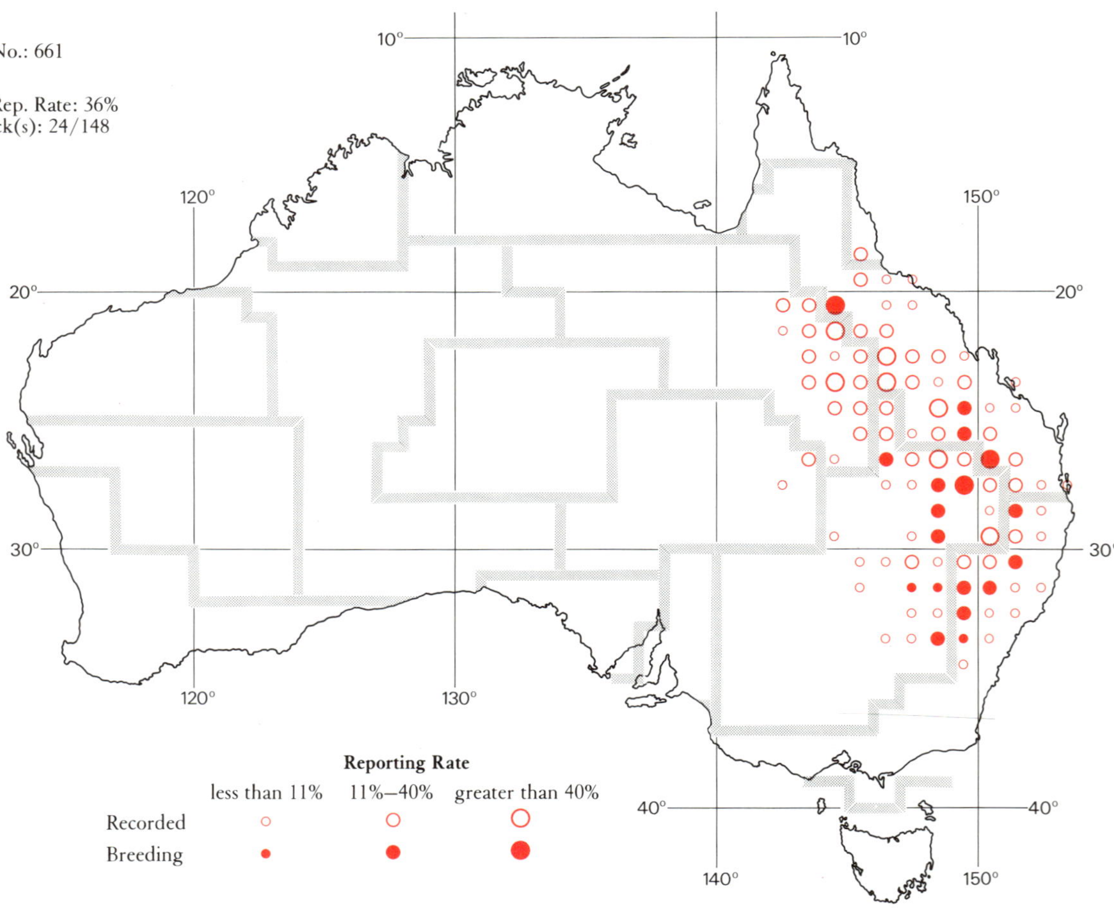

Pictorella Mannikin

Lonchura pectoralis

There is no evidence of geographical variation in the endemic Pictorella Mannikin,[901] though the Field Atlas suggests that there may now be two isolated populations. Records before 1950 from the Barkly Tableland (the western portion of the Rolling Downs Region) fill the gap indicating that the species' range was once probably continuous. At the turn of the century the species bred as far east as Homestead (20/145) but was already becoming rare there in 1904.[1673] The Atlases contain no records from the district after 1906. The Field Atlas sighting from Edward R (14/141) is a northward extension of range in the Cape York Region; flocks of about 20, including birds in juvenile plumage, were seen there in May 1980.

The Pictorella Mannikin moves inland during the wet summer season. At Kununurra (15/128) flocks of up to 200 congregate round water as the dry winter season progresses. They disperse when the rains come, leaving only pairs and small flocks.[901] In the Darwin (12/130), Mt Isa (20/139) and Richmond (20/143) districts large numbers of Pictorella Mannikins are seen only infrequently.[168,423,862] The Field Atlas suggests a movement of the population to coastal districts in the dry winter season with no records at the end of the dry season (spring) south of 19°S in the east and 17°S in the west.

The Pictorella Mannikin lives in acacia scrub where there is a grassy understorey and in spinifex.[1661] It is usually near watercourses.[982] It visits irrigated crops on the Ord R (15/128) but tends to avoid human habitation.[901] It feeds on the ground on the seeds of grasses, occasionally those of rice, and takes insects, including ants and flying termites.[901] During the Field Atlas a small group of birds was observed to feed on green algae from a shallow pool.

The bird's sources of food may be damaged by frequent fires. The country has been increasingly burnt with the development of cattle stations and this may be responsible for the reduction in range. Licensed trappers still catch some birds in the Kimberleys for the finch trade.

There are no measures of abundance.

Field Atlas Records: 107

1° blocks–with records: 44 (5%)
–with breeding records: 1 (2%)

Atlas No.: 659

Max. Rep. Rate: 14%
In block(s): 15/128

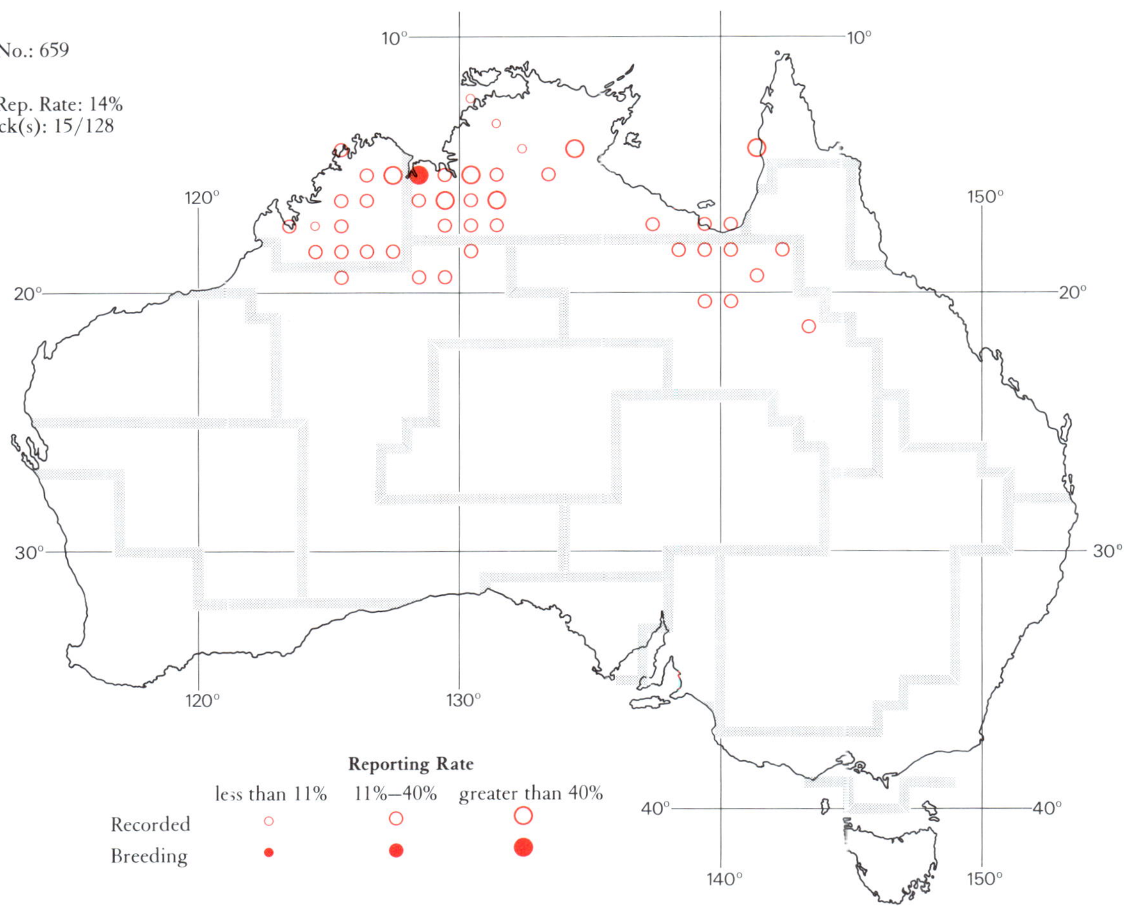

Chestnut-breasted Mannikin

Lonchura castaneothorax

The Chestnut-breasted Mannikin is native to New Guinea and adjacent islands and Australia; it was introduced to New Caledonia (21S/165E), Tahiti (17S/149W) and the Society Is (17S/153W). In Australia two subspecies are provisionally recognized: *castaneothorax* east of the Gulf of Carpentaria and *assimilis* to the west.[901] Historical records south of the Gulf of Carpentaria, e.g. Moonlight Ck (17/139),[748] suggest that there is no gap between them. The species' range in the Kimberley Region extends south-west of that mapped, with a record from Mt Anderson (18/123) in 1968. Presumed excapees have been seen in Melbourne (37/145) since 1967 and probably breed.[917] A population in Perth (31/115) is expanding slowly and presumably is also breeding.

Where the Yellow-rumped Mannikin also lives, the two species form mixed flocks that roost and breed together, producing hybrids that are probably fertile. Near Kununurra (15/128) the proportion of Chestnut-breasted to Yellow-rumped is 2:1.[901]

The Field Atlas shows no evidence of large-scale seasonal movement. At Murphy's Ck (27/152) the bird arrives occasionally for short periods but does not breed;[1099] at Darwin (12/130) many appear in some years and few in others.[423] Nomadism has been reported in the Kimberley Region taking the birds farther inland in the wet summer season.[901] The longest movement recorded is 50 km in New Guinea.[1441]

The Chestnut-breasted Mannikin lives in swamps, reed-beds, mangroves, canefields and irrigated crops of rice, sorghum and millet.[901] It breeds colonially and in the non-breeding season forms flocks of several hundreds. The bird feeds from the heads of grass in seed, rarely taking seeds from the ground. Early in the breeding season it will catch termites. In northern Australia the species has benefited from the cultivation of irrigated crops and may have increased.[901] Opinions differ whether the spread of the introduced Nutmeg Mannikin has caused a decline of the Chestnut-breasted in northern Qld.[657,901] Licensed trappers still catch birds in the Kimberley Region for the finch trade.

Density in New Guinea in eucalypt woodland near Port Moresby (9S/147E) was 0.04 birds/ha in the breeding season to 0.8 birds/ha when flocking in the non-breeding season.[158]

Field Atlas Records: 1736

1° blocks–with records: 109 (13%)
–with breeding records: 30 (28%)

Atlas No.: 657

Max. Rep. Rate: 47%
In block(s): 19/147

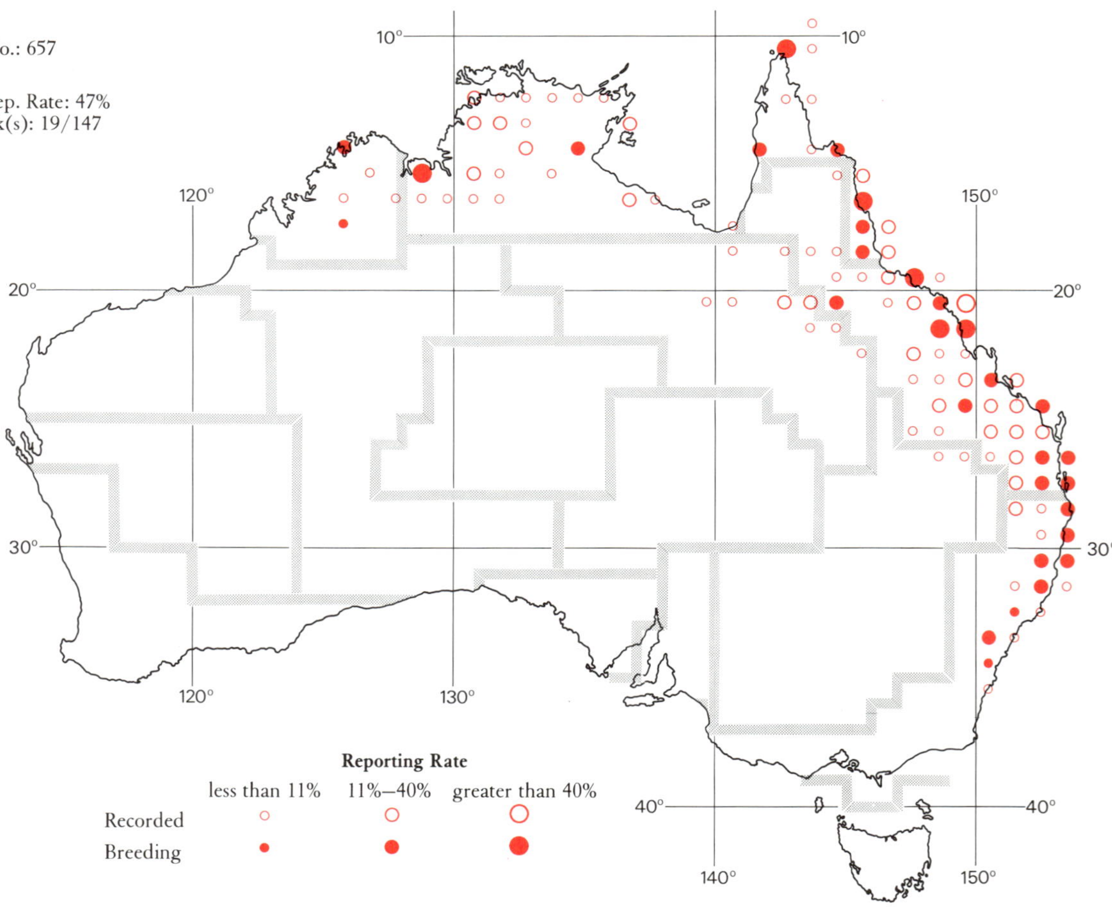

Yellow-rumped Mannikin

Lonchura flaviprymna

The endemic Yellow-rumped Mannikin feeds, roosts and nests with the Chestnut-breasted Mannikin and the two species have produced hybrids, probably fertile.[901] About 10% of the pairs observed at Kununurra (15/128) were mixed and the proportion of Yellow-rumped to Chestnut-breasted was 1:2.[901]

Nomadism has been reported near Kununurra[901] and, round Darwin (12/130), the Yellow-rumped Mannikin appears in good numbers in some years and low numbers in others.[423] It is said to move to the coast in very dry seasons.[982] At Kidman Springs (16/130) pairs and flocks of up to seven are present only in the wet summer season.

The Yellow-rumped Mannikin lives in flocks in swamps, reedbeds, mangroves and farmland. It feeds on half-ripe and ripe seeds of at least five species of grass.[901] It is trapped illegally in the Kimberley Region.

There are no measures of abundance.

Field Atlas Records: 67

1° blocks-with records: 12 (1%)
-with breeding records: 1 (8%)

Atlas No.: 658

Max. Rep. Rate: 15%
In block(s): 15/128

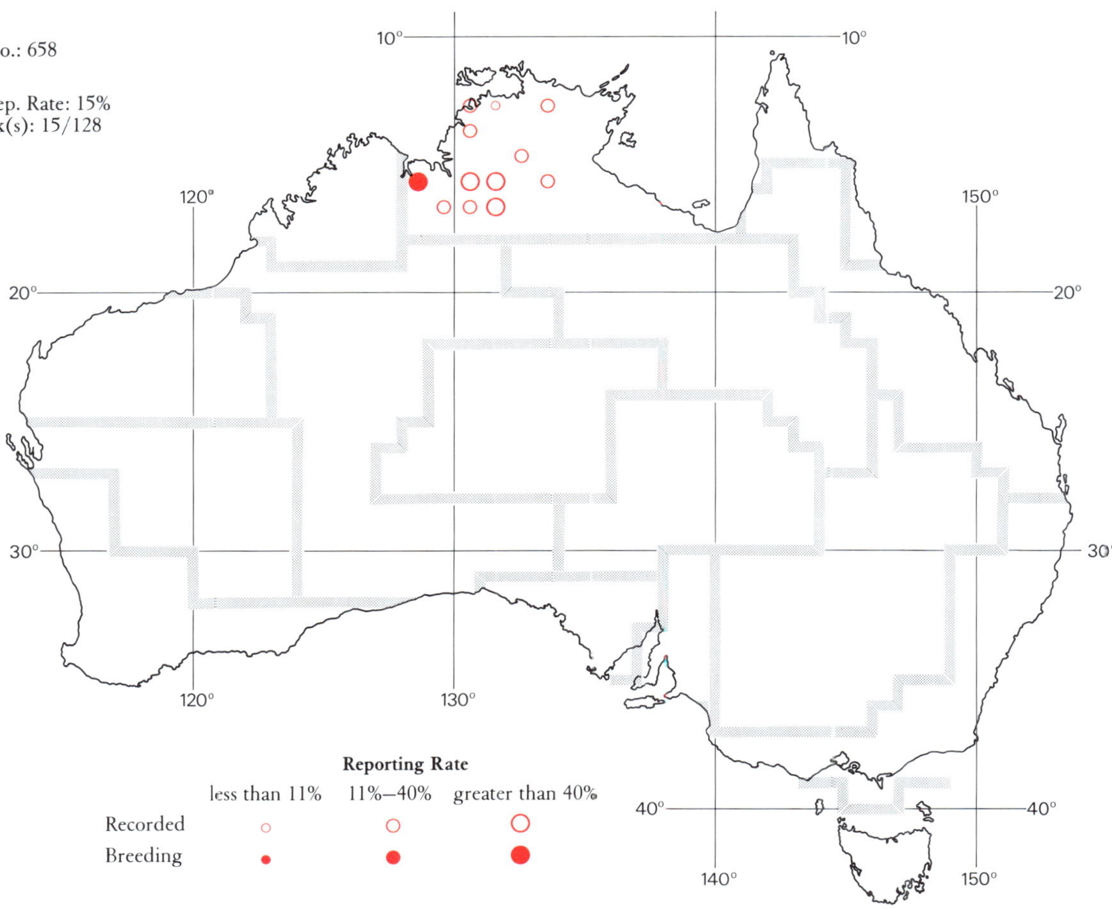

Nutmeg Mannikin

Lonchura punctulata

The Nutmeg Mannikin, a native of south-eastern Asia, has been introduced or escaped in several centres along the eastern Australian coast (p. 669). The earliest Atlas record is from Brisbane (27/153) in 1939; that population, well-established by the 1940s, is thought to be derived from birds that escaped about 1930.[1085] The Atlases and other sources suggest at least three other independent releases: Townsville (19/146) in 1950;[1085] Rockhampton (23/150), where the first Atlas records were in 1954; and Sydney (33/151), where the species was certainly established by 1947.[1133,1734] If, as dated records suggest, Nutmeg Mannikins in the Atherton and adjacent regions are all derived from the Townsville releases, spread has been rapid. By 1960 the species was well established in Townsville, Innisfail (17/146) and Cairns (16/145) and in the country between them[901] and had reached Cooktown (15/145) by 1961.[1085] To the south it was in Mackay (21/149) by 1959–60.[1085] The Rockhampton, Brisbane and Sydney populations seem to be expanding more slowly. Birds, presumably from Brisbane, had reached Grafton (29/152) by 1960[722] but have not since moved much farther south.

The Nutmeg Mannikin generally inhabits disturbed areas near settlements, including canefields and farmland in the Atherton Region, and seems to be rare in undisturbed native vegetation. It will live in the introduced lantana.[135] Because this is such a common weed of many otherwise native forests, it may allow the bird access to forest areas. The Nutmeg Mannikin often nests colonially. It feeds almost entirely on seeds,[901] half-ripe and ripe, and will peck at carcasses of animals killed on roads, probably taking 'meat meal' partly cooked on the warm, black roads.

In Qld the Nutmeg Mannikin breeds for 11 months of the year.[1047] This capacity for increase matches its rapid spread in some areas. At Murphy's Ck (27/152) four were seen one year, a small flock the next and about 50 in the third year.[1099] The species has been most successful in the Atherton Region where the warmer climate stimulates a longer growing season for grasses. Opinion differs as to whether the Nutmeg Mannikin has displaced native finches in some places.[657,901]

There are no measures of abundance.

Field Atlas Records: 1004

1° blocks–with records: 41 (5%)
–with breeding records: 19 (46%)

Atlas No.: 983

Max. Rep. Rate: 31%
In block(s): 17/146

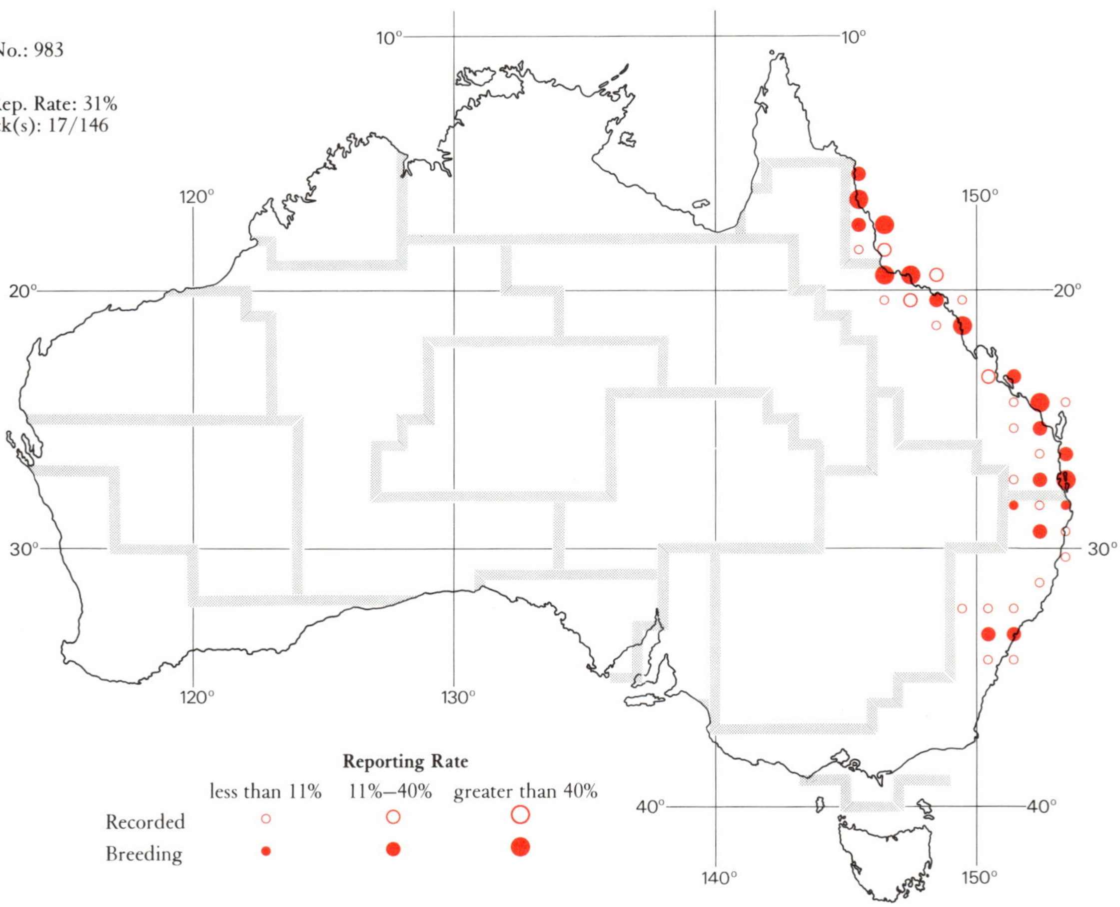

Blue-faced Finch

Erythrura trichroa

The Blue-faced Finch lives in Indonesia, the Pacific islands and New Guinea, with a small population in Australia.[901] The few Australian birds collected are regarded as synonymous with the New Guinean subspecies *sigillifera*. There is a single published report from the Cape York Region: Lloyd I. (12/143) in 1914;[1147] all other Australian records are from the Atherton Region between Helenvale (15/145) and Ravenshoe (17/145). Since 1950 there have been no records south of 17°S.

Resident observers in the Atherton Region see the species throughout the year, suggesting that it is sedentary, but the record from Lloyd I. has been used to suggest that there may be migration between New Guinea and Australia.

The Blue-faced Finch inhabits rainforest from sea level to at least 1000 m[801] moving at times into clearings and adjacent eucalypt forest. One sighting was close to school buildings.[212] The bird on Lloyd I. was near mangroves. In New Guinea the species is only a vagrant to the lowlands but is common in the mountains where it is seldom seen though often caught in mist-nets.[150] In Australia the species is usually seen singly and sometimes in pairs or small groups of 4–6 in thick foliage where it mostly forages. The Blue-faced Finch gathers much of its food in rainforest trees but it has also been recorded feeding on seeds of casuarinas and of grasses including the cultivated Guinea grass. Seeds and insects have been found in its guts.[748] The Blue-faced Finch is not often seen even by resident observers familiar with the species. It probably has not declined except perhaps where rainforest has been cleared at the southern end of its range.

There are no measures of abundance.

Field Atlas Records: 31
1° blocks–with records: 2 (1%)
–with breeding records:

Atlas No.: 660

Max. Rep. Rate: 17%
In block(s): 15/145

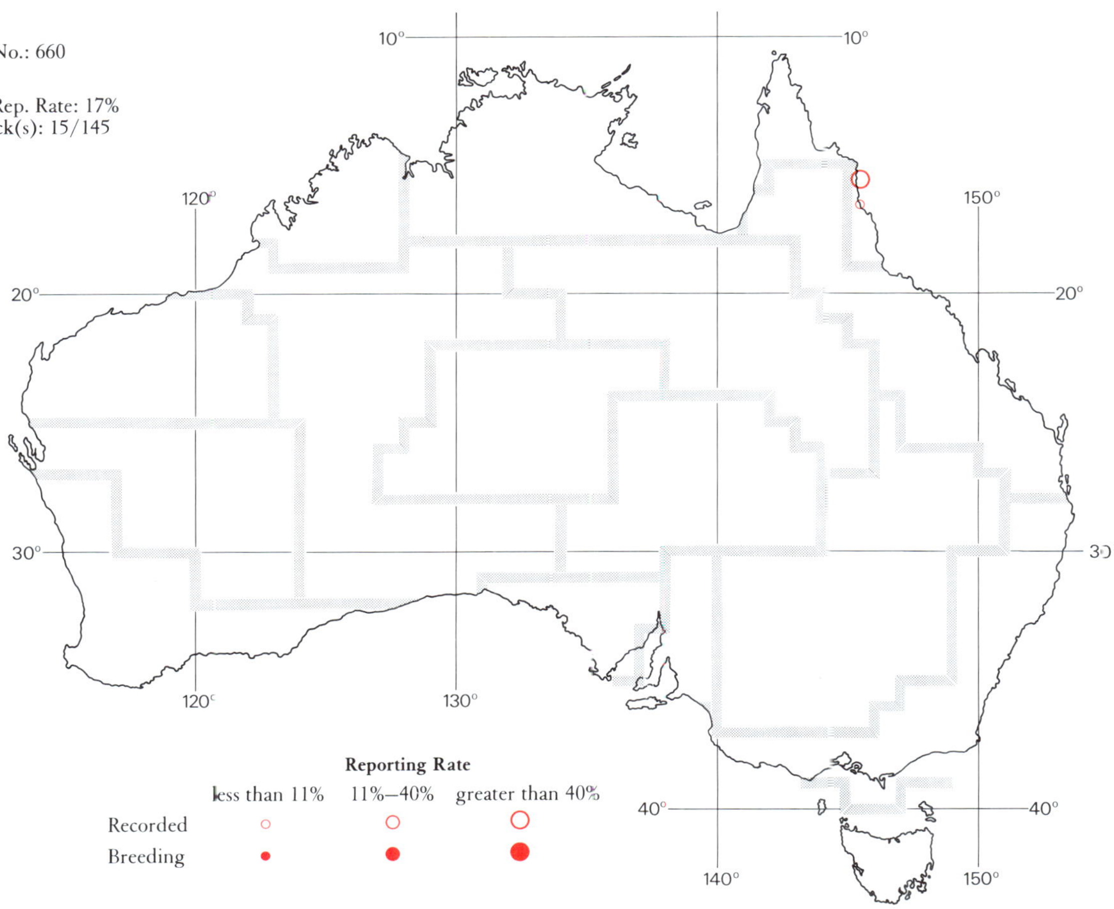

Gouldian Finch

Erythrura gouldiae

Both range and numbers of the endemic Gouldian Finch have declined, specially populations south and east of the Gulf of Carpentaria. Before 1900 the species was regularly reported as far south as Homestead (20/145)[1673] and was in small numbers in the Cloncurry (20/140) district.[1143] In the Cape York Region it ranged as far north as the Archer R (13/142).

Since 1950 there have been reports only from the central parts (15–19°S) of this original range and the Field Atlas report of a single bird from the Leichhardt R (18/139) in June 1980 is the only one from this district since 1910.[1709] In the remainder of the Top End and in the Kimberley Region Field Atlas reports cover most of the historical range of the species.

The population is polymorphic, about a quarter having red heads, three-quarters black and a small number yellow.[901] There are no geographical differences in the colour ratios.

The Gouldian Finch is at least partly migratory, moving south in the wet summer season.[167,901] There are too few records in the Field Atlas to determine the extent of the movement and too few breeding records to confirm whether, as is suggested,[901] breeding takes place mainly in the southern parts of its range.

The Gouldian Finch is never far from water. It lives in eucalypt woodland where there is a grassy understorey, in adjacent clearings and on the edges of mangroves.[901] It feeds on grass seeds, ripe and half-ripe, taking them from seeding heads and rarely from the ground. It also eats insects.[901]

Though the range of the Gouldian Finch has altered little in the Top End and Kimberley Regions, numbers are greatly reduced. In the non-breeding season in 1967 a flock exceeding 300 was reported at Warlock Ponds (15/133)[423] but, at waterholes in the Top End where thousands were once recorded, flocks have been reduced to fewer than 30 birds. Round Kununurra (15/128) where the Gouldian Finch was still legally trapped under licence until 1981, the number declared in trappers' returns fell steadily over the ten years before. Widespread annual burning of grass in northern Australia may be a factor causing its decline.

There are no measures of abundance.

Field Atlas Records: 81

1° blocks–with records: 33 (4%)
–with breeding records: 4 (12%)

Atlas No.: 670

Max. Rep. Rate: 9%
In block(s): 14/132

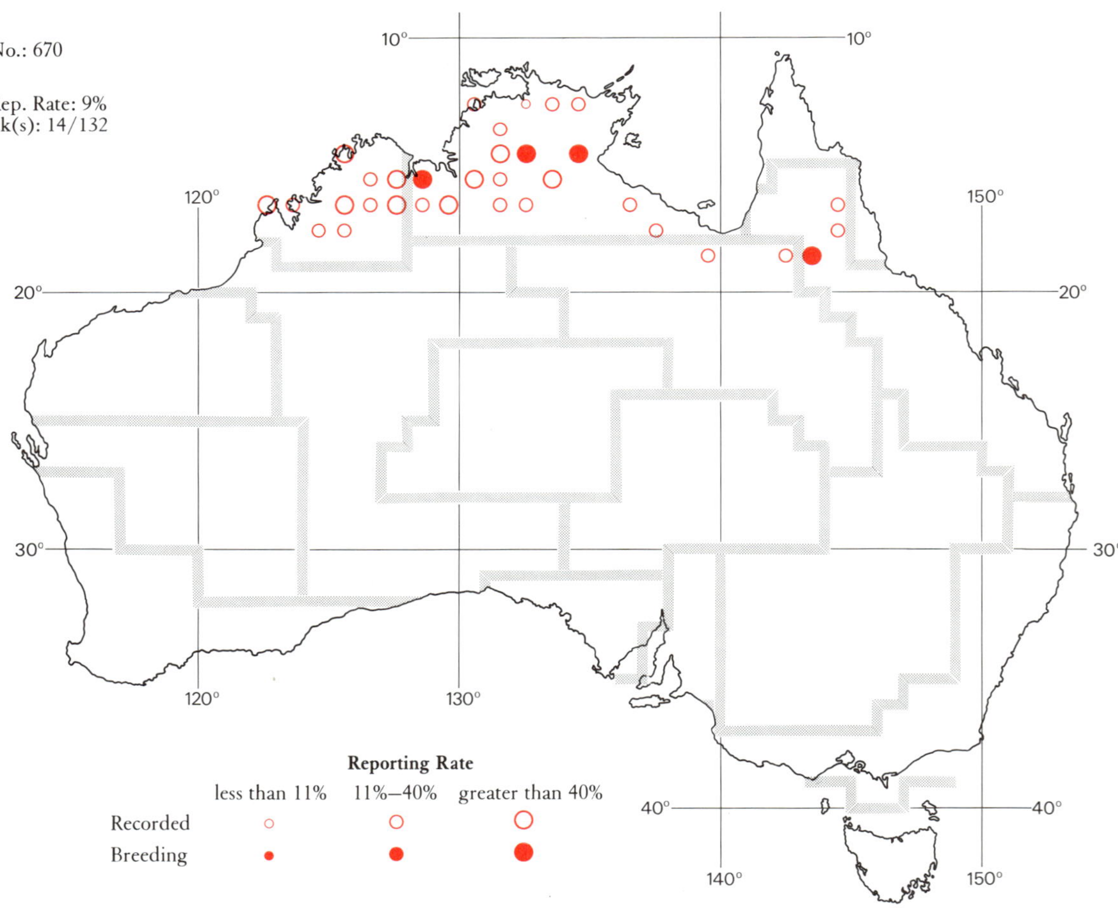

Metallic Starling

Aplonis metallica

The Metallic Starling ranges from New Guinea and adjacent islands to Australia.[1422] It is the only native member of its family in Australia. Historical records of the Metallic Starling from Peel I. (27/153)[8] are unconfirmed and it is uncertain whether two from Sydney (33/151) were wild birds.[816] Breeding has been recorded historically in the same 1° blocks as during the Field Atlas.

Most of the Atherton population migrates, presumably out of Australia, Mar.-Aug.[99,104,231,303,698] Reporting Rates for the Region were 32% in summer and 8% in winter. Whether the population of the Cape York Region also migrates is not known but the Field Atlas suggests it may (Reporting Rates: summer 24%; winter 7%). The species is regarded as nomadic throughout Torres Strait and probably migratory[503] but no passage has been observed.

The Metallic Starling lives in rainforest, parks and gardens. It roosts and nests colonially, one tree housing 188 nests.[1552] Flocks seen in the Atherton Region after the breeding season have been estimated to contain 3000–5000 birds on several occasions.[698]

The Metallic Starling feeds on fruit and insects and in New Guinea is known to take much nectar.[129] Sometimes birds travel a considerable distance to feed, those from the colony on Lloyd I. (12/143) flying each day to the mainland and returning at dusk in tightly packed flocks to roost in the mangroves.[1147] Factors reducing breeding success are the breaking of branches under the weight of the nests and attacks by hawks, which sometimes dismember nests to reach the young.[99] Snakes also prey upon the colonies.

Density in New Guinea in lowland rainforest near Port Moresby (9S/147E) was 0.2 birds/ha, though this may be an underestimate because birds are centred on a nest colony.[129]

Field Atlas Records: 377

1° blocks–with records: 16 (2%)

–with breeding records: 10 (63%)

Atlas No.: 674

Max. Rep. Rate: 45%
In block(s): 12/143

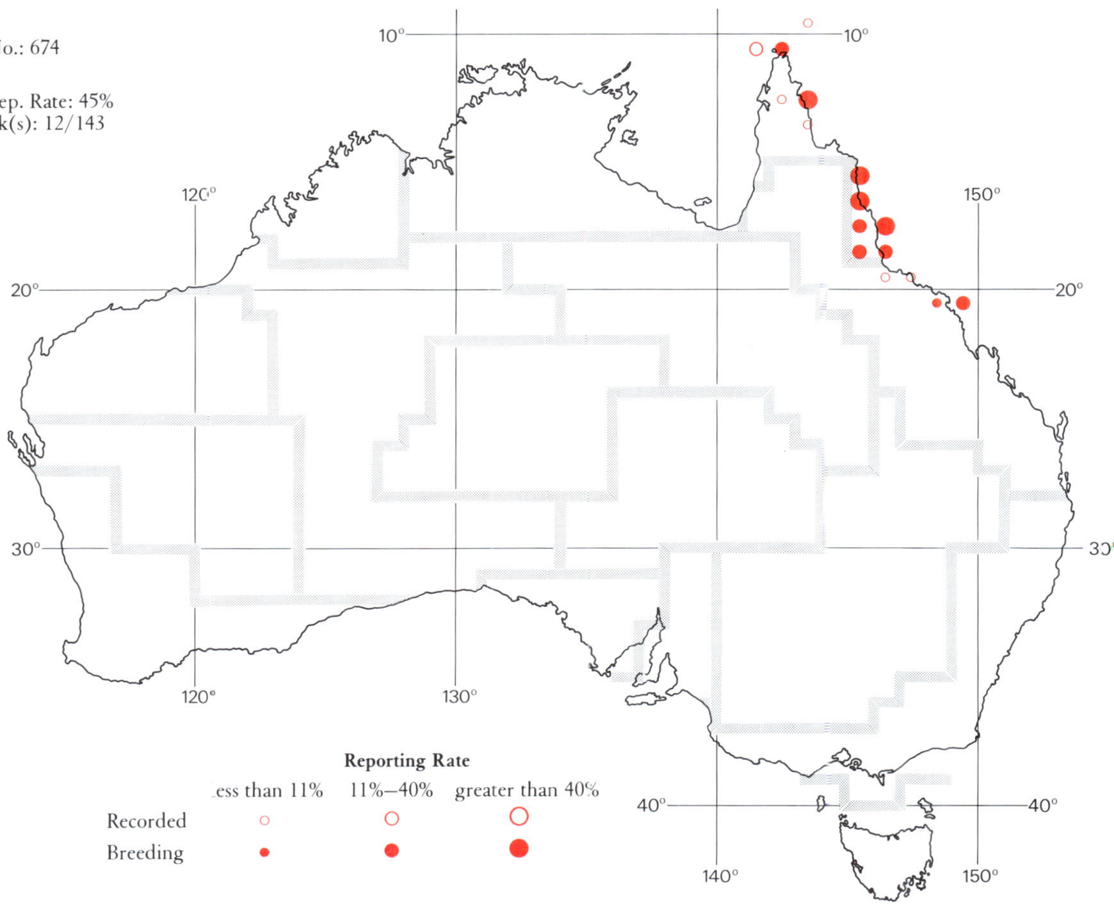

Common Starling

Sturnus vulgaris

The Common Starling, native to Europe and western Asia, was introduced to Australia by European settlers and its spread has been associated with settlement. It was released near Melbourne (37/145) in the late 1850s, Hobart (42/147) in 1860, Brisbane (27/153) in 1869–70, Sydney (33/151) in 1880 and Adelaide (34/138) in 1881.[1085] Spread was rapid (p. 669). Between 1880 and 1900 it became established at Mildura (34/142), Kangaroo I. (35/137) and Eyre Pen. It reached Qld from the south in 1919, the earlier release having apparently failed, and by 1926 had occupied all settled areas of NSW. The spread from Adelaide to Eyre Pen. (500 km) took 20 years (25 km/year) and from Sydney to Brisbane (800 km) 40 years (20 km/year). North of Brisbane along the coast colonization has slowed. There was an isolated record from Port Douglas (16/145) in 1944[1650] and sporadic sightings near Innisfail (17/146) from 1965 onwards.[698] The main spread now seems to be westward. In 1976 the Starling reached the SA/WA border[1085], a rate of spread of about 9 km/year from Eyre Pen. There were isolated earlier sightings in the South-West at Albany (35/117) in 1917, Gingin (31/115) in 1936 and Esperance (33/121) in 1971–72, 1976 and 1982. Authorities in WA destroy any birds reported, e.g. 2000 in 1980–81 at Nullarbor Station (31/130). In the NT also, attempts are being made to halt the Starling's spread.

In Tas. the Starling has recently become established in the far south-west, expanding from Strathgordon (42/146) into communities along the coast.

The Starling can move far, confirmed by the recovery of a bird in Brisbane banded a month before in Tas.[1709] Flocks live in pasture, farmland, swamps and suburbia but the bird is absent from wet forest. It competes with native species for nest hollows and forms large communal roosts in autumn and winter.[653] One on a Hobart bridge held about 25 000 birds.[1803] The Starling is omnivorous[1085] and requires fresh water.

Densities were: Armidale (30/151) in partly-cleared, grazed woodland 0.31 birds/ha;[582] and Rotamah I. Bird Observatory (37/147) in eucalypt woodland and forest 2 breeding birds/ha.[284] In an irrigated fruit growing area at Mildura (34/142) the estimated population was 100 000 in 1954,[1753] about 2.5 birds/ha.

Field Atlas Records: 31 764

1° blocks–with records: 246 (30%)
–with breeding records: 157 (64%)

Atlas No.: 999

Max. Rep. Rate: 86%
In block(s): 34/137

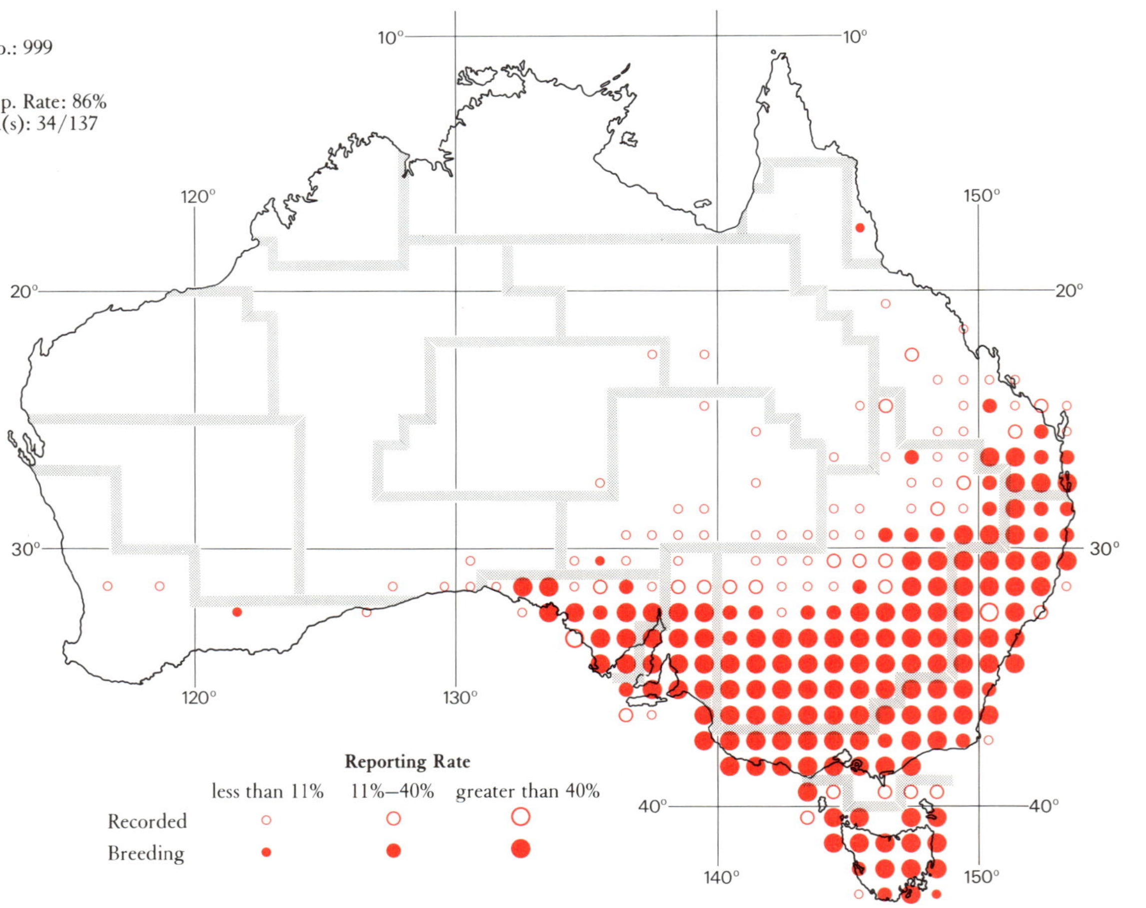

Common Mynah

Acridotheres tristis

The Common Mynah is native to India and southern China. The earliest known release in Australia was in Melbourne (37/145) in 1862.[1085] In 1883 birds from Melbourne were taken to Qld and released in canefields on the Herbert R (18/146) and Johnstone R (17/145) and in Townsville (19/146)[343] where they quickly established themselves. Some of these birds were taken to Cairns (16/145) and Toowoomba (27/151) in 1918.[1085] In Sydney (33/151), after a few small introductions along the harbour and Parramatta R last century, the Mynah did not expand its range greatly until the 1940–50 period, by the end of which it had occupied the greater part of the metropolitan area.[854,1085] It was released in Tas. in 1900 or shortly after but failed to survive. A small population became established in Launceston (41/147) about 1955. It has gone but a probable sighting of one bird was made there during the Field Atlas.[1533] The Mynah was reported in Adelaide (34/138) in 1957[1085] but did not expand greatly. Active steps have been taken to exterminate it there and any strays that remain are shot. The species was released in Canberra (35/149) in the late 1960s and has established itself. The historical maps (p. 670) show its gradual spread.

No seasonal or other movements have been reported.

The Mynah lives in pairs or flocks and roosts communally, usually in trees.[403] It inhabits urban areas and spreads out from them mainly along roads.

The Common Mynah is omnivorous, feeding on refuse and insects.[853] Most of its food is collected on the ground. It is aggressive to other species competing for the same food source and has been recorded killing fledgling sparrows and eating their brains.

The Common Mynah was introduced to northern Qld to combat insects in the sugarcane fields[657] but gathered in the environs of settlements and was not a useful predator of insects.

There are no measures of abundance.

Field Atlas Records: 6289

1° blocks–with records: 55 (7%)
–with breeding records: 20 (36%)

Atlas No.: 998

Max. Rep. Rate: 65%
In block(s): 38/145

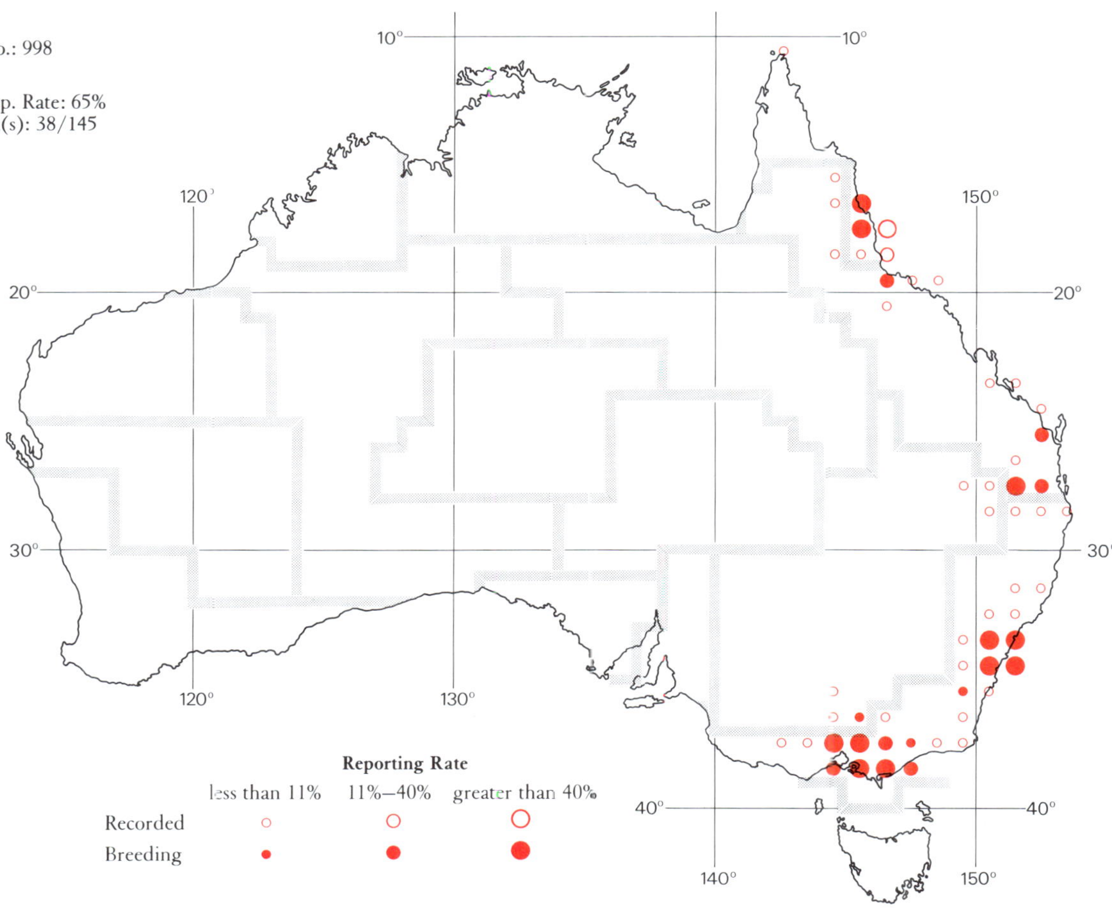

Yellow Oriole

Oriolus flavocinctus

The Yellow Oriole ranges from New Guinea and adjacent islands to Australia. The Atlases show the populations in Qld and in the Top End and Kimberley Regions to be separate but they are not morphologically distinct and all belong to the subspecies *flavocinctus*.[608,748] The Kimberley population has been reported as far east as 127°E. The species' range appears unchanged from the historical periods.

The Field Atlas does not suggest a large-scale seasonal pattern of movement. The species is present all year in the Innisfail (17/146) district[698] but is only occasionally recorded on the highlands of the Atherton Region.[231]

The Yellow Oriole lives mainly in rainforest and adjacent mangroves. Sometimes it inhabits orchards, fruiting trees in clearings and rainforest along streams, even when the surrounding country has been cleared.[698] The bird is generally solitary and seldom in parties of more than three. It repeats its loud bubbling calls from the forest trees for up to two hours at a time[926] but is difficult to see in the dense canopy.[1758] The Yellow Oriole feeds on native and cultivated fruit[303] and guts have contained only seeds and fruit.[748]

There are no measures of abundance.

Field Atlas Records: 1045

1° blocks–with records: 55 (7%)
–with breeding records: 13 (24%)

Atlas No.: 672

Max. Rep. Rate: 60%
In block(s): 13/130

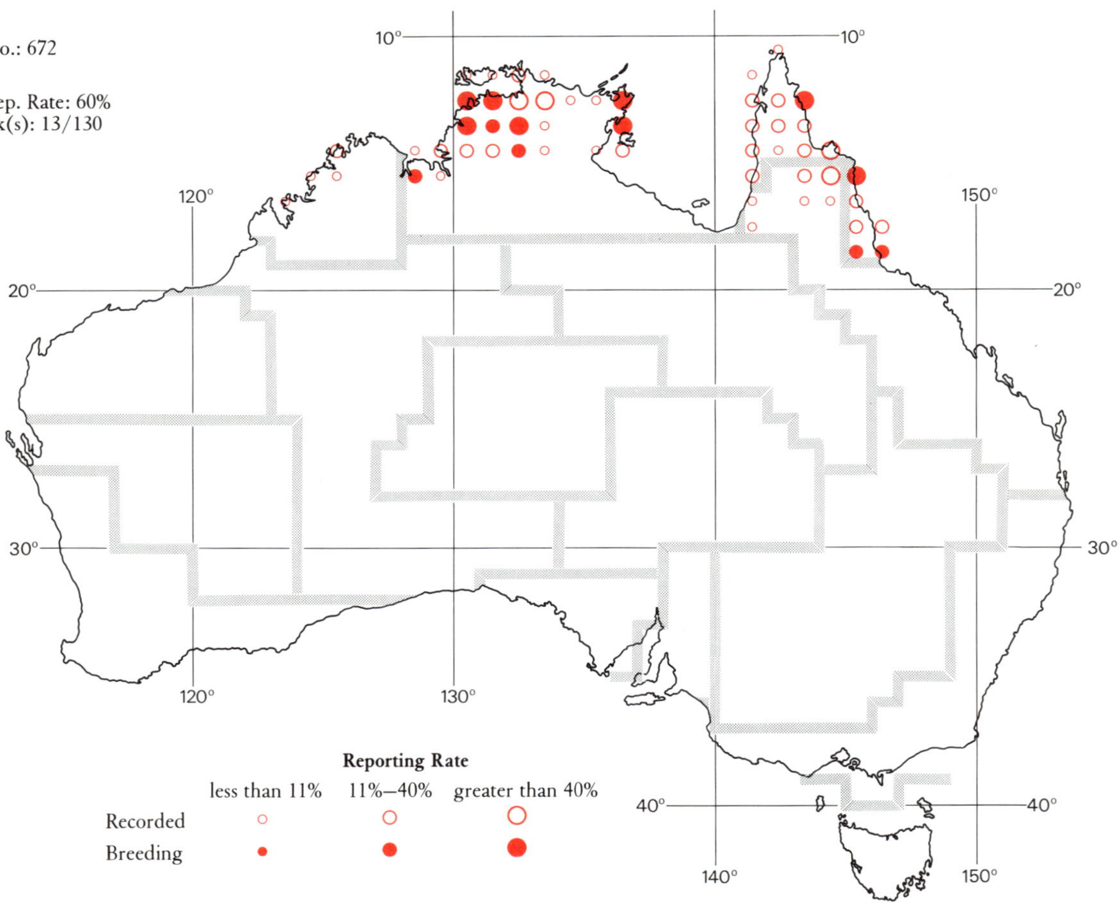

Olive-backed Oriole

Oriolus sagittatus

The Olive-backed Oriole ranges from New Guinea and adjacent islands to Australia. In Australia three subspecies are recognized: *magnirostris* in the Cape York Region; *sagittatus* to the south and *affinis* west of the Leichhardt and Norman Rivers (141°E).[1422] The species is an uncommon visitor to the SA Ra. Region recorded in all months except May.[436] A single pair is reported to have bred there in 1968 and 1969.

The eastern population, *sagittatus*, is partly migratory. Observers in south-eastern Australia have recorded its almost complete absence in winter at Murphy's Ck (27/152),[1099] Wollomombi (30/152),[129] Tocumwal (35/145),[833] Canberra (35/149),[154] Moruya (35/150)[1192] and Jamieson (37/146).[1469] Movements related to abundances of food have also been noted. In Torres Strait the Oriole is an uncommon visitor at the end of the dry winter season when fruit is ripe[503] and at Menangle Park (34/150) up to 50 arrived each year in Apr. to feed on infestations of caterpillars.[516] In the Top End the Reporting Rate round Darwin (12/130) was higher in winter (17%) than in summer (6%) and this suggestion of movements is confirmed by observation.[423] The longest movement recorded is that of an immature from Lynch's Ck (33/150) to Minyon Falls (28/153), 644 km.[1437]

The Olive-backed Oriole lives in eucalypt forest and woodland and occasionally on the edges of rainforest. Round Brisbane (27/153) ones or twos are often among flocks of Figbirds.[128]

The Olive-backed Oriole feeds on insects as well as fruit, native and introduced,[748,1049] and gathers at plagues of caterpillars.[516,1099] During the Field Atlas it was observed to feed on nestling honeyeaters.

Density at Atherton (17/45) was 0.5 birds/ha.[231] In eucalypt woodland densities were: Armidale (30/151) 0.06 birds/ha,[582] Wollomombi 0.02–0.28 birds/ha[129] and Moruya 0.2 birds/ha;[1197] in eucalypt forest at Jamieson it was 0.5–1.0 bird/ha[1469] and at Boola Boola (38/146) 0.16–0.70 birds/ha.[1113]

Field Atlas Records: 6688

1° blocks–with records: 276 (34%)
–with breeding records: 68 (25%)

Atlas No.: 671

Max. Rep. Rate: 57%
In block(s): 27/149

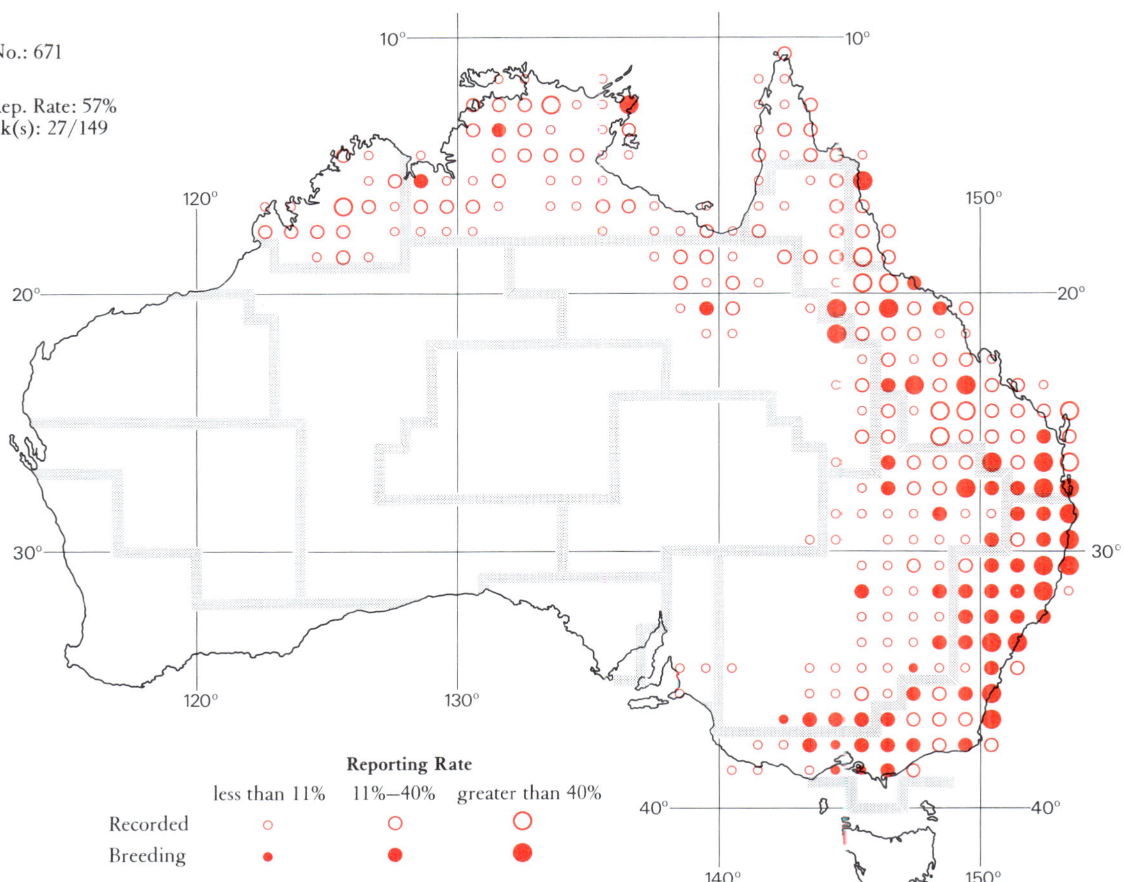

Figbird

Sphecotheres viridis

All figbirds are now included in one species that is divided into several subspecies, two of which are in Australia: *flaviventris* north of 19°S and in the Cape York Region and regions to the west; and *vieilloti* south of 19°S.[604] A hybrid zone between the two exists in the Townsville (19/146) area.[623] The population of *flaviventris* in the Cape York Region extends south to the Norman R (17/140) on the western coast and the Top End and Kimberley populations are isolated. Vagrant Figbirds have been recorded near Berwick (38/145)[1207] and there is an unconfirmed report from Rutherglen (36/146).[1128]

On the Atherton Tableland *flaviventris* is said to be resident and *vieilloti* to visit Apr.-June[231] but the occurrence of *vieilloti* in this district, where hybrids are known to exhibit several different plumages, has yet to be confirmed.[623] At Murphy's Ck (27/152) *vieilloti* is absent Apr.-Aug.[1099] but banded birds at Murwillumbah (28/153) seldom moved far and most were sedentary.[1431] The species winters in Sydney (33/151) but at Nowra (34/150) it appears only Aug.-Apr.[396] At Berwick, far south of their usual range, birds occurred Jan.-Mar.[1207] In the Batchelor (13/131) and Adelaide R (12/131) districts the Figbird appears to be nomadic.[1602] These seasonal visits have been interpreted as local wanderings to keep in contact with food.[623] The longest movement recorded is that of a bird from Pialba (25/152) to Kilkivan (26/152), 105 km.[1439]

The Figbird forms flocks of 100 or more in winter. It lives in and on the edges of rainforest, in eucalypt forest and in parks and gardens and is often an urban bird.[350,953] Round Brisbane (27/153) flocks are often joined by a few Olive-backed Orioles.[128] Communal breeding has once been observed.[1913] The Figbird's food includes fruit of many native shrubs and trees, specially figs. It also takes fruit of exotic species, including lantana, and will raid cultivated fruit.[1205]

In New Guinea density at Port Moresby (9S/147E) in eucalypt woodland was 0.2 birds/ha.[158]

Field Atlas Records: 3819

1° blocks–with records: 104 (13%)
–with breeding records: 32 (31%)

Atlas No.: 432

Max. Rep. Rate: 70%
In block(s): 20/149

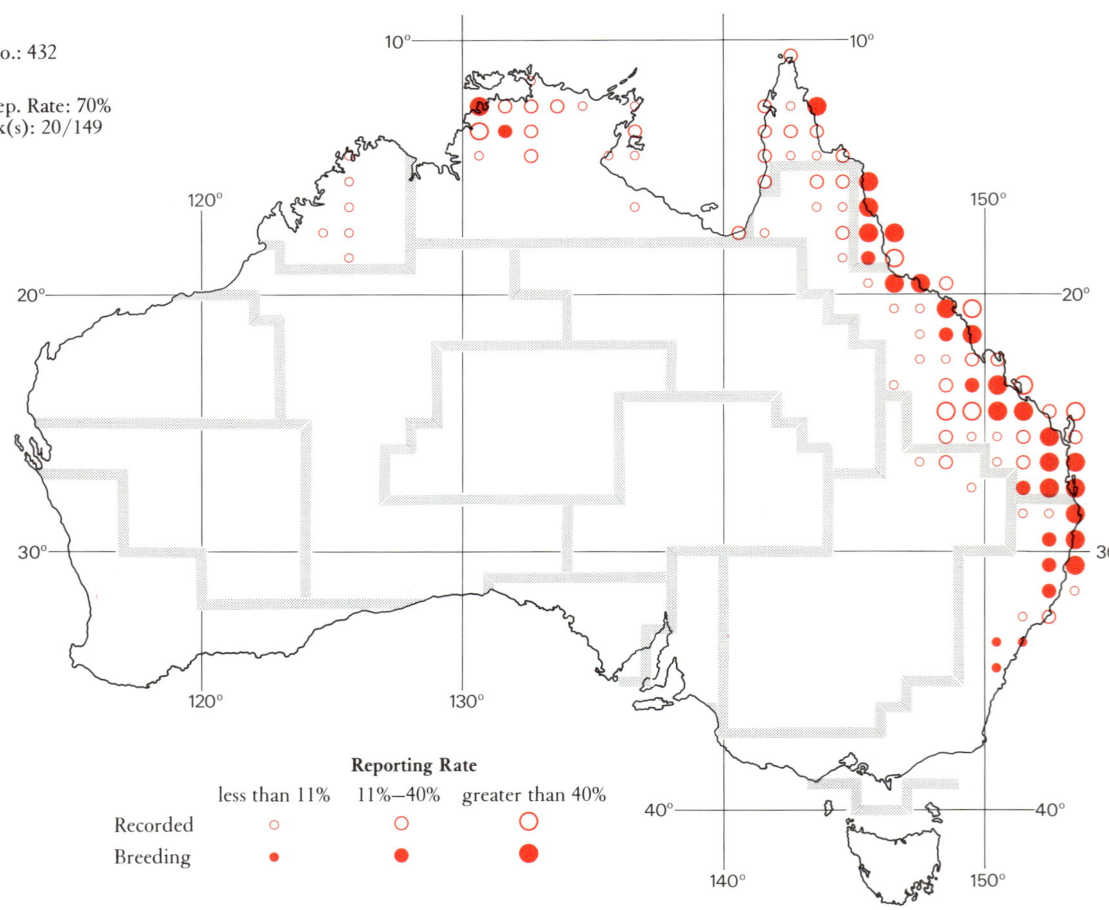

Spangled Drongo

Dicrurus hottentottus

The Spangled Drongo ranges from southern Asia to New Guinea and Australia. The Atlases show gaps in the distribution of the Australian subspecies *bracteatus* between the Qld, Top End and Kimberley populations but there is little morphological difference between them.[1780] The Drongo rarely breeds south of 31°S.

In eastern Australia the Drongo is partly migratory, some birds moving north in winter, others south, possibly some moving altitudinally and some apparently not migrating.[1211] Australian birds winter in parts of New Guinea[1235] though no seasonal change in numbers was detected in south-eastern New Guinea.[160] In Torres Strait flocks of up to about 20 move north in Mar.-Apr. and south in Oct.-Nov.[503] On 6 Nov. 1978 more than 100 per minute passed over Thursday I. (10/142) for 30 minutes. Birds migrate southward through the Atherton Region Oct.-Nov.[231,698] and a summer influx is reported at Rockhampton (23/150)[1089] and occasionally at Murphy's Ck (27/152).[1099]

From Torres Strait to Rockhampton, in addition to the migrants, some birds are present all year. Farther south moderate numbers of Drongos remain all year, e.g. Cooloola (26/153)[1495] (Reporting Rates 24–31°S: summer 20.0%, winter 15.6%). South of 31°S (the southern breeding limit) the Drongo is a winter visitor[1292] (Reporting Rates: summer 2%, winter 7%). Tas. records were in May, June and July.

In the Top End and Kimberley Regions, the Drongo is present all year.[423,1211]

The Drongo lives where there are high lookout perches. Rainforest, including that along streams, eucalypt and paperbark woodland, eucalypt forest and mangroves are all used. It feeds mainly on insects such as beetles, grasshoppers, dragonflies, mantids, cicadas[98,748,1211] and insects emerging from the ground at dusk, catching them as they start to fly. One fed on a colony of paper wasps for several weeks until the wasps were almost exterminated.[716] Sometimes the Drongo robs food from, or eats, other birds.[141]

Density in New Guinea in lowland rainforest near Port Moresby (9S/147E) was c.6 birds/ha.[160]

Field Atlas Records: 3949

1° blocks–with records: 132 (16%)

–with breeding records: 30 (23%)

Atlas No.: 673

Max. Rep. Rate: 66%
In block(s): 13/130

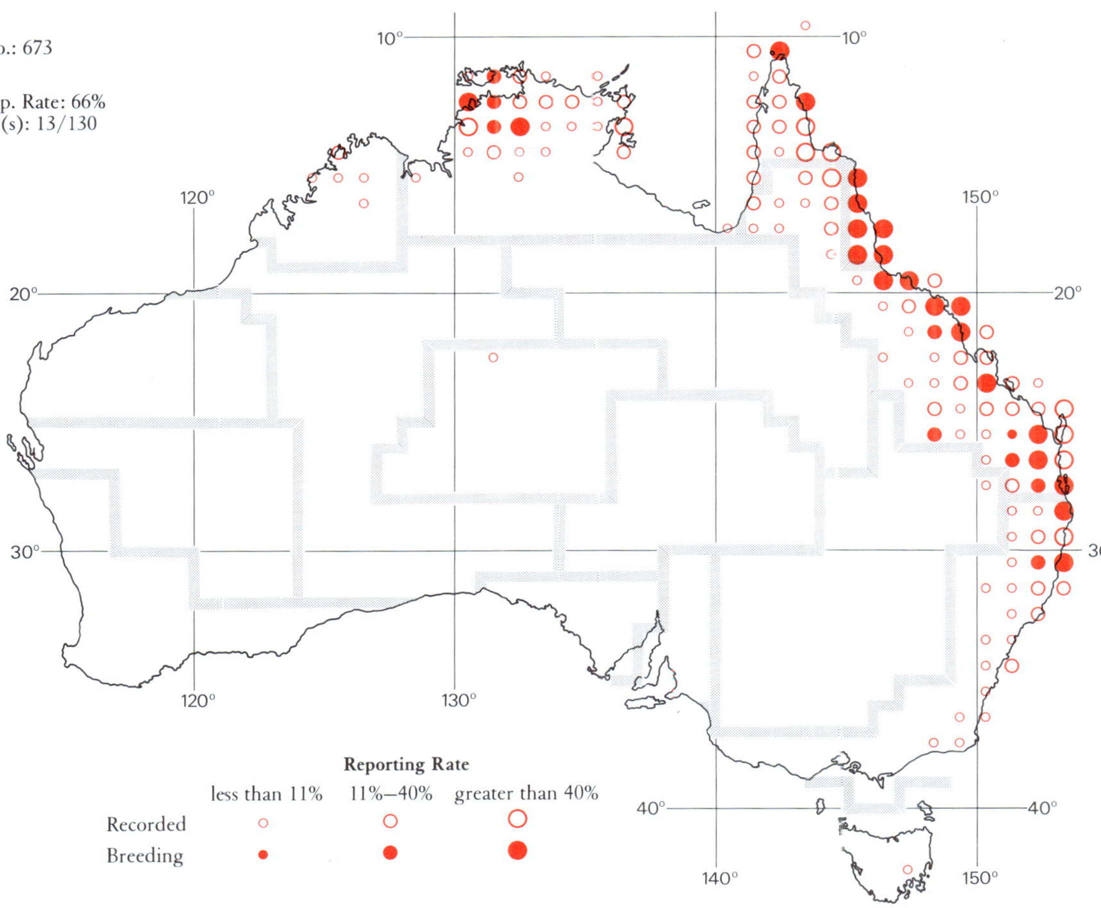

Golden Bowerbird

Prionodura newtoniana

The endemic Golden Bowerbird was recorded in the Field Atlas in most 1° blocks where it was recorded historically. There are recent historical records at 18/145 in the ranges behind Ingham.

Playgrounds (bowers) are visited occasionally or regularly throughout the year by golden-plumaged males and less frequently and irregularly by female-plumaged males.[643,1812] The species is present all year near Atherton (17/145) on the tablelands[231] but in winter a few birds are said to appear at lower altitudes than in summer.[1709] The Golden Bowerbird lives at altitudes between 900 and 1500 m.[1709] The Reporting Rate above 1000 m was 32.6% compared with 0.02% below. It inhabits rainforest and is the only Australian representative of a small group of bowerbirds, otherwise restricted to New Guinea, which builds a maypole type of bower.[643] The Golden Bowerbird eats mainly fruit but it also takes beetles and cicadas.[643]

In a 50-ha study area at Paluma (19/146) over five years 17 bower sites have been observed (one bower site or complex to 2.94 ha). Ten sites were permanently maintained and defended by golden-plumaged males (1 male/5 ha) and the remainder attended by one to several female-plumaged males on an irregular to permanent basis.[643]

Field Atlas Records: 97

1° blocks–with records: 7 (1%)
–with breeding records: 1 (14%)

Atlas No.: 685

Max. Rep. Rate: 13%
In block(s): 18/146

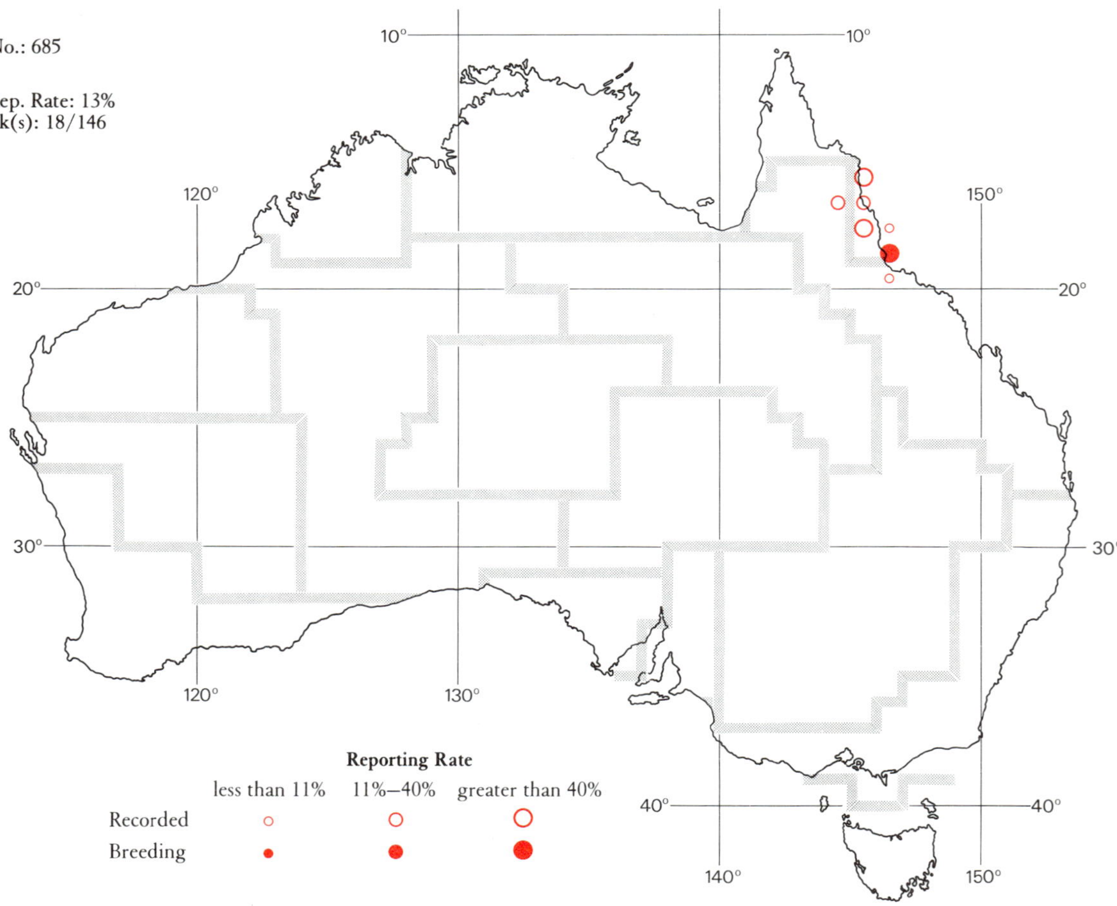

Satin Bowerbird

Ptilonorhynchus violaceus

Two subspecies of the endemic Satin Bowerbird are recognized: *minor* in the Atherton Region and *violaceus* to the south.[400] A population in the Otway Ra. (38/143) is cut off from the main distribution by the city of Melbourne (37/145).

At Leura (33/150), most males leave their playgrounds (bowers) in Nov. and return in July.[1782] In the Brindabella Ra. (35/148) the bird is absent during the winter from some localities, apparently moving to lower altitudes.[1016] During autumn and early winter flocks of more than 30 birds roam locally.[400] The longest movement recorded is of a banded adult from Bogong (36/147) to Tallangatta (36/147), 60 km.[51]

The population *minor* lives in rainforest above 900 m and *violaceus* in rainforest and eucalypt forest below 1000 m.[400] The Bowerbird will visit suburban gardens and parks, specially in winter. The male builds a bower on the ground that is visited by females and immatures. The Satin Bowerbird feeds on leaves, grass, fruit and insects and their larvae.[132,400,1042] Nomadic flocks sometimes damage cultivated fruit.

At Leura dominant males maintain bowers some distance apart interspersed with those of subsidiary males. The distances from the bower of one dominant male to those of its three nearest dominant neighbours were 2.9 km, 4.3 km and 6.4 km.[1782,1783] Fifteen different females were courted at the bower of one of the dominant males in one season and 600 birds were banded in a garden within its territory during three years. At Tooloom Scrub (28/152), there was an average of five birds on a 3.5 km transect through rainforest walked at least six times.[1342]

Field Atlas Records: 3310

1° blocks–with records: 59 (7%)
–with breeding records: 29 (49%)

Atlas No.: 679

Max. Rep. Rate: 52%
In block(s): 36/150

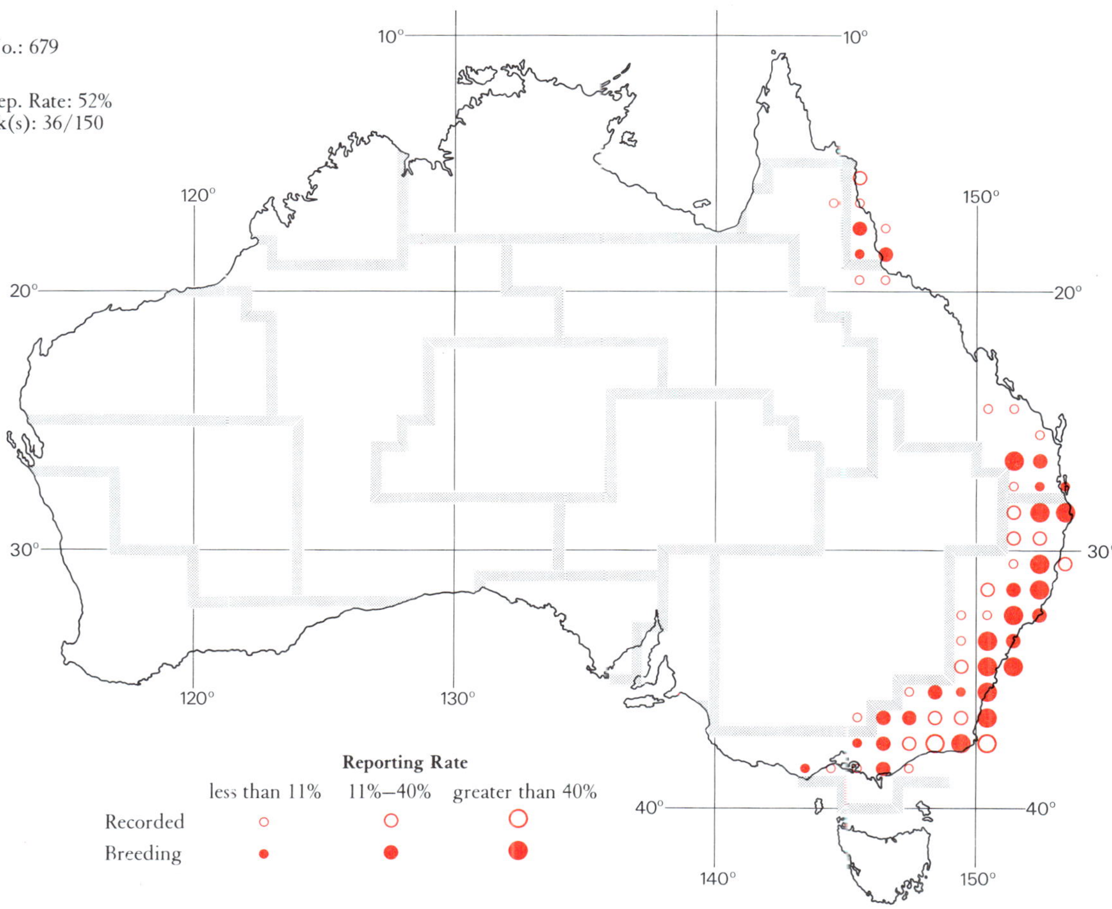

Regent Bowerbird

Sericulus chrysocephalus

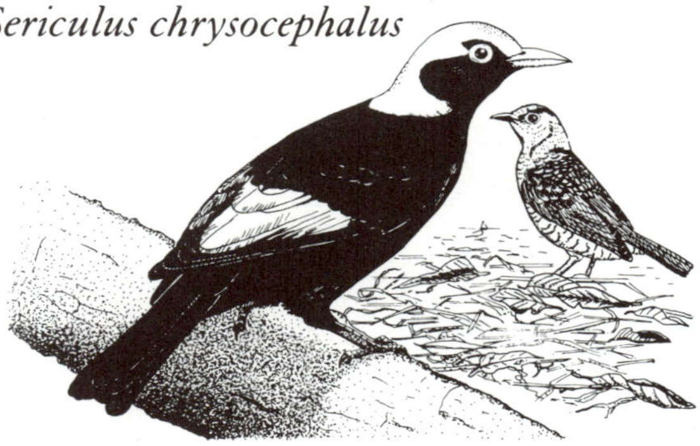

Two subspecies of the endemic Regent Bowerbird have been described, though some doubts have been expressed about their validity.[400] They are *rothschildi* in the Eastern Qld Region and *chrysocephalus* in the South-East Region. The species still occupies the range it inhabited during the historical periods. There are historical records linking the Dawes Ra. (24/150) population with that farther south but the population in the Eungella (21/148) district appears isolated.

In the northern South-East Region the birds desert the high altitudes in winter to visit fruiting trees, including those in orchards, on the lowlands.[400] Reporting Rates in the New England (30/152) district were higher below 200 m (58%) in autumn-winter than in spring-summer (22%).

The Regent Bowerbird lives in rainforest but will forage in eucalypt forest and even in the suburbs of Brisbane (27/153) during the non-breeding season. The playground (bower) is well concealed on the forest floor. The bird feeds on fruit, native and cultivated, including berries of the introduced lantana, and insects.[400]

Only two Regent Bowerbirds were seen on at least six 3.5 km transects through Tooloom Scrub (28/152) in Dec. 1962.[1342] There are no other measures of abundance.

Field Atlas Records: 518

1° blocks–with records: 24 (3%)
–with breeding records: 6 (25%)

Atlas No.: 684

Max. Rep. Rate: 18%
In block(s): 30/152

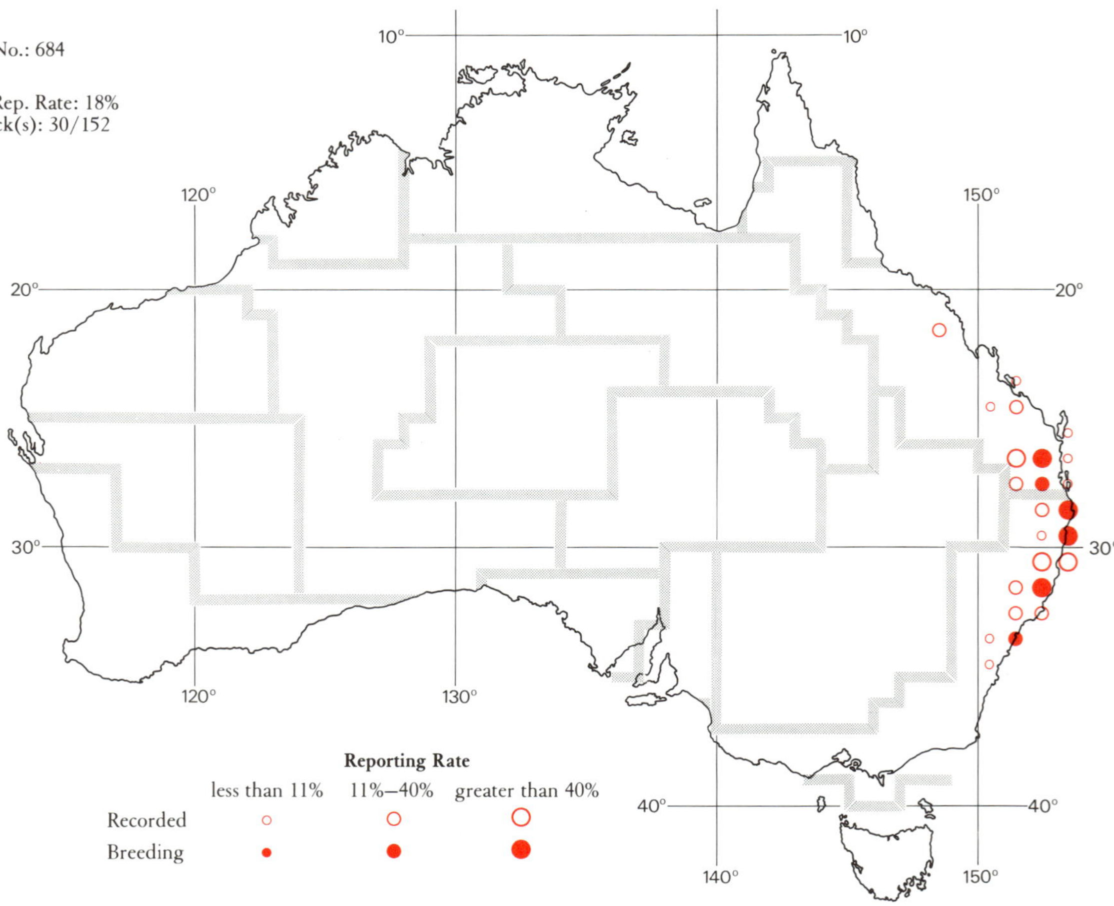

Spotted Bowerbird

Chlamydera maculata

Two subspecies of the endemic Spotted Bowerbird are recognized: *maculata* east of about 138°E and *guttata* in the Central Ra. Region and areas to the west.[400,1382] The Atlases show that the Central Ra. population of *guttata* approaches the eastern *maculata* closely though there have been no records of the Spotted Bowerbird between 137°E and 138°E. Neither do historical records close the gap between the Central Ra. and western populations of *guttata*. The range of the Spotted overlaps that of the Great Bowerbird in the Eastern Qld and Rolling Downs Regions.

The range of *maculata* has declined in the south where formerly it extended to the Murray R (36°S). In the SA Ra. Region it was last reported near Swan Reach (34/139) in 1929[1763a] and Field Atlas records in the southern Murray-Darling Region came from only two isolated sites near the Murray R.

There is no information about movement but sightings such as that in the Field Atlas, of a single bird in Brisbane (27/153) in Dec. 1981, indicate that it can travel far. There is also a record from Sydney (33/151) in 1902 made during a drought.[816] The Field Atlas does not suggest any large-scale seasonal movement.

The Spotted Bowerbird lives in dense thickets of low scrub. It will also enter gardens. The male constructs a bower which is visited by more than one female. The Spotted Bowerbird feeds on fruit, seeds, berries, insects and their larvae, nectar and spiders.[400,748] It is usually near fresh water.[458,1509,1881a]

The only measures of density are: on 12 000 ha at 25/146 near Charleville there were at least 20 bowers; and near Wiluna (26/120) 80 birds were shot illegally in three months in one orchard.

Field Atlas Records: 1988
1° blocks–with records: 185 (23%)
–with breeding records: 39 (21%)

Atlas No.: 680

Max. Rep. Rate: 72%
In block(s): 27/147

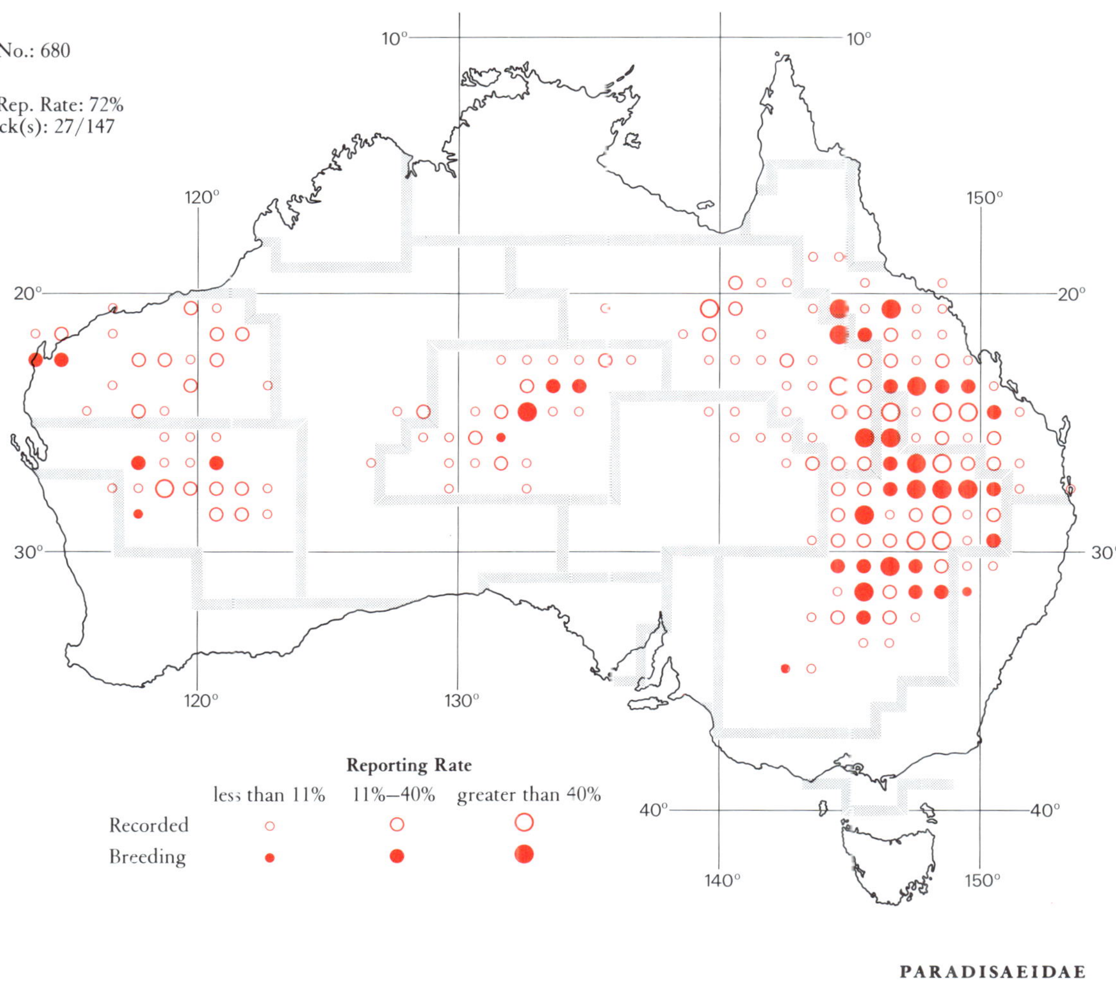

Great Bowerbird

Chlamydera nuchalis

Four subspecies of the endemic Great Bowerbird are recognized: *yorki* in the Cape York Region; *orientalis* to the south; *nuchalis* in the Top End; and *oweni* in the Kimberley Region. There are intermediates where they meet[400] and no breaks in the range are suggested by the Atlases. The distribution of the Great Bowerbird overlaps that of the Fawn-breasted in the Cape York Region and the Spotted in the Eastern Qld and Rolling Downs Regions.

There is no information about movement. In Qld the Great Bowerbird lives on the edges of rainforest, including that along streams and in isolated patches; it will enter towns. In the Top End it lives in eucalypt forest and eucalypt woodland; and in the Kimberley Region in rainforest, eucalypt woodland, mangroves and gardens. Groups of up to 20 birds have been recorded in the non-breeding season. Usually it lives round playgrounds (bowers) which are close to permanent water, sometimes hidden under bushes. A new bower is constructed by the male each year not far from the previous one. It feeds on fruit, native and cultivated, berries, nectar, seeds, insects and spiders.[400] Food is gathered in trees and from the ground.

Near Edward R (14/141) in the Cape York Region four bowers were in use one season within 1.6 km of a camp, two within 100 m of it.[1758] Five active bowers were seen along 16 km of vehicle track on the Adelaide R plain (12/131).[456] There are no other measures of abundance.

Field Atlas Records: 1908

1° blocks–with records: 136 (17%)
–with breeding records: 13 (10%)

Atlas No.: 683

Max. Rep. Rate: 66%
In block(s): 16/130

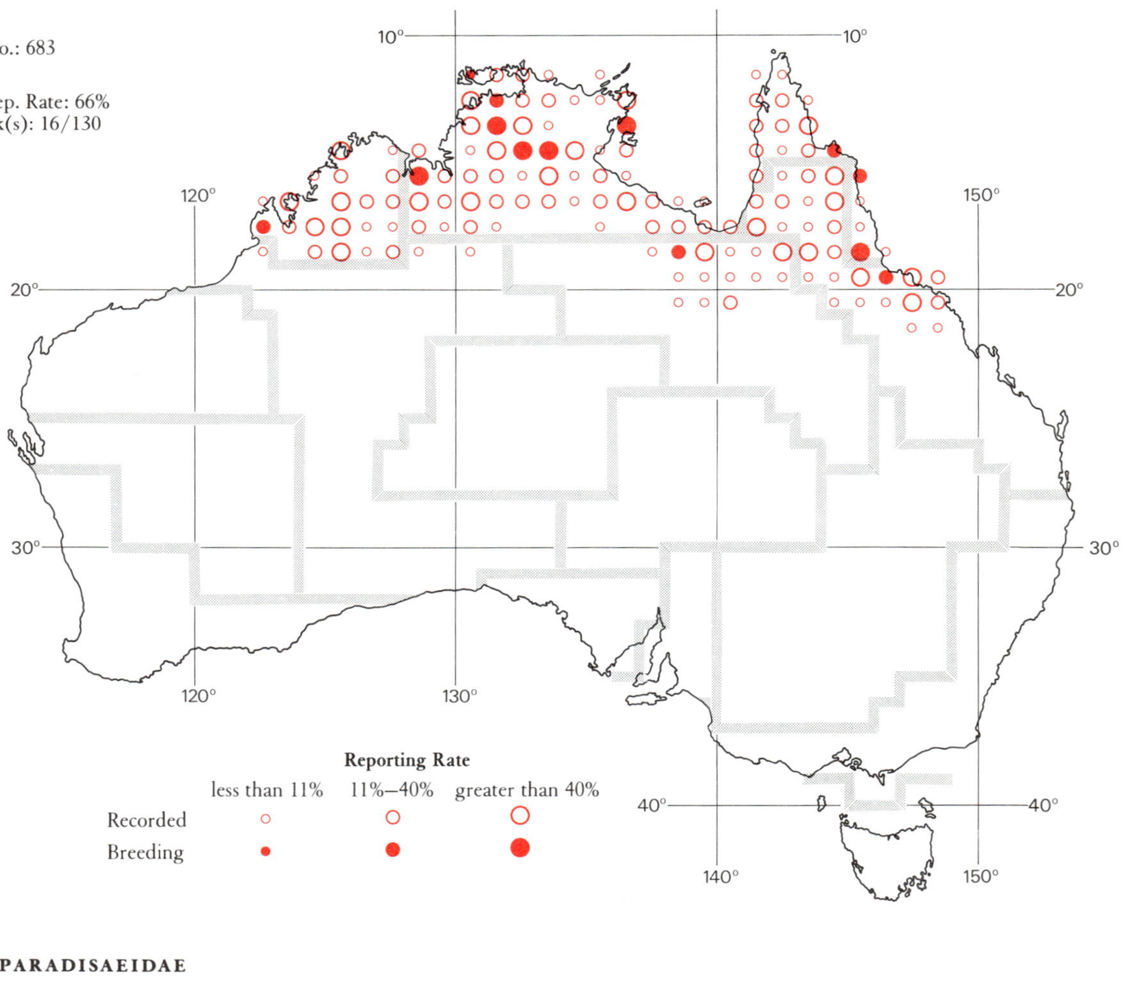

Fawn-breasted Bowerbird

Chlamydera cerviniventris

The Fawn-breasted Bowerbird is predominantly a New Guinean species reaching the southern limit of its Australian range at Silver Plains (13/143) in the Cape York Region. Though said to inhabit islands in Torres Strait, no record has been located and resident observers did not find it during the Field Atlas.[503] Records from the western side of the Cape York Region are probably erroneous[400] because the Great Bowerbird is also present in the Cape York Region; the two species overlap in distribution there.[104] There are breeding records in the Historical Atlas though none during the Field Atlas.

There is no information about movements. The Fawn-breasted Bowerbird lives on the coastal lowlands in eucalypt and paperbark woodland, rainforest regrowth and mangroves but specially in the clearings and margins of rainforest. In New Guinea it is abundant in urban areas.[129] The male builds a playground (bower) on the forest floor, even in mangroves and, on one occasion, below high tide mark. Food includes berries, fruit and insects.[400]

In New Guinea density near Port Moresby (9S/147E) in eucalypt woodland was 0.15 birds/ha but would probably be greater in urban areas.[158] At Bautama near Port Moresby there were 11 bowers in 6 sq. km.[1410]

Field Atlas Records: 35

1° blocks–with records: 6 (1%)
–with breeding records:

Atlas No.: 682

Max. Rep. Rate: 37%
In block(s): 12/143

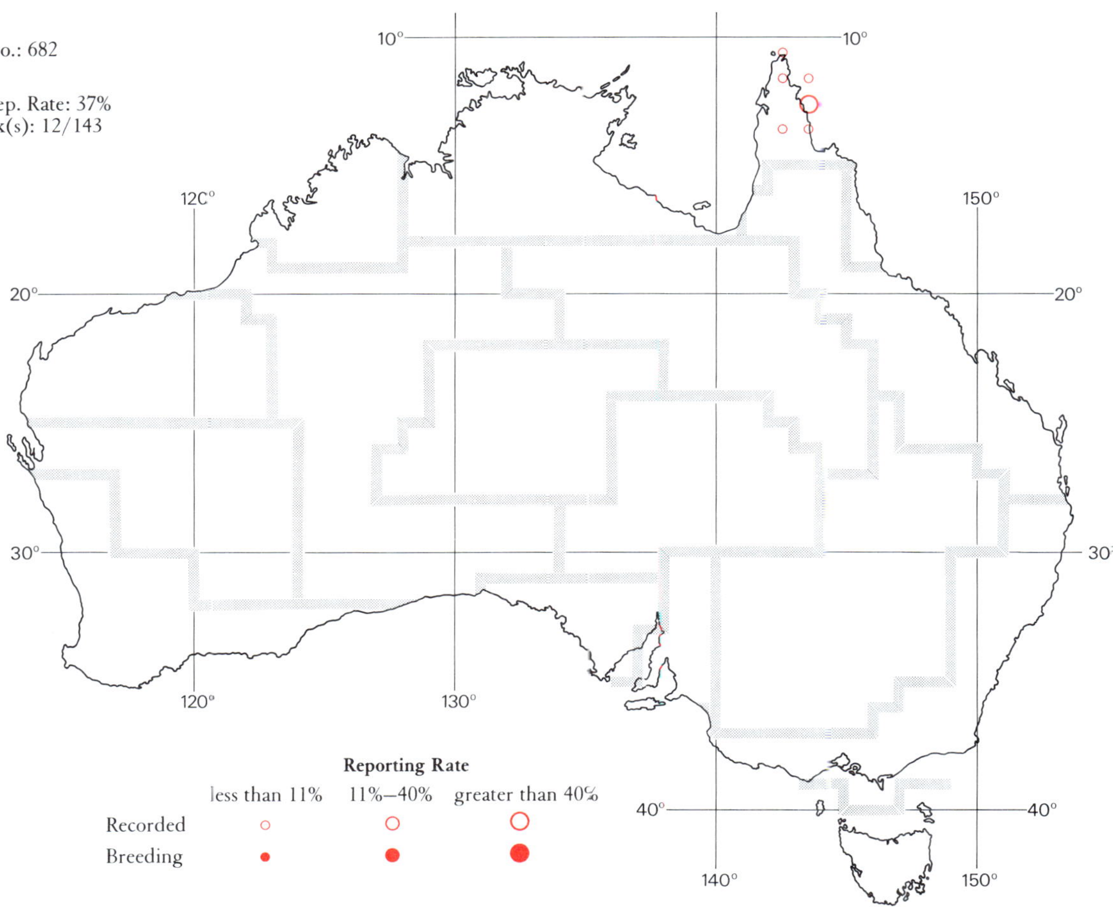

Tooth-billed Catbird

Ailuroedus dentirostris

The endemic Tooth-billed Catbird was recorded during the Field Atlas in all the 1° blocks where it was known historically.

It is sedentary. When adult, the male Tooth-bill appears to live permanently in a territory. The species lives at altitudes between 600 m and 1400 m. Sometimes a few may be in country as low as 350 m[698] but the Field Atlas did not, in the 10′ blocks sampled for altitudinal movement, record any below 500 m. Above 500 m the Reporting Rate for spring-summer (37.1%) was higher than that for autumn-winter (26.8%), probably as a result of seasonal differences in conspicuousness. The species inhabits rainforest including small remnant pockets bordering agricultural land.[400] On the forest floor the male prepares a playground (court), a cleared area round one or several saplings, carpeted with leaves placed face down.[643] The Tooth-bill feeds on fruit, berries, seeds, insects and their larvae.[400] It uses its toothed bill to tear or cut the edges of leaves, e.g. olive-berry tree and shoots, and for eating leaf fragments.[644] At some times of the year it may subsist mainly on leaves.[1042]

There are no measures of abundance.

Field Atlas Records: 174

1° blocks–with records: 8 (1%)
–with breeding records: 2 (25%)

Atlas No.: 678

Max. Rep. Rate: 17%
In block(s): 18/146

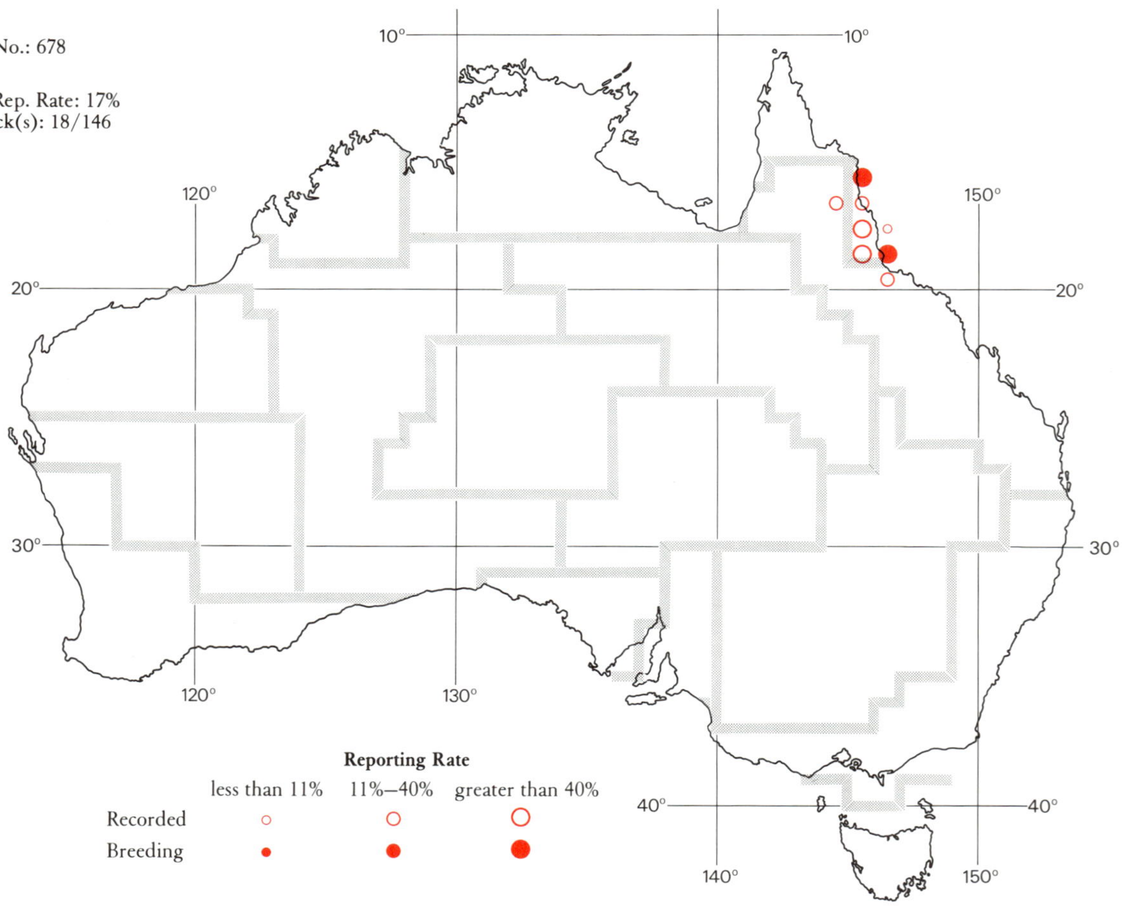

Spotted Catbird

Ailuroedus melanotis

The Spotted Catbird lives in New Guinea and adjacent islands and in Australia. In the Cape York Region the subspecies *joanae* extends from near the Claudie R (12/143), where it was recorded in the Field Atlas, to the Rocky R (13/143). The subspecies *maculosus* inhabits the Atherton Region.[606]

Breeding adults are sedentary in permanent territories.[231,643,698] . The Field Atlas suggests that there may be some dispersal to lower altitudes in winter. Below 500 m, the Reporting Rate rose from 7.1% in spring-summer to 21.7% in autumn-winter. Above 500 m there was little change (spring-summer 48.6%, autumn-winter 41.5%).

The Spotted Catbird lives in pairs, mainly at higher altitudes. It inhabits rainforest but is more easily observed foraging along road verges.[643] No display arena has been recorded. The Spotted Catbird feeds on fruit, specially figs, buds, shoots, flowers, seeds, insects and other invertebrates, birds' eggs and nestlings and was once recorded taking a lizard.[400,643,1268,1812]

There are no measures of abundance.

Field Atlas Records: 413

1° blocks-with records: 8 (1%)

-with breeding records: 7 (88%)

Atlas No.: 677

Max. Rep. Rate: 41%
In block(s): 12/143

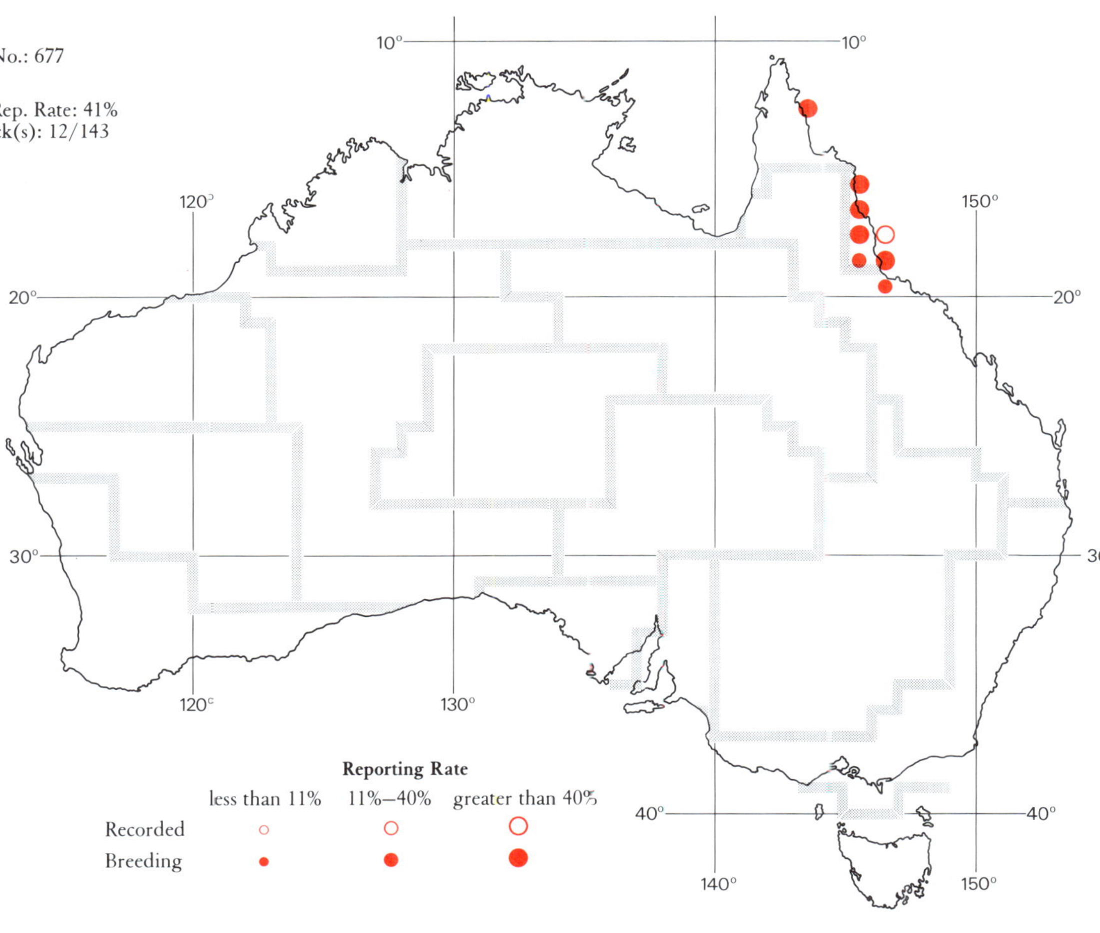

Green Catbird

Ailuroedus crassirostris

The endemic Green Catbird is generally sedentary, birds banded at Tumbi Umbi (33/151) and Kincumber (33/151) having been recaptured there up to nine years later.[1285] Individuals may move some distance. One appeared on Wreck I. (23/151) in Sept.-Oct. 1970[205] and in Aug. 1935 a few visited open farming country near Raymond Terrace (32/151) where none had been seen for more than 60 years.[527] Another, near Bermagui (36/150), was in eucalypt forest near the coast for a short period in autumn 1978.

The Green Catbird lives singly or in pairs in rainforest and occasionally in eucalypt forest.[536,744] Up to 20 birds have been recorded in foraging flocks in the non-breeding season. The species has been observed to build a rudimentary playground (display arena) of leaves with undersides turned upwards on the ground.[400]

The Green Catbird feeds on insects and larvae, bird's eggs and nestlings, fruit, both native and cultivated, flower heads and large seeds.[324,519,1049,1205] Because it eats cultivated fruit and is reputed to be good eating, it has been shot illegally.[974] Where rainforest survives it is locally common.[400]

There are no measures of abundance.

Field Atlas Records: 643

1° blocks–with records: 27 (3%)

–with breeding records: 4 (15%)

Atlas No.: 676

Max. Rep. Rate: 22%
In block(s): 30/152

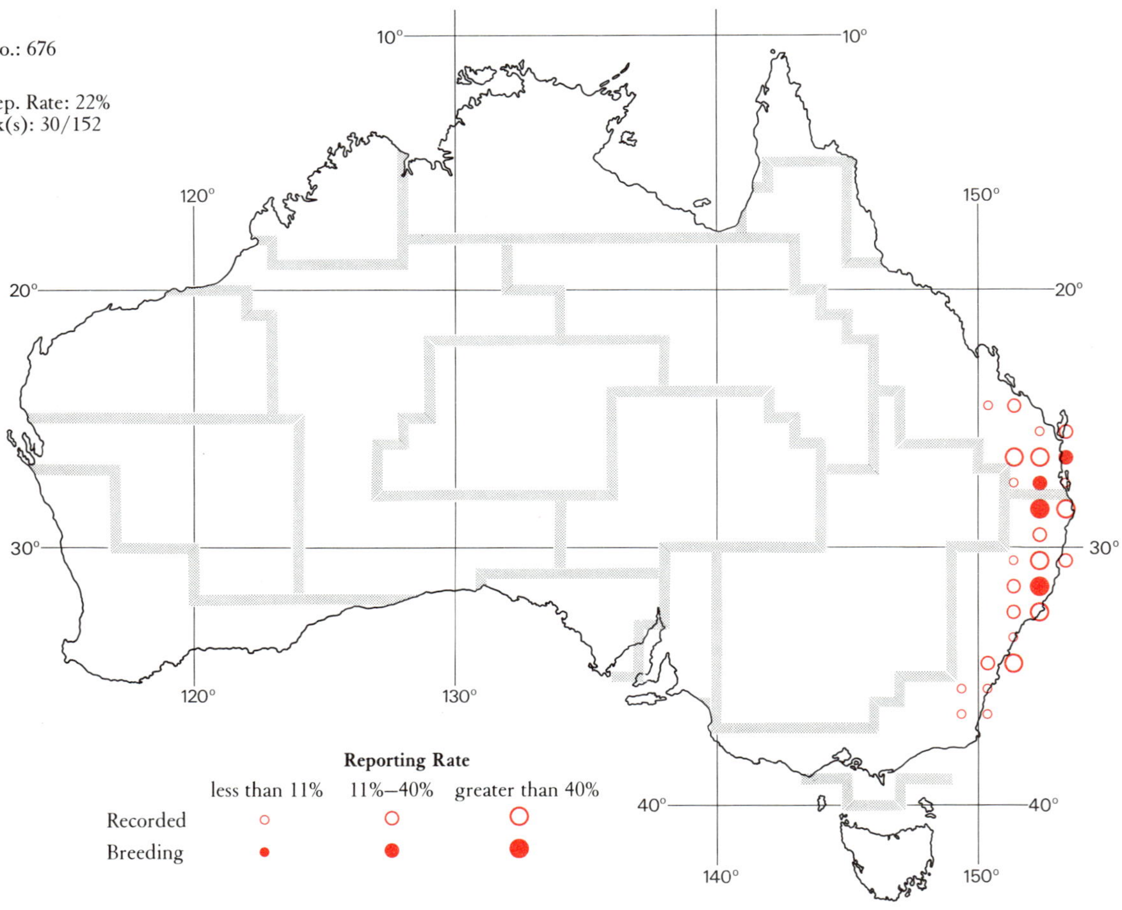

Paradise Riflebird

Ptiloris paradiseus

Records in the Field Atlas in the Calliope Ra. (24/150, 24/151) are of an apparently isolated population north of the previously accepted range of the species. From the Bunya Mts (26/151) southward, distribution is more or less continuous in the highlands of the Great Dividing Ra. The species is said to be sedentary[1292] but individuals will wander. There are records from high country (23/151) near Rockhampton in 1888 and 1930.[1089] One bird was reported near Murrurundi (31/150) in 1978.[1292]

The Paradise Riflebird lives in rainforest including that dominated by nothofagus and adjacent eucalypt forest. It is now largely confined to the highlands but formerly also occupied tracts of lowland rainforest most of which have been cleared. The Paradise Riflebird eats fruit, berries, spiders, insects and larvae, probing under bark like a treecreeper and stripping it from the trees.

On 3.5 km transects walked twice daily over ten days in Tooloom Scrub (28/152) an average of two Paradise Riflebirds was seen on each of at least six counts.[1342]

Field Atlas Records: 160

1° blocks–with records: 14 (2%)
–with breeding records: 3 (21%)

Atlas No.: 686

Max. Rep. Rate: 9%
In block(s): 26/151

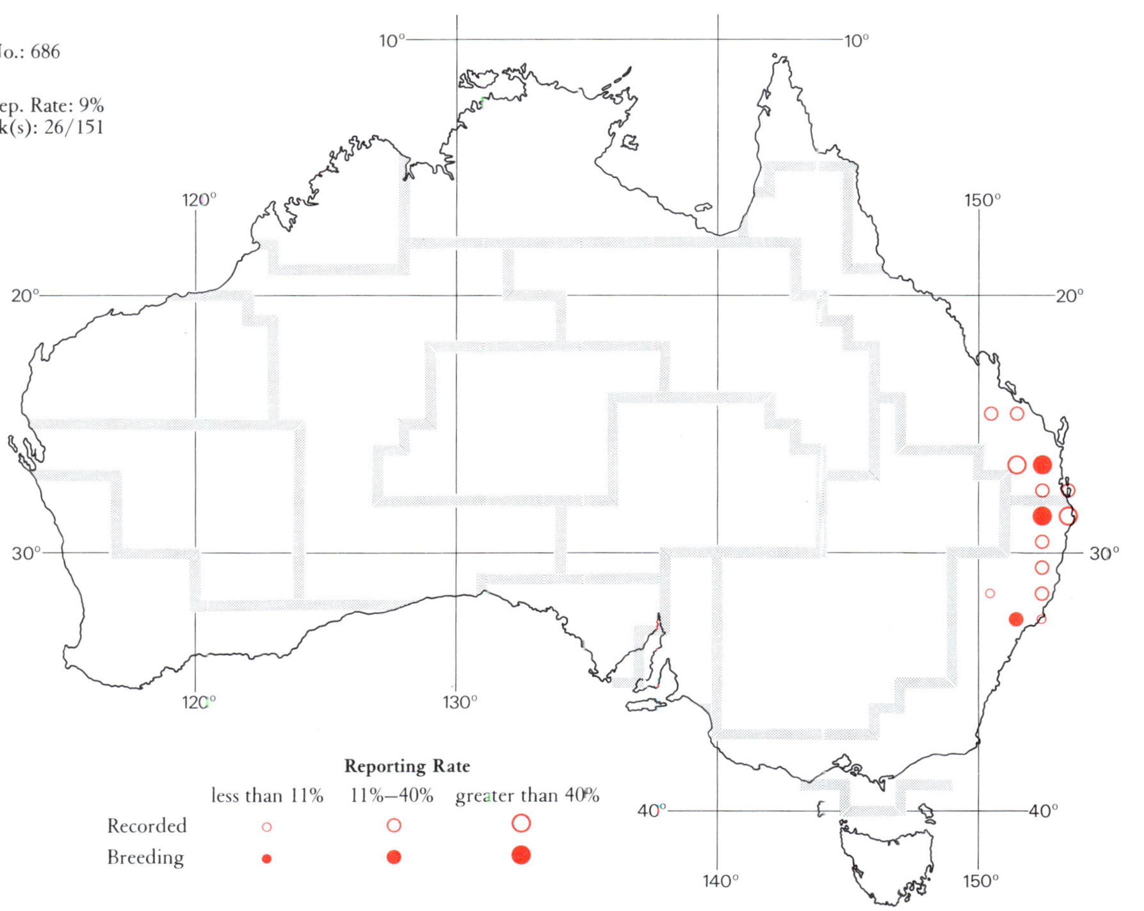

Victoria's Riflebird

Ptiloris victoriae

The endemic Victoria's Riflebird lives in the Atherton Region, inhabiting some coastal islands as well as the mainland. It was recorded during the Field Atlas in all the 1° blocks where it was recorded historically.

The species is present all year at some localities, the male appearing to keep to a territory, while the female may wander.[231,698] On the Bloomfield R (15/145) it was regarded as more common July-Jan. than at other times.[1343] Non-breeding flocks move out into eucalypt and paperbark woodland and flocks of up to 50 birds have been recorded in an orchard in winter.[1469]

Victoria's Riflebird lives in and on the fringes of rainforest where each male dominates all the vantage points over about 2 ha.[212] It feeds on fruit, native and cultivated, berries and insects and their larvae. The bird forages in many ways: using its bill to wrench off large pieces of bark in search of crevice-dwelling insects, chiselling into dead branches, probing into rotten wood on the forest floor and spiralling about the trunks of trees like a treecreeper.[12,212] It also feeds close to the ground on fruiting shrubs and on fallen fruit.[400] Because it damages cultivated fruit it is sometimes shot illegally.

There are no measures of abundance.

Field Atlas Records: 367

1° blocks–with records: 7 (1%)
–with breeding records: 3 (43%)

Atlas No.: 687

Max. Rep. Rate: 32%
In block(s): 15/145

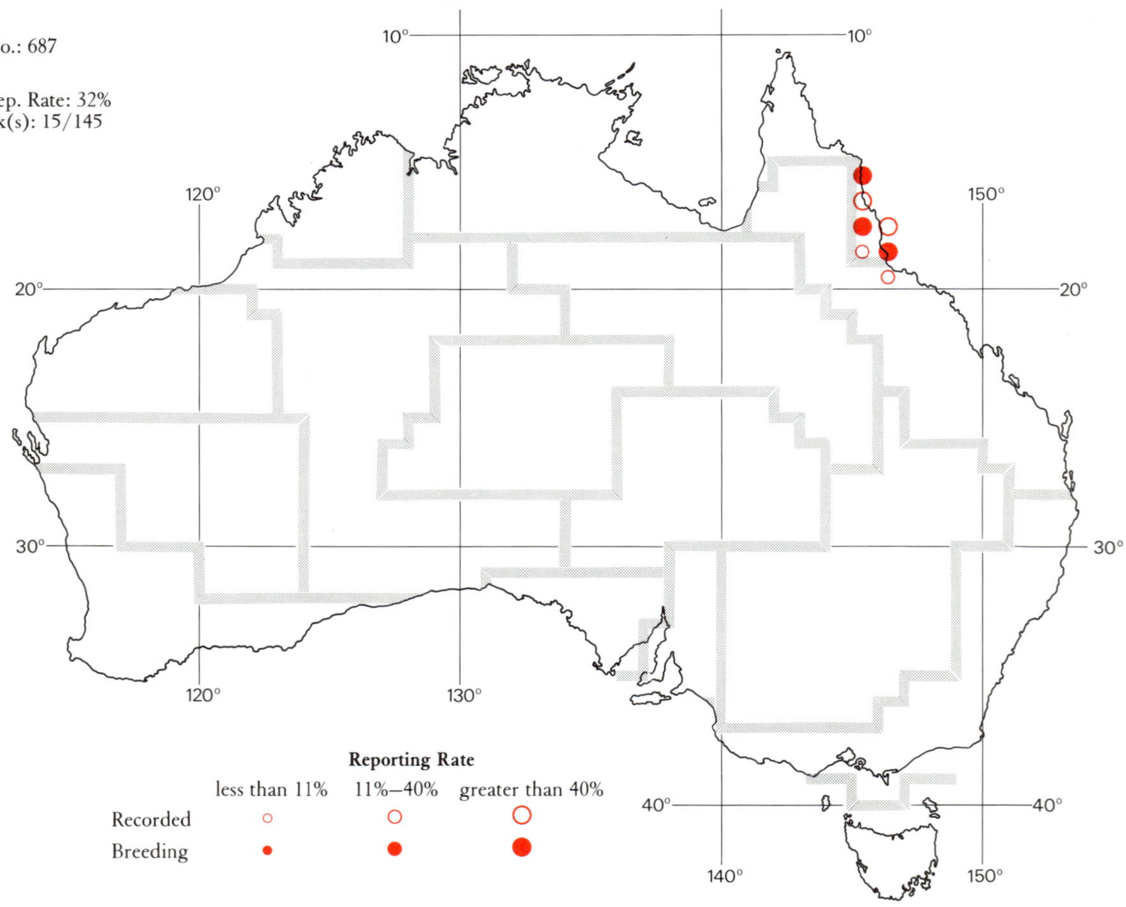

Magnificent Riflebird

Ptiloris magnificus

The Magnificent Riflebird lives in New Guinea[856] and in the Cape York Region. It has been recorded on Albany I. (10/142) in Torres Strait.[1813]

There is no information about movement.

The Riflebird lives in rainforest, including that bordering watercourses, and is usually heard before it is seen. Most foraging is in the mid- to upper levels of rainforest trees but birds descend to bathe in forest pools[400] and the male usually displays close to the ground.[129]

The Magnificent Riflebird feeds on insects and their larvae exposed by chiselling into dead branches, specially the dead canes of the lawyer vine[129] or by prising bark from the tree trunk and base of large branches with bill and foot. It also feeds on berries, fruit and seeds.[400]

One male ranged over a territory more than 300 m across.[1343] Density in New Guinea in lowland rainforest near Port Moresby (9S/147E) was 0.6 birds/ha.[129]

Field Atlas Records: 56

1° blocks-with records: 5 (1%)
-with breeding records: 1 (20%)

Atlas No.: 688

Max. Rep. Rate: 54%
In block(s): 12/143

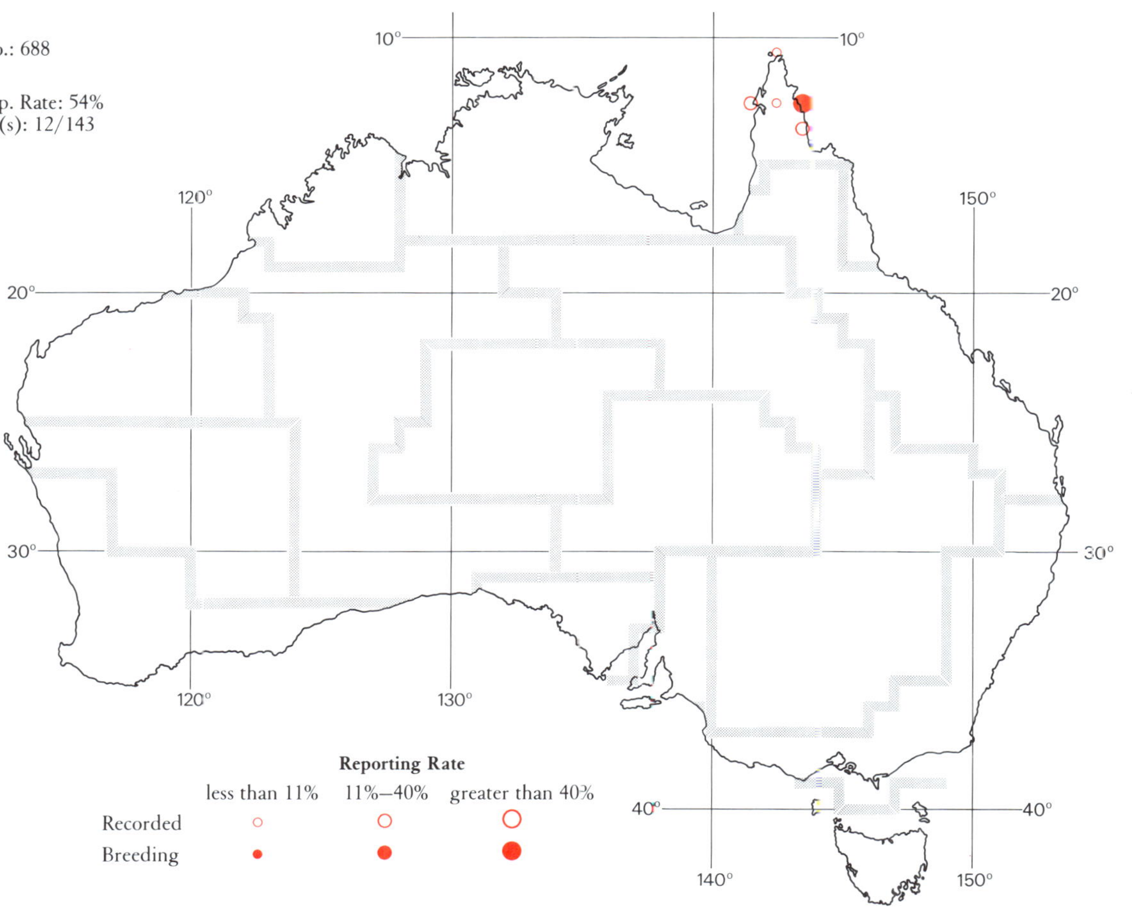

Trumpet Manucode

Manucodia keraudrenii

The Trumpet Manucode lives in New Guinea and adjacent islands, in the Cape York Region and on several islands in Torres Strait.[503] It has not previously been known from the western coast of the Cape York Region.

It has been suggested that the Manucode might be a partial migrant to New Guinea.[140,400] One early observer stated that the bird seemed to arrive in Australia early Aug. and depart at the end of Mar. when it gathered into flocks[1145] but another regarded it as sedentary near Somerset (10/142).[1343] Recent observers resident in Torres Strait doubt that migration takes place[503] and there is no evidence of a change in numbers in south-eastern New Guinea.[160] Reporting Rates in Australia were higher in summer (13.2%) than in winter (6.2%).

In Australia and much of New Guinea the Manucode is a lowland species. It lives singly, in pairs or, rarely, in small parties in the canopy on the margins of rainforest.[1147] The nest is often near that of a Black Butcherbird.[104,1145] In New Guinea the Manucode has been implicated as a brood parasite[775] but there are no field observations of Trumpet Manucode chicks being reared by other species. The species feeds on fruit, berries, seeds and insects and their larvae.[400]

Density in New Guinea in lowland rainforest near Port Moresby (9S/147E) was 0.5 birds/ha.[160]

Field Atlas Records: 47

1° blocks-with records: 6 (1%)
-with breeding records: 1 (17%)

Atlas No.: 689

Max. Rep. Rate: 40%
In block(s): 12/143

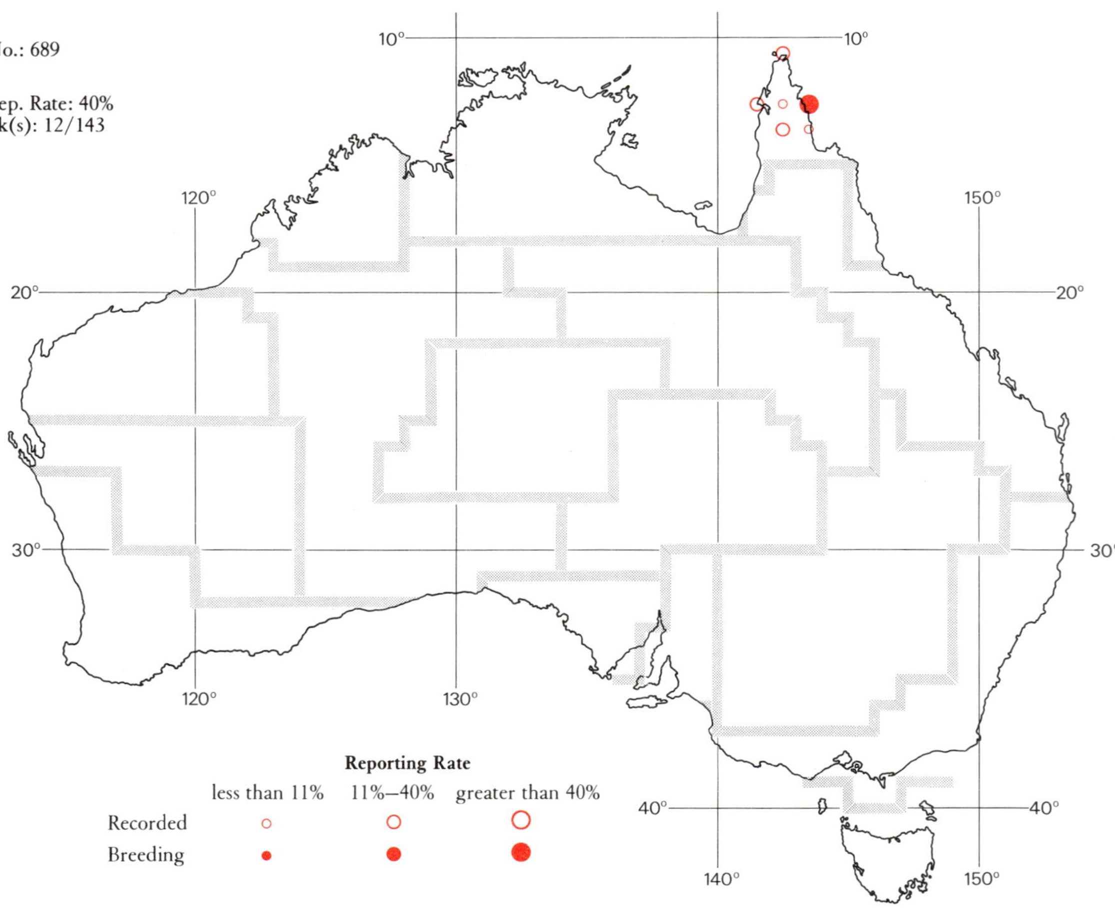

White-winged Chough

Corcorax melanorhamphos

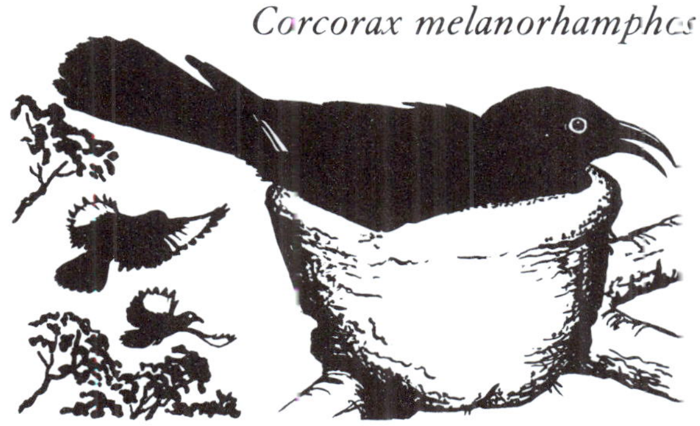

The endemic White-winged Chough has no subspecies and its range is unchanged from the historical periods. It is absent from Kangaroo I. (35/137).

At Canberra (35/149) groups wander within a home range of 800–1000 ha when not breeding;[1548] the species behaves similarly near Inverell (29/151).[94] The longest movement recorded is 10 km.[85]

The Chough lives in eucalypt woodland, forest and mallee in groups of 2–20. At Canberra more than 80% of the birds studied were in groups of 4–8.[1548] Flocks numbering 100 or more, sometimes reported, are temporary congregations of several groups. The species breeds communally, sometimes more than one female laying in the nest.

The Chough feeds on insects and other invertebrates for which it searches in the litter on the woodland floor and by probing into soft ground. It will also break open and eat freshwater mussels.[337] All members of a group join an individual that finds an abundant source of food. At Inverell birds followed a stereotyped pattern of searching for food daily, 30 birds covering 3 ha systematically for eight days then moving elsewhere.[94] The birds require water daily for drinking and, in the breeding season, to build their mud nests.

At Canberra each group defends a territory of about 20 ha in the breeding season.[1548]

Field Atlas Records: 11 805

1° blocks–with records: 207 (25%)
–with breeding records: 133 (64%)

Atlas No.: 693

Max. Rep. Rate: 80%
In block(s): 27/149

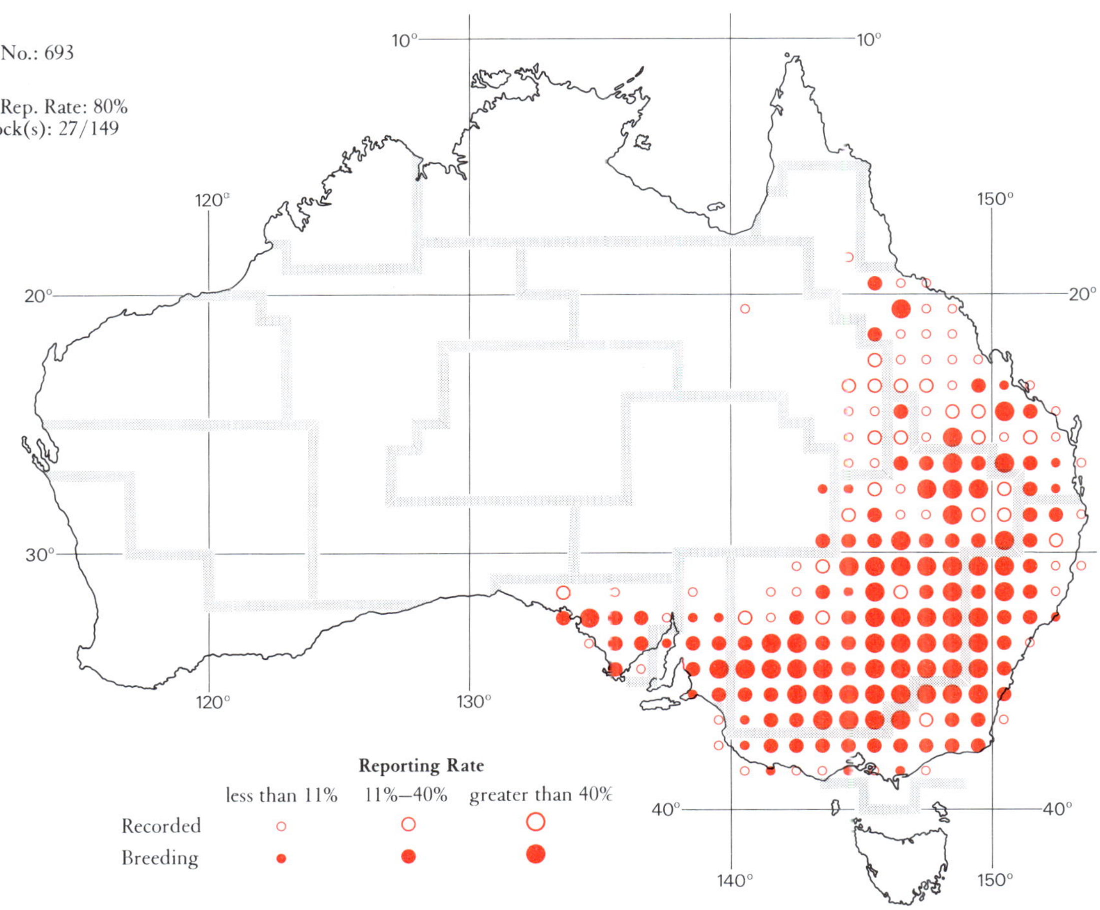

Apostlebird

Struthidea cinerea

No subspecies of the endemic Apostlebird are recognized.[1422] The species' distribution has apparently changed since 1900 (p. 670). In the Top End an isolated population has spread since it was first recorded in 1922 and in Eastern Qld the bird has established itself on the coast since 1900.[1709] It has also moved into the SA Ra. Region since 1950. Only in the South-East Region have declines been reported: a small population has disappeared from near Gilgai (29/151) since 1975;[93] and another, apparently isolated, population has gone from between Dimboola (36/142) and Naracoorte (36/140) since 1935.

Banding has shown the Apostlebird to be sedentary in some places[334] but near the edge of its range it may desert its breeding territory in winter to form large flocks in places where food, water and shelter are available.

The Apostlebird lives in eucalypt woodland, including that lining streams in the inland, and mallee. It is never far from water and needs mud to build its nest. It lives and breeds communally,[501] the group not always numbering the 12 individuals that its name suggests. There is some interchange of members in a group[93] and sometimes 20 or more birds will stay together for several years though 8–18 are more usual sizes.[501] Groups do not often conflict with each other; each is focused on a source of water and tends to avoid other groups.[653]

The Apostlebird is an opportunistic feeder.[93] In summer it eats mainly insects taken on the ground and in winter mainly small seeds.[653] During plagues of house mice it will catch and kill them.[840] It drinks frequently, specially in summer. The settlement of the inland, with the accompanying provision of stockwater, may have benefited the species and enabled it to expand its range.

At Inverell (29/151) group territories were about 9.5 ha in extent.[95]

Field Atlas Records: 6749

1° blocks–with records: 218 (27%)
–with breeding records: 117 (54%)

Atlas No.: 675

Max. Rep. Rate: 92%
In block(s): 27/149

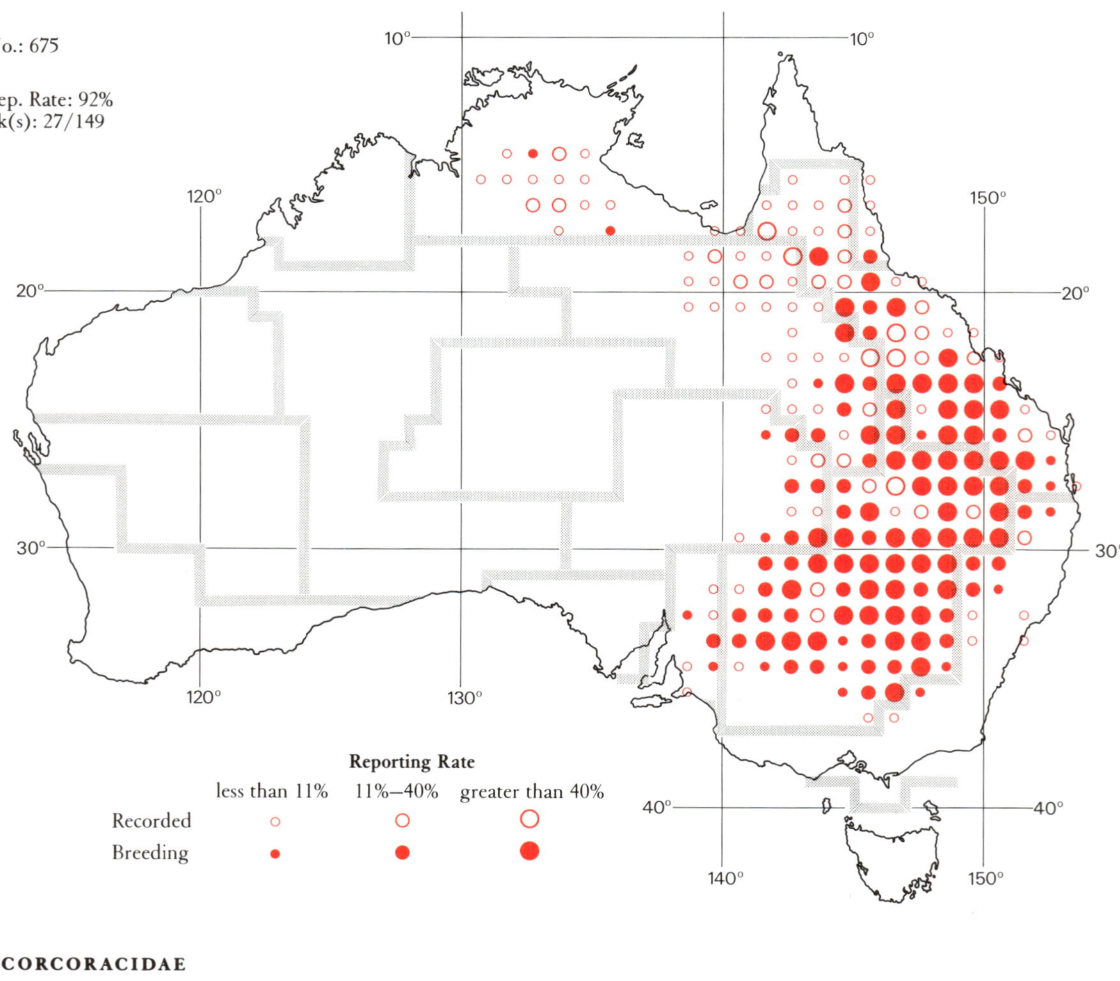

Australian Magpie-lark

Grallina cyanoleuca

The Australian Magpie-lark also lives in southern New Guinea.[550] It was either overlooked there or has recently invaded because the first New Guinea record was not made until 1969.[42] In Australia the species is a vagrant in the Tas. Region, more often reported from King (39/144) and Flinders (40/148) Is than from Tas. itself. The only Tas. breeding record is of eggs collected in 1903. The Magpie-lark has become established on Kangaroo I. (35/137) since 1950.[121]

In three areas in the South-West Region banding showed that breeding adults were sedentary and territorial throughout the year.[1764] After breeding, young birds from the local area formed flocks but there was no evidence of regular seasonal movement. In other places apparent seasonal movements have been reported. The Magpie-lark is a winter visitor to the tip of Cape York (10/142) but has an irregular pattern of movement within Torres Strait.[503] At Inverell (29/151) only males remained and defended territories in winter and flocks were observed flying north in autumn.[96] At Darwin (12/130) there were flocks in the dry winter season (May-Sept.) but the birds were scarce for the rest of the year[423] (Reporting Rates: summer 32%, winter 66%). At Coolup (32/115) part of the population migrated in summer, after breeding, perhaps to wetter areas.[1511] Except near Darwin the Field Atlas does not suggest any substantial seasonal shift of the population. Records in the Tas. Region were made at all times of year.

The Magpie-lark lives on plains, pasture and cleared areas, provided that surface water is available to construct its mud nest. The post-breeding flocks are sometimes large (20–3000 birds) and roost together at night.[698,792,1511] The bird feeds mainly on small insects from pasture and muddy areas.[1511] The provision of permanent water for stock enables it to remain all year in places formerly vacated during dry weather. Its establishment on Kangaroo I. has followed clearing for agriculture.[121] At Northam (31/116), where many natural waters are becoming saline, the species seems to have declined in abundance in farmland.[1216]

At Jamieson (37/146) in eucalypt forest density was 0.1 birds/ha.[1469] In pasture in the South-West most pairs held permanent territories of 8–32 ha (average 15 ha, 0.1 birds/ha).[1511,1764]

Field Atlas Records: 47 200

1° blocks–with records: 674 (83%)
–with breeding records: 340 (50%)

Atlas No.: 415

Max. Rep. Rate: 94%
In block(s): 28/148,29/148

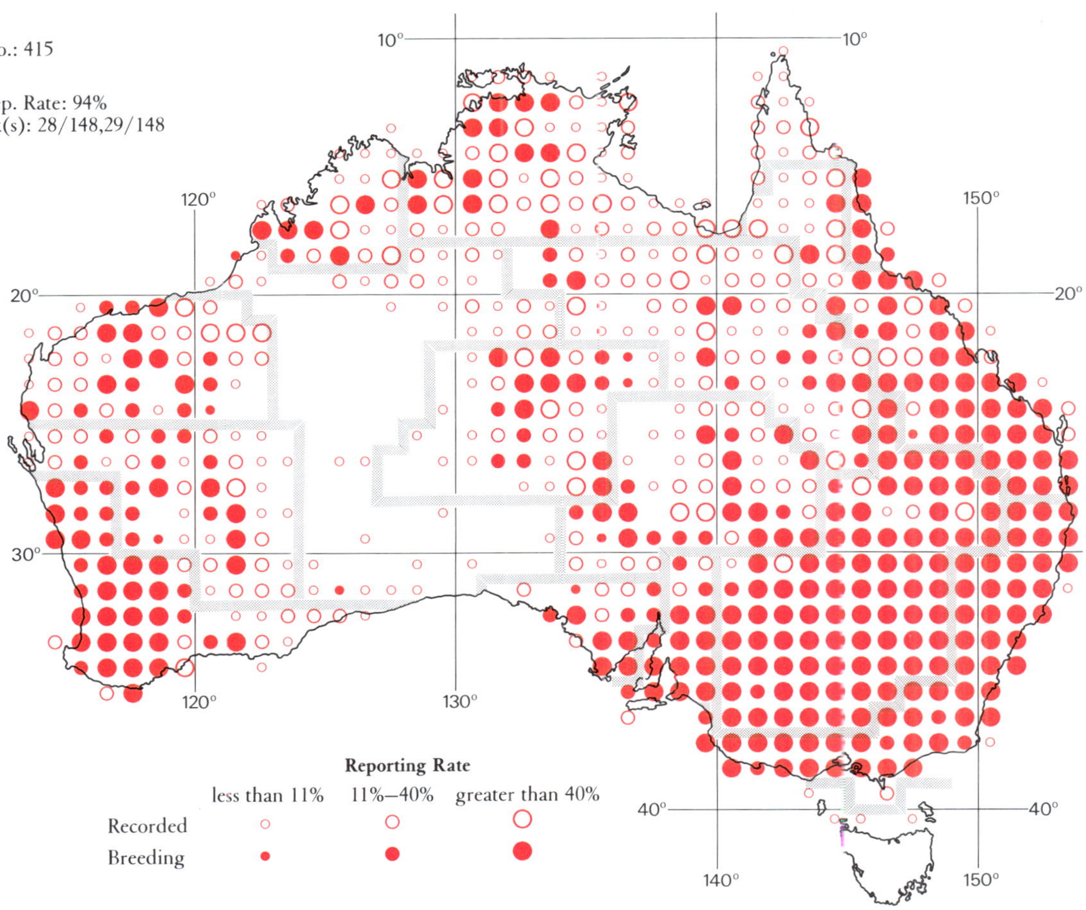

White-breasted Woodswallow

Artamus leucorhynchus

The White-breasted Woodswallow ranges from western Pacific islands through south-eastern Asia and New Guinea to Australia. Views about subspecies in Australia differ.[981,1239] The species may be increasing its range in the inland: before 1977 there was only one record from the Central Ra. Region in the Atlases; most of the Field Atlas records were in 1978–79 and may represent an irruption but there was a small number of sightings later. In the Tas. Region the only record was of four birds on King I. (39/144) in June 1961.[734]

Most White-breasted Woodswallows from southern Australia are thought to migrate north in winter.[981] Reporting Rates south of 26°S support the suggestion of a general exodus (summer 5.1%, winter 1.8%). In the southern Murray-Darling the bird is present only Sept.-Apr.[833] Observers in the Atherton Region regard the local population as sedentary though more birds are present in winter[134,231] but the Field Atlas does not suggest any winter augmentation of the population (Reporting Rates: summer 30.0%, winter 28.1%). Near Darwin (12/130) the reverse is true with the Field Atlas suggesting a winter influx (summer 6.0%, winter 25.0%) though a local observer noted no seasonality in abundance.[423] It was thought that Australian birds could migrate overseas[981] but no passage has been observed through Torres Strait or in New Guinea, where the species is considered sedentary.[158,503] Some specimens taken in the L. Eyre Region in winter resemble northern breeding birds, which suggests that, as well as the northward movement of southern breeding birds, there may be some southward movement of northern populations.[748]

Always closely associated with water, the White-breasted Woodswallow lives in forest, woodland and mangroves along the coast;[104,748,1876,] inland in eucalypt woodland along watercourses; along roads in sugarcane-growing areas; and on islands of the Great Barrier Reef.[1732,1813] Birds cluster and 200 have been observed going to roost.[134] Sometimes breeding is colonial. Overseas, birds breed communally[379] and in Australia up to five adults have been observed feeding young.[900] The bird eats small flying insects.[900]

In New Guinea density was 0.1 birds/ha in eucalypt woodland near Port Moresby (9S/147E) but greater in urban areas.[158]

Field Atlas Records: 5369

1° blocks–with records: 386 (48%)
–with breeding records: 114 (30%)

Atlas No.: 543

Max. Rep. Rate: 60%
In block(s): 24/153

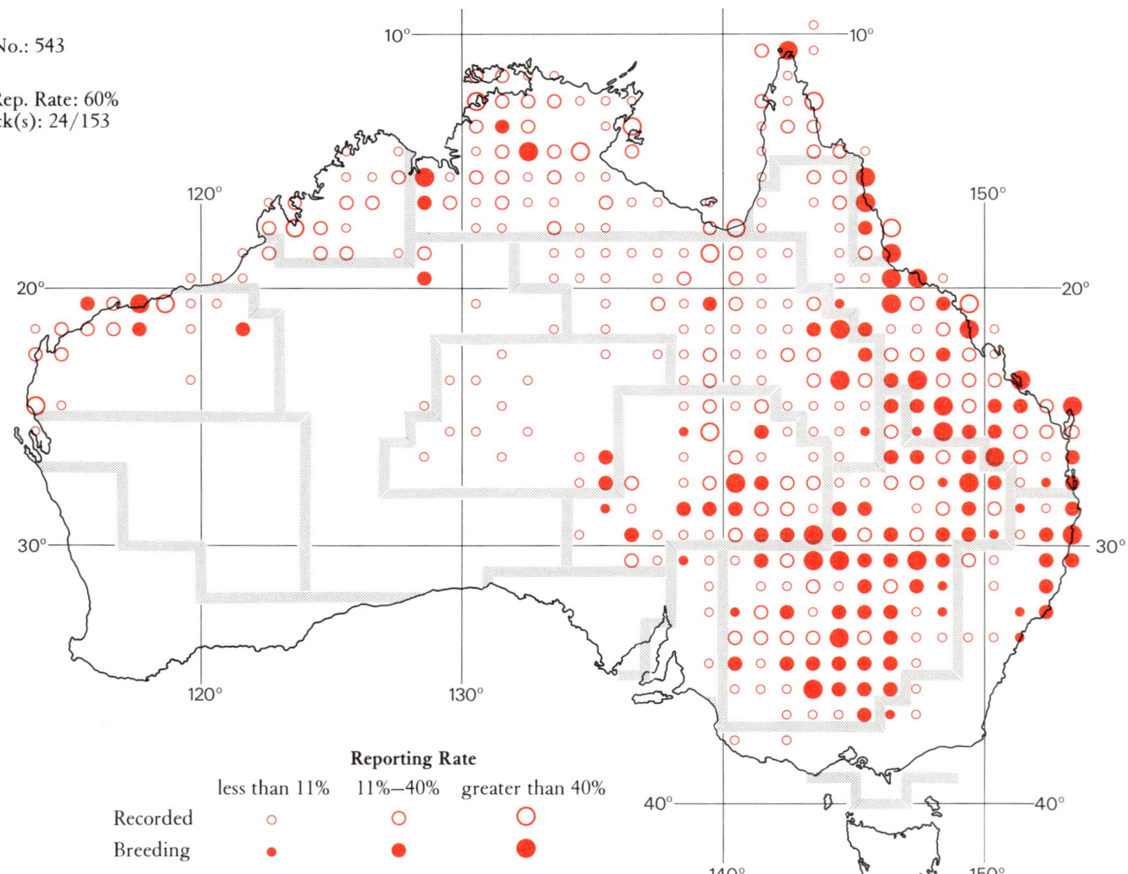

Masked Woodswallow

Artamus personatus

No subspecific variation of the endemic Masked Woodswallow is recognized.[981] The species breeds in mixed colonies with the White-browed Woodswallow and there are several reports of apparent hybridization in the field.[1736]

Small numbers of the Masked Woodswallow occasionally reach King I. (39/144) and it bred there in 1972.[717] North of 20°S breeding is rarely reported.

Large flocks characteristically appear suddenly at a locality. These influxes have sometimes been associated with the arrival or approach of low-pressure weather systems[246,1512,1835] but are elsewhere identified as a regular seasonal phenomenon.[189,627,833,981,1508] Reporting Rates south of 26°S were higher in summer (2.9%) than in winter (1.4%), which suggests a regular pattern of movement though the bird remains widespread in southern Australia in winter.

The Masked Woodswallow inhabits open wooded country. Usually it is observed in flocks, frequently in company with the White-browed Woodswallow.[692,1835] Nesting is colonial but communal breeding has not been reported. The Masked Woodswallow eats insects, mainly caught in the air, and sometimes nectar from flowering trees and shrubs.[456] Large numbers of the Woodswallow have been reported to congregate round swarms of grasshoppers[522] and ants[111] as well as flowering eucalypts[1602] and legumes.[1872] The bird may be important in the pollination of mistletoe and other plant species.[1403]

There are no measures of abundance.

Field Atlas Records: 2834

1° blocks–with records: 518 (64%)

–with breeding records: 106 (20%)

Atlas No.: 544

Max. Rep. Rate: 43%
In block(s): 29/143

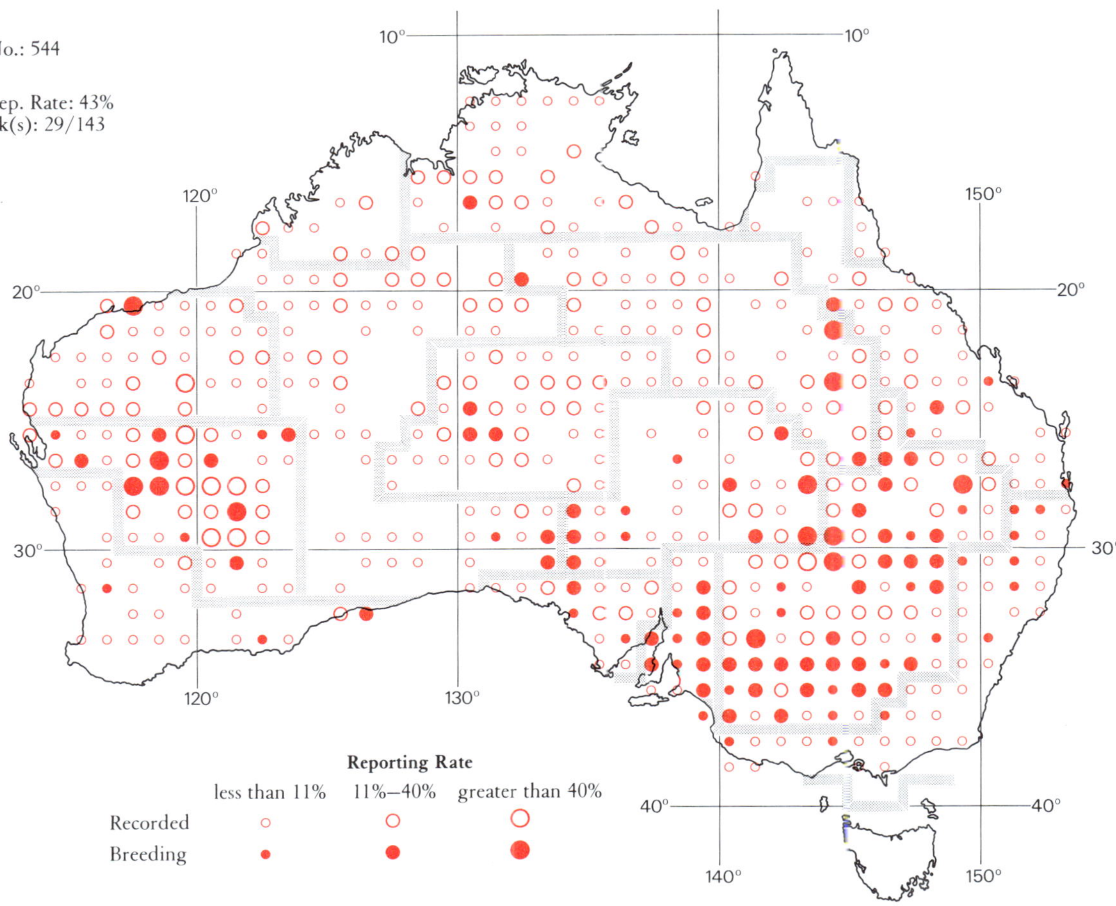

White-browed Woodswallow

Artamus superciliosus

The endemic White-browed Woodswallow shows no subspecific variation. It breeds in mixed colonies with the Masked Woodswallow and there are several reports of apparent hybridization in the field.[1736] There are occasional irruptions into the Tas. Region: in Tas. itself, the bird was recorded breeding in 1972–73 and again in 1982 but was not sighted between these years; on King I. (39/144) it may be a more regular visitor.[717,739] On the Australian mainland breeding has not been reported north of 19°S. Unlike the Masked Woodswallow, only a few White-browed Woodswallows visit, or breed in, WA.[307]

A north-south pattern of movement is suggested by the Field Atlas. Reporting Rates in summer in the South-East (6.9%), Murray-Darling (25.6%) and SA Ra. (7.1%) Regions were much higher than in winter (0.3%, 3.3%, 1.5% respectively). Most Tas. records were Nov.-Apr. In WA during the Field Atlas records were made in all seasons except summer, which probably reflects lack of observers inland in summer rather than a real seasonal pattern.

The White-browed Woodswallow inhabits woodlands.[981] It is usually in large flocks, often in company with the Masked Woodswallow .[692,1835] In the breeding season the unexpected arrival of a flock may be followed by the speedy establishment of a breeding colony.[490,1460] Such irruptions rarely bring birds to nest in the same site in successive years. The White-browed Woodswallow feeds on insects, either in the air or on the ground.[748] It has often been observed arriving to feed on grasshoppers when they are abundant.[189,1099] It also consumes nectar from flowering shrubs and trees.[356]

Colonies of 17 and 38 nests were observed in a study site of about 12 ha at Bombala (36/149).[1460] During an irruption at Jamieson (37/146) in 1977 more than 50 birds were present, most of them breeding, in a small section of the 20 ha under observation.[1469] In eucalypt woodland at Armidale (30/151) density was 0.25 birds/ha.[582]

Field Atlas Records: 4653

1° blocks–with records: 348 (43%)
–with breeding records: 141 (41%)

Atlas No.: 545

Max. Rep. Rate: 54%
In block(s): 27/149

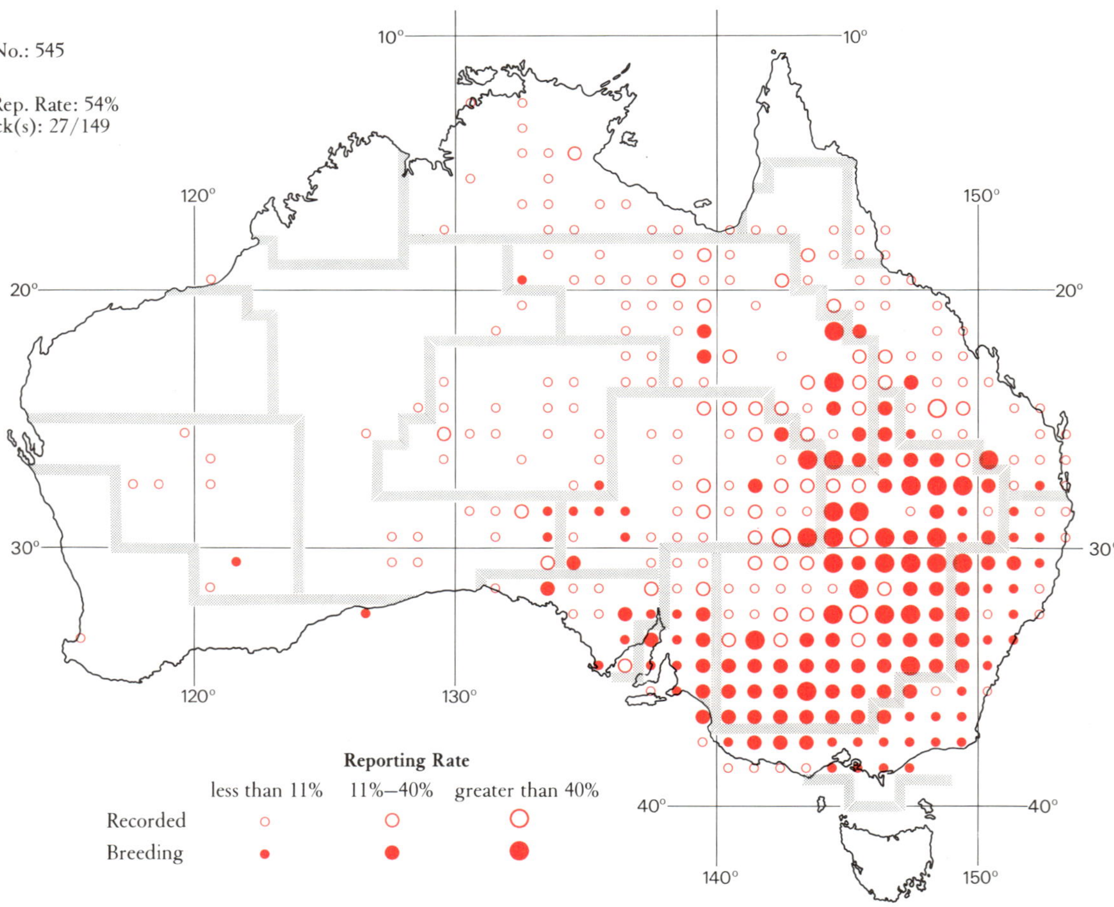

Black-faced Woodswallow

Artamus cinereus

The range of the Black-faced Woodswallow extends to Timor (9S/125E) and New Guinea.[612] The two subspecies recognized in Australia lie east and west of the Great Dividing Ra. with a hybrid zone between them; there are differing opinions about their nomenclature.[612,1250]

The species ranks as one of the most commonly reported birds in the Field Atlas though not recorded in the Tas. Region.

In most parts of Australia the Black-faced Woodswallow appears to be sedentary and near Rawlinna (31/125) banding has shown it to be so: individuals remained near their banding place, one for 26 months.[246] Fewer are observed round Darwin (12/130) during the wet summer season than in the dry winter season[423] but Reporting Rates showed a reverse trend (summer 16.6%; winter 10.2%). Changes in numbers have been noticed at several places in Qld though with no seasonal pattern.[512]

The Black-faced Woodswallow lives in eucalypt woodland, acacia scrub and spinifex usually where there are shrubs and dead trees on which to perch. It exhibits clustering behaviour[1701] and has been recorded breeding communally.[501] The Woodswallow feeds on insects mostly caught in flight but sometimes taken from the ground.[748,900] Near Kununurra (15/128) the bird fed almost exclusively on grasshoppers for several weeks at the end of the wet summer season. Nectar from flowering shrubs is also consumed.

Following clearing of the forest and woodland in the South-West, the range of the Black-faced Woodswallow has extended into the far south-western corner (south of 33°S and west of 117°E).[1638] Clearing in south-eastern Australia appears not to have resulted in a similar expansion. On the Nullarbor Plain density was 0.25 birds/ha.[246]

Field Atlas Records: 12 675
1° blocks–with records: 672 (83%)
–with breeding records: 180 (27%)

Atlas No.: 546

Max. Rep. Rate: 94%
In block(s): 21/122

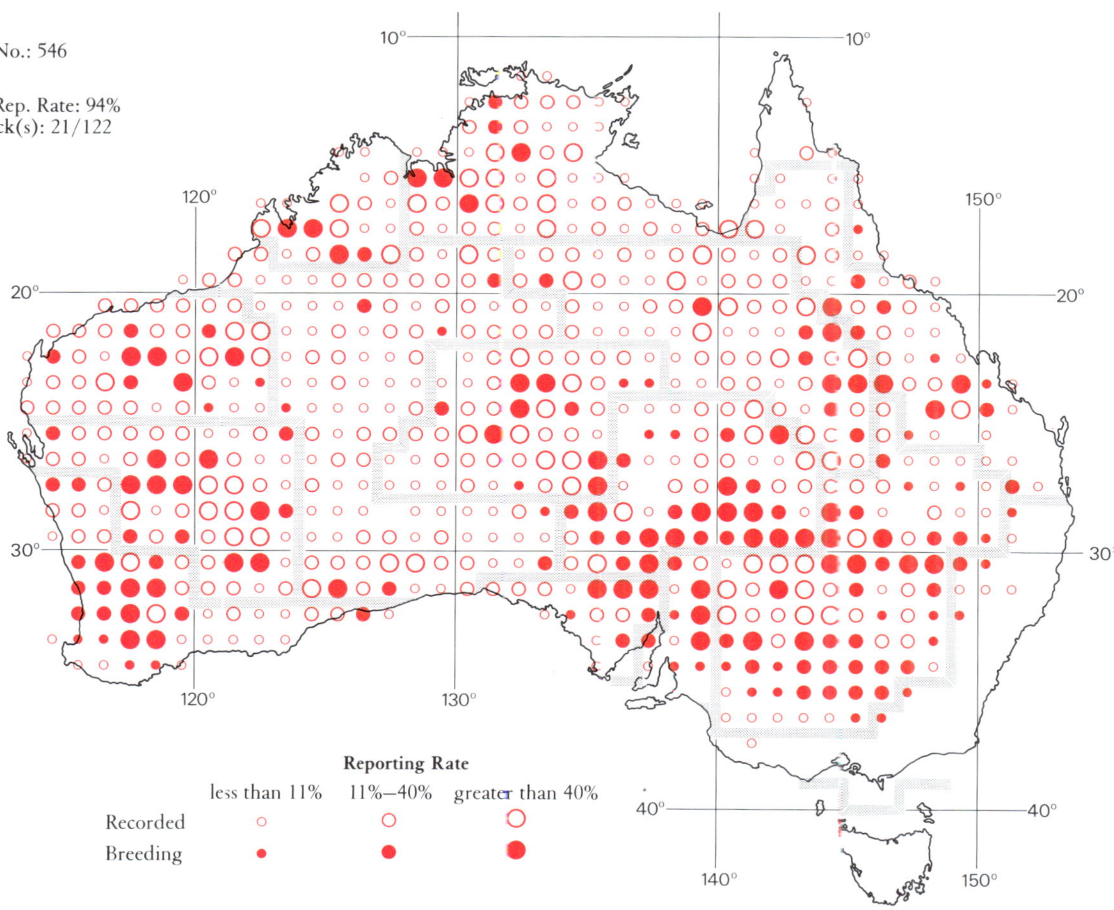

Dusky Woodswallow

Artamus cyanopterus

Two subspecies of the endemic Dusky Woodswallow are recognized: *cyanopterus* in the east and *perthi* in the west. Differentiation between them is not great and neither is the gap between their ranges.[981] There are historical breeding records in the Atherton Region, then a gap to 23°S, which suggests that the northern breeding population may be isolated. Though breeding in Tas. it does not do so on Flinders I. (40/148).

In parts of its range the Dusky Woodswallow is migratory and in others sedentary. In the Atherton Region it is present all year[698] though there were no Field Atlas reports in summer north of 21°S; the Atherton population must be small. In the South-East and Tas. Regions Reporting Rates were higher in summer (28.5% and 22.5% respectively) than winter (10.5% and 0.3%), which reflects migration north in winter. In Canberra (35/149) birds arrive in spring in flocks of 10–30, breed and flock again at the end of summer before migrating.[653] At Jamieson (37/146) birds are present only Sept.-Apr.[1469] In Tas. migrating flocks of 10–200 Woodswallows have been observed in Apr.[1273] Records such as those on Cooper Ck (27/140) and the Paroo R (28/144) show that birds will move inland along watercourses. The first was a group of 12 in June 1978 and the second 5-6 adults in Aug. 1980. In the South-West no movement has been detected.

The Dusky Woodswallow inhabits woodland and clearings in forest. Consequently it is almost completely absent from western Tas. except on the coast. It often lives in flocks, some estimated at 200–300 birds[748] and is known to roost in clusters at night[1206] up to 200 individuals hanging on to the trunk of a tree.[653] It has been recorded breeding communally.[410,501] The members of a flock remain in contact, roaming widely over the surrounding countryside once breeding is over. The Dusky Woodswallow feeds mainly on insects, mostly catching them in the air, sometimes on the ground.[653] It also seeks them in bark crevices like a treecreeper[1120] and feeds in the leafy branches of eucalypts.[777] Grasshoppers and bees are often taken and the bird also consumes nectar.[129,356]

At Armidale (30/150) in eucalypt woodland density was 0.01–0.17 birds/ha[582] and at Jamieson in eucalypt forest 1.0–2.5 birds/ha.[1469]

Field Atlas Records: 14 229

1° blocks–with records: 291 (36%)
–with breeding records: 153 (53%)

Atlas No.: 547

Max. Rep. Rate: 56%
In block(s): 32/134

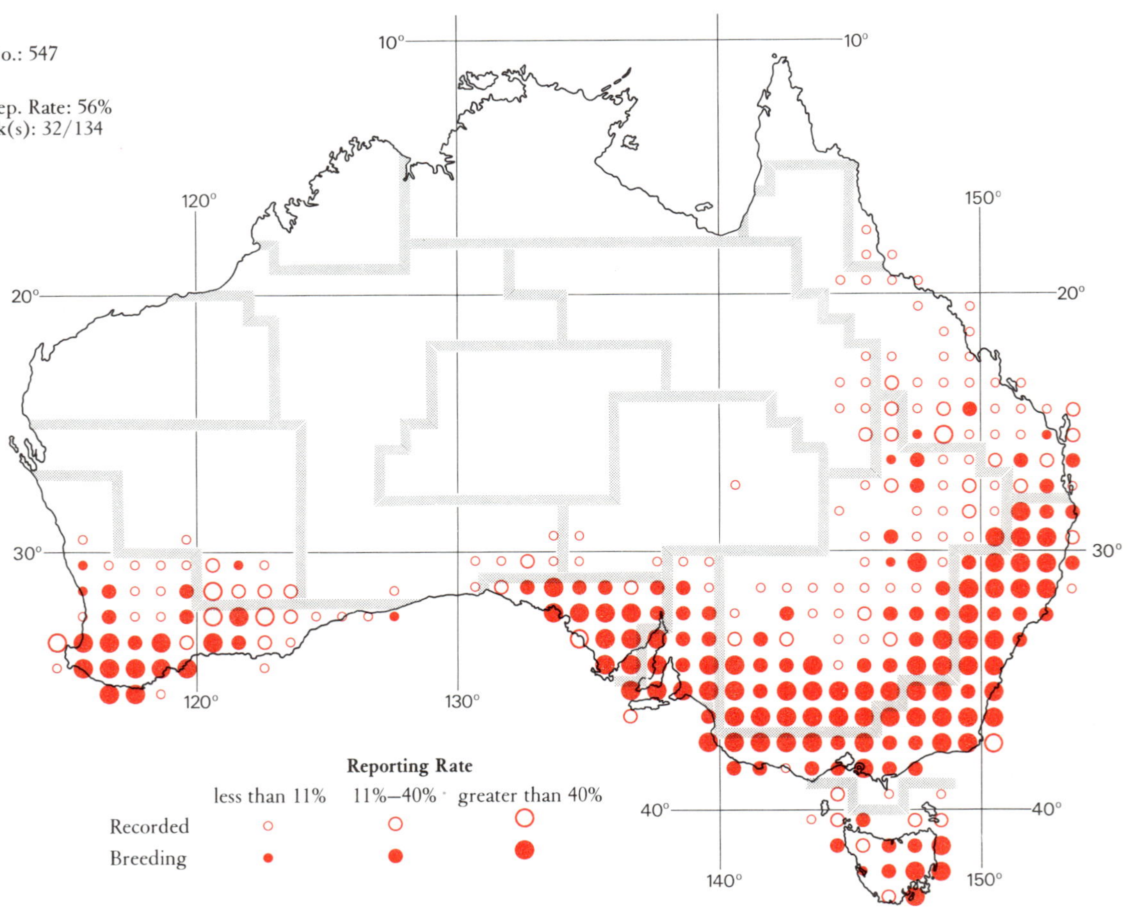

Little Woodswallow

Artamus minor

No subspecific variation is recognized in the endemic Little Woodswallow.[981] The populations at the northern end of the Flinders Ra. (39/138) and in the Gawler Ra. (32/136) appear to be isolated and were overlooked until the 1960s.[387,547] There are occasional records of vagrants farther south, e.g. L. Merreti (34/140) in Jan. 1966,[705] Edenhope (37/141) in Sept. 1941.[382]

The Little Woodswallow is considered resident in some parts of its range and nomadic rather than migratory in others.[423,787,981,1508,1871,1880] Reporting Rates were higher in summer (1.7%) than winter (0.4%) in the Murray-Darling Region and higher in winter (23.4%) than in summer (12.9%) in the Top End, which suggests that there may be some movement south in summer and north in winter.

The Little Woodswallow usually inhabits acacia scrub and spinifex where there are gorges in rocky country[748,981,1871] but also lives in tussock grassland in the eastern parts of its range. It is known to cluster in roosts at night. Generally flocks number 20 or fewer though more than 200 have been recorded in a communal roost.[748] Communal breeding has been recorded.[529]

The Little Woodswallow takes most of its insect food in flight. Ants are sometimes found in guts[748] and the bird feeds on the nectar of flowering shrubs, sometimes in company with other species of woodswallow.

There are no measures of abundance.

Field Atlas Records: 2112

1° blocks-with records: 339 (42%)
-with breeding records: 41 (12%)

Atlas No.: 548

Max. Rep. Rate: 57%
In block(s): 13/132

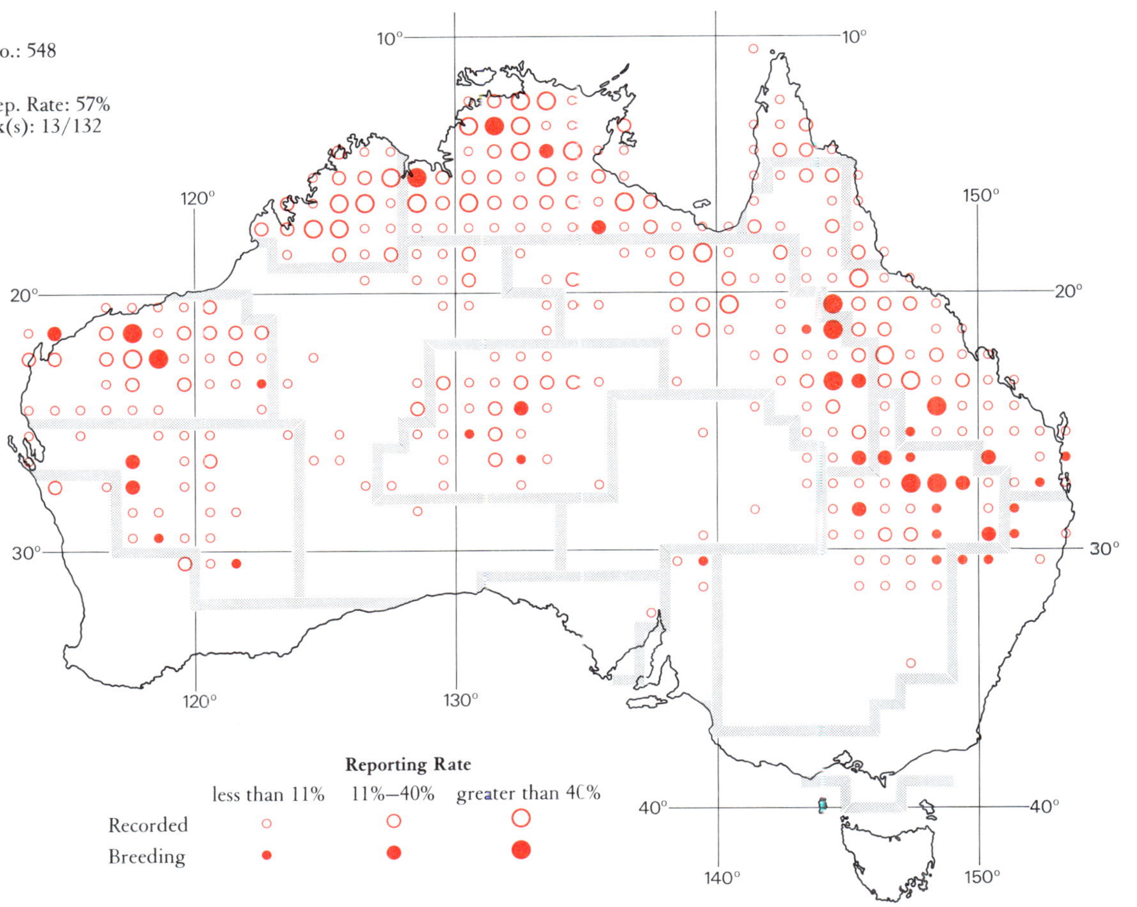

Black Butcherbird

Cracticus quoyi

The Black Butcherbird lives in New Guinea and Australia. Birds of the Cape York Region and Top End are sometimes considered the same subspecies and sometimes separated. The birds in the Cape York Region, *quoyi*, show only one type of plumage; those between the Endeavour R (15/145) and the Herbert R (18/146) are *rufescens*, distinguished by having a brown plumage type, thought by some to be immatures, as well as the usual black.[748,1422,1703,1709,1711] A small, unnamed population between Proserpine (20/148) and Port Clinton (22/150) has only a black plumage type.

There are no records from 1° blocks round the Gulf of Carpentaria other than those shown. The Top End population has been reported as far south as Port Keats (14/129).[71]

The Black Butcherbird is sedentary in rainforest in New Guinea[129] and is present all year in the Atherton Region[698] and round Darwin (12/130).[423] It lives in mangroves and rainforest, inland and coastal. In the Cape York Region the Trumpet Manucode often builds its nest nearby[104,1145] apparently for protection. The Black Butcherbird feeds on crabs,[748,1859] fruit[856] and many large insects, small vertebrates and, specially, nestling birds.[429,1194]

Density in New Guinea in lowland rainforest near Port Moresby (9S/147E) was 0.3 birds/ha.[160]

Field Atlas Records: 557

1° blocks-with records: 34 (4%)

-with breeding records: 8 (24%)

Atlas No.: 701

Max. Rep. Rate: 51%
In block(s): 12/143

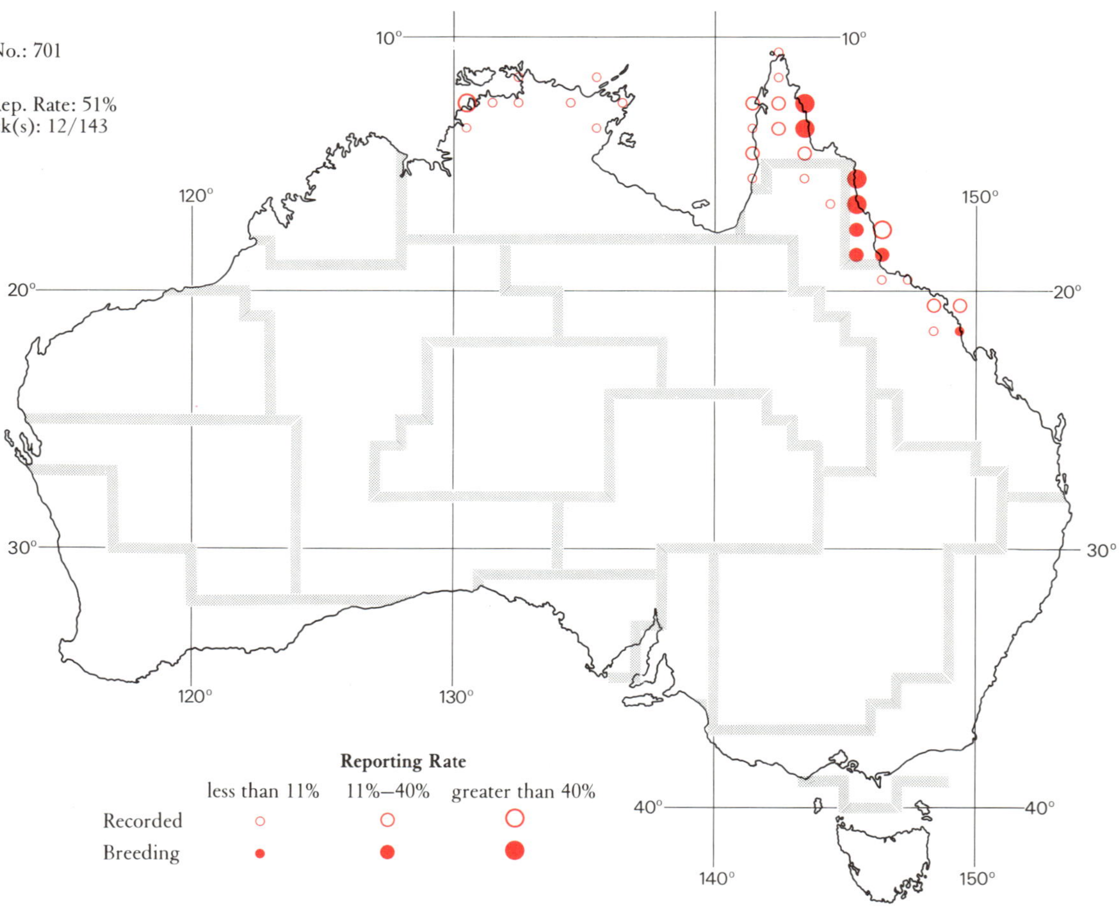

Grey Butcherbird

Cracticus torquatus

There are four subspecies of the endemic Grey Butcherbird: *argenteus* in the Top End; *latens* in the Kimberley; *cinereus* in Tas.; and *torquatus* everywhere else.[614] The Atlases show that the subspecies *argenteus* and *latens* are separated by a gap of no more than 110 km, the closest records being in the 10′ blocks 14°25′/129°55′ and 15°25′/129°55′; the area between was scarcely visited. The species is absent from the mid-west and south-west of Tas., King I. (39/144), the Furneaux Group (39/148), Maria I. (42/148) and Kangaroo I. (35/137).

Near Sydney (33/151) the Grey Butcherbird is sedentary and territorial throughout the year.[919] In Darwin (12/130), where wet and dry season conditions contrast, it retreats from flooded creeks during the wet summer season.[426] The Field Atlas suggests no large-scale seasonal movement and the longest recorded is 11 km.[85]

The Grey Butcherbird lives singly or in pairs in eucalypt woodland, forest margins and suburban parks. At Wollomombi (30/152) it occupies cleared land adjacent to woodland but only immatures have been observed to enter the woodland.[129] In the lower South-East Region and Tas. it is absent from high altitudes.[734] Communal breeding has been reported.[501] The Grey Butcherbird feeds mainly on insects[1049] though it also takes birds including nestlings.[908,919,1602a] Small reptiles and mice are regularly collected by the bird and vegetable matter, seeds and berries have been found in its guts.[748]

The species has slowly disappeared from some places since European settlement. At one locality in the Orange (33/149) district it had gone by 1920 and was absent from most of the district by 1940, with only occasional sightings since then.[784]

Density at Armidale (30/151) in eucalypt woodland was 0.22 birds/ha;[582] at Eyre Bird Observatory (32/126) in mallee in Aug. 0.23 birds/ha;[464] and on the Nullarbor Plain 0.01 birds/ha.[246]

Field Atlas Records: 19 780

1° blocks–with records: 517 (64%)

–with breeding records: 184 (36%)

Atlas No.: 702

Max. Rep. Rate: 86%
In block(s): 27/149

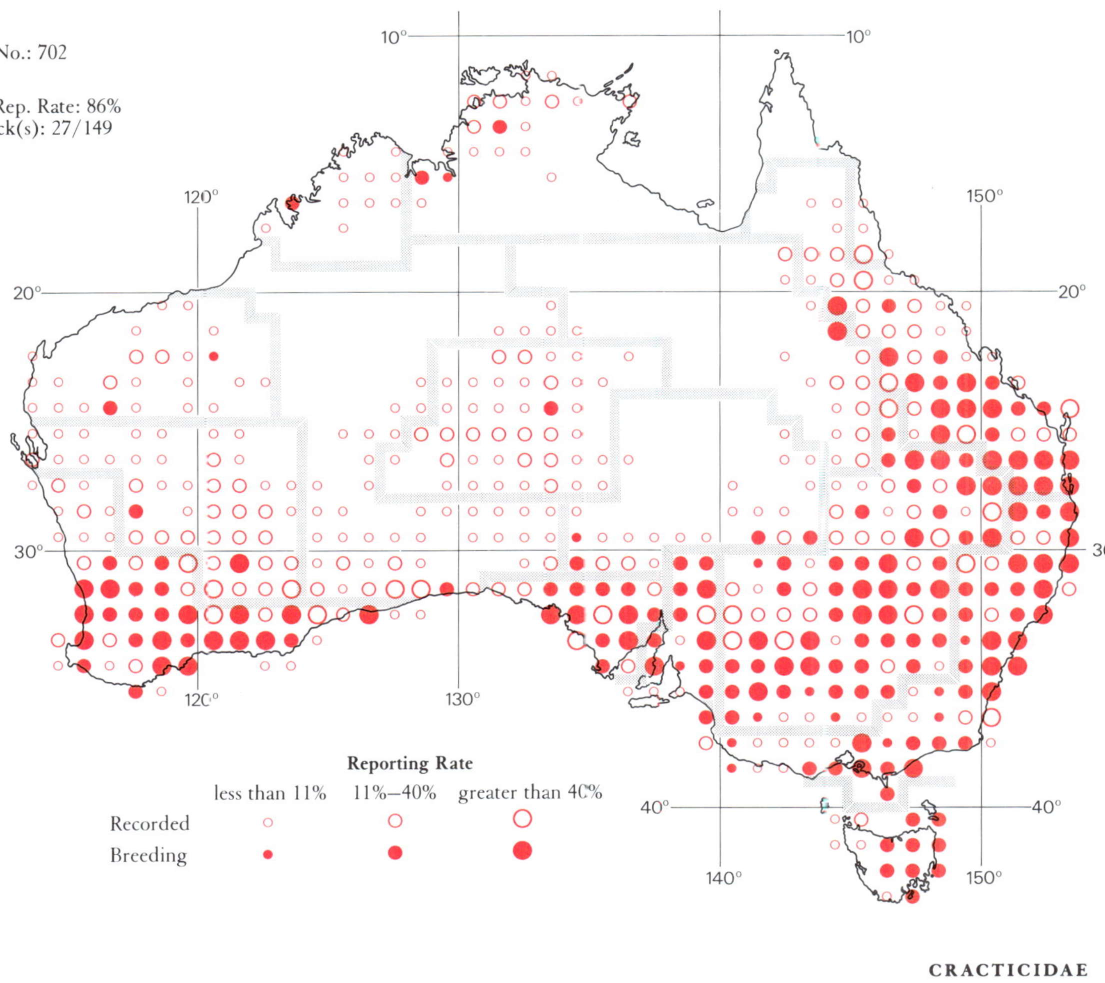

Black-backed Butcherbird

Cracticus mentalis

The Black-backed Butcherbird lives in southern New Guinea and Australia. Its distribution in Australia extends north to the tip of Cape York Pen. (10/142) but not to the islands of Torres Strait. The southern limits of its distribution are on the Palmer R (16/144) in the east and the Coleman R (15/141) in the west.

At Port Moresby (9S/147E) in New Guinea, where it is an urban bird, it can be seen all year.[129] It lives in eucalypt woodland or forest and at Cooktown (15/145) is common and conspicuous near houses.[1703] It eats many kinds of food: large insects, lizards from the branches of trees in New Guinea[1449] and other vertebrates including small birds.[1758] It has been seen to take acacia seed.[1861]

In New Guinea density in eucalypt woodland at Port Moresby was about 0.3 birds/ha.[158]

Field Atlas Records: 117

1° blocks–with records: 17 (2%)
–with breeding records: 1 (6%)

Atlas No.: 704

Max. Rep. Rate: 36%
In block(s): 15/144

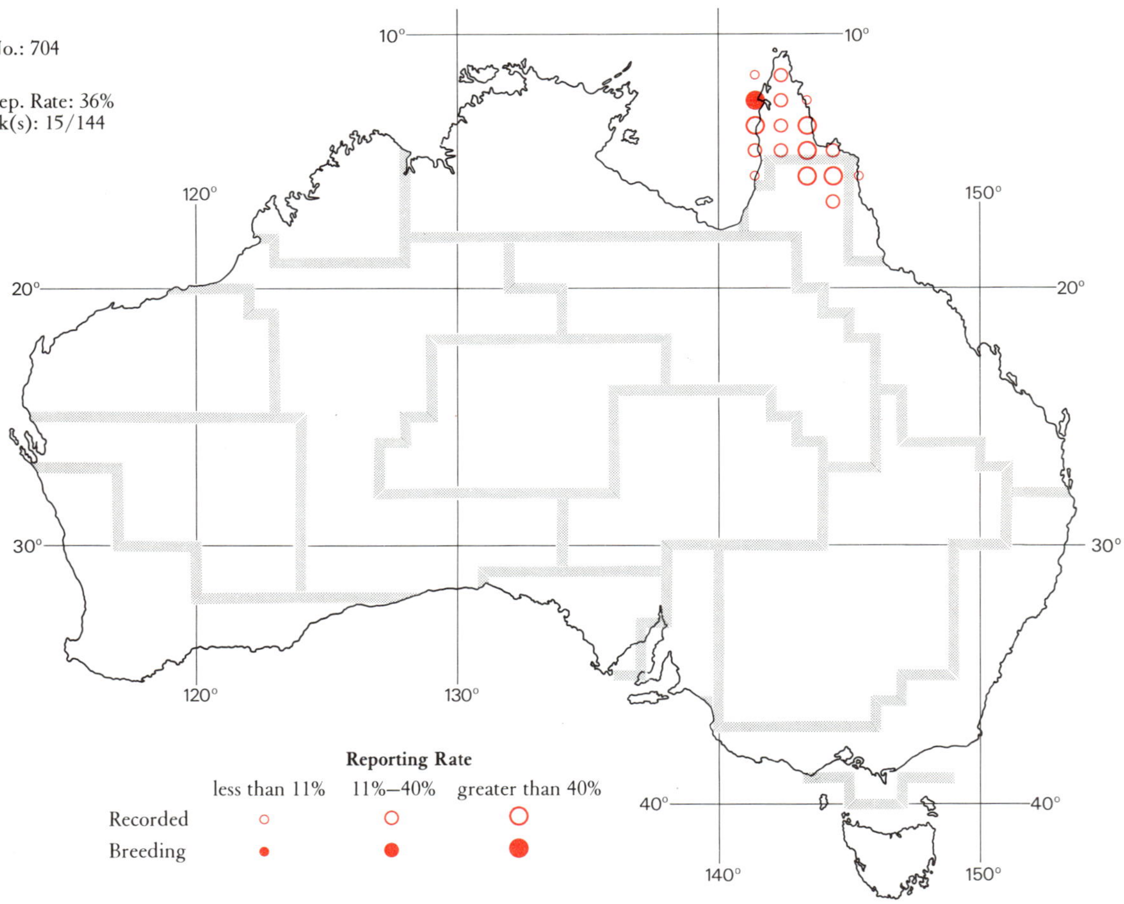

Three subspecies of the endemic Pied Butcherbird are recognized: *picatus* in the north; *nigrogularis* in the south-east; and *kalgoorli* in the south-west.[748,1422]

In the South-West Region the range of the Pied Butcherbird is expanding slowly southward as settlement reduces the area of forest.[1216,1634] The Atlases contain no records from the South-West before 1900 and by 1950 only one south of 33°S. There is no suggestion of a similar increase in range in the east.

The Pied Butcherbird is recorded as present throughout the year by many observers[231,833,862,1089] but is a vagrant on the Nullarbor Plain.[246]

The species lives in pairs or groups of 3–4 in acacia scrub, mallee and eucalypt woodland and has been recorded breeding communally.[410,501,1330,1752] It feeds chiefly on large insects taken on the ground but will take nestlings, mice,[723] snakes[752] and lizards.[748] It often drinks at mills and dams in the inland.[458]

There are no measures of abundance.

Pied Butcherbird

Cracticus nigrogularis

Field Atlas Records: 17 334

1° blocks–with records: 578 (71%)
–with breeding records: 170 (29%)

Atlas No.: 700

Max. Rep. Rate: 93%
In block(s): 27/149

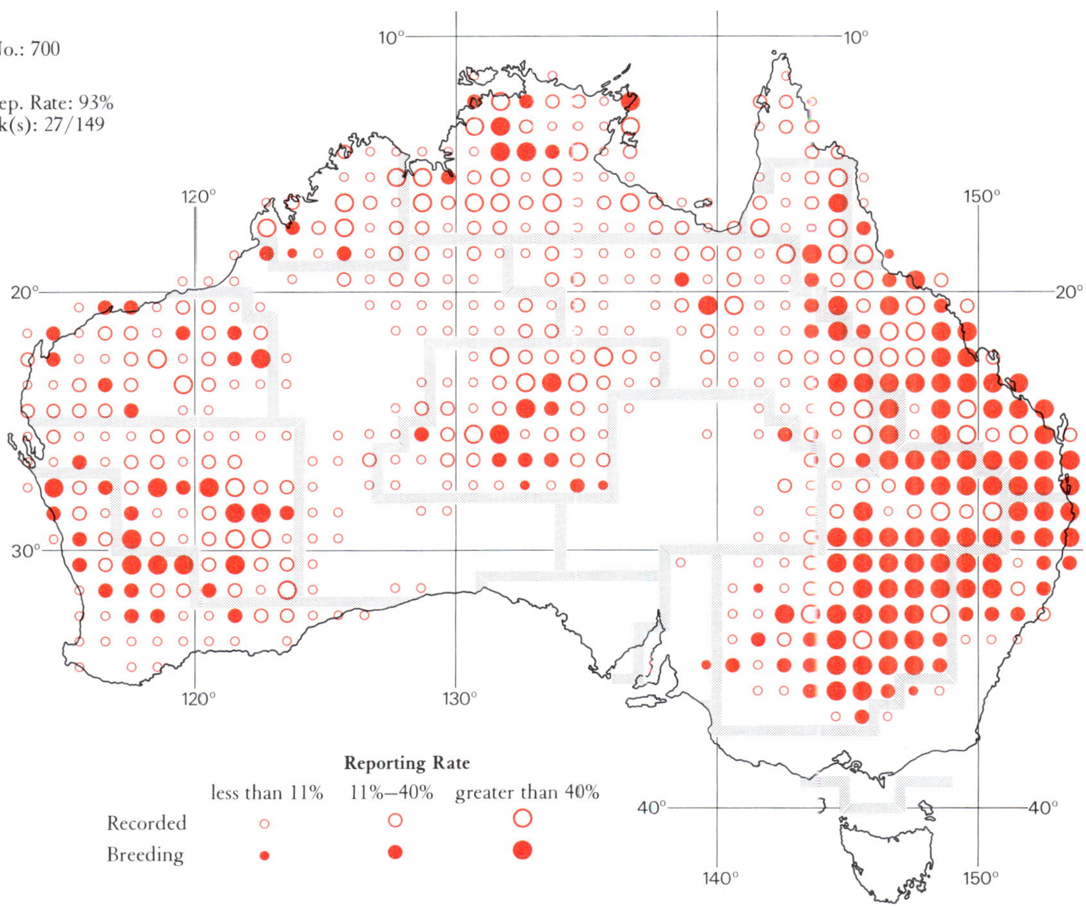

Australian Magpie

Gymnorhina tibicen

The Australian Magpie lives in New Guinea and Australia and has been introduced successfully to New Zealand. Eight subspecies have been described in Australia.[1422] The white-backed subspecies are distributed in the lower South-East Region west to Eyre Pen. and north to the Central Ra. Region, in the Tas. Region and in the South-West Region. The black-backed subspecies range throughout northern Australia.[1124] Within the overlap zones of the subspecies there are many hybrid types varying from fully white- to fully black-backed.[287,1370]

Throughout its range the Magpie is sedentary, the longest movement recorded being 67 km.[85] Groups of 2–20 birds defend a permanent territory within which they obtain all their food and select nesting and roosting sites.[311] The species inhabits eucalypt woodland, mallee, acacia scrub and woodland along watercourses. It does not inhabit forest and therefore is absent from western Tas.[1513]

The Magpie feeds on ground-dwelling insects, soft-bodied invertebrates and seeds throughout the year.[1794] Inland it is shy of people and flies out of sight when disturbed, a behaviour completely different from that of Magpies in southern Australia.[456]

In agricultural land near Canberra (35/149) the Magpie's density was 0.18 birds/ha.[311] In 37 territories groups included on average 2.8 adults and occupied areas of 1.3–6.3 ha/bird. At Rotamah I. Bird Observatory (37/147) groups of 2–4 Magpies occupied territories of 7–10 ha.[285] Near Adelaide (34/138) groups of 2–8 birds occupied territories of about 3.0–6.3 ha.[1664] At Coolup (32/115) in agricultural land groups of 4–20 birds occupied territories of 12–40 ha.[1513]

Field Atlas Records: 52 945

1° blocks–with records: 662 (82%)
–with breeding records: 305 (46%)

Atlas No.: 705

Max. Rep. Rate: 94%
In block(s): 29/147,34/149

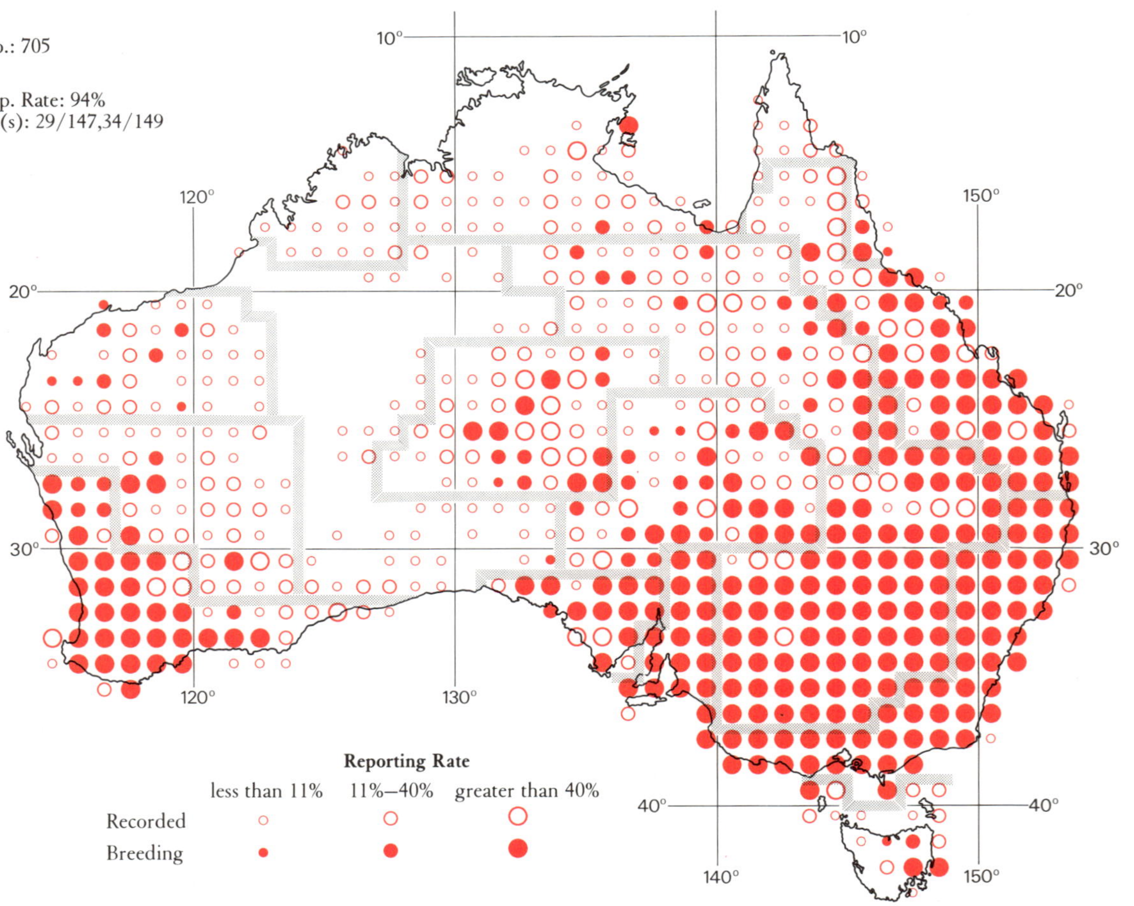

Pied Currawong

Strepera graculina

The endemic Pied Currawong shows a gradient of morphological change from north to south from which subspecies have been described but there is doubt about their ranges.[1422,1709] The Field Atlas 10′ maps indicate that the population in the Grampians (37/142) is isolated and possibly also that in the south-western corner of the species' range (38/141).

The Pied Currawong is present all year on the tablelands of the Atherton Region;[231] on the coast at Innisfail (17/146) a few visit May-Sept.[698] Birds in Eastern Qld are said to move north in winter or to lower altitudes.[1709] In the South-East Region most movements are local, usually less than 80 km, taking birds away from high country in winter to flock in lower areas.[1454] The distance moved depends on how far it is from the bird's breeding ground to an abundance of the stick insects to which it is attracted. The Pied Currawong is also attracted to settlements, near which one-third of the winter flocks were found during a study in the South-East Region.[1454] This perhaps accounts for the higher Reporting Rates in winter compared with summer. In the Eastern Qld and South-East Regions differences were similar, which suggests that there is no large-scale northward movement in winter (Eastern Qld: summer 20.4%, winter 25.7%; South-East: summer 33.2%, winter 47.5%).

The difference in the Murray-Darling Region was greater (summer 5.2%, winter 15.7%), which supports the observation[833] that the Pied Currawong is a winter visitor to the region.

Some birds are always present in forest and woodland. In winter flocks visit suburbs and feed on the fruit of orchard and ornamental trees, lawn insects[653] and other food that they scavenge. During breeding insects and small vertebrates are eaten.[1454]

Density in a suburb of Armidale (30/151) without dense housing was 0.39 birds/ha.[582] In eucalypt woodland densities were: Armidale 0.02–0.13 birds/ha,[582] Wollomombi (30/151) 0.07–0.22 birds/ha[129] and Moruya (35/150) 0.13 birds/ha.[1197] In eucalypt forest they were: Jamieson (37/146) 0.1 birds/ha[1469] and Boola Boola (38/146) 0.10–0.46 birds/ha.[1113] In 1964 it was estimated that there were 1 000 000 individuals occupying the part of the South-East Region lying between Sydney (33/151) and Melbourne (37/145).[1454]

Field Atlas Records: 15 082

1° blocks–with records: 160 (20%)

–with breeding records: 66 (41%)

Atlas No.: 694

Max. Rep. Rate: 82%
In block(s): 25/148

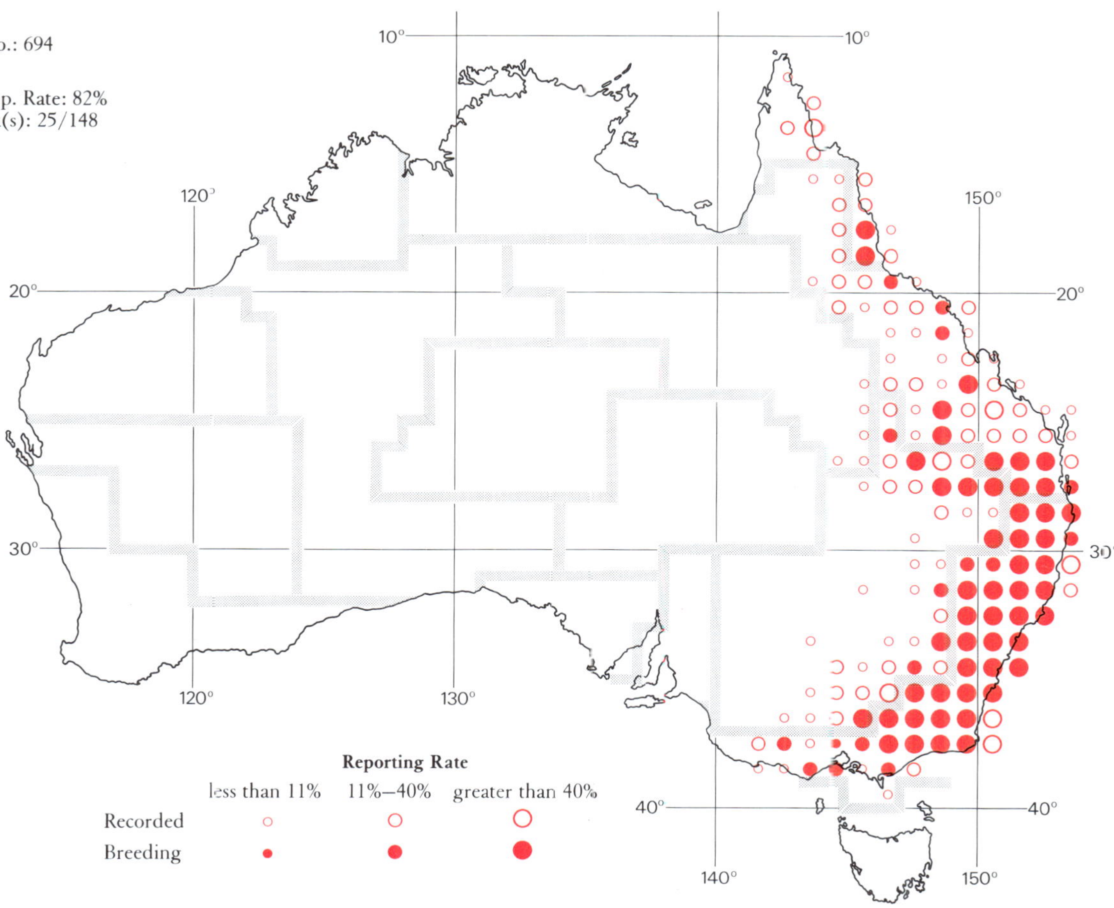

Black Currawong

Strepera fuliginosa

The Black Currawong is endemic to the Tas. Region. Some birds descend from high country to lower altitudes in winter.[740,1486] The longest movement recorded is that of an immature bird from Cascades (42/147) to Bronte Park (42/146), 105 km.[823]

The Black Currawong lives in rainforest, eucalypt forest and heath as well as parks. In eastern Tas. the dry forest and woodland are inhabited by the Grey Currawong but patches of wet vegetation allow the Black Currawong to live in these areas as well. In addition, in winter the Black Currawong will move into drier vegetation. On the Bass Strait islands, where the Grey Currawong is absent, the Black Currawong is ubiquitous in heath and woodland as well as in the damper areas.[734] The Black Currawong forms flocks in autumn and winter, often round settlements.[573] Flocks of up to 400 are regular in the central highlands.[546] Those at Ferntree (42/147) contain a high proportion of immatures (up to 90%).[1409] A mixed flock of Black and Grey Currawongs has been observed on Maria I. (42/148).[1534] The Black Currawong feeds on large insects, carrion, fruit and debris along the shore.[572,740,1409] Native berries, such as those of the native cranberry, are particularly important.[1533] It takes small vertebrates, including nestling birds and young ducks, sometimes placing them in cracks in logs to hold them while they are butchered.[573] The Currawong will also enter hay barns in winter to hunt mice[572] and scavenge for scraps in farmyards, picnic grounds and small settlements.

The Black Currawong appears in an area once human settlement is established, e.g. at the Great Lake (41/146).[572]

There are no measures of abundance.

Field Atlas Records: 1909

10′ blocks–with records: 295 (53%)
–with breeding records: 37 (13%)

Atlas No.: 696

Max. Rep. Rate: 95%
In block(s): 41°55′/146°55′

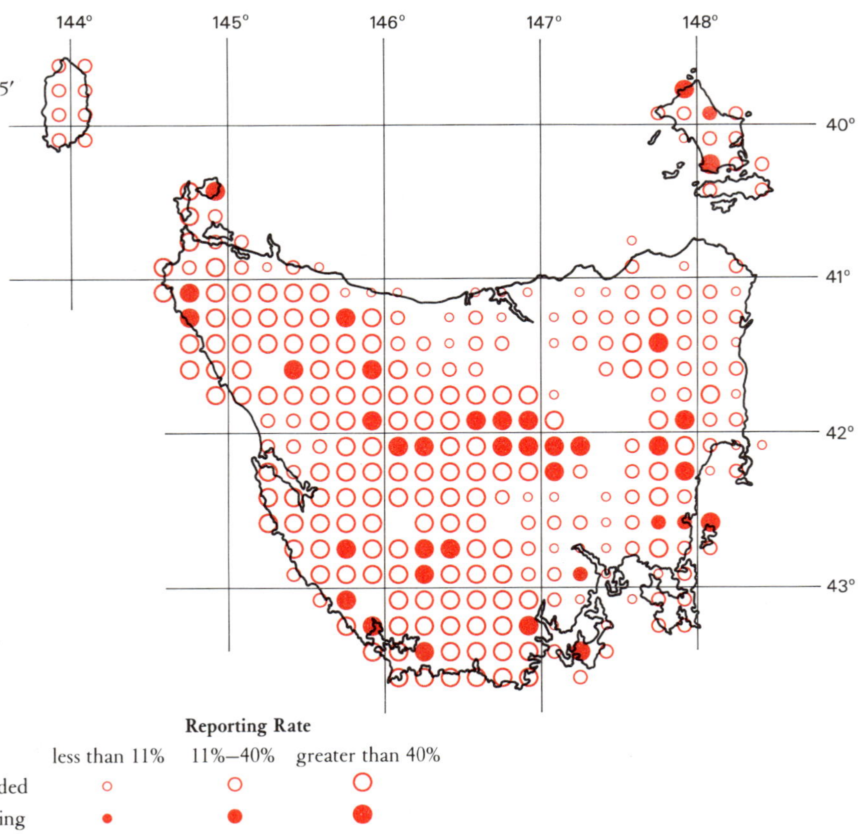

Grey Currawong

Strepera versicolor

The endemic Grey Currawong is composed of four distinct groups of subspecies some of which were formerly regarded as species.[599,1422,1576] Nominate *versicolor* occupies most of the South-East Region; the second group is *melanoptera* (Black-winged) (SA Ra. and western South-East Regions); the third group includes *plumbea* (Central Ra., Western Deserts, Goldfields and South-West), *howei* (Murray-Darling) and *intermedia* (foot of Yorke Pen. and Eyre Pen. Region); the fourth is the Clinking Currawong *arguta* of Tas. confined to the eastern side of the state and absent from the Bass Strait islands. The population in the Central Ra. Region was not recorded at all during the Field Atlas but historical records show it extending through the Western Deserts to the Goldfields Region and south probably to Eyre Pen., absent only from the Nullarbor Plain.[599]

Altitudinal migration is suggested in the South-East Region.[1128,1292] At Moruya (35/150) a few birds arrive in winter.[1192] In the Snowy Mts (36/148) Reporting Rates were higher in autumn-winter both above and below 500 m, possibly because the bird forages in more open country in autumn-winter[653] making it more conspicuous. The differences do not suggest large-scale altitudinal movement (above 500 m: spring-summer 26%, autumn-winter 33%; below 500 m: spring-summer 11%, autumn-winter 20%).

The Grey Currawong inhabits eucalypt forest, woodland, mallee and heath near the coast. In Tas. it lives in the dry forest and woodland and the Black in wetter country. The Grey Currawong is usually solitary or in pairs on the mainland[1343] but loose flocks of up to 40 birds form in Tas. in the non-breeding season.[1409] It feeds on large insects, often gathered on the ground, and will raid soft fruit and apple orchards.[1638] In Tas. the berries of native plants such as the native cranberry are important.[1533]

The Grey Currawong was reported to be retreating from the wheatbelt of the South-West Region in the 1940s but has since begun to return.[1638] Density at Boola Boola (38/146) in eucalypt forest was 0.1 birds/ha[1113] and at Eyre Bird Observatory (32/126) in mallee 0.05–2.58 (usually 0.25) birds/ha.[464]

Field Atlas Records: 8770

1° blocks–with records: 180 (22%)
–with breeding records: 77 (43%)

Atlas No.: 697

Max. Rep. Rate: 69%
In block(s): 34/119

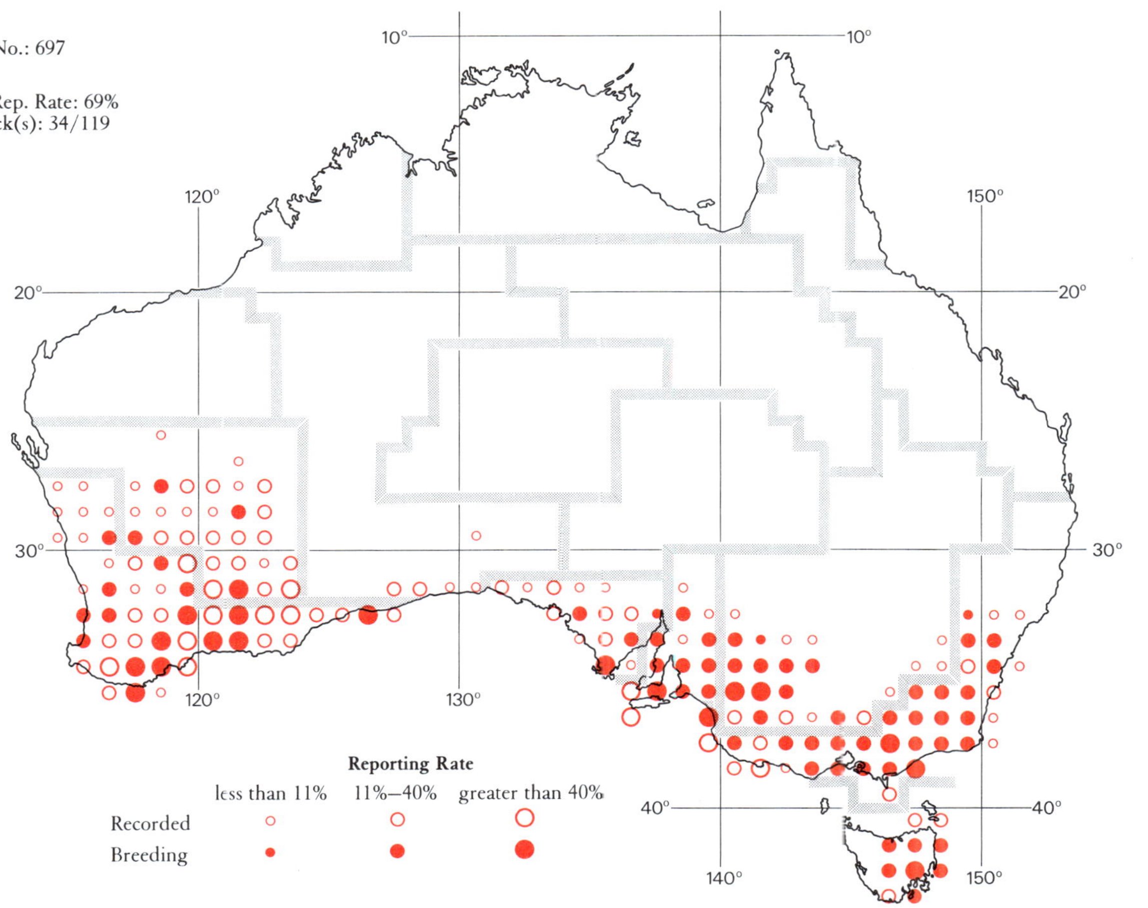

Australian Raven

Corvus coronoides

The crows and ravens proved difficult to identify in the field and the map shown here may not represent the true distribution of the endemic Australian Raven. It was derived by querying all records outside the range previously accepted by three people experienced with corvids. Only extralimital records accepted by all three are shown on the map. The map differs from one published in 1973[1546] which did not show that the Australian Raven reached the Gulf of Carpentaria, any of the Eastern Qld Region or the northern part of the South-East Region.

The Australian Raven varies morphologically on a cline of increasing size from west to east but no subspecies are recognized.[1543] The species' range overlaps that of the other four corvids in places.

Though breeding pairs of the Australian Raven are sedentary and occupy a permanent territory, young birds form flocks, sometimes with Little Ravens. The flocks move out of the dry inland in summer and return there in winter.[1544]

The Australian Raven inhabits a variety of country provided there is water and a tree or tall structure for nesting. It has been recorded breeding in the centre of Sydney (33/151).[470] In arid country, territories are centred on richer sites, such as watercourses, with large unoccupied areas between.[1546] The species maintains a territory for the exclusive use of the breeding pair and its young, except when there is an abundance of food. Then immature flocks, joined by other locally resident pairs, may temporarily 'swamp' the territorial defence of the pair. The Australian Raven is omnivorous, one study showing its diet by volume to be 34% carrion, 42% insect and 24% plant material.[1551] It has been recorded taking eggs from a coastal colony of Pied Cormorants.[1369] It has been predicted that agricultural development will make available more food and allow the Australian Raven to increase its range.[1543]

The average size of territories measured near Jandowae (26/151), Toganmain (34/145) and Canberra (35/149) was 112 ha,[1545] which gives a breeding density in good areas of 0.02 birds/ha.

Field Atlas Records: 27 155

1° blocks–with records: 419 (52%)
–with breeding records: 196 (47%)

Atlas No.: 930

Max. Rep. Rate: 82%
In block(s): 32/117

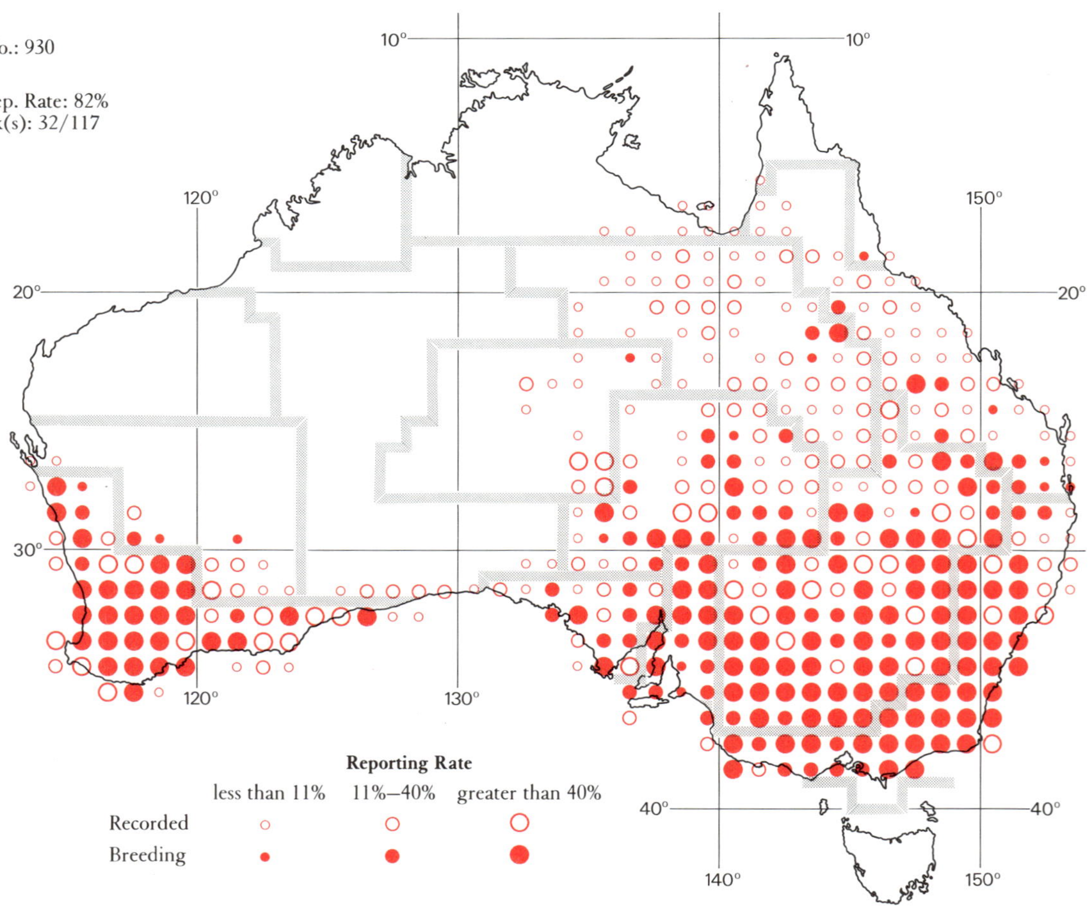

Forest Raven

Corvus tasmanicus

The endemic Forest Raven has two subspecies: *boreus* in the upper South-East Region and *tasmanicus* in the lower South-East and Tas. Regions.[1543] The population in southern SA[1278] was not discovered until 1970 and the species was confirmed to be on King I. (39/144) during the Field Atlas. The Forest Raven overlaps the range of the Torresian Crow near Glen Innes (29/151), Armidale (30/151) and Barrington Tops (32/151) and those of the Australian and Little Ravens over a wide range.

Breeding pairs occupy permanent territories. Non-breeding birds flock: 30–40 were recorded at a rubbish tip near Lorne (38/143)[1545] and flocks of up to 100 are common in winter in Tas.[1533]

In Tas. the Forest Raven is the only corvid; it inhabits eucalypt forest and woodland and often forages on beaches. On the mainland it sometimes occupies other kinds of country. Near Barrington Tops it lives in eucalypt forest and farmland; on the adjacent coast, between South-west Rocks (30/153) and Port Stephens (32/152), there is an apparently isolated population occupying a narrow belt of heath growing on dunes, eucalypt forest and paper-bark swamp.[471] At Armidale (30/151) Forest and Australian Ravens mutually exclude each other from their territories; the Forest Raven and Torresian Crow also appear to occupy exclusive territories.[1543] At Wilsons Prom. (39/146) and Cape Otway (38/143) the Forest Raven is the only corvid and inhabits eucalypt forest. The population in SA inhabits mature plantations of introduced pine and open pasture.[1278] Near Naracoorte (36/140) mixed feeding flocks of Forest, Australian and Little Ravens have been seen.[1278]

The Forest Raven is omnivorous, one study showing its diet by volume to be 31% carrion, 41% insects and 28% plant material.[1551] In seabird colonies the Forest Raven takes eggs and chicks.[250] In Tas. it is a pest in vegetable crops, damaging seedlings, potatoes and maize.[1533]

The distribution of the Forest Raven has been interpreted as the fragmented range of a declining species but because mainland populations were overlooked until recently, their status cannot be assessed by comparison with earlier records. The breeding territory of a pair of Forest Ravens covers more than 40 ha.[1545]

Field Atlas Records: 4586

1° blocks–with records: 44 (5%)
–with breeding records: 17 (39%)

Atlas No.: 868

Max. Rep. Rate: 76%
In block(s): 42/146

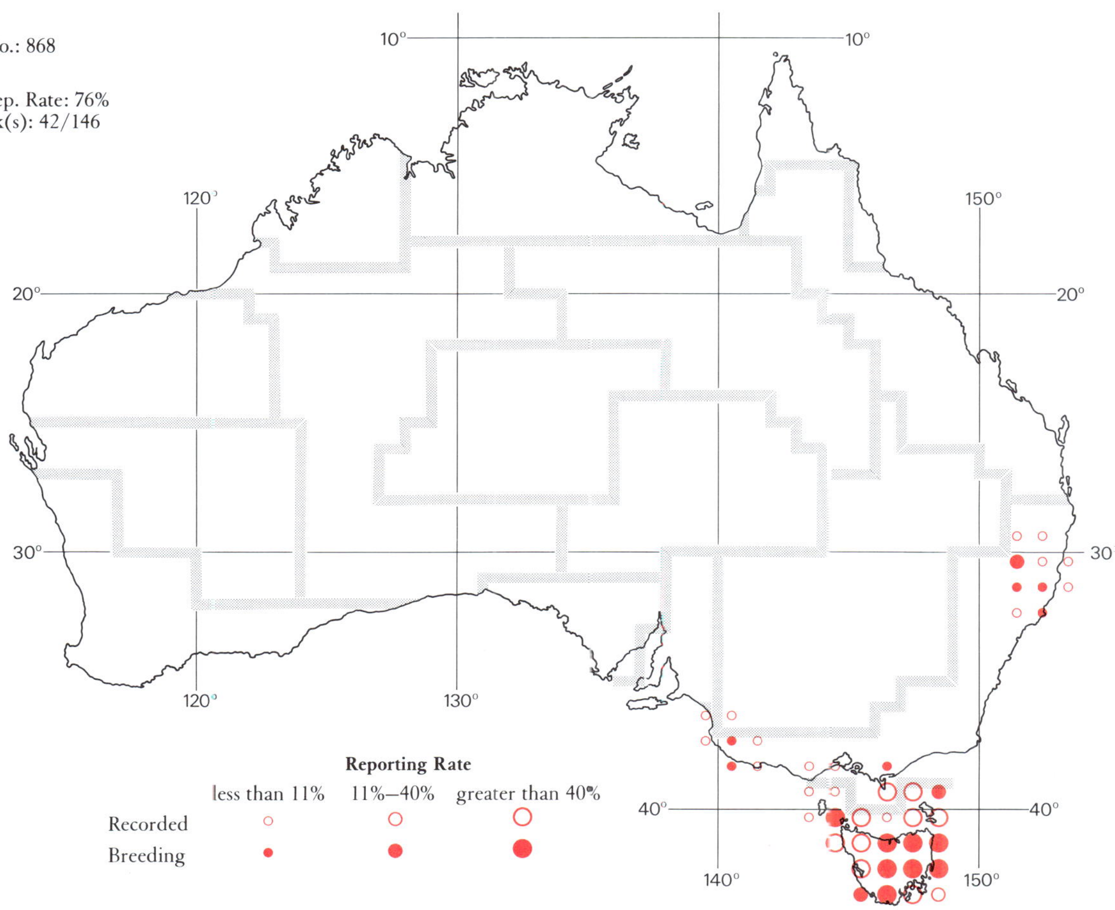

Little Raven

Corvus mellori

The crows and ravens proved difficult to identify in the field and the map shown here may not represent the true distribution of the endemic Little Raven. It was derived by querying all records outside the range previously accepted by three people experienced with corvids. Only extralimital records accepted by all three are shown on the map. The map shows the Little Raven to be on King I. (39/144) but does not otherwise differ greatly from one published in 1973.[1546] The range of the Little Raven meets that of the Torresian Crow only near Barrington Tops (32/151). Overlap with the Australian and Forest Ravens and the Little Crow is extensive.

Adult and immature Little Ravens are nomadic.[1544] In the southern Murray-Darling Region birds tend to move south-east in summer to better-watered areas, returning in autumn. Most movements in excess of 15 km involve non-breeding birds, which far outnumber the breeding individuals, and account for major fluctuations in abundance in a district. Breeding adults tend to return to the same areas each year.[1544] Banded adults and juveniles have been recovered up to 309 km distant from Mt Mary Plains (34/139)[195] and the longest movement recorded of a banded bird is 352 km.[85] Numbers fluctuate on King I. where the Little Raven seems to be a common visitor from mainland Australia during the non-breeding season.[734,1543]

The Little Raven lives in eucalypt forest, woodland and acacia scrub. It also enters and breeds in heavily built-up parts of Melbourne (37/145). Breeding is semi-colonial, the territories clumped together though the birds forage far beyond their boundaries. Breeding occupies about three months after which the birds again become nomadic. Flocks of up to 300 have been recorded.

The Little Raven is omnivorous; one study showed that its diet by volume was 14% carrion, 61% insects and 25% plant material.[1551] It will take stick insects in forest but more commonly forages in open sites.

The average territory size measured at Mt Kosciusko (36/148) and Toganmain (34/145) was 2.6 ha per pair.[1546]

Field Atlas Records: 8313

1° blocks–with records: 108 (13%)
–with breeding records: 60 (56%)

Atlas No.: 954

Max. Rep. Rate: 60%
In block(s): 34/138

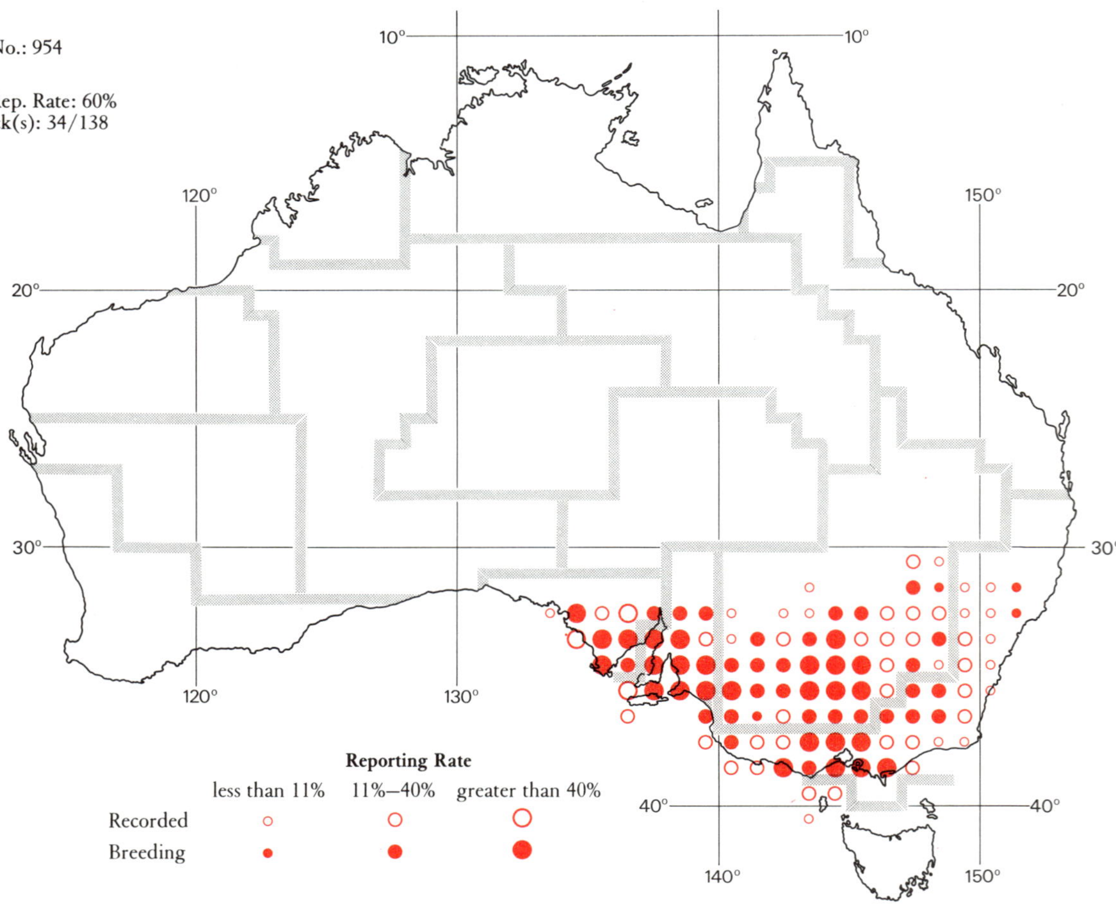

Little Crow

Corvus bennetti

The crows and ravens proved difficult to identify in the field so the map shown here may not represent the true distribution of the endemic Little Crow. It was derived by querying all records outside the range previously accepted by three people experienced with corvids. Only extralimital records accepted by all three are shown on the map. The map differs from one published in 1973 based on specimens,[1546] which showed that the Little Crow was absent from the eastern part of the Rolling Downs, the eastern and southern Murray-Darling, the southern SA Ra. and the lower South-West Regions; and that it was present in the Kimberley Region. The Field Atlas shows that the range of the Little Crow overlaps that of the Torresian Crow and the Little and Australian Ravens.

The Little Crow wanders during the non-breeding season. The most distant recovery of a banded bird up to 1982 was from Jiggalong (23/120) to 60 km south of Menzies (29/121), 691 km.[1074]

The Little Crow is a bird of the arid zone, inhabiting all types of country. It is the urban bird of Kalgoorlie (30/121) and other WA country towns, even entering shop doorways; at the fishing port of Denham (25/113) it scavenges along the waterfront.[1638] It is usually in flocks and breeds semi-colonially.[1545] Breeding birds must be within reach of water because mud is used in the nest. It is omnivorous; its diet by volume in one study was 26% carrion, 48% insect and 26% plant material.[1551] It benefits from the activities of larger corvids, which are better able to open carcasses. When the larger ones leave, the Little Crow picks out the remaining flesh that is accessible to its small bill. It is suggested that, when conditions are less favourable inland, the Little Crow moves into the Eastern Qld and Murray-Darling Regions.[1543] In the Pinnaroo (35/140) area clearing for agriculture may have enabled the Little Raven to displace the Little Crow.[1543]

At Ivanhoe (32/144) size of territory round the nest was 1.1 ha but the birds foraged over a much greater area.[1546]

Field Atlas Records: 3323

1° blocks–with records: 414 (51%)
–with breeding records: 117 (28%)

Atlas No.: 691

Max. Rep. Rate: 78%
In block(s): 28/121

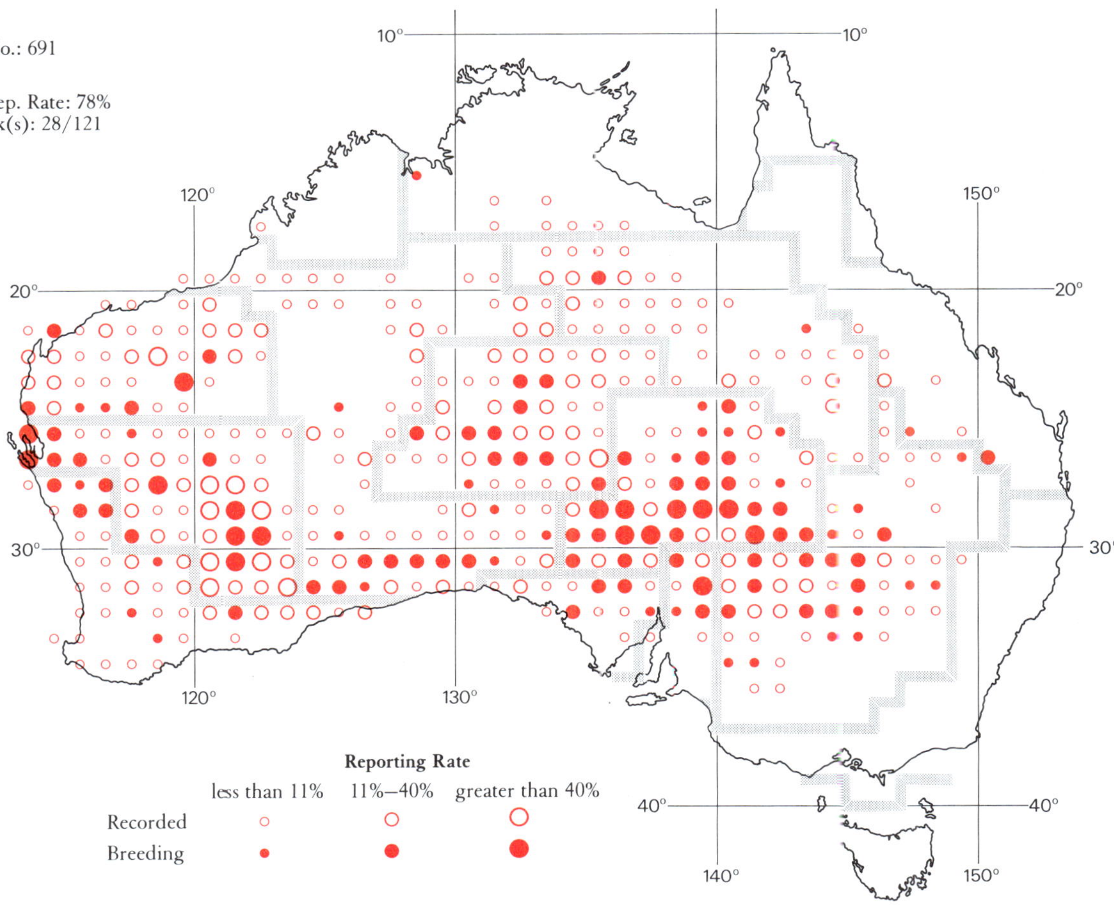

Torresian Crow

Corvus orru

The Torresian Crow lives in New Guinea and Australia, the Australian subspecies being *cecilae.*[1546] The crows and ravens proved difficult to identify in the field. The map shown here may not represent the true distribution of the Torresian Crow. It was derived by querying all records outside the range accepted by three people experienced with corvids. Only extralimital records accepted by all three are shown on the map.

The Field Atlas map differs from one published in 1973 based on specimens,[1546] which showed that the Torresian Crow occurred in the northern Western Deserts but was absent from a large part of inland eastern Australia including all but the western portion of the Rolling Downs, the northern Murray-Darling and northern L. Eyre Regions.

The Field Atlas shows that the species contacts the Little Raven at Barrington Tops (32/151). It also contacts the Forest Raven near Glen Innes (29/151), Armidale (30/151) and Barrington Tops and the Australian Raven and Little Crow over a wide range.

The Torresian Crow is mostly sedentary, the longest movement recorded being 22 km.[85] Adults are permanently resident in a territory and summer sightings of pairs at inland localities are likely to be of this species. In the better-watered areas of the north and on the eastern coast flocks form during autumn and winter, probably of non-territorial birds.[1543,1546] Mixed flocks with the Little Crow have been reported. The Torresian Crow occupies farmland and woodland, including that in ranges and along watercourses in the inland. The bird is omnivorous, its diet by volume being 26% carrion, 43% insect and 31% plant material.[1551] It can land on a standing crop of grain and feed from grain in ear; cultivation has therefore increased the area suitable for the Torresian Crow. The species will also enter forest for foraging.

At Jandowae (27/151), the Crow's density was 0.02 birds/ha; 19 pairs held territories averaging 131 ha, which contained all their requirements.[1546]

Field Atlas Records: 8037

1° blocks–with records: 408 (50%)
–with breeding records: 71 (17%)

Atlas No.: 692

Max. Rep. Rate: 76%
In block(s): 24/153

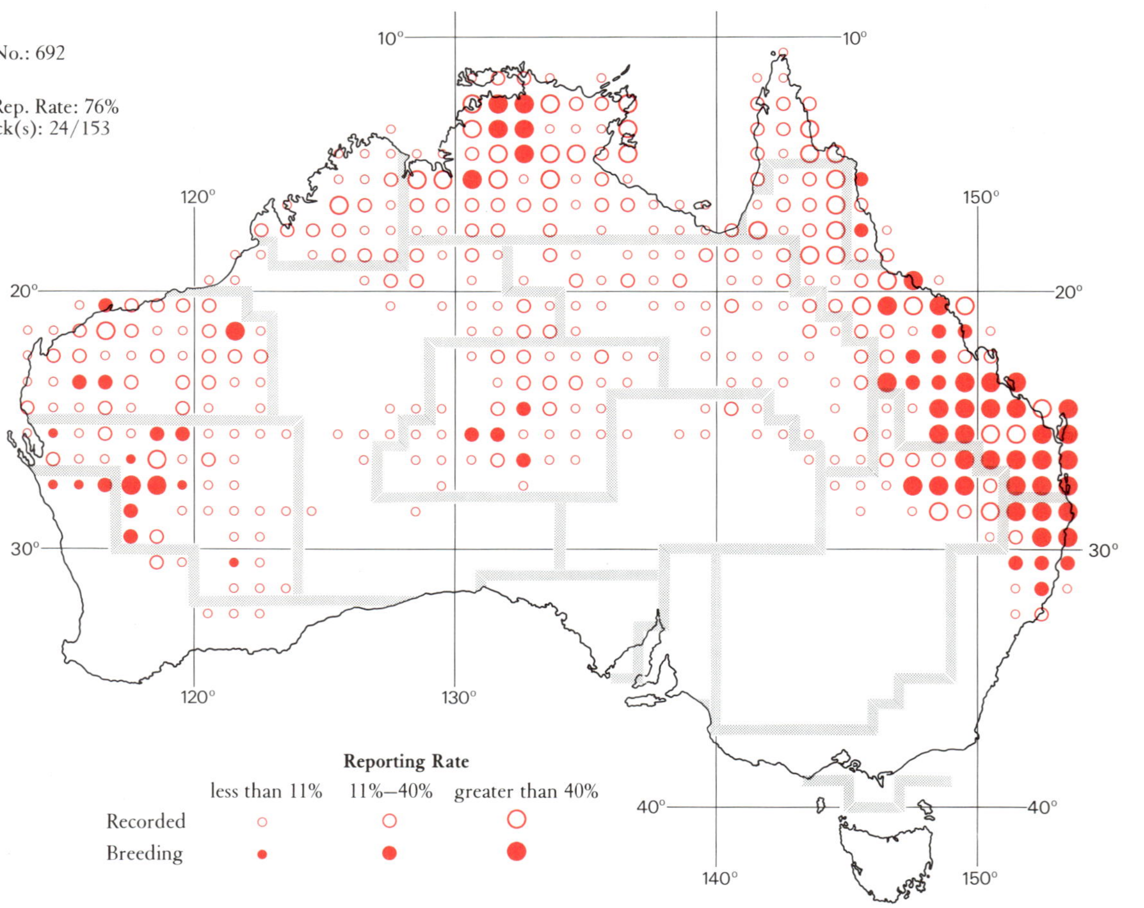

UNCOMMON AND VAGRANT SPECIES

All species on the Australian list[1581] not covered by main texts (pp. 2–648) are mentioned, as well as species added since its compilation. Species known only from escaped or released birds with no evidence of breeding in the wild, are omitted.

Confirmed records are listed in 1° grids in the four Atlas time periods (p. xvi). These include only records up to Dec. 1981 unless otherwise specified. The letter 'B' following the grid reference indicates a breeding record.

Unconfirmed records are those not supported by conclusive evidence but which there is no reason to doubt. These are referred to in the preamble to each species but not listed in the time periods below.

STRUTHIONIDAE

928 Ostrich
Struthio camelus

Native to Africa. Released from farms: at 35/139 on The Coorong; Morgan (34/139); Port Augusta (32/137);[1085] near Swan Hill (35/143); and Murgah (35/144).
1977–81 32/137B.
1951–76 32/137B, 33/139B, 35/139B.
1901–50 35/144B.

SPHENISCIDAE

927 King Penguin
Aptenodytes patagonicus

Breeds on sub-Antarctic islands including Macquarie (54S/158E) and Heard (53S/73E) Is. Accidental to Australia.
1977–81 33/115, 37/148, 43/147.
1951–76 34/115, 38/142, 42/145, 43/147.
1901–50 42/148.

926 Gentoo Penguin
Pygoscelis papua

Breeds on sub-Antarctic islands including Macquarie (54S/158E) and Heard (53S/73E) Is. Accidental to Tas.
1977–81 41/145.
1951–76 42/147.
1901–50 43/147.

925 Chinstrap Penguin
Pygoscelis antarctica

Breeds in Antarctica including Heard I. (53S/73E). Accidental to Tas.
1977–81 43/146.
1951–76 43/146.

924 Adelie Penguin
Pygoscelis adeliae

Breeds on the Antarctic continent and adjacent islands; rare y strays as far north as Macquarie I. (54S/158E). Two records of doubtful provenance: Vic. 1933 and WA 1937.[388]

003 Rockhopper Penguin
Eudyptes chrysocome

Breeds on sub-Antarctic islands including Macquarie (54S/158E) and Heard (53S/73E) Is; single birds sometimes reach Australia.
1977–81 34/118, 36/139, 38/141, 41/146, 43/147.
1951–76 32/115, 34/115, 37/140, 38/141, 38/142, 38/14[illegible], 38/145, 39/143, 39/146, 40/143, 42/145.
1901–50 32/115, 32/133, 35/117, 37/139, 37/140, 38/14[illegible], 38/142, 38/143, 39/144.
Before 1901 34/115, 39/144.

004 Fiordland Penguin
Eudyptes pachyrhynchus

Breeds on the South I. of New Zealand and adjacent islands; single birds sometimes reach Australia.
1977–81 35/138, 36/139, 37/140, 38/140, 38/141, 40/14[illegible], 41/148, 42/145, 43/147.

1951–76 32/124, 35/138, 36/139, 38/141, 38/144, 38/145, 39/143, 39/144, 40/143, 40/147, 41/145, 41/148, 42/145, 43/147.
1901–50 33/151, 34/150, 38/141, 43/147.
Before 1901 42/147, 43/147.

923 Snares Penguin

Eudyptes robustus

Breeds on Snares I. (48S/166E), New Zealand; rarely reaches Australia.
1977–81 42/147, 43/147.
1951–76 42/147.
1901–50 37/140.

959 Erect-crested Penguin

Eudyptes sclateri

Breeds on islands to the south of New Zealand. Accidental to Australia.
1977–81 38/145.
1951–76 33/121, 35/138, 38/145.
Before 1901 38/142.

968 Royal Penguin

Eudyptes schlegeli

Breeds on Macquarie I. (54S/158E). Accidental to Australia.
1977–81 38/140, 41/148, 43/147.
1951–76 41/148, 43/146, 43/147.
1901–50 35/138, 42/148, 43/147.

858 Magellanic Penguin

Spheniscus magellanicus

A South American penguin. One Australian record: Phillip I. (38/145), Mar. 1976.
1951–76 38/145.

DIOMEDEIDAE

931 Buller's Albatross

Diomedea bulleri

Breeds on Solander (46S/166E), Snares (48S/166E) and Chatham (44S/176W) Is. Few but regular sightings of live birds off the coasts of the South-East and Tas. Regions since 1967.
1977–81 33/151, 35/138, 36/150, 37/139, 37/150, 38/141, 41/144, 43/147.
1951–76 30/153, 33/151, 34/150, 34/151, 37/150, 38/144.

090 Grey-headed Albatross

Diomedea chrysostoma

Breeds on sub-Antarctic islands including Macquarie I. (54S/158E). Regular winter visitor to Australian waters.
1977–81 33/120, 34/135, 35/137, 35/138, 36/150, 37/139, 37/140, 37/147, 37/150, 38/140, 38/141, 38/142, 38/144, 38/145, 41/146, 42/144.
1951–76 27/153, 28/153, 29/153, 31/115, 32/151, 33/115, 33/151, 34/138, 34/150, 35/136, 35/137, 35/138, 35/139, 36/136, 37/149, 37/150, 38/141, 38/142, 38/145, 38/146, 38/148, 39/143, 39/144, 39/147, 40/143, 40/147, 40/148, 42/145.
1901–50 25/113, 31/115, 31/152, 32/115, 33/115, 33/151, 34/138, 34/150, 35/138, 35/139, 36/137, 37/137, 38/141, 38/144, 38/145, 40/147.
Before 1901 34/115, 35/138, 36/133, 38/140, 40/144.

092 Sooty Albatross

Phoebetria fusca

Breeds on islands in the Atlantic and Indian Oceans, mainly sub-Antarctic. Rare but regular visitor to Australian waters.
1977–81 32/115, 34/151, 35/138, 37/150, 40/148, 41/148, 42/148, 43/148.
1951–76 23/152, 27/153, 32/151, 34/150, 34/151, 35/138, 35/150, 37/149, 38/142, 40/143, 40/147, 41/150, 43/146.
1901–50 30/153, 31/130, 32/115, 35/116, 35/137, 38/140.
Before 1901 37/144.

093 Light-mantled Sooty Albatross

Phoebetria palpebrata

Breeds on sub-Antarctic islands including Macquarie (54S/158E) and Heard (53S/73E) Is; less often seen in Australian waters than the Sooty.
1977–81 32/126, 34/138, 37/140, 38/141, 43/148.
1951–76 27/153, 29/153, 33/151, 34/138, 34/150, 35/138, 38/141.
1901–50 38/141, 42/148.

PROCELLARIIDAE

949 Antarctic Petrel

Thalassoica antarctica

Breeds on the Antarctic continent and adjacent islands. All Atlas records are of beachwashed birds.
1977–81 35/138, 38/140, 38/141, 38/144, 42/145.

971 Providence Petrel

Pterodroma solandri

Breeds on Lord Howe I. (31/159). A former breeding colony on Norfolk I. (29/167) is extinct. Regular visitor to waters off central and northern NSW.
1977–81 27/153, 33/151, 43/148.
1951–76 25/153, 27/153, 28/153, 30/153, 31/152, 32/151, 33/151, 34/150, 34/151, 35/150, 37/149.
1901–50 33/151, 34/151.

922 Kermadec Petrel

Pterodroma neglecta

Breeds on Lord Howe I. (31/159), Kermadec I. (29S/178W) and islands in eastern South Pacific Ocean. Status in Australian waters unknown.
1951–76 28/153, 33/151, 35/150.

921 Herald Petrel

Pterodroma arminjoniana

Breeds between 10°S and 30°S in Atlantic, Indian and Pacific Oceans. A few birds present on Raine I. (11/144) when visited in 1959 and 1980–83; breeding confirmed July 1982.[1000]
1977–81 11/144, 28/153.
1951–76 11/144, 28/153.

920 Tahiti Petrel

Pterodroma rostrata

Breeds on islands of Pacific Ocean south of the equator. Recorded most years in small numbers in waters off central eastern Australia.[851]
1977–81 9/133, 10/145, 1[illegible]/145, 14/146.
1951–76 27/153, 30/153, 31/154.

076 Soft-plumaged Petrel

Pterodroma mollis

Breeds on islands in Atlantic and Indian Oceans south to about 50°S. Rarely recorded in Australian waters. Two sight records off south-eastern Tas. (43/148); remaining records are of beachwashed specimens.
1977–81 35/138, 36/139, 43/148.
1951–76 34/135, 38/141, 39/146.
1901–50 31/115, 32/115, 33/115.

919 Mottled Petrel

Pterodroma inexpectata

Breeds on islands of the New Zealand region; migrates to northern Pacific Ocean as far as the Arctic and occurs south to the sub-Antarctic in summer. Accidental to Australia.
1977–81 34/150, 37/139, 37/140, 37/147, 38/145, 41/145.
1951–76 28/153, 33/151, 34/150, 35/138, 35/150, 37/147, 37/149, 38/142, 38/144, 38/145, 42/145.
1901–50 38/145.
Before 1901 40/145.

955 Black-winged Petrel

Pterodroma nigripennis

Breeds on islands of southern Pacific Ocean including Lord Howe I. (31/159); ranges widely in tropical and sub-tropical Pacific Ocean. One inland record: at the You Yangs (37/144), Apr. 1974.[1677]
1977–81 23/151, 30/153.
1951–76 23/151, 23/152, 25/153, 27/153, 29/153, 30/153, 32/151, 34/151, 35/150, 37/144, 38/144.

918 Cook's Petrel

Pterodroma cookii

Known to breed only on Little Barrier (36S/175E), Great Barrier (36S/175E) and Codfish (46S/167E) Is off New Zealand;[776] migrates to northern Pacific Ocean. Rare in Australian waters.[1636]
1951–76 32/151, 34/150, 34/151.

984 Snow Petrel

Pagodroma nivea

An Antarctic species, rarely straying out of the pack-ice.
1951–76 35/138, 38/144.

081 Blue Petrel

Halobaena caerulea

Breeds on many sub-Antarctic islands[776] including islets near Macquarie I. (54S/158E).[936] Most records in the Field Atlas were of beachwashed birds.
1977–81 31/115, 34/135, 35/138, 37/140, 38/141, 38/142, 38/144, 38/145, 38/146, 40/147, 40/148, 41/145, 41/146, 42/145, 42/147, 43/147.
1951–76 33/151, 34/138, 34/150, 34/151, 35/138, 38/141, 38/142, 38/144, 38/145, 39/143, 39/146, 40/143, 41/147.
1901–50 31/115, 32/115, 33/115, 35/138, 38/145.
Before 1901 38/145.

073 Grey Petrel

Procellaria cinerea

Breeds in winter on many sub-Antarctic islands. A former colony on Macquarie I. (54S/158E) is now believed exterminated, probably by cats.[936] Rarely near the Australian coast.
1977–81 34/135, 35/138, 37/140, 38/144, 43/147.
1951–76 35/137, 38/141, 38/142, 39/143, 42/147.
1901–50 33/115.

917 Black Petrel

Procellaria parkinsoni

Breeds on Little Barrier (36S/175E) and Great Barrier (36S/175E) Is, New Zealand. One specimen: Sydney (33/151), May 1875.[388]
Before 1901 33/151.

916 Westland Petrel

Procellaria westlandica

Breeds only behind Barrytown (42S/171E), New Zealand. Rare in Australian waters.
1977–81 43/147.
1951–76 28/153, 30/153, 33/151, 34/150, 34/151.

915 White-chinned Petrel

Procellaria aequinoctialis

Breeds on islands mostly in the sub-Antarctic, perhaps including Macquarie I. (54S/158E).[776] Regular visitor to Australian waters in small numbers.
1977–81 28/153, 37/140, 37/150, 38/141, 38/142, 38/144, 38/148, 40/144, 42/145, 43/146, 43/147, 43/148.
1951–76 28/153, 31/115, 32/150, 32/151, 34/150, 38/141, 41/148.
1901–50 28/112.
Before 1901 35/117.

975 Buller's Shearwater

Puffinus bulleri

Known to breed only on Poor Knights (35S/174E) and Whale (37S/177E) Is off northern New Zealand;[1359] migrates to northern Pacific Ocean. One live bird found in a burrow on Cabbage Tree I. (32/152) in 1960 but no breeding confirmed.[493] Regular visitor to Australian waters in small numbers.[847]
1977–81 28/153, 32/152, 36/150, 37/150, 43/147.
1951–76 28/153, 29/153, 30/153, 32/151, 32/152, 33/151, 34/150, 34/151, 35/150, 37/149, 43/147.

914 Manx Shearwater

Puffinus puffinus

Breeds on islands in Atlantic Ocean; migrates to south-western Atlantic. One Australian record: Venus Bay (33/134), Nov. 1961.[388] See also Fluttering/Hutton's Shearwater p. 29.
1951–76 33/134.

912 Audubon's Shearwater

Puffinus lherminieri

One alleged record: bird collected by Sir Joseph Banks in 1770 'off Cape Upstart' (19/147) known only from description in Solander's diaries.[388]

OCEANITIDAE

064 Grey-backed Storm-Petrel

Oceanites nereis

Breeds on sub-Antarctic islands. Records in all seasons from waters off south-eastern Australia.
1977–81 37/139, 37/150, 38/140, 38/141, 38/142, 41/144, 42/144, 43/147, 43/148.
1951–76 33/151, 34/150, 38/142, 40/148, 41/148, 43/146.

066 Black-bellied Storm-Petrel

Fregetta tropica

Breeds on islands near the Antarctic Convergence. A skin in the British Museum (Nat. Hist.) labelled 'Port Essington N.W. Australia',[1709] provenance doubtful. Status in Australian waters unknown.
1977–81 28/153, 29/153.
1951–76 27/153, 35/150.
Before 1901 43/140.

944 White-bellied Storm-Petrel

Fregetta grallaria

Breeds on islands in the southern Atlantic, Pacific and Indian Oceans including Lord Howe I. (31/159). Rarely strays near the Australian coast.
1951–76 30/153.
1901–50 43/147.

911 Leach's Storm-Petrel

Oceanodroma leucorhoa

Breeds on islands off continents in northern Pacific and Atlantic Oceans. Accidental to Australia.
1977–81 31/115, 32/115.
1951–76 38/142.

830 Sooty Storm-Petrel

Oceanodroma matsudeirae

Breeds on islands south-east of Japan. Rarely recorded in Australian waters.
1977–81 16/121.
1951–76 13/120, 15/118, 19/116, 19/118.

PELECANOIDIDAE

910 South Georgian Diving-Petrel

Pelecanoides georgicus

Breeds on islands off New Zealand and on sub-Antarctic islands. One Australian record: Bellambi (34/150), Dec. 1958.
1951–76 34/150.

SULIDAE

825 Cape Gannet

Morus capensis

Breeds southern Africa. A single bird, first reported in the colony of Australasian Gannets breeding on the Wedge Light (38/144), Port Phillip Bay, in the 1980–81 breeding season,[295] has been observed at the same site during the 1981–82 and 1982–83 seasons.
1977–81 38/144.

FREGATIDAE

909 Christmas Frigatebird

Fregata andrewsi

Breeds on Christmas I. (10S/105E). Only Australian record: Darwin (12/130), Jan. 1974.[1170]
1951–76 12/130.

PHAETHONTIDAE

108 White-tailed Tropicbird

Phaethon lepturus

Breeds on islands in Pacific, Indian and Atlantic Oceans mainly in the tropics. Regular but rare visitor to eastern Australian waters. A wreck followed a cyclone in Mar. 1978 when many birds were found inland in the South-East and Murray-Darling Regions.[1288]
1977-819/143, 10/145, 11/144, 11/145, 13/145, 14/146, 28/153, 30/153, 31/148, 31/150, 32/150, 34/145, 34/150, 34/151, 35/117, 37/147.
1951-76 13/143, 16/149, 22/154, 27/153, 28/153, 30/153, 32/152, 33/151, 34/150, 34/151.
1901-50 11/142, 16/145, 16/149, 27/152, 33/151, 35/150.
Before 1901 28/113, 33/151, 34/151.

ARDEIDAE

908 Grey Heron

Ardea cinerea

Breeds Africa, the Palearctic, Indian subcontinent and south-eastern Asia. Australian records doubtful.[388]

907 Yellow Bittern

Ixobrychus sinensis

Breeds eastern and southern Asia, western Pacific islands and New Guinea. One unconfirmed record: a specimen from 'Queensland'.[1709] One confirmed record: Kalgoorlie (30/121), Jan. 1967.[1638]
1951–76 30/121.

ANATIDAE

906 Mute Swan

Cygnus olor

Native to Eurasia. Introduced to several sites in Australia as an ornamental bird. Only known established population in the wild is at Northam (31/116) but breeding has once been reported at Coolbellup (32/115).[1638]
1977–81 31/116B.
1951–76 31/116B, 32/115B.
1901–50 31/116B.

905 Northern Shoveler

Anas clypeata

Breeds northern hemisphere. Does not normally move south of the equator. Specimen collected in NSW in 1839, exact location unknown.[723] Unconfirmed record: Yandilla (27/151), before 1900.[299]
1977–81 27/140.
1951–76 30/145.

209 Garganey

Anas querquedula

Breeds in the Palearctic; migrates south. Accidental to Australia but Atlas records in most months.
1977–81 12/130, 15/128, 23/133, 24/119.
1951–76 12/130, 13/143, 27/151, 31/115, 36/142.
Before 1901 38/143, 38/144.

ACCIPITRIDAE

904 White-eyed Buzzard

Butastur teesa

Breeds southern Asia. One doubtful record: Lithgow (33/150), Nov. 1899.[388]

PHASIANIDAE

903 Peafowl

Pavo cristatus

Native to India and Sri Lanka. Widely kept in Australia, usually free ranging.[1085] Only established feral populations are listed.
1977–81 40/144B, 42/147B.
1951–76 31/115B, 32/115B, 35/145B, 39/143B.

902 Feral Chicken

Gallus gallus

Native to south-eastern Asia. Introduced to many parts of Australia but established only on Heron I. (23/151).[1085,1116]
1977–81 23/151B.
1951–76 23/151.
1901–50 23/151B.

950 Common Pheasant

Phasianus colchicus

Native to Asia. Introductions and escapees, some recent, at many places in the South-East, Tas. and South-West Regions. Only established feral populations are listed.
1977–81 32/115B, 39/144B, 39/148B, 40/148B.
1951–76 32/115B, 34/139B.
1901–50 32/115.

901 California Quail

Lophortyx californicus

Native to western USA. Introduced south-eastern Australia. Known to be established only on King I. (39/144)[1085] but 25+ reported near Newcastle (32/151), Mar. 1981.[1080]
1977–81 39/143, 39/144B, 40/144.
1951–76 39/143, 39/144, 40/144.

RALLIDAE

900 Red-legged Crake

Rallina fasciata

Widely distributed in south-eastern Asia and known to travel considerable distances.[1117] One record in Australia: Broome (17/122), July 1958.[1628]
1951–76 17/122.

899 Corncrake

Crex crex

Breeds Europe and northern Asia; migrates to Africa. Accidental to Australia.[388]
1901–50 30/115.
Before 1901 33/151.

JACANIDAE

897 Pheasant-tailed Jacana
Hydrophasianus chirurgus

Breeds southern Asia; part of population migrates south. One unconfirmed Australian record: Paraburdoo (23/117), Dec. 1974.[1710]

CHARADRIIDAE

896 Golden Plover
Pluvialis apricaria

Breeds northern Europe and Asia; migrates south, usually no farther than the equator. Two Australian specimens of doubtful origin.[388]

851 Little Ringed Plover
Charadrius dubius

Breeds from the Palearctic to tropical Eurasia. Now considered that small numbers regularly visit Australia in summer, specially the Top End.[1164a]
1977–81 12/130, 32/115, 34/137, 37/140.
1951–76 12/130, 20/119.

895 Ringed Plover
Charadrius hiaticula

Breeds in the Arctic; migrates to Europe, Africa and Asia mostly north of the equator. Accidental to Australia.
1977–81 12/130, 33/151, 36/150, 38/144, 38/146.
1951–76 32/151, 34/138.
1901–50 33/151.
Before 1901 32/152.

894 Caspian Plover
Charadrius asiaticus

Breeds southern Europe and Asia; migrates to Africa. One Australian record: Pine Ck (13/131), 1896.[388] Unconfirmed record: Darwin (12/130), Oct. 1974.[1175]
Before 1901 13/131.

SCOLOPACIDAE

893 Eurasian Curlew
Numenius arquata

Breeds northern Europe and Asia; migrates to Africa and south-eastern Asia. One unconfirmed record: near Darwin (12/130), 1948.[475]

892 Upland Sandpiper
Bartramia longicauda

Breeds southern Alaska and Canada to USA; migrates to southern South America. One Australian specimen collected Sydney (33/151), 1848.[388]
Before 1901 33/151.

850 Spotted Greenshank
Tringa guttifer

Breeds in the Palearctic; migrates to south-eastern Asia. Unconfirmed sighting: Darwin (12/130), Mar. 1974.[1175]

891 Redshank
Tringa totanus

Breeds northern Palearctic; migrates to Africa and southern Asia, regularly as far south as Indonesia. Some birds reach Australia, probably most years.
1977–81 12/130, 14/125.
1951–76 12/130, 20/118, 23/113.

832 Green Sandpiper
Tringa ochropus

Breeds northern Eurasia; migrates south to Africa and southern Asia. One unconfirmed record: Rookery Pt (12/132), Sept. 1979.

827 Stilt Sandpiper
Micropalama himantopus

A North American wader, breeding in the Arctic tundras. One record for Australia: Darwin (12/130), Sept. 1980.
1977–81 12/130.

939 Asian Dowitcher
Limnodromus semipalmatus

Breeds Siberia; migrates to south-eastern Asia. A specimen from Darwin (12/130), Oct. 1971, was first confirmed record for Australia. Large numbers were discovered in 1982: at Port Hedland (20/118) 130 in Mar.-Apr., 35 in Sept. and 20 in Nov.; and at Bush Pt (18/122) 100 in Sept.[913,1017] This, together with Atlas records, suggests that the species reaches Australia regularly, though probably only a few localities, mainly in the north.
1977–81 12/130, 16/145, 27/153, 34/150.
1951–76 12/130, 16/145, 20/118, 27/153, 38/144.

890 Baird's Sandpiper
Calidris bairdii

Breeds Siberia and arctic North America; migrates to South America. First record for Australia was a specimen from Lauderdale (42/147), Oct. 1966. Unconfirmed record: Botany Bay (33/151), Nov. 1976.[394]
1977–81 32/126.
1951–76 35/143, 42/147.

889 Western Sandpiper

Calidris mauri

Breeds northern Alaska and north-eastern Siberia; migrates to South America. In Australia three published records considered doubtful: Sandford (42/147), Sept. 1969;[1744] Ivanhoe (32/144), May 1974;[839] Homebush Bay (33/151), Nov. 1978.[1416]

849 White-rumped Sandpiper

Calidris fuscicollis

Breeds arctic North America; migrates to southern South America. Accidental to Australia.
1977–81 32/115, 33/150, 38/144.
1951–76 38/143.

857 Little Stint

Calidris minuta

Breeds northern Europe and north-western Asia; migrates south. First recorded in Australia at Pt Wilson (38/144), Jan. 1977.[1773] Three unconfirmed records: Darwin (12/130), Sept.-Nov. 1979, Jan. 1980 and Sept. 1981. Vagrant to Australia.
1977–81 34/138, 35/143, 38/144, 42/147, 42/148, 43/147.

888 Dunlin

Calidris alpina

Breeding range nearly circumpolar in the Arctic; migrates south, seldom south of the equator. In Australia, unconfirmed specimens and records. Recent records in Vic. were withdrawn in 1981[1679] following recognition of the new species *C. paramelanotos*, see species 822.

822 Cox's Sandpiper

Calidris paramelanotos

A newly recognized wader visiting Australia in summer.[1390,1680] Known from two specimens: Mosquito Pt (35/139), Feb. 1975, and Price Saltfields (34/138), Mar. 1977, and a number of unconfirmed sightings. See also species 888.
1977–81 34/138.
1951–76 35/139.

887 Buff-breasted Sandpiper

Tryngites subruficollis

Breeds in arctic north-western America and eastern Siberia; migrates to South America, mainly Argentina. Accidental to Australia.
1977–81 32/151.
1951–76 34/151, 35/143, 37/144, 38/144, 38/145.

PHALAROPODIDAE

835 Grey Phalarope

Phalaropus fulicarius

Breeds in Arctic regions farther north than the other two species of *Phalaropus*; pelagic outside the breeding season. A single record for Australia: near Swan Hill (35/143), Feb. 1976.[1678]
1951–76 35/143.

932 Red-necked Phalarope

Phalaropus lobatus

Breeds in the Holarctic; normally spends the non-breeding season at sea in tropical and sub-tropical Pacific and Indian Oceans. Vagrant to Australia.
1977–81 30/151, 31/115, 32/115, 32/126, 34/138, 35/143, 38/144.
1951–76 12/130, 32/137, 34/138, 35/139, 37/144, 38/144.

886 Wilson's Phalarope

Phalaropus tricolor

Breeds interior North America; migrates to South America. Accidental to Australia. Unconfirmed record: Moolap (38/144), Dec. 1981.
1977–81 35/143, 37/144.
1951–76 37/144, 38/143.

STERCORARIIDAE

933 Long-tailed Jaeger

Stercorarius longicauda

Breeds mainly within the Arctic Circle; migrates to waters off the Pacific and Atlantic shores of South America. Vagrant to Australia.
1977–81 30/153, 33/151, 37/149, 38/142.
1951–76 27/153, 28/153, 33/151, 37/139, 38/144.

LARIDAE

856 Black-tailed Gull

Larus crassirostris

Breeds coastal Japan and China; ranges to coasts of eastern Asia One unconfirmed record in Vic.: single bird seen at Pt Lonsdale (38/144), Mar.-Apr. 1978, and Ricketts Pt (38/145), Sept 1978—both localities in Port Phillip Bay.

855 Lesser Black-backed Gull

Larus fuscus

A northern hemisphere species. One record on the Australian list: Melville Bay (12/136), which is now considered doubtful.[376]

885 Franklin's Gull

Larus pipixcan

Breeds North America; migrates to South America. Accidental to Australia.[1638] Unconfirmed record: Jurien Bay (30/115), Mar. 1974.
1977–81 27/153, 33/151.
1951–76 28/114, 33/115.

884 Black Tern

Chlidonias niger

Breeds North America, Europe and western Asia; migrates to South America and western Africa. Accidental to Australia.
1951–76 32/151, 33/151.

952 Arctic Tern

Sterna paradisaea

Breeds in the Holarctic; migrates to the Antarctic. Regular in small numbers in Australian waters.
1977–81 16/145, 31/115, 32/115, 32/126, 33/151, 36/150, 37/149, 37/150, 38/141, 38/144, 38/145, 40/148, 41/145, 41/146, 42/145, 43/148.
1951–76 16/145, 27/153, 31/115, 32/115, 32/151, 33/115, 33/151, 34/115, 34/150, 34/151, 35/138, 35/150, 36/149, 37/143, 37/145, 37/149, 38/142, 38/143, 38/145, 38/147, 39/143, 39/144, 39/146, 42/147, 43/147.
1901–50 33/115, 35/138, 38/145.

883 Saunders's Tern

Sterna saundersi

A specimen in the National Museum, Vic., labelled as this species is considered doubtful.[508]

982 Grey Ternlet

Procelsterna albivittata

Breeds on Pacific Ocean islands including Lord Howe (31/159) and Norfolk (29/167) Is. Small numbers off eastern Australia Dec.-May.[848]
1977–81 27/153, 32/151, 38/149.
1951–76 27/153, 28/153, 30/153, 31/152, 32/152, 33/151, 34/150.
1901–50 33/151, 34/151.

972 White Tern

Gygis alba

A tropical tern breeding on Norfolk (29/167) and Kermadec (29S/178W) Is and on other oceanic islands remote from Australia.
1977–81 25/152, 33/151, 34/150.
1951–76 17/146, 18/146, 19/147, 25/152, 25/153, 27/153, 29/153, 31/152, 34/150.

STRIGIDAE

823 Brown Hawk-Owl

Ninox scutulata

Breeds south-eastern Asia. One record for Australia: a specimen found dead on Ashmore Reef (12/122), Jan. 1973.[1592]
1951–76 12/122.

APODIDAE

882 Glossy Swiftlet

Collocalia esculenta

Ranges across Indian Ocean from Andaman (12N/93E) and Nicobar (8N/93E) Is through New Guinea to New Caledonia and rarely visits Australia. One specimen: Cape York (10/142), before June 1929;[388] confirmed record: Iron Ra. (12/143), Jan. 1966; and unconfirmed records: Julatten (16/145), May 1978, and Iron Ra., Aug. 1980.
1951–76 12/143.
1901–50 10/142.

881 Uniform Swiftlet

Collocalia vanikorensis

Ranges from Sulawesi through New Guinea to New Caledonia and the Cape York Region (rarely). First Australian record: a specimen taken in northern Qld (10/142), Sept. 1913.[388] Unconfirmed record: Sue I. (10/142), Feb. 1978.
1901–50 10/142.

828 House Swift

Apus affinis

Breeds eastwards from Spain and Africa to south-eastern Asia. Only Australian record: an injured bird found at Pt Stuart (12/131), Mar. 1979.[1504]
1977–81 12/131.

844 Swift spp

Two records of small, grey swifts: Warrego (19/133), Mar. 1980, 3–4 birds; Alice Springs (23/133), Dec. 1981, several hundred birds. One record of a small dark swift in a flock of Fork-tailed Swifts: False Bay (33/137), Dec. 1973.[548] None identified to specific level; probably all of Asian origin.

PITTIDAE

880 Blue-winged Pitta

Pitta moluccensis

Breeds south-eastern Asia; migrates south, normally as far as the Malay Pen. Two specimens from Australia of birds found dead: Mandora Station (19/120), Nov. 1927, and Derby (17/123), about the end of 1930.[1629]
1901–50 17/123, 19/120.

HIRUNDINIDAE

878 Pacific Swallow

Hirundo tahitica

Said to occur in Torres Strait but the evidence is inadequate.[1709]

MOTACILLIDAE

876 Yellow-headed Wagtail

Motacilla citreola

Breeds north-central Asia; migrates to the Himalaya, Burma and Thailand. One record in Australia: Woolooware Bay (34/151), July 1962.[1139]
1951–76 34/151.

875 Grey Wagtail

Motacilla cinerea

A Palearctic species migrating south to Africa, Asia and New Guinea. Accidental to Australia. All Atlas records Nov.-Mar.
1977–81 12/130, 17/145, 18/124, 34/138.
1951–76 17/145.

874 White Wagtail

Motacilla alba

Breeds Eurasia and northern Africa; migrates to central and southern Africa and south-eastern Asia. One specimen: Geraldton (28/114), Mar. 1981.[1718]
1977–81 28/114.

PYCNONOTIDAE

873 Red-vented Bulbul

Pycnonotus cafer

Native to south-eastern Asia; temporarily established in Melbourne (37/144–145) about 1917; last reported Oct. 1942.[1057] Unconfirmed report: Sydney (33/151), 1917.[1065]
1901–50 37/144, 37/145B.

MUSCICAPIDAE

831 Arctic Warbler

Phylloscopus borealis

Breeds in northern Eurasia and Alaska; nominate race migrates to southern Asia. Only Australian record: a specimen of the nominate race, Scott Reef (14/121), Nov. 1979.[1168]
1977–81 14/121.

SYLVIIDAE

872 Great Reed-Warbler

Acrocephalus arundinaceus

Breeds in the Palearctic; migrates south as far as south-eastern Asia and, rarely, New Guinea. One specimen from Melville I. (11/130), Jan. 1912.[1711] Unconfirmed record: Darwin (12/130), Mar. 1979. Two birds, sighted and tape-recorded, Kununurra (15/128), Jan. 1982.[1518]
1901–50 11/130.

PLOCEIDAE

870 Black-headed Mannikin

Lonchura malacca

Native to south-eastern Asia. Small populations have been established from escapees in Sydney (33/151) from 1929; not recorded there since 1960. A single bird sighted at Nowra (34/150), Aug. 1977.[1525]
1977–81 34/150.
1951–76 33/151B.
1901–50 33/151B.

951 White-winged Wydah

Euplectes albonotatus

Native to Africa. Introduced to Australia at Sydney (33/150) about 1931.[1085] Not recorded since early 1970s and population is believed no longer to exist.[1131a]
1951–76 33/150B.
1901–50 33/150B.

869 Red Bishop

Euplectes orix

Native to Africa. A breeding population established near Adelaide (34/138) in 1926 but not recorded during the Field Atlas. Considered extinct in 1976.[1693]
1951–76 35/138, 35/139.
1901–50 34/138B, 34/140 35/138B, 35/139B.

CORVIDAE

867 House Crow

Corvus splendens

Breeds Indian subcontinent. Individual birds have occasionally reached Fremantle (32/115) on visiting ships and the species was once found breeding there; all birds found in WA have been destroyed.[922] Three birds left a ship at Geelong (37/144), May 1960;[689] one bird seen at 38/144 near Werribee, Mar. 1963.[1675]
1951–76 30/116, 31/115, 32/115B, 32/116, 33/115, 33/116, 37/144, 38/144.
1901–50 32/115.

HISTORICAL MAPS

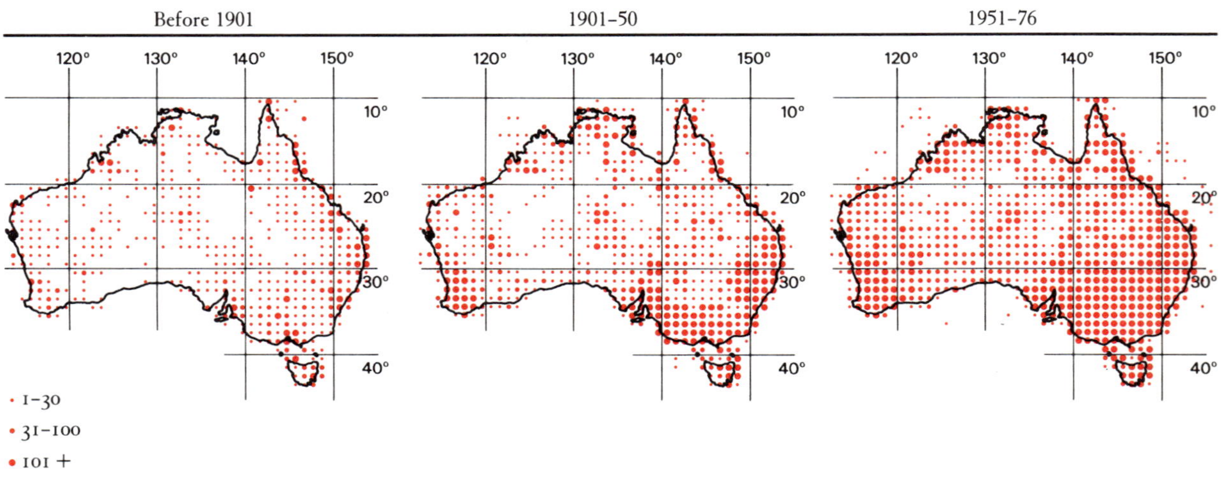

Number of species recorded in 1° blocks

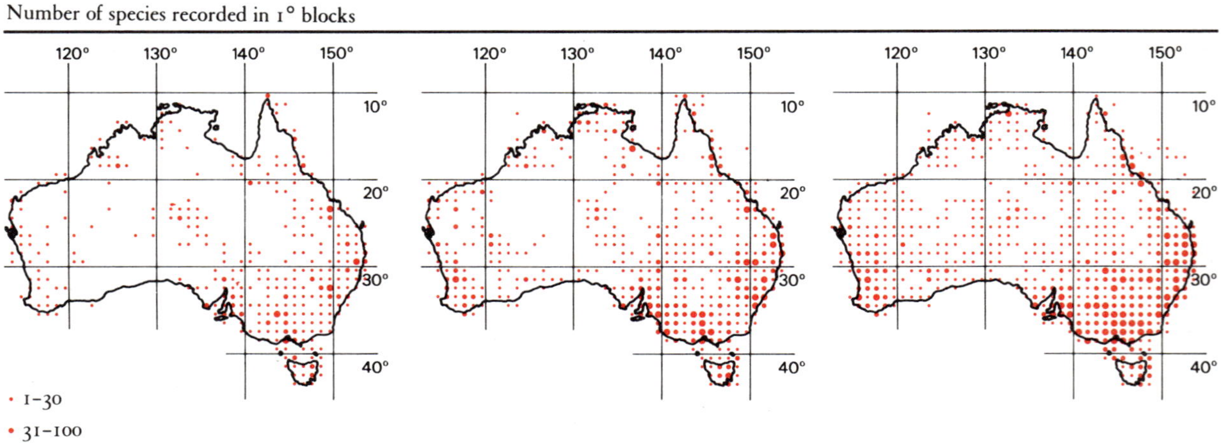

Number of species recorded breeding in 1° blocks

Before 1901	1901–50	1951–76

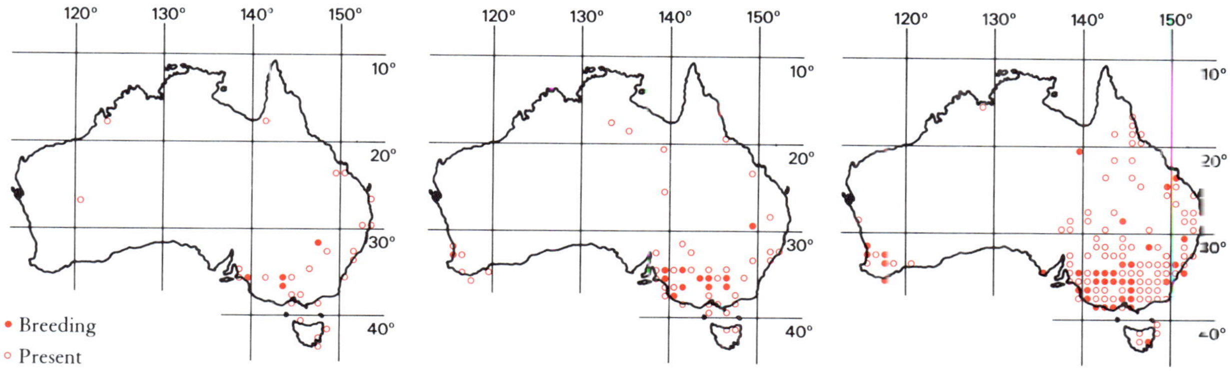

060 Great Crested Grebe *Podiceps cristatus* p. 4

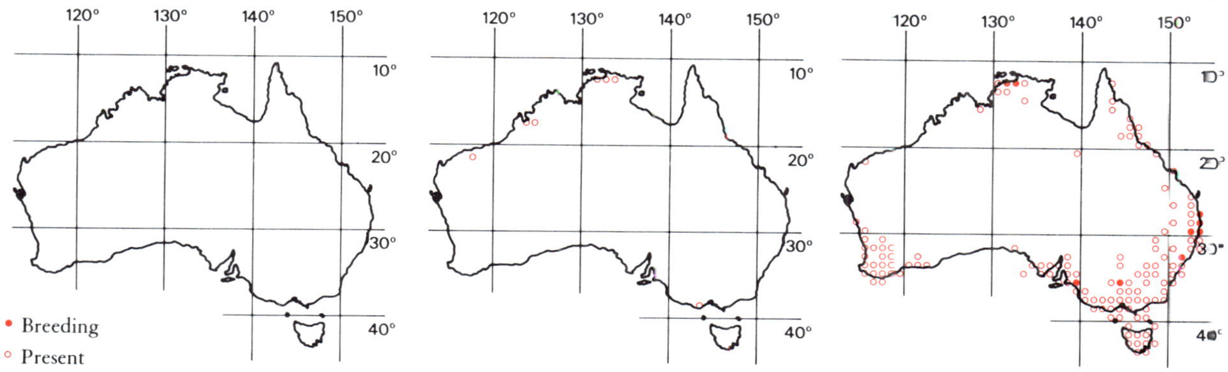

977 Cattle Egret *Ardeola ibis* p. 52

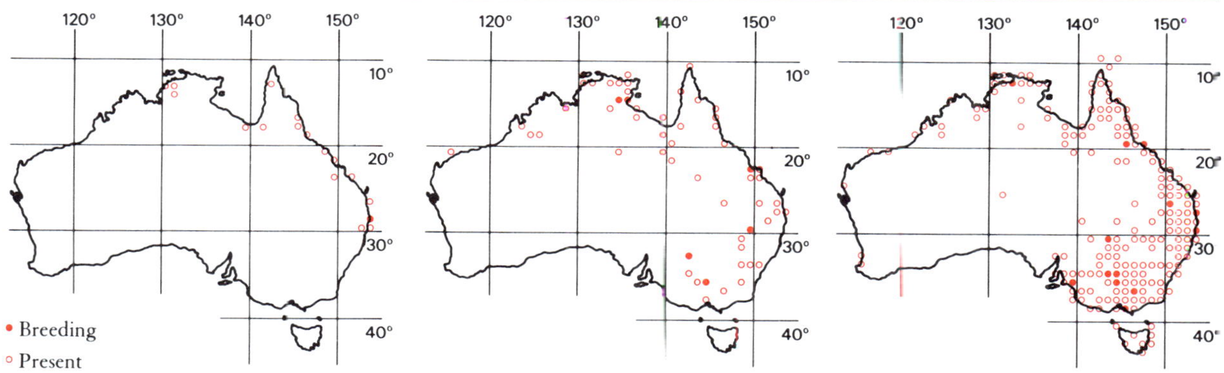

185 Little Egret *Egretta garzetta* p. 54

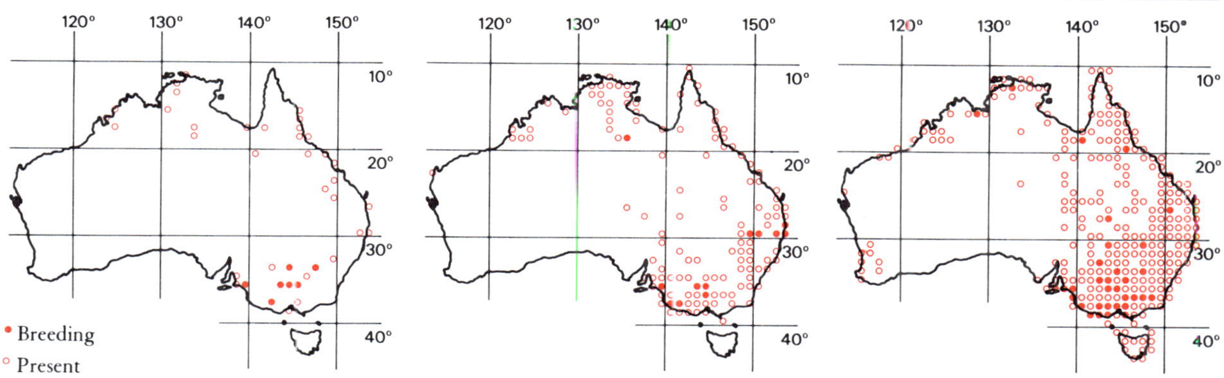

179 Sacred Ibis *Threskiornis aethiopica* p. 64

Before 1901 | 1901–50 | 1951–76

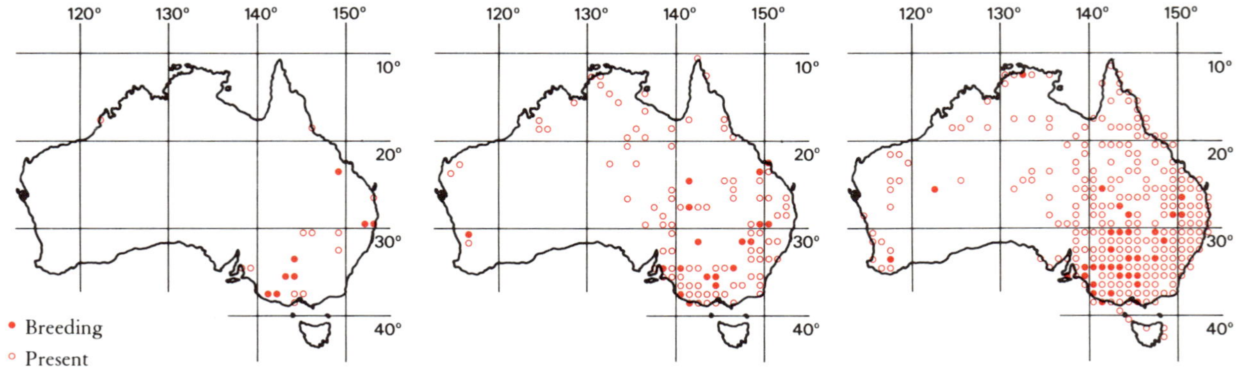

182 Yellow-billed Spoonbill *Platalea flavipes* p. 67

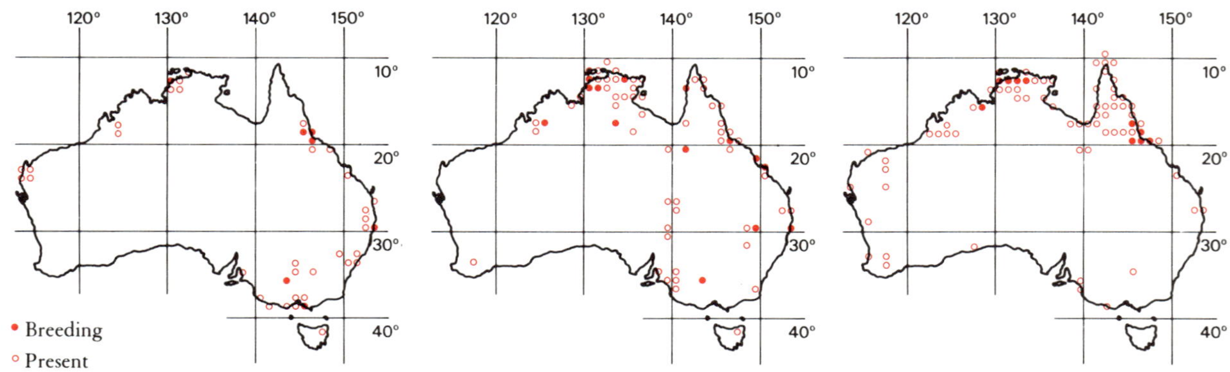

199 Magpie Goose *Anseranas semipalmata* p. 68

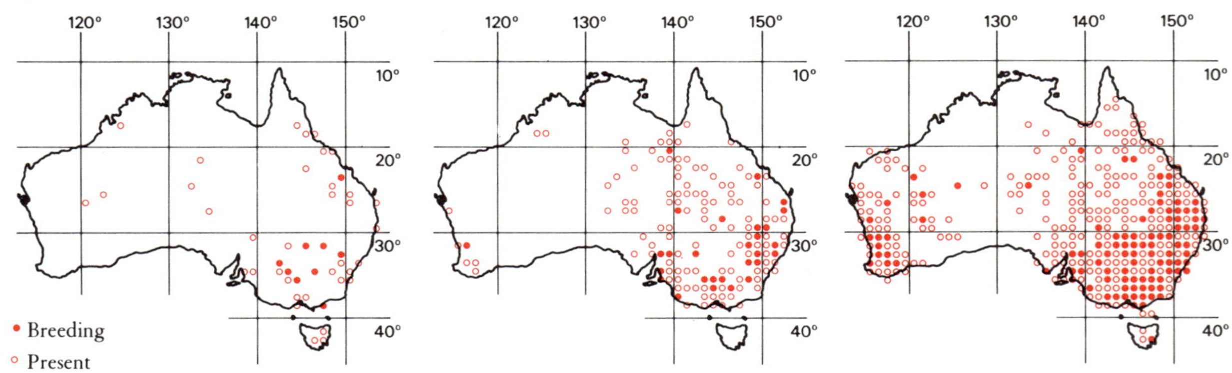

202 Maned Duck *Chenonetta jubata* p. 83

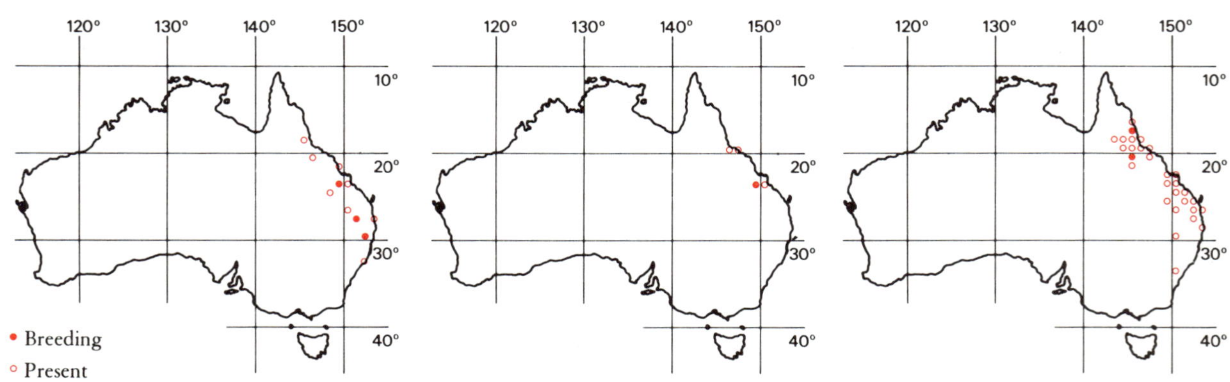

200 Cotton Pygmy-Goose *Nettapus coromandelianus* p. 84

Before 1901 | 1901–50 | 1951–76

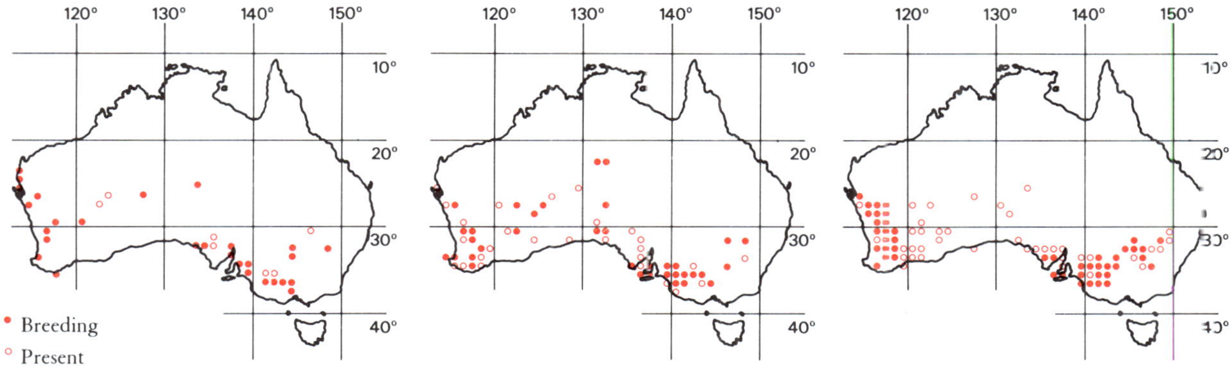

007 Malleefowl *Leipoa ocellata* p. 113

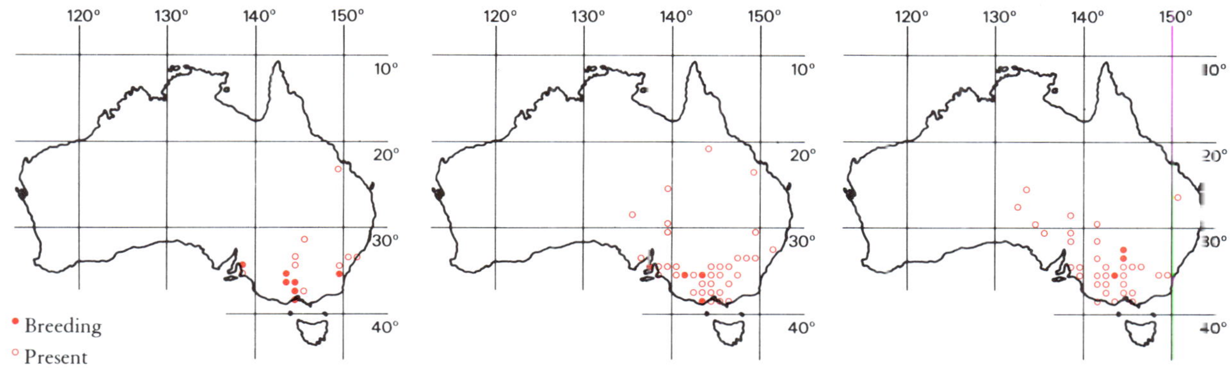

020 Plains-wanderer *Pedionomus torquatus* p. 125

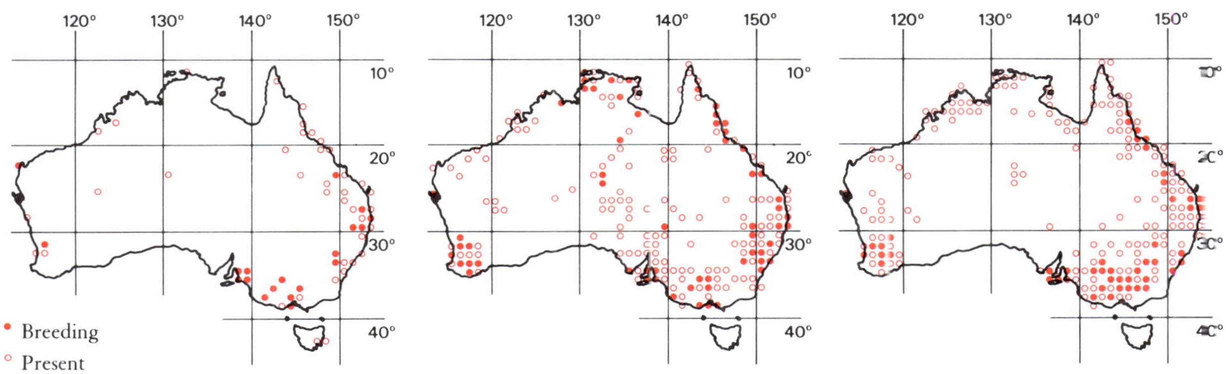

174 Bush Thick-knee *Burhinus magnirostris* p. 144

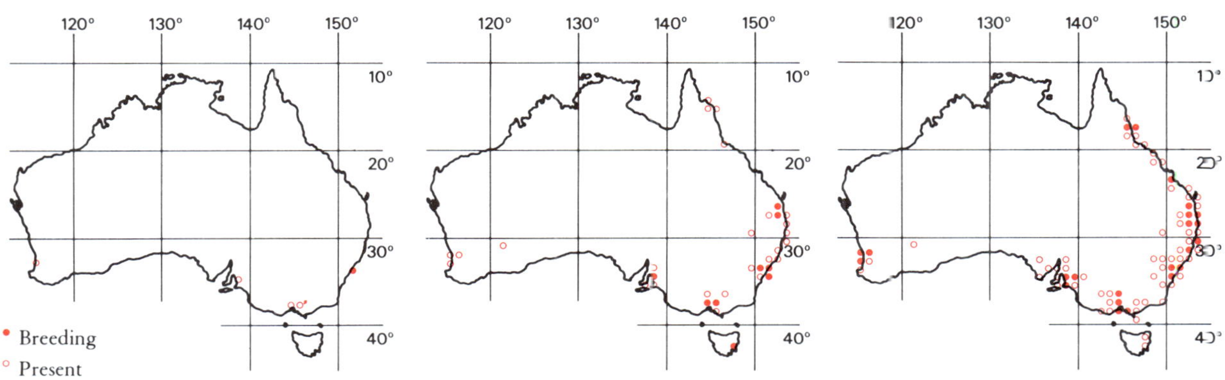

989 Spotted Turtle-Dove *Streptopelia chinensis* p. 222

Before 1901	1901–50	1951–76

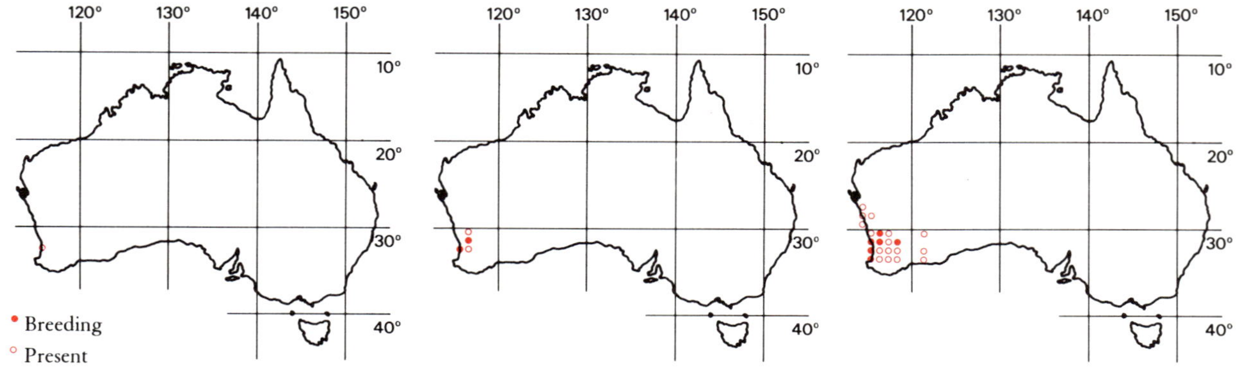

988 Laughing Turtle-Dove *Streptopelia senegalensis* p. 223

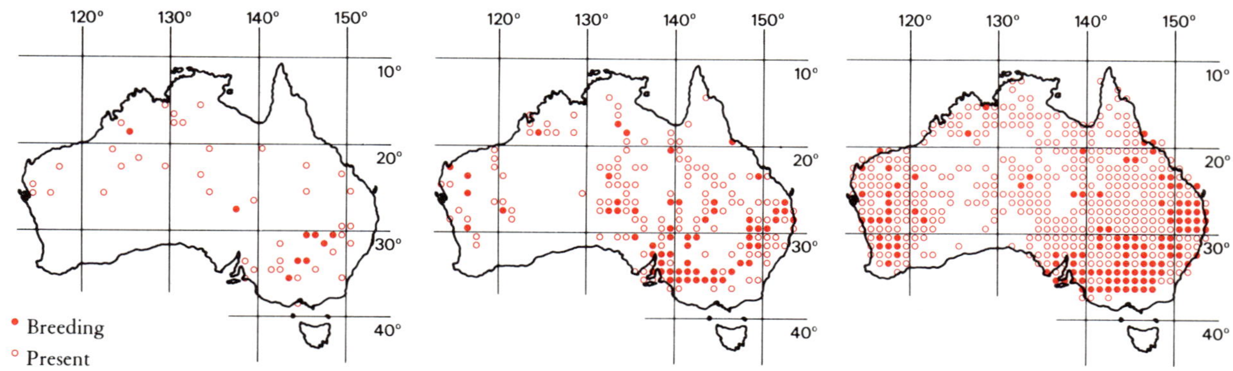

043 Crested Pigeon *Ocyphaps lophotes* p. 232

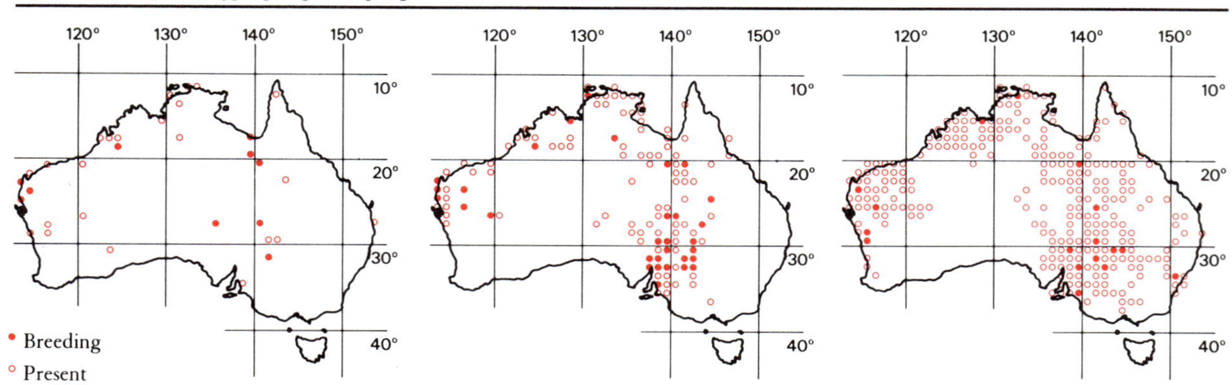

271 Little Corella *Cacatua sanguinea* p. 247

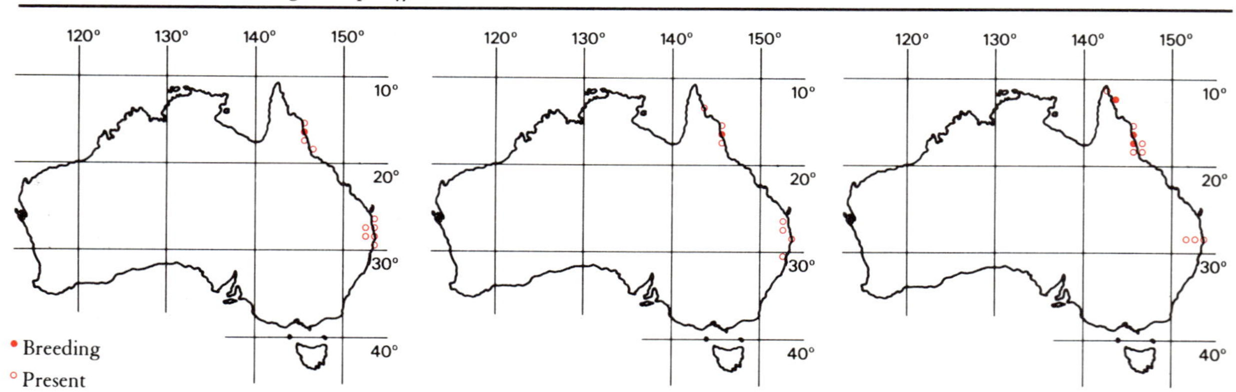

261 Double-eyed Fig-Parrot *Psittaculirostris diophthalma* p. 259

Before 1901 | 1901–50 | 1951–76

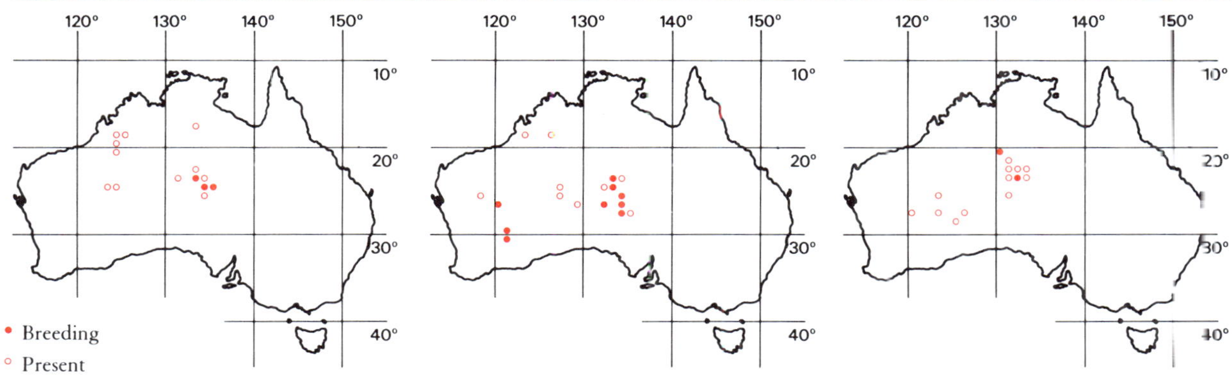

279 Alexandra's Parrot *Polytelis alexandrae* p. 264

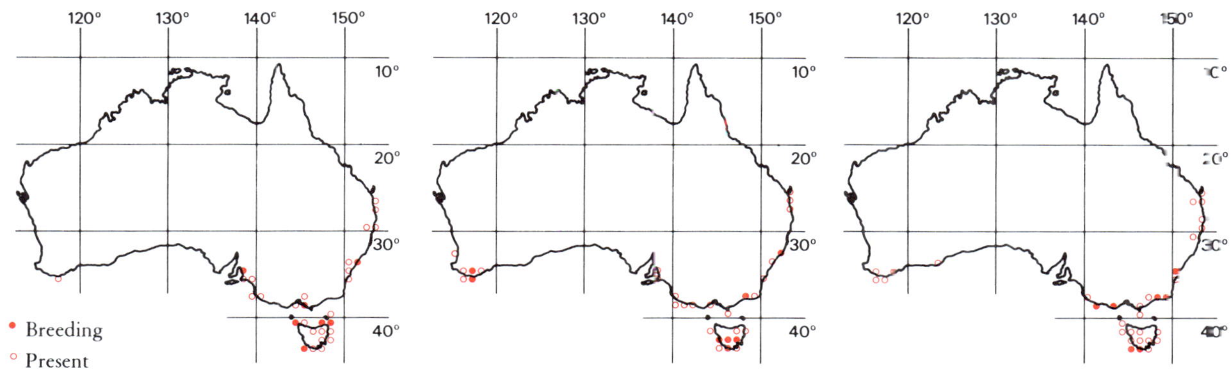

311 Ground Parrot *Pezoporus wallicus* p. 266

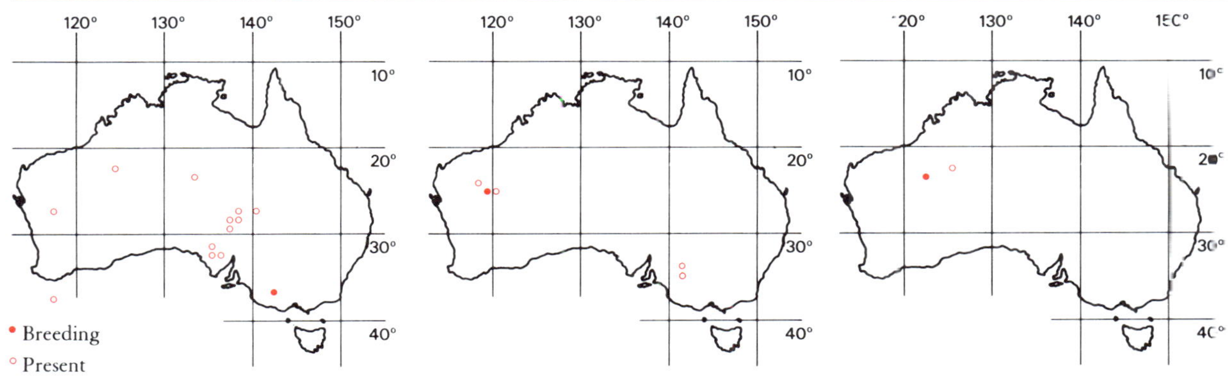

312 Night Parrot *Geopsittacus occidentalis* p. 267

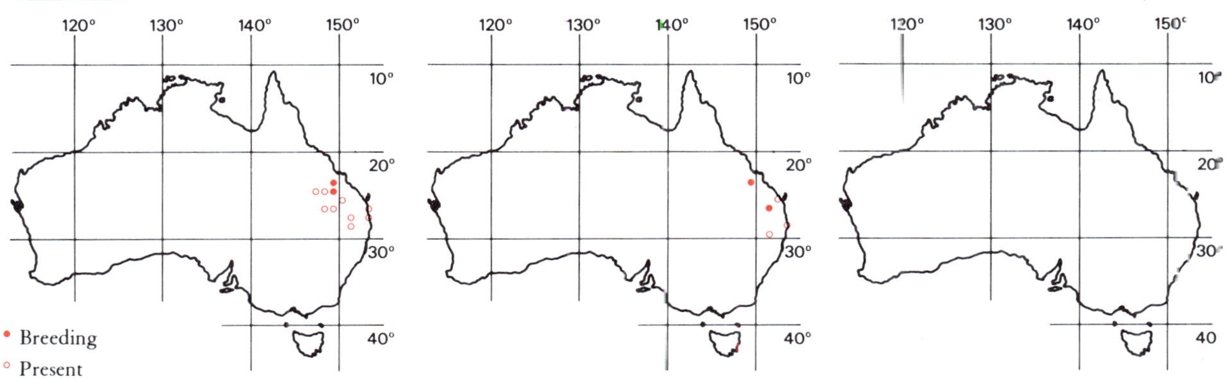

299 Paradise Parrot *Psephotus pulcherrimus* p. 284

Before 1901 | 1901–50 | 1951–76

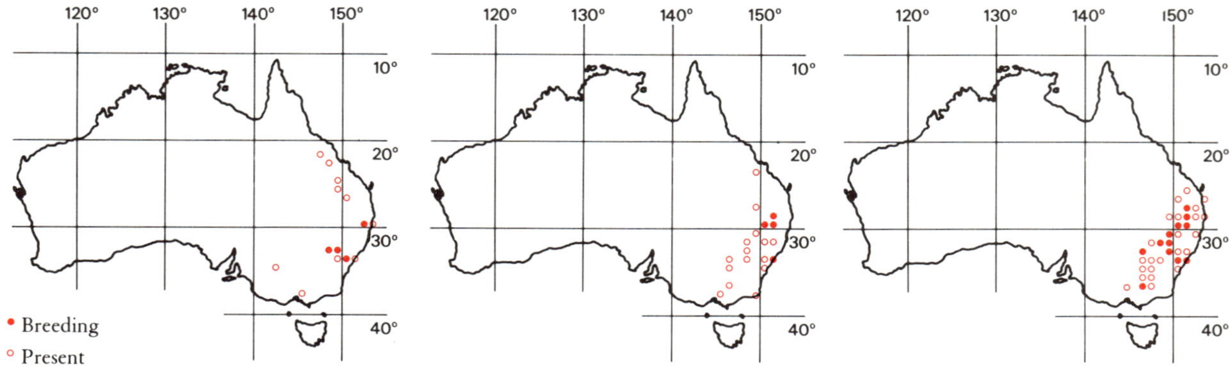

302 Turquoise Parrot *Neophema pulchella* p. 291

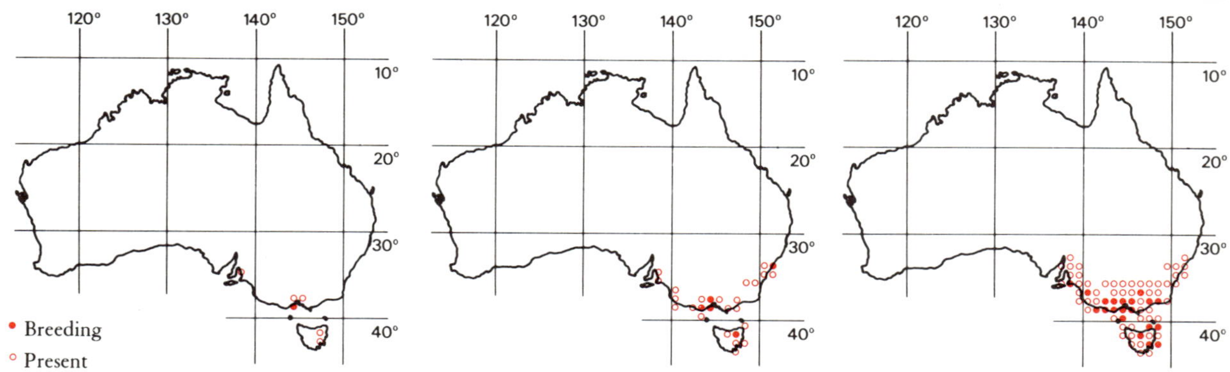

993 Skylark *Alauda arvensis* p. 343

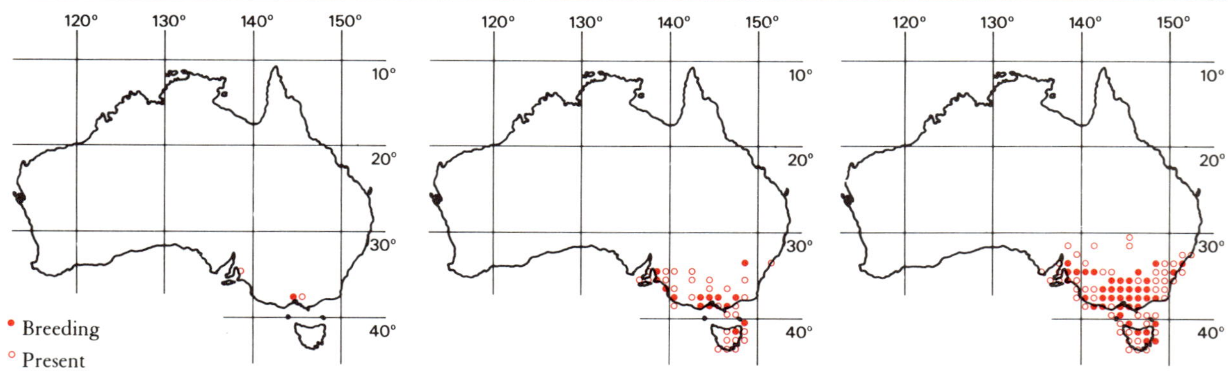

991 Blackbird *Turdus merula* p. 360

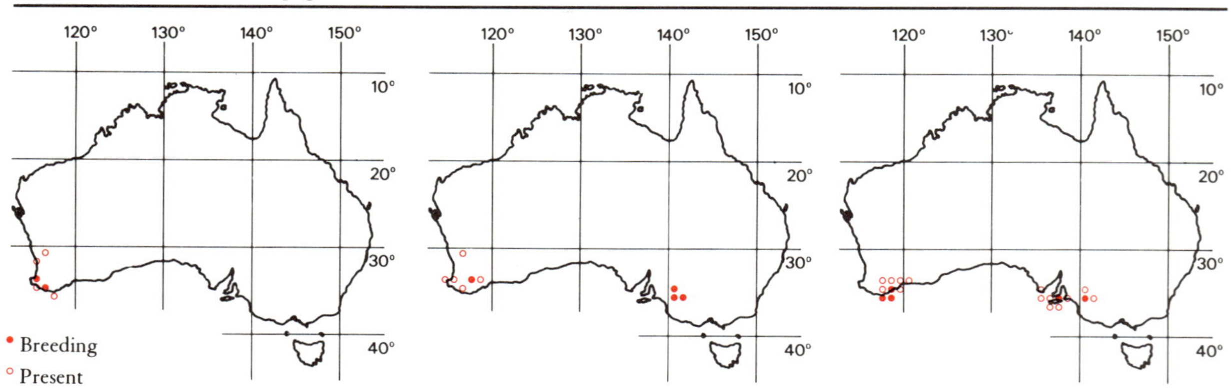

422 Western Whipbird *Psophodes nigrogularis* p. 416

Before 1901 | 1901–50 | 1951–76

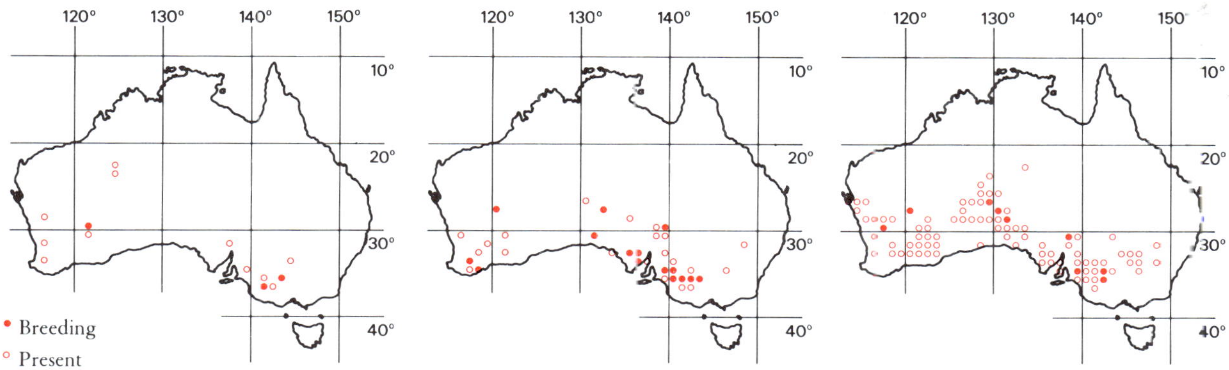

437 Chestnut Quail-thrush *Cinclosoma castanotum* p. 420

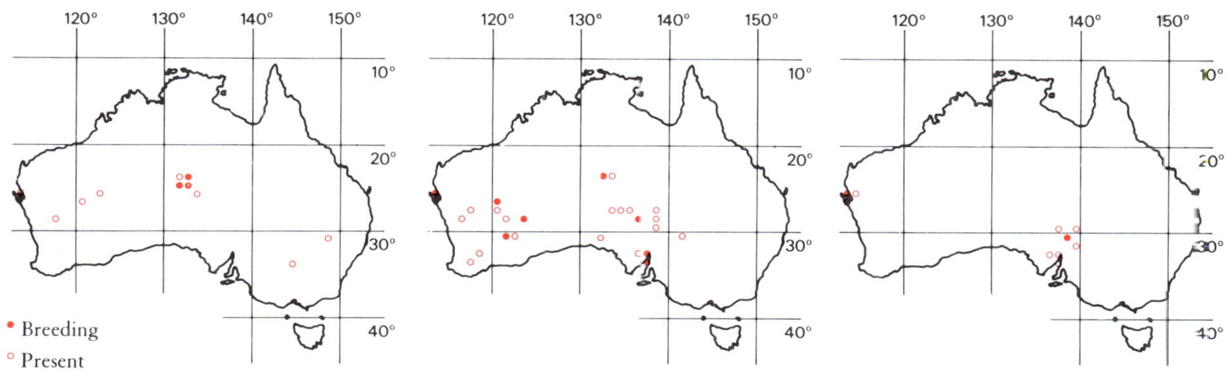

512 Thick-billed Grasswren *Amytornis textilis* p. 451

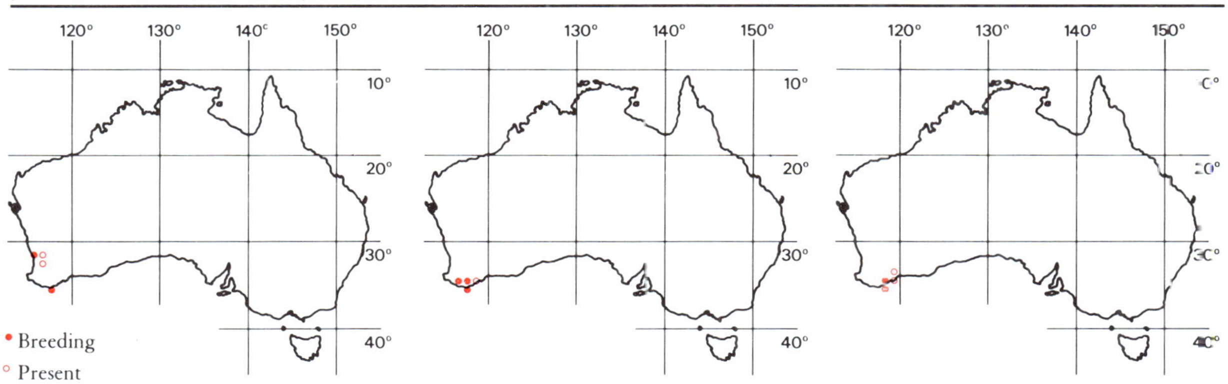

520 Western Bristlebird *Dasyornis longirostris* p. 454

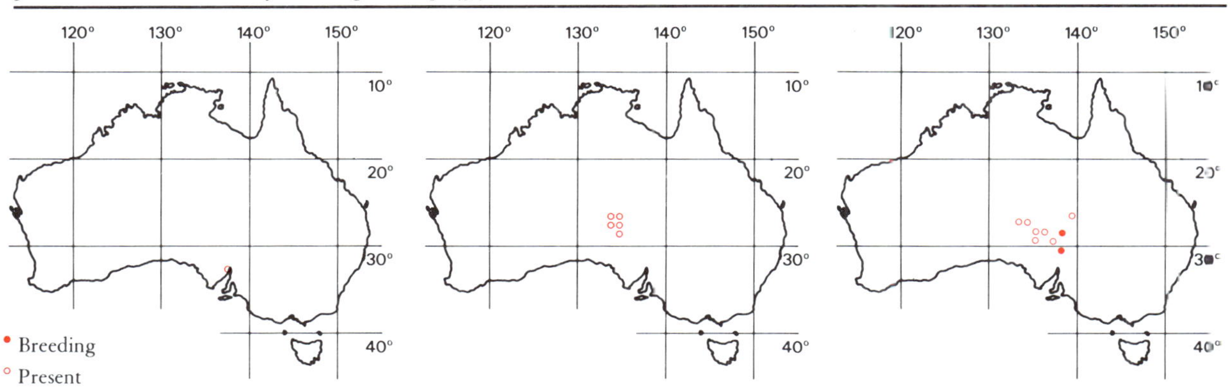

468 Chestnut-breasted Whiteface *Aphelocephala pectoralis* p. 492

Before 1901 | 1901–50 | 1951–76

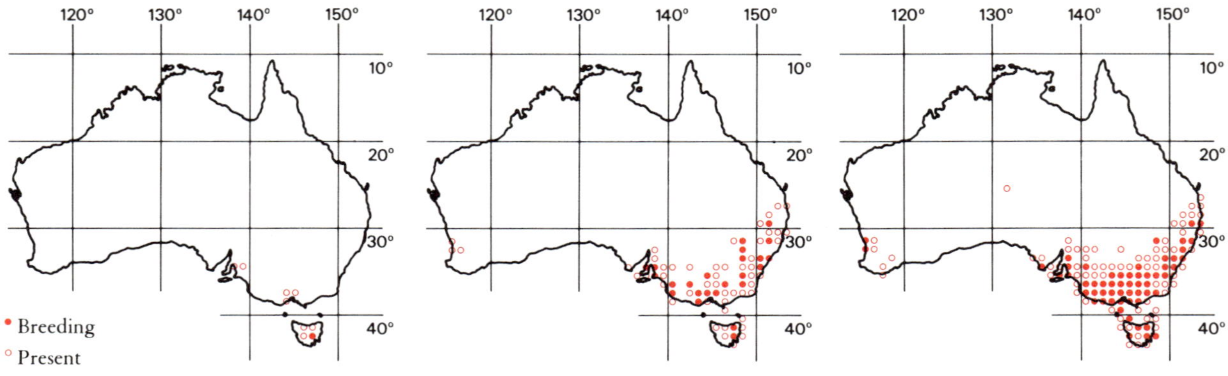

996 European Goldfinch *Carduelis carduelis* p. 584

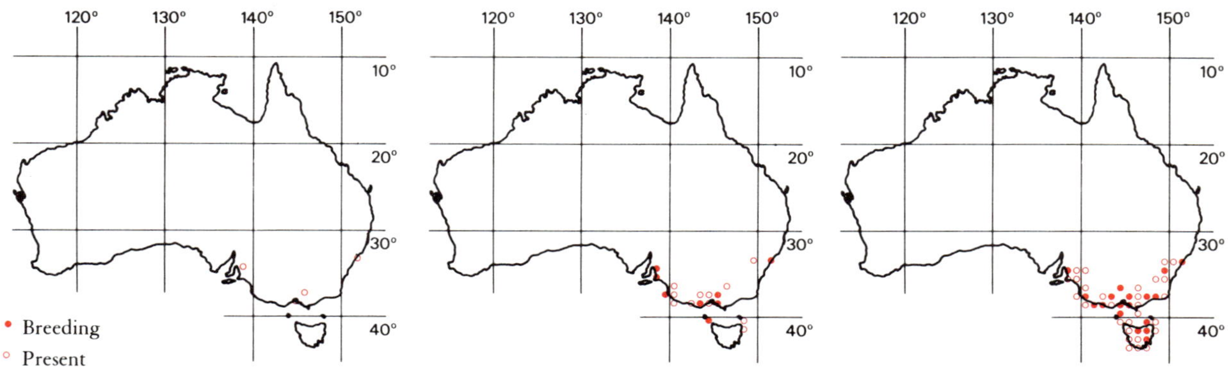

997 European Greenfinch *Carduelis chloris* p. 585

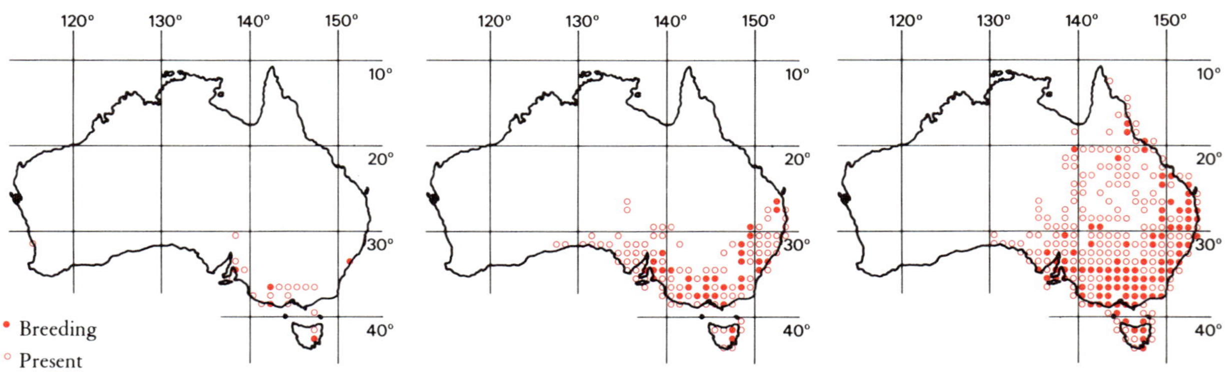

995 House Sparrow *Passer domesticus* p. 586

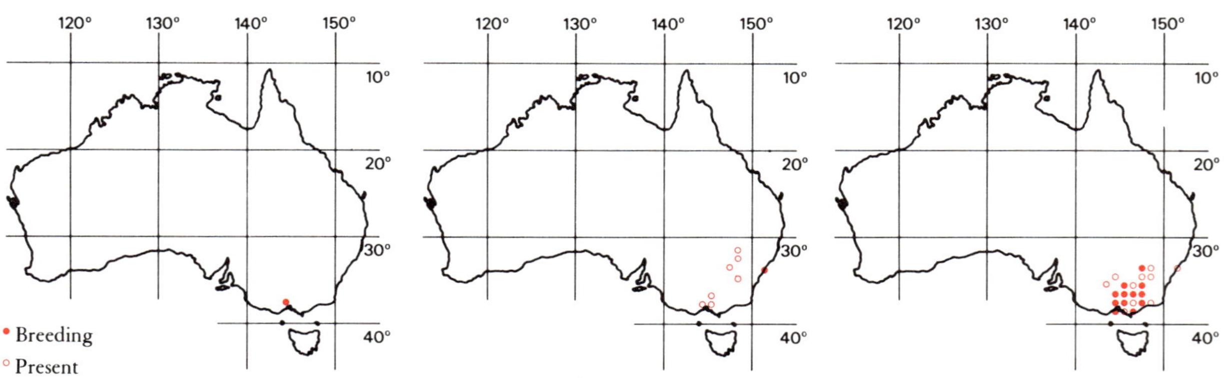

994 Tree Sparrow *Passer montanus* p. 587

Before 1901 | 1901–50 | 1951–76

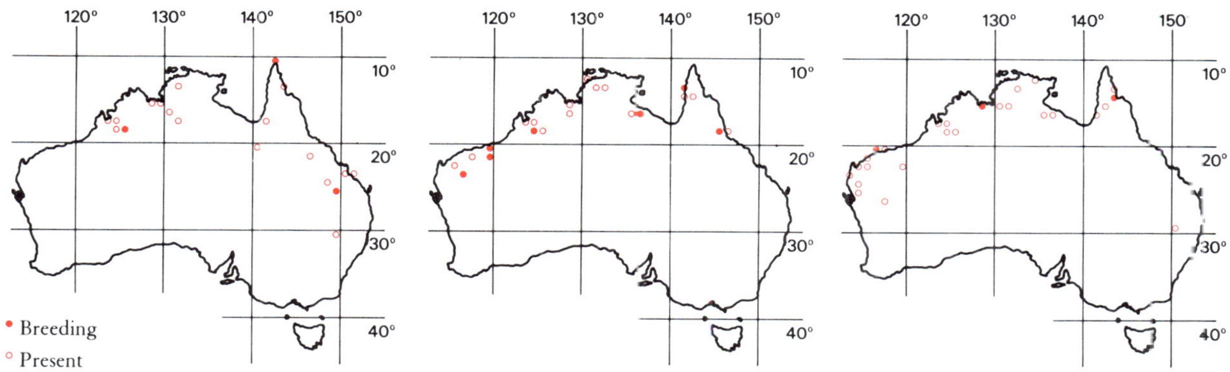

663 Star Finch *Neochmia ruficauda* p. 593

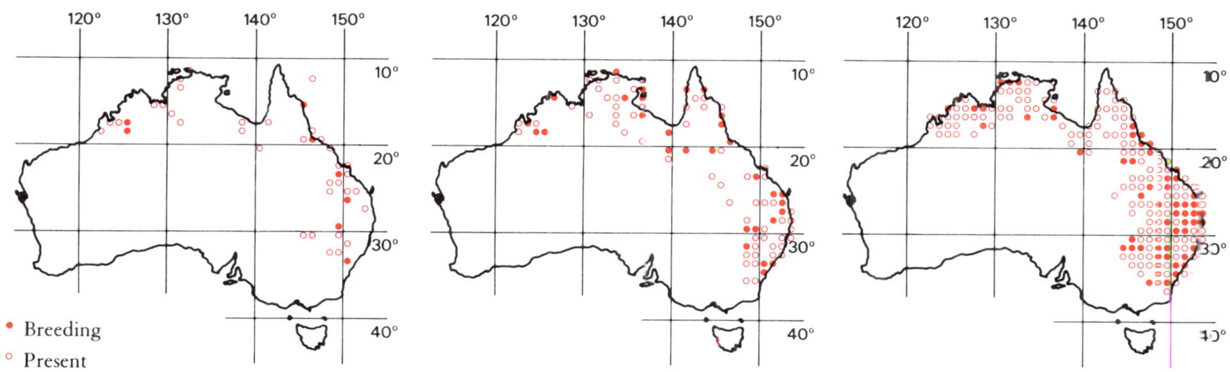

655 Double-barred Finch *Poephila bichenovii* p. 596

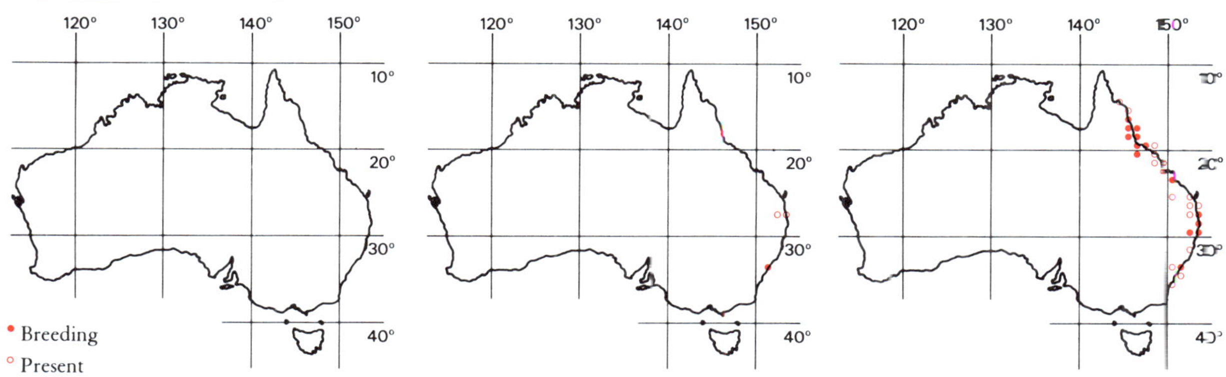

983 Nutmeg Mannikin *Lonchura punctulata* p. 604

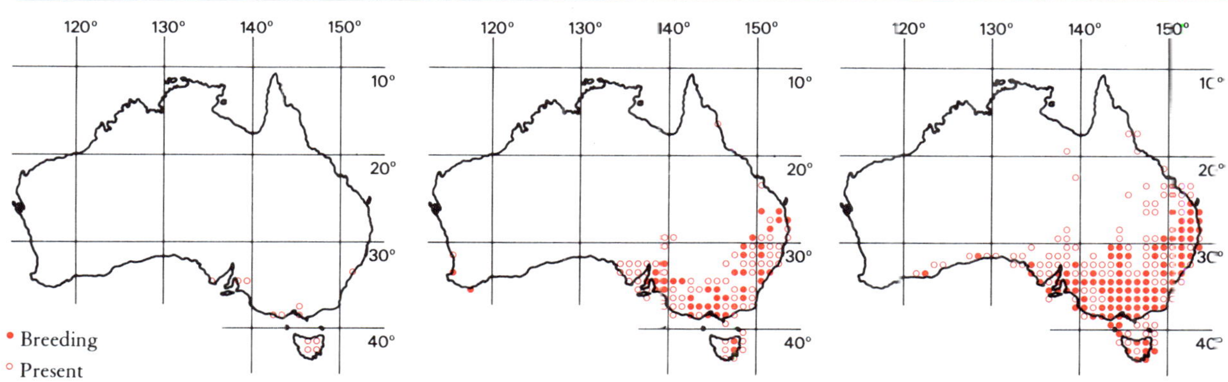

999 Common Starling *Sturnus vulgaris* p. 608

Before 1901 | 1901–50 | 1951–76

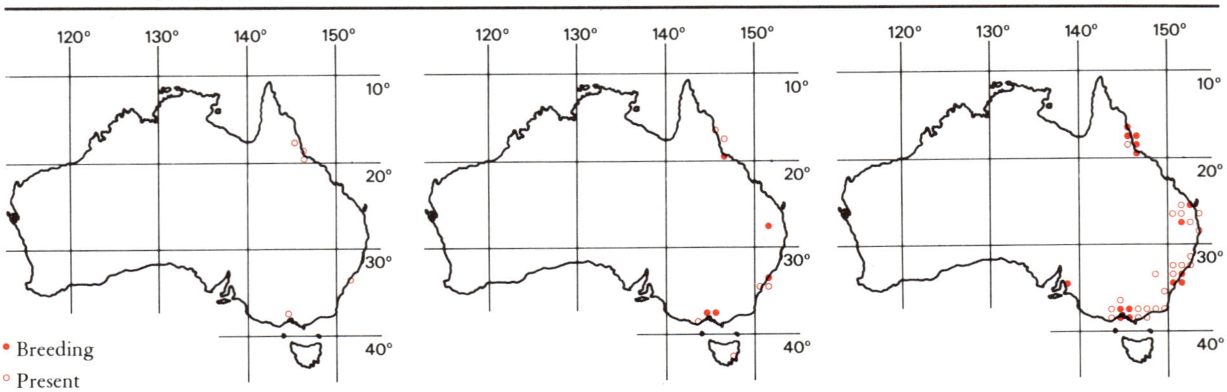

998 Common Mynah *Acridotheres tristis* p. 609

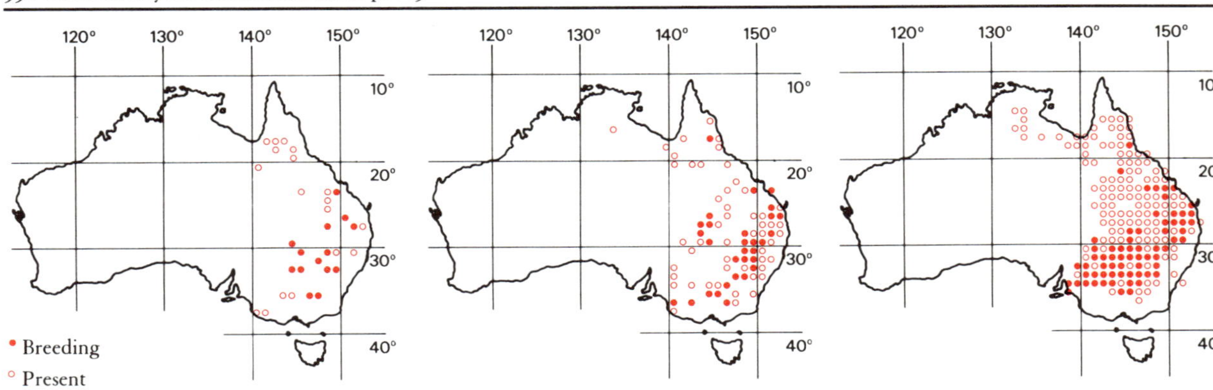

675 Apostlebird *Struthidea cinerea* p. 628

SOURCES OF RECORDS FOR THE HISTORICAL ATLAS

Data for the Historical Atlas were gathered from published and unpublished documents and from collections. The main ones are listed below. Many people contributed records from their own notebooks or recollections but these were too numerous to list separately. Issues of journals published later than 1976 were not searched systematically.

Collections

*AMERICAN MUSEUM OF NATURAL HISTORY. New York, U.S.A.

*AUSTRALIAN MUSEUM. 'A' and 'B' Registers. H. G. Bray egg register. MacGillivray egg data books. Register of the Moore egg collection. Sydney, Australia.

BONNIN, J. M. Specimen collection. Adelaide, Australia.

*BRITISH MUSEUM OF NATURAL HISTORY. London, U.K.

*CAMBRIDGE DEPARTMENT OF ZOOLOGY MUSEUM. Cambridge, U.K.

CLELAND, J. B. Egg collection. Adelaide, Australia.

*DEPARTMENT OF FORESTRY AND NATURAL RESOURCES MUSEUM. West Lafayette, U.S.A.

*FIELD MUSEUM OF NATURAL HISTORY. Chicago, U.S.A.

GREEN, R. H. Oological collection. Launceston, Australia.

*LOS ANGELES COUNTY MUSEUM OF NATURAL HISTORY. Los Angeles, U.S.A.

*MERSEYSIDE COUNTY MUSEUM. Liverpool, U.K.

*MUSEUM OF COMPARATIVE ZOOLOGY. Cambridge, U.S.A.

*MUSEUM OF ZOOLOGY. Ann Arbor, U.S.A.

*NATIONAL MUSEUM OF VICTORIA. General collection. R. Hall collection. F. C. Morse collection. D. Thompson collection. H. L. White skin collection. H. L. White egg collection. Melbourne, Australia.

NATIONAL PARKS AND WILDLIFE SERVICE COLLECTION. Sydney, Australia.

*NATIONAL WILDLIFE COLLECTION, Canberra, Australia.

PATON, J. B. Specimen collection. Adelaide, Australia.

*PEABODY MUSEUM OF NATURAL HISTORY. New Haven, U.S.A.

QUEEN VICTORIA MUSEUM AND ART GALLERY. Launceston, Australia.

*QUEENSLAND MUSEUM. Brisbane, Australia.

RAGLESS, G. Egg collection. Adelaide, Australia.

*ROYAL SCOTTISH MUSEUM. Edinburgh, U.K.

*SOUTH AUSTRALIAN MUSEUM. Adelaide, Australia.

TASMANIAN MUSEUM. Hobart, Australia.

*UNIVERSITY OF CALIFORNIA. Berkeley, U.S.A.

*UNIVERSITY OF FLORIDA. Gainsville, U.S.A.

*UNIVERSITY OF WESTERN ONTARIO. London, Canada.

WHITE, S. A. Skin collection register. Adelaide, Australia.

* The source was not completely searched for records.

Unpublished References

*AGRICULTURE PROTECTION BOARD OF WESTERN AUSTRALIA. Files.

ALEXANDER, W. B. Notebooks, 1912–1925. Held by C.S.I.R.O. Div. Wildlife Res., Helena Valley, WA.

AMIENS, J. H. Notes. Held by R. W. Walter, Qld.

ANGOVE, W. T. Personal notes in the margins of: Gould, J. 1898. *Handlist of Australian Birds*. Held by R. Angove, Adelaide.

ANGOVE, W. T. Notebooks. Held by R. Attwood, SA.

ANONYMOUS. Drawings of birds chiefly from Australia 1791–92. 101 watercolours in various sizes. Held by Mitchell Library, Sydney.

ANONYMOUS. List of specimens collected on Flinders Island, 1924. Held by South Australian Museum, Adelaide.

ANONYMOUS. Notes 1925–26. Held by South Australian Museum, Adelaide.

AUSTRALIAN NATIONAL UNIVERSITY, DEPT OF ZOOLOGY. Report of the 8th Field Excursion to western New South Wales, May 11–19, 1974. Copy held by R.A.O.U., Melbourne.

BLIGH, R. Photographs of original sketches in scrap book. Held by Mitchell Library, A565, Sydney.

BOSS-WALKER, H. B. Personal diaries. Held by Hobart Library, Hobart.

BROTHERS, N. Diaries. Held by N. Brothers, Tasmania.

CARTER, T. Notes from south-west and mid-west Australia. 1922. Held by R.A.O.U., Melbourne.

CLARK, M. S. Extracts from his diaries made by J. Sutton. Held by South Australian Museum, Adelaide.

COLCLOUGH, M. J. and E. D. FRIZELLE. Notes from the Roper River, 1910. Copy held by R.A.O.U., Melbourne.

COLES, W. List of native and introduced birds noted at Melbourne Botanic Gardens. Sept. 1884-Nov. 1937. Held by B. J. Fielding, Vic.

DAVIS, A. D. and H. E. BRENTON. Checklist of the Birds of Mt. Tomah. Copy held by R.A.O.U., Melbourne.

DICK, ALBERT. Diaries. Held by C.S.I.R.O. Div. Wildlife Res., Canberra.

ECKERT, J. Bird List, 1963. Held by J. B. Paton, Adelaide.

ELSEY, J. R. A letter to John Gould. Gould Collection MS. 1217. Held by National Library, Canberra.

ERICKSON, R. Observations on birds at Fairlea Farm, Bolgart, Western Australia. 1949–1960. Held by S. Davies, Perth.

FINLAYSON, H. H. Bird list, 1937. Held by J. B. Paton, Adelaide.

FORD, J. Diaries. Copy held by R.A.O.U., Melbourne.

GARDNER, W. Manuscript. Held by Armidale College of Advanced Education Library (Historical Resources Centre), NSW.

GIBLIN, W. W. Diary. Giblin egg collection in the Tasmanian Museum; no data; diary has dates and places. Held by Tasmanian Museum, Hobart.

GILBERT, J. Port Essington. Expedition Diaries 1844–45. Vol. 1–3. Held by Mitchell Library, FM4/749, Sydney.

HANN, F. H. Diaries of trips, 1903, 1906, 1907. Copies held by D. Reid, Adelaide.

HARRIS, G. P. 1806. Australian painting. Held by Mitchell Library, PXD 388, Sydney.

HARRIS J. 1791. Papers of Dr John Harris. Held by Mitchell Library, CYA 1597, Sydney.

HARRIS, M. P. 1979. The seabirds of the Victorian islands. A report to the Ministry for Conservation, Victoria. ITE Project Report No. 588.

HELMS, R. 1891. Diary of work done and notes thereon recorded by Richard Helms, naturalist, during the time of exploration beginning 25th April, 1891. Held by Mitchell Library, CYA2575, Sydney.

HOLDEN, DR LONSDALE. Tasmanian Ornithological Notes. Two notebooks. Held by Australian Museum, Sydney.

HUNTER, J. Birds and flowers of New South Wales drawn on the spot in 1788, 89 and 90. Paintings. Held by Mitchell Library, B1428, Sydney.

JACKSON, S. Field trip books 1908–1922. Held by National Library, Canberra.

JAMES, M. and F. JAMES. Bird lists. Held by R.A.O.U., Melbourne.

KOLICHIS, N. Field notes. Held by N. Kolichis, WA.

KREFFT, G. 1856–57. Narrative of the Exploring Expedition led by Blandowski to the Lower Murray and Darling Rivers (Krefft, G. Papers, Vol 8). Held by Mitchell Library, A268, Sydney.

LASHMAR, A. Notebooks. Held by A. Lashmar, SA.

LEARMONTH, N. Private journals. Held by Portland Field Naturalists Club, Vic.

LEARMONTH, N. Bird notes, 1956–67. Held by Portland Field Naturalists Club, Vic.

LEICHHARDT, F. W. L. Notebooks, 1842–47. Held by Mitchell Library, Manuscript C154, Sydney.

LENDON, A. Notebooks. Held by C. Lendon, Perth.

LOARING, W. H. Manuscript notes. Held by family, WA.

McGILL, A. Diary. Held by A. McGill, Sydney (original); R.A.O.U., Melbourne (copy).

MacGILLIVRAY, J. 1844–49. Notebooks containing copies of letters and notes written while on H.M.S. Rattlesnake, 1844–48. (a) Correspondence (including to J. Gould). (b) Notebook containing notes on natural history of Cape York and north Australia. Held by Mitchell Library, Microfilm (neg.) FM4/2231, Sydney.

McLENNAN, W. Notebooks. (1) Pascoe River 1913. (2) Arnhem Land 1915–16. (3) Moa (Banks) Island 1919–20. (4) Groote Eylandt 1921. (5) Coen District 1921. (6) Rocky Scrub 1922. Held by R.A.O.U., Melbourne.

MENZIES, A. 1791. Journal of A. Menzies, Botanist, with George Vancouver at King George's Sound, 27 Sept. 1791–13 Oct. 1791. Held by Mitchell Library, B1135, Sydney.

MESTON, A. L. Marginal notes in his copy of: Littler, F. M. 1910. A Handbook of Tasmania and its Dependencies. Held by Public Library, Launceston.

MILLEDGE, D. Tasmanian notebooks. Held by Tas. National Parks and Wildlife Service, Hobart.

MORGAN, A. M. Notebooks. Held by South Australian Museum, Adelaide (original); R.A.O.U., Melbourne (copy).

MORIARTY, T. K. Birds of Wanjarri. (Published Emu 72: 1–7). Original manuscript held by R.A.O.U., Melbourne.

MORSE, F. G. Field notes. Held by R.A.O.U., Melbourne.

MURPHY, J. 1844. Journal kept by J. Murphy on the Port Essington Expedition. Held by Mitchell Library, ML MS 2193, Sydney.

NATIONAL PARKS & WILDLIFE SERVICE (TAS). Folder on the Japanese Snipe. 1976.

*NEST RECORD SCHEME. Held by R.A.O.U., Melbourne.

PAYNES, H. Description of the Birds of Creswick 1928–30. Held by School of Forestry Library, Creswick, Vic.

PIZZEY, G. Notebooks. Held by G. Pizzey, Melbourne.

PIZZEY, G., A. ROBINSON, V. N. SERVENTY and D. L. SERVENTY. Queen Victoria Spring, Jan. 1964. Held by D. L. Serventy, Perth.

RAMSEY, E. P. Letters received and other papers, 1865–91, from collectors in the field. Held by Mitchell Library, MLMSS 563/4, Sydney.

RAPER, G. 1789. Bird and Flower of Port Jackson. Painting. Held by Mitchell Library, PXD 184, Sydney.

RAPER, G. Copies of watercolour drawings of birds in the British Museum. Held in Mitchell Library, FM5, Sydney.

ROBERTSON, J. S. Personal Notes 1938–76. Held by J. Robertson, Qld.

ROSE, A. B. Birds of the Gloucester District, 1954–55. Copy held by R.A.O.U., Melbourne.

ROSE, A. B. A Preliminary Checklist of the Birds of Ku-ring-gai Chase. Copy held by R.A.O.U., Melbourne.

ROSE, A. B. Flora and fauna of north Wahroonga Bushland. Copy held by R.A.O.U., Melbourne.

RUSSELL, R. and P. DEVONSHIRE. Bird list. Held by School of Forestry Library, Creswick, Vic.

RYAN, J. V. Diaries. (1) Mainly northern Melbourne suburbs, 1924–49. (2) Bendigo district, northern suburbs of Melbourne 1957–61. (3) Mittyack 1935–38, Noorcongong 1939–41, Atherton Tableland 1943–44, Goornong 1941–49. (4) Mainly Bendigo area, 1950–56. Held by J. V. Ryan, Melbourne.

SEDGWICK, E. H. Notebooks. Held by E. H. Sedgwick, Perth.

SERVENTY, D. L. Journal of Sep.-Oct. 1958: North-west Barrow Island Kimberley. Held by D. L. Serventy, Perth.

SERVENTY, D. L., V. N. SERVENTY, A. R. MAIN, W. H. BUTLER and R. STEWART. Great Victoria Desert Expedition 1960. Held by D. L. Serventy, Perth.

SHAW, W. and E. 1829–33. Notes (copies of letters to England) describing life on a farm and flora and fauna of WA. Held by Mitchell Library, B164, Sydney.

SOUTH AUSTRALIAN ORNITHOLOGICAL ASSOCIATION. Correspondence. Held by S.A.O.A., Adelaide.

SOUTH AUSTRALIAN ORNITHOLOGICAL ASSOCIATION. National Parks Survey. Held by S.A.O.A., Adelaide.

SPRATT, F. A. Notes. Held by D. M. Sandstrom, WA.

SUTTON, J. Notebooks. Held by South Australian Museum, Adelaide.

SWEATMAN, J. Journal of a Voyage in H.M.S. 'Bramble', 1842–47. Held by Mitchell Library, Posit. Microfilm FM4/2486, Sydney.

TAYLOR, W. Diary of Wilfred Taylor of Bugaldie, wildlife photographer. Held by A. Morris, NSW.

TINGAY, A. and S. R. TINGAY. The Lakes Regional Open Space Biological Survey for the Shire of Rockingham, 1977. Held by A. Tingay, WA.

TOBIN, G. 1792. Sketches on H.M.S. Providence. Paintings. Held by Mitchell Library, 2PX A563, Sydney.

TREVOR, M. 1894. 1800 Miles on Australian Rivers. Held by H. Gibson, Melbourne.

VICTORIAN MINISTRY FOR CONSERVATION. National Parks Files. Held by Ministry for Conservation, Melbourne.

VICTORIAN ORNITHOLOGICAL RESEARCH GROUP. Pilot Atlas Project field cards 1/12/1975–31/12/1976. Held by R.A.O.U., Melbourne. Results published in: ASTON, H. I. and R. A. BALMFORD. 1978. A Bird Atlas of the Melbourne Region. Melbourne: Victorian Ornithological Research Group.

WEINDORFER, G. Log, June 1927-Feb. 1931. Held by R.A.O.U., Melbourne.

WHEELER, R. Birds of the lower north of South Australia. Held by R. Wheeler, Victoria.

WILCOX, J. F. 1848–49. Journal 7th Jan.-21st Dec. 1848; 8th May–2nd July 1849. Original Manuscript. Held by Mitchell Library, FM3/453, Sydney.

WORLD BIRD DAY LISTS. District bird lists 1958–59, 1960–61 (Victoria and New South Wales). Held by R.A.O.U., Melbourne. Results published in: WHEELER, W. R. 1962. World Bird Day, district bird lists, 1958–59. The Bird Observer, Apr. supplement.

Published References

ANGAS, G. F. 1847. Savage Life & Scenes in Australia and New Zealand, being an artist's impressions of countries and people at the Antipodes. 2 vols. London: Elder Smith & Co.

ANONYMOUS. 1856. The Birds of Victoria. The Argus Newspaper, 14 July 1856. p. 4.

ANONYMOUS. 1890. Tasmanian Steam Navigation Company's guide for visitors to Tasmania or how to spend my holiday. 3rd ed. Hobart: Mercury.

AUSTIN, K. A. 1964. The voyage of the Investigator, 1801–1803—Commander Matthew Flinders, R.N. London: Angus & Robertson.

*AUSTRAL AVIAN RECORD. Journal. Ed. G. M. Mathews. London: Witherby.

THE AUSTRALIAN BIRD BANDER. Journal of the Bird Banders' Association of Australia. Sydney.

*AUSTRALIAN BIRD WATCHER. Journal of the Bird Observers Club. Melbourne.

AUSTRALIAN BIRDS. (Formerly 'Birds'). Journal of the New South Wales Field Ornithologists' Club. Sydney.

*AUSTRALIAN ZOOLOGIST. Journal of the Royal Zoological Society of New South Wales. Sydney.

BANFIELD, E. D. 1908. The Confessions of a Beachcomber. Sydney: Angus & Robertson.

BATES, D. 1944. The Passing of the Aborigines; a lifetime spent among the natives of Australia. London: Murray.

BATES, H. W. (Ed.). 1875. Journey Across the Western Interior of Australia by Colonel Peter Egerton Warburton. London: Sampson Low.

BEGG, C. and N. BEGG. 1979. The World of John Boultree. (Sealing in Australia and New Zealand). Christchurch, N.Z.: Whitcoulls Ltd.

BELCHER, C. F. 1914. The Birds of the District of Geelong, Australia. Geelong: W. J. Griffiths.

BENNETT, G. 1834. Wanderings of a Naturalist in New South Wales, Batavia, Pedir Coast, Singapore and China being the journal of a Naturalist in those countries during 1832, 1833 and 1834. 2 vols. London: Richard Bently.

BIRD OBSERVERS CLUB. 1965. Birds of the Mallacoota Area, Victoria. Melbourne: Bird Observers Club.

BOLAM, A. G. 1925. The Trans Australian Wonderland. 4th ed. Melbourne: Modern Printing Co.

BOSS-WALKER, I. R. 1950. Peaks and High Places. Hobart: Scenery Reservation Board.

BROADBENT, J. and S. CLARK (Eds). 1976. A Faunal Survey of East Australian Rainforests. Sydney: Australian Museum.

BROADBENT, K. 1888. Birds of the central part of Queensland. Proc. Roy. Soc. Queensl. 5: 14–31.

BROCK, D. G. 1975. To the Desert with Sturt; a diary of the 1844 expedition. K. PEAKE-JONES (Ed.). Adelaide: Roy. Geogr. Soc. of Aust. (SA Branch).

BUNCE, D. 1857. Australasiatic Reminiscences of Twenty-three Years Wanderings in Tasmania and the Australias; including Travels With Dr. Leichhardt in North or Tropical Australia. Melbourne: Hendy.

BURNS, T. E. and J. R. SKEMP. 1961. Van Diemens Land Correspondents. Letters from R. C. Gunn, R. W. Lawrence and J. Jorgensen, Sir F. Franklin and others to Sir W. Hooker. 1827–1849. Launceston: Queen Victoria Museum.

BYERLEY, F. J. (Ed.). 1867. Narrative of the Overland Expedition of the Messrs Jardine from Rockhampton to Cape York, Northern Queensland. Compiled from the Journals of the Brothers. Brisbane: Buxton.

CAMPBELL, A. J. 1901. Nests and Eggs of Australian Birds. 2 vols. Melbourne: The Author.

CARNEGIE, D. W. 1898. Spinifex and Sand. London: C. A. Pearson. 1973, Ringwood: Penguin.

CIRCULAR HEAD SCIENTIFIC JOURNAL. An MS journal on ornithology. Vol. 1, No. 1, 21 June 1836, with supplement and contributions. Mitchell Library, Manuscript 1180, Sydney.

CLELAND, J. B. 1918. The food of Australian birds. An investigation into the character of the stomach and crop contents. Dept. of Agric. NSW, Science Bull. 15: July.

CORELLA. Journal of the Australian Bird Study Association. Sydney.

COPPINGER, R. W. 1883. Cruise of the Alert. Four years in Patagonian, Polynesian and Mascarene Waters (1878–82). London: Sonnenschein & Co.

COX, J. D. and A. G. HAMILTON. 1889. List of the birds of the Mudgee district. Proc. Linn. Soc. NSW 4: 395–424.

DAHL, K. 1925. In Savage Australia. An account of a hunting and collecting Expedition in Arnhem Land and Dampier Land. London: Philip Allan.

DAVIES, S. J. J. F. 1969. An aerial survey of inland Western Australia. C.S.I.R.O. Div. Wildl. Res. Tech. Memo 1: 1–27.

DAVIS-WRIGHT, W. 1923. Canberra. Sydney: John Andrew & Co.

DEIGNAN, H. G. 1964. Birds of the Arnhem Land Expedition. In SPECHT, R. L. (Ed.). Records of the American-Australian Scientific Expedition to Arnhem Land, Vol. 4. Melbourne: M.U.P.

DIGGLES, S. 1866–70. The Ornithology of Australia. Brisbane: The Author.

EARL, G. W. 1846. Enterprise in Tropical Australia. London: Madden & Malcolm.

EDEN, C. H. 1872. My Wife and I in Queensland. London: Longmans, Green & Co.

EMU. Journal of the Royal Australasian Ornithologists Union. Melbourne.

EVANS, G. W. 1822. A geographical, historical and topographical description of Van Diemen's Land. London: J. Souter.

EYRE, E. J. 1845. Journals of Expeditions of Discovery into Central Australia and Overland from Adelaide to King Georges Sound 1840–41, including an Account of the Manners and Customs of the Aborigines and the State of their Relations with Europeans. 2 vols. London: T. & W. Boone.

FENTON, J. 1891. Bush Life in Tasmania Fifty Years Ago. London: Hazel Watson & Viney Ltd.

FIELD, BARRON. 1825. Geographical Memoirs on New South Wales by various hands. London: Murray.

FINSCH, O. 1872. Uber die von Frau Amalie Dietrich in Australien gesammelten Vogel. Verhandl. zool.-bot. Gesellschaft in Wien 22: 315–340.

FLEAY, D. 1968. Nightwatchmen of Bush and Plain. Brisbane: Jacaranda Press.

FORBES, W. A. 1878. Reports of the Collections of Birds made during the Voyage of H.M.S. 'Challenger'.—No. VII. On the Birds of Cape York and the Neighbouring Islands (Raine, Wednesday and Booby Islands). Proc. Zool. Soc. London for 1878: 120–128.

FOUNTAIN, P. 1907. Rambles of an Australian Naturalist. From the notes and journals of Thomas Ward. London: Murray.

FRANKLIN, M. 1865. Childhood at Brindabella. 2nd ed. Sydney: Angus & Robertson.

FRASER, M. A. C. 1903. Notes on the Natural History of Western Australia. Perth: W. M. Watson, Govt Printer.

GALE, J. 1927. Canberra; history of and legends relating to the Federal capital territory of the Commonwealth of Australia. Queanbeyan: Fallick.

GEELONG NATURALIST. Journal of the Geelong Field Naturalists' Club. Geelong.

GIBLIN, R. W. 1928. The Early History of Tasmania; the geographical era, 1642–1804. Vol. 1 & 2. London: Methuen & Co.

*GOULD, J. 1840–69. The Birds of Australia. 7 vols London: The Author.

GOULD, J. 1865. Handbook to the Birds of Australia. London: The Author.

*GOULD LEAGUE NOTES. Journal of the Gould League of Bird Lovers of New South Wales. Sydney.

GRANT, J. 1803. The Narrative of a Voyage of Discovery performed in His Majesty's Vessel 'The Lady Nelson' of sixty tons burthen with sliding keels, in the years 1800, 1801 and 1802 to New South Wales. London: T. Egerton, Military Library.

GREEN, R. H. 1959. Birds of Tasmania. Launceston: The Author. Reprinted and enlarged 1977.

GREGORY, A. C. and F. T. GREGORY. 1884. Journals of Australian Explorations. Brisbane: James C. Beal. Facsimile edition 1969, Adelaide: Libraries Board of South Australia.

HALL, B. P. (Ed.). 1974. Birds of the Harold Hall Australian Expeditions. London: Brit. Mus. (Nat. Hist.).

HALL, R. 1907. The Useful Birds of Southern Australia. Melbourne: T. C. Lothian.

HALL, R. 1922. Australian Bird Maps. Hobart: The Author.

HAVARD, WARD & OLIVE. 1938. A Frenchman sees Sydney in 1819. J. and Proc. R. Aust. Hist. Soc. 24: 17–42.

HAY, J. G. 1906. The Visit of Mr Charles Fraser, Colonial Botanist of New South Wales, with Captain Stirling, in H.M.S. 'Success', to the Swan River in 1827, with his Report on the Botany, Soil and Capabilities of the Locality. Perth: J. G. Hay.

HENDERSON, J. 1851. Excursions and Adventures in New South Wales with Pictures of Squatting and of life in the Bush. 2 vols. London: W. Shoberl.

HERMAN, M. (Ed.). 1965. Annabella Boswell's Journal. Sydney: Angus & Robertson.

HILL, E. 1943. The Great Australian Loneliness. Melbourne: Robertson & Mullens.

HILL, M., D. ROGERS and M. L. C. BAKER. 1978. Merrimu Reservoir, Stage III. State Rivers & Water Supply Commission of Vic., Environmental Report. June.

HINDWOOD, K. A., K. KEITH and D. L. SERVENTY. 1963. Birds of the south-west Coral Sea. C.S.I.R.O. Div. Wildl. Res. Tech. Paper 3.

*HUNTER NATURAL HISTORY. Journal of the Newcastle Flora and Fauna Protection Society.

*JOURNAL OF THE ROYAL SOCIETY OF WESTERN AUSTRALIA. Perth.

KING, P. P. 1826. Narrative of a Survey of the Intertropical and Western Coasts of Australia. Performed between the years 1818 and 1822. 2 vols. London: Murray.

LEARMONTH, N. F. 1966. Birds of Portland (Victoria) District. Portland: Portland Field Nat. Club.

LEE, I. 1925. Early Explorers in Australia from the Log-Books and Journals, including the diary of Allan Cunningham, Botanist, from March 1, 1817 to November 19, 1818. London: Methuen.

LE SOUEF, W. H. D. 1907. Wild life in Australia. Christchurch: Whitcombe and Tombs Ltd.

LEWIN, J. W. 1822. A natural history of the birds of New South Wales collected, engraved and faithfully painted after nature. 1st Edition 1808. London: G. Schulze.

LITCHFIELD, J. S. 1930. Far North Memories, being the account of ten years spent on the diamond drills, and of things that happened in those days. Sydney: Angus & Robertson.

LITTLER, F. M. 1910. A Handbook of the Birds of Tasmania and its Dependencies. Launceston: The Author.

LORD, C. E. and H. H. SCOTT. 1924. Synopsis of the Vertebrate Animals of Tasmania. Hobart: no publisher.

LOYN, R. H. 1975. Westernport Bay Environmental Study 1973–74, Project Report 4.4.8. Melbourne: Min. Cons. Vict.

LUMHOLTZ, K. 1889. Among Cannibals: an account of Four Years' Travels in Australia and of Camp-Life with the Aborigines of Queensland. London: Murray.

LYCETT, J. 1825. Views in Australia, New South Wales and Van Diemen's Land Delineated. London: J. Souter. Facsimile edition 1971, Australia: Thomas Nelson Ltd. Reprinted 1977.

McEVEY, A. 1974. John Cotton's Birds of the Port Phillip District of New South Wales. Sydney & London: Collins.

MacGILLIVRAY, J. 1852. Narrative of the Voyage of H.M.S. 'Rattlesnake' during the years 1846–1850. 2 vols. London: T. and W. Boone.

MACKANESS, G. 1947. Some Private Correspondence of Sir J. & Lady J. Franklin. Part 1 & 2. Aust. Hist. Monograph No. 15.

MACKANESS, G. (Ed.). 1965. Fourteen Journeys Over the Blue Mountains of New South Wales, 1813–1841. Sydney: Horwitz-Grahame.

MADIGAN, C. T. 1946. Crossing the Dead Heart. Melbourne: Georgian House.

MARKHAM, E. 1952. Voyage to Van Diemen's Land, 1833. Launceston: Telegraph.

MARSHALL, A. J. and R. DRYSDALE. 1962. Journey Among Men. London: Hodder & Stoughton.

MARTIN, J. 1864. Journals and Reports of two voyages to the Glenelg River and the North-west Coast of Australia 1863–64. Perth: A. Shenton.

MATHEWS, G. M. 1910–27. The Birds of Australia. 12 vols. London: Witherby.

MESTON, A. L. 1889. Report on the Government Scientific Expedition to the Bellenden-Ker Range (Wooroonooran), North Queensland. Queensl. Parl. Pap. C.A.95.Fo.

MILLETT, MRS EDWARD. 1872. An Australian Parsonage; or, the settler and the savage in Western Australia. London: Stanford.

MITCHELL, T. L. 1838. Three Expeditions into the Interior of Eastern Australia, with descriptions of the recently explored region of Australia Felix, and of the present Colony of New South Wales. 2 vols. London: T. & W. Boone.

MITCHELL, T. L. 1848. Journal of an Expedition into the Interior of Tropical Australia in search of a route from Sydney to the Gulf of Carpentaria. London: Longman, Brown, Green & Longman.

MOSCAL, T. 1979. Natural History Case Studies from south-west Tasmania. Hobart: Steering Committee South-West Tasmania Resources Survey.

MOSELEY, H. N. 1879. Notes by a Naturalist on the 'Challenger', being an account of various observations made during the voyage of H.M.S. Challenger round the World in the years 1872–76. London: Macmillan.

MURRAY, A. S. 1898. Twelve Hundred Miles on the River Murray. London: J. S. Virtue.

NAPIER, F. 1876. Notes of a Voyage from New South Wales to the north coast of Australia, from the Journal of the late Francis Napier. Glasgow: no publisher.

*NEW SOUTH WALES FIELD ORNITHOLOGISTS CLUB NEWSLETTER. Sydney.

NIX, H. A. 1972. Fauna of the Shoalwater Bay area. Part 7. In GUNN, R. A. (Ed.). Shoalwater Bay Area, Queensland. C.S.I.R.O. Div. Land Res. Tech. Memo. 72/10.

NORTH, A. J. 1896. Aves. In SPENCER, W. B. (Ed.). Report on the work of the Horn Scientific Expedition to Central Australia. London: Dulau & Co.

NORTH, A. J. 1898. List of Birds collected by the Calvert

Exploring Expedition in Western Australia. Trans. Roy. Soc. S. Aust. for 1897–98. 22 Dec: 125–192.

NORTH, A. J. 1901–14. Nests and Eggs of Birds found breeding in Australia and Tasmania. Aust. Museum Sydney: Special Catalogue No. 1.

OXLEY, J. 1820. Journals of Two Expeditions into the interior of New South Wales. London: Murray. Facsimile Edition 1964, Adelaide: Libraries Board of South Australia.

*PAPERS AND PROCEEDINGS OF THE ROYAL SOCIETY OF VAN DIEMEN'S LAND (TRANSACTIONS). Hobart.

*QUEENSLAND JOURNAL OF AGRICULTURAL AND ANIMAL SCIENCES. Brisbane.

*QUEENSLAND NATURALIST. Journal of the Queensland Naturalists' Club.

*QUEENSLAND ORNITHOLOGICAL SOCIETY NEWSLETTER. Brisbane.

RAMSEY, E. P. 1888. Tabular List of All the Australian Birds at present known to the Author. Sydney: The Author.

RATCLIFFE, F. N. 1947. Flying Fox and Drifting Sand. Sydney: Angus & Robertson.

RECORDS OF THE QUEEN VICTORIA MUSEUM. Launceston.

*RECORDS OF THE WESTERN AUSTRALIAN MUSEUM. SUPPLEMENTS. Perth.

RIX, C. E. 1978. Companion to Cooper's Birds of a Saltfield. Melbourne: ICI Australia Ltd.

ROGERS, K. 1976. Notes on changes in mountain areas. The Clematis 15: Nov. 1976.

ROUNSEVELL, D. E., S. A. BLACKHALL and D. G. THOMAS. 1977. Birds of Maria Island: results of a field survey and literature review. N.P.W.S. Tas., Tech. Rep. 77/3.

ROYAL SOCIETY OF SOUTH AUSTRALIA. TRANSACTIONS AND PROCEEDINGS. Adelaide.

*ROYAL SOCIETY OF TASMANIA. PROCLAMATIONS AND PAPERS. Hobart.

*ROYAL SOCIETY OF VAN DIEMEN'S LAND. PAPERS AND PROCEEDINGS (TRANSACTIONS). Hobart.

SCOTT, E. (Ed.). 1929. Australian Discovery. 2 vols. London: Dent.

SENIOR, W. 1880. Travel and Trout in the Antipodes: an angler's sketches in Tasmania and New Zealand. Melbourne: Beccles.

SERVENTY, D. L. and A. J. MARSHALL. 1958. A Natural History Reconnaissance of Barrow and Montebello Islands. C.S.I.R.O. Div. Wildl. Res. Tech. Paper 6.

SERVENTY, D. L. and H. M. WHITTELL. 1976. Birds of Western Australia. 5th ed. Perth: Univ. of Western Australia Press.

SHARLAND, M. 1945. Tasmanian Birds: How to Identify Them. Hobart: Oldham, Beddome & Meredith.

SHARLAND, M. 1958. Tasmanian Birds: a field guide to the birds inhabiting Tasmania and adjacent islands, including the sea birds. 3rd ed. Sydney: Angus & Robertson.

SHARLAND, M. 1964. A Territory of Birds. Adelaide: Rigby.

SHARP, A. 1963. The Discovery of Australia. Oxford: Clarendon Press.

SHUMACH, S. 1967. An Autobiography; or Tales and Legends of Canberra Pioneers. Canberra: ANU Press.

SKEMP, J. R. 1971. My Birds. Devonport: C. L. Richmond & Sons.

SMITH, G. 1909. A Naturalist in Tasmania. Tasmania: Clarendon Press.

SMITH, J. 1886. *In* GARREN, A. (Ed.). Picturesque Atlas of Australasia, Vol. 2. Sydney: The Picturesque Atlas Pub. Co.

SMITH, S. 1903. Parrakeets: A Handbook to Imported Species. London: R. H. Porter.

SOUTH AUSTRALIAN NATURALIST. Journal of Field Naturalists' Society of South Australia. Adelaide.

*SOUTH AUSTRALIAN ORNITHOLOGICAL ASSOCIATION NEWSLETTER. Adelaide.

THE SOUTH AUSTRALIAN ORNITHOLOGIST. Journal of the South Australian Ornithological Association. Adelaide.

SPENCER, W. B. (Ed.). 1890. Handbook of Melbourne. Melbourne: Spectator Publishing Co.

STOKES, J. L. 1846. Discoveries in Australia; with an account of the coasts and rivers explored and surveyed in the years 1837–38–39–40–41–42–43 during the Voyage of *H.M.S. Beagle*. 2 vols. London: T. & W. Boone.

STUART, J. M. 1864. The Journals of J. McDouall Stuart during the years 1858–62 W. HERDMANN (Ed.). London: Saunders, Otley & Co.

STUART, J. M. 1863. Explorations Across the Continent of Australia, 1861–1862. Melbourne: F. F. Bailliere. Facsimile edition 1962, Adelaide: Libraries Board of South Australia.

THE SUNBIRD. Journal of the Queensland Ornithological Society.

TASMANIAN GOVERNMENT. 1937. Tasmania—The Jewel of the Commonwealth.

*TASMANIAN JOURNAL OF NATURAL SCIENCE. Hobart.

TASMANIAN NATURALIST. Journal of the Tasmanian Field Naturalists Club. Hobart.

TAYLOR, N. M. (Ed.). 1966. The Journal of Ensign Best, 1837–43. Wellington: Govt Printer.

THOMSON, D. F. 1935. Birds of Cape York Peninsula. Melbourne: Govt Printer.

TOKYO UNIVERSITY OF FISHERIES. 1968. 34th Cruise of Umitaka Maru. Tokyo: Tokyo Univ. of Fisheries.

*TOOWOOMBA FIELD NATURALISTS CLUB JOURNAL. Toowoomba.

TOWNSEND, J. P. 1849. Rambles and observations in New South Wales with sketches of men and manners, notices of the aborigines, glimpses of scenery, and some hints to emigrants. London: Chapman & Hall.

TUCKEY, J. H. 1805. An Account of a Voyage to Establish a Colony at Port Phillip in Bass's Strait. London: Longman, Hurst, Rees & Orme.

*VICTORIAN NATURALIST. Journal of the Field Naturalists' Club of Victoria. Melbourne.

VIGORS, N. A. and T. HORSFIELD. 1827. A description of the Australian Birds in the Collection of the Linnean Society; with an Attempt at arranging them according to their natural Affinities. Trans. Linn. Soc. London. 15. Feb. 15: 170–331.

*VORG NOTES. Journal of the Victorian Ornithological Research Group. Melbourne.

WALCH, G. 1871. Huddart, Parker and Co.'s guide to Tasmania. Melbourne: Huddart Parker & Co.

WEEDING, J. S. 1976. A History of the Lower Midlands. Mary Fisher Bookshop.

WELLS, K. F., P. N. GOWLAND and A. P. SPATE. 1977. Natural resources of the Buckland Training Area, Tasmania. C.S.I.R.O. Div. Land Use Res. Tech. Memo 77/14.

*WESTERN AUSTRALIAN DEPARTMENT OF FISHERIES AND WILDLIFE. Reports.

WESTERN AUSTRALIAN NATURALIST. Journal of the Western Australian Naturalists' Club. Perth.

WHEELER, R. (Ed.). 1959. World Bird Day, district bird lists, 1957–58. The Bird Observer, Dec. supplement.

WHEELER, R. (Ed.). 1962. World Bird Day, district bird lists, 1958–61. The Bird Observer, Apr. supplement.

WHEELER, R. and D. MORGAN (Eds). 1954. World Bird Day, district bird lists, 1953. The Bird Observer, Apr. supplement.

WHEELER, R. and D. MORGAN (Eds). 1955. World Bird Day, district bird lists, 1954. The Bird Observer, Aug. supplement.

WHEELRIGHT, H. 1861. Bush Wanderings of a Naturalist; or Notes on the Field Sports and Fauna of Australia Felix. By An Old Bushman. London: Routledge, Warne, Routledge.

WHITWORTH, R. P. 1877. Bailliere's Tasmanian gazetteer and road guide, Hobart Town. Bailliere.

WILDLIFE. (1938–54). Melbourne.

*WILDLIFE RESEARCH BULLETINS. Published by WA Wildlife Research Centre, Dept of Fisheries and Wildlife. Perth.

WINNECKE, C. G. A. 1897. Journal of the Horn Scientific Exploring Expedition, 1894. Adelaide: Govt Printer.

CALCULATION OF REPORTING RATES FOR FIELD ATLAS MAPS

1° Maps of Australia. The number of record sheets submitted in each 1° block was taken as a measure of coverage (see p. xxxii). In a sample of 11 well-covered grid blocks at least 50% of the total number of species had been recorded by the time 50 or more record sheets were received. This number of record sheets was considered an acceptable criterion for the calculation of Reporting Rates. In 1° blocks containing 50 or fewer record sheets, an averaging process was used that took into account the Reporting Rates in surrounding 1° blocks.

The procedure was as follows where

RR = the Reporting Rate,

x = the number of records of a species and

y = the number of record sheets in a 1° block.

1) 1° blocks with more than 50 record sheets. Calculate the Reporting Rate directly:

$$RR = \frac{x}{y}$$

2) 1° blocks with 50 or fewer record sheets. The Reporting Rate for grid block A is to be calculated:

B	C	D
E	A	F
G	H	I

$$RR = \frac{\frac{\Sigma x_{B-I}}{\Sigma y_{B-I}}(50 - y_A) + x_A}{50}$$

10′ Maps of Tasmania. The procedure was the same as for the 1° maps except that a threshold of 25 record sheets for a 10′ block was used because few 10′ blocks contained as many as 50 record sheets.

ACKNOWLEDGEMENTS

The RAOU owes its thanks to the following people and institutions. They are given in chronological order, not order of priority.

PAULINE REILLY. The idea of an Atlas was conceived while she was President of the RAOU; as Chairman of the Atlas Committee, she carried it through.

JOHN DISNEY followed by STEPHEN DAVIES, as Chairmen of the RAOU's Field Investigation Committee, guided the project from beginning to end.

THE RAOU COUNCIL. In particular, Secretary TOMMY GARNETT performed the early spadework and Treasurer IAN SINCLAIR made the financial resources stretch to unbelievable lengths.

GERALD HOREY and JOHN BROADBENT conducted, for the RAOU, the first pilot atlas[488] that began the training of atlassers.

BOAT (The Bird Observers Association of Tasmania),[1747] SAOA (The South Australian Ornithological Association)[1693] and VORG (Victorian Ornithological Research Group)[75] conducted pilot atlases of great value in their own right and as training for the national Atlas.

The AUSTRALIAN GOVERNMENT made grants through the Australian National Parks and Wildlife Service and the Bureau of Flora and Fauna. Without this financial support the Atlas could not have been carried out.

Atlas Field Officers PETER CURRY and CHRIS RILEY began organizing the Atlas in 1976. MARGARET BLAKERS, in Oct. 1976, picked up the project and carried it through to publication.

The Atlas Committee: SIMON BENNETT, MARGARET BLAKERS, DAVID PETERS and PAULINE REILLY laboured in hectic haste to start the Field Atlas on 1 January 1977.

PETER TRUSLER designed the Atlas logo which so well represented the project.

BOB ROWLANDS did all the computer programming as a volunteer. His dedication was essential to the success of the Atlas.

The STAFF, full- and part-time, laboured far beyond their rewards: Liz Bennet, Simon Bennett, Margaret Blakers, Kay Busija, Margaret Considine, Peter Curry, Rob Drummond, Burna Dunn, David Eades, Pat Emery, Ian Harris, Aileen Hickey, Judy McVicar, Kerrie Milburn-Clark, Penny Paton, Chris Riley, Julie Strudwick, Rod Waterman and typists Anne Chew and Tricia Merrett.

VOLUNTEER STAFF at the Atlas office worked faithfully for love and cups of tea: Cecily Allen, Kevin Bartram, Liz Bennet, Bob Cross, Mike Cullen, Frank Davison, Alexandra Djurovich, Rob Drummond, Pauline Duncan, Pat Emery, Alf Finch, Barbara Gleeson, Jack Jones, Betty Joyce, Ann and Bruce McGregor, Judy McVicar, Arthur Reilly, Charlie Silveira, Tom Smith, Elizabeth Turnbull, Rod Waterman, John Withell. Many more helped for a few hours or a day.

REGIONAL ORGANIZERS. They were the backbone of the Atlas and sacrificed themselves endlessly. Armies of helpers assisted them.

Western NSW: GRAEME CLARK and later CEDRIC BEAR. They were helped by Sub-ROs Barry Baker, Rodney Cox, Bob Digan, Neil Hermes, Trisha and Philip Maher, Neil Macfarlane and John Penhallurick. Others who assisted: Mike Doyle.

Eastern NSW: DICK COOPER. He was helped by Sub-ROs John Brooke, David Geering, Alan Morris, Richard Noske, Neville Schrader and David Thompson. Others who assisted: Robin Bigg, Stephen Debus, Barbara Howie, Morton Kaveney, Julie Strudwick and Ian McAllan.

Vic.: DAVID PETERS. He did not have Sub-ROs but was helped by Richard and Margaret Alcorn, Helen Aston, Rosemary Balmford, Simon Bennett, Marie Boyd, Margaret Cameron, Pat Emery, Sarah Fitzherbert, Don Franklin, Marc Gottsch, Murray Gunn, Kath Hough, Joy Johnstone, Neil Macfarlane, Ruth McLay, Harley Mawdsley, Joan Naarding, Frank Noelker, Ward Potts, Pauline

Reilly, Ken Sandy, Garry Tirrell, Barry Traill and Max Yardy.

Qld: Bill Horton and later Sue Elmes. They were helped by Sub-ROs Suzanne Cameron, Marion Crouther, Andree Griffin, Roy Hando, Richard Noske, David Stewart, John Walter and Graham Waterman. Others who assisted: Chris Corben, Doug Dow, Alf Finch, Les Hall, Helen Horton, Barry Jahnke, Jiro Kikkawa, Billie Smallgange, Ian Venables, Don Vernon, John and Ruth Walter and Denis Watson.

SA: Helen Laybourne-Smith and later John Bransbury. They were helped by Sub-ROs Frank Badman, Jack Bourne, Colin Gill, Garth Jackson, Margaret Killick, Bob Lines, Joe Mack, Jeff Nottage, John Mengersen, Lynn Pedler, Dulcie Rowley, John Seymour, Richard Smyth, Kent Treloar and Graham White. The local vetting committee was: Rod Attwood, Andrew Black, David Close, Brian Glover, Roger Jaensch, Leo Joseph, Joan Paton and David Vincent. Others who assisted: Tom Bradley, Edie Bransbury, Wally Cunningham, Joan Dewhirst, Lindy Dewhirst, Lena Dunkley, Rex Ellis, Clinton Garrett, Stephanie Hocking, Phillipa Horton, Cathy Houston, Bill and Di Lade, Helen Laybourne-Smith, Bronwyn Mart, Bill Matheson, Jane Needle, Shane Parker, Penny Paton, Joan Ramsey, Margaret Rennison, Peter Taylor, Kevin Thiele, John Truran, Bob Whatmough and Ken Woodcock.

WA: Stephen Davies. He was helped by Sub-ROs Kym Brennan, Tony Bush, Peter Congreve, Liz Edwards, Wally Klau, John Martindale, Jim Masters, Don Reid, Dick Rule, Don Shepherd and Doug Watkins; and his local vetters and co-ordinators: Graeme Chapman, Peter Curry, Tom Delaney, Roger Jaensch, Jim Lane and Dom Serventy.

Tas.: Mike Newman. He also did not have Sub-ROs but was helped by Peter and Pat Duckworth, Peter Fielding, Bob Green, Ken Harris, Sue and Warren Jones, Priscilla Park, Bob Patterson, Paul Rosevear, David Rounsevell and Len Wall.

Southern NT: Colin and Robin Lendon, Mike Fisher, Gary Wilson and John Erlich. They were helped by Chris Benz, David Bird, Dennis Chinner, Mike Fleming, Ken Johnson, Dick Kimber and Audrey Turner.

Northern NT: Hilary Thompson and later Mike Reed. They were helped by John McKean and numerous others.

The Families of Regional Organizers and of many atlassers provided support and encouragement.

The RAOU Finance Advisory Committee under the indefatigable Chairmanship of Norman Wettenhall bridged the gap between income and expenditure.

CSIRO permitted its officers to collaborate in various aspects of the organization of the Atlas.

The Historical Atlas Helpers scoured the literature and other sources for records: Leigh Adamson, Wilma Barden, Cedric Bear, Liz Bennet, Cecil Bird, Christine Bonser, Stephen Boyd, John Bransbury, Andrew Burbidge, Sue Cameron, Melanie Carron, Bob Cross, Jeff Davies, Stephen Debus, Carmen Drake, Dan Drakes, Rob Drummond, Peter Duckworth, Sue Elmes, Norman Favaloro, Peter Fielding, Heather and Don Fleming, Berrice Forrest, Don Franklin, Barbara Gleeson, Murray Gunn, Margaret Hendry, Katharine Highman, Cathy Houston, Illawarra Bird Observers Club, Barry Jahnke, Sue and Warren Jones, Ethel Lisser, Ian McAllan, Macleay Museum volunteers, Mr and Mrs McPhie, Bill Matheson, Rosemary Metcalf, Alan Morris, Brenda Murliss, Jean Neaves, Priscilla Park, Joan Paton, Graeme Phipps, David Purchase, Noel Quinn, Don Reid, Margaret Rennison, A. B. Rose, Alison Rowell, John Seymour, Don Shepherd, Billie Smallegange, Lola Smith, David Tanner, Ron Thoday, Lyndal Thorburn, John Truran, Elizabeth Turnbull, Ian Veacock, Len Wall, John Walter, Rod Waterman, Dennis Watson and Ross Wrigley.

Bird groups, naturalists clubs and other organizations trained atlassers, conducted field trips and promoted the Atlas to their members: Armidale College of Advanced Education, Arnhem Bird Club, Bingara atlassers, Bird Observers Association of Tasmania, Bird Observers Club of Australia, Canberra Ornithologists Group, Coonabarabran Atlas Group, NSW Field Ornithologists Club, Hunter Bird Observers Club, Illawarra Bird Observers Club, Mid-Murray Field Naturalists Club, Murdoch University Wildlife Association, Queensland Ornithological Society, South Australian Ornithological Association, University of New England, Victorian Ornithological Research Group, WA Group of the RAOU.

Property owners permitted access to their land and provided many facilities.

Aboriginal communities in Qld, NT, SA and WA permitted access to their land.

The Governments of Qld, NSW, Vic., Tas., NT, SA and WA, particularly their fauna authorities and the Forestry Commission of Tas., provided financial and practical assistance.

Vignettes Committee. Sue Ebury, Charles McCubbin, Pauline Reilly, Will Rolland and Kaye Turner had the unenviable task of selecting the artists and checking the work.

Artists. Many assisted and those who contributed received only a small fee. They are listed on pp. 692-3.

The Vetting Committee—Helen Aston, Kevin Bartram, Rob Drummond, David Eades, Howard Jarman, Richard Loyn and Julie Strudwick—agonized to reach the highest standards of accuracy.

Ken Norris, Belinda Schoefell and Gary Wilson led expeditions to the NT and Andy Buchanan to south-western Tas.

Pilots Don Jeans, Bos Parsons and Vic Smith flew atlassers all over Australia.

Duncan McKellar and the staff of the Australian Survey Office accompanied expeditions to the Great Sandy Desert and the Tanami Desert.

The following museums and their staff permitted access to their collections and answered numerous queries: American Museum of Natural History, Mary LeCroy; Australian Museum, Wally Boles, Wayne Longmore; British Museum of Natural History, Derek Read; Cambridge Dept of Zoology, Con Benson; Los Angeles County Natural History Museum, Ralph Schreiber; Macleay Museum, Graeme Phipps; Merseyside County Museum, Clem Fisher; National Museum Melbourne, Allan McEvey and Belinda Gillies; National Wildlife Collection, Dick Schodde; Queen Victoria Museum Launceston, Bob Green and Kathy Bulman; Queensland Museum, Don Vernon; Royal Scottish Museum, I. Lyster; South Australian Museum, Shane Parker; Tasmanian Museum, Len Wall.

Henry Nix helped throughout with everything from vetting maps to advice on data analysis and biogeographical regions.

David Purchase provided information from the Australian Bird Banding Scheme and was custodian of the bird numbers.

Harry Bell, Allan Burbidge, Shane Parker and Julie Strudwick read and commented on all the texts, correcting our mistakes and misinterpretations and providing much new information.

Many others did the same for selected texts and maps: Stephen Ambrose, Helen Aston, David Baker-Gabb, Merle Baldwin, Cedric Bear, Simon Bennett, Gavin Blackman, Wally Boles, Pat Bourke, Wayne Braithwaite, John Bransbury, Kim Brennan, Sue Briggs, Michael Brooker, Nigel Brothers, Dick and Molly Brown, Peter Brown, Murray Bruce, Andrew Burbidge, Margaret Cameron, Christine Cannon, Mike Carter, Graeme Chapman, Graeme Clark, Mike Clarke, David Close, Peter Congreve, Dick Cooper, Chris Corben, David Crawford, Frank Crome, Marion Crouther, Mike Cullen, Peter Curry, Sid Curtis, Peter Dann, Stephen Debus, Xenia Dennett, Doug Dow, David Eades, John Eckert, Sue Elmes, Bill Emison, John Erlich, Simon Ferrier, Win Filewood, Kate Fitzherbert, Mike Fleming, Hugh Ford, Julian Ford, Joseph Forshaw, Don Franklin, Clifford and Dawn Frith, Harry Frith, Peter Fullagar, Phil Fuller, Stephen Garnett, Tommy Garnett, Billie Gill, Merv Goddard, Peter Gowland, Bob Green, Andree Griffin, John Hobbs, Glenn Holmes, Ern Hoskin, Stephen Hopper, Kees Hulsman, Tim Hunt, Roger Jaensch, Ken Johnson, Gavin Johnstone, Ron Johnstone, Darryl Jones, Jack Jones, Richard Jordan, Leo Joseph, Jiro Kikkawa, Brian King, Bill Lane, Brett Lane, Jim Lane, Hugh Lavery, Allan Lill, Terry Lindsey, John Long, Wayne Longmore, Kim Lowe, Richard Loyn, Arnold McGill, John McKean, Henry Nix, Ian Mason, Charles Meredith, Bill Middleton, Nick Mooney, Alan Morris, Durno Murray, Hans Naarding, Mike Newman, Ian Norman, Richard Noske, Sue Noske, Penny Olsen, Priscilla Park, David Paton, Joan Paton, Bob Patterson, Leo Pecotich, Trevor Pescott, David Peters, Jack Pettigrew, Graeme Phipps, Gordon Ragless, Harry Recher, Mike Reed, Barry Reville, Mike Ridpath, Bruce Robertson, Norman Robinson, Win Rohan-Jones, Ian Rooke, David Rounsevell, Ian Rowley, Denis Saunders, Dick Schodde, Neville Schrader, Eric Sedgwick, John Seebeck, Dom Serventy, Don Seton, Ken Simpson, Grahame Smith, Glen Storr, Mike Tarburton, Gerry van Tets, Mr and Mrs Thorsborne, Sonia Tidemann, Sue Tingay, Bob Tomkins, Don Vernon, Joan Vincent, Len Wall, Doug Watkins, M. Watson, Chris Weaver, Gary White, Jo Wieneke, Kent Williams, Gary Wilson, Steve Wilson, John Woinarski, Ron Wooller, Boyd Wykes, Ed Wyndham, Richard Zann.

Stephen Marchant read all the texts and attempted to make them comprehensible.

David Ride commented on the Introduction.

Leigh-Mardon Graphics produced the dot maps used in the *Atlas*.

John las Gourgues, Chief Drafting Officer of the Victorian Ministry for Conservation, drew the outline map of Australia and the maps used in the introduction of the *Atlas*. The staff of the Australian Survey Office and Mike Bamford also assisted with the preparation of the maps.

Leo Joseph, Tess Kloot and Stephen Marchant proofread the *Atlas*.

B. A. and A. G. Wells donated the colour transparency used on the case of the Atlas.

The RAOU apologizes to anyone whose name may inadvertently have been omitted.

CONTRIBUTORS

The following list includes the names of all those known to have contributed to the Atlas. There are probably many more of whom we have no record. We apologize to anyone whose name has been omitted and thank all who helped.

Contributors are listed in the State of their last known address. People living in the Australian Capital Territory are included under New South Wales.

New South Wales

D. C. C. Aartsen, N. Aartsen, D. Albertson, B. Allen, L. Allen, W. Allan, M. D. Allom, R. F. Almond, P. Alsop, P. Alsop, J. Anderson, N. Anderson, R. Anderson, A. E. Andrew, D. Andrew, R. Annels, K. L. Anway, A. C. Appleby, B. F. Archer, J. S. Armstrong.

I. Baird, D. B. Baker, G. B. Baker, M. Baldwin, V. R. Baldwin, D. Balfour, W. P. Barden, A. L. Barker, L. M. Barnes, K. G. Barrett, D. Barton, D. Bates, G. Bath, G. D. Batten, J. L. Baulch, C. A. Bear, A. S. Beard, J. Belcher, S. Bellchambers, J. Benham, M. Benjamin, Mr and Mrs S. Benn, D. Best, N. Best, E. K. Bigg, D. L. Binns, R. Bird, A. W. Bischoff, W. Blackman, G. A. Blackwell, C. Blakers, G. E. Blakers, G. L. Blaxland, B. J. Blaylock, R. W. Boden, W. E. Boles, S. C. Bolliger, A. G. Bomford, C. M. Bonser, P. Boreham, M. Boswell, J. M. Bourne, P. Bowe, C. C. Bower, G. Bradley, L. W. Braithwaite, G. C. Breen, M. Brewer, J. Brickhill, J. W. Bromby, W. N. Bromby, A. J. Brown, I. Brown, M. R. Brown, G. Browne, B. Brownlee, Mrs Brummell, E. Buckle, D. Bull, E. Burce, J. E. Burfoot, N. Byrne.

T. J. Callaghan, A. Cameron, J. Cameron, M. Cameron, R. Cameron, I. L. Campbell, M. Carey, J. Carson, C. Carter, D. Carter, G. Carter, M. V. Carthew, Mr Catchpole, H. Catling, J. Chalmers, Mr and Mrs L. Chapple, S. C. Chittick, R. J. Church, G. Clancy, J. Clancy, G. S. Clark, N. E. Clark, G. Clarke, N. Clarke, M. Clayton, E. S. Clissold, C. Clout, E. R. Clout, G. K. Clout, A. Clucas, M. I. Cochrane, L. R. Cole, M. Cole, M. Cole, A. W. Colemane, A. J. Collett, R. Condon, P. W. Connors, C. Cooper, E. H. Cooper, R. M. Cooper, V. M. Corp, M. C. Cort, P. A. Cosier, J. W. Costello, A. N. Cowan, G. Cowlishaw, A. Cox, P. J. Cox, R. J. Cox, K. Crawford, M. Crawford, C. Crayford, F. Creer, I. F. Crowe, P. Cull, P. M. Curling.

A. Dampney, J. Darnell, C. C. Davey, K. Davey, D. Davidson, P. N. Davidson, L. W. Davies, L. De Silva, S. J. S. Debus, E. Denny, M. Denny, J. Densley, M. Densley, G. Derrin, E. B. Dettman, B. Dibley, M. Dibley, P. A. Disher, H. J. Disney, T. Dixon, E. Doody, Mr and Mrs J. H. Dorrington, E. Dovey, H. A. Dowd, R. W. Dowd, H. M. Doyle, J. Doyle, R. Draffan, V. A. Drake, C. C. Duncan, T. J. Dunlea, S. Dunlop, J. Duranti.

C. M. Edwards, E. C. Edwards, Mr and Mrs K. Edwards, M. D. B. Eldridge, R. A. Elford, E. S. Elphick, B. F. Emery, W. Emery, M. Eves, G. A. Ewers, M. W. Ewers.

A. Faulks, E. Ferrari, S. Ferrier, R. H. Fielding, L. W. C. Filewood, H. Firth, M. E. Fisher, S. Fletcher, O. Florin, H. A. Ford, J. C. Forshaw, J. M. Forshaw, R. Forster, J. Frankenburg, B. Fraser, G. C. Fraser, G. Friend, G. R. Fry, G. Fry, P. J. Fullagar.

M. Gallen, A. M. Garrard, B. A. Geddes, M. Geddes, D. J. Geering, M. H. George, A. Gibson, C. Gibson, H. B. Gill, R. R. Gill, S. Gill, B. J. Gilligan, A. M. Gilmore, I. Glasson, S. D. Glyde, N. Godfrey, Mrs Gold, H. G. Goldstein, K. Goodrich, P. J. Gooley, C. H. Gorely, R. Gould, W. K. Goulding, T. J. Gourlay, D. R. Grace, B. Graham, M. Graham, W. Graham, C. Green, T. W. Green, R. Gregory-Smith, K. Groat, M. Guppy, G. E. Guy.

D. Haig, B. Haigh, D. Hair, J. Hall, E. Hamonet, S. Hamonet, A. Hansen, M. Harding, F. J. Hardwick, J. W. Hardy, G. N. Harrington, J. Harris, P. Harris, B. Harrison, M. Hauswirth, W. Hawes, H. R. Hawkeswood, J. Hawkins, D. Hayne, J. Haynes, R. W. Hedgecoe, A. Heinrich, J. A. Henderson, G. Herbert, N. L. Hermes, J. Higgins, H. G. Hiller, M. D. Hinchey, R. B. Hinchley, J. Hinton, S. M. Hnatiuk, J. N. Hobbs, D. Hobcroft, P. Hogan, H. Holland, J. S. Holland, G. Holloway, F. Holmes,

K. Holmes, M. N. Holmes, N. G. Holmes, M. Holst, L. J. Hone, W. Horrigan, T. Hosking, J. M. Houghton, D. Howarth, E. Howe, B. Howie, A. Huggett, C. P. Humphries, A. C. Hunt, J. W. Hunt, G. Hutchings, C. E. Hutchinson, Dr Huxtable, B. R. Huxtable, J. F. Hyslop.

R. Imisides, J. Imrie, K. Imrie, S. J. Inwards, J. Izzard.

A. Jackson, B. G. James, J. E. R. James, N. L. Jeffery, S. Jeisman, R. Jenkins, V. Jenkins, B. Jenkinson, B. Jenkinson, G. Johnson, I. G. Johnson, M. D. Johnson, Prof. and Mrs Johnson, D. M. Johnston, F. G. Johnston, J. Johnston, J. Johnston, W. Jollie, D. Jones, D. Jones, E. W. Jones, R. Jordan, T. Jordan.

M. Kala, R. P. Kavanagh, M. Kaveney, G. Kearns, R. T. Keene, T. Kendall, A. N. Kennedy, J. Kennedy, B. Kenyon, O. Kershaw, A. King, N. L. Kirby, T. G. Korn, J. R. Kupke, L. G. Kupke, N. Kurtz.

W. Laing, B. Lake, D. Lamm, P. D. Lamont, W. J. Land, S. G. Lane, G. M. Langdale-Smith, R. A. Langdale-Smith, D. Larkins, A. J. Latham, G. C. Law, J. E. Law, A. R. Lawson, D. G. Lawson, W. Lawson, S. M. Learmonth, G. Lenehan, S. Leslie, P. A. Lester, M. Lewis, A. J. Ley, Mr and Mrs Lightowler, B. D. Lindenmayer, A. Lindsay, A. Lindsey, T. R. Lindsey, K. Lineham, P. Lipscombe, E. Lisser, K. Lisser, P. Liston, R. A. Littlejohn, A. A. Locke, J. A. Lockley, P. Lonergan, D. Long, R. Long, J. M. Low, F. Lucas.

I. McAllan, A. McBride, S. McCann, C. McClelland, A. McClure, A. O. McCutcheon, I. McDonald, M. McDonald, R. A. McDonald, R. MacFarlane, A. R. McGill, A. McGowan, A. J. McHugh, L. J. F. McHugh, J. McKinnon-Matthews, W. A. McKinnon-Matthews, D. K. McNally, N. McNally, J. McVicar, A. McWhirter.

R. K. Madely, R. Magrath, J. Maguire, P. Maher, P. N. Maher, B. Mannes, B. Marchant, R. Marchant, S. Marchant, G. Marchell, J. D. Marthick, R. J. Martin, W. R. Martin, I. G. Matthias, N. J. Maxwell, G. J. Mendel, Dr and Mrs D. J. Merrall, J. R. Merritt, P. B. Metcalf, R. Metcalf, A. H. Metson, N. Metson, A. Mewett, J. Miles, D. R. Milledge, D. Miller, L. P. Miller, J. Mills, K. G. Mills, V. Minter, F. H. Mitchell, R. Moffatt, A. Montgomery, L. Moodie, T. Moore, E. Morgan, M. Morphett, A. K. Morris, J. Morris, N. H. Morrison, E. Mulholland, B. F. Mulvogue, W. Murphy, D. Murray, G. B. Murray, G. M. Murray.

P. A. Nagle, L. Nancarrow, N. Neasby, J. Nevinson, Mr and Mrs H. Nicholson, H. A. Nix, K. Norris, R. A. Noske, J. E. A. Noyce.

M. O'Brien, D. M. O'Grady, G. Y. O'Grady, P. I. Ormay, C. M. Orr, E. M. Osborne, The Venerable and Mrs C. Osborne, J. R. Ovenden, R. Owen.

Mr and Mrs J. Page, R. Page, H. C. Panter, T. Paramore, I. K. Parker, L. Parry, B. W. Patch, J. Patience, O. Patterson, M. C. Paul, L. J. Payne, J. Pearse, R. A. Pearson, B. Peden, J. M. Pegler, J. M. Penhallurick, B. E. Pennefather, R. Perry, A. Peters, F. Peters, N. A. Phillips, W. T. Pigg, J. Piggin, L. Pitt, R. Poile, B. Pointer, M. Pointer, E. Poole, R. Poulton, V. Powys, E. Pratt, J. H. Pratt, C. H. Pratten, A. R. Preston, W. Price, F. A. Prichard, H. N. Pryor, J. Prytherch, D. Purchase.

T. Quested.

K. Ranson, J. G. Raymond, D. Read, V. Read, R. Rehwinkel, J. M. Reinschmidt, E. G. Reynolds, J. Reynolds, J. Rice, D. Richards. I. Richards, C. T. Richardson, B. Riches, V. Riggs, J. Riley, V. Robb, J. Roberts, G. Roberts, W. Roberts, J. R. Robertson, S. Robson, I. Rodriques, F. Rodwell, A. E. Rogers, D. Rogers, N. C. Rogers, S. J. Rooke, A. B. Rose, A. D. Ross, F. W. D. Rost, R. B. Rostron, A. Rottenbury, R. Routley, M. T. Rowe, A. Rowell, D. Ruming, A. G. Russell, J. Russill, N. D. Russill, W. A. Ryan, N. L. Ryrie.

P. D. Salisbury, V. Sandeman, J. A. Sanders, J. Saw, K. Schaefer, D. Schlunke, B. L. Schmidt, R. Schodde, B. Schoefell, W. H. Schofield, N. W. Schrader, J. Scollay, M. L. Scrivenor, J. Scrivens, M. Seamer, D. Secomb, J. B. See, P. Seely, J. Seymour, R. E. Sharrock, N. Sheppard, N. R. Shoebridge, G. Siepen, R. O. Sims, D. Sinclair, M. P. Sinclair, J. Skinner, A. Slee, A. I. Sloane, Mr and Mrs J. Slobbe, D. I. Smedley, A. G. Smith, B. Smith, D. Smith, E. L. Smith, G. Smith, G. Smith, L. Smith, N. Smith, P. G. Smith, P. J. Smith, R. G. Smith, C. N. Smithers, J. Sparks, M. Stein, G. E. Stevens, D. A. Stewart, M. Stokes, R. C. Stokes, D. E. Stokes, A. R. Strachan, D. S. Stringfellow, I. Stuart, P. C. Stubbings, C. C. Styles, M. Sutherland, F. G. Swindley, E. A. Synnot.

S. Tarleton, J. N. S. Tarr, T. Tasoulis, R. Telford, R. M. Thackway, E. B. Thomas, G. D. Thomas, M. Thomas, I. Thompson, L. Thompson, P. Thompson, D. H. Thomson, L. J. Thorburn, P. A. Thorburn, C. R. Thouless, S. C. Tidemann, M. Tiernan, K. Timmins, B. Tinsey, N. Town, D. Turner, R. J. Turner.

R. Van Der Boon, F. W. C. Van Gessel, Vanderlaan Family, G. F. van Tets, N. Vaughan, P. A. Veerman, D. M. Vickery, Mr and Mrs B. Virtue.

N. Wakeling, J. Walker, Mr and Mrs R. Walker, E. Walmsley, A. G. Walters, M. G. Watson, T. Watson, J. D. Watt, J. W. Waugh, G. Weatherstone, R. B. Webster. R. C. Weston, J. Whatman, C. Wheeler, E. Wheeler, W. M. Wheeler, F. White, G. White, I. White, M. F. White, S. Wicksteed, J. Wilkinson, A. Williams, A. Williams, E. Williams, Mr and Mrs L. A. Williams, M. Williams, M. Williams, N. Williams, J. Williamson, K. Wilson, M. Wilson, S. J. Wilson, B. M. Wimborne, J. Winter, D. M. Winterbottom, M. Withers, G. C. Wolff, R. E. Wollstein, J. Wombey, D. Worland, A. F. Wright, J. Wright McKinney, E. Wyndham.

C. H. Young.

Victoria

E. Abel, M. J. Adam-Smith, M. Adams, L. R. Adamson, C. H. Agar, J. Agst, E. Aickin, L. V. Akehurst, D. Akers, R. T. Alcorn, J. Alderson, H. Alexander, S. W. Alford, C. Allamby, J. G. Allan, P. A. M. Allan, C. M. E. Allen, G. J. Ambrose, C. Anderson, P. Anton, M. C. Appleby, G. V. E. Apted, P. Arblaster, A. Aston, H. I. Aston, L. Aston, T. Aumann, C. N. Austin, E. R. Austin.

F. Baggaley, C. J. Baillie, J. R. Bain, K. M. Bain, R. C. Baird, D. J. Baker-Gabb, H. Bakker, J. W. Baldock, C. Balmford, P. Balmford, R. A. Balmford, M. A. Bardwell, B. Barlow, G. Barnes, K. Bartram, L. H. Bassett, I. M. Bateman, A. Batten, P. M. Beacom, C. Beardsall, R. S. Beasley, J. Beckman, R. Beckman, G. W. Bedggood, H. Beggs, S. R. Beggs, K. Bell, E. Bennet, W. Bennet, M. Bennett, S. C. Bennett, J. D. Berry, C. M. Biggs, L. R. Billing, N. R. Billing, P. Bingham, G. W. Binns, G. Birch, M. Bird, E. Bishop, D. F. Blackford, M. Blakers, C. J. Blyth, G. Blyth, P. Blyth, K. Bode, C. Boekel, H. K. Bohnke, P. Booth, K. Bottcher, J. R. Bowen, G. Bowker, D. L. Bowrey, G. Box, H. J. Boyd, M. Boyd, S. Boyd, H. A. Bradley, K. H. Bradshaw, R. W, Bradshaw, M. Brady, E. Brennan, E. P. Brewster, O. G. Brewster, D. E. Britton, M. Brooke, G. Brown, H. G. Brown, R. Brown, C. Browne, S. J. Bruton, B. Buchanan, R. Buckingham, D. Bullock, C. Bunn, A. H. Burbidge, K. Burrage, J. Burston, M. Burston, A. Bushell, M. L. Bushell, F. Bussat, E. Butler, D. M. Byrne, Mr and Mrs G. Byrne, M. Byrne.

D. Cameron, D. Cameron, M. A. Cameron, R. G. Cameron, B. Cane, K. D. Carlyle, M. Carron, M. Carter, D. Carver, D. Cass, Castlemaine Field Naturalists, R. Cavey, M. Cavill, I. Chadwick, A. Chambers, A. E. Chapman, M. P. Chapman, P. Chapple, R. Chatto, P. Cheal, C. Cheers, E. M. Chevalier, P. Chick, I. Christensen, B. Christophers, D. Clark, J. Clark, G. Clarke, I. Clarke, M. Clarke, E. Coggan, R. S. Coghill, J. Collins, M. Collister, M. Colson, C. Compton, L. Conole, M. Considine, Mr and Mrs P. Conte, D. Cooke, A. K. Cookson, R. V. Cooling, E. J. Cope, R. E. Corbould, G. Cornish, A. H. Corrick, M. E. Coulson, V. Cousins, J. Cowell, R. D. Cowley, S. J. Cowling, T. Crafter, N. Craig, R. Crocker, C. Crook, D. L. Cross, R. W. Cross, B. Crossett, J. Crouch, G. M. Crowl, J. M. Cullen, D. Cullimore, E. A. Cuming, L. Cupper, A. J. Curry, V. Curtis.

I. Dalrymple, G. Daly, S. Danaher, F. Dann, P. M. Dann, L. Dart, J. A. Davies, J. N. Davies, M. Davies, P. Davies, R. L. Davies, R. Davis, F. T. Davison, G. Daws, J. G. Dawson, B. L. Day, J. N. Day, N. K. J. Day, G. R. Deason, D. M. Deerson, X. Dennett, D. M. Dennis, P. C. Dennis, F. R. B. Denton, M. Denton, D. Dickson, M. V. Dixon, N. Dogger, T. Dominelli, A. Dowling, S. L. Drape, J. L. Druce, C. Drum, R. I. Drummond, F. Du Guesclin, A. Dugan, J. Dullard, P. E. Duncan, H. Dunn, I. Dunn, K. Durden.

D. Eades, A. Eames, Mr and Mrs K. Eden, E. L. Edwards, L. B. Edwards, G. Ellis, M. Ellis, P. Emery, W. B. Emison, J. Endacott, I. D. Endersby, S. Endersby, W. R. Eury, R. Evans, S. Evans.

A. Faithful, F. Farr, N. J. Favaloro, P. J. Fell, D. Fester, C. Field, B. J. Fielding, B. Files, R. Fishenden, A. Fisher, M. Fisher, M. Fishley, S. Fitzherbert, A. Fitzpatrick, E. W. Fitzpatrick, P. Flahavin, A. M. Fleming, B. R. Fleming, D. S. Fleming, W. M. Flentje, A. Forbes, J. Forrester, D. Forte, C. Foster, D. Foster, M. A. Foster, J. L. Fox, D. Franklin, D. C. Franklin, J. L. Fransen, C. Fraser, M. Fraser, N. A. Fraser, J. Freitag, G. Fulton.

A. H. Gadsden, J. Galbraith, M. Galbraith, T. R. Garnett, B. Garrett, P. A. Gell, R. L. Gell, T. A. Gellert, Mrs Geysen, J. G. Gibb, K. Gibbs, H. Gibson, L. A. Gibson, R. C. Gibson, R. J. Gibson, D. Giddy, F. Gladstone, J. Gladstone, M. B. Gladstone, B. Gleeson, K. Gleeson, C. Gloury, G. Glover, A. L. Godfrey, C. Godman, S. D. Goldsworthy, G. Goods, M. Gottsch, C. N. Gove, A. W. Graham, E. Graham, M. Graham, R. Graham, A. Graham-Higgs, D. Grant, J. Grant, N. Grant, A. H. Green, C. Green, V. Green, G. Griffin, N. L. Grimwade, G. Grogan, R. J. Gully, M. O. Gunn.

E. Haarhoff, M. Hackleton, R. S. Hadley, B. Hahnel, B. Hall, M. W. Halliday, K. Hammond, R. E. Hancock, S. J. Hanley, J. Harding, J. Hardy, F. E. G. Harrap, A. M. Harris, I. Harris, K. Harris, M. Hartland, I. Hastings, K. Hastwell, J. Hauptmann, K. Hawkey, P. Hawking, A. Haylock, J. L. Hayward, I. S. Hazelcine, J. Heard, K. J. Hegarty, A. Heislers, M. E. Hendry, S. Henry, J. Hermans, P. Hermans, M. Herrmann, D. Hewett, C. A. Hibbert, A. Hickey, B. Hill, R. Hill, L. Hilton, W. Hilton, F. Hirst, G. J. Hirth, J. Hirth, H. H. Hobson, D. M. Hodges, P. Hodson, D. G. W. Hollands, S. Holliday, K. Holmes, P. Homan, A. Hone, N. Hooper, B. Horler, G. F. B. Horrocks, M. Horton, R. T. Horwood, G. A. Hosken, K. Hough, A. Howard, P. E. Howden, V. Howell, P. Howes, D. C. Howlett, F. L. Howlett, W. J. Howlett, H. R. Hudson, M. Huggins, M. Hughes, N. Hunter, M. Huntley.

E. Incoll, J. C. Ipsen, R. Ireland, J. Irvine, L. Irving, A. C. Isles.

M. Jackson, P. Jackson, S. Jackson, V. Jacobs, R. Jakobi, U. Jakubowski, Mr and Mrs F. G. James, I. G. James L. James, B. Jarman, H. Jarman, L. A. Jarman, B. W. Jarrett, D. Jeans, W. Johns, H. R. Johnson, P. G. Johnson, S. Johnson, J. Johnstone, J. Jones, P. Jones, T. A. Jordon, K. Joseph, A. Jubb.

P. J. Kadwell, B. Kane, J. Kean, J. W. Kellam, D. Keller, B. Kenna, M. D. Kester, N. Kilfoyle, D. King, J. E. King, Mr and Mrs Kirton, J. Klapste, K. Kleesh, E. Kleingeerts, T. Kloot, P. Knell, D. Kuzmic.

W. Labbett, P. Lade, A. Ladsden, R. K. Laidlaw, D. Lakey, F. Lancaster, V. Lancaster, B. A. Lane, R. Lang, P. Lansley, M. K. Laverick, G. Lawrie, M. Le Grew, V. R. Leaton, C. Lendon, B. H. Leorke, L. Leorke, W. Lester, D. G. Lewis, Dr Lipshutz, E. B. Littlejohn, L. Living, J. D. Lobb, D. Lockwood, G. Lockwood, K. Lowe, T. G. Lowe, N. Lowton, R. Loyn, M. J. P. Lubcke, L. F. Lumsden, E. Lyndon, B. Lyon, J. A. Lyons.

P. G. McCallum, G. McCarthy, J. McCarthy,

E. S. McCauley, J. McClusky, P. McCulloch, F. McDonald, M. McDonald, P. McDonald, S. McEvey, D. B. MacFarlane, N. MacFarlane, A. M. McGregor, B. McGregor, I. McIlraith, B. Mackintosh, I. M. McIntyre, L. McKay, T. C. McKellar, R. McLay, D. MacLean, L. McMillan, E. McNabb, J. A. McPherson, Mr and Mrs F. McPhie, G. McRae.

R. Magrath, J. Mailer, A. Malcolm, R. Malseed, J. B. Mangan, I. Mansergh, V. Marriner, J. Marshall, A. J. Marsland, A. Martin, B. R. Martin, J. Martindale, H. Mason, G. R. Mathison, J. A. H. Mattiske, H. S. Mawdsley, I. Mawson, E. F. May, G. Meadows, R. Medlings, J. Meier, P. Menkhorst, C. W. Meredith, M. Meredith, C. Merlin, D. Merrick, R. N. Middleton, W. G. D. Middleton, M. S. Miles, L. Millsom, P. R. Milne, C. D. T. Minton, R. M. Missen, K. Mitchell, P. Mitchell, Mr and Mrs A. E. Monger, K. Monk, R. A. Moon, E. Moore, G. Moore, G. L. J. Moors, I. L. Morgan, A. Morison, C. Morley, D. Moroney, I. C. Morris, W. Morris, J. Morrish, J. L. Muller, J. Murphy, M. J. Murphy, S. Murphy, N. Murray.

L. Naismith, J. Neaves, J. Ness, D. S. Nicholas, F. Noelker, T. Noone, E. Norris, R. C. Norris.

G. Oakley, D. J. Oataway, R. M. O'Brien, W. J. Obst, P. O'Connor, P. O'Dell, A. O'Callaghan, H. R. Officer, M. O'Sullivan, J. Ozols.

J. Pagon, M. Pagon, G. Park, K. Parker, G. Parkes, J. Parry, N. Pascall, B. Pascoe, D. Pascoe, L. H. Paton, K. A. Patterson, A. Paulin, P. Peake, R. Pearce, M. Pengelly, Penguin Study Group, J. R. Petering, D. E. Peters, J. Peters, V. Pettigrove, D. Phillips, M. Pietsch, D. Plant, W. Platt, J. Pollock, W. Potts, A. J. Poynton, J. Pratt, I. Preston, G. Price, P. Pritchard, N. J. Purbrick, B. Pye, T. Pye.

D. Rae, K. Raff, R. M. Raid, J. Raines, C. Ralph, V. Rawlings, K. Raymond, A. G. Redfern, G. A. Reid, J. Reid, S. Reid, P. N. Reilly, R. Renn, P. V. Rich, T. H. Rich, J. Richardson, R. Ridge, M. Rigg, R. J. Ritchie, A. D. Roadley, C. Roberts, D. G. Roberts, M. A. Roberts, H. B. Robertson, M. B. Robertson, D. Robinson, J. Robinson, M. Rollason, R. H. M. Ross, N. Rossiter, R. Rowlands, K. Ruby, A. D. Rudolph, M. P. Ruff, K. C. Runciman, P. Rush, M. C. Russell, G. P. Ryan.

S. Salmon, R. Sanders, K. C. Sandy, F. Sargent, N. Scarfe, R. Scheffer, R. M. Scholes, M. Schultz, R. Schulz, T. Schurman, V. Schurmann, J. H. Seebeck, A. Selzer, R. B. Semmens, J. Sendy, P. Senten, M. J. Seyfort, S. Seymour, M. Shackleford, R. Shackleford, B. Shalit, N. Shaw, R. Shaw, L. L. Sheary, H. Sherritt, C. E. Silveira, E. M. Simpson, K. N. G. Simpson, A. I. Sinclair, T. Sinclair, G. Slater, A. D. Smith, B. Smith, B. Smith, L. E. Smith, M. Smith, M. L. Smith, N. Smith, P. C. Smith, T. Smith, B. C. Snape, D. Snape, L. Solly, H. B. Somerset, C. R. Sonter, D. E. Souter, R. J. Southey, K. Sowter, C. Spiker, I. D. Spokes, J. Spry, E. L. Stabb, I. Stanyer, J. Starks, P. Starks, S. Steele, L. Stephens, R. Stephens, R. H. Stephens, P. Stewart, M. W. Stewart, A. Stirling, P. Stoddart, S. Stone, R. E. Storey, J. Stott, P. Stretton, I. A. Stringer, R. W. H. Strong, J. Strudwick, A. Sudbury, A. Summer, W. Sutherland, A. Swan, R. Symons, M. F. Sympson, R. T. Sympson.

A. C. Talbot, M. E. Talmage, C. C. Taylor, G. Taylor, M. Taylor, P. Taylor, B. W. Teague, I. D. Temby, Mr and Mrs R. V. Terry, A. Thiele, K. Thiele, V. Thiele, R. G. Thoday, O. L. Thomas, R. Thomas, M. Thompson, L. Thomson, Mr and Mrs I. E. Tilbury, G. Tirrell, N. Todd, S. E. Townsend, B. J. Traill, I. Trevaskis, L. Trewenack, G. Tribe, T. Trickett, B. E. Triggs, R. Tripp, J. A. Tubb, E. M. Tucker, P. E. Tucker, E. Turnbull, N. H. Turnbull, K. M. Turner, D. G. Twaits, B. M. Tyers, J. Tyers.

S. B. Unthank.

A. Vanzet. F. R. Vasey, K. M. Vaughan, J. Vince, M. J. Vincent, M. Vye-Taylor.

J. Waayers, S. Wadsworth, J. W. Wainer, B. Wait, A. Walker, C. M. Walker, R. Walpole, L. Waterman, R. Waterman, F. Watts, M. Watts, D. J. Waugh, R. Weatherly, W. Weatherly, J. R. Webb, N. W. Webb, S. Webb, F. J. Webley, E. Webster, I. C. Webster, L. R. Webster, M. Wellings, I. C. Werner, P. D. West, H. N. B. Wettenhall, W. R. Wheeler, T. Wheller, A. White, A. P. White, M. White, M. Whitelaw, C. Wild, D. Wilkie, J. Wilkinson, R. Willig, I. Willis, F. Wilmink, A. Wilson, E. M. Wilson, G. W. Wilson, N. Witham, W. R. Witham, P. M. Withington, J. C. Z. Woinarski, G. W. Wolfe, E. L. Wood, M. Wood, J. P. Woods, L. P. Woodward, M. Woodward, D. J. Woolcock, J. Wright, K. Wright, A. E. Wylie.

M. C. Yardy, H. Young.

R. A. Zann, J. Zitta.

Queensland

A. W. Adams, J. M. Adamson, C. S. Aiken, R. A. Aisbett, Mr and Mrs R. Aldridge, H. Alexander, D. R. Allen, R. Allingham, F. Allison, E. R. Anderson, J. P. H. Arden, E. Arthur, B. G. Ashley, M. Atzehi.

D. Baker, B. Ballingall, A. Barnaart, A. E. Barnett, G. H. Barr, A. J. Beale, J. E. Beardmore, J. Becker, J. Bell, J. M. Bell, O. K. Benn, J. Bennie, P. W. Bensted, G. Beruldsen, H. Beste, J. Beste, H. Betts, S. R. Biddulp, E. B. Birt, D. Bjelke-Petersen, S. Blaber, J. Blackwood, M. Blake, S. W. Blandford, C. Bond, J. Bond, L. Bone, M. J. Borg, R. J. Boyes, U. M. Bradley, R. Bradshaw,

Mr and Mrs R. Brashaw, R. Bridges, P. K. Bridle, P. Broadhead, C. H. Broadley, T. R. Brodribb, E. C. Brown, L. Bryden, L. Buchan, R. R. Burdon.

C. Cameron, S. Cameron, A. Campbell, G. H. Campbell, E. G. Carne, E. Carroll, W. R. Carter, M. L. Cassels, J. Chandler, R. Chandler, P. Chapman, Mr and Mrs G. A. Christien, J. Chudleigh, G. Claridge, E. Coates, J. Colvin, P. Comerford, M. J. C. Conolly, C. J. Corben, I. A. Cowan, C. I. Cox, L. M. Crouther, M. M. Crowther, E. Cruden, G. Crump, G. Cumming, G. J. Curran.

R. Dahlstrom, J. Dalton, C. E. Daniell, A. H. Davidson, M. H. Dawes, M. Dean, R. Delforce, T. Denny, M. Dingwall, N. Dorman, R. N. Dorman, A. Dow, D. D. Dow, A. Dowd, H. R. Downward, J. Dunlop.

J. Eales, R. J. Eales, K. Edington, C. Eley, R. Eley, C. Elliott, I. Elliott, S. Elliott, P. Elmes, S. Elmes, R. A. Elvish, R. W. Etheridge, L. E. Euston, L. Evans, M. A. Evans, E. J. Eves.

B. S. Falkiner, M. Falvey, Mr and Mrs J. Fearnley, J. Fearnside, F. Ferguson, P. Ferguson, C. Ferris, G. J. Field, A. K. Finch, E. M. Finley, G. N. Fishburn, A. C. Fisher, W. J. Fisher, J. Fletcher, J. Flory, L. Foord, E. J. Foran, A. E. Forbes, H. Fox, L. Frahm, Mr and Mrs D. Fraser, P. L. Freier, J. French, J. E. Friend, K. Fritz.

W. A. Gadsby, A. Galbraith, J. Galeano, S. T. Garnett, C. Gasteen, L. R. Gibson, N. R. Gibson, W. F. Giles, P. L. Gillan, A. J. Gillanders, M. C. Glasman, G. Glass, G. J. Glass, G. B. Goadby, M. M. Goadby, G. Godsell, D. N. Gordon, E. A. Gordon, M. Gordon, M. Gough, A. Graham, M. Grant, R. J. Graydon, C. R. Greet, G. E. Greet, J. Gregor, A. C. M. Griffin, J. Grillmeier, C. Grillo.

A. Haffenden, I. F. Haig, J. G. Hale, L. Hall, R. N. Halsall, R. A. Hando, D. Hanlon, K. Hansen, V. R. J. Hansen, L. Harpham, P. L. Harris, H. P. Haselgrove, M. I. Hawken, L. J. Hawkins, R. Hawkins, R. Henderson, P. J. Hendry, M. Higgs, B. K. Hills, M. J. Hirst, D. W. Hiscock, N. Hopkins, R. E. Hopkinson, H. Horton, W. Horton, L. Hotchin, P. Hughes, Mr and Mrs J. Hunter.

K. A. Ireland, J. Irving, G. M. Ivett.

W. L. Jago, B. R. Jahnke, C. Jakku, E. Jakku, R. C. James, S. James, J. Johnson, J. Johnson, J. A. Johnson, D. N. Jones, E. Jones, G. K. Jones, K. Jones, M. V. Jones, N. Jones, J. Josefski, I. Joyce.

J. R. Kane, S. Keliher, N. J. Kelly, B. Kemp, M. Kentwell, J. E. Kerle, E. A. Kerr, I. M. Kerr, K. Kerr, B. R. King, D. King, T. King, M. Kingsford, T. H. Kirkpatrick, J. N. Kleinschmidt, C. T. Kratzmann.

D. Lade, W. Lade, R. L. Lake, P. A. Lander, B. Laver, H. J. Lavery, B. Laws, G. J. Leach, N. S. Lee, S. Lee, J. L'Estrange, J. F. Liddy, C. G. Lloyd, D. Lobageier, J. Lucas, P. K. Lund.

K. McArthur, G. I. McBurney, J. McCabe, B. McDarra, K. R. McDonald, K. R. McGuigan, K. McKay, A. MacKee, P. M. MacKenzie, R. MacKenzie, D. McLaughlin, J. McLean, F. McNamara, A. MacRae, J. MacTaggart.

D. Magarry, R. Mahoney, D. Mainwaring, F. J. Manwaring, N. Marr, H. Mason, I. J. Mason, J. Mason, J. J. Maxwell, W. T. May, D. Merrall, C. Messer, C. Meyers D. Meyers, S. H. Midgley, M. S. Miles, D. Miller, B. K. Mills, P. Milne, A. F. Minching, D. Mitchell, A. Moloney, E. G. Moodie, J. Moon, M. Moore, M. Morgan, A. Morice, A. Morton, S. Mostachetti, J. H. Moverley, E. M. P. Murray, J. M. Murray, S. J. Murray.

G. P. Neilson, K. J. Neilson, L. New, S. A. Nicolosi, D. C. Niland, J. Nolan, T. Norris.

J. F. O'Donnell, B. J. O'Neill, G. Opit, M. R. O'Reilly.

T. Pacey, G. Page, R. Page, A. M. Palliser, Mr and Mrs Parr, E. A. Parr, M. D. Passmore, J. Paterson, S. M. Patterson, D. H. Pearse, M. Peers, N. M. Pegler, L. Perry, A. Podlich, R. J. Poile, G. L. Porter.

T. Querl, N. Quinn.

R. W. Radnell, V. J. Radnell, A. Reeves, D. Regan, S. Reilly, S. D. Reilly, T. Reis, D. Reynolds, I. S. Reynolds, W. G. Richards, B. Richardson, N. Richardson, M. Rigg, E. M. Roberts, G. Roberts, L. Roberts, C. Robertshaw, C. Robertshaw, B. Robertson, J. S. Robertson, Y. Robertson, T. Robinson, E. M. Robson, D. Rogers, A. C. Rohde, M. L. Rohde, K. Rowe, J. Russell, J. N. Russell, L. M. Russell, B. J. Ryan.

R. E. Sandaver, J. Savage, P. Schmidtchen, D. A. Schulz, J. B. Schwennesen, J. Scott, C. Seawright, A. A. Sellers, D. H. C. Seton, O. Seymour, P. A. Shannon, B. L. Shelley, D. Shepherdson, R. Slater, D. M. Slattery, M. B. Sloan, B. M. Smallgange, L. N. Smith, R. Smith, R. G. Smith, W. H. Smith, A. Smyth, C. Soppa, J. A. Sorley, G. Spearritt, G. Spencer, M. J. Spreadborough, A. R. Steginga, M. R. Stevenson, D. A. Stewart, D. J. Stewart, E. Sticklen, R. Sticklen, R. Stokie, T. Stuart, G. I. Surman, M. Sutherland, P. H. Sutton, J. Sweeney, G. Sweetzer, N. Sweetzer, F. G. Swindley.

J. Tadgell, D. R. Tait, D. Tanner, R. Taplin, J. M. Taylor, J. M. Temple-Watts, M. T. Templeton, H. Tew, J. Thompson, P. Thompson, A. Thorsborne, M. Thorsborne, P. Tierney, G. G. Tiffany, R. J. Treloar, J. Trezise, J. D. Turnbull, R. Turnbull, R. C. Turnbull.

R. Underwood.

P. S. Valentine, H. D. Vander Have, C. H. Van Der Held, J. Veacook, E. J. Venables, I. F. Venables, P. Venables, M. A. Verroen, S. M. Verroen, N. J. Versack.

T. J. Wainwright, L. N. Waldron, K. B. Walker, R. M. Walker, E. M. Wall, J. Wall, L. Wallace, R. Wallace, J. C. Walter, R. Ward, G. J. Waterman, B. Watson, D. J. Watson, J. Waugh, P. Weaver, S. R. Weigall, I. Weston, T. Weston, M. Weymouth, C. M. Wheeler, J. Wheeler, D. White, K. J. White, J. Wieneke, W. Wilkinson, J. Williams, K. A. Williams, K. A. W. Williams, B. G. Wilson, E. M. T. Wilson, M. Wilson, C. Wingate, J. W. Winter, R. Wolcott, L. B. Woodall, P. F. Woodall, A. Woods, J. S. Wren, J. R. Wright, R. Wright, R. A. Wyatt, J. Wyndham.

M. Young.

South Australia

C. J. Absalom, H. Adams, J. A. Ahrens, R. I. Albury, J. Alcock, R. Aldersey, D. Altmann, B. Anderson, R. Andrew, E. Arthur, J. Arthur, C. B. Ashton, S. Atkinson, R. Attwood.

F. J. Badman, J. Badman, T. V. Bampton, The Venerable and Mrs I. F. Barlow, L. A. Barmuta, D. E. Barnes, N. Basham, Mr and Mrs I. R. Baumann, C. Baxter, J. Bell, N. Bennett, N. Benson, T. V. Bentley, R. F. Beverley, L. S. Bigg, P. G. Bigg, P. Bird, P. L. Bird, N. N. Birks, C. M. Bishop, A. B. Black, R. Blacker, E. F. Boehm, G. W. Bond, J. M. Bonnin, S. Bosch, J. Bourne, J. M. Bourne, P. G. Bowie, D. C. Bowman, T. J. Bradley, D. A. Bragg, S. L. Braithwaite, A. J. Bransbury, E. Bransbury, R. Bray, L. E. Brice, A. Brinkworth, E. K. Bristowe, A. Brockmeyer, D. Brookes, R. A. Buckley, R. T. Burford, W. H. Burgess, R. F. Burke, R. S. Burke.

D. L. Catford, G. A. Carpenter, P. Caust, P. D. Cawthorne, P. M. Champion, C. Chennells, C. L. Chick, D. Churchill, D. Clark, M. Clarke, P. Clarke, J. Clegg, D. H. Close, D. Cole, M. J. Collier, R. W. Collier, D. A. Colliver, M. Commane, M. Constable, G. Cook, R. Cooper, B. J. Copley, P. Copley, D. M. Cornish, H. Cosh, T. J. Cox, A. Crooks, H. W. Crouch, M. Crouch, G. M. Cundell, R. G. Cundell, R. Cunningham, W. H. Cunningham, M. Curnow, D. W. Cuttle.

D. Davis, D. C. Daw, F. A. Day, B. M. De Courcy-Ireland, B. Delchau, P. Dennis, J. E. Dewhirst, R. M. Downer, B. J. Downes, D. W. Dowsett, J. Drummond, R. E. Dudley, G. Duggan, A. H. Dunkley, E. Dupree.

G. H. Eatts, H. J. Eckert, K. Edwards, B. Eggins, J. R. Eley, G. Ellis, R. Ellis, G. Elson, G. Esau, D. Evans, J. Evans, R. Evans, R. E. Evans.

E. Fennell, H. Ferguson, R. Ferguson, G. L. Fewsdale, J. Firman, C. H. Fleming, A. Fletcher, B. Foote, R. Fox, T. Fraser.

B. J. Gabell, E. M. Gare, G. C. Garrett, Mr and Mrs L. R. Garrett, M. J. Garrett, C. G. Gatehouse, N. Gemmell, B. Gerschwitz, A. M. Geyer, N. Gibbons, P. Gilchrist, C. L. Gill, J. Gloster, B. Glover, Mr and Mrs G. J. Goldfinch, L. Y. Golding, J. Goode, Y. Goode, R. J. Goodfellow, E. J. Gough, L. M. Gough, D. B. Gray, H. J. Gredley, W. B. Greenwood, R. Grimshaw, D. Grocock.

J. Haines, K. Haines, J. R. Hammat, T. W. Hampel, M. V. Hann, A. C. Hardy, D. Harper, D. C. Harris, W. Harris, A. Hartnett, J. H. Hatch, W. R. Hawkes, B. J. Head, B. Heininger, M. Heron, M. R. Hewett, K. M. Heyne, S. M. Hocking, D. J. R. Hodge, W. Hole, C. Hollamby, I. M. Holman, D. C. Hopkins, P. Horton, C. Houston, H. Howell, N. Hudson, T. Hunt, Mr and Mrs R. Hutt.

G. P. Jackson, W. Jenner, M. R. Johnson, G. Johnston, L. G. Joseph.

R. G. F. Kalberer, J. C. Kay, E. Kelsey, L. Kennett, E. M. Kentish, Mr and Mrs A. S. Kernick, R. Kernot, M. A. Killick, M. J. Kinesbury, D. Kraehenbuehl, Kuitpo Bird Study Group.

D. Lade, D. Lade, J. Laird, A. J. Lance, P. A. Langdon, V. R. Langley, A. Lashmar, H. Laybourne-Smith, C. Leckonby, A. Lees, A. F. Lees, M. Lees, D. L. Lillyman Jnr, S. A. Lillywhite, A. R. Litchfield, K. Lloyd, G. Lonzar, E. E. Lucas, N. F. Lynch.

G. M. McArthur, A. McCallum, M. McCallum, I. McClure, K. J. McConnell, M. P. McCormack, N. C. McCrie, I. MacDonald, J. C. McEntee, A. McFarlane, J. McFarlane, R. M. McFarlane, P. F. McGillivray, A. McIntyre, A. McKenzie, A. McMahon, J. A. McNamara, J. McPherson.

K. J. Mack, G. B. Markey, B. J. Marsh, D. J. Marsters, B. K. Mart, W. E. Matheson, G. H. Matulich, I. A. May, K. Maynard, L. M. Maynard, J. Mengersen, J. E. Menz, R. D. Milde, L. Mitchell, R. M. Moorhouse, T. D. Morgan, T. L. Mort, E. A. Mount, J. Mount, P. D. Munchenberg, M. J. Murdock, R. J. Murdock, A. Murray, J. A. Murray.

S. C. Nairn, Mr and Mrs D. A. Nankivell, J. Needle, P. R. Needle, M. Newton, W. M. Noske, J. K. Nottage.

M. Obst, J. H. O'Connor, R. Ogilvie, W. R. O'Malley, M. J. Opie, K. E. Oster, H. F. Oswald.

C. J. Palin, D. C. Palmer, S. Parker, D. R. Parslow, S. D. Parsons, D. C. Paton, J. B. Paton, P. Paton, C. K. Pawsey, D. Pearce, N. A. Peart, L. P. Pedler, K. R. Perkin, J. S. Peterson, D. A. Pfeiffer, J. M. Pfeiffer, R. G. Pickett, S. Pikusa, S. G. Plane, M. L. Possingham, H. T. Pritchard, Mr and Mrs A. L. Proud.

G. B. Ragless, J. Ramsey, C. Rasby, V. G. Rattley, R. E. Read, D. R. Reid, Mr and Mrs J. Reid, J. R. Reid, N. C. Reid, M. E. Rennison, D. Riches, J. S. Riddell, I. Roberts, B. Robertson, D. G. Robertson, T. Robinson, J. F. Roe, J. H. Roeters, R. I. Rohrlach, S. R. Rohrlach, E. M. Rolfe, N. P. Rose, B. Rosier, S. Rosier, E. Rothe, B. M. Rowley, T. Russ, D. J. Ryan, P. F. Rymill.

P. Sadler, D. A. Sando, J. Saunders, H. M. Scarborough, Mr and Mrs J. C. Schiller, D. Searle, J. Seymour, I. Sharp, R. A. Sharpe, P. D. Shaughnassy, B. M. Sherstone, J. Shutes, M. Sigston, S. Simon, H. Sizer, D. J. Skinner, M. Skinner, S. Skinner, F. N. Smeaton, J. Smith, J. A. Smith, K. H. Smith, K. W. Smith, M. Smith, N. Smith, O. M. Smith, T. S. Smith, C. E. Smyth, J. D. Smyth, R. F. Smyth, R. J. Smyth, A. N. Somerville, G. Spanton, J. F. H. Steer, G. S. Stephens, E. B. Stewart, P. F. Stuchbery, B. A. Swaby, R. J. Swaby, N. Swanson.

E. Taylor, F. Taylor, G. Taylor, P. A. Taylor, P. B. Taylor, P. W. Taylor, R. Taylor, T. E. Taylor, J. Tedder, A. C. Thomas, P. M. Thomas, W. J. Thomas, D. N. Thompson, J. Thorn, F. R. Thurston, J. R. Tilbrook, K. Tilbrook, A. Tiver, P. Tonkin, K. W. Treagus, A. Treloar, K. Treloar, J. M. Truran, T. Tyson.

L. Underwood.

W. A. Vaitkus, D. Vincent.

L. Walford, M. Walker, J. E. Waller, W. M. Walter, J. Wandel, Mr and Mrs L. Warren, Mr and Mrs V. K. Warwick, M. Waugh, A. T. Webber,

N. M. Wedding, T. Wendland, M. West, I. R. Westley, G. B. Wetherall, P. A. Wetherall, R. J. Whatmough, M. Wheaton, D. Whidley, G. K. White, K. White, B. Whitford, Mr and Mrs D. Wilkin, D. Williams, D. Wilson, H. C. Wilson, V. A. Winen, J. Winslet, K. Woodcock, B. Wright, C. W. Wright, M. Wright.

Mr and Mrs C. H. Yerbury.

Western Australia

G. F. Agar, E. I. Aitken, R. A. Aitken, S. J. Ambrose, A. Anderson, G. Arnold, H. Arnold, E. Atkins.

J. Baines, G. F. U. Baker, P. Baker, S. D. Baker, M. Bamford, M. E. Barber, A. Barker, P. Barker, P, Barr, H. Barrett, E. Bell, B. Bellairs, B. Bellanger, V. Bennett, M. Beurteaux, S. Bilich, D. Blagden, G. Blagden, J. Blair, F. Blakeway, R. Boase, A. D. Bornman, N. Bougher, L. Boyle, J. M. Bradley, R. A. Breedon, M. H. Bremner, P. Briggs, E. Briggs, J. Brimson, L. Broadhurst, N. F. Brockman, W. J. L. Brooke, M. G. Brooker, L. Brown, M. Brown, A. A. Burbidge, E. Burfein, R. Burking, J. Burt, T. E. Bush, H. Butler.

Calingiri School, S. M. Callahan, M. B. Callanan, D. O. Cameron, D. R. Carlson, A. Carr, C. Chapman, G. S. Chapman, R. Chapman, M. Charlton, J. Chilvers, P. R. Choate, P. Christensen, Dr and Mrs G. M. Clarke, G. D. Clarkson, B. Clay, P. Clay, H. Clow, G. Cocks, P. Cole, S. Colegate, D. Collings, P. Congreve, S. Cook, C. Cowell, K. Creed, S. Cromack, M. A. Cross, G. E. Curedale, P. J. Curry, W. M. Cusack.

J. Dart, J. E. Dart, S. J. J. F. Davies, G. Davis, A. Daw, C. G. Dawe, P. Dean, A. De Haan, T. J. Delaney, J. Dell, P. De Long, A. E. Dempster, R. J. Denny, T. C. Dercksen, A. M. De Rebeira, C. P. S. De Rebeira, L. C. Dixon, A. P. Dolling, J. Donohue, J. Douglas, D. G. Drakes, M. Dunne, J. Dunsmore, P. Durack, D. Dureau.

D. Edinger, K. Edwards, J. Erickson, R. Erickson, L. Eva, Mr and Mrs J. D. Ewert.

J. Falconer, L. Falconer, J. Faulkner, S. Fawcett, B. V. Fewings, J. G. Fewings, A. Fewster, B. Fewster, E. L. Finlay, J. R. Ford, R. Forrest, S. D. Forsyth, G. Fraser, H. N. Fretwell, Mr and Mrs G. Frusher, P. J. Fuller.

L. E. Gardiner, A. Garstone, J. Garstone, R. Garstone, J. A. T. Gibb, H. B. Gibbs, S. Giles, R. Gill, P. J. Gouldthorp, A. Graham, A. H. Green, D. G. Green, G. Green, N. Green, O. Green, M. Gregson, P. Griffin, Mr and Mrs K. J. Grimes, Mrs Grimes, G. Guise.

B. Hale, H. G. Hall, A. W. Hamersley, G. Hancock, P. K. Hardey, G. Harewood, T. Harris, D. J. Heard, A. H. Hele, H. Henderson, R. Hewitt, A. K. Highman, A. Hill, P. S. Hogarth, A. Hope, T. L. Hordacre, M. J. Howard, T. Howard, B. J. Hume, C. Hume-Wright, R. Hunt, P. Hussy.

J. A. Ingham.

F. Jacob, R. P. Jaensch, R. M. Jakob, D. F. James, K. James, F. C. Jeanes, L. E. Jefferis, D. Johnson, N. Johnson, G. Jones, L. Jones, P. Jones, H. L. Jordan.

E. D. Kabay, G. Keall, D. Keast, C. Keen, N. P. Keen, K. F. Kenneally, P. Kimber, Kimberley School Of The Air, B. King, C. Kirby, R. P. Kirkpatrick, W. L. Klau, F. V. Knapp, B. Kneebone, S. Kopij, P. Kovalevs.

P. Lambert, E. Lammonby, Mr and Mrs P. Lammondy, M. F. Lamperd, J. A. K. Lane, G. N. Langfield, R. D. Lawer, H. Lay, L. Lebel, I. Leggat, P. Leighton, E. Lindsay, M. Lloyd, P. J. Loeper, P. London, J. L. Long, A. F. Longbottom, I. R. Lutz, A. J. Lynam.

R. McCarthy, W. A. McCarthy, F. G. McColl, E. R. McCrum, A. M. McDonald, M. B. McGregor, N. L. McKenzie, F. H. McKinnell, B. McMeikan, N. MacNeil.

P. Mack, T. Maddocks, H. Malone, J. L. Malone, F. Mansfield, G. Martin, B. K. Masters, J. N. Masters, J. R. Masters, A. G. Mathews, K. V. Mathews, K. J. Mathwin, R. May, I. Michael, A. L. Milhinch, R. M. Milhinch, M. T. Millard, K. R. Mills, E. I. Milne, R. Moir, J. Monaghan, L. A. Moore, A. Morgan, M. E. Moroney, F. Morris, N. Morse, J. Moyes, O. Mueller, J. R. Muhling, R. Munster, B. Mussen.

J. Nankivell, C. J. Napier, M. E. Nash, R. J. Nash, S. J. Nevill, B. J. Newbey, O. G. Nichols, C. J. Nicholson, J. W. Nilson, C. Nock, J. E. Nock, R. Nottingham, J. Nurick.

M. O'Halloran, W. A. Ckell, E. J. Oliver.

R. Palmer, B. Parker, V. M. Patterson, R. Peakall, A. C. Pegg, C. Pegg, P. A. J. Pennings, K. D. Perry, Penrhos College Club, H. M. Pierce, M. J. Pinner, J. Platt, E. Pole, R. S. Price, B. Purdue.

S. G. Quartly.

J. Radys, T. Ratcliffe, N. A. Rea, V. Read, B. Reaveley, W. Richards, K. C. Richardson, L. Richmond, C. M. Richter, A. Riegen, J. Ritchie, G. Roberts, J. Roberts, S. Roberts, J. D. Robertson, M. Robertson, L. Robins, F. E. Robinson, F. N. Robinson, P. A. Rodgers, Mr and Mrs R. J. Rogers, G. Rogerson, I. J. Rooke, A. M. Rose, S. Rose, E. Rosenberg, D. D. Row, I. Rowley, B. Rudeforth, R. Rule, P. J. Ryan.

W. P. J. Sanderson, D. M. Sandstrom, D. A. Saunders, Mr and Mrs B. Scheer, R. Schulz, J. Scott, L. V. Scott,

L. Scrivener, J. Seabrook, E. Sedgwick, A. B. Selkirk, D. Serventy, M. H. Seymour, T. G. D. Shannon, E. M. W. Sharp, I. Sheeler, D. Shepherd, R. Shepherd, A. D. Sieber, D. Sieber, A. R. Silberstein, L. Silvester, E. Singleton, E. Singleton, G. T. Smith, P. Smith, R. Smith, V. W. Smith, R. E. S. Sokolowski, E. Solomon, I. Solomon, D. W. Stallard, A. Standring, G. B. Standring, A. N. Start, K. Stewart, A. Stokes, C. Story, A. L. Stout, R. Stranger, J. Street, G. Strong, L. Styles.

J. N. Talbot, J. Taylor, F. X. Thomas, K. Thomas, N. Thomas, D. Thompson, H. A. Thompson, P. C. Thomson, B. Tiddy, R. A. Tilt, A. Tingay, C. Towers, J. Towers, N. Trigwell, A. Turner, J. Turner.

M. Vaughan, L. Vernon.

K. Wallace, T. R. Walters, G. B. Walton, B. Ward, P. Waterhouse, D. G. Watkins, E. G. Watts, A. G. Wells, R. C. Wemm, W. J. West, D. Westwood, S. Whatley-Dale, I. Wheeler, M. White, J. Wilson, H. C. Wolf, F. W. Wood, E. Woodrow, F. S. B. Woods, J. Woods, R. Wooller, B. J. Wykes, E. Wykes.

K. V. York, L. Young, K. Youngson.

Tasmania

J. Allenby, L. Andrews, N. Andrews, J. Atkins, P. Atkins, P. Atkinson.

J. Bayley-Stark, R. S. Bell, S. J. Berrigan, G. Best, G. Biggs, J. Biggs, D. Binns, C. Bird, S. A. Blackall, B. J. Blake, R. Blakers, P. Bosworth, M. Boyden, G. Bromfield, L. Brothers, N. P. Brothers, P. Brown, R. Brown, R. M. Brozek, C. Bulman.

G. D. Campbell, R. J. Carr, B. R. Champion, D. T. Chester, B. Chetwynd, C. Chin, M. Chin, M. A. Chladil, R. W. Chynoweth, M. Clougher, J. Cook. G. Cooke, R. Cooper, G. Copson, G. Coulson, K. Cox, P. Craven, S. Cronin, J. Cunningham.

J. C. Davey, G. A. Davis, P. Davis, S. Davis, K. R. Dean, S. M. Dent, H. L. Dodson, M. Douglas, P. Duckworth, B. J. Dutton.

J. M. Elliott, V. Essam.

K. Felton, W. Ferrar, P. Fielding, A. Fletcher.

N. Gellie, S. Gray, R. H. Green, M. Gregory, J. A. Grimm.

R. Hamilton, R. W. Hamilton, C. Hanwood, J. G. K. Harris, S. Harris, W. Harris, P. Hay, J. Hedley, D. Henderson, J. Hinton, R. J. Hoare, M. Holdsworth, E. Holt, R. Hooper, S. W. Hunniford, M. Hunt, I. Hutchinson.

T. D. Johns, Mr and Mrs M. D. Johnson, K. Johnston, G. Johnstone, S. Jones, W. Jones.

C. Kaye.

G. Leaman, C. G. Leary, L. Lewis, K. Littlejohn, M. K. Lord, R. L. Lucas.

M. J. McBain, A. McClure, A. McClure, A. McGifford.

P. Marker, B. R. Miller, N. Mooney, P. Mooney, I. Mormion, M. Mykytowycz.

J. Naarding, J. R. Napier, O. M. G. Newman.

M. Orchard.

P. Park, N. Parker, V. Pattemore, E. Patterson, R. M. Patterson, L. J. Payne, R. J. Pearce, D. Pinner.

S. M. Rackham, A. V. Ratkowski, D. Richmond, C. Riley, G. Robertson, R. Rose, P. Rosevear, D. E. Rounsevell, P. Rowe.

J. Saddington, A. Sather, G. Searle, H. P. Shaw, P. Shepherd, T. Singline, I. J. Skira, M. Smith, M. J. Smith, P. Smith, S. Smith, D. Sparks.

C. H. Taylor, D. G. Thomas, W. Thompson, G. Towney, B. Treadwell, A. Trichanos, D. Tucker, R. Tyson.

A. Upcher, R. Upcher.

O. Vaughan, P. Vaughan.

T. Waite, W. C. Wakefield, L. E. Wall, R. Walters, E. Wapstra, M. Wapstra, J. B. Warren, G. Webb, D. Weston, E. Wheeler, J. S. Whinray, C. White, T. White, M. M. Wilson, R. I. Wilson, M. J. Windley, E. J. Woehler.

K. Young.

D. Ziegler.

Northern Territory

C. Banks, P. R. Barnett, C. A. Benz, G. A. Bertrand, D. C. Bird, J. E. Bosley, R. W. Braithwaite, M. Breen, K. G. Brennan, P. S. Brown.

J. Chalmers, D. Chinner, M. Chuk, H. Clow, C. Cox.

C. Devonport, R. E. B. Don, L. R. Downes.

R. Ebdon, R. Edwards, J. Erlich, J. A. Estbergs.

K. Fisher, M. R. Fleming, J. Fowler, G. R. Friend.

B. Gall, D. Gibson, K. Grant, G. Griffin.

G. R. Haddon, J. M. Hanna, J. Harris, S. Hendy, C. Herbert, A. Hertog, M. Hewett, B. Hill, H. C. Howell.

D. Jacobs, L. Johannsen, K. Johnson, K. A. Johnson, P. Jurkijevic, S. Jurkijevic.

M. J. Karga, A. Kerle, W. J. Kerr, R. G. Kimber, P. T. King.

N. G. C. Lau, T. Lillingstone, R. M. Lines, R. Lipscombe, M. W. J. Locke.

R. B. McElroy, K. McIntosh, J. L. McKean.

W. R. Mason, D. Matthews, B. Milburn, G. Miles.

D. O'Byrne, D. H. Odling-Smee.

M. Passmore, D. Peacock, C. Peterson, J. C. Pfitzner, C. Preece, J. S. Preston, E. W. Pretty.

M. Reed, Mr and Mrs G. A. Robson, D. Roff, P. Rower.

K. Shurcliff, R. Southgate, D. G. Steele, R. J. Steele, B. Strong.

H. A. F. Thompson, B. G. Thomson, D. Tuke, A. Turner.

J. Waddy, M. Wallen, R. Whiteford, J. Whitlow, R. Williams, H. Wilson, F. Woerle, B. E. Woods, M. Wright.

D. Zoellner.

Overseas

W. Bartlett, J. Brouwer, B. Davies, S. A. Halse, C. Harwood, R. W. Howe, P. Kasper, G. L. MacLean, D. McMillan, J. D. Moffatt, C. Radcliffe, S. Russell, C. Stoops, M. K. Tarburton, D. C. Tucker, M. J. Whitmore.

ARTISTS

The name of each artist is followed by a list of the pages on which his or her work appears.

GLOSSARY

acacia scrub	vegetation dominated by shrubs of the genus *Acacia*, specially *A. aneura* (p. xlii)
accidental	a bird outside its usual range, see 'vagrant'
allopatric	occupying mutually exclusive geographical areas
altitudinal movement	movement from one elevation to another seasonally, generally to lower country in winter
ANARE	Australian National Antarctic Research Expedition
Antarctic	of latitudes south of the Antarctic Convergence
Antarctic Convergence	zone, moving with the seasons, where warm tropical currents meet cold Antarctic currents
anvil	stone on which snail shells are beaten until they break
aphid	Hemiptera, family Aphididae
arid zone	regions where the annual average rainfall is less than 250 mm
autumn	1 Mar. to 31 May
bandicoot	Marsupialia, family Peramelidae; terrestrial native mammal (600–1000 g)
banding	application of a numbered band or ring, usually to the leg
block	1°: area bounded by lines of latitude and longitude one degree apart 10′: area bounded by lines of latitude and longitude ten minutes apart (p. xvii)
bore	hole in the ground through which underground water is pumped to a surface tank and reticulated to stock water troughs by gravity
breakaway	small cliff in inland Australia
broad-acre grazing	system in which domestic stock are kept on uncleared native pasture throughout the year
brood parasite	bird that lays in the nest of another species, the host species raising the chick of the brood parasite
census	count of individual birds in a defined area
claypan	shallow intermittently filled pool
clear-fell	forestry operation in which all trees on a site are cut down
cline	continuous gradation in characters among members of a species
coarse grains	cereals other than wheat
coccid	Hemiptera, family Coccidae
colonial	breeding in a colony
commensal	of two different species of organism living in close association without being interdependent
communal breeding	breeding unit containing more than one male and one female
conspecific	of the same species
crayfish-pots	cages to catch crayfish *Jasus* (Crustacea, Decapoda)
creche	flock of young birds, not yet independent, from more than one brood
creek	watercourse; often carries water only seasonally and is usually a tributary of a larger river; in the inland some creeks discharge into salt lakes; in coastal areas 'mangrove' creeks discharge into the sea
dam	excavation to hold water
density	measure of the number of individual birds in a unit area, usually a hectare
desert	arid area usually sparsely vegetated with spinifex and shrubs, see 'arid zone'

dingo — Mammalia, Carnivora *Canis familiaris*

dog-whelk — Mollusca, Gastropoda

donga — depression where water accumulates after rain; vegetated with acacia scrub

dredge — to feed in shallow water with head and bill wholly or partly submerged and sometimes the body upended

dry winter season — in tropical areas that time of the year when rain does not fall, usually Apr. to Nov.

East Australian Current — warm ocean current flowing south along the eastern coast

eclipse — post-breeding plumage in which males look like females

emergent vegetation — plants emerging above the canopy or water surface

endemic — organism restricted to an area, e.g. Australia

escapee — bird that has escaped from aviculture

European settlement — 1788

Field Atlas — contemporary data on the distribution of Australian birds collected between 1 Jan. 1977 and 31 Dec. 1981, during the Atlas of Australian Birds

firefly moth — Arthropoda, Insecta, Coleoptera, family Lampyridae

firing — deliberately burning vegetation, native or introduced

flying fox — Mammalia, Megachiroptera, family Pteropidae

forest — vegetation of trees, usually over 10 m high, with projective foliage cover of more than 50% (p. xlii)

gibber plain — level land covered with pebbles, usually in arid areas; little perennial vegetation

glean — to search leaves, twigs and bark for small insects, spiders and other invertebrates

glider — Mammalia, Marsupialia, family Petauridae; greater *Schoinobates volans*, sugar *Petaurus breviceps*

goanna — Reptilia, family Varanidae *Varanus*

guano — compacted mass of faeces of birds, generally colonial species, accumulated over years

guts — digestive system; includes crop, proventriculus, gizzard and intestines; samples (gut contents) for analysis may be taken from any or all of these regions

hawk — to hunt in the air, usually without hovering

headland — strip of land along the edge of an arable field left unploughed to allow space for machines

heath — vegetation dominated by shrubs (p. xlii)

herb — non-woody plant

Historical Atlas — collection of data on distribution from observations and specimens before 1 Jan. 1977 (p. xvi)

Holarctic — northern America, Europe, Africa north of the Sahara and Asia north of the Himalaya

honeydew — sugary substance excreted by aphids, coccids and psyllids

host — bird in whose nest a cuckoo deposits its egg

hybrid — animal or plant resulting from a cross between individuals of two distinguishable populations, usually subspecies or species

immature — stage of a bird's life from the time it fledges (flies) until it is mated and reproducing

improved pasture — pasture to which fertilizer has been applied

introduced — animal or plant, not endemic to a region, which has been brought, deliberately or accidentally, into it

isohyet — line joining points of equal rainfall

juvenile — bird in its first plumage of true feathers

krill — Arthropoda, Crustacea, order Euphausiacea, *Euphausia* and/or *Nyctiphanes*

lek — area, usually small, used by the males of some species as a display arena; mating takes place on or near it

lerp — protective covering of many psyllids; mostly carbohydrate

litter — layer of partly decomposed leaves, twigs and debris on the ground

mallee — multi-stemmed eucalypt growing from a subterranean rhizome; also the vegetation in which mallee is dominant (p. xlii)

manna — sugary fluid exuded from damaged plant material

marsupial — Mammalia, sub-class Metatheria

midden — accumulation of inedible food remains, specially shells

migrant — bird that moves regularly from one area to another, often from a breeding area to a non-breeding area and back again

migration	movement of all or part of a population of migrants
myxomatosis	virus introduced for the eradication of rabbits
native cat	Marsupialia, family Dasyuridae *Dasyurus*
nectar	sugary fluid produced in the flowers of some species of plant
nestling	young bird that has not yet left the nest in which it hatched
nomad	bird whose movements are unpredictable with regard to timing and direction
nominate subspecies	subspecies bearing same name as the species
nuptial flight	flight of the fertile individuals of some insects during which mating takes place
opportunistic	of behaviour which takes advantage of an unpredicted or unexpected event
paddock	land enclosed by a fence; in the inland may be many hundred sq. km
Palearctic	Europe, Africa north of the Sahara and Asia north of the Himalaya
parasite	animal or plant that lives in or on another (the host) from which it obtains nourishment
passage migrant	migrant observed at a place between its departure point and its destination
pelagic	of or relating to the open sea
plankton	small drifting plants and animals inhabiting the surface of a sea or lake
playground	small area used, and often modified, by one or more male birds for display
polymorphism	two or more readily distinguishable forms within an interbreeding population
possum	Mammalia, Marsupialia; brush-tailed, family Phalangeridae *Trichosurus vulpecula*; ring-tailed, family Petauridae *Pseudocheirus peregrinus*
pre-breeder	bird that has not begun to breed
psyllid	Arthropoda, Insecta, Hemiptera, family Psyllidae; small insects that suck the juices of plants; immature stages build a lerp under which to shelter; sometimes mistakenly called scale
rabbit	Mammalia, Lagomorpha, family Leporidae *Oryctolagus cuniculus*, introduced
rainforest	dense forest growing in areas of heavy rainfall; trees are evergreen and predominantly broadleaved (p. xlii)
rat	Mammalia, Rodentia, family Muridae; there are at least six native *Rattus* species as well as introduced *Rattus norvegicus* and *R. rattus*
rat-kangaroo	Mammalia, Marsupialia, family Macropodidae *Bettongia*
recapture	banded bird caught alive and released
record	documented observation of a bird including place and time
recovery	banded bird found dead
relict	remnant of a formerly widely distributed group
Reporting Rate	see p. xxxv
resident	species present in the same locality throughout the year
roost	place where birds rest or sleep
salt glands	glands in the head or bill of some species that secrete a salty fluid to rid the body of excess salt
saltbush	vegetation in which chenopods are dominant (p. xlii)
saltmarsh	areas dominated by samphire and other salt tolerant plants; the ground is often soft and wet but not flooded
scale	Arthropoda, Insecta, Hemiptera, family Coccidae; the insects feed on leaves, secreting a protective scale around themselves
seasonal movement	migration
sedentary	individuals staying in the same locality throughout the year
semi-colonial	birds nesting near each other as opposed to breeding in isolation
shrub	woody perennial plant, less than 10 m high, with several major branches arising from near the base of the main stem
soak	small pool of water or swampy ground, usually recharged from an underground spring
sodium fluoroacetate	poison, also called 1080
species	groups of actually or potentially interbreeding populations reproductively isolated from other such groups
spinifex	vegetation association in which spinifex is dominant (p. xlii)
spring	1 Sept. to 30 Nov.
station	large farm
sub-Antarctic	of latitudes immediately north of the Antarctic Convergence
subspecies	taxonomic subdivision of a species

succulent	plant having thick fleshy leaves or stems
summer	1 Dec. to 29 Feb.
swamp	low-lying, wet or flooded ground, partly or wholly vegetated; some hold water permanently, others dry out seasonally or intermittently
sympatric	existing in the same geographical area
Tasmanian devil	Marsupialia, Dasyuridae *Sarcophilus harrisii*
taxonomy	science or practice of classification
termitaria	nest of a termite colony; often large earth structures rising from the ground; conical are cone shaped; meridian are flat-sided and narrow, the long axis orientated north and south
termite	Arthropoda, Insecta, Isoptera
tuatara	Reptilia, Rhyncocephalia *Sphenodon*
tundra	treeless zone in the holarctic with a permanently frozen subsoil
tussock grassland	vegetation association dominated by grasses (p. xlii)
unconfirmed record	not supported by conclusive evidence but no reason for doubt
understorey	vegetation below the canopy in a forest or woodland
vagrant	bird wandering outside the normal range of the species
watercourse	channel, bed or route along which water flows after rain or during flooding
wet summer season	in tropical areas, that time of the year when the rain falls, usually Dec. to Mar.
wetland	permanently or temporarily flooded ground
wheatbelt	areas where wheat and other cereals are grown; mainly in the Murray-Darling, Eyre Pen. and South-West Regions
winter	1 June to 31 Aug.
woodland	vegetation association of well-spaced trees less than 30 m high (p. xlii)
wreck	simultaneous death of unusual numbers of birds, e.g. caused by storms, starvation or pollution

PLANT LIST

Names of families follow *A Dictionary of Flowering Plants and Ferns*.[1898]

acacia	Leguminosae, division Mimosoideae *Acacia*
angophora	Myrtaceae *Angophora*
astroloma	Epacridaceae *Astroloma*
austral mulberry	Monimiaceae *Hedycarya angustifolia*
bamboo	Gramineae *Arundo donax*, introduced
banksia	Proteaceae *Banksia*
bauhinia	Leguminosae, division Papilionoideae *Bauhinia*
baumea	Cyperaceae *Baumea*
belah	Casuarinaceae *Casuarina cristata*
blackberry	Rosaceae *Rubus fruticosus*, introduced
blackwood	Leguminosae, division Mimosoideae *Acacia melanoxylon*
bloodwood	Myrtaceae *Eucalyptus terminalis*
bluebush	Chenopodiaceae *Maireana*; usually *M. sedifolia*
bluegum	Myrtaceae *Eucalyptus leucoxylon* (SA), *E. globulus* (Tas.)
boobialla	Myoporaceae *Myoporum insulare*
boronia	Rutaceae *Boronia*
bottlebrush	Myrtaceae *Callistemon*
boxthorn	Solanaceae *Lycium*, some introduced
brachysema	Leguminosae, division Papilionoideae *Brachysema*
bracken	Dennstaedtiaceae *Pteridium aquilinum*
broombush	Myrtaceae *Melaleuca uncinata*
bulkuru sedge	Cyperaceae *Eleocharis dulcis*
buloke	Casuarinaceae *Casuarina luehmannii*
buttongrass	Cyperaceae *Gymnoschoenus sphaerocephalus*
camphor laurel	Lauraceae *Cinnamomum camphora*, introduced
canegrass	Gramineae; sandhill *Zygochloa*; swamp *Eragrostis* or *Leptochloa*
capeweed	Compositae *Arctotheca calendula*, introduced
cassia	Leguminosae, division Caesalpinioideae *Cassia*
cassytha	Lauraceae *Cassytha*
casuarina	Casuarinaceae *Casuarina*
chenopod	Chenopodiaceae, includes saltbush, bluebush and samphire
cider gum	Myrtaceae *Eucalyptus gunnii*
clover	Leguminosae, division Papilionoideae *Trifolium*, introduced
composite	Compositae; daisy
coral tree	Leguminosae, division Papilionoideae *Erythrina*, some introduced
cotoneaster	Rosaceae *Cotoneaster*, introduced
couch	Gramineae *Sporobolus virginicus*
cumbungi	Typhaceae *Typha*
cutting grass	Cyperaceae *Gahnia*
daviesia	Leguminosae, division Papilionoideae *Daviesia*
double-gee	Polygonaceae *Emex australis*, introduced
dryandra	Proteaceae *Dryandra*
emu-bush	Myoporaceae *Eremophila*
epacris	Epacridaceae *Epacris*
eucalypt	Myrtaceae *Eucalyptus*
fig	Moraceae *Ficus*
gidgee	Leguminosae, division Mimosoideae, often *Acacia cambagei*
grass	Gramineae, some introduced
grass tree	Xanthorrhoeaceae *Xanthorrhoea*
grevillea	Proteaceae *Grevillea*

guinea grass	Gramineae *Panicum maximum*
gum	Myrtaceae *Eucalyptus*
hakea	Proteaceae *Hakea*
hawthorn	Rosaceae *Crataegus*, introduced
hoop pine	Araucariaceae *Araucaria cunninghamii*
hydrilla	Hydrocharitaceae *Hydrilla*
inkweed	Phytolaccaceae *Phytolacca octandra*
ironbark	Myrtaceae; common name for several species of *Eucalyptus*: *E. sieberi* (Tas.), *E. creba*, *E. staigerana*, *E. fibrosa*, *E. melanophloia*, *E. paniculata*, *E. caleyi*, *E. sideroxylon*
jarrah	Myrtaceae *Eucalyptus marginata*
kangaroo paw	Haemodoraceae *Anigozanthos*
karri	Myrtaceae *Eucalyptus diversicolor*
lantana	Verbenaceae *Lantana*, introduced
laurel	Lauraceae *Laurus*
lawyer vine	Palmae *Calamus*
legume	Leguminosae
lemonwood	Rutaceae *Acronychia*
lignum	Polygonaceae *Muehlenbeckia*
loquat	Rosaceae *Eriobotrya japonica*, introduced
mango	Anacardiaceae *Mangifera indica*, introduced
mangrove	Rhizophoraceae; many genera in Australia
manna gum	Myrtaceae *Eucalyptus viminalis*
marri	Myrtaceae *Eucalyptus calophylla*
mirrorbush	Rubiaceae *Coprosma hirtella*
mistletoe	Loranthaceae and Vitaceae, often *Amyema* and related genera
mitchell grass	Gramineae *Astrebla*
mulga	Leguminosae, division Mimosoideae, usually *Acacia aneura*
mulla mulla	Amaranthaceae *Ptilotus*
myrtle	Myrtaceae
narrow-leaved peppermint	Myrtaceae *Eucalyptus radiata*
native fuchsia	Rutaceae *Correa*
native pine	Cupressaceae *Callitris*
nothofagus	Fagaceae *Nothofagus*
olive-berry tree	Elaeocarpaceae *Elaeocarpus*
onion weed	Liliaceae *Asphodelus fistulosus*, introduced
paddymelon	Curcurbitaceae *Cucumis myriocarpus*
palm	Palmae
pandanus	Pandanaceae *Pandanus*
paperbark	Myrtaceae *Melaleuca*
parrot bush	Proteaceae *Dryandra sessilis*
passion flower	Passifloraceae *Passiflora*
pea	Leguminosae, division Papilionoideae
peppercorn	Anacardiaceae *Schinus (areira) molle*, introduced
pigface	Aizoaceae *Carpobrotus aequilaterus*; *C. edulis* is introduced
pine	Pinaceae *Pinus*, introduced
pisonia	Nyctaginaceae *Pisonia*
pittosporum	Pittosporaceae *Pittosporum*
prickly pear	Cactaceae *Opuntia inermis* and *O. stricta*, introduced
privet	Oleaceae *Ligustrum vulgare*, introduced
pyracantha	Rosaceae *Pyracantha*, introduced
reed	Gramineae *Phragmites communis*
river red gum	Myrtaceae *Eucalyptus camaldulensis*
rush	Juncaceae
saltbush	Chenopodiaceae *Atriplex* often specifically *A. numularia*
samphire	Chenopodiaceae *Sarcocornia*
sassafrass	Atherospermataceae *Atherosperma moschata* or *Doryphora* sp.
saw sedge	Cyperaceae *Gahnia*
sea-rocket	Cruciferae *Cakile maritima*
sea-tassel	Ruppiaceae *Ruppia*
sedge	Cyperaceae and some other families
sheep's burr	Rosaceae *Acena novaezeelandiae* and related species
sheoke	Casuarinaceae, usually *Casuarina stricta*
silver wattle	Leguminosae, division Mimosoideae *Acacia dealbata*
snowgum	Myrtaceae *Eucalyptus pauciflora*
spinifex	Gramineae *Triodia* and *Plectrachne*
storksbill	Geraniaceae *Erodium*, introduced
stringybark	Myrtaceae; brown *Eucalyptus baxteri*
sugar gum	Myrtaceae *Eucalyptus cladocalyx*
teatree	Myrtaceae, usually *Leptospermum*
thistle	Compositae *Cirsium* and related genera, mostly introduced
Townsville lucerne	Leguminosae, division Papilionoideae *Stylosanthes*
turpentine	Myrtaceae *Syncarpia*
wandoo	Myrtaceae *Eucalyptus wandoo*; *E. redunca* is sometimes regarded as a subspecies of it
waterlily	Nymphaeaceae
willow	Salicaceae *Salix*, introduced
wind grass	Gramineae *Aristida contorta*
wireweed	Polygonaceae *Polygonum*
wongai	Sapotaceae *Manilkara kauki*
yate	Myrtaceae *Eucalyptus cornuta*

REFERENCES

1 ABBOTT, I. 1974. The avifauna of Kangaroo Island and causes of its impoverishment. Emu 74: 124–134.

2 ABBOTT, I. 1978. Breaksea Island, Western Australia. Corella 2: 24–25.

3 ABBOTT, I. 1978. Green Island, Western Australia. Corella 2: 34–35.

4 ABBOTT, I. 1978. Hamelin Island, Western Australia. Corella 2: 38–39.

5 ABBOTT, I. 1978. Lancelin Island, Western Australia. Corella 2: 40–42.

6 ABBOTT, I. 1979. The distribution and abundance of seabirds at sea and on islands near the mid and north-western coasts of Australia. Corella 3: 93–102.

7 ABBOTT, I. 1981. Sandy Island, Western Australia. Corella 5: 69–70.

8 AGNEW, N. V. I. 1913. The Birds of Peel Island. Emu 13: 93–97.

9 ALEXANDER, W. B. 1918. White-tailed Black Terns in Western Australia. Emu 18: 134–135.

10 ALEXANDER, W. B. 1919. Notes from Western Australia. Emu 19: 59–60.

11 ALEXANDER, W. B. 1920. Australian species of Tubinares (Petrels and Albatrosses). Emu 20: 14–24.

12 ALEXANDER, W. B. 1926. Notes on a visit to North Queensland. Emu 25: 245–260.

13 ALI, S. and S. D. RIPLEY. 1980. Handbook of the Birds of India and Pakistan. Vol. 2. 2nd Ed. Delhi: O.U.P.

14 ALTHOFER, G. W. 1934. Birds of the Wellington District, N.S.W. Emu 34: 105–112.

15 AMADON, D. 1941. Notes on some Australian birds of prey. Emu 40: 365–384.

16 AMADON, D. 1978. Remarks on the taxonomy of some Australasian raptors. Emu 78: 115–118.

17 AMBROSE, S. J. Pers. comm.

18 AMIET, L. 1956. Re-appearance of the White-winged Black Tern on the Queensland coast. Emu 56: 95–99.

19 AMIET, L. 1957. A wader survey on some Queensland coastal localities. Emu 57: 236–254.

20 AMIET, L. 1958. The distribution of *Diomedea* in Eastern Australian waters: North of Sydney. Notornis 7: 219–230.

21 ANON. 1902. Exhibits. Vict. Nat. 18: 175.

22 ANON. 1906. Obituary notice—Fletcher. Emu 5: 226–230.

23 ANON. 1923. Obituaries—Chubb. Emu 23: 154–155.

24 ANON. 1963. Recovery round-up: White-plumed Honeyeater. Aust. Bird Bander 1: 105.

25 ANON. 1964. Recovery round-up: Welcome Swallow. Aust. Bird Bander 2: 84.

26 ANON. 1964. Recovery round-up: Australian Gannet. Aust. Bird Bander 2: 115.

27 ANON. 1965. Recovery round-up: Red-capped Dotterel. Aust. Bird Bander 3: 18.

28 ANON. 1966. Recovery round-up: Black-faced Cormorant. Aust. Bird Bander 4: 81.

29 ANON. 1966. Recovery round-up: Eastern Silvereye. Aust. Bird Bander 4: 82.

30 ANON. 1967. Recovery round-up: Spiny-cheeked Honeyeater. Aust. Bird Bander 5: 41.

31 ANON. 1967. Recovery round-up: Sooty Tern. Aust. Bird Bander 5: 65.

32 ANON. 1968. Recovery round-up: Common Bronzewing. Aust. Bird Bander 6: 18.

33 ANON. 1968. Recovery round-up: White-cheeked Honeyeater. Aust. Bird Bander 6: 41.

34 ANON. 1968. Recovery round-up: Brown Gannet. Aust. Bird Bander 6: 87.

35 ANON. 1969. Recovery round-up: Crested Tern. Aust. Bird Bander 7: 16.

36 ANON. 1969. Recovery round-up: Blackbird. Aust. Bird Bander 7: 17.

37 ANON. 1969. Recovery round-up: Reed Warbler. Aust. Bird Bander 7: 65.

38 ANON. 1969. Recovery round-up: Yellow-faced Honeyeater. Aust. Bird Bander 7: 66.

39 ANON. 1969. Recovery round-up: New Holland Honeyeater. Aust. Bird Bander 7: 88.

40 ANON. 1970. Recovery round-up: Stubble Quail. Aust. Bird Bander 8: 41.

41 ANON. 1970. Recovery round-up: White-eyed Duck. Aust. Bird Bander 8: 88.

42 ANON. 1970. Additions to Rand & Gilliard. PNG Bird Soc. Newsl. 51: 2.

43 ANON. 1972. Recovery round-up: Australian Shelduck. Aust. Bird Bander 10: 82.

44 ANON. 1972. Recovery round-up: White-breasted Sea Eagle. Aust. Bird Bander 10: 82.

45 ANON. 1973. Recovery round-up: Silver Gull. Aust. Bird Bander 11: 41.

46 ANON. 1973. Recovery round-up: Black-shouldered Kite. Aust. Bird Bander 11: 63.

47 ANON. 1974. Recovery round-up: Caspian Tern. Aust. Bird Bander 12: 82.

48 ANON. 1975. Recovery round-up: Cape Barren Goose. Aust. Bird Bander 13: 63.

49 ANON. 1976. Recovery round-up: Little Black Cormorant. Aust. Bird Bander 14: 55.

50 ANON. 1976. Recovery round-up: Nankeen Kestrel. Aust. Bird Bander 14: 56.

51 ANON. 1976. Recovery round-up: Satin Bowerbird. Aust. Bird Bander 14: 57.

52 ANON. 1976. Recovery round-up: Plumed Egret. Aust. Bird Bander 14: 77.

53 ANON. 1977. Recovery round-up: Nankeen Night-Heron. Corella 1: 21.

54 ANON. 1977. Recovery round-up: Little Tern. Corella 1: 22.

55 ANON. 1977. Recovery round-up: Grey-tailed Tattler. Corella 1: 40.
56 ANON. 1977. Recovery round-up: Little Pied Cormorant. Corella 1: 40.
57 ANON. 1980. Recovery round-up: Sharp-tailed Sandpiper. Corella 4: 21.
58 ANON. 1980. Recovery round-up: Torresian Imperial Pigeon. Corella 4: 42.
59 ANON. 1980. Recovery round-up: Spotted Pardalote. Corella 4: 63.
60 ANON. 1980. Recovery round-up: Noisy Friarbird. Corella 4: 63.
61 ANON. 1980. Recovery round-up: Cattle Egret. Corella 4: 133.
62 ANON. 1980. Recovery round-up: Whistling Kite. Corella 4: 134.
63 ANON. 1981. Recovery round-up: White-winged Tern. Corella 5: 47.
64 ANON. 1981. Recovery round-up: Australian Hobby. Corella 5: 118.
65 ANON. 1981. Recovery round-up: Terek Sandpiper. Corella 5: 118.
66 ANON. 1981. Recovery round-up: Red-necked Stint. Corella 5: 118.
67 ANON. 1982. Recovery round-up: Masked Booby. Corella 6: 21.
68 ANON. 1982. Recovery round-up: Spotted Harrier. Corella 6:22.
69 ANON. 1982. Recovery round-up: Bar-tailed Godwit. Corella 6: 126.
70 APPLEBY, R. 1976. Brown Falcons and grasshoppers. S. Aust. Orn. 27: 159.
71 ASHBY, E. 1906. Northern notes. Emu 6: 72–74.
72 ASHBY, E. 1917. Field notes on *Acanthornis magnus* (Gld) Scrub-Tit or Great-Tit. S. Aust. Orn. 3: 10–12.
73 ASHBY, E. 1918. Notes on some birds met with in the neighbourhood of Pungonda. Emu 17: 219–220.
74 ASHBY, E. 1933. Detailed observations of the nesting habits of the Black-throated Grebe. Emu 32: 250–259.
75 ASTON, H. I. and R. A. BALMFORD. 1978. A Bird Atlas of the Melbourne Region. Melbourne: VORG.
76 ATHERTON, R. G., P. A. MATTHEW and J. W. WINTER. 1980. Notes on the crop contents and locality records of the Marbled Frogmouth *Podargus ocellatus marmoratus* Gould from Cape York Peninsula. Sunbird 11: 71–72.
77 ATTIWILL, A. R. 1954. Field Notes on Red-capped Robin. S. Aust. Orn. 21: 35–36.
78 AUSTIN, C. N. 1950. Further notes on the birds of Dunk Island, Queensland. Emu 49: 225–231.
79 AUSTIN, C. N. 1953. The Black Falcon and some other raptors in south-west Victoria. Emu 53: 77–80.
80 AUSTIN, C. N. 1956. Stray feathers—range extension of three bird species. Emu 56: 80–81.
81 AUSTIN, C. N. 1969. Extension of range of the Chestnut-breasted Cuckoo. Emu 69: 187.
82 AUSTIN, C. N., A. G. BROWN, H. R. OFFICER and H. N. B. WETTENHALL. 1963. Notes on the Black-banded Pigeon. Emu 62: 238–239.
83 AUSTIN, T. B. 1907. Field notes on birds from Talbragar River, New South Wales. Emu 7: 74–79.
84 AUSTRALASIAN WADER STUDIES GROUP.
85 AUSTRALIAN BIRD BANDING SCHEME. Division of Wildlife Research, CSIRO, Canberra.

86 BADMAN, F. J. 1979. Birds of the southern and western Lake Eyre drainage. S. Aust. Orn. 28: 29–56; 57–81.
87 BADMAN, F. J. 1981. Bourke's Parrot in the western Lake Eyre drainage. S. Aust. Orn. 28: 161–164.
88 BADMAN, F. J. and I. A. MAY. 1983. Waders in northern South Australia. S. Aust. Orn. 29: 29–39.
89 BAKER-GABB, D. J. Pers. comm.
90 BAKER-GABB, D. J. 1982. Comparative ecology and behaviour of Swamp Harriers *Circus approximans*, Spotted Harriers *C. assimilis* and other raptors in Australia and New Zealand. Ph.D. thesis. Monash University.
91 BAKER-GABB, D. J. and J. D. PETTIGREW. 1982. Non-breeding diet of the Letter-winged Kite in north-eastern South Australia. Corella 6: 19–20.
92 BALDA, R. P. and J. L. BROWN. 1977. Observations on the behaviour of Hall's Babbler. Emu 77: 111–114.
93 BALDWIN, M. Pers. comm.
94 BALDWIN, M. 1971. Group movements of the White-winged Chough. Aust. Bird Watcher 4: 69–77.
95 BALDWIN, M. 1974. Studies of the Apostle Bird at Inverell. Part I. General behaviour. Sunbird 5: 77–88.
96 BALDWIN, M. 1975. Birds of Inverell District, N.S.W. Emu 75: 113–120.
97 BALMFORD, R. 1978. Early introductions of birds to Victoria. Aust. Bird Watcher 7: 237–248; 262–265.
98 BANFIELD, E. J. 1908. The Spangled Drongo-Shrike. Emu 7: 178–181.
99 BANFIELD, E. J. 1912. Metallic Starlings (*Calornis*). Emu 12: 22–26.
100 BANFIELD, E. J. 1913. Megapode mounds and pits. Emu 12: 281–283.
101 BARKER, G. H. 1949. Queensland (Branch Report). Emu 48: 208–210.
102 BARNARD, C. A. 1925. A review of the bird life on Coomooboolaroo Station, Duaringa District, Queensland, during the past fifty years. Emu 24: 252–265.
103 BARNARD, C. A. 1934. Notes from "Coomooboolaroo". Emu 33: 251–252.
104 BARNARD, H. G. 1911. Field Notes from Cape York. Emu 11: 17–32.
105 BARNARD, H. G. 1914. Search for *Amytornis woodwardi* in the Northern Territory. Emu 13: 188–190.
106 BARNARD, H. G. 1914. Birds of the Brunette Downs (N.T.). Emu 13: 205–210.
107 BARNARD, H. G. 1914. Northern Territory birds. Emu 14: 39–57.
108 BARNARD, H. G. 1926. Birds of the Cardwell District, Queensland. Emu 26: 1–13.
109 BARNARD, H. G. 1927. Effects of droughts on bird-life in Central Queensland. Emu 27: 35–37.
110 BARNARD, H. G. 1934. Notes on the Square-tailed Kite and the Red Goshawk. Emu 34: 25–28.
111 BARRETT, C. 1920. Wood-Swallows and Ants. Emu 19: 246.
112 BARTON, D. 1978. The Streaked Shearwater in Australian and Japanese waters. Aust. Seabird Group Newsl. 11: 27–30.
113 BARTON, D. 1979. Albatrosses in the western Tasman Sea. Emu 79: 31–35.
114 BARTON, D. 1982. Notes on Skuas and Jaegers in the Western Tasman Sea. Emu 82: 56–59.
115 BARTRAM, K. Pers. comm.
116 BATEMAN, M. 1975. Report on wintering Flame Robins in the Cobram district, Victoria. Aust. Bird Bander 13: 47–52.
117 BATEMAN, M. I. 1978. Report on Flame Robin banding—Cobram area 1978. VORG Notes 14: 75–77.
118 BATEMAN, M. I. 1982. Flame Robins—Cobram area. VORG Notes 18: 47.
119 BATES, G. L. 1930. Handbook of the Birds of West Africa. London: John Bale, Sons & Danielsson Ltd.
120 BATEY, I. 1907. On fifteen thousand acres: its bird-life sixty years ago. Emu 7: 1–17.
121 BAXTER, C. I. 1981. Some remarks on the birds of Kangaroo Island. S. Aust. Orn. 28: 172–174.
122 BAXTER, C. and S. A. PARKER. 1981. The status of the Elegant Parrot and the Rock Parrot on Kangaroo Island. S. Aust. Orn. 28: 164–165.
123 BECK, J. R. and D. W. BROWN. 1972. The biology of Wilson's Storm Petrel *Oceanites oceanites* (Kuhl) at Signey Is., South Orkney Is. Br. Antarctic Surv. Sci. Rep. No. 68.
124 BEDGGOOD, G. M. 1959. Some notes on the birdlife at Bungil, Victoria. Emu 59: 159–163.
125 BEDGGOOD, G. W. 1958. Parrots and cockatoos of the Mooroopna district, Victoria. Emu 58: 72–73.
126 BEDGGOOD, G. W. 1960. Square-tailed Kite at Coomboona, Victoria. Emu 60: 38.
126a BEEHLER, B. 1978. Historical changes in the avifauna of the Wau Valley, Papua New Guinea. Emu 78: 61–64.

127 BEEHLER, B. McP. 1980. Black Swan *Cygnus atratus*, a new species for the New Guinea region. PNG Bird Soc. Newsl: 173–174.
128 BELAND, P. 1977. Mimicry in orioles of south-eastern Queensland. Emu 77: 215–218.
129 BELL, H. L. Pers. comm.
130 BELL, H. L. and R. A. NOSKE. Pers. comm.
131 BELL, H. L. 1959. Some observations on the Black-headed Pardalote in the Brisbane area. Emu 59: 131–135.
132 BELL, H. L. 1960. Observations on the habits of bower-birds in captivity. Emu 60: 8–10.
133 BELL, H. L. 1960. Nesting notes of the Mangrove King-fisher. Emu 60: 64–65.
134 BELL, H. L. 1961. Notes on migratory birds in north Queensland. Emu 61: 56–57.
135 BELL, H. L. 1961. The introduced Spice Finch in north-eastern Queensland. Emu 61: 94–96.
136 BELL, H. L. 1962. The Green-winged Pigeon in Victoria. Emu 61: 328–330.
137 BELL, H. L. 1963. Distribution of the Jabiru in south-eastern Australia. Emu 63: 201–206.
138 BELL, H. L. 1965. Communal nesting by the White-winged Triller. Emu 65: 133–138.
139 BELL, H. L. 1966. A population study of heathland birds. Emu 65: 295–304.
140 BELL, H. L. 1967. Bird-life of the Balimo Sub-district, Papua New Guinea. Emu 67: 57–79.
141 BELL, H. L. 1967. An association of two New Guinea bird species. Emu 67: 95–98.
142 BELL, H. L. 1968. Some distribution notes on New Guinea highland birds. Emu 67: 211–214.
143 BELL, H. L. 1968. The Noisy Pitta in New Guinea. Emu 68: 92–94.
144 BELL, H. L. 1968. Silver-crowned Friarbird—extension of range. Emu 68: 126.
145 BELL, H. L. 1969. Recent Papuan breeding records. Emu 69: 235–237.
146 BELL, H. L. 1969. Status of the White-winged Triller in New Guinea. Emu 69: 243–244.
147 BELL, H. L. 1970. Field notes on the birds of Amazon Bay, Papua. Emu 70: 23–26.
148 BELL, H. L. 1970. Field notes on birds of the Nomad River Sub-district, Papua. Emu 70: 97–104.
149 BELL, H. L. 1970. Additions to the avifauna of Goodenough Island, Papua. Emu 70: 179–182.
150 BELL, H. L. 1971. Field notes on birds of Mt. Albert Edward, Papua. Emu 71: 13–19.
151 BELL, H. L. 1975. Avifauna of the Ninigo and Hermit Islands, New Guinea. Emu 75: 77–84.
152 BELL, H. L. 1976. Painted Finch at Woomera, S. Aust. Orn. 27: 111.
153 BELL, H. L. 1979. The effect on rainforest birds of plantings of teak, *Tectona grandis*, in Papua New Guinea. Aust. Wildl. Res. 6: 305–318.
154 BELL, H. L. 1980. The effects of a power-line clearing on birds of dry sclerophyll forest at Black Mountain Reserve, Australian Capital Territory. Corella 4: 8–19.
155 BELL, H. L. 1980. Composition and seasonality of mixed-species feeding flocks of insectivorous birds in the Australian Capital Territory. Emu 80: 227–232.
156 BELL, H. L. 1981. Information on New Guinean Kingfishers Alcedinidae. Ibis 123: 51–61.
157 BELL, H. L. 1982. Sexual differences in the foraging ecology of the Frill-necked Flycatcher *Arses telescophthalmus* in New Guinea. Aust. J. Ecol. 7: 137–148.
158 BELL, H. L. 1982. Abundance and seasonality of the savanna avifauna at Port Moresby, Papua New Guinea. Ibis 124: 252–274.
159 BELL, H. 1982. Social organization of Thornbills *Acanthiza*. R.A.O.U. 80th Annual Congress, Armidale, N.S.W. 1982.
160 BELL, H. L. 1982–1983. A bird community of lowland rainforest in New Guinea. I-V. Emu 82: 24–41; 65–74; 143–162; 217–224; 256–275.
161 BELL, H. L. 1983. Co-operative breeding by the White-browed Scrub-wren *Sericornis frontalis*. Emu 82: 315–316.
162 BELL, H. L. 1983. Resource-partitioning among three syntopic species of Thornbills (Aves: *Acanthiza*) at Wollomombi, New South Wales. Ph.D. Thesis, University of New England, Armidale.
163 BELLCHAMBERS, T. P. 1916. Notes on the Mallee Fowl. S. Aust. Orn. 11: 134–140.
164 BENNETT, S. Pers. comm.
165 BENNETT, S. 1983. A review of the distribution, status and biology of the Plains-wanderer *Pedionomus torquatus*, Gould. Emu 83: 1–11.
166 BENNETT, S. 1983. The Northern Scrub-robin *Drymodes superciliaris* in the Northern Territory. Emu 83: 105–107.
167 BERNEY, F. L. 1903. North Queensland notes on some migratory birds. Emu 2: 210–213.
168 BERNEY, F. L. 1905–1907. Field notes on birds of the Richmond District, North Queensland. II-V. Emu 5: 72–78; 6: 41–47; 106–115; 155–159.
169 BERNEY, F. L. 1936. The Bustard in Queensland. Emu 36: 4–9.
170 BERNEY, F. L. 1938. A note on the Black Honeyeater. Emu 37: 318.
171 BERULDSEN, G. 1975. The Bush-hen in south-eastern Queensland. Aust. Bird Watcher 6: 75–76.
172 BERULDSON, G. R. 1978. Observations on the White-eared Flycatcher *Monarcha leucotis*. Aust. Bird Watcher 7: 249–255.
173 BERULDSEN, G. 1980. A Field Guide to Nests and Eggs of Australian Birds. Adelaide: Rigby.
174 BESTE, H. 1982. Observations at a nest of Eastern Grass Owls. Aust. Bird Watcher 9: 154–157.
175 BILNEY, R. J. Pers. comm.
176 BILNEY, R. J. and W. B. EMISON. 1983. Breeding of the White-bellied Sea-eagle in the Gippsland Lakes Region of Victoria, Australia. Aust. Bird Watcher 10: 61–68.
177 BINNS, G. 1953. Birds of Terang, south-western Victoria. Emu 53: 211–221.
178 BIRD, C. L. 1980. Fork-tailed Swifts in Tasmania. Tas. Bird Rep. 10: 22–24.
179 BLACK, A. B. Pers. comm.
180 BLACK, A. 1984. A biological survey of the Eastern Great Victoria Desert. Nature Conservation Society of South Australia.
181 BLACKBURN, A. 1970. Unusual records of Spine-tailed and Fork-tailed Swifts. Emu 70: 90.
182 BLACKMAN, J. G. 1971. Distribution of the Sarus Crane in northern Queensland. Emu 71: 137–138.
183 BLACKMAN, J. G. and D. K. LOCKE. 1978. Nocturnal flying height of a Dollarbird. Emu 78: 163.
184 BLIGH, A. C. 1919. Bustard and mice. Emu 18: 205.
185 BODEN, R. W. 1980. A note on the diet of the Cape Barren Goose. W. Aust. Nat. 14: 238.
186 BOEHM, E. F. 1934. Ornithological observations in the Sutherlands and Mount Mary District, S.A. S. Aust. Orn. 12: 154–163.
187 BOEHM, E. F. 1939. The Fork-tailed Swift (*Micropus pacificus*, Latham 1801): with special reference to its occurrence in South Australia. S. Aust. Orn. 15: 54–58.
188 BOEHM, E. F. 1944. The Fork-tailed Swift in South Australia. S. Aust. Orn. 17: 21.
189 BOEHM, E. F. 1957. Perching birds (Passeriformes) of the Mount Mary Plains, South Australia. Emu 57: 311–324.
190 BOEHM, E. F. 1959. Parrots and Cockatoos of the Mount Mary Plains, South Australia. Emu 59: 83–87.
191 BOEHM, E. F. 1960. Notes on some South Australian Waders. Part I. Emu 60: 211–218.
192 BOEHM, E. F. 1961. Indian Turtle-dove extends range in South Australia. Emu 61: 55.
193 BOEHM, E. F. 1964. Notes on some South Australian Waders. Part II. Emu 63: 276–282.
194 BOEHM, E. F. 1974. Golden Bronze-Cuckoo in dry country. S. Aust. Orn. 26: 167.
195 BOEHM, E. F. 1977. Results from banding Little Ravens. Corella 1: 29–32.
196 BOEKEL, C. 1979. Notes on the status and behaviour of the Purple-crowned Fairy-Wren *Malurus coronatus* in the

Victoria River Downs area, Northern Territory. Aust. Bird Watcher 8: 91–97.
197 BOLES, W. E. Pers. comm.
198 BOLES, W. E. and R. SCHODDE. Pers. comm.
199 BOLES, W. E. 1976. The "rediscovery" of the first record of the Bush-Hen for Australia. Aust. Birds 11: 21–25.
200 BOLES, W. E. 1979. Comments on *Sericornis beccarii* in southern Cape York Peninsula. Sunbird 10: 70–72.
201 BOLES, W. E. 1980. A notable record of the Yellow-legged Flycatcher. Aust. Birds 15: 12–13.
202 BOLES, W. E. 1983. A taxonomic revision of the Brown Thornbill *Acanthiza pusilla* (Shaw) 1790 with description of a new subspecies. Emu 83: 51–58.
203 BOLGER, P. F. and L. E. WALL. 1959. Egrets in Tasmania. Emu 59: 184–188.
204 BONNIN, M. and C. E. RIX. 1972. Meeting 16.11.71. Bird Talk 1: 4.
205 BOOTH, J. 1970. Birds of Fairfax Island (Bunker Group) and Wreck Island (Capricorn Group), Great Barrier Reef. Sunbird 1: 85–92.
206 BOURKE, P. A. Pers. comm.
207 BOURKE, P. A. 1947. Notes on the Horsfield Bush-Lark. Emu 47: 1–7.
208 BOURKE, P. A. 1949. The Breeding Population of a thirty-five acre 'Timber Paddock'. Emu 49: 73–83.
209 BOURKE, P. A. 1954. The Gilbert Whistler in New South Wales. Emu 54: 163–168.
210 BOURKE, P. A. 1956. Notes on the Whiskered Tern. Emu 56: 339–348.
211 BOURKE, P. A. 1960. The Purple-crowned Lorikeet in New South Wales. Emu 60: 134.
212 BOURKE, P. A. and A. F. AUSTIN. 1947. The Atherton Tablelands and its avifauna. Emu 47: 87–116.
213 BOURKE, P. A., V. T. LOWE and T. G. LOWE. 1973. Notes on the Gull-Billed Tern. Aust. Bird Watcher 5: 69–79.
214 BOURNE, J. M. 1982. Remarks on the status of the Painted, Little and Red-chested Button-Quails in the south-east of South Australia. S. Aust. Orn. 29: 5–6.
215 BOURNE, W. R. P. 1967. Long-distance vagrancy in the petrels. Ibis 109: 141–167.
216 BOURNE, W. R. P. 1968. Notes on the Diving-petrel. Bull. Br. Orn. Club 88: 77–85.
217 BOURNE, W. R. P. and J. WARHAM. 1966. Geographical variation in the Giant Petrels of the genus *Macronectes*. Ardea 54: 45–67.
218 BRADLEY, E. and J. BRADLEY. 1958. Notes on the behaviour and plumage of colour-ringed Blue Wrens. Emu 58: 313–326.
219 BRAITHWAITE, L. W. Pers. comm.
220 BRAITHWAITE, L. W. 1970. The Black Swan. Aust. Nat. Hist. 16: 375–379.
221 BRAITHWAITE, L. W. 1976. Notes on the breeding of the Freckled Duck in the Lachlan River Valley. Emu 76: 127–132.
222 BRAITHWAITE, L. W. 1976. Breeding seasons of waterfowl in Australia. Proc. 16th Intern. Ornith. Congr.: 235–247.
223 BRAITHWAITE, L. W. 1981. Ecological studies of the Black Swan. II. Colour and plumage changes, growth rates, sexual maturation and timing and frequency of breeding. Aust. Wildl. Res. 8: 121–133.
224 BRAITHWAITE, L. W. 1982. Ecological studies of the Black Swan. IV. The timing and success of breeding on two nearby lakes on the southern tablelands of New South Wales. Aust. Wildl. Res. 9: 261–75.
225 BRANDON, T. 1936. The birds of Wilmington district. S. Aust. Orn. 13: 228–230.
226 BRANSBURY, A. J. Pers. comm.
227 BRAVERY, J. A. 1962. Notes on *Mirafra javanica* in the Atherton District, North Queensland. Emu 62: 55–58.
228 BRAVERY, J. A. 1964. Waders of Tinaroo Dam, North Queensland. Emu 64: 61–64.
229 BRAVERY, J. A. 1967. Field notes on the Oriental Cuckoo. Emu 66: 267–271.
230 BRAVERY, J. A. 1969. The Sarus Crane in north-eastern Queensland. Emu 69: 52–53.
231 BRAVERY, J. A. 1970. The birds of Atherton Shire, Queensland. Emu 70: 49–63.
232 BRENT, A. E. 1905. The Black-headed Honeyeater *Melithreptus melanocephalus*. Emu 5: 12–15.
233 BRIDGEWATER, A. E. 193·. A list of the birds of Mansfield, Victoria. Emu 31: 280–285.
234 BRIGGS, S. V. Pers. comm.
235 BRIGGS, S. V. 1982. Food habits of the Freckled Duck and associated water-fowl in north-western N.S.W. Wildfowl 33: 88–93.
236 BRIGHT, J. 1935. Notes on a few birds of the Rochester District. Emu 34: 293–302.
237 BRIGHT, J. and A. R. TAYSOM. 1932. Birds of Lake Cooper, Victoria and surroundings. Emu 32: 42–48.
238 BROADBENT, K. 1888. Birds of the central part of Queensland. Proc. Roy. Soc. Queensland 5: 14–31.
239 BROADBENT, K. 1910. Birds of Cardwell and Herbert River districts, (N.Q.). Emu 10: 233–245.
240 BROINOWSKI, G. J. 1885. Birds and Mammals of Australia. Sydney: G. Murray.
241 BROOKER, M. G. 1969. The nesting of the Chestnut-breasted Quail-Thrush in south-western Queensland. Emu 69: 47.
242 BROOKER, M. G. 1974. The Satin Flycatcher—a new record for W.A. W. Aust. Nat. 12: 181.
243 BROOKER, M. G. 1976. A record of the Eastern Grass Owl from the Northern Territory. Emu 76: 154.
244 BROOKER, M. G. 1977. Some notes on the mammalian fauna of the Western Nullarbor Plain, Western Australia. W. Aust. Nat. 14: 2–15.
245 BROOKER, M. G. and A. J. ESTBERGS. 1976. A survey of terrestrial vertebrates in the Carnarvon Region, W.A. W. Aust. Nat. 13: 160–170.
246 BROOKER, M. G., M. G. RIDPATH, A. J. ESTBERGS, J. BYWATER, D. S. HART and M. S. JONES. 1979. Bird observations on the north-western Nullarbor Plain and neighbouring regions, 1967–78. Emu 79: 176–190.
247 BROOKER, M. G. and M. G. RIDPATH. 1980. The diet of the Wedge-tailed Eagle, *Aquila audax*, in Western Australia. Aust. Wildl. Res. 7: 433–452.
248 BROTHERS, N. P. Pers. comm.
249 BROTHERS, N. P. 1979. Further notes on the birds of Maatsuyker Island, Tasmania. Emu 79: 89–91.
250 BROTHERS, N. P. 1979. Tasman Island, Tasmania. Corella 3: 55–57.
251 BROTHERS, N. P. 1979. Pedra Branca, Tasmania. Corella 3: 58–60.
252 BROTHERS, N. P. 1979. Mewstone, Tasmania. Corella 3: 68–69.
253 BROTHERS, N. P. 1979. Breeding seabirds of Eddystone Rock, Tas. Corella 3:83.
254 BROTHERS, N. P. 1980. The Nuggets, Tasmania. Corella 4: 110–112.
255 BROWN, A. G. 1949. Notes on some birds of the Whitsunday Group, Queensland. Emu 49: 44–49.
256 BROWN, A. G. 1950. The birds of 'Turkeith', Victoria. Emu 50: 105–113.
257 BROWN, C. B. and G. W. ARCHIBALD. 1977. Captive Brolgas and Sarus Cranes prey on wild mice. Emu 77: 39–40.
258 BROWN, J. L., D. D. DOW, E. R. BROWN and S. D. BROWN. 1978. Effects of helpers on feeding of nestlings in the Grey-crowned Babbler (*Pomatostomus temporalis*). Behav. Ecol. Sociobiol. 4: 43–59.
259 BROWN, P. B. Pers. comm.
260 BROWN, P. B. 1979. The status of Parrot species in western Tasmania. Tas. Bird Report 9: 4–12.
261 BROWN, P. B. and R. I. WILSON. 1981. A survey of the Orange-bellied Parrot *Neophema chrysogaster* in Tasmania, Victoria and South Australia. A Report for WWF (Aust.), NPWS Tas. November 1981.
262 BROWN, R. J. and M. N. BROWN. Pers. comm.
263 BROWN, R. J. and M. N. BROWN. 1980. Co-operative breeding in robins of the genus *Eopsaltria*. Emu 80: 89.
264 BROWN, R. J. and M. N. BROWN. 1980. Eurasian Coots breeding on irrigation dams near Manjimup, Western Australia. Corella 4: 33–36.

265 BROWN, R. J. and M. N. BROWN. 1981. Reports of the Middlesex Field Study Centre I to IV 1976–81. Perth: RAOU.
266 BROWN, R. S. and D. M. DEERSON. Pers. comm.
267 BROWN, R. S. and D. M. DEERSON. 1982. RAOU Conservation Group—Duck survey. RAOU Newsl. June 1982.
268 BROWNSMITH, C. B. 1978. Sarus Cranes dominate over Brolgas in captivity. Emu 78: 98.
269 BRUCE, M. D. Pers. comm.
270 BRUCE, M. D. 1974. A review of the Red-backed Quail in south-eastern Australia. Aust. Birds 9: 17–19.
271 BRYANT, C. E. 1934. The camp-out at Moree, N.S.W. and the birds observed. Emu 33: 159–173.
272 BRYANT, C. E. 1938. The Fantail-Warbler at home. Emu 37: 262–264.
273 BRYANT, C. E. 1940. A note on the Australian Dotterel. Emu 39: 153–155.
274 BRYANT, C. E. 1940. Photography in the swamps: The Eastern Swamphen. Emu 39: 288–292.
275 BRYANT, C. E. 1940. Photography in the swamps: The Little Grassbird. Emu 40: 162–164.
276 BRYANT, C. E. 1942. Photography in the swamps: The Marsh Crake. Emu 42: 31–35.
277 BRYANT, C. E. and E. AMOS. 1949. Notes on Crakes of the genus *Porzana* around Melbourne, Victoria. Emu 48: 249–275.
278 BULLER, K. G. 1945. A new record of the Western Bristle-Bird. Emu 34: 78–80.
279 BULLER, W. L. 1905. Supplement to 'The Birds of New Zealand'. London: The Author.
280 BURBIDGE, A. A. Pers. comm.
281 BURBIDGE, A. A. 1980. Additional records of the White-backed Swallow south of the Swan River. W. Aust. Nat. 14: 199.
282 BURBIDGE, A. A. and P. J. FULLER. 1982. Banded Stilt breeding at Lake Barlee, Western Australia. Emu 82: 212–216.
283 BURBIDGE, A. A., P. J. FULLER, J. A. K. LANE and S. A. MOORE. The breeding status of Boobies and Frigate Birds in Western Australia. Submitted to Emu.
284 BURBIDGE, A. H. Pers. comm.
285 BURBIDGE, A. H. 1982. Birds. Rotamah Island Bird Observatory Report 1980–81: 8–36.
286 BURGER, A. E. 1978. Interspecific breeding attempts by *Macronectes giganteus* and *M. halli*. Emu 78: 234–235.
287 BURTON, T. C. and A. A. MARTIN. 1976. Analysis of hybridization between Black-backed and White-backed Magpies in south-eastern Australia. Emu 76: 30–36.
287a BUSBY, J. R. 1979. Australian Biotaxonomic Information System. Introduction and Data Interchange Standards. Canberra: A.G.P.S.
288 BUSH, T. E. and G. A. LODGE. 1977. Birds of Bedout Island—A visit in May 1972. W. Aust. Nat. 13: 189–190.
289 BUTLER, W. H. 1977. Notes on Kimberley birds. W. Aust. Nat. 13: 194–195.

290 CADE, T. J. 1982. The Falcons of the World. London: Collins.
291 CALABY, J. H. 1951. Notes on the Little Eagle; with particular reference to rabbit predation. Emu 51: 33–56.
292 CAMERON, A. C. 1932. Notes on the Ground Cuckoo-Shrike. Emu 31: 306–307.
293 CAMERON, A. C. 1952. At what height do birds fly? Emu 52: 86–88.
294 CAMERON, A. C. 1968. Feeding habits of the Spine-tailed Swift. Emu 68: 217–219.
295 CAMERON, M. 1981. Cape Gannet photographed at Wedge Light, Port Phillip Bay. Geelong Nat. 18: 34–35.
296 CAMPBELL, A. G. 1902. Birds of north-eastern Victoria. Emu 2: 9–18.
297 CAMPBELL, A. G. 1927. The genus *Amytornis*: a review. Emu 27: 23–35.
298 CAMPBELL, A. G. 1935. The genus *Sericornis* in Australia, with notes on four monotypic genera. Emu 34: 249–274.
299 CAMPBELL, A. J. 1901. Nests and eggs of Australian birds. Melbourne: The Author.
300 CAMPBELL, A. J. 1917. The Yellow-breasted Bush Chat (*Ephthianura crocea*). Emu 17: 61–62.
301 CAMPBELL, A. J. 1918. Notes on some additions to the H. L. White Collection. Emu 18: 2–8.
302 CAMPBELL, A. J. 1920. Notes on additions to the 'H. L. White Collection". Emu 20: 49–66.
303 CAMPBELL, A. J. and H. G. BARNARD. 1917. Birds of the Rockingham Bay District, North Queensland. Emu 17: 2–38.
304 CAMPBELL, A. J. and A. G. CAMPBELL. 1926. A review of the genus *Cinclosoma*. Emu 26: 26–40.
305 CAMPBELL A. J. and D. LE SOUEF. 1918. Cyclones and bird destruction. Emu 18: 72–73.
306 CANNON, C. E. Pers. comm.
307 CARNABY, I. C. 1965. The White-browed Wood-Swallow in Western Australia. Emu 65: 74.
308 CARNABY, I. C. 1933. The birds of the Lake Grace District, W.A. Emu 33: 103–109.
309 CARRICK, R. 1959. The food and feeding habits of the Straw-necked Ibis *Threskiornis spinicollis* (Jameson), and the White Ibis *T. molucca* (Cuvier), in Australia. CSIRO Wildl. Res. 4: 69–92.
310 CARRICK, R. 1962. Breeding, movements and conservation of ibises (Threskiornithidae) in Australia. CSIRO Wildl. Res. 7: 71–88.
311 CARRICK, R. 1972. Population Ecology of the Australian Black-backed Magpie, Royal Penguin and Silver Gull. Wildl. Research Report (Bureau of Sport, Fisheries and Wildlife, Washington), 2: 41–99.
312 CARRUTHERS, R. K. 1968. Notes on an influx of Oriental Pratincoles at Mount Isa. Emu 68: 216–217.
313 CARRUTHERS, R. K. 1969. Co-operative fish drives. Emu 68: 280–282.
314 CARRUTHERS, R. K., W. HORTON and D. P. VERNON. 1970. Distribution, habits and seasonal dimorphism of the Western Grass Wren *Amytornis textilis ballarae* Condon in northwestern Queensland. Mem. Queensl. Mus. 15: 335–341.
315 CARTER, M. J. Pers. comm.
316 CARTER, M. J. 1983. Streaked Shearwaters in northern Australia. Aust. Bird Watcher 10: 113–122.
317 CARTER, T. 1904. Birds occurring in the region of the North-west Cape. Emu 3: 207–213.
318 CARTER, T. 1923–24. Birds of the Broome Hill District. Emu 23: 125–142; 223–235; 306–318.
319 CAWKELL, E. M. and J. E. HAMILTON. 1961. The birds of the Falkland Islands. Ibis 103a: 1–27.
320 CAYLEY, N. W. 1931. What Bird is That? (5th Ed. Rev. 1972). Sydney: Angus & Robertson.
321 CAYLEY, N. W. 1949. The Fairy Wrens of Australia. Sydney: Angus & Robertson.
322 CAYLEY, N. W. 1973. (Ed. A. H. LENDON). Australian Parrots in Field and Aviary. Sydney: Angus & Robertson.
323 CHAFFER, N. 1930. Notes on the Brown Warbler. Emu 30: 58–60.
324 CHAFFER, N. 1931. The Catbird. Emu 31: 61–62.
325 CHAFFER, N. 1931. The gorgeous Regent Honeyeater. Emu 31: 140–141.
326 CHAFFER, N. 1933. The Bulbul. Emu 33: 136–137.
327 CHAFFER, N. 1945. Noisy Friar-bird as a nest robber. Emu 44: 287.
328 CHAFFER, N. 1954. The Eastern Bristle-bird. Emu 54: 153–162.
329 CHALMERS, C. E. 1972. Cattle Egrets in the Gippsland area. Emu 72: 180–181.
330 CHANDLER, L. G. 1910. Notes on pardalotes. Emu 10: 113–118.
331 CHANDLER, L. G. 1913. Bird-life of Kow Plains, Victoria. Emu 13: 33–45.
332 CHANDLER, L. G. 1944. Birds and vineyards. Emu 44: 93–96.
333 CHAPMAN, F. R. H. 1963. The Pelican in South Australia. S. Aust. Orn. 24: 6–1[illegible].
334 CHAPMAN, G. S. Pers. comm.
335 CHAPMAN, G. 1981. White-fronted Honeyeaters. W.A. Bird Notes 17: 2.

336 CHAPMAN, G. and I. ROWLEY. 1978. Co-operative breeding by Red-winged Wrens *Malurus elegans*. W. Aust. Nat. 14: 74.
337 CHENEY, G. M. 1915. Birds of Wangaratta district, Victoria. Emu 14: 199–213.
338 CHESHIRE, N. 1982. Birds seen at sea off Australia. Aust. Seabird Group Newsl. 17: 10–13.
339 CHESHIRE, N., J. JENKINS and P. NESFIELD. 1979. Distribution of the Cape Pigeon in the Tasman Sea and south-west Pacific. Notornis 26: 37–46.
340 CHINNER, D. W. 1977. Observations on the effect of increased rainfall on birdlife in central Australia. S. Aust. Orn. 27: 188–192.
341 CHISHOLM, A. H. 1915. Notes on the Yellow-bellied Shrike-Tit *Falcunculus frontatus*. Emu 15: 78–85.
342 CHISHOLM, A. H. 1918. Two singing species of *Gerygone*. Emu 17: 150–153.
343 CHISHOLM, A. H. 1919. Introduced birds in Queensland. Emu 19: 60–62.
344 CHISHOLM, A. H. 1922. The "lost" Paradise Parrot. Emu 22: 4–17.
345 CHISHOLM, A. H. 1932. Remarkable strays. Emu 32: 63–64.
346 CHISHOLM, A. H. 1932. The White-tailed Kingfisher. Emu 32: 81–83.
347 CHISHOLM, A. H. 1944–45. Birds of the Gilbert Diary. 1–2. Emu 44: 131–150; 183–200.
348 CHISHOLM, A. H. 1946. Observations and reflections on birds of the Victorian mallee. Emu 46: 168–186.
349 CHISHOLM, A. H. 1948. Some birds of central Victoria. Emu 47: 203–205.
350 CHISHOLM, A. H. 1948. Bird Wonders of Australia. 3rd Ed. Sydney: Angus & Robertson.
351 CHISHOLM, A. H. 1951. The story of the Scrub-birds. Emu 51: 89–112; 285–297.
352 CHISHOLM, A. H. 1957. The Albert Lyrebird—a puzzle in distribution. Emu 57: 25–30.
353 CHISHOLM, A. H. 1960. Extension of range in three species. Emu 60: 145–146.
354 CHISHOLM, A. H. 1960. Remarks on robins. Emu 60: 221–236.
355 CHISHOLM, A. H. 1965. Further remarks on vocal mimicry. Emu 65: 57–64.
356 CHISHOLM, A. H. 1971. Various bird notes. Aust. Bird Watcher 4: 41–43.
357 CHISHOLM, E. C. 1926. Birds introduced into New South Wales. Emu 25: 276–279.
358 CHISHOLM, E. C. 1934. Birds of the Comboyne Plateau, 1923–34. Emu 34: 8–23.
359 CHISHOLM, E. C. 1936. Birds of the Pilliga Scrub. Emu 36: 32–38.
360 CHISHOLM, E. C. 1938. The birds of Barellan, New South Wales. Emu 37: 301–313.
361 CHRISTENSEN, V. 1954. Dates of arrival of three bird species. Emu 54: 72–73.
362 CLANCEY, P. 1978. On the nominate race of the Wandering Albatross *Diomedea exulans* Linnaeus, 1758. Durban Mus. Novitate XI(19): 309–311.
363 CLANCY, G. P. 1977. Boobook Owls in the Sydney district. Aust. Birds 12: 12–13.
364 CLARKE, M. F. 1982. The evolution of co-operative breeding in the Bell Miner *Manorina melanophrys* Latham. B.Sc. (Hons) Thesis, University of Melbourne.
365 CLELAND, J. B. 1910. Examination of contents of stomachs and crops of some Australian birds. Emu 9: 219–226.
366 CLELAND, J. B. 1911. Examination of contents of stomachs and crops of Australian birds. Emu 11: 79–95.
367 CLELAND, J. B. 1912. Examination of contents of stomachs and crops of Australian birds. Emu 12: 8–18.
368 CLELAND, J. B. 1919. The birds of the Pilliga Scrub, New South Wales. Emu 18: 272–285.
369 CLELAND, J. B. 1920. The Spinebill as a flower fertilizer. Emu 19: 247–248.
370 CLELAND, J. B. 1926. Monthly proceedings. S. Aust. Orn. 8: 292.
371 CLELAND, J. B. 1965. Fork-tailed Swifts in South Australia before a cyclone. Emu 65: 144–148.
372 CLELAND, J. B. and E. B. SIMS. 1968. Food of the Glossy Black Cockatoo. S. Aust. Orn. 25: 47.
373 CLOSE, D. H. 1981. The Kelp Gull in South Australia. S. Aust. Orn. 28: 155–156.
374 CLOSE, D. H. 1982. Recent records of the Oriental Plover. S. Aust. Orn. 28: 205–206.
375 CLOSE, D. H. 1982. Offshore sightings of Procellariiformes in South Australia. S. Aust. Orn. 28: 210–213.
376 CLOSE, D. H., C. CORBEN and J. B. COX. 1979. Comment on a record of the Lesser Black-backed Gull. Aust. Bird Watcher 8: 135–136.
377 CLOSE, D. H. and O. M. G. NEWMAN. Pers. comm.
378 CLOSE, D. H. and D. TEESE. 1978. Yellow-tinted Honeyeater at Weipa. Sunbird 9: 59–60.
379 CLUNIE, F. 1976. Behaviour and nesting of Fijian White-breasted Woodswallows. Notornis 23: 61–75.
380 COLES, H. J. 1905. Notes and notices. Emu 5: 104.
381 COLLINS, H. 1934. Some notes on the birds of the Nullarbor Plain. S. Aust. Orn. 12: 198–201.
382 COLLINS, H. 1942. Little Wood-Swallow in Victoria. Emu 42: 16.
383 COLLISON, F. L. 1959. Observations on Regent Honeyeaters. S. Aust. Orn. 23: 30–31.
384 CONDON, H. T. 1942. The Ground Parrot (*Pezoporus wallicus*) in South Australia. S. Aust. Orn. 16: 4–7.
385 CONDON, H. T. 1951. Variation in the Brown Hawk. Emu 50: 152–174.
386 CONDON, H. T. 1966. The Western Whipbird: Preliminary notes on the discovery of a new subspecies on Southern Yorke Peninsula, South Australia. S. Aust. Orn. 24: 79–91.
387 CONDON, H. T. 1968. A Handlist of the Birds of South Australia. (2nd Ed.). Adelaide: S.A.O.A.
388 CONDON, H. T. 1975. Checklist of the Birds of Australia. Part I. Non-Passerines. Melbourne: RAOU.
389 CONGREVE, D. and P. CONGREVE. 1982. The beach census. Eyre Bird Obs. Rep. 1979–81: 28–43.
390 CONGREVE, P. Pers. comm.
391 CONGREVE, P. 1972. Torpidity in the White-backed Swallow. Emu 72: 32–33.
392 CONGREVE, P. 1981. East to west movement of White-fronted Honeyeaters. W.A. Bird Notes 18: 2.
393 CONSIDINE, M. T. 1979. The feeding behaviour of the Sooty Oystercatcher (*Haematopus fuliginosus*) on rocky shores in Victoria. B.Sc.(Hons) Thesis, Monash University.
394 COOK, J. E. and J. W. WAUGH. 1977. A Baird's Sandpiper at Botany Bay. Aust. Birds 12: 32–35.
395 COOPER, R. 1981. Birds of Cape Portland. Tas. Bird Rep. 11: 21–24.
396 COOPER, R. M. Pers. comm.
397 COOPER, R. M. 1969. Burdekin Duck in NSW. Emu 69: 242–243.
398 COOPER, R. P. 1962. A revision of the distribution of the Brown Pigeon. Emu 61: 253–269.
399 COOPER, R. P. 1974–1975. The avifauna of Wilson's Promontory. Aust. Bird Watcher 5: 137–174; 205–233; 253–276; 6: 17–34; 47–60; 87–102; 103–116.
400 COOPER, W. T. and J. M. FORSHAW. 1977. The Birds of Paradise and Bower Birds. Sydney: Collins.
401 CORBEN, C. J. Pers. comm.
402 CORFE, B. 1977. A sighting of the Fig Parrot in south-east Queensland. Sunbird 8: 44.
403 CORNWALL, E. M. 1909. Notes on birds found breeding near Mackay, North Queensland. Part II. Emu 8: 138–141.
404 CORNWALL, E. M. 1910. Notes on the Great-billed Heron *Ardea sumatrana*. Emu 9:138–141.
405 CORNWALL, E. M. 1921. The Sunbird in Mackay. Emu 21: 69.
406 CORRICK, A. H. 1980. Freckled Duck on lakes in the Western District, Victoria. Aust. Bird Watcher 8: 254–255.
407 CORRICK, A. H. 1982. Records of Freckled Duck in Victoria from Dec. 1980 to July 1981 and of the number shot on opening day of the 1981 duck season. Aust. Bird Watcher 9: 260–268.
408 COUNSILMAN, J. J. 1974. Waking and roosting behaviour

of the Indian Myna. Emu 74: 135–148.
409 COURTNEY, J. 1971. Breeding of the Red-rumped Kingfisher at Swan Vale, N.S.W. Emu 71: 174–175.
410 COURTNEY, J. and S. MARCHANT. 1971. Breeding details of some common birds in south-eastern Australia. Emu 71: 121–133.
411 COWLES, G. S. 1964. A new Australian Babbler. Emu 64: 1–5.
412 COWLING, S. J. and K. W. LOWE. 1981. Studies of Ibises in Victoria, I: Records of breeding since 1955. Emu 81: 33–39.
413 COX, J. B. 1973. The identification of the smaller Australasian *Diomedea* and the status of the *Diomedea* in South Australia. S. Aust. Orn. 26: 67–75.
414 COX, J. B. 1974. Some birds of north-eastern Eyre Peninsula, S.A. S. Aust. Orn. 26: 142–144.
415 COX, J. B. 1976. A review of the Procellariiformes occurring in South Australian waters. S. Aust. Orn. 27: 28–82.
416 COX, J. B. 1976. Grey Grasswrens and Grass Owls at Goyder's Lagoon, South Australia. S. Aust. Orn. 27: 96–100.
417 COX, J. B. 1978. The distribution and seasonal occurrence of Gannets off the South Australian coast. S. Aust. Orn. 28: 18–20.
418 COX, J. B. 1978. Short-tailed Shearwater colonies of the South-East. S. Aust. Orn. 28: 16–17.
419 COX, J. B. 1980. Some remarks on the breeding distribution and taxonomy of the Prions (Procellariidae: *Pachyptila*). Rec. S. Aust. Mus. 18: 92–121.
420 COX, J. B. and D. H. CLOSE. 1977. Interbreeding of Little and Fairy Terns. Emu 77: 28–32.
421 COX, J. B. and L. P. PEDLER. 1977. Birds recorded during three visits to the far north-east of South Australia. S. Aust. Orn. 27: 231–250.
422 CRAMP, S. and K. E. L. SIMMONS. 1977–83. Handbook of the Birds of Europe, the Middle East, and North Africa: The Birds of the Western Palearctic. Vols 1–3. Oxford: O.U.P.
423 CRAWFORD, D. N. 1972. Birds of Darwin area, with some records from other parts of Northern Territory. Emu 72: 131–148.
424 CRAWFORD, D. N. 1977. Notes on the feeding of two *Chlidonias* terns. Emu 77: 146–147.
425 CRAWFORD, D. N. 1979. Effects of drought on Green Pygmy-Geese and Comb-crested Jacanas in Northern Territory. Emu 79: 37–39.
426 CRAWFORD, D. N. 1979. Effects of grass and fires on birds in the Darwin area, Northern Territory. Emu 79: 150–152.
427 CRAWFORD, D. N. 1980. Seasonal fluctuations in numbers of three species of tern in Northern Territory. Emu 80: 166–169.
428 CROME, F. H. J. 1973. The relationship of the Helmeted and Yellow-tufted Honeyeaters. Emu 73: 12–18.
429 CROME, F. H. J. 1975. Notes on the breeding of the Purple-crowned Pigeon. Emu 75: 172–174.
430 CROME, F. H. J. 1975. Breeding, feeding and status of the Torres Strait Pigeon at Low Isles, north-eastern Queensland. Emu 75: 189–198.
431 CROME, F. H. J. 1975. The ecology of fruit pigeons in tropical northern Queensland. Aust. Wildl. Res. 2: 155–185.
432 CROME, F. H. J. 1976. Some observations on the biology of the Cassowary in northern Queensland. Emu 76: 8–14.
433 CROME, F. H. J. 1976. Breeding, moult and food of the Squatter Pigeon in north-eastern Queensland. Aust. Wildl. Res. 3: 45–59.
434 CROME, F. H. J. 1976. Macleay's Honeyeater. *In* Reader's Digest Complete Book of Australian Birds: 473.
435 CROME, F. H. J. and H. E. BROWN. 1979. Notes on social organization and breeding of the Orange-footed Scrubfowl *Megapodius reinwardt*. Emu 79: 111–119.
436 CROUCH, H. W. 1970. The Olive-backed Oriole in South Australia. S. Aust. Orn. 25: 195–200.
437 CROUTHER, M. M. 1983. Observations on White-rumped Swiftlets breeding at Finch Hatton Gorge 1981–82. Aust. Bird Watcher 10: 1–11.
438 CULLEN, M. J. Pers. comm.
439 CUPPER, J. and L. CUPPER. 1981. Hawks in Focus. Mildura: Jaclin.
440 CURRY, P. J. Pers. comm.
441 CURRY, P. J. 1979. Long-toed Stints, Sanderling and other waders at Lake Violet, Central W.A. in midsummer 1977–78. W. Aust. Nat. 14: 109–113.
442 CURRY, P. 1981. A survey of the birds of Herdsman Lake 1980–81. W.A. Dept. of Conservation & Environment Bull. 105: 1–35.
443 CURRY, P. J. and P. MENKHORST 1979. Breeding records of the Grey Honeyeater in the upper Lyons River district. W. Aust. Nat. 14: 162–163.
444 CURTIS, H. S. Pers. comm.
445 CURTIS, H. S. 1972. The Albert Lyrebird in display. Emu 72: 81–84.
446 CYRUS, D. and N. ROBSON. 1980. Bird Atlas of Natal. Pietermaritzburg: Univ. Natal Press.
447 CZECHURA, G. V. 1979. Observations on quail-hunting strategies in some Australian raptors (Aves: Falconiformes). Sunbird 10: 59–66.
448 CZECHURA, G. V. 1980. Carrion utilization by two species of Australian Goshawks. Raptor Research 14: 62–63.
449 CZECHURA, G. V. 1981. Predation on a small colony of bats by an Australian Hobby, *Falco longipennis*. Vict. Nat. 98: 200–203.

450 DANN, P. Pers. comm.
451 DANN, P. 1980. Vic. Wader Study Group Bull. 3.
452 DANN, P. 1981. Resource allocation among three congeneric species of Sandpiper. Stilt 1: 3.
453 DANN, P. 1981. Breeding of the Banded and Masked Lapwings in southern Victoria. Emu 81: 121–127.
454 DANN, P. M. and K. W. LOWE. Pers. comm.
455 DAVIES, J. 1982. Daylight landings of White-throated Needletails. Bird Observer 609: 76.
456 DAVIES, S. J. J. F. Pers. comm.
457 DAVIES, S. J. J. F. 1963. Aspects of the behaviour of the Magpie Goose *Anseranas semipalmata*. Ibis 105: 76–98.
458 DAVIES, S. J. J. F. 1972. Results of 40 hours' continuous watch at five waterpoints in an Australian desert. Emu 72: 8–12.
459 DAVIES, S. J. J. F. 1976. The natural history of the Emu in comparison with that of other ratites. Proc. 16th Intern. Ornith. Congr.: 109–120.
460 DAVIES, S. J. J. F. 1976. Environmental variables and the biology of Australian arid zone birds. Proc. 16th Intern. Ornith. Congr.: 418–482.
461 DAVIES, S. J. J. F. 1977. Man's activities and birds' distribution in the arid zone. Emu 77: 169–172.
462 DAVIES, S. J. J. F. 1977. The timing of breeding by the Zebra Finch *Taeniopygia castanotis* at Mileura, Western Australia. Ibis 119: 369–372.
463 DAVIES, S. J. J. F. 1978. The food of Emus. Aust. J. Ecol. 3: 411–422.
464 DAVIES, S. J. J. F. 1982. Counting bush birds at Eyre. Eyre Bird Obs. Rep. 1979–81: 43–49.
465 DAVIES, S. J. J. F., M. W. R. BECK and J. P. KRUISKAMP. 1971. The results of banding 154 Emus in Western Australia. C.S.I.R.O. Wildl. Res. 16: 77–79.
466 DAVIES, S. J. J. F. and G. S. CHAPMAN. 1975. The status of birds on Peron Peninsula and Dirk Hartog Island, Shark Bay, W.A. Emu 75: 55–61.
467 DAVIS WRIGHT, W. 1923. Canberra. Sydney: John Andrew & Co.
468 DAWSON, W. R. and C. D. FISHER. 1969. Responses to temperature by the Spotted Nightjar (*Eurostopodus guttatus*). Condor 71: 49–53.
469 DEAN, E. W. 1903. Ground-larks in Tasmania. Emu 2: 174.
470 DEBUS, S. J. S. Pers. comm.
471 DEBUS, S. J. S. 1980. Little and Forest Ravens in New South Wales. Aust. Birds 15: 7–12.
472 DEBUS, S. J. S. 1982. Range and status of the Red Goshawk in N.S.W. Aust. Birds 16: 41–45.
473 DEBUS, S. J. S. 1982. Seasonal changes in raptor numbers near Sydney, N.S.W. Aust. Birds 16: 60–61.

474 DEBUS, S. J. S. 1983. The Square-tailed Kite as a migrant in south-eastern Australia. Aust. Birds 17: 56–58.
475 DEIGNAN, H. G. 1964. Birds of the Arnhem Land Expedition. *In* SPECHT, R. L. (Ed.). Records of the American-Australian Scientific Expedition to Arnhem Land, Vol. 4: 345–425. Melbourne: M.U.P.
476 DELL, J. Pers. comm.
477 DELL, J. 1965. The Red-browed Finch, *Aegintha temporalis* in Western Australia. W. Aust. Nat. 9: 160–169.
478 DELL, J. 1980. White-backed Swallow in the south-west. W. Aust. Nat. 14: 200.
479 DELROY, L. B. 1974. The food of waterfowl (Anatidae) in the southern Coorong saltwater habitat of South Australia. S. Aust. Orn. 26: 157–163.
480 DENNETT, X. 1982. Flame Robin survey. VORG Notes 18: 31–34.
481 DE REBEIRA, P. and P. CONGREVE. 1982. Bird banding at Eyre. Eyre Bird Obs. Report 1979–81: 20–27.
482 DEVILLIERS, P. 1977. The Skuas of the north American Pacific coast. Auk 94: 417–429.
483 DE WARREN, J. J. 1928. The avifauna of the upper reaches of the Macleay River, N.S.W. Emu 28: 111–120.
484 DIAMOND, J. M. 1972. Avifauna of the Eastern Highlands of New Guinea. Cambridge, Massachusetts: Nuttall Orn. Club.
485 DIAMOND, J. M. and A. G. MARSHALL. 1977. Niche shifts in New Hebridean birds. Emu 77: 61–72.
486 DICKISON, D. 1930. The Brown Thornbill. Emu 29: 210.
487 DIETRICH, A. Specimens in Godeffroy Museum, Hamburg. See FINSCH, D. 1872. Uber die von Frau Amalie Dietrich in Australien gesammelten Vogel. Verhandl. zool.-bot. Gesellschaft in Wien 22: 315–340.
488 DISNEY, H. J. de S. 1979. Royal Australasian Ornithologists Union Pilot Atlas Scheme. Corella 2: 97–163.
489 DISNEY, H. J. de S. and A. STOKES. 1976. Birds in pine and native forests. Emu 76: 133–138.
490 D'OMBRAIN, A. F. 1934. The White-browed Wood-Swallow. Emu 33: 292–297.
491 D'OMBRAIN, A. F. 1947. Death of 'Old Tom'. Emu 47: 34.
492 D'OMBRAIN, A. F. 1955. Notes on young Black Bitterns. Emu 55: 216–218.
493 D'OMBRAIN, A. F. and A. GWYNNE. 1962. Buller's Shearwater on Cabbage Tree Island, New South Wales. Emu 61: 274–276.
494 DOMM, S. 1977. Seabirds and waders of Lizard Island area. Sunbird 8: 1–8.
495 DOMM, S. and H. F. RECHER. 1973. The birds of One Tree Island with notes on their yearly cycle and feeding ecology. Sunbird 4: 63–86.
496 DOVE, H. S. 1910. The Dusky Robin *Petroica vittata*. Emu 10: 127–131.
497 DOVE, H. S. 1937. Migrants and others. Emu 37: 61–62.
498 DOW, D. D. Pers. comm.
499 DOW, D. D. 1977. Indiscriminate interspecific aggression leading to almost sole occupancy of space by a single species of bird. Emu 77: 115–121.
500 DOW, D. D. 1978. Breeding biology and development of the young of *Manorina melanocephala*, a communally breeding honeyeater. Emu 78: 207–222.
501 DOW, D. D. 1980. Communally breeding Australian birds with an analysis of distributional and environmental factors. Emu 80: 121–140.
502 DOW, D. D., M. A. DOW, P. G. FISK, B. J. GILL, C. E. MOFFATT, J. D. MOFFATT and M. J. WHITMORE. 1980. Breeding biology of Hall's Babbler. RAOU 78th Annual Congress.
503 DRAFFAN, R. D. W., S. T. GARNETT and G. MALONE. 1983. Birds of the Torres Strait. Emu 83: 207-34.
504 DUCKWORTH, P. E. Pers. comm.
505 DUNLOP, J. N. and G. M. STORR. 1981. Carnac Island, Western Australia. Corella 5: 71–74.
506 DUNLOP, R. R. 1970. Behaviour of the Banded Rail. Sunbird 1: 3–15.
507 DYBBRO, T. 1976. De danske ynglefugles udbredelse. Copenhagen: Dansk Ornithologisk Forening.
508 EADES, D. Pers. comm.
509 EBERHARD, I. H. and R. J. PEARSE. 1981. Conservation of Cape Barren Geese *Cereopsis novaehollandiae* in Tasmania. Aust. Wildl. Res. 8: 147–162.
510 ECKERT, H. J. Pers. comm.
511 ECKERT, J. 1969. The Common Tern in South Australia. S. Aust. Orn. 25: 87–89.
512 ECKERT, J. 1971. Birds of the Franklin Islands and Eyre Island, South Australia. Emu 71: 61–64.
513 ECKERT, J. 1972. Further notes on the Little Quail. S. Aust. Orn. 26: 34–36.
514 ECKERT, J. 1977. The distribution of the Emu Wrens *Stipiturus malachurus* and *S. ruficeps mallee* in South Australia. S. Aust. Orn. 27: 186–187.
515 EDDY, R. J. 1961. Twenty years of Painted Honeyeaters. Aust. Bird Watcher 1: 122–128.
516 EDWARDS, E. O. 1946. Notes from Menangle Park. Emu 46: 232- 234.
517 EDWARDS, E. O. 1948. Have the White-backed Swallows their 'territories'? Emu 47: 314.
518 EDWARDS, H. V. 1921. A south-coastal Selborne. Emu 21: 42–48.
519 EDWARDS, H. V. 1925. Birds of a New South Wales garden. Emu 24: 282–286.
520 EDWARDS, H. V. 1926. The Rock Warbler. Emu 26: 143–144.
521 ELLIOTT, A. J. 1935. Notes on the White-throated Nightjar. Emu 35: 129–132.
522 ELLIOTT, A. J. 1938. Birds of the Moonie River District adjacent to the border of New South Wales with Queensland. Emu 38: 30–49.
523 ELLIOTT, M. W. 1916. Bird life at Dumbleyung, W.A. Emu 15: 179–182.
524 EMISON, W. B. Pers. comm.
525 EMISON, W. B. and W. M. BREN. 1981. Banding of Peregrine Falcon chicks in Victoria, Australia. Emu 80: 288–291.
526 EMISON, W. G., P. B. DUGUESCLIN and I. D. TEMBY. 1981. Management of the Long-billed Corella. *In* RINEY, THANE (Ed.). Wildlife Management in the '80s. Melbourne: Graduate School of Environmental Science, Monash University.
527 ENWRIGHT, W. J. 1936. Random notes. Emu 36: 55–56.
528 ERICKSON, R. 1951. Notes on Rufous Whistlers. Emu 51: 1–10; 153–165.
529 ESTBERGS, J. 1975. Co-operative nesting behaviour in Little Wood-Swallows. W. Aust. Nat. 13: 63–64.
530 ESTBERGS, J. A., R. GARSTONE and A. HERTOG. 1978. Observations of the Eastern Grass Owl near Darwin, N. T. Emu 78: 93–94.
531 EY, A. 1977. Notes on the Streaked Grass-Warbler. Sunbird 8: 20–21.
532 EYLMANN, E. 1911. Die Vogelwelt der Kolonie Sudaustralien. Z. fur Ornith. 59: 93–178; 259–299.

533 FAVALORO, N. J. Pers. comm.
534 FAVALORO, N. J. 1931. Two Australian Flycatchers. Emu 30: 161–168.
535 FAVALORO, N. J. 1931. Notes on *Arses kaupi* and *Arses lorealis*. Emu 30: 241–242.
536 FAVALORO, N. J. 1931. Notes of a trip to the Macpherson Range, south-eastern Queensland. Emu 31: 48–59.
537 FAVALORO, N. J. 1941. Notes on the Blue-and-White Wren. Emu 40: 260–265.
538 FAVALORO, N. 1942. The usurpation of nests, nesting sites and materials. Emu 41: 268–276.
539 FAVALORO, N. 1944. Notes on the two resident Victorian plovers. Emu 43: 145–153.
540 FAVALORO, N. 1944. The White-breasted Sea-Eagle along the Murray Valley. Emu 43: 233–242.
541 FAVALORO, N. 1949. Notes on the Red-capped Dotterel in inland localities. Emu 49: 13–18.
542 FAVALORO, N. J. 1981. The Red Goshawk. Aust. Bird Watcher 9: 44–53.
543 FAVALORO, N. J. and A. McEVEY. 1968. A new species of Australian Grass-Wren. Mem. Nat. Mus. Vic. 28: 1–9.

544 FERRIER, S. Pers. comm.
545 FERRIER, S. 1982. Habitat requirements of the Rufous Scrub-Bird. R.A.O.U. Congr. Abstracts.
546 FIELDING, P. Pers. comm.
547 FINCH, B. W. 1974. Southerly breeding of Little Wood Swallows. S. Aust. Orn. 26: 141.
548 FINCH, B. W. 1975. Sight-record of a species of Swift new for Australia. Emu 75: 44.
549 FINCH, B. 1979. The status of the Barn Swallow *Hirundo rustica* in the Port Moresby district. PNG Bird Soc. Newsl. 151: 5-6.
550 FINCH, B. W. 1980. Bensbach. PNG Bird Soc. Newsl. 171-172: 10-37.
551 FINCH, B. W. 1980. Discovery of the Zitting Cisticola *Cisticola juncidis* in the Bensbach Floodplain (Bulla-Bullamuk), Western Province. PNG Bird Soc. Newsl. 173-174: 13-17.
552 FINCH, B. W. 1981. Sight record of a probable Grey Falcon *Falco hypoleucos* on the Aroa Plains (Central Province). PNG Bird Soc. Newsl. 175-176: 11-13.
553 FINCH, B. W. 1982. Discovery of the Mangrove Robin *Eopsaltria pulverulenta* on the Middle Sepik—a new habitat adaptation. PNG Bird Soc. Newsl. 187-188: 7-8.
554 FINCH, B. W. 1982. Notes on the migration patterns of some common migrants in the Port Moresby area. PNG Bird Soc. Newsl. 189-190: 3-6.
555 FINCH, B. W. and L. and A. HOWELL. 1982. Observations June-August, 1982. PNG Bird Soc. Newsl. 193-194: 37.
556 FINLAYSON, H. H. 1951. Further notes of some birds seen upon islands of the Eyre Peninsula coasts, South Australia, with remarks on the soil erosion of North Gambier Island. S. Aust. Orn. 20: 3-5.
557 FILEWOOD, L. W. C., K. HOUGH, I. C. MORRIS and D. E. PETERS. 1978. Helpers at the nest of the Rainbow Bee-eater. Emu 78: 43-44.
558 FISHER, J. and R. M. LOCKLEY. 1954. Seabirds. London: Collins.
559 FITZHERBERT, K. Pers. comm.
560 FLEAY, D. 1948. Notes on the White-breasted Sea-Eagle. Emu 48: 20-31.
561 FLEAY, D. 1950. Random notes on the White Goshawk. Emu 50: 1-4.
562 FLEAY, D. 1951. The Little Eagle in the Healesville district, Victoria. Emu 51: 57-65.
563 FLEAY, D. 1968. Nightwatchmen of Bush and Plain. Brisbane: Jacaranda Press.
564 FLETCHER, A. W. J. Pers. comm.
565 FLETCHER, A. W. J. and O. M. G. NEWMAN. 1979. Breeding of Fairy Terns at Mortimer Bay, South Arm Peninsula. Tas. Bird Rep. 9: 17-19.
566 FLETCHER, J. A. 1904. Bird notes from Wilmot Tasmania. Part III. Emu 4: 14-17.
567 FLETCHER, J. A. 1911. Honey-eaters of the Cleveland district, Tasmania. Emu 11: 105-109.
568 FLETCHER, J. A. 1913. Occurrence of Cisticola in Tasmania. Emu 12: 171-172.
569 FLETCHER, J. A. 1915. Further field notes on the Emu-Wren *Stipiturus malachurus*. Emu 14: 213-217.
570 FLETCHER, J. A. 1915. Nesting of the Black Cormorant (*Phalacrocorax carbo*) in Tasmania. Emu 15: 114-118.
571 FLETCHER, J. A. 1916. Notes from Springfield, Tasmania. Emu 15: 193-194.
572 FLETCHER, J. A. 1918. The Black Bell-Magpie *Strepera fuliginosa*. Emu 17: 227-228.
573 FLETCHER, J. A. 1922. Field notes on the Black Bell-Magpie *Strepera fuliginosa*. Emu 22: 60-63.
574 FLETCHER, J. A. 1924. Birds of the Steppes. Emu 24: 107-117.
575 FLETCHER, J. A. 1932. Bird Notes from Tas. Emu 32: 119.
576 FLETCHER, T. I. 1965. White-throated Nightjar in Otway Ranges, Victoria. Emu 64: 229-230.
577 FORD, H. A. Pers. comm.
578 FORD, H. 1976. The honeyeaters of Kangaroo Island. S. Aust. Orn. 27: 134-138.
579 FORD, H. A. 1977. The ecology of honey-eaters in South Australia. S. Aust. Orn. 27: 199-203.
580 FORD, H. A. 1978. The Black Honeyeater: nomad or migrant? S. Aust. Orn 27: 263-269.
581 FORD, H. A. 1981. A comment on the relationships between miners *Manorina* spp. in South Australia. Emu 81: 247-250.
582 FORD, H. A. and H. BELL. 1981. Density of birds in eucalypt woodland affected to varying degrees by dieback. Emu 81:202-208.
583 FORD, H. A. and R. HOWE. 1980. The future of birds in the Mount Lofty Ranges. S. Aust. Orn. 28: 85-89.
584 FORD, H. A. and D. C. PATON. 1975. Impoverishment of the avifauna of Kangaroo Island. Emu 75: 155-156.
585 FORD, H. A. and D. C. PATON. 1976. The value of insects and nectar to honeyeaters. Emu 76: 83-84.
586 FORD, H. W. 1918. Birds about the Tanjil River and Ranges, Victoria. Emu 17: 221-223.
587 FORD, J. R. Pers. comm.
588 FORD, J. R. 1958. Seasonal variation in populations of Anatidae at the Bibra Lake District, Western Australia. Emu 58: 31-41.
589 FORD, J. 1961. The increase in abundance of the Bourke Parrot in Western Australia, 1938-60. Emu 61: 211-217.
590 FORD, J. R. 1963. Breeding behaviour of the Yellow-tailed Thornbill in south-western Australia. Emu 63: 185-200.
591 FORD, J. 1965. An early occurrence of the Kelp Gull in south-western Australia. W. Aust. Nat. 9: 172.
592 FORD, J. 1965. New information on the distribution of birds of south-western Australia. W. Aust. Nat. 10: 7-12.
593 FORD, J. 1966. First occurrence of the Little Egret in south-western Australia. Emu 65: 208.
594 FORD, J. 1966. Taxonomy and variation of the Chestnut-shouldered Wrens of Western Australia. Emu 66: 47-57.
595 FORD, J. 1967. Occurrences of the Double-banded Dotterel in Western Australia. W. Aust. Nat. 10: 84-85.
596 FORD, J. 1968. Distribution of the Masked Owl in Western Australia. W. Aust. Nat. 11: 21-22.
597 FORD, J. R. 1970. Distribution of Quail-Thrushes in the Northern Territory and their taxonomic relations. Emu 70: 135-139.
598 FORD, J. 1970. Variation in the *Sericornis frontalis maculatus* sub-species group. Emu 70: 168-172.
599 FORD, J. 1971. Distribution and taxonomy of southern birds in the Great Victoria Desert. Emu 71: 27-36.
600 FORD, J. 1971. Distribution, ecology and taxonomy of some Western Australian passerine birds. Emu 71: 103-120.
601 FORD, J. 1974. Taxonomic significance of some hybrid and aberrant-plumaged Quail-Thrushes. Emu 74: 80-90.
602 FORD, J. 1974. Speciation in Australian birds adapted to arid habitats. Emu 74: 161-168.
603 FORD, J. 1975. Hybridization of Splendid and Turquoise Wrens. Emu 75: 153-154.
604 FORD, J. 1975. Systematics and hybridization of Figbirds *Sphecotheres*. Emu 75: 163-171.
605 FORD, J. 1976. Systematics and speciation in the Quail-Thrushes of Australia and New Guinea. Proc. 16th Intern. Ornith. Congr. 542-556.
606 FORD, J. 1977. Taxonomic status of the Spotted Catbird on Cape York Peninsula. Sunbird 8: 61-64.
607 FORD, J. 1977. Sympatry in Hall's and White-browed Babblers in New South Wales. Emu 77: 40.
608 FORD, J. 1978. Geographical isolation and morphological and habitat differentiation between birds of the Kimberley and the Northern Territory. Emu 78: 25-35.
609 FORD, J. 1978. Intergradation between the Varied and Mangrove Honeyeaters. Emu 78: 71-74.
610 FORD, J. 1978. Conspecificity of the Black-throated and Fairy Warblers. Emu 78: 75-79.
611 FORD, J. 1978. Subspeciation in the White-throated Warbler of Australia and New Guinea. Emu 78: 90-92.
612 FORD, J. 1978. Hybridization between the White-vented and Black-vented forms of the Black-faced Wood-Swallow. Emu 78: 105-114.
613 FORD, J. 1979. Speciation or subspeciation in the Yellow Robins? Emu 79: 103-106.

614 FORD, J. 1979. A new subspecies of Grey Butcherbird from the Kimberley, Western Australia. Emu 79: 191–194.
615 FORD, J. 1979. Subspeciation, hybridization and relationships in the Little Shrike-thrush *Colluricincla megarhyncha* of Australia and New Guinea. Emu 79: 195–210.
616 FORD, J. 1980. Hybridization between contiguous subspecies of the Varied Sittella in Queensland. Emu 80: 1–12.
617 FORD, J. 1980. Morphological and ecological divergence and convergence in isolated populations of the Red-tailed Black-Cockatoo. Emu 80: 103–120.
618 FORD, J. 1981. Morphological and behavioural evolution in populations of the *Gerygone fusca* complex. Emu 81: 57–81.
619 FORD, J. 1981. Evolution, distribution and stage of speciation in the *Rhipidura fuliginosa* complex in Australia. Emu 81: 128–144.
620 FORD, J. 1981. Geographical variation in *Cinclosoma castanotum* and its historical significance. Emu 81: 185–192.
621 FORD, J. 1981. Hybridization and migration in Australian populations of the Little and Rufous-breasted Bronze-cuckoos. Emu 81: 209–222.
622 FORD, J. 1982. Origin, evolution and speciation of birds specialized to mangroves in Australia. Emu 82: 12–23.
623 FORD, J. R. 1982. Hybrid phenotypes in male Figbirds *Sphecotheres viridis* in Queensland. Emu 82: 126–130.
624 FORD, J. R. 1983. Speciation in the Ground Thrush complex *Zoothera dauma* in Australia. Emu 83: 152–172.
625 FORD, J. and S. A. PARKER. 1973. A second species of Wedgebill? Emu 73: 113–118.
626 FORD, J. and S. A. PARKER. 1974. Distribution and taxonomy of some birds from south-western Queensland. Emu 74: 177–194.
627 FORD, J. and P. S. STONE. 1957. Birds of the Kellerberrin/Kwolyin District, Western Australia. Emu 57: 9–21.
628 FORD, J. and B. V. TEAGUE. 1959. Extension of range in the White-breasted Robin. Emu 59: 37–39.
629 FORD, J. and B. V. TEAGUE. 1959. Observations between the Moore and Murchison Rivers, Western Australia. Emu 59: 89–92.
630 FORDYCE, J. C. 1978. Flame Robin banding in the Shepparton district, Victoria. Corella 2: 85–88.
631 FORSHAW, J. M. Pers. comm.
632 FORSHAW, J. M. 1964. Some field observations on the Great Palm Cockatoo. Emu 63: 327–331.
633 FORSHAW, J. M. 1979. Eggs of the Double-eyed Fig Parrot. Aust. Bird Watcher 8: 98–99.
634 FORSHAW, J. M. and W. T. COOPER. 1981. Parrots of the World. (2nd (Rev.) Ed.). Melbourne: Lansdowne.
635 FORSHAW, J. M. and W. T. COOPER. 1981. Australian Parrots. (2nd (Rev.) Ed.). Melbourne: Lansdowne.
636 FOSTER, F. P. Pers. comm.
637 FRANKLIN, D. C. Pers. comm.
638 FRASER, G. C. and G. J. MENDEL. 1976. The Bush-Hen in New South Wales. Aust. Birds 11: 25–27.
639 FRASER, T. 1982. Black Falcons feeding gregariously. Aust. Raptor Assoc. Newsl. 3(2): 10.
640 FRASER, T. 1982. Black Falcons eating carrion. Aust. Raptor Assoc. Newsl. 3(2): 12.
641 FREEMAN, D. J. 1970. The Rediscovery of the Black Grass Wren *Amytornis housei* with additional notes on this species. Emu 70: 193–195.
642 FREEMAN, D. J. 1971. The nest of the White-lined Honeyeater. Emu 71: 86–87.
643 FRITH, C. B. and D. W. FRITH. Pers. comm.
644 FRITH, C. B. and D. W. FRITH. 1979. Leaf-eating by Birds-of-Paradise and Bowerbirds. Sunbird 10: 21–23.
645 FRITH, H. J. 1952. Notes on the pigeons of the Richmond River, N.S.W. Emu 52: 89–99.
646 FRITH, H. J. 1952. A record of the Turquoise Parrakeet. Emu 52: 99–101.
647 FRITH, H. J. 1956. Breeding habits in the family Megapodiidae. Ibis 98: 620–640.
648 FRITH, H. J. 1957. Food habits of the Topknot Pigeon. Emu 57: 341–345.
649 FRITH, H. J. 1959. The ecology of wild ducks in inland New South Wales: II. Movements. III. Food habits. IV. Breeding. CSIRO Wildl. Res. 4: 108–130; 131–155; 156–181.
650 FRITH, H. J. 1962. The Mallee Fowl. Sydney: Angus & Robertson.
651 FRITH, H. J. 1963. Movements and mortality rates of the Black Duck and Grey Teal in south-eastern Australia. CSIRO Wildl. Res. 8: 119–131.
652 FRITH, H. J. 1965. Ecology of the Freckled Duck *Stictonetta naevosa* (Gould). CSIRO Wildl. Res. 10: 125–139.
653 FRITH, H. J. (Ed.) 1969. Birds in the Australian High Country. Sydney: Reed.
654 FRITH, H. J. 1970. The Japanese Snipe, *Gallinago hardwickii*: a preliminary report. CSIRO Wildl. Res. Tech. Mem. No. 6: 1–26.
655 FRITH, H. J. 1972. Nesting of the Black-banded Pigeon and the Australian Rock Pigeons. Emu 72: 13–16.
656 FRITH, H. J. 1977. Band recoveries of the Magpie Goose *Anseranas semipalmata*. Aust. Wildl. Res. 4: 81–84.
657 FRITH, H. J. 1979. Wildlife Conservation. (Rev. Ed.). Sydney: Angus & Robertson.
658 FRITH, H. J. 1982. Waterfowl in Australia. (Rev. Ed.). Sydney: Angus & Robertson.
659 FRITH, H. J. 1982. Pigeons and Doves of Australia. Adelaide: Rigby.
660 FRITH, H. J. and R. D. BARKER. 1975. Food of the Plumed Pigeons *Geophaps plumifera* and *G. ferruginea*. Aust. Wildl. Res. 2: 63–76.
661 FRITH, H. J., L. W. BRAITHWAITE and J. L. McKEAN. 1969. Waterfowl in an inland swamp in New South Wales. II. Food. CSIRO Wildl. Res. 14: 17–64.
662 FRITH, H. J., B. K. BROWN and A. K. MORRIS. 1977. Food habits of the Stubble Quail, *Coturnix pectoralis*, in south-eastern Australia. CSIRO Wildl. Res. Tech. Pap. 32: 1–70.
663 FRITH, H. J. and J. H. CALABY. 1953. The Superb Parrot in southern New South Wales. Emu 53: 324–330.
664 FRITH, H. J. and J. H. CALABY. 1974. Fauna survey of the Port Essington District, Cobourg Peninsula, Northern Territory of Australia. CSIRO Wildl. Res. Tech. Pap. 28: 1–208.
665 FRITH, H. J. and S. M. CARPENTER. 1980. Breeding of the Stubble Quail, *Coturnix pectoralis*, in south-eastern Australia. Aust. Wildl. Res. 7: 117–137.
666 FRITH, H. J., F. H. J. CROME and B. K. BROWN. 1977. Aspects of the biology of the Japanese Snipe *Gallinago hardwickii*. Aust. J. Ecol. 2: 341–368.
667 FRITH, H. J., F. H. J. CROME and T. O. WOLFE. 1976. Food of Fruit-pigeons in New Guinea. Emu 76: 49–58.
668 FRITH, H. J. and S. J. J. F. DAVIES. 1961. Ecology of the Magpie Goose, *Anseranas semipalmata*, Latham (Anatidae). CSIRO Wildl. Res. 6: 91–141.
669 FRITH, H. J. and J. L. McKEAN. 1975. Races of the introduced Spotted Turtle-dove *Streptopelia chinensis* (Scopoli), in Australia. Aust. J. Zool. 23: 295–306.
670 FRITH, H. J., J. L. McKEAN and L. W. BRAITHWAITE. 1976. Sexual cycles and food of the doves *Streptopelia chinensis* and *S. senegalensis* in Australia. Emu 76: 15–24.
671 FRITH, H. J. and M. H. WATERMAN. 1977. Movements of Stubble Quail *Coturnix pectoralis* from South Australian grainfields. Aust. Wildl. Res. 4: 85–90.
672 FRITH, H. J., T. O. WOLFE and R. D. BARKER. 1976. Food of eight species of Columbidae in the genera *Geopelia*, *Phaps*, *Geophaps* and *Petrophassa*. Aust. Wildl. Res. 3: 159–171.
673 FULLAGAR, P. J. 1976. Cabbage Tree Island, N.S.W. Aust. Bird Bander 14: 94–97.
674 FULLAGAR, P. 1977. Beach mortality of Short-tailed Shearwaters. Aust. Seabird Group Newsl. 8: 46.
675 FULLAGAR, P. J. 1978. Eclipse Island, Western Australia. Corella 2: 21–23.
676 FULLAGAR, P. J., J. L. McKEAN and G. F. VAN TETS. 1974. Report on the birds. *In* RECHER, H. F. and S. S. CLARK (Eds). Environmental survey of Lord Howe Island. A report to the Lord Howe Island Board. Sydney: Australian Museum viii 86: 55–72.

677 FULLAGAR, P. J. and G. F. VAN TETS. 1976. Bird notes from a winter visit to Eclipse Island, Western Australia. W. Aust. Nat. 13: 136–144.
678 FULLER, P. J. and A. A. BURBIDGE. 1981. The birds of Pelsart Island, Western Australia. W.A. Dept of Fisheries and Wildl. Rep. 44: 1–41.

679 GALBRAITH, I. C. J. 1967. The Black-tailed and Robust Whistlers *Pachycephala melanura* as a species distinct from the Golden Whistler. Emu 66: 289–292.
680 GALBRAITH, I. C. J. 1969. The Papuan and Little Cuckoo-shrikes *Coracina papuensis* and *robusta* as races of a single species. Emu 69: 9–29.
681 GALBRIATH, I. C. J. and S. A. PARKER. 1969. The Atherton Scrub-wren *Sericornis keri* Mathews: a neglected Australian species. Emu 69: 212–232.
682 GALL, B. C. and N. W. LONGMORE. 1978. Avifauna of the Thredbo Valley, Kosciusko National Park. Emu 78: 189–196.
683 GANNON, G. R. 1962. Distribution of the Australian honeyeaters. Emu 62: 145–166.
684 GANNON, G. R. 1966. The influence of habitat on the distribution of Australian birds. Emu 65: 241–253.
685 GARNETT, S. T. Pers. comm.
686 GARNETT, S. T. 1980. The social organization of the Dusky Moorhen *Gallinula tenebrosa* Gould (Aves: Rallidae). Aust. Wildl. Res. 7: 103–112.
687 GENELLY, R. E. 1978. Observations of the Australian Kestrel on northern tablelands of New South Wales, 1975. Emu 78: 137–144.
687a GIBSON, A. 1981. Great-billed Heron at Yamba. Aust. Birds 16: 32–33.
688 GIBSON, C. G. 1908. Notes on some birds of the Abrolhos Islands (W.A.). Emu 8: 64–66.
689 GIBSON, J. D. 1961. Colombo Crows in Victoria. Emu 61: 244–245.
690 GIBSON, J. D. 1977. The birds of the County of Camden. Aust. Birds 11: 41–45.
691 GIBSON, J. D. and A. SEFTON. 1953. Notes on the desertion of nests by Fantail-Warblers. Emu 53: 74–77.
692 GILBERT, P. A. 1934–35. The seasonal movements and migrations of birds in eastern New South Wales. Emu 34: 101–105; 200–209; 35: 17–27.
693 GILBERT, P. A. 1936. The Topknot Pigeon. Emu 35: 301–312.
694 GILBERT, P. A. and H. KEANE. 1913. The Black-eared Cuckoo *Mesocalius palliolatus*, Lath. Emu 13: 80–82.
695 GILL, H. B. Pers. comm.
696 GILL, H. B. 1964. The White-tailed Kingfisher. Emu 63: 273–276.
697 GILL, H. B. 1969. First record of the Sarus Crane in Australia. Emu 69: 49–52.
698 GILL, H. B. 1970. Birds of Innisfail and hinterland. Emu 70: 105–116.
699 GILL, H. B. 1971. Further records of Sarus Cranes in northern Queensland. Emu 71: 140–141.
700 GILL, R. G. 1965. Some observations on the Red-necked Rail. Emu 64: 321–322.
701 GIVENS, T. V. 1948. Notes on the White-browed Crake. Emu 48: 141–147.
702 GIVENS, T. V. and W. B. HITCHCOCK. 1953. *Cisticola juncidis* (Raf.) in the Northern Territory. Emu 53: 193–200.
703 GLAUERT, L. 1945. A Western Australian Grass Owl. Emu 44: 229–230.
704 GLOVER, B. 1965. Bird Report 1964. S. Aust. Orn. 24: 55–68.
705 GLOVER, B. 1966. Little Wood-Swallow at Lake Merreti. S. Aust. Orn. 24: 111.
706 GLOVER, B. 1966. Burdekin Duck—report from south-east. S. Aust. Orn. 24: 116.
707 GLOVER, B. 1968. Bird Report 1966–67. S. Aust. Orn. 25: 27–45.
708 GLOVER, B. 1971. Bird Report 1968–69. S. Aust. Orn. 25: 219–233.
709 GLOVER, B. 1972. Bird Report 1969–70. S. Aust. Orn. 26: 26–33.
710 GLOVER, B. 1973. Bird Report 1971–72. S. Aust. Orn. 26: 121–125.
711 GLOVER, B. 1975. Bird Report 1972–73. S. Aust. Orn. 26: 175–180.
712 GLOVER, B. 1976. The Little Bittern in South Australia. S. Aust. Orn. 27: 100–101.
713 GLOVER, B. and R. SCHODDE. 1956. Additional South Australian Records. S. Aust. Orn. 22: 20–22.
714 GODDARD, M. T. Pers. comm.
715 GODDARD, M. T. 1955. Notes on the breeding of the Cattle Egret in north-eastern New South Wales. Emu 55: 275–277.
716 GOGERLEY, J. F. H. 1922. Drongo Shrike and hornets. Emu 21: 315.
717 GOOD, R., A. M. McGARVIE, O. M. G. NEWMAN and D. PINNER. 1973. White-browed and Masked Wood-Swallows in Tasmania. Tas. Nat. 1973: 2–4.
718 GOODFELLOW, R. J. 1966. Burdekin Duck in northern South Australia. S. Aust. Orn. 24: 116.
719 GOODWIN, D. 1967. Notes on behaviour of some Australian birds. Emu 66: 237–251.
720 GOODWIN, D. 1982. Estrildid Finches of the World. Oxford: O.U.P.
721 GOSPER, D. 1964. Observations on the breeding of the Koel. Emu 64: 39–41.
722 GOSPER, D. G. 1976. The Nutmeg Mannikin on the north coast of N.S.W. Aust. Birds 11: 8–11.
723 GOULD, J. 1865. Handbook to the Birds of Australia. London: Gould.
724 GOWLAND, P. N. 1983. A guide to the ecology and management of bird pests of commercial agriculture in the Ord River Irrigation area. No. 2. Waterbirds. Unpub. MS.
725 GRAY, J. M. 1933 Notes on the Eastern Grass-Owl. Emu 33: 139–140.
726 GREEN, G. A. 1972. Fifth Abrolhos Expedition, 1970. Aquinas College, Manning, W. A.
727 GREEN, R. H. Pers. comm.
728 GREEN, R. 1956. Notes on the inland breeding of the Red-capped Dotterel. Emu 56: 140–143.
729 GREEN, R. H. 1959. A Tasmanian nesting note on the White-breasted Sea-Eagle. Emu 59: 215–217.
730 GREEN, R. H. 1965. Impact of rabbit on predators and prey. Tas. Nat. 3: 1–2.
731 GREEN, R. H. 1966. Gut contents of some Tasmanian birds. Emu 66: 105–110.
732 GREEN, R. H. 1969. The birds of Flinders Island. Rec. Queen Vict. Mus. 34: 1–32.
733 GREEN, R. H. 1973. Albatross Island, Tasmania. Aust. Bird Bander 11: 81–83.
734 GREEN, R. H. 1977. Birds of Tasmania. Launceston: The Author.
735 GREEN, R. H. 1980. The food of the White-browed Scrubwren. Tas. Nat. 60: 9–10.
736 GREEN, R. H. 1981. Barn Owls in Tasmania. Bird Watcher 9: 94–95.
737 GREEN, R. H. 1982. A Red-tipped *Pardalotus striatus* at Maggs Mountain. Tas. Nat. 70: 6–7.
738 GREEN, R. H. and D. MacDONALD. 1963. The Black Pyramid gannetry. Emu 63: 177–184.
739 GREEN, R. H. and A. M. McGARVIE. 1971. The birds of King Island. Rec. Queen Vict. Mus. 40: 1–42.
740 GREEN, R. H. and B. C. MOLLISON. 1961. Birds of Port Davey and south coast of Tasmania. Emu 61: 223–236.
741 GRIFFIN, A. C. M. Pers. comm.
742 GRIFFIN, A. C. M. 1974. Birds of Mount Spec. Sunbird 5: 29–39.
743 GUILER, E. R. 1967. The Cape Barren Goose, its environment, numbers and breeding. Emu 66: 211–235.
744 GWYNNE, A. J. 1937. Notes on the Green Catbird. Emu 37: 76–78.
745 GWYNNE, A. J. 1948. Notes on the Brown Honeyeater. Emu 47: 161–162.

746 HAINES, L. C. 1943. Unusual nest of King Quails. Emu 42: 252.
747 HAINES, L. C. 1968. Field Notes. Birds 3:13–14.

748 HALL, B. P. (Ed.) 1974. Birds of the Harold Hall Australian Expeditions. London: Brit. Mus. (Nat. Hist.).
749 HALL, L. Pers. comm.
750 HALL, R. 1907. Notes on a collection of birds from the Townsend River, north-western Australia. Emu 7: 25–27.
751 HALL, R. 1909. Notes on ducks. Emu 9: 77–79.
752 HALL, R. and J. P. ROGERS. 1908. Notes on a collection of birds from north-west Australia. Emu 7: 138–142.
753 HALSE, S. A. 1978. Feeding habits of six species of honeyeater in south-western Australia. Emu 78: 145–148.
754 HALSE, S. A. 1981. Migration by Hutton's Shearwater. Emu 81: 42–44.
755 HAMILTON-SMITH, E. 1965. Scrub Turkey at Chillagoe, Queensland. Emu 65: 118.
756 HANCOCK, J. and H. ELLIOTT. 1978. The Herons of the World. London: London Editions.
757 HANKS, E. S. 1959. Cuckoos and the parasitic habit. Emu 59: 250–258.
758 HANKS, P. 1930. Notes on the Eastern Whipbird. Emu 30: 84–87.
759 HANSCOMBE, S. A. 1915. Observations on the Nankeen Night-Heron (*Nycticorax caledonicus*). Emu 15: 132–134.
760 HARDY, A. D. 1928. Particulars regarding acclimatization experiments with the Nightingale and Skylark in Australia. Argus (Melbourne). 7 January 1928.
761 HARPER, P. C. 1972. The field identification and distribution of the Thin-billed Prion (*Pachyptila belcheri*) and the Antarctic Prion (*Pachyptila desolata*). Notornis 19: 140–175.
762 HARPER, P. C. 1980. The field identification and distribution of the Prions (genus *Pachyptila*), with particular reference to the identification of storm-cast material. Notornis 27: 235–286.
763 HARPER, P. C., G. E. WATSON and J. P. ANGLE. 1972. New records of the Kerguelen Petrel (*Pterodroma brevirostris*) in the south Atlantic and Pacific Oceans. Notornis 19: 56–60.
764 HARRIS, A. 1981. Brolgas in danger, says study. W. Australian, 7 Dec. 1981.
765 HARRIS, J. G. K. and O. M. G. NEWMAN. 1974. Communal nesting behaviour in the genus *Sericornis*. Tas. Nat. 37: 4–5.
766 HARRIS, M. P. and K. G. BODE. 1981. Populations of Little Penguins, Short-tailed Shearwaters and other seabirds on Phillip Island, Victoria, 1978. Emu 81: 20–28.
767 HARRIS, M. P. and D. M. DEERSON. 1980. Notch Island, Seal Islands Group, Victoria. Corella 4: 71–72.
768 HARRIS, M. P. and F. I. NORMAN. 1981. Distribution and status of coastal colonies of seabirds in Victoria. Mem. Nat. Mus. Vict. 42: 89–106.
769 HARRISON, C. J. O. 1968. Some comparative notes on the Common and Yellow Silver-eyes. Emu 68: 127–131.
770 HARRISON, C. J. O. 1969. Emu-Wrens and the forest zone in south-western Australia. W. Aust. Nat. 11: 66–67.
771 HARRISON, C. J. O. 1970. The relationship of the Brown and Black-backed Treecreepers. Emu 70: 9–11.
772 HARRISON, C. J. O. 1973. The origin of the apparent speciation in eastern central Australia. Emu 73: 138–139.
773 HARRISON, C. J. O. 1975. The Australian subspecies of Lewin's Rail. Emu 75: 39–40.
774 HARRISON, C. J. O. 1976. Some aspects of adaptation and evolution in Australian Fan-tailed Flycatchers. Emu 76: 115–119.
775 HARRISON, C. J. O. and M. P. WALTERS. 1973. Use of nests of other species by the Trumpetbird. Emu 73: 189–190.
776 HARRISON, P. 1983. Seabirds. Beckenham, Kent: Croom Helm.
777 HEATHCOTE, W. 1931. Notes on Dusky Wood-Swallows. Emu 31: 168.
778 HEATHCOTE, W. 1933. Black-faced Cuckoo-Shrike hovering. Emu 33: 66.
779 HEATHER, B. D. 1978. The Cattle Egret in New Zealand in 1977. Notornis 25: 218–234.
780 HEATHER, B. D. 1982. The Cattle Egret in New Zealand, 1978–1980. Notornis 29: 241–268.
781 HENDERSON, D. G. 1981. The Banded Stilt in Tasmania. Tas. Bird Report 1981, 11: 14–15.
782 HENDERSON, R. C. 1977. Breeding records of the Scarlet-chested Parrot in the Great Victoria Desert, and further records in the lower north-east. S. Aust. Orn. 27: 222–223.
783 HERON, S. J. 1970. The birds of Middle Swan, Western Australia. Emu 70: 155–158.
784 HERON, S. J. 1973. Birds of the Orange District, N.S.W. Emu 73: 1–8.
785 HEUMAN, G. A. 1926. Birds in the Northern Territory and the new finch. Emu 25: 134–136.
786 HEWITT, J. M. 1960. The Cattle Egret in Australia. Emu 60: 99–102.
787 HILL, G. F. 1911. Field notes on the birds of Kimberley, north-west Australia. Emu 10: 258–290.
788 HILL, G. F. 1913. Ornithological notes, Barclay Expedition. Emu 12: 238–262.
789 HINDWOOD, K. A. 1931. Historical associations and early records of the Emu-Wren. Emu 31: 99–110.
790 HINDWOOD, K. A. 1933. The Green-backed Mangrove Heron. Emu 33: 27–43.
791 HINDWOOD, K. A. 1934. The Spine-tailed Logrunner *Orthonyx temminckii*. Emu 33: 257–267.
792 HINDWOOD, K. A. 1935. Birds inhabiting mangroves in the neighbourhood of Sydney. Emu 34: 181–189.
793 HINDWOOD, K. A. 1937. Painted Quail. Emu 37: 60–61.
794 HINDWOOD, K. A. 1940. Notes on the distribution and habits of the Jacana or Lotus-bird. Emu 39: 261–267.
795 HINDWOOD, K. A. 1946. The White-fronted Tern (*Sterna striata*) in Australia. Emu 45: 179–200.
796 HINDWOOD, K. A. 1946. The Mangrove Warbler—extension of range. Emu 45: 311–314.
797 HINDWOOD, K. A. 1947. The Tawny Grass-bird. Emu 47: 29–34.
798 HINDWOOD, K. A. 1947. Occurrence of the Red-tailed Tropic-bird in south-eastern Australia and Tasmania. Emu 47: 57–58.
799 HINDWOOD, K. A. 1947. Occurrence of the Crested Pigeon near Sydney, N.S.W. Emu 47: 63–64.
800 HINDWOOD, K. A. 1948. The Rufous Fantail in city buildings. Emu 48: 31–32.
801 HINDWOOD, K. A. 1948. The Blue-faced Finch in Australia. Emu 48: 53–56.
802 HINDWOOD, K. A. 1949. Waders at Boat Harbour near Sydney. Emu 48: 323–324.
803 HINDWOOD, K. A. 1950. Breeding of the White-plumed Honeyeater near Sydney. Emu 49: 211–213.
804 HINDWOOD, K. A. 1953. Channel-billed Cuckoo in New Caledonia. Emu 53: 334–335.
805 HINDWOOD, K. A. 1954. The distribution of the Coucal *Centropus phasianinus* in Australia. Emu 54: 1–6.
806 HINDWOOD, K. A. 1956. The Mangrove Honeyeater in New South Wales. Emu 56: 353–356.
807 HINDWOOD, K. A. 1956. The migration of the White-naped and Yellow-faced Honeyeaters. Emu 56: 421–425.
808 HINDWOOD, K. A. 1959. The nesting of birds in the nests of social insects. Emu 59: 1–36.
809 HINDWOOD, K. A. 1959. The Mangrove Honeyeater—an extension in range. Emu 59: 57–58.
810 HINDWOOD, K. A. 1959. The Purple-crowned Pigeon in south-eastern Australia. Emu 59: 219–220.
811 HINDWOOD, K. A. 1965. Brown Songlark at sea. Emu 65: 56.
812 HINDWOOD, K. A. and E. S. HOSKIN. 1954. The waders of Sydney (County of Cumberland), New South Wales. Emu 54: 217–255.
813 HINDWOOD, K. A. and A. R. McGILL. 1951. The 'Derra Derra' 1950 camp-out of the RAOU. Emu 50: 217–238.
814 HINDWOOD, K. A. and A. R. McGILL. 1951. Tawny Grass-bird *Megalurus timoriensis*: A correction. Emu 51: 173–174.
815 HINDWOOD, K. A. and A. R. McGILL. 1953. The Wood-Sandpiper in Australia. Emu 53: 1–13.
816 HINDWOOD, K. A. and A. R. McGILL. 1958. The Birds of Sydney: Sydney: Royal Zool. Soc. NSW.
817 HINDWOOD, K. A. and E. MAYR. 1946. A revision of

the Striped-crowned Pardalotes. Emu 46: 49–67.
818 HINDWOOD, K. A. and M. SHARLAND. 1964. The Swift Parrot. Emu 63: 310–326.
819 HINES, M. P. 1962. Seabird remains on Stradbroke Island, Qld. Emu 62: 62–63.
820 HITCHCOCK, W. B. 1952. New and rare Victorian records. Emu 52: 273–284.
821 HITCHCOCK, W. B. 1954. Occurrence of the Arctic Tern in South Australia. S. Aust. Orn. 21: 45–46.
822 HITCHCOCK, W. B. 1959. A review of the "Least" Terns in Australian waters. S. Aust. Orn. 22: 87–106.
823 HITCHCOCK, W. B. 1964. Ninth Annual Report of the Australian Bird-banding Scheme. CSIRO Wildl. Res. Tech. Pap. 7.
824 HITCHCOCK, W. B. 1965. Geography and seasonal movements of the Common Tern in Australia. Emu 64: 157–171.
825 HITCHCOCK, W. B. 1966. Tenth Annual Report of the Australian Bird-banding Scheme. CSIRO Wildl. Res. Tech. Pap. 11.
826 HITCHCOCK, W. B. and R. CARRICK. 1960. Fifth Annual Report of the Australian Bird-banding Scheme. CSIRO Wildl. Res. 5: 52–84.
827 HITCHCOCK, W. B. and J. L. McKEAN. 1969. Square-tailed Kite in the Northern Territory. Emu 69: 115.
828 HOBBS, J. N. Pers. comm.
829 HOBBS, J. N. 1956. Tree Sparrow in the Riverina. Emu 56: 434.
830 HOBBS, J. N. 1957. Tree Sparrow in the Riverina. Emu 57: 200.
831 HOBBS, J. N. 1958. Some notes on Grebes. Emu 58: 129–132.
832 HOBBS, J. N. 1959. Migratory movements of the Swamp Harrier. Emu 59: 87–88.
833 HOBBS, J. N. 1961. The birds of south-west New South Wales. Emu 61: 21–55.
834 HOBBS, J. N. 1967. Honeyeaters and emu bush. Emu 66: 386–388.
835 HOBBS, J. N. 1970. Southern Emu-Wren at high altitude. Emu 70: 142.
836 HOBBS, J. N. 1971. A plague of mice at Warren. Aust. Bird Watcher 4: 43–46.
837 HOBBS, J. N. 1971. Use of tools by the White-winged Chough. Emu 71: 84–85.
838 HOBBS, J. N. 1972. Breeding of Red-capped Dotterel at Fletcher's Lake, Dareton, NSW. Emu 72: 121–125.
839 HOBBS, J. N. 1974. A Western Sandpiper in New South Wales. Aust. Birds 9: 21–24.
840 HOBBS, J. N. 1976. Birds feeding on house mice. Emu 76: 219–220.
841 HOBBS, J. N. 1979. An irruptive extension of range of the Golden-headed Cisticola. Aust. Birds 14: 25–29.
842 HOBBS, J. N. and M. KAVENEY. 1962. Notes on the birds of the central coast area of New South Wales. Emu 61: 295–300.
843 HOGAN, J. 1925. Bird notes from Willis Island. Emu 24: 266–275.
844 HOLLAND, L. M. 1972. Field notes on the Green-backed Honeyeater. Aust. Bird Watcher 4: 226–233.
845 HOLLANDS, D. L. G. 1979. The Letter-winged Kite irruption of 1976–77. Aust. Bird Watcher 8: 1–18.
846 HOLMES, G. Pers. comm.
847 HOLMES, G. 1975. The Australian status of the Grey-backed Shearwater. Aust. Birds 9: 98.
848 HOLMES, G. 1976. Post-breeding dispersal of the Grey Noddy. Aust. Birds 10: 50–51.
849 HOLMES, G. 1977. The ecology of Petrels in mid-northern NSW. Aust. Seabird Group Newsl. 8: 20–35.
850 HOLMES, G. 1981. Field identification of the Marbled Frogmouth. Aust. Birds 15: 74–75.
851 HOLMES, G. 1981. Occurrence and identification of the Tahiti Petrel in central eastern Australia. Aust. Birds 16: 24.
852 HOLSWORTH, W. N. 1965. Nesting success of the Osprey on Rottnest Island, Western Australia. W. Aust. Nat. 10: 13–15.
853 HOLYOAK, D. T. 1979. Notes on the birds of Viti Levu and Taveuni, Fiji. Emu 79: 7–18.
854 HONE, J. 1978. Introduction and spread of the Common Myna in New South Wales. Emu 78: 227–230.
855 HOOGERWERF, A. C. 1962. Enkele gegevens over het vogelschade-onderzoek bij het Rijstproefbedrijf "Koembe" te Koerik (nabij Merauke) Zuid Nieuw-Guinea. Bull. Agr. Res. Station Manokwari Agr. Series. No. 7.
856 HOOGERWERF, A. 1971. On a collection of birds from the Vogelkop, near Manokwari, north-western New Guinea. Emu 71: 1–12; 73–83.
857 HOPKINS, N. 1948. Birds of Townsville and district. Emu 47: 331–347.
858 HOPKINS, N. 1949. Sea-Eagles harrying flying foxes. Emu 49: 52.
859 HOPKINS, N. 1957. Blue-winged Kookaburra and snake. Emu 57: 145–146.
860 HOPPER, S. D. 1980. Pollen and nectar feeding by Purple-crowned Lorikeets on *Eucalyptus occidentalis*. Emu 80: 239–240.
861 HOPPER, S. D. and A. A. BURBIDGE. 1979. Feeding behaviour of a Purple-crowned Lorikeet on flowers of *Eucalyptus buprestium*. Emu 79: 40–42.
862 HORTON, W. 1975. The birds of Mt. Isa. Sunbird 6: 49–69.
863 HOSKIN, E. S. Pers. comm.
864 HOSKIN, E. S. and K. A. HINDWOOD. 1964. Fairy Tern near Sydney, N.S.W. Emu 63: 335–336.
865 HOUSTON, C. 1980. Recent records of the Grey-crowned Babbler. S. Aust. Orn. 28: 159–160.
866 HOWCHIN, W. 1926. Some references to the literature concerning the extinct Emus of Kangaroo Island and elsewhere. S. Aust. Orn. 8: 244–253.
867 HOWE, D. and W. DODD. 1973. Common Noddy in Northern Territory. Emu 73: 140.
868 HOWE, F. E. 1915. Notes on the genus *Pycnoptilus*. Emu 14: 192–199.
869 HOWE, F. E. 1917. Observations on the genus *Hylacola* (Ground Wrens). Emu 17: 87–94.
870 HOWE, F. E. 1921. The genus *Climacteris* (Tree-creepers). Emu 21: 32–41.
871 HOWE, F. E. 1928. Notes on some Victorian birds. Emu 27: 252–265.
872 HOWE, F. E. 1932. Observations on the genus *Petroica* with notes on nidification. Emu 32: 39–41.
873 HOWE, F. E. 1933. The Mallee Emu-Wren *Stipiturus mallee*. Emu 32: 266–269.
874 HOWE, F. E. and W. BURGESS. 1942. Ornithologists in the mallee. Emu 42: 65–73.
875 HOWE, R. W., T. D. HOWE and H. A. FORD. 1981. Bird distributions on small rainforest remnants in NSW. Aust. Wildl. Res. 8: 637–651.
876 HOWE, R. W. and R. A. NOSKE. 1980. Co-operative feeding of fledgelings by Crested Shrike-tits. Emu 80: 40.
877 HUDDY, L. 1979. Social behaviour and feeding ecology of Scarlet Robins *Petroica multicolor*. B.Sc. (Hons) Thesis. University of New England.
878 HUDSON, R. and J. W. H. CONROY. 1975. Southern Giant-petrel in New Guinea. Emu 75: 43–44.
879 HUGHES, B. and P. HUGHES. 1980. Notes on Monarch Flycatchers at Widgee. Sunbird 11: 22–23.
880 HULSMAN, K. Pers. comm.
881 HULSMAN, K. 1976. The robbing behaviour of terns and gulls. Emu 76: 143–149.
882 HULSMAN, K. 1977. Breeding success and mortality of terns at One Tree Island, Great Barrier Reef. Emu 77: 49–60.
883 HULSMAN, K. 1979. One Tree Island, Queensland. Corella 3: 37–40.
884 HUMPHRIES, C. P. 1947. Among the birds at Melville Bay. Emu 47: 130–136.
885 HUNT, A. C. 1977. Luci-jet kills Whistling Eagles. Bird Obs. 552: 85–86.
886 HUNT, T. J. Pers. comm.
887 HUNT, T. J. and R. F. KENYON. 1970. The rediscovery of the Mallee Whipbird in Victoria. Aust. Bird Watcher 3: 222–226.
888 HUNTER, 1983. Interspecific breeding in Giant Petrels at South Georgia. Emu 82: 312–314.

889 HYEM, E. L. 1930. Yellow Robin's eggs. Emu 29: 277.
890 HYEM, E. L. 1933. Glossy Black Cockatoo. Emu 33: 56–58.
891 HYEM, E. L. 1936–37. Notes on the birds of "Mernot", Barrington, N.S.W. Parts I & II. Emu 36: 109–127; 262–272.
892 HYEM, E. L. 1953. Notes on the nesting of the Spotted Quail-thrush. Emu 53: 273–276.
893 HYEM, E. L. 1957. Thornbills huddling together. Emu 57: 145.
894 HYEM, E. L. 1979. Observations on owls in the Upper Manning River District, N.S.W. Corella 3: 17–25.

895 IMBER, M. J. and J. A. F. JENKINS. 1981. The New Caledonian Petrel. Notornis 28: 149–160.
896 IMBER, M. J. and R. RUSS. 1975. Some foods of the Wandering Albatross (*Diomedea exulans*). Notornis 22: 27–36.
897 IMMELMANN, K. 1960. Behavioural observations on several species of Western Australian birds. Emu 60: 237–244.
898 IMMELMANN, K. 1961. Beitrage zur biologie und ethologie Australischer Honigfresser. Z. fur. Orn. 102: 164–207.
899 IMMELMANN, K. 1966. Notes on the breeding biology of the White-winged Triller *Lalage sueurii tricolor*, Swainson, in north-western Australia. Emu 66: 1–15.
900 IMMELMANN, K. 1966. Beobachtungen an Schwalbenstaren. Z. fur. Orn. 107: 37–69.
901 IMMELMANN, K. 1982. Australian Finches. Sydney: Angus & Robertson.
902 INGLE, A. P. 1910. Notes on the birds seen around Inglenook, Merriman's Creek, South Gippsland. Emu 10: 121–127.
903 INGRAM, C. 1907. On the birds of the Alexandra District, Northern Territory of South Australia. Ibis 1 (9th Series): 387–415.
904 IRBY, F. M. 1927. Further notes on rare parrots. Emu 27: 13–16.
905 IRBY, F. M. 1933. Notes on times of arrival of migratory birds to Casino District, N.S.W. Emu 32: 171–172.
906 IRBY, F. M. 1934. Letter-winged Kites in the east. Emu 33: 268–271.
906a IVES, N. L. 1973. Overnight torpidity in Australian arid country birds. Emu 73: 140.
907 IVES, N. 1982. Bush Birds in Captivity. Perth: W.A. Avicultural Soc.

908 JACK, N. 1958. Queensland. Emu 58: 45–47.
909 JACK, N. 1971. Dominican Gull in northern Queensland. Emu 71: 181.
910 JACKSON, S. W. 1919. Haunts of the Letter-winged Kite *Elanus scriptus*, Gould. Emu 18: 160–172.
911 JAENSCH, R. P. Pers. comm.
912 JAENSCH, R. P. 1980. White-bellied Cuckoo-shrike in the Upper Murray. S. Aust. Orn. 28: 109.
913 JAENSCH, R. 1982. Port Hedland Saltworks shows its colours. RAOU Newsl. 52: 1, 4.
914 JAHNKE, B. R. 1977. Masthead Island, Queensland. Corella 1: 48–50.
915 JARMAN, H. 1973. The Turquoise Parrot. Aust. Bird Watcher 4: 239–250.
916 JARMAN, H. 1974. The Rock Parrot. Aust. Bird Watcher 5: 237–245.
917 JARMAN, H. A. 1981. Notes on introduced finches occurring in the Heidelberg area, Victoria. Aust. Bird Watcher 9: 78–79.
918 JASPER, T. 1962. Nesting notes on the Red-browed Finch. Emu 62: 177–180.
919 JASPER, T. 1964. Predation by the Grey Butcherbird in an orchard area. Emu 63: 413–414.
920 JASPER, T. 1965. Further notes on the nesting of the Red-browed Finch. Emu 64: 145–146.
921 JEHL, J. R. 1973. The distribution of marine birds in Chilean waters in winter. Auk 90: 114–135.
922 JENKINS, C. F. H. 1959. Introduced birds in Western Australia. Emu 59: 201–207.
923 JENKINS, C. F. H. 1959. The Cattle Egret in Western Australia. W. Aust. Nat. 7: 19–21.
924 JENKINS, C. F. H. 1961. Notes on some waders in south-western Australia. W. Aust. Nat. 7: 201–204.
925 JENKINS, J. A. F. 1979. Observation on the Wedge-tailed Shearwater (*Puffinus pacificus*) in the south-west Pacific. Notornis 26: 331–348.
926 JERRARD, C. H. H. 1931. Vocal powers of the Yellow Oriole. Emu 31: 42–43.
927 JESSOP, A. E. Pers. comm.
928 JOHNSON, C. N. 1983. Courtship feeding and mating in the Channel-billed Cuckoo. Corella 7: 44.
929 JOHNSTON, D. 1983. Nesting of the Square-tailed Kite in the Baradine area. Aust. Birds 17: 35–37.
930 JOHNSTONE, G. W. Pers. comm.
931 JOHNSTONE, G. W. 1971. Giant Petrels. Aust Bird Bander 9: 86–87.
932 JOHNSTONE, G. W. 1974. Field characters and behaviour at sea of Giant Petrels in relation to their oceanic distribution. Emu 74: 209–218.
933 JOHNSTONE, G. W. 1977. Comparative feeding ecology of the Giant Petrels *Macronectes giganteus* (Gmelin) and *M. halli* (Mathews). *In* LLANA, G. A. (Ed.). Proc. 3rd SCAR Symp. Antarct. Biol.: 647–668. Houston, Texas: Gulf Publ.
934 JOHNSTONE, G. W. 1978. Interbreeding by *Macronectes halli* and *M. giganteus* at Macquarie Island. Emu 78: 235.
935 JOHNSTONE, G. W. 1979. Agonistic behaviour of the Giant Petrels *Macronectes giganteus* and *M. halli* feeding at seal carcasses. Emu 79: 129–132.
936 JOHNSTONE, G. W. 1980. Current studies of seabirds in Antarctic waters. Aust. Seabird Group Newsl. 14: 17–18.
937 JOHNSTONE, G. W. and K. R. KERRY. 1976. Ornithological observations in the Australian sector of the Southern Ocean. Proc. 16th. Intern. Ornith. Congr.: 725–738.
938 JOHNSTONE, G. W., D. J. LUGG and D. A. BROWN. 1973. The biology of the Vestfold Hills, Antarctica. ANARE Sci. Rep. Pen. B (1) Zoology. Antarctic Div. Dept of Science. Aust. Publ. No. 123.
939 JOHNSTONE, G. W., D. MILLEDGE and D. F. DORWARD. 1975. The White-capped Albatross of Albatross Island: numbers and breeding behaviour. Emu 75: 1–11.
940 JOHNSTONE, R. E. 1975. Distribution and taxonomic status of the Dusky Warbler *Gerygone tenebrosa*. Emu 75: 185–188.
941 JOHNSTONE, R. E. 1977. An Asian Gull-billed Tern in Western Australia. W. Aust. Nat. 14: 26–27.
942 JOHNSTONE, R. E. 1978. North Fisherman Island, Western Australia. Corella 2: 43–45.
943 JOHNSTONE, R. E. 1978. South Fisherman Island, Western Australia. Corella 2: 46–47.
944 JOHNSTONE, R. E. 1979. Three more records of the Kerguelen Diving-Petrel in Western Australia. W. Aust. Nat. 14: 133.
945 JOHNSTONE, R. E. 1981. Notes on the distribution, ecology and taxonomy of the Red-crowned Pigeon (*Ptilinopus regina*) and Torres Strait Pigeon (*Ducula bicolor*), in Western Australia. Rec. W. Aust. Mus. 9: 7–22.
946 JOHNSTONE, R. E. 1981. Notes on the distribution, ecology and taxonomy of the Partridge Pigeon (*Geophaps smithii*) and Spinifex Pigeon (*G. plumifera*) in Western Australia. Rec. W. Aust. Mus. 9: 49–64.
947 JOHNSTONE, R. E. 1982. Streaked Shearwaters *Calonectris leucomelas* off mid-west coast of Western Australia. W. Aust. Nat. 15: 58.
948 JOHNSTONE, R. E. 1982. The Yellow Wagtail *Motacilla flava* in Australia. W. Aust. Nat. 15: 61–66.
949 JOHNSTONE, R. F. and W. J. KLITZ. 1977. Variation and evolution in a granivorous bird: the House Sparrow. *In* PINOWSKI, J. and S. C. KENDEIGH (Eds). Granivorous birds in ecosystems. International Biological Programme 12: 15–51. London: Cambridge University Press.
950 JONES, B. 1973. Studies of Giant Petrels at Malabar, New South Wales. Aust. Bird Bander 11: 47–50.
951 JONES, D. N. Pers. comm.
952 JONES, D. N. 1981. Temporal changes in the suburban avifauna of an inland city. Aust. Wildl. Res. 8: 109–119.

953 JONES, D. N. 1983. The suburban bird community of Townsville, a tropical city. Emu 83: 12–18.

954 JONES, E. and I. J. SKIRA. 1979. Breeding distribution of the Great Skua at Macquarie Island in relation to numbers of rabbits. Emu 79: 19–23.

955 JONES, J. 1940. A Victorian breeding record of the Lesser Egret. Emu 39: 300–301.

956 JONES, J. 1943. The Wood-Sandpiper. Emu 43: 107–112.

957 JONES, J. 1945. The Banded Stilt. Emu 45: 1–36.

958 JONES, J. 1946. Australian distribution of the Pigmy-Geese. Emu 46: 128–132.

959 JONES, J. 1952. The Hattah Lakes camp-out, October 1951. Emu 52: 225–254.

960 JONES, S. G. and W. M. BREN. 1978. Observations on the wintering behaviour of Victorian Peregrine Falcons. Aust. Bird Watcher 7: 198–203.

961 JONES, W. 1981. Bird Observations 1981—Systematic List. Tas. Bird Report 11: 30–45.

962 JORDAN, R. and P. J. FULLAGAR. Pers. comm.

963 JORDAN, P. and R. JORDAN. Pers. comm.

964 JOSEPH, L. 1975. Southern record of the Painted Finch. S. Aust. Orn. 27: 5.

965 JOSEPH, L. 1976. Recent records of the Scarlet-chested Parrot. S. Aust. Orn. 27: 144–145.

966 JOSEPH, L. 1978. Range and movements of the Regent Parrot in South Australia. S. Aust. Orn. 28: 26–27.

967 JOSEPH, L. 1982. The Red-tailed Black-Cockatoo in south-eastern Australia. Emu 82: 42–45.

967a JOSEPH, L. 1982. The Glossy Black-Cockatoo on Kangaroo Island. Emu 82: 46–49.

968 JOSEPH, L. 1982. A further population of the Grey Grasswren. Sunbird 12: 51–53.

969 JOSEPH, L. and J. REID. 1981. The Crested Shrike-tit on the Darling and Murray Rivers. S. Aust. Orn. 28: 157–159.

970 KABAY, E. D. and A. A. BURBIDGE. 1977. A biological survey of the Drysdale River National Park, North Kimberley, Western Aust. Wildl. Res. Bull., W.A. No. 6: 1–133.

971 KAIGLER, C. G. 1968. Red-necked Avocet in Westland. Notornis 15: 123.

972 KAVENEY, M. 1958. Notes on the Brush Turkey. Emu 58: 152–153.

973 KAVENEY, M. T. 1958. Protective flocking of Regent Honeyeater. Emu 58: 266.

974 KEAST, J. A. 1944. A winter list from the Tweed River District, NSW with remarks on some nomadic species. Emu 43: 177–187.

975 KEAST, J. A. 1949. Field notes on the Grey-tailed Tattler. Rec. Aust. Mus. 22: 207–211.

976 KEAST, J. A. 1956. Variation in the genus *Megalurus*. Proc. Roy. Zool. Soc. N.S.W. 1954–55: 25–28.

977 KEAST, J. A. 1957. Variation in the Australian Emu-Wrens (*Stipiturus*). Proc. Roy. Soc. NSW 1955–56: 47–53.

978 KEAST, A. 1957. Variation and speciation in the genus *Climacteris* Temminck (Aves: Sittidae). Aust. J. Zool. 5: 474–495.

979 KEAST, A. 1957. Variation in the Australian Kingfishers (Aves: Alcedinidae). Rec. Aust. Mus. 24: 61–72.

980 KEAST, A. 1958. The influence of ecology on variation in the Mistletoe-bird *Dicaeum hirundinaceum*. Emu 58: 195–206.

981 KEAST, A. 1958. Seasonal movements and geographic variations in the Australian Wood-Swallows (Artamidae). Emu 58: 207–218.

982 KEAST, A. 1958. Infraspecific variation in the Australian finches. Emu 58: 219–246.

983 KEAST, A. 1958. The genus *Psophodes* Vigors and Horsfield, and its significance in demonstrating a possible pathway for the origin of Eyrean species from Bassian ones. Emu 58: 247–255.

984 KEAST, J. A. 1958. The relationship between seasonal movements and the development of geographic variation in the Australian Chats *Ephthianura* Gould and *Ashbyia* North (Passeres: Muscicapidae, Malurinae). Aust. J. Zool. 6: 53–68.

985 KEAST, A. 1958. Variation and speciation in the Australian Campephagidae (Passeres). Aust. J. Zool. 6: 248–267.

986 KEAST, A. 1958. Variation and speciation in the Australian Flycatchers (Aves: Muscicapinae). Rec. Aust. Mus. 24: 73–108.

987 KEAST, A. 1968. Seasonal movements in the Australian Honeyeaters (Meliphagidae) and their ecological significance. Emu 67: 159–209.

988 KEAST, A. J. 1970. Adaptive evolution and shifts in niche occupations in island birds. Biotropica 2: 61–75.

989 KEAST, A. 1978. The status of the Mountain Thornbill *Acanthiza katherina*. Emu 78: 7–10.

990 KEAST, A. 1978. Zoogeography and evolutionary history of the Scrubwren genus *Sericornis*. Emu 78: 119–125.

991 KENYON, R. F. 1972. Polygyny among Superb Lyrebirds in Sherbrooke Forest Park, Kallista, Victoria. Emu 72: 70–76.

992 KERRY, K. R., R. S. C. HORNE and D. F. DORWARD. 1983. Records of the Short-tailed Shearwater *Puffinus tenuirostris* in Antarctic waters. Emu 83: 35–37.

993 KIKKAWA, J. 1973. The status of Silvereyes *Zosterops* on the islands of the Great Barrier Reef. Sunbird 4: 30–37.

994 KIKKAWA, J. 1976. The birds of Cape York Peninsula. Sunbird 7: 25–41; 81–106.

995 KIKKAWA, J. 1980. Seasonality of nesting by Zebra Finches at Armidale, N.S.W. Emu 80: 13–20.

996 KIKKAWA, J. and W. BOLES. 1976. Heron Is., Queensland. Aust. Bird Bander 14: 3–6.

997 KIKKAWA, J., L HORE-LACY and J. LE GAY BRERETON. 1965. A preliminary report on the birds of the New England National Park. Emu 65: 139–143.

998 KIKKAWA, J., G. B. MONTEITH and G. INGRAM. 1981. Cape York Peninsula: Major region of faunal interchange. *In* KEAST, A. (Ed.) Ecological Biogeography of Australia: 1695–1741. The Hague: Junk.

999 KILGOUR, J. F. 1904. A trip to the Ord River, (N.-W.A.). Emu 4: 37–43.

1000 KING, B. R. Pers. comm.

1001 KING, B. R. 1980. Social organization and behaviour of the Grey-crowned Babbler *Pomatostomus temporalis*. Emu 80: 59–76.

1002 KINGHORN, J. R. 1924. Notes on a trip to Upper Colo, N.S.W. Emu 24: 134–141.

1003 KINSKY, F. C. 1970. Annotated Checklist of the Birds of New Zealand. Wellington: Reed.

1004 KLAPSTE, J. 1976. Spotted Crake as the prey of the White Egret. Aust. Bird Watcher 6: 164.

1005 KLAPSTE, J. 1977. The Barn Swallow (*Hirundo rustica*) in Australia. Aust. Bird Watcher 7: 25–34.

1006 KLAPSTE, J. 1981. Where to go to see the Barn-Swallow. The Bird Observer 599: 101.

1007 KOLICHIS, N. 1976. New breeding records of the Banded Stilt in Western Australia. W. Aust. Nat. 13: 114–119.

1008 KOLICHIS, N. 1977. Birds of Bedout Island—a visit in May 1975. W. Aust. Nat. 13: 191–194.

1009 KOLICHIS, N. and I. ABBOTT. 1978. Gull Rock, King George Sound, Western Australia. Corella 2: 28–29.

1010 KURTZ, N. K. 1981. Notes on the nesting behaviour of the Yellow-rumped Thornbill in the Mudgee District. Aust. Birds 15: 58–62.

1011 LABUTIN, Y. V., V. V. LEONOVITCH and B. N. VEPRINTSEV. 1982. The Little Curlew *Numenius minutus* in Siberia. Ibis 124: 302–319.

1012 LACK, D. 1956. Swifts in a Tower. London: Methuen.

1013 LAMM, D. W. 1965. Seasonal counts of birds at Lake George, New South Wales. Emu 64: 114–128.

1014 LAMM, D. W. and J. H. CALABY. 1950. Seasonal variation of bird populations along the Murrumbidgee in the Australian Capital Territory. Emu 50: 114–122.

1015 LAMM, D. W. and D. WHITE. 1950. The changing status of avifauna in the Australian Capital Territory. Emu 49: 199–204.

1016 LAMM, D. W. and S. J. WILSON. 1966. Seasonal fluctuations of birds in the Brindabella Range, Australian Capital Territory. Emu 65: 183–205.

1017 LANE, B. A. Pers. comm.
1018 LANE, B. A. 1981. A preliminary report on the status of the Fairy Tern *Sterna nereis*—Breeding colonies in Victoria. Unpublished report.
1019 LANE, B. 1981. The Hooded Plover survey, October 1980. Vict. Wader Study Group Bull. 3: 6–8.
1020 LANE, B. and A. JESSOP. 1983. Distribution and habitat of the Red-capped Plover in Victoria. Vict. Wader Study Group Bull. 7: 35–40.
1021 LANE, J. A. K. Pers. comm.
1022 LANE, J. A. K. 1978. Saint Alouarn Island, Western Australia. Corella 2: 36–37.
1023 LANE, S. G. 1964. Breeding sea-birds on Bird Island, Norah Head, N.S.W. Emu 64: 317–319.
1024 LANE, S. G. 1965. Seasonal banding of Fairy Martins. Aust. Bird Bander 3: 67–70.
1025 LANE, S. G. 1966. Plum-headed Finches near Richmond, New South Wales. Emu 66: 111–112.
1026 LANE, S. G. 1968. Some aspects of banding. Emu 67: 231–233.
1027 LANE, S. G. 1972. Tasmanian type Silvereyes in New South Wales. Aust. Bird Bander 10: 33–34.
1028 LANE, S. G. 1972. White-phased Wedge-tailed Shearwater on North Solitary Island, NSW. Emu 72: 184–185.
1029 LANE, S. G. 1979. Summary of the breeding sea-birds on New South Wales coastal islands. Corella 3: 7–9.
1030 LANE, S. G. 1982. Frederick Island, Archipelago of the Recherche, Western Australia. Corella 6: 61–62.
1031 LANE, S. G. 1982. Remark Island, Archipelago of the Recherche, Western Australia. Corella 6: 65–66.
1032 LANE, S. G. 1982. Figure of Eight Island, Archipelago of the Recherche, Western Australia. Corella 6: 75–76.
1033 LANE, S. G. and H. BATTAM. 1971. Silvereye movement in eastern Australia. Aust. Bird Bander 9: 80–82.
1034 LANE, S. G. and G. WHITE. 1983. Nesting of the Sooty Shearwater in Australia. Emu 83: 117–118.
1035 LANG, C. L. 1930. Fairy Martins nesting. Emu 29: 251.
1036 LAVERY, H. J. Pers. comm.
1037 LAVERY, H. J. 1966. Pygmy Geese in Australia. Qld J. Agric. May 1966: Plant Industry Advisory Leaflet No. 861: 1–7.
1038 LAVERY, H. J. 1969. Studies of waterfowl (Anatidae) in north Queensland. 3. Harvests. Qld J. Agric. Animal Sci. 26: 677–690.
1039 LAVERY, H. J. 1970. The comparative ecology of waterfowl in North Queensland. Wildfowl 21: 69–77.
1040 LAVERY, H. J. (Ed.). 1978. Exploration North. Richmond: Richmond Hill.
1041 LAVERY, H. J. and R. J. GRIMES. 1974. Purple-crowned Pigeon at Townsville, Queensland. Emu 74: 53–54.
1042 LAVERY, H. J. and R. J. GRIMES. 1974. The functions of the bill of the Tooth-billed Bowerbird. Emu 74: 255–256.
1043 LAVERY, H. J. and R. J. GRIMES. 1974. The Yellow Silvereye in north-east Queensland. Sunbird 5: 42–43.
1044 LAVERY, H. J. and N. HOPKINS. 1963. Birds of the Townsville district of North Queensland. Emu 63: 242–252.
1045 LAVERY, H. J. and P. M. JOHNSON. 1979. Population behaviour of the Black Kite *Milvus migrans* (Boddaert) in north-eastern Australia. RAOU Scientific Day, 24 May 1979.
1046 LAVERY, H. J. and D. SETON. 1967. *Cisticola juncidis* (Rafinesque) in north-east Queensland. Emu 67: 125–132.
1047 LAVERY, H. J., D. SETON and J. A. BRAVERY. 1968. Breeding seasons of birds in north-eastern Australia. Emu 68: 133–147.
1048 LAWRENCE, C. C. 1948. Nests of Little Grass-birds. Emu 47: 370–372.
1049 LEA, A. M. and J. T. GRAY. 1935–36. The food of Australian birds. Parts 1–5. Emu 34: 275–292. 35: 63–98; 145–178; 251–280; 335–347.
1050 LEACH, J. A. 1911. An Australian Bird Book. (9th Ed. Rev. 1958). Melbourne: Whitcombe & Tombs.
1051 LEARMONTH, N. 1950. Bristle-birds on the seashore. Emu 50: 83.
1052 LEARMONTH, N. F. 1953. Bird movement in the Portland District, Victoria, in 1951. Emu 53: 86–88.
1053 LEARMONTH, N. F. 1966. Birds of Portland (Victoria) District. Portland: Portland Field Nat. Club.
1054 LEICESTER, M. 1960. Some notes on the Lewin Rail. Emu 60: 20–24.
1055 LENDON, A. 1944. Bird notes: Bulbuls in Adelaide. S. Aust. Orn. 17: 1.
1056 LENDON, A. 1951. Unidentified Egret. Emu 51: 120.
1057 LENDON, A. 1952. Bulbuls in Melbourne. Emu 52: 67–68.
1058 LENDON, A. H. 1973. Neville W. Cayley's Australian Parrots in Field and Aviary. (Rev. Ed.). Sydney: Angus & Robertson.
1059 LEOPOLD, A. S. and T. O. WOLFE. 1970. Food habits of nesting Wedge-tailed Eagles *Aquila audax*, in south-eastern Australia. CSIRO Wildl. Res. 15: 1–17.
1060 LE SOUEF, A. S. 1923. How settlement and other factors are affecting native birds. Emu 23: 105–108.
1061 LE SOUEF, A. S. 1924. Rabbit poisoning and birds. Emu 23: 247.
1062 LE SOUEF, D. 1901. A visit to the Riverina district, N.S.W. Vict. Nat. 17: 179–186.
1063 LE SOUEF, D. 1913. Bird migration. Emu 12: 195–197.
1064 LE SOUEF, E. A. 1931. The effect of our civilization on the bird life of Australia. Emu 30: 188–192.
1065 LE SOUEF, W. H. D. 1918. Red-vented Bulbul. Emu 17: 236.
1066 LIDDY, J. 1959. Notes on the Black Kite in north-west Queensland. Emu 59: 268–274.
1067 LIDDY, J. 1960. Waders of the interior of north-west Queensland. Emu 60: 203–210.
1068 LIDDY, J. 1962. Honeyeaters of north-west Queensland. Emu 61: 285–291.
1069 LIDDY, J. 1966. Autumnal migration of the Yellow-faced Honeyeater. Emu 66: 87–103.
1070 LIDDY, J. 1970. A summary of Blackbird banding in Tasmania. Aust. Bird Bander 8: 57–58.
1071 LILL, A. 1979. An assessment of male parental investment and pair bonding in the polygamous Superb Lyrebird. Auk 96: 489–498.
1072 LIMPUS, C. J. 1980. Birds of Maclennan Cay, Great Barrier Reef. Corella 4: 58–59.
1073 LIMPUS, C. J. and B. J. LYON. 1981. Seabirds breeding on the Swain Reefs, Queensland. Corella 5: 101–105.
1074 LINDGREN, E. 1961. Natural history notes from Jigalong, north-western Australia. III. The birds. W. Aust. Nat. 7: 169–176.
1075 LINDGREN, E. 1971. Records of new and uncommon species for the island of New Guinea. Emu 71: 134–136.
1076 LINDGREN, E. 1973. Studies in the ecology and physiology of three species of Grass Parrots (*Neophema* Aves: Psittacidae). Ph.D. Thesis. Univ. of Western Australia.
1077 LINDSEY, A. I. G. 1979. A feeding association between Australian Fernwren and Orange-footed Scrubfowl. Sunbird 10: 47.
1078 LINDSEY, T. 1979. More on Tattlers. Atlas Newsl. 10: 2.
1079 LINDSEY, T. R. 1981. NSW Bird Report for 1980. Aust. Birds 16: 1–23.
1080 LINDSEY, T. R. 1982. NSW Bird Report for 1982. Aust. Birds 17: 1–26.
1081 LITTLER, F. M. 1903. Notes on some birds peculiar to Tasmania. Emu 3: 81–89.
1082 LITTLER, F. M. 1910. A Handbook of the Birds of Tasmania. Launceston: The Author.
1083 LLOYD, A. 1982. Communal feeding in Chestnut-rumped Thornbills. Sunbird 12: 40.
1084 LONG, J. L. Pers. comm.
1085 LONG, J. L. 1981. Introduced Birds of the World. Sydney: Reed.
1086 LONGMORE, N. W. Pers. comm.
1087 LONGMORE, W. 1974. Field notes from the east-central Queensland region. Q. Orn. Soc. Newsl. 5(4): 4.
1088 LONGMORE, N. W. 1976. Squatter Pigeon near West Wyalong, N.S.W. Aust. Birds 11: 18.

1089 LONGMORE, N. W. 1978. Avifauna of the Rockhampton area, Queensland. Sunbird 9: 25–53.

1090 LONGMORE, N. W. and W. E. BOLES. 1983. Description and systematics of the Eungella Honeyeater *Meliphaga hindwoodi*, a new species of honeyeater from central Queensland. Emu 83: 59–65.

1091 LORD, C. E. 1918. On the occurrence of the Australian Roller (*Eurystomus pacificus*) in Tasmania. Emu 18: 136.

1092 LORD, C. E. 1919. A trip to the National Park of Tasmania at Mount Field. Emu 19: 98–106.

1093 LORD, E. A. R. 1933. Movements of migratory birds in the Murphy's Creek district, Queensland. Emu 32: 207–210.

1094 LORD, E. A. R. 1934. The Crested Hawk. Emu 34: 31–32.

1095 LORD, E. A. R. 1935. Notes on the Red-backed Kingfisher. Emu 35: 59–61.

1096 LORD, E. A. R. 1936. Notes on Swifts. Emu 35: 216–218.

1097 LORD, E. A. R. 1943. Migratory notes (1941–1942) from the Murphy's Creek district, Queensland. Emu 43: 18–23.

1098 LORD, E. A. R. 1952. Field notes on the Red Goshawk. Emu 52: 23–24.

1099 LORD, E. A. R. 1956. The birds of the Murphy's Creek district, southern Queensland. Emu 56: 100–128.

1100 LOVE, J. R. B. 1917. Notes on *Ashbyia lovensis*. S. Aust. Orn. 3: 40–41.

1101 LOWE, K. W. Pers. comm.

1102 LOWE, K. W. 1978. The foraging strategies of the Royal Spoonbill, Sacred Ibis and White-faced Heron in Westernport Bay. B.Sc.(Hons) Thesis. Univ. of Melbourne.

1103 LOWE, K. W. 1981. Habitat requirements of herons, egrets, ibises and spoonbills in the Westernport region. Project Report: Westernport Bay Environmental Study. Melbourne: Min. Cons. Vict.

1104 LOWE, K. W. 1982. The feeding behaviour and diet of Royal Spoonbills *Platalea regis* in Westernport Bay. Emu 82: 163–168.

1105 LOWE, K. W. 1983. Egg size, clutch size and breeding success of the Glossy Ibis *Plegadis falcinellus*. Emu 83: 31–34.

1106 LOWE, V. T. 1956. The way of a dove. Emu 56: 167–182.

1107 LOWE, V. T. 1958. Notes on ducks and duck-shooting. Emu 58: 26–30.

1108 LOWE, V. T. 1963. Observations on the Painted Snipe. Emu 62: 221–237.

1109 LOWE, V. T. 1971. Field Observations on the Australian Dotterel. Aust. Bird Watcher 4: 35–40.

1110 LOYN, R. H. Pers. comm.

1111 LOYN, R. H. 1975. Westernport Bay Environmental Study 1973–74, Project Report 4.4.8. Melbourne: Min. Cons. Vict.

1112 LOYN, R. H. 1978. A survey of birds in Westernport Bay, Victoria, 1973–74. Emu 78: 11–19.

1113 LOYN, R. H. 1980. Bird populations in a mixed eucalypt forest used for production of wood in Gippsland, Victoria. Emu 80: 145–156.

1114 LYNES, H. 1930. Review of the genus *Cisticola*. Ibis (12th Ser.) 6 Suppl.: 1–673.

1115 McARTHUR, K. 1974. Movements of Australian Gannets past Caloundra. Sunbird 5: 99–100.

1116 McBRIDE, G., I. P. PARER and F. FOENANDER. 1969. The social organization and behaviour of the feral domestic fowl. Anim. Behav. Monogr. 2: 127–181.

1117 McCLURE, H. E. 1974. Migration and survival of the birds of Asia. Bangkok: Applied Scientific Research Corporation of Thailand.

1118 McCLURE, H. E. 1975. Bird migration in eastern Asia and its relation to that of Australia. Emu 74: 315.

1119 McCULLOCH, E. M. 1966. Swifts and bushfires. Emu 65: 290.

1120 McCULLOCH, E. M. 1970. Feeding of Dusky Woodswallows. Emu 70: 200.

1121 MacDONALD, J. D. 1968. Speciation in the *Colluricincla harmonica* complex. Emu 67: 215–223.

1122 MacDONALD, J. D. 1969. Albert Lyrebird in Blackall Range. Emu 69: 118–119.

1123 MacDONALD, J. D. 1971. Validity of the Buff-breasted Quail. Sunbird 2: 1–5.

1124 MacDONALD, J. D. 1973. Birds of Australia. Sydney: Reed.

1125 MacDONALD, J. D. and P. R. COLSTON. 1966. J. R. Elsey and his bird observations on Gregory's Overland Expedition, Australia, 1856. Emu 65: 255–278.

1126 McEVEY, A. 1949. Notes on the Australian Pipit and its territory. Emu 49: 35–43.

1127 McEVEY, A. 1952. Further notes on the Australian Pipit and its territory. Emu 52: 117–120.

1128 McEVEY, A. 1965. The birds of the Rutherglen District. Emu 65: 1–56.

1129 McEVEY, A. 1974. John Cotton's Birds of the Port Phillip District of New South Wales 1843–1849. Sydney: Collins.

1130 McFARLAND, D. C. Pers. comm.

1131 McGARVIE, A. M. and M. T. TEMPLETON. 1974. Additions to the birds of King Island, Bass Strait. Emu 74: 91–96.

1131a McGILL, A. R. Pers. comm.

1132 McGILL, A. R. 1947. Unravelling the Knots. Emu 47: 137–146.

1133 McGILL, A. R. 1948. The Asiatic Ring Dove as an escapee. Emu 47: 232–233.

1134 McGILL, A. R. 1955. The two larger gulls in eastern Australia. Emu 55: 90–98.

1135 McGILL, A. R. 1959. New South Wales. Emu 59: 52–54.

1136 McGILL, A. R. 1959. The northern range limits of the Purple-crowned Lorikeet. Emu 59: 264–266.

1137 McGILL, A. R. 1960. Parrots of the genus *Neophema* in New South Wales. Emu 60: 39–46.

1138 McGILL, A. R. 1960. The Little Whimbrel. Emu 60: 89–94.

1139 McGILL, A. R. 1963. A Record of the Yellow-headed Wagtail for Australia. Emu 63: 66–71.

1140 McGILL, A. R. 1970. Australian Warblers. Melbourne: Bird Observers' Club.

1141 McGILL, A. R. and S. G. LANE. 1955. The Mt. Keira camp-out. Emu 55: 49–71.

1142 McGILL, M. 1978. An assessment of known information on the Plumed Frogmouth. Aust. Birds 12: 37–40.

1143 MacGILLIVRAY, W. 1901. Some north-west Queensland birds. Vict. Nat. 18(3): 45–47.

1144 MacGILLIVRAY, W. 1910. Along the Great Barrier Reef. Emu 10: 216–233.

1145 MacGILLIVRAY, W. 1914. Notes on some North Queensland birds. Emu 13: 132–186.

1146 MacGILLIVRAY, W. 1915. A new honeyeater: *Macgillivrayornis claudi*. Emu 15: 77–78.

1147 MacGILLIVRAY, W. 1917–18. Ornithologists in North Queensland. Emu 17: 63–87; 145–148; 180–212.

1148 MacGILLIVRAY, W. 1926. Birds of the Capricorn Islands. Emu 25: 229–238.

1149 MacGILLIVRAY, W. 1927. The charming Bourke Parrot. Emu 27: 65–67.

1150 MacGILLIVRAY, W. 1927. Birds from a coastal boat. Emu 27: 92–101.

1151 MacGILLIVRAY, W. D. K. 1929. Through a drought-stricken land. Emu 29: 113–129.

1152 MacGILLIVRAY, W. D. K. 1931. The Shining Flycatcher. Emu 31: 81–82.

1153 MacGILLIVRAY, W. D. K. 1932. The Flock Pigeon. Emu 31: 169–174.

1154 MacGILLIVRAY, W. D. K. 1932. The Yellow-billed Kingfisher. Emu 32: 1–2.

1155 McGILP, J. N. 1920. *Eudromias australis* (Australian Dotterel). S. Aust. Orn. 5: 49–52.

1156 McGILP, J. N. 1921. The Black-banded Whiteface. Emu 21: 49–50.

1157 McGILP, J. N. 1921. Black Kites. Emu 21: 311.

1158 McGILP, J. N. 1922. The drinking habits of *Peltohyas australis* (Australian Dotterel). S. Aust. Orn. 6: 155–156.

1159 McGILP, J. N. 1923. Birds of Lake Frome district, South Australia. Emu 22: 237–243; 274–287.

1160 McGILP, J. N. 1932. Wedgebill *Sphenostoma cristatum*. S. Aust. Orn. 11: 191–193.

1161 McGILP, J. N. 1934. The nesting of the Painted Snipe *Rostratula australis*. S. Aust. Orn. 12: 167–169.
1162 McGILP, J. N. 1934. The hawks of South Australia. S. Aust. Orn. 12: 261–292.
1163 McGILP, J. N. 1935. Birds of the Musgrave Ranges. Emu 34: 163–176.
1164 McINNES, R. S. and P. B. CARNE. 1978. Predation of cossid moth larvae by Yellow-tailed Cockatoos causing losses in plantations of *Eucalyptus grandis* in north coastal New South Wales. Aust. Wildl. Res. 5: 101–121.
1164a McKEAN, J. L. Pers. comm.
1165 McKEAN, J. L. 1959. Blue Wren at Kulkyne Forest, Vict. Emu 59: 63.
1166 McKEAN, J. L. 1960. White-fronted Terns in South-eastern Australia in 1959. Emu 60: 262–264.
1167 McKEAN, J. L. 1980. Oceanic Ornithological Survey between Derby, W.A. and Darwin, N.T. NT Nat. 1: 8–13.
1168 McKEAN, J. L. 1980. The First Record of the Arctic Warbler *Phylloscopus borealis* from Australia. W. Aust. Nat. 14: 200.
1169 McKEAN, J. L. 1981. The status of gulls and terns (Laridae) in the Darwin area 1974–80. Aust. Seabird Group Newsl. 15: 11–16.
1170 McKEAN, J. L., M. C. BARTLETT and C. M. PERRINS. 1975. New Records from the Northern Territory. Aust. Bird Watcher 6: 45–46.
1171 McKEAN, J. L. and D. GRAY. 1973. Unusual sea-bird records from the Northern Territory. Emu 73: 184.
1172 McKEAN, J. L. and W. B. HITCHCOCK. 1969. The taxonomic status of *Acanthiza katherina* De Vis. Emu 69: 113.
1173 McKEAN, J. L., I. J. MASON and L. W. O'CONNOR. 1975. Birds not previously recorded from Timor. Emu 75: 62–64.
1174 McKEAN, J. L. and M. REED. 1979. Further sightings of the Bush-hen *Gallinula olivacea* from sub-coastal Northern Territory. Sunbird 10: 73–74.
1175 McKEAN, J. L., H. A. F. THOMPSON and J. A. ESTBERGS. 1976. Records of Uncommon Migrant Waders near Darwin, N.T. Aust. Bird Watcher 6: 143–148.
1176 McKEOWN, K. C. 1944. Notes on the food of Australian birds. Emu 43: 188–191.
1177 McKILLIGAN, N. G. 1975. Breeding and movements of the Straw-necked Ibis in Australia. Emu 75: 199–212.
1178 MacLAINE, J. D. 1908. Clarke Island (Bass Strait) notes. Emu 7: 191–193.
1179 MacLEAN, G. L. 1976. A field study of the Australian Pratincole. Emu 76: 171–182.
1180 MacLEAN, G. L. 1976. A field study of the Australian Dotterel. Emu 76: 207–215.
1181 MacLEAN, G. L. 1977. Comparative notes on Black-fronted and Red-kneed Dotterels. Emu 77: 199–207.
1182 McMANUS, T. J. 1979. St. Helens Island, Tasmania. Corella 3: 52–54.
1183 McNAMARA, E. 1937. Birds of the blackberries. Emu 37: 99–102.
1184 McNAMARA, E. 1946. Field notes on the Eastern Bristle-bird. Emu 45: 260–265.
1185 McNAMARA, J. A. 1980. Nocturnal feeding of the Inland Dotterel. Emu 80: 39–40.
1186 MACE, B. 1981. ARA field trip to the Strzelecki Creek, north-eastern S.A. Aust. Raptor Assoc. Newsl. 2(4): 3.
1187 MACK, G. 1934. Grey-headed Robin. Emu 33: 157–158.
1188 MACKNESS, B. 1979. The ecology of the Pheasant Coucal *Centropus phasianinus* (Latham) in Australia. Sunbird 10: 1–8.
1189 MAIN, A. R. Pers. comm.
1190 MAKIN, D. 1961. Mass migration. Emu 61: 139–141.
1191 MARCHANT, J. and K. TAYLOR. 1981. Bird population changes for the years 1979–80. Bird Study 28: 147–152.
1192 MARCHANT, S. Pers. comm.
1193 MARCHANT, S. 1972. Evolution of the genus *Chrysococcyx*. Ibis 114: 219–233.
1194 MARCHANT, S. 1974. Analysis of nest-records of the Willie Wagtail. Emu 74: 149–160.
1195 MARCHANT, S. 1977. A seawatch on the southern coast of new South Wales. Emu 77: 9–18.
1196 MARCHANT, S. 1979. Nesting of the Cicadabird *Coracina tenuirostris*. Ibis 121: 80–84.
1197 MARCHANT, S. 1979. The birds of forest and woodland near Moruya, N.S.W. Aust. Birds 13: 59–68.
1198 MARCHANT, S. 1982. The sedentary nature of passerine birds in woodland at Moruya, N.S.W. Corella 6: 87–88.
1199 MARCHANT, S. and E. O. HOHN. 1980. Field notes on the Fan-tailed Cuckoo. Emu 80: 77–80.
1200 MARSHALL, A. J. 1930. The Yellow-throated Scrub-wren: a monograph. Emu 30: 3–9.
1201 MARSHALL, A. J. 1931. The Azure Kingfisher. Emu 31: 44–47.
1202 MARSHALL, A. J. 1932. In search of a 'lost' bird. Emu 32: 99–103.
1203 MARSHALL, A. J. 1933. The Mangrove Robin. Emu 32: 233–235.
1204 MARSHALL, A. J. 1933. Notes from the Leichhardt River, Qld. Emu 32: 263–266.
1205 MARSHALL, A. J. 1935. On the birds of the Mcpherson Ranges, Mt. Warning, and Contiguous Lowlands. Part I. Emu 34: 189–196.
1206 MARSHALL, A. J. 1957. On the function of 'clustering' in Wood-Swallows. Emu 57: 53–54.
1207 MARSHALL, J. 1964. Southern Figbird near Bass Strait. Emu 63: 339.
1208 MARSHALL, A. J. 1965. The recent occurrence of the Flock Pigeon in North Australia. Emu 64: 150–151.
1209 MARTINDALE, J. 1980. Warden's report: 1977–79. Eyre Bird Observatory Report 1977–79: 4–40.
1210 MARTINDALE, J. 1983. Aspects of counts of Freckled Duck *Stictonetta naevosa* in eastern Australia, Jan.-Feb. 1983. Melbourne: RAOU.
1211 MASON, I. J. Pers. comm.
1212 MASON, I. 1981. Letter to the Editor. R.A.O.U. Newsl. 50: 10.
1213 MASON, I. J., H. B. GILL and J. H. YOUNG. 1981. Observations on the Red-necked Crake. Aust. Bird Watcher 9: 69–77.
1214 MASON, I. J., J. L. McKEAN and M. L. DUDZINSKI. 1984. Geographic variation in the Pheasant Coucal (*Centropus phasianinus*) (Latham) and a description of a new subspecies from Timor. Emu 84: in press.
1215 MASON, I. J. and T. O. WOLFE. 1975. First record of the Marsh Crake for the Northern Territory. Emu 75: 235.
1216 MASTERS, J. R. and A. L. MILHINCH. 1974. Birds of the shire of Northam, about 100 km east of Perth, W.A. Emu 74: 228–244.
1217 MATHESON, W. E. 1978. A further irruption of native-hens in 1975. S. Aust. Orn. 27: 270–273.
1218 MATHEWS, G. M. 1909–10. On the birds of North-West Australia. Emu 9: 1–16; 53–69; 103–110.
1219 MATHEWS, G. M. 1910–27. The Birds of Australia. London: Witherby.
1220 MATHEWS, G. M. 1914. Additions and corrections to my list of the Birds of Australia. Aust. Av. Rec. 2: 83–107.
1221 MATHEWS, G. M. 1930. Some interesting links with John Gould. Emu 30: 141–143.
1222 MATTHIESSEN, J. 1973. Observations on the food of some birds of south-western Western Australia. Emu 73: 191–193.
1223 MAY, I. A. 1977. Recent re-discovery of the Eyrean Grasswren. Emu 77: 230–231.
1224 MAYO, L. M. 1923. Nest of the White-eared Fly-catcher. Emu 23: 30–31.
1225 MAYO, L. M. 1931. Bird notes from here and there in Queensland. Emu 30: 288–289.
1226 MAYO, L. M. 1931. A Queensland 'Sanctuary'. Emu 31: 71–76.
1227 MAYO, L. M. 1933. Random bird notes from Queensland. Emu 32: 300–301.
1228 MAYO, L. M. 1935. Rufous Shrike-Thrushes. Emu 34: 302.
1229 MAYR, E. 1940. Notes on Australian birds, I: the genus *Lalage*. Emu 40: 111–117.
1230 MAYR, E. 1944. Timor and the colonization of Australia by birds. Emu 44: 113–130.

1231 MAYR, E. 1951. Notes on some pigeons and parrots from Western Australia. Emu 51: 137–145.

1232 MAYR, E. 1953. Geographic and individual variation in the Shrike-Tit (*Falcunculus frontatus*). Emu 53: 249–252.

1233 MAYR, E. 1953. Taxonomic notes on *Oreoica gutturalis*. Emu 53: 252–253.

1234 MAYR, E. and A. McEVEY. 1960. The distribution and variation of *Mirafra javanica* in Australia. Emu 60: 155–192.

1235 MAYR, E. and A. L. RAND. 1937. Results of the Archbold Expeditions. 14. Birds of the 1933–1934 Papuan Expedition. Bull. Amer. Mus. Nat. Hist. 73: 1–248.

1236 MAYR, E. and D. L. SERVENTY. 1938. A review of the genus *Acanthiza* Vigors and Horsfield. Emu 38: 245–292.

1237 MEAD, C. J. 1970. The winter quarters of British Swallows. Bird Study 17: 229–240.

1238 MEDWAY, Lord and D. R. WELLS. 1976. The Birds of the Malay Peninsula. Vol. 5. London: Witherby.

1239 MEES, G. F. 1961. An annotated catalogue of a collection of bird skins from West Pilbara, W.A. J. Roy. Soc. W.A. 44: 97–143.

1240 MEES, G. F. 1963. Status and distribution of some species of owls in Western Australia. W. Aust. Nat. 8: 166–169.

1241 MEES, G. F. 1964. A revision of the Australian owls (Strigidae and Tytonidae). Zool. Verhandl., Leiden 65: 1–62.

1242 MEES, G. F. 1965. The status of *Pardalotus ornatus* Temminck. Emu 65: 97–102.

1243 MEES, G. F. 1967. A note on the pollination of the Kangaroo Paw. W. Aust. Nat. 10: 149–151.

1244 MEES, G. F. 1968. Colour of orbital ring in the Partridge-Pigeon *Geophaps smithii*. Emu 67: 294.

1245 MEES, G. F. 1969. A systematic review of the Indo-Australian Zosteropidae (Part III). Zool. Verhandl., Leiden 102: 1–390.

1246 MEES, G. F. 1974. The migration of the Tasmanian race of the Silvereye. Aust. Bird Bander 12: 51–54.

1247 MEES, G. F. 1975. Some birds recorded from Timor by Salomon Muller in 1828–29. Emu 76: 150–151.

1248 MEES, G. F. 1977. Geographical variation of *Caprimulgus macrurus* Horsfield (Aves, Caprimulgidae). Zool. Verhandl., Leiden 155: 1–47.

1249 MEES, G. F. 1977. The subspecies of *Chlidonias hybridus* (Pallas), their breeding distribution and migration (Aves, Laridae, Sternidae). Zool. Verhandl., Leiden 157: 1–64.

1250 MEES, G. F. 1982. Birds of the lowlands of southern New Guinea (Merauke and Koembe). Zool. Verhandl., Leiden 191: 1–188.

1251 MELLOR, J. W. 1908. Birds identified in N.S.W. during the seventh session of the A.O.U. Emu 7: 184–186.

1252 MEREDITH, C. W. Pers. comm.

1253 MEREDITH, C. W. 1983. R.A.O.U. Conservation Statement No. 1. The Ground Parrot. RAOU Newsl. 55: 6–11.

1254 MICHIE, R. H. 1959. Welcome Swallows nesting in Northland—a new breeding bird for New Zealand. Notornis 8: 61–62.

1255 MIDDLETON, A. L. A. 1970. Foods and feeding habits of the European Goldfinch near Melbourne. Emu 70: 12–16.

1256 MIDDLETON, W. G. D. Pers. comm.

1257 MILES, J. M. and A. A. BURBIDGE. 1975. A biological survey of the Prince Regent River Reserve, north-west Kimberley, Western Australia. W.A. Dept of Fisheries & Wildl. Wildl. Res. Bull. 3: 1–114.

1258 MILHINCH, A. L. 1983. Crimson Chats in the south-west of W.A. in 1981. W. Aust. Nat. 15: 109–111.

1259 MILLEDGE, D. R. Pers. comm.

1260 MILLEDGE, D. R. 1971. Systematic list for 1971. Fairy Tern. Tas. Bird Report 1: 6.

1261 MILLEDGE, D. 1972. The birds of Maatsuyker Island, Tasmania. Emu 72: 167–170.

1262 MILLEDGE, D. 1977. One year's observations of seabirds in continental shelf waters off Sydney, N.S.W. Corella 1: 1–12.

1263 MILLEDGE, D. and N. BROTHERS. 1976. Maatsuyker Island, Tasmania. Aust. Bird Bander 14: 32–34

1264 MILLER, R. S. 1932. Some notes on the Little Kingfisher *Alcyone pusilla*. Emu 31: 257–259.

1265 MILLER, R. S. 1932. Some remarks on the nesting of the Brown-backed Honeyeater. Emu 32: 110–112.

1266 MILLER, R. S. 1937. The Mangrove Kingfisher. Emu 36: 149–152.

1267 MILLER, R. S. 1937. The Helmeted Friarbird. Emu 36: 249–253.

1268 MILLER, R. S. 1937. The Spotted Catbird. Emu 37: 73–76.

1269 MILLER, R. S. 1938. Notes on the Stubble Quail and the Brown Quail in Victoria. Emu 37: 285–299.

1270 MILLIGAN, A. W. 1901. Description of a New Bristle Bird *Sphenura*. Emu 1: 67–69.

1270a MILLIGAN, A. W. 1902. Field observations on Western Australian birds Emu 2: 68–77.

1271 MILNE, H. V. 1938. The Black Honeyeater. Emu 37: 247–250.

1272 MOFFATT, R. 1978. Sight record of a White-headed Petrel in central New South Wales. Aust. Birds 12: 55–56.

1273 MOLLISON, B. C. 1962. Notes on Tasmanian birds, IV-VI. Emu 62: 112–114.

1274 MOLLISON, B C. and R. H. GREEN. 1962. Mist Netting Tree-Martins on charcoal patches. Emu 61: 277–280.

1275 MONTAGUE, T. L. 1982. The food and feeding ecology of the Little Penguin *Eudyptula minor* at Phillip Island, Victoria, Australia. M.Sc. Thesis, submitted Monash University.

1276 MOONEY, N. Pers. comm.

1277 MOONEY, N. 1981. Reasons for size differences within forest hawks. Tas. Bird Report 11: 4–11.

1278 MOORE, L. A. 1977. The Forest Raven *Corvus tasmanicus*—a new record for South Australia. S. Aust. Orn. 27: 251–253.

1279 MORGAN, B. and J. MORGAN. 1964. Cattle Egrets in Brisbane. Emu 64: 230–232.

1280 MORGAN, B. and J. MORGAN. 1968. The Bush-hen in south-eastern Queensland. Emu 68: 150.

1281 MORGAN, D. G., L. R. MORGAN, L. N. ROBINSON, P. A. ROBINSON and D. H. ASHTON. 1961. The Eyrean Grass-Wren (*Amytornis goyderi*). Aust. Bird Watcher 1: 161–171.

1282 MORIARTY, T. K. 1972. Birds of Wanjarri, W.A. (27°25′S, 120°40′E). Emu 72: 1–7.

1283 MORRIS, A. K. Pers. comm.

1284 MORRIS, A. K. 1971. The Red-chested Quail in New South Wales. Emu 71: 178–180.

1285 MORRIS, A. K. 1975. The birds of Gosford, Wyong and Newcastle (County of Northumberland). Aust. Birds 9: 37–43.

1286 MORRIS. A. K. 1976. Endangered fauna—birds. Parks & Wildlife 1: 151–154.

1287 MORRIS, A. K. 1979. The declining status of the Little Tern in N.S.W. Corella 3: 105–110.

1288 MORRIS, A. K. 1979. The inland occurrence of Tropic-Birds in N.S.W. during March, 1978. Aust. Birds 13: 51–54.

1289 MORRIS, A. K. 1979. The spread of the Cattle Egret in N.S.W. Aust. Birds 13: 72–74.

1290 MORRIS, A. K. 1980. The status and distribution of the Turquoise Parrot in N.S.W. Aust. Birds 14: 57–67.

1291 MORRIS, A. K. and N. KURTZ. 1977. Red-chested and Little Button-quail in the Mudgee District of N.S.W. Corella 1: 77–79.

1292 MORRIS, A. K., A. R. McGILL and G. HOLMES. 1981. Handlist of Birds in New South Wales. Sydney: N.S.W. Field Orn. Club.

1293 MORRIS, C. Pers. comm.

1294 MORRIS, F. T. 1978. Feeding association between Little Egret and Sacred Ibis. Emu 78: 164.

1295 MORSE, F. C. 1918. Quail and mice. Emu 18: 74.

1296 MORSE, F. C. 1922. Birds of the Moree district. Emu 22: 24–36.

1297 MORTON, H. J. 1953. Notes on Spotless Crake (*Porzana plumbea*). S. Aust. Orn. 21: 5.

1298 MORTON, S. R. 1975. The diet of the Barn Owl *Tyto alba* in southern Victoria. Emu 75: 31–34.

1299 MORTON, S. R. and A. A. MARTIN. 1979. Feeding ecology of the Barn Owl, *Tyto alba* in arid Southern Australia. Aust. Wildl. Res. 6: 191–204.

1300 MOSEY, H. 1956. Birds observed on a visit to New Guinea during June to August, 1950. Emu 56: 357–366.

1301 MULDER, J. F. 1898. Birds of Geelong and Cape Otway District. Geelong Nat. 6: 48–52.

1302 MURPHY, R. C. 1936. Oceanic Birds of South America. New York: Macmillan.

1303 MURRAY, M. D. and R. CARRICK. 1964. Seasonal movements and habitats of the Silver Gull, *Larus novaehollandiae* Stephens, in south-eastern Australia. C.S.I.R.O. Wildl. Res. 9: 160–188.

1304 MURTON, R. K. 1971. Man and Birds. London: Collins.

1305 NAARDING, J. A. 1981. Population estimates of the Short-tailed Shearwater *Puffinus tenuirostris* in Tasmania. Tas. Nat. 66: 3–4.

1306 NAARDING, J. A. 1983. Latham's Snipe (*Gallinago hardwickii*) in southern Australia. Tas. Wildl. Div. Tech. Rep. 83/1: 1–89.

1307 NAPIER, J. R. 1973. Crested Grebe and Coot nesting in Tasmania. Aust. Bird Watcher 5: 7–10.

1308 NAPIER, J. R. 1979. Bird Island, Georges Rocks, Tasmania. Corella 3: 48–49.

1309 NAPIER, J. R. 1979. Georges Island, Georges Rocks, Tasmania. Corella 3: 50–51.

1310 NAPIER, J. R. 1982. Note on the breeding of the White-fronted Tern in Franklin Sound, Furneaux Group, Tasmania. Aust. Bird Watcher 9: 210–212.

1311 NELSON, J. B. 1975. The breeding biology of Frigatebirds—a comparative review. The Living Bird 1975: 113–156.

1312 NELSON, J. B. 1978. The Sulidae Gannets and Boobies. London: O.U.P.

1313 NETHERSOLE-THOMPSON, D. 1951. The Greenshank. London: Collins.

1314 NEWMAN, O. M. G. Pers. comm.

1315 NEWMAN, O. M. G. 1973. Systematic List for 1973. Tas. Bird Rep. 3: 2–21.

1316 NEWMAN, O. M. G. 1976. Systematic List for 1975. Tas. Bird Rep. 5: 2–26.

1317 NEWMAN, O. M. G. 1979. The Ground Parrot—census, calls, nest and movement. Tas. Bird Rep. 8: 4–6.

1318 NEWMAN, O. M. G. 1980. Successful breeding of Fairy Terns at Mortimer Bay, South Arm Peninsula. Tas. Bird Rep. 10: 24–26.

1319 NEWMAN, M. 1981. Wader count highlights—Eastern Curlew decline. R.A.O.U. Newsl. 48: 11.

1320 NEWMAN, O. M. G. 1982. Dispersal of immature Pied Oystercatchers in the Hobart Area. An Occasional Stint 1: 51–58.

1321 NEWMAN, O. M. G. and G. C. BRATT. 1976. The Pink Robin—an avian lichen buff. Search 7: 487–488.

1322 NEWMAN, O. M. G. and A. W. J. FLETCHER. 1980. Fluctuations in Hobart area wader populations 1964 to 1981. Tas. Bird Rep. 10: 4–11.

1323 NEWMAN, O. M. G. and N. MOONEY. Pers. comm.

1324 NEWMAN, O. M. G. and P. PARK. 1982. Sooty Oystercatcher: evaluation of Hobart area count data 1964–1982. An Occasional Stint 1: 65–72.

1325 NEWMAN, O. M. G. and L. E. WALL. Pers. comm.

1326 NICHOLLS, C. A. 1962. Recovery of a ringed Sharp-tailed Sandpiper in Siberia. Emu 61: 328.

1327 NICHOLS, O. G., D. WATKINS and E. D. KABAY. 1982. The distribution of the Red-eared Firetail *Emblema oculata* in relation to bauxite mining in the northern jarrah forest. Emu 82: 169–172.

1328 NIELSEN, L. 1959. Albert Lyrebird in the Blackall Ranges. Emu 59: 249.

1329 NIELSEN, L. 1964. Movements of the Willie Wagtail. Emu 64: 27.

1330 NIELSEN, L. 1965. Double clutch of Pied Butcher-bird. Emu 64: 113.

1331 NIELSEN, L. 1966. Migration of the Blue-faced Honeyeater. Emu 65: 305–309.

1332 NIX, H. 1982. Environmental determinants of biogeography and evolution in Terra Australis. *In* BARKER, W. R. and P. J. M. GREENSLADE (Eds). Evolution of the Flora and Fauna of Arid Australia: 47–66. Adelaide: Peacock.

1333 NIX, H. A. and M. P. AUSTIN. 1973. Mulga: a bioclimatic analysis. Tropical Grasslands 7: 9–21.

1334 NOELKER, F. Pers. comm.

1335 NORMAN, F. I. Pers. comm.

1336 NORMAN, F. I. 1970. Notes on the birds of Big Green Island, Furneaux Group, Tas. 1965–1968. Vict. Nat. 87: 131–141.

1337 NORMAN, F. I. 1971. Movement and mortality of Wood Ducks banded in Victoria. Emu 71: 57–60.

1338 NORMAN, F. I. 1977. Granite Island, Corner Inlet, Victoria. Corella 1: 54–55.

1339 NORMAN, F. I. 1979. Results from banding Eurasian Coots in Victoria, 1953–77. Corella 3: 73–76.

1340 NORMAN, F. I. and L. MUMFORD. 1982. Food of the Chestnut Teal, *Anas castanea*, in the Gippsland Lakes region of Victoria. Aust. Wildl. Res. 9: 151–155.

1341 NORMAN, F. I. and K. C. NORRIS. 1982. Some notes on Freckled Duck shot in Victoria, Australia, 1981. Wildfowl 33: 81–87.

1342 NORRIS, A. Y. 1964. Observations on some birds of the Tooloom Scrub, northern N.S.W. Emu 63: 404–412.

1343 NORTH, A. J. 1901–14. Nests and Eggs of Birds Found Breeding in Australia and Tasmania. Aust. Museum Sydney: Special Catalogue No. 1.

1344 NORTON, S. P. W. 1922. Bird notes from Boree (New England Plateau). Emu 22: 39–44.

1345 NOSKE, R. A. Pers. comm.

1346 NOSKE, R. 1975. The Common Sandpiper in N.S.W. Aust. Birds 10: 1–7.

1347 NOSKE, R. 1978. Range extensions of the Grey-headed Honeyeater and Fan-tailed Cuckoo. Sunbird. 9: 12.

1348 NOSKE, R. A. 1979. Co-existence of three species of Treecreepers in north-eastern New South Wales. Emu 79: 120–128.

1349 NOSKE, R. 1980. Co-operative breeding by Treecreepers. Emu 80: 35–36.

1350 NOSKE, R. A. 1980. Co-operative breeding and plumage variation in the Orange-winged (Varied) Sittella. Corella 4: 45–53.

1351 NOSKE, R. A. 1982. Comparative behaviour and ecology of some Australian bark-foraging birds. Ph.D. Thesis, University of New England.

1352 NOSKE, R. A. 1983. Communal breeding of Brown-headed Honeyeaters. Emu 83: 38–41.

1353 NOSKE, R. A. and R. STICKLEN. 1979. Nest and eggs of the Yellow-legged Flycatcher. Emu 79: 148–149.

1354 OFFICER, H. R. 1964. Australian Honeyeaters. Melbourne: Bird Observers' Club.

1355 OFFICER, H. R. 1967. In the footsteps of McLennan and Barnard at Cape York. Emu 66: 279–287.

1356 OFFICER, H. R. 1969. Australian Flycatchers. Melbourne: Bird Observers' Club.

1357 OGDEN, J. 1979. Estimates of the population sizes of the Black Noddy and Wedge-tailed Shearwater at Heron Island in 1978. Sunbird 10: 33–39.

1358 O'GORMAN, B. 1981. A prolonged field study of the Diamond Firetail. Aust. Aviculture 35: 14–27.

1359 OLIVER, W. 1974. New Zealand Birds. Wellington: Reed.

1360 OLSEN, P. Pers. comm.

1361 OLSEN, P., W. J. M. VESTJENS and J. OLSEN. 1979. Observations on the diet of the Australian Kestrel *Falco cenchroides*. Emu 79: 133–138.

1362 OLSEN, S. L. 1980. The significance of the distribution of the Megapodiidae. Emu 80: 21–24.

1363 ONIANS, G. 1933. The Brown Bittern. Emu 32: 311–312.

1364 OPIE, M. J. 1974. A further sighting of the Oriental Pratincole. S. Aust. Orn. 26: 166.

1365 ORTON, C. L. E. and P. T. SANDLAND. 1913. Birds of Moora (W.A.) and district. Emu 13: 75–80.

1366 OSBORNE, R. E. P. 1931. Report on bird life in central Australia and northern Australia. S. Aust. Orn. 11: 4–7.
1367 PARK, P. Pers. comm.
1368 PARK, P. 1981. A colour-banding study of Welcome Swallows breeding in southern Tasmania. Corella 5: 37–41.
1369 PARKER, S. A. Pers. comm.
1370 PARKER, S. 1969. New and interesting distribution records of central Australian birds. S. Aust. Orn. 25: 59–71.
1371 PARKER, S. A. 1970. Taxonomy of the populations of *Sericornis beccarii* inhabiting Cape York Peninsula. Emu 70: 69–72.
1372 PARKER, S. A. 1970. Critical notes on the status of some Northern Territory birds. S. Aust. Orn. 25: 115–125.
1373 PARKER, S. A. 1971. Taxonomy of the Northern Territory Friarbirds known as *Philemon buceroides gordoni*. Emu 71: 54–56.
1374 PARKER, S. A. 1971. Critical notes on the status of some central Australian birds. Emu 71: 99–102.
1375 PARKER, S. A. 1971. Distribution of *Meliphaga flavescens* and *M. fusca* in northern Queensland. Sunbird 2: 41–47.
1376 PARKER, S. A. 1972. Remarks on distribution and taxonomy of the grass wrens *Amytornis textilis, modestus* and *purnelli*. Emu 72: 157–166.
1377 PARKER, S. A. 1973. The identity of *Microeca brunneicauda* Campbell, 1902. Emu 73: 23–25.
1378 PARKER, S. A. 1974. Review of Storr, G. M. 1973. List of Queensland Birds. S. Aust. Orn. 26: 146–148.
1379 PARKER, S. A. 1977. Notes on the distribution of the Chiming and Chirruping Wedgebills in South Australia. S. Aust. Orn. 27: 175–176.
1380 PARKER, S. A. 1977. Records of the Barking Owl from South Australia. S. Aust. Orn. 27: 204–206.
1381 PARKER, S. A. 1977. The distribution and occurrence in South Australia of owls of the genus *Tyto*. S. Aust. Orn. 27: 207.
1382 PARKER, S. A. 1979. The Spotted Bowerbird in southwestern Queensland. Sunbird 10: 67–68.
1383 PARKER, S. A. 1979. Birds. *In* Flinders Ranges National Park. Draft Management Plan: 139–151. Adelaide: SA Nat. Parks and Wildl. Service.
1384 PARKER, S. A. 1980. The records of the Speckled Warbler from South Australia. S. Aust. Orn. 28: 102–103.
1385 PARKER, S. A. 1980. Samuel White's ornithological explorations in northern South Australia in 1863. S. Aust. Orn. 28: 113–119.
1386 PARKER, S. A. 1980. An early breeding record of the Sooty Tern from Kangaroo Island. S. Aust. Orn. 28: 139–140.
1387 PARKER, S. A. 1980. Birds and conservation parks in the north-east of South Australia. S.A. Parks and Conservation 3: 11–18.
1388 PARKER, S. A. 1981. Prolegomenon to further studies in the *Chrysococcyx 'malayanus'* group (Aves, Cuculidae). Zool. Verhandl., Leiden 187: 1–56.
1389 PARKER, S. A. 1982. Notes on *Amytornis striatus merrotsyi* Mellor, a subspecies of the Striated Grasswren inhabiting the Flinders Ranges. S. Aust. Orn. 29: 13–16.
1390 PARKER, S. A. 1982. A new Sandpiper of the Genus *Calidris*. S. Aust. Nat. 56: 63.
1391 PARKER, S. A., H. J. ECKERT, G. B. RAGLESS, J. B. COX and N. C. H. REID. 1979. List of species. An Annotated Checklist of the Birds of South Australia. Part 1: emus to spoonbills. Adelaide: SAOA.
1392 PARKER, S. A., I. A. MAY and W. HEAD. 1978. Some observations on the Eyrean Grasswren *Amytornis goyderi* (Gould 1875). Records of S. Aust. Mus. 17: 361–371.
1393 PARKER, S. A. and I. A. MAY. 1982. Additional notes on seabirds recorded in South Australia. S. Aust. Orn. 28: 213–216.
1394 PARKER, S. A. and N. REID. 1979. Remarks on the status of some Australian passerines. *In* TYLER, M. J. (Ed.). The Status of Endangered Australasian Wildlife. Proc. Centenary Symp. R. Zool. Soc. S. Aust.: 109–115.
1395 PARKES, K. C. 1958. Extra-limital records of the Australian Whiskered Tern. Emu 58: 288.
1396 PARKES, K. C. 1980. A new subspecies of the Spiny-cheeked Honeyeater *Acanthagenys rufogularis*, with notes on generic relationships. Bull. Br. Orn. Club 100: 143–147.
1397 PARRY, V. A. 1970. Kookaburras. Melbourne: Lansdowne.
1398 PARRY, V. 1973. The auxiliary social system and its effect on territory and breeding in Kookaburras. Emu 73: 81–100.
1399 PARSONS, F. E. 1935. Bird notes. S. Aust. Orn. 13: 19–20.
1400 PARSONS, F. E. and J. N. McGILP. 1935. The two red-throated Whistlers. Emu 35: 113–126.
1401 PATERSON, A. 1928. Notes made during a visit to the Bogong High Plains, Victoria. Emu 28: 66–69.
1402 PATON, D. C. 1980. The importance of manna, honeydew and lerp in the diets of honeyeaters. Emu 80: 213–226.
1403 PATON, D. C. and H. A. FORD. 1977. Pollination by birds of native plants in South Australia. Emu 77: 73–85.
1404 PATON, D. C. and B. J. WYKES. 1978. Re-appraisal of moult of Red-necked Stints in southern Australia. Emu 78: 54–60.
1405 PATON, J. B. 1981. The Grey Honeyeater: a review of distribution and variation. S. Aust. Orn. 28: 185–190.
1406 PATON, J. B. and D. C. PATON. 1977. West Island, South Australia. Corella 1: 65–67.
1407 PATON, J. B. and D. C. PATON. 1977. Wright Island, South Australia. Corella 1: 68–69.
1408 PATON, J., D. PATON and M. WATERMAN. 1968 The Pink and Rose Robins in South Australia. S. Aust. Orn. 25: 19–20, 22.
1409 PATTERSON, R. M. Pers. comm.
1410 PECKOVER, W. S. 1969. The Fawn-breasted Bowerbird *Chlamydera cerviniventris*. Proc. PNG Sci. Soc. 21: 23–25.
1411 PECKOVER, W. S. and L. W. C. FILEWOOD. 1976. Birds of New Guinea and Tropical Australia Sydney: Reed.
1412 PECOTICH, L. 1982. Speciation of the Grey Swiftlet *Aerodramus spodiopygius* in Australia. Tower Karst 4: 53–56.
1413 PEDLER, L. P. 1975. Quails in mid-northern South Australia. S. Aust. Orn. 27: 3–5.
1414 PEDLER, L. P. 1976. Breeding and other notes on Black-shouldered Kites. S. Aust. Orn. 27: 139–141.
1415 PEDLER, L. 1982. Northern records of the Painted Button-quail in South Australia. S. Aust. Orn. 29: 1–5.
1416 PEGLER, J. and A. R. McGILL. 1979. Identification of a Western Sandpiper *Calidris mauri* and a comparison of this species with other small calidrids. Aust. Birds 14: 30–32.
1417 PENNY, M. 1974. The Birds of Seychelles and the Outlying Islands. London: Collins.
1418 PERRING, F. H. and S. M. WALTERS. 1962. Atlas of the British Flora. London: Botanical Society of the British Isles.
1419 PESCOTT, T. W. 1980. Lawrence Rocks, Victoria. Corella 4: 107–109.
1420 PETERS, D. E. Pers. comm.
1421 PETERS, D. E. 1979. Some evidence for a decline in population status of the Regent Honeyeater. Aust. Bird Watcher 8: 117–123.
1422 PETERS, J. L. 1931–79. Checklist of the Birds of the World. Vols I-XV. Cambridge, Mass: Museum of Comparative Zoology.
1423 PHILLIPS, J. H. 1963. The pelagic distribution of the Sooty Shearwater *Procellaria grisea*. Ibis 105: 340–353.
1424 PHILLIPS, R. E. 1980. Behaviour and systematics of New Zealand Plovers. Emu 80: 177–197.
1425 PIERCE, R. J. 1974. Presumed attempted breeding of the White-winged Black Tern in New Zealand. Notornis 21: 129–134.
1426 PINNER, D. and L. A. BIRD. 1970. White-winged Triller breeding in north-western Tasmania. Emu 70: 32.
1427 PIZZEY, G. 1969. Reclaiming the Bustard. Animals 12: 254–256.

1428 PIZZEY, G. 1980. A Field Guide to the Birds of Australia. Sydney: Collins.
1429 PLUMB, W. 1948. A nesting season in a Wellard paddock. Emu 47: 291–303.
1430 POORE, G. C. B., A. H. CORRICK and F. I. NORMAN. 1979. Food of three waders at Lake Reeve, Victoria. Emu 79: 228–229.
1431 PRATT, E. K. 1968. Notes on the Southern Figbird. Aust. Bird Bander 6: 56–57.
1432 PRATT, E. 1973. Progress notes on the banding of Bar-shouldered Doves and Crested Pigeons. Aust. Bird Bander 11: 8–10.
1433 PRATT, E. 1979. Red-tailed Black Cockatoo in north-eastern New South Wales. Aust. Birds 14: 36–37.
1434 PRUETT-JONES, S. G., C. M. WHITE and W. R. DEVINE. 1981. Breeding of the Peregrine Falcon in Victoria, Australia. Emu 80: 253–269.
1435 PRUETT-JONES, S. G., C. M. WHITE and W. B. EMISON. 1981. Eggshell thinning and organochlorine residues in eggs and prey of Peregrine Falcons from Victoria, Australia. Emu 80: 281–287.
1436 PURCHASE, D. 1969. Twelfth Annual Report of the Australian Bird-banding Scheme. CSIRO Wildl. Res. Tech. Pap. 16: 1–54.
1437 PURCHASE, D. 1969. Thirteenth Annual Report of the Australian Bird-banding Scheme. CSIRO Wildl. Res. Tech. Pap. 18: 1–56.
1438 PURCHASE, D. 1969. Fourteenth Annual Report of the Australian Bird-banding Scheme. CSIRO Wildl. Res. Tech. Pap. 19: 1–64.
1439 PURCHASE, D. 1970. Fifteenth Annual Report of the Australian Bird-banding Scheme. CSIRO Wildl. Res. Tech. Pap. 21: 1–79.
1440 PURCHASE, D. 1971. Sixteenth Annual Report of the Australian Bird-banding Scheme. CSIRO Wildl. Res. Tech. Pap. 22: 1–78.
1441 PURCHASE, D. 1972. Seventeenth Annual Report of the Australian Bird-banding Scheme. CSIRO Wildl. Res. Tech. Pap. 25: 1–87.
1442 PURCHASE, D. 1973. Eighteenth Annual Report of the Australian Bird-banding Scheme. CSIRO Wildl. Res. Tech. Pap. 27: 1–93.

1443 QUESTED, T. 1982. Spine-tailed Swift landing in tree. Aust. Birds 16: 64.

1444 RAGLESS, G. B. Pers. comm.
1445 RAGLESS, G. B. 1969. Nest and eggs of the Chestnut-breasted Whiteface. S. Aust. Orn. 25: 98–99.
1446 RAGLESS, G. B. 1977. The Chestnut Rail at Darwin. S. Aust. Orn. 27: 254–255.
1447 RAMSAY, E. P. 1888. Tabular List of all the Australian Birds. Sydney: The Author.
1448 RAMSAY, J. 1919. Notes on birds observed in the Upper Clarence River district, N.S.W. Sept.-Dec. 1918. Emu 19: 2–9.
1449 RAND, A. L. and E. T. GILLIARD. 1967. Handbook of New Guinea Birds. London: Weidenfeld & Nicolson.
1450 RATCLIFFE, D. A. 1963. The status of the Peregrine in Great Britain. Bird Study 10: 56–90.
1450a RATKOWSKY, A. V. Pers. comm.
1451 RATKOWSKY, A. V. and D. A. RATKOWSKY. 1977. The birds of the Mt. Wellington Range, Tasmania. Emu 77: 19–22.
1452 RATKOWSKY, A. V. and D. A. RATKOWSKY. 1978. A survey of the birds of the Mt. Wellington Range, Tasmania, during the non-breeding months. Emu 78: 223–226.
1453 RATKOWSKY, A. V. and D. A. RATKOWSKY. 1980. A survey of the birds of two areas of sclerophyll bushland in southern Tasmania. Emu 80: 169–170.
1454 READSHAW, J. L. 1968. The distribution, abundance and seasonal movements of the Pied Currawong, *Strepera graculina* (Shaw), an important bird predator of Phasmatidae, in eastern Australia. Aust. J. Zool. 16: 37–47.
1454a RECHER, H. F. Pers. comm.
1455 RECHER, H. F. 1971. Sharing of habitat by three congeneric honeyeaters. Emu 71: 147–152.
1456 RECHER, H. F. 1972. Territorial and agonistic behaviour of the Reef Heron. Emu 72: 126–130.
1457 RECHER, H. F. 1977. Ecology of co-existing White-cheeked and New Holland Honeyeaters. Emu 77: 136–142.
1458 RECHER, H. F. and R. T. HOLMES. 1982. The foraging behaviour of herons and egrets on the Magela Creek floodplain, Northern Territory. Canberra: A.G.P.S.
1459 RECHER, H. F. and J. A. RECHER. 1972. The foraging behaviour of the Reef Heron. Emu 72: 85–90.
1460 RECHER, H. F. and M. SCHULZ. 1983. Observations on the breeding of White-browed Wood-Swallows. Corella 7: 1–6.
1461 RECHER, H. F., D. G. THOMAS and D. R. MILLEDGE. 1971. Census of dry sclerophyll. Emu 71: 40–43.
1462 REED, J. 1952. Common Sandpiper at Sorrento, Vic. Emu 52: 215–216.
1463 REED, M. and J. L. McKEAN. 1982. A Northern Territory specimen of Hutton's Shearwater. Emu 82: 230.
1464 REESE, L. 1927. Bird notes. S. Aust. Orn. 9: 69.
1465 REESE, L. R. 1932. Bird notes. S. Aust. Orn. 11: 229.
1465a REID, J. 1980. Bird Report 1976. S. Aust. Orn. 28: 127–137.
1466 REID, N. 1975. The Western Warbler on Eyre Peninsula. S. Aust. Orn. 27: 6–8.
1467 REID, N. 1980. Australia-wide notes. The massacre at Bool Lagoon, S.A.—Requiem for the Freckled Duck. Bird Observer 581: 44.
1468 REID, N. and J. B. COX. 1978. The status of Satin and Leaden Flycatchers in South Australia with a note on female plumages. S. Aust. Orn. 27: 277–279.
1469 REILLY, P. N. Pers. comm.
1470 REILLY, P. N. 1966. Predation by Grey Shrike-thrush. Emu 65: 318.
1471 REILLY, P. N. 1970. Nesting of the Superb Lyrebird in Sherbrooke Forest, Victoria. Emu 70: 73–78.
1472 REILLY, P. N. 1974. Breeding of Little Penguins along the Great Australian Bight. Emu 74: 198–200.
1473 REILLY, P. N. 1977. Gabo Island, Victoria. Corella 1: 51–53.
1474 REILLY, P. N. 1978. Birds of Gabo Island. Corella 2: 73–75.
1475 REILLY, P. N. 1978. Little Penguins at sea. Aust. Seabird Group Newsl. 10: 20–21.
1476 REILLY, P. N. 1981. A Thirteen-month Study of Birds in a Suburban Coastal Environment. Aust. Bird Watcher 9: 124–129.
1477 REILLY, P. N., M. G. BROOKER and G. W. JOHNSTONE. 1975. Birds of the south-western Nullarbor Plain. Emu 75: 73–76.
1478 REILLY, P. N. and J. M. CULLEN. 1981. The Little Penguin *Eudyptula minor* in Victoria, II: Breeding. Emu 81: 1–19.
1479 REILLY, P. N. and J. M. CULLEN. 1982. The Little Penguin *Eudyptula minor* in Victoria, III: Dispersal of chicks and survival after banding. Emu 82: 137–142.
1480 REILLY, P. N. and W. J. GARRETT. 1973. Nesting of Fairy Martins in culverts. Emu 73: 188–189.
1481 REILLY, P. N. and J. A. KERLE 1981. A study of the Gentoo Penguin *Pygoscelis papua*. Notornis 28: 189–202.
1482 REYNOLDS, I. S., J. C. WALTER and P. F. WOODALL. 1982. Observations on Yellow Chats *Ephthianura crocea* in Western Queensland. Sunbird 12: 21–29.
1483 RHODES, L. J. 1944. Birds of the Adelaide River District, Northern Territory. Emu 44: 87–93.
1484 RICHARDS, K. A. 1982. In search of Flock Pigeons. Bird Observer 607: 59.
1485 RIDPATH, M. G. 1972. The Tasmanian Native Hen, *Tribonyx mortierii*. III. Ecology. CSIRO Wildl. Res. 17: 91–118.
1486 RIDPATH, M. G. and R. E. MOREAU. 1966. The birds of Tasmania: ecology and evolution. Ibis 108: 348–393.
1487 RIGGERT, T. L. 1977. The biology of the Mountain Duck

on Rottnest Island, Western Australia. Wildlife Monographs 52: 1–67.

1488 RIPLEY, S. D. 1977. Rails of the World. Toronto: M. F. Feheley.

1489 RIX, C. E. 1940. *Pachycephala rufogularis* (Gould) Red-lored Whistler. Extension in distribution. S. Aust. Orn. 15: 96–98.

1490 RIX, C. E. 1970. Birds of the Northern Territory. S. Aust. Orn. 25: 147–191.

1491 ROBERTS, G. J. Pers. comm.

1492 ROBERTS, G. 1977. Queensland records of Sooty Albatross, Light-mantled Albatross and White-headed Petrel. Sunbird 8: 69–71.

1493 ROBERTS, G. J. 1981. Field observations of the Grey Honeyeater in central Australia. S. Aust. Orn. 28: 190.

1494 ROBERTS, G. 1982. Some seabird observations in Western Australia. Aust. Seabird Group Newsl. 17: 3–5.

1495 ROBERTS, G. J. and G. J. INGRAM. 1976. An annotated list of the land birds of Cooloola. Sunbird 7: 1–20.

1496 ROBERTS, G. J. and G. J. INGRAM. 1978. Marbled Frogmouth in the Conondale Range, south-eastern Queensland. Emu 78: 41–42.

1497 ROBERTS, N. L. 1935. Spinebill Honeyeater. Emu 35: 99.

1498 ROBERTS, P. E. 1957. Notes on birds of the Cumberland Islands. Emu 57: 303–310.

1499 ROBERTSON, B. I. Pers. comm.

1500 ROBERTSON, B. I. 1977. Identification of Pacific and Dominican Gulls. Aust. Bird Watcher 7: 5–10.

1501 ROBERTSON, B. I. 1981. Chalky Island, Furneaux Group, Tasmania. Corella 5: 49–52.

1502 ROBERTSON, C. J. R. 1974. Birds in New Zealand. Wellington: Reed.

1503 ROBERTSON, C. J. R. and F. C. KINSKY. 1972. The dispersal movements of the Royal Albatross (*Diomedea epomophora*). Notornis 19: 289–301.

1504 ROBERTSON, D. G. 1980. First Record of the House Swift *Apus affinus* (Apodidae) in Australia. Aust. Bird Watcher 8: 239–242.

1505 ROBERTSON, J. S. 1946. The Eastern Bristlebird in Queensland. Emu 45: 265–270.

1506 ROBERTSON, J. S. 1962. Mackay Report. Emu 61: 270–274.

1507 ROBERTSON, J. S. 1962. Range of the Rufous-throated Honeyeater. Emu 61: 332–333.

1508 ROBINSON, A. 1933. Notes on Wood-Swallows and Swallows of the Barlee Range, W.A. Emu 33: 95–96.

1509 ROBINSON, A. 1936. Notes on the Western Bower-bird. Emu 35: 229–231.

1510 ROBINSON, A. 1939. Birds of the Barlee Range. Emu 38: 461–466.

1511 ROBINSON, A. 1947–48. Magpie-Larks: A study in behaviour. Emu 46: 265–281; 382–391; 47: 11–28; 147–153.

1512 ROBINSON, A. 1949. The biological significance of bird song in Australia. Emu 48: 291–315.

1513 ROBINSON, A. 1956. The annual reproductory cycle of the Magpie, *Gymnorhina dorsalis* Campbell, in south-western Australia. Emu 56: 233–336.

1514 ROBINSON, A. 1960. The importance of the marri as a food source to south-western Australian birds. W. Aust. Nat. 7: 109–115.

1515 ROBINSON, F. N. 1975. Vocal mimicry and the evolution of bird song. Emu 75: 23–27.

1516 ROBINSON, F. N. 1977. Environmental origins of the Menurae. Emu 77: 167–168.

1517 ROBINSON, F. N. and H. J. FRITH. 1981. The Superb Lyrebird *Menura novaehollandiae* at Tidbinbilla, A.C.T. Emu 81: 145–157.

1518 ROBINSON, P. D. Pers. comm.

1519 ROGERS, A. 1971. N.S.W. Bird Report for 1970. Birds 5: 66–76.

1520 ROGERS, A. 1973. N.S.W. Bird Report for 1972. Birds 7: 89–107.

1521 ROGERS, A. 1974. N.S.W. Bird Report for 1973. Birds 8: 97–118.

1522 ROGERS, A. 1975. N.S.W. Bird Report for 1974. Aust. Birds 9: 77–96.

1523 ROGERS, A. E. F. 1975. Movements of Wedge-tailed Shearwaters off the coast of New South Wales. Emu 75: 178–180.

1524 ROGERS, A. 1977. N.S.W. Bird Report for 1976. Aust. Birds 11: 81–104.

1525 ROGERS, A. 1978. N.S.W. Bird Report for 1977. Aust. Birds 13: 1–21.

1526 ROHAN-JONES, W. G. Pers. comm.

1527 ROHAN-JONES, W. G. and H. F. RECHER. Pers. comm.

1528 ROOKE, I. J. Pers. comm.

1529 ROOKE, I. J. 1979. The social behaviour of the Honeyeater *Phylidonyris novaehollandiae*. Ph.D. Thesis, Univ. of W. Aust.

1530 ROOKE, I. J. 1983–84. The Silvereye (*Zosterops lateralis*: Aves, Zosteropidae). A review. J. Roy. Soc. W. Aust.: in press.

1531 ROSE, A. B. 1973. Food of some Australian birds. Emu 73: 177–183.

1532 ROSS, J. A. 1926. The excursion to Murrayville. Emu 25: 175–183.

1533 ROUNSEVELL, D. E. Pers. comm.

1534 ROUNSEVELL, D. E., S. A. BLACKHALL and D. G. THOMAS. 1977. Birds of Maria Island. Tas N.P.W.S. Wildl. Div. Tech. Rep. No. 77/3.

1535 ROUNSEVELL, D. E. and N. P. BROTHERS. Pers. comm.

1536 ROUNSEVELL, D. E., N. P. BROTHERS, S. A. BLACKHALL, C. RILEY and R. J. WILSON. 1979 Some new or unusual records of birds for Tasmania during 1977. Emu 79: 139–140.

1537 ROUNSEVELL, D. E. and N. P. BROTHERS. The status and conservation of seabirds at Macquarie Island. *In* Proc. 18th ICBP World Conf. Seabird Workshop, 1982. In press.

1538 ROUNSEVELL, D. E. and J. C. Z. WOINAESKI. 1983. Status and distribution of the Forty-spotted Pardalote, *Pardalotus quadragintus*. Aust. Wildl. Res. 10: 343–349.

1539 ROWAN, M. K. 1972. Winter quarters of Swallows. Bird Study 19: 54–55.

1540 ROWLEY, I. C. R. Pers. comm.

1541 ROWLEY, I. 1965. The life history of the Superb Blue Wren *Malurus cyaneus*. Emu 64: 251–297.

1542 ROWLEY, I. 1968. Unusual feeding technique of Bald Coot. Emu 67: 295–296.

1543 ROWLEY, I. 1970. The genus *Corvus* (Aves: Corvidae) in Australia. CSIRO Wildl. Res. 15: 27–71.

1544 ROWLEY, I. 1971. Movements and longevity of ravens in south-eastern Australia. CSIRO Wildl. Res. 16: 49–72.

1545 ROWLEY, I. 1973. The comparative ecology of Australian Corvids. II. Social organization and behaviour. CSIRO Wildl. Res. 18: 25–65.

1546 ROWLEY, I. 1973. The comparative ecology of Australian Corvids VI. Why five species? CSIRO Wildl. Res. 18: 157–169.

1547 ROWLEY, I. 1976. Co-operative breeding in Australian birds. Proc. 16th Intern. Orn. Congr.: 657–666.

1548 ROWLEY, I. 1978. Communal activities among White-winged Choughs *Corcorax melanorhamphus*. Ibis 120: 178–197.

1549 ROWLEY, I. 1981. A relic population of Blue-breasted Wrens *Malurus pulcherrimus* in the central wheatbelt. W. Aust. Nat. 15: 1–8.

1550 ROWLEY, I. 1981. The communal way of life in the Splendid Wren *Malurus splendens*. Z. fur. Tierpsychol. 55: 228–267.

1551 ROWLEY, I. and W. J. M. VESTJENS. 1973. The comparative ecology of Australian Corvids. V. Food. CSIRO Wildl. Res. 18: 131–155.

1552 ROYAL AUSTRALASIAN ORNITHOLOGISTS UNION. Nest Record Scheme.

1553 RYAN, C. S. 1906. On European and other birds liberated in Victoria. Emu 5: 110–119.

1554 RYAN, J. V. 1951. Honeyeaters in the Bendigo district. Emu 51: 175–176.

1555 RYAN, J. V. 1952. Notes on the Rufous Song-Lark. Emu 52: 158–161.
1556 RYAN, J. V. 1953. Some fruit-loving birds. Emu 53: 331.
1557 RYAN, J. V. 1954. Further notes on the Rufous Song-Lark. Emu 54: 128–131.
1558 RYAN, J. V. 1955. Ground-Thrush in northern Victoria. Emu 55: 309.
1559 RYAN, J. V. 1979. Nearly forty years with *Hylacola* spp. Aust. Bird Watcher 8: 42–47.

1560 SAGE, B. L. 1956. Remarks on the racial status, history and distribution of the Tree Sparrow introduced into Australia. Emu 56: 137–140.
1561 SALMON, H. A. 1965. Distribution of the Jabiru in central and northern coastal New South Wales. Emu 65: 149–151.
1562 SANDLAND, P. T. 1937. Notes on birds on Pelsart Island. Emu 37: 144–149.
1563 SANDLAND, P. T. and O. L. E. ORTON. 1922. List of birds found breeding in and around the swamps near Moora, W.A., during 1921. Emu 22: 134–137.
1564 SARGENT, O. H. 1928. Reactions between birds and plants. Emu 27: 185–192.
1565 SAUNDERS, D. A. Pers. comm.
1566 SAUNDERS, D. A. 1976. Simultaneous use of a nesting hollow by a cockatoo and a duck. Emu 76: 223–224.
1567 SAUNDERS, D. A. 1977. Red-tailed Black Cockatoo breeding twice a year in the south-west of Western Australia. Emu 77: 107–110.
1568 SAUNDERS, D. A. 1977. The effect of agricultural clearing on the breeding success of the White-tailed Black Cockatoo. Emu 77: 180–184.
1569 SAUNDERS, D. A. 1979. Distribution and taxonomy of the White-tailed and Yellow-tailed Black Cockatoos *Calyptorhynchus* spp. Emu 79: 215–227.
1570 SAUNDERS, D. A. 1979. The availability of tree hollows for use as nest sites by White-tailed Black Cockatoos. Aust. Wildl. Res. 6: 205–216.
1571 SAUNDERS, D. A. 1980. Food and movements of the short-billed form of the White-tailed Black Cockatoo. Aust. Wildl. Res. 7: 257–269.
1572 SAUNDERS, D. A. 1982. The breeding behaviour and biology of the short-billed form of the White-tailed Black Cockatoo *Calyptorhynchus funereus*. Ibis 124: 422–455.
1573 SCHMIDT, B. L. 1978. Birds of the Cobar region, New South Wales. Aust. Birds 12: 61–86.
1574 SCHODDE, R. Pers. comm.
1575 SCHODDE, R. 1965. Observations on new distribution and habitat of five Australian land birds. Emu 64: 204–208.
1576 SCHODDE, R. 1975. Interim list of Australian songbirds, Passerines. Melbourne: RAOU.
1577 SCHODDE, R. 1977. The identity of the Powerful Owl *Ninox strenua* from Fulham, S.A. S. Aust. Orn. 27: 184–185.
1578 SCHODDE, R. 1982. The Fairy-wrens. Melbourne: Lansdowne Editions.
1579 SCHODDE, R. and R. de NAUROIS. 1982. Patterns of variation and dispersal in the Buff-banded Rail (*Gallirallus philippensis*) in the south-west Pacific, with description of a new subspecies. Notornis 29: 131–142.
1580 SCHODDE, R., P. FULLAGAR and N. HERMES. 1983. A review of Norfolk Island birds past and present. Canberra: ANPWS Spec. Pub. 8.
1581 SCHODDE, R., B. GLOVER, F. C. KINSKY, S. MARCHANT, A. R. McGILL and S. A. PARKER. 1978. Recommended English Names for Australian Birds. Emu 77: 245–307.
1582 SCHODDE, R. and J. L. McKEAN. 1976. The relations of some monotypic genera of Australian Oscines. Proc. 16th Intern. Ornith. Congr.: 530–541.
1583 SCHODDE, R. and I. J. MASON. 1974. Common Tern in the Northern Territory. Emu 74: 103–4.
1584 SCHODDE, R. and I. J. MASON. 1975. Occurrence, nesting and affinities of the White-throated Grass-wren *Amytornis woodwardi* and White-lined Honeyeater *Meliphaga albilineata*. Emu 75: 12–18.
1585 SCHODDE, R. and I. J. MASON. 1976. A new subspecies of *Colluricincla megarhyncha* Quoy and Gaimard from the Northern Territory. Emu 76: 109–114.
1586 SCHODDE, R. and I. J. MASON. 1976. Infra-specific variation in *Alcedo azurea* Latham (Alcedinidae). Emu 76: 161–166.
1587 SCHODDE, R. and I. J. MASON. 1979. Revision of the Zitting Cisticola *Cisticola juncidis* (Rafinesque) in Australia, with description of a new subspecies. Emu 79: 49–53.
1588 SCHODDE, R. and I. J. MASON. 1980. Nocturnal Birds of Australia. Melbourne: Lansdowne.
1589 SCHODDE, R., I. J. MASON, M. L. DUDZINSKI and J. L. McKEAN. 1980. Variation in the Striated Heron *Butorides striatus* in Australasia. Emu 80: 203–212.
1590 SCHODDE, R., G. T. SMITH, I. J. MASON and R. G. WEATHERLEY. 1979. Relationships and speciation in the Australian Corellas (Psittacidae). Bull. Br. Orn. Club 99: 128–137.
1591 SCHODDE, R., G. F. VAN TETS, C. R. CHAMPION and G. S. HOPE. 1975. Observations on birds at glacial altitudes on the Carstensz Massif, western New Guinea. Emu 75: 65–72.
1592 SCHODDE, R. and G. F. VAN TETS. 1981. First record of the Brown Hawk-Owl *Ninox scutulata* from Australasia. Emu 81: 171.
1593 SCHRADER, N. W. Pers. comm.
1594 SCHRADER, N. W. 1974. Invasion of Black-tailed Nativehens. Aust. Bird Watcher 5: 234.
1595 SCHRADER, N. W. 1976. Gibber-bird feeding on blowfly larvae. Aust. Bird Watcher 6: 179.
1596 SCHRADER, N. W. 1980. A review of the distribution of the Superb Parrot in central New South Wales. Aust. Birds 14: 45–50.
1597 SCHREIBER, R. W. and D. A. HENSLEY. 1976. The diets of *Sula dactylatra, Sula sula* and *Fregata minor* on Christmas Island, Pacific Ocean. Pac. Sci. 30: 241–248.
1598 SCHULZ, M. Pers. comm.
1599 SCHULZ, M. 1983. Fluctuations in Hooded Plover numbers at Venus Bay in 1981 and 1982. Bull. Vict. Wader Study Group 7: 11–12.
1600 SEDGWICK, E. H. 1936. Notes on the Chestnut-tailed Thornbill. Emu 35: 231–232.
1601 SEDGWICK, E. 1940. Birds of the Rockingham district. Emu 40: 129–152; 237–245.
1602 SEDGWICK, E. H. 1947. Northern Territory bird notes. 1–2. Emu 46: 294–308; 349–378.
1602a SEDGWICK, E. H. 1947. Feeding of Butcher-birds. Emu 47: 68–69.
1603 SEDGWICK, E. H. 1949. Observations on the Lower Murchison R.A.O.U. camp, September 1948. Emu 48: 212–242.
1604 SEDGWICK, E. H. 1949. Mixed associations of small birds in the south-west of Western Australia. Emu 49: 9–13.
1605 SEDGWICK, E. H. 1952. Bird life at Leonora, Western Australia. Emu 52: 285–296.
1606 SEDGWICK, E. H. 1956. Further notes on mixed associations in Western Australia. Emu 56: 425–429.
1607 SEDGWICK, E. 1958. The introduced Turtledoves in Western Australia. W. Aust. Nat. 6: 92–100; 112–127.
1608 SEDGWICK, E. H. 1971. Movements of *Pardalotus substriatus* Red-tipped Pardalote, in Western Australia. W. Aust. Nat. 12: 1–4.
1609 SEDGWICK, E. H. 1973. Birds of the Benger Swamp. W. Aust. Nat. 12: 147–155.
1610 SEDGWICK, E. H. 1977. Movements of the Spotted Pardalote. W. Aust. Nat. 13: 176.
1611 SEDGWICK, E. H. 1978. A population study of the Barrow Island avifauna. W. Aust. Nat. 14: 85–108.
1612 SEDGWICK, L. E. 1964. Birds of the Stirling Ranges, Western Australia. Emu 64: 7–19.
1613 SEEBECK, J. H. Pers. comm.
1614 SEEBECK, J. H. 1976. The diet of the Powerful Owl, *Ninox strenua* in Western Victoria. Emu 76: 167–170.
1615 SEFTON, A. R. 1958. Eastern Australian records of the Little Shearwater. Emu 58: 374.

1616 SEFTON, A. R. 1958. Bird notes from the Illawarra District, New South Wales. Emu 58: 393–394.
1617 SEFTON, A. R. and J. A. DEVITT. 1962. Additions to the birds recorded from the Illawara District. Emu 62: 184–187.
1618 SELBY, A. D. 1952. Photographing some birds of the Gunbower District, Victoria. Emu 52: 145–146.
1619 SERVENTY, D. L. 1932. *Calidris tenuirostris* in Western Australia. Emu 31: 309.
1620 SERVENTY, D. L. 1936. Feeding methods of *Podargus*. Emu 36: 74–90.
1621 SERVENTY, D. L. 1937. Local migration in the Perth District, Western Australia. Emu 37: 90–94.
1622 SERVENTY, D. L. 1948. The birds of the Swan River District, Western Australia. Emu 47: 241–286.
1623 SERVENTY, D. L. 1951. The evolution of the chestnut-shouldered wrens (*Malurus*). Emu 51: 113–120.
1624 SERVENTY, D. L. 1951. Observations on migration in the Timor Sea in 1949. Emu 51: 134.
1625 SERVENTY, D. L. 1952. The bird islands of the Sahul Shelf. Emu 52: 33–59.
1626 SERVENTY, D. L. 1952. Movements of the Wilson Storm-Petrel in Australian seas. Emu 52: 105–116.
1627 SERVENTY, D. L. 1953. The southern invasion of northern birds during 1952. W. Aust. Nat. 3: 177–196.
1628 SERVENTY, D. L. 1958. A new bird for the Australian list—the Malay Banded Crake. Emu 58: 415–418.
1629 SERVENTY, D. L. 1968. Wanderings of the Blue-winged Pitta to Australia. Bull. Br. Orn. Club 88: 160–162.
1630 SERVENTY, D. L. 1970. Torpidity in the White-backed Swallow. Emu 70: 27–28.
1631 SERVENTY, D. L. 1970. Torpidity in Australian birds. Emu 70: 201–202.
1632 SERVENTY, D. L. 1972. The Shearwaters of Shark Bay, W.A. Emu 72: 175–177.
1633 SERVENTY, D. L. 1973. 'Wrong-way' migration of the Koel in Western Australia. Emu 73: 138.
1634 SERVENTY, D. L. 1977. The use of data on the distribution of birds to monitor climatic change. Emu 77: 162–166.
1635 SERVENTY, D. L. and N. KOLICHIS. Pers. comm.
1636 SERVENTY, D. L., V. SERVENTY and J. WARHAM. 1971. The Handbook of Australian Seabirds. Sydney: Reed.
1637 SERVENTY, D. L. and H. M. WHITTELL. 1948. Birds of Western Australia. Perth: Paterson Brokensha.
1638 SERVENTY, D. L. and H. M. WHITTELL. 1976. (5th Edn). Birds of Western Australia. Perth: Univ. of W.A. Press.
1639 SERVENTY, V. N. 1943. Notes on nesting birds of the Abrolhos Islands. Emu 42: 235–241.
1640 SERVENTY, V. 1959. The birds of Willis Island. Emu 59: 167–176.
1641 SERVENTY, V. 1979. A new dawn for the Night Parrot. The Bulletin, 26 June 1979: 41.
1642 SETON, D. H. C. 1973. Observations on breeding of the Great-billed Heron in northern Queensland. Emu 73: 9–11.
1643 SETON, D. and H. J. LAVERY. 1974. Nest and eggs of the Atherton Scrub Wren. Emu 74: 53.
1644 SHANKS, D. 1949. Observations from the Upper King River District, Victoria. Emu 49: 132–141.
1645 SHARLAND, M. S. R. 1925. Tasmania's indigenous birds. Emu 25: 94–103.
1646 SHARLAND, M. S. R. 1936. Photographing the Black-throated Grebe. Emu 35: 354–356.
1647 SHARLAND, M. S. R. 1943. Channel-bill Cuckoo in Tasmania. Emu 43: 75.
1648 SHARLAND, M. S. R. 1943. Nesting habits of the Tree-Martin. Emu 43: 126–130.
1649 SHARLAND, M. 1952. Banded Plover increasing. Emu 52: 256.
1650 SHARLAND, M. 1953. Starling in north Queensland. Emu 53: 35.
1651 SHARLAND, M. 1953. Field-Wren's habitat. Emu 53: 92–93.
1652 SHARLAND, M. 1954. The Tasmanian Scrub-tit. Emu 54: 81–88.
1653 SHARLAND, M. 1957. Egrets at Ulmarra, N.S.W. Emu 57: 295–301.
1654 SHARLAND, M. 1958. Tasmanian Birds. Sydney: Angus and Robertson.
1655 SHARLAND, M. 1958. The Swamp Harrier as a migrant. Emu 58: 75–80.
1656 SHARLAND, M. 1959. Queensland's rare quail. Emu 59: 179–183.
1657 SHARLAND, M. 1960. Tasmanian Coot mystery. Emu 60: 95–98.
1658 SHARLAND, M. 1981. A Guide to the Birds of Tasmania. Hobart: Drinkwater Publishing.
1659 SHARLAND, M. S. R. and S. W. CRANE. 1922. Birds of Hobart. Emu 22: 127–133.
1660 SHARROCK, J. T. R. 1976. The Atlas of Breeding Birds in Britain and Ireland. England: British Trust for Ornithology, Irish Wildbird Conservancy.
1661 SHILLING, D. 1948. The birds of upper Liveringa Station, Western Australia. Emu 48: 64–72.
1662 SHIPWAY, A. K. 1969. The numbers of terns and shearwaters nesting on Heron Island in 1965. Emu 69: 108–109.
1663 SHRADER, K. 1956. White phase in Tree-creepers. Emu 56: 143.
1664 SHURCLIFF, A. and K. SHURCLIFF. 1973. Territory in the Australian Magpie *Gymnorhina tibicen*: an analysis of its size and change. S. Aust. Orn. 26: 127–132.
1665 SIBLEY, F. C. and R. B. CLAPP. 1967. Distribution and dispersal of Central Pacific Lesser Frigatebirds *Fregata ariel*. Ibis 109: 328–337.
1666 SIMMS, E. 1978 British Thrushes. London: Collins, New Naturalist.
1667 SIMPSON, K. G. and D. J. NOONAN. 1967. Diving display flights of the Spine-tailed Swift. Emu 67: 27–31.
1668 SINGLINE, T. A. 1981. Nesting notes on Lewin's Rail *Rallus pectoralis* and Spotless Crake *Porzana tabuensis* in north-east Tasmania. Tas. Nat. 65: 6–7.
1669 SLATER, P. 1959. Breeding periods of birds in the Kimberley Division, Western Australia. W. Aust. Nat. 7: 35–41.
1670 SLATER, P. 1961. Nesting of the Australian Goshawk (*Accipiter fasciatus*). Emu 61: 134–136.
1671 SLATER, P. 1961. Eastern Barn Swallow *Hirundo rustica gutturalis* at Derby. W. Aust. Nat. 8: 33–36.
1672 SMEDLEY, D. I. Pers. comm.
1673 SMEDLEY, J. H. 1904. Finches in northern Queensland. Emu 4: 68–69.
1674 SMITH, A. J. and B. I. ROBERTSON. 1978. Social organization of Bell Miners. Emu 78: 169–178.
1675 SMITH, F. T. H. 1967. A sighting of the Colombo Crow (*Corvus splendens*) in Victoria. Aust. Bird Watcher 3: 49.
1676 SMITH, F. T. H. 1968. The Long-toed Stint *Erolia subminuta* in southern Victoria. Aust. Bird Watcher 3: 132–140.
1677 SMITH, F. T. H. 1975. A Victorian Record of the Black-winged Petrel, *Pterodroma nigripennis*. Aust. Bird Watcher 6:40.
1678 SMITH, F. T. H. 1976. An Australian record of the Grey Phalarope. Aust. Bird Watcher 6: 292–299.
1679 SMITH, F. T. H. 1981. A retraction of Victorian Dunlin records. Aust. Bird Watcher 9: 43.
1680 SMITH, F. T. H. 1982. Dunlins undone. The Stilt 2: 10–11.
1681 SMITH, F. T. H., R. SWINDLEY and R. SYMPSON. 1974. Little Shearwater near Barwon Heads, Victoria. Aust. Bird Watcher 5: 187.
1682 SMITH, G. T Pers. comm.
1683 SMITH, G. T. 1976. Ecological and behavioural comparisons between the Atrichornithidae and Menuridae. Proc. 16th Intern. Ornith. Congr.: 125–136.
1684 SMITH, G. T 1977. The effect of environmental change on six rare birds. Emu 77: 173–179.
1685 SMITH, G. T. 1979. The status of Australian parrots. *In* TYLER, M. J. (Ed.). The Status of Endangered Australian Wildlife: 101–108. Adelaide: Roy. Zool. Soc. S. Aust.
1686 SMITH, G. T. and R. I. FORRESTER. 1981. The status

of the Noisy Scrub-bird *Atrichornis clamosus*. Biol. Conserv. 19: 239–254.
1687 SMITH, G. T. and N. KOLICHIS. 1981. Coffin Island, Western Australia. Corella 5: 66–68.
1688 SMITH, G. T. and F. N. ROBINSON. 1976. The Noisy Scrub-bird: an interim report. Emu 76: 37–42.
1689 SMITH, L. A. and R. E. JOHNSTONE. 1974. First record of White-lined Honeyeater in northern Kimberley, W.A. Emu 74: 58.
1690 SMITH, L. A. and R. E. JOHNSTONE. 1977. Status of the Purple-crowned Wren *Malurus coronatus* and Buff-sided Robin *Poecilodryas superciliosa* in Western Australia. W. Aust. Nat. 13: 185–188.
1691 SMYTH, D. M. 1980. Notes on the diet of the Grey Swiftlet *Aerodramus spodiopygius*. Sunbird 11: 20–21.
1692 SMYTH, D. M., L. PECOTICH and J. R. ROBERTS. 1980. Notes on the distribution and breeding of the Grey Swiftlet *Aerodramus spodiopygius*. Sunbird 11: 1–19.
1693 SOUTH AUSTRALIAN ORNITHOLOGICAL ASSOCIATION. 1977. A Bird Atlas of the Adelaide Region. Adelaide: SAOA.
1694 SOUTHERN, H. N. and D. L. SERVENTY. 1947. The two phases of *Astur novae-hollandiae* (Gm.) in Australia. Emu 46: 331–347.
1695 SPECHT, R. L. 1981. Major vegetation formations in Australia. *In* KEAST, A. (Ed.). Ecological Biogeography of Australia: 163–297. The Hague: Junk.
1696 SPECHT, R. L., E. M. ROE and V. H. BROUGHTON (Eds). 1974. Conservation of major plant communities in Australia and Papua New Guinea. Aust. J. Bot. Suppl. 7: 1–667.
1697 SPELLERBERG, I. F. 1971. Arrival and departure of birds at McMurdo Sound, Antarctica. Emu 71: 167–171.
1698 SPENCE, T. Pers. comm.
1699 STEWART, D. 1977. A second breeding record of the Chestnut-breasted Whiteface. S. Aust. Orn. 27: 258.
1700 STOKES, T. 1975. The effect of a bushfire on the banding of Flame Robins in the Brindabella Ranges. Aust. Bird Bander 13: 75–76.
1701 STOKES, T. and N. HERMES. 1979. Cluster roosting in the Black-faced Woodswallow. Emu 79: 84–86.
1702 STORR, G. 1952. On the distribution of some South Australian birds. S. Aust. Orn. 20: 92.
1703 STORR, G. M. 1953. Birds of the Cooktown and Laura Districts, North Queensland. Emu 53: 225–248.
1704 STORR, G. M. 1960. Migration and breeding season in *Sterna nereis* and *S. albifrons*. Emu 60: 135–137.
1705 STORR, G. M. 1964. Zonation and seasonal occurrence of marine birds in the seas off Fremantle, Western Australia. Emu 63: 297–303.
1706 STORR, G. M. 1965. The avifauna of Rottnest Island, Western Australia. Parts 1–3. Emu 64: 48–60; 105–113; 172–180.
1707 STORR, G. M. 1966. Birds of the northern islands of the Houtman Abrolhos. Emu 65: 209–221.
1708 STORR, G. M. 1967. List of Northern Territory birds. Spec. Publ. W. Aust. Mus. 4: 1–90.
1709 STORR, G. M. 1973. List of Queensland birds. Spec. Publ. W. Aust. Mus. 5: 1–177.
1710 STORR, G. M. 1975. An observation of the Pheasant-tailed Jacana in W.A. W. Aust. Nat. 13: 60–61.
1711 STORR, G. M. 1977. Birds of the Northern Territory. Spec. Publ. West. Aust. Mus. 7: 1–130.
1712 STORR, G. M. 1980. The western sub-species of the Cape Barren Goose. W. Aust. Nat. 14: 202–203.
1713 STORR, G. M. 1980. Birds of the Kimberley Division, Western Australia. Sp. Publ. W. Aust. Mus. 11: 1–117.
1714 STORR, G. M. 1981. Birds of the north eastern interior of Western Australia. Rec. W. Aust. Mus. 9: 65–99.
1715 STORR, G. M. and R. E. JOHNSTONE. 1977. The Short-tailed Shearwater in south-western Western Australia. W. Aust. Nat. 13: 203.
1716 STORR, G. M. and R. E. JOHNSTONE. 1977. First Australian Record of the Pintail Snipe (*Gallinago stenura*). W. Aust. Nat. 13: 207–208.
1717 STORR, G. M. and R. E. JOHNSTONE. 1979. Field Guide to the Birds of Western Australia. Perth: W. Aust. Mus.
1718 STORR, G. M., R. E. JOHNSTONE and L. A. SMITH. 1982. A specimen of the White Wagtail (*Motacilla alba*) from Geraldton W.A. W. Aust. Nat. 15: 55–56.
1719 STRANGER, R. H. 1967. Field notes on the White-breasted Robin. W. Aust. Nat. 10: 115–116.
1720 STRANGER, R. H. 1968. A study of homing performance in the Senegal Dove. W. Aust. Nat. 11: 4–14.
1721 STRONOCH, N. R. H. 1980. Chestnut Teal *Anas castanea* at Bensbach (Western Province)—a new species for the New Guinea region. PNG Bird Soc. Newsl. 173–174: 3.
1722 STRONOCH, N. 1981. Grey Falcon *Falco hypoleucos* at Bensbach, Western Province—a new species for the New Guinea region. P.N.G. Bird Soc. Newsl. 178–179: 5.
1723 STRONOCH, N. R. 1981. Pallid Cuckoo *Cuculus pallidus*—a new species for Papua New Guinea. P.N.G. Bird Soc. Newsl. 181–182: 23.
1724 STRUDWICK, J. E. Pers. comm.
1725 SULLIVAN, C. 1927. A bird-lover in the mallee of north-eastern Victoria. Emu 27: 69–75.
1726 SULLIVAN, C. 1961. Bar-shouldered Dove near Sydney. Emu 61: 170.
1727 SUTTON, J. 1924. An ornithological trip around Eyre Peninsula. S. Aust. Orn. 7: 118–159.
1728 SWAINSON, G. W. 1970. Co-operative rearing in the Bell Miner. Emu 70: 183–188.
1729 SWALES, M. K. 1965. The sea-birds of Gough Island. Ibis 107: 17–42; 215–229.

1730 TARBURTON, M. K. 1977. Nesting of the Red-tailed Tropicbird at Sugarloaf Rock, W.A. Emu 77: 122–126.
1731 TARBURTON, M. K. 1981. Seabirds nesting at Norfolk Island. Notornis 28: 209–219.
1732 TARR, H. E. 1948. Birds of Dunk Island, north Queensland. Emu 48: 8–13.
1733 TARR, H. E. 1948. Notes on the King Quail. Emu 48: 103–106.
1734 TARR, H. E. 1949. Another introduced bird making headway in Australia. Emu 49: 142–143.
1735 TARR, H. E. 1950. The distribution of foreign birds in Australia. Emu 49: 189–198.
1736 TARR, J. N. S. 1981. An apparent nesting by a hybrid Woodswallow. Aust. Birds 15: 76.
1737 TEMPLETON, M. T. 1976. Movement of Welcome Swallows through Loorana, King Island, Tasmania. Emu 76: 86–88.
1738 THOMAS, D. G. 1966. Observations on the arrival of the Welcome Swallows and Tree-martins in the Hobart area, spring 1964–65. Emu 66: 33–38.
1739 THOMAS, D. G. 1968. Bird movements in southern Victoria and Tasmania 1965–66. First Report of the Individual Observation Points Scheme. Emu 68: 42–84.
1740 THOMAS, D. G. 1968. Waders of Hobart. Emu 68: 95–125.
1741 THOMAS, D. G. 1969. Individual observation points Second Annual Report, August 1966-July 1967. Emu 68: 249–271.
1742 THOMAS, D. G. 1969. Breeding biology of the Australian Spur-winged Plover. Emu 69: 81–102.
1743 THOMAS, D. G. 1969. Composition of Flame Robin flocks wintering in Tasmania. Emu 69: 240–241.
1744 THOMAS, D. G. 1970. Western Sandpiper in Tasmania. Emu 70: 88–89.
1745 THOMAS, D. G. 1970. Wader migration across Australia. Emu 70: 145–154.
1746 THOMAS, D. G. 1974. The Scrub-tit *Acanthornis magnus*—status and ecology. Tas. Nat. 38: 1–8.
1747 THOMAS, D. 1979. Tasmanian Bird Atlas. Fauna of Tasmania Handbook No. 2. Hobart: Univ. of Tasmania.
1748 THOMAS, D. G. 1980. Foraging of Honeyeaters in an area of Tasmanian schlerophyll forest. Emu 80: 55–58.
1749 THOMAS, D. G. and A. J. DARTNALL. 1971. Ecological aspects of the feeding behaviour of two Calidritine Sandpipers wintering in south-eastern Tasmania. Emu 71: 20–26.

1750 THOMAS, D. G. and A. J. DARTNALL. 1971. Moult of the Curlew Sandpiper in relation to its annual cycle. Emu 71: 153–158.
1751 THOMAS, D. G. and L. E. WALL. 1966. A probable record of the Ruff (*Philomachus pugnax*) and notes on other waders at Cape Portland, Tasmania. Aust. Bird Watcher 2: 238–241.
1752 THOMAS, H. 1951. Notes on the Pied Butcher-bird. Emu 51: 165–168.
1753 THOMAS, H. F. 1957. The Starling in the Sunraysia District, Victoria. Part IV. Roost sites. Emu 57: 269–284.
1754 THOMPSON, H. A. F. 1977. Dusky Moorhen in the Northern Territory. Sunbird 8: 24.
1755 THOMPSON, H. A. F. 1977. Notes on birds in the Darwin and northern areas of the Northern Territory. Sunbird 8: 83–91.
1756 THOMPSON, M. C. 1973. Migratory patterns of Ruddy Turnstones in the central Pacific region. Living Bird 12: 5–23.
1757 THOMSON, A. L. 1964. A New Dictionary of Birds. London: Nelson.
1758 THOMSON, D. F. 1935. Birds of Cape York Peninsula. Melbourne: Govt Printer.
1759 TICKELL, W. L. N. 1976. The distribution of Black-browed and Grey-headed Albatrosses. Emu 76: 64–68.
1760 TICKELL, W. L. N. and J. D. GIBSON. 1968. Movements of Wandering Albatrosses *Diomedea exulans*. Emu 68: 6–20.
1761 TICKELL, W. L. N. and C. D. SCOTLAND. 1961. Recoveries of ringed Giant Petrels *Macronectes giganteus*. Ibis 103a: 260–266.
1762 TIDEMANN, S. C. 1980. Notes on breeding and social behaviour of the White-winged Fairy-wren *Malurus leucopterus*. Emu 80: 157–161.
1763 TILLEY, S. 1982. The diet of the Powerful Owl, *Ninox strenua*, in Victoria. Aust. Wildl. Res. 9: 157–175.
1764 TINGAY, S. R. 1981. The social behaviour and vocalisations of Australian Magpie Larks. Ph.D. Thesis. Univ. of W. Aust.
1765 TINGAY, A. and S. R. TINGAY. 1982. Middle Island, Archipelago of the Recherche, Western Australia. Corella 6: 49–50.
1766 TINGAY, A. and S. R. TINGAY. 1982. Hood Island, Archipelago of the Recherche, Western Australia. Corella 6: 59.
1767 TUBB, J. A. 1945. Field notes on some New Guinea birds. Emu 44: 249–273.
1768 TWAITS, D. 1982. Birds of the Emerald district. Unpublished report.
1769 TYERS, J. and B. I. ROBINSON. Pers. comm.

1770 VALENTE, A. 1981. Vertebrate remains in pellets of the Barn Owl, *Tyto alba*, from Planet Downs Station, south-western Queensland. Aust. Wildl. Res. 8: 181–185.
1771 VAN TETS, G. F. 1965. Eastern Swamphen takes a downy from a pair of Chestnut Teal. Emu 64: 100.
1772 VAN TETS, G. F. 1977. Wright and Egg Islands, Horseshoe Reef, Tasmania. Corella 1: 63–64.
1773 VAN TETS, G. F. 1978. Second Amendments to the 1975 RAOU Checklist. Emu 78: 80–87.
1774 VAN TETS, G. F., A. H. D'ANDRIA and E. SLATER. 1967. Nesting distribution and nomenclature of Australasian Vanelline Plovers. Emu 67: 85–93.
1775 VAN TETS, G. F. and B. J. MARLOW. 1977. Dangerous Reef, South Australia. Corella 1: 70–71.
1776 VAN TETS, G. F. and P. A. VAN TETS. 1967. A report on the resident birds of the territory of Christmas Island. Emu 66: 309–317.
1777 VAN TETS, G. F., W. J. M. VESTJENS, A. H. D'ANDRIA and R. BARKER. 1977. Guide to the recognition and reduction of aerodrome bird hazards. Canberra: A.G.P.S.
1778 VAN TETS, G. F., M. H. WATERMAN and D. PURCHASE. 1976. Dispersal patterns of Cormorants banded in South Australia. Aust. Bird Bander 14: 43–46.
1779 VAN TETS, J. 1978. Transmarine dispersal of Australian Pelicans. Aust. Seabird Group Newsl. 11: 5–6.
1780 VAURIE, C. 1949. A revision of the bird family Dicruridae. Bull. Amer. Mus. Nat. Hist. 93: 199–342.
1781 VEITCH, D. 1981. Wader banding near Auckland. Stilt 1: 1.
1782 VELLENGA, R. E. 1970. Behaviour of the male Satin Bowerbird at the bower. Aust. Bird Bander 8: 3–11.
1783 VELLENGA, R. E. 1980. Distribution of bowers of the Satin Bowerbird at Leura, N.S.W. with notes on parental care, development and independence of the young. Emu 80: 97–102.
1784 VERNON, D. P. Pers. comm.
1785 VERNON, D. P. 1978. Food of Prions with particular reference to Fairy Prion, *Pachyptila turtur*. Sunbird 9: 7–9.
1786 VERNON, D. P. and D. H. FLEAY. 1975. One Queensland and two northern NSW specimen records of the Kerguelen Petrel. Sunbird 6: 90–92.
1787 VERNON, D. P. and J. H. D. MARTIN. 1974. A Queensland specimen record of Silver-grey Petrel (Antarctic Fulmar). Sunbird 5: 68–69.
1788 VERNON, D. P. and J. H. D. MARTIN. 1974. A Queensland specimen record of Thin-billed Prion. Sunbird 5: 73–74.
1789 VERNON, D. P. and J. H. D. MARTIN. 1975. Birds of Stradbroke Island. Proc. Roy. Soc. Q. 86: 61–72.
1790 VESTJENS, W. J. M. 1975. Breeding behaviour of the Darter at Lake Cowal, N.S.W. Emu 75: 121–131.
1791 VESTJENS, W. J. M. 1975. Feeding behaviour of Spoonbills at Lake Cowal, N.S.W. Emu 75: 132–136.
1792 VESTJENS, W. J. M. 1977. Breeding behaviour and ecology of the Australian Pelican, *Pelecanus conspicillatus* in N.S.W. Aust. Wildl. Res. 4: 37–58.
1793 VESTJENS, W. J. M. 1977. Status, habitats and food of vertebrates at Lake Cowal. CSIRO Wildl. Res. Tech. Mem. 12: 1–87.
1794 VESTJENS, W. J. M. and R. CARRICK. 1974. Food of the Black-backed Magpie, *Gymnorhina t. tibicen*, at Canberra. Aust. Wildl. Res. 1: 71–83.
1795 VINCENT, M. J. Pers. comm.
1796 VINCENT, R. T. 1972. Fairy Martin in southern Tasmania. Emu 72: 115–116.
1797 VIRTUE, R. M. 1947. Birds observed at Torokina, Bougainville Island. Emu 46: 324–331.

1798 WAKEFIELD, N. A. 1958. The Yellow-tufted Honeyeater with a description of a new subspecies. Emu 58: 163–194.
1799 WAKEFIELD, W. C. Pers. comm.
1800 WALKINSHAW, L. 1973. Cranes of the World. New York: Winchester.
1801 WALL, L. E. Pers. comm.
1802 WALL, L. E. 1965. The Whistling Eagle in Tasmania. Emu 65: 75–77.
1803 WALL, L. E. 1973. Starling roost on the Tasman Bridge, Hobart. Tas. Nat. 32: 4–6.
1804 WALL, L. E. 1977. Spoonbills in Tasmania. Tas. Nat. 48: 5.
1805 WALL, L. E. 1979. Australasian Grebe (*Tachybaptus novaehollandiae*) in Tasmania. Tas. Nat. 56: 1–2.
1806 WALLACE, B. J. Pers. comm.
1807 WALLACE, B. and D. GOSPER. 1967. Starlings and Falcon eat flying ants. Emu 66: 317.
1808 WARHAM, J. 1956. Observations on the birds of Pelsart Island. Emu 56: 83–93.
1809 WARHAM, J. 1957. Cockatoo Island birds. Emu 57: 225–231.
1810 WARHAM, J. 1960. Bird notes from the Carnarvon Ranges, Queensland. Emu 60: 1–7.
1811 WARHAM, J. 1961. The birds of Raine Island, Pandora Cay and Murray Island Sandbank, north Queensland. Emu 61: 77–93.
1812 WARHAM, J. 1962. Field notes on Australian Bower-birds and Cat-birds. Emu 62: 1–30.
1813 WARHAM, J. 1962. Bird islands within the Barrier Reef and Torres Strait. Emu 62: 99–111.

1814 WARHAM, J. 1967. The White-headed Petrel *Pterodroma lessoni* at Macquarie Island. Emu 67: 1–22.
1815 WARHAM, J. 1977. Raine Island, Queensland. Corella 1: 45–47.
1816 WARHAM, J. 1979. Cat Island, Tasmania. Corella 3: 43–45.
1817 WARHAM, J. 1981. Does Hutton's Shearwater circumnavigate Australia? Emu 81: 44.
1818 WARHAM, J. and B. D. BELL. 1979. The birds of Antipodes Island, New Zealand. Notornis 26: 121–169.
1819 WATERMAN, M. H. and L. C. LLEWELLYN. 1968. Further notes on White-backed Swallows and Rainbowbirds. Aust. Bird Bander 6: 8–9.
1820 WATSON, G. E. 1975. Birds of the Antarctic and Sub-Antarctic. Washington: Amer. Geophysical Union.
1821 WATSON, G. E., J. P. ANGLE, P. C. HARPER, M. A. BRIDGE, R. P. SCHLATTER, W. L. N. TICKELL, J. C. BOYD and M. M. BOYD. 1971. Birds of the Antarctic and Sub-Antarctic. Antarctic Map Folio Series 14: 1–18. New York: Amer. Geog. Soc.
1822 WATSON, I. M. 1955. Some species seen at the Laverton Saltworks, Victoria, 1950–53, with notes on seasonal changes. Emu 55: 224–248.
1823 WATSON, J. B. 1910. Further data on the homing sense of Noddy and Sooty Terns. Science 32: 470–473.
1824 WATSON, J. D., W. R. WHEELER and E. WHITBOURN. 1962. With the RAOU in Papua-New Guinea, October 1960. Emu 62: 31–50; 67–98.
1825 WATSON, M. 1969. Significance of antiphonal song in the Eastern Whipbird *Psophodes olivaceus*. Behaviour 35: 157–178.
1826 WEAVER, C. M. Pers. comm.
1827 WEAVER, C. M. 1982. Breeding habitats and status of the Golden-shouldered Parrot *Psephotus chrysopterygius* in Queensland. Emu 82: 2–6.
1828 WEBB, T. T. 1933. Aboriginal bird names in East Arnhem Land. Emu 33: 18–23.
1829 WEBSTER, H. O. 1947. Tawny-crowned Honeyeaters in a heavily forested area. Emu 47: 63.
1830 WEBSTER, H. O. 1948. Field notes on *Malurus elegans*, the Red-winged Wren of Western Australia. Emu 47: 287–290.
1831 WELLS, A. G. and B. A. WELLS. 1977. Breeding records of the Grey Honeyeater. W. Aust. Nat. 13: 209–210.
1832 WHEELER, J. R. 1960. Blue-billed Ducks at Lake Wendouree, Ballarat. Emu 60: 281–284.
1833 WHEELER, J. R. 1962. Observations on the Cattle Egret in eastern Australia. Emu 62: 192–193.
1834 WHEELER, J. 1965. Purple-crowned Wren: habitat and call-note. Emu 64: 219.
1835 WHEELER, R. 1947. Wood-Swallows and north winds. Emu 47: 55.
1836 WHEELER, R. 1947. Winter abundance of birds in Port Phillip Bay. Emu 47: 84–86.
1837 WHEELER, R. 1950. Further observations from Fishermen's Bend, Melbourne. Emu 50: 73–83.
1838 WHEELER, R. 1955. Charadriiformes at the Laverton Saltworks, Victoria, 1950–53. Emu 55: 279–295.
1839 WHEELER, R. 1957. Altona Survey Group. Report No. 4. Emu 57: 185–190.
1840 WHEELER, R. 1959. The RAOU Camp-out at Noosa Heads, Queensland, 1958. Emu 59: 229–249.
1841 WHEELER, R. 1963. Accipitriformes at the Laverton Saltworks, Victoria. 1950–53. Emu 63: 8–13.
1842 WHEELER, W. R. 1967. The birds of Cairns, Cooktown and the Atherton Tablelands. Aust. Bird Watcher 3: 55–76.
1843 WHEELRIGHT, H. W. 1861. Bush Wanderings of a Naturalist (by an old bushman). London: Routledge, Warne, Routledge.
1844 WHINRAY, J. S. 1980. The first Australian breeding record of the White-fronted Tern. Aust. Bird Watcher 8: 137–146.
1845 WHITE, C. M., S. G. PRUETT-JONES and W. B. EMISON. 1981. The status and distribution of the Peregrine Falcon in Victoria, Australia. Emu 80: 270–280.
1846 WHITE, C. M. N. 1975. Migration of Palaearctic Waders in Wallacea. Emu 75: 37–39.
1847 WHITE, C. M. N. 1976. Migration of Palaearctic nonpasserine birds in Wallacea. Emu 76: 79–82.
1848 WHITE, C. M. N. 1977. Migration of Palaearctic passerine birds in Wallacea. Emu 77: 37–38.
1849 WHITE, G. Pers. comm.
1850 WHITE, G. 1979. Flat Witch Island, Tasmania. Corella 3: 66–67.
1851 WHITE, G. 1979. Flat Island, Tasmania. Corella 3: 70–72.
1852 WHITE, G. 1980. Islands of south-west Tasmania. Sydney: The Author.
1853 WHITE, G. 1981. Ile du Golfe, Tasmania. Corella 5: 53–55.
1854 WHITE, G. 1981. Round Top Island, Maatsuyker Group, Tasmania. Corella 5: 58–59.
1855 WHITE, H. L. 1910. Description of two new nests and eggs from north-west Australia. Emu 10: 132–134.
1856 WHITE, H. L. 1913. Notes on the Cassowary *Casuarius australis*, Wall. Emu 12: 172–178.
1857 WHITE, H. L. 1915. Australian Cuckoos. Emu 14: 144–154.
1858 WHITE, H. L. 1915. Descriptions of nests and eggs new to science. Emu 15: 35–36.
1859 WHITE, H. L. 1917. North Australian birds. Emu 16: 117–158; 205–231.
1860 WHITE, H. L. 1917. Descriptions of new or rare eggs. Emu 16: 159–164.
1861 WHITE, H. L. 1922. A collecting trip to Cape York Peninsula. Emu 22: 99–116.
1862 WHITE, S. A. 1913. Field Ornithology in South Australia. Emu 13: 16–32.
1863 WHITE, S. A. 1914. A long-lost bird. Emu 14: 99–100.
1864 WHITE, S. A. 1915. An expedition to the Musgrave and Everard Ranges. Emu 14: 181–191.
1865 WHITE, S. A. 1915. The birds of Mallacoota. Emu 14: 135–144.
1866 WHITE, S. A. 1915. On some pellets or casts of a Screech Owl *Tyto alba delicatula*, Gould. S. Aust. Orn. 2: 90–95.
1867 WHITE, S. A. 1917. Further notes upon the Arctic Skua (*Stercorarius parasiticus*). S. Aust. Orn. 3: 38–39.
1868 WHITE, S. R. 1940. Note on the Western Spinebill. Emu 40: 169–170.
1869 WHITE, S. R. 1946. Notes on the bird life of Australia's heaviest rainfall region. Emu 46: 81–122.
1870 WHITE, S. R. 1950. Domestic co-operation among Chestnut-tailed Thornbills. W. Aust. Nat. 2: 93–94.
1871 WHITLOCK, F. L. 1909. Notes on birds observed on the Pilbara Goldfield, north-western Australia. Emu 8: 173–194.
1872 WHITLOCK, F. L. 1910. On the East Murchison. Emu 9: 181–219.
1873 WHITLOCK, F. L. 1911. In the Stirling Ranges, Western Australia. Emu 10: 305–317.
1874 WHITLOCK, F. L. 1912. Further notes from the Stirling Ranges, W.A. Emu 11: 239–243.
1875 WHITLOCK, F. L. 1912. Notes on *Megalurus striatus* (Milligan). Emu 11: 244–245.
1876 WHITLOCK, F. L. 1918. Notes on north-western birds. Emu 17: 166–179.
1877 WHITLOCK, F. L. 1921. Notes on Dirk Hartog Island and Peron Peninsula, Shark Bay, Western Australia. Emu 20: 168–186.
1878 WHITLOCK, F. L. 1922. Notes from the Nullarbor Plain. Emu 21: 170–187.
1879 WHITLOCK, F. L. 1923. A trip to the Fortescue River and Hamersley Ranges, north-west Australia. Emu 22: 259–273.
1880 WHITLOCK, F. L. 1924. Journey to central Australia in search of the Night Parrot. Emu 23: 248–281.
1881 WHITLOCK, F. L. 1925. Ten months on the Fitzroy River, north-western Australia. Emu 25: 69–89.
1881a WHITLOCK, F. L. 1928. Peculiarities in the distribution of birds in Western Australia. Emu 27: 179–184.
1882 WHITLOCK, F. L. 1940. Notes on the Thick-billed Thornbill. Emu 39: 176–178.

1883 WHITMORE, M. J. and D. D. DOW. 1982 Noisy Miners—the girls have it made. Aust. Nat. Hist. 20: 290–294.
1884 WHITTELL, H. M. 1933. The birds of the Bridgetown District, south-west Australia. Emu 32: 182–189.
1885 WHITTELL, H. M. 1933. Notes on the White-breasted Robin. Emu 32: 236–240.
1886 WHITTELL, H. M. 1938. The birds of the Bridgetown District, south-western Australia. II. Emu 38: 54–59.
1887 WHITTELL, H. M. 1941–42. A review of the work of John Gilbert in Western Australia. Emu 41: 112–129; 216–242; 289–305.
1888 WHITTELL, H. M. 1944. Black-shouldered Kite in Western Australia. Emu 43: 294–296.
1889 WIENEKE, J. E. Pers. comm.
1890 WIGAN, L. 1929. Lyrebird and Pilot-Bird. Emu 28: 231.
1891 WILLIAMS, C. K. Pers. comm.
1892 WILLIAMS, C. K. 1979. Ecology of Australian Chats (*Ephthianura* Gould): reproduction in aridity. Aust. J. Zool. 27: 213–229.
1893 WILLIAMS, C. K. and A. R. MAIN. 1976. Ecology of Australian chats (*Ephthianura* Gould): seasonal movements, metabolism and evaporative water loss Aust. J. Zool. 24: 397–416.
1894 WILLIAMS, C. K. and A. R. MAIN. 1977. Ecology of Australian chats (*Ephthianura* Gould): aridity, electrolytes and water economy. Aust. J. Zool. 25: 673–691.
1895 WILLIAMS, K. 1977. "Brinebool" the Beautiful. Q. Orn. Soc. Newsl. 8: 10.
1896 WILLIAMS, R. F. 1955. Vegetation regions. *In* Atlas of Australian Resources. Canberra: Dept Nat. Resources.
1897 WILLIG, R. 1982. A preliminary assessment of the status and distribution of terns in Victoria. Report to Fisheries & Wildlife Div., Vict. Min. for Cons.: 1–63.
1898 WILLIS, J. C. 1966. A Dictionary of the Flowering Plants and Ferns. 7th Ed. Rev. by H. K. Airy Shaw. Cambridge: Cambridge University Press.
1899 WILSON, H. 1937. Notes on the Night Parrot, with references to recent occurrences. Emu 37: 79–87.
1899a WILSON, R. I. Pers. comm.
1900 WILSON, R. 1980. The birds of Lake Dulverton. Part 1: Grebes to Ducks. Tas. Bird Report 10: 11–13.
1901 WILSON, R. I. 1981. The woodchip industry and Tasmanian birds. Tas. Bird Report 11: 11–14.
1902 WILSON, R. I. 1981. Tawny Grassbird, King Island. Tas. Bird. Rep. 11: 17–20.
1903 WILSON, S. J. 1965. New information on some birds of New South Wales. Emu 64: 209–213.
1904 WINTER, J. Pers. comm.
1905 WITHERBY, H. F., F. C. R. JOURDAIN, N. F. TICEHURST and B. W. TUCKER. 1938–41. The Handbook of British Birds. Vols 1–5. London: Witherby.
1906 WODZICKI, K. and P. STEIN. 1958. Migration and dispersal of New Zealand Gannets. Emu 58: 289–312.
1907 WOINARSKI, J. C. Z. Pers. comm.
1908 WOINARSKI, J. C. Z., D. F. DORWARD and J. M. CULLEN. 1983 Variation in the *Pardalotus striatus* complex in south-eastern Australia. Emu 83: 82–93.
1909 WOINARSKI, J. C. Z. and A. R. McEVEY. 1983. Mortality in pardalotes in Melbourne, 1960–1980. Corella 7: 55–59.
1910 WOINARSKI, J. C. Z. and D. E. ROUNSEVELL. 1983. Comparative ecology of pardalotes, including the Forty-spotted Pardalote, *Pardalotus quadragintus* (Aves: Pardalotidae) in south-eastern Tasmania. Aust. Wildl. Res. 10: 351–361.
1911 WOLSTENHOLME, H. 1925. Notes on birds observed during the Queensland Congress and camp-out, 1924. Emu 24: 230–236; 243–251.
1912 WOLSTENHOLME, H. 1930. Triple nesting of the Scarlet Honeyeater Emu 29: 264–266.
1913 WOODALL, P. F. 1980. Communal breeding in the Figbird. Sunbird 11: 73–75.
1914 WOOD JONES, F. 1937. The breeding of the White-faced Storm Petrel (*Pelagodroma marina*) on South Australian islands. S. Aust. Orn: 35–41.
1915 WOOLLER, R. D. 1981. Seasonal changes in a community of honeyeaters in south-western Australia. Emu 81: 50–51.
1916 WOOLLER, R. D. and J. S. BRADLEY. 1981. Consistent individuality in the calls of Spinifexbirds *Eremiornis carteri* on Barrow Island, W.A. Emu 81: 40.
1917 WOOLLER, R. D. and M. C. CALVER. 1981. Diet of three insectivorous birds on Barrow Island, W.A. Emu 81: 48–50.
1918 WYKES, B. J. Pers. comm.
1919 WYKES, B. J. 1982. Resource partitioning and the role of competition in structuring *Lichenostomus* honeyeater (and *Manorina melanophrys*) communities in southern Victoria. Ph.D. Thesis, Monash Univ.
1920 WYNDHAM, E. Pers. comm.
1921 WYNDHAM, E. 1978. Birds of the Milparinka district and Cooper Creek basin. Emu 78: 179–187.
1922 WYNDHAM, E. 1980. Environment and food of the Budgerigar *Melopsittacus undulatus*. Aust. J. Ecol. 5: 47–61.
1923 WYNDHAM, E., R. E. HUTCHISON and B. F. BROCKWAY-FULLER. 1981. Gonadal condition of nonbreeding wild and domesticated Budgerigars *Melopsittacus undulatus*. Ibis 123: 511–518.

1924 YEATMAN, L. 1976. Atlas des oiseaux nicheurs de France. Paris: Societe Ornithologique de France.

1925 ZANN, R. A. Pers. comm.
1926 ZANN, R. 1976. Distribution, status and breeding of Black-throated Finches *Poephila cincta* in northern Queensland. Emu 76: 201–206.
1927 ZILLMAN, E. E. 1964. Observations on the Winking or Barking Owl. Aust. Bird Watcher 2: 102–104.
1928 ZUSI, R. L. 1978. Notes on song and feeding behaviour of *Orthonyx spaldingii*. Emu 78: 156–157.
1929 ZWART, M. H. 1973. Breeding and behaviour of Pilotbirds. Emu 73: 124–128.

INDEX